AF324296

Computational Fluid Dynamics Review 1998

Computational Fluid Dynamics Review 1998

Vol. I

Editors

M. Hafez
University of California, Davis

K. Oshima
University of Tokyo

World Scientific
Singapore • New Jersey • London • Hong Kong

Published by

World Scientific Publishing Co. Pte. Ltd.

P O Box 128, Farrer Road, Singapore 912805

USA office: Suite 1B, 1060 Main Street, River Edge, NJ 07661

UK office: 57 Shelton Street, Covent Garden, London WC2H 9HE

British Library Cataloguing-in-Publication Data
A catalogue record for this book is available from the British Library.

COMPUTATIONAL FLUID DYNAMICS REVIEW 1998

ISBN 981-02-3959-9 (Vol. I)
ISBN 981-02-3960-2 (Vol. II)
ISBN 981-02-3564-X (Set)

Printed in Singapore by Uto-Print

PREFACE

The first volume of *CFD Review* was published in 1995. The purpose of this new publication is to present comprehensive surveys and review articles which provide up-to-date information about recent progress in computational fluid dynamics, on a regular basis. Because of the multidisciplinary nature of CFD, it is difficult to cope with all the important developments in related areas. There are at least ten regular international conferences dealing with different aspects of CFD, including

* Grid Generation and Adaptation
* Artificial Viscosity and Upwind Schemes for Hyperbolic Systems
* Boundary Elements and Vortex Methods
* Finite Elements in Fluids
* Spectral Methods and High Order Schemes
* Iterative Algorithms and Preconditioning Techniques
* Domain Decomposition Methods
* Multigrid Methods
* Parallel Computation
* Flow Visualization

This list does not include optimization; a field bigger than all CFD areas combined.

Besides these specialized conferences, one should mention:

* International Conference on Numerical Methods in Fluid Dynamics
* International Symposium on CFD
* American CFD Meetings (AIAA, ASME, SIAM)
* European Conferences on CFD (ECCOMAS)
* Asian Conferences on CFD (ACFD)

as well as the British, Canadian, Australian, Japanese, Chinese conferences on CFD.

There are also the general conferences on computational methods and on fluid mechanics and aerodynamics in both engineering and applied mathematics circles.

It is a real challenge to keep up with all these activities and to be aware of essential and fundamental contributions in these areas. It is hoped that *CFD Review* will help in this regard by covering the state-of-the-art in this field.

The present book contains sixty-two articles written by authors from US, Europe Japan and China, covering the main aspects of CFD. There are five sections: general topics, numerical methods, flow physics, interdisciplinary applications, parallel computation and flow visualization. The section on numerical methods includes grids, schemes and solvers, while that on flow physics includes incompressible and compressible flows, hypersonics and gas kinetics as well as transition and turbulence. This book should be useful to all researchers in this fast-developing field.

Finally, we would like to thank Ben Ransom, Janeen Curtis, and Susan Torguson of UCD for their assistance with the computer work and the editing process.

Editors

CONTENTS OF VOLUME I

CONTENTS OF VOLUME II

IIIc. Hypersonics and Gas Kinetics

IIId. Transition And Turbulence

Computational Fluid Dynamics Review 1998

HISTORICAL REMARK ON

FLUIDDYNAMICS AND COMPUTERS AT

I S A S , UNIVERSITY OF TOKYO

Koichi Oshima

Professor Emeritus of University of Tokyo

Abstract

After a short interruption due to the defeated war, old Aeronautical Research Institute, Imperial University of Tokyo revived as Aeronautical Research Institute, University of Tokyo at 1953, which has been growing with the aerospace science and technology. In 1964, it renamed as Institute of Space and Aeronautical Science, University of Tokyo to expanding its activity into space. In 1981, it still expands its activity into astronautics under new name of Institute of Space and Astronautical Science. Here is an overview concerning to the history of fluiddynamics and numerical fluiddynamics and their interaction with digital technologies taking place in this Institute, and, in some extent, in Japan and Asia from 1950 to present. It is discussed, how fluid dynamics as a modern science has been introduced and developed, and how digital technology as modern engineering tools has grown up to modern industry and how it has influenced to fluiddynamics in Japan.

Contents

1.FLUIDDYNAMICS BEFORE COMPUTER AGE

Japan opened its door to the modern western science around the end of 18th century, which is also the case of other Asian countries. Of course, Asian countries have their own culture over 4,000 years, nevertheless western culture, particularly its science and technology, has strongly influenced to the academic as well as social structure of all the Asian countries. Also it should be noted that, when Japan opened its door, all the fundamental structure of modern sciences have been already established.

1.1 Pre-historical Days of Fluiddynamics

Fluiddynamics as modern science has started during 17th century in the western world, the Euler equation and the Navier-Stokes equations were formulated and major concern of the fluid-dynamicists at these days is to looking for analytical (approximate) solutions, which exist quite limited cases only, such geometrically simple flows as parallel flows, source flows or at most flows around two-dimensional circular cylinder in infinite, uniform flow fields. They are basically for academic interests only.

In 1903 Prandtl has introduced the boundary layer theory. Based on this idea flowfields around real objects became tractable, that is,

Flow Field = Outer, Inviscid Flow

+ Thin, Boundary layer

Thus 20th century was the highlight time of fluiddynamics when it grows from mere academic interests onto the real world. Here, *Fluiddynamics as Science became Engineering Tool'*.

1.2 Last Frontier

Around 1940's transonic and supersonic flight became a reality, which is new environment we have not experienced before. Thus fluidynamics of compressible fluid emerged. That is, nonlinear, hyperbolic equations was introduced into fluiddynamics, which is a frontier toward the new field of science, which we, Asian scientists, could actually observe, and are even involved.

Development of supersonic aerodynamic engineering was quite swift. Quick progress of aeronautics had strong impetus from it. Based on deep insight of natural laws, Sedov, Taylor, Imai, *et al.* work out the similarity laws in compressible fluiddynamics around these days. Various kinds of expansions or asymptotic solutions were introduced, and open up wide fields of applications of fluiddynamics to aeronautics, fluid machinery, and atmospheric sciences. Although emergence of computers we have to wait till 1960's.

1.3 Fluiddynamics in Japan - 1950's

Only as my personal view, I like to recall briefly the historical background of fluiddynamics in Japan. Modern fluid dynamics was started by Prof. Prandtl, and inherited by Prof. von Karman, who left Germany to Cal.Tech., where a large group of fluiddynamicists gathered, headed by Prof. Liepman. Prof.Tani (my teacher) was there, and he established the Japanese aeronautics group at Aeronautical Research Institute, University of Tokyo, which was later renamed as Institute of Space and Aeronautical Science. On the other hand, there has been strong group of fluiddynamicists who mostly originated from Faculty of Physics, Profs. Tomotika, Kyoto University, and Imai, University of Tokyo, were group leader and they have strongly influenced from UK group headed by Profs. Taylor and Batchelor. Thus there are two groups of fluiddynamicists in Japan: one from Faculty of Technology, the other from Faculty of Science. They were collaborating and competing, resulting prosperous progress both in fundamental research and practical applications of fluiddynamics. Prof. Tamaki, my supervisor, came from Faculty of Science and became a professor of Faculty of Technology. I graduated Faculty of Technology and got the degree of Science. This multiplicity is a characteristic of Japanese fluiddynamics group and probably the main reason of our success in this field of science.

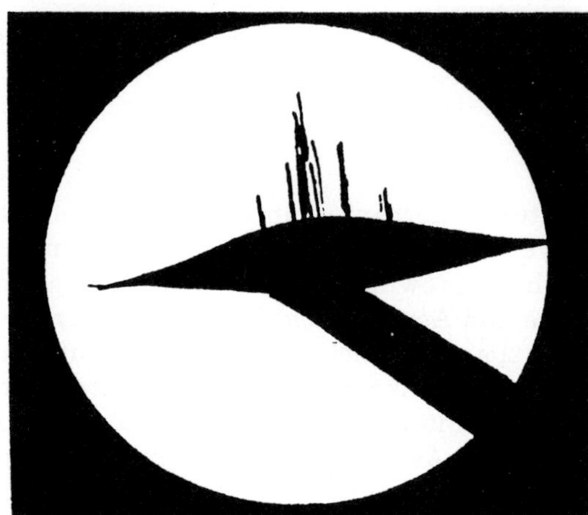

Fig.1: Moving shocks on Tamada profile

In 1950 I joined Prof. Tamaki's laboratory (University of Tokyo) as a graduate student. He was one of the youngest professors in the Faculty, and constructed tiny induction tunnel. Figure 1 shows a schlieren photograph taken by that tunnel, which clearly shows moving shocks on the surface of so-called shock-free transonic airfoil [1]. This profile, Tamada profile, was proposed by Prof. Tamada, Kyoto University, based on analytical solution of the flow around an particular airfoil which he found, and supposed to have shock-free supersonic region on the surface in subsonic uniform flow. This type of *shock-free, transonic* airfoil was rediscovered later and now widely used for aircraft, and the fact that this shock free region does in reality consist of moving weak shocks became well-known, 20 years later after this picture was taken.

Under his supervision, we have constructed small shock tube and assembled measuring devices, which consists of high speed time counter and instantaneous optical measuring system.

Modern digital technologies at that days were amply applied. Systematic survey of the transonic flows around NACA0012 airfoil has been done [2]. Also we still extended this line of work, and constructed divergent-type shock tube, world first try at that time. Whole time history of interaction of plane shock wave with circular cylinder was measured, and approach to the steady state was discovered precisely [3]. After 40 years later, even today, this process cannot be dissolved numerically even using today's high-end computers.

At the later part of 1950's, several electron-tube computers imported from USA were operating in Japan. But they are too expensive to be installed in research institutions, and software systems for scientific applications were practically none. (Only exception was Weather Bureau of Japan, which had IBM704.) They offered computing service which includes the programming, computing and printing out the result. I had used once such service in order to solve a fluiddynamical problem [4]. Since we did not touch upon the programming, there was no way to check the result directly. It was pity to have found some bugs later on [5]. Somehow this is the first case to utilize an electronic computer to numerically solve fluiddynamical problems in Japan. Late 1960's, Fortran compiler became common in Japan also, and rental services for research purposes were all die out.

1.4 Aerodynamics as National Concern

During World War I, air forces well demonstrated its decisive power at the battle field, and every nations want to have this technological products and established national center of aerodynamics research. NACA was established 1915 to build air power for US. ONERA (France), DVL (Germany), NPL (UK), TsAGI (USSR) followed. In Japan also Aeronautical Research Institute, Imperial University of Tokyo was established 1918.

World War II even confirmed the importance of air power against the national security. US Government sent a mission headed by Prof. von Karman to Germany to investigate their research status on aerodynamics. The mission report recommended construction of research facilities for future hypersonic flight. Accordingly, Arnold Engineering Development Center (AEDC), a huge complex of wind tunnels, was constructed. Wind tunnels then became status symbol of a nation. China constructed China Aerodynamics Research and Development Center, a huge complex of aerodynamics R/D facilities distributed over wide mountain site in Sichuan, in order to protect the nation against Pacific Power. Right after their independence, India established the National Aeronautics Laboratory at Bangalore and built the aerodynamics research as well as aerospace industry. Even Japan, right after the peace treaty, constructed a large transonic wind tunnel, even though there was absolutely no possibility to have air power in this nation.

2. DEVELOPMENT OF DIGITAL TECHNOLOGY

Counting is one of the most fundamental culture of human beings, but counting by machine is relatively new technology, which only appeared in this century. Once it was proved that business uses of electronic computers are successful, they invite many business people, which brought mass production and low cost, which ensured large customer. Thus computers became big business enterprise.

2.1 Emerging of CFD

Applications of computers to scientific calculations was rather slow. Because it need to develop numerical schemes and wrote down programs to each problem. Before computers came out, so-called Richardson scheme appeared at 1910, which solves by time marching the heat conduction equation, and eventually was found that it diverges. Later Courant-Friedrichs-Levy (CFL) condition of convergence criteria of time marching process was reported in 1927.

It is said that the first programmable, electronic computer, ENIAC was build on 1943 - 1946, which consists of 18,000 electron tubes and operates at 100kHz. Fluiddynamical problems were firstly solved by UNIVAC at 1953 of Courant Institute, which has stored programs.

After computers have introduced into CFD community, their intensive uses resulted in many, various techniques for solving the Navier-Stokes equations. Courant Institute of New York University has been most active in CFD; Conservation Law scheme for Reversible Process; Entropy Condition for Irreversible Process; and concept of Weak Solution for flows containing shock waves, etc came out from this group. Prof. Peter D. Lax, Director of this Institute proudly wrote in [4];

Why Courant Institute was Successful in

CFD ?.

He answered himself;

Experimental Computing is the Royal Road in CFD.

However he also complains that

CFDist was an orphan inside the Academia,

and

Orthodox mathematicians simply consider CFD as a service function to engineers, resorting to brutal force.

It should be emphasized here that major parts of numerical fluiddynamic theory have already completed from 1950's to early 1970's, during which fluiddynamicists directly involved to construct machine itself.

2.2 Digital Technology for Business

It is said that the first computer as a business tool is UNIVAC, which uses electron tube processor and mercury delay-line memory. UNIVAC was introduced 1951 and mainly used as business tool at up to about 40 offices. This was the most successful system of the 1st generation machine (era of electron tube processor).

The second generation is the era of transistor processor, IBM7070, IBM1401 introduced at late 1958, are representative machine throughout 1960's. Top end machines such as IBM7090, CDC6600 are still too expensive to be commodity item and only for research institutes, but middle class machine like IBM1401 became a standard item for business office and has been used more than 15,000 sets by 1965

The 3rd Generation is of LSI's era, which started early 1960's and continuously prospering. Its high reliability invited business use, resulted in mass production, then low cost. IBM360 or Hitac8800 are some of names to be recalled.

The 3.5th Generation - era of VLSI follows them at the end of 1960's to 70's. Time sharing system appeared, and big names such as IBM, FACOM, NEC, HITAC provided powerful center computers to every institutions. Business applications include accounting, ticketing, documentation, etc.

Massive users —≫ mass-production

—≫ high reliability —≫ low cost

brought many success story and

Performance is proportional to square of price.

was a guiding rule of this era. Every users try to combine their funds to buy single, powerful computers. That is, this is the era of computing center. Monopoly by one big business enterprise -IBM- was also established.

Snow beauty and seven dwarfs

was a word heard at those days.

Computers have been progressing with high pace, which is well predictable, since they are based on the same physics. Once a mass production line is established, constant flow of products (CPU's, Memories) has been poured into market, which ensure the low price and invited many office workers as business machines, so-called SOHO (Small Office, Home Office) become common.

Commodity Computers on every desk, commodity servers in every room,

is now a reality.

2.3 Digital Technology for Laboratory

In parallel to the business use, digital equipment to laboratory automation has come out at the same time, As mentioned previously, time counter was an indispensable tool for shock tube experiment, which was made by research people by themselves using diode tubes at that time.

Late 1960's, mini-computer, such as Hitac10, appeared in the market, whose price was affordable by single user (laboratory). Then they enthusiastically welcomed almost all the fluiddynamics laboratory for laboratory automation. Soon LSI's for analog-to-digital converter appeared in market and LC's for digital controller also became available. Time resolution of the measuring data was close to several microsecond, and measuring operation was fully automated. This was particularly useful for taking ensemble average of turbulent flow data.

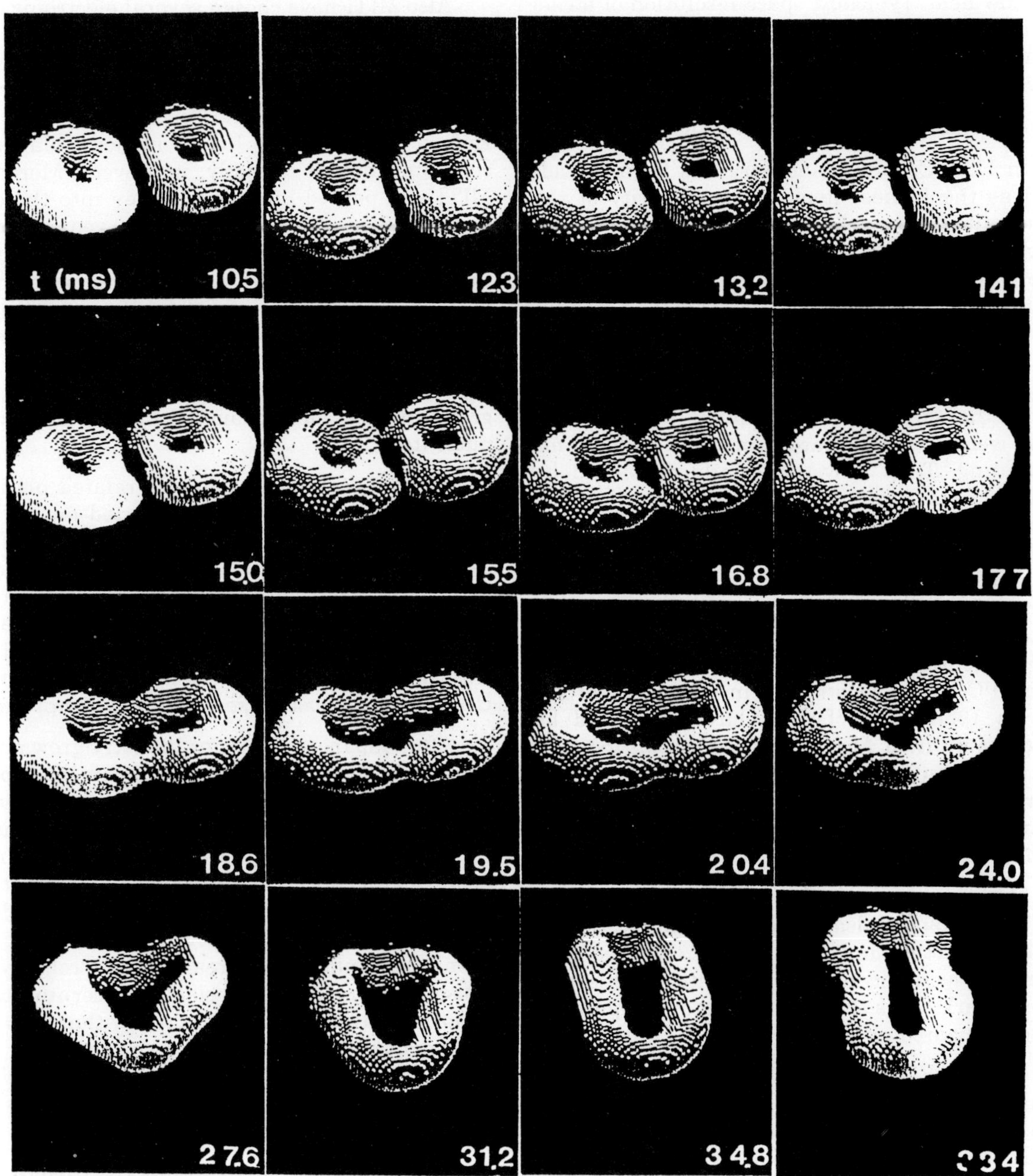

Fig.2: Cut and connect of two vortex rings

Figure 2 [7] shows a simulation of time history of two-interacting vortex rings, which produced by ejecting air through two holes side by side. The vortex rings are ejected periodically and a hot-wire anemometer measure the velocity field at a point, repeating this time-sequential data are stored, and ensemble averaging of several ten's data are derived. Then the anemometer traverses the flow field. Typically, space resolution of measuring point is a fraction of 1 mm, time resolution 5 microsecond, and the ejection period is 5 sec. Total measuring time is well over several days.

3. JAPANESE GOVERNMENT POLICY

In USA Computer industry has been considered as the highest item for the security of the nation. R/D money for it has been continuously flowing in from DOD, AEC or NASA. In Japan also governmental money comes with strong governmental control over all the aspects of this industry.

3.1 Ministry of International Trade and Industry

At the early days of computer age - 1950's, there were several research programs of electronic computing. TAC, a system jointly developed by University of Tokyo and Toshiba, was a machine using about 7,000 electron tubes, which started early 1950's and completed in 1959. Mr. Eiichi Goto, a young student of University of Tokyo invented Parametron, a logic element, in 1954, which have attracted much interest among computer scientists. Soon a computer using this element, PC-1, was constructed in University of Tokyo in 1958. Both of them have supported by funds came from Ministry of Education, Science and Culture (MOE) of Japan as fundamental research activity. Although they were not successful as industry, just like many other ideas appeared during those days (1st and 2nd generations) .

Ministry of International Trade and Industry (MITI) of Japan also support a transistor computer project taking place at its own research branch, Electronic Research Institute. This machine, Mark III, is the first transistor computer in Japan appeared in 1956.

Computer application to large business systems also started. A seat selection system of Japan National Railway was introduced in 1960 using MARS-1, a system made by Hitachi. This is the world first seat assignment system.

Computer industry begins to explosively develop in the middle of 1950's and industrial capacity of Japan at that time was extremely weak. Then MITI advised several Japanese electronic company to make joint venture and contracted out a big computer project at 0.35 billion yen for 3 years started at 1962, which successfully completed.

Also MITI have chosen several electronic companies and guide them to have technology exchange contract with chosen US companies, Hitachi with RCA, Mitsubishi with TRW, NEC with Honeywell, Oki with RR, Toshiba with GE, etc.. Helped by advanced US technology, Japanese computer output steadily improved. Moreover, MITI still poured in project money into the computer industry, 22 billion yen for 10 years starting 1970, which was a project to construct advanced image processing system. Development of main part of computers were joint program with US partners, then this kind of work has to be infrastructure of computer industry and so many human resources were spent for engineering program and less for fundamental research on computer itself.

By 1980, export of computer-related products from Japan exceeded those imported ones. However, this policy disturbs free enterprise of Japanese computer technology, and, particularly, resulted in weak software development in Japan. Under strong governmental support, all the Japanese computer companies, like Fujitsu, Toshiba, Hitachi etc., continue to be prospering contrary to USA, where almost all *seven dwarfs* disappeared.

After Japanese computer industry grew strong enough, MITI still looking for government support project, and started the *5th generation computer project*, which started 1982 for 13 years spending 5.7 billion yen. Its purpose is to construct AI (Artificial Intelligence) software system to promote AI application from general public. It was completed and released to public, however, there has not been much influence to real world. Probably because that fundamental science concerning to AI is not matured yet. Also it is noted that this project spent so much money for so long time, and many young able scientists were involved, which makes the Japanese computer community too much AI-minded.

3.2. Ministry of Education, Science and Culture

As stated above, there were several research activities in Japanese Universities during the beginning stage of computer, such as TAC. However, later on, Japanese Government MOE has had a policy that don't do research about computer in University laboratory, but buy them from Japanese farms, which have heavily supported by MITI. This is even the case of classroom that let them use educational softwares, but don't let them write programs by themselves. That is, computers are not calculator but stationary. Thus, in a math class, for example, students have no means to find a value of, say, a transcendental function but just watch and enjoy beautiful graphic display which computers produce at one key touch.

Around 1966 - 1968, MOE established computing centers to several major universities in Japan. Hitac5020E to University of Tokyo, Hitac5020F to several others including ISAS. Fujitsu230/60 followed them. These machines are series products of each computer company and users (research people) usually were not allowed to touch inside. In the universities, there are, naturally, many people working in various fields of science. They want to solve various problems by various means. Then computing center can only provide general, basic libraries for scientific computing.

3.3 Science and Technology Agency

Somehow, CFD has grown to be applied to various fields of industry and various industrial codes appeared in the market.

Qualification procedures to such industrial codes;

Verification - Code Against Theory:

Validation - Code Against Experiment: and

Certification - Code Against Experience:

are also established. These industrial, application softwares, such as FLOW3D, PHOENICS, NASTRAN, SINDA, are all originated from foreign countries. Probably because that MITI has concerned with computer system only, and because that MOE has been reluctant to have joint activity with the industry.

R and D of nuclear power for power plant has been carried out in Nuclear Research Institute of Japan which is supervised by Atomic Energy Commission. Space-related projects have been conducted by National Space Development Agency of Japan and research programs related to aerospace have been carried out at National Aerospace Laboratory. Weather Bureau of Japan, which is a branch of Ministry of Transportation, has been utilizing the highest-end computers since very beginning of computer age and still keeps that position. Recent environmental crisis call attention of policy makers in Japan, and several research organizations were established. They are the

biggest user of supercomputers for scientific calculation, and much governmental R/D money has been flowing through these organizations to computers.

Since the computers in R/D budget are one of the easiest item to convince the policymakers and general public in our government system, much money tends to flowing in without much caution about its cost effectiveness. Also there is public temptation that the computers can do anything. Japan has spent multi-billion yen every year over last 20 years for *Earthquake Prediction Program*. Now questions about the capability of scientific analysis, including use of computers, for this problem are raised.

4. COMPUTER AS RESEARCH TOOL

Once LSI technology established around mid 1960', the development of computing machines becomes purely technological problem. That is, its progress is predictable, in other words, linear extrapolation is applicable.

4.1 Era of Computing Center - 1970'

1960' through 1970' manufacturing technology of LSI was still in infant stage. It means that the computing speed of LSI (Flops) packed in a computer was roughly proportional to the cost, and the memory capacity (Byte) contained in it was also proportional to the its cost. Thus the computing performance, which is, roughly say, proportional to the product of the computing speed and the memory capacity, is proportional to square of the price of a computer. If we pay twice of money for a computer, its performance may be four times larger than a computer of half price.

Due to this reasoning, every institutions allocated all-combined big money to a single computer, and established one computing center. In 1968, the Computing Center of University of Tokyo was established, and other universities and research organizations followed. By the end of 1970' almost all Japanese Universities had computing centers provided with high-end machines, and they are regularly renewed. Since the research people who are supposed to use them were rarely involved procurement planning, loading rates of these machines were generally not busy.

Early 1970, I tried to solve large linear simultaneous equation for thermal design of our satellite project. Center library provided on 50 order solver, then I have had to write own soft for our project, which can solve125 order equations on the same hardware [9]. About the same time we made Monte-Carlo simulation, in which, in order to save the CPU time, the interaction process has to be written by machine language [10].

By the end of 1970's, almost all Japanese universities had computing center equipped with highest end computers. Japanese university people at those days could enjoy world highest accessibility to high-end computers.

4.2 Era of Supercomputers - 1980'

Around 1980, computational fluid dynamics proved itself to be useful for aircraft design, NASA started its National Aerospace Simulation Program, which was supposed to compliment wind tunnel testing, which became too expensive to maintain. At the First Computational Fluid Dynamics Symposium held in Tokyo, 1985 [11], comprehensive reports on this program were presented, which provide strong impetus to Japanese aerospace community and soon they established powerful supercomputing center at National Aerospace Laboratory, Tokyo. We like to call this time as Supercomputer Era.

Generally say, every researchers want to make their own computer according to their particular research purposes. This is particularly true, when the multi-processor system came out. Hoshino, Tsukuba University, has proposed a parallel system computer - PAX series [8].

Now we discuss why supercomputers were so useful for aircraft design ? An answer is found in [12]. He said that

CFD treats

Simple Configuration *such as streamlined body in uniform flow, and*

Simple Physics *such fluids of ideal fluid, having linear constitutive equation.*

Above all CFD in aircraft design is in nature complimentary to wind tunnel testing, that is, some extension of known solutions.

Use of supercomputer for R&D work, in contrary, needs skilled personnel. Supercomputers could buy at reasonable cost, but researchers were not, thus many rental business of supercomputers went to bankrupt during those days.

MOE has a program to promote basic scientific research in the universities in Japan – *Grant-in-Aid for Scientific Research.* In 1987 – 1989, Grant: Scientific Research on Priority Area was awarded to *Numerical Fluiddynamics.* This was a big group of University professors gathering from various fields of science and engineering, aerospace, mechanical and civil engineering and physics departments, *etc.* Also notable is that this Grand is awarded to mere tools, not to objects as usual cases. As one of the outcome of this group activity, Japan Society of Computational Fluid Dynamics (JSCFD) was inaugurated, and Computational Fluid Dynamics Journal has been started .

4.3 Era of Network Computing - 1990'

At late 1980's cold war ends, and excess expenditure of government money ceased to pour in to the big project-type institutions. On the other hand, we have seen that network computing and super parallel computers are coming out.

A Grant-in-Aid for Scientific Research was awarded to Mathematical Modelling of Turbulent Flows for 1993-1995, which was successfully over.

Japanese government still continues to pour in big money to R/D, such as nuclear energy, aerospace, environmental problems. Then various super-parallel computer projects are running in parallel. In order to make a success on a technological project, it is necessary to have thorough scientific knowledge of the object and to have good tools capable to handle the problem encountered. The Japan's government supports tend to go tools, software like the 5th Generation AI Computing System or hardware like computer itself. And fundamental research, which needs genuine human power, are likely discouraged. Today's situation in Japan may perhaps say that

Too many computing environment, too less human resources to use it.

Computers have been very powerful to solve technological problems in all aspects of fluiddynamics. Then every expert knowledge thus obtained are put all together in blackbox, and anyone can access to such expert knowledge through worldwide network.

This is world of technology, and an era of dilettante has begun.

Those who don't know aerodynamics design airplane. Those who don't know fluid dynamics design fluid machinery.

5. CONCLUDING REMARKS

1. When Japan join to the modern western science world at late 19th century, science as modern culture has been already established. In order to catch up modern world, Japan has been just busy to develop science and technology altogether. Aeronautical Research Institute, University of Tokyo was started 1918 to develop aeronautics in this country.

2. New frontier of modern science, compressible fluiddynamics, has emerged late 1940's. Fluiddynamicists in Japan, including ARI, had some contributions theoretically as well as experimentally to this new field of science.

3. Computers has emerged 1950's, and its fundamental physics has established late 1960's as 3.5th generation machine. Fundamentals of CFD also has established during those days, but Japanese fluiddynamics has little contribution probably because we are too busy to catch up technological progress of Computers.

4. Throughout 1970's (Computing Center Era) and 1980's (Supercomputer Era) Japanese computer industry has continuously progressing. Japanese fluiddynamicists, including of ISAS, have enjoyed world highest accessibility the computing environment by 1980's.

5. MITI of Japan has provided strong support and control to the Japanese computer industry, but less interests on the industrial softwares.

6. MOE of Japan has installed and regularly updated high-end computers for general purpose research tools at every university campuses, including ISAS, and distributed computers as stationary to classrooms, which is not supposed to be used as math-calculator.

7. Japanese policymaker, and perhaps general public, have a day dream that the modern computers can solve any problem, and put in large budget to the problems whose solution does not likely exist.

REFERENCES

[1] F.Tamaki; Measurement of the pressure distribution of the aerofoil in the sub-sonic flow by Mach-Zehnder interferometer, Report of the Aeronautical Research Institute, Tokyo Imperial University, No.283 Feb. 1944 (Vol.XX,16)

[2] K.Oshima; Shock tube study of the flow past circular-arc airfoils using a Mach-Zehnder interferometer, J. Phys. Soc. Jpn, vol.10 (1955), pp.571-577

[3] Chul-Soo Kim, Experimental studies of supersonic flow past a circular cylinder, J. Phys. Soc. Jpn, vol.11 (1956), pp.439-445

[4] K.Oshima; Quasisimilar solution of blast waves, Aero. Res. Inst. Univ. Tokyo, Rep.386, 1964

[5] C.H.Lewis; Plane, cylindrical, and spherical blast waves bsed upon Oshima's quai-similarity model, AEDC-TN-61-157 (1961)

[6] P.D.Lax, Computational fluid dynamics at the Courant Institute, CFD Review 1995 ed. M.Hafez & K.Oshima, John Wiley & Sons Ltd. pp.1-5

[7] Y.Oshima, T.Noguchi and K.Oshima; Numerical study of interaction of two vortex rings, Fluid Dyn. Research vol.1 pp.215-227

[8] T.Hoshino; Parallel compuers and parallel computing in scientific simulations, 11th ICNMFD, Lecture Notes in Physics 323, Springer 1990, pp.31-39

[9] K.Oshima and Y.Kuriyama; A thermal network correction method and its application to some Japanese stellites, ESA SP-139 (1978)

[10] M.Murakami and K.Oshima; Kinetic approach to the evaporation and condensation problem, ISAS Report no.518 (1974)

[11] K.Oshima; Numerical methods in fluid mechanics I and II, JSCFD (1985)

[12] C.A.J.Fletcher; "Computational Techniques For Fluid Dynamics I & II", Springer-Verlag 1989.

PROGRESS IN CFD FOR TURBOMACHINE CASCADE FLOW PROBLEMS EMPHASIZING INVESTIGATIONS IN JAPAN

Hisaaki DAIGUJI[†]

1. INTRODUCTION

1.1 CFD before Computer

Although the computational fluid dynamics(CFD) has progressed as the development of large-scale computer, the investigations of CFD began in the time that the computer did not popularize yet. The first CFD investigation in Japan at the dawn was probably a simulation of twin vortices behind a circular cylinder by Kawaguti(1953). In the simulation the vorticity transport equation and the Poisson equation of stream function were manually solved using the relaxation method by Southwell(1946). Figure 1 shows the calculated flow pattern. In regard to the cascade flow problem, a numerical study by Seya(1957) is the oldest one where the two-dimensional plane cascade flow is obtained by solving the Laplace equation of stream function by the relaxation method improved by Fox(1947). At that time the analytical methods using conformal mapping were conventionally used, but it was difficult to deal with the cascades composed of largely cambered blades or with high solidity. On the contrary, the numerical methods have a merit that they are applicable to every kind cascade geometries. The writer began in 1960 cascade flow simulations, and proposed proper block relaxations for the upstream and downstream regions, treatments of the blade boundary condition and a calculation of flow past a edge(Daiguji 1967). The fundamental equation to be able to calculate by the Sowthwell's relaxation method is

$$\frac{\partial^2 \psi}{\partial x^2} + \frac{\partial^2 \psi}{\partial y^2} = (\ln b\rho)_x \psi_x + (\ln \rho)_y \psi_y + 2\omega r^2 b\rho \sin \delta$$

where ψ is the stream function defined as $\partial\psi/\partial\theta = rb\rho w_m$, $\partial\psi/\partial m = -b\rho w_u$, m, θ are the coordinates on a revolutional flow surface, $dx = dm/r$, $dy = d\theta$, and b, ρ, ω, δ and w_m, w_u are the breadth of impeller passage, fluid density, angular velocity of impeller, inclined angle of revolutional surface and relative velocities. The equation was solved on the graded square net using the block relaxation. Figure 2 shows the pressure profiles on the blade surfaces of a radial-flow impeller experimented by Morelli(1951), calculated by the relaxation method.

---

† Emeritus professor of Tohoku University

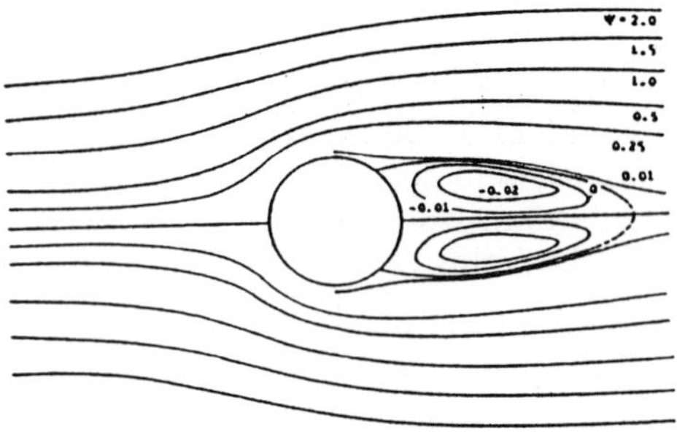

Fig.1: Flow pattern aroud a circular cylinder, $Re = 40$ [Kawaguti(1953)]

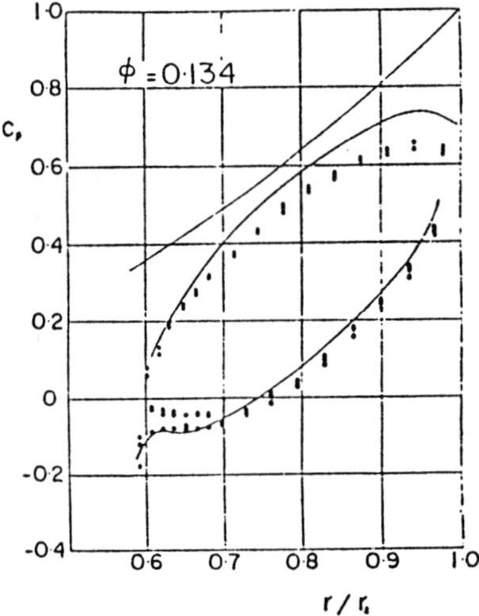

Fig.2: Calculated pressure profiles on blade surfaces of centrifugal impeller o • experimental data [Daiguji(1967)]

1.2 Quasi-Three-Dimensional Analyses

Since the time computer appeared until the time of so-called supercomputer, there was a severe limitation of computer time, and the flow problems to be solved, number of computational points and the numerical methods were also restricted according to the available computer. The well-known quasi-three-dimensional analysis by Wu(1952) is that the 3D flow is reduced to a number of 2D flows with less computational efforts.

That is, the flow through an impeller of turbomachine is obtained by mutual iteration of three cascade flows on blade-to-blade flow surfaces(S1-surface) and three through flows on hub-to-casing flow surfaces(S2-surface) taken nearly parallel to the blades. However, this analysis has some defects that since the S1-surfaces do not always close in the downstream region the periodic boundary conditions do not exactly hold, and that for the flow with strong channel vortex the flow surfaces extremely twist. Therefore, the quasi-3D analysis combined some cascade flows on revolutional flow surfaces with a meridional through flow has been used. The development on the quasi-3D analyses in the 1970s is well described in the reviews of Adler(1980a, b). The cascade flows were calculated by the methods of lines(MOL, streamline curvature method, velocity gradient method)(Katsanis 1964, Wilkinson 1969, Bindon 1973), the shock capturing finite-difference methods (Gopalakrishnan-Bozzola 1972, Kurzrock-Novick 1975, Veuillot 1977) and the shock capturing finite area(volume) methods(McDonald 1971, Denton 1974). Figure 3 shows the streamlines and Mach number contours of the flows of a diagonal-flow compressor cascade calculated by the MOL and the FEM. The MOL is less computer time and applicable to the transonic flow without shock wave. It seems that these results of inviscid subsonic flows are not so different by the numerical methods.

The shock capturing relaxation methods were widely used at the time before supercomputer. In the original method by Murman-Cole(1971) the transonic small perturbation equation is solved using the central-difference scheme for the subsonic region and the upwind-difference scheme for the supersonic region. The switching from central-difference to upwind-difference can also be realized by an addition of artificial viscosity. The method was applied to the full potential equation and extended to the FVM and the FEM. Dodge(1976) pointed out that too large supersonic region in the early stage methods is a result of the fact that the discretized equation of supersonic region contains the data outside of the domain of dependence, and proposed an improved method. The relaxation methods solving the full potential equation were applied to the cascade flow problems having weak shock waves(Ives-Liutermoza 1977, Caspar *et al.* 1980, Ecer-Akay 1981). The CPU-time is relatively small.

In regards to the through flow on the S2-surface, an averaged flow with respect to θ should be taken than the flow on a special S2-surface, e.g. camber surface of blade(Katsanis-McNally 1977), and mass flow weighted average rather than arithmetic mean(Hirsch-Warzee 1976). The flow is usually inviscid, and terms due to inclined blade and sometimes to total pressure loss and friction are taken into account. The finite-difference method(Bosman-El-Shaarawi 1976, Katsanis-McNally 1977) and the FEM(Hirsch-Warzee 1976, 79, Oates *et al.* 1976) are used for the subsonic flow, and the MOL(Senoo-Nakase

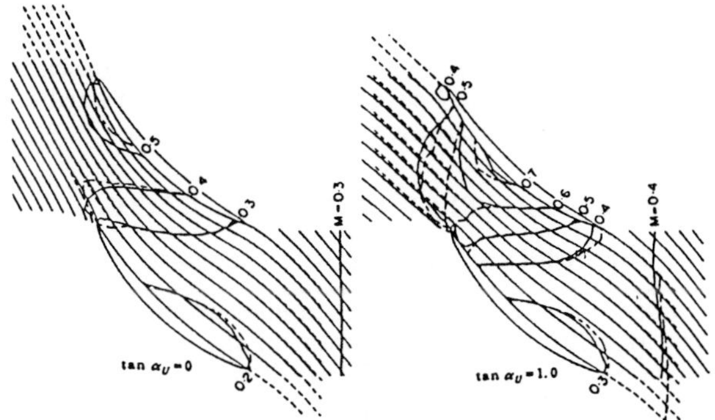

Fig.3: Streamlines and Mach number contours in diagonal-flow compressor cascade ——— FEM, −−−− method of streamlines [Shirahata-Daiguji (1981)]

1972, Novak-Hearsey 1976, Katsanis-McNally 1977) and the time-marching method(Spurr 1976) for the transonic flow.

1.3 Supercomputer and CFD

As utilizing the supercomputer the trends of studies on the cascade flow problems largely changed to the inviscid Euler analyses, and soon to the turbulent flow analyses by the Reynolds-averaged Navier-Stokes(NS) equations. The change from the Euler solver to the NS solver raised the practicality of CFD in the cascade flow problems, and at the same time released from troublesome problems concerning the solid wall boundary and the trailing vortex sheet in the Euler solver. For a reasonable computation of the turbulent flows, especially using the algebraic or the low Reynolds number turbulence model, great many computational points must be taken. However, in the early stage computations, since to take such many computational points was generally impossible it was difficult to evaluate appropriate values of the drag force of airfoil and the fluid-dynamic loss of cascade flow.

The recent development of the large scale computer has made possible to simulate the stall propagation in the compressor stage and even to construct the turbulent flow database by the direct numerical simulation(DNS) and the large eddy simulation(LES). On the other hand, the universal turbulence model does not complete yet, though the applicability of the turbulence models has been gradually extended. The flows having large separation zone and accompanying wide wake region should at present be calculated as unsteady flows by DNS or LES. But, for the transition to turbulent flow and the development of turbulent boundary layer which are difficult to simulate directly, completion of the more accurate models is expected. The evolution of large scale computer once saturated about twenty years ago, but after that development of parallel computer has continuously accelerated as shown in Fig.4. In the present review, the fundamental equations, the numerical methods and their applications of the 3D cascade flows, the turbulent cascade flows and the unsteady cascade flows shall be described emphasizing the Japanese investigations.

2. 3D POTENTIAL CASCADE FLOWS

2.1 Formulation of Subsonic Cascade Flow

The absolute flow through an impeller which originates in the rest fluid is usually irrotational. Therefore, the velocity potential ϕ can be defined as

$$\boldsymbol{v} = \boldsymbol{w} + \boldsymbol{\omega} \times \boldsymbol{r} = -\nabla\phi \tag{1}$$

where $\boldsymbol{v}$, $\boldsymbol{w}$, $\boldsymbol{\omega}$ are the absolute velocity, the relative velocity and the angular velocity of impeller, respectively. The governing equation for the incompressible and subsonic flows is(Diguji 1983b)

$$\nabla^2\phi = \begin{cases} 0 & \text{(incompressible)} \\ \boldsymbol{w}\cdot\nabla h/c^2 \equiv B & \text{(subsonic)} \end{cases} \tag{2}$$

where c is the velocity of sound and h is the specific enthalpy. For such homentropic flow, the rothalpy defined as $I = h + v^2/2 - \omega r\, w_u$ is constant throughout the flow field.

In the far-upstream and far-downstream regions the flows are uniform in circumferential direction, and the following relations hold.

$$v_n = 0$$

$$\frac{1}{\rho}\frac{\partial p}{\partial n} = \frac{1}{r}v_u^2\cos\delta$$

$$\frac{\partial w_m}{\partial n} = \frac{1}{w_m}\left\{\omega\frac{\partial}{\partial n}\frac{\Gamma}{2\pi} - \frac{1}{2\,r^2}\frac{\partial}{\partial n}\left(\frac{\Gamma}{2\pi}\right)^2\right\} \tag{3}$$

where subscripts m, n, u mean the meridional, quasi-normal(to hub and casing) and circumferential components, respectively. δ is the inclined angle of meridian velocity to the rotational axis, and the second equation of Eq.(3) is the generalized condition of radial equilibrium for the mixed-flow machine. $\Gamma = 2\pi r\, v_u$ is the circulation, and from the third equation of Eq.(3) if the flow is free vortex ($\Gamma(n) = $ const.), then the flow has uniform meridian velocity ($w_m(n) = $ const.). The third equation is derived from the condition of uniform rothalpy.

The boundary conditions are imposed as

$$\phi(x,\theta,r) = \phi(x,\theta_0,r) - \Gamma(\theta - \theta_0)/2\pi$$
$$\text{(upstream)}$$

$$\frac{\partial\phi}{\partial n} = f \equiv \begin{cases} -w_m & \text{(downstream)} \\ \pm\boldsymbol{n}\cdot\boldsymbol{\omega}\times\boldsymbol{r} & \text{(blades)} \\ 0 & \text{(hub \& casing)} \end{cases}$$

$$\phi(x, \theta_0 + \Delta\theta, r) = \phi(x, \theta_0, r) - \Gamma/N$$

$$\frac{\partial\phi(x, \theta_0 + \Delta\theta, r)}{\partial n} = -\frac{\partial\phi(x,\theta_0,r)}{\partial n} \quad \text{(periodic)} \tag{4}$$

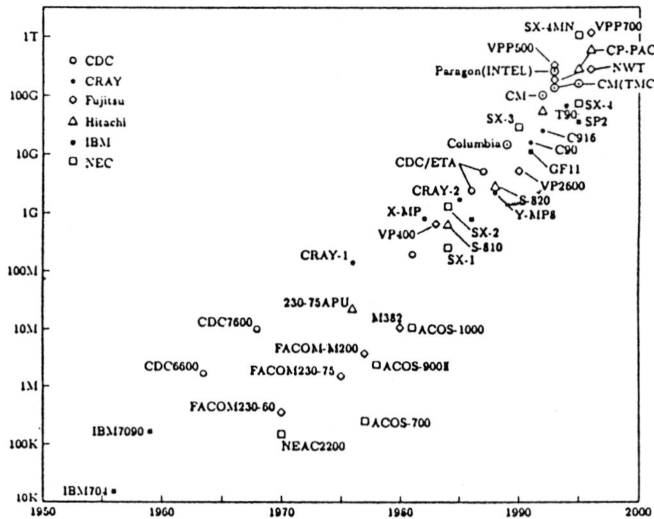

Fig.4: Evolution of large scale computers (Peak or effective value)

where $\partial/\partial n = \boldsymbol{n}\cdot\nabla$, $\boldsymbol{n}$ is the outward unit normal vector on the boundaries and N is the number of blades. The boundary value problem of Eq.(2) has a unique solution under these boundary conditions.

2.2 Finite Element Methods for Subsonic Cascade Flow

Eqs.(2) and (4) can be relatively easily solved by the variational FEM and the Galerkin method. The functional is

$$J(\phi) = \int_\Omega\left\{\frac{1}{2}(\nabla\phi)^2 + B\phi\right\}d\boldsymbol{x} - \int_{\partial\Omega_2}f\phi\,dS$$
$$= \min. \tag{5}$$

where $\partial\Omega_2$ is the boundary imposed the Neumann condition of the computational domain Ω. In the calculation the hexahedral trilinear elements, 6-node triangular cylinder elements or hexahedral Lagrangian quadratic elements are used. By the FEM the incompressible cascade flows of water turbines were calculated by Nagafuji(1976, 1980), and the compressible cascade flows of axial-flow turbine by Laskaris(1978).

The far-downstream flow angle in the inviscid cascade flow must be determined to satisfy the Kutta condition that the fluid smoothly flows out from the trailing edge of the blade. For the wing flow analyses a trailing vortex sheet due to shedding vortices from the trailing edge is usually introduced behind the wing. Similarly, in the cascade flow analyses trailing vortex sheets should be considered for ordinary loadings—the constant reaction, semi-forced vortex and $\beta_2 = $ const. designs, except for the free vortex design.

A computational method considering the trailing vortex sheets was proposed by the writer(1982). The calculation in the downstream region needs a mesh discretized to fit to the trailing vortex sheet, and the values of Γ_D and w_{mD} at the downstream boundary. They are determined in the computational process by the predictor-corrector-method. The predictors are given by(Daiguji 1983b)

$$\frac{r\,d\theta}{dm} = \frac{1}{w_m}\left(\frac{\Gamma_D}{2\pi r} - \omega r\right)$$
$$\Gamma_D^{(0)} = 2\pi r(\omega r + w_m \tan\beta_T) \tag{6}$$
$$w_m^{(0)} = G/2\pi r\,b\,\rho_U$$

where G is the mass flow rate, b the breadth of passage and β_T blade exit angle. $B^{(0)} = 0$. The velocities on both sides of the trailing vortex sheet have the same magnitude but generally different directions. A vortex filament in the trailing vortex sheet is an envelope of the mean velocities $\overline{w}$, and the value of Γ_D is constant along a vortex filament. The corrector of the mesh is rediscretized to fit the j-surfaces to the trailing vortex sheet and the k-surfaces to the vortex filaments. The corrector of Γ_D is

$$\Gamma_D^{(n+1)} = \Gamma_D^{(n)} + \alpha\,2\pi r(\overline{w}_m \tan\beta_T - \overline{w}_u) \tag{7}$$

where $\alpha \approx 0.4$ is the damping factor. The corrector of w_m is obtained from the third equation of Eq.(3) considering $G = \iint \rho w \cdot n\,rd\theta dn$. The converged solution is usually obtained within five iterations, but for the high Mach number subsonic flow is in need of more iterations.

The solution ϕ of the Laplace equation by the FEM using 8-node cubic elements has better accuracy than the solution by the 7-point central-difference scheme. The accuracy of ϕ for the distorted hexahedral element mesh depends on the smoothness and the orthogonality, but does not suddenly deteriorate by these factors. On the other hand, the accuracy of the derivatives of ϕ, i.e. the velocities would largely depends on the distortion of mesh without the greatest possible care. The binomial satisfying a nodal value $\phi(x_0)$ can be expressed as(Daiguji 1983b)

$$\phi(x) = \phi_0 + \sum_{\ell=1}^{9} c_\ell \psi_\ell(\bar{x}) \tag{8}$$

where $\psi_i = \bar{x}_i$, $(i = 1, 2, 3)$, $\psi_{i+3} = \bar{x}_i^2$, $(i = 1, 2, 3)$, $\psi_7 = 2\bar{x}_2\bar{x}_3$, $\psi_8 = 2\bar{x}_3\bar{x}_1$, $\psi_9 = 2\bar{x}_1\bar{x}_2$ and $\bar{x} = x - x_0$. The 9 unknown coefficients c_ℓ are determined by a weighted least square method such that the conditions $\phi_k - \phi_0 = \sum_{\ell=1}^{9} c_\ell \psi_{\ell k}$ at the 14 nodal points shown in Fig.5 are satisfied as well as possible. That is, the values of c_ℓ are determined from the simultaneous linear equations

$$\sum_{\ell=1}^{9} a_{n\ell} c_\ell = f_n \qquad (n = 1, 2, \cdots, 9) \tag{9}$$

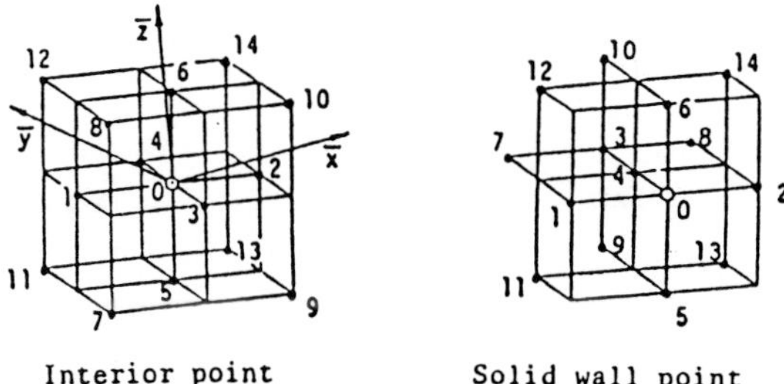

Fig.5: 14 reference points of weighted least square approximate binomial

where

$$a_{n\ell} = \sum_{k=1}^{14} W_k^2 \psi_{nk}\psi_{\ell k},$$
$$f_n = \sum_{k=1}^{14} W_k^2 \psi_{nk}(\phi_k - \phi_0) \tag{10}$$

W_k is the weight function introduced in order to regard the nodal point conditions near the point x_0 as important, and put that $W_k^2 = \{(x - x_0)^2\}^{-3}$. It is recommended to select the 14 nodal poits such that sets of three points lying on a line homogeneously distribute around the nodal point x_0. Then the velocities are

$$v_i = -\partial\phi/\partial x = -c_i \qquad (i = 1, 2, 3) \tag{11}$$

2.3 Numerical Examples

Figure 6 shows the calculated pressure profiles on the blade surfaces for the radial compressor impeller with the experimental data by Mizuki *et al.*(1975). The flow is incompressible and the impeller has almost constant sectional area passage. It is found from this figure that the calculated results of the three-dimensional analysis by the so-far mentioned FEM(Daiguji 1983a) and the quasi-3D analysis by Krimerman-Adler(1978) well coincide with the experimental data, but that the calculated results of the quasi-3D analysis using one through flow(Katsanis 1964), which has been practically used in the design of turbomachines, do not always coincide with the experimental data, especially near leading edge of near front shroud($k = 6$). In the quasi-3D analysis by Krimerman-Adler the flows on the seven S1-surfaces and on the seven S2-surfaces are iteratively calculated by the FEM.

The hexahedral trilinear elements can be well applied to the radial compressor impeller, but for the centrifugal pump impeller having small blade angle(for example, exit blade angle $\beta_2 = 22.5°$) to make use of 6-node triangular cylinder elements is more convenient. For such an impeller, the skewness of the meshes in the

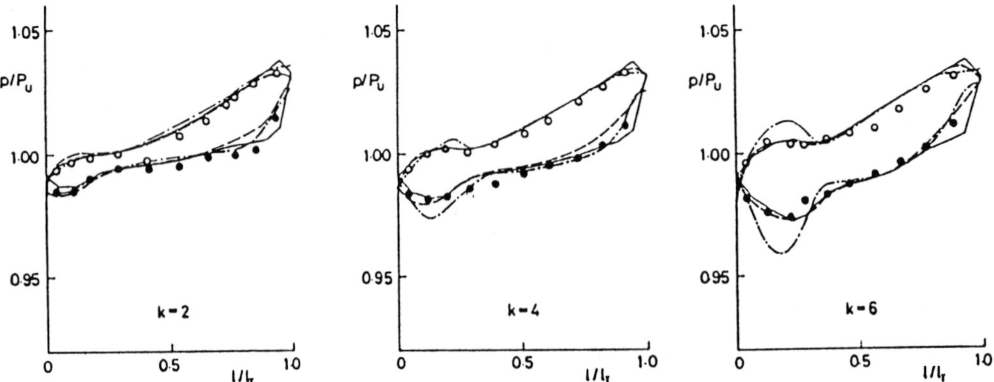

Fig.6: Pressure profiles on blade surfaces of radial compressor
——— 3D potential(Daiguji 1983),
– – – – Quasi-3D(Krimerman-Adler 1978),
– · – · – Quasi-3D using one through flow(Katsanis 1964),
○ ● Experiment by Mizuki *et al.*(1974, 75)

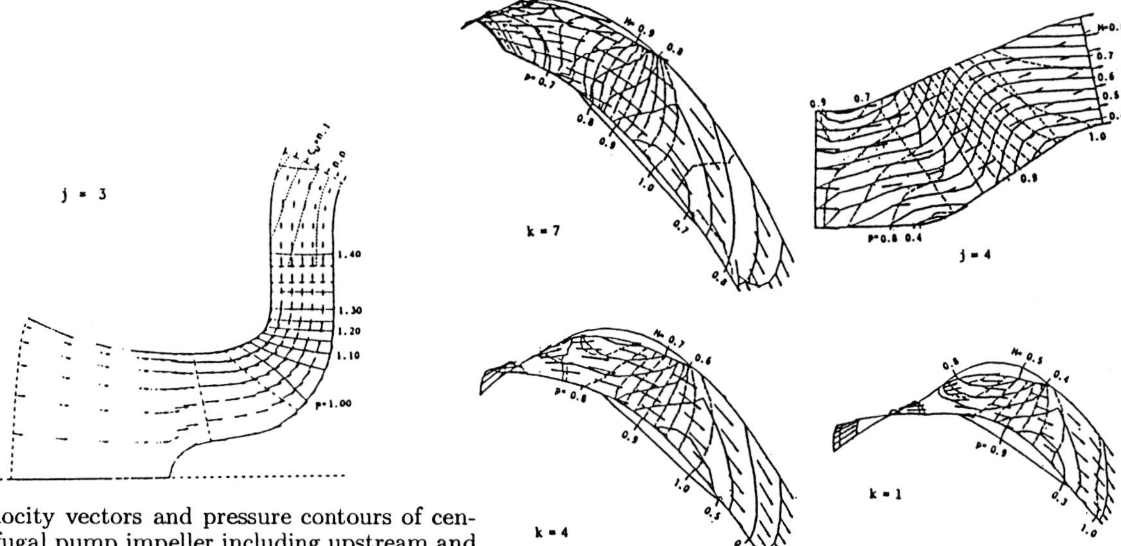

Fig.7: Velocity vectors and pressure contours of centrifugal pump impeller including upstream and downstream regions – – – – boundaries of axisymmetric regions and blades [Daiguji-Takano (1986)]

Fig.8: Velocity vectors, pressure contours and Mach number contours in transonic diagonal-flow compressor rotor [Daiguji(1983)]

upstream and downstream regions is rapidly magnified as the distances from the leading and trailing edges. To avoid instability due to this skewness, a 3D flow analysis connected with 2D axisymmetric flows is proposed(Daiguji-Takano 1986). The governing equation of the 2D axisymmetric flow is

$$\frac{\partial}{\partial x}\left(\frac{1}{r}\frac{\partial \psi}{\partial x}\right) + \frac{\partial}{\partial r}\left(\frac{1}{r}\frac{\partial \psi}{\partial r}\right) = \zeta_u \tag{12}$$

where ψ is the stream function and ζ_u is the vorticity which is zero in the upstream region and

$$\zeta_u = -\omega r \frac{d}{d\psi}\frac{\Gamma}{2\pi} + \frac{1}{2r}\frac{d}{d\psi}\left(\frac{\Gamma}{2\pi}\right)^2 \tag{13}$$

in the downstream region. Figure 7 shows the velocity vectors and pressure contours of a centrifugal pump impeller considering axisymmetric upstream and downstream flow fields. Figure 8 shows the velocity vectors, static pressure contours and Mach number contours of a subsonic diagonal-flow compressor rotor.

3. FUNDAMENTAL EQUATIONS OF COMPRESSIBLE FLOW

3.1 Calculations of Turbulent Flow

The computational methods of turbulent flows can be divided into three categories—the direct numerical simulation(DNS), the large eddy simulation(LES) and the methods using the time(ensemble)-averaged Navier-Stokes equations(Reynolds equations). The DNS is applicable to all kinds of the turbulent flows, but in order to obtain reliable results by the DNS it is necessary to use very fine grid near the Kolmogorov scale. Therefore, the appropriate results have been obtained only for simple turbulent flows at relatively low Reynolds numbers. Some cascade flows have been obtained by the DNS, but they do not give sufficiently reasonable results and are no more than challenges to the Navier-Stokes solutions of the cascade flows.

In the calculation of the LES the grid finer than the Taylor microscale must be used and the large scale eddies are directly simulated by solving the spatial-average Navier-Stokes equations. The effects of the remaining subgrid scale eddies are considered statistically in the equations. As the subgrid scale turbulence model, the Smagorinsky(1963) model has been used widely, but for not so fine grid the recently developed dynamic models(Horiuti 1996) mixed the Smagorinsky model and some scale similarity model according to the fineness of the computational grid are recommended. The computational grid, efforts and applicability of LES lie between the DNS and the methods using turbulence model mentioned below. Computation of cascade flow by the LES is scarcely performed.

In the calculations of the turbulent cascade flows the Reynolds equations have been solved with few exceptions. The turbulence models are classified as follows:

- eddy viscosity model
  - 0-eqn. model (algebraic)
    - Cebeci-Smith(1974)
    - Baldwin-Lomax(1978)
    - Johnson-King(1985)
  - 2-eqn. model
    - k-ε model
      - Launder-Spalding(1974)
    - low Re k-ε model
      - Jones-Launder(1972)
      - Chien(1982)
      - Launder-Spalding(1983)
      - Myon-Kasagi(1988)
      - Abe *et al.*(1994)
    - k-ω model—Wilcox(1988)
- algebraic stress(ASM) model—Rodi(1976)
- Reynolds stress transport equation (SM) model
  - Hanjalić-Launder(1972)
  - Launder-Reece-Rodi(1975)
  - Launder-Shima(1989)
  - Hanjalić *et al.*(1997)

3.2 Eddy Viscosity Models

In this model the eddy viscosity approximation

$$-\rho\left(\overline{u_i'u_j'} - \frac{2}{3}k\,\delta_{ij}\right) = \mu_t\left(\frac{\partial u_i}{\partial x_j} + \frac{\partial u_j}{\partial x_i} - \frac{2}{3}\frac{\partial u_k}{\partial x_k}\delta_{ij}\right)$$

$$(14)$$

that is, an assumption that anisotropic components of Reynolds stress tensor are proportional to the corresponding anisotropic components of rate of strain tensor is imposed. Under this assumption the Reynolds equations can be treated in the same way as the laminar flow, and therefore most of the computations of cascade flow problems have been implemented employing the eddy viscosity models.

The simplest turbulence model is the zero equation model in which the eddy viscosity ν_t is given by algebraic forms. The Cebeci-Smith(CS) model is a skillfully arranged algebraic model. In this model, the boundary layer is divided into two layers, ν_t in the inner layer is expressed based on the Prandtl mixing length theory and ν_t in the outer layer in terms of the displacement thickness δ^*. The CS model is applicable to the boundary layer flows where the local equilibrium of turbulence energy is satisfied, and contains the effects of low Reynolds number, transition, pressure gradient and solid wall curvatures. But it requires a large number of grid points, because the neighboring grid points of the solid wall must be taken $y^+ = 1 \sim 4$ inside viscous sublayer.

The Baldwin-Lomax(BL) model which succeeds to the idea of the CS two layer model is widely used for the compressible flows. In this model, ν_t in the outer layer is given using the vortex function $F(y) = y|\omega|(1 - \exp(-y^+/26))$ instead of δ^* to be able to compute more easily, and the transition point is determined from the condition $\mu_{t\,\mathrm{max}} \geq 14\mu$. The BL model is simple and has been successfully used for the subsonic wing flows, and also applied to many cascade flow problems. However, this does not always mean this model is superior compared with other models. For the separated boundary layers with two peaks of $F(y)$, the Degani-Schiff(1986) model modified the BL model is recommended.

The algebraic models so far mentioned only depend on the data on some transversal(normal) to the wall, and are independent of the upstream region. The upstream effects can be considered by solving the transport equations of turbulence quantities. In the Johnson-King(JK) model, an ordinary differential equation of the variable $g = (-\overline{u_1'u_2'}\,_{\mathrm{max}})^{-1/2}$, which is the maximum Reynolds shear stress across the boundary layer, is solved and the result is considered in the BL model. This model gives better results than the CS and BL models for the wing flow with small separation bubble or embedded shock wave.

The k-ε model is the most popular transport equation model. In this model the eddy viscosity is expressed as $\nu_t = 0.09 f_\mu k^2/\varepsilon$, two transport equations of the turbulent kinetic energy k and its rate of dissipation ε are solved, and for the wall turbulent flow the law of the wall is employed. The turbulence is fundamentally characterized by k and the length scale ℓ. However, in this model the equation of $\varepsilon(\sim k^{3/2}/\ell)$, which has clearer meaning than the ℓ and is constructed after the equation of k, is adopted. The many validations of the k-ε model were performed by Spalding, Launder, et $al.$ for the diverse flows including combustion, and the empirical constants by Launder-Spalding(1974) are generally used.

For the separated boundary layer it is unsuitable to use the law of the wall, and the low Reynolds number models were developed. The Chien model containing the distance from the wall explicitly can be applied to the curvilinear coordinate grid, the Myon-Kasagi model shows the correct asymptotic behaviors to the wall, the Grasso-Falconi(1993) model contains the compressibility effects, and the Abe-Kondoh-Nagano(1994) model is improved to be applicable to the separated and reattached flow. When using such models, very small values of k and ε must be given at the upstream boundary, and then these values suddenly amplify near the physical transition point. Recently, a model paid a special attention to the transition was proposed. The neighboring grid points should be taken close to the wall like the algebraic models.

In the two-equation turbulence models there are the k-ω model by Kolmogorov(1942) and Saffman-Wilcox(1974), the k-ω^2 model by Wilcox-Rubesin (1980), in addition to the k-ε model. Where $\omega(\sim k^{1/2}/\ell \sim \varepsilon/k)$ is called as the psudo-vorticity or specific dissipation rate. The k-ω model by Wilcox(1988a) can be well applied to the boundary layer with adverse pressure gradient.

3.3 Turbulence Models without Eddy Viscosity Approximation

Under the assumption (14) of eddy viscosity model, when a flow begins to subject to a uniform mean shear strain, it needs too long time for homogeneous turbulence, and when the mean shear strain is removed, the flow instantaneously returns to isotropy. The multiscale model by Wilcox(1988b) overcomes these defects of the eddy viscosity model, keeping the excellent applicability of Wilcox(1988a) model to the boundary layer with adverse pressure gradient.

The algebraic stress model(algebraic second-moment closure, ASM) placed between the eddy viscosity model and the Reynolds stress transport equation model was proposed by Launder(1971) and Rodi(1972) separately, and the Rodi's ASM has been widely used. In the computation, first the values of k and ε are determined from their transport equations, then the Reynolds stresses are determined from algebraic equations deduced from a comparison of the transport equations of the Reynolds stress and k. The model can also be used for the diverging annular duct flows.

The Reynolds stress transport equation(second-moment closure, SM) model enable us to treat more complicated strain and stress fields on the turbulence structure. The equations are derived by taking turbulent velocity-weighted moments of the NS equations. Launder-Reece-Rodi(1975) model is often referred in the investigations of turbulent flows as a standard SM model, Launder-Shima model is that emphasizes on the low Reynolds number effects, and recently developed Hanjalić-Jakirlić-Hadžić model has extended limits. The coefficients of the LRR model are determined from many experimental data of simple standard turbulent flows — nearly homogeneous shear flow, near-wall turbulence, decay of grid turbulence, plane strain distortion, axisymmetric contraction, plane mixing layer, plane jet, wake behind plate, channel flow, asymmetric channel flow and flat-plate boundary layer. Therefore, for the flows similar to such flows fairly good simulation results will be obtained. The HJH model is further extended to be able to apply to the boundary layer in adverse pressure gradient, with transverse shear, by-pass transition, oscillating and pulsating flows, developing Couette flow and backward-facing step flow.

The turbulence models have been proposed from the simple mixing length theory to the complicated SM closures, and recently the limits of the models are gradually extending. However, it is not easy to develop a turbulence model that is applicable to the flows having large coherent structures — diverging duct flow, having large separation zone and wake behind a blunt body, and some compressible effects. For example, for the subcritical flow past a circular cylinder a very large value of the drag force is evaluated from the 2D computation, and the appropriate values can be barely obtained by the 3D computation of the longitudinal vortices in the wake. For the transonic wing flow with an embedded shock and a plateau in the pressure profile behind the shock, the calculated pressure profile by the low Reynolds number k-ε model is better than by the BL model, but largely different from the experiment. The reason is that the calculated separation bubble behind the shock is too small. If the value of turbulence diffusion ν_t is reduced to 50 %, then the calculated results will close to the experiment, but it will be difficult to settle using the SM model or by the LES(Takakura 1989).

3.4 Governing Equations for Cascade Flow Problems

The governing equations for the absolute flow through a stator can be written as

$$\rho_t + \nabla \cdot \rho v = 0$$
$$(\rho v)_t + \nabla \cdot \rho v v = -\nabla p + \nabla \cdot \boldsymbol{\Pi} \qquad (15)$$
$$e_t + \nabla \cdot (e + p)v = -\nabla \cdot q + \nabla \cdot (\boldsymbol{\Pi} \cdot v)$$

where ρ is the density, v the absolute velocity, $e = \rho(\epsilon + v^2/2)$ the stagnation internal energy par unit volume and p the static pressure. $\boldsymbol{\Pi}$ and $\boldsymbol{q} = -k\nabla T$ are the viscous stress tensor and the heat flux, respectively, and these components are

$$
\begin{aligned}
\tau_{ij} &= \mu\left[\left(\frac{\partial v_i}{\partial x_j} + \frac{\partial v_j}{\partial x_i}\right) - \frac{2}{3}\delta_{ij}\frac{\partial v_k}{\partial x_k}\right] - \overline{\rho v_i' v_j'} \\
q_i &= \left(\frac{\mu}{Pr} + \frac{\mu_t}{Pr_t}\right)\frac{1}{\gamma-1}\frac{\partial c^2}{\partial x_i}
\end{aligned}
\tag{16}
$$

which contain the terms of the Reynolds stress $-\overline{\rho v_i' v_j'}$ and the eddy viscosity μ_t for the turbulent flow. $e + p = \rho H$, $H = h + v^2/2$ is the stagnation enthalpy, which is constant along an absolute streamline for the steady isentropic flow through a stator.

The governing equations for the relative flow past a rotor are

$$
\begin{aligned}
\rho_t + \nabla \cdot \rho \boldsymbol{w} &= 0 \\
(\rho \boldsymbol{w})_t + \nabla \cdot \rho \boldsymbol{w}\boldsymbol{w} &= -\nabla p + \nabla \cdot \boldsymbol{\Pi} + \boldsymbol{f} \\
e_t + \nabla \cdot (e+p)\boldsymbol{w} &= -\nabla \cdot \boldsymbol{q} + \nabla \cdot (\boldsymbol{\Pi} \cdot \boldsymbol{w})
\end{aligned}
\tag{17}
$$

where $\boldsymbol{w} = \boldsymbol{v} - \boldsymbol{\omega} \times \boldsymbol{r}$ is the relative velocity, $\boldsymbol{f} = \rho(\omega^2 \boldsymbol{r} - 2\boldsymbol{\omega} \times \boldsymbol{w})$ is the centrifugal and coriolis forces and in this case

$$
\begin{aligned}
e &= \rho\left(\epsilon + \frac{\boldsymbol{v}^2}{2} - \omega r v_u\right) = \frac{p}{\gamma-1} + \frac{\rho}{2}\{w^2 - (\omega r)^2\} \\
\tau_{ij} &= \mu\left[\left(\frac{\partial w_i}{\partial x_j} + \frac{\partial w_j}{\partial x_i}\right) - \frac{2}{3}\delta_{ij}\frac{\partial w_k}{\partial x_k}\right] - \overline{\rho w_i' w_j'}
\end{aligned}
\tag{18}
$$

$e + p = \rho I$, $I = H - \omega r v_u = h + w^2/2 - (\omega r)^2/2$ is called the rothalpy, which is constant along a relative streamline for the steady isentropic flow through a rotor.

3.5 Formulations for Finite-Difference Method and Finite Volume Method

The governing equations of the absolute flow and the relative flow have a similar form and can be expressed in the vector form as

$$
U_t + \frac{\partial F_i}{\partial x_i} + D + g = 0
\tag{19}
$$

where

$$
U = \begin{bmatrix} \rho \\ \rho \boldsymbol{u} \\ e \end{bmatrix}, \quad
F_i = \begin{bmatrix} \rho u_i \\ \rho u_i \boldsymbol{u} + p\mathbf{1}_i \\ (e+p)u_i \end{bmatrix}
\tag{20}
$$

D is the diffusion term and g is the force term for the relative flow. These definitions are somewhat abbreviated and for example, U for the 3D flow in cartesian coordinates is

$$
U = \begin{bmatrix} \rho \\ \rho u_1 \\ \rho u_2 \\ \rho u_3 \\ e \end{bmatrix}
$$

In the finite volume method the equations expressed in the integral form for each computational cell are also used.

$$
\int_\Omega U_t \, dV + \int_{\partial\Omega} \boldsymbol{n} \cdot \boldsymbol{H} \, dS + \int_\Omega g \, dV = 0
\tag{21}
$$

where

$$
\boldsymbol{H} = \begin{bmatrix} \rho \boldsymbol{u} \\ \rho \boldsymbol{u}\boldsymbol{u} + (p\boldsymbol{I} - \boldsymbol{\Pi}) \\ (e+p)\boldsymbol{u} - \boldsymbol{\Pi} \cdot \boldsymbol{u} - k\nabla T \end{bmatrix}, \quad
g = \begin{bmatrix} 0 \\ \boldsymbol{f} \\ 0 \end{bmatrix}
\tag{22}
$$

$\boldsymbol{I}$ is the unit matrix and $\boldsymbol{n}$ is the outward unit normal vector to the surface $\partial\Omega$ of a cell Ω.

In the finite-difference method the curvilinear coordinate grid generated so as to fit the impeller configuration and the fundamental equations written in general curvilinear coordinates are generally used. Such equations can be directly derived from the equations in cartesian coordinates not via the cylindrical coordinates.

$$
\hat{U}_t + \frac{\partial \hat{F}_i}{\partial \xi_i} + \hat{D} + \hat{g} = 0
\tag{23}
$$

where

$$
\begin{aligned}
\hat{U} &= J\begin{bmatrix} \rho \\ \rho \boldsymbol{u} \\ e \end{bmatrix}, \quad
\hat{F}_i = J\begin{bmatrix} \rho U_i \\ \rho U_i \boldsymbol{u} + p\nabla\xi_i \\ (e+p)U_i \end{bmatrix}, \\
\hat{D} &= -\frac{\partial}{\partial \xi_i} J\begin{bmatrix} 0 \\ \nabla\xi_i \cdot \boldsymbol{\Pi} \\ \nabla\xi_i \cdot \boldsymbol{\Pi} \cdot \boldsymbol{u} + k\,g^{ij}\partial T/\partial \xi_j \end{bmatrix}, \\
\hat{g} &= J\begin{bmatrix} 0 \\ \boldsymbol{f} \\ 0 \end{bmatrix}
\end{aligned}
\tag{24}
$$

$J = \partial(x_1, x_2, x_3)/\partial(\xi_1, \xi_2, \xi_3)$, $g^{ij} = \nabla\xi_i \cdot \nabla\xi_j$. $U_i = \boldsymbol{u} \cdot \nabla\xi_i$ are the contravariant velocities.

Taking a linear combination of the momentum equations of Eq.(23) we obtain the equations of the flow in transformed space $\boldsymbol{\xi}$.

$$
\begin{aligned}
B&\left(\hat{U}_t + \frac{\partial \hat{F}_i}{\partial \xi_i} + \hat{D} + \hat{g}\right) \\
&= \tilde{U}_t + \frac{\partial \tilde{F}_i}{\partial \xi_i} + \tilde{R} + \tilde{D} + \tilde{g} = 0
\end{aligned}
\tag{25}
$$

where

$$
\tilde{U} = J\begin{bmatrix} \rho \\ \rho U \\ e \end{bmatrix}, \quad
\tilde{F}_i = J\begin{bmatrix} \rho U_i \\ \rho U_i U + p\nabla\xi_i \cdot \nabla\xi \\ (e+p)U_i \end{bmatrix}
\tag{26}
$$

$\tilde{R}$ is a sum of additional terms introduced to express the flux terms in divergence form and generally small for the smooth computational grid, though the values do not need to evaluate in the actual computation.

4. NUMERICAL METHODS FOR COMPRESSIBLE FLOW

4.1 Grid Generation

The computational grid of the meridional through flow in the quasi-3D analysis is usually generated by lines dividing between the hub and casing and nearly orthogonal lines to them. The grid of cascade flow is mainly H-type, but C-type for the turbine blade with large blunt nose. In the downstream region of the cascade with large stagger, nearly orthogonal lines to the streamlines instead of the circumferential lines must be taken to keep well orthogonality of grid.

In the 1980s it was possible to compute directly the 3D flows. The 3D computational grid is usually generated by the following procedure. (i) some revolutional surfaces are made in the space between hub to casing; (ii) on each revolutional surfaces the H-type, C-type or other type grids having the same topology are generated and (iii) if necessary, interpolation, blending or smoothing are performed. For the largely staggered cascade, triangular grid, triangular grid combined with quadrilateral grids around the blades(Fig.9) or chimera grid are also used. The hexahedral or triangular cylinder cells should be mounted nearly normal to the revolutional surfaces if possible. The 2D and 3D multi-black adaptive grid whose mesh size is finer across the shock wave is generated by analytical method(Fig.10).

4.2 Delta-Form Implicit Schemes

The cascade flow problems has been solved in the early stage by the explicit methods such as the finite-difference scheme by MacCormack(1969), MacCormack-Lomax(1979), and the finite volume methods by MacCormack-Paulley(1972) and Denton(1974). The Denton's method was improved in 1983, and succeeded to the finite volume method by Dawes(1983,86) containing the viscous effects. These methods have been utilized widely for the research and development of turbomachines, but are not always sufficient as CFD techniques for the complicated flows with shock waves and vortices.

The implicit method is stable for the relatively large Courant number compared to the explicit method, and therefore is suitable to obtain the steady state solution by the time-marching method. Further, in the computation using the low Reynolds number turbulence model, since the adjacent grid points to the solid wall must be taken in the range of $y^+ = 1 \sim 4$, the condition $CFL = c\Delta t/\Delta y < 1$ becomes very severe in the explicit method, and to use the implicit method is recommended. CFL is the Courant number and c is the velocity of sound. The delta-form approximate-factorization(AF) implicit scheme by Beam-Warming(1978) had a great influence on the schemes in the 80's.

The fundamental equations of the compressible flow can be written as

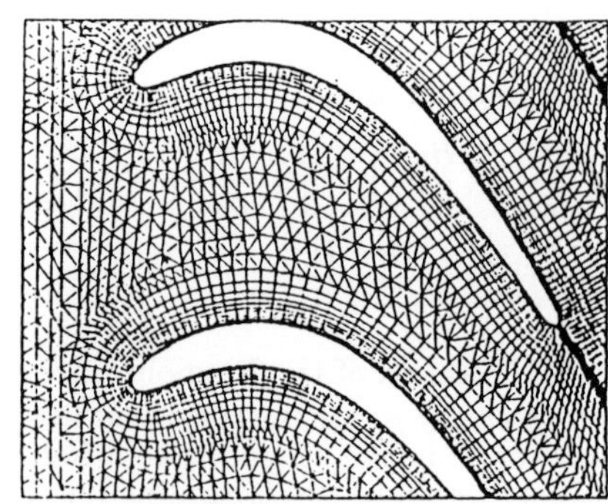

Fig.9: Patched grid for FDM/FEM zonal approach [Nakahashi(1986)]

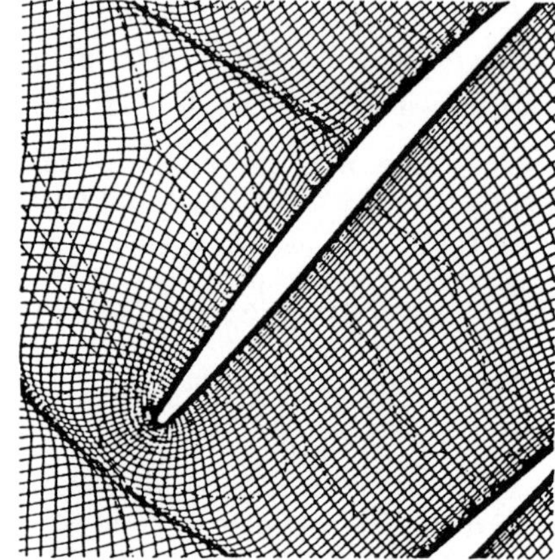

Fig.10: Multi-block solution adaptive grid and Mach number contours [Yamamoto-Engel(1997)]

$$u_t + \frac{\partial F_i}{\partial x_i} + d = 0 \tag{27}$$

Applying the trapezoidal law to Eq.(27) and rewriting to the delta-form implicit scheme, we get

$$\left(I + \Delta t\,\theta \frac{\partial}{\partial x_i} A_i \right) \Delta u^n = -\Delta t \left(\frac{\partial F_i}{\partial x_i} + d \right)^n \equiv RHS^n \tag{28}$$

where $F_i = A_i u$, $A_i = \partial F_i / \partial u$ are Jacobian matrices and $u^{n+1} = u^n + \Delta u^n$. The accuracy of the steady state solution obtained form Eq.(28) by the time-marching method only depends on the right hand side, RHS and is independent of the left hand side operator. Therefore, in the left hand side it is possible to omit the diffusion terms and to use the AF method and the first-order upwind scheme. For the flow having large flow separation only main parts of the diffusion terms are considered in the left hand side to converge the asymptotic solution correctly and to accelerate the convergency.

For the unsteady flows it is more convenient to use the following delta-form implicit scheme in which the solution at each time step is obtained by the Newton iteration

$$\left(I + \frac{1}{2}\Delta t \frac{\partial}{\partial x_i} A_i\right)\Delta u^{(m)} = -(u^{(m-1)} - u^n)$$

$$- \frac{1}{2}\Delta t\left\{\left(\frac{\partial F_i}{\partial x_i} + d\right)^n + \left(\frac{\partial F_i}{\partial x_i} + d\right)^{(m-1)}\right\} \quad (29)$$

where $u^{(0)} = u^n$, $u^{(m)} = u^{(m-1)} + \Delta u^{(m)}$. The solution converges within a few iterations, then $u^{n+1} = u^{(m)}$. Using the second-order backward-difference scheme instead of the Crank-Nicholson method in Eq.(29), we get

$$\left(I + \frac{2}{3}\Delta t \frac{\partial}{\partial x_i} A_i\right)\Delta u^{(m)} = -\frac{1}{3}(u^{n-1} - 4u^n$$

$$+ 3u^{(m-1)}) - \frac{2}{3}\Delta t\left\{\left(\frac{\partial F_i}{\partial x_i}\right)^{(m-1)} + d^n\right\} \quad (30)$$

4.3 Efficient Algorithms

Eq.(28), (29) or (30) is simultaneous linear equations with a very large coefficient matrix, and can be usually solved by the fractional step method. In the Beam-Warming scheme using the AF by Yanenko(1963, 71) the equation becomes as

$$[I + \theta(\nabla_1 C_1^+ + \Delta_1 C_1^-)][I + \theta(\nabla_2 C_2^+ + \Delta_2 C_2^-)]$$
$$[I + \theta(\nabla_3 C_3^+ + \Delta_3 C_3^-)]\Delta u = RHS \quad (31)$$

where $C_i^{\pm} = \Delta t A_i^{\pm}/\Delta x_i$ are the Courant numbers, $A_i = A_i^+ + A_i^-$, A_i^+, A_i^- are the Jacobian matrices only containing positive and negative eigenvalues of A_i, respectively, and ∇_i, Δ_i are the backward- and forward-difference operators, respectively. Eq.(31) can be divided into three steps, and in each step simultaneous linear equations with a 5×5 block tridiagonal matrix are solved. Further, applying the diagonalization by Pullium-Chaussee(1981), the linear equations are further reduced to five sets of linear equations with a scalar tridiagonal matrix and the computational efforts decrease still more.

The lower-upper successive Gauss-Seidel(LU-SGS) method proposed by Jameson-Yoon(1987) is recently used as a more efficient algorithm(Yoon-Kwak 1992). In this method Eq.(28), (29) or (30) is facterized as

$$(D + L)D^{-1}(D + U)\Delta u = RHS \quad (32)$$

where

$$L = \theta\sum_i(\nabla_i C_i^+ - C_i^+), \quad D = I + \theta\sum_i|C_i|,$$

$$U = \theta\sum_i(\Delta_i C_i^- + C_i^-)$$

Eq.(32) is solved dividing into two steps. Putting $D^{-1}(D + U)\Delta u = \Delta u^*$, the equations become as

$$D\,\Delta u^* = RHS - L\,\Delta u^*$$
$$D\,\Delta u = D\,\Delta u^* - U\,\Delta u \quad (33)$$

D is diagonalized, and these equations can be solved explicitly. The LU-SGS method can be improved more efficient algorithm by combining with the multigrid method(Anderson-Thomas 1988, Yokota-Caughey 1988).

The errors caused by the AF are order of $(C_i)^2$ for 2D and $(C_i)^3$ for 3D. Therefore, for the large Courant number the convergence of the solution rather deteriorates. On the other hand, in the LU-SGS method the errors caused by the factorization are order of C_i in spite of the dimension, and therefore it becomes possible to take the large Courant number.

The diagonalization in the LU-SGS method makes worse the convergence of the solution. Therefore, further putting $D = R_i\Sigma L_i$, (i=1, 2 or 3), Σ becomes a semi-diagonalized matrix. Where L_i, $R_i = L_i^{-1}$ are the matrices composed of left and right eigenvectors, respectively. In the LU-SGS-GE method(Yuan-Daiguji 1995b), each equations of Eq.(33) are solved dividing into three steps, and simultaneous linear equations with the matrix Σ can be solved by the Gaussian elimination. The method is apparently complicated, but the computer time can be considerably saved than the original LU-SGS method.

4.4 Flux Splitting Methods

In the right hand side of the delta-form implicit method and in the explicit method the high-accuracy, high-resolution upwind schemes have to be used. The compressible Euler equations are transformed by the theory of characteristics to the equations of shear and entropy waves propagating along a path line and of sound waves propagating along bicharacteristics on a sonic conoid. The upwinding should be made separately for these waves. The oldest method utilized such proper upwinding is the finite volume method by Godunov(1959). To simplify the explanation, we here consider the 1D Euler equations

$$u_t + F_x = u_t + Au_x = 0 \quad (34)$$

In the Godunov scheme, first the flow at t^n is approximated by constant functions in each cell, next the flux functions at the cell boundaries $F_{i\pm1/2}$ are determined from the Riemann problem, and then the flow at t^{n+1} is obtained from the equation

$$u_i^{n+1} = u_i^n - \frac{\Delta t}{\Delta x}(F_{i+1/2} - F_{i-1/2}) \quad (35)$$

In this scheme the values of $F_{i\pm1/2}$ are taken at the upstream side of all the three waves emanated from the cell boundaries.

In the fluctuation splitting method proposed by Roe(1982) and later called the flux-difference splitting method, the finite-difference equation of Eq.(34)

is written as

$$u_i^{n+1} = u_i^n - \frac{\Delta t}{\Delta x}(\Delta F_{i-1/2}^+ + \Delta F_{i+1/2}^-) \tag{36}$$

$$\Delta F^\pm = \bar{A}^\pm \Delta u = \sum_k \lambda_k^\pm \alpha_k R_k \tag{37}$$

where $\Delta u_{i+1/2} = u_{i+1} - u_i$, $\bar{A} = \bar{R}\bar{\Lambda}\bar{L}$ and λ_k, α_k, R_k are the k-components of $\bar{\Lambda}$, $\alpha(= \bar{L}\Delta u)$, $\bar{R}$, respectively. In this method $\bar{A}$ has to be taken a special form by Roe(1980) to satisfy the conservative condition of Eq.(37).

In the higher-order scheme it is recommended to make use of the conservative from expressed as

$$\left(\frac{\partial F}{\partial x}\right)_i = \frac{1}{\Delta x}(\hat{F}_{i+1/2} - \hat{F}_{i-1/2}) \tag{38}$$

where $\hat{F}_{i\pm1/2}$ are called the numerical flux functions, and not always the values of fluxes at points $x_{i\pm1/2}$. In the flux vector splitting method by Steger-Warming (1981), the upwind differences are taken for F^+ and F^- separately, where $F = F^+ + F^-$, $F^\pm = R\Lambda^\pm Lu$. However, at that time the solution often oscillated near the sonic point where the direction of propagation of a wave changes, because that the finite-difference approximation was ruined there. In order to overcome this difficulty, a method using smoothly varying F^+ and F^- was proposed by van Leer(1982). And the method below mentioned can also avoid this difficulty [see Eq.(41)].

4.5 Higher-Order TVD Schemes

In the former higher-order schemes some artificial viscosity was added uniformly to stabilize the solution, whereas in the TVD(total variation diminishing) schemes a necessary viscosity is added only a region, where the solution becomes unstable, by introducing limiter functions. The numerical flux function of the Chakravarthy-Osher(1985) TVD scheme is

$$\hat{F}_{i+1/2} = \hat{F}_{i+1/2}^{(1)}$$
$$+ \frac{1-\kappa}{4}\Delta\tilde{F}_{i-1/2}^+ + \frac{1+\kappa}{4}\Delta\tilde{\tilde{F}}_{i+1/2}^+$$
$$- \frac{1-\kappa}{4}\Delta\tilde{F}_{i+3/2}^- - \frac{1+\kappa}{4}\Delta\tilde{\tilde{F}}_{i+1/2}^- \tag{39}$$

where

$$\hat{F}_{i+1/2}^{(1)} = \frac{1}{2}(F_i + F_{i+1}) - \frac{1}{2}|A_{i+1/2}|\Delta u$$
$$\Delta\tilde{F}_{j+1/2}^\pm = \mathrm{minmod}(\Delta F_{j+1/2}^\pm, b\,\Delta F_{j-1/2}^\pm)$$
$$(j = i,\, i+1)$$
$$\Delta\tilde{\tilde{F}}_{j+1/2}^\pm = \mathrm{minmod}(\Delta F_{j+1/2}^\pm, b\,\Delta F_{j+3/2}^\pm)$$
$$(j = i-1,\, i) \tag{40}$$

The numerical flux function of Eq.(39) consists of the flux of the first-order upwind scheme $\hat{F}_{i+1/2}^{(1)}$ and correction terms that make the scheme higher order. In order to avoid the destruction of finite-difference form, $\Delta F_{j+1/2}^\pm$ have to be taken as

$$\Delta F_{j+1/2}^\pm = A_{i+1/2}^\pm \Delta u_{j+1/2}$$
$$\text{or}\quad \Delta F_{j+1/2}^\pm = |\mathrm{sign}(\Lambda_{i+1/2}^\pm)|\Delta F_{j+1/2}$$
$$(j = i,\, i\pm1) \tag{41}$$

$\lambda_k^\pm = (\lambda_k \pm |\lambda_k|)/2$, $\mathrm{sign}(\lambda_k^+) = 1\,(\lambda_k > 0)$, $= 0\,(\lambda_k \le 0)$. The accuracy of the CO scheme is in general second-order, and third-order only for $\kappa = 1/3$, though decreases to first-order if the minmod functions act. The conditions that the scheme is TVD stable are

$$1 \le b \le (3-\kappa)/(1-\kappa)$$
$$\text{and}\quad (1-\theta)b_1 C \le 1,$$
$$b_1 = \{5 - \kappa + (1+\kappa)b\}/4$$

that is, if $\kappa = 1/3$ then $b \le 4$, and the solution for the fully implicit method($\theta = 1$) is stable in spite of the Courant number C, but the explicit method($\theta = 0$) is stable under the condition $C \le 0.4$.

Eq.(39) becomes the Leonard(1979) scheme for $\kappa = 1/2$, and to the second-order central difference scheme (Yee 1987) for $\kappa = 1$.

$$F_{i+1/2}^{(2)} = F_{i+1/2}^{(1)} + \frac{1}{2}\{A_{i+1/2}^+\Psi(r_{i+1/2}^-)$$
$$- A_{i+1/2}^-\Psi(r_{i+1/2}^+)\}\Delta u_{i+1/2} \tag{42}$$

where $r_{i+1/2}^- = \Delta u_{i-1/2}/\Delta u_{i+1/2}$, $r_{i+1/2}^+ = \Delta u_{i+3/2}/\Delta u_{i+1/2}$. The various kinds of the limiters $\Psi(r)$ have been used, and putting

$$\Psi(r) = \mathrm{minmod}[3r,\, (r+2)/3,\, 2]$$

then Eq.(42) becomes the third-order CO scheme ($\kappa = 1/2$, $b = 4$, $r_1 = 1/4$). The limiters are generally used to stabilize the solution, but some have a function to improve the accuracy as well as the stability.

The numerical flux of the third-order scheme is

$$\hat{F}_{i+1/2} = \hat{F}_{i+1/2}^{(1)} + \frac{1}{6}\Delta F_{i-1/2}^+ + \frac{1}{3}\Delta F_{i+1/2}^+$$
$$- \frac{1}{6}\Delta F_{i+3/2}^- - \frac{1}{3}\Delta F_{i+1/2}^-$$

And the numerical flux of the fourth-order scheme is in general

$$\hat{F}_{i+1/2} = \hat{F}_{i+1/2}^{(1)} + \frac{1}{6}\Delta F_{i-1/2}^+ - \frac{1-\phi}{24}\Delta^3 F_{i-1/2}^+$$
$$+ \frac{1}{3}\Delta F_{i+1/2}^+ - \frac{1+\phi}{24}\Delta^3 F_{i+1/2}^+$$
$$- \frac{1}{6}\Delta F_{i+3/2}^- + \frac{1-\phi}{24}\Delta^3 F_{i+3/2}^-$$

$$-\frac{1}{3}\Delta F^{-}_{i+1/2} + \frac{1+\phi}{24}\Delta^3 F^{-}_{i+1/2}$$

$$(43)$$

where $\Delta^3 F_{j+1/2} = \Delta F_{j-1/2} - 2\Delta F_{j+1/2} + \Delta F_{j+3/2}$. Choosing $\phi = 1/3$ in Eq.(43) and rewriting it to the TVD scheme, we get(Yamamoto-Daiguji 1993)

$$\hat{F}_{i+1/2} = \hat{F}^{(1)}_{i+1/2} + \frac{1}{6}D\tilde{\tilde{F}}^{+}_{i-1/2} + \frac{1}{3}D\tilde{F}^{+}_{i+1/2}$$
$$- \frac{1}{6}D\tilde{F}^{-}_{i+3/2} - \frac{1}{3}D\tilde{\tilde{F}}^{-}_{i+1/2} \qquad (44)$$

where

$$D\tilde{F}^{\pm}_{j+1/2} = \mathrm{minmod}(DF^{\pm}_{j+1/2},\, b\, DF^{\pm}_{j-1/2})$$
$$(j = i,\, i+1)$$
$$D\tilde{\tilde{F}}^{\pm}_{j+1/2} = \mathrm{minmod}(DF^{\pm}_{j+1/2},\, b\, DF^{\pm}_{j+3/2})$$
$$(j = i-1,\, i)$$
$$DF^{\pm}_{j+1/2} = \Delta F^{\pm}_{j+1/2} - \frac{1}{6}(\Delta^2 \tilde{F}^{\pm}_{j+1} - \Delta^2 \tilde{\tilde{F}}^{\pm}_{j})$$
$$(j = i,\, i\pm1)$$
$$\Delta^2 \tilde{F}^{\pm}_{j+1} = \mathrm{minmod}(\Delta^2 F^{\pm}_{j+1},\, b_2 \Delta^2 F^{\pm}_{j})$$
$$\Delta^2 \tilde{\tilde{F}}^{\pm}_{j} = \mathrm{minmod}(\Delta^2 F^{\pm}_{j},\, b_2 \Delta^2 F^{\pm}_{j+1})$$

$$(45)$$

The fourth-order scheme is in general higher resolution than the third-order scheme, and can capture clearly not only the shock wave but also the weak discontinuities — shear layer and contact surface. However, the fourth-order scheme is inclined to amplify the numerical disturbances, and needs to stabilize more robustly. The first minmod functions of DF restrict the magnitude of F to prevent the occurrence of new extremum points of F, and accordingly the condition $1 < b \leq 4$ must be satisfied. Similarly, the second minmod functions for $\Delta^2 F$ suppress the occurrence of new inflection points of F, and the condition $1 < b_2 \leq 4$ must be satisfied(Daiguji *et al.* 1997). The instability seems first to appear as an inflection point and then to grow extremum points. The first limiters are the same as the CO scheme, and the second limiters stabilize by limiting the magnitude of the fourth-order correction terms, but nothing to do for the part of the third-order CO TVD scheme. The computer time by this fourth-order scheme is little different from the CO scheme.

Next, choosing $\phi = 1/5$ in Eq.(43), the resultant flux vector of the fifth-order scheme has the same forms as Eqs.(44) and (45), except for

$$DF^{+}_{i-1/2} = \Delta F^{+}_{i-1/2} - \frac{1}{5}(\Delta^2 \tilde{F}^{+}_{i} - \Delta^2 \tilde{\tilde{F}}^{+}_{i-1})$$
$$DF^{\pm}_{i+1/2} = \Delta F^{\pm}_{i+1/2} - \frac{3}{20}(\Delta^2 \tilde{F}^{\pm}_{i+1} - \Delta^2 \tilde{\tilde{F}}^{\pm}_{i}) \qquad (46)$$
$$DF^{-}_{i+3/2} = \Delta F^{-}_{i+3/2} - \frac{1}{5}(\Delta^2 \tilde{F}^{-}_{i+2} - \Delta^2 \tilde{\tilde{F}}^{-}_{i+1})$$

The actual calculated results by these fourth- and fifth-order schemes are almost the same, since the values of ϕ are not so different.

4.6 Higher-Order MUSCL Type Schemes

The first-order Godunov method was extended to the second-order MUSCL(monotone upstream-centered schemes for conservation laws) by van Leer(1979), and further to the higher-order ENO(essentially non-oscillatory) schemes by Harten *et al.*(1987). In these finite volume methods u_i is defined as the mean value of u in one cell, and in the second-order scheme the left and right hand side values of u at the cell boundary are given as

$$u^{L}_{i+1/2} = u_i + \frac{1-\kappa}{4}\Delta\tilde{u}_{i-1/2} + \frac{1+\kappa}{4}\Delta\tilde{u}_{i+1/2}$$
$$u^{R}_{i+1/2} = u_{i+1} - \frac{1-\kappa}{4}\Delta\tilde{u}_{i+3/2} - \frac{1+\kappa}{4}\Delta\tilde{u}_{i+1/2}$$

$$(47)$$

where

$$\Delta\tilde{u}_{j+1/2} = \mathrm{minmod}(\Delta u_{j+1/2},\, b\Delta u_{j-1/2})$$
$$(j = i,\, i+1)$$
$$\Delta\tilde{\tilde{u}}_{j+1/2} = \mathrm{minmod}(\Delta u_{j+1/2},\, b\Delta u_{j+3/2})$$
$$(j = i-1,\, i)$$

$$(48)$$

$\Delta u_{j+1/2} = u_{j+1} - u_j$. Eq.(47) consists of u_j of the first-order Godunov scheme and higher-order correction terms, and the accuracy is in general second-order and third-order only for $\kappa = 1/3$, if the limiters do not act. At the boundary there is the small jump

$$u^{R}_{i+1/2} - u^{L}_{i+1/2} \equiv \Delta u = -\frac{1-\kappa}{4}\Delta^3 u_{i+1/2}$$

The numerical flux at the boundary is determined from the approximate Riemann solver by Roe(1980) as

$$\hat{F}_{i+1/2} = \frac{1}{2}\{F(u^{R}_{i+1/2}) + F(u^{L}_{i+1/2})\}$$
$$- \frac{1}{2}\sum_k |\lambda_k|\alpha_k R_k \qquad (49)$$

where $F(u^{R}_{i+1/2}) - F(u^{L}_{i+1/2}) = \bar{A}\Delta u$, $\bar{A} = \bar{R}\bar{\Lambda}\bar{L}$, $\alpha = \bar{L}\Delta u$ and λ_k, α_k, R_k are the k-components of $\bar{\Lambda}$, α, $\bar{R}$, respectively.

In the fourth-order scheme the numerical fluxes can be determined from Eq.(49) and (Yamamoto-Daiguji 1993)

$$u^{L}_{i+1/2} = u_i + \frac{1}{6}D\tilde{u}_{i-1/2} + \frac{1}{3}D\tilde{u}_{i+1/2}$$
$$u^{R}_{i+1/2} = u_{i+1} - \frac{1}{6}D\tilde{u}_{i+3/2} - \frac{1}{3}D\tilde{u}_{i+1/2} \qquad (50)$$
$$Du_{j+1/2} = \Delta u_{j+1/2} - \frac{1}{6}(\Delta^2 \tilde{u}_{j+1} - \Delta^2 \tilde{\tilde{u}}_{j})$$

and the jump at the cell boundary is

$$u^{R}_{i+1/2} - u^{L}_{i+1/2} = \Delta^5 u_{i+1/2}/36$$

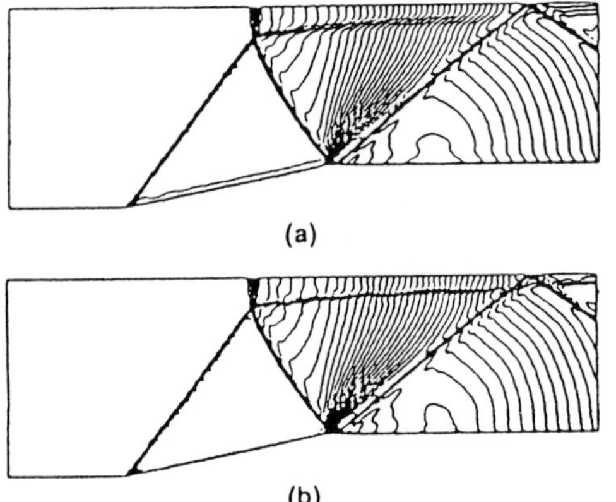

Fig.11: Mach number contours of 2D supersonic inviscid duct flow (a) third-order MUSCL TVD
scheme, (b) fourth-order compact MUSCL
TVD scheme [Yamamoto-Daiguji(1993)]

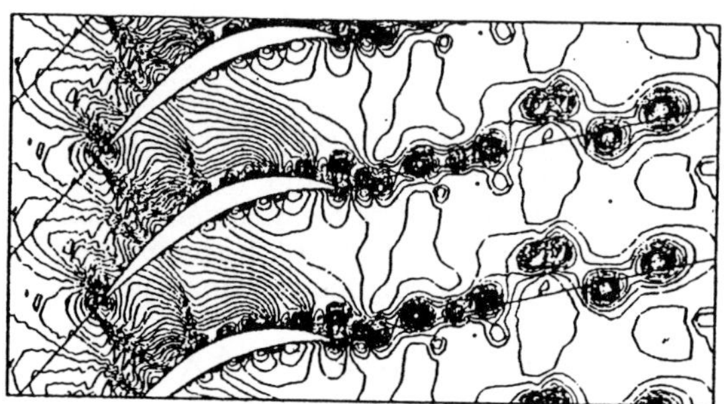

Fig.12: Instantaneous density contours of transonic
compressor cascade flow by laminar analysis
[Fukuda *et al.*(1990)]

The computational results by this fourth-order scheme
can be obtained in almost the same computer time
as the third-order scheme. Figure 11 shows a comparison of the Mach number contours calculated using the third- and fourth-order schemes. These results
were obtained by solving the Euler equations using the
grid points of 241×81. It is found from the figure that
the slip lines calculated by the fourth-order scheme are
substantially improved, though the shock waves do not
so differ.

Harten(1989) was proposed the ENO schemes with
subcell resolution for 1D flow which can capture the
discontinuities with excellent resolvability. However, it
seems very difficult to extend such schemes straightforwardly to the multi-dimensional flow. On the other
hand, the multi-dimensional upwinding schemes(van
Leer 1993) have been developed to capture properly
the discontinuities, especially cross the grid diagonally.
But these schemes have not been applied to the cascade
flow problems yet.

5. COMPUTATIONAL RESULTS OF COM-PRESSIBLE CASCADE FLOWS

5.1 2D Simulations

The simulation of the compressor cascade flow is not so
easy because of the decelerating flow subject to adverse
pressure gradient. Figure 12 shows a result of DNS
of transonic compressor cascade flow by Fukuda *et
al.*(1990). In this computation great many grid points
of 801 × 201 were used. It is found from the figure that
many vortices occur successively on the blade surfaces
and flow down into the wake, and that the simulation
as a steady flow using the turbulence model is serious.

On the other hand, the simulation of turbine cascade
flow is relatively easy because of the accelerating flow
with favorable pressure gradient. Recently, fairly good
estimations of fluid-dynamic losses of turbine cascade
have been possible. Figure 13 shows a simulation result
of transonic turbine cascade flow which was reported
by Kiock *et al.*(1986). In the computation Eq.(25) with
the Chien's low Reynolds number k-ε model, 181×61
H-type grid with improved orthogonality in the down-
stream region, the fourth-order compact TVD scheme
of Eqs.(44) and (45), and the LU-SGS-GE algorithm
were used. $Re = 8.5 \times 10^5$, max $CFL \approx 20$. The contours of velocity component w_m in Fig.13 indicate occurrence of vortex streets and shock wave/vortex interactions. The detailed computation of the unsteady
vortices brings the satisfactory loss coefficients and exit
flow angles to coincide well with the experimental data
as shown in Fig.14. Figure 15 shows another simulation
result of transonic turbine cascade flow with a Schlieren
photograph which was experimented in Qinghe Laboratory of Beijing. The flows are steady, but the loss
coefficients and the exit flow angles change largely and
delicately according to the Mach number(shock wave
pattern) as shown in Fig.16. The coincidence between
the computations and experiments is not always well
especially in the high Mach number range.

5.2 3D Simulations

The first example is transonic fan rotor flow where two
shock waves occur from the blade trailing edge only
near the tip and one shock wave impinges on the neighboring blade pressure side(Nozaki *et al.* 1987). In the
computation the BL algebraic model, overlapped C-
type grid with $151 \times 33 \times 50$ points, the LU-ADI scheme
by Fujii-Obayashi(1986) were used. Figure 17 shows
the calculated oilflow pattern on the suction side and
the detail at the hub corner the flow separates. The
second example is transonic compressor rotor flow with
channel shocks(Daiguji-Yamamoto 1989). The boundary layer flow separates behind the shock and the fluid
in the separation region flows towards the tip as shown
in Fig.18. The rotor flows having large flow separation
due to incidence angle, wall curvature, shock wave, corner flow and tip clearance have in general large 3D
effects.

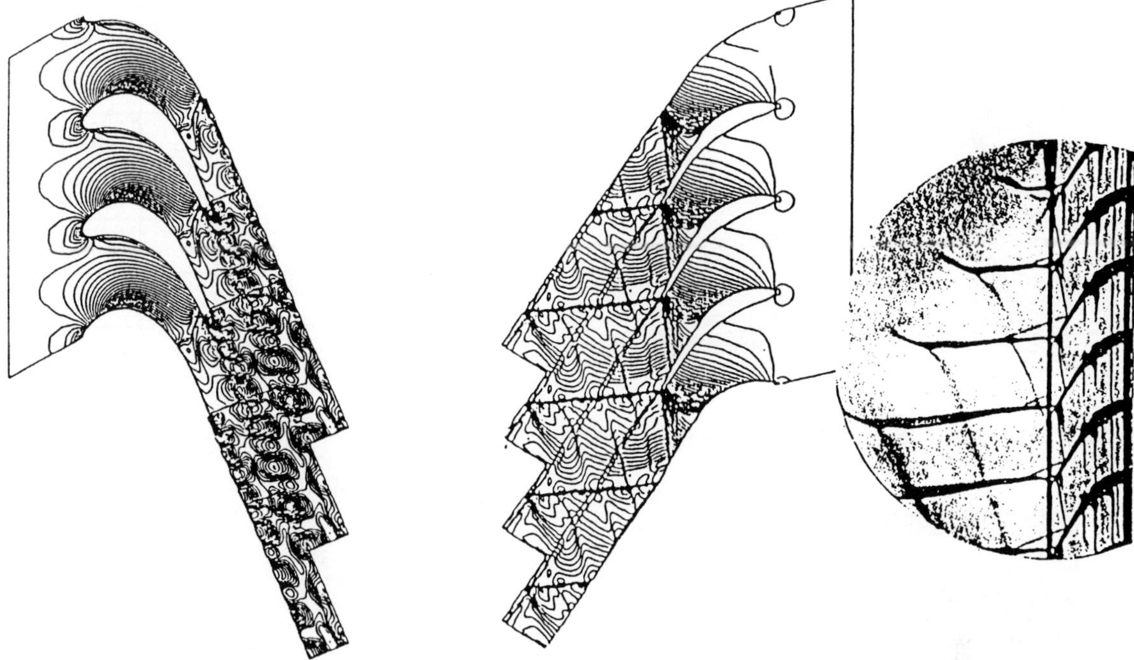

Fig.13: Instantaneous velocity contours for transonic turbine cascade composed of VKI LS-59 blades [Daiguji *et al.*(1997)]

Fig.15: Instantaneous Mach number contours and Schlieren photograph for transonic turbine cascade composed of GE-851 blades [Yuan-Daiguji (1993)]

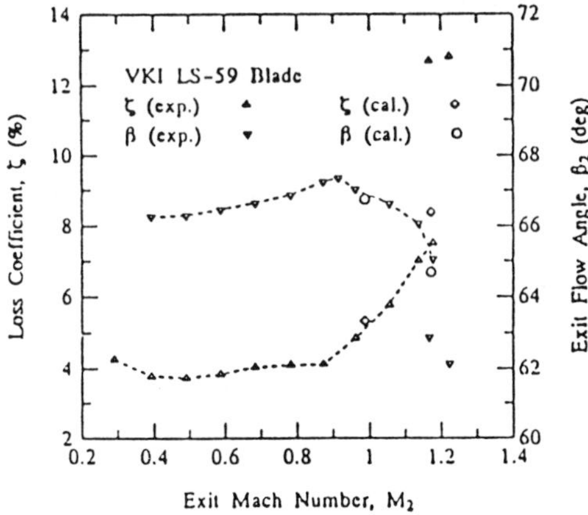

Fig.14 Loss coefficients and exit flow angles *vs* exit Mach numbers for VKI LS-59 blades [Yuan-Daiguji(1995)]

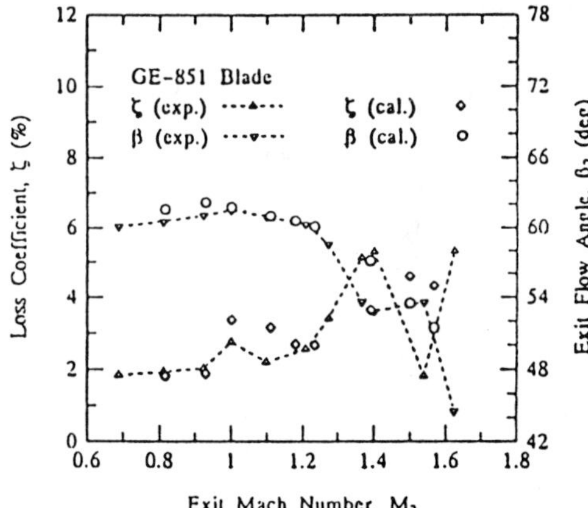

Fig.16: Loss coefficients and exit flow angles *vs* exit Mach numbers for GE-851 blades [Yuan-Daiguji(1995)]

The final example is a simulation of the transonic axial-flow fan NASA Rotor 67 experimented by Strezisar *et al.*(1989)(Arima *et al.* 1997). The simulation was implemented using the Chien's low Reynolds number k-ε model, $156 \times 61 \times 81$ grid points, the diagonalized Beam-Warming delta-form AF method and the third-order CO TVD scheme, considering the tip clearance. The calculated adiabatic efficiencies (Fig.19)

and pressure ratios, and the static pressure, temperature and flow angle distributions at the exit generally coincide well with the experiments. And the calculated relative Mach number contours also coincide well with the experiments(Fig.20). However, the simulation results of Rotor 37 implemented at the same time do not well coincide with the experiments.

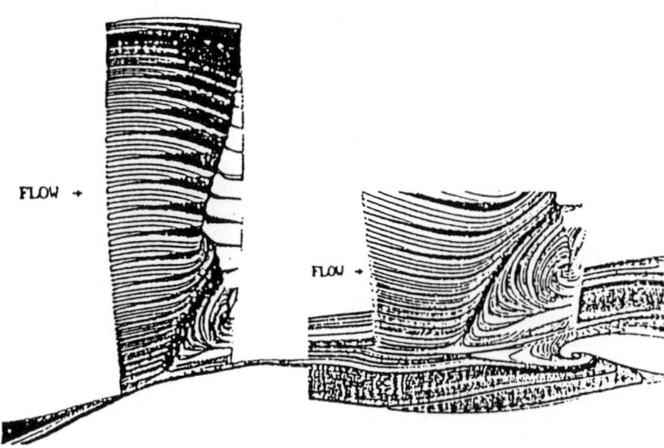

Fig.17: Computed oilflow patterns on suction side of fan blade [Nozaki *et al.*(1987)]

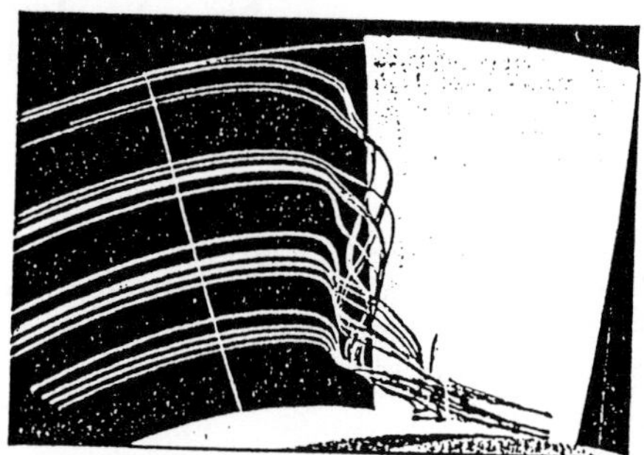

Fig.18: Computed path lines in separation zone of transonic compressor rotor flow [Daiguji-Yamamoto(1989)]

5.3 Tip Clearance Effects and Leakage Vortex

3D analyses of the flow through an axial flow turbomachine rotor with tip clearance have been performed by Pouagare-Delaney(1986), Watanabe *et al.* (1991), Kunz-Lakshminarayana(1993) and Basson-Lakshminarayana(1995). In the last paper, the flow through an axial-flow turbine cascade is simulated employing the Chien's low Reynolds number k-ε model, using main grid of $81 \times 57 \times 57$ points together with $41 \times 21 \times 15$ points in the tip gap and considering the second- and fourth-order dissipation terms in the right hand side.

The leakage vortex of the compressor rotor occurs from the somewhat large tip clearance. Figure 21 shows a simulation result of leakage vortex visualized using some relative path lines(Daiguji-Yamamoto 1989). In the computation, the grid with $81 \times 25 \times 25$ points whose 4 points are allotted in the clearance, Eq.(25) with the Chien's low Reynolds number k-ε model, the diagonalized delta-form AF method, and the third-order CO TVD scheme are used. $Re = 1.0$

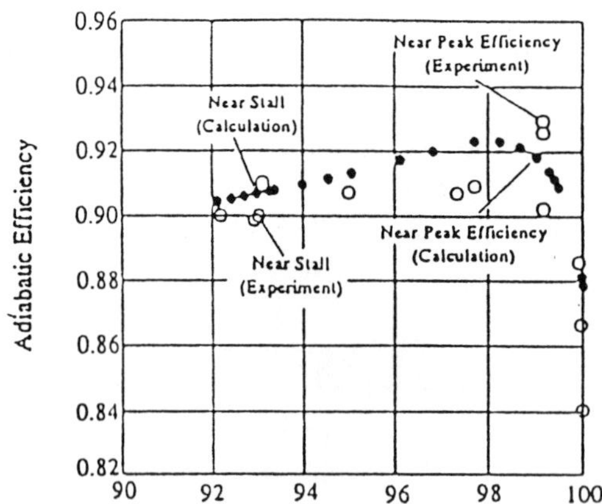

Fig.19: Adiabatic efficiencies *vs* mass flow rates for transonic fan NASA Rotor 67 [Arima *et al.*(1997)]

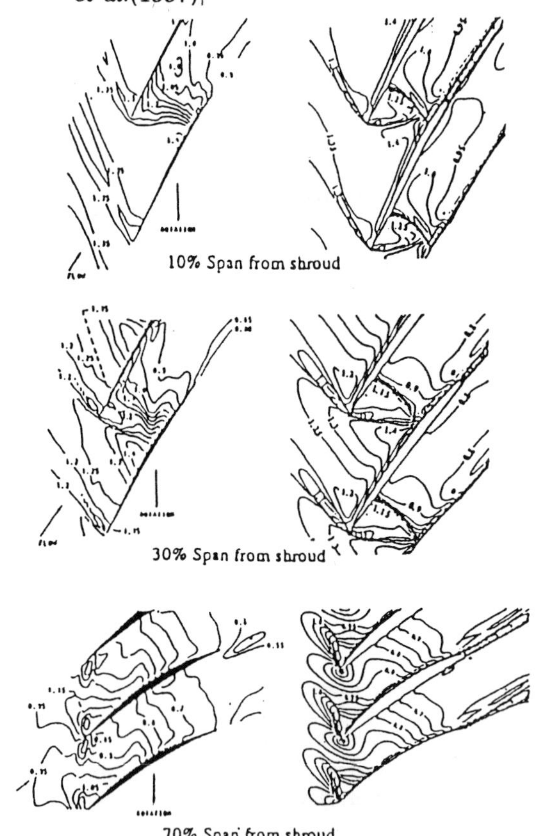

Fig.20: A comparison of relative Mach number contours by experiment(left) and computation(right) for Rotor 67 [Arima *et al.*(1997)]

$\times 10^6$. Although the calculated results are qualitative due to the relatively small grid points, the leakage vortex occurs from the casing corner on the blade suction

Fig.21: Leakage vortex visualized using path lines in transonic compressor rotor flow [Daiguji-Yamamoto(1989)]

side where backward flow essentially exists. The backward flow stretches and strengthens the vorticity in the flow shed from the clearance, and thus contributes to the growing of the leakage vortex. The leakage vortex extends along the relative flow, adheres to the casing wall and has oval cross-sections. A local pressure drop on the blade suction side appears near the occurrence point of the leakage vortex.

5.4 Wet Steam Flow through a Turbine Cascade

The fundamental equations of the wet steam flow consists of the usual compressible NS equations and the conservative equations of the mass of liquid phase and the number density of droplets as

$$(\rho\beta)_t + \nabla \cdot \rho\beta u = \Gamma$$
$$(\rho n)_t + \nabla \cdot \rho n u = \rho I \tag{51}$$

where β is the liquid mass faction, Γ the mass generation rate due to evaporation/condensation, n the number density of droplets and I the nucleation rate. The values of Γ and I are determined from the classical nucleation and condensation theories.

The homogeneous gas-liquid two-phase flow is assumed considering the very fine droplets at a rapid steam expansion process in the low pressure region near the final stage of steam turbine. Then, $p_g = p_\ell = p$ and $T_g = T_\ell = T$, where subscripts g and ℓ mean the gas and liquid phases. The specific volume $1/\rho$, specific internal energy ϵ, specific enthalpy h and specific heat C_p of the two-phase flow can be expressed as

$$f_m = (1 - \beta)f_g + \beta f_\ell \tag{52}$$

The equation of state and the velocity of sound of the two-phase flow are(Ishizaka *et al.* 1995)

$$p = (1-\beta)\rho RT + \beta\rho(AT^2 + BT + C)\frac{p}{p+p_c}$$
$$c^2 = \left.\frac{dp}{d\rho}\right|_s = \left(\frac{C_0}{R}\frac{\rho}{\rho_g} - \frac{C_1}{C_{pm}}\right)^{-1}T \tag{53}$$

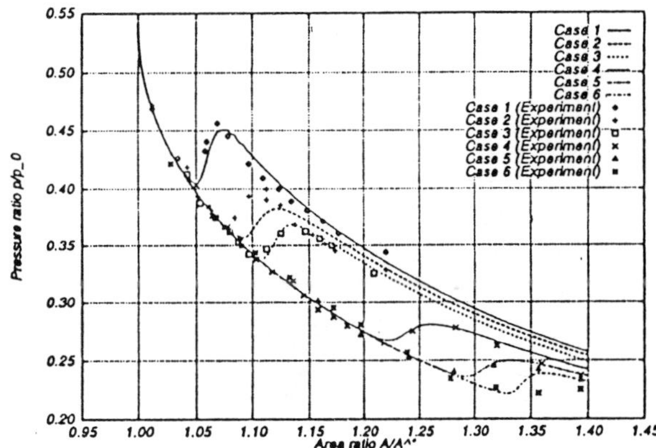

Fig.22: Pressure distributions of wet steam diverging duct flow [Ishizaka *et al.*(1995)]

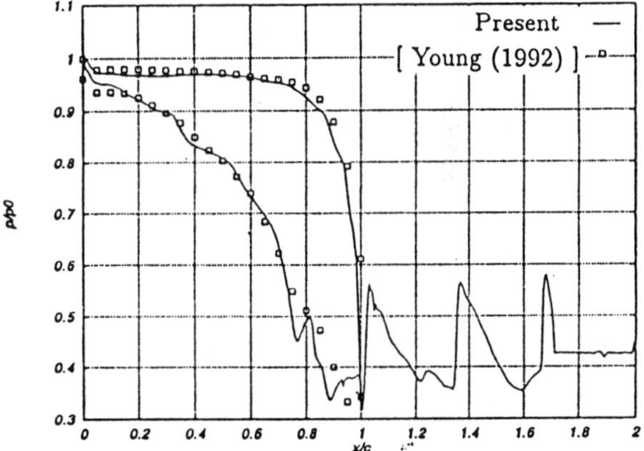

Fig.23: Pressure distributions on turbine blade surface and in wake [Ishizaka *et al.*(1995)]

where $C_0 = 1 - \beta(\rho/\rho_\ell)p_c/(p + p_c)$, $C_1 = 1 + \beta\rho(AT^2 - C)/(p + p_c)$ and $\gamma_m = C_p/\{C_p - (1 - \beta)R\}$. For the sufficiently small liquid mass fraction β the above equations are reduced to

$$p = (1 - \beta)\rho RT$$
$$= (\gamma_m - 1)(e - \rho u_i u_i/2 - \rho h_{0m}) \tag{54}$$
$$c^2 = \gamma_m(1 - \beta)RT = \gamma_m p/\rho$$

Using these relations in the fundamental equations of the two-phase flow, the linearization, diagonalization and upstreaming techniques and therefore the usual numerical methods for the compressible flow can be applied to these equations.

In order to examine the validity of the method, first the wet steam flows through a converging-diverging nozzle used in the experiment by Bennie-Green(1943) were computed as 1D flows. The computation was

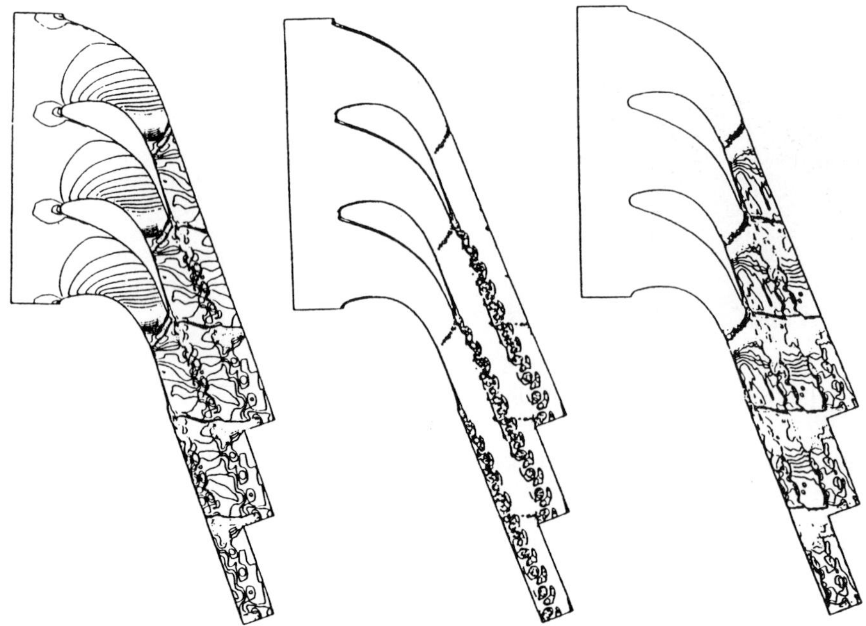

Fig.24: Instantaneous contours of (a) Mach number, (b) vorticity, (c) liquid mass fraction in turbine cascade wet steam flow (Ishizaka *et al.* 1995)

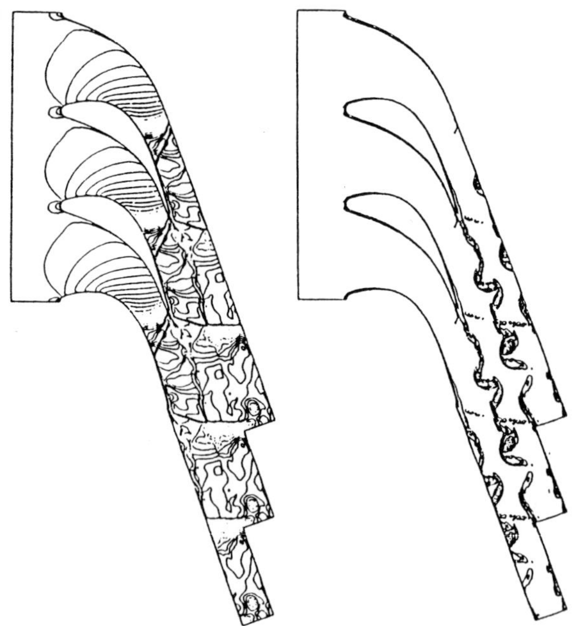

Fig.25: Instantaneous contours of (a) Mach number, (b) vorticity in turbine cascade flow under dry assumption (Ishizaka *et al.* 1995)

performed by the third-order CO type MUSCL TVD scheme using 201 points with uniform spacing. Figure 22 shows the pressure distributions in the diverging duct of nozzle. As is evident from the figure condensation shocks whose locations depend on the upstream stagnation conditions (p_0, T_0) occur in the duct, and the simulation results well coincide with the experimental data. Therefore, that the assumption of the homogeneous two-phase flow is appropriate is found. What is not usual shock wave is clear from the fact that the flow behind the shock is supersonic and the pressure is descending.

Next, the flows with nonequilibrium condensation in the low pressure stage of transonic steam turbine calculated by Young(1988, 92) were computed. In the computation the fundamental equations of the wet steam flow with the Chien's low Reynolds number k-ε model are discretized by the fourth-order MUSCL TVD scheme and the Crank-Nicholson method using the C-type grid with 221×61 points, and they are solved by the LU-SGS algorithm efficiently. That a condensation shock on the suction side of the blade is captured by the high resolution scheme is found from the pressure profiles on the blade surfaces in Fig.23. Figure 24 shows the instantaneous contours of the Mach number M, the vorticity ζ and the liquid mass fraction β. The vortex street is formed in the wake and interacts with the shock waves. When the inlet temperature T_0 is relatively low and the liquid mass fraction β is large, the vortex street becomes larger, the fluid dynamic losses also enlarge, and the outlet flow angle decreases. Figure 25 shows the instantaneous contours of M and ζ calculated under the dry condition for the same inlet conditions as in Fig.24.

6. INCOMPRESSIBLE CASCADE FLOWS

6.1 Fundamental Equations

The fundamental equations of the incompressible flow are the equation of continuity and the Reynolds-averaged Navier-Stokes equations

$$\nabla \cdot u = 0 \tag{55}$$

$$u_t + \nabla \cdot uu = -\nabla p + \nabla \cdot \Pi + f \tag{56}$$

where $u = v$ and $f = 0$ for the absolute flow, and $u = w$ and $f = \omega^2 r - 2\omega \times w$ for the relative flow. The components of the viscous stress tensor Π are

$$\tau_{ij} = \nu \left(\frac{\partial u_i}{\partial x_j} + \frac{\partial u_j}{\partial x_i} \right) - \overline{u_i' u_j'} \tag{57}$$

where $-\rho \overline{u_i' u_j'}$ is the Reynolds stress. These equations are dimensionless, and $Re = 1/\nu$ is the Reynolds number.

In the case of the incompressible flow, different from the compressible flow, u^{n+1} can be obtained from Eq.(56), but p^{n+1} of the remaining variable can not be done from the remaining Eq.(55). In order to obtain p the pseudo-compressibility method by Chorin(1967) introduces the modified continuity equation

$$p_t + \beta \nabla \cdot u = 0 \tag{58}$$

where $\beta(= \rho c^2)$ is the pseudo-compressibility coefficient. In the case obtaining the steady state solution by the time-marching method, when once the solution converges, $p_t = 0$, hence the solution of Eqs.(55) and (56) can be obtained from Eqs.(58) and (56).

The weak formulation of Eqs.(56) and (55) by the Galerkin method becomes as

$$\int_\Omega \{u_t v + (\nabla \cdot uu)v - p\nabla v + \nu \nabla u \cdot \nabla v$$
$$- fv\}dx = 0 \tag{59}$$

$$\int_\Omega (\nabla \cdot u)\, q\, dx = 0 \tag{60}$$

where v, q are the shape functions. Using the abbreviations of Eqs.(59) and (60) as

$$a(u , v) - (p , \nabla v) = (f , v)$$
$$(q , \nabla \cdot u) = 0$$

the basic equations of the penalty function method which contain some small perturbations can be expressed as

$$a(u_\epsilon , v) - (p_\epsilon , \nabla v) = (f , v) \tag{61}$$

$$(q , \nabla \cdot u_\epsilon) = -\epsilon(q , p_\epsilon) \tag{62}$$

where ϵ is a small positive parameter(Hughes *et al.* 1979, Bercovier-Engelman 1979). If ϵ approaches to zero, then the solutions u_ϵ and p_ϵ of Eqs.(61) and (62) approach to solutions u and p of Eqs.(59) and (60) to be required. From Eq.(62)

$$p_\epsilon = -\epsilon^{-1}\nabla \cdot u_\epsilon \tag{63}$$

and substituting this expression into Eq.(61)

$$a(u_\epsilon , v) + \epsilon^{-1}(\nabla \cdot u_\epsilon , \nabla v) = (f , v) \tag{64}$$

In the penalty function method, u is first obtained from Eq.(64) and then p is determined from Eq.(63). In the psudo-compressibility method the time-variation of p depends on $\nabla \cdot u$ to be originally zero, whereas in this method p itself directly on $\nabla \cdot u$.

By taking divergence of Eq.(56) the Poisson equation of p is derived as

$$\nabla^2 p = -\nabla \cdot (\nabla \cdot uu) + \nabla \cdot f$$
$$+ [-(\nabla \cdot u)_t + \nabla \cdot (\nabla \cdot \Pi)] \tag{65}$$

where $\nabla \cdot f = 2(\omega^2 + \omega \zeta_x)$ for relative flow. In the MAC scheme(Harlow-Welch 1965, Amsden-Harlow 1970), u is determined as a solution of the Cauchy problem of Eq.(56), and p is found from the boundary value problem of Eq.(65). Similarly, in the explicit FEM, u is first calculated from Eq.(59), then p is calculated from the weak formulation of Eq.(65) as

$$\int_\Omega (\nabla p + \nabla \cdot uu - f) \cdot \nabla q\, dx = 0 \tag{66}$$

Further, these MAC type methods require a special measure to be satisfied the continuity condition (55).

6.2 Pseudo-Compressibility Methods

The equations of the pseudo-compressibility method in general curvilinear coordinates can be written as

$$\hat{U}_t + \frac{\partial \hat{F}_i}{\partial \xi_i} + \hat{D} + \hat{g} = 0 \tag{67}$$

where

$$\hat{U} = J \begin{bmatrix} p \\ u \end{bmatrix}, \quad \hat{F}_i = J \begin{bmatrix} \beta U_i \\ U_i u + p\, \nabla \xi_i \end{bmatrix},$$
$$\hat{D} = -\frac{\partial}{\partial \xi_i} J \begin{bmatrix} 0 \\ \nabla \xi_i \cdot \Pi \end{bmatrix}, \quad \hat{g} = J \begin{bmatrix} 0 \\ f \end{bmatrix} \tag{68}$$

These equations have the similar characters to the equations of compressible flow, and it is possible to apply to them many computational techniques developed for the compressible flow. The flux vectors can be linearized as $\hat{F}_i = \hat{A}_i \hat{U}$. Using the linearization Eq.(67) can be rewritten to the delta-form implicit scheme for the steady flow

$$\left(I + \Delta t\, \theta \frac{\partial}{\partial \xi_i} \hat{A}_i \right) \Delta \hat{U}^n = RHS^n \tag{69}$$

where

$$\hat{U}^{n+1} = \hat{U}^n + \Delta \hat{U}^n$$
$$RHS = -\Delta t(\partial \hat{F}_i / \partial \xi_i + \hat{D} + \hat{g})$$

The pseudo-compressibility method is applicable to the unsteady flow problem provided that a sufficiently large value of β is taken. Since ten years, the pseudo-compressibility methods for the unsteady flow have been developed by Merkle-Athavale(1987), Soh-Goodrich(1988), Rogers-Kwak(1990) and Rogers *et al.*(1991). Using Eq.(30)

$$\left(I + I_t^{-1}\frac{\partial}{\partial \xi_i}\hat{A}_i\right)\Delta\hat{U}^{(m)} = RHS \qquad (70)$$

where

$$\hat{U}^{(m)} = \hat{U}^{(m-1)} + \Delta\hat{U}^{(m)}$$
$$RHS = -I_m(\hat{U}^{n-1} - 4\hat{U}^n + 3\hat{U}^{(m-1)})$$
$$- I_t^{-1}\left(\frac{\partial\hat{F}_i}{\partial\xi_i} + \hat{D} + \hat{q}\right)^{(m-1)}$$
$$I_t^{-1} = \begin{bmatrix} \Delta\tau & 0 \\ 0 & 2\Delta t\,\boldsymbol{I}/3 \end{bmatrix}, \; I_m = \begin{bmatrix} 0 & 0 \\ 0 & \boldsymbol{I}/3 \end{bmatrix}$$

$\beta = 1 \sim 100$ for the steady flow and $\Delta\tau = 10^{12}$, $\beta = \text{a}$ few hundreds $\sim$ thousands for the unsteady flow.

6.3 MAC Type Schemes

The MAC scheme can obtain the solution that completely satisfies the continuity condition and does not occur any spurious error by using the staggered grid (u, v, w, p are defined at different points) and by taking account of the terms $[\quad]$ to be originally zero in Eq.(65) with the condition $\nabla \cdot \boldsymbol{u}^{n+1} = 0$. The scheme has been improved and extended to the curvilinear coordinate grid by introducing the contravariant velocity(Nakamura-Takemoto 1985, Shyy *et al.* 1985), volume flux(Shin *et al.* 1988, Rosenfeld *et al.* 1988), contravariant physical velocity(Pope 1978, Demirdzic *et al.* 1987, Koshizuka *et al.* 1989) or covariant physical velocity(Karki-Patankar 1988). The equations for the steady flow of the implicit SMAC scheme using volume flux are written as(Ikohagi-Shin 1991)

$$J\{1 + \Delta t\,\theta(U_i\frac{\partial}{\partial\xi_i} - \tilde{D}_k)\}\Delta U_k^* = RHS_k^n \qquad (71)$$

$$JU_k^{n+1} = JU_k^* - \Delta t\,Jg_{ki}\frac{\partial\phi}{\partial\xi_i} \qquad (72)$$

$$\frac{\partial}{\partial\xi_i}\left(Jg_{ij}\frac{\partial\phi}{\partial\xi_j}\right) = \frac{1}{\Delta t}\left(\frac{\partial JU_k^*}{\partial\xi_k}\right) \qquad (73)$$
$$p^{n+1} = p^n + \phi$$

where

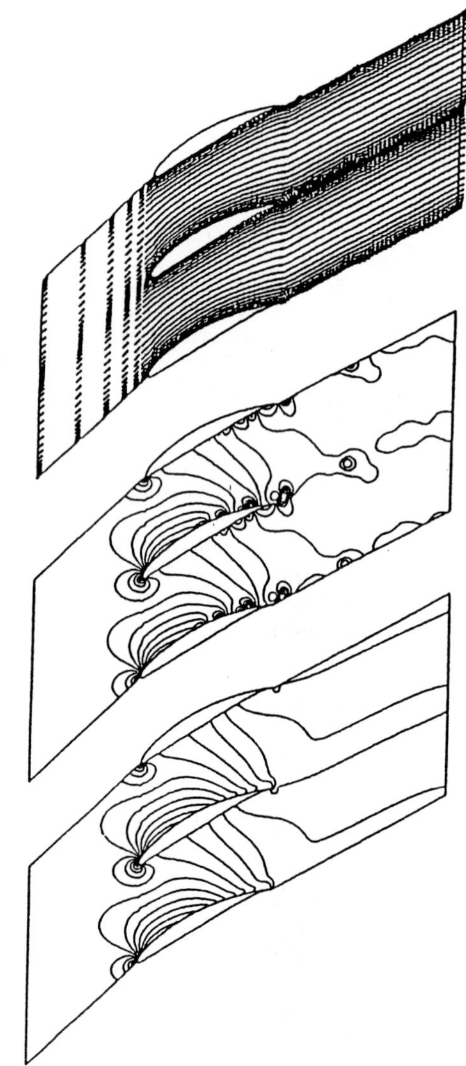

Fig.26: Instantaneous velocity vectors, pressure contours and time-averaged pressure contours of decelerating cascade flow [Shin *et al.*(1993)]

$$U_k^* = U_k^n + \Delta U_k^*$$
$$RHS_k = -\Delta t\left\{\frac{\partial}{\partial\xi_i}JU_iU_k - A_k + Jg_{ki}\frac{\partial p}{\partial\xi_i} + D_k\right\}$$
$$A_k = \boldsymbol{u}\cdot JU_i\frac{\partial}{\partial\xi_i}\nabla\xi_k$$
$$D_k = \nabla\xi_k\cdot\frac{\partial}{\partial\xi_i}\left\{(\nu+\nu_t)Jg_{ij}\frac{\partial\boldsymbol{u}}{\partial\xi_j}\right\}$$
$$+ Jg_{kj}\left(\nabla\xi_i\cdot\frac{\partial\nu_t}{\partial\xi_i}\frac{\partial\boldsymbol{u}}{\partial\xi_j} - \frac{2}{3}\frac{\partial k}{\partial\xi_j}\right)$$

U_k^* from Eq.(71), ϕ from Eq.(73) and U_k^{n+1} from Eq.(72) are calculated in order. The equations for the

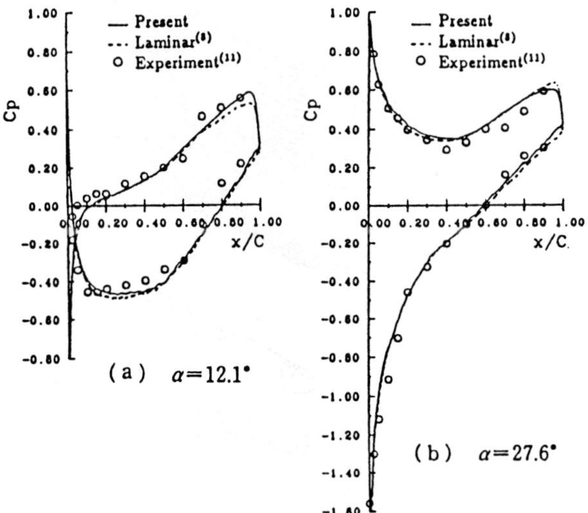

(a) $\alpha = 12.1°$

(b) $\alpha = 27.6°$

Fig.27 Pressure profiles on blade surfaces of decelerating cascade flow ———— turbulent, ———— laminar, ○ experiment by Emery et al. [Shin et al.(1993)]

unsteady flow are(Shin *et al.* 1993a)

$$J\{1 + \frac{1}{2}\Delta t(U_i\frac{\partial}{\partial\xi_i} - \tilde{D}_k)\}\Delta U_k^{*(m)}$$
$$= -J(U_k^{(m-1)} - U_k^n) + \frac{1}{2}(RHS_k^n + RHS_k^{(m-1)})$$
$$\tag{74}$$

$$U_k^{*(m)} = U_k^{(m-1)} + \Delta U_k^{*(m)}$$

$$JU_k^{(m)} = JU_k^{*(m)} - \frac{1}{2}\Delta t\, Jg_{ki}\frac{\partial\phi^{(m)}}{\partial\xi_i} \tag{75}$$

$$\frac{\partial}{\partial\xi_i}\left(Jg_{ij}\frac{\partial\phi^{(m)}}{\partial\xi_j}\right) = \frac{2}{\Delta t}\frac{\partial JU_k^{*(m)}}{\partial\xi_k} \tag{76}$$

The iterative calculation is performed at each time step.

In the psudo-compressibility methods and the implicit SMAC schemes the computation of the right hand side is implemented using the higher-order TVD scheme, and the time-integration is performed using the approximate-factorization(AF), the LU-SGS algorithm or the following improved AF(Shin *et al.* 1995). For the scalar equation

$$(1 + c_i^+\nabla_i + c_i^-\Delta_i)\,u = rhs$$

the equations of the improved AF method are expressed as

$$c_0\bar{u}_{0,0,0} - c_1^+\bar{u}_{-1,0,0} + c_1^-\bar{u}_{1,0,0} = rhs$$
$$c_0\bar{\bar{u}}_{0,0,0} - c_2^+\bar{\bar{u}}_{0,-1,0} + c_2^-\bar{\bar{u}}_{0,1,0} = c_0\bar{u}_{0,0,0}$$
$$c_0 u_{0,0,0} - c_3^+ u_{0,0,-1} + c_3^- u_{0,0,1} = c_0\bar{\bar{u}}_{0,0,0}$$
$$\tag{77}$$
$$c_0 = 1 + |c_1| + |c_2| + |c_3|$$

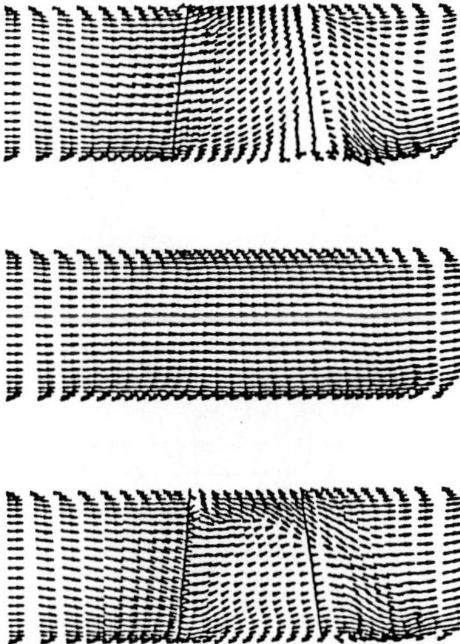

Fig.28: Velocity vectors of axial-flow turbo-fan near suction side of blade, mid-passage, near pressure side of blade from top [Shin(1991)]

The simultaneous linear equations with a tridiagonal matrix are implicitly solved by the Gaussian elimination, and the computer time of the improved AF is rather less than the explicit LU-SGS algorithm.

6.4 Applications to Cascade Flows

The calculated results of the flows through a decelerating cascade composed of NACA 65(12)10 blades are first shown(Shin *et al.* 1993a). The Reynolds number $Re = 2.45 \times 10^6$ and the incidence angle $\alpha = 12.1°$. The computation is performed by the implicit SMAC scheme using the grid points 165×55. Figure 26 shows the calculated instantaneous velocity vectors, instantaneous pressure contours and time-averaged pressure contours. As increasing α, small-scale vortices emerge successively on the suction side of the blade, and these vortices shedding into the wake cause large pressure fluctuation on the blade. Figure 27 shows a comparison of the time-averaged pressure profiles on the blade surfaces with the experimental data by Emery et al.(1954). Here, the low Reynolds number k-ε model by Myon-Kasagi was used.

Next, the calculated results of the flows through an impeller of axial-flow turbo-fan are shown(Shin 1991). The computation is also performed by the implicit SMAC scheme using the grid points of $61 \times 25 \times 25$. Figure 28 shows the calculated velocity vectors on the hub-to-casing surfaces of near suction side of blade, mid-passage and near pressure side of blade. From these figures a separation bubble at the hub corner on the suction side wall, and the flow toward tip in the separation zone near the trailing edge and the reaction flow in the wake are observed.

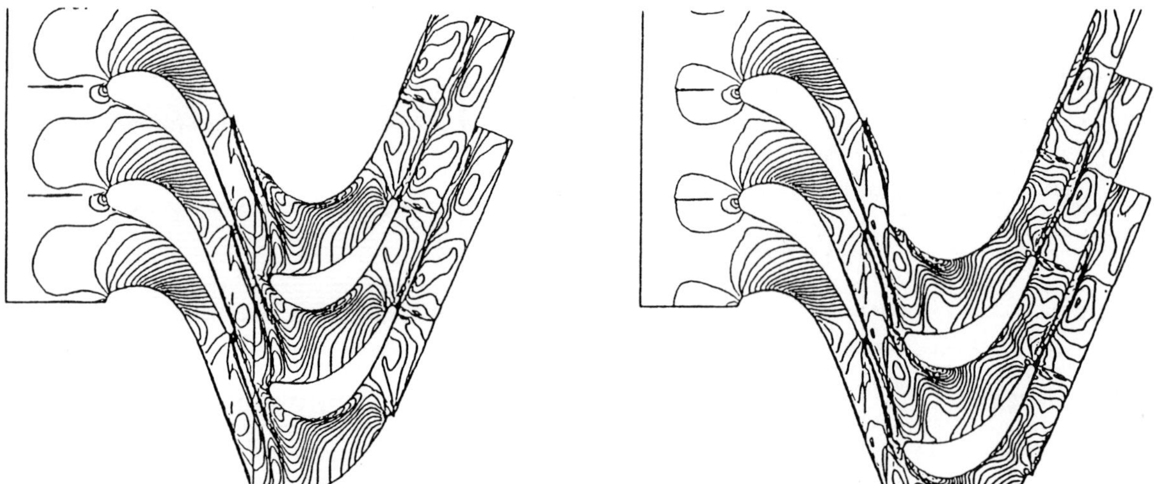

Fig.29: Instantaneous absolute Mach number contours of turbine stage(Yamamoto-Daiguji 1990)

7. UNSTEADY CASCADE FLOW PROBLEMS

7.1 Stator/Rotor Interactions

Most turbomachines have some stages. The stage consists of a stator(diffuser vanes, guide vanes) and a rotor(impeller, runner). The flows of the stator and the rotor mutually interact and periodically oscillate due to their relative motion. In the case that the stator and rotor have the same pitch, the flows can be completely obtained by computing each one passage under the periodic conditions. And in the case of a simple pitch ratio like 3:2 the flows can be determined from the computation of two stator passages and three rotor passages. Figure 29 shows the calculated Mach number contours of absolute flow through a transonic axial-flow turbine stage(Yamamoto-Daiguji 1990). In the computation the Chien's low Reynolds number k-ε model, the grids points of 71×41 and 91×41 and the third-order Chakravarthy-Osher type MUSCL TVD scheme are used. It is found that the wakes of the stator smoothly pass through the interface and reach to the rotor, and that the flow in the rotor behaves fairly unsteady and the shock waves periodically appear at the rotor exit. One can not obtain such a satisfactory result until the use of the high resolution scheme and an adequate numerical connection at the interface.

In the case of the general pitch ratio, there are two strategies in which the flows of stator and rotor can be determined from the computation of each one passage. The first strategy was proposed by Erdos-Alzner(1977a, b). The periodic boundary and interface conditions are explained as follows. Defining the pitches of stator and rotor by p_S and p_R, then the periods of the stator and rotor flows become $\tau_S = rp_R/u = p_R/\omega$ and $\tau_R = p_S/\omega$, respectively, where $u = \omega r$ is

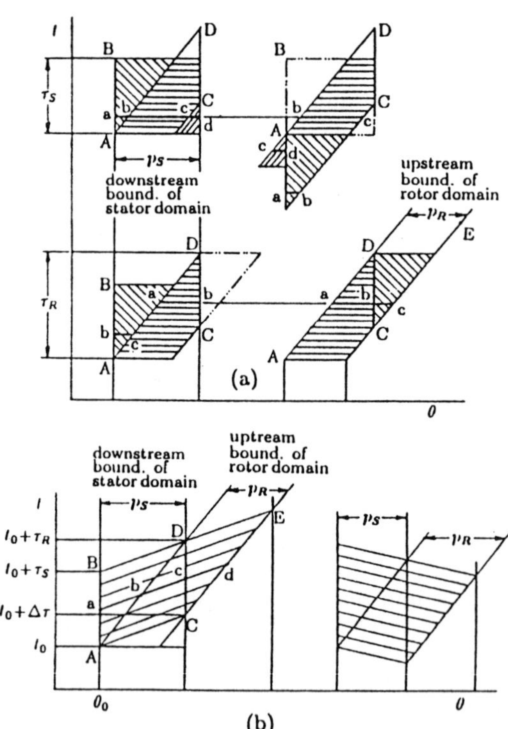

Fig.30 Interface conditions in stator/rotor interaction
(a) Erdos-Alzner method, (b) Giles method

the circumferential velocity of the rotor. Therefore, the flows at points A, B, C, D, E in Fig.30 are just the same. On the periodic boundary conditions, the right boundary condition of the stator is given as the values at the time $\Delta\tau = \tau_R - \tau_S$ before on the corresponding

Fig.31: Calculated static pressure distribution of modified DFVLR transonic fan stage [Naka-mura *et al.*(1996)]

left boundary[Fig.30(a) top]. Similarly, the remaining boundary conditions can also be given by the past values on the corresponding boundaries. On the interface conditions, the downstream boundary condition of the stator domain(a—d) is given as the values on the corresponding upstream boundary of the rotor domain at the present(b—c), the time $\Delta\tau$ before(c—d) and the time τ_S before(a—b)[Fig.30(a) top]. Similarly, the upstream boundary condition of the rotor domain can be given by the present and past values on the downstream boundary of the stator domain[Fig.30(a) bottom]. In this method the explicit method and the dividing streamline fit grid must be used.

In the theory of partial differential equations, the initial surface of the Cauchy problem is possible to take not only $t = $ const. but also $\phi = \phi(x, t)$ surface(Courant 1962). However, no one has noticed till quite recently that this initial surface is successfully utilized to the stator/rotor interaction problem. Giles(1988, 90) proposed the second strategy using the time-inclined computational domain. The domain is taken parallel to the line AC in Fig.30(b) as

$$t = t_0 + \frac{\Delta\tau}{p_S}(\theta - \theta_0) = t_0 + \left(1 - \frac{p_R}{p_S}\right)\frac{\theta - \theta_0}{\omega} \qquad (78)$$

In such domain both the flows of stator and rotor change in cyclic with the same wave length p_S, irrespective the different pitches. Therefore, the periodic boundary conditions can be in simple imposed like the case of the same pitch. On the interface conditions, for an interface a—d in Fig.30(b) the part b—c is a true interface of both domains, and the downstream boundary a—b of the stator domain contacts to the upstream boundary c—d of the rotor domain. The computation during one cycle starts at AC and ends at BDE, and

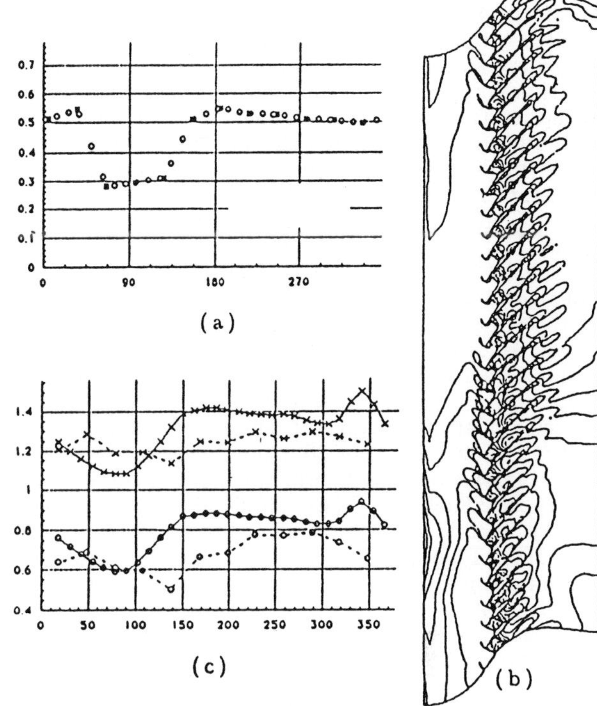

Fig.32: (a) Inlet Mach number distributions of axial-flow compressor stage o input data in computation, ■ experimental data, (b) Calculated Mach number distribution on developed mid-span plane of axial-flow compressor stage and (c) Exit total pressure(×) and Mach number(o) distributions of axial-flow compressor stage ——— computation, — — — experimental data [Kikuchi *et al.*(1997)]

further is repeated until the complete and periodic solution is obtained. The 3D analyses of the stator/rotor interaction have been performed by the Giles' method(Jung *et al.* 1995).

7.2 Large-Scale Computations of Whole Cascade Passages Utilized NWT

The Numerical Wind Tunnel(NWT) is a parallel computer installed in National Aerospace Laboratory of Japan in 1993, which had 140 processing elements(PE), the peak speed of 236 GFLOPS and the main memory of 35.8 GB, and was reinforced 166 PE and 280 GFLOPS in 1996.

In the analyses of the fan blade flow of a turbo-fan engine with inlet distortion and of the stall propagation of a multistage axial-flow compressor it is necessary to compute the whole passages of the blade row simultaneously. A computational method that each passage (or three passages) is allotted to a PE has been developed. In the method the Reynolds-averaged Navier-Stokes equations in curvilinear coordinates with the Baldwin-Lomax algebraic turbulence model are solved using the approximate-factorization, the time-integration by the Newton iteration and the Chakravarthy-Osher TVD scheme. This numerical method was validated for the counter rotating propeller by Matsuo(1993) such that the thrust coefficients are in satisfactory coincidence with the experimental data. And the method to allot the passages to the PEs was also verified by Nakamura *et al.*(1996) for the DFVLR transonic fan stage. The computational grid has 112×51×21 points for the rotor and 92×30×21 points for the stator. Figure 31 shows the calculated static pressure distribution of the fan stage without inlet distortion.

Recently, unsteady flow through a compressor rotor with inlet distortion is analyzed by Kikuchi *et al.* (1997). The rotor has 30 blades and each passage has 119×51×21 grid points, therefore the total grid points are 3.8 million. The computation was performed for the inlet Mach number distributions as shown in Fig.32(a). Figures 32(b) and (c) show the calculated instantaneous Mach number contours on the developed mid-span plane and the total pressure and Mach number distributions of the mid-span at the rotor exit. The computational results do not well coincide with the experiment yet.

Recently the interaction of centrifugal impeller with diffuser vanes is also solved(Dawes 1995, Yamane-Nagashima 1995). Yamane-Nagashima calculated the flows of two passages of the impeller and three passages of the diffuser vanes under the usual periodic boundary conditions, though the Giles method is applicable to this flow. The computational grid with 100×48×49 points is employed for a passage of both the impeller and diffuser vanes, and the numerical method is modified by the flux vector splitting by Anderson-Thomas-van Leer(1986) and by the LU-ADI factorization algorithm by Obayashi *et al.*(1986). Figure 33 shows the calculated instantaneous density contours on the mid-span in the beginning of choking.

7.3 Unsteady Incompressible Cascade Flow Analyses Using Stream Function

The equation of the diagonal cascade flow on a revolutional surface is

$$\frac{\partial}{\partial m}\left(\frac{r}{b}\frac{\partial \psi}{\partial m}\right) + \frac{\partial}{\partial \theta}\left(\frac{1}{br}\frac{\partial \psi}{\partial \theta}\right) = -r\zeta_r \tag{79}$$

where b is the breadth of passage, ζ_r the vorticity in the relative flow and ψ the stream function defined as

$$\frac{\partial \psi}{\partial \theta} = brw_m, \quad \frac{\partial \psi}{\partial m} = -bw_u \tag{80}$$

The cascade flow on the revolutional surface can be transformed to the straight axis cascade flow on xy-plane using the mapping functions

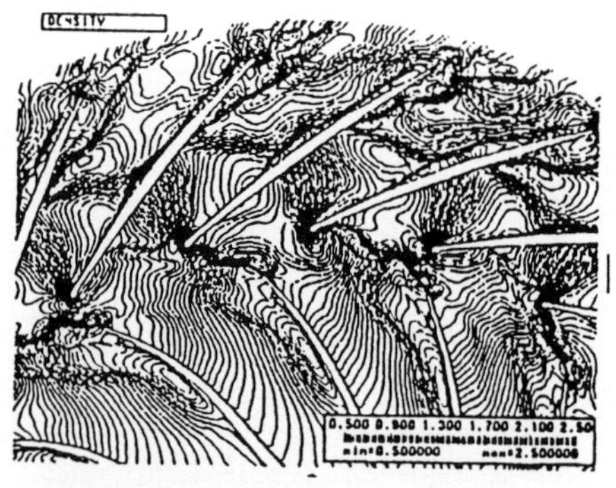

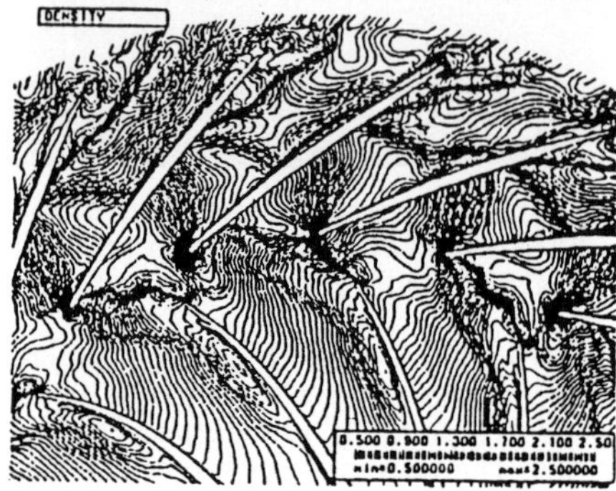

Fig.33: Instantaneous pressure contours on mid-span in choked flow of centrifugal compressor stage [Yamane-Nagashima(1995)]

$$dx = \frac{dm}{r}, \quad dy = d\theta \tag{81}$$

Then, the equation of ψ is rewritten as

$$\frac{\partial}{\partial x}\left(\frac{1}{b}\frac{\partial \psi}{\partial x}\right) + \frac{\partial}{\partial y}\left(\frac{1}{b}\frac{\partial \psi}{\partial y}\right)$$
$$= 2\omega r^2 \sin\delta - r^2\zeta - r\gamma\,\delta(y - \bar{y}) \tag{82}$$

where ω is the angular velocity of impeller, δ the inclined angle of revolutional surface($\sin\delta = dr/dm$), γ the strength of trailing vortex and $\delta(y - \bar{y})$ the δ-function. The right hand side of Eq.(82) means the vorticity, and the first term is the vorticity appearing in the rotating frame and the second term the vorticity of inflow prescribed at the upstream boundary. The boundary conditions are

$$\frac{\partial \psi}{\partial x} = \begin{cases} br\{\omega r - w_m \tan\alpha\} & \text{(upstream)} \\ -brw_m \tan\beta & \text{(downstream)} \end{cases}$$
$$\psi = 0 \text{ and } \Phi \qquad \text{(blades)}$$
$$\begin{aligned} \psi(x, y_1) &= \psi(x, y_0) + \Phi \\ \partial\psi(x, y_1)/\partial y &= \partial\psi(x, y_0)/\partial y \end{aligned} \quad \text{(periodic)}$$
$$\tag{83}$$

where $w_m(y, t)$ is the meridian velocity and given considering the wake of the inlet guide vanes, and $\Phi = \int_0^1 w_m br\, d\theta$ is the flow rate through a passage.

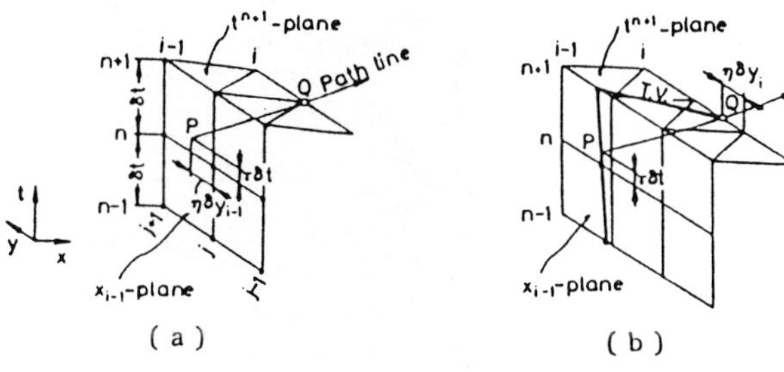

Fig.34: Calculations of ζ and γ by convective-difference method

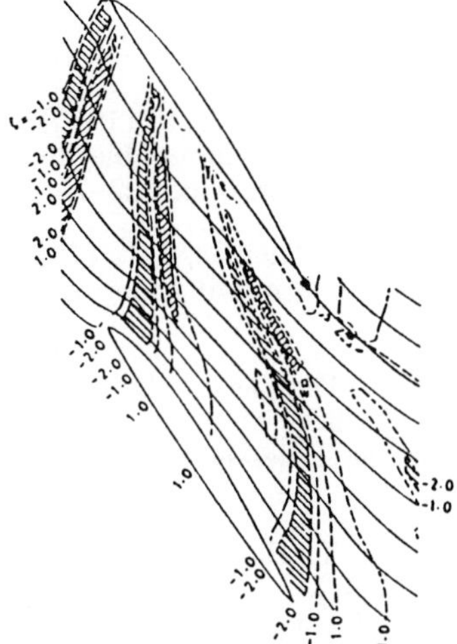

Fig.35: Instantaneous streamlines and equi-vorticity lines on revolutional flow surface [Daiguji-Shirahata(1980)]

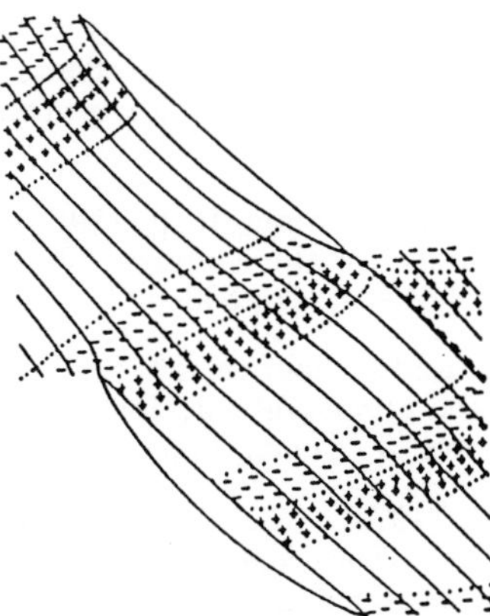

Fig.36: Instantaneous streamlines and a part of positive(+), negative(−) and zero(·) discrete vertex points of diagonal-flow pump [Shirahata-Daiguji(1982)]

The transport equation of vorticity ζ is

$$\frac{d}{dt}\left(\frac{\zeta}{b}\right) = 0 \qquad (84)$$

The initial value is

$$\zeta_U = -\frac{1}{r}\sec^2\alpha\,\frac{\partial}{\partial y}w_m(y,t) \qquad (85)$$

And the transport equation of trailing vortex γ is

$$\frac{d}{dt}\left(\frac{\bar{w}_m\gamma}{\Phi}\right) = 0 \qquad (86)$$

The initial value of γ, that is, the strength of shed vortex is

$$(\bar{w}_m\gamma)_T = -\int_0^1\left(\frac{\partial}{\partial t}w_u r\,d\theta + \frac{\zeta}{b}d\psi\right) \qquad (87)$$

β_D is determined from the Kutta condition. The values of γ_T and β_D must be calculated by the predictor-corrector-method.

The equation of ψ can be solved by the Galerkin FEM using the triangular linear elements, and the transport equations of ζ and γ by a convective-difference scheme. That is, the values of ζ/b and $w_m\gamma/\Phi$ are determined by interpolation on the upstream surfaces shown in Fig.34. The quadratic interpolation formula by Matsumoto-Daiguji(1992) is

$$\begin{aligned}
f_i^* = \ &f_i + \alpha^-\Delta f_{i-1/2} + \alpha^+\Delta f_{i+1/2}\\
&+ \frac{1}{2}(|\alpha| - \alpha^2)\,\mathrm{minmod}(\Delta f_{i-1/2}, b\Delta f_{i+1/2})\\
&- \frac{1}{2}(|\alpha| - \alpha^2)\,\mathrm{minmod}(\Delta f_{i+1/2}, b\Delta f_{i-1/2})
\end{aligned} \qquad (88)$$

where $1 < b \le 3$ and $\alpha = (x_i^* - x_i)/\Delta x$. This interpolation formula is very stable.

The unsteady 2D cascade flows on a revolutional flow surface in a diagonal-flow water turbine are analyzed(Daiguji-Shirahata 1980). Figure 35 shows the calculated instantaneous streamlines of relative flow and equi-vorticity lines of absolute flow. In the calculation by the convective-difference scheme somewhat large numerical diffusion of the vorticity is inevitable. On the other hand, in the calculation by the discrete vortex method the permanence of the vorticity can be completely kept. Next, the unsteady 2D cascade flows of a diagonal-flow pump are analyzed(Shirahata-Daiguji 1982). Figure 36 shows the calculated instantaneous streamlines of relative flow and the distribution of a part of singular points having positive, negative or zero value of the vorticity, and Fig.37 shows the time-variation of the relative velocity profiles on the blade surfaces during one period.

The simulation of the rotating stalls in the 2D cascade flow using the discrete vortex method has been performed by Nishizawa-Takata(1990, 1994). The flow in this simulation consists of outer flow and boundary layer flows around the blades.   The boundary layer

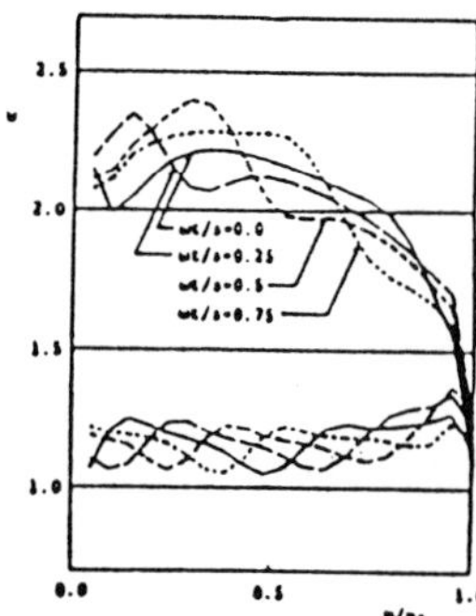

Fig.37: Instantaneous relative velocity profiles on blade surfaces of diagonal-flow pump [Shirahata-Daiguji(1982)]

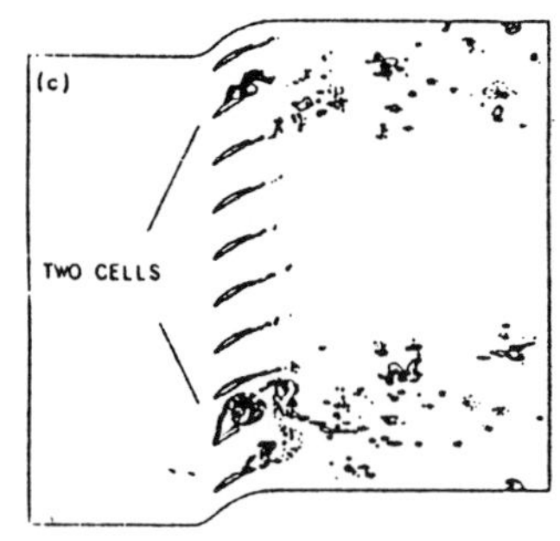

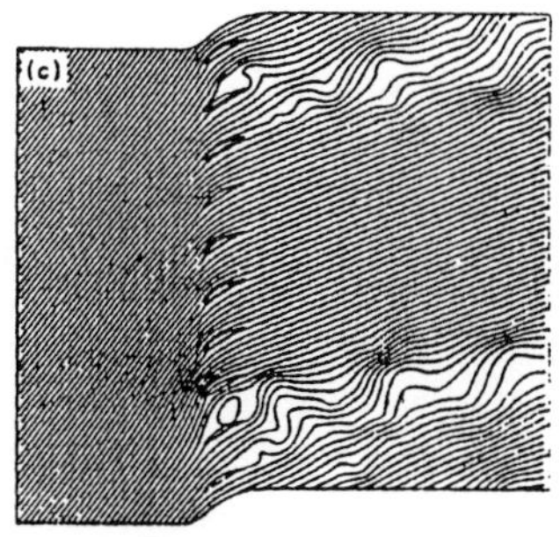

Fig.38: Vorticity contours and streamlines in rotating stall cascade flow with finite pitch [Nishizawa-Takata(1994)]

flows are obtained by solving the momentum integral equations using the pressure supplied from the outer flow computation. On the other hand, the outer flow is obtained by solving the Poisson equation of stream function and the vorticity transport equation in integral form using the displacement thickness and the shed vortices at separation points supplied from the boundary layer computation.

$$\frac{\partial^2 \psi}{\partial x^2} + \frac{\partial^2 \psi}{\partial y^2} = -\zeta \tag{89}$$

$$\frac{d}{dt}\int_\Omega \zeta \, dx = \oint_{\partial\Omega} \frac{\partial}{\partial n}\{(\nu + \nu_t)\zeta\} \, ds$$
$$+ \oint_{\partial\Omega}\left(\frac{\partial \nu_t}{\partial n}\frac{\partial w_n}{\partial s} + \frac{\partial \nu_t}{\partial s}\frac{\partial w_s}{\partial s}\right) ds \tag{90}$$

where n, s are the normal and tangential directions on $\partial\Omega$, respectively. Eq.(89) is solved by the FEM using the values of ψ on the blade surfaces determined from the one-valued pressure condition. The vorticity is calculated by the discrete vortex method, and the ν_t is given by the algebraic turbulence model. The calculated results of instantaneous vorticity contours and streamlines are shown in Fig.38.

Bibliography

Abe, K., Kondoh, T. and Nagano, Y. (1994). A new turbulence model for predicting fluid flow and heat transfer in separating and reattaching flows. — Part 1. Flow field calculations. *Int. J. Heat Mass Transfer*, **37**, 139–51.

Adler, D. (1980a). Status of centrifugal impeller internal aerodynamics, Part I: Inviscid flow prediction methods. *Trans. ASME, J. Engng. Power*, **102**, 728–37.

Adler, D. (1980b). Part II: Experiments and influence of viscosity. *Trans. ASME, J. Engng. Power*, **102**, 738–46.

Amsden, A. A. and Harlow, F. H. (1970). A simplified MAC technique for incompressible fluid flow calculations. *J. Comput. Phys.*, **6**, 322–25.

Anderson, W. K. and Thomas, J. L. (1988). Multigrid acceleration of the flux split Euler equations. *AIAA J.*, **26**, 649–54.

Anderson, W. K., Thomas, J. L. and Van Leer, B. (1986). Comparison of finite volume flux vector splittings for the Euler equations. *AIAA J.*, **24**, 1453–60.

Arima, T., Sonoda, T., Shirotori, M., Tamura, A. and Kikuchi, K. (1997). A numerical investigation of transonic axial compressor rotor flow using a low Reynolds number k-ε turbulence model.

Baldwin, B. S. and Lomax, H. (1978). Thin layer approximation and algebraic model for separated turbulent flows. AIAA Paper 78-275.

Basson, A. and Lakshminarayana, B. (1995). Numerical simulation of tip clearance effects in turbomachinery. *Trans. ASME, J. Turbomachinery*, **117**, 348–59.

Beam, R. M. and Warming, R. F. (1978). An implicit factored scheme for the compressible Navier-Stokes equations. *AIAA J.*, **16**, 393–402.

Bercovier, M. and Engelman, M. (1979). A finite element for the numerical solution of viscous incompressible flows. *J. Comp. Phys.*, **30**, 181–201.

Bindon, J. P. (1973). Stability and convergence of streamline curvature flow analysis procedure. *Int. J. num. Meth. Engng.*, **7**, 69–83.

Caretto, L. S., Gosman, A. D., Patankar, S. V. and Spalding, D. B. (1972). Two calculation procedures for steady, three-dimensional flows with recirculation. Lecture Notes in Physics, Vol.II, 60–68,

Springer-Verlag.

Caspar, J. R., Hobbs, D. E. and Davis, R. (1980). Calculation of two-dimensional potential cascade flow using finite area methods. *AIAA J.*, **18**, 103–109.

Cebeci, T. and Smith, A. M. O. (1974). *Analysis of Turbulent Boundary Layers*. Academic Press.

Chakravarthy, S. R. and Osher, S. (1985). A new class of high accuracy TVD schemes for hyperbolic conservation law. AIAA Paper 85-0363.

Chorin, S. J. (1967). A numerical method for solving incompressible viscous flow problems. *J. Comput. Phys.*, **2**, 12–26.

Chien, K. Y. (1982). Predictions of channel and boundary-layer flows with a low-Reynolds-number turbulence model. *AIAA J.*, **23**, 1308–19.

Courant, R. (1962). *Methods of Mathematical Physics, Vol.II Partial Differential Equations*. Chapter IV, Interscience Publishers.

Daiguji, H. (1967). A simplified analysis of cascade flow in mixed-flow pumps and turbines. *Bull. JSME*, **10**, 496–506.

Daiguji, H. (1983a). Numerical analysis of 3-D potential flow in centrifugal turbomachines. *Bull. JSME*, **26**, 1495–1501.

Daiguji, H. (1983b). Finite element analysis for 3-D compressible potential flow in turbomachinery. Proc. of 1983 Tokyo Int. Gas Turbine Congr., 455–62.

Daiguji, H. and Takano, T. (1986). A finite element analysis for 3-D potential cascade flow including upstream and downstream effects[in Japanese]. *Trans. JSME*, Ser.B, **52**, 3139–45.

Daiguji, H. and Shirahata, H. (1980). Upstream cascade effects in diagonal cascade flow analysis. Proc. of 10th Symp. on International Association for Hydraulic Research, Tokyo, Vol.1, 571–82.

Daiguji, H. and Yamamoto, S. (1989). Numerical methods for fluid flows[in Japanese]. *J. Gas Turbine Soc. Japan*, **17**, 4–10.

Daiguji, H., Yuan, X. and Yamamoto, S. (1997). Stabilization of higher-order high resolution schemes for the compressible Navier-Stokes equations. *Num. Meth. Heat & Fluid Flow*, **7**, 250–74.

Dawes, W. N. (1983). An efficient implicit algorithm for the equations of 2D viscous compressible flow. *Int. J. Heat and Fluid Flow*, **4**.

Dawes, W. N. (1986). Computation of off-design flows in a transonic compressor rotor. *Trans. ASME, J. Engng. Power*, **108**, 144–50.

Dawes, W. N. (1995). A simulation of the unsteady interaction of a centrifugal impeller with its vaned diffuser: Flow analysis. *ASME J. Turbomachinery*, **117**, 213–22.

Degani, D. and Schiff, L. B. (1986). Computation of Turbulent supersonic flows around pointed bodies having crossflow separation. *J. Comp. Phys.*, **66**, 173–96.

Demirdzic I., Gosman, A. D., Issa, R. I. and Peric, M. (1987). A calculation procedure for turbulent flow in complex geometrics. *Comput. & Fluids*, **15**, 251–73.

Denton, J. D. (1974). A time-marching method for two- and three-dimensional blade to blade flow. ARC R & M 3775.

Denton, J. D. (1983). An improved time-marching method for turbomachinery flow calculation. *Trans. ASME, J. Engng. Power*, **105**, 514–24.

Dodge, P. R. (1977). *Trnsonic Flow Problems in Turbomachinery* (Adamson, Jr., T. C. and Platzer, M. F., ed.), 253– , Hemisphere Pub. Co.

Ecer, A. and Akay, H. U. (1981). Investigation of transonic flow in a cascade using the finite element method. *AIAA J.*, **19**, 1174–82.

Erdos, J. I. and Alzner, E. (1977). Computation of unsteady transonic flows through rotating and stationary cascades. Vol.I—Method of analysis. NASA CR 2900.

Erdos, J. I., Alzner, E. and McNally, W. (1977). Numerical solution of periodic transonic flow through a fan stage. *AIAA J.*, **15**, 1559–68.

Fox, L. (1947). Some improvements in the use of relaxation methods for the solution of ordinary and partial differential equations. *Proc. Roy. Soc. London*, Ser.A, **190**, 31–59.

Fujii, K. and Obayashi, S. (1986). Practical applications of new LU-ADI scheme for the three-dimensional Navier-Stokes computations of transonic flows. AIAA Paper 86-0513.

Fukuda, M., Kikuchi, K., Tamura, A., Hashimoto, K. and Matsuoka, A. (1990). Unsteady flow analysis in two-dimensional compressor cascade[in Japanese]. Porc. of 8th NAL Symp. on Aircraft Computational Aerodynamics, NAL SP-14, 221–26.

Giles, M. B. (1988). Calculation of unsteady wake/rotor interaction. *AIAA J. Propulsion Power*, **4**, 356–62.

Giles, M. B. (1990). Stator/rotor interaction in a transonic turbine, *AIAA J. Propulsion Power*, **6**, 621–27.

Grasso, F. and Faleoni, D. (1993). High-speed turbulence modeling of shock-wave/boundary-layer interaction. *AIAA J.*, **31**, 1199–1206.

Hanjalić, K., Jakirlić, S. and Hadžić, I. (1997). Expanding the limits of "equilibrium" second-moment turbulence closures. *Fluid Dynamics Research*, **20**, Special Issue, 25–41.

Hanjalić, K. and Launder, B. E. (1972). A Reynolds stress model of turbulence and its application to thin shear flows. *J. Fluid Mech.*, **52**, 609–38. Harlow, F.H. and Welch, J. E. (1965). Numerical calculation of time dependent viscous incompressible flow of fluid with free surface. *Phys. Fluids*, **8**, 2182–89.

Harten, A. (1989). ENO schemes with subcell resolution, *J. Comp. Phys.*, **83**, 148–84.

Harten, A., Engquist, B., Osher, S. and Chakravarthy, S. R. (1987). Uniformly high order accurate essentially non-oscillatory schemes, III. *J. Comp. Phys.*, **71**, 231–303.

Hirsch, Ch. and Warzee, G. (1976). A finite-element method for through flow calculations in turbomachines. *Trans. ASME, J. Fluids Engng.*, **98**, 403–21.

Hirsch, Ch. and Warzee, G. (1979). An integrated quasi-3-D finite element calculation program for turbomachinery flows. *Trans. ASME, J. Engng. Power*, **101**, 141–48.

Horiuti, K. (1996). A new dynamic two-parameter mixed model for large-eddy simulation. *Phys. Fluids*,

Hughes, T. J. R., Liu, W. K. and Brooks, A. (1979). Finite element analysis of incompressible viscous flows by the penalty function formulation. *J. Comput. Phys.*, **30**, 1–6.

Ikohagi, T. and Shin, B. R. (1991). Finite-difference scheme for steady incompressible Navier-Stokes equations in general curvilinear coordinates. *Comput. & Fluids*, **19**, 479–88.

Ishizaka, K., Ikohagi, T. and Daiguji, H. (1995). A high-Resolution numerical method for transonic non-equilibrium condensation flows through a steam turbine cascade. 6th Int. Symp. on CFD, Lake Tahoe, Vol.I, 479–84.

Ives, D. C. and Liutermoza, J. F. (1977). Analysis of transonic cascade flow using conformal mapping and relaxation techniques. *AIAA J.*, *15*, 647–52.

Jameson, A. and Yoon, S. (1987). Lower-upper implicit schemes with multiple grids for the Euler equations. *AIAA J.*, **25**, 929–35.

Johnson, D. A. and King, L. S. (1985). A mathematically simple turbulence closure model for attached and separated turbulent boundary layers. *AIAA J.*, **23**, 1684–92.

Jones, W. P. and Launder, B. E. (1972). The prediction of laminarization with a two-equation model of turbulence, *Int. J. Heat Mass Transfer*, **15**, 301–14.

Jung, A., Merz, R., Mayer, J. F. and Stetter, H. (1995). Calculation of stator/rotor interaction in a transonic turbine stage using multiple blade passages and time-inclining for three-dimensional flows. 6th Int. Symp. on CFD, Lake Tahoe, Vol.II, 552–59.

Karki, K. C. and Patankar, S. V. (1988). Calculation procedure for viscous incompressible flow in complex geometrics. *Numerical Heat Transfer*, **14**, 295–307.

Katsanis, T. (1964). Use of arbitrary quasi-orthogonals for calculating flow distribution in the meridional plane of a turbomachine. NASA TN D-2546.

Katsanis, T. and McNally, W. D. (1977). Revised FORTRAN program calculating velocities and streamlines on the hub-shroud mid channel stream surface of an axial, radial, or mixed flow turbomachine of annular duct. NASA TN D-8430, 8431.

Kawaguti, M. (1953). Numerical solution of the Navier-Stokes equations for the flow around a circular cylinder at Reynolds number 40. *J. Phys. Soc. Japan*, **8**, 747–57.

Kikuchi, K., Nozaki, O., Matsuo, Y., Mtsunaga, K., Takeuchi, H. and Hirai, K. (1997). Unsteady flow analysis of a compressor blade row with inlet distortion[in Japanese]. NAL SP.

Koshizuka, S., Oka, Y. and Kondo, S. (1989). A staggered differencing technique on boundary-fitted curvilinear grid for incompressible Navier-Stokes equations written with physical components. Proc. of 3rd Int. Symp. on CFD, Sendai, Vol.I, 163–69.

Krimerman, Y. and Adler, D. (1978). The complete three-dimensional calculation of the compressible flow field in turbo impellers. *J. Mech. Engng. Sci.* **20**, 149–58.

Kunz, R. F., Lakshminarayana, B. and Basson, A. H. (1993). Investigation of tip clearance phenomena in an axial compressor cascade using Euler and Navier-Stokes procedures. *ASME J. Turbomachinery*, **115**, 453–67.

Laskaris, T. E. (1978). Finite-element analysis of three-dimensional potential flow in turbomachines. *AIAA J.*, **16**, 717–22.

Launder, B. E., Reece, G.J. and Rodi, W. (1975). Progress in the development of a Reynolds-stress turbulence closure, *J. Fluid Mech.*, **68**, 537–66.

Launder, B. E. and Shima, N. (1989). Second-moment closure for the near wall sublayer: development and application. *AIAA J.*, **27**, 1319–25.

Launder B. E. and Spalding, D. B. (1974). The numerical computation of turbulent flow. *Comp. Meth. in Appl. Mech. Engng.*, **3**, 269–89.

Leonard, B. P. (1979). A stable and accurate convective modelling procedure based on quadratic upstream interpolation. *Computer Meth. Appl. Meth. Engng.*, **19**, 59–98.

MacCormack, R. W. (1969). The effect of viscosity in hypervelocity impact cratering, AIAA Paper 69-354.

MacCormack, R. W. and Lomax, H. (1979). Numerical solution of compressible viscous flows. *Ann. Rev. Fluid Mech.*, Vol.11, 289–316.

MacCormack, R. W. and Paulley, A. J. (1972). Computational efficiency achieved by time splitting of finite difference operators, AIAA Paper 72-154.

McDonald, P. W. (1971). The computation of transonic flow through two-dimensional gas turbine cascades. ASME Paper, 71-GT-89.

Matsumoto, Y. and Daiguji, H. (1992). Convective-difference scheme using a general curvilinear coordinate grid for incompressible viscous flow problems. *JSME Int. J.*, Ser.B, **35**, 559–64.

Matsuo, Y. (1993). An efficient numerical method for simulating unsteady viscous flows through a counter rotating propeller. Proc. of 5th Int. Symp. on Comput. Fluid Dyn., Sendai, Vol.II, 200–205.

Merkle, C. L. and Athavale, M. (1987). Time-accurate unsteady incompressible flow algorithms based on artificial compressibility. AIAA Paper 87-1137.

Murman, E. M. and Cole, J. D. (1971). Calculation of plane steady transonic flows. *AIAA J.*, **9**, 114–21.

Myong, H. K. and Kasagi, N. (1990). A new approach to the improvement of k-ε turbulence model for wall-bounded shear flow. *JSME Int. J.*, Ser.B, **33**, 63–72.

Nagafuji, T. (1976). Numerical analyses of internal flows of fluid machinery. JSME Papers 760-18, 211–16.

Nagafuji, T. and Morii, H. (1980). A flow study in Francis turbine runner. Proc. of 10th Symp. on International Association for Hydraulic Research, Tokyo, Vol.1, 583–95.

Nakahashi, K. (1986). FDM-FEM zonal approach for computations of compressible viscous flows. Lecture Notes in Physics, **264**, 494–98, Springer-Verlag, Berlin.

Nakahashi, K., Nozaki, O., Kikuchi, K. and Tamura, A. (1989). Navier-Stokes computations of two- and three-dimensional cascade flowfields. *AIAA J. Propulsion Power*, **5**, 320–26.

Nakamura, T., Iwamiya, T., Yoshida, M., Matsuo, Y. and Fukuda, M. (1996). Simulation of the 3 dimensional cascade flow with numerical wind tunnel(NWT). Proc. of Super Computing,

Nakamura, Y. and Takemoto, Y. (1985). Solutions of incompressible flows using a generalized QUICK method. Proc. Int. Symp. Comput. Fluid Dyn., Vol.II, 285–96.

Nishizawa, T. and Takata, H. (1990). Numerical analysis of rotating stall by a vortex model[in Japanese]. *Trans. JSME*, Ser.B, **56**, 618–27.

Nishizawa, T. and Takata, H. (1994). Numerical study on rotating stall in finite pitch cascades. ASME Paper, 94-GT-258.

Nozaki, O., Nakahashi, K. and Tamura, A. (1987). Numerical analysis of three dimensional cascade flow by solving Navier-Stokes equations. Proc. of 1987 Tokyo Int. Gas Turbine Congress, Vol.II, 325–31.

Obayashi, S., Matsushima, K., Fujii, K. and Kuwahara, K. (1986). Improvements in efficiency and reliability for Navier-Stokes computations using the LU-ADI factorization algorithm. AIAA Paper 86-0338.

Pouagare, M. and Delaney, R. A. (1986). Study of three-dimensional viscous flows in an axial compressor cascade including tip leakage effects using a SIMPLE-based algorithm. *Trans. ASME, J. Turbomachinery*, **108**, 51–58.

Pulliam, T. H. and Chaussee, D. S. (1981). A diagonal form of an implicit approximate factorization algorithm. *J. Comp. Phys.*, **39**, 347–63.

Rodi, W. (1976). A new algebraic relation for calculating the Reynolds stresses. *ZAMM*, **56**, 219–21.

Roe, P. L. (1980). The use of the Riemann problem in finite-difference schemes, Lecture Notes in Physics, **141**, 354–59.

Roe, P. L. (1982). Fluctuations and signals, a framework for numerical evolution problems. *Numerical Methods in Fluid Dynamics*, 219–57.

Roe, P. L. (1986). Discrete methods for the numerical analysis of time-dependent multidimensional gas dynamics. *J. Comp. Phys.*, **63**, 458–76.

Rogers, S. E. and Kwak, D. (1990). Upwind difference scheme for the time-accurate incompressible Navier-Stokes equations. *AIAA J.*, **28**, 253–62.

Rogers, S. E., Kwak, D. and Kiris, C. (1991). Steady and unsteady solutions of the incompressible Navier-Stokes equations. *AIAA J.*, **29**, 603–10.

Rosenfeld, M., Kwak, D. and Vinokur, M. (1991). A fractional step solution method for the unsteady incompressible Navier-Stokes equations in general coordinate systems. *J. Comput. Phys.*, **94**, 102–39.

Smagorinsky, J. (1963). General circulation experiments with the primitive equations . I. The basic experiment. *Mon. Weather Rev.*, **91**, 99–164.

Senoo Y. and Nakase, Y. (1972). An analysis of flow through a mixed flow impeller, *Trans. ASME, J. Engng. Power*, **94**, 43–50.

Seya, K. (1957). A numerical method for flow performances through a cascade with large camber and large solidity[in Japanese]. *Memo. Institute of High Speed Mechanics, Tohoku Univ.*, **13**, 13–73.

Shin, B. R. (1991). Numerical schemes for the incompressible Navier-Stokes equations and their applications. Ph.D. Thesis, Tohoku University.

Shin, B. R., Ikohagi, T. and Daiguji, H. (1993a). An unsteady implicit SMAC scheme for the two-dimensional incompressible Navier-Stokes equations. *JSME Int. J.*, Ser.B, **36**, 598–606.

Shin, B. R., Ikohagi, T. and Daiguji, H. (1993b). A finite-difference scheme for two-dimensional incompressible turbulent flows using curvilinear coordinates. *JSME Int. J.*, Ser.B, **36**, 607–11.

Shin, B. R., Ikohagi, T. and Daiguji, H. (1995). An implicit finite-difference scheme for the incompressible Navier-Stokes equations using an improved factored scheme. *CFD J.*, **4**, 191–208.

Shin, B. R., Yamamoto, S., Ikohagi, T. and Daiguji, H. (1988). Fractional step finite-difference scheme for solving the incompressible Navier-Stokes equations in curvilinear coordinates. Proc. of Soviet Union-Japan Symp. on CFD—Khabarovsk, Vol.2, 91-97.

Shirahata, H. and Daiguji, H. (1981). Subsonic cascade flow analysis by a finite element method. *Bulletin of the JSME*, **24**, 29–36.

Shirahata, H. and Daiguji, H. (1982). A numerical method of unsteady flow through a mixed flow impeller using finite element method and discrete vortex method. Prof. of 4th Int. Symp. on FEM in Flow Problems, Tokyo, 1027–34.

Shyy, W., Tong, S. S. and Correa, S. M. (1985). Numerical recirculating flow calculation using a body-fitted coordinate system. *Numerical Heat Transfer*, **8**, 99–113.

Soh, W. Y. and Goodrich, J. W. (1988). Unsteady solution of incompressible Navier-Stokes equations. *J. Comput. Phys.*, **79**, 113–34.

Southwell, R. V. (1946). *Relaxation Methods in Theoretical Physics*. Vol.II, Oxford Clarendon Press.

Steger, J. L. and Warming, R. F. (1981). Flux vector splitting of the inviscid gas-dynamic equation with applications to finite difference methods. *J. Comp. Phys.*, **40**, 263–93.

Takakura, Y., Ogawa, S. and Ishiguro, T. (1989). Turbulence models for 3D transonic viscous flows II. Proc. of 3rd Int. Symp. on CFD, Nagoya, Vol.II, 680–85.

van Leer, B. (1979). Towards the ultimate conservative difference scheme. V. A second-order sequel to Godunov's method. *J. Comp. Phys.*, **32**, 101–38.

van Leer, B. (1982). Flux vector splitting for the Euler equations. Lecture Notes in Physics, **170**, 507–512, Springer-Verlag.

van Leer, B. (1993). Progree in multi-dimensional upwind differencing. Lecture Notes in Physics, **414**, 1–26, Springer-Verlag.

Veuillot, J.-P. (1977). Calculation of the quasi three-dimensional flow in a turbomachine blade row. *Trans. ASME, J. Engng. Power*, **99**, 53–62.

Watanabe, T., Nozaki, O., Kikuchi, K. and Tamura, A. (1991). Numerical simulations of the flow through cascade with tip clearance. Numerical Simulation in Turbomachinery, ASME FED-Vol.20, 131–39.

Wilcox, D. C. (1988a). Reassessment of the scale determining equation for advanced turbulent models. *AIAA J.*, **26**, 1299–1310.

Wilcox, D. C. (1988b). Multiscale model for turbulent flows. *AIAA J.*, **26**, 1311–20.

Wilkinson, D. H. (1969). Stability, convergence and accuracy fo two-dimensional streamline curvature methdos using quasi-orthogonals. *Proc. I. Mech. Engrs.*, **184**, Part 3G(I).

Wu, C. H. (1952). A general theory of three-dimensional flow in subsonic and supersonic turbomachines of axial, radial, and mixed-flow types. *Trans. ASME*, **74**, 1363–80.

Yamamoto, K. and Engel K. (1997). Multi-block grid generation using an elliptic differential equation. AIAA Paper 97-0201.

Yamamoto, S. and Daiguji, H. (1990). Numerical simulation of unsteady turbulent flow through transonic and supersonic cascades. Lecture Notes in Physics, **371**, 485–89, Springer-Verlag.

Yamamoto, S. and Daiguji, H. (1991). A numerical method for the transonic cascade flow problem. (1991). *Comp. & Fluids*, **19**, 461–78.

Yamamoto, S. and Daiguji, H. (1993). Higher-order-accurate upwind schemes for solving the compressible Euler and Navier-Stokes equations. *Comp. & Fluids*, **22**, 259–70.

Yamane, T. and Nagashima, T. (1995). Flow choking and shock wave structure at diffuser vanes in a high speed centrifugal compressor. Proc. of Int. Symp. on Air Breathing Engines.

Yanenko, N. N. (1963). On the implicit difference computing methods for solving multidimensional heat conduction equations [in Russian]. *Izv. Uchebn. Zaved., Mathematika*, **4**, 148–57.

Yanenko, N. N. (1971). *The Method of Fractional Steps*. Springer-Verlag.

Yee, H. C. (1987). Construction of explicit and implicit symmetric TVD schemes and their applications. *J. Comp. Phys.*, **68**, 151–79.

Yokota, J. W. and Caughey, D. A. (1988). LU implicit multigrid algorithm for the three-dimensional Euler equations. *AIAA J.*, **26**, 1061–69.

Yoon, S. and Kwak, D. (1992). Implicit Navier-Stokes solver for three-dimensional compressible flows. *AIAA J.*, **30**, 2653–59.

Young, J.B. (1988). An equation of state for steam for turbomachinery and other flow calculations. *Trans. ASME, J. Engng. Gas Turbines and Power*, **110**, 1–7.

Young, J. B. (1992). Two dimensional, nonequilibrium, wet-steam calculations for nozzles and turbine cascades. *Trans. ASME, J. Turbomachinery*, **114**, 569–79.

Yuan, X. and Daiguji, H. (1993). Navier-Stokes simulations of transonic cascade flows considering unsteady wake influences. *CFD J.*, **2**, 307–18.

Yuan, X. and Daiguji, H. (1995a). Development of efficient Navier-Stokes solver for application to general large-scale turbine blade designs. Proc. of CSPE-JSME-ASME Int. Conf. on Power Engng., Shanghai, Vol.2, 573–83.

Yuan, X. and Daiguji, H. (1995b). A new LU-type implicit scheme for three-dimensional compressible Navier-Stokes equations. 6th Int. Symp. on CFD, Lake Tahoe, Vol.III, 1473–78.

CONTEMPORARY AUSTRALIAN CFD SCENE

K. SRINIVAS[1]   **C. A. J. FLETCHER**[2]

(1)   Senior Lecturer, Department of Aeronautical Engineering, University of Sydney, NSW 2006, Australia
(2)   Director, CANCES (Centre for Advanced Numerical Computation in engineering and Science), University of New South Wales, NSW 2052, Australia

Abstract

The paper presents a summary of some of the major activity in the area of Computational Fluid Dynamics within Australia. The summary is divided into various sections - Industrial CFD, DSMC, High Speed Flow, Multi Phase Flow, Application of CFD to study fundamental fluid flow processes and Algorithms. In every section the important research work undertaken is reviewed from the point of view of its aim and achievements.

1. Introduction

Computational Fluid Dynamics is perhaps the most versatile and powerful of the approaches to fluid dynamics, the others being the experimental and the theoretical. The relative merits of these methods have been a topic of debate in the past. But today the power of CFD and the benefits it can provide in analysis and design have been very well realised. Consequently, CFD is now a well developed discipline in many countries. Australia is no exception despite its geographical location. Active work in the area has been in progress in many of the universities and other research centres for many years. It is a formidable task to trace back to the beginning of CFD in Australia. Perhaps it is also not necessary. On the other hand it is interesting and useful to review the current scene of CFD activity in Australia. The present review attempts to highlight the major research projects, the achievements and the possible future directions CFD is likely to take. It is hoped that the reader will get a bird's eye view of the trends of CFD in Australia and will be able to appreciate the rapid and interesting developments taking place.

University, research institutions and industry are some of the key centres in any country that promote CFD activity. Australian universities are contributing substantially to the progress through many of the research projects with participation from students, research fellows and the university staff. It may also be noted that the industry is now just beginning to employ CFD in its ventures. As a result major share of this review is taken up by the work in the universities.

The review is divided into sections - Industrial CFD, DSMC, High Speed Flow, Multi Phase Flow, Application of CFD to study fundamental fluid flow processes and Algorithms. This classification does not adhere to any standards and is by no means unique. The only justification for it is that these classes form some of the major areas in which active work is being carried out in Australia today. It is believed that this structure of the review may provide the reader information about CFD in Australia in a ready manner.

As the authors of the review we have limited ourselves to collating all the information from various researchers and presenting them in an order for the reader to appreciate. We have not checked the validity of any claims made by researchers about their work.

Due to constraints on the length the review is limited to developments in Mechanical and Aeronautical Engineering disciplines with the exception of Metal Processing from University of Melbourne. We do acknowledge the significant progress made in the related areas like non - Newtonian flows, meteorological flows, geophysical flows etc. However, reviews of such work should be found in the specialised journals from such disciplines. For example a good review of work in the computation of Non - Newtonian fluids is found in Huilgol and Phan-Thien (1997).

Further, even from the disciplines included in the present review, some of the researchers were unable to provide summaries of their work. It is with regrets that we have to omit their work.

2. Industrial CFD

It is industry application that provides the real testing ground for the many concepts, algorithms and codes that are developed in research institutions. Industry offers the real life situation which is involved and far more complicated than the simple flows com-

---

Received on July 21, 1997.

puted to benchmark the codes in the laboratories. Complication may arise from too many physical processes being involved or because of complex geometries being encountered. Australian CFD has taken up the challenge provided by the local industry ; some of the important areas being Power Production, Building Ventilation and Metal Processing. In this section we review the work which employ CFD to compute some of the flows encountered in these fields.

2.1 Flow Through Industrial Boilers

Computation of flow through an industrial boiler from the point of view of analysis and design has received considerable attention at CANCES (Centre for Advanced Numerical Computation in Engineering and Science, University of New South Wales). The main application of this work is in the area of power production. As is well known, a boiler configuration is very complex with many heat transfer surfaces like heat exchanger, condenser, economiser etc. Further, the processes taking place inside a boiler are varied and complicated. Accordingly, the overall project includes many smaller ones addressing specific issues. Some of the major activity in progress at the centre is described in the following subsections.

2.1.1 Tube Bank Heat Exchangers

In this project, computational investigation of laminar and turbulent flow and heat transfer over tube banks in crossflow is being carried out. The steady-state Reynolds-Averaged Navier-Stokes equations are discretised by means of a cell-centred finite-volume algorithm. An Additive-Correction Multigrid-SIMPLEC algorithm (see Section 7 on Algorithms) with dynamic tuning of the relaxation coefficients has been developed to accelerate the convergence of the iterative solver. The computational domain is defined to take advantage of the symmetry conditions of the flow, resulting in sections of the domain being modelled independently and progressing the simulation deeper into the tube bank, using the conditions at the outlet of the previous section as inlet boundary conditions for the next domain, until the row(s) of interest are reached. It is found that a relatively fine grid is required to be able to predict heat transfer behaviour accurately. Two- dimensional results have shown good agreement with experiments (see Fig.1). Details can be found in references - Zdravistch, Fletcher, and Srinivas, 1994, Zdravistch, Fletcher, and Behnia, 1994, Zdravistch, Fletcher, and Behnia, 1995.

2.1.2 The Momentum and Energy Sources Formulation

Computation of flow and heat transfer over tube banks in crossflow as the one described above, can be demanding in terms of CPU time. As a remedy, a momentum algorithm with a band energy sources formulation capable of simulating the global flow and heat transfer behaviour has been developed (Zdravistch, Fletcher, and Reizes, 1996). It is assumed that the resistance to flow created by the presence of the tubes can be simulated as a drag force. The drag force

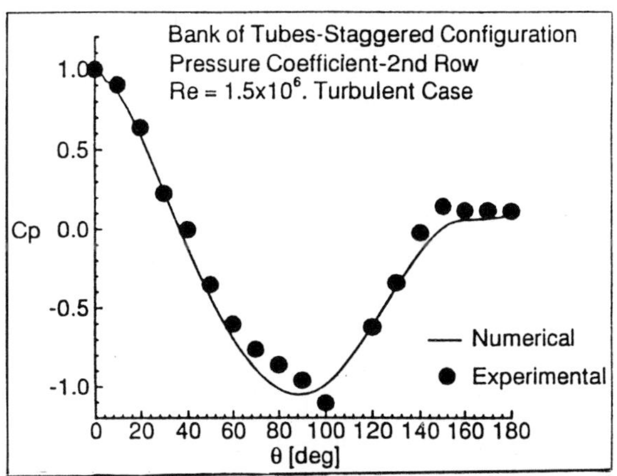

Fig.1: This figure shows the computed distribution of pressure coefficient over the second row of a staggered tube bank configuration. The Reynolds number is 1.5×10^6. Experimental results are due to Achenbach (1969).

is added as a "sink" in the momentum equations. Similarly, the amount of heat extracted from the hot fluid into the steam flowing inside the tubes can be represented as a "sink" added to the energy equation. Radiative heat transfer effects are accounted for using empirical relations. The strength of the momentum sinks is determined using a drag coefficient, which is set to match experimental values of pressure drop across the tube bank. The heat transfer coefficients are obtained using empirical relations, and the energy sinks are calibrated iteratively to maintain a heat balance over the tube bank. The formulation has been implemented over channel flows with isolated obstructions and as a distributed formulation to simulate tube bank effects in coal fired boiler configurations. Using this methodology, the global character of the velocity, pressure and temperature fields over complete tube banks can be obtained, without the need for simulating the detailed flow over the individual tubes.

2.1.3 Electrostatic Precipitation

Electrostatic precipitation is the industrial process by which particulate waste is removed from a stream of noxious gas. The process uses very high voltages and electric fields to charge the particulate waste, so that the electrical forces upon the particles cause them to follow different trajectories from the gas streamlines.

Electrostatic precipitation has very complex interaction mechanisms between the electric field, gas flow and particulate flow. The electrostatic body force acting on the gas by the electric field causes a secondary gas flow and contributes to the generation of turbulent kinetic energy. Ions generated by the corona discharge charge the particles in the gas stream. Charged dust particles migrate to the collecting plate by Coulomb force due to the electric field, but are also under the influence of the momentum interaction with the gas flow through aerodynamic drag. The presence of particle space charge affects the electric field and works to decrease corona current from the discharge electrode.

Complete modelling of the electrostatic precipitator (ESP) is a formidable task, as a number of different and interacting physical mechanisms govern its behaviour. In particular, the electrical behaviour of an ESP must be adequately described and linked to the fluid and particle dynamics. CANCES is developing comprehensive models that incorporate the detailed electrical, fluid and particle dynamics, and coupling these models to obtain new insights into ESP physics and design.

The governing equations for this model are the incompressible viscous Navier-Stokes equations for the gas, the Lagrangian particle equation of motion, and electrical potential and current continuity equations for the electrical corona. Further, a turbulence model is included for the gas phase. Special attention has been paid to the electrical corona in ESP's. The governing equations for the corona are a Poisson equation for the electrical potential and the current continuity equation.

A highly coupled model requires frequent, and therefore efficient, solution of these equations. This efficiency is realised by borrowing numerical techniques from analogous problems in computational fluid dynamics. The equations are written in a pseudotransient form and discretised using the finite volume method on a structured grid (Medlin, Fletcher and Morrow, 1996a, 1996b). The resulting equations are relaxed to a steady state by advancing in time using the approximate factorisation methodology. The Additive Correction Multigrid method is used to accelerate iterative convergence.(see Section 7 on Algorithms). Figure 2 shows the spatial electric field strength that forces the particles towards the walls of the duct.

Modelling the fluid and particle dynamics is the other main computational challenge in the ESP problem. Turbulent gas flow and particle trajectories under the influence of the electric field are calculated by using the commercial CFD package FLUENT combined with the finite volume solver for the electric potential and ion charge density. This enables one to solve the strongly coupled governing equations describing the motion of ions, gas and particles to get a detailed understanding of electrohydrodynamic interactions in an ESP (Choi and Fletcher, 1997). Figure 3 shows the predicted particle mass concentration from a coupled electrohydrodynamic flow simulation.

The next stage of ESP modelling at CANCES will involve progression to increasingly coupled steady state models and time-dependent models of the cou-

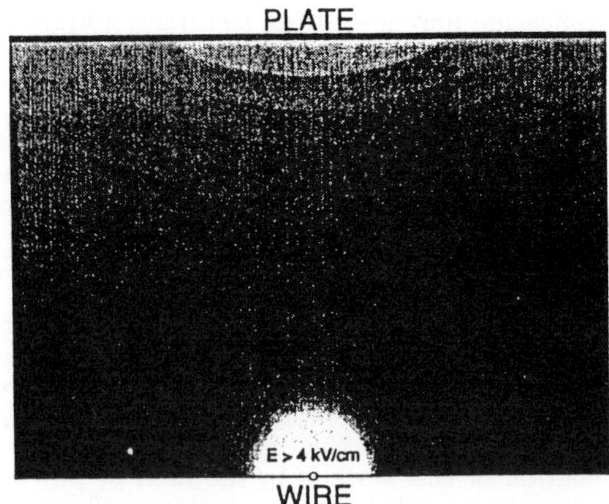

Fig.2: Contours of electric field strength in section of precipitator.

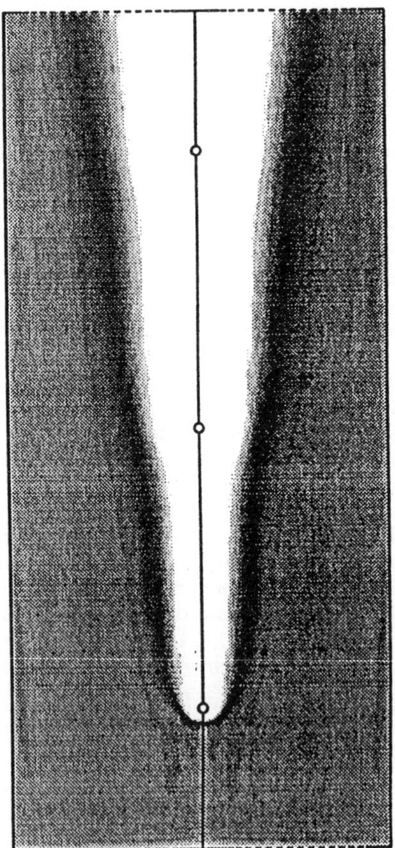

Fig.3: Particle mass concentration for 2.0 mm particles with inlet mass concentration of 1.0 g/m³.

pled electrical-fluid-particle dynamics. With reliable predictive models at hand, numerical experiments on novel ESP geometries will be conducted and techniques for improving overall ESP efficiency explored.

Some of the related work in progress at CANCES is described in Section 5 on Multiphase Flows and Section 7 on Algorithms.

2.2 CFD in Built Environment

CFD research interests at the Advanced Thermal Technologies Laboratory, CSIRO, lie in the development of numerical prediction techniques for practical building ventilation air flows involving heat transfer and/or dust dispersion. Recent work has been basically concerned with

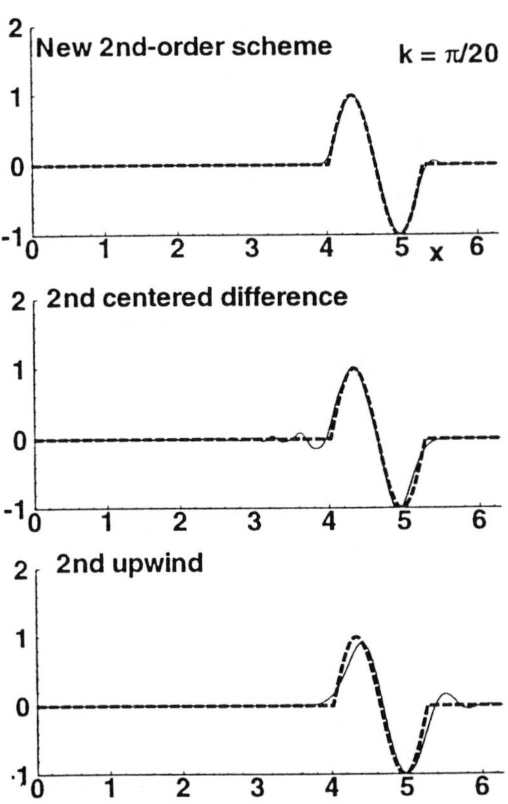

Fig.4: (a) Comparison of three convection schemes for a propagating sinusoidal wave packet.

1. Development and demonstration of an efficient multi-grid method with local grid refinement technique (Li and Fuchs, 1996) and higher-order numerical schemes (Li and Murray, 1995, Li and Baldacchino, 1995, Li, 1996).

2. Analysis of basic flow elements found in building air flows such as colliding free-convection boundary layers (Li et al., 1996).

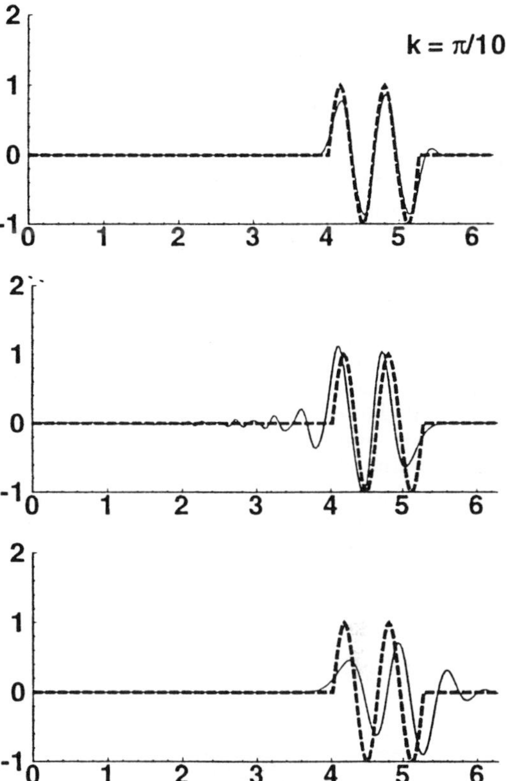

Fig.4: (b) Comparison of three convection schemes for a propagating sinusoidal wave packet.

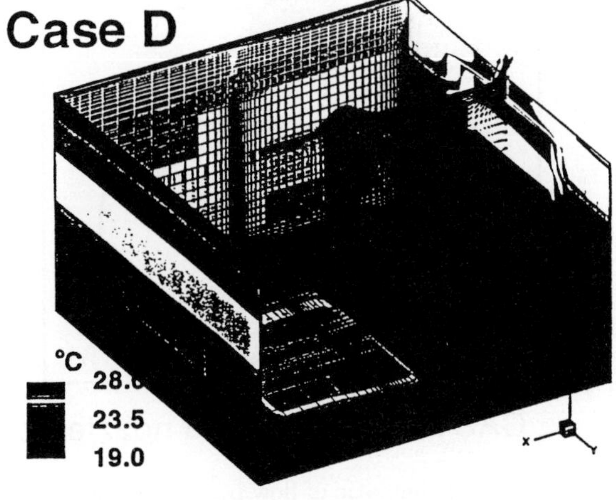

Fig.5: Surface temperature field and flow pattern in a room ventilated by displacement. The simulation considers the effect of surface thermal radiation among surface elements.

3. Application of developed numerical methods for simulating practical ventilation air flows and developing new design concepts in ventilation (Li and Delsante, 1996).

New wave number-extended high-order upwind-biased schemes have been developed (Li, 1996) using some additional constraints from the classical Fourier analysis. Both the wave number- and direction-dependent dispersion and dissipation characteristics are analysed for the conventional high-order upwind schemes up to 11th-order. This has provided a better overall quantitative control of numerical dispersion and dissipation. Figure 4 shows computed propagation of a sinusoidal wave packet. The fourth-order Runge-Kutta time stepping scheme was used and the time step so chosen that there is almost no error resulting from the temporal term discretisation. As the wave number increases, the advantages of the second-order wave number-extended scheme are clearly seen.

Future research topics envisaged in this laboratory include development of Large-Eddy Simulation methods for time dependent turbulent flows, accurate simulation of dusty two-phase flows and near-body natural convection flows, and analysis of thermal instability of enclosure flows (see also Fig.5)

2.3 CFD for Mineral and Metal processing

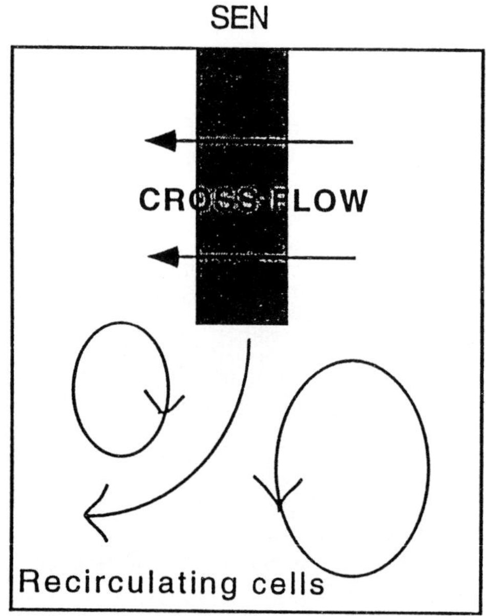

Fig.6: (a) Schematic of flow in a mould showing a single exit submerged entry nozzle (SEN).

The Department of Chemical Engineering, University of Melbourne concentrates on the application of CFD to various aspects of Mineral and Metal process-ing. One such is the process of liquid metal delivery during thin slab casting. It is intended to develop a computational model of submerged jet oscillation, oscillatory free surface flow and heat transfer for liquid injection into a model of a thin slab casting mould (Gebert, Davidson and Rudman, 1997). Of particular interest is the jet oscillation observed at higher casting speeds. Finite volume numerical techniques, based on established methods involving the SIMPLE algorithm are used to solve the transient flow equations, including the transport equations for the turbulence model. The scalar advection equation which defines the free surface is solved by a Volume of Fluid based method. The predicted oscillation is manipulated by the use of an effective resistance force in the model (see Figs.6(a) and (b)).

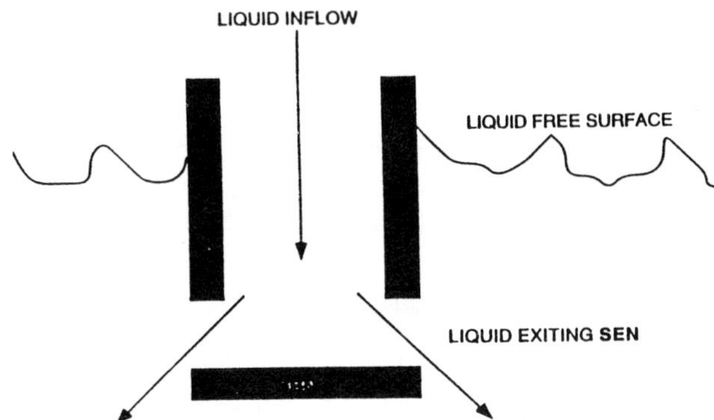

Fig.6:(b) Schematic of the region near a double exit submerged entry nozzle (SEN).

Some of the other investigations undertaken at this centre are summarised in Section 5 on Multi Phase Flows.

2.4 Heat Transfer and Fluid Flow in Cavities

The School of Mechanical and Manufacturing Engineering, University of New South Wales is considering various aspects of natural convection in cavities as well as cooling of electronic components. Combined convection, conduction and radiation heat transfer in an open cavity with a heat source (Dehghan and Behnia, 1996a) has been modelled numerically. Surface emissivity has been varied and its effect on the flow determined. It is found that radiation influences the flow significantly. The authors (Dehghan and Behnia, 1996b) have also modelled natural convection air cooling of two heat generating devices mounted on a conducting substate.

A Direct Numerical Simulation (DNS) of onset of unsteadiness, the route to chaos in natural convection of air in a cavity has been carried out by Le Quere and Behnia (1997). The numerical algorithm used integrates the Boussinesque type Navier - Stokes equations in velocity-pressure formulation with Chebyshev spatial approximation and a finite difference second

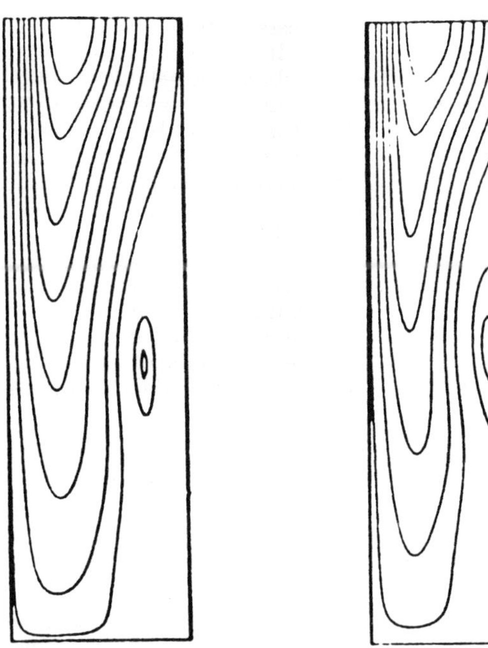

(a) Numerical

(b) Experimental

Fig.7: Comparisons between the predicted and observed flow patterns for cavity of aspect ratio 3.57 with bakelite plate: (a) $Ra^* = 2.08 \times 10^5$ (b) $Ra^* = 3.0 \times 10^5$.

order scheme for time marching. Very accurate solutions for air with Rayleigh numbers upto 10^{10} have been obtained. Internal gravity waves are shown to play an important role.

Numerical simulation of natural convection in liquid metals has also been performed (Esposito and Behnia 1994) and transition to oscillatory flow and chaos computed. Also the effect of free surfaces and thermocapillary on convection in liquid metals has been brought out by Behnia, Stella and Guj (1995a, 1995b). They have shown that thermocapillary action can drastically change the flow pattern and heat transfer behaviour due to natural convection.

The natural convection flow and heat transfer in an inclined open cylindrical thermosyphon representing an evacuated solar collector tube have been numerically simulated by Behnia, Gaa, Leong and Morrison (1996). It is shown that heat traction efficiency of the thermosyphon is strongly affected by the angle of inclination of the tube. The numerical predictions are validated by LDA velocity measurements.

Flow in the intake manifold of an engine has been numerically simulated by Milton, Behnia, Chen and Blakeley (1994). Fuel drops are injected into the carrier gas and by tracking them the formulation of the fuel film on the side walls is studied. In the study by Kim, Milton, Chen and Behnia (1995) the evaporation of the particles as well as the fuel film is simulated. Chen, Behnia and Milton (1995) have also included a butterfly valve in the manifold. It is shown that this obstruction significantly affects the behaviour of both the gas flow and the fuel flow.

A number of problems related to the cooling of electronic components and computers have been studied. Copeland, Behnia and Nakayama (1996a) have simulated the laminar flow of a dielectric liquid in a manifold microchannel heat sink. They have identified the location of maximum and minimum heat transfer coefficients for the design and optimisation of these heat sinks. In a companion study, Copeland, Behnia and Nakayama (1996 b) also took into account the effect of conduction in the walls of the microchannel. Simulations were performed for different flow rates and wall thermal conductivities. Mizunuma, Behnia and Nakayama (1996) simulated turbulent flow in microchannels etched in electronic components. They showed that the heat transfer performance of grooved channels is better than that of flat channels. The simulation results were validated with experimental data obtained in the laboratory. Nakyama and Behnia (1996) have discussed the role of CFD simulation in the design of heat sinks drawing examples from plate finned array cooled by planar impinging air flow and a simulated chip cooled by a dielectric coolant. CFD has also been exploited to clarify details not obtainable through experiments. The local heat transfer coefficients on a chip being cooled by air are shown in Fig.7.

3. DSMC (Direct Simulation Monte Carlo)

One of the success stories of Australian CFD is the pioneering work of Bird who has been involved in the development and application of the DSMC method for a number of years. Following his work, the activity in the area is now widespread throughout the world.

3.1 Development and Application of the DSMC Method

The direct simulation Monte Carlo (or DSMC) method is a computer simulation of the gas flow at the molecular level and does not depend on the conventional mathematical models such as the continuum Navier- Stokes or the discrete particle Boltzmann equations. The DSMC procedures and molecular models have been described in detail by Bird (1994).

The transport terms in the Navier-Stokes equations break down at sufficiently high Knudsen numbers and the DSMC method is widely used for basic studies and engineering applications that involve rarefied flows or extremely small physical dimensions. Reductions in computing costs mean that the method can now be applied at lower Knudsen numbers and to flows that involve small disturbances. Recent studies have included the development with time of the flow instabilities in axially symmetric Taylor-Couette flow (Bird 1996). Figure 8 is for the geometry as shown with the end-walls as planes of symmetry and a peripheral speed ratio of 3. It shows the relatively high Knudsen number at which vortices first appear and the much lower number (corresponding to a Reynolds number of 4,300) at which the flow becomes permanently chaotic.

Separate runs with different random number sequences were made for most of the Knudsen number cases. The development process was different in each run but, for Knudsen numbers just below the upper limit for vortex formation, the only difference was in a delay preceding the initiation of the vortex formation process. The average delay has been found (Bird, 1997) to increase with the number of simulated molecules in the calculation. The statistical fluctuations evidently play a role, possibly the entire role in the initiation of the disturbances. The level of fluctuations in a real gas depends on the number of molecules in a cubic mean free path which is inversely proportional to the square of the gas density. The fluctuations in a typical DSMC calculation correspond to those in a real gas at about three or four atmospheres. In this study, the DSMC method is used to obtain information that is inaccessible not only to the Navier-Stokes equations but also to the Boltzmann equation.

The study of flow instabilities is continuing and other calculations are concerned with thermal creep. This is a physical effect that can be important in practical flows, for example it can lead to convection in a zero-gravity situation, but is absent at the level of the Navier-Stokes model.

A graphical DSMC demonstration program and a general program and for either two-dimensional or axially symmetric flows are among files that can be downloaded from http://ourworld.compuserve.com/homepages/gabird.

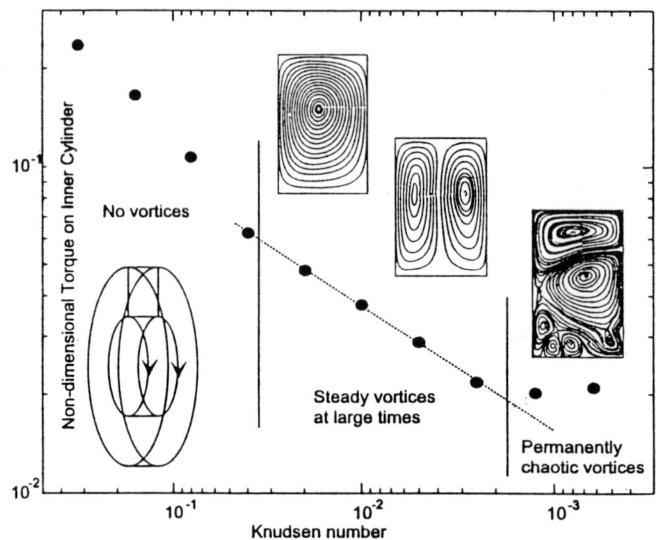

Fig.8: Flow regimes in axially-symmetric Taylor–Couette flow.

3.2 Application of DSMC to Nozzle Calculations for Near Continuum Flow

At the University of Queensland, the DSMC method has been modified to predict the flow in a new rarefied gas dynamics wind tunnel under development. In this tunnel, the hypervelocity flow produced by an expansion tube is to be expanded in a nozzle to produce a rarefied flow with a flow velocity in excess of 10 km/s. Typical Knudsen numbers will be of the order of 0.01; the product of Knudsen number and Mach number will be of $O(0.1)$ so that continuum fluid mechanics cannot be expected to be valid. Calculations are being performed with Navier-Stokes codes and Direct Simulation Monte-Carlo (DSMC) codes and will be compared with experimental measurements in order to determine the operating regime of the new facility.

The DSMC calculations are particularly difficult since a large portion of the flow in the nozzle is characterised by a very small Knudsen number (large collision rate) which makes the calculations prohibitively expensive. To make the calculations feasible, a modification of the DSMC method is used, originally due to Pullin (1980), called the equilibrium particle simulation method. This method is invoked in those parts of the flow where the expected collision rate is so large that it can be confidently predicted that the local state is characterised by a Maxwell-Boltzmann distribution of the simulator particles. The new method is described in an ICASE report (Macrossan 1995); in essence, local equilibrium amongst the simulator particles is enforced, without resorting to the collision calculations where the collision rate is extremely high. This method, which is a hybrid of continuum & rarefied methods, uses particle Monte-Carlo simulation throughout the flow and thus avoids some problems of statistical noise transfer which arise when continuum and Monte-Carlo particle based methods are amalgamated into a hybrid code.

4. High Speed Flow

It may be said that most of the CFD work carried out in Australia is in the incompressible flow area. But a few exceptions do occur. One such is the Department of Mechanical Engineering, University of Queensland which is known for its work on Shock Tubes. The other is the Department of Aeronautical Engineering, University of Sydney which is summarised in Sections 7.1 and 7.2. The activity at the University of Queensland is highlighted in the following sections.

4.1 Hypervelocity chemical reacting flows

A long term CFD research effort at the Department of Mechanical Engineering, University of Queensland has been directed to hypervelocity chemical reacting flows. Initial interest in this area was sparked by the experience of the first Shuttle Orbiter reentry flight, where the Shuttle experienced a nose-up pitching moment larger than expected. Although, non-inclusion of viscous effects in the pre-flight design calculations could be responsible for this, Stalker (1989) suggested that chemical non-equilibrium effects in the dissociated air could cause a significant forward shift in the centre of pressure and he used a simple analytical model to demonstrate this effect. This effect has been observed in CFD calculations for an ideal dissociating gas in a series of papers (Macrossan & Stalker 1987, Macrossan 1989, Macrossan 1990, Macrossan & Pullin 1994, Macrossan & Eckett 1996) which looked at the effect for a two dimensional flow about a blunt-nosed flat plate at an angle of attack, and the three dimensional flow about sharp and blunted cones at an angle of attack. The results show that, indeed, the center of pressure for a chemical non-equilibrium reacting flow can lie outside the limits set for the case of chemical frozen or chemical equilibrium flow; that is, the pressure distribution for the non-reacting case is not intermediate case between those found for no chemistry and equilibrium chemistry. This result highlights the need for CFD calculations in the design stage of hypervelocity flight vehicles.

As part of this work, the team at the University of Queensland, in collaboration with Caltech, has led the development of kinetic theory methods (as an alternative to Riemann solver methods) for hypervelocity flows. Starting from the original work by Pullin (1980) which introduced the Equilibrium Flux Method (EFM) the kinetic theory method has been extended (Macrossan & Oliver 1993) and its dissipative properties have been investigated (Macrossan 1989, Macrossan & Hancock 1996). The conclusion is that the kinetic theory based methods contain a natural dissipation, that is a dissipation which arises naturally from physically based modelling of molecular motions, and that the dissipation is larger than the true physical dissipation because the methods, in effect, contain a mean free path which scales with the local cell size. The dissipation decreases as the Mach number increases and EFM (and its derivatives) have been used as the basis of Navier-Stokes solvers by Macrossan & Oliver (1993) and Mallett, Pullin and Macrossan (1995). In the later work, the viscous flow over a blunt-nosed delta wing at an angle of attack was studied, with particular emphasis on the leeward flow. Separation was found to occur just inboard of the leading edge and the vortex which forms in the leeward flow causes intense local heating.

4.2 Applications of CFD

A suite of computer programs have been developed at the Department of Mechanical Engineering, University of Queensland for the simulation of transient gas-dynamic flows and for the design of shock tube nozzles and expansion tubes. Depending on the application, flows in one-, two- or three-dimensional geometries can be considered (see Figs.9 and 10).

For the detailed engineering design of a pulse-flow facilities such as a free-piston shock tube, a model of the flow in the entire facility is useful. A one-dimensional code was written for the purpose and has become a standard tool in the design of high enthalpy flow facilities and light-gas guns at the University of Queensland.

In situations where multidimensional effects are important, simulations are performed with a two-dimensional (axisymmetric) time-dependent code. This code started out as a single-block Navier-Stokes solver and has been extended to include multiple-block geometries. When trying to capture viscous effects with high resolution, run times can become large (on the order of hundreds of CPU hours) and some effort has been directed towards producing versions of the code that have been parallelised for the (shared-memory architecture) Silicon Graphics Power Challenge and also for the (distributed-memory) IBM SP-2.

Other two-dimensional flow codes that have been developed by the group include an unstructured-grid Euler solver and a shock-fitting Navier-Stokes solver.

Applications of the two-dimensional codes have included

(i) the simulation of the flows over models being tested in a shock tunnel or expansion tube;

(ii) identification of noise in expansion tubes ;

(iii) examination of flow starting problems in shock tunnel ; (iv) an investigation of spurious pressure measurements in a shock tube and

(v) a study of the effects of finite-opening times on the performance of shock tubes and expansion tubes

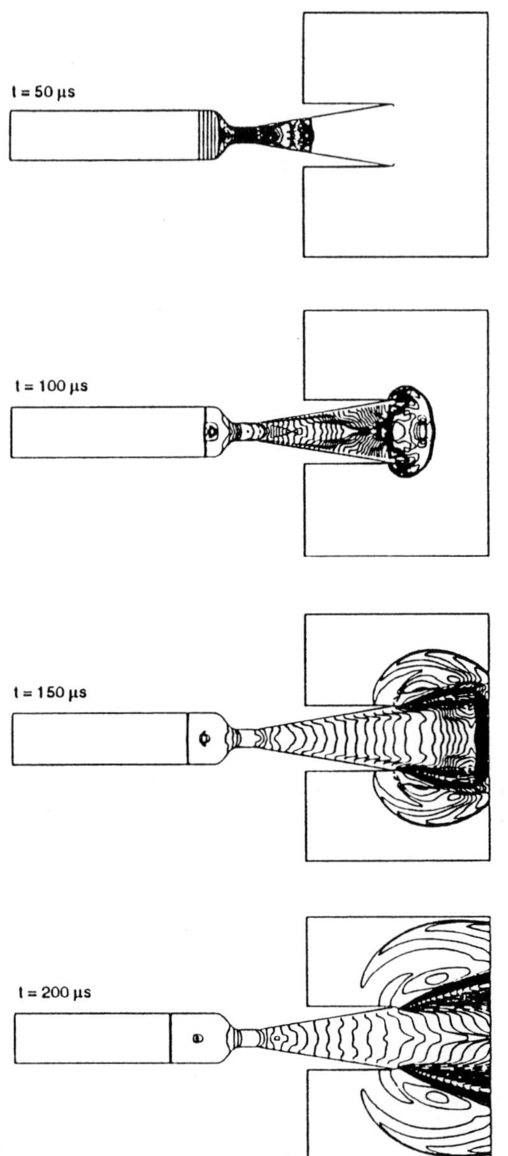

Fig.9: The starting flow in a shock tunnel nozzle. The simulation is used to determine test flow conditions that cannot be measured easily.

Construction of a three-dimensional code was motivated by the need to analyse reasonably complex flow geometries that were being studied in a scramjet scaling project. The flow solver was coupled to an optimisation algorithm to aid the design of scramjet thrust nozzles in the presence of large cross-stream variations of flow properties and has demonstrated that shape optimisation can be done, at least for relatively simple geometries. This code will also be one of the main design tools for the a project on building better shock tunnel nozzles. (see Jacobs, 1994, Jacobs 1991, Kendall, Morgan, and Jacobs 1997, Doolan and Jacobs 1996, McGhee and Jacobs 1996, Petrie-Repar and Jacobs 1996, Jacobs 1993, Smith, Johnston, and Austin 1996, Jacobs 1994, Jacobs 1991, Kendall, Morgan, and Petrie-Repar 1997, Jacobs and Craddock 1995)

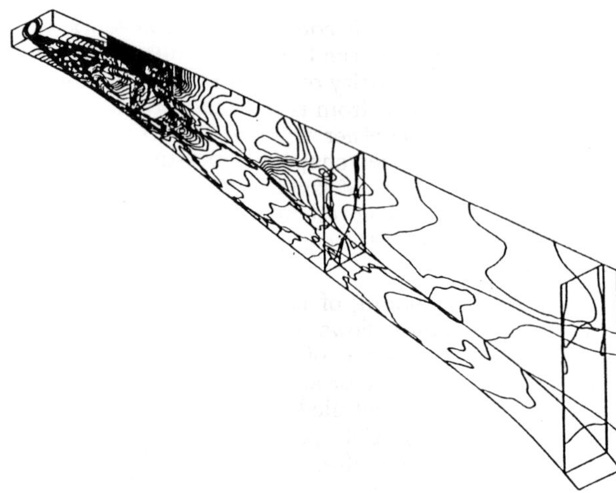

Fig.10: Pressure contours over a scramjet thrust surface for circular-jet heat addition.

Work is in progress in the computation of high speed flows at other centres in Australia. One such is the Department of Aeronautical Engineering, University of Sydney. This work is detailed in the more appropriate Section 7 on Algorithms.

5. Multi Phase Flow

Multiphase flows are commonly encountered in the Process Industry and are quite complicated to handle. CFD efforts are in progress in Australia to develop suitable methods for the purpose.

5.1 Fundamental Gas Particle Research

This is another extensive investigation in progress at CANCES, University of New South Wales and its goals are summarised as follows.

Although significant progress has been made in understanding particle motion due to fluid turbulence, new insights have shown that there are fundamental research issues for particle interactions that have to be resolved for continued model development (Crowe, 1993). The theoretical description of particle-laden flows is still in its infancy (Crowe et al., 1993), thereby further limiting current models of more complex flows such as in the presence of heat and mass transfer and chemical reactions. A comprehensive review of available mathematical models for particle-laden flows is provided by Elghobashi (1994). There are two approaches commonly used for modelling particulate flows, namely the Eulerian (continuum) and the Lagrangian (particle trajectory) methods. In the Lagrangian formulation, the motion of individual particles is considered and relevant variables are calculated along the particle trajectories. The Lagrangian formulation is a very fundamental procedure to describe the individual particle interacting with the fluid and wall surface, and can yield a detailed physical description of the individual particle motion in a flow domain (Crowe, 1982).

However, a drawback common to most numerical implementations using the Lagrangian method is that the instantaneous velocity of the fluid surrounding the particle is determined from the time-averaged Navier-Stokes equations together with a turbulence closure model, and an assumed shape of the probability density function (p.d.f.) of the velocity. Furthermore, major simplifying assumptions are needed to prescribe the residence time of the particle in a large-scale turbulence eddy.

Another shortcoming of using the Lagrangian approach for complex flows in most engineering applications, is that some of the particulate features rely on a deterministic or stochastic model which requires unreasonably detailed and expensive computations. Also, the particle paths which result from a Lagrangian prediction often do not contain enough information for design engineers. From an engineering design perspective, the mean particulate flow fields need to be known in terms of mean particulate velocities and the distribution of particle concentration, to facilitate quantifiable decisions. Therefore, the continuum model (Eulerian approach) becomes attractive as an alternative way of representing the particulate phase to obtain the mean particulate flow fields in complex domains.

The need for effective turbulence modelling (Hetsroni, 1989) is an important reason for preferring an Eulerian to a Lagrangian description of the particulate behaviour. The difficulties of adding turbulence modelling to a Lagrangian formulation are lucidly set out by Elghobashi (1994). Given the complexity of the flows being studied, quantitative accuracy from a Lagrangian formulation would be prohibitively expensive in computer time. Recent developments of accurate continuum particulate boundary conditions at solid surfaces (Tu and Fletcher, 1995a), now permit the particulate behaviour to be predicted with much greater accuracy than before, at least for flows with low to moderate particulate loading. The Eulerian (continuum) formulation for the particulate phase also allows turbulence effects to be modelled more efficiently than does the Lagrangian (trajectory) approach.

Another research need is to improve understanding of particle-wall interaction for geometrically constrained flows. The particle-wall interaction is the primary mechanism for transfer of momentum, kinetic energy and heat from a surface to both the fluid and particulate phase flow field. Murray and Humphrey (1994) also reported that, although heat transfer in the wake of a tube was reduced by the particles, a significant enhancement of heat transfer was measured over the front of the tubes, which is considered to result mainly from the increased thermal capacity of the gas-particle mixture. Recent developments of accurate particle-wall collision models (Tu, Lee and Fletcher, 1996) show that a significantly higher concentration of particles in the front of a tube has been predicted and substantial influence on both the fluid turbulence and temperature field can be expected.

When an obstacle is presented in a particle-laden flow, the particulate concentration in the front of the obstacle will be locally increased due to the reflection from the wall. For many intermediate loadings it is expected that locally the loading will be increased sufficiently that particle-particle interactions allow some additional upstream influence. The extension of the modelling of the particulate phase for higher loading situations requires the modelling of particle-particle interaction which may be the dominant mechanism of particulate flow in this particular region where the particles reflected from the wall move opposite to the incident particles.

The purpose of the present research is to build a generic two-phase modelling capability in continuum framework, focused on particle-fluid, particle-wall and particle-particle interactions, and their influence on heat transfer behaviour.

Recent analysis (Tu, Lee and Fletcher, 1996) indicated that there exists a particle-wall rebounding layer within which the mean particulate flow fields are significantly influenced by both the particle-wall collision process and the aerodynamic drag associated with the rebounding particles. Modelling the mean particulate flow fields within this region requires the consideration of the momentum balance between the drag and particle inertia for the incident and reflected particles (see Fig.11).

Due to the reflection from the wall, the particulate concentration will be locally increased. For many intermediate loadings it is expected that locally the loading will be increased sufficiently that particle-particle interactions allow some additional upstream influence. In the present research a local particle-wall momentum exchange mechanism representing the particulate pressure (Tu and Fletcher, 1993) will be used. Current research (Tu, Lee and Fletcher, 1996) indicates that this is necessary in order to predict geometrically complex gas particle interactions accurately.

The influence of the particulate field on the turbulent temperature fluctuations of the carrier fluid has been analysed by Derevich et al (1989) and Yarin and Hetsroni (1994). For small Stokes number and using a Prandtl mixing length theory simple dependencies of the particulate temperature fluctuations are obtained on the carrier fluid temperature fluctuations and on the particulate velocity fluctuations. The rela-

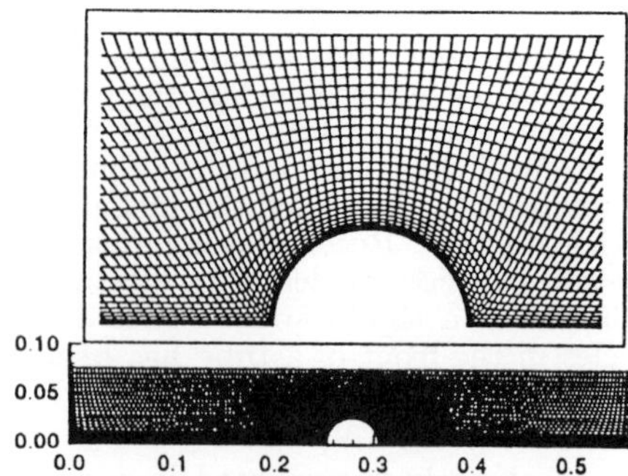

(a) Computational domain and grids

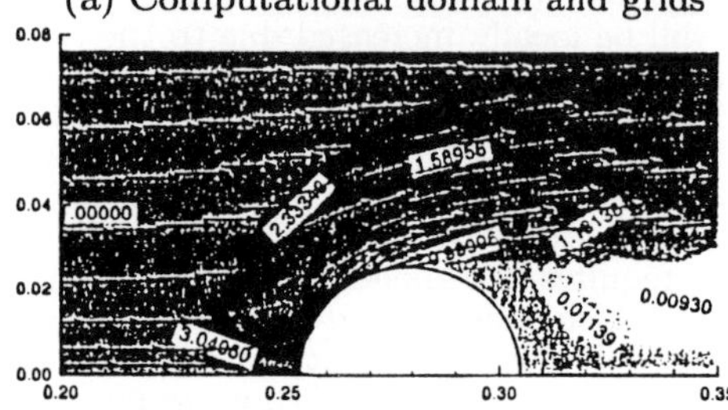

(b) Particle concentration contour and velocity vector plots

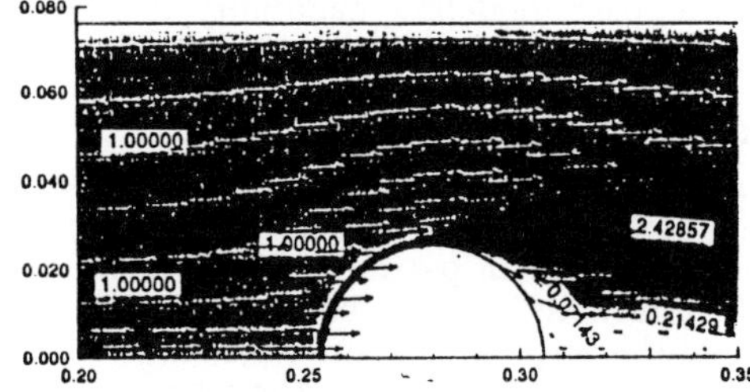

(c) Particle concentration contour and velocity vector plots

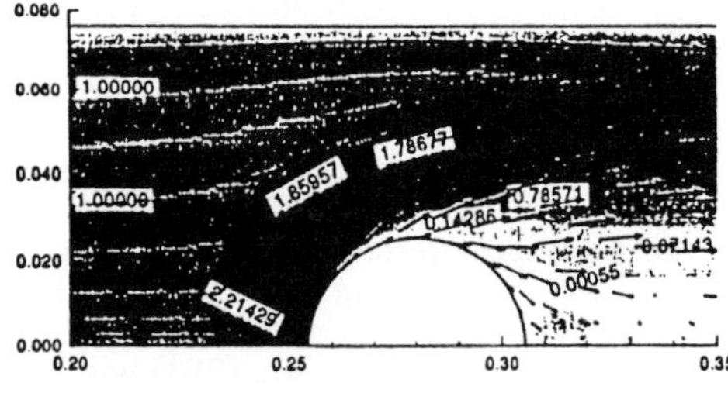

(d) Particle concentration contour and velocity vector plots

Fig.11: Numerical solutions of mean particulate flow near an isolated single tube.

tive magnitudes of the fluid and particulate temperature fluctuations are dependent on the loading, the relative specific heats and particularly radiation, if present. It is expected that the transport of thermal energy by rebounding particles will be responsible for much of the increase in heat transfer.

5.2 Multiphase Pipe Flow

FLAIR, the Fluid-dynamics Laboratory for Aeronautical and Industrial Research, Monash University addresses multiphase (liquid-gas, liquid- solid) fluid flow that occur in many industrial processes, from slurry transport to spray droplet combustion. Although complicated adjusted single- phase models have been developed in the past for multiphase fluids, the full two-phase model is the only one able to handle all cases found to date. Based on separate Navier-Stokes type equations for each phase, along with non-linear coupling and interactions terms, it has proven to give particularly good comparisons in test problems. The model has been applied to the prediction of pressure drop within multiphase pipe systems, and the study of how the addition of non-passive solid particles within the fluid affects the flow structures and hence the mixing characteristics (see Morris, Hourigan and Thompson, 1997)

5.3 Multi Phase Flow in Metal Processing

At the Department of Chemical Engineering, University of Melbourne, considerable work has been undertaken to develop computational models for some of the interesting flows that occur in Mineral and Metal Processing. One of them is the CFD application to the smelting processes where pelletised feed solids are frequently introduced as a continuous stream into a bath of molten liquid. It is intended to develop a computational model of the dispersion of solids falling into a flowing liquid stream (Smith, Davidson, Fletcher, 1997) to gain an understanding of the mixing of the solids in flowing liquids as a function of particle properties and operating conditions. An Eulerian multiphase formulation with a non diffusive convective scheme is used. Predictions have been obtained for a range of values of solid densities, inlet particle volume fractions, impact velocities and particle diameters (see Fig.12).

Another project in progress is relevant to enhancement of spray dryer performance. The operation of a spray dryer involves a complex gas-droplet flow coupled with heat and mass transfer. Understanding of these is essential for making the process more efficient and economic. Both the standard $k - \varepsilon$ and Reynolds Stress Transport (RST) turbulence models have been applied with mixed success. The $k-\varepsilon$ model is known to give poor predictions in many strongly swirling flows but its implementation is less computer intensive than the more accurate RST model. A modified form of the $k - \varepsilon$ model which seeks to account for the anisotropy of turbulence in swirling flow has been used with success for flow in cyclone separators while retaining the computational advantage of the $k - \varepsilon$ model. It is found that, although the modifications improved the prediction of cyclone separator and other flows, there is no consistent improvement for the spray dryer flow (Hoveden and Davidson 1997).

Gas-Particle Flash Furnace Jets forms one other application of CFD at this centre. In flash smelting of

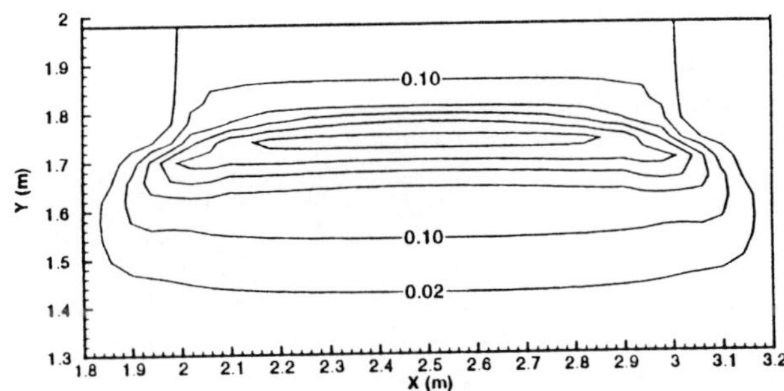

Fig.12: Predicted contours of solids volume fraction formed when neutrally buoyant particles of diameter 5cm are injected for 0.15 secs into a stationary liquid. The particles of density 7000 kg/m3 and volume fraction 0.05 enter at 15 m/s through a 1 m inlet into a liquid contained in a two dimensional rectangular tank 5 m wide and 2 m deep. Volume fraction contours shown include 0.02 and then increase from 0.1 to 0.5 in steps of 0.1.

nickel or copper, sulphide concentrate and oxygen- enriched air are injected via burners into a reaction shaft in which sulphur is removed from the particles by oxidation. Essential to the efficiency of this process is the flow and dispersion of the particles within the gas jet. Objectives of this project are to evaluate $k-\varepsilon$ two-way coupling models of particle-turbulence interaction for confined jet flows under isothermal, non-reacting conditions using computational techniques and to computationally model the effect of mass exchange on turbulent dispersion of small particles in flash smelting, assuming hot isothermal flow (Davidson, 1996a, 1997). Flow predictions based on three different two way coupling models of gas-particle interaction are compared for confined jets in the present context in Davidson (1996 b).

6. STUDY OF FUNDAMENTAL FLUID FLOW PROCESSES

Application of CFD to explain and understand some of the basic fluid flow phenomena is now wide spread. Australia has not lagged behind the rest of the world in this regard as the description in the following subsections show.

6.1 Computer model for a pulsating vapour bubble

The CFD project of Soh at the University of Wollongong is aimed at gaining a further understanding of the thermodynamic processes inside a pulsating bubble; and how the change in the thermodynamic state will affect the behaviour of the bubble. The computer model consists of three domains. The first is the hydrodynamics domain. The method of computing for the deformation of the bubble is based on the bound-

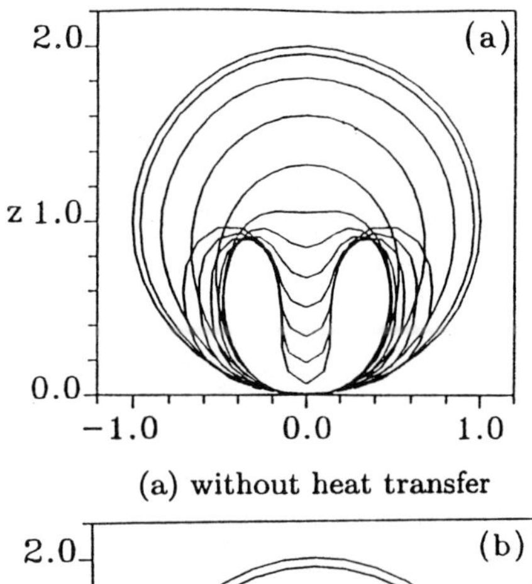

(a) without heat transfer

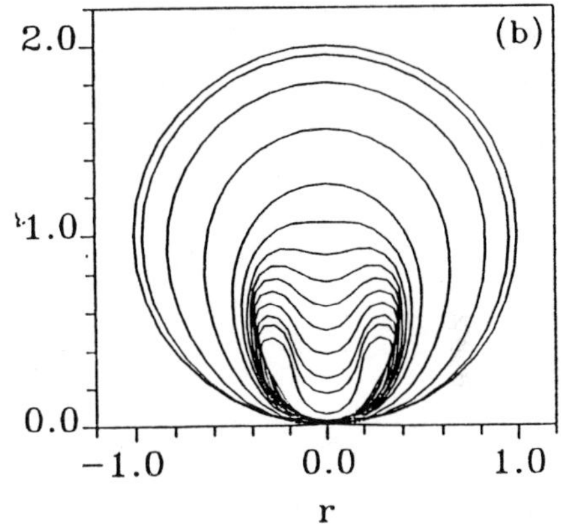

(b) with heat transfer

Fig.13: Collapse of bubbles at various nondimensional times: r and z are multiples of the initial bubbles' radii. The horizontal axis represents a rigid wall.

ary integral scheme developed in Australia by Best and Kucera (1992). The second domain is the temperature domain of the incompressible liquid water in which the force convection of heat is calculated. A dual reciprocal boundary integral scheme was developed by Karimi and Soh (1994) to perform the calculations in this domain. The third domain is the vapour inside the bubble. An analysis for the thermodynamic process in a bubble, including vaporisation and condensation, has been developed by Soh (1994) to formulate the transfer of heat between the bubble (third domain) and the temperature domain and the transfer of works between the bubble and the hydrodynamic domain.

The computer model is being refined by comparisons with experimental data. An experiment at Wollongong uses an electric spark discharge to generate a vapour bubble in water. The images of the bubble are captured by high speed cinematography at a framing rate of 6000 frames per second. The images of the bubble are digitised for comparison with computational results. Computations show that the behaviour of the bubble depends on the thermodynamic equation of state of the vapour (or gas) in it. The transfer of heat is found to be a significant factor which contributes to the damping of bubble rebound. The study has provided a plausible explanation for the mechanism in which the bubble transfers external energy into the surrounding liquid (see Figs.13 and 14)

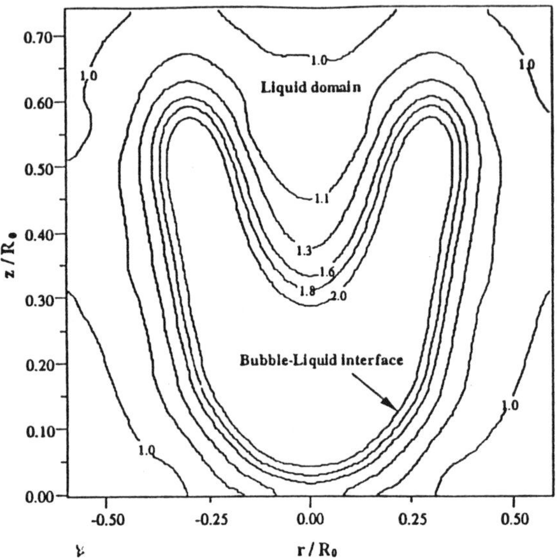

Fig.14: Isotherms of a bubble near the end of its collapse.

In parallel with the work mentioned above, a boundary integral technique was developed by Harvey, Best and Soh (1996) to calculate the bubble's surface velocity potential and pressure from the experimental data (see also Soh & Karimi 1996, Karimi and Soh 1994, Soh and Shervani Tabar 1994, Karimi & Soh 1996, Soh and Karimi 1996, Karimi & Soh 1995).

6.2 FLAIR, Fluid-dynamics Laboratory for Aeronautical and Industrial Research, Monash University)

A number of projects important from a basic fluid dynamic point of view are in progress at this centre and are described below.

6.2.1 Turbulence Mixing

In Australia as in other countries, the availability of faster computers with larger memories has enabled researchers to perform massive computations like Large Eddy Simulation (LES). Figure 15 shows the structures within a turbulent channel flow that have been captured using the technique due to Morris and Ferziger (1996). Visualisation of these fields have allowed greater understanding of the structure interaction and the benefits of local high turbulence intensity regions.

The numerical method uses a global spectral discretisation along with a fourth-order Runge-Kutta

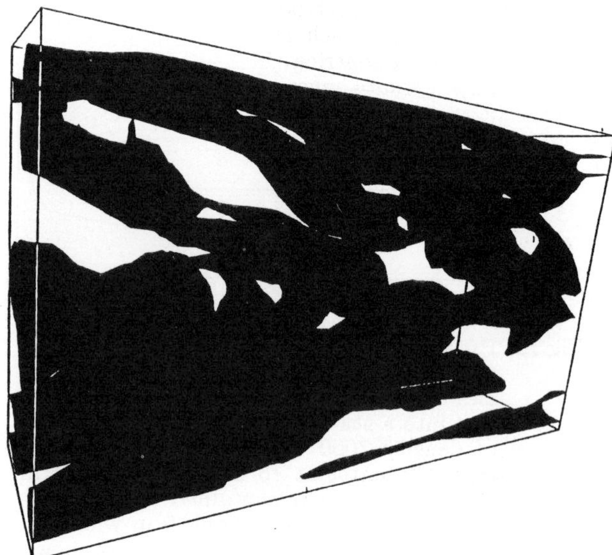

Fig.15: Turbulent structures within a channel mixing flow as predicted by a Large Eddy Simulation.

time stepping scheme. The turbulence modelling employed was developed in collaboration with the Stanford University and involves a new hyperviscosity term. This leads to improved smoothing within the flow field and so eliminates most of the unstable behaviour that might destroy the computations.

6.2.2 Computation of Three-Dimensional Cylinder Wake Flows

A spectral/spectral-element code has been used to compute the transition to three-dimensional flow for a circular cylinder wake. Shedding frequency, base suctions, transition Reynolds number and the instability wavelength are predicted to within experimental error. The study has led to important observations about the three-dimensional instability modes and their interaction. There are indications that the interaction of the two modes may suppress, or at least mask, the period-doubling found in previous simulations by the group (see Thompson and Hourigan, 1992, Thompson, Hourigan and Sheridan, 1994, 1995, 1996).

The developed numerical code uses a higher order Galerkin method which attempts to strike a balance between versatility and geometrical adaptability of low order h-type methods (like finite-difference and finite element) and the superior convergence characteristics of higher order p-type schemes (such as full spectral methods). The method has been shown to exhibit exponential convergence rates characteristic of spectral schemes. The current implementation uses the classical three step splitting scheme for time stepping. Second order temporal accuracy is achieved by applying higher order pressure boundary conditions.

Various techniques are introduced to increase the computational efficiency such as static condensation to reduce the size of matrices involved in the implicit diffusion and pressure substeps and the use of Gauss-Lobatto quadrature for the finite element integrations leading to diagonal mass matrices (see Fig.16).

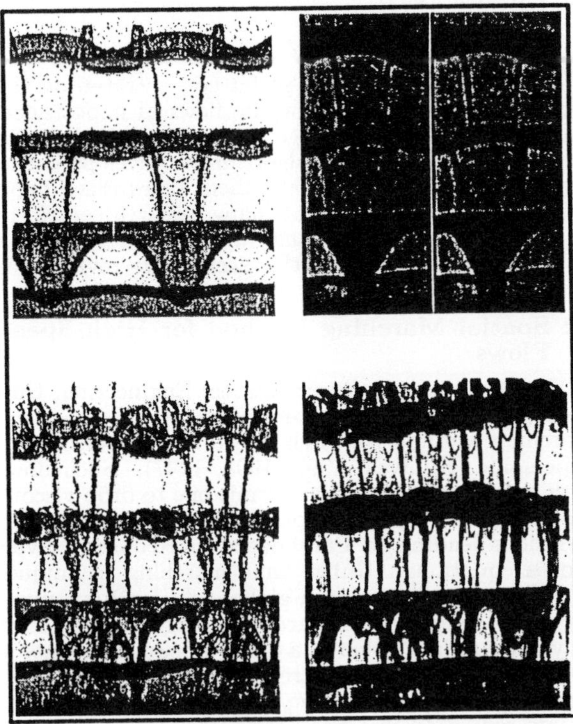

Fig.16: Comparison of hydrogen bubble (left) and experimental dye visualisations (right) of the flow structures in a circular cylinder wake. The cylinder is at the bottom of each diagram and the flow is upwards. The figures at the top are visualisations at Reynolds number of 210 showing mode A streamwise vortex loops connecting the predominantly two dimensional Strouhal vortices parallel to the cylinder. The structures are periodic at this Reynolds number. The lower figure shows the situation at a Reynolds number of 250. Here a different mode (Mode B) predominates which has a significantly shorter wavelength. The flow is no longer periodic. Experimental results reproduced here are due to Williamson, Cornel University.

6.2.3 Simulation of Acoustically Forced Flows Around Long Rectangular Plates

Previous experimental work has shown that the base drag experienced by a long flat plate under cross- stream acoustic forcing can depend sensitively on plate length. The numerical study, using the spectral element code described above, attempts to quantify this behaviour in terms of the phasing of the vortices originating from the leading edge of the plate as they pass the trailing edge. The simulations do mimic the experimental results and reveal the influence of phasing of vortices on the base suction. This research has implications for the design of low drag heat exchangers. (see Hourigan, Welch, Thompson, Cooper and Welsh, 1991, Rudman, Thomson and Hourigan, 1992, Thomson, Hourigan, Mills and Sheridan, 1995, Thompson, Hourigan and Sheridan, 1995, Thomson, Hourigan, Mills and Sheridan, 1995)

6.2.4 Vortex Breakdown in Swirling Flows

In this study vortex breakdown in swirling flow has been investigated numerically using a Finite Element code. It is demonstrated that the approach to numerical convergence is strongly dependent on whether grid compression near the rotating lid is included. In the case of a uniform mesh, for the Reynolds number considered, increasing the grid resolution leads to the appearance of initially fewer recirculation regions and then an increase. This non-monotonic route to convergence was not observed when the Ekman layer was better resolved using grid compression at the rotating lid. Measurement uncertainties associated with rheometers leads to an uncertainty in the viscosity and hence the Reynolds number of the flow. It is demonstrated that comparison of predicted flow patterns at nominally the same Reynolds number can be misleading; what is termed validation in these circumstances is at best coincidental (see Rudman, Hourigan and Thompson, 1994, Hourigan, Graham and Thompson, 1995a and 1995 b, Graham, Morris, Hourigan, and Thompson, 1996).

6.2.5 Free Surface Flows

Free surface flows occur in a wide variety of situations in both industry and the environment. However, due to the problems associated with the implementation of appropriate boundary conditions on a surface whose position is constantly changing, such flows are often arduous to model numerically. In this research, application of Smooth Particle Hydrodynamics (S.P.H.) is considered as a remedy.

This scheme is Lagrangian, and as a result it is able to automatically track the free surface while also handling any free surface gradient, thus overcoming some of the problems which often plague alternative schemes. S.P.H.'s application to incompressible flows has only occurred during more recent times, with Monaghan(1994), Thompson, Hourigan and Monaghan(1994) and Takeda et. al.(1995) considering its application to a number of test problems.

In order to test the merits of this technique in the modeling of free surface flows, the current work has considered the case of an impinging plane jet at various angles to the vertical. The heights of the two streams were then measured and the results of these simulations were then compared with those obtained from experiments. Good agreement between both sets of results was obtained for most angles.

The case of flow past a cylinder is currently being investigated using S.P.H. in order to determine its ability to deal with boundary layer flows. Concurrent studies have also included the investigation of different types of boundary conditions.

It is anticipated that such examinations will lay the basis for the investigation of flow past a cylinder close to a free surface, for which extensive experimental results exist, and also for the flow associated with the controlled oscillation of a cylinder through a free surface. (see Thompson, Hourigan and Monaghan, 1994, Monaghan, Thompson and Hourigan, 1994)

6.2.6 Instabilities in a supersonic impinging jets

The instabilities in a supersonic impinging jet are investigated by solving the two-dimensional Euler equations using the piecewise parabolic method (PPM) and Roe's linearised Riemann-solver. The predicted shock cell spacing agrees well with the observed and theoretical value. The frequency and nature of the dominant instabilities are found to be a function of the impingement distance. Two instability modes are possible, a symmetric (or varicose mode) and an asymmetric (or sinuous) mode. For two given jet exit Mach numbers (M = 0.98 and 1.29), the energy in, and frequency of, these modes are a function of impingement distance, leading to an integral staging due to an acoustic feedback loop. The predicted frequencies of the fundamental axisymmetric and asymmetric instabilities agree with the theoretically allowed values. Staging of predicted frequencies, that occurs in experiments, is also predicted. (see Rudman, Hourigan and Brocher, 1994 and 1996).

6.2.7 Acoustic Resonant Flows

Numerical models employing the vortex method have been used to investigate a number of different flows in which a resonant acoustic field is driven by vortex shedding.

For example, the separated flow around a trip rod placed upstream of a resonator tube. The acoustic power generated by the flow is calculated using Howe's theory of aerodynamic sound. The results are consistent with previous experimental observations using a free surface hydraulic analogy and provide insight into the flow structures responsible for the transfer of energy from the flow field to the resonant acoustic field.

In another case, experimental and numerical investigations of the generation of resonant sound by flow in a duct containing two sets of baffles and the 'feedback' of the sound on the vortex shedding process are reported. The experiments are conducted in a wind tunnel and the numerical simulations are used to predict the sources of resonant sound in the flow. Comparison is made between the predicted time-dependent structures and the observed flow structures using smoke visualisation.

The vortex cloud model predicts the flow conditions under which net acoustic energy is generated by the flow and therefore when resonance can be sustained; the results are consistent with the occurrence of peaks in the observed resonant sound pressure levels. (see Welsh, Hourigan, Welch, Downie, Thompson, and Stokes, 1990, Hourigan, Welsh, Thompson and Stokes, 1990 and 1991, Hourigan, Thompson, Welsh and Brocher, 1992).

7. ALGORITHMS

The present review has so far concentrated on the application of CFD to compute flows of importance. But one has to acknowledge the fact that in order to do so one needs efficient algorithms. Development of algorithms and research into them takes place hand in hand with application at many research centres in other countries. But in Australia it takes place on a somewhat lower key. In this section we review this activity within Australia.

7.1 Spatial Marching Method for High Speed Flows

The Department of Aeronautical Engineering, University of Sydney has been developing a Spatial Marching Method for High Speed flow supersonic or hypersonic (Srinivas, 1992, 1995, 1997). Such flows with a dominant direction permit one to drop viscous terms in the flow direction from the Navier Stokes equations leading to Parabolised or Reduced Navier Stokes equations. Unlike time marching techniques where one iterates upon a global solution for a number of time steps, with space marching one iterates at only one spatial station at a time. Once convergence is obtained the next downstream station is considered. Such a procedure can give almost an order of magnitude reduction in CPU time compared to the time marching methods.

The method being developed is an explicit method and is Finite Volume based. The Space Marching procedure, in principle, demands that upstream influence not be present, a condition true only for inviscid flows. But when one considers viscous flows there are boundary layer regions of the flow where the signals do propagate upstream. If these are not taken into account what are called Departure Solutions (Davis et. al. 1986) may result. A remedy is to split pressure at the interface between Finite Volumes based on Vigneron's strategy (Vigneron et. al. 1978). Effectively the pressure at the interface is calculated as a weighted mean of the pressures from adjacent cells. In the present method the contribution from the down stream cell is neglected.

For marching the explicit modified Runge-Kutta method due to Jameson is employed. The Space Marching method has gone through various versions. In the early versions (Srinivas 1992a, 1992b) of the method the traditional dissipation terms due to Jameson, Schmidt and Turkel were used. Subsequently the scalar dissipation terms were replaced by matrix type dissipation terms (Srinivas 1995). Recently the method has been recast in the Local Extremum Diminishing (LED) framework (Srinivas 1995, 1997).

The developed Spatial Marching Method has been validated with many test cases. First one is a laminar flow past an isothermal flat plate at Mach number of 2. Computed heat transfer coefficients along the plate are shown in Fig.17 in comparison with the results of Lawrence et. al. (1989). The method was also used to compute one of the test cases for the Workshop on Hypersonic Flows for Reentry Problems organised by INRIA, France at Antibes. The third example considered is the turbulent flow at Mach number of 5 past an adiabatic flat plate.

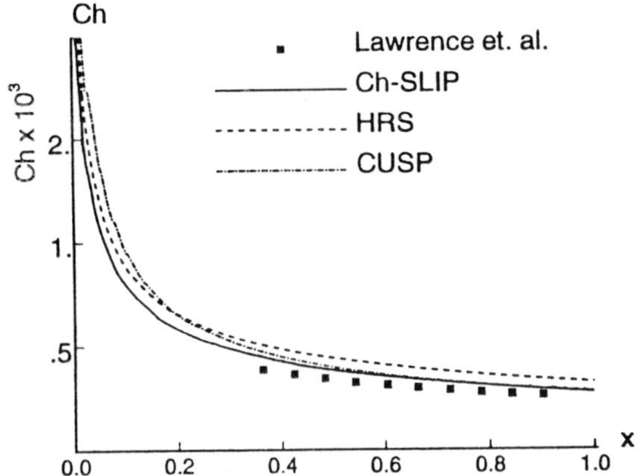

Fig.17: Computed Distribution of Heat Transfer Coefficient along a Flat Plate for a Laminar Flow using three of the LED schemes. Discrete points are due to Lawrence et. al. (1989).

The other test case is of the interaction of a hypersonic flow at Mach number of 14.1 with a 15° compression corner involving many interesting features - leading edge shock, compression fan, resulting shock, expansion fan, slip stream etc. Figure 18 from Srinivas (1997 a) shows the pressure and Mach number contours for the flow. The results were obtained using the most recent version of the method which employs a characteristic Symmetric Limited Positive Scheme SLIP. A sharp definition of features comparable to the ones produced by more complicated schemes like the TVD is evident.

7.2 Research into recent algorithms for Transonic Flows

This work is carried out by Srinivas and his group and considers some of the recent methods proposed for the computation of transonic flows. Most important of these are the LED schemes and Matrix form of dissipation terms. The work essentially uses the Runge - Kutta time stepping for steady state solution of Navier Stokes equations. Matrix form of dissipation terms in place of the popular scalar ones are shown

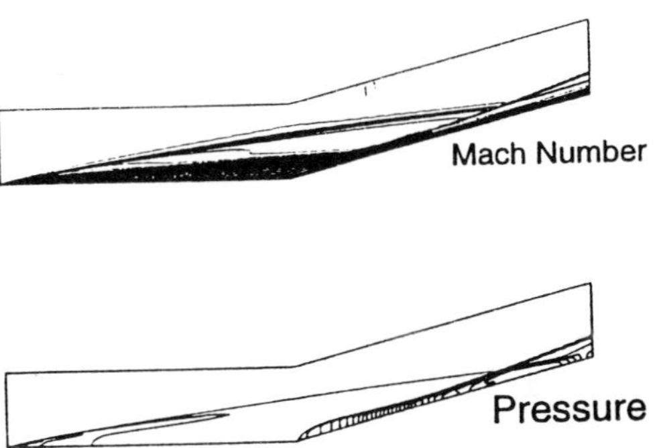

Fig.18: Pressure and Mach number contours for a Hypersonic Flow negotiating a Compression Corner for the Characteristic SLIP scheme.

to produce better definition of shock and other features both in Spatial Marching for high speed flows and time marching for transonic flows (Srinivas, 1995 a, Fard and Srinivas 1995 a).

Recently what are called Local Extremum Diminishing schemes have been proposed by Jameson (1995) and these seem to be promising from many angles. The LED schemes control dissipation more precisely than the conventional forms of dissipation used with Runge Kutta time stepping. It is shown that single point shocks are possible when some of the members of the family are employed. A very desirable feature of LED schemes is that they are easily incorporated into existing Runge Kutta codes. It is expected that these schemes will be ideal for viscous flow computation where a good control of dissipation is essential.

Many of the LED schemes have been tested in this work. Fard and Srinivas (1995c, 1996) consider the application of the characteristic SLIP scheme to compute transonic flow past a projectile and a bump in the wind tunnel. As a part of the study a modified algebraic ReNormalised Group (RNG) model of turbulence has also been proposed and tested (Fard and Srinivas, 1995b, 1995b, 1996). The modification consists of solving an ordinary differential equation to account for the upstream history of turbulence in a manner similar to Johnson and King model.

In parallel Srinivas (1996, 1997b) considers the computation of transonic turbulent flow past cascades. As shown in Fig.19 the results are encouraging. The study is incomplete and is still in progress.

7.3 The Additive-Correction Multigrid-SIMPLEC Dynamic Tuning algorithm

The use of CFD on a design mode for engineering applications requires for the numerical algorithm to be efficient and robust. To accelerate the convergence of the iterative numerical method, and Additive- Correction Multigrid-SIMPLEC Dynamic Tuning procedure was developed. In this methodology, an Additive-

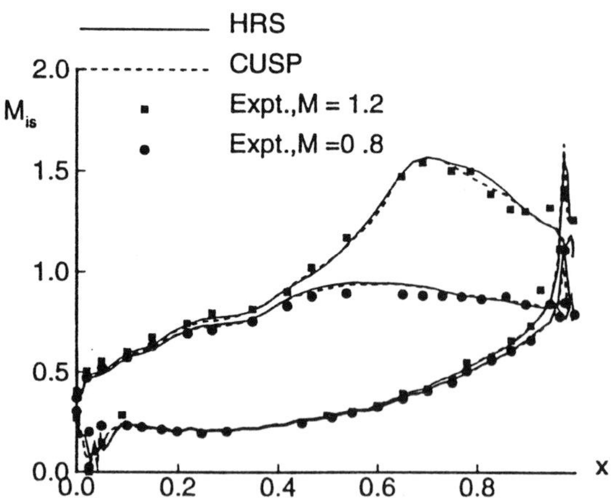

Fig.19: Computed Surface Mach number distribution for the VKI 1 cascade using two of the LED schemes.

Correction Multigrid algorithm is used to accelerate the convergence of the momentum and velocity potential (mass conservation) equations. A velocity potential method is used as a solver on the fine (solution) grid, while the corrections from the velocity potential correction equation are directly introduced into the momentum correction equations as a source term on the coarse grid. For the turbulent transport and energy equations, the relaxation coefficients are dynamically tuned to enhance the convergence rate. Two methods, the relative error based residual and the cell Peclet-Reynolds based residual are used. This methodology leads to savings up to 35 % of CPU time when compared with the original SIMPLEC algorithm, being more effective for problems with a large number of grid points (see Zdravistch, Fletcher and Srinivas,, 1995, Zdravistch and Fletcher, 1996, Zdravistch, and Fletcher, 1996).

7.4 Iterative Methods for Computation of Incompressible Flows

This work is due to Mellen and Srinivas at the Department of Aeronautical Engineering, University of Sydney. The study uses the primitive variable formulation on a non staggered grid (Mellen and Srinivas, 1994). A coupled Gauss - Seidel type point solver and a coupled Gauss Seidel type Line Solver have been developed (Mellen and Srinivas, 1996). Computations were performed on many geometries including driven cavity and backward facing step both on single and multigrids The coupled line solver performs significantly better than the point solver. The following table gives the required CPU times for the driven cavity problem using a 90×90 grid. Comparisons are made with the SIMPLEC procedure as well.

Reynolds number	SIMPLE	Point Solver	Line Solver
100	2520 sec	3720 sec	1362 sec
1000	1560 sec	5700 sec	1216 sec
10000	2460 sec	9840 sec	752 sec

8. Concluding Remarks

A short review of the major current CFD activity within Australia in the Mechanical and Aeronautical Engineering disciplines is presented. It appears that Industrial CFD is one of the strong interests in the universities and other research centres. There is significant progress in computing single phase as well as multiphase flows in varied applications like industrial boiler, metal and mineral processing, buildings, cooling of electronic components, etc.

Australia has established itself as a world leader in the development and application of DSMC. This field is still breaking new grounds. Now DSMC is being applied to compute near-continuum flows.

As in other countries there is considerable interest within Australia in exploiting CFD to study basic fluid flow phenomena. Interesting phenomena like flow in a pulsating bubble, three dimensional wake behind a cylinder and acoustically forced flows have been studied.

Large Eddy Simulation (LES) and Direct Numerical Simulation have also been undertaken solely in Australia or sometimes in collaboration with the Universities in other countries.

High Speed and High Temperature Flows do attract attention of researchers in Australia, but to a lesser extent than low speed flows. It is mostly the interest in flows like the one through shock tubes that keeps this research active.

Regarding the development of algorithms to compute fluid flows it seems that the activity is in a low key.

Acknowledgements

We would like to thank all our colleagues who provided us the summaries of their work and also answered some of the queries we had. Thanks are also due to Mr. S. Wishart, Mr. B. Grocholsky and Mr. R. Pinto for their help in the preparation of the manuscript.

REFERENCES

Section 1

Huilgol, R.R. and Phan-Thien, N. (1997) Fluid Mechanics of Viscoelasticity: General Pronciples, Constitutive Modelling, Analytical and Numerical Techniques for recent advances in Viscoelastic Computational Methods (Chap 7), Elsevier, Amsterdam.

Section 2.1

Achenbach, E. (1969) Investigations on the Flow

through a Staggered Tube Bundle at Reynolds numbers up to $Re = 10^7$, Waerme und Stoffubertragung, 2, 47-52.

Choi, B.S. and Fletcher, C.A.J. (1997) Computation of Particle Transport in an Electrostatic Precipitator, submitted to the 8th International Conference on Electrostatics (Poitiers, 4-6 June, 1997)

Crowe, C.T. (1982) Review - Numerical Models for Dilute Gas-Particle Flows, J. Fluid Engrg., Vol.104 pp.297-303.

Crowe, C.T. (1993) Modelling Turbulence in Multiphase Flows, Engineering Turbulence Modelling and Experiments 2 (Eds. Rod,W. and Martelli,F.), Elsevier Science Publ.

Crowe, C.T., Chung, J.N. & Troutt, T.R. (1993) Particle dispersion by organized turbulent structures. In Particulate Two-phase Flows (Edited by Roco, M. C.), Chap. 18. Butterworth-Heinemann, New York.

Derevich, I.Y., Yeroshenko, V.M. and Zaichik, L.l. (1989) Hydrodynamics and Heat Transfer of turbulent Gas Suspension Flows in Tubes - 2. Heat Transfer, Int. J. Heat Mass Transfer, 32, 2341-2350.

Medlin, A.J., Fletcher, C.A.J and Morrow,R. (1996a) An Efficient Pseudo-Transient Solution Method for Monopolar Corona with Charge Advection and Diffusion, Proceedings of the 6th International Conference on Electrostatic Precipitation, Budapest, 17-22 June.

Medlin, A.J., Fletcher, C.A.J. and Morrow, R. (1996b) A Pseudotransient Approach to Space-Charge Coupled Computations, submitted to J. Comput. Phys.

Zdravistch, F. and Fletcher, C.A.J. (1996) Convergence Acceleration of SIMPLE Type Algorithms using a Dynamic Tuning Additive Correction Multigrid Strategy, Second Asian Computational Fluid dynamics Conference, Tokyo.

Zdravistch, F. and Fletcher, C.A.J. (1996) Multigrid Acceleration of the Full SIMPLEC Algorithm, Computational Techniques and Applications Conference: CTAC-95, R.L.May and A.K.Easton (Editors), World Scientific, pp 839-845.

Zdravistch, F., Fletcher, C.A.J. and Behnia, M. (1993) Laminar and Turbulent Heat Transfer Predictions in Tube Banks, Preprints of Papers of the Fifth Australasian Heat and mass Transfer Conference, Brisbane, pp 65-1 65-6.

Zdravistch, F., Fletcher, C.A.J. and Behnia, M. (1995) Numerical Laminar and Turbulent Fluid Flow and Heat Transfer Predictions in Tube Banks, Int. J. Num. Meth. Heat Fluid Flow, No. 8, 717-733.

Zdravistch, F., Fletcher, C.A.J. and Reizes, J.A. (1996) Numerical Simulation of Fluid Flow and Heat Transfer over Tube Bank Heat Exchangers in Power Station Boilers Using a Momentum and Energy Sources Formulation, 9th Int. Symposium on Transport Phenomena in Thermal-Fluids Engineering, Singapore.

Zdravistch, F., Fletcher, C.A.J. and Srinivas, K. (1994) Flow Prediction in Tube Banks with a Finite Volume Procedure, Computational Techniques and Applications Conference: CTAC-93, D. Stewart, H. Gardner and D. Singleton (Editors), World Scientific, pp 479-483.

Zdravistch, F., Fletcher, C.A.J. and Srinivas, K., (1995) A Multigrid Additive Correction Strategy for Highly Complex Flows, Sixth International Symposium on Computational Fluid Dynamics: ISCFD-6, Lake Tahoe, Nevada.

Section 2.2

Li, Y (1996) Wave number extended high-order upwind-biased finite difference schemes for convective scalar transport. Accepted for publication in Journal of Computational Physics.

Li, Y. and Baldacchino, B. (1995) Implementation of some higher-order schemes on non-uniform grids. International Journal of Numerical Methods in Fluids, vol.21, pp.1201-1220.

Li, Y. and Delsante, A. (1996) Derivation of capture efficiency of residential kitchen range hoods in a confined space. Building and Environment, Vol. 31, No.5, pp. 461-468.

Li, Y. and Fuchs, L. (1995) An anisotropic local grid refinement method for fluid flow simulation. Numerical Heat Transfer, Part B: Fundamentals, vol. 30, pp. 195-215.

Li, Y. and Rudman, M.(1995), Assessment of high-order upwind schemes incorporating FCT for convection dominated problems. Numerical Heat Transfer, Part B: Fundamentals, vol. 27, pp. 1-21.

Li, Y., Vidakovic and Symons, J (1996), Some numerical observations on three-dimensional vortices and oscillations in colliding free convection boundary layers. In Proceedings of Roomvent'96, Yokohama, Japan, 17-19 July.

Section 2.3

Gebert, B.M., Davidson, M.R. and Rudman, M.J. (1997), Oscillations of a Confined Submerged Liquid Jet in preparation for CFD in Mineral and Metal Processing and Power Generation Conf., 3-4, July, Melbourne.

Section 2.4

Behnia, M, Gaa, F.O., Leong, S.S. and Morison, G.L. (1996) Numerical Study of Inclined thermosyphons, Computational Fluid Dynamics Journal, Vol. 5, No. 3, pp 319-340.

Behnia, M., Stella, and Guj, G., (1995a) Numerical Study of Three Dimensional Low Pr Buoyancy and Thermocapillary Convection, J Num. Heat Transfer, Part A, Vol. 27, pp 73-88.

Behnia, M., Stella and Guj, G., (1995b), A Numerical Study of Three Dimensional Combined Buoyancy and Thermocapillary Convection, Int. J Multiphase Flow, Vol.21, pp 529-542.

Chen, P.Y.P, Behnia, M. and Milton, B.E. (1995) Computation of Air Fue Droplet Flows in S I Engine Manifolds, Presented at the 14th International Conference Methods in Fluid Dynamics, published in Lecture Notes in physics, Elsevier Publishers, pp 447-453.

Copeland, D., Behnia, M. and Nakayama, W. (1996a) Manifold Microchannel Heat Sinks: Isothermal Analysis, Proc. 5th Intersociety Conf. On Thermal Phenomena in Electronic Systems (IEEE), pp 251-257.

Copeland, D., Behnia, M. and Nakayama, W. (1996b) Manifold, Microchannel Heat Sinks, Conjugate and Extended Analysis, Presented at the Ninth International Symposium on Transport Phenomena in thermal-Fluid Engineering, (ISTP-9), pp 498-503.

Dehghan, A.A. and Behnia, M., (1996a) Combined Natural Convection, Conduction and Radiation Heat Transfer in a Discretely Heated Open cavity, Transaction of ASME, Vol 118, pp 56 - 64.

Dehghan, A.A. and Behnia, M. (1996b) Numerical Investigation of Natural Convection in a vertical Slot with two Heat Source Elements, Int. J Heat and Fluid Flow, Vol 17, pp 474-482.

Esposito, P.G. and Behnia, M. (1994) Transition to unsteadiness in Three Dimensional Low Prandtl Natural convection, Heat Transfer, Vol. 7, pp 43 - 48.

Kim, C.H., Milton, B.E., Chen, P.Y.P. and Behnia, M. (1995) Modelling of vaporised Two Phase Induction Flow in a Fuel injection System, Proc. 6th Asian Congress of Fluid Mechanics, Vol II, pp 1118 - 1121.

Le Quere, P. and Behnia, M. (1996) From onset of unsteadiness to chaos in a differentially heated square cavity, (accepted J. Fluid Mechanics).

Milton, B.E., Behnia, M., Chen, P.Y.P. and Blakeley, (1994) Simulation of Fuel Film Formation in Ducts, Int. J Applied Finite Elements and Computer Aided Engineering, Vol 18, pp 349-363.

Mizunuma, H., Behnia, M. and Nakayama, W. (1996) Heat Transfer from Grooved Surfaces to Flow of Fluorinert Coolant in Reduced Size Channels, Proc. 5th Intersociety Conf. On Thermal Phenomena in Electronic Systems (IEEE), pp 274-283.

Nakayama, W. and Behnia, M. (1996) CFD Simulations in the Design and Evaluation of Heat Sinks, ASME- EPP, Vol 18, pp 11-21.

Section 3.1

Bird, G.A. (1994) Molecular Gas Dynamics, Oxford University Press.

Bird, G.A. (1996) Recent Advances and Current Challenges for DSMC, Computers and Mathematics with Applications,

Bird, G.A. (1997) The Initiation of Centrifugal Instabilities in an Axially Symmetric Flow, Proc. 20th Int. Symp. on Rarefied Gas Dynamics, in press.

Section 3.2

Macrossan, M.N. (1995) 'Some Developments of the Equilibrium-Particle Simulation Method for the Direct Simulation of Compressible Flows', NASA CR-187613, ICASE Interim Rep 27.

Pullin, D. I. (1980) "Direct simulation methods for compressible inviscid ideal-gas flow", J. Comput Phys, v34, pp231-244.

Section 4.1

Macrossan, M.N. (1989) "The equilibrium flux method for the calculation of flows with non-equilibrium chemical reactions", J Comput Phys, v80, pp204-231.

Macrossan, M.N. (1990) "Hypervelocity flow of dissociating nitrogen downstream of a blunt nose", J Fluid Mech, v207, pp167-202.

Macrossan, M.N. and Eckett, C. (1996) "Chemical non- equilibrium inviscid flow over a blunt cone at incidence", AIAA J, v34 pp2054-2061.

Macrossan, M.N. and Hanock, M. (1996) "A method of assessing the effects of artificial dissipation in Navier-Stokes codes", Int J Num Method Fluids, v22, pp 979-985.

Macrossan, M.N. and Oliver, R.I. (1993) "A kinetic theory solution method for the Navier-Stokes equations", Int J Num Meth Fluids, v17, pp177-193.

Macrossan, M.N. and Pullin, D.I. (1994) "A computational investigation of inviscid hypervelocity flow of a dissociating gas past a cone at incidence", J. Fluid Mech, v226, pp69-92.

Macrossan, M.N. and Stalker, R.J. (1987) "After-body flow of a dissociating gas down-stream of a blunt nose", AIAA Paper 87-407.

Mallett, E.R., Pullin, D.I. and Macrossan, M.N. (1995) "Numerical study of hypersonic leeward flow over a blunt nosed delta wing", AIAA J v33, pp1626-1633.

Pullin, D.I. (1980) 'Direct Simulation Methods for Compressible Inviscid Ideal-Gas Flow', J. Comput. Phys. v. 34(2), pp. 231-244.

Stalker, R.J. (1989) "Approximations for non-equilibrium hypervelocity aerodynamics", Ann Rev Fluid Mech, v21, p37.

Section 4.2

Doolan, C.J. and Jacobs, P,A. (1996) Modeling mass entrainment in a quasi-one-dimensional shock tube code. A.I.A.A. Journal, 34(6) pp.1291-1293.

Jacobs, P.A. (1991) Simulation of transient flow in a shock tunnel and a high Mach number nozzle.In 4th International Symposium on Computational Fluid Dynamics.

Jacobs P.A. (1991) Single-block Navier-Stokes integrator, ICASE Interim Report 18.

Jacobs, P.A. (1993) Simulation of transient flows in a shock tunnel, In Computational Applications and Techniques Conference '93, Canberra, Australia., pages 311-319. World Scientific.

Jacobs, P.A. (1994) Numerical simulation of transient

hypervelocity flow in an expansion tube, Computers and Fluids, 23(1) pp.77-101.

Jacobs, P.A. (1994) Quasi-one-dimensional modelling of a free-piston shock tunnel, A.I.A.A. Journal, 32(1):137-145.

Jacobs, P.A. and Craddock, C.S. (1995) Simulation and optimisation of three-dimensional flows in scramjet ducts. In Twelfth International Symposium on Air Breathing Engines., Washington, DC., September, A. I. A. A.

Kendall, M.A. Morgan, R.G. and Jacobs, P.A, (1997) A compact, shock-assisted free-piston driver for impulse facilities. Shock Waves, Accepted.

Kendall, M.A., Morgan, R.G. and Petrie-Repar, P. (1997) A study of free-piston double-diaphragm drivers for expansion tubes A.I.A.A. Paper 97-0985, January.

McGhee, A.M. and Jacobs, P.A. (1996) Parallel computation of hypervelocity flow, In R.L.May and A.K.Easton, editors, Computational Techniques and Applications: CTAC95, pages 549-556, World Scientific, 1996.

Petrie-Repar, P.J. and Jacobs, P.A. (1996) A computational study of shock speeds in high performance shock tubes Shock Waves, Submitted.

Smith, A.L., Johnston, I.A. and Austin, K.J. (1996) Comparison of numerical and experimental drag measurement in hypervelocity flow. The Aeronautical Journal of the Royal Aeronautical Society, 100(999) pp.385-388.

Section 5.1

Elghobashi, S. (1994) On Predicting Particle-Laden Turbulent Flows, Applied Scientific Research, Vol. 52, pp. 309-329.

Hetsroni, G. (1989) Particle-turbulence interaction, Int. J. Multiphase Flow, Vol. 15, pp.735-746.

McLaughlin, J.B. (1994) Numerical computation of particle- turbulence interaction, Int. J. Multiphase Flow, Vol.20, pp.211-232.

Murray, D.B. and Humphreys, I. (1994) Local enhancement of heat transfer in a particulate cross flow, 2. Experimental data and predicted trends, Int. J. Multiphase Flow, Vol. 20, pp.170- 179.

Tu, J.Y., Eghlimi, A. and Fletcher, C.A.J. (1995) Novel Turbulence Model for Gas-Particle Flows, in Proceedings of the 6th International Symposium of CFD, pp.1310-1315.

Tu, J.Y. and Fletcher, C.A.J. (1994) An improved model for particulate turbulence modulation in confined two-phase flows, Int. Comm. Heat and Mass Transf., 21(6), pp. 775-783.

Tu, J.Y. and Fletcher, C.A.J. (1995a) Numerical computation of turbulent gas-solid particle flow in a 90 bend, AIChE Journal, 41 (10), pp. 2187-2197.

Tu, J.Y. and Fletcher, C.A.J. (1995b) Continuum Hypothesis in the Computation of Gas-Solid Flows, in Computational fluid Dynamics, (Eds., Leutloff, D. and Srivastava, R.C.) Springer- Verlag, Heidelberg, pp.1-10.

Tu, J.Y. and Fletcher, C.A.J. (1995c) Computational analysis of turbulent gas-solid particle flows over tube banks, ASME Gas-Solid Flow FED-Vol.228, pp. 309.

Tu J.Y. and Fletcher, C.A.J. (1996a) Eulerian Modelling of Dilute Particle-Laden Gas Flows Past Tubes, Journal of Computational Fluid Dynamics, 5, pp.45-70.

Tu, J.Y. and Fletcher, C.A.J. (1996b) Computational Modelling of Particle Dispersion in a Backward-Facing Step Flow, to appear in Proceedings of The 2nd International Symposium Numerical Methods for Multiphase Flows, ASME-FED, San Diego, CA, USA, July 7-11.

Tu, J.Y., Fletcher, C.A.J. and Behnia, M. (1996) Numerical modelling of three-dimensional fly ash flow in power utility boilers, Int. J. Numer. Methods Fluids, in press.

Tu, J.Y., Fletcher, C.A.J., Behnia, M. and Reizes, J.A. (1996) Prediction of Flow and Erosion in Coal Fired Power Boilers and Comparison with Measurement, Transactions ASME J. Eng. Gas Turbines & Power, accepted.

Tu, J.Y., Lee, B.E. and Fletcher, C.A.J. (1996) Computational Modelling of Particle-Wall Interactions for Tube Bank Erosion Prediction, to appear in Proceedings of The Symposium on Erosion Processes, ASME-FED, San Diego, CA, USA, July 7-11.

Yarin, L.P. and Hetsroni, G. (1994) Temperature fluctuations in particle-laden flows dilute flows, Int. J. Multiphase Flow, 20, pp. 17-25.

Section 5.2

Morris, P., Hourigan, K. and Thompson, M.C. (1997) The flow of a multiphase fluid through piping. ASME fluids engineering summer meeting (in preparation).

Section 5.3

Davidson, M.R. (1996a) Comparison of Two-way Coupling Models for Confined Turbulent Gas-Particle Jets in Flash Smelting, Submitted to Appl. Math. Modelling.

Davidson, M.R. (1996b) Turbulence Modulation by Particles in Models of Flash Furnace Jets. 6th Australasian Heat and Mass Transfer Conference, 9-12 Dec, 1996. Begell House Inc, in press.

Davidson, M.R. (1997) Turbulence Modulation by Mass Transfer in a Model of a Flash Furnace Gas-Particle Jet, in preparation for CFD in Mineral and Metal Processing and Power Generation Conf., 3-4, July, Melbourne.

Hovenden, R. and Davidson, M.R. (1997) Particle-Turbulence Interaction in a Model of Two-phase Flows in Spray Drying In preparation for CFD in Mineral and Metal Processing and Power Generation Conf., 3-4, July, Melbourne.

Smith, K.M., Davidson, M.R. and Fletcher D.F. (1997) Dispersion of a Solids Stream Falling into a Stationary Liquid, in preparation for CFD in Mineral and Metal Processing and Power Generation Conf., 3-4, July, Melbourne.

Section 6.1

Best, J.P. and Kucera, A. (1992) "A numerical investigation of non-spherical rebounding bubbles", J. Fluid Mech. Vol. 247, pp 137-154. 012

Harvey, S.B., Best, J.P. & Soh, W.K. (1996) "Vapour bubble measurement using image analysis", Meas. Sci. Technology, vol. 7, pp 592-604

Karimi, A.A. and Soh, W.K. (1994) 'The application of dual reciprocity boundary element technique to the study of bubble dynamics', Computational Fluid Mechanics Journal, Vol. 3, No. 3, October, pp. 367 - 378.

Karimi, A. & Soh, W.K. (1995) "The effects of heat transfer on a pulsating cavity bubble", Proceedings of the Sixth Asian Congress of Fluid Mechanics, Chew T.Y. & Tso C.P., Centre for continuing education, Nanyang Technological University, Singapore, May, pp361-354.

Karimi, A. & Soh, W.K. (1996) "Adiabatic and non-adiabatic cavity bubbles near rigid boundaries", The Ninth International Symposium on Transport Phenomena (ISTP-9) in Thermal-Fluids Engineering, National University of Singapore, editors Winoto, S. H., Chew, T. T. & Wijeysundera, N. E., National University of Singapore, June, pp. 1048-1053.

Karimi, A.A. and Soh, W.K. (1996) "Transient heat transfer of a non-spherical bubble near a rigied boundary", Submitted to The Journal of Computational Fluid Mechanics.

Soh, W.K. & Karimi, A.A. (1996a), "On the calculation of heat transfer in a pulsating bubble", Applied Mathematical Modelling, Vol. 20, September, 638-645.

Soh, W.K. and Karimi, A.A. (1996b) "The thermodynamic process and heat transfer of a pulsating bubble in liquid" 6th Australasian Heat and Mass Transfer Conference, University of New South Wales, Australia.

Soh, W.K. and ShervaniTabar, M. (1994) 'Computer model for a pulsating vapour bubble near a rigid surface', The Journal of Computational Fluid Mechanics, Vol 3, No. 1, pp. 223-236.

Section 6.2

Graham, L.J. W., Morris, P., Hourigan, K. and Thompson, M.C. (1996) Convergence and Validation of Vortex Breakdown. (In preparation.)

Hourigan, K., Graham, L.J.W. and Thompson, M.C. (1995a) Spiral Streaklines in Pre-Vortex Breakdown Regions of Axisymmetric Swirling Flows, Physics of Fluids, Vol. 7 (12), 3126-8.

Hourigan, K., Graham, L.J.W and Thompson, M.C. (1995b), The Torsionally Driven Cavity: A Test Case for Comparison between Experiment and Prediction, Proceedings of 12th Australasian Fluid Mechanics Conference, Vol. 1, 295-298, Sydney, December 10-15.

Hourigan, K., Thompson, M.C., Welsh, M.C. and Brocher, E. (1992), Acoustic Sources in a Tripped Flow Past a Resonator Tube, AIAA Journal, Vol. 30(2), 1484-1491.

Hourigan, K., Welch, L.W., Thompson, M.C., Cooper, P.I. and Welsh, M.C. (1991), Augmented Forced Convection Heat Transfer in Separated Flow Around a Blunt Flat Plate, Exp. Thermal and Fluid Science, 4, 182-191.

Hourigan, K., Welsh, M.C., Thompson, M.C. and Stokes, A.N. (1990) Aerodynamic Sources of Acoustic Vibration in a Duct with Baffles, Journal of Fluids and Structures, 4, 345-370.

Hourigan, K., Welsh, M.C., Thompson, M.C. and Stokes, A.N. (1991), Sources of Resonant Sound in Separated Duct Flows, Inter-noise, 91, 2-4 December, Sydney, Australia, Vol. 1, 601-604.

Monaghan, J.J., Thompson, M.C. and Hourigan, K. (1994) A simulation of fixed and moving jets, International Colloquium on Jets, Wakes and Shear Layers, 18-20 April, Melbourne, 30.1-30.12.

Morris, P. and Ferziger, J.H. (1996) The 2-4 model for large eddy simulations. Submitted to Physics of Fluids A.

Rudman, M., Hourigan, K. and Brocher, E. (1994) Instabilities in an impinging jet, International Colloquium on Jets, Wakes and Shear Layers, 18-20 April, Melbourne, 29.1-29.8.

Rudman, M., Hourigan, K. and Brocher, E. (1996) The Feedback Loop in an Impinging Supersonic Jet, International Journal of Experimental Heat Transfer, Thermodynamics, and Fluid Mechanics, Vol. 12.

Rudman, M., Hourigan, K. and Thompson, M.C. (1994) Particle Shear Rate History in a Taylor-Couette Column, Liquid-Solid Flows, The 1994 ASME Fluids Engineering Division Summer Meeting, Lake Tahoe, Nevada, June 19-23.

Rudman, M., Thompson, M.C. and Hourigan, K. (1992) An Adaptive Fast Multipole Moment Technique for Calculating Flow around Two-Dimensional Bluff Bodies, Computational Fluid Dynamics '92, Vol. 1, 449-445.

Thompson, M.C. and Hourigan, K. (1992) Prediction of Vortex Junction Flow Upstream of a Surface Mounted Obstacle, 11th Australasian Fluid Mechanics Conference, 14-18 December, Hobart, Australia, Vol I, pp. 111-114

Thompson, M.C., Hourigan, K., Mills, R. and Sheridan, J. (1995a) The effect of acoustic forcing on base pressures, ASME International Mechanical Engineering Congress, San Francisco, Nov.

Thompson, M.C., Hourigan, K., Mills, R. and Sheridan, J. (1995b), Simulation of Acoustically Forced

Flows Around Long Rectangular Plates, Proceedings of 12th Australasian Fluid Mechanics Conference, Sydney, Vol. 1, 513-516, December 10-15.

Thompson, M.C., Hourigan, K. and Monaghan, J.J. (1994) The simulation of free-surface flows with SPH, Advances in Computational Methods in Fluid Mechanics, The 1994 ASME Fluids Engineering Division Summer Meeting, Lake Tahoe, Nevada, June 19-23.

Thompson, M., Hourigan, K. and Sheridan, J. (1995) Three-Dimensional Mode Development in a Low Reynolds Number Flow Over a Cylinder, Proceedings of 12th Australasian Fluid Mechanics Conference, Vol. 1, 247-250, Sydney, December 10-15.

Thompson, M., Hourigan, K. and Sheridan, J. (1994) Three-Dimensional structures in Cylinder Wakes, Fifth European Turbulence Conference, 214, Siena, Italy, July 5-8.

Thompson, M.C., Hourigan, K. and Sheridan, J. (1995) The Development of Three-Dimensionality in the Wake of a Circular Cylinder, Turbulent Shear Flows 5, Kluwer Press, Ed. E. Benzi.

Thompson, M.C., Hourigan, K. and Sheridan, J. (1996) Three-Dimensional Instabilities in the Wake of a Circular Cylinder, International Journal of Experimental Heat Transfer, Thermodynamics, and Fluid Mechanics, 190-196, Vol. 12.

Welsh, M.C., Hourigan, K., Welch, L.W., Downie, R.J., Thompson, M.C. and Stokes, A.N. (1990) Acoustics and experimental methods: the influence of sound on flow and heat transfer. Experimental Thermal and Fluid Science, 3, 138-152.

Section 7.1

Davis, R.T., Barnet, M. and Rakich, J.V. (1986) The Calculation of Supersonic Viscous Flows using the Parabolise Navier-Stokes Equations, Computers and Fluids, Vol. 14, pp 197-224.

Lawrence, S.L., Tannehill, J.C. and Chausee, (1989) Upwind Algorithm for the Parabolised Navier-Stokes Equations, AIAA J, 27, 1175-1183.

Srinivas, K. (1992a) A Spatial Marching Algorithm for Navier Stokes Equations, Computers and Fluids, Vol. 21, No. 2, pp 291 - 299

Srinivas, K. (1992b) Computation of Hypersonic Flow past a Compression Corner by a Spatial Marching Scheme, Workshop - Hypersonic Flows for Reentry Problems, Part2, organised by INRIA, France at Antibes, France, April 15-19, 1991 (See proceedings edited by Abgrall,R, Desideri,J.A., Glowinski,R., Mallet,M and Periaux,J. Published by springer-Verlag, p 338-347).

Srinivas, K. (1992c) Computation of High Speed Flows using an Explicit Spatial Marching Algorithm, presented at the 11th Australasian Fluid Mechanics Conference, Hobart, Australia, 14 - 18 December, pp 711-714.

Srinivas, K. (1995a) An Explicit Finite Volume Spatial Marching Method for Reduced Navier Stokes Equations, Proceedings, 8th International conference on Numerical Methods in Laminar and Turbulent Flow, Int. J Num. Methods in Fluids, Vol 22, pp 121-135.

Srinivas, K. (1995b) An Explicit Finite Volume Spatial Marching Method for Reduced Navier Stokes Equations -2, Proceedings Ninth International Conference on Laminar and Turbulent Flows (Atlanta, USA, July 10 -14), Vol 9, Part 1, pp 516-527, edited by C Taylor and P Durbetaki, Pineridge Press.

Srinivas, K. (1997a) LED Based Spatial Marching Method for High Speed Flows, (to appear Computational Fluid Dynamics Journal)

Vigneron, Y.C., Rakich, J.V. and Tanehill, J.C. (1978) Calculation of supersonic viscous flows over delta wings with sharp subsonic leading edges, NASA TM 78500, June.

Section 7.2

Fard, M.P. and Srinivas, K. (1993) Multigrid Algorithms and Residual Averaging for Transonic Flow, presented at the Computational Techniques and Applications conference, CTAC93, Australian National University, Canberra, July 5th - 9th.

Fard, M.P. and Srinivas, K. (1995a) Validation of Matrix form of Artificial Dissipation Terms for an Axisymmetric, Transonic Flow, Proceedings, Sixth Asian Congress of Fluid Mechanics, (May 22-26, Singapore),Vol2, pp 1328-1331, edited by Y T Chew and C P Tso.

Fard, M,P. and Srinivas, K. (1995b) An Investigation of Renormalisation Group Based Algebraic turbulence Model presented at the 6th ISCFD conference at lake Tahoe, Nevada, USA, Sept. 4-8, pp 287-292.

Fard, M.P. and Srinivas, K. (1995c) An Investigation of Renormalisation Group Based Algebraic turbulence Model-2 presented at the Twelfth Australasian Fluid Mechanics Conference, Sydney, Australia, 10-15 December, pp 763-766.

Fard, M.P. and Srinivas, K. (1996) Computation of Complex Flows with SLIP scheme and a Half equation Model of Turbulence. (presented at the 15th International conference on Numerical Methods in fluid Dynamics, June 24 - 28, Monterey, California.

Jameson, A. (1995) Analysis and Design of Numerical Schemes for Gas Dynamics, 2 : Artificial Diffusion and Discrete Shock Structure, Int. J Computational Fluid Dynamics, Vol. 5, pp 1-38.

Srinivas, K. (1996) Computation of Cascade Flows by a High Resolution Switched Scheme, Proceedings, the second Asian Computational Fluid Dynamics Conference, Tokyo, 15-18 December, edited by Kubota H and Aso S, pp 177-182.

Srinivas, K. (1997b) Computation of Cascade Flows by LED Schemes (to appear Computational Fluid Dynamics Journal)

Section 7.2 (See References for Section 7.1)

Section 7.3 (See References for Section 2.1)

Section 7.4

Mellen, C.P. and Srinivas, K. (1994) Treatment of Branch Cut Lines in the computation of Incompressible Flow with C-grids, presented at the 14th International conference on Numerical Methods for Fluid Dynamics,(held in Bangalore, India, July 11-15), Procedings, Edited by S M Deshpande, S S Desai and R Narasimha pp 201-205, Springer-Verlag.

Mellen, C.P. and Srinivas, K. (1996) A Comparison of three different Implicit Solution Schemes for the Incompressible Navier-Stokes Equation, Proceedings, the second Asian Computational Fluid Dynamics Conference, Tokyo, 15-18 December, edited by Kubota H and Aso S, pp 301-306.

SOLUTION OF THE EULER EQUATIONS ON SOLUTION-ADAPTIVE CARTESIAN GRIDS

Kenneth G. POWELL

W. M. Keck Foundation CFD Laboratory Department of Aerospace Engineering University of Michigan Ann Arbor, MI 48109-2118

Abstract

The appeal of the solution-adaptive Cartesian Grid approach to CFD is the degree to which the grid-generation procedure can be automated. In most grid-generation approaches , the starting point is a discretized surface geometry, and the quality of the volume grid that is ultimately generated is highly dependent on the discretization of the surface geometry. In the Cartesian approach, the quality of the volume grid is much less dependent on the original surface geometry description; in fact, the two are almost entirely decoupled. The Cartesian approach to grid generation introduces difficulties in the development of a solver, however. For proper resolution, adaptive gridding is a virtual necessity, and, even with adaptation, the number of cells necessary for a high-fidelity Cartesian flow solution is higher than that of a solution on a body-fitted grid, whether structured or unstructured. The Cartesian solver must also be able to handle the small cut cells that occur in the Cartesian grid-generation procedure. This paper describes approaches for Cartesian grid generation, and for flow solvers that work well on adaptive Cartesian grids.

1. Introduction

1.1 The Appeal of the Cartesian Approach

One of the fervent wishes of almost all CFD users and researchers is for "fully automatic" grid generation. However, there is a fundamental trade-off in the grid-generation process: between automation of the process, and control over the properties of the grid. Most users of CFD see grid generation as a means unto an end, and would opt for as automatic

a grid-generation as possible, but for one very important problem — the CFD results obtained on a grid are dependent, sometimes extremely so, on the "quality" of the grid. The quality of a grid is a fairly loosely defined measure of grid skewness, grid stretching, grid smoothness, and grid resolution. Generally, flow solvers behave the best, giving best results and best convergence, when grids:

- have relatively small cells in regions of the flow with relatively large gradients, and vice versa;

- have regular cells, that are not highly

Received on October 12, 1997.

stretched or skewed;

- have smooth variations in cell properties.

Sometimes these properties are in conflict with each other: for example, achieving efficient resolution of boundary-layer flows requires highly stretched cells.

If flow solvers can be made as insensitive as possible to variations in cell sizes, grid stretching and grid skewness, this opens the door to greater automation of the grid-generation process. This approach has been central to the increasing popularity of unstructured-grid schemes for CFD. Triangle/tethrahedron unstructured grid generation, which can be much more highly automated than quadrilateral/hexahedron structured grid generation, has gained popularity in direct proportion to progress made in accuracy and robustness of solvers for unstructured grids. The Cartesian approach is one step farther along this path: Cartesian quadrilateral/hexahedron grid generation can be more highly automated than unstructured triangle/tetrahedron grid generation, but puts even greater demands on the flow solver.

In a Cartesian-based solver, a body is "cut" out of a background Cartesian grid. Generation of a non-body-conforming Cartesian grid is a purely algebraic process, and even the cutting of the body out of the grid, although it requires extreme care in implementing, is fundamentally a straightforward process. A simple example of a Cartesian grid is shown in Figure 1. Cells that are outside the body are marked as active cells on which the flow solver operates; cells that are inside the body are marked as inactive cells; cells that straddle the body are marked as cut cells, and require some modified algorithm. It is the treatment of these cut cells that posed one of the primary challenges in development of Cartesian solvers for the Euler equations,

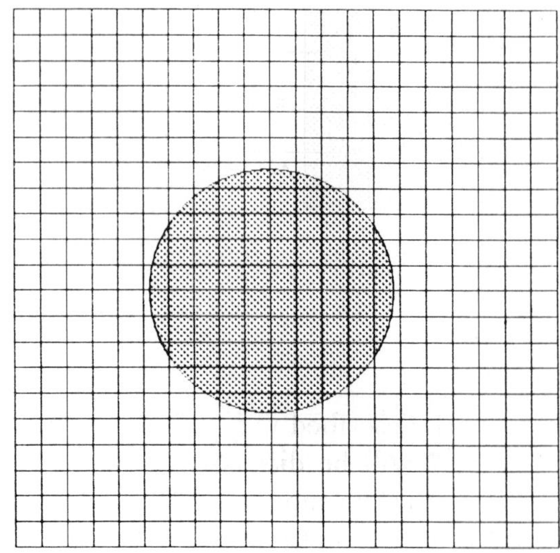

Figure 1: Simple Cartesian Grid for Flow Past a Circle

and that differentiates between the various Cartesian-based schemes in use today.

Aftosmis et al [1] have grouped the boundary treatments for cut cells into three levels of sophistication:

Level 1 The volume fraction of the cut cell that is inside the flow is computed, and thus the fraction of the cell that is fluid and that that is not is accounted for in the solution procedure in that cell.

Level 2 In addition to computing the volume fraction for the cut cell, the intersection between the cell and the body is computed and stored. In this approximation, the cell volume is accounted for correctly, and so are the areas of the cell faces. The portion of the body that lies within the cell is treated as a flat facet.

Level 3 In addition to having the volume and intersection data, sub-cell information about the geometry of the portion of the body that lies in the cell is taken into account. This allows curved surfaces to be represented on a sub-cell level.

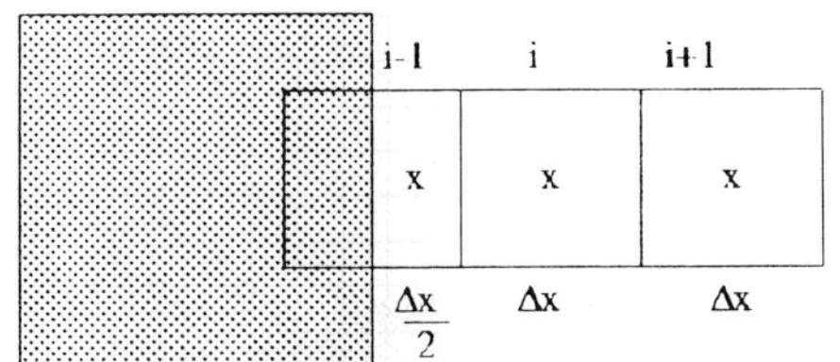

Figure 2: A Simple Cut Cell in One Dimension

In the work described in this paper, only levels 2 and 3 will be discussed, as this is the state of the art in modern Cartesian solvers. A level 1 approximation can lead to inconsistent difference formulas.

1.2 Cut Cells: The Primary Difficulty in Development of Cartesian Schemes

There are two primary difficulties raised by the cut cells generated in a Cartesian grid. One is accuracy: many of the most common numerical schemes for solving partial differential equations, particularly those derived from a finite-difference point of view, are based on the assumption that the grid spacing is constant. This is a dangerous assumption in a Cartesian-based scheme; cut cells require special treatment to ensure consistency, and some cut-cell treatments are more accurate than others. Because cut cells can be much smaller than their uncut neighbor cells, the solver must be based on a numerical approximation to the spatial derivatives that is accurate even for a tiny cell next to a large one.

The second difficulty introduced by the cut cells is stability: many schemes for solving partial differential equations have stability constraints that are tied to the smallest cell in the grid — the smaller the cell, the smaller the maximum stable time step. Using the simple grid in Figure 2 to make the point, discretization of a linear advection equation on this grid

$$\frac{\partial u}{\partial t} + a\frac{\partial u}{\partial x} = 0 \qquad (1)$$

by a first-order upwind scheme with forward-Euler time stepping leads to a CFL constraint based on the smallest cell, cell $i-1$. That constraint is that

$$\Delta t \le \frac{\Delta x}{2a} \,, \qquad (2)$$

which is twice as strict as the constraint for the uncut cells. In a more general grid, there is no control over how small the cut cells may be, and it is not uncommon for cut cells to be six or more orders of magnitude smaller than their uncut neighbors, leading to time-step penalties of a factor of a million or greater.

The schemes that have been developed for dealing with small cut cells fall into two basic categories, described below.

Approaches Based on Wave Propagation

The first approach, due to Chern and Colella [2], is one that primarily addresses the stability problem raised by cut cells. It was developed as an approach for front-tracking, but was later applied to the cut-cell problem for Cartesian grids, with the surface of the body that is cut out of the grid treated as the front.

The basic idea of the redistribution approach can be understood by considering a first-order upwind discretization of the linear advection equation, Equation 1, on the grid of Figure 2. On cell i, which is uncut, the update scheme is derived as follows:

$$\frac{\partial u}{\partial t} = -a\frac{\partial u}{\partial x}$$

$$\int_{\text{cell } i} \frac{\partial u}{\partial t} = -\int_{\text{cell } i} \frac{\partial u}{\partial x}$$

$$\frac{\partial \bar{u}_i}{\partial t}\Delta x = -a\left(u_{i+1/2} - u_{i-1/2}\right) \,, \quad (3)$$

where the bar denotes a cell average quantity and the $i + 1/2$ and $i - 1/2$ subscripts denote the cell interfaces. Using a forward Euler temporal discretization,

$$\frac{\partial \bar{u}_i}{\partial t} = \frac{\bar{u}_i^{n+1} - \bar{u}_i^n}{\Delta t} \tag{4}$$

and using a first-order upwind spatial discretization, assuming $a > 0$ gives

$$u_{i+1/2}^n - u_{i-1/2}^n = \bar{u}_i^n - \bar{u}_{i-1}^n . \tag{5}$$

Thus, the update scheme for cell i is

$$\bar{u}_i^{n+1} \Delta x = \bar{u}_i^n \Delta x + a \Delta t \left(\bar{u}_{i-1}^n - \bar{u}_i^n \right) . \tag{6}$$

The stability constraint is that the coefficient of $\bar{u}_i^n$ is non-negative, which gives

$$\Delta x - a \Delta t \geq 0$$

$$\Delta t \leq \frac{\Delta x}{a} \tag{7}$$

An update formula for the cut cell is derived similarly, giving

$$\bar{u}_{i-1}^{n+1} \Delta x_{cut} = \bar{u}_{i-1}^n \Delta x_{cut} +$$

$$a \Delta t \left(u_W^n - \bar{u}_{i-1}^n \right) , \tag{8}$$

where u_W^n is the wall-boundary value of u, but this leads to a stability problem. The stability constraint for this cell is

$$\Delta t \leq \frac{\Delta x_{cut}}{a} . \tag{9}$$

For the simple grid shown, this means the maximum allowable time step in the cut cell is one half of that allowed in the uncut cell. On a more general grid, Δx_{cut} may well be several orders of magnitude smaller than Δx, leading to an unreasonable stability constraint.

The basic idea behind the redistribution approach to this problem is to modify the update formula for the cut cell. Instead of the term

$$\delta = a \Delta t \left(u_W^n - \bar{u}_{i-1}^n \right) \tag{10}$$

on the right-hand side of Equation 8, this contribution is rescaled, to

$$\delta' = \frac{\Delta x_{cut}}{\Delta x} \delta \tag{11}$$

so that the update formula for the cut cell is

$$\bar{u}_{i-1}^{n+1} \Delta x_{cut} = \bar{u}_{i-1}^n \Delta x_{cut} +$$

$$\frac{\Delta x_{cut}}{\Delta x} a \Delta t \left(u_W^n - \bar{u}_{i-1}^n \right) \tag{12}$$

which has the stability constraint given in Equation 7.

This rescaling, however, introduces an error; it is a violation of conservation. The original form for δ came directly from the conservation of u; rescaling δ means that u is not conserved in cell $i+1$. In order to fix this, the redistribution approach is to take the quantity

$$\delta - \delta' = \left(1 - \frac{\Delta x_{cut}}{\Delta x} \right) a \Delta t \left(u_W^n - \bar{u}_{i-1}^n \right) \tag{13}$$

and redistribute it to another cell or cells on the grid.

A sophisticated approach for redistributing this quantity was developed by Leveque [3], and applied to the cut-cell Cartesian problem by Berger and LeVeque [4]. The wave emanating from the wall will, in a time step Δt, sweep through cell $i - 1$ and into cell i (see Figure 2. Thus, in this one-dimensional example, the redistribution is fairly simple — a change δ' is made to the state in cell $i - 1$ and a change $\delta - \delta'$, due to the cut cell, is added to the change computed for the uncut cell i. Cell $i + 1$ is not affected by the cut cell, since the wave from the wall does not cross into that cell during the time step. This same idea carries over to two dimensions, as depicted in Figure 3. The direction in which the wave will travel, and the distance it will travel in time Δt, are used to compute the fraction of the quantity $\delta - \delta'$ that must

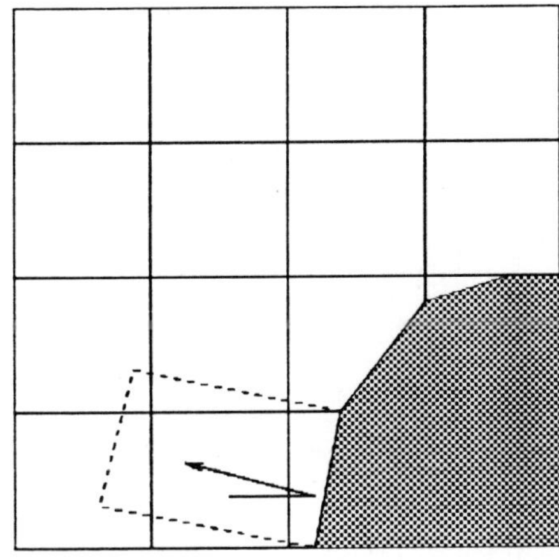

Figure 3: A Simple Cut Cell in Two Dimensions

be distributed to each of the cells neighboring the cut cell. Berger and Leveque refined further on this approach by introducing rotated Riemann solvers [5, 6]. Boxes like that of Figure 3 are drawn not just in the wave-propagation direction, but also in the direction normal to the propagation. On each face of the cut cell, Riemann problems for both directions are solved, and the results combined to form a flux normal to the face. This approach, combining the redistribution and rotated Riemann solvers, leads to an accurate and robust treatment of cut cells.

The Cell-Merging Approach

The philosophy of the cell-merging approach is akin to that of public-health vaccination programs — if the small cut cells are eradicated, there will be no small-cut-cell problem. In the merging approach, cut cells that are less than some user-defined fraction of the area of their uncut version are "merged" with a neighboring uncut cell. This process is depicted in Figure 4. The original grid, on the left, is examined for cut cells whose area is less than one half of their uncut area. Those small cut cells are merged with neighboring uncut cells, resulting in the grid on the

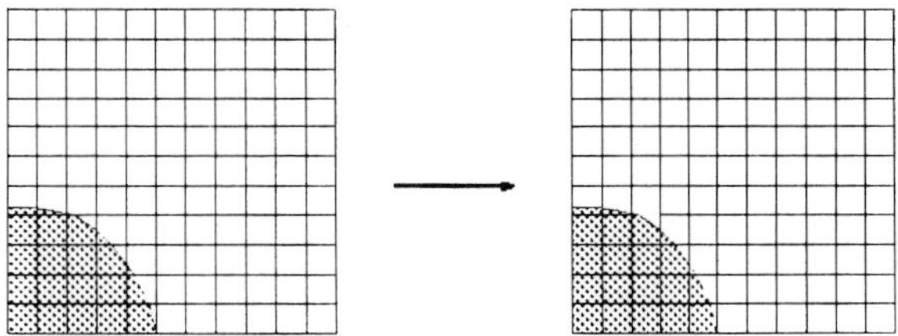

Figure 4: The Cell-Merging Process

right. For steady flows, this is a simple pre-processing step on the grid. The solver never sees the original grid; it simply acts on the pre-processed grid, which has, by definition, no small cut cells.

The cell-merging approach basically solves the stability problem associated with small cut cells. There still must be a modified approach in the remaining cut cells (those that were big enough that they did not need to be merged, and those that result from the merging process) to ensure accuracy of the scheme on those cells. Standard finite-volume scheme for unstructured grids can be used for this purpose. These will be described more fully in a later section of this paper. The merging process itself is also described in more detail in a later section.

1.3 The Need for Adaptive Refinement

One difficulty with the grids generated by the Cartesian approach, outside of the difficulties raised by small cut cells, is the problem of resolution. A simple Cartesian grid treats all portions of the flow equally; all cells in the grid are squares/cubes, and all are the same size. Flows of interest are almost never that isotropic, however. To resolve real geometries, and real flows, it is virtually necessary to allow the grid to adapt in some way.

One way to adapt the grid is to keep the grid Cartesian, i.e. successive points lie along a line of constant x (or y or z), but the spac-

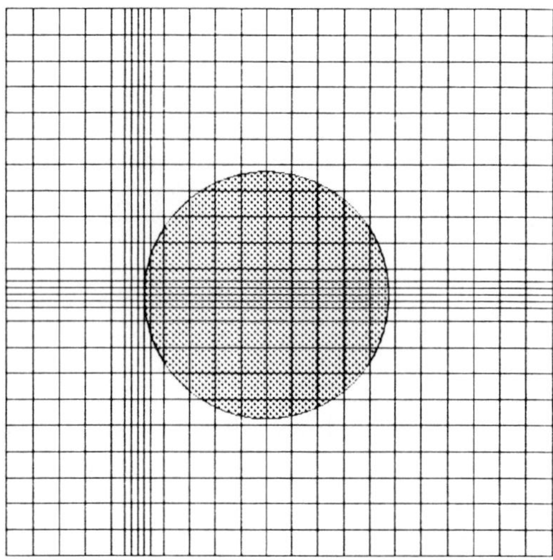

Figure 5: Adaptive Refinement by Stretching

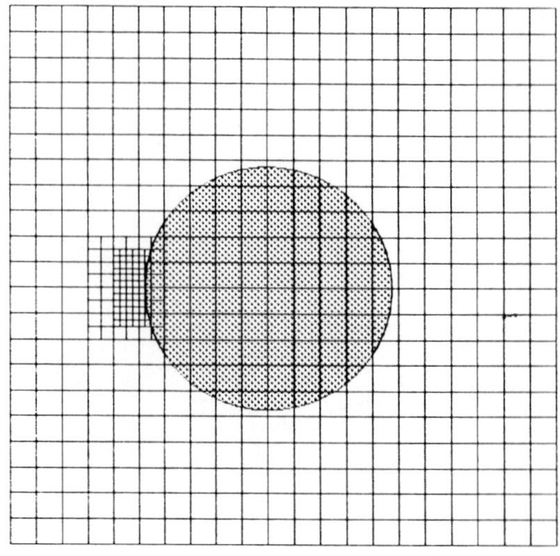

Figure 6: Adaptive Refinement by Patches

ing is allowed to vary so that certain features of the geometry or flow are better resolved than others. This complicates the flow solver somewhat, as non-constant grid spacing must be accounted for, but the basic structure of the code is unchanged. Figure 5 shows a simple example; the grid of Figure 1 is refined in the vicinity of the forward stagnation point. It is clear from the figure that there is an inefficiency associated with this type of refinement. Clustering grid points near the stagnation point leads to clustering away from the body as well.

Another approach to adaptive refinement is to select a rectangular region in which refinement is desired, and overlay the grid in that region with a patch of finer grid. This approach, depicted in Figure 6, leads to Cartesian grid away from the patch, and Cartesian grid inside the patch, but requires special treatment on both sides of the boundary of the patch. This patch-overlay can be done recursively, so that large variations in cell size can be achieved. In the figure, a 3×4 patch of the base grid has been refined one level. Following that, a 3×6 sub-patch of that patch has been refined by one level. The resulting grid resolves the forward stagnation point, and the area of refinement is kept local.

A third approach is cell-by-cell refinement, as depicted in Figure 7. This approach can be viewed as the small-patch (i.e. single-cell patch) limiting case of the patch refinement. Its advantage over the patch refinement is economical use of cells; its disadvantage is that the grid does not have the (i, j, k) structure that the unstretched or stretched Cartesian approach has globally, and the patch-refinement approach has locally. Data structures that work well for this refinement scheme, as well as the others, are described below.

2 The Basic Elements of an Adaptive Cartesian Scheme

2.1 Data Structures for Cartesian Schemes

Any CFD code has at its heart a basic data structure, used to store the geometric quan-

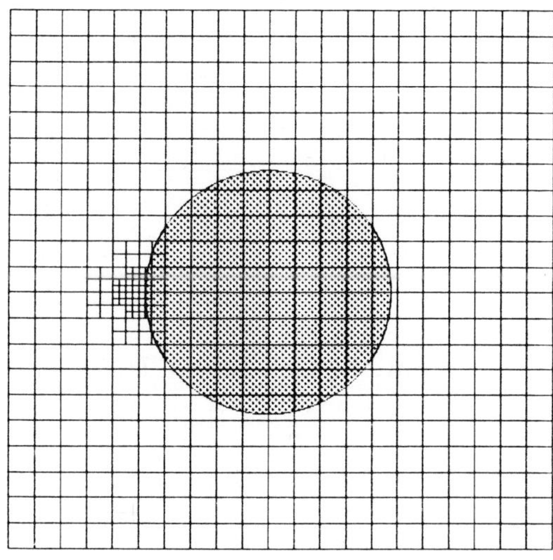

Figure 7: Adaptive Refinement by Cells

This data structure is that used for most structured-grid CFD codes, and is thus very familiar to most people in CFD. It is also very computationally efficient; finding a neighboring cell simply requires incrementing or decrementing an i or j index. Updates for the cells take place by double loops (a loop over i from 1 to i_{max} inside a loop over j from 1 to j_{max}, or vice versa), with cells that are inside the flow updated in one way, cells that are cut by a body updated another way, and cells that are inside a body skipped entirely. The shortcoming of this approach is that it is inherently limited to structured grids such as that of Figure 1: solution-adaptive Cartesian grid codes need a more sophisticated data structure.

tities (cell areas, vertex locations, etc.) and the flow quantities (density, momenta, energy etc.). There is more than one Cartesian-based approach to solving the Euler equations, and the various approaches can be categorized by the data structure used. There are four basic data structures commonly used in Cartesian approaches for compressible flow. Each will be described below for a two-dimensional Cartesian code; all carry over to three dimensions (some more easily than others).

Array-Based Storage

The simplest data structure for a Cartesian approach, and the one most easily coded (particularly in FORTRAN), is the array. Each cell is given an i and a j index. Cell-based quantities, such as flow variables, are stored as two-dimensional arrays, with the arrays dimensioned $i_{max} \times j_{max}$, where i_{max} is the maximum number of cells in the i direction and j_{max} is the maximum number of cells in the j direction. Vertex-based quantities, such as x and y coordinates of the vertices, are stored in two-dimensional arrays that are dimensioned $(i_{max} + 1) \times (j_{max} + 1)$.

Linked-List Data Structure

One way to deal with adapted Cartesian grids is to treat the grid as being totally unstructured, and use a data structure suitable for an unstructured-grid code. This is typically done via linked lists. A list of cells is set up, i.e. a vector of length n_{cells}. A neighbor list can then be created, for instance, by setting up an array that is $n_{cells} \times 4$, which stores, for the ith cell, the indices corresponding to the neighbors to the north, east, west and south of the cell. Since some operations (such as flux calculations) take place on a face-by-face basis, a vector of faces, n_{faces} long, is also typically stored. Similarly, a vertex-based list, $n_{vertices}$ long is stored. Flow variables and cell areas are typically stored in cell-based lists; fluxes and face normals are stored in face-based lists; vertex coordinates are stored in vertex-based lists. The number of connectivity lists stored is a trade-off between the computation time saved by having the connectivity stored versus the memory used in storing it. Cell-to-neighbor, cell-to-face, face-to-cell, face-to-vertex and cell-to vertex lists are ones that are often stored.

This type of data structure has the advantage of being quite general. Unlike the array-based structure, solution-adaptive grids are easily represented by this type of structure. That generality comes at a cost, however. First, there is a memory penalty — the connectivity lists take up space. These lists are not necessary in an array-based structure because connectivity is implied there; the neighboring cells, or a cell's vertices, can be found by incrementing and decrementing array indices. A second penalty is efficiency. The linked-list approach makes use of indirect addressing. Compilers produce code for addressing an element of an array-based structures that is often as much as five times faster than that produced for addressing a linked-list structure. A third penalty of the linked-list data structure is that adapting the grid requires updating the connectivity lists. Primarily for this reason, this approach, which was used in some early Cartesian work, is not being used in more recent work by the same authors [1, 7, 8].

Cell-Based Trees

Another data structure that lends itself well to the Cartesian approach, and particularly to the cell-refinement approach depicted in Figure 7, is a cell-based tree. The basic unit of this data structure is a cell, and the cells are arranged in a tree. In two dimensions. each cell has one parent cell and four children cells; in three dimensions, each cell has one parent cell and eight children cells. A two-dimensional cell-based tree is depicted in Figure 8.

The grid is represented as a tree in which cells in the grid correspond to nodes of the tree. The root of the tree is a large cell that covers the entire solution domain. This root cell is then divided, yielding (in two dimensions) four children cells.

Tree-based structures are well-suited to an adaptive Cartesian-grid approach for several

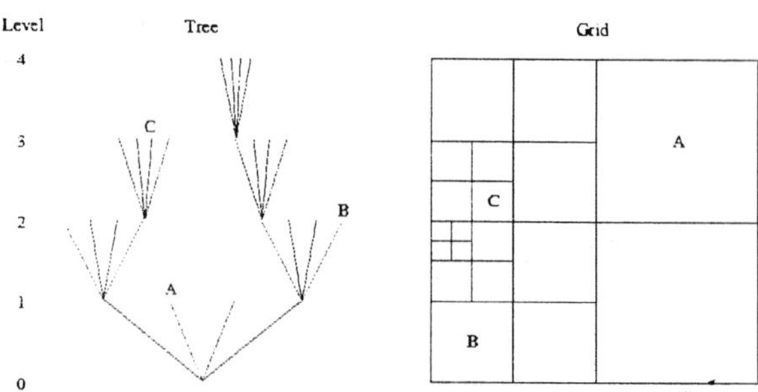

Figure 8: Cell-Based Tree

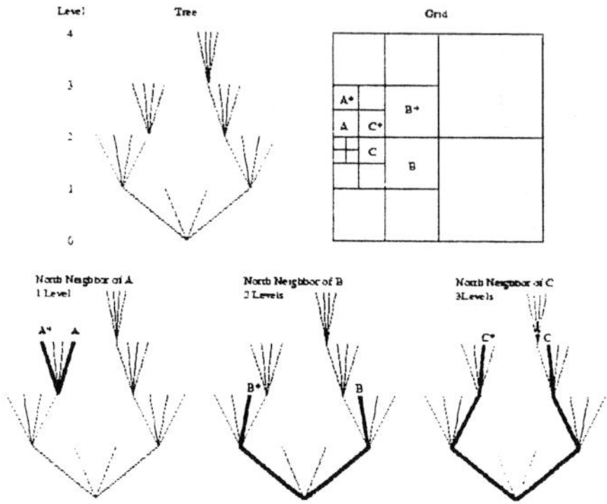

Figure 9: Finding a Neighboring Cell in a Tree

reasons. One reason is that a tree-based structure is memory-efficient; connectivity information relating cells to neighboring cells is unnecessary, as this information can be inferred from the tree structure. This neighbor-finding process is depicted in Figure 9. The northward neighbors A*, B* and C* of three cells A, B and C are desired. The figure shows the path traversed through the tree in each case. Case A is the simplest; querying A's parent cell yields the answer. In case C, the tree must be traversed all the way to the root and back again to find the neighbor.

Another strength of this structure is the ease with which the grid can be adapted: local refinement simply adds children cells to one of the nodes of the tree; local coarsening simply deletes the children cells of one of the nodes of the tree.

This data structure has two advantages over the linked-list approach. First, the storage required is less. Memory usage is heavily dependent on implementation, but the connectivity in a 2D cell-based tree can be described minimally with five pointers (to the parent and four children) and an integer (for the level of the tree); a typical linked-list structure requires, at a minimum, pointers from edges to cells, edges to nodes, and cells to edges, leading to approximately ten pointers per cell. A more important advantage is that the tree structure can be adapted (i.e. refined and/or coarsened) much more simply. Adaptation affects the tree structure very locally; the pointers in the linked-list structure must be redirected each time a cell is adapted. These two advantages come at a cost; connectivity information is not as explicitly stored in a tree-based structure, and so tasks such as finding neighboring cells must be done by traversing the tree, as depicted in Figure 9. In a typical tree-based Cartesian code, between 10% and 20% of the time is spent in tree traversals. This is actually a double cost; besides the CPU time spent on tree traversals, the extra levels of indirection mean substantially slower memory access than in a structured code.

Cartesian codes developed by the Boeing TRANAIR group (a full-potential solver) [9], Melton et al [10], DeZeeuw and Powell [7] and Bayyuk [11] are based on this approach.

Patch-Based Trees
A patch-based tree is a middle ground between the cell-based tree and the simple array data structures. The root of the tree is a relatively coarse grid that covers the entire domain. An arbitrary number of patches are then formed, which are one level (or two levels, in the work of Quirk) finer than the "parent" grid. Each of these patches can have an arbitrary number of finer patches inside them, and so on, building up a tree of successively finer patches. As opposed to the cell-based tree, in which the number of children cells is known (four in 2D, eight in 3D), the data structure for the patch-based approach must be general enough to allow an arbitrary number of children patches.

The patch-based approach was pioneered by Berger, and is the most frequently used data structure for adaptive Cartesian solvers. The work of Berger and Colella et al [12] and that of Quirk [13, 14] are based on this approach. Quirk has shown that this data structure lends itself particularly well to parallelization [15]. In addition, Berger [16] has shown that refining by patches does not have as much overhead associated with it as it might appear to have. For typical three-dimensional cases, the patch-refinement method typically leads to approximately twice as many cells as the cell-by-cell refinement method. However, due to the lower memory-per-cell cost of the patch-based approach, and the higher computational efficiency of a solver on the patch-based tree data structure, the patch-based approach may come out as the winner in the long run, even for steady flows.

2.2 Grid Generation and Adaptation

Geometry-Based Adaptive Refinement
In the geometry-based adaptive-refinement scheme used to generate the initial grids for calculations, both the body-surface discretization and the solution-domain volume discretization are carried out automatically, based on a minimum of user input. The inputs to this step are:

- A length scale for the root cell (typical value 100 chords);

- A maximum length scale for cells that intersect a body (typical value 0.01 chords);

- A maximum angle deviation between successive faces on a body (typical value $5°$);

- A minimum length scale of interest (typical value 0.001 chords).

In addition, a location for the centroid of the root cell (typically the origin) is required, as is a representation of the surfaces of the bodies in the flow.

The grid-generation algorithm makes use of these input parameters by carrying out the following steps;

1. A root cell is constructed based on the input size and location;

2. The root cell is recursively refined until each body has at least one cell intersecting it;

3. Cells that intersect bodies are recursively refined until the maximum body length-scale constraint is met (or the minimum length-scale is hit);

4. Cells that intersect bodies are recursively refined until the angle-deviation constraint is met (or the minimum length-scale is hit).

This approach is robust, and relatively straightforward to code, although it relies on recursion. Most of the time is spent querying the representations of the bodies, in order to determine whether Cartesian cell vertices are in or out of the flow, and computing intersections of the Cartesian cells with the bodies. Efficient and robust coding of those intersection calculations is extremely important.

Figure 10: The Split-Cell Problem

Quirk [14] uses integer arithmetic to compute these intersections, which guarantees robustness and can resolve arbitrarily fine geometric details. Aftosmis et al [17] have implemented an extremely efficient and robust technique, that uses finite-precision arithmetic when possible, and infinite-precision arithmetic when necessary.

One added difficulty is the so-called split-cell problem, depicted (in two dimensions) in Figure 10. The trailing edge shown splits the cell in two, and without grossly over-refining the trailing edge, the cell at the trailing edge will need two separate fluid states. This problem, which occurs occasionally in two dimensions, is endemic in three dimensions. Allowing cells to have two fluid states is virtually necessary to remedy the problem.

Solution-Based Adaptive Refinement
Because adaptation is necessary in Cartesian codes just to resolve the geometry, solution-based adaptive refinement comes virtually for free. All the machinery is in place to refine or coarsen a cell (or a patch) in the grid, and

the only piece necessary to make solution-based adaptive refinement work is an adaptation criterion. Many criteria have been put forward for solution-based adaptation [18]. Some are based on estimating truncation error, others are more heuristic. A typical approach, based on two heuristic criteria, one for detecting compressions/expansions and one for detecting shears, is given here. The criteria are scaled in the manner suggested by Warren et al [19] so as to avoid over-refining high-gradient regions at the cost of smooth regions. The criteria are

$$\tau_c = |\nabla \cdot \mathbf{u}| \, h_i^{1.5} \qquad \sigma_c = \sqrt{\frac{1}{n} \sum_{i=1}^{n} \tau_{c_i}^2}$$

$$\tau_v = |\nabla \times \mathbf{u}| \, h_i^{1.5} \qquad \sigma_c = \sqrt{\frac{1}{n} \sum_{i=1}^{n} \tau_{v_i}^2} \quad (14)$$

where n is the number of cells in the grid, and h_i is a length-scale associated with cell i. Based on these criteria, cells are flagged for refinement if

$$|\tau_c| > \sigma_c \text{ or } |\tau_v| > \sigma_v \qquad (15)$$

and flagged for coarsening if

$$|\tau_c| < \frac{1}{10}\sigma_c \text{ and } |\tau_v| < \frac{1}{10}\sigma_v . \qquad (16)$$

For cell-by-cell refinement, any cell that is flagged for coarsening or refining is immediately coarsened or refined. For patch-based refinement, an intermediate step, fitting patches around regions of refined or coarsened cells, is necessary. Quirk describes this patch-fitting process in his thesis [13].

2.3 Flow-Solver Pieces

Limited Linear Reconstruction

In order for the scheme to be more than first-order accurate, a local reconstruction must be done; in order for the scheme to yield oscillation-free results, the reconstruction must be limited. The limited linear reconstruction described here is due to Barth [20]. A least-squares gradient is calculated, using neighboring cells, by locally solving the following non-square system for the gradient of the primitive variable vector W by a least-squares approach

$$\mathcal{L}\nabla W^{(k)} = f \qquad (17)$$

$$\mathcal{L} = \begin{pmatrix} \Delta x_1 & \Delta y_1 \\ \vdots & \vdots \\ \Delta x_N & \Delta y_N \end{pmatrix} \qquad f = \begin{pmatrix} \Delta W_1^{(k)} \\ \vdots \\ \Delta W_N^{(k)} \end{pmatrix} \qquad (18)$$

where

$$\Delta x_i = x_i - x_0$$
$$\Delta y_i = y_i - y_0$$
$$\Delta u_i = u_i - u_0 \qquad (19)$$

and the points are numbered so that 0 is cell in which the gradient is being calculated, and i is one of N neighboring cells used in the reconstruction.

The gradients calculated in this manner must be limited in order to avoid overshoots. Any limiter suited for an unstructured grid can be used. A typical choice is an unstructured-grid implementation of the minmod limiter, due to Barth [20]. The reconstructed values are limited by a quantity ϕ in the following way

$$\mathbf{W}(\mathbf{x}) = \bar{\mathbf{W}} + \phi(\mathbf{x} - \bar{\mathbf{x}}) \cdot \nabla \mathbf{W} , \qquad (20)$$

where ϕ is given by

$$\phi = \min \left\{ \begin{array}{l} 1 \\ \min_k \left(\frac{|\bar{W}^{(k)} - \max_{neighbors}(W^{(k)})|}{|\bar{W}^{(k)} - \max_{cell}(W^{(k)})|} \right) \\ \min_k \left(\frac{|\bar{W}^{(k)} - \min_{neighbors}(W^{(k)})|}{|\bar{W}^{(k)} - \min_{cell}(W^{(k)})|} \right) \end{array} \right. . \qquad (21)$$

Flux Function

In a finite-volume approach, the governing equations, are integrated over a cell in the mesh, giving

$$\frac{\partial \mathbf{U}}{\partial t} + \sum_{faces} \mathbf{F} A_{face} = 0 \qquad (22)$$

where V is the volume of the cell, A_{face} is the area of a face of the cell, and $\mathbf{U}$ and $\mathbf{F}$ are the state vector and flux vector, given by (in two dimensions)

$$\mathbf{U} = \begin{pmatrix} \rho \\ \rho u \\ \rho v \\ E \end{pmatrix} \qquad \mathbf{F} = \begin{pmatrix} \rho u_n \\ \rho u_n u + p \\ \rho u_n v \\ u_n (E + p) \end{pmatrix} \qquad (23)$$

where

$$E = \frac{p}{(\gamma - 1)} + \rho \frac{u^2 + v^2}{2} . \qquad (24)$$

A number of different flux functions can be used to calculate the fluxes through faces of the cells. While Roe's scheme is the most popular, and is the one described here, there are arguments for using other flux functions (see, for example, Quirk's contribution to the Riemann-solver debate [21]). In Roe's scheme, the interface fluxes $\mathbf{F}$ are computed from a combination of the average flux and the flux difference. The average flux comes from evaluating the flux vector $\mathbf{F}$ at the states on the left and right sides of the cell interface; these states come from the limited reconstruction described above, applied in the cells to the left and right of the interface. The flux-difference terms are then added, giving

$$\mathbf{F}(\mathbf{U}_L, \mathbf{U}_R) = \frac{1}{2}(\mathbf{F}(\mathbf{U}_L) + \mathbf{F}(\mathbf{U}_R)) -$$

$$\frac{1}{2}\sum_{k=1}^{k=4} \alpha_k |\lambda_k| \mathbf{r}_k , \qquad (25)$$

where the r_k are the right eigenvectors of the Euler system, the λ_k are the eigenvalues, and the α_k are the wave strengths. Roe's scheme is well-documented enough at this point that these quantities are not reproduced here; any modern CFD text (such as Hirsch [22]) has the formulas and derivation of Roe's scheme.

Time-Stepping Schemes

With the exception of Boeing's TRANAIR code [9], which uses a direct solver, most Cartesian work is based on explicit time stepping. Typically, simple multi-stage schemes are used, with anywhere from two to six stages. Optimally damping multi-stage schemes [23] work well in the Cartesian framework. Multigrid acceleration has also been implemented in some of the explicit codes [24].

2.4 Assessment of the Approach

For two-dimensional steady solutions of the Euler equations, adaptive Cartesian approaches have proven to be competitive with unstructured triangle-grid approaches. An ICASE/NASA Langley workshop on solution-adaptive techniques in CFD was held in the fall of 1994 to assess the state-of-the-art of various approaches. Benchmark cases for steady and unsteady flows were run with various codes, and compared. Results for the AGARD-1 and AGARD-3 benchmark cases [25] are shown in Figures 11–17. The first was run on a grid of 11,366 cells; the second on a grid of 15,234 cells. As can be seen, the adaptation criteria capture the shocks and wakes, without over-resolving other flow regions. The C_p on the wing for the first case, and the Mach number on the wing for the second case, show that the cut cells on the wing do not lead to non-smooth values of the flow variables there. The shock locations, C_ℓ

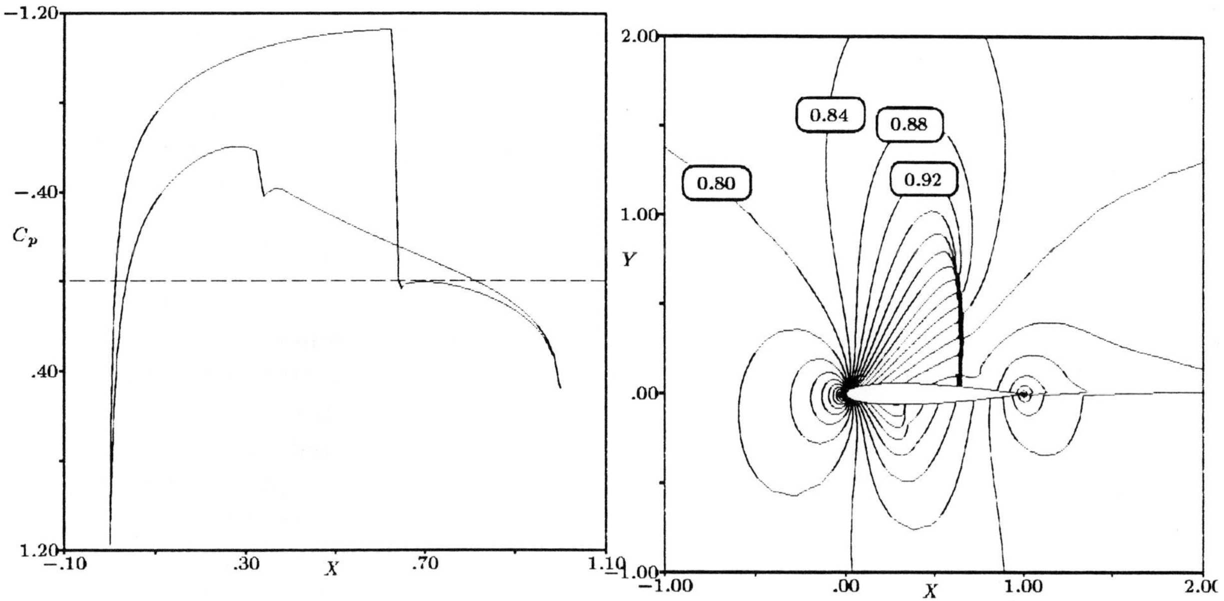

Figure 11: AGARD 01 Benchmark Case — C_p

Figure 12: AGARD 01 Benchmark Case — M contours

and C_d fall within the range of those given by competing techniques, including ultra-fine structured grids.

Another way of assessing accuracy of a code is comparison with an exact solution. Computations of Ringleb flow on three grids

- a structured, body-fitted grid

- a uniform Cartesian grid with no adaptation

- a solution-adapted Cartesian grid

were carried out by Coirier [26, 27], and error norms are shown in Figures 18 and 19. In the weaker norm, the L_1 norm plotted in Figure 18, it is clear that the structured and Cartesian grids are the same order of accuracy; a least-squares curve fit gives a slope of 2.08 for the structured grid and 2.02 for the uniform Cartesian grid. The structured grid has, for the same number of cells, a slightly lower error, as expected. In the stronger

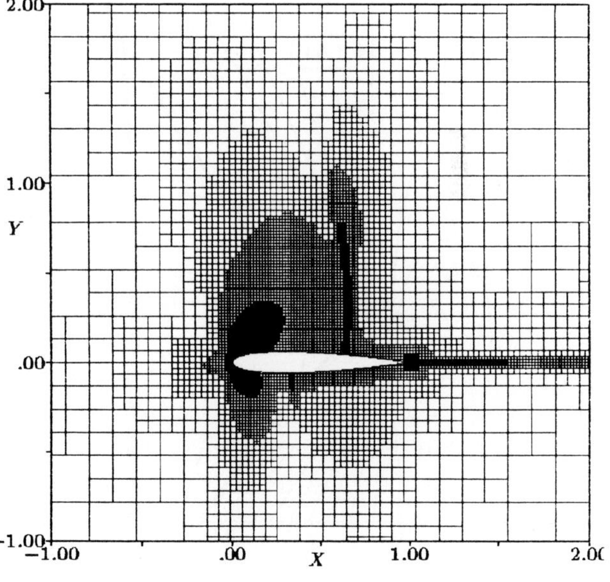

Figure 13: AGARD 01 Benchmark Case — Grid

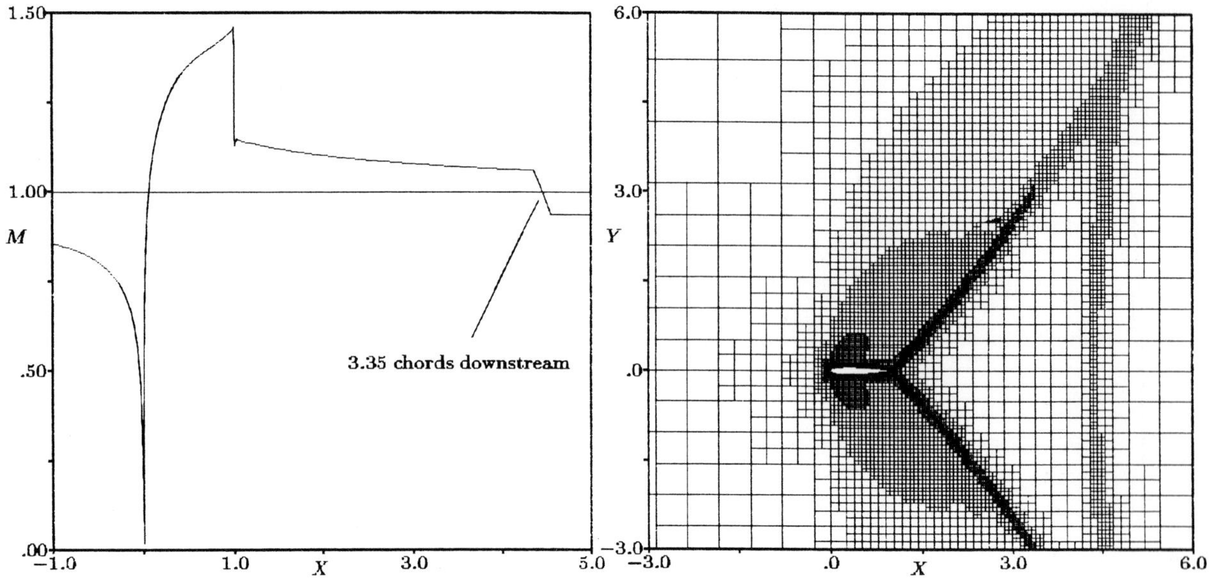

Figure 14: AGARD 03 Benchmark Case — Mach Number on axis

Figure 16: AGARD 03 Benchmark Case — Grid

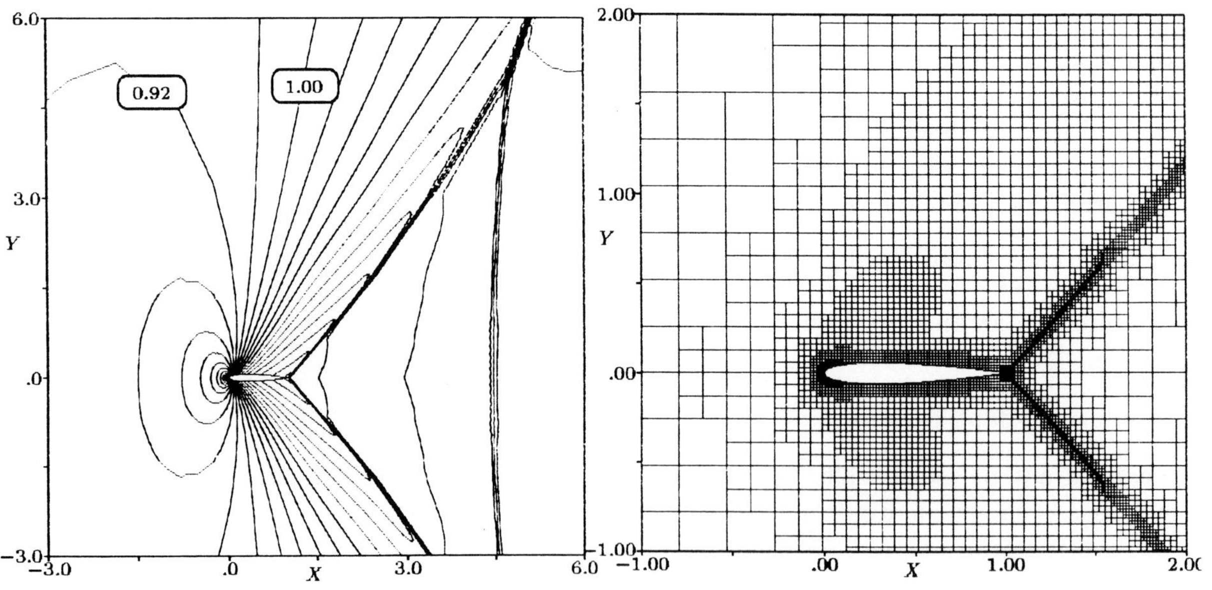

Figure 15: AGARD 03 Benchmark Case — M contours

Figure 17: AGARD 03 Benchmark Case — Grid Close-up

norm, the L_∞ norm shown in Figure 19, it is clear that the Cartesian grid is not fully second-order accurate. The slope of the line for the structured grid is 1.97; for the Cartesian it is 1.40. Examination of the cells in which the error is the largest shows, unsurprisingly, that it is small cut cells, and their neighbors, that contribute to this degradation in accuracy.

The net result of the Ringleb study is that, in two dimensions, a Cartesian approach based on cell-merging and linear reconstruction requires about 40% more cells than an equivalent structured grid to achieve the same accuracy. This result, while not really disappointing, is not the whole story — recent work by Berger [28] indicates that use of quadratic reconstruction in cut cells can yield truly second-order accuracy for those cells as well as interior cells.

3 Beyond Steady Flow: A Promising Approach for Moving-Geometry Problems

One class of problems in which grid-generation issues dominate are problems in which bodies move relative to each other and/or deform. If the relative motions are small, an initial structured grid can be generated, and can be stretched and sheared as the bodies move and deform. In cases for which the relative motions are large, or the gaps separating bodies become relatively small, however, this grid-movement approach leads to highly stretched and sheared grids, compromising solution accuracy. Using overset grids has been used with a great deal of success to overcome the large relative motion problem, but even that technique is problematic when gaps between bodies become small. For

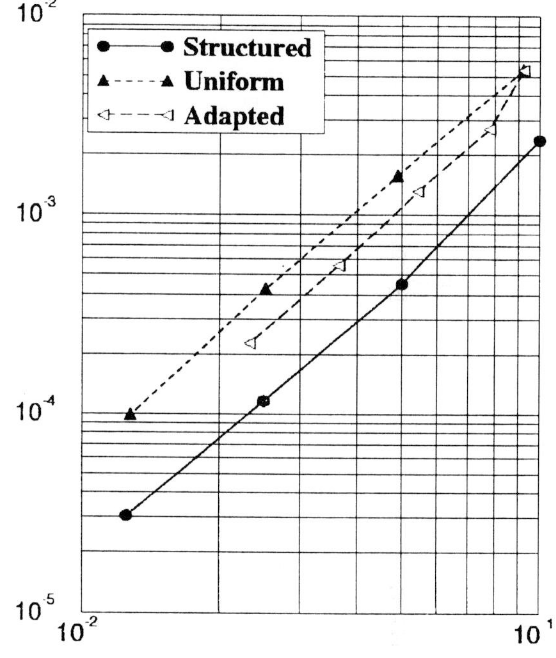

Figure 18: L_1 Norm of Error for Ringleb Calculation

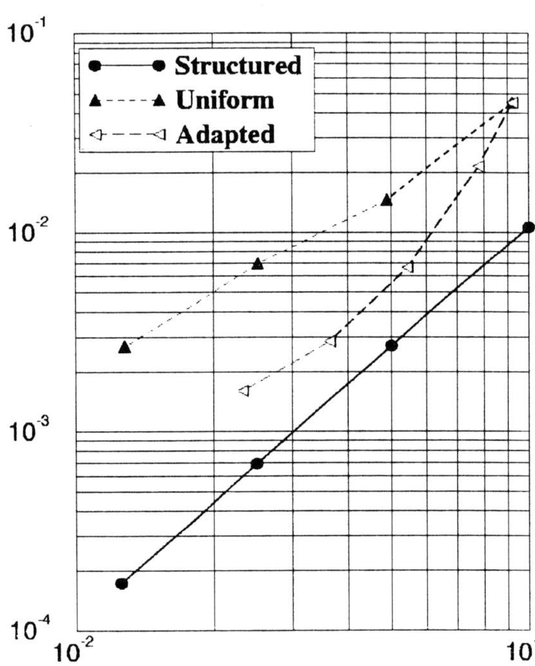

Figure 19: L_∞ Norm of Error for Ringleb Calculation

these cases, in which relative motions of bodies are large, and gaps between bodies can become small, it is virtually necessary to adopt a remeshing approach, in which an entirely new grid, at least in the vicinity of the bodies, is generated at each time step.

A promising alternative, at least in the case where inviscid physics are a useful model, is to use an adaptive Cartesian grid, and allow the bodies to move across the grid. At each time step, the intersection of each body with the grid is calculated, so that the geometry of the grid in the vicinity of the body changes. Away from the bodies, however, the grid does not change. This gives the adaptive Cartesian approach the advantages of both the remeshing and the grid-movement approach: the grid shearing and stretching are virtually eliminated (as in the remeshing approach) and yet the changes to the grid due to the moving body are localized (as in the grid-movement approach).

Implementing this approach requires confronting the following problems:

1. small cut cells on boundaries;

2. vacation of cells by boundaries during a motion step; and

3. invasion of cells by boundaries during a motion step.

As will be described, the process of cell merging alleviates all three of these problems. The three problems are illustrated in Figure 20, which shows the initial and final positions of a deforming boundary that moves on a grid of, for simplicity, uniform refinement. The figure also shows the motion envelope which is defined as the boundary of the set of all cells that are intersected by the interface at any time during the motion step. Only the cells that define the motion envelope are drawn in the figure.

The first problem is inherent to the Cartesian approach in general, regardless of the

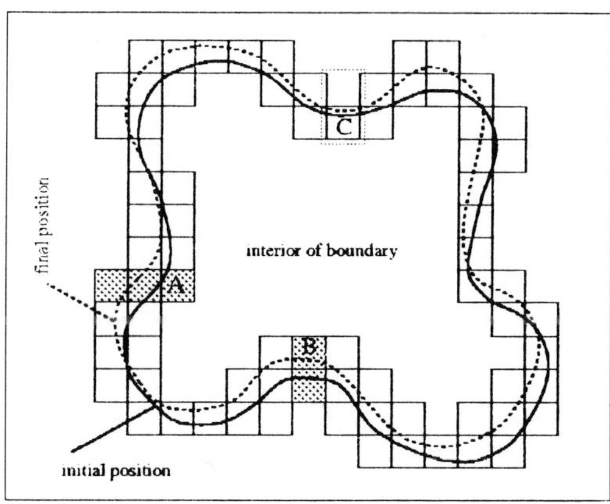

Figure 20: The three problems solved by cell merging.

body motion, and has already been discussed. An example of a small cell is indicated by the "C" in Figure 20. There, the topologic status of the cell is invariant during the boundary motion step, i.e. the cell begins the iteration as a cut cell, and ends it as a cut cell, and each vertex that started inside the body ends inside the body, and each vertex that started outside the body ends outside the body. At both the initial and final time, however, the way that the boundary cuts the cell leads to a small cut cell in the flow domain. Cell merging would combine this cell with the one immediately above it (which is not drawn since it does not lie in the motion envelope) to form the composite cell that lies within the dashed rectangle. The composite cell shown has acceptable initial and final areas.

The second and third problems are encountered only if boundary motion is present, since they are related to change in topologic status. Cell "A" in Figure 20 is vacated by the boundary during the motion step, beginning the time step as a cut cell, and ending it as an inactive cell. Cell "B" is invaded by the boundary, starting the time step as an inactive cell, and ending it as a cut cell.

In both cases, combining the given cell with other neighboring cells, shown in the enclosing shaded rectangles, to form composite cells of variable volume enables the problems to be easily and simply overcome within the finite-volume method, since the method allows a straightforward solution for cells with variable volume and translating boundary conditions. For cell "A" the exterior part of the composite cell contracts during the motion step while for cell "B" the exterior part expands during the motion step.

Summarizing, merging cells to form composite cells provides a simple means to overcome the problems of irregular cell geometry due to small cut cells, and the problems associated with changes in topologic status caused by motion of a boundary on a stationary grid. The goal of the merging is to provide new composite cells, comprising small groups of cells on the original grid, that remain topologically invariant over the time step. A cell (or composite cell) is topologically invariant if and only if each vertex that started outside the body ends outside it, and each vertex that started inside the body ends inside it.

The merging process is carried out for each body in the flow, one body at a time. The cells in the motion envelope of each body are merged as follows:

1. Find a suitable line segment at which to initiate the merging process for this body. This segment has to lie entirely within the motion envelope of the body, and it is best if this segment is not near a sharp corner in the body, or near a region of large variation in refinement level, or in a region where the body motion is large relative to the local cell size.

2. Starting from the initiation site, construct a topologically invariant rectangle that straddles the boundary. The next invariant rectangle is formed using the edge from which the boundary

exits the previous rectangle as the entry edge. The formation of invariant rectangles continues, marching along the boundary until the initiation site is re-encountered, signaling the completion of the loop around the boundary.

3. Expand each merged rectangle generated in the above step to satisfy as far as possible the targets set for minimum and maximum areas and for area ratios. If expansion is not sufficient, combination of merged groups and local remerging are also attempted.

Figure 21, which shows a portion of a stationary boundary, gives an example of the above procedure. Cells that straddle the body are numbered; other cells are not shown unless they are called into play by the merging. Cell 1 is chosen to be the initiation site. Since the boundary is stationary, every intersected cell is by definition topologically invariant. Assuming that the minimum initial and final areas are required to be no less than half the smallest uncut cell in volume, several cells must be merged with neighbors to satisfy these constraints. In particular, cell 3 will be combined with its shaded neighbor, cell 4 must be merged with cell 5, cell 6 with cell 7, cell 8 with cell 9, cells 11 and 12 with cell 10, cell 21 with the shaded cell directly above it, and cell 22 likewise with the shaded cell directly above it.

An illustration of the merging process for the same geometry, but this time in motion, is given in Figure 22. Composite cells are shown by shading; cells that do not require merging are unshaded.

Bayyuk [11] has supplied a proof to a theorem about the properties of the merging process for two-dimensional grids. The ultimate implication of the theorem is that the merging will succeed provided the refinement everywhere in the vicinity of a boundary "adequately" resolves the local geometry and mo-

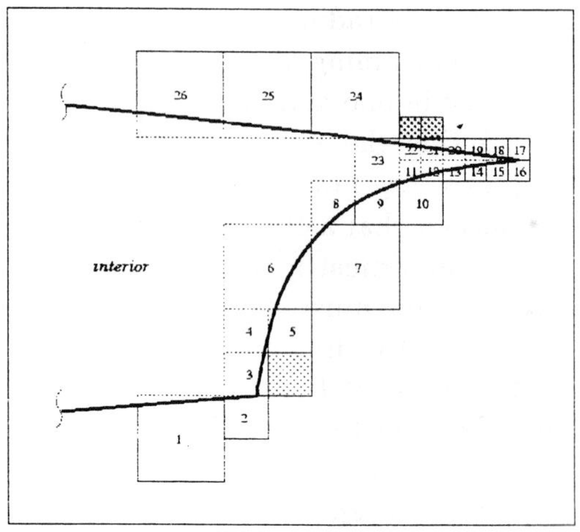

Figure 21: Cell merging around a stationary boundary.

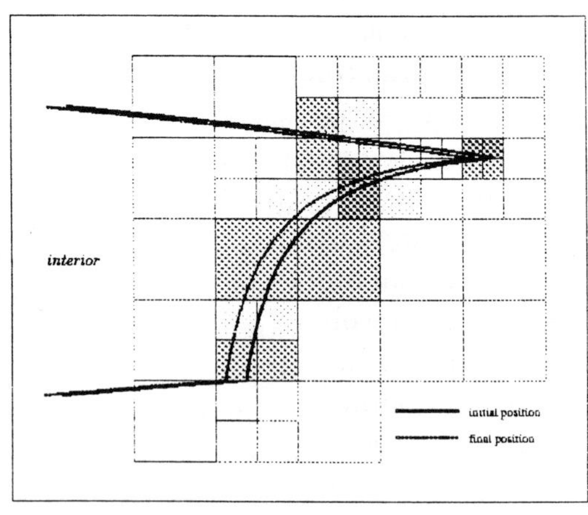

Figure 22: Cell merging around a moving boundary.

tion. The resolution is defined in terms of three length scales: the length scale of the local discretization; the length scale of the local geometry (a radius of curvature); and a length scale pertaining to the local motion of the boundary in one time step.

Implementing the solver on the resulting grids requires very few changes to the steady solver. The fact that a cell volume can change in one time step must be accounted for, and the flux function must be modified slightly to account for the moving boundary faces. A two-stage explicit time-stepping is convenient for these unsteady problems, and it is important to use a limiter that is not too diffusive, in order to capture the unsteady shock physics. The full scheme for the solver is described by Bayyuk in his thesis [11].

Some of the capabilities of this approach for unsteady flows and moving geometries are shown in the following three computations. In the first one, depicted in Figure 3, a shock wave of Mach number 2.44 collides with a half-diamond wedge, leading to a complex flow pattern after diffraction at the apex, with interacting shock, shear, and contact surfaces, as well as roll-up vortices and an expansion fan. The computation shows excellent quantitative agreement (in terms of density ratios) with the results of the corresponding experiment [29], and good agreement with the results obtained using different grid types and computational methods, e.g. [30]. All the flow features are resolved, even those that are difficult to discern in the interferogram.

The second computation, depicted in Figures 3 and 3, shows two bodies that are initially at rest. The gas in the enclosure of the C-shaped body, up to the mid-plane of the orifice, is initially at high pressure. The centroid of the cylindrical body is initially coincident with the centroid of the orifice. Upon release, the compressed gas expands through the orifice, propelling the cylindrical body along with it and propelling the larger

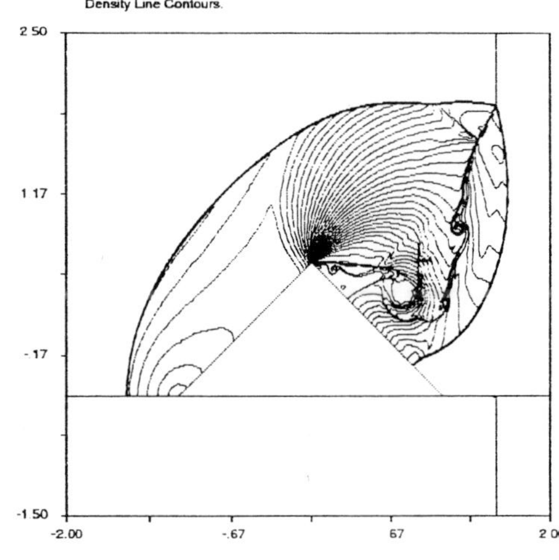

Figure 23: Density Contours for Shock Collision with a Half-Diamond.

body in the opposite direction. Both bodies are treated as rigid and the motions are fully determined by the aerodynamic forces. The larger body quickly attains supersonic speed, as evidenced by the bow shock in-front of it. The shock, shear, and contact waves that develop in the flow-field are all well-resolved. The figures show the grid and the flow details at two times during the computation.

In the third computation, shown in Figure 3, a projectile collides with a deformable solid, and the solid deforms and fractures. A two-stage projectile is accelerated from rest under the aerodynamic forces imparted by the pressurized gas in its rear cavity. Upon attaining a pre-selected velocity, the pressure and temperature of the gas in the projectile's front cavity instantaneously rise causing the projectile to fracture and its front and rear parts to accelerate apart. The front part subsequently collides with the target. Except during the fracture of the projectile, its two parts are treated as rigid bodies. The motion of the target's solid wall is computed from the applied aerodynamic and impact forces using a simplified finite-element computation uti-

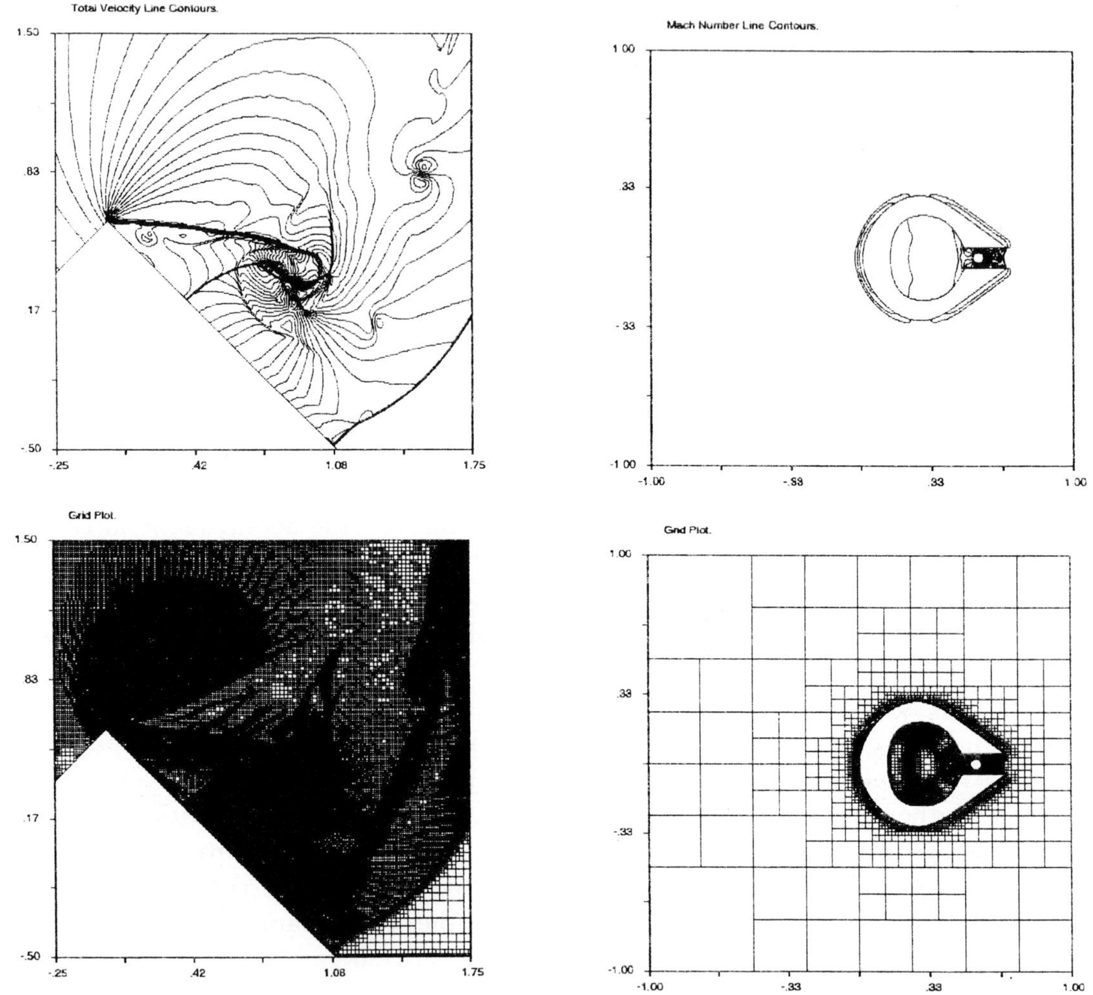

Figure 24: Speed Contours and Grid in the Roll-up Region of the Half-Diamond Problem.

Figure 25: Mach Contours and Grid at Early Time for Propulsion Problem.

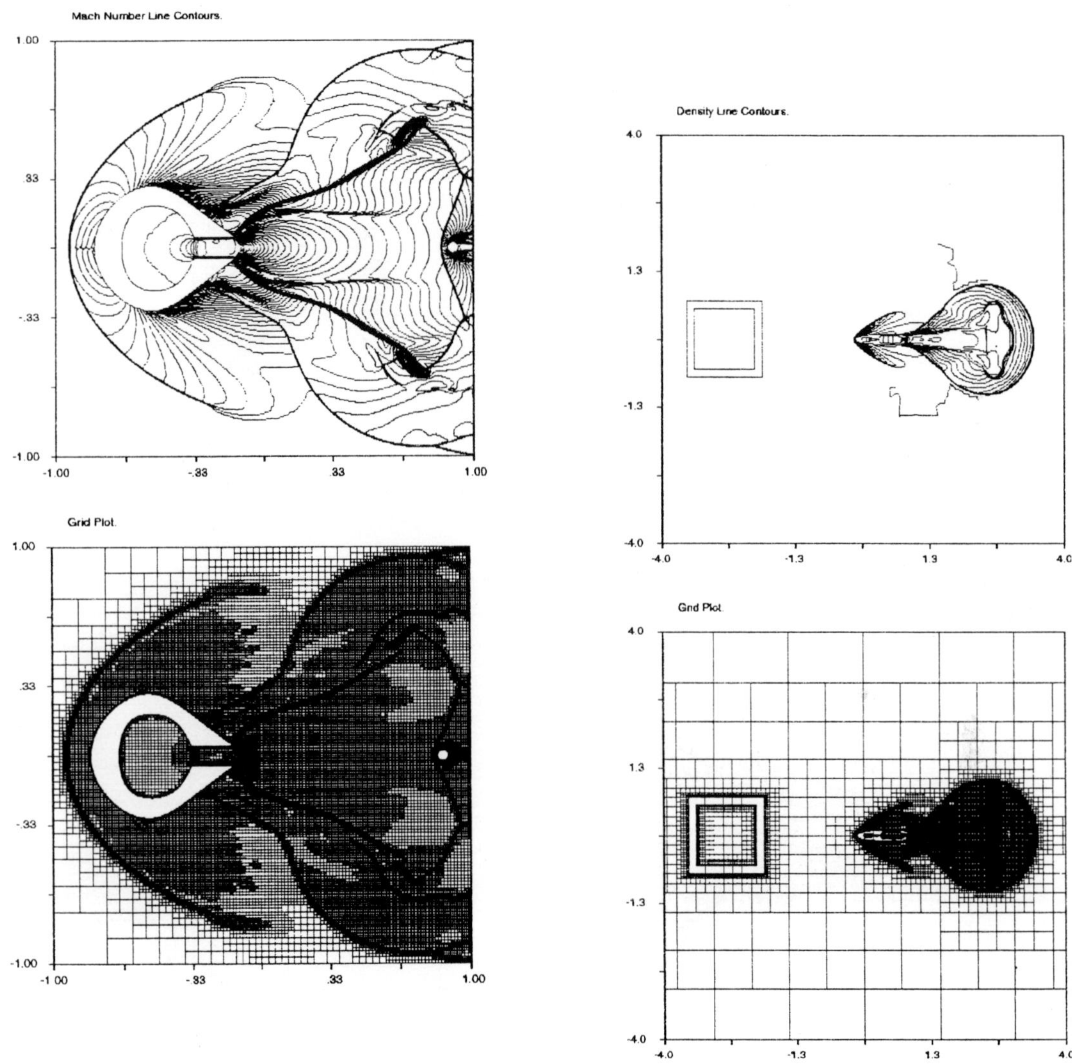

Figure 26: Mach Contours and Grid at Late Time for Propulsion Problem.

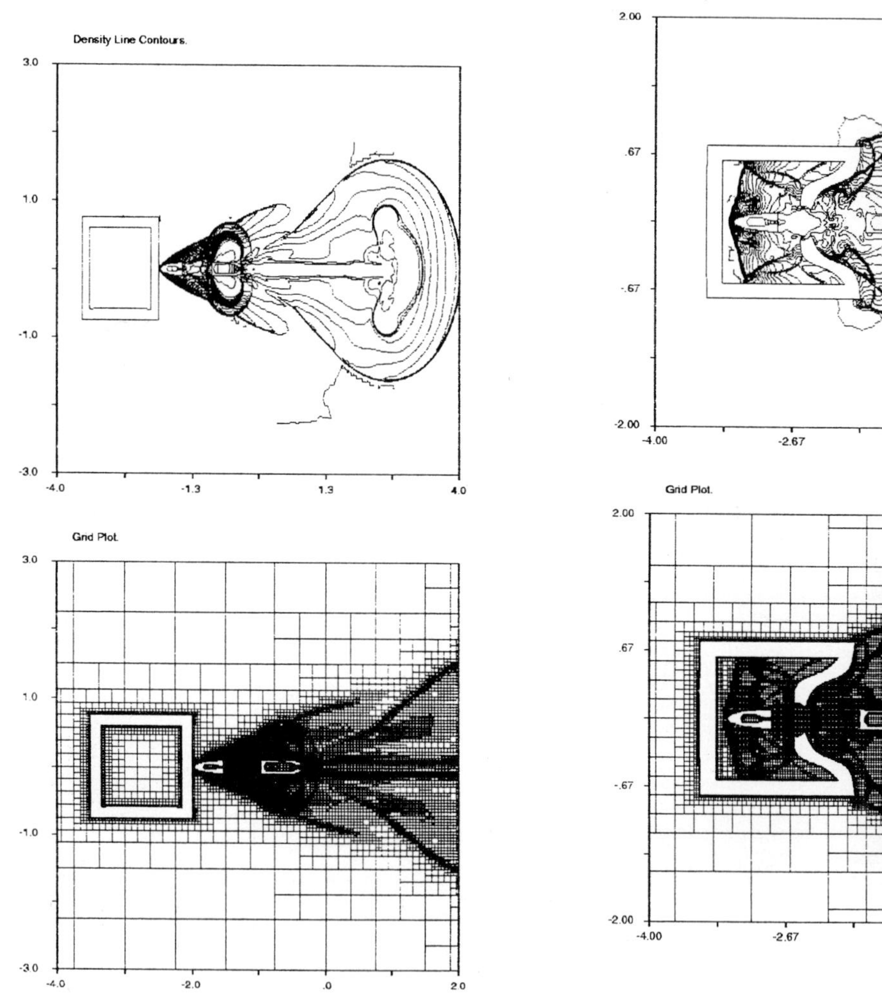
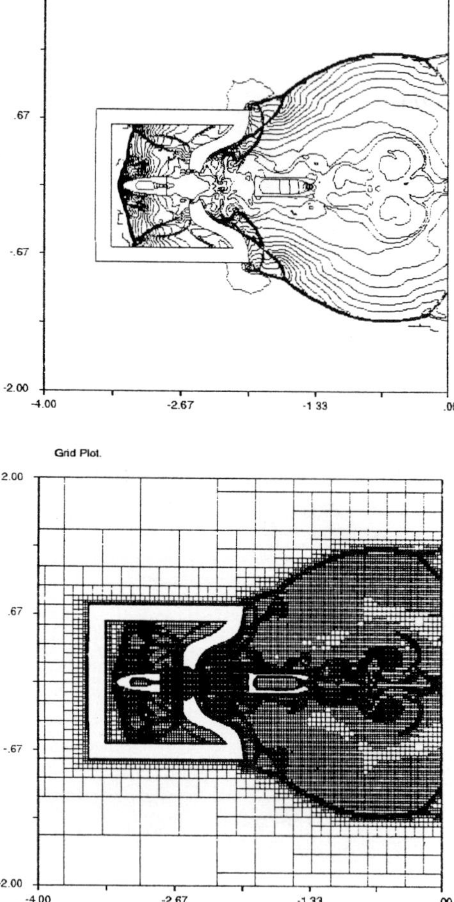

Figure 27: Density Contours and Grids for The Projectile Penetration Problem.

lizing an elasto-plastic non-strain-hardening material model with the yield strength of mild steel. A simple model for plastic fracture was included. This example demonstrates the automatic handling of boundaries that move, deform, and undergo topologic transformation. It also demonstrates the ability to handle separation, impact, and fluid-structure interaction problems.

4 Beyond Two Dimensions: Coupling Cartesian to CAD

If the real promise of the Cartesian approach is the automation of the grid-generation process, then the real payoff is for three-dimensional flows. With the exception of the moving-boundary problems described above, grid generation for two-dimensional problems, whether via unstructured grids of triangles or automated multi-block structured grid generation, has advanced to the point where grid generation does not seriously hamper work. For three-dimensional complex geometries, however, multi-blocking has not been totally automated, and even unstructured tetrahedron generation is not as automatic as desired. This is primarily because of the dependence of the volume grid of tetrahedra on the surface mesh of triangles.

The Cartesian approach has the advantage of decoupling the volume grid description from the surface grid description. However the body is described (e.g. by a solid-model in a CAD system, or by NURBS patches, or by triangular facets), the volume grid is generated, and the discretization of the body is set by the intersection of cells of the volume grid with the body geometry.

Cartesian grid generation in three dimensions comes down to determining whether the Cartesian cells are inside a body, outside all of the bodies, or straddling a body, and, for those straddling a body, computing the intersection of the cells with the body. Theoretically, this could be done however the bodies were defined. In practice, bodies defined by NURBS patches cause the grid-generation code to be unacceptably slow, since computing the intersection of a Cartesian cell with a NURBS patch is an expensive process. It is much more efficient to triangulate the surface of the bodies in the flow, and then compute the intersection of Cartesian cells with the triangular facets. The triangulation must be suitably fine to resolve the geometry of the body. However, the quality of the final cut-Cartesian mesh is not dependent on the surface triangulation in the way that a body-conforming mesh would be. This is because, in a body-conforming mesh, the surface triangles become a part of the volume mesh; in the Cartesian case, a cut cell may contain only part of one surface triangle, or may contain several surface triangles.

The time-consuming portions of 3D Cartesian grid generation really come down to two operations:

- determining whether a point is inside or outside a body;

- "clipping" a Cartesian cell against a triangular facet.

The first operation is carried out by using an appropriate reference point, and seeing whether a line segment connecting the reference point and the point of interest crosses an even or odd number of body facets. The second is carried out by a classic clipping algorithm such as that of Sutherland and Hodgmann [31].

The speed of a Cartesian grid generation scheme is primarily set by determining when to carry out these operations, and when they can be ignored. If every Cartesian cell were to be clipped against every surface triangle, a *very* slow code would result. Instead, if cells that are clearly far from any body are ruled out before carrying out the clipping, and if cells in the vicinity of the body are only clipped against surface triangles that they actually intersect, the code is sped up by orders of magnitude, since the complexity of the code is substantially lower. Aftosmis et al [17] discuss some of the ways in which the complexity of a Cartesian grid generator can be kept in control by use of appropriate data structures and bounding-box information. The robustness of the grid generation is primarily set by careful scaling and exception handling, described in the same paper.

A result for a generic business jet with is shown in Figures 28 and 29. The geometry for the run was specified parametrically [32], and is similar to that of the Cessna *Citation X*. The discretized geometry has 460,000 polygons on the surface. The volume mesh has 370,000 nodes, nearly 140,000 of which are cut cells. The boundary conditions at the nacelle inlets and outlets were chosen so as to specify a mass-flow rate and thrust provided by the engine. The flight Mach number was taken to be $M_\infty = 0.86$.

Figure 30 shows the results for a photo-reconnassance (Lockheed *U-2*) geometry. The grid has 270,000 cells, 98,000 of which are cut, with 278,000 surface polygons. The landing gear and the engine inlet and exhaust were the dominant features caught by the adaptation. The code used for the business jet and reconnaissance plane calculations is described more fully elsewhere [33].

A penalty associated with the Cartesian approach that appears in three dimensions is related to anisotropy. Cartesian cells are, by definition, isotropic — they have the same Δx, Δy and Δz. Problems with non-trivial geometries are rarely well-suited to this isotropy. High-aspect ratio swept wings, for example, require more resolution in the chordwise direction than the spanwise direction, and neither direction necessarily lines up with one of the coordinate directions. Thus, in any three-dimensional Cartesian calculation, resolving flow or geometry features in one direction typically means over-resolving it in the other two directions. Thus, even if Cartesian approaches were as accurate as body-conforming approaches, more cells would be required in the Cartesian calculation. Given that an accuracy penalty is also paid for the Cartesian approach, substantially more cells will typically be needed in the Cartesian approach than in body-conforming approaches in order to give the same accuracy. The more anisotropic the flow, the larger this penalty.

The trade-off is clear — substantially simplified grid generation at the cost of substantially more grid points in the calculation. For many calculations, particularly preliminary calculations for a given configuration or configurations, this is a trade-off that many would willingly make.

5   Concluding Remarks

The appeal of the Cartesian approach is the automation of the grid-generation process. The process comes down to computing intersections of Cartesian cells with the body surfaces, however they may be represented. There are some difficulties in designing robust and fast schemes for computing these intersections, but rapid progress is being made on this front, in the work of Aftosmis, Melton and Berger and in commercial codes such as Boeing's TRANAIR code [9], Lockheed's SPLITFLOW code [34], and the work at CF-DRC [35].

In addition to its promise for three-dimensional flows, the Cartesian approach is very appealing for two-dimensional unsteady flows with moving boundaries. These problems, which also have serious grid-related

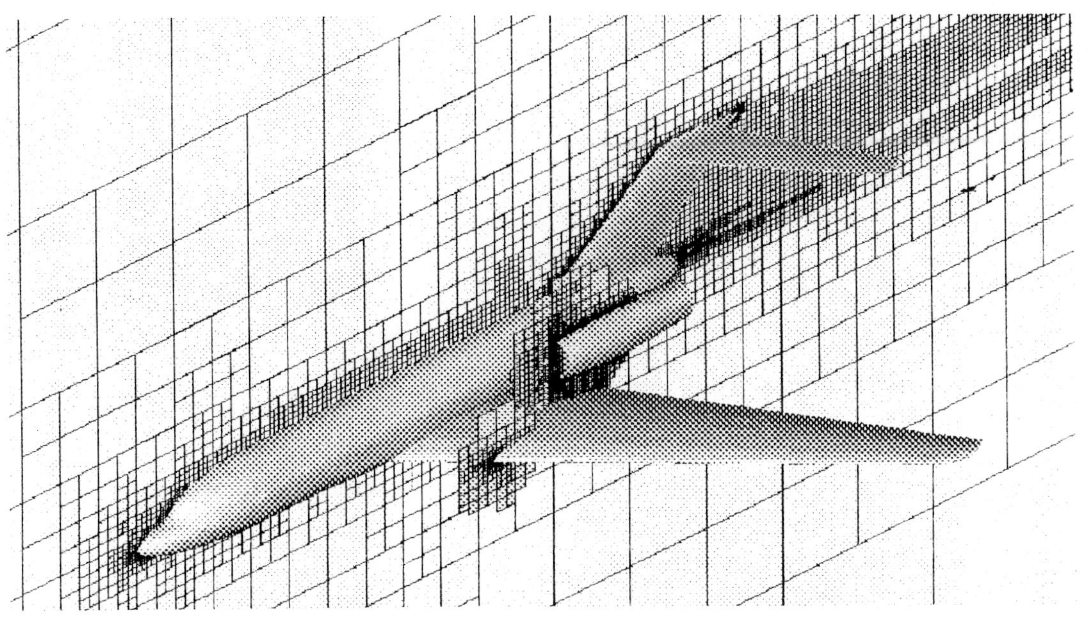

Figure 28: Generic Business Jet — Grid

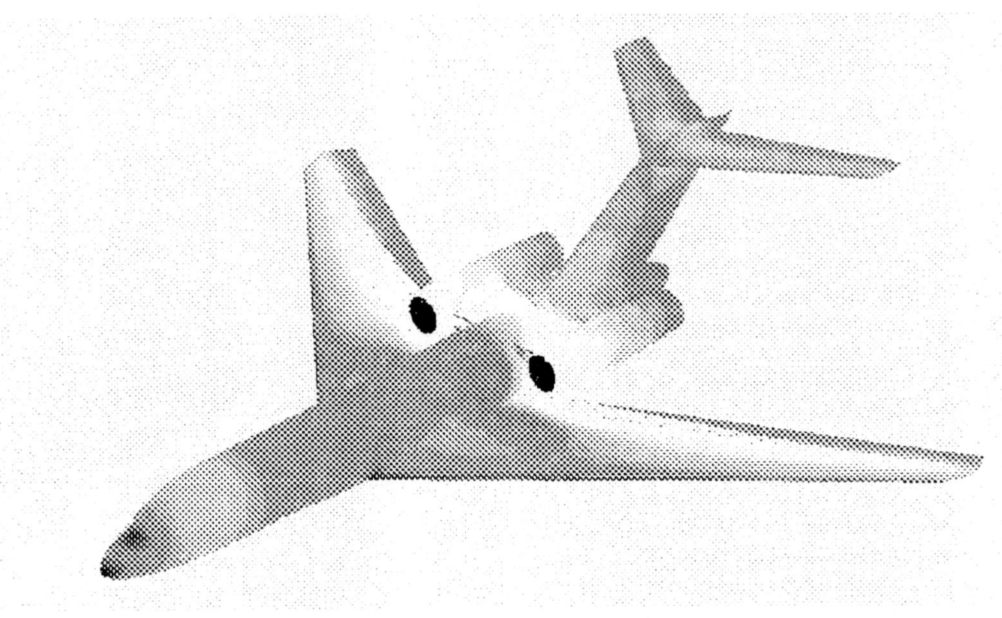

Figure 29: Generic Business Jet — Surface Pressure Contours

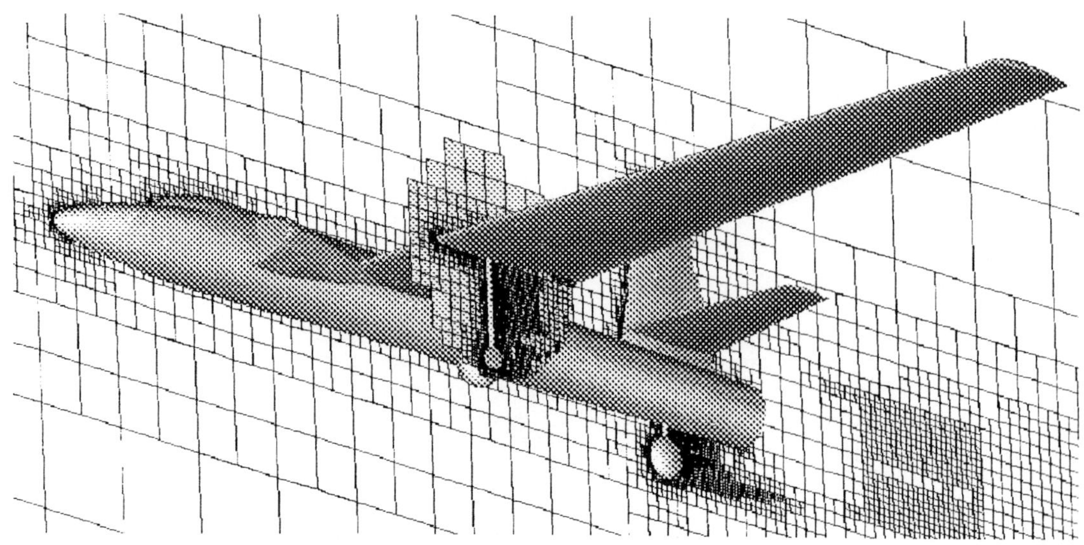

Figure 30: Photo-Reconnaissance Aircraft — Pressure Contours and Grid

problems to surmount when more conventional approaches are used, are particularly well-suited to the Cartesian approach.

In addition, while not discussed in this paper, the Cartesian approach has also been shown to be easily extensible to other equation sets, allowing solution of MHD, multifluid, and reacting flows, just as a few examples.

These successes of the Cartesian approach do not come without a cost, however. Current treatments for cut cells, while stable and better than first-order accurate, are not as accurate as treatments for boundary cells in body-conforming approaches. The isotropy of the Cartesian approach means that, for anisotropic problems, there can be a substantial resolution penalty. Cartesian approaches will always require more cells than body-conforming approaches in order to get as good a result.

The most glaring weakness of the Cartesian approach is the difficulties that arise in implementing viscous solvers. The small cut cell problem raises its head again, in an even uglier way, when viscous operators need to be constructed [26]. And the isotropy problem leads to impractical numbers of cells, even at moderate Reynolds numbers. Cartesian approaches for high Reynolds number will require special boundary-layer treatment — either a viscous/inviscid interaction approach, or a layer of body conforming prismatic cells to capture the boundary-layer physics.

These difficulties simply imply that the Cartesian approach will never be the only tool in a CFD user's toolbox. However, Cartesian codes should play a growing role in CFD as more work is done on them. They are ideally suited to preliminary design work — the automation of the grid generation procedure is a very powerful lever for decreasing the turnaround time on calculating the flow around a new geometry.

References

[1] M. J. Aftosmis, J. E. Melton, and M. J. Berger, "Adaptation and surface modeling for Cartesian mesh methods," AIAA Paper 95-1725, 1995.

[2] I. L. Chern and P. Colella, "A conservative front tracking method for hyperbolic conservation laws," Tech. Rep. UCRL-97200, Lawrence Livermore National Laboratory, 1987.

[3] R. J. LeVeque, "High resolution finite volume methods on arbitrary grids via wave propagation." ICASE Report 87-68, 1987.

[4] M. J. Berger and R. J. LeVeque, "An adaptive Cartesian mesh algorithm for the Euler equations in arbitrary geometries," in *AIAA 9th Computational Fluid Dynamics Conference*, 1989.

[5] S. F. Davis, "A rotationally-biased upwind difference scheme for the Euler equations," *Journal of Computational Physics*, vol. 56, 1984.

[6] D. Levy, K. G. Powell, and B. van Leer, "Use of a rotated Riemann solver for the two-dimensional Euler equations," *Journal of Computational Physics*, vol. 106, pp. 201–214, 1993.

[7] D. De Zeeuw and K. G. Powell, "An adaptively-refined cartesian mesh solver for the Euler equations," *Journal of Computational Physics*, vol. 104, no. 1, pp. 56–68, 1993.

[8] S. Bayyuk, K. G. Powell, and B. van Leer, "A simulation technique for two-dimensional unsteady inviscid flows around arbitrarily moving and deforming bodies of arbitrary geometry," AIAA Paper 93-3391-CP, 1993.

[9] D. P. Young, R. G. Melvin, M. B. Bieterman, F. T. Johnson, and S. S. Samant, "A locally refined rectangular grid finite element method: Application to computational fluid dynamics and computational physics," *Journal of Computational Physics*, vol. 62, pp. 1–66, 1991.

[10] J. E. Melton, M. J. Berger, M. J. Aftosmis, and M. D. Wong, "3D applications of a Cartesian grid Euler method," aiaa paper, 1995.

[11] S. Bayyuk, *Euler Flows with Arbitrary Geometries and Moving Boundaries*. PhD thesis, University of Michigan, 1996.

[12] M. J. Berger and P. Colella, "Local adaptive mesh refinement for shock hydrodynamics," Tech. Rep. UCRL-97196, Lawrence Livermore National Laboratory, 1987.

[13] J. Quirk, *An Adaptive Grid Algorithm for Computational Shock Hydrodynamics*. PhD thesis, Cranfield Institute of Technology, 1991.

[14] J. J. Quirk, "An alternative to unstructured grids for computing gas dynamic flows around arbitrariliy complex two-dimensional bodies." ICASE Report 92-7, 1992.

[15] J. J. Quirk, "A parallel adaptive algorithm for computational shock hydrodynamics." To appear in *Applied Numerical Mathematics*, 1996.

[16] M. J. Berger. Private Communication, 1996.

[17] M. J. Aftosmis, J. E. Melton, and M. J. Berger, "Robust and efficient Cartesian mesh generation for component-based geometry," AIAA Paper 97-0196, 1997.

[18] K. G. Powell, P. L. Roe, and J. Quirk, "Adaptive-mesh algorithms for computational fluid dynamics," in *Algorithmic Trends for the 1990's*, 1991.

[19] G. Warren, W. K. Anderson, J. Thomas, and S. Krist, "Grid convergence for

adaptive methods," in *AIAA 10th Computational Fluid Dynamics Conference*, 1991.

[20] T. J. Barth, "On unstructured grids and solvers," in *Computational Fluid Dynamics*, Von Kármán Institute for Fluid Dynamics, Lecture Series 1990-04, 1990.

[21] J. J. Quirk, "A contribution to the great Riemann solver debate." ICASE Report 92-64, 1992.

[22] C. Hirsch, *Numerical Computation of Internal and External Flows*. John Wiley & Sons, 1988.

[23] B. van Leer, W. T. Lee, P. L. Roe, K. G. Powell, and C. H. Tai, "Design of optimally-smoothing schemes for the Euler equations," *Communications in Applied Numerical Mathematics*, vol. 8, pp. 761–769, 1992.

[24] D. D. Zeeuw, *A Quadtree-Based Adaptively Refined Cartesian-Grid Algorithm for Solution of the Euler Equations*. PhD thesis, University of Michigan, 1993.

[25] AGARD Subcommittee C, "Test cases for inviscid flow field methods." AGARD Advisory Report 211, 1986.

[26] W. L. Coirier, *Simulation of Steady Viscous Flow on an Adaptively Refined Cartesian Grid*. PhD thesis, University of Michigan, in preparation, 1994.

[27] W. J. Coirier and K. G. Powell, "An accuracy assessment of Cartesian-mesh approaches for the Euler equations," *Journal of Computational Physics*, vol. 117, pp. 121–131, 1995.

[28] M. J. Berger and J. E. Melton, "An accuracy test of a Cartesian grid method for steady flow in complex geometries,"

in *Proceedings of the Fifth International Conference on Hyperbolic Problems*, 1994.

[29] D. Zhang and I. Glass, "An interferometric investigation of the diffraction of planar shock waves over a half-diamond cynlinder in air," Tech. Rep. 322, UTIAS, 1988.

[30] I. Lottati and S. Eidelman, "A second-order Godunov scheme on a spatial adapted triangular grid," *Applied Numerical Mathematics*, vol. 14, 1994.

[31] J. D. Foley, A. van Dam, S. K. Feiner, and J. F. Hughes, *Computer Graphics Principles and Practice*. Addison Wesley, second ed., 1987.

[32] E. Charlton, *An Octree Solution to Conservation-Laws over Arbitrary Regions (OSCAR) with Applications to Aircraft Aerodynamics*. PhD thesis, University of Michigan, 1997.

[33] E. Charlton and K. Powell, "An octree solution to conservation laws over arbitrary regions (OSCAR)," AIAA Paper 97-0198, 1997.

[34] J. S. L. Karman, "SPLITFLOW: A 3d unstructured Cartesian/prismatic grid CFD code for complex geometries," AIAA Paper 95-0343, 1995.

[35] Y. Jiang, Z. J. Wang, and A. J. Przekwas, "Pressure-based high order accuracy flow solver on adaptive, mixed type unstructured grids," AIAA Paper 96-0417, 1996.

ANISOTROPIC CARTESIAN GRID METHOD FOR VISCOUS FLOW COMPUTATION

Zi-Niu WU

National Laboratory for Computational Fluid Dynamics, Institute of Fluid Dynamics, Beijing University of Aeronautics and Astronautics; Beijing 100083, P.R.China, ziniu@c5.sebuaa.ac.cn

Abstract

An anisotropic Cartesian grid (ACG) method is developed for viscous flow computations. This method is more compatible with the anisotropic feature of flows near the body than the traditional isotropic Cartesian grid (ICG) method and is therefore grid saving. An efficient algorithm for ACG generation is presented and analyzed. A bilinear interpolation is applied to construct an easily implementable second-order accurate solid wall condition. The stability of this solid wall treatment is established using the GKS-stability theory. The ACG method along with the solid wall condition is finally validated by computing viscous flows around airfoils. The ACG method is compared with the ICG method. It is found that the ACG method can significantly save the number of grid points without jeopadizing the accuracy. Such a gain can be expected to be substantially more important in three-dimensional flow computations.

Key words: Cartesian grid, anisotropic grid, solid wall condition, stability, viscous flows

1. Introduction

In recent years, the Cartesian grid method was studied for external flow computation with aribitrarily shaped solid walls, see for instance [2, 4, 8, 10, 12, 14, 20, 22, 24] for inviscid flow computation and [11, 21, 29] for viscous flow computations. A Cartesian grid involves simple cells in interior regions and cutcells near solid walls. This method requires a mesh refinemnt strategy [1, 2, 3, 10, 24] to resolve the geometry and/or the flow and a special treatment of the solid boundary conditions [2, 8, 19, 22, 24, 25]. This method can be classified into two categories: the block-structured (BS) approach and the globally-unstructured (GU) approach. The BS method, developed in [1, 2, 24] for hyperbolic problems, uses a nested sequence of structured grids obtained by collecting subcells of the same level into a rectangular while the GU method [10, 11, 12, 21] refines a cell whenever it is needed and does not cluster the subcells of the same level to form a rectangular grid. In both methods, the finest grid (the highest level grid) is connected to the coarsest grid by a series of grids of intermediate levels.

Received on December 10, 1996.

This work has been supported by Chinese National Science Foundation.

A common feature of the existing Cartesian grid method is that it uses an isotropic refinement strategy. The grid generation starts with a background Cartesian grid with squared cells. The final grid is generated by the recursive subdivision of a single cell (parent) into four squared cells (children) when it is necessary. The hierarchical relation between the children and parents is stored and used for cell neighbour searching. The resulting grid thus involves cells of various sizes. A subgrid of level l with $l = 0, 1, \ldots, L$ refers to all the cells having the same mesh size $\vec{h}_l = (h_l^{(x)}, h_l^{(y)})$. The mesh size satisfies the relation $\vec{h}_l = r^l \vec{h}_0$ where $r = \frac{1}{2}$ is the the refinement ratio and $\vec{h}_0$ is the mesh size on the coarse grid. Since a cell is refined equally in both directions independently of the geometry and the flow gradient, the resulting Cartesian grid is isotropic.

Traditional ICG method uses a finite-volume approach to define solid boundary condition by solving a conservative difference equation on each polygone resulting from cell cutting. Cell cutting is expensive in general. The finite-volume approach may lead to instability for small cutcells. One either merges small cutcells to form large ones [12, 24] or modifies the scheme near the wall[18] to maintain the stability. This unavoidably reduces the accuracy or complicates the interior scheme. The advantage of the finite-volume approach is that it ensures conservation near the wall. This is important when solving the Euler equations for which discontinuous solutions may exist near the wall. A conceptually rather different method was presented in [25] where a hybrid Cartesian/curvilinear grid is applied. The grid is Cartesian in interior regions and is structured cuvilinear near the solid wall. This ensures a good precision both at interior regions and near the wall. Difference equations on the two grids can be matched in a stable way. Such a method is further developed in [19] where the Cartesian grid in interior region is matched to a body-fitted-structured grid through a thin unstructured grid.

The geometry encountered in many problems is often highly anisotropic (wings, missiles, high-speed airplanes, etc). For such problems, the isotropic Cartesian grid (ICG) method does not take into account the anisotropy of the geometry and thus leads to too many extra grid points near the body where the flow is strongly anisotropic. When solving the Navier-Stokes equations, cutcells should be completely merged inside the viscous boundary layer. In this case a conservative treatment of the solid wall by a finite-volume approach is not necessary and makes the algorithm too complicated.

In order to make the Cartesian grid method more efficient, we propose in this paper an anisotropic Cartesian grid (ACG) method and a finite-difference solid wall treatment. The ACG method can also be automatically switched to the traditional isotropic method and thus includes the ICG method as a special case. The ACG is expected to reduce the number of cells without lossing the accuracy in comparison with the ICG method. We restrict this paper to the case of geometry-oriented Cartesian grid generation. The finite-difference solid wall treatment is expected to simplify the boundary condition and to ensure stability and accuracy. The present method provides a good basis for solution adaptive Cartesian grid generation much studied by others, see for instance [1, 2, 3, 12, 24]. It is also easily extendible to three dimensions. Using Cartesian grid method to solve three dimensional problems is still rare, see [21] for an exception.

The ACG method divides a cell into subcells independently in each direction. Near the wall with small curvature, the cells are essentially refined in the normal direction. When there is a great curvature, the cells are refined in both directions. The ACG method not only provides an efficient resolution of the geometry but also fits to the anisotropic nature of the viscous boundary layer near the body where the flow gradient is essentially in the direction normal to the wall. Furthermore, the ACG method will also be quite useful in the interior flow region with strong gradient.

For example, near an essentially vertical or horizaontal shock, the use of an ACG will substantially reduce the number of grid points since a shock only needs to be refined in the normal direction. In Figs. 1 and 2, we display an anisotropic Cartesian grid and an isotropic Cartesian grid constructed and applied in the present study.

The finite-difference solid wall condition is based on the nonslip condition on the wall and the known solution at several interior points. These solutions are then interpolated to the center of the boundary cell cut by the solid wall. The finite-difference approach does not require the boundary cells to be really cut. For convenience, we still use the term of cutcell for those cells having an intersection with the solid wall. Such a treatment is not conservative. Since we compute viscous flows, the conservation is not necessary at the wall if the cutcells are completely inside the boundary layer. We point out that a similar but slightly different method is proposed in [21] for solid wall condition.

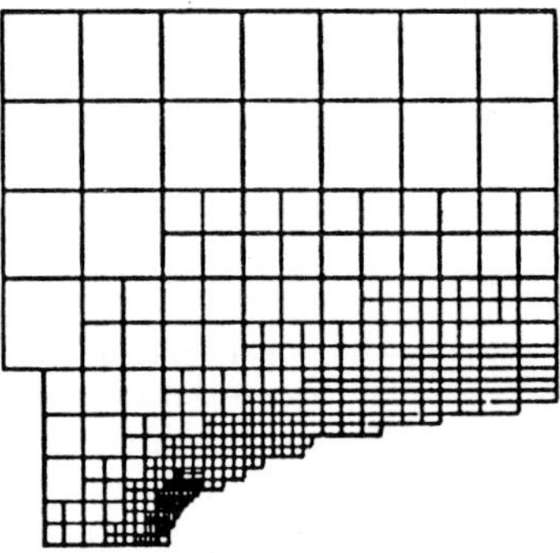

Fig.1: Anisotropic Cartesian grid near a part of NACA0012 airfoil

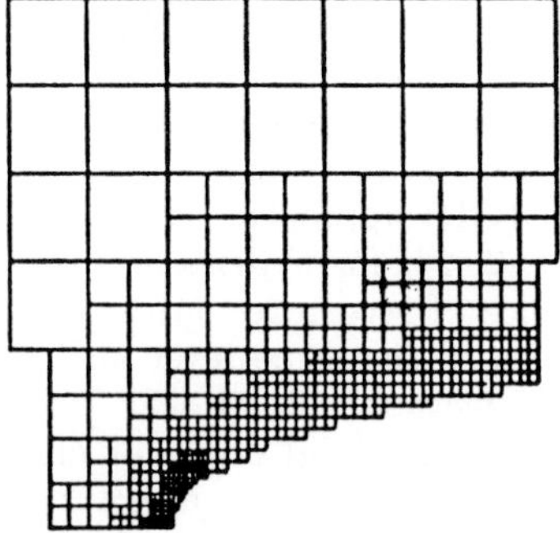

Fig.2: Isotropic Cartesian grid near a part of NACA0012 airfoil

This paper will be presented as follows.

In Section 2, an efficient algorithm for ACG generation is described and validated. The refinement criterion will be based on the local deviation from a straight line combined with a prescribed parameter although other criteria can also be similarly used. The local deviation is projected into both directions based on the normal direction of the wall. The algorithm is made efficient even though multi-element bodies are involved. The statistical advantage of the anisotropic Cartesian grid over the isotropic grid is analyzed.

Section 3 is devoted to boundary treatment near solid walls. We use bilinear interpolation to construct an easily implementable second-order accurate solid wall condition. By using the GKS-stability[15] theory to a model problem, we prove that the present solid wall condition does not have instability due to small cutcells.

The ACG method along with the solid wall condition is finally validated by computing viscous flows around airfoils in Section 4. We first apply a scheme from Lerat [16] to construct a partially discrete scheme which is especially suitable for a Cartesian grid with anisotropic refinement interfaces. In one dimension and for the inviscid case, the Lerat's scheme reduces to the classical Lax-Wendroff scheme. We then compute several test cases. The ACG method is compared with the ICG method. It is found that the ACG method can significantly save the number of grid points without jeopadizing the accuracy. In all the computations, only the geometry-oriented algorithm is applied for grid generation. The gain of ACG over ICG can be expected to be substantially more important when a solution adaptive AMR is used or when the problem is three dimensional.

We also provide two appendices which are related to the difference scheme on a Cartesian grid. In Appendix A, we show how to obtain simple interpolation formulas for computing averages and derivatives at cell interfaces. In Appendix B, we give an analysis of the stability of the Lax-Wendroff scheme on a refined Cartesian grid with a series of interfaces.

2. Anisotropic grid generation

2.1 General requirement

Algorithms for isotropic Cartesian grid generation can be found in [24, 12]. All the algorithms are basically algebraic and involve logical correspondance between cells of the same level and cells of different levels. In consequence, there would be no end to elaborate the algorithm. The algorithm proposed here brings much from the work in [12] but is made efficient for anisotropic refinement and for interaction between separated bodies. An elaborated algorithm can save the CPU time by one order in magnitude than a nonelaborated algorithm. The algorithm presented here is the fast among those we have tested. For a typical two-dimensional geometry, the present algorithm generates an anisotropic Cartesian grid with ten thousands of grid points in less than one minute on a personal computer. A worse algorithm would need more than ten minutes while a curvilinear grid generator by solving PDE would require half an hour on the same computer.

The cells in a Cartesian grid can be divided into three types: fluid cells, solid cells and cutcells. See Fig.3. Fluid cells are cells completely inside the flow field. Solid cells are cells inside the body and are not used in computations[1]. Cutcells are cells which have an intersection with the wall.

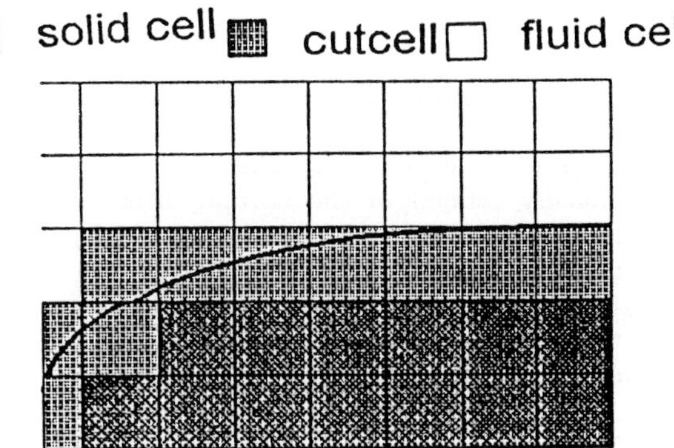

Fig.3: Fluid Cells, solid cells and cutcells.

The present grid generation algorithm is divided into five steps: 1) construction of the coarse grid, 2) determination of the refinement criterion, 3) refinement of each cell when necessary, 4) location of cutcells, and 5) suppress of solid cells. Each step seems to be quite direct in implementation but the efficiency of the global algorithm depends much on how to elaborate each step, especially steps 2, 4 and 5. We will elaborate each step so that the algorithm has the following properties:

- the hierachical relation between cells of various levels can be clearly defined for anisotropic cells
- the interaction between the refinement requirement issued from separated bodies can be easily taken into account
- the algorithm can be directly extended to three dimensions.

2.2 Refinement criterion

This is the most essential part of the algorithm. Since we restrict ourselves to geometry-oriented refinement, the refinement criterion is based on the geometry of the body(ies). We first need to compute the

---

[1]If one needs to know the detailed temperature evolution inside the body, then solid cells may be used to discretise the heat equation.

refinement criterion on the body and then determine the refinement criterion in any interior point.

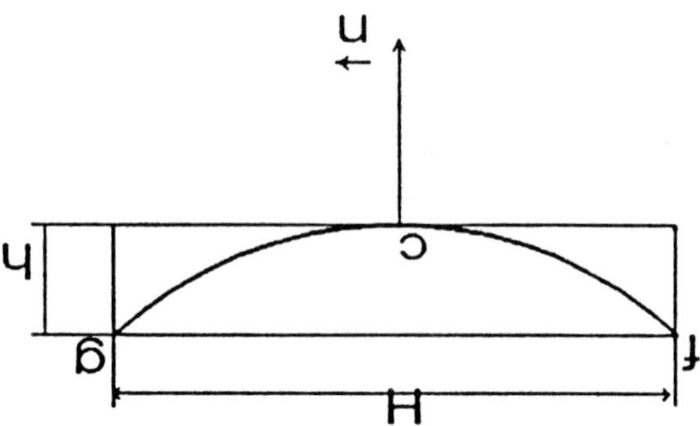

Fig.4: Rectangle including the curve fcg.

Body refinement criterion. The grid near the body is essentially refined in the normal direction. The refinement in the tangent direction depends on the curvature of the curve or its local deviation from a straight line. We use a statistical method [9] to estimate the local deviation. Such a method can be easily extended to three dimensions and was applied in [3] to cluster cells of the same level to form a rectangular grid.

Consider the curve fcg displayed in Fig.4. Assume that on this curve there are a set of several discrete points $Z = \{(x_i, y_i)\}_{i=1,2,...,P}$ around the point c where the normal is $\vec{n}$. The set Z includes the points f, c and g. We need to find a rectangle with the smallest area and containing the curve fcg. The width and length of this rectangle are respectively h and H. The ratio $C = h/H$ provides a good estimate of the deviation of the set Z from a best fitted straight line. This rectangle can be efficiently determined as follows.

Let A be the $P \times 2$ matrix of the coordinates of the P points in the set Z, and A_m the matrix of the same dimension with the x and y coordinates of the mean of the set (x_m, y_m). The symmetric matrix $M = (A - A_m)^t (A - A_m)$, i.e.,

$$M = \begin{pmatrix} \sum_{i=1}^{P} x_i^2 - x_m^2 & \sum_{i=1}^{P} x_i y_i - x_m y_m \\ \sum_{i=1}^{P} x_i y_i - x_m y_m & \sum_{i=1}^{P} y_i^2 - y_m^2 \end{pmatrix},$$

contains the second moments of the set Z about their mean. The symmetric matrix M has two eigenvalues e_1 and e_2 and two real eigenvectors $\vec{E}_1$ and $\vec{E}_2$. For convenience we assume $| e_1 | \le | e_2 |$. The ratio $| e_1 |^{\frac{1}{2}} / | e_2 |^{\frac{1}{2}}$ provides a good estimate of h/H. Furthermore, the eigenvalues E_1 and E_2 provide a good estimation of the orientation of the rectangle. In consequence, we will use the following relations to esti-

mate the local deviation C and the normal n:

$$C = \left(\frac{| e_1 |}{| e_2 |} \right)^{\frac{1}{2}}, \quad \vec{n} = E_1. \tag{1}$$

This algorithm is invariant under rotations and translations. Since the computation of the eigenvalues and eigenvectors of a 2×2 matrix is very simple, the above method is quite efficient in practice. Normally we only need 3 points ($P = 3$) in two dimensions. But the described method is valid for $P > 3$ which would be useful for three dimensions.

Define the local refinement degree by

$$\vec{R} = (R^{(x)}, R^{(y)}) = (\frac{1}{l_x}, \frac{1}{l_y}) \tag{2}$$

where l_x and l_y denote the mesh level in x and y. When the local refinement degree is $\vec{R}$, then the mesh should be refined such that the local mesh size $\vec{h} = (h^{(x)}, h^{(y)})$ satisfies

$$h^{(x)} = h_0^{(x)} R^{(x)}(\vec{x}_w), \quad h^{(y)} = h_0^{(y)} R^{(y)}(\vec{x}_w). \tag{3}$$

or

$$h^{(x)} = \frac{h_0^{(x)}}{l_x}, \quad h^{(y)} = \frac{h_0^{(y)}}{l_y}. \tag{4}$$

The integers l_x and l_y are computed as follows.

Let l_{min}, l_{max} and l_{iso} be three integers. The integer l_{min} is the minimal number of refinements on the wall and l_{max} is the maximal number of refinements on the wall. The integer l_{iso} is the number of anisotropic refinements counting from the finest subgrid. If $l_{iso} = 0$, then the refinement is totally isotropic. As a result, the present method can be easily switched to the traditional isotropic method or includes the isotropic Cartesian grid algorithm as a special case. Let C_{min} and C_{max} be the minimaum and maximum local deviations on the wall. At any point $\vec{x}_w$ on the wall, define the local relative deviation $c(\vec{x}_w)$ by

$$c(\vec{x}_w) = \frac{C(\vec{x}_w) - C_{min}}{C_{max} - C_{min}} \tag{5}$$

Then the number of refinements in the directions normal and tangent to the wall are computed as:

$$l_n = (1 - c)l_{min} + c l_{max}, \quad l_t = l_n - l_{iso}.$$

Finally, l_x and l_y are computed by

$$l_x = \sqrt{(l_n n_x)^2 + (l_t n_y)^2},$$

$$l_y = \sqrt{(l_n n_y)^2 + (l_t n_x)^2}. \tag{6}$$

Refinement criterion in interior points. Remember that we do not consider refinement based on the solution. The grid refinement in interior regions originates from the refinement near the walls. The refinement criterion should be determined by requiring the cells to vary gradually from the wall to the far flow field. The refinement parameter should decay gradually from the wall to the far flow field. The simplest choice is to ensure approximatively a constant subgrid width p in terms of grid points in the direction $\vec{n}(\vec{x}_{w,i})$. This requires the refinement degree to decay linearly in function of the distance. Let $\vec{R}(\vec{x}_i)$ be the refinement criterion at any interior point $\vec{x}_i$ (a cell center). Assume $\vec{x}_w$ be a point on the wall nearest to $\vec{x}_i$. Their distance is denoted d_w. Then by computing the refinement criterion at $\vec{x}_i$ as

$$\begin{cases} R^{(x)}(\vec{x}_i) = R^{(x)}(\vec{x}_w) + (\frac{1}{r} - 1)\frac{d_w}{2ph_0^{(x)}} \\ R^{(y)}(\vec{x}_i) = R^{(y)}(\vec{x}_w) + (\frac{1}{r} - 1)\frac{d_w}{2ph_0^{(y)}} \end{cases}, \quad (7)$$

we obtain a constant subgrid width p.

When there are a multielement body as shown in Fig.5, the refinement degree at an interior point $\vec{x}_i$ computed from one body may differ from that from another body. Consider for instance the interior point I in Fig.5. The vector $\vec{R}(\vec{x}_I)$ computed using the refinement criterion defined at point A of body 1 is in the vertical direction. The vector $\vec{R}(\vec{x}_I)$ computed using the refinement criterion defined at point B of body 2 is in the lower-left or upper-right direction. A unique value of $\vec{R}(\vec{x}_I)$ can be defined by using the minimum components computed from all the bodies. Assume that there are N separated bodies $1, 2, \ldots, N$. Let $\vec{R}^{(n)}(\vec{x}_I) = (R_x^{(n)}, R_y^{(n)})$ be the refinement degree computed from body n. Then the minimum component method yields:

$$\vec{R}(\vec{x}_I) = (R^{(x)}, R^{(y)}), \quad \begin{cases} R^{(x)} = \min_{1 \leq n \leq N}[R_x^{(n)}] \\ R^{(y)} = \min_{1 \leq n \leq N}[R_y^{(n)}] \end{cases} \quad (8)$$

The computation of $R^{(x)}$ and $R^{(y)}$ is costly if for each point I we need to find the nearest point on each wall. The following simplification would reduce the cost substantially.

For the coarse cells, the exact algorithm is not costly since the number of coarse cells is small. On the higher level grid, if a fine cell is close to the wall, we still use the exact algorithm to compute $R^{(x)}$ and $R^{(y)}$, if the cell is sufficiently far away from the wall, we interpolate the value $R^{(x)}$ and $R^{(y)}$ from the known values on the coaser cells.

2.3 Anisotropic refinement

The refinement starts from grid level $l = 0$. At each grid level l, we compute $\vec{R}(\vec{x}_c)$ for each cell where $\vec{x}_c$ is the cell center. Compute

$$l_x = \left[\frac{1}{R^{(x)}}\right], \quad l_y = \left[\frac{1}{R^{(y)}}\right] \quad (9)$$

where $[\cdot]$ denotes the integer part of a real number.

If $l < l_x$ and $l < l_y$, then the current cell is divided into four squared cells. For example, this occurs for cell 1 in Fig.6-a,b.

If $l \geq l_x$ and $l < l_y$, then the current cell is divided into two rectangle cells in the vertical direction. For example, this occurs for cell 2 in Fig.6-a,b.

If $l < l_x$ and $l \geq l_y$, then the current cell is divided into two rectangle cells in the horizontal direction. For example, this occurs for cell 3 in Fig.6-a,b.

If $l \geq l_x$ and $l \geq l_y$, then the current cell will not be divided.

Let us consider the refinement for the geometry in Fig.6. At level $l = L - l_{iso}$ where $l_{iso} = 2$, the grid is still isotropic (Fig.6-a). At two steps of anisotropic refinement, the anisotropic grids are respectively given by Fig.6-c and Fig.6-d.

2.4 Search of cutcells and solid cells

We need to search cutcells before searching solid cells.

Cutcells. The boundary condition that will be studied uses a finite-difference approach so that we do not need to find the intersection between the wall and the cutcells. In order the search to be quite efficient, a cell (rectangle or square) will be considered to be a cutcell once the nearest point on the wall lies inside the $(1 + \eta)$-amplified cell. The $(1 + \eta)$-amplified cell has the same shape and the same center as the original cell except that its width and length are amplified by $(1 + \eta)$. Consider the cells displayed in Fig.7. The original cells have no intersection with the wall BB' but the $(1+\eta)$-amplified cells have an intersection with the wall. The parameter η is a small positive number. For example, we can take $\eta = 0.1$. Such a search leads to cutcells of two types: genuine cutcells and pseudo cutcells. A genuine cutcell has intersection with the wall while a pseudo cutcell is disjoint from the wall and its amplified cell intersects the wall. The cutcells can be further classified into f-cutcells and s-cutcells. The f-cutcells have fluid cells as neighbours but have no solid neighbours. The s-cutcells have solid cells as neighbours. See Fig.8-a.

After all the cutcells and solid cells are searched (see

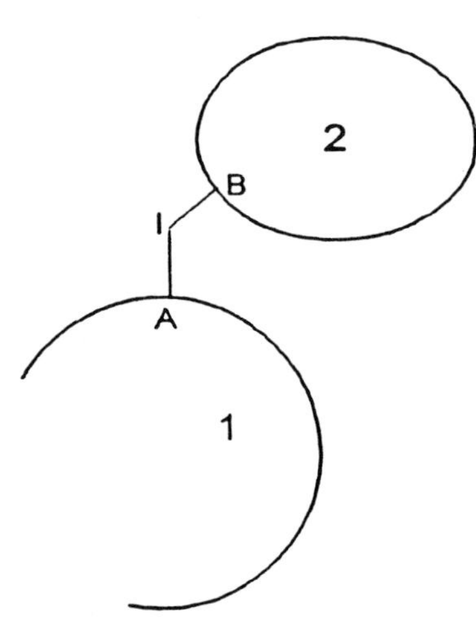

Fig.5: Multibody interaction.

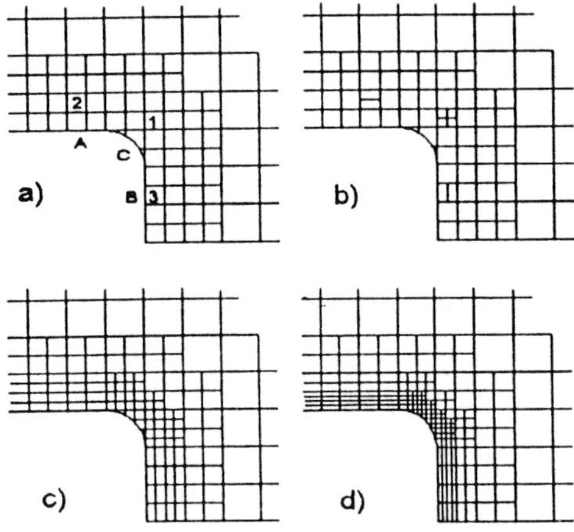

Fig.6: Anisotropic refinement.

next paragraph), only the s-cutcells are kept and the f-cutcells are transformed into fluid cells. See Fig.8-b.

Such a method, which broadens the search and keeps the s-cutcells, is extremely efficient. This treats without ambiguity the case of cell 1 and cell 2 displayed in Fig.8-a. The wall lies at the interface of cells 1 and 2. A different algorithm would face the difficulty of deciding which of cells 1 and 2 intersects with the wall.

The case of cell 3 in Fig.8 is also important. Cell 3 is a genuine cutcell but it is finally transformed into a fluid cell. Since cell 3 is essentially inside the fluid, it

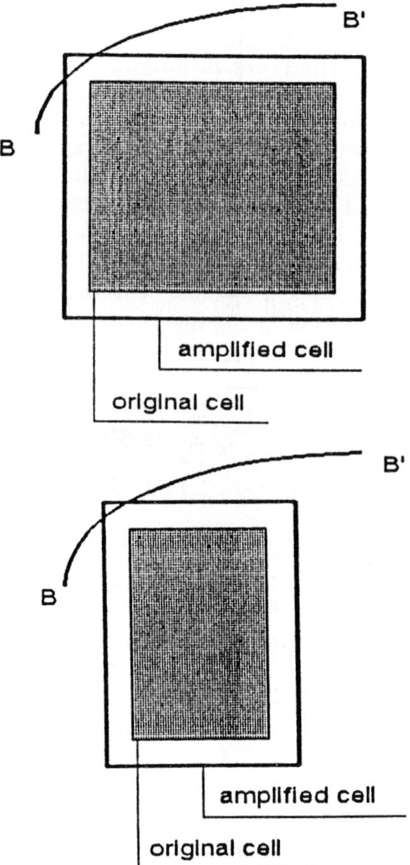

Fig.7: Definition of amplified cells

is better to treat it as a fluid cell instead of defining a boundary condition on it.

<u>Search of solid cells</u>. Once all the cutcells are searched, the search of solid cells is relatively simple. Consider for instance the body displayed in Fig.9. In order to suppress all the solid cells inside the oven, a reference point P is given. The closest cell, denoted as cell P, is searched and then suppressed, see Fig.9-1. Once cell P is suppressed, its four neighbours are suppressed, see Fig.9-2. The neighbours of all the suppressed cells are suppressed unless they are cutcells, see Fig.9-3,4,5. This process is repeated untill all the solid cells are suppressed, see Fig.9-6.

2.5 Summary of the algorithm

The separate steps of the algorithm presented in the previous paragraphs should be properly assembled in order the global algorithm to be efficient. The proposed global algorithm is given below:

Step 1. Generate the coarse grid ($l = 0$).

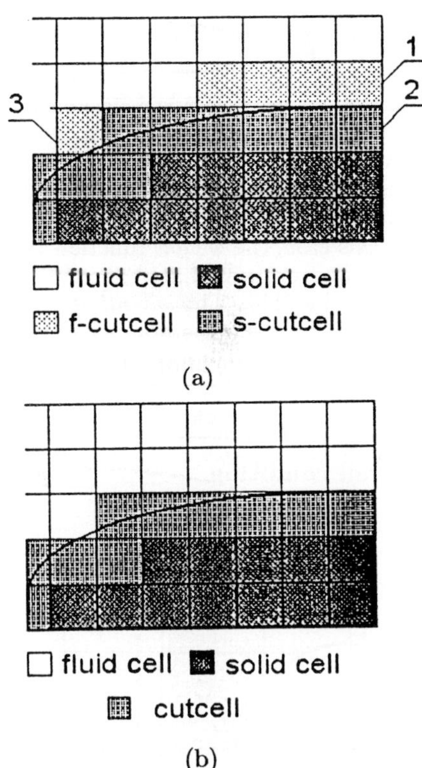

(a)

(b)

Fig.8: Generation and transformation of cutcells.

Step 2. Compute the refinement degree associated with each solid wall by using (3)-(6).

Step 3. Compute the refinement degree at each coarse cell using (8).

Step 4. Search all the cutcells for all the solid walls and keep only the s-cutcells.

Step 5. Suppress the solid cells inside all the solid bodies.

Step 6. Compute the refinement degree for cutcells using (7) and (8).

Step 7. Compute refinement degree for fluid cells by using (7) and (8).

Step 8. Refine each cell by dividing it into rectangular or squared subcells by comparing l with l_x and l_y where l_x and l_y are computed with (9).

Step 9. Set $l = l + 1$. If $l \leq L$, go to Step 4. If $l = L$, stop.

2.6 Comparison between ACG and ICG

Consider a straight line defined by

$$y = x \tan(\alpha), \quad 0 \leq x \leq \cos \alpha \qquad (10)$$

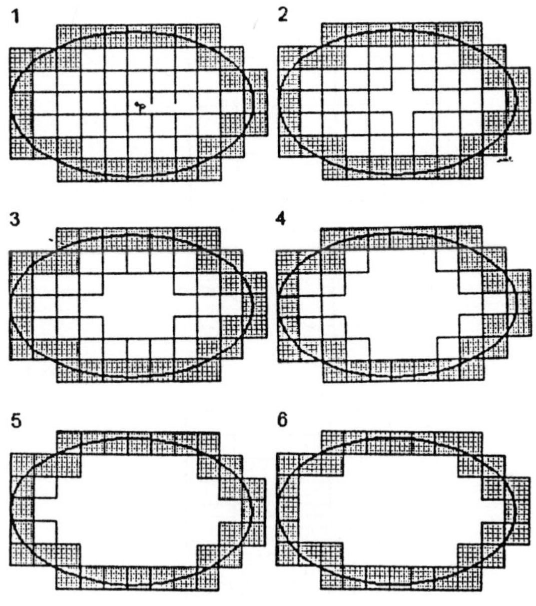

Fig.9: Search of solid cells.

where $0 \leq \alpha \leq \pi$. The grid is refined in the normal direction of the line starting from an isotropic initial grid as displayed in Fig.10-a. The anisotropic grid and isotropic grid after two levels of refinement are displayed in Fig.10-b and Fig.10-c for two typical angles α. It is clear that the anisotropic grid has much fewer points than the isotropic one especially for $\alpha \approx 0$ and $\alpha \approx \pi/2$.

Let $\alpha = 0$. Assume that the grid is refined L times with respect to the initial grid. Furthermore, assume the number of grids counted in the direction normal to the refined line and corresponding to the level l is p. Then each line of coarse cells in the normal direction is divided into $J_{aniso} = pL$ cells for the anisotropic refinement and $J_{iso} = pL^2$ cells for the isotropic refinement. Thus, the ratio between new cells by anisotropic refinement and isotropic refinement is given by $r_{\alpha=0} = \frac{1}{L}$

Now let $0 \leq \alpha \leq \pi$. By symmetry we only need to consider $0 \leq \alpha \leq \pi/4$. Here we are interested in the statistical average $\bar{r} = \frac{4}{\pi} \int_0^{\frac{\pi}{4}} W(\alpha) r_\alpha d\alpha$ where $W(\alpha)$ is a weight function. By assuming α to have a uniform distribution we can take $W(\alpha) = 1$. For any α, the grid is refined $L \cos \alpha$ times in y and $L \sin \alpha$ times in x for the anisotropic refinement. Then each line of coarse cells in the normal direction is divided into

$$J_{aniso} = [pL^2 \sin^2 \alpha + pL(\cos \alpha - \sin \alpha)]$$

$$\approx 0.27 + \frac{0.67}{L}. \tag{11}$$

Expression (11) shows that $\bar{r}$ is a decreasing function of the number of refinements L but it is bounded by 0.27. In practice, the most probabal value of α is near 0 if the line (10) represents the wall or the viscous boundary layer, or near $\pi/2$ if the line (10) represents a shock. In this case, the weight function $W(\alpha)$ is different from 1 and ensures that $\bar{r}$ is smaller than the value predicted by (11). The saving of grid points of the anisotropic refinement method in comparison with the traditional isotropic method is therefore rather important.

3. Solid wall treatment

3.1 Solid wall condition

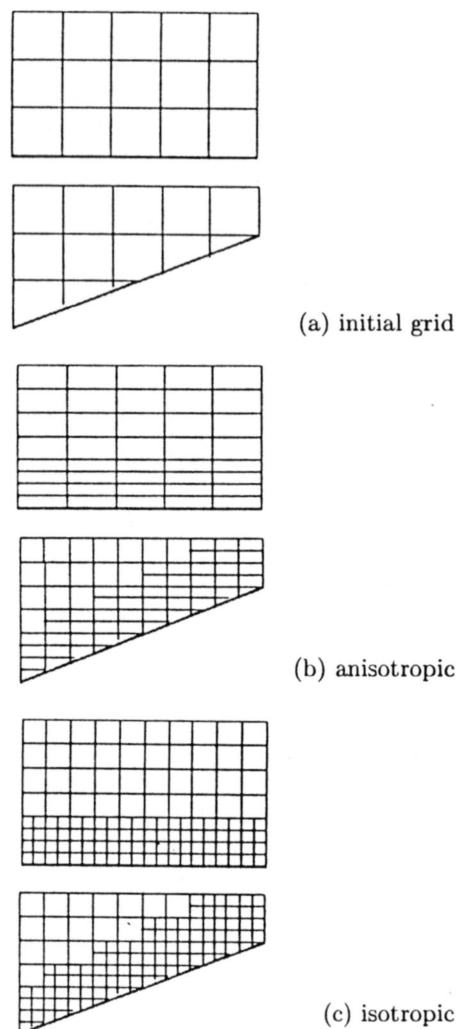

(a) initial grid

(b) anisotropic

(c) isotropic

Fig.10: Anisotropic refinement and isotropic refinement near a straight line.

cells where $[\cdot]$ denotes the integer part of a real value. For isotropic refinement, we have $J_{iso} = [pL^2 \cos^2 \alpha]$ $(0 \leq \alpha \leq \pi/4)$. Then by $\bar{r} = \frac{4}{\pi} \int_0^{\pi/4} W(\alpha) \frac{J_{aniso}}{J_{iso}} d\alpha$ and $W(\alpha) = 1$ we have

$$\bar{r} = \frac{4}{\pi} \int_0^{\frac{\pi}{4}} \frac{[pL^2 \sin^2 \alpha + pL(\cos \alpha - \sin \alpha)]}{[pL^2 \cos^2 \alpha]} d\alpha$$

$$\approx \frac{4}{\pi} \int_0^{\frac{\pi}{4}} \frac{pL^2 \sin^2 \alpha + pL(\cos \alpha - \sin \alpha)}{pL^2 \cos^2 \alpha} d\alpha$$

$$= \frac{4}{\pi} \left(1 - \frac{\pi}{4} + \frac{1}{L} (\ln \tan \frac{3\pi}{8} + \ln \cos \frac{\pi}{4}) \right)$$

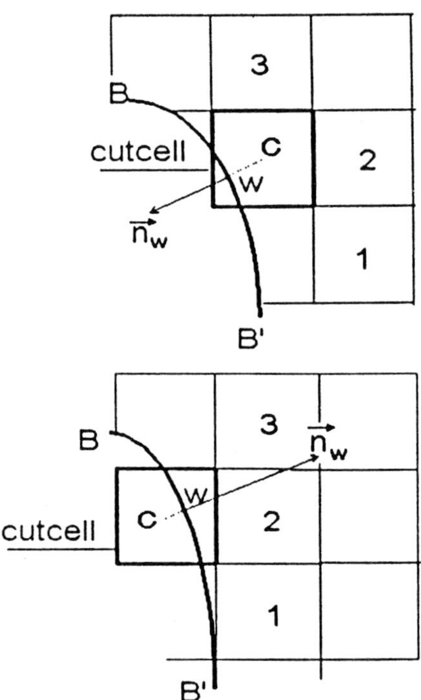

Fig.11: Cutcell definition on the wall.

When a finite-volume approach is applied to define the boundary condition, one needs to find the possible intersection in order to define the polygone lying inside the flow field for each cutcell. In the present finite-difference approach, the cutcell will not be cut. This avoids the use of a costly and complicated algorithm to find the intersection. The boundary condition here is directly defined at the center of the cutcell. For the cutcell in Fig.11, the boundary condition is defined at its center c. Cutcells can be divided into two classes: inner cutcell and outer cutcell. An inner cutcell has its cell center inside the flow field, see the cutcell in Fig.11-a). An outer cutcell has its cell center inside the solid, see the cutcell in Fig.11-b). The present boundary condition will be invariant whether the cutcell is inner or outer.

The boundary condition is defined by using the following bilinear interpolation easily extendible to three dimensions:

$$\phi_c = L(\phi, x_c, y_c),$$
$$L(\phi, x, y) = d_0^{(\phi)} + d_1^{(\phi)} x + d_2^{(\phi)} y + d_3^{(\phi)} xy$$
$$(12)$$

where ϕ refers to the two velocity components u and v, the density ρ and the pressure p.

The interpolation coefficients $d_i^{(\phi)}$ with $i = 1, 2, 3, 4$ are determined using known solutions at three adjacent fluid cells ϕ_1, ϕ_2, ϕ_3. This yields:

$$d_0^{(\phi)} + d_1^{(\phi)} x_1 + d_2^{(\phi)} y_1 + d_3^{(\phi)} x_1 y_1 = \phi_1 \quad (13)$$
$$d_0^{(\phi)} + d_1^{(\phi)} x_2 + d_2^{(\phi)} y_2 + d_3^{(\phi)} x_2 y_2 = \phi_2 \quad (14)$$
$$d_0^{(\phi)} + d_1^{(\phi)} x_3 + d_2^{(\phi)} y_3 + d_3^{(\phi)} x_3 y_3 = \phi_3 \quad (15)$$

Furthermore, we need the nonslip condition at the point $\mathbf{w}$ on the wall. The location of the point $\vec{x}_w$ is crucial for accuracy. Based on some numerical experiments, we find that **w should be the wall point nearest to the cutcell center** $\vec{x}_c$ in order to have a good accuracy. The nonslip condition is of Dirichlet type for some components and of Neumann type for others. The Dirichlet condition $\phi_w = g$ leads to:

$$d_0^{(\phi)} + d_1^{(\phi)} x_w + d_2^{(\phi)} y_w + d_3^{(\phi)} x_w y_w = g. \quad (16)$$

For the Neumann condition, we need to know the unit normal $\vec{n}_w$ at $\vec{x}_w$. The Neumann condition can be written as:

$$\frac{\partial \phi}{\partial n_w} = g. \quad (17)$$

Remark that $\frac{\partial \phi}{\partial n_w} = (\frac{\partial \phi}{\partial x}, \frac{\partial \phi}{\partial y})_w \bullet (n_w^{(x)}, n_w^{(y)})$, and $\frac{\partial \phi}{\partial x}|_w = d_1 + y_w d_3$, $\frac{\partial \phi}{\partial y}|_w = d_2 + x_w d_3$. These relations along with (17) lead to:

$$n_w^{(x)} d_1 + n_w^{(y)} d_2 + (n_w^{(x)} y_w + n_w^{(y)} x_w) d_3 = g. \quad (18)$$

For the velocity components u and v, the relations (13)-(15) and (16) with $g = 0$ determine the coefficients $d_i^{(u)}$ and $d_i^{(v)}$ uniquely. For the pressure, the relations (13)-(15) and (18) with $g = 0$ (high Reynolds number) determine the coefficients $d_i^{(p)}$ uniquely.

For the density, it depends on whether the wall is adiabatic or has a fixed temperature. If the wall is adiabatic, the coefficients $d_i^{(\rho)}$ are computed similarly as for the pressure. If the wall has a fixed temperature T_w, then we first need to compute the pressure on the wall by (12):

$$p_w = d_0^{(p)} + d_1^{(p)} x_w + d_2^{(p)} y_w + d_3^{(p)} x_w y_w \quad (19)$$

Then the coefficients $d_i^{(\rho)}$ are computed similarly as for the velocity with $g = p_w/(R_g T_w)$ where R_g is the gas constant.

3.2 Stability analysis

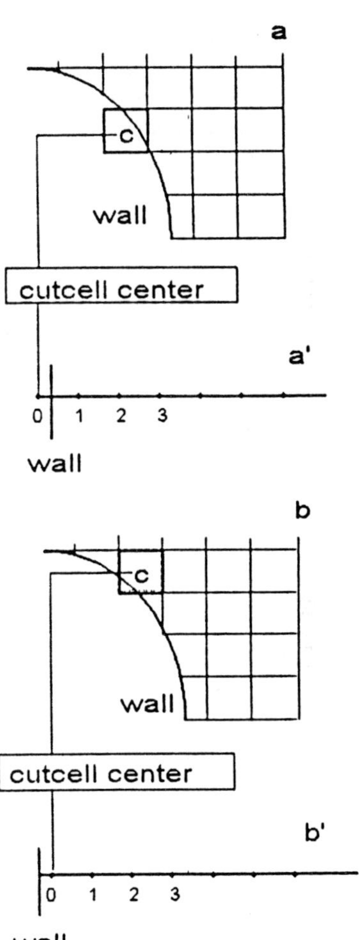

Fig.12: Correspondance between the two-dimensional boundary configuration and the one dimensional model. Upper: outer cutcell. Lower: inner cutcell

The solid wall condition based on a finite-volume approach has the difficulty of instability due to small cutcells. Here we want to know whether the present finite-difference approach has the same instability problem. To achieve such a purpose, it is sufficient to analyze the stability for the corresponding one-dimensional model problem. A two-dimensional analysis, though not impossible, will not say more than the corresponding one dimensional analysis.

The correspondance between the two-dimensional grid near the wall and the one-dimensional model grid for analysis is displayed in Fig.12. The boundary condition is defined at the boundary cell center $0(C)$. The point 0 may be outside or inside the wall, see Fig.12-a or Fig.12-b$'$.

The   Navier-Stokes equations form a mixed hyperbolic-parabolic system. Near the wall, the parabolic part is dominated by the dissipation term. It is therefore reasonable to analyze the stability separately for a hyperbolic model and a parabolic model defined as below:

$$u_t + au_x = 0, \ x > 0 \qquad \text{hyperbolic} \qquad (20)$$

$$u_t = u_{xx}, \ x > 0 \qquad \text{parabolic} \qquad (21)$$

The stability of boundary treatment for the hyperbolic equation can be studied using the stability theory of GKS [15]. That for the parabolic equation can be studied using the stability theory of Varah [27]. A stability theory has also been developed by Osher [23] when the parabolic equation includes a lower order term, that is, when the equation becomes

$$u_t + u_x = u_{xx}.$$

Consider a Cauchy-stable three-point scheme in interior points. The wall is assumed to locate at $x = 0$. The boundary condition is divided into two classes: Dirichlet condition (e.g. the velocity) and Neumann condition[2] (e.g. the pressure).

For the Dirichlet condition, the value at the wall is fixed, that is $u_w = g$. Using linear interpolation leads to:

$$\alpha u_0^n + (1 - \alpha)u_1 = g,$$
$$\frac{1}{2} < \alpha \le 1 \quad \text{inner cutcell} \qquad (22)$$
$$\alpha u_0^n + (\alpha - 1)u_1 = g,$$
$$\frac{3}{4} < \alpha \le 1 \quad \text{outer cutcell} \qquad (23)$$

Here α is the interpolation coefficient depending on the location of the wall.

---

[2] For the scalar hyperbolic equation, there is a Dirichlet condition for a inflow boundary while there is no condition for a outflow boundary. But for the outflow boundary, we may need a numerical boundary condition often defined by extrapolation. A boundary condition defined by extrapolation is similar to a Neumann condition.

For the Neumann condition, the derivative at the wall is fixed, that is $\frac{du}{dx}\big|_w = g$. Using linear interpolation yields:

$$u_0^n = u_1 + (x_0 - x_1)g \quad \text{inner or outer cutcell} \qquad (24)$$

Boundary conditions (22), (23) and (24) correspond to the boundary condition previously defined for the Navier-Stokes equations.

We use normal mode analysis [15] to study the stability. Let u_j^n be the numerical solution defined at the grid point j and at instant n. In the normal mode analysis the solution takes the following form:

$$u_j^n = \hat{u}z^n\kappa^j \qquad (25)$$

where z and κ are complex. We also express the boundary data as $g = \hat{g}z^n$. Introduce (25) into the interior difference scheme leads to $D(z,\kappa)\hat{u} = 0$. The characteristic equation $\det D(z,\kappa) = 0$ determines $\kappa = \kappa(z)$ for each z. Introduce (25) into the boundary condition gives:

$$M(z,\kappa)\hat{u} = \hat{g} \qquad (26)$$

First consider the hyperbolic equation. For $|z| \ge 1$, the characteristic equation for a three-point scheme has two roots: an inner root κ_1 which is compatible with the Cauchy stability and an outer root κ_2 which is not used in stability analysis. For a scheme dissipative in the sense of Kreiss, we have [15, 13] the following inequalities:

$$|\kappa_1| < 1, \ |\kappa_2| > 1, \ \forall |z| \ge 1, z \ne 1 \qquad (27)$$
$$\kappa_1 = 1, \ |\kappa_2| > 1, \ \text{if } z = 1, a > 0 \qquad (28)$$
$$\kappa_2 = 1, \ |\kappa_1| < 1, \ \text{if } z = 1, a < 0 \qquad (29)$$

If $\det M(z,\kappa_1) \ne 0 \ \forall \ |z| \ge 1$, then the boundary treatment excludes the solution with $|z| \ge 1$ and the problem is said to be GKS-stable [15]. If $\det M(z,\kappa_1) = 0$ for at least one z with $|z| \ge 1$, then the problem is GKS-unstable. The GKS-stability is rather strong since it excludes the marginal case of $|z| = 1$. In some special case, a GKS-unstable method may be stable in a weak sense, see [28] and the references cited therin.

Now consider the parabolic equation. For $|z| \ge 1$, the characteristic equation for a three-point scheme has an inner root κ_1 and an outer root κ_2 satisfying

$$|\kappa_1| < 1, \ |\kappa_2| > 1, \ \forall |z| > 1 \qquad (30)$$
$$\kappa_1 = 1 - O(\sqrt{z - 1}),$$
$$\kappa_2 = 1 + O(\sqrt{z - 1}), \ z \to 1 \qquad (31)$$

If $\det M(z,\kappa_1) \ne 0$ for $|z| > 1$ and $\det M(z) = O[(z - 1)^q]$ with $q \le \frac{1}{2}$ for $z \to 1$, then the problem is stable [27].

Now we analyze the stability for each boundary condition.

Theorem 3.1 *Consider the hyperbolic equation (20). Let the three-point difference scheme be dissipative in the sense of Kreiss. Then, independently of the wall position, the problem is GKS-stable if 1) for an inflow boundary with $a > 0$, the boundary condition is given by (22) or (23); 2) for an outflow boundary with $a < 0$, the numerical boundary condition is given by (24).*

Proof. First consider $a > 0$. For an inner cutcell with the condition (22), $M(z)$ is found to be

$$M(z) = \alpha + (1 - \alpha)\kappa_1, \quad \frac{3}{4} < \alpha \le 1.$$

By assumption of dissipation, we obtain from (27)-(28) that $\mid \kappa_1 \mid \le 1$ for $\mid z \mid \ge 1$. Furthermore, from $\frac{3}{4} < \alpha \le 1$ we have $0 \le 1 - \alpha < \frac{1}{4}$. In consequence, the inequality $\mid (1 - \alpha)\kappa_1 \mid < \alpha$ holds. This means $\mid M(z) \mid \ne 0$ for all $\mid z \mid \ge 1$ so that the problem is GKS-stable. For an outer cutcell with the condition (23), $M(z)$ is found to be

$$M(z) = \alpha + (\alpha - 1)\kappa_1, \quad \frac{1}{2} < \alpha \le 1$$

and we still find that $\mid M(z) \mid \ne 0$ for all $\mid z \mid \ge 1$ so that the problem is GKS-stable.

Now consider $a < 0$ for which the boundary condition is defined by (24). In this case, $M(z)$ is given by:

$$M(z) = 1 - \kappa_1$$

Due to dissipation, we obtain from (27) and (29) that $\mid \kappa_1 \mid < 1$ for all $\mid z \mid \ge 1$. In consequence, $\mid M(z) \mid \ne 0$ and the problem is GKS-stable. $\square$.

Theorem 3.2 *For the parabolic equation (21), the problem is stable for each of the condition (22), (23) and (24).*

Proof. For $\mid z \mid > 1$, the root properties (30)-(31) imply that $\mid \kappa_1 \mid < 1$. This ensures $\det M(z) \ne 0$ as for the hyperbolic equation.

Now consider $z \to 1$. For the condition (22) we have $M(z) \mid_{z \to 1} = \alpha + (1 - \alpha)\kappa_1 = 1$ so that the problem is stable. Similarly, for the condition (23) the problem is also stable. For the condition (24) we have $M(z) = 1 - \kappa_1$. Since $\kappa_1 = 1 - O(\sqrt{z - 1})$ by (31), we obtain $M(z) = O(\sqrt{z - 1})$ so that the problem is also stable. $\square$.

In conclusion, the present solid wall condition based on linear interpolation has no problem of instability independently of the position of the wall or size of the cutcell lying inside the fluid. This is the advantage of such a boundary condition over the condition based on a finite-volume approach.

4. Numerical experiments

4.1 Difference approximation for viscous flows

4.1.1 The Navier-Stokes equations

We apply the anisotropic Cartesian grid method to solve the following compressible Navier-Stokes equations:

$$w_t + f_x + g_y = 0 \tag{32}$$

where $f = f^{(E)} - f^{(V)}$, $g = g^{(E)} - g^{(V)}$,

$$w = \begin{pmatrix} \rho \\ \rho u \\ \rho v \\ \rho E \end{pmatrix}, \quad f^{(E)} = \begin{pmatrix} \rho u \\ \rho u^2 + p \\ \rho uv \\ \rho uH \end{pmatrix},$$

$$g^{(E)} = \begin{pmatrix} \rho v \\ \rho vu \\ \rho v^2 + p \\ \rho vH \end{pmatrix}, \quad \begin{aligned} p &= (\gamma - 1)\rho e \\ e &= E - \tfrac{1}{2}(u^2 + v^2) \\ H &= E + p/\rho \end{aligned} \tag{33}$$

$$f^{(V)} = \frac{1}{R} \begin{pmatrix} 0 \\ 2u_x - \frac{2}{3}(u_x + v_y) = \tau_{xx} \\ u_y + v_x = \tau_{xy} \\ u\tau_{xx} + v\tau_{xy} + \frac{\gamma}{Pr}e_x \end{pmatrix},$$

$$g^{(V)} = \frac{1}{R} \begin{pmatrix} 0 \\ u_y + v_x = \tau_{xy} \\ 2v_y - \frac{2}{3}(u_x + v_y) = \tau_{yy} \\ u\tau_{yx} + v\tau_{yy} + \frac{\gamma}{Pr}e_y \end{pmatrix} \tag{34}$$

Here γ denotes the ratio of specific heats, ρ, p, E and H are respectively the density, the pressure, the total energy and the total enthalpy, $\vec{U} = u\vec{i} + v\vec{j}$ is the velocity vector with the components u and v, R and Pr are the Reynolds number and the Prandtl number. With the assumption of a perfect gas, the following relations hold:

$$H = E + \frac{p}{\rho}, \quad p = (\gamma - 1)\rho(E - \frac{1}{2}\vec{U}^2).$$

4.1.2 A partially discrete difference scheme

For the inviscid part, we apply the Lerat's scheme [16] and for the viscous part we use the second-order central difference formula. Due to the particular feature of the Cartesian grid, any difference approximation can be simplified if it is put into a partially discrete form. The idea is as follows. The scheme is put into the traditional conservative form. Since the cell is in the form of a square or rectangle, the formula to compute the numerical flux is quite simple. But the numerical flux still involves derivatives. These derivatives are not discretized but will be computed elsewhere using the user-prefered method such as interpolation or method of least squares. This is why the scheme is called partially discrete. Such an idea is extremely useful for anisotropic Cartesian grid where the cell-to-cell relation is of unstructured nature.

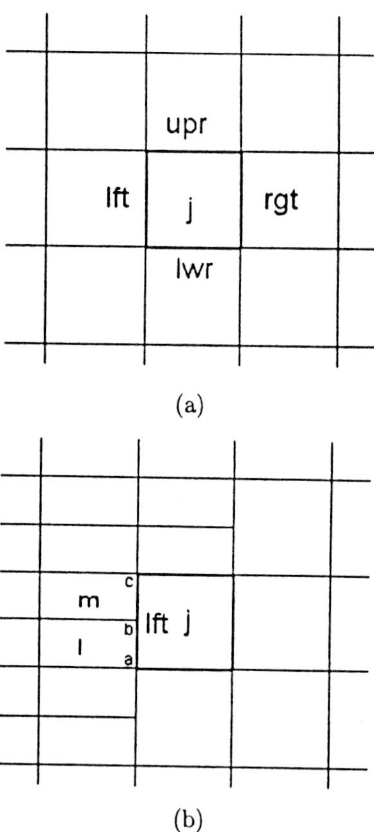

(a)

(b)

Fig.13: Cell for definition of a conservative scheme.

Now we use the cell displayed in Fig.13 to write the partially discrete difference scheme in the following form:

$$\Delta w_j = -\sigma_j(h^*_{rgt} - h^*_{lft} + h^*_{upr} - h^*_{lwr}) \qquad (35)$$

Here $\sigma_j = \Delta t_j/S_j$, Δt_j being the time-step and S_j the area of the cell. Let Δx_j and Δy_j be the cell size in

x and y, then the numerical fluxes h^*_{rgt} and h^*_{upr} are given by:

$$\frac{h^*_{rgt}}{\Delta y_j} = f^{(E)}(\bar{w}^{n+\frac{1}{2}}_{rgt})$$
$$- f^{(V)}(\bar{w}_{rgt}, \frac{\partial \bar{w}}{\partial x}\big|_{rgt}, \frac{\partial \bar{w}}{\partial y}\big|_{rgt}) \qquad (36)$$
$$\frac{h^*_{upr}}{\Delta x_j} = g^{(E)}(\bar{w}^{n+\frac{1}{2}}_{upr})$$
$$- g^{(V)}(\bar{w}_{upr}, \frac{\partial \bar{w}}{\partial x}\big|_{upr}, \frac{\partial \bar{w}}{\partial y}\big|_{upr}) \qquad (37)$$

and the fluxes h^*_{lft}, h^*_{lwr} are similarly defined.

In (36) and (37), $\bar{w}^{n+\frac{1}{2}}_{rgt}$ and $\bar{w}^{n+\frac{1}{2}}_{rgt}$ are computed as

$$\bar{w}^{n+\frac{1}{2}}_{rgt} = \bar{w}^n_{rgt}$$
$$- \frac{1}{2}\Delta t_{rgt}(\bar{A}_{rgt}\frac{\partial \bar{w}}{\partial x}\big|_{rgt} + \bar{B}_{rgt}\frac{\partial \bar{w}}{\partial y}\big|_{rgt}) \qquad (38)$$
$$\bar{w}^{n+\frac{1}{2}}_{upr} = \bar{w}^n_{upr}$$
$$- \frac{1}{2}\Delta t_{upr}(\bar{A}_{upr}\frac{\partial \bar{w}}{\partial x}\big|_{upr} + \bar{B}_{upr}\frac{\partial \bar{w}}{\partial y}\big|_{upr}) \qquad (39)$$

Here, $\bar{A} = \frac{df^{(E)}}{dw}(\bar{w})$, $\bar{B} = \frac{dg^{(E)}}{dw}(\bar{w})$.

Excepting $\bar{w}$, $\frac{\partial \bar{w}}{\partial x}$, and $\frac{\partial \bar{w}}{\partial y}$, the equations (35)-(39) completely define the difference scheme we need. This is the partial discrete scheme we use in our computations.

Remark 1. For the inviscid part, the scheme is linearly equivalent with the scheme of Lerat [16]. Lerat's scheme is of Lax-Wendroff type but it is totally dissipative even in two dimensions.

Remark 2. One can also replace (38) by

$$\bar{w}^{n+\frac{1}{2}}_{rgt} = \bar{w}^n_{rgt} - \frac{1}{2}\Delta t_{rgt}(\frac{\partial f^{(\bar{E})}}{\partial x}\big|_{rgt} + \frac{\partial g^{(\bar{E})}}{\partial y}\big|_{rgt}) \qquad (40)$$

The latter does not involve computation of Jacobian matrix. However, since for any function ϕ, we need to compute its average and derivatives by interpolation or other methods, formula (38) is cheaper than (40) since the latter requires to compute the expensive fluxes f and g in a cloud of points to define their derivatives at each point. The same is true for (39).

Remark 3. The average $\bar{w}$ and derivatives $\frac{\partial \bar{w}}{\partial x}$, $\frac{\partial \bar{w}}{\partial y}$ can be computed by various methods such as linear interpolation or method of least squares. In the present computation, we have used linear interpolation. The interpolation formulas are defined in such a way that they have the same form everywhere independently of the type of interface. See Appendix A in the end of this paper.

Remark 4. When the flow has discontinuity solutions, it will be important to ensure conservation at grid interfaces such as the left face of cell j displayed in Fig.13-b). This is done as follows. The computation starts with the finest grid and finishes with the coarsest grid. The numerical fluxes computed for a fine grid are projected to the coarse grid at their interfaces. For example, for the left face of cell j displayed in Fig.13-b), we compute the numerical flux by $h_{lft}^{(*)} = h_{ab}^{(*)} + h_{bc}^{(*)}$ where $h_{ab}^{(*)}$ and $h_{bc}^{(*)}$ are numerical fluxes at the right face of the finer cells l and m.

Remark 5. The Lerat's scheme for a linear hyperbolic equation is stable for $0 < CFL \leq 1$ in one dimension and $0 < CFL \leq 1/2$ in two dimensions. On a Cartesian grid with refinement interfaces, we find numerically that this stability domain is slightly reduced. Such a result may have a common sense for other schemes on a refined Cartesian grid and for other mesh refinement problems. We thus provide a stability analysis for the Lax-Wendroff scheme on a model Cartesian grid in Appendix B.

4.2 Numerical results

4.2.1 Symmetric flow around a NACA0012 airfoil

Consider a symmetric flow around a NACA0012 airfoil. The free stream Mach number is $M_\infty = 0.85$ and the Reynolds number based on the chord length is $R_e = 2000$. The open boundary extends to a distance of 10 times the chord length. The corresponding Cartesian grids obtained by the isotropic algorithm and the anisotropic algorithm are displayed in Fig.14 and Fig.15. Both grids have 12 levels. The isotropic refinement algorithm leads to 12702 cells while the anisotropic algorithm reduces the total number of cells to 6682. In comparison with the isotropic algorithm, the anisotropic algorithm has saved 47% of the cells. It remains to see whether the two grids lead to the same solution.

The computed Mach contours for both grids are displayed in Fig.16 and Fig.17. The viscous boundary layer is well resolved on both grids. The difference between the solutions on both grids is not visible.

In Fig.18 we display the distribution of pressure coefficient on the chord. The curve obtained from anisotropic Cartesian grid is not distinguishible from that obtained by isotropic Cartesian grid.

The skin friction is most sensitive to the accuracy of the treatment. In Fig.19 we display the distribution of skin friction on the chord. As for the pressure distribution, the curve obtained from anisotropic Cartesian grid is not distinguishible from that obtained by isotropic Cartesian grid.

As a result, the anisotropic Cartesian grid method saves considerably the number of cells while maintaining the same accuracy as the corresponding isotropic Cartesian grid method.

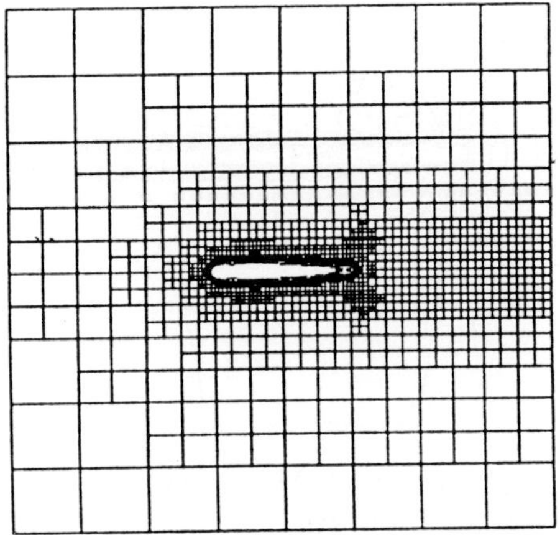

Fig.14: NACA0012 airfoil. Isotropic Cartesian grid with 12702 points.

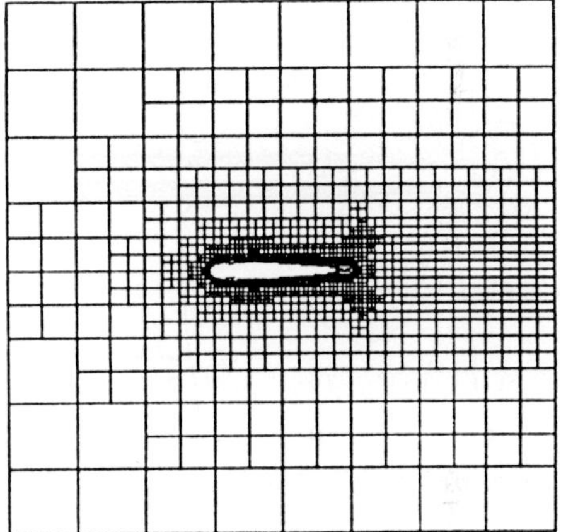

Fig.15: NACA0012 airfoil. Anisotropic Cartesian grid with 6682 points.

4.2.2 Flow around a Bi-NACA airfoil

Consider a flow around a Bi-NACA airfoil. The Bi-NACA airfoil is formed by two parallel NACA0012 airfoils. They are shifted by a distance equal to half a chord in both the parallel and perpendicular directions. The free stream Mach number is $M_\infty = 0.85$ and the Reynolds number based on the chord length is $R_e = 500$. The flow has now incidence. The open boundary extends to a distance of 10 times the chord length. The corresponding Cartesian grids obtained by the isotropic algorithm and the anisotropic algorithm are displayed in Fig.20 and Fig.21. The isotropic and anisotropic grids have respectively 9827 and 5817 cells.

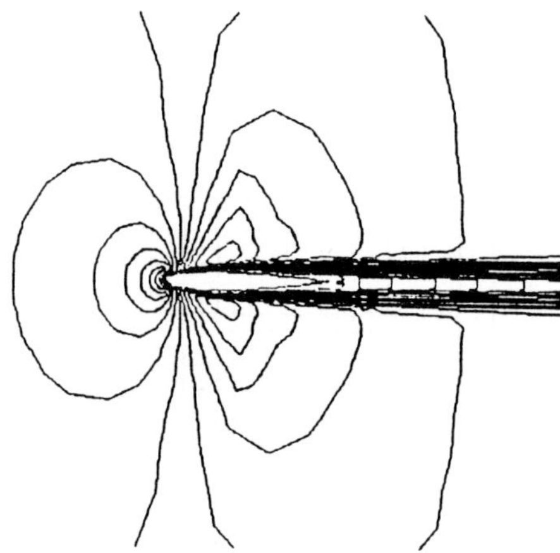

Fig.16: NACA0012 airfoil. Mach contours with isotropic Cartesian grid

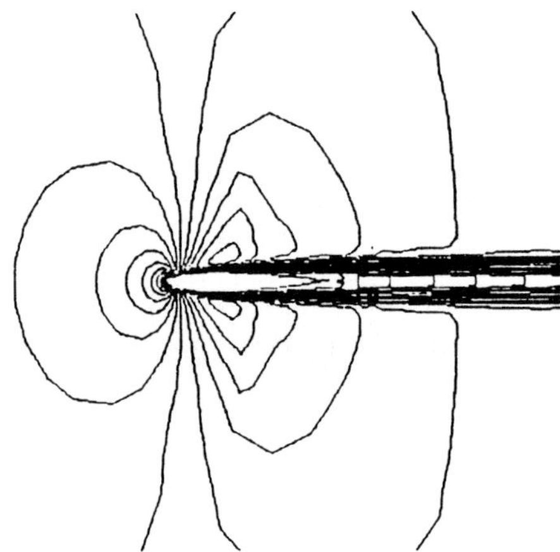

Fig.17: NACA0012 airfoil. Mach contours with anisotropic Cartesian grid

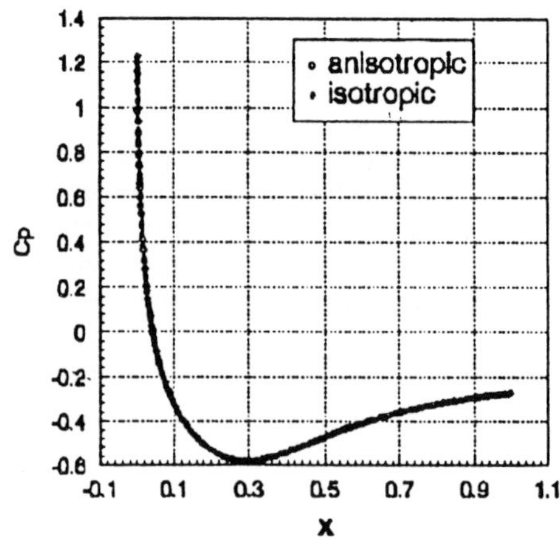

Fig.18: NACA0012 airfoil. Pressure distribution along the airfoil.

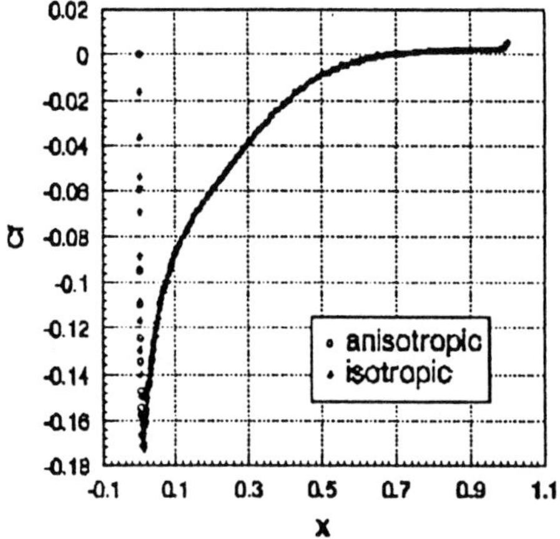

Fig.19: NACA0012 airfoil. Skin-friction coefficient along the airfoil.

The computed Mach contours for both grids are displayed in Fig.22 and Fig.23. We observe no difference between the solutions on the two grids. The agreement between the two solutions is more convincingly shown in Figs. 24 and 25 where the distributions of pressure coefficients and skin-friction coefficients are displayed.

In consequence, the anisotropic Cartesian grid method saves considerably the number of cells while maintaining the same accuracy as the corresponding isotropic Cartesian grid method even for a multi-element body.

4.2.3 Flow with incidence

Finally we consider a flow around a NACA0012 airfoil with angle of attack. The free stream Mach number is $M_\infty = 0.85$ and the Reynolds number based on the chord length is $R_e = 500$. The incidence of the flow is equal to 10°. Only the anisotropic grid method is used to compute the flow. The anisotropic grid is diplayed in Fig.26. It has 8579 cells in total. The boundary layer is aligned with the flow direction so that a geometry-oriented anisotropic refinement does not fit exactly to the anisotropy of the solution in the boundary layer. We therefore specifiy two boundary-layer lines near which the grid is refined using the anisotropic refinement method. This is solution adaptive but not yet fully automatic. Since the purpose of this work is not to present a solution adaptive algorithm, we are presently satisfied with such an artificial solution adaptive refinement.

The computed Mach contours is given in Fig.27. The distributions of pressure coefficients and skin-friction coefficients are displayed in Fig.28 and Fig.29.

This solution compares well with the known solution

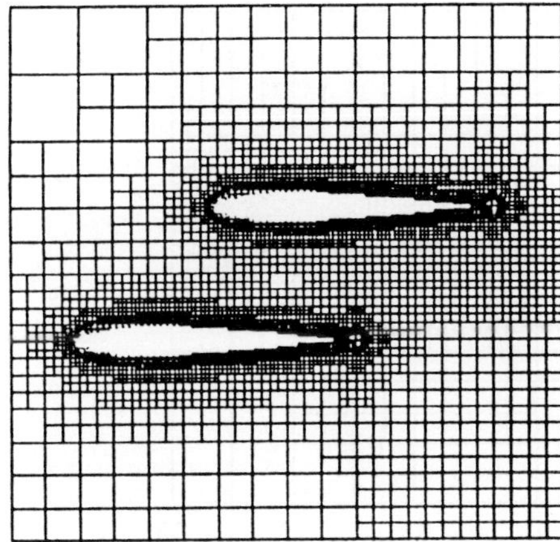

Fig.20: Bi-NCACA airfoil. Isotropic grid with 9827 points.

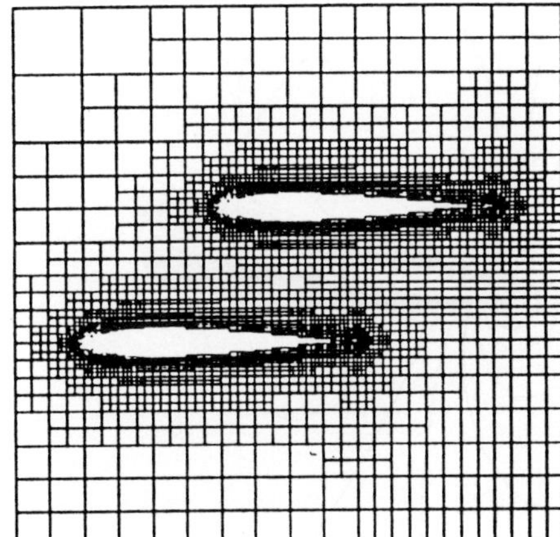

Fig.21: Bi-NCACA airfoil. Anisotropic grid with 5817 points.

computed with other methods [5].

Acknowledgments

This work was supported by Chinese National Science Foundation under Contract Number 19502002. This study has benefited from discussions with Professor N. Satofuka of Kyoto Institute of Technology and Dr. S.A. Bayyuk of CFD Coporation-USA during the 15th ICNMFD, and from remarks by Professor A. Lerat of SINUMEF Laboratory-ENSAM Paris and Professor Z.Q. Zhu at Institute of Fluid Dynamics of BUAA.

Appendix

A.1 Interpolation formulas with uniform structure

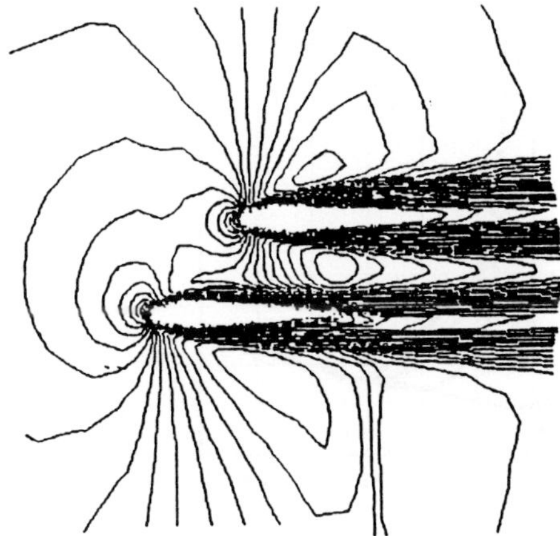

Fig.22: BI-NACA. Mach contours with isotropic Cartesian grid

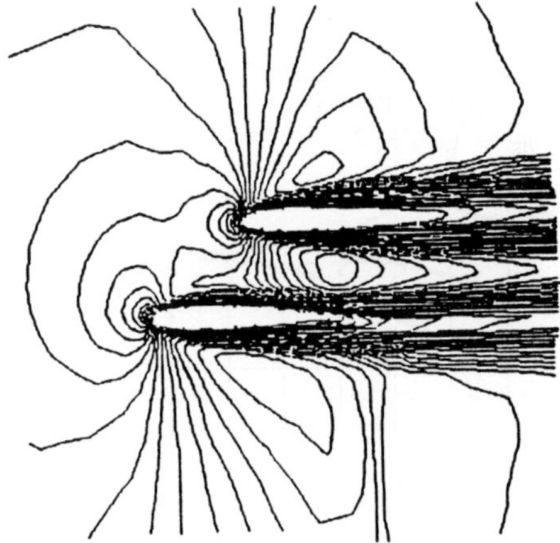

Fig.23: BI-NACA. Mach contours with anisotropic Cartesian grid

Let ϕ be a function of $\vec{x}$. We use the right face GF of the cell 1 displayed in Fig.30 and construct linear interpolation formulas to compute the average $\bar{\phi}$ and derivatives $\frac{\partial\bar{\phi}}{\partial x}$, $\frac{\partial\bar{\phi}}{\partial y}$ defined at the center m of GF. We want the formulas to have a uniform structure independently of whether there is refinement across the face GF or no.

If there is no mesh refinement at the face GF (Fig.30-1), we have the following simple formulas:

$$\bar{\phi} = \frac{\phi_1 + \phi_2}{2} \tag{41}$$

$$\frac{\partial\bar{\phi}}{\partial x} = \frac{\phi_2 - \phi_1}{x_2 - x_1} \tag{42}$$

$$\frac{\partial\bar{\phi}}{\partial y} = \frac{\phi_3 + \phi_4 - \phi_5 - \phi_6}{x_3 + x_4 - x_5 - x_6} \tag{43}$$

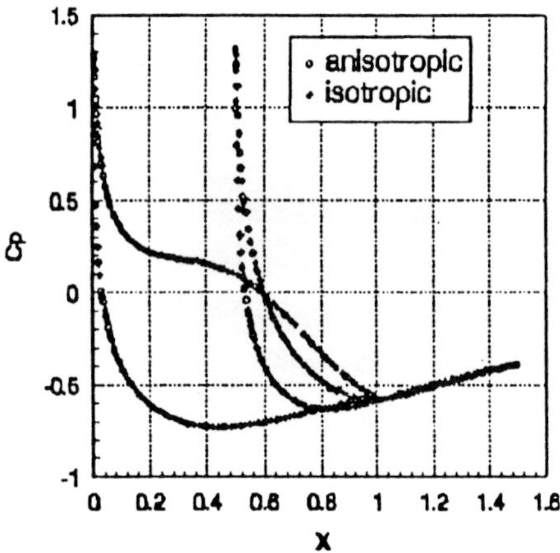

Fig.24: BI-NACA. Pressure distribution along the airfoil.

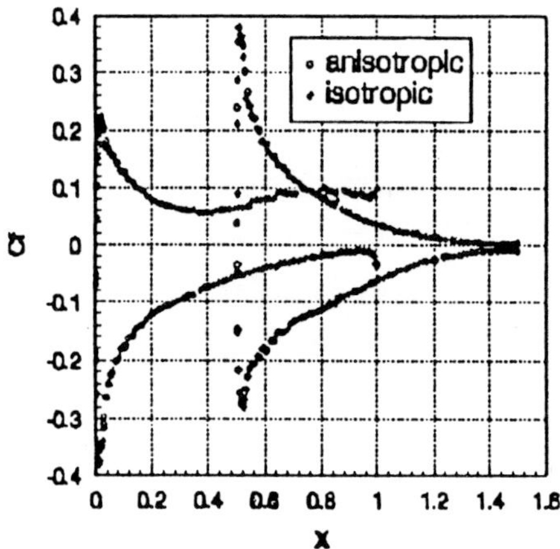

Fig.25: BI-NACA. Skin-friction coefficient along the airfoil.

If there is mesh refinement at the face GF (Fig.30-2), the interpolation formulas can be devised with some flexibility. The following formulas are independent of the type of interface and can uniform all the formulas.

For convenience, define $L(\phi : x_1, x, x_2)$ by

$$L(\phi : x_1, x, x_2) = \phi_1 + \frac{x - x_1}{x_2 - x1}(\phi_2 - \phi_1).$$

Compute $\bar{\phi}$. Let d be the intersection between the segment $\overline{23}$ and the horizontal line passing the point 1. The value at d is computed as $\phi_d = L(\phi : y_2, y_d, y_3)$ where $y_d = y_1$. The value $\bar{\phi}$ is then computed by linear interpolation using ϕ_1 and ϕ_d:

$$\bar{\phi} = L(\phi : x_1, x_m, x_d) \tag{44}$$

If the right face of 1 is not a refinement interface, then d and 2 are at the same location so that the

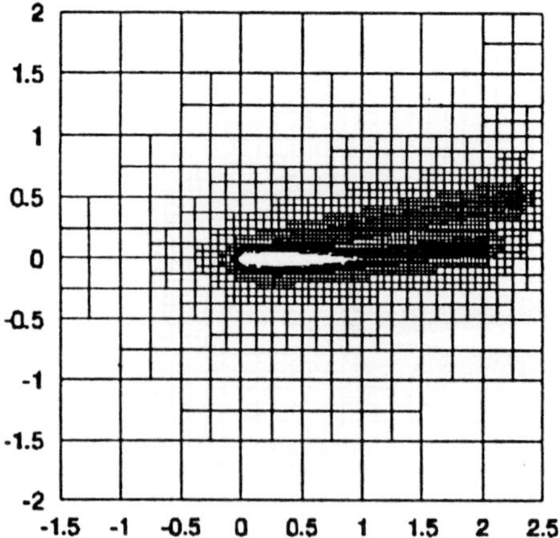

Fig.26: NACA0012 with $\alpha = 10^o$. Anisotropic grid.

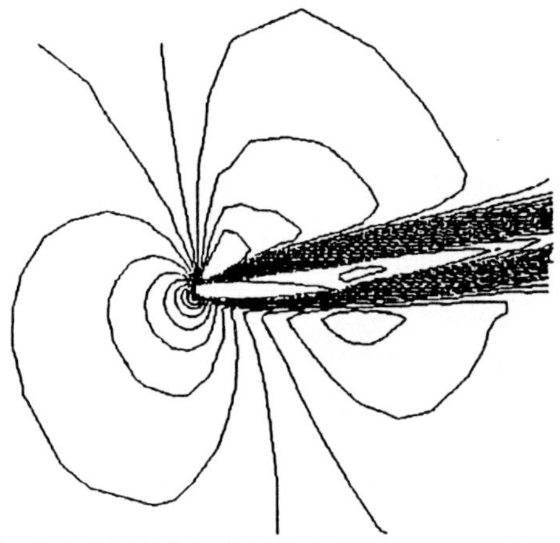

Fig.27: NACA0012 with $\alpha = 10^o$. Mach contours.

general formula (44) reduces to the simple formula (41). This unifies the formulas in both refinement and non-refinement faces.

Compute $\frac{\partial \bar{\phi}}{\partial x}$. The simplest formula is given by

$$\frac{\partial \bar{\phi}}{\partial x} = \frac{\phi_d - \phi_1}{x_d - x_1} \tag{45}$$

This formula has a second-order accuracy at the middle of the segment $1d$ but its accuracy drops to first order at the middle m of GF. We cannot do better if we want the scheme to be three point. Due to supraconvergence that often exits in mesh refinement problem, the global accuracy is expected to be still second order. This is indeed so, see [29].

If the right face of 1 is not a refinement interface, then d and 2 are at the same location so that the general formula (45) reduces to the simple formula

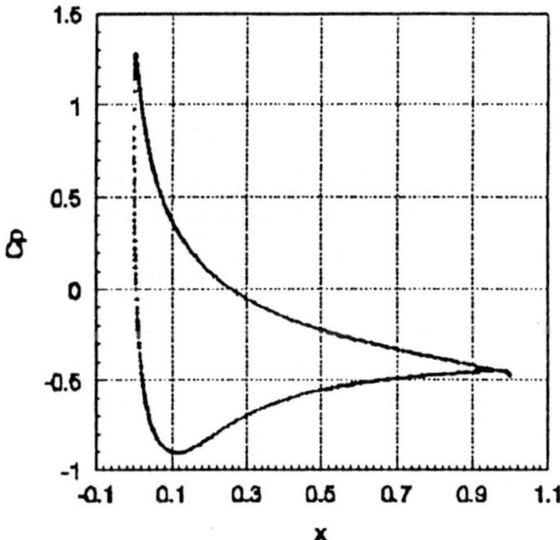

Fig.28: NACA0012 with $\alpha = 10°$. Pressure distribution.

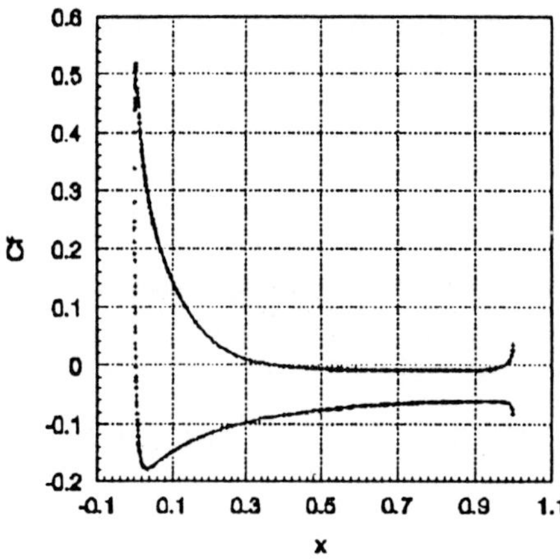

Fig.29: NACA0012 with $\alpha = 10°$. Skin-friction coefficient.

(42). This unifies the formulas for both refinement and non-refinement faces.

Compute $\overline{\dfrac{\partial \bar{\phi}}{\partial y}}$. First compute the intersections $\overline{(x_a, y_a)}, (x_b, y_b), (x_c, y_c)$ between the segment EF and the segments 52, 12, 43:

$$x_a = x_1 + \frac{1}{2}\Delta x_1, \quad y_a = L(y : x_5, x_a, y_2)$$

$$x_b = x_1 + \frac{1}{2}\Delta x_1, \quad y_b = L(y : x_1, x_b, x_2)$$

$$x_c = x_1 + \frac{1}{2}\Delta x_1, \quad y_c = L(y : x_4, x_c, x_3)$$

The values at a, b and c are computed as:

$$\phi_a = L(\phi : x_5, x_a, y_2),$$
$$\phi_b = L(\phi : x_1, x_b, x_2),$$
$$\phi_c = L(\phi : x_4, x_c, x_3) \qquad (46)$$

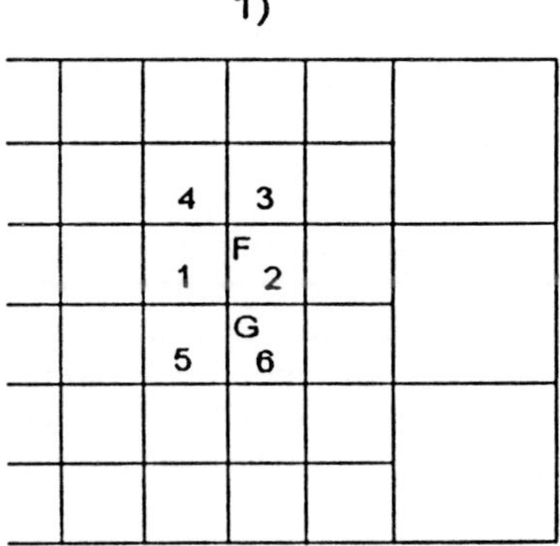

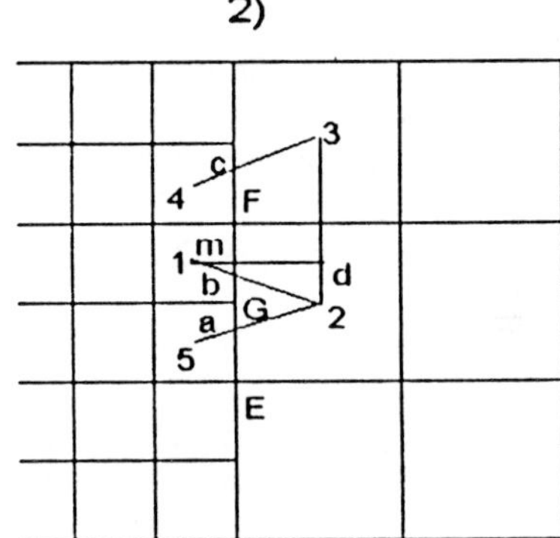

Fig.30: Interpolation at cell faces.

The y-derivatives at the center of ab and bc are calculated by the following formulas:

$$\frac{\partial \bar{\phi}}{\partial y}\Big|_{bc} = \frac{1}{x_c - x_b}(\phi_c - \phi_b),$$

$$\frac{\partial \bar{\phi}}{\partial y}\Big|_{ab} = \frac{1}{x_b - x_a}(\phi_b - \phi_a) \qquad (47)$$

The derivative at m is finally computed by:

$$\frac{\partial \bar{\phi}}{\partial y} = L\left(\frac{\partial \bar{\phi}}{\partial y} : y_{ab}, y_m, y_{bc}\right),$$

$$y_{ab} = \frac{1}{2}(y_a + y_b),$$

$$y_{bc} = \frac{1}{2}(y_c + y_b). \qquad (48)$$

If the cell 1 is not close to a refinement interface, then it is easy to see that the general formula (48) reduces to the simple formula (43). This unifies the formulas for both refinement and non-refinement faces.

A.2 Stability of the Lax-Wendroff scheme on a refined Cartesian grid

2.1 Model problem for analysis

Consider the scalar hyperbolic equation

$$u_t + au_x = 0 \tag{49}$$

The model grid that we consider is displayed in Fig.31. This grid models a refined Cartesian grid in one dimension. The grid is composed of L subgrids G_l, $l = 1, 2, \ldots, L$, and $L - 1$ interfaces $I_{l+\frac{1}{2}}$, $l = 1, 2, \ldots, L-1$. The subgrid G_l contains p interior meshes $j = 1, 2, \ldots, p$ and two interface points $j = 0, p+1$ with a uniform mesh size h_l satisfying $h_l = rh_{l-1} = h_0 r^l$. The refinement ratio r is equal to $\frac{1}{2}$ for a Cartesian grid. The subgrid width p is kept constant here.

The numerical solution at the point j of subgrid l is denoted $u_{j,l}$. The Lax-Wendroff scheme for a scalar and linear hyperbolic equation can be written as:

$$u_{j,l}^{n+1} = u_{j,l}^n - \frac{\lambda_l}{2}(u_{j+1,l}^n - u_{j-1,l}^n)$$
$$+ \frac{1}{2}\lambda_l^2(u_{j+1,l}^n - 2u_{j,l}^n + u_{j-1,l}^n), \tag{50}$$

where $\lambda_l = a\sigma_l$, $\sigma_l = k_l/h_l$, k_l being the time step. We use a local time stepping for steady state computation so that both λ_l and σ_l become constant. We therefore let $\lambda_l = \lambda$ and $\sigma_l = \sigma$.

On the model grid, the Lax-Wendroff scheme written on a Cartesian grid using interpolation defined in Appendix A reduces to (50) plus the following interface condition for the interface values $u_{0,l}$ and $u_{p+1,l}$:

$$\frac{u_{p,l-1} + u_{p+1,l-1}}{2} = \frac{u_{0,l} + u_{1,l}}{2},$$
$$\frac{u_{p+1,l-1} - u_{p,l-1}}{h_{l-1}} = \frac{u_{1,l} - u_{0,l}}{h_l}, \quad 2 \leq l \leq L. \tag{51}$$

Now we want to study the stability of the multiple interface problem (50)-(51) for long time integration. Precisely, we want to know if the existence of multiple refined interfaces changes the stability of the Lax-Wendroff scheme. Since a hyperbolic system can be diagonalised and the interface condition (51) is invariant under any linear transformation, the result based on a scalar analysis remains valid for a system of hyperbolic equations. We know that (51) is stable for a large class of dissipative and nondissipative schemes if only one interface exists [6, 17]. We also know that a stable single interface problem may become unstable when several interfaces are put together [26].

2.2 Necessary stability condition by energy method

The purpose here is to show that the existance of multiple interfaces indeed reduces the stability range of the Lax-Wendroff scheme. For this purpose, we define the l_2-norm of the solution by $\| u \|^2 = \sum_{l=1}^{L} \sum_{j=1}^{p} h_l u_{j,l}^2$. In order to obtain the result in a rapid way, we consider a subset $\mathbf{U}$ of all the possible numerical solutions. If $\| u \|_{u \in \mathbf{U}}^n$ increases unboudedly in time, then the problem is unstable. This allows us to obtain necessary but not sufficient condition for stability.

The subset $\mathbf{U}$ is defined as follows. Denote $\Phi_{j,l} = \overline{(u_{j,l} - u_{j+1,l})}^2$ and $\psi_l = \overline{(u_{1,l} - u_{p,l-1})}^2$ where the overline defines an average with respect to a short time. The subset $\mathbf{U}$ is defined by solutions for which $\overline{\Phi}_{j,l}$ is constant in each subgrid, that is, $\mathbf{U} \overset{def}{=} \{u_{j,l} : \overline{\Phi}_{j,l} = \overline{\phi}_l \ \forall j\}$ where ψ_l does not depend on j.

Using the interface condtion (51), we obtain the following recursive relations:

$$\overline{\psi}_l = \frac{(1+r)^2}{4}\overline{\phi}_{l-1} = \frac{(1+r)^2}{4r^2}\overline{\phi}_l. \tag{52}$$

Multiply the scheme by $2h_l u_{j,l}^n$, use (52) and the inequality $\sum h_{j,l}u_{j,l}^n u_{j,l}^{n+1} \leq \frac{1}{2}(\| u^{n+1} \| + \| u^n \|^2)$, and perform summation by parts, we obtain

$$\| u^{n+1} \|^2 - \| u^n \|^2 \geq \lambda(I_1 + I_2 + B + B_b) \tag{53}$$

where

$$I_1 = -\lambda \sum_{l=0}^{L} \sum_{j=1}^{p-1}(u_{j,l} - u_{j+1,l})^2$$

$$I_2 = -\lambda \sum_{l=1}^{L}[(u_{p,l-1} - u_{1,l})^2$$
$$+ \frac{1}{1+r}(u_{p,l-1}^2 - u_{1,l}^2)] \tag{54}$$

$$B = \sum_{l=1}^{L} \frac{a(1-r)}{1+r}(u_{p,l-1} - u_{1,l})^2$$

$$B_b = (\frac{\lambda}{h_0} + a)u_{0,0}u_{1,0}$$
$$+ (\frac{\lambda}{h_L} - a)u_{p,L}u_{p+1,L} \tag{55}$$

In the above expression, B_b is due to boundary treatment. When the number of subgrid is large, then the influence of B_b is negligible so that we simply assume $B_b = 0$.

Let $u_{j,l} \in \mathbf{U}$, then it is straightforward to verify that $\overline{I_1 + I_2 + B} > 0$ if

$$a > 0 \quad \text{and} \quad \sigma < \frac{1}{p}\frac{1-r}{1+r}\frac{p^{\frac{(1+r)^2}{4}}}{p-1+\frac{(1+r)^2}{4}}, \qquad (56)$$

This means that $\overline{\|u\|}^n$ is an increasing function in time even for a subset of the solution if (56) is satisfied. As a result, the l_2-norm will grow unboundedly after a long time integration so that the problem becomes unstable.

Condition (56) is only a necessary condition for long time stability. It only means that the lower bound of the stability range of the Lax-Wendroff scheme has been increased by multiple mesh refinements. We know that on a uniform grid the Lax-Wendroff scheme is stable for $0 \leq \sigma \mid a \mid \leq 1$.

2.3 Sufficient stability condition by eigenvalue analysis

Let $u_{j,l}^n = z^n \phi_{j,l}$ where $z \in \mathbf{C}$. Introduce this solution into the scheme (50) and the interface condition (51) for all j and all l, we obtain the following system

$$z M_1 Y = M_2 Y \qquad (57)$$

where M_1 and M_2 are two real matrices, Y is a column-vector with components $\phi_{j,l}$, $1 \leq l \leq L$, $1 \leq j \leq p$. System (57) is closed by a Dirichlet condition at the inflow boundary and a first-order extrapolation condition at the outflow boundary.

System (57) has in total $(p+1)L$ eigenvalues z_σ. The spectral radius $\rho = \max(\mid z_\sigma \mid)$ characterises the stability for long time integration. If $\rho \leq 1$, then the solution will be bounded; if $\rho > 1$, then the solution will be unbounded after long time integration; if $\rho < 1$, then the solution will converge to a steady state.

In Fig.32 we display the spectral radii for $\sigma = 0.1$ and $\sigma = 0.2$. We see that the stability also depends on the subgrid width p. When $\sigma = 0.1$, the problem becomes unstable for $1 \leq p \leq 6$. When $\sigma = 0.2$, the problem becomes stable for $p > 1$. It is interesting to compare the sufficient condition with the necessary condition (56). Let $p = 2$, then the necessary condition (56) requires $\sigma \geq 0.06$ for stability. While Fig.32 shows that σ should ly between 0.1 and 0.2 for stability. Let $p = 6$, then the necessary condition requires $\sigma \geq 0.0337$ and the sufficient condition is $\sigma > 0.1$. In consequence, the necessary condition (56) is only qualitatively significant and should not be used as a sufficient condition for stability.

In Fig.33 we display the spectral radii in function of σ. We obtain the following results: a) the lower bound of the stability range is near $\sigma = 0.12$, b) the

upper bound of the stability range is near 0.9, c) the minimal spectral radius, thus maximum convergence speed, occurs near $\sigma = 0.8$.

2.4 Conclusion

In consequence, the multiple mesh refinement interfaces not only change the lower bound but also the upper bound of the stability range. However, this change is relatively small.

In the above analysis, we have assumed that $a > 0$, that is, the wave moves in the direction of mesh refinement. If the wave moves in the direction of mesh coarsening, then the stability range is not reduced.

We also remark that the concept of using a subset of all the possible solutions to find a necessary stability condition seems to be new. It would be possible to extend such a concept to more general stability problems by proceeding in the following way: construct subsets for which the stability can be easily analyzed, prove that a linear combination of the subsets converges to the whole solution space and show that the problem is stable for each subset.

REFERENCES

[1] M. J. Berger and P. Colella, *Local adaptive mesh refinement for shock hydrodynamics*, J. Comput. Phys., **82** (1989), pp.64-84.

[2] M. J. Berger and R. LeVeque, *An adaptive Cartesian mesh algorithm for the Euler equations in arbitrary geometries*, AIAA Paper-89-1930 (1989).

[3] M. J. Berger and J. Oliger, *Adaptive mesh refinement for hyperbolic partial differential equations*, J. Comput. Phys., **53** (1984), pp.484-512.

[4] S.A. Bayyuk, K.G. Powell, and B. van Leer, *An algorithm for the simulation of 2-D unsteady inviscid flows around moving and deforming bodies of aribitrary geometry*, AIAA Paper-93-3391 (1993).

[5] M.O. Bristeau, R. Glowinski, J. Periaux, and H. Viviand (Eds), Notes on Numerical Fluid Mechanics, Vol. 18, 1987 (Numerical Simulation of Compressible Navier-Stokes Flows, GAMM Workshop, Nice, December 1985).

[6] G. Browning, H.-O. Kreiss and J. Oliger, *Mesh refinement*, Math. Comp., **27** (1973), pp.29-39.

[7] M. Ciment, *Stable difference schemes with uneven mesh spacings*, Math. Comp., **114** (1971), pp.219-226.

[8] D. K. Clarke , M. D. Salas and H. A. Hassan, *Euler calculations for multi-element airfoils using Cartesian grids*, AIAA J., **24**, pp.353-358

(1986).

[9] H. Cramer, Mathematical methods of statistics, Princeton Univ. Press, Princeton, N.Y., 1951.

[10] W.J. Coirier and K. G. Powell, *An accuracy assessment of Cartesian-mesh approaches for the Euler equations*, J. Comput. Phys., **117** (1995), pp. 121-131.

[11] W.J. Coirier and K. G. Powell, *Solution adaptive Cartesian cell approach for viscous and inviscid flows*, AIAA J., **34**, (1996), pp. 938-945.

[12] D. De Zeeuw and K. G. Powell, *An adaptively refined Cartesian mesh solver for the Euler equations*, J. Comput. Phys., **104** (1993), pp. 56-68.

[13] M. Goldberg and E. Tadmor, *Scheme-independent stability critiria for difference approximations of hyperbolic initial-boundary value problems. II, Math. Comp.*, **36** (1981), pp. 603-626.

[14] W. D. Gropp, *Local uniform mesh refinement with moving grids*, SIAM J. Sci. Stat. Comput., **8** (1987), pp. 292-304.

[15] B. Gustafsson, H.-O. Kreiss, and A. Sundström, *Stability theory of difference approximations for initial boundary value problems II*, Math. Comp., **26**(1972), pp.649-686.

[16] A. Lerat, *Multidimensional centered schemes of the Lax-Wendroff type*, CFD Review 1995 J. Wiley, pp.124-140.

[17] A. Lerat and Z. N. Wu, *Stable conservative multidomain treatments for implicit Euler solvers*, J. Comput. Phys, **123** (1996), pp. 45-64.

[18] R.J. LeVeque, *Cartesian grid methods for flow in irregular regions*, In Num. Meth. Fl. Dyn. III, K. W. Morton and M.J. Baines, Eds., Clarendon Press, (1988), pp. 375-382.

[19] M. S. Liou and K. H. Kao, *Progress in grid generation: from chimeria to dragon grids*, Frontiers of Computational Fluid Dynamics, Eds A. Caughey and M.M. Hafez, John Wiley & Sons, (1994), pp.385-411.

[20] R. A. Mitcheltree, M. D. Salas and H. A. Hassan, *Grid embedding technique using Cartesian grids for Euler solutions, AIAA J.*, **26** (1988), pp. 754-756.

[21] N. Satofuka, A. Nakano, and N. Shimomura, *Numerical simulations of compressible viscous flows using hierarchical Cartesian grid*, in the 15th Conf. Numer. Meth. Fl. Dyn., Monterey, USA, June 24-28, 1996.

[22] K. Morinishi, *A finite difference solution of the Euler equations on non-body-fitted Cartesian grids, Computers Fluids*, **21** (1992), pp.331-344.

[23] S. Osher, *Stability of parapolic difference approximations to certain mixed initial boundary value problems*, Math. Comp., **26** (1972), pp. 13-39.

[24] J. Quirk, *An alternative to unstructured grids for computing gas dynamic flows around arbitrarily complex two-dimensional bodies*, Computers Fluids, **23** (1994), pp.125-142.

[25] G. Starius, *On composite mesh difference methods for hyperbolic differential equations*, Numer. Math., **35** (1980), pp.241-255.

[26] L. N. Trefethen, *Stability of finite-difference models containing two boundaries or interfaces*, Math. Comp., **45**(1985), pp.279-300.

[27] J. M. Varah, *Stability of difference approximations to the mixed initial boundary value problems for parabolic systems*, SIAM J. Numer. Anal., **8**, (1971) pp.598-615.

[28] Z. N. Wu, *Uniqueness of steady state solutions for difference equations on overlapping grids*, SIAM J. Numer. Anal., **33** (1996) pp.1335-1357.

[29] Z. N. Wu, *A genuinely second-order accurate method for viscous flow computations around complex geometries*, in the 15th Conf. Numer. Meth. Fl. Dyn., Monterey, USA, June 24-28, 1996.

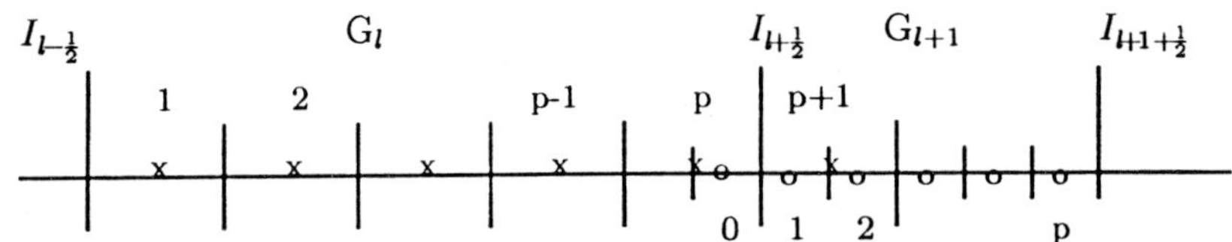

Fig.31: Multi-refined grid in one dimension.

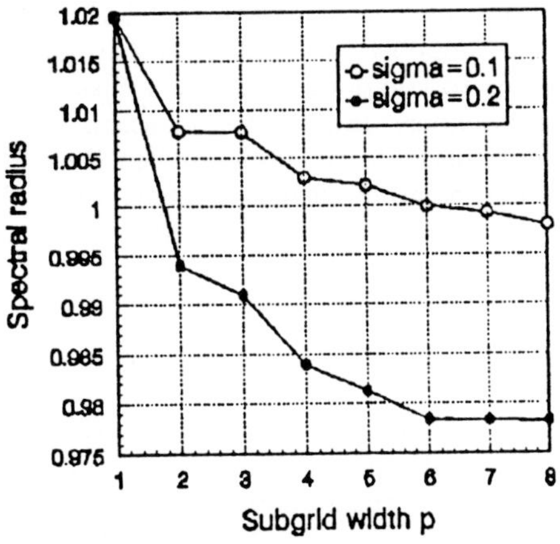

Fig.32: Influence of the subgrid width p on the spectral radius: $\sigma = 0.1, 0.2$.

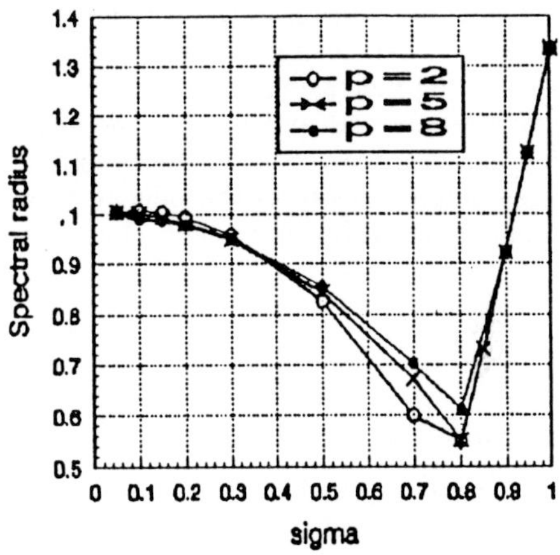

Fig.33: Influence of σ on the spectral radius.

NUMERICAL SIMULATION OF INCOMPRESSIBLE FLOWS AND ANALYSIS OF THE SOLUTIONS

Charles-Heri BRUNEAU

Mathématiques Appliquées de Bordeaux, Université Bordeaux 1;
351 cours de la Libération, 33405 Talence cedex, France
bruneau@math.u-bordeaux,fr

Abstract

The aim of this survey is to discuss some of the difficulties one can encouner both when solving Navier-Stokes equations for compressible flows by an obstacle and analyzing the approximate solutions. Far to be exhaustive, some main aspects of the numerical simulation are deliberately pointed out, in addition to the way the obstacle is taken into account and to the far field boundary conditions. Then, using one of the robust methods it is possible to simulate the transition to turbulence for increasing Reynolds numbers. That means to compute transient solutions which need to be analyzed and here is the second topic of this paper. Indeed, the classical tools like Fourier analysis are very efficient as long as the solutions is periodic but useless when the solution is more complex. Despite the development of wavelets and new algorithms it seems still difficult to distinguish quasi-periodic and chaotic solutions.

1   Introduction

It is nowadays quite impossible to review all the ways the researchers have found out all around the world and for thirty years to solve the Navier-Stokes equations for incompressible flows. There are now classical books devoted to these equations and their approximations [6,11,12,18,25,30,31]. There are also international conferences focusing globally or partially on this topic [17,29,23,9,16]. This shows the success of Navier-Stokes equations among the computational fluid dynamics community. Success that gives rise to a tremendous research activity and to so many papers the reader is overwhelmed. Therefore this paper does not pretend to give an exhaustive review of the field but only some comments on some aspects of the formulations, the boundary conditions, the approximations, the solving methods and also the analysis of the solutions. Indeed, using a method robust enough on a fine mesh at least in the boundary layer area, it is now possible to compute transient solutions quite easily in 2D and even in 3D when making the best of the new computers and the new computational techniques. That means that one has to use appropriate tools of analysis to qualify the computed solutions. As long as the solutions are periodic this is very easy by Fourier analysis but when they are more complex it is very difficult to analyze precisely the solution even in laminar cases.

Received on September 10, 1997.

2 Navier-Stokes models

2.1 The equations

From the mass and momentum conservation laws, it is easy to derive Navier-Stokes equations for an incompressible Newtonian viscous fluid in a domain $\Omega \subset \mathbb{R}^N$ with $N \leq 3$

$$\partial_t U + (U \cdot \nabla) U - \frac{1}{Re} \Delta U + \nabla p = F$$
$$in \; \Omega_T = \Omega \times (0, T) \quad (2.1)$$

$$div \, U = 0 \; in \; \Omega_T. \quad (2.2)$$

The first equation can be rewritten as

$$\partial_t U + (U \cdot \nabla) U - div \, \tilde{\sigma}(U, p) = F \; in \; \Omega_T \quad (2.3)$$

or

$$\partial_t U + (U \cdot \nabla) U - div \, \sigma(U, p) = F \; in \; \Omega_T \quad (2.4)$$

where $\tilde{\sigma}(U, p)$ and $\sigma(U, p)$ are respectively the pseudostress tensor and the stress tensor defined by :

$$\tilde{\sigma}(U, p) = \frac{1}{Re} \nabla U - p \, I \quad (2.5)$$

$$\sigma(U, p) = \frac{2}{Re} D(U) - p \, I$$
$$with \; D(U)_{ij} = \frac{1}{2} \left(\frac{\partial u_i}{\partial x_j} + \frac{\partial u_j}{\partial x_i} \right) \quad (2.6)$$

with $U = (u_i)_i$ the velocity vector, p the pressure, Re the dimensionless Reynolds number and F the external forces. Most often $F = 0$ and the motion is given through a non homogeneous Dirichlet boundary condition imposed on a part Γ_D of the boundary $\partial \Omega$.

These equations for the primitive variables can be transformed by introducing the vorticity $\omega = \nabla \wedge U$. The general form in 3D of the velocity-vorticity equations is

$$\partial_t \, \omega + (U \cdot \nabla)\omega - \frac{1}{Re} \Delta \omega = (\omega \cdot \nabla)U$$
$$+ \nabla \wedge F \; in \; \Omega_T$$

$$\nabla \wedge U = \omega \; in \; \Omega_T$$

$$div U = 0 \; in \; \Omega_T$$
$$\quad (2.7)$$

where the second part of the non linear term is treated as a source term in the first equation. In 2D this term vanishes. There are other models like the stream function-vorticity model which is valid only in 2D (see [14] for more details).

2.2 The initial datum

The evolution problem (2.1) (2.2) requires an initial condition

$$U(x, 0) = U_0(x) \; in \; \Omega \quad (2.8)$$

and from a mathematical point of view this initial datum must belong to the right space, in particular U_0 must a priori satisfy both the divergence-free condition and the boundary conditions. In practice it is not so easy to check the divergence-free condition and the numerical experience shows it is not compulsory. The first time steps will produce the good initial solution.
An other question related to the initial condition is the use of a high-order scheme in time requiring several initial data $U^{-j}(x), 0 \leq j \leq J$ for the first time steps. This can be solved either by setting $U^{-j}(x) = U_0(x)$ for $0 \leq j \leq J$ or by using an Euler scheme for these first time steps.

2.3 The boundary conditions

It is well-known that the boundary conditions constitute one of the main difficulties we can encounter. Here there are three types : inflow, no-slip and outflow or open boundary conditions. The first two correspond to non homogeneous and homogeneous Dirichlet conditions, the last one is much more difficult to find out in order to get a well-posed problem and a realistic approximate solution. These conditions are gathered for instance when computing the flow behind an obstacle in a channel (figure 1) where the domain Ω has for boundary $\partial\Omega = \Gamma_D \cup \Gamma_0 \cup \Gamma_1 \cup \Gamma_N$.

On Γ_D the flow at infinity (U_∞, p_∞) is imposed, that is a Poiseuille flow is set at the entrance section. On Γ_1 there is a no-slip condition $U = 0$ as well as on Γ_0 if the mesh is adapted to the limit of the obstacle. We shall see in the next section that there are other ways to take into account the obstacle. But the condition to set on Γ_N is far to be so easy. Indeed if Γ_D is not too close to the obstacle Ω_0, the Dirichlet condition is relevant at the entrance section and does not produce any perturbation. On the contrary, even when Γ_N is not so close to Ω_0 some boundary conditions can produce strong reflections when vortices are convected through the artificial limit. The treatment of the open boundary conditions for Navier-Stokes equations is itself a large field of research as nothing tell us what to do to get on the truncated domain the restriction of the solution on the infinite domain. There are essentially two ways of dealing with this difficulty which are either to use a buffer region outside of Ω in wich the equations are modified [8, 27] or to impose the best condition known on Γ_N. Many researchers have find out good

open boundary conditions and we refer to [28] and references therein for more details. One of the most used is probably the zero-stress boundary condition $\sigma(U, p)n = 0$ we generalize in [3] by

$$\tilde{\sigma}(U, p)n = \tilde{\sigma}(U_\infty, p_\infty)n$$

or

$$\sigma(U, p)n = \sigma(U_\infty, p_\infty)n$$

for Stokes flow and by for instance

$$\tilde{\sigma}(U, p)n + \frac{1}{2}(U \cdot n)^-(U - U_\infty)$$
$$= \tilde{\sigma}(U_\infty, p_\infty)n \quad (2.9)$$

or

$$\sigma(U, p)n + \frac{1}{2}(U \cdot n)^-(U - U_\infty)$$
$$= \sigma(U_\infty, p_\infty)n \quad (2.10)$$

for Navier-Stokes flows with the notation $a = a^+ - a^-$. On Γ_N, the new term is equal to zero except if $U.n$ is negative to ease the convection of vortices and avoid reflections. From a mixed formulation we can show by energy estimates that conditions (2.9) or (2.10) yield a well-posed problem [2].

2.4 The obstacle

To take into account the obstacle there are essentially two ways, either the mesh is constructed so that Γ_0 is approximated by the sides of some cells or an immersion method is used. In the first case a no-slip boundary condition $U = 0$ is imposed on Γ_0 and the computation is done on the unstructured mesh via a finite elements or a finite volumes approximation. In the second case a cartesian mesh is applied on $D = \Omega \cup \Gamma_0 \cup \Omega_0$ and the approximation is achieved by means of finite differences or spectral methods. Of

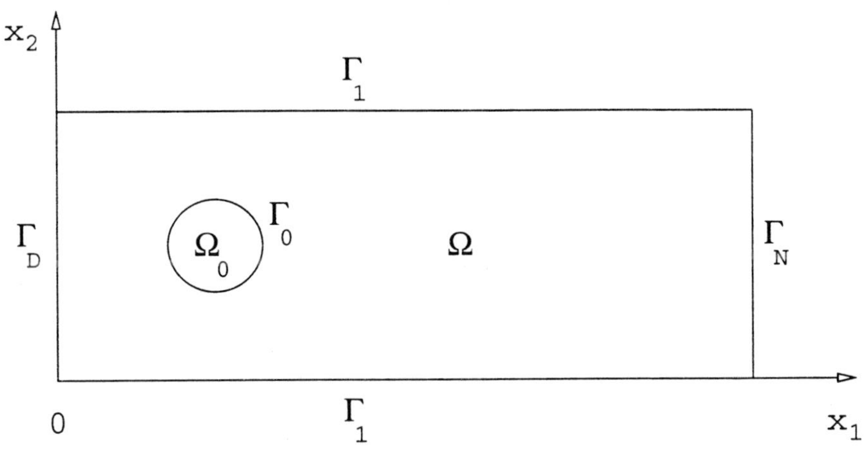

Figure 1: Computing domain

course, this needs an additional tool to represent the obstacle. One is to force the no-slip condition at the surface of the body by adding a feedback forcing function to the momentun equations. The points defining the surface are chosen by the user, they can be either the closest vertices or the intersection points between the body surface and the mesh. In this second case several interpolations can be used [13, 27]. Another tool is to consider the obstacle as a porous medium with a very small permeability coefficient K. This yields a fluid-solid formulation in the domain D by solving Navier-Stokes equations in the fluid and Darcy equations in the solid. One way to do that is to set $K = 1$ at every point in Ω and to set $K = K_{\Omega_0} \ll 1$ at every point in Ω_0. Then equations (2.1),(2.2) are replaced by

$$\partial_t U + (U \cdot \nabla)U - \frac{1}{Re}\Delta U + \frac{U}{ReD_a K^3}$$

$$+\frac{1}{K}\nabla p = F \ in \ D_T = D \times (0, T) \qquad (2.11)$$

$$divU = 0 \ in \ D_T \qquad (2.12)$$

where Darcy number is given by $D_a = \dfrac{1}{ReK_{\Omega_0}}$. It is clear that this adds an extra work as it requires to solve equations (2.11), (2.12) in Ω_0. But the cartesian mesh simplifies the computation and allows to use the spectral or multigrid methods. Moreover, the pressure in Ω_0 permits to compute the drag and lift forces [5]. Finally the velocity in Ω_0 is of the same order than K_{Ω_0}.

3 Numerical simulation

3.1 The approximation

It is obvious that we can not give here even an outline of the numerous types of approximations used to solve the Navier-Stokes equations. Indeed, this needs several books [6, 11, 12, 18, 25, 30, 31]. But , we can give a taste of the main difficulties which lay on one hand on the equilibrium between the convection and the diffusion terms and on the other hand on the divergence-free condition. It is now clear that we have to treat the convection term explicitly to avoid artificial numerical diffusion in time. Let us say that this term must be expressed at time $n\delta t$ when computing the solution at time $(n+1)\delta t$. On the contrary the other terms can be discretized implicitely at time $(n+1)\delta t$. The discretization in space is subjected to the mesh and thus to the way the obstacle is taken into account. The more used is probably the finite volumes approximation on unstructured meshes [24]. But with one of the immersion procedures it is possible to benefit of the spectral or finite differences methods ([19, 27] or [4]). As the convection term is put in the second member of the momentum equation (2.1) or (2.11), a centered discretisation of the other terms yields a well-conditioned matrix easy to invert.

But the convection term needs some more work. A good discretization is needed to guarantee the success of the simulation. Indeed, one has to be very careful dealing with this term as every extra diffusion brought up by the scheme is added to $-\frac{1}{Re}\Delta U$ and changes the real value of the Reynolds number. For instance, the discretization of

$$u\frac{\partial u}{\partial x}$$

at point j in one dimension by a first-order upwind scheme

$$u_j^n\left(u_j^n - u_{j-1}^n\right)/\delta x \qquad (3.1)$$

if u_j^n is positive corresponds to a second-order approximation of

$$u\frac{\partial u}{\partial x} - \frac{\delta x}{2}\frac{\partial}{\partial x}\left(|u|\frac{\partial u}{\partial x}\right)$$

and thus adds a viscosity term that alters the Reynolds number. Consequently the simulation can be qualitatively correct but not quantitatively. It is well-known that the

critical Reynolds number corresponding to the first Hopf bifurcation for the driven cavity problem is not yet determined for sure. Because, for this problem, this first bifurcation occurs at high Reynolds number and therefore it is not easy to achieve a good accuracy. Then it is necessary to use a less diffusive scheme ; a possible choice is to re-place (3.1) by

$$u_{j-1/2}^n\left(4u_j^n - 5u_{j-1}^n + u_{j-2}^n\right)/3\delta x$$
$$- u_{j+1/2}^n\left(4u_j^n - 5u_{j+1}^n + u_{j+2}^n\right)/3\delta x \qquad (3.2)$$

if $u_{j-1/2}^n$ is positive and $u_{j+1/2}^n$ is negative. The results presented in this paper are obtained with such a scheme. For the driven cavity problem, it yields the hopf bifurcation for Re around 7500 which appears in recent work to be a good value of the critical Reynolds number for this problem [23, 16]. Other results are generally obtained with high order compact schemes.

Another key to the success is the approximation of the divergence-free condition and here again there are numerous methods to do it. For a finite elements approximation, many ways were developped to find a good approximate space and often the equation (2.2) is satisfied only in a weak sense on each element [30]. With finite differences, a easy way to approximate (2.2) is to use a centered discretization on a staggered grid (figure 2) so that $divU = 0$ can be written at the pressure point directly without interpolation [4].

But probably the most famous way to force the incompressibility is given by the duality method as described in the next subsection.

3.2   The convergence methods

The last point is to find the whole method of resolution which insures a good performance

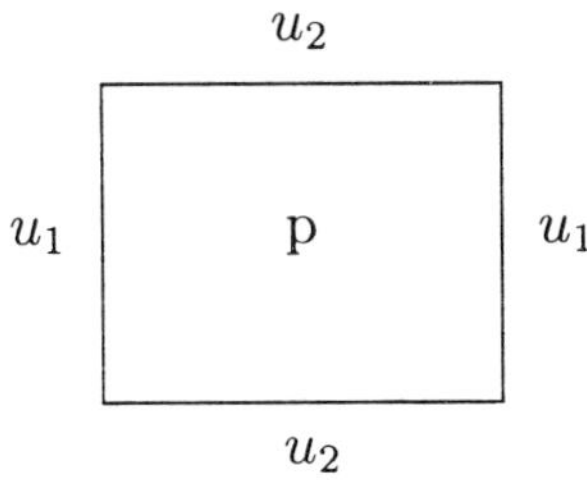

Figure 2: A staggered cell

necessary to observe the long time behaviour of the solution. As already pointed out, the duality method is one choice. Then, the pressure plays the role of the Lagrange multiplier and is computed by Uzawa's algorithm. Coupling this to a good gradient type method to invert the linear system a good performance can be achieved.

Another choice is to use the decomposition of the solution in its different scales. The new and now well-known nonlinear Galerkin method consists in cutting these scales into two parts, the large and the small ones. Then, the original Navier-Stokes equations are splitted into two parts to better represent the relative behaviour of the two types of scale. The result is an improvement of the performance obtained with a classical method whatever the approximation is [21]. Linked also to the different scales, the multigrid method is a very strong tool [1, 15]. Indeed, by using successive grids it is possible to capture very fast the scales related to each grid. On one hand the solution is computed on a really coarse mesh to get the large scales and on the other hand the finer the grid is the smaller the scales can be reached [4, 33].

This choice of method is decisive. Indeed to make a direct simulation of the transition to turbulence it is necessary both to use a very fine mesh in the boundary layer and to compute the solution for a long time. Even in 2D, this can require several days of computing time on the best work stations.

4 Numerical tests and analysis of solutions

The numerical tests presented here correspond to the domain Ω of figure 1. The channel is the rectangle $(0, L) \times (0, 1)$ with $L = 3$ or 4 and the obstacle is a circle of radius 0.2 which center is located at point $(1, 0.5)$. We recall Re is a dimensionless Reynolds number. To get a meaningful Reynolds number, one has to multiply Re by the diameter of Ω_0 which is here $d = 0.4$. So a solution at $Re = 100$ corresponds to a solution at real Reynolds number 40. For low Reynolds numbers the numerical experiments are performed on a uniform grid of 256×64 cells which is fine enough to describe the solution. For instance, at $Re = 100$ there is a symmetric steady solution with a recirculating bubble behind the cylinder as we can see on the stream function isolines (figure 3).

Then, as Re increases, the steady solution looses its stability to the benefit of a purely periodic solution very stable for higher values of Re. The recirculation zone alternates from the top to the bottom of the

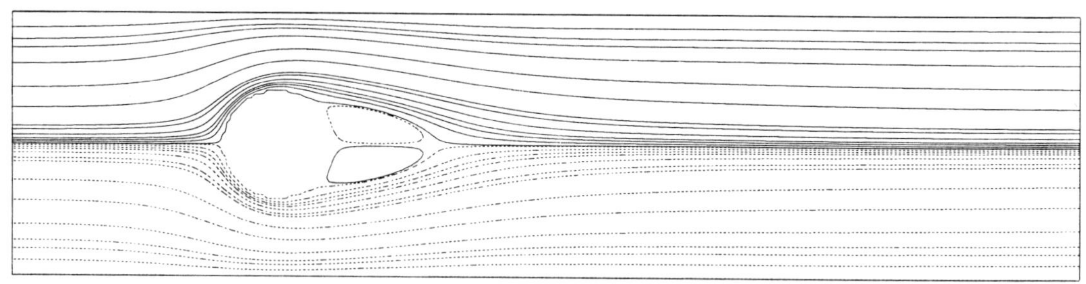

Figure 3: Solution at $Re = 100$

cylinder. We can detect quite accurately the critical Reynolds number corresponding to the first Hopf bifurcation. Indeed, for this simple geometry it corresponds to the loss of symmetry. The question is : Are we sure it is quantitatively correct ? To answer this question we can make another numerical test on the same geometry with $d = 0.2$ by applying an open boundary condition on Γ_1 instead of the no-slip condition. In this case a constant flow $U = (1, 0)$ is imposed on Γ_D instead of the Poiseuille flow of flowrate 1.

We then have a very well-known physical test and can compare the results with physical experiments. The values of the Strouhal number S_t for various real Reynolds numbers are in very good agreement with the physics [32] and assert the accuracy of the method (table 1). Nevertheless this comparison is possible only for low Reynolds numbers.

Coming back to the initial problem and using a finer mesh, we increase Re to reach other regimes. For $Re = 1000$, for instance, there is still a periodic solution but this time there are strong alternate vortices convected through the domain. We can then control that the open conditions (2.10) do not affect the solution computed on a shorter domain as it can be seen on the isolines of the vorticity field (figure 4). Indeed both solutions

are computed with exactly the same parameters. The only difference is the length of the domain $L = 4$ and $L = 3$. We see in particular that there is no reflections induced by the artificial boundary and that the computed solution on the shorter domain corresponds to the restriction of the solution computed on the larger domain at the same time (figure 4).

Then, increasing Re, there is still a periodic solution until $Re = 3700$ but with various behaviours. Indeed, a classical Fourier analysis reveals that approximately from $Re = 200$ to $Re = 3700$ the flow is periodic and exhibits the same main frequency $f_m \simeq 1$ (this value depends on the various parameters). But from $Re = 2200$ to $Re = 3600$ it appears two subharmonics corresponding to $f \simeq \frac{f_m}{3}$ and $\frac{2f_m}{3}$ and for $Re = 3700$ it appears seven subharmonics corresponding to $f \simeq \frac{f_m}{8}$ and its multiples (figure 5). Let us note that this qualitative behaviour, in particular the number of subharmonics, changes with the geometry of the obstacle. For instance, a square of side length 0.4 does not give the same subharmonics.

Until now, the Fourier analysis is a very efficient tool that gives very accurately the frequencies of a time signal corresponding to the value of one component of the velocity at a chosen point of the domain behind the

Re	Reynolds number	Computed S_t	Value of S_t in [32]
300	60	0.130	0.136
500	100	0.164	0.164
800	160	0.188	0.186

Table 1: Comparison of the Strouhal number

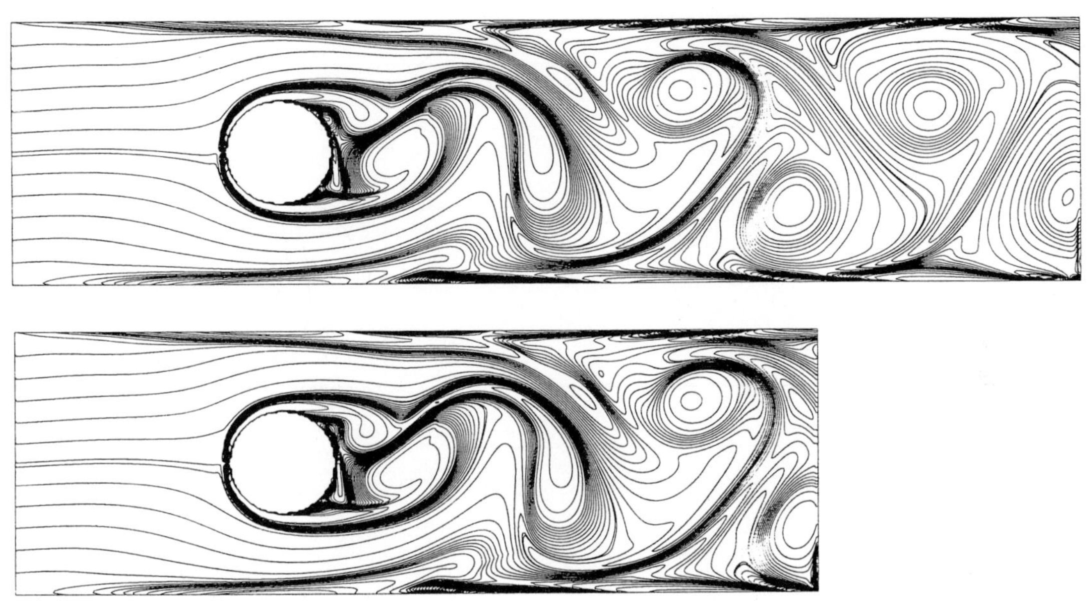

Figure 4: Comparison of the solutions obtained on the domain Ω with $L = 4$ and $L = 3$

cylinder. Of course, the behaviour of the solution and thus the spectrum does not depend of the point. We can complete the analysis with a phase portrait that corroborates the presence of subharmonics as the same curve is drawn several times. We see on figure 6 that a wavelets analysis is not so accurate even if the main frequency f_m and its subharmonics are detected at $Re = 3700$.

At $Re = 3800$, a long time simulation shows us the solution alternates between two states. It is well-known that the wavelets are very efficient to detect a dis-continuity [7, 22, 26] but here there is a slow transition between the two states much more difficult to analyze as it contains a large part of the spectrum. On figure 7 are represented the time-frequency analysis obtained with both a windowed Fourier transform and an adapted wavelets transform. We see the two methods provide about the same informations but as soon as there is a transition the analysis is spoiled.

When the Reynolds number increases we get more complex solutions. We can see the field becomes more complex and looks chaotic (figure 8). Nevertheless the different

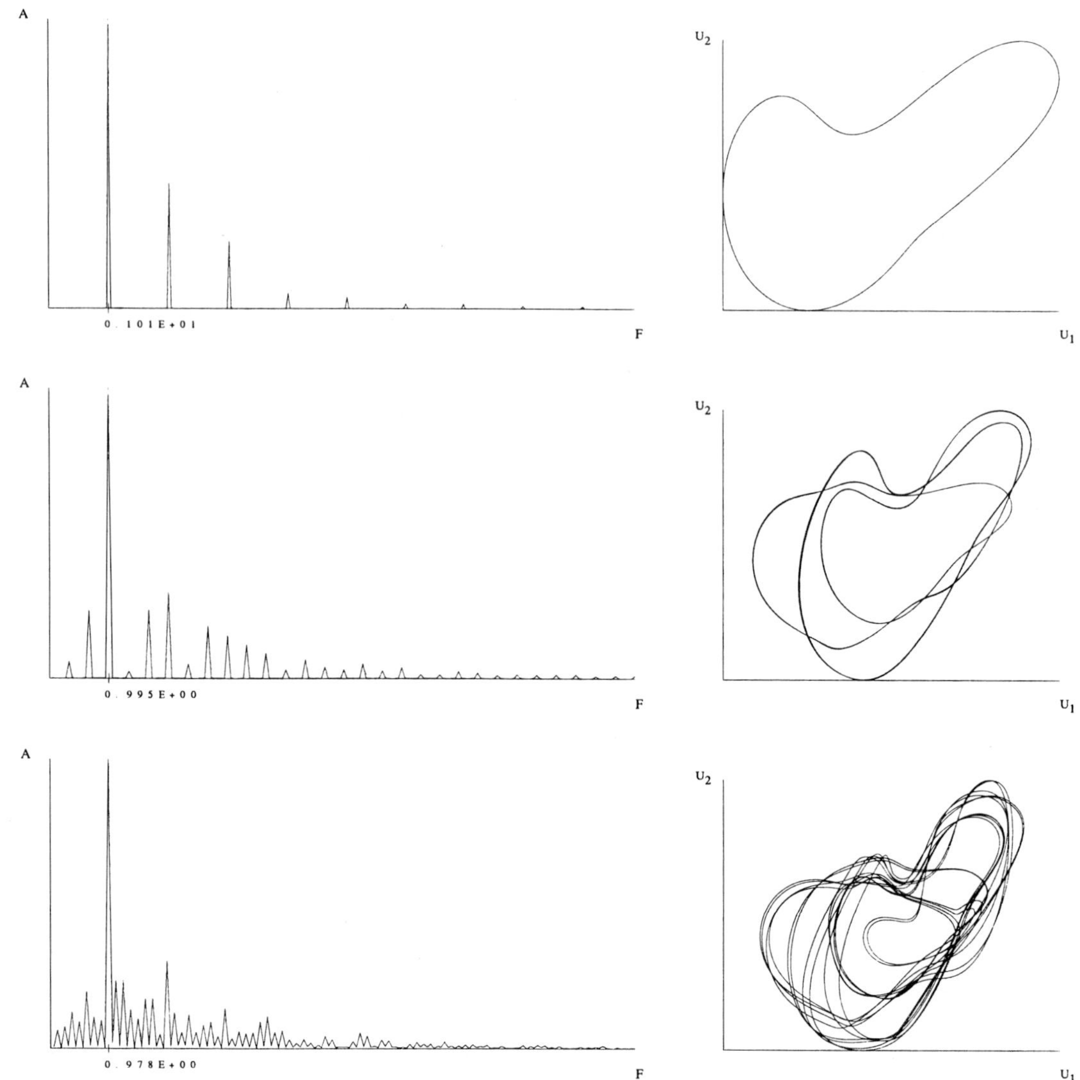

Figure 5: Spectrum and phase portrait for $Re = 1000$, 2500 and 3700

analysis give very few informations except the presence of the main frequency which changes with time. We think that a matching pursuit procedure [20] can give some more informations providing a good dictionary adapted to the kind of signal we analyze is used.

Another question is to understand the behaviour of the vortices, how they are convected through the domain for high Reynolds numbers, if they can merge or be divided into two parts ? Here again, some work has been done by means of wavelets essentially to compress two-dimensional turbulent flows [10]. The result shows that a 2D field can be analyzed with this tool and well represented. It is now necessary to see if such an analysis from time to time can

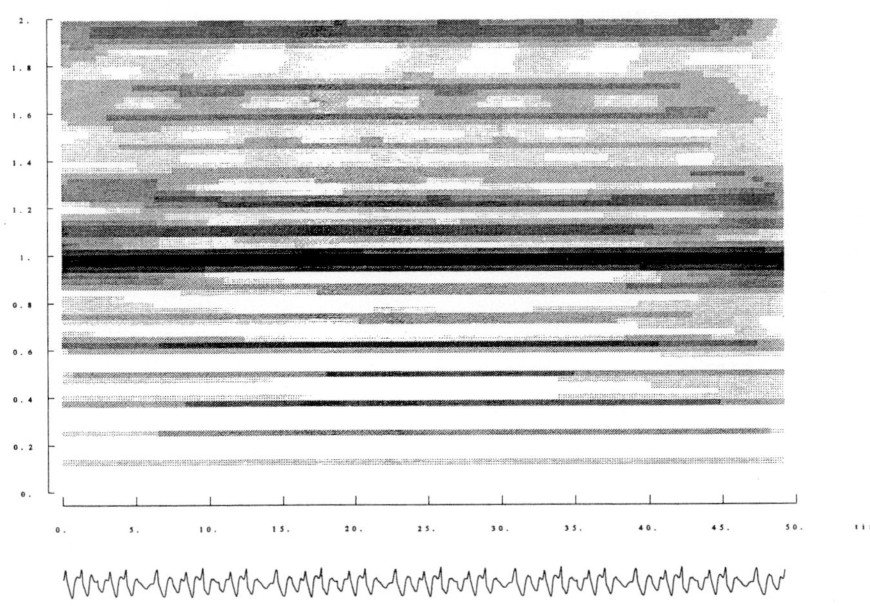

Figure 6: Wavelets analysis of the signal at $Re = 3700$

give enough informations to understand the behaviour of the whole field.

5 Conclusion

With the improvement of both the numerical techniques and the computing power it is now possible to compute directly from the Navier-Stokes equations good transient solutions in route to turbulence. However it is still difficult to qualify the computed solutions with the existing tools of analysis.

6 Acknowledgements

We would like to thank Roger Gay, Bruno Torrésani and Jean-François Delage for fruitful discussions and the numerical tests on wavelets analysis.

References

[1] A. Brandt, *Multigrid techniques Guide with applications to fluid dynamics*, GMD-Studien 85 (1984).

[2] C.-H. Bruneau, P. Fabrie, *New efficient boundary conditions for incompressible Navier-Stokes equations : A well-posedness result*, M^2AN 30 (1996).

[3] C.-H. Bruneau, P. Fabrie, *Effective downstream boundary conditions for incompressible Navier-Stokes equations*, Int. J. Num. Meth. Fluids 19 (1994).

[4] C.-H. Bruneau, C. Jouron, *An efficient scheme for solving steady incom-*

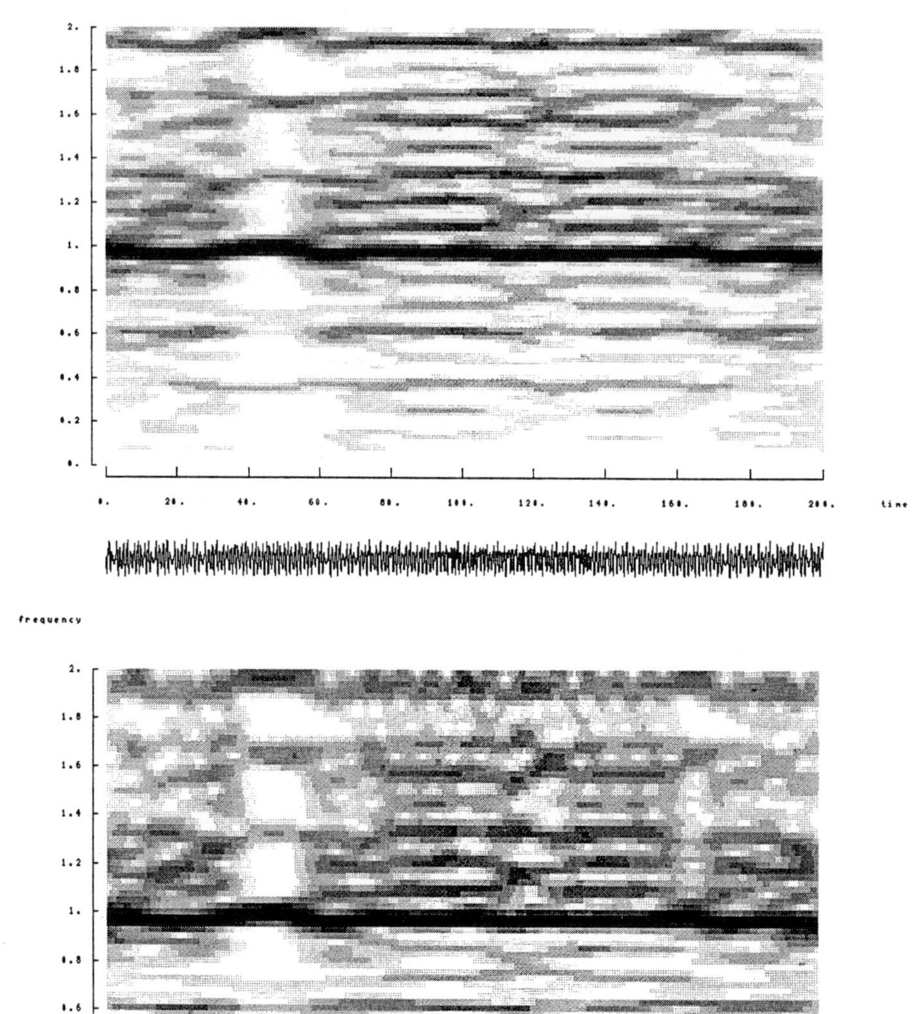

Figure 7: Windowed Fourier and wavelets analysis of the signal at $Re = 3800$

pressible Navier-Stokes equations, J. Comp. Phys. 89, $n^0 2$ (1990).

[5] J.-P. Caltagirone, *Sur l'interaction fluide-milieu poreux : Application au calcul des efforts exercés sur un obstacle par un fluide visqueux*, CRAS 318,

Figure 8: Solution at $Re = 10000$

série 2 (1994).

[6] C. Canuto, M.Y. Hussaini, A. Quarteroni, T.A. Zang, *Spectral methods in fluid dynamics*, Springer-Verlag (1988).

[7] C. Chui, *Wavelets : A tutorial in theory and applications*, Academic Press (1992).

[8] G. Danabasoglu, S. Biringen, C.L. Streett, *Application of the spectral multidomain method to the Navier-Stokes equations*, J. Comp. Phys. 113, $n^0 2$ (1994).

[9] S.M. Deshpande, S.S. Desai, R. Narasimha (Eds), *Numerical methods in fluid dynamics*, Proceedings of the 14th Int. Conf. , Lecture notes in physics 453 (1995).

[10] M. Farge, E. Goiraud, Y. Meyer, F. Pascal, M.V. Wickerhauser, *Improved predictability of two-dimensional turbulent flows using wavelet packet compression*, Fluid Dyn. Res. 10 (1992).

[11] C.A. Fletcher, *Computational techniques for fluid dynamics*, Vol.1 & 2, Springer-Verlag (1991).

[12] V. Girault, P.-A. Raviart, *Finite element methods for Navier-Stokes equations : Theory and algorithms*, Springer-Verlag (1986).

[13] D. Goldstein, R. Handler, L. Sirovich, *Modeling a no-slip flow boundary with an external force field*, J. Comp. Phys. 105, $n^0 2$ (1993).

[14] P.M. Gresho, *Incompressible fluid dynamics : Some fundamental formulation issues*, Ann. Rev. Fluid Mech. 23 (1991).

[15] W. Hackbush, *Multigrid methods and applications*, Springer-Verlag (1985).

[16] M. Hafez (Ed), *Computational fluid dynamics*, Proceedings of the 6th Int. Symp. (1995).

[17] J.G. Heywood, K. Masuda, R. Rautwann, S.A. Solarnikov (Eds), *The Navier-Stokes equations - Theory and numerical methods*, Proceedings of the 2nd Conf. , Lecture notes in math. 1530 (1992).

[18] Ch. Hirsch, *Numerical Computation of internal and external flows*, Vol. 1 & 2 (1990).

[19] P. Le Quere, T. Alziary de Roquefort, *Computation of natural convection in two-dimensional cavities with Chebyshev polynomials*, J. Comp. Phys. 57 (1985).

[20] S. Mallat, Z. Zhang, *Matching pursuits with time-frequency dictionaries*, Report 619 Courant Institute (1992).

[21] M. Marion, R. Temam, *Navier-Stokes equations : Theory and approximation*, Handbook of Numerical Analysis (to appear).

[22] Y. Meyer, S. Roques (Eds), *Progress in wavelet analysis and applications*, Proceedings of the Int. Conf. (1992).

[23] M. Napolitano, F. Sabetta (Eds), *Numerical methods in fluid dynamics*, Proceedings of the 13th Int. Conf. , Lecture notes in physics 414 (1993).

[24] S. V. Patankar *Calculation procedure for viscous incompressible flows in complex geometries*, J. Numer. Heat Transfer 14, $n^0 3$ (1988).

[25] R. Peyret, T.D. Taylor, *Computational methods for fluid flow*, Springer-Verlag (1983).

[26] M.B. Ruskai, G. Beylkin, R. Coifman, I. Daubechies, S. Mallat, Y. Meyer, L. Raphael (Eds), *Wavelets and their applications*, Jones and Barlett Publishers (1992).

[27] E.M. Saiki, S. Biringen, *Numerical simulation of a cylinder in uniform flow : Application of a virtual boundary method*, J. Comp. Phys. 123 (1996).

[28] R.L. Sani, P.M. Gresho, *Résumé and remarks on the open boundary condition minisymposium*, Int. J. Num. Meth. in Fluids, 18 (1994).

[29] C. Taylor, J.H. Chin, G.M. Homsy (Eds), *Numerical methods in laminar and turbulent flow*, Vol.7, Part 1 & 2, Proceedings of the 7th Int. Conf (1991).

[30] F. Thomasset, *Implementation of finite elements methods for Navier-Stokes equations*, Springer-Verlag (1981).

[31] R. Temam, *Navier-Stokes equations*, North-Holland (1984).

[32] C.H. Williamson, *Oblique and parallel modes of vortex shedding in the wake of a circular cylinder at low Reynolds numbers*, J. Fluid Mech. 206 (1989).

[33] N.G. Wright, P.H. Gaskell *An efficient multigrid approach to solving highly recirculating flows*, Comput. Fluids 24, $n^0 1$ (1995).

SOLUTION-ADAPTIVE METHODS
FOR ELLIPTIC GRID GENERATION

Kenichi MATSUNO

Dept.of Mech. and System Engng.; Kyoto Institute of Technology,
Matsugasaki, Sakyoku, Kyoto 606, JAPAN

Abstract

Adaptive grid methods of control function approach in elliptic grid generation developed by the author are described in this article. A use of a modified solution instead of actual solution is shown to be effective for estimation of a weight function related to the control function. In the development of the methods, a use of upwind formulation and a locally-variable relaxation parameter method have been introduced for the elliptic equation, which are described in detail in this article. An orthogonality of the grid and grid adaption cannot be strictly applied simultaneously. The adaptive method of elliptic type produces badly-skewed grid in many cases. To overcome this problem, an overset adaptive-grid method is introduced, which combines the overset grid method with adaptive-grid method. In this article, the developments of the adaptive grid methods are presented with numerical examples.

1. Introduction

Developments of computers and numerical methods have made it possible to simulate practical flow problems with reasonable accuracy[1]. However, since the computer memory and processor speed are still limited we cannot use enough grid points for complex flow problems. It is also well known that the accuracy and resolution of the numerical solutions are strongly dependent on the grid generated in the flowfield. Thus it is very important in Computational Fluid Dynamics to generate the grid of high quality. In generating the grid of high quality, there exist two important issues. The one is how to efficiently fit the grid to the body boundaries of complex geometry.

The other important issue is how to concentrate the grid in the high-gradient regions of the flowfield. A solution-adaptive method is the one which make it possible to concentrate the grid automatically in such regions. There are two approaches of the grid adaption: grid refinement and moving grid. The grid refinement approach[2] is the method which alters the number of the grid point by locally adding or removing points in the initial grid when they are needed according to the flow physics. As a consequence of local point addition and subtraction, the overall grid structure becomes complicated and hard to implement the existing flow solvers. In contrast, the moving grid approach is the method which moves the grid point according to the flow physics retaining grid structure. The overall grid structure is never changed, and thus the implementation of the any existing flow solver is achieved without any change if it is written on general curvilinear coordinate system. The moving grid method is simpler in the data structure and easier to implement the existing flow solvers than the grid refinement method. The one of the drawbacks of moving grid method is that the local resolution is achieved with the expense of depreciating other regions because of the fixed number of the grid points. However, this drawback is not serious if the adaptive gridding works perfectly, because the depreciated region is inherently the region of very small gradient of flow physics.

There are many solution-adaptive methods of the moving-grid approach[3]. Among them, the adaptive methods using a system of elliptic equations are relatively popular since the elliptic equation generates smooth grid distribution due to the inherent maximum principle. The solution-adaptive grid method of elliptic type uses the source term to control the grid distribution so that the grid density is large at the large-gradient region. The one of the drawbacks of the adaptive grid method of elliptic type is the grid skewness. The solution adaptive grid method generates possibly highly skewed grid, for example, in the case that the shock wave stands in the diagonal direction of the mesh lines. This often happens in the single grid structure.

Received on March 15, 1997.

"

There exist some review papers on the moving grid approach, for example,[3][4]. In this article, the adaptive method developed by the author[5] is mainly described. Some development strategies, modified solution method, use of upwind formulation, and locally variable relaxation parameter method, are also described as for the adaptive gridding. These methods will be able to be applied to general elliptic equation. To overcome the drawback of the grid skewness in elliptic adaptive method the author's recent approach, the overset adaptive-grid method, is described in the final part of this article.

Here we describe the method briefly using one-dimensional space and after that we extend the method to three dimensions.

2. Adaptive-Methods for Elliptic Grid Generation: Basic Method of Approach

Now we consider the one-dimensional space, whose coordinate is denoted by s. The starting point has been the Poisson equation for grid generation that has been well described in Ref.[6]. This elliptic grid generation method has been applied to various types of geometry. The elliptic equation proposed by Thomas and Middlecoff[7] is written in the one-dimensional form as:

$$\xi_{ss} = (\xi_s)^2 P(\xi). \tag{1}$$

Here P is a so-called "control function", which controls grid point distribution. This equation can be transformed to ξ coordinate by interchanging the role of dependent and independent variables. This yields following elliptic equation for one-dimensional grid generation,

$$s_{\xi\xi} + Ps_\xi = 0. \tag{2}$$

The purpose of grid adaption is to find functions for P which adapt the grid on properties of the solution and still retain the known attributes of the above Poisson equation. It is necessary for the adaptive method to generate the grid such that the grid spacing is small when the derivative of flow physics is large and vice versa. This property is described by the equation, defining a weight function w,

$$w\Delta s_i = C \,(\text{Constant}). \tag{3}$$

Here, $\Delta s_i = s_{i+1} - s_i$ is the grid spacing at grid index i. The equation (3) may be consider as a finite-difference equation of the following differential equation when we regard the term Δs_i as $\Delta s_i = \Delta s_i / \Delta \xi$ with $\Delta \xi = 1$.

$$ws_\xi = const. \tag{4}$$

To avoid the estimation of the constant in the right-hand side of Eq.(4), we differentiate Eq.(4) with respect to ξ. Then we get the following one-dimensional equation:

$$s_{\xi\xi} + \left(\frac{w_\xi}{w}\right) s_\xi = 0. \tag{5}$$

Equation (5) is the differential form for Eq.(3).

Comparing Eq.(5) with Eq.(2), the control function P relates with weight function w as follows:

$$P = w_\xi / w. \tag{6}$$

Thus we get the elliptic equation for one-dimensional adaptive-grid generation,

$$s_{\xi\xi} + Ps_\xi = 0, \qquad P = w_\xi / w. \tag{7}$$

In order to be able to solve an equation such as (7) for realistic conditions there must be restrictions put on P which depend on the choice of w.

A central difference approximation is usually employed in numerical solution of the Poisson equation. If the central deferential approximation is employed with Eq.(7), the resulting difference equation can be written

$$r_i - 1 = -P_i \frac{1}{2}(r_i + 1) \tag{8}$$

where

$$r_i = \frac{\Delta s_{i+1}}{\Delta s_i}, \quad \Delta s_i = s_i - s_{i-1}. \tag{9}$$

The most basic constraint on the grid spacing ratio, r_i in Eq.(8), is that it must be positive for any value of i,

$$0 < r_i, \tag{10}$$

and this leads directly to

$$|P_i| = |(w_\xi / w)_i| < 2. \tag{11}$$

This restriction, (11) can be relaxed when an upwind-like formulation is used, which is described in the next section. For flow-solution methods, it is unfavorable from the viewpoint of solution-accuracy that the grid-spacing ratio of three consecutive points changes suddenly. Thus, it is natural that the user specifies an allowable limit of grid-spacing ratio of the adaptive grid. Now, let K be the maximum of the allowable grid-spacing ratio, then we get following inequality,

$$\frac{1}{K} \le r_i \le K. \tag{12}$$

Combining Eqs.(11) and (12) yields the following constraint on the differnce equation:

$$|P_i| = |(w_\xi / w)_i| \le \frac{2(K-1)}{K+1}. \tag{13}$$

One of the most useful forms for the adaptive weight function w is that form where it depends on the first derivative of the variable, f, to be adapted. A useful form of first derivative adaption is the total arclength of the variable f, and it can be written as

$$w = \sqrt{1 + b_1(f_s)^2}. \tag{14}$$

Here, f is any flow property such as pressure p, density ρ, Mach number M, and so on. The coefficient b_1 is a user-specified free parameter, which is related to grid spacing due to the effect of flow change.

Once we specify the allowable limit of grid-spacing ratio K, Eq.(12), the parameter b in Eq.(14) is no more free. The value of b_1 must satisfy the inequality at grid point i from Eq.(13),

$$(b_1)_i \leq \left[\frac{4\frac{K-1}{K+1}}{|\{(f_s)^2\}_\xi| - 4\frac{K-1}{K+1}(f_s)^2} \right]_i. \tag{15}$$

We have only one value of b_1. Therefore b_1 is determined by taking minimum in the lagest $(b_1)_i$, that is,

$$b_1 = \min\left[(b_1)_i : i = 1, 2, \ldots, imax\right] \tag{16}$$

where,

$$(b_1)_i = \left[\frac{4\frac{K-1}{K+1}}{|\{(f_s)^2\}_\xi| - 4\frac{K-1}{K+1}(f_s)^2} \right]_i. \tag{17}$$

If both the first and second derivatives of the flow property are considered in grid adaption, a useful formulation containing both the first and second derivatives of the function f is

$$w = w_1 w_2 = \sqrt{1 + b_1(f_s)^2}\sqrt{1 + b_2(f_{ss})^2} \tag{18}$$

When this weight function is used in the Poisson equation, the control function is

$$P = \frac{w_\xi}{w} = \frac{w_{1\xi}}{w_1} + \frac{w_{2\xi}}{w_2} \tag{19}$$

or, with the use of the grid spacing ratio restriction,

$$|P| = |P_1 + P_2| \leq \frac{2(K-1)}{K+1}. \tag{20}$$

The relative importance of first and second derivative adaption can be controlled by introducing the two constant K_1 (first derivative) and K_2 (second derivative) and the relative weight parameter θ, which have the properties

$$K_1 + K_2 = \frac{2(K-1)}{K+1}, \tag{21}$$

where

$$K_1 = \theta \frac{2(K-1)}{K+1} \quad \text{and} \quad K_2 = (1-\theta)\frac{2(K-1)}{K+1}. \tag{22}$$

The values of b_1 and b_2 in w follow in the same way as the minimum values obtained from equations

$$\begin{aligned}
(b_1)_i &\leq \frac{2K_1}{\left[|\{(f_s)^2\}_\xi| - 4\frac{K-1}{K+1}(f_s)^2\right]_i} \\
(b_2)_i &\leq \frac{2K_2}{\left[|\{(f_{ss})^2\}_\xi| - 4\frac{K-1}{K+1}(f_{ss})^2\right]_i}.
\end{aligned} \tag{23}$$

More generally, when we consider N kinds of properties in grid adaption simultaneously, then the weight function f is

$$w = w_1 w_2 \cdots w_N, \tag{24}$$

and the control function becomes

$$P = P_1 + P_2 + \cdots + P_N \tag{25}$$

The restrictions become, with relative importance θ_ℓ,

$$\begin{aligned}
K_1 + K_2 + \cdots &+ K_N \\
= \theta_1 \frac{2(K-1)}{K+1} &+ K_2 \frac{2(K-1)}{K+1} + \cdots \\
&+ K_N \frac{2(K-1)}{K+1} \\
= \frac{2(K-1)}{K+1}&,
\end{aligned} \tag{26}$$

where

$$\sum_{\ell=1}^{N} \theta_\ell = 1. \tag{27}$$

If the weight function w_ℓ is written as

$$w_\ell = \sqrt{1 + b_\ell \phi_\ell^2} \tag{28}$$

then the value b_ℓ can be determined by

$$(b_\ell)_i \leq \frac{2K_\ell}{\left[|\{\phi^2\}_\xi| - 4\frac{K-1}{K+1}\phi^2\right]_i}. \tag{29}$$

It is theoretically possible to reflect any kind of flow properties in the adaptive gridding through Eq.(??): first derivative, second derivative, pressure gradient, density gradient, or everything.

The use of adaptive methods described above does not preclude the use of purely geometric control terms in the Poisson equation for grid generation. We define the control function such that it include the purely geometrical part P_g and solution-adaptive part P_a as a linear combination

$$|P| = |P_g + P_a| \leq \frac{2(K-1)}{K+1}, \tag{30}$$

or, if we use the weight between P_g and P_a then

$$P = (1 - \theta)P_g + \theta P_a, \quad (0 \leq \theta \leq 1) \tag{31}$$

where θ is again a parameter which controls the weight between the geometrical and adaptive grids properties. This will be discussed in detail at the later section.

3. Use of Central-Upwind Hybrid Formulation for Elliptic Grid Generation

The use of the central difference approximation is the standard approximation for the elliptic equation. To ensure the positive grid spacing, $\Delta s_i > 0$, the control function must satisfy the restriction (11) when the central finite difference approximation is used in Eq.(7). As long as this restriction is satisfied, the negative grid spacing does not occur. However, the coefficient P is the function of the grid position s_i, and thus the Eq.(7) is a kind of nonlinear equation. We have to estimate the value P_i using s_i at the previous iteration step. The restriction (11) could happen to be destroyed in the Jacobi or Gauss-Seidel iteration process. In such a case, once the negative grid spacing is calculated the iteration may not converge and results in divergence. The equation (7) is a kind of advection-diffusion equation. The occurrence of negative value is actually the "wiggle" in the advection-diffusion equation. Therefore the use of the upwind difference scheme is effective[8]. The first order upwind scheme is, however, too dissipative. Thus the combination of the central and upwind scheme is preferable in the present situation. Here, we introduce the use of a central-upwind hybrid method for the elliptic adaptive-grid generation[9].

When we apply the central-upwind hybrid approximation[10] to the first derivative term and central difference approximation to the second derivative term, the finite difference approximation to Eq.(7) becomes

$$s_{i+1} - 2s_i + s_{i-1} \\ + P_i \left\{ \phi_i \left(s_{i+1} - s_i \right) + \left(1 - \phi_i \right) \left(s_i - s_{i-1} \right) \right\} = 0 \tag{32}$$

where ϕ_i is a parameter which shifts the finite difference approximation from central to upwind side according to the value P_i at the grid s_i and evaluated as

$$\phi_i = \begin{cases} \frac{1}{2} & \text{if } | P_i | \leq 2 - \epsilon \\ -\frac{1}{P_i} + \frac{1}{1 - \exp(-P_i)} & \text{otherwise.} \end{cases} \tag{33}$$

Here, ϵ is a small positive number and given by 0.01. Note that $\phi_i = \phi(P_i)$ and $\phi(-P) = 1 - \phi(P)$. The present central-upwind hybrid method approximates the advection-diffusion equation so as to minimise the numerical error without the wiggles[10]. Once the value of ϕ_i is determined with Eq.(33), the solution of Eq.(32) is assured to be $s_{i-1} < s_i < s_{i+1}$ even if the value of $| P_i |$ is larger than 2.

Combining Eq.(32) with Eq.(12) gives the restriction on P_i,

$$| P_i | \leq \frac{K - 1}{\tilde{\phi}_i + (1 - \tilde{\phi}_i)K}, \tag{34}$$

where

$$\tilde{\phi}_i = \max \left(\phi_i, 1 - \phi_i \right). \tag{35}$$

Combination the weight function

$$w = \sqrt{1 + b(f_s)^2}, \quad 0 < b, \tag{36}$$

and the restriction on the grid spacing ratio K, given by Eq.(12), gives the condition on b. The maximum value of b at the i grid point is

$$b_i = \begin{cases} \infty & \text{if } A_i \leq 0 \\ M_i / A_i & \text{otherwise} \end{cases} \tag{37}$$

where

$$A_i = \{ | f_s(f_s)_\xi | - M_i(f_s)^2 \}_i , \\ M_i = \frac{K - 1}{\tilde{\phi}_i + (1 - \tilde{\phi}_i)K}. \tag{38}$$

We choose b as $b = \min(b_1, b_2, \ldots, b_{imax})$. When we use the N combination of the flow properties for the weight function w

$$w = w_1(\varphi_1)w_2(\varphi_2) \cdots w_N(\varphi_N) \tag{39}$$

with

$$w_\ell(\varphi_\ell) = \sqrt{1 + b_\ell \varphi_\ell^2}, \quad \ell = 1, 2, \ldots, N, \tag{40}$$

then the control function becomes

$$P = \frac{w_\xi}{w} = \frac{w_{1\xi}}{w_1} + \frac{w_{2\xi}}{w_2} + \cdots + \frac{w_{N\xi}}{w_N} \tag{41}$$

The control function P must satisfy the following restriction at every grid point i,

$$| P_i | \leq \frac{K - 1}{\tilde{\phi}_i + (1 - \tilde{\phi}_i)K} = M_i. \tag{42}$$

Let P_ℓ be

$$P_\ell = \frac{w_{\ell\xi}}{w_\ell}, \tag{43}$$

then

$$P = P_1 + P_2 + \cdots + P_N \tag{44}$$

and

$$| P_i | \leq | P_{1i} | + | P_{2i} | + \cdots + | P_{Ni} |$$

$$\leq \theta_1 M_i + \theta_2 M_i + \cdots + \theta_N M_i$$
$$= M_i, \tag{45}$$

with

$$\sum_{\ell=1}^{N} \theta_\ell = 1, \quad 0 \leq \theta_\ell,$$
$$| P_{\ell i} | \leq \theta_\ell M_i. \tag{46}$$

Here θ_ℓ is a weight parameter which controls the effect of the property φ_ℓ. Again, the coefficient b_ℓ can be determined with

$$b_\ell = \min\left(b_{\ell 1}, b_{\ell 2}, \ldots, b_{\ell imax}\right),$$
$$b_{\ell i} = \begin{cases} \infty & \text{if } A_i \leq 0 \\ \theta_\ell M_i / A_{\ell i} & \text{otherwise,} \end{cases}$$
$$A_{\ell i} = \left\{ | \varphi_\ell(\varphi_\ell)_\xi | - M_i(\varphi_\ell)^2 \right\}_i, \tag{47}$$

Now we can specify the values K and θ_ℓ. The values P_ℓ and b_ℓ are determined automatically according to the specified K and θ_ℓ.

4. Modified Flow Solution for Weight Function

When the high-resolution Euler solver is applied to get the flow solution, the shock wave is captured in one or two grid spacings[12]. For that situation, the control function has large value at only two or three grid points and zero value at the other grid points. Such control function cannot realize smooth concentration of the grid points at the shock region. The control function works for at most these non-zero value grid points. To avoid such situation, the use of the modified distribution of the flow properties has been motivated by the consideration of making the variation of the control function wider and smoother[13]. The modification of the distribution of the flow variables means here to smear the obtained flow solution. Suppose f_i is an actual solution at the grid point i, then the modified solution $\tilde{f}_i$ is obtained by averaging,

$$\tilde{f}_i = \psi f_i + \frac{1-\psi}{N} \sum_{p=1}^{N} f_p, \quad 0 \leq \psi \leq 1, \tag{48}$$

where f_p is a value of f at the pth grid point of N points surrounding i, and ψ is the weight at the averaging. It is effective to get the modified smooth solution that the above averaging procedure is repeated several times. The control function P_g is estimated using the distribution of the modified $\tilde{f}_i$.

Now we show an example of the result which use modified flow solution for the estimation of the weight function. The test problem is the nonlinear Burgers' equation,

$$u_t + (u^2/2)_x = 0, \quad 0 \leq x \leq 1, \tag{49}$$

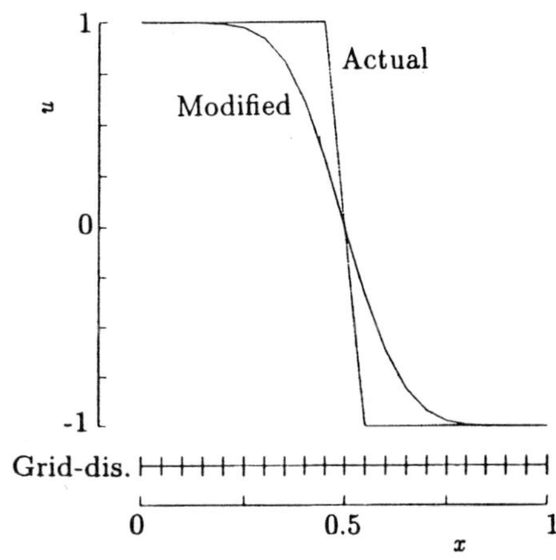

Fig.1: Actual and modified solution for Burgers' equation.

with the boundary condition,

$$x = 0 : u = 1 \text{ and } x = 1 : u = -1. \tag{50}$$

Initial condition and exact steady-state solution with $t \to \infty$ are

$$u(x, 0) = \begin{cases} 1 & 0 \leq x \leq 0.25 \\ 1 - 2\frac{(x-0.25)}{0.5} & 0.25 \leq x \leq 0.75 \\ -1 & 0.75 \leq x \leq 1 \end{cases}$$
$$u(x, \infty) = \begin{cases} 1 & 0 \leq x \leq 0.5 \\ -1 & 0.5 \leq x \leq 1 \end{cases} \tag{51}$$

The solution is solved by the Roe's flux-difference splitting scheme[11] of first order accuracy. Note that the Roe's flux-difference splitting scheme of first order accuracy captures discontinuity solution of steady state in one grid spacing in this case. The adaptive grid is generated for the steady-state solution. The modified solution for estimating weight function is determined by applying the smoothing procedure, Eq(48), ten times with $\psi = 0.5$. Figure 1 shows the actual solution and its modified solution. Figure 2(a) shows the adaptive grid distribution, where the weight function is estimated using actual solution. Figure 2(b) shows the adaptive grid distribution, where the modified solution is used to estimate the weight function. The first derivative is used to estimate the weighting function for adaptive gridding. From these figures, it is shown that the use of modified solution for estimation of weight function is very promising for adaptive gridding.

5. Three-Dimensional Formulation

The extension of the method described above to three dimensions is done to apply the above mentioned one-dimensional formulation straightforward to the three arclength directions defined by the generalized coordinates ξ, η, and ζ.

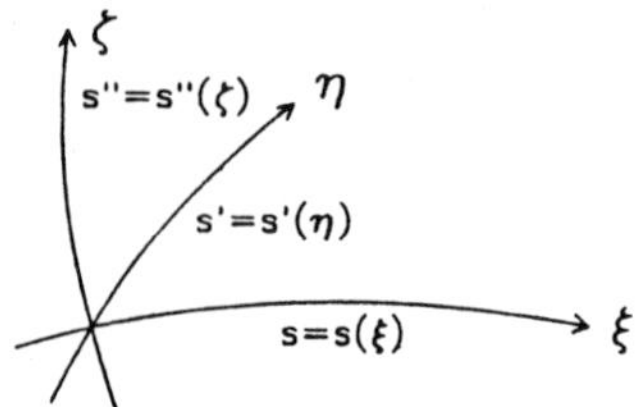

Fig.3 Primary arclengths and generalized coordinates

Let s, s', and s'' be the arclengths along the each constant grid lines shown in Fig.1. The basic criterion for grid adaption is

$$
\begin{aligned}
ws_\xi &= const., \quad s_\xi^2 = x_\xi^2 + y_\xi^2 + z_\xi^2 \\
w's'_\eta &= const., \quad s'^2_\eta = x_\eta^2 + y_\eta^2 + z_\eta^2 \\
w''s''_\zeta &= const., \quad s''^2_\zeta = x_\zeta^2 + y_\zeta^2 + z_\zeta^2
\end{aligned}
\tag{52}
$$

The three-dimensional forms of the Poisson equations for grid generation due to Thomas and Middlecoff[7] are

$$
\begin{aligned}
\xi_{xx} + \xi_{yy} + \xi_{zz} &= P(\xi,\eta,\zeta)\left(\xi_x^2 + \xi_y^2 + \xi_z^2\right) \\
\eta_{xx} + \eta_{yy} + \eta_{zz} &= Q(\xi,\eta,\zeta)\left(\eta_x^2 + \eta_y^2 + \eta_z^2\right) \\
\zeta_{xx} + \zeta_{yy} + \zeta_{zz} &= R(\xi,\eta,\zeta)\left(\zeta_x^2 + \zeta_y^2 + \zeta_z^2\right)
\end{aligned}
\tag{53}
$$

where P, Q, and R are control functions and defined by weight functions w, w', and w'' estimated respectively along the arclengths $s(\xi)'$, $s'(\eta)$, and $s''(\zeta)$.

$$
\begin{aligned}
P &= w_\xi/w \\
Q &= w'_\eta/w' \\
R &= w''_\zeta/w''.
\end{aligned}
\tag{54}
$$

Equations (53) can be inverted to give form which can be used for numerical solution to obtain the grid

$$
\begin{aligned}
&\alpha(\mathbf{r}_{\xi\xi} + P\mathbf{r}_\xi) + \beta(\mathbf{r}_{\eta\eta} + Q\mathbf{r}_\eta) + \gamma(\mathbf{r}_{\zeta\zeta} + R\mathbf{r}_\zeta) \\
&+2\kappa\mathbf{r}_{\xi\eta} + 2\lambda\mathbf{r}_{\eta\zeta} + 2\mu\mathbf{r}_{\zeta\xi} = 0,
\end{aligned}
\tag{55}
$$

where, $\mathbf{r} = [x, y, z]^T$ and

$$
\begin{aligned}
\alpha &= t_{11}^2 + t_{12}^2 + t_{13}^2 \\
\beta &= t_{21}^2 + t_{22}^2 + t_{23}^2 \\
\gamma &= t_{31}^2 + t_{32}^2 + t_{33}^2 \\
\kappa &= t_{11}t_{21} + t_{12}t_{22} + t_{13}t_{23} \\
\lambda &= t_{21}t_{31} + t_{22}t_{32} + t_{23}t_{33} \\
\mu &= t_{31}t_{11} + t_{32}t_{12} + t_{33}t_{13}
\end{aligned}
$$

$$
\begin{aligned}
t_{11} &= y_\eta z_\zeta - y_\zeta z_\eta, \\
t_{12} &= z_\eta x_\zeta - z_\zeta x_\eta, \\
t_{13} &= x_\eta y_\zeta - x_\zeta y_\eta, \\
t_{21} &= y_\zeta z_\xi - y_\xi z_\zeta, \\
t_{22} &= z_\zeta x_\xi - z_\xi x_\zeta, \\
t_{23} &= x_\zeta y_\xi - x_\xi y_\zeta, \\
t_{31} &= y_\xi z_\eta - y_\eta z_\xi, \\
t_{32} &= z_\xi x_\eta - z_\eta x_\xi, \\
t_{33} &= x_\xi y_\eta - x_\eta y_\xi.
\end{aligned}
\tag{56}
$$

The weight functions w, w', and w'' are defined the same manner as described in one-dimensional formulation. The first and second derivative formulations along the three arclength directions are

$$
\begin{aligned}
w &= \sqrt{1 + b_1(f_s)^2}\sqrt{1 + b_2(f_{ss})^2} \\
w' &= \sqrt{1 + b'_1(f_{s'})^2}\sqrt{1 + b'_2(f_{s's'})^2} \\
w'' &= \sqrt{1 + b''_1(f_{s''})^2}\sqrt{1 + b''_2(f_{s''s''})^2}.
\end{aligned}
\tag{57}
$$

The parameters b_1, b'_1, ..., and b'' are determined in a fashion similar the one-dimensional case by specifying the restrictions on the grid spacing ratios K, K', and K'' in three directions, respectively.

The additional question of grid orthogonality also must addressed for multi-dimensional case, and this places a significant restriction on adaption. This issue can be avoided by a new overset adaptive-grid method which will be discussed on the later section.

In some cases, especially in boundary layer flows, the grid orthognality near the wall boundary is very important and the orthogonal condition has been geometrically forced. The adaptive grid also must keep this geometric constraint. The way of inclusion of the geometric condition in control function has already been described in section 2.

Now, let the control functions P, Q, and R include the geometrical condition such as orthogonality near the wall and adaptive condition as the linear combination of the geometric part P_g and adaptive part P_a.

$$
P = (1-\theta)P_g + \theta P_a, \quad (0 \le \theta \le 1)
\tag{58}
$$

where θ is a parameter which controls the weight between the geometrical and adaptive grids property. The question is how to get P_g, Q_g, and R_g, which are usally the control function of the initial (original) grid for adaptive grid.

If the initial grid has been generated by another type of grid generation method, then we need to incorporate the information of the initial grid distribution in the control P_g, Q_g, and R_g. To obtain the information of the initial grid distribution, we inversely solve the

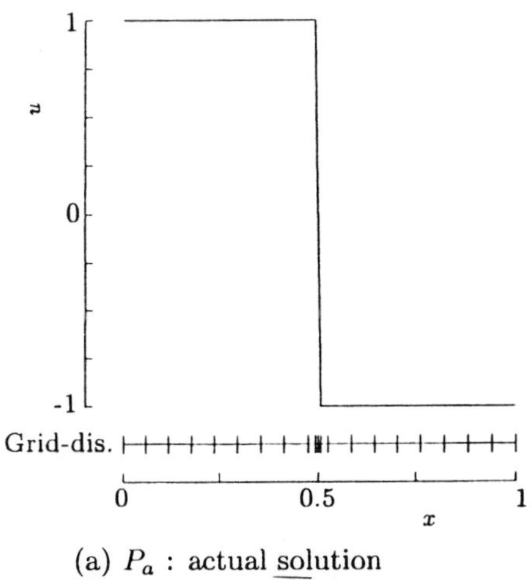

(a) P_a : actual <u>solution</u>   (b) P_a : modified solution

Fig.2: Adaptive grid for the Burgers' equation (21 points)

control function $P_{i,j,k}$, $Q_{i,j,k}$ and $Q_{i,j,k}$ from Eq(55).

$$(P_g)_{i,j,k} = -\frac{1}{\alpha J}\left[d_1 t_{11} + d_2 t_{12} + d_3 t_{13}\right]$$

$$(Q_g)_{i,j,k} = -\frac{1}{\beta J}\left[d_1 t_{21} + d_2 t_{22} + d_3 t_{23}\right]$$

$$(R_g)_{i,j,k} = -\frac{1}{\gamma J}\left[d_1 t_{31} + d_2 t_{32} + d_3 t_{33}\right] \quad (59)$$

where,

$$J = \frac{\partial(x,y,z)}{\partial(\xi,\eta,\zeta)} = \begin{vmatrix} x_\xi & x_\eta & x_\zeta \\ y_\xi & y_\eta & y_\zeta \\ z_\xi & z_\eta & z_\zeta \end{vmatrix}$$

$$\begin{aligned}
d_1 &= \alpha x_{\xi\xi} + \beta x_{\eta\eta} + \gamma x_{\zeta\zeta} \\
&\quad + 2\kappa x_{\xi\eta} + 2\lambda x_{\eta\zeta} + 2\mu x_{\zeta\xi}, \\
d_2 &= \alpha y_{\xi\xi} + \beta y_{\eta\eta} + \gamma y_{\zeta\zeta} \\
&\quad + 2\kappa y_{\xi\eta} + 2\lambda y_{\eta\zeta} + 2\mu y_{\zeta\xi}, \\
d_1 &= \alpha z_{\xi\xi} + \beta z_{\eta\eta} + \gamma z_{\zeta\zeta} \\
&\quad + 2\kappa z_{\xi\eta} + 2\lambda z_{\eta\zeta} + 2\mu z_{\zeta\xi}.
\end{aligned} \quad (60)$$

The derivatives in the above equations are estimated at the grid point (i,j,k) with the central difference approximations.

6. Locally Variable Relaxation Parameter

The adaptive-grid generation equation (53) can be solved by relaxation method, usually the line Jacobi method, and written symbolically as

$$\mathcal{L}\delta r = -\mathcal{R}(r), \qquad r^{m+1} = r^m + \omega \delta r \quad (61)$$

where m is the index of iteration, and ω is relaxation parameter of usually $1 \leq \omega \leq 2$.

The under-relaxation parameter is effective for robustness in the adaptive method when the elliptic Poisson equation is solved iteratively. It decreases the grid movement in each iteration process, especially in the case that the extremely high-gradient regions such as the shock wave presents. However, the under-relaxation parameter decreases the convergence rate. Hence in Ref.[9] the under-relaxation parameter has been used locally and the decrease of the total convergence rate has been able to be avoided. This is similar to the use of the local-time step techniques in the compressible flow solution method. In Ref.[13], the relaxation parameter ω is determined according to the local values of control functions, $P_{i,j,k}$, $Q_{i,j,k}$, and $R_{i,j,k}$ as,

$$\begin{aligned}
\mathcal{L}\delta r = -\mathcal{R}(r), \qquad r^{m+1} = r^m + \omega_{i,j,k}\delta r, \\
\omega_{i,j,k} = \text{func}(P_{i,j,k}, Q_{i,j,k}, R_{i,j,k}), \\
0 \leq \omega_{i,j,k} \leq 2,
\end{aligned} \quad (62)$$

The value $\omega_{i,j,j}$ is determined with

$$\omega_{i,j,k} = \begin{cases} 0.8 & \text{if } 1.9 \leq \max(P_{i,j,k}, Q_{i,j,k}, R_{i,j,k}) \\ 1.6 & \text{if } \max(P_{i,j,k}, Q_{i,j,k}, R_{i,j,k}) \leq 1/4 \\ 1 & \text{otherwise} \end{cases}$$

$$(63)$$

7. Two- and Three-Dimensional Applications

The procedure of the adaptive method is as follows:

1. Generate initial grid by a conventional method. Estimate the initial control function P_g, Q_g, and R_g, if necessary.
2. Solve flow equations on the initial grid and get f.

133

3. Calculate modified distribution $\tilde{f}$ from actual solution f.
4. Estimate weight function w, w', and w''.
5. Determine control function P, Q, and R.
6. Solve the elliptic equation (55) by a line relaxation method, and get adaptive grid.
7. Solve flow equations on generated adaptive grid.
8. If necessary, repeat **3** to **7**.

Pressure is selected as flow property f.

Here we show some examples of the application of the method described above. The applications were focused on capturing shock waves sharply.

The first example is the supersonic compression duct flow shown in Fig.4. Inflow Mach number is 2.0. The inviscid flowfield governed by the Euler equation written by conservation law form is solved by a cell-centered finite-volume scheme. The scheme employs the Roe's flux-difference splitting scheme combined with third-order MUSCL interpolation with minmod-limiter[12]. Figure 5 shows the initial grid of 127×44 which was made with equi-spacing grid. Figure 6 shows the pressure contours calculated on the initial grid. Figures 7 and 8 show the adaptive grid and the solution on it. The grid points are effectively concentrated at the high gradient region. The shock waves are calculated more sharply on the adaptive grid.

The next example is a three-dimensional application to the supersonic duct flow shown in Fig.9. The slope angle of the floor is 6°. At the ceiling of the duct the slope angle changes linearly from 7° to 9°. Inflow Mach number is 2.7. Boundary conditions include a supersonic Mach 2.7 uniform inflow, exit outflow extrapolation, and solid-wall tangency. The number of the grid points is $61 \times 24 \times 38$. Figure 10 shows the comparison of (a) nonadaptive grid and (b) generated adaptive grid. Comparison of the non-adaptive grid solution and adaptive-grid solution is shown in Fig.11 (a) and (b). In these figures, the computed pressure distributions are drawn. The interval of the isobars is 0.05, nondimensionalized by $\rho_{in} c_{in}^2$ (c: speed of sound) at the duct inlet. In the figures 10 and 11, grid and pressure contours on the three boundary surfaces are drawn. There exists only one plain interaction of two shock waves in the flowfield. Adaptive grid solution shows sharper shock waves than the non-adaptive grid solution. In this case, the shock waves cross diagonally grid cells. It is well known that the Roe's flux difference scheme with one dimensional MUSCL interpolation technique described above smears shock wave relatively wide range when the shock wave crosses diagonally the grid cells. Thus, the shock waves of the non-adaptive grid solution are highly smeared as can be seen in Fig. 11 (a). However, as can be seen in

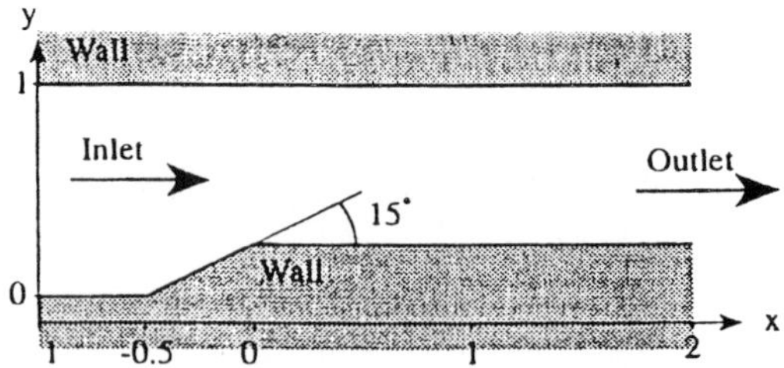

Fig.4: 2D supersonic compression duct geometry

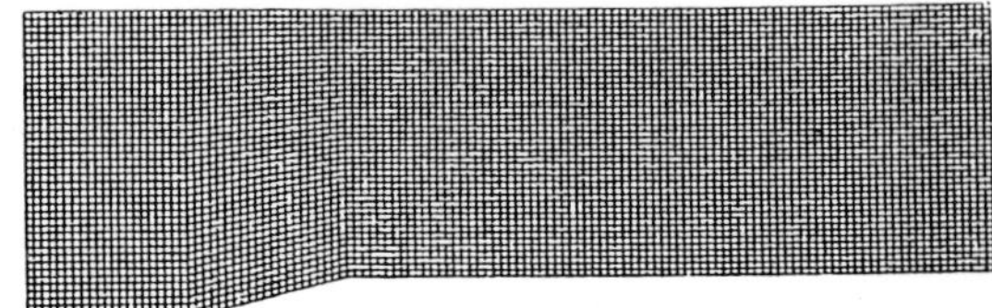

Fig.5: Initial (non-adapted) grid

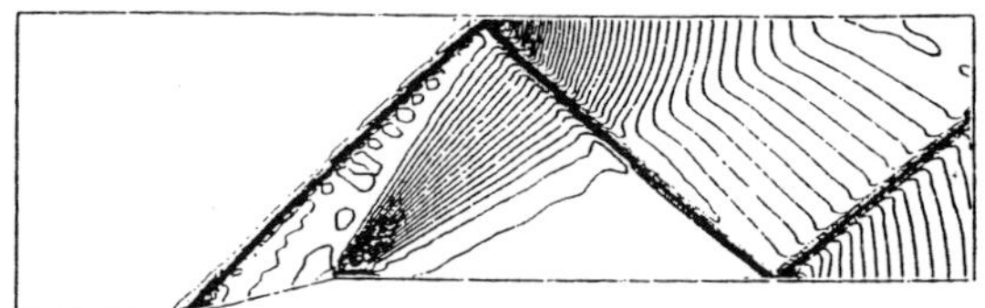

Fig.6: Non-adaptive grid solution, pressure contours($\Delta p = 0.05$)

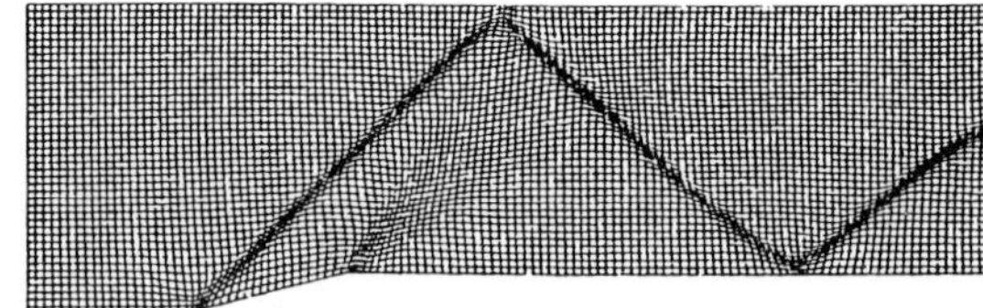

Fig.7: Adaptive grid

Fig. 11 (b), the adaptive-grid solution shows excellent shock wave thickness, under the half of the non-adaptive grid shock.

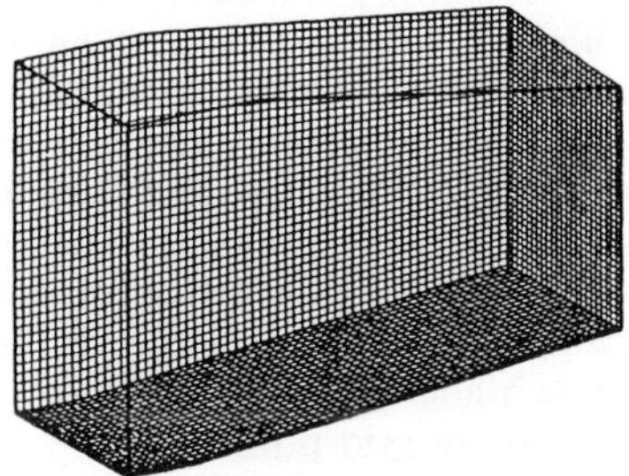

(a) Non-adaptive grid   (b) Adaptive grid
Fig.10: Grids of the three-dimensional compression duct, $61 \times 24 \times 38$

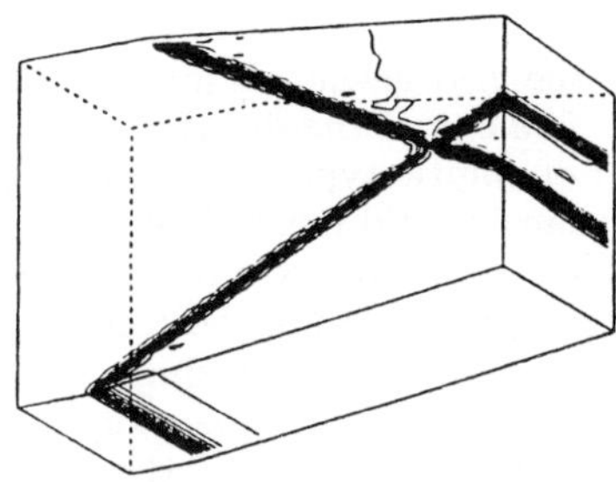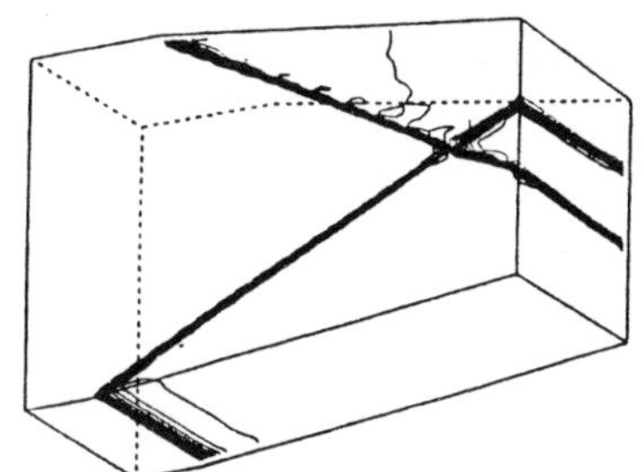

(a) Non-adaptive grid solution   (b) Adaptive-grid solution
Fig.11: Pressure contours of the supersonic compression duct flow

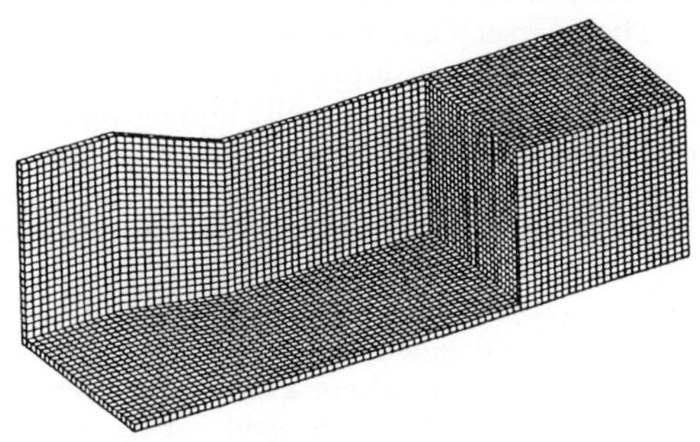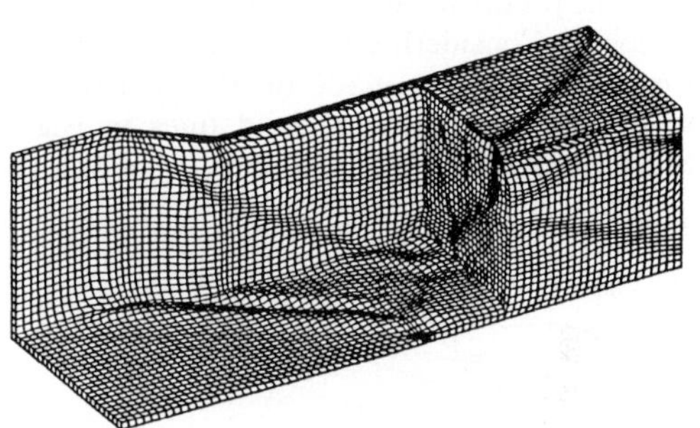

(a) Non-adaptive grid   (b) Adaptive grid
Fig.13: Grids of the three-dimensional squre duct, $79 \times 26 \times 26$

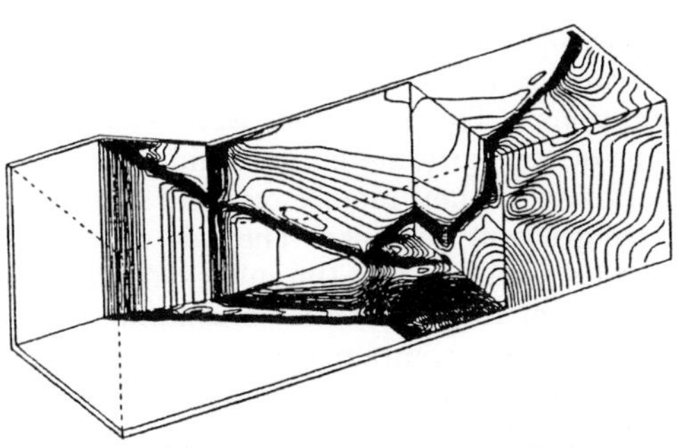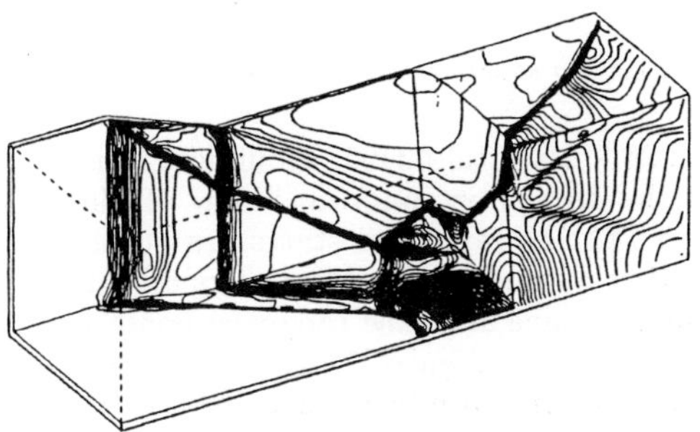

(a) Non-adaptive grid solution   (b) Adaptive-grid solution
Fig.14: Pressure contours of shock-shock interaction in the squre duct flow

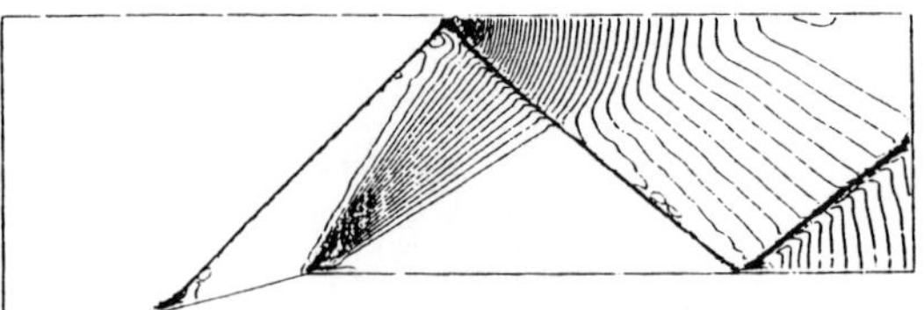

Fig.8: Adaptive grid solution, pressure contours($\Delta p = 0.05$)

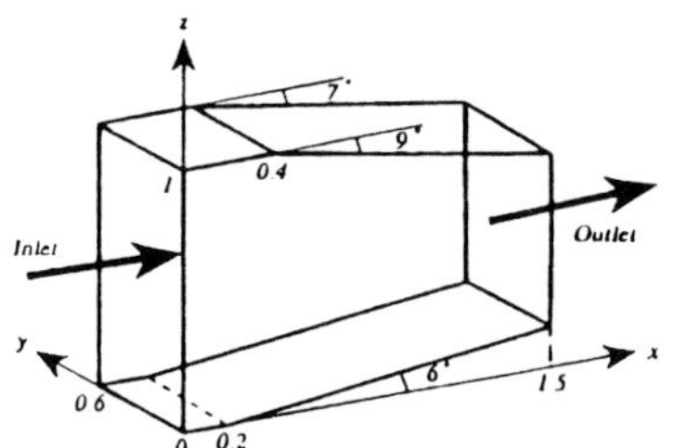

Fig.9: 3D supersonic compression duct geometry

Next application of the method is a simulation of the complex interactions of several shock waves. The geometry of the problem is shown in Fig.12. Four walls have 15° slopes. Considering the geometrical symmetry, the computation was made on the quarter part of the full geometry shown by solid lines in Fig.12. Inflow Mach number is 2.7. The number of the grid

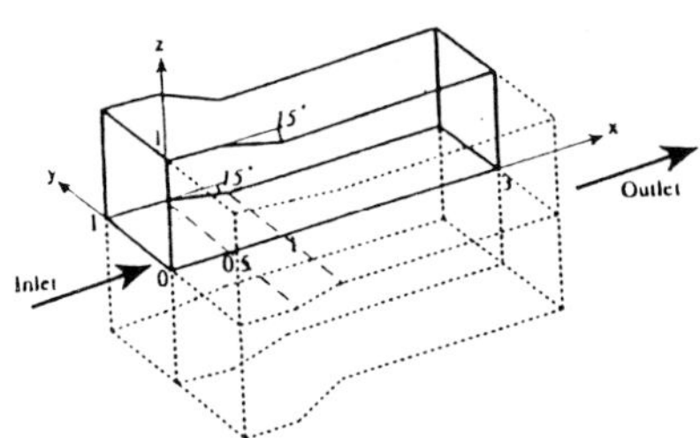

Fig.12: 3D square duct geometry

points is $79 \times 26 \times 26$, total 53404. Boundary conditions are; At the duct entry, uniform inflow of Mach 2.7 is specified. The extrapolated states are given as outflow condition at duct exit. Mirror condition is used at symmetry plane and flow tangency condition is specified at the walls. Figures 13 and 14 show the results. The solution is shown by pressure distribution in Fig.14, where the isobars are drawn with 0.1 interval. The figures include a partial cut of the region. In this case, several shock waves exist and interact each other. Again, the adaptive grid solution represents

shock waves sharply. As can be seen, the adaptive grid shows high skewness. This problem can be severe for some situations since orthogonality and adaption cannot be applied simultaneously. As for the efficiency, the necessary computing time is at least twice of the conventional method, since the method generates the grid and solves the flow twice for initial grid and solution and adaptive ones. However, the quality of the solution is higher than the non-adaptive grid solution of 8 times more grid points.

8. Overset Adaptive-Grid Methods

The solution-adaptive grid method of elliptic type uses the source term to control the grid distribution so that the grid density is large at the large-gradient region. The one of the drawback of the adaptive grid method of elliptic type is the grid skewness. The solution adaptive grid method generates possibly highly skewed grid, for example, in the case that the shock wave stands in the diagonal direction of the mesh lines. This often happens in the single grid structure. The orthogonality and grid adaption cannot be strictly applied simultaneously, and they must be traded off against each other. The one of the ideas to realize both the orthogonality and grid adaption simultaneously is to abandon the single grid topology[13].

The overset(Chimera) grid method[14] was originally developed to adjust the grid smoothly to the body surface in complex multi-body geometry by oversetting local subgrid around bodies on the main grid covering whole of the flow field. The overset grid method has been shown to be very useful to adjust the grid geometry locally to the complex body geometry[15]. The overset-grid method can be used to get fine grid locally. That is, the subgrid is overlaid on the region where gradient of the flow property is large on the main grid. This idea is to use the overset-grid method in solution-adaptive manner. Similar idea was used by Fujii[16] for tracking blast wave. Kao et al[17] proposed the idea that the use of the overset Chimera grid in the regions of large gradients to capture the salient features accurately during computations. These authors were successfully applied the overset-grid method to fluid dynamics problems. However, they merely overlaid the subgrid in the large-gradient regions. They do not apply any moving-grid techniques to the subgrids and main grid.

The idea of Ref.Mat95b is to combine the overset grid method with solution-adaptive grid method described above in order to generate adaptive grid of high quality and of high efficiency. That is, "overset adaptive-grid method"[13] is as follows. Firstly, the subgrids are overlaid in the high-gradient regions of the flowfield. Next the adaptive-grid method described above is applied to these overlaid subgrids. Thus, the overset adaptive-grid method adapts the grid twice to the flow solution. In other word, the overset-grid method is combined with the adaptive-grid method in order to get the solution adaptive grid of totally high quality and efficiency. The adaptive-grid method on single grid structure is often bothered by grid skewness. It is hard to generate orthogonal adaptive grid in multi-dimensions. However, the overset adaptive-grid method can avoid such bad skewness by overlaying subgrid adjusting to the flow direction or gradient. The overset adaptive-grid generation procedure is as follows. The body-fitted coarse main grid is firstly generated by the elliptic grid generation method in usual manner. The flow is solved on the coarse main grid as a base flow for overset gridding and adaptive gridding. The regions with large gradients of flow properties, such as shock waves, are detected. The subgrids which cover the large-gradient regions are generated and put on the large-gradient regions in the main grid. In this step, the each subgrid is not adapted to the flow solution but merely put on the large-gradient region. Next, the solution-adaptive method using elliptic equation described above is applied to the subgrids. Finally the flowfield is solved again on the main grid and subgrids simultaneously. The data transfer between the main grid and subgrid is done by bi- or tri-linear interpolation. To get the desirable adaptive properties, the solution-adaptive grid generation procedure is usually repeated or iterated twice or three times.

9. Overset Adaptive-Grid Applications

The first example is the same flow as in Fig.4, the two-dimensional supersonic compression duct flow. In this case, seven subgrids were overlaid in the regions of the shock waves and expansion fan, illustrated in Fig.15. Figures 16 and 17 show a "simple solution-

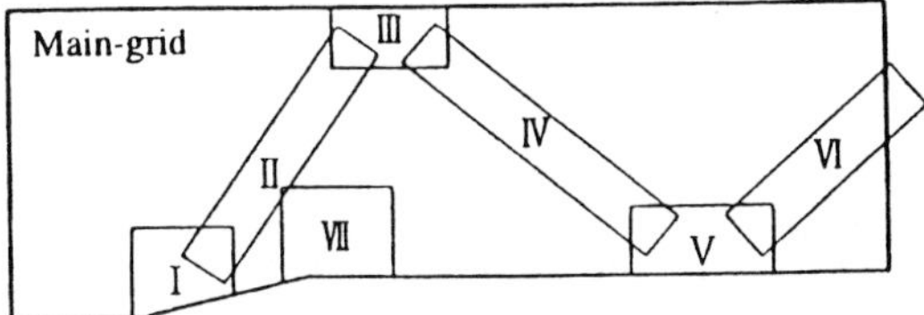

Fig.15: Layout of the overlaid subgrids

adapted overset" grid and its flow solution, where the adaptive grid was not applied to the subgrids. The subgrids were merely overlaid in the high-gradient regions. Figures 18 and 19 are the results of the overset

adaptive-grid, where the subgrids were not only overlaid on the high-gradient region but also applied the adaptive-grid method to. As can be seen from Fig.15, the subgrids II, IV, and VI are overlaid parallel to the shock waves. In consequence, the adaptive grids of that subgrids have excellent orthogonality, seen in Fig.18. It is also seen from the figures that the resolution of the shock wave is excellent in the overset adaptive-grid solution.

Three-dimensional overset grid applied to the same problem in Figs.9-11 is illustrated in Fig.20. The related overset adaptive grid is shown in Fig.21 and the pressure solution on it is shown in Fig.22. Again, the solution shows the excellent shock resolution.

For the comparison sake, the total number of the grid points of the above examples are the same as the single grid structure cases, respectively.

10. Concluding Remarks

As for the adaptive methods using the elliptic equation, the series of the research efforts by the author have been described in this paper.

Even if we can use enormous grid points in numerical simulations by the benefit of today's decline of memory price, the adaptive gridding is key issue in the CFD. The grid position itself is a part of the solution. It is the present author's opinion that adaptive methods offer the hope of significant improvement today and in the future for numerical simulation of fluid flow.

The use of the upwind formulation and the locally-variable relaxation parameter method have been developed for the elliptic grid generation in the studies. They, of cause, can be applied to numerical methods for general elliptic problems and will be valid and effective.

REFERENCES

[1] R.L.Meakin , 'Moving Body Overset Grid Methods for Complete Aircraft Tiltrotor', Simulation, AIAA-93-3350-CP, (1993).

[2] M.J.Berger and P.Colella, 'Local Adaptive Mesh Refinement for Shock Hydrodynamics' J.Comp.Phys., Vol.82, 1989, pp.64-84.

[3] J.F.Thompson, 'A Survay of Dynamically Adaptive Grids in the Numerical Solution of Partial Differential Equations, Appl. Numer. Math, Vol.1, 1985, pp.3-27.

[4] P.R. Eiseman, 'Adaptive Grid Generation', Comput. Meth. Appl. Mech. Engng, 1987, **64**, 321-376.

[5] K. Matsuno and H.A. Dwyer, 'Adaptive

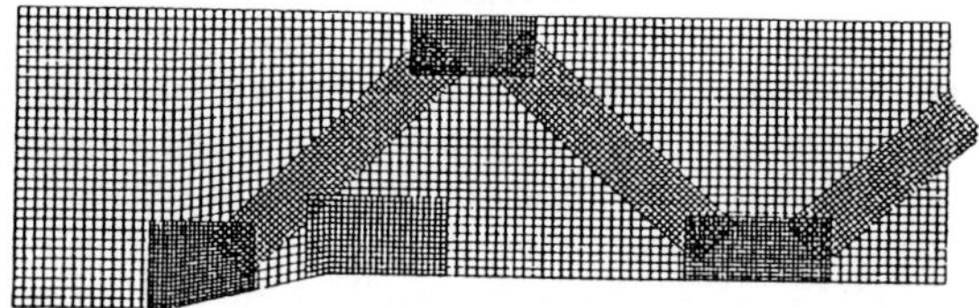

Fig.16: Simple adapted overset grid

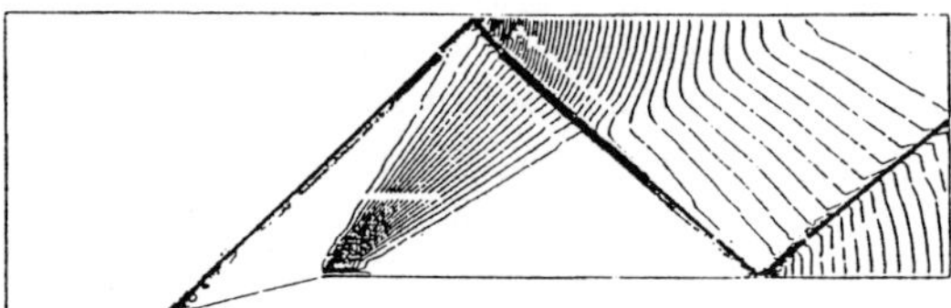

Fig.17: Pressure contours on the simple adapted overset grid

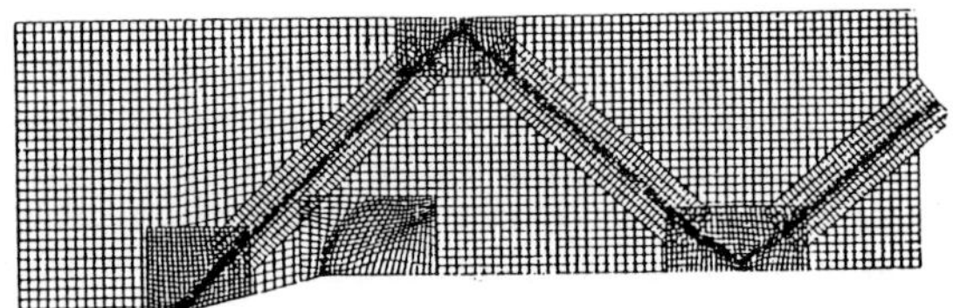

Fig.18: Overset adaptive grid

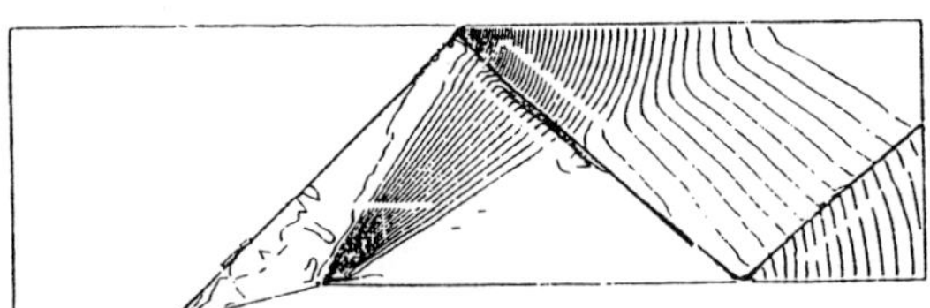

Fig.19: Pressure contours on the overset adaptive grid

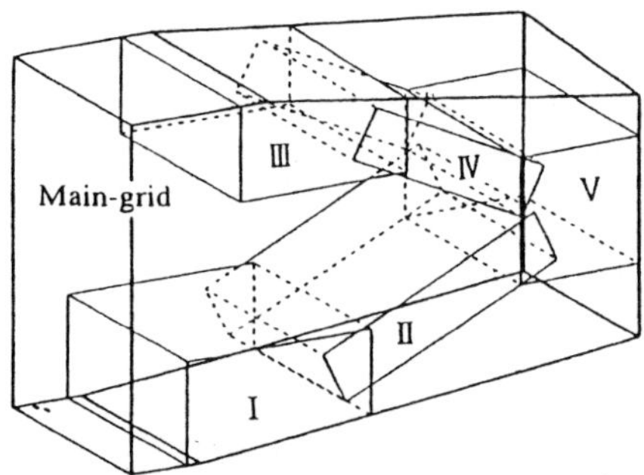

Fig.20: Layout of the overlaid subgrids

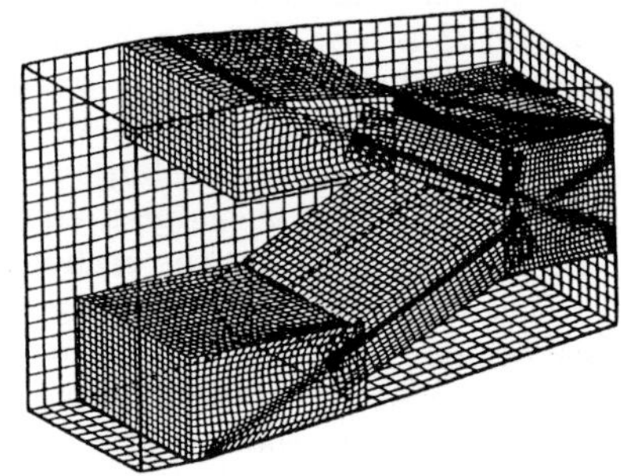

Fig.21: Three-dimensional overset adaptive grid

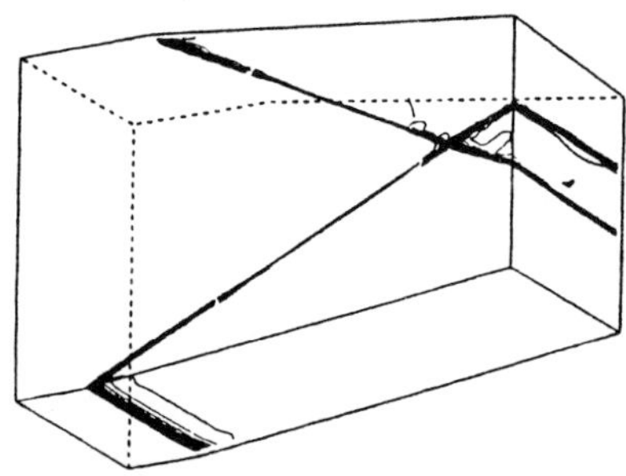

Fig.22: Pressure contours on the overset adaptive grid

Methods for Elliptic Grid Generation', J. Comp. Phys., Vol. 77, 1988, pp.40-52.

[6] J.F. Thompson, Z.U.A. Warsi, and C.W. Mastin, *Numerical Grid Generation*, 1985, Elsevier, NewYork.

[7] P.D. Thomas and J.F. Middlecoff, 'Direct Control of the Grdi Point Distribution in Meshes Generated by Elliptic Equations', AIAA J., Vol.18, 1980, pp.652-656.

[8] P.J. Roache, *Computational Fluid Dynamics*, Hermosa, 1982.

[9] K. Matsuno, 'Solution-Adaptive Grid Method Using Elliptic Equation', Jour. of Japan Society for Aeronautics ans Space Sciences, Vol.43, 1995, pp.26-31, (in Japanese).

[10] K. Matsuno, "Wiggles" and Central-Upwind

Hybrid Methods, Jour. of Japan Society for
Aeronautics ans Space Sciences, Vol.38, 1990,
pp.386-389, (in Japanese).

[11] P.L. Roe, 'Approximate Riemann Solvers,
Parameter Vectors and Difference Scheme'.
J.Comp. Phys., Vol.43, 1981, pp.357-372.

[12] C. Hirsch, *Numerical Computation of Internal
and External Flows*, Vol.2, John Wiley & Sons,
1988.

[13] K. Matsuno, M. Yamakawa, and N. Satofuka,
'Overset Adaptive-Grid Method with
Applications to Compressible Flows', Proc. of
the 6th International Symposium on CFD,
Lake Tahoe, (1995), pp.788-793.

[14] J.L.Steger and J.A.Benek, On the Use of
Composite Grid Schemes in Computational
Aerodynamics, Comp. Methods Appl. Mech.
Engng., 1987, Vol.64, 301-320.

[15] R.L.Meakin, Moving Body Overset Grid
Methods for Complete Aircraft Tiltrotor
Simulation, AIAA-93-3350-CP, 1993

[16] K.Fujii, 'A Zonal method for Practical Fluid
Dynamics Problems', Proc. of 4th
ISCFD-Davis, (1991), 353-358.

[17] K.-H.Kao, M.-S. Liou, and C.-Y.Chow, 'Grid
Adaption Using Chimera Composite
Overlapping Meshes', AIAA-93-3389-CP, 1993.

[18] M.Yamakawa, K. Matsuno, and N. Satofuka,
'Adaptive Grid Method Using Elliptic Equation
for Compressible Flows'. Trans. of JSME(B),
Vol.62, 1996, pp.2640-2645, (in Japanese).

[19] J.L.Steger and J.A.Benek, 'On the Use of
Composite Grid Schemes in Computational
Aerodynamics', Comp. Methods Appl. Mech.
Engng., Vol.64, (1987), 301-320.

ADVANCING-FRONT/LOCAL-RECONNECTION (AFLR) UNSTRUCTURED GRID GENERATION

David L. MARCUM

Mechanical Engineering Department, NSF Engineering Research Center for
Computational Field Simulation
P.O. Box 9627, Mississippi State University, MS 39759
e-mail: marcum@erc.msstate.edu

Abstract

Methods for generation of unstructured planar, surface, and volume grids using the advancing-front/local-reconnection (AFLR) procedure are presented. AFLR uses an iterative point creation and insertion scheme wherein points are created using advancing-front, -point, or -normal type point placement. Initially, the connectivity for these generated points is obtained by direct subdivision, without regard to connectivity quality. The connectivity is then improved by iteratively using local reconnection subject to a quality criterion. A min-max type (minimize the maximum angle) criterion is used. The overall procedure is applied repetitively until a complete field grid is generated. Field point distribution is controlled by a point distribution function based on the boundary point spacing. This function is propagated through the field by interpolation or specified growth. Procedures for generating edge and surface grids which are fully compatible with the volume grid generation are presented. Results for a variety of configurations are presented. The results demonstrate that the AFLR procedure consistently produces grids of very-high quality with minimal user input. Efficiency is such that standard PCs or workstations can be used to generate three-dimensional unstructured grids for complex configurations.

1 Introduction

Unstructured grid generation procedures for triangular and tetrahedral elements have typically been based on either an octree [1], advancing-front [2,3], or Delaunay [4-7] approach. Efficiency is the primary advantage of the octree approach. The advancing-front approach offers advantages of high-quality elements and integrity of the boundary. And, the Delaunay approach offers advantages of efficiency and a sound mathematical basis. Methods using a combined approach with advancing-front type point placement and a Delaunay connectivity have been developed for triangular elements [8-10]. These methods can produce grids with quality similar to that of traditional advancing-front methods. However, efficiency has not been substantially improved. Alternative approaches have been developed using variations of the edge-swapping algorithm of Lawson [11]. In this approach, the grid is repetitively reconnected to locally satisfy a desired criterion. Barth [12] has implemented this approach with a Delaunay criterion and circumcenter point placement. Marcum [13] has developed a very efficient local reconnection procedure using advancing-front point placement and an effective combined Delaunay/min-max type criterion for generation of triangular or tetrahedral element grids. This procedure differs substantially from the previously cited methods in that it makes effective use of the existing grid as a search data structure, point insertion is performed using direction subdivision, a Delaunay criterion is only used locally in an intermediate step within the local-reconnection process, and the combined Delaunay/min-max reconnection criterion is the only criteria developed to-date which allows effective optimization of a three-dimensional triangulation. This methodology has also been extended for generation of high-aspect ratio elements [14], right-angle elements [15], and solution-adapted grids [16,17]. High-quality grids have been generated about geometrically complex configurations in two and three dimensions for a variety of applications using this method. Various point placement strategies and connectivity criteria have been implemented and compared within this procedure. Results verify that advancing-front point placement with a min-max connectivity criterion consistently produces the highest quality grid in an efficient manner [18]. Fully compatible edge and surface grid generation components using this procedure have also been developed [19]. In this paper, an overview of this method for planar, surface, and volume grid generation is presented. Several application examples are presented which demonstrate the capabilities, consistency, efficiency, and quality of this approach.

Received on July 29, 1997.

"

AFLR Unstructured Grid Generation Procedure

The triangular/tetrahedral grid generation procedure used in the present work is a combination of automatic point creation, advancing type ideal point placement, and connectivity optimization schemes. A valid grid is maintained throughout the grid generation process. This provides a framework for implementing efficient local search operations using a simple data structure. It also provides a means for smoothly distributing the desired point spacing in the field using a point distribution function. This function is propagated through the field by interpolation from the boundary point spacing or by specified growth normal to the boundaries. Points are generated using either advancing–front type point placement for isotropic elements, advancing–point type point placement for isotropic right angle elements, or advancing–normal type point placement for high–aspect ratio elements. The connectivity for new points is initially obtained from direct subdivision and then improved by iteratively using local–reconnection subject to a quality criterion. A min–max type (minimize the maximum angle) type criterion is used. The overall procedure is applied repetitively until a complete field grid is obtained.

The basic steps in the procedure are briefly outlined below. More complete details and results are presented in Refs. 13 and 14.

1) Specify point spacing on the boundary surface.

2) Generate a boundary surface grid.

3) Generate a valid triangulation of the boundary surface points and recover all boundary surfaces. An example initial triangulation is shown in Fig. 1 a.

4) Assign a point distribution function to each boundary point based on the local point spacing. Also, optionally assign geometric growth rates normal to the boundary surface.

5) For isotropic elements, generate points using advancing–front point placement. Points are generated by advancing from from the edge/face that satisfies the point distribution function of elements that only satisfy the point distribution function on one edge/face. An example triangulation generated using advancing–front point placement is shown in Fig 2 a.

6) For right angle elements, generate points using advancing–point point placement. Points are generated by advancing as in step 5, except two points are created by advancing along edge/face

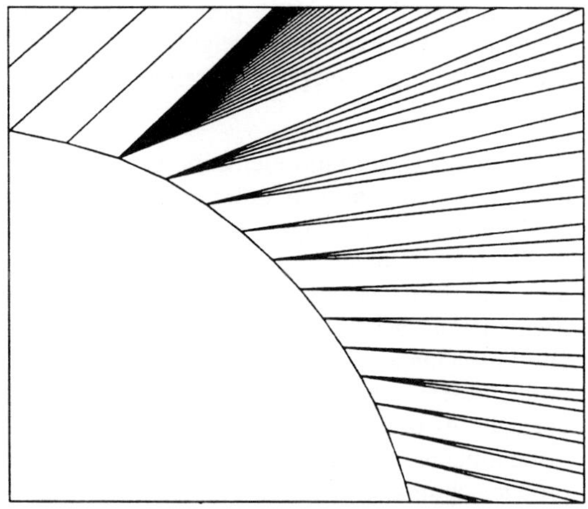

a. Initial triangulation.

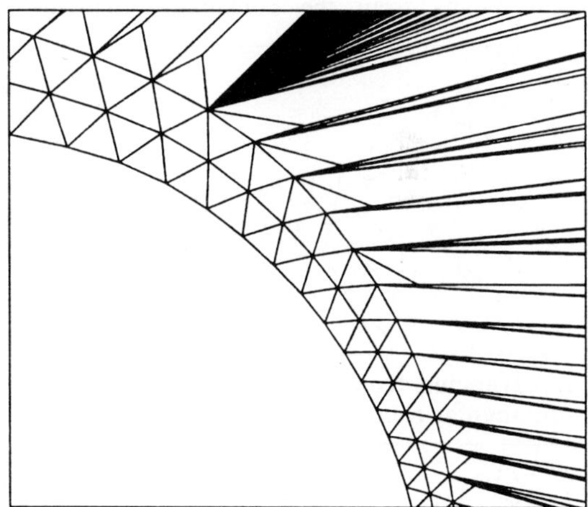

b. Triangulation after direct point insertion on third grid generation iteration.

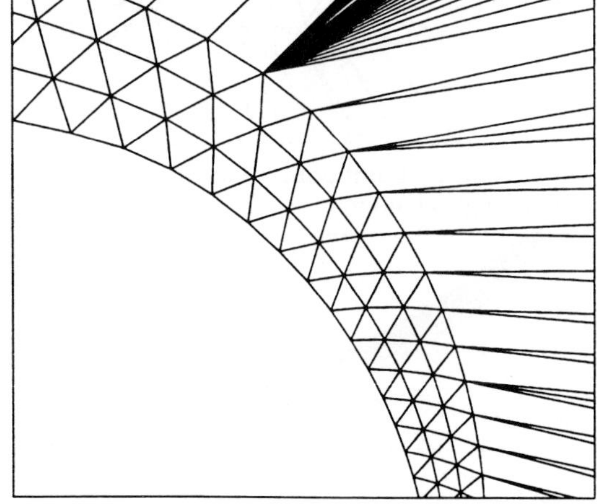

c. Triangulation after local reconnection on third grid generation iteration.

Fig. 1 Unstructured grid generation process.

normals from the two/three points of the satisfied edge/face. An example triangulation generated using advancing–point point placement is shown in Fig 2 b.

7) For high–aspect ratio elements, generate points using advancing–normal point placement. Points are generated one layer at a time from the boundaries by advancing along normals dependent upon the boundary surface geometry. An example triangulation generated using advancing–normal point placement is shown in Fig 2 c. A key aspect of the present implementation is the use of multiple normals. At points where the boundary surface is discontinuous, multiple normals are assigned to produce optimal grid quality.

8) Interpolate the point distribution function for new points from the containing elements. If

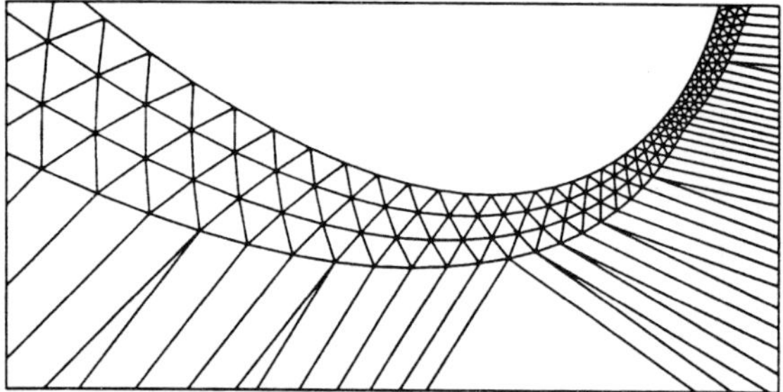

a. Advancing–front point placement for isotropic equiangular elements.

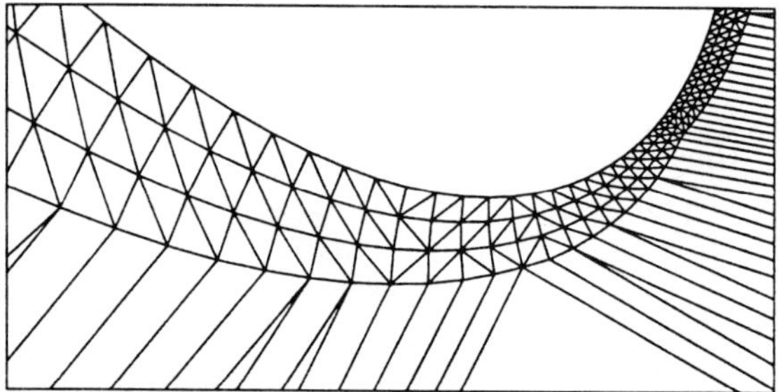

b. Advancing–point point placement for isotropic right–angle elements.

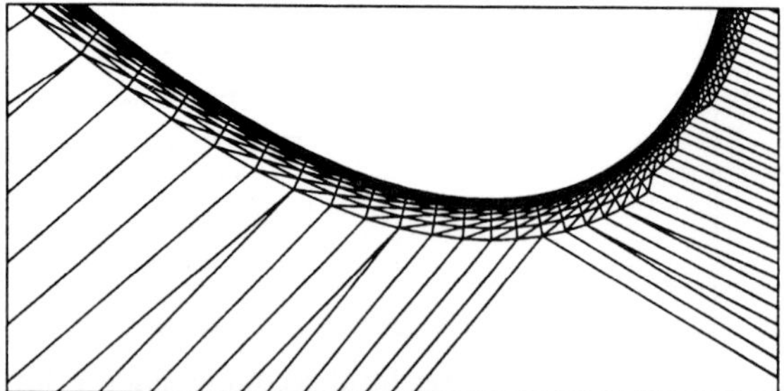

c. Advancing–normal point placement for high–aspect–ratio right–angle elements.

Fig. 2   Different point placement strategies.

geometric growth is specified then the distribution function is determined from an approximate distance to the nearest boundary and the specified geometric growth from that boundary.

9) Reject new points that are too close to other new points. Rejected points are averaged with advancing–point point placement for generation of right angle elements.

10) Insert the accepted new points by directly subdividing the elements which contain them. A triangulation after direct insertion is shown in Fig. 1 b.

11) Optimize the connectivity using local–reconnection. For each element pair, compare the reconnection criterion for all possible connectivities and reconnect using the most optimal one. Possible triangulations in two and three dimensions are shown in Fig. 3. Repeat this local–reconnection process until no elements are reconnected. In three–dimensions a combined Delaunay/min–max type criterion is used[13]. In this process a Delaunay criterion is used initially and then the min–max criterion is applied. This improves the overall grid quality substantially and partially overcomes the problems associated with optimum local states that prohibit a global optimum from being obtained. A triangulation after local–reconnection is shown in Fig 1 c.

12) Repeat the point generation and local–reconnection process, steps 5 through 11, until no new points are generated.

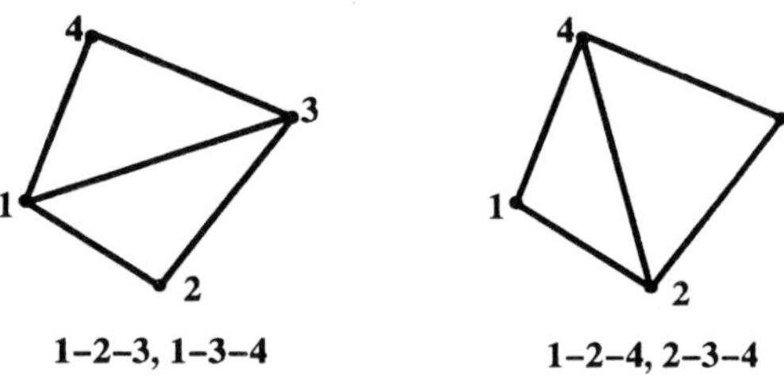

a. Four reconnectable points in two dimensions.

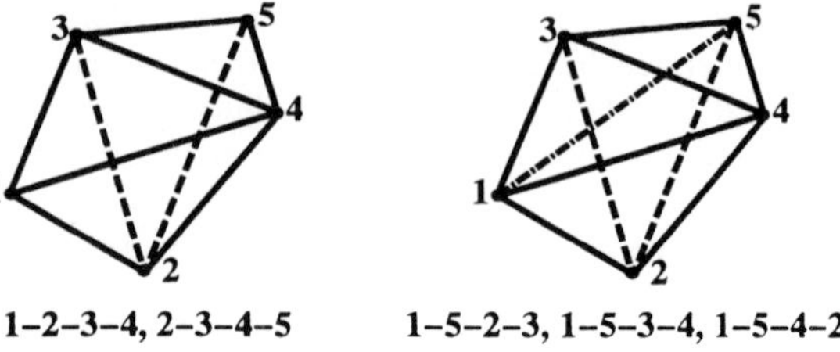

b. Five reconnectable points in three dimensions.

Fig. 3   Possible triangulations for reconnectable element pairs.

13) Optionally combine elements to form quadrilaterals in two dimensions or hexahedral, prism, or pyramid elements in three dimensions. Elements are combined by advancing from boundary surfaces and selecting combinations based on alignment and quality.

14) Smooth the coordinates of the field grid.

15) Optimize the connectivity using the local–reconnection process (step 11).

The procedure described in the above steps allows complete control over the type and quality of grid to be generated with minimal user interaction. In generating the boundary surface grid, user input is required to specify point spacings at selected control points. Further control over the spacing of the field points can be obtained using specified geometric growth[15], fixed field points[15], embedded boundary surfaces[15] or adaptation sources[16]. Once a boundary surface grid is generated no further user input or adjustment of parameters are required other than selecting desired options such as point placement type or geometric growth.

With high–aspect–ratio elements and advancing–normal type point placement, the procedure described above does produce sliver elements in three dimensions. These elements are generated only in regions of high–aspect–ratio elements with a very structured alignment. The slivers are generated between prism element pairs as shown in Fig. 4. Elimination of these elements with local–reconnection is not feasible. There may be no nearby optimization path which produces a better connectivity. The problem is inherently due to the very structured nature of the grid in these regions. There are only a limited set of possible triangulations for a set of tetrahedra aligned in prismatic groups. One alternative is to combine the tetrahedra in high–aspect–ratio regions into prismatic elements. In this case the slivers are not a problem as they can be directly eliminated. If a fully tetrahedral grid is desired,

then the connectivity within the prismatic elements can be specified in a manner similar to that done with pseudo structured methods[20]. This issue remains a research topic as the possible solutions have not been advanced to the state of generality and robustness of the isotropic generator.

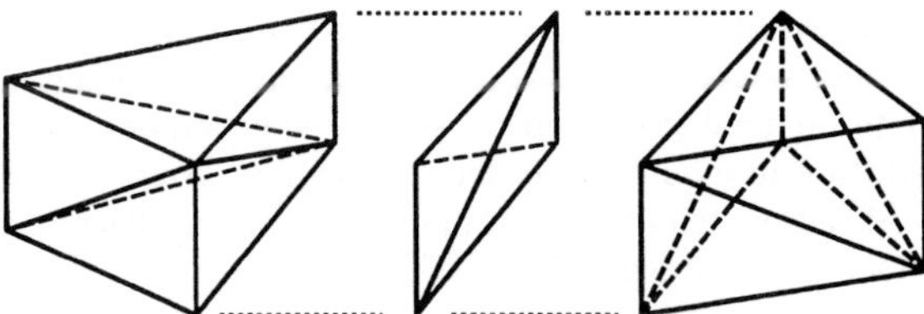

Fig. 4 Trapped sliver elements between prism groups of tetrahedral elements.

Two–Dimensional Application Examples

Selected application examples are presented here to demonstrate the capabilities of the present procedure for generation of two–dimensional unstructured grids. A summary of grid quality and CPU time required for the primary examples is presented in Table 1.

Grid quality distributions and statistics are presented for each primary example. Element angle is used as the grid quality measure. The complete set of grid quality data consists of the three corner angles for all triangles. Maximum and standard deviation values are presented along with distribution plots in 5 deg. increments. The results for the examples presented are representative of those obtained for a variety of configurations. Typically, for an isotropic grid, the maximum element angle is 120 deg. or less, the standard deviation is 7 deg. or less, and 99.5% or more of the elements have angles between 30 and 90 deg. The minimum angle is usually dictated by the geometry and boundary point spacing. Standard deviation is not applicable for grids with high–aspect–ratio elements as there should be peak distributions at a small angle, 60 deg., and 90 deg. The minimum angle is set by the maximum aspect–ratio.

2D Case	Max. Angle (deg)	Std. Dev. Angle (deg)	CPU Time (sec)		
			Pentium 120 Toshiba *Tecra 500* 128 MB SunSolaris, g77	Pentium Pro 200 Gateway 2000 *G6–200* 128 MB SunSolaris, g77	MIPS R10k SGI *Indigo^2* 512 MB IRIX64, f77
Multi–Element Airfoil 29,564 triangles	116	6.9	5.7	2.4	1.0
Wake–Adapted Multi–Element Airfoil 140,609 triangles	127	n/a	40	16	9.5
Mediterranean 213,323 triangles	118	7.2	61	27	13

Table 1 Summary of grid quality and CPU requirements for two–dimensional example cases.

CPU time required on a laptop PC, desktop PC, and workstation is presented for each primary example. Computer routines for the two–dimensional grid generator are written in Fortran. All floating–point calculations are performed using 64 bit precision with 8 byte data. The CPU times reported include all I/O and generation of grid quality data. A discretized boundary edge grid file is the input and a FAST unformatted grid file and a quality data file are the output. The efficiency of the overall procedure is such that generation of a typical grid requires only seconds on any current PC or workstation. All of the cases presented can be generated on a PC with at least 16 Mb of RAM.

User input required to generate a complete grid is minimal and includes specifying the point spacing at selected control points on boundary curves and selection of options such as growth from boundaries or generation of high–aspect–ratio elements. There are no user adjustable parameters that need to be changed from case to case. Specification of point spacings is minimized by automatic reduction of the boundary point spacing in regions where the spacing is greater than the distance between nearby boundaries. The present code is very robust and thoroughly tested. It does not fail to produce a valid grid, given a set of boundary curves that are valid and have a reasonable discretization.

Multi–Element Airfoil

A grid suitable for inviscid CFD analysis was generated for a multi–element airfoil. The grid near the airfoil surface is shown in Fig. 5. The grid generated contains 15,217 points and 29,564 elements. Element size varies smoothly within the grid. Grid quality distribution for the grid is shown in Fig. 6. Element angle distribution, maximum and minimum values, and standard deviation verify that the grid is of very high quality. Required CPU time is listed in Table 1. For comparison a similar grid was generated using the advancing–point point placement to generate right–angle isotropic triangles. A mixed element grid is automatically generated by combining suitable triangle pairs to form quadrilaterals. The resulting grid is shown in Fig. 7. Near the surface the elements are aligned in a very structured manner.

Mediterranean Sea

A grid was generated for the geometrically complex coastline of the Mediterranean Sea. The boundary curves for the computational domain are shown in Fig. 8. The initial boundary curve discretization was nearly uniform. Automatic point spacing reduction was used to reduce the point spacing near points of high–curvature

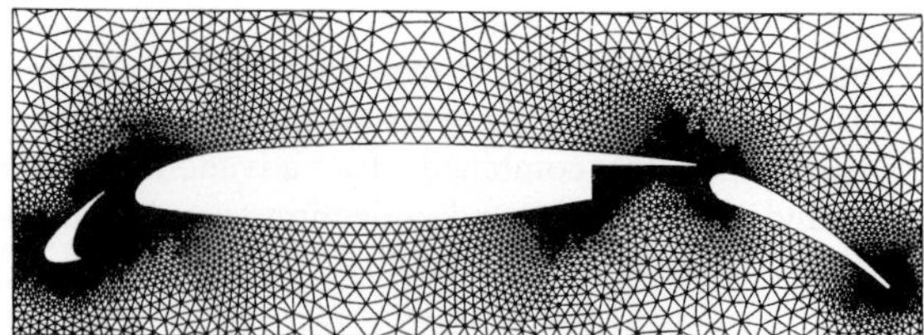

Fig. 5 Unstructured grid for multi–element airfoil.

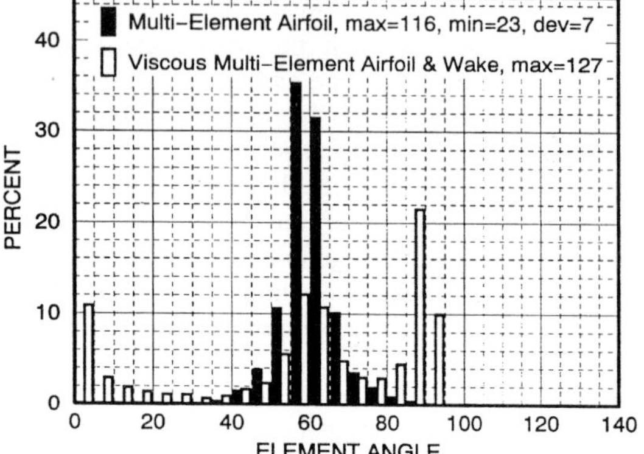

Fig. 6 Grid quality distributions for multi–element airfoil grids.

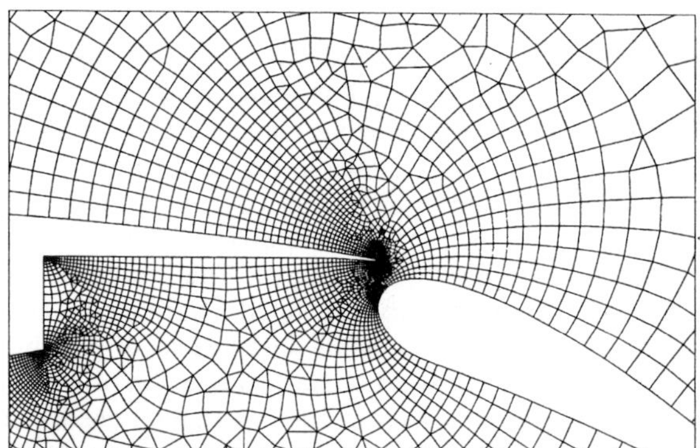

Fig. 7 Mixed quadrilateral and triangular element grid generated using advancing–point point placement for multi–element airfoil.

and in regions where boundaries are close to one another. Views of the grid near the Aegean Sea and Sea of Crete are shown in Figs. 9 a and b, respectively. The grid generated contains 111,612 points and 213,323 elements. Point distribution function growth was used to increase the element size away from the coastline. Element size varies smoothly within the grid. Grid quality distribution for the grid is shown in Fig. 10. Element angle distribution, maximum and minimum values, and standard deviation verify that the grid is of very high quality. Required CPU time is listed in Table 1.

Fig. 8 Boundary edges of Mediterranean Sea grid.

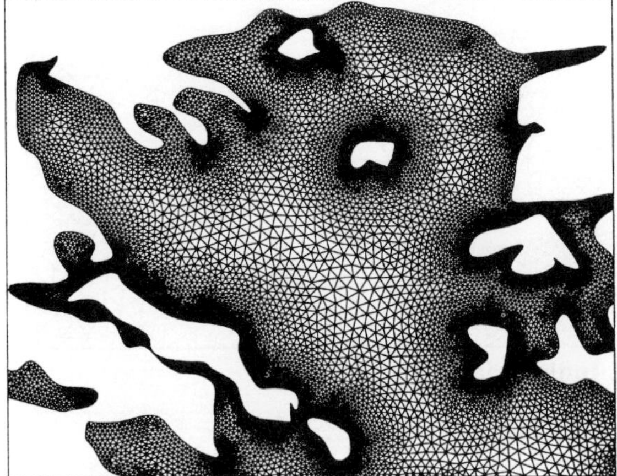

a. Grid near Aegean Sea.

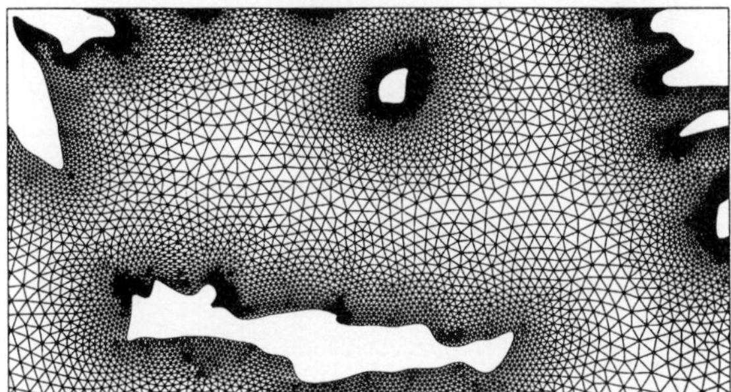

b. Grid near Sea of Crete.

Fig. 9 Mediterranean Sea grid generated using point distribution function growth.

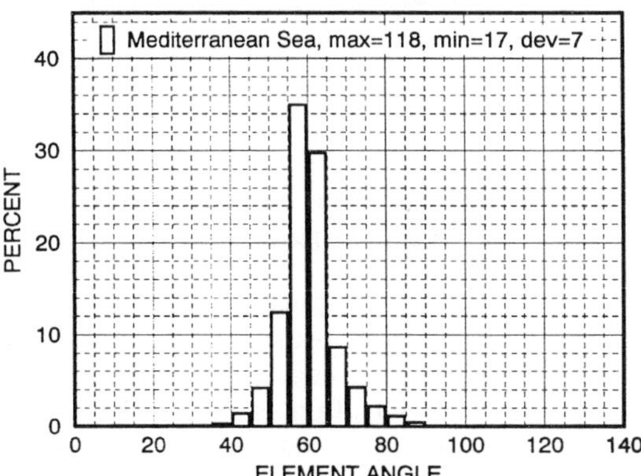

Fig. 10 Grid quality distributions for Mediterranean Sea grid.

Multi–Element Airfoil with Wake–Adaptation

A solution–adapted grid suitable for viscous CFD analysis was generated for a multi–element airfoil. For this case an initial grid without adaptation was generated and a viscous solution was obtained. Selected streamlines were tracked in the wake regions. These streamlines were then treated as embedded boundary surfaces for generation of aligned high–aspect ratio elements in wake regions. Aligned high–aspect ratio elements produce optimal resolution and grid quality. The boundary edges and embedded wakes are shown in Fig. 11. The final solution–adapted grid contains 71,032 points and 140,609 elements and is shown Fig. 12. Grid quality distribution for this grid is shown in Fig. 6. Element angle distribution and maximum angle verify that the grid is of very high quality. The maximum element angle is generated within the high–aspect–ratio element region adjacent to one of the corners of the blunt leading edges. Required CPU time is listed in Table 1. Details of solution–adaptation for viscous flow fields are presented in Ref. 16.

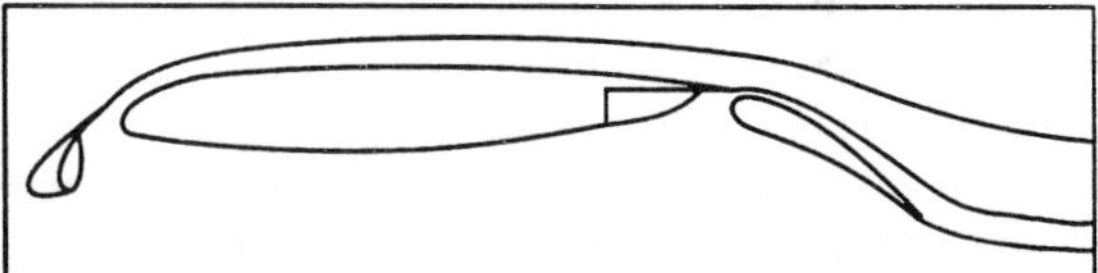

Fig. 11 Boundary edges for multi–element airfoil with multiple wakes.

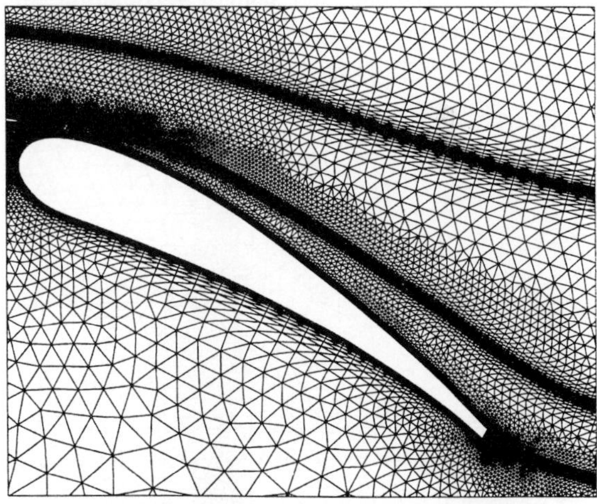

Fig. 12 Final solution–adapted grid for multi–element airfoil with multiple wakes.

Airfoil with Anisotropic Solution–Adaptation

A solution–adapted grid suitable for inviscid CFD analysis was generated for a NACA 0012 airfoil. Free stream conditions were a Mach number of 0.8 and an angle of attack of 1.25 deg. Solution–adaptation was obtained using feature detection combined with pattern recognition. Physically based feature detectors were used to identify regions with physically important features. Regions identified by feature detection are shown in Fig. 13. For this case, detection of slip line type features in the wake region was intentionally turned off. Pattern recognition was then used to reduce those regions to simple geometric entities representative of the physical features. For example, stagnation regions are reduced to singular points and shock waves are reduced to curves, as shown in Fig. 13. Within the grid generator, singular points are converted to adaptation point sources which locally reduce the point distribution function. Curves are treated as embedded boundaries for generation of high–aspect ratio elements aligned with flow features. The concept is the same as that for the previous wake adaptation case. An initial grid, one isotropic adapted grid, and two additional anisotropic adapted grids and solutions were generated. The initial and final solution–adapted grid are shown in Figs. 14 a and b. The initial grid contains 1,426 points and the final solution–adapted grid contains 6,655 points. For the final grid, the maximum aspect–ratio in the adapted regions is about 100:1. Computed pressure contours for the initial and final solutions are shown in Figs. 15 a and b. The final grid provides very good resolution of the flow features. An equivalent resolution grid using isotropic solution adaptation contains 108,610 points. The grid used here with anisotropic adaptation offers a significant reduction in the total number of grid points required. Details of anisotropic solution–adaptation using feature detection and pattern recognition are presented in Ref 17.

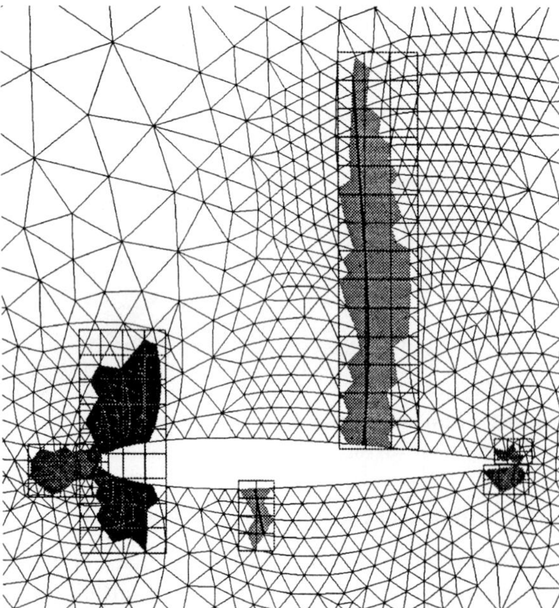

Fig. 13 Regions identified by feature detection and representative lines created by pattern

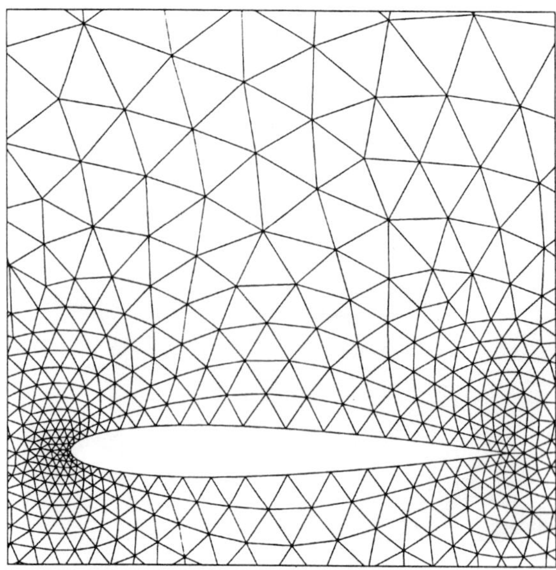

a. Initial grid.

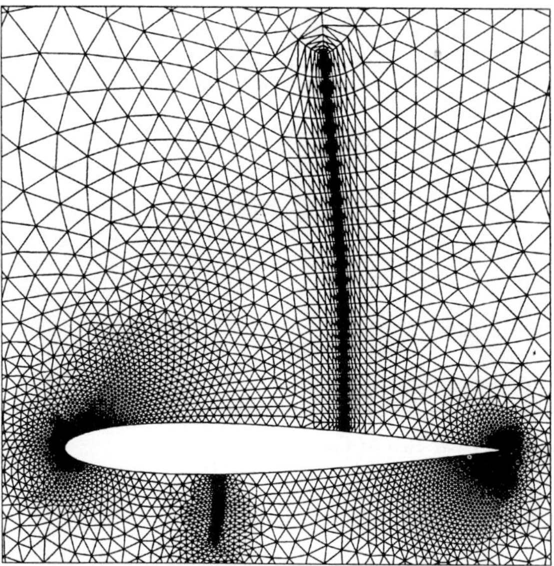

b. Final anisotropic solution–adapted grid.

Fig. 14 Grids for transonic NACA 0012 airfoil.

Three–Dimensional Surface Grid Generation

For grid generation with the present methodology, the grid point distribution is automatically propagated from specified control points to edge grids, from edge to surface grids, and finally from surface grids to the volume grid. Surface patches, edges, and corner points for a fighter geometry definition are shown in Fig. 16. The first step in the grid generation process is to initially set the desired point spacing to a global value at all edge

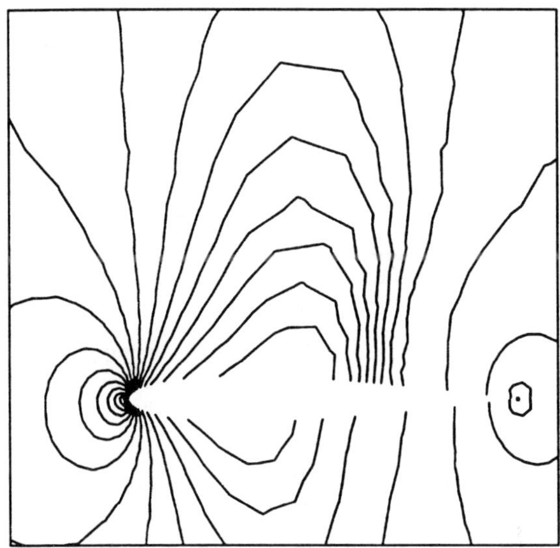

a. Contours for initial grid solution.

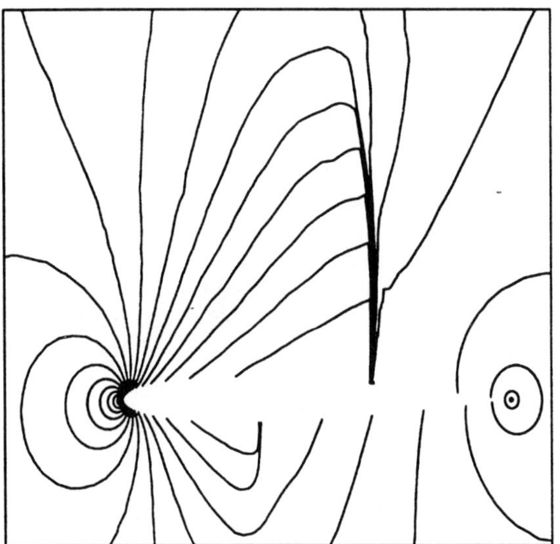

b. Contours for final anisotropic solution–adapted grid.

Fig. 15 Computed pressure contours for NACA0012 transonic airfoil.

end points. Point spacings are then set to different values at desired control points on edges in specific regions requiring further resolution. For example, end points along leading edges and trailing edges would typically be set to a very fine point spacing. Point spacings can be set anywhere along an edge. For example, a point in the middle of a wing section would typically be set to a larger point spacing than at the leading or trailing edges. As control point spacings are set, a discretized edge grid is created for each edge. Specification of desired control point spacings is typically the only user input required in the overall grid generation process.

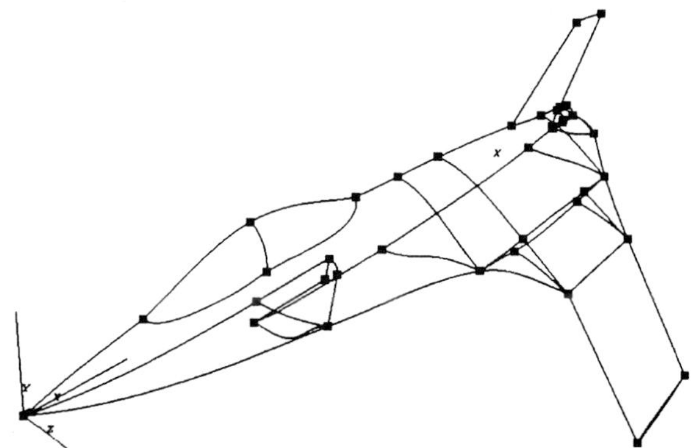

Fig. 16 Surface patches, edges, and corner points for fighter geometry definition.

A CAD system is used to define and evaluate the surface geometry. Edge and surface grid generation requires use of CAD surface evaluation routines and access to the CAD geometry database. Surface topology is extracted from the CAD database and a separate data structure is used for grid generation. The grid generation procedures used have been designed to isolate CAD geometry access. All access to CAD geometry is outside the grid generation routines. This approach produces a very clean interface between the grid generation and CAD system. It also makes it very easy to use different CAD systems with very little modification. The edge grid generation and subsequent surface grid generation procedures are described in the following sections. Additional information can be found in Ref. 19. The primary CAD system routine required for the present edge and surface grid generation is one that determines physical space coordinates, x,y,z, given mapped space coordinates, u,v. This routine is labeled CAD routine xyz_from_uv in the following sections.

Edge Grid Generation Procedure

Edge grids are created using a one–dimensional version of the standard grid generation procedure. This insures that point distribution and growth rates are fully compatible for optimal final grid quality. For each edge or segment the point spacing is specified at both ends, as shown in Fig 17a. Edge grid generation is then used to produce the point distribution shown in Fig. 17b. The basic steps in the edge grid procedure are outlined below.

1) Create an interpolation table for the physical space coordinates and arc–length versus mapped space coordinates using the CAD routine xyz_from_uv.

2) Advance from each end of the edge segment and create two new points. The point spacings for these points are interpolated from the exposed end points of the edge.

Fig. 17 Specified point spacings and final point distribution for a surface edge.

3) Reject a new point if it is too close to the other new point.

4) Repeat the edge grid point generation process steps 2 and 3 until no new points are created.

5) Smooth the arc–length coordinates of the edge grid.

6) Interpolate at the generated arc–length coordinates for mapped space coordinates u,v.

7) Obtain the physical space coordinates, x,y,z, at the interpolated mapped space coordinates, u,v, using the CAD routine, xyz_from_uv.

The edge grid generation routine consists of steps 2 through 6 above. All generation parameters for details such as interpolation, limiting, rejection, and smoothing are identical to those used in the standard planar and volume grid generation procedures.

Surface Grid Generation Procedure

Given a geometry definition that uses a surface mapping, one can generate a surface grid in either mapped or physical space. With a mapped space approximation (MSA), the standard two–dimensional grid generator can be used as is. The advantage of this approach is efficiency. However, for realistic configurations, the mappings are often distorted in physical space and an MSA approach produces a poor quality surface grid. The two–dimensional grid generation procedure can be modified to generate near optimal grids on a surface using a physical space approximation (PSA). In this approach, an approximate surface definition is used within the surface grid generation procedure to determine point placement such that ideal physical space surface triangles are created. The approximate surface definition provides an efficient means of iterating on the surface an allows separation of the grid generation and CAD systems within the grid procedure. A valid mapped and physical space grid is maintained throughout the procedure and all searching is done in mapped space.

Local–reconnection is performed in both mapped and physical space. The physical space reconnection can not be used for elements which are considerably larger than the desired element size. These elements exist early in the process. In physical space these elements can be composed of highly curved edges that are not accounted for in the reconnection criteria. The output grid is a mapped space grid which corresponds to an approximately optimal surface grid when mapped back to the actual surface definition. The basic steps in the overall procedure are listed below.

1) Generate a surface grid entirely in mapped space using the standard two–dimensional procedure. This grid is used as the physical space approximation.

2) Obtain the physical space coordinates, x,y,z, at the mapped space coordinates, u,v, using the CAD routine xyz_from_uv.

3) Generate a valid triangulation in mapped space of the edge points and recover all discrete edges.

4) Assign a point distribution function to each edge point based on the local point spacing in physical space.

5) Generate points using advancing–front point placement by advancing from from satisfied edges. These points are generated to obtain approximately optimal elements in physical space. Iteration and interpolation of physical space coordinates from mapped space coordinates are required. The grid from steps 1 and 2 is used as a locally linear approximation to the surface definition.

6) Interpolate in mapped space the point distribution function for new points from the containing elements.

7) Reject new points that are too close to other new points in physical space.

8) For each accepted new point, search in mapped space for the containing element and directly inset the point.

9) Optimize the connectivity using local–reconnection in mapped space.

10) Optimize the connectivity using local–reconnection in physical space. Only elements which are close to satisfying the distribution function are allowed to be reconnected.

11) Repeat the point generation and local–reconnection process, steps 5 through 10, until no new points are generated.

12) Smooth the mapped space coordinates of the surface grid using physical space edge length

weighting. This is equivalent to smoothing directly in physical space.

13) Interpolate for the smoothed physical space coordinates using the grid from steps 1 and 2.

14) Optimize the connectivity using physical space local–reconnection .

15) Obtain the physical space coordinates, x,y,z, on the true surface at the generated mapped space coordinates, u,v, using the CAD routine, xyz_from_uv.

The PSA surface grid generation routine consists of steps 3 through 14 above. All generation parameters for details such as interpolation, limiting, rejection, and smoothing are identical to those used in the standard planar and volume grid generation procedures. For both the edge and surface grid generation procedures, the final physical space grid is located on the actual surface defined by the CAD data base. The approximate physical space edge and surface is used only within the grid generation procedures.

Three–Dimensional Surface Grid Generation Application Examples

Two selected application examples are presented here to demonstrate the capabilities of the present procedure for generation of three–dimensional unstructured surface grids. Grid quality distributions and statistics are presented for each example. Element angle is used as the grid quality measure. The complete set of grid quality data consists of the three corner angles for all surface triangles. Maximum, minimum, and standard deviation values are presented along with distribution plots in 5 deg. increments. The results for the examples presented are representative of those obtained for a variety of surfaces. Typically, the resulting grid quality is the same as that expected for the two–dimensional grid generator. Required CPU times are about three times that required of the the two–dimensional grid generator.

Generic Shell

The first case is a generic shell that was derived from a circular surface patch with a circular hole that is distorted in physical space. The surface mapping and geometry definition is shown in Fig. 18. Surface grids were generated for this case with two different procedures. One was generated with the mapped space approximation (MSA) approach. With MSA the standard two–dimensional grid generator is used in mapped space. The other grid was generated using the physical space approximation (PSA) approach. Both

grids are shown in Figs. 19 a and b in mapped space. The MSA grid is optimal in mapped space while the PSA grid is not. The grids in physical are shown in Figs. 20 a and b. The MSA grid contains distorted elements in physical space while the PSA grid is of very high–quality. Grid quality distributions in physical space for these grids are shown in Fig. 21. Element angle distribution, maximum and minimum values, and standard deviation verify that the PSA surface grid is of very high quality and the MSA surface grid is not.

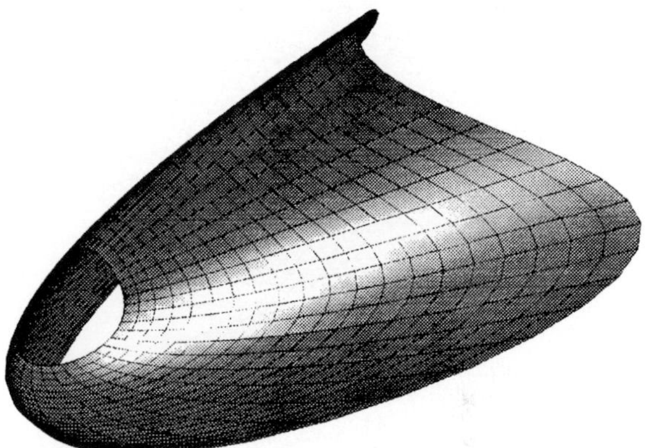

Fig. 18 Surface mapping for geometry definition of generic shell.

Hawaiian Islands

A surface grid was generated on the geometrically complex ocean bottom around the Hawaiian Islands. For this case, a quality grid can only be obtained using some form of physical space grid generation. The surface grid generated using the PSA approach is shown in Fig. 22. A nearly uniform point spacing was specified and as shown in Fig. 22 a nearly uniform grid is generated. Grid quality distribution in physical space for this case are shown in Fig. 23. Element angle distribution, maximum and minimum values, and standard deviation verify that the PSA surface grid is of very high quality.

Three–Dimensional Application Examples

Selected application examples are presented here to demonstrate the capabilities of the present procedure for generation of three–dimensional unstructured grids. All surface grids used were generated using the previously described PSA surface grid generation procedure. A summary of grid quality and CPU time required for the primary examples is presented in Table 2.

Grid quality distributions and statistics are presented for each primary example. Element angle is used as the grid quality measure. The complete set of grid quality data consists of the six dihedral angles for all tetrahedra.

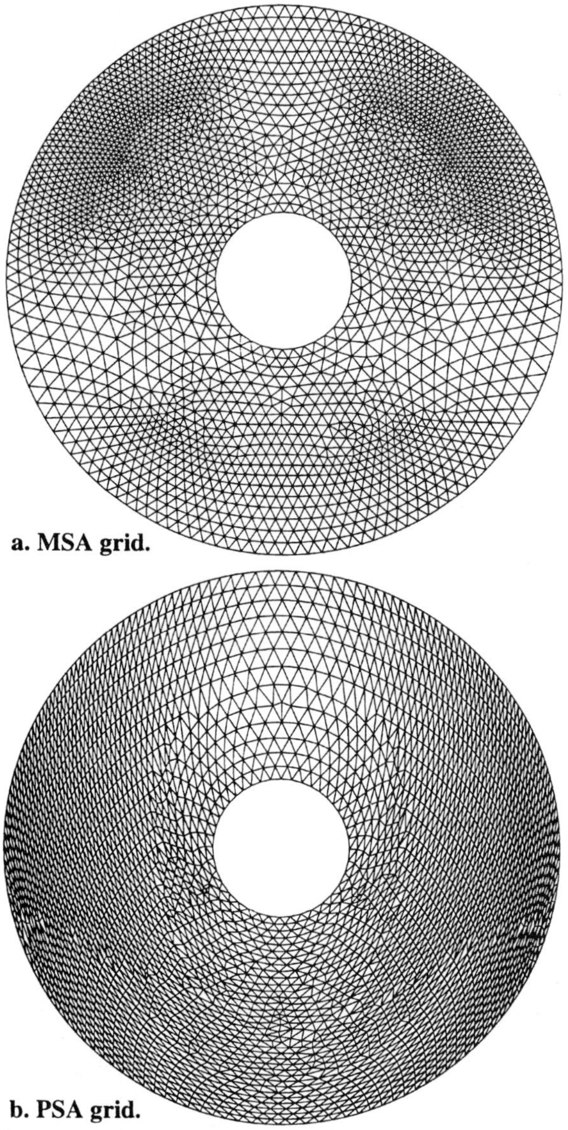

a. MSA grid.

b. PSA grid.

Fig. 19 Surface grids in mapped space for generic shell.

Maximum and standard deviation values along with distribution plots in 5 deg. increments are presented for both the surface and volume grids. The results for the examples presented are representative of those obtained for a variety of configurations. Typically, for an isotropic grid, the maximum element angle is 160 deg. or less, the standard deviation is 17 deg. or less, and 99.5% or more of the elements have angles between 30 and 120 deg. The minimum angle is usually dictated by the geometry and boundary point spacing. Standard deviation typically increases when geometric growth is used to increase the field point spacing. Surface grid quality is typically the same as that obtained for two–dimensional cases.

a. MSA grid.

b. PSA grid.

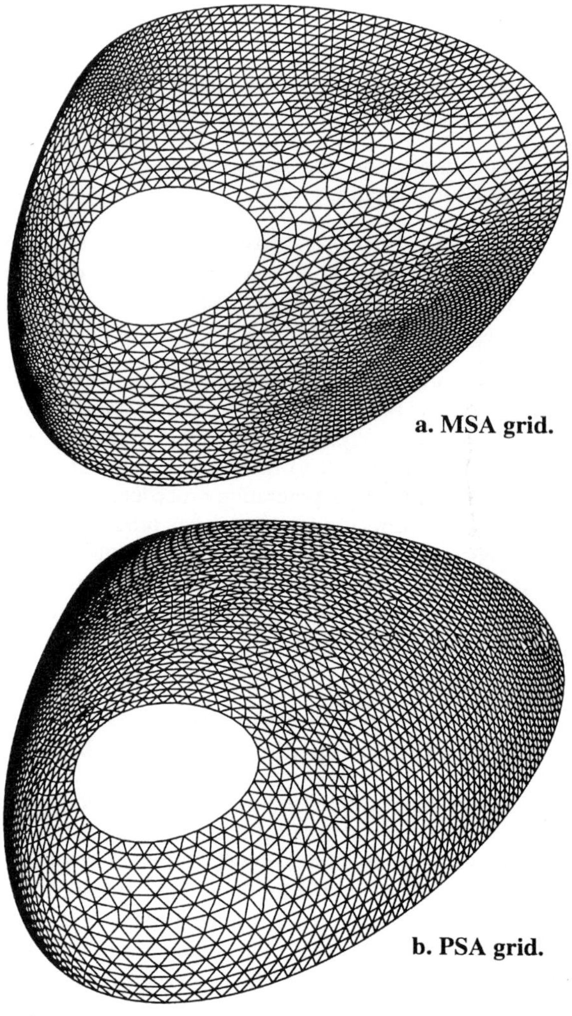

Fig. 20 Surface grids in physical space for generic shell.

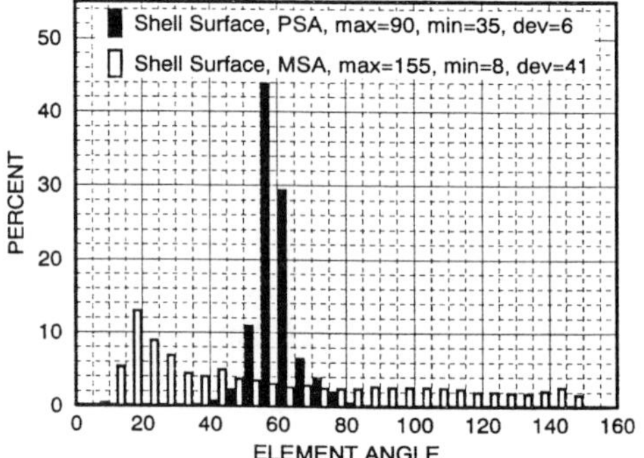

Fig. 21 Grid quality distributions for generic shell surface grids.

CPU time required on a laptop PC, desktop PC, and workstation is presented for each primary example. Computer routines for the three–dimensional grid

Fig. 22 PSA surface grid for ocean bottom near Hawaiian Islands.

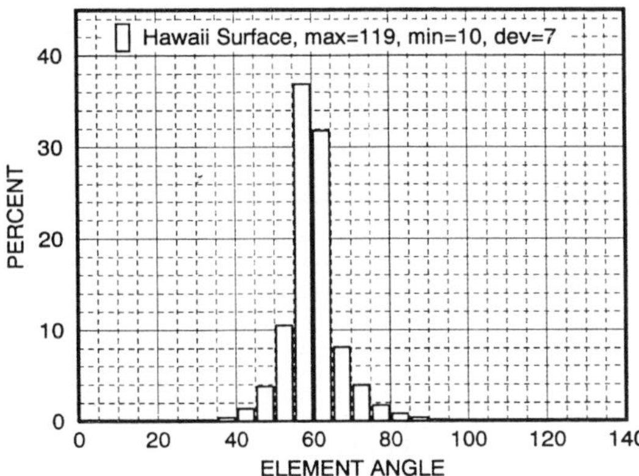

Fig. 23 Grid quality distributions for Hawaiian Island surface grid.

generator are written in C with dynamic memory that is automatically reallocated based upon actual requirements. All floating–point calculations are performed using 64 bit precision with 8 byte data. The CPU times reported include all I/O and generation of grid quality data. A boundary surface grid file is the input and a FAST binary grid file and a quality data file are the output. The efficiency of the overall procedure is such that generation of a typical grid requires only minutes on many current PCs or workstations. Generation of a typical surface grid requires only seconds (approximately three times the CPU time required for a two–dimensional grid). Memory required is about 100 bytes per element generated.

User input required to generate a complete grid is minimal and includes specifying the point spacing at selected control points on the boundary curves for surface grid generation. Selection of options such as growth from boundaries is the only required user input for volume grid generation. There are no user adjustable parameters that need to be changed from case to case. The present code is very robust and thoroughly tested. It does not fail to produce a valid volume grid, given a set of boundary surface triangulations that are valid and have a reasonable discretization.

Fighter

A complete grid suitable for inviscid CFD analysis was generated for a generic fighter. The surface grid on the fighter surface only is shown in Fig. 24. The complete surface grid contains 24,322 total boundary faces. Distribution of grid points within the volume grid can be visualized using a tetrahedral field cut as shown in Fig. 25. The complete volume grid contains 65,020 points and 349,516 elements. Element size varies smoothly on the surface and within the volume grid. Grid quality distributions for the surface and volume grids are shown in Figs. 26 a and b. Element angle distributions, maximum values, and standard deviations verify that the surface and volume grids are of very high quality. Required CPU time is listed in Table 2. For comparison, a tetrahedral field cut through a fighter wing tip of a high–aspect–ratio element grid generated using advancing–normal point placement is shown in Fig. 27.

3D Case	Max. Angle (deg)	Std. Dev. Angle (deg)	CPU Time (min)		
			Pentium120 Toshiba *Tecra 500* 128 MB SunSolaris,gcc	Pentium Pro 200 Gateway 2000 *G6–200* 128 MB SunSolaris, gcc	MIPS R10k SGI *Indigo²* 512 MB IRIX64, cc
Fighter 349,516 tetrahedra	148	17	15	6.4	2.9
Combustor (w/growth) 643,484 tetrahedra	156	17	26	12	7.6
Launch Vehicle 877,103 tetrahedra	155	17	33	15	12
Ducted Fan (w/growth) 1,214,018 tetrahedra	155	18	58	28	23
Ducted Fan 2,144,516 tetrahedra	154	17	n/a	n/a	40

Table 2 Summary of grid quality and CPU requirements for three–dimensional example cases.

The elements within the viscous regions are aligned with the surface in a very structured manner. Multiple normals are used at discontinuous edge points to improve element quality.

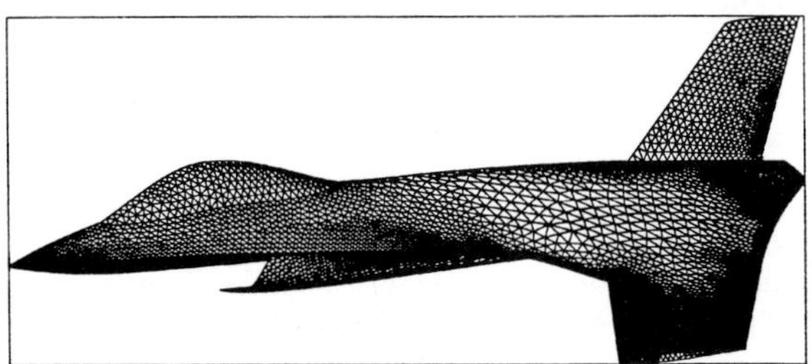

Fig. 24 Fighter surface grid.

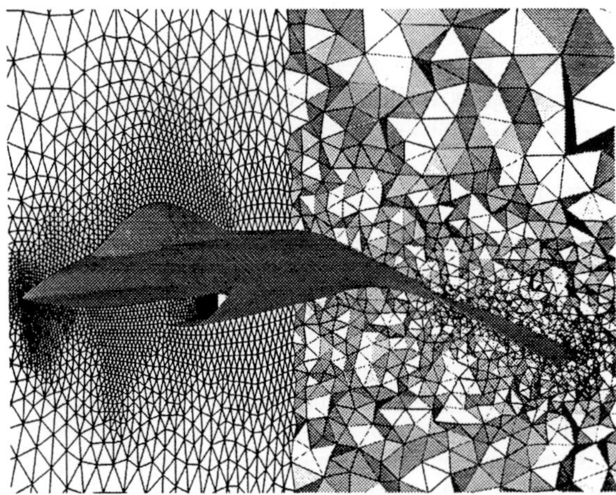

Fig. 25 Tetrahedral field cuts for fighter grid.

Combustor

A complete grid was generated for a geometrically complex turbine engine combustor. A cut–away view of the combustor surface is shown in Fig. 28. A cross–section view of the surface grid is shown in Fig.

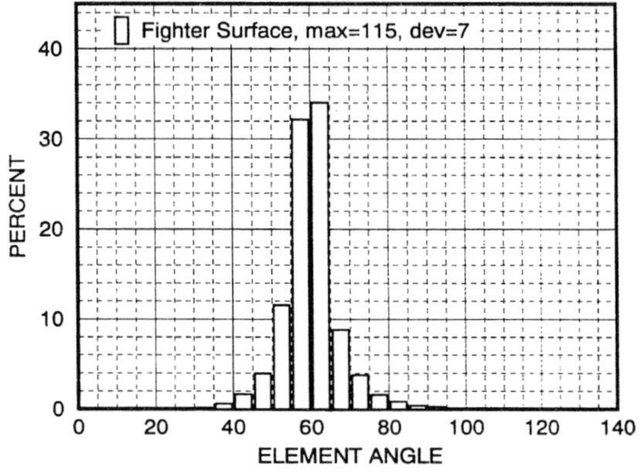

a. Surface grid quality.

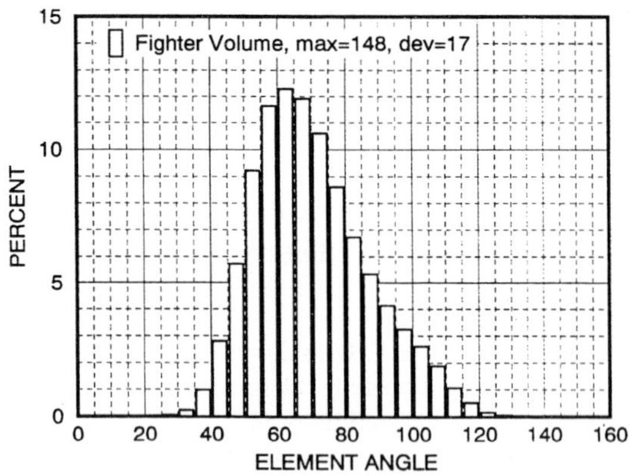

b. Volume grid quality.

Fig. 26 Grid quality distributions for fighter grid.

29. The complete surface grid contains 75,440 total boundary faces. A tetrahedral field cut near the injector is shown in Fig. 30. The complete volume grid contains 127,135 points and 644,305 elements. Element size

Fig. 27 Tetrahedral field cut through wing tip of fighter grid with high–aspect–ratio elements near surface.

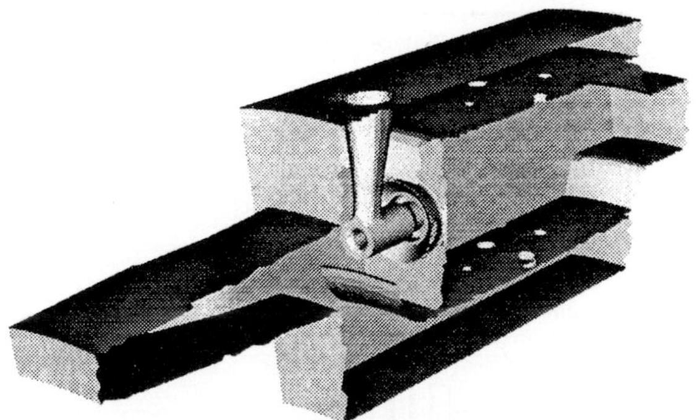

Fig. 28 Sectioned surface of combustor.

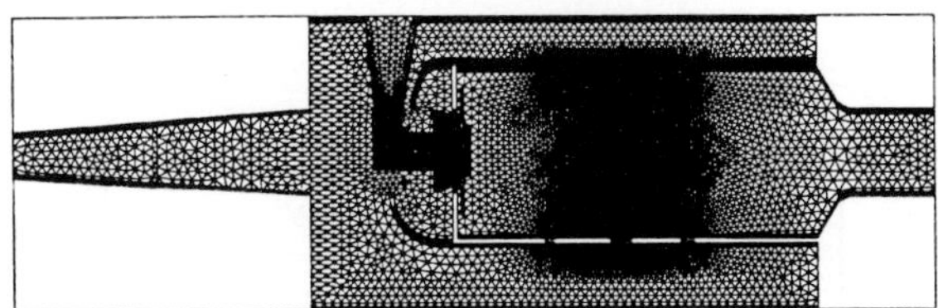

Fig. 29 Combustor surface grid.

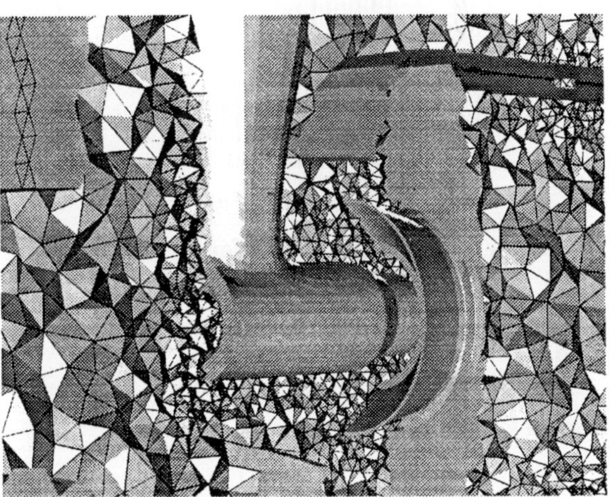

Fig. 30 Tetrahedral field cut for combustor grid generated using point distribution function growth.

varies smoothly on the surface and within the volume grid. Grid quality distributions for the surface and volume grids are shown in Figs. 31 a and b. Element angle distributions, maximum values, and standard deviations verify that the surface and volume grids are of very high quality. Required CPU time is listed in Table 2.

Point distribution function growth was used to automatically increase element size in the interior regions. Use of growth is beneficial in cases where available computer memory for the solver is limited. And, in cases where an increased point spacing in the interior is compatible with expected solution gradients. Without growth the resulting volume grid for the combustor contains 170,335 points and 898,979 elements. Overall quality does degrade slightly with the use of geometric growth. Usually, the result is a higher standard deviation in element angles. A comparison of the grid quality for a growth rate of 1.2 and 1.5 is shown in Fig. 31b. The volume grid shown in Fig. 30 and listed in Table 2 was generated with a rate of 1.2. Using a growth rate of 1.5 the resulting volume grid contains 57,644 points and 224,718 elements. Increasing the growth rate results in a further reduction in the total number of elements at the expense of overall grid quality. Degradation in quality is typically not a significant factor for growth rates of 1.5 or less.

Launch Vehicle

A complete grid suitable for inviscid CFD analysis was generated for a launch vehicle with strap–on boosters. The computational domain is a symmetric slice of the complete configuration. Partial–views of the surface grid and geometry for one symmetry plane, the main booster, and the strap–on booster are shown in Figs. 32 a and b. The complete surface grid contains 121,860 total boundary faces. A tetrahedral field cut near the strap–on booster rocket motor is shown in Fig. 33. The complete volume grid contains 178,099 points and 877,103 elements. Element size varies smoothly on the surface and within the volume grid. Grid quality distributions for the surface and volume grids are shown in Figs. 34 a and b. Element angle distributions, maximum values, and standard deviations verify that the surface and volume grids are of very high quality. Required CPU time is listed in Table 2.

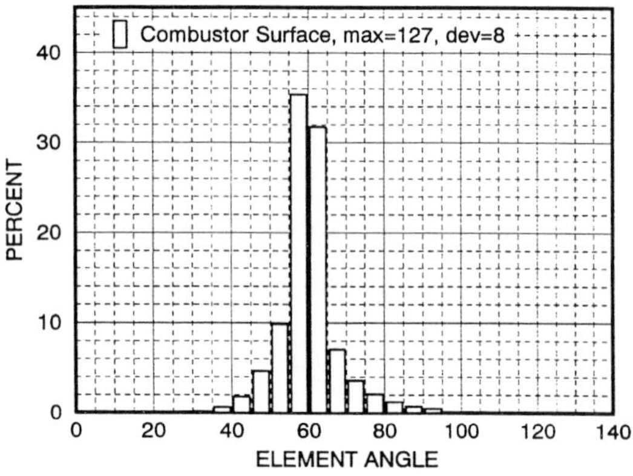

a. Surface grid quality.

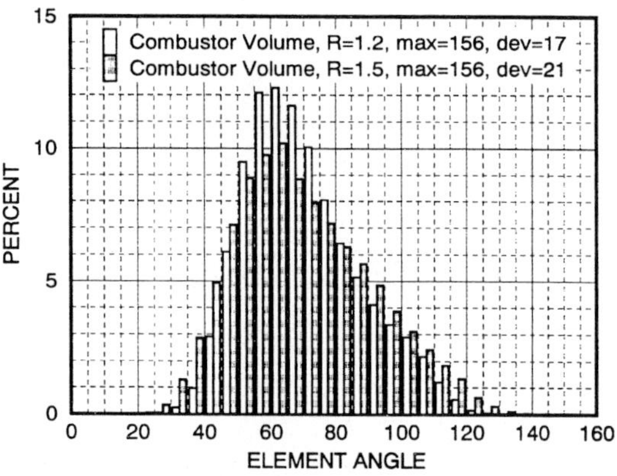

b. Volume grid quality with different growth rates.

Fig. 31 Grid quality distributions for combustor grid.

Ducted Fan

A complete grid was generated for a geometrically complex two–stage ducted fan. The surface grid on the nacelle, fan blades, and hub are shown in Figs. 35 a and b. The complete surface grid contains 150,848 total boundary faces. A tetrahedral field cut is shown in Fig. 36. The complete volume grid generated without growth contains 397,381 points and 2,144,516 elements. Using point distribution function growth with a growth rate of 1.2, the volume grid generated contains 240,780 points and 1,214,018 elements. Tetrahedral field cuts between the blade rows are shown in Figs. 37 a and b for both grids. The effect of point distribution function growth on the point spacing within the volume grid is evident in the region between the blade rows. Element size varies smoothly on the surface and within both of the volume grids. Grid quality distributions for the surface and both volume grids are shown in Figs. 38 a and b. Element angle distributions, maximum values, and standard deviations verify that the surface and both

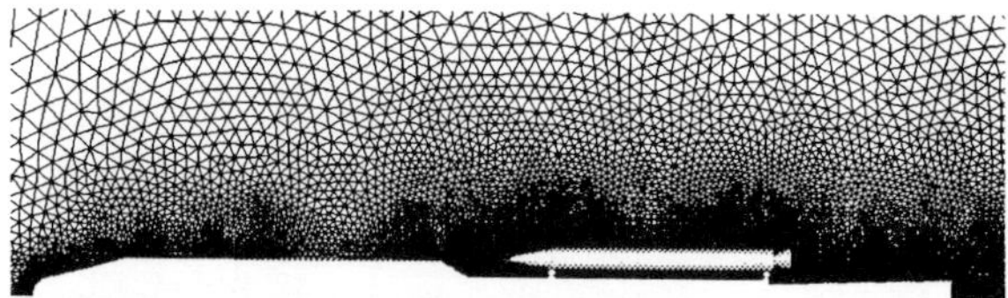

a. Symmetry plane, main booster, and strap–on booster.

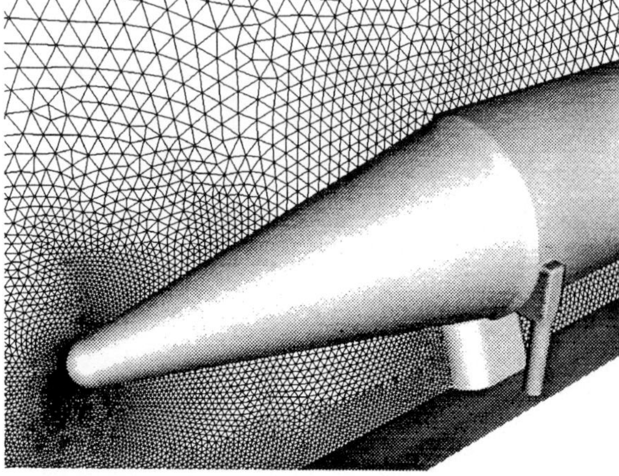

b. Nose of strap–on booster

Fig. 32 Launch vehicle surface grid.

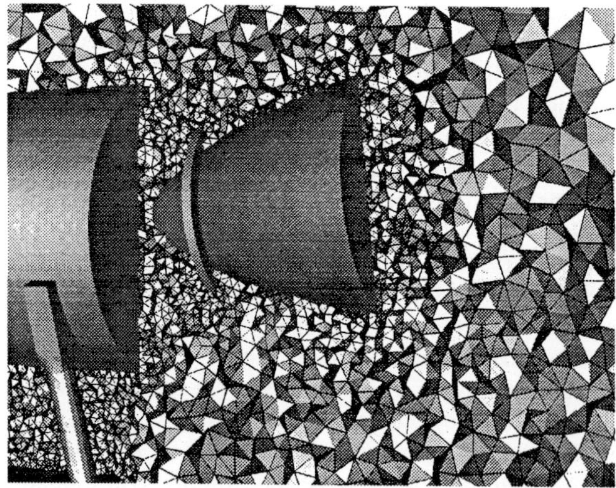

Fig. 33 Tetrahedral field cut for launch vehicle grid.

volume grids are of very high quality. Required CPU times are listed in Table 2. For the grid generated without growth, required CPU times are not available for the PCs tested as they each are configured with 128 Mb of RAM and this case requires about 215 Mb of RAM.

Summary

Methods for generation of unstructured planar, surface, and volume grids using the AFLR procedure have been

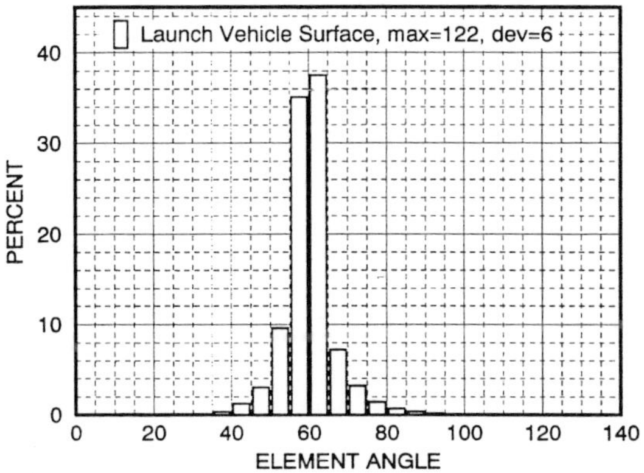

a. Surface grid quality.

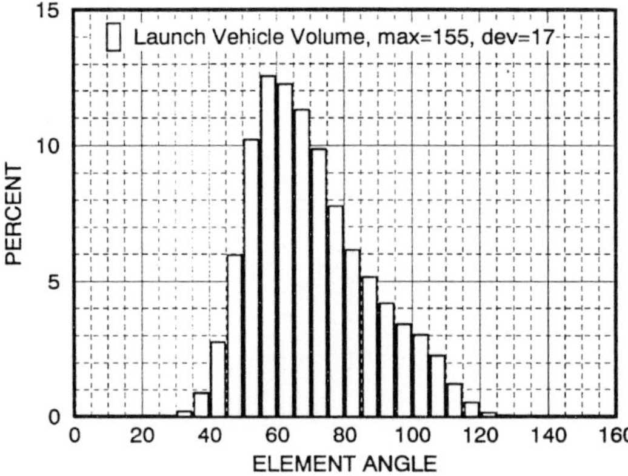

b. Volume grid quality.

Fig. 34 Grid quality distributions for launch vehicle grid.

presented. The AFLR procedure is based on an iterative point insertion scheme using advancing–front, –point, or –normal type point placement and local–reconnection. Results for a variety of configurations have been presented. The results demonstrate that the AFLR procedure consistently produces grids of very–high quality. Efficiency is such that standard PCs or workstations can be used to generate three–dimensional unstructured grids for complex configurations. The combined quality and efficiency of the AFLR procedure represents the current state–of–the–art in unstructured tetrahedral grid generation.

Acknowledgements

The author would like to acknowledge the efforts of Adam Gaither at the MSU ERC for preparing the CAD geometry definitions, generating the surface grids, and integrating the software used to produce the results presented in the Three–Dimensional Application Examples section and McDonnell Douglas Aerospace for preparing and supplying the launch vehicle geometry. The author would also like to acknowledge

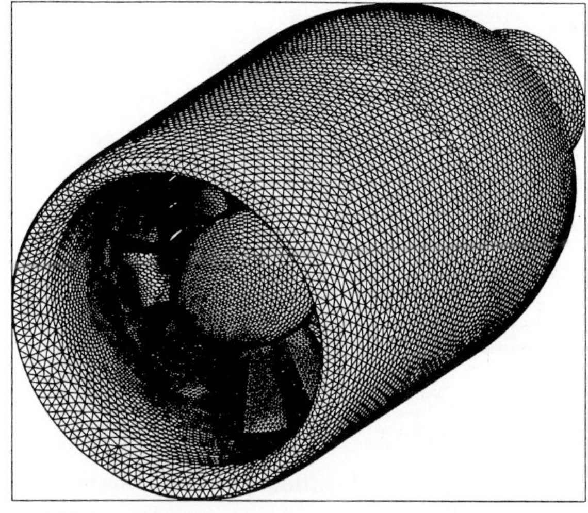

a. With nacelle.

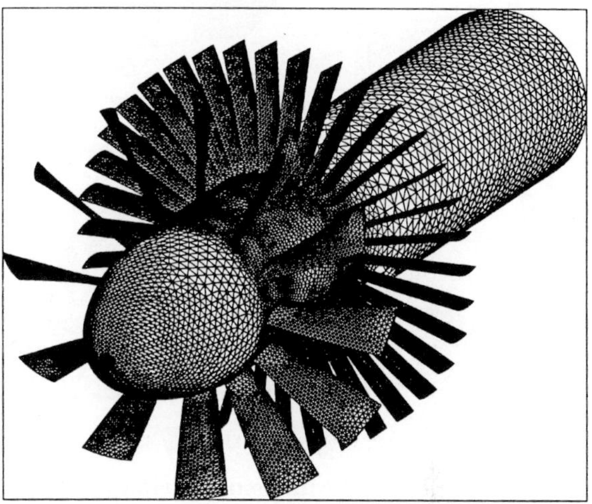

b. Fan blades and hub only.

Fig. 35 Ducted fan surface grid.

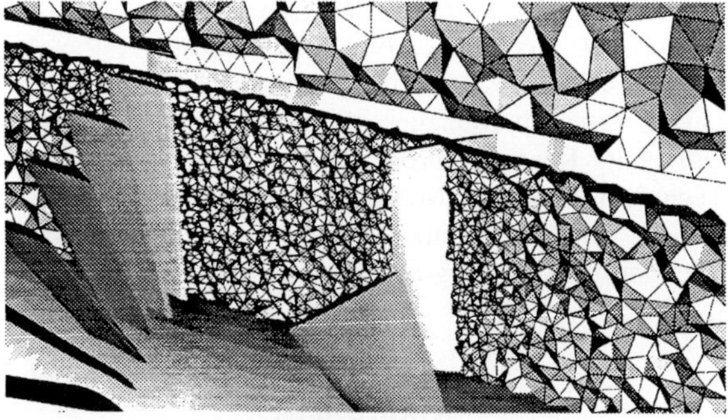

Fig. 36 Tetrahedral field cut for ducted fan grid without point distribution function growth.

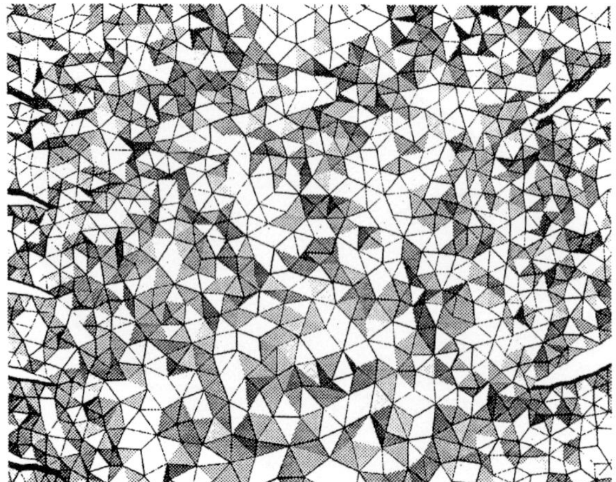

a. Fan grid generated without point distribution function growth.

b. Fan grid generated with point distribution function growth.

Fig. 37 Tetrahedral field cuts for ducted fan grids.

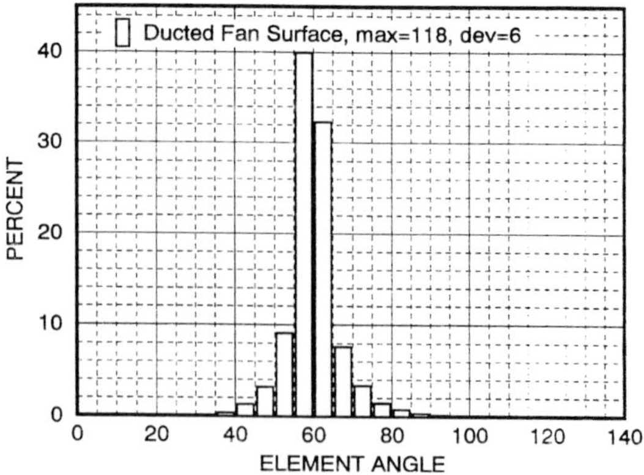

a. Surface grid quality.

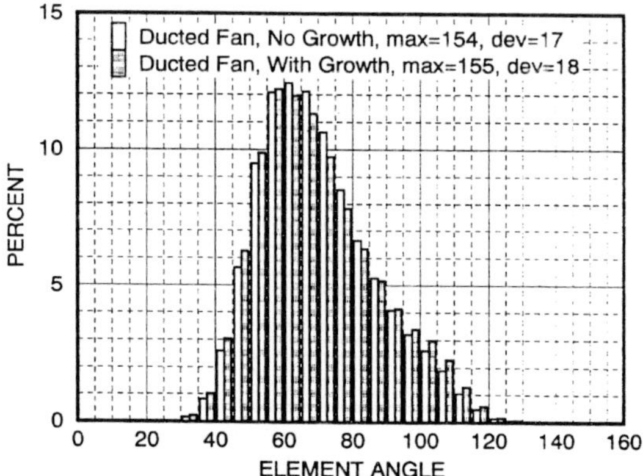

b. Volume grid quality with and without point distribution function growth.

Fig. 38 Grid quality distributions for ducted fan grid.

support for this work from Altair Software, the Air Force Office of Scientific Research, Dr. Leonidas Sakell, Program Manager, the Ford Motor Company, University Research Program, Dr. Thomas P. Gielda, Technical Monitor, McDonnell Douglas Aerospace, Dan L. Pavish, Technical Monitor, National Science Foundation, ERC Program, Dr. George K. Lea, Program Director.

References

[1] Shepard, M.S. and Georges, M.K., "Automatic Three–Dimensional Mesh Generation by the Finite Octree Technique," *International Journal of Numerical Methods in Engineering*, **32**, 709–749, 1991.

[2] Peraire, J., Peiro, J., Formaggia, L., Morgan, K., and Zienkiewicz, O.C., "Finite Element Euler Computations in Three–Dimensions," *International Journal of Numerical Methods in Engineering*, **26**, 2135–2159, 1988.

[3] Lohner, R. and Parikh, P., "Three–Dimensional Grid Generation by the Advancing–Front Method," *International Journal of Numerical Methods in Fluids*, **8**, 1135–1149, 1988.

[4] Weatherill, N.P., "A Method for Generation of Unstructured Grids Using Dirichlet Tessellations," Princeton University, *MAE Report No. 1715*, 1985.

[5] Baker, T.J., "Three–Dimensional Mesh Generation by Triangulation of Arbitrary Point Sets," *AIAA Paper 87–1124*, 1987.

[6] Holmes, D.G. and Snyder, D.D., "The Generation of Unstructured Meshes Using Delaunay Triangulation," *Proceedings of the Second International Conference on*

Numerical Grid Generation in Computational Fluid Dynamics, Eds. Sengupta, S., Hauser, J. Eisman, P.R., and Thompson, J.F., Pineridge Press Ltd., 1988.

[7] George, P.L., Hecht, F., and Saltel, E. "Fully Automatic Mesh Generator for 3D Domains of any Shape," *Impact of Computing in Science and Engineering*, **2**, 187–218, 1990.

[8] Mavriplis, D.J., "An Advancing Front Delaunay Triangulation Algorithm Designed for Robustness," *AIAA Paper 93–0671*, 1993.

[9] Rebay, S. "Efficient Unstructured Mesh Generation by Means of Delaunay Triangulation and Bowyer–Watson Algorithm," *Journal of Computational Physics*, **106**, 125–138, 1993.

[10] Muller, J.D., Roe, P.L., and Deconinck, H., "A Frontal Approach for Internal Node Generation in Delaunay Triangulations," *International Journal of Numerical Methods in Fluids*, **17**, 241–256, 1993.

[11] Lawson, C.L., "Properties of n–dimensional Triangulations," *Computer Aided Geometric Design*, **3**, 231–246, (1986).

[12] Barth, T.J., "Steiner Triangulation for Isotropic and Stretched Elements," *AIAA Paper 95–0213*, 1995.

[13] Marcum, D. L. and Weatherill, N. P., "Unstructured Grid Generation Using Iterative Point Insertion and Local Reconnection," *AIAA Journal,* **33**, 1619–1625, 1995.

[14] Marcum, D. L., "Generation of Unstructured Grids for Viscous Flow Applications," AIAA–95–0212, *33rd AIAA Aerospace Sciences Meeting*, Reno, NV, January 1995.

[15] Marcum, D. L., "Generation of High–Quality Unstructured Grids for Computational Field Simulation," *6th International Symposium on Computational Fluid Dynamics*, Lake Tahoe, NV, September 1995.

[16] Marcum, D.L., "Adaptive Unstructured Grid Generation for Viscous Flow Applications," *AIAA Journal*, **34**, No. 11, p. 2440–2443, 1996.

[17] Marcum, D. L. and Gaither, K.P., "Solution Adaptive Unstructured Grid Generation Using Pseudo–Pattern Recognition Techniques," AIAA–97–1869, *13th AIAA Computational Fluid Dynamics Conference*, Snowmass, CO, June 1997.

[18] Marcum, D.L., "Control of Point Placement and Connectivity in Unstructured Grid Generation Procedures," *IX International Conference on Finite Elements in Fluids*, Venice, Italy, October 1995.

[19] Marcum, D. L., "Unstructured Grid Generation Components for Complete Systems," *5th International Conference on Grid Generation in Computational Fluid Simulations*, Starkville, MS, April 1996.

[20] Lohner, R., "Matching Semi–Structured and Unstructured Grids for Navier–Stokes Calculations," AIAA Paper 93–3348, *11th AIAA Computational Fluid Dynamics Conference*, Orlando, FL, July 1993.

ANISOTROPIC MESH ADAPTION BY METRIC CONTROL FOR SYSTEMS OF PDE INVOLVING MULTISCALE PHENOMENA

Frederic HECHT [1]   Bijan MOHAMMADI [1]

(1) INRIA, BP 105, 78 153 Le Chesnay, France
Email: 1stname.lastname@inria.fr

Abstract

Several new ideas for anisotropic adaption of un- structured triangular grids are presented with par- ticular emphasis to fluid flows computations in- volving multiscale phenomena.

KEY WORDS: Mesh, Anisotropic, Adaptation, Delaunay, Metric, CFD.

1. Introduction

From finite element theory, it is well known that for isotropic problems, meshes with equilateral triangles are more suitable. But, equilaterality notion involves lengths through scalar product in a given metric. Therefore, anisotropic meshes might be seen as isotropic with respect to a different metric. Now, if we define this metric by an a posteriori error estimation, we can do adaption of the mesh following the solution.

Of course, an unstructured grid environment is the natural framework for the introduction of general adaptivity and anisotropy concepts ([7], [8], [14]). However, our experience shows that in adaption procedure based on local metric changes, three major difficulties remain:

1) As the metric is defined from the interpolation error of some quantity, the extension to systems of PDE is not clear. Especially when phenomena of different nature interact (for instance in a shock-boundary layer interactions).
2) Boundary layers are not correctly resolved and wall coefficients (especially the friction coefficient) obtained over adapted meshes are useless. This is due to the fact that the distance of the first layer points to the wall is not uniform. This is one of the major weakness of anisotropic adaption in CFD.
3) Multiscale phenomena are hard to capture by the adaption (for instance small eddies in a turbulent flow).

In this paper, we present three idea to remove the above difficulties.

The first idea is the intersection of the different metrics obtained for the different variables of a PDE which are in our case the conservation variables. This remove also the difficulty we had until now for the 'right' variable to choose for building the metric from its interpolation error. Indeed, we used to choose for instance the pressure or the density for Euler computations and for instance the local Mach number or the entropy for viscous computations. Of course, this is a dead-end as none of these choices could give satisfaction because one variable cannot encapsulate all the physics of the system.

The second idea is motivated by two requirements coming in fact from structured meshes which are known to be more suitable near the walls:

- we want the meshes to be as orthogonal as possible near the body,
- we want each nodes layer to be at a uniform distance of the wall.

The idea consists of a modification of our local metric near the wall. To enforce the orthogonality of the mesh in this area, we change the eigenvectors of our metric. The second requirement is satisfied by giving to the user the possibility of prescribing the normal size of the elements along the wall and to propagate this through the domain by a relaxation procedure to obtain a quasi-orthogonal mesh in near-wall regions with each nodes layer at a uniform distance from the wall.

The third ingredient helps the capture of multiscale phenomena by introducing a relative rather than global definition of the metric. This also removes the dimensional incompatibilities between the metrics coming from different variables when we use the first ingredient.

These ingredients are extensible to three dimensional cases. In fact, to generate 2D meshes, we have used our 3D surface grid generator after set-

Received on August 1, 1997.

ting z to zero. The generation procedure is fast (an average being 30000 triangles by minute by MegaFlops).

Our adaption loop can be defined as follows:
Start from an initial mesh.
Adaptation loop.

a) Solve the PDE (in this case the fluid dynamics system);

b) Build a metric from the multi-variable solution obtained in **a)**;

c) Build an equilateral mesh with respect to the metric of **b)** using a Delaunay type mesh generator.

End of the loop.

We will discuss in particular the point **b)** where we introduce the ingredients presented above. We also give a brief description of the fluid solver and the mesh generator.

2 Navier-Stokes solver

We use the NSC2KE fluid solver for these computations. More details can be found in [10]. A Finite-Volume-Galerkin formulation of Navier-Stokes equations in conservation form has been considered. A four stages Runge-Kutta scheme is used for time integration. The Roe Riemann solver [6] has been used for the Euler part together with a MUSCL type reconstruction and Van Albada limiters [15] for second order accuracy. P^1 finite element has been used for the viscous part of the operators. A Stegger-Warming [11] flux splitting has been used at the inflow and outflow boundaries while non penetration or slip boundary conditions are applied to solid walls depending on the nature of the flow. Turbulent modeling is done using the classical $k - \varepsilon$ [12] model with special wall-laws enabling for separated and unsteady flows computations [13, 16].

3 Mesh generation

We give a brief description of our Delaunay type mesh generator, more details can be found in [3] and [4]. For mesh adaption, two strategies are possible. Local optimization of the mesh or global regeneration at each adaption ([7], [8]). In this work, we use the first approach. Both approach have advantages and disadvantages often from the implementation point of view.

Once the metric is given (see below for the metric definition), the mesh generation procedure is applied. This algorithm uses five different local tools: *edge suppression, vertex suppression, vertex addition, edge swapping and vertex reallocation (barycentering step)*. We need three different grids: one defining the domain geometry (*'C.A.D. mesh'*, $\mathcal{T}_G$) which is fixed during the mesh adaption loop, one that contains the control space informations (i.e. the metric tensor definition) denoted by $\mathcal{T}_0$ and the adapted mesh $\mathcal{T}_1$. We can say that this algorithm solves the following problem:

Find "an optimal mesh" $\mathcal{T}_1$ adapted from $\mathcal{T}_0$ with respect to the criterion imposed by the metric tensor $\mathcal{M}$ and compatible with $\mathcal{T}_G$.

So, we suppose that the meshes $\mathcal{T}_G$ and $\mathcal{T}_0$ are given together with the metric tensor $\mathcal{M}$ obtained from a finite element solution over $\mathcal{T}_0$.

Initially, we have $\mathcal{T}_1 = \mathcal{T}_0$, but $\mathcal{T}_G$ can of course be different of $\mathcal{T}_0$.

Different steps can be distinguished in our mesh adaption procedure:

Data structure creation and verification. Prior to grid adaption, it is necessary to create some auxiliary data: mesh connectivity arrays, boundary and inter-sub-domain (geometrical and physical)[5] edges with their associated tangent vectors, fixed point localization, etc. A topological verification step together with a triangle orientation is made in order to guarantee that the given meshes are correct and oriented.

Geometry reconstruction. Generalized Farin's algorithm [3] is used in order to define a 'G^1' Bézier surface over $\mathcal{T}_G$.

Initial regularization. Depending on user choices, an initial optimization step (edge swapping) is made in order to improve the initial mesh quality. Numerical experiences show that the final result is better if this initial optimization is made.

Grid adaption. The triangles of $\mathcal{T}_1$ are ordered in a particular data structure called *double dynamic list (DDL)*. Three double lists are considered (vertices, edges and triangles). Suppose that the first $(i-1)$ edges $a_j, j = (1, .., i-1)$ are treated, then the edge a_i is taken. The procedure that we propose is the following one:

1. Let d_i be the length of a_i computed with the metric tensor $\mathcal{M}$. We have three possibilities:

   - If $d_i > l_{max}$ ($l_{max} \approx 1.4 L_{ref}$) then a_i must be cut in two edges using the add-

---

vertex procedure. The new mesh elements (vertices, edges and triangles) are added at the end of their corresponding DDL. Now, a_i has changed and we again compute its length, d_i. If $d_i > l_{max}$ then we repeat the same process until its associated length $d_i \leq l_{max}$.

- If $d_i < l_{min}$ ($l_{min} \approx 0.6 L_{ref}$) then a_i is suppressed, this leads to new edges creations. We check if the length of these new edges is bigger than l_{max}, in which case the previous process is applied, or smaller than l_{min} where this process is repeated.

- if $l_{min} \leq d_i \leq l_{max}$, a_i is kept.

2. If we have done a local mesh modification, a local optimization step must be considered. In this case, edge swapping is applied to the new or modified edges.

This process is repeated over all the edges.

Final optimization. Finally, a global optimization process is considered. All non prescribed interior vertices with less that 4 elements connected to it are suppressed and a barycentering step is applied: all non fixed vertices are moved to the center of gravity of their neighbors with respect to the metric $\mathcal{M}$.

Thus, a final mesh $\mathcal{T}_1$, adapted from $\mathcal{T}_0$ with respect to the metric tensor $\mathcal{M}$ is obtained and it satisfies the geometrical and topological constraints. After each adaption step, an initial solution is obtained over the new mesh by interpolation of the solution on the previous mesh. A good interpolation is crucial for unsteady computations.

Remarks on mesh generation:

1) The remeshing algorithm always works locally. This strategy allows us to remesh 3D surfaces with only a local parameterization.

2) As based on a Delaunay algorithm, this technique avoids the possibility of any overlapping in the mesh from the moment that the metric is well defined (i.e. definite positive).

4 Metric computations

During automatic mesh generation it is necessary to have as much informations as possible on the nature and the local behavior of the solution. These indications constitute *the control space* which governs the grid generation. As we are interested by informations on the elements shape and size, we need to convert somehow what we known on

the solution to something having the dimension of a length and containing directional informations. This might be done by giving at each point $x \in \Omega \subset \mathbb{R}^2$ three parameters (six in $\mathbb{R}^3$). This kind of control is equivalent to an isotropic control with a change of the metric tensor all over the domain[1].

It is easy to prove that if K_0 is a non degenerated triangle then, there exists a unique metric, where K_0 is equilateral with unity edge lengths. So, in order to obtain triangles with a given stretching and size over a sub-domain, it is sufficient to construct an isotropic mesh using the metric tensor $\mathcal{M}$ with an associated refinement function h equal to 1 everywhere. $\mathcal{M}$ is given at every point $x \in \Omega \subset \mathbb{R}^2$ by

$$\mathcal{M}(x) = \mathcal{R}(x) \begin{pmatrix} \lambda_1(x) & 0 \\ 0 & \lambda_2(x) \end{pmatrix} \mathcal{R}(x)^{-1}, \quad (1)$$

where $\lambda_1(x) > 0$ and $\lambda_2(x) > 0$ are the $\mathcal{M}(x)$ eigen-values and $\mathcal{R}(x)$ is the rotation matrix of angle $\alpha(x)$ that maps the $\mathbb{R}^2$ canonical basis over the $\mathcal{M}(x)$ unit eigen-vectors.

Elementary differential geometry says that the length of a parametric curve $\Gamma(t)$ where $t \in [0, 1]$ in the new metric is defined by:

$$L(\Gamma) = \int_0^1 \sqrt{\Gamma'(t)^T \mathcal{M}(\Gamma(t)) \Gamma'(t)} \, dt. \quad (2)$$

We propose the following discretization of the metric tensor $\mathcal{M}$. Let $\mathcal{M}$ be given at vertices $\{\mathcal{M}(x_i), \ i = 1, \ldots, n_v\}$, then over each mesh triangle, K_0, the $\mathcal{M}$ coefficients are linearly interpolated and a continuous discretization of the metric tensor $\mathcal{M}$ is obtained.

Let $\Gamma = [x_0, x_1]$ be a segment, $\Gamma \subset K_0$. Computing its length using equation (2), we obtain

$$L(\Gamma) = \int_0^1 \sqrt{(\Gamma')^T \mathcal{M}(\Gamma(t)) \Gamma'} \, dt \quad (3)$$

$$= \int_0^1 \sqrt{l_0^2 + t(l_1^2 - l_0^2)} \, dt \quad (4)$$

$$= \frac{2}{3} \frac{l_0^2 + l_0 l_1 + l_1^2}{l_0 + l_1}, \quad (5)$$

where $l_i = \sqrt{(\Gamma')^T \mathcal{M}(x_i) \Gamma'}$, $i = 0, 1$.

We now discuss how the metric tensor $\mathcal{M}$ is defined in order to satisfy an adaption criterion. Suppose that we only work with one variable denoted

---

[1] For the sake of simplicity, we will consider in this section that $\Omega \subset \mathbb{R}^2$. We obtain the same results if Ω defines a surface or if $\Omega \subset \mathbb{R}^3$. In the latest case, 'triangles' are replaced by 'tetrahedron'.

η. We are going to determine the metric tensor in order to equilibrate the interpolation error.

Assume that a first solution has been computed over a given mesh and let η be continuous piecewise linear interpolated, then (see [1] and [2]) the interpolation error depends on the Hessian matrix of η:

$$\mathcal{E} = |\eta - \Pi_h \eta|_0 \leq c_0 h^2 |\mathcal{H}(\eta)|_0, \tag{6}$$

where $\Pi_h \eta$ is the P^1 interpolation of η and

$$\mathcal{H} = \begin{pmatrix} \partial^2 \eta / \partial x^2 & \partial^2 \eta / \partial x \partial y \\ \partial^2 \eta / \partial x \partial y & \partial^2 \eta / \partial y^2 \end{pmatrix} \tag{7}$$

$$= \mathcal{R} \begin{pmatrix} \lambda_1 & 0 \\ 0 & \lambda_2 \end{pmatrix} \mathcal{R}^{-1},$$

and the metric tensor $\mathcal{M}$ is defined by

$$\mathcal{M} = \mathcal{R} \begin{pmatrix} |\lambda_1| & 0 \\ 0 & |\lambda_2| \end{pmatrix} \mathcal{R}^{-1}. \tag{8}$$

The error over a mesh edge a_i can be computed as

$$E_i \approx |a_i^T H a_i| \leq \sqrt{c_0} a_i^T M a_i. \tag{9}$$

Now, the mesh is equilateral and of edge length $\sqrt{c_0}$ in the metric defined by $\mathcal{M}$.

We can associate an ellipse $\mathcal{E}_\mathcal{M}$ to the metric tensor $\mathcal{M}$ having axes d_1 and d_2 (eigen-vectors of $\mathcal{M}$) and with lengths $1/\sqrt{\lambda_1}$ and $1/\sqrt{\lambda_2}$, respectively.

Remarks on metric computations:

1) In the metric definition, we have to introduce the maximum and minimum edge lengths in the mesh to avoid unrealistic metrics. This is not really a restriction as usually we have a good idea on what these quantities should be. More precisely, the eigenvalues of the metric are limited as follow:

$$\tilde{\lambda}_{1,2} = min(max(|\lambda_{1,2}|, \frac{1}{h_{max}^2}), \frac{1}{h_{min}^2}),$$

with h_{min} and h_{max} being the minimal and maximal edge lengths allowed in the mesh.

2) As the key point in the metric definition is the second derivatives. But, in our application the solver is a P1 finite element solver. Therefore, a weak formulation (by Green formula) has to be used to compute the Hessians.

4.1 Extension to systems

Suppose now, that several variables $\eta_1, \eta_2, \ldots, \eta_r$ are given. The problem becomes: *find the metric so that the maximum interpolation error is minimized for all the variables.* It is clear, from the geometrical identification ellipse–metric that the solution to the previous minimization problem is to find the biggest ellipse contained in the intersection of all the ellipses $\mathcal{E}_1, \mathcal{E}_2, \ldots, \mathcal{E}_r$ corresponding to metrics $\mathcal{M}_1, \mathcal{M}_2, \ldots, \mathcal{M}_r$ computed from the variables $\eta_1, \eta_2, \ldots, \eta_r$. It is not easy in general to find the optimal solution of this problem. However the following algorithm seems to be suitable enough (see fig. 1). Suppose that only two variables (η_1 and η_2) are provided. We find an approximation of the optimal intersection ellipse by the following procedure:

- if the two ellipses $\mathcal{E}_1, \mathcal{E}_2$ do not intersect (this happens when one ellipse is contained into another), the one with smallest area is taken as intersection and the suitable metric is the associated one;

- otherwise, let λ_i^j and $v_i^j, i, j = 1, 2$ the eigenvalues and eigen-vectors of $\mathcal{M}_j, j = 1, 2$. The intersection metric $\hat{\mathcal{M}}$, is defined by

$$\hat{\mathcal{M}} = \frac{\hat{\mathcal{M}}_1 + \hat{\mathcal{M}}_2}{2}. \tag{10}$$

where $\hat{\mathcal{M}}_1$ (resp. $\hat{\mathcal{M}}_2$) has the same eigenvectors of $\mathcal{M}_1, (v_1^1, v_2^1)$ and with eigen-values

$$\tilde{\lambda}_i^1 = max(\lambda_i^1, v_i^{1^T} \mathcal{M}_2 v_i^1), \quad i = 1, 2. \tag{11}$$

Now, if n variables are given, the final intersection metric is computed as follows:

- $\hat{\mathcal{M}} = \text{intersection}(\mathcal{M}_1, \mathcal{M}_2)$.

- for $i = 3, \ldots, n$
  $\hat{\mathcal{M}} = \text{intersection}(\hat{\mathcal{M}}, \mathcal{M}_i)$.

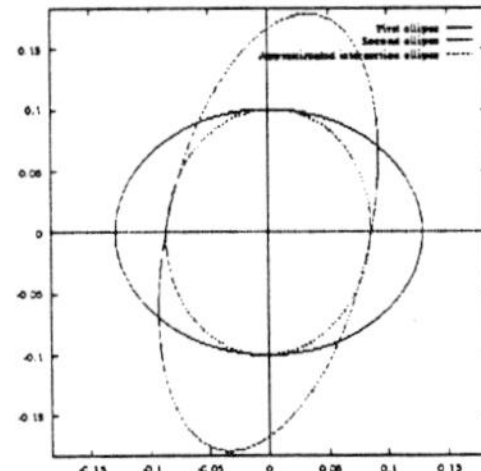

Figure 1: Approximated optimal ellipse of two-ellipse intersection.

Here a dimensional problem appears due to the fact that the different metrics are based on different variables having different dimension. We will see that a relative rather than the global error estimation (6) permits to avoid this problem.

4.2 Boundary layer improvement

As said previously, metric evaluation is done by Green's formula with homogenous Neumann boundary condition on all boundaries. This is quite logical and means that elements do not change in size and shape near boundaries. But this also means that informations on normal mesh sizes are not available at the walls and as a result they are generally overestimated. Our experience shows that this is not suitable for viscous computation in boundary layers. This is why structured meshes are usually more suitable for these regions.

We have tried to give a new boundary condition for the metric on the walls to improve the mesh definition in near-wall area by approaching structured meshes. More precisely, along the wall the previous metric $\mathcal{M}(x)$ is replaced by a new metric $\hat{\mathcal{M}}(x)$:

$$\hat{\mathcal{M}}(x) = T \Lambda T^{-1},$$

where

$$\Lambda = diag(\frac{1}{h_n^2}, \lambda_\tau) \quad \text{and} \quad T = (\vec{n}(x), \vec{\tau}(x)).$$

In other words, along solid walls the eigen-vectors are now the unit normal and tangent vectors at the wall with $1/h_n^2$ and λ_τ the corresponding eigen-values. The refinement along the wall can now be monitored through h_n. More precisely, in the adapted mesh the distance from the wall of the first layer nodes will be h_n. This quantity is given by the user and depends on the Reynolds number. On the other hand, λ_τ is computed taking the maximum of the two projection lengths on $\vec{\tau}(x)$ of the $\mathcal{M}$ unit eigen-vectors times the corresponding eigen-values.

It is necessary to keep the information coming from the original metric on λ_τ to be able to recognize situations such as shock-boundary layer interactions.

A relaxation procedure can now be applied to propagate this orthogonality property further in the mesh. In this paper, this has been done for only two nodes layers.

4.3 Multiscale phenomena

As we said, our experience shows that the previous approach is not satisfactory to capture multiscale phenomena. For instance, when several eddies with variable energy are present, the mesh adaption has difficulty to capture the weaker ones, especially if there are shocks involved in the flow. We notice that (6) leads to a global error while we would like to have a relative one. Indeed, for the backward facing step for instance, the mean flow velocities at the inflow and in the main and in the secondary recirculations are different by several order of magnitude ($10^4 - 10^5$). Therefore, when using a global criteria to evaluate the metric, the secondary recirculation is difficult to capture. We propose the following estimation which takes into account not only the dimension of the variables but also their magnitude:

$$\mathscr{E} = |\frac{\eta - \Pi_h \eta}{\max(|\Pi_h \eta|, \epsilon)}|_0 \qquad (12)$$

$$\leq c_0 h^2 |\frac{D^2 \eta}{\max(|\Pi_h \eta|, \epsilon)}|_0,$$

where we have introduced the local value of the variable in the norm. ϵ is a cut-off to avoid numerical difficulties and also to define the difference between the orders of magnitude of the smallest and largest scales we try to capture. Indeed, when a phenomena falls bellow ϵ, it will not be captured. This can also be seen as looking for a more precise estimation in regions where the variable is small (i.e. a variable c_0 in (6)). Another important consequence of this estimation is that it removes the dimensional problems when intersecting metrics coming from different quantities.

5 Numerical examples

In this section we describe several configurations of compressible inviscid and viscous flows. For all these cases, the initial meshes have been generated using EMC2 [9].

For the steady cases, the stop criteria for the adaption loop has been the independence with respect of the mesh of the results (especially wall coefficients).

These techniques have been validated on several other configurations like multi-element airfoils. These computations can be found in [5].

5.1 Turbulent flow over a backward step

This is the classical backward-step (ratio 2 between the step and the canal heights) at $Re_{\infty/H} = 44000$ and inflow Mach number of 0.1. The aim here is to show the difference between the relative and global error estimation for the capture of multiscale phenomena. The flow has been computed on a fine regular mesh. The predicted separation length is seven. We capture surprisingly well the secondary

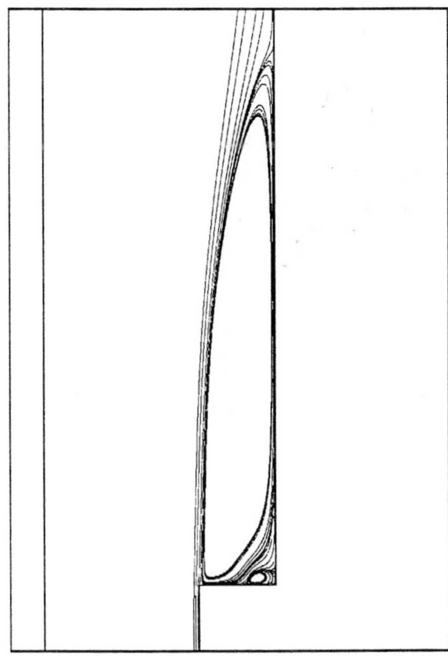

Figure 2: *Backward facing step: particle tracking for the computation using the relative error estimation.*

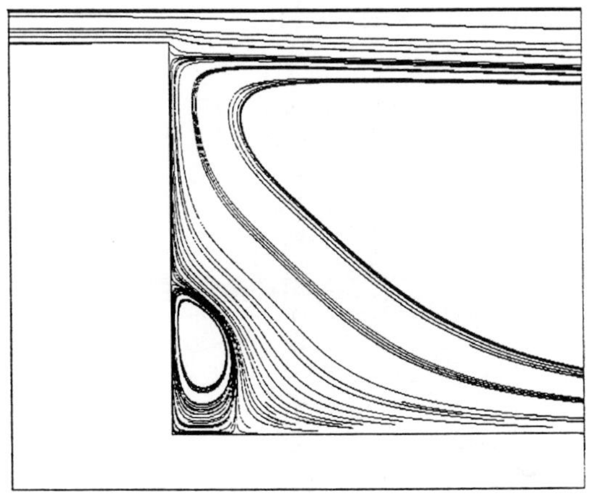

Figure 3: *Backward facing step: particle tracking for the computation using the relative error estimation (zoom).*

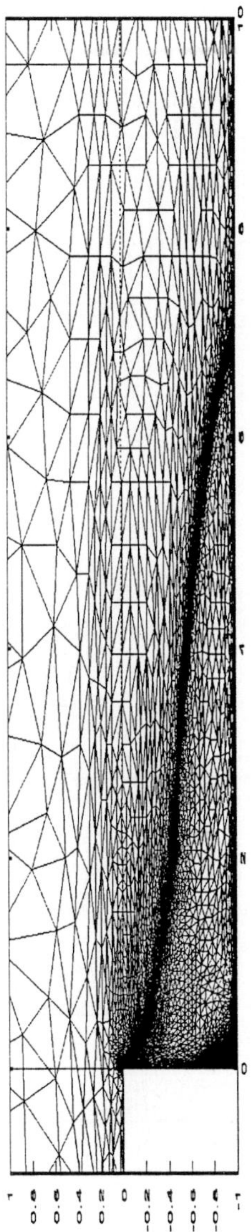

Figure 4: *Backward facing step: partial view of the mesh obtained with the relative criteria. The main and secondary recirculations are correctly identified.*

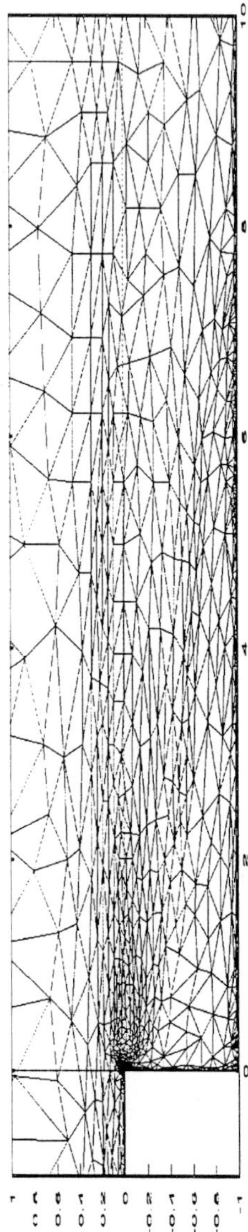

Figure 5: *Backward facing step: partial view of the mesh obtained with the global criteria. The main recirculation is weakly detected and the secondary one has not been captured.*

recirculation with our wall-laws. This is an important point as it is usually stated that this is an impossible task with eddy viscosity models and wall-laws and that we need more sophisticated models for this. The secondary recirculation has a length of about 0.2 H and a height of about 0.35 H.

From this mesh and solution we do one adaption using the global and relative criteria with the same $c_0\mathcal{E} = 1.e - 2$. We can see that when using the global estimation, the secondary bubble area (despite it exists) is not detected by the adaption. The relative estimation however clearly capture the bubble. The relative estimation also leads to a much finer mesh in regions of interest. With the global criteria, to have a similar refinement, we need to ask for a global error of several order of magnitude less but this will lead to a uniform refinement everywhere. Therefore, if the global criteria is not able to detect a phenomena which is already present, there is no hope for it to predict it at all.

5.2 Transonic Turbulent Flow over a RAE2822

This is a transonic flow at inflow Mach number of 0.734 and 2.79 degrees of incidence. The chord based Reynolds number is $Re_C = 6.510^6$. Experimental data are available for the pressure and friction coefficients distribution [17]. One difficulty here is to correctly predict the shock position. The aim here is to show the impact of the ingredients described above on mesh generation for boundary layers. The parameters h_{min}, h_{max} and h_n are respectively $0.01C$, $3.C$ and $0.0002C$. The interpolation error is $c\mathcal{E} = 510^{-3}$.

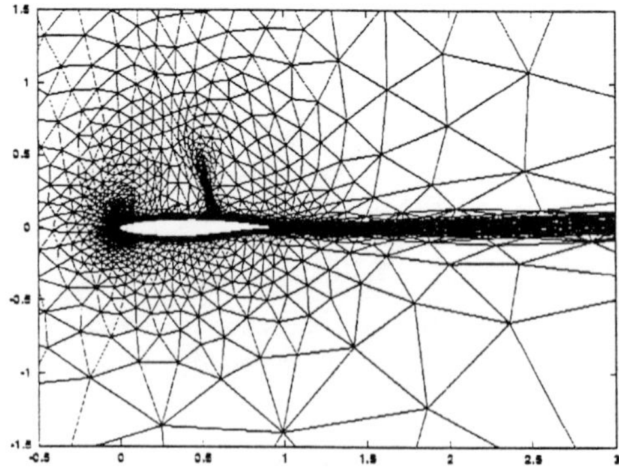

Figure 6: *RAE 2822: Adapted mesh, about 7000 nodes (partial view).*

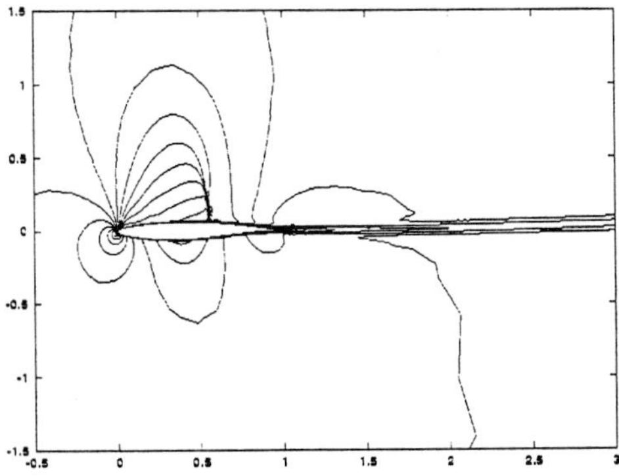

Figure 7: *RAE 2822: Iso-Mach contours (partial view).*

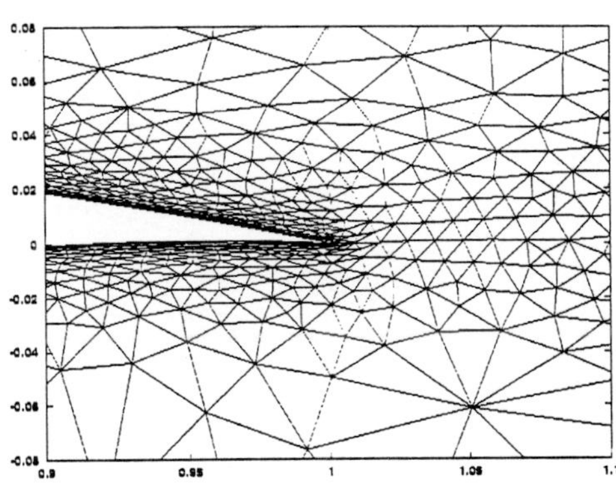

Figure 10: *RAE 2822: zoom around the trailind edge.*

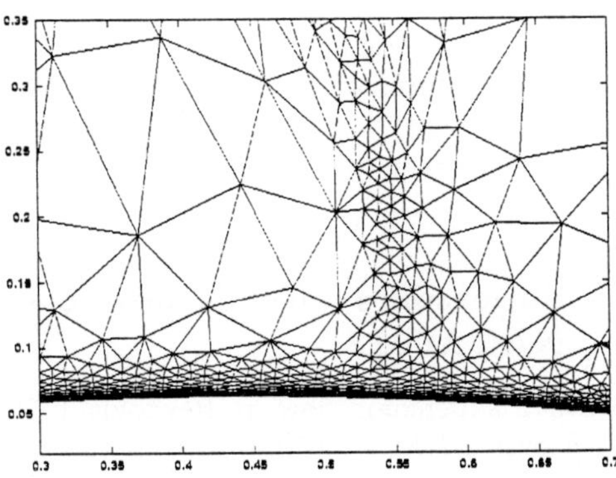

Figure 8: *RAE 2822: zoom over the region of shock-boundary layer interaction.*

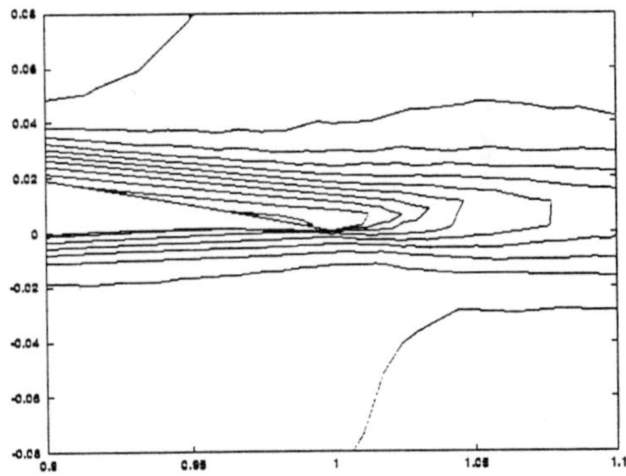

Figure 11: *RAE 2822: Iso-Mach contours around the trailind edge.*

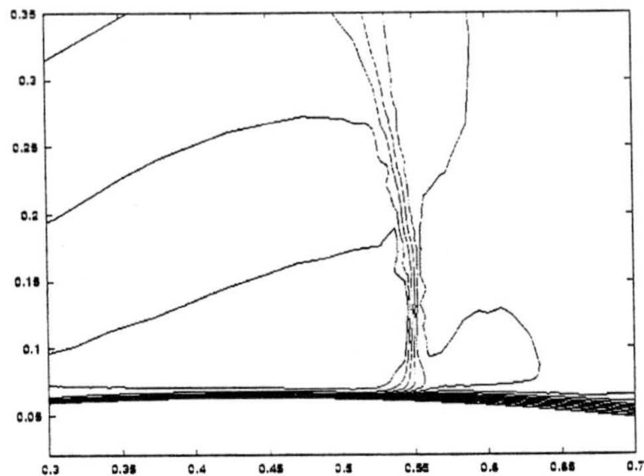

Figure 9: *RAE 2822: Iso-Mach contours in the region of shock-boundary layer interaction.*

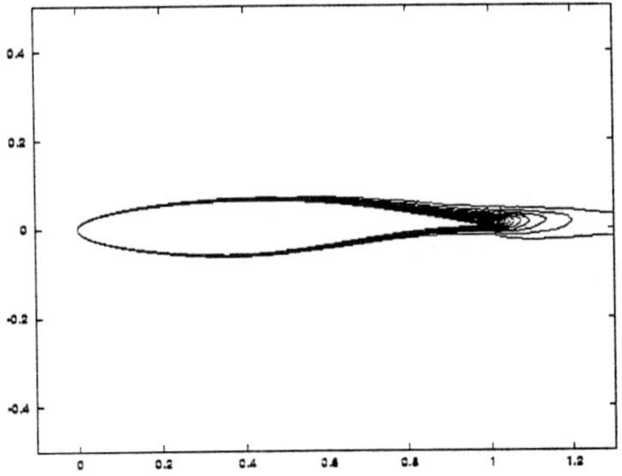

Figure 12: *RAE 2822: Iso-k contours.*

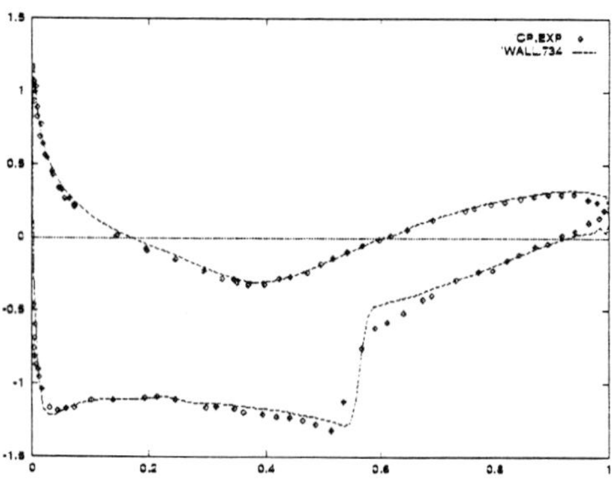

Figure 13: *RAE 2822: pressure coefficient distribution.*

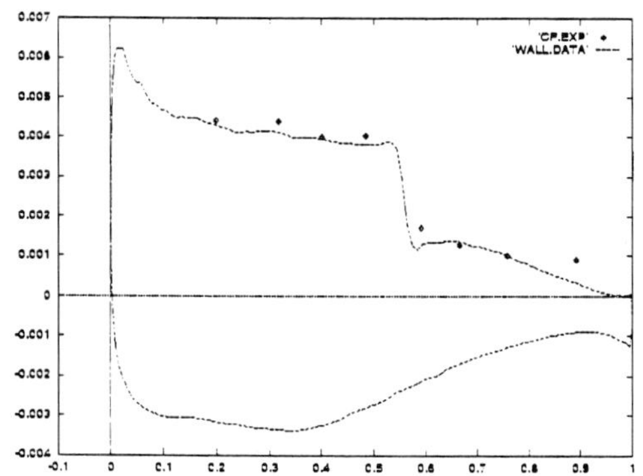

Figure 14: *RAE 2822: friction coefficient distribution.*

5.3   Supersonic scramjet inlet

This case consists of an internal supersonic flow at Mach 3 in a scramjet inlet. Despite the configuration being symmetric, the whole domain has been computed to see if the solution remains symmetric or not. This case is devoted to an evaluation of our multi-variable strategy for the metric definition. The second modification is not used here as it is an inviscid flow.

Five automatic adaptions have been applied. The accuracy of the resolution of the discontinuities for this flow at step 5 is surprising. It is interesting to notice that there is no oscillation in the solution and that the fluid solver seems to nicely accept such a meshes. Moreover, we see

that the solution stays perfectly symmetric even on a clearly non-symmetric mesh.

The initial mesh has around 8000 triangles and the final mesh around 40000. This gives an idea of the number of elements we might need if we want an equivalent resolution on a structured grids.

We compare the normalized residual evolution of the adapted computation to a direct computation where we have started on the last adapted mesh from an uniform solution using the same time integration procedure. The residual is based on the norm of the right-hand-side of the equations and not on the time derivative terms. In this way, the time step size do not influence the convergence history. This gives an idea of the cost of similar computations on a large uniform grid. In fact, for the same resolution we would need much more grid points in an uniform mesh but this just enforces our conclusions.

The convergence is accelerated by the adaption technique. In term of CPU, as in the adaption loop most of the work is done in coarser levels, the computation cost is reduced by at least a factor of 20.

5.4   Supersonic viscous flow over a NACA0012

A Mach 2 supersonic flow at Reynolds 10000 is studied over a NACA 0012 airfoil. In this configuration, the two new ingredients we propose here are used. The quality of wall coefficients (especially the friction coefficient) is studied too. In particular, we want to see how much they depend on the mesh.

The adaption loop length is 6 for this case. As expected, for viscous computations the convergence acceleration of the adaption strategy is more important.

We can also see that the perturbation due to the adaption on the residual is more important that in the last case. This comes from the fact that the flow is viscous and that we have our new boundary condition for the metric along the wall.

The multi-variable intersection procedure works well for the identification of shock, boundary layer and wake regions, even the weak shocks at the trailing edge have been detected. In the same way, are mixed boundary condition for the metric along the wall implies a good repartition of the first and second nodes layers. For these computations we have just propagated the property of being at the same distance of the wall only to the two first layers but the same procedure can be applied on the other layers. We can also see what we obtain without this boundary condition.

5.5 Unsteady flow around a cylinder

This is a sub-sonic flow at Mach 0.4 and Reynolds 80. Our aim here is to see how the mesh adaption procedure follows an unsteady phenomena.

We make a first computation on a initial mesh having around 6000 triangles until the unsteady flow is established. The adimensional period of the movement being around 7, mesh adaption has been carried out eight times in a period.

There is a major difficulty in unsteady computations compared to the steady cases. Indeed, as we saw in the previous convergence history, at each adaption step, the residual was perturbated. This is due to the fact that the linear interpolation of the solution on the new mesh cannot be compatible with the nonlinear nature of the solution and the equations need time to recover their level of resolution. Basically, we need either a relaxation or subcycling procedure after each adaption for the solution to be coherent with the previous resolution level, or an interpolation being consistent with the equations we solve. In this work, nothing has been done for this purpose, therefore the time evolution of quantities like the drag or lift coefficients presents the same behavior than a convergence history in steady computations. The mesh adaption procedure however follows the unsteadiness of the flow while the number of nodes remains almost constant. Current work concerns the improvement of these techniques for unsteady flows.

5.6 CPU and Memory considerations

As NSC2KE is an explicit code, it does not require so much memory. The global memory required is about (1500 times the number of nodes) bytes. The speed is about 10^{-4} second by node by iteration on a 1 Megaflop computer for second order viscous computations using the Hybrid Upwind Scheme.

The metric computation and mesh generation modules are in C++ and have been optimized for memory requirement. An average speed for the generation of the adapted mesh is 500 triangles by second on a 1 Megaflop computer.

6 Concluding remarks

Anisotropic mesh adaption has been applied with success to compressible viscous flows in a wide range of the Reynolds and Mach number. In particular, three new idea have been described to remove the basic difficulties with adapted meshes for flows computations concerning multiple physical interactions, boundary layers and multiscale phenomena. It has been shown that these modifications greatly improve the quality of the adapted meshes for these situations.

These idea are simple to take into account in any adaption procedure because they involve only local and simple modifications of codes. Therefore, they can be easily applied by other users. In the same way, their extension to 3D configurations is straightforward.

We also saw that our upwind solver did not particularly suffer from the fact that meshes had poor quality in the Euclidean sense. We could also expect difficulties in the evaluation of second derivatives on such meshes. But we saw that in weak form the obtained Hessian, despite not being perfect, identifies rather correctly the regions of interest.

Another important conclusion is that for steady computations, convergence of the explicit flow solver is improved when using mesh adaption. Indeed, as we saw only a few thousand of iterations were necessary to converge for both the Euler and Navier-Stokes cases on quite large meshes with only a few of them on the finer levels. We recover here some kind of multi-grid behavior but with a set of independent refined levels where only interpolations (Coarse to Fine) are used but not projections (Fine to Coarse). Therefore, the global CPU is significantly reduced compared to a direct computation on a uniform fine mesh.

Developments are under progress for the extension of these techniques to 3D applications. This includes four steps:

- the metric evaluation as in 2D, obtained by a laplacian assembling (done);

- the intersection of the solution metric with the geometrical metric of the 3D surface representing the boundary (done);

- the adaption of the corresponding 3D surface mesh (done)

- existing 3D Delaunay mesh generator [7] modification to take into account a local metric defined on a background mesh (in progress).

References

[1] E. F. D'AZEVEDO AND R. B. SIMPSON, 'On Optimal Interpolation Triangle Incidences'. *SIAM'S Journal on Scientific and Statistical Computing*, no. 6: pp. 1063–1075, (1989).

[2] E. F. D'AZEVEDO AND R. B. SIMPSON, 'On Optimal Triangular Meshs for Minimizing the Gradient Error'. *Numerische Mathematik* **59**, no. 4: pp. 321–348, (1991).

[3] M.J. CASTRO DÍAZ, 'Mesh refinement over surfaces'. *Rapport de recherche INRIA, Rocquencourt* no. RR-2462, (1994).

[4] M. J. CASTRO DÍAZ AND F. HECHT, 'Anisotropic surface mesh generation'. *INRIA Research Report, Rocquencourt* no. RR-2672, (1995).

[5] M.J. CASTRO DÍAZ, F. HECHT AND B. MOHAMMADI, 'New Progress in Anisotropic Mesh Adaption for Inviside and Viscous Flow Simulations'. *INRIA Research Report, Rocquencourt* no. RR-2671, (1995).

[6] P.L.ROE, 'Approximate Riemann Solvers, Parameters Vectors and Difference Schemes' *JCP. Vol. 43* (1981).

[7] P.L. GEORGE, *Automatic mesh generation, application to Finite Element Methods*. Wiley & Sons, Chichester, 1991.

[8] P.L. GEORGE, F. HECHT AND E. SALTEL, 'Automatic mesh generation with specified boundary'. *Computer methods in applied mechanics and engineering*. **92**: pp. 269–288 (1991).

[9] F. HECHT AND E. SALTEL, 'EMC^2 un logiciel d'édition de maillages et de contours bidimensionnels'. *Technical report INRIA, Rocquencourt*, no. RT-0118 (1990).

[10] B. MOHAMMADI, 'CFD with NSC2KE: An User-Guide'. *Technical report INRIA, Rocquencourt* no. RT-0164 (1994).

[11] J. STEGER AND R.F. WARMING , 'Flux Vector Splitting for the Inviscid gas dynamic with Applications to Finite-Difference Methods'. *Journal Comp. Phys.* no. 40, pp: 263–293 (1982).

[12] B.E. LAUNDER AND D.B. SPALDING , 'Mathematical Models of Turbulence'. *Academic Press* (1972).

[13] B. MOHAMMADI AND O. PIRONNEAU , ' Analysis of the K-Epsilon Turbulence Model'. *J. Wiley and Masson Pub.* (1994).

[14] M.G. VALLET, *Génération de maillages Éléments Finis anisotropes et adaptatifs*. Université Paris VI, Paris, 1992.

[15] G.D. VAN ALBADA AND B. VAN LEER, 'Flux Vector Splitting and Runge-Kutta Methods for the Euler Equations'. *ICASE Report*, no. 84-27, (1984).

[16] F. HECHT, B. MOHAMMADI, 'Mesh Adaption by Metric Control for Multi-scale Phenomena and Turbulence'. *AIAA paper*, no. 97-0859, (1997).

[17] P.H. COOK, M. A. MCDONALD AND M.C. FIRMIN, 'Pressure Distributions, Boundary Layer and Wake Measurements'. *AGARD AR138*, no. A6-1 to A6-77, (1979).

AN APPLICATION OF THE DIFFERENCE POTENTIALS METHOD TO SOLVING EXTERNAL PROBLEMS IN CFD

Victor S. RYABEN'KII [1]   Semyon V. TSYNKOV [1]

(1) Keldysh Institute for Applied Mathematics, Russian Academy of Science; Miusskaya sq. 4, 125047 Moscow, Russia. Fax: (7-095)972-0737; E-mail: ryab@applmat.msk.su

(2) NASA Langley Research Center, Hampton, VA, U.S.A. Permanent address(Correspondence author): Dept. of Appl. Mathematics, School of Mathematical Sciences, Tel Aviv University; Ramat Aviv, Tel Aviv 69978, Israel. Fax: (972-3)640-9357, E-mail: tsynkov@math.tau.ac.il; URL: http://fmad-www.larc.nasa.gov:80/~tsynkov/

Abstract

Numerical solution of infinite-domain boundary-value problems requires some special techniques that would make the problem available for treatment on the computer. Indeed, the problem must be discretized in a way that the computer operates with only finite amount of information. Therefore, the original infinite-domain formulation must be altered and/or augmented so that on one hand the solution is not changed (or changed slightly) and on the other hand the finite discrete formulation becomes available.

One widely used approach to constructing such discretizations consists of truncating the unbounded original domain and then setting the artificial boundary conditions (ABC's) at the newly formed external boundary. The role of the ABC's is to close the truncated problem and at the same time to ensure that the solution found inside the finite computational domain would be maximally close to (in the ideal case, exactly the same as) the corresponding fragment of the original infinite-domain solution.

Let us emphasize that the proper treatment of artificial boundaries may have a profound impact on the overall quality and performance of numerical algorithms. The latter statement is corroborated by the numerous computational experiments and especially concerns the area of CFD, in which external problems present a wide class of practically important formulations.

In this paper, we review some work that has been done over the recent years on constructing highly accurate nonlocal ABC's for calculation of compressible external flows. The approach is based on implementation of the generalized potentials and pseudodifferential boundary projection operators analogous to those proposed first by Calderon. The difference potentials method (DPM) by Ryaben'kii is used for the effective computation of the generalized potentials and projections. The resulting ABC's clearly outperform the existing methods from the standpoints of accuracy and robustness, in many cases noticeably speed up the multigrid convergence, and at the same time are quite comparable to other methods from the standpoints of geometric universality and easiness of implementation.

1 Basic Ideas

1.1 Preliminaries

Let us consider two domains: the finite computational domain Dnx and its infinite complement to the entire plane (entire space) Dx; the domains are separated by the artificial boundary Γ, which is supposed to be a simple closed curve (surface). We originally formulate our problem on $\bar{D}_{in} \cup D_{ex}$; specifically, we are looking for a vector-function $\mathbf{u} = \mathbf{u}(\mathbf{r}, t)$, $\mathbf{r}$ represents the space coordinates and t is the time, that satisfies

some (generally speaking, nonlinear) system of partial differential equations (PDE's)

$$\Phi\left(\mathbf{u}, \frac{\partial \mathbf{u}}{\partial t}, \nabla \mathbf{u}, \ldots\right) = \mathbf{g}(\mathbf{r}, t), \tag{1}$$

$$\mathbf{r} \in \bar{D}_{in} \cup D_{ex},$$

as well as some boundary conditions at infinity

$$\phi(\mathbf{u}) \longrightarrow \mathbf{0}, \quad \text{as } r \equiv |\mathbf{r}| \longrightarrow \infty. \tag{2}$$

We will always assume that the problem (1) is uniquely solvable and well-posed. Later on, we will think of $\mathbf{u}$ as of the hydrodynamic variables, e.g., velocity components, density, and pressure. In

Received on July 1, 1997.
This paper was prepared while the second author held a National Research Council Research Associateship at NASA Langley Research Center, Hampton, VA 23681-2199, U.S.A.

this case, system (1a) is one of the relevant systems of PDE's used in fluid dynamics, e.g., the Navier-Stokes equations, and boundary conditions (1b) may reflect, e.g., the fact that as the coordinate approaches infinity the local flow parameters approach the free-stream ones. We will also assume that in the case of an external flow problem the computational domain D_{in} entirely contains the immersed body(ies), and equations (1a) include the proper boundary conditions at the solid surface(s). In this case, we may think (for simplicity) that $\bar{D}_{in} \cup D_{ex} = R^n$, $n = 2$ or $n = 3$.

The next step is to assume that in the domain D_{ex} system (1a) is either linear and homogeneous from the very beginning or admits replacement by some linear homogeneous system of PDE's under certain conditions. Analogously, we will assume that boundary conditions (1b) can also be thought of as some linear homogeneous relations. Then, instead of (1) we consider

$$\Phi\left(\mathbf{u}, \frac{\partial \mathbf{u}}{\partial t}, \nabla\mathbf{u}, \dots\right) = \mathbf{g}(\mathbf{r}, t), \quad \mathbf{r} \in D_{in}, \qquad (2a)$$

$$\mathbf{L}\tilde{\mathbf{u}} = 0, \quad \mathbf{r} \in \bar{D}_{ex}, \qquad (2b)$$

$$\mathbf{l}\tilde{\mathbf{u}} \longrightarrow 0, \quad \text{as } r \longrightarrow \infty. \qquad (2c)$$

In (2b) and (2c), $\mathbf{L}$ and $\mathbf{l}$ are some linear operators, and $\tilde{\mathbf{u}} = \tilde{\mathbf{u}}(\mathbf{r}, t)$ may either coincide with $\mathbf{u}$ or be some function of $\mathbf{u}$. For example, system (2b) can be obtained on the basis of (1a) using linearization in D_{ex}. (Note, linearization of the governing equations in the far field against the constant free-stream background is relevant to many external flows.) Then, $\tilde{\mathbf{u}} = \mathbf{u} - \mathbf{u}_0$, where $\mathbf{u}_0$ represents the corresponding free-stream parameters. Analogously to (1), we will assume that problem (2) is uniquely solvable and well-posed.

Our final goal will be to find the solution $\mathbf{u}$ to (2a) on D_{in}. However, we cannot do it directly since system (2a) is not closed if considered separately. On the other hand, we also cannot directly solve on the computer neither (1) nor (2) because D_{ex} is infinite. Consequently, we will need to supplement system (2a) by the special artificial boundary conditions (ABC's) at Γ, so that the resulting problem

$$\Phi\left(\mathbf{u}, \frac{\partial \mathbf{u}}{\partial t}, \nabla\mathbf{u}, \dots\right) = \mathbf{g}(\mathbf{r}, t), \quad \mathbf{r} \in D_{in}, \qquad (3a)$$

$$\mathbf{B}_\Gamma\left(\mathbf{u}, \frac{\partial \mathbf{u}}{\partial t}, \nabla\mathbf{u}, \dots\right) = 0, \quad \mathbf{r} \in \Gamma, \qquad (3b)$$

is closed and its solution on D_{in} is in a certain sense close to the solution of (2). To construct the ABC's (3b), we will use the linearity of (2b)-(2c).

Before proceeding further, let us define here an important concept of exact ABC's, which we will refer to henceforth. Namely, if the ABC's (3b) are constructed so that the solution to (3) and (2) coincide on D_{in}, then we will call such boundary conditions the exact ABC's. Note, we always assume that the solutions of (1) and (2) are close enough and always treat the exactness of ABC's within the accuracy of replacement of (1) by (2) (e.g., within the accuracy of far-field linearization). Clearly, we can reformulate the definition of exact ABC's as follows: one should be able to uniquely complement the solution of (3) from D_{in} to D_{ex} so that the resulting function on $\bar{D}_{in} \cup D_{ex}$ solves (2).

An extensive work has been done by many authors over the recent years towards constructing the exact ABC's for different problems, as well as towards developing the effective approximate boundary conditions. A survey of this work can be found, e.g., in the recent review papers by Tsynkov [1, 2], as well as in the comprehensive reviews by Givoli [3, 4].

1.2 Boundary Equations with Projections

Clearly, the idea of exact ABC's is that the relation (3b) should equivalently replace (2b)-(2c). Since the equations (2b)-(2c) are linear we can use the apparatus of generalized potentials and boundary projections in order to obtain the desirable boundary conditions. In so doing, $\mathbf{B}_\Gamma$ from (3b) appears to be some linear pseudodifferential operator.

The generalized potentials and boundary projection operators that we employ for constructing the ABC's were first proposed by Calderon [5] and then also studied by Seeley [6]. Ryaben'kii [7, 8] (see also the recent paper [9] and the book by Mikhlin, et al. [10]) had extended and modified the original approach, and also proposed an effective numerical technique for calculation of potentials and projections, this technique is called the difference potentials method (DPM). Additionally, Ryaben'kii in [11] had for the first time shown how the generalized potentials and projections could be used for constructing the ABC's for unsteady finite-difference problems.

We, however, will for the moment restrict ourselves by considering the steady-state problems only, which means $\mathbf{u} = \mathbf{u}(\mathbf{r})$ and $\tilde{\mathbf{u}} = \tilde{\mathbf{u}}(\mathbf{r})$. To simplify the notations, we will further omit the tilde that distinguishes between the "linear" and "nonlinear" solutions; it should not cause misunderstandings since hereafter we will mostly concentrate on the "linear part" of (2).

On $\bar{D}_{in} \cup D_{ex}$, we introduce the following auxiliary problem (AP)

$$\mathbf{L}\mathbf{u} = \mathbf{f}, \quad \mathbf{r} \in \bar{D}_{in} \cup D_{ex}, \quad \text{supp} \, \mathbf{f}(\mathbf{r}) \subset D_{in}, \qquad (4a)$$

$$\mathbf{lu} \longrightarrow 0, \quad \text{as} \quad r \longrightarrow \infty, \tag{4b}$$

which is the key element of our further construction. In the theory of generalized potentials, the solution of the AP (4) plays approximately the same role as the convolution with the fundamental solution plays in the classical potential theory. We will assume that problem (4) is uniquely solvable for any compactly supported right-hand side (RHS) $\mathbf{f(r)}$ and well-posed; the corresponding inverse (Green's) operator is denoted $\mathbf{G}$. For any function $\mathbf{u(r)}$, $\mathbf{r} \in \bar{D}_{ex}$, we also introduce its clear trace ξ_Γ [7, 8] on Γ,

$$\xi_\Gamma = \mathbf{Tr}_\Gamma \mathbf{u}. \tag{5}$$

For the applications considered below ($\mathbf{L}$ in (2b) and (4a) is a second-order operator), ξ_Γ is chosen as a vector-function with the components $\left(\mathbf{u}, \dfrac{\partial \mathbf{u}}{\partial \zeta} \right)\Big|_\Gamma$, ζ is the normal to Γ. The corresponding space of vector-functions ξ_Γ is called the space of clear traces. For any element ξ_Γ of this space, we define the generalized potential with the density ξ_Γ:

$$\mathbf{P}\xi_\Gamma = \left(\mathbf{G} \left((\mathbf{Lw})\big|_{D_{in}} \right) \right)\Big|_{\bar{D}_{ex}}. \tag{6}$$

Here, $\mathbf{w} = \mathbf{w(r)}$ is an arbitrary function with the only requirement of that it has the clear trace ξ_Γ, $\mathbf{Tr}_\Gamma \mathbf{w} = \xi_\Gamma$, and the RHS for the AP, i.e., the function that the Green operator $\mathbf{G}$ operates on in equation (6), is given by

$$(\mathbf{Lw})\big|_{D_{in}} = \begin{cases} \mathbf{Lw}, & \text{on } D_{in}, \\ 0, & \text{on } \bar{D}_{ex}. \end{cases} \tag{7}$$

We also define the operator $\mathbf{P}_\Gamma$,

$$\mathbf{P}_\Gamma \xi_\Gamma = \mathbf{Tr}_\Gamma \mathbf{P} \xi_\Gamma, \tag{8}$$

as the trace (5) of the generalized potential (6); this operator obviously maps the space of clear traces ξ_Γ onto itself. It turns out that $\mathbf{P}_\Gamma$ in (8) is a projection, $\mathbf{P}_\Gamma^2 = \mathbf{P}_\Gamma$, it is called the Calderon boundary projection. This operator will play a fundamental role in our further analysis. It is possible to show [7, 8] that those and only those ξ_Γ that belong to the image of the projection, $\xi_\Gamma \in \mathrm{Im}\mathbf{P}_\Gamma$, i.e., satisfy the boundary equation with projection (BEP)

$$\mathbf{P}_\Gamma \xi_\Gamma = \xi_\Gamma, \tag{9}$$

can be complemented from Γ to D_{ex} so that the complement $\mathbf{u} = \mathbf{u(r)}$, $\mathbf{r} \in \bar{D}_{ex}$ solves (2b)–(2c). In other words, BEP (9) provides for a complete classification of those and only those densities ξ_Γ of generalized potentials that are actually the traces

of some solution $\mathbf{u} = \mathbf{u(r)}$ to (2b)–(2c) on $\bar{D}_{ex}$. When the BEP (9) is satisfied, the corresponding complement on $\bar{D}_{ex}$ can be restored as the potential (6). Clearly, since $\mathrm{Im}\mathbf{P}_\Gamma$ contains all those and only those ξ_Γ that can be complemented on D_{ex} so that the complement solves (2b)–(2c), then equation (9) equivalently replaces system (2b)–(2c), which means the BEP (9) serves as a desirable exact ABC of type (3b). We emphasize that these ABC's are usually nonlocal; on the other hand, the algorithm for constructing the ABC's does not impose any essential limitations on the shape of Γ.

BEP (9) provides for a general recipe to construct the exact ABC's. However, any specific problem requires its own special approaches. First, certain assumptions in regard to the behavior of solution have to be done in order to pass from (1) to (2). For example, to linearize the governing equations one has to assume that the flow perturbations in the far field are small; this assumption can usually be verified a posteriori. Second, the AP (4), which needs to be actually solved for calculating $\mathbf{P}_\Gamma$, is still formulated on the infinite domain. Special techniques must be employed to replace it by some finite-domain AP. Third, in practice the ABC's are to be constructed for the discrete rather than for the continuous formulation of the problem, which means that some additional steps may be required for the actual implementation of BEP (9). Finally, the numerical process of the DPM [7, 8], which is used for the actual calculation of ABC's, may also vary in some details from one problem to another. In Section 2, we will delineate the corresponding algorithm for the Navier-Stokes equations. Here, we will briefly describe a simpler case of the two-dimensional external inviscid flows.

1.3 An Inviscid Example

Tsynkov [12, 13] and Sofronov and Tsynkov [14] consider a finite body (airfoil) immersed in an infinite flow of inviscid compressible fluid. The flow is governed by the Euler equations and is assumed to be subsonic at infinity; for the purpose of numerical solution, the equations are discretized on a finite-difference O-type grid that is generated around the body. The computational domain D_{in} in [12, 13, 14] is formed by the grid; no special assumptions in regard to the shape of its external boundary Γ are done. Outside the computational domain, i.e., in D_{ex}, the Euler equations are linearized against the free-stream background. Moreover, we assume the existence of the velocity potential in the far field and split the linearized system into elliptic (velocity) and advection (entropy) parts. After the term associated with the circulation of flow around the airfoil is subtracted, the regular part of the potential of velocity perturbations can be described by the Prandtl-Glauert equation

$$(1 - M_0{}^2) \frac{\partial^2 \varphi}{\partial x^2} + \frac{\partial^2 \varphi}{\partial y^2} = 0 \qquad (10a)$$

with zero boundary condition at infinity:

$$\varphi(x, y) \longrightarrow 0, \quad \text{as} \quad x^2 + y^2 \longrightarrow \infty. \qquad (10b)$$

Equation (10a) can be easily reduced to the Laplace equation by means of an affine coordinate transform; in so doing, boundary condition (10b) obviously remains intact. To obtain the ABC's for velocity components, we represent the solution to (10) in the form of a generalized potential and then construct the corresponding BEP. We use the AP formulated on a ring-shaped domain $\{R_1 \leq r \leq R_0\}$; the external circle $\{r = R_0\}$ encompasses Γ and the internal circle $\{r = R_1\}$ lies inside D_{in}. At $r = R_1$, we specify homogeneous Dirichlet boundary conditions; at $r = R_0$, we specify boundary conditions

$$\left. \frac{d\hat{\varphi}_k}{dr} \right|_{r=R_0} + \left. \frac{|k|}{r} \hat{\varphi}_k \right|_{r=R_0} = 0, \qquad (11)$$

$$k = 0, \pm 1, \pm 2, \ldots,$$

where $\hat{\varphi}_k$, $k = 0, \pm 1, \pm 2, \ldots$, are the Fourier components of the potential (10) after the transform with respect to the polar angle. It is possible to make sure that this AP is uniquely solvable and well-posed for any RHS concentrated inside the ring. Moreover, it is easy to see that once complemented to the exterior of the ring, the solution to this AP vanishes at infinity (compare to (4b)). Numerically, the AP is easy to solve by means of the discrete Fourier transform in polar coordinates.

Analogously to the continuous formulations, the ABC's in the discrete form should simply close the system that is solved inside D_{in}. For a second-order scheme employed inside the computational domain, we can always assume that the velocity components on the penultimate coordinate row Γ of the O-type grid are known, whereas the corresponding values on the outermost row Γ_1 should be determined using the ABC's. Therefore, we consider the penultimate coordinate line Γ as the actual artificial boundary. Referring the reader to the original work [12, 13, 14], as well as to Section 2, in which the Navier-Stokes case is analyzed, we skip here the details of the numerical procedure for calculating generalized potentials and projections and setting the inviscid ABC's. We only mention that after the finite-difference BEP

is obtained, it is solved so that the velocity on Γ provided from inside D_{in} coincides with the gradient of the potential (10) in a certain generalized sense. Once the BEP is solved, we can find the discrete potential density for any data (velocity components) specified on Γ. When this grid density is known, we calculate the generalized potential and find the trace of its gradient on the outermost coordinate line Γ_1 by means of interpolation; this procedure yields the ABC's for velocities. Finally, the ABC's for thermodynamic parameters are obtained using local relations, specifically, the Bernoulli equation and the entropy advection equation.

The technique of [12, 13, 14] for constructing the ABC's was combined with an iterative method [15] by Sofronov for calculating steady solutions to the Euler equations. A few subsonic and transonic airfoil flows have been numerically studied; in the transonic case, local supercritical regions were always kept inside D_{in} so that one could treat the exterior linearized flow in D_{ex} as purely subsonic. Numerical results presented in [14] demonstrate clear superiority of the nonlocal DPM-based ABC's over the standard local techniques based on quasi-one-dimensional characteristic analysis. For a fixed computational domain, nonlocal ABC's [12, 13, 14] provide for better accuracy and faster convergence rate than local techniques; the entire numerical algorithm also appears to be more robust. Additionally, when the artificial boundary approaches the airfoil, the solution obtained with the technique of [12, 13, 14] appears to be essentially less influenced by the decrease of the size of computational domain than the solution obtained on the basis of the local boundary conditions. In other words, ABC's [12, 13, 14] allow one to maintain good accuracy of computations for much smaller computational domains than the standard boundary conditions do. These results, along with the geometric universality of ABC's [12, 13, 14], make this approach useful for calculating external Euler flows.

Analyzing this Euler example, we can see that the principle favorable circumstance that allowed us to relatively easy obtain the ABC's is the availability of a finite-domain AP. This, in turn, is due to the fact that the Poisson equation admits variables separation in polar coordinates. However, one obviously cannot rely on such circumstances in constructing the ABC's for more general cases. For example, the linearized Navier-Stokes equations (see Section 2) and even the linearized Euler equations (without introducing the potential) most likely do not admit the separation of variables in any other coordinate system except in the Cartesian one. Therefore, we need to develop some technique that would generally allow us to approximate the solutions of the infinite-domain AP of type (4) by the solutions of a new finite-domain AP. In so doing, we should be able to control the corresponding approximation error so that the resulting ABC's be as close to the exact ones as desired. Below, the procedure for constructing appropriate finite-domain approximations to the AP's of type (4) is described for a particular class of systems of PDE's with constant coefficients. For wider classes of systems, the possibility to use the same procedure is corroborated by the numerical experiments.

1.4 <u>Systems of Simple Structure</u>

Hereafter in this subsection, we assume that all functions depend on only two space variables, i.e., $\mathbf{r} = (x, y)$, x and y being scalars. This assumption allows us to simplify the presentation and at the same time does not imply any loss of generality since the analysis of the case $\mathbf{r} = (x, \mathbf{y})$, $\mathbf{y}$ being a vector, is straightforward.

Let us consider the system of PDE's

$$\frac{\partial \mathbf{u}}{\partial x} = \mathbf{A}\left(\frac{\partial}{\partial y}\right)\mathbf{u} + \mathbf{f}(x, y), \qquad (12)$$

$$-\infty < x, y < \infty,$$

with respect to a n-component vector-function $\mathbf{u} = \mathbf{u}(x, y)$. In (12), $\mathbf{A}\left(\dfrac{\partial}{\partial y}\right)$ is a $n \times n$ matrix, each entry of this matrix is a symbolic polynomial of $\dfrac{\partial}{\partial y}$. The RHS $\mathbf{f}(x, y)$ is a compactly supported n-component vector-function, $\mathbf{f}(x, y) \equiv \mathbf{0}$ for $|x - a| > a$, $|y| > a$, a being some positive constant. System (12) can obviously be thought of as a particular class of systems of type (4a); consideration of only the first-order derivatives with respect to x presents no loss of generality since the higher derivatives can be eliminated by introducing additional variables.

We designate by $\mathbf{A}(i\alpha)$ the matrix that is obtained by substituting $i\alpha$ into $\mathbf{A}\left(\dfrac{\partial}{\partial y}\right)$ instead of $\dfrac{\partial}{\partial y}$. The entries of the matrix $\mathbf{A}(i\alpha)$ are, therefore, some polynomials of the real variable α (the coefficients of these polynomials may, generally speaking, be complex).

Definition 1 *System (12) is called the system of simple structure if the roots $\lambda_j(\alpha)$ of the equation* $\det \|\mathbf{A}(i\alpha) - \lambda\mathbf{I}\| = 0$ *(I is the $n \times n$ identity matrix) satisfy the inequalities*

$$|\Re\lambda_j(\alpha)| \geq \delta(\alpha) \geq \delta > 0, \qquad (13)$$

where $\delta > 0$ does not depend on α.

Let now U be a class of n-component vector-functions, in which we will be looking for the solutions $\mathbf{u}(x, y)$ of system (12). The inclusion $\mathbf{u}(x, y) \in U$ holds for all those and only those vector-functions that satisfy the following conditions.

- Each component of the vector-function $\mathbf{u}(x, y)$ has continuous derivatives of all those orders that are involved in system (12).

- Vector-function $\mathbf{u}(x, y)$ can be represented as a Fourier integral:

$$\mathbf{u}(x, y) = \frac{1}{\sqrt{2\pi}} \int\limits_{-\infty}^{\infty} \hat{\mathbf{u}}(x, \alpha)e^{i\alpha y}\, d\alpha, \qquad (14a)$$

$$\hat{\mathbf{u}}(x, \alpha) = \frac{1}{\sqrt{2\pi}} \int\limits_{-\infty}^{\infty} \mathbf{u}(x, y)e^{-i\alpha y}\, dy. \qquad (14b)$$

- The derivatives involved in (12) can be represented as

$$\frac{\partial \mathbf{u}}{\partial x} = \frac{1}{\sqrt{2\pi}} \int\limits_{-\infty}^{\infty} \left[\frac{d}{dx}\hat{\mathbf{u}}(x, \alpha)\right] e^{i\alpha y}\, d\alpha, \qquad (15a)$$

$$\mathbf{A}\left(\frac{\partial}{\partial y}\right)\mathbf{u} = \frac{1}{\sqrt{2\pi}} \int\limits_{-\infty}^{\infty} \mathbf{A}(i\alpha)\hat{\mathbf{u}}(x, \alpha)e^{i\alpha y}\, d\alpha. \qquad (15b)$$

- Fourier transformation $\hat{\mathbf{u}}(x, \alpha)$ of the function $\mathbf{u}(x, y)$ is bounded for each α,

$$|\hat{\mathbf{u}}(x, \alpha)| < \infty. \qquad (16)$$

Note, $|\hat{\mathbf{u}}(x, \alpha)|$ in (16) can be thought of as the sum of the absolute values of components.

Further, for any real $Y > 2a$ we associate with (12) another system of PDE's

$$\frac{\partial \mathbf{u}_Y}{\partial x} = \mathbf{A}\left(\frac{\partial}{\partial y}\right)\mathbf{u}_Y + \mathbf{f}_Y(x, y), \qquad (17)$$

$$-\infty < x, y < \infty,$$

with the unknowns $\mathbf{u}_Y = \mathbf{u}_Y(x, y)$. In (17), $\mathbf{f}_Y(x, y)$ is a n-component vector-function that is periodic in the y direction with the period Y and that coincides with the RHS $\mathbf{f}(x, y)$ of system (12) for $|y| < Y/2$.

Let U_Y be a class of n-component vector-functions, in which we will be looking for the solutions $\mathbf{u}_Y(x, y)$ of system (17). The function $\mathbf{u}_Y(x, y)$ belongs to U_Y if and only if it satisfies the following conditions.

- Each component of $\mathbf{u}_Y(x, y)$ has continuous derivatives of all those orders that are involved in system (17); moreover, $\mathbf{u}_Y(x, y \pm Y) = \mathbf{u}_Y(x, y)$.

- Vector-function $\mathbf{u}_Y(x, y)$ can be represented as a Fourier series:

$$\mathbf{u}_Y(x, y) = \sum_{k=-\infty}^{k=\infty} \hat{\mathbf{u}}_{Y_k}(x) e^{iky\frac{2\pi}{Y}}, \qquad (18a)$$

$$\hat{\mathbf{u}}_{Y_k}(x) = \frac{1}{Y} \int_{-Y/2}^{Y/2} \mathbf{u}_Y(x, y) e^{-iky\frac{2\pi}{Y}} dy. \qquad (18b)$$

- The derivatives involved in (17) can be represented as

$$\frac{\partial \mathbf{u}_Y}{\partial x} = \sum_{k=-\infty}^{k=\infty} \left[\frac{d}{dx} \hat{\mathbf{u}}_{Y_k}(x) \right] e^{iky\frac{2\pi}{Y}}, \qquad (19a)$$

$$\mathbf{A}\left(\frac{\partial}{\partial y}\right) \mathbf{u}_Y = \qquad (19b)$$

$$\sum_{k=-\infty}^{k=\infty} \mathbf{A}\left(i\frac{2\pi k}{Y}\right) \hat{\mathbf{u}}_{Y_k}(x) e^{iky\frac{2\pi}{Y}}.$$

- Fourier coefficients $\hat{\mathbf{u}}_{Y_k}(x)$ of the function $\mathbf{u}_Y(x, y)$ are bounded for all k,

$$|\hat{\mathbf{u}}_{Y_k}(x)| < \infty. \qquad (20)$$

Theorem 1 *The system of simple structure (12) may have at most one solution $\mathbf{u}(x, y) \in U$, and system (17) may have at most one solution $\mathbf{u}_Y(x, y) \in U_Y$.*

Proof. Clearly, it is sufficient to show that the homogeneous system

$$\frac{\partial \mathbf{u}}{\partial x} = \mathbf{A}\left(\frac{\partial}{\partial y}\right) \mathbf{u}$$

has only trivial solution $\mathbf{u}(x, y) \equiv 0$ in the class U. Let $\mathbf{u}(x, y) \in U$ be some solution to this homogeneous system. We represent this solution as a Fourier integral in accordance with (14a) and, using (15), obtain

$$\frac{1}{\sqrt{2\pi}} \int_{-\infty}^{\infty} \left[\frac{d}{dx} \hat{\mathbf{u}}(x, \alpha) - \mathbf{A}(i\alpha)\hat{\mathbf{u}}(x, \alpha) \right] e^{i\alpha y} d\alpha = 0.$$

Therefore, for any α

$$\frac{d\hat{\mathbf{u}}(x, \alpha)}{dx} = \mathbf{A}(i\alpha)\hat{\mathbf{u}}(x, \alpha), \qquad (21)$$

and $\hat{\mathbf{u}}(x, \alpha)$ is bounded. On the other hand, all roots of the characteristic equation of system (21) always have non-zero real parts since (12) is a system of simple structure. It is well-known that in this case system (21) has a unique bounded solution $\hat{\mathbf{u}}(x, \alpha)$, which is identically zero. Consequently,

$$\mathbf{u}(x, y) = \frac{1}{\sqrt{2\pi}} \int_{-\infty}^{\infty} \hat{\mathbf{u}}(x, \alpha) e^{i\alpha y} d\alpha \equiv \mathbf{0}.$$

The statement about $\mathbf{u}_Y(x, y)$ can be proven analogously. $\square$

To formulate the next theorem, we introduce the following notations. Let $\varphi(x, y)$ be some n-component vector-function, $\varphi(x, y) = (\varphi_1(x, y), \varphi_2(x, y), \ldots, \varphi_n(x, y))^T$. Designate

$$|\varphi(x, y)| = \sum_{j=1}^{n} |\varphi_j(x, y)|; \qquad (22a)$$

$$\left| \frac{\partial^{p+q} \varphi(x, y)}{\partial x^p \partial y^q} \right| = \sum_{j=1}^{n} \left| \frac{\partial^{p+q} \varphi_j(x, y)}{\partial x^p \partial y^q} \right|; \qquad (22b)$$

$$\|\varphi(x, y)\|_k = \sum_{p+q=0}^{k} \sup_{x, y} \left| \frac{\partial^{p+q} \varphi(x, y)}{\partial x^p \partial y^q} \right|. \qquad (22c)$$

Theorem 2 *Consider an arbitrary bounded domain $D \subset R^2$. Let z be an arbitrary non-negative integer number. Then, one always can find a sufficiently large number $k = k(z, D)$ such that for any compactly supported function $\mathbf{f}(x, y)$, $\mathbf{f}(x, y) \equiv \mathbf{0}$ for $|x - a| > a$ and $|y| > a$, that satisfies $\|\mathbf{f}\|_k < \infty$ (see (22)) the solutions $\mathbf{u} \in U$ and $\mathbf{u}_Y \in U_Y$ of systems (12) and (17), respectively, exist (Y > 2a is arbitrary) and*

$$\|\mathbf{u} - \mathbf{u}_Y\|_{z, D} \leq \frac{const}{Y^\kappa}, \qquad (23)$$

where $\kappa = \kappa(z, k, D)$, and $\kappa \longrightarrow \infty$ as $k \longrightarrow \infty$.

The importance of Theorem 2 is that it allows us to approximate the solution of the infinite-domain AP by the solutions of another problem, which is periodic and, therefore, finite, in all but one Cartesian directions. Recall, to calculate the ABC's we need to know the projection $\mathbf{P}_\Gamma$ (see (8)), which is the superposition of the trace $\mathbf{Tr}_\Gamma$ (see (6)) and the potential $\mathbf{P}$ (see (5)). We, therefore, conclude that in order to obtain the ABC's we do not need to know the potential, i.e., the solution of the AP, everywhere on D_{ex}; we need to know it only on some neighborhood of Γ. Consequently, the approximation of $\mathbf{u}$ by $\mathbf{u}_Y$ on any finite domain D as the period Y increases (see (23)) is sufficient for achieving our goal of constructing the ABC's.

The proof of Theorem 2 is essentially based on the estimates for one-dimensional fundamental solutions obtained by Godunov and Gordienko in [16]. Specifically, let

$$\frac{d\mathbf{v}}{dx} = \mathbf{A}\,\mathbf{v} + \mathbf{f}(x) \qquad (24)$$

be an abstract system of n ordinary differential equations (ODE's) with constant coefficients, and $\mathbf{f}(x)$ be a continuous bounded function. Let also the roots λ of the characteristic equation $\det|\mathbf{A} - \lambda\mathbf{I}| = 0$ of system (24) satisfy the inequality

$$|\Re\lambda| \geq \delta > 0. \qquad (25)$$

We supplement system (24) by the condition of boundedness of its solution $\mathbf{v}(x) = (v_1(x), v_2(x), \ldots, v_n(x))^T$ on the entire line $-\infty < x < \infty$,

$$\|\mathbf{v}(x)\| \leq const. \qquad (26)$$

Because of the equivalence of norms on a finite-dimensional space for a fixed n, the choice of the specific norm $\|\mathbf{v}(x)\|$ in (26) is not essential. Hereafter, we will always use the l_1 norm of vectors, i.e., consider as a norm of a vector the sum of absolute values of its components. We now formulate the following result, see [16]. Fundamental solution $\mathbf{G}(x)$ of the corresponding linear differential operator from (24) exists and is unique in the class of functions that meet boundary condition (26). This fundamental solution is actually a $n \times n$ matrix that satisfy the inequality

$$\|\mathbf{G}(x)\| \leq \Omega(n) \left(\frac{\|\mathbf{A}\|}{\delta}\right)^{n-1} e^{-\delta\frac{|x|}{2}}, \qquad (27)$$

where $\Omega(n)$ is some number that depends only on n, and the norms of the matrices $\mathbf{G}(x)$ and $\mathbf{A}$ are consistent with the chosen vector norm. Note, according to the definition of the fundamental solution, one can represent the solution to (24), (26) as

$$\mathbf{v}(x) = \int_{-\infty}^{\infty} \mathbf{G}(x - t)\mathbf{f}(t)\,dt. \qquad (28)$$

We will be looking for the solution of system (12) in the form of the Fourier integral (14a), where the Fourier transformation $\hat{\mathbf{u}}(x, \alpha)$ is subject to the determination. Analogously to (14), we represent the RHS $\mathbf{f}(x, y)$ of system (12) as

$$\mathbf{f}(x, y) = \frac{1}{\sqrt{2\pi}} \int_{-\infty}^{\infty} \hat{\mathbf{f}}(x, \alpha)e^{i\alpha y}\,d\alpha, \qquad (29a)$$

$$\hat{\mathbf{f}}(x, \alpha) = \frac{1}{\sqrt{2\pi}} \int_{-\infty}^{\infty} \mathbf{f}(x, y)e^{-i\alpha y}\,dy. \qquad (29b)$$

Substituting (14a) and (29a) into (12) and using (15), we formally obtain for $\hat{\mathbf{u}}(x, \alpha)$ the following system of ODE's with constant coefficients

$$\frac{d\hat{\mathbf{u}}(x, \alpha)}{dx} = \mathbf{A}(i\alpha)\hat{\mathbf{u}}(x, \alpha) + \hat{\mathbf{f}}(x, \alpha). \qquad (30)$$

System (30) depends on α as on the parameter. We will supplement this system by boundary condition (16).

To make sure that the function $\mathbf{u}(x, y)$ obtained using (14a) (provided that $\hat{\mathbf{u}}(x, \alpha)$ satisfies (30), (16)) does belong to U and does solve (12), we will first have to make certain assumptions in regard to the smoothness of the RHS $\mathbf{f}(x, y)$ (recall, this function is always supposed to have compact support). Then, we estimate from above the Fourier transformation $\hat{\mathbf{f}}(x, \alpha)$ (see (29a)) and its derivatives by certain rational functions of α. Using the estimates for $\hat{\mathbf{f}}(x, \alpha)$, inequality (27), and representation (28), we can estimate from above the solution $\hat{\mathbf{u}}(x, \alpha)$ of (30), (16) and its derivatives with respect to x by the (22)-type norm of $\mathbf{f}(x, y)$ multiplied by a certain rational function of α and a certain exponential function of x. Further, the differentiability of $\hat{\mathbf{u}}(x, \alpha)$ with respect to α and the estimates for the derivatives (in particular, the rate of decay for big $|\alpha|$) can be obtained by induction with respect to the order of differentiation. Finally, using the aforementioned estimates we make sure that the inverse Fourier transformation (14a) does exist in U and therefore, provides for the solution to (12); using the same estimates, we also make sure that inequality (23) does hold, first for $z = 0$ and then for $z > 0$.

This brief outline will be transformed into the full proof of Theorem 2 in a future paper. Here, we will show how to handle the last remaining infinite direction, x, and to therefore finally replace the infinite-domain AP by a new problem formulated on some bounded domain.

Since we have already replaced the original infinite-domain formulation of the AP by the periodic formulation with respect to y, then, instead of solving systems (30), (16) for each α, $-\infty < \alpha < \infty$, we will need to solve the systems

$$\frac{d}{dx}\hat{\mathbf{u}}_{Y_k}(x) = \mathbf{A}\left(i\frac{2\pi k}{Y}\right)\hat{\mathbf{u}}_{Y_k}(x) + \hat{\mathbf{f}}_{Y_k}(x) \qquad (31)$$

with boundary condition (20) for each k, $k = 0, \pm 1, \pm 2, \ldots$ Note, system (31) is obtained on the basis of representation

$$\mathbf{f}_Y(x, y) = \sum_{k=-\infty}^{k=\infty} \hat{\mathbf{f}}_{Y_k}(x)e^{iky\frac{2\pi}{Y}}, \qquad (32a)$$

$$\hat{\mathbf{f}}_{Y_k}(x) = \frac{1}{Y}\int_{-Y/2}^{Y/2} \mathbf{f}_Y(x, y)e^{-iky\frac{2\pi}{Y}}dy \qquad (32b)$$

by substituting (18a) and (32a) into (17) and using (19).

For each k, $k = 0, \pm 1, \pm 2, \ldots$, system (31) formally needs to be solved on the entire line $-\infty < x < \infty$. However, since we need to know the solution of the AP only on some finite neighborhood of Γ, we can always choose a sufficiently large but still finite segment on the x axis, on which it would be sufficient to calculate the solution to (31). Without loss of generality, we may think that it is the segment $(0, X)$, $X > 2a$. Since the RHS $\mathbf{f}(x, y)$ is compactly supported, system (31) is homogeneous for $x \leq 0$ and $x \geq X$. To ensure that boundary condition (20) is met, we have to prohibit for $x \geq X$ all the eigensolutions of the homogeneous counterpart to (31) that increase as $x \longrightarrow +\infty$ and to prohibit for $x \leq 0$ all the eigensolutions that increase as $x \longrightarrow -\infty$. This can be done by imposing the boundary conditions

$$\left[\prod_{\Re\lambda_j(k)>0} (\mathbf{A}^k - \lambda_j(k)\mathbf{I})\right]\hat{\mathbf{u}}_{Y_k}(0) = \mathbf{0} \qquad (33a)$$

and

$$\left[\prod_{\Re\lambda_j(k)<0} (\mathbf{A}^k - \lambda_j(k)\mathbf{I})\right]\hat{\mathbf{u}}_{Y_k}(X) = \mathbf{0} \qquad (33b)$$

at $x = 0$ and $x = X$, respectively. In (33), $\lambda_j(k)$ are the eigenvalues of $\mathbf{A}^k \equiv \mathbf{A}\left(i\frac{2\pi k}{Y}\right)$ and the matrix products are computed taking into account the multiplicities of these eigenvalues. Clearly, since for the system of simple structure $|\Re\lambda| \geq \delta > 0$, boundary conditions (33) actually imply that the solution

$\hat{\mathbf{u}}_{Y_k}(x)$ vanishes at infinity (which is a stronger property than simply boundedness (20)).

Thus, we have shown how to replace the infinite-domain AP for a system of simple structure by the new problem formulated on the rectangle $(0, X) \times (-Y/2, Y/2)$ for the same RHS so that the solutions of these two problems are arbitrarily close to one another on a fixed finite neighborhood of D_{in}. We only have to mention, that the "asymmetric treatment" of the Cartesian directions x and y is caused mostly by the physical concerns. Specifically, typical external flow formulations (see Section 2) would bear some natural non-isotropy when we treat the Cartesian direction x as a streamwise and the Cartesian direction y as a cross-stream. In so doing, exact analytic treatment (33) of the Cartesian direction x seems to be most relevant for practical computations, since the periodic treatment for x may result in much larger values of period required for achieving the same accuracy than the periodic treatment for y. On the other hand, for the purposes other than the practical computation of projections, we may consider Fourier representation of the solution to the AP in all Cartesian directions. In particular, this has been done by Tsynkov in [17] when proving the solvability of the AP for the linearized thin-layer equations in the sense of tempered distributions.

To conclude the introductory part of the paper, we will also briefly comment on the relation between the systems of simple structure and other equations that can often be encountered in practice.

1.5 Wider Classes of Systems

Many systems of equations that are frequently solved in mathematical physics may not appear to be the systems of simple structure. For example, consider the system

$$\frac{\partial u}{\partial x} = v \qquad (34)$$

$$\frac{\partial v}{\partial x} = -\frac{\partial^2 u}{\partial y^2} + \mu u + f(x, y),$$

$$\mu \in R, \quad \mu = \text{const},$$

which is obtained when replacing the equation

$$\frac{\partial^2 u}{\partial x^2} + \frac{\partial^2 u}{\partial y^2} - \mu u = f(x, y) \qquad (35)$$

by the first-order system with respect to x. For (34), the matrix $\mathbf{A}(i\alpha)$ becomes

$$\mathbf{A}(i\alpha) = \begin{bmatrix} 0 & 1 \\ \alpha^2 + \mu & 0 \end{bmatrix},$$

and its eigenvalues are $\lambda(\alpha) = \pm\sqrt{\alpha^2 + \mu}$. When $\mu > 0$ we have $|\Re\lambda(\alpha)| \geq \sqrt{\mu} > 0$, and system

(34) is a system of simple structure. However, when $\mu = 0$ (in this case, equation (35) is the Poisson equation) or when $\mu < 0$ (equation (35) is the Helmholtz equation), the real part of $\lambda(\alpha)$ for some α's is zero, which means that system (34) is not a system of simple structure. Accordingly, the Poisson equation on R^2 not always has a bounded solution. As for the Helmholtz equation, the condition of vanishing of its solution as $r \equiv (x^2 + y^2)^{1/2} \longrightarrow \infty$ does not ensure the uniqueness. To select the unique solution, one has to impose a more fine Sommerfeld radiation condition, which, in particular, also implies vanishing of the solution at infinity.

As shown above, if the linear differential operator $\mathbf{L}$ from (2b) being represented in the form of (12) implies the inequality (13), i.e., the corresponding system of PDE's meets the Definition 1 of the systems of simple structure, and if the boundary conditions at infinity (2c) are actually the conditions of a sufficiently fast decay, then the ABC's at the artificial boundary $\Gamma = \partial D_{in}$ can be constructed using the finite-domain AP on the rectangle $(0, X) \times (-Y/2, Y/2)$. Indeed, under these conditions the infinite-domain AP (4) formulated for the RHS (7) can be replaced by the finite-domain AP using Theorem 2 and boundary conditions (33). If, however, system (2b) represented in the form of (12) is not a system of simple structure, then we can sometimes approximate its solutions by the solutions of a specially chosen system of simple structure. (Note, the representation in the form of (12), if possible at all, may require the solution with respect to $\partial \mathbf{u}/\partial x$ after the Fourier transform; in so doing, the entries of $\mathbf{A}(i\alpha)$ may become rational rather than polynomial functions of α).

Let $\mathbf{B} = \mathbf{B}\left(\dfrac{\partial}{\partial y}, \varepsilon\right)$ be a $n \times n$ matrix, which approaches the zero matrix as $\varepsilon \longrightarrow +0$, and let the roots $\lambda(\alpha, \varepsilon)$ of the equation

$$\det \| \mathbf{A}(i\alpha) + \mathbf{B}(i\alpha, \varepsilon) - \lambda(\alpha, \varepsilon)\mathbf{I} \| = 0 \tag{36}$$

satisfy for each ε the inequality

$$|\Re\lambda(\alpha, \varepsilon)| \geq \delta(\varepsilon) > 0. \tag{37}$$

Let, in addition, the problem

$$\Phi(\mathbf{u}, \nabla\mathbf{u}, \ldots, \varepsilon) = \mathbf{g}(x, y), \quad (x, y) \in D_{in} \tag{38a}$$

$$\frac{\partial \mathbf{u}(x, y, \varepsilon)}{\partial x} = \left[\mathbf{A}\left(\frac{\partial}{\partial y}\right) + \mathbf{B}\left(\frac{\partial}{\partial y}, \varepsilon\right)\right]\mathbf{u}, \tag{38b}$$

$(x, y) \in \bar{D}_{ex}$

$$\mathbf{u}(x, y, \varepsilon) \longrightarrow 0, \quad \text{as} \quad r \longrightarrow \infty \tag{38c}$$

be uniquely solvable and let its solution $\mathbf{u}(x, y, \varepsilon)$ approach the solution to the steady-state counterpart of problem (2) as $\varepsilon \longrightarrow +0$ on any finite subdomain of R^2. Then, for constructing the ABC's on Γ one obviously can use the AP

$$\frac{\partial \mathbf{u}(x, y, \varepsilon)}{\partial x} = \tag{39a}$$

$$\left[\mathbf{A}\left(\frac{\partial}{\partial y}\right) + \mathbf{B}\left(\frac{\partial}{\partial y}, \varepsilon\right)\right]\mathbf{u} + \mathbf{f}(x, y),$$

$(x, y) \in R^2, \quad \text{supp} \mathbf{f}(x, y) \subset D_{in}$

$$\mathbf{u}(x, y, \varepsilon) \longrightarrow 0, \quad \text{as} \quad r \longrightarrow \infty, \tag{39b}$$

where ε is chosen sufficiently small.

For example, if $\mu = 0$ in system (34), then the correction

$$\mathbf{B}\left(\frac{\partial}{\partial y}, \varepsilon\right) = \left[\begin{array}{cc} 0 & 0 \\ \varepsilon & 0 \end{array}\right].$$

will transform (34) into a system of simple structure. In the case of the Helmholtz equation ($\mu < 0$ in (34)) supplemented by the Sommerfeld radiation conditions at infinity, we choose

$$\mathbf{B}\left(\frac{\partial}{\partial y}, \varepsilon\right) = \left[\begin{array}{cc} 0 & 0 \\ i\varepsilon & 0 \end{array}\right].$$

This choice corresponds to the well-known principle of limitary absorption, which allows one to approximate both the Helmholtz equation and the Sommerfeld radiation conditions with the increasing accuracy as $\varepsilon \longrightarrow +0$. We should note that another choice of $\mathbf{B}$,

$$\mathbf{B}\left(\frac{\partial}{\partial y}, \varepsilon\right) = \left[\begin{array}{cc} 0 & 0 \\ -i\varepsilon & 0 \end{array}\right].$$

also provides for the system of simple structure (39a). However, in this case we approximate (as $\varepsilon \longrightarrow +0$) the solution composed of incoming waves rather than the one that meets the radiation conditions.

Finally, we mention that for a given system of PDE's it is, generally speaking, not always easy to find a good analogue to the principle of limitary absorption so that the solutions of this system can be approximated by the solutions of a certain system of simple structure. However, the replacement of an infinite-domain AP by the periodic problem may still be possible, the validity of such a replacement can be verified a posteriori by means of the numerical experiments as done, e.g., in our work [18].

2 ABC's for External Viscous Flows

2.1 Governing Equations

We consider a flow of compressible viscous fluid over a finite body or configuration of bodies (e.g., single-element or multi-element airfoil in two space dimensions). The flow is supposed to be uniform and subsonic at infinity. As mentioned above, we intend to actually calculate the flow only on some finite computational domain D_{in} (which contains the immersed body(ies)). The infinite exterior D_{ex} of the domain D_{in} is truncated, and the equations there along with the boundary conditions at infinity are to be replaced by the ABC's at $\Gamma = \partial D_{in}$.

To construct the ABC's, we linearize the governing equations in D_{ex} against the constant free-stream background. Generally, to justify the possibility of linearization we have to make sure that first, the flow perturbations in the far field are small, and second, the differential equations that describe these small perturbations are linear. As concerns the first assumption, it does hold for the external viscous flows, at least when the size of D_{in} is much larger than the size of the immersed body. The linearity of the governing equations for small perturbations is easy to establish for the subsonic flows, i.e., when the free-stream Mach number M_0 is not too close to one. For the transonic flows, it basically requires a special study (see our work [19] and also the recent paper [20]). The situation appears different for three space dimensions, when the far field is essentially linear, and for two space dimensions, when the consideration of a simplified potential model may formally require to introduce some nonlinear corrections in the transonic limit, i.e., as $M_0 \longrightarrow 1$. We, however, always consider the full flow system rather than the potential equation, we also never approach the transonic limit too closely. In so doing, we use the far-filed linearization for two space dimensions as well. In practice, the validity of the far-field linearization is always verified by an a posteriori numerical check.

Either one of the two following systems of PDE's

$$\frac{\partial \rho}{\partial x} + \frac{\partial u}{\partial x} + \frac{\partial v}{\partial y} = 0 \tag{40}$$

$$\frac{\partial u}{\partial x} + \frac{\partial p}{\partial x} - \frac{1}{Re}\left[\frac{4}{3}\frac{\partial^2 u}{\partial x^2} + \frac{1}{3}\frac{\partial^2 v}{\partial x \partial y} + \frac{\partial^2 u}{\partial y^2}\right] = 0$$

$$\frac{\partial v}{\partial x} + \frac{\partial p}{\partial y} - \frac{1}{Re}\left[\frac{4}{3}\frac{\partial^2 v}{\partial y^2} + \frac{1}{3}\frac{\partial^2 u}{\partial x \partial y} + \frac{\partial^2 v}{\partial x^2}\right] = 0$$

$$\frac{\partial p}{\partial x} - \frac{1}{M_0^2}\frac{\partial \rho}{\partial x} - \frac{\gamma}{Re\,Pr}\left[\frac{\partial^2 p}{\partial x^2} + \frac{\partial^2 p}{\partial y^2} - \right.$$

$$\left. \frac{1}{\gamma M_0^2}\left(\frac{\partial^2 \rho}{\partial x^2} + \frac{\partial^2 \rho}{\partial y^2}\right)\right] = 0$$

and

$$\frac{\partial \rho}{\partial x} + \frac{\partial u}{\partial x} + \frac{\partial v}{\partial y} = 0 \tag{41}$$

$$\frac{\partial u}{\partial x} + \frac{\partial p}{\partial x} - \frac{1}{Re}\frac{\partial^2 u}{\partial y^2} = 0$$

$$\frac{\partial v}{\partial x} + \frac{\partial p}{\partial y} - \frac{1}{Re}\frac{4}{3}\frac{\partial^2 v}{\partial y^2} = 0$$

$$\frac{\partial p}{\partial x} - \frac{1}{M_0^2}\frac{\partial \rho}{\partial x} - \frac{\gamma}{Re\,Pr}\left[\frac{\partial^2 p}{\partial y^2} - \frac{1}{\gamma M_0^2}\frac{\partial^2 \rho}{\partial y^2}\right] = 0$$

can be used (and has actually been used) for constructing the ABC's in two space dimensions for steady-state flows. Hereafter, we will concentrate mostly on this specific case (2D steady-state flows); in the end of the paper, we will also present some recent three-dimensional results (see our work [19, 20], as well as the earlier papers [21, 22]) and comment on the possible approaches to treating the time-dependent problems.

System (40) represents the linearized dimensionless full Navier-Stokes equations, and system (41) represents the linearized dimensionless thin-layer equations. In (40) and (41), u, v, p, and ρ are the perturbations of the Cartesian velocity components, pressure, and density with respect to the corresponding free-stream parameters, M_0 is the free-stream Mach number, Re is the Reynolds number, Pr is the Prandtl number, and γ is the ratio of specific heats. In both cases we assume that the direction of flow at infinity coincides with the positive x direction, and that the gas is perfect. To obtain the dimensionless quantities, we use the following scales: u_0, for velocity components; ρ_0, for density; $\rho_0 u_0^2$, for pressure; μ_0, for viscosity; characteristic size L (typically, airfoil chord), for all distances. Here, the subscript "0" denotes the free-stream parameters.

As mentioned above, to construct the ABC's we need to be able to solve the AP for a nonhomogeneous counterpart to either (40) or (41) driven by a certain compactly supported RHS. In our work [17] we have, in particular, shown that when supplemented by a compactly supported RHS, system (41) is always solvable in the sense of tempered distributions (see, e.g., book [23] by Hörmander or [24] by Vladimirov for a detailed description of the concept of distributions and its application to solving the PDE's). We have also shown that if this solution satisfies the boundary condition

$$(u, v, p, \rho) \longrightarrow (0, 0, 0, 0), \tag{42}$$

$$\text{as} \quad r \equiv (x^2 + y^2)^{1/2} \longrightarrow +\infty,$$

then it is unique in the class of distributions vanishing at infinity. Note, the solvability in $\mathcal{S}'$ (space of tempered distributions) does not necessarily mean the fulfillment of (42). To make sure condition (42) does hold, we need to do certain assumptions about the RHS $\mathbf{f}(x, y)$. From the very beginning, we always think that $\mathbf{f}$ is compactly supported and that $\mathbf{f} \in L^1(R^2)$, which is no loss of generality. If we additionally assume that $\mathbf{f} \in L^2(R^2)$ (which basically presents no loss of generality as well), then the solution $\mathbf{u} \equiv (u, v, p, \rho)$ to the infinite-domain AP (of type (4)) can be represented as $\mathbf{u} = \mathbf{u}^{(1)} + \mathbf{u}^{(2)}$, where $\mathbf{u}^{(1)}$ satisfies (42) and $\mathbf{u}^{(2)} \in L^2(R^2)$; the latter inclusion can be treated as "a generalized decay at infinity". If, however, we require that $\mathbf{f}(x, y)$ be sufficiently smooth on R^2 so that its Fourier transformation with respect to both x and y belongs to $L^1(R^2)$, then we can show (see [17]) that the solution $\mathbf{u}$ to the AP for system (41) meets boundary condition (42). Similar results for three space dimensions have been obtained in our work [20]. Let us also note that the work [17] is devoted to studying a wider class of formulations than only the steady-state flows, specifically, we study there the flows that oscillate in time. The aforementioned solvability results for the steady-state thin-layer equations can be obtained as one consequence of the considerations of [17]. We will briefly review the general results of [17] in the end of this paper.

2.2 Geometric Setup

Let us now introduce the geometric setup typical for external flow problems in two space dimensions; an example is shown in Figure 1. We are interested in calculating the flow around an airfoil, the single-element configuration presented on the figure does not imply any loss of generality since the treatment of external boundary for the multiple immersed bodies would basically be the same. To calculate the flow, we first generate the grid around the airfoil; for this specific example it will be a C-type curvilinear grid, which actually forms the computational domain D_{in}. The nonlinear flow equations are discretized and solved on this grid inside D_{in}. However, analogously to the continuous case (see Section 1) the discrete system inside D_{in} is subdefinite unless we supplement it by some ABC's. Indeed, the stencil of the finite-difference operator used inside D_{in} cannot, generally speaking, be applied to those nodes of the C-grid that are located near the external boundary, e.g., it cannot be applied to any node of the outermost coordinate row of this grid, since in so doing the part of the stencil may simply "fall out" of the domain. Therefore, the discrete

system inside D_{in} without ABC's would merely have less equations than it has unknowns. Consequently, unlike the continuous case, for which to close the system inside D_{in} means to set the ABC's exactly at the continuous external boundary, to close the system inside D_{in} in the discrete framework means to provide for some additional relations between the values of the solution in the nodes located in a certain external part of the grid. For example, if the scheme employed inside D_{in} is written on the 3×3 stencil, which, in particular, corresponds to a widely used second-order central-difference approach, then the ABC's should provide for the missing relations between the values of the solution on the penultimate and outermost coordinate rows of the C-grid. On Figure 1, the penultimate coordinate row is designated Γ and the outermost row is designated Γ_1.

Henceforth, we will treat the penultimate coordinate row Γ of the C-grid as a formal continuous artificial boundary. This, in particular, means that the outermost curve Γ_1 belongs already to the area, in which we linearize the governing equations. It also means that if we were looking for the continuous solution on D_{in}, then we would need to construct the ABC's exactly at the penultimate coordinate line Γ. For the discrete formulation, however, we need to obtain missing relations between the values of the solution on Γ and Γ_1 (see above). To do that, we will first formulate the ABC's of type (9) exactly at Γ and then use the generalized potential (6) for complementing the boundary data from Γ to D_{ex}, the trace of this complement on Γ_1 will provide us with the unknown values of the solution at the outermost coordinate line of the C-grid. In so doing, we not only obtain the desirable missing relations that close the discrete system in D_{in}, but at the same time automatically make sure that these relations are right in the sense that they properly take into account the structure of the solution in the far field; the latter is true because the complement we construct on D_{ex} is obtained on the basis of (9).

2.3 Computation of the Potentials and Projections

In fact, we, of course, cannot calculate directly the continuous generalized potentials (6) and boundary projections (8); instead, we calculate their discrete counterparts called the difference potentials and the difference boundary projections, respectively. The corresponding numerical procedure is based on application of the DPM [7, 8]. The issues of consistency and convergence for the difference potentials and their continuous prototypes have been studied by Ryaben'kii in [8] and Reznik in [25].

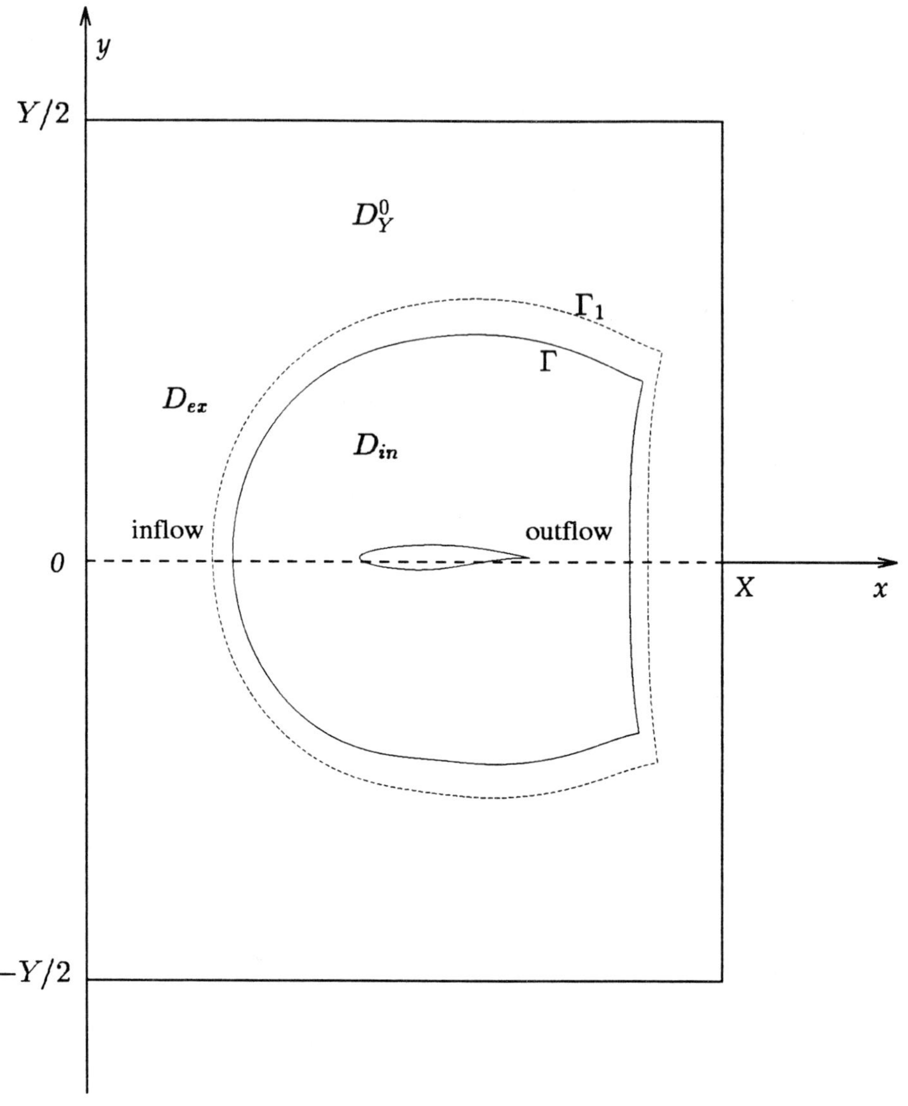

Figure 1. Configuration of domains.

As described in Section 1, the AP for the nonhomogeneous version of either system (40) or (41) is first formulated on the entire plane and then truncated so that one needs to solve it only on the rectangular domain $D_Y^0 = (0, X) \times (-Y/2, Y/2)$; this auxiliary domain should fully contain both Γ and Γ_1, see Figure 1. We will now describe the finite-difference formulation of the AP and the DPM-based algorithm for calculating the difference potentials and projections.

Let us introduce in D_Y^0 two Cartesian grids, $\mathcal{M}^0$ and $\mathcal{N}^0$. The grid $\mathcal{M}^0$ will be used for specifying the RHS for the finite-difference AP, the grid $\mathcal{N}^0$ is the one, on which the solution to this AP will be defined. We also introduce the space $F^0 \ni \mathbf{f}^0 \equiv \mathbf{f}_{\mathcal{M}^0}$ of the RHS's for the AP, the space $U^0 \ni \mathbf{u}^0 \equiv \mathbf{u}_{\mathcal{N}^0}$ of its solutions, and the finite-difference operator $\mathbf{L}_h : U^0 \longmapsto F^0$, which can be a discretization of the left-hand side of either (40) or (41). Note, in the work [18, 26, 27] we have used the operator $\mathbf{L}_h$ obtained by the second-order central-difference approximation of (40), in so doing the grids $\mathcal{M}^0$ and $\mathcal{N}^0$ coincide except on the lines $x = 0$ and $x = X$, where the RHS grid $\mathcal{M}^0$ is simply not defined. In a later work [17, 28, 29], we have used the operator $\mathbf{L}_h$ obtained by the second-order approximation of (41) with the central differences along y and the first-order differences along x. In so doing, the RHS grid $\mathcal{M}^0$ is shifted in the x direction with respect to $\mathcal{N}^0$ by the half of the grid size, and the finite-difference equations of the AP are written as

$$\mathbf{L}_h \mathbf{u}^0 \equiv \mathbf{D} \frac{\mathbf{u}_{m+1,j} - \mathbf{u}_{m,j}}{h_x} + \tag{43}$$

$$\frac{1}{2}\mathbf{F}\left(\frac{\mathbf{u}_{m,j+1} - \mathbf{u}_{m,j-1}}{2h_y} + \frac{\mathbf{u}_{m+1,j+1} - \mathbf{u}_{m+1,j-1}}{2h_y}\right) +$$

$$\frac{1}{2}\mathbf{H}\left(\frac{\mathbf{u}_{m,j+1} - 2\mathbf{u}_{m,j} + \mathbf{u}_{m,j-1}}{h_y^2} + \right.$$

$$\left.\frac{\mathbf{u}_{m+1,j+1} - 2\mathbf{u}_{m+1,j} + \mathbf{u}_{m+1,j-1}}{h_y^2}\right) = \mathbf{f}_{m+1/2,j},$$

$$m = 0, \ldots, M - 1, \quad j = 0, \ldots, 2J,$$

where h_x and h_y are the Cartesian grid sizes; subscript m corresponds to the Cartesian direction x, and $M + 1$ is the total number of nodes of the grid $\mathcal{N}^0$ in this direction; subscript j corresponds to the Cartesian direction y, and $2J + 2$ is the total number of nodes of the grid $\mathcal{N}^0$ (and $\mathcal{M}^0$) in this direction. The 4×4 matrices $\mathbf{D}$, $\mathbf{F}$, and $\mathbf{H}$ are given by

$$\mathbf{D} = \begin{bmatrix} 1 & 0 & 0 & 1 \\ 1 & 0 & 1 & 0 \\ 0 & 1 & 0 & 0 \\ 0 & 0 & 1 & -M_0^{-2} \end{bmatrix}, \tag{44}$$

$$\mathbf{F} = \begin{bmatrix} 0 & 1 & 0 & 0 \\ 0 & 0 & 0 & 0 \\ 0 & 0 & 1 & 0 \\ 0 & 0 & 0 & 0 \end{bmatrix},$$

$$\mathbf{H} = -\frac{1}{Re} \begin{bmatrix} 0 & 0 & 0 & 0 \\ 1 & 0 & 0 & 0 \\ 0 & 4/3 & 0 & 0 \\ 0 & 0 & \gamma Pr^{-1} & -Pr^{-1}M_0^{-2} \end{bmatrix}.$$

Note, we omit the superscript "0" for the solutions and the RHS's of the AP when referring these functions to specific grid nodes, see (43).

Following the considerations of Section 1, we require that any function $\mathbf{u}^0 \in U^0$ be periodic in the y direction with the period Y, i.e.,

$$\mathbf{u}_{m,0} = \mathbf{u}_{m,2J+1}, \quad m = 0, \ldots, M, \tag{45}$$

$$\mathbf{u}_{m,-1} = \mathbf{u}_{m,2J}, \quad m = 0, \ldots, M.$$

As concerns the RHS's $\mathbf{f}^0 \in F^0$, they may, generally speaking, differ from zero only for those nodes of the grid $\mathcal{M}^0$ that belong to D_{in}. Instead of (18)

and (32), we now use the standard discrete Fourier transform (see, e.g., [17, 18, 26] for details), and obtain the following family of systems of ordinary difference equations

$$\mathbf{A}_k \hat{\mathbf{u}}_{m+1,k} + \mathbf{B}_k \hat{\mathbf{u}}_{m,k} = \hat{\mathbf{f}}_{m+1/2,k}, \tag{46}$$

$$m = 0, \ldots, M - 1, \quad k = -J, \ldots, J,$$

which represent the finite-difference analogue of equations (31). In system (46), subscript k is the dual discrete Fourier variable that corresponds to the Cartesian direction y, $\hat{\mathbf{u}}_{m+1,k}$ and $\hat{\mathbf{f}}_{m+1/2,k}$ are the discrete Fourier transformations of the solution and the RHS, respectively; $\mathbf{A}_k$ and $\mathbf{B}_k$ are the square coefficient matrices, which depend on k but do not depend on m. These matrices are obviously determined by the structure of the finite-difference operator $\mathbf{L}_h$. For system (43), $\mathbf{A}_k$ and $\mathbf{B}_k$ are given by

$$\mathbf{A}_k = \frac{1}{h_x}\mathbf{D} + \frac{r_k}{2}\mathbf{F} + \frac{t_k}{2}\mathbf{H}, \tag{47}$$

$$\mathbf{B}_k = -\frac{1}{h_x}\mathbf{D} + \frac{r_k}{2}\mathbf{F} + \frac{t_k}{2}\mathbf{H},$$

where the matrices $\mathbf{D}$, $\mathbf{F}$, and $\mathbf{H}$ are defined in formula (44), $r_k = i\sin\left(kh_y\frac{2\pi}{Y}\right)/h_y$, and $t_k = -4\sin^2\left(\frac{1}{2}kh_y\frac{2\pi}{Y}\right)/h_y^2$. Specific expressions for $\mathbf{A}_k$ and $\mathbf{B}_k$ that correspond to system (40) can be found in [18] and those that correspond to a more general case of time-periodic flows can be found in [17]. Note, the matrices $\mathbf{A}_k$ and $\mathbf{B}_k$ from (47) have order 4, whereas for system (40) these matrices would have order 8 (see [18]), this is caused by the necessity to introduce additional variables when reducing the order of differencing with respect to x from the second to the first.

To complete the formulation of the finite-difference AP on the rectangle D_Y^0, we have to specify the boundary conditions at $x = 0$ and $x = X$, i.e., at $m = 0$ and $m = M$. Analogously to how it is done in Section 1 for the continuous formulation, we first note that system (46) formally considered on the infinite mesh $-\infty < m < \infty$ is homogeneous for $m \leq 0$ and $m \geq M$. Then, to meet boundary condition (42), we have to prohibit all the growing eigensolutions of (46) for $m \longrightarrow -\infty$, as well as for $m \longrightarrow +\infty$. This can be done by imposing for each k, $k = -J, \ldots, J$, the boundary conditions

$$\left[\prod_{|\lambda_r(k)|>1} (\mathbf{Q}_k - \lambda_r(k)\mathbf{I})\right]\hat{\mathbf{u}}_{0,k} = \mathbf{0}, \tag{48a}$$

$$\left[\prod_{|\lambda_r(k)| \leq 1} (\mathbf{Q}_k - \lambda_r(k)\mathbf{I}) \right] \hat{\mathbf{u}}_{M,k} = \mathbf{0}, \qquad (48b)$$

where $\mathbf{Q}_k = (\mathbf{A}_k)^{-1}\mathbf{B}_k$, $\lambda_r(k)$ are the eigenvalues of $\mathbf{Q}_k$, $\mathbf{I}$ is the identity matrix, and the matrix products are calculated in accordance with the multiplicities of eigenvalues. Discrete boundary conditions (48) are analogous to the continuous boundary conditions (33). The only difference is that unlike (33a) and (33b), equations (48a) and (48b) are not symmetric. In formula (48b) (downstream boundary condition), we admit the eigenvalues $|\lambda(k)| = 1$, which in the continuous case would correspond to $|\Re\lambda(\alpha)| = 0$. The reason for this asymmetry is that none of the systems (40) and (41) is actually a system of simple structure. For $\alpha = 0$ ($k = 0$ in the discrete formulation), we have (multiple) eigenvalue with $|\Re\lambda(\alpha)| = 0$ ($|\lambda(k)| = 1$) for both (40) and (41). Unfortunately, neither for the linearized Navier-Stokes nor for the linearized thin-layer equations we are unaware of any good analogues of the principle of limitary absorption (see Section 1) that could have helped us to approximate the solutions of these systems by the solutions of systems of simple structure. We have, therefore, to entirely rely on the numerical experiments as the means to justify the possibility to replace the original AP by the periodic one. The computations that do corroborate the possibility to introduce the periodic formulation are reported in [18]. Moreover, it is also possible to make sure that the matrix $\mathbf{Q}_k$ for $|\lambda(k)| = 1$ ($k = 0$) still has a full system of eigenvectors, which means that the corresponding eigensolutions of the homogeneous one-dimensional system are at most oscillatory, but never increasing. Using this fact, one can show (see [18]) that the resulting solution obtained by means of an inverse Fourier transform still vanishes at infinity, even if for a selected finite number of α's (k's) we require only the boundedness (48b) rather than the true decay of the one-dimensional solution in the Fourier space.

Thus, we have finally completed the formulation of the finite-difference AP, and have, therefore, defined its Green (i.e., inverse) operator $\mathbf{G}_h$, $\mathbf{G}_h : F^0 \longmapsto U^0$. It is easy to see that this AP is uniquely solvable for any compactly supported RHS and well-posed, the well-posedness can be established on the basis of the considerations of [30]. An effective numerical algorithm for solving the AP (more precisely, for solving one-dimensional systems (46) with boundary conditions (48)) is described in our work [31]. This algorithm can be referred to as a version of the well-known successive substitution technique, but with-

out its "inverse" or "resolving stage". The particular efficacy of the approach of [31] is based on the fact that boundary conditions (48) are formulated in terms of eigen subspaces of the operator $\mathbf{Q}_k$.

Let us now split the nodes of the grid $\mathcal{M}^0$ into two groups, $\mathcal{M}_{in} = \mathcal{M}^0 \bigcap \bar{D}_{in}$ and $\mathcal{M}_{ex} = \mathcal{M}^0 \backslash \mathcal{M}_{in}$. Then, we apply the stencil of the finite-difference operator $\mathbf{L}_h$ to each node of the set $\mathcal{M}_{in}$ and call the union of all these stencils $\mathcal{N}_{in}$. Analogously, we apply the stencil of $\mathbf{L}_h$ to each node of $\mathcal{M}_{ex}$ and obtain $\mathcal{N}_{ex}$. The intersection of the two sets $\mathcal{N}_{in}$ and $\mathcal{N}_{ex}$ is called the grid boundary γ, $\gamma = \mathcal{N}_{in} \bigcap \mathcal{N}_{ex}$. The subset $\gamma \subset \mathcal{N}^0$ is actually a multi-layered fringe composed of those nodes of the grid $\mathcal{N}^0$ that are located in a certain sense near the continuous artificial boundary Γ.

For any function $\mathbf{u}^0 \in U^0$ we introduce its difference clear trace ξ_γ on the grid boundary γ as merely a contraction,

$$\xi_\gamma = \mathbf{Tr}_\gamma \mathbf{u}^0 \equiv \mathbf{u}_{\mathcal{N}^0}\big|_\gamma. \qquad (49)$$

Since γ is a multi-layered set of nodes, ξ_γ of (49) in a certain sense models ξ_Γ of (5). Let us now introduce the space of difference clear traces that would contain all grid vector-functions of dimension 4 defined on γ. For each element ξ_γ of this space, we can construct the generalized difference potential

$$\mathbf{P}_{ex}\xi_\gamma = \left(\mathbf{G}_h \left((\mathbf{L}_h\mathbf{w}^0)\big|_{\mathcal{M}_{in}} \right) \right)\Big|_{\mathcal{N}_{ex}}, \qquad (50)$$

where $\mathbf{w}^0 \in U^0$ is chosen so that it has the trace ξ_γ, $\mathbf{Tr}_\gamma\mathbf{w}^0 = \xi_\gamma$, and is arbitrary in the rest. For example, one always may choose $\mathbf{w}^0 = \begin{cases} \xi_\gamma, & \text{on } \gamma \\ \mathbf{0}, & \text{on } \mathcal{N}^0 \backslash \gamma \end{cases}$. As concerns the RHS for the difference AP, i.e., the function that the discrete Green operator $\mathbf{G}_h$ operates on in equation (50), it is defined as

$$(\mathbf{L}_h\mathbf{w}^0)\big|_{\mathcal{M}_{in}} = \begin{cases} \mathbf{L}_h\mathbf{w}^0, & \text{on } \mathcal{M}_{in}, \\ \mathbf{0}, & \text{on } \mathcal{M}_{ex}. \end{cases} \qquad (51)$$

Clearly, to calculate the difference potential $\mathbf{P}_{ex}\xi_\gamma$ (50), which is analogous to the continuous generalized potential (6), we need to actually solve the difference AP.

Finally, we define the operator $\mathbf{P}_\gamma$,

$$\mathbf{P}_\gamma\xi_\gamma = \mathbf{Tr}_\gamma\mathbf{P}_{ex}\xi_\gamma, \qquad (52)$$

as the composition of potential (50) and trace (49). This operator obviously maps the space of difference clear traces ξ_γ onto itself. As in the continuous case (compare (52) with (8)), $\mathbf{P}_\gamma$ appears to be a projection, $\mathbf{P}_\gamma^2 = \mathbf{P}_\gamma$, it is called the difference boundary projection. It is possible to show [7, 8] that those and only those ξ_γ that satisfy the BEP

$$\mathbf{P}_\gamma \xi_\gamma = \xi_\gamma, \tag{53}$$

i.e., belong to the image of the boundary projection $\mathbf{P}_\gamma$, $\xi_\gamma \in \mathrm{Im}\mathbf{P}_\gamma$, can be complemented on $\mathcal{N}_{ex}$ so that the complement $\mathbf{u}_{\mathcal{N}_{ex}}$ satisfies the homogeneous equation $\mathbf{L}_h \mathbf{u}_{\mathcal{N}_{ex}} = \mathbf{0}_{\mathcal{M}_{ex}}$ and boundary conditions (45), (48) of the difference AP. In other words, BEP (53) provides for an exhaustive classification of those and only those ξ_γ's that can be represented as a trace of some solution $\mathbf{u}_{\mathcal{N}_{ex}}$ of the equation $\mathbf{L}_h \mathbf{u}_{\mathcal{N}_{ex}} = \mathbf{0}_{\mathcal{M}_{ex}}$ supplemented by boundary conditions (45), (48). Provided that the BEP (53) is satisfied, the aforementioned complement $\mathbf{u}_{\mathcal{N}_{ex}}$ can be obtained in the form of the generalized potential (50), $\mathbf{u}_{\mathcal{N}_{ex}} = \mathbf{P}_{ex}\xi_\gamma$. Difference boundary projection (52) and difference BEP (53) are analogous to the continuous boundary projection (8) and continuous BEP (9), respectively.

Recall, we have formulated the periodic AP so that its solution approaches the solution of the infinite-domain AP on any finite neighborhood of Γ as the period Y increases. This gives us grounds (see Section 1) for setting the continuous ABC's in the form of BEP (9), in which the Calderon projection $\mathbf{P}_\Gamma$ is calculated using the new finite-domain rather than the original infinite-domain AP. Here, we, in turn, approximate the solution of the continuous periodic AP by the solution of the finite-difference periodic AP, which presents the next key step of the entire procedure (we mean the next one after introducing the periodic formulation). This two-step scheme leads us to a somewhat non-standard concept of convergence for the solutions of the difference AP. Namely, we will consider convergence of the difference solution to the solution of the infinite-domain continuous AP on some finite fixed domain (e.g., on any rectangle $|x - a| < a$, $|y| < a$, where $a \le X/2$, $a < Y/2$) as $(h_x, h_y, Y) \longrightarrow (0, 0, +\infty)$. We have already discussed the reasons and consequences of considering the convergence on a fixed finite subdomain only. We should also emphasize that the convergence is considered not only as the grid size vanishes but also as the period Y synchronously grows. Note, to achieve some initially prescribed accuracy, one should increase the period and decrease the grid size consistently. Some estimates connecting the grid size, the period, and the desired accuracy can be found in [18].

Following the considerations of Section 1, one can conclude that the convergence of the foregoing type is sufficient for the purpose of constructing the ABC's. We will, therefore, use the difference analogues (50), (52), and (53) to the continues potentials (6), projections (8), and BEP's (9), respectively, to set the ABC's in the discrete framework. Later on, we will comment on how to choose the specific values of the grid size and the period; in practice, this choice is always done on the basis of the numerical experience.

2.4 An Application of the Difference Potentials and BEP's for Setting the ABC's

Let us denote by ν the set of those nodes of the C-grid that actually determine the penultimate coordinate line Γ (see Figure 1); analogously, nodes ν_1 will correspond to Γ_1. Additionally, we introduce on Γ the set of collocation points ω, which is also called the collocation grid. This collocation grid will be used for specifying the unknowns; typically, it is coarser than the grid ν. The size of the collocation grid ω is not arbitrary, it is connected to the size of the Cartesian grid $\mathcal{N}^0$, some relevant estimates can be found in [8]. For the practical purposes, we often take the total number $|\omega|$ of nodes of the collocation grid ω proportional to the square root of the total number $|\gamma|$ of nodes that constitute the grid boundary γ. We should also note that the collocation grid is usually not uniform; as a rule, it is more concentrated towards the wake region.

We will approximate the space of clear traces ξ_Γ (see Section 1) by the finite-dimensional space ξ_ω. Specifically, ξ_ω are the eight-component vector-functions defined at the nodes ω, the components of ξ_ω contain the trace of the solution $\mathbf{u}$ and the trace of its normal derivative $\dfrac{\partial \mathbf{u}}{\partial \zeta}$, $\xi_\omega = \left(\mathbf{u}, \dfrac{\partial \mathbf{u}}{\partial \zeta} \right)\Big|_\omega$. We assume that there is an operation $\mathbf{R}_{\Gamma\omega}$ of interpolation along Γ so that as the collocation grid ω is refined the continuous function $\mathbf{R}_{\Gamma\omega}\xi_\omega$ approaches the corresponding ξ_Γ in the sense of a sufficiently strong norm.

We also introduce the operation of continuation of the boundary data from the continuous artificial boundary Γ to the grid boundary γ. Assuming that the size of the Cartesian grid $\mathcal{N}^0$ is reasonably small, one can say that all the nodes γ are located in some small neighborhood of Γ. Therefore, considering ξ_Γ as the given data, we can drop normal from each node γ to Γ and then use the first two terms of the Taylor expansion to calculate ξ_γ. We will designate this operation of continuation by $\pi_{\gamma\Gamma}$, $\pi_{\gamma\Gamma}\xi_\Gamma = \xi_\gamma$. Combining $\pi_{\gamma\Gamma}$ with the previously introduced interpolation $\mathbf{R}_{\Gamma\omega}$, we obtain the operation of continuation of the discrete data ξ_ω from Γ to γ, $\pi_{\gamma\omega}\xi_\omega \equiv \pi_{\gamma\Gamma}\mathbf{R}_{\Gamma\omega}\xi_\omega = \xi_\gamma$.

Finally, recall that the actual data on the curve Γ is calculated numerically at the nodes ν. Therefore, we will further need the operation $\mathbf{R}_{\omega\nu}$ of one-dimensional interpolation along Γ that for a specified $\xi_\nu = \left(\mathbf{u}, \dfrac{\partial \mathbf{u}}{\partial \zeta}\right)\bigg|_\nu$ would give ξ_ω, $\xi_\omega = \mathbf{R}_{\omega\nu}\xi_\nu$. Since both ξ_ω and ξ_ν are vector-functions of the same dimension, the interpolation $\mathbf{R}_{\omega\nu}$ is implemented componentwise.

The finite-difference boundary projection $\mathbf{P}_\gamma$ of (52) and BEP (53) can be used for setting the ABC's differently. We will begin with the brief description of the algorithm of [18], which chronologically has come first.

Substituting the expression $\xi_\gamma = \pi_{\gamma\omega}\xi_\omega$ into the BEP (53), we obtain the following equation

$$(\mathbf{I}_\gamma - \mathbf{P}_\gamma)\,\pi_{\gamma\omega}\xi_\omega = \mathbf{0}_\omega \tag{54}$$

with respect to ξ_ω. Equation (54) can be treated as a certain implicit relation between the solution $\mathbf{u}$ and its normal derivative $\dfrac{\partial \mathbf{u}}{\partial \zeta}$ at the nodes ω. For the difference boundary projection $\mathbf{P}_\gamma$ obtained on the basis of the central-difference approximation of system (40), we actually solve equation (54) with respect to the normal derivatives $\dfrac{\partial \mathbf{u}}{\partial \zeta}\bigg|_\omega$ (see [18]) and therefore express the normal derivatives explicitly in terms of $\mathbf{u}_\omega$. The method we employ for solving equation (54) is based on the application of a certain variational approach, see [18] for more details. Note that in the literature, the operators that express boundary values of normal derivatives of the solution to a PDE or system of PDE's in terms of boundary values of the solution itself are called the Poincaré-Steklov operators [32] or Dirichlet-to-Neumann maps [3, 4].

Having obtained the discrete Dirichlet-to-Neumann map $\mathbf{S}_\omega$ by solving (54), we are then able to calculate ξ_γ for any $\mathbf{u}_\nu$ provided from inside the computational domain D_{in}:

$$\xi_\gamma = \pi_{\gamma\omega}\left(\mathbf{R}_{\omega\nu}\mathbf{u}_\nu,\ \mathbf{S}_\omega\mathbf{R}_{\omega\nu}\mathbf{u}_\nu\right). \tag{55}$$

Finally, we use the function ξ_γ of (55) as the density of the generalized difference potential (50) and interpolate this potential from the grid $\mathcal{N}_{ex}$ to the nodes $\nu_1 \subset \Gamma_1$ using some local formulas of sufficiently high order [18]. In so doing, we obtain the desirable ABC's in the form

$$\mathbf{u}_{\nu_1} = \mathbf{R}_{\nu_1\mathcal{N}_{ex}}\mathbf{P}_{ex}\pi_{\gamma\omega}\left(\mathbf{R}_{\omega\nu}\mathbf{u}_\nu,\ \mathbf{S}_\omega\mathbf{R}_{\omega\nu}\mathbf{u}_\nu\right) \equiv \tag{56}$$

$$\mathbf{T}\,\mathbf{u}_\nu,$$

where $\mathbf{R}_{\nu_1\mathcal{N}_{ex}}$ is the aforementioned interpolation from $\mathcal{N}_{ex}$ to ν_1. Boundary conditions (56) obviously

close the discrete system solved inside D_{in} because they provide for the missing relations between the values of the solution at the penultimate and outermost coordinate rows of the C-grid. Moreover, we are guaranteed that this closure is right from the standpoint of the far-field asymptotic behavior of the solution since the operator $\mathbf{T}$ of (56) is constructed using the resolved form $\mathbf{S}_\omega$ of the boundary projection and the potential (50). We should also note that since all the operators involved in (56) are linear we can actually calculate the matrix of the operator $\mathbf{T}$ in some appropriately chosen basis. This makes the practical implementation of boundary conditions (56) particularly easy even in spite of their nonlocal nature because this implementation is, in fact, reduced to a matrix-vector multiplication. Note, the simplicity of practical implementation of the DPM-based ABC's is not affected by the shape of artificial boundary Γ; the matrix form (56) of the ABC's always remains the same although for the different shapes of Γ and Γ_1 the matrices $\mathbf{T}$ are also different. Moreover, as could be seen from our previous considerations these different matrices $\mathbf{T}$ are themselves calculated by means of one and the same (i.e., geometrically universal) numerical algorithm, which simply uses the shape of the artificial boundary (more precisely, the actual locations of nodes ν and ν_1) as the input data. This algorithm requires one solution of the difference AP per basis vector, as well as some special numerical procedure that includes matrix inversion and multiplication for obtaining $\mathbf{S}_\omega$. Note, the basis, in which we actually calculate the matrices of all operators, is actually chosen in the space of ξ_ω's, and the interpolation $\mathbf{R}_{\omega\nu}$ is applied afterwards. It is also important to emphasize that the RHS's for the difference AP are always concentrated near Γ (see (51)), and the solution of the AP also needs to be known only near Γ and Γ_1. Therefore, the solution of the difference AP (direct and inverse Fourier transforms and the solution of systems (46), (48) for all k) requires only $\mathcal{O}(M \cdot J)$ floating-point operations; a detailed justification of this estimate is contained in [18], see also [31].

Further delineation of the foregoing numerical algorithm for calculating the matrix $\mathbf{T}$ from (56) can be found in our work [18]. The results of implementation of boundary conditions (56) for flow computations are reported in [26, 27], some of these results are reproduced and discussed in Section 3 of this paper.

Another approach to setting the DPM-based ABC's is based on the direct implementation of boundary projections, this approach has recently been proposed in [28, 29], see also [17]. The main purpose of introducing the new approach was to reduce the computational cost of the ABC's. In the new methodology, the difference boundary projection $\mathbf{P}_\gamma$ of (52) is constructed on the basis of the thin-layer system (41) using the operator $\mathbf{L}_h$ defined in (43); the operation $\pi_{\gamma\omega}$ that we use for the continuation of boundary data from Γ to γ is the same as described above (it is based on the Taylor expansion). However, unlike the previous case [18, 26, 27], in which the boundary conditions are driven only by $\mathbf{u}_\nu$ (see (56)), we now consider both $\mathbf{u}_\nu$ and $\left.\frac{\partial \mathbf{u}}{\partial \zeta}\right|_\nu$ as the input data for the ABC's. Assuming that these data are provided from inside the computational domain D_{in}, we then obtain the density of the generalized difference potential by first continuing the boundary data from Γ to γ and then applying the difference projection $\mathbf{P}_\gamma$,

$$\xi_\gamma = \mathbf{P}_\gamma \pi_{\gamma\omega} \mathbf{R}_{\omega\nu} \left.\left(\mathbf{u}, \frac{\partial \mathbf{u}}{\partial \zeta} \right)\right|_\nu . \qquad (57)$$

In other words, we project the arbitrary boundary data provided from inside D_{in} onto the "right manifold" in the sense that ξ_γ of (57) already belongs to the image of the boundary projection $\mathbf{P}_\gamma$, $\xi_\gamma \in \mathrm{Im}\mathbf{P}_\gamma$, and can therefore be represented as a trace of some $\mathbf{u}_{\mathcal{N}_{ex}}$ that solves the equation $\mathbf{L}_h \mathbf{u}_{\mathcal{N}_{ex}} = 0_{\mathcal{M}_{ex}}$ and satisfies boundary conditions (45), (48).

To actually find the aforementioned complement $\mathbf{u}_{\mathcal{N}_{ex}}$ of ξ_γ on $\mathcal{N}_{ex}$, we need to compute the difference potential $\mathbf{P}_{ex}\xi_\gamma$ (see (50)) for the density ξ_γ of (57) (which requires the solution of the difference AP). Then, interpolating this potential from $\mathcal{N}_{ex}$ to $\nu_1 \subset \Gamma_1$, we obtain $\mathbf{u}_{\nu_1}$. Combining the foregoing steps, we can write the desirable ABC's in the form

$$\mathbf{u}_{\nu_1} = \mathbf{R}_{\nu_1 \mathcal{N}_{ex}} \mathbf{P}_{ex} \mathbf{P}_\gamma \pi_{\gamma\omega} \mathbf{R}_{\omega\nu} \left.\left(\mathbf{u}, \frac{\partial \mathbf{u}}{\partial \zeta} \right)\right|_\nu . \qquad (58)$$

Let us now consider an arbitrary function ξ_γ. By definition (see formula (50)), the generalized potential $\mathbf{P}_{ex}\xi_\gamma$ with the density ξ_γ satisfies on $\mathcal{N}_{ex}$ the equation $\mathbf{L}_h \mathbf{P}_{ex}\xi_\gamma = 0_{\mathcal{M}_{ex}}$ and boundary conditions of the AP (45), (48). In turn, one can easily show (see [8]) that any function $\mathbf{u}_{\mathcal{N}_{ex}}$ that solves the equation $\mathbf{L}_h \mathbf{u}_{\mathcal{N}_{ex}} = 0_{\mathcal{M}_{ex}}$ with boundary conditions (45), (48) can be represented as $\mathbf{u}_{\mathcal{N}_{ex}} = \mathbf{P}_{ex}\eta_\gamma$, where $\eta_\gamma = \mathrm{Tr}_\gamma \mathbf{u}_{\mathcal{N}_{ex}}$. In our case $\eta_\gamma = \mathrm{Tr}_\gamma \mathbf{P}_{ex}\xi_\gamma = \mathbf{P}_\gamma \xi_\gamma$ and therefore $\mathbf{P}_{ex}\xi_\gamma = \mathbf{u}_{\mathcal{N}_{ex}} = \mathbf{P}_{ex}\eta_\gamma = \mathbf{P}_{ex}\mathbf{P}_\gamma \xi_\gamma$. In other words, for any ξ_γ the following relation

$$\mathbf{P}_{ex}\xi_\gamma = \mathbf{P}_{ex}\mathbf{P}_\gamma \xi_\gamma$$

is true. Consequently, instead of (58) we can write the ABC's as follows

$$\mathbf{u}_{\nu_1} = \mathbf{R}_{\nu_1 \mathcal{N}_{ex}} \mathbf{P}_{ex} \pi_{\gamma\omega} \mathbf{R}_{\omega\nu} \left.\left(\mathbf{u}, \frac{\partial \mathbf{u}}{\partial \zeta} \right)\right|_\nu \equiv \qquad (59)$$
$$\mathbf{T} \left.\left(\mathbf{u}, \frac{\partial \mathbf{u}}{\partial \zeta} \right)\right|_\nu$$

(the matrices $\mathbf{T}$ in equations (56) and (59) are obviously different). Clearly, boundary conditions (59) provide for the missing relations between the values of the solution on the penultimate and outermost coordinate rows of the C-grid and therefore close the finite-difference system that we solve inside D_{in}. Moreover, this closure is consistent with the desirable far-field behavior of the solution because of the projection $\mathbf{P}_\gamma$ incorporated in (58).

Comparing boundary conditions (56) and (59) we see that they are essentially different. Direct usage of the boundary projection $\mathbf{P}_\gamma$ in (58), (59) allows us to completely avoid the entire resolving stage of the algorithm that is inherent for the approach of the first type summarized in formula (56). Recall, the resolving stage in (56) is associated with the computation of $\mathbf{S}_\omega$ (resolved form of the boundary projection) on the basis of equation (54). Elimination of this stage implies an essential simplification of the algorithm, as well as noticeable economy of computer resources, i.e., the reduction of the computational cost of boundary conditions (59) in comparison with boundary conditions (56).

Of course, another part of this cost reduction, which is even more essential, is accounted for by the reduction of order n of one-dimensional finite-difference system (46) from the eighth to the fourth. Indeed, we recall that boundary conditions (56) were obtained on the basis of the second-order central-difference discretization of system (40) and boundary conditions (59) were obtained on the basis of the discretization (43) of system (41). In so doing, the matrices $\mathbf{A}_k$ and $\mathbf{B}_k$ in system (46) have order $n = 8$ for boundary conditions (56) (see [18]) and order $n = 4$ for boundary conditions (59) (see (44), (47)). According to [31], the solution of one-dimensional problem (46), (48) for each k costs $\mathcal{O}(M \cdot n^2)$ floating-point operations. Therefore, the solution of the entire AP costs $\mathcal{O}(M \cdot J \cdot n^2)$ operations. For both boundary conditions (56) and (59) we have to solve the AP repeatedly, one time per basis vector. Consequently, one can expect that for the same geometry of the discrete sets ν, ν_1, and ω, for the same basis in the space of ξ_ω's, and for the same grid $\mathcal{N}^0$, the computational cost of the matrix $\mathbf{T}$ from (59) will be at least four times less that the cost of matrix $\mathbf{T}$ from (56). Taking into account the foregoing elimination of $\mathbf{S}_\omega$ from the structure of ABC's, we conclude that the overall improvement of the computational efficacy when going from (56) to (59) will be even more drastic. Our computational experiments do corroborate this theoretical expectations. In fact, we could gain up to a factor of five in the reduction of cost of the ABC's (56) in comparison with ABC's (59). Moreover, our preliminary estimates show that in the case of three space dimensions this gain may increase.

As mentioned above, along with being computationally cheaper than the original technique (56) the new methodology (59) also appears much simpler from the algorithmic standpoint. At the same time, boundary conditions (59) do posses all the aforementioned favorable properties that are relevant to boundary conditions (56). Namely, they are geometrically universal, easy to implement in practice, and of course, they perform as good as (or even better than) (56) from the standpoints of accuracy, overall efficacy, and robustness (see Section 3).

2.5 Implementation of DPM-based ABC's

The ABC's of both types (56) and (59) are designed to close the finite-difference system solved inside D_{in}, i.e., to make sure that the number of equations solved inside D_{in} is equal to the number of unknowns. Both relations (56) and (59) are spatially nonlocal, which in practical terms means that the matrices $\mathbf{T}$ are dense. Although these matrices are, in fact, structural, we have not used this property for practical computations yet. (Qualitative structure of the matrices $\mathbf{T}$ can be understood from physical considerations. If the vectors $\mathbf{u}_\nu$ and $\mathbf{u}_{\nu_1}$ are arranged properly, then we can consider $\mathbf{T}$ as being composed of several blocks. Each block of $\mathbf{T}$ would correspond to one physical variable (u, v, p, or ρ) row-wise, i.e., for all nodes ν, and one physical variable column-wise, i.e., for all nodes ν_1, and would have a kind of "diagonal dominance" in the sense that the entries located near the main diagonal will be greater that those located far away from this diagonal. In physical terms, it merely means that each specific node influences its close neighbors stronger than it influences the nodes located on the other side of the computational domain.)

So far, we have been discussing the ABC's only from the viewpoint of closing the system solved inside D_{in} so that the closure is consistent with the desirable far-field behavior of the solution. In practice, however, the construction of a formal closure of the finite-difference system solved inside D_{in} is not sufficient, we also have to combine this closing procedure with the specific solver. The majority of solvers currently used in CFD for calculating the steady-state viscous flows on the basis of finite-difference discretizations employ various types of pseudo-time iterations. In most cases, the iterations are explicit in time and may be enhanced by different techniques for the purpose of accelerating the convergence, e.g., by multigrid.

We have to emphasize that both boundary conditions (56) and (59) are particularly well fitted for the combined usage with explicit iterative solvers. Indeed, let us assume that on some time level (which, in particular, may be zero) the solution is already known on the entire C-grid. Then, advancing one time step by means of some explicit technique we obviously cannot obtain the next-level solution also on the entire C-grid. The reason for that is exactly the same as why the original steady-state finite-difference system inside D_{in} would be subdefinite without the ABC's — the stencil applied to some external nodes of the C-grid may partially "fall out" of the domain. In other words, when using solely the procedure employed inside D_{in}, we can obtain the solution on the upper time level only at the "internal" nodes of the C-grid. For the case of the stencil 3×3, this "internal" set includes all nodes of the C-grid except for the outermost coordinate row $\nu_1 \subset \Gamma_1$. The values of the solution on this outermost coordinate row should therefore be provided by the ABC's so that the solution on the upper time level becomes available everywhere, which makes the next iteration feasible. Looking at boundary conditions (56) and (59) one can see that the desirable complement of the solution on the upper time level can easily be obtained by means of either one of these techniques. When using boundary conditions (56), we simply take $\mathbf{u}_\nu$ that is already computed on the upper time level and, applying the operator $\mathbf{T}$ (matrix-vector multiplication), obtain $\mathbf{u}_{\nu_1}$. This operation is repeated on every iteration; if the relaxation procedure requires evaluation of the residuals more than once per iteration (e.g., multi-stage Runge-Kutta), then boundary conditions (56) are used as many times per iteration as the residuals need to be evaluated.

The implementation of boundary conditions (59) is analogous. After making one iteration of the Navier-Stokes solver inside D_{in} we know the solution on the upper time level everywhere in the interior of the curve Γ. Therefore, we can take $\mathbf{u}_\nu$ on the upper time level as done above and also can easily calculate $\left.\dfrac{\partial \mathbf{u}}{\partial \zeta}\right|_\nu$ using the available data. Then, applying the matrix $\mathbf{T}$ from (59), we obtain $\mathbf{u}_{\nu_1}$ on the upper time level and therefore make it possible to advance another time step. One can see that in both cases (56) and (59) practical implementation of the DPM-based ABC's is very easy since it is reduced to a matrix-vector multiplication on each iteration. Moreover, the implementation is in no way affected by the shape of artificial boundary, although the operators $\mathbf{T}$ themselves would, of course, depend on the geometry.

The implementation of the DPM-based boundary conditions would obviously require some changes if the solver employed inside D_{in} uses multigrid for the acceleration of convergence. Namely, for the multigrid iterative solver the values of the solution on the outermost coordinate row of the grid should be provided every time the residuals need to be evaluated on every level of multigrid. Since the operators $\mathbf{T}$ do depend on the geometry, we may formally need to calculate a separate operator for each subsequent grid. It, however, turns out that at least for some multigrid strategies the numerical process appears to be sensitive only to the ABC's specified at the finest level of multigrid, whereas the sensitivity of the numerical process to the boundary conditions on all the coarser levels is negligible.

In all the computations reported in Section 3, we have used the algorithm [33, 34, 35] by Swanson and Turkel for obtaining the steady-state solutions of the Navier-Stokes equations inside D_{in}. This algorithm is based on the second-order central-difference approximation in space and Runge-Kutta relaxation in time. In practice, we have always used five-stage Runge-Kutta time stepping. The code, in which the algorithm of [33, 34, 35] is implemented, allows one to use different multigrid strategies for the acceleration of convergence. Depending on the specific computational variant, we used from three to five levels of multigrid with W-cycles, when one iteration is done on the finest level and two iterations are done on each of the coarser levels of multigrid. As the additional means of convergence acceleration, the algorithm of [33, 34, 35] includes local time stepping and residual smoothing.

The standard treatment of the external boundary in the code [33, 34, 35] is based on the locally-one-dimensional analysis of characteristics for the inflow part of the boundary and the boundary conditions of extrapolation for the outflow part of the boundary. This approach is actually one of the most well-known and widely used in CFD, its different versions have many times been described in the literature, see, e.g., the reviews [1, 2, 3, 4]. The purely local characteristics/extrapolation treatment may or may not be enhanced in the code [33, 34, 35] by the point-

vortex correction. This lift-based correction [36] usually improves the results provided by the original local ABC's.

We have experimentally made certain that for the different flow regimes (see Section 3, as well as [26, 27]) computed on the basis of the foregoing multigrid strategy, neither the convergence history nor the calculated solution depend on whether we specify the standard local ABC's (see above) on all levels of multigrid or we specify these boundary conditions on the finest level only and for all the coarser levels simply retain the boundary values provided from the finest level. This gave us reasons to expect that the nonlocal DPM-based ABC's (56) or (59) can also be set on the finest level of multigrid only, whereas the boundary values for all the coarser levels will be provided from the finest one. In all the computations reported below, we used exactly this scheme of implementation of the nonlocal DPM-based ABC's. Numerical results (see Section 3) corroborate the possibility Y of doing so, at least for those multigrid strategies that we used for our computations.

Finally, we should note that the implementation of the DPM-based ABC's with implicit iterative solvers also seems feasible, at least from the formal standpoint. Although we have never tried to run any numerical experiments, theoretically this implementation would simply mean that the nonlocal relations of type (56) or (59) are incorporated in the system that is solved on the upper time level on each step of the implicit iteration process.

2.6 <u>Miscellaneous Issues Important for Calculation of DPM-based ABC's</u>

First, we comment here on the choice of the discretization parameters for the difference AP. The unknowns for this problem are specified on the Cartesian grid $\mathcal{N}^0$. The cell size of this grid should be chosen so that the distance between the curves Γ and Γ_1 (see Figure 1) is resolved. In practical computations we usually take the cell size of the grid $\mathcal{N}^0$ to be 3–5 times smaller than the minimal distance between Γ and Γ_1.

As concerns another important parameter of the difference AP, the period Y (see Figure 1), it should be chosen so that to ensure the sufficient accuracy of computations. This choice, of course, cannot be done without taking into account the previous computational experience. However, we first have to choose the unit for measuring Y. It is reasonable to expect that the average diameter of the computational domain D_{in} would make a better unit for measuring the period Y than the characteristic size of the immersed body(ies) (e.g., airfoil chord, see Figure 1). Indeed, the final accuracy will depend on the "extent of convergence", i.e., on how close the solution of the periodic AP has approached the solution of the original nonperiodic AP. Since the size of the computational domain D_{in} actually determines the size of that subdomain in D_Y^0, on which we consider the convergence (e.g., the rectangle $|y| < a$, $|x - a| < a$), then exactly the size of D_{in} becomes a natural unit for measuring the period Y as the variable that controls the convergence.

As mentioned above, the value of the period Y is one of the factors that determine the accuracy of computations. The accuracy, of course, also depends on the actual size of D_{in}. Generally, the DPM-based ABC's allow one to use much smaller computational domains than other available methods do. The specific numerical results will be reported in Section 3; here, we provide only for a qualitative picture. In [26, 27], we have conducted a series of computations for the relatively small values of the period Y, about 2–4 diameters of D_{in}. It turns out that for a small computational domain (2–3 chords of the airfoil for transonic flow regions), the solution obtained on the basis of the DPM-based ABC's differs within 2–4% from the asymptotic solution obtained in a very large computational domain (about 30–50 chords depending on the specific variant). (Note, when comparing the two solutions, we actually compare the corresponding force coefficients: lift, drag, skin friction). These results (see [26, 27]) can be satisfactory for some cases; however, when the accuracy needs to be improved one has to choose larger periods Y. For example, to keep the force coefficients within 0.01%–0.1% of the corresponding asymptotic values, one may need to choose the period Y of about 50 diameters of the computational domain or more.

That big periods Y, along with the small cell size of the grid $\mathcal{N}^0$ prescribed by the distance between Γ and Γ_1 may, generally speaking, result in an expensive numerical procedure for computation of the operators $\mathbf{T}$. To reduce this cost, we introduce nonuniform grids with respect to y. For example, the cell size of such a grid may remain constant (the same as it would be for the uniform grid) in the neighborhood of D_{in} and then enlarge as $|y|$ increases. Clearly, the discrete Fourier transform in the y direction that we have formerly used for the separation of variables does not apply to the stretched grids. Consequently, we will need some other procedure to separate the variables and to therefore reduce the difference AP to a family of one-dimensional systems (46) with boundary conditions (48). The separation of variables for a finite-difference counterpart

of system (41) means that the discrete transform should simultaneously diagonalize the difference approximations to both the first and the second derivatives with respect to y. We have not studied thoroughly the question of whether or not one can find a special distribution of nodes in the y direction and some consistent with this distribution discretization so that the diagonalizing transform appears orthogonal. In practice, it turns out that usage of the skew bases solves the problem of simultaneous diagonalization of the first and the second derivatives and at the same time allows one to still maintain high accuracy of the final results.

Following this idea we first introduce some second order discretization of the first derivative with respect to y on the stretched grid, we also take into account the periodicity conditions (45). In the matrix form, this operation can be represented as a certain $(2J + 1) \times (2J + 1)$ matrix $\mathcal{D}$, which has three non-zero diagonals, $j - 1 \leq i \leq j + 1$, and also non-zero off-diagonal corner entries. Then, the matrix $\mathcal{D}^2$ represents the second difference derivative with respect to y. Both these matrices can obviously be diagonalized by the same transform, which is computed in practice with the help of the standard library eigenvalues/eigenvectors subroutines (IMSL). The rest of the algorithm remains the same as described above. The only difference is that instead of r_k and t_k in (47) one should plug in the eigenvalues of $\mathcal{D}$ and their squares, respectively. Our experiments show that usage of the stretched grids can drastically reduce the computer effort required for calculating the nonlocal DPM-based ABC's. An essential part of the results reported in Section 3 is obtained on the basis of the stretched-grid algorithm.

Another means for reducing the computational cost of the DPM-based ABC's is implementation of the algorithm on parallel platforms. There are, generally speaking, a few ways to parallelize the computation of operators $\mathbf{T}$. Since we anyway calculate the matrix of the linear operator in a certain basis, it is possible to calculate independently and, therefore, simultaneously, the columns of this matrix, where each column corresponds to its own basis vector. We, however, have chosen another way of parallelization, which, in our opinion, requires minimal modifications of the already developed sequential code. Namely, one-dimensional systems (46) with boundary conditions (48) are obviously independent because they are obtained by the separation of variables (k is a parameter). Therefore, these systems can be solved in parallel on the different processors of a multi-processor computer. This approach has been implemented on an eight-processor CRAY Y-

MP. Numerical experiments show the reduction of the "wallclock" time required for calculating the operator $\mathbf{T}$ of (59) by up to a factor of five in comparison with the standard single-processor implementation.

Finally, we should note that the treatment of turbulence in the far field requires special attention. Indeed, the parameter Re in systems (40) and (41) represents the true molecular Reynolds number, therefore these forms of the governing equations apply directly only to the laminar flows. The approximate treatment of turbulence in the far field for the purpose of constructing the ABC's was proposed in the work [27]. Below, we reproduce some results of this work.

To numerically simulate turbulent flows, we use an algebraic turbulence model (Baldwin–Lomax) incorporated in the code [33, 34, 35]. This model is relevant to describing the flow in the vicinity of the immersed body(ies). In the far field, we use simpler approach based on the concept of effective turbulent viscosity [37]. The idea is to qualitatively describe turbulent flow (i.e., the process of turbulent mixing) as a laminar flow of model fluid having some new "turbulent" viscosity.

To obtain the relation between the molecular and effective turbulent viscosity, we use the following considerations. First, we refer to the incompressible case and consider laminar flow. Here, we have the following distribution of u–velocity (perturbation with respect to the far-field value u_0) [37]:

$$u = \frac{W}{2\sqrt{\pi \rho^2 \nu u_0 x}} \, exp\left(-\frac{u_0 y^2}{4\nu x}\right). \qquad (60)$$

We recall [37, 38] that formula (60) is obtained under the natural assumption that the far-field solution actually depends neither on the shape of the immersed body nor on the type of flow in its close vicinity but only on one constant W, which is the total drag. (Note that the dimension of W in (60) is that of force per unit length.)

For the turbulent case we assume [37, 38] that the mixing length l for the wake flow is proportional to the local width b of the wake, $l/b = const$. Then, approximately replacing the value of the derivative in the expression for turbulent viscosity, $\nu_t = l^2 \left|\dfrac{d\,u}{d\,y}\right|$, by the ratio u_{max}/b, where u_{max} is the maximal deviation of the actual velocity from u_0 (which corresponds to the middle of the wake), we easily obtain $\nu_t = const \cdot b \cdot u_{max}$, which, in particular, implies that ν_t does not depend on y. We may also assume [37] that analogously to the laminar case the wake-type

solution for the turbulent flow is self-similar. Then, we have $u = u_{max} \cdot \phi(y/b)$, which immediately yields,

$$W = \rho u_0 \int_{-b}^{b} u \, dy = \rho u_0 b u_{max} \int_{-1}^{1} \phi\left(\frac{y}{b}\right) d\left(\frac{y}{b}\right),$$

and consequently,

$$\nu_t = \frac{kW}{\rho u_0}, \qquad (61)$$

where $k = const$. Once can easily see from (61) that $\nu_t = const$ throughout the whole wake region and therefore, the structure of turbulent wake behind the body (i.e., the profile of mean velocity) appears to be the same as the structure of laminar wake [37, 38] since we actually have the same solution as (60),

$$u = \frac{W}{2\sqrt{\pi \rho^2 \nu_t u_0 x}} \, exp\left(-\frac{u_0 y^2}{4\nu_t x}\right) = \qquad (62)$$

$$\frac{1}{2\sqrt{\pi}} \sqrt{\frac{W}{k\rho x}} \, exp\left(-\frac{\rho u_0{}^2 y^2}{4kWx}\right),$$

only for some other constant ν_t instead of the true molecular viscosity ν.

Recall, our purpose is to find such ν_t that, being substituted into (60), will provide the same wake solution as we would have for the turbulent case. Clearly, ν_t from (61) satisfies this requirement (see (62)), so, let us now determine the specific value of k. Since the effective viscosity is constant throughout the whole wake region, we assume that in the far field it preserves the same value as it has in the outer part of the boundary layer near the trailing edge of the immersed airfoil. Using the Clauser conjecture [37] to calculate the latter quantity, and restricting ourselves (for qualitative consideration) by the case of not high longitudinal pressure gradients, we can obtain,

$$\nu_t = \frac{k_C W}{\rho u_0},$$

which means that the value of the unknown constant k in (61) may be chosen the same as the value of the Clauser constant, $k_C = 0.0168$.

To independently determine W, we recall that in the case of turbulent flow past a flat plate the drag is given by [37]

$$W = 0.0307 Re^{-1/7} \rho u_0^2 L, \qquad (63)$$

where Re is the actual Reynolds number based on the molecular viscosity and L is the characteristic length. To take into account the compressibility, we multiply equation (63) by $\left(\dfrac{2}{2+0.5(\gamma-1)\,M_0^2}\right)^{5/7}$ in accordance with the Tucker conjecture [37]. Then, we introduce an empirical constant Θ, which is the ratio of the pressure and viscous contributions to the total drag (obviously, $\Theta = 0$ for the flat plate); for the transonic flows computed below, $\Theta \approx 5/2$. Finally, we obtain the following relation,

$$Re_t = \frac{u_0 L}{\nu_t} =$$

$$\frac{Re_0^{1/7}}{0.0307 k_C(\Theta+1)}\left(\frac{2}{2+0.5(\gamma-1)\,M_0^2}\right)^{-5/7},$$

which is used for determining the effective turbulent Reynolds number in all the turbulent computations presented in the next section. We emphasize that the proposed treatment of turbulence in the far field is only qualitative. However, this approach seems to be be justified by the numerical experiments.

3 Numerical Results

3.1 Acceleration of Convergence

One of the most important aspects of implementation of any ABC's is the influence that the boundary conditions exert on the convergence to steady state. Our numerical experiments for both two [26, 27] and three [19, 20] space dimensions show that the nonlocal DPM-based ABC's can essentially speed up the convergence of the multigrid iterations compared to the standard characteristics-based boundary conditions.

This positive influence on the convergence rate is, however, not a general situation. It appears that the acceleration of convergence typically occurs only when the interior iterative solver involves multigrid. Otherwise, the nonlocal highly accurate ABC's either do not influence the convergence at all or may even slow it down. For example, the observation of the latter kind was done by Ferm in [39] for the nonlocal ABC's [39, 40] (by Gustafsson and Ferm) that are constructed for the Euler flows in ducts using Fourier transform in the cross-stream direction. The same phenomenon also occurs for the external inviscid flows as shown by Ferm in the work [41], in which he studies the convergence of pseudo-time iterations for the Euler equations supplemented by the nonlocal ABC's [42] (boundary conditions [42] are constructed analogously to [39, 40] for elliptic artificial boundaries). To accelerate the convergence of pseudo-time iterations with nonlocal boundary conditions, Ferm in [39, 41] employs the technique of [43] by Engquist and Halpern (or its modification), which

allows him to make the convergence at least as fast as it is for the simplest locally-one-dimensional non-reflecting boundary conditions that are based on the analysis of characteristics. On the other hand, when boundary conditions [42] are implemented along with some multigrid Euler solver inside the computational domain they no longer slow down the convergence and, therefore, no longer require special acceleration procedures (like the one from [43]). This has been demonstrated by Ferm in the work [44], in which he shows that in order to reduce the initial error by a prescribed factor one needs roughly the same number of multigrid cycles for both nonlocal ABC's [42] and the characteristic non-reflecting boundary conditions.

As mentioned above, when implemented along with the multigrid algorithm [33, 34, 35] (see Section 2), the DPM-based ABC's are capable of even speeding up the convergence to steady state compared to the standard boundary conditions. Let us reproduce here several graphs from [26] that represent the convergence history for different subsonic and transonic laminar flows around the airfoil NACA0012. In the captions to the figures below, α denotes the angle of attack.

From Figures 2, 3, 4, and 5, one can easily see that usage of the DPM-based ABC's can increase the convergence rate of the multigrid iterations by up to a factor of three depending on the specific variant of computations. Note, the subcritical (i.e., fully subsonic) laminar cases that correspond to Figures 2, 3, 4, and 5 have been computed on the grids with small stretching ratios because near the airfoil surface those grids could be chosen relatively coarse. As a result, we have used global rather than local Courant step for iterations in time. In this respect, one can say that Figures 2, 3, 4, and 5 demonstrate the influence exerted by the DPM-based ABC's on a "pure" multigrid (augmented only by residual smoothing).

For the case of two-dimensional turbulent flows that are computed on the grids with much higher stretching ratio and with the local (i.e., chosen cell by cell) Courant step in time, we have not been able to obtain as drastic convergence speedup as for the foregoing laminar cases. The history of convergence for two different two-dimensional transonic turbulent cases is presented in Figures 6 and 7. We however, mention, that for many three-dimensional transonic turbulent cases, the DPM-based ABC's have been able to produce the increase of the convergence rate about as big as shown above for the two-dimensional laminar flows. The corresponding results are reported in our work [19, 20] and will also

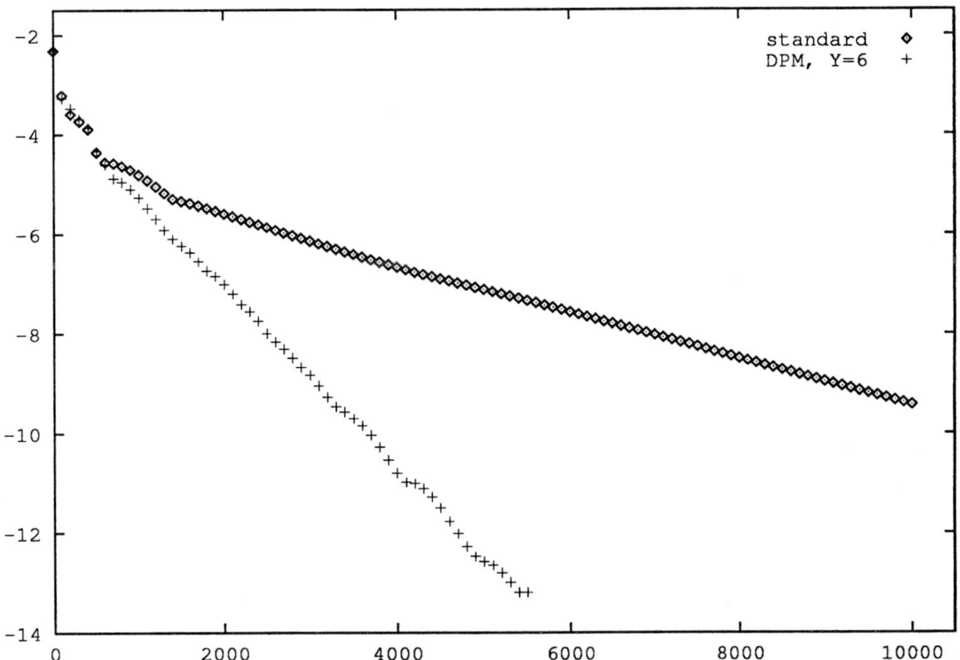

Figure 2. Convergence history: $\log\|\rho_{residual}\|_\infty$ versus number of cycles; NACA0012, $M_0 = 0.63$, $\alpha = 2^\circ$, $Re = 400$. Grid 256×64, average radius of D_{in} about 15 chords.

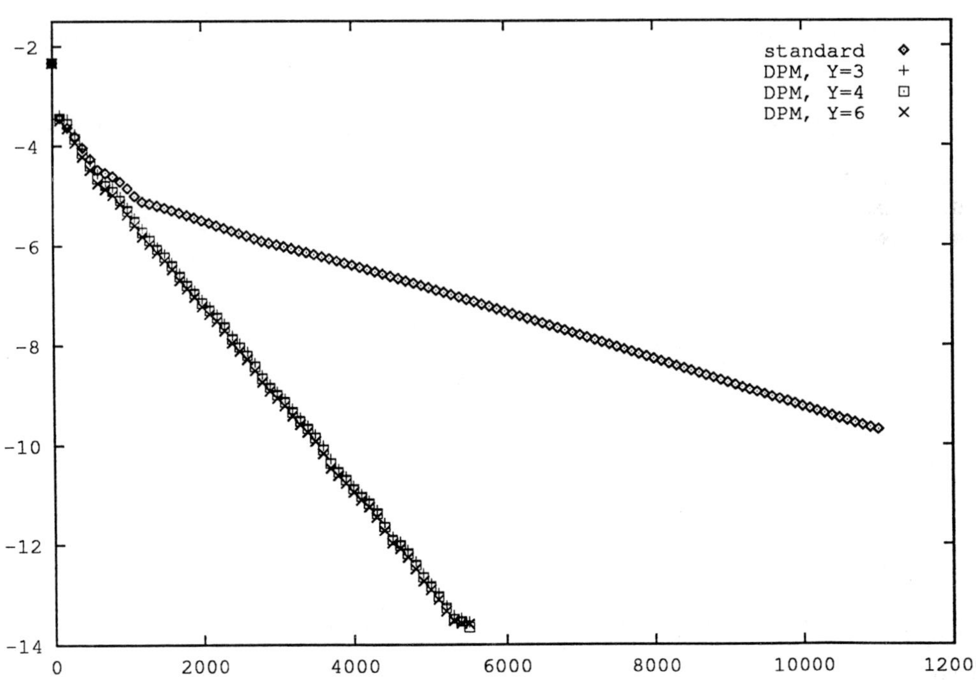

Figure 3. Convergence history: $\log\|\rho_{residual}\|_\infty$ versus number of cycles; NACA0012, $M_0 = 0.63$, $\alpha = 2^\circ$, $Re = 400$. Grid 256×64, average radius of D_{in} about 5.5 chords.

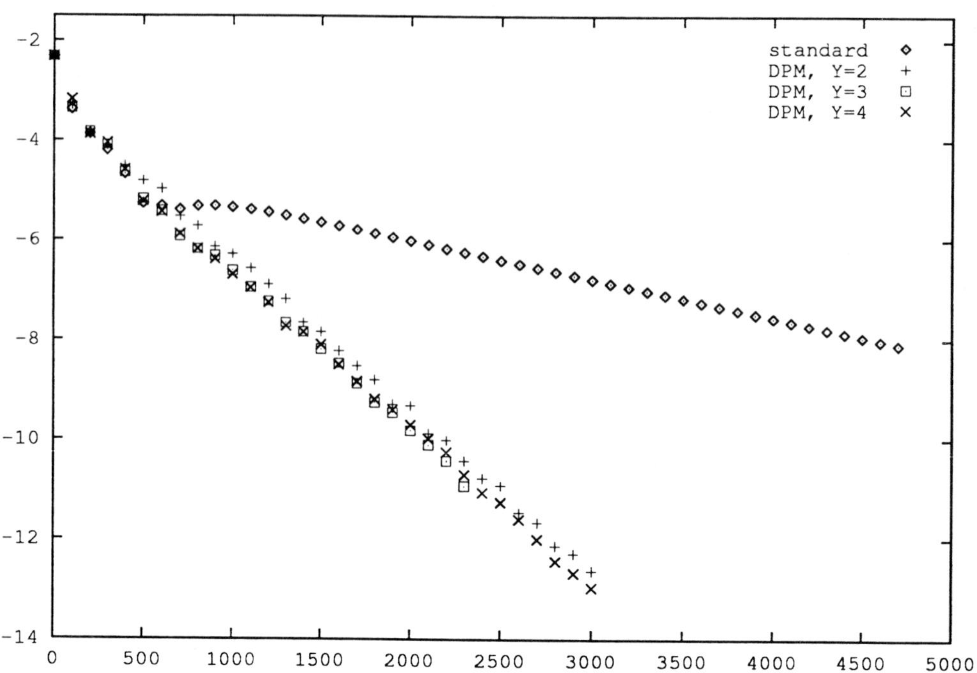

Figure 4. Convergence history: $\log\|\rho_{residual}\|_\infty$ versus number of cycles; NACA0012, $M_0 = 0.63$, $\alpha = 2°$, $Re = 4000$. Grid 256×64, average radius of D_{in} about 5.5 chords.

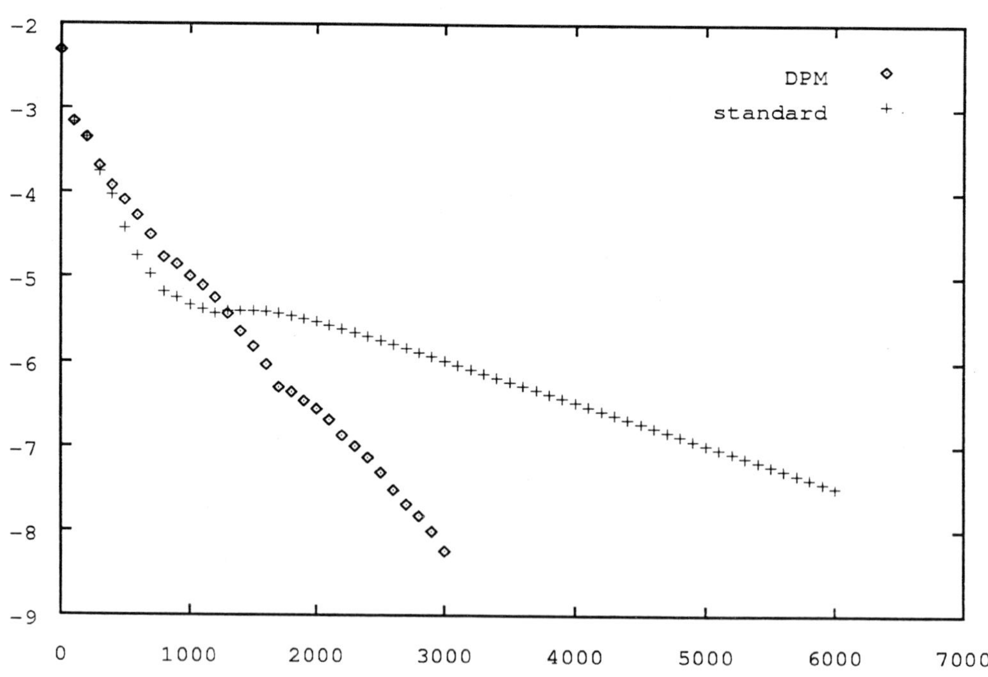

Figure 5. Convergence history: $\log\|\rho_{residual}\|_\infty$ versus number of cycles; NACA0012, $M_0 = 0.63$, $\alpha = 2°$, $Re = 5000$. Grid 256×64, average radius of D_{in} about 10 chords.

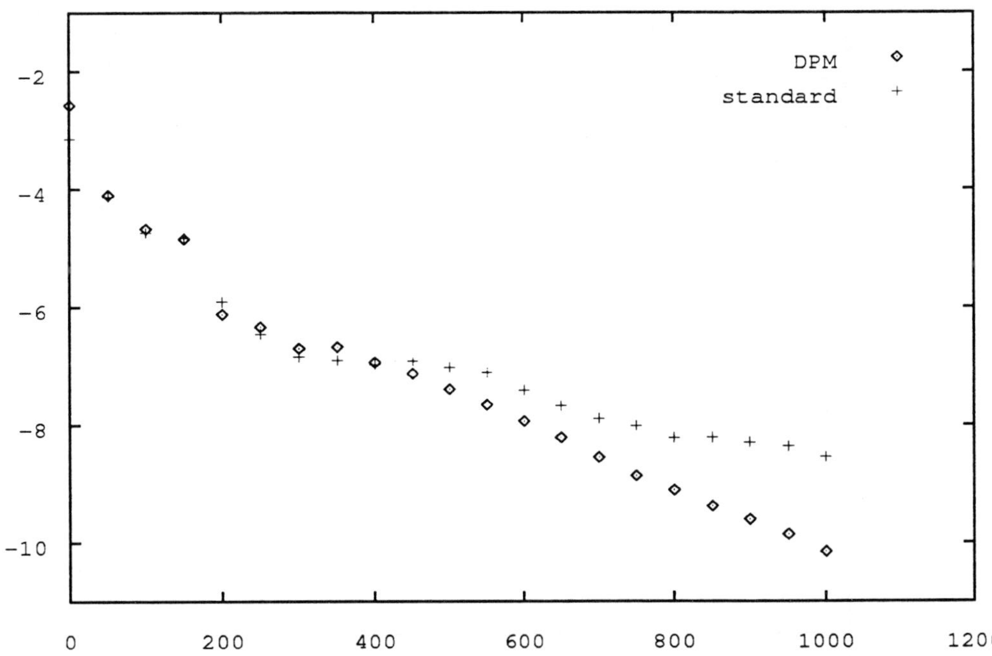

Figure 6. Convergence history: $\log\|\rho_{residual}\|_\infty$ versus number of cycles; RAE2822, $M_0 = 0.73$, $\alpha = 2.79°$, $Re = 6.5 \cdot 10^6$. Average radius of D_{in} about 11 chords, normal spacing near the airfoil 10^{-4}.

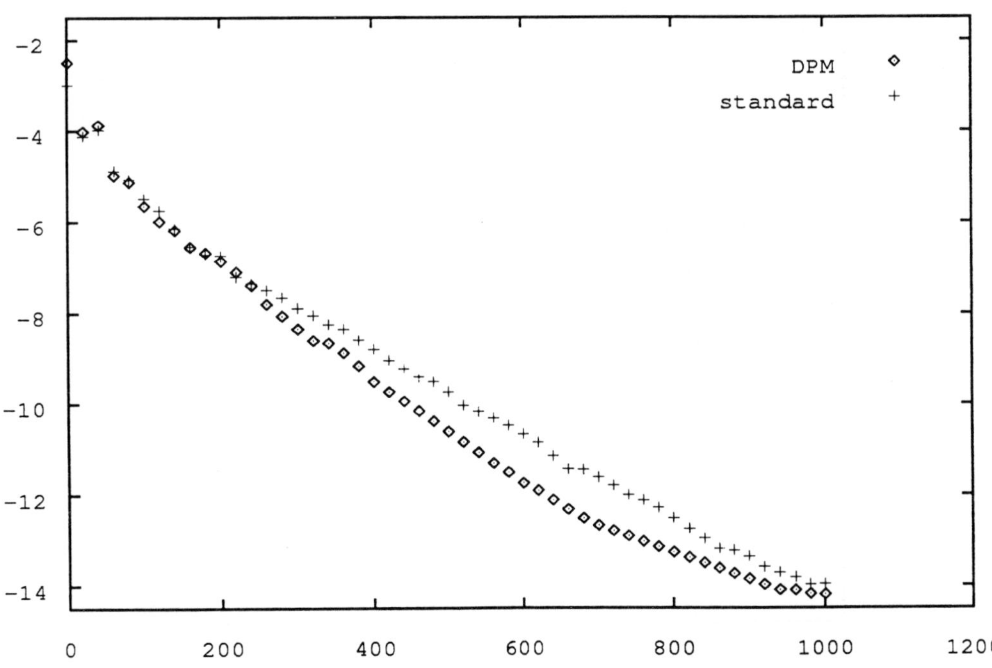

Figure 7. Convergence history: $\log\|\rho_{residual}\|_\infty$ versus number of cycles; RAE2822, $M_0 = 0.73$, $\alpha = 2.79°$, $Re = 6.5 \cdot 10^6$. Average radius of D_{in} about 11 chords, normal spacing near the airfoil 10^{-5}.

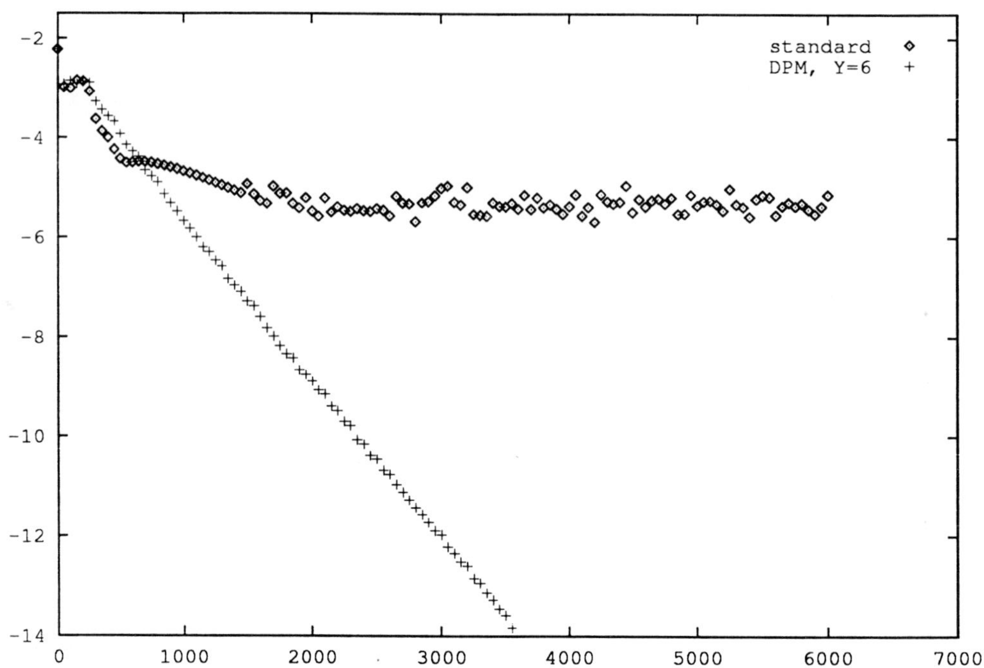

Figure 8. Convergence history: $\log\|\rho_{residual}\|_\infty$ versus number of cycles; NACA0012, $M_0 = 0.85$, $\alpha = 1°$, $Re = 4000$. Grid 256×64, average radius of D_{in} about 5.5 chords.

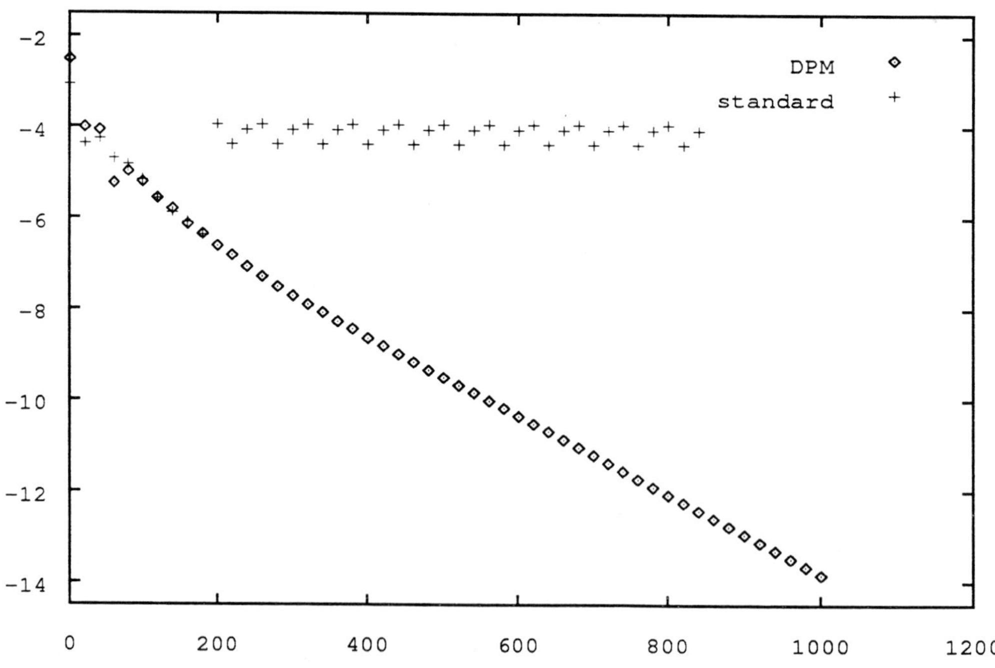

Figure 9. Convergence history: $\log\|\rho_{residual}\|_\infty$ versus number of cycles; RAE2822, $M_0 = 0.73$, $\alpha = 2.79°$, $Re = 6.5 \cdot 10^6$. Average radius of D_{in} about 6 chords, normal spacing near the airfoil $.33 \cdot 10^{-4}$.

be briefly discussed later in this paper.

Returning to Figures 2, 3, 4, and 5, we see that the convergence rates for two different types of ABC's are the same on the initial stage of the iteration process; then, for the DPM-based ABC's the convergence rate remains the same all the time and for the standard boundary conditions it drastically decreases. Therefore, it would be reasonable to assume that the ABC's start to actually influence the convergence only after the numerical perturbations caused by the immersed body reach the external boundary. In other words, the DPM-based ABC's become most effective from the standpoint of convergence acceleration on the so-called asymptotic stage of the multigrid. The similar type of behavior can be observed for the three-dimensional computations as well (see below).

We, however, have to say that although the acceleration of multigrid convergence provided by the DPM-based ABC's is extremely important for applications, the mechanism of interaction of the nonlocal DPM-based ABC's with multigrid may require an additional study. In fact, neither rigorous mathematical explanation of the convergence speedup nor a definite experimental conclusion of why and when it happens is available as of yet. For our two- and three-dimensional computations, we have used different multigrid strategies (W and V cycles, respectively), also all the three-dimensional cases have been computed with the local time step, and the results are also different. Whereas for the two-dimensional transonic turbulent flows we did not see much of an increase in the convergence rate, in three dimensions the strongest speedup occurs right for the transonic turbulent cases (see the work [19, 20] and also below). At the same time, the convergence rates for the subsonic turbulent flows in three space dimensions are the same for the ABC's of different types [19, 20]. As for the laminar flows, the experiments have been conducted in two space dimensions only.

Analyzing the influence that nonlocal boundary conditions may exert on the convergence of multigrid iterations, we should also note that many modern multigrid solvers are not optimal themselves. A massive effort is currently underway towards constructing the new finite-difference schemes, for which the convergence characteristics of multigrid methods would essentially improve. For example, the work in this direction has been done by Ta'asan [45] and Sidilkover [46]. In [45], Ta'asan devised an essentially optimal multigrid solver for the Euler equations in subsonic regime. Due to the separate treatment of the elliptic and advection parts of the system, this approach allows one to achieve in subsonic regime the convergence rates similar to those that can be obtained when solving the full potential equation. Sidilkover in [46] proposed the so-called genuinely multidimensional high-resolution scheme. This scheme has stability properties much superior to those that are relevant to the standard methods and, therefore, facilitates the construction of a very simple and efficient multigrid algorithm (using Gauss-Seidel relaxation as a smoother) that would apply to the entire range of Mach number. Particular efficacy demonstrated by the DPM-based ABC's when implemented in combination with multigrid gives us reasons to hope that these boundary conditions may essentially contribute in developing the new generation of effective multigrid-based algorithms.

Finally, we should mention that the DPM-based ABC's generally improve the robustness of the entire numerical procedure. In conducting our computational experiments [26, 27], we have noticed that sometimes the multigrid iterations [33, 34, 35] supplemented by the standard characteristic/extrapolation boundary conditions simply fail to converge, which never happens if these standard ABC's are replaced by the nonlocal DPM-based boundary conditions. In Figures 8 and 9, we show the history of convergence for the corresponding computations. The similar phenomenon has been observed in three dimensions as well. Namely, for a transonic turbulent flow with separation (see [20]), the multigrid iteration procedure with standard boundary conditions failed to converge, whereas the DPM-based ABC's have still been able to ensure a fast convergence to steady state.

3.2 Accuracy

Another most important outcome of usage of the DPM-based ABC's is the essential increase of accuracy that these boundary conditions provide for in computing the external viscous flows. Below, we compare some numerical results obtained on the basis of boundary conditions (59) for a certain transonic turbulent flow around the airfoil RAE2822 with the results obtained for the same flow regime on the basis of the standard local ABC's (characteristics/extrapolation) enhanced by the point-vortex correction [36]. In Table 1, we present the results for three different grids, 640×128, 608×112, and 600×104 nodes, that correspond to the computational domains of the average radii of 50, 8, and 2.5 chords of the airfoil, respectively. It is important that each subsequent (smaller) grid is obtained here by cutting off several external coordinate lines

Table 1. Comparison with the point-vortex (p.-v.) model for RAE2822 airfoil; $M_0 = 0.73$; $Re_0 = 6.5 \cdot 10^6$; $\alpha = 2.79°$; normal grid spacing near the airfoil $0.5 \cdot 10^{-5}$.

Domain "radius"	3 chords		8 chords		50 chords	
Grid	600×104		608×112		640×128	
Type of ABC's	p.-v.	(59)	p.- v.	(59)	p.-v.	(59)
C_l	0.8653	0.8591	0.8624	0.8589	0.8603	0.8593
relative error	0.58%	0.02%	0.24%	0.04%	0%	0%
$C_d \times 10$	0.1203	0.1263	0.1209	0.1261	0.1255	0.1260
relative error	4.14%	0.24%	3.67%	0.08%	0%	0%
$C_D \times 10$	0.1755	0.1816	0.1762	0.1815	0.1810	0.1815
relative error	3.04%	0.05%	2.65%	0%	0%	0%

Table 2. Comparison with the point-vortex (p.-v.) model for RAE2822 airfoil; $M_0 = 0.73$; $Re_0 = 6.5 \cdot 10^6$; $\alpha = 2.79°$; normal grid spacing near the airfoil $0.5 \cdot 10^{-5}$.

Domain "radius"	2.5 chords		50 chords			
Grid	320×64		320×64		640×128	
Type of ABC's	p.-v.	(59)	p.-v.	(59)	p.-v.	(59)
C_l	0.8688	0.8560	0.8504	0.8492	0.8603	0.8593
relative error	2.15%	0.38%	1.15%	1.17%	0%	0%
$C_d \times 10$	0.1123	0.1259	0.1260	0.1265	0.1255	0.1260
relative error	10.5%	0.07%	0.40%	0.39%	0%	0%
$C_f \times 100$	0.5469	0.5492	0.5478	0.5480	0.5543	0.5544
relative error	1.34%	0.94%	1.17%	1.15%	0%	0%
$C_D \times 10$	0.1670	0.1808	0.1808	0.1814	0.1810	0.1815
relative error	7.73%	0.39%	0.11%	0.05%	0%	0%

of the preceding (bigger) grid. This is done in order to completely avoid any possible influence that the change of the grid near the airfoil surface may exert on the solution.

From Table 1 one can see that the corresponding asymptotic values of the force coefficients (lift C_l, wave drag C_d, total drag C_D), i.e., the values obtained for the large (50 chords) computational domain, are very close to one another for the different types of ABC's. However, as the artificial boundary approaches the airfoil the discrepancy between the corresponding values increases, and the force coefficients obtained on the basis of boundary conditions (59) deviate from their asymptotic values much less than the coefficients obtained using local ABC's. In other words, the nonlocal DPM-based ABC's allow one to use much smaller computational domains than the standard boundary conditions do and to still maintain high accuracy of computations. Moreover, from Table 1 one can see that unlike the ABC's (59), which perform well for all coefficients, the point-vortex boundary conditions perform much better for the lift coefficient C_l than they do for the drag coefficients C_d and C_D. This behavior seems reasonable since the point-vortex model is a purely lift-based treatment and does not take into account drag at all.

In Table 2 we also compare the results obtained using the two aforementioned types of ABC's; however, the computations presented in this table were conducted on the different grids. One can see that boundary conditions (59) outperform the point-vortex ABC's in these cases as well (C_f in Table 2 is the skin friction).

We should also emphasize that the benefit of using smaller computational domains and, as a consequence, smaller grids, i.e., the grids with lesser number of nodes, is not only the direct reduction of the computational work because of the grid shrinkage but also the improvement of convergence because the grids may be chosen less stretched.

3.3 Entry-Wise Interpolation

The computational overhead associated with the usage of nonlocal ABC's (56) or (59) consists of two parts. The first part is accounted for by the matrix-vector multiplications that we do every time we need to evaluate the residuals (see Section 2). These operations add about 1–2% to the total cost of computations (more precisely, to the cost of the same number of multigrid cycles but without the nonlocal ABC's). Generally, we estimate this additional expense as low (taking into account the benefits provided by the DPM-based ABC's), and we do not think that any special effort towards reducing this cost is currently required. Nonetheless, we note that there may still be some room for reducing this cost by using the multiresolution-based techniques (see our work [19, 22]). Moreover, we should mention that the operation of matrix-vector multiplication is, as a rule, fully vectorizeable, which provides for an additional advantage of using the DPM-based ABC's on the CRAY machines. Furthermore, we expect that the new discretizations that are currently under development (see [45, 46]) will allow one to use **relaxation** procedures other than the methods of Runge-Kutta type that are most widely used now. As a consequence, the overhead associated with the matrix-vector multiplications (56) or (59) may be reduced since, for example, the methods of Gauss-Seidel type would require only one such operation per iteration, whereas the multi-stage Runge-Kutta methods require as many multiplications (56) or (59) as the number of stages is. Finally, we mention that the DPM-based ABC's can be implemented directly, i.e., the operation $\mathbf{T}$ can be computed explicitly (without first calculating the matrix in a basis) every time $\mathbf{u}\big|_{\nu_1}$ needs to be updated. This strategy has, in fact, been chosen for all our three-dimensional computations [19, 20, 21, 22]; the corresponding results will be briefly commented on later in this paper.

If the matrix-based strategy is used, then the second part of the total overhead due to the ABC's is accounted for by the computational cost of the operators $\mathbf{T}$ themselves (see (56) and (59)). In the case of two space dimensions, we could always keep the corresponding cost at a level of about 10% of the total work required to calculate the steady-state solution using the algorithm [33, 34, 35] with the specific multigrid strategy discussed in Section 2. Basically, this additional expense can be regarded low as well.

As mentioned above, in three space dimensions we have so far been using another strategy that did not require the calculation of matrices $\mathbf{T}$ at all. However, in the future we may need it, especially in the view of possible multiple runs. Our prelimi-

nary estimates show that in the case of three space dimensions the cost of the operator $\mathbf{T}$ may appear higher (in relative terms) than it is for two space dimensions. Therefore, we propose a special approach to decreasing the computational cost of the nonlocal DPM-based ABC's in the framework of massive computations.

Very often in CFD, one needs to calculate different flows around the same configuration of bodies; in so doing, the same grid is likely to be used. In particular, this may be the case in three space dimensions, especially as the grid generation for this case typically requires a much more substantial effort than for two space dimensions. As can be seen from our previous considerations, the operators $\mathbf{T}$ depend on the geometry, which does not change as long as the grid remains the same. These operators also depend on the coefficients of system (40) or (41), for example, on the Mach number M_0, as well as on the angle of attack α. Note, the angle of attack α formally appears neither in (40) nor in (41) since we always assume that the free-stream velocity is aligned with the positive x direction. We, however, take into account the angle of attack α by rotating the entire computational domain, see Figure 1.

Suppose now that all the computational experiments that we are going to carry out for some chosen geometry belong to a certain range of the parameters involved, say $\alpha_{min} \leq \alpha \leq \alpha_{max}$, $M_{0_{min}} \leq M_0 \leq M_{0_{max}}$. Then, we can pick up several points $\alpha^{(p)}$ within the initially prescribed range for the angle of attack and several points $M_0^{(q)}$ within the initially prescribed range for the Mach number and calculate the operator $\mathbf{T}$ for each resulting pair $\left(\alpha^{(p)}, M_0^{(q)}\right)$. Note, the same is possible for the triplets (α, M_0, Re); however, the far-field effective Reynolds number has been noticed to influence the results at a much less extent than the other two parameters. Finally, to obtain the matrix $\mathbf{T}$ for any specific pair of values (α, M_0) (within the corresponding range), we simply interpolate each entry of $\mathbf{T}$ independently with respect to the angle of attack and the Mach number between the known values for $\left(\alpha^{(p)}, M_0^{(q)}\right)$.

We have implemented this approach numerically for the same transonic turbulent flow around RAE2822 as studied above. To simplify our task on the preliminary stage, we used one-dimensional interpolation separately for α and M_0 instead of using the two-dimensional interpolation on the mesh $\left(\alpha^{(p)}, M_0^{(q)}\right)$. The results shown in Table 3 corroborate usefulness of the approach based on the interpolation of coefficients of the operators $\mathbf{T}$. From

Table 3. Interpolation of coefficients of the operator $\mathbf{T}$ for RAE2822 airfoil; $M_0 = 0.73$; $Re_0 = 6.5 \cdot 10^6$; $\alpha = 2.79°$; normal grid spacing near the airfoil $0.5 \cdot 10^{-5}$.

Domain "radius"	3 chords				50 chords	
Grid	600×104				640×128	
Type of ABC's	p.-v.	(59)	Int. M_0	Int. α	p.-v.	(59)
C_l	0.8653	0.8591	0.8593	0.8587	0.8603	0.8593
relative error	0.58%	0.02%	0.0%	0.07%	0%	0%
$C_d \times 10$	0.1203	0.1263	0.1257	0.1252	0.1255	0.1260
relative error	4.14%	0.24%	0.24%	0.6%	0%	0%
$C_D \times 10$	0.1755	0.1816	0.1811	0.1805	0.1810	0.1815
relative error	3.04%	0.05%	0.22%	0.55%	0%	0%

this table, we see that the accuracy of the solutions obtained on the basis of the interpolated matrices is almost not worse than the accuracy that we get using the genuine operator $\mathbf{T}$.

Of course, the results of Table 3 are only preliminary, and the issue of the entry-wise interpolation of the matrices $\mathbf{T}$ with respect to α and M_0 requires a thorough further study. In particular, the approach based on interpolation may have certain limitations on the size of the computational domain, as well as on the actual admissible ranges of the parameters involved. However, the initial results are encouraging. For the repeated computations, this approach can drastically decrease the overall cost of the DPM-based ABC's since it requires one substantial initial effort for calculating the matrices $\mathbf{T}$ on the mesh $\left(\alpha^{(p)}, M_0^{(q)}\right)$, and then the boundary conditions for each subsequent variant of computations (within the prescribed range) will come for almost no extra computational cost because the cost of interpolation itself is virtually negligible.

3.4 Low Mach Number Flows

We finally address another interesting aspect of implementation of the DPM-based ABC's. It is well-known that many standard explicit solvers for compressible flows encounter difficulties when directly applied to calculating the flows with low Mach numbers. The difficulties are caused by the "different scales" of eigenvalues u and $u \pm c$ (u is the flow velocity and c is the speed of sound, $|u| \ll c$), and result in the severe Courant-type limitations on the time step. One possible cure for this problem is based on the so-called local preconditioning techniques. The idea of these techniques is to change the time-evolving system (multiplying it by some non-singular matrix-preconditioner) so that the gap between the eigenvalues is narrowed but at the same time the steady state remains unchanged. An approach of this type has been recently proposed by Turkel, Fiterman, van Leer, and Vatsa, see [47, 48, 49], and has already been implemented in practice on the basis of the code [33, 34, 35].

It, however, turns out that the standard ABC's incorporated in the code [33, 34, 35] (characteristics/extrapolation) perform poorly for the case of low Mach number flows. On the other hand, boundary conditions (59) in this case demonstrate the same good performance as they show in the case of transonic flows. In Table 4, we compare numerical results obtained using two different types of ABC's for a low Mach number turbulent flow around the airfoil RAE2822. One can see that as in the previous cases, the DPM-based ABC's allow us to maintain high accuracy of computations for small computational domains.

According to the authors of [47, 48, 49], their preconditioning technique actually performs better if the governing equations are written with respect to some other equivalent set of unknowns rather than (u, v, p, ρ). As concerns the DPM-based ABC's, we do not rewrite equations (40) or (41). For those cases presented in Table 4, we have been able to obtain accurate results without any changes (except in the input data) in the boundary conditions algorithm. We, however, note that in so doing the system matrices (44) become strongly non-symmetric. The accuracy of the results from Table 4 in this case is probably due to the fact that we solve the difference AP by a direct method. However, for the small free-stream Mach numbers the use of the symmetrizers for the system matrices (for example, those presented in work [50]) may still be recommended. Moreover, in our work [20] we have constructed the three-dimensional DPM-based ABC's for the true incompressible case and then implemented these boundary

Table 4. Low Mach number turbulent flow around RAE2822 airfoil; $M_0 = 0.01$; $Re_0 = 6.5 \cdot 10^6$; $\alpha = 2.79°$; normal grid spacing near the airfoil $0.6 \cdot 10^{-5}$.

Domain "radius"	2.5 chords		20 chords	
Grid	320×64		320×64	
Type of ABC's	p.-v.	(59)	p.-v.	(59)
C_l	0.5708	0.5387	0.5419	0.5390
relative error	5.3%	0.05%	0%	0%
C_d	-0.0005	0.0016	0.0014	0.0016
relative error	???	0%	0%	0%
$C_D \times 100$	0.5676	0.7761	0.7551	0.7764
relative error	24.8%	0.03%	0%	0%

conditions along with the compressible solver for a very low Mach number ($M_0 = 0.01$). In this case, the DPM-based boundary conditions performed as well as they do for the higher subsonic and transonic Mach numbers (see [20]).

3.5 Three-Dimensional Flows

So far, we have been mostly describing the two-dimensional algorithms and the two-dimensional results. In our work [19, 20] (see also [21, 22]), we have constructed and implemented the nonlocal DPM-based ABC's for three-dimensional steady-state viscous flows (perhaps, the most important case from the standpoint of current computational practice).

As can be seen from Section 2, the ABC's algorithm basically consists of two parts. The first part may be called geometrical, it comprises the construction of the grid sets $\mathcal{M}_{in}$, $\mathcal{M}_{ex}$, $\mathcal{N}_{in}$, $\mathcal{N}_{ex}$, γ, collocation grid ω, all the necessary interpolation operations ($\mathbf{R}_{\Gamma\omega}$, $\mathbf{R}_{\omega\nu}$, $\mathbf{R}_{\nu_1 \mathcal{N}_{ex}}$), and continuation $\pi_{\gamma\omega}$. The second part is computational, it consists of the cross-stream transforms (e.g., discrete Fourier) and the solution of systems (46), (48).

The principle differences between the two-dimensional case and the three-dimensional case are mostly concentrated in the first (geometrical) part, whereas the second (numerical) part changes only quantitatively. The computational geometry is obviously more cumbersome for the case of three space dimensions than it is for the case of two space dimensions. Moreover, there is a qualitative difference between interpolation along the one-dimensional curve, which constitutes the artificial boundary in two dimensions and which does not differ from the straight line when it comes to the issue of internal geometry, and interpolation along the two-dimensional curvilinear surface, which constitutes the artificial boundary in three dimensions and which has a non-

trivial internal geometry. In practice, our computations [19, 20] show that although the geometrical part of the algorithm for three space dimensions is more complicated, it is still universal in the sense that the same procedure serves a variety of artificial boundaries with different shapes; moreover, this geometrical part also turns out numerically cheap.

The major difference between the two- and three-dimensional cases for the computational part of the algorithm is that we basically add another cross-stream direction (in the case of a three-dimensional wing, for example, it may be a span-wise direction). Then, we separate the variables in the AP by transforming in both cross-stream and span-wise directions and obtain a family of one-dimensional systems of type (46) with boundary conditions (48) that formally looks exactly the same although the matrices are of order five and depend on a pair of parameters (wavenumbers) rather than on only one parameter k. The numerical part of the ABC's algorithm in three space dimensions also appears universal in the sense that it does not depend on the shape of the specific computational domain. The details of the three-dimensional DPM-based algorithm, as well as various computational results, can be found in our recent paper [20]. Here, we reproduce only one numerical example.

We consider a steady-state flow of viscous compressible gas around the ONERA M6 wing. We use the NASA-developed code by Vatsa, et al. [51] to integrate the thin-layer equations on a one-block curvilinear C-O type grid generated around the wing (the geometric setup for three space dimension is delineated in our work [19, 20]). The code [51] is similar to the one [33, 34, 35] that we used for two-dimensional computations, it is based on the central-difference finite-volume discretization in space with the first- and third-order artificial dissipation. Pseudo-time

Table 5. Comparison with the standard characteristics-base boundary conditions for a turbulent flow around the ONERA M6 wing: $M_0 = 0.84$; $Re_0 = 11.7 \cdot 10^6$; $\alpha = 3.06°$.

Domain "radius"	3 root chords		10 root chords	
Grid	$197 \times 49 \times 33$		$209 \times 57 \times 33$	
Type of ABC's	standard	DPM	standard	DPM
Full lift, C_L	0.298±0.004	0.2798	0.2805	0.2786
Relative error	6.24%±1.43%	0.43%	0%	0%
Full drag, $C_D \times 10$	0.168±0.008	0.1537	0.1542	0.1531
Relative error	8.95%±5.19%	0.39%	0%	0%

iterations are used for obtaining the steady-state solution; the integration in time is done by the five-stage Runge-Kutta algorithm (with the Courant number calculated locally) supplemented by the residual smoothing. For the purpose of accelerating the convergence, the multigrid methodology is implemented; in our computations we used three subsequent grid levels with V cycles; the full multigrid methodology (FMG) could be employed as well. In addition, we use the preconditioning technique [52] to improve the convergence to steady state. We implement the DPM-based ABC's on the finest level of multigrid on the final FMG stage; the boundary data for coarser levels are provided by the coarsening procedure. As mentioned above, the implementation of the ABC's was direct, i.e., it did not require the calculation of the matrices $\mathbf{T}$. Moreover, even on the finest level we implement the DPM-based ABC's only on the first and the last Runge-Kutta stages, which seems to make very little difference compared to the implementation on all five stages; the boundary data for the three intermediate stages are provided from the DPM-based ABC's on the first stage. Unlike the two-dimensional case, the standard treatment of the external boundary in three dimensions is based on merely the locally one-dimensional characteristics analysis and extrapolation (the point-vortex model is not applicable).

We have conducted the computations for a standard three-dimensional transonic test case for the ONERA M6 wing: $M_0 = 0.84$; $Re_0 = 11.7 \cdot 10^6$; $\alpha = 3.06°$. The solution was calculated for two different computational domains of the average radii of approximately 10 and 3 root chords of the wing, respectively. The results summarized in Table 5 clearly demonstrate that for the small computational domains the DPM-based ABC's generate much more accurate solutions that the standard (characteristics-based) boundary conditions

do. Note, the grids for the different domains have different dimensions, and the smaller $197 \times 49 \times 33$ grid (3 root chords) is now an exact subset of the bigger $203 \times 57 \times 33$ grid (10 root chords). This is done in order to eliminate any influence that the change of the grid in the near field could possibly exert on the solution.

Besides the improvement of accuracy, the application of the DPM-based ABC's to transonic flow computations on the small (3 root chords) computational domain yielded much higher convergence rate of the residual (continuity equation), as well as much faster convergence of other quantities, including those deemed as sensitive, e.g., the number of supersonic points in the domain. In Figures 10a and 10b, we show the convergence history for this supercritical flow variant. One can see that the convergence for the standard boundary conditions is poor; therefore the corresponding force coefficients in Table 5 are given with the error bands indicated.

For the 10 root chords domain, the DPM-based ABC's also provide for some convergence speedup, although the difference between the two ABC's techniques is less dramatic here. This is reasonable because one could generally expect that the bigger the computational domain, the smaller is the influence that the external boundary conditions exert on the numerical procedure. The convergence history for the 10 root chords computations is shown in Figures 11a and 11b.

Note, from Figure 10b one can conclude that on the small domain the two algorithms converge to quite different solutions, whereas Figure 11b allows one to assume that on the big domain the final solutions are close to one another. The data from Table 5 corroborate these conclusions. This behavior again fits into the aforementioned concept that the impact of the ABC's decreases as the domain size increases.

As concerns the computational cost of the three-dimensional DPM-based ABC's, we note that by ap-

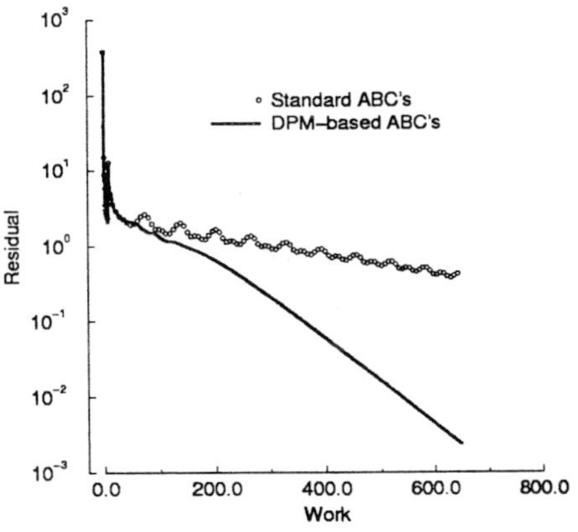

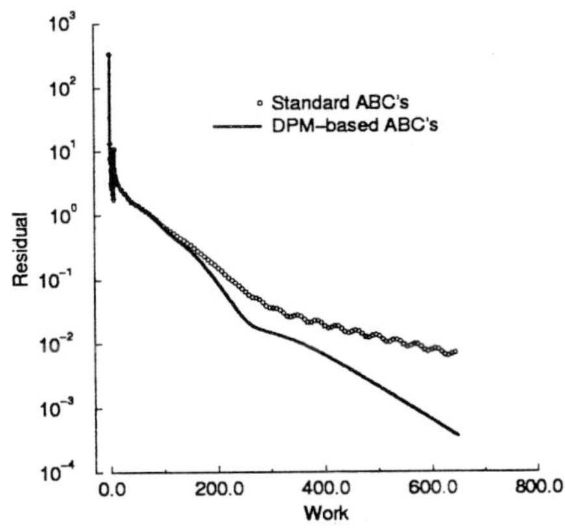

Figure 10a. ONERA M6: $M_0 = 0.84$, $Re_0 = 11.7 \cdot 10^6$, $\alpha = 3.06°$. Convergence history for the residual of the continuity equation. Average domain "radius" is 3 root chords of the wing; grid $197 \times 49 \times 33$.

Figure 11a. ONERA M6: $M_0 = 0.84$, $Re_0 = 11.7 \cdot 10^6$, $\alpha = 3.06°$. Convergence history for the residual of the continuity equation. Average domain "radius" is 10 root chords of the wing; grid $209 \times 57 \times 33$.

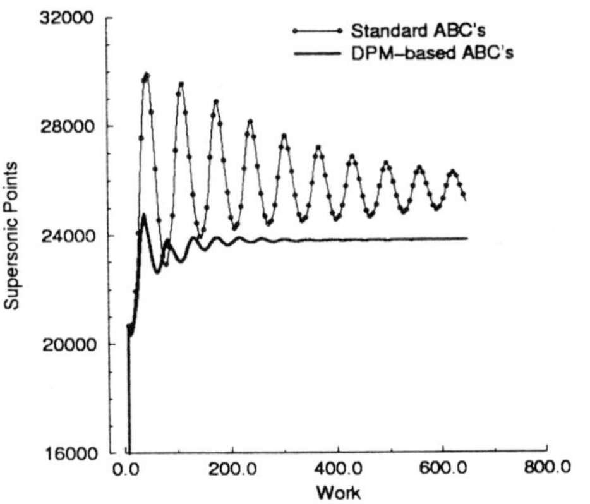

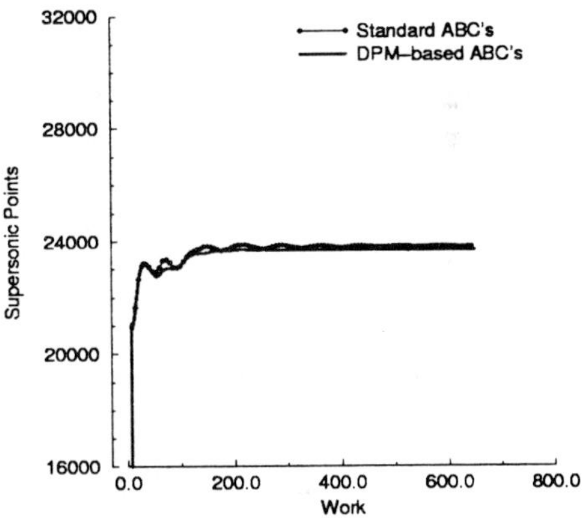

Figure 10b. ONERA M6: $M_0 = 0.84$, $Re_0 = 11.7 \cdot 10^6$, $\alpha = 3.06°$. Convergence history for the number of supersonic nodes in the domain. Average domain "radius" is 3 root chords of the wing; grid $197 \times 49 \times 33$.

Figure 11b. ONERA M6: $M_0 = 0.84$, $Re_0 = 11.7 \cdot 10^6$, $\alpha = 3.06°$. Convergence history for the number of supersonic nodes in the domain. Average domain "radius" is 10 root chords of the wing; grid $209 \times 57 \times 33$.

plying this procedure only on the first and the last Runge-Kutta stages and only on the finest multigrid level, the total number of the required calculations of generalized potential has been brought to a minimum. In so doing, the average cost of application of the DPM-based ABC's adds about 20–25% of the CPU time to the cost of the same procedure with the standard (characteristics-based) boundary conditions. This extra expense is not high (taking into account the improvement of accuracy); moreover, it can often be compensated for and even noticeably prevailed over by the convergence acceleration and the reduction of the domain size. Besides, to ex-

plicitly decrease the computational cost associated with the DPM-based ABC's the entry-wise interpolation of boundary operators (see above) and/or the multiresolution-based methodologies (see [19, 22]) can be used. We expect that the latter can also be employed when implementing the DPM-based ABC's for multi-block grids.

4 Concluding Remarks

The DPM-based approach provides for a geometrically universal and robust means to set the ABC's for steady-state external flow computations. These ABC's enable one to essentially decrease the size of the computational domain (for the case of steady-state viscous flows) in comparison with the domains that can be used with standard boundary conditions. This implies the possibility to improve the accuracy of computations without increasing their cost or to reduce the total cost of computations without decreasing the accuracy. Moreover, the total computational cost may also be essentially reduced because of the convergence speedup that is accounted for by usage of the DPM-based ABC's. We also emphasize that the DPM-based ABC's apply to the computational domains of irregular shape with equal ease and that the practical implementation of these boundary conditions is easy from the algorithmic standpoint. In particular, it takes a relatively little effort to supplement some already existing code by the new DPM-based ABC's. Altogether, these properties make the DPM-based ABC's a very attractive tool for external flow computations.

The next big challenge would, of course, be extending the DPM-based approach to the case of time-dependent problems. As mentioned above, a particular class of such problems, namely, the fluid flows oscillating in time, has been studied in [17]. In practice, this formulation originates, e.g., from the well-known problem of pitching airfoil. Analogously to the steady-state case, the ABC's of [17] also connect the values of the solution at the penultimate and outermost rows of the grid (e.g., C-grid); however, it is done for the entire time interval T equal to one period. To obtain the ABC's in [17], we first approximate the data on $\Gamma \times [t_0, t_0 + T]$ (t_0 is arbitrary) by some periodic in time vector-function in the sense of least squares. The approximant obviously appears to be the Fourier series of the actual data on $\Gamma \times [t_0, t_0 + T]$. Assuming that the period T is known in advance, and that there are no other essential time-dependent effects in the model, we implement the Fourier transform in time and end up with the family of "steady-state" systems. Each member of this family has, generally speaking, complex coefficients but can nevertheless be treated analogously to how it is done above, so that the resulting boundary conditions in the frequency domain are obtained in the form (59). Then, implementing the inverse Fourier transform we obtain the ABC's in time domain. When the problem is discretized in time, both direct and inverse transforms are discrete as well, and the aforementioned family of "steady-state" systems is finite, which gives us a way to practically calculate the ABC's. Clearly, the DPM-based ABC's of [17] appear to be nonlocal in both space and time. However, the nonlocality in time is limited by the interval T equal to one period.

As concerns the case of general time-dependent flows, for which we do not do any initial assumptions (like periodicity) about the behavior of the solution, it is, of course, more complicated and more difficult to handle. The main obstacle here is the nonlocality of the ABC's in time. Unlike the periodic case, the exact ABC's for the general time-dependent problem would formally require storing all the preceding information on the artificial boundary from $t = 0$ to $t = t_{final}$. As the solution develops in time, such boundary conditions would become more and more expensive from the standpoints of both memory and computer time, which is required for processing the constantly growing amount of boundary data. The estimated high computational cost severely limits the possibilities of developing the exact ABC's for time-dependent problems and makes most of the available constructions of such boundary conditions practically infeasible for any long-term runs.

Clearly, the crucial issue for any time-dependent ABC's algorithm is how to effectively restrict the nonlocality of the boundary conditions in time. (In time-periodic formulation of [17], this restriction was incorporated in the formulation of the problem from the very beginning.) A promising approach based on implementation of the Laplace transform in time and then on usage of special recursion relations for calculating the convolutions with the kernels of nonlocal operators has recently been proposed by Sofronov in [53]. The limitation of this technique is the requirement that the boundary should be of some regular, e.g., linear, shape. There are several other approaches to this problem, none of which has been studied thoroughly yet, although each one may in principle appear useful. One approach is based on the idea of an artificial periodic formulation in time. It is analogous to what we do for the cross-stream space coordinates in the steady-state problem. Basically, we introduce some error, which is controlled by the value of the period, and in so doing obtain finite formulation of the problem which is available for the numerical treatment (see Section 2). Note, the idea of an artificial periodic formulation in time was earlier proposed by Lax in [54]. Other possible approaches could be based on some properties of solutions relevant to particular classes of equations (systems). For example, one can make use of lacunas that exist in the solutions of hyperbolic equations, or exploit the exponential decay of coefficients of the Green operator as the time interval increases, which is relevant to parabolic equations. More details on these and analogous approaches can be found in the work [11], in which they have been first proposed, as well as in the reviews [1, 2].

We emphasize that the issue of ABC's for time-dependent problems is important not only in CFD, but in many other areas of scientific computing. For example, the problems of computational acoustics and aeroacoustics (Helmholtz, wave, or linearized Euler equations), as well as the problems of electromagnetic waves propagation (Maxwell equations), present significant mathematical interest and also attract more and more attention of the practitioners. Effective algorithms for handling unbounded domains are required for such problems in both single-mode (Helmholtz-type) and wide-band (as a rule, time domain) formulations. For the time domain algorithms, the restriction of nonlocality of the ABC's in time obviously acquires particular importance and presents a major challenge for the future research. Among many interesting links that connect this problem to other areas of computational mathematics we would like to point out one. Namely, one of the principle elements of the construction of highly accurate and numerically efficient DPM-based ABC's for time-dependent problems would be to calculate the Calderon projections for the schemes that are able of clear capturing the lacunas relevant to the solutions of hyperbolic equations (systems). In turn, the ideas for constructing such schemes seem to be closely related to another group of ideas and techniques associated with the genuinely multidimensional methods. The latter are intensively studied now, especially in CFD, where these methods present a major hope for the drastic increase of efficiency of the multigrid solvers (see [46]).

Acknowledgments

The authors are very thankful to Saul Abarbanel of Tel-Aviv University, Alvin Bayliss of Northwestern University, David Gottlieb of Brown University, Bertil Gustafsson of Uppsala University, Dmitrii Kamenetskii of Keldysh Institute for Applied Mathematics, Russian Academy of Sciences, David Sidilkover of ICASE, NASA Langley Research Center, Ivan Sofronov of Keldysh Institute for Applied Mathematics, Russian Academy of Sciences, R. Charles Swanson of NASA Langley Research Center, James L. Thomas of NASA Langley Research Center, Eli Turkel of Tel-Aviv University, and Veer Vatsa of NASA Langley Research Center for the numerous fruitful discussions on the subject of this paper and for their most helpful suggestions and comments. Our special thanks to Eli Turkel, R. Charles Swanson, and Veer Vatsa for giving us an opportunity to use their multigrid Navier-Stokes codes and to Edward Parlette of Vigyan, Inc. for his most valuable help in generating the three-dimensional grids.

We are most grateful to the Editor of the volume Mohamed Hafez of UC Davis for inviting and encouraging us to write this review.

References

1. Tsynkov, S. V., "Artificial Boundary Conditions Based on the Difference Potentials Method," NASA Technical Memorandum No. 110265, Langley Research Center, July 1996.

2. Tsynkov, S. V., "Numerical Solution of Problems on Unbounded Domains and Application of the Difference Potentials Method," submitted to *Applied Numerical Mathematics*.

3. Givoli, D., "Non-reflecting Boundary Conditions," *Journal of Computational Physics*, 94 (1991) pp. 1–29.

4. Givoli, D., *Numerical Methods for Problems in Infinite Domains*, Elsevier, Amsterdam, 1992.

5. Calderon, A. P., "Boundary-Value Problems for Elliptic Equations," in *Proceedings of the Soviet-American Conference on Partial Differential Equations at Novosibirsk*, Fizmatgiz, Moscow, 1963, pp. 303–304.

6. Seeley, R. T., "Singular Integrals and Boundary Value Problems," *American Journal of Mathematics*, 88 (1966) pp. 781–809.

7. Ryaben'kii, V. S., "Boundary Equations with Projections," *Russian Mathematical Surveys*, 40 (1985) pp. 147–183.

8. Ryaben'kii, V. S., *Difference Potentials Method for Some Problems of Continuous Media Mechanics*, Nauka, Moscow, 1987 (in Russian).

9. Ryaben'kii, V. S., "Difference Potentials Method and its Applications," *Math. Nachr.*, 177 (1996) pp. 251–264.

10. Mikhlin, S. G., Morozov, N. F., and Paukshto, M. V., "The Integral Equations of the Theory of Elasticity," B. G. Teubner Verlagsgesellschaft, Stuttgart, 1995.

11. Ryaben'kii, V. S., "Exact Transfer of Boundary Conditions," *Computational Mechanics of Solids,* Issue 1 (1990) pp. 129–145 (in Russian).

12. Tsynkov, S. V., "Boundary Conditions at the External Boundary of the Computational Domain for Subsonic Problems in Computational Fluid Dynamics," Keldysh Institute of Applied Mathematics, U.S.S.R. Academy of Sciences, Preprint 108, Moscow, 1990 (in Russian).

13. Tsynkov, S. V., "An Implementation of the Potential Flow Model in Setting the External Boundary Conditions for the Euler Equations. Part I," Keldysh Institute of Applied Mathematics, U.S.S.R. Academy of Sciences, Preprint 40, Moscow, 1991 (in Russian).

14. Sofronov, I. L., and Tsynkov, S. V., "An Implementation of the Potential Flow Model in Setting the External Boundary Conditions for the Euler Equations. Part II," Keldysh Institute of Applied Mathematics, U.S.S.R. Academy of Sciences, Preprint 41, Moscow, 1991 (in Russian).

15. Sofronov, I. L., "A Rapidly Converging Method for Solving the Euler Equation," *Computational Mathematics and Mathematical Physics*, 31 (1991) pp. 66–78.

16. Godunov, S. K., and Gordienko, V. M., "The Green Matrix of a Boundary-Value Problem for Ordinary Differential Equations," *Russian Mathematical Surveys*, 39 (1984) pp. 45–85.

17. Tsynkov, S. V., "Artificial Boundary Conditions for Computation of Oscillating External Flows," to appear in *SIAM Journal on Scientific Computing* (1997). Also see: NASA Technical Memorandum No. 4714, Langley Research Center, August 1996.

18. Ryaben'kii, V. S., and Tsynkov, S. V., "Artificial Boundary Conditions for the Numerical Solution of External Viscous Flow Problems," *SIAM Journal on Numerical Analysis*, 32 (1995) pp. 1355–1389.

19. Vatsa, V. N., and Tsynkov, S. V., "An Improved Treatment of External Boundary for Three-Dimensional Flow Computations," AIAA Paper No. 97-2074, June 1997.

20. Tsynkov, S. V., "External Boundary Conditions for Three-Dimensional Problems of Computational Aerodynamics," NASA Technical Memorandum No. 110337, Langley Research Center, March 1997; also submitted to *SIAM Journal on Scientific Computing*.

21. Tsynkov, S. V., "Nonlocal Artificial Boundary Conditions for Computation of External Viscous Flows," in: Computational Fluid Dynamics'96, Proceedings of the Third ECCOMAS CFD Conference, September 9–13, 1996, Paris, France, J.-A. Desideri, C. Hirsch, P. Le Tallec, M. Pandolfi, and J. Périaux, eds., John Wiley & Sons, 1996, pp. 512–518.

22. Tsynkov, S. V., "Artificial Boundary Conditions for Infinite-Domain Problems," to appear in Proceedings of the ICASE/LaRC Workshop on Barriers and Challenges in Computational Fluid Dynamics, Hampton, August 5–7, 1996, M. D. Salas et al., eds., Kluwer Academic Publishers, 1997.

23. Hörmander, L., *Linear Partial Differential Operators*, Springer-Verlag, Berlin, 1963.

24. Vladimirov, V. S., *Equations of Mathematical Physics*, Dekker, New York, 1971.

25. Reznik, A. A., "Approximation of Surface Potentials of Elliptic Operators by Difference Potentials," *Soviet Mathematics Doklady*, 25 (1982) pp. 543–545.

26. Tsynkov, S. V., "An Application of Nonlocal External Conditions to Viscous Flow Computations," *Journal of Computational Physics*, 116 (1995) pp. 212–225.

27. Tsynkov, S. V., Turkel, E., and Abarbanel, S., "External Flow Computations Using Global Boundary Conditions," *AIAA Journal*, 34 (1996) pp. 700-706. Also see: AIAA Paper No. 95-0562, January 1995.

28. Tsynkov, S. V., "Construction of Artificial Boundary Conditions Using Difference Potentials Method," to appear in *Proceedings of the International Conference on Optimization of Finite-Element Approximations*, June 25–29, 1995, St.-Petersburg, Russia.

29. Tsynkov, S. V., "Nonlocal Artificial Boundary Conditions Based on the Difference Potentials Method," in *Proceedings of the Sixth International Symposium on Computational Fluid Dynamics*, IV, pp. 114–119, September 4–8, 1995, Lake Tahoe, Nevada.

30. Ryaben'kii, V. S., "Necessary and Sufficient Conditions for Good Definition of Boundary Value Problems for Systems of Ordinary Difference Equations," *U.S.S.R. Computational Mathematics and Mathematical Physics*, 4 (1964) pp. 43–61.

31. Ryaben'kii, V. S., and Tsynkov, S. V., "An Effective Numerical Technique for Solving a Special Class of Ordinary Difference Equations," *Applied Numerical Mathematics*, 18 (1995) pp. 489–501.

32. Agoshkov, V. I., "Domain Decomposition Techniques for Problems in Mathematical Physics,"

in *Computational Processes and Systems,* issue 8, G. I. Marchuk, ed., Nauka, Moscow, 1990, pp. 3–51 (in Russian).

33. Swanson, R. C., and Turkel, E., "A Multistage Time-Stepping Scheme for the Navier-Stokes Equations," AIAA Paper 85-0035, January 1985.

34. Swanson, R. C., and Turkel, E., "Artificial Dissipation and Central Difference Schemes for the Euler and Navier-Stokes Equations," AIAA paper 87-1107, June 1987.

35. Swanson, R. C., and Turkel, E., "Multistage Schemes with Multigrid for the Euler and Navier-Stokes Equations. Volume I: Components and Analysis," NASA Technical Paper No. 3631, Langley Research Center, 1997.

36. Thomas, J. L., and Salas, M. D., "Far-Field Boundary Conditions for Transonic Lifting Solutions to the Euler Equations," AIAA Paper 85-0020, January 1985.

37. Loytsyansky, L. G., *Mechanics of Fluid and Gas*, Nauka, Moscow, 1987 (in Russian).

38. Schlichting, H., *Boundary Layer Theory*, McGraw-Hill, New York, 1968.

39. Ferm, L., "Non-Reflecting Accurate Open Boundary Conditions for the Steady Euler Equations," Technical Report 143, Department of Scientific Computing, Uppsala University, Uppsala, Sweden, September 14, 1992.

40. Ferm, L., and Gustafsson, B., "A Downstream Boundary Procedure for the Euler Equations", *Computers and Fluids*, 10 (1982) pp. 261–276.

41. Ferm, L., "Modified External Boundary Conditions for the Steady Euler Equations," Technical Report 153, Department of Scientific Computing, Uppsala University, Uppsala, Sweden, August 25, 1993.

42. Ferm, L., "Open Boundary Conditions for External Flow Problems," *Journal of Computational Physics*, 91 (1990) pp. 55–70.

43. Engquist, B., and Halpern, L., "Far Field Boundary Conditions for Computation over Long Time," *Applied Numerical Mathematics*, 4 (1988) pp. 21–45.

44. Ferm, L., "Multigrid for External Flow Problems," Technical Report, Department of Scientific Computing, Uppsala University, Uppsala, Sweden, September 9, 1993.

45. Ta'asan, S., "Canonical-Variables Multigrid Method for Steady-State Euler Equations," ICASE Report 94-14, NASA Langley Research Center, Hampton, VA, U.S.A., 1994.

46. Sidilkover, D., "A Genuinely Multidimensional Upwind Scheme and Efficient Multigrid Solver for the Compressible Euler Equations," ICASE Report No. 94-84, NASA Langley Research Center, Hampton, VA, U.S.A., 1994.

47. Turkel, E., "Review of Preconditioning Methods for Fluid Dynamics," *Applied Numerical Mathematics*, 12 (1993) pp. 257–284.

48. Turkel, E., Fiterman, A., and van Leer, B., "Preconditioning and the Limit of the Compressible to the Incompressible Flow Equations for Finite Difference Schemes," in *Frontiers of Computational Fluid Dynamics 1994*, Caughey, D. A., and Hafez, M. M., eds., John Wiley and Sons, 1994, pp. 215–234.

49. Fiterman, A., Turkel, E., and Vatsa, V., "Pressure Updating Methods for Steady-State Fluid Equations," AIAA paper 95-1652, June 1995.

50. Abarbanel, S. and Gottlieb, D., "Optimal Time Splitting for Two- and Three-Dimensional Navier-Stokes Equations with Mixed Derivatives," *J. Comput. Phys.*, 41 (1981) pp. 1–33.

51. Vatsa, V. N., Sanetrik, M. D., and Parlette, E. B., "Development of a Flexible and Efficient Multigrid-Based Multiblock Flow Solver," AIAA Paper 93-0677, January 1993.

52. Turkel, E., Vatsa, V. N., and Radespiel, R., "Preconditioning Methods for Low-Speed Flows," AIAA Paper 96-2460-CP, June 1996.

53. Sofronov, I. L., "Transparent Boundary Conditions for Unsteady Transonic Flow Problems in Wind Tunnel," Universität Stuttgart, Mathematisches Institut A, Preprint 95-21, Stuttgart, Germany, 1995.

54. Lax, P. D., "On Cauchy's Problem for Hyperbolic Equations and the Differentiability of Solutions of Elliptic Equations," *Communications of Pure and Applied Mathematics*, 8 (1955) pp. 615–633.

THE METHOD OF SPACE-TIME CONSERVATION ELEMENT AND SOLUTION ELEMENT — A NEW PARADIGM FOR NUMERICAL SOLUTION OF CONSERVATION LAWS

Sin-Chung CHANG [1]   Sheng-Tao YU [2]   Ananda HIMANSU [3]

Xiao-Yen WANG [4]   Chuen-Yen CHOW [4]   Ching-Yuen LOH [3]

(1) NASA Lewis Res. Center,   Cleveland, OH 44135

(2) NYMA Technology, Inc.,   NASA Lewis Research Center

(3) Cleveland Telecomm. Corp., NASA Lewis Research Center

(4) University of Colorado, Boulder, CO 80309

Abstract

In this paper, we review the basic principles of the method of space-time conservation element and solution element for solving the conservation laws in one and two spatial dimensions. The present method is developed on the basis of local and global flux conservation in a space-time domain, in which space and time are treated in a unified manner. In contrast to the modern upwind schemes, the approach here does not use the Riemann solver and the reconstruction procedure as the building blocks. Therefore, the logic and rationale are considerably simpler. The present approach has yielded high-resolution shocks, rarefaction waves, acoustic waves, vortices, ZND detonation waves and shock/acoustic waves/vortices interactions. Moreover, since no directional splitting is employed, numerical resolution of two-dimensional calculations is comparable to that of the one-dimensional calculations.

1   Introduction

Recently, Chang and coworkers [1–9] have reported a new numerical framework for solving conservation laws, namely, the Method of Space-Time Conservation Element and Solution Element, or the CE/SE method for short. This method is distinguished by the simplicity of its conceptual basis – flux conservation in space and time. The method has been illustrated in the context of conservation laws in fluid dynamics. In describing the present method, the original reports took the most direct approach. That is, starting from the basic integral equations, the algebraic details and the mathematical analyses were presented in a systematic way, such that all information needed to implement a computer program was provided. That approach made it clear that the present method was developed from fundamentals. It is not an incremental improvement of a previously existing method.

The CE/SE method, however, can also be described in a more intuitive manner by focusing on its unique space-time discretization. The essence of the method may be grasped from a simple delineation of the inherent space-time geometry. With it, the rigorous algebraic details and mathematical proofs, though not superfluous, take on more of the flavor of confirming the intuitively obvious facts. In this review, the CE/SE method will be illustrated by the geometric approach. For mathematical details, the reader is referred to the original papers.

In addition, we wish to take a somewhat comparative approach in this review, to clarify the differences between the CE/SE method and traditional schemes. In particular, we wish to pique the interest of readers familiar with the modern upwind schemes. Toward this end, we shall place the present method into the context of the projection-upwinding-evolution approach advocated by van Leer [10,11].

In this introduction, to assist the reader in getting a flavor of the present method, particularly in its role as tool to solve initial-value problems, several remarks which will be fully explained later given here: (i) Although the difference form is also considered, the main emphasis of the present

Received on July 1, 1997.

method is on solving the integral form of the transport equations in the space-time domain. (ii) The present method was designed to conserve space-time flux locally and globally. Conventional finite-volume schemes, however, concentrate on the spatial flux calculation; temporal evolution is usually treated by finite-differencing or semi-integration. The unified treatment of space and time in the present method cannot be overemphasized. (iii) For isentropic flows, the present method can be used to construct explicit solvers that are non-dissipative (neutrally stable) for all Courant numbers ≤ 1. Also, these solvers are two-way time-marching schemes, i.e., each forward marching scheme can be inverted to become a backward marching scheme. In other words, the marching variables at the $(n-1)$th time level can be determined in terms of those at the nth time level. (iv) A zigzagging marching strategy in the space-time domain is employed, such that flow information at each interface separating two conservation elements can be evaluated without interpolation or extrapolation. In particular, no Riemann solver is needed in calculating interfacial fluxes. (v) The flow solution structure is not calculated through a reconstruction procedure. Instead, the gradients of flow variables are treated as independent unknowns, and they are not influenced by the flow properties in neighboring elements at the same time level. This is in full compliance with the flow physics of the initial value problem. (vi) For flows in multiple spatial dimensions, no directional splitting is employed. The two and three-dimensional spatial meshes employed by the present method are built from triangles and tetrahedrons, respectively. Note that triangles and tetrahedrons, respectively, are also the simplest building blocks for two- and three-dimensional unstructured meshes.

At first glance, the special features of the CE/SE method, such as the staggered mesh and the treatment of flow gradient values as unknowns, may seem cumbersome. Closer examination, however, shows that they are requirements for a faithful discrete counterpart to the conservation laws. The resulting schemes are simple and yield accurate results. As a concrete illustration of these features, a FORTRAN program is listed in the appendix. It is a CE/SE solver for an extended Sod's shock tube problem [12], in which the shock tube problem is extended by imposing a non-reflecting boundary condition at each end of the computational domain.

The challenge of the non-reflective boundary condition is no less difficult than that of capturing the shock and the contact discontinuity. First, the flow under consideration is subsonic throughout and the treatment of the non-reflecting boundary condition for a subsonic flow is more difficult than that for a supersonic flow. Second, this difficulty is exacerbated by the existence of a shock and a contact discontinuity, which must be allowed to exit the domain without reflection. We remark that traditional non-reflective boundary conditions, such as the characteristics-based, the radiation (asymptotic), and the buffer-zone conditions, are based on the assumption that the flow is continuous. In spite of these difficulties, the present solver is capable of generating highly accurate non-reflecting solutions using a uniform time-step size from the beginning of time-marching (refer to Sec. 6 for a full discussion of the numerical result). The main loop in this CE/SE program contains only 39 FORTRAN statements. A single "if" is used to identify the time-levels at which to activate the boundary conditions. The program contains no other Fortran conditional statements such as "if", "amax", and "amin" that are used so often in programs of modern upwind methods.

Computer programs based on the CE/SE method have been developed for calculating flows in one and two spatial dimensions. Numerous results were obtained [13–22], including various shock tube problems, the ZND detonation waves, the implosion and explosion problem, shocks over a forward-facing step, acoustic waves, and shock/acoustic wave interactions. The method can clearly resolve shock/acoustic wave interactions wherein the difference of the magnitude between acoustic wave and shock could be up to six orders. In two-dimensional flows, the reflected shock is as crisp as the leading shock. From the evidence of these results, the CE/SE method has proved to be a promising numerical framework for solving fluid dynamics problems. In addition, Scott [23-27] has developed an implicit steady-state version of the space-time method for simulating steady-state incompressible flows, including viscous boundary layers and developing pipe flows.

The remainder of the paper is organized as follows. In Sec. 2, we present the general aspects of the space-time methods. A new space-time integral form of conservation laws will be described. In Sec. 3, a brief review at the conceptual level of the modern upwind schemes is presented. In Sec. 4, we present the CE/SE method for solving flow equations in one spatial dimension. In Sec. 5, the extensions of the CE/SE method in two and three spatial dimensions, without directional-splitting, are illustrated. In Sec. 6, numerical examples calculated by the present method are shown. We then offer the concluding remarks.

2 The Space-Time Methods

The conventional finite volume methods for simulating conservation laws were formulated according to flux balance over *a fixed spatial domain*. As such, the conservation laws state that the rate of change of the total amount of a substance contained in a fixed spatial domain V is equal to the flux of that substance across the boundary of V. Let the density of the substance be u and its spatial flux be $\vec{f}$; then the convectional conservation law can be expressed as:

$$\frac{d}{dt} \int_V u\, dV = - \int_{S(V)} \vec{f} \cdot \vec{ds}, \qquad (2.1)$$

where $S(V)$ is the boundary of V, and $\vec{ds} = d\sigma\, \vec{n}$ with $d\sigma$ and $\vec{n}$, respectively, being the area and the outward unit normal vector of a surface element on $S(V)$. The finite volume methods concentrate on the evaluation of the right hand side of Eq. (2.1). The left hand side of Eq. (2.1) is usually discretized by a finite difference method, such as the Runge Kutta method.

Alternatively, an integration method for space-time flux balance could be employed:

$$\int_V u\, dV \bigg|_{t_1}^{t_2} = - \int_{t_1}^{t_2} dt \int_{S(V)} \vec{f} \cdot \vec{ds}. \qquad (2.2)$$

This integration method, however, bears strong resemblance to the method of finite-differencing the temporal derivative term. Algebraically, one can always find its counterpart in the finite difference methods. Therefore, we refer to this method as semi-integration. Perhaps, the usefulness of this method lies in depicting a clearer picture of the space-time flux balance as compared to the methods of finite-differencing the time derivative term.

As shown in Fig. 2.1(a), due to the fixed spatial domain, the shape of the space-time Conservation Elements (CEs) in one spatial dimension must be rectangular. In addition, these elements must stack up exactly on top of each other in the time direction, i.e., no staggering of these elements in time is allowed. For equations in two spatial dimensions, as depicted in Fig. 2.1(b), a conservation element is

a uniform-cross-section cylinder in space-time, and again no staggering in time is allowed. This arrangement results in vertical interfaces extended in the direction of time evolution between adjacent CEs. Across these interfaces, flow information travels in both directions. Therefore, in calculating the interfacial flux, upwind biasing (or a Riemann solver) becomes necessary.

In the following subsections, we shall first state a non-conventional integral equation for conservation laws, in which space and time are treated in a unified manner. This space-time integral equation allows a space-time geometry of the CEs in the present CE/SE method which is different from that used in the conventional finite volume methods and shown in Fig. 2.1. As a result, the present CE/SE method is able to capture shocks without using a Riemann solver.

In Secs. 2.2 and 2.3, we shall discuss the non-dissipative property of the isentropic conservation laws. In Sec. 2.4, the breakdown of the non-dissipative property due to the entropy increasing processes will be discussed. We then conclude this section by posing several criteria for an ideal numerical analogue of the conservation laws. For convenience, the discussions in this section are based on one-dimensional flow equations.

2.1 Governing Equations

The one-dimensional unsteady Euler equations of a perfect gas can be expressed as

$$\mathbf{U}_t + \mathbf{F}_x = 0, \qquad (2.3)$$

where

$$\mathbf{U} = \begin{pmatrix} u_1 \\ u_2 \\ u_3 \end{pmatrix} = \begin{pmatrix} \rho \\ \rho v \\ \rho E \end{pmatrix}, \qquad (2.4)$$

$$\mathbf{F} = \begin{pmatrix} f_1 \\ f_2 \\ f_3 \end{pmatrix} = \begin{pmatrix} \rho v \\ \rho v^2 + p \\ (\rho E + p) v \end{pmatrix}, \qquad (2.5)$$

with ρ, v, p, and E being the density, velocity, static pressure, and specific total energy, respectively. By definition, $E = v^2/2 + e$, where e is the specific internal energy. The equation of state is $p = (\gamma - 1)\rho e$ with γ being the specific heat ratio.

As shown in Fig. 2.2, let $x_1 = x$, and $x_2 = t$ be the coordinates of a two-dimensional Euclidean space E_2. Then Eq. (2.3) can be expressed as the divergence-free conditions

$$\nabla \cdot \vec{h}_m = 0, \qquad m = 1, 2, 3, \qquad (2.6)$$

where $\vec{h}_m = (f_m, u_m)$, $m = 1, 2, 3$, are the *space-time* mass, momentum, and energy current-density vectors, respectively. Equation (2.6) is valid everywhere in E_2 for continuous and isentropic flow solutions. For flows with shock waves, we must use the more fundamental form of the conservation laws:

$$\oint_{S(R)} \vec{h}_m \cdot \vec{ds} = 0, \qquad m = 1, 2, 3. \qquad (2.7)$$

Here $S(R)$ is the boundary of a space-time region R, and $\vec{ds} = d\sigma \, \vec{n}$ with $d\sigma$ and $\vec{n}$, respectively, being the area and the outward unit normal vector of a surface element on $S(R)$. Note that (i) because $\vec{h}_m \cdot \vec{ds}$ is the space-time flux of $\vec{h}_m$ leaving R through $\vec{ds}$, Eq. (2.7) simply states that the total space-time flux of $\vec{h}_m$ leaving R through its boundary vanishes; (ii) all mathematical operations can be carried out as though E_2 were an ordinary two-dimensional Euclidean space; (iii) because $S(R)$ is a simple closed curve in E_2, the "surface" integration in Eq. (2.7) can be converted into a line integration [6], i.e.,

$$\oint_{S(R)} \vec{h}_m \cdot \vec{ds} = \oint_{S(R)}^{c.c.} (f_m \, dt - u_m \, dx) = 0 \qquad (2.8)$$

where the notation *c.c.* indicates that the line integration is carried out in the counterclockwise direction; and (iv) Eq. (2.6) is valid only for smooth flows, and it can be derived from Eq. (2.7) using Gauss' divergence theorem.

Here, we offer two remarks about Eq. (2.7): (i) Unlike Eqs. (2.1) and (2.2), the present formulation does not impose any constraint on the shape of the CEs in the space-time domain. This is a crucial difference that, at the conceptual level, separates the present method from the conventional finite volume methods. (ii) Equation (2.7) must be satisfied over any bounded sub-domain R of E_2. As a result, if the total space-time fluxes leaving R through a subset of the boundary $S(R)$ are known, the total space-time fluxes leaving through the rest of $S(R)$ can be deduced. Perhaps, a marching scheme can be built by a careful selection of the space-time geometry of the CEs. In the following discussions, we offer some thoughts about this envisioned marching scheme.

2.2　Marching Forward and Backward

A time-marching scheme for solving the Euler equations is valid because the Jacobian matrix $\mathbf{A} = \partial \mathbf{F}/\partial \mathbf{U}$ has the real eigenvalues v and $v \pm c$, where c is the sonic speed. Because $\mathbf{F}_x = \mathbf{A}\mathbf{U}_x$, Eq. (2.3) can be diagonalized by multiplying it from the left by the left eigenvector matrix $\mathbf{M}$ of $\mathbf{A}$, i.e.,

$$\mathbf{M}\mathbf{U}_t + \mathbf{M}\mathbf{A}\mathbf{M}^{-1}\mathbf{M}\mathbf{U}_x = 0, \qquad (2.9)$$

where $\mathbf{M}^{-1}$ is the inverse of $\mathbf{M}$. Let $\partial\hat{\mathbf{U}} = \mathbf{M}\partial\mathbf{U}$ be the characteristic variable vector. Then we have

$$\hat{\mathbf{U}}_t + \mathbf{\Lambda}\hat{\mathbf{U}}_x = 0, \qquad (2.10)$$

where $\mathbf{\Lambda} = \mathbf{M}\mathbf{A}\mathbf{M}^{-1}$ is a diagonal matrix composed of the eigenvalues a_m ($m = 1, 2, 3$) of $\mathbf{A}$. Eq. (2.10) is the classical system of characteristic compatibility equations, from which one can derive the Riemann invariants along the characteristics.

From the solution of the Method Of Characteristics (MOC), one knows that the flow solution has a finite domain of dependence at a previous time and is completely determined by it. One can calculate the flow solution along the characteristics for the whole space-time domain if the initial conditions are specified at every point in space. Moreover, because the Riemann invariant is constant along any characteristic without regard to the marching direction, given the flow solution at a certain time, one can *trace backward* along the characteristics to obtain the solution in the past. Thus, an ideal numerical analogue of the Euler equations should be able to march forward to obtain the flow solution at a new time level, and from there march backward to recover the solution at the starting time.

Moreover, because the marching procedure can be initiated from any initial-data curve in the space-time domain (as long as it nowhere has a characteristic direction), the nature of the forward-and-backward marching of the Euler equations is not limited to the time evolution. Therefore, an ideal numerical analogue of the isentropic convection equations should possess no preferred marching direction in the E_2 space, except insofar as grid-lines are selected.

For flows in multiple spatial dimensions, one can also derive the divergence-free condition, similar to Eq. (2.6), and its integral form similar to Eq. (2.7), as the governing equations. Although the equations cannot be diagonalized as in the one-dimensional case, the equation set is nonetheless hyperbolic. Therefore, the abovementioned forward-and-backward marching nature is still valid.

2.3 Space-Time Inversion

Another important property of Eq. (2.3) is its invariance under space-time inversion. Let (x_o, t_o) be a fixed point in E_2, and let $x' = 2x_o - x$ and $t' = 2t_o - t$. Note that (x', t') is the image of (x, t) and vice-versa under the space-time inversion with respect to (x_o, t_o). By the coordinate transformation, Eq. (2.3) is equivalent to

$$\mathbf{U}_{t'} + \mathbf{F}_{x'} = 0. \qquad (2.11)$$

Let $\mathbf{U}^o(x, t)$ be a smooth function of x and t, and let $\mathbf{U} = \mathbf{U}^o$ satisfy Eq. (2.3). Then the above invariance property of Eq. (2.3) and the fact that $\mathbf{F}$ is function of $\mathbf{U}$ imply that $\mathbf{U} = \mathbf{U}^o(2x_o - x, 2t_o - t)$ is also a solution to Eq. (2.3). This property of the smooth solutions to Eq. (2.3) should be shared by an ideal numerical analogue of Eq. (2.3).

We remark that a numerical analogue that is stable and also preserves the above invariance property under space-time inversion must be non-dissipative, or neutrally stable. This non-dissipative nature of a numerical analogue of the convection equations reflects the fact that a smooth solution to Eq. (2.3) does not dissipate with time. For the one and two-dimensional scalar convection equations, this conclusion is established in [2, 5].

For flow equations in multiple spatial dimensions, the invariance property under space-time inversion is still valid. An ideal numerical analogue should preserve this property, and it should be neutrally stable.

2.4 The Shock Capturing Scheme

Two issues need to be addressed by a shock capturing scheme : (i) the flow properties are not continuous across the shock, and (ii) the flow entropy increases across the shock.

To address the first issue, an integral, rather than differential, form of the governing equations must be employed. Based on the conservation of mass, momentum, and energy, the flow properties on one side of a shock are determined by the flow properties on the other side. Hence, flux conservation should be the only principle used to relate the two distinct flow states. Other methods, based on the differential equations such as the MOC, should not be considered.

The second issue pertains to the fact that the evolution of a flow with shocks is an entropy increasing process, and it is irreversible. However, there is no constraint built into the Euler equations for flow solutions to obey the second law of thermodynamics. Therefore, an additional constraint must be imposed when flows with shocks are modeled using the Euler equations. In computational fluid dynamics, this additional constraint can be conveniently implemented by numerical treatment. For example, in a finite-differencing setting, Lax [28] showed that the discretized Euler equations must satisfy an entropy condition in order to capture shocks successfully. Essentially, an even-order artificial damping should be added to the discretized equation such that spurious oscillations near the shock can be suppressed and a physically sensible solution can be obtained.

In Sec. 4, we shall present the method of adding the artificial damping in the present scheme for capturing shocks. We shall show that within any one time step, the added artificial damping only influences the calculation of the flow gradient; the calculation of the flow properties is not changed. In addition, a specific parameter is employed to allow direct control of the amount of the artificial damping. When the simulations of isentropic flows are of interest, the artificial damping can be turned off.

2.5 An Ideal Numerical Analogue

From the above discussions, it is seen that a smooth solution to the Euler equations, Eq. (2.3), has the following important properties: (i) it does not dissipate with time; (ii) its value at any point (x, t) has a finite domain of dependence at an earlier time; and (iii) it is completely determined by the initial data at a given time. In the light of these properties, we remark that (i) a solution to a dissipative numerical scheme will dissipate with time; (ii) the value of a solution to an implicit scheme at any point (x, t) depends on all the initial data and all the boundary data up to the time t; and (iii) a scheme involving more than two time levels requires the specification of the initial data at more than one time level. Therefore, we conclude that an ideal numerical analogue to Eq. (2.3), in addition to having the properties discussed in Secs. 2.2 and 2.3, should be *neutrally stable, explicit,* and involving only *two time levels.*

By adding an artificial dissipation term, an ideal solver of Eq. (2.3) can be extended to model flows with shocks. We want to emphasize that the artificial dissipation in an ideal numerical method should occur only in shock capturing; without added artificial damping, there should be no other source of numerical dissipation.

Furthermore, in an ideal Navier Stokes solver, the above guidelines of modeling the Euler equations should be applied to the discretization of the convective terms of the Navier Stokes equations. We note that, stripped of any added artificial terms, traditional numerical schemes are in general not free from inherent numerical dissipation. For flows at large Reynolds numbers, numerical dissipation may overwhelm the physical dissipation and cause a complete distortion of the solution. Because an ideal analogue of Eq. (2.3) has no numerical dissipation, when it is applied to discretize the convective terms of the Navier Stokes equations, the Navier-Stokes solver has the property that the numerical dissipation of its solutions approaches zero as the physical dissipation approaches zero. æ

3 Modern Upwind Schemes in the Space-Time Domain

In this section, we consider the modern upwind schemes, namely those developed based on the Godunov scheme. In this context, we want to clarify the differences between the CE/SE method and the conventional shock-capturing schemes. The structure of the modern upwind schemes will be described using van Leer's projection-upwinding-evolution formulation. In this regard, Huynh [29] has given a particularly lucid description. In general, the upwind schemes consists of three steps: (i) a projection or reconstruction step, in which the flow property distribution within each cell is approximated by polynomial curve fitting; (ii) an upwind step involving the solution of a Riemann problem to calculate the spatial fluxes at cell interfaces; and (iii) the temporal evolution step, in which the flow properties at the next time step are determined by either finite-differencing or by space-time flux conservation. Note that a space-time splitting form of the conservation laws, i.e., either Eq. (2.1) or Eq. (2.2), is employed

in the upwind schemes.

The original Godunov scheme employs the simplest reconstruction – the piecewise constant function, the most computationally expensive upwind scheme – the exact Riemann solver, and a simple time-marching method – an averaging procedure. Since the Godunov method is only first-order accurate, all modern upwind schemes are endeavors to improve the efficiency and accuracy of the method. To this end, modern upwind schemes utilize higher order reconstruction methods, more efficient Riemann solvers, and various time marching schemes. The basic structure of the modern upwind approach, however, has not changed.

3.1 The Evolution Step

In the present paper, a second-order upwind scheme presented by Huynh [29] is used to represent this class of schemes. As shown in Fig. 3.1, the E_2 space is divided into rectangular CEs. In each CE, the known column matrix of flow properties $\mathbf{U}_j^n$ is located at the center of the bottom boundary, and the unknown to be solved for, $\mathbf{U}_j^{n+1}$, is at the center of the top boundary. The space-time flux conservation over the CE using Eq. (2.2) can be expressed as

$$\mathbf{U}_j^{n+1}\Delta x - \mathbf{U}_j^n\Delta x$$
$$+\mathbf{F}_{j+1/2}^{n+1/2}\Delta t - \mathbf{F}_{j-1/2}^{n+1/2}\Delta t = \mathbf{0}. \qquad (3.1)$$

Since a second-order scheme is used, a linear distribution of $\mathbf{U}$ and $\mathbf{F}$ along the respective boundary segments is assumed. Therefore, as shown in Eq. (3.1), the integration of the flow properties along the respective boundary segment is equal to the values at the midpoint multiplied by the length of the line segment. In Eq. (3.1), the value of $\mathbf{U}_j^n$ is known, and we need to evaluate $\mathbf{F}_{j-1/2}^{n+1/2}$ and $\mathbf{F}_{j+1/2}^{n+1/2}$ to determine $\mathbf{U}_j^{n+1}$. This equation serves to track the temporal evolution of the solution, and thus forms the evolution step in van Leer's formulation. However, in order to calculate $\mathbf{F}_{j-1/2}^{n+1/2}$ and $\mathbf{F}_{j+1/2}^{n+1/2}$, the use of Eq. (3.1) must be preceded by a reconstruction step and an upwind step.

Temporal evolution schemes other than Eq. (3.1) could be used, in which, however, the picture of the space-time flux conservation is less clear. For instance, the Runge-Kutta (RK) method could be used for time marching. In each intermediate RK step, a reconstruction step and an upwind step are performed to calculate the spatial fluxes. The unbalanced spatial flux is then used as the inhomogeneous term in the RK method. After several intermediate steps of the RK method, the flow solution at the new time level is obtained.

In the space-time setting of Eq. (3.1), a possibility of solution discontinuity must be allowed between $\mathbf{U}_j^{n+1}$ and $\mathbf{U}_j^n$. Its exact location, however, is unclear. In Huynh's approach, the solution is assumed to be continuous within the lower half of each CE. As such, if $(\mathbf{U}_x)_j^n$ is known, $(\mathbf{U}_t)_j^n$ can be obtained by assuming

$$(\mathbf{F}_x)_j^n = \mathbf{A}_j^n (\mathbf{U}_x)_j^n \tag{3.2}$$

and

$$(\mathbf{U}_t)_j^n = - (\mathbf{F}_x)_j^n . \tag{3.3}$$

Here (i) $\mathbf{A}_j^n$ is the matrix $\mathbf{A}$ (which is a function of $\mathbf{U}$) evaluated with $\mathbf{U} = \mathbf{U}_j^n$, and (ii) Eqs. (3.2) and (3.3) are the numerical analogues of the relation $\mathbf{F}_x = \mathbf{A}\mathbf{U}_x$ and the differential equation, Eq. (2.3), respectively. It follows from Eqs. (3.2) and (3.3) that one can use the second-order Taylor polynomial to calculate $\mathbf{U}_{j\pm1/2}^{n+1/2}$:

$$\mathbf{U}_{j\pm1/2}^{n+1/2} = \mathbf{U}_j^n + (\mathbf{U}_t)_j^n \frac{\Delta t}{2} \pm (\mathbf{U}_x)_j^n \frac{\Delta x}{2} \tag{3.4}$$

if the value of $(\mathbf{U}_x)_j^n$ is known. This leads to the reconstruction step in the upwind schemes.

3.2   The Reconstruction Step

As a common practice in upwind schemes, flow properties at neighboring nodes are employed for a polynomial curve-fit to calculate $(\mathbf{U}_x)_j^n$. The difficulty with such a procedure has always lain in fitting a polynomial curve across a possible solution discontinuity, which results in spurious oscillations. To suppress the spurious overshoot of $(\mathbf{U}_x)_j^n$, scheme developers introduced *predetermined constraints* to the flow distribution, based on the expected nature of the solution. Van Leer [10, 11] suggested that the reconstruction procedure must preserve the monotonicity of the flow property distribution. An extension of this idea is the Total Variational Diminishing (TVD) method [30]. Another approach is the Essentially Non-Oscillatory (ENO) method [31] based on a strategy of stencil nodal selection. In general, these complex procedures are combinations of limiting the slopes of the flow properties and sharpening the jump conditions whenever a shock is detected.

Usually, these methods are effective in suppressing oscillations near shocks.

The drawback of these complex procedures is that the special properties, such as monotonicity, are not universal, and they are not prescribed by the conservation laws. For example, due to the existence of a source term, the flow field of a detonation wave is not monotonic nor does it display a TVD property. In this case, one must resort to other ideas to constrain the slopes.

3.3   The Upwind Step

Once $(\mathbf{U}_x)_j^n$ is determined, the flow property distribution inside a CE can be expressed by the Taylor polynomial, Eq. (3.4). As such, flow properties at the immediate two sides of a cell interface, $\left(\mathbf{U}_{j+1/2}^{n+1/2}\right)_R$ and $\left(\mathbf{U}_{j+1/2}^{n+1/2}\right)_L$, can be calculated. Refer to Fig. 3.2.

Note that $\left(\mathbf{U}_{j+1/2}^{n+1/2}\right)_R$ and $\left(\mathbf{U}_{j+1/2}^{n+1/2}\right)_L$ are distinct states, and one must reconcile the two states to yield a unique interfacial flux for the temporal evolution calculation described by Eq. (3.1). The simplest approach is to average the two states. This approach, however, leads to central differencing the spatial flux terms in Eq. (3.1), and results in numerical oscillations for flows with shocks.

In the Godunov scheme, flow properties in each cell at $t = t^n$ are assumed constant. By using $\mathbf{U}_j^n$ and $\mathbf{U}_{j+1}^n$ as the initial values, one can set up a Riemann (shock tube) problem at the interface of two adjacent CEs. With the aid of the known solution (a function of x/t) to this classical problem, the flux passing through the interface can be obtained by an integration over the time interval $t^n \leq t \leq t^{n+1}$ at $x = x_{j+1/2}$.

In the second-order modern upwind schemes, a distribution of the flow property inside each CE exists. Thus, inside a CE, the characteristics associated with the varying flow properties interact with one another. As such, the flow physics at the interface is much more complex than that of the Godunov scheme. In this case, there is no known analytical solution. As a result, it becomes a formidable task to calculate the evolution of the interfacial flux. When faced with this difficulty, the usual approach is to consider only the left and right interface midpoint values, $\left(\mathbf{U}_{j+1/2}^{n+1/2}\right)_L$ and $\left(\mathbf{U}_{j+1/2}^{n+1/2}\right)_R$, and use them to yield a unique value for the flux at that same point in the space-time domain. The calculation is usually based on the combination of left and right fluxes with upwind biasing, and the procedure is usually termed as an approximate Riemann solution. The idea of tracking the temporal evolution of flow solution at the interface, such as is used in the Godunov scheme, is discarded.

What is actually involved in practice is the use of the characteristic-based techniques. In the classical MOC, to march one step forward along a characteristic, one has to assume that the characteristic can be approximated locally as a straight line segment with its slope (an eigenvalue of $\mathbf{A}$) either explicitly determined by the initial flow states (a first-order linearization), or iteratively determined by the combination of the initial flow state and the final flow state (a second-order linearization).

In the approximate Riemann solver, a similar procedure is performed, in which the coefficient matrix $\mathbf{A}$ and its eigenvalues are evaluated based on a specially chosen "averaged flow state." Following this, the flux calculation can be performed based on the characteristic-value splitting. In the modern upwind schemes, the efficiency and accuracy of the approximate Riemann solver hinge on a careful choice of this averaged flow state. As an example, we refer the reader to [29] for a detailed discussion of the linearization procedure in Roe's flux splitting scheme [32] based on the "Roe state."

The drawback of this treatment is that the order of accuracy of the resultant scheme is usually less than the order of the polynomial employed in the reconstruction. This is because the flux calculation is based on an approximated Riemann problem, in which the flow property variation inside each CE is not directly taken into account in calculating the interfacial flux. In addition, the linearization process in the flux splitting also involves approximation. For these reasons, modern upwind schemes are often referred to as *high-resolution* instead of high-order, even when a third- or fourth-order polynomial was used in the reconstruction.

Moreover, the logic of the upwind biasing is largely based on the characteristic-value splitting, which is sensible only when it is used to describe the space-time evolution of a smooth flow. The application of this method in a situation without space-time evolution and with the possible existence of a shock is difficult to justify.

In addition, the above characteristics-based technique is applicable only to flows in one spatial dimension. For flows in multiple spatial dimensions, the MOC is more complex and has its own inherent space-time geometries. Therefore, it is not amenable to any numerical method using a predetermined lattice stencil in the space-time domain. As a result, directional splitting is employed for solving flow equations in multiple spatial dimensions. This practice causes deterioration of numerical resolution in multi-dimensional flows. Furthermore, because source terms have no preferred direction, the directional splitting approach poses significant difficulties in constructing an approximate Riemann solver.

3.4 Summary of Distinguishing Features

To recapitulate the distinguishing features of the modern upwind schemes, we provide the following remarks: (i) In the reconstruction step, the flow structure is estimated by a curve fit among neighboring cells. The use of a polynomial curve fit over a possible solution discontinuity at cell interfaces is theoretically unjustifiable, and also leads to spurious oscillations. A *priori* knowledge of the flow solution, such as a TVD property, is then used to suppress these oscillations. (ii) The need to solve the Riemann problem in the modern upwind schemes is due to the choice of a fixed spatial mesh in the space-time domain as shown in Fig. 2.1. As a result of the choice, flow information propagates in both directions through the interface, and the upwind-biasing becomes necessary in calculating the interfacial flux. (iii) Only the flow properties immediately adjacent to an interface are used to calculate the interfacial flux. Due to the variation of the flow properties inside each CE, the actual evolution of the interfacial flux is much more complex. (iv) Given two distinct flow states in contact, the calculation of the interfacial flux is carried out by using characteristics-based splitting, which is valid only for describing the space-time evolution of a smooth flow. (v) For flows in multiple spatial dimensions, directional splitting is used to implement one-dimensional characteristic flux splitting. This practice causes deterioration of numerical resolution and difficulties in solving conservation laws with source terms.

4 The CE/SE Method in One Spatial Dimension

In the setting of evolution and reconstruction, the following distinguishing features of the CE/SE method result in a simpler and more consistent numerical flow model: (i) A space-time discretization is chosen for the flux conservation such that there is no Riemann problem to be solved at the cell interface. (ii) To avoid imposing predetermined constraints on the flow solutions such as monotonicity and TVD, the flow properties and their gradients are treated as unknowns. For smooth flows, the unknowns are completely determined by flux conservation, and the resultant numerical procedure can march forward and backward in time. (iii) For flows with shocks, an adjustable artificial damping is added to the discretized equations such that the numerical entropy condition is satisfied.

In the following, we shall first discuss the preliminaries of the CE/SE method. We then present the evolution step for calculating flow properties, the reconstruction step for obtaining the flow property gradient, and the shock capturing method for resolving shocks.

4.1 Preliminaries

In Fig. 4.1, we illustrate the nodes, denoted by dots (filled circles), where the unknowns are located. The space and time intervals between neighboring nodal lines are respectively denoted by $\Delta x/2$ and $\Delta t/2$. There is a Solution Element (SE) associated with each node (j, n). Let the $\mathrm{SE}(j, n)$ be the interior of the space-time region bounded by a dashed line depicted in Fig. 4.2. It includes a horizontal line segment, a vertical line segment, and their immediate neighborhood. Between SEs, discontinuities are allowed. As will be shown immediately, for the Euler equations, Eq. (2.3), which have no source terms, the actual size of the neighborhood does not matter. However, for other equations with source terms, because these sources may be distributed over the entire computational domain, solution elements must be constructed such that they can fill the entire domain. In this case the neighborhood may be chosen such that a SE looks like the rhombus depicted in Fig. 4.8(b). Inside a SE, the flow properties are assumed continuous.

Within a SE, the flow property vector $\mathbf{U}$ and the flux vector $\mathbf{F}$ are approximated by their discretized counterparts $\mathbf{U}^*$ and $\mathbf{F}^*$. Since a second-order scheme is desired, piecewise linear distributions $\mathbf{U}^*$ and $\mathbf{F}^*$ are assumed. For (x, t) in $\mathrm{SE}(j, n)$, we assume that

$$\mathbf{U}^*(x, t; j, n) = \mathbf{U}_j^n + (\mathbf{U}_x)_j^n (x - x_j) + (\mathbf{U}_t)_j^n (t - t^n) \tag{4.1}$$

and

$$\mathbf{F}^*(x, t; j, n) = (\mathbf{F})_j^n + (\mathbf{F}_x)_j^n (x - x_j) + (\mathbf{F}_t)_j^n (t - t^n). \tag{4.2}$$

Here $\mathbf{F}_j^n$ is the column matrix $\mathbf{F}$ (which is a function of $\mathbf{U}$) evaluated with $\mathbf{U} = \mathbf{U}_j^n$.

The expansion coefficients $(\mathbf{U}_t)_j^n$, $(\mathbf{F}_x)_j^n$ and $(\mathbf{F}_t)_j^n$ in Eqs. (4.1) and (4.2) will be expressed as functions of the independent unknowns $\mathbf{U}_j^n$ and $(\mathbf{U}_x)_j^n$ of the present scheme by assuming Eqs. (3.2) and (3.3), and

$$(\mathbf{F}_t)_j^n = \mathbf{A}_j^n (\mathbf{U}_t)_j^n, \tag{4.3}$$

which is the numerical analogue of the relation $\mathbf{F}_t = \mathbf{A}\mathbf{U}_t$. Furthermore, because $\vec{h}_m = (f_m, u_m)$, $m = 1, 2, 3$, we shall assume that for $m = 1, 2, 3$,

$$\vec{h}_m^*(x, t; j, n) = (f_m^*(x, t; j, n), u_m^*(x, t; j, n)), \tag{4.4}$$

where u_m^* and f_m^*, $m = 1, 2, 3$, are the components of the column matrices $\mathbf{U}^*$ and $\mathbf{F}^*$, respectively.

At this juncture, note that: (i) In an alternative approach to be presented in Sec. 4.3, the differential condition Eq. (3.3) is not assumed. Rather it arises naturally as a result of flux conservation and Eqs. (4.1) and (4.2). (ii) Hereafter, the components of the column matrices $\mathbf{U}_j^n$, $(\mathbf{U}_x)_j^n$, $(\mathbf{U}_t)_j^n$, $\mathbf{F}_j^n$, $(\mathbf{F}_x)_j^n$ and $(\mathbf{F}_t)_j^n$, will be denoted by $(u_m)_j^n$, $(u_{mx})_j^n$, $(u_{mt})_j^n$, $(f_m)_j^n$, $(f_{mx})_j^n$ and $(f_{mt})_j^n$, $m = 1, 2, 3$, respectively.

4.2 The Evolution and Reconstruction Steps

For smooth flows, the calculations of $\mathbf{U}_j^n$ and $(\mathbf{U}_x)_j^n$ are determined by requiring fluxes to be conserved over space-time Conservation Elements (CEs). As depicted in Figs. 4.3(a) and 4.3(b), two CEs, denoted by $\mathrm{CE}_-(j, n)$ and $\mathrm{CE}_+(j, n)$, are associated with every mesh point (j, n). A glance over Figs. 4.1, 4.2, and 4.3 reveals that the set of $\mathrm{CE}_\pm(j, n)$ over all mesh points (j, n) do not overlap among themselves and can fill the entire space-time computational domain.

For each (j, n), the following discrete analogue to the space-time flux conservation, Eq. (2.7), is imposed:

$$\oint_{S(CE_-(j,n))} \vec{h}_m^* \cdot \vec{ds} = 0, \quad m = 1, 2, 3, \qquad (4.5)$$

and

$$\oint_{S(CE_+(j,n))} \vec{h}_m^* \cdot \vec{ds} = 0, \quad m = 1, 2, 3. \qquad (4.6)$$

According to Figs. 4.2, 4.3a, and 4.3b, we have the following observations: (i) The edges $\overline{CB}$ and $\overline{CD}$ of $CE_-(j, n)$ lie in $SE(j - 1/2, n - 1/2)$; (ii) The edges $\overline{AB}$ and $\overline{AD}$ of $CE_-(j, n)$ lie in $SE(j, n)$; (iii) The edges $\overline{ED}$ and $\overline{EF}$ of $CE_+(j, n)$ lie in $SE(j + 1/2, n - 1/2)$; (iv) The edges $\overline{AD}$ and $\overline{AF}$ of $CE_+(j, n)$ lie in $SE(j, n)$. As a result, with the aid of the numerical counterpart of Eq. (2.7), and Eqs. (3.2), (3.3) and (4.1)–(4.4), we conclude that Eq. (4.5) leads to three relations involving the independent unknowns $\mathbf{U}_j^n$, $(\mathbf{U}_x)_j^n$, $\mathbf{U}_{j-1/2}^{n-1/2}$, $(\mathbf{U}_x)_{j-1/2}^{n-1/2}$, and Eq. (4.6) leads to another three relations, involving $\mathbf{U}_j^n$, $(\mathbf{U}_x)_j^n$, $\mathbf{U}_{j+1/2}^{n-1/2}$, and $(\mathbf{U}_x)_{j+1/2}^{n-1/2}$. Assuming that the unknowns at the mesh points $(j - 1/2, n - 1/2)$ and $(j + 1/2, n - 1/2)$ are given, the six components of $\mathbf{U}_j^n$ and $(\mathbf{U}_x)_j^n$ can be determined by the above six relations. In the following, it will be shown that the procedure for solving these relations can be divided into two sequential steps: an evolution step followed by a reconstruction step.

Note that the space-time flux of $\vec{h}^*$ leaving $CE_-(j, n)$ through $\overline{AD}$ and that leaving $CE_+(j, n)$ through $\overline{AD}$ are evaluated using the same unknowns, i.e., $\mathbf{U}_j^n$ and $(\mathbf{U}_x)_j^n$. Thus, these two space-time fluxes are each the negative of the other. As a result, a combination of Eq. (4.5) and Eq. (4.6) imply that

$$\oint_{S(CE(j,n))} \vec{h}_m^* \cdot \vec{ds} = 0, \quad m = 1, 2, 3, \qquad (4.7)$$

where $CE(j, n)$ (see Fig. 4.3(c)) is the union of $CE_-(j, n)$ and $CE_+(j, n)$. Here, as explained above, the fluxes of $\vec{h}_m^*$ leaving $CE(j, n)$ through $\overline{CD}$, $\overline{CB}$, $\overline{ED}$, and $\overline{EF}$ can be evaluated in terms of $\mathbf{U}_{j\pm1/2}^{n-1/2}$ and $(\mathbf{U}_x)_{j\pm1/2}^{n-1/2}$.

The flux of $\vec{h}_m^*$ leaving $\overline{BF}$, however, is a function of $\mathbf{U}_j^n$ only. This conclusion can be reached from the following observations: (i) $dt = 0$ along $\overline{BF}$, (ii) $\overline{BF}$ lies in $SE(j, n)$, (iii) u_m^*, $m = 1, 2, 3$, is linear in x on $\overline{BF}$, and (iv) the mesh point (j, n), i.e., point

A, is the mid-point of $\overline{BF}$. As a matter of fact, with the aid of Eq. (2.8), it can be shown that the flux of $\vec{h}_m^*$ leaving $CE(j, n)$ through $\overline{BF}$ is simply $(u_m)_j^n \Delta x$. In other words, Eq. (4.7) implies that $\mathbf{U}_j^n$ can be determined explicitly in terms of $\mathbf{U}_{j\pm1/2}^{n-1/2}$ and $(\mathbf{U}_x)_{j\pm1/2}^{n-1/2}$. This is the *evolution* step for the present marching scheme.

After obtaining $\mathbf{U}_j^n$, $\mathbf{F}_j^n$ and $\mathbf{A}_j^n$ can be determined because they are functions of $\mathbf{U}_j^n$ only. As a result, by applying either Eq. (4.5) or Eq. (4.6) (only one of these two equations is independent after Eq. (4.7) is used), one can obtain a system of three linear equations with the three unknowns being the three components of $(\mathbf{U}_x)_j^n$. In other words, $(\mathbf{U}_x)_j^n$ can be determined in terms of $\mathbf{U}_j^n$, $\mathbf{U}_{j\pm1/2}^{n-1/2}$, and $(\mathbf{U}_x)_{j\pm1/2}^{n-1/2}$ by solving either Eq. (4.5) or Eq. (4.6). This is the *reconstruction* step of the present marching scheme.

Recall that the reason leading to the decoupling of the evaluation of $\mathbf{U}_j^n$ from that of $(\mathbf{U}_x)_j^n$ is that the mesh point (x_j, t^n) is located at the mid-point of $\overline{BF}$. For a space-time mesh with non-uniform spatial mesh intervals, the above decoupling can still be achieved, with some modifications in the procedure. We shall explain these modifications using a special nonuniform-spatial-interval mesh. From this description, the reader can easily infer the modifications required for a general nonuniform-spatial-interval mesh.

Consider the space-time mesh depicted in Fig. 4.4. Here, we assume that the spatial mesh intervals to the right and the left of the vertical mesh line $j = 0$ are separately uniform and have the lengths Δx and $\Delta x'$, respectively. Spatial mesh intervals are separated from one another by solid vertical mesh lines. Each spatial mesh interval is divided into two subintervals of equal length by a dashed vertical line. For this space-time mesh, non-uniformity of spatial intervals occurs around the mesh line $j = 0$. Thus, we are interested in the marching scheme solving for $\mathbf{U}_0^n$ and $(\mathbf{U}_x)_0^n$. Note that the mesh point $(0, n)$ is not located at the intersection of mesh lines. Instead, it is located at the mid-point of $\overline{BF}$ (see Fig. 4.5). Refer to Figs. 4.4-4.6 for the mesh point $(0, n)$ and its associated SEs and CEs (for $n = 1/2, 3/2, ...$). As a result, in this case too, $\mathbf{U}_0^n$ can be calculated by using Eq (4.7) only. Once $\mathbf{U}_0^n$ is obtained, $(\mathbf{U}_x)_0^n$ can be calculated by either Eq. (4.5) or Eq. (4.6).

4.3 An Alternative Construction

As described in [6], the above CE/SE scheme for isentropic flows can be constructed from a different flux-balance perspective. In this construction, the locations of mesh points (dots in Fig. 4.7) are identical to that in the original construction (refer to Fig. 4.1). The computation domain, however, is divided into rhombic regions in the interior and triangular regions at the boundary. Each region is associated with a mesh point (j, n) and serves as a conservation element, as denoted by $CE'(j, n)$ (refer to Fig. 4.8(a)). The solution element associated with point (j, n) is the interior of $CE'(j, n)$, and is denoted by $SE'(j, n)$ (refer to Fig. 4.8(b)).

For any (x, t) in $SE'(j, n)$, $\mathbf{U}(x, t)$, $\mathbf{F}(x, t)$ and $\vec{h}_m(x, t)$ are approximated by their discretized counterparts: $\mathbf{U}^*(x, t; j, n)$, $\mathbf{F}^*(x, t; j, n)$, and $\vec{h}_m^*$ as defined in Eqs. (4.1), (4.2), and (4.4), respectively. Here, the marching scheme requires that for each mesh point (j, n), the space-time flux of $\vec{h}_m^*$ leaving the boundary of the space-time rhombus $CE'(j, n)$ vanishes, i.e.,

$$\oint_{S(CE'(j,n))} \vec{h}_m^* \cdot d\vec{s} = 0, \quad m = 1, 2, 3. \quad (4.8)$$

By using Eqs. (4.1), (4.2) and (4.4), we have, for $m = 1, 2, 3$,

$$\nabla \cdot \vec{h}_m^*(x, t; j, n) = (f_{mx})_j^n + (u_{mt})_j^n, \quad (4.9)$$

for any (x, t) in $SE'(j, n)$. By using Gauss' divergence theorem and the fact that the right-hand-side of Eq. (4.9) is a constant within $SE'(j, n)$, Eq. (4.8) implies that

$$(f_{mx})_j^n + (u_{mt})_j^n = 0, \quad m = 1, 2, 3, \quad (4.10)$$

i.e., $(\mathbf{U}_t)_j^n + (\mathbf{F}_x)_j^n = \mathbf{O}$, which is Eq. (3.3). Here, however, Eq. (4.8) is not an imposed assumption as it was in the original construction. Instead, it is the result of the flux conservation condition Eq. (4.8).

In the present construction, the interaction between two neighboring SE's lies in the balance of space-time flux over the oblique interface that separates them. Refer to Fig. 4.9. Note that distinct flow states exist along the two sides of the interface. As would be done in the derivation of the Rankine-Hugoniot relation, we impose the space-time-flux-balance condition across the interface, i.e., the space-time flux leaving $CE'(j - 1/2, n - 1/2)$ through $\overline{B'C'}$ should be the space-time flux entering $CE'(j, n)$ through $\overline{BC}$.

In order to avoid the ambiguity of a double-valued function, SE's are defined such that an interface separating two rhombic CE's does not belong to any of the two adjacent SE's. As depicted in Fig. 4.9, the values of $\vec{h}_m^*$ along $\overline{BC}$ and $\overline{B'C'}$ are evaluated using information from $SE'(j, n)$ and $SE'(j-1/2, n-1/2)$, respectively. Moreover, by using Eqs. (4.9) and (4.10), and Gauss' divergence theorem, we have

$$\oint_{S(\triangle ABC)} \vec{h}_m^* \cdot d\vec{s} = 0, \quad m = 1, 2, 3, \quad (4.11)$$

and

$$\oint_{S(\triangle A'B'C')} \vec{h}_m^* \cdot d\vec{s} = 0, \quad m = 1, 2, 3. \quad (4.12)$$

Equation (4.5) follows immediately from Eqs. (4.11) and (4.12) and the interface flux balance condition. Similarly, Eq. (4.6) can also be obtained. As a result, the marching scheme constructed using the alternative approach is identical to that constructed with the original approach.

Note that, as a result of Eq. (4.11), the flux of $\vec{h}_m^*$ leaving $\triangle ABC$ through the oblique line segment $\overline{BC}$ is balanced by the flux of $\vec{h}_m^*$ leaving $\triangle ABC$ through the horizontal line segment $\overline{AB}$ (where $dt = 0$) and the vertical line segment $\overline{AC}$ (where $dx = 0$). It follows that, in general, flux integration along an oblique line segment can be turned into an equivalent and simpler problem that involves only flux integration along a horizontal line segment and a vertical line segment.

4.4 Special Features of the CE/SE Method for Isentropic Flows

We conclude the discussion of the CE/SE method for isentropic flows by the following remarks: (i) Since the space-time fluxes at the interfaces of CEs cancel each other, the local conservation conditions, Eqs. (4.5) and (4.6), lead to a global conservation condition

$$\oint_{S(V')} \vec{h}_m^* \cdot d\vec{s} = 0, \quad m = 1, 2, 3, \quad (4.13)$$

where V' is the union of any combination of CE_+s and CE_-s. Similarly, as a result of Eq. (4.8), Eq. (4.13) is also valid in the alternative construction if V' is the union of any combination of CE's. (ii) The present marching scheme is a two-level explicit scheme. Only one set of initial conditions at a single time level is needed to start the computation, and the numerical solution is completely determined by the initial conditions. This is in full compliance with the flow physics. (iii) Backward marching schemes can also be constructed using Eqs. (4.5) and (4.6). For each mesh point (j, n), they also imply

$$\oint_{S(CE_+(j-1/2,n+1/2))} \vec{h}_m^* \cdot \vec{ds} = 0, \qquad (4.14)$$

and

$$\oint_{S(CE_-(j+1/2,n+1/2))} \vec{h}_m^* \cdot \vec{ds} = 0. \qquad (4.15)$$

Note that the rectangular space-time regions to the left and the right of the mesh point (j, n) depicted in Fig. 4.10 are $CE_+(j - 1/2, n + 1/2)$ and $CE_-(j + 1/2, n + 1/2)$, respectively. Previously, we noted that Eqs. (4.5) and (4.6) show that $\mathbf{U}_j^n$ and $(\mathbf{U}_x)_j^n$ are determined in terms of $\mathbf{U}_{j\pm1/2}^{n-1/2}$ and $(\mathbf{U}_x)_{j\pm1/2}^{n-1/2}$. Here, Eqs. (4.14) and (4.15) show that $\mathbf{U}_j^n$ and $(\mathbf{U}_x)_j^n$ can be determined in terms of $\mathbf{U}_{j\pm1/2}^{n+1/2}$ and $(\mathbf{U}_x)_{j\pm1/2}^{n+1/2}$. In other words, the same set of local conservation conditions that was used to construct the forward-marching scheme can also be used to construct the backward-marching scheme. (iv) Recall the invariance property of the Euler equations under space-time inversion that was discussed in Sec. 1.2. Let

$$\mathbf{U} = \mathbf{U}^o(x, t) \qquad (4.16)$$

be a smooth solution to Eq. (2.3). For this solution, we have

$$\mathbf{U}_x = \mathbf{U}_1^o(x, t) \equiv \frac{\partial \mathbf{U}^0(x, t)}{\partial x}, \qquad (4.17)$$

for any x and t. Previously, we showed that

$$\mathbf{U} = \mathbf{U}^o(2x_o - x, 2t_o - t) \qquad (4.18)$$

is also a solution to the Euler equations, Eq. (2.3). For the above solution, Eq. (4.18), the spatial gradients of the flow properties can be expressed as

$$\begin{aligned}
\mathbf{U}_x &= \frac{\partial \mathbf{U}^o(2x_o - x, 2t_o - t)}{\partial(2x_o - x)} \frac{d(2x_o - x)}{dx}, \\
&= -\mathbf{U}_1^o(2x_o - x, 2t_o - t). \qquad (4.19)
\end{aligned}$$

The present numerical analogue to the Euler equations, Eq. (2.3), also possesses the invariance property under space-time inversion. Let $(\mathbf{U}^o)_j^n$, and $(\mathbf{U}_1^o)_j^n$ be given at all mesh points (j, n). Let

$$\mathbf{U}_j^n = (\mathbf{U}^o)_j^n \quad ; \qquad (\mathbf{U}_x)_j^n = (\mathbf{U}_1^o)_j^n \qquad (4.20)$$

be a solution to Eqs. (4.5) and (4.6). With the aid of the equations provided in the previous sections, it can be shown that

$$\begin{aligned}
\mathbf{U}_j^n &= (\mathbf{U}^o)_{2j_o-j}^{2n_o-n} \quad ; \\
(\mathbf{U}_x)_j^n &= -(\mathbf{U}_1^o)_{2j_o-j}^{2n_o-n} \qquad (4.21)
\end{aligned}$$

are also a solution to Eqs. (4.5) and (4.6). Here, $(2j_o - j, 2n_o - n)$ is the image of (j_o, n_o) and vice-versa, under space-time inversion with respect to (j_o, n_o) (see Fig. 4.11). Note that Eqs. (4.20) and (4.21) are the numerical counterparts of Eqs. (4.16–17) and (4.18–19), respectively.

It has been shown by numerical experiments that the present scheme is neutrally stable in the interior of the computational domain up to at least a thousand time steps when the Courant number does not exceed unity. In these numerical experiments simulating a shock-tube problem, the computational domain was allowed to grow with time, so that the undisturbed fluid state could always be prescribed at the computational boundaries as the exact solution. As a matter of fact, by using an analysis similar to that given at the end of Sec. 6 in [6], one can show that the linearized form of the present numerical analogue is neutrally stable when the Courant number does not exceed unity. The above conclusions are consistent with a remark made in Sec. 2.3, i.e., a numerical analogue of the Euler equations that is stable and also preserves the invariance property under space-time inversion must be neutrally stable. The Courant number mentioned above, denoted by CFL, is defined at each time level n for 1D flow computations as

$$CFL = \max_j \left[\frac{\left(|v_j^n| + |c_j^n| \right) \Delta t}{\Delta x} \right],$$

where c is the sonic speed. Analogous Courant numbers can be defined for 2D flow computations.

To summarize, the present numerical scheme preserves the forward-backward marching nature and the space-time inversion invariance property of the Euler equations. The present scheme also meets the requirements of an ideal numerical analogue set forth in Sec. 2.4, i.e., it should be neutrally stable, explicit, and involving only two time levels. The present scheme can be considered as a nonlinear extension of the a scheme, i.e., the inviscid ($\mu=0$) version of the a-μ scheme described in [6]. Because the a scheme is neutrally stable, generally one would expect that a nonlinear extension of such a scheme is numerically unstable. The present scheme appears to be an exception to this common wisdom.

4.5 The Shock-Capturing Scheme

The above marching scheme for isentropic flows can be expressed as

$$\mathbf{U}_j^n = \mathbf{H}\left(\mathbf{U}_{j-1/2}^{n-1/2}, \mathbf{U}_{j+1/2}^{n-1/2}, (\mathbf{U}_x)_{j-1/2}^{n-1/2}, (\mathbf{U}_x)_{j+1/2}^{n-1/2}\right), \quad (4.22)$$

and

$$(\mathbf{U}_x)_j^n = \mathbf{H}_x\left(\mathbf{U}_{j-1/2}^{n-1/2}, \mathbf{U}_{j+1/2}^{n-1/2}, (\mathbf{U}_x)_{j-1/2}^{n-1/2}, (\mathbf{U}_x)_{j+1/2}^{n-1/2}\right) \quad (4.23)$$

Here $\mathbf{H}$ and $\mathbf{H}_x$ are column-matrix functions. Their explicit forms can be obtained from Eqs. (5.20)-(5.29) in [6] with the assumption that the viscosity $\mu = 0$. In the construction of the shock-capturing scheme, the local conservation condition, Eq. (4.7), is again assumed. Because Eq. (4.22) follows directly from Eq. (4.7), the former is incorporated into the shock-capturing scheme without modification. As a result, given the same $\mathbf{U}_{j\pm 1/2}^{n-1/2}$ and $(\mathbf{U}_x)_{j\pm 1/2}^{n-1/2}$, the shock-capturing scheme shares with the non-dissipative scheme the same zero-order terms on the right sides of Eqs. (4.1) and (4.2). In addition, the shock-capturing scheme observes a global conservation condition that is also a direct result of Eq. (4.7), i.e., for any space-time region that is the union of any combination of the CEs of the type depicted in Fig. 4.3(c), the total flux of $\vec{h}_m^*$, $m = 1, 2, 3$, leaving its boundary vanishes.

The shock-capturing scheme is obtained by modifying Eq. (4.23). To proceed, let

$$(\mathbf{U}')_{j\pm 1/2}^n = \mathbf{U}_{j\pm 1/2}^{n-1/2} + \frac{\Delta t}{2}(\mathbf{U}_t)_{j\pm 1/2}^{n-1/2}, \quad (4.24)$$

i.e., $(\mathbf{U}')_{j\pm 1/2}^n$ is a first-order Taylor's approximation of $\mathbf{U}$ at the mesh point $(j \pm 1/2, n)$. Thus,

$$(\mathbf{U}_x^c)_j^n = \frac{(\mathbf{U}')_{j+1/2}^n - (\mathbf{U}')_{j-1/2}^n}{\Delta x} \quad (4.25)$$

is a central-difference approximation of $\mathbf{U}_x$ at the mesh point (j, n). In the shock-capturing scheme, Eq. (4.23) is replaced by

$$(\mathbf{U}_x)_j^n = (1 - 2\epsilon)(\mathbf{H}_x)_j^n + 2\epsilon(\mathbf{U}_x^c)_j^n, \quad (4.26)$$

where $(\mathbf{H}_x)_j^n$ denotes the expression on the right side of Eq. (4.23), and ϵ is a real number. Note that $(\mathbf{U}_x^c)_j^n$ is defined in terms of a central-difference approximation. Generally, numerical dissipation is introduced as a result of using such an approximation. On the other hand, $(\mathbf{H}_x)_j^n$ represents the

solution from a non-dissipative scheme. The right side of Eq. (4.26) is a weighted averaged of $(\mathbf{H}_x)_j^n$ and $(\mathbf{U}_x^c)_j^n$ with the weight factor of $1 - 2\epsilon$ and 2ϵ, respectively. Therefore, one may heuristically conclude that the numerical dissipation associated with the shock-capturing scheme can be increased by increasing the value of ϵ. This conclusion is verified by numerical experiments. As shown in [6], the stability domain of the shock-capturing scheme is defined by

$$CFL \leq 1 \quad \text{and} \quad 0 \leq \epsilon \leq 1. \quad (4.27)$$

Note that Eq. (4.26) can also be expressed as

$$(\mathbf{U}_x)_j^n = (\mathbf{H}_x)_j^n + 2\epsilon(D\mathbf{U})_j^n, \quad (4.28)$$

or

$$(\mathbf{U}_x)_j^n = (\mathbf{U}_x^c) + (2\epsilon - 1)(D\mathbf{U})_j^n, \quad (4.29)$$

where

$$(D\mathbf{U})_j^n \equiv (\mathbf{U}_x^c)_j^n - (\mathbf{H}_x)_j^n. \quad (4.30)$$

According to Eq. (4.28), $(\mathbf{U}_x)_j^n$ for the shock-capturing scheme is the sum of the non-dissipative term $(\mathbf{H}_x)_j^n$ and the dissipative term $2\epsilon(D\mathbf{U})_j^n$. The latter provides the necessary entropy-increasing condition within the stability domain defined by Eq. (4.27). Also it is seen from Eqs. (4.28) and (4.29) that $(\mathbf{U}_x)_j^n$ reduces to $(\mathbf{H}_x)_j^n$ and $(\mathbf{U}_x^c)_j^n$ in the cases of $\epsilon = 0$ and $\epsilon = 0.5$, respectively.

We remark that there is a slight difference between the shock-capturing scheme defined above and that defined by Eqs. (4.24)-(4.28) in [6]. Hereafter, the latter is referred to as the simplified shock-capturing scheme. Both schemes assume Eq. (4.22). However, Eq. (4.26) is replaced by

$$(\mathbf{U}_x)_j^n = (1 - 2\epsilon)(\mathbf{H}_x')_j^n + 2\epsilon(\mathbf{U}_x^c)_j^n \quad (4.31)$$

in the simplified scheme. Here,

$$(\mathbf{H}_x')_j^n = (\mathbf{U}_x^c)_j^n - (d\mathbf{U})_j^n \quad (4.32)$$

with

$$(d\mathbf{U})_j^n = \frac{1}{2}\left[(\mathbf{U}_x)_{j+1/2}^{n-1/2} + (\mathbf{U}_x)_{j-1/2}^{n-1/2}\right] - \frac{\mathbf{U}_{j+1/2}^{n-1/2} - \mathbf{U}_{j-1/2}^{n-1/2}}{\Delta x}. \quad (4.33)$$

Note that $(d\mathbf{U})_j^n$ is the difference of two numerical approximations to $(\mathbf{U}_x)_j^{n-1/2}$: (i) an average of $(\mathbf{U}_x)_{j+1/2}^{n-1/2}$ and $(\mathbf{U}_x)_{j-1/2}^{n-1/2}$, and (ii) a central-difference approximation in terms of $\mathbf{U}_{j+1/2}^{n-1/2}$ and $\mathbf{U}_{j-1/2}^{n-1/2}$. Also, as a result of Eqs. (3.2), (3.3), (4.24),

(4.25), (4.32), and (4.33), $(\mathbf{H}'_x)^n_j$ is a simple explicit function of $\mathbf{U}^{n-1/2}_{j\pm1/2}$ and $(\mathbf{U}_x)^{n-1/2}_{j\pm1/2}$. On the other hand, as explained earlier, the evaluation of $(\mathbf{H}_x)^n_j$ requires the solution of a system three linear equations at each mesh point (j,n).

It follows from Eqs. (4.31) and (4.32) that, in the simplified scheme, Eqs. (4.28) and (4.29) should be replaced by

$$(\mathbf{U}_x)^n_j = (\mathbf{H}'_x)^n_j + 2\epsilon(d\mathbf{U})^n_j \qquad (4.34)$$

and

$$(\mathbf{U}_x)^n_j = (\mathbf{U}^c_x)^n_j + (2\epsilon - 1)(d\mathbf{U})^n_j, \qquad (4.35)$$

respectively.

The reader is reminded that Eq. (4.31) was originally derived [6] from a "natural generalization" of the $a - \epsilon$ scheme, which is a solver for a scalar convection equation with a constant convection speed a. In a paper to be published, the conditions under which $(\mathbf{H}_x)^n_j$ can be approximated by $(\mathbf{H}'_x)^n_j$ will be given.

By comparison with the shock-capturing scheme defined by Eqs. (4.22) and (4.26), the simplified scheme is more computationally efficient. Yet the numerical results are almost identical to those generated by the former except when ϵ is very small (≤ 0.03). Note that the stability domain of the simplified scheme is approximately

$$CFL \leq 1 \quad \text{and} \quad 0.03 \leq \epsilon \leq 1. \qquad (4.36)$$

The two shock-capturing schemes described above generally can capture shocks with high resolution and without generating substantial numerical oscillations near shock if $0.3 \leq \epsilon \leq 0.8$. To further damp out these oscillations, $(\mathbf{U}^c_x)^n_j$ in Eq. (4.29) (which is equivalent to Eq. (4.26)) and Eq. (4.35) (which is equivalent to Eq. (4.31)) can be modified using a weighting procedure [6]. Let

$$(\mathbf{U}_{x\pm})^n_j = \pm\frac{(\mathbf{U}')^n_{j\pm1/2} - \mathbf{U}^n_j}{\Delta x/2}. \qquad (4.37)$$

Because $(\mathbf{U}')^n_{j\pm1/2}$ and $\mathbf{U}^n_j$ are the numerical analogues of $\mathbf{U}$ at the mesh points $(j\pm1/2, n)$ and (j, n), respectively, $(\mathbf{U}_{x+})^n_j$ and $(\mathbf{U}_{x-})^n_j$ are two numerical analogues of $\mathbf{U}_x$ at the mesh point (j, n), with one being evaluated from the right and another from the left. It follows from Eqs. (4.25) and (4.37) that

$$(\mathbf{U}^c_x)^n_j = \frac{1}{2}\left[(\mathbf{U}_{x+})^n_j + (\mathbf{U}_{x-})^n_j\right], \qquad (4.38)$$

i.e., $(\mathbf{U}^c_x)^n_j$ is the simple average of $(\mathbf{U}_{x+})^n_j$ and $(\mathbf{U}_{x-})^n_j$. The nonlinear weighting function is defined as

$$W(x_-, x_+; \alpha) = \frac{|x_+|^\alpha x_- + |x_-|^\alpha x_+}{|x_+|^\alpha + |x_-|^\alpha} \qquad (4.39)$$

where x_-, x_+, and α are real variables with $|x_+| + |x_-| > 0$ and $\alpha \geq 0$. Note that (i) $W(x_-, x_+; 0)$ is the simple average of x_- and x_+, and (ii) $W(x_-, x_+; 1)$ and $W(x_-, x_+; 2)$ are used in the slope-limiters proposed by van Leer [11] and van Albada et al. [33], respectively.

Recall that $(u_{mx\pm})^n_j$ denotes the m-th component of $(\mathbf{U}_{x\pm})^n_j$. Let

$$(u^w_{mx})^n_j = W\left[(u_{mx-})^n_j, (u_{mx+})^n_j, \alpha\right] \qquad (4.40)$$

Then, as shown in [6], numerical oscillations near shocks can be suppressed very efficiently if $(\mathbf{U}^c_x)^n_j$ is replaced by $(\mathbf{U}^w_x)^n_j$, i.e., the column matrix formed by $(u^w_{mx})^n_j$, $m = 1, 2, 3$, if $\alpha \geq 1$.

4.6 Concept of Dual Space-Time Mesh and Development of New Implicit Solvers

In addition to being used in the Euler solvers described in this section, the uniform space-time mesh depicted in Fig. 4.1 was also used in the simplest CE/SE scheme, i.e., the a scheme referred to earlier. The latter is a solver for the pure convection equation

$$\frac{\partial u}{\partial t} + a\frac{\partial u}{\partial x} = 0, \qquad (4.41)$$

where a is a constant convection speed.

The mesh depicted in Fig. 4.1 is staggered in time, i.e., the mesh points that have the same spatial location appear only at alternating time levels. In Fig. 4.12, the mesh depicted in Fig. 4.1 is superimposed on another staggered mesh, with the mesh points of the latter being marked by filled triangular symbols. Hereafter, in this subsection, (i) the mesh depicted in Fig. 4.1, with its mesh points marked by dots, is referred to as the original mesh, (ii) the mesh with its mesh points marked by triangles is referred to as the alternate mesh, and (iii) the combination of the above two meshes, i.e., that depicted in Fig. 4.12, is referred to as the dual mesh. As shown in Fig. 4.13, a CE of a mesh point marked by a triangle may coincide with a CE of another mesh point marked by a dot.

Obviously, through a similar procedure, one can construct the a scheme or any CE/SE Euler solver using the alternate mesh. As a matter of fact, one can even combine two independent a schemes, one constructed on the original mesh, and the other on the alternate mesh. This combined explicit scheme, referred to as the dual a scheme, has two completely decoupled solutions.

In [8], the construction of two implicit solvers for the convection-diffusion equation

$$\frac{\partial u}{\partial t} + a\frac{\partial u}{\partial x} - \mu\frac{\partial^2 u}{\partial x^2} = 0 \qquad (\mu \geq 0) \qquad (4.42)$$

was described. These two solvers are referred to as the a-$\mu(I1)$ and a-$\mu(I2)$ schemes, respectively. Here "I" stands for "implicit", and is used to distinguish these schemes from the explicit a-μ scheme described in [1, 4, 6].

In the case that $\mu = 0$, both the a-$\mu(I1)$ and the a-$\mu(I2)$ schemes reduce to the non-dissipative (explicit) dual a scheme. As a result, these two schemes have the important property that their numerical dissipation approaches zero as the physical dissipation approaches zero. This property provides better simulation of nearly inviscid flows (i.e., large-Reynolds-number flows), and ensures that numerical dissipation will not overwhelm physical dissipation. Furthermore, as can be inferred from several discussions in [6], the amplification factors arrived at by a von Neumann analysis of the dual a scheme are identical to those of the Leapfrog scheme. In other words, for the special case $\mu = 0$, the amplification factors of the a-$\mu(I1)$ and the a-$\mu(I2)$ schemes reduce to those of the Leapfrog scheme.

In case that $\mu > 0$, both a-$\mu(I1)$ and a-$\mu(I2)$ schemes become implicit. This is consistent with the fact that, for $\mu > 0$, the value of a solution to Eq. (4.42) at any point (x, t) depends on the initial data and all the boundary data up to the time t. In other words, an implicit scheme should be used to solve an initial/boundary-value problem, such as one involving Eq. (4.42). This requirement becomes more important as the diffusion term in Eq. (4.42) becomes more dominant.

Furthermore, for both a-$\mu(I1)$ and a-$\mu(I2)$ schemes, the solution at the mesh points marked by dots is coupled with that at the mesh points marked by triangles if $\mu > 0$. Also, it was shown in [8] that, in the pure diffusion case (i.e., when $a = 0$), the principal amplification factors of both a-$\mu(I1)$ and a-$\mu(I2)$ schemes are identical to the amplification

factor of the Crank-Nicolson scheme. Note that the latter has only one amplification factor.

Finally, note that both a-$\mu(I1)$ and a-$\mu(I2)$ schemes are stable if the Courant number $|\nu| \equiv |a|\,\Delta t\,/\,\Delta x \leq 1$. Also, both schemes have second-order accuracy in space and time if ν is held constant. This is numerically confirmed in [8].

5 The Euler Solvers for Multiple Spatial Dimensions

5.1 The 2D Euler Solvers

In Sec. 4, it was established that there were only *two* sets of independent marching variables, i.e., (i) $(u_m)^n_j$, $m = 1, 2, 3$, and (ii) $(u_{mx})^n_j$, $m = 1, 2, 3$, at each mesh point (j, n), if Eqs. (3.2), (3.3) and (4.1)–(4.4) are assumed. As a result, it requires *two* sets of conservation conditions, i.e., Eqs. (4.5) and (4.6) to construct the 1D non-dissipative Euler scheme. As a prerequisite to Eqs. (4.5) and (4.6), *two* CEs, i.e., $CE_-(j, n)$ and $CE_+(j, n)$ are defined for each mesh point (j, n).

The 2D CE/SE non-dissipative Euler solver [5, 7] was constructed using the same set of design principles that was used to construct its 1D counterpart. The differences between them stem entirely from the fact that there is one more spatial dimension to be considered in the 2D solver. In this section, only the basic geometric structures of the 2D solver will be described. For other details, the reader is referred to [5].

The 2D unsteady Euler equations of a perfect gas [5, 7] consist of *four* independent equations, $m = 1$, 2, 3, 4, instead of the *three* equations applicable to 1D flow. Also, in the 2D case, there are two spatial components of the gradient of each u_m (i.e., u_{mx} and u_{my}, where x and y are Cartesian coordinates for the 2D space). This is in contrast to the 1D case, in which, for each u_m, there is only one spatial component of the gradient of u_m (i.e., u_{mx}).

In the development of the 2D non-dissipative Euler solver and its extensions [5], a set of equations that is a natural 2D extension of Eqs. (3.2), (3.3) and (4.1)–(4.4) is assumed. As a result, there are *three* sets of independent marching variables at each mesh point (j, k, n) (see Figs. 5.1 and 5.2 for the locations of the mesh points. The reader is referred to [5, 7] for the definitions of the spatial mesh indices j and k). They are $(u_m)_{j,k}^n$, $(u_{mx})_{j,k}^n$ and $(u_{my})_{j,k}^n$, $m = 1, 2, 3, 4$. It follows that it requires *three* sets of conservation conditions (each set comprises four conditions, corresponding to $m = 1, 2, 3, 4$) at each mesh point to construct the 2D non-dissipative solver. Therefore, as a prerequisite, one must define *three* conservation elements for each mesh point. The construction of these CEs, which is the most intriguing part of the development of the 2D CE/SE Euler solver, will be described in what immediately follows.

Consider a spatial domain formed by congruent triangles (see Fig. 5.1). The center of each triangle is marked by either an empty circle or a filled circle. The distribution of these empty and filled circles is such that if the center of a triangle is marked by a filled (empty) circle, then the centers of the three neighboring triangles with which the first triangle shares a side are marked by empty (filled) circles. As an example, point G, the center of the triangle $\triangle BDF$, is marked by a filled circle while points A, C and E, the centers of the triangles $\triangle BFM$, $\triangle BJD$ and $\triangle DLF$, respectively, are marked by empty circles. These centers are the spatial projections of the space-time mesh points used in the 2D solver [5, 7].

To specify the exact locations of the mesh points in space-time, one must also specify their temporal coordinates. In the 2D CE/SE development, again we assume that the mesh points are located at the time levels $n = 0$ $\pm 1/2$, ± 1, $\pm 3/2$, ..., with $t = n \Delta t$ at the nth time level. Furthermore, we assume that the spatial projections of the mesh points at a whole-integer (half-integer) time level are the points marked by empty (filled) circles in Fig. 5.1.

Let the triangles depicted in Fig. 5.1 lie on the time level $n = 0$. Then those points marked by empty circles are the mesh points at this time level. On the other hand, those points marked by filled circles are not the mesh points at the time level $n = 0$. They are only the spatial projections of the mesh points at half-integer time levels.

Points A, C and E, which are depicted in Figs. 5.1 and 5.2(a), are three mesh points at the time level $n = 0$. Point G', which is depicted in Fig. 5.2(a), is a mesh point at the time level $n = 1/2$. Its spatial projection at the time level $n = 0$ is point G. Because point G is not a mesh point, it is not marked by a filled circle in the space-time plots given in Figs. 5.2(a)–(c). Hereafter, only a mesh point, e.g., point

G', will be marked by a filled or empty circle in a space-time plot.

The conservation elements associated with point G' are defined to be the space-time quadrilateral cylinders $GFABG'F'A'B'$, $GBCDG'B'C'D'$, and $GDEFG'D'E'F'$ that are depicted in Fig. 5.2(a). Here (i) points B, D and F are the vertices of the triangle with point G being its center (centroid) (see also Fig. 5.1), and (ii) points A', B', C', D', E' and F' are on the time level $n = 1/2$ with their spatial projections on the time level $n = 0$ being points A, B, C, D, E and F, respectively.

Point G' is a mesh point at a half-integer time level. For a mesh point at a whole-integer time-level, the conservation elements associated with it can be constructed in a similar fashion. As an example, consider Fig. 5.2(b). Here points B' (B''), I' (I''), J' (J''), K' (K''), D' (D''), G' (G'') and C' (C'') are on the time level $n = 1/2$ ($n = 1$) with their spatial projections on the time level $n = 0$, respectively, being the points B, I, J, K, D, G and C that are depicted in Fig. 5.1. Point C'' is a mesh point at the time level $n = 1$. By definition, the conservation elements associated with point C'' are the quadrilateral cylinders $C'J'K'D'C''J''K''D''$, $C'D'G'B'C''D''G''B''$ and $C'B'I'J'C''B''I''J''$.

The CEs associated with point G' and those associated with point C'' are depicted in Figs. 5.2(a) and 5.2(b), respectively. The relative positions of these six CEs in the global space-time mesh are depicted in Fig. 5.2(c).

Recall that, in the development of the 1D non-dissipative Euler solver, a pair of diagonally opposite vertices of each $\text{CE}_\pm(j, n)$ (see Figs. 4.3(a) and (b)) are assigned as mesh points. Furthermore, the boundary of each $\text{CE}_\pm(j, n)$ is a subset of the union of the SEs associated with the two diagonally opposite mesh points of this CE. In the 2D development, as seen from Figs. 5.2(a)–(c), two diagonally opposite vertices of each CE are also assigned as mesh points. In the following, we shall define the SEs such that even in the 2D case, the boundary of a CE is again a subset of the union of the SEs associated with the two diagonally opposite mesh points of this CE.

As an example, the SE associated with point G' is depicted in Fig. 5.3(a). It is the union of three vertical rectangles (i.e., $G''B''BG$, $G''D''DG$ and $G''F''FG$), a horizontal hexagon (i.e., $A'B'C'D'E'F'$) and their immediate neighborhood.

Note that points G'', B'', D'' and F'' are on the time level $n = 1$ and their spatial projections on the time level $n = 0$ are points G, B, D and F, respectively. As another example, the SE associated with point C'' is depicted in Fig. 5.3(b). Again, it is the union of three vertical rectangles (i.e., $C'''D'''D'C'$, $C'''B'''B'C'$ and $C'''J'''J'C'$), a horizontal hexagon (i.e., $K''D''G''B''I''J''$) and their immediate neighborhood. Note that points C''', D''', B''' and J''' are on the time level $n = 3/2$, and their spatial projections on the time level $n = 0$ are points C, D, B and J, respectively. The definition of the SE of any mesh point at a half-integer (whole-integer) time level is similar to the definition of the SE of the point $G'(C'')$.

As depicted in Fig. 5.2(b), one of the CEs associated with point C''' is the space-time quadrilateral cylinder $C'D'G'B'C''D''G''B''$. Among the vertices of this CE, only points C'' and G' are mesh points. From Figs. 5.3(a) and (b), it is seen that (i) three of the faces of this CE, i.e., $G'G''D''D'$, $G'B'B''G''$ and $G'D'C'B'$ are subsets of the SE of point G', and (ii) the other three faces, i.e., $C''B''B'C'$, $C''C'D'D''$ and $C''D''G''B''$ are subsets of the SE of point C''. As a result, by assuming that the flux of each $\vec{h}_m^*$ ($m = 1, 2, 3, 4$) is conserved over this CE, one can impose four conditions involving only the independent marching variables at the mesh points C'' and G'. Similarly, by using the flux conservation conditions over the other two CEs associated with point C'', one can obtain eight other conditions that relate the independent marching variables at the mesh points C'', I' and K'. Using the above 12 conditions, the 12 independent marching variables, i.e., u_m, u_{mx} and u_{my}, $m = 1, 2, 3, 4$, at the mesh point C'' can be determined in terms of the independent marching variables at the mesh points G', I' and K'. Similarly, the independent marching variables at the mesh point G' (see Fig. 5.2(a)) can be determined in terms of those at the mesh points A, C and E. By considering the mesh point C'' (G') as a typical mesh point at a whole-integer (half-integer) time level, the reader can understand how the 2D non-dissipative Euler solver was constructed [5, 7].

The non-dissipative Euler solver is only one of several 2D solvers described in [5, 7]. The latter document includes the 2D extensions of all but one of the 1D solvers described in [6]. The only exception is the 2D extension of the 1D Navier-Stokes solver. This will be dealt with in a separate paper. Also, because of the similarity in their design, each of the 2D extensions shares with its 1D version virtually the same

fundamental characteristics. As an example, the 2D non-dissipative Euler solver is neutrally stable, explicit, and involves only two time levels during a single time step. It also preserves the forward-backward marching nature and the space-time inversion invariance property of the 2D unsteady Euler equations. These are the same properties that characterize the 1D non-dissipative Euler solver.

The discussion of the 2D Euler solvers is concluded with the following remarks:

1. Because (i) the spatial geometric structure embedded in the CE/SE 2D space-time mesh is constructed from triangles, and (ii) triangles are the simplest polygon in the 2D space, the CE/SE solvers described in [5] can easily be modified and extended to solve flow problems with complex geometries.

2. Several 2D CE/SE solvers using nonuniform mesh have been developed [18–22]. Some of the numerical results generated with these solvers will be presented in Sec. 6.

3. Recall that the 1D non-dissipative Euler solver can be constructed using an alternative perspective in which the CEs and SEs have the shape of a rhombus (see Figs. 4.7–4.9). A similar alternative perspective can also be used to construct the 2D non-dissipative Euler solver. In this construction, the 2D counterpart of the rhombus referred to above is a space-time region with 12 faces.

To visualize the region referred to in remark (c), above, consider Figs. 5.3(a) and 5.4. Let points A' and G'' be joined by a line segment $\overline{A'G''}$. Because (i) points F' and A' are at one time level, and (ii) points F'' and G'' are at another time level, for any point P on the line segment $\overline{A'G''}$ (see Fig 5.4), there is one and only one point P' on $\overline{F'F''}$ such that P and P' have the same temporal coordinates. One face of the space-time region is generated by the line segment $\overline{PP'}$ as point P moves from one end of $\overline{A'G''}$ to the other end. Note that, by its definition, the intersection of this face and a plane of constant t is a straight line. However, it can be shown that the intersection of the face and a plane of constant x (constant y) is a hyperbola on a y-t (x-t) plane. Thus, in general, this face is not a plane. It is a "hyperbolic" surface. The same conclusion also follows from the fact that (i) the face contains four points A', G'', F'' and F', and (ii) generally one can not

222

find a plane in a 3D space-time that contains four arbitrary points in this space-time.

Similarly, one can generate the other eleven faces which, respectively, contain the following quads of points (i) A', G'', B'', B', (ii) C', G''', B'', B', (iii) C', G'', D'', D', (iv) E', G'', D'', D', (v) E', G''', F'', F', (vi) A', G, F, F', (vii) A', G, B, B', (viii) C', G, B, B', (ix) C', G, D, D', (x) E', G, D, D' and (xi) E', G, F, F'. Because the boundary of the space-time region is formed by 12 "hyperbolic" surfaces, it may be referred to as a "hyperbolic dodecahedron".

Point G' is a mesh point at a half-integer time level. Similarly, for a point at a whole-integer time level, eg., the point C'' depicted in Fig. 5.3(b), one can construct a corresponding "hyperbolic dodecahedron". For the current space-time mesh, the hyperbolic dodecahedron associated with a mesh point at a half-integer level and that at a whole-integer level are images of each other under space-time inversion. They are also congruent to each other. Note that these hyperbolic dodecahedra are constructed such that they can fill the entire computational domain without gaps or overlaps. Each of them plays the roles of both a CE and a SE in the 2D case, just as is done by a rhombus in the 1D case.

5.2 The Basis for a 3D Euler Solver

We indicate the discretization of space-time which forms the basis for a 3D Euler solver currently under development. The extension of the CE/SE method to three spatial dimensions follows reasoning similar to that used when extending the 1D solver to the 2D case (see the previous subsection). In the 3D case, the unsteady Euler equations of a perfect gas consist of five independent equations, $m = 1, 2, 3, 4, 5$. There are *three* spatial components of the gradient of each u_m (i.e., u_{mx}, u_{my} and u_{mz}, where x, y and z are Cartesian coordinates for the 3D space). When piecewise linear variation with space and time are assumed for the numerical solution, as is done in the 1D and 2D cases, and after the differential equation is assumed valid at each mesh point, there remain *four* sets of independent marching variables at each mesh point. It follows that *four* sets of conservation conditions are required at each mesh point to construct the non-dissipative 3D solver. Hence, *four* conservation elements must be defined for each mesh point. Just as a triangle was the polygon sharing its bounding edges with three neighbors, so a tetrahedron is the polyhedron sharing its bounding surfaces with *four* neighbors.

In the 2D case, referring to Figs. 5.1 and 5.2(a), $GFAB$, $GBCD$ and $GDEF$ are the spatial projections of the CEs associated with G'. The CEs in the 3D case can be constructed in analogous fashion. Consider the tetrahedron $ABCD$ with centroid G, and the tetrahedron $ABCP$ with centroid H, depicted in Fig. 5.5. They share the face ABC. The polyhedron $GABCH$ is then defined as the spatial projection of a CE associated with a point G'. The CE is thus a right cylinder in space-time, with $GABCH$ as its spatial base. The point G is the spatial image of the mesh point G', which is displaced temporally from G by half a time step.

In similar fashion, three additional CEs associated with the mesh point G' can be constructed by considering in turn three tetrahedra that share with $ABCD$ one of its other three faces. Thus the numerical solution at G' can be determined from a knowledge of the solution at the four mesh points (one of which is H) which are the centroids of the tetrahedra sharing a face with $ABCD$. This forms the basis of a non-dissipative 3D Euler solver.

Just as the structured mesh of Fig. 5.1 is obtainable by sectioning the parallelograms of Fig. 5.1 into triangles, so it is possible to construct a structured mesh of tetrahedra by sectioning a mesh of parallelepipeds. Details of the construction will be given in a future paper.

6 Computational Examples

6.1 Shock Tube with Non-Reflecting Computational Boundaries

The CE/SE computational results for the extended shock-tube problem described in Sec. 1 are presented in this subsection.

Flow of an ideal gas with specific-heat ratio $\gamma = 1.4$ is considered in an infinite shock-tube. The initial condition, at time $t = 0$, is $(\rho, v, p) = (1, 0, 1)$ if $x < 0$, and $(\rho, v, p) = (0.125, 0, 0.1)$ if $x > 0$. Here, ρ, v, p denote the density, velocity, and pressure of the fluid, respectively. A uniform space-time mesh with $\Delta x = 0.01$ and $\Delta t = 0.004$ (corresponding to a maximum Courant number of about 0.88) is used over the computational domain defined by $-0.5 \leq x \leq 0.5$ and $t \geq 0$.

The numerical results are generated using the FORTRAN program listed in the Appendix. The program implements the 1D simplified shock-capturing scheme (see Sec. 4.5). The settings $\epsilon = 0.5$ and $\alpha = 1$ are used for the artificial dissipation parameters. Note that the results are obtained without the need of any local mesh-refinement techniques or any time-step tuning.

The non-reflecting boundary conditions used are (i) $\mathbf{U}_j^n = \mathbf{U}_{j-1/2}^{n-1/2}$ and $(\mathbf{U}_x)_j^n = (\mathbf{U}_x)_{j-1/2}^{n-1/2}$ if (j, n) is a mesh point on the right boundary, and (ii) $\mathbf{U}_j^n = \mathbf{U}_{j+1/2}^{n-1/2}$ and $(\mathbf{U}_x)_j^n = (\mathbf{U}_x)_{j+1/2}^{n-1/2}$ if (j, n) is a mesh point on the left boundary. The reasons why such trivial extrapolations can serve so well as non-reflecting boundary conditions in the CE/SE method are explained in a separate paper [40].

Figs. 6.1–6.3 show the numerical solution (triangular data points) compared with the analytical solution (unbroken line) at three different times, namely, $t = 0.2$, 0.4 and 0.6. It is seen that excellent agreement is obtained between the numerical results and the analytical solution. In particular, as seen in Fig. 6.1, the shock wave discontinuity is resolved almost within one mesh interval and the contact discontinuity is resolved in four mesh intervals. Fig. 6.2 shows that by $t = 0.4$, the numerically computed shock wave has passed cleanly through the right boundary, with no spurious reflections. Similarly, Fig. 6.3 shows that by $t = 0.6$, the contact discontinuity has passed through the right boundary, while the expansion region has partially passed through the left boundary. Agreement with the exact solution continues to be excellent.

6.2 Convection-Diffusion Examples

The CE/SE computations described in this subsection were originally presented in [8], where an implicit CE/SE solver for the convection-diffusion equation $u_t + au_x - \mu u_{xx} = 0$ ($\mu \geq 0$) was developed. The solver, termed the a-$\mu(I1)$ solver and previously referred to in Sec. 4.6, is an extension of the a scheme, which is the solver for $u_t + au_x = 0$ referred to at the end of Sec. 4.4. See Sec. 4.6 for a brief discussion of how the construction of the a-$\mu(I1)$ solver ensures that numerical dissipation does not overwhelm physical dissipation in large-Reynolds-number flow computations. The examples below help show that the scheme is accurate over the whole Reynolds number range, from pure diffusion

to convection-dominated solutions.

Pure Diffusion. We consider a special case of the convection-diffusion equation with $a = 0$ and $\mu = 1$, in the domain $0 \leq x \leq 1$ and $t \geq 0$. The initial/boundary conditions completing the problem specification are (i) $u(0, t) = u(1, t) = 0$ for $t \geq 0$, (ii) $u(x, 0) = 2x$ for $0 \leq x \leq 0.5$, and (iii) $u(x, 0) = 2(1 - x)$ for $0.5 \leq x \leq 1$. The solution $u(x, t)$ exhibits the diffusive decay of the initial saw-tooth shape. An exact series solution is available, see for e.g. p.15 of [34]. For the CE/SE computation, uniform mesh intervals $\Delta x = 0.02$ and $\Delta t = 0.005$ are used. Fig. 6.4 shows the time-slice at $t = 0.05$, comparing numerical and exact solutions, and also showing the error scaled with the peak exact value at that time level. The maximum error magnitude is seen to be about 0.5% of the peak solution value. At $t = 1$ (not shown), when the peak solution value has dwindled to about 4×10^{-5}, the maximum error magnitude is about 0.15% of the peak solution value.

Boundary Layer, Re $= 100$. We next consider the problem defined for the convection-diffusion equation in the domain $0 \leq x \leq 1$ and $t \geq 0$ by the conditions (i) $u(0, t) = 0$ for $t \geq 0$, (ii) $u(1, t) = 1$ for $t \geq 0$, and (iii) $u(x, 0) = x$ for $0 \leq x \leq 1$. The 'steady-state' or time-asymptotic limit of the solution is $u(x, \infty) = [\exp(ax/\mu) - 1] \, / \, [\exp(a/\mu) - 1]$. The case $a = 1$, $\mu = 0.01$ (i.e. $Re = 100$) is considered, which leads to a steady-state boundary layer at $x = 1$. Uniform mesh intervals $\Delta x = 0.0025$ and $\Delta t = 0.002$ are used, so that the Courant number is 0.8. Fig. 6.5 shows the computed and exact steady-state limits, together with the error. The boundary layer is seen to be well resolved, with the maximum magnitude of the error being about 1% of the solution peak.

6.3 Shock Wave in a Constant-Temperature Bath

We next take up a flow problem proposed by Pember [35]. Consider the one-dimensional Euler equations with a special heat transfer term in the energy equation:

$$\mathbf{U}_t + \mathbf{F}_x + \mathbf{H} = 0, \qquad (6.1)$$

where the flow properties $\mathbf{U}$ and the flux $\mathbf{F}$ are as defined in Eqs. (2.4) and (2.5). The source term is defined as $\mathbf{H} = (0, 0, K\rho(T - T_o))^T$. The function of the source term is to force a constant temperature T_o upon the whole flow field. The equilibrium counterpart of the relaxation system is

$$\frac{\partial \rho}{\partial t} + \frac{\partial \rho v}{\partial x} = 0, \qquad (6.2)$$

$$\frac{\partial \rho v}{\partial t} + \frac{\partial \rho v^2 + p^*}{\partial x} = 0, \qquad (6.3)$$

where the pressure $p^*(\rho) = (\gamma - 1)\rho e_o$, with e_o as the internal energy of the gas at $T = T_o$. We conducted the calculation with 201 grid nodes over the x interval $0 \leq x \leq 1$. The conductivity K in the flow system was set be to 10^8, 10^{12}, and 10^{16}. The relaxation time $\epsilon = 1/K$ is under-resolved for all three cases. Essentially, we get the same result for different Ks. Figures 6.6(a)–(d) show the flow properties at $t = 0.3$. The flow field contains a right-moving shock and a left-moving rarefaction. The numerical solution is in agreement with that presented by Jin [36]. For details of the treatment of source terms in the present CE/SE method, we refer the reader to [16].

6.4 Diffraction of a Shock Wave around a Wedge

According to the experimental shadowgraph results shown in [37], when a plane shock wave of $M_s = 1.3$ is moving over the beginning of a finite wedge of semi-vertex angle $\theta = 26.565°$, an ordinary Mach reflection is generated. As the shock wave passes the base, the flow separates to form vortex sheets at the sharp corners. Further interaction produces an increasingly elaborate pattern of shock waves, slip lines and vortices.

As reported in [18, 21], this flowfield is simulated using the CE/SE Euler solver. By virtue of the symmetry in the solution, attention is restricted to the upper half of the domain. The extent of the computational domain is set based on an estimation from Fig. 522 in [37]. The shock wave is at $x = -0.5$ at $t = 0$. The numerical boundary condition imposed on the vertical wall of the wedge is described in [20]. Numerical solutions at eight time levels ($t = 0.725, 0.9075, 1.2125, 1.55, 1.825, 2.1375, 2.4875,$ and 2.9475), obtained by using two subdomains with 321×89 and 209×34 mesh points, and with $\Delta t = 0.0025$, are shown in Fig. 6.7. It should be pointed out that the upper and lower walls of the channel shown in the shadowgraphs of [37] are actually further apart than the top and bottom edges of the shadowgraphs. Therefore some flow phenomena that are seen in Fig. 6.7, in the region near the upper wall, are lost in shadowgraphs, especially at the 4th time level. Comparisons of the computed solutions with experimental pictures of [37] have shown an excellent agreement in general flow features except for those phenomena induced by the effect of viscosity. The shock waves, slip lines and vortices are captured very well.

6.5 Implosion/Explosion of Polygonal Shock Waves in a Box

The 2-D CE/SE Euler solver has been used in [22] to solve a problem studied in [38], concerning the implosion/explosion of a polygonal shock wave in a square box. In addition to the early stage of the implosion/explosion process, the later development of the process, which was not studied in [38], is also simulated in [22]. The computation further demonstrates the robustness of the CE/SE Euler scheme in handling discontinuous flows.

A uniformly distributed 241x241 grid is utilized in the computational domain, which is a square defined by $-2 \leq x \leq 2$ and $-2 \leq y \leq 2$. The initial shock wave configuration is a polygon, the geometric center of which coincides with that of the square. Inside the polygon is a low pressure region, with a pressure ratio of 10 across the shock. The radius of the circumscribed circle of the polygon is selected to be $0.8\sqrt{3}$ for all shapes of the polygon. In the numerical scheme, the two parameters ϵ and α are set to be 0.5 and 1 respectively, everywhere in the computational domain for all cases, and the maximum Courant number is always kept at a value of 0.9.

In one set of computations, the early flowfield is studied for polygonal shock waves with initial shapes of an equilateral triangle, a square, and a pentagon. The density contour plots at different time levels are shown in Fig. 6.8. Wave patterns similar to those captured in Figs. 1–5 of [38] using a TVD method on a 359x359 grid are clearly shown in the CE/SE solutions, displaying detailed features such as Mach stems and the newly-developed smaller polygons.

In another computation, the implosion/explosion of a hexagonal shock wave is simulated until the re-implosion of the shock wave is observed in the box. More complex flow phenomena can be seen in the density contour plots of Fig. 6.9, including the reflections of shock waves, shock-shock interaction, and shock-contact surface interaction. It is interesting to note that the shape of the contact surface centered at the origin of the box remains unchanged even after the passage of shock waves.

6.6 Examples from Computational Aeroacoustics

The CE/SE computational examples we describe in the next two subsections were reported in [9] and [15]. The investigations in [9] and [15] found the CE/SE Euler scheme to be capable of handling the complete spectrum of flows, from small-amplitude linear acoustic waves, all the way to nonlinear or even discontinuous waves (shocks). Through numerical experiments in computational aeroacoustics, the following salient properties of the CE/SE Euler scheme emerge: (i)The CE/SE scheme possesses very low dispersion error and yields high resolution results comparable to that of a high order compact difference scheme, although nominally the CE/SE scheme is only of 2nd order accuracy. (ii) In general, the numerical non-reflecting boundary condition applicable to the CE/SE scheme is genuinely multi-dimensional, and can be implemented in a simple and elegant way without resorting to the complexities of characteristic forms or buffer zones. (iii) The CE/SE scheme is both a CFD (Computational Fluid Dynamics) and a CAA (Computational Aeroacoustics) scheme, capable of handling continuous and discontinuous flows. It thus represents a unique numerical technique for flows where sound waves and shocks and their interactions are important.

It is well-known that in CAA, the non-reflecting boundary condition plays a dominant role in the final numerical results. In general, there are three ways to impose the non-reflecting boundary conditions, namely,

(i) to apply 1-D characteristic variables (Riemann invariants) in the direction normal to the boundary,

(ii) to minimize spurious numerical reflections from the boundaries by inserting a buffer zone with increased numerical damping,

(iii) to apply an asymptotic analytical solution at the boundaries.

In the new CE/SE scheme, none of the above complex treatments of non-reflecting boundary conditions is needed. Instead, a simple non-reflecting

boundary condition is:

$$(u_{mx})^n_{j,k} = (u_{my})^n_{j,k} = 0, \qquad m = 1, 2, 3, 4,$$

while $(u_m)^n_{j,k}$ is defined by simple extrapolation from the interior neighbors. In general, the consequent reflection amounts to about 1% of the strength of the incident waves.

6.7 Acoustic Waves Generated by a Flat-Plate Loudspeaker

In our first example, a continuous wave problem is considered. The physical domain is a rectangle, as shown as in Fig. 6.10. A mesh of 200×200 nodes is employed over a domain of length 300 in x and 200 in y. The background mean flow is a uniform one in the x direction, with Mach number $M = 0.5$. The mean flow is given by

$$u_0 = M, \quad v_0 = 0, \quad \rho_0 = 1, \quad p_0 = \frac{1}{\gamma}, \quad \gamma = 1.4$$

where the subscript 0 identifies mean flow quantities. At the center of the inlet boundary, a small vertical flat plate of size 1/10 of the height of the domain is vibrating in the x direction, generating a perturbation to the mean flow that is given by

$$u^* = 0.01 \sin\left(\frac{\pi t}{5}\right)$$

Fig. 6.10 illustrates the isobars at time $t = 600$. For clear display, 800 uniform contour levels are used. In Fig. 6.10, the well-known acoustic directivity pattern of a flat plate loudspeaker is clearly seen, and practically no reflection at the four boundaries is observed.

6.8 Acoustic Pulse / Shock-Wave Interaction

In order to demonstrate the capability of the CE/SE scheme to handle interactions of acoustic waves and shock waves, we describe an example of a weak acoustic pulse wave passing through a strong shock. We use a mesh of 200×200 nodes. The domain is centered at the origin $(0, 0)$, with an extent of $-100 \leq x \leq 100$ and $-100 \leq y \leq 100$. A steady oblique shock at a position along a diagonal of the computational domain is precalculated to form part of the initial condition for the computation. The initial conditions of an isolated acoustic pulse are superimposed on this precalculated shock to form the initial condition of the given problem. The data upstream and downstream of the shock are respectively

$$u_0 = 2.378056, \; v_0 = 0,$$
$$\rho_0 = 1 \quad \text{and} \quad p_0 = 0.7142857;$$

and

$$u_0 = 2.1017481, \; v_0 = 0.4062729,$$
$$\rho_0 = 1.5807555 \quad \text{and} \quad p_0 = 1.3713613.$$

A weak acoustic pulse propagating across a strong shock is considered. An acoustic pulse, initially centered at $(x_0, y_0) = (-75, 0)$, with initial data

$$u^* = v^* = 0, \qquad p^* = \rho^* = \epsilon e^{-a\left[(x-x_0)^2 + (y-y_0)^2\right]}$$

is superimposed on the mean flow, where the initial pulse amplitude $\epsilon = 0.001$ and $a = (\ln 2)/9$. It is observed that the oblique shock strength is three orders of magnitude larger than the initial amplitude of the acoustic pulse. The pulse propagates in all directions with the speed of sound, while being carried downstream by the mean flow. During the computation, the non-reflecting boundary condition described above is enforced at all the four sides of the computational domain.

For such an interaction between a weak (linear) wave and a discontinuous wave, the theoretical exact solution is not available. However, the numerical results obtained with the CE/SE scheme demonstrate physically plausible phenomena. Fig. 6.11 illustrates the isobars at various time steps. At first, the acoustic pulse is blown downstream and propagates freely. As the pulse collides with the strong oblique shock, the shock is practically unaffected, while the acoustic pulse ring is distorted in its passage through the shock, due to different speeds of sound and flow velocities on either side of the oblique shock.

In other examples described in [9] and [15], the interactions of a strong (i.e. nonlinear) acoustic pulse, and of weak and strong vortical and entropy pulses with a strong shock were computed. Currently, the CE/SE method is being applied to benchmark problems in CAA, and has proved to be exceptionally accurate.

7 Summary and Conclusions

In the present article, we reviewed the method of space-time conservation element and solution element (the CE/SE method, for short) for the numerical solution of conservation laws. We described several CE/SE schemes for computing fluid flows, and touched upon other CE/SE schemes and extensions. Our descriptions emphasized the geometry of the space-time discretization.

We examined salient general properties of solutions to the Euler equations for time-dependent compressible flow. From this, we arrived at requirements for a scheme to be an ideal Euler solver. An ideal solver for smooth flows must be neutrally stable, explicit and two-level, and must be such that the discrete equations are invariant under space-time inversion. The CE/SE non-dissipative Euler solvers for isentropic flows meet all these requirements. In the present article, we described the non-dissipative 1D and 2D Euler solvers in terms of the conservation of piecewise linear space-time fluxes over discrete space-time volumes. Thus, given the space-time discretization, the schemes have a simple specification in terms of flux conservation. When shock waves are present in the solution, numerical dissipation must be introduced into numerical schemes in a controllable fashion, to model the irreversibility in the exact solution. We described the shock-capturing 1D Euler solver, which is a modification of the non-dissipative solver. The added numerical dissipation has a simple geometric description and a straightforward generalization to the 2D case. The Navier-Stokes solvers, not described here, reduce to the non-dissipative solvers when the physical viscosity vanishes, and hence the latter is never overwhelmed by numerical dissipation.

We contrasted the structure and properties of the CE/SE 1D Euler solvers with those typical of high-resolution finite-volume upwind schemes. The key strategies that enable the CE/SE schemes to avoid the limitations of the upwind schemes are: (i) The more general form of the conservation laws, i.e., the integral form, is cast in a form in which space and time are treated on an equal footing. This gives flexibility in the shape of the space-time conservation elements, which is useful for defining CEs when, for e.g., sources are present in the CE. (ii) A staggered space-time mesh is employed. This results in the simplest stencil. It also obviates the need for interpolation of fluxes at the interface between CEs. Thus, there is no need for an approximate Riemann solver. Hence, characteristics-based upwind-biasing methods, which are complicated and strictly valid only for smooth solutions, are avoided. There is thus also no compromise in the symmetry of treatment of the spatial fluxes. This also has implications for flows in multiple spatial dimensions. For the computation of such flows, upwind techniques must use directional splitting with its attendant difficulties. The CE/SE method in multiple spatial dimensions, on the other hand, does not involve any directional splitting. (iii) The flow property gradient is treated as an additional unknown in the CE/SE schemes. Therefore, there is no need for reconstruction of the flow gradient by polynomial curve fitting over neighboring mesh points, and for the subsequent use of complicated flux limiters. (iv) Space-time fluxes are conserved at both the local and global level. The condition of flux conservation, rather than any extrapolation, links the solution at a mesh point with its neighbors at the previous time level. This emphasis on the integral conservation law is critical for accurate flow simulations, particularly if they involve long marching times and/or regions of rapid change (e.g., boundary layers and shocks).

We reproduced here several numerical results obtained with various CE/SE flow solvers. The results included a demonstration of extremely simple yet highly effective non-reflecting boundary conditions for the extended Sod's shock-tube problem. The CE/SE solver for the scalar convection-diffusion equation was shown to be accurate in all Reynolds number regimes. The CE/SE solver for the Euler equations with source terms representing heating/cooling was also shown to be accurate, even when the source term is of the order of 10^{16}. We reproduced numerical solutions obtained with the 2D CE/SE Euler solver, including the process of diffraction of a shock wave around a wedge and the implosion/explosion of a polygonal shock wave in a box, as well as computational aeroacoustic phenomena involving the interaction of strong shocks and weak acoustics. The results reproduced here are only some of the difficult problems readily solved with CE/SE schemes; see [1–9] and [13–22] for more examples.

We remark here that the CE/SE schemes developed thus far are characterized by simplicity, generality of applicability and second-order accuracy in space and time. The simplest possible stencils are employed. The 2D spatial mesh is constructed from triangles, and the 3D spatial mesh will be constructed from tetrahedra. Triangles and tetrahedra are the simplest polytopes in 2D and 3D, respec-

tively. The 1D and 2D Euler solvers bear a remarkable resemblance to the solvers of the 1D and 2D scalar convection-diffusion equations, respectively, with the discrete equations in the former two being matrix versions of the scalar equations in the latter two. All of the above schemes are characterized by virtually the same properties. Furthermore, the viscous flow solvers are designed to reduce to the respective non-dissipative solvers when the physical viscosity vanishes. The CE/SE method thus represents a new unified framework for the numerical solution of conservation laws. The concept of the dual space-time mesh, that was explained in Sec. 4.6, plays a key role in the development of implicit viscous solvers. It is also essential in the application of the CE/SE method to 2D unstructured meshes, when the triangles in space can not be segregated into two sets which are dual to each other. A 2D Euler solver which uses an unstructured mesh, and a 2D implicit Navier-Stokes solver are currently under development.

Acknowledgement

The authors are grateful to Dr. H.T. Hyunh of NASA Lewis Research Center for several insightful conversations regarding upwind techniques.

Appendix

```
c     The CE/SE 1D Euler Solver solves
c     the extended Sod's shock-tube problem
c
      implicit real*8(a-h,o-z)
      parameter (nxd=1000)
      dimension  q(3,nxd), qn(3,nxd),
     *           qx(3,nxd), qt(3,nxd),
     *           s(3,nxd), vxl(3), vxr(3),
     *           xx(nxd)
c
c     nx must be an odd integer.
      nx = 101
      it = 100
      dt = 0.4d-2
      dx = 0.1d-1
      ga = 1.4d0
      rhol = 1.d0
      ul = 0.d0
      pl = 1.d0
      cl = 1.d0
      rhor = 0.125d0
      ur = 0.d0
      pr = 0.1d0
      cr = 1.d0
      ia = 1
c
      nx1 = nx + 1
      nx2 = nx1/2
      hdt = dt/2.d0
      tt = hdt*dfloat(it)
      qdt = dt/4.d0
      hdx = dx/2.d0
      qdx = dx/4.d0
      dtx = dt/dx
      a1 = ga - 1.d0
      a2 = 3.d0 - ga
      a3 = a2/2.d0
      a4 = 1.5d0*a1
      u2l = rhol*ul
      u3l = pl/a1 + 0.5d0*rhol*ul**2
      u2r = rhor*ur
      u3r = pr/a1 + 0.5d0*rhor*ur**2
      do 5 j = 1,nx2
      q(1,j) = rhol
      q(2,j) = u2l
      q(3,j) = u3l
      q(1,nx2+j) = rhor
      q(2,nx2+j) = u2r
      q(3,nx2+j) = u3r
      do 5 i = 1,3
      qx(i,j) = 0.d0
      qx(i,nx2+j) = 0.d0
5     continue
c
      open (unit=8,file='for008')
      write (8,10) tt,it,ia,nx
      write (8,20) dt,dx,ga
      write (8,30) rhol,ul,pl,cl
      write (8,40) rhor,ur,pr,cr
c
      do 400 i = 1,it
      m = nx + i - (i/2)*2
      do 100 j = 1,m
      w2 = q(2,j)/q(1,j)
      w3 = q(3,j)/q(1,j)
      f21 = -a3*w2**2
      f22 = a2*w2
      f31 = a1*w2**3 - ga*w2*w3
      f32 = ga*w3 - a4*w2**2
      f33 = ga*w2
      qt(1,j) = -qx(2,j)
      qt(2,j) = -(f21*qx(1,j) + f22*qx(2,j)
     *           + a1*qx(3,j))
      qt(3,j) = -(f31*qx(1,j) + f32*qx(2,j)
     *           + f33*qx(3,j))
      s(1,j) = qdx*qx(1,j) + dtx*(q(2,j)
     *           + qdt*qt(2,j))
      s(2,j) = qdx*qx(2,j) + dtx*
     *           (f21*(q(1,j) + qdt*qt(1,j))
     *           + f22*(q(2,j)+ qdt*qt(2,j))
     *           + a1*(q(3,j) + qdt*qt(3,j)))
      s(3,j) = qdx*qx(3,j) + dtx*
     *           (f31*(q(1,j) + qdt*qt(1,j))
     *           + f32*(q(2,j) + qdt*qt(2,j))
     *           + f33*(q(3,j) + qdt*qt(3,j)))
100   continue
      if (i.ne.(i/2)*2) goto 150
      do 120 k = 1,3
      qx(k,1)= cl*qx(k,1)
      qx(k,nx1) = cr*qx(k,nx)
      qn(k,1) = q(k,1)
      qn(k,nx1) = q(k,nx)
120   continue
150   j1 = 1 - i + (i/2)*2
      mm = m - 1
      do 200 j = 1,mm
      do 200 k = 1,3
      qn(k,j+j1) = 0.5d0*(q(k,j) + q(k,j+1)
     *             + s(k,j) - s(k,j+1))
      vxl(k) = (qn(k,j+j1) - q(k,j)
     *         - hdt*qt(k,j))/hdx
      vxr(k) = (q(k,j+1) + hdt*qt(k,j+1)
     *         - qn(k,j+j1))/hdx
      qx(k,j+j1) =
     *       (vxl(k)*(dabs(vxr(k)))**ia
     *       + vxr(k)*(dabs(vxl(k)))**ia)/
```

229

```fortran
     *            ((dabs(vxl(k)))**ia
     *            + (dabs(vxr(k)))**ia +1.d-60)
200   continue
      m = nx1 - i + (i/2)*2
      do 300 j = 1,m
      do 300 k = 1,3
      q(k,j) = qn(k,j)
300   continue
400   continue
c
      m = nx1 -it + (it/2)*2
      mm = m - 1
      xx(1) = -0.5d0*dx*dfloat(mm)
      do 500 j = 1,mm
      xx(j+1) = xx(j) + dx
500   continue
      do 600 j = 1,m
      x = q(2,j)/q(1,j)
      z = a1*(q(3,j) - 0.5d0*x**2*q(1,j))
      y = x/dsqrt(ga*z/q(1,j))
      write (8,50) xx(j),q(1,j),x,y,z
600   continue
c
      close (unit=8)
10    format(' t = ',f8.4,' it = ',i4,
     *        ' ia = ',i4,' nx = ',i4)
20    format(' dt = ',f8.4,' dx = ',f8.4,
     *        ' gamma = ',f8.4)
30    format(' rhol = ',f8.4,' ul = ',f8.4,
     *        ' pl = ',f8.4,' cl = ',f8.4)
40    format(' rhor = ',f8.4,' ur = ',f8.4,
     *        ' pr = ',f8.4,' cr = ',f8.4)
50    format(' x =',f8.4,' rho =',f8.4,
     *        ' u =',f8.4,' M =',f8.4,
     *        ' p =',f8.4)
      stop
      end
```

References

[1] S.C. Chang and W.M. To, "A New Numerical Framework for Solving Conservation Laws – The Method of Space-Time Conservation Element and Solution Element", NASA TM 104495, August 1991.

[2] S.C. Chang, "On an Origin of Numerical Diffusion: Violation of Invariance Under Space-Time Inversion", Proceedings of the 23rd Modeling and Simulation Conference, April 30 - May 1, 1992, Pittsburgh, PA, William G. Vogt and Marlin H. Mickle eds., Part 5, pp. 2727-2738. Also published as NASA TM 105776.

[3] S.C. Chang and W.M. To, "A Brief Description of a New Numerical Framework for Solving Conservation Laws – The Method of Space-Time Conservation Element and Solution Element", Proceedings of the 13th International Conference on Numerical Methods in Fluid Dynamics, July 6-10, 1992, Rome, Italy, M. Napolitano and F. Sabetta, eds. Also published as NASA TM 105757.

[4] S.C. Chang, "New Developments in the Method of Space-Time Conservation Element and Solution Element – Applications to the Euler and Navier-Stokes Equations", Presented at the Second U.S. National Congress on Computational Mechanics, August 16-18, 1993, Washington D.C. Published as NASA TM 106226.

[5] S.C. Chang, X.Y. Wang and C.Y. Chow, "New Developments in the Method of Space-Time Conservation Element and Solution Element – Applications to Two-Dimensional Time-Marching Problems", NASA TM 106758, December 1994.

[6] S.C. Chang, J. Comput. Phys., 119 (1995), pp. 295-324.

[7] S.C. Chang, X.Y. Wang and C.Y. Chow, "The Method of Space-Time Conservation Element and Solution Element – Applications to One-Dimensional and Two-Dimensional Time-Marching Flow Problems", AIAA Paper 95-1754, in A Collection of Technical Papers, 12th AIAA CFD Conference, June 19-22, 1995, San Diego, CA, pp. 1258-1291. Also published as NASA TM 106915.

[8] S.C. Chang, X.Y. Wang, C.Y. Chow and A. Himansu, "The Method of Space-Time Conservation Element and Solution Element – Development of a New Implicit Solver", Proceedings of the Ninth International Conference on Numerical Methods in Laminar and Turbulent Flow, July 10-14, 1995, Atlanta, GA. Also published as NASA TM 106897.

[9] S.C. Chang, C.Y. Loh and S.T. Yu, "Computational Aeroacoustics via a New Global Conservation Scheme", to appear in the Proceedings of the 15th International Conference on Numerical Methods in Fluid Dynamics, June 24-28, 1996, Monterey, CA.

[10] B. van Leer, J. Comput. Phys., 23 (1977), pp. 263-275.

[11] B. van Leer, *J. Comput. Phys.*, **23** (1977), pp. 276-299.

[12] G.A. Sod, *J. Comput. Phys.*, Vol. 27 (1978), pp. 1-31.

[13] C.Y. Loh, S.C. Chang, J.R. Scott and S.T. Yu, "Application of the Method of Space-Time Conservation Element and Solution Element to Aeroacoustics Problems", Proceedings of the 6th International Symposium of CFD, September 1995, Lake Tahoe, NV.

[14] C.Y. Loh, S.C. Chang, J.R. Scott and S.T. Yu, "The Space-Time Conservation Element Method – A New Numerical Scheme for Computational Aeroacoustics", AIAA Paper 96-0276, presented at the 34th AIAA Aerospace Sciences Meeting, January 15-18, 1996, Reno, NV.

[15] C.Y. Loh, S.C. Chang and J.R. Scott, "Computational Aeroacoustics via the Space-Time Conservation Element / Solution Element Method", AIAA Paper 96-1687, presented at the 2nd AIAA/CEAS Aeroacoustics Conference, May 6-8, 1996, State College, PA.

[16] S.T. Yu and S.C. Chang, "Treatments of Stiff Source Terms in Hyperbolic Conservation Systems by the Method of Space-Time Conservation Element and Solution Element", AIAA Paper 97-0435, presentated at the 35th AIAA Aerospace Sciences Meeting, January 1997, Reno, NV.

[17] X.Y. Wang, C.Y. Chow and S.C. Chang, "Application of the Space-Time Conservation Element and Solution Element Method to Shock-Tube Problem", NASA TM 106806, December 1994.

[18] X.Y. Wang, "Computational Fluid Dynamics Based on the Method of Space-Time Conservation Element and Solution Element", Ph.D. Dissertation, 1995, Department of Aerospace Engineering, University of Colorado, Boulder, CO.

[19] X.Y. Wang, C.Y. Chow and S.C. Chang, "Application of the Space-Time Conservation Element and Solution Element Method to Two-Dimensional Advection-Diffusion Problems", NASA TM 106946, June 1995.

[20] X.Y. Wang, C.Y. Chow and S.C. Chang, "High Resolution Euler Solvers Based on the Space-Time Conservation Element and Solution Element Method", AIAA Paper 96-0764, presented at the 34th AIAA Aerospace Sciences Meeting, January 15-18, 1996, Reno, NV.

[21] X.Y. Wang, C.Y. Chow and S.C. Chang, "Numerical Simulation of Flows Caused by Shock-Body Interaction", AIAA Paper 96-2004, presented at the 27th AIAA Fluid Dynamics Conference, June 17-20, 1996, New Orleans, LA.

[22] X.Y. Wang, C.Y. Chow and S.C. Chang, "An Euler Solver Based on the Method of Space-Time Conservation Element and Solution Element", to appear in the Proceedings of the 15th International Conference on Numerical Methods in Fluid Dynamics, June 24-28, 1996, Monterey, CA.

[23] J.R. Scott and S.C. Chang, "A New Flux Conserving Newton's Method Scheme for the Two-Dimensional, Steady Navier-Stokes Equations", NASA TM 106160, June 1993.

[24] J.R. Scott, "A New Flux-Conserving Numerical Scheme for the Steady, Incompressible Navier-Stokes Equations", NASA TM 106520, April 1994.

[25] J. R. Scott and S.C. Chang, *Comp. Fluid Dyn.*, Vol. **5** (1995), pp. 189-212.

[26] J.R. Scott and S.C. Chang, "The Space-Time Solution Element Method – A New Numerical Approach for the Navier-Stokes Equations", AIAA Paper 95-0763, presented at the 33rd AIAA Aerospace Sciences Meeting, January 9-12, 1995, Reno, NV.

[27] J. R. Scott, "Further Development of a New, Flux-Conserving Newton Scheme for the Navier-Stokes Equations", NASA TM 107190, March 1996.

[28] P. Lax, "Hyperbolic Systems of Conservation Laws and the Mathematical Theory of Shock Waves," *SIAM,* Philadelphia, Pennsylvania (1973).

[29] H.T. Huynh, *SIAM J. Numer. Anal.*, **32**, 5 (1995), pp. 1565-1619.

[30] A. Harten, *J. Comput. Phys.*, Vol. 49 (1983), pp. 357-393.

[31] A. Harten, B. Engquist, S. Osher, and S. Chakravarthy, *J. Comput. Phys.*, Vol. 71 (1987), pp. 231-303.

[32] P.L. Roe, *Ann. Rev. Fluid Mech.*, Vol. 18 (1986), pp. 337-365.

[33] G.D. van Albada, B. van Leer, and W.W. Roberts, *Astronom. Astrophys.*, Vol. 108 (1982), p.76.

[34] G.D. Smith, *Numerical Solution of Partial Differential Equations: Finite Difference Methods*, 3rd ed., 1985. Oxford Univ. Press, New York.

[35] R.B. Pember, *SIAM J. Appl. Math.*, **53**, 5 (1993), pp. 1293-1330.

[36] S. Jin, *J. Comput. Phys.*, **122** (1995), pp. 51-67.

[37] H. Oertel, Sr., 1966 *Stossrohre*, Vienna: Springer-Verlag.

[38] T. Aki and F. Higashino, AIP Conference Proceedings 208, 1989.

[39] S.T. Yu, Y.-L.P. Tsai and K.C. Hsieh, "Simulating Waves in Flows by Runge-Kutta and Compact Difference Schemes", AIAA Journal, Vol. **33**, No. 3, pp. 421-429. Also presented as AIAA Paper 92-3210.

[40] S.C. Chang, A. Himansu, C.Y. Loh, X.Y. Wang, S.T. Yu and P.C.E. Jorgenson, "Robust and Simple Non-Reflecting Boundary Conditions for the Space-Time Conservation Element and Solution Element Method", AIAA Paper 97-2077, presented at the 13th AIAA CFD Conference, June 29-July 2, 1997, Snowmass, CO.

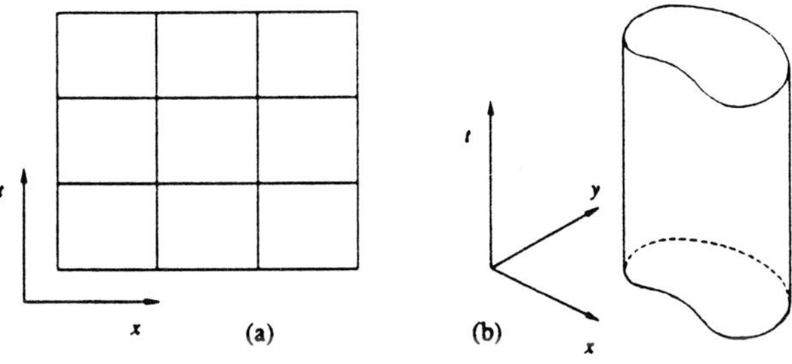

Figure 2.1 — Space-time conservation elements for methods using a fixed
spatial domain: (a) one space dimension (b) two space dimensions

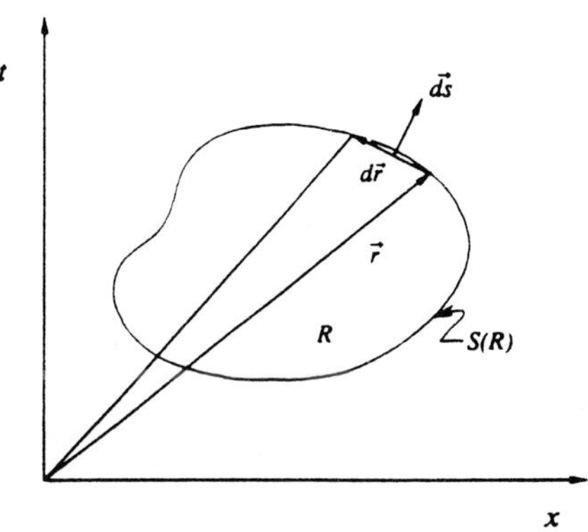

Figure 2.2 — A space-time conservation element with an
arbitrary space-time domain

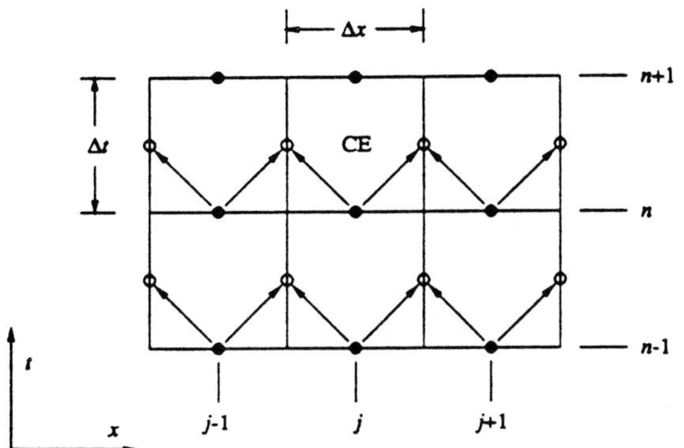

Figure 3.1 — The direction of information transfer (shown by arrows)
during spatial flux estimation in an upwind scheme

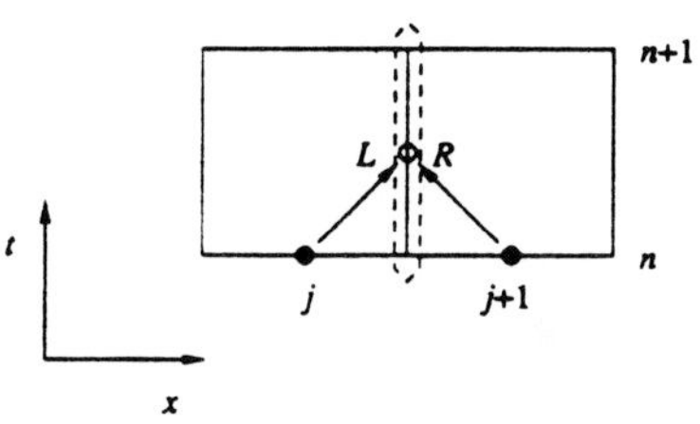

Figure 3.2 — Estimating interface solution value from left (L)
and right (R), in an upwind scheme

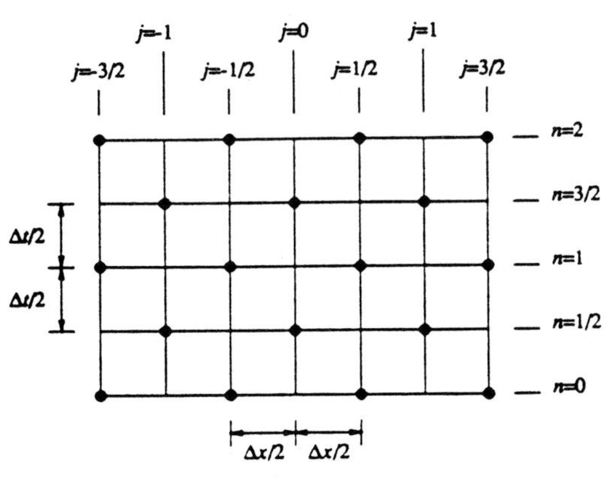

Figure 4.1 — The space-time mesh

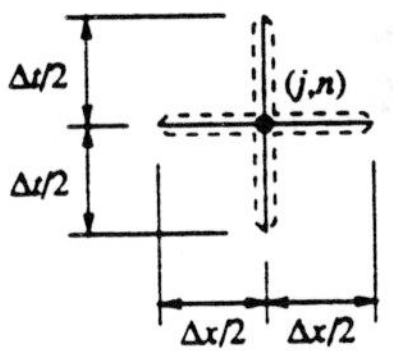

Figure 4.2 — SE(j,n)

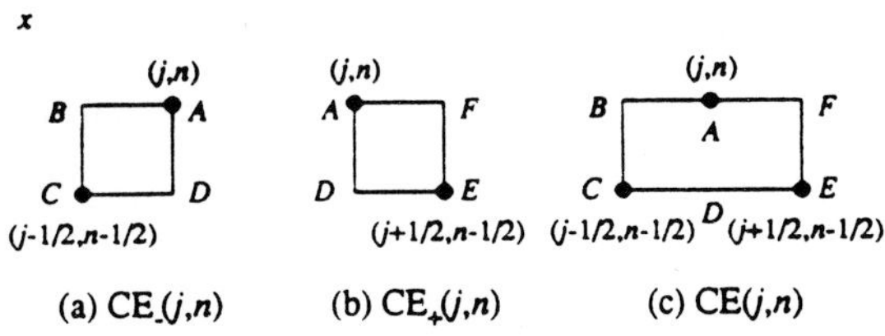

(a) CE_(j,n)   (b) CE_+(j,n)   (c) CE(j,n)

Figure 4.3 — The CEs associated with the mesh point (j,n)

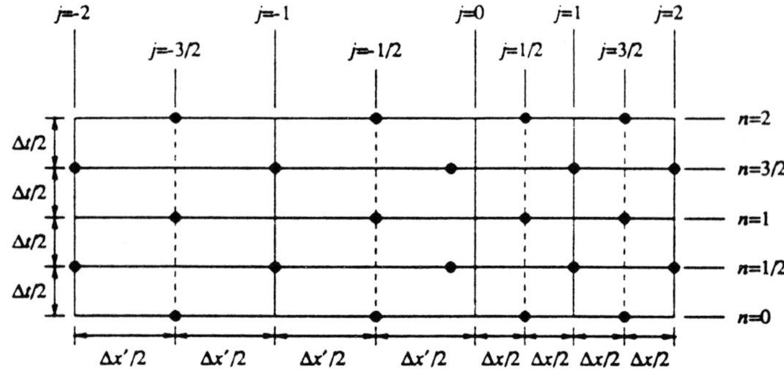

Figure 4.4 — A space-time mesh with nonuniform spatial mesh intervals

Figure 4.5 — The mesh point $(0,n)$ (denoted by a dot) and SE$(0,n)$ for $n=1/2, 3/2, \ldots$

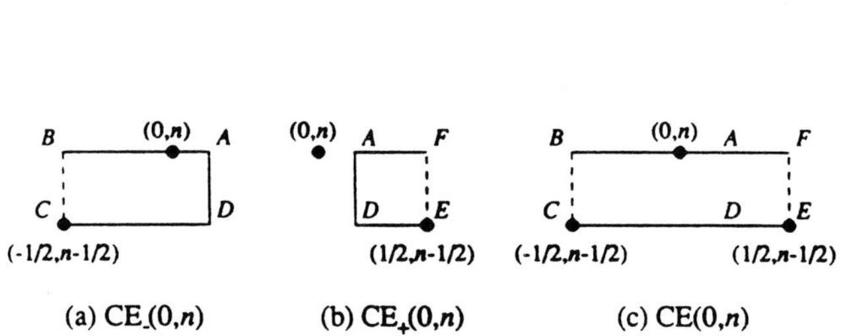

(a) $CE_-(0,n)$　　(b) $CE_+(0,n)$　　(c) $CE(0,n)$

Figure 4.6 — The CEs associated with the mesh point $(0,n)$, $n=1/2, 3/2, \ldots$

Figure 4.7 — The alternative type of SEs and CEs

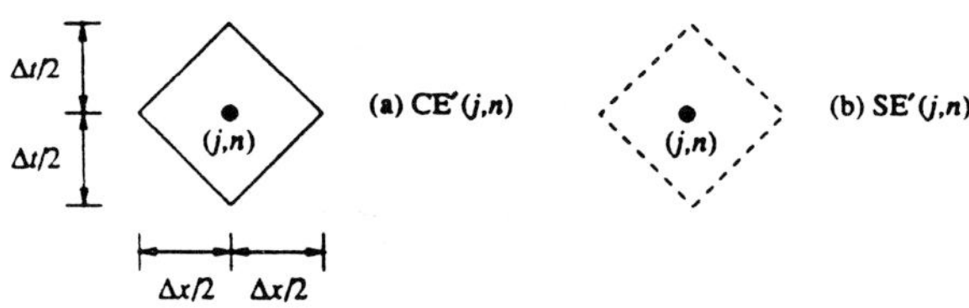

(a) $CE'(j,n)$　　(b) $SE'(j,n)$

Figure 4.8 — $CE'(j,n)$ and $SE'(j,n)$ for an interior mesh point (j,n)

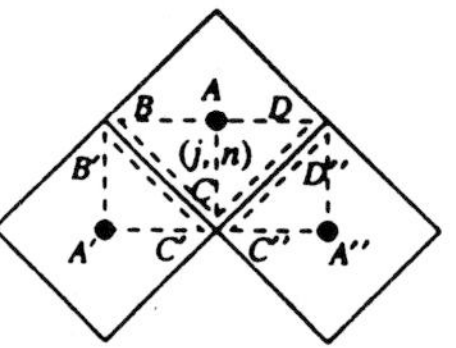

Figure 4.9 — Three neighboring CEs (SEs) of alternative type

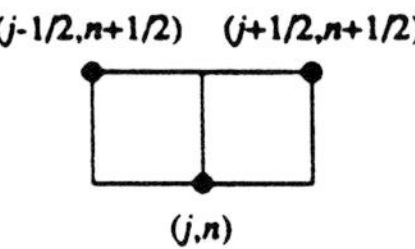

Figure 4.10 — The backward-marching conservation conditions

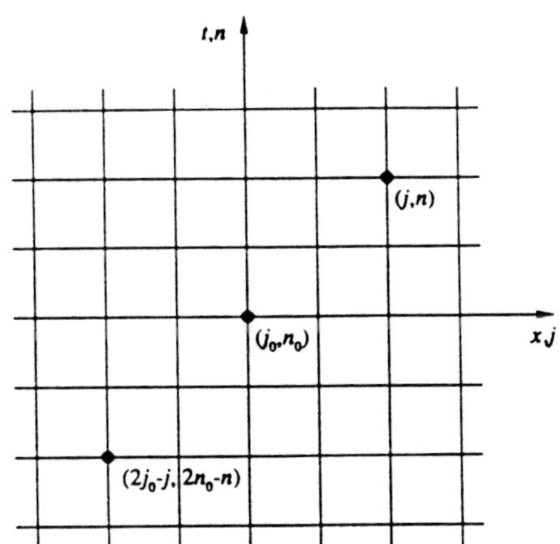

Figure 4.11 — The mesh points (j,n) and $(2j_0-j, 2n_0-n)$ are images of each other under space-time inversion with respect to (j_0,n_0)

Figure 4.12 — The dual space-time mesh

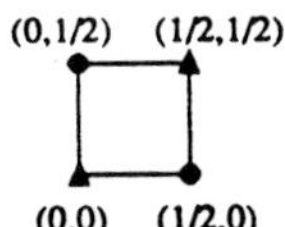

Figure 4.13 — A rectangular space-time region shared by $CE_-(1/2,1/2)$ and $CE_+(0,1/2)$

Figure 5.1 — A spatial domain formed from congruent triangles, showing the spatial projections of the mesh points

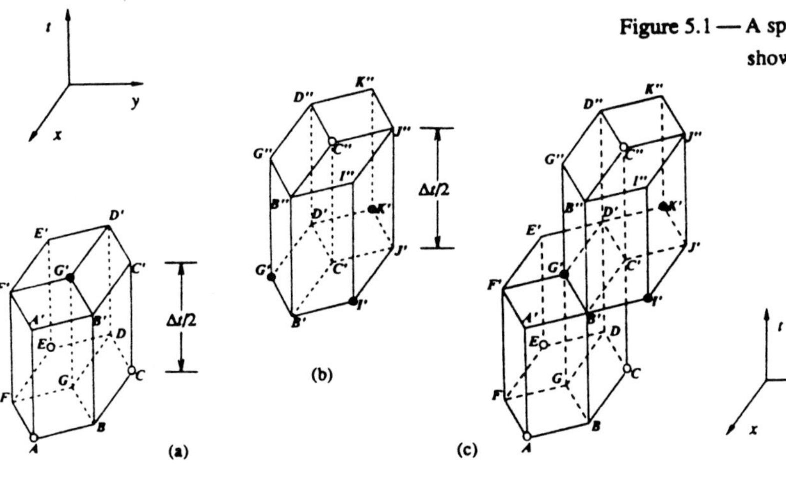

Figure 5.2 — (a) The CEs associated with G' (b) the CEs associated with C'' (c) The relative positions of the CEs of successive time steps

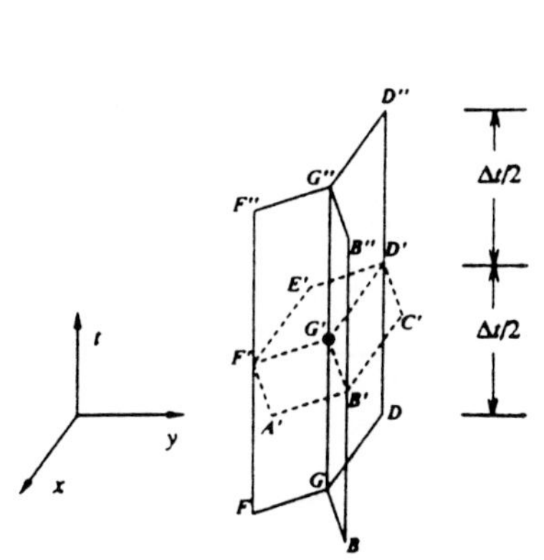

Figure 5.3(a) — The SE associated with the point G'

235

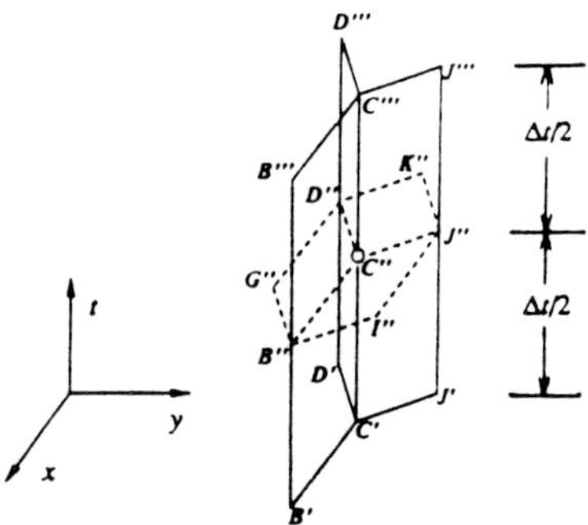

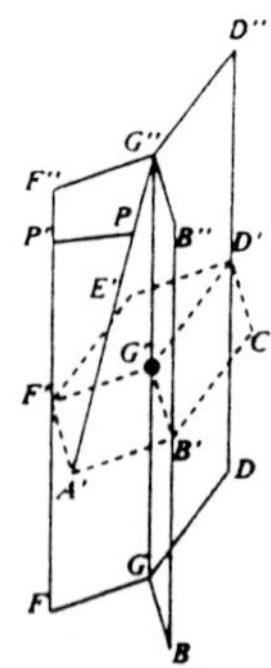

Figure 5.3(b) — The SE associated with the point C''

Figure 5.4 — The SE of alternative type associated with G'

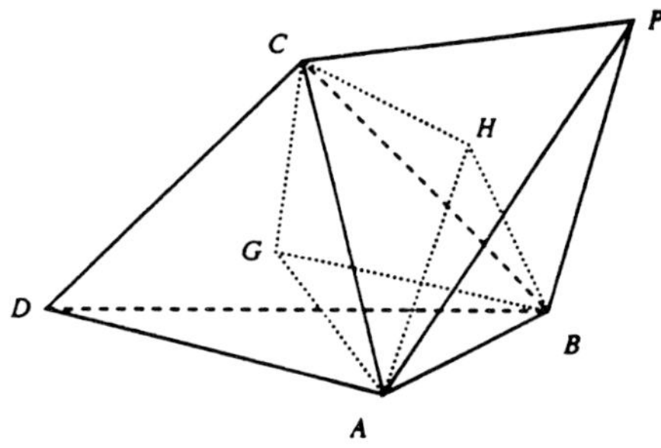

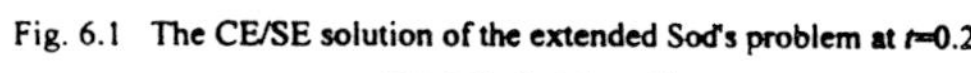

Figure 5.5 — Spatial projection of part of a 3D space-time mesh,
showing the construction of a CE

Fig. 6.1 The CE/SE solution of the extended Sod's problem at t=0.2

Fig. 6.2 The CE/SE solution of the extended Sod's problem at t=0.4

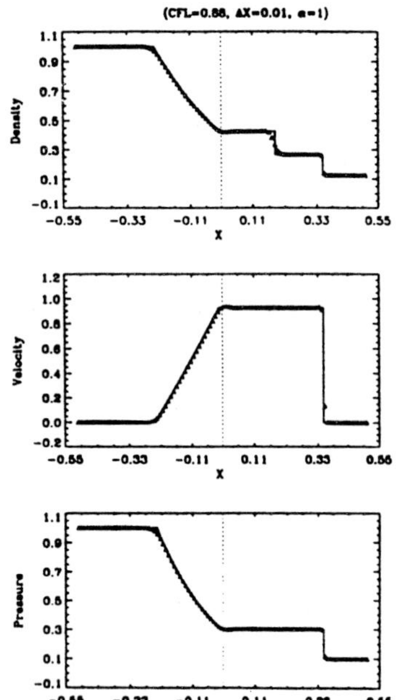

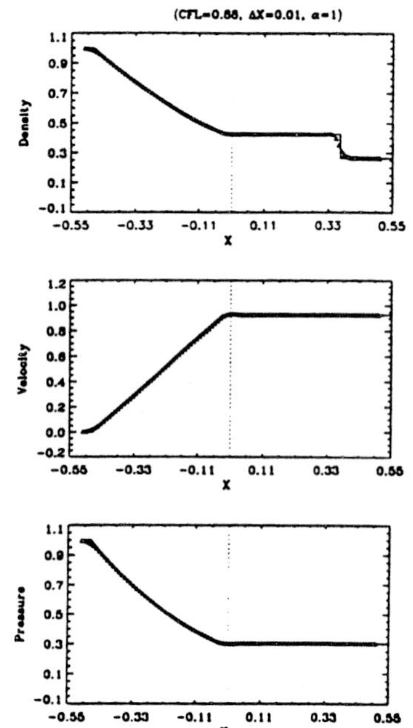

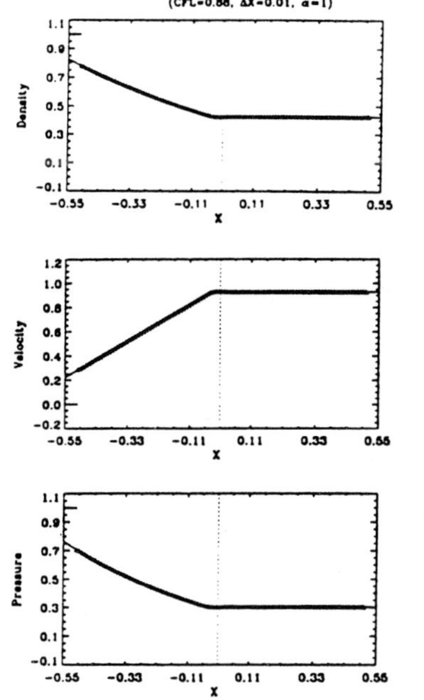

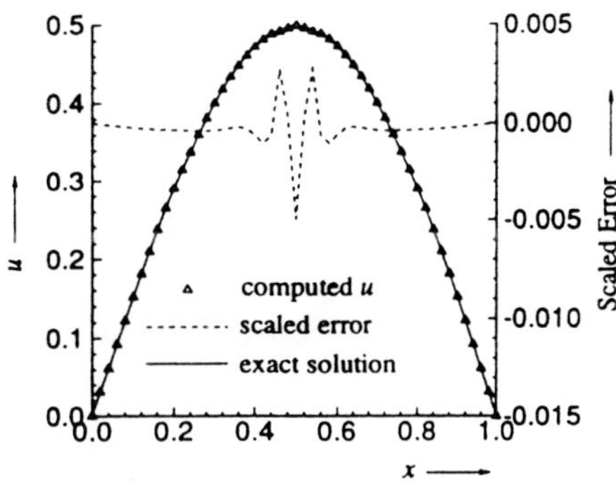

Figure 6.4 — Implicit a-μ Scheme : Pure Diffusion

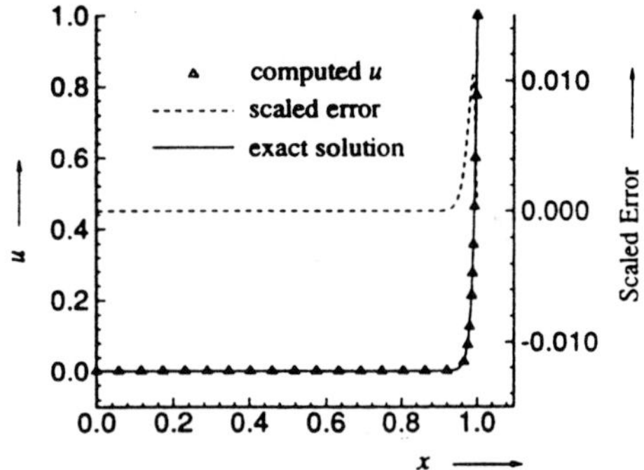

Figure 6.5 — Implicit a-μ Scheme : Boundary Layer, Re = 100

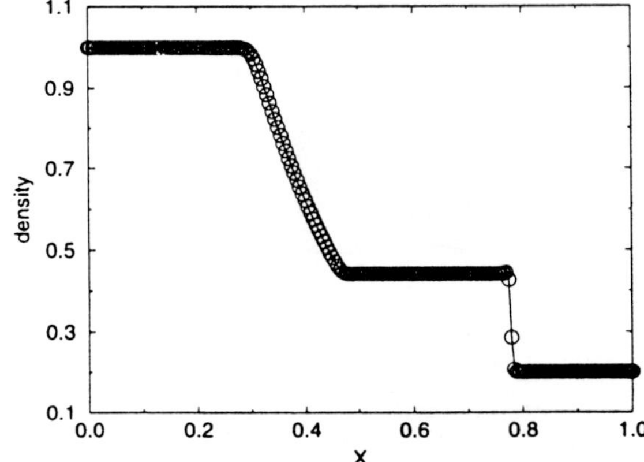

Figure 6.6(a) — Shock in a constant temperature bath with $K = 10^{16}$: density distribution at $t = 0.3$

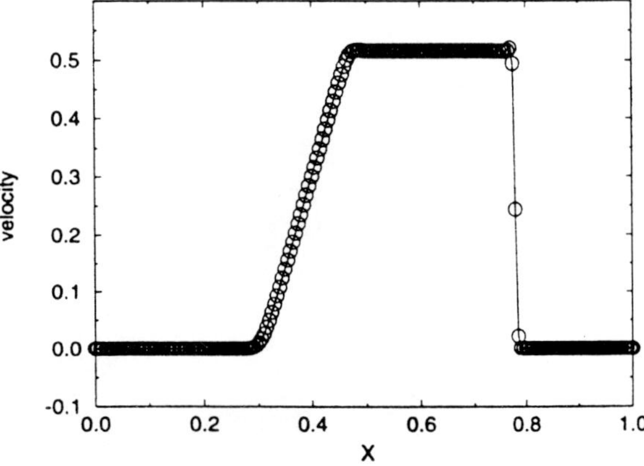

Figure 6.6(b) — Shock in a constant temperature bath with $K = 10^{16}$: velocity distribution at $t = 0.3$

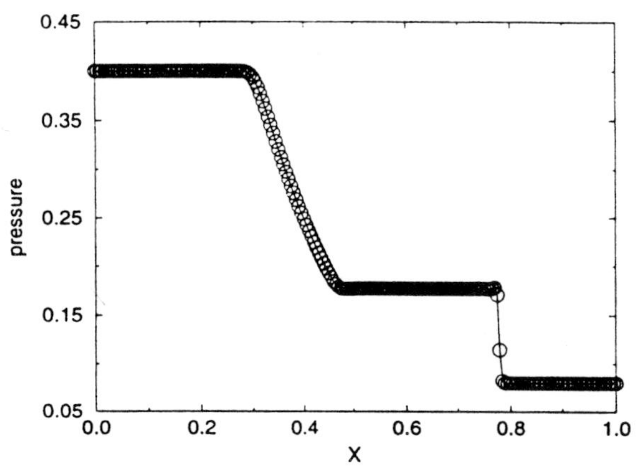

Figure 6.6(c) — Shock in a constant temperature bath with $K = 10^{16}$: pressure distribution at $t = 0.3$

Figure 6.6(d) — Shock in a constant temperature bath with $K = 10^{16}$: temperature distribution at $t = 0.3$

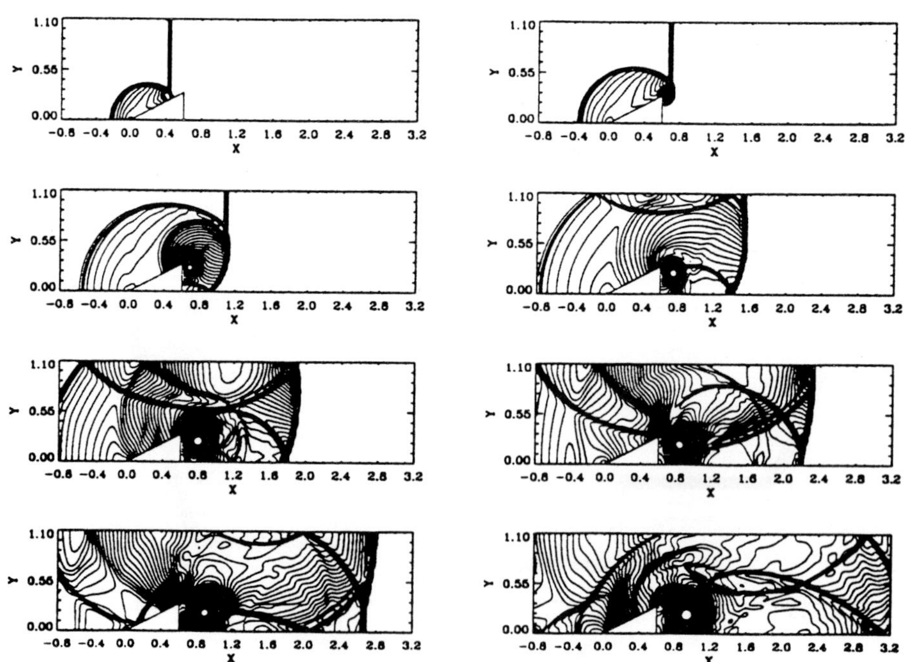

Fig. 6.7 Diffraction of a shock wave around a wedge:
Density contours of the CE/SE solution

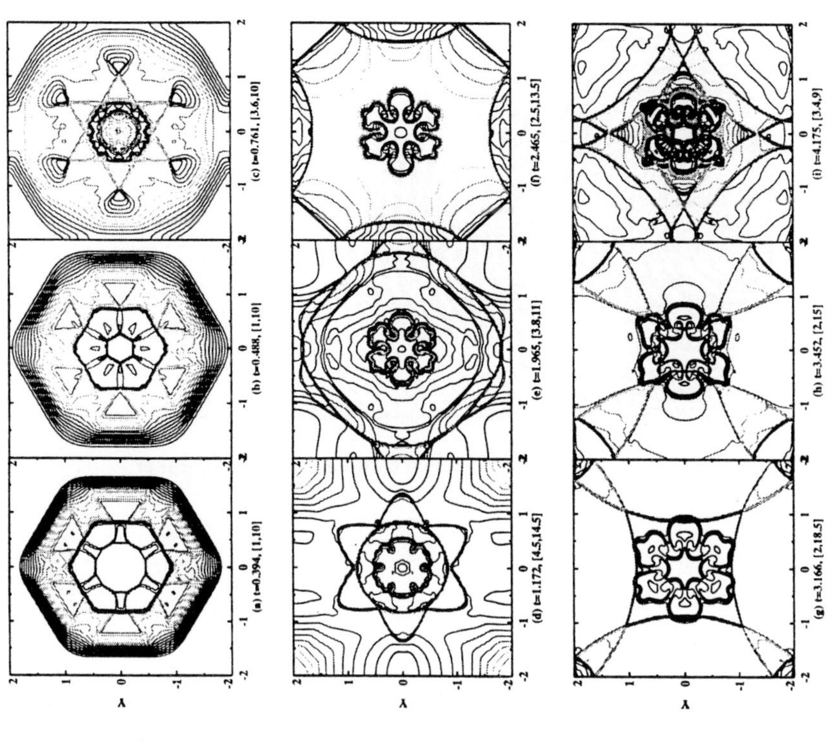

Fig. 6.9 Density contours for implosion/explosion of a hexagonal shock in a square box

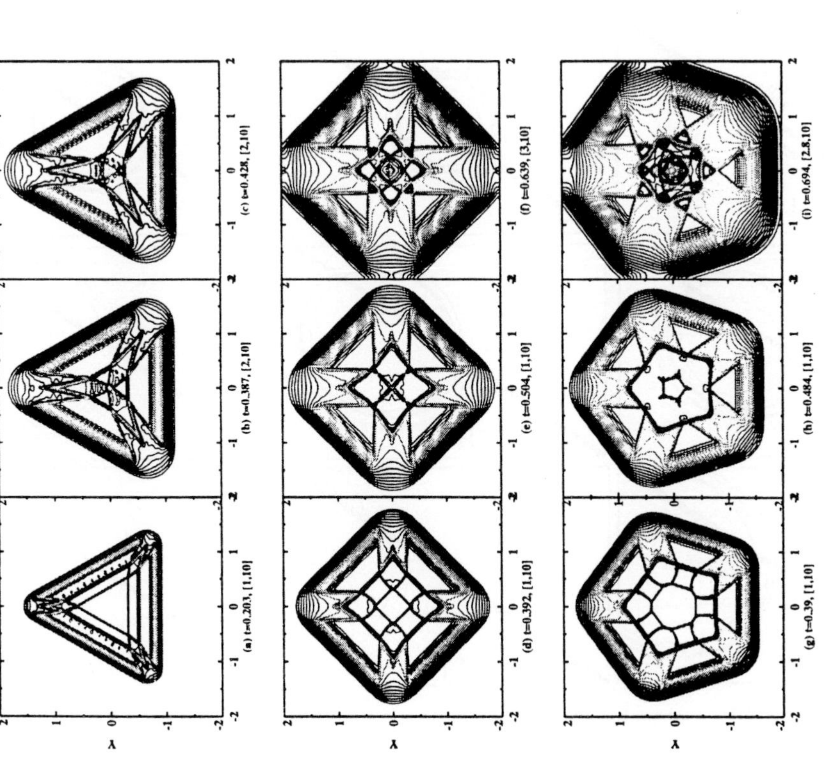

Fig. 6.8 Density contours for implosion/explosion of a polygonal shock in a square box

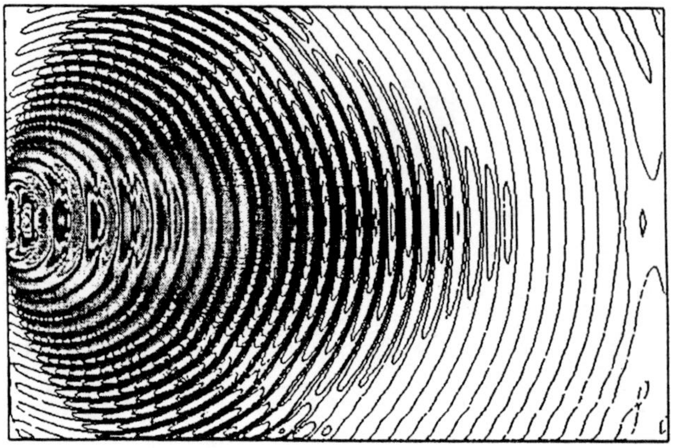

Fig. 6.10 Isobars for acoustic waves generated by a flat-plate loudspeaker

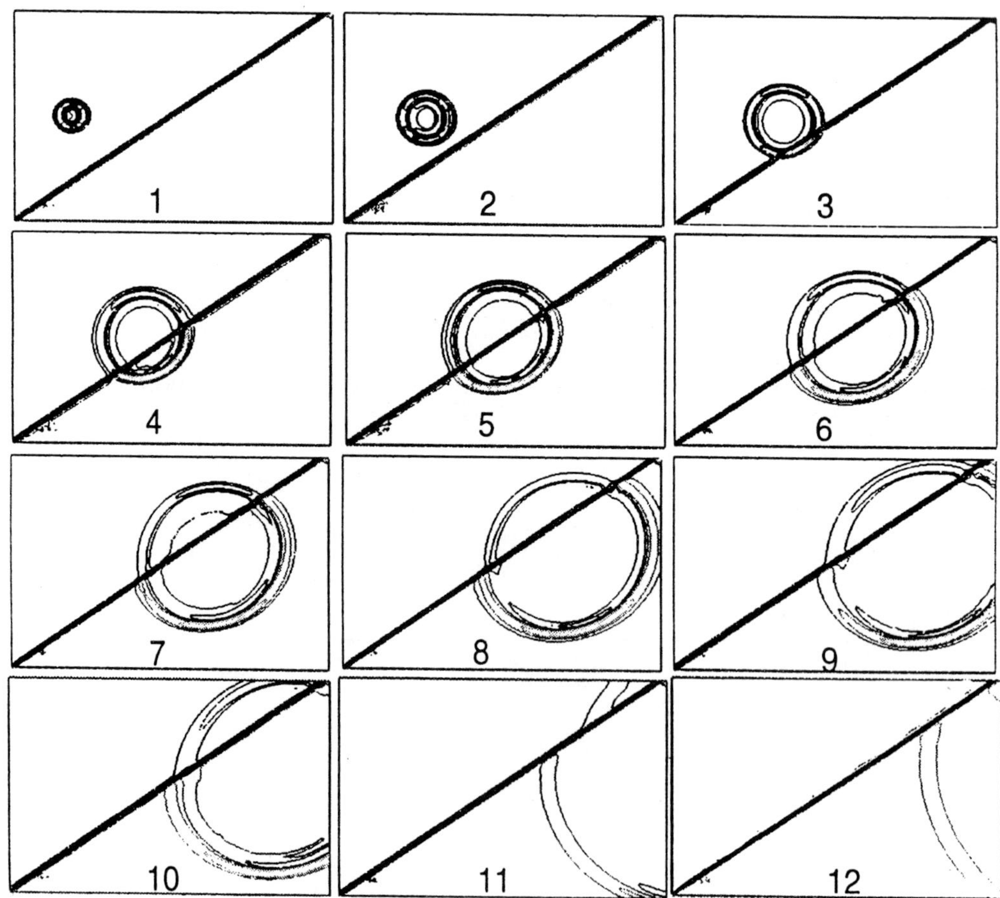

Fig. 6.11 Isobars for interaction of an acoustic pulse with a shock wave

DISCONTINUOUS FINITE ELEMENTS AND FINITE VOLUME VERSIONS OF THE LAX-FRIEDRICHS AND NESSYAHU-TADMOR SCHEMES FOR COMPRESSIBLE FLOWS ON UNSTRUCTURED GRIDS

Paul ARMINJON [1]   Marie-Claude VIALLON [2]

Aziz MADRANE [1]   Lahcen KADDOURI [3]

(1) Dép. de mathématiques et statistique, Universitde Montréal; C.P.6128, Succ. centre-ville, Montréal, Québec, Canada, H3C 3J7

(2) Dép. de mathématiques appliquées, Université du Littoral; F-62228 Calasis Cedex France

(3) Équipe d'analyse numérique, Université de Saint-Étienne, 23 rue Paul Michlon, 42023 Saint-Étienne Cedex 2, Francce

Abstract

We present a new finite volume version ([1], [2], [3]) of the 1-dimensional Lax-Friedrichs and Nessyahu-Tadmor schemes ([5]) for nonlinear hyperbolic equations on unstructured grids, and compare it to a recent discontinuous finite element method ([6], [23]) in the computation of some typical test problems for compressible flows.

The non-oscillatory central difference scheme of Nessyahu and Tadmor, in which the resolution of Riemann problems at the cell interfaces is by-passed thanks to the use of the staggered Lax-Friedrichs scheme, is extended here to a two-step, two-dimensional non-oscillatory centered scheme in finite volume formulation. The construction of the scheme rests on a finite volume extension of the Lax-Friedrichs scheme, in which the finite volume cells are the barycentric cells constructed around the nodes of an FEM triangulation, for even time steps, and some quadrilateral cells associated with this triangulation, for odd time steps. Piecewise linear cell interpolants using least-squares gradients combined with a van Leer-type slope limiting allow for an oscillation-free second-order resolution.

The discontinuous finite element method consists of two steps. We first perform a finite element computation which includes calculation of the fluxes across the edges of the triangular elements using 1-D Riemann solvers with a modification to satisfy the entropy condition. We then proceed to a truly multidimensional slope limitation performed on the physical variables.

Numerical applications to several test problems show the accuracy and stability of the finite volume method.

1. INTRODUCTION

We present a new finite volume method for nonlinear hyperbolic systems on unstructured triangular grids inspired by the Lax-Friedrichs and Nessyahu-Tadmor one-dimensional difference schemes [5] , and compare it to a recent discontinuous finite element method introduced by Jaffré, Kaddouri and Gowda ([6], [23]) in the computation of some typical two-dimensional test problems for compressible flows.

Discontinuous finite elements were first introduced independently, for the neutron transport equation and applications to nuclear engineering, by Reed and Hill [7] and Lesaint and Raviart [8], and then adapted to the equations of hydrodynamics and elasticity ([9], [10], [11], [12]) or even to such applications as the motion of a load on an ice layer ([13], in joint work with P.Jamet), reservoir simulation both without slope limiters ([14], [15]) and with limitation ([16], [17], [18]), and many other applications where they proved to be very successful.

Numerical analysis of discontinuous finite element methods can be found e.g. in [8], with improvements in [19] for scalar hyperbolic equations, and in [20].

In the finite element method used in this paper, the solution is approximated by discontinuous piecewise linear polynomials; numerical fluxes are calculated, on the triangle edges, at appropriate integration points through the use of one-dimensional Riemann solvers. To prevent spurious oscillations, we use a multidimensional version of van Leer's limiter borrowed from ([6], [23]).

Our finite volume method ([1], [2], [3], [4]) is a two-dimensional finite volume scheme inspired by an elegant difference scheme proposed, for one-dimensional problems, by Ness yahu and Tadmor [5], which is itself a high resolution non-oscillatory Godunov-type method for hyperbolic systems of conservation laws constructed on the principle of the staggered Lax-Friedrichs scheme.

It is decomposed in two time steps for second order accuracy in time, performed on alternate, staggered grids, thus allowing a complete by-pass of the (usually expensive) detailed exact or approximate solution of the local Riemann problems generated at the cell interfaces, thanks to the use of staggered form of the Lax-Friedrichs scheme.

In [4], we have presented an extension of the Nessyahu-Tadmor difference scheme to the simpler case of **rectangular grids**; several applications (linear advection, Burgers' equation, diffraction of a planar shock wave around a 90° corner for the Euler equations [21] , Mach 3 wind tunnel with forward facing step [22]) showed the feasibility of the method, which led to good results on regular grids without any mesh adaptation.

In ([1], [2], [3]), we had constructed a finite volume scheme for unstructured triangular grids inspired by the Lax-Friedrichs and Nessyahu-Tadmor one-dimensional difference schemes; some numerical experiments (Supersonic flow around a NACA 0012 airfoil, supersonic flow around a double ellipse) showed the high accuracy of the method, but did not provide a systematic comparison with an already well established method.

In this paper we attempt to present such a comparison, and introduce for that purpose in section 3 the discontinuous finite element, second order accurate method recently proposed, and successfully tested, by Jaffré, Kaddouri and Gowda ([6], [23], [25]).

Notice that in this paper, we used Roe's Riemann solver with an entropy correction recently proposed by Dubois and Mehlman, while the solvers used in references ([6], [23], [25]) are the Osher and Osher-Solomon solvers, respectively.

In section 2 we describe the mathematical modelling of the problem; in section 3 we give a detailed presentation of the discontinuous finite element method of Jaffré, Kaddouri and Gowda.

In section 4, we present the first two authors' construction [1], for unstructured triangular grids, of a finite volume version of the staggered Lax-Friedrichs scheme, and then describe their second order accurate non-oscillatory finite volume method [1] inspired by the Nessyahu-Tadmor one-dimensional central difference scheme [5].

In section 5, we present applications to some typical test problems for supersonic flows (supersonic flow past a blunt body;supersonic flow around a double ellipse [35]).

Finally, section 6 is an appendix describing the solution of the projection problems generated by the slope limitation process.

2. MATHEMATICAL MODELLING

2.1. Governing equations.

We consider the two-dimensional Euler equations for compressible flows, written in conservation form as:

$$\frac{\partial}{\partial t}U(x,y;t) + \frac{\partial}{\partial x}F(U(x,y;t)) + \frac{\partial}{\partial y}G(U(x,y;t)) = 0$$

$$\text{for } (x,y;t) \in \Omega \times \mathbb{R}_+ \quad (2.1)$$

where Ω is a closed bounded domain of the plane,

$$U = \begin{pmatrix} \rho \\ \rho u \\ \rho v \\ \rho E \end{pmatrix} \quad F(U) = \begin{pmatrix} \rho u \\ \rho u^2 + p \\ \rho uv \\ (\rho E + p)u \end{pmatrix},$$

$$(2.2)$$

$$G(U) = \begin{pmatrix} \rho v \\ \rho uv \\ \rho v^2 + p \\ (\rho E + p)v \end{pmatrix}$$

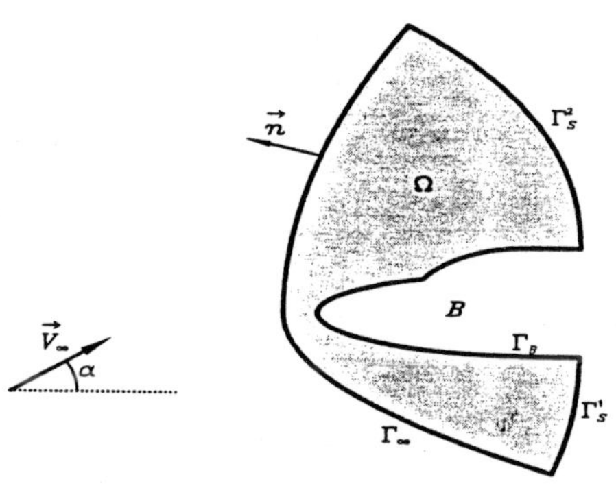

FIGURE 1. Boundary of the computational domain

ρ is the density, $\vec{V} = (u,v)$ is the velocity vector, E is the total energy by unit mass, and p is the pressure. We assume that the fluid satisfies the perfect gas law :

$$p = (\gamma - 1)(\rho E - \frac{1}{2}\rho(u^2 + v^2)) \qquad (2.3)$$

where γ, the ratio of specific heats, is taken equal to 1.4 for air.

2.2. Boundary conditions :

In the sequel, we consider domains of computation related to external flows around bodies; in fig.1 the body is represented by a double-ellipse [40] which limits the domain of computation by its wall Γ_B . In order to deal with a bounded computational domain, a second (artificial) farfield boundary $\Gamma_\infty \cup \Gamma_S$ is introduced, with $\Gamma_S = \Gamma_S^1 \cup \Gamma_S^2$.

The flow is assumed to be uniform at infinity, and we prescribe

$$\rho_\infty = 1, \quad V_\infty = \begin{pmatrix} cos\alpha \\ sin\alpha \end{pmatrix}, \quad p_\infty = \frac{1}{\gamma M_\infty^2} \qquad (2.4)$$

where α is the angle of attack and M_∞ denotes the free-stream Mach number.

On the wall Γ_B we use the usual "no normal velocity" condition: $\vec{V}.\vec{n} = 0$, where $\vec{n} \in R^2$ is the outer normal vector to Γ_B.

Finally, for unsteady calculations, an initial flow is prescribed :

$$U(x,y;0) = U_0(x,y) \quad \forall (x,y) \in \Omega. \qquad (2.5)$$

3. A TWO-DIMENSIONAL DISCONTINUOUS FINITE ELEMENT METHOD

In this section, we give a detailed description of the discontinuous finite element method proposed by Jaffré, Kaddouri and Gowda ([6], [23], [25]).

We consider here a (P_1)-piecewise linear spatial discretization based on unstructured triangular elements (the number of elements sharing a common node is not constant); the computational domain Ω is subdivided into triangles by a triangulation $\mathcal{T}_h$:

$$\Omega = \bigcup_{i=1}^{N_e} K_i \; ; \; K_i \in \mathcal{T}_h,$$

where the K_i's are the elements of the triangulation, N_e is the total number of triangles, and h is the largest diameter of all elements.

Let $U_{j,i} = U(A_{j,i};t)$ denote the value of the dependent variable vector U, at time t and at the jth vertex $A_{j,i}$ of element $K_i \in \mathcal{T}_h$ $(i = 1,\ldots,N_e; j = 1,2,3)$.

The degrees of freedom are the components of the vector U at the vertices of all elements of the triangulation (fig. 2). Let W denote the approximation space formed by the

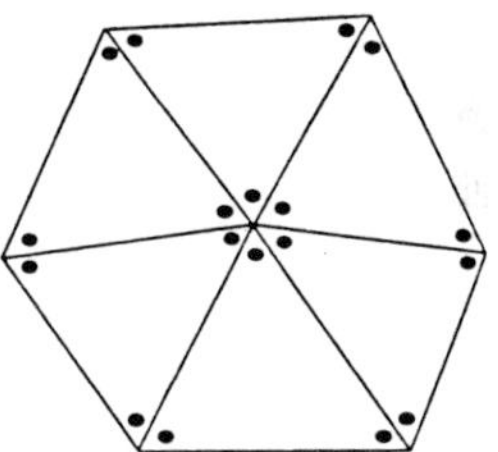

FIGURE 2. Degrees of freedom of the triangulation

piecewise continuous functions which are linear on each triangle $K_i \in \mathcal{T}_h$ (P_1-approximation).

The vector U of conservative variables ρ, ρu, ρv, ρE is approximated by functions in the product space W^4 and the corresponding approximations will again be denoted by ρ, ρu, ρv, ρE, elements of W, for simplicity.

For each element $K_i(i = 1,\ldots,N_e)$ and each node $A_{j,i} \in K_i$ there exists a unique basis or shape function $(\varphi_{j,i})$ with the property

$$\forall\, i = 1,\ldots,N_e\,, \quad \varphi_{j,i}(A_{k,i}) = \begin{cases} 1 & \text{if } j = k \\ 0 & \text{otherwise} \end{cases}$$

The functions $(\varphi_{j,i})_{i=1,\ldots,N_e}^{j=1,\ldots,3}$ form a basis of the approximation space W.

In the present formulation we shall use an explicit Euler time discretization.

We now consider a Galerkin discontinuous finite element approximation, which proceeds from a variational formulation of the Euler equations.

Multiplying (2.1) by a shape function $\varphi_{j,i}$ and integrating by parts the terms with spatial derivatives, we obtain the following system for the piecewise linear vector U^{n+1}

243

of approximate dependent variables to be computed at time t^{n+1}:

$$\begin{cases} \text{Find } U^{n+1} \in W^4 \\ \displaystyle\int_{K_i} \frac{U^{n+1} - U^n}{\Delta t^n} \varphi_{j,i} \, dS = \\ \displaystyle\int_{K_i} \left(F(U^n) \frac{\partial \varphi_{j,i}}{\partial x} + G(U^n) \frac{\partial \varphi_{j,i}}{\partial y} \right) dS \\ \displaystyle - \int_{\partial K_i} \left(\vec{\mathbb{F}}(U^n) \cdot \boldsymbol{n} \right) \varphi_{j,i} \, d\sigma \end{cases} \tag{3.1}$$

where $U^n = U(x,y;t^n) \in W^4$, $\Delta t^n = t^{n+1} - t^n$ is the time step and $\vec{n} = \begin{pmatrix} n_x \\ n_y \end{pmatrix}$ is the unit outer normal (directed towards the exterior of K_i).

In (3.1), $\vec{\mathbb{F}}(U^n) \cdot \vec{n} = F(U^n) \cdot n_x + G(U^n) \cdot n_y \equiv F_n$ is the (outward) numerical flux across an edge $\mathcal{A}$ of ∂K_i; it will be computed with the help of an approximate Riemann solver for the Riemann problem generated, in the direction normal to the edge $\mathcal{A}$, by the limits of the values of the dependent variables on both sides of $\mathcal{A}$ as one tends to $\mathcal{A}$ along $\vec{n}$.

In this paper, we have used Roe's Riemann solver ([28], [24], [34]) with an entropic correction due to Dubois and Mehlman [24].

Remark 3.1. *Replacing the above approximation space by the space of piecewise constant functions (constant on each triangle $K_i \in \mathcal{T}_h$) reduces (3.1) to the equation governing the standard finite volume scheme*

$$\int_{K_i} \frac{U_i^{n+1} - U_i^n}{\Delta t^n} \, dS + \int_{\partial K_i} \left(\vec{\mathbb{F}}(U^n) \cdot \boldsymbol{n} \right) d\sigma = 0, \quad K_i \in \mathcal{T}_h.$$

3.1. Numerical integration.

To complete the description of the spatial discretization, we must specify the quadrature formulas which will be used to compute the integrals appearing in (3.1). Numerical experiments for scalar equation ([25], [18], [6]) have suggested the following quadrature techniques.

In equation (3.1) the terms containing the spatial derivatives are computed with the help of the values at the centroid M of mesh K_i :

$$\int_{K_i} \left(F(U^n) \frac{\partial \varphi_{j,i}}{\partial x} + G(U^n) \frac{\partial \varphi_{j,i}}{\partial y} \right) dS \simeq$$

$$\text{Area}(K_i) \left(F(\overline{U}_i^n) \frac{\partial \varphi_{j,i}}{\partial x}(M_i) + G(\overline{U}_i^n) \frac{\partial \varphi_{j,i}}{\partial y}(M_i) \right)$$

where M_i is the centroid of K_i and $\overline{U}_i^n$ is the average value of U^n on K_i :

$$\overline{U}_i^n = \frac{1}{3} \sum_{j=1}^{3} U(A_{j,i})$$

since U^n is linear on K_i.

For the integral associated with the outward flux

$$\int_{\partial K_i} \left(\vec{\mathbb{F}}(U^n) \cdot \boldsymbol{n} \right) \varphi_{j,i} \, d\sigma = \sum_{\mathcal{A} \in \partial K_i} \int_{\mathcal{A}} \left(\vec{\mathbb{F}}(U^n) \cdot \boldsymbol{n} \right) \varphi_{j,i} \, d\sigma$$

we use either the values at the midpoints of the edges $\mathcal{A}$, or the values at both Gauss points of each edge

1st choice :

$$\int_{\mathcal{A}} \left(\vec{\mathbb{F}}(U^n) \cdot \boldsymbol{n} \right) \varphi_{j,i} \, d\sigma \simeq \mathrm{l}(\mathcal{A}) \cdot \mathcal{F}_{|\mathcal{A}}^n(U_l, U_r) \cdot \varphi_{j,i}(M)$$

where $\mathrm{l}(\mathcal{A})$ and M denote the length and midpoint of edge $\mathcal{A}$, respectively, and $\mathcal{F}_{|\mathcal{A}}^n(U_l, U_r)$ is the <u>numerical flux</u> across edge $\mathcal{A}$ which separates the states U_l and U_r, obtained by taking the limits of U along the normal to $\mathcal{A}$ at M; this numerical flux will be computed with Roe's Riemann solver, as described below.

2nd choice :

$$\int_{\mathcal{A}} \left(\vec{\mathbb{F}}(U^n) \cdot \boldsymbol{n} \right) \varphi_{\mathcal{A}j,i} \, d\sigma \simeq$$

$$\frac{\mathrm{l}(\mathcal{A})}{2} \left(\mathcal{F}_{|\mathcal{A}}^n(U_l, U_r)(G_1)\varphi_{j,i}(G_1) + \right.$$

$$\left. \mathcal{F}_{|\mathcal{A}}^n(U_l, U_r)(G_2)\varphi_{j,i}(G_2) \right)$$

where G_1, G_2 are the Gauss points of edge $\mathcal{A}$.

Numerical experiments have shown that the computation at the edge midpoint (1st choice) is sufficient.

For the first integral in (3.1) (time derivative) we use the values at the three vertices:

$$\int_{K_i} \frac{U^{n+1} - U^n}{\Delta t^n} \varphi_{j,i} \, dS \simeq \frac{\text{Area}(K_i)}{3} \frac{U_{K_i,A_j}^{n+1} - U_{K_i,A_j}^n}{\Delta t^n}$$

by the properties of $\varphi_{j,i}$.

3.2. Multidimensional slope limitation.

We will describe a multidimensional extension of a slope limitation procedure which has been successfully used for scalar equations ([17], [6], [23]).

When dealing with the Euler equations, it has been widely recognized that one should limit the physical variables ρ, u, v, p rather than the conservative variables ρ, ρu, ρv, ρE.

Let $U^n \in W^4$ denote the solution previously computed at time t^n, and $U^\star \in W^4$ the solution predicted at t^{n+1} by solving system (3.1).

We want to modify $U^\star$ and obtain a corrected vector of conservative variables U^{n+1}, by the following procedure.

For each triangle $K_i \in \mathcal{T}_h$, let

- $w_{K,A_i} = w_{|K}(A_i):\quad i = 1,\ldots,3\quad$ be the value of $w_{|K}$ at node i,

- $\overline{w}_K = \dfrac{1}{3}\sum_{i=1}^{3} w_{K,A_i}$, the mean value of $w_{|K}$ in element K,

- $T(A)$ be the set of element $K \in \mathcal{T}_h$ such that vertex $A \in K$ (Fig. 3).

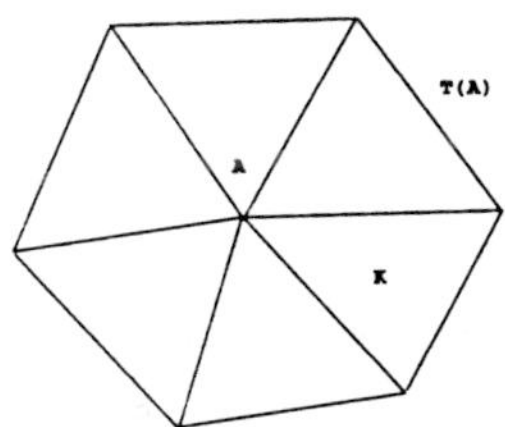

FIGURE 3. The set $T(A)$

For each element K, we compute the mean values of the conservative variables, noted $\overline{\rho}_K^{\star}, \overline{(\rho u)}_K^{\star}, \overline{(\rho v)}_K^{\star}, \overline{(\rho E)}_K^{\star}$, which are simply the arithmetic means of these variables at the three vertices of K.

In order to obtain a conservative scheme, the vectors U^{n+1} and $U^{\star}$ must have the same mean value on each element.

We then compute the mean values of the physical variables, $\overline{\rho}_K^{\star}, \overline{u}_K^{\star}, \overline{v}_K^{\star}, \overline{p}_K^{\star}$ (the mean value $\overline{\rho}_K^{\star}$ of the density has already been calculated).

For pressure we also take the arithmetic mean of p at the vertices of K. In contrast, the mean value of the velocity components u, v are defined by

$$\overline{u}_K^{\star} = \frac{\overline{(\rho u)}_K^{\star}}{\overline{\rho}_K^{\star}} \ , \quad \overline{v}_K^{\star} = \frac{\overline{(\rho v)}_K^{\star}}{\overline{\rho}_K^{\star}} \ . \tag{3.2}$$

For the components of momentum at time t^{n+1}, we will use the following value

$$\overline{(\rho u)}_K^{n+1} = \frac{1}{3}\sum_{i=1}^{3}\rho_{K,A_i}^{n+1}u_{K,A_i}^{n+1} \ ,$$

$$\overline{(\rho v)}_K^{n+1} = \frac{1}{3}\sum_{i=1}^{3}\rho_{K,A_i}^{n+1}v_{K,A_i}^{n+1} \ . \tag{3.3}$$

Observe that these values are different from $\overline{u}_K^{\star}$, $\overline{v}_K^{\star}$, as they use the nodal values of density and velocity instead of those of the momentum.

Formulas (3.2)-(3.3) have been chosen to ensure existence and uniqueness of the solution of the minimisation problems to be defined below.

For every node A of the grid we compute the minimum and maximum of the mean values of the physical variables in the elements sharing node A :

$$w_{min}(A) = \min_{K \in T(A)} \overline{w}_K^{\star}, \quad w_{max}(A) = \max_{K \in T(A)} \overline{w}_K^{\star},$$

$$\text{for}\quad w = \rho, u, v, p \tag{3.4}$$

The slopes of the physical variables ρ, u, v, p will be limited, in this order, in the following way.

Let V denote the vector $(\rho, u, v, p)^T$ of physical variables.

In each element K with vertices A_i $(i = 1,\ldots,3)$, $V_{|K}^{n+1}$ is defined by :

$$\begin{cases} (i) \quad \overline{w}_K^{n+1} = \overline{w}_K^{\star}, \quad \text{for}\quad w = \rho, \ p, \ \text{and} \\[2ex] \overline{(\rho w)}_K^{n+1} = \overline{(\rho w)}_K^{\star}, \text{for}\quad w = u, \ v, \\[2ex] (ii) \quad \text{For } i = 1,\ldots,3, \ w = \rho, \ u, \ v, \ p: \\[2ex] A_1 \leq w_{K,A_i}^{n+1} \leq A_2 \ , \ 0 \leq \alpha \leq 1, \\[2ex] \text{where } A_1 = (1 - \alpha)\overline{w}_K^{\star} + \alpha w_{min}(A) \\[2ex] \text{and } A_2 = (1 - \alpha)\overline{w}_K^{\star} + \alpha w_{max}(A) \\[2ex] (iii) \quad \text{For } w = \rho, \ u, \ v, \ p \text{ the distance in } \mathbb{R}^3 \\[2ex] \text{between } w^{n+1} = (w_{K,A_i}^{n+1})_{i=1,\ldots,3} \\[2ex] \text{and } w^{\star} = (w_{K,A_i}^{\star})_{i=1,\ldots,3} \text{ is minimum .} \end{cases}$$

The computation of $V_{|K}^{n+1}$ from $V_{|K}^{\star}$ thus amounts to four projection problems in $\mathbb{R}^3$

(one for each physical variable); as (i) defines a plane, and (ii) a cube, we look for the projection, on their intersection, of the corresponding variable

$$w^{\star} = (w_{K,A_i}^{\star})_{i=1,\ldots,3}.$$

Condition (i) allows for mass conservation, (ii) limits the variation of ρ, u, v, p (in that order), and (iii) guarantees uniqueness of the solution.

After computing the vector of physical variables $V^{n+1} = \begin{pmatrix} \rho^{n+1} \\ u^{n+1} \\ v^{n+1} \\ p^{n+1} \end{pmatrix}$ we return to the conservative variables according to

$$U^{n+1} = \begin{pmatrix} \rho^{n+1} \\ \rho^{n+1}u^{n+1} \\ \rho^{n+1}v^{n+1} \\ \dfrac{p^{n+1}}{(\gamma - 1)} + \dfrac{1}{2}\rho^{n+1}\left[(u^{n+1})^2 + (v^{n+1})^2\right] \end{pmatrix}.$$

The slope limitation therefore requires the solution of a series of local minimization problems in 3 dimensional space, with the constraints (i) and (ii).

These projection problems can easily be solved by duality, as will be shown in the **appendix.**

In order to ensure the existence of a solution for the projection problems, we have to make sure that the intersection of the corresponding plane and cube is not empty. For density and pressure, it is easily seen that if we let $\rho_{K,A_i}^{n+1} = \overline{\rho}_K^{\star}$ and $p_{K,A_i}^{n+1} = \overline{p}_K^{\star}$ for $i = 1, 2, 3$, conditions (i) and (ii) are then satisfied, so that the relevant intersection is not empty.

For the velocity components, we can easily check, applying (3.2) and (3.3), that if we let $u_{K,A_i}^{n+1} = \overline{u}_K^{\star}$ ($i = 1, 2, 3$), then

$$\overline{(\rho u)}_K^{n+1} = \frac{1}{3} \sum_{i=1}^{3} \rho_{K,A_i}^{n+1} \overline{u}_K^{\star} = \frac{1}{3} \left(\sum_{i=1}^{3} \rho_{K,A_i}^{n+1} \right) \frac{\overline{(\rho u)}_K^{\star}}{\overline{\rho}_K^{\star}}$$

$$= \overline{(\rho u)}_K^{\star} \, .$$

with a similar result for the second velocity component.

The parameter α controls the extent of the slope limitation process. For $\alpha = 0$, we get the most stringent limitation: the solution V^{n+1} (and therefore U^{n+1}) is piecewise constant, thus reducing the method to the usual (spatially) first order accurate scheme.

In our numerical experiments, we have usually chosen $\alpha = 0.5$, a value which led to optimal results in the scalar case (cf. [25], [23], [6]).

For one-dimensional problems, the limitation procedure reduces to the usual van Leer solpe limitation ([26], [41]).

3.3. The numerical scheme.

The above description of the Jaffré-Kaddouri-Gowda discontinuous finite element method can be summarized within the frame of a two-step scheme.

Assuming for simplicity that we use an explicit Euler time discretization, let $U^n \in W^4$ be the solution obtained at time $t = t^n$. In the first step (predictor), we compute an approximation $U^{\star} \in W^4$ of the solution at time t^{n+1}. This predictor step consists in a finite element calculation, but features the use of Riemann solvers.

In the second step, which can be viewed as a correction step, we limit the vector $U^{\star}$ to obtain an approximate solution U^{n+1}.

1 - Predictor step : Finite element calculation

$$\begin{cases} \text{Compute } U^{\star} \in W^4 \text{ such that} \\ \displaystyle\int_{K_i} \frac{U^{\star} - U^n}{\Delta t^n} \varphi_{j,i} \, dS = \\ \displaystyle\int_{K_i} \left(F(U^n) \frac{\partial \varphi_{j,i}}{\partial x} + G(U^n) \frac{\partial \varphi_{j,i}}{\partial y} \right) dS \\ \displaystyle - \int_{\partial K_i} \left(\vec{\mathbb{F}}(U^n) \cdot n \right) \varphi_{j,i} \, d\sigma \\ \text{for each } K_i \in \mathcal{T}_h \text{ and } \varphi_{j,i} \ (j = 1, 2, 3) \end{cases} \quad (3.5)$$

2 - Limitation step

This step limits the variation range of components of the

vector of physical variables $V^{\star} = \begin{pmatrix} \rho^{\star} \\ u^{\star} \\ v^{\star} \\ p^{\star} \end{pmatrix}$ obtained in the predictor step. It leads to V^{n+1} and then U^{n+1}, the final approximation of the vector of conservative variables at time t^{n+1}.

We observe that the two steps are independent from each other, and the limitation process is distinct from the flux calculation, contributing to the originality of the method.

3.4. Riemann solver.

In this subsection we present a short description of Roe's approximate Riemann solver, which will be used here in conjunction with an entropic correction recently proposed by Dubois and Mehlman ([24]) described below.

Let $\mathcal{A}$ be an edge in the grid, and $\vec{n}_{\mathcal{A}}$ a unit vector normal to $\mathcal{A}$; let $F_n(U)$ be the flux in the direction of $\vec{n}_{\mathcal{A}}$, with $U \in W^4$. Let U_l and U_r be the limiting values of U obtained when approaching the edge $\mathcal{A}$ along $\vec{n}_{\mathcal{A}}$ from the upwind and downwind side, respectively, with respect to the direction of $\vec{n}_{\mathcal{A}}$; for the description of Roe's numerical flux, we will provisionally assume that U_l and U_r are two constant states along each side of $\mathcal{A}$ (while we will later use limits of piecewise linear functions, see below).

Let $\widetilde{F}_{\mathcal{A}}(U_l, U_r)$ denote the numerical flux of a specific Riemann solver. It satisfies the <u>consistency</u> relation $\widetilde{F}_{\mathcal{A}}(U, U) = F_n(U)$, for all $U \in W^4$, where

$$F_n(U) = n_1 F(U) + n_2 G(U) = \vec{\mathbb{F}}(U).\vec{n} \, . \quad (3.6)$$

is the physical flux across $\mathcal{A}$, and

$$\vec{n} = \vec{n}_{\mathcal{A}} = \begin{pmatrix} n_1 \\ n_2 \end{pmatrix} = \begin{pmatrix} \cos\theta \\ \sin\theta \end{pmatrix} \in \mathbb{R}_{\star}^2 \ (\vec{n} \neq \vec{0})$$

If $\vec{V} = \begin{pmatrix} u \\ v \end{pmatrix}$ denotes the velocity, we can write it in the <u>edge-normal to edge</u> local basis formed by $\vec{n}$ and a unit <u>vector</u> $\vec{a}$ along edge $\mathcal{A}$ (fig.4) as

$$\vec{V}_n = \begin{pmatrix} u_n = u\cos\theta + v\sin\theta \\ v_n = -u\sin\theta + v\cos\theta \end{pmatrix} \, . \quad (3.7)$$

We can then define an invertible linear transformation T in $\mathbb{R}^4$:

$$T = \begin{bmatrix} 1 & 0 & 0 & 0 \\ 0 & \cos\theta & \sin\theta & 0 \\ 0 & -\sin\theta & \cos\theta & 0 \\ 0 & 0 & 0 & 1 \end{bmatrix} \, ,$$

This relation allows for the computation of $F_n(U)$ with the help of the first (vector) component F of the flux $\vec{\mathbb{F}}$ only, without resorting to G.

This leads to a reduction of the computing time, since the calculation of the matrices T and T^{-1} and the flux component F requires fewer operations than the calculation of the global flux F_n from its original definition.

In the sequel, the flux calculations will therefore be performed for $n_1 = 1$, $n_2 = 0$ (corresponding to $\theta = 0$).

3.5. Numerical flux of Roe's original scheme.

In Roe's scheme for the one-dimensional Euler equations $U_t + F(U)_x = 0$, one computes the exact solution of the linearized Riemann problem defined at the cell interface $x_{i+\frac{1}{2}}$ between cells C_i and C_{i+1} with corresponding constant states U_i, U_{i+1}, by replacing the Jacobian matrix $\frac{\partial F(U)}{\partial U} = A(U)$, computed at $x_{i+\frac{1}{2}}$, by a special constant matrix $\bar{A}_{i+\frac{1}{2}} = \bar{A}(U_i, U_{i+1})$ called Roe' average matrix, with the properties

(i) $F(U_{i+1}) - F(U_i) = \bar{A}_{i+\frac{1}{2}}(U_{i+1} - U_i)$

(ii) For $U_i = U_{i+1} = U$, we have
$\bar{A}(U,U) = A(U) \equiv \frac{\partial F(U)}{\partial U}$

(iii) The eigenvalues λ_k of $\bar{A}$ are real and its eigenvectors r_k are linearly independant ($k = 1, \ldots, 3$).

It can be shown that $\bar{A}_{i+\frac{1}{2}} = \bar{A}(U_i, U_{i+1})$ is in fact equal to $A(\bar{U})_{i+\frac{1}{2}} \equiv \frac{\partial F(U)}{\partial U}|_{U=\bar{U}_{i+\frac{1}{2}}}$ (see [27]) where $\bar{U}_{i+\frac{1}{2}} = (\bar{\rho}_{i+\frac{1}{2}}, \bar{u}_{i+\frac{1}{2}}, \bar{H}_{i+\frac{1}{2}})$ is a specific average of the vectors U_i, U_{i+1} called Roe's average, defined by

$$\bar{\rho}_{i+\frac{1}{2}} = \sqrt{\rho_i \rho_{i+1}}, \quad \bar{u}_{i+\frac{1}{2}} = \frac{(u\sqrt{\rho})_i + (u\sqrt{\rho})_{i+1}}{\sqrt{\rho_i} + \sqrt{\rho_{i+1}}},$$

$$\bar{H}_{i+\frac{1}{2}} = \frac{(H\sqrt{\rho})_i + (H\sqrt{\rho})_{i+1}}{\sqrt{\rho_i} + \sqrt{\rho_{i+1}}}$$

In the present context of the two-dimensional Euler equations, and owing to the previous remarks at the end of section 3.4, we will consider two adjacents states U_l, U_r separated by a triangle edge $\mathcal{A}$. Roe's average matrix will be given by (see [27] for a description of this matrix)

$$\bar{A}(U_l, U_r) = \frac{\partial F(U)}{\partial U}|_{U=\bar{U}_{lr}} \qquad (3.13)$$

where Roe's average vector

$$\bar{U}_{lr} = (\bar{\rho}, \bar{u}, \bar{v}, \bar{H}) \qquad (3.14)$$

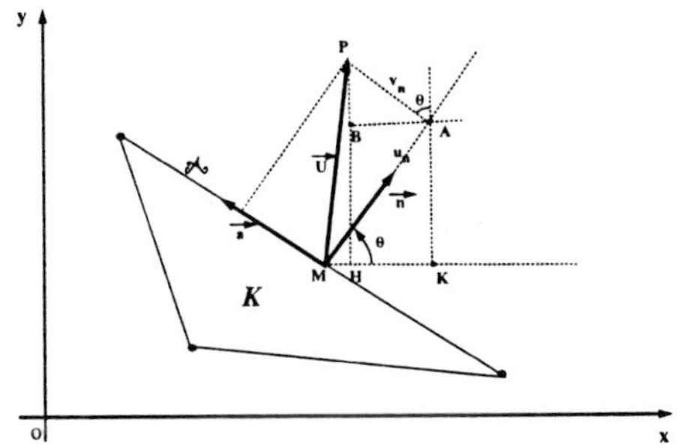

FIGURE 4. Local basis for the calculation of the flux

$$T^{-1} = \begin{bmatrix} 1 & 0 & 0 & 0 \\ 0 & \cos\theta & -\sin\theta & 0 \\ 0 & \sin\theta & \cos\theta & 0 \\ 0 & 0 & 0 & 1 \end{bmatrix}. \qquad (3.8)$$

An arbitrary state vector $U = \begin{pmatrix} \rho \\ \rho\vec{V} \\ \rho E \end{pmatrix}$ is then mapped into $\hat{U}$:

$$U \longrightarrow T \cdot U = \begin{pmatrix} \rho \\ \rho\vec{V_n} \\ \rho E \end{pmatrix} = \hat{U}. \qquad (3.9)$$

The flux in the direction $\vec{n}_\mathcal{A}$

$$F_n(U) = n_1 F(U) + n_2 G(U).$$

can be written in the edge-normal to edge local basis as

$$F_n(U) = \cos\theta \begin{pmatrix} \rho u \\ p + \rho u^2 \\ \rho uv \\ (\rho E + p)u \end{pmatrix} + \sin\theta \begin{pmatrix} \rho v \\ \rho uv \\ p + \rho v^2 \\ (\rho E + p)v \end{pmatrix}$$

$$= \begin{pmatrix} \rho u_n \\ (\rho u)u_n + p\cos\theta \\ (\rho v)u_n + p\sin\theta \\ (\rho H)u_n \end{pmatrix}, \qquad (3.10)$$

where $H = E + \frac{p}{\rho}$ is the specific total enthalpy (per unit mass).

It can be easily verified that if

$$\hat{U} = TU = \left(\rho, \rho\vec{V_n}, \rho E\right)^T,$$

then

$$F_n(U) = T^{-1} \circ F(\hat{U}) \qquad (3.11)$$

where

$$F(\hat{U}) = \left(\rho u_n, p + \rho u_n^2, \rho u_n v_n, (\rho E + p)u_n\right)^T \qquad (3.12)$$

247

is defined by

$$\begin{cases} \bar{\rho} = \sqrt{\rho_l \rho_r} \\[2mm] \text{and for } \ a = u, \ v \text{ or } H \\[2mm] \bar{a} = \frac{a_l \sqrt{\rho_l} + a_r \sqrt{\rho_r}}{\sqrt{\rho_l} + \sqrt{\rho_r}} \end{cases} \qquad (3.15)$$

The Roe averages of the remaining dependent variables E, p and c are then computed from $\bar{\rho}, \bar{u}, \bar{v}, \overline{H}$:

$$\begin{cases} \bar{E} = \frac{1}{\gamma}\bar{H} + \frac{\gamma-1}{2\gamma}(\bar{u}^2 + \bar{v}^2) \\[2mm] \bar{c} = \{(\gamma - 1)(\bar{H} - \frac{1}{2}(\bar{u}^2 + \bar{v}^2))\}^{\frac{1}{2}} \\[2mm] \bar{p} = \frac{\bar{\rho}\bar{c}^2}{\gamma} \end{cases}$$

Properties (i), (ii) above now read

(i) $\quad F(U_l) - F(U_r) = \bar{A}(U_l, U_r)(U_r - U_l)$

(ii) $\quad \bar{A}(U, U) = dF(U) \equiv \frac{\partial F(U)}{\partial U}$

We write $\lambda_k(U_l, U_r)(k = 1, \ldots, 4)$ and $r_k(U_l, U_r)$ for the eigenvalues and eigenvectors of $\bar{A}(U_l, U_r)$ (see [27]) and let $\triangle U \equiv U_r - U_l$.

Since the $r_k(U_l, U_r)$ are linearly independent, there exists $(\alpha_k)_{k=1,\ldots,4} \in R^4$ such that

$$\triangle U \equiv U_r - U_l = \sum_{k=1}^{4} \alpha_k r_k(U_l, U_r) \qquad (3.16)$$

the numerical flux of Roe's scheme can then be written ([28], [29])

$$\tilde{F}_A(U_l, U_r) = \frac{1}{2}(F(U_l) + F(U_r)) - \frac{1}{2}d(U_l, U_r) \qquad (3.17)$$

where $d(U_l, U_r)$ is the viscous term, given by

$$d(U_l, U_r) = |\bar{A}(U_l, U_r)|(U_r - U_l) =$$

$$\sum_{k=1}^{4} \alpha_k |\lambda_k(U_l, U_r)| r_k(U_r, U_l) \qquad (3.18)$$

Unfortunately, Roe's scheme does not satisfy an entropy inegality ([27], [29], [30]) and allows non-physical expansion shocks in the vicinity of sonic points ([21]), violating the Lax-Oleinik entropy condition; it is not an E-scheme ([43], [30], [27]), so that the numerical solution does not necessarily converge to the (unique) entropy solution of the original nonlinear initial value problem.

To modify Roe's scheme so as to satisfy an entropy inequality, several ideas have been proposed (e.g. [29], [32], [33], [44], [45]), which are based on the representation of sonic rarefaction waves, and are equivalent to introducing a certain amount of artificial viscosity, the exact amount depending on a parameter which requires a case-dependent adjustment.

Dubois and Mehlman ([24]) have recently introduced a parameter-free modification of Roe's scheme based on nonlinear Hermite interpolation of an approximate flux function, leading to an entropic scheme.

3.6. Dubois-Mehlman's entropic correction of Roe's scheme. This approach is based on a nonlinear modification of the flux function defined by (3.17) , in the neighbourhood of sonic points only.

For a given state U we write $U - U_l$ and $U_r - U_l$ as linear combinations of the eigenvectors.

Denoting by $(w_k)_{k=1,\ldots,4}$ and $(\alpha_k)_{k=1,\ldots,4}$ the characteristic variables associated[1] with $U - U_l$ and $U_r - U_l$, respectively, we have

$$U - U_l = \sum_{k=1}^{4} w_k \cdot r_k(U_l, U_r) \qquad (3.19)$$

$$U_r - U_l = \sum_{k=1}^{4} \alpha_k \cdot r_k(U_l, U_r) \qquad (3.20)$$

Following [29], we define the intermediate states

$$\begin{cases} U_0 & = \quad U_l \\ \quad\vdots \\ U_k & = \quad U_{k-1} + \alpha_k r_k(U_l, U_r) \qquad (3.21) \\ \quad\vdots \\ U_4 & = \quad U_r \end{cases}$$

Let $\mathbb{S}$ denote the set of sonic indices

$$\mathbb{S} = \{i, \ \lambda_i(U_{i-1}) < 0 < \lambda_i(U_i)\} \qquad (3.22)$$

We now introduce the Dubois-Mehlman modified flux function, parameterized by the states U_l and U_r

$$F^{\mathrm{dm}}(U_l, U_r; U) = F(U_l) + \sum_{k=1}^{4} g_k(w_k) \cdot r_k(U_l, U_r) \qquad (3.23)$$

where the $w'_k s$ are the characteristic variables introduced in (3.19), and the $g'_k s$ are parameterized by the state $(U_j)_{j=1,\ldots,4}$, and defined as follows:

if $\ k \notin \mathbb{S} \quad$ then $\quad g_k(w) = \lambda_k(U_l, U_r) \cdot w \quad \forall w$

if $\ k \in \mathbb{S}$

$$g_k(w) = \begin{cases} p_k(w) & \text{for } 0 < w < \alpha_k \\[3mm] \lambda_k(U_l, U_r) \cdot w & \text{for } w \notin (0, \alpha_k) \end{cases} \qquad (3.24)$$

---

[1]In this quasi one-dimensional context (due to the remarks following (3.12)) we can assume the existence of characteristic variables as in the purely one-dimensional case (cf. [27] Vol II p.155).

where $p_k(w)$ is the unique Hermite polynomial of degree 3 determined by the conditions

$$p_k(0) = 0, \ p_k(\alpha_k) = \lambda_k(U_l, U_r) \cdot \alpha_k,$$

$$p_k^{'}(0) = \lambda_k(U_{k-1}), \ p_k^{'}(\alpha_k) = \lambda_k(U_k)$$

We recall that $\lambda_k(U_j)$ denotes the k-th eigenvalue of the jacobian matrix $A(U) \equiv \frac{\partial F}{\partial U}$ calculated at the intermediate state U_j, while $\lambda_k(U_l, U_r)$ is the k-th eigenvalue of the Roe matrix $\hat{A}(U_l, U_r)$.

It can be verified that away from sonic points, the modified flux F^{dm} coincides with Roe's linearized flux:

$$F^R(U) = F(U_l) + \hat{A}(U_l, U_r)(U - U_l) \quad (3.25)$$

If the original physical flux $F(U)$ is at least continuously differentiable and if $\hat{A}(U_l, U_r)$ is a continuous function of U_l and U_r, F^{dm} is a continuous function of U_l, U_r and U.

Lemma 3.1.

The Riemann problem

$$\begin{cases} \dfrac{\partial U}{\partial t} + \dfrac{\partial F^{dm}(U_l, U_r; U)}{\partial x} = 0 \\[4mm] U(x; 0) = \begin{cases} U_l & x < 0 \\[3mm] U_r & x > 0 \end{cases} \end{cases}$$

has a unique entropy solution.

See [24] for the proofs of lemmas 3.1 and 3.2.

Lemma 3.2. *For $k \in \mathbb{S}$, the Hermite polynomial $p_k(w)$ is defined by*

$$p_k(w) = a_k w^3 + b_k w^2 + c_k w \ ,$$

with

$$a_k = \frac{\lambda_k(U_k) + \lambda_k(U_{k-1}) - 2\lambda_k(U_l, U_r)}{\alpha_k^2}$$

$$b_k = \frac{3\lambda_k(U_l, U_r) - 2\lambda_k(U_{k-1}) - \lambda_k(U_k)}{\alpha_k}$$

$$c_k = \lambda_k(U_{k-1})$$

The modified numerical flux can be written as

$$\tilde{F}_A^{dm}(U_l, U_r) = F(U_l) +$$

$$\sum_{k \notin \mathbb{S}, \lambda_k(U_l, U_r) < 0} \alpha_k \cdot \lambda_k(U_l, U_r) \cdot r_k + \sum_{k \in \mathbb{S}} g_k(w_k^\star) \cdot r_k \quad (3.26)$$

where

$$w_k^\star = \frac{-\lambda_k(U_{k-1}) \cdot \alpha_k}{D} \quad (3.27)$$

where

$$D = 3\lambda_k - 2\lambda_k(U_{k-1}) - \lambda_k(U_k) + \sqrt{E}$$

and

$$E = (3\lambda_k - \lambda_k(U_{k-1}) - \lambda_k(U_k))^2 - \lambda_k(U_{k-1}) \cdot \lambda_k(U_k)$$

is the argument of the unique extremum of $g_k(w)$ in the interval $0 < w < \alpha_k$, and λ_k in (3.27) denotes $\lambda_k(U_l, U_r)$.

Description of the eigenvectors $r_k(U_l, U_r)$ and characteristic variables α_k

To complete our description of the Dubois-Mehlman entropy correction to Roe's scheme, we must define the eigenvectors $r_k(U_l, U_r)$ and coefficients α_k appearing in (3.19), (3.20), (3.27).

From [28], we have

$$r_1 = \begin{pmatrix} 1 \\ \hat{u} - \hat{c} \\ \hat{v} \\ \hat{H} - \widehat{uc} \end{pmatrix}, \ r_2 = \begin{pmatrix} 0 \\ 0 \\ \hat{v} \\ \hat{v}^2 \end{pmatrix},$$

$$r_3 = \begin{pmatrix} 1 \\ \hat{u} \\ \hat{v} \\ \frac{1}{2}(\hat{u}^2 + \hat{v}^2) \end{pmatrix}, \ r_4 = \begin{pmatrix} 1 \\ \hat{u} + \hat{c} \\ \hat{v} \\ \hat{H} + \widehat{uc} \end{pmatrix}$$

For $w = \rho, \rho u, \rho v$ et ρE, we set $\Delta w = w_r - w_l$.

Equation (3.20) defines a linear system in the 4 unknowns $\alpha_1, \ldots, \alpha_4$. For $\hat{v} \neq 0$, the unique solution is given by

$$\begin{cases} \alpha_1 = \dfrac{\gamma - 1}{2\hat{c}^2}\left\{\Delta(\rho E) - \hat{v}\Delta(\rho v) - \left(\hat{u} + \dfrac{\hat{c}}{\gamma - 1}\right)\Delta(\rho u) + \right. \\ \qquad \left. \left(\dfrac{\hat{c} \cdot \hat{u}}{\gamma - 1} + \dfrac{1}{2}(\hat{u}^2 + \hat{v}^2)\right)\Delta\rho\right\} \\[4mm] \alpha_2 = \dfrac{1}{\hat{v}}\Delta(\rho v) - \Delta\rho \\[4mm] \alpha_3 = \left(1 + \dfrac{\hat{u}}{\hat{c}}\right)\Delta\rho - \dfrac{1}{\hat{c}}\Delta(\rho u) - 2\,\alpha_1 \\[4mm] \alpha_4 = \alpha_1 + \dfrac{1}{\hat{c}}\left(\Delta(\rho u) - \hat{u}\Delta\rho\right) \end{cases}$$

If $\hat{v} = 0$, α_2 is indeterminate. One can set it equal to 0, in which case there exists a unique $(\alpha_k)_{k=1,\ldots,4} \in \mathbb{R}^4$

satisfying (3.20), given by

$$
\begin{cases}
\alpha_1 = \dfrac{\gamma - 1}{2\,\widehat{c}^2}\left\{\Delta(\rho E) - \left(\widehat{u} + \dfrac{\widehat{c}}{\gamma - 1}\right)\Delta(\rho u) + \right. \\
\qquad \left.\left(\dfrac{\widehat{c}\cdot\widehat{u}}{\gamma - 1} + \dfrac{\widehat{u}^2}{2}\right)\Delta\rho\right\} \\[2mm]
\alpha_2 = 0 \\[2mm]
\alpha_3 = \left(1 + \dfrac{\widehat{u}}{\widehat{c}}\right)\Delta\rho - \dfrac{1}{\widehat{c}}\Delta(\rho u) - 2\,\alpha_1 \\[2mm]
\alpha_4 = \alpha_1 + \dfrac{1}{\widehat{c}}\left(\Delta(\rho u) - \widehat{u}\Delta\rho\right)
\end{cases}
$$

The Dubois-Mehlman correction has proved to be useful for explicit schemes, but difficult to implement for implicit schemes, owing to the difficulty of the linearization process.

3.7. Note on the implementation of the time discretization.

In this paper, we limit our applications to the computation of stationary solutions, and will therefore use a local time stepping process described below.

Introducing the diagonal matrix

$$
\Sigma = \operatorname{diag}\left[\frac{\operatorname{Area}(K_i)}{\Delta t_i^n}\right]_{K_i \in \mathcal{T}_h},
$$

we can write (3.5) as follows:

$$
\begin{cases}
\text{Find } U^\star \in W^4 \quad \text{such that} \\[2mm]
U^\star - U^n = -\Sigma^{-1}\,\mathcal{R}(U^n)
\end{cases}
\tag{3.28}
$$

where $\mathcal{R}(U^n)$ is the residual defined from the right-hand side of (3.5)

$$
\mathcal{R}(U^n) = \int_{\partial K_i}\left(\vec{\mathbb{F}}(U^n)\cdot\boldsymbol{n}\right)\phi_{j,i}\,d\sigma -
$$
$$
\int_{K_i}\left(F(U^n)\frac{\partial\phi_{j,i}}{\partial x} + G(U^n)\frac{\partial\phi_{j,i}}{\partial y}\right)dS.
\tag{3.29}
$$

The scheme (3.28) is stable under an appropriate $CFL-$ *condition*.

Let μ denote the $CFL - number$ (assumed to be uniform on the whole grid). For each element $K_i \in \mathcal{T}_h$ we note

- ν_i : mean value, in element K_i, of the characteristic speed corresponding to the largest eigenvalue

$$
\nu_i = \sqrt{\overline{u}_i^2 + \overline{v}_i^2} + \overline{c}_i\,,
$$

- h_i : ratio of the area of K_i by its perimeter

$$
h_i = \frac{\operatorname{Area}(K_i)}{\operatorname{L}(K_i)}\,.
$$

The local time step is then chosen so that

$$
\Delta t_i^n \le \mu\frac{h_i}{\nu_i}\,.
\tag{3.30}
$$

In most cases we have used a $CFL - number$ $\mu = 0.5$.

In the numerical experiments presented in this paper, we study the distribution on the body B (fig.1) and the isolines of the Mach number M, the pressure p, or the pressure coefficient C_p, defined by

$$
Cp = \frac{p_\infty - p}{\dfrac{1}{2}\rho_\infty \|\boldsymbol{V}_\infty\|^2}
\tag{3.31}
$$

To study the convergence of the method, we will present graphs of the $\mathbb{L}_2$-norm of the residual $\mathcal{R}(U^n)$ as a function of the number of iterations.

4. Finite volume methods on unstructured triangular grids

4.1. A two-dimensional finite volume method inspired by the Lax-Friedrichs scheme.

We consider the solution $U(x, y, t)$ of the two-dimensional Euler equations (2.1)-(2.2)

$$
U_t + F(U)_x + G(U)_y = 0
\tag{4.1}
$$

in some region Ω of the $x - y$ plane. In one space dimension [5], both the staggered form of the Lax-Friedrichs scheme and the Nessyahu-Tadmor scheme use two alternate grids $\{x_j\}$ and $\{x_{j+1/2}\}$ at odd and even time steps, respectively. In two dimensions, we proceed in a similar way, starting from an arbitrary FEM triangular grid $\mathcal{T}_h$ such that

$$
\Omega = \bigcup_{T \in \mathcal{T}_h} T \text{ and } T \cap T' =
$$

$$
\begin{cases}
\phi \\
\text{one vertex for any } T, T' \in \mathcal{T}_h \\
\text{one side}
\end{cases}
\tag{4.2}
$$

The nodes of the FEM triangulation are the vertices a_i of the triangles, and in this subsection the degrees of freedom are the vector values of U at the nodes, which can also be considered as cell average values for the cell C_i centered at each individual node a_i (defined below).

For the first grid associated with our finite volume extension of the Lax-Friedrichs scheme, the nodes are the vertices a_i of $\mathcal{T}_h$ while the finite volume cells are the barycentric cells C_i associated with these nodes, obtained by joining the midpoints M_{ij} of the sides originating in a_i to the centroids G_{ij} of the triangles of $\mathcal{T}_h$ which meet at a_i (fig.5).

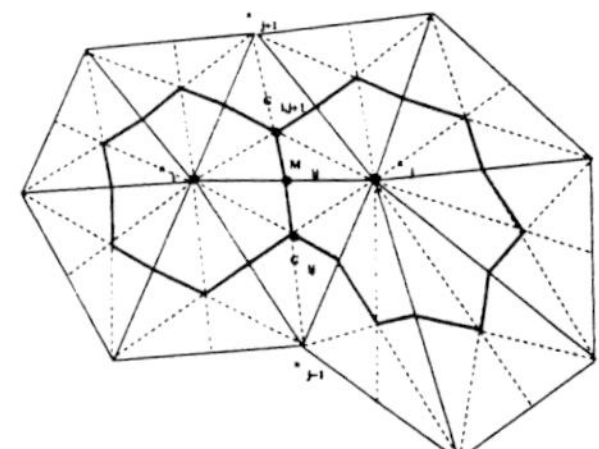

FIGURE 5. **Barycentric cells around nodes a_i, a_j; quadrilateral cell $a_i G_{ij} a_j G_{i,j+1}$.**

For the second grid the nodes are the midpoints M_{ij} of the sides of the original triangulation, while the cells are the quadrilaterals of the form $L_{ij} = a_i G_{ij} a_j G_{i,j+1}$ having M_{ij} as midpoint of one diagonal, obtained by joining two adjacent nodes a_i, a_j to the centroids of the two triangles of $\mathcal{T}_h$ of which $a_i a_j$ is a side.

Let $U_i^n \cong U(a_i, t^n)$ and $U_{ij}^{n+1} \cong U(M_{ij}, t^{n+1})$ denote the nodal (or cell average) vector values in the first and second grid at time $t = t^n$ and $t = t^{n+1}$, respectively (n even).

For the barycentric cell C_i, let $\vec{v}_{ij}^1$ and $\vec{v}_{ij}^2$ denote the unit outer normal vectors to $G_{ij} M_{ij}$ and $M_{ij} G_{i,j+1}$ respectively, pointing out of cell C_i (fig.6) , and for the quadrilateral cell L_{ij}, let $\vec{n}^1{}_{ij}, \ldots, \vec{n}_{ij}^4$ be the normal vectors to the cell edges $a_i G_{ij}$, $G_{ij} a_j$, $a_j G_{i,j+1}$ and $G_{i,j+1} a_i$, respectively, pointing out of cell L_{ij} (fig.7).

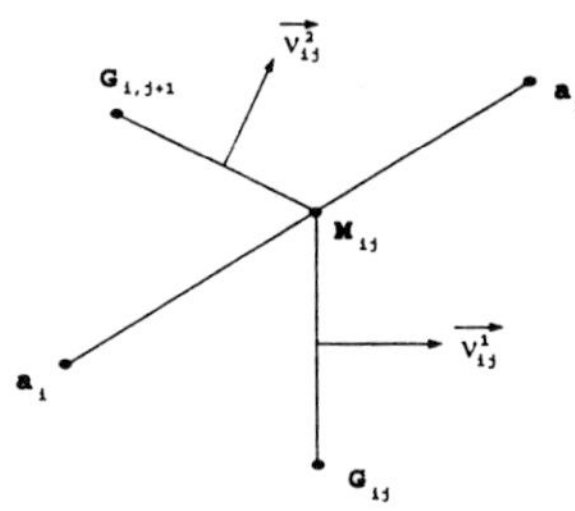

FIGURE 6. **Barycentric cell boundary element $\Gamma_{ij} = G_{ij} M_{ij} \cup M_{ij} G_{i,j+1}$.**

We must also define the elementary flux vectors

$$\vec{\eta}_{ij} = \int_{\Gamma_{ij} = G_{ij} M_{ij} G_{i,j+1}} \vec{v}\, d\sigma = |\overrightarrow{G_{ij} M_{ij}}| \vec{v}_{ij}^1 + |\overrightarrow{M_{ij} G_{i,j+1}}| \vec{v}_{ij}^2 \quad (4.3)$$

and

$$\begin{cases} \vec{\theta}_{ij} = |\overrightarrow{a_i G_{ij}}| \vec{n}_{ij}^1 + |\overrightarrow{a_i G_{i,j+1}}| \vec{n}_{ij}^4 \\ \vec{\theta}_{ji} = |\overrightarrow{a_j G_{ij}}| \vec{n}_{ij}^2 + |\overrightarrow{a_j G_{i,j+1}}| \vec{n}_{ij}^3. \end{cases} \quad (4.4)$$

We write furthermore

$$\vec{\nu}_{ij}^k = \begin{pmatrix} \nu_{ijx}^k \\ \nu_{ijy}^k \end{pmatrix} \text{ for } k = 1, 2,$$

$$\vec{n}_{ij}^k = \begin{pmatrix} n_{ijx}^k \\ n_{ijy}^k \end{pmatrix} \text{ for } k = 1, \ldots, 4$$

and

$$\vec{\eta}_{ij} = \begin{pmatrix} \eta_{ijx} \\ \eta_{ijy} \end{pmatrix}, \quad \vec{\theta}_{ij} = \begin{pmatrix} \theta_{ijx} \\ \theta_{ijy} \end{pmatrix}. \quad (4.5)$$

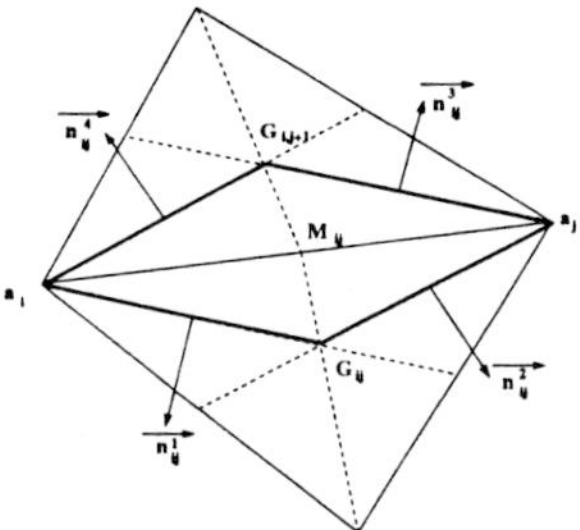

FIGURE 7. **Quadrilateral cells L_{ij}.**

An advantage of using a finite volume formulation where the degrees of freedom are values of the unknown function at the triangulation vertices lies in the possibility to couple (4.1) with an elliptic equation. This can be very convenient in the case of a scalar conservation law where

$$\begin{cases} f(u) = w_1\, h(u) \\ g(u) = w_2\, h(u) \end{cases} \quad (4.6)$$

i.e. when (4.1) can be written as $u_t + \mathrm{div}\big(\vec{W}\, h(u)\big) = 0$ with $\vec{W} = (w_1, w_2)^T$ and $\vec{W}$ stems from an elliptic problem. This situation arises in the study of polyphase flows in porous media, where fluid mechanical and thermodynamical considerations are combined, leading to coupled hyperbolic and elliptic equations [42].

The first step of the two-dimensional finite volume extension of the Lax-Friedrichs scheme is defined by integrating (4.1) on the 3-dimensional cell $L_{ij} \times [t^n, t^{n+1}]$, assuming that the (barycentric) cell values U_i^n at the vertices a_i of the original triangulation are known:

$$\int_{t^n}^{t^{n+1}} \int_{L_{ij}} \big(U_t + F(U)_x + G(U)_y\big)\, dx\, dy\, dt = 0. \quad (4.7)$$

Applying the divergence theorem and observing that $L_{ij} = (L_{ij} \cap C_i) \cup (L_{ij} \cap C_j)$ we get

$$\int_{L_{ij}} U(x, y, t^{n+1})\, dA - \int_{L_{ij} \cap C_i} U(x, y, t^n)\, dA - \int_{L_{ij} \cap C_j} U(x, y, t^n)\, dA$$

$$+ \int_{t^n}^{t^{n+1}} \int_{\partial L_{ij}} \big(F(U) n_x + G(U) n_y\big)\, d\sigma\, dt = 0. \quad (4.8)$$

Since $U(x,t)$ is approximated by U_i^n in C_i, U_j^n in C_j, we can choose the approximation U_i^n on $\partial L_{ij} \cap C_i$ and U_j^n on $\partial L_{ij} \cap C_j$, whence the first step of our finite volume Lax-Friedrichs scheme:

$$A(L_{ij})U_{ij}^{n+1} - A(L_{ij} \cap C_i) \cdot U_i^n - A(L_{ij} \cap C_j) \cdot U_j^n$$
$$+ \Delta t\big(F(U_i^n)\theta_{ijx} + G(U_i^n)\theta_{ijy}\big) +$$
$$\Delta t\big(F(U_j^n)\theta_{jix} + G(U_j^n)\theta_{jiy}\big) = 0. \quad (4.9)$$

We note that this approximation corresponds to choosing the approximate time derivative

$$U_t \cong$$
$$\left(U_{ij}^{n+1} - \frac{A(L_{ij} \cap C_i) \cdot U_i^n + A(L_{ij} \cap C_j) \cdot U_j^n}{A(L_{ij})}\right)\bigg/ \Delta t. \quad (4.10)$$

For the second step we proceed similarly with the help of the barycentric cells C_i:

$$A(C_i)U_i^{n+2} - \sum_{j \text{ neighbour of } i} A(L_{ij} \cap C_i)U_{ij}^{n+1}$$
$$+ \Delta t \sum_{j \text{ neighbour of } i} \big(F(U_{ij}^{n+1})\eta_{ijx} + G(U_{ij}^{n+1})\eta_{ijy}\big) = 0 \quad (4.11)$$

where the value of $U(x, y, t^{n+1})$ on the boundary $\partial C_i = \bigcup_{j \text{ neighbour of } i} \Gamma_{ij}$ is approximated locally, on $\partial C_i \cap L_{ij} = \Gamma_{ij}$, by U_{ij}^{n+1} (i.e. the approximate value of U^{n+1} on the quadrilateral cell L_{ij}. We thus alternately define an approximate solution U_{ij}^{n+1} which is piecewise constant on the quadrilateral cells L_{ij}, at odd time steps ($n = 0, 2, \ldots$), and a solution U_i^{n+2} constant on the barycentric cells C_i, at even time steps.

4.2. A two-dimensional finite volume inspired by the Nessyahu-Tadmor scheme.

The construction of a finite volume method inspired by [5] implies the computation of gradients of piecewise linear functions, and the limitation of the gradients. To simplify the presentation, we shall describe our finite volume method in the case of scalar conservation laws in this subsection. The extension to the vector case is obtained from a field by field decomposition (see, e.g. [47] , [46], [48]).

We consider a scalar nonlinear conservation equation

$$(4.1') \qquad u_t + f(u)_x + g(u)_y = 0$$

At the beginning of the $(n+1)^{\text{st}}$ time step (n even), we have obtained approximate barycentric cell values u_i^n (a_i : a vertex of $\mathcal{T}_h$). We must now, in order to follow the van Leer MUSCL approach used by Nessyahu and Tadmor, construct a piecewise linear profile on the barycentric cells C_i; this can be achieved as follows.

We first construct a piecewise linear approximant on each triangle T of the original triangulation, continuous on the whole computational domain Ω_h, with the help of the barycentric cell/nodal values u_i^n: if $T \in \mathcal{T}_h$ is

a triangle with vertices a_{i_j}, ($j = 1, 2, 3$), we construct $p_T \in P^1$ such that

$$p_T(a_{i_j}) = u_{i_j}^n \ (j = 1, 2, 3)$$

$p_T(x, y)$ is easily obtained from the barycentric coordinates of (x, y) with respect to the vertices of T:

$$p_T(\vec{x}) = \sum_{j=1}^{3} \lambda_j(\vec{x})u_j^n$$

where the vertices of T have been relabelled $1, 2, 3$ or a_1, a_2, a_3, and $\vec{x} = (x, y) \in T$.

The gradient of the (barycentric) cellwise *piecewise linear interpolant* $L(x, y, t^n)$ to be defined will now be chosen (as e.g. in [39] p.28), for cell C_i, as the arithmetic average of the gradients of the polynomials p_T for all triangles T such that $a_i \in T$: on C_i we take

$$L = L_i(x, y, t^n) = u_i^n + (x - x_i)P_i^n + (y - y_i)Q_i^n$$
$$(x, y) \in C_i \quad (4.12)$$

where

$$\begin{pmatrix} P_i^n \\ Q_i^n \end{pmatrix} = \operatorname*{Average}_{a_i \in T} \{\operatorname{grad} p_T\}. \quad (4.13)$$

Contrary to what prevailed in the one-dimensional case, where the average value $\bar{u}_j^n$ of the piecewise linear interpolant $L_j(x, t^n)$ was also its value at the node x_j, we can no longer identify the average value of the piecewise linear interpolant (4.12) , on cell C_i, with its nodal value u_i^n at the "center" a_i of C_i, since a_i need not be the centroid of C_i, and $\frac{1}{A(C_i)} \int_{C_i} L_i(x, y, t^n)dA \neq u_i^n$ in general.

The new cell values at t^{n+1} and t^{n+2} will nevertheless again be *defined* by formulas similar to (4.9) (first step), and (4.11) (second step), obtained by integrating (4.1') on $L_{ij} \times [t^n, t^{n+1}]$ for the first step, and on $C_i \times [t^n, t^{n+1}]$ for the second step:

$$A(L_{ij})u_{ij}^{n+1} \equiv \text{numerical approximation of}$$
$$\int_{L_{ij}} u(x, y, t^{n+1})dA \quad (4.14)$$

$$A(C_i)u_i^{n+2} \equiv \text{numerical approximation of}$$
$$\int_{C_i} u(x, y, t^{n+2})dA. \quad (4.15)$$

For the first step of our scheme we write

$$\int_{t^n}^{t^{n+1}} \int_{L_{ij}} (u_t + f(u)_x + g(u)_y)dA\,dt = 0 \quad (4.16)$$

which leads to

$$\int_{L_{ij}} u(x,y,t^{n+1})dA =$$

$$\int_{L_{ij}\cap C_i} L(x,y,t^n)dA + \int_{L_{ij}\cap C_j} L(x,y,t^n)dA$$

$$- \int_{t^n}^{t^{n+1}} \int_{\partial L_{ij}} (f(u)n_x + g(u)n_y)d\sigma dt. \quad (4.17)$$

The numerical approximation of the right-hand side, and (4.14), will thus lead to u_{ij}^{n+1}, which will be our cell value for the quadrilateral cell L_{ij}.

4.3. Approximation of $\int_{L_{ij}\cap C_i} L(x,y,t^n)dA$.

$L(x,y,t^n)$ is the piecewise linear function defined, on cell C_i, by (4.12) . Let $A_i A_{ij} B_{ij}\ A_{i,j+1}$ be the points of the plane defined by the linear function L_i on C_i which correspond to the four vertices of $L_{ij}\cap C_i = [a_i G_{ij} M_{ij} G_{i,j+1}]$ where $[...]$ denotes the quadrilateral generated by the corresponding vertices (fig. 8).

The integral of L on $L_{ij}\cap C_i$ is equal to the total volume of the two prisms with triangular base $a_i G_{ij} M_{ij} A_i A_{ij} B_{ij}$ and $a_i M_{ij} G_{i,j+1} A_i B_{ij} A_{i,j+1}$, constructed on the triangular bases $L_{ij}^r\cap C_i$ and $L_{ij}^\ell\cap C_i$ where $L_{ij}^r = $ triangle $(a_i G_{ij} a_j)$ and $L_{ij}^\ell = $ triangle $(a_i a_j G_{i,j+1})$; r,ℓ stand for right, left (for an observer at a_i).

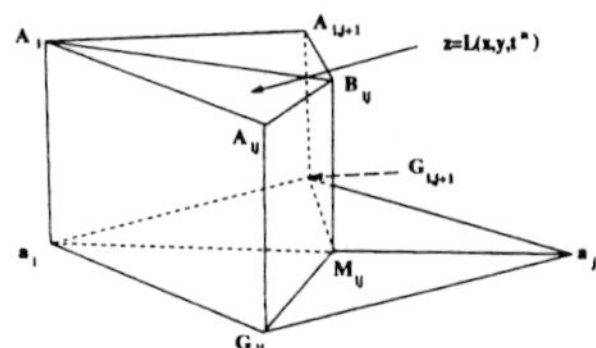

FIGURE 8. Prismatic regions for the computation of $\int_{L_{ij}\cap C_i} L(x,y,t^n)dA$.

The volume of the first prism, for instance, is given by

$$\text{Vol}\{a_i G_{ij} M_{ij} A_i A_{ij} B_{ij}\}$$

$$= \frac{1}{3}\,\text{Area}\,(a_i G_{ij} M_{ij}) \cdot [a_i A_i + G_{ij} A_{ij} + M_{ij} B_{ij}]$$

$$= \frac{1}{3}\,\text{Area}\,(L_{ij}^r\cap C_i)\{u_i^n + u_i^n + (x_{G_{ij}} - x_i)P_i^n$$

$$+ (y_{G_{ij}} - y_i)Q_i^n + u_i^n + (x_{M_{ij}} - x_i)P_i^n$$

$$+ (y_{M_{ij}} - y_i)Q_i^n\}$$

$$= \text{Area}\,(L_{ij}^r\cap C_i)\{u_i^n + \frac{1}{3}(x_{G_{ij}} + x_{M_{ij}} - 2x_i)P_i^n$$

$$+ \frac{1}{3}(y_{G_{ij}} + y_{M_{ij}} - 2y_i)Q_i^n\}$$

where

$$\text{Area}\,(L_{ij}^r\cap C_i) \equiv \text{Area triangle}\,(a_i G_{ij} M_{ij})$$

$$= \frac{1}{2}\{(x_{G_{ij}} - x_i)(y_{M_{ij}} - y_i) - (x_{M_{ij}} - x_i)(y_{G_{ij}} - y_i)\}. \quad (4.19)$$

Similarly, the volume of the second prism is

$$\text{Vol}\{a_i M_{ij} G_{i,j+1}\ A_i B_{ij} A_{i,j+1}\} = \text{Area}(L_{ij}^\ell\cap C_i) \cdot$$

$$\{u_i^n + \frac{1}{3}(x_{M_{ij}} + x_{G_{i,j+1}} - 2x_i)P_i^n + \frac{1}{3}(y_{M_{ij}} + y_{G_{i,j+1}} - 2y_i)\} \quad (4.20)$$

Summing (4.18) and (4.20), we get

$$\int_{L_{ij}\cap C_i} L(x,y,t^n)dA =$$

$$\text{Area}(L_{ij}\cap C_i)\{u_i^n + \frac{1}{3}(x_{M_{ij}} - x_i)P_i^n + \frac{1}{3}(y_{M_{ij}} - y_i)Q_i^n\}$$

$$+ \frac{1}{3}\text{Area}(L_{ij}^r\cap C_i)\{(x_{G_{ij}} - x_i)P_i^n + (y_{G_{ij}} - y_i)Q_i^n\}$$

$$+ \frac{1}{3}\text{Area}(L_{ij}^\ell\cap C_i)\{(x_{G_{i,j+1}} - x_i)P_i^n + (y_{G_{i,j+1}} - y_i)Q_i^n\}$$

with

$$\text{Area}(L_{ij}^\ell\cap C_i) =$$

$$\frac{1}{2}\{(x_{M_{ij}} - x_i)(y_{G_{i,j+1}} - y_i) - (x_{G_{i,j+1}} - x_i)(y_{M_{ij}} - y_i)\}. \quad (4.22)$$

4.4. Approximation of $\int_{L_{ij}\cap C_j} L(x,y,t^n)dA$.

Proceeding in the same way we find

$$\int_{L_{ij}\cap C_j} L(x,y,t^n)dA =$$

$$\text{Area}(L_{ij}\cap C_j)\{u_j^n + \frac{1}{3}(x_{M_{ij}} - x_j)P_j^n + \frac{1}{3}(y_{M_{ij}} - y_j)Q_j^n\}$$

$$+ \frac{1}{3}\text{Area}(L_{ij}^r\cap C_j)\{(x_{G_{ij}} - x_j)P_j^n + (y_{G_{ij}} - y_j)Q_j^n\}$$

$$+ \frac{1}{3}\text{Area}(L_{ij}^\ell\cap C_j)\{(x_{G_{i,j+1}} - x_j)P_j^n + (y_{G_{i,j+1}} - y_j)Q_j^n\}$$

where

$$\text{Area}(L_{ij}^r\cap C_j) =$$

$$\frac{1}{2}\{(x_{M_{ij}} - x_j)(y_{G_{ij}} - y_j) - (x_{G_{ij}} - x_j)(y_{M_{ij}} - y_j)\}. \quad (4.24)$$

4.5. Approximation of $\int_{t^n}^{t^{n+1}} \int_{\partial L_{ij}} \{f(u)n_x + g(u)n_y\}d\sigma dt$.

This is achieved with the midpoint rule for integration with respect to time:

$$\int_{t^n}^{t^{n+1}} \int_{\partial L_{ij}} \big(f(u)n_x + g(u)n_y\big)d\sigma dt \cong$$

$$\Delta t \int_{\partial L_{ij}} \Big\{ f\big(u(x,y,t^{n+1/2})\big)n_x + g\big(u(x,y,t^{n+1/2})\big)n_y \Big\} d\sigma \tag{4.25}$$

where a first order Taylor expansion is used for $u(x,y,t^{n+1/2})$; using (4.1') we write

$$u(x,y,t^{n+1/2}) \cong$$

$$u(x,y,t^n) - \frac{\Delta t}{2}\Big\{ f'\big(u(x,y,t^n)\big)u_x(x,y,t^n) +$$

$$g'\big(u(x,y,t^n)\big)u_y(x,y,t^n) \Big\} \tag{4.26}$$

On $L_{ij} \cap C_i$ we have chosen $u_x \equiv P_i^n$ and $u_y \equiv Q_i^n$, but we must find an approximate value of $u(x,y,t^n)$ on the line segments $a_i G_{ij}$ and $a_i G_{i,j+1}$ (and similarly on $a_j G_{ij}$ and $a_j G_{i,j+1}$). One possible choice consists in choosing the value of $L(x,y,t^n)$, our linear interpolant, at the midpoints of these segments; we then take, for any (x,y) on $a_i G_{ij}$:

$$u(x,y,t^n) \cong$$

$$u_i^n + \frac{1}{2}(x_{G_{ij}} - x_i)P_i^n + \frac{1}{2}(y_{G_{ij}} - y_i)Q_i^n \equiv u_{a_i,G_{ij}}^n \tag{4.27}$$

thus defining our value $u_{a_i,G_{ij}}^n$ for the side $a_i G_{ij}$ of L_{ij}.

In view of (4.26), we can now define an approximate average value of $u(x,y,t^{n+1/2})$ along the side $a_i G_{ij}$ to be used in (4.25):

$$u_{a_i G_{ij}}^{n+1/2} =$$

$$u_{a_i,G_{ij}}^n - \frac{\Delta t}{2}\big\{ f'(u_{a_i,G_{ij}}^n)P_i^n + g'(u_{a_i,G_{ij}}^n)Q_i^n \big\}. \tag{4.28}$$

Introducing these values in (4.25), we finally get

$$\frac{1}{\Delta t}\int_{t^n}^{t^{n+1}} \int_{\partial L_{ij}} \big\{ f(u)n_x + g(u)n_y \big\} d\sigma dt \cong$$

$$f(u_{a_i,G_{ij}}^{n+1/2})n_{ijx}^1 \cdot |a_i G_{ij}| + f(u_{a_i,G_{i,j+1}}^{n+1/2})n_{ijx}^4 \cdot |a_i G_{i,j+1}|$$

$$+ f(u_{a_j,G_{ij}}^{n+1/2})n_{ijx}^2 \cdot |a_j G_{ij}| + f(u_{a_j,G_{i,j+1}}^{n+1/2})n_{ijx}^3 \cdot |a_j G_{i,j+1}|$$

$$+ g(u_{a_i,G_{ij}}^{n+1/2})n_{ijy}^1 \cdot |a_i G_{ij}| + g(u_{a_i,G_{i,j+1}}^{n+1/2})n_{ijy}^4 \cdot |a_i G_{i,j+1}|$$

$$+ g(u_{a_j,G_{ij}}^{n+1/2})n_{ijy}^2 \cdot |a_j G_{ij}| + g(u_{a_j,G_{i,j+1}}^{n+1/2})n_{ijy}^3 \cdot |a_j G_{i,j+1}|. \tag{4.29}$$

4.6. First step of the finite volume extension of the Nessyahu-Tadmor scheme.

Collecting our approximations (4.21), (4.23), (4.29) of the three terms appearing in the R.H.S. of (4.17) and taking (4.14) into account, we obtain the following approximation u_{ij}^{n+1} for the first (odd) time step of our scheme:

$$\text{Area}(L_{ij})u_{ij}^{n+1} =$$

$$\text{R.H.S.}(4.21) + \text{R.H.S.}(4.23) - \Delta t \cdot \{\text{R.H.S.}(4.29)\} \tag{4.30}$$

where u_{ij}^{n+1} can be considered as a cell value for cell L_{ij} at time t^{n+1}, or as a nodal value at the midpoint M_{ij}, at time t^{n+1}.

In preparation for the second (even) time step, we now construct a piecewise linear approximation of u on the quadrilaterals L_{ij}:

$$u(x,y,t^{n+1}) \cong$$

$$L^{(o)}(x,y,t^{n+1}) \equiv u_{ij}^{n+1} + (x - x_{M_{ij}})P_{ij}^{n+1} + (y - y_{M_{ij}})Q_{ij}^{n+1} \tag{4.31}$$

where the slopes P_{ij}^{n+1}, Q_{ij}^{n+1} can be computed as follows.

First we construct a piecewise linear approximate function defined on the triangles $T \in \mathcal{T}_h$ of the original triangulation. On triangle $T = a_i a_j a_k$, we can use for that purpose the newly obtained values u_{ij}^{n+1} at the midpoints of the sides of T. We then compute the average of the slopes of the linear interpolants in the two triangles $T, T' \in \mathcal{T}_h$ sharing $a_i a_j$ as a common side (fig. 7), and use these averages in (4.31).

4.7. Second step of the finite volume Nessyahu-Tadmor scheme.

The second step is obtained by integrating (4.1') on the cylindric region $C_i \times [t^{n+1}, t^{n+2}]$, using the same finite volume approach as for the first step, to define a cell average value u_i^{n+2} on cell C_i:

$$\text{Area}(C_i)u_i^{n+2} -$$

$$\sum_{j \text{ neighbour of } i} \int_{L_{ij} \cap C_i} u(x,y,t^{n+1})dxdy$$

$$= -\Delta t \sum_{j \text{ neighbour of } i} \int_{\Gamma_{ij}} \big\{ f\big(u(x,y,t^{n+3/2})\big)\nu_x +$$

$$g\big(u(x,y,t^{n+3/2})\big)\nu_y \big\} d\sigma \tag{4.32}$$

where $u(x,y,t^{n+1})$ is approximated, on $L_{ij} \cap C_i$, by the piecewise linear interpolant (4.31). Its integral on $L_{ij} \cap C_i$ is computed as described in section 4.3.

To obtain an approximate value of $u(x,y,t^{n+3/2})$ we use a Taylor expansion with respect to time combined with (4.1'), and we subdivide the cell-boundary element Γ_{ij} into $G_{ij}M_{ij} \cup M_{ij}G_{i,j+1}$.
On $G_{ij}M_{ij}$ (resp. $M_{ij}G_{i,j+1}$), $u(x,y,t)$ is then approximated by its value of the midpoint of the line segment $G_{ij}M_{ij}$ (resp. $M_{ij}G_{i,j+1}$).

4.8. Approximation of the slopes.

In order to compute the gradient (P_i^n, Q_i^n) of the piecewise linear interpolant $L(x, y, t^n)$ for the cell C_i, we must first compute the gradient of the first degree polynomials P_T for all triangles $T \in \mathcal{T}_h$ such that $a_i \in T$. Although we could then in principle directly take the average of the gradients of the polynomials P_T, obtained as described at the beginning of section 4.2, we shall consider here a least-squares technique (cf. [40]). For simplicity, we shall describe it for the case of triangular (finite volume) cells.

Let T be a triangle with centroid G, and let T_j, $j = 1, 2, 3$ be the neighbouring triangles, with centroids G_j $(j = 1, \ldots, 3)$ (fig.9); assume the values of the numerical approximation of the solution u at the four points $\{G, G_j\}_{j=1}^{3}$ are known at time t^n, equal to u_T^n, $u_{T_j}^n$ $(j = 1, \ldots, 3)$ (these values can be considered as cell values playing for the triangular cells T, T_j the same role as u_i^n, u_{ij}^n for the cells C_i, L_{ij}).

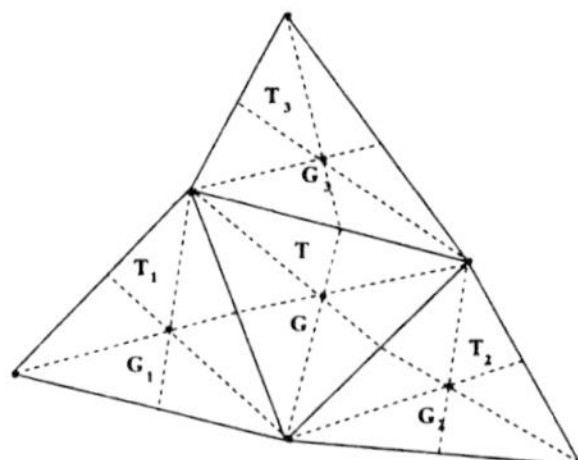

FIGURE 9. **Computation of the least-squares gradient for a triangular cell T.**

The least-squares gradient $\widetilde{(grad\, u)}_T^n = (\tilde{a}_T^n, \tilde{b}_T^n)$ for triangle T will then be chosen such as to minimize the functional

$$I = \sum_{j=1}^{3} \{u_T^n + \overrightarrow{GG_j} \cdot (grad\, u)_T^n - u_{T_j}^n\}^2 \quad (4.33)$$

where

$$(grad\, u)_T^n = (a_T^n, b_T^n)$$

The minimum is obtained when

$$\frac{\partial I}{\partial a_T^n} = \frac{\partial I}{\partial b_T^n} = 0 \quad (4.34)$$

and is shown in [40] to lead to the following least-squares gradient:

$$\tilde{a}_T^n = \frac{1}{D} \sum_{j=1}^{3} (y_{G_j} - y_G)^2 \sum_{j=1}^{3} (u_{T_j}^n - u_T^n)(x_{G_j} - x_G)$$

$$- \frac{1}{D} \sum_{j=1}^{3} (x_{G_j} - x_G)(y_{G_j} - y_G) \sum_{j=1}^{3} (u_{T_j}^n - u_T^n)(y_{G_j} - y_G) \quad (4.35)$$

$$\tilde{b}_T^n = \frac{1}{D} \sum_{j=1}^{3} (x_{G_j} - x_G)^2 \sum_{j=1}^{3} (u_{T_j}^n - u_T^n)(y_{G_j} - y_G)$$

$$- \frac{1}{D} \sum_{j=1}^{3} (x_{G_j} - x_G)(y_{G_j} - y_G) \sum_{j=1}^{3} (u_{T_j}^n - u_T^n)(x_{G_j} - x_G) \quad (4.36)$$

where the denominator

$$D = \sum_{j=1}^{3} (x_{G_j} - x_G)^2 \sum_{j=1}^{3} (y_{G_j} - y_G)^2 -$$

$$\left[\sum_{j=1}^{3} (x_{G_j} - x_G)(y_{G_j} - y_G)\right]^2 \quad (4.37)$$

is strictly positive for any non-degenerated triangle.

For the barycentric cells C_i or the quadrilateral cells L_{ij}, the procedure is quite similar to the one described above for triangular cells. Alternately, for a barycentric cell C_i with center a_i, we could first compute the least squares gradient $grad\, u|_{T_j} = (\tilde{a}_{T_j}^n, \tilde{b}_{T_j}^n)$ of each neighbouring triangle T_j (such that $a_i \in T_j$), and then take the cell gradient $grad\, u|_{C_i} = \text{average}\{grad\, u|_{T_j}\}$, with a similar procedure for a quadrilateral cell L_{ij}.

Unfortunately, this procedure does not preserve monotonicity of the data in the usual van Leer sense described below, and allows the creation of local extremas between the nodes; this phenomenon may lead to (or amplify already existing) spurious oscillations, with the associated loss of stability and convergence difficulties in the case of steady flows. We have therefore introduced some slope limitation in the computation of the gradients.

4.8.1. *Slope limitation.*

To ensure the stability of the scheme and prevent the generation of oscillations in regions of strong gradients, we must perform a slope correction. Following van Leer's approach ([26], [41]), in which the value at some interface point $x_{i+1/2}$ (in the one dimensional case) must fall within the range of values spanned by the adjacent grid averages, u_{i-1} and u_{i+1}, we limit the slopes of the linear interpolant L defined by (4.12) (resp. $L^{(o)}$ defined by (4.31) to ensure that its value at the boundary points $G_{ij}, M_{ij}, G_{i,j+1}$ of ∂C_i (resp. at the vertices $a_i, a_j, G_{ij}, G_{i,j+1}$ of ∂L_{ij}) are bounded by the values at the cell center u_i^n (resp. u_{ij}^{n+1}) and the value u_j^n at the corresponding neighbouring node a_j (resp $u_{i,j-1}^{n+1}$ and $u_{i,j+1}^{n+1}$ at the adjacent quadrilateral cell "midpoints" $M_{i,j-1}$ and $M_{i,j+1}$).

The limitation procedure is implemented on each cell as follows. Let

$$(grad\, u)_i = (P_i, Q_i)^T$$

denote the gradient for cell i, where $P_i \approx u_x$, $Q_i \approx u_y$ at node i. If u satisfies the van Leer requirement we choose

$$P_i^{\text{lim}} = \quad \min_{j \in \mathcal{N}(i)} \text{mod} \ \{P_j\}$$

$$= \begin{cases} \min_{j \in \mathcal{N}(i)} |P_j|.(\text{common sign of all values } P_j) \\ \qquad \text{if all the values } P_j \ (j \in \mathcal{N}(i)) \\ \qquad \text{have the same sign} \\ \\ 0 \qquad\qquad \text{otherwise} \end{cases} \quad (4.38)$$

where $\mathcal{N}(i)$ is the set of nodes j adjacent to node i. If u

does not satisfy the van Leer requirement, we set $P_i^{\text{lim}} = 0$. The computation of Q_i^{lim} is done in the same manner. For quadrilateral cells L_{ij} we proceed in a similar way.

5. Numerical Experiments

In order to assess the relative advantages of both methods , we applied them to several typical test problems; we present results for

- Euler flow around a Double-ellipse at supersonic regime.
- Supersonic flow past a blunt body at 0° of angle of attack and $M_\infty = 2$

Both methods, which are second-order accurate and non-oscillatory thanks to the use of limiters, have been applied on the same grids, to compute the steady flows by stationarization, with the help of grid adaptation to improve the resolution. For the grid adaptation, we have used a procedure developed by M.J.Castro Diaz and F.Hecht at INRIA (France) ([37], [38]).For the finite volume method, the extension of section 4.2 to two-dimensional systems of conservation laws is achieved by the procedure described in ([46], [47], [48]).

On the whole, the finite volume version of the Nessyahu-Tadmor scheme has proved to be less time-consuming and significantly more accurate, for given grids. . The discontinuous finite element method required more computing time to reach convergence. Nevertheless, both methods seem to provide a reliable alternative to other well established schemes, and lend themselves to an extension to three-dimensional problems for unstructured grids.

Example 1 Supersonic flow past a double-ellipse
at 20° of angle of attack and $M_\infty = 2$

For this problem inspired by ([35]), but with Mach number $M_\infty = 2$ instead of the range of hypersonic Mach numbers considered there, and 20° of angle of attack, the geometry is a double ellipse ([35]), defined by

$$\begin{cases} x \leq 0 & \begin{cases} z \leq 0 & \left(\frac{x}{0.06}\right)^2 + \left(\frac{z}{0.015}\right)^2 = 1 \\ z \geq 0 & \left(\frac{x}{0.035}\right)^2 + \left(\frac{z}{0.025}\right)^2 = 1 \end{cases} \\ 0 \leq x \leq 0.016 & \begin{cases} z \geq 0 & z = 0.025 \\ z \leq 0 & z = -0.015 \end{cases} \end{cases}$$

For this steady flow problem we used the same three grids with both methods.We present the results obtained with the initial and final grid only.

For the initial mesh (1558 vertices), both methods give fairly comparable results; notice that the C_p curves can be nearly superposed, which is an indication that both methods are indeed doing some reasonable calculation. The same is true for the pressure contours of both methods, with perhaps a very small advantage for our finite volume method (FV) which gives slightly sharper shocks.

The final grid (5055 vertices) shows a clear advantage for the FV method, which gives a nearly perfect shock resolution with very smooth contours, while the DFE method shows a serious breach of monotonicity in the lower part of the bow shock.

As was the case with the initial grid, the C_p curves can again be nearly exactly superposed.

The major difference between the two methods appears to lie in the convergence history and computing times. Fig. 12 shows a clear advantage for our finite volume method, for the initial grid (1558 vertices). Computing times for the initial grid (CPU : 3564 for FV and 48288 for DFE) confirm the advantage of the proposed Finite Volume Method.

Example 2 Supersonic flow past a blunt body
at 0° of angle of attack. see ([36]).

This is a standard test problem which has been considered by many authors. The problem definition is borrowed from [36].The Mach number is $M_\infty = 2.0$ with an incidence $\alpha = 0^o$. The initial and final grid had 2737 and 7039 nodes, respectively. Fig.16 and 17 show the corresponding pressure contours. Our results, obtained with grid adaptation ([37], [38]), sustain comparison with other results for which grid adaptation has also been used.

Method	CPU	nodes	elements
F.V	2701	2737	5244
D.F.E.M	11700	2737	5244
F.V	16905	7039	13801
D.F.E.M	46886	7039	13801

TABLE 1. Characteristics of F.V and D.F.E.M (Initial and Final Grids, Blunt Body)

Fig. 15 , which presents the residuals for both methods, shows a strong advantage in favour of our finite volume method, considering the results on fig.16 and 17, a fact which is confirmed by table 1: both the CPU-times and the speed of convergence are substantially better.

This test case displays the robustness and the accuracy of the finite volume method, given the fact that the discontinuous finite element method is already an excellent and well established method.

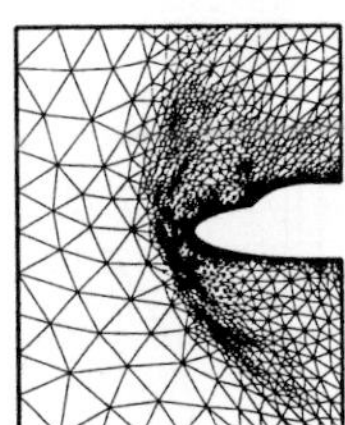

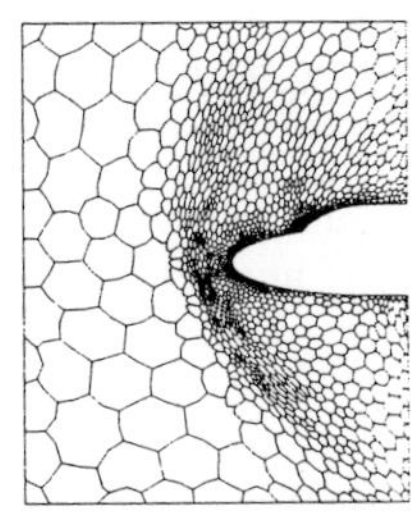 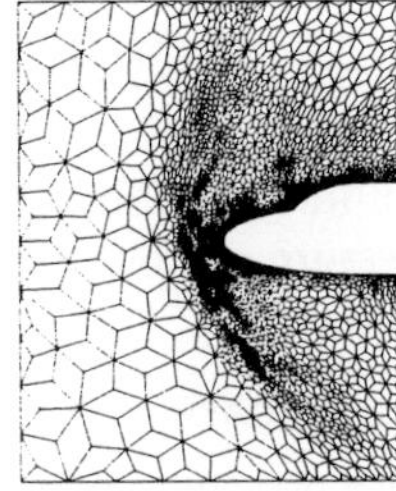

FIGURE 10. Euler flow around a double ellipse. Original grid, barycentric cells C_i and quadrilateral cells L_{ij}

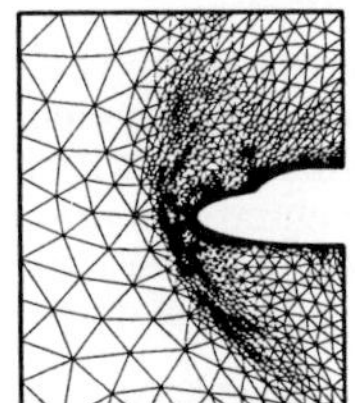

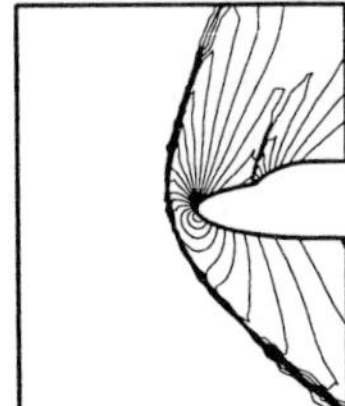 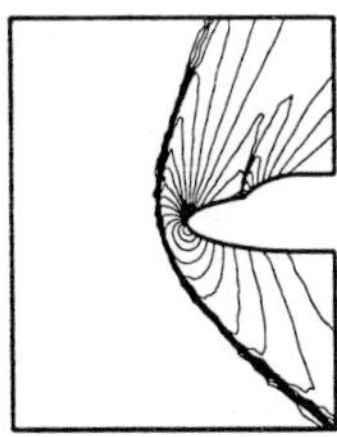

FIGURE 11. Double ellipse : Initial grid (1558 vertices) and solution (pressure contours) (FV, left) and (DFE, right)

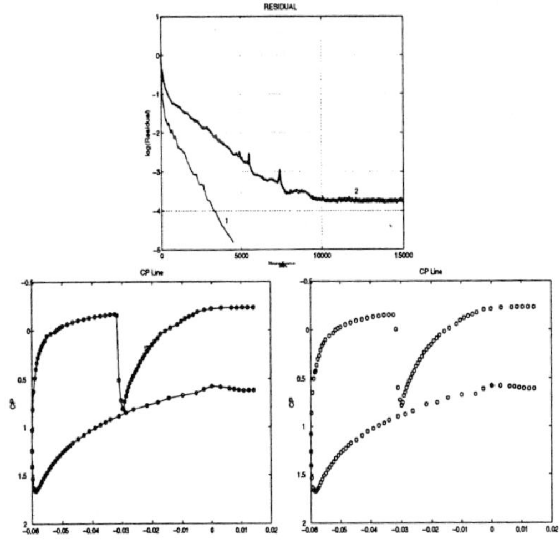

FIGURE 12. Residual for initial grid (1558 vertices)(1=FV, 2=DFE) and C_p body cuts (FV, left) and (DFE, right) (Double ellipse)

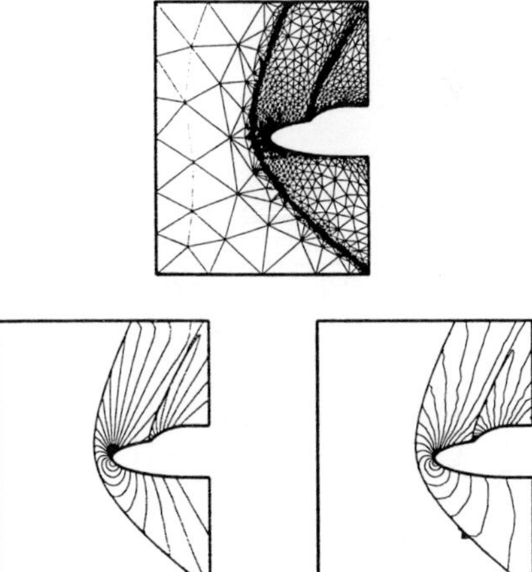

FIGURE 13. Double ellipse : Final grid (5055 vertices) and solution (pressure contours) (FV, left) and (DFE, right)

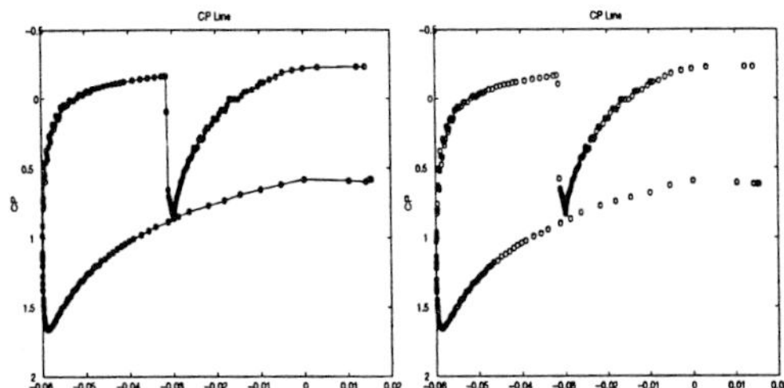

FIGURE 14. C_p body cuts for the final grid (5055 vertices) (FV, left) and (DFE, right) (Double ellipse)

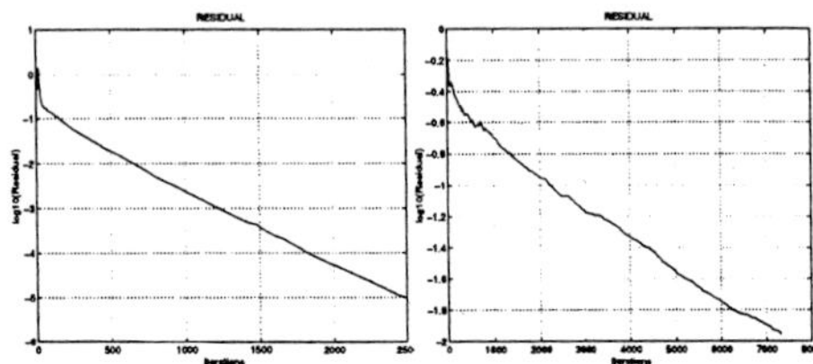

FIGURE 15. Convergence for Finite Volume scheme (left) and Discontinuous finite element method (right) (Blunt body)

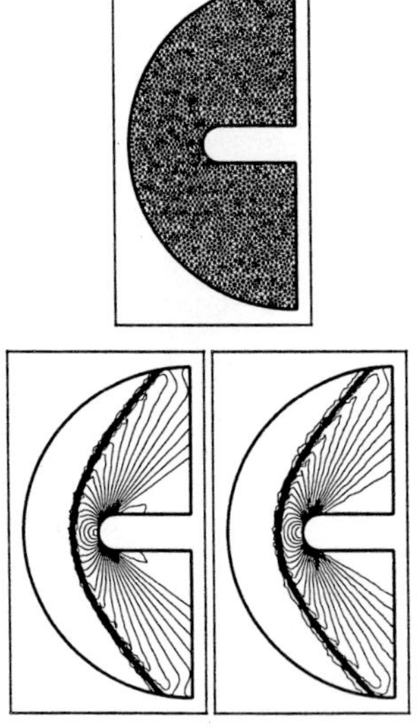

FIGURE 16. Initial grid and solution (pressure contours) FV (left) and DFE (right)

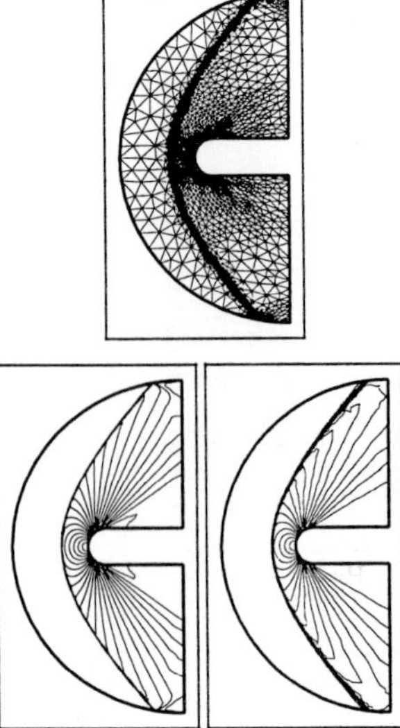

FIGURE 17. Final grid and solution (pressure contours) FV (left) and DFE (right)

Concluding remarks

The examples presented here suggest that the finite volume method proposed by the authors is capable of a very high resolution, with a clear advantage, as compared with the discontinuous finite element method studied here, at least at the level of the convergence histories, computing times, smoothness of the pressure contours, and monotonicity enforcement.

We are presently adapting the finite volume method to a mixed finite volume/Galerkin finite element method for the Navier-Stokes equations ([49]) where the convective part of the Navier-Stokes equations are treated with the finite volume method presented here while the diffusive term will be handled by a finite element method.

In another paper ([50], [51]) we consider the particular case of the linear equation $u_t + div(u\vec{V}) = 0$ where $\vec{V} = \left(V_1(x,y), V_2(x,y)\right)^T$ with $div\,\vec{V} = 0$; we obtain an L^∞ bound for the numerical solution, from which we deduct the existence of a subsequence $\{u_{\mathcal{T}_k, \Delta t_k}\}$ which converges weakly to some function u in $L^\infty(\mathbb{R}^2 \times \mathbb{R}^+)$-weak. Using another set of inequalities, we obtain a total variation-type bound, slightly weaker than a classical bound on the total variation of the numerical solution, called a bound on the "weighted total variation", from which we are then able to show that the above weak limit function u is indeed a weak solution of $u_t + div(u\vec{V}) = 0$.

REFERENCES

1. P. ARMINJON ET M.C. VIALLON , Généralisation du schéma de Nessyahu-Tadmor pour une équation hyperbolique à deux dimensions d'espace, Comptes Rendus de l'Acad. des Sciences, Paris, t.320, série I , 85-88, January 1995.

2. P. ARMINJON, M.C. VIALLON, A. MADRANE , From Lax-Friedrichs to a Multidimensional Finite Volume Extension of the Nessyahu-Tadmor Scheme for Compressible Flows, Proc. Int. Conf. on Numerical Meth. for the Euler and Navier-Stokes Equations. Centre de Rech.Math. Univ de Montréal, P.Arminjon and A.Dervieux, editors, september 1995, to appear in the Am.Math.Soc-CRM series.

3. P. ARMINJON, M.C. VIALLON AND A. MADRANE (1997), A Finite Volume Extension of the Lax-Friedrichs and Nessyahu-Tadmor Schemes for Conservation Laws on Unstructured Grids, Int.J. Comp.Fluid Dynamics, in press.

4. P.ARMINJON, D.STANESCU AND M.C. VIALLON , A two-dimensional finite volume extension of the Lax-Friedrichs and Nessyahu-Tadmor schemes for compressible flows, Proc. of the 6 th.Int.Symp. on Comp. Fluid Dynamics, Lake Tahoe (Nevada), September 4-8, 1995, M. Hafez and K.Oshima, editors, vol.4.

5. H. NESSYAHU AND E. TADMOR , Non-oscillatory Central Differencing for Hyperbolic Conservation Laws, J. Comp. Physics 87, No. 2, 408–463, 1990.

6. G.D.V GOWDA AND J.JAFFRÉ , A discontinuous finite element method for non linear scalar conservation laws, IN-RIA Res. Rep. No. 1848, Rocquencourt, 78153 Le Chesnay, France, 1993.

7. W. REED AND T. HILL , Triangular mesh methods for the neutron transport equation, Tech.Rep. LA-UR-73-479, Los Alamos Scientific Laboratory, 1973.

8. P. LESAINT AND P. RAVIART , On a finite element method for solving the neutron transport equations, in Mathematical Aspects of Finite Elements in Partial Differential Equations, Academic Press, New York, 1974.

9. P.JAMET AND R.BONNEROT, Numerical solution of the Eulerian equations of compressible flow by a finite element method which follows the free boundary and the interfaces, J. Comput. Phys., 18, 21-45, 1975.

10. J.T.ODEN , Mathematical Aspect of Finite Element Approximations in Theoretical Mechanics, in Nemat-Nasser, Ed., Mechanics Today, Vol II, Pergamon Press, Oxford, pp. 159-250, 1975.

11. T.J.R. HUGHES ET AL., A new finite element formulation for computational fluid dynamics 1, 2, 3, 4, Computer Methods in Applied Mechanics and Engineering, 54 (1986), pp. 223-234 and 341-355 and 58, (1986), pp.305-336.

12. P.ARMINJON AND A.ROUSSEAU, Discontinuous finite elements and Godunov-type methods for the Eulerian equations of gas dynamics. Comput.Meths.Appl.Mech.Engrg., 49, pp.17-36, 1985.

13. P.ARMINJON, Finite element analysis of moving loads on a floating ice sheet, in Topics in Numerical Analysis III, Proc.Royal Irish Acad. Conf. on Num. Anal., 1976, J.J.H.Miller, editor, Academic Press, 1977.

14. G. CHAVENT AND G. SALZANO, A finite element method for the 1-D water flooding problem with gravity, J. Comp. Phys., 45, pp. 1-21, 1982.

15. G. CHAVENT, G. COHEN, AND J.JAFFRÉ, Discontinuous upwinding and mixed finite elements for two-phase flow in reservoir simulation. Comp. Meth. Appl. Mech. Eng., 47 , pp. 93-118, 1984.

16. G. CHAVENT, B. COCKBURN, G. COHEN, AND J.JAFFRÉ, A discontinuous finite element method for nonlinear hyperbolic equations, in Innovative Numerical Methods in Engineering, R.P.Schaw, J.Periaux, A.Chaudouet, J.Wu, C.Marino and C.A. Bredia, Eds, Berlin, 1986.

17. G. CHAVENT AND J. JAFFRÉ, Mathematical Models and Finite Elements for Reservoir Simulation, North Holland, Amsterdam, 1986.

18. G. CHAVENT, J. JAFFRÉ, R. EYMARD, D. GUÉRILLOT AND L. WEILL, Discontinuous and mixed finite elements for two-phase incompressible flow, SPE Reservoir Engineering, 5 , pp. 567-575, 1990.

19. C. JOHNSON AND J.PITKARANTA, An analysis of the discontinuous Galerkin method for a scalar hyperbolic equation, Math. Comp., 46 , pp. 1-26, 1986.

20. G. RICHTER, An optimal order error estimate for the discontinuous Galerkin method, Math. Comp., 50 , pp. 75-88, 1988.

21. J.J.QUIRK, A Contribution to the Great Riemann Solver Debate, NASA-ICASE Rep. No. 92-64, Nov. 1992, and Int.J.Num.Meth.Fluids, Vol.18, pp.555-574, 1994.

22. P.WOODWARD AND P.COLELLA, The Numerical Simulation of Two-Dimensional Fluid Flow with Strong Shocks, J.Comp.Phys.54, p.115-173, 1984.

23. J. JAFFRÉ AND L. KADDOURI, Discontinuous finite elements for the Euler equations, Proc. Third Int. Conf. Hyperbolic Problems, Uppsala 1990, B.Engquist and B.Gustafsson, Eds. Studentlitteratur (Lund, Sweden)- Chartwell-Bratt (Bromley, England), Vol.III, pp.602-610.

24. F. DUBOIS AND G. MEHLMAN, A non-parameterized Entropy Correction for Roe's Approximate Riemann Solver, rapport CMAP-Ecole Polytechnique, n° 248, 1991; see also Proc. 10th AIAA CFD Conf., Honolulu, June 1991, and AIAA Journal.

25. G.D.V. GOWDA, Discontinuous finite elements for nonlinear scalar conservation laws, Thèse de Doctorat, Université Paris IX-Dauphine, 1988.

26. B.V. LEER, Towards the ultimate conservative difference scheme: IV. A new approach to numerical convection, J.Comput.Physics, 23, 276–299, 1977.

27. C. HIRSCH, Numerical Computation of Internal and External Flows, Vol.2 : Computational Methods for Inviscid and Viscous Flows, Wiley Series in Numerical Methods in Engineering, J. Wiley & Sons, 1990.

28. P.L. ROE , Approximate Riemann solvers, parameter vectors, and difference schemes, J. Comp. Phys. 43, 357–372, 1981.

29. A. HARTEN, P.D. LAX AND B. V. LEER, On Upstream Differencing and Godunov Type Schemes for Hyperbolic Conservation Laws, SIAM Review, 25(1), 35–61, 1983.

30. E.GODLEWSKI AND P.A.RAVIART , Hyperbolic Systems of Conservation Laws. Soc.Math. Appl.Ind., Ellipses, Paris, 1990.

31. S. OSHER AND F. SOLOMON, Upwind difference schemes for hyperbolic systems of conservation laws, Math. Comp, 38, 339–374, 1982.

32. A. HARTEN AND J.M. HYMAN, Self-Adjusting Grid Methods for One-Dimensional Hyperbolic Conservation Laws, J. Comput. Physics, 50, 235–269, 1983.

33. HARTEN A., High resolution schemes for hyperbolic conservation laws, J. Comp Phys., 49, pp. 357-393, 1983.

34. D.W.LEVY, K.G.POWELL AND B.VAN LEER, An implementation of a grid independent upwind scheme for the Euler equations, AIAA paper, 89-1931, 1989.

35. INRIA and GAMNI-SMAI , Workshop on hypersonic flows for reentry problems, Problem 6 : Flow over a double ellipse, test case 6.1 : Non-Reactive Flows. Antibes, France, January 22-25, 1990.

36. J.PERAIRE, M.VAHDATI, K.MORGAN AND O.C.ZIENKIEWICZ , Adaptive Remeshing for Compressible Flow Computations, J.Comp.Phys.72, 449-466, 1987.

37. M.J.CASTRO DIAZ AND F.HECHT, Anisotropic Surface Mesh

Generation, INRIA Res. Rep. No. 2672, October 1995, INRIA, Rocquencourt, 78153 Le Chesnay, France.

38. M.J.Castro Diaz, Mesh Refinement over Triangulated Surfaces, INRIA Res. Rep. No. 2462, June 1994, INRIA, Rocquencourt, 78153 Le Chesnay, France.

39. F. Fezoui , Résolution des équations d'Euler par un schéma de van Leer en élément finis. Res.Rep. INRIA No. 358, Rocquencourt, 78153 Le Chesnay, France, 1985.

40. S. Champier , Convergence de schémas numériques type Volumes Finis pour la résolution d'équations hyperboliques. Thèse, Univ. de St-Etienne, 1992.

41. B. van Leer , Towards the Ultimate Conservative Difference Scheme.II. Monotonicity and conservation combined in a second order scheme, J. Comp. Phys., Vol.14, 361–370, 1974.

42. R.Eymard and T.Gallouët, Convergence d'un schéma de type éléments finis-volumes finis pour un système d'une équation elliptique et d'une équation hyperbolique, M^2AN , Vol.27 , No.7, 843-861, 1993.

43. S. Osher, Riemann solvers, the entropy condition and difference approximations, SIAM J.Num.Analysis, 21, 217-235, 1984

44. P.L.Roe, Some contributions to the modelling of discontinuous flows, in Lectures in Applied Mathematics, vol.22, pp.163-193, (B.Engquist, S.Osher and R.C.J.Somerville, Editors), SIAM, Philadelphia, 1985.

45. P.L.Roe and J.Pike, Efficient construction and utilisation of approximate Riemann solutions, in Computing Meth. in Appl.Sciences and Engineering, VI (R.Glowinski and J.L.Lions, Editors) North-Holland (1984), pp. 499-518.

46. F.Angrand, A. Dervieux, L. Loth and G. Vijayasundaram , Simulations of Euler transonic flows by means of explicit finite-element type schemes, INRIA Res. Rep. No. 250, Rocquencourt, 78153 Le Chesnay, France, 1983.

47. F. Angrand and A. Dervieux , Some explicit triangular finite element schemes for the Euler equations, Int. J. Num. Meth. in Fluids, Vol. 4, 749–764, 1984.

48. P. Arminjon and A. Dervieux , Construction of TVD-like Artificial Viscosities on Two-Dimensional Arbitrary FEM Grids, J. Comp. Phys., 106, No.1, 176–198, 1993.

49. F.Beux, S.Lanteri, A.Dervieux and B.Larrouturou, Upwind stabilization of Navier-Stokes solvers, INRIA Res.Rep.No. 1885, Rocquencourt, 78153 Le Chesnay, France, 1993.

50. M.C. Viallon et P. Arminjon , Convergence du schéma de Nessyahu-Tadmor sur des maillages non structurés pour une équation hyperbolique linéaire bidimensionnelle, Rapport de recherche No. C.N.R.S. U.R.A. 740, Équipe d'Analyse Numérique, Universités de Lyon et Saint-Étienne, septembre 1994.

51. M.C. Viallon and P. Arminjon, Convergence of a finite volume extension of the Nessyahu-Tadmor scheme on unstructured grids for a two-dimensional linear hyperbolic equation, Rapport de recherche, No. 2239, Centre de Recherches mathématiques, Université de Montréal, january 1995, to appear in SIAM J.Num.Anal.

6. APPENDIX

Solution of the projection problem for the slope limitation

The slope limitation process consists in solving a series of local minimization problems of dimension $nv(K) = 3$, with two constraints.

In this appendix we will show how these minimization problems can be solved; we follow the description given in [6] for the scalar case.

For $w = \rho,\ u,\ v,\ p$: we denote by

- P_K, the plane with equation

$$\sum_{i=1}^{3} X_i = 3\overline{w}_K^\star = w_{K,A_1} + w_{K,A_2} + w_{K,A_3}$$

- Q_K, the cube

$$\prod_{i=1}^{3} [wmin(A_i), wmax(A_i)] \quad \text{with}$$

$$\begin{cases} wmin(A_i) &=& (1-\alpha)\overline{w}_K^\star + \alpha w_{min}(A_i) \\ wmax(A_i) &=& (1-\alpha)\overline{w}_K^\star + \alpha w_{max}(A_i) \end{cases}$$

where $w_{min}(A_i), w_{max}(A_i)$ are defined by (3.4).

For $0 \le \alpha \le 1$, $\overline{w}_K^\star$ satisfies the inequalities

$$wmin(A_i) \le \overline{w}_K^\star \le wmax(A_i) \ , \ i = 1, \dots , 3 \quad (6.1)$$

- $J(X) = \dfrac{1}{2}\|X - w_K^\star\|^2$ where $w_K^\star = (w_{K,A_i})_{i=1,\dots,3}$

- $V = \begin{pmatrix} 1 \\ 1 \\ 1 \end{pmatrix}$ is a vector normal to P_K

With these notations, the computation of w_K^{n+1} amounts to the following minimization problem:

$$\begin{cases} \text{Find } w_K^{n+1} \in P_K \cap Q_K \quad \text{such that} \\[2mm] J(w_K^{n+1}) = \min_{X \in P_K \cap Q_K} J(X) \end{cases} \quad (6.2)$$

This is a convex minimization problem which has a unique solution since $P_K \cap Q_K \ne \emptyset$ (the point $X = (X_i)_{i=1,\dots,3}$ with $X_i = \bar{w}_K^\star, i = 1, \dots , 3$, belongs to $P_K \cap Q_K$ by (6.1)) .

In order to solve the minimization problem (6.2), we dualize the constraint $X \in P_K$.

We introduce the Lagrangian L defined in $\mathbb{R}^3 \times \mathbb{R}$ by

$$L(X,\mu) = J(X) + \mu \left\{ \left(\sum_{i=1}^{3} X_i \right) - 3\overline{w}_K^\star \right\} \quad (6.3)$$

Problem (6.2) is then equivalent to the following associated saddle point problem:

$$\left\{ \begin{array}{l} \text{Find the saddle point} \\[4pt] (\, w_K^{n+1}, \lambda\,) \in Q_K \times \mathbb{R} \text{ such that} \\[4pt] L(w_K^{n+1}, \lambda) = \\[2pt] \min_{X \in P_K} \max_{\mu \in \mathbb{R}} L(X, \mu) = \\[2pt] \max_{\mu \in \mathbb{R}} \min_{X \in P_K} L(X, \mu) \end{array} \right. \qquad (6.4)$$

To solve problem (6.4), we first solve, for any given $\mu \in \mathbb{R}$, the minimization problem

$$\left\{ \begin{array}{l} \text{Find } \widehat{X}(\mu) \in Q_K \text{ such that} \\[6pt] L(\widehat{X}(\mu), \mu) = \min_{X \in Q_K} L(X, \mu) \end{array} \right. \qquad (6.5)$$

We observe that $w_K^\star = (w_{K,A_i})_{i=1,\dots,3} \in P_K$, so that $< V, w_K^\star > = 3\overline{w}_K^\star$.

The Lagrangian can therefore be written as

$$L(X, \mu) = \frac{1}{2}\left\{ \|X - w_K^\star\|^2 + 2\mu < V, X - w_K^\star > \right\}$$

or

$$L(X, \mu) = \frac{1}{2}\|X - (w_K^\star - \mu V)\|^2 - \frac{\mu^2}{2}\|V\|^2 \quad (6.6)$$

The minimization problem (6.5) is thus equivalent to minimizing the distance between $w_K^\star - \mu V$ and the cube Q_K; the solution $\widehat{X}(\mu)$ is therefore given by the projection of $w_K^\star - \mu V$ on Q_K : $\widehat{X}(\mu) = \mathrm{Proj}_{(Q_K)}(w_K^\star - \mu V)$.

$\widehat{X}(\mu)$ can be obtained by truncation of the components of $w_K^\star - \mu V$:

$$\left\{ \begin{array}{ll} \text{if } w_{K,A_i}^\star - \mu \in [wmin(A_i), wmax(A_i)] & \text{then} \\[2pt] \left(\widehat{X}(\mu)\right)_i = w_{K,A_i}^\star - \mu & \\[8pt] \text{if } w_{K,A_i}^\star - \mu > wmax(A_i) & \text{then} \\[2pt] \left(\widehat{X}(\mu)\right)_i = wmax(A_i) & \\[8pt] \text{if } w_{K,A_i}^\star - \mu < wmin(A_i) & \text{then} \\[2pt] \left(\widehat{X}(\mu)\right)_i = wmin(A_i) & \end{array} \right. \qquad (6.7)$$

To complete the solution of problem (6.4), we must now find the real number λ which maximizes the real function

$$\mu \longmapsto F(\mu) = L\left(\widehat{X}(\mu), \mu\right)$$

The desired limited vector $w_K^{n+1} = (w_{K,A_i}^{n+1})_{i=1,\dots,3}$ will then be given by

$$w_K^{n+1} = \widehat{X}(\lambda) \quad \text{where} \quad \lambda \text{ yields } \max_{\mu \in \mathbb{R}} F(\mu)$$

To find the maximum of F, let us write its first and second derivatives:

$$F'(\mu) = < V, \widehat{X}(\mu) - w_K^\star > = \sum_{i=1}^{3}\left(\widehat{X_i}(\mu) - w_{K,A_i}^\star\right)$$

$$F''(\mu) = \sum_{i=1}^{3} \widehat{X_i}'(\mu)$$

where $\widehat{X}(\mu)$ is given by (6.7), so that

$$\left\{ \begin{array}{l} \text{if } w_{K,A_i}^\star - \mu \in [wmin(A_i), wmax(A_i)], \quad \text{then} \\[2pt] \widehat{X}'(\mu) = -1 \\[8pt] \text{otherwise,} \quad \widehat{X}'(\mu) = 0. \end{array} \right.$$

We therefore get

$$F''(\mu) = - \text{ card } \{i \in \{0, 1, \dots, 3\}$$
$$w_{K,A_i}^\star - \mu \in [wmin(A_i), wmax(A_i)]\} \leq 0.$$

F is thus a continuous and differentiable, piecewise quadratic concave function.

F'' has $2 \times 3 = 6$ points of discontinuity which arise when $\widehat{X}(\mu)$ crosses a face of the cube Q_K.

The function F is therefore unambiguously defined and readily available, so that one can easily find the value λ for which is attains its maximum.

Fig.18 shows how the duality method works in the case of the slope limitation of a variable w in the one-dimensional case $(nv(K) = 2)$.

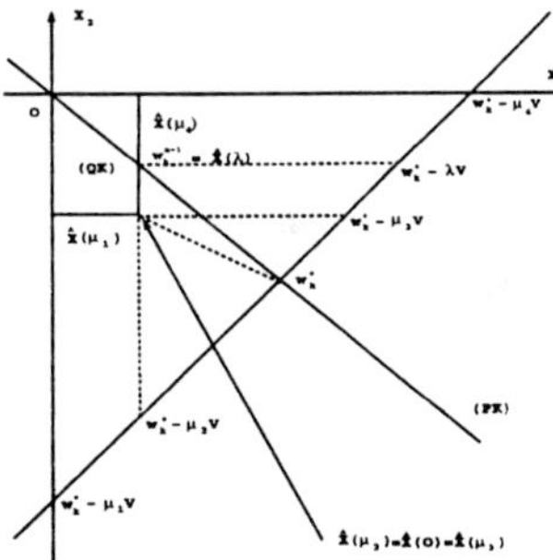

FIGURE 18. Determination of w_K^{n+1} by the duality method in the one-dimensional case $(nv(K) = 2)$

Conclusion

The limitation process, for the slopes of the physical variables, reduces to the maximization, for each element K of the grid, of a piecewise quadratic function of one real variable, which is differentiable and easy to calculate; as the limitation is only active in the neighborhood of strong gradients or fronts of the vector of physical variables, it turns out to be a simple, relatively inexpensive part of the Jaffré-Kaddouri-Gowda discontinuous finite element method.

ARTIFICIAL DISSIPATION AS A FEEDBACK CONTROL, WITH APPLICATION TO THE CELL VERTEX METHOD

K. W. MORTON [1]   S. M. STRINGER [1]

(1)   ICFD, Oxford University Computing Laboratory Wolfson Building, Parks Road, Oxford, England

1. Introduction

The cell vertex method as presently applied to the steady, compressible Reynolds-averaged Navier–Stokes equations on a structured mesh has three main algorithmic components: cell residuals, distribution matrices, and simple mixed second and fourth order artificial dissipation terms — see [5]. The residuals R_α for each cell Ω_α are mapped onto the vertices or nodes of the cell by means of the distribution matrices $D_{\alpha,j}$ to form a system of node-based equations of the form

$$N_j(W)_j \equiv N_j^0(W) + A_j(W)$$

$$:= \frac{\sum_{\alpha=1}^{m} V_\alpha \left(D_{\alpha,j} R_\alpha + A_{\alpha,j} \right)}{\sum_{\alpha=1}^{m} V_\alpha} \qquad (1)$$

where $A_{\alpha,j}$ and its summation at a node A_j is the numerical dissipation, V_α is the measure of cell Ω_α, and m is the number of cells meeting at node j. Letting δ_h^2 denote an undivided Laplacian grid operator, 'standard' artificial dissipation terms, consisting of a combination of second and fourth order differences adapted from the cell centre design in [13], take the form

$$A_{\alpha,j}(W) := \tau_\alpha^{(2)} \left(W_j - W_\alpha \right)$$

$$- \tau_\alpha^{(4)} \left(\delta_h^2 W_j - \delta_h^2 W_\alpha \right) \qquad \forall j \in \aleph_\alpha, (2)$$

where $\aleph_\alpha$ is the set of nodes of Ω_α, W_α and $\delta_h^2 W_\alpha$ are average cell center values, and

$$\tau-(2)_\alpha = \frac{\varepsilon^{(2)}}{\Delta t_\alpha} \max_{i \in \aleph_\alpha} \kappa_j, \tau_\alpha^{(4)} = \max \left(0, \frac{\varepsilon^{(4)}}{\Delta t_\alpha} - \tau_\alpha^{(2)} \right),$$

$$\kappa_j = \frac{\left| \delta_h^2 p_j \right|}{(8 + \delta_h^2) p_j}. \qquad (3)$$

Received on August 8, 1997.
(The work of the second author has been funded by DRA Research Contract CB/FRN1c/U/67.)

These are the forms used for inviscid Euler and viscous laminar Navier-Stokes flow problems in 2D, but for turbulent flow, $\varepsilon^{(2)}$ and $\varepsilon^{(4)}$ are often scaled by the factor $\min(1, M/M_\infty)$; and in 3D the denominator in κ_j is of course $(12 + \delta_h^2) p_j$. In some codes, too, the effect of $\delta_h^2 p$ is spread further by introducing extra averaging over neighbouring cells into the definition of $\tau_\alpha^{(2)}$.

The artificial dissipation is needed for several purposes. Second order dissipation, which is motivated by physical viscosity, is used at shocks because the distribution matrices which perfectly model the ingoing and outgoing waves in one dimension (as employed for instance in [24]) do not have their counterpart for multidimensional systems; in addition, cell residuals calculated using the simple trapezoidal rule have large errors for cells which are crossed by shocks, so their effect needs to be smoothed out. In this connection it should be noted here that the use of $\delta_h^2 p_j$ in (3) for calculating the coefficient κ_j of the second order dissipation serves two purposes: it concentrates the dissipation in the neighbourhood of shocks, although of course it does switch on in other regions of high pressure gradients such as near the leading edge of an aerofoil; and it acts as a smooth coefficient to spread the effect in that neighbourhood.

The fourth order dissipation is operative in smooth regions of flow and plays a number of roles: it needs to damp the spurious chequerboard mode that is present in the cell residuals, even in the Navier-Stokes case; it also has to damp other spurious modes (typically washboard modes) which are introduced by the averaging inherent in the use of distribution matrices; and it has to mitigate the effects of any mismatch between the number of cell residuals and the number of nodal unknowns. The first and second are essentially multi-dimensional phenomena, but the last can be met even in one dimension, for example at a sonic point where it is well known that difficulties due to solution indeterminancy occur; a thorough analysis of this situation in [18] shows how the addition of local fourth order artificial dissipation restores the coercivity of the system of equations while maintaining their accuracy.

All these issues of ill-determinancy, stabilisation and model inaccuracy are typical of those addressed in state space control theory by the application of feedback control. In this paper we will interpret artificial dissipation as a feedback control mechanism; and we will use this viewpoint to derive modified distribution matrices and improved artificial dissipation models, whose effectiveness we shall demonstrate on a range of standard test problems.

In general, spurious oscillatory modes (either the chequerboard modes allowed in the cell residuals or the washboard modes in the nodal residuals) signal the occurrence of these difficulties, and are distinguished from the target solution by their lack of smoothness and their grid dependence. This provides the basis for the feedback control. So let us consider a typical problem in state space control theory: suppose the target solution $\mathbf{W}_T$ is mainly governed by the relation

$$(I - A)\mathbf{W}_T \approx \mathbf{f}, \tag{4}$$

which does not fully determine $\mathbf{W}_T$ and is also subject to error; but, on the other hand, suppose the target solution is distinguished by a relation

$$C\mathbf{W}_T = \mathbf{g}. \tag{5}$$

Then the iteration that is used may take the form

$$\mathbf{W}^{n+1} = A\mathbf{W}^n + \mathbf{f} + G(\mathbf{g} - C\mathbf{W}^n), \tag{6}$$

where G is designed to give effective feedback control through use of the relation (5). Now let us compare this with the usual update algorithm based on the nodal residual in (1), namely

$$\mathbf{W}_j^{n+1} = \mathbf{W}_j^n - \omega_j^n[\mathbf{N}_j^0(\mathbf{W}^n) + \mathbf{A}_j(\mathbf{W}^n)]. \tag{7}$$

Thus the nodal residual equations $\mathbf{N}_j^0(\mathbf{W}) = \mathbf{0}$, without artificial dissipation, play the rôle of the relation (4); and the artificial dissipation acts as a feedback control mechanism. This framework implies that we ideally want

$$\mathbf{A}_j(\mathbf{W}_T) = \mathbf{0}, \tag{8}$$

and it is this that motivates "solution transparent" dissipation models which have the target solution $\mathbf{W}_T$ in their null space.

In the close neighbourhood of a shock, the magnitude of the second order artificial dissipation terms can be used to identify the target solution, but application of such terms across the shock either requires very delicate tuning or it causes undue smearing of the shock. Hence we shall make use of a shock detection algorithm,

either to switch off the dissipation in shocked cells, or to combine a local modification of the distribution matrix with a minimal application of dissipation. On the other hand, in a boundary layer the solution is smooth if the chosen mesh is adequate to represent it. Thus we shall use fourth and higher order artificial dissipation terms, which we call "smooth solution transparent" or SST terms, that allow smooth boundary layers into their null space. In both cases, it is the comparison of divided differences on meshes of differing fineness that provides the basis for distinguishing the target solution from spurious oscillations and designing the artificial dissipation.

In the next section we summarise the key features of the cell vertex method, and then in section 3 we describe the test cases that we shall use in the development of the new artificial dissipation approach. The core of the paper is contained in the three subsections of section 4: in the first, the choice of target solution is described together with the principles behind the design of the artificial dissipation terms; then in section 4.2 these are applied to shocked flows by means of a shock detection algorithm; and in section 4.3 higher order smooth solution transparent artificial dissipation terms are developed for use in laminar and turbulent boundary layers. In both cases, some local modifications to the standard Lax-Wendroff distribution matrices are introduced, so that the artificial dissipation terms do not have to be used to define the basic scheme that determines the target solution, but only to distinguish this from various spurious modes.

2 The cell vertex method

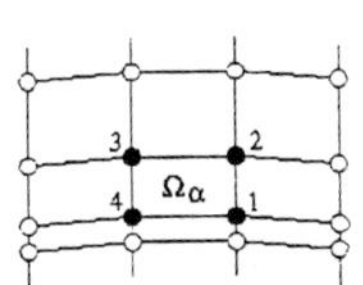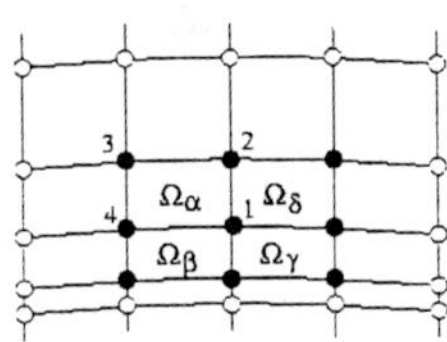

Figure 1: Mesh layout and notation for 2D cell vertex: on the left, nodes to define $\mathbf{R}_\alpha$ and on the right nodes needed for $\mathbf{N}_1$; bold dots correspond to the Euler system, with Navier-Stokes needing the extra open dots.

The specific form of the cell vertex algorithm that we shall use is well documented in [5], [20], [21] and [24]; but for convenience we summarise the key components here, using the notation indicated on Fig. 1. In two dimensions on a quadrilateral mesh, the system of conservation laws $\nabla.(\mathbf{f}, \mathbf{g}) = \mathbf{0}$ is integrated over a typical cell Ω_α and the trapezoidal rule applied to the line integrals arising from application of Gauss' Theorem, to give the cell residual $\mathbf{R}_\alpha$ as follows:

$$\int_{\Omega_\alpha} \nabla\cdot(\mathbf{f},\mathbf{g})\mathrm{d}\Omega = \int_{\partial\Omega_\alpha} (\mathbf{f},\mathbf{g})\cdot\mathbf{n}\mathrm{d}\Omega = \int_{\partial\Omega_\alpha} [\mathbf{f}\mathrm{d}y - \mathbf{g}\mathrm{d}x]$$

$$\approx \tfrac{1}{2}[(\mathbf{F}_1 - \mathbf{F}_3)\delta y_{24} + (\mathbf{F}_2 - \mathbf{F}_4)\delta y_{31}$$

$$- (\mathbf{G}_1 - \mathbf{G}_3)\delta x_{24} - (\mathbf{G}_2 - \mathbf{G}_4)\delta x_{31}]$$

$$\equiv V_\alpha\mathbf{R}_\alpha, \tag{9}$$

where V_α is the measure of cell Ω_α, δx_{ij} denotes $x_i - x_j$, δy_{ij} denotes $y_i - y_j$ and $\mathbf{F}_j$, $\mathbf{G}_j$ are the approximate fluxes at node j. For the Euler equations, if $\mathbf{W}$ is the approximation to the vector of conserved variables $\mathbf{w}$ we have simply $\mathbf{F}_j = \mathbf{f}(\mathbf{W}_j)$, $\mathbf{G}_j = \mathbf{g}(\mathbf{W}_j)$; but for the Navier-Stokes equations the fluxes are also functions of the gradients of velocity and temperature, or more generally $\mathbf{f} \equiv \mathbf{f}(\mathbf{w},\nabla\mathbf{w})$, $\mathbf{g} \equiv \mathbf{g}(\mathbf{w},\nabla\mathbf{w})$. The consistent use of (9) in this case requires us to define gradients of variables at the nodes. This is done by first using the formula (9) to recover gradient approximations at the centroid of each cell, and then interpolating between these centroids:

$$\left(\frac{\partial \mathbf{W}}{\partial x}\right)_\alpha \approx \frac{1}{V_\alpha} \int_{\partial\Omega_\alpha} \mathbf{W}\mathrm{d}y,$$

$$\left(\frac{\partial \mathbf{W}}{\partial y}\right)_\alpha \approx \frac{1}{V_\alpha} \int_{\partial\Omega_\alpha} (-\mathbf{W})\mathrm{d}x, \tag{10a}$$

$$(\nabla\mathbf{W})_j =$$
$$\frac{V_\gamma(\nabla\mathbf{W})_\alpha + V_\delta(\nabla\mathbf{W})_\beta + V_\alpha(\nabla\mathbf{W})_\gamma + V_\beta(\nabla\mathbf{W})_\delta}{V_\alpha + V_\beta + V_\gamma + V_\delta}.$$
$$\tag{10b}$$

The latter interpolation is simple, robust, and in one dimension corresponds to taking the derivative of an interpolating quadratic.

The standard distribution matrices are formed from a generalised Lax-Wendroff update, as originated by Ni [31] and developed by Hall [9], Morton and Paisley [21], Morton [17] and others. For both the Euler equations and the Navier-Stokes equations, the second order update is based on the approximation

$$\frac{\partial^2 w}{\partial t^2} \approx \nabla.(A\mathbf{R}, B\mathbf{R}), \tag{11}$$

where A and B are the Jacobians of the inviscid fluxes $\mathbf{f}^I$, $\mathbf{g}^I$ only and $\mathbf{R}$ is the cell residual given by (9). Integrating this over a node-centred quadrilateral formed by the diagonals of the cells meeting at the node, gives the second order nodal update which is combined with the first order update obtained from the volume-averaged cell residuals. The result is a nodal residual of the form given in (1), where the distribution matrices for the cell Ω_α shown in Fig. 1 have the form

$$D_{\alpha,1} = I - \nu_C\Delta t_\alpha(\delta y_{42}A_\alpha - \delta x_{42}B_\alpha)/V_\alpha \tag{12a}$$
$$D_{\alpha,2} = I - \nu_C\Delta t_\alpha(\delta y_{13}A_\alpha - \delta x_{13}B_\alpha)/V_\alpha \tag{12b}$$
$$D_{\alpha,3} = I - \nu_C\Delta t_\alpha(\delta y_{24}A_\alpha - \delta x_{24}B_\alpha)/V_\alpha \tag{12c}$$
$$D_{\alpha,4} = I - \nu_C\Delta t_\alpha(\delta y_{31}A_\alpha - \delta x_{31}B_\alpha)/V_\alpha. \tag{12d}$$

Here ν_C is a parameter which corresponds to a global CFL number and Δt_α is a maximal cell-based local timestep. Fourier analysis of the update for the one-dimensional convection-diffusion equation and the two-dimensional convection equation, which is given in [5], leads to defining the inviscid and viscous timesteps as

$$\Delta t_\alpha^I = V_\alpha\{(|u| + c)^2(\Delta y_\alpha)^2 + (|v| + c)^2(\Delta x_\alpha)^2\}^{-\frac{1}{2}}$$
$$\tag{13a}$$

$$\Delta t_\alpha^V = \tfrac{1}{2}\frac{\rho\mathrm{Re}}{\max(\tfrac{4}{3}\mu, \kappa\gamma/\mathrm{Pr})}\frac{V_\alpha^2}{(\Delta x_\alpha)^2 + (\Delta y_\alpha)^2} \tag{13b}$$

and then setting

$$\Delta t_\alpha = \min\left(\Delta t_\alpha^I, \Delta t_\alpha^V\right). \tag{13c}$$

Combined with the update (7) in which

$$\omega_j^n = \nu_N\Delta t_j, \text{ where } \Delta t_j = \min(\Delta t_\alpha, \Delta t_\beta, \Delta t_\gamma, \Delta t_\delta),$$
$$\tag{14}$$

and ν_N is a nodal CFL number, we have the generalised Lax-Wendroff scheme. Variation of the two CFL numbers ν_N and ν_C gives greater smoothness in the converged solution when ν_C is increased, and faster convergence when ν_N is increased; however, the model problems already referred to show that convergence requires

$$\nu_N \leq \nu_C \;, \; \nu_N\nu_C < 1. \tag{15}$$

These conditions generalise the usual Lax-Wendroff stability condition $\nu \leq 1$.

This basic algorithm can be readily generalised to three dimensions (see [4], [28]) and essentially the same scheme can be applied to an unstructured triangular or tetrahedral mesh (see [3]). When combined with a full approximation multigrid procedure as in [5], it gives a reasonable rate of convergence which is comparable with that obtained for alternative methods. However, two points need to be made about the convergence to a steady state solution. Firstly, our concern so far in the development of the method has been concentrated on the quality of the converged solution; and undoubtedly there are substantial improvements to be obtained in convergence rates, particularly by improving the choice of distribution matrices. Secondly, although convergence is to a solution of the nodal residual equations (1), on a quadrilateral mesh there are roughly the same number of cell residuals as unknowns so that our objective is to drive most of these close to zero — the main exceptions are the cells crossed by shocks for which the cell residuals are inaccurate and should not be driven to zero. Thus, in assessing the accuracy of the scheme in later sections we shall plot two sets of residual balances. Firstly, as in Fig. 9 through a boundary layer, we plot the contributions to the nodal residuals from the inviscid and viscous fluxes and from the artificial dissipation terms: at convergence these are always in overall balance, and we wish to check that the last contribution is insignificant. Secondly as in Fig. 10, we plot the inviscid and viscous flux contributions to the cell residuals: these will not always be in balance, both because of the artificial dissipation terms, but also because of the action of the distribution matrices; it is the effect of the latter that we here want to demonstrate, and to minimise.

3 Test problems

To develop and demonstrate the effectiveness of any new CFD algorithm entails the use of carefully selected and widely known test cases. This is particularly true in the development of artificial dissipation terms for such as the cell vertex method: this is because in this case simple problems, like scalar convection-diffusion, do not require such terms at all: but, on the other hand, more complex problems do not have analytic solutions, and published comparative numerical solutions are often not well documented, especially for turbulent flows modelled by specific turbulence models embodying numerous model constants. Very often, then, the tests have to be aimed at internal consistency — in regard to mesh refinement (globally or locally) or variation of numerical parameters — and response to changes in the problem specification, such as the Reynolds number. And overall a balance has to be struck between final accuracy and computation time, especially the convergence rate of the iteration. All the computations described below will be carried out in the multigrid framework used in [5]: so it is often the case that increased amounts of a simple linear artificial dissipation will give rapid convergence to a smooth solution, at the expense of loss of detail in flow predictions. Bearing in mind these general points, we have therefore selected the following problems for our test calculations.

3.1 Transonic inviscid flow past NACA0012 aerofoil

The test problem with $M_\infty = 0.8$ and angle of attack $\alpha = 1.25°$ is one of the most widely used and documented — see [32], [34], [38], [11], [12], [25]. The strong shock on the upper surface and the weak shock on the lower surface together provide a good test of an algorithm's shock capturing capability. However, the lift and the drag are largely determined by the accuracy around the leading and trailing edges.

Thus the results that are presented consist of: (i) *Mach no. plots* around the surface of the aerofoil; (ii) *entropy loss* around the aerofoil; and (iii) *lift and drag coefficients*. In addition, for the cell vertex scheme a significant quantity to note is (iv) *the magnitude of the cell residuals* at convergence, other than those crossed by shocks: it is the nodal residuals that are driven to zero by the iteration (7), but it is the cell residuals that determine the local accuracy.

3.2 Subsonic laminar viscous flow past NACA0012 aerofoil

A good test case for SST dissipation models is laminar viscous flow past a NACA0012 aerofoil at $M_\infty = 0.5$, $Re_\infty = 5000$ and $\alpha = 0.0°$ — see [5], [16], [26], [27], [35], and [15]. Here, the challenge for the design of artificial dissipation terms is to damp spurious modes whilst preserving the natural variation through the boundary layer: and the strength and position of the recirculation region needs to be predicted correctly. Hence, diagnostics consist of: (i) *Mach no. plots* around the aerofoil; (ii) *y-momentum plots* along the aerofoil; (iii) *lift and drag coefficients*; and (iv) *nodal and cell budget plots*.

3.3 Subsonic turbulent viscous flow past NACA0012 aerofoil

A more demanding test for SST dissipation models that involves the more severe physical conditions of turbulent flow, and the greater numerical challenges posed by extra model terms, higher mesh aspect ratios etc., is flow past a NACA0012 aerofoil at $M_\infty = 0.5$, $Re_\infty = 2.89\times10^6$ and $\alpha = 0.0°$ — see [36], [27]. Due to high natural variation, turbulent calculations are far more susceptible than laminar problems to severe degradation by artificial viscosity. Hence, numerical results include: (i) *y-momentum plots* near the leading edge; (ii) *lift and drag coefficients*; and (iii) *nodal and cell budget plots*. We have used the simple Baldwin-Lomax algebraic turbulence model [2] with two modifications adopted from [33] and used in [5] — namely, use of the maximum local shear stress through the boundary layer and setting the constant $C_{WK} = 1.0$.

3.4 Transonic turbulent viscous flow past RAE5225 aerofoil

A well-documented test case used for checking the capabilities of a shock-capturing algorithm under turbulent flow is for the RAE5225 aerofoil at $M_\infty = 0.735$, $Re_\infty = 6 \times 10^6$ and $\alpha = 1.57°$ — see [10], [7], [8]. The main feature of the flow is a strong shock on the upper surface at about 55% chord. Diagnostic tests consist of: (i) *Mach no. plots around the aerofoil*; (ii) *lift and drag coefficients*; and (iii) *nodal and cell budget plots*.

4 Improved artificial dissipation models for feedback control

Practical use of the standard artificial viscosity (AVIS) model of (2) and (3) shows that it has several limitations. As already remarked, the coefficient in (3) allows too much second order dissipation at the aerofoil leading edge, so that we shall use a shock detection switch instead to control the switching between second and fourth order dissipation. Studies for the linear convection-diffusion equation at a normal boundary layer (see [35]), which simulates a shock for the Navier-Stokes equations, show that fourth order dissipation is ineffective in damping the washboard oscillatory mode that is present in the nodal residual equations, and it distorts the true solution; second order dissipation is much more effective here, so we shall still use the new switch to combine the second and fourth order dissipation as in (3). On the other hand, it is easy to have too much dissipation across the shock; and the matrix-valued dissipation of Swanson & Turkel [37] essentially introduces a factor proportional to the wavespeed $|\lambda_s|$ associated with the shocked characteristic component, so this is nearly zero at the shock. In section 4.2 we shall therefore develop a combination of modified distribution matrices and second order AVIS to deal with shocks, based on trying to ensure that the discrete target solution lies in the null space of the nodal residual equations, with the AVIS terms omitted.

With the use of a shock detection switch, only the fourth order dissipation is switched on through most of the flow domain, and its simple constant coefficient form embodied in (3) is quite inadequate: if it is sufficient to damp typical washboard modes, it has too great an effect on the flux balance through the boundary layer of a Navier-Stokes calculation. We shall therefore present in section 4.3 a variety of improved models, finally selecting one which makes use of a sixth order operator.

4.1 The target solution and the choice of dissipation operators

Away from shocks and other discontinuities the discrete target solution should be smooth and only rather weakly dependent on the mesh — through the truncation error. But near a shock it will depend on the mesh, and it is important that we can characterise its behaviour fairly precisely. *The target solution is the best representation of the true solution that is possible on the given mesh, and using the given discrete scheme*; we should therefore expect that it will depend on the scheme, or at least our interpretation of it will depend on the scheme. Consider a simple example, of a one-dimensional scalar conservation law $u_t + f_x = 0$. If the unsteady problem is approximated by a conservative scheme using piecewise constants on a set of intervals or cells $(x_{i-\frac{1}{2}}, x_{i+\frac{1}{2}})$, the target solution is typically the projection of the exact solution onto piecewise constants. So a moving shock between two constant states is represented at a given time t_n by an intermediate state U_s^n equal to an average of u^n across the shock; thus the value U_s^n gives the precise position of the shock. Now, alternatively, consider the cell vertex approximation to the steady problem, with $U_0 = u_L$ and $U_J = u_R$ prescribed at the left and right boundaries. The approximation is here regarded as a continuous piecewise linear function, and the cell residual equations are $f(U_i) - f(U_{i-1}) = 0$ for $i = 1, 2, ..., J$. Clearly, these equations cannot all be satisfied because there are only $J - 1$ unknowns; and across the shock the best that one can achieve in general is $f(U_{s-1}) = f(U_{s+1})$, with the intermediate $f(U_s)$ being rather arbitrary and determined by the distribution matrices (or factors) and artificial dissipation terms that are used. Thus in these two cases, although the resulting one-point shock solutions may look superficially similar, their interpretation is quite different; hence the design of the update scheme to achieve convergence to the target solution will be different.

Application of the cell vertex method to give a one point shock in the case of the Euler equations in one dimension is fully described in [24] using a nozzle problem example with flow from left to right. A *shock cell* can be identified as lying between a set of supersonic states at nodes to its left, and a set of subsonic states at nodes to the right. On the other hand, in the iteration process the distribution matrix for a cell is determined from an average state for the cell. Thus, according to whether the shock cell at any stage is considered to be subsonic or supersonic, it will or will not influence the update of the supersonic state to its left; it is clear that in either case the final target approximation does not set to zero all the components of the shock cell residual. We can similarly characterise the corresponding situation in two dimensions, for example for flow past an aerofoil: near the aerofoil, where the true shock roughly follows one mesh direction, the situation will be similar to that in one dimension; but as the shock curves away across the mesh its normal direction must be estimated for each cell so that whether the cell is a shock cell can be calculated. The details will be given in the next section, but a typical situation is illustrated in Fig. 2 in the mesh cordinate system.

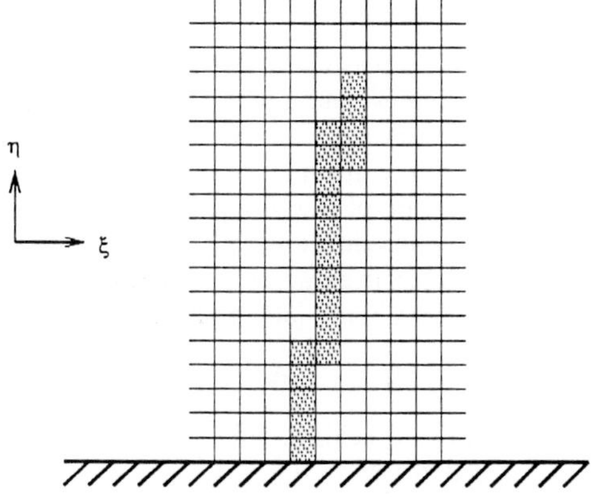

Figure 2: Shock cells (shaded) on upper surface of NACA0012 aerofoil for transonic inviscid flow.

Let us consider now the choice of dissipation operator, on a structured but nonuniform mesh, which may have a very high aspect ratio in the case of Navier-Stokes calculations. Starting with the Euler equations, we expect first derivatives to be sufficiently well represented on the mesh that they are also close to those calculated on a coarse mesh obtained by deleting alternate mesh points in each coordinate direction. In a typical direction, suppose the mesh interval is h_- to the left and h_+ to the right of mesh point j, so that we can define backward and forward divided differences by

$$D_- U_j := \frac{U_j - U_{j-1}}{h_-} \quad , \quad D_+ U_j := \frac{U_{j+1} - U_j}{h_+}. \quad (16)$$

Then on the coarse mesh, with point j omitted, we ob-

tain the corresponding central difference operator

$$D_0 U_j := \frac{U_{j+1} - U_{j-1}}{h_- + h_+}. \quad (17)$$

The grid independence of these two sets of values is conveniently measured by

$$\begin{aligned} \delta_h^2 U_j &:= \tfrac{1}{2}(h_- + h_+)(D_+ - D_-)U_j \quad (18) \\ &\equiv \tfrac{1}{2}\frac{(h_- + h_+)^2}{h_-}(D_+ - D_0)U_j \\ &\equiv \tfrac{1}{2}\frac{(h_- + h_+)^2}{h_+}(D_0 - D_-)U_j. \quad (19) \end{aligned}$$

This motivates our choice of (18) as the basic one-dimensional second order dissipation operator. Note that it is a linearly transparent, second order undivided difference: thus it is $O(h^2)$ for a smooth function; and for the washboard mode equal to ± 1 on alternate mesh points, it yields $(h_- + h_-)^2/h_-h_+$ which is always greater than or equal to 4. So it is a very effective discriminant on any mesh.

Now let us consider how this might be generated by the second order AVIS terms in (1) — (3). We wish to form $\mathbf{A}_{\alpha,j}$ from the difference $\mathbf{W}_j - \mathbf{W}_\alpha$, as this imposes global conservation by ensuring that

$$\sum_{j \in \mathcal{N}_\alpha} \mathbf{A}_{\alpha,j} = \mathbf{0} \quad \forall \alpha; \quad (20)$$

but the coefficients of (3) do not give the correct smooth mesh dependence. In the model case of (18), $V_\alpha = h_\pm$ and we can take $\Delta t_\alpha = (h/a)_\pm$ so that the factors $h_\pm^{-1}$ in the $D_\pm U_j$ have to come from the κ coefficients. A brief calculation shows that we need to replace the $\delta_h^2 p$ dependence of (3) by

$$\tau_\pm^{(2)} = \frac{\epsilon^{(2)}}{\Delta t_\pm} \frac{N^{-3}}{h_\pm},$$

where the average mesh length $1/N$ is introduced to obtain the correct order of magnitude. Such a choice is incorporated into a smoothly varying AVIS scheme in a 1D problem in the next section, but this is more difficult to accomplish for general multidimensional problems.

However, this is not particularly significant since we shall use second order dissipation only in the neighbourhood of shocks. The linear transparency of (18) is much more important when it is incorporated in the $\tau^{(4)}(\delta_h^2 \mathbf{W}_j - \delta_h^2 \mathbf{W}_\alpha)$ term of (2), since this is used more widely. We shall in fact define $\delta_h^2 \mathbf{W}_j$ by applying the scheme (18) in each local coordinate direction and adding the results. The justification for this scaling, rather than using a consistent discretisation of $\nabla^2 \mathbf{W}$, lies in the assumption that a high aspect ratio mesh would have been generated and designed to roughly balance the important differential terms in a boundary layer. It is worth noting, too, that when these third order difference fluxes are combined at a node to give the fourth order AVIS term, on a uniform mesh this can also be interpreted as a fine/coarse mesh difference through the well-known identity $\delta_h^4 V \equiv \delta_{2h}^2 U - 4\delta_h^2 U$.

4.2 Transonic problems: artificial viscosity and the use of shock detection

We start the reconsideration of cell vertex shock modelling with the one dimensional inviscid and transonic nozzle problem used in [24], before going on to the test problems of sections 3.1 and 3.4. We select as the target solution of the nozzle problem on a uniform mesh of 17 points the approximation obtained in [24] by the use of upwind distribution matrices, which is plotted against the exact solution in the top figure of Fig. 3. One sees that even on a coarse mesh the nodal values are very accurate upstream of the shock, which lies in a shock cell bounded by an accurate supersonic nodal value and an inaccurate subsonic value; the latter is the only seriously inaccurate "shock point", all other subsonic nodes on the right having good accuracy. The upwind distribution matrices are obtained by spectrally decomposing the Jacobian matrix A and setting the cell CFL number $\nu_C = 1$; in effect, a matrix-valued time step Δt_α is used in the one-dimensional version of (12). In [24] the resulting nodal equations were solved in 10-20 Gauss-Seidel sweeps through the mesh, using a few approximate Newton iterations at each node.

In two dimensions, the simultaneous decomposition of the Jacobian matrices A and B is not possible, upwind distribution matrices cannot always be defined, and the Gauss-Seidel sweeping is much less effective. So we seek to devise a simpler and more generally applicable modification to the standard cell vertex scheme described in sections 1 and 2. In [26], [27], the 2nd order AVIS term was focused around the shock, whilst it was scaled down in the shock cell itself; i.e. equation (2) was replaced by

$$\begin{aligned} \mathbf{A}_{\alpha,j}(\mathbf{W}) &:= S_\alpha \tau_\alpha^{(2)} (\mathbf{W}_j - \mathbf{W}_\alpha) \\ &- \tau_\alpha^{(4)} \left(\delta_h^2 \mathbf{W}_j - \delta_h^2 \mathbf{W}_\alpha \right) \quad \forall j \in \mathcal{N}_\alpha; \end{aligned} \quad (21)$$

where S_α is small in the shock cell, but elsewhere $S_\alpha = 1$. This approach arises very naturally from our control objective of designing dissipation terms that damp only spurious oscillatory behaviour whilst having the target solution in their null space. In the smooth region of flow outside the shock, the AVIS terms can more easily distinguish the smooth target solution from

the spurious mode, and so we can ensure that the target solution, and in particular the shock structure, lies further within their null space. On the other hand, with a small value of S_α the shock is protected from the smearing effects of the dissipation. In the second figure of Fig. 3 are plotted three solutions with standard dissipation, but where the 2nd order AVIS is scaled down in the shock cell by (i) 1.0, (ii) 0.3 and (iii) 0.1. It can be seen that as the amount of AVIS is reduced in the shock cell, the shock becomes far better resolved. However, at least a small level of AVIS is always required in the shock cell to balance the error transmitted to the upstream supersonic node from the shock cell residuals (see below). Moreover, because the AVIS terms are far from transparent to the shock structure, if the AVIS coefficient is not well chosen in the shock cell there will either be an overshoot or the shock will be excessively smeared. However, we shall now extend this approach, employing improved distribution matrices in the shock cell to ameliorate this difficulty.

Shock detection is reasonably straight forward even in two and three dimensions, and for the nozzle problem reduces to merely recognising the last supersonic value $\mathbf{W}_{s-1}$ and the first subsonic value $\mathbf{W}_s$. Then a key point in the definition of the target solution is the recognition that $\mathbf{W}_{s-1}$ should be wholly defined by information coming from upstream, as in the ground-breaking Murman and Cole scheme [30]; that is, we need

$$D_{s-\frac{1}{2},s-1}\mathbf{R}_{s-\frac{1}{2}} + \mathbf{A}_{s-\frac{1}{2},s-1} = \mathbf{0}. \quad (22)$$

The simplest way to achieve this is to distribute all of the shocked cell residual $\mathbf{R}_{s-\frac{1}{2}}$ to the subsonic shock point value $\mathbf{W}_s$, and to set the artificial dissipation terms for this cell to zero; this results in having

$$D_{s-\frac{1}{2},s-1} = 0 \quad ; \quad D_{s-\frac{1}{2},s} = 2I \quad (23a)$$

$$A_{s-\frac{1}{2},s-1} = -A_{s-\frac{1}{2},s} = 0. \quad (23b)$$

For weak shocks this strategy may need to be modified; but for the moment we suppose that this is the only modification made to the distribution matrices. It ensures that the error committed in the shock cell residual is not transmitted upstream, and the 2nd order AVIS is needed only to damp the spurious washboard mode allowed by the standard distribution matrices in the smooth flow downstream of the shock. Moreover, since no AVIS is now needed in the shock cell itself to prevent overshoots, sensitivity to the AVIS levels is much reduced.

Let us now consider how this AVIS term should be designed. We can suppose that fourth order dissipation is used to overcome the solution indeterminancy of the cell residual equation system at the sonic point, as already described; and we shall suppose that as in two dimensional problems, where it is needed to control chequerboard modes, it is applied quite generally. However, in order to push the target solution further into the null space of the dissipation operator, differences across the shock cell should not be used in its calculation, and in any case we want to observe (23b). Thus $\tau_\alpha^{(4)}$ is set to zero in the shock cell and its two neighbours. The second order AVIS needed only in the subsonic region immediately downwind of the shock cell, is scaled by $O(h^2)$ as when using $\delta_h^2 p$ in (3) and is switched off smoothly by a damping factor $\gamma < 1$; so we set

$$\tau_\alpha^{(2)} = \frac{\hat{\epsilon}^{(2)} N^{-3}}{\Delta t_\alpha h_\alpha} \gamma^{|s-\alpha|} \text{ if } M_\alpha < 1,$$
$$\tau_\alpha^{(2)} = 0 \text{ if } M_\alpha \geq 1, \tag{24}$$

with $\hat{\epsilon}^{(2)} = O(1)$. The results shown in the bottom figure of Fig. 3 show excellent agreement with the target solution and only a weak dependence on the parameter $\hat{\epsilon}^{(2)}$. Indeed, it is clear that a reasonable solution can be obtained in the complete absence of the second order AVIS.

By contrast, results obtained with standard second order AVIS which are shown in the third figure of Fig. 3 are heavily dependent on the value of $\epsilon^{(2)}$. Yet is worth noting that the best value of $\epsilon^{(2)}$ in this case, and in the second set of graphs obtained from the scheme (21), can be deduced from the relation (22). If we set $\Delta \mathbf{F} = A \Delta \mathbf{W}$ and neglect the source term in the shock cell residual, with $\nu_C = 1$ in the standard distribution matrix, the first term in (22) is given by

$$-(I - (u+c)^{-1}A)A\Delta\mathbf{W}/\Delta x =$$
$$-(\Delta x)^{-1}[\Delta\rho - (u+c)^{-1}\Delta(\rho u)] \begin{bmatrix} u-c \\ (u-c)^2 \end{bmatrix} ;$$
$$\tag{25}$$

while the standard AVIS term from this cell, which gives the dominant contribution, is

$$-\tfrac{1}{2}S\tau^{(2)} \begin{bmatrix} \Delta\rho \\ \Delta(\rho u) \end{bmatrix} =$$
$$-\tfrac{1}{2}S(\Delta x)^{-1}\epsilon^{(2)}(u+c)\max(\kappa_+, \kappa_-) \begin{bmatrix} \Delta\rho \\ \Delta(\rho u) \end{bmatrix} .$$
$$\tag{26}$$

Because $(u-c)$ is very small in the cell, and even in this shock cell $\Delta(\rho u)$ is much smaller than $\Delta\rho$, the first components of these vectors dominate; thus we have a

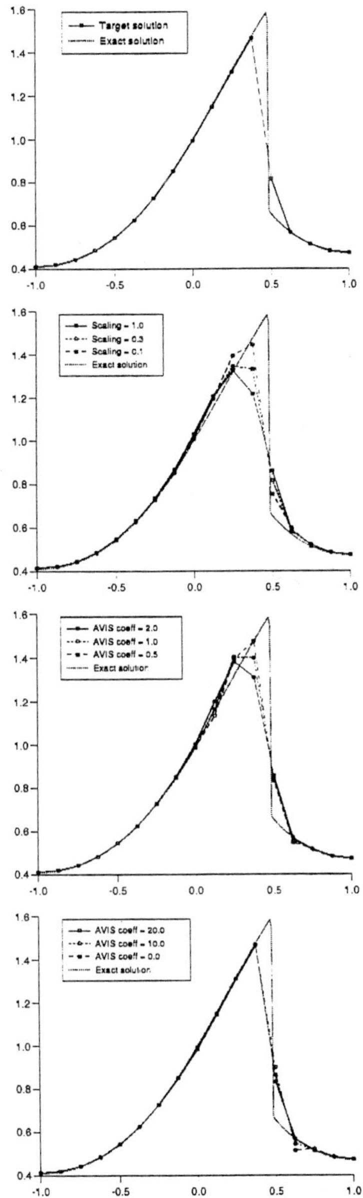

Figure 3: Mach No. for cell vertex approximations to 1D nozzle problem: from the top, target solution; standard AVIS with $\epsilon^{(2)} = 4.0$ reduced in shock cell; standard AVIS with $\epsilon^{(2)} = 2, 1, \tfrac{1}{2}$; new scheme with $\hat{\epsilon}^{(2)} = 20, 10, 0$.

matching of these terms if

$$S\epsilon^{(2)}\max(\kappa_+, \kappa_-) = 2(u - c)/(u + c).$$

This gives an optimal value of $S\epsilon^{(2)}$ of around 0.2, which corresponds closely with best values obtained in both the second and third set of graphs in Fig. 3

In moving to two dimensions the first step is the shock detection. The technique used in [28], [26], [27], and developed from that in [22] and [23], is based on recovering the shock normal from

$$\mathbf{n}_\alpha = -(\nabla M)_\alpha \tag{27}$$

as in (10a), where M is the Mach number. Then the decision on whether a cell Ω_α is judged to be crossed by a shock is based on three sets of conditions being satisfied:

(i) an initial test that $M_j > 1$ holds for at least one node in $\mathcal{N}_\alpha$;

(ii) the flow is decelerating and the pressure is rising through the shock, i.e. $\mathbf{M}_\alpha.\mathbf{n}_\alpha > 0$, $(\nabla p)_\alpha.\mathbf{n}_\alpha > 0$, where $\mathbf{M}_\alpha = \mathbf{v}_\alpha/c_\alpha$;

(iii) there are a pair of nodes $i, j \in \mathcal{N}_\alpha$ which are respectively upstream supersonic and downstream subsonic, and between which there is a substantial fractional rise in pressure and decrease in normal velocity component.

In [28] it was shown how this technique can be used to replace the $\delta_h^2 p$ switch of (3) in the application of the standard artificial dissipation terms, in both two and three dimensional problems. Here we will also use it as a basis for changing the standard distribution matrices as in (23).

For this purpose, it is useful and convenient to modify stage (iii) of the procedure. Firstly, as compared with the one-dimensional situation, a shock may cross the corner of a cell so that the cell may easily slip in and out of the selected set as the iteration proceeds; thus the selection needs to be made more stable and its consequences modified. Secondly, the distribution matrices as given in (12) can be seen to form diagonal pairs, with each pair summing to 2I. Thus in stage (iii) the tests are applied to the two diagonal pairs only; and the distribution matrix modification of (23a) is replaced by the following. Consider a diagonal pair of nodes u, d of cell Ω_α that are upstream and downstream of the shock respectively. The new shock cell distribution matrices are taken as

$$D_{\alpha,d} = D_{\alpha,d}^{LW} + \phi D_{\alpha,u}^{LW}, \quad D_{\alpha,u} = (1 - \phi)D_{\alpha,u}^{LW}. \tag{28}$$

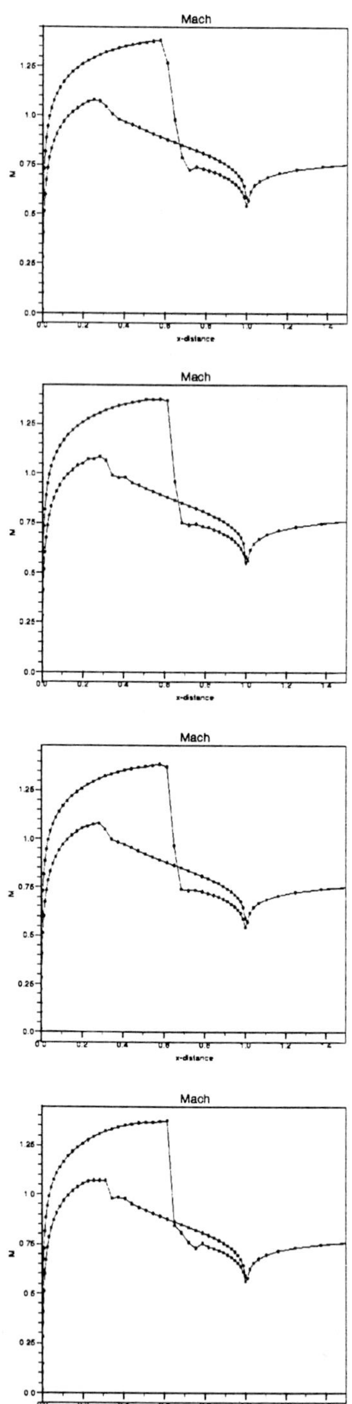

Figure 4: Mach No. for Euler test case 3.1: from the top, standard AVIS; matrix AVIS; $|\lambda_s|$ scaling; new scheme with shock-switched dissipation.

Grid	Lift	Drag
129×33	0.3724	0.0230
257×65	0.3673	0.0228
513×129	0.3626	0.0227

Grid	Lift	Drag
129×33	0.3722	0.0231
257×65	0.3645	0.0227
513×129	0.3610	0.0226

Table 1: Lift and drag coefficients for calculations of Fig. 4: standard AVIS (top), matrix AVIS (bottom).

where D^{LW} are the standard Lax-Wendroff distribution matrices given by (12). Ideally, we would like to set $\phi = 1$; however, in practice we have found this leads to limit cycles with certain cells oscillating between shocked and unshocked states. Hence we have found it necessary to set

$$\phi = \epsilon_\phi \max(\frac{P_d - P_u}{P_d}) \quad \text{with} \quad \epsilon_\phi = 2. \tag{29}$$

For the same reason we have found the δ_h^2 scaling of $\tau^{(2)}$ preferable to generalisations of (24), though its lack of smoothness leaves the same glitch downwind of the shock as with the standard AVIS — see Fig. 4.

In numerical tests, four solutions have been calculated for the transonic inviscid NACA0012 aerofoil test case of section 3.1 on a coarse 129×33 grid, and are shown in Fig. 4. The first is calculated with the standard artificial viscosity model, the second is calculated with matrix dissipation as in [37], the third is calculated with the scalar dissipation scaling $|\lambda_s|$, and the fourth is calculated with the new shock cell distribution matrices (28) and no 2nd order AVIS in the shock cell. The new distribution matrices (28) offer good shock resolution with comparatively little sensitivity to the value of $\epsilon^{(2)}$. Fig. 5 is reproduced from [25], and shows entropy plots around the aerofoil for coarse, medium and fine mesh solutions with standard and matrix artificial dissipation. These results are to be compared with a coarse mesh solution calculated with the new distribution matrices (28) shown in Fig. 6, for which the lift and drag coefficients are 0.3672 and 0.0225 respectively. It can be seen that the values of the entropy jump through the shock, and the lift and drag coefficients, are dramatically improved on a coarse mesh with the new scheme.

Finally, numerical results are presented for the tran-

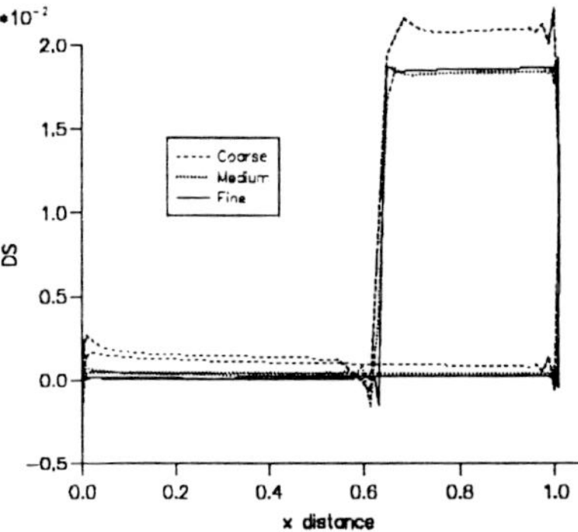

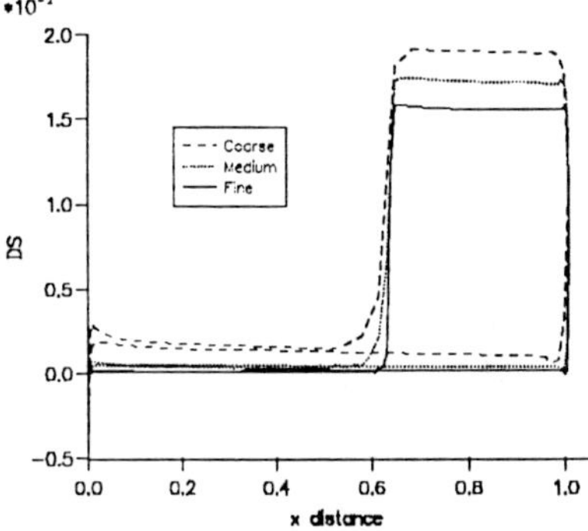

Figure 5: Entropy variable plots for standard AVIS (top) and matrix AVIS (bottom) applied to test case 3.1 on three meshes.

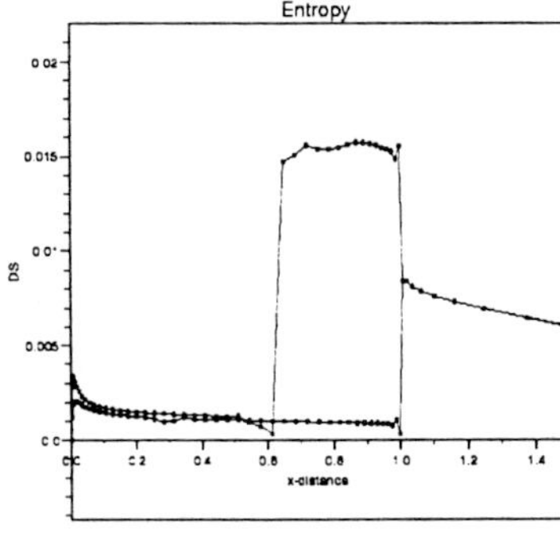

Figure 6: Entropy variable plot for coarse mesh case of Fig. 5 with new scheme.

washboard modes are generated by sharp flow changes, such as at the leading and trailing edges of an aerofoil; and if sufficient artificial dissipation is included to control these, then the boundary layers are distorted. The acid test of a scheme and the key practical requirement is good prediction of lift and drag; but detailed analysis of the flow prediction is needed to diagnose and correct deficiencies. We shall use the subsonic laminar test case of section 3.2 to carry out most of the analysis and development of new artificial dissipation schemes, while testing the developments on the other more demanding turbulent flow test case of section 3.3.

The results obtained with the standard cell vertex method for the well documented test case 3.2 are presented in detail in [5] and [20]. They show that the drag coefficient, boundary layer separation point, boundary layer profiles, skin friction coefficient etc. are reliably and accurately predicted with $\epsilon^{(2)} = 0$ and $\epsilon^{(4)}$ chosen over a wide range of values. But detailed examination of the recirculating flow region — particularly the velocity vectors, the Mach contours and the nodal and cell residual budget plots — show that too little AVIS leads to spurious mode oscillations and too much to cell residual imbalance. For example, Fig. 8 shows a plot of y-momentum tangential to the aerofoil in the recirculation region for a solution calculated with no artificial dissipation, where a spurious washboard mode can be plainly seen. The effect of these phenomena become more pronounced with more severe flow regimes, and we hope to largely eliminate them for this test case. Thus, our aim is to design new dissipation models that properly damp spurious modes, whilst protecting the natural variation in boundary layers.

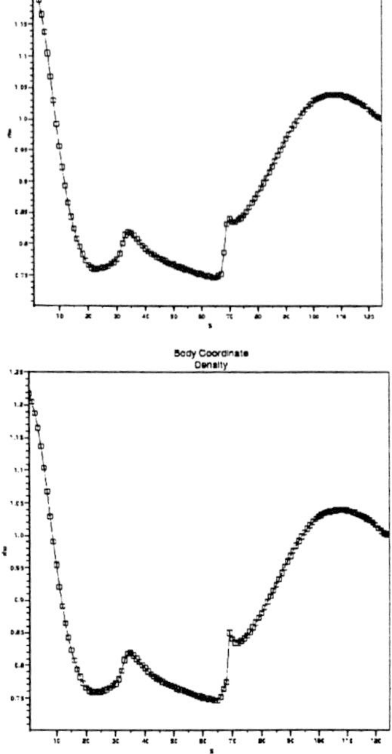

Figure 7: Density plots tangential to RAE5225 aerofoil for Navier-Stokes test case 3.4: standard AVIS (top), new scheme (bottom).

sonic turbulent Navier–Stokes RAE5225 aerofoil test case of section 3.4 on a 257×65 grid. Fig. 7 shows density plots tangential to the aerofoil, forty coordinate lines out from body, for solutions calculated with standard 2nd order dissipation, and shock switched dissipation with the new distribution matrices (28). It is clear that the new scheme has led to similarly good results for this turbulent test case. Furthermore, multigrid convergence rates remain good.

4.3   Subsonic viscous flows: smooth solution transparent (SST) artificial dissipation

With the subsonic Navier-Stokes equations, the emphasis in their numerical approximation is on the accurate prediction of boundary layers — their development, their separation, their effect on drag etc.; and then in the transonic case their interaction with shocks is an important predictive requirement. The principal difficulties with the standard cell vertex scheme, as with many other schemes, in meeting these targets is that spurious

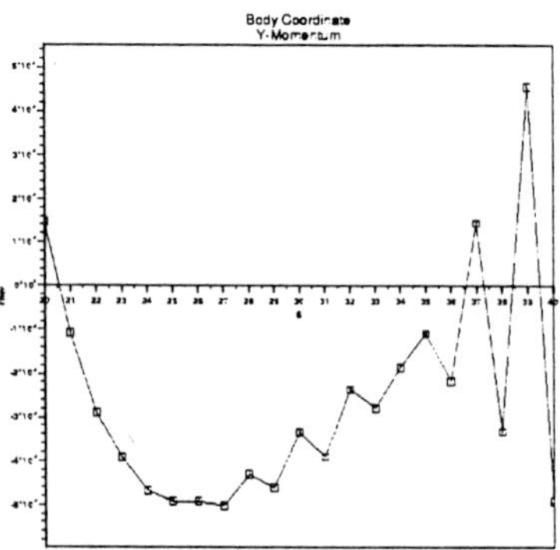

Figure 8: Test case 3.2 with no AVIS, showing spurious washboard mode in y-momentum tangential to the aerofoil in the recirculation region.

Numerical tests with both matrix and linearly transparent 4th order dissipation have been disappointing by these criteria. On a well scaled mesh, standard dissipation terms are already nearly linearly transparent, and so linear transparency offers very little improvement. With matrix dissipation, the wavespeed scaling reduces the amount of dissipation close to the aerofoil, which can lead to the presence of a spurious mode there. Furthermore, the reduction in the AVIS contributions to nodal budget plots can be understood in terms of the effective streamlining of the dissipation due to the wavespeed scaling. In fact, for the laminar NACA0012 test case, far better results can be obtained by a simple scalar dissipation model in the streamwise ξ direction only. Thus our general approach now is to design dissipation terms that have the local smooth solution further within their null space, but not the spurious modes. We term such dissipation models *smooth solution transparent* (SST). A number of different approaches to smooth solution transparency have been explored in [26], [27], [35], e.g. generalised SST 4th difference terms, and SST coefficients. However, most of these designs lead to erratic behaviour in nodal and cell budget plots due to non-smooth switches. Thus, our preferred approach involves enlarging the stencil to form higher order differences that have a more linear character.

A cell-based nth order dissipation model has terms of the form

$$\mathbf{A}_{\alpha,j}^{(n)}(\mathbf{W}) := -(-1)^{n/2}\tau_\alpha^{(n)}\left(\delta_h^{n-2}\mathbf{W}_j - \delta_h^{n-2}\mathbf{W}_\alpha\right)$$
$$\forall j \in \Omega_\alpha, \tag{30}$$

where n is even, $\delta_h^{n-2}\mathbf{W}_\alpha$ is an average cell centre value, and where

$$\tau_\alpha^{(n)} = \max\left\{0, \frac{\epsilon^{(n)}}{\Delta t_\alpha} - \tau_\alpha^{(2)}\right\}. \tag{31}$$

It is clearly conservative and the undivided differences $\delta_h^{n-2}\mathbf{W}_j$ may be constructed in a simple manner: firstly linearly transparent 2nd differences $\delta_h^2\mathbf{W}_j$ are calculated as in (18) and stored at each node of the mesh; then these are used to calculate $\delta_h^4\mathbf{W}_j$ (2nd differences of 2nd differences), which in turn are used to calculate $\delta_h^6\mathbf{W}_j$, and so on. At boundaries the differences $\delta_h^{n-2}\mathbf{W}_j$ incorporate a tangential component only; the normal contribution is taken as zero. This approach was generally found to give the best experimental results in [6]. In fact, the whole process may be viewed as successive iterations of a global operator which constructs nth order differences from $(n-2)$th order differences.

Consider now the size of the higher differences for a chequerboard mode: it is clear that for unit values

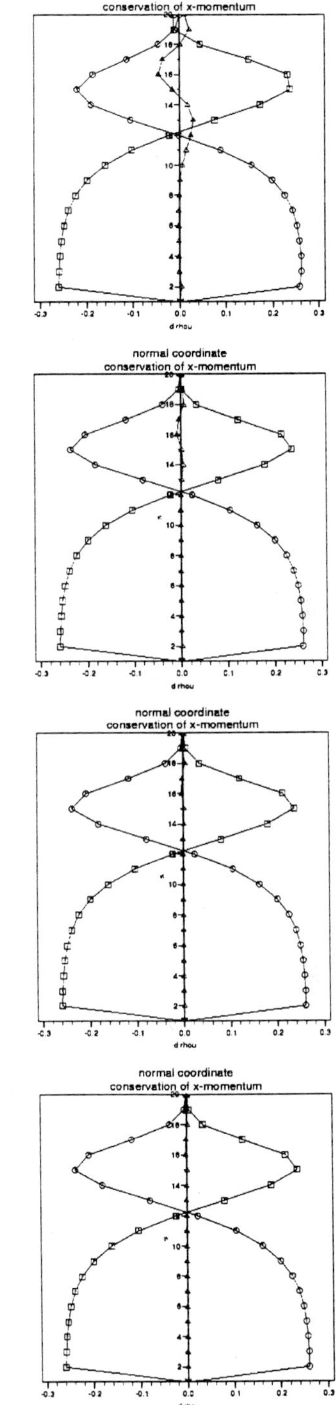

Figure 9: Nodal budget plots for test case 3.2 with, from the top, 4th, 6th, 8th and 10th order artificial viscosity [□, inviscid flux; o, viscous flux; △ artificial dissipation].

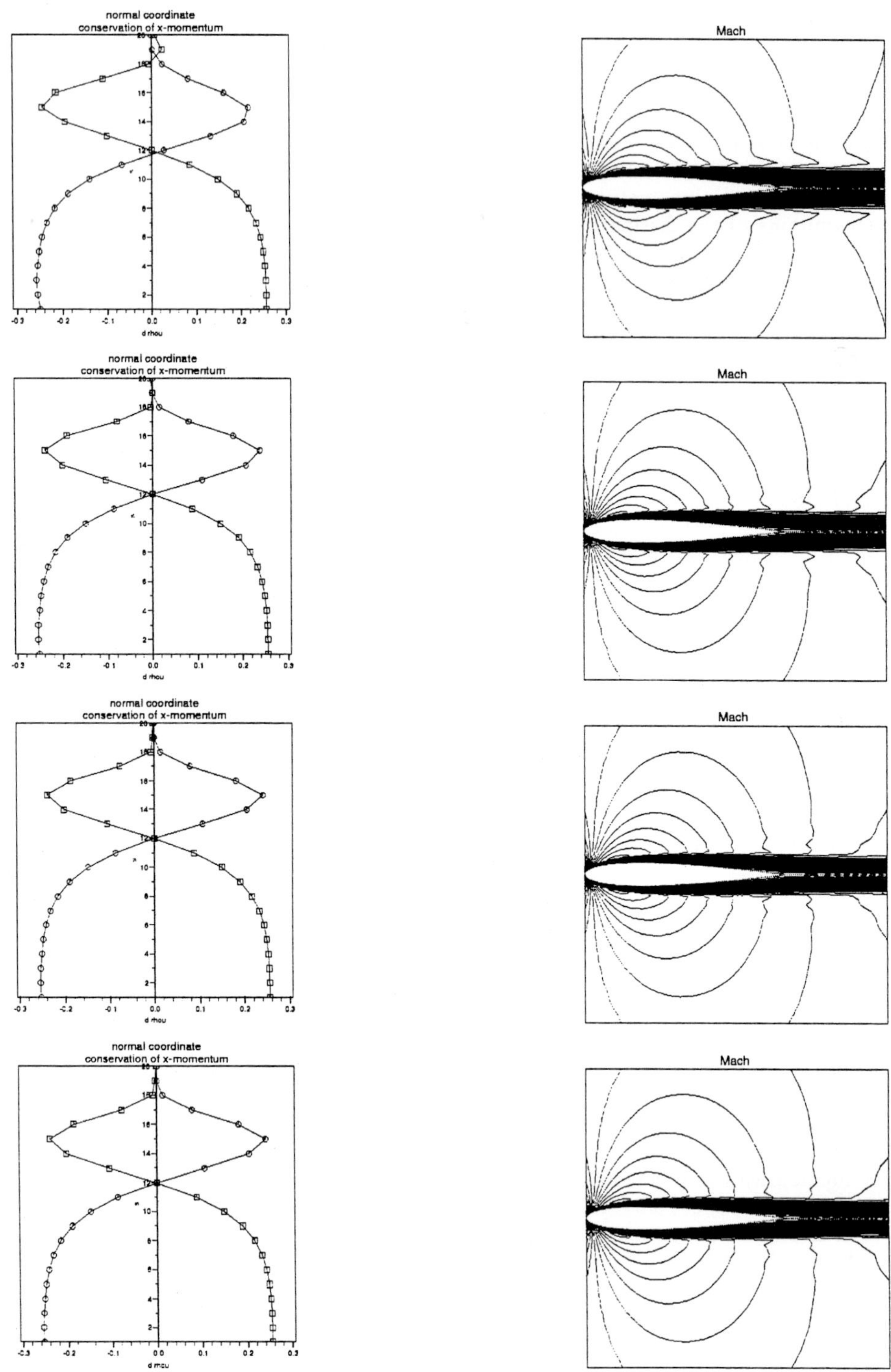

Figure 10: Cell residual budget plots as for Fig. 9.   Figure 11: Plots of Mach No. contours as for Fig. 9.

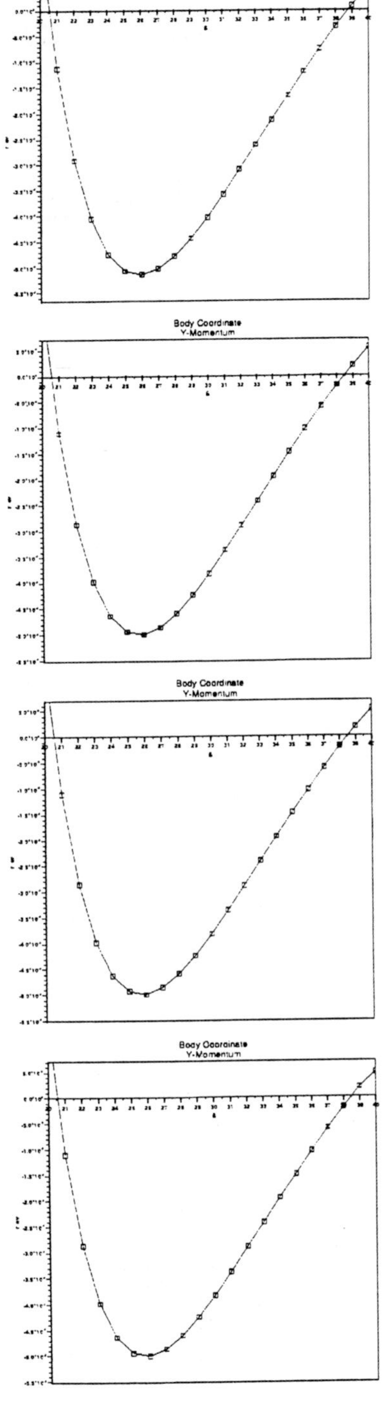

Figure 12: Plots of y-momentum tangential to the aerofoil in the recirculation region as for Fig. 9.

$\delta_h^2 \mathbf{W}_j \geq 8$, with equality holding on a uniform mesh, and hence generally we have

$$\delta_h^{n+2}\mathbf{W}_j \approx 8 \times \delta_h^n \mathbf{W}_j.$$

Thus, in numerical tests we have found that $\epsilon^{(n)}$ needs to be reduced as n increases in order to maintain convergence of the pseudo-timestepping solution procedure. For each value of n, the coefficient $\epsilon^{(n)}$ was tuned to an optimal value through experimentation.

Four solutions of the laminar Navier–Stokes NACA0012 test problem 3.2 are presented in Figs. 9, 10, 11 and 12 with: (i) 4th, (ii) 6th, (iii) 8th and (iv) 10th order artificial viscosity. The artificial viscosity coefficients used were: $\epsilon^{(4)} = 0.05$, $\epsilon^{(6)} = 0.005$, $\epsilon^{(8)} = 0.001$, $\epsilon^{(10)} = 0.0002$. Fig. 9 shows nodal budget plots normal to the aerofoil at 50 percent chord; Fig. 10 shows the correspondingly budget plots for the cell residuals; Fig. 11 shows the Mach No. contours; and Fig. 12 shows the y-momentum in the recirculation region, as in Fig. 8. It is seen that the schemes are extremely effective, with even the cell residuals well balanced throughout the boundary layer while from Fig. 12 it is seen that there is no visible trace of a spurious mode. Furthermore, the linear character of these higher order dissipation models helps to avoid limit cycles and fast multigrid convergence is maintained. The Mach No. contours continue to show some oscillations for the higher order schemes; but the viscous drag coefficients shown below show steady improvement.

	inviscid drag coeff	viscous drag coeff
4th order	0.022563	0.032935
6th order	0.022574	0.032651
8th order	0.022573	0.032645
10th order	0.022572	0.032644

It is our view such higher order dissipation models will automatically damp spurious modes wherever they occur without serious smoothing of the underlying solution. Moreover, it is the linear character of these higher order dissipation models that ensures good convergence rates and 'smooth' behaviour in nodal and cell budget plots. The rather erratic behaviour seen with other SST designs may well be partly due to their non-linearity. Lastly, it seems natural to expect that enlarging the stencil of the dissipation model will allow more precise discrimination between natural variation and spurious modes, but these are usually inconvenient in practice near boundaries; here the iteration of the global second order operator overcomes this problem.

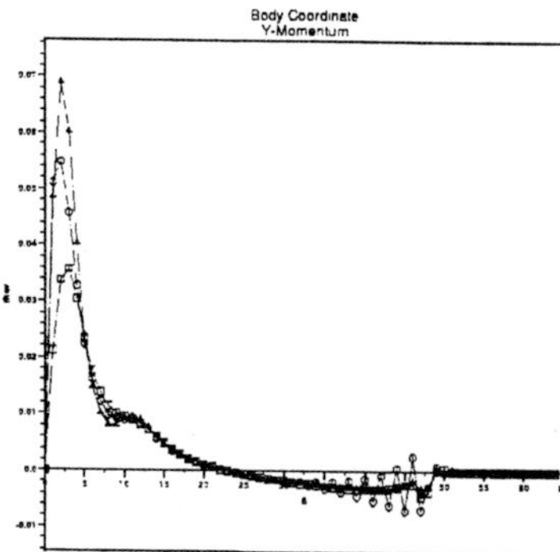

Figure 13: Plots of y-momentum tangential to the aerofoil for test case 3.3: Mach No. scaled standard 4th order dissipation (o); standard 4th order dissipation without Mach No. scaling (□); and 6th order dissipation without Mach No. scaling (△).

Finally, numerical results are given in Fig. 13 for the subsonic turbulent Navier–Stokes NACA0012 aerofoil test case 3.3 on a 129×65 grid. We compare standard 4th order dissipation with our favoured SST 6th order dissipation. Fig. 13 shows plots of y-momentum tangential to the aerofoil, two coordinate lines out from body, for solutions calculated with (i) Mach No. scaled standard 4th order dissipation (o); (ii) standard 4th order dissipation without Mach No. scaling (□) and (iii) 6th order dissipation without Mach No. scaling (△). In these experiments we set $\epsilon^{(4)} = 0.05$ and $\epsilon^{(6)} = 0.01$. The Mach No. scaling takes the form $\min\{1, M/M_\infty\}$ described in section 1. The introduction of this scaling reduces the level of 4th order dissipation close to the aerofoil, which results in less smoothing of the underlying physical solution but also allows the occurrence of a spurious washboard mode near the trailing edge, that can be seen clearly in the figure. So in practice a balance must be struck between damping of spurious modes and protecting the natural solution, and because turbulent calculations involve high natural variation, they are far more susceptible than laminar problems to severe degradation by artificial viscosity. From Fig. 13 it can be seen that with smooth solution transparent 6th order dissipation the physical behaviour near the leading edge is far less smeared, whilst the spurious mode is properly damped. Moreover, in the table below for the nonoscillatory cases (ii) and (iii) we see a very significant improvement in the accuracy of lift and drag coefficients which should be compared with experimental data from [1] for Re=3.0×10⁶, where the total drag is given as 0.0058±0.0002.

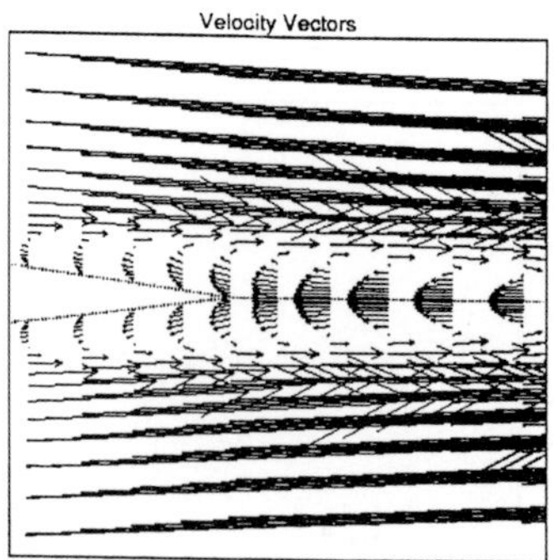

Figure 14: Velocity vectors near trailing edge for test case 3.2.

	4th order	6th order
inviscid drag coeff.	0.0021721	0.0018188
viscous drag coeff.	0.0077162	0.0075530

However, it should be noted that convergence is less robust with 6th order dissipation than with 4th order; convergence was achieved for only a relatively small range of $\epsilon^{(6)}$.

4.4 Improved distribution matrices

Lastly, we consider briefly whether any modifications to the standard distribution matrices of (12) would be appropriate. Theoretical studies of the convection-diffusion equation show that at low Péclet or Reynolds numbers one should not base the choice of distribution matrices solely on the convective (i.e. inviscid) fluxes; thus some change in the boundary layer could be beneficial. This would also reflect the differences in numerical behaviour arising from the Navier-Stokes boundary conditions. However, these have their greatest effect in the recirculating region near the trailing edge in test case 3.2. The effect there is that the flow has a component directed towards the body as in flow incident on a wedge, which could easily give rise to the oscillations obtained with an inadequately upwinded difference scheme and a normal boundary layer. Now for the cell vertex method applied to the convection diffusion equation, these oscillations can be completely avoided, as shown in [14] and [19]. This is achieved by not using the cell residuals adjacent to the boundary to update the interior nodal values, but rather to define the normal derivative terms on the boundary.

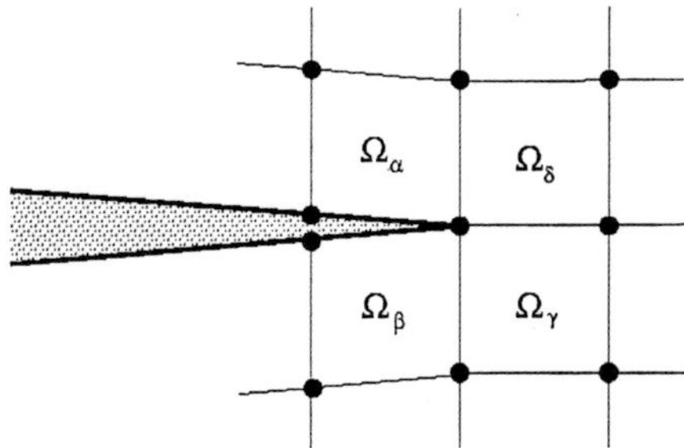

Figure 15: Four cells at a trailing edge.

Fig. 14 shows the velocity vectors near the trailing edge for the laminar Navier-Stokes NACA0012 test case 3.2, showing the recirculation region. In particular, the flow in the four cells adjacent to the trailing edge shown in Fig. 15 is either towards or parallel to the aerofoil surface, where there are Dirichlet boundary conditions. As a simple modification we therefore set to zero the distribution matrices for these four cells. The result is shown in Fig. 16 for standard AVIS and the relatively small value for $\epsilon^{(4)}$ of 0.0025, that would normally allow the occurrence of the spurious washboard mode near the trailing edge. It is clear that this simple local change has largely eliminated the spurious washboard mode; and fast multigrid convergence is also maintained. This is an indication that there is considerable scope for developing the cell vertex scheme by improving the distribution matrices.

5   Conclusions

Widely differing views are held by algorithm developers on the rôle of artificial dissipation terms: some are prepared to start with a simple central difference scheme and use artificial dissipation as the major tool to knock it into shape; others, principally those who advocate upwind schemes, abhor its direct use at all, while acknowledging that the difference between their schemes and related central difference schemes can be interpreted as artificial dissipation.

The cell vertex scheme that is the subject of the present paper gives a central difference scheme — but one centred on a cell rather than a node. As is now well documented (e.g. [14], [29], [19]), it can thus give accurate solutions to scalar convection-diffusion problems, at any Péclet number, without the addition of artificial dissipation. On the other hand, these terms have been required when solutions to the Euler and Navier-Stokes systems of equations are sought and simple Lax-Wendroff distribution matrices have to be used.

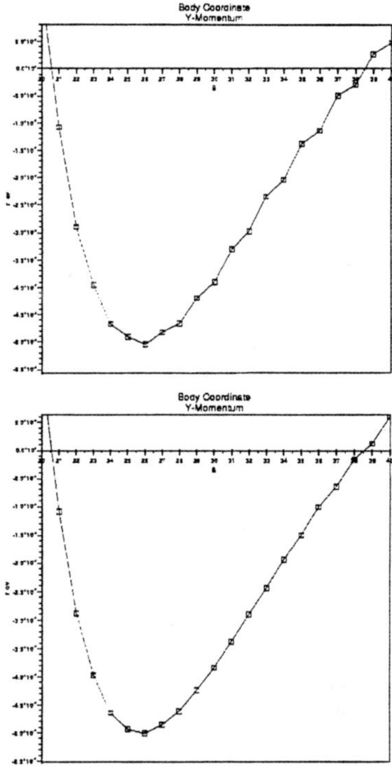

Figure 16: Plots of y-momentum tangential to the aerofoil in test case 3.2: on the left with standard distribution matrices; on the right, with the trailing edge modification.

In the present paper, we have made elementary use of a state space control theory viewpoint to design artificial dissipation terms for fluid flow computations that are not used to define the target solution, but merely to eliminate spurious solution modes. In practice this refinement to the basic scheme is relatively minor, but the improvement to its performance quite marked. Thus we feel that the new viewpoint is helpful and we have barely started its exploitation; it could, for instance, be equally well used for other schemes such as the vertex-centred schemes of [10].

References

[1] I.H. Abbott and A.E. von Doenhoff. *Theory of Wing Sections*. Dover, 1959.

[2] B.S. Baldwin and H. Lomax. Thin layer approximation and algebraic model for separated turbulent flows. AIAA Paper 78-257, 1978.

[3] P. Crumpton and M.B. Giles. Aircraft computations using multigrid and an unstructured parallel library. AIAA Paper 95-0210, 1995.

[4] P.I. Crumpton. A cell vertex method for 3d Navier Stokes solutions. Technical Report NA93/09, Oxford University Computing Laboratory, Wolfson Building, Parks Road, Oxford, OX1 3QD., 1993.

[5] P.I. Crumpton, J.A. Mackenzie, and K.W. Morton. Cell vertex algorithms for the compressible Navier-Stokes equations. *J. Comput. Phys.*, 109(1):1–15, 1993.

[6] M.R. Field. *The setting up and solution of the cell vertex equations*. PhD thesis, Oxford University Computing Laboratory, Wolfson Building, Parks Road, Oxford, OX1 3QD., 1994.

[7] A. Gould. Results of the CVTCC-VII test case. Unpublished Report of the Cell Vertex Technology Coordination Centre, Sowerby, April 1993.

[8] A. Gould. Results of the CVTCC-VIII test case. Unpublished Report of the Cell Vertex Technology Coordination Centre, Sowerby, July 1993.

[9] M.G. Hall. Cell vertex schemes for the solution of the Euler equations. In K.W. Morton and M.J. Baines, editors, *Proceedings of the Conference on Numerical Methods for Fluid Dynamics, University of Reading, U.K.*, pages 303–345. Oxford University Press, 1986.

[10] M.G. Hall. On the reduction of artificial viscosity in viscous flow solutions. In D.A. Caughey and M.M. Hafez, editors, *Frontiers of Computational Fluid Dynamics*, Computational Methods in Applied Science, pages 303–317. John Wiley & Sons, 1994.

[11] A. Jameson. Analysis and design of numerical schemes for gas dynamics: Part 1 artificial diffusion, upwind biasing, limiters and their effect on accuracy and multigrid convergence. *Internat. J. Comput. Fluid Dynamics*, 4(3-4):171–218, 1995.

[12] A. Jameson. Analysis and design of numerical schemes for gas dynamics, part 2 artificial diffusion and discrete shock structure. *Internat. J. Comput. Fluid Dynamics*, 5(1-2):1–38, 1995.

[13] A. Jameson, W. Schmidt, and E. Turkel. Numerical solutions of the Euler equations by finite volume methods with Runge–Kutta time stepping schemes. AIAA Paper 81-1259, 1981.

[14] J.A. Mackenzie and K.W. Morton. Finite volume solutions of convection-diffusion test problems. *Math. Comp.*, 60(201):189–220, 1992.

[15] L. Martinelli. *Calculations of Viscous Flows with a multigrid method*. PhD thesis, Dept. of Mech. and Aerospace Eng., Princeton University, 1987.

[16] D.J. Mavriplis, Jameson A., and L. Martinelli. Multigrid solution of the Navier–Stokes equations on trianguler meshes. AIAA Paper 89-0120, 1989.

[17] K.W. Morton. Finite volume and finite element methods for the steady Euler equations of gas dynamics. In J.R. Whiteman, editor, *The Mathematics of Finite Elements and Applications VI MAFELAP 1987*, pages 353–378. Academic Press, 1988.

[18] K.W. Morton. Coercivity for one-dimensional cell vertex approximations. In D.F. Griffiths and G.A. Watson, editors, *Numerical Analysis: A R Mitchell 75th Birthday Volume*, pages 189–206. World Scientific, 1996.

[19] K.W. Morton. *Numerical Solution of Convection-Diffusion Problems*, volume 12 of *Applied Mathematics and Mathematical Computation*. Chapman & Hall, 1996.

[20] K.W. Morton, P.I. Crumpton, and J.A. Mackenzie. Cell vertex methods for inviscid and viscous flows. *Comput. & Fluids*, 22(2/3):91–102, 1993.

[21] K.W. Morton and M.F. Paisley. A finite volume scheme with shock fitting for the steady Euler equations. *J. Comput. Phys.*, 80:168–203, 1989.

[22] K.W. Morton and M.A. Rudgyard. Shock recovery and the cell vertex scheme for the steady Euler equations. In D.L. Dwoyer, M.Y. Hussaini, and R.G. Voigt, editors, *Proceedings of 11th International Conference on Numerical Methods in Fluid Dynamics, Williamsburg, U.S.A. 1988*, volume 323 of *Lecture Notes in Physics*, pages 424–428. Springer-Verlag, 1989.

[23] K.W. Morton and M.A. Rudgyard. Finite volume methods with explicit shock representation. In L. Fezoui, J.C.R. Hunt, and J. Periaux, editors, *Computational Aeronautical Fluid Dynamics*, IMA Conference Series, pages 103–121. Clarendon Press, 1994. Proceedings of GAMNI/SMAI - IMA Conference on Computational Aeronautical Fluid Dynamics, Antibes, 1989.

[24] K.W. Morton, M.A. Rudgyard, and G.J. Shaw. Upwind iteration methods for the cell vertex scheme in one dimension. *J. Comput. Phys.*, 114(2):209–226, 1994.

[25] K.W. Morton and S.M. Stringer. Finite volume methods for inviscid and viscous flows, steady and unsteady. In H. Deconinck, editor, *Computational Fluid Dynamics*, Lecture Series 1995-02, pages 1–63. von Karmen Institute, 1995.

[26] K.W. Morton and S.M. Stringer. Recent developments of the cell vertex method. In *Sixth International Symposium on Computational Fluid Dynamics. A Collection of Technical Papers*, volume II, pages 857–872, 1995. Lake Tahoe, September.

[27] K.W. Morton and S.M. Stringer. Artificial viscosity and the cell vertex method. To appear in the Proceedings of the 15th ICNMFD Conference, Monterey, 24-28 June, 1996.

[28] K.W. Morton, S.M. Stringer, and M.A. Woodgate. A cell vertex method for 3D Navier-Stokes equations with matrix artificial dissipation and shock detection. In W.H. Hui, Yue-Kuen Kwok, and J.R. Chasnov, editors, *First Asian Computational Fluid Dynamics Conference*, volume One, pages 41–50. Hong Kong Univ. of Sci. & Tech., 1995.

[29] K.W. Morton, M. Stynes, and E. Süli. Analysis of cell-vertex methods for convection-diffusion problems. Technical Report NA94/06, Oxford University Computing Laboratory, Wolfson Building, Parks Road, Oxford, OX1 3QD., 1994. Submitted for publication.

[30] E.M. Murman and J.D. Cole. Calculation of plane steady transonic flows. *AIAA Journal*, 9:114–121, 1971.

[31] R.-H. Ni. A multiple grid scheme for solving the Euler equations. *AIAA Journal*, 20(11):1565–1571, Nov 1981.

[32] T.H. Pulliam. Artificial dissipation models for the Euler equations. *AIAA J.*, 24:1931–1940, 1986.

[33] R. Radespiel and R.C. Swanson. An investigation of cell centered and cell vertex multigrid schemes for the Navier–Stokes equations. AIAA Paper 89-0548, 1989.

[34] P.L. Roe. Characteristic–based schemes for the Euler equations. *Annual Review of Fluid Mechanics*, 18:337–365, 1986.

[35] S.M. Stringer and K.W. Morton. Artificial viscosity for the cell vertex method. Technical Report NA96/08, Oxford University Computing Laboratory, Wolfson Building, Parks Road, Oxford, OX1 3QD., 1996.

[36] R.C. Swanson and E. Turkel. A multistage timestepping scheme for the Navier-Stokes equations. AIAA Paper 85-0035, 1985.

[37] R.C. Swanson and E. Turkel. On central-difference and upwind schemes. Technical report, ICASE, NASA, 1990. Report no. 90-44.

[38] R.C. Swanson and E. Turkel. On central-difference and upwind schemes. *J. Comput. Phys.*, 101:297–306, 1992.

[39] N.D. Wash. *Upwind Iteration Techniques for Compressible Flow Computations*. PhD thesis, Oxford University Computing Laboratory, 1995.

A GENERAL CLASS OF DIFFERENCE APPROXIMATION FOR SCALAR CONSERVATION LAWS REALIZING BOTH HIGH RESOLUTION AND THE CONVERGENCE TO ENTROPY SOLUTION
(An Analysis from the Viewpoint of Numerical Viscosity)

AISO, Hideaki

Computational Sciences Division, National Aerospace Laboratory; Jindaiji-Higashi-Machi 7-44-1, Chofu, TOKYO 182, JAPAN

Abstract

In this article we are concerned with difference approximation for conservation law. We mainly analyze the convergence property from the viewpoint of numerical viscosity. In the main theorem we obtain a rather wide class of difference approximations converging to the entropy solution. The main theorem is one of the best possible results of this kind.

INTRODUCTION

Conservation law, a class of partial differential equation, is mathematical description of the concept of conservation that is essential in physics, and plays an important role in fluid dynamics. It still provides us with interesting mathematical problems as well. It is well known that the motion of inviscid compressible fluid is described by a system of conservation laws called Euler equation and that physical phenomena (propagation of small perturbations, shocks, etc.) of the fluid gives us good examples of mathematical concepts like characteristics, weak solution, Rankine-Hugoniot relation etc. Even in the case of compressible viscous fluid governed by Navier-Stokes equation, a large part of the fluid motion follows the mechanism of conservation if Reynolds number is large.

Now CFD (Computational Fluid Dynamics) is a popular and important technology. Difference approximation is a numerical method popularly used in CFD. In this situation we need a reliable theory of difference approximation for conservation law which helps us improve numerical schemes of difference approximation, while we have used difference approximation as a tool for the theoretical analysis. In other words, it is important to construct a good scheme of difference approximation which provides us with numerical results accurate enough for the practical use as well as

converges to the entropy solution (physically relevant solution), *i.e.* the unique weak solution that satisfies the entropy condition.

But it is not easy to construct a difference scheme satisfying both requirements of the convergence and the accuracy because the two requirements are contradictory in some sense. Generally speaking, some numerical viscosity must be added to guarantee the convergence to entropy solution as well as to suppress harmful numerical instability like oscillation. On the other hand the numerical viscosity comes along with the inevitable side effect of smearing numerical solution. Then naturally arises an essential and important problem: Determine the lower limit of numerical viscosity that guarantees the convergence to entropy solution. The complete solution to problem would provide us with a strong and useful tool to construct the ultimate schemes of difference approximation, but the problem is not yet completely solved. One of the reason for the difficulty of problem is that almost all the system of conservation laws remain still unsolved, *i.e.* the existence and uniqueness of entropy solution is not proved. Another is that the discretized equation not only inherits the property of original differential equation of conservation law but also receives secondary property coming from discretization and that our analysis tools are not enough to discriminate the secondary one.

Here we restrict ourselves into the case of scalar con-

Received on March 20, 1997.

servation law, where the property of exact solution is well known. We discuss the convergence of difference approximation to entropy solution from the viewpoint of numerical viscosity, and finally we determine a general class of difference approximations converging to the entropy solution in terms of numerical viscosity.

Let us present the background of our discussion reviewing a brief history of research in the field.

The famous work by Oleĭnik[17] may be one of the earliest mathematical works that treat difference approximation for conservation law. In the work the existence and uniqueness of entropy solution to scalar conservation law is proved with Lax-Friedrichs scheme of difference approximation[10]. It implies that a numerical solution by Lax-Friedrichs scheme should be accurate enough if the mesh of difference is fine enough. But, in practical use we can not avoid strong smearing of numerical solution thanks to the large numerical viscosity. Another scheme of Lax-Wendroff[12] was proposed to attain the second order accuracy, where the order of accuracy is understood in the sense of Taylor expansion. In fact Lax-Wendroff scheme has much less numerical viscosity than Lax-Friedrichs scheme and it may give more accurate numerical solutions. On the other hand, it is well known that the numerical instability may cause spurious oscillation to prevent the numerical solution from the convergence to entropy solution.

Regarding the property of characteristics, it is a natural idea to switch the algorithm of difference according to the direction of characteristics. The idea is called upwind difference (in a weak sense). Several difference schemes have been proposed based on the idea. Murmann-Roe scheme [15, 19, 20], so called the upwind scheme (in a strong sense), is the simplest upwind difference. It has a little more numerical viscosity than Lax-Wendroff scheme and is stable enough to converge to a weak solution. But the convergence to entropy solution is not guaranteed. Also Godunov scheme [7], that is based on the exact solution to Riemann problem, is an upwind difference. Godunov scheme has numerical viscosity between those of Murmann-Roe scheme and Lax-Friedrichs scheme, and converges to the entropy solution. We also mention Engquist-Osher scheme [6] as another example of upwind difference. The numerical viscosity of scheme is between those of Godunov scheme and Lax-Friedrichs scheme and the convergence to entropy solution is guaranteed. Observing the fact above, we easily imagine that some relation would exist between the numerical viscosity and the convergence property of difference approximation.

It was proved by A.Y. LeRoux[14] and Harten[8, 9]

that a difference approximation with any numerical viscosity between those of Murmann-Roe scheme and Lax-Friedrichs scheme is stable in the sense of TVD (Total Variation Diminishing) and then converges to a weak solution[1], although the convergence to entropy solution was not proved. It is an innovative result determining a class of difference approximations that satisfy some particular property in terms of numerical viscosity. We also mention Harten's entropy-fix, which is a procedure to make Murmann-Roe scheme converge to the entropy solution by adding a small numerical viscosity when the numerical viscosity vanishes or is very small. But the theoretical proof for the entropy-fix remained still open at the time. The proof was given later in [2].

Tadmor and Osher [18] determined a class of E-scheme. A difference approximation is called an E-scheme if the numerical viscosity is between those of Godunov scheme and Lax-Friedrichs scheme. In [18] it is proved that any E-scheme converges to the entropy solution with some CFL restriction a little more strict than the natural CFL condition.

In [2] two main theorems are proved. The first theorem tells that any E-scheme converges to the entropy solution only with the natural CFL condition supposed, which is an extension of [18]. In the second, assuming that the flux function of conservation law is strictly convex, one determines a wider class of difference approximations converging to the entropy solution. The class includes difference approximations improved from Murmann-Roe scheme by Harten's entropy-fix, *i.e.* the theorem gives a proof for Harten's entropy fix. The class may be the widest if we consider 3-stencil difference approximations. The work [18] was followed by some extensions [22] that determines a class of higher order-accurate difference approximations converging to the entropy solution.

In this article we follow the line above and finally reach the main theorem (Theorem 9) to determine a general class of difference approximations converging to the entropy solution. The class that the main theorem determines includes high order-accurate difference approximations some of which are well known in practical use.

This article is organized as follows.

1. Scalar Conservation Law.
2. Difference Approximation in Viscous Form.
3. Examples and TVD-Property.
4. Consistency with Entropy Condition.
5. Consistency with Entropy Condition and Small Numerical Viscosity.

---

[1]Strictly speaking, the fact is a little more complicated. See Theorem 2 , Theorem 3 and Corollary 1 in Section 3.

6. Understanding Modified Flux Functions.
7. Consistency with Entropy Condition and Higher Order Accuracy.
8. Conclusion.

In Section 1 we review the property of exact solution to scalar conservation law. In Section 2 we prepare general concepts used in our discussion on difference approximation. We review several well known schemes of difference approximation and TVD property in Section 3.

Section 4 is devoted to a theorem and its proof which tell the sufficiency condition for the convergence to entropy solution in the case of general flux functions. In Section 5 we give an extension of the theorem in Section 4 by restricting the flux functions into strictly convex ones. In Section 6 we consider the concept of modified flux function. The concept is employed to give the proof of theorems in Section 4 and 5, but it is not only a technical tool but also seems to have some essential meaning. We give the main theorem in Section 7.

1. SCALAR CONSERVATION LAW.

We begin with a review of the theory for scalar conservation law.

The initial value problem for scalar conservation law is written in the form

$$
\begin{cases}
u_t + f(u)_x = 0, & -\infty < x < \infty, t > 0 \\
u(x,0) = u_0(x), & -\infty < x < \infty.
\end{cases}
\tag{1}
$$

We discuss the problem in BV, the category of functions of bounded variation. Then we suppose that the initial data u_0 is of bounded variation and we consider only weak solutions of bounded variation. The flux function f is general or strictly convex.

It is well known that the global existence of smooth solution may be violated because of the nonlinearity of flux function. Then we introduce the concept of weak solution to allow discontinuities in the solution. We define that $u = u(x,t)$ is a weak solution if the equality

$$
\int_0^\infty \int_{-\infty}^\infty \{\phi_t(x,t)u(x,t) + \phi_x(x,t)f(u(x,t))\}\,dxdt
$$

$$
+ \int_{-\infty}^\infty \phi(x,0)u_0(x)dx = 0
\tag{2}
$$

is satisfied for any test function $\phi \in C_0^\infty(\mathbf{R} \times [0,\infty))$.

The disadvantage of weak solution is the loss of uniqueness. In fact infinitely many weak solution may

exists to a problem and only one weak solution is physically relevant. Therefore we usually impose an additional condition on weak solutions to select the physically relevant one. The condition is called the entropy condition. We define the entropy condition following Lax[11].

To define the entropy condition we need the concept of entropy pair.

Definition 1 *A pair of function (U,F) is called an entropy pair if the pair satisfies*

1. U is convex,
2. $F' = U'f'$.

The functions U and F are called an entropy function and a corresponding entropy flux function, respectively.

We easily observe that infinitely many entropy pairs exist. In fact every convex function U can be an entropy function with the corresponding entropy flux function F determined by $F(s) = \int^s U'(\sigma)f'(\sigma)d\sigma$. Then we define the entropy condition.

Definition 2 *We say that a weak solution $u = u(x,t)$ to the problem (1) satisfies the entropy condition if u satisfies the entropy inequality*

$$
U(u)_t + F(u)_x \le 0
\tag{3}
$$

for every entropy pair (U,F). The entropy inequality (3) is understood in the sense of distribution, i.e.

$$
\int_0^\infty \int_{-\infty}^\infty \{\phi_t(x,t)U(u(x,t))
$$

$$
+ \phi_x(x,t)F(u(x,t))\}\,dxdt \ge 0
\tag{4}
$$

for any non-negative test function $\phi \in C_0^\infty(\mathbf{R} \times (0,\infty))$.

A weak solution satisfying the entropy solution is called the entropy solution. The existence and uniqueness of entropy solution is already known (See [17], also [21]).

The flux function f is restricted to be strictly convex later. When the flux function is strictly convex, discussion on the entropy condition is simplified because of the following theorem [5].

Theorem 1 *When the flux function f is strictly convex, the entropy condition is satisfied if the entropy inequality (3) is satisfied for one entropy pair (U,F) whose entropy function U is strictly convex.*

2. DIFFERENCE APPROXIMATION IN VISCOUS FORM

We mainly consider difference approximation in viscous form

$$u_i^{n+1}$$
$$= u_i^n - \frac{\lambda}{2}\left\{f(u_{i+1}^n) - f(u_{i-1}^n)\right\}$$
$$+ \frac{\lambda}{2}\left\{a_{i+\frac{1}{2}}^n(u_{i+1}^n - u_i^n) - a_{i-\frac{1}{2}}^n(u_i^n - u_{i-1}^n)\right\},$$
$$i, n \in Z, n \leq 0 \tag{5}$$

where $\lambda = \dfrac{\Delta t}{\Delta x}$ is the ratio of the widths of difference Δt and Δx in time and space, respectively, and each u_i^n is an approximate value at the node $(i\Delta x, n\Delta t)$. The second parenthesis of right-hand side is called the numerical viscosity, and each $a_{i+\frac{1}{2}}^n$ is called the numerical viscosity coefficient. Throughout the article we suppose that Δx and Δt satisfies the natural CFL condition

$$\lambda \cdot \sup_{\inf_{i,n} u_i^n \leq s \leq \sup_{i,n} u_i^n} |f'(s)| \leq 1. \tag{6}$$

The viscous form (5) is written also in conservative form

$$u_{i+1}^n = u_i^n - \lambda\left\{\bar{f}_{i+\frac{1}{2}}^n - \bar{f}_{i-\frac{1}{2}}^n\right\} \tag{7}$$

with the relation

$$\bar{f}_{i+\frac{1}{2}}^n = \left\{f(u_i^n) + f(u_{i+1}^n)\right\} - \frac{1}{2}a_{i+\frac{1}{2}}^n(u_{i+1}^n - u_i^n) \tag{8}$$

supposed. Each $\bar{f}_{i+\frac{1}{2}}^n$ is called the numerical flux. The conservative form is closely related with the finite volume concept. Each u_i^n is regarded to be a representative value of u over the interval (finite volume) $\left((i - \frac{1}{2})\Delta x, (i + \frac{1}{2})\Delta x\right)$ at the of $t = n\Delta t$ and each numerical flux $\bar{f}_{i+\frac{1}{2}}^n$ is regarded to represent the time-average of the flux function $f(u)$ at the point $x = (i + \frac{1}{2})\Delta x$ over the time-interval $[n\Delta t, (n+1)\Delta t]$. The conservative form is convenient when we apply the idea from physics to difference approximation.

We expect that the difference approximation would approach the real solution as the difference mesh is getting finer. Therefore it is important to analyze the convergence of difference approximation as Δ tends to $(0, 0)$. A difference approximation is understood to assign a set $\{u_i^n\}_{i,n \in \mathbf{Z}, n \geq 0}$ of approximate values over the nodes $(i\Delta x, n\Delta t)$, $i, n \in Z, n \geq 0$ to each $\Delta = (\Delta x, \Delta t)$ which satisfies $\dfrac{\Delta t}{\Delta x} = \lambda$, where we suppose the natural CFL condition (6).

But there is still ambiguity because the set $\{u_i^n\}_{i,n \in \mathbf{Z}, n \geq 0}$ is determined only over the discrete set $\{(i\Delta x, n\Delta t)\}_{i,n \in \mathbf{Z}, n \geq 0}$ of nodes of difference mesh but not over on the half space $\mathbf{R} \times [0, \infty)$. Then, for each Δ, we define a function u_Δ determined over the half space $\mathbf{R} \times [0, \infty)$ from $\{u_i^n\}_{i,n \in \mathbf{Z}, n \geq 0}$. Such a function is called an approximate function.

Finally we discuss convergence of difference approximation in the following manner.

1. Fix the ratio $\lambda = \dfrac{\Delta t}{\Delta x}$ so that the natural CFL condition (6) should be satisfied, and suppose that a set $\{u_i^n\}_{i,n \in \mathbf{Z}, n \geq 0}$ of approximate values is determined for each $\Delta = (\Delta x, \Delta t)$.

2. For each Δ, we define an approximate function u_Δ by

$$u_\Delta(x, t) = u_i^n$$
$$\text{if } \begin{cases} (i - \frac{1}{2})\Delta x < x < (i + \frac{1}{2})\Delta x \\ n\Delta t \leq t < (n + 1)\Delta t. \end{cases} \tag{9}$$

3. The initial data $\{u_i^0\}_{i \in \mathbf{Z}}$ should be determined so that the following conditions 1) and 2) should be satisfied.

   1) The total variation of $\{u_i^0\}_{i \in \mathbf{Z}}$ is uniquely bounded irrespective of Δ, i.e.

   $$\sum_{i \in \mathbf{Z}} |u_{i+1}^n - u_i^n| \leq M \tag{10}$$

   for some constant M irrespective of Δ.

   2) The initial value $u_\Delta(\cdot, 0)$ of approximate function u_Δ converges to the original initial data u_0 of the problem (1) in $L_{loc}^1(\mathbf{R})$. as Δ tends to $(0, 0)$.

4. Then we discuss the convergence of the family $\{u_\Delta\}_\Delta$ of approximate functions in $L_{loc}^1(\mathbf{R} \times [0, \infty))$ as Δ tends to $(0, 0)$.

In this article the word "the convergence of difference approximation " denotes the convergence of the family $\{u_\Delta\}_\Delta$ of approximate functions. We note that we may sometimes consider the convergence of subsequence from the family of approximate functions.

3. EXAMPLES AND TVD-PROPERTY

We begin the discussion on convergence with reviewing several examples of schemes written in viscous form and their convergence property.

1. Lax-Friedrichs scheme.

The scheme's popular form is

$$u_i^{n+1} = \frac{u_{i-1}^n + u_{i+1}^n}{2} - \frac{\lambda}{2}\left\{f(u_{i+1}^n) - f(u_{i-1}^n)\right\}. \tag{11}$$

The scheme is written in viscous form (5) with each numerical viscosity coefficient $a_{i+\frac{1}{2}}^n$ being a constant $\frac{1}{\lambda}$, $i.e.$

$$a_{i+\frac{1}{2}}^n = a^{LF} = \frac{1}{\lambda}. \tag{12}$$

The scheme is important in the theoretical analysis. Oleĭnik [17] proved the existence and uniqueness using the scheme. It means also that the convergence of scheme to the entropy solution is proved at the same time. On the other hand, numerical solutions by the scheme suffers from smearing effect because of the large numerical viscosity and especially the discontinuities are strongly damped.

2. Murmann-Roe scheme.
Murmann-Roe scheme [15, 19, 20] is the simplest upwind difference and is called also the upwind scheme without additional viscosity. The idea of upwind scheme is easily explained in the conservative form (7). The numerical flux $\bar{f}_{i+\frac{1}{2}}^n$ is switched to be $f(u_i^n)$ or $f(u_{i+1}^n)$ according to the sign of average characteristic speed $\dfrac{f(u_{i+1}^n) - f(u_i^n)}{u_{i+1}^n - u_i^n}$, $i.e.$

$$\bar{f}_{i+\frac{1}{2}}^n = \begin{cases} f(u_{i+1}^n) & \text{if } \dfrac{f(u_{i+1}^n) - f(u_i^n)}{u_{i+1}^n - u_i^n} < 0 \\[2mm] f(u_i^n) & \text{otherwise.} \end{cases} \tag{13}$$

In other words the numerical flux $\bar{f}_{i+\frac{1}{2}}^n$, which is regarded as flux at the contact between the two finite volumes $((i - \frac{1}{2})\Delta x, (i + \frac{1}{2})\Delta x)$ and $((i + \frac{1}{2})\Delta x, (i + \frac{3}{2})\Delta x)$ where the values of variable u are represented by u_i^n and u_{i+1}^n, respectively, is determined to be $f(u_i^n)$ or $f(u_{i+1}^n)$ according to the direction of wind, where the wind means the propagation of information by characteristics.
The scheme is written also in viscous form (5) with each numerical viscosity given by

$$a_{i+\frac{1}{2}}^n = a^{MR}(u_i^n, u_{i+1}^n)$$
$$\equiv \left| \int_0^1 f'(u_i^n + (u_{i+1}^n - u_i^n)\theta)d\theta \right|. \tag{14}$$

The scheme is good at capturing shock waves. But it is known that numerical solutions by the scheme sometimes converge to physically irrelevant solutions which include discontinuities violating the entropy condition.

3. Godunov scheme.
Godunov scheme [7], one of upwind differences, is derived from the concept of Riemann solution. Originally, the scheme is obtained from the following procedure. Assuming a step function

$$u^n(x) = u_i^n \text{ if } (i - \tfrac{1}{2})\Delta x < x < (i + \tfrac{1}{2})\Delta x, \tag{15}$$

from the difference approximation $\{u_i^n\}_{i\in\mathbf{Z}}$ at each time step $t = n\Delta t$, one solves the initial value problem

$$\begin{cases} w_t + f(w)_x = 0, & -\infty < x < \infty, n\Delta t < t < \infty \\[2mm] w(x, n\Delta t) = u^n(x), & -\infty < x < \infty \end{cases} \tag{16}$$

exactly up to the next time step $t = (n + 1)\Delta t$. Then one obtains each $u_i^{n+1}, i \in Z$ by averaging the exact entropy solution $w = w(x, t)$ over the interval $((i - \frac{1}{2})\Delta x, (i + \frac{1}{2})\Delta x)$ at the time $t = (n + 1)\Delta t$;

$$u_i^{n+1} = \frac{1}{\Delta x} \int_{(i+\frac{1}{2})\Delta x}^{(i-\frac{1}{2})\Delta x} w(x, (n + 1)\Delta t)dx. \tag{17}$$

Godunov scheme is written in the viscous form as well. The numerical viscosity is the following.

$$a_{i+\frac{1}{2}}^n$$
$$= a^G(u_{i+1}^n, u_i^n)$$
$$\equiv \max_{(s-u_i^n)(s-u_{i+1}^n)\leq 0} \frac{f(u_i^n) + f(u_{i+1}^n) - 2f(s)}{u_{i+1}^n - u_i^n} \tag{18}$$

It is known that Godunov scheme converges to the entropy solution. The fact sounds quite natural if we remember the original procure above consists of exact evolution procedure and averaging.

4. Enquist-Osher scheme.
The difference approximation (5) with each numerical viscosity given by

$$a_{i+\frac{1}{2}}^n = a^{EO}(u_i^n, u_{i+1}^n)$$
$$\equiv \int_0^1 \left| f'(u_i^n + (u_{i+1}^n - u_i^n)\theta) \right| d\theta, \tag{19}$$

is called Engquist-Osher scheme [6]. In [6] the scheme is proved to converge to the entropy solution. It should be mentioned that the main part of proof is direct calculation to verify an numerical entropy inequality, which is a discrete version of entropy inequality. The calculation we have in the proof of Theorem 5 is developed from that in the work.

5. Harten scheme (first order-accurate).

Harten [8, 9] offered an idea called the entropy-fix to improve Murmann-Roe scheme by removing the fault that Murmann-Roe scheme may form non-physical discontinuities which violate the entropy condition. The entropy-fix is a procedure to add small numerical viscosity to the numerical viscosity of Murmann-Roe scheme if it is very small or 0. A few example of numerical viscosity coefficients are shown below.

$$a^n_{i+\frac{1}{2}} = \max\left\{a^{MR}(u^n_i, u^n_{i+1}), \epsilon\right\} \qquad (20)$$

or

$$a^n_{i+\frac{1}{2}} = \begin{cases} \dfrac{1}{2\epsilon}\left\{a^{MR}(u^n_i, u^n_{i+1})\right\}^2 + \dfrac{\epsilon}{2} \\ \qquad \text{if } a^{MR}(u^n_i, u^n_{i+1}) < \epsilon \\ a^{MR}(u^n_i, u^n_{i+1}) \\ \qquad \text{otherwise,} \end{cases} \qquad (21)$$

where ϵ is a positive number less that $\dfrac{1}{\lambda}$. Many practical examples show that this additional viscosity works effectively to obtain the convergence to entropy solution. But the proof of convergence to the entropy solution has not been obtained until [2].

6. Lax-Wendroff scheme [12].

From the viewpoint of order of accuracy in the sense of Taylor expansion, the examples 1-5 above are only of the first order accuracy. When each numerical viscosity coefficient $a^n_{i+\frac{1}{2}}$ is a function of u^n_i and u^n_{i+1}, the difference approximation (5) is of the second order accuracy if and only if the relation

$$a^n_{i+\frac{1}{2}} = \lambda\left\{a^{MR}(u^n_i, u^n_{i+1})\right\}^2 + O(\Delta) \qquad (22)$$

is satisfied. Lax-Wendroff scheme is the case. But Lax-Wendroff scheme may cause spurious oscillation in the numerical solution and violate the convergence.

We note the relation

$$a^{LW}(u^n_i, u^n_{i+1}) \leq a^{MR}(u^n_i, u^n_{i+1}) \leq a^G(u^n_i, u^n_{i+1})$$
$$\leq a^{EO}(u^n_i, u^n_{i+1}) \leq a^{LF} \equiv \frac{1}{\lambda} \qquad (23)$$

among numerical viscosity coefficients of the examples above, where a^{LW} denotes the numerical viscosity coefficient of Lax-Wendroff scheme.

Here we note a basic concept, the consistency of difference approximation with the original differential equation, which means that each convergent subsequence from the family $\{u_\Delta\}_\Delta$ of approximate functions converges to a weak solution to the problem (1). In the case of viscous form (5) we have the following theorem.

Theorem 2 *If the numerical viscosity coefficients $a^n_{i+\frac{1}{2}}$ are uniquely bounded with upper and lower bounds irrespective of i, n and Δ, any convergent subsequence from the family $\{u_\Delta\}_\Delta$ of approximate functions converges to a weak solution to (1) as Δ tends to $(0, 0)$.*

Usually the analysis of convergence to entropy solution consists of two parts of discussion. First we discuss the compactness of family $\{u_\Delta\}_\Delta$ of approximate functions, which means the existence of subsequence from $\{u_\Delta\}$ converging to a weak solution. We note that the stability in an appropriate norm may imply the compactness in some topology. Then we consider the consistency with entropy condition, *i.e.* we examine if the convergence limit, which is already a weak solution, satisfies the entropy condition. What is mainly discussed in this article is the consistency with entropy condition.

In the remaining part of this section, we mention the concept of TVD (Total Variation Diminishing) and related results. We define the concept of TVD as follows.

Definition 3 *A difference approximation $\{u^n_i\}_{i, n\in\mathbf{Z}, n\geq 0}$ is said to be TVD if the inequality*

$$\sum_{i\in\mathbf{Z}} |u^{n+1}_{i+1} - u^{n+1}_i| \leq \sum_{i\in\mathbf{Z}} |u^n_{i+1} - u^n_i| \qquad (24)$$

is satisfied for any $n \geq 0$.

The following theorem [8, 9] determines a class of TVD difference approximations in terms of numerical viscosity.

Theorem 3 *The difference approximation (5) is TVD if each numerical viscosity $a^n_{i+\frac{1}{2}}$ satisfies the inequality*

$$a^{MR}(u^n_i, u^n_{i+1}) \leq a^n_{i+\frac{1}{2}} \leq a^{LF} \equiv \frac{1}{\lambda} \qquad (25)$$

We note the following fact also.

Theorem 4 *If the condition (25) is satisfied, the following inequality holds.*

$$\inf_{i \in \mathbf{Z}} u_i^n \leq \inf_{i \in \mathbf{Z}} u_i^{n+1} \leq \sup_{i \in \mathbf{Z}} u_i^{n+1} \leq \sup_{i \in \mathbf{Z}} u_i^n \qquad (26)$$

The property of TVD guarantees the existence of convergent subsequence from the family $\{u_\Delta\}$ of approximate functions. (See, for example, [16]) By Theorem 2, we have the following corollary.

Corollary 1 *If the condition (25) is satisfied, any convergent subsequence from the family $\{u_\Delta\}_\Delta$ of approximate functions given by difference approximation (5) converges to a weak solution to (1).*

But the uniqueness of convergence limit is not guaranteed. The condition (25) does not guarantee the convergence to entropy solution either. In fact Murmann-Roe scheme satisfies the condition (25) and may converge to a weak solution violating the entropy condition. Therefore we analyze the consistency with entropy condition in the following sections.

4. CONSISTENCY WITH ENTROPY CONDITION.

In this section and the next, we consider the convergence of difference approximation (5) to entropy solution. Throughout the sections we assume the TVD condition (25) and fixed real constants m and M satisfying

$$m \leq u_i^0 \leq M, \quad i \in \mathbf{Z} \qquad (27)$$

irrespective of Δ. We note that (27) and TVD condition (25) automatically gives $m < u_i^n < M$, $i, n \in \mathbf{Z}, n \geq 0$.

We call the inequality

$$U(u_i^{n+1}) - U(u_i^n) + \frac{\lambda}{2} \left\{ F(u_{i+1}^n) - F(u_{i-1}^n) \right\}$$
$$- \frac{\lambda}{2} \left\{ A_{i+\frac{1}{2}}^n (u_{i+1}^n - u_i^n) - A_{i-\frac{1}{2}}^n (u_i^n - u_{i-1}^n) \right\} \leq 0 \qquad (28)$$

a numerical entropy inequality, where (U, F) is an entropy pair and each coefficient $A_{i+\frac{1}{2}}^n$ is an appropriate real number. If a difference approximation satisfies the numerical entropy inequality (28) with the coefficients $A_{i+\frac{1}{2}}^n$ uniquely bounded by some upper and lower bounds irrespective of i, n and Δ, then any weak solution which is the convergence limit of convergent subsequence from the family $\{u_\Delta\}$ of approximate functions given by the difference approximation satisfies the entropy inequality (3).

Therefore, in order to prove the convergence to entropy solution, we construct uniquely bounded coefficients $A_{i+\frac{1}{2}}^n$ and verify prove the numerical entropy

inequality (28) for every entropy pair (U, F) (if the flux function f is a general one) or for a particular entropy pair (U, F) with the entropy function U strictly convex (if the flux function f is strictly convex).

In this section we consider the case that the flux function f is general. In this case we have the following theorem. (See also [2].)

Theorem 5 *If each numerical viscosity coefficient $a_{i+\frac{1}{2}}^n$ in difference approximation (5) satisfies*

$$a^G(u_i^n, u_{i+1}^n) \leq a_{i+\frac{1}{2}}^n \leq \frac{1}{\lambda}, \qquad (29)$$

then the family $\{u_\Delta\}_\Delta$ of approximate functions determined by (9) converges to the entropy solution as $\Delta \longrightarrow (0,0)$.

Tadmor and Osher [18] called a difference approximation satisfying (29) an E-scheme. In [18] the convergence of E-schemes to entropy solution is proved with some CFL restriction. The theorem above tells that any E-scheme converges to the entropy solution under the natural CFL condition (6).

In the remaining part of the section we give the proof of theorem following the proof in [2].

To conduct the proof we employ the concept of modified flux function, which was first introduced in [18]. In brief, we use the concept of modified flux function to describe the numerical viscosity in a different form, and we compare the modified flux function with the original flux function f to know whether the numerical viscosity is enough to attain the numerical entropy inequality defined later.

Definition 4 *A function g is called a modified flux function associated with the numerical viscosity coefficient $a_{i+\frac{1}{2}}^n$ at $(i + \frac{1}{2}, n)$ if it satisfies the following conditions (30),(31),(32).*

$$g(s) = f(s) \text{ if } (s - u_i^n)(s - u_{i+1}^n) \geq 0. \qquad (30)$$

$$|g'(s)| \leq \frac{1}{\lambda}, \quad m \leq s \leq M. \qquad (31)$$

$$\int_{u_i^n}^{u_{i+1}^n} |g'(s)| \, ds = a_{i+\frac{1}{2}}^n (u_{i+1}^n - u_i^n). \qquad (32)$$

We note that the condition (32) is equivalent to

$$\int_0^1 \left| g'(u_i^n + (u_{i+1}^n - u_i^n)\theta) \right| d\theta = a_{i+\frac{1}{2}}^n. \qquad (33)$$

As for the existence of modified flux function we have the following lemma.

Lemma 1 *There exists a modified flux function associated with $a^n_{i+\frac{1}{2}}$ at $(i+\frac{1}{2},n)$ if and only if the numerical viscosity coefficient $a^n_{i+\frac{1}{2}}$ satisfies TVD condition (25) ;*

$$a^{MR}(u^n_i, u^n_{i+1}) \leq a^n_{i+\frac{1}{2}} \leq a^{LF} \equiv \frac{1}{\lambda}.$$

Proof. If $g(s)$ is a modified flux function associated with a numerical viscosity coefficient $a^n_{i+\frac{1}{2}}$, the relation (33) gives

$$a^n_{i+\frac{1}{2}} \geq \left| \int_0^1 g'(u^n_i + (u^n_{i+1} - u^n_i)\theta)d\theta \right|$$

$$\geq \left| \frac{g(u^n_{i+1}) - g(u^n_i)}{u^n_{i+1} - u^n_i} \right|$$

$$= \left| \frac{f(u^n_{i+1}) - f(u^n_i)}{u^n_{i+1} - u^n_i} \right|$$

$$= a^{MR}(u^n_i, u^n_{i+1})$$

and

$$a^n_{i+\frac{1}{2}} \leq \int_0^1 \frac{1}{\lambda}d\theta = \frac{1}{\lambda}.$$

Therefore TVD condition (25) is a necessity condition for the existence of modified flux function.

Suppose that the condition (25) is satisfied. We define a function $g^n_{i+\frac{1}{2}}$ as follows. (See Figures 1 and 2.)

If $u^n_i < u^n_{i+1}$,

$$g^n_{i+\frac{1}{2}}(s)$$

$$= \begin{cases} f(s), & s \leq u^n_i \\ \max\left\{ -\frac{1}{\lambda}(s - u^n_i) + f(u^n_i), \right. \\ \qquad \left. \bar{f}^n_{i+\frac{1}{2}}, \frac{1}{\lambda}(s - u^n_{i+1}) + f(u^n_{i+1}) \right\}, \\ \qquad\qquad u^n_i \leq s \leq u^n_{i+1} \\ f(s), & s \geq u^n_{i+1}. \end{cases}$$

$$(34)$$

If $u^n_i > u^n_{i+1}$,

$$g^n_{i+\frac{1}{2}}(s)$$

$$= \begin{cases} f(s), & s \leq u^n_{i+1} \\ \min\left\{ \frac{1}{\lambda}(s - u^n_{i+1}) + f(u^n_{i+1}), \right. \\ \qquad \left. \bar{f}^n_{i+\frac{1}{2}}, -\frac{1}{\lambda}(s - u^n_i) + f(u^n_i) \right\}, \\ \qquad\qquad u^n_{i+1} \leq s \leq u^n_i \\ f(s), & s \geq u^n_i. \end{cases}$$

$$(35)$$

If $u^n_i = u^n_{i+1}$,

$$g^n_{i+\frac{1}{2}}(s) = f(s), -\infty < s < \infty. \qquad (36)$$

Figure 1.
The graph of $g^n_{i+\frac{1}{2}}(s), u^n_i \leq s \leq u^n_{i+1}$ (if $u^n_i < u^n_{i+1}$)

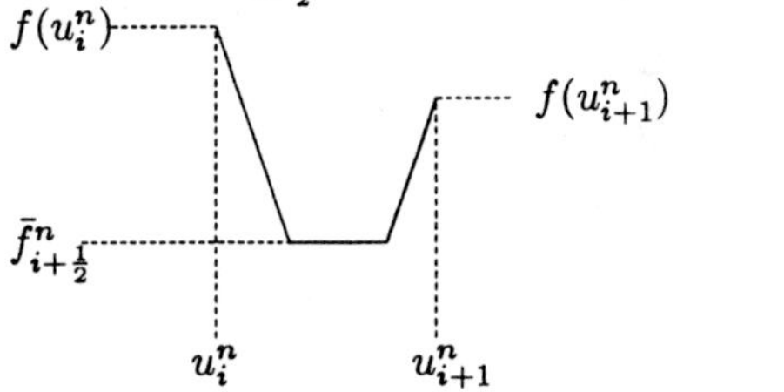

Figure 2.
The graph of $g^n_{i+\frac{1}{2}}(s), u^n_{i+1} \leq s \leq u^n_i$ (if $u^n_i > u^n_{i+1}$)

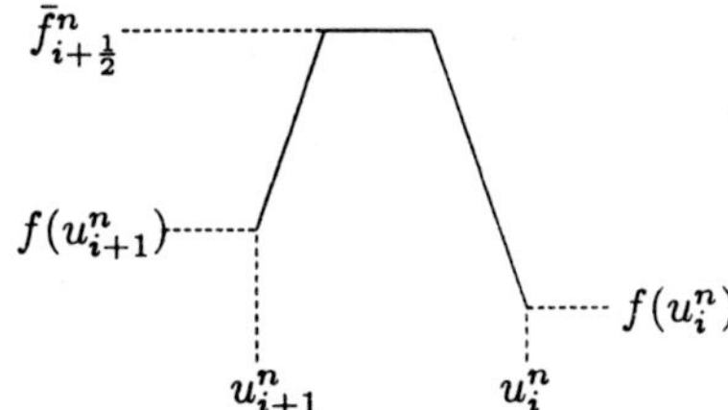

$\bar{f}^n_{i+\frac{1}{2}}$ in the definition denotes the numerical flux defined by (8). It is easy to observe that the function $g^n_{i+\frac{1}{2}}$ is a modified flux function associated with $a^n_{i+\frac{1}{2}}$.

$$(\text{q.e.d.})$$

We call the function $g^n_{i+\frac{1}{2}}$ defined by (34)-(36) the standard modified flux function associated with $a^n_{i+\frac{1}{2}}$. We construct each coefficient $A^n_{i+\frac{1}{2}}$ using the standard modified flux function $g^n_{i+\frac{1}{2}}$. We note the following relation as well.

$$\bar{f}^n_{i+\frac{1}{2}} = \frac{1}{2}\left\{f(u^n_i) + f(u^n_{i+1})\right\}$$
$$-\frac{1}{2}\int_{u^n_i}^{u^n_{i+1}}\left|(g^n_{i+\frac{1}{2}})'(s)\right|ds. \tag{37}$$

Restricting ourselves into the standard modified flux functions, we have the following lemma.

Lemma 2 *Suppose that the numerical viscosity coefficient $a^n_{i+\frac{1}{2}}$ satisfies*

$$a^G(u^n_{i+\frac{1}{2}}) \le a^n_{i+\frac{1}{2}} \le a^{LF} \equiv \frac{1}{\lambda}. \tag{38}$$

Then the standard modified flux function $g^n_{i+\frac{1}{2}}$ associated with $a^n_{i+\frac{1}{2}}$ satisfies

$$g^n_{i+\frac{1}{2}}(s) \le f(s), u^n_i \le s \le u^n_{i+1}$$

or

$$g^n_{i+\frac{1}{2}}(s) \ge f(s), u^n_{i+1} \le s \le u^n_i$$

if

$$u^n_i < u^n_{i+1} \quad or \quad u^n_i > u^n_{i+1},$$

respectively.

Proof. First we consider the case of $u^n_i < u^n_{i+1}$. It is clear that

$$-\frac{1}{\lambda}(s - u^n_i) + f(u^n_i) \le f(s)$$

and

$$\frac{1}{\lambda}(s - u^n_{i+1}) + f(u^n_{i+1}) \le f(s)$$

are valid if s is between u^n_i and u^n_{i+1}. The condition (38) gives

$$\bar{f}^n_{i+\frac{1}{2}} \le \frac{1}{2}\left\{f(u^n_i) + f(u^n_{i+1})\right\} - \frac{1}{2}a^G(u^n_i, u^n_{i+1})(u^n_{i+1} - u^n_i).$$

We obtain

$$a^G(u^n_i, u^n_{i+1}) = \frac{f(u^n_i) + f(u^n_{i+1}) - 2\min_{u^n_i \le s \le u^n_{i+1}} f(s)}{u^n_{i+1} - u^n_i}$$

from the relation (18). Then we see

$$\bar{f}^n_{i+\frac{1}{2}} \le \min_{u^n_i \le s \le u^n_{i+1}} f(s).$$

Therefore we conclude

$$g^n_{i+\frac{1}{2}}(s) \le f(s), \ u^n_i \le s \le u^n_{i+1}.$$

The same discussion applies to the case of $u^n_i > u^n_{i+1}$.

Now we come to the stage where we construct uniquely bounded coefficients $A^n_{i+\frac{1}{2}}$ of the numerical entropy inequality. To simplify the discussion, we define a family $\{(U(\,\cdot\,;k), F(\,\cdot\,;k))\}_{k\in\mathbf{R}}$ of entropy pairs parametrized by a real number k;

$$\begin{cases} U(s;k) = (s-k)^+ = \chi^+(s-k)\cdot(s-k) \\ F(s;k) = \chi^+(s-k)\left\{f(s) - f(k)\right\}, \end{cases} \tag{39}$$

where

$$s^+ = \max\{0, s\} = \frac{s + |s|}{2} \tag{40}$$

and

$$\chi^+(s) = \begin{cases} 1 & \text{if } s > 0 \\ 0 & \text{otherwise.} \end{cases} \tag{41}$$

We obtain the following lemma.

Lemma 3 *If coefficients $A^n_{i+\frac{1}{2}}(k)$, $i, n \in Z, n \ge 0, m \le k \le M$ exist and satisfy the numerical entropy inequality*

$$U(u^{n+1}_i; k) - U(u^n_i; k)$$
$$+\frac{\lambda}{2}\left\{F(u^n_{i+1}; k) - F(u^n_{i-1}; k)\right\}$$
$$-\frac{\lambda}{2}\left\{A^n_{i+\frac{1}{2}}(k)(u^n_{i+1} - u^n_i) - A^n_{i-\frac{1}{2}}(k)(u^n_i - u^n_{i-1})\right\}$$
$$\le 0, \tag{42}$$

then the numerical entropy inequality (28) holds for an arbitrary entropy pair (U, F) with each coefficient $A^n_{i+\frac{1}{2}}$ given by

$$A^n_{i+\frac{1}{2}} = \int_m^M A^n_{i+\frac{1}{2}}(k)U''(k)dk. \tag{43}$$

Proof. Let $E^n_i(k)$ denote the left-hand side of numerical entropy inequality (42), *i.e.*

$$E^n_i(k)$$
$$= U(u^{n+1}_i; k) - U(u^n_i; k)$$
$$+\frac{\lambda}{2}\left\{F(u^n_{i+1}; k) - F(u^n_{i-1}; k)\right\}$$
$$-\frac{\lambda}{2}\left\{A^n_{i+\frac{1}{2}}(k)(u^n_{i+1} - u^n_i) - A^n_{i-\frac{1}{2}}(k)(u^n_i - u^n_{i-1})\right\}. \tag{44}$$

For an arbitrary entropy pair function (U, F), we easily observe

$$U(x) = (x - m)U'(m) + U(m) + \int_m^x (x - s)U''(s)ds \tag{45}$$

and

$$F(x) = (x - m)F'(m) + F(m) + \int_m^x (x - s)U''(s)ds. \tag{46}$$

Noting that $m \le u_i^n \le M$ for some constants m and M, we here restrict ourselves into the case that $m \le x \le M$ to obtain

$$U(x) = (x - m)U'(m) + U(m) + \int_m^M U(x; s)U''(s)ds$$

$$F(x) = (x - m)F'(m) + F(m) + \int_m^M F(x; s)U''(s)ds. \tag{47}$$

We note $U''(s)$ is not negative because U is convex. Then we obtain

$$\int_m^M E_i^n(k)dk$$
$$= U(u_i^{n+1}) - U(u_i^n) + \frac{\lambda}{2}\left\{F(u_{i+1}^n) - F(u_{i-1}^n)\right\} \tag{48}$$
$$- \frac{\lambda}{2}\left\{A_{i+\frac{1}{2}}^n(u_{i+1}^n - u_i^n) - A_{i-\frac{1}{2}}^n(u_i^n - u_{i-1}^n)\right\}$$

and the proof is complete.

$$(\text{q.e.d.})$$

The lemma above shows that we may analyze the numerical entropy inequality only for entropy pairs $(U(\cdot; k), F(\cdot; k))$, $m \le k \le M$. We define each constant $A_{i+\frac{1}{2}}^n$ by

$$A_{i+\frac{1}{2}}^n(k)$$
$$= \int_0^1 \chi^+(u_i^n + (u_{i+1}^n - u_i^n)\theta - k) \tag{49}$$
$$\times \left|(g_{i+\frac{1}{2}}^n)'(u_i^n + (u_{i+1}^n - u_i^n)\theta - k)\right| d\theta,$$

where $g_{i+\frac{1}{2}}^n$ is the standard modified flux function associated with $a_{i+\frac{1}{2}}^n$.

Then we have some calculation. First we calculate $u_i^{n+1} - k$ and $U(u_k^{n+1})$;

$$u_i^{n+1} - k$$
$$= u_i^n - k - \frac{\lambda}{2}\int_{u_{i-1}^n}^{u_{i+1}^n} f'(s)ds$$
$$+ \frac{\lambda}{2}\int_{u_i^n}^{u_{i+1}^n} \left|(g_{i+\frac{1}{2}}^n)'(s)\right| ds$$
$$- \frac{\lambda}{2}\int_{u_{i-1}^n}^{u_i^n} \left|(g_{i-\frac{1}{2}}^n)'(s)\right| ds$$

$$= u_i^n - k$$
$$+ \frac{\lambda}{2}\int_k^{u_{i-1}^n} \left\{f'(s) + \left|(g_{i-\frac{1}{2}}^n)'(s)\right|\right\} ds$$
$$+ \frac{\lambda}{2}\int_k^{u_i^n} \left\{-\left|(g_{i-\frac{1}{2}}^n)'(s)\right| - \left|(g_{i+\frac{1}{2}}^n)'(s)\right|\right\} ds$$
$$+ \frac{\lambda}{2}\int_k^{u_{i-1}^n} \left\{-f'(s) + \left|(g_{i+\frac{1}{2}}^n)'(s)\right|\right\} ds$$
$$= \int_k^{u_{i-1}^n} \frac{\lambda}{2}\left\{\left|(g_{i-\frac{1}{2}}^n)'(s)\right| + (g_{i-\frac{1}{2}}^n)'(s)\right\} ds$$
$$+ \frac{\lambda}{2}\left\{g_{i-\frac{1}{2}}^n(k) - f(k)\right\}$$
$$+ \int_k^{u_i^n} \left\{1 - \frac{\lambda}{2}\left|(g_{i-\frac{1}{2}}^n)'(s)\right| - \frac{\lambda}{2}\left|(g_{i+\frac{1}{2}}^n)'(s)\right|\right\} ds$$
$$+ \int_k^{u_{i+1}^n} \frac{\lambda}{2}\left\{\left|(g_{i+\frac{1}{2}}^n)'(s)\right| - (g_{i+\frac{1}{2}}^n)'(s)\right\} ds$$
$$- \frac{\lambda}{2}\left\{g_{i+\frac{1}{2}}^n(k) - f(k)\right\},$$

$$U(u_i^{n+1}; k)$$
$$= \chi^+(u_i^{n+1} - k)(u_i^{n+1} - k)$$
$$= \chi^+(u_i^{n+1} - k)$$
$$\times \int_k^{u_{i-1}^n} \frac{\lambda}{2}\left\{\left|(g_{i-\frac{1}{2}}^n)'(s)\right| + (g_{i-\frac{1}{2}}^n)'(s)\right\} ds$$
$$+ \chi^+(u_i^{n+1} - k)\frac{\lambda}{2}\left\{g_{i-\frac{1}{2}}^n(k) - f(k)\right\}$$
$$+ \chi^+(u_i^{n+1} - k)$$
$$\times \int_k^{u_i^n} \left\{1 - \frac{\lambda}{2}\left|(g_{i-\frac{1}{2}}^n)'(s)\right| - \frac{\lambda}{2}\left|(g_{i+\frac{1}{2}}^n)'(s)\right|\right\} ds$$

$$+\chi^+(u_i^{n+1} - k)$$

$$\times \int_k^{u_{i+1}^n} \frac{\lambda}{2}\left\{\left|(g_{i+\frac{1}{2}}^n)'(s)\right| - (g_{i+\frac{1}{2}}^n)'(s)\right\} ds$$

$$-\chi^+(u_i^{n+1} - k)\frac{\lambda}{2}\left\{g_{i+\frac{1}{2}}^n(k) - f(k)\right\}.$$

We also have

$$U(u_i^n; k) - \frac{\lambda}{2}\left\{F(u_{i+1}^n; k) - F(u_{i-1}^n; k)\right\}$$

$$+\frac{\lambda}{2}\left\{A_{i+\frac{1}{2}}^n(k)(u_{i+1}^n - u_i^n) - A_{i-\frac{1}{2}}^n(k)(u_i^n - u_{i-1}^n)\right\}$$

$$= \chi^+(u_{i-1}^n - k)$$

$$\times \int_k^{u_{i-1}^n} \frac{\lambda}{2}\left\{\left|(g_{i-\frac{1}{2}}^n)'(s)\right| + (g_{i-\frac{1}{2}}^n)'(s)\right\} ds$$

$$+\chi^+(u_{i-1}^n - k)\frac{\lambda}{2}\left\{g_{i-\frac{1}{2}}^n(k) - f(k)\right\}$$

$$+\chi^+(u_i^n - k)$$

$$\times \int_k^{u_i^n} \left\{1 - \frac{\lambda}{2}\left|(g_{i-\frac{1}{2}}^n)'(s)\right| - \frac{\lambda}{2}\left|(g_{i+\frac{1}{2}}^n)'(s)\right|\right\} ds$$

$$+\chi^+(u_{i+1}^n - k)$$

$$\times \int_k^{u_{i+1}^n} \frac{\lambda}{2}\left\{\left|(g_{i+\frac{1}{2}}^n)'(s)\right| - (g_{i+\frac{1}{2}}^n)'(s)\right\} ds$$

$$+\chi^+(u_{i+1}^n - k)\frac{\lambda}{2}\left\{f(k) - g_{i+\frac{1}{2}}^n(k)\right\}.$$

Then we obtain

$$E_i^n(k)$$

$$= \left\{\chi^+(u_i^{n+1} - k) - \chi^+(u_{i-1}^n - k)\right\}$$

$$\times \int_k^{u_{i-1}^n} \frac{\lambda}{2}\left\{\left|(g_{i+\frac{1}{2}}^n)'(s)\right| + (g_{i+\frac{1}{2}}^n)'(s)\right\} ds$$

$$+ \left\{\chi^+(u_i^{n+1} - k) - \chi^+(u_i^n - k)\right\}$$

$$\times \int_k^{u_i^n} \left\{1 - \frac{\lambda}{2}\left|(g_{i-\frac{1}{2}}^n)'(s)\right| - \frac{\lambda}{2}\left|(g_{i+\frac{1}{2}}^n)'(s)\right|\right\} ds$$

$$+ \left\{\chi^+(u_i^{n+1} - k) - \chi^+(u_{i+1}^n - k)\right\}$$

$$\times \int_k^{u_{i+1}^n} \frac{\lambda}{2}\left\{\left|(g_{i+\frac{1}{2}}^n)'(s)\right| - (g_{i+\frac{1}{2}}^n)'(s)\right\} ds$$

$$+ \left\{\chi^+(u_i^{n+1} - k) - \chi^+(u_{i-1}^n - k)\right\}$$

$$\times \frac{\lambda}{2}\left\{g_{i-\frac{1}{2}}^n(k) - f(k)\right\}$$

$$+ \left\{\chi^+(u_i^{n+1} - k) - \chi^+(u_{i+1}^n - k)\right\}$$

$$\times \frac{\lambda}{2}\left\{f(k) - g_{i+\frac{1}{2}}^n(k)\right\}.$$

$$(50)$$

The sign of $\chi^+(u_i^{n+1} - k) - \chi^+(u_{i-1}^n - k)$ is not the same as that of $u_{i-1}^n - k$ unless $u_{i-1}^n - k = 0$. The sign of integral

$$\int_k^{u_{i-1}^n} \frac{\lambda}{2}\left\{\left|(g_{i+\frac{1}{2}}^n)'(s)\right| + (g_{i+\frac{1}{2}}^n)'(s)\right\} ds$$

is not opposite to that of $u_{i-1}^n - k$ because the integrand is not negative. Therefore the first term of right-hand side of equality (50) is not positive. Similar argument is applied to the second and third terms. Then we obtain the following estimate.

$$E_i^n(k)$$

$$\leq \left\{\chi^+(u_i^{n+1} - k) - \chi^+(u_{i-1}^n - k)\right\}$$

$$\times \frac{\lambda}{2}\left\{g_{i-\frac{1}{2}}^n(k) - f(k)\right\}$$

$$+ \left\{\chi^+(u_i^{n+1} - k) - \chi^+(u_{i+1}^n - k)\right\}$$

$$\times \frac{\lambda}{2}\left\{f(k) - g_{i+\frac{1}{2}}^n(k)\right\}.$$

$$(51)$$

Again the sign of $\chi^+(u_i^{n+1} - k) - \chi^+(u_{i-1}^n - k)$ is not the same as that of $u_{i-1}^n - k$ unless $u_{i-1}^n - k = 0$. LEMMA 2 implies that the sign of $g_{i-\frac{1}{2}}^n(k) - f(k)$ is not opposite to that of $u_i^n - k$. Then we obtain

$$\{\chi^+(u_i^{n+1} - k) - \chi^+(u_{i-1}^n - k)\}$$

$$\times \frac{\lambda}{2}\left\{g_{i-\frac{1}{2}}^n(k) - f(k)\right\} \leq 0.$$

By similar discussion, we also obtain

$$\{\chi^+(u_i^{n+1} - k) - \chi^+(u_{i+1}^n - k)\}$$

$$\times \frac{\lambda}{2}\left\{f(k) - g_{i+\frac{1}{2}}^n(k)\right\} \leq 0.$$

Therefore we conclude that, if each numerical viscosity coefficient $a_{i+\frac{1}{2}}^n$ satisfies the condition (2), then the numerical entropy inequality

$$U(u_i^{n+1}; k) - U(u_i^n; k)$$

$$+\frac{\lambda}{2}\left\{F(u_{i+1}^n; k) - F(u_{i-1}^n; k)\right\}$$

$$-\frac{\lambda}{2}\left\{A_{i+\frac{1}{2}}^n(k)(u_{i+1}^n - u_i^n) - A_{i-\frac{1}{2}}^n(k)(u_i^n - u_{i-1}^n)\right\}$$

$$\leq 0 \tag{52}$$

is satisfied for $k \in [m, M]$, with each coefficient $A_{i+\frac{1}{2}}^n(k)$ determined by (49). This completes the proof of Theorem 5 .

5. CONSISTENCY WITH ENTROPY CONDITION AND SMALL NUMERICAL VISCOSITY.

In this section we restrict our interest into the case that the flux function f is strictly convex. In this case we apply Theorem 1 to have the advantage that it is enough to prove the entropy inequality (3) or the numerical entropy inequality (28) only for a particular entropy pair (U, F) with the entropy function U strictly convex. Of course the entropy pair should be fixed irrespective of Δ. Then we may expect that the consistency with entropy condition might be guaranteed with some weaker condition than that in Theorem 5 . In fact we have the following theorem.

Theorem 6 *Suppose that the flux function f is strictly convex. Let ϵ be an arbitrary positive number less than 1. If each numerical viscosity coefficient $a_{i+\frac{1}{2}}^n$ satisfies*

$$\max\left\{a^{MR}(u_i^n, u_{i+1}^n), \frac{\epsilon}{\lambda} sgn(u_{i+1}^n - u_i^n)\right\}$$

$$\leq a_{i+\frac{1}{2}}^n \leq a^{LF} \equiv \frac{1}{\lambda}, \tag{53}$$

the family $\{u_\Delta\}_\Delta$ of approximate functions determined by difference approximation (5) converges to the entropy solution as $\Delta \longrightarrow (0,0)$.

The proof of theorem is more complicated than that of Theorem 5 . We note that the condition (53) is too weak to guarantee the numerical entropy inequality (42) for all $k \in \mathbf{R}$. Therefore we construct a particular entropy pair (U, F) with the entropy function U strictly convex and coefficients $A_{i+\frac{1}{2}}^n$ for which the numerical entropy inequality (28) is satisfied. For that purpose we need more precise estimate of $E_i^n(k)$, which is determined by (44) as the left-hand side of the numerical entropy inequality (42) for entropy pairs $(U(\,\cdot\,;k), F(\,\cdot\,;k))$ parametrized by $k \in \mathbf{R}$. We define $E_{i,-}^n(k)$ and $E_{i,+}^n(k)$ as follows.

$$E_{i,-}^n(k)$$

$$= \{\chi^+(u_i^{n+1} - k) - \chi^+(u_i^n - k)\}$$

$$\times \int_k^{u_i^n} \frac{1}{2}\left\{1 - \lambda\left|(g_{i-\frac{1}{2}}^n)'(s)\right|\right\} ds$$

$$+ \{\chi^+(u_i^{n+1} - k) - \chi^+(u_{i-1}^n - k)\} \tag{54}$$

$$\times \int_k^{u_{i-1}^n} \frac{\lambda}{2}\left\{\left|(g_{i-\frac{1}{2}}^n)'(s)\right| + (g_{i-\frac{1}{2}}^n)'(s)\right\} ds$$

$$+ \{\chi^+(u_i^{n+1} - k) - \chi^+(u_{i-1}^n - k)\}$$

$$\times \frac{\lambda}{2}\left\{g_{i-\frac{1}{2}}^n(k) - f(k)\right\},$$

$$E_{i,+}^n(k)$$

$$= \{\chi^+(u_i^{n+1} - k) - \chi^+(u_i^n - k)\}$$

$$\times \int_k^{u_i^n} \frac{1}{2}\left\{1 - \lambda\left|(g_{i+\frac{1}{2}}^n)'(s)\right|\right\} ds$$

$$+ \{\chi^+(u_i^{n+1} - k) - \chi^+(u_{i+1}^n - k)\} \tag{55}$$

$$\times \int_k^{u_{i+1}^n} \frac{\lambda}{2}\left\{\left|(g_{i+\frac{1}{2}}^n)'(s)\right| - (g_{i+\frac{1}{2}}^n)'(s)\right\} ds$$

$$+ \{\chi^+(u_i^{n+1} - k) - \chi^+(u_{i+1}^n - k)\}$$

$$\times \frac{\lambda}{2}\left\{f(k) - g_{i+\frac{1}{2}}^n(k)\right\}.$$

It is clear that $E_i^n(k)$ splits into the sum of $E_{i,-}^n(k)$ and $E_{i,+}^n(k)$;

$$E_i^n(k) = E_{i,-}^n(k) + E_{i,+}^n(k).$$

For the proof of Theorem 6 it suffices to prove the following lemma.

Lemma 4 *There exists a strictly convex function U irrespective i, n, Δ that satisfies the inequality*

$$\int_m^M E_{i,\pm}^n(k)U''(k)dk \le 0, \ \forall i, n, \qquad (56)$$

where M and m are determined so that they satisfy $m \le u_i^n \le M, \ \forall i, n$.

In fact, once we obtain such a strictly convex function U which is supposed to be an entropy function, we immediately define a corresponding entropy flux function F and each coefficient $A_{i+\frac{1}{2}}^n, i, n \in \boldsymbol{Z}, n \ge 0$ by $F' = U'f'$ and (43). Then the relation (48) shows that the numerical entropy inequality (28) is satisfied.

In the remaining part of this section we give a sketch of proof of LEMMA 4 .

If $E_{i,\pm}^n(k) \le 0$ holds for any $k \in \boldsymbol{R}$, it is clear that the inequality (56) holds for any convex function U. We easily observe the following fact.

Lemma 5 *The inequalities $E_{i,+}^n(k) \le 0$ and $E_{i+1,-}^n(k) \le 0$ hold for any $k \in \boldsymbol{R}$ when $u_i^n \ge u_{i+1}^n$. Even when $u_i^n < u_{i+1}^n$, the inequalities $E_{i,+}^n(k) \le 0$ and $E_{i+1,-}^n(k) \le 0$ are still valid for any $k \in \boldsymbol{R}$ if the modified flux function $g_{i+\frac{1}{2}}^n$ satisfies $g_{i+\frac{1}{2}}^n(s) \le f(s), u_i^n < s < u_{i+1}^n$. (Remember that $g_{i+\frac{1}{2}}^n(s) = f(s)$ if $s \le u_i^n$ or $s \ge u_{i+1}^n$.)*

Otherwise, $E_{i,+}^n(k)$ and $E_{i+1,-}^n(k)$ might be positive for some $k \in \boldsymbol{R}$. and it is necessary to estimate them more precisely. Therefore we continue the estimate under the assumption that $u_i^n < u_{i+1}^n$ and that some $s \in (u_i^n, u_{i+1}^n)$ violates the inequality $g_{i+\frac{1}{2}}^n(s) \le f(s)$. For simplicity we also assume $f(0) = f'(0) = 0$ without loss of generality.

We define v_- and v_+ by

$$\begin{aligned} v_- &= u_i^n + \lambda\left\{f(u_i^n) - \bar{f}_{i+\frac{1}{2}}^n\right\}, \\ v_+ &= u_{i+1}^n - \lambda\left\{f(u_{i+1}^n) - \bar{f}_{i+\frac{1}{2}}^n\right\}, \end{aligned} \qquad (57)$$

and define w_- and w_+ so that the condition

$$f(w_-) = f(w_+) = \bar{f}_{i+\frac{1}{2}}^n, \ w_- < w_+ \qquad (58)$$

is satisfied. v_- and v_+ mean the both ends of horizontal part of the graph of modified flux function $g_{i+\frac{1}{2}}^n$. w_- and w_+ mean the intersection points of the horizontal part of graph of $g_{i+\frac{1}{2}}^n$ and the graph of flux function f. We easily observe that $v_- \le w_- < 0 < w_+ \le v_+$. More precisely, the equalities $v_- = w_-$ and $w_+ = v_+$ never occur at the same time.

We follow only the discussion for $E_{i,+}^n(k)$ because a similar discussion is valid for $E_{i,-}^n(k)$.

First we estimate $E_{i,+}^n(k)$ as the following lemma.

Lemma 6

$$E_{i,+}^n(k)$$

$$\le \begin{cases} \dfrac{\lambda}{2}\left\{g_{i+\frac{1}{2}}^n(k) - f(k)\right\} = \dfrac{\lambda}{2}\left\{\bar{f}_{i+\frac{1}{2}}^n - f(k)\right\} \\ \qquad \text{if } k \in \left[\dfrac{v_- + w_-}{2}, \dfrac{w_+ + v_+}{2}\right]. \\ 0 \quad \text{otherwise.} \end{cases} \qquad (59)$$

Proof. We consider only the case that $k \in [\frac{v_-+w_-}{2}, \frac{w_++v_+}{2}]$, because the estimate is trivial if $k < \frac{v_-+w_-}{2}$ or $k > \frac{w_++v_+}{2}$.

If $k > u_i^{n+1}$, the definition (54) of $E_{i,+}^n(k)$ directly gives

$$\begin{aligned} E_{i,+}^n(k) &= \chi^+(u_{i+1}^n - k)\frac{\lambda}{2}\left\{g_{i+\frac{1}{2}}^n(k) - f(k)\right\} \\ &= \frac{\lambda}{2}\left\{g_{i+\frac{1}{2}}^n(k) - f(k)\right\}. \end{aligned}$$

If $k \le u_i^{n+1}$, we obtain

$$E_{i,+}^n(k) \le \int_k^{u_i^n} \frac{1}{2}\left\{1 - \lambda\left|(g_{i+\frac{1}{2}}^n)'(s)\right|\right\}ds = \frac{1}{2}(v_- - k).$$

We compare $g_{i+\frac{1}{2}}^n(k) - f(k)$ with $\frac{1}{\lambda}(v_- - k)$ to obtain

$$\frac{1}{\lambda}(v_- - k) \le g_{i+\frac{1}{2}}^n(k) - f(k).$$

This completes the proof.

$$(\text{q.e.d.})$$

$E_{i,+}^n(k)$ may have the positive sign if $k \in (w_-, w_+)$, and it is strictly negative if $\frac{v_-+w_-}{2} < k < w_-$ or $w_+ < k < \frac{w_++v_+}{2}$. To make the integral $\int_m^M E_{i,+}^n(k)U''(k)dk$ negative or 0, we have to cancel the positive part $\int_{w_-}^{w_+} E_{i,+}^n(k)U''(k)dk$ with

$$\int_{\frac{v_-+w_-}{2}}^{w_-} E_{i,+}^n(k)U''(k)dk \text{ or } \int_{w_+}^{\frac{w_++v_+}{2}} E_{i,+}^n(k)U''(k)dk$$

either of which is strictly negative. If there exists a positive function P that satisfies the inequality

$$\int_m^M E_{i,+}^n(k)P(k) \leq 0,$$

we can complete the proof by determining U by $U'' = P$. In fact it is possible to construct such a positive function P.

We outline how to construct the function P. Let functions h_- and h_+ be the inverse functions of $f(s), s \leq 0$ and $f(s), s \geq 0$, respectively, i.e. the functions h_- and h_+ should satisfy

$$f(h_-(\tau)) = f(h_+(\tau)) = \tau, \quad h_-(\tau) < 0 < h_+(\tau) \tag{60}$$
$$\text{for } \tau > 0$$

and

$$h_-(0) = h_+(0) = 0. \tag{61}$$

Then we define another function h by

$$h(\tau) = h_+(\tau) - h_-(\tau), \tag{62}$$

where we note that $w_+ - w_- = h(\bar{f}_{i+\frac{1}{2}}^n)$. Using these three function, we determine a real number K_-, K_+ and K by

$$K_- = \inf_{A_-} \frac{h_-(\alpha) - h_-(\alpha + \frac{\epsilon}{2\lambda}h(\alpha))}{h(\alpha)} - \frac{\epsilon}{2}$$
$$K_+ = \inf_{A_+} \frac{h_+(\alpha + \frac{\epsilon}{2\lambda}h(\alpha)) - h_+(\alpha)}{h(\alpha)} - \frac{\epsilon}{2} \tag{63}$$

and

$$K = \min\{K_+, K_-\}, \tag{64}$$

where the domains A_+ and A_- are determined by

$$A_- = \left\{\alpha \geq 0 \,\Big|\, f'\left(h_-(\alpha + \tfrac{\epsilon}{2\lambda}h(\alpha))\right) \geq -\frac{1}{\lambda}\right\},$$
$$A_+ = \left\{\alpha \geq 0 \,\Big|\, f'\left(h_+(\alpha + \tfrac{\epsilon}{2\lambda}h(\alpha))\right) \leq \frac{1}{\lambda}\right\}.$$

While we skip the proof, we claim the following 1)-2) which implies the role of K and functions h_-, h_+, h in the proof of LEMMA 4.

1) K is a positive number (neither 0 nor negative), and the following estimate is valid,

$$\max\{w_- - v_-, v_+ - w_+\} \geq Kh(\bar{f}_{i+\frac{1}{2}}^n), \tag{65}$$

where we note that K is irrespective of i, n, Δ.

2) When P is a non-negative function, $\int_m^M E_i^n(k)P(k)dk$ is estimated as follows.

$$\int_m^M E_i^n(k)P(k)dk$$
$$\leq \int_{h_-(\bar{f}_{i+\frac{1}{2}}^n)}^{h_+(\bar{f}_{i+\frac{1}{2}}^n)} \left\{\bar{f}_{i+\frac{1}{2}}^n - f(k)\right\} P(k)dk$$
$$-\frac{1}{8}\min\left\{P(h_-(\bar{f}_{i+\frac{1}{2}}^n)), P(h_+(\bar{f}_{i+\frac{1}{2}}^n))\right\}$$
$$\times \frac{\bar{f}_{i+\frac{1}{2}}^n}{h(\bar{f}_{i+\frac{1}{2}}^n)}\left\{Kh(\bar{f}_{i+\frac{1}{2}}^n)\right\}^2 \tag{66}$$

We define P by

$$P(s) = \{h(f(s))\}^{\frac{1}{\gamma}},$$

where $\gamma = \frac{1}{8}K^2$. It is easy to see that the function P makes the right-hand side of inequality (66) not positive. Then any function U satisfying $U'' = P$ attains the inequality (56). Strictly speaking, P is not always positive (i.e. $P(s) > 0, s \neq 0$ and $P(0) = 0$), but there is no problem because the function U is really strictly convex.

This completes the sketch of proof of lemma 4.

The entropy function U that we construct to satisfy the numerical entropy inequality (28) is getting more convex as the positive constant ϵ in the condition (53) tends to 0. If our purpose is only to prove the convergence to entropy solution, we do not care how convex the entropy function U is. In fact it suffices that U is strictly convex because of Theorem 1. But, from the viewpoint of numerical results, we need to observe the effect of numerical viscosity coefficients upon the numerical solution.

We consider numerical viscosity coefficients taking the smallest value that satisfies the condition (53) with small $\epsilon > 0$ and entropy functions of the form $U(s) = \{h(f(s))\}^\alpha \, (\alpha > 0)$. The sketch of proof implies that $\frac{1}{\gamma} = \frac{8}{K^2}$ is almost the smallest value of α that validates the numerical entropy inequality (28) with the entropy function $U(s) = \{h(f(s))\}^\alpha \, (\alpha > 0)$. In other words, the numerical entropy inequality is not satisfied with an entropy function $U(s) = \{h(f(s))\}^\alpha, 0 < \alpha \ll \frac{1}{\gamma}$. It is clear that K and γ tend to 0 as $\epsilon \to 0$. Then we understand that, if ϵ is very small, entropy functions for which the numerical solution satisfies the numerical entropy inequality (28) may be very particular and far from an entropy function which seems to be natural from the viewpoint of physics. In fact we easily observe the fact above by calculating numerical solutions for decay of inverse shock[2]: The numerical solution shows some unnatural decay at the beginning of decay of inverse shock, but the unnatural decay satisfies the numerical entropy inequality with the particular entropy function $U(s) = \{h(f(s))\}^{\frac{1}{7}}$. Of course the numerical solution finally converges to the entropy solution as $\Delta x, \Delta t \longrightarrow 0$, which Theorem 6 means.

It is also well known that Harten scheme (5), (20) or (5), (21) shows similar phenomena if ϵ is very small.

The discussion above implies the difficulty in relating numerical viscosity coefficients with the quality of numerical solution. But Theorem 6 is important both theoretically and practically because it gives one of the best possible solution if the maximum and minimum of each numerical viscosity coefficient $a_{i+\frac{1}{2}}^n$ are determined in terms of u_i^n and u_{i+1}^n.

6. UNDERSTANDING MODIFIED FLUX FUNCTIONS

In this section we give some more understanding of modified flux function.

First we review Riemann solver and Godunov scheme briefly.

The Riemann problem is an initial value problem whose initial value has two constant states with a discontinuity between the two states;

$$
\begin{cases}
u_t + f(u)_x = 0 \\
u(x,0) = \begin{cases} u_L, x < 0 \\ u_R, x > 0. \end{cases}
\end{cases}
\tag{68}
$$

For a fixed flux function f, the solution $u = u(x,t)$ to Riemann problem is determined only by u_L and u_R, and has the similarity which means that the value of u at (x,t) depends on x/t. Then the solution is written in the form

$$
u(x,t) = w(x/t; u_L, u_R).
\tag{69}
$$

$w(x/t; u_L, u_R)$ is called the (exact) Riemann solver.

We remember that Godunov scheme is obtained by repeating the procedure (15)-(17). The definition (17)

---

[2]For example, such decay happens in solving the following initial value problem

$$
\begin{cases}
u_t + f(u)_x = 0 \\
u(x,0) = \begin{cases} u_L, x < 0 \\ u_R, x > 0, \end{cases}
\end{cases}
\tag{67}
$$

where $u_L < u_R$.

for u_i^{n+1} and Green's formula implies that

$$
\begin{aligned}
u_{i+\mp}^n &= u_i^n \\
&- \int_{n\Delta t}^{(n+1)\Delta t} f(v((i + \tfrac{1}{2})\Delta x, t))dt \\
&+ \int_{n\Delta t}^{(n+1)\Delta t} f(v((i - \tfrac{1}{2})\Delta x, t))dt,
\end{aligned}
\tag{70}
$$

where $v = v(x,t)$ is the exact solution to the problem (16). Because of the property of finite propagation speed and CFL condition, we easily see that

$$
\begin{aligned}
v((i + \tfrac{1}{2})\Delta x, t) &= w(0; u_i^n, u_{i+1}^n), \\
n\Delta t &\leq t < (n + 1)\Delta t.
\end{aligned}
\tag{71}
$$

Then Godunov scheme is written in conservative form (7) with each numerical viscosity $\bar{f}_{i+\frac{1}{2}}^n$ given by

$$
\bar{f}_{i+\frac{1}{2}}^n = f(w(0; u_i^n, u_{i+1}^n)).
\tag{72}
$$

We naturally understand the fact remembering the finite volume concept and the conservation concept in physics.

We consider a difference approximation in viscous form (5) with each numerical viscosity coefficient $a_{i+\frac{1}{2}}^n$ satisfying the condition (25);

$$
a^{MR}(u_i^n, u_{i+1}^n) \leq a_{i+\frac{1}{2}}^n \leq a^{LF} \equiv \frac{1}{\lambda}
$$

and the standard modified flux functions $g_{i+\frac{1}{2}}^n$, $(i, n \in \mathbf{Z}, i \geq 0)$ defined by (34)-(36). Note that we can really take the standard modified flux functions thanks to the condition (25). We already know that the difference approximation is written in conservative form (7) with each numerical flux $\bar{f}_{i+\frac{1}{2}}^n$ given by

$$
\begin{aligned}
\bar{f}_{i+\frac{1}{2}}^n &= \frac{1}{2}\left\{f(u_i^n) + f(u_{i+1}^n)\right\} - \frac{1}{2}a_{i+\frac{1}{2}}^n(u_{i+1}^n - u_i^n) \\
&= \frac{1}{2}\left\{f(u_i^n) + f(u_{i+1}^n)\right\} - \frac{1}{2}\int_{u_i^n}^{u_{i+1}^n} \left|(g_{i+\frac{1}{2}}^n)'(s)\right| ds.
\end{aligned}
\tag{73}
$$

Here we give another way to write the numerical flux using the standard modified flux function.

Theorem 7 *Suppose that each numerical viscosity $a_{i+\frac{1}{2}}^n$ a difference approximation (7) satisfies the condition (25). Then the difference approximation is* *written in conservative form (7) with each numerical flux $\bar{f}^n_{i+\frac{1}{2}}$ given by*

$$\bar{f}^n_{i+\frac{1}{2}} = g^n_{i+\frac{1}{2}}(\bar{w}^n_{i+\frac{1}{2}}(0; u^n_i, u^n_{i+1})), \qquad (74)$$

where $g^n_{i+\frac{1}{2}}$ is the standard modified flux function defined by (34)-(36) and $w^n_{i+\frac{1}{2}}(x/t; u_L, u_R)$ is the exact Riemann solver to the Riemann problem

$$\begin{cases} u_t + g^n_{i+\frac{1}{2}}(u)_t = 0 \\ u(x,0) = \begin{cases} u_L, x < 0 \\ u_R, x > 0. \end{cases} \end{cases} \qquad (75)$$

We skip the proof. We note that we need to consider solution to conservation law whose flux function does not have smooth derivative when we prove the theorem. But the physical understanding of conservation law brings us to natural acceptance of the theorem.

When we discretize a differential equation to construct a difference approximation, we are based on the consistency concept, which means that a difference approximation tends to the original differential equation. But the concept of consistency is not enough to give a deterministic algorithm but it allows some ambiguity or variety in discretization and difference approximation. In our discussion on difference approximation in viscous form or conservative form, the variety appears in definition of each numerical viscosity coefficient or numerical flux. To analyze some class of difference approximation with such variety we need to develop a concept or tool which describes the variety completely only by itself and removes it from the other part of discussion. The modified flux function is employed in our discussion as one of such concepts. Its advantage is that the problem of consistency with entropy condition is converted to an easier problem of the comparison between each modified flux function with the original flux function.

7. CONSISTENCY WITH ENTROPY CONDITION AND HIGHER ORDER ACCURACY.

We again suppose that the function f in the problem (1) is strictly convex. Theorem 6 in section 5 includes difference approximations only of first order accuracy everywhere. Therefore we extend Theorem 6 to obtain a stronger result which include difference approximation of the higher order accuracy. We also give a sketch of the proof in the latter half of this section.

We already review Lax-Wendroff scheme and that the difference approximation (5) has the second order accuracy (in the sense of Taylor expansion) if and

only if each numerical viscosity coefficient satisfies the condition (22) which contradicts the condition (53) in Theorem 6. Therefore we have to give up the second order accuracy in some situation (for example, shock, extrema or some other singularity etc.) to guarantee the convergence to entropy solution. But we might expect the higher order accuracy in the region where the solution is regular enough. We point out that the any numerical viscosity coefficient satisfying the condition (53) in Theorem 6 is still too large to acquire the higher order accuracy even if we restrict our discussion into the region where the solution is smooth.

On the other hand, there is no stronger result than Theorem 6 if the lower limit of each numerical viscosity coefficient $a^n_{i+\frac{1}{2}}$ that guarantees the convergence to entropy solution is described in terms of u^n_i and u^n_{i+1}. Then, naturally arises an idea to describe the lower limit of $a^n_{i+\frac{1}{2}}$ in terms of more than two node values, for example, four node values $u^n_{i-1}, u^n_i, u^n_{i+1}$ and u^n_{i+2}, *i.e.* the condition is written in the form

$$a_m(u^n_{i-1}, u^n_i, u^n_{i+1}, u^n_{i+2}) \leq a^n_{i+\frac{1}{2}} \leq \frac{1}{\lambda}. \qquad (76)$$

But the idea is difficult to be realized with mathematical proof.

To resolve the inconvenience above we employ the method of flux modification which is proposed by Harten [8, 9]. We consider difference approximation in the following form;

$$\begin{aligned} u^{n+1}_i &= u^n_i - \frac{\lambda}{2}\left[\{f(u^n_{i+1}) + \phi^n_{i+1}\} - \{f(u^n_{i-1}) + \phi^n_{i-1}\}\right] \\ &\quad + \frac{\lambda}{2}\left\{\tilde{a}^n_{i+\frac{1}{2}}(u^n_{i+1} - u^n_i) - \tilde{a}^n_{i-\frac{1}{2}}(u^n_i - u^n_{i-1})\right\}. \end{aligned} \qquad (77)$$

Each ϕ^n_i is called the flux modification and $\tilde{f}^n_i = f(u^n_i) + \phi^n_i$ is called the modified flux. For keeping the consistency with original conservation law, each flux modification ϕ^n_i should satisfy the condition;

$$|\phi^n_i| \leq C \cdot \min\left\{|u^n_i - u^n_{i-1}|, |u^n_{i+1} - u^n_i|\right\} \qquad (78)$$

for a fixed constant C independent of i, n, Δ. The coefficients $\tilde{a}^n_{i+\frac{1}{2}}$ should be uniquely bounded irrespective of i, n, Δ as well. We note that the difference approximation (77) is written also in the ordinary viscous form (5) with each $a^n_{i+\frac{1}{2}}$ given by

$$a^n_{i+\frac{1}{2}} = \tilde{a}^n_{i+\frac{1}{2}} - \frac{\phi^n_i + \phi^n_{i+1}}{u^n_{i+1} - u^n_i}. \qquad (79)$$

Note that each $a^n_{i+\frac{1}{2}}$ is finite because of the condition (78).

We begin with the discussion on stability. The condition (25) for TVD property is no more directly applied to the form (79). But we easily obtain the following result.

Theorem 8 *The difference approximation (77) is TVD and satisfies*

$$\inf_{i\in\mathbf{Z}} u_i^n \leq \inf_{i\in\mathbf{Z}} u_i^{n+1} \leq \sup_{i\in\mathbf{Z}} u_i^{n+1} \leq \sup_{i\in\mathbf{Z}} u_i^n, \ n \in Z, n \geq 0, \tag{80}$$

if each coefficient $\tilde{a}_{i+\frac{1}{2}}^n$ satisfies

$$\left| \frac{\tilde{f}_{i+1}^n - \tilde{f}_i^n}{u_{i+1}^n - u_i^n} \right| \leq \tilde{a}_{i+\frac{1}{2}}^n \leq \frac{1}{\lambda}, \tag{81}$$

where $\tilde{f}_i^n = f(u_i^n) + \phi_i^n$.

Theorem 8 extends the condition (25) in Theorem 3 . If $\sigma_i^n = \phi_{i+1}^n = 0$, they are exactly the same. The essential part of proof is almost similar to Theorem 3 . We may add the observation of the fact $\left| \tilde{f}_{i+1}^n - \tilde{f}_i^n \right| \leq \frac{1}{\lambda} \left| u_{i+1}^n - u_i^n \right|$.

Next we consider the consistency with entropy condition We have the following theorem, which is our main theorem.

Theorem 9 *Let $\bar{\phi}_i^n$ be defined by*

$$\bar{\phi}_i^n = \frac{1}{2} \left\{ sgn\Delta u_{i-\frac{1}{2}}^n + sgn\Delta u_{i+\frac{1}{2}}^n \right\}$$

$$\times \min \left\{ \frac{1}{2} q_{i-\frac{1}{2}}^n \left(1 - \lambda q_{i-\frac{1}{2}}^n \right) \left| \Delta u_{i-\frac{1}{2}}^n \right|, \right. \tag{82}$$

$$\left. \frac{1}{2} q_{i+\frac{1}{2}}^n \left(1 - \lambda q_{i+\frac{1}{2}}^n \right) \left| \Delta u_{i+\frac{1}{2}}^n \right| \right\},$$

*where $\Delta u_{i+\frac{1}{2}}^n = u_{i+1}^n - u_i^n$, $\Delta f_{i+\frac{1}{2}}^n = f(u_{i+1}^n) - f(u_i^n)$ and $q_{i+\frac{1}{2}}^n = \left| \dfrac{\Delta f_{i+\frac{1}{2}}^n}{\Delta u_{i+\frac{1}{2}}^n} \right|$ $(q_{i+\frac{1}{2}}^n = f'(u_i^n)$ if $\Delta u_{i+\frac{1}{2}}^n = 0)$.
Suppose that each flux modification ϕ_i^n satisfies*

$$(\phi_i^n - \bar{\phi}_i^n)\phi_i^n \leq 0. \tag{83}$$

Then the difference approximation (77) converges to the entropy solution if each coefficient $\tilde{a}_{i+\frac{1}{2}}^n$ satisfies

$$\max \left\{ \left| \frac{\tilde{f}_{i+1}^n - \tilde{f}_i^n}{u_{i+1}^n - u_i^n} \right|, \frac{\epsilon}{\lambda} sgn\Delta u_{i+\frac{1}{2}}^n \right\} \leq \tilde{a}_{i+\frac{1}{2}}^n \leq \frac{1}{\lambda}, \tag{84}$$

where ϵ is an arbitrary fixed positive number such that $0 < \epsilon \leq 1$.

A brief sketch of proof is given later.

The theorem is an extension of Theorem 6 . It is possible to rewrite the condition in the theorem above into the form of inequalities determining the range of each numerical viscosity coefficient $a_{i+\frac{1}{2}}^n$ using (79) and other relations. If we do so, however, the statement of theorem would be more confusing. The statement of theorem 9 is better both theoretically and practically.

We also mention that the natural CFL condition (6) yields

$$\left| \frac{\tilde{f}_{i+1}^n - \tilde{f}_i^n}{u_{i+1}^n - u_i^n} \right| \leq \frac{1}{\lambda}$$

and then the condition (84) always makes sense under the assumption of natural CFL condition.

We obtain the following corollary about the order of accuracy.

Corollary 2 *If we add the assumption that each ϕ_i^n satisfies*

$$\phi_i^n = \bar{\phi}_i^n + O(\Delta^2) \tag{85}$$

to the condition of theorem 9 , then the difference approximation (77) is also second order-accurate in the smooth region (i.e. except around the discontinuities).

Before Theorem 9 a few works [1, 18] treat higher-order difference approximations to get some results for the consistency with entropy condition. But the results seem to have some artificial restriction which comes from the method of proof. Theorem 9 seems to have the least of such artificial restriction and is a very general result including [1] and [18]. The theorem includes also several higher order-accurate difference approximations, for example, Harten's second order-accurate TVD scheme [8, 9] and some difference approximation (van Leer's MUSCL[13] etc.) obtained by the method of reconstruction within each cell (see [1, 4] also [13]).

The sketch of proof is given in the remaining part of the section.

We again note that the assumption of strict convexity of flux function f is essential for Theorem 6 and Theorem 9 . Otherwise we can not prove them. Theorem 8 is still valid even if the flux function f is not strictly convex.

The basic argument is similar to Theorem 6 . We easily recognize that the difference approximation is TVD from Theorem 8 . We construct a particular entropy pair (U, F) with the entropy function U strictly convex and a family of coefficients $\{A_{i+\frac{1}{2}}^n\}_{i,n\in\mathbf{Z},n\geq 0}$ which satisfy the numerical entropy inequality (28), where the entropy pair (U, F) should be independent of $\Delta = (\Delta x, \Delta t)$.

First we observe that every entropy pair is a convex linear combination of some family of entropy pairs parametrized by $k \in \mathbf{R}$ and prepare modified flux functions. The family is a little different from that in the proof of Theorem 5 and 6, and the modified flux functions are given in a little difference definition as well.

Take two constants m and M so that $m \leq u_i^0 \leq M$ be satisfied for all $i \in \mathbf{Z}$. Then we easily see that $m \leq u_i^n \leq M$ is satisfied for all n and i because of Theorem 8 and the condition (84) We define a family of entropy pairs $\{U(\,\cdot\,; k), F(\,\cdot\,; k)\}_{k \in \mathbf{R}}$ by

$$
\begin{aligned}
&U(s; k) \\
&= \chi^-(k)(s-k)^- + \chi^+(k)(s-k)^+ \\
&= \begin{cases} (s-k)^- & \text{if } k < 0 \\ (s-k)^+ & \text{if } k > 0 \end{cases}
\end{aligned}
\tag{86}
$$

and

$$
\begin{aligned}
&F(s; k) \\
&= -\chi^-(k)\chi^-(s-k)\{f(s) - f(k)\} \\
&\quad + \chi^+(k)\chi^+(s-k)\{f(s) - f(k)\} \\
&= \begin{cases} -\chi^-(s-k)\{f(s) - f(k)\} & \text{if } k < 0 \\ \chi^+(s-k)\{f(s) - f(k)\} & \text{if } k > 0 \end{cases}
\end{aligned}
\tag{87}
$$

where $s^+ = \max\{|s|, 0\}$, $s^- = \max\{-|s|, 0\}$,

$$
\chi^+(s) = \begin{cases} 0 \text{ if } s < 0 \\ 1 \text{ if } s > 0 \end{cases}
$$

and

$$
\chi^-(s) = \begin{cases} 1 \text{ if } s < 0 \\ 0 \text{ if } s > 0. \end{cases}
$$

We do not define $\chi^+(0)$, $\chi^-(0)$, $U(\,\cdot\,; 0)$ nor $F(\,\cdot\,; 0)$ in the statement above. Then we suppose that they have some finite values. It does not make any inconvenience. We have the following relation for each entropy pair (U, F) with U strictly convex and of C^2-class.

$$
\begin{cases} U(s) = \displaystyle\int_{-\infty}^{\infty} U(s; k)U''(k)dk + sU'(0) + U(0) \\[2mm] F(s) = \displaystyle\int_{-\infty}^{\infty} F(s; k)U''(k)dk + sF'(0) + F(0) \end{cases}
\tag{88}
$$

We suppose that $U(0) = F(0) = U'(0) = F'(0) = 0$ without loss of generality. We also note that the interval of integral in (88) could be $[m, M]$ instead of $[-\infty, \infty]$ if we restrict variable s within the interval $[m, M]$.

Next we define modified flux functions, which consist of two kinds. Modified flux functions of the first kind are defined as follows. A function f_i^n is called a modified flux function of the first kind associated with $\tilde{f}_i^n$ at (i, n) if it satisfies that

$$
|(f_i^n)'(s)| \leq \frac{1}{\lambda} \quad \text{if } s \in [m, M],
$$

$$
f_i^n(u_i^n) = \tilde{f}_i^n
$$

and

$$
f_i^n(s) = f(s) \ \text{ if } \begin{cases} s \leq \min\left\{u_{i-1}^n, u_i^n, u_{i+1}^n\right\} \\ \text{or} \\ s \geq \max\left\{u_{i-1}^n, u_i^n, u_{i+1}^n\right\}. \end{cases}
$$

Modified flux functions of the second kind is defined as follows. A function $g_{i+\frac{1}{2}}^n$ is called a modified flux function of the second kind associated with $\bar{a}_{i+\frac{1}{2}}^n$ at $(i + \frac{1}{2}, n)$ if it satisfies that

$$
\left|(g_{i+\frac{1}{2}}^n)'(s)\right| \leq \frac{1}{\lambda} \quad \text{if } s \in [m, M],
$$

$$
\bar{a}_{i+\frac{1}{2}}^n = \int_0^1 \left|(g_{i+\frac{1}{2}}^n)'(u_i^n + (u_{i+1}^n - u_i^n)\theta)\right| d\theta
$$

and

$$
g_{i+\frac{1}{2}}^n = \begin{cases} f_i^n(s) & \text{if } (s - u_i^n)(u_{i+1}^n - u_i^n) \leq 0 \\ f_{i+1}^n(s) & \text{if } (s - u_{i+1}^n)(u_i^n - u_{i+1}^n) \leq 0. \end{cases}
$$

A modified flux function f_i^n or $g_{i+\frac{1}{2}}^n$ is not uniquely determined. To conduct the proof we choose particular modified flux functions, which are like the standard modified flux functions in the proof of Theorem 5 and 6. But we skip the procedure to choose them because the procedure is too complicated to be written down here. We continue the sketch supposing that the particular modified flux functions f_i^n or $g_{i+\frac{1}{2}}^n$ are already chosen.

We define functions $F_i^n(\,\cdot\,; k)$, $k \in \mathbf{R}$ by

$$
\begin{aligned}
&F_i^n(s; k) \\
&= -\chi^-(k)\chi^-(s-k)\{f_i^n(s) - f_i^n(k)\} \\
&\quad + \chi^+(k)\chi^+(s-k)\{f_i^n(s) - f_i^n(k)\}
\end{aligned}
\tag{89}
$$

and $A^n_{i+\frac{1}{2}}(k)$ ($k \in \mathbf{R}$ is a parameter) by the equation

$$A^n_{i+\frac{1}{2}}(k)(u^n_{i+1} - u^n_i)$$
$$= F(u^n_i; k) + F(u^n_{i+1}; k)$$
$$\quad - F^n_i(u^n_i; k) - F^n_{i+1}(u^n_{i+1}; k) \qquad (90)$$
$$\quad + \int_{u^n_i}^{u^n_{i+1}} \left| (g^n_{i+\frac{1}{2}})'(s) \right| ds.$$

Here we mention that each $A^n_{i+\frac{1}{2}}(k)$ is determined to be a finite value and to be uniquely bounded, *i.e.* $\left| A^n_{i+\frac{1}{2}}(k) \right| \leq M$ for a constant M independent of i, n, k, Δ. Then we define $E^n_i(k)$ by

$$E^n_i(k)$$
$$= U(u^{n+1}_i; k) - U(u^n_i; k)$$
$$\quad + \frac{\lambda}{2} \left\{ F(u^n_{i+1}; k) - F(u^n_{i-1}; k) \right\}$$
$$\quad - \frac{\lambda}{2} \left\{ A^n_{i+\frac{1}{2}}(k)(u^n_{i+1} - u^n_i) - A^n_{i-\frac{1}{2}}(k)(u^n_i - u^n_{i-1}) \right\}.$$
$$(91)$$

$E^n_i(k)$ is the left-hand side of numerical entropy inequality for the entropy pair $(U(\cdot\,; k), F(\cdot\,; k))$. $E^n_i(k)$ is not always negative or zero, and changes the sign depending on k. Through discussion almost similar to that in the proof of Theorem 6 we can construct a fixed strictly convex function U satisfying the inequality

$$\int_m^M E^n_i(k) U''(k) dk \leq 0. \qquad (92)$$

U is irrespective of i, n, Δ. We easily see that the relation

$$\int_m^M E^n_i(k) U''(k) dk$$
$$= U(u^{n+1}_i) - U(u^n_i) + \frac{\lambda}{2} \left\{ F(u^n_{i+1}) - F(u^n_{i-1}) \right\} \qquad (93)$$
$$\quad - \frac{\lambda}{2} \left\{ A^n_{i+\frac{1}{2}}(u^n_{i+1} - u^n_i) - A^n_{i-\frac{1}{2}}(u^n_i - u^n_{i-1}) \right\}$$

holds, where

$$F(s) = \int_m^M F(s; k) U''(k) dk \qquad (94)$$

and

$$A^n_{i+\frac{1}{2}} = \int_m^M A^n_{i+\frac{1}{2}}(k) U''(k) dk. \qquad (95)$$

It is clear that the functions U and F above make an entropy pair (U, F).

Therefore we prove the numerical entropy inequality (28) for the entropy pair (U, F) with U strictly convex. This complete the proof.

8. CONCLUSION

In this article we analyze difference approximation for scalar conservation law from the viewpoint of numerical viscosity and finally obtain the main theorem determining a general and wide class of difference approximation converging to the entropy solution. The class includes a wide subclass of difference approximations of the second order accuracy (except around the singularities), to which several well known second order-accurate schemes of difference approximation belong. The main theorem gives one of the best solutions to the natural and classical problem "Determine a class of difference approximation good enough to solve the original differential equation " in the case of scalar conservation law.

From the theoretical interest, the word "a good difference approximation" means a difference approximation converging to the proper solution (which is entropy solution in our case) to the original differential equation. Then the main theorem is one of the ultimate results theoretically.

From the viewpoint of numerics, the class of main theorem is too wide to guarantee the quality of numerical solutions given by difference approximations belonging to the class. Of course we have the criterion of the order of accuracy in the sense of Taylor expansion, but even the criterion is not enough to select difference approximations giving numerical solutions of good quality. Then we still need to develop some new criteria to measure the quality of numerical solutions in order to approach the ultimate difference approximations. The main theorem however should be a good landmark to construct better difference approximations.

REFERENCES

[1] H. Aiso. Second order TVD difference approximations of scalar conservation laws and the consistency with the entropy condition. *Report of Uppsala University, Dep. of Scientific Computing.* No.152/1993.

[2] H. Aiso. Admissibility of difference approximation for scalar conservation laws. *Hiroshima Math. J.*, 23(1):15–61, 1993.

[3] H. Aiso. Higher Order-Accurate Difference Approximation for Scalar Conservation Laws and the Consistency with Entropy Condition. *Collection of Technical Papers, 6th International Symposium on Computational Fluid Dynamics*, pages 7–12, 1995.

[4] H. Aiso and T. Iwamiya. Construction and Analysis of Difference Approximation for Scalar Conservation Laws and Its Consistency with Entropy Condition Using Boltzmann-like Approach. *Special Publication of National Aerospace Laboratory JAPAN, SP-27*, pages 123–128, 1994.

[5] R. J. DiPerna. Uniqueness of solutions to hyperbolic conservation laws. *Indiana Univ. Math. J.*, 28:244–257, 1979.

[6] B. Engquist and S. Osher. Stable and entropy satisfying approximations for transonic flow calculations. *Math. Comp.*, 34:45–75, 1980.

[7] S. K. Godunov. Finite difference method for numerical computation of discontinuous solutions of the equations of fluid dynamics (in Russian). *Mat. Sb. (N.S.)*, 47:251–306, 1959.

[8] A. Harten. High resolution schemes for hyperbolic conservation laws. *J. Comput. Phys.*, 49:357–393, 1983.

[9] A. Harten. On a class of high resolution total-variation-stable finite-difference schemes. *SIAM J. Numer. Anal.*, 21(1):1–23, 1984.

[10] P. D. Lax. Weak solution of nonlinear hyperbolic equations and their numerical computation. *Comm. Pure Appl. Math.*, 7:159–193, 1954.

[11] P. D. Lax. Shock waves and entropy. In *Contributions to nonlinear functional analysis*, pages 603–634. Academic Press, NewYork, 1971.

[12] P. D. Lax and B. Wendroff. Systems of conservation laws. *Comm. Pure Appl. Math.*, 13:217–237, 1960.

[13] B. van Leer. Towards the ultimate conservative difference scheme V. A second order sequel to Godunov's method. *J. Comput. Phys.*, 32:101–136, 1979.

[14] A. Y. LeRoux. A numerical conception of entropy for quasi-linear equations. *Math. Comp.*, 31:848–872, 1977.

[15] E. M. Murmann. Analysis of embedded shock waves calculated by relaxation methods. *AIAA J.*, 12:626–633, 1974.

[16] I. P. Natanson. *Theory of functions of a real variable*. Frederick Ungar, 1955. Translated from Russian by L. F. Boron.

[17] O. Oleĭnik. Discontinuous solutions of nonlinear differential equations. *Uspekhi Mat. Nauk.(N.S.)*, 12:3–73, 1957. English transl. in *Amer. Math. Soc. Transl., Ser. 2, vol. 26*, 95-172.

[18] S. Osher and E. Tadmor. On the convergence of difference approximations to scalar conservation laws. *Math. Comp.*, 50:19–51, 1988.

[19] P. L. Roe. Approximate Riemann solvers, parameter vectors and difference schemes. *J. Comput. Phys.*, 43:357–372, 1981.

[20] P. L. Roe. The use of Riemann Problem in finite difference schemes. In *Lecture Notes in Physics, 141*, volume 141, pages 354–559. Springer-Verlag, NewYork, 1981.

[21] J. Smoller. *Shock waves and reaction-diffusion equations*. Springer-Verlag, NewYork, 1982.

[22] E. Tadmor. Numerical viscosity and the entropy condition for conservative difference schemes. *Math. Comp.*, 43:369–382, 1984.

NON-EXISTENCE OF THIRD ORDER MUSCL SCHEMES UNIFIED CONSTRUCTION OF ENO SCHEMES AND A NEW DISCONTINUITY SHARPENNING TECHNIQUE - STIFF SOURCE TERM APPROACH

Huamo WU [1]   **Lier WANG** [2]   **Geng SUN** [2]

(1)   Institute of Computational Mathematics and
      Scientific/Engineering Computing, Chinese Academy of Sciences, Beijing,
      100080, China
      State Key Laborator of Scientific/Engineering Computing
(2)   Institute of Mathematics, Chinese Academy of Sciences, Beijing, 100080,
      China

Abstract

In this paper the non-existence of third order accurate MUSCL schemes for genuinely nonlinear conservation laws is proved. A new class of non-MUSCL highly accurate ENO schemes is constructed in an unified way which is useful for parallel computing. A new discontinuity sharpening technique -artificial stiff source term approach is suggested. Numerical results are presented.

Keywords: ENO schemes   Non-MUSCL schemes   Unified Structure   Parallel
            computing   Discontinuity Sharpenning

1. INTRODUCTION

Various high resolution difference schemes have been proposed in recent years for solving hyperbolic conservation laws

$$\frac{\partial u}{\partial t} + \frac{\partial f}{\partial x} = 0. \qquad (1.1)$$

Among these methods the semi-discrete ENO schemes

$$\left(\frac{\partial v}{\partial t}\right)_j + \frac{1}{\Delta x}(\hat{f}_{j+1/2} - \hat{f}_{j-1/2}) = 0 \qquad (1.2)$$

attract active applications in CFD by their efficiency in preventing accuracy degeneration near extrema and by the simplicity in use in multidimensional cases. There are two important classes of semi-discrete ENO schemes of (1.2). One is the class of ENO schemes of MUSCL type

$$\hat{f}_{j+1/2} = \hat{f}(v_{j+1/2}), \quad v_{j+1/2} = \sum_{i=-k}^{k} \alpha_{j+i}^{(k)} v_{j+i} \qquad (1.3)$$

Another is the class of ENO schemes of non-MUSCL type

$$\hat{f}_{j+1/2} = \sum_{i=-k}^{k} \beta_{j+i}^{(k)} f(v_{j+i}). \qquad (1.4)$$

It will be shown in this paper that MUSCL-type approximation (1.3) can not achieve third (or higher) order of accuracy in the case of genuinely nonlinear $f(u)$. In contrast with this pessimistic point of MUSCL schemes, the non-MUSCL type approximation (1.4) can always provide arbitrary high order of accuracy. This is one of the reasons why ENO schemes of this kind are frequently used for flow field computations of high accuracy . However, the construction of many used schemes of this kind is not unified in the sence that calculations involve many branches and is not suitable for parallel computing. Hence the unified way of constructing high accurate and high resolution schemes is worthy to be developed.

Received on March 1, 1997.

Project supported by The National Science Foundation of China and The National Climbing Project of China

It is well known that the high resolution of the interfaces of the multi-phase flow motion has significant importance. In the most of recently developed Level Set (LS) methods[20] the interfaces of the flow are described by the equation $\phi(x,y) = 0$, where ϕ, called as the level set function, is assumed to be sufficiently smooth and satisfies a Hamilton-Jacobi equation

$$\frac{\partial \phi}{\partial t} + H\left(\frac{\partial \phi}{\partial x}\right) = 0 \qquad (1.5)$$

or

$$\frac{\partial \phi}{\partial t} + H\left(\frac{\partial \phi}{\partial x}, \frac{\partial \phi}{\partial y}\right) = 0 \qquad (1.6)$$

in two dimensions.

The location of the zero of $\phi(x,y)$ is captured using high order accurate ENO methods[24,25]. Many complicated multi-phase flows are resolved excellently by LS method, although there are spaces for improving the resolution of the interfaces.

Another possible choice of LS function is to use piece-wise constant function[5,22], the characteristic function $\chi(x,y)$ of the flow domain, i.e., in the case of $K+1$ phase flows, $\chi(x,y) = k$, for k-phase, $0 \le k \le K$. Then the discontinuities (k;k+1) are captured by well developed high resolution methods. However, it is well known that the resolution of contact discontinuities using TVD or ENO methods is much worse than the resolution of strong shocks. In the process of improvement in overcoming this difficulty one has designed several efficient methods such as the method of artificial compression [7,33], method of successive viscosity reduction [15,17], essentially conservative or nonconservative schemes [6,15,29], subcell resolution method [9], super compact finite difference methods[18], high order accurate methods[39,40] and variants of front-tracking methods [12,19,41]. In this paper we introduce a stiff source term approach, i.e., instead of solving equations in one dimension

$$\frac{\partial u}{\partial t} + \frac{\partial u}{\partial x} = 0 \qquad (1.7)$$

we solve the following equation

$$\frac{\partial u}{\partial t} + \frac{\partial u}{\partial x} = \psi(u), \qquad (1.8)$$

with Naguma type source term $\psi(u) = \alpha\, u(1-u)(u - 0.5)$, $\alpha >> 1$ in the case of one dimensional two-phase flow with $u = 0$ for phase A and $u = 1$ for phase B. We have an attemption to get stable and good resolution of interfaces.

The paper is organized as follows. In section 2 some semi-discrete schemes are given for linear equations and conservation laws. We give the proof of the fact that semi-discrete MUSCL type schemes of accuracy order higher than second do not exist for genuinely nonlinear conservation laws. In section 3 several useful second order and third order high resolution schemes are presented to illustrate the fact of non-unified construction of some highly accurate schemes. Section 4 describes our unified construction of the second order high resolution ENO schemes. Higher order schemes are presented in section 5. Recurrent formulae for high order accuracy correction of semi-discrete non-MUSCL schemes are given as well. Section 6 is devoted to the numerical solution of stiff ordinary differential equations. We derive a highly fidelity algorithm (HFA) for logistic model in mathematical biology and then present HFA for Naguma equation which has important applications in this paper. In section 7 we give an application of HFA as an ODE solver for numerical solution of PDE with Naguma stiff source term. Similar technique is applied for numerical solution of (1;0) discontinuity for advection equation (1.7). In section 8 we present numerical results for unified ENO methods to see the resolution and the results for (1;0) discontinuity sharpenning using artificial stiff terms approach. Section 9 concludes the paper.

2. SEMI-DISCRETE SCHEMES

Consider the following high order accurate semi-discrete schemes for conservation laws

$$\left(\frac{\partial v}{\partial t}\right)_j + \frac{1}{\Delta x}\left(\hat{f}_{j+1/2} - \hat{f}_{j-1/2}\right) = 0. \qquad (2.1)$$

In the linear case $f = av$ with constant $a > 0$ we have the following unlimited scheme

$$\hat{f}_{j+\frac{1}{2}} = a\, v_{j+\frac{1}{2}} \qquad (2.2)$$

$$v_{j+\frac{1}{2}} = v_j + \frac{1}{4}(1+k)\Delta_{j+\frac{1}{2}}v + \frac{1}{4}(1-k)\Delta_{j-\frac{1}{2}}v \qquad (2.3)$$

$$\Delta_{j+\frac{1}{2}}v = v_{j+1} - v_j \qquad (2.4)$$

This scheme is second order accurate if $k = -1$ (upwind) or $k = 1$ (central), or $k = 0$ (Fromm). The scheme is of third order accuracy if $k=\frac{1}{3}$.

In the non-linear case: $f = f(u), f'_u \ge 0$, we have the following two kinds of generalizations.

i Non-MUSCL type Scheme [21]:

$$\hat{f}_{j+\frac{1}{2}} = f_j + \tfrac{1}{4}(1+k)\tilde{\Delta}_{j+\frac{1}{2}}f$$
$$+\tfrac{1}{4}(1-k)\bar{\Delta}_{j-\frac{1}{2}}f$$
$$\tilde{\Delta}_{j+\frac{1}{2}}f = mm(\Delta_{j+\frac{1}{2}}f, b\Delta_{j-\frac{1}{2}}f),$$
$$\bar{\Delta}_{j+\frac{1}{2}}f = mm(\Delta_{j-\frac{1}{2}}f, b\Delta_{j+\frac{1}{2}}f),$$
$$1 \le b \le 3$$
$$\Delta_{j+\frac{1}{2}}f = f_{j+1} - f_j \qquad (2.5)$$

where the limiter function $mm(x, y)$ is defined as follows

$$mm(x,y) = \begin{cases} sign(x)\ min(|x|, |y|), & if\ x\ y \ge 0 \\ 0, & elsewhere \end{cases} \qquad (2.6)$$

It is clear that $mm(x, y)$ is the usual *minmod* function in the literatures.

ii MUSCL type Scheme [1]:

$$\tilde{f}_{j+\frac{1}{2}} = f(v_{j+\frac{1}{2}})$$
$$v_{j+\frac{1}{2}} = v_j + \tfrac{1}{4}(1+k)\tilde{\Delta}_{j+\frac{1}{2}}v$$
$$+\tfrac{1}{4}(1-k)\bar{\Delta}_{j-\frac{1}{2}}v$$
$$\tilde{\Delta}_{j+\frac{1}{2}}v = mm(\Delta_{j+\frac{1}{2}}v, b\Delta_{j-\frac{1}{2}}v),$$
$$\bar{\Delta}_{j+\frac{1}{2}}v = mm(\Delta_{j-\frac{1}{2}}v, b\Delta_{j+\frac{1}{2}}v),$$
$$1 \le b \le 3 \qquad (2.7)$$
$$\Delta_{j+\frac{1}{2}}v = v_{j+1} - v_j$$

The MUSCL type schemes with similar limitations are widely used with success in fluid dynamical problems [1,23]. However, it was mentioned in [1] that the third order scheme (k=$\frac{1}{3}$) and second order schemes ($k = -1$ or 1) give very similar results. The reason for this will be clear later.

Next we consider the following fourth or fifth order schemes for linear equations:

$$\hat{f}_{j+\frac{1}{2}} = av_{j+\frac{1}{2}}$$
$$v_{j+\frac{1}{2}} = v_j + \tfrac{1}{6}(\Delta_{j-\frac{1}{2}}v - \tfrac{1-k_1}{4}\Delta^3_{j-\frac{1}{2}}v)$$
$$+\tfrac{1}{3}(\Delta_{j+\frac{1}{2}}v - \tfrac{1+k_1}{8}\Delta^3_{j+\frac{1}{2}}v),$$
$$-1 \le k_1 \le 1. \qquad (2.8)$$
$$\Delta_{j+\frac{1}{2}}v = v_{j+1} - v_j$$
$$\Delta^3_{j+\frac{1}{2}}v = \Delta_{j-\frac{1}{2}}v - 2\Delta_{j+\frac{1}{2}}v + \Delta_{j+\frac{3}{2}}v$$

The scheme is fourth order accurate if $k_1 = -1$ (upwind) or $k_1 = 1$ (central) or $k_1 = \frac{1}{3}$ (compact). It is of fifth order accuracy if $k_1 = \frac{1}{5}$.

There are two kinds of generalizations of this scheme to non-linear case.

i Non-MUSCL type flux limited scheme:

$$\tilde{f}_{j+\frac{1}{2}} = f_j + \tfrac{1}{6}mm(\tilde{\Delta}_{j-\frac{1}{2}}f, b\tilde{\tilde{\Delta}}_{j+\frac{1}{2}}f)$$
$$+\tfrac{1}{3}mm(\tilde{\tilde{\Delta}}_{j+\frac{1}{2}}f, b\tilde{\Delta}_{j-\frac{1}{2}}f) \qquad (2.9)$$

$$\tilde{\Delta}_{j-\frac{1}{2}}f = \Delta_{j-\frac{1}{2}}f - \frac{1-k_1}{4}\bar{\Delta}^3_{j-\frac{1}{2}}f \qquad (2.10)$$

$$\tilde{\tilde{\Delta}}_{j+\frac{1}{2}}f = \Delta_{j+\frac{1}{2}}f - \frac{1+k_1}{8}\bar{\Delta}^3_{j+\frac{1}{2}}f \qquad (2.11)$$

$$\Delta_{j+\frac{1}{2}}f = f_{j+1} - f_j,$$
$$\bar{\Delta}^3_{j+\frac{1}{2}}f = \bar{\Delta}_{j-\frac{1}{2}}f - 2\bar{\Delta}_{j+\frac{1}{2}}f + \bar{\Delta}_{j+\frac{3}{2}}f$$
$$\bar{\Delta}_{j+\frac{1}{2}}f = mm(\Delta_{j-\frac{1}{2}}f, b_1\Delta_{j+\frac{1}{2}}f, b_2\Delta_{j-\frac{3}{2}}f) \qquad (2.12)$$

ii MUSCL type flux limited scheme [13]:

$$\tilde{f}_{j+\frac{1}{2}} = f(v - j + \tfrac{1}{2})$$
$$v_{j+\frac{1}{2}} = v_j + \tfrac{1}{6}mm(\tilde{\Delta}_{j-\frac{1}{2}}v, b\tilde{\tilde{\Delta}}_{j+\frac{1}{2}}v)$$
$$+\tfrac{1}{3}mm(\tilde{\tilde{\Delta}}_{j+\frac{1}{2}}v, b\tilde{\Delta}_{j-\frac{1}{2}}v)$$
$$\tilde{\Delta}_{j-\frac{1}{2}}v = \Delta_{j-\frac{1}{2}}v - \tfrac{1-k_1}{4}\bar{\Delta}^3_{j-\frac{1}{2}}v$$
$$\tilde{\tilde{\Delta}}_{j+\frac{1}{2}}v = \Delta_{j+\frac{1}{2}}v - \tfrac{1+k_1}{8}\bar{\Delta}^3_{j+\frac{1}{2}}v$$
$$\Delta_{j+\frac{1}{2}}v = v_{j+1} - v_j$$
$$\bar{\Delta}^3_{j+\frac{1}{2}}v = \bar{\Delta}_{j-\frac{1}{2}}v - 2\bar{\Delta}_{j+\frac{1}{2}}v + \bar{\Delta}_{j+\frac{3}{2}}v$$
$$\bar{\Delta}_{j+\frac{1}{2}}v = mm(\Delta_{j-\frac{1}{2}}v, b_1\Delta_{j+\frac{1}{2}}v, b_2\Delta_{j-\frac{3}{2}}v) \qquad (2.13)$$

This scheme is used recently with success in CFD [13].

Here we prove a theorem on the possible order of accuracy of the semi-discrete MUSCL schemes, namely we prove the non-existence of third order accurate semi-discrete MUSCL schemes for genuinely nonlinear conservation laws.

Consider the nonlinear conservation law

$$\frac{\partial u}{\partial t} + \frac{\partial f(u)}{\partial x} = 0. \quad f''_{uu} \ne 0 \qquad (2.14)$$

Semi-discrete scheme of MUSCL type has the form:

$$\left(\frac{\partial v}{\partial t}\right)_j + \frac{1}{\Delta x}(\hat{f}_{j+1/2} - \hat{f}_{j-1/2}) = 0 \qquad (2.15)$$

$$\hat{f}_{j+1/2} = \hat{f}(v_{j+1/2}) \qquad (2.16)$$

$$v_{j+1/2} = \sum_{i=-k}^{k} \alpha^{(k)}_{j+i} v_{j+i}. \qquad (2.17)$$

Third order (spatial) accuracy of (2.1) means

$$\hat{f}(v_{j+1/2}) - \hat{f}(v_{j-1/2}) = \Delta x \left(\frac{\partial f}{\partial x}\right)_j + O(\Delta x^4) \qquad (2.18)$$

The coefficients $\alpha_{j+i}^{(k)}$ in (2.17) are complicated functions of v depending on the limiters. The final expression for v_{j+i} is a linear combination of $v_{j+\frac{1}{2}}$ with coefficients $\alpha_{j+i}^{(k)}$. In the case of sufficiently smooth solution v, it suffices to consider the case of constant $\alpha_{j+i}^{(k)}$ to check the possible accuracy of the scheme.

Expanding $v_{j+\frac{1}{2}}$ and $v_{j-\frac{1}{2}}$ into Taylor series:

$$v_{j+\frac{1}{2}} = v_j + \alpha\Delta x\, v'_j + \beta\Delta x^2\, v''_j + \gamma\Delta x^3\, v'''_j + O(\Delta x^4) \tag{2.19}$$

$$v_{j-\frac{1}{2}} = v_{j-1} + \alpha\Delta x\, v'_{j-1} + \beta\Delta x^2\, v''_{j-1} + \gamma\Delta x^3\, v'''_{j-1} + O(\Delta x^4) \tag{2.20}$$

Substituting (2.19) and (2.20) into (2.18), and expanding the LHS of (2.18) into Taylor series, we get

$$\begin{aligned}
\hat{f}(v_{j+1/2}) - \hat{f}(v_{j-1/2}) &= \Delta x\, f'_u\, v'_j \\
&+ \Delta x^2[f'_u\, v''_j\,(\alpha - \tfrac{1}{2}) + \tfrac{1}{2}f''_{uu}\, v'_j(\alpha^2 - (\alpha-1)^2)] \\
&+ \Delta x^3[f'_u\, v'''_j\,(\tfrac{1}{6} - \tfrac{1}{2}\alpha + \beta) \\
&+ \tfrac{1}{2}f''_{uu}\, v'_j\, v''_j(2\alpha\beta - (\alpha-1)(\tfrac{1}{2} - \alpha + \beta)) \\
&+ \tfrac{1}{6}f'''_{uuu}\, v'^3_j\,(\alpha^3 - (\alpha-1)^3)] + O(\Delta x^4)
\end{aligned} \tag{2.21}$$

Second order accuracy requires that the coefficient with Δx^2 of the RHS of (2.21) vanishes, *i.e.*

$$\Delta x^2: \quad \alpha = \frac{1}{2}, \quad \Rightarrow \quad \alpha^2 - (\alpha-1)^2 = 0 \tag{2.22}$$

For the third order accuracy the following condition should be satisfied

$$\Delta x^3: \quad f'_u: \quad \frac{1}{6} - \frac{1}{2}\alpha + \beta = 0, \quad \Rightarrow \quad \beta = \frac{1}{12}. \tag{2.23}$$

It is clear that $f''_{uu} = 0$, $f'''_{uuu} = 0$ for linear $f(u)$, hence in this case $\alpha = 1/2$ and $\beta = 1/12$ are sufficient for the third order accuracy. While for the genuinely nonlinear $f(u)$, $f''_{uu} \neq 0$, it is necessary for the third order accuracy that at least, the coefficients in the terms with f''_{uu} should equal to zero in addition. This means

$$2\alpha\beta - (\alpha-1)(\frac{1}{2} - \alpha + \beta) = 0, \tag{2.24}$$

But in the same time we have $\alpha = 1/2$, $\beta = 1/12$ which will lead to the following

$$2\alpha\beta - (\alpha-1)(\frac{1}{2} - \alpha + \beta) = \frac{1}{8}(\neq 0) \tag{2.25}$$

The contradiction of (2.24) with (2.25) proves the following

Theorem *No MUSCL type schemes with accuracy order higher than second exist for genuinely nonlinear conservation laws.*

In the following sections we are interested in the non-MUSCL semi-discrete ENO schemes.

3. NON-MUSCL ENO SCHEMES

In this section we show some examples of high resolution schemes of unified and non-unified constructions. For the sake of simplicity we only consider the conservation laws with $f = f^+$, $f_u^+ \geq 0$

$$\left(\frac{\partial f^+}{\partial v}\right)_j + \frac{1}{\Delta x}\left(\hat{f}^+_{j+\frac{1}{2}} - \hat{f}^+_{j-\frac{1}{2}}\right) = 0 \tag{3.1}$$

Introduce the following notations of difference quantities:

$$\begin{aligned}
\Delta_{j+\frac{1}{2}}^{(1)} f^+ &= f^+_{j+1} - f^+_j \\
\Delta_j^{(2)} f^+ &= \Delta_{j+\frac{1}{2}}^{(1)} f^+ - \Delta_{j-\frac{1}{2}}^{(1)} f^+ \\
D_{j+\frac{1}{2}}^{(2)} f^+ &= mm(\Delta_j^{(2)} f^+, \Delta_{j+1}^{(2)} f^+) \\
\Delta_{j+\frac{1}{2}}^{(3)} f^+ &= \Delta_{j+1}^{(2)} f^+ - \Delta_j^{(2)} f^+
\end{aligned} \tag{3.2}$$

Consider the second order NND scheme [38,42]:

$$\hat{f}^+_{j+\frac{1}{2}} = f^+_j + \frac{1}{2}mm\left(\Delta_{j+\frac{1}{2}}^{(1)} f^+, \Delta_{j-\frac{1}{2}}^{(1)} f^+\right) \tag{3.3}$$

$$\Delta_{j+\frac{1}{2}}^{(1)} f^+ = f^+_{j+1} - f^+_j \tag{3.4}$$

This is a TVD scheme which may degenerate the accuracy near local extrema. Two modifications of this scheme to be UNO or ENO are as follows.

i   Semi-discrete version of second order UNO scheme [10]

$$\begin{aligned}
\hat{f}^+_{j+\frac{1}{2}} = f^+_j + \tfrac{1}{2}mm\Big(&\Delta_{j+\frac{1}{2}}^{(1)} f^+ - \tfrac{1}{2}D_{j+\frac{1}{2}}^{(2)} f^+, \\
&\Delta_{j-\frac{1}{2}}^{(1)} f^+ + \tfrac{1}{2}D_{j-\frac{1}{2}}^{(2)} f^+\Big)
\end{aligned} \tag{3.5}$$

and

ii   Semi-discrete version of Jin's second order monotonicity preserving scheme [15]

$$\begin{aligned}
\hat{f}^+_{j+\frac{1}{2}} = f^+_j + \tfrac{1}{2}mm\Big(&\Delta_{j-\frac{1}{2}}^{(1)} f^+ + D_{j-\frac{1}{2}}^{(2)} f^+, \\
&\Delta_{j+\frac{1}{2}}^{(1)} f^+, \Delta_{j+\frac{3}{2}}^{(1)} f^+ - D_{j+\frac{3}{2}}^{(2)} f^+\Big)
\end{aligned} \tag{3.6}$$

One notes that the $mm(x,y)$ function can be calculated without any branching functions (See section 4). Hence the above NND scheme and its two modifications are constructed in an unified way and are suitable for parallel computing.

Now consider the case of third order modifications of NND scheme, the ENN scheme [39] and Z33 scheme[4]. Introduce the following function $ms(x,y)$:

$$ms(x,y) = \begin{cases} x, if\, |x| < |y| \\ y, if\, |y| < |x| \\ x, if\, |x| = |y|,\, and\, xy > 0 \\ 0, if\, |x| = |y|,\, and\, xy \le 0 \end{cases} \qquad (3.7)$$

Roughly speaking, function $ms(x,y)$ chooses the argument with minimum modulus.

In the ENN scheme the numerical flux $\hat{f}^+_{j+\frac{1}{2}}$ is defined as follows.

i. If $|\Delta^{(1)}_{j+\frac{1}{2}} f^+| \le |\Delta^{(1)}_{j-\frac{1}{2}} f^+|$ then

$$\hat{f}^+_{j+\frac{1}{2}} = f^+_j + \tfrac{1}{2} ms \left(\Delta^{(1)}_{j+\frac{1}{2}} f^+, \Delta^{(1)}_{j-\frac{1}{2}} f^+ \right) \\ - \tfrac{1}{6} ms \left(\Delta^{(2)}_{j+1} f^+, \Delta^{(2)}_{j} f^+ \right) \qquad (3.8)$$

ii. If $|\Delta^{(1)}_{j+\frac{1}{2}} f^+| > |\Delta^{(1)}_{j-\frac{1}{2}} f^+|$ then

$$\hat{f}^+_{j+\frac{1}{2}} = f^+_j + \tfrac{1}{2} ms \left(\Delta^{(1)}_{j+\frac{1}{2}} f^+, \Delta^{(1)}_{j-\frac{1}{2}} f^+ \right) \\ + \tfrac{1}{3} ms \left(\Delta^{(2)}_{j} f^+, \Delta^{(2)}_{j-1} f^+ \right) \qquad (3.9)$$

Another third order scheme is the Z33 scheme [4]. In this case f^+ is defined as follows:

i. If $|\Delta^{(1)}_{j+\frac{1}{2}} f^+| \le |\Delta^{(1)}_{j-\frac{1}{2}} f^+|$ then

(i1). If $|\Delta^{(2)}_{j+1} f^+| \le |\Delta^{(2)}_{j} f^+|$ then

$$\hat{f}^+_{j+\frac{1}{2}} = f^+_j + \tfrac{1}{2} ms \left(\Delta^{(1)}_{j+\frac{1}{2}} f^+, \Delta^{(1)}_{j-\frac{1}{2}} f^+ \right) \\ - \tfrac{1}{6} ms \left(\Delta^{(2)}_{j+1} f^+, \Delta^{(2)}_{j} f^+ \right) \\ + \tfrac{1}{12} ms \left(\Delta^{(3)}_{j+\frac{3}{2}} f^+, \Delta^{(3)}_{j+\frac{1}{2}} f^+ \right) \qquad (3.10)$$

(i2). If $|\Delta^{(2)}_{j+1} f^+| > |\Delta^{(2)}_{j} f^+|$ then

$$\hat{f}^+_{j+\frac{1}{2}} = f^+_j + \tfrac{1}{2} ms \left(\Delta^{(1)}_{j+\frac{1}{2}}, \Delta^{(1)}_{j-\frac{1}{2}} f^+ \right) \\ - \tfrac{1}{6} ms \left(\Delta^{(2)}_{j+1} f^+, \Delta^{(2)}_{j} f^+ \right) \\ - \tfrac{1}{12} ms \left(\Delta^{(3)}_{j+\frac{1}{2}} f^+, \Delta^{(3)}_{j-\frac{1}{2}} f^+ \right) \qquad (3.11)$$

ii. If $|\Delta^{(1)}_{j+\frac{1}{2}} f^+| > |\Delta^{(1)}_{j-\frac{1}{2}} f^+|$ then

(ii1). If $|\Delta^{(2)}_{j} f^+| \le |\Delta^{(2)}_{j-1} f^+|$ then

$$\hat{f}^+_{j+\frac{1}{2}} = f^+_j + \tfrac{1}{2} ms \left(\Delta^{(1)}_{j+\frac{1}{2}} f^+, \Delta^{(1)}_{j-\frac{1}{2}} f^+ \right) \\ + \tfrac{1}{3} ms \left(\Delta^{(2)}_{j} f^+, \Delta^{(2)}_{j-1} f^+ \right) \\ - \tfrac{1}{12} ms \left(\Delta^{(3)}_{j+\frac{1}{2}} f^+, \Delta^{(3)}_{j-\frac{1}{2}} f^+ \right) \qquad (3.12)$$

(ii2). If $|\Delta^{(2)}_{j} f^+| > |\Delta^{(2)}_{j-1} f^+|$ then

$$\hat{f}^+_{j+\frac{1}{2}} = f^+_j + \tfrac{1}{2} ms \left(\Delta^{(1)}_{j+\frac{1}{2}} f^+, \Delta^{(1)}_{j-\frac{1}{2}} f^+ \right) \\ + \tfrac{1}{3} ms \left(\Delta^{(2)}_{j} f^+, \Delta^{(2)}_{j-1} f^+ \right) \\ + \tfrac{1}{4} ms \left(\Delta^{(3)}_{j-\frac{1}{2}} f^+, \Delta^{(3)}_{j-\frac{3}{2}} f^+ \right) \qquad (3.13)$$

We can see heavy branching calculations of the $\hat{f}^+_{j+\frac{1}{2}}$ are involved in these last two schemes. Too many 'if' and '$then$' are contained in the switch functions, although the $ms(x,y)$ function may be calculated without branching (see section 4). Hence, the way of the construction of these schemes is not suitable for parallel computing. Similar situations take place in cases of usual high order ENO schemes in choosing the stencils of the schemes [24,25].

4. UNIFIED CONSTRUCTION OF NON-MUSCL SEMI-DISCRETE ENO SCHEMES

We suppose

$$f = f^+ + f^-, \quad \frac{\partial f^+}{\partial u} \ge 0, \quad \frac{\partial f^-}{\partial u} \le 0, \qquad (4.1)$$

and

$$\hat{f} = \hat{f}^+ + \hat{f}^- \qquad (4.2)$$

are numerical fluxes for f^+ and f^- respectively.

The construction principles of the schemes for f^+ and f^- are similar. Hence, for the sake of simplicity we only construct schemes for f^+. Our constructions of the unified schemes are based on the $mm(x,y)$ and $ms(x,y)$ functions (see (2.6, 3.7)). Here we give modified expressions for these functions

$$mm(x,y) = \tfrac{1}{4} \left(sign(x) + sign(y) \right) \\ \left(|x| + |y| - ||x| - |y|| \right) \qquad (4.3)$$

$$ms(x,y) = \tfrac{1}{2} sign((x + y)xy) \\ \left(|x| + |y| - ||x| - |y|| \right) \qquad (4.4)$$

One can see that the calculations of $mm(*)$ and $ms(*)$ using (4.3) and (4.4) are not branching provided the function $sign(*)$ and $|*|$ are calculated in computer without bifurcations.

Let

$$\Delta^{(k)}_* \equiv \Delta^{(k)}_* f^+, \quad k = 0, 1, 2, ... \\ \Delta^{(0)} \equiv f^+, \\ \Delta^{(2k-1)}_{j+\frac{1}{2}} = \Delta^{(2k-2)}_{j+1} - \Delta^{(2k-2)}_{j}, \\ \Delta^{(2k)}_{j} = \Delta^{(2k-1)}_{j+\frac{1}{2}} - \Delta^{(2k-1)}_{j-\frac{1}{2}}, \\ (k = 1, 2, ...) \qquad (4.5)$$

Denote the exact value of the flux $f^+(u(x,t))$ of (1.1) at $x_{j+\frac{1}{2}}$ by $f^+_{j+\frac{1}{2}}$. $\tilde{f}^+_{j+\frac{1}{2}}$ is the approximation of $f^+_{j+\frac{1}{2}}$. Then the numerical flux $\hat{f}^+_{j+\frac{1}{2}}$ of the m th order approximation of (3.1) is as follows.

$$f^+_{j+\frac{1}{2}} = \tilde{f}^+_{j+\frac{1}{2}} + O\left(\Delta x^{m+1}\right) \tag{4.6}$$

and

$$\hat{f}^+_{j+\frac{1}{2}} = \tilde{f}^+_{j+\frac{1}{2}} + \sum_{s=1}^{[(m-1)/2]} \alpha_{2s} \left(\frac{\partial^{2s} f^+}{\partial x^{2s}}\right)_{x_{j+\frac{1}{2}}}$$
$$+ O\left(\Delta x^{m+1}\right) \tag{4.7}$$

with some constant coefficients α_{2s} ([24,25] and section 5 below).

We give two new ways of construction of m-th order accurate ENO scheme by upwind and central approximations of $f^+(x)$ in certain intervals.

In the upwind case (cf.[14]) $f^+(x)$ is approximated in the interval $x_{j-\frac{1}{2}} = x_j - \frac{\Delta x}{2} \leq x \leq x_j + \frac{\Delta x}{2} = x_{j+\frac{1}{2}}$ by a polynomial $P(x)$ of degree m

$$P(x) = \sum_{k=0}^{m} \frac{(x - x_j)^k}{k! \Delta x^k} D_j^{(k)}, \tag{4.8}$$

with

$$D_j^{(k)} = \Delta x^k \left(\frac{\partial^k f^+}{\partial x^k}\right)_{x_j} + O\left(\Delta x^{m+1}\right),$$
$$(k = 0,1,2,...), \quad m \geq 1 \tag{4.9}$$

and

$$\tilde{f}^+_{j+\frac{1}{2}} = P(x_{j+\frac{1}{2}}) = \sum_{k=0}^{m} \frac{1}{k! 2^k \Delta x^k} D_j^{(k)}, \tag{4.10}$$

(2) In the central case $f^+(x)$ is approximated in the interval $x_j \leq x \leq x_{j+1}$ by a polynomial $P(x)$ of degree m

$$P(x) = \sum_{k=0}^{m} \frac{(x - x_{j+\frac{1}{2}})^k}{k! \Delta x^k} D_{j+\frac{1}{2}}^{(k)}, \tag{4.11}$$

with

$$D_{j+\frac{1}{2}}^{(k)} = \Delta x^k \left(\frac{\partial^k f^+}{\partial x^k}\right)_{x_{j+1}} + O\left(\Delta x^{m+1}\right),$$
$$(k = 0,1,2,...), \quad m \geq 1 \tag{4.12}$$

we have

$$\tilde{f}^+_{j+\frac{1}{2}} = P\left(x_{j+\frac{1}{2}}\right) = D_{j+\frac{1}{2}}^{(0)} \tag{4.13}$$

The final numerical flux $\hat{f}^+_{j+\frac{1}{2}}$ is obtained by accuracy correcting $\tilde{f}^+_{j+\frac{1}{2}}$ according to (4.7).

The ENO schemes constructed in this unified way is named as UENO schemes. Now we present second order accurate UENO schemes. We difine

$$D_{j+\frac{1}{2}}^{(2)} = ms\left(\Delta_{j+1}^{(2)}, \Delta_j^{(2)}\right),$$
$$D_j^{(2)} = \tfrac{1}{2}\left(D_{j+\frac{1}{2}}^{(2)} + D_{j-\frac{1}{2}}^{(2)}\right)$$
$$D_j^{(1)} = ms\left(\Delta_{j+\frac{1}{2}}^{(1)} - \tfrac{1}{2} D_{j+\frac{1}{2}}^{(2)}, \Delta_{j-\frac{1}{2}}^{(1)} + \tfrac{1}{2} D_{j-\frac{1}{2}}^{(2)}\right)$$
$$D_{j+\frac{1}{2}}^{(1)} = \tfrac{1}{2}\left(D_{j+1}^{(1)} + D_j^{(1)}\right)$$
$$D_j^{(0)} = \Delta_j^{(0)} (\equiv f_j^+)$$
$$D_{j+\frac{1}{2}}^{(0)} = \Delta_j^{(0)} + \tfrac{1}{2} D_{j+\frac{1}{2}}^{(1)} - \tfrac{1}{8} D_{j+\frac{1}{2}}^{(2)}. \tag{4.14}$$

In the upwind case (scheme N22U) we have the expression for flux $\hat{f}^+_{j+\frac{1}{2}}$:

$$\hat{f}^+_{j+\frac{1}{2}} = f_j^+ + \frac{1}{2} D_j^{(1)} + \frac{1}{8} D_j^{(2)} \tag{4.15}$$

In the central case, (scheme N22C) we have

$$\hat{f}^+_{j+\frac{1}{2}} = f_j^+ + \frac{1}{2} D_{j+\frac{1}{2}}^{(1)} - \frac{1}{8} D_{j+\frac{1}{2}}^{(2)} \tag{4.16}$$

It is observed that N22U and N22C are generalizations of the schemes (3.5) and (3.6) respectively. Our experience shows that changing function ms(*) to the minmod function mm(*) in calculating $D^{(k)}$ we have almost identical numerical results. But this change may lead to better theoretical results in numerical analysis.

5. HIGHER ORDER ACCURATE UENO SCHEMES

The above idea for second order UENO scheme construction can be extended to third, fourth and higher order schemes. Let us consider the third order scheme of upwind type. Defining

$$D_j^{(3)} = ms\left(\Delta_{j+\frac{1}{2}}^{(3)}, \Delta_{j-\frac{1}{2}}^{(3)}\right),$$
$$D_{j+\frac{1}{2}}^{(3)} = \tfrac{1}{2}\left(D_{j+1}^{(3)} + D_j^{(3)}\right)$$
$$D_{j+\frac{1}{2}}^{(2)} = ms\left(\Delta_{j+1}^{(2)} - \tfrac{1}{2} D_{j+1}^{(3)}, \Delta_j^{(2)} + \tfrac{1}{2} D_j^{(3)}\right),$$
$$D_j^{(2)} = \tfrac{1}{2}\left(D_{j+\frac{1}{2}}^{(2)} + D_{j-\frac{1}{2}}^{(2)}\right)$$
$$D_j^{(1)} = ms\left(\Delta_{j+\frac{1}{2}}^{(1)} - \tfrac{1}{2} D_{j+\frac{1}{2}}^{(2)} + \tfrac{1}{12} D_{j+\frac{1}{2}}^{(3)},\right.$$
$$\left.\Delta_{j-\frac{1}{2}}^{(1)} + \tfrac{1}{2} D_{j-\frac{1}{2}}^{(2)} + \tfrac{1}{12} D_{j-\frac{1}{2}}^{(3)}\right),$$
$$D_{j+\frac{1}{2}}^{(1)} = \tfrac{1}{2}\left(D_{j+1}^{(1)} + D_j^{(1)}\right)$$
$$D_{j+\frac{1}{2}}^{(0)} = \Delta_j^{(0)} + \tfrac{1}{2} D_j^{(1)} + \tfrac{1}{8} D_j^{(2)} + \tfrac{1}{48} D_j^{(3)}. \tag{5.1}$$

we have an third order upwind approximation of $f_{j+\frac{1}{2}}$

$$\tilde{f}^+_{j+\frac{1}{2}} = f_j^+ + \frac{1}{2} D_j^{(1)} + \frac{1}{8} D_j^{(2)} + \frac{1}{48} D_j^{(3)}. \tag{5.2}$$

and the third order central approximation

$$\tilde{f}^+_{j+\frac{1}{2}} = f^+_j + \frac{1}{2}D^{(1)}_{j+\frac{1}{2}} - \frac{1}{8}D^{(2)}_{j+\frac{1}{2}} + \frac{1}{48}D^{(3)}_{j+\frac{1}{2}} \quad (5.3)$$

Now consider the fourth order scheme. Defining

$$
\begin{aligned}
D^{(4)}_{j+\frac{1}{2}} &= ms\left(\Delta^{(4)}_{j+1}, \Delta^{(4)}_j\right), \\
D^{(4)}_j &= \frac{1}{2}\left(D^{(4)}_{j+\frac{1}{2}} + D^{(4)}_{j-\frac{1}{2}}\right) \\
D^{(3)}_j &= ms\left(\Delta^{(3)}_{j+\frac{1}{2}} - D^{(4)}_{j+\frac{1}{2}}, \Delta^{(3)}_{j-\frac{1}{2}} + D^{(4)}_{j-\frac{1}{2}},\right), \\
D^{(3)}_{j+\frac{1}{2}} &= \frac{1}{2}\left(D^{(3)}_{j+1} + D^{(3)}_j\right) \\
D^{(2)}_{j+\frac{1}{2}} &= ms\left(\Delta^{(2)}_{j+1} - \frac{1}{2}D^{(3)}_{j+1} + \frac{1}{12}D^{(4)}_{j+1},\right. \\
&\qquad \left. \Delta^{(2)}_j + \frac{1}{2}D^{(3)}_j + \frac{1}{12}D^{(4)}_j\right), \\
D^{(2)}_j &= \frac{1}{2}\left(D^{(2)}_{j+\frac{1}{2}} + D^{(2)}_{j-\frac{1}{2}}\right) \\
D^{(1)}_j &= ms(\Delta^{(1)}_{j+\frac{1}{2}} - \frac{1}{2}D^{(2)}_{j+\frac{1}{2}} + \frac{1}{12}D^{(3)}_{j+\frac{1}{2}} \\
&\qquad - \frac{1}{48}D^{(4)}_{j+\frac{1}{2}}, \Delta^{(1)}_{j-\frac{1}{2}} + \frac{1}{2}D^{(2)}_{j-\frac{1}{2}} \\
&\qquad + \frac{1}{12}D^{(3)}_{j-\frac{1}{2}} + \frac{1}{48}D^{(4)}_{j-\frac{1}{2}}), \\
D^{(1)}_{j+\frac{1}{2}} &= \frac{1}{2}\left(D^{(1)}_{j+1} + D^{(1)}_j\right)
\end{aligned}
\quad (5.4)
$$

we have fourth order upwind approximation

$$\tilde{f}^+_{j+\frac{1}{2}} = f^+_j + \frac{1}{2}D^{(1)}_j + \frac{1}{8}D^{(2)}_j + \frac{1}{48}D^{(3)}_j + \frac{1}{384}D^{(4)}_j \quad (5.5)$$

and the central approximation of the fourth order

$$\tilde{f}^+_{j+\frac{1}{2}} = f^+_j + \frac{1}{2}D^{(1)}_{j+\frac{1}{2}} - \frac{1}{8}D^{(2)}_{j+\frac{1}{2}} + \frac{1}{48}D^{(3)}_{j+\frac{1}{2}} - \frac{1}{384}D^{(4)}_{j+\frac{1}{2}} \quad (5.6)$$

We note that in the case of third or fourth orders of accuracy the numerical fluxes $\tilde{f}^+_{j+\frac{1}{2}}$ obtained have to be modified in the following way

$$\hat{f}^+_{j+\frac{1}{2}} = \tilde{f}^+_{j+\frac{1}{2}} - \frac{1}{24}D^{(2)}_{j+\frac{1}{2}} \quad (5.7)$$

to match the third or fourth order accuracy of the scheme.

In a similar way we can construct upwind and central approximations $\tilde{f}^+_{j+\frac{1}{2}}$ with arbitrary order of accuracy. In the general case of m-th order accuracy of the scheme the final numerical fluxes $\hat{f}^+_{j+\frac{1}{2}}$ is defined as

$$\hat{f}^+_{j+\frac{1}{2}} = \tilde{f}^+_{j+\frac{1}{2}} + \sum_{s=1}^{[(m-1)/2]} \alpha_{2s}\left(\frac{\partial^{2s} f^+}{\partial x^{2s}}\right)_{x_{j+\frac{1}{2}}} \quad (5.8)$$

where $\tilde{f}^+_{j+\frac{1}{2}}$ is the approximation of $f^+_{j+\frac{1}{2}}$ like (4.8) or (4.11)and the coefficients α_{2s} are defined by recurrent

formulae:

$$\frac{1}{2^{2s}(2s+1)!} + \sum_{k=1}^{s} \frac{\alpha_{2k}}{2^{2s-2k}(2s-2k+1)!} = 0$$

$$s = 1, 2, ..., [(m-1)/2]$$

In particular we have

$$
\begin{aligned}
&\alpha_2 = -\frac{1}{24}, \quad \alpha_4 = \frac{7}{5760}, \\
&\alpha_6 = -\frac{31}{967680}, \quad \alpha_8 = \frac{381}{464486400}, \quad \cdots
\end{aligned}
\quad (5.9)
$$

6. HIGH FIDELITY ALGORITHMS OF STIFF ODEs

In the previous sections we constructed UENO schemes in an unified way which are of high order accurate. However as was mentioned in the Introduction the resolution of discontinuities of UENO schemes of linear equations will not be sharp compairing with the one for conservation laws. We propose here a new discontinuity sharpenning technique- adding stiff source term approach. In solving the new linear equation with stiff source terms the ODE solver and the way of adding source term are very important. Sections 6 and 7 will devoted for these subjects.

The research of LeVeque and Yee [32] show us that many often used numerical methods will to physically spurous solutions for stiff ODEs. For example solution of the Cauchy problem of logistic equation

$$\frac{du}{dt} = \alpha u(1-u), \quad \alpha > 0, \quad u(0) = u^0(> 0), \quad (6.1)$$

is

$$u(t) = \frac{u^0}{u^0 + (1 - u^0)e^{-\alpha t}} \quad (6.2)$$

which is monotone, $u(t) \to 0$ as $t \to \infty$. However, classical numerical methods such as explicit Euler method, Adams-Bashforth method, modified Euler method, Runge-Kutta methods, predictoe-corrector methods, linear multistep methods(LMM) etc. may give correct asymptote ($u = 1$) only in the case when the stiffness parameter $A(= \Delta t\alpha)$ is taken within the region of linear stability of the algorithm under consideration. Hence, number A can not be large, otherwise, the numerical solution will have period-doubling bifurcation or chaotic structures. In this section we derive the highly fidelity algorithms (HFA) which can give correct solution for any large A. Then, the HFA methods will be given for the Naguma equation:

$$\frac{du}{dt} = \alpha u(1-u)(u - 0.5), \quad \alpha > 0, \quad (6.3)$$

Let $\tau = \Delta t$ be the stepsize in time, $u = u(n\tau), \hat{u} = u((n+1)\tau)$ denote the values of the solution $u(t,x)$

at the time level $n\tau$ and $(n+1)\tau$. Using Taylor's expantion of u, we have

$$\hat{u} - u = \tau \frac{du}{dt} + \frac{\tau^2}{2!}\frac{d^2u}{dt^2} + O(\tau^3)$$
$$= A(u - u^2) + \frac{\tau}{2!}A(1 - 2u)\frac{du}{dt} + O(\tau^3) \qquad (6.4)$$

The key step in the derivation of HFA methods is to replace the term $A(u - u^2)$ by $A(u - u\hat{u})$. We have

$$\hat{u} - u$$
$$= A(u - u\hat{u}) + Au(\hat{u} - u)$$
$$\quad + \frac{\tau}{2!}A(1 - 2u)\frac{du}{dt} + O(\tau^3)$$
$$= A(u - u\hat{u}) + Au(\tau\frac{du}{dt} + O(\tau^2))$$
$$\quad + \frac{\tau}{2!}A(1 - 2u)\frac{du}{dt} + O(\tau^3)$$
$$= A(u - u\hat{u}) + Au(Au(1 - u) + O(\tau^2)) \qquad (6.5)$$
$$\quad + \frac{A^2}{2!}(1 - 2u)u(1 - u) + O(\tau^3)$$
$$= A(u - u\hat{u}) + \frac{A^2}{2!}(u - u^2) + O(\tau^3)$$
$$= A(u - u\hat{u}) + \frac{A^2}{2!}(u - u\hat{u})$$
$$\quad + \frac{A^2}{2!}u(\hat{u} - u) + O(\tau^3)$$

Expanging (6.5) further we have

$$\hat{u} - u = A(u - u\hat{u}) + \frac{A^2}{2!}(u - u\hat{u}) + \frac{A^3}{3!}u(1 - \hat{u}) + O(\tau^3) \qquad (6.6)$$

Truncating the RHS of (6.6) we have the expression

$$\hat{u} = u + \sum_{k=1}^{m}\frac{A^k}{k!}u(1 - \hat{u}), \ (m \leq 1)$$

or after solving for $\hat{u}$:

$$\hat{u} = u + \frac{(1 + \sum_{k=1}^{m}\frac{A^k}{k!})u}{1 + \sum_{k=1}^{m}\frac{A^k}{k!}u} \qquad (6.7)$$

It can be shown (using the expression (6.2)) that the formula (6.7) is valid for all natural number m, and is of m-th order of accuracy (of cause in the sence of small time stepsize), and it is important that we can prove [31] that $\hat{u} \to 1$ monotonically as $t \to \infty$ for any $A > 0$. These important properties coin the name of the method (6.7) as **H**ighly **F**idelity **A**lgorithm of the logistic equation.

For the derivation of the HFA methods for Naguma equation (6.3) we have to mention the following special features: solution $u = 0$ and $u = 1$ are stable when $t \to +\infty$, while $u = 0.5$ is unstable. This means that for all starting value $u < 0.5$ the solution $u(t)$ will tend to 0 , and for all $u > 0.5$ $u(t)$ will tend to 1 quickly. In the construction of HFA methods for Naguma equation we have to consider two cases: (1). starting value $0 < u < 0.5$ and (2). $u > 0.5$

After a similar but more complex calculations we arrive at the following HFA methods for Naguma equation.

The first order accurate HFA method:

$$\hat{u} = \begin{cases} \dfrac{[1 + A(u - 0.5)]u}{1 + A(u - 0.5)u}, & \text{if } u > 0.5 \\[2ex] \dfrac{u}{1 + A(0.5 - u)(1 - u)}, & \text{if } u < 0.5 \end{cases} \qquad (6.8)$$

and second order accurate HFA method:

$$\hat{u} = \begin{cases} \dfrac{\{1 + [A + \frac{A^2}{2!}(u(1 - u) + u - 0.5)](u - 0.5)\}u}{1 + [A + \frac{A^2}{2!}(u(1 - u) + u - 0.5)](u - 0.5)u}, \\[1ex] \qquad\qquad \text{if } u > 0.5 \\[2ex] \dfrac{u}{1 + [A + \frac{A^2}{2!}(0.5 - u^2)](1 - u)(0.5 - u)}, \\[1ex] \qquad\qquad \text{if } u < 0.5 \end{cases}$$
$$(6.9)$$

The above HFA methods of Naguma equation will be used as the ODE solver for (1;0) discontinuity sharpenning in the next section.

7. NEW DISCONTINUITY SHAPENNING METHOD, ARTIFICIAL STIFF SOURCE TERM APPROACH

The numerical solution of the intitial value problem

$$\frac{\partial u}{\partial t} + \frac{\partial u}{\partial x} = 0 \qquad (7.1)$$

$$u(0, x) = u^0(x) = \begin{cases} 1, if \ x \leq 0 \\ 0, if \ x > 0 \end{cases} \qquad (7.2)$$

is the basic test problem for the study of high resolution methods. Figs.8(a), 9(a) and 10(a) show numerical results of three high resolution schmes, namely, the TVD method using so called 'subbee' and 'superbee' limiters [26] and the uniformly second order accurate non-oscillatory method[10]. We can see the resolutions of the interfaces are quite different. In order to improve the resolution of the discontinuity we have mentioned several methods. Here we try to introduce add artificial stiff term $\psi(u) = \alpha u(1 - u)(u - 0.5)$ to equation(7.1), i.e. we are solving the equation

$$\frac{\partial u}{\partial t} + \frac{\partial u}{\partial x} = \psi(u), \ \psi(u) = \alpha u(1 - u)(u - 0.5), \qquad (7.3)$$

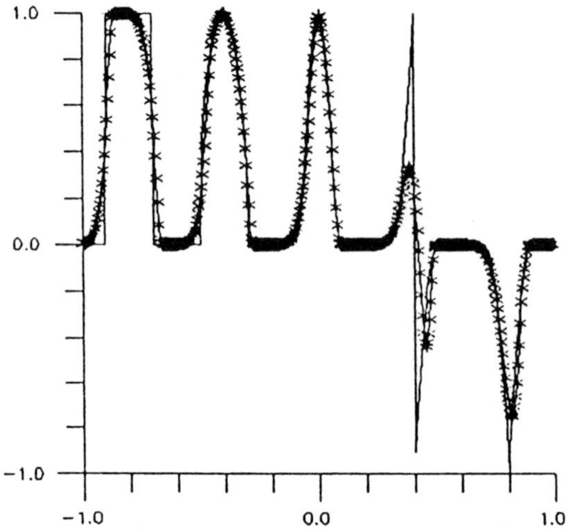

Fig.1: Second order method UNO (3.4) Equation
(8.1)   Initial data (8.2) NX=400   CFL=0.5   T=2

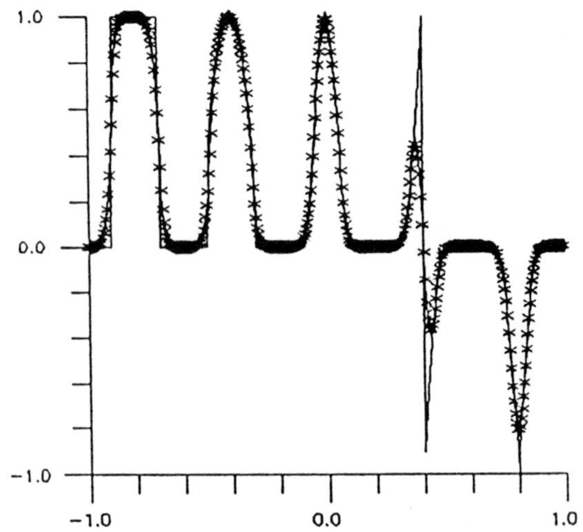

Fig.3: Third order method ENN (3.8)-(3.9) Equation
(8.1)   Initial data (8.2) NX=400   CFL=0.5   T=2

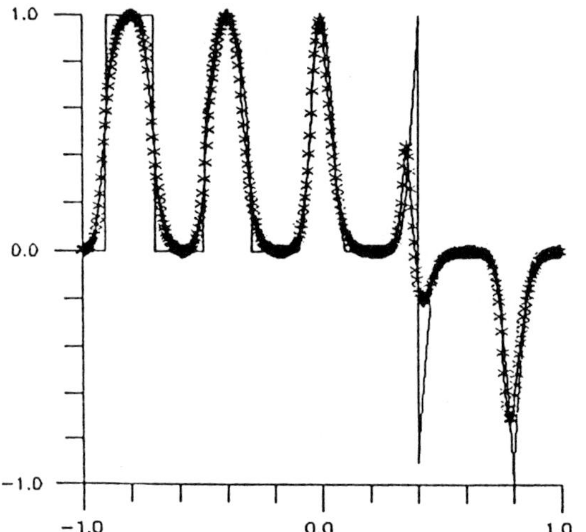

Fig.2: Second order method JIN (3.5) Equation (8.1)
Initial data (8.2) NX=400   CFL=0.5   T=2 Problem

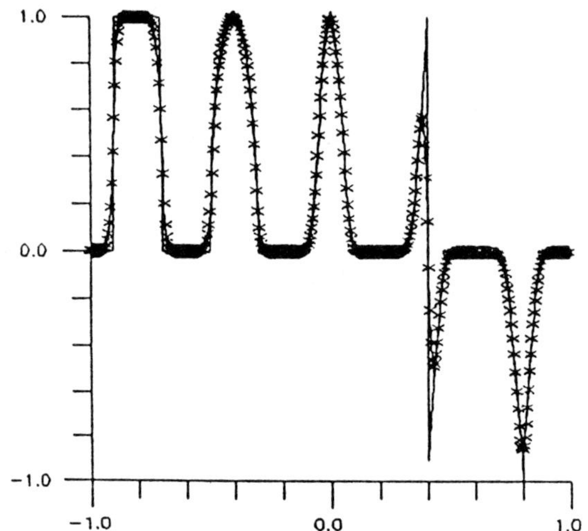

Fig.4: Third order method Z33 (3.10)-(3.13)
Equation (8.1)   Initial data (8.2) NX=400
CFL=0.5   T=2

with initial value (7.2) We use the important properties of $\psi(u)$: it pulls all $u < 0.5$ to 0, and pulls all $u > 0.5$ to 1. We expect that in this way the resolution of the discontinuity of the linear equations will become better. On the other hand, as we can find from the work [16], LeVeque and Yee pointed out the severe difficulties in solve (7.3).

Consider the following initial value problem

$$\frac{\partial u}{\partial t} + \frac{\partial u}{\partial x} = \psi(u), \ \psi(u) = \alpha u(1 - u)(u - 0.5), \quad (7.4)$$

$$u(0, x) = u^0(x) = \begin{cases} 1, \text{if } x \leq 0 \\ 0, \text{if } x > 0 \end{cases} \quad (7.5)$$

Along the charactristic line $x = x_0 + t$, the solution of (7.4) satisfies the following ODE

$$\frac{d}{dt}u(x_0 + t, t) = \psi(u(x_0 + t, t)) \quad (7.6)$$

For an arbitrary initial value $u(0, x)$ if α is large $u(t, x)$ tends to the following running wave solution $w(x - t)$

for t large, where

$$w(x) = \begin{cases} 1, & \text{if } x < 0 \\ \frac{1}{2}, & \text{if } x = 0 \\ 0, & \text{if } x > 0 \end{cases} \qquad (7.7)$$

Especially the solution of the IVP (7.3) must be the simple running wave $u(t,x) = u^\circ(x - t)$, in other words, $u(t,x)$ must be the solution of (7.1),(7.2). The successful solution methods of (7.4) must give correct running wave with sharp numerical wave front, keep correct moving speed D of the numerical wave front ($D = 1$). Another requirement is the admissible used time stepsize $\tau = \Delta t$ (satisfies the CFL condition) must not be small when the stiffness parameter $A = \tau\alpha$ is large. Because τ and Δx are quantities of same order, Δx must not be small, otherwise small Δx will lead to increase the scale of the mesh for problem solving. The trapezoidal linear implicit discretization of source term and the HFA methds may satisfy the above reqirement,i.e., the time step size used is large enough.

Another difficulty is to keep right speed of the numerical wave propogation.

In [16] LeVeque and Yee present two kinds of difference schemes. The first is a implicit discretization of two step second order MacCormack type algorithm with trapezoidal implicit treatment of the source term:

$$[1 - \frac{1}{2}\tau\psi'(u_j^n)]\Delta u_j^{(1)} = -\frac{\tau}{\Delta x}(u_j^n - u_{j-1}^n) + \tau\psi(u_j^n)$$

$$u_j^{(1)} = u_j^n + \Delta u_j^{(1)}$$

$$[1 - \frac{1}{2}\tau\psi'(u_j^n)]\Delta u_j^{(2)} = -\frac{\tau}{\Delta x}(u_{j+1}^{(1)} - u_j^{(1)}) + \tau\psi(u_j^n)$$

$$u_j^{(2)} = u_j^n + \frac{1}{2}(\Delta u_j^{(1)} + \Delta u_j^{(2)})$$

$$u_j^{n+1} = u_j^{(2)} \qquad (7.8)$$

In addition they also make a TVD like flux limitation modification for $u_j^{(2)}$

$$u_j^{n+1} = u_j^{(2)} + (R_{j+\frac{1}{2}}\phi_{j+\frac{1}{2}} - R_{j-\frac{1}{2}}\phi_{j-\frac{1}{2}})$$

where R, ϕ are limitations of *minmod* sort. Second one proposed in [16] is a natural second order splitting method:

$$\hat{u} = S_\psi(\Delta t/2)\, S_f(\Delta t)\, S_\psi(\Delta t/2)\, u \qquad (7.9)$$

S_ψ is the implicit trapezoidal linear ODE operator S_f is the MacCormack-type operator without or with flux limitation:

$$S_\psi(\Delta t/2) : [1 - \frac{1}{4}\tau\psi'(u_j^n)]\Delta u_j^* = \frac{\tau}{2}\psi(u_j^n),$$
$$u_j^* = u_j^n + \Delta u_j^*$$
$$S_f(\Delta t) : \Delta u_j^{(1)} = -\frac{\tau}{\Delta x}(u_j^* - u_{j-1}^*),$$
$$u_j^{(1)} = u_j^* + \Delta u_j^{(1)}$$
$$\Delta u_j^{(2)} = -\frac{\tau}{\Delta x}(u_{j+1}^{(1)} - u_j^{(1)}), \qquad (7.10)$$
$$u_j^{(2)} = u_j^* + \frac{1}{2}(\Delta u_j^{(1)} + \Delta u_j^{(2)})$$
$$u_j^{**} = u_j^{(2)} + (R_{j+\frac{1}{2}}^*\phi_{j+\frac{1}{2}}^* - R_{j-\frac{1}{2}}^*\phi_{j-\frac{1}{2}}^*)$$
$$S_\psi(\Delta t/2) : [1 - \frac{1}{4}\tau\psi'(u_j^{**})]\Delta u_j^{**} = \frac{\tau}{2}\psi(u_j^{**}),$$
$$u_j^{n+1} = u_j^{**} + \Delta u_j^{**}$$

where R^*, ϕ^* are some sort of limitations.

As is pointed out in [16], if $A = \Delta t\alpha$ is small,the numerical wave speed D is alway correct ($D = 1$), but if A becomes sufficiently large (for example, $A=15$) all calculations with or without flux limitations give incorrect wave speed. The speed D may < 1 even $D = 0$ or $D > 1$ depending on the scheme and the way of flux limitations. We also tried different well known TVD schemes to see whether there are any exceptional schemes to give correct wave propogation speed, but fail.

A.Bermudez and M.E.Vazquez [2] tested first order upwind difference method as S_f ,explicit Euler method as ODE solver S_ψ , the source term ψ_j treated as an upwind average $\beta\psi_j + (1 - \beta)\psi_{j-1}$. The comparision made in [2] for $A \leq 3$ shows that $\beta = \frac{1}{2}$ gives the best numerical wave speed $D \approx 1$. However one may find from the figures in [2] if A becomes larger the mean speed of the numerical wave speed D will become smaller and smaller. So this problem seems have not resolved completely.

In order to simulate the wave motion correctly we modify the way of adding stiff source term. We introduce

$$\tilde{\psi}_j = (1 - \theta_j)\,\psi(u_j)$$

where θ_j is some kind of switch function which will detect the interval with numerical 'shock'.

Let $\varepsilon = 10^{-6}$ be a small number. We define θ_j as the following

$$\theta_j = max(\theta_j^{(-1)}, \theta_j^{(0)}, \theta_j^{(1)})$$

where

$$\theta_j^{(-1)} = \left(\frac{|D_0|}{|D_0| + |D_{-1}| + \varepsilon}\right)^m,$$
$$\theta_j^{(0)} = \left(\frac{|D_1 - D_0|}{|D_1| + |D_0| + \varepsilon}\right)^m,$$
$$\theta_j^{(1)} = \left(\frac{|D_2 - D_1|}{|D_2| + |D_1| + \varepsilon}\right)^m,$$

$$\theta_j^{(0)} = \left(\frac{|D_1 - D_0|}{|D_1| + |D_0| + \varepsilon} \right)^m ,$$

$$\theta_j^{(1)} = \left(\frac{|D_1|}{|D_2| + |D_1| + \varepsilon} \right)^m ,$$

Remark 2. We may change all D_i in the expressions $\theta_j^{(\ell)}$ to $A_i = |D_i|$, $(\ell = -1, 0, 1)$.

We can observe from the expressions $\theta_j^{(\ell)}, \ell = -1, 0, 1$ that there will be a term with some j which equals or is close to 1 near the wave front. Hence, with this j, $\theta_j = 1$ or ≈ 1 . In other words, for this particular j, $\tilde{\psi}_j$ is zero or is close to zero, and therefore, actually we are perform calculations with very small or even without source terms and so we can expect correct numerical wave speed.

This approach of adding modified stiff source term is used to solve linear equation (7.1) with initial data (7.2) The comparitive results are presented in the next section.

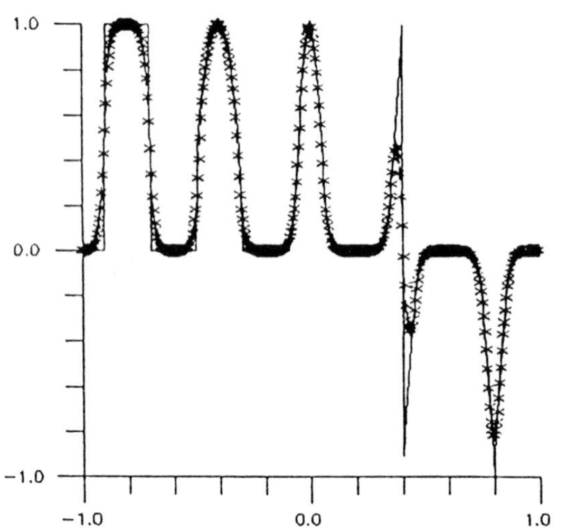

Fig.5: Second order method N22U (4.18) Equation (8.1)   Initial data (8.2) NX=400   CFL=0.5   T=2

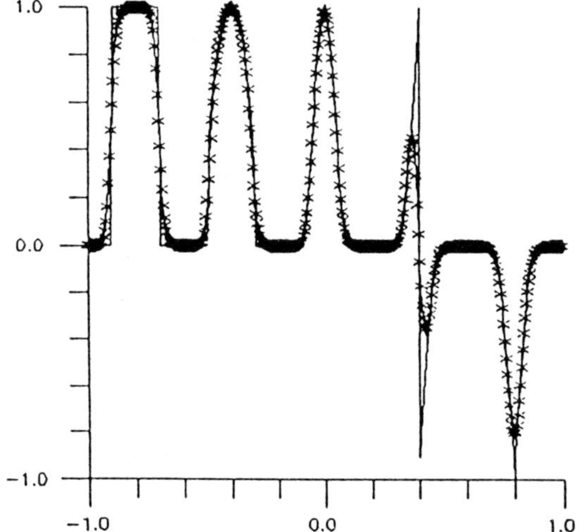

Fig.6: Second order method N22C (4.22) Equation (8.1)   Initial data (8.2) NX=400   CFL=0.5   T=2

$$(m \geq 1)$$
$$D_i = u_i - u_{i-1}, \quad i = -1, 0, 1, 2$$

Remark 1. In the case of the equation with $D < 0$, for example:

$$\frac{\partial u}{\partial t} - \frac{\partial u}{\partial x} = \psi(u), \quad \psi(u) = \alpha u(1 - u)(u - 0.5),$$

the definition of $\theta_j^{(-1)}$ and $\theta_j^{(1)}$ should be exchanged.

$$\theta_j^{(-1)} = \left(\frac{|D_0 - D_{-1}|}{|D_0| + |D_{-1}| + \varepsilon} \right)^m ,$$

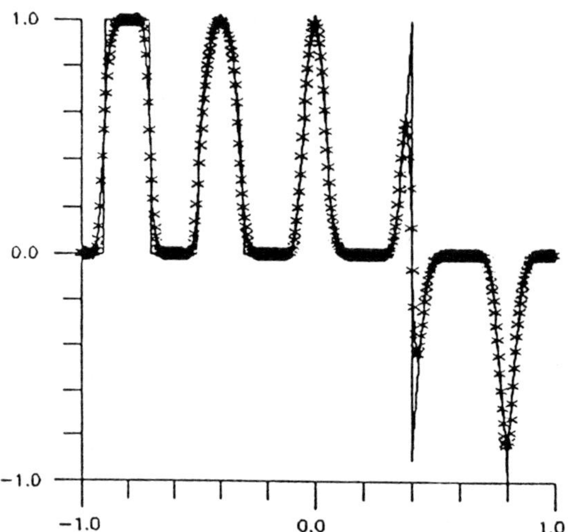

Fig.7: Third order method N33U (5.2) Equation (8.1)   Initial data (8.2) NX=400   CFL=0.5   T=2

8. NUMERICAL RESULTS

8.1 Test of Unified ENO Methods

For the practical use the semi-discrete UENO schemes constructed are solved by well known TVD Runge-Kutta time-steping methods[24].

Numerical tests of the new UENO schemes are performed for linear advection problem

$$\frac{\partial u}{\partial t} + \frac{\partial u}{\partial x} = 0, \tag{8.1}$$

Initial data is a combination of five profiles (square, extended semi-circle, $\sin^2 x$, parabolic sharp shock, and parabolic sharp corner)(See the solid lines in figures bellow):

$$u(x,0) = \begin{cases} 1, & \\ \quad if -0.9 < x < -0.7 & \\ (1 - 100\,(x+0.4)^2)^{1/2}, & \\ \quad if -0.5 < x < -0.3 & \\ (sin\,5\pi(x+0.1))^2, & \\ \quad if -0.1 < x < 0.1 & \\ 100\,(x-0.3)^2, & \\ \quad if \quad 0.3 < x < 0.4 & \\ -100\,(x-0.5)^2, & \\ \quad if \quad 0.4 < x < 0.5 & \\ -100\,(x-0.7)^2, & \\ \quad if \quad 0.7 < x < 0.8 & \\ -100\,(x-0.9)^2, & \\ \quad if \quad 0.8 < x < 0.9 & \\ 0, & \\ \quad elsewhere & \end{cases} \tag{8.2}$$

The problem is solved by the high resolution schemes and the schemes ENN and Z33 using 400 grids in the interval [-1,1] with CFL=0.5 and T=2. The results are presented in Figs.1-7. Solid lines are exact solutions of the problem, marks indicate the numerical results. The notations N22U,N22C stand for new UENO second order upwind and central schemes. The notations UNO2 and JIN stand for the schemes using (3.5), (3.6) respectively, ENN and Z33 for schemes from [39] and [4]. N33U stands for UENO third order upwind scheme. One might find that the results of the uniformly second order accurate schemes N22U and N22C improve the results of UNO2 and JIN, and have resolution close to that of third order schemes.

8.2 Test of Artificial Stiff Source Term Technique

Here we present numerical results of solving initial value problems (7.1),(7.2) without source term and (7.3),(7.2) with source term. The tests are performed using well known ubified TVD schemes with subbee and superbee flux limiters [26] and the uniformly second order accurate scheme UNO2 [10] and first order upwind method. Figures 8-11 show the results by using Subbee TVD, Superbee TVD, UNO2 and upwind schemes respectively. The symble 'a' means the results using equation without source term, The symble 'b' means the results using the equation with

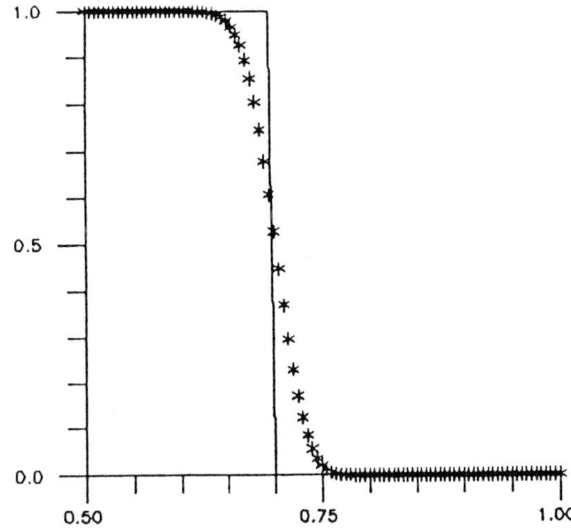

Fig.8(a): First order upwind method Initial data(7.2). Equation (7.1) NX=200 CFL=0.75 T = 0.5

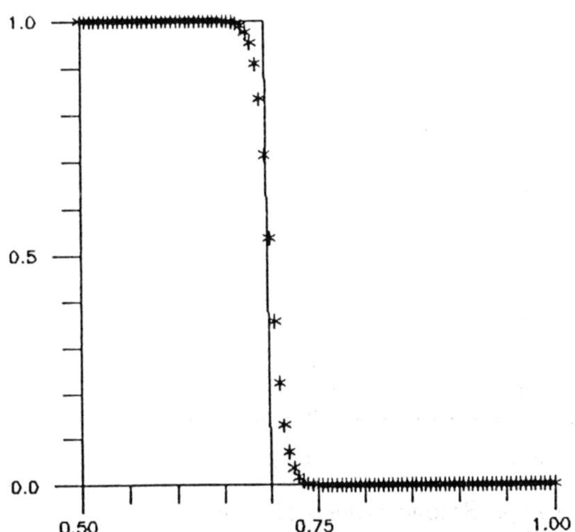

Fig.9(a): TVD Subbee method [26] Initial data(7.2). Equation (7.1) NX=200 CFL=0.75 T = 0.5

source term without modification. The symble 'c' corresponds to the results using the equation with source term with modification. The parameters in the calculations are as follows. The interval of the calculation is $(0 < x < 1)$, $\Delta x = 0.005$, CFL =0.75, T = 0.5, $A = 1000$. The parameter A for first order upwind scheme is changed to $A = 500$. We also tried the calaulations using different CFL numbers, namely, CFL=0.5, CFL= 0. 25 , and found that there are some influences of CFL on the resolution of wave front. Our

311

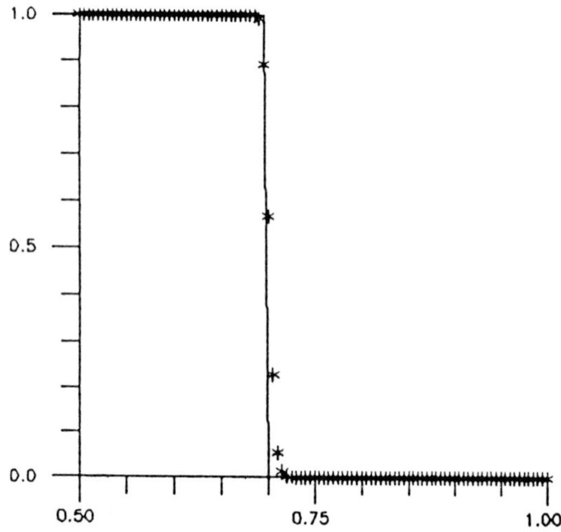

Fig.10(a) TVD Superbee method [26] Initial data(7.2). Equation (7.1) NX=200 CFL=0.75 T = 0.5

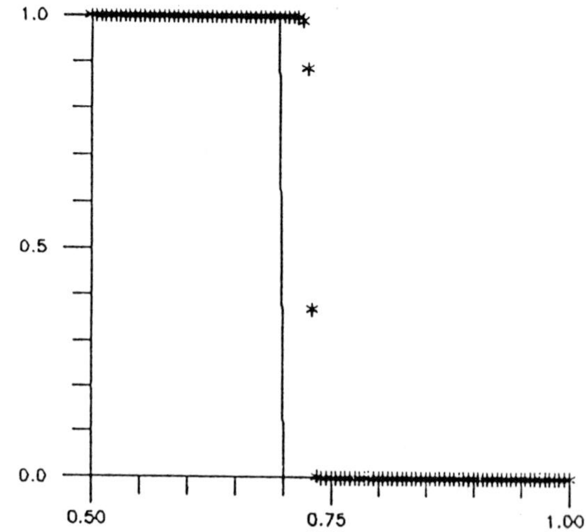

Fig.8(b): First order upwind method Initial data(7.2). Equation (7.4) without source term modification NX=200 CFL=0.75 T = 0.5 A = 500

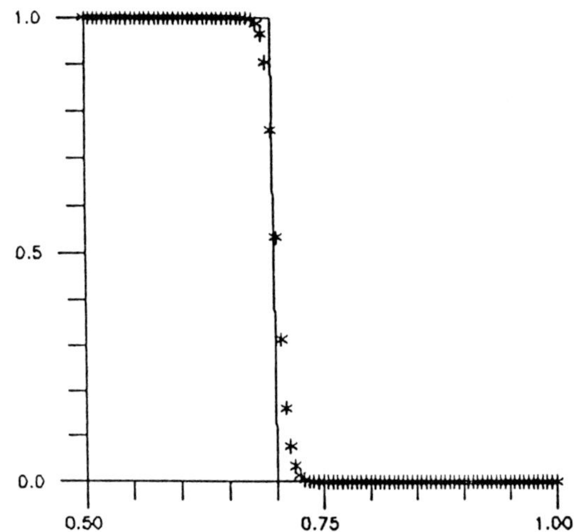

Fig.11(a) Second order UNO method [10] Initial data(7.2). Equation (7.1) NX=200 CFL=0.75 T = 0.5

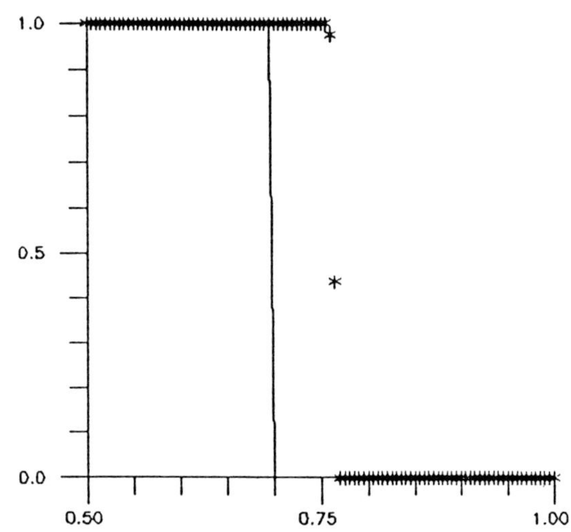

Fig.9(b) TVD Subbee method [26] Equation (7.4) without source term modification Initial data(7.2) NX=200 CFL=0.75 T = 0.5 A = 1000

experience shows that a good choice of A is to match the relation $A/\text{CFL} \approx 1000$.

8.3 Test of ODE Solvers for Naguma Stiff ODE

In this subsection we compare three ODE solvers for Naguma equation

$$\frac{du}{dt} = \alpha\ \psi(u), \quad \psi(u) = u(1-u)(u-0.5), \quad \alpha > 0$$

The first method is the explicit Euler method

$$\hat{u} - u = A\ \psi(u) = Au(1-u)(u-0.5)$$

Second one is the linearlized trapezoidal role which is derived from the fully implicit trapezoidal role as

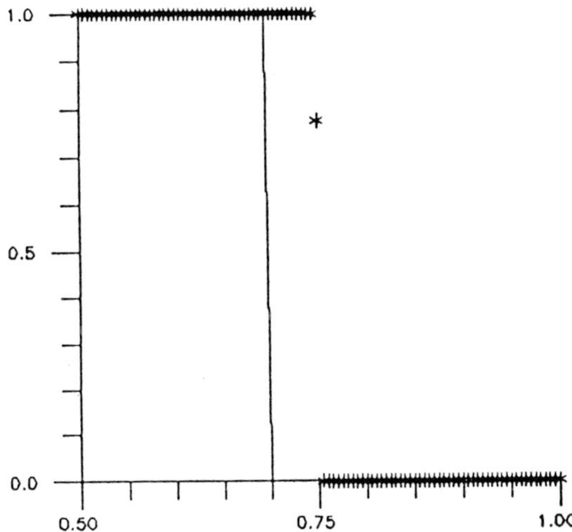

Fig.10(b): TVD Superbee method [26] Equation (7.4) without source term modification Initial data(7.2)   NX=200   CFL=0.75   T = 0.5   A = 1000

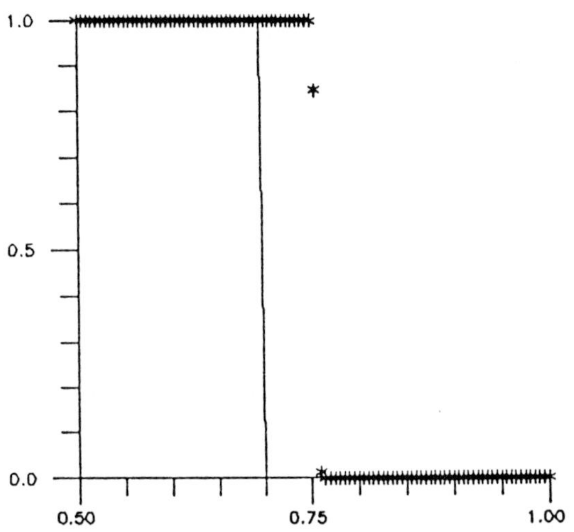

Fig.11(b): Second order UNO method [10] Equation (7.4) without source term modification Initial data(7.2)   NX=200   CFL=0.75   T = 0.5   A = 1000

follows.

$$\hat{u} - u = \frac{A}{2}\left(\psi(u) + \psi(\hat{u})\right)$$
$$= A\left(\psi(u) + \frac{1}{2}\psi'(u)(\hat{u} - u)\right) + O((\hat{u} - u)^2)$$
$$\tag{8.3}$$

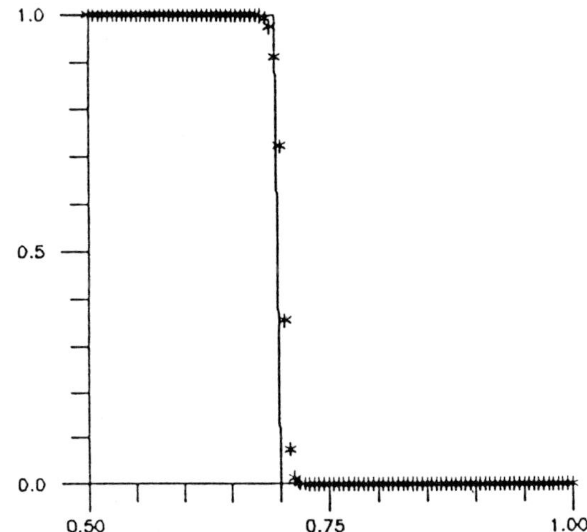

Fig.8(c): First order upwind method Equation (7.4) without source term modification Initial data(7.2)   NX=200   CFL=0.75   T = 0.5   A = 1000

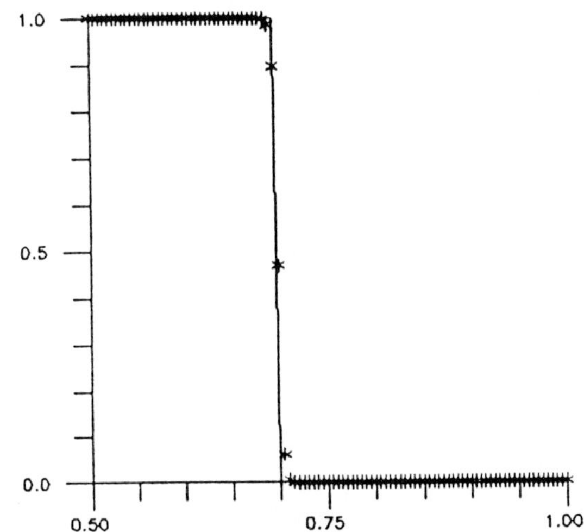

Fig.9(c): TVD Subbee method [26] Equation (7.4) with modification of source term Initial data(7.2)   NX=200   CFL=0.75   T = 0.5   A = 1000

Trancating (8.3) we have

$$\hat{u} - u = A\left(\psi(u) + \frac{1}{2}\psi'(u)(\hat{u} - u)\right)$$

or

$$\hat{u} - u = A\left(1 - \frac{A}{2}\psi'(u)\right)^{-1}\psi(u)$$

which is the desired linearly trapezoidal role. The third one is our highly fidelity algorithm (HFA) of

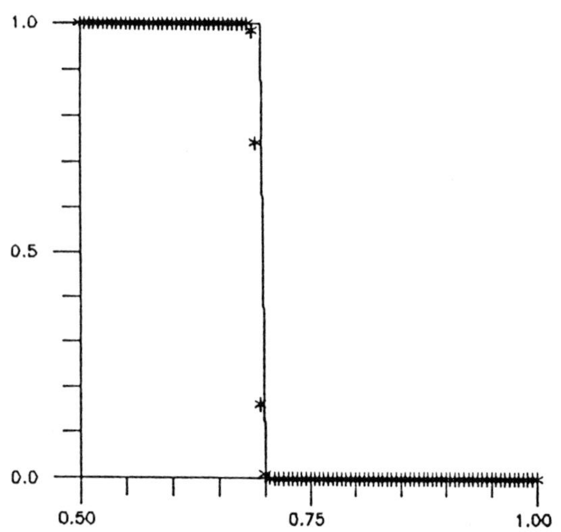

Fig.10(c): TVD Superbee method [26] Equation (7.4)
with modification of source term Initial data(7.2)
NX=200   CFL=0.75   T = 0.5   A = 1000

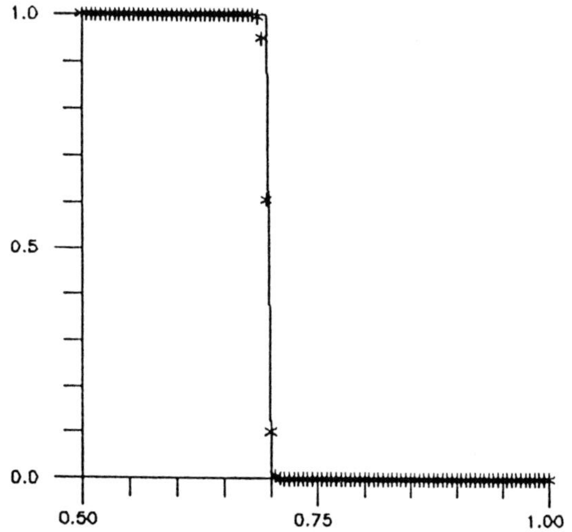

Fig.11(c): Second order UNO method [10] Equation
(7.4) with modification of source term Initial
data(7.2) NX=200   CFL=0.75   T = 0.5   A = 1000

second order accuracy

$$\hat{u} =$$

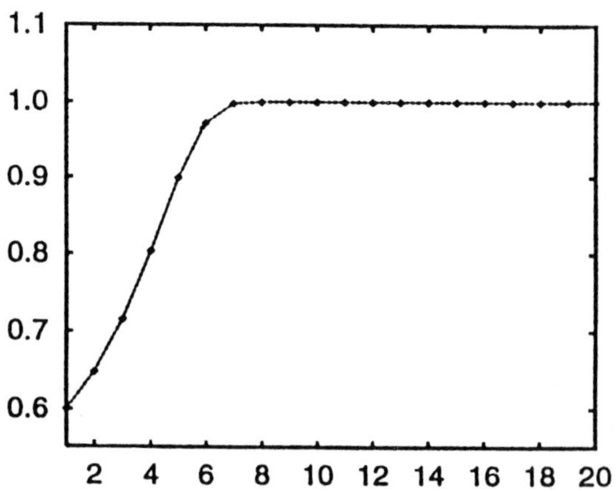

Fig.12(a): Explicit Euler method A = 2 Naguma
equation $du/dt = \alpha u(1 - u)(u - 0.5)$

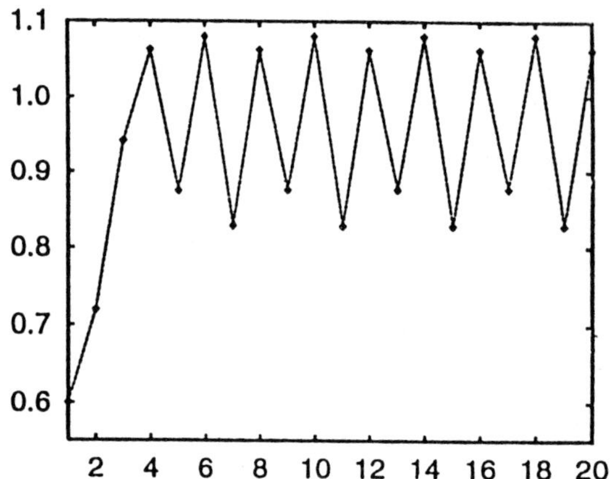

Fig.12(b): Explicit Euler method A = 5 Naguma
equation $du/dt = \alpha u(1 - u)(u - 0.5)$

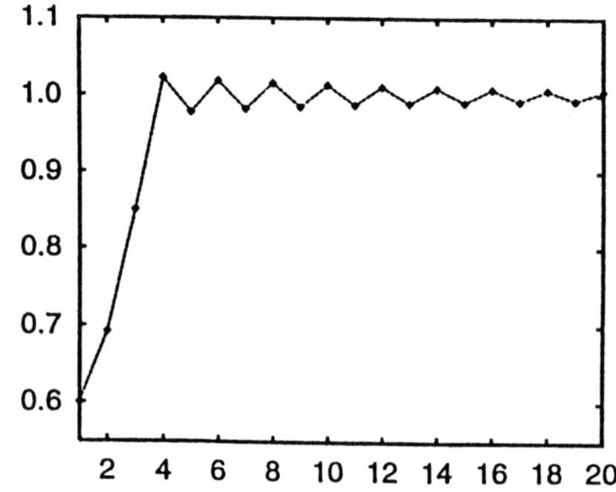

Fig.13(a): Trapezoidal role A=100 Naguma equation
$du/dt = \alpha u(1 - u)(u - 0.5)$

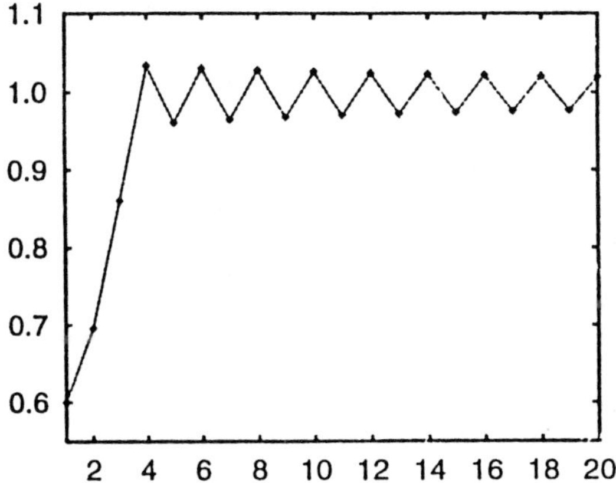

Fig.13(b): Trapezoidal role A=1000 Naguma
equation $du/dt = \alpha u(1-u)(u-0.5)$

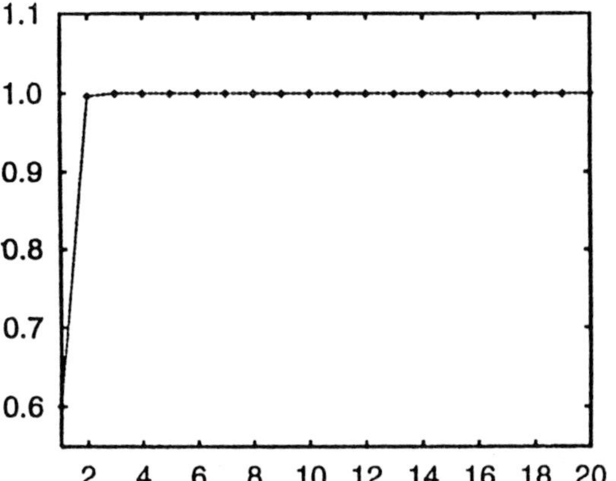

Fig.14(a): HFA method A=100 Naguma equation
$du/dt = \alpha u(1-u)(u-0.5)$

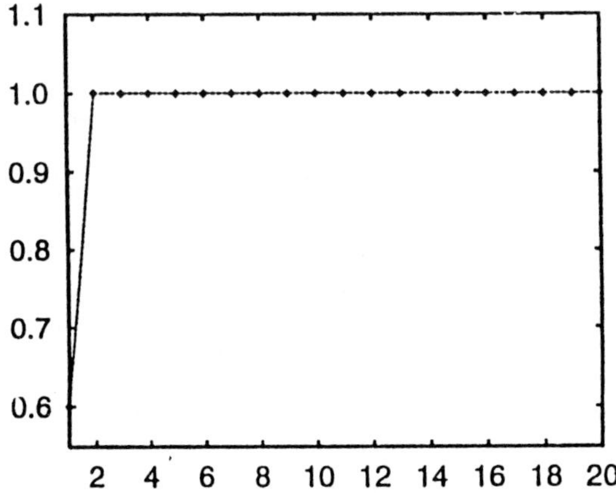

Fig.14(b): HFA method A=1000 Naguma equation
$du/dt = \alpha u(1-u)(u-0.5)$

$$
\begin{cases}
\dfrac{\left\{\left\{1+[A+\dfrac{A^2}{2!}(u(1-u)+u-0.5)]\right\}_u (u-0.5)\right\}}{\left\{1+[A+\dfrac{A^2}{2!}(u(1-u)+u-0.5)]\times(u-0.5)u\right\}}, & \text{if } u > 0.5 \\[2em]
\dfrac{u}{\left\{1+[A+\dfrac{A^2}{2!}(0.5-u^2)]\times(1-u)(0.5-u)\right\}}, & \text{if } u < 0.5
\end{cases}
\tag{6.9}
$$

Here are the numerical results obtained by using the Euler method (Figs.12(a) and (b)), linear implicit trapezoidal role (Figs.13(a) and (b)) and highly fidelity algorithm (HFA) of second order (Figs.14(a) and (b)). The parameters A in computations are $A = 2, A = 5$ for Euler method, $A = 100, 1000$ for linearly trapezoidal role and HFA. The numerical results show that Euler method can not use stiffness parameter $A > 2$ for keeping monotonicity of the solution; although the linearly implicit trapezoidal role is implicit large number $A \geq 100$ for trapezoidal role may cause oscillations in the solution. The results of HFA method show the efficiency in rapid transition of the starting value $u(0)$ to the stable asymptote in only one time step keeping the monotonicity of the solution. This is why we use HFA as the ODE solver for solving Naguma PDE. This excellent feature of HFA was firstly observed in solving the logistic equation [31].

9. CONCLUSIONS

In this paper we give a proof of the fact that semi-discrete MUSCL-like schemes of accuracy higher than second do not exist for genuinely nonlinear conservation laws.

In order to overcome the un-uniformality of the widely used ENO schemes we presented two kind of unified constructions of ENO schemes using upwind and central approximations of the fluxes. The scheme constructed in this way is suitable for parallel computing especially for vectorized computers.

In order to improve the resolution of the interfaces of multi-phase flows in this paper we suggested a new discontinuity sharpenning technique - adding modified artificial stiff source term approach. We also suggested use highly fidelity algorithms (HFA) for solving stiff ODEs having stable and unstable asymptotes as

$t \to +\infty$. The HFA method for ODE with Naguma source term has an important applications as the ODE solver for solving PDE with stiff source terms. Our numerical test gives satisfactory results of discontinuity sharpenning. It may be applied for solving interface tracking problems using piecewise constant level set functions. The new discontinuity sharpenning technique can be extend to multidimensional case using usual space-splitting approach.

It should noted, by the way, one of the purposes of this paper is to study the possibility of use artificial stiff source terms to sharpen the numerical interfaces. For the case of the problem with real chemical reaction terms the described numerical procedure using modification of the source term will be reasonable or not requires further investigations.

REFERENCES

[1] K.W. Anderson, J.L. Thomas and B. van Leer, 'A Comparison of Finite Volume Flux-Vector Splittings for the Euler Equations', *AIAA J.* **24**,1453-1460, (1986)

[2] A. Bermudez and M. E. Vazquez, ' Upwind methods for hyperbolic concervation laws with source terms', *Computers and Fluids*, **23**, 1049-1071, (1994)

[3] S. Chakravarthy and S. J. Osher, 'A new class of high accuracy TVD schemes for hyperbolic conservation laws', *AIAA* 85-0363,(1985)

[4] Y.S. Chen, *Master Degree Thesis*, Computing Center, Academia Sinica, (1994).

[5] S. Fu, 'Numerical simulation of blast wave-free surface interaction with level -set function method', *Computational Fluid Dynamical Journal.*, **4**, 451-462,(1996)

[6] E. Harabetian, ' Nonconservative hybrid shock capturing schemes', *Journal of Computational Physics*,**105**, 1-13, (1993)

[7] A. Harten, ' The artificial compression method for computation of shocks and contact discontinuities. I: Single conservation laws'. *Communication on Pure and Applied Mathematics*,**30**, 611-638,(1977)

[8] A. Harten, ' High Resolution Schemes for Hyperbolic Conservation Laws', *Journal of Computational Physics*, **49**, 357-393, (1983)

[9] A. Harten, 'ENO schemes with subcell resolution', *Journal of Computational Physics*, **83**, 148-184, (1989)

[10] A. Harten, S. Osher, 'Uniformly High Order Accurate Non-Oscillatory Schemes I', *SIAM Journal on Numerical Analysis*, **24**, 279-309, (1987)

[11] A. Harten, B. Engquist, S. Osher, S.R. Chakravarthy ' Uniformly High Order Accurate Essentially Non-Oscillatory Schemes III', *Journal of Computational Physics*, **71**, 231-303, (1987)

[12] W. H. Hui, C. Y. Loh, ' A new Lagrangian method for steady supersonic flow computation, Part II Slip-line resolution ', *Journal of Computational Physics*, **103**, 450-464,(1992)

[13] K. Ishizaka, T. Ikohagi and H. Daiguji, ' A High Resolution Scheme for Compressible Gas-liquid Phase Flows ',*Proceedings of Fifth International Symposium on CFD*, Sandai, **I**,352-357, (1993)

[14] K. Itoh, H. Tanno, M. Takahashi, ' A pointwise non-oscillatory shock capturing scheme', *Proceeeding of Fifth International Symposium on CFD*, Sendai I, pp.370-375, (1993)

[15] B.X. Jin, ' On an essentially conservative schemes for hyperbolic conservation Laws', *Journal of Computational Physics*,**112**, 308-315, (1994)

[16] R.J. LeVeque and H.C. Yee, ' A study of numerical method for hyperbolic conservaion laws with stiff source terms', *Journal of Computational Physics*, **86** , 187-210,(1990)

[17] X.L. Li , B.X. Jin and J. Glimm , 'Numerical study for the three-dimensional Rayleigh-Taylor instability through the TVD/AC scheme and parallel computation', *Journal of Computational Physics*,**126**, 343-355,(1996)

[18] Y. W. Ma and D. X. Fu, ' Super compact finite difference method (SCFDM) with arbitrarily high accuracy', *Computational Fluid Dynamics Journal*, **5**, 259-276,(1996)

[19] D. K. Mao, ' A treatment of discontinuities for finite difference methods in the two-dimensional case', *Journal of Computational Physics*, **104**, 377-397,(1993)

[20] B. Merriman, J. K. Bence and S. J. Osher, 'Motion of multiple junctions: A level set approach', *Journal of Computational Physics*, **112**, 334-363,(1994)

[21] S. Osher, S.R. Chakravarthy, 'Very High Order Accurate TVD Schemes',*IMA Volume in Mathematics and Its Applications.*, **2**, *Osci. Theory, Computation and Methods of Compensated Compactness*, 229-274, (1987)

[22] P.L. Roe, 'Some contributions to the modelling of discontinuous flows', *Lecture on Applied Mathematics.*,**22**,(1985)

[23] S.K. Saxena, 'Thin-layer Navier-Stokes Solution of Supersonic / Hypersonic Viscous Flows',

Computational Fluid Dynamics Journal,2, 103-119(1993)

[24] C.W. Shu and S. Osher, ' Efficient implementation of essentially non-oscillatory shock-capturing schemes', *Journal of Computational Physics*,77, 439-471, (1988)

[25] C.W. Shu and S. Osher, 'Efficient implementation of essentially non-oscillatory shock-capturing schemes, II', *Journal of Computational Physics*, 83, 32-78, (1989)

[26] P.K. Sweby, High resolution schemes using flux limiters for hyperbolic conservation laws, *SIAM Journal on Numerical Analysis*,21, 995-1011,(1984)

[27] B. van Leer, 'Towards the ultimate conservative difference scheme V. A second-order sequel to Godunov's method', *Journal of Computational Physics*, 32,101-136,(1979)

[28] H.M. Wu, ' MmB - A New Class of Accurate High Resolution Schemes in Two Dimensions - Analysis and Applications, *Proceedings of Third International Symposium on CFD*, Nagoya, 64-72, (1990)

[29] H. M. Wu, 'Conservation laws and entropy for high resolution schemes', *"Integral Methods in Science and Engineering"*, (Edited by C. Constanda,*Pitman Research Notes in Mathematical Series 317*, 95-108, (1994)

[30] H. Wu, S.L. Yang, 'MmB-A new class of accurate high resolution schemes for conservation laws in two dimensions', *Impact of Computing on Science and Engineering*, 1, No.4, (1989)

[31] H.M. Wu and L.E. Wang, ' Linearly Implicit Methods for One-Dimensional Steady Navier-Stokes Shock Layers', *Proc. 7th International Conference Boundary and Interior Layers Computational and Asymptotic Methods (BAIL VII)*, Beijing, (1994)

[32] T. Yabe and F. Xiao, 'A simulation technique that simultaneously solves liquid and gas', *Proceedings of Fifth International Symposium on CFD-Sendai*, III,358-362, (1993)

[33] Huanan Yang, 'An artificial compression method for ENO schemes: The slope modification method', *Journal of Computational Physics*, 89, 125-160, (1990)

[34] H. C. Yee, 'On the implimention of a class of upwind schemes for systems of hyperbolic conservation laws'*NASA TM86839.*

[35] H. Yee, 'Upwind and Symmetric shock capturing schemes', *NASA TM 89464,* May (1987)

[36] H. Yee, 'Construction of explicit and implicit symmetric TVD schemes and their applications' *Journal of Computational Physics*, 68, 151-179,(1987)

[37] H.C. Yee, P.K. Sweby and D.F. Griffiths, ' Dynamical approach study of spurious steady-state numerical solutions of nonlinear difference equations. I. The dynamics of time discretization and implications for algorithm development in compurional fluid dynamics.',*Journal of Computational Physics*, 97, 249-310, (1991)

[38] H.X. Zhang, ' Non-oscillatory No-free Parameters and Dissipative Schemes', *Acta Aerodynamica Sinica,,* 2, 143-165,(1988)

[39] H.X. Zhang, G.H. He, L. Zhang, 'Some Important Problems for the High Accurate Difference Scheme Solving Gas Dynamic Equations', *Acta Aerodynamica Sinica,* 11,347-356, (1993)

[40] H.X. Zhang, F. G. Zhuang, 'On the constructions of high order accurate difference schemes', *Proceeding of Eighth Natioanal Conference on CFD*, 1-11,(1996)

[41] Y.L. Zhu, X.C. Zong, B.M. Chen, Z.M. Zhang, *Initial and Boundary Value Problems and Flows around Bodies*, Science Press, Beijing, (1980),

[42] F.G. Zhuang, H.X. Zhang, ' Non-oscillatory No-free Parameters and Dissipative Schemes',*Advances in Applied Mechanics*, 29, 193-256, (1992)

PROBING NUMERICAL FLUXES: MASS FLUX, POSITIVITY, AND ENTROPY-SATISFYING PROPERTY

Meng-Sing LIOU

NASA Lewis Research Center, MS 5-11, Cleveland, OH 44135

Abstract

An analysis of inviscid numerical fluxes is presented with emphasis on their capability for accurately capturing shock and contact discontinuities, preserving the positivity for density and pressure, and maintaining shock stability in multidimensions. We investigate in detail the cause of certain catastrophic breakdowns in some numerical flux schemes. In particular, the terms responsible for the odd-even decoupling and the so-called "carbuncle phenomenon" are identified, leading to a proposed conjecture and hence a cure. The validity of this conjecture is confirmed by examining the mass flux of existing upwind schemes. Therefore, the conjecture is very useful to flux function development as to whether the flux scheme will be afflicted with these failings. Study of mass flux is important because the numerical diffusivity introduced in it can be identified and it is the term common to all conservation equations. It also bears a direct consequence to the prediction of contact (stationary and moving) discontinuities which are considered to be the limiting case of the boundary layer. Specifically, the recent AUSM-family schemes are examined and we prove the positivity property of these schemes for both density and pressure. Furthermore, we show the scheme not only exactly captures the 1D shock condition, but in addtion satisfies the entropy condition, thus yielding only the physically admissible solution.

1 Introduction

Numerical representation of inviscid fluxes, namely the numerical flux function, has been a subject of intensive effort by many researchers during the last three decades. Despite the enormous progress that has been achieved, deficiencies have also been experienced, as they may lack robustness, accuracy, or efficiency. In this paper we will attempt to construct a set of tools for the analysis of numerical fluxes, with the following goals: (1) establish basic requirements needed for a reliable scheme, (2) provide a clear and systematic way for designing flux function, and (3) identify causes of failures reported in the literature. We will examine such fundamental properties as capturing of contact and shock discontinuities, entropy condition, and the positivity-preserving property. These properties are believed to be of great relevance to a scheme's robustness and accuracy. We can identify the root of the odd-even decoupling and "carbuncle" phenomena, which in turn can provide a quick check against a scheme, as to its predictability for these problems. The pressure term appearing in the mass flux of existing schemes is identified to be the cause and this is verified in the numerical tests as the symptoms will disappear with the removal of this pressure term in the original scheme.

Received on July 31, 1997.

To demonstrate the properties analyzed, we will consider mostly the unsteady calculations because they generally shed more light on the capability of the scheme in terms of robustness and accuracy.

The paper is organized as follows. A formulation of the numerical flux will be presented in Section 2, a study of mass fluxes in Section 3, the positivity property in Section 4, and finally we will discuss the shock capturing and entropy property in Section 5. In each section, numerical examples are included for illustration. Finally, we will summarize the present study with concluding remarks.

2 NUMERICAL FLUX

For illustrative purposes, we shall begin by considering the 1D flux for ideal gas. For the AUSM-family schemes [1-4,13], the inviscid numerical flux function is written as a sum of separately treated convective and pressure fluxes:

$$\boldsymbol{f}_{1/2} = \rho u \begin{bmatrix} 1 \\ u \\ H \end{bmatrix}_{1/2} + \begin{bmatrix} 0 \\ p \\ 0 \end{bmatrix}_{1/2} = \boldsymbol{f}^{(c)}_{1/2} + \boldsymbol{p}_{1/2}. \quad (1)$$

Heerreafter the subsrcipt "1/2" is used to denote quantities evaluated at the interface stradding the j-th and $(j+1)$-th cells. Here the convective flux is defined as follow,

$$\boldsymbol{f}^{(c)}_{1/2} = \dot{m}_{1/2} \begin{cases} \Phi_j, & \text{if } \dot{m}_{1/2} \geq 0, \\ \Phi_{j+1}, & \text{otherwise,} \end{cases} \quad (2)$$

where $\Psi = (1, u, H)^T$. From the above equations, it is clear that the crux of the matter in defining the numerical flux lies in the two interface quantities: (1) the mass flux $\overset{\bullet}{m}_{1/2}$, and (2) the pressure flux $p_{1/2}$. The pressure flux contains nothing but the pressure term,

$$\mathbf{p}_{1/2} = (0, p_{1/2}, 0)^T. \tag{3}$$

Of these two fluxes, the pressure flux is the less complicated one because the conditions that must be fulfilled at various speed regimes are easier to deal with. Nevertheless, different formulas will lead to results with somewhat different quality, certainly not at a catastrophic level. Thus one might conceive some enhancement to improve the performance of the scheme (see for example AUSM$^+$-W [13]).

3. MASS FLUX

In this paper, we will argue that the mass flux holds an important key in explaining certain catastrophic failings associated with existing schemes. The mass flux plays a central role because it is common for all equations and through which its effects perpetuate.

In what follows, we write the mass flux in the most general form:

$$\overset{\bullet}{m} = \langle \overset{\bullet}{m} \rangle - \frac{1}{2}\mathcal{D}(\mathbf{U}_j, \mathbf{U}_{j+1}). \tag{4}$$

where $\langle \overset{\bullet}{m} \rangle$ is a some centrally weighted average and is required to meet the condition, $\langle \overset{\bullet}{m} \rangle \in (\overset{\bullet}{m}_j, \overset{\bullet}{m}_{j+1})$, $\overset{\bullet}{m}_j = \rho_j u_j$. The simplest form is

$$\langle \overset{\bullet}{m} \rangle = \frac{1}{2}[(\rho u)_j + (\rho u)_{j+1}].$$

The dissipative term $\mathcal{D}(\mathbf{U}_j, \mathbf{U}_{j+1})$ can be further expanded in terms of differences of primitive variables $\mathbf{U} = (\rho, u, p)^T$,

$$\mathcal{D} = \mathcal{D}^{(\rho)}(\overline{\mathbf{U}})\Delta\rho + \mathcal{D}^{(u)}(\overline{\mathbf{U}})\Delta u + \mathcal{D}^{(p)}(\overline{\mathbf{U}})\Delta p, \tag{5}$$

where $\overline{\mathbf{U}}(\mathbf{U}_j, \mathbf{U}_{j+1})$, are some mean quantities whose precise definitions are immaterial for the present discussion, and the difference operator $\Delta(\bullet) = (\bullet)_{j+1} - (\bullet)_j$.

Needless to say, different schemes will have different representations for each individual element in $\mathcal{D}$ and as such they contribute to a variety of numerical behaviors. In what follows, we shall give mass flux formulas for the AUSM$^+$ and AUSMDV. Let $a_{1/2} = a(\mathbf{U}_j, \mathbf{U}_{j+1})$ be a common speed of sound defined at the interface $1/2$,

the mass flux of both schemes can be cast in a single expression,

$$\overset{\bullet}{m}_{1/2} = a_{1/2}(\rho_j m^+_{1/2} + \rho_{j+1} m^-_{1/2}), \tag{6}$$

but the following definitions for $m^\pm_{1/2}$.

<u>AUSM$^+$ [3]</u>:

$$m_{1/2} = \mathcal{M}^+_{(4,\beta)}(M_j) + \mathcal{M}^-_{(4,\beta)}(M_{j+1}), \tag{7a}$$

and

$$m^\pm_{1/2} = \frac{1}{2}(m_{1/2} \pm |m_{1/2}|). \tag{7b}$$

<u>AUSMDV [4]</u>:

$$m^+_{1/2} = \omega^+_{1/2}\mathcal{M}^+_{(4,\beta)}(M_j) + (1 - \omega^+_{1/2})\mathcal{M}^+_{(1)}(M_j), \tag{8a}$$

and

$$m^-_{1/2} = \omega^-_{1/2}\mathcal{M}^-_{(4,\beta)}(M_{j+1}) + (1 - \omega^-_{1/2})\mathcal{M}^-_{(1)}(M_{j+1}), \tag{8b}$$

together with the blending functions,

$$\omega^+_{1/2} = \frac{2f_j}{f_j + f_{j+1}}, \quad \omega^-_{1/2} = \frac{2f_{j+1}}{f_j + f_{j+1}}, \quad f = p/\rho. \tag{8c}$$

The Mach number split polynomials $\mathcal{M}^\pm$ are given in [3,4]. Their precise expressions are not important insofar as this paper is concerned and thus omitted.

Equation (6) can be rewritten by respectively substituting corresponding formulas for $m^\pm_{1/2}$. We get for AUSM$^+$

$$\begin{aligned} \langle \overset{\bullet}{m} \rangle &= (\rho_j + \rho_{j+1})u_{1/2}/2, \quad u_{1/2} = a_{1/2}m_{1/2}, \\ \mathcal{D}^{(\rho)} &= |u_{1/2}|, \\ \mathcal{D}^{(p)} &= 0; \end{aligned} \tag{9}$$

and the corresponding formulas for AUSMDV

$$\begin{aligned} \langle \overset{\bullet}{m} \rangle &= (\rho_j u_j + \rho_{j+1} u_{j+1})/2, \\ \mathcal{D}^{(\rho)} &= (|u_j| + |u_{j+1}|)/2, \\ \mathcal{D}^{(p)} &= 2a_{1/2}\frac{g_j + g_{j+1}}{f_j + f_{j+1}}, \quad g = \mathcal{M}^+_{(4,\beta)} - \mathcal{M}^+_{(1)} \geq 0. \end{aligned} \tag{10}$$

For both schemes, the coefficient $\mathcal{D}^{(u)}$ is somewhat involved, but fortunately is not important within the scope of the present investigation.

Similarly, we can write the dissipation coefficients in $\overset{\bullet}{m}$ for the Roe schemes:

$$\begin{aligned} \langle \overset{\bullet}{m} \rangle &= (\rho_j u_j + \rho_{j+1} u_{j+1})/2, \\ \mathcal{D}^{(\rho)} &= |\lambda_2|, \\ \mathcal{D}^{(p)} &= [|\lambda_1| + |\lambda_3| - 2|\lambda_2|]/2a^2, \\ \mathcal{D}^{(u)} &= \rho[|\lambda_3| - |\lambda_1|]/2a, \end{aligned} \tag{11}$$

where $\{\lambda_k, k = 1, 2, 3\} = \{u - a, u. u + a\}$ and the quantities are understood to be evaluated using the usual Roe averages.

It is worth mentioning that in all of the above-mentioned schemes the coefficients $\mathcal{D}^{(p)}$ and $\mathcal{D}^{(\rho)}$ are nonnegative and AUSM$^+$ is the only scheme yielding $\mathcal{D}^{(p)} = 0\forall\mathbf{U}$. We note that this way of classifying the dissipation terms gives an interesting approach for studying flux functions. Specifically, it provides a tool for discriminatingly detecting causes of failures, because each individual term can be probed pathologically, as it will become clear in the following **Lemma 1** and **Conjecture**.

Lemma 1. The exact solution of the Riemann problem for a contact discontinuity moving with speed u_c requires that

$$\mathcal{D}^{(\rho)} = |u_c|, \qquad \text{as } (u, p)_j = (u, p)_{j+1} = (u_c, p),$$
$$\text{and } \rho_j \neq \rho_{j+1}.$$

$$(12)$$

This is shown to be a sufficient and necessary condition. This result is useful for checking the accuracy of a numerical flux functions and can provide us with a new insight about their behavior.

Since we are interested in the Euler as well as Navier-Stokes solutions, a major criterion concerns the ability of accurately and stably resolving shock and contact discoutinuities under a wide variety of conditions in both steady and unsteady flows. Since a contact discontinuity is thought of as the limiting case of a viscous layer, a scheme's ability of accurately resolving the contact discontinuity is an indication how well the viscous layer can be computed with the same scheme. **Lemma 1** is then useful in this sense. Hence, we included in this study schemes of Roe [7], Osher and Solomon [11], modified HLLE [15] and some recent AUSM-family [3,4]. It can be checked easily that they all satisfy Eq. (12). Noting that while the earlist AUSM scheme [1] results in $\mathcal{D}^{(\rho)} = 0$ for a stationary contact, it however does not satisfy (12) for a moving contact. The consequences are what is shown in Fig.1, in which a glitch is clearly seen in a slowly-moving contact discontinuity. The dramatic improvement by the AUSM$^+$ scheme is due to the use of a common speed of sound $a_{1/2}$ in Eq. (6).

The concept of using the common speed of sound turns out to be useful also in the formulation of the AUSMDV scheme. It is a necessary ingredient to blend two Mach number distributions such that stationary and moving contact discontinuities can be accurately

predicted, according. As a result, the following relationship,

$$\rho_j \omega^+_{1/2} = \rho_{j+1} \omega^-_{1/2}, \qquad (13)$$

is required and satisfied in (8c).

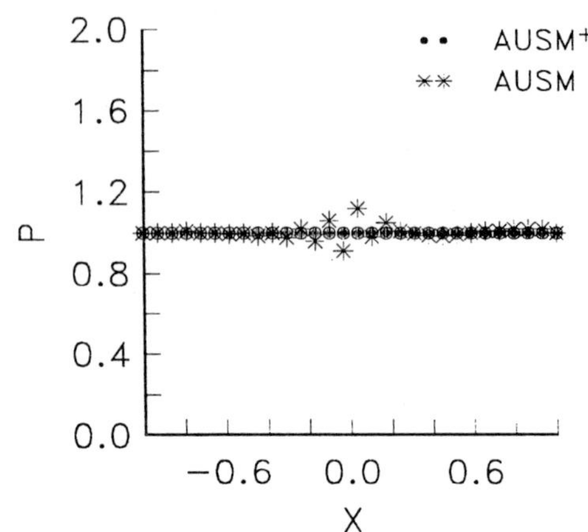

Fig. 1 Slowly moving contact; comparison of AUSM and AUSM$^+$ solutions.

Next we will consider the performance of each mass flux specifically with respect to the odd-even decoupling and "carbuncle" phenomenon.

3.1 Shock Instability

The catastrophic failings in the supersonic blunt body flow (dubbed the "carbuncle" phenomenon) and the odd-even grid perturbation problem [9] can be described as transverse numerical instability associated with a shock wave [4], which we shall classify as the *shock instability* because a small disturbance near the shock results in an explosive amplification. These seemingly benign problems prove to be rather daunting for many upwind schemes, including the exact Riemann solver by Godunov [10] and approximate ones by Roe [7] and Osher-Solomon [11]. Even to our disappointment, the recent HUS scheme [12] and the AUSMDV [4] can not escape the difficulty. By contrast, the AUSM$^+$ has been shown to yield solutions completely free of this instability [4]. A heuristic explanation of this instability is described in [4] in which an effective cure called *"Shock Fix"* is prescribed. It is shown that this shock fix is quite general to be fitted into any scheme, working equally well as in the AUSMDV scheme and the Roe scheme. Despite the success of this shock fix, it is still not so desirable to have since it needs a detection and it might fail when problem involves interactions of complex features. In [4] we give a somewhat general statement of the shock fix as to the replacement choice of the underlying scheme with a more dissipative partner scheme. With the present study, we now have the criterion that can specifically suggest which are the alternative replacement schemes in the shock fix.

Moreover, further study suggests that the term $\mathcal{D}^{(p)}$ plays a key role in causing the shock instability. As a result we have reached the following conjecture.

Conjecture. The condition $\mathcal{D}^{(p)} \neq 0$, $\forall M$ in the mass flux is necessary for a scheme to develop, as t increases, the shock instability as manifested by the odd-even decoupling and carbuncle phenomenon.

A rigorous proof for the **Conjecture** is difficult, if not impossible. But a heuristic argument for a system involving a set of simplifying assumptions can be constructed and leads to the result that the amplification factor is greater than unity, thus growth of disturbances. A more general analysis would be preferred. Nevertheless, we will resort to the numerical proof with evidences to substantiate the validity of the condition and counter-prove it by modifying those schemes afflicted with the instability in such a way that they have $\mathcal{D}^{(p)} = 0$, $\forall M$.

Among the schemes that satisfy **Lemma 1**, AUSM$^+$ (original AUSM is of its special case) is the only scheme in its original form that gives rise to $\mathcal{D}^{(p)} = 0, \forall M$, while the AUSMDV, Roe schemes (as shown above) and others yield nonvanishing $\mathcal{D}^{(p)}$ when $|M| < 1$. For Example, the Roe splitting in (11) becomes

$$\mathcal{D}^{(p)} = \begin{cases} 0, & \text{if } |M| \geq 1, \\ (1 - |M|)/a, & \text{otherwise.} \end{cases} \quad (14)$$

While the expression of $\mathcal{D}^{(p)}$ for the other schemes (such as AUSMDV in Eqs. (8) and (10)) may be more complicated, the detail however is not essential for the discussion of shock instability. Most important of all is to recognize whether the term $\mathcal{D}^{(p)}$ exists or not. A summary of the pressure coefficient of various popular shock capturing schemes is presented in Table 1, along with a column indicating shock instability. This clearly displays the close correlation of the role of $\mathcal{D}^{(p)}$ term and the occurrence of shock instability. We note that the AUSM$^+$-W [13] differs from the AUSM$^+$ only in the momentum flux for the pressure term, hence both schemes behave identically (not shown) insofar as the conjecture is concerned.

Noticing that in the framework of the AUSM schemes the only difference between AUSM$^+$ and AUSMDV is in the definition of the mass flux $\overset{\bullet}{m}$ (see Eqs.

(7) and (8)). Thus, it is natural to believe that the mass flux in Eq. (2) can be alternatively substituted with the Roe mass flux, as suggested by Shima and Jyonouchi [8], denoted here as AUSM$^+$-R for it falls in the form of AUSM scheme.

Table 1. Summary of $\mathcal{D}^{(p)}$ and shock instability.

Scheme	$\mathcal{D}^{(p)} = 0$	Shock instability	Lemma 1
AUSM$^+$	Yes	No	Yes
AUSMDV	No	Yes	Yes
HUS	No	Yes	Yes
Roe	No	Yes	Yes
Osher	No	Yes	Yes
HLLE	Yes	No	No
HLLEM	No	Yes	Yes
AUSM$^+$-R	No	Yes	Yes

We display the Mach contours in Figs. 2 for the odd-even grid perturbation problem. It is evident that the **Conjecture** is confirmed by the fact that the AUSM$^+$ predicts correct solution and the AUSMDV, Roe, and AUSM$^+$-R are afflicted with anomalies. Conversely, if these schemes are modified in such a way that $\mathcal{D}^{(p)} = 0, \forall M$, then the instability that is triggered and amplified by the pressure term is now completely gone, as is evident in the figures. The trick is achieved by fiddling with the linear field, i.e.,

$$|\lambda_2| = \max(a, |u|). \quad (15)$$

In other words, the *linear* field now is propagated with the speed of sound, instead of fluid speed. This is in contrast to the HLLE scheme in which the largest values are adopted for th *nonlinear* fields. We must note that this shock instability can not be remedied in the Roe flux by tuning the entropy fix because the fix is applied in the contribution associated with the nonlinear field, while the root really lies in the linear field. We must also point out that this modification unfortunately fouls up the property in **Lemma 1**. We shall call it the modified Roe scheme, or denoted as Roe$_{\text{mod}}$ in the caption. It is evident the modification given by (15) for the Roe scheme cures the carbuncles.

It is very interesting that when the Roe mass flux is used instead in the AUSM$^+$ (referred to as AUSM$^+$-R), we get the carbuncles. However, with the modification in (15), the carbuncles are pleasantly removed, even with a second-order calculation as shown in Fig.2. This further convinces the validity of the **Conjecture**

The original blending function in the AUSMDV scheme is chosen by including the pressure effect which leads to a nonvanishing $\mathcal{D}^{(p)}$, thereby resulting in the shock instability. Fortunately, this can be easily corrected by retaining only the density effect, but without sacrificing accurate capturing of contact discontinuity. This is achieved by defining $f = 1/\rho$, instead of $f = p/\rho$ in Eq. (8) (also denoted in the caption of Fig. 2), while Eq. (13) still holds.

The modified HLLE (or HLLEM) scheme is also designed to reduce the numerical dissipation for accurately resolving the contact discontinuity, by inserting a term resulting from the linear field of the Roe scheme. Thus a $\mathcal{D}^{(p)}$ term is inserted along with the $\mathcal{D}^{(\rho)}$ term, which can be easily seen from Eq. (11). It is exactly due to this process that the HLLEM inadvertently produces the shock instability, while the original HLLE scheme does not, see Obayashi and Wada [16].

We include in Figs. 3 the density contours and the profiles along the stagnation streamline of the supersonic blunt body problem. The AUSM$^+$ again yields clean and smooth solutions; the original AUSMDV has tiny nonmonotone contours near the stagnation streamline, which are now regularized after redefining the variable f in the mass flux. Though the "carbuncle" phenomenon produced by the Roe scheme is catastrophic, it however can be completely cured with the modification, Eq. (15), that ensures $\mathcal{D}^{(p)} = 0 \; \forall M$ in $\dot{m}$.

4. POSITIVITY

In the previous studies on the AUSM$^+$ and AUSMDV schemes [3,4], we give the positivity-preserving property only for the density. In this paper, we complete the positivity analysis by adding the proof of positivity for the pressure, hence positivity of internal energy also.

Lemma 2. Let the initial condition be $(-u, p, \rho)_L = (u, p, \rho)_R$, $u > 0$. Using the Euler one step explicit time integration, the AUSM$^+$ and AUSMDV schemes guarantee the following results. (a) pressure flux is positive if

$$0 \leq u\Delta t/\Delta x \leq 1. \qquad (16a)$$

(b) pressure in a cell remains positive as $t > 0$ if

$$0 \leq u\Delta t/\Delta x \leq 1/\gamma. \qquad (16b)$$

That is, under the necessary condition, Eq. (16b), for $p_j^0 > 0, \forall j$, we have

$$p_j^n > 0, \quad n > 0, \quad \forall j. \qquad (17)$$

Equations (16a) and (16b) are called the *positivity condition*. The proof of (b) involves lengthy algebraic manipulations and will be omitted here. Einfeldt et al [6] show that under the above initial conditions, the Roe scheme will reach negative pressure and density when the streams are separating relative to each other at a speed exceeding (roughly) twice the speed of sound (or $u \geq 0.945a$). In (b) we show the evolution of the flow in which the pressure and density remain positive under the stated time step constraint. In other words, **Lemma 2** states that both AUSM$^+$ and AUSMDV are not limited by any initial conditions and they will preserve positivity of pressure and density during the evolution of the flow.

In Fig. 4 we show the comparison of results by the Van Leer, HLLE, AUSM$^+$, and AUSMDV schemes, for a receding flow in which two streams are moving away from each other at $M = 2$, specifically $(\rho, -u, a)_L = (\rho, u, a)_R = (1, 2, 1)$. They all behave correctly, except that the HLLE result displays a totally smeared velocity profile at the center, which is consistent with the result shown in [6], indicating excessive dissipation at vanishing speed. In Fig. 5 we demonstrate the effectiveness of the positivity-preserving property for the case at a receding speed $M = 25$. All schemes but the HLLE (albeit positive in pressure and density) reaches a machine-zero density (vacuum condition) without difficulty. For this condition, aside from the Roe scheme, the Osher-Solomon scheme [11] will immediately encounter negative density. Noting that the initial condition considered in this case is well beyond the Osher-Solomon scheme's limit for preserving positive sonic speed at the contact discontinuity.

Two dimensional results for the case with initial conditions, $(-u, -v, p, \rho)_L = (u, v, p, \rho)_R$, $u/a = 20\sin 45°, v/a = 20\cos 45°$, are presented in Figs. 6. Again the calculation went robustly and smoothly, preserving positivity in both pressure and density (noting the logarithmic scale for p and ρ).

5. SHOCK CAPTURING AND ENTROPY-SATISFYING PROPERTY

5.1 Shock Capturing

We now come to the issues concerning the capturing of a shock discontinuity. Three questions arise: (1) can a shock wave be supported numerically with no intermediate points? (2) can the entropy condition be satisfied as well? and furthermore, (3) will the result be free of the shock instability? An ideal scheme should be designed to achieve positive answers to these questions. The last question has already been addressed above and in what follows we shall only focus on the remaining two issues.

A significant ingredient in the AUSM$^+$ [2] and AUSMDV [4] is the concept of a common speed of sound $a_{1/2} = a(\mathbf{U}_j, \mathbf{U}_{j+1})$. As a result, this guarantees the satisfaction of **Lemma 1**. We can then take a step further to quantify the definition for $a_{1/2}$ by meeting the following exact property.

Lemma 3. Let a stationary shock be described by: $\mathbf{U} = \{\mathbf{U}_L, x \leq x_j; \mathbf{U}_R, x > x_j\}$, where the normal shock relation is satisfied by the states $\mathbf{U}_L$ and $\mathbf{U}_R$, where $u_L > a_L$. The shock is exactly preserved if we define the numerical speed of sound as:

$$a_{1/2} = a_L^{*2}/u_L, \tag{18}$$

where a_L^* is the critical speed of sound evaluated at $\mathbf{U}_L$ and for ideal gas $a^{*2} = 2(\gamma - 1)/(\gamma + 1)h_t$.

Note that the formula (18) is valid for $u_L > a_L^*$ (also equivalently $u_L > a_L$). The definition can be extended to cover entire speed regimes of (u_L, u_R):

$$a_{1/2} = \min(\tilde{a}_L, \tilde{a}_R), \quad \tilde{a}_L = a_L^{*2}/\max(u_L, a_L^*) \text{ and}$$
$$\tilde{a}_R = a_R^{*2}/\max(-u_R, a_R^*). \tag{19}$$

We bring the reader's attention to the arguments in the max functions in Eq. (19), which now contain signs rather than the absolute value proposed in [3]. It is this change that recognizes the physically-admissable solution. Figure 7 shows the numerical results for a single stationary shock discontinuity. The present AUSM$^+$ and AUSMDV are seen to yield perfect shock solution for this case. Moreover, the Van Leer splitting (with or without Hanel's modification for the energy equation [14]) now dramatically removes the two intermediate shock points [5] by simply incorporating this specially-chosen speed of sound.

5.2 Entropy-Satisfying Property

The question of whether the AUSM-family schemes have this property has not been previously investigated. With the exact shock-capturing property obtained above, it was doubted whether the scheme now would behave like the Roe scheme in that no distinction can be detected between the shock wave and physically inadmissible expansion shock. While theoretical analysis to determine the entropy-satisfying property is difficult, we shall resort to numerical examination of it for some benchmark problems. One of which is the so-called inverse shock problem. The initial condition prescribes an acceleration of flow in the direction of a sudden pressure drop such that the fluxes across the jump are equal. This corresponds to an exact reversal of a stationary shock for which no stationary states can hold, setting an expansion wave to begin the motion. In Fig. 8 the entropy-satisfying property is evident for the AUSM$^+$ and AUSMDV. The results compare well with that of the Godunov scheme, only the AUSM$^+$ showing a somewhat better resolution of the plateau in temperature.

To further examine the accuracy issue for unsteady flow, we calculate a slowly-moving shock flow, moving at 50 time steps per cell to the left. Dispersive errors introduced by numerical schemes can be measured with the integral of the Riemann variables, i.e., $\int dp \pm \rho a\, du$ and $\int dp - a^2 d\rho$. We show in Fig. 9 the results in terms of the discrete representation of these Riemann variables. The Roe solution clearly shows the wave-like error trailing behind the shock for the two downstream-running characteristics; the AUSM$^+$ solution maintains nearly constant values, except with a jump at the shock, as it should be. It is interesting to note that the modified Roe scheme give a slightly less wavy result.

Lastly we consider the diffraction of a supersonic moving shock over a 90° bend. In contrast to the documented failings by several prominent flux schemes [9], the AUSM$^+$ flux demonstrates robust shock-capturing capability, as displayed in Fig. 10. Although the entropy fix is sufficient for the Roe scheme to overcome the expansion fan at the corner, it is clearly useless in dealing with the shock instability anomaly, which is now avoided with the use of Eq. (15). The AUSM$^+$ solutions on two grids are shown to be robust and free of the shock instability.

6. CONCLUDING REMARKS

The mass flux, which is a common factor to all inviscid fluxes of conservation laws, is claimed to be more important than what has been treated in the past. Schemes that have paid attention to the construction of the mass flux, such as the AUSM-family schemes, have demonstrated remarkable successes in dealing with flows having various types of conditions. A detailed analysis of mass flux has been attempted, especially with regard to the role of the dissipative pressure term. It is confirmed that this term is the root for causing the celebrated "carbuncle" and the odd-even decoupling phenomena which we classify as the shock instability. We have also proposed a fix to free the Roe scheme from developing these instabilities.

In this paper, we have completed the positivity analysis on the AUSM$^+$ and AUSMDV schemes by adding the proof for the pressure. Several cases, including a two-dimensional one, have been shown to display the schemes' effectiveness/robustness in treating fast expanding flows.

Continuing the great saga of the upwind shock-capturing schemes, we now have developed a scheme that (1) is free from the shock instability, (2) has the positivity property, (3) can reproduce the exact stationary shock, and (4) gives rise to the entropy-satisfying solution.

ACKNOWLEDGMENTS

The research has been supported by the Computing and Interdisciplinary Systems Office, Lewis Research Center and monitored by Dr. Chuck Lawrence.

REFERENCES

1. M.-S. Liou and C. J. Steffen, *J. Comput. Phys.* **107**, 23 (1993).
2. M.-S. Liou, *Lecture Notes in Physics* **414**, 115 (1993).
3. M.-S. Liou, NASA TM 106524 (1994); also *J. Comput. Phys.* **129**, 364 (1996).
4. Y. Wada and M.-S. Liou, AIAA paper 94-0083 (1994); to appear in *SIAM J. Sci. Comput.* (1997).
5. B. van Leer, *Lecture Notes in Physics* **170**, 507 (1982).
6. B. Einfedt, C. D. Munz, P. L. Roe, and B. Sjögreen, *J. Comput. Phys.* **92**, 273 (1991).
7. P. L. Roe, *J. Comput. Phys.* **43**, 357 (1981).
8. E. Shima and T. Jounouchi, (preprint, private communication) (1996).
9. J. J. Quirk, ICASE Report 92-64 (1992).
10. S. K. Godunov, *Mat. Sb.* **47**, 271 (1959).
11. S. Osher and F. Solomon, *Math. Comp.* **38**, 339 (1982).
12. F. Coquel and M.-S. Liou, NASA TM 106843 (1995).
13. M.-S. Liou, AIAA paper 95-1701-CP (1995).
14. D. Hänel et al, AIAA paper 87-1105-CP (1987).
15. B. Einfeldt, *SIAM J. Numer. Anal.* **25**, 294 (1988).
16. S. Obayashi and Y. Wada, *AIAA J.* **32**, 1093 (1994).

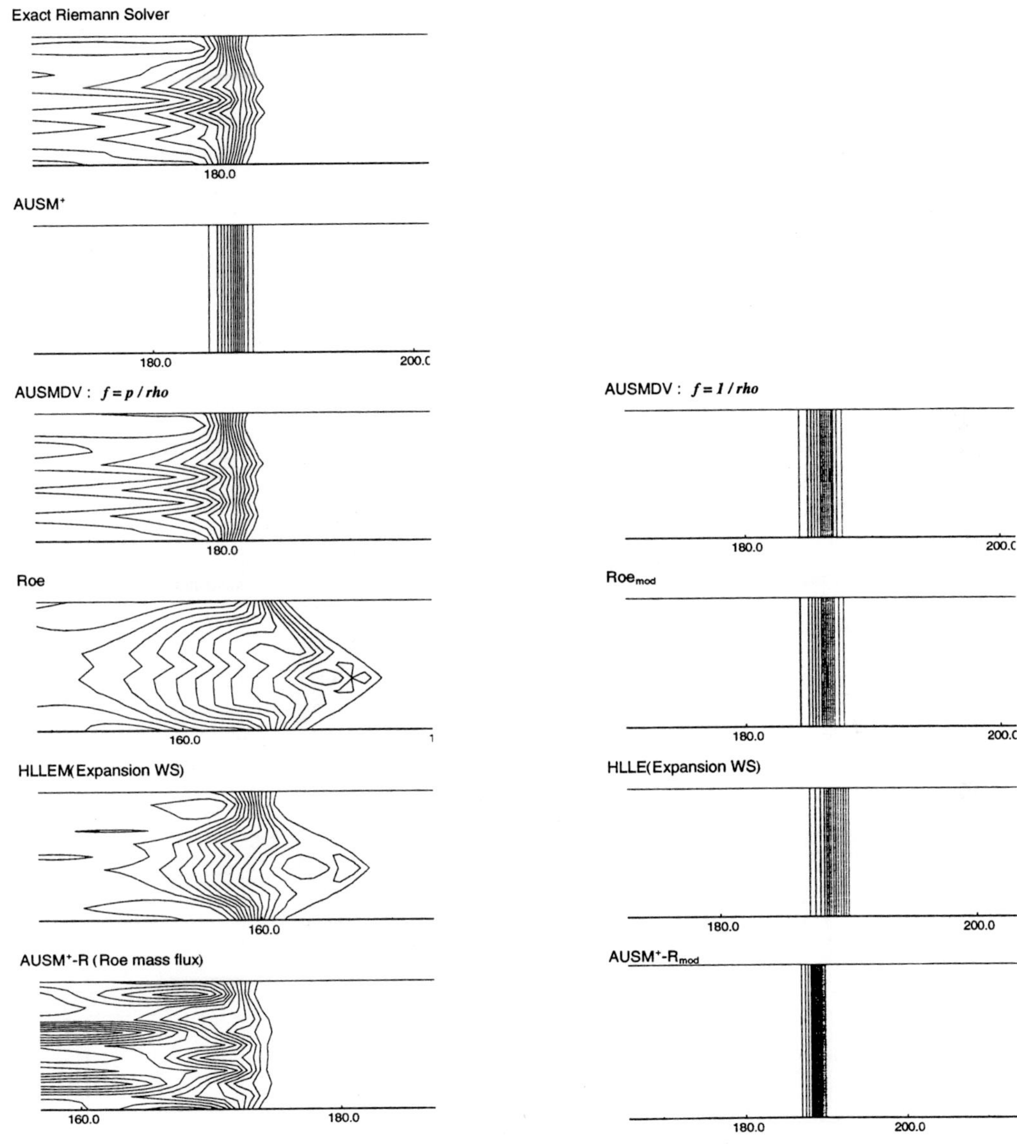

Fig. 2 Odd-even grid perturbation problem; Left: original schemes, Right: modification with $\mathcal{D}^{(p)} = 0$. All, but the modified AUSM$^+$-R, are O(h).

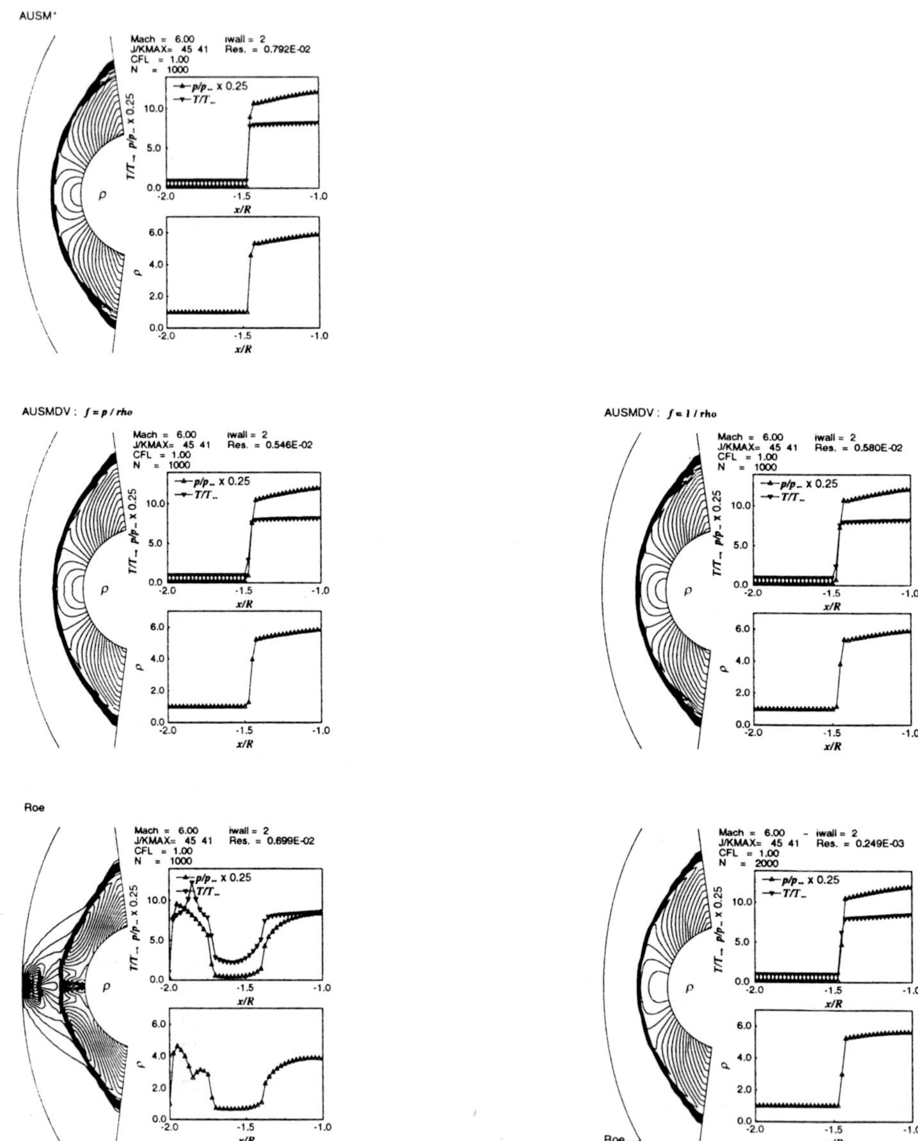

Fig. 3 Supersonic ($M_\infty = 6$) blunt body problem; profiles along the stagnation streamline.

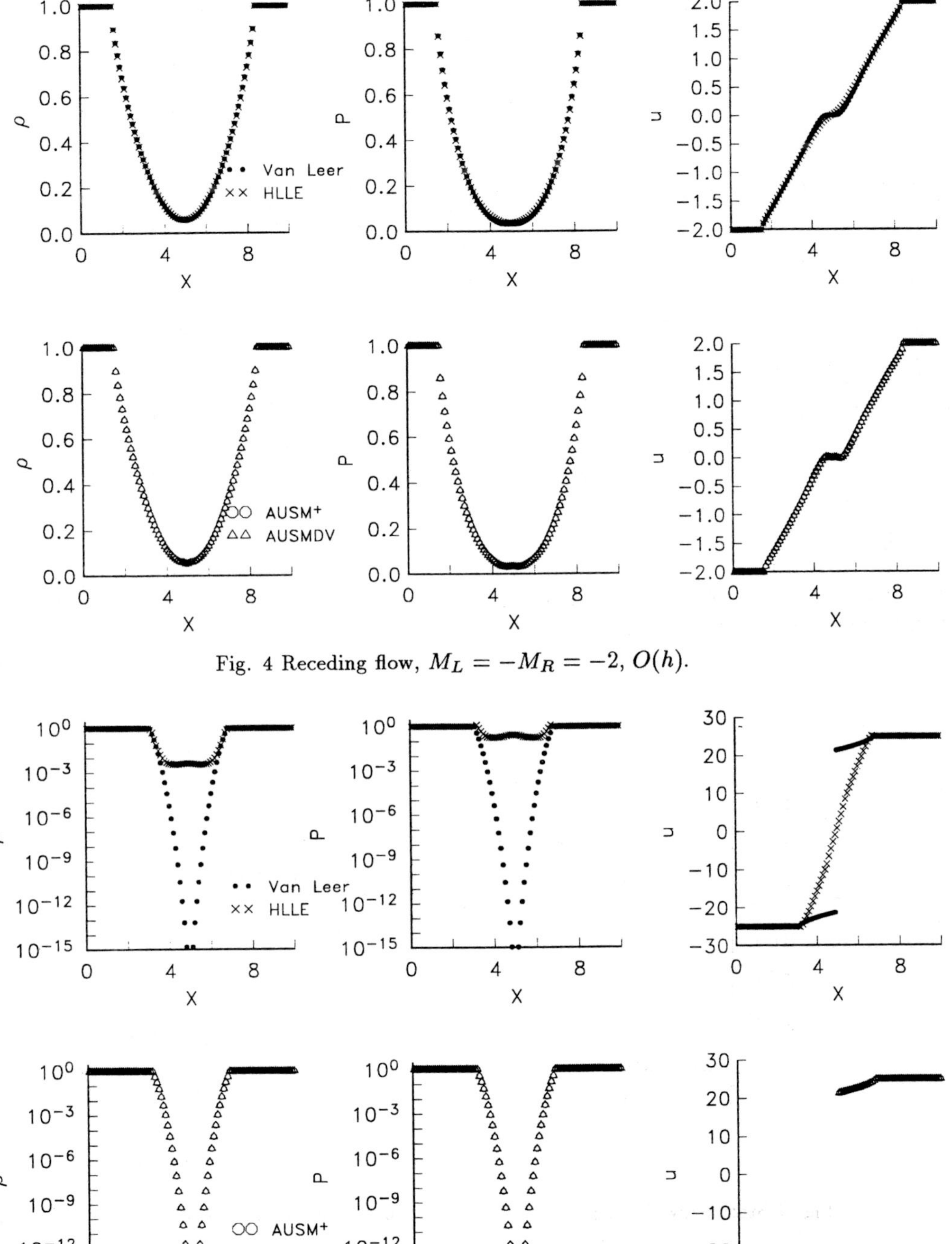

Fig. 4 Receding flow, $M_L = -M_R = -2$, $O(h)$.

Fig. 5 Receding flow, $M_L = -M_R = -25$, $O(h)$.

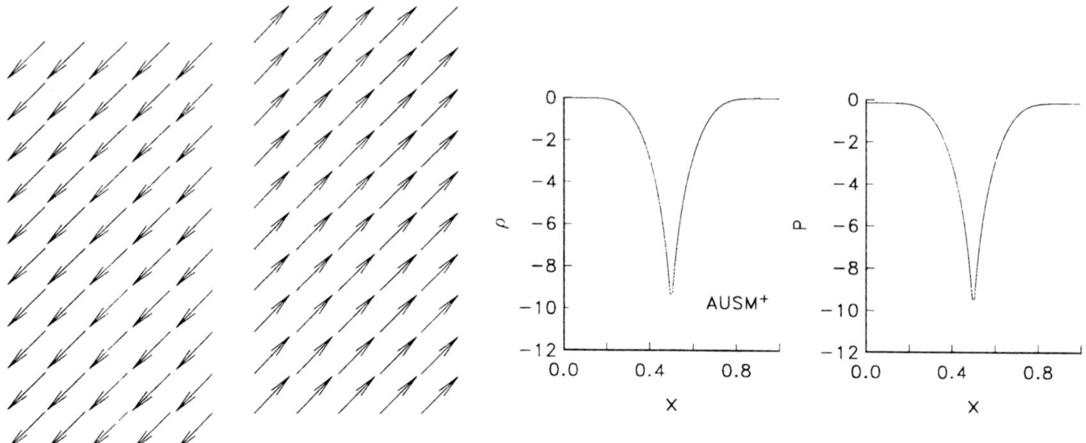

Fig. 6 2D receding flow, $(M_x, M_y)_L = -(M_x, M_y)_R = -(20/\sqrt{2}, 20/\sqrt{2})$, CFL =1.0.; Left: velocity vectors, Right: density and pressure profiles along $y = 0.5$.

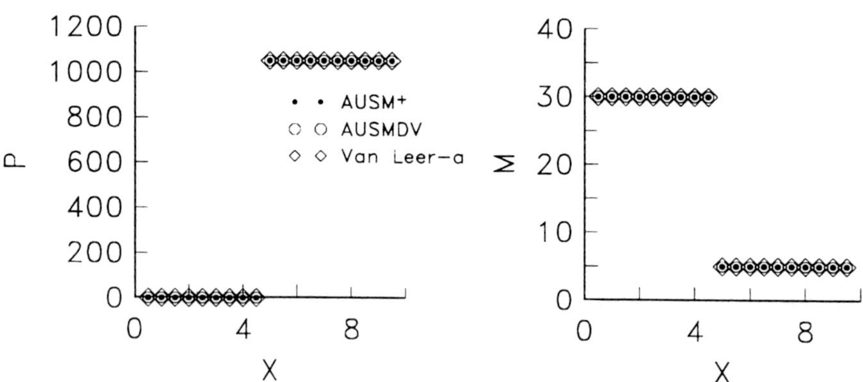

Fig. 7 Stationary shock problem; Van Leer-a uses common speed of sound.

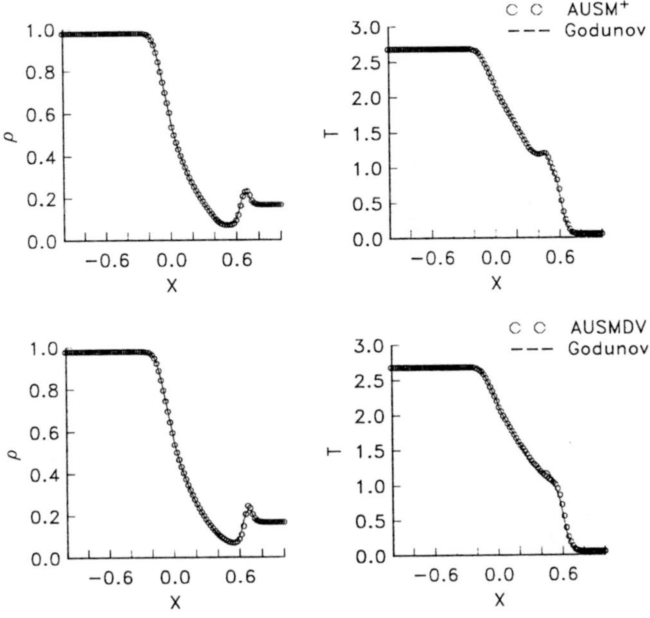

Fig. 8 Reverse shock problem; by AUSM$^+$ (top), and AUSMDV (bottom).

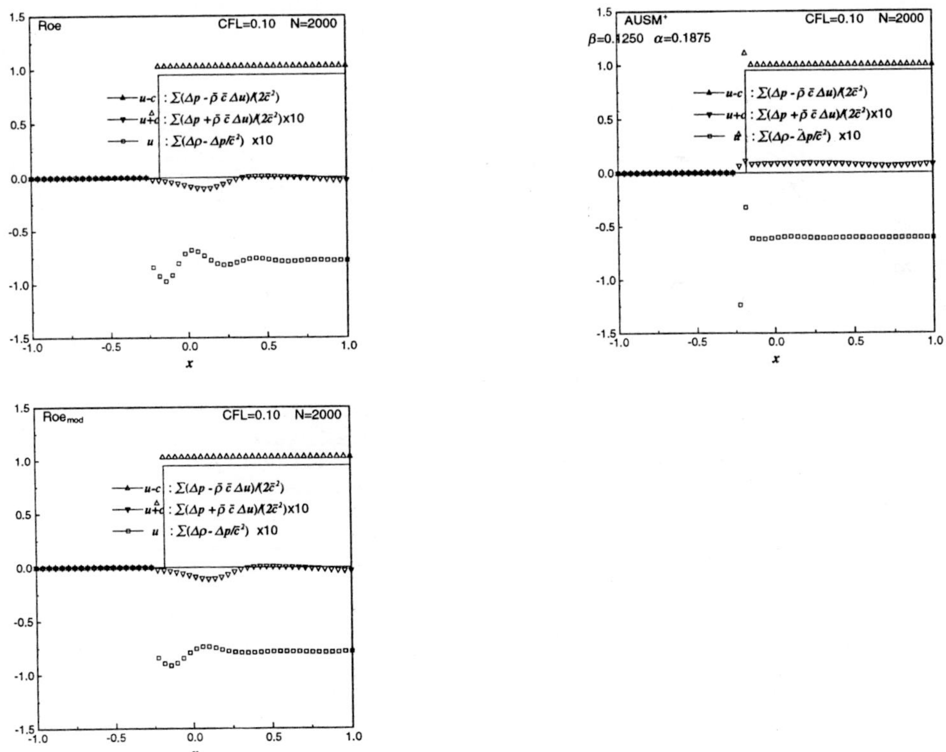

Fig. 9 Slowly moving shock; comparison of the first-order Roe (left) and AUSM$^+$ (right) solutions.

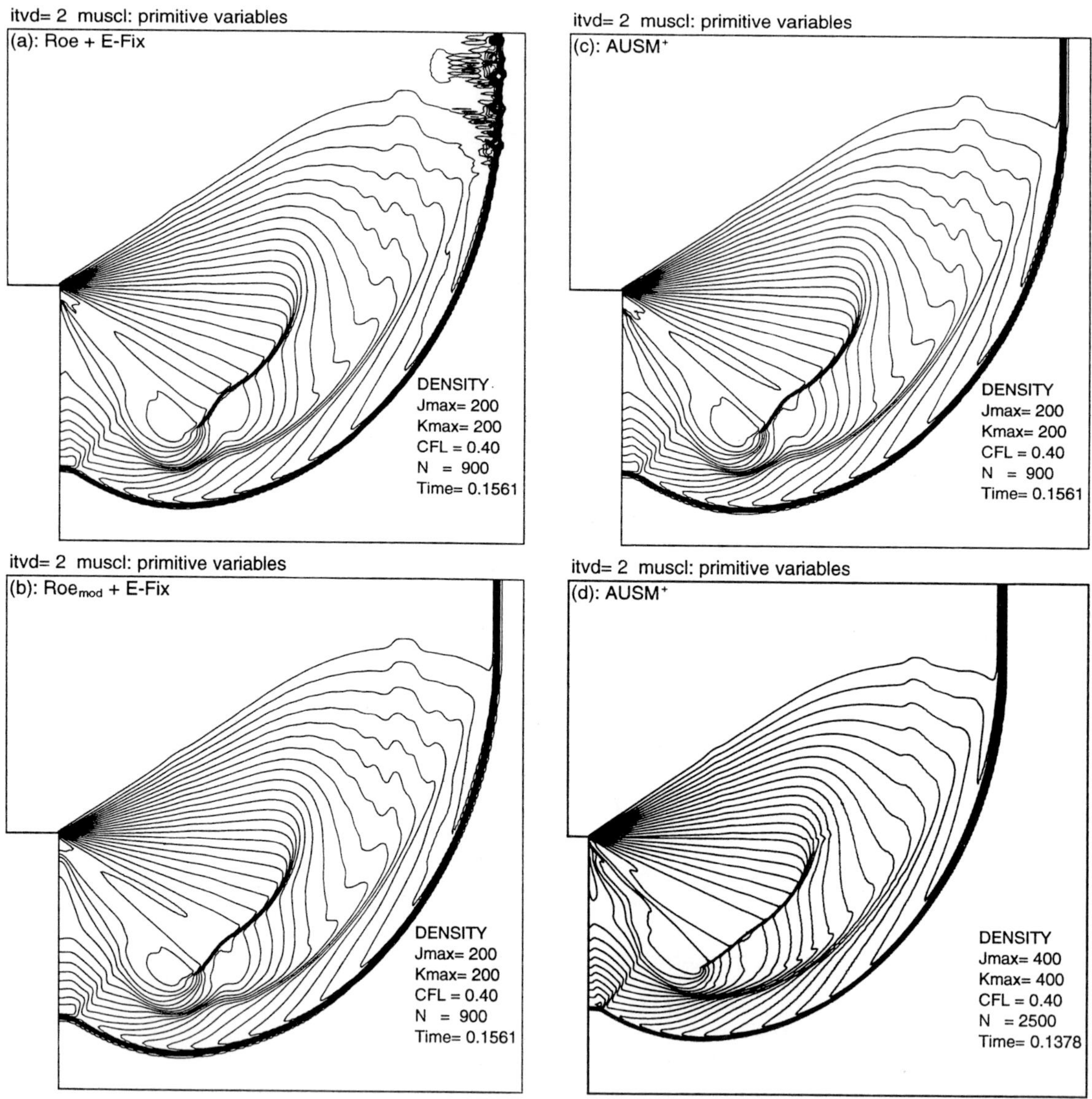

Fig. 10 Supersonic corner problem. The Roe flux with entropy fix produces shock instability which disappears after the modification enforcing $\mathcal{D}^{(p)} = 0$. A fine grid solution (at different instant) obtained with the AUSM$^+$ scheme is included for comparison, showing sharper resolution of shocks. All calculations were made with linear extrapolation of primitive variables using the Van Albada limiter

A FRACTIONAL STEP METHOD FOR THE SOLUTION OF THE COMPRESSIBLE NAVIER–STOKES EQUATIONS

R. CODINA [1]   M. VÁZQUEZ [1]   O.C. ZIENKIEWICZ [2]

1) International Center for Numerical Methods in Engineering, Universitat Polit'ecnica de Catalunya, Gran Capit'a s/n, Edifici C1, 08034 Barcelona, Spain
2) Institute of Numerical Methods in Engineering, University College of Swansea, Swansea SA2 8PP, UK.

Abstract

In this paper we present a summary of the splitting technique for both compressible and incompressible flows previously proposed in [22, 23, 7]. Also, we extend it to the case of a fully implicit treatment of the viscous and convective terms of the momentum equations. For incompressible flows, this scheme reduces to classical fractional step methods, except for a non-standard treatment of the boundary conditions. For compressible flows the continuity equation involves two variables which must be related through the equation of state. Convective terms of the conservation equations to be solved are stabilized by means of a Characteristic - Galerkin scheme. Also, in the presence of shocks some additional dissipation is needed. Both numerical techniques are explained here taking the transport of a scalar quantity as a model problem.

1. Introduction

The algorithm here described is designed aiming to deal equally well with a broad spectrum of problems, all of them physically modeled by the Navier - Stokes equations of fluids (either viscous or inviscid, compressible or incompressible, laminar or turbulent and so on). A fractional step or splitting, a technique first thought for incompressible flows and proposed by Chorin [3] and Temam [20], allows the use of equal interpolation spaces for pressure and velocity fields. Further, the split of the linear momentum equation produces a stabilizing effect on the pressure, which eliminates the need for special interpolation when the incompressible limit is reached. Therefore not only separate incompressible or compressible problems can be solved using the same algorithm, but compressible problems with regions of very low Mach number can be properly solved too.

In [7] we discuss the semi-implicit version of the algorithm, in which only the pressure gradient term is treated implicitly. Here we extend this scheme to the case in which both the viscous and the convective terms of the momentum equation may be treated implicitly. In principle, this introduces a splitting error in the scheme that will be discussed below. However, we propose a method for eliminating the splitting error due to the implicit treatment of the convective term, and therefore only that due to the viscous term will remain. This is applicable to all types of flow and in particular to incompressible cases, when the algorithm described here reduces to the classical fractional step or projection methods, except for the treatment of the boundary conditions.

The reformulation of the equations in a characteristics co-moving frame provides a consistent artificial diffusion when space discretization is done. This diffusion, which is similar to that of some other classical methods, like SUPG, can handle the purely numerical oscillations that usually appear when the Galerkin method is applied to equations with dominant convective terms. But if shock waves are present in the solution (very likely to appear in compressible flows), the artificial diffusion supplied by the method itself is not enough to eliminate spurious localized oscilations produced around the shock in the numerical solution. Additional "shock capturing" diffusion is needed there. We use in these cases a numerical diffusion proportional to the element residual of each of the scalar transport equations to be solved. We have organized this paper as follows. In the following section we state the problem and introduce some notation. Next we describe briefly the explicit version of the characteristic Galerkin method and the shock capturing technique used here for each of the scalar equations of the Navier–Stokes equations. In section 4 we describe the splitting method applied to the continuous equations, and then we obtain their weak form incorporating boundary conditions. The discrete problem is analyzed in section 5, where the choice of the variables in the case of compressible flows is discussed. Finally, we present some simple numerical examples and draw some conclusions.

Received on August 2, 1997.

2 NAVIER–STOKES EQUATIONS

Let us picture a fluid contained in a given domain Ω. Its velocity $u = u(x,t)$ can be described by means of a vectorial function of position x, within Ω, and time t, within $[0,\infty)$. To complete the dynamic description, two of the following three variables are needed: $\rho = \rho(x,t)$, $p = p(x,t)$ and $T = T(x,t)$ which describe its thermodynamic state variables density, pressure and temperature respectively. The linear momentum

$$U_i := \rho u_i,$$

and the total energy

$$E := \rho e,$$

where the total energy per unit mass

$$e := e_o + \frac{1}{2} u_i u_i,$$

can be defined as above. Here, the first term e_o is the internal energy, and the second, the kinetic energy.

Different types of fluids (or flows) are described by their equation of state. In particular, we shall consider the cases of incompressible flows, barotropic flows and the flow of ideal gases. In this last case, the state variables are related according to the following state law:

$$p = \rho RT, \tag{1}$$

where R is the universal gas constant. If, besides, the gas is polytropic, the internal energy is dependent only on T linearly

$$e_o = C_v T,$$

with the constant of proportionality C_v, the specific heat at constant volume.

The equations that model the behaviour of a fluid are known as the Navier-Stokes equations, a problem that is widely described in many books, for instance, [1, 16, 9]. This set of differential equations is derived from general conservation principles of mass, energy and momentum. In its pure conservative form, they can be written

$$\frac{\partial V}{\partial t} + \frac{\partial C_i}{\partial x_i} + \frac{\partial D_i}{\partial x_i} + S = 0, \tag{2}$$

where the conservative variables

$$V^T = (\rho, \rho u_1, \rho u_2, \rho u_3, \rho e)$$

are transported by means of convection, through the convective fluxes:

$$C_i^T = (\rho u_i, \rho u_i u_1 + \delta_{i1}p, \rho u_i u_2 + \delta_{i2}p,$$
$$\rho u_i u_3 + \delta_{i3}p, u_i(\rho e + p))$$

and of diffusion, through the diffusive fluxes:

$$D_i^T = (0, -\tau_{i1}, -\tau_{i2}, -\tau_{i3}, q_i - \tau_{ij}u_j).$$

The source is

$$S^T = (0, \rho g_1, \rho g_2, \rho g_3, \rho(g_i u_i + r)),$$

where g_i is the acceleration due to gravity, which points vertically downwards, and r is the heat source per unit mass.

The thermal flux is assumed proportional to temperature gradients, i.e. it is assumed the Fourier law:

$$q_i = -k(x)\frac{\partial T}{\partial x_i},$$

where $k(x)$ is the thermal conductivity.

The deviatoric (i.e. excluding pressure isotropic term) stress tensor τ_{ij} is related linearly to velocity gradients as is usual in newtonian fluids:

$$\tau_{ij} = \mu\left(\frac{\partial u_i}{\partial x_j} + \frac{\partial u_j}{\partial x_i} - \frac{2}{3}\frac{\partial u_k}{\partial x_k}\delta_{ij}\right), \tag{3}$$

where $\mu = \rho\nu$ is the viscosity, and ν is the kinematic viscosity. For some types of flow, kinematic viscosity is found to be temperature dependent through the Sutherland law:

$$\mu = AT^\alpha/(T + B),$$

where A,B,α are given constants.

3 A MODEL PROBLEM: CONVECTION–DIFFUSION– REACTION EQUATION

In this section we describe the numerical formulation employed to stabilize the convective terms of the transport equations to be solved. If this is done, some local overshoots and undershoots may appear in the vicinity of sharp gradients of the unknowns, in particular near shocks. A further numerical dissipation needs to be introduced in these cases. We describe both numerical techniques using the convection–diffusion–reaction equation as model problem.

3.1 Characteristic based schemes

In orther to unveil one possible solution to the problem arised by the convective fluxes, consider a general convection–diffusion equation. In it, these fluxes can be written as $C_i = u_i V$. For that reason

$$\frac{\partial V}{\partial t} + \frac{\partial(u_i V)}{\partial x_i} + \frac{\partial D_i(V)}{\partial x_i} + S = 0$$

can be re written as

$$\frac{\partial V}{\partial t} + u_i\frac{\partial V}{\partial x_i} + V\frac{\partial u_k}{\partial x_k} + \frac{\partial D_i(V)}{\partial x_i} + S = 0. \tag{4}$$

As we shall see, the fractional momentum introduced in section 4 is solution of a time-discrete problem of this form.

Now consider each component of (4) separately and suppose that the diffusive terms are uncoupled. The first two terms of (4) compose the material derivative of V (now a scalar):

$$\frac{dV}{dt} := \frac{\partial V}{\partial t} + u_i\frac{\partial V}{\partial x_i}.$$

A material derivative of variable V means the rate of change of V, as it is observed from a reference system in which the fluid is locally and instantaneously at rest: the *co-moving reference frame*. In it, the convective terms disappear. As was said early, equations containing convective terms are often satisfied by non-smooth solutions. In this case, the usual Galerkin finite element method loses accuracy and can produce a numerical solution with spurious effects, very far away from the physical one (cf. [14]). But if we want to solve the transport equation for the variable V taking proffit of the fact that the convective term vanishes in a different coordinate frame, the equation must be wholly reformulated in the co-moving system. If we note

$$L(V) := \frac{\partial D_i(V)}{\partial x_i} + V \frac{\partial u_k}{\partial x_k},$$

then (4) is simplified to

$$\frac{dV}{dt} + L(V) + S = 0. \tag{5}$$

Until the end of the section we consider the source S as zero to simplify the algebra. Let us label a particle and follow it as it wanders within the fluid. Then, its motion can be described by the *characteristics equation*

$$\frac{d\tilde{x}(t)}{dt} = u(\tilde{x}(t)), \tag{6}$$

where the tilde means "the trajectory of a particle of fluid that was at a reference point x_{ref} at a refence time t_{ref}". This statement is in fact the initial condition for the equation (6):

$$\tilde{x}(t_{\text{ref}}) = x_{\text{ref}}.$$

Time integration of this ordinary differential equation with this initial condition would solve the problem of tracking particles of fluid, the "carriers" of co-moving frames. In this form, equation (5) can be restated in that frame. Let S be the fixed reference spatial system, with origin pinned at x_{ref}. Let S' be the co-moving frame. If the origin of S' coincides at time t_{ref} with that of S, i.e., x_{ref}, then

$$\left.\frac{dV}{dt}(\tilde{x}(t),t)\right|_{x_{\text{ref}},t_{\text{ref}}} = \left.\left[\frac{\partial V}{\partial t} + u_i \frac{\partial V}{\partial x_i}\right]\right|_{x_{\text{ref}},t_{\text{ref}}},$$

so that

$$\frac{dV}{dt}(\tilde{x}(t),t) + L(V(\tilde{x}(t),t)) = 0.$$

This equation can be time-discretized using the trapezoidal rule:

$$\frac{1}{\Delta t}(V(\tilde{x}^{n+1},t^{n+1}) - V(\tilde{x}^n,t^n)) + \tag{7}$$

$$\theta L(V(\tilde{x}^{n+1},t^{n+1})) + (1-\theta)L(V(\tilde{x}^n,t^n)) = 0,$$

where $\Delta t = t^{n+1} - t^n$ (assumed constant, for simplicity of notation) and $\tilde{x}^k$ is an approximation to $\tilde{x}(t^k)$, $k = n$ and $k = n+1$.

Once the time-discretization is done, it is necessary to choose the pair x_{ref}, t_{ref} according to it. If S' coincides with $\tilde{x}^n$, the trajectory of our particle of fluid is integrated forwards in time using the values at $\tilde{x}^n$. And backwards if $x_{\text{ref}} := \tilde{x}^{n+1}$, expanding from the values at $\tilde{x}^{n+1}$. A point between them can be used too. We choose $x_{\text{ref}} := \tilde{x}^{n+1}$ (fig. 1). As was said above, Galerkin method is going to be used to discretize the space because in the co-moving frame the convective terms disappear. For that reason, the space convergence for Galerkin method is optimal, but in frame S' (i.e. $(\Delta x')^2$). If $\theta = 1/2$ is chosen (*Crank-Nicholson*) in (8), the trapezoidal rule gives the highest possible order in time: $O(\Delta t^2)$. Then, as $(\Delta x')^2 = (\Delta x)^2 + (u\Delta t)^2$, second order in space, but in S (i.e. $O(\Delta x^2)$) is reached in the streamline direction.

Through four successive steps, a second order time-discretized transport equation can be obtained in the co-moving frame, but with values obtained at the same point in the same, fixed, reference frame:

1. Integrate trajectory backwards, with $O(\Delta t^2)$:

$$\tilde{x}^n \approx \tilde{x}^{n+1} - \Delta t u(\tilde{x}^{n+1},t^n)$$
$$\approx x - \Delta t u^n.$$

2. Approximate velocity of this particle at $(\tilde{x}^n,t^n)$:

$$u(\tilde{x}^n,t^n) \approx u\,(x - \Delta t u^n\,,\,t^n)$$
$$\approx u^n - \Delta t u_i^n \frac{\partial u^n}{\partial x_i}.$$

3. Integrate trajectory backwards, now with $O(\Delta t^3)$:

$$\tilde{x}^n \approx \tilde{x}^{n+1} - \frac{\Delta t}{2}\left[u(\tilde{x}^{n+1},t^{n+1}) + u(\tilde{x}^n,t^n)\right]$$
$$\approx x - \frac{\Delta t}{2}u^{n+1} - \frac{\Delta t}{2}u^n + \frac{\Delta t^2}{2}u_i^n \frac{\partial u^n}{\partial x_i}$$
$$\approx x - \Delta t u^{n+1/2} + \frac{\Delta t^2}{2}u_i^n \frac{\partial u^n}{\partial x_i},$$

where $u^{n+1/2} = \left(u^{n+1} + u^n\right)/2$.

4. Calculate the variable V at $(\tilde{x}^n,t^n)$:

$$V(\tilde{x}^n,t^n) = V\left(x - \Delta t u^{n+1/2}\right.$$
$$\left. + \frac{\Delta t^2}{2}u_i^n \frac{\partial u^n}{\partial x_i} + O(\Delta t^3), t^n\right)$$
$$= V^n - \Delta t u_j^{n+1/2} \frac{\partial V^n}{\partial x_j}$$
$$+ \frac{\Delta t^2}{2}\left[u_i^n \frac{\partial u_j^n}{\partial x_i}\frac{\partial V^n}{\partial x_j}\right.$$
$$\left. + u_i^{n+1/2}u_j^{n+1/2}\frac{\partial^2 V^n}{\partial x_i \partial x_j}\right] + O(\Delta t^3).$$

As $u^{n+1/2} = u^n + O(\Delta t)$ and

$$u_i^n u_j^n \frac{\partial^2 V^n}{\partial x_i \partial x_j} =$$
$$u_i^n \frac{\partial}{\partial x_i}\left(u_j^n \frac{\partial V^n}{\partial x_j}\right) - u_i^n \frac{\partial u_j^n}{\partial x_i}\frac{\partial V^n}{\partial x_j},$$

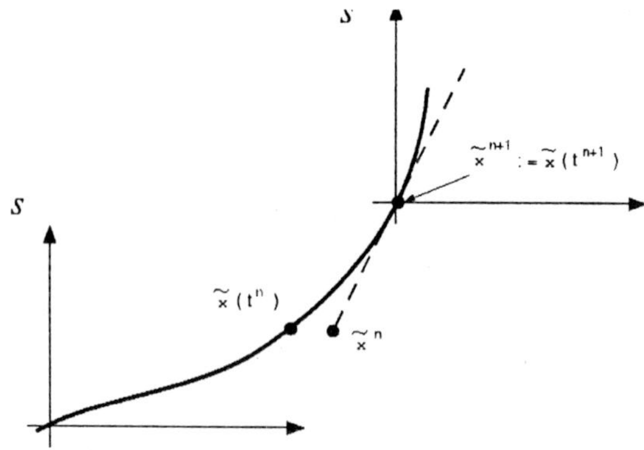

Figure 1. The co-moving reference frame.

we have for the variable V:

$$V(\tilde{x}^n, t^n) = V^n - \Delta t u_j^{n+1/2} \frac{\partial V^n}{\partial x_j}$$

$$+ \frac{\Delta t^2}{2} u_i^n \frac{\partial}{\partial x_i} \left(u_j^n \frac{\partial V^n}{\partial x_j} \right) + O(\Delta t^3). \qquad (8)$$

To find an approximation for $L(V(\tilde{x}^n, t^n))$, it is done

$$L(V(\tilde{x}^n, t^n)) = L\left(V^n - \Delta t u_j^n \frac{\partial V^n}{\partial x_j} + O(\Delta t^2) \right)$$

$$= L(V^n) - \Delta t u_j^n \frac{\partial L(V^n)}{\partial x_j} + O(\Delta t^2). \qquad (9)$$

Finally, using (8) and (9) in (8)

$$V^{n+1} - V^n = -\Delta t \left[u_j^{n+1/2} \frac{\partial V^n}{\partial x_j} + L(V^{n+1/2}) \right]$$

$$+ \frac{\Delta t^2}{2} \left[u_i \frac{\partial}{\partial x_i} \left(u_j \frac{\partial V}{\partial x_j} + L(V) \right) \right]^n ,$$

which can be now approximated evaluating all the right hand side explicitly or implicitly.

In brief, if we note

$$R(V) := -u_i \frac{\partial V}{\partial x_i} - L(V)$$

the continuum equation

$$\frac{\partial V}{\partial t} = R(V) \qquad (10)$$

is discretized in time, according to the methodology hitherto exposed, as:

$$\Delta V = \Delta t R(V)^{n+\theta} - \frac{\Delta t^2}{2} u_i^n \frac{\partial R(V)^n}{\partial x_i} \qquad (11)$$

with $\theta \in [0, 1]$, that is: *time - discretization of transport equation (10) using this method has led us to conclude*

that the temporal variation of V is controlled by both the residual of the equation (at first order) and the convective derivative of it (at second order).

Consider now the operator $L(V)$ purely as diffusive, only to fix the main ideas, and also take $\theta = 0$. Thus

$$R(V) := u_i \frac{\partial V}{\partial x_i} - k \frac{\partial^2 V}{\partial x_i \partial x_i},$$

with $k > 0$. Assuming that $R(V)$ vanishes on the boundary Γ we have that

$$\int_\Omega W u_i \frac{\partial R(V)}{\partial x_i} d\Omega$$

$$\overbrace{}^{0}$$

$$= \int_\Gamma n_i W u_i R(V) d\Gamma - \int_\Omega \frac{\partial}{\partial x_i} (W u_i) R(V) d\Omega$$

$$= -\int_\Omega \left(u_i \frac{\partial W}{\partial x_i} \right) R(V) d\Omega - \int_\Omega \left(W \frac{\partial u_i}{\partial x_i} \right) R(V) d\Omega$$

for all test functions W. Using this fact, the weak form of the time-discrete equation to be solved is

$$\int_\Omega W V^{n+1} d\Omega = \int_\Omega W V^n d\Omega$$

$$- \Delta t \left[\int_\Omega W u_j \frac{\partial V}{\partial x_j} d\Omega + \int_\Omega k \frac{\partial W}{\partial x_j} \frac{\partial V}{\partial x_j} d\Omega \right]^n$$

$$- \frac{\Delta t^2}{2} \left[\int_\Omega \left(u_i \frac{\partial W}{\partial x_i} \right) R(V) d\Omega \right.$$

$$\left. + \int_\Omega \left(W \frac{\partial u_i}{\partial x_i} \right) R(V) d\Omega \right]^n + \Delta t \int_{\Gamma_N} W g d\Gamma (12)$$

for all test functions W vanishing on Γ_D, where Γ_D and Γ_N means respectively contours with a Dirichlet's condition and with a Neumann's condition (normal derivative $= g$).

The main features of this last equation are two. First, the terms linear in Δt are integrals evaluated over all the domain and boundary terms coming from boundary conditions. And second, the terms quadratic in Δt are a SUPG-like term (see [2, 12, 13]), different only in that the *intrinsic time* (τ) of SUPG is here replaced by a linear function of time, plus a term which depends on velocity divergence, and for that reason, mainly active in the case of compressible flows. Although the final outcome of SUPG is very similar, its starting point is quite different: in its simpler formulation, the test functions are slightly modified by adding to them a term linear in its convective derivative. See [4] for a further discussion about different stabilization techniques for transport equations.

3.2 A DISCONTINUITY CAPTURING TECHNIQUE

Although the method proposed adds artificial diffusion needed when the physical diffusion vanishes in the differential equations of the type here considered, localized overshoots and undershoots can appear around the strong discontinuities that could be present at the solution. For that reason, some techniques have been developed in order to deal consistently with them.

One of these techniques (proposed in [5, 6]) consists of adding an anisotropic diffusion tensor in that particular places where it is not enough the streamline diffusion of the Characteristic–Galerkin method. It is based on two concepts. First, to preserve consistency, this diffusion tensor must be proportional to the residual of the equation evaluated within each of the elements. Second, it must be small where the convection is small.

If a convection–diffusion–reaction (CDR) equation

$$\frac{\partial V}{\partial t} + u_i \frac{\partial V}{\partial x_i} - k \frac{\partial^2 V}{\partial x_k \partial x_k} + sV = Q, \qquad (13)$$

is considered, to the artificial elementary diffusion due to the Characteristic–Galerkin (CG) method, given by

$$k_{\mathrm{cg}} = \frac{1}{2} \frac{\Delta t}{2} u^2,$$

where $u^2 = u_i u_i$, we propose to add a numerical, discontinuity capturing elementary diffusion

$$k_{\mathrm{dc}} = \frac{1}{2} \alpha_{\mathrm{dc}} h \frac{|R(V^{\mathrm{h}})|}{|\nabla V^{\mathrm{h}}|},$$

where $R(V^{\mathrm{h}})$ is the residual (only spatial terms considered) of (13) and

$$\alpha_{\mathrm{dc}} = \max \left(0, C - \frac{2k|\nabla V^{\mathrm{h}}|}{hR(V^{\mathrm{h}})} \right).$$

C is a constant depending on the interpolation order (0.7 if linear and 0.35 if quadratic). Supraindex "h" means that it belongs to the discretized usual FEM space.

But while the former only acts along the streamlines, the latter do it in all directions, taking into account both of the previous concepts. Like the streamline CG diffusion, the final discontinuity capturing tensor diffusion is not diagonal: in the streamline direction it is compared to the CG diffusion. So, the terms added to the right hand side of equation (12), once space discretization in N_{el} elements is done, are:

$$\sum_{e=1}^{N_{\mathrm{el}}} \int_{\Omega^e} \left[k_{\mathrm{dc}} \frac{\partial W^{\mathrm{h}}}{\partial x_i} \delta_{ij} \frac{\partial V^{\mathrm{h}}}{\partial x_j} \right.$$
$$\left. + (k_{\mathrm{sl}} - k_{\mathrm{dc}}) \frac{\partial W^{\mathrm{h}}}{\partial x_i} \left(\frac{u_j u_i}{u^2} \right) \frac{\partial V^{\mathrm{h}}}{\partial x_j} \right] \mathrm{d}\Omega, \quad (14)$$

where $k_{\mathrm{sl}} = \max(0, k_{\mathrm{dc}} - k_{\mathrm{cg}})$.

As in the preceding section, in order to apply these concepts to Navier-Stokes equations, they must be rewritten in a CDR form. Then, those same identifications can be done.

4 FRACTIONAL STEP METHOD FOR THE COMPRESSIBLE NAVIER–STOKES EQUATIONS

In this section the basic algorithm is presented. We start with the formulation of the splitting technique for the continuous equations and then we obtain the weak form of each of the equations to be solved incorporating boundary conditions. This weak form is the starting point for the finite element discretization.

4.1 Splitting

Let us write the conservation equations for the momentum $U_i := \rho u_i$ and the density ρ (continuity equation) as

$$\frac{\partial U_i}{\partial t} = M_i - \frac{\partial p}{\partial x_i} =: R_i, \qquad (15)$$

$$\frac{\partial \rho}{\partial t} = -\frac{\partial U_i}{\partial x_i}. \qquad (16)$$

where R_i is the i-th component of the steady-state residual and we have used the abbreviation

$$M_i := -\frac{\partial}{\partial x_j} (\rho u_i u_j - \tau_{ij}) - \rho g_i. \qquad (17)$$

The convective contribution $u_j \partial(\rho u_i)/\partial x_j$ appearing in M_i could lead to numerical instabilities if the standard Galerkin formulation is used to discretize the space. In order to stabilize this effect, we first discretize (15) in time along the characteristics of the total derivative $\partial/\partial t + u_j \partial/\partial x_j$ as explained in section 2. This leads to the following equations:

$$\frac{\Delta U_i^n}{\Delta t} = M_i^{n+\theta_3} - \frac{\partial p^{n+\theta_2}}{\partial x_i} - \frac{\Delta t}{2} u_k^n \frac{\partial R_i^n}{\partial x_k}, \quad (18)$$

$$\frac{\Delta \rho^n}{\Delta t} = -\frac{\partial U_i^{n+\theta_1}}{\partial x_i}, \qquad (19)$$

where Δt is the time step size (assumed to be constant for simplicity), the superscripts denote time step level, $\theta_1, \theta_2, \theta_3 \in [0,1]$ and we use the notation $f^{n+\theta} = \theta f^{n+1} + (1-\theta) f^n$, $\Delta f^n = f^{n+1} - f^n$ for any function f and $\theta \in [0,1]$.

Observe that in (18) the term coming from the discretization along the characteristics has been treated explicitly, but the rest are allowed to be treated implicitly. In [7] we described the same algorithm as here but with $\theta_3 = 0$, that is, treating explicity the contribution from the viscous and convective terms. The case $\theta_3 > 0$ will introduce an important difference, as will be shown below.

A deeper insight of the implicit treatment of M_i can be achieved by separating its convective and viscous parts. We define them respectively as

$$M_{\mathrm{c},i} := -\frac{\partial}{\partial x_j} (u_j U_i), \qquad (20)$$

$$M_{\mathrm{v},i} := \frac{\partial}{\partial x_j} \tau_{ij}, \qquad (21)$$

so that

$$M_i = M_{c,i} + M_{v,i} - \rho g_i.$$

In order to avoid the need for solving a nonlinear problem within each time step, we take

$$M_{c,i}^{n+\theta_3} = -\frac{\partial}{\partial x_j}\left(u_j^n U_i^{n+\theta_3}\right), \qquad (22)$$

that is, the convective velocity is evaluated at the previous time step. This approach is used for example in [19] for incompressible flows.

Let

$$\Delta \bar{U}_i^n := \Delta U_i^n + \Delta t \frac{\partial p^{n+\theta_2}}{\partial x_i}. \qquad (23)$$

Having introduced this new variable, (18) and (19) can be written as

$$\frac{\Delta \bar{U}_i^n}{\Delta t} = M_i^{n+\theta_3} - \frac{\Delta t}{2} u_k^n \frac{\partial R_i^n}{\partial x_k}, \qquad (24)$$

$$\frac{\Delta \rho^n}{\Delta t} = -\frac{\partial}{\partial x_i}\left(U_i^n + \theta_1 \Delta \bar{U}_i^n - \theta_1 \Delta t \frac{\partial p^{n+\theta_2}}{\partial x_i}\right), \quad (25)$$

$$\frac{\Delta U_i^n}{\Delta t} = \frac{\Delta \bar{U}_i^n}{\Delta t} - \frac{\partial p^{n+\theta_2}}{\partial x_i}. \qquad (26)$$

Hereafter, we shall refer to $\bar{U}_i^{n+1} := U_i^n + \Delta \bar{U}_i^n$ as the *fractional momentum*.

In principle, the term $M_i^{n+\theta_3}$ in (24) must be computed using U_j^{n+1}. If this is done, (24), (25) and (26) are exactly equivalent to (18) and (19) for the continuous in space problem that we consider for the moment. However, the use of U_j^{n+1} in (24) prevents from the possibility of computing directly the fractional momentum from this equation. This can be avoided by replacing $M_i^{n+\theta_3}$ by $\bar{M}_i^{n+\theta_3}$, which is obtained by computing M_i with $\bar{U}_j^{n+1}$ instead of U_j^{n+1}. This introduces of course a splitting error.

There is the possibility of eliminating the error coming from the fact that the convective contribution to M_i is computed with the fractional momentum and not with the momentum itself. Using (22) and (23) it is found that, up to second order accuracy for the pressure term,

$$M_{c,i}^{n+\theta_3} = -\frac{\partial}{\partial x_j}\left(u_j^n U_i^n\right)$$

$$- \theta_3 \frac{\partial}{\partial x_j}\left(u_j^n \Delta \bar{U}_i^n\right) + \theta_3 \Delta t \frac{\partial}{\partial x_j}\left(u_j^n \frac{\partial p^n}{\partial x_i}\right)$$

$$= \bar{M}_{c,i}^{n+\theta_3} + \theta_3 \Delta t \frac{\partial}{\partial x_j}\left(u_j^n \frac{\partial p^n}{\partial x_i}\right). \qquad (27)$$

The last term corrects the splitting error in the convective fluxes and only that corresponding to the viscous fluxes will remain.

If in (24) $M^{n+\theta_3}$ is replaced by $\bar{M}^{n+\theta_3}$, using the correction given by (27) or not, we obtain an equation for the fractional momentum alone, which can be solved. Once this is done, (25) may be used to compute either

ρ^{n+1} if $\theta_2 = 0$ or p^{n+1} if $\theta_2 > 0$. In this last case, the equation of state is needed to express ρ^{n+1} in terms of p^{n+1}. This point is treated in the following section.

Finally, (26) can be used to compute the momentum U_i^{n+1}. The important point is the substitution of ΔU_i^n in (25) using (23), all this at the continuous level. This will lead to a stabilizing pressure dissipation term in the discrete finite element scheme that allows to use this scheme for incompressible flows with the same velocity-pressure finite element interpolation if the semi-implicit version of the algorithm is employed.

4.2   Fractional momentum equation

Let us obtain now the weak form of (24), (25) and (26). Considering first (24), let $\bar{W}_i$ be the i-th component of the test function for the fractional momentum. We shall compute it in the problem domain Ω and also on its boundary $\Gamma = \partial \Omega$, and therefore $\bar{W}_i$ is subject to no conditions. Multiplying (24) by $\bar{W}_i$, integrating over Ω and integrating the viscous term and the term coming for the discretization along the characteristics by parts we get

$$\int_\Omega \bar{W}_i \frac{\Delta \bar{U}_i^n}{\Delta t} \mathrm{d}\Omega = \int_\Omega \bar{W}_i \bar{M}_{c,i}^{n+\theta_3} \mathrm{d}\Omega$$

$$- \theta_3 \Delta t \int_\Omega \frac{\partial \bar{W}_i}{\partial x_j} u_j^n \frac{\partial p^n}{\partial x_i} \mathrm{d}\Omega + \theta_3 \Delta t \int_\Gamma n_j u_j^n \bar{W}_i \frac{\partial p^n}{\partial x_i} \mathrm{d}\Gamma$$

$$+ \int_\Omega \bar{W}_i \rho g_i - \int_\Omega \frac{\partial \bar{W}_i}{\partial x_j} \bar{\tau}_{ij}^{n+\theta_3} \mathrm{d}\Omega + \int_\Gamma \bar{W}_i n_j \bar{\tau}_{ij}^{n+\theta_3} \mathrm{d}\Gamma$$

$$+ \frac{\Delta t}{2} \int_\Omega \frac{\partial}{\partial x_k}(u_k^n \bar{W}_i) R_i^n \mathrm{d}\Omega, \qquad (28)$$

where n is the unit outward normal to Γ and we have assumed that $R_i^n = 0$ on Γ. Observe that the two terms in the second row correspond to the modification introduced in (27).

If in (28) the viscous stresses defined in (3) are all computed at $n + \theta_3$, this equation couples all the components of the fractional momentum, that is to say, it is a system of d coupled scalar equations and d unknowns, $d = 2$ or 3 being the number of space dimensions. In order to avoid this coupling, we evaluate the part of the viscous stresses that couples the d equations explicitly. Also, in order to avoid the need for using the density at $n + \theta_3$ (which is unknown at the moment of solving (28)) we also evaluate it explicitly. After doing this, the viscous stresses are approximated by

$$\bar{\tau}_{ij}^{n+\theta_3} \approx \frac{\mu}{\rho^n} \frac{\partial \bar{U}_i^{n+\theta_3}}{\partial x_j} - \frac{\mu}{(\rho^n)^2} \bar{U}_i^{n+\theta_3} \frac{\partial \rho^n}{\partial x_j}$$

$$+ \mu \left(\frac{\partial u_j^n}{\partial x_i} - \frac{2}{3} \frac{\partial u_k^n}{\partial x_k} \delta_{ij}\right). \qquad (29)$$

Boundary conditions expressed in terms of traction can be (weakly) prescribed in (28). Apart from the prescription of the momentum itself (directly or by imposition of the velocity), we consider the following possibilities of boundary conditions:

a) The whole traction prescribed on Γ_T: $-pn_i + n_j\tau_{ij} = t_i$ (given).
b) Only the pressure component of the traction prescribed on Γ_P: $-pn_i = t_i^P$ (given).
c) Free part of the boundary Γ_F.

Conditions a) and b) are standard, especially a). However, condition c) is not as clear as the others. The idea is to leave Γ_F free, without any prescription neither on the velocity nor on the traction or part of it. This approach has been commonly used in compressible flow problems at supersonic outflows, but can be used as an outflow boundary condition for other types of flow (see [17]).

The prescription of boundary conditions a) in (28), taking the pressure at time step n, leads to replace

$$\int_\Gamma \tilde{W}_i n_j \tilde{\tau}_{ij}^{n+\theta_3} \mathrm{d}\Gamma$$

by

$$\int_{\Gamma-\Gamma_T} \tilde{W}_i n_j \tilde{\tau}_{ij}^{n+\theta_3} \mathrm{d}\Gamma + \int_{\Gamma_T} \tilde{W}_i (t_i + p^n n_i) \mathrm{d}\Gamma. \quad (30)$$

It is observed that boundary integrals have to be evaluated if the fractional momentum is to be computed also on the boundary.

4.3 Continuity equation

Let us consider now (25) and weight it by a test function W_p. We have that

$$\int_\Omega W_p \frac{\Delta\rho^n}{\Delta t} \mathrm{d}\Omega = -\int_\Omega W_p \frac{\partial U_i^n}{\partial x_i} \mathrm{d}\Omega$$
$$+ \theta_1 \int_\Omega \frac{\partial W_p}{\partial x_i} \left(\Delta\bar{U}_i^n - \Delta t \frac{\partial p^{n+\theta_2}}{\partial x_i} \right) \mathrm{d}\Omega$$
$$- \theta_1 \int_\Gamma W_p n_i \left(\Delta\bar{U}_i^n - \Delta t \frac{\partial p^{n+\theta_2}}{\partial x_i} \right) \mathrm{d}\Gamma. \quad (31)$$

As a boundary condition, we impose that the normal component of (26) be also verified on Γ, a condition equivalent to impose that the normal component of the momentum equation (18) be verified on Γ. This leads to

$$n_i \left(\Delta\bar{U}_i^n - \Delta t \frac{\partial p^{n+\theta_2}}{\partial x_i} \right) = n_i \Delta U_i^n. \quad (32)$$

on the part of the boundary Γ_C where the test function for the continuity equation W_p doesn't vanish. Observe that if $\theta_2 > 0$ both the pressure and the density appear in (31). Either of these can be chosen as the variable for the continuity equation, as will be discussed in the following section. Thus, Γ_C is the part of Γ where either p or ρ are free, depending on which variable is used. Suppose for example that the choice is p. According to the type of boundary conditions above, we have that

a) On Γ_T: $p = n_i \tau_{ij} n_j - n_i t_i$.

b) On Γ_P: $p = n_i t_i^P$.

In both cases, we have a Dirichlet type of boundary condition for the pressure, so that $W_p = 0$ on that part of Γ and $\Gamma_C = \Gamma - \Gamma_T - \Gamma_P$. On the other hand, ΔU_i^n is also known on the part of the boundary where the momentum is given. The problem arises on Γ_F, that is, for condition c) stated above. In this case, neither $W_p = 0$ nor ΔU_i^n is known. If (32) is used in the boundary integral of (31), we obtain an equation that involves U_i^{n+1}, which is not yet known. Therefore, this equation becomes coupled with the weak form of (26) discussed next. In order to avoid this coupling, we take $n_i \Delta U_i^n$ as 0. For transient calculations, if the normal component of the momentum varies on Γ_F this will be an approximation of order Δt. In any case, the steady-state solution (if reached) will be correct. Recall that this approximation is needed only when Γ_F is not empty, that is, when the non-standard boundary condition c) is used.

Let Γ_D be the part of Γ where the momentum is known. Using (32) and the approximation just described, (31) can be written as

$$\int_\Omega W_p \frac{\Delta\rho^n}{\Delta t} \mathrm{d}\Omega = -\int_\Omega W_p \frac{\partial U_i^n}{\partial x_i} \mathrm{d}\Omega$$
$$+ \theta_1 \int_\Omega \frac{\partial W_p}{\partial x_i} \left(\Delta\bar{U}_i^n - \Delta t \frac{\partial p^{n+\theta_2}}{\partial x_i} \right) \mathrm{d}\Omega$$
$$- \theta_1 \int_{\Gamma_D} W_p n_i \Delta U_i^n \mathrm{d}\Gamma. \quad (33)$$

This is the weak form of the continuity equation that we use, either if the unknown is the pressure or the density. In the second case, the pressure may be considered known where the density is given by using the equation of state and a guess for the temperature, if required.

4.4 Momentum equation

Finally, for (26) we have that

$$\int_\Omega W_i \frac{\Delta U_i^n}{\Delta t} \mathrm{d}\Omega = \int_\Omega W_i \frac{\Delta\bar{U}_i^n}{\Delta t} \mathrm{d}\Omega$$
$$- \int_\Omega W_i \frac{\partial p^{n+\theta_2}}{\partial x_i} \mathrm{d}\Omega, \quad (34)$$

where W_i is the i-th component of the test function. In this equation all the components of the momentum can be prescribed. This is possible due to the fact that the fractional momentum has been computed precisely by imposing that (26) be also satisfied on the boundary.

In summary, the equations that we have now are (28), (31) and (34), and the boundary conditions that have been introduced are the traction conditions and (32), that can be considered as the normal component of the momentum equations. Moreover, since the fractional momentum is also computed on the boundary, all the components of the momentum itself can be prescribed on it. However, the momentum is usually not directly fixed for compressible flows, but instead the velocity is given as boundary condition. We use the common approach of taking the momentum as prescribed using the given velocity values and the density computed in the current time step. This prescription is performed at the end of this step.

4.5 Energy equation

Once (28), (31) and (34) are solved, we have the momentum and either the pressure or the density at the current time step. It remains to compute the total energy. For that, one can solve explicitly or implicitly the last scalar equation in the vector equation (2). To simplify the exposition, we use here the former option with a discretization along the characteristics, which leads to

$$\frac{\Delta E^n}{\Delta t} = R_{\mathrm{E}}^n - \frac{\Delta t}{2} u_k^n \frac{\partial R_{\mathrm{E}}^n}{\partial x_k}, \tag{35}$$

where R_{E} is defined as

$$R_{\mathrm{E}} := -\frac{\partial}{\partial x_i} \left[u_i(E+p) - k\frac{\partial T}{\partial x_i} - \tau_{ij}u_j \right].$$

Weighting this equation by a test function W_{E}, integrating the diffusion and heat production terms by parts, setting $R_{\mathrm{E}} = 0$ on the boundary and prescribing the total heat flux (from production and from conduction) to H on a part of the boundary Γ_{H} we get

$$\int_{\Omega} W_{\mathrm{E}}\frac{\Delta E^n}{\Delta t}\mathrm{d}\Omega =$$

$$- \int_{\Omega} W_{\mathrm{E}}\frac{\partial}{\partial x_i} \left[u_i \left(E+p \right) \right]^n \mathrm{d}\Omega$$

$$- \int_{\Omega} \frac{\partial W_{\mathrm{E}}}{\partial x_i} \left(k\frac{\partial T}{\partial x_i} + \tau_{ij}u_j \right)^n \mathrm{d}\Omega$$

$$+ \frac{\Delta t}{2} \int_{\Omega} \frac{\partial}{\partial x_k} \left(u_k^n W_{\mathrm{E}} \right) R_{\mathrm{E}}^n \mathrm{d}\Omega$$

$$+ \int_{\Gamma_{\mathrm{H}}} W_{\mathrm{E}}H\mathrm{d}\Gamma. \tag{36}$$

On $\Gamma - \Gamma_{\mathrm{H}}$ we assume that $W_{\mathrm{E}} = 0$, that is, the energy is known there. As for the momentum, the total energy is not normally prescribed, but instead of this the temperature is given. In this case, we prescribe the total energy using the values already known of velocity and density and the prescribed temperatures.

If the solution of the flow equations has no shocks, instead of the energy equation written in conservation form one can solve the heat equation

$$\frac{\partial T}{\partial t} = R_{\mathrm{T}} := -u_i\frac{\partial T}{\partial x_i} + \frac{1}{C_v\rho}\frac{\partial}{\partial x_i} \left(k\frac{\partial T}{\partial x_i} \right)$$

$$+ \frac{1}{C_v\rho}\sigma_{ij}\frac{\partial u_i}{\partial x_j}. \tag{37}$$

Usually, this equation is written with the heat capacity $C_v\rho$ multiplying the temporal derivative of the temperature. However, this would prevent the possibility of

using a constant diagonal approximation to the mass matrix (via nodal numerical quadrature, for example) in the case of variable densities.

If an explicit time approximation along the characteristics is used for (37) we get

$$\frac{\Delta T^n}{\Delta t} = R_{\mathrm{T}}^n - \frac{\Delta t}{2} u_k^n \frac{\partial R_{\mathrm{T}}^n}{\partial x_k}. \tag{38}$$

Let us weight now this equation by a test function W_{T}, integrate the diffusion term by parts, set $R_{\mathrm{T}} = 0$ on the boundary and prescribe the conduction heat flux to H on a part of the boundary Γ_{H}. The result is

$$\int_{\Omega} W_{\mathrm{T}}\frac{\Delta T^n}{\Delta t}\mathrm{d}\Omega =$$

$$\int_{\Omega} W_{\mathrm{T}} \left[\left(-u_i + \frac{k}{C_v\rho^2}\frac{\partial \rho}{\partial x_i} \right) \frac{\partial T}{\partial x_i} \right]^n \mathrm{d}\Omega$$

$$+ \int_{\Omega} W_{\mathrm{T}} \left[\frac{1}{C_v\rho}\sigma_{ij}\frac{\partial u_i}{\partial x_j} \right]^n \mathrm{d}\Omega$$

$$- \int_{\Omega} \frac{\partial W_{\mathrm{T}}}{\partial x_i} \left(\frac{k}{C_v\rho}\frac{\partial T}{\partial x_i} \right)^n \mathrm{d}\Omega$$

$$+ \frac{\Delta t}{2} \int_{\Omega} \frac{\partial}{\partial x_k} \left(u_k^n W_{\mathrm{T}} \right) R_{\mathrm{T}}^n \mathrm{d}\Omega$$

$$+ \int_{\Gamma_{\mathrm{H}}} \frac{1}{C_v\rho^n} W_{\mathrm{T}}H\mathrm{d}\Gamma. \tag{39}$$

The temperature is assumed to be known on $\Gamma - \Gamma_{\mathrm{H}}$.

4.6 Time increment calculation

Based on stability criteria, the time increment can be calculated for Convection-diffusion-reaction (CDR) equations [10, 5]. Then, to be used here, Navier-Stokes terms must be identified with the analogue terms in a CDR equation. In this case, for the whole set of equations we use the same time increment. This is evaluated for each node using the following:

$$\Delta t = \frac{F^{\mathrm{TI}}}{\left(\dfrac{1}{\Delta t_c} + \dfrac{1}{\Delta t_u} \right)} \tag{40}$$

where Δt_c is the "crosswind" time increment, calculated using the diffusive limit for the 1-D CDR equation and Δt_u is the "upwind" one, calculated using the general form of for the 1-D CDR equation, which depends on a ratio between diffusion and convection (through Péclet or Reynolds number). F^{TI} is a factor that for explicit advance can be considered as a safety factor, always lower than 1.0. For implicit treatment this factor can be larger than 1.0. In the numerical diffusion which comes from Characteristic - Galerkin method the Δt which appears divided by two is evaluated using $F^{\mathrm{TI}} = 1.0$ in all cases.

5 DISCRETE PROBLEM AND SOLUTION STRATEGIES

With the weak form of the differential equations already established, we can proceed to discretize the space. We do this using the standard Galerkin method, since the term coming from the discretization in time along the characteristics will stabilize the convective terms. This means that we take all the test functions $\tilde{W}_i$, W_p, W_i, W_E and W_T equal to the shape functions. Also, some additional shock-capturing viscosity will be needed in the presence of discontinuities or sharp gradients of the solution, as explained in Section 3.2.

Let us consider first the equations for the fractional momentum (28) and for the end-of-step momentum (34). For the sake of simplicity, we take $\theta_3 = 0$ in what follows. Once the spatial discretization has been performed, the discrete version of these equations can be written in matrix form, the structure of which is

$$M \frac{\Delta \tilde{\tilde{U}}^n}{\Delta t} = F_1 - K\bar{U}^n, \tag{41}$$

$$M_0 \frac{\Delta \bar{U}_0^n}{\Delta t} = M_0 \frac{\Delta \tilde{\tilde{U}}_0^n}{\Delta t} - G_0 \bar{p}^{n+\theta_2} + F_2. \tag{42}$$

Vectors of nodal unknowns have been indicated by a boldface character and an overbar. Matrices M, K and G are the standard mass matrix for vector fields, the matrix coming from the viscous and convective terms in the equation for the fractional momentum and the matrix coming from the gradient operator, respectively. Subscript naught in the previous equations refers to not prescribed degrees of freedom for the momentum (in the sense indicated above), and F_2 contains precisely the contribution from $\Delta \tilde{\tilde{U}}^n$ and $\Delta \bar{U}^n$ corresponding to the prescribed degrees of freedom for the latter. Here and below we use F with subscripts to denote a vector which is known at the moment of solving a particular equation.

The discrete version of the energy equation written in conservation form (36) or the heat equation (39) can be solved at the beginning or at the end of the time step. These equations have the structure

$$M_{s,0} \frac{\Delta \bar{T}^n}{\Delta t} = F_T \quad \text{and} \quad M_{s,0} \frac{\Delta \bar{E}^n}{\Delta t} = F_E, \tag{43}$$

where M_s is the mass matrix for scalar unknowns and $M_{s,0}$ its modification to account for Dirichlet boundary conditions.

It remains to write the discrete version of the continuity equation (33). We consider different cases according to the type of flow being analyzed. We will see that it is useful to introduce the matrices M_α and L_β, of components

$$M_{\alpha,ij} = \int_\Omega \alpha N_i N_j \, d\Omega,$$

$$L_{\beta,ij} = \int_\Omega \beta \frac{\partial N_i}{\partial x_k} \frac{\partial N_j}{\partial x_k} \, d\Omega,$$

where N_i is the shape function associated to the i-th node of the finite element mesh with which we assume

that all the variables are interpolated and α and β are functions that depend on the type of flow.

5.1 Incompressible and slightly compressible flows

These two types of flows can be defined by the relation

$$\Delta \rho^n = \alpha \Delta p^n, \tag{44}$$

with $\alpha = 0$ for fully incompressible flows and $\alpha = 1/c^2$ (a positive constant) for slightly compressible flows. In this case, (33) can be written as

$$\int_\Omega \alpha W_p \frac{\Delta p^n}{\Delta t} d\Omega + \theta_1 \Delta t \int_\Omega \frac{\partial W_p}{\partial x_i} \frac{\partial p^{n+\theta_2}}{\partial x_i} d\Omega =$$

$$- \int_\Omega W_p \frac{\partial U_i^n}{\partial x_i} d\Omega + \theta_1 \int_\Omega \frac{\partial W_p}{\partial x_i} \Delta \tilde{U}_i^n d\Omega$$

$$- \theta_1 \int_{\Gamma_D} W_p n_i \Delta U_i^n d\Gamma. \tag{45}$$

Once the finite element discretization of this equation has been done, the matrix form of the discrete problem is

$$M_\alpha \frac{\Delta \bar{p}^n}{\Delta t} + \theta_1 \Delta t L_\beta \bar{p}^{n+\theta_2} = F_C, \tag{46}$$

with α the parameter appearing in (44) and $\beta = 1$ in this case. In (46) we have introduced

$$F_C := -D\bar{U}^n + \theta_1 G^t \Delta \bar{U}^n + F_D,$$

where F_D is the vector coming from the last term in (45) that is, from the boundary values of the momentum, and D is the matrix coming from the divergence operator. Dirichlet boundary conditions for the pressure are assumed to be included in (46).

Of special interest is the case of fully incompressible flows, that is, $\alpha = 0$. It is well known that in this case the velocity and pressure finite element interpolations must satisfy the Babuška-Brezzi conditions when the classical U-p approach is used. This is not the case using the type of fractional step methods that we are considering. We justify this in the following. To simplify the discussion, we assume that U is prescribed to zero on the whole boundary Γ.

Omitting the subscript β for a moment (it is 1), the matrix form of (41), (42) and (46) can be written as

$$M_0 \frac{\Delta \bar{U}_0^n}{\Delta t} = F_1^* - K_0 \bar{U}_0^n, \tag{47}$$

$$\theta_1 \Delta t L \bar{p}^{n+\theta_2} = -D_0 \bar{U}_0^n + \theta_1 G_0^t \Delta \bar{U}_0^n + F^*, \tag{48}$$

$$M_0 \frac{\Delta \bar{U}_0^n}{\Delta t} = M_0 \frac{\Delta \tilde{\tilde{U}}_0^n}{\Delta t} - G_0 \bar{p}^{n+\theta_2} + F_2. \tag{49}$$

Now subscript naught refers to degrees of freedom of interior nodes. Matrices D_0 and G_0^t are the submatrices of D and G^t corresponding to these nodes. They are related by $D_0 = -G_0^t$. Vectors F_1^* and F^* have been introduced to take into account the boundary values of the fractional momentum.

From (49) we get that

$$\Delta \bar{\bar{U}}_0^n = \Delta \bar{U}_0^n + \Delta t M_0^{-1} G_0 \bar{p}^{n+\theta_2} - \Delta t M_0^{-1} F_2,$$

and using this in (47) and (48) we obtain

$$M_0 \frac{\Delta U_0^n}{\Delta t} + K_0 \bar{U}_0^n + G_0 \bar{p}^{n+\theta_2} = F_1^* + F_2,$$
$$D_0 \bar{U}_0^{n+\theta_1} + \theta_1 \Delta t \left(L - G_0^t M_0^{-1} G_0 \right) \bar{p}^{n+\theta_2} = F_C^*, \tag{50}$$

with

$$F_C^* := F^* - \theta_1 \Delta t G_0^t M_0^{-1} F_2.$$

Clearly, we must have $\theta_1 > 0$ and $\theta_2 > 0$ in order to have a solvable problem.

The important point in (50) is the presence of the matrix $B := L - G_0^t M_0^{-1} G_0$, that can be understood as the difference between two discrete Laplacian operators. This matrix provides additional stability and, in particular, allows to use equal velocity pressure finite element interpolations in the incompressible case, as it had been noticed for example in [18] and [15]. This is so because this matrix is *positive semidefinite*. The proof of this fact can be found in [7].

5.2 Barotropic flows

Let us consider now the flow of compressible barotropic fluids, that is to say, fluids for which there is an equation of state that involves only the density and the pressure, and not the temperature. In general, we write this equation as $p = p(\rho)$, but we will particularize it to the case

$$p = A\rho^\gamma, \tag{51}$$

where A and γ, the *adiabatic exponent*, are physical constants. This situation is found for example in the case of isentropic flow of perfect gases.

In the case of incompressible or slightly compressible flows we have formulated the continuity equation in terms of the pressure only. However, now we have the possibility of choosing either the density or the pressure as unknown of the problem. Let us start with the former option:

Density as variable If we choose to write the continuity equation (33) using the density we have to express the pressure gradient in terms of the density. For this we use the approximation

$$\frac{\partial p^{n+\theta_2}}{\partial x_i} = \left(\frac{\mathrm{d}p}{\mathrm{d}\rho} \right)^n \frac{\partial \rho^{n+\theta_2}}{\partial x_i} = \frac{\gamma p^n}{\rho^n} \frac{\partial \rho^{n+\theta_2}}{\partial x_i}. \tag{52}$$

The approximation relies on the fact that we evaluate the derivative of p with respect to ρ (the square of the speed of sound) at n instead of at $n + \theta_2$. This may be thought of as a linearization of the problem.

Using (52) in (33), it is found that the discrete continuity equation can be written in this case as:

$$M_\alpha \frac{\Delta \bar{p}^n}{\Delta t} + \theta_1 \Delta t L_\beta \bar{p}^{n+\theta_2} = F_C, \tag{53}$$

now with $\alpha = 1$ and $\beta = \gamma p^n / \rho^n$. Observe that this equation has the same structure as (46) but with the density being the unknown instead of the pressure.

Pressure as variable If instead of using the density we use the pressure, the approximation that we employ is

$$\Delta \rho^n = \left(\frac{\mathrm{d}\rho}{\mathrm{d}p} \right)^n \Delta p^n$$

which is of order $O((\Delta p^n)^2)$. This approximation leads to

$$\frac{\Delta \rho^n}{\Delta t} = \left(\frac{\mathrm{d}\rho}{\mathrm{d}p} \right)^n \frac{\Delta p^n}{\Delta t} = \frac{\rho^n}{\gamma p^n} \frac{\Delta p^n}{\Delta t},$$

and the discrete continuity equation can now be written again as

$$M_\alpha \frac{\Delta \bar{p}^n}{\Delta t} + \theta_1 \Delta t L_\beta \bar{p}^{n+\theta_2} = F_C, \tag{54}$$

that is, exactly as (46), but now with $\alpha = \rho^n / (\gamma p^n)$ and $\beta = 1$.

For this type of flows, unlike the incompressible case, the continuity equation (53) or (54) can be solved explicitly ($\theta_2 = 0$) also. The fully explicit form of the algorithm allows very fast calculations at each time step, for matrix inversions are avoided. On the other hand, smaller time increments are to be used which leads to a poorer convergence rate to stationary states.

5.3 Perfect gases

In this case the equation of state involves not only the pressure and the density, but also the temperature. This equation is (1). The appearence of the temperature in it complicates a little the treatment of the continuity equation. As before, we may use either the density or the pressure as variables.

Density as variable Again, if $\theta_2 > 0$ we need to relate the pressure gradient to the density. We have that

$$\frac{\partial p^{n+\theta_2}}{\partial x_i} = \frac{\partial \rho^{n+\theta_2}}{\partial x_i} R T^{n+\theta_2} + \rho^{n+\theta_2} R \frac{\partial T^{n+\theta_2}}{\partial x_i}. \tag{55}$$

Clearly, if $\theta_2 = 0$ the pressure gradient term is on the right hand side of the continuity equation, and it is entirely evaluated in the previous time step explicitly. But for $\theta_2 > 0$, if we use directlty this expression in (33) the continuity equation will be coupled to the energy (or heat) conservation equation. In order to avoid this, for the implicit solution of this equation we use an iterative strategy based on assuming that $T^{n+\theta_2}$ is known and then correcting it. There is also another aspect that is computationally inconvenient. If we take

$\rho^{n+\theta_2}$ as unknown in the second term of the RHS of (55) this will lead to a non-symmetric matrix (see (33)). This can be circumvented if we also assume that $\rho^{n+\theta_2}$ is known and then we correct it.

Let then T_g be a guess for $T^{n+\theta_2}$ within the time step under consideration and ρ_g a guess for $\rho^{n+\theta_2}$. Equation (55) may be replaced by

$$\frac{\partial p^{n+\theta_2}}{\partial x_i} = \frac{\partial \rho^{n+\theta_2}}{\partial x_i} RT_g + \rho_g R \frac{\partial T_g}{\partial x_i}$$

The second term in this equation contributes to the RHS of the discrete continuity equation. If we denote by F_ρ this contribution, this discrete equation is

$$M_\alpha \frac{\Delta \bar{\rho}^n}{\Delta t} + \theta_1 \Delta t L_\beta \bar{p}^{n+\theta_2} = F_C + F_\rho. \qquad (56)$$

with $\alpha = 1$ and $\beta = RT_g$. This equation is similar to (53). Apart from the coefficients α and β, the only difference is the term F_ρ, which comes from the spatial derivative of the temperature.

Pressure as variable As for the case of barotropic flows, we may also use the pressure as the unknown of the continuity equation. For that we only need to use the equation of state (1), from which we have

$$\rho^{n+1} = \frac{p^{n+1}}{RT^{n+1}}. \qquad (57)$$

As in the previous case, we need to guess the value of T^{n+1} by T_g in order to uncouple the resulting continuity equation and the energy equation. We may then write

$$\Delta \rho^n = \frac{\Delta p^n}{RT_g} + \left[\frac{p^n}{RT_g} - \frac{p^n}{RT^n} \right].$$

The bracketed term term contributes to the RHS of the discrete continuity equation with a vector F_p. This equation can be written as

$$M_\alpha \frac{\Delta \bar{p}^n}{\Delta t} + \theta_1 \Delta t L_\beta \bar{p}^{n+\theta_2} = F_C + F_p, \qquad (58)$$

with $\alpha = 1/(RT_g)$ and $\beta = 1$. Again, this equation has the same structure as (54) with a modification of the RHS due to the variation (now in time) of the temperature.

From numerical experiments we have found that this approach doesn't work well in the presence of strong shocks, in the sense that we haven't been able to obtain a converged steady state solution in these cases. We attribute this to the appearence of the temperature as a denominator in the function α. This makes the coefficients of M_α difficult to evaluate numerically and with possibly high variations from one time step to the other in the vicinity of shocks.

For the case $\theta_2 = 0$, in general compressible cases, the algorithm scheme is:

1. Solve for the fractional momentum (41).
2. Advance the rest of the variables, all at a time.

Step 2 is different for barotropic flow because no energy or heat equation is needed. Besides, the fully explicit algorithm is not correct for incompressible flow.

On the other hand $\theta_2 > 0$ leads to a more complex although more general scheme. If we use either the pressure or the density as unknown, within each time step we need an iterative scheme to correct the temperature that has been guessed. This iterative scheme is:

1. Solve the energy equation or the heat equation (43).
2. Solve for the fractional momentum (41).
3. Guess a temperature T_g.
4. Solve the continuity equation (58) for p^{n+1} (or (56) for ρ^{n+1}).
5. Obtain ρ^{n+1} from the equation of state (57) (or p^{n+1}, if ρ^{n+1} has been used in 4).
6. Solve for the end-of-step momentum (42).
7. Correct T_g using T^{n+1} and ρ_g using ρ^{n+1}, if needed.
8. Check convergence. If not, go to 4.

In the case of barotropic or incompressible flow at constant temperature, only steps 2, 4 and 5 are to be done in this order. Let us make some remarks about this algorithm. The first concerns the use of the heat equation in step 1. If this is done, we already have T^{n+1} and therefore there is *no need to iterate* at all. However, we have found that this approach may yield wrong results in the presence of shocks, with a wrong location for them and/or without satisfying the jump conditions. It is well known that in these situations it is necessary to use the energy equation written in conservation form. By doing this, after step 1 we have E^{n+1}. A natural way to compute T_g is to use this and the density and velocity of the previous time step.

There is the possibility of not checking convergence, that is to say, take T_g computed as indicated before as T^{n+1} in the continuity equation and also ρ_g as ρ^{n+1} in the vector F_ρ if the density is used as unknown. This is an approximation of order $O(\Delta t)$ that works well if only the steady-state is of interest.

The steady-state is reached slightly faster and time steps slightly larger can be used if a couple of iterations of the previous scheme are performed. We have found almost no difference neither in the numerical results nor in the convergence behavior if more than two iterations are done.

Again, if $\theta_2 = 0$ and the explicit scheme is used, smaller time steps are allowed and convergence rates to steady states can be worsened. However, due to the smaller amount of arithmetic operations per time step this is the best option for compressible flow, unless more terms are implicitly treated in the equations (viz. convective or diffusive terms, using $\theta_3 > 0$).

5.4 General expression of the continuity equation

For all the type of flows considered we have written the continuity equation in a very similar way. Using the pressure as variable the general form is

$$M_\alpha \frac{\Delta \bar{p}^n}{\Delta t} + \theta_1 \Delta t L_\beta \bar{p}^{n+\theta_2} = F'_\mathrm{C},$$

with $\beta = 1$ and

$$\alpha = \begin{cases} 0 & \text{for incompressible flows} \\ \frac{1}{c^2} & \text{for slightly compressible flows} \\ \frac{\rho^n}{\gamma p^n} & \text{for barotropic flows} \\ \frac{1}{RT_\mathrm{g}} & \text{for perfect gases} \end{cases}$$

and $F'_\mathrm{C} = F_\mathrm{C}$, except in the case of perfect gases, for which $F'_\mathrm{C} = F_\mathrm{C} + F_p$.

The density can be used as variable only for barotropic fluids and perfect gases. In this case the discrete continuity equation is

$$M_\alpha \frac{\Delta \bar{\rho}^n}{\Delta t} + \theta_1 \Delta t L_\beta \bar{\rho}^{n+\theta_2} = F'_\mathrm{C},$$

now with $\alpha = 1$ and

$$\beta = \begin{cases} \frac{\gamma p^n}{\rho^n} & \text{for barotropic flows} \\ RT_\mathrm{g} & \text{for perfect gases} \end{cases}$$

and $F'_\mathrm{C} = F_\mathrm{C}$ for barotropic fluids and $F'_\mathrm{C} = F_\mathrm{C} + F_\rho$ for perfect gases.

In all the cases, the matrix of the algebraic system of equations to be solved is symmetric and positive-definite (for incompressible confined flows a pressure needs to be specified). We use the conjugate gradient method to solve it. In general, very few iterations are needed to converge, since the unknown at the previous time step is a good initial guess for its value at the current one.

6 NUMERICAL EXAMPLES

In this section some numerical results are shown. These few examples are presented in order to show the correct behavior of the algorithm under different regimes of flow. More examples can be seen in [21, 22, 23, 7]. Even though some of these examples are very simple, we have considered convenient to include them, since the algorithm presented herein has several original features.

6.1 Incompressible flows

In these two examples no volume forces are considered and the problems are solved keeping the temperature constant in all the domain. Parameters θ_1 and θ_2 are fixed to one in both cases, whereas θ_3 varies according to the case analyzed.

6.1.1 Inviscid flow passing a wing profile

In this example inviscid flow passing a NACA 0012 airfoil, placed at an attack angle of $5°$ is modeled. It is well known (see, for instance, [11]) that although no condition is imposed in the circulation around the airfoil (the so called *Kutta condition*) the final stationary solution

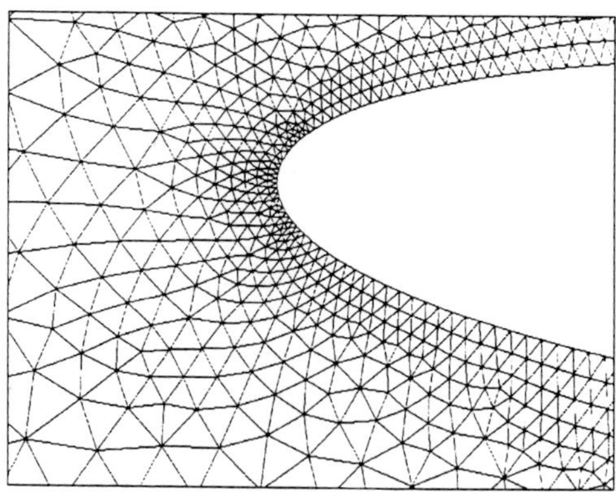

Figure 2. NACA 0012 profile. Detail of the mesh around stagnation point.

reached is that corresponding to the viscous problem, viz. with the downstream stagnation point at the very trailing edge. This fact seems to violate Kelvin's theorem, because if the initial condition is circulation free then if no diffusion is present, the final stationary state must be also circulation free. However, this is not the case, and a circulation which is different from zero appears around the airfoil. The mechanism that triggers this process is the artificial diffusion added by the numerical method, which in turn is supposed to be small. Therefore, the convergence rate to the final state could be very poor unless the equations are treated implicitly. For that reason, this is a good problem for testing the behaviour and possible advantages of the implicit form of the scheme. As no energy equation must be solved, time factors F^{TI} (see (40)) much larger than 1.0 can be used improving the convergence to the steady state.

The 2-D domain is discretized using a mesh made of 15075 P1 elements (7838 nodal points) slightly refined from the exterior, far from the wing, to the profile itself (fig. 2). The velocity is fixed to 1.0 at the inflow and the pressure to 0.0 at the outflow. Normal component of the velocity is prescribed to 0.0 at the airfoil and at the upper and lower boundaries.

Parameters θ_2 and θ_3 are in this case 1.0 (both fractional momentum and pressure equations are solved implicitly). If this is done, the time factor F^{TI} can be between two and three orders of magnitude greater than in the case $\theta_2 = 1.0$ and $\theta_3 = 0.0$. Although an additional linear system of equations needs to be solved per time step, the time step size may be taken much larger, making the total CPU time needed much smaller. This fact strongly favours the use of an implicit method for solving this equation in most of the types of flow. The stationary state reached is shown in fig. 3.

As time increments are here so large, our suggestion for correcting the splitting error (see (27) and (28))

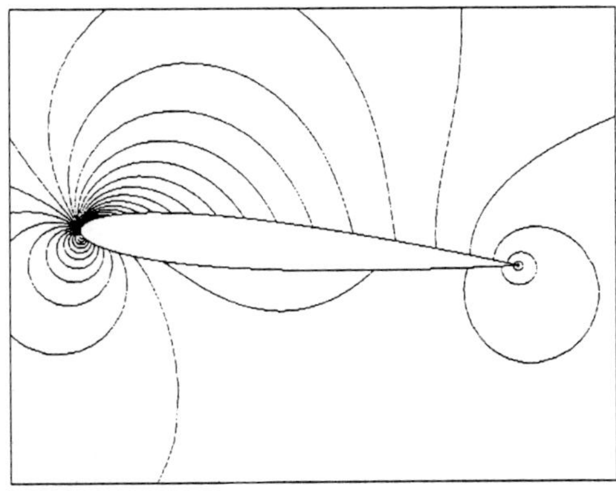

Figure 3. NACA 0012 profile. Pressure contours around
airfoil.

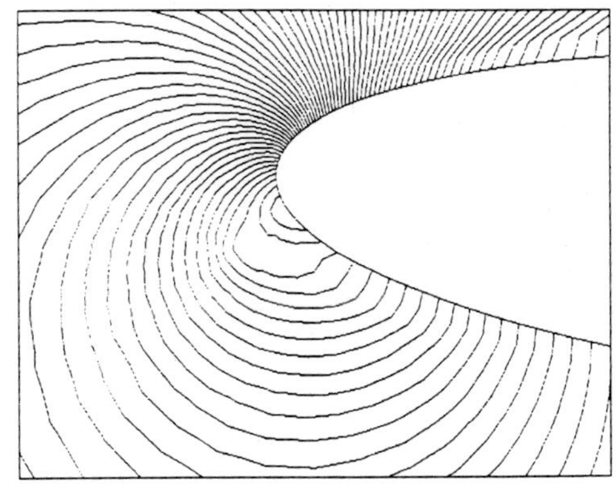

Figure 5. NACA 0012 profile. Pressure contours around
stagnation point. Splitting error of stationary state
produces an overdiffusive result.

is also tested: fig. 4 and fig. 5 show the difference in pressure level contours around the stagnation point either if it is taken into account or not. The analytical value for the pressure at the stagnation point is 0.5. If the correction is done, the value is 0.515. If not, the pressure value reached there is 0.309 being much more diffusive due to the splitting error. This becomes more apparent when larger F^{TI} are used, due to the fact that this error is $O(\Delta t)$.

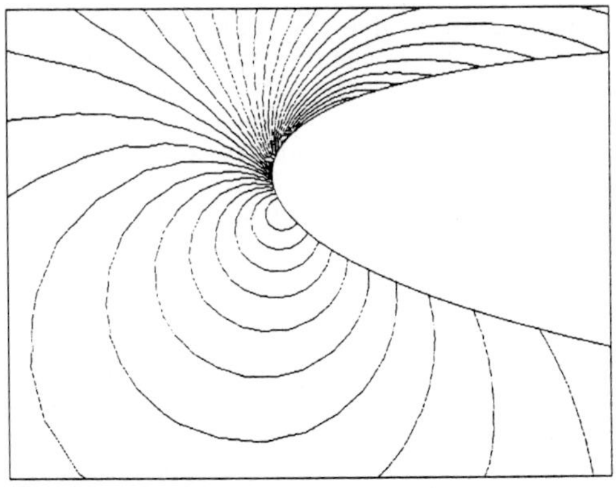

Figure 4. NACA 0012 profile. Pressure contours around
stagnation point. Splitting error of stationary state
corrected.

6.1.2 Flow passing a cylinder at Re=100

In this case the mesh is made of 2000 Q1 elements (2100 nodal points). An inlet horizontal velocity of norm 1 is prescribed, whereas the outlet is left free. There, pressure is fixed to $n_j \tau_{ij} n_i$. The no slip condition is prescribed on the cylinder. At this Reynolds number, the stationary state is oscillatory: a trail of vortexes is left behind the cylinder. The onset of the oscillatory behaviour was produced by a small initial perturbation on the velocity. Fully developed flow is shown in fig. 6.

In this problem we have used $\theta_3 = 0$, i.e. the semi-implicit form of the algorithm, which has no splitting error.

The period obtained is around 5.7, which is very close to that obtained in [19]. The period can be evaluated in many ways: through the evolution of a variable at a point behind the cylinder or the net force over it, or through the evolution of the error norm, etc. In the example shown here, one period of time is covered in approximately 270 time steps.

6.2 Compressible inviscid flow passing a cylinder

In this example, flow at Mach 3 reaches a cylinder and a steady shock is formed upstream of it. At the inflow, velocity, density and temperature are prescribed, for it is a supersonic inlet. The normal velocity on the cylinder is fixed to zero. The domain is discretized using a uniform mesh of 5351 P1 elements (2772 nodal points).

Pressure coefficient

$$P_c = \frac{-2(p - p_{\mathrm{ref}})}{\rho v_{\mathrm{ref}}^2},$$

where reference values are those of the inflow, and Mach number level contours are shown in fig. 7. Also

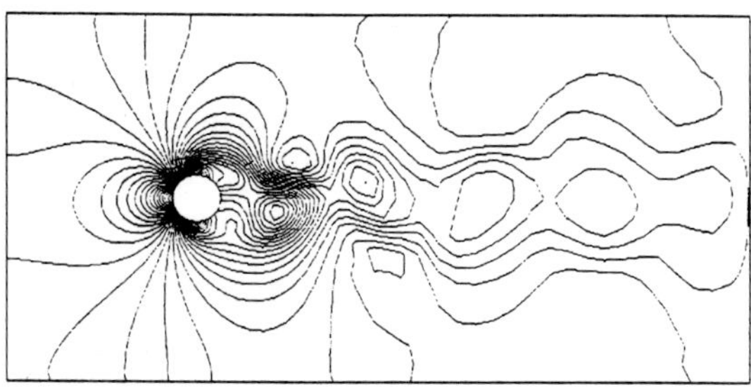

Figure 6. Flow passing a cylinder, Re=100. Streamlines
and pressure contours.

their profile along a horizontal cut through the mid
point of the domain is shown in fig. 8. It can be seen
in this last graph that at the stagnation point (which
correspond to x coordinate 3.0) the Mach number is
slightly larger than zero, being the reason for that both
a non-symmetric mesh and the interpolating procedure
of the variables along the cut.

6.3 Compressible viscous flow over a flat plate

The supersonic flow over a plate *(Carter's Flat Plate
Problem)* develops many different features that can appear when solving the complete Navier-Stokes equations, like boundary layers and shocks, and the interaction between them.

The Mach number at the inflow is $M_\infty = 3.0$. The
viscosity μ depends on the temperature according to
Sutherland's law:

$$\mu = \frac{0.0906T^{1.5}}{T + 0.0001406}.$$

Prandtl number ($\mathrm{Pr} = \mu C_p/k$) is in this case 0.72,
where $C_p = \gamma C_v$ is the specific heat at constant pres-

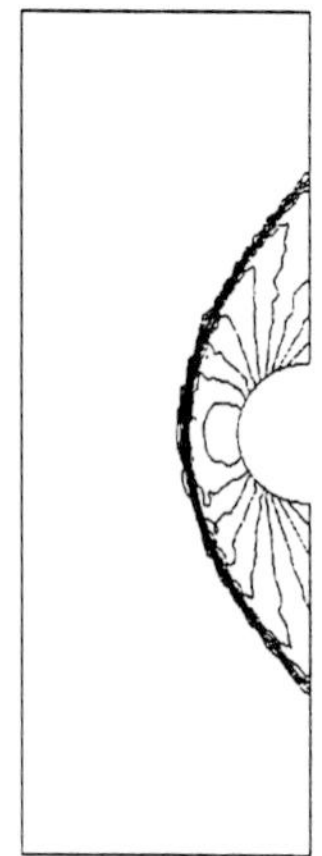

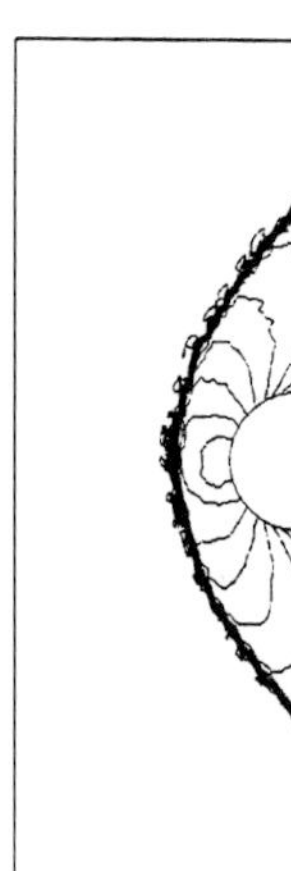

Figure 7. Supersoninc flow passing a cylinder, Mach 3.
Left: pressure coefficient. Right: Mach number.

sure, $C_v = 715$ and $\gamma = 1.4$. The state law is that of
an ideal gas.

The domain is divided using a uniform mesh of $112 \times$
64 (7168) Q1 elements, corresponding to 7345 nodal
points. If the coordinates origin is at the left bottom
corner, the domain goes from 0.0 to 0.8 vertically and
from 0.0 to 1.4 horizontally. Density, temperature and
velocity are prescribed at the inflow, because this inlet
is supersonic. The values prescribed at the inflow are
1.0 and 2.8×10^{-4} for the first two and $(1.0; 0.0)$ for
the horizontal and vertical components of the velocity.
The non-slip condition is imposed at the floor of the
plate, which starts at $x = 0.25$.

The stagnation temperature is calculated according
to:

$$T_{\text{stag}} = T_\infty \left(1 + \frac{\gamma - 1}{2} M_{\infty}^2 \right)$$

which is the prescription of this variable along the
plate. No prescriptions are made at the outflow. This
point must be remarked, because most of the outlet
is subsonic, eventually requiring a prescription on the
density. Nevertheless, the only prescribed node of the
outflow is that of the right bottom corner which is considered belonging to the plate, with its boundary conditions on temperature and velocity.

In fig. 9, fig. 10, fig. 11 and fig. 12 the results obtained for this example are shown. Note the sharpness
of the shock and the gradual change of the variables
along the boundary layer. In fig. 11 and fig. 12 a comparison with the original results of Carter (as appearing

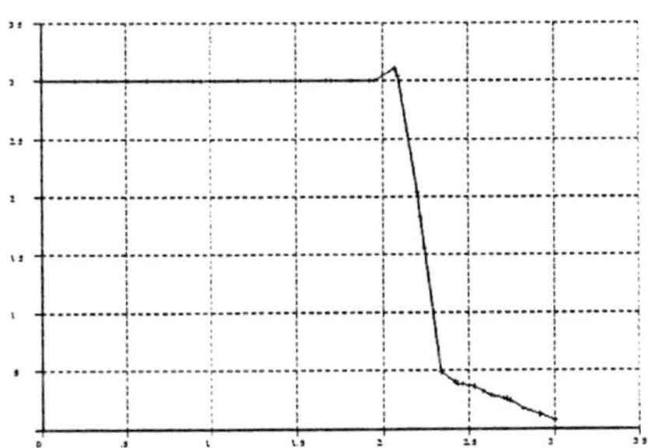

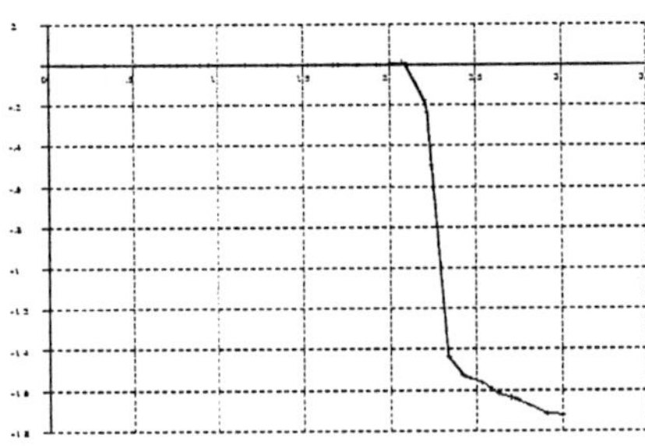

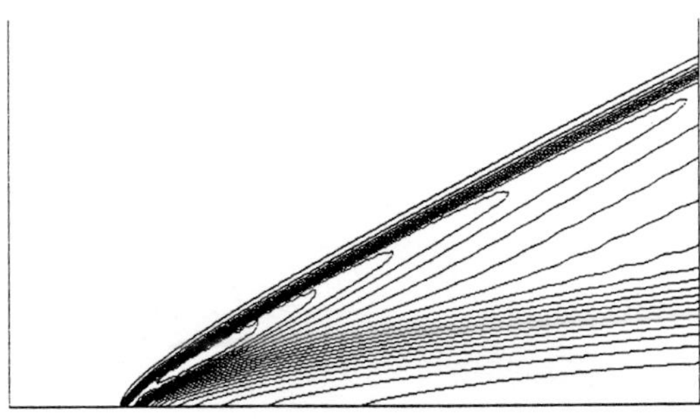

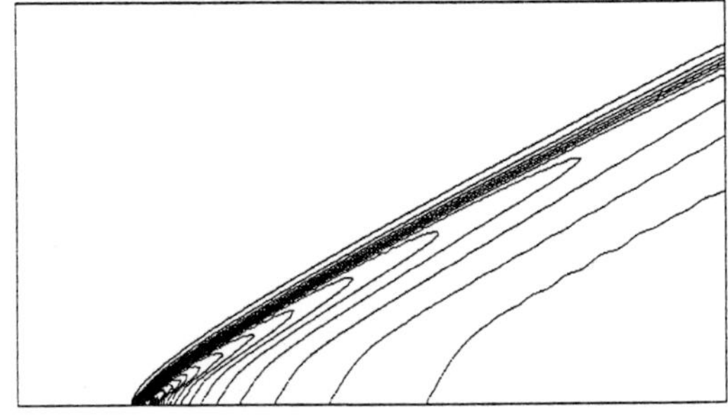

Figure 8. Supersoninc flow passing a cylinder, Mach 3. Top: Mach number. Bottom: pressure coefficient. The cut is done horizontally at half of domain, normal to the shock.

Figure 9. Flow over a flat plate. Contour levels. Top: density. Bottom: pressure.

in [8]) is made, showing a good agreement with them. These figures correspond to the profiles of some variables along a cut at $x = 1.25$. Density, pressure and temperature are normalized using their inflow values.

For the velocity, density and temperature, the only and slight difference is in the very maximum value at the shock. Carter's pressure profile is not shown because it presents some oscillations. In this problem, shock capturing diffusion is artificially put according to section 3.2. It is activated for all the equations. The algorithm works equally well for both the strong and low compression regions of the domain, viz. the shock and the boundary layer respectively.

7 CONCLUSIONS

In this paper we have presented an overview of the splitting technique that we have developed for both compressible and incompressible flow problems. We have also extended it to the case in which the fractional momentum equation is treated implicitly and discussed some implementation issues related to the implicit treatment of the viscous and convective terms.

Besides the split of the momentum equation, the basic ingredients of the numerical formulation are the use of an explicit version of the Characteristic-Galerkin scheme together with the introduction of a nonlinear shock capturing diffusion for each scalar equation depending on the spatial residual.

Numerical examples have shown that these techniques are effective to solve convection dominated flows. Shocks have been reproduced with their correct strength and without local oscillations.

Two important aspects of the model presented are the treatment of boundary conditions and the correction of the splitting error coming from the convective term when it is treated implicitly. Both can be applied to incompressible cases as well, thus improving the original fractional step method.

From the numerical examples presented it can be concluded that the formulation proposed here works well in very different flow regimes, compressible and incompressible, viscous and inviscid.

ACKNOWLEDGEMENTS

This research has been partially supported by the NASA grant NAGW/2127, Ames Control Number 90-

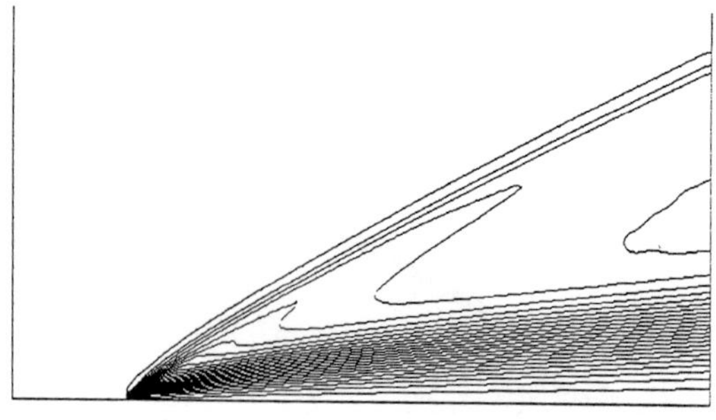

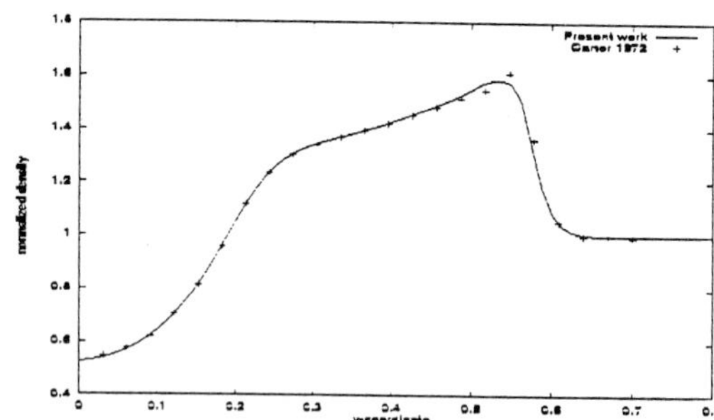

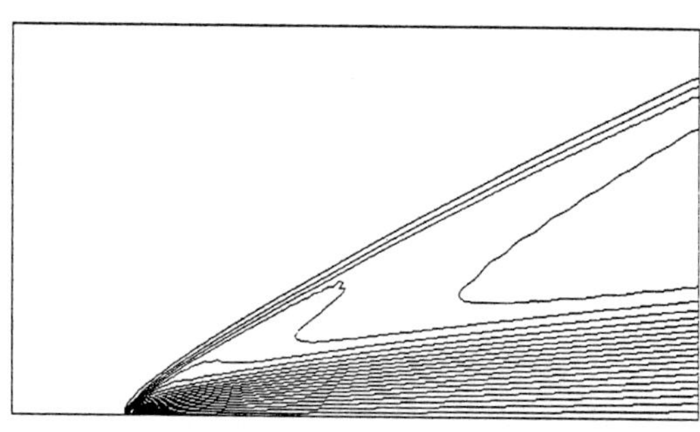

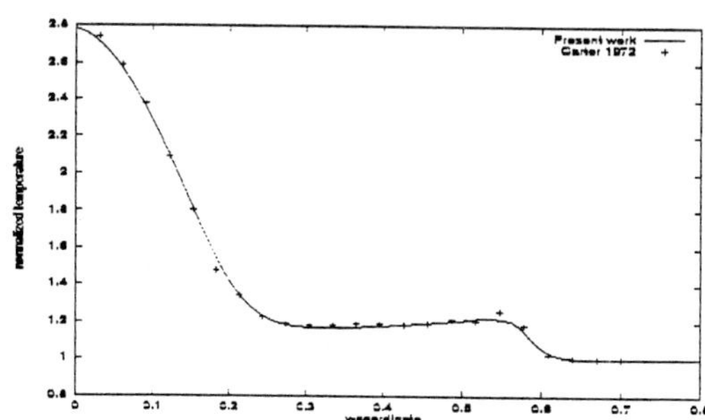

Figure 10. Flow over a flat plate. Contour levels. Top: temperature. Bottom: Mach number.

Figure 11. Flow over a flat plate. Profile along $x = 1.25$. Top: density. Bottom: temperature. (Normalized using inflow values)

144.

REFERENCES

[1] G. BATCHELOR, *An Introduction to Fluid Dynamics*, Cambridge University Press, 1967.

[2] A. BROOKS AND T. HUGHES, *Streamline upwind / Petrov-Galerkin formulations for convection dominated flows with particular emphasis on the incompressible Navier-Stokes equation*, Comp. Meth. Appl. Mech. Eng., 32 (1982), pp. 199–259.

[3] A. CHORIN, *A numerical method for solving incompressible viscous problems*, J. Comput. Phys., 2 (1967), pp. 12–26.

[4] R. CODINA, *Comparison of some finite element methods for solving the diffusion – convection – reaction equation*, Comp. Meth. Appl. Mech. Eng. (submitted).

[5] ———, *A discontinuity-capturing crosswind – dissipation for the finite element solution of the convection – diffusion equation*, Comp. Meth. Appl. Mech. Eng., 110 (1993), pp. 325–342.

[6] ———, *A shock-capturing anisotropic diffusion for the finite element solution of the diffusion – convection – reaction equation*, in Proc. VIII International Conference on Finite Elements in Fluids, vol. Part I, Barcelona, Spain, September 1993, Ed. CIMNE ./ Pineridge Press, pp. 67–75.

[7] R. CODINA, M. VÁZQUEZ, AND O. ZIENKIEWICZ, *A general algorithm for compressible and incompressible flow–Part III. The semi-implicit form*, Int. J. Num. Meth. Fluids. (to appear).

[8] L. DEMKOWICZ, J. ODEN, AND W. RACHOWICZ, *A new finite element method for solving compressible Navier-Stokes equations based on an operator splitting method and h-p adaptivity*, Comp. Meth. Appl. Mech. Eng., 84 (1990), pp. 275–326.

[9] T. FABER, *Fluid Dynamics for Physicists*, Cambridge University Press, 1995.

[10] A. HINDMARSH, P. GRESHO, AND D. GRIFFITHS, *The stability of explicit Euler time-integration for certain finite-difference approximation of the multidimensional advection-diffusion equations*, Int. J. Num. Meth. Fluids., 4 (1984), pp. 853–897.

[11] C. HIRSCH, *Numerical Computation of Internal and External Flows - Vol. 2*, John Wiley & sons, 1990.

[12] T. HUGHES AND M. MALLET, *A new finite element method for computational fluid dynamics: III. The generalized streamline operator for multidimensional advective-diffusive systems*, Comp. Meth. Appl. Mech.

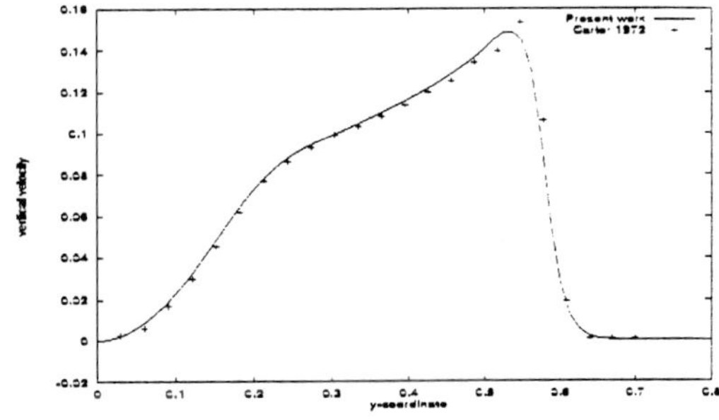

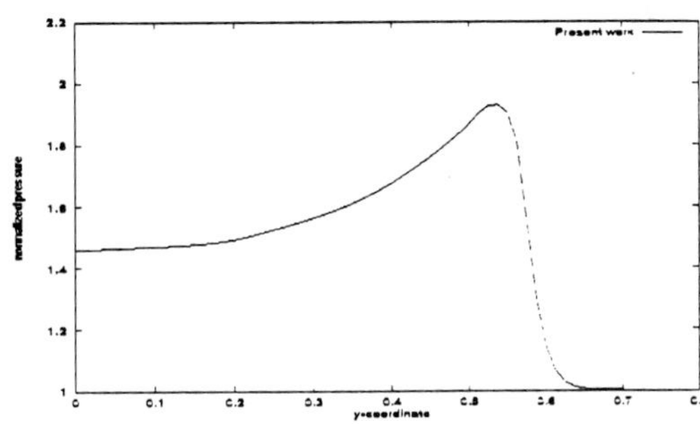

Figure 12. Flow over a flat plate. Profile along $x = 1.25$.
Top: vertical component of velocity. Bottom: pressure.
(Normalized using inflow values)

Eng., 58 (1986), pp. 305–328.

[13] T. HUGHES AND T. TEZDUYAR, *Finite element methods for first - order hyperbolic systems with particular emphasis on the compressible euler equation*, Comp. Meth. Appl. Mech. Eng., 45 (1984), pp. 217–284.

[14] C. JOHNSON, *Numerical Solution of Partial Differential Equations by the Finite Element Method*, Cambridge University Press, 1987.

[15] M. KAWAHARA AND K. OHMIYA, *Finite element analysis of density flow using the velocity correction method*, Int. J. Num. Meth. Fluids, 5 (1985), pp. 981–993.

[16] L. LANDAU AND E. LIFSHITZ, *Mecánica de Fluidos*, Editorial Reverté, 1986.

[17] T. PAPANASTASIOU, N. MALAMATARIS, AND K. ELLWOOD, *A new outflow boundary condition*, Int. J. Num. Meth. Fluids., 14 (1992), pp. 587–608.

[18] G. SCHNEIDER, G. RAITHBY, AND M. YOVANOVICH, *Finite element analysis of incompressible fluid flow incorporating equal order pressure and velocity interpolation*, in Numerical Methods in Laminar and Turbulent Flow (C. Taylor, K. Morgan, C.A. Brebbia eds), Pentech Press, Plymouth, 1978.

[19] J. SIMO AND F. ARMERO, *Unconditional stability and long term behavior of transient algorithms for the incompressible Navier-Stokes equations*, Comp. Meth. Appl. Mech. Eng., 111 (1994), pp. 111–154.

[20] R. TEMAM, *Sur l'approximation de la solution des équations de Navier–Stokes par la méthode des pas fractionaires (I)*, Arch. Rat. Mech. Anal., 32 (1969), pp. 135–153.

[21] M. VÁZQUEZ, R. CODINA, AND O. ZIENKIEWICZ, *A fractional step method for the solution of the Navier-Stokes equations*, Tech. Rep. 103, CIMNE, December 1996.

[22] O. ZIENKIEWICZ AND R. CODINA, *A general algorithm for compressible and incompressible flow - Part I. The Split, Characteristic-Based Scheme*, Int. J. Num. Meth. Fluids., 20 (1995), pp. 869–885.

[23] O. ZIENKIEWICZ, K. MORGAN, B. SATYA SAI, R. CODINA, AND M. VÁZQUEZ, *A general algorithm for compressible and incompressible flow - Part II. Tests on the Explicit Form*, Int. J. Num. Meth. Fluids., 20 (1995), pp. 887–913.

COMPRESSIBLE TURBULENT FLOW COMPUTATIONS ON UNSTRUCTURED TRIANGULAR GRIDS

M.T. MANZARI   K. MORGAN   O. HASSAN

Department of Civil Engineering, University of Wales Swansea, SA2 8PP, U.K.

Abstract

The simulation of two dimensional compressible turbulent flow problems is considered. The turbulence is represented via a two–equation eddy viscosity model and both the $k - \epsilon$ and the $k - \omega$ models are employed. The mass–averaged compressible Navier–Stokes equations, coupled with the turbulence–transport equations, are solved using a Galerkin finite element method. Consistent numerical fluxes are constructed using Roe flux–difference splitting. A MUSCL scheme is employed to achieve higher–order accuracy for both the mean flow and the turbulence variables. A general unstructured triangular grid is employed to represent the computational domain and the solution algorithm is implemented using an edge–based data structure. Several test cases, involving a range of flow conditions, are solved to show the applicability of the proposed procedure.

1 Introduction

Unstructured grid methods are now well–established for the simulation of compressible inviscid flows. With the approach, it has been demonstrated that it is currently possible to achieve a rapid prediction of the nature of flows involving complex geometries. Recent interest has concentrated upon demonstrating that the approach is capable of being extended to deal with the simulation of both laminar and turbulent viscous flow. The new problems which have to be overcome in this area are the construction of appropriate mesh generation procedures for thin viscous layers and the implementation of suitable, efficient turbulent flow solvers. The mesh generation problem has received considerable attention in the literature and major strides have been made in providing a general capability [15,5]. In this paper, we will be concerned with flow solver development with a view to implementing turbulence models and investigating their numerical performance over a wide spectrum of flow conditions. In this context, a range of turbulence modelling options are currently available. Algebraic and one–equation models have been shown to perform well for certain transonic flow applications. Such models are not considered further here as it is felt that two–equation turbulence models offer the minimum basis for a general turbulent flow computational capability. Among the existing two–equation models, the $k - \epsilon$ and $k - \omega$ models have proved to be the most popular over a range of different flow regimes. The original $k - \epsilon$ model was proposed by Jones and Launder [6] and many variants are now in existence. The $k - \omega$ model was developed by Wilcox [22] and has been demonstrated to be an excellent alternative to the $k - \epsilon$ model for many applications. Both these models suffer from well–known drawbacks, but also enjoy certain advantages. From the standpoint of aerodynamics, the two major shortcomings of the $k - \epsilon$ model can be regarded as (1) its lack of sensitivity to adverse pressure gradients and (2) the numerical stiffness of the model equations when integrated through the viscous sub–layer, due to the need for damping functions. In the context of $k - \epsilon$ model, Grasso and Falconi [4] developed a low–Reynolds number model which includes terms which are designed to correctly account for compressibility effects arising due to high Mach numbers. This model is adopted in this work. The $k - \omega$ model does not suffer from these deficiencies, but it does exhibit strong dependency on the adopted free stream values of ω [11]. Although Menter has shown how this drawback can be removed by blending the $k - \omega$ model with the high Reynolds number $k - \epsilon$ model [?], the original $k - \omega$ model as proposed by Wilcox is investigated here. This model is especially attractive for unstructured grid computations since there is no need to calculate variables such as the distance from the wall or the wall shear stress. This requirement is common in the computation of damping functions in $k - \epsilon$ models.

In the following, the unsteady mass–averaged Navier–Stokes equations are presented and the two turbulence models which will be employed are described. The resulting equations are discretised in space using a Galerkin finite element method on a grid of linear triangular elements. With computational efficiency aspects in mind, the grid is represented in terms of an edge based data structure [1,10,14]. To achieve a practical algorithm, the actual flux function on each edge of the grid is replaced by a consistent numerical flux function, constructed by the application of Roe's flux difference splitting along the edge. The MUSCL approach is used to achieve a high resolution extension. Explicit time stepping is employed to ad-

Received on July 30, 1997.

vance the solution to steady state. The predictive performance of the resulting numerical algorithm is demonstrated by considering the simulation of a number of test cases covering a wide range of aerodynamic flow regimes.

2. GOVERNING EQUATIONS

For unsteady 2D compressible turbulent flow, the Favre (mass)–averaged form of the conservation laws for the mean flow variables is obtained by introducing the Favre and conventional time averages into the Navier–Stokes equations [3]. The Favre average is used for the velocity components and the total energy, while the conventional time average is employed for the density and the pressure. Here, non–dimensional variables are used, with the non–dimensionalisation based upon the density, velocity and molecular viscosity of the free stream and a characteristic length scale of the problem. The resulting system of equations is expressed in the non–dimensional conservation form

$$\frac{\partial \mathbf{U}}{\partial t} + \frac{\partial \mathbf{F}^j}{\partial x_j} - \frac{\partial \mathbf{G}^j}{\partial x_j} = \mathbf{S} \qquad (1)$$

where the summation convention is employed and $\mathbf{U} = [\rho \ \rho u_1 \ \rho u_2 \ \rho E]^T$, $\mathbf{S} = [0 \ 0 \ 0 \ 0]^T$ and

$$\mathbf{F}^j = \begin{bmatrix} \rho u_j \\ \rho u_1 u_j + p\delta_{1j} \\ \rho u_2 u_j + p\delta_{2j} \\ (\rho E + p)u_j \end{bmatrix} \quad \mathbf{G}^j = \begin{bmatrix} 0 \\ \tau_{1j} \\ \tau_{2j} \\ u_i \tau_{ij} - q_j + D_{kj} \end{bmatrix} \qquad (2)$$

for i and $j = 1, 2$. Here t denotes time, x_i the coordinates relative to a Cartesian coordinate system Ox_1x_2, u_i the component of the velocity vector in direction x_i, p the pressure and δ_{ij} the Kronecker delta. The total energy per unit mass is defined as $E = e + u_i u_i/2 + k$, where e is the mass–averaged specific internal energy and k is the turbulence kinetic energy. The equation of state for an ideal gas is employed in the form

$$p = (\gamma - 1)\rho(E - 0.5u_j u_j - k) = \frac{\gamma - 1}{\gamma}\rho T \quad (3)$$

where γ is the ratio of the specific heats. The stress tensor, τ_{ij}, is regarded as being made up of contributions from both the molecular and the eddy viscosity and is written as

$$\tau_{ij} = \frac{2\mu}{Re}\left(S_{ij} - \frac{1}{3}\frac{\partial u_k}{\partial x_k}\delta_{ij}\right) + \tau_{ij}^t \qquad (4)$$

Here τ_{ij}^t denotes the turbulent stress tensor, whose components are defined by

$$\tau_{ij}^t = \frac{2\mu_t}{Re}\left(S_{ij} - \frac{1}{3}\frac{\partial u_k}{\partial x_k}\delta_{ij}\right) - \frac{2}{3}\rho k \delta_{ij} \qquad (5)$$

and $S_{ij} = (\partial u_i/\partial x_j + \partial u_j/\partial x_i)/2$. In these equations, μ is the molecular dynamic viscosity, whose variation is defined by Sutherland's law, μ_t is the turbulent viscosity and Re is the free–stream Reynolds number. The components of the heat flux vector q_j are defined by

$$q_j = \frac{-1}{Re}\left(\frac{\mu}{Pr} + \frac{\mu_t}{Pr_t}\right)\frac{\partial T}{\partial x_j} \qquad (6)$$

so that both molecular and eddy contributions are considered. In this expression, Pr and Pr_t denote the molecular and turbulent Prandtl numbers respectively and T is the temperature. In the examples considered here, Pr_t will be assumed to be constant ($= 0.9$). Finally, D_{kj} is defined by

$$D_{kj} = \frac{1}{Re}\left(\mu + \frac{\mu_t}{\sigma_k}\right)\frac{\partial k}{\partial x_j} \qquad (7)$$

where σ_k is a closure constant.

3. TURBULENCE MODEL

Before we can solve the system of equations (1), the turbulent viscosity μ_t must be evaluated. In the context of two–equation turbulence modelling, the turbulent viscosity is related to the turbulence kinetic and dissipation energies. For each of these quantities an additional differential equation is solved along with the conservation laws. Here we consider both the $k - \epsilon$ and the $k - \omega$ models. In this case, the turbulent transport equations for k and ϕ (ϵ or ω) can be written in the form of equation (1) with $\mathbf{U} = [\rho k \ \rho \phi]^T$ and

$$\mathbf{F}^j = \begin{bmatrix} \rho k u_j \\ \rho \phi u_j \end{bmatrix} \quad \mathbf{G}^j = \begin{bmatrix} D_{kj} \\ D_{\phi j} \end{bmatrix} \quad \mathbf{S} = \begin{bmatrix} P_k - D_k \\ P_\phi - D_\phi \end{bmatrix} \qquad (8)$$

for i and $j = 1, 2$. Fuller details of these models are outlined below.

The $k - \epsilon$ model

The $k-\epsilon$ model proposed by Grasso and Falconi [4] is considered. This model is designed to account for compressibility and low Reynolds–number effects in high Mach number flows. For this model, we define

$$D_{\epsilon j} = \frac{1}{Re}\left(\mu + \frac{\mu_t}{\sigma_\epsilon}\right)\frac{\partial \epsilon}{\partial x_j}$$

$$P_k = \tau_{ij}^t \frac{\partial u_i}{\partial x_j} \qquad P_\epsilon = C_{\epsilon_1}f_1\frac{\epsilon}{k}P_k$$

$$D_k = \rho\epsilon - \Pi_{c,1} - \Pi_{c,2} - \Pi_{c,3} \qquad D_\epsilon = C_{\epsilon_2}f_2\,\rho\epsilon\frac{\epsilon}{k} \qquad (9)$$

and the turbulent viscosity is then written as

$$\mu_t = C_\mu f_\mu \rho k \left(\frac{k}{\epsilon}\right) Re \qquad (10)$$

The closure constants are assigned the values $\sigma_k = 1.55$, $\sigma_\epsilon = 2.0$, $C_\mu = 0.09$, $C_{\epsilon_1} = 1.60$ and

$$C_{\epsilon_2} = 1.83 \left[1 - \frac{2}{9} \exp\left(\frac{-Re_t{}^2}{36} \right) \right]. \tag{11}$$

The damping functions are defined as

$$
\begin{aligned}
f_1 &= 1.0 \qquad f_2 = \left[1 - \exp\left(\frac{-y^+}{A} \right) \right]^2 \\
f_\mu &= \left(1 + \frac{3.45}{\sqrt{Re_t}} \right) \tanh\left(\frac{y^+}{80} \right)
\end{aligned}
\tag{12}
$$

where $A = 4.9$, $y^+ = u_\tau y Re/\nu_w$, the friction velocity $u_\tau = \sqrt{\tau_w/(\rho Re)}$ and the turbulent Reynolds number $Re_t = k^2 Re/(\nu \epsilon)$. In the definition of y^+, the quantity y denotes the distance from the wall. In addition, the compressibility effects $\Pi_{c,i}$ are computed as

$$\Pi_{c,1} = (-0.4 P_k + 0.2 \rho \epsilon) M_t^2$$

$$\Pi_{c,2} = \frac{-\mu_t}{\rho^2} \frac{1}{\sigma_p} \frac{\partial \rho}{\partial x_j} \frac{\partial p}{\partial x_j}$$

$$\Pi_{c,3} = -0.4 \rho \epsilon \left\{ 1 - \exp\left[-\left(\frac{M_t - 0.3}{0.8} \right)^2 \right] \right\} \tag{13}$$

where the turbulent Mach number is defined as $M_t = \sqrt{2k/(\gamma R T)}$ and $\sigma_p = 0.5$.

To address the problem of overprediction of the peak heat transfer rate, we follow Marvin and Coakley [9] and redefine the turbulent viscosity, so that it is based on a limited turbulent length scale l_m, according to

$$\mu_t = C_\mu f_\mu \rho l_m \sqrt{k} Re \qquad l_m = \min\left(C_l y, \frac{k^{\frac{3}{2}}}{\epsilon} \right) \tag{14}$$

where $C_l = 2.25$ and y denotes the distance from the wall.

The $k - \omega$ model

The $k - \omega$ model as proposed by Wilcox [22] is a two–equation model which does not require the use of damping functions and behaves well at any solid wall. For this model, we have

$$D_{\omega j} = \frac{1}{Re} (\mu + \frac{\mu_t}{\sigma_\omega}) \frac{\partial \omega}{\partial x_j} \qquad \mu_t = \alpha^* \frac{\rho k}{\omega} Re \tag{15}$$

$$P_\omega = \frac{\alpha \omega}{k} P_k \qquad D_k = \beta^* \rho \omega k \qquad D_\omega = \beta \rho \omega^2$$

where the closure constants have the values

$$
\begin{aligned}
\beta &= 0.075 \qquad \beta^* = 0.09 \qquad \alpha = 5/9 \\
\alpha^* &= 1 \qquad \sigma_k = 2 \qquad \sigma_\omega = 2
\end{aligned}
\tag{16}
$$

An interesting feature of this model is its ability to deal with high adverse pressure gradients and separated flows. When the solution of high Mach number flows is considered, the implemented model incorporates the following modifications, proposed by Wilcox [23], which are intended to account for compressibility and viscous effects.

Compressibility Modifications

The closure coefficients are written as functions of the turbulence Mach number $M_t^2 = 2k/a^2$, where a is the local speed of sound, as

$$
\begin{aligned}
\beta^* &= \beta_0^* [1 + \xi^* F(M_t)] \\
\beta &= \beta_0 \left[1 - \frac{\beta_0^*}{\beta_0} \xi^* F(M_t) \right]
\end{aligned}
\tag{17}
$$

Here β_0^* and β_0 are the corresponding incompressible values of β^* and β as given in (16), and

$$\xi^* = \frac{3}{2} \qquad F(M_t) = (M_t^2 - M_{t_0}^2) H(M_t - M_{t_0})$$

where $M_{t_0} = 1/4$ and $H(x)$ denotes the Heaviside step function.

Viscous Modifications

To predict the sub–layer features more accurately near transition, a laminar Prandtl number σ_l is added to the diffusion term in the ω–equation so that this term is considered in the form

$$\frac{\partial}{\partial x_j} \left[\left(\sigma_l \mu + \frac{\mu_t}{\sigma_\omega} \right) \frac{\partial \omega}{\partial x_j} \right] \tag{18}$$

with $\sigma_l = 5/18$. In addition, certain closure coefficients are assumed to vary with the turbulent Reynolds number $Re_t = Re k/\omega \nu$ as

$$\alpha^* = \frac{\alpha_0^* + Re_t/R_k}{1 + Re_t/R_k}$$

$$\alpha = \frac{5/9}{\alpha^*} \frac{\alpha_0 + Re_t/R_\omega}{1 + Re_t/R_\omega} \tag{19}$$

$$\beta^* = \frac{9}{100} \frac{5/18 + (Re_t/R_\beta)^4}{1 + (Re_t/R_\beta)^4}$$

where

$$
\begin{aligned}
\alpha_0 &= 0.014 \qquad \alpha_0^* = 0.014 \qquad \beta = \tfrac{3}{40} \\
R_k &= 20 \qquad R_\omega = 14.55 \qquad R_\beta = 8
\end{aligned}
\tag{20}
$$

4. INITIAL/BOUNDARY CONDITIONS

To develop a numerical solution procedure, we consider a flow domain, Ω, bounded by a closed boundary curve, Γ, with unit outward normal vector n_j. An initial condition for the system of six differential equations (1) can be written as

$$\mathbf{U}(\mathbf{x}, t_0) = \mathbf{U}_0(\mathbf{x}) \tag{21}$$

for all $\mathbf{x}$ in Ω at time $t = t_0$, where $\mathbf{U}_0$ is a specified function. In practice, free stream conditions are initially imposed everywhere in Ω. The turbulence kinetic energy is set at a value equal to 0.1% of the mean flow kinetic energy, i.e. ($k_\infty = 0.001$). The free–stream turbulent viscosity is assumed to be equal to the free–stream molecular viscosity and, based on this assumption, the initial (free–stream) value for ϵ (or ω) is estimated from the definition of μ_t. On a solid wall, the no slip condition $u_i = 0$ is imposed. For an isothermal wall, a prescribed temperature condition of the form $T = T_w$ is imposed, while for an adiabatic wall the condition $n_j(\partial T/\partial x_j) = 0$ is weakly imposed. Both k and ϵ are assumed to vanish at a solid wall, while ω approaches $60\mu/\rho\beta\Delta y^2 Re$, where Δy is the distance between the wall and the first grid point away from the wall.

At far–field computational boundaries, a characteristic analysis, based upon the solution of a 1D Riemann problem in the direction normal to the boundary, is used to determine the appropriate form of the normal fluxes [7].

5. FINITE ELEMENT DISCRETISATION

Consider a trial function space $\mathcal{T}$ and a weighting function space $\mathcal{W}$, which are initially defined to consist of all suitably smooth functions. A weak variational formulation of the problem outlined above can then be stated as: find $\mathbf{U} \subset \mathcal{T}$ such that $\forall\, W \in \mathcal{W}$ and $\forall t > t_0$

$$\int_\Omega \frac{\partial \mathbf{U}}{\partial t} W d\Omega = \int_\Omega (\mathbf{F}^j - \mathbf{G}^j)\frac{\partial W}{\partial x_j} d\Omega - \int_\Gamma (\mathbf{F}^j - \mathbf{G}^j)n_j W d\Gamma + \int_\Omega \mathbf{S} W d\Omega \tag{22}$$

The domain Ω is discretised into a general assembly of linear triangular elements, with the nodes numbered from 1 to p. Subsets $\mathcal{T}^{(p)}$ and $\mathcal{W}^{(p)}$ of the trial and weighting function sets $\mathcal{T}$ and $\mathcal{W}$ can then be defined as

$$\mathcal{T}^{(p)} = \{\mathbf{U}^{(p)}(\mathbf{x}, t) | \mathbf{U}^{(p)} = \textstyle\sum_{J=1}^{p} \mathbf{U}_J(t)N_J(\mathbf{x});$$

$$\mathbf{U}_J(t_0) = \mathbf{U}_0(\mathbf{x}_J)\}$$

$$\mathcal{W}^{(p)} = \left\{ W(\mathbf{x}) | W = \sum_{J=1}^{p} a_J N_J(\mathbf{x}) \right\} \tag{23}$$

where N_J is the standard linear finite element shape function associated with node J, located at

$\mathbf{x} = \mathbf{x}_J$, and $\mathbf{U}_J$ is the nodal value of $\mathbf{U}^{(p)}$ at node J while a_J is a constant.

An approximate solution of the problem can be obtained from the Galerkin statement: find $\mathbf{U}^{(p)}$ in $\mathcal{T}^{(p)}$ such that

$$\sum_{E \in I} \int_{\Omega_E} \frac{\partial \mathbf{U}^{(p)}}{\partial t} N_I d\Omega = \sum_{E \in I} \int_{\Omega_E} \mathbf{S}^{(p)} N_I d\Omega +$$
$$\sum_{E \in I} \int_{\Omega_E} (\mathbf{F}^{j(p)} - \mathbf{G}^{j(p)})\frac{\partial N_I}{\partial x_j} d\Omega -$$
$$\sum_{B \in I} \int_{\Gamma_B} (\mathbf{F}^{j(p)} - \mathbf{G}^{j(p)})n_j N_I d\Gamma \tag{24}$$

for $I = 1, 2, ..., p$ and for all $t > t_0$. The integrals appearing in the Galerkin variational statement can be evaluated in the standard finite element form by summing individual element and boundary edge contributions. The integral appearing on the left hand side of this equation can be evaluated exactly, to give

$$\sum_{E \in I} \int_{\Omega_E} \frac{\partial \mathbf{U}^{(p)}}{\partial t} N_I d\Omega =$$
$$\sum_{E \in I} \left[\int_{\Omega_E} N_I N_J d\Omega \right] \frac{d\mathbf{U}_J}{dt} = [\mathbf{M}]_{IJ} \frac{d\mathbf{U}_J}{dt} \tag{25}$$

where $\mathbf{M}$ denotes the finite element consistent mass matrix. For the steady flow analysis which is of interest here, this matrix is replaced by the lumped (diagonal) mass matrix $\mathbf{M}_L$. This allows the use of truly explicit time marching procedures. The last term on the right hand side in equation (24) arises because of the existence of source terms in the governing equations. Assuming that the source terms vary linearly over each element, this term can be evaluated as

$$\sum_{E \in I} \int_{\Omega_E} \mathbf{S} N_I d\Omega =$$
$$\sum_{E \in I} \left[\int_{\Omega_E} N_I N_J d\Omega \right] \mathbf{S}_J = [\mathbf{M}_L] \mathbf{S}_J \tag{26}$$

Here, the consistent mass matrix has been replaced by the lumped mass matrix, following the approach advocated by Vemaganti and Prabhu [19].

In the integrals involving the inviscid fluxes, we assume that

$$\mathbf{F}^j = \sum_{I=1}^{p} \mathbf{F}_I^j N_I(x_j) \tag{27}$$

where $\mathbf{F}_I^j = \mathbf{F}^j(\mathbf{U}_I^{(p)})$. It then follows that

$$\int_{\Omega_E} \mathbf{F}^j \frac{\partial N_I}{\partial x_j} d\Omega = \frac{\Omega_E}{3} \left[\frac{\partial N_I}{\partial x_j} \right]_E (\mathbf{F}_I^j + \mathbf{F}_J^j + \mathbf{F}_K^j) \tag{28}$$

$$\int_{\Gamma_B} \mathbf{F}^j n_j N_I d\Gamma = \frac{\Gamma_B}{6}(2\mathbf{F}_I^j n_j + \mathbf{F}_J^j n_j) \qquad (29)$$

In these equations, Ω_E is the area of the element E, which has nodes I, J, K, and, assuming nodes I and J lie on the boundary, Γ_B is the distance between nodes I and J.

The viscous fluxes are approximated in the piecewise linear form

$$\mathbf{G}^j = \sum_{I=1}^{p} \mathbf{G}_I^j N_I(x_j) \qquad (30)$$

where $\mathbf{G}_I^j = \mathbf{G}^j \left[\mathbf{U}_I, (\mathbf{U}_{,x_1}^{(p)})_I, (\mathbf{U}_{,x_2}^{(p)})_I \right]$. The gradients of the variables at node I are computed using a standard variational recovery procedure [7]. Performing the integration, we obtain

$$\int_{\Omega_E} \mathbf{G}^j \frac{\partial N_I}{\partial x_j} d\Omega = \frac{\Omega_E}{3} \left[\frac{\partial N_I}{\partial x_j} \right]_E (\mathbf{G}_I^j + \mathbf{G}_J^j + \mathbf{G}_K^j) \qquad (31)$$

$$\int_{\Gamma_B} \mathbf{G}^j n_j N_I d\Gamma = \frac{\Gamma_B}{6}(2\mathbf{G}_I^j n_j + \mathbf{G}_J^j n_j) \qquad (32)$$

It has been found that, in practice, this approach is identical numerically to the standard finite element approach for evaluating these terms [24].

An alternative approach for discretising these equations uses an edge–based data structure for the triangular mesh [1, 10, 14]. With the element–based data structure, the discrete equations are formed by summing individual element contributions to nodes, but in the edge–based implementation the discrete equations are formed by summing edge contributions to the nodes. The adoption of this data structure can lead to significant reductions in both cpu and memory requirements [1] for 3D simulations. It can be readily shown that the edge–based data–structure counterpart of our Galerkin finite element formulation reads [14]

$$\left[\mathbf{M}_L \frac{d\mathbf{U}}{dt} \right]_I =$$
$$-\sum_{s=1}^{m_I} \frac{C_{II_s}^j}{2}\{(\mathbf{F}_I^j + \mathbf{F}_{I_s}^j) - (\mathbf{G}_I^j + \mathbf{G}_{I_s}^j)\} +$$
$$\langle \sum_{f=1}^{2} D_f \{ (4\bar{\mathbf{F}}_I^n + 2\bar{\mathbf{F}}_{J_f}^n + \mathbf{F}_I^n - \mathbf{F}_{J_f}^n) -$$
$$(4\bar{\mathbf{G}}_I^n + 2\bar{\mathbf{G}}_{J_f}^n + \mathbf{G}_I^n - \mathbf{G}_{J_f}^n)\}\rangle_I + [\mathbf{M}_L]_I \, \mathbf{S}_I \qquad (33)$$

where edge s joins nodes I and I_s and J_1 and J_2 are the boundary nodes which are connected to boundary node I. The weights $C_{II_s}^j$ and D_f are

defined as

$$C_{II_s}^j = -\sum_{E \in II_s} \frac{2\Omega_E}{3} \left[\frac{\partial N_I}{\partial x_j} \right]_E + \langle \frac{\Gamma_f}{6} n_{jf} \rangle_{II_s}$$

$$D_f = -\Gamma_f / 12$$
$$\qquad (34)$$

where n_{jf} is the component in the x_j direction of the unit normal to the boundary edge, of unit length Γ_f, which joins the nodes I and J_f. Note that the bracketed terms are only non–zero if II_s is a boundary edge.

Roe Flux–Difference Splitting

The basic inviscid flux contribution on the right hand side of equation (33) can be expressed in the form

$$-\sum_{s=1}^{m_I} \frac{C_{II_s}^j}{2}(\mathbf{F}_I^j + \mathbf{F}_{I_s}^j) = -\sum_{s=1}^{m_I} \mathbf{F}_{II_s} \qquad (35)$$

This is a central difference type approximation to the inviscid flux term. To obtain a practical numerical scheme, we replace the actual edge flux $\mathbf{F}_{II_s}$ by a consistent numerical flux $\mathcal{F}_{II_s}$ based upon Roe's first order upwind flux–difference splitting [16], reformulated for turbulence modelling by Morrison [13], as

$$\mathcal{F}_{II_s} = \frac{1}{2}\{(\mathbf{F}_I^j + \mathbf{F}_{I_s}^j)C_{II_s}^j - |\mathbf{A}_{II_s}|(\mathbf{U}_{I_s} - \mathbf{U}_I)\} \qquad (36)$$

Here the Jacobian matrix $\mathbf{A}_{II_s}$, defined by

$$\mathbf{A}_{II_s} = \mathbf{A}(\mathbf{U}_I, \mathbf{U}_{I_s}) = C_{II_s}^j \left(\frac{d\mathbf{F}^j}{d\mathbf{U}} \right)_{II_s} \qquad (37)$$

is evaluated using Roe's averaging procedure [16, 13]. To construct a higher–order algorithm we employ a MUSCL scheme [8] in which equation (36) is replaced by the expression

$$\mathcal{F}_{II_s} = \frac{1}{2}\{(\mathbf{F}_l^j + \mathbf{F}_r^j)C_{II_s}^j - |\mathbf{A}(\mathbf{U}_l, \mathbf{U}_r)|(\mathbf{U}_r - \mathbf{U}_l)\} \qquad (38)$$

where $\mathbf{U}_l$ and $\mathbf{U}_r$ are interface values of the state variables for edge II_s. Figure 1 illustrates the formation of a one–dimensional stencil to allow for the reconstruction of the interface values for an edge. Dummy nodes, I_L and I_R, are located by extending the edge II_s, defined by nodes I and I_s, from the left and right equidistantly. These two points lie inside elements E_L (with nodes L_1, L_2 and L_3) and E_R (with nodes R_1, R_2 and R_3) and once their location is specified, the value of any desired variable can be computed from interpolation between the element nodal values. Special care must be taken near boundaries, as in this case, a dummy node may fall outside the domain. In this case, a linear extrapolation of the variables is employed. The interface values are computed from the nodal values of this one–dimensional stencil using reconstruction plus limiting. The limiter introduced by Venkatakrishnan [20] has been employed.

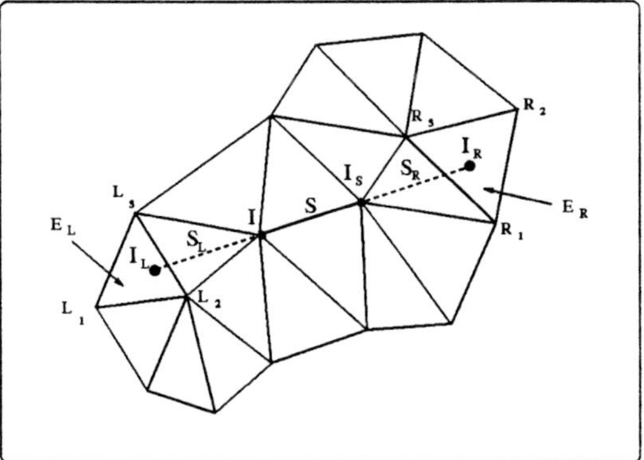

Figure 1: Location of Dummy nodes and corresponding elements.

Time integration

The forward difference approximation is employed to construct the time marching method

$$\mathbf{U}_I^{m+1} = \mathbf{U}_I^m + \Delta t\,[M_L]_I^{-1}\,\mathbf{R}_I \qquad (39)$$

where $\mathbf{U}_I^m$ denotes the solution at node I at time $t = t^m$, $\Delta t = t^{m+1} - t^m$, and $\mathbf{R}_I$ is the right hand side of equation (33). A local time stepping approach is used to accelerate the convergence rate towards the steady–state.

6. NUMERICAL RESULTS

In this section, three test cases involving different flow regimes are studied to examine the numerical performance of the procedure. The solutions are started from free–stream conditions and no special treatment has been found to be required to obtain a stable converged solution.

Supersonic flow over a flat plate

This is a standard test case which is used to examine the performance of any turbulence model. The problem consists of a supersonic flow ($M = 2$) at a Reynolds number of $1 \times 10^7 (1/m)$. The wall is isothermal with $T_w = 222^o K$ and is one metre long. The free–stream temperature is $222^o K$, $Pr_t = 0.9$ and $Pr = 0.72$. Numerical results for this problem are presented in references [13, 19].

The mesh employed is a triangulation of structured grid, with 101×101 nodes and $20,000$ elements. The first layer of nodes is located at a distance 3×10^{-6} metre above the wall ($y^+ < 0.5$). The aspect ratio of the elements close to wall varies from 1,600 at the leading edge to 5,200 at the trailing edge. Figure 2 compares the velocity profiles at the section $x_1 = 0.93$ metre, for both the $k - \epsilon$ and the $k - \omega$ models, with Prandtl's (1/7)–th power law. It is observed that both schemes give good agreement with the power law. More details about the velocity profiles can be obtained from the (u^+, y^+) plots of figure 3, which show an enlargement of the inner region. The results obtained by the $k - \omega$ model show very good agreement with the standard inner region profile. It should be noted that the presented profiles are Van Driest's effective velocity profiles, obtained after applying the Van Driest transformation [18]. In figure 4, the wall skin friction distributions computed by the $k - \omega$ model and the $k - \epsilon$ implementations are compared with the White–Christoph skin friction distribution [21].

Subsonic flow over a NACA0012 aerofoil

This problem involves a subsonic turbulent flow ($M = 0.502$) over an adiabatic NACA0012 aerofoil. The Reynolds number, based upon the aerofoil chord, is $Re = 2.91 \times 10^6$ and the angle of attack is 1.77^o. The Prandtl numbers are $Pr = 0.72$ and $Pr_t = 0.9$ and the free–stream temperature is $528^o R$.

The problem is solved using the $k - \omega$ model on a fully unstructured grid consisting of 27,939 elements and 14,097 nodes. The first layer of nodes is 10^{-5} units above the wall. In the Roe approximate Riemann solver, an entropy correction parameter of 0.1 was employed in this case. In figure 5, the computed wall pressure coefficient is compared with the experimental results obtained by Thibert *et al* [17]. The computed results are in good agreement with experiment. The pressure coefficient and Mach contour plots are shown in figure 5.

Hypersonic flow over a compression corner

Flow over compression corner is simulated to demonstrate the capability of the method to predict wall pressure, skin friction and heat transfer coefficients at high Mach numbers.

Compression corner $\theta = 15^0$

In the first example, the ramp angle is 15 degrees and the flow conditions are $M = 9.22$, $Re = 4.7 \times 10^7$/metre, $T_\infty = 64.5$ K and $T_w = 295$ K. The flat plate length is 56 cm. For this problem, experimental results have been presented by Coleman and Stollery [2]. A triangulation of a structured 200×30 grid was used with the first layer of nodes located at a distance 10^{-5} metre above the wall. The computed normalised pressure p/p_∞, skin friction $C_f = 2\tau_w/\rho_\infty u_\infty^2$ and heat transfer q W/cm^2 from the $k-\omega$ and the $k-\epsilon$ models are shown in figures 6 to 8. The experimental results for both pressure and heat transfer rate are also plotted for comparison.

It is seen from the figures that the levels of the pressure and the heat transfer are computed well with both of the turbulence models. However, in the reattachment zone, while the $k-\epsilon$ model shows close agreement with experiment for the heat transfer rate, the $k-\omega$ model shows a variation which is too rapid. This problem may be due either to the model itself or to the requirement for a more refined mesh to capture the proper physics of the problem.

Compression corner $\theta = 34^0$

The ramp angle was then increased to 34^0. The resulting flow shows extremely high pressure gradients and heat transfer rates and the flow separates on the flat plate. The computed results for pressure coefficient, skin friction and heat transfer rate are shown in figures 9 to 11 for the $k-\omega$ and the $k-\epsilon$ models. Experimental results [2] are also shown for comparison where available. It is seen from the figures that the levels of the pressure and the heat transfer are not computed correctly. Again, both models give similar results for the pressure distribution. For the heat transfer rates, the $k-\epsilon$ model is closer to the experiment than the $k-\omega$ model. Both models predict the position of the peak heat transfer rate slightly downstream of the correct point. Again, the difference between the computed and the experimental results can be attributed to the turbulence models and to the lack of mesh resolution close to the corner.

7. CONCLUSIONS

An unstructured grid flow solver, employing two–equation turbulence models has been implemented. It has been shown that the algorithm provides a capability for the solution of turbulent flow problems on general triangular grids. The $k-\omega$ model seems to be better suited for unstructured grid implementation because of the simplicity of its transport equations. It has been shown that the method can handle high pressure gradients and shock–wave boundary–layer interaction and can predict practical quantities such as pressure coefficient, skin friction and heat transfer with reasonable accuracy. The extension of the method to include modified turbulence models and to the modelling of three–dimensional applications is an area of current research.

ACKNOWLEDGEMENTS

The first author would like to acknowledge the support received from the Iranian Ministry of Culture and Higher Education.

References

[1] T.J. Barth. Numerical aspects of computing viscous high Reynolds number flows on unstructured meshes. Technical Report 91-0721, AIAA, 1991.

[2] G.T. Coleman and J.L. Stollery. Heat transfer from hypersonic turbulent flow at a wedge compression corner. *J. Fluid Mech.*, 56:741-752, 1972. part 4.

[3] A. Favre. Equations des gas turbulents compressibles. *J. Mecanique*, 4(3):361-390, 1965.

[4] F. Grasso and D. Falconi. High speed turbulence modeling of shock–wave/boundary-layer interaction. *AIAA J.*, 31(7):1199-1206, 1993.

[5] O. Hassan, K. Morgan, E.J. Probert, and J. Peraire. Unstructured tetrahedral mesh generation for three–dimensional viscous flows. *International Journal for Numerical Methods in Engineering*, 39:549-567, 1996.

[6] W.P. Jones and B.E. Launder. The prediction of laminarization with a two–equation model of turbulence. *Int. J. Heat Mass Transfer*, 15(2):301-314, 1972.

[7] P.R.M. Lyra, M.T. Manzari, K. Morgan, O. Hassan, and J. Peraire. Upwind side-based unstructured grid algorithms for compressible viscous flow computations. *International Journal for Engineering Analysis and Design*, 2:197-211, 1995.

[8] M.T. Manzari, P.R.M. Lyra, K. Morgan, and J. Peraire. An unstructured grid FEM/MUSCL algorithm for the compressible Euler equations. In *Proc. of VIII International Conference on Finite Elements in Fluids: New Trends and Applications*, pages 379-388, Barcelona, Spain, 1993.

[9] J.G. Marvin and T.J. Coakley. Turbulence modeling for hypersonic flows. Technical Report TM-101079, NASA, 1989.

[10] D.J. Mavriplis. Three dimensional unstructured multigrid for the Euler equations. Technical Report 91-1549, AIAA, 1991.

[11] F.R. Menter. Influence of free–stream values on $k-\omega$ turbulence model predictions. *AIAA J.*, 30(6):1657–1659, 1992.

[12] F.R. Menter. Two–equation eddy–viscosity turbulence models for engineering applications. *AIAA J.*, 32(8):1598–1605, 1994.

[13] J. Morrison. Flux–difference split scheme for turbulent transport equations. Technical Report 90–5251, AIAA, 1990.

[14] J. Peraire, J. Peiró, and K. Morgan. A 3–D finite element multigrid solver for the Euler equations. Paper 92–0449, AIAA, 1992.

[15] S. Pirzadeh. Progress toward a user–oriented unstructured viscous grid generator. Technical Report 96–0031, AIAA, Nevada, Reno, 1996.

[16] P.L. Roe. Approximate Riemann solvers, parameter vectors and difference schemes. *J. Comp. Phys.*, 43:357–372, 1981.

[17] J.J. Thibert, M. Granjaques, and L. Ohman. Naca0012 airfoil pressure distribution measurements. AGARD Advisory Report AR–138, AGARD, 1979.

[18] E.R. Van Driest. Turbulent boundary layer in compressible fluids. *J. Aero. Sciences*, 18(3):145–160,216, 1951.

[19] G.R. Vemaganti and R.K. Prabhu. Application of a two–equation turbulence model for high speed compressible flows using unstructured grids. Technical Report 93–3029, AIAA, 1993.

[20] V. Venkatakrishnan. Preconditioned conjugate gradient methods for the compressible Navier–Stokes equations. *AIAA J.*, 29(7):1092–1100, 1991.

[21] F.M. White. *Viscous Fluid Flow*. McGraw–Hill, 1974.

[22] D.C. Wilcox. Reassessment of the scale determining equation for advanced turbulence models. *AIAA J.*, 26(11):1299–1310, 1988.

[23] D.C. Wilcox. Progress in hypersonic turbulence model. Technical Report 91–1785, AIAA, 1991.

[24] O.C. Zienkiewicz and K. Morgan. *Finite Elements and Approximation*. John Wiley & Sons, 1983.

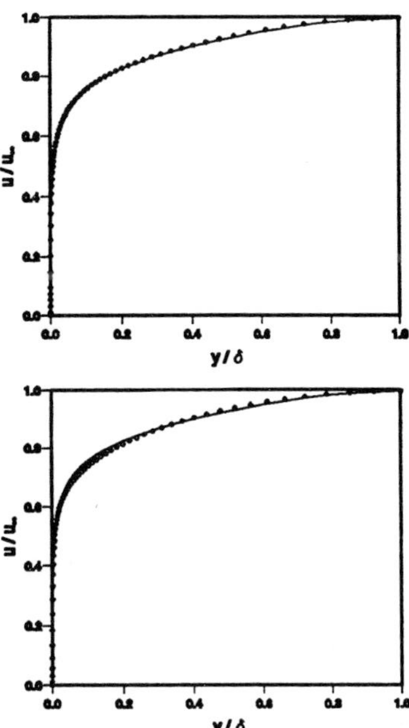

Figure 2: Supersonic flow over flat plate; velocity profile for the $k-\omega$ (top) and the $k-\epsilon$ (bottom) models. (solid line: power law, circle: computed)

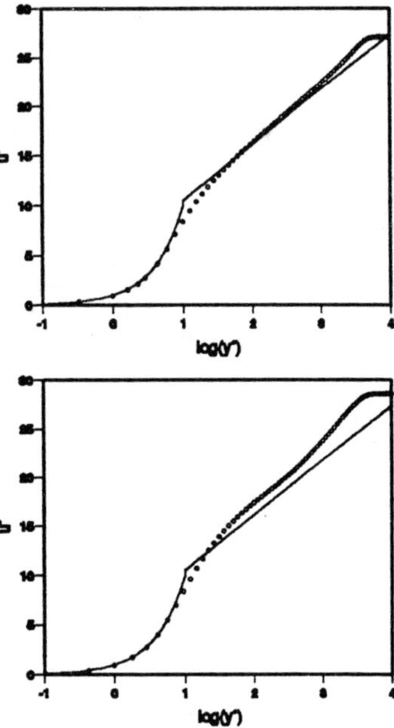

Figure 3: Supersonic flow over flat plate; inner region velocity profile for the $k-\omega$ (top) and the $k-\epsilon$ (bottom) models. (solid line: Standard inner region profile, circle: computed)

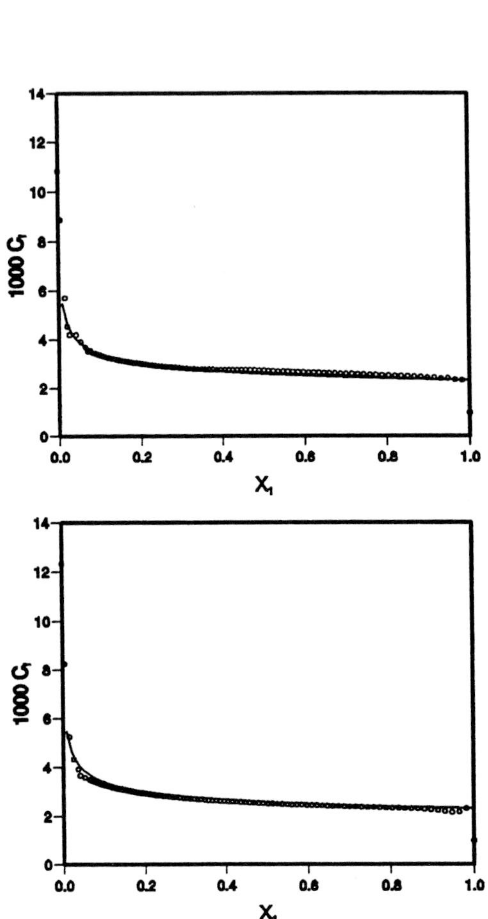

Figure 4: Supersonic flow over flat plate; wall skin friction for the $k - \omega$ (top) and the $k - \epsilon$ (bottom) models. (solid line: White–Christoph, circle: computed)

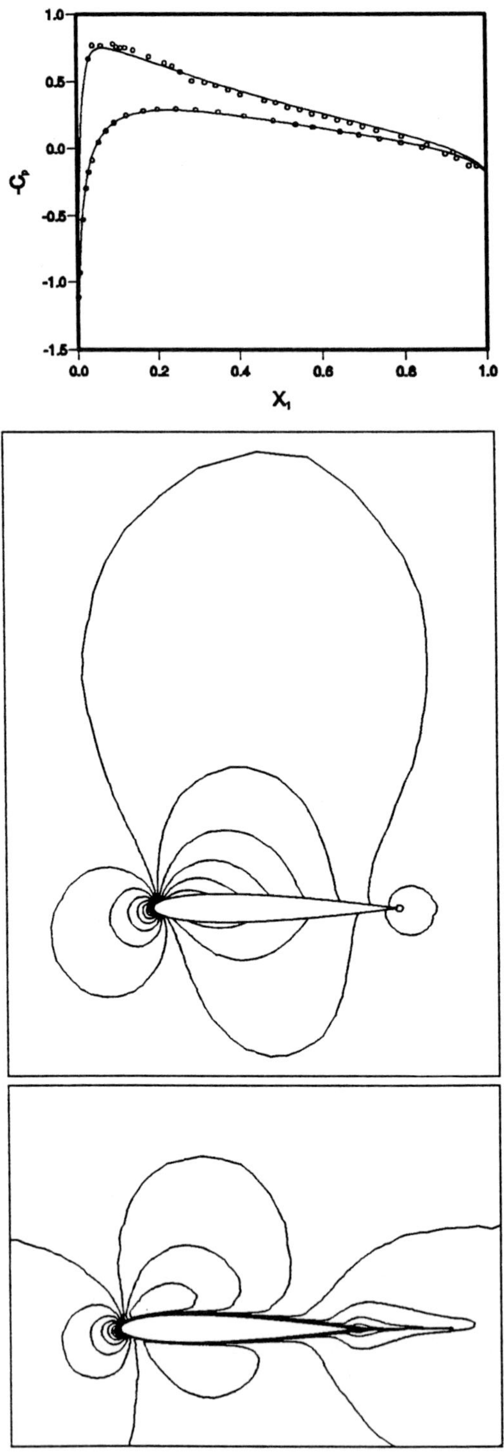

Figure 5: Subsonic flow over NACA0012 aerofoil; wall pressure coefficient (circle: experiment, solid line: computed) (top), pressure coefficient (middle), Mach number (bottom) contour plots.

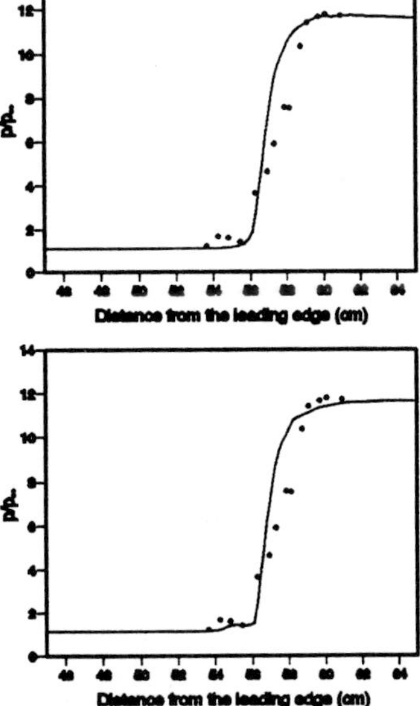

Figure 6: Hypersonic flow over compression corner ($\theta = 15^o$); wall pressure coefficient distribution obtained by the $k - \omega$ (top) and the $k - \epsilon$ (bottom) models. (solid line: computed, circle: experiment)

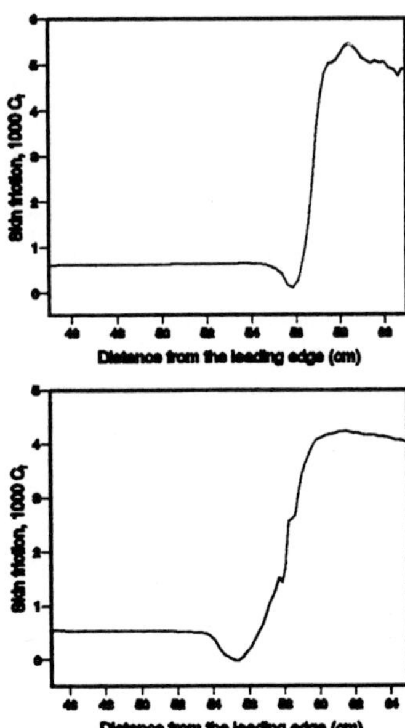

Figure 7: Hypersonic flow over compression corner ($\theta = 15^o$); wall skin friction coefficient distribution obtained by the $k - \omega$ (top) and the $k - \epsilon$ (bottom) models.

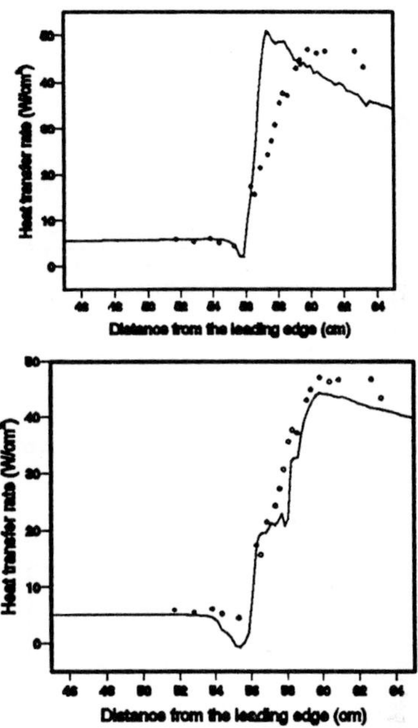

Figure 8: Hypersonic flow over compression corner ($\theta = 15^o$); wall heat transfer coefficient distribution obtained by the $k - \omega$ (top) and the $k - \epsilon$ (bottom) models. (solid line: computed, circle: experiment)

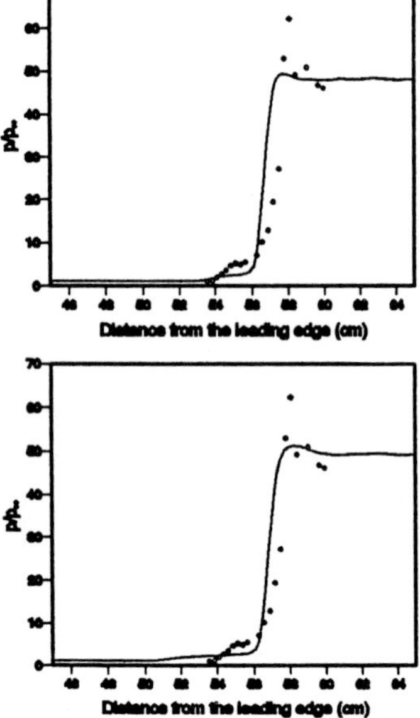

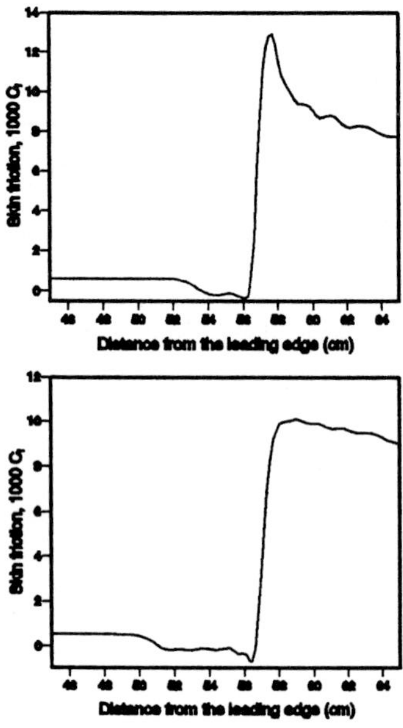

Figure 9: Hypersonic flow over compression corner ($\theta = 34^o$); wall pressure coefficient distribution obtained by the $k - \omega$ (top) and the $k - \epsilon$ (bottom) models. (solid line: computed, circle: experiment)

Figure 10: Hypersonic flow over compression corner ($\theta = 34^o$); wall skin friction coefficient distribution obtained by the $k - \omega$ (top) and the $k - \epsilon$ (bottom) models.

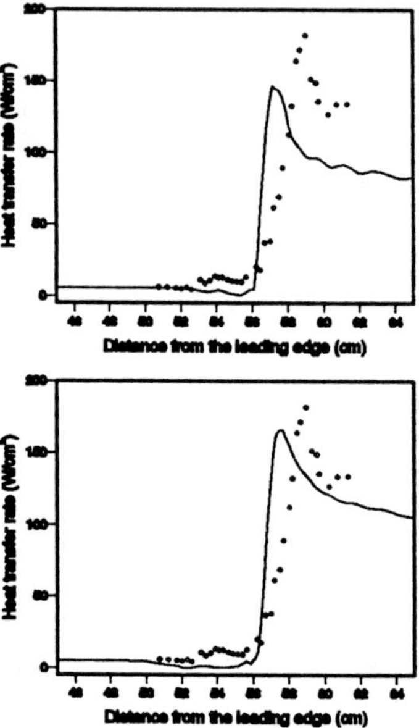

Figure 11: Hypersonic flow over compression corner ($\theta = 34^o$); wall heat transfer coefficient distribution obtained by the $k - \omega$ (top) and the $k - \epsilon$ (bottom) models. (solid line: computed, circle: experiment)

NEW HYBRID-STREAMLINE-UPWIND FINITE-ELEMENT METHOD IN A DUAL SPACE
(Natural Convection at High Rayleigh Numbers)

Takahiko TANAHASHI   **Takafumi MAKIHARA**

Faculty of Science and Technology, Keio University; 3-14-1 Hiyoshi, Kohoku-ku, Yokohama 223, Japan

Abstract

In the present paper, unsteady natural convection flows of air in a square cavity with constant heated and cooled walls are investigated numerically using the hybrid-GAMAC (generalized and simplified marker and cell) finite-element method. A direct-numerical simulation of two- dimentional turbulence is shown in the present paper without introducing any random forces. The present result at $Ra = 1.0 \times 10^8$ is compared with the other numerical results. The present result of calculation is in good agreement with that of the others. This study allows us to say that the hybrid-GSMAC method is very stable at high Rayleigh numbers. In addition, selected frames from an animation generated from the computational results at $Ra = 1.0 \times 10^9$ show very clearly the formation of large-scale eddies via the following sequence: initial instability, proceeding through transition, and eventually the statistical steady state. Using this method, it is possible to investigate natural convection and heat transfer phenomena at high Rayleigh number up to 10^{10}.

Key Words   Natural Convection, High Rayleigh Number, Hybrid-GSMAC-FEM, Square Cavity, Dual Space

1. Introduction

The recent development of computer and Computational Fluid Dynamics has become possible to analize numerically the more complex phenomena of fluid flows including the thermal effect. Analysis of natural convection in a square cavity is accepted as a bench mark problem to examine validation of calculating scheme on thermal fluid problems, because of simplicity of shape and boundary conditions. So many authors have analized this problem for verification of scheme. The subject of almost all these analysis is flows of Rayleigh number $10^6 \sim 10^8$. And all analytical results are agreed with each other. However, if Rayleigh number increases higher, flow becomes more unstable and hard to analize [4]. And also transition process from laminer to turbulent flow has not been completely understood yet. As in the case of forced convection at high Reynolds number flows, strong nonlinearity of convection term in high Rayleigh number flows [5] causes numerical diffusivity and pressure oscillation, which makes analysis more difficult. It is also difficult to do highly accurate experiment in these high Rayleigh number flows. So it is necssary to compare accurate numerical analysis

with good experiment. When we apply an analitical method of imcompressible fluid to high Rayleigh number natural convection, in many cases accuracy and stability of numerical solution depends highly on the energy equation rather than the momenum equation. Therefore it is much necessary to develop stable and high accurate numerical scheme at high Rayleigh numbers. In the previous report [6 ~ 8], we have studied the effectiveness of hybrid-streamline-upwind finite-element-method in a dual space on the two dimentional advection-diffusion equation. As the result, we could stably analize two dimentional high Peclet number flows with high accuracy using both streamline-upwind weight function and high accurate interpolation function.

In this present paper, we construct a new numerical analitical method suitable for high Rayleigh number natural convection problem as a more practical problem using hybrid-streamline upwind FEM in a dual space. Then we study about the accuracy of calculation by comparing the present result with others. And this method makes it possible to investigate natural convection at high Rayleigh number up to 10^{10}.

Nomenclature

Gr : Grashof number$(= g\beta\Delta T L_r^3/\nu^2)$

g : acceleration due to gravity

H : Bernoulli function

Received on February 20, 1997.

j_e : element mean value of Jaccobian
$K_{\alpha\beta}$: advection diffusion matrix
L_r : representative length
L_α : interpolation function
(upwind shape function)
L_α^* : weight function (adjoint
upwind shape function)
$M_{\alpha\beta}$: mass matrix
Nu : Nusselt number
Pr : Prandtl number($= \nu/\alpha$)
Ra : Rayleigh number($= Gr \cdot Pr$)
T : dimensionless temperature
$\boldsymbol{u}$: dimensionless velocity vector
α : thermal diffusivity
β : volumetric coefficient of expansion
ΔL_{min} : minimum mesh width
Δt : time increment
ε : convergence criterion
λ : half value of cell Peclet number
in x direction
μ : half value of cell Peclet number
in y direction
ν : kinematic viscosity
ϕ : modified velocity potential
$\boldsymbol{\omega}$: dimensionless vorticity vector

Subscripts

α, β : local node number in element
$\langle\ \ \rangle_e$: element mean value

2. Basic Equations

In this section, we will mention about the algorithm of time marching and hybrid-streamline-upwind FEM in a dual space.

2.1 Algorithm of Time Marching

Fundamental equations are continuity equation, Navier-Stokes equation in rotational form and energy equation with Boussinesq approximation. Dimensionless basic equations are as follows:

$$\nabla \cdot \boldsymbol{v} = 0 \tag{1}$$

$$\frac{\partial \boldsymbol{u}}{\partial t} = -\nabla H + \boldsymbol{v} \times \boldsymbol{\omega} + \frac{1}{Gr}(\nabla^2 \boldsymbol{u} + T\boldsymbol{e}) \tag{2}$$

$$\frac{\partial T}{\partial t} + (\boldsymbol{u} \cdot \nabla)T = \frac{1}{Ra}\nabla^2 T \tag{3}$$

where $\boldsymbol{e}$ is unit vector in the direction of gravity and $e_1 = 0, e_2 = -1$. And representative velocity is decided on condition that buoyancy force and viscous force balance each other.

In the present paper, the algorithm of GS-MAC method is used as time marching method. Equation(2) is discretizated explicitly for velocity and implicitly for Bernoulli function and simultaneous relaxation is applied considering imcompressibility of fluid. Discretization of the basic equations by GS-MAC method is as follows:

Predicter step ($\boldsymbol{u}^n$, H^n and T^n are known)

$$\boldsymbol{\omega}^n = \nabla \times \boldsymbol{u}^n \tag{4}$$

$$\frac{\tilde{\boldsymbol{u}} - \boldsymbol{u}^n}{\Delta t} = -\nabla H^n + \boldsymbol{u}^n \times \boldsymbol{\omega}^n + \frac{1}{Gr}\nabla^2(\boldsymbol{u}^n + T^n\boldsymbol{e}) \tag{5}$$

Repeat simultaneous relaxation ($\tilde{\boldsymbol{u}}^{(0)}$, $H^{(0)} = H^n$ are initial value)

$$\nabla^2 \phi^{(m)} = \nabla \cdot \tilde{\boldsymbol{u}}^{(m)} \tag{6}$$

$$H^{(m+1)} = H^{(m)} + \phi^{(m)}/\Delta t \tag{7}$$

$$\tilde{\boldsymbol{u}}^{(m+1)} = \tilde{\boldsymbol{u}}^{(m)} - \nabla\phi^{(m)} \tag{8}$$

Energy equation

$$\frac{T^{n+1} - T^n}{\Delta t} = -(\boldsymbol{u}^{n+1} \cdot \nabla)T^n + \frac{1}{Ra}\nabla^2 T^n \tag{9}$$

where subscripts n is time step number and m is repeat level. Time marching algorithm of the present scheme is as follows:

step 1) Velocity predicter $\tilde{\boldsymbol{v}}$ is obtained explicitly from equations(4) and (5)
step 2) Divergence velocity $\nabla \cdot \tilde{\boldsymbol{v}}$ is calculated for the velocity predicter.
step 3) Modified velocity potential ϕ is got from equation(6).
step 4) Bernoulli function is recorrected in equations(7) and (8).
step 5) After repeating calculation from step 2 to step 4 until Bernoulli function and velocity satisfy the imcompressible condition, we get finally $\boldsymbol{u}^{n+1} = \tilde{\boldsymbol{u}}^{m+1}, H^{n+1} = H^{m+1}$.
step 6) T^{n+1} is obtained explicitly from equation(9).

2.2 Descritization by Hybrid-Stleamline-Upwind FEM

Equations(4)$\sim$(8) are discretized by the Galerkin method. But by use of discrete del operater, discretization of the governing equations becomes easy and amount of the calculation is decreased. Equation(9) is discretized by hybrid-streamline-upwind FEM. In this section, we inspect discretization of only equation(9).

Energy equation(9) is multiplied with weight function and integrated in each element. Mean value of velocity vector is defined in element as follows:

$$u = \langle u \rangle_e \quad v = \langle v \rangle_e \tag{10}$$

T is interpolated using the upwind shape function L_α as follows:

$$T = L_\alpha T_\alpha \tag{11}$$

where subscript α follows summation convention. Further, weight function is described using the adjoint upwind shape function L_α^* as follows:

$$\delta T^* = L_\alpha^* \delta T_\alpha \tag{12}$$

where L_α is an upwind shape function which is obtained from locally linearized steady advection-diffusion equation. And L_α^* is an adoint upwind shape function which is got from the dual space and described as follows:

$$L_\alpha(\lambda, \mu) = L_\alpha^*(-\lambda, -\mu) \tag{13}$$

So the discretization of energy equation becomes as follows:

$$\overline{M_{\alpha\beta}} \frac{T_\beta^{n+1} - T_\beta^n}{\Delta t} + K_{\alpha\beta}^x T_\beta^n + K_{\alpha\beta}^y T_\beta^n = 0 \tag{14}$$

$$\overline{M_{\alpha\beta}} = \int L_\alpha^* L_\beta d\Omega \approx J_e \delta_{\alpha\beta} = \frac{S_e}{4} \delta_{\alpha\beta} \tag{15}$$

$$K_{\alpha\beta}^x = \int \left(\langle u \rangle_e L_\alpha^* + \frac{1}{Ra} \frac{\partial L_\alpha^*}{\partial x} \right) \frac{\partial L_\beta}{\partial x} d\Omega \tag{16}$$

$$K_{\alpha\beta}^y = \int \left(\langle v \rangle_e L_\alpha^* + \frac{1}{Ra} \frac{\partial L_\alpha^*}{\partial y} \right) \frac{\partial L_\beta}{\partial y} d\Omega \tag{17}$$

where $\delta_{\alpha\beta}$ is Kronecker's delta. In fact, when we transform these equations from physical space to calculation space, mass matrix is lumped twice row and column. And advection-diffusion matrix $K_{\alpha\beta}^x$ is upwinded in x direction and $K_{\alpha\beta}^y$ upwinded in y direction. In other word, we have the following formula as upwinded only in x direction:

$$L_\alpha = \frac{\xi_\alpha \left(e^{\lambda\xi} - e^{-\lambda\xi_\alpha} \right)}{e^\lambda - e^{-\lambda}} \cdot \frac{1}{2} (1 + \eta_\alpha \eta) \tag{18}$$

as upwinded only in y direction:

$$L_\alpha = \frac{1}{2} (1 + \xi_\alpha \xi) \cdot \frac{\eta_\alpha \left(e^{\mu\eta} - e^{-\mu\eta_\alpha} \right)}{e^\mu - e^{-\mu}} \tag{19}$$

If mesh is orthogonal, coordinate axes of physical space and calculation space are coincident each other, so the advection-diffusion matrix is approximated as follows:

$$K_{\alpha\beta}^x = \frac{\Delta y u_e \xi_\alpha \xi_\beta e^{\lambda\xi_\alpha}}{8 \sinh \lambda} \tag{20}$$

$$K_{\alpha\beta}^y = \frac{\Delta x v_e \eta_\alpha \eta_\beta e^{\mu\eta_\alpha}}{8 \sinh \mu} \tag{21}$$

But if cell Peclet nember is equal 0 or positive infinity or negative infinity, it is not possible to get solution. Moreover extraordinary digit is not desirable because it increases calculation time. So in calculation, decresing calculation effectiveness is defended by using approximate equation of limite value to part of exponential function.

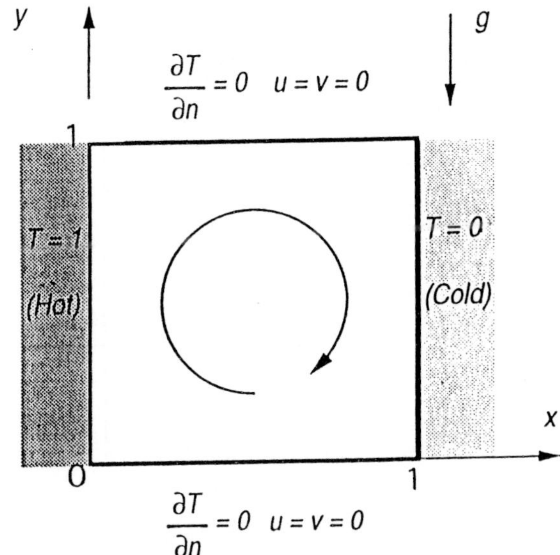

Fig.1: Flow geometry and coordinate system

3. Analytical Model and Calculation Condition

Analytical model and the boundary conditions are shown in Fig.1. And as initial condition, velocity is given as $u = v = 0$ and pressure H is also $H = 0$ in all domain. Temparature is $T = 0$ except hot wall ($T = 1$). Working fluid is air and $Pr = 0.71$. When the difference of mean Nusselt number between hot and cold walls becomes under 10^{-2}, it is accepted as a steady state condition. Mean Nusselt number is defined by the following equation:

$$\overline{Nu_{av}} = \int_0 \left[\frac{\partial T}{\partial x} \right]_{x=0} dx \tag{22}$$

Calculation conditions are shown in Tab.1, and calculation meshes in Fig.2. In proportion to the increase of Rayleigh number, it is necessary to use more fine meshs. But calculation mesh which uses at $Ra = 10^6$ is regular mesh of 20×20, and it is not shown in here. In the case of square cavity, if the rate of irregular decomposition is a, function $f(x)$ which decides irregular decomposition is as follows:

$$f(x) = \begin{cases} \dfrac{1}{2} \cdot \dfrac{e^{ax} - 1}{e^a - 1} & 0 \le x \le 0.5 \\[3ex] 1 - \dfrac{1}{2} \cdot \dfrac{e^{a(1-x)} - 1}{e^a - 1} & 0.5 \le x \le 1 \end{cases} \tag{23}$$

In the present paper, the rate of irregular decomposition is $a = 2$.

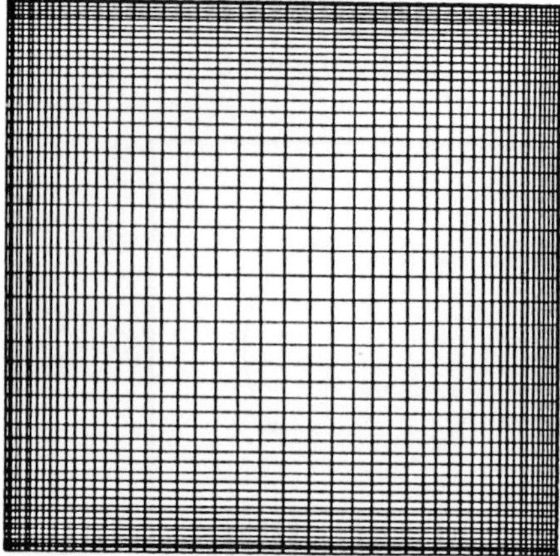

(a) 50×50 irregular

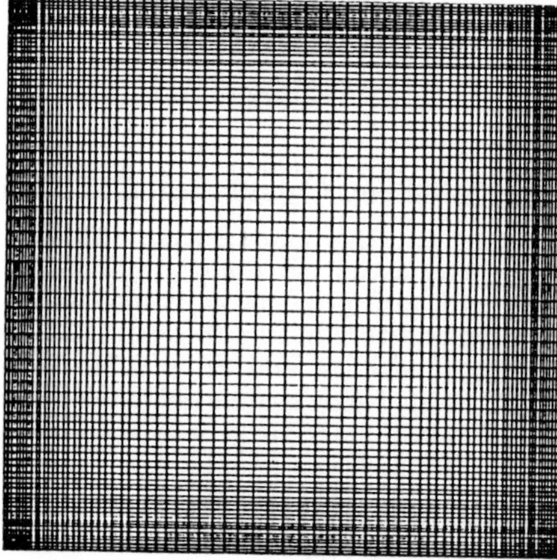

(b) 80×80 irregular

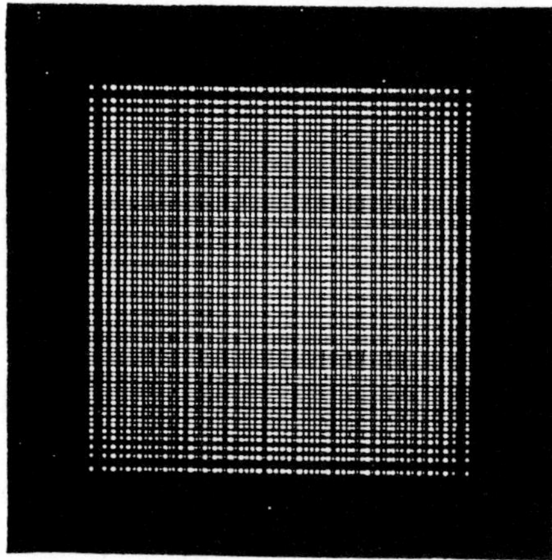

(c) 200×200 irregular

Fig.2: Finite element meshes

4. Results and Discussions

4.1 Result of $Ra = 10^6$ Flow

Firstly, to confirm effectiveness of the present scheme we compared results of this scheme with analytical result of the conventional GSMAC method at $Ra = 10^6$. Data size of coefficient matrix at that time and memory size used during calculation are written in Tab.2. Data size of coefficient matrix decreases 75.6%, and memory size at calculation decreases 12.0% with respect to the conventional scheme. Therefore, large coefficient matrix in the energy equation is reduced to small one, then it becomes possible to apply FEM to practical problems of imcompressible flow.

Isothermal lines and streamlines at $Ra = 10^6$ are shown in Fig.3. The results of these two schemes do not change qualitatively and are agree each other. Both of them give stable solutions without numerical oscillation and diffusion.

4.2 Result of $Ra = 10^8$ Flow

Next, we analize the natural convection of $Ra = 10^8$ in which many analytical results exist comparatively . It is possible for hybrid-streamline-upwind FEM to get stable solutions without numerical oscillation even if using more rough mesh of 50×50 elements. On the other hand, it is impossible for the conventional GS-MAC to analize with same mesh structure. Even if using 1/10 time step, it is impossible to get solutions. As a result, this hybrid-streamline-upwind EEM using a local exact solution is comfirmed to be stable and highly accurate. At steady state, streamlines and temparature distribution (Fig.4) show that the secandary vortex occures as it is explained in references[2]~[4]. Comparison with others about maximum value of velocity on the center axis and maximum and minimum values of local Nusselt number is shown in Tab.3. And when mean nusselt number at steady state is compared with other results, in agreement of results the effectiveness of the present scheme is confirmed.

4.3 Results of $Ra = 10^9$ and $Ra = 10^{10}$ Flow

Next, we will analize $Ra = 10^9$ and $Ra = 10^{10}$ flows which are turbulent in generally. Because the present paper is two dimensional calculation, numerical result is considered to be an approximation of mean value of space at near the center about three dimentional phenomena. By refernce[4] the boundary layer thickness of temperature in the square cavity is approximated as follows:

$$\delta \approx \left(\frac{2}{PrRa}\right)^{\frac{1}{4}} \tag{24}$$

so when the domain is divided, the minimum mesh width is decided as in Tab.1.

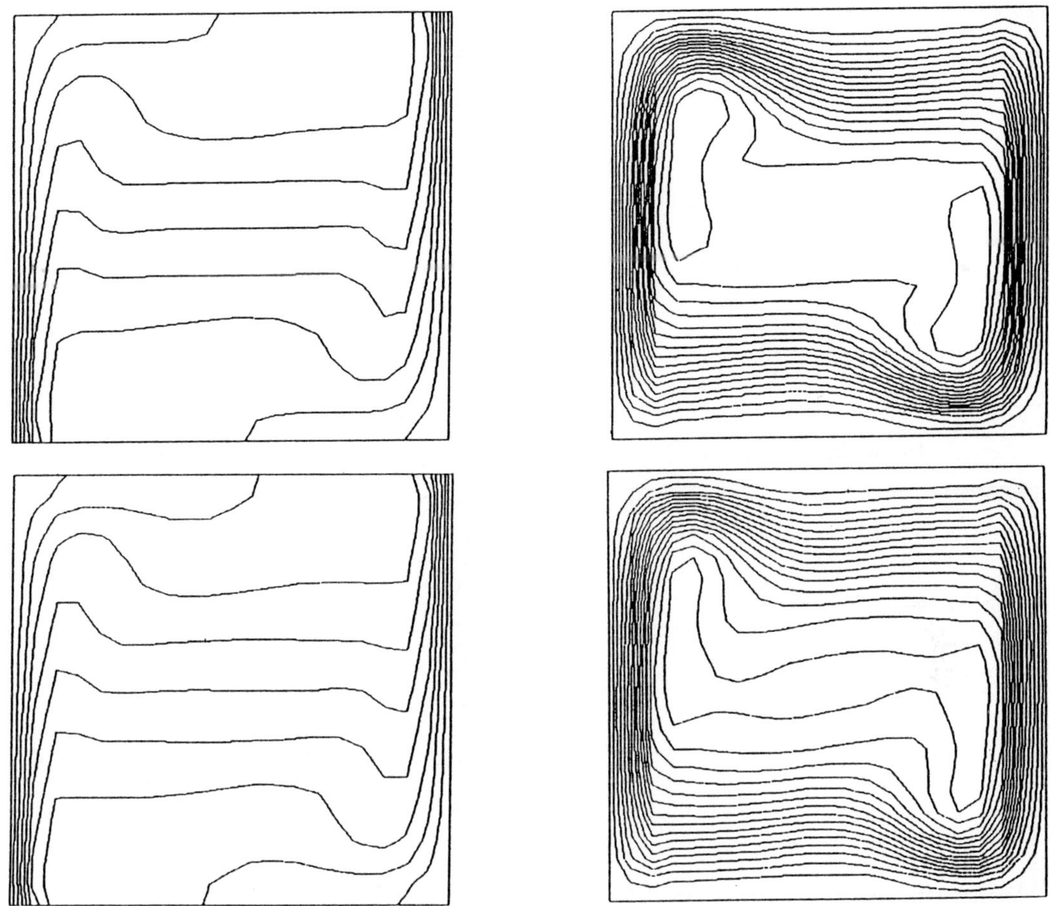

$(Ra = 10^6,\ \Delta T = 0.1, \psi = 1.0 \times 10^{-6}$ (upper panels: Hybrid, lower panels: GSMAC)

Fig.3: Comparison of isotherms and stream lines in a steady state

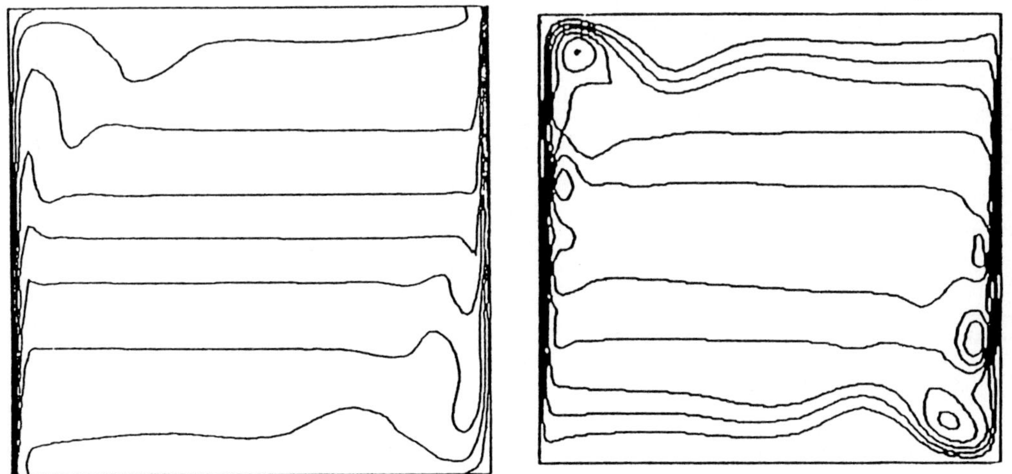

Fig.4: Temperature distributions and streamlines at $Ra = 10^8$ ($\Delta\psi = 1.0 \times 10^{-7}$)

Fig.5: Temperature distributions in transient states $(Ra = 10^9, \; Pr = 0.71)$

Fig.6: Temperature distributions in transient states $(Ra = 10^{10}, \; Pr = 0.71)$

$$t = 1.0 \times 10^5$$

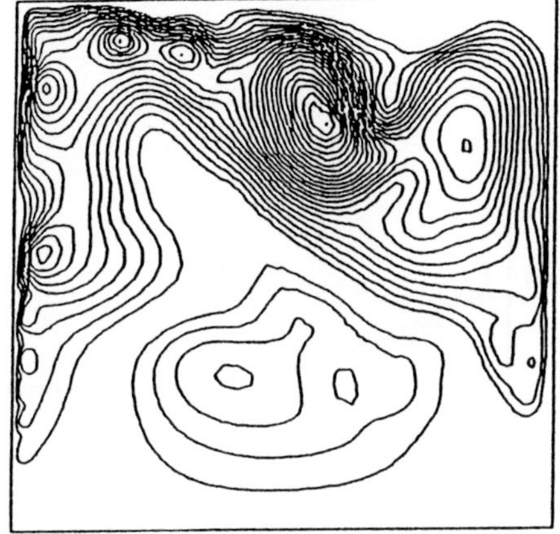

$$t = 20.0 \times 10^5$$

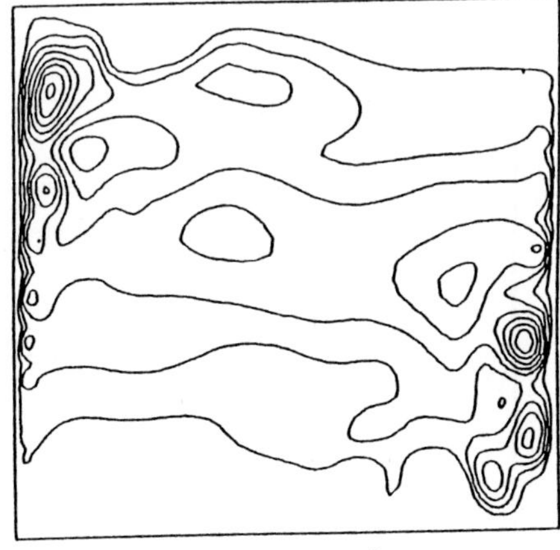

$$t = 150.0 \times 10^5$$

Fig.7: Streamlines in transient states
($Ra = 10^9$, $Pr = 0.71$, $\Delta\psi = 3.0\times10^{-8}$)

$$t = 5.0 \times 10^5$$

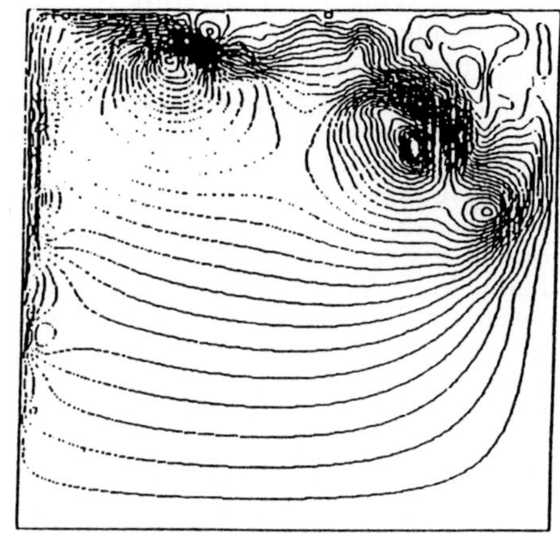

$$t = 16.0 \times 10^5$$

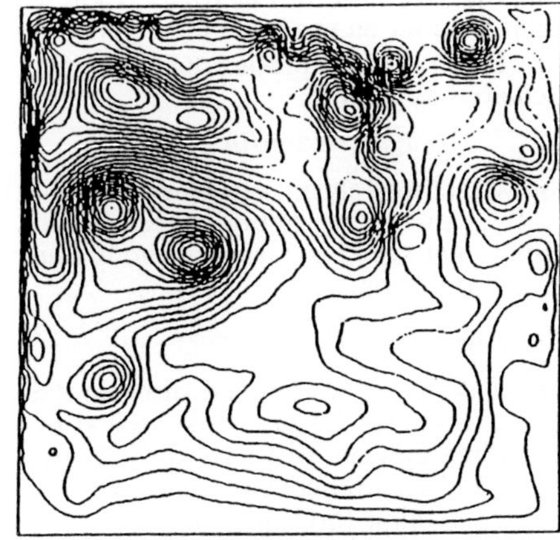

$$t = 75.0 \times 10^5$$

Fig.8: Streamlines in transient states
($Ra = 10^9$, $Pr = 0.71$, $\Delta\psi = 3.0\times10^{-8}$)

Looking at the temparature distribution at $Ra = 10^9$ (Fig.5), that is in agree with the result at $Ra = 10^8$ qualitatively. In other word, vortex which occurs on the hot wall reaches the cold wall, which is moving along the top wall. After that, high temperature field spreads the whole cavity from top to bottom. But motion of the initial vortex is more complex than at $Ra = 10^8$, and at the first unstable wave which occurs in the laminer boundary layer grows up gradually while rising up by convection. Afterward it is growing up while forming vortex, and flow becomes turbulent flow after that vortex breaks down. Looking at streamlines in Fig.6, many small vortex occurs on near the hot wall at about dimensionless time $t = 20.0 \times 10^5$. It is impossible to observe this phenomena at under $Ra = 10^8$. The occurrence of small vortex is thought to be by reason that various sizes of vortex of high wave number affects flow by strong nonlinerity of convection term.

When we are looking at temperature distribution (Fig.7) and streamlines (Fig.8) at $Ra = 10^{10}$, the motion of transient state coincides with that at $Ra = 10^9$ qualitatively. But many small vortices are observed at first time on the bottom of the hot wall. These vortices are appeared in the analysis of Paolucci[4], so it can confirm that the present scheme is effective. Time history of mean Nusselt number at every Rayleigh number, is shown in Fig.9. At $Ra = 10^8$ it goes to steady state gradually, while $Ra = 10^9$ and 10^{10} it is observed overshoot near $t = 5.0 \times 10^5$. When we observe the temparature distribution and streamlines at this time, vortex which occurs on the hot wall runs into the top wall. Namely at over $Ra = 10^9$, because of the influence of strong buoyancy force, turbulence occurs in the boundary layer, and the phenomena is thought that collision of up stream onto the top wall disturbes flow in the boundary layer.

Mean Nuselt number on the hot wall at every Rayleigh number is shown in Fig.10. It is known that mean Nusselt number on the hot wall is generally proportional to $Ra^{1/4}$. In the present analysis, this relation is calculated as follows:

$$\overline{Nu_{av}} = 0.286 Ra^{1/4} \tag{25}$$

This is almost completely in agreement with others. The present scheme using hybrid-streamline-upwind FEM is able to get stable solution without numerical oscillation and diffusion.

5. Conclusions

In the present paper, we applied discrete del operater in order to solve the Navier-Stokes equation using hybrid-streamline-upwind FEM to the energy equation. We analized natural convection flow problem in a square cavity at high Rayleigh number and studied accuracy of numerical solution. As examples of calculation, we analized $Ra = 10^6 \sim 10^{10}$ flows and got conclusions as follows:

In the flow at $Ra = 10^6$ we aimed at more efficient calcultion and studied about decreasing calcu-

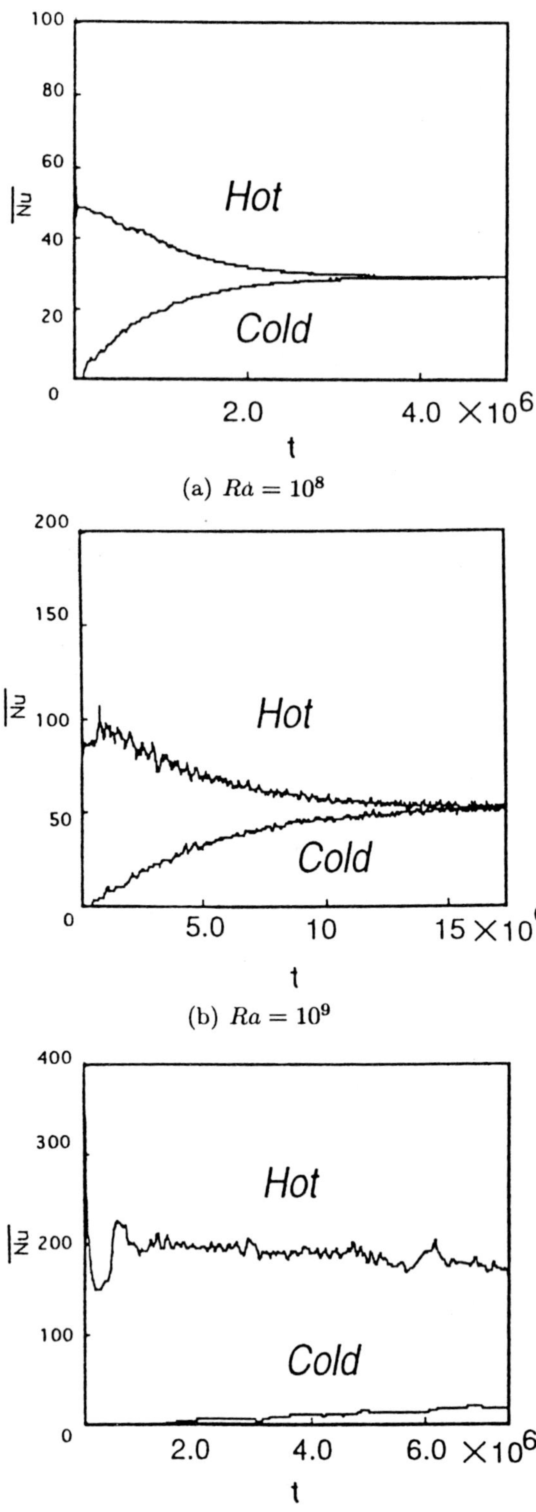

Fig.9: Time dependence of mean Nusselt numbers

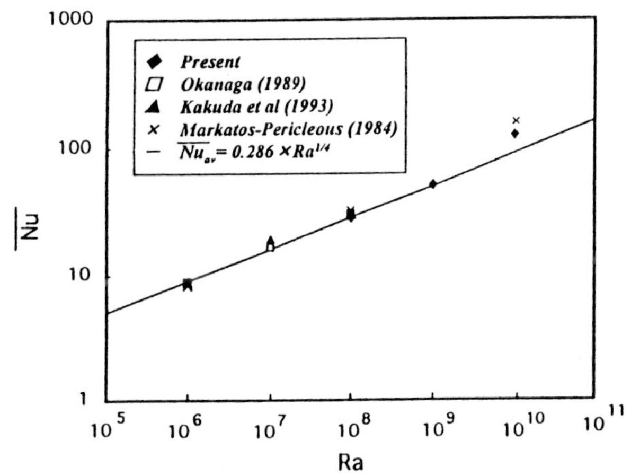

Fig.10: Correlation of heat transfer coefficient

lation capacity. As a result, we have found that the discretization of advection-diffusion matrix using discrete del operator caused 75.6% decreases about data size of coefficient matrix, and 12.0% decreases about memory size at calculation in comparison with conventional discretization. This showed that discretization of the present scheme is suitable for practical imcompressible flow problems. At $Ra = 10^8$ flow we compared numerical result with others and have showed that the present scheme is effective. At $Ra = 10^9$ and $Ra = 10^{10}$, using the present scheme we could get stable solution of complex phenomena in the coupled fields of velocity and temparature. In particularly, computed result of natural convection at $Ra = 10^{10}$ agree with others qualitatively. From the above mentioned, we believe that useful, stable and high accurate scheme has constructed by using FEM in a dual space.

REFERENCES

[1] De vaul Davis (1983) Natural convection in a square cavity: A bench mark numerical solution; Int. J. Num. Methods Fluids, vol.3 p.249

[2] Kawahara,Y. *et al.* (1992) Finite element simulation for natural convection at high Rayleigh numbers; 6th CFD Symposium 1992, p.423

[3] Okanaga,H. (1989) Study on high accurate numerical analysis for natural convection at high Rayleigh numbers; Doctor Thesis of Keio University 1989

[4] Paolucci,S. (1990) Direct numerical simulation of two-dimensional turbulent natural convection in an enclosed cavity; J. Fluid Mech., vol.215 p.229

[5] Kato,Y. and Tanahashi,T. (1991) Finite -element method for three-dimensional incompressible viscous flow using simultaneous relaxation of velocity and Bernoulli function (2nd report, Natural convection in horizontal concentric annuli); Bull of JSME, (in Japanese), 58-551, B, 160

[6] Tanahashi,T. and Oki,Y. (1993) New hybrid-steamline-upwind finite-element method in a dual space (1st Report; Fundamental theory and discrete Del operator); Bull of JSME, (in Japanese), 59-568, B, 3823

[7] Oki,Y. and Tanahashi,T. (1994) New hybrid-streamline-upwind finite-element method in a dual space (2nd Report; Approach to unsteady one-dimensional Burger's equation); Bull of JSME, (in Japanese), 60-573, B, 1639

[8] Ikeda,H. and Tanahashi,T. (1994) New hybrid-streamline-upwind finite-element method in a dual space (3rd Report; Verification for two-dimensional advection-diffusion equation); Bull of JSME, (in Japanese), 60-578, B, 3311

[9] Tunoda,K. *et al.* (1993) Petrov-Galerkin finite element analysis using exponential functions for natural convection in a square cavity; Bull of JSME, (in Japanese), 59-564, B, 123

[10] Mizukami,A. (1992) Finite element analysis for natural convection in a square cavity at high Rayleigh numbers; Numerical Analysis on Structure Engineering, 16, 433

[11] Markatos,N. *et al.* (9184) Laminar and turbulent natural convection in an enclosed cavity; J. Heat Mass Transf., 27, 755

Table 1: Numerical conditions

Ra	10^6	10^8	10^9	10^{10}
Pr	0.71	0.71	0.71	0.71
Δt	100	50	50	50
Mesh size	20×20	50×50	80×80	200×200
$\Delta L_{\min}$	5.0×10^{-2}	4.7×10^{-3}	4.0×10^{-3}	1.58×10^{-3}
ε	1.0×10^{-6}	1.0×10^{-8}	1.0×10^{-9}	1.0×10^{-9}

Tabble 2: Comparison of computer memories at $Ra = 10^6$

	Data Size(bytes)	Memory Size(Mbytes)
Present	394750	1.248
GSMAC	1622812	1.344

Table 3: Comparison of numerical results at $Ra = 10^8$

	U_{max} x_2	V_{max} x_1	Nu_{av}	Nu_{max} x_2	Nu_{min} x_2
Present	269.7	2039		74.40	1.950
(DS-FEM)	0.9413	0.0138	28.60	0.013	1.000
Kakuda &	306.5	2036		66.24	1.510
Tosaka[9]	0.9381	0.0116	30.40	0.0231	1.000
	320.7	2221		84.93	1.904
Okanaga[3]	0.9240	0.0136	30.06	0.0100	1.000
Mizukami[10]			28.93		
Markatos &	514.3	1812		61.06	5.224
Pericleous[11]	0.9410	0.0135	32.04	0.0075	0.097

KEY TO PROBLEMS IN SPECTRAL METHODS

Jian-Ping WANG

State Key Laboratory for Turbulence Research; Department of Mechanics and
Engineering Sciences, Peking University; Peking 100871

1. Introduction

In the past quarter century, great progress has been made in developing spectral methods in computational fluid dynamics. Various schemes have been proposed for dealing with various problems. Nowadays, spectral methods are recognized as one of the main methods in CFD.

Researches in experimental fluid dynamics can be divided into three categories: measuremental techniques, fluid properties and industrial applications. Similarly, those in computational fluid dynamics are the same except that the measuremental techniques are replaced with numerical techniques. The measuremental and numerical techniques themselves are interesting research themes, and they serve as tools for the researches of fluid properties and industrial applications. In fact, spectral methods have been widely used for researches on fluid properties rather than on industrial applications, since the latter needs more robustness and flexibility. This is the obvious difference in the uses between conventional spectral method with finite difference method and finite element method. However, spectral methods are still eagerly studied because of their excellent properties, for instance, the properties of no phase error, exponentially convergence rate, efficient spectral form *etc.*

The requirement for a good numerical method is that any problem can be calculated easily, accurately and efficiently. Spectral methods can be divided into two types in terms of the discrete form of the governing equations, i.e., the spatial spectral form and the form. The former is denoted as Galerkin method and the latter is denoted as collocation method. The one with both forms is called as pseudospectral method. Spectral methods are much more accurate than other methods when approximating smooth functions. The linear terms of governing equations can be easily transformed into spectral form, and it is much more efficient than the spatial form, and needs less operations than finite difference method and finite element method. The spectral method was developed by Orszag (1971a) for calculating nonlinear terms with collocation method instead of complicated operations of spectral expansion. Spectral methods can also be classified by the orthogonal functions such as trigonometric polinomials, Chebyshev polinomials and Legendre polinomials. Using Fourier expansion and Fast Fourier Transform (FFT) in a simple geometry with periodic boundary condition, the computation process is both easy and efficient.

For non-periodic problems, Chebyshev polynomials or Legendre polynomials are suitable eigenfunctions, but their formulae are complicated. It is clear that spectral methods are suitable for wavy flows with various scales of vortices such as turbulence. This is why spectral methods are used in most direct numerical simulations of turbulence. Although various problems have been solved by spectral methods, it is still far from solving arbitrary problem due to their global properties.

The most important characteristics of spectral methods is the global property (Solomonoff and Turkel 1989). Global eigenfunctions are used for approximating functions in the whole region, domain or element. Inversely, the expansion coefficients are calculated by global integration in terms of the orthogonal relations. This on one hand provides high accuracy and superior efficiency, but on the other hand owns little flexibility, preventing the progress of spectral methods.

Overlooking the existing spectral schemes, one can distinguish them into two kinds: extending the advantages or overcoming the disadvantages of spectral methods. On the progress of spectral methods in fluid dynamics before 1987, Hussaini and Zang (1987) made summarization in their review paper. The present review focuses on the numerical techniques of spectral methods, especially on the difficulties encountered such as treatments of discontinuities, boundary conditions, complex geometries and time discretizations.

Our aim is to find a key to solving these problems.

Received on March 19, 1997.

This research was supported by National Aerospace Laboratory of Japan and National Science Foundation of China.

2. Discontinuities

Many problems of fluid dynamics include both discontinuities and wave components, for example, the turbulent shear flows, the turbulent flames, and the interaction of shock wave with vortices or with turbulent flows. These problems emerge from industrial needs such as various jet engines. Since it is still difficult to measure out the detail data of these flows in present experimental techniques, numerical studies take an important role in these fields. The numerical method should satisfy two requirements, the high resolution at discontinuities and the high accuracy in smooth areas. Although various TVD schemes have been proposed for dealing with shock waves, smearing of wavy solutions cannot be avoided. The ENO schemes (Harten *et al.* 1987) have taken improvements in calculating this kind of problems but still possess phase errors. Therefore, spectral schemes for treating shock waves or contact surfaces are desired to solve this contradiction.

A well-known phenomenon named Gibbs phenomenon arises in using eigenfunctions to approximate step functions. The truncated eigenfunctions show the characteristic oscillatory behavior in the neighborhood of the discontinuity point (Canuto *et al.* 1987). The reason for that can be explained in terms of Dirichlet kernel, which is a global weighted function in the integration of representing a point value. The oscillation near a step does not disappear even with large numbaer of truncated terms.

There exist two categories in treating shock waves numerically, i.e., the shock fitting techniques and the shock capturing techniques. The former regards shock waves as outer or inner boundaries, where the Rankine-Hugoniot relations are used for setting the boundary conditions. The latter treats shock waves as continuum as smooth area, while detecting discontinuities in terms of conservation laws. These two kinds of techniques have been also applied to spectral methods.

Hussaini *et al.* (1956b) first introduced shock fitting technique to spectral methods. Chebyshev collocation method was used in calculating Euler equations on one side or two sides of two-dimensional shock waves. Results of shock/acoustic wave interaction, shock/vortex interaction and blunt-body problems show that spectral shock-fitting techniques are superior in accuracy. Wang *et al.* (1990a) developed shock-fitting codes for three-dimensional Euler and Navier-Stokes equations with applications to blunt-bodies. Kopriva *et al.* (1992) proposed a multidomain spectral collocation method to solve both steady and unsteady inviscid supersonic flows over blunt-bodies.

Another approach is the shock-capturing technique.

Since discontinuities are approximated in global eigenfunctions, the Gibbs phenomenon becomes an important problem. The disadvantages appear in two aspects, i.e., the spurious oscillations near discontinuities and their amplification in time. The amplification is caused by calculating space derivatives and it depends on the wave number of spectral modes. The larger of the wave number, the larger of the amplification. For instance, the n-th component of Fourier series is multiplied by n in calculating its first derivative. A number of numerical techniques have been proposed for eliminating or controlling these oscillations. All of them can be divided into two groups, one is global filter in spectral space, another is local limiter in spatial space. The former is based on the characteristics of spectral methods, while the latter comes from finite difference schemes. The two kinds of techniques are sometimes used together.

Gazdag (1973) obtained the first numerical results for the Burgers equation with a viscosity term. Gottlieb *et al.* (1981) utilized a post-processing Shuman filter and a spectral low-pass filter to calculate shock waves with one-dimensional Euler equations. The shock wave location was determined by examining the spectral coefficients, and the solution oscillated by the spectral filter is smoothed by the local post-processing filter on both sides of the shock wave. Cornille (1982) developed a scheme reducing the oscillations due to Gibbs phenomenon. Sakell (1984) tested the effects of low pass filtering and second- and fourth-order artificial viscosity schemes. Hussaini *et al.* (1985a) compared the results of the astrophysical model obtained by spectral collocation methods and by MacCormack finite difference scheme. Several spectral filters and a nonlinear artificial viscosity were tested. The comparison leads to a pessimistic conclusion that the accuracy of spectral methods was deteriorated to first order, which is worse than finite difference method. Cai *et al.* (1992) proved the existence of one-side filters for Fourier approximations of discontinuous functions, which can preserve spectral accuracy up to the discontinuity from one side. However the process of calculating three filtered solution on each point at each time step seems too complicated.

In contract to the above global operations, local operations can be implemented in flux conservative forms. The fluxes are reconstructed locally, such as the FCT and ENO schemes. Zalesak (1979) for the first time introduced the flux corrected transport (FCT) algorithm of Boris and Book (1976) to spectral methods. Taylor *et al.* (1981) applied FCT algorithm to similar problems. McDonald (1989) showed that the finite difference phase error causes more artificial oscillation than Gibbs error. Cai and Shu (1993) proposed an uniform high-order spectral method. Four techniques were taken into the algorithm, i.e., the method of adding step function, the ENO polynomial interpolation, the filtering technique and the uniform spectral approximation. Uniform convergence and spectral accuracy in the smooth region was recognized in one and two-dimensional test problems, though small oscillations remain near the discontinuities. The essence of adding step function is to subtract the Dirichlet kernel from the truncated Fourier series of discontinuous functions. Since the remainder is only C_0 continuous, spectral accuracy can not be achieved near the discontinuity. This is the reason of the small oscillations and they are smoothed by filtering. This method was utilized to several application problems by Wai (1994). Sidilkover and Karniadakis (1993), Giannakouros and Karniadakis (1994) introduced the uniform high-order spectral method and FCT method into the spectral element method, which relaxed the restriction of treating shock waves in wide region.

All of this amounts to saying that the treatment of discontinuities is still limited by the global properties of spectral methods because they are local phenomena. In order to solve this problem, Wang (1995) proposed a finite spectral ENO scheme that the smoothest stencil is searched by comparing divided differences, and the non-periodic Fourier interpolation is used instead of Newton interpolation. An error limiter was developed by Wang (1997) to determine the length of the stencil adaptively. This scheme can be treated cellwise, and it is much simpler than the previous ones. Figure 1 is the result of using this scheme for the shock tube problem.

3. Boundary Conditions

Any one who has experience on calculating CFD problems with non-periodic boundary conditions by spectral methods might meet the difficulties of treating boundary conditions. For periodic boundary conditions, Fourier series are the suitable eigenfunctions due to their inherent periodic nature, while Chebyshev and Ledgendre polynomials are used for approximating non-periodic problems.

Generally, the troubles of treating boundary conditions in finite difference method stem from coordinate singularity, non-characteristic treatment and extrapolation accuracy. For spectral methods, however, additional difficulties come from their global properties, because local variations affect the accuracy and stability of the whole region. Since a solution depends on the treatment of boundary conditions as well as on the scheme for inner region, it is important to impose boundary conditions stably and accurately. The ways

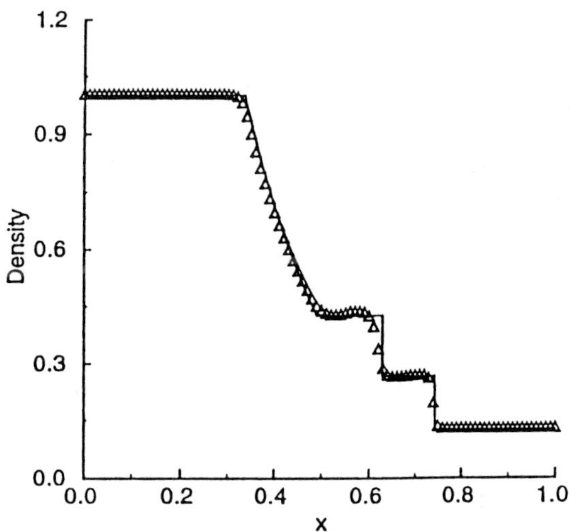

Fig.1: Density distribution of Shock tube problem using finite spectral ENO scheme with five-point stencil and with error limiter.

for treating boundary conditons in spectral form and in spatial form are different.

In spectral forms, the most used way is to expand boundary conditions in spectral forms, and then solve them together with those of governing equations. Let N be the number of the truncation modes, and N_b be the number of the boundary conditions. The N relations between the N coefficients are constituted by $N - N_b$ equations derived from governing equations and N_b equations from boundary conditions. This technique was proposed as tau method by Lanczos (1964) for the first time, and was applied to the calculations of incompressible viscous flows by Orszag (1971b). However, as discussed by Gottlieb and Orszag (1977), the tau method generates spurious eigenvalues for eigenvalue problems, which lead to instability. Gottlieb and Lustman (1983) proposed the characteristic variable operating method for linear systems. Canuto and Quarteroni (1987) suggested to use implicit time-advancing schemes to avoid spurious instability in linear and nonlinear systems. Funnaro (1990) gave a varivational formulation for Chebyshev collocation methods with Neumann conditions. Gardner et al. (1989) presented a modified tau method for eliminating spurious eigenvalues, but it is difficult to be generalized to CFD problems.

In spatial forms, solutions are obtained at collocation points including those at boundaries every time step. The Dirichlet conditions can be easily imposed on the boundaries. The Neumann conditions, however, are troublesome to deal with. Street et al. (1985) imposed Neumann conditions on the intermediate solutions of the boundary value problems in spatial space. However, this led to a substantial deterioration of the convergence rate of iteration. Hussaini et al. (1985b) adjusted the highest Chebyshev coefficient to fit the Neumann condition in their explicit time-advancing code of Euler equation. Kopriva et al. (1991) presented a characteristic treatment for shock fitting techniques. Carcione and Wang (1993) derived the characteristic modes of the elastodynamic equation, which are used to the wave propagation problem.

Although various methods have been attempted, the origin of the difficulties has been neglected, i.e., combining boundary conditions with governing equations and adjusting boundary conditions globally. Once the boundary conditions can be imposed independently and locally, the situation was greatly changed. Wang *et al.* (1993) proposed a global coefficient adjustment method for treating boundary conditions. This method based on the idea of imposing the boundary condition by only changing the boundary value. This results a formula of adjusting all Chebyshev coefficient, unlike the highest coefficient adjustment method. The quantity added to the boundary or to each coefficient can be easily derived both for Dirichlet and Neumann conditions. Figure 2 illustrates the spatial variation by global coefficient adjustment method. The influence is restricted to the neighborhood of the boundary, and it is similar to the Dirichlet kernel. At each time step, the boundary conditions can be imposed by adjusting the intermediate solutions, independently from the process of solving the governing equations. This method can be used both for spectral form and spatial form. In the former, the coefficients are adjusted, while in the latter the boundary values are adjusted. Since it is a one-dimensional operation, there is no problem to use it in multi-dimensional problems. Applications to the calculations of hypersonic viscous flows over a sphere show that it imposes the boundary conditions both on the body surface and the coordinate singular line easily, accurately and stably.

4. Complex Geometries

Although spectral methods have been one of the main numerical methods in CFD, there exists still a hurdle of treating complex geometries, which prevents them from competing with finite difference method and finite element method. However spectral methods are required for dealing with complex geometries due to their high accuracy which can not be obtained with other methods. This necessity was pointed out by Orszag (1980a), and some progress has been made since then. The efforts of extending the scope of spectral methods have been done in two aspects: the co-

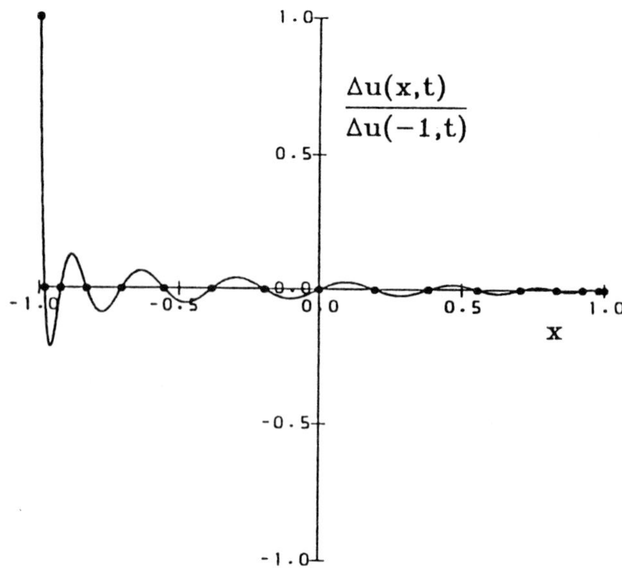

Fig.2: Spatial variation by global coefficient adjustment method

ordinate transformation techniques and the domain decomposition techniques.

Orszag (1980a) suggested an idea of mapping the complex geometry to a simpler one by smooth stretching transformation. This type of mapping can be found in Hussaini *et al.* (1985b). However these are limited to the cases that the coordinates fit the body because the directions of the mesh should be close to perpendicular. In order to relieve these restrictions, Wang *et al.* (1990c) for the first time introduced generalized coordinates into spectral methods, and applied it to the calculations of the head of a shuttle-like body (Fig.3). The derivatives in metric are also calculated by spectral methods to keep consistency in accuracy. This technique was also taken into the cooperation of Carcione and Wang (1990) for the wave propagation problem. Further later, the generalized coordinates were utilized by Ku (1995).

The domain decomposition technique was first described by Orszag (1980a) as a patching idea. Morchoisne (1982) first proposed a multi-domain technique with which an airfoil was calculated. The spectral element method proposed by Patera (1984) has given a great influence on spectral methods. It combines the varivational formulation used in finite element methods and Chebyshev polynomials used in spectral methods. This method was applied to the laminar back-step flow by Patera (1984), cylindrical annulus flow by Korczak and Patera (1986), and three-dimensional laminar horseshoe vortex flow by

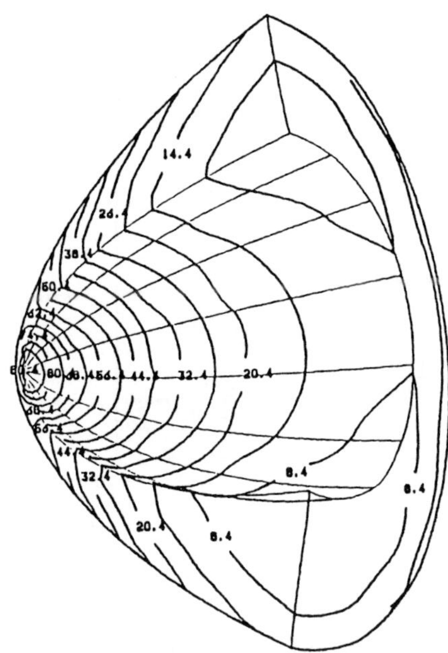

Fig.3: Pressure contours of hypersonic flows over
a shuttle-like body by 3-D Euler equations

Tan (1989). The detail of spectral element method is summarized by Maday and Patera (1989) in a chapter of a book. Another approach is the pseudospectral matrix element method developed by Ku (1989). Wang *et al.* (1988) obtained the results of hypersonic viscous flows by using a multi-domain method imposing inner boundary conditions with global coefficient adjustment method. Kopriva (1991) presented an interface treatment technique which allows patched or overlapping subdomains.

The use of generalized coordinates and domain decomposition techniques has relaxed the limitation of treating complex geometries to a certain extent. However, the fundamental problem, i.e., the contradiction between the local property of flow fields with the global nature of spectral methods, has not been solved yet. Two difficulties can be considered when representing a variable through a complicated region. One is the difficulty of grid generation over suddenly changed coners. Another is the averaging effect for each spectral mode in the integration across the wide region with different scale features. Domain decomposition techniques reduce the global effects to each subdomain or element, but they are still not local schemes. In other words, they possess not only structured mesh, but also structured domains. Three disadvantages of the domain decomposition methods can be indicated. First, the distribution of domains is generally not uniform. Secondly, the collocation points in each domain or element are not uniform. Thirdly, it is difficult to obtain spectral accuracy at inner boundaries. Therefore, pointwise or cellwise schemes are required for dealing with arbitrary geometries.

5. Time Discretizations

Generally, the time discretization for spectral methods are those transplanted from other methods. For incompressible flows, Orszag (1971a, 1971b, 1971c, 1971d) proposed a first order Galerkin scheme, which was used for the numerical simulations on Taylor-Green vortex-decay problem, flows with slabs, spheres, and cylinders, and Orr-Sommerfeld stability equation. Gottlieb and Turkel (1980) proposed leapfrog and Runge-Kutta methods. Moin and Kim (1980), Orszag and Kells (1980b) yield semi-implicit schemes by combining the Adams-Bashforh method and Crank-Nicolson method. These numerical schemes have played an important role in later simulations on turbulence. Hasegawa *et al.* (1989, 1992) used fifth order Runge-Kutta scheme to simulate the turbulent flames (Fig.4). The schemes proposed by Moser *et al.* (1983), Malik et al. (1985), Quere *et al.* (1985), Ku *et al.* (1987) own their originalities respectively.

The stability of spectral methods is as important as other methods. In fact, most stability analyses in other methods utilize Fourier expansion after von Neumann. Two means are used to make spectral schemes stable, the implicit treatment and the filtering technique. The typical semi-implicit scheme is the Adams-Bashforth-Crank-Nicolson scheme mentioned above, which need no filtering for calculating smooth flows. In explicit schemes, however, filtering techniques are necessary for suppressing the amplification of high frequency modes. Lanczos (1938) suggested using filters for the first time. Majda *et al.* (1978) proved that using a cutoff function can stabilize the nonsmooth initial data problems. Gottlieb and Turkel (1980) proposed a filter to achieve unconditionally stable. However Fulton and Taylor (1984) pointed out that this filter could lead to absolute instability. Mulholland and Sloan (1991) investigated the effect of filtering on the pseudospectral solution of the advection equation, the KdV equation and Burgers' equation. Wang proposed a residual dependent filter for increasing the accuracy in time (Yasuhara et al. 1988, Wang *et al.* 1890). Unlike the conventional fixed filters, the critical wave number of the filter is determined by comparing the residual with the coefficients one by one. This technique provides a compromise to the contradiction between the accuracy and stability, with which good resolution was obtained at viscous boundary layer. Figure 5 shows the temperature contours using fixed and residual dependent filter respectively. The difference

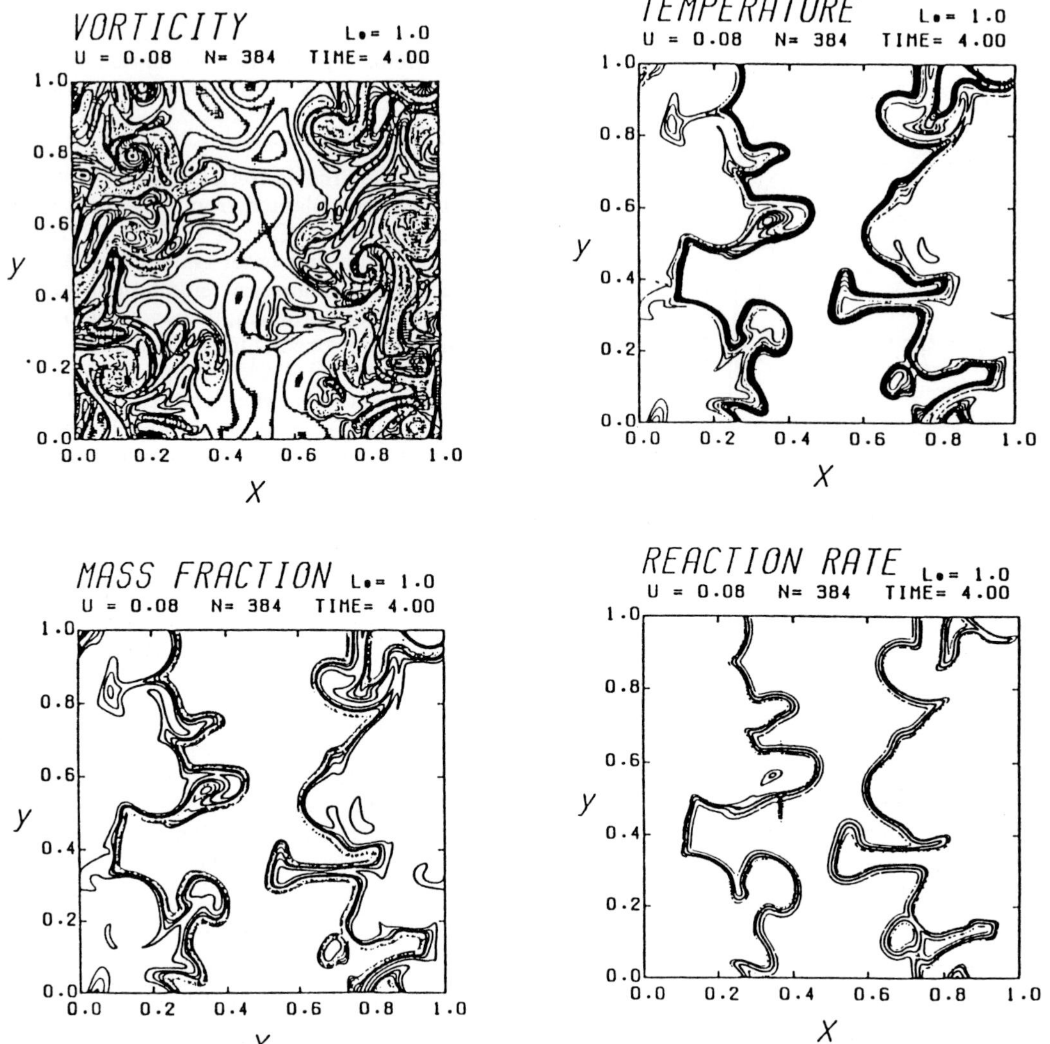

Fig.4: Propagation of flame in two dimensional turbulent premixed gas

of the resolution at the boundary layer is obvious.

Another branch of time discretization is convergence acceleration techniques, i.e., the multigrid methods. Orszag (1980a) illustrated several iterative schemes combining spectral and finite difference approximations. Zang *et al.* (1982,1984) introduced multigrid technique into spectral methods. Since the number of truncated eigenfunctions in spectral space correspond with their collocation number in spatial space, the interpolation processes can be done through the spectral modes, which is rather convenient than usual multigrid methods. Brandt *et al.* (1985) gave several modification on that algorithm. Street *et al.* (1985) applied spectral multigrid method to solving

transonic potential flows over an airfoil. Heinrichs (1988,1992) improved the multigrid convergence by using line relaxation techniques and preconditioning techniques.

In numerical simulations of unsteady flows such as turbulent flows, the accuracy of time discretization is as important as the accuracy in space, because wrong intermediate solutions may lead to entirely different results due to the nonlinear properties of hydrodynamic equations. An attempt worth noticing is using spectral approximation in time. The pioneers are Ierley *et al.* (1992), who introduced a Fourier ex-

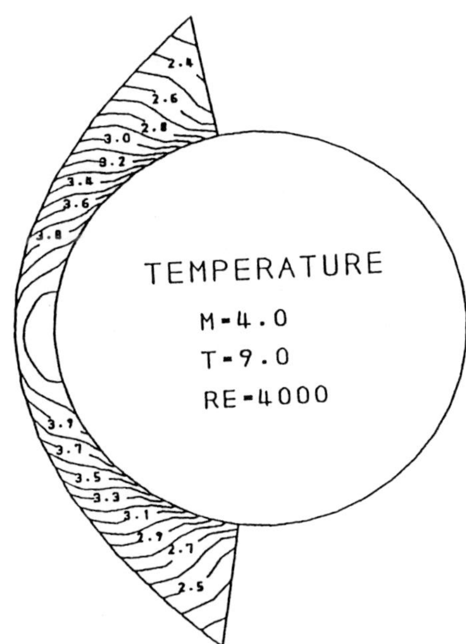

(a) by time-fixed filter

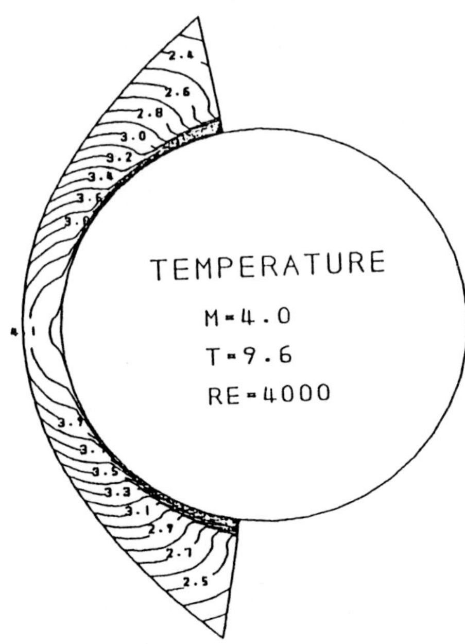

(b) by residual dependent filter

Fig.5: Temperature contours of supersonic flows over a sphere by axisymmetric Navier-Stokes Equations

pansion in spatial direction and Chebyshev expansion in time direction. The equation of spectral form is solved by iterative schemes. Then the solutions at the end time boundary are used as initial data at the start time boundary of the next time-marching step. This time multi-domain method yields some results for Kuramoto-Sivashinsky problem and Burgers equation. However, at first glance, it remains the weak points similar to those explained in spectral multi-domain methods and spectral element methods.

6. Conclusions

Spectral methods are still in developing because the problems discussed above have not been completely solved. To make spectral methods to be competitive with finite difference method and finite element method, we must show that they are strong not only on simulations of fluid properties, but also on those of industrial applications. In other words, it is necessary to make spectral methods from special ones to normal ones. The key is to overcome the contradiction between the global property of the methods and the local property of the objects. In fact, almost all efforts of developing spectral methods are toward finding this key. The way is obvious, that is, to construct local spectral methods suited to local properties of fluids and geometries. These methods may be treated pointwise or cellwise rather than domainwise or elementwise. To do this, eigenfunctions with both non-periodic and uniform mesh are necessary conditions. According to conventionality, unfortunately, neither Fourier series nor Chebyshev and Ledendre polynomials fit this need. A breakthrough is being attempted by Wang (1995) recently. The fundamental idea is using Fourier series to approximate non-periodic functions. The Fourier coefficients are calculated by solving an interpolation system. The non-periodic Fourier transform makes it possible to construct various local spectral schemes (Wang and Fujiwara (1992)). Wang also proposed a central spectral scheme and a finite spectral ENO scheme, with which the wave propagation problem and the shock tube problem (Fig.1) were successfully simulated. A further effort of achieving the free transform between the discrete form in physical space and that in spectral space is undergoing.

REFERENCES

[1] Boris, J.P. and Book, D.L. (1976): Solution of the Continuity Equation by the Method of Flux-Corrected Transport, Methods in Comput. Phys. 16, 85-129.

[2] Cai, W., Gottlieb, D. and Shu, C.-W. (1992): On One-Sided Filters for Spectral Fourier Approximations of Discontinuous Functions, SIAM J. Numer. Anal. 29, 905-916.

[3] Cai, W., Gottlieb, D. and Shu C.-W. (1993): Uniform High-Order Spectral Methods for One- and Two-Dimensional Euler Equations, J. Comput. Phys. 104, 427-443.

[4] Canuto, C., Hussaini, M.Y., Quarteroni, A. and Zang, T.A. (1987): Spectral Methods in Fluid Dynamics, Springer-Verlag.

[5] Canuto, C. and Quarteroni, A. (1987): On the Boundary Treatment in Spectral Methods for Hyperbolic Systems, J. Comput. Phys. 71, 100-110.

[6] Carcione, J.M. and Wang, J.P. (1993): A Chebyshev Collocation Method for the Elastodynamic Equation in Generalized Coordinates, Comput. Fluid Dynamics J. 2, 269-290.

[7] Cornille, P. (1982): A Psudospectral Scheme for the Numerical Calculation of Shocks, J. Comput. Phys. 47, 146-159.

[8] Don, W.S. (1994): Numerical Study of Psudospectral Methods in Shock Wave Applications, J. Comput. Phys. 110, 103-111.

[9] Fulton, S.R. and Taylor, G.D. (1984): On the Gottlieb-Turkel Time Filter for Chebyshev Spectral Methods, J. Comput. Phys. 55, 302-312.

[10] Funaro, D. (1990): A Variational Formulation for the Chebyshev Pseudospectral Approximation of Neumann Problems, SIAM J. Numer. Anal. 27, 695-703.

[11] Gardner, D.R., Trogdon, S.A. and Douglass, R.W. (1989): A Modified Tau Spectral Method That Eliminates Spurious Eigenvalues, J. Comput. Phys. 80, 137-167.

[12] Gazdag, J. (1973): Numerical Convective Schemes Based on Accurate Computation of Space Derivatives, J. Comput. Phys. 13, 100-113.

[13] Giannakouros, J. and Karniadakis, G.E. (1994): A Spectral Element-FCT Method for the Compressible Euler Equations, J. Comput. Phys. 115, 65-85.

[14] Gottlieb, D. and Orszag, S.A. (1977): Numerical Analysis of Spectral Methods: Theory and Applications, SIAM-CBMS, Phyladelphia.

[15] Gottlibe, D. and Turkel, E. (1980): On Time Discretization for Spectral Methods, Studies in Applied Mathematics 63, 67-86.

[16] Gottlieb, D., Lustman, L. and Orszag, S.A. (1981): Spectral Calculations of One-Dimensional Inviscid Compressible Flows, SIAM J. Sci. Stat. Comput. 2, 296-310.

[17] Gottlieb, D. and Lustman, L. (1983): The Spectrum of the Chebyshev Collocation Operator for the Heat Equation, SIAM J. Numer. Anal. 20, 909-921.

[18] Harten, A., Engquist, B., Osher, S. and Chakravarthy, S. (1987): J. Comput. Phys., 71, 231-232.

[19] Hasegawa, T., Yamaguchi, S., Wang, J.P. and Jounouchi, T. (1989): Propagation of Premixed Flames in Vortical Flows, Proc. 27th Sympo. Combustion, 49-51 (in Japanese).

[20] Hasegawa, T., Arai, T., Kadowaki, S. and Yamaguchi, S. (1992): Autoignation of a Turbulent Premixed Gas, Combust. Sci. and Tech. 84, 1-13.

[21] Heinrichs, W. (1988): Line Relaxation for Spectral Multigrid Methods, J. Comput. Phys. 77, 166-182.

[22] Heinrichs, W. (1992): A Spectral Multigrid Method for the Stokes Problem in Streamfunction Formulation, J. Comput. Phys. 102, 310-318.

[23] Hussaini, M.Y., Kopriva, D.A., Salas, M.D. and Zang, T.A. (1985a): Spectral Methods for the Euler Equations: Part I-Fourier Methods and Shock Capturing, AIAA J. 23, 64-70.

[24] Hussaini, M.Y., Kopriva, D.A., Salas, M.D. and Zang, T.A. (1985a): Spectral Methods for the Euler Equations: Part II-Chebyshev Methods and Shock Fitting, AIAA J. 23, 234-240.

[25] Hussaini , M.Y. and Zang, T.A. (1987): Spectral Methods in Fluid Dynamics, Ann. Rev. Fluid Mech. 19, 339-367.

[26] Ierley, G., Spencer, B. and Worthing, R. (1992): Spectral Methods in Time for a Class of Parabolic Partial Differential Equations, J. Comput. Phys. 102, 88- 97.

[27] Kopriva, D.A., Zang, T.A. and Hussaini, Y. (1991): Spectral Methods for the Euler Equations: the Blunt Body Problem Revisited, AIAA J. 29, 1458-1462.

[28] Kopriva, D.A. (1992): Spectral Solution of Inviscid Supersonic Flows over Wedges and Axisymmetric Cones, Computers Fluids 21, 247-266.

[29] Kopriva, D.A. (1994): Multidomain Spectral Solution of Compressible Viscous Flows, J. Comput. Phys. 115, 184-199.

[30] Korczak, K.Z. and Patera, A.T. (1984): An Isoparametric Spectral Element Method for Solution of the Navier-Stokes Equations in Complex Geometry, J. Comput. Phys. 62, 361-382.

[31] Ku, H.-C., Hirsh, R.S. and Talor T.D. (1987): A Pseudospectral Method for Solution of the Three-Dimensional Incompressible Navier-Stokes Equations, J. Comput. Phys. 70,

439-462.

[32] Ku, H.-C., Hirsh, R.S and Talor T.D. (1989): A Pseudospectral Matrix Element Method for Solution of Three-Dimensional Incompressible Flows and Its Parallel Implementation, J. Comput. Phys. 83, 260-291.

[33] Ku, H.-C. (1995): Solution of Flow in Complex Geometries by the Psudospectral Element Method, J. Comput. Phys. 117, 215-227.

[34] Lanzos, C. (1938): Trigonometric Interpolation of Empirical and Analytical Functions, J. Math. Phys. 17, 123-199.

[35] Lanzos, C. (1964): Applied Analysis, Prentce-Hall, Englewood Gliffs, NJ, 464- 517.

[36] Maday, Y. and Patera, A.T. (1989): Spectral Element Methods for the Incompressible Navier-Stokes Equations, Chapter 3, State-Of-The-Art Surveys on Computational Mechanics, Edited by A.K. Noor and J.T.Oden, SIAM, 1989.

[37] Mailk, M.G. Zang, T.A. and Hussaini, M.Y. (1985): A Spectral Collocation Method for the Navier-Stokes Equations, J. Comput. Phys. 61, 64-88.

[38] Majda, A., McDonough, J. and Osher, S. (1978): The Fourier Method for Nonsmooth Initial Data, Mathematics of Computation 32, 1041-1081.

[39] McDonald, B.E. (1989): Flux-Corrected Pseudospectral Method for Scalar Hyperbolic Conservation Laws, J. Comput. Phys. 82, 413-428.

[40] Moin, P. and Kim, J. (1980): On the Numerical Solution of Time-Dependent Viscous Incompressible Fluid Flows Involving Solid Boundaries, J. Comput. Phys. 35, 381-392.

[41] Morchoisne, Y. (1982): Inhomogeneous Flow Calculations by Spectral Methods: Mono-Domain and Multi-Domain Techniques, ONERA TP-1982-67.

[42] Moser, R.D., Moin, P. and Leonard, A. (1983): A Spectral Numerical Method for the Navier-Stokes Equations with Applications to Taylor-Couette Flow, J. Comput. Phys. 52, 524-544.

[43] Mulholland, L.S. and Sloan, D.M. (1991): The Effect of Filtering on the Pseudospectral Solution of Evolutionary Partial Differential Equations 96, 369-390.

[44] Nakamura, Y., Wang, J.P. and Yasuhara, Y. (1991): A Calculation of the Compressible Navier-Stokes Equations in Generalized Coordinates by the Spectral Method, Lecture Notes Physic. 323, 127-131.

[45] Orszag, S.A. (1971a): Garlerkin Appoximations to Flows within Slabs, Spheres and Cylinders, Phys. Rev. Lett. 26, 1100-1103.

[46] Orszag, S.A. (1971b): Accurate Solution of the Orr-Sommerfeld Stability Equation, J.Fluid Mech. 50, 689-703.

[47] Orszag, S.A. (1971c): Numerical Simulation of Incompressible Flows Within Simple Boundaries. I Galaerkin (Spectral) Representation, Stud. Appl. Math. 50, 293-327.

[48] Orszag, S.A. (1971d): Numerical Simulations of Incompressible Flows within Simple Boundaries: II Accuracy, J. Fluid Mech. 49, 75-112.

[49] Orszag, S.A. (1980a): Spectral Methods for Problems in Complex Geometries, J. Comput. Phys. 37, 32-51.

[50] Orszag, S.A. and Kells, L.C. (1980b): Transition to Turbulence in Plane Poiseuille and Plane Couette Flow, J. Fluid Mech. 96, 159-205.

[51] Patera, A.T. (1984): A Spectral Element Method for Fluid Dynamics: Laminar Flow in a Channel Expansion, J. Comput. Phys. 54, 468-488.

[52] Quere, P. and Roquefort, T.A. (1985): Computation of Natural Convection in Two-Dimensional Cavities with Chebyshev Polynomials, J. Comput. Phys. 57, 210-228.

[53] Sakell, L. (1984): Pseudospectral Solutions of One- and Two-Dimensional Invisid Flows with Shock Waves, AIAA J. 22, 929-934.

[54] Sidilkover, D. and Karniadakis, G.E. (1993): Non-oscillator Spectral Element Chebyshev Method for Shock Wave Calculations, J. Comput. Phys. 107, 10-22.

[55] Street, C.L., Zang, T.A. and Hussaini, M.Y. (1985b): Spectral Multigrid Methods with Applications to Transonic Potential Flow, J. Comput. Phys. 57, 43-76.

[56] Talor, T.D., Myers, R.B. and Albert, J.H. (1981): Pseudo-Spectral Calculations of Shock Waves, Rarefaction Waves and Contact Surfaces, Computers Fluids 9, 469-473.

[57] Tan, C.S. (1989): A Multi-Domain Spectral Computation of Three-Dimensional Laminar Horseshoe Vortex Flow Using Incompressible Navier-Stokes Equations, J. Comput. Phys. 85, 130-158.

[58] Wang, J.P., Nakamura, Y. and Yasuhara, M. (1990a): Several Improvements of Spectral Method in Compressible Flow Calculation, Numerical Methods in Fluid Dynamics II, Proc. 3rd Int. Sympo. Comput. Fluid Dynamics, Nagoya, Japan Society of Computational Fluid Dynamics, 1035-1042.

[59] Wang, J.P., Nakamura, Y. and Yasuhara, M. (1990b): Accurate Computations of Compressible Navier-Stokes Equations by the Spectral Collocation Method, J. Jpn. Soc. Aerosp. Sci. 38, 232-240.

[60] Wang, J.P, Nakamura, Y. and Yasuhara, M. (1990c): A Chebyshev Collocation Method for the Compressible Navier-Stokes Equations in Generalized Coordinates, T. Japan Soc. Aero Spa. 33, 120-134.

[61] Wang, J.P. and Fujiwara, T. (1992): A method transforming Aperiodic Functions in Fourier Series, Proc. 2nd Int. Conf. Spectral High-Order Methods, Montpellier.

[62] Wang, J.P, Nakamura, Y. and Yasuhara, M. (1993): Global Coefficient Adjustment Method for Neumann Condition in Explicit Chebyshev Collocation Method and its Application to Compressible Navier-Stokes Equations, J. Comput. Phys. 107, 160-175.

[63] Wang, J.P. (1995): Non-Periodic Fourier Transform and Finite Spectral Method, Proc. Sixth Int. Sympo. Computational Fluid Dynamics, Lake Tahoe, 1339-1344.

[64] Wang, J.P. (1997): Finite Spectral Method based on Non-Periodic Fourier Transform, Computers Fluids (to apear).

[65] Yasuhara, M., Nakamura, Y. and Wang, J.P. (1988): Computations of the Hypersonic Flow by Spectral Methods, J. Jpn Soc. Aero. Spa. Sci. 36, 542-549.

[66] Yasuhara, M., Nakamura, Y. and Wang, J.P. (1989): Numerical Calculation of Hypersonic Flow by the Spectral Method, Lecture Notes in Physics 323, Springer-Verlag, 607-611.

[67] Zalesak, S. (1979): Fully Multidimensional Flux-Corrected Transport, J. Comput. Phys. 31, 335-362.

[68] Zang, T.A., Wong, Y.S. and Hussaini, M.Y. (1982): Spectral Multigrid Methods for Elliptic Equations, J. Comput. Phys. 48, 485-501.

[69] Zang, T.A., Wong, Y.S. and Hussaini, M.Y. (1984): Spectral Multigrid Methods for Elliptic Equations II, J. Comput. Phys. 54, 489-507.

EFFECTIVENESS OF A SPECTRAL FINITE DIFFERENCE SCHEME

Yoshihiro MOCHIMARU [1]

Dept. of International Development Engineering, Tokyo Institute of Technology, Tokyo 152, Japan

Abstract

A newly developed spectral finite difference scheme (Fourier expansion, Dini expansion, Legendre expansion) applicable to numerical analysis of heat and fluid flows is reviewed. Concrete solution procedures applied to different (mathematical or physical) types of problems are given for analyses corresponding to two-dimensional external flows, axisymmetric heat and fluid flows, two-dimensional natural convection in an elliptic enclosure, two-dimensional natural convection with a free surface, supplemented with obtained thermal or flow fields.

1. Introduction

Methods recently applied in numerical analysis in the field of fluid dynamics are divided into mainly finite difference schemes and finite element schemes in a point of ways for formulation of a system of equations from the original differential equations supplemented with boundary conditions and / or initial conditions. The former (FDM) is mostly based on finite difference approximation for derivatives, whereas the latter (FEM) is based on a principle of weighted residuals. Conventional finite element method adopt polynominals as a base for expanding (or approximating) unknowns in a finite series. On the other hand, spectral finite element methods, which have been developed recently and mostly called just spectral methods [1] and [2], adopt a finite part of a complete set of periodic functions such as sine and cosine functions as a base. Although the spectral finite difference scheme reviewed in this article uses a spectral function as a base, where in most cases Fourier series is selected (not necessarily restricted), the algorism for formulation of the system of equations from the original governing differential equations is completely different from the so-called finite element methods or from any one of weighted residual methods. The key point of the said scheme is that it introduces no approximation in leading the resulting system of equations in principle, but just introduces discretization to the resulting system of equations by a finite difference approximation only for getting numerical solutions.

Received on February 3, 1997.

2. Features of a Current Spectral Finite Difference Method

2.1 Equations Applicable

In the current spectral finite difference scheme, it is assumed that the original governing equations can be decomposed into each component of the corresponding base functions (spectral functions) if unknowns are expressed in terms of the base functions. Thus, any rational equations (differential, integral, algebraic, or their mixed ones) are to be applied. Among these there are:

Incompressible Navier-Stokes equations of constant viscosity (a laminar flow model); the energy equation for incompressible fluid with negligible dissipation; substantially incompressible Navier-Stokes equations with buoyancy under Boussinesq approximation; incompressible non-Newtonian fluid flow with a rational constitutive equation of constant coefficients.

Application of the said method is not restricted to thermal or fluid flow phenomena, but hereafter treatment leading to examples is limited to thermal/fluid flow fields.

2.2 Applicable Flow Fields (Geometry)

Basically the said spectral finite difference method is restricted to apply to a simply-connected or doubly-connected region in a view point of mapping domains. However, extention of usage is possible.

2.3 Applicable Boundary Conditions

Basically along each boundary, Neumann condition, Dirichlet condition, or any other rational condition with respect to unknowns may be accepted. Quantities appearing in the equations for these conditions are assumed to be at least piecewise continuous. Extention to the situation where Neumann condition/Dirichlet condition/the other types of condition are mixed along each boundary is possible.

2.4 Applicable Initial Conditions

Quantities at an initial time are assumed to be at least piecewise continuous.

3. Spectral Finite Difference Formulation

In case of using periodic base functions, it is necessary to set up a coordinate system such that physical quantities are periodic at least along one coordinate. Conversely, if a coordinate system is given so that physical quantities are periodic at least along one coordinate, then at least some restriction that the base functions are at least smooth (for the second order partial differential equations) at the point corresponding to the ends of the interval of the period will apply. To support boundry-value problems for spatial finite difference analysis, a boundary-fitted coordinate system is recommended to be formed at least locally. In case quantities are periodic in space, a coordinate system may be formed in such a way that one direction of the coordinates coincides with that in periodicity. Alternatively, a coordinate system may be formed so that any specified point on the physical space are assigned multiply to coordinates with a uniform difference (period) like a Rieman surface. In a two-dimensional case, an example is a polar coordinate system over a cylinder, which has a period of 2π in a circumferential coordinate. However, use of a conformal mapping system for a two-dimensional case or for an axisymmetric case in the section including the axis will be strongly suggested for getting simpler expressions. For example, used are an exponential function for a field around a single cylinder (or a single sphere) and hyperbolic or trigonometric functions for a field around a single elliptic cylinder (or a single ellipsoid of revolution).

Consider a two-dimensional case or an axisymmetric case. Let $u_i (i = 1, 2, \cdots, N)$ be unknowns, and let u_i be decomposed into $u_i = \sum_n U_{in} F_n$, where $\{F_n\}(n = 1, 2, \cdots, \infty)$ is a complete set for base functions. If one of the coordinate, say β, possesses a period of B in space (natural period), then as complete base functions the following is possible:

$$\left\{ \sin 2\pi \frac{n\beta}{B} \mid n : \text{integer} \geq 1 \right\}$$

$$+ \left\{ \cos 2\pi \frac{n\beta}{B} \mid n : \text{integer} \geq 0 \right\} .$$

This corresponds to a Fourier spectral finite difference method. (Dini expansions and Legendre expansions are also possible.) Substituting these expressions into the original governing equations which are assumed to be rational with respect to u_i's and decomposing them into each component F_n gives a system of (possibly nonlinear) differential equations. Decomposition (i.e. inversion) itself may be executed analytically,

and the results may be expressed in terms of elementary or special functions or just be evaluated in terms of numerical integration. In such a way one spatial variable can be separated to give a system of simultaneous partial differential equations (with respect to one spatial variable and time for unsteady-state problems) or a system of simultaneous ordinary differential equations (with respect to one spatial variable). In a three-dimensional case, one spatial variable or two spatial variables can be separated depending on using one kind of base functions or two kinds of base functions for two spatial variables. In these analyses it is assumed that the series converges at least asymptotically as a number of terms tends to infinity. Then the resulting system of simultaneous partial or ordinary differential equations can be solved by truncating the series up to a certain order and by discretizing them in space (and in time if necessary) by a finite difference approximation which accepts non-uniform grid spacing a great deal. The discretized equations can be easily integrated semi-implicitly (for nonlinear cases) with respect to time for unsteady-state problems.

In some cases, an artificial boundary corrsponding to the edge of the mapped domain will be produced in the interior of the analyzing domain due to setting up a boundary fitted coordinate system, where additional auxiliary boundary condition(s) have to be supplied along the newly formed apparent boundary. These additional boundary condition(s) may be continuity of the values of the unknowns in space and that of their gradients if necessary along the said artificial boundary. The number of required condition(s) depends on the order of the differential operators for the unknowns.

4. Examples of Application of a Spectral Finite Difference Method

4.1 Analysis of a Two-dimensional Steady-state Non-Newtonian Fluid Flow

Steady-state characteristics of the flow past a cicular cylinder placed normally to a uniform flow of a polymer solution is analyzed using a spectral finite difference method with a time marching method[3]. The equation of motion and the equation of continuity can be expressed as

$$\rho(D/Dt)\mathbf{V} = -\nabla p + \nabla \cdot \tau + \rho \mathbf{g} \quad , \qquad (1)$$

$$\nabla \cdot \mathbf{V} = 0 \qquad (2)$$

respectively, τ: extra stress tensor, p: pressure, V: velocity vector, ρ: density. The following constitutive equation is adopted:

$$\tau^{ij} = \mu d^{ij} - \lambda \frac{\delta}{\delta t} d^{ij} \quad , \qquad (3)$$

where d^{ij}: rate of deformation tensor, λ : a material constant (≥ 0), $\delta/\delta t$: a convective time derivative operator, and $\lambda = 0$ corresponds to Newtonian fluids.

Then the equation of motion in the stream function-vorticity formulation in a Cartesian coordinate can be given in a dimensionless form by

$$\frac{D}{Dt}\left\{\zeta + \epsilon\left(\frac{\partial^2}{\partial x^2} + \frac{\partial^2}{\partial y^2}\right)\zeta\right\}$$

$$= \frac{1}{Re}\left(\frac{\partial^2}{\partial x^2} + \frac{\partial^2}{\partial y^2}\right)\zeta \quad , \qquad (4)$$

where $Re \equiv \rho U_\infty a/\mu, \epsilon \equiv \lambda/(\rho a^2), a$: radius of the cylinder, ζ : dimensionless vorticity, ψ: a dimensionless stream function; reference length $= a$, reference velocity $= U_\infty$ (free stream velocity). Coordinate transformation:

$$x + i\,y = \exp(\alpha + i\beta) \quad . \qquad (5)$$

Then Eq.(4) leads to

$$\left\{\exp(2\alpha)\frac{\partial}{\partial t} + \frac{\partial\psi}{\partial\beta}\frac{\partial}{\partial\alpha} - \frac{\partial\psi}{\partial\alpha}\frac{\partial}{\partial\beta}\right\}\left\{\zeta + \epsilon\exp(-2\alpha)\right.$$

$$\left.\times\left(\frac{\partial^2}{\partial\alpha^2} + \frac{\partial^2}{\partial\beta^2}\right)\zeta\right\} = \frac{1}{Re}\left(\frac{\partial^2}{\partial\alpha^2} + \frac{\partial^2}{\partial\beta^2}\right)\zeta \quad . \quad (6)$$

Relationship between the stream function and the vorticity is given by

$$\zeta\exp(2\alpha) = -\left(\frac{\partial^2}{\partial\alpha^2} + \frac{\partial^2}{\partial\beta^2}\right)\psi \quad . \qquad (7)$$

Boundary conditions on the surface ($\alpha = 0$) are

$$\psi(\alpha = 0, \beta) = \frac{\partial}{\partial\alpha}\psi(\alpha = 0, \beta) = 0 \quad . \qquad (8)$$

Far-away conditions under a steady-state (designated with a sufix s):

$$\psi_S(\alpha = \alpha_\infty, \beta) = \exp(\alpha_\infty)\sin\beta$$

$$+ \frac{1}{2}C_D\left\{\frac{\beta}{\pi} - \operatorname{erf} Q\right\} \quad , \qquad (9)$$

$$\zeta_S(\alpha = \alpha_\infty, \beta) = -\frac{C_D Re}{2\sqrt{\pi}}\frac{Q}{\exp(\alpha_\infty)}\exp\left(-Q^2\right) \quad , \qquad (10)$$

$$Q = \sqrt{Re}\exp\left(0.5\,\alpha_\infty\right)\sin\left(\frac{\beta}{2}\right) \quad , \qquad (11)$$

$$C_D = \frac{2}{Re}\int_0^\pi \sin\beta\left(\frac{\partial}{\partial\alpha} - 1\right)\zeta(\alpha = 0, \beta)\,d\beta$$

$$- \epsilon\int_0^\pi \cos\beta\ \zeta^2(\alpha = 0, \beta)\,d\beta \quad , \qquad (12)$$

where C_D is a drag coefficient based on $\rho U_\infty^2 a$. Spectral formulation:

$$\begin{bmatrix}\psi\\\zeta\end{bmatrix} = \sum_{n=1}^\infty \begin{bmatrix}\psi_n\\\zeta_n\end{bmatrix}\sin n\beta \quad . \qquad (13)$$

(Even functions can be neglected from the symmetrical configulation.) The n-th components of Eqs.(6) and (7) lead to

$$\frac{\partial}{\partial t}\left\{\exp(2\alpha)\,\zeta_n + \theta\epsilon\left(\frac{\partial^2}{\partial\alpha^2} - n^2\right)\zeta_n\right\}$$

$$- \frac{1}{2}\left\{\sum_{m=1}^{n-1} m\left(\psi'_{n-m}\eta_m - \psi_m\eta'_{n-m}\right)\right.$$

$$+ \sum_{m=1}^\infty m\left(\psi'_{m+n}\eta_m - \psi_m\eta'_{m+n}\right)$$

$$\left.- \sum_{m=n+1}^\infty m\left(\psi'_{m-n}\eta_m - \psi_m\eta'_{m-n}\right)\right\}$$

$$= \frac{1}{Re}\left(\frac{\partial^2}{\partial\alpha^2} - n^2\right)\zeta_n \quad , \qquad (14)$$

$$\eta_n \equiv \zeta_n + \epsilon\,\exp(-2\alpha)\left(\frac{\partial^2}{\partial\alpha^2} - n^2\right)\zeta_n \quad , \qquad (15)$$

$$' \equiv \frac{\partial}{\partial\alpha} \quad ,$$

$$\exp(2\alpha)\ \zeta_n = -\left(\frac{\partial^2}{\partial\alpha^2} - n^2\right)\psi_n \quad , \qquad (16)$$

where θ is a nonpositive parameter introduced for getting higher numerical stability, which does not alter steady-state characters. [Whether a relevant steady-state solution will be attained is a different matter.] Eq.(12) reduces to

$$C_D = \frac{\pi}{Re}\left(\frac{\partial}{\partial\alpha}\zeta_1 - \zeta_1\right)_{\alpha = 0}$$

$$- \epsilon\cdot\frac{\pi}{2}\sum_{n=1}^\infty \zeta_n(\alpha = 0)\zeta_{n+1}(\alpha = 0) \quad . \qquad (17)$$

Numerical integration of the resulting system of equations are based on a semi-implicit method. As initial values for ψ and ζ, the following (a nearly potential flow field $+$ a developing boundary layer) may be used:

$$\psi_i = \left(r - \frac{1}{r}\right)\sin\beta\ \operatorname{erf}\left(\frac{r-1}{2T_0}\right)$$

$$- \frac{4T_0}{\sqrt{\pi}}\sin\beta\left[1 - \exp\left\{-\frac{(r-1)^2}{4T_0^2}\right\}\right] \quad ,$$

$$\zeta_i = -\frac{2}{\sqrt{\pi}T_0}\sin\beta\,\exp\left\{-\frac{(r-1)^2}{4T_0^2}\right\}\ ,\qquad (19)$$

$$T_0 \equiv \sqrt{(t_0/Re)}\ ,\qquad (20)$$

$$r = \exp(\alpha)\ \ ,t_0 \sim 1\ \ (\text{ arbitrary })\ .\qquad (21)$$

As an example, analyzed flow patterns are shown in Fig.1[3].

Similarly, examples of an analysis of a heat

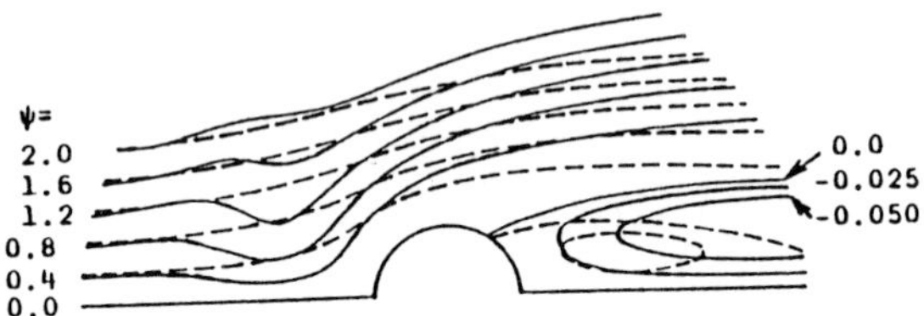

Fig.1: Streamlines in the upper half plane for $Re = 20[3]$. ———: $Re \times \epsilon = 4$, $- - -$: $Re \times \epsilon = 0$. Fluid flows from left to right.

and/or fluid flow past an elliptic cylinder are found in Refs.[4],[5],and [6].

4.2 Analysis of Natural Convection in an Elliptical Cavity Heated from below

Treated is transient natural laminar convection followed by a steady natural laminar convection in an elliptical cavity heated from below[7]. Fluid (air) is assumed to be at rest up to time $t = 0$ in a horizontal elliptic cylinder with a uniform temperature, and at time $t = 0$ the lower half part of the cylinder is assumed to be subject to a step change of temperature. Coordinate transformation:

$$x + i\,y = \exp(i\kappa)\,\sinh(\alpha + i\beta)/\cosh\alpha_0 \qquad (22)$$

where the y-axis is the vertical upward, and $\alpha = \alpha_0$ corresponds to the cylinder surface; κ is the angle between the direction of the major axis of the elliptic section and the vertical upward ($0 \le \kappa \le \pi/2$). Dimensionless governing equations:

$$K\frac{\partial\zeta}{\partial t} + \frac{\partial(\zeta,\psi)}{\partial(\alpha,\beta)} = \frac{1}{\sqrt{Gr}}\left(\frac{\partial^2}{\partial\alpha^2} + \frac{\partial^2}{\partial\beta^2}\right)\zeta + \frac{\partial(T,y)}{\partial(\alpha,\beta)},$$
$$(23)$$

$$K\zeta = -\left(\frac{\partial^2}{\partial\alpha^2} + \frac{\partial^2}{\partial\beta^2}\right)\psi\ ,\qquad (24)$$

$$K\frac{\partial T}{\partial t} + \frac{\partial(T,\psi)}{\partial(\alpha,\beta)} = \frac{1}{Pr\sqrt{Gr}}\left(\frac{\partial^2}{\partial\alpha^2} + \frac{\partial^2}{\partial\beta^2}\right)T\ ,\qquad (25)$$

$$K \equiv \frac{\partial(x,y)}{\partial(\alpha,\beta)} = \left(\frac{\partial y}{\partial\alpha}\right)^2 + \left(\frac{\partial y}{\partial\beta}\right)^2$$
$$= \frac{1}{2\cosh^2\alpha_0}(\cosh 2\alpha + \cos 2\beta)\ ,\qquad (26)$$

where ζ,ψ,T,Gr,Pr denote dimensionless vorticity, a dimensionless stream function, dimensionless temperature [$\equiv (T^* - T_u)/(T_l - T_u)$], Grashof number (based on a and $(T_l - T_u)$), and Prandtl number respectively. T^*: local temperature, T_l: temperature at the lower half of the cylinder surface, T_u: temperature at the upper half cylinder surface. Reference length and reference velocity is defined as the semi-major axis a and $(\nu/a)\sqrt{Gr}$ respectively. (ν: kinematic viscosity.) Boundary conditions at the cylinder surface are:

$$\psi(\alpha_0,\beta,t) = 0\ \ (t > 0)\ ,\qquad (27)$$

$$(\partial/\partial\alpha)\psi(\alpha_0,\beta,t) = 0\ \ (t > 0)\ ,\qquad (28)$$

$$T(\alpha_0,\beta,t) = 0\ \ (-\beta_0 < \beta < \pi - \beta_0)\ ,\qquad (29)$$

$$T(\alpha_0,\beta,t) = 1$$
$$(-\pi \le \beta \le -\beta_0\ \text{ or }\ \pi - \beta_0 \le \beta \le \pi)\ ,\qquad (30)$$

$$\beta_0 = \tan^{-1}(\tanh\alpha_0 \times \tan\kappa)\ \ (0 \le \beta_0 \le \pi/2)\ .\ (31)$$

Virtual boundary conditions at $\alpha = 0$ give

$$\phi(0,\beta,t) = \phi(0,\pi - \beta,t)\ ,\qquad (32)$$

$$(\partial/\partial\alpha)\phi(0,\beta,t) = -(\partial/\partial\alpha)\phi(0,\pi - \beta,t)\ ,\qquad (33)$$

where ϕ stands for ψ,ζ, and T. Initial conditions are:

$$\psi(\alpha,\beta,0) = \zeta(\alpha,\beta,0) = 0\ \ (0 \le \alpha \le \alpha_0)\ ,\qquad (34)$$

$$T(\alpha,\beta,0) = 0\ \ (0 \le \alpha < \alpha_0)\ .\qquad (35)$$

Spectral expressions of primary variables:

$$\begin{bmatrix}\psi\\\zeta\\T\end{bmatrix} = \sum_{n=1}^{\infty}\begin{bmatrix}\psi_{sn}(\alpha,t)\\\zeta_{sn}(\alpha,t)\\T_{sn}(\alpha,t)\end{bmatrix}\sin n\beta$$
$$+ \sum_{n=0}^{\infty}\begin{bmatrix}\psi_{cn}\\\zeta_{cn}\\T_{cn}\end{bmatrix}\cos n\beta\ .\qquad (36)$$

The following are examples corresponding to the zeroth cosine component of Eqs.(23),(24),and (25):

$$\frac{1}{2\cosh^2\alpha_0}\left(\cosh 2\alpha\,\frac{\partial}{\partial t}\zeta_{c0} + \frac{1}{2}\frac{\partial}{\partial t}\zeta_{c2}\right)$$

$$+\frac{1}{2}\sum_{n=1}^{\infty}n\frac{\partial}{\partial\alpha}(\zeta_{cn}\psi_{sn} - \zeta_{sn}\psi_{cn})$$

$$= \frac{1}{\sqrt{Gr}}\frac{\partial^2}{\partial\alpha^2}\zeta_{c0} + \frac{1}{2\cosh\alpha_0}\left\{\cos\kappa\,\frac{\partial}{\partial\alpha}(T_{c1}\cosh\alpha)\right.$$

$$-\sin\kappa\,\frac{\partial}{\partial\alpha}(T_{s1}\sinh\alpha)\Biggr\}\quad,\qquad(37)$$

$$\frac{1}{2\cosh^2\alpha_0}\left(\cosh 2\alpha\,\zeta_{c0}+\frac{1}{2}\zeta_{c2}\right)=-\frac{\partial^2}{\partial\alpha^2}\psi_{c0}\ ,$$
$$(38)$$

$$\frac{1}{2\cosh^2\alpha_0}\left(\cosh 2\alpha\,\frac{\partial}{\partial t}T_{c0}+\frac{1}{2}\frac{\partial}{\partial t}T_{c2}\right)$$
$$+\frac{1}{2}\sum_{n=1}^{\infty}n\frac{\partial}{\partial\alpha}\left(T_{cn}\psi_{sn}-T_{sn}\psi_{cn}\right)$$
$$=\frac{1}{Pr\sqrt{Gr}}\frac{\partial^2}{\partial\alpha^2}T_{c0}\qquad(39)$$

respectively. A part of the boundary conditions, Eq.(28), can be expressed in an implicit way by

$$\frac{2}{h^2}\psi(\alpha_0-h,\beta,t)+\frac{\cosh 2\alpha_0+\cos 2\beta}{2\cosh^2\alpha_0}\zeta(\alpha_0-h,\beta,t)$$
$$=0\ ,\qquad(40)$$

where α_0-h is the coordinate of the grid point adjacent to the surface. Eqs.(29) and (30) reduce to

$$T_{sn}(\alpha_0,t)+i\,T_{cn}(\alpha_0,t)=-2\exp(in\beta_0)/(n\pi)$$
$$(n:\text{odd})\quad,\qquad(41)$$

$$T_{sn}(\alpha_0,t)=T_{cn}(\alpha_0,t)=0$$
$$(n:\text{even}>0)\quad,\qquad(42)$$

$$T_{c0}(\alpha_0,t)=1/2\quad.\qquad(43)$$

If ϕ stands for $\psi,\zeta,$ or T, then likewise Eqs.(32) and (33) reduce to

$$\phi_{sn}(0,t)=(\partial/\partial\alpha)\phi_{cn}(0,t)=0\ (n:\text{even})\quad,\qquad(44)$$

$$\phi_{cn}(0,t)=(\partial/\partial\alpha)\phi_{sn}(0,t)=0\ (n:\text{odd})\quad.\qquad(45)$$

Decomposed initial conditions are not reproduced here. Mean Nusselt number Nu_m (based on a length a) along the heated surface (except for small portion designated by ϵ from the discrete surface temperature point) can be given by

$$Nu_m=\frac{2}{L}\int_{-\beta_0-\pi+\epsilon}^{-\beta_0-\epsilon}\frac{\partial}{\partial\alpha}T(\alpha_0,t)\,d\beta$$
$$\cong\frac{2}{L}\Biggl\{\pi\frac{\partial}{\partial\alpha}T_{c0}(\alpha_0,t)$$
$$-\sum_{n:\text{odd}}^{\lfloor\pi/\epsilon\rfloor}\frac{2}{n}\sin n\beta_0\,\frac{\partial}{\partial\alpha}T_{cn}(\alpha_0,t)$$
$$-\sum_{n:\text{odd}}^{\lfloor\pi/\epsilon\rfloor}\frac{2}{n}\cos n\beta_0\,\frac{\partial}{\partial\alpha}T_{sn}(\alpha_0,t)\Biggr\}\ ,$$
$$(46)$$

where L is the dimensionless perimeter of the elliptic section. Although mathematically $|T|<+\infty$, $Nu_m\propto\ln\epsilon$ as $\epsilon\to+0$. Numerical integration schemes are the same as in the previous section. Analyzed examples of flow/thermal fields are shown in Figs.2 and 3 [7]. Similar treatment on an artificially introduced boundary

is also found in Ref.[8].

Fig.2: Streamlines at a nearly steady state for $b/a=1/5, Gr=10^6, Pr=0.72, \kappa=\pi/2$. Absolute values of ψ inside are 0.015[7].

Fig.3: Isotherms at a nearly steady state for $b/a=1/5, Gr=10^6, Pr=0.72, \kappa=\pi/2$. Differences of temperature between isotherms are 0.2 [7].

4.3 Analysis of a Steady-state Axisymmetric Heat and Fluid Flow Field

Forced convection heat transfer problems from an ellipsoid of revolution is analyzed, where the ellipsoid is placed in a uniform flow such that it is parallel to the axis of revolution[9]. Material properties are assumed to be constant. Under this circumstance the flow field itself is uncoupled with the energy equation, so that the steady-state flow field can be obtained in a time marching way as in the previous cases. The steady-state thermal field can be obtained afterward. Coordinate transformation:

$$z+i\,r=\begin{cases}\cosh(\alpha+i\beta)/\cosh\alpha_0 & (\text{prolate spheroid})\\ \exp(\alpha+i\beta)/\exp\alpha_0 & (\text{sphere})\\ \sinh(\alpha+i\beta)/\sinh\alpha_0 & (\text{oblate spheroid})\end{cases}$$
$$(47)$$

where the z-axis coincides with the symmetrical axis, to which the r-axis is perpendicular. The direction of the positive z-axis is assumed to coincide with that of the flow direction. Dimensionless governing equations are:

$$K\frac{\partial\zeta}{\partial t}+\frac{\partial(\zeta,\psi/r)}{\partial(\alpha,\beta)}$$

$$+\frac{\partial}{\partial\alpha}\left(\frac{1}{r}\frac{\partial r}{\partial\beta}\zeta\frac{\psi}{r}\right)-\frac{\partial}{\partial\beta}\left(\frac{1}{r}\frac{\partial r}{\partial\alpha}\zeta\frac{\psi}{r}\right)$$

$$=\frac{1}{Re}\left\{\left(\frac{\partial^2}{\partial\alpha^2}+\frac{\partial^2}{\partial\beta^2}\right)\zeta+\frac{\partial}{\partial\alpha}\left(\frac{1}{r}\frac{\partial r}{\partial\alpha}\zeta\right)\right.$$

$$\left.+\frac{\partial}{\partial\beta}\left(\frac{1}{r}\frac{\partial r}{\partial\beta}\zeta\right)\right\}\quad,\tag{48}$$

$$-K\zeta=\left(\frac{\partial^2}{\partial\alpha^2}+\frac{\partial^2}{\partial\beta^2}\right)\left(\frac{\psi}{r}\right)$$

$$+\frac{\partial}{\partial\alpha}\left(\frac{1}{r}\frac{\partial r}{\partial\alpha}\frac{\psi}{r}\right)+\frac{\partial}{\partial\beta}\left(\frac{1}{r}\frac{\partial r}{\partial\beta}\frac{\psi}{r}\right)\quad,$$

$$\tag{49}$$

$$\frac{\partial(T,\psi/r)}{\partial(\alpha,\beta)}+\frac{\psi}{r}\frac{1}{r}\frac{\partial(T,r)}{\partial(\alpha,\beta)}$$

$$=\frac{1}{PrRe}\left\{\left(\frac{\partial^2}{\partial\alpha^2}+\frac{\partial^2}{\partial\beta^2}\right)T\right.$$

$$\left.+\left(\frac{1}{r}\frac{\partial r}{\partial\alpha}\frac{\partial}{\partial\alpha}+\frac{1}{r}\frac{\partial r}{\partial\beta}\frac{\partial}{\partial\beta}\right)T\right\}\quad,\tag{50}$$

$$K\equiv\frac{\partial(z,r)}{\partial(\alpha,\beta)}=\left(\frac{\partial r}{\partial\alpha}\right)^2+\left(\frac{\partial r}{\partial\beta}\right)^2\quad,\tag{51}$$

where ψ,ζ,T are a diemsionless stream function, dimensionless vorticity, and dimensionless temperature respectively. $T\equiv(T^*-T_\infty)/(T_m-T_\infty)$, T^* : local temperature. Eqation (50) is a form valid only for a steady-state. In Eq.(48) $Re\equiv U_\infty a/\nu$, U_∞ : free stream velocity, a : one half of the symmetric axis, ν: kinematic viscosity, Pr : Prandtl number, $\alpha_0=\tanh^{-1}(b/a)$ (prolate spheroid), $\alpha_0=0$ (sphere), $\alpha_0=\tanh^{-1}(a/b)$ (oblate spheroid), b: 1/2 of the length of the axis (of the ellipsoid) normal to the symmetrical axis, reference length $=a$, reference velocity $=U_\infty$. Boundary conditions for the velocity field are

$$\psi(\alpha_0,\beta)=0\quad,\tag{52}$$

$$\frac{\partial}{\partial\alpha}\psi(\alpha_0,\beta)=0\quad,\tag{53}$$

$$\psi(\alpha_\infty,\beta)=\frac{1}{2}r^2-\frac{C_D}{2\pi}\left\{\frac{1+\cos\beta}{2}\right.$$

$$\left.-\exp\left(-ReR\sin^2\frac{\beta}{2}\right)\right\}\quad,\tag{54}$$

$$\zeta(\alpha_\infty,\beta)=-\frac{Re^2C_D}{8\pi R}\sin\beta\exp\left(-ReR\sin^2\frac{\beta}{2}\right),$$

$$\tag{55}$$

$$C_D=\frac{\pi}{Re}\left[\int_0^\pi r^3\frac{\partial}{\partial\alpha}\left(\frac{\zeta}{r}\right)d\beta\right]_{\alpha=\alpha_0}\quad,\tag{56}$$

where C_D is a drag coefficient based on $\rho U_\infty^2 a^2$, $R=r/\sin\beta$ at $\alpha=\alpha_\infty$ and $\exp\alpha_\infty\gg 1$. A boundary condition for a thermal field far away is

$$T(R=R_\infty,\beta)$$

$$=\frac{A}{4\pi}Nu_m\left\{\frac{1}{R_\infty}\exp\left(-PrReR_\infty\sin^2\frac{\beta}{2}\right)\right.$$

$$\left.+C_D\frac{Re}{R_\infty^2}G\left(2\sqrt{PrReR_\infty}\sin\frac{\beta}{2}\right)\right\}\quad,\tag{57}$$

where $G()$ is a function[9], and Nu_m stands for a mean Nusselt number based on the temperature difference T_m-T_∞ and a length a ; T_m : mean surface temperature, T_∞ : free stream temperature, A: dimensionless surface area of the body, $R_\infty\equiv R(\alpha=\alpha_\infty)$, R_∞ may be independent of that for the flow field. On the surface ($\alpha=\alpha_0$)in case of a uniform surface temperature

$$T=1\quad,\tag{58}$$

$$A\,Nu_m=-2\pi\int_0^\pi r\frac{\partial T}{\partial\alpha}\,d\beta\quad,\tag{59}$$

and in case of uniform heat flux

$$-\frac{\partial T}{\partial\alpha}=\sqrt{K}\,Nu_m\quad,\tag{60}$$

$$A=2\pi\int_0^\pi rT\sqrt{K}\,d\beta\quad.\tag{61}$$

Spectral expressions for the unknowns:

$$\frac{\psi}{r}=\sum_{n=1}^\infty\psi_n(\alpha,t)\sin n\beta\quad,\tag{62}$$

$$\zeta=\sum_{n=1}^\infty\zeta_n(\alpha,t)\sin n\beta\quad,\tag{63}$$

$$T=\sum_{n=0}^\infty T_n(\alpha)\cos n\beta\quad.\tag{64}$$

Throughout the decomposition the following identity may be used:

$$\sin n\beta\,\cot\beta=\begin{cases}\cos n\beta+2\displaystyle\sum_{k=1}^{(n-1)/2}\cos(2k-1)\beta\\\qquad\qquad(n{:}\text{odd})\\[2mm]\cos n\beta+1+2\displaystyle\sum_{k=1}^{n/2-1}\cos 2k\beta\\\qquad\qquad(n{:}\text{even})\end{cases}$$

$$\tag{65}$$

Eq.(58) reduces to

$$T_0(\alpha_0) = 1 \quad , \qquad (66)$$

$$T_n(\alpha_0) = 0 \quad (n \geq 1) \quad . \qquad (67)$$

Alternatively Eq.(60) reduces to

$$-\frac{\partial}{\partial \alpha} T_0(\alpha_0) = \frac{Nu_m}{\pi} \int_0^{\pi} \left(\sqrt{K}\right)_{\alpha = \alpha_0} d\beta \quad , \qquad (68)$$

$$-\frac{\partial}{\partial \alpha} T_n(\alpha_0) = \frac{2Nu_m}{\pi} \int_0^{\pi} \left(\sqrt{K}\right)_{\alpha = \alpha_0} \cos n\beta \, d\beta$$
$$n \geq 1 \quad . \qquad (69)$$

Discretization in space may be made in the following

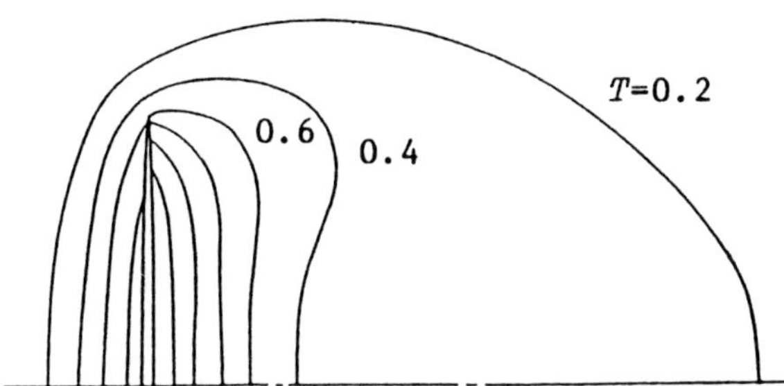

Fig.4: Isotherms in the section of a meridian for an oblate spheroid of $b/a = 50$ with a uniform heat flux at $Re_D[(\equiv 2(b/a)Re)] = 50$ and $Pr = 0.72$. Temperature differences between isotherms are 0.2 . Fluid flows from left to right[9].

non-uniform grid spacing, that is, the coordinate α_n for the n-th grid point is given by

$$\alpha_n = \alpha_0 + h \left\{ \frac{\sinh(n-1)\gamma}{\sinh \gamma} + 1 \right\} \quad , \qquad (70)$$

where γ is a suitable positive constant. This results in substantially doubly-exponentially transformation in the physical space. In the limit $\gamma \to 0$ it produces a uniform grid spacing in α. The initial flow field adopted (a developing boundary-layer flow field superimposed on a potential flow) is similar in a type to, but a little different from Eqs.(18) and (19) for a two-dimensional field. Numerical integration schems are the same as before. However, the thermal field T_n itself is linear, once the flow field is fixed. An example of the thermal field of the numerical analysis is shown in Fig.4 [9].

Other examples treating on the quantities on the symmetrical axis are also found in Refs.[10], [11], [12], and [13].

4.4 Analysis of the Flow Fields Bounded by Two Boundaries

Examples (concerning on, a fluid flow between rotating eccentric cylinders) are found in Refs.[14] and [15]; in these analyses possibly one auxiliary condition that pressure p itself is a scalar function shall be introduced.

5. Application of a Fourier Spectral Finite Difference Method in a Non-Periodic System

Analyzed is a natural transient convection followed by a steady-state natural convection of liquid metals filled in a two-dimensional cut circular cavity consisting of a chord (free surface) and a part of a circle (solid wall) in the section of a cylinder[16]. The following analysis can be applied also to liquid bath. Coordinate transformation in a dimensionless form:

$$x + i\,y = -i \, \tan\left(\frac{\alpha + i\beta}{2}\right) \quad , \qquad (71)$$

where the y-axis is upward and coincident with the symmetrical axis of the circular section, and the x-axis is perpendicular to the y-axis in the section. The range of coordinates is

$$-\infty < \beta < +\infty, \quad 0 \leq \alpha \leq \alpha_0 \quad , \qquad (72)$$

where $\alpha_0 = 2\tan^{-1}(b/a)$, a : one half of the chord length at the free surface, b : depth of the liquid metal bath. The free surface corresponds to $\alpha = 0$ and the solid wall corresponds to $\alpha = \alpha_0$. In terms of dimensionless vorticity ζ, a dimensionless stream function ψ, and dimensionless temperature T, governing equations are:

$$K\frac{\partial \zeta}{\partial t} + \frac{\partial(\zeta, \psi)}{\partial(\alpha, \beta)} = \frac{1}{\sqrt{Gr}} \left(\frac{\partial^2}{\partial \alpha^2} + \frac{\partial^2}{\partial \beta^2}\right)\zeta + \frac{\partial(T, y)}{\partial(\alpha, \beta)},$$
$$(73)$$

$$K\zeta = -\left(\frac{\partial^2}{\partial \alpha^2} + \frac{\partial^2}{\partial \beta^2}\right)\psi \quad , \qquad (74)$$

$$K\frac{\partial T}{\partial t} + \frac{\partial(T, \psi)}{\partial(\alpha, \beta)} = \frac{1}{Pr\sqrt{Gr}} \left(\frac{\partial^2}{\partial \alpha^2} + \frac{\partial^2}{\partial \beta^2}\right)T \quad , \qquad (75)$$

$$K \equiv \frac{\partial(x, y)}{\partial(\alpha, \beta)} = \left(\frac{1}{\cos \alpha + \cosh \beta}\right)^2 \quad , \qquad (76)$$

where length, velocity, and time are made dimensionless with respect to $a, (\nu/a)\sqrt{Gr}$, and $a^2/(\nu\sqrt{Gr})$ respectively; $T \equiv (T^* - T_g)/(T_g - T_a)$; T^* : local liquid metal temperature, Gr: Grashof number $\equiv (T_g - T_a)g\beta^* a^3/\nu^2$; ν: kinematic viscosity of a liquid metal; T_g: gas temperature above the free surface; T_a: room temperature; β^*: coefficient of volume expansion of a liquid metal; Pr : Prandtl number. As boundary conditions, no slip flow on the wall and no shear stress on the free surface (the effect of the gas atmosphere can be neglected) are used, whereas $T = 0$ on the free surface (from the definition) and the profile of the dimensionless heat flux q on the wall (based on a thermal conductivity of the liquid metal and the length a) as well as a mean Nusselt number Nu averaged over the wall of the cylinder, based on the thermal conductivity of the liquid metal and $(T_g - T_a)$ are specified independently of time, and are expressed in terms of T as

$$Nu = \frac{1}{L} \int_{-\infty}^{\infty} q \sqrt{K} \, d\beta \quad , \tag{77}$$

$$q = - \left(\frac{1}{\sqrt{K}} \frac{\partial T}{\partial \alpha} \right)_{\alpha = \alpha_0} \quad , \tag{78}$$

where L is dimensionless wetted length of the liquid metal bath along the wall. On the free surface

$$\psi(\alpha = 0, \beta) = 0 \quad , \tag{79}$$

$$\zeta(\alpha = 0, \beta) = 0 \quad . \tag{80}$$

On the cylinder surface

$$\psi(\alpha = \alpha_0, \beta) = 0 \quad , \tag{81}$$

$$\frac{\partial}{\partial \alpha} \psi(\alpha = \alpha_0, \beta) = 0 \quad . \tag{82}$$

Initial conditions are

$$\psi(\alpha, \beta, t = 0) = \zeta(\alpha, \beta, t = 0) = 0 \quad , \tag{83}$$

$$T(\alpha, \beta, t = 0) = 0 \quad . \tag{84}$$

Spectral formulation of the unknowns:

$$\begin{bmatrix} \psi \\ \zeta \end{bmatrix} = \sum_{n=1}^{\infty} \begin{bmatrix} \psi_n(\alpha, t) \\ \zeta_n(\alpha, t) \end{bmatrix} \sin n\pi\xi \quad , \tag{85}$$

$$T = \sum_{n=1}^{\infty} T_n(\alpha, t) \sin n\pi\xi \quad , \tag{86}$$

$$\xi = \frac{1}{2} (1 + \tanh \beta) \quad . \tag{87}$$

Eq.(83) reduces to

$$\psi_n(0, t) = \zeta_n(0, t) = 0 \quad (n = 1, 2, \cdots) \quad . \tag{88}$$

Also

$$T_n(0, t) = 0 \quad . \tag{89}$$

Eq.(81) reduces to

$$\psi_n(\alpha_0, t) = 0 \quad (n = 1, 2, \cdots) \quad . \tag{90}$$

To support high Grashof number cases, introduced is a non-uniform grid spacing in α, where the n-th coordinate α_n numbered from the wall is specified, using a suitable positive constant γ (h : a constant to be determined to satisfy such a condition that $\alpha = 0$ is on a grid), by

$$\alpha_n = \alpha_0 - h \left\{ \frac{\sinh(n - 1)\gamma}{\sinh \gamma} + 1 \right\} \quad . \tag{91}$$

If Gr or α_0 gets larger, γ can be as large as 0.1 . In general, Eq.(82) can be replaced by

$$-\frac{2}{h^2} \psi(\alpha_0 - h, \beta, t) = (K\zeta)_{\alpha = \alpha_0 - h} \quad , \tag{92}$$

which gives

$$-\frac{1}{h^2} \psi_n(\alpha_0 - h, t)$$
$$= \int_0^1 \sin n\pi\xi \sum_{m=1}^{\infty} (K\zeta_m \sin m\pi\xi)_{\alpha = \alpha_0 - h} \, d\xi$$
$$(n = 1, 2, \cdots) \quad . \tag{93}$$

Eq.(78) reduces to

$$\frac{\partial}{\partial \alpha} T_n(\alpha_0, t) = -2 \int_0^1 \left(\sqrt{K} \right)_{\alpha = \alpha_0}$$
$$\times q \sin n\pi\xi \, d\xi$$
$$(n = 1, 2, \cdots) \quad . \tag{94}$$

Also Eq.(77) is converted to

$$Nu = \frac{1}{2L} \int_0^1 \frac{q}{\xi(1 - \xi)} \left(\sqrt{K} \right)_{\alpha = \alpha_0} d\xi \quad . \tag{95}$$

Decomposition of Eqs.(83) and (84) is not reproduced here. Numerical integration schemes are the same as before. Two examples of steady-state streamlines applied to a case of water at a moderate temperature (not a case of liquid metals) are shown in Figs.5 and 6.

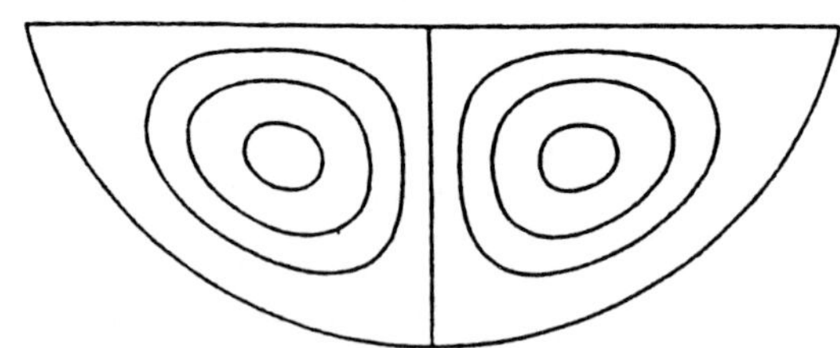

Fig.5: Steady-state streamlines at $Gr = 10^4, b/a = 0.8, Pr = 3, Nu = 0.01, q =$ const. For the left half, $\psi = 6 \times 10^{-5}$ (inside), $4 \times 10^{-5}, 2 \times 10^{-5}, 0$ (wall, the central axis, free surface).

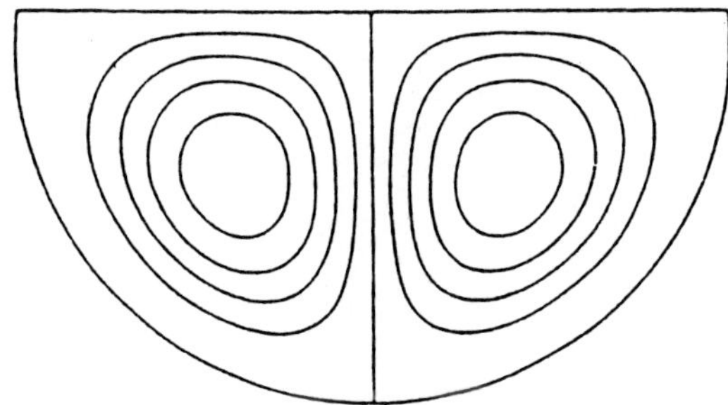

Fig.6: Steady-state streamlines at $Gr = 10^4, b/a = 1.1, Pr = 3, Nu = 0.01, q =$ const. For the left half, $\psi = 8 \times 10^{-4}$ (inside), $6 \times 10^{-4}, 4 \times 10^{-4}, 2 \times 10^{-4}, 0$ (wall, the central axis, free surface).

6. A Spectral Finite Difference Method, using a Dini Expansion

6.1 General Feature

Among many possibilities, the following is a complete set of base functions possessing a period of 2π. For odd functions

$$\mathrm{J}_1\left(\lambda_n \frac{\beta}{\pi}\right) \quad ; |\beta| \leq \pi, \ n \geq 1$$

For even functions

$$\mathrm{J}_0\left(\lambda_n \frac{\beta}{\pi}\right) \quad ; |\beta| \leq \pi, \ n \geq 0$$

where J_0 and J_1 are the zero-th and first order of the first kind of Bessel functions respectively, and λ_n is the n-th positive zero of $\mathrm{J}_1(x)$ if $n > 0$, and $\lambda_0 = 0$.

6.2 An Example of Analysis, Using a Dini Expansion

Treated is transient two-dimensional natural laminar convection followed by a steady natural laminar convection in an elliptic cavity placed horizontally and heated from below [17]. Fluid (air) is assumed to be at rest up to time $t = 0$ with a uniform temperature, and at $t = 0$ the lower half part of the cylinder is assumed to be subject to a step change of a uniform temperature (a Dirichlet type condition) or at $t = 0$ the wall of the cavity is assumed to be subject to a given heat flux profile (a Neumann type condition). Governing equations are the same as Eqs.(23) − (25), where in case of a Dirichlet type condition

$$T \equiv (T^* - T_u)/(T_l - T_u) ,$$

T^* : local temperature, T_u : upper wall temperature. T_l : lower wall temperature, or in case of a Neumann type condition

$$T \equiv (T^* - T_i)/\Delta T$$

T_i : initial fluid temperature, ΔT is so defined that Gr (Grashof number based on ΔT) = 1. The Cartesian coordinate (x, y) and the elliptic one are related by

$$x + iy = -i \sinh(\alpha + i\beta)/\cosh \alpha_0$$

if the major axis is horizontal, or

$$x + iy = -i \cosh(\alpha + i\beta)/\sinh \alpha_0$$

if the major axis is vertical. In these expressions a cavity surface is given by $\alpha = \alpha_0 (> 0)$, and the range of coordinates is

$$0 \leq \alpha \leq \alpha_0 ,$$

$$-\pi < \beta \leq \pi .$$

Under these circumstances, the dimensionless stream function ψ, vorticity ζ, temperature T are assumed to be of the form

$$\begin{bmatrix} \psi \\ \zeta \end{bmatrix} = \sum_{n=1}^{\infty} \begin{bmatrix} \psi_n(\alpha, t) \\ \zeta_n(\alpha, t) \end{bmatrix} \mathrm{J}_1\left(\lambda_n \frac{\beta}{\pi}\right) . \tag{96}$$

$$T = \sum_{n=0}^{\infty} T_n(\alpha, t) \mathrm{J}_0\left(\lambda_n \frac{\beta}{\pi}\right) , \tag{97}$$

For decomposing the governing equations, the following identities may be used:

$$\int_0^{\pi} \beta \mathrm{J}_0\left(\frac{\lambda_n \beta}{\pi}\right) \mathrm{J}_0\left(\frac{\lambda_m \beta}{\pi}\right) d\beta$$

$$= \begin{cases} \frac{\pi^2}{2} \mathrm{J}_0^2(\lambda_n) , & (m = n) \\ 0 , & (m \neq n) \end{cases} , \tag{98}$$

$$\int_0^{\pi} \beta \mathrm{J}_1\left(\frac{\lambda_n \beta}{\pi}\right) \mathrm{J}_1\left(\frac{\lambda_m \beta}{\pi}\right) d\beta$$

$$= \begin{cases} \frac{\pi^2}{2} \mathrm{J}_0^2(\lambda_n) , & (m = n[> 0]) \\ 0 , & (m \neq n, \text{or } m = n = 0) \end{cases} .$$

$$\tag{99}$$

Dynamical boundary conditions lead to

$$\psi_n(\alpha_0, t) = 0 \ (n \geq 1) , \tag{100}$$

$$\frac{2}{h^2} \psi_n(\alpha_0 - h, t)$$

$$= \frac{-2}{\pi^2 \mathrm{J}_0^2(\lambda_n)} \int_0^{\pi} K(\alpha_0, \beta) \zeta(\alpha_0 - h, \beta, t)$$

$$\times \beta \mathrm{J}_1\left(\lambda_n \frac{\beta}{\pi}\right) d\beta , \ (n \geq 1) . \tag{101}$$

The thermal boundary condition in case of a Dirichlet type) leads to

$$T_n\left(\alpha_0, t\right) = \begin{cases} \dfrac{J_1\left(\lambda_n/2\right)}{\lambda_n J_0^2\left(\lambda_n\right)} \, , & (n \geq 1) \\[2mm] 1/4 \, , & (n = 0) \end{cases} \, .$$

(102)

whereas in case of a Neumann type q (heat flux) $= Nu\cos\beta$ (Nu : const.)

$$\frac{\partial}{\partial\alpha}T_n\left(\alpha_0, t\right)$$

$$= -\frac{2Nu}{\pi^2 J_0^2(\lambda_n)}$$

$$\times \int_0^\pi \sqrt{K(\alpha_0,\beta)}\,\beta J_0\left(\lambda_n\frac{\beta}{\pi}\right)\cos\beta d\beta \, ,$$

$$(n \geq 0) \, . \qquad (103)$$

In case the major axis is horizontal, auxiary boundary conditions (between foci) lead to

$$\int_0^\pi \phi(0,\beta,t)\sin 2n\beta \, d\beta = 0 \, ,$$

$$(n: \text{ integer}) \ (\phi = \psi \text{ or } \zeta) \, , \qquad (104)$$

$$\int_0^\pi \frac{\partial}{\partial\alpha}\phi(0,\beta,t)\sin(2n-1)\beta \, d\beta = 0 \, ,$$

$$(n: \text{ integer}) \ (\phi = \psi \text{ or } \zeta) \, , \qquad (105)$$

$$\int_0^\pi T(0,\beta,t)\cos(2n-1)\beta \, d\beta = 0 \, ,$$

$$(n(\text{integer}) \geq 1) \, , \qquad (106)$$

$$\int_0^\pi \frac{\partial}{\partial\alpha}T(0,\beta,t)\cos 2n\beta \, d\beta = 0 \, ,$$

$$(n(\text{integer}) \geq 0) \, . \qquad (107)$$

whereas, in case the major axis is vertical, they lead to

$$\psi_n(0,t) = 0 \, , \ (n \geq 1) \, , \qquad (108)$$

$$\zeta_n(0,t) = 0 \, , \ (n \geq 1) \, , \qquad (109)$$

$$\frac{\partial}{\partial\alpha}T_n(0,t) = 0 \, , \ (n \geq 0) \, . \qquad (110)$$

Figures 7 and 8 show steady-state streamlines and isotherms (cellular flow pattern) for $Pr = 0.7, Gr = 6000, b/a = 0.5$, in case of a Dirichlet condition [Nu_m(mean Nusselt number) $= 2.10$] respectively. Figures 9 and 10 show steady-state streamlines and isotherms respectively for $Pr = 0.7, Nu = -3400$(heated from below), $b/a = 0.5$ in case of a Neumann condition [ΔT_m (mean temperature difference) $= 2039$]; mean Nusselt number based on the temperature difference ΔT_m under a steady-state is 1.18 .

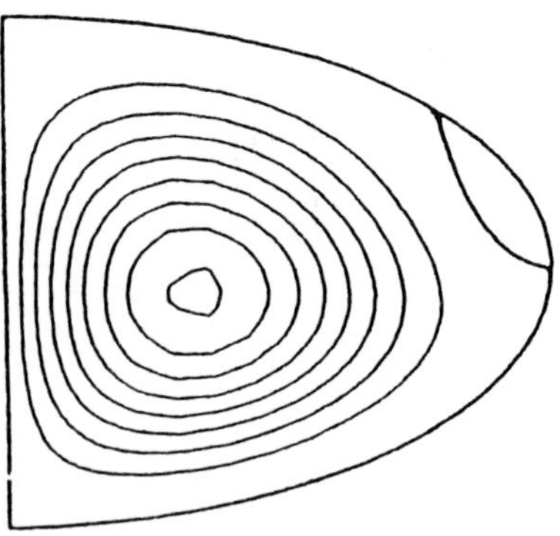

Fi.7: Steady-state streamlines in the right half plane for $Pr = 0.7, b/a = 0.5$ at $Gr = 6000$(Dirichlet problem). Centerline velocity is downward, $\delta\psi = 0.005$.

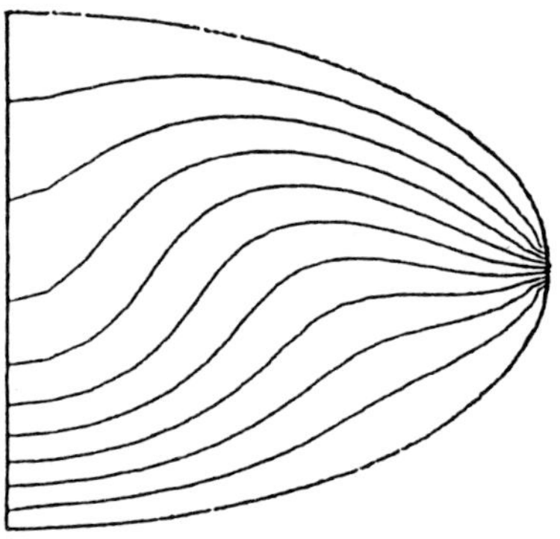

Fig.8: Steady-state isotherms in the right half plane for $Pr = 0.7, b/a = 0.5$ at $Gr = 6000$(Dirichlet problem). δT (temperature difference) $= 0.1$.

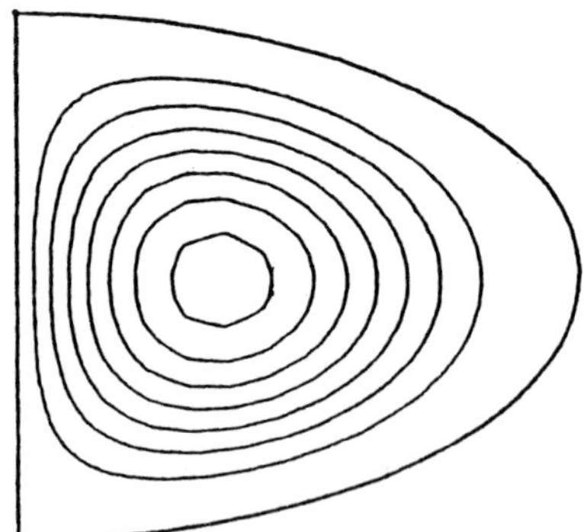

Fig.9: Steady-state streamlines in the right half plane for $Pr = 0.7, b/a = 0.5$ at $Nu = -3400$ (Neumann problem). Centerline velocity is upward, $\delta\psi = 0.1$.

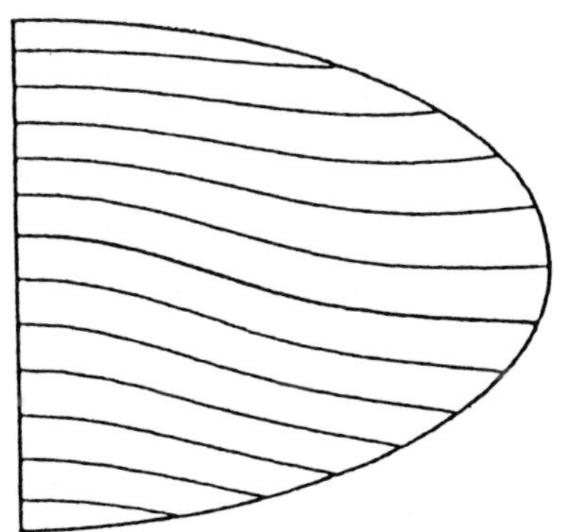

Fig.10: Steady-state isotherms in the right half
plane for $Pr = 0.7, b/a = 0.5$ at
$Nu = -3400$ (Neumann problem).
$\delta T = 250$.

7. A Spectral Finite Difference Method, using A Legendre Expansion

7.1 General Feature

Among many possibilities, the following is a complete set of base functions for functions possessing a period of 2π. For odd functions

$$P_{2n-1}\left(\frac{\beta}{\pi}\right) - P_{2n+1}\left(\frac{\beta}{\pi}\right) \equiv \Phi_{2n-1};$$

$$|\beta| \le \pi,\ n \ge 1 \tag{111}$$

For even functions

$$\begin{cases} P_0\left(\beta/\pi\right) \equiv \Phi_0; \\ \qquad\qquad\qquad\qquad |\beta| \le \pi \\ \\ \dfrac{P_{2n}\left(\beta/\pi\right)}{2n^2+n} - \dfrac{P_{2n+2}\left(\beta/\pi\right)}{2\left(n+1\right)^2+n+1} \equiv \Phi_{2n}; \\ \qquad\qquad\qquad\qquad |\beta| \le \pi,\ n \ge 1 \end{cases}$$

$$\tag{112}$$

where $P_n()$ denotes a Legendre polynomial. The n-th odd component of the equation $f(\beta) = 0$ can be separated through

$$\int_{-\pi}^{\pi} F_{on}(\beta) f(\beta) d\beta = 0 \ , \tag{113}$$

where

$$F_{on} = \frac{1}{2\pi} \sum_{k=1}^{n} (4k-1) P_{2k-1}\left(\frac{\beta}{\pi}\right) \ . \tag{114}$$

if unknowns are expressed in linear combination of the base functions (111) and (112). The n-th even component of the equation $f(\beta) = 0$ can be separated through

$$\int_{-\pi}^{\pi} F_{en}(\beta) f(\beta) d\beta = 0 \ , \tag{115}$$

where

$$F_{en} = \frac{1}{2\pi} \sum_{k=1}^{n} (4k+1)(2k^2+k) P_{2k}\left(\frac{\beta}{\pi}\right) \ , (n \ge 1) \tag{116}$$

$$F_{e0} = \frac{1}{2\pi} \ . \tag{117}$$

The following identity may be used:

$$\int_{-\pi}^{\pi} P_{2k-1}\left(\frac{\beta}{\pi}\right) \frac{\partial^2}{\partial\beta^2} P_{2n-1}\left(\frac{\beta}{\pi}\right) d\beta$$

$$= \begin{cases} 2(n-k)(2n+2k-1)/\pi & (k \le n) \\ 0 & (k > n) \end{cases} \ , \tag{118}$$

$$\int_{-\pi}^{\pi} P_{2k}\left(\frac{\beta}{\pi}\right) \frac{\partial^2}{\partial\beta^2} P_{2n}\left(\frac{\beta}{\pi}\right) d\beta$$

$$= \begin{cases} 2(n-k)(2n+2k+1)/\pi & (k \le n) \\ 0 & (k > n) \end{cases} \ , \tag{119}$$

$$\int_{-\pi}^{\pi} F_{ek} \frac{\partial^2}{\partial\beta^2} \Phi_{2n} d\beta$$

$$= \begin{cases} -\dfrac{(4n+3)(4n^2+15n-1)}{6\pi^2(n+1)} \ , \\ \qquad\qquad\qquad (1 \le n \le k) \\ \\ -\dfrac{(4n+3)k(2k+1)(2k+3)(4k^2+15k-1)}{6\pi^2 n(n+1)(2n+1)(2n+3)} \ , \\ \qquad\qquad\qquad (n > k \ge 1) \end{cases}$$

$$\tag{120}$$

Fig.11: Steady-state streamlines for
$Pr = 0.7, Gr = 10, \delta\psi = 0.00005$.
Centerline velocity near the bottom is
downward.

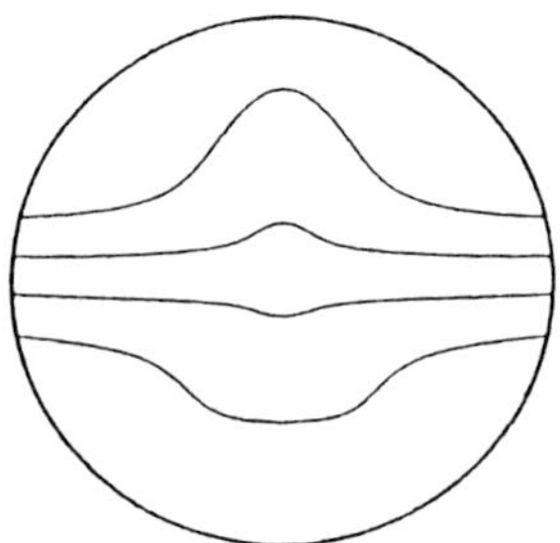

Fig.12: Steady-state isotherms for
$Pr = 0.7, Gr = 10, \delta T = 0.2$.

A numerical example is given for a steady-state two-dimensional natural convecton in a circular cavity heated from below (upper and lower half parts are kept at a constant temperature), where the stream function and vorticity are decomposed into odd order functions, and temperature is decomposed into even order functions.

In Figs.11 and 12, the first three terms for the stream function and vorticity, and the first six terms for temperature are used.

8. Treatment of Different Kinds of Boundary Conditions

8.1 Reduction of Boundary Conditions Expressed Separately to Single Form(s)

One possibility to find a single functional / function valid throughout a closed contour of a boundary where a boundary condition is specified piecewise in a different form for each interval has been given in Ref.[18]. Extended is the following [19] : it is assumed that an unknown quantity ϕ on the said boundary $\alpha = \alpha_0$ (a physical boundary or an artificially equivalent boundary) is given by $\phi(\alpha = \alpha_0, \beta, t) = F_i(\beta, t)$ if $\beta \in [\beta_{i-1}, \beta_i]$ $(i = 1, 2, \cdots, N)$, where $F_i()$ is a discretized form (in space and time) of the boundary condition corresponding to the interval, [type(s) ⟨ open or close ⟩ of the ends of which may be arbitrary], and F_i may be a functional of unknowns such that $(\partial/\partial\phi(\alpha = \alpha_0, \beta; t))\, F_i(\beta; t) \not\equiv 1$; the unknown ϕ is assumed to be piecewise continuous, and $F_i(\beta; t)$ continuous . Then $\phi(\alpha = \alpha_0, \beta, t)$ can be reduced to a single form by

$$\phi(\alpha = \alpha_0, \beta; t)$$

$$- \sum_{i=1}^{N} \{H(\beta - \beta_{i-1}) - H(\beta - \beta_i)\} F_i(\beta; t) = 0 \ .$$

(121)

Hereafter (α, β) is assumed to be a boundary fitted coordinate (produced in a conformal mapping connected to a Cartesian coordinate (x, y)) such that $\alpha = \alpha_0$ corresponds to the boundary and that the space is periodic in β with a period of 2π and

$\beta_N - \beta_0 = 2\pi$; $H()$ denotes a Heviside step function. The form of Eq.(121) can be decomposed into Fourier components, and thus clearly the Fourier components of $\phi(\alpha = \alpha_0, \beta; t)$ is expressed in terms of the functional of F_i's. If F_i is rational with respect to $\phi(\alpha_k, \beta; t)(k = 0, 1, \cdots)$, also the Fourier components of $\phi(\alpha = \alpha_0, \beta; t)$ are rational with respect to the Fourier components of $\phi(\alpha_k, \beta; t)(k = 0, 1, \cdots)$. [Fourier spectral methods are used in the proposed scheme.] Although the form F_i is not necessarily unique (as shown later in a finite difference approximation), it is available if the validity of the interchange of the order of operators (differentiation, integration, discretization) is assumed for the quantities encountered. One possible concrete choice of F_i's for thermal boundary conditions to the unknown T (temperature) is as follows:

$$F_i = \text{const.} \qquad \text{(for a Dirichlet condition)} \quad ,$$

(122)

$$F_i = -\left\{ \sum_{k=1}^{M} A_k T(\alpha_k, \beta; t) \right.$$
$$\left. + q(\beta, t)\sqrt{K(\alpha = \alpha_0, \beta)} \right\} / A_0 \ , \quad (A_0 \neq 0)$$

(for a Neumann condition) , (123)

where

$$\sum_{k=0}^{M} A_k T(\alpha_k, \beta; t) \Big/ \sqrt{K(\alpha = \alpha_0, \beta)} + q(\beta; t) = 0$$

(124)

is assumed to be a discretized form of a Neumann condition [$(\partial/\partial\alpha)T(\alpha = \alpha_0, \beta; t)/\sqrt{K(\alpha = \alpha_0, \beta)} + q(\beta; t) = 0$] and $K \equiv [\partial(x, y)/\partial(\alpha, \beta)]$. Another one possible concrete choice of F_i's for dynamical boundary conditions in a two-dimensional heat and fluid flow field for the unknown ζ (vorticity, which is evaluated at $\alpha = \alpha_0 - h$ just inside the flow region, but not at the boundary) with a free surface facing gaseous atmosphere, resulting in neglecting shearing stresses, is as follows:

$$F_i = 0 \quad \text{(along a free surface)} \ , \qquad (125)$$

$$F_i = - \frac{1}{K(\alpha = \alpha_0 - h, \beta)} \frac{2}{h^2} \psi(\alpha = \alpha_0 - h, \beta; t)$$

(along a solid wall) , (126)

(ψ : stream function). Boundary condition(s) for other quatities if any, which can be specified independently in a single Neumann or Dirichlet condition, will not be affected by the said boundary condition(s).

8.2 Examples of Boundary Conditions

Transient natural laminar convection followed by steady-state convection in an elliptic cavity, the plane of which is placed vertically, with or without open free surface (in a two-dimensional flow model) is considered, where the major axis of the elliptic section is assumed to be horizontal. Let (x, y) be a Cartesian coordinate system such that the origin is located at the center of the elliptic section and that the y-axis is vertically upward:

$$x + iy = -i\,\frac{\sinh(\alpha + i\beta)}{\cosh \alpha_0} \;\; ; \;\; \alpha_0 = \tanh^{-1}(b/a) \;,$$

$$(127)$$

$2a$: major axis, $2b$: minor axis; the location ($\alpha = \alpha_0, \beta = 0$) corresponds to the bottom point. Three types of boundary conditions are considered:

$\langle\langle 1 \rangle\rangle$ Without open surfaces, cooled or heated from below (all boundaries $\alpha = \alpha_0$ consist of solid walls). Half of the wall temperature ($y > 0$) is kept constant, i.e. without loss of generality

$$T(\alpha = \alpha_0, \tfrac{\pi}{2} < |\beta| \le \pi; t) = 0 \;. \qquad (128)$$

Along the lower half of the wall, dimensionless heat flux profile q is specified:

$$q\left(|\beta| < \tfrac{\pi}{2}; t\right) \;\propto\; \cos\beta \qquad (129)$$

along with a mean Nusselt number Nu_m (in the lower half part) based on $\Delta T, a$, and the thermal conductivity of the fluid. It is assumed that $Nu_m > 0$ stands for cooling and $Nu_m < 0$ heating. Dynamical conditions are not affected by the thermal boundary conditions, that is, along all the solid wall

$$\psi(\alpha = \alpha_0, \beta; t) = \frac{\partial}{\partial \alpha}\psi(\alpha = \alpha_0, \beta; t) = 0 \;. \qquad (130)$$

$\langle\langle 2 \rangle\rangle$ Without open surfaces, cooled from above. The lower half wall ($y < 0$) is held at a constant temperature. Under a specified mean Nusselt number Nu_m along the wall ($y > 0$), different thermal boundary conditions are given:

$$T(\alpha = \alpha_0, |\beta| < \tfrac{\pi}{2}; t) = 0 \;, \qquad (131)$$

$$q\left(\tfrac{\pi}{2} < |\beta| \le \pi; t\right) \;\propto\; -\left(1 - \tfrac{1}{2}\cos 4\beta\right)\cos\beta \;. \qquad (132)$$

$\langle\langle 3 \rangle\rangle$ With an open surface, cooled from below. The open surface is assumed to occupy $\pi \ge |\beta| > \beta_1 (> \tfrac{\pi}{2})$ and $\alpha = \alpha_0$ (mathematical modeling, approximation for a physical problem, $(\pi - \beta_1)/\pi \ll 1$). For $\beta_1 < |\beta| \le \pi, \alpha = \alpha_0$:

$$T = 0 \;, \;\; \psi = \zeta = 0 \;. \qquad (133)$$

For $|\beta| < \beta_1, \alpha = \alpha_0$:

$$q\,(\beta; t) \;\propto\; \cos\left(\frac{\pi\beta}{2\beta_1}\right) \;,\;\; \psi = \frac{\partial}{\partial \alpha}\psi = 0 \;. \qquad (134)$$

Also mean Nusselt number along the solid wall is specified.

Concrete choice of F_i corresponding to Eq.(123) is based on the following approximation:

$$\frac{\partial}{\partial \alpha}T(\alpha_0, \beta; t) \cong \{T(\alpha_0, \beta; t) + cT(\alpha_0 - h, \beta; t)$$

$$-(1 + c)T(\alpha_0 - 2h, \beta; t)\}$$
$$/\{(2 + c)h\} \;, \qquad (135)$$

where c is a constant (arbitrarily chosen in principle, $\neq -2$). Also for analytical estimation under Eq.(127), instead of a finite series expansion, the following may be useful:

$$\int_0^{\beta_1} \frac{\cos m\beta}{K(\alpha = \alpha_0 - h, \beta)}\, d\beta$$

$$= \frac{2\cosh^2 \alpha_0}{\sinh 2\alpha_1}\left[(-1)^m \exp(2m\alpha_1)\right.$$

$$\times \sum_{k=m}^{\infty}(-1)^k \exp(-2k\alpha_1)\,\frac{\sin k\beta_1}{k}$$

$$+ \sum_{k=1}^{m-1}(-1)^k \exp(-2k\alpha_1)\,\frac{\sin(m-k)\beta_1}{m-k}$$

$$+(-1)^m \exp(-2m\alpha_1)\left\{\frac{\pi}{2}\right.$$

$$\left.\left.- \tan^{-1}\left(\frac{\exp(-2\alpha_1) + \cos\beta_1}{\sin\beta_1}\right)\right\}\right] \;,$$

$$(136)$$

where $\alpha_1 \equiv \alpha_0 - h(> 0)$, $0 < \beta_1 < \pi$, and m is any positive integer.

8.3 Numerical Results

Thermal fields and flow patterns obtained as a steady-state are shown in Fig.13 $-$ Fig.18

9. Analysis Leading to Direct Integration

Analyzed is a steady slow flow of an incompressible viscous fluid past a circular cylinder placed normally to a uniform flow. Governing equations are the same as in the section 4.1 except for $\epsilon = 0$. Under the assumption of a slow flow, the following estimation may be made: i.e. it is assumed that only the first two terms of the Fourier components are predominant.

Fig.13: Steady-state isotherms (case $\langle\langle 1\rangle\rangle$) for $b/a = 0.1$ at $Pr = 0.7, Gr = 10^8, Nu_m = -5, \delta T$ (temperature difference between isotherms) $= 0.1$.

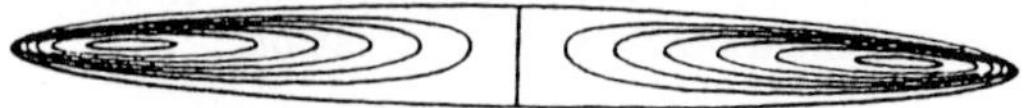

Fig.14: Steady-state streamlines (case $\langle\langle 1\rangle\rangle$) for $b/a = 0.1$ at $Pr = 0.7, Gr = 10^8, Nu_m = -5, \delta\psi$ (stream function difference between streamlines) $= 10^{-4}$, Centerline velocity is upward.

Fig.15: Steady-state isotherms (case $\langle\langle 2\rangle\rangle$) for $b/a = 0.1$ at $Pr = 0.7, Gr = 10^6, Nu_m = 5, \delta T = 0.1$.

Then the governing equations are simplified for the sin β and sin 2β components as

$$\frac{d\psi_1}{d\alpha}\zeta_2 - \psi_2\frac{d\zeta_1}{d\alpha} - \frac{1}{2}\frac{d\psi_2}{d\alpha}\zeta_1 + \frac{1}{2}\psi_1\frac{d\zeta_2}{d\alpha}$$

$$= \frac{1}{Re}\left(\frac{d^2\zeta_1}{d\alpha^2} - \zeta_1\right) \quad , \tag{137}$$

$$\frac{1}{2}\psi_1\frac{d\zeta_1}{d\alpha} - \frac{1}{2}\frac{d\psi_1}{d\alpha}\zeta_1 = \frac{1}{Re}\left(\frac{d^2\zeta_2}{d\alpha^2} - 4\,\zeta_2\right) \quad , \tag{138}$$

$$e^{2\alpha}\zeta_1 = \psi_1 - \frac{d^2\psi_1}{d\alpha^2} \quad , \tag{139}$$

$$e^{2\alpha}\zeta_2 = 4\,\psi_2 - \frac{d^2\psi_2}{d\alpha^2} \quad . \tag{140}$$

If Oseen approximation, which is not necessarily uniformly valid, is applied to Eqs.(137) and (138), i.e.

Fig.16: Steady-state streamlines (case $\langle\langle 2\rangle\rangle$) for $b/a = 0.1$ at $Pr = 0.7, Gr = 10^6, Nu_m = 5, \delta\psi = 10^{-4}$, Centerline velocity is upward.

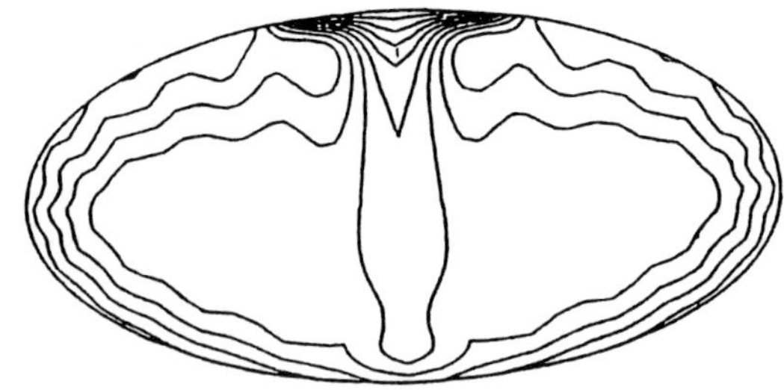

Fig.17: Steady-state isotherms (case $\langle\langle 3\rangle\rangle$) for $b/a = 0.5, \beta_1 = 2.9$ at $Pr = 3, Gr = 10^6, Nu_m = 0.1, \delta T = 0.005$.

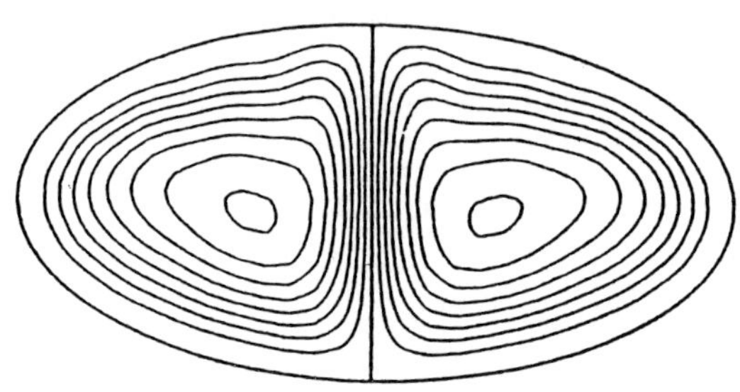

Fig.18: Steady-state streamlines (case $\langle\langle 3\rangle\rangle$) for $b/a = 0.5, \beta_1 = 2.9$ at $Pr = 3, Gr = 10^6, Nu_m = 0.1, \delta\psi = 0.001$, Centerline velocity is upward.

$\psi_1 = \exp\alpha$ and $\psi_2 = 0$ are substituted in the left hand sides, the following are obtained:

$$e^{\alpha}\zeta_2 + \frac{1}{2}e^{\alpha}\frac{d\zeta_2}{d\alpha} = \frac{1}{Re}\left(\frac{d^2\zeta_1}{d\alpha^2} - \zeta_1\right) \quad , \tag{141}$$

$$-\frac{1}{2}e^{\alpha}\zeta_1 + \frac{1}{2}e^{\alpha}\frac{d\zeta_1}{d\alpha} = \frac{1}{Re}\left(\frac{d^2\zeta_2}{d\alpha^2} - 4\,\zeta_2\right) \quad . \tag{142}$$

Boundary conditions are:

$$\psi_n(0) = \frac{\partial}{\partial\alpha}\psi_n(0) = 0 \quad (n = 1, 2) \quad , \tag{143}$$

$$\psi_1 \to e^{\alpha}, \quad \zeta_1 \to 0 \quad \text{as} \quad \alpha \to \infty \quad , \tag{144}$$

$$|\psi_2| < +\infty, \quad \zeta_2 \to 0 \quad \text{as} \quad \alpha \to \infty \quad . \tag{145}$$

Then the following are obtained:

$$\psi_1 = r - \left\{1 + \frac{4}{Re}\frac{K_1(Re/2)}{K_0(Re/2)}\right\}\frac{1}{r} + \frac{4}{Re}\frac{K_1(Re\,r/2)}{K_0(Re/2)} ,$$

$$\tag{146}$$

$$\psi_2 = -\frac{K_1(Re/2)}{K_0(Re/2)} + \left[\frac{4}{Re} + \left\{1 + \left(\frac{4}{Re}\right)^2\right\}\right.$$
$$\left.\times\frac{K_1(Re/2)}{K_0(Re/2)}\right]\frac{1}{r^2} - \frac{4}{Re}\frac{K_2(Re\,r/2)}{K_0(Re/2)} \quad , \tag{147}$$

$$\zeta_1 = -Re\,\frac{K_1(Re\,r/2)}{K_0(Re/2)} \quad , \tag{148}$$

$$\zeta_2 = Re\,\frac{K_2(Re\,r/2)}{K_0(Re/2)} - \frac{K_1(Re/2)}{K_0(Re/2)}\frac{4}{r^2} \quad , \tag{149}$$

where $r \equiv \exp\alpha$. Then the drag coefficient C_D becomes

$$C_D = \frac{\pi}{Re}\left\{\frac{d}{d\alpha}\zeta_1(0) - \zeta_1(0)\right\}$$
$$= \frac{\pi Re}{2}K_2(Re/2) \quad , \tag{150}$$

where $K_n()[(n = 0, 1, 2)]$ is a modified Bessel function of order n. The vorticity on the surface corresponding to the above solution is given by

$$\zeta(0, \beta) = Re\,\sin\beta\,\{2\cos\beta$$
$$- K_1(Re/2)/K_0(Re/2)\} \quad , \tag{151}$$

which predicts the minimum Reynolds number $Re_{\min}$ for producing a pair of standing vortices:

$$2K_0(Re_{\min}/2) = K_1(Re_{\min}/2) \quad , \tag{152}$$

from which $2Re_{\min} = 1.52$, whereas the experimental one ($2Re_{\min}$) is 6.23 as in Ref.[20]. Figure 19 shows the relation between the drag coefficient C_D and Re. Although characteristics obtained by linearized equations do not necessarily show sufficiently quantitative agreement with experimental ones, numerical treatment of a full set of equations for a low or moderate Reynolds number case by a spectral finite difference scheme produces good estimation such as shown in Fig.1 ($\epsilon = 0$).

10. Discussions

10.1 Number of Terms Used for Numerical Analysis

For regular coordinate systems, as far as the values of parameters are moderate, $10-50$ terms can be used. However, if the coordinate system possesses apparent singular points such as in Fig.(5), more terms will be required.

10.2 Coordinate Systems

In case of a little more complicated geometry of the body, it is recommended to use a regular (complex) function that can produce one coordinate surface which is nearly fitting to the body surface and also to the far away condition if any.

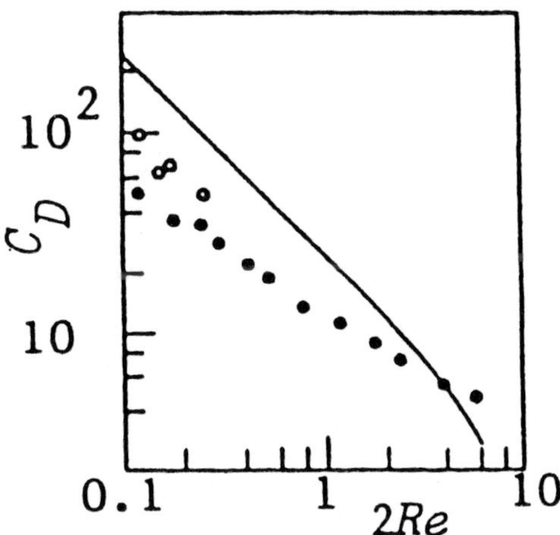

Fig.19: C_D vs. Re for a circular cylinder. ——— : Eq.(150) for a solution of a two-component Oseen type approximation; $\bullet$: experimental data by Finn[21]; $\circ$: experimental data by Jones and Knudsen, cited from Ref.[22].

10.3 Time Integration Schemes

For solving a nonlinear simultaneous equations in a semi-implicit way, it is recommended that the terms appering in an accelerating term and having different indices from those in the Laplacian terms are regarded to be known quantities, which produces a first order accuracy in time. This scheme produces a high speed computation.

11. Conclusions

It is found that the proposed spectral difference method gives fine numerical simulation results for a variety of combination of parameters and for a variety of geometry of heat and fluid flow if non-uniform grid spacing is introduced in case of an extreme combination of physical or geometrical parameters, and that it gives high speed of computation due to the reduction of dimensions.

REFERENCES

[1] J.P.Boyd, *Chebyshev & Fourier Spectral Methods (Lect. Notes in Eng. 49)*, Springer-Verlag, Berlin, (1989).

[2] C.Canuto, M.Y.Hussani, A.Quarteroni, & T.A.Zaug, *Spectral Methods in Fluid Dynamics*, Springer-Verlag, New York, (1988).

[3] Y.Mochimaru, 'Numerical Simulation of a Flow past a Circular Cylinder with Moderate Weissenberg Numbers', *Xth Internat. Congr. Rheology*, vol.2, 131-133(1988).

[4] Y.Mochimaru, 'Numerical Simulation of a Flow past an Elliptic Cylinder at Low and Moderate Reynolds Numbers', *2nd KSME-JSME Fluids Eng. Sym.*, vol.1, 424-429(1990).

[5] Y.Mochimaru, 'Numerical Simulation of Heat Transfer from an Elliptic Cylinder at Moderate Reynolds Numbers', *Computational Mechanics '91*, 714-717(1991).

[6] Y.Mochimaru, 'Numerical Simulation of a Flow past an Elliptic Cylinder at Moderate and High Reynolds Numbers, Using a Spectral Method', *Eleventh Australasian Fluid Mech.*, vol.1, 99-102(1992).

[7] Y.Mochimaru, 'Numerical Simulation of Natural Convection in an Elliptical Cavity Heated from Below, Using a Spectral Finite Difference Method', *2nd Internat. Sym. on Experimental and Computational Aerodynamics of internal Flows*, vol.2, 387-392(1993).

[8] Y.Mochimaru, Y.Gotoh, & M.Gonda, 'Numerical Simulation of Sedimentation Control in a Furnace', *Proc. 1st Int. Conf. Flow Interaction cum Exhibition/Lect. on Interaction of Science & Art*, 278-281(1994).

[9] Y.Mochimaru, 'Numerical Simulation of Forced Convection Heat Transfer from an Ellipsoid of Revolutiom, Using a Fourier Spectral Method', *Transport Phenomena in Thermal Engineering*, vol.1, 161-166(1993).

[10] Y.Mochimaru, 'Transient Natural Convection Heat Transfer in a Spherical Cavity', *Heat Transfer Japanese Research*, vol.18, 9-19(1989).

[11] Y.Mochimaru, 'Numerical Simulation of a Steady Flow Behind a Body of Revolutiom at Moderate and High Reynolds Numbers, Using a Spectral Method', *IUTAM Sym. Bluff-Body Wakes, Dynamics and Instabilities*, 189-192(1992).

[12] Y.Mochimaru, 'Numerical Simulation of a Steady Non-Newtonian Fluid Flow past a Body of Revolution at MOderate Reynolds Numbers, Using a Spectral Method', *Proc. 5th Int. Sym. Computational Fluid Dynamics*, vol.2, 273-278(1993).

[13] Y.Mochimaru, 'Numerical Simulation of Natural Convection in a Cavity of an Ellipsoid of Revolution, Using a Spectral Finite Difference Method', *14th Int. Conf. Numerical Methods in Fluid Dynamics/ Lect. Notes Physics*, Vol.453, 485-489(1994).

[14] Y.Mochimaru, 'Direct Numerical Simulation on Flows in a Complete Journal Bearing, Using a Spectral Method', *Int. Sym. Advanced Computers and Design '89*, 339-344(1989).

[15] Y.Mochimaru, 'Numerical Simulation of a Second Order Fluid Flow between Rotating Eccentric Cylinders, Using a Spectral Method', *China-Japan Int. Conf. Rheology*, 93-96(1991).

[16] Y.Mochimaru, 'Numerical Simulation of Natural Convection in a Cut Circular Cavity', *Internat. Conf. on Fluid and Thermal Energy Conversion '94*, Vol.1, 269-274(1994).

[17] Y.Mochimaru, 'Analysis of Natural Convection in an Elliptical Cavity, Using a Dini Expansion', *3rd Int. Sym. Experimental and Computational Aerothermodynamics of Internal Flows*, 909-916(1996).

[18] Y.Mochimaru, 'A Spectral Finite Difference Scheme and its Effectiveness', *Proc. 26th Ann. Iranian Mathematics Conference*, Vol.2, 245-254(1995).

[19] Y.Mochimaru, 'Extension of Spectral Finite Difference Schemes for Computational Fluid Dynamics', *6th Int. Sym. Computational Fluid Dynamics*, Vol.2, 845 - 850 (1995).

[20] S.W.Churchill, *Viscous Flows : The Practical Use of Theory*, Butterworths, 334-335(1988).

[21] R.K.Finn, *J. Applied Phys.*, vol.24, 771 - 773 (1953).

[22] D.Sucker and H.Brauer, *Wärme und Stoffübertragung*, vol.8, 149 (1975).

SPECTRAL METHODS FOR THE VORTICITY-STREAMFUNCTION EQUATIONS

Roger PEYRET

CNRS; Laboratoire J.A. Dieudonné, Université de Nice-Sophia-Antipolis,
Parc Valrose, 06000 Nice, France,   and
INRIA; 2004, route des Lucioles, 06902 Sophia-Antipolis, France

Abstract

The solution of the Navier-Stokes equations within the vorticity-streamfunction formulation by means of spectral methods is presented and discussed. By using semi-implicit time-discretization schemes, the Navier-Stokes problem amounts to the solution of a Stokes-type problem. The solution of this problem makes use of the influence matrix technique for prescribing the boundary conditions in the Chebyshev collocation approximation. The method is direct and reduces to matrix products to be performed at each time-cycle. Then, the method is adapted to the domain decomposition approach. Finally Fourier-Chebyshev methods for calculating free-surface flows with the vorticity-streamfunction equations are described.

Contents

Received on July 1, 1997.

1 Introduction

For a large variety of problems in fluid mechanics the accuracy is a crucial issue. This is true, for example, when looking for bifurcating solutions where the precise determination of the threshold is of importance. The appearance of unsteady behaviours may be occulted because of the numerical diffusion associated with a low-order approximation. That happened very often in the past, when the numerical methods needed to be highly dissipative to work (e. g. first order upwind finite-differences). With the progress observed in the numerical analysis of the Navier-Stokes equations and with the gain in efficiency of modern computers, a number of flows which were believed to be steady are revealed to be effectively unsteady. So, for numerical studies in fundamental or applied fluid mechanics, the quest for accuracy becomes one of the main issues.

There exist several manners to obtain highly accurate approximations. Increasing the stencil in a finite - difference method is the more easily imaginable way but not necessarily the most efficient. In the context of finite-differences it seems preferable to consider compact Hermitian methods (see [1]).

However. the most accurate results are obtained by means of spectral methods. These methods constitute the ultimate step in the conquest of accuracy since they produce an exponentially small error for infinitely differentiable functions. Contrary to finite-difference or finite-element methods which are local approximations, the spectral methods furnish a global approximation. As a result, these latter are very sensitive to a local perturbation or error which does not remain localized and may pollute the solution everywhere. In particular, the presence of discontinuities introduces Gibbs phenomenon extending in the whole domain. This is the reason why spectral methods should be restricted to the calculation of smooth solutions. Nevertheless, some boundary singularities can be handled by substracting the most singular part from the solution [2]. [3] or by using domain decomposition techniques [4]. Also, efficient procedures for removing the Gibbs phenomenon are presented in [5]. A brief introduction to spectral methods will be presented in Section 2.

For the numerical calculation of two-dimensional incompressible viscous flows in simply connected domains, the use of the vorticity-streamfunction equations (exposed in Section 3) presents undeniable advantages with respect to the velocity-pressure equations. These advantage are well known : the velocity field is automatically divergence-free ; the mathematical properties of the vorticity-stream-function equations allow the construction of robust and efficient numerical methods ; at last. computing time is saved due to the reduced number of equations.

All these arguments, which were put forward as early as the first finite-difference solutions of the Navier-Stokes equations appear, were again valid and even more justified for spectral methods. The difficulty to obtain a divergence-free velocity field as well as a good accuracy for the pressure field (possible existence of spurious modes) made the vorticity-streamfunction formulation more attractive. However, we must recognize that significant advances have been made in the development of spectral methods for the velocity-pressure equations during the last years and that efficient spectral codes, based on these equations, are now available. If the advantages of the vorticity-streamfunction formulation are celebrated, on the other hand its drawbacks cannot be masked. The main drawback, often put forward to justify the lack of interest for the formulation, is that it is restricted to two- dimensional, plane or axisymmetric. flows. The natural extension of the vorticity-streamfunction equations to three-dimensional flows consists in the vorticity-vector potential formulation. Some interesting attemps have been made in [6] and [7] to develop spectral methods for solving the three-dimensional vorticity-vector potential equations. However, some of the arguments in favour for the two-dimensional case do not apply now. The main difficulty is to get solenoidal vorticity and vector potential fields. Therefore, to our opinion, the velocity-pressure formulation remains the best one for the calculation of three-dimensional flows, as far as the vorticity-velocity equations (see e.g. [8]) will not be investigated in the field of spectral methods.

The classical difficulty associated with the vorticity-streamfunction formulation is the lack of boundary conditions for the vorticity ω at a noslip wall, while the streamfunction ψ and its normal derivative $\partial_n\psi$ are prescribed. In finite-difference methods, this difficulty is generally surmounted by using the definition of the vorticity (as the Laplacian of the streamfunction) at the boundary and then by including the Neumann condition on the streamfunction. This local way to remove the difficulty can be replaced by a more global technique based on the use of the Green formula [1], [9] for deriving boundary conditions for the vorticity.

A little in the same spirit, one can look for an operator acting on the boundary values of the vorticity which ensures (in a weak or strong sense) the satisfaction of the Neumann condition on the streamfunction . Such a method has been proposed in [10] in association with a finite-element approximation.

A similar approach can be developed for spectral Chebyshev approximations : the so-called *influence matrix technique*. The influence matrix (IM) is the Chebyshev version of the boundary operator above mentioned. When associated with the collocation method, the IM technique can be interpreted as a special algorithm for the solution of the full discrete problem obtained by prescribing the vorticity and the streamfunction equations to be satisfied at every inner collocation points with the addition of the boundary conditions for the streamfunction and its normal derivative. Therefore, the IM technique is a method of solution rather than a method of approximation. However, the influence matrix approach can be used also for theoretical studies, for example the stability of the time-discretization scheme as done in [11], [12]. The IM technique, also called discrete Green function method, has been introduced in [13] and [14] for one-dimensional problems. In the two-dimensional configuration, it has been developed in [12], [15] and [16].

The time-discretization of the Navier-Stokes equations by means of a semi-implicit scheme will be presented in Section 4. The IM technique used for solving the resulting Stokes-type problem and applications to Navier-Stokes flows will be described in Section 5.

The presentation of the domain decomposition method associated with the vorticity-streamfunction equations makes the object of Section 6. Domain decomposition is useful, and even indispensable, in various circumstances. First, it is the best way, if not the only one, to deal with non-rectangular geometries. Then, it gives an accurate and economical manner to calculate solutions with steep gradients and even with singularities. Moreover, it allows to deal with relatively low order matrices, that is beneficial if recalling the ill-conditioning of Chebyshev differentiation matrices. Finally, the domain decomposition approach is the natural way to use modern parallel computers. The method which will be exposed in the sequel is again based on the extensive use of the influence matrix technique [4],[17]. Some applications of the method will be briefly described.

Finally, Section 7 will be devoted to the calculation of free-surface flows in the frame of the Fourier-Chebyshev approximation. The use of the vorticity-streamfunction equations is not classical for such flows because the pressure is involved in the dynamical free-surface conditions. Therefore, after elimination of the pressure thanks to the momentum equation, one obtains a system of equations for the vorticity, the streamfunction and the height of the free surface. Note that such a formulation has been used with success in [18] in association with finite-difference methods. Two time-discretization schemes will be described. The semi-implicit scheme considered in [19] leads to a direct solution method but may suffer of instabilities when the slope of the free surface becomes large. Therefore, a fully implicit scheme is developed leading to a nonlinear problem solved by an iterative procedure [20], [21]. The extension of the method to special physical situations like stratification and surfactant will be mentioned.

2 FUNDAMENTALS OF SPECTRAL METHODS

The starting point of *spectral methods* is the well known Fourier method that consists in looking for the solution of a given differential problem as a truncated Fourier series expansion. The Fourier method belongs to the general class of Galerkin methods. It is easily applicable when the problem is linear with constant coefficients and is really accurate only if the solution is periodical. The extension without too much complication of the Fourier method to nonlinear problems is obtained in the case of unsteady equations by evaluating the nonlinear part of the equations in an explicit manner. Then, efficiency is preserved if these nonlinear terms are evaluated by means of the so called *pseudospectral technique* that consists in performing differentiations in the spectral space (the space of Fourier coefficients), the products in the physical space (the space of the function values at collocation points) and the link between both spaces being made with fast Fourier transform.

For nonperiodical problems the Fourier basis is no longer adapted and more suitable basis functions have to be considered, for example orthogonal polynomials like Chebyshev or Legendre polynomials are well in favour. The Chebyshev polynomials are advantaged by the possibility to apply fast Fourier transforms.

For the basic principles of spectral methods and associated algorithms we refer to the books by Gottlieb and Orszag [22], Canuto *et al.* [23] and by Boyd [24]. Informations can also be found in [25] and in the book by Peyret and Taylor [1]. In the following, we want to briefly describe the solution of linear simple elliptic equations, in order to give to the reader the notion of what is a spectral method and what is the computational effort required by its implementation.

Let us consider the one-dimensional boundary value problem

$$Lu = f, \quad -1 < x < 1 \tag{2.1}$$

$$B_- u(-1) = 0, \quad B_+ u(1) = 0 \tag{2.2}$$

where L is a linear second-order differential operator with constant coefficients. The solution is looked for in the form of the Chebyshev polynomial expansion

$$u_N(x) = \sum_{k=0}^{N} \hat{u}_k T_k(x) \tag{2.3}$$

where $T_k(x) = \cos(k \cos^{-1} x)$ is the kth Chebyshev polynomial. The coefficients $\hat{u}_k$, $k = 0,..,N$ can be calculated by means of a Galerkin-type method (the *tau method*) in which the equations are obtained by annulating the scalar products

$$(L u_N - f, T_j)_w = 0, \quad j = 0,..., N - 2 \tag{2.4}$$

and associating the boundary conditions (2.2) to close the system :

$$B_- u_N(-1) = 0, \quad B_+ u_N(1) = 0 \tag{2.5}$$

In Eq.(2.4) w refers to the Chebyshev weight $w = (1 - x^2)^{-1/2}$, so that the scalar product writes as:

$$(u, v)_w = \int_{-1}^{1} \frac{u\,v}{\sqrt{1 - x^2}}\,dx. \tag{2.6}$$

Equations (2.4), (2.5) furnish an algebraic system to determine the Chebyshev coefficients $\hat{u}_k$, $k = 0,...,N$. The tau method has been very often used in the past but, mainly for its difficulty to be applied to variable-coefficient equations, it is more and more replaced by the collocation method.

In the Chebyshev collocation method, the collocation points $x_i = 0,..,N$ are generally the extrema of the polynomial of higher degree $T_N(x)$; in other words, they are the zeros of $(1-x^2)\,T_N'(x)$ and their expression is $x_i = \cos(\pi i/N)$, $i = 0,...,N$. The collocation method consists in forcing the equation (2.1) to be satisfied at the inner collocation points and the system is closed by adding the boundary conditions (2.2). Let

$$L u_N(x_i) = f(x_i), \quad i = 1,...N - 1 \tag{2.7}$$

$$B_- u_N(-1) = 0, \quad B_+ u_N(1) = 0. \tag{2.8}$$

Therefore, the method produces an algebraic system for the Chebyshev coefficients $\hat{u}_k$, $k = 0,..,N$. However, the usual way to apply the collocation method is to consider the values of u_N at the collocation points x_i as unknowns rather than the coefficients $\hat{u}_k$. For that, it is necessary to express the derivatives of u_N at a given collocation point in terms of the values of u_N itself. For example, the expression of the first-order derivative is :

$$\partial_x u_N(x_i) = \sum_{j=0}^{N} d_{i,j}^{(1)} u_N(x_j) \tag{2.9}$$

where the coefficients $d_{i,j}^{(1)}$ (given in Appendix) can be calculated from the expansion (2.3). differentiated, evaluated at x_i and with $\hat{u}_k$ defined by the inverse expansion :

$$\hat{u}_k = \frac{1}{2\,\bar{c}_k\,N} \sum_{j=0}^{N} \bar{c}_j\,u_N(x_j)\,T_k(x_j) \tag{2.10}$$

with $\bar{c}_0 = \bar{c}_N = 2$, $\bar{c}_k = 1$ for $1 \leq k \leq N - 1$. The expression of the second-order derivative is accurately obtained by the repeated application of (2.9), that is by calculating the square of the differentiation matrix $\mathcal{D} = [d_{i,j}^{(1)}]$.

Another manner to consider the collocation method and, consequently to calculate the coefficients $d_{i,j}^{(1)}$, is to regard the expansion (2.3) as the Lagrange interpolation polynomial based on the values of u_N at the collocation points x_i, that is :

$$u_N(x) = \sum_{j=0}^{N} p_j(x)\,u_N(x_j) \tag{2.11}$$

with

$$p_j(x) = \frac{(-1)^{j+1}(1 - x^2)\,T_N'(x)}{\bar{c}_j\,N^2(x - x_j)} \tag{2.12}$$

Then, by differentiating (2.11) we get the coefficients $d_{i,j}^{(1)}$ of (2.9).

Generally, problems like (2.1), (2.2) appear, at each time level of an unsteady calculation, with the same operator L. In this case, the rule is to use a method of solution of the algebraic system resulting from (2.7), (2.8) such that the largest part of the calculations is performed in a preprocessing stage executed before to start the time-integration. In these respects the solution of the system by LU decomposition is not necessarily the must economical although very accurate, while the method based on the inversion of the associated matrix can be very efficient (in the present one-dimensional case). The matrix is inverted in the preprocessing stage and its inverse is stored. Then, at each time-cycle, the computational effort reduces to a matrix-vector product which is very efficiently done on a vector computer.

Because the Chebyshev differentiation matrices are ill-conditioned, it is recommended for the inversion of the matrix associated to (2.7), (2.8) to use a routine whose accuracy is very reliable (e.g. FO4AEF from NAG or LINRG from IMSL) or to apply a preconditioning before to perform the inversion. The preconditioning operator can be the finite-difference operator (a tridiagonal matrix easily inverted) approximating the same problem (see [12] for test calculations).

Let us consider now the solution of the two-dimensional Helmholtz equation :

$$\nabla^2 u - \sigma u = f \quad \text{in } D \qquad (2.13)$$

$$B\,u = g \quad \text{on } \Gamma = \partial D \qquad (2.14)$$

Where D is the square $-1 < x, y < 1$, σ is constant and B is a linear boundary operator. We introduce the Chebyshev collocation points in both directions:

$$x_i = \cos \frac{\pi i}{N}, \quad i = 0, .., N,$$
$$\qquad\qquad\qquad\qquad\qquad (2.15)$$
$$y_j = \cos \frac{\pi j}{M}, \quad j = 0, .., M.$$

The approximate solution is looked for as a polynomial $u_{N,M}(x, y)$ of degree at most N in x and M in y, respectively. Then, using differentiation formulas of type (2.9) and after elimination of the boundary values of $u_{N,M}$ thanks to the boundary conditions (2.14), we get the following algebraic system

$$\mathcal{D}_x U_{N,M} + U_{N,M}\,\mathcal{D}_y^T - \sigma U_{N,M} = \mathcal{F} \qquad (2.16)$$

where $U_{N,M}$ is the matrix of inner unknowns $u_{N,M}$ and the matrices $\mathcal{D}_x$ and $\mathcal{D}_y$ are constructed from the differentiation matrices of second order. If the system (2.16) has to be solved at each time-cycle of an unsteady process, an efficient way of solution is to use a full diagonalization procedure [12]. The efficiency of the method lies in the fact that the calculations of eigenvalues and eigenvectors as well as the inversion of the eigenvector matrices are

done once and for all in the preprocessing stage performed before to start the time-integration. So that, at each time-cycle, the computational effort reduces to 4 matrix-matrix products.

As mentioned in the Introduction, the main interest of spectral methods lies in their high accuracy. For example, the error associated with the Fourier as well as the Chebyshev approximation is $O(1/N^\alpha)$ where N refers to the truncation and α is directly connected to the number of continuous derivatives of the function under consideration (see [23]). In particular, for infinitely differentiable functions, α is larger than any integer and then exponential (or "spectral" or "infinite") accuracy is obtained. Such a behaviour has to be compared to the $O(1/N^p)$ error of a finite-difference approximation where $1/N$ is the mesh size and p, which depends on the scheme, is essentially finite and even relatively small.

3   VORTICITY-STREAMFUNCTION EQUATIONS

The Navier-Stokes equations within the vorticity-streamfunction formulation write as :

$$\partial_t \omega + \vec{V}.\nabla \omega - \nu \nabla^2 \omega = f, \qquad (3.1)$$

$$\nabla^2 \psi + \omega = 0, \qquad (3.2)$$

where ω is the vorticity, ψ is the streamfunction. ν the kinematic viscosity and f a given forcing term. In the case of a plane flow, the velocity vector $\vec{V} = (u, v)$ is connected to the streamfunction ψ by the relations :

$$u = \partial \psi_y , \quad v = -\partial \psi_x, \qquad (3.3)$$

and to the vorticity ω by

$$\omega = \partial_x v - \partial_y u . \qquad (3.4)$$

The vorticity-streamfunction formulation is restricted to two-dimensional flows : plane or axisymmetric. In this latter case, the system (3.1)-(3.2) is supplemented by an equation for the azimuthal velocity.

In the following, the description of numerical methods will be for plane flows in two situations : (1) flow in an infinite duct assuming periodicity in the horizontal direction, (2) flow in a simply connected square domain. In both situations, the boundary conditions derive from those prescribed for the velocity field. In the case where the value of the velocity vector is given on the boundary, we get boundary conditions of the general following type :

$$\psi = g, \quad \partial_n \psi = h \quad \text{on } \Gamma \qquad (3.5)$$

where ∂_n refers to the normal derivative to the boundary Γ. The initial condition for ω is deduced from the initial condition for the velocity

$$\vec{V} = \vec{V}_0 \quad \text{at } t = 0, \qquad (3.6)$$

(with $\nabla . \vec{V}_0 = 0$) so that

$$\omega = \partial_x v_0 - \partial_y u_0 = \omega_0 \quad \text{at } t = 0. \qquad (3.7)$$

We refer to [1], [12] and [26] for the general conditions which must be satisfied by g, h and $\vec{V}_0$. A theoretical proof on the equivalence between vorticity-streamfunction and velocity-pressure formulations is given in [26].

Equations (3.1) and (3.2) are partial differential equations of second-order with respect to space for the two dependent variables ω and ψ. Therefore, one should expect to have boundary conditions associated to each of these variables, while two conditions are available for ψ and none for ω. This difficulty has longtime been considered as a real drawback of the vorticity-streamfunction formulation. This difficulty can be, however, completely removed thanks to the influence matrix technique which will described later.

4 TIME-DISCRETIZATION

The equations are discretized in time with a finite-difference scheme. The basic principles of the time-discretization are : implicit treatment of diffusion, explicit for convection and strongly implicit connection between vorticity and streamfunction. A class of schemes fulfilling these requirements is constructed by combining backward-differentiation formula (BDF) for time/diffusion and an Adams-Bashforth extrapolation for convection. The second-order scheme (AB/2BE) of this class has been considered in [15] and its third-order version (AB/3BE) in [27].

The interest of the BDF discretization compared to the well-known Crank-Nicolson scheme appears in its better stability properties. This has been shown in [28] for the advection-diffusion equation. Let us now consider the stability of the

time- discretization of the unsteady Stokes equations :

$$\partial_t \omega - \nu \nabla^2 \omega = 0 \qquad (4.1)$$

$$\nabla^2 \psi + \omega = 0 \qquad (4.2)$$

to be solved in the domain $0 < x < 2\pi$, $-1 < y < 1$. assuming 2π-periodicity in the x-direction and noslip boundary conditions $\psi = 0$. $\partial_y \psi = 0$ at $y = \pm 1$. The initial condition $\omega = \omega_0(x,y)$ is assumed to be periodic in x-direction. The solution is expanded in truncated Fourier series

$$\phi(x,y,t) = \sum_{k=-K+1}^{K} \hat{\phi}_k(y,t)\, e^{i k x} \qquad (4.3)$$

where $\phi = \omega, \psi$. Then, we construct, for each Fourier component $(\hat{\omega}_k, \hat{\psi}_k)$ the following time-discretized problem :

$$\frac{(1+\varepsilon)\,\hat{\omega}_k^{n+1} - 2\varepsilon\,\hat{\omega}_k^{n} - (1-\varepsilon)\,\hat{\omega}_k^{n-1}}{2\,\Delta t}$$

$$-\nu(\partial_{yy} - k^2)\,[\theta\,\hat{\omega}_k^{n+1} + (1-\theta)\,\hat{\omega}_k^{n}] = 0 \qquad (4.4)$$

$$(\partial_{yy} - k^2)\,\hat{\psi}_k^{n+1} + \hat{\omega}_k^{n+1} = 0 \qquad (4.5)$$

to be solved in $-1 < y < 1$ with the boundary and initial conditions :

$$\hat{\psi}_k^{n+1}(\pm 1) = 0. \quad \partial_y \hat{\psi}_k^{n+1}(\pm 1) = 0,$$
$$\hat{\omega}_k^{0}(y) = \hat{\omega}_{0k}. \qquad (4.6)$$

where $(\hat{\omega}_k^n, \hat{\psi}_k^n)$ refers to the approximation at time $n\,\Delta t$. The above scheme is first-order accurate in time whatever the values of the parameters ε and θ . It is second-order accurate if $\varepsilon = 2\,\theta$. The AB/2BE scheme corresponds to $\varepsilon = 2$, $\theta = 1$ and the Crank-Nicolson scheme is obtained for $\varepsilon = 1$, $\theta = 1/2$.

Then, the solution of the above one-dimensional problem in y is approximated with the Chebyshev-collocation method described in Section 2. The stability of the resulting discrete approximation has been studied in [11] and [29], the main results being reported in [12]. In particular, for $\varepsilon = 1$ (two-level schemes) or $\varepsilon = 2\,\theta$ (second-order schemes) it can be shown that a necessary condition for stability is $\theta \geq 1/2$ (numerical calculations show that this condition is also sufficient). For $\theta = 1/2$ (Crank-Nicolson scheme) the scheme is marginally stable and instabilities have been experienced in the case of the Navier- Stokes equations solved with the tau method developed in [15].

The unconditional instability found if $\theta < 1/2$ is remarkable when compared to the conditional stability $(\Delta t < C(\theta)/N^4)$ obtained for the only diffusion equation.

The scheme generally used is the second-order scheme defined by $\varepsilon = 2\theta = 2$. In the case of the Navier-Stokes equations, the nonlinear convective term is explicitly evaluated through the Adams-Bashforth extrapolation

$$(\vec{V}.\nabla\omega)^{n+1} \cong$$
$$2(\vec{V}.\nabla\omega)^n - (\vec{V}.\nabla\omega)^{n-1} \equiv \{\vec{V}.\nabla\omega\}^{n,n-1}$$
$$(4.7)$$

Therefore, the time-discretization of the Navier-Stokes equations (3.1), (3.2) writes as :

$$\frac{3\omega^{n+1} - 4\omega^n + \omega^{n-1}}{2\Delta t} +$$
$$\{\vec{V}.\nabla\omega\}^{n,n-1} - \nu\nabla^2\omega^{n+1} = f^{n+1} \qquad (4.8)$$

$$\nabla^2\psi^{n+1} + \omega^{n+1} = 0 \qquad (4.9)$$

The limitation on the size of the time-step (see [28] for the advection-diffusion equation) due to the explicit evaluation of the nonlinear terms can be avoided by using a fully implicit BDF scheme as done in [30] for the calculations of Rayleigh-Bénard convection. For solving the nonlinear resulting problem these authors use an iterative procedure with preconditioning and relaxation. The interest of such a fully implicit method essentially depends on the possible compromise between the resulting gain in the size of the time-step and the extra-cost induced by the internal iterations. This is highly problem-dependent but is also largely directly connected with the efficiency of the iterative procedure. In these respects, in some situations, it can be more efficient to evaluate dynamically at each iteration the relaxation parameter [20],[21],[31] rather than to keep it constant as done in [30].

Note that second-order accuracy in time could be considered as insufficient when compared to the high accuracy of the spatial approximation. As a matter of fact, for stability reasons, the allowable time-step is often sufficiently small so that the time-accuracy becomes comparable to the space-accuracy. Anyway, higher-order time-accurate schemes can be constructed as mentioned above and exposed in Section 7.

Finally, the implementation of the semi-implicit scheme with the Adams-Bashforth extrapolation

(4.7) necessitates the evaluation of the nonlinear term $\vec{V}.\nabla\omega$ at level n and $n-1$. When dealing with Fourier or Chebyshev approximation this evaluation can be done by the pseudospectral technique whose basic principles have been stated in Section 2. However, in the Chebyshev case, the differentiations can be done directly in the physical space using (2.9) : a matrix product can be less time-consuming than a Fast Fourier Transform if the number of collocation points does exceed a bound (of the order of 64) depending on the computer and the subroutine.

5 INFLUENCE MATRIX METHOD FOR BOUNDARY CONDITIONS

In the context of spectral methods, the influence matrix method has been introduced in [32] for the determination of the values of the pressure at a boundary ensuring the solenoidal character of the velocity field. The same idea was used in [12]-[16] for prescribing the noslip conditions in the vorticity-streamfunction formulation. Therefore, the influence matrix method constitutes an efficient way to remove the classical difficulty associated with the boundary conditions for the vorticity - streamfunctions equations. We shall now describe the method.

5.1 Fourier-Chebyshev case

Coming back to the Navier-Stokes equations (3.1) and (3.2) discretized in time with the semi-implicit scheme (4.8), (4.9), we get a Stokes-type problem to be solved at each time-cycle. Assuming again the periodicity of the solution in the x-direction, the following one-dimensional Stokes-type problem has to be solved for each Fourier component of wave-number k (see eq. (4.3)) :

$$(\partial_{yy} - \sigma)\omega = f, \quad -1 < y < 1, \qquad (5.1)$$

$$(\partial_{yy} - k^2)\psi + \omega = 0, \quad -1 < y < 1, \qquad (5.2)$$

$$\psi(\pm 1) = g_\pm, \qquad (5.3)$$

$$\partial_y\psi(\pm 1) = h_\pm, \qquad (5.4)$$

where $\sigma = k^2 + 3/(2\nu\Delta t)$. For the sake of simplicity, we have posed $\hat{\omega}_k^{n+1} \equiv \omega$ and $\hat{\psi}_k^{n+1} \equiv \psi$ in the above equations. The solution (ω, ψ) is looked for as the combination :

$$\omega = \hat{\omega} + \bar{\omega}, \quad \psi = \hat{\psi} + \bar{\psi}. \qquad (5.5)$$

The part $(\tilde{\omega}, \tilde{\psi})$ satisfies to the equations (5.1) and (5.2) with the boundary conditions

$$\tilde{\omega}(\pm 1) = 0, \quad \tilde{\psi}(\pm 1) = g_{\pm}. \qquad (5.6)$$

Then, the part $(\bar{\omega}, \bar{\psi})$ is solution of the homogeneous equations deduced from (5.1) and (5.2) by setting $f = 0$, with the boundary conditions

$$\bar{\omega}(-1) = \lambda_1, \quad \bar{\omega}(1) = \lambda_2, \quad \bar{\psi}(\pm 1) = 0, \qquad (5.7)$$

where the constants λ_1 and λ_2 have to be found such that the Neumann condition (5.4) is satisfied. This is obtained by considering the decomposition

$$\bar{\omega} = \lambda_1 \bar{\omega}_1 + \lambda_2 \bar{\omega}_2, \quad \bar{\psi} = \lambda_1 \bar{\psi}_1 + \lambda_2 \bar{\psi}_2 \qquad (5.8)$$

where the elementary solution $(\bar{\omega}_l, \bar{\psi}_l)$, $l = 1, 2$, satisfy to the homogeneous equations with the boundary conditions :

$$\begin{array}{lll} \bar{\omega}_1(-1) = 1, & \bar{\omega}_1(1) = 0, & \bar{\psi}_1(\pm 1) = 0, \\ \bar{\omega}_2(-1) = 0, & \bar{\omega}_2(1) = 1. & \bar{\psi}_2(\pm 1) = 0. \end{array}$$

Finally, the constants λ_1 and λ_2 are solution of the algebraic system obtained by prescribing (5.4), let:

$$\partial_y \psi(\pm 1) = \partial_y \tilde{\psi}(\pm 1) + \partial_y \bar{\psi}(\pm 1) = h_{\pm}.$$

The matrix M of this system is the influence matrix (IM).

For each Fourier mode, the algorithm necessitates the solution of six Helmholtz equations with Dirichlet conditions. These problems are solved by means of the Chebyshev (tau or collocation) method briefly described in Section 2. It must be noticed that the Helmholtz problems determining $(\bar{\omega}_l, \bar{\psi}_l)$, $l = 1, 2$, are time-independent and, hence, can be solved once and for all in a preprocessing stage performed before to start the time-integration. Therefore, the matrix M (which is here a simple 2×2 matrix) is constructed and inverted in the preprocessing stage, the inverse M^{-1} as well as the elementary solutions $(\bar{\omega}_l, \bar{\psi}_l)$, $l = 1, 2$ are stored. Consequently, the computational effort at each time-cycle, and for each Fourier mode, for solving the Stokes-type problem reduces to the solution of two Helmholtz problems, that is to say to two matrix-vector products.

It seems important to observe that the influence matrix method is nothing else than a special technique for solving the fully discretized version (tau or collocation) of the system (5.1)-(5.4), where the unknowns are, in the collocation case,

$\omega(y_j)$, $j = 0, \dots M$ and $\psi(y_j)$, $j = 1, .., M - 1$ with $y_j = \cos(\pi j / M)$. The solution of this system would be then obtained by the superposition technique described in [22], p.119 for the tau method. but which applies straightforwardly to the collocation method. However, the equivalence between the full discrete system and the influence matrix technique is true provided that all the differential problems involved in this latest technique are solved by using the same discrete differential operators that those used in the construction of the full discrete system. In particular, the analytical solution of the elementary problems for $(\bar{\omega}_l, \bar{\psi}_l)$, $l = 1, 2$ has to be proscribed because it detroys the equivalence and, hence, can reduce the accuracy of the final solution as experimented in [29].

5.2 Chebyshev-Chebyshev case

The influence matrix method has been extended to the two-dimensional case without periodicity using the Chebyshev tau method in [15] and the Chebyshev collocation method in [12] and [16]. In the two-dimensional case, the Stokes-type problem analogous to (5.1)-(5.4) to be solved is written as :

$$\nabla^2 \omega - \sigma \omega = f \quad \text{in } D, \qquad (5.9)$$

$$\nabla^2 \psi + \omega = 0 \quad \text{in } D, \qquad (5.10)$$

$$\psi = g \quad \text{on } \Gamma = \partial D, \qquad (5.11)$$

$$\partial_n \psi = h \quad \text{on } \Gamma = \partial D, \qquad (5.12)$$

where $D =]-1, 1[\times]-1, 1[$; g and h are assumed to satisfy some conditions of continuity and compatibility at the corners of the domain (see [12]). We shall describe now the collocation method developed in [12], [16]. The solution is again decomposed according to (5.5), let

$$\omega = \tilde{\omega} + \bar{\omega}, \quad \psi = \tilde{\psi} + \bar{\psi}. \qquad (5.13)$$

The part $(\tilde{\omega}, \tilde{\psi})$ is solution of the equations (5.9) and (5.10) with the boundary condition (5.11) supplemented with homogeneous Dirichlet conditions for $\tilde{\omega}$, let

$$\left\{ \begin{array}{ll} \nabla^2 \tilde{\omega} - \sigma \tilde{\omega} = f & \text{in } D, \\ \tilde{\omega} = 0 & \text{on } \Gamma, \end{array} \right. \qquad (5.14)$$

$$\left\{ \begin{array}{ll} \nabla^2 \tilde{\psi} = -\tilde{\omega} & \text{in } D, \\ \tilde{\psi} = g & \text{on } \Gamma. \end{array} \right. \qquad (5.15)$$

These Dirichlet problems are solved by the collocation method described in Section 2. Then, the part $(\bar{\omega}, \bar{\psi})$ is looked for as the linear combination

$$\begin{pmatrix} \bar{\omega} \\ \bar{\psi} \end{pmatrix} = \sum_{l=1}^{L} \lambda_l \begin{pmatrix} \bar{\omega}_l \\ \bar{\psi}_l \end{pmatrix} \qquad (5.16)$$

where the pairs $(\bar{\omega}_l, \bar{\psi}_l)$, $l = 1, .., L$ are solution of the discrete problems :

$$\begin{cases} \nabla^2 \bar{\omega}_l - \sigma\, \bar{\omega}_l = 0 & \text{in } D_c \\ \bar{\omega}_l(\eta_k) = \delta_{kl} & \text{at } \eta_k \in \Gamma_c \end{cases} \qquad (5.17)$$

and

$$\begin{cases} \nabla^2 \bar{\psi}_l = -\bar{\omega}_l & \text{in } D_c \\ \bar{\psi}_l(\eta_k) = 0 & \text{at } \eta_k \in \Gamma_c \end{cases} \qquad (5.18)$$

where D_c is the discretized domain D using the collocation points (2.15), Γ_c is the boundary of D_c, the corners being excluded and η_k refers to the collocation points belonging to Γ_c. Note that the values of ω and ψ at the corners are not involved in the collocation solution. In (5.17), δ_{kl} is the Kronecker symbola.

From (5.16) and the second equation (5.17) it is seen that the constants λ_l, $l = 1, .., L$, are the values of ω at the boundary Γ_c. These values are determined by asking the streamfunction $\psi = \tilde{\psi} + \bar{\psi}$ to satisfy to the Neumann condition (5.12) at $\eta_k \in \Gamma_c$. We obtain the algebraic system :

$$M\,\Lambda = \tilde{F} \qquad (5.19)$$

where $\Lambda = (\lambda_1, .., \lambda_L)^T$ and M is the influence matrix constructed in terms of the elementary solutions.

The remaining question is to determine the proper value of L. For the Chebyshev collocation method based on polynomials of degree N and M in the x and y direction respectively, the number of collocation points (excepted the corners) on the boundary is $2(N + M - 2)$, so that the total number of elementary solutions should be $L = 2(N + M - 2)$. However, for this value of L, the matrix M is found to have four null eigenvalues [12], [15], [16], therefore it is not invertible. This behaviour can be explained in the following manner. As above mentioned, the influence matrix method is nothing else than a special algorithm for solving the full discrete system. Thus, one may think that the singular nature of the influence matrix simply reflects the singular nature of the full discrete system itself. This singular nature comes from the properties of any polynomial $p_{N,M}(x, y)$ to which one asks to be zero at collocation points belonging to the boundary, as well as its normal derivative (the corners are excluded).

To be more precise, let us consider the simpler case where the noslip conditions ($\psi = 0$, $\partial_n \psi = 0$) are prescribed on two adjacent sides : $x = 1$ and $y = 1$, for example. On the other sides ($x = -1$ and $y = -1$), stress-free conditions are assumed, i.e. $\psi = 0$, $\omega = 0$. Then. we consider the streamfunction which is approximated with a polynomial $\psi_{N,M}$ and we write the equations expressing that $\psi_{N,M}$ satisfies to the boundary conditions at all collocation points on Γ_c. Because the values of $\psi_{N,M}$ are given on Γ_c, the only remaining equations are those which express the Neumann condition. We find that one (anyone but one only) of these equations can be expressed as a linear combination of the others. The meaning of this result is : if $\psi_{N,M}$ satisfies to the Dirichlet conditions at all collocation points on Γ_c and the Neumann condition at these points excepted one, then the Neumann condition will be automatically satisfied at the removed point. This behaviour is a direct consequence of the equality of the cross-derivatives of $\psi_{N,M}$ at the corner, i.e. $\partial_x(\partial_y \psi_{N,M}) = \partial_y(\partial_x \psi_{N,M})$ at ($x = 1$, $y = 1$), which furnishes a relationship between the values of the normal derivative on the two sides under consideration. Obviously, the above results do not apply if the two sides where $\partial_n \psi$ is prescribed are no longer adjacent but opposite. In that case, the above equations are found to be linearly independent and the corresponding influence matrix is non singular. This reasoning leads to a simple rule for counting the number of null eigenvalues of the influence matrix according to the type of boundary conditions (prescription of $\partial_n \psi$ or ω) : considering a corner, there is a zero eigenvalue attached to this corner if and only if $\partial_n \psi$ is prescribed on both adjacent sides.

Coming back to the case where ψ and $\partial_n \psi$ are prescribed on the four sides, then the influence matrix has four eigenvalues equal to zero. Therefore, 4 collocation points on the boundary have to be removed when the influence matrix is constructed so that $L = 2(N + M)$. The choice of these points is not completely arbitrary and must follow some rules as shown in [33]. One possible choice [12] is constituted by the set of points $(x_1, -1)$, $(x_{N-1}, -1)$, $(x_1, 1)$, $(x_{N-1}, 1)$. Another possible choice is $(x_1, 1)$, $(1, y_{N-1})$, $(x_{N-1}, -1)$, $(-1, y_1)$. The location of the removed points can have an effect on the conditioning of the influence matrix as found in [4]. With the first set of points above proposed the condition number of the influence matrix behaves like $a(\sigma) N^{1.96}$, with $a(0) = 0.46$ and $a(1000) = 0.03$, this behaviour being determined in [12] from numerical calculations assuming $N = M$.

As above discussed, the Neumann condition on ψ is automatically satisfied at the removed points, although it has not been explicitly prescribed when constructing the IM. Consequently the calculated approximate solution satisfies to the equations at inner collocation points and to the boundary conditions, the approximate solution for the streamfunction is determined in a unique way but the vorticity is uniquely determined only at the inner collocation points. The values of ω at the boundary are not given by the solution method : to get these values, it suffices to employ the definition of the vorticity, that is to evaluate $\omega = -\nabla^2 \psi$ at the boundary.

Therefore, by using the IM technique, the Stokes-type problem is reduced to the solution of a number of Dirichlet problems for the Helmholtz (or Poisson) equation. These problems are easily solved by means of the Chebyshev collocation method associated with the matrix diagonalization procedure mentioned in Section 2. The largest part of the Helmholtz problems (those determining the elementary solutions $\bar{\omega}_l$, $\tilde{\psi}_l$) are time-independent and then they are solved in a preprocessing stage performed before to start the time-integration.

Therefore, the influence matrix M is constructed and inverted during this preprocessing stage ; however, contrary to the one-dimensional case, it is now more efficient to do not store the elementary solutions. At each time-cycle, the computational task is the following : first, two Helmholtz problems have to be solved for determining $(\tilde{\omega}, \tilde{\psi})$: the knowledge of $\tilde{\psi}$ allows us to construct $\tilde{F}$ in Eq. (5.19). Then, the vector Λ is determined from $\Lambda = M^{-1} \tilde{F}$. At last, the final solution (ω, ψ) is calculated by solving again two Helmholtz equations with Dirichlet conditions : $\omega(\eta_k) = \lambda_k$, $k = 1,..,L$ (and $\omega = 0$ at the removed points) and $\psi = 0$.

5.3   Applications

The method has been applied to the calculation of convective flows in various situations. The flow of low Prandtl number fluids submitted to an horizontal temperature gradient has been calculated in [34] by using the Chebyshev-tau influence matrix method developed in [15]. The same method was used in [35] and [36] for the study of the double-diffusive convection induced by the lateral heating of a fluid in a tank, stably stratified by a linear concentration gradient (Fig.1). The case where the heating is applied at the bottom has been considered in [12] using the collocation IM method above described. Note that, in this last reference, the regularized driven cavity flow has been calculated as test case. A study of time-dependent flows in the Bridgman-Stockbarger system for crystal growth has been done in [37] using the same method. Finally the method has been extended to axisymmetric geometry in [38] and [39] for the calculation of rotating flows in an annular configuration.

6   DOMAIN DECOMPOSITION METHOD

The domain decomposition (or "multidomain") method consists in dividing the computational domain in several subdomains, so that the solution is calculated in each subdomain and matched at the interfaces separating two adjacent subdomains. There exist various types of domain decomposition methods. The features of the Chebyshev method [4], [40], [41] exposed in the present section are :

(i) the subdomains are not overlapping.

(ii) the matching condition at an interface are prescribed in a strong sense.

(iii) the solution method is direct, avoiding iterations.

The interest of the domain decomposition method has been evoked in the Introduction. It is presently the most used technique for dealing with nonrectangular domains. Its application to domains which can be divided in rectangles is relatively easy [42]. It becomes more complex when the original domain is more irregular. In such a situation the technique draws nearer to the finite-element method and constitutes the so-called spectral-element method based on a weak formulation and developed in [43], [44], for example.

From the numerical point of view, the domain decomposition method presents also the advantage to deal with matrices of relatively low dimensions, that prevents from inaccuracies associated with the ill-conditioning of the Chebyshev differentiation matrices. Also it is well known that the domain decomposition approach is the most efficient way to solve large scale problems using parallel computers.

At last, the technique presents a great interest for the calculation of solutions exhibiting not only large inner gradients but also thin boundary layers as shown in [45], or even singularities at a boundary due to the local change of boundary conditions [4] or to geometrical singularities [42].

6.1 Fourier-Chebyshev case

Let us briefly describe the two-domain method for the one-dimensional Stokes-type problem (5.1)-(5.4). The domain D $(-1 < y < 1)$ is divided into two sub-domains D_m, $m = 1, 2$, separated by the interface $y = y_I$. Let (ω_m, ψ_m) the solution in D_m and f_m the restriction of f to D_m. Each pair (ω_m, ψ_m) is solution of the equations (5.1), (5.2) with the corresponding conditions (5.3), (5.4) at $y = -1$ or $y = 1$. The matching at the interface is obtained by requiring the continuity of the functions and of the first-order derivatives :

$$\omega_1(y_I) = \omega_2(y_I), \quad \psi_1(y_I) = \psi_2(y_I) \qquad (6.1)$$

$$\partial_y \omega_1(y_I) = \partial_y \omega_2(y_I), \quad \partial_y \psi_1(y_I) = \partial_y \psi_2(y_I) \qquad (6.2)$$

Note that, in some methods like the spectral-element method [43], [44] , the matching conditions can be weaker. In the following, we consider only the case of strong matching conditions (6.1), (6.2).

An efficient solution method is again based on the influence matrix technique [40], [41]. The principle is similar to the one applied for the treatment of the boundary conditions developed in Section 5. Now we are looking for the values of the unknowns at the interface which ensure the continuity of the derivatives. The solution in D_m is looked for as the decomposition

$$\omega_m = \tilde{\omega}_m + \bar{\omega}_m, \quad \psi_m = \tilde{\psi}_m + \bar{\psi}_m, \quad m = 1, 2. \qquad (6.3)$$

The pair $(\tilde{\omega}_m, \tilde{\psi}_m)$, $m = 1, 2$ is solution of

$$\begin{cases} (\partial_{yy} - \sigma)\tilde{\omega}_m = f_m \ \ \text{in} \ \ D_m, \\ \tilde{\omega}_m(y_m^0) = 0, \quad \tilde{\omega}_m(y_I) = 0, \end{cases} \qquad (6.4)$$

$$\begin{cases} (\partial_{yy} - k^2)\tilde{\psi}_m = -\tilde{\omega}_m \ \ \text{in} \ \ D_m, \\ \tilde{\psi}_m(y_m^0) = g_m, \quad \tilde{\psi}_m(y_I) = 0, \end{cases} \qquad (6.5)$$

where $y_m^0 = 2\,m - 3$ such that $y_1^0 = -1$ and $y_2^0 = 1$. and $g_1 = g_-$, $g_2 = g_+$.

Then, the pair $(\bar{\omega}_m, \bar{\psi}_m)$, $m = 1, 2$ is solution of the problem :

$$(\partial_{yy} - \sigma)\bar{\omega}_m = 0 \ \ \text{in} \ \ D_m \qquad (6.6)$$

$$(\partial_{yy} - k^2)\bar{\psi}_m + \bar{\omega}_m = 0 \ \ \text{in} \ \ D_m \qquad (6.7)$$

$$\bar{\omega}_m(y_m^0) = \beta_m, \quad \bar{\omega}_m(y_I) = \lambda_I, \qquad (6.8)$$

$$\bar{\psi}_m(y_m^0) = 0, \quad \bar{\psi}_m(y_I) = \mu_I, \qquad (6.9)$$

Where the 4 constants β_1, β_2, λ_I and μ_I have to be found such that :
(i) the Neumann conditions (5.4) are satisfied, and
(ii) the continuity conditions (6.2) are satisfied. The solution method exactly follows the technique developed in the previous Section. We introduce the linear combinations :

$$\begin{cases} \bar{\omega}_1 = \beta_1 \bar{\omega}_{1,1} + \lambda_I \bar{\omega}_{1,2} + \mu_I \bar{\omega}_{1,3} \\ \bar{\psi}_1 = \beta_1 \bar{\psi}_{1,1} + \lambda_I \bar{\psi}_{1,2} + \mu_I \bar{\psi}_{1,3} \end{cases}$$

and

$$\begin{cases} \bar{\omega}_2 = \beta_2 \bar{\omega}_{2,1} + \lambda_I \bar{\omega}_{2,2} + \mu_I \bar{\omega}_{2,3} \\ \bar{\psi}_2 = \beta_2 \bar{\psi}_{2,1} + \lambda_I \bar{\psi}_{2,2} + \mu_I \bar{\psi}_{2,3} \end{cases}$$

where $(\bar{\omega}_{m,l}, \bar{\psi}_{m,l})$, $m = 1, 2$ and $l = 1, 2, 3$ satisfies to the equations (6.6), (6.7) with the boundary conditions shown in Table 1.

		D_1		D_2	
y	-1		y_I	y_I	1
$\bar{\omega}_{m,1}$	1		0	0	1
$\bar{\psi}_{m,1}$	0		0	0	0
$\bar{\omega}_{m,2}$	0		1	1	0
$\bar{\psi}_{m,2}$	0		0	0	0
$\bar{\omega}_{m,3}$	0		0	0	0
$\bar{\psi}_{m,3}$	0		1	1	0

Table 1

We note that $\bar{\omega}_{1,3} = \bar{\omega}_{2,3} = 0$. Now, by imposing the solution (6.3) to satisfy to the Neumann conditions (5.4) and the continuity conditions (6.2), we obtain an algebraic system which determines the constants β_1, β_2, λ_I and μ_I. The matrix A of this system is the influence matrix.

As in Section 5, the problem reduces to the solution of Helmholtz equations with Dirichlet conditions, which are solved by means of the Chebyshev (collocation or tau) method. As previously, and for the same reasons, some of the Helmholtz problems are solved in the preprocessing stage, and the matrix A is calculated and inverted once and for all.

It seems worthwile to observe that, here again, the IM technique is, in fact, only a special algorithm for solving the full discrete system which could be constructed by discretizing the equations, boundary conditions and matching conditions.

The extension of the method to an arbitrary number of subdomains is straightforward [40] (see also [45] for the three-domain decomposition).

6.2 Chebyshev-Chebyshev case

For the one-dimensional problem above discussed the influence matrix concerns simultaneously the boundary values (β_1, β_2) of ω and the interface values (λ_I, μ_I) of (ω, ψ). It is obvious that such a technique would lead to a very large matrix in the two-dimensional case. This is avoided by separating the problem of boundary conditions from the problem of matching at the interface. In this way, we construct boundary influence matrices (BIM) and a continuity influence matrix (CIM) which are uncoupled ones from the others [4].

Let us now describe this technique for the two-dimensional Stokes-type problem (5.9)-(5.12). The domain D is divided into two subdomains D_1 and D_2, whose interface γ is defined by $x = x_I$ (see Fig.2).

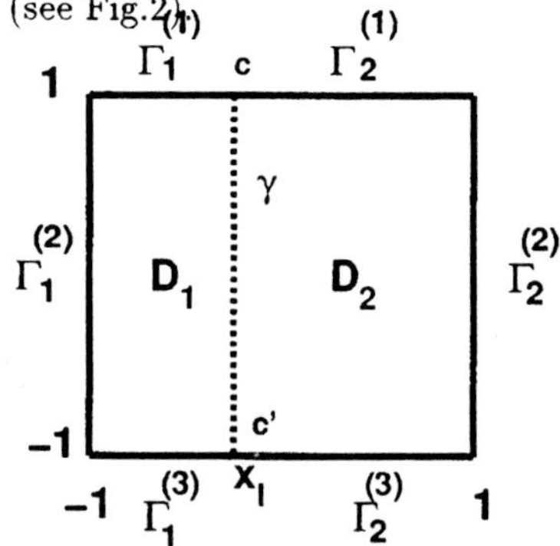

Figure 2: Two-domain decomposition configuration.

The global problem is replaced by a set of two problems (one in each subdomain D_m, $m = 1, 2$) which are coupled by the conditions of continuity at the interface :

$$
\begin{aligned}
\phi_1 &= \phi_2 \\
\partial_x \phi_1 &= \partial_x \phi_2
\end{aligned}
\qquad (6.10)
$$

where $\phi = \omega, \psi$. Let (ω_m, ψ_m) the solution of the Stokes-type problem in D_m, $m = 1, 2$. The decomposition (6.3) is again considered :

$$
\omega_m = \tilde{\omega}_m + \bar{\omega}_m, \quad \psi_m = \tilde{\psi}_m + \bar{\psi}_m, \quad m = 1, 2.
\qquad (6.11)
$$

The pair $(\tilde{\omega}_m, \tilde{\psi}_m)$, $m = 1, 2$ is solution of the problem :

$$
\nabla^2 \tilde{\omega}_m - \sigma \tilde{\omega}_m = f_m \quad \text{in } D_m
\qquad (6.12)
$$

$$
\nabla^2 \tilde{\psi}_m + \tilde{\omega}_m = 0 \quad \text{in } D_m
\qquad (6.13)
$$

$$
\tilde{\psi}_m = g_m, \quad \partial_n \tilde{\psi}_m = h_m \quad \text{on } \Gamma_m
\qquad (6.14)
$$

$$
\tilde{\psi}_m = 0, \quad \tilde{\omega}_m = 0 \quad \text{on } \gamma
\qquad (6.15)
$$

where $\Gamma_m = \Gamma_m^{(1)} \cup \Gamma_m^{(2)} \cup \Gamma_m^{(3)}$ (see Fig.2). This Stokes-type problem is solved by the Chebyshev collocation method associated with the IM technique described in Section 6. The Neumann condition on $\tilde{\psi}_m$ is prescribed only on Γ_m (and not on γ) so that two collocation points on Γ_m have to be excluded when constructing the BIM. A convenient choice is constituted by the two first points near the corners on the horizontal boundaries $\Gamma_m^{(1)}$ and $\Gamma_m^{(3)}$.

The pair $(\bar{\omega}_m, \bar{\psi}_m)$, $m = 1, 2$ is solution of the problem

$$
\nabla^2 \bar{\omega}_m - \sigma \bar{\omega}_m = 0 \quad \text{in } D_m
\qquad (6.16)
$$

$$
\nabla^2 \bar{\psi}_m + \bar{\omega}_m = 0 \quad \text{in } D_m
\qquad (6.17)
$$

$$
\bar{\psi}_m = 0, \quad \partial_n \bar{\psi}_m = 0 \quad \text{on } \Gamma_m
\qquad (6.18)
$$

$$
\bar{\psi}_m = \mu, \quad \bar{\omega}_m = \lambda \quad \text{on } \gamma
\qquad (6.19)
$$

where the values at the interface λ and μ have to be determined such that the continuity of the normal derivatives of ω and ψ is satisfied (see Eq. (6.10)). This is done by using again the IM technique. The pair $(\bar{\omega}_m, \bar{\psi}_m)$ is sought as the linear combination

$$
\begin{pmatrix} \bar{\omega}_m \\ \bar{\psi}_m \end{pmatrix} = \sum_{l=1}^{L} \lambda_l \begin{pmatrix} \bar{\omega}'_{m,l} \\ \bar{\psi}'_{m,l} \end{pmatrix} + \sum_{l=1}^{L} \mu_l \begin{pmatrix} \bar{\omega}''_{m,l} \\ \bar{\psi}''_{m,l} \end{pmatrix}
\qquad (6.20)
$$

where L is the number of inner collocation points η_l belonging to the interface γ.

The pairs $(\bar{\omega}'_{m,l}, \bar{\psi}'_{m,l})$ and $(\bar{\omega}''_{m,l}, \bar{\psi}''_{m,l})$ satisfy to equations (6.16) and (6.17) with boundary conditions (6.18) but λ and μ in (6.19) are replaced by characteristic functions. More precisely, the boundary conditions on γ are :

$$\bar{\omega}'_{m,l}(\eta_l) = \delta_{l,m}, \qquad \bar{\psi}'_{m,l}(\eta_l) = 0$$

$$\bar{\omega}''_{m,l}(\eta_l) = 0, \qquad \bar{\psi}''_{m,l}(\eta_l) = \delta_{l,m}$$

with $m = 1, 2$, $l = 1, .., L$. Therefore, the parameters λ_l and μ_l in (6.20) are respectively the values of $\bar{\omega}_m$ and $\bar{\psi}_m$ at points η_l on γ. These values are solution of the algebraic system obtained by imposing the normal derivatives of ω_m and ψ_m to be continuous across γ. This system writes as :

$$\bar{A}\,X = \tilde{F}$$

where $X = (\lambda_1, .., \lambda_L, \mu_1, .., \mu_L)^T$. The continuity influence matrix (CIM) $\bar{A}$ is also often called "capacitance matrix" or "Schur complement" in the field of domain decomposition methods [46]. In some situations, the matrix $\bar{A}$ may be ill-conditioned. Possible remedies are to consider a suitable preconditioning [4] or, more simply, to invert $\bar{A}$ by means of a conjugate gradient solver [47].

The Stokes-type problems determining the elementary solutions $(\bar{\omega}'_{m,l}, \bar{\psi}'_{m,l})$ and $(\bar{\omega}''_{m,l}, \bar{\psi}''_{m,l})$ are time-independent. As previously, they are solved in the precalculation phase, but they are not stored. The matrix $\bar{A}$ is constructed and inverted during this phase, and its inverse is stored. Therefore, at each time cycle, the computational effort is the following :

1 - Calculation of $(\tilde{\omega}_m, \tilde{\psi}_m)$ by solution of (6.12)-(6.15), and evaluation of $\tilde{F}$.

2 - Calculation of (λ_l, μ_l), $l = 1, .., L$ from $X = \bar{A}^{-1}\tilde{F}$.

3 - Calculation of the final solution (ω_m, ψ_m) by solving a new Stokes-type problem of type (6.12)-(6.15) where the homogeneous boundary conditions (6.15) are replaced by $\omega_m(\eta_l) = \lambda_l$, $\psi_m(\eta_l) = \mu_l$.

Il is interesting to observe that all the Stokes-type problems (whose solution amounts to matrix products) can be solved independently in each subdomain. That leads to an algorithm well adapted to parallel computers, communications between processors being needed only when the interfacial values are calculated.

In case of numerous subdomains, the CIM may become very large. Nevertheless, if the original domain is a long rectangle divided into successive adjacent rectangles separated by parallel interfaces the CIM is block-tridiagonal, each block corresponding to an interface. The resulting block-tridiagonal system is susceptible to be easily solved by classical methods, using in particular the cyclic reduction algorithm. An alternate way is to introduce a dichotomy procedure. This procedure consists by dividing first the original domain into two subdomains ; then the various Stokes problems in each of these two subdomains are solved by performing a new decomposition into two subdomains, and so on. Although the algorithm may become complicated if the number of dichotomies is large, its systematic logic is, however, a valuable advantage. This dichotomy technique has been recently applied in [42].

6.3   Applications

The present domain decomposition method has been employed for studying various problems in fluid mechanics.

The Fourier-Chebyshev version of the method has been used for the calculation of viscous flows with inner layers in which the solution varies rapidly. In such a situation, the domain is decomposed so that the large-gradient region is included in a subdomain in which the degree of the local Chebyshev approximation is high. If the location of the steep gradient is not known or if it is moving with time, the domain decomposition technique is used here in the aim to achieve a more regular distribution of collocation points. It is however possible to develop domain decomposition methods with moving interfaces [45], possibly self-adapted to the stiffness of the solution as recently done in [48].

The evolution of the double-diffusive interface has been studied in [49]. The convective motion of a fluid stratified by concentration and heated by a hot spot located in its bottom has been considered in [41] and [45]. The case where the fluid is heated by an internal heat source was addressed in [50].

Calculations have been done in [51] for studying the creation of convective cells in a fluid stratified by concentration and subjected to a horizontal temperature gradient.

Figure 3 illustrates the development of the temporal instability of a wake submitted to a non-uniform vertical temperature gradient.

In all these applications, the flow is periodic in one direction (Fourier approximation) and the computational domain is divided into three subdomains in the other direction (Chebyshev approximation).

In the two-dimensional case without periodicity the domain decomposition method has been employed with success for calculating flows with singularities at a boundary. An example is the flow in the melt of the Czochralski process for crystal growth considered in [4], [52], in which the vorticity is singular at the contact point between the crystal and the free surface above the melt. A mono-domain solution exhibits large Gibbs oscillations, because it amounts to look for a polynomial approximation of a function having a singularity at an inner point on the line crystal-free surface. In the two-domain solution, the interface is located at the contact point, so that the singularity is rejected at a corner which is not explicitly involved in the Chebyshev collocation technique. Moreover, and principally, every dependent variable is approximated with two different polynomials, one in each subdomain. The result is free of oscillations, its accuracy is largely acceptable although not being exponential.

Another examples of crystal-growth problems have been studied in [47] and [53]. The change in thermal boundary conditions at the wall of the Bridgman-Stockbarger system induces a singularity in the temperature. In that case, a six-domain solution combining flow in the melt and heat conduction in the walls has given physically realistic results [53]. A six-domain decomposition has also been introduced in [47] for calculating the motion of two superposed fluid separated by an horizontal interface assumed to be fixed (multi-layer floating zone system). The fluids are bounded laterally by heated walls in which the conduction for the temperature is taken into account.

Finally, a nonrectangular geometry is considered in [42] and [54]. More precisely the axisymmetric configuration concerns a rotating system made of a cylindrical duct with a lateral cavity (Fig.4). The four-domain technique (three domains in the duct, one domain for the cavity) gives a smooth solution in spite of the presence of the corner singularity for the vorticity.

7 FREE-SURFACE FLOWS

The usual difficulty associated with the calculation of free-surface flows lies in the fact that the computational domain is not rectangular and that its shape is an unknown which varies with time. These features are not easily accepted by spectral methods. However, the actual need of high accuracy in a number of physical problems connected with free surfaces, and more generally with interfaces, makes valuable the efforts for developping efficient spectral methods for such problems. We shall describe now the Fourier-Chebyshev method developed in [19], [20], [21] and [31].

The method is again based on the vorticity-streamfunction equations. The use of such a formulation is not classical for free-surface problems. Generally, one prefers to deal with the primitive variables because the pressure naturally enters into the dynamical free-surface conditions. This apparent drawback is easily removed by using the momentum equation for eliminating the pressure from the free-surface conditions.

7.1 Equations of free-surface flows

We consider a layer of liquid such that $y = h(x, t)$ represents the dimensionless height of the free surface measured from the bottom $y = 0$ (the reference length is the mean height at the initial time). The flow is assumed to be periodic (of dimensionless period A) in the horizontal x-direction. Thanks to a coordinate transformation based on the height of the free surface, the computational domain becomes a fixed rectangle D defined by $0 < X < 2\pi$, $-1 < Y < 1$ where $X = 2\pi x/A$ and $Y = 2y/h - 1$. Let $\Omega(X, Y, t)$, $\Psi(X, Y, t)$ and $H(X, t)$ respectively the vorticity, streamfunction and height of the free surface, the Navier-Stokes equations in dimensionless form write as

$$\partial_t \Omega - \frac{1}{Re} L(\Omega, H) + C(\Omega, \Psi, H) = 0 \qquad (7.1)$$

$$L(\Psi, H) + \Omega = 0 \qquad (7.2)$$

where Re is the Reynolds number ; $L(\phi, H)$ is the transformed Laplacian which can be written as

$$L(\phi, H) = \alpha^2 \partial_{XX}\phi + \frac{4}{H^2}\partial_{YY}\phi + \\ \alpha\partial_X H.N(\phi, H) + \alpha^2 \partial_{XX} H.M(\phi, H) \qquad (7.3)$$

where $\alpha = 2\pi/A$, N and M are two nonlinear differential operators. The nonlinear term C in (7.1) contains the convective term $\vec{V}.\nabla\Omega$ and a term coming from the time-derivative of Ω in the transformed plane (see [19]). The equations (7.1) and (7.2) have to be solved in the domain D with boundary conditions at the bottom $Y = -1$ and the free surface $Y = 1$.

At the bottom $Y = -1$, two types of boundary condition can be considered, noslip :

$$\Psi = 0, \quad \partial_Y \Psi = 0, \qquad (7.4)$$

or stress-free :

$$\Psi = 0, \quad \Omega = 0. \qquad (7.5)$$

At the free surface $Y = 1$, we have to consider the kinematic condition :

$$\partial_t H + \alpha \partial_X \Psi = 0 \qquad (7.6)$$

and the dynamical conditions

$$\Omega + 2\alpha^2 \partial_{XX} \Psi + Q(\Omega, \Psi, H) = 0 \qquad (7.7)$$

$$2\partial_{tY} \Psi + \frac{2}{Re}\left(\partial_Y \Omega - 2\alpha^2 \partial_{XXY} \Psi\right) + R(\Omega, \Psi, H) = 0 \qquad (7.8)$$

where Q and R are two nonlinear differential operators. In the term R appear the two dimensionless parameters: Fr (Froude number) and We (Weber number characteristic of the surface tension).

The equations (7.6)-(7.8) come from the classical free-surface conditions [55] after expressing it in the vorticity-streamfunction variables in the X, Y coordinate system. Moreover, the equation (7.8) is obtained after elimination of the pressure thanks to the momentum equation projected tangentially to the free surface. Finally, the problem is closed by adding the initial conditions at $t = 0$:

$$\vec{V} = \vec{V}^0, \quad \text{then} \quad \Omega = \Omega^0 \qquad (7.9)$$

and

$$H = H^0. \qquad (7.10)$$

7.2 Numerical method of solution

The time-discretization of the motion equations based on the semi-implicit scheme of type (4.8), (4.9) presents a difficulty due to the nonlinearity of the transformed Laplacian (7.3). The fully implicit evaluation of $L(\phi, H)$ would leave the problem nonlinear. This is avoided [19] by considering the second-order approximation :

$$L(\phi^{n+1}, H^{n+1}) \simeq L(\phi^{n+1}, H_0) + \{L(\phi, H) - L(\phi, H_0)\}^{n,\,n-1} \qquad (7.11)$$

with $\{\,.\,\}^{n,\,n-1}$ is the Adams-Bashforth extrapolation defined by (4.7). The constant H_0 must satisfy the condition $H_0^2 < 4H^2/3$ in order to ensure the linear stability of the (Fourier-Chebyshev) approximation. In practice, H_0 is taken equal (or close) to the mean value of the initial height H^0.

Bringing the approximation (7.11) into the Navier-Stokes equations (7.1), (7.2) and after introducing the semi-implicit time-discretization of type (4.8), (4.9) in (7.1), (7.2) as well as in the dynamical free-surface conditions (7.7), (7.8), we get the following problem which defines Ω^{n+1}, Ψ^{n+1} :

$$\alpha^2 \partial_{XX} \Omega^{n+1} + \frac{4}{H_0^2} \partial_{YY} \Omega^{n+1} - \sigma \, Re\,\Omega^{n+1} = f_1 \quad \text{in } D \qquad (7.12)$$

$$\alpha^2 \partial_{XX} \Psi^{n+1} + \frac{4}{H_0^2} \partial_{YY} \Psi^{n+1} + \Omega^{n+1} = f_2 \quad \text{in } D \qquad (7.13)$$

with the free-surface conditions at $Y = 1$

$$\Omega^{n+1} + 2\alpha^2 \partial_{XX} \Psi^{n+1} = f_3 \qquad (7.14)$$

$$2\sigma \partial_Y \Psi^{n+1} + \frac{2}{Re} \partial_Y \Omega^{n+1} - \frac{4\alpha^2}{Re} \partial_{XXY} \Psi^{n+1} = f_4 \qquad (7.15)$$

and the boundary conditions (7.4) or (7.5) at the bottom $Y = -1$. In (7.12) and (7.15) the constant σ is defined by $\sigma = 2\,\Delta t/3$. Then, the kinematic condition (7.6), at $Y = 1$, is discretized in time according to :

$$\frac{1}{2\Delta t}(3H^{n+1} - 4H^n + H^{n-1}) + \alpha \partial_X \Psi^{n+1} = 0 \qquad (7.16)$$

gives directly the height H^{n+1}. Note that a little more implicit version of the scheme for the condition (7.15) is also considered in [19].

The above linear problem is solved by means of Fourier series expansion in the X-direction and Chebyshev collocation in the Y-direction. The free-surface conditions and the noslip condition at the bottom are simultaneously prescribed thanks to the influence matrix technique.

The method, which has been tested on various academic examples in [19] may be subject to nonlinear instabilities when the slope of the free surface becomes too large. Therefore, a fully implicit method has been developed in [20] and [21]. The method allows us to calculate larger deformations of the free surface but at the price of an iterative procedure. For the description of the implicit method, we write the full system of equa-

409

tions, free-surface and boundary conditions as

$$\mathcal{A}\partial_t\Phi + \mathcal{B}\Phi = 0 \qquad \text{in } D$$

$$\bar{\mathcal{A}}\partial_t\bar{\Phi} + \bar{\mathcal{B}}\bar{\Phi} = 0 \qquad \text{at } Y = 1 \qquad (7.17)$$

$$\mathcal{B}_0\Phi = 0 \qquad \text{at } Y = -1$$

where $\Phi = (\Omega, \Psi)^T$ and $\bar{\Phi} = (\Omega, \Psi, H)^T$. This system (7.17) is discretized in time with a fully implicit scheme where the time derivatives are approximated with

$$(\partial_t\Phi)^{n+1} \simeq \frac{1}{\Delta t}\sum_{j=0}^{J} a_j\Phi^{n+1-j} \qquad (7.18)$$

with $a_0 = 3/2$, $a_1 = -2$, $a_2 = 1/2$ for the three-level second-order scheme and $a_0 = 11/6$, $a_1 = -3$, $a_2 = 3/2$, $a_3 = -1/3$ for the four-level third-order scheme. All the other terms in (7.17) are evaluated at level $n+1$. The resulting semi-discrete system writes as

$$\mathcal{C}\Phi^{n+1} = \mathcal{F} \qquad \text{in } D$$

$$\bar{\mathcal{C}}\bar{\Phi}^{n+1} = \bar{\mathcal{F}} \qquad \text{at } Y = 1 \qquad (7.19)$$

$$\mathcal{B}_0\Phi^{n+1} = 0 \qquad \text{at } Y = -1$$

where $\mathcal{C}$ and $\bar{\mathcal{C}}$ are nonlinear differential operators. The problem (7.19) is solved by means of the following preconditioned iterative procedure

$$\mathcal{C}_0\delta\Phi^{n+1,m+1} \quad = \mathcal{F} - \mathcal{C}\Phi^{n+1,m} \qquad \text{in } D$$

$$\bar{\mathcal{C}}_0\delta\bar{\Phi}^{n+1,m+1} \quad = \bar{\mathcal{F}} - \bar{\mathcal{C}}\bar{\Phi}^{n+1,m} \qquad \text{at } Y = 1$$

$$\mathcal{B}_0\delta\Phi^{n+1,m+1} \quad = 0 \qquad\qquad\quad \text{at } Y = -1$$

$$(7.20)$$

and

$$\bar{\Phi}^{n+1,m+1} = \bar{\Phi}^{n+1,m} + \gamma\,\delta\bar{\Phi}^{n+1,m+1} \qquad (7.21)$$

where the superscript m refers to the iteration. The iteration is started with $\bar{\Phi}^{n+1,0} = \bar{\Phi}^n$. The preconditioning operators $\mathcal{C}_0$ and $\bar{\mathcal{C}}_0$ are linear operators respectively deduced from $\mathcal{C}$ and $\bar{\mathcal{C}}$ by replacing H by H_0 and by neglecting all nonlinear terms. As a matter of fact, $\mathcal{C}_0$ and $\bar{\mathcal{C}}_0$ are the operators appearing in the semi-implicit scheme (7.12)-(7.15). Again, the linear problem (7.20) is solved by the Fourier-Chebyshev method above mentioned in the case of the semi-implicit scheme. The relaxation parameter γ is either a constant or dynamically calculated at each iteration. For

$\gamma =$ constant, a linear study [21] gives the necessary condition of convergence $\gamma < 2(H/H_0)^2$ provided the polynomial degree N is sufficiently large. In some cases, the use of a constant relaxation parameter cannot ensure convergence. Therefore it is calculated at each iteration from the residual of the vorticity equation

$$\mathcal{C}_{\Omega_0}\delta\Omega^{n+1,m+1} = \mathcal{F}_\Omega - \mathcal{C}_\Omega\Omega^{n+1,m} = \mathcal{R}_\Omega^m.$$

The parameter $\gamma = \gamma_m$ can be defined by the Minimal Residual technique (see [23]) or, better, by the following expression [20], [21]

$$\gamma_m = \frac{\|\,\mathcal{R}_\Omega^m\,\|}{\|\,\mathcal{C}_\Omega\mathcal{C}_{\Omega_0}^{-1}\mathcal{R}_\Omega^m\,\|} \qquad (7.22)$$

where $\|\,\phi\,\|$ is the Euclidean norm of the vector consisting of the values of ϕ at the collocation points in both spatial directions.

7.3 Applications

In [19], the authors have employed the second-order time-accurate semi-implicit method for the calculation of academic free-surface flows without surface tension : damping of a wave and oscillatory injection-withdrawal at the bottom of the liquid layer. The second problem, which can be rather stiff, has shown the limitation of the semi-implicit scheme concerning its stability. Therefore, the fully implicit method was used in [21], [31], and [56] for the study of the interaction of a vortex-pair with a free surface. The method was based on the third-order time-accurate scheme (7.18).

The presence of surface tension is an important feature of the problem, physically as well as numerically. From the numerical point of view, the presence of the third-order derivative $\partial_{XXX}H$ in the R term of the free-surface condition (7.8) may be the cause of large inaccuracies in the Fourier representation of this derivative leading to instabilities. This is avoided by evaluating the derivative $\partial_{XXX}H$ by means of a compact sixth-order finite-difference formula, associated with a very high-order filter [57].

In the study of the interaction vortex-pair/free-surface presented in [21], [31], the initial layer of liquid is stably stratified by the presence of a vertical temperature gradient (Fig.6). The evolution of the temperature is governed by a transport-diffusion equation which has to be included in the motion equation (7.17) and solved iteratively by the processus (7.20).

In [31], [56] the interaction vortex-pair/free surface in presence of surfactant has been addressed. The variations of the concentration of the surfactant on the free surface induce variations in the surface tension which must be taken into account in the dynamical free-surface conditions [58], [59]. The evolution of the concentration of the surfactant is governed by an equation of transport-diffusion type [60] to be solved on the free surface. The iterative implicit method allows the correct representation of very stiff concentration fronts.

8 CONCLUSION

Spectral methods for solving the Navier-Stokes equations within the vorticity-streamfunction formulation have been presented and discussed. The boundary conditions as well as the matching conditions in the domain decomposition method are imposed through the influence matrix technique. This technique, which can also be considered in the algebraic point of view as an algorithm for solving the full discrete system, can be applied to various problems associated with partial differential equations. Its application to the vorticity-streamfunction equations leads to efficient numerical codes, well adapted to vectorization and parallelization.

As for any spectral method, the consideration of complex geometries is a real difficulty. Domain decomposition is a good way to surmount such a difficulty. Another way, which begins to be investigated and merits to be deepened is the embedding method. The method (also called "fictitious domain") consists in embedding the original domain, with arbitrary shape, in to a larger rectangular domain in which the spectral approximation is applied. Recently, a method of that type based on the combination of Fourier spectral and boundary-element approximations has been proposed in [61] for solving the Navier-Stokes equations.

Appendix : Coefficients of the differentiation matrix $\mathcal{D}$

$$d_{i,j}^{(1)} = \frac{\bar{c}_i(-1)^{i+j+1}}{2\bar{c}_j \sin \dfrac{(i+j)\pi}{2N} \sin \dfrac{(i-1)\pi}{2N}},$$

$$0 \le i,j \le N, \quad i \ne j$$

$$d_{i,i}^{(1)} = -\frac{\cos \dfrac{i\pi}{N}}{2\sin^2 \dfrac{i\pi}{N}}, \quad 1 \le i \le N-1,$$

$$d_{0,0}^{(1)} = -d_{N,N}^{(1)} = \frac{2N^2+1}{6}$$

where $\bar{c}_0 = \bar{c}_N = 2$, $\bar{c}_i = 1$ for $1 \le i \le N-1$.

References

[1] R. Peyret and T.D. Taylor, Computational Methods for Fluid Flow, Springer-Verlag, New York, 1983.

[2] W.W. Schultz, N.Y. Lee and J.P. Boyd, Chebyshev pseudospectral method of viscous flows with corner singularities, *J. Sci. Comput.*, **4**, pp. 1-24, 1989.

[3] O. Botella and R. Peyret, The Chebyshev approximation for the solution of singular Navier-Stokes problems, *to be published*.

[4] I. Raspo, J. Ouazzani and R. Peyret, A spectral multidomain technique for the computation of the Czochralski melt configuration, *Int. J. Numer. Methods Heat Fluid Flow*, **6**, pp. 31-58, 1996.

[5] D. Gottlieb and C.-W. Shu, On the Gibbs phenomenon and its resolution, *to be published in SIAM Review*.

[6] R. Pasquetti, R. Peyret, J.M. Lacroix and R. Bwemba, Spectral method and vorticity-potential vector formulation for three-dimensional cylindrical calculations, In 2nd Internat. Forum on Expert Systems and Computer Simulation in Energy Engineering, Paper 8-2, Erlangen, 17-20 March, 1992.

[7] R. Pasquetti and R. Bwemba, A spectral algorithm for the Stokes problem in vorticity-vector potential formulation and cylindrical geometry, *Comput. Methods Appl. Mech. Engng*, **117**, pp. 71-90, 1994.

[8] W.Z. Shen and T.P. Loc, Numerical method for unsteady 3D Navier-Stokes equations in velocity-vorticity form, *Computers and Fluids*, **26**, pp.193-216, 1997.

[9] J.C. Wu and M.M. Wahbah, Numerical solution of viscous flow equations using integral representations, In Proc. Fifth Int. Conf. Numer. Methods in Fluid Dynamics, Lecture Notes in Physics, A.I. Van de Vooren and P.J. Zandbergen editors, Vol. 59, pp. 448-453, Springer-Verlag, Berlin, 1976.

[10] R. Glowinski and O. Pironneau, Numerical methods for the first biharmonic equation and for the two-dimensional Stokes problem, *SIAM Review*, **21**, pp.167-212, 1979.

[11] U. Ehrenstein, The spectrum of a Chebyshev-Fourier approximation for the Stokes equations, *J. Sci. Comput.*, **5**, pp.55-84, 1990.

[12] U. Ehrenstein and R. Peyret, A Chebyshev collocation method for the Navier-Stokes equations with application to double-diffusive convection, *Int. J. Numer. Methods Fluids*, **9**, pp. 427-452, 1989.

[13] L. Tuckerman, Formation of Taylor vortices in spherical Couette flow, PhD thesis, MIT, Cambridge, Mass., 1983.

[14] S.C.R. Dennis and L. Quartapelle, Direct solution of the vorticity-stream function ordinary differential equations by a Chebyshev approximation, *J. Comput. Phys.*, **52**, pp. 448-463, 1983.

[15] J.M. Vanel, R. Peyret and P. Bontoux, A pseudo-spectral solution of vorticity-stream function equations using the influence matrix technique, In Numerical Methods for Fluid Dynamics II, pp. 463-475, K.W. Morton and M.J. Baines editors, Clarendon Press, Oxford, 1986.

[16] U. Ehrenstein and R. Peyret, A collocation Chebyshev method for solving Stokes-type equations, In Proc. 6th Int. Symp. Finite Elements in Flow Problems, M.O. Bristeau *et al.* editors, pp. 213-218, Antibes, June 16-20, 1986.

[17] J.P. Pulicani, A spectral multidomain method for the solution of 1D-Helmholtz and Stokes-type equations. *Computers and Fluids*, **16**, pp. 207-215, 1988.

[18] I.S. Kang and L.G. Leal, Numerical solution of axisymmetric unsteady free-boundary problems at finite Reynolds number. I. Finite-difference scheme and its application to the deformation of a bubble in a uniaxial straining flow, *Phys. Fluids*, **30**, pp. 1929-1940, 1987.

[19] A. Garba, A. Mofid and R. Peyret, Spectral solution of free surface flows, *Finite Elements in Analysis and Design*, **16**, pp. 191-206, 1994.

[20] A. Garba, Méthodes spectrales en domaines non-rectangulaires. Application au calcul d'écoulements à surface libre, Thèse de Doctorat, Université de Nice-Sophia-Antipolis, 1993.

[21] A. Garba, T. Pagès and R. Peyret, Spectral calculations of the interaction vortex pair-free surface in presence of a thermal gradient, *to be published*.

[22] D. Gottlieb and S.A. Orszag, Numerical Analysis of Spectral Methods: Theory and Applications, SIAM, Philadelphia, 1977.

[23] C. Canuto, M.Y. Hussaini, A. Quarteroni and Zang, T.A., Spectral Methods in Fluid Dynamics, Springer-Verlag, New-York, 1988.

[24] J.P. Boyd, Chebyshev and Fourier Spectral Methods, Springer-Verlag, Berlin, 1989.

[25] R. Peyret, Introduction to Spectral Methods, Computational Fluid Dynamics Lecture Series 1986-04, Von Karman Institute for Fluid Dynamics, Rhode-St-Genèse, Belgium, 1986.

[26] J.L. Guermond and L. Quartapelle, Equivalence of u-p and ζ-ψ formulations of the time-dependent Navier-Stokes equations, *Inter. J. Numer. Methods Fluids*, **18**, pp. 471-487, 1994.

[27] P. Le Quéré, Transition to unsteady natural convection in a tall water-filled cavity, *Phys. Fluids*, **A2**, pp. 503-515, 1990.

[28] J. Ouazzani, R. Peyret and A. Zakaria, A., Stability of collocation-Chebyshev schemes

with application to the Navier-Stokes equations, In Proc. Sixth GAMM Conference on Numerical Methods in Fluid Mechanics, Notes on Numerical Fluid Mechanics, Vol. 13, pp. 287-294, D. Rues and W. Kordulla editors, Vieweg, Braunschweig, 1986.

[29] U. Ehrenstein, Méthodes spectrales de résolution des équations de Stokes et de Navier-Stokes. Application à des écoulements de convection double-diffusive, Thèse de Doctorat, Université de Nice-Sophia-Antipolis, 1986.

[30] J. Fröhlich, T. Gerhold, J.M. Lacroix and R. Peyret, Fully implicit spectral methods for convection, In High Performance Computing II, pp. 585-596, M. Durand and F. El Dabaghi editors, North-Holland, Amsterdam, 1991.

[31] T. Pagès, R. Pasquetti and R. Peyret, Interaction of vortices with free surfaces, In ICIAM 95, Hamburg, 3-7 July, 1995.

[32] L. Kleiser and U. Schumann, Treatment of incompressibility and boundary conditions in 3-D numerical spectral simulations of plane channel flows, In Proc. Third GAMM Conf. in Numerical Methods in Fluids Dynamics, Notes on Numerical Fluid Mechanics, Vol. 2, pp. 165-173, E.H. Hirschel editor, Vieweg, Braunschweig, 1980.

[33] R. Bwemba and R. Pasquetti, On the influence matrix in the spectral solution of the 2D Stokes problem (vorticity-stream function formulation), *Appl. Numer. Math.*, **16**, pp. 299-315, 1995.

[34] J.P. Pulicani, E. Crespo del Arco, A. Randriamampianina, P. Bontoux and R. Peyret, Spectral simulations of oscillatory convection at low Prandtl number, *Inter. J. Numer. Methods Fluids*, **10**, pp. 481-517, 1990.

[35] Y. Demay, J.M. Lacroix, R. Peyret and J.M. Vanel, Numerical experiments on stratified fluids subject to heating, In Proc. Third Inter. Symp. on Stratified Flows, Pasadena, CA, 1987, pp. 588-597, E.J. List and G.H. Jirka editors, American Society of Civil Engineers, New York, 1990.

[36] R. Peyret and J.M. Vanel, Numerical experiments in double-diffusive convection, Computational Fluid Dynamics, pp.33-52, D. Leut-loff and R.C. Srivastava editors, Springer-Verlag, Berlin, 1995.

[37] P. Larroudé, J. Ouazzani, J.I.D. Alexander and P. Bontoux, Symmetry breaking flow transitions and oscillatory flows in a 2D directional solidification model, *Eur. J. Mech. B/Fluids*, **13**, pp. 353-383, 1994.

[38] A. Chaouche, Une méthode de collocation-Chebyshev pour la simulation des écoulements axisymétriques dans un domaine annulaire en rotation, *La Recherche Aérospatiale*, **1990-5**, pp. 1-13, 1990.

[39] A. Chaouche and A. Randriamampianina, A collocation method based on the influence matrix technique for Navier-Stokes problems in annular domains, *Comput. Methods Appl. Mech. Engng*, **80**, pp. 237-244, 1990.

[40] J.P. Pulicani, A spectral multi-domain method for the solution of 1-D Helmholtz and Stokes-type equations, *Computers and Fluids*, **9**, pp. 499-515, 1988.

[41] J.M. Lacroix, R. Peyret and J.P. Pulicani, A pseudospectral multi-domain method for the Navier-Stokes equations with application to double-diffusive convection, In Proc. Seventh GAMM Conf. on Numerical Methods in Fluid Mechanics, Notes on Numerical Fluid Mechanics, Vol. 20, pp. 167-174, M. Deville editor, Vieweg, Braunschweig, 1988.

[42] I. Raspo, Méthodes spectrales et de décomposition de domaine pour les écoulements complexes confinés en rotation, Thèse de Doctorat, Université d'Aix-Marseille II, 1996.

[43] A.T. Patera, A spectral element method for fluid dynamics: laminar flow in a channel expansion, *J. Comput. Phys.*, **54**, pp. 468-488, 1984.

[44] Y. Maday and A.T. Patera, Spectral element methods for the incompressible Navier-Stokes equations, State-Of-The-Art Survey, pp. 71-143, A.K. Noor and J.T. Oden editors, ASME, 1989.

[45] R. Peyret, The Chebyshev multidomain approach to stiff problems in fluid mechanics, *Comput. Methods Appl. Mech. Engng*, **80**, pp. 129-145, 1990.

[46] P.F. Chan and B.C. Resasco, A framework for the analysis and construction of domain decomposition preconditioners, In 1st Inter. Symp. Domain Decomposition Methods for PDE, R. Glowinski *et al.* editors, SIAM, Philadelphia, 1988.

[47] P. Larroudé, Approximations pseudo-spectrales des phénomènes convectifs instationnaires en configurations complexes. Application à la croissance cristalline, Thèse de Doctorat, Université d'Aix-marseille II, 1995.

[48] S. Gauthier, H. Guillard, T. Lumpp, J.M. Malé, R. Peyret and F. Renaud, A spectral domain decomposition technique with moving interfaces for viscous compressible flows, In Computational Fluid Dynamics'96, Proc. Third ECCOMAS Computational Fluid Dynamics Conf., Paris, pp. 839-844, J.A. Désidéri *et al.* editors, pp. 839-844, J. Wiley & Sons, New York, 1996.

[49] J.P. Pulicani, Modélisation isotherme d'une interface diffusive en convection de double-diffusion, *Inter. J. Heat Mass Transfer*, **37**, pp. 2835-2858, 1994.

[50] A. Mofid, Application des méthodes spectrales à l'étude de l'effet d'une source de chaleur dans un fluide stratifié, Thèse de Doctorat, Université de Nice-Sophia-Antipolis, 1992.

[51] C. Sabbah, Etude numérique de la formation et l'évolution de cellules convectives dans un fluide stratifié chauffé latéralement, Stage de DEA "Turbulence et Systèmes Dynamiques", Université de Nice-Sophia-Antipolis, 1994.

[52] I. Raspo, J. Ouazzani and R. Peyret, A direct Chebyshev multidomain method for flow computation with application to rotating systems, *Contemporary Mathematics*, **180**, pp.533-538, 1994.

[53] Ph. Larroudé, I. Raspo, J. Ouazzani and R. Peyret, Application of the spectral multidomain method to flow in crystal growth system, In Computational Fluid Dynamics'94, Proc. Second ECCOMAS Computational Fluid Dynamics Conf., Stuttgart, pp. 964-970, S. Wagner *et al.*editors, J. Wiley & Sons, New York, 1994.

[54] I. Raspo, A. Randriamampianina, E. Crespo del Arco and P. Bontoux, Phénomènes d'instabilité convective dans les systèmes "canal-cavité" en rotation différentielle en présence d'un flux forcé, *to be published in C. R. Acad. Sci. Paris*.

[55] J.V. Wehausen and E.V. Laitone, Surface Waves, Handbuch der Physik, **9**, pp. 446-478, 1960.

[56] T. Pagès and R. Peyret, Interaction vortex pair-free surface in presence of surfactant, *to be published.*

[57] T.J. Hou, J.S. Lowengrub and M.J. Shelley, Removing the stiffness from interfacial flows with surface tension, *J. Comput. Phys.*, **114**, pp. 312-338, 1994.

[58] L. Landau and E.M. Lifshitz, Fluid Mechanics. Addison-Wesley, Reading, 1960.

[59] V.G. Levich and V.S. Krylov, Surface-tension-driven phenomena, *Annual Review of Fluid Mechanics*, Vol. 1, pp. 293-316, 1969.

[60] H.A. Stone, A simple derivation of the time-dependent convective-diffusion equation for surfactant transport along a deforming surface, *Phys. Fluids*, **A2**, pp. 111-112, 1990.

[61] M. Elghaoui and R. Pasquetti, A spectral embedding method for the incompressible Navier-Stokes equations, In Computational Fluid Dynamics'96, Proc. Third ECCOMAS Computational Fluid Dynamics Conf., Paris, pp. 124-130, J.A. Désidéri *et al.*editors, J. Wiley & Sons, New York, 1996.

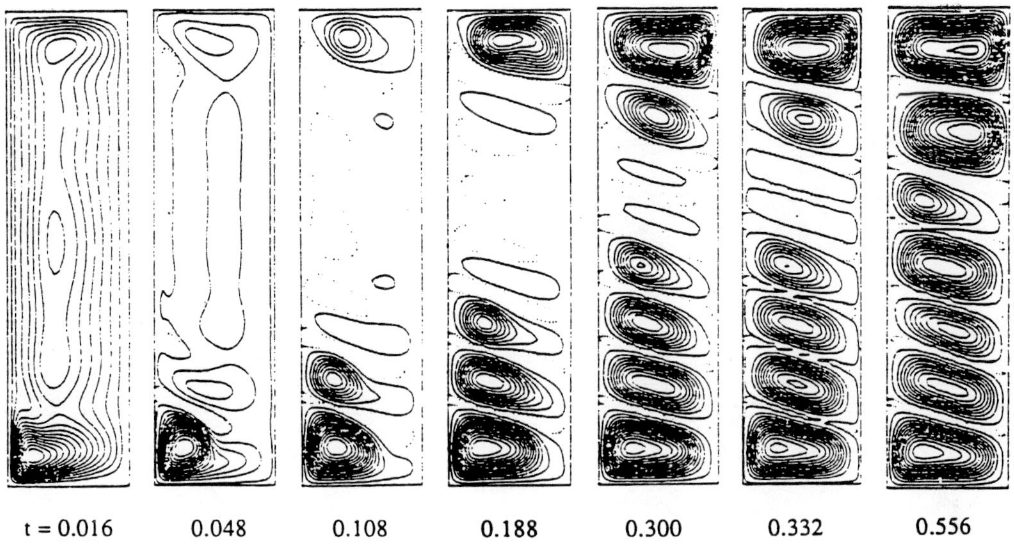

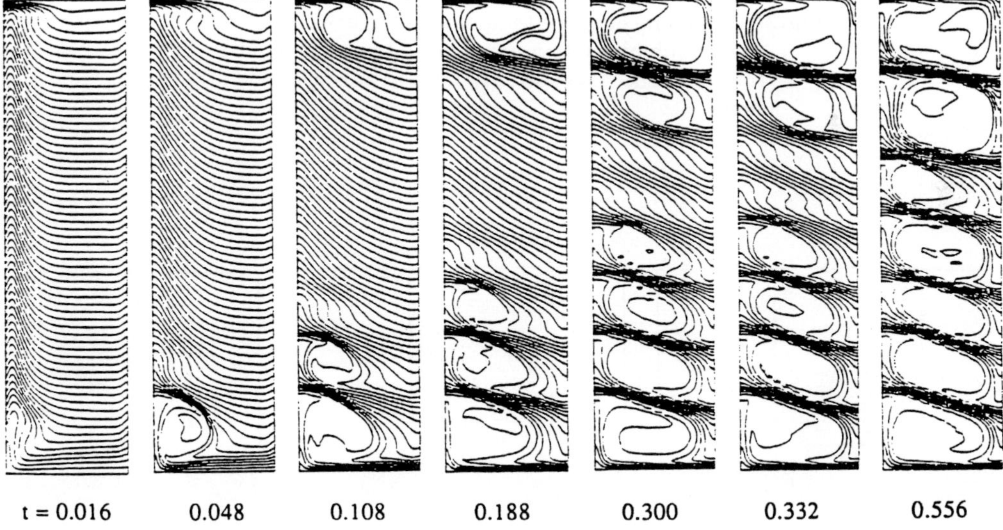

Fig.1. Formation of double-diffusive convective layers in the lateral heating of a fluid stratified by concentration: instantaneous streamlines (above) and iso-concentration lines (below) at various times (ref.[12]).

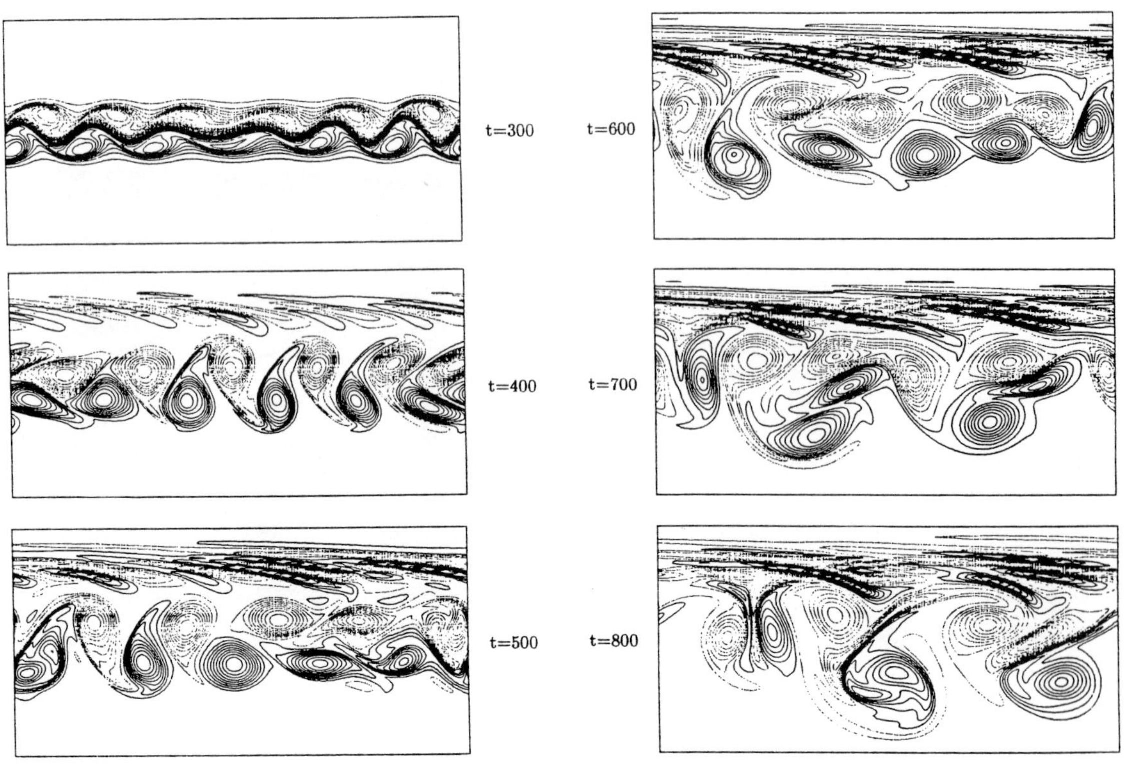

Fig.3. Development of the temporal instability of a wake submitted to a nonuniform temperature gradient : iso-vorticity lines at various times.

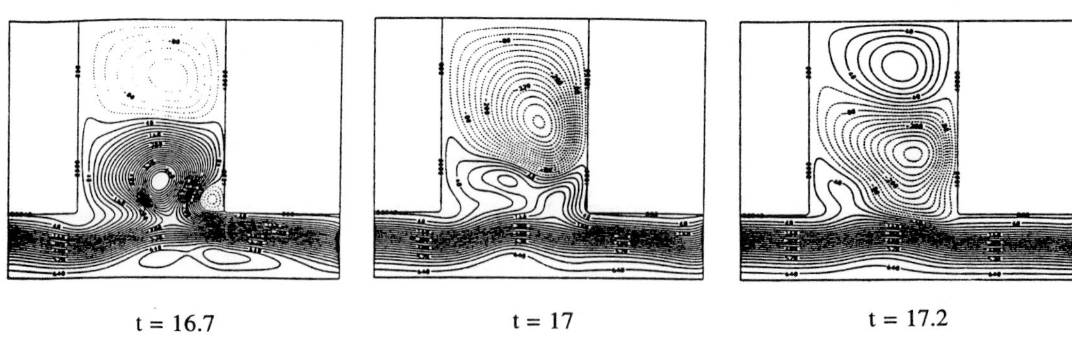

Fig.4. Unsteady flow in a duct-cavity system in rotation : instantaneous streamlines at various times (ref.[54]).

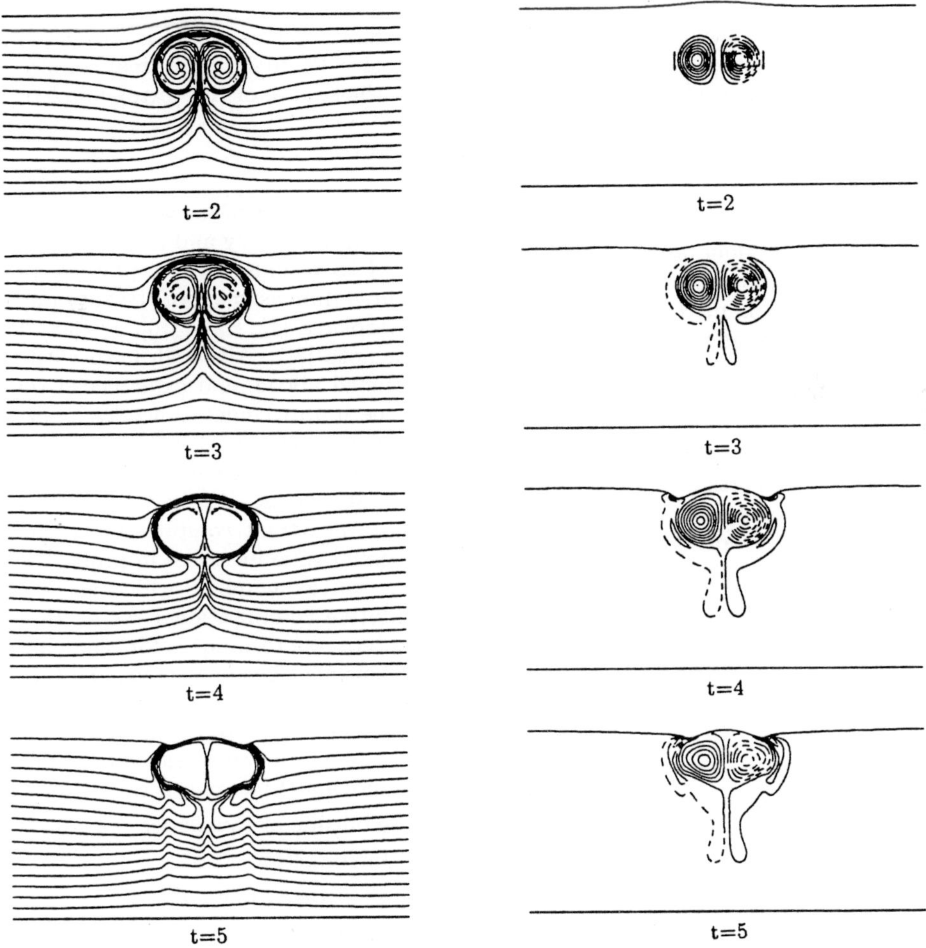

Fig.5. Interaction of a vortex-pair and the free surface of a layer of fluid stratified by a constant temperature gradient : isothermal lines (left) and iso-vorticity lines (right) (ref.[21], [31]).

THREE RESTRICTIONS ON THE VORTEX METHODS

Yoshifumi OGAMI

Department of Mechanical Engineering, Ritsumeikan University, 1-1-1
Noji-higashi, Otsu-city, Shiga 525-77, Japan

Abstract

This paper gives an overview of the advances made in the recent vortex methods for removing three
restrictions on the methods, i.e, (1) inability to treat viscosity, (2) inability to treat compressibility,
and (3) prohibitive increase of computational costs with the square of the vortex number.

1. Introduction

The vortex methods are unique among numerous
computational methods for fluid dynamics. They are
referred to as the Lagrangian methods in contrast
to the Eulerian methods such as the finite difference
methods or the finite element methods which have
been extensively used for studying dynamics and mechanics of various types of fluid motions. In the vortex methods, the continuous vorticity distribution is
divided into particles (point vortices, vortex blobs or
vortex sheets), which then interact with one another
just like stars or electrons. The vortices are traced in a
Lagrangian scheme by solving a set of ordinary differential equations derived from the Biot-Savart law like
the stars are from the Newton's law and the electrons
from the Coulomb's law (unlike stars and electrons
the vortices do not produce accelerations but velocities). The moving vortices then represent unsteady
rotational (but inviscid) fluid motions, which approximate a solution to the Euler's motion equations (the
inviscid vorticity transport equation).

The advantages of the vortex methods are (1) they
do not require grids which are indispensable to the Eulerian methods, (2) the ordinary differential equations
which approximate solutions to the Euler's equations
are feasible to deal with compared to the partial differential equations, (3) computation is performed only
where vortices exist, (4) the boundary condition at infinity is naturally satisfied, (5) the physical meanings
of the computational method and of its results are
easily understood via the vortex dynamics, and (6)
computing of the pressure is not required to proceed
with the simulations.

Since the first attempt was made by Rosenhead in
1931 [1] to simulate a flow by a vortex method, numerous variations of the vortex method have been presented, theoretically and mathematically studied, and
applied to various types of physical problems including three dimensions (see the surveys by Leonard [2],
[3], Sarpkaya [4], Hald [10], Puckett [6], Sethian [7],
Kamemoto [8]). Although born with the advantages
mentioned above, and experienced the developments
and the improvements since their births until today,
the vortex methods still lag considerably behind the
grid-based Eulerian methods because of the serious inborn restrictions: (1) they do not treat viscosity, (2)
they do not treat compressibility, and (3) the computational time increases in proportion to the square
of the vortex number, which consequently leads to
prohibitive computational costs. However, in recent
years, notable techniques for removing these three restrictions have been presented and adopted from other
areas (e.g. astrophysics, plasma physics) into the vortex schemes. This paper gives an overview of the advances made in the recent vortex methods especially
for removing these inborn restrictions.

The outline of this paper is as follows.

1. Introduction

2. The vortex methods

3. Viscosity

    3.1 Core spreading method

    3.2 Random vortex method

    3.3 Diffusion velocity method

    3.4 Strength exchange method

4. Fast methods

    4.1 Far-field potential

        4.1.1 Monopole method

        4.1.2 Multipole method

    4.2 Rapidly decaying potential

5. Compressibility

Received on February 27, 1997.

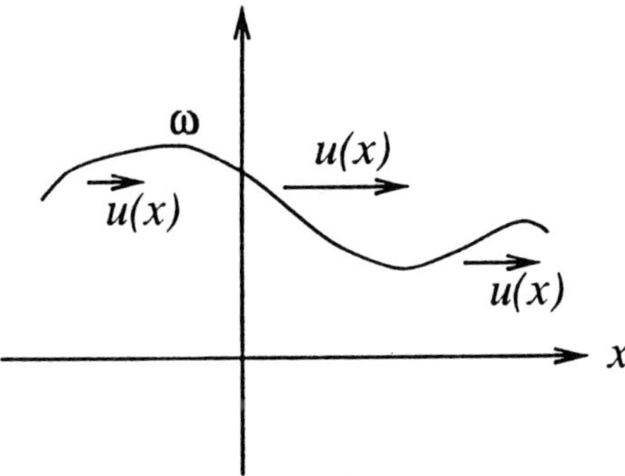

Fig.1: A vorticity distribution ω moving in the
x-direction with the velocity $u(x)$

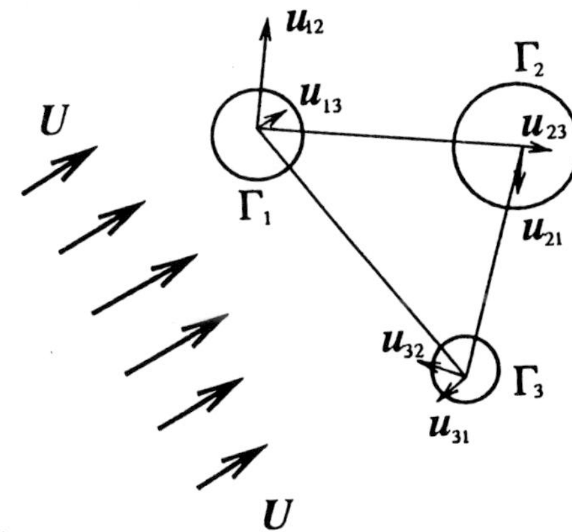

Fig.2: Vortices induce velocity vectors each other

6. Concluding remarks

2. THE VORTEX METHODS

In this section, we briefly review the vortex methods
for two dimensional incompressible inviscid fluids.

The aim of the vortex methods is to give approximate solutions to the following vorticity transport
equation

$$\frac{d\omega}{dt} = 0 \qquad (1)$$

This equation can be rewritten in a conservation form

$$\frac{\partial \omega}{\partial t} + \mathrm{div}(\omega \boldsymbol{u}) = 0 \qquad (2)$$

where the equation of continuity $\mathrm{div}\boldsymbol{u} = 0$ has been
used. The physical interpretations of this motion
equation are that the function ω is moving with the
fluid velocity — the convection velocity $\boldsymbol{u}$ (for one-
dimensional case, see Fig.2), and that the circulation
$\Gamma = \iint \omega dx dy$ is conserved. It is analogous to the fact
that the equation of continuity

$$\frac{\partial \rho}{\partial t} + \mathrm{div}(\rho \boldsymbol{u}) = 0 \qquad (3)$$

is derived from the assumptions that the function of
the fluid density, ρ, is carried with the fluid velocity
$\boldsymbol{u}$, and that the mass $m = \iint \rho dx dy$ is conserved.

Mathematically, solving equation (2) is obtaining
analytical solutions (if possible), or numerical solutions with bounded errors, which are proved to converge to the exact solutions, with given initial and
boundary conditions. In a Lagrangian scheme, however, solving equation (2) is equal to moving the function ω with the velocity vector $\boldsymbol{u}$ according to the
physical interpretations of the motion equation men-
tioned above. The vortex methods realize this by
tracking the particles which are created by discretizing
the continuous function ω and are to reproduce this
function. The velocity vector induced on a particle by
the rest of them is obtained through the use of well
known law — the Biot–Savart law.

Each particle is provided with the circulation Γ
which is

$$\Gamma = \iint_S \omega dx dy \qquad (4)$$

where S is the region this particle occupies. The region S is infinitesimal when the particle is a point vortex and is infinite when the particle has, for example,
the Gaussian structure. This circulation is constant
in time as shown by

$$\frac{d\Gamma}{dt} = \iint_S \frac{d\omega}{dt} dx dy = 0 \qquad (5)$$

even when the shape of the region S changes due to a
characteristic of the velocity field.

The continuous vorticity distribution is produced by
an assembly of the particles each with the circulation
Γ_i located at $\boldsymbol{x}_i = (x_i, y_i)$ as expressed by

$$\omega(\boldsymbol{x}, t) = \sum_{i=1}^{N} \Gamma_i K(\boldsymbol{x} - \boldsymbol{x}_i) \qquad (6)$$

where N is the particle number and K is a kernel of
a particle. The time variation of the vorticity distribution is then given by the evolution of the assembly. Here we should realize that if the kernel is that

of a point vortex, discretization is done only space-wise, however if the kernel is that of a vortex blob, discretization is done both space-wise and depth-wise (namely these vortex blobs overlap).

To sum up, solving Eq.(2) by the vortex scheme is tracking the particles with constant circulations which move with the convection velocities given by the Biot–Savart law (Fig.2). This is done by solving computationally the following system of ordinary differential equations (Leonard [2])

$$\frac{dx_i}{dt} = -\frac{1}{2\pi} \sum_{j=1}^{N} \frac{(x_i - x_j) \times \hat{e}_z \Gamma_j g(|x_i - x_j|/\sigma_j)}{|x_i - x_j|^2}$$

$$(7)$$

where g is given by

$$g(y) = 2\pi \int_0^y K(z)z\,dz \qquad (8)$$

$\hat{e}_z$ is the unit vector in the z-direction, and both no interior boundaries in the flow field and zero velocity at infinity are assumed.

Hald and del Prete [9] gave a short-time convergence proof of the vortex methods with several types of smoothed kernels in two dimensions. Later, Hald established long-time convergence [10] and then showed convergence for a wide class of two-dimensional methods [11]. Beale and Majda [12], [13] presented convergence proofs of two and three dimensions. Beale [14] and Greengard [15] also gave proofs of the three-dimensional vortex method. For a survey of the practical applications of the methods as well as the theoretical foundations, see Sarpkaya [4], Sethian [7].

In the inviscid flow as considered here, the vorticity is carried only with the convection velocity, which essentially is the velocity of the fluid itself, so that it looks as if the vorticity is attached to the fluid. Of course it is not true. The vorticity is allowed to deviate from the paths of the fluid particles. This fact leads to various techniques for simulating the viscous diffusion in a Lagrangian scheme as shown in the next chapter.

3. VISCOSITY

The first restriction on the vortex methods we consider in this paper is inability to treat viscosity of a fluid.

The vorticity transport equation for viscous flows includes the diffusion term as

$$\frac{d\omega}{dt} = \nu \Delta \omega \qquad (9)$$

The problem is how to incorporate the diffusion term into the vortex scheme. In the following subsections

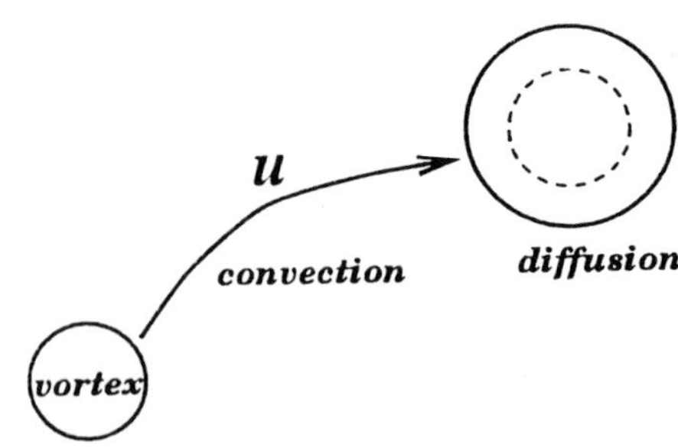

Fig.3: A vortex is convected and the core size of it expands

we will see how it is achieved. The diffusion techniques we are particularly interested in are such that the process of diffusion is treated in a Lagrangian manner, and that solutions of the flows with an object in them have been obtained. They are

1. Core spreading method (Kuwahara and Takami [16], Zhang and Ghoniem [17], Cheng et al. [18], Nourazar and Milane [19])
2. Random vortex method (Chorin [20], Najm and Ghoniem [21], Taehyeung and Flynn [22], Clarke and Tutty [29], Kida and Kurita [26])
3. Diffusion velocity method (Ogami and Akamatsu [27], Shintani and Akamatsu [28], Clarke and Tutty [29], Ogami and Ayano [30], Strickland et al. [31])
4. Strength exchange method (Degond and Mas-Gallic [33], [34], Leonard and Koumoutsakos [35], [36], Subramaniam and van Dommelen [37])

(As a rule, the first and the recent papers are listed as references)

Although our interest is focused on the recent vortex methods, we briefly see the first two methods, which were introduced more than twenty years ago, because they are simple, easy to implement and still used.

For another approaches to the modelling of diffusion, see Fishelov [38],[39], where the vorticity is approximated by convolving with a cutoff function, Lu [40], and Chang and Chern [41], where the diffusion equation for the vorticity is solved by a finite difference method, Russo [42], where the diffusion equation is solved on an irregular grid formed by the particles, Nakanishi and Kamemoto [43], where the diffusion is considered by changing the size of the cross section of a vortex blob, and Najm [44], where the diffusion is treated by a hybrid Lagrangian-Eulerian implementation.

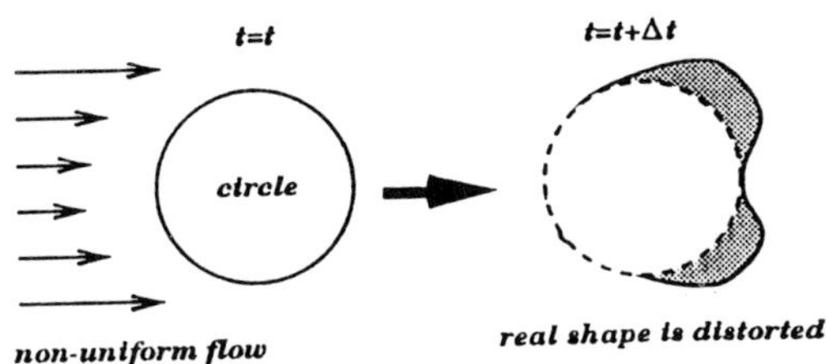

Fig.4: The shape of vortex is distorted in a non-uniform flow

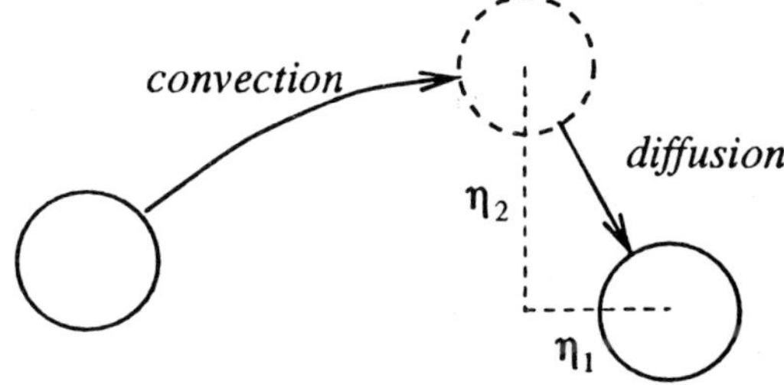

Fig.6: A vortex is convected and displaced

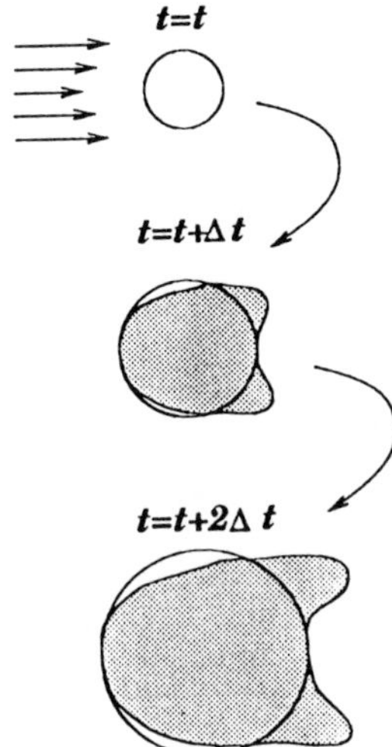

Fig.5: The real shape of a vortex (shaded area) in a viscous flow is expanded due to diffusion and is distorted due to convection but the shape is kept circular in the core spreading algorithm

3.1 Core spreading method

The simplest idea one may come up with to incorporate the diffusion effect into the vortex methods would be to use the fundamental solution of the diffusion equation as a kernel in (6), that is

$$K = \frac{1}{4\pi\nu t} \exp\left(-\frac{|x - x_i|^2}{4\nu t}\right) \qquad (10)$$

In this scheme, a vortex travels with the convection velocity and diffuses (namely the core of a particle spreads with time as shown in Fig.3. This technique is called *the core spreading method* ([16], [17], [18], [19]).

The major numerical error for using a kernel of fi-nite core arises from the fact that the shape of the core, circle, is not necessarily the real shape after even a short period of time. In a non-uniform flow as is often the case, the vortex shape which is a circle at first is distorted actually but it is not in a computational algorithm (Fig.3.). Namely, the vorticity is not correctly convected and thus the gap between the two shapes (shaded area in Fig.3.) becomes a numerical error. This error will be infinitesimal in the limit of infinitely many vortex blobs with fixed core radius. Unfortunately, in the core spreading method, the core radius, $\delta = \sqrt{2\nu t}$, expands with time and consequently the error grows although the diffusion process is correctly simulated (Fig.3.) (since viscosity is assumed constant everywhere in a flow, a vortex by nature expands radially via the diffusion process). In short, *vorticity is correctly diffused, but incorrectly convected* (Greengard [45]). This leads to the conclusion that *the core spreading method approximates the wrong equation* [45].

3.2 Random vortex method

The Brownian motion is the core of the next method. The probability of existence, at a certain position and time, of a particle moving at random is given as a solution of the diffusion equation. A pollen floating on a water surface moving chaotically due to the collision with water molecules, known as the Brownian motion, or a person walking drunkenly can be treated by the same computational technique — the random walk method (the Monte Calro method).

Naturally, it is expected that the diffusion term of Eq.(9) can be incorporated into the vortex scheme by adding random walks to the vortex movements. This has been proposed by Chorin [20] and the scheme is

$$x_i^{n+1} = x_i^n + \Delta t u^n + \eta_1 \qquad (11)$$

$$y_i^{n+1} = y_i^n + \Delta t v^n + \eta_2 \qquad (12)$$

where (x_i^n, y_i^n) is the position of ith vortex at time $n\Delta t$, and η_1 and η_2 are Gaussianly distributed random

variables with zero mean and variance $2\Delta t\nu$ (Fig. 3.1
). This is called *the random vortex method* ([21], [22],
[29]).

The theoretical basis for applying the random walks
to viscous flow simulations as well as a proof of weak
convergence of the method is given by Marchioro and
Pulvirenti [23]. Following this, a stronger proof of
convergence is provided by Goodman [24]. The most
precise result is presented by Long [25]. However, all
of these deal with the case of no boundaries, and thus
*there is still no convergence proof for the original vor-
tex algorithm applied to the flow of a slightly viscous
fluid around a cylinder* (Hald [10]).

Although the random vortex method has been used
extensively for many years because of its simplicity,
easiness to implement and applicability to complicated
geometries, several drawbacks have been pointed out
as follows:

1. the accuracy is very poor because of the use of
   random numbers
2. the convergence is very slow
3. the physical quantities such as velocity and pressure have to be averaged over several time-steps
   to obtain smooth distributions
4. the Reynolds number cannot be very low
5. the boundary layer equations have to be solved to
   produce vortices on boundaries

For a more detailed criticism of this method, see Sarpkaya [4]. It is desirable to have such a method as-is
deterministic and capable of treating strong viscosity.

3.3 Diffusion velocity method

Such a method as is deterministic and capable of
treating strong viscosity has been presented by Ogami
and Akamatsu [27]. They introduced new concept of
a diffusion velocity.

Suppose that many particles floating in a vacant
space with no external force and that these particles
start to move due to pairwise interactions of the particles in a manner that the density of the particle distribution satisfies the diffusion equation (Fig.3.)β This
pairwise interaction may be force-wise or velocity-wise
or else. The task is to find out the law of this interaction (if one ever exists) and define it. The concept of
the diffusion velocity is the realization of this task. It
should be noted that a numerical model is to solve existent equations numerically and a physical model is to
derive equations which govern physical phenomena or
to explain physical phenomena. The two models look
different due to the different purposes but essentially
they may be the same.

The physical concept of the diffusion velocity may

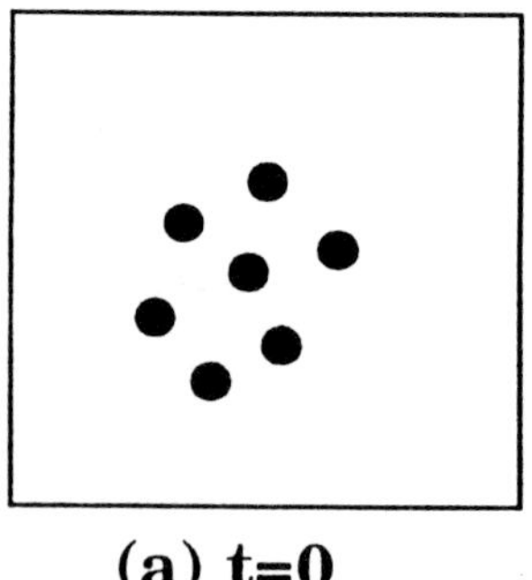

(a) t=0

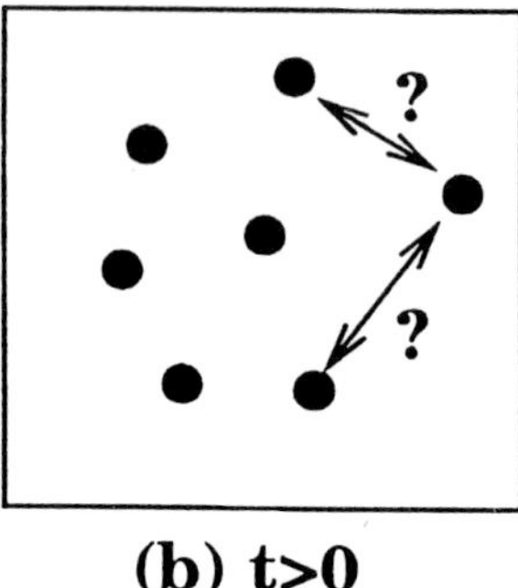

(b) t>0

Fig.7: (a) Particles floating in a space.
(b) Particles diffuse due to pairwise
interactions

be imaginary at first as described above but the mathematical definition, which will make the physical concept substantial, of the diffusion velocity is not. This
definition is easily found directly from Eq.(9) by transforming it into a conservation form, i.e.,

$$\frac{\partial\omega}{\partial t} + \text{div}(\omega\boldsymbol{u} + \omega\boldsymbol{u}_d) = 0 \qquad (13)$$

where $\boldsymbol{u}$ is the convection velocity vector and $\boldsymbol{u}_d$ is
the diffusion velocity vector defined by

$$\boldsymbol{u}_d = -\frac{\nu}{\omega}\nabla\omega \qquad (14)$$

Similar to Eq.(2), we can obtain the physical interpretations of Eq.(13), which are that the function ω is
moving both with the convection velocity $\boldsymbol{u}$ and with
the diffusion velocity $\boldsymbol{u}_d$ (Fig.3.)β and that the circulation $\Gamma = \iint \omega dx dy$ is conserved (namely $d\Gamma/dt = 0$ in
the co-ordinate system moving both with the convec-
tion velocity and the diffusion velocity, Shintani and
Akamatsu [28]). These interpretations enable us to
understand the physical concept of the diffusion velocity and most importantly to create a vortex tech-

nique for viscous flow simulations within a frame work
of pure Lagrangian scheme. The technique based on
this is called *the diffusion velocity method* ([27], [28],
[29], [30]). A simple convergence proof of this method
is given by Shintani and Akamatsu [28] following the
results for an inviscid vortex method given by Beale
and Majda [32].

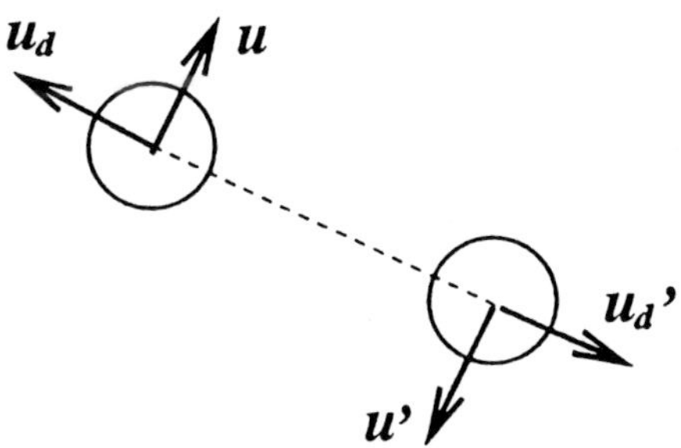

Fig.8: Convection velocities u, u' and diffusion
velocities u_d, u_d'

The diffusion velocity is calculated simply as a sum
of pairwise interactions of vortices just like the convection velocity is. In this method, the fluid particles are
transferred only with the convection velocity while the
vorticity/vortex is both with the convection velocity
and with the diffusion velocity. Therefore the vortex blobs deviate from the paths of the fluid particles
(Fig.3.9).

Also it should be noted that the diffusion velocity
vector does not always satisfy the relation

$$\mathrm{div}\, u_d = 0 \tag{15}$$

while the convection velocity satisfies

$$\mathrm{div}\, u = 0 \tag{16}$$

which is known as the continuity equation for incompressible fluids. This significant difference may be easily explained by the physical fact that the vorticity distribution ω generally varies with time in a Lagrangian
frame, i.e.

$$\frac{d\omega}{dt} \neq 0 \tag{17}$$

but the fluid density ρ does not, i.e.

$$\frac{d\rho}{dt} = 0 \tag{18}$$

Here are examples of the flows around an impulsively started circular cylinder calculated by the diffusion velocity technique. Figure 3.9 shows the vortex

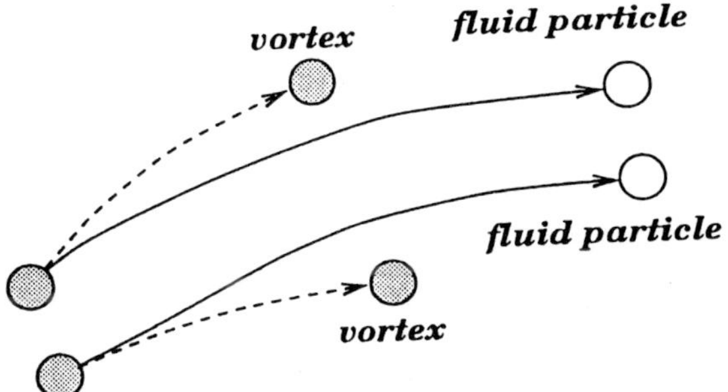

Fig.9: Vortices deviate from the paths of fluid
particles

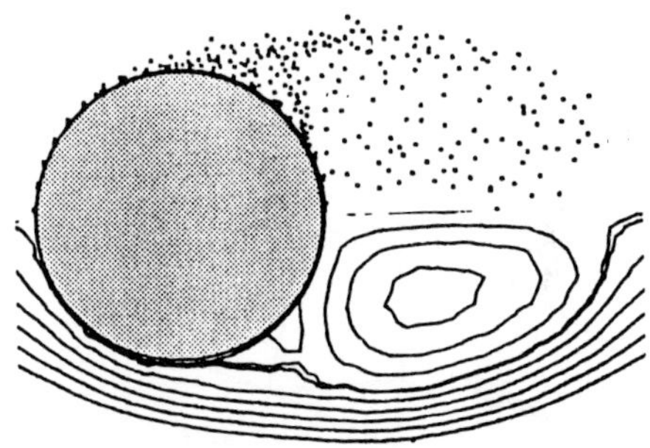

Fig.10: Vortex distribution (upper half) and
streamlines (lower half) at $Re = 1200$ and
time 2.9

distribution (upper half) and the stream lines (lower
half) at $Re = 1200$ (time 2.9), and Fig.3.9 shows the
vortex distribution (upper half) and the contours of
vorticity (lower half) at $Re = 40$ (time 3.0). Though
the resolution in these figures is not very high because
of the use of the limited number of vortices, we may
say that this is the first successful simulations of the
unsteady separated flows with strong viscosity by a
Lagrangian scheme which is both pure and deterministic. This technique was applied to the flows of a
wide range of Reynolds numbers ($Re = 0.1 \sim 10^6$)
[30]. Note that the lower limit of the applicability of
the random vortex method is $Re = 100$ (Chorin [20]),
and that there are numerous vortex techniques presented which are said to be deterministic, but most of
them are grid-based (e.g. [40], [41], [42], [44]).

The advantages of the diffusion velocity technique
are

1. this method is deterministic and purely Lagrangian
2. the solution is much smoother than that of the

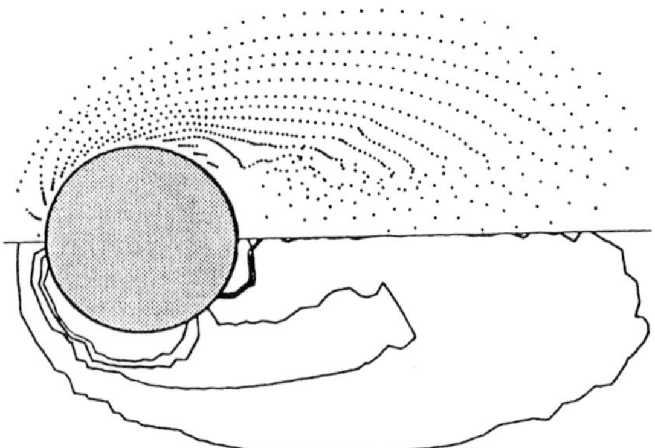

Fig.11: Vortex distribution (upper half) and contours of vorticity (lower half) at $Re = 40$ and time 3.0

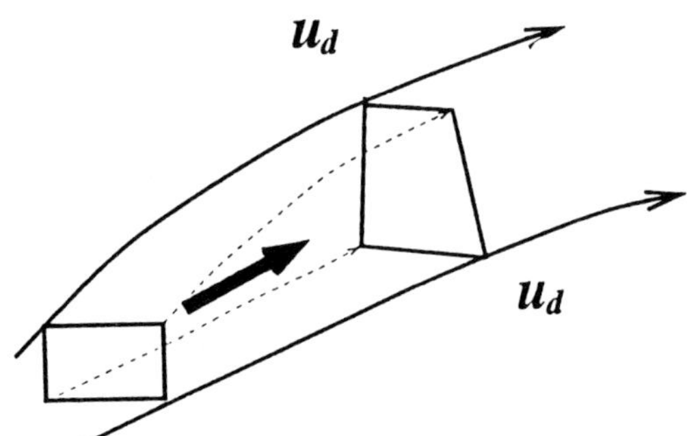

Fig.12: The shape and the size vary

random vortex method
3. the diffusion velocity can be computed as a summation of small supports from other vortices just like the convection velocity, and thus it can be easily implemented within the vortex methods
4. this method is valid even when $Re = 0$
5. the separation of the boundary layer occurs spontaneously and consequently the separation point can be found with no help of the boundary layer theory
6. the results are clearly dependent on the Re number and thus comparable with those of the finite difference methods

Also some problems are pointed out, i.e.,

1. the diffusion velocity field is non-solenoidal so that the vortex radius should vary
2. when the vortices are located at some distance apart they no longer interact to produce a finite

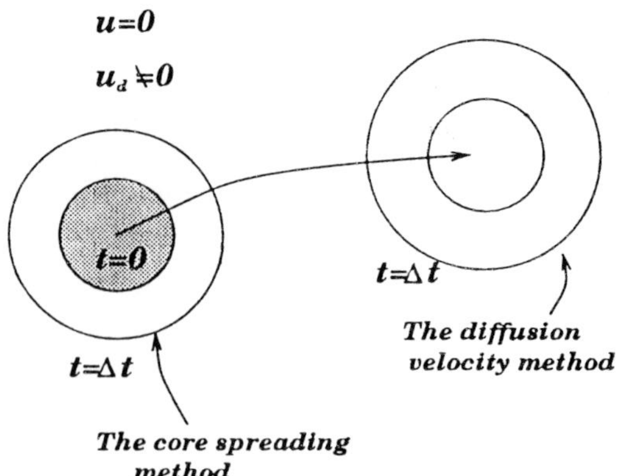

Fig.13: The core spreading method and the diffusion velocity method

value of the diffusion velocity

Some explanations may be necessary for these problems.

Problem 1   Since the diffusion velocity does not satisfy Eq.(15), both the shape and the size of the small region moving with the *diffusion* velocity should vary (Fig.3)β(as is well known, in incompressible flows satisfying Eq.(16), the shape of the small region moving with the *convection* velocity varies but the size of it does not. On the contrary, in compressible flows, which do not always satisfy Eq.(16), both the shape and the size of the region vary similar to the diffusion velocity case).

Therefore, as shown by Kempka and Strickland [46], the size of the vortex core has to change according to

$$\frac{1}{\delta^2}\frac{d(\delta^2)}{dt} = \mathrm{div}\boldsymbol{u}_d \qquad (19)$$

The core radius increases or decreases depending on the sign of $\mathrm{div}\boldsymbol{u}_d$.

This concept is completely different from that of the core spreading method. In the zero convection velocity field, for example, a vortex remains at the initial position and spreads in the core spreading method while this vortex travels with the diffusion velocity and spreads according to the law (19) in the diffusion velocity method (Fig.3)β

Problem 2   The Gaussian distribution employed by Ogami and Akamatsu for the vortex core decays exponentially so does the diffusion velocity (Fig.3)β

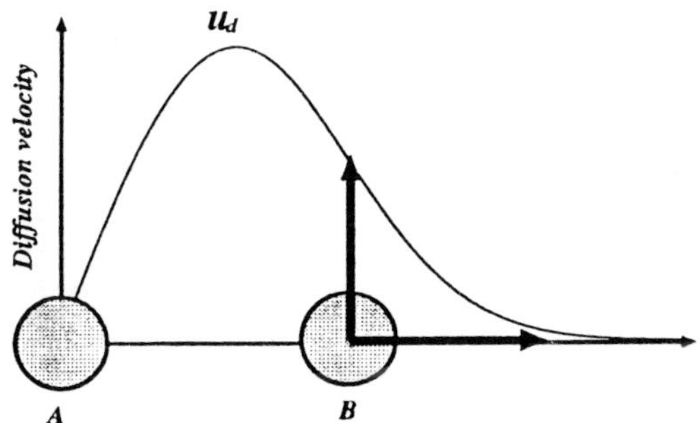

Fig.14: The diffusion velocity decays exponentially. The diffusion velocity induced on vortex B by vortex A fixed at the origin vs. the distance between the two vortices is plotted.

It is a drawback that the diffusion process does not continue where the vortices are sparsely distributed.

This suspension of the diffusion process happens simply because a vorticity structure smaller than a vortex blob cannot be expressed. However, there may be some ways to overcome this restriction as follows:

a. if the circulation of each vortex is set small enough initially to give the resolution one wishes, there is no need for the diffusion process to continue in the region where the vortices are sparse and the diffusion velocity is negligibly small. This is because it is considered that the goal — the required resolution (the minimum value of vorticity set in advance) — has been reached there (see Fig.3.3 for one-dimension). With the use of appropriate computational techniques, it would be possible to completely stop the diffusion process in the region where it is not needed anymore while maintaining the accuracy and the smoothness of the vorticity distribution.

b. if the diffusion process still has to continue, there are two ways, (1) spread the core radius according to Eq.(19) so that the vortex blobs overlap sufficiently to produce the diffusion velocities of finite value (the accuracy will be better but the resolution will be worse), and (2) split a vortex into multiple vortices (the accuracy and the resolution will be better but the vortex number increases).

3.4 Strength exchange method

Leonard and Koumoutsakos ([35], [36]) have applied the technique of particle strength exchange, pre-

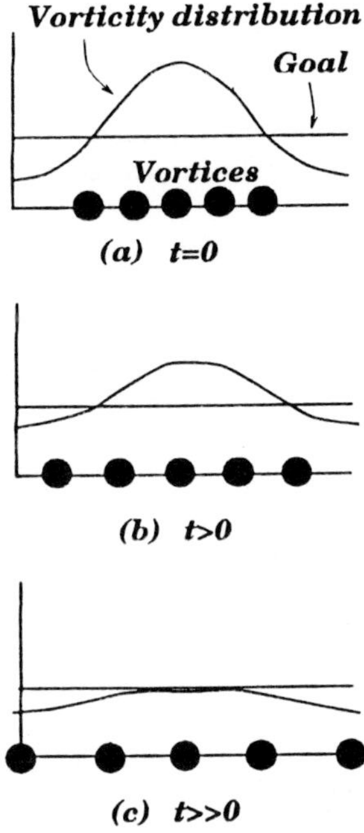

Fig.15: Simulation of one-dimensional diffusion. (a) the vortices, the vorticity distribution and the goal — the desired resolution — at initial stage, (b) the vorticity distribution decreases as the vortices spread, (c) the vortices stop to move when the goal has been reached

sented by Degond and Mas-Gallic ([33], [34]), to the vortex methods and performed unprecedented high-resolution simulations. They provided benchmark quality simulations for the early stages of the flow around an impulsively started cylinder using up to a million computational elements. This high resolution was realized probably because the strength exchange was treated within a frame work of the Eulerian scheme rather than the Lagrangian one as explained later in this section.

Similar techniques to change the strengths of vortices are presented also by Subramaniam and van Dommelen [37], Fishelov [38], and Cottet and Mas-Gallic [47].

As indicated by the vorticity transport equation (9), the vorticity distribution ω moving with the convection velocity varies with time due to diffusion. As a result the strength of a vortex, $\Gamma = \omega h^2$ where h^2 is the vortex area, varies when this vortex is being convected as shown by

$$\frac{d\Gamma}{dt} = h^2 \nu \Delta\omega \qquad (20)$$

This indicates that the diffusion of vorticity can be treated by changing the strengths of the moving vortices.

In [35] and [36], the term $\Delta\omega$ is approximated as

$$\Delta\omega \sim \frac{1}{\sigma^2} \int K\left(\frac{|x-y|}{\sigma}\right) \{\omega(y) - \omega(x)\} dy \qquad (21)$$

where

$$K\left(\frac{z}{\sigma}\right) = \frac{4}{\pi\sigma^2} \exp\left(-\frac{|z|^2}{\sigma^2}\right) \qquad (22)$$

Using Eq.(21) and discretizing the integration over the particles (vortices), the following equation is obtained

$$\frac{d\Gamma_i}{dt} = \frac{\nu h^2}{\sigma^2} \sum_j K\left(\frac{|x_i - x_j|}{\sigma}\right) (\Gamma_j - \Gamma_i) \qquad (23)$$

This shows that i^{th} vortex is given $\nu h^2/\sigma^2 K \cdot (\Gamma_j - \Gamma_i)dt$ amount of circulation by j^{th} vortex during the time period dt (Fig.3.%. The same amount of circulation is taken from j^{th} vortex so that the total circulation is conserved (i.e., $\sum_i d\Gamma_i/dt = 0$). This method is called *the strength exchange method*.

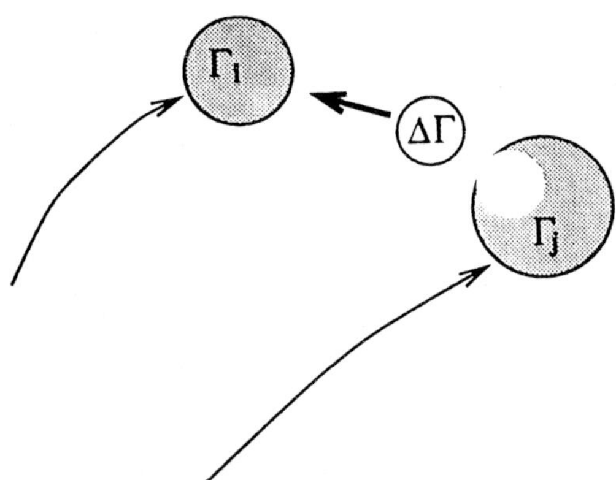

Fig.16: A portion of circulation is transported from particle to particle

Unlike the random vortex method and the diffusion velocity method, the strength exchange method does not work by itself. It needs grids. Suppose a vortex created on a body surface and see if this vortex starts to diffuse by itself. The random vortex method enables this vortex to diffuse via random walks and the

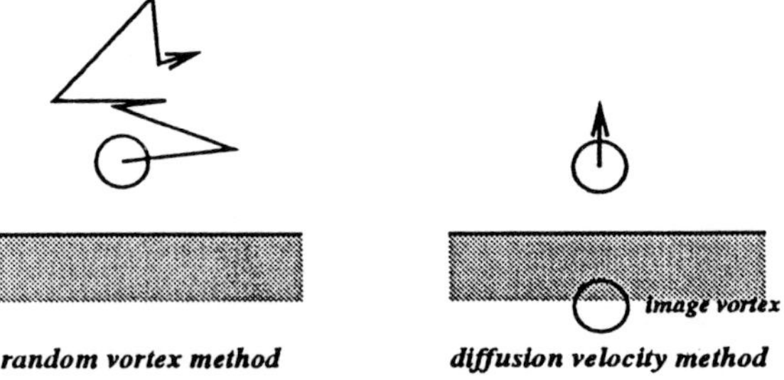

Fig.17: Diffusion models

diffusion velocity method via the diffusion velocity induced by the image vortex placed for the boundary condition [27] (Fig.3.%. As for the strength exchange method, this vortex just remains here without diffusing (Fig.3.%a)) because it needs partners with which it exchanges the vortex strength. These partners, with no strength initially, have to be positioned beforehand at every time step in the region where the vorticity is zero but will be non-zero at the next time step (i.e., $\Gamma_i = 0$ and $d\Gamma_i/dt \neq 0$ in Eq.(23)). The simplest way of doing this is to use grids (Fig.3.%b)).

In [35] and [36], it is said that the particle locations have to be re-initialized (remeshed) onto a uniform grid while interpolating the old vorticity on the new particle locations. The reason for doing this is, according to the papers, to reform "particle clustering in one direction accompanied by an expansion in another direction." However, the more important explanation for the necessity of grids may be the one just mentioned above.

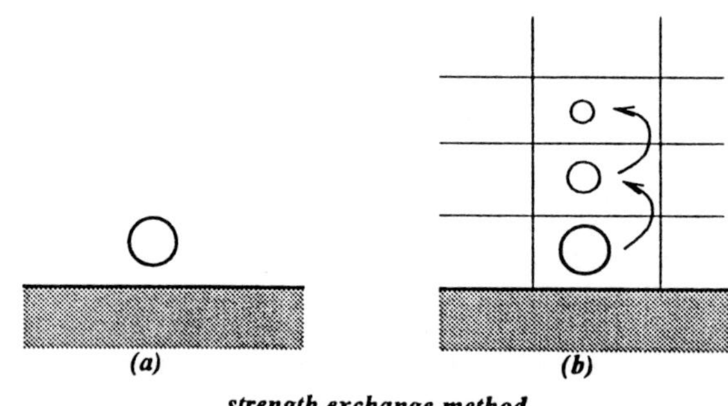

Fig.18: Circulation travels from grid to grid

It is not sufficient to regard the mechanism of diffusion only as strength exchange between particles. Based on the concept of the strength exchange method, this mechanism is considered to consist both of strength exchange between the existent particles and of creation of new particles (Fig.3.%. It is be-

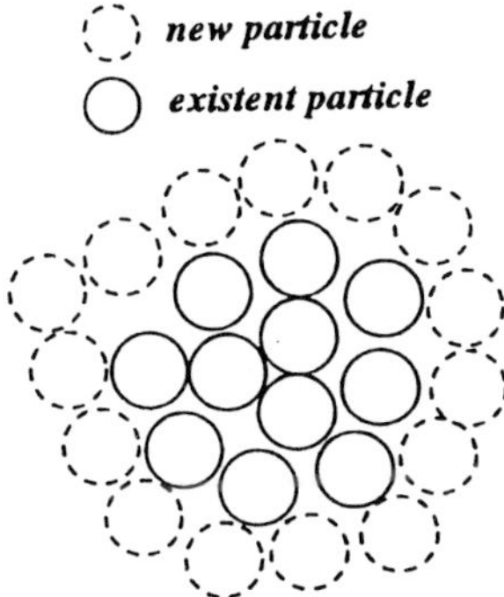

Fig.19: Circulation travels between the existent particles (solid line). New particles (dashed line) are necessary for further diffusion

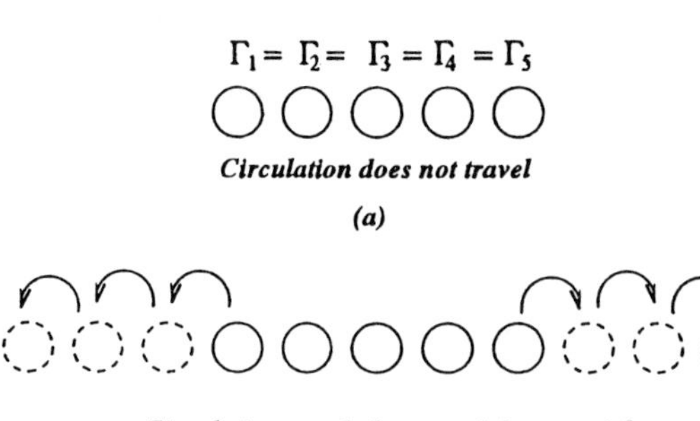

Fig.20: Circulation travels from particle to particle

cause the diffusive substance (circulation/vorticity) seeps out of the particles (vortices) not only into the existent particles (vortices) but also into the vacant (irrotational) area. Note that this creation does not mean the creation of vorticity caused by viscosity on a body surface but means fission of a vortex into multiple vortices to simulate spreading of vorticity field due to diffusion.

As for the random vortex method and the diffusion velocity method, the mechanism of diffusion can be regarded simply as transfer of particles. In these methods the particles work as carriers of vorticity/circulation while they work as mediums through which vorticity/circulation travels in the strength exchange method.

It seems that the strength exchange method is classified as an Eulerian method as far as diffusion is concerned. To see this, let us consider the one-dimensional diffusion equation in an infinite domain and simulate the solution to this equation with an ini-

tial condition of a concentrated vorticity distribution. This initial condition may be approximated by using particles with the same strength placed at the same distance apart. What happens next is that the particles remain at the initial positions because there is no velocity at all, and that the strengths of them also remains as they are because the strength of each particle is the same, i.e., $d\Gamma_i/dt = 0$ (Eq.(23)). No diffusion process is simulated (Fig.3.4(a)).

If an array of particles (with zero strength) is placed at each end of the initial particle array, the particles remain again at the initial positions but a certain amount of circulation travels from particle to particle (Fig.3.4(b)). Thus the diffusion process is simulated. The particles, referred to as the Lagrangian grids, remain at the fixed positions and the diffusive substance is transfered from particle to particle (grid to grid). This is a significant feature of an Eulerian method.

The strength exchange method theoretically can be a Lagrangian method if the creation of particles (spreading of vorticity field due to diffusion) is treated in a Lagrangian scheme. Practically it is a mixed Lagrangian-Eulerian method because the creation of particles is performed using grids.

A vortex mania would wonder if simulations of high resolution are possible through a pure Lagrangian scheme, or an Eulerian scheme is the only way to achieve high resolution.

4. FAST METHODS

The second restriction on the vortex method we consider in this paper is the increase of the computational time in proportion to N^2 where N is the number of the computational elements (e.g. vortices, vortex sheets and vortex blobs). The computation of pairwise interactions costs prohibitively. To reduce this computational time without losing the accuracy of the solutions, efficient techniques have been presented. We will see those techniques in the following subsections.

4.1 Far-field potential

The decay of the convection velocity produced by a vortex is so slow that all pairwise interactions must be accounted for to calculate the total velocity induced on a vortex by the rest of the vortices. The same problem occurs when calculating the gravitational force in celestial mechanics or the Coulombic force in plasma physics. Those forces are obtained from the gradient of a potential function called *far-field potential*. In this subsection, we will see time reduction techniques, which do not use a finite difference grid, for the far-field potential.

Such methods as use a finite difference grid and solve the Poisson's equation on it, instead of direct calculation of the pairwise interactions, in order to obtain the velocity field are presented by e.g. Hockney and Eastwood [55], Baden [56], Baden and Puckett [57].

4.1.1 Monopole method

Two things which are located close enough to each other and, at the same time, are located far enough from an observer look like one thing to the observer. This experience may help us to understand intuitively the physical basis of the method cited in this subsection (Barnes and Hut [48]).

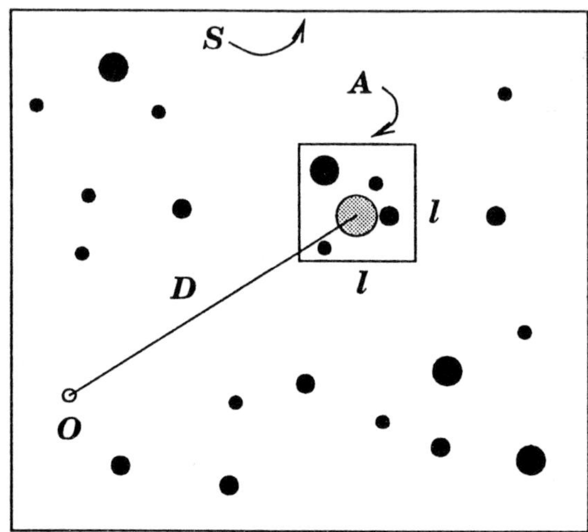

Fig.21: The system S and a small cell A. Black circles represent vortices of different circulation. The shaded circle in A represents a pseudo-vortex, with the total circulation in A, placed at the center of vorticity of A.

Figure 4.1.1 shows a cell, S, large enough to contain the entire system of the vortex distribution and a small cell, A, which contains several vortices of different circulation (black circles). Both the width and the height of the small cell are l, and the distance between the center of vorticity (Batchelor [49]), *monopole*, of the small cell (shaded circle) and the observer O (white circle) is D.

If the distance D is large enough and, at the same time, the size of the small cell l is small enough, the total velocity induced at the position O by all the vortices in the small cell is approximated by the velocity induced by one vortex, located at the center of vorticity, which has the total circulation in the small cell. Thus the computation required reduces from the number of the vortices in the small cell to one. If

the subdivided cell system for the entire region is well constructed, the computational time will reduce dramatically. Such a technique, as efficiently constructs a tree structured hierarchical subdivision of the entire system into cells, avoids ambiguity and tangling, allows rigorous analysis of errors, and most importantly reduces the computational time from $O(N^2)$ to $O(N \log N)$ is presented by Barnes and Hut [48]. This technique is based on the works by Appel [50] and Jernigan [51]. A different subdivision algorithm is presented by Van Dommelen and Rudensteiner [52], and then it is restructured for implementation on a parallel machine by Clarke and Tutty [29].

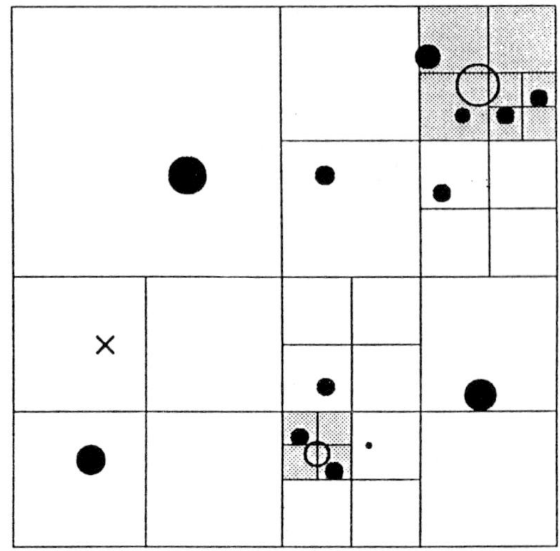

Fig.22: Hierarchical cells. Black circles represent vortices. × indicates the observation point. The vortices in a shaded cell are lumped together when calculating the velocity at the observation point. White circles represent pseudo-vortices in the shaded cells

Briefly we see how a hierarchical tree structured cell system is constructed, the velocity is calculated and the computation time is reduced in the case of Barnes and Hut's technique [48] considering a simple example of two dimension.

The entire system is divided into square cells so that each cell contains only one vortex or none in the following manner. Begin with an empty square cell large enough to contain the entire system. One by one, load vortices into this 'root' cell. If any two vortices fall into the same cell, divide that cell into four square subcells (*daughter* cells). Figure 4.1.1 shows a result of this cell division method.

Suppose that we calculate the total velocity induced, by all the vortices, at the position indicated by $\times$ in Fig.4.1.1The core of this time reduction technique is to lump the vortices close enough to each other and far enough from the observation point ($\times$). The criterion used in [48] for doing this is that, if the width of a *mother* cell, which contains four or more cells, is smaller than the distance between the center of vorticity of this mother cell and the observation point, the vortices in this mother cell are lumped and regarded as one vortex (pseudo-vortex) which contains the total circulation in this cell, and is located at the center of vorticity of this cell. There are two such mother cells in Fig.4.1.1as indicated by the shaded cells. The velocity is then calculated by using these pseudo-vortices and the *single* (non-lumped) vortices with an appropriate book-keeping work of dealing with the necessary information on all the cells.

The order of computing work of this monopole technique is estimated as follows. The average size of a mother/daughter cell $\bar{l}$ is the order of the inter-vortex spacing $D_s/\sqrt{N}$, where D_s is the size of the root cell and N is the vortex number. The number of subdivisions, n, required to reach the average cell size $\bar{l}$ staring at the root is obtained from the relation

$$\bar{l} = \frac{D_s}{\sqrt{N}} = D_s \left(\frac{1}{2}\right)^n \qquad (24)$$

as

$$n = \frac{1}{2} \log_2 N \qquad (25)$$

The time required to construct the whole cell system is order of nN, i.e., $O(N \log N)$. The book-keeping work of tagging the subdivided cells with the total circulation and center-of-vorticity position of the vortices they contain, and finally the work of calculating velocities using the necessary information may also require a computation time of $O(N \log N)$.

4.1.2 Multipole method

The core of the multipole method is based on a mathematical technique rather than a physical consideration. Briefly we see the time reduction technique following Greengard and Rokhlin [53].

The complex function F of a vortex of circulation Γ located at $z_0 = x_0 + iy_0$ is

$$F = \frac{i\Gamma}{2\pi} \log(z - z_0) \qquad (26)$$

For any z such that $|z| > |z_0|$, this function is equal to the multipole expansion — the Laurent expansion,

$$F = \frac{i\Gamma}{2\pi} \left(\log(z) - \sum_{k=1}^{\infty} \frac{1}{k} \left(\frac{z_0}{z}\right)^k \right) \qquad (27)$$

Similarly the complex function of m vortices of strengths $\{q_i, i = 1, \cdots, m\}$ located at points $\{z_i, i = 1, \cdots, m\}$ with $|z_i| < r$ is given by

$$F = Q \log(z) + \sum_{k=1}^{\infty} \frac{a_k}{z^k} \qquad (28)$$

for any z with $|z| > r$, where

$$Q = \sum_{k=1}^{\infty} q_i \qquad \text{and} \qquad a_k = \sum_{k=1}^{\infty} \frac{-q_i z_i^k}{k} \qquad (29)$$

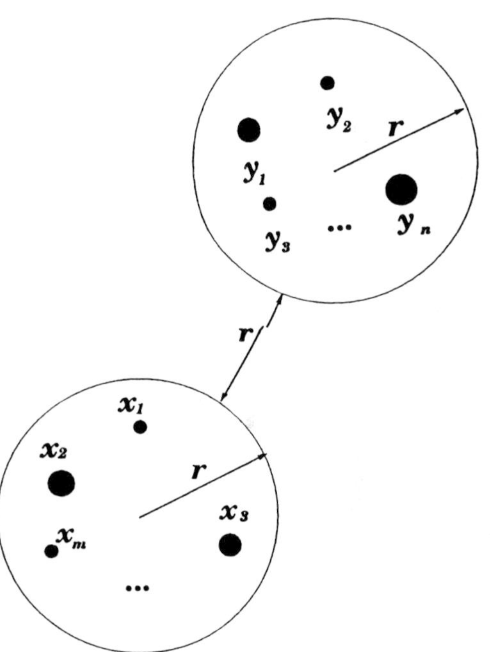

Fig.23: Two separated sets of vortices

Suppose m vortices at the points $\{x_i\}$ in a circle of radius r and obtain the complex function due to these vortices to calculate the velocities at the points $\{y_j\}$ (Fig.4.1.2 Traditionally this requires work of $O(mn)$. However, if we first compute the coefficients of a $p-$term multipole expansion of the complex function due to m vortices at the points $\{x_i\}$ using Eqs.(28), (29), then the order of the computation becomes $O(mp)$. Calculating the velocities at the points $\{y_j\}$ using the resulting multipole expansion requires work of $O(np)$. The total amount of computation is of the order $O(mp) + O(np)$. The value of the expansion term, p, is determined by the required precision and once the precision is specified, the amount of computation reduces to $O(m) + O(n)$, which is significantly smaller than mn for large m and n. For more details of this technique, see [53]. Another time reduction method which uses the Tayler expansion instead of the Laurent expansion is presented by Draghicescu [54].

Leonard and Koumoutsakos ([35], [36]) have used both the monopole method and the multipole method taking care with regard to vectorization, and obtained the theoretical time reduction. More specifically, "A timestep requiring one evaluation of all velocities for a million particles requires about one minute on single processor of a CRAY YMP while the N^2 algorithm would require roughly 24 hours [35]." Clarke and Tutty [29] also used both methods and restructured them for the suitable use of a parallel machine.

4.2 Rapidly decaying potential

A thing far from an observer is invisible to the observer. This may be a physical basis for the intuitive idea of the time reduction technique for the rapidly decaying potential (e.g., Van der Waals).

The vorticity distribution of a Gaussian-cored vortex or the diffusion velocity distribution induced by this kind of vortex blob decays so rapidly that the blobs far from the observation point do not have to be considered to calculate the total value of those distributions at this observation point. Ogami et al. [59] presented a time reduction technique for this kind of rapidly decaying potential. We briefly see this technique. (This method is so simple that a similar method might have been presented sometime somewhere)

Suppose that we calculate the density distribution of Eq.(6) at the locations of N particles with the Gaussian kernel Eq.(22). The amount of the computation is $O(N^2)$ without the reduction method described below.

First, we divide the computational area into square blocks labelled $B_{i,j}$ as in Fig.4.2 Next, we register the number of each particle and the label of the block which this particle belongs to. For example, this registration can be done by using an array $a(N)$,

$$a(1), a(2), ..., a(n_{i,j}), ..., a(N) = \overbrace{2, 6, 8,}^{1,1} \overbrace{10, 3, 9, 20, 21,}^{1,2}$$
$$\overbrace{44}^{1,3}, ..., \overbrace{26, 67, 33,}^{i,j} ...$$

This indicates that the 2nd, the 6th and the 8th particles belong to the block $B_{1,1}$, the 10th, the 3rd, the

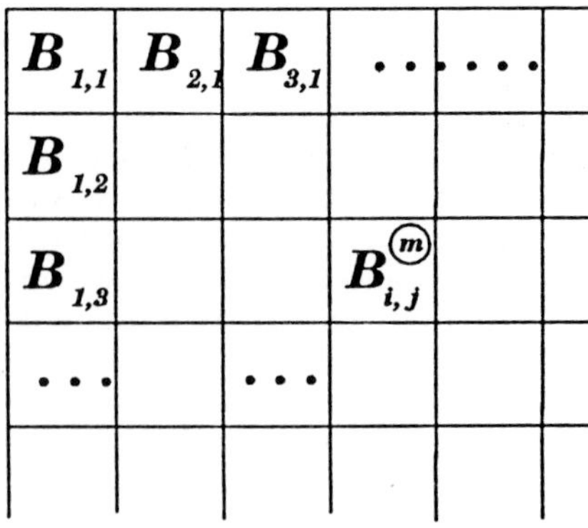

Fig.24: The labelled blocks.

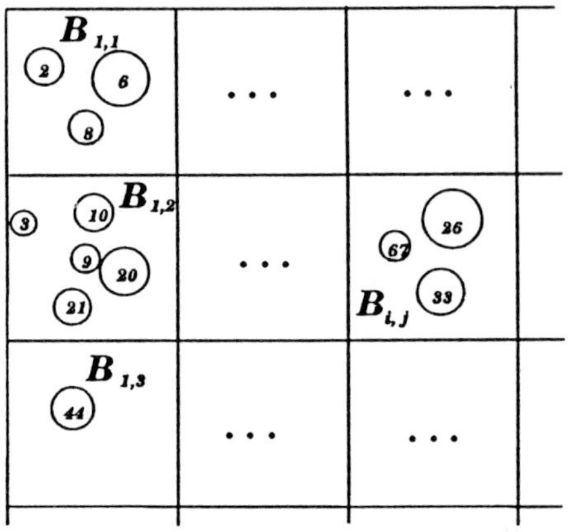

Fig.25: The registered particles.

9th, the 20th and the 21st particles belong to the block $B_{1,2}$, etc (Fig.4.2 Additionally, an array $b(i_B, j_B, 2)$, where i_B is the maximum block number in i-direction and j_B is that in j-direction,

$$b(1,1,1)=3, \qquad b(1,1,2)=1$$
$$b(1,2,1)=5, \qquad b(1,2,2)=4$$
$$b(1,3,1)=1, \qquad b(1,4,2)=9$$
$$\vdots \qquad\qquad \vdots$$
$$b(i,j,1)=3, \qquad b(i,j,2)=n_{i,j}$$
$$\vdots \qquad\qquad \vdots$$

should be provided. This indicates that the number of the particles in the block $B_{i,j}$ is $b(i,j,1)$ and that the numbers given to these particles begin with $b(i,j,2)$-th number in $a(N)$. For example, there are five particles in $B_{1,2}$ as indicated by $b(1,2,1) = 5$ and the numbers given to these particles are from 4th number in $a(N)$, which is 10, as indicated by $b(1,2,2) = 4$.

Now suppose mth particle exists in the block $B_{i,j}$ If the width of each block is large enough, the particles which contribute to the calculation of the density at the location of this particle exist only in the nine blocks, i.e., the block $B_{i,j}$ and the surrounding eight blocks. Considering the effective particles in these nine blocks only, we can accurately calculate Eq.(6). These effective particles are easily identified by the use of arrays $a(N)$ and $b(i_B, j_B, 2)$ explained above. The computing time greatly reduces because it is now proportional to N instead of N^2. This technique may be called *block-registration method*. This will be easily applied to three dimensions.

5. COMPRESSIBILITY

The last restriction on the vortex method we consider in this paper is inability to treat compressible fluids.

Though it may not be possible to directly apply the vortex methods to compressible flow simulations, some of the techniques developed for them can be used or modified for the simulations.

Ogami et al. presented Lagrangian formulations for solving the Burgers equation, the linear wave equation, and one– and two–dimensional isentropic compressible equations ([58], [59]). Similar to the vortex techniques, the compressible fluid density ρ is represented by an assembly of particles and the movements of the particles are calculated by the pairwise interactions of the particles. Ogami also applied his techniques to the full Navier–Stokes equations of one dimension ([60] in Japanese).

Hui has developed a *Generalized Lagrangian Formulation* which uses streamlines and their orthogonals as coordinate lines. The fluid particles are followed in their direction of motion but not with their speeds (for more details see [61]).

A particle method called *Smoothed Particle Hydrodynamics*, originally invented to simulate phenomena in astrophysics (Monaghan [62]), has been applied to one– and two– dimensional shock simulations (Monaghan and Gingold [63], Monaghan [64], Henneken and Icke [65]). This method is similar to that presented by Ogami except that it treats an inviscid compressible flow with the use of an artificial viscosity.

Lagrangian-Eulerian formulations for compressible flow simulations are presented by Katz and Shaughnessy [66], [67], and Konstantinov and Orszag [68]. A mixed method which uses the vortex method for the boundary layer and the random choice method, which is not deterministic, for compressible fluid regions is

presented by Sod [69].

Briefly we see the numerical method, originated from the vortex methods, and some results for the one-dimensional compressible viscous N–S equations presented by Ogami [60]. The three governing equations are (1) the continuity equation,

$$\frac{\partial \rho}{\partial t} + \frac{\partial (\rho u)}{\partial x} = 0 \tag{30}$$

where ρ is the fluid density and u the fluid velocity, (2) the momentum equation,

$$\frac{\partial (\rho u)}{\partial t} + \frac{\partial (\rho u^2 + p)}{\partial x} = (\lambda + 2\mu)\frac{\partial^2 u}{\partial x^2} \tag{31}$$

where p is the pressure, λ the heat conductivity and μ the viscosity, and (3) the energy equation,

$$\frac{\partial e}{\partial t} + \frac{\partial (e + p)u}{\partial x} = (\lambda + 2\mu)\frac{\partial}{\partial x}\left(u\frac{\partial u}{\partial x}\right) + \frac{\gamma\mu}{P_r}\frac{\partial^2 \varepsilon}{\partial x^2} \tag{32}$$

where e is the total energy, γ the gas specific heat ratio, ε the internal energy and P_r the Prandtl number.

The density at x_i, $\rho(x_i)$, is given as a summation of the Gaussian cored particles located at x_j as follows.

$$\rho(x_i) = \sum_{j=1}^{N} \frac{\rho_j \Delta x_j}{\sqrt{\pi}\sigma_j} \exp\left[-\frac{(x_i - x_j)^2}{\sigma_j^2}\right] \tag{33}$$

where ρ_j is the initial density at x_j, Δx_j the initial width of jth particle and σ_j the core radius of jth particle.

Using Eqs.(30) and (31), the acceleration is obtained as

$$\frac{du}{dt} = -\frac{1}{\rho}\frac{\partial p}{\partial x} + \frac{\lambda + 2\mu}{\rho}\frac{\partial^2 u}{\partial x^2} \tag{34}$$

This equation gives the acceleration of a moving particle.

Equation (32) becomes

$$\frac{d(e\Delta x_i)}{dt} = -\Delta x_i \frac{\partial (up)}{\partial x}$$
$$+\Delta x_i \left[(\lambda + 2\mu)\frac{\partial}{\partial x}\left(u\frac{\partial u}{\partial x}\right) + \frac{\gamma\mu}{Pr}\frac{\partial^2 \varepsilon}{\partial x^2}\right] \tag{35}$$

The total energy of ith particle varies with time according to this equation when this particle is moving with the velocity u.

Using the relation [46]

$$\frac{1}{\sigma_i}\frac{d\sigma_i}{dt} = \frac{\partial u}{\partial x} \tag{36}$$

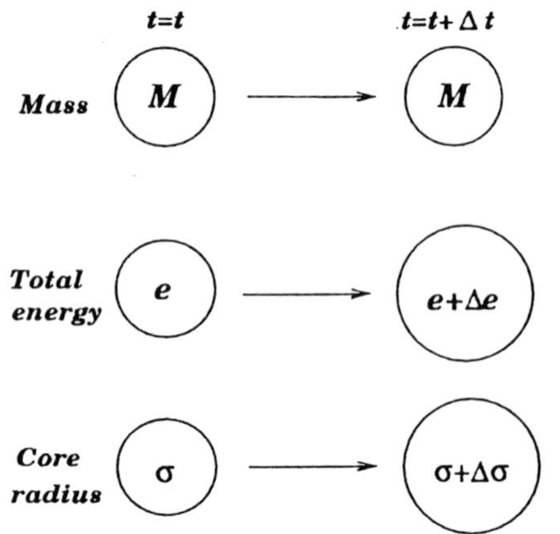

Fig.26: The mass is conserved while the total energy and the core radius vary with time

Eq.(30) becomes

$$\frac{d(\sigma_i \rho)}{dt} = 0 \qquad (37)$$

This indicates that the mass of ith particle is conserved even when it is moving with the velocity u and the size of the particle, σ_i, varies.

To sum up, the compressible viscous fluid motion can be simulated by the particles with constant mass which are accelerated according to Eq.(34) and whose total energy and core radius vary according respectively to Eq.(35) and Eq.(36) (Fig.5). As can be observed, a new blob species – fluid blob – is used instead of a vortex blob.

The fluid density ρ and its derivatives are calculated from Eq.(33), and the term $\partial p/\partial x$, $\partial u/\partial x$, $\partial^2 u/\partial x^2$ and $\partial^2 \varepsilon/\partial x^2$ are obtained by the interpolation techniques using the Spline function. The pressure is given by the relation $p = \rho R T = (\gamma - 1)(e - \rho/2u^2)$ where R is the gas constant and T the temperature.

Here are some results. Figure 5. shows a shock wave and an expansion wave produced in a shock tube filled with argon. The solid line indicates the initial status of the density distribution where the left side of the pressure is 50 times higher than the right side of it. This can be created by placing particles with $\Delta x_i/50$ apart at $x < x_0$ and with Δx_i apart at $x > x_0$ where x_0 is the position of a diaphragm. The rest of the lines plot the density distributions at the times indicated in this figure. We see that an expansion wave is travelling to the left and a shock wave to the right. The parameters used for this simulation are $\Delta x_i = 0.04/7999$cm, $\Delta x_i/\sigma_i = 1$,

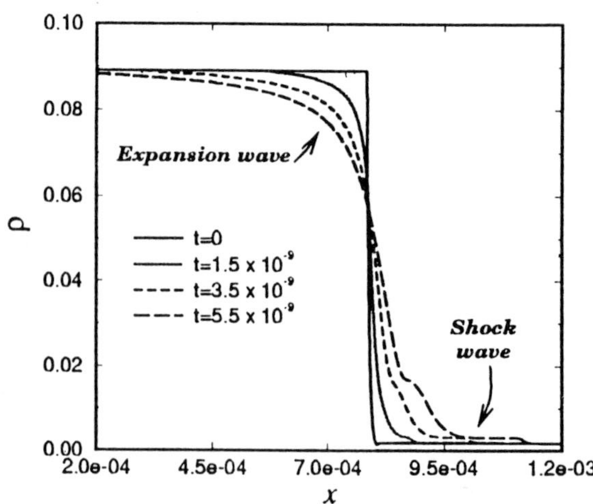

Fig.27: Simulation of a shock wave and an expansion wave produced in a shock tube. Density distributions are plotted

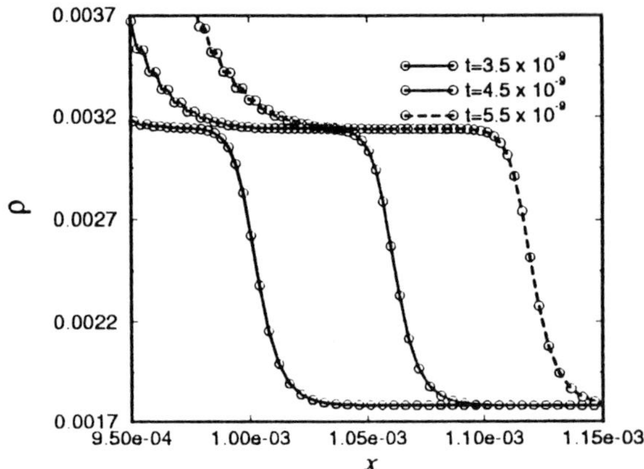

Fig.28: The shock wave reaches the steady state

$\rho_i = 1.784 \times 10^{-3}$g/cm³, $\mu = 2.1 \times 10^{-4}$g/cm·s, $\gamma = 1.67$, $R = 208.2 \times 10^4$J/kgK, the initial temperature $T_0 = 300$K, the time step $\Delta t = 5 \times 10^{-15}$s and the particle number $N = 8000$.

The travelling shock wave is magnified in Fig.5. The velocity of this wave is calculated as $M = 1.823$ where M is the Mach number. The shape of the wave does not vary, which means the wave has reached the steady state. The x coordinates of the symbol marks o in Fig.5. correspond to those of the particles. We see that the structure of the shock wave is represented by almost ten particles and that interestingly enough the distance between the particles is of order of the mean free path (i.e. 10^{-6}cm). This may indicate that our numerical model is physically consistent.

Figures 5, 5.and 5.show the time variations of the pressure distribution, the velocity distribution and the energy distribution.

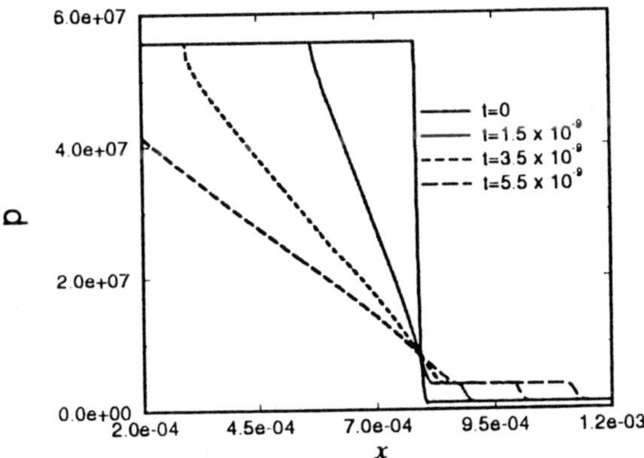

Fig.29: Pressure distribution

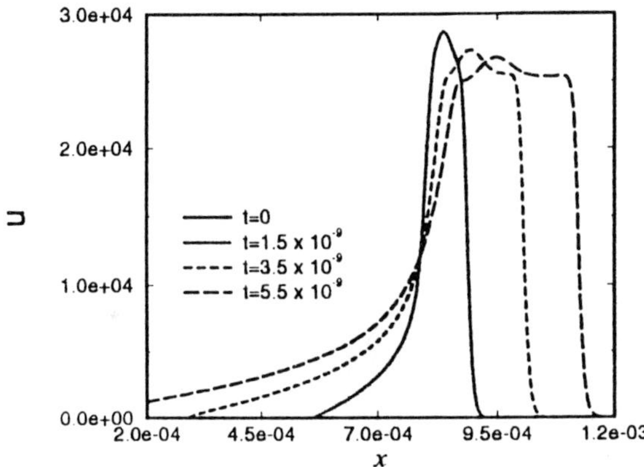

Fig.30: Velocity of the particles. At time=0, the velocity is zero everywhere

6. CONCLUDING REMARKS

We have seen how the three inborn restrictions on the vortex methods, i.e. (1) inability to treat viscosity, (2) inability to treat compressibility and (3) prohibitive growth of the computational costs, are eased or removed. We apologize here if we have missed important work. Our interest has been focused on a Lagrangian technique rather than an Eulerian one because we believe that there are still many unknown ways to make the best use of the advantages of the Lagrangian technique for many known and unknown purposes, though we all admit that a grid-based Eulerian method also is a powerful computational tech-

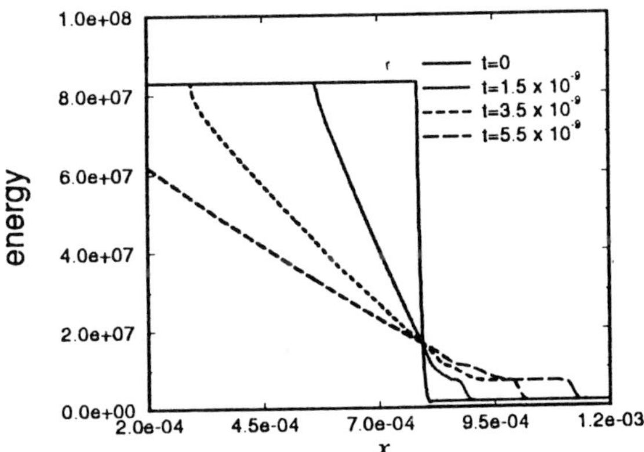

Fig.31: Energy distribution

nique itself and that it can compensate for the drawbacks of the vortex methods.

Four types of diffusion model for the vortex scheme have been discussed. The core spreading method and the random vortex method are classic whose advantages and disadvantages are well known. The diffusion velocity method is a pure deterministic Lagrangian technique. This provided the first successful simulations of the unsteady separated flows with strong viscosity. This new concept needs more numerical experiments and analyses to reveal its characteristics. The strength exchange method requires a mixed Lagrangian-Eulerian formulation. The remeshing technique and the fast summation algorithms are practical ways to tune up the vortex methods so that they can provide benchmark quality simulations with up to a million computational elements. We wonder if a pure Lagrangian scheme also is able to provide simulations of high resolution, or an Eulerian scheme is the only way to achieve high resolution. Incidentally, the Lagrangian technique of handling the diffusion of vorticity will be applicable to the diffusion of heat. If this is true, then the vortex methods will be able to solve heat transfer problems where vorticity and heat interestingly interact. A new blob species — heat blob — needs to be created.

For time reduction techniques of the far-field potential, the monopole method and the multipole method were viewed. The combination of the two methods would be most practical and efficient to reduce the computation time of the convection velocity. For the rapidly decaying potential such as the vorticity distribution or the diffusion velocity, the block-registration method was described.

One way of treating a compressible flow by a pure Lagrangian technique was introduced. Since a new blob species (fluid blob) has been used in stead of a vortex blob, this method is no longer classified under a vortex method but under a particle method (is it because the vortex method evolves into a particle method or because the vortex method is kin to it after all?).

For further development of the vortex methods, it may be necessary to remove the borderlines of the related computational methods or to adopt useful techniques from other fields, as well as to create new concepts, techniques and blob species, of course. With these it may not be impossible to apply the vortex techniques to the heat transfer problems, the natural convection flows, the flows involving chemical reactions and so on. To widen the range of application to the physical problems definitely would be a challenging, creative and pioneering work.

On the other hand, the ultimate goal of the present vortex methods may be simulations of realistic three dimensional flows with objects in them, which the Eulerian methods are skillful at. With the efficient algorithms tuned for a vector processor or a parallel processor, and powerful computer equipment capable of handling millions of vortices, these simulations will be realized before long.

Almost all the recent vortex simulations have been done using vortex blobs rather than point vortices. This is because the vortex blobs exclude the singularity of the velocity and they conveniently produce a continuous vorticity distribution. However, Sarpkaya [4] offered crucial problems about using blobs as follows.

1. The rigid-blob idealization is not dynamically consistent either and violates Euler's equations and Helmholtz's laws!
2. A linear sum of vortices with finite cores does not constitute an exact solution of the nonlinear equations of motion.
3. The overlapping blobs must be regarded as mathematical artifices (many souls in one body!) since vorticity carrying non-deforming fluid particles cannot occupy the same space at the same time (the exclusion principle).

We could answer this way. The purpose of the vortex methods is not to give an exact solution but to give an approximate solution, which, if possible, is proved to converge to the exact solution. The violation of Euler's equations and Helmholtz's laws can be tolerated if the degree of this violation, which is to be estimated mathematically, is as small as the error due to the approximation.

We note that the fluid particles carry the vorticity but the vorticity does not carry the fluid particles. The overlapping blobs do not mean the overlapping fluid particles but mean the overlapping vorticity distributions. It is not the fluid itself but the vorticity that is discretized into blobs (as mentioned in Section 2, discretization is done both space-wise and depth-wise). In an incompressible flow, the fluid density is constant everywhere, which would prohibit the overlapping of fluid particles (overlapping causes a change of fluid density). However, the vorticity is not generally constant, which would permit overlapping of vortex blobs (without overlapping the spatial variation of the vorticity could not be treated). On the other hand, in a compressible flow, the fluid density is spatially variable so that the technique to overlap the fluid particles is essential.

The position where the density is ρ may be expressed by two particles of density $\rho/2$ or by three particles of $\rho/3$. The possible combinations are infinity so are the possible models for the particle representation. It may be important to realize that a numerical model is to solve existent equations numerically and a physical model is to derive equations which govern physical phenomena or to explain physical phenomena, and that essentially the two models may be the same, even though they look different due to the different purposes.

This is analogous to the fact that we can approximate the integral $\int_a^b f(x)dx$ by the Riemann sum

$$\int_a^b f(x)dx \approx \sum_{i=0}^{n-1} f(c_i)(x_{i+1} - x_i)$$

where $a = x_0 < x_1 < \cdots < x_n = b$ of $[a,b]$ and $c_i \in [x_i, x_{x+1}]$. We all know that a limit of the Riemann sums is equal to $\int_a^b f(x)dx$ even though the curve of the function $f(x)$ evidently differs from the outline of the rectangles of $f(c_i) \times (x_{i+1} - x_i)$ which approximate this curve (Fig.6). This concept might apply to the case where the numerical models simulate the physical phenomena. Namely, if a numerical model of discrete representation is conceptionally and technically (physically and mathematically) correct in modelling the real physical phenomena, the following would hold

Physical phenomena $\approx$ Numerical simulations

(e.g. Helmholtz's laws $\approx$ Rigid-blob idealization)

even when the numerical model used for these simulations looks inconsistent to the physical phenomenon.

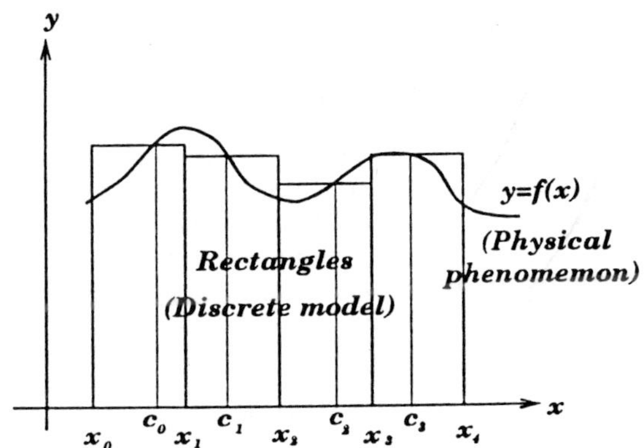

Fig.32: The outline of the rectangles does not fit the curve of $f(x)$. However, this does not mean a Riemann sum approximation is wrong

Simulations are not real. There always will be a gap between the simulations and the real physical flows. For instance, a vortex blob simulation is like a printed picture. If we magnify a picture of Marilyn Monroe, we only see scattered dots. We do not see a protein structure of her skin. All we try to do is to re-create a thing which by nature contains an infinite amount of physical data by using a finite number of numerical data. However, the computational simulations will undoubtedly continue to bring a better understanding and a new insight of the real physical flows.

REFERENCES

[1] L. Rosenhead, 'The formulation of vortices from a surface of discontinuity', *Proc. Roy. Soc., Series A*, **134**, 170-192, (1931).

[2] A. Leonard, 'Vortex methods for flow simulation', *J. Comput. Physics*, **37**, 289-335, (1980).

[3] A. Leonard, 'Computing three-dimensional flows with vortex elements', *Ann. Rev. Fluid Mech.*, **17**, 523-559, (1985).

[4] T. Sarpkaya, 'Computational methods with vortices — the 1988 Freeman scholar lecture', *J. Fluids Engineering*, **111**, 5-52, (1989).

[5] O.H. Hald, *Convergence of Vortex Methods*, 33-58, Vortex Methods and Vortex Motion, SIAM, Philadelphia, 1991.

[6] E.G. Puckett, *Vortex Methods: An Introduction and Survey of Selected Research Topics*, Incompressible Computational Fluid Dynamics – Trends and Advances, Cambridge University Press, Cambridge , 1991.

[7] J.I. Sethian, *A Brief Overview of Vortex Methods*, 1-32, Vortex Methods and Vortex Motion, SIAM, Philadelphia, 1991.

[8] K. Kamemoto, *On Attractive Features of the Vortex Methods*, 334-353, Computational Fluid Dynamics Review 1995, Wiley, Chichester, 1995

[9] O.H. Hald and V.M. Del Prete, 'Convergence of vortex methods for Euler's equations', *Math. Comp.*, **32**, 791-809, 1978.

[10] O.H. Hald, 'Convergence of vortex methods for Euler's equations, II', *SIAM J. Numer. Anal.*, **16**, 726-755, 1979.

[11] O.H. Hald, 'Convergence of vortex methods for Euler's equations, III', *SIAM J. Numer. Anal.*, **24**, 538-582, 1987.

[12] J.T. Beale and A. Majda, 'Vortex methods I: convergence in three dimensions', *Math. Comp.*, **39**, 1-27, 1982.

[13] J.T. Beale and A. Majda, 'Vortex methods II: higher order accuracy in two and three dimensions', *Math. Comp.*, **39**, 29-52, 1982.

[14] J.T. Beale, 'A convergent 3-D vortex method with grid-free stretching', *Math. Comp.*, **46**, 401-424, 1986.

[15] C. Greengard, 'Convergence of the vortex filament method', *Math. Comp.*, **47**, 387-398, 1986.

[16] K. Kuwahara and H. Takami, 'Numerical studies of two-dimensional vortex motion by a system of point vortices', *J. Phys. Soc. Japan*, **34**, 247-253, (1973).

[17] X. Zhang and A.F. Ghoniem, 'A computational model for the rise and dispersion of wind-blown, buoyancy-driven plumes —I. neutrally stratified atmosphere', *Atmospheric Environment*, **27A**, 2295-2311, (1993).

[18] M. Cheng, Y.T. Chew, and S.C. Luo, 'Discrete vortex simulation of the separated flow around a rotating circular cylinder at high Reynolds number', *Finite Elements in Analysis and Design*, **18**, 225-236, (1994).

[19] S. Nourazar and R. Milane, 'Comparison of the effect of two subgrid-scale models in the simulation of a two-dimensional spatially growing mixing layer using the vortex method', *Turbulent Flows*, **188**, 47-52, (1994).

[20] A.J. Chorin, 'Numerical study of slightly viscous flow', *J. Fluid Mech.*, **57**, 785-796, (1973).

[21] H.N. Najm and A.F. Ghoniem, 'Coupling between vorticity and pressure oscillations in combustion instability', *J. Propulsion and Powers*, **10**, 769-776, (1994).

[22] K. Taehyeung and M.R. Flynn, 'Numerical simulation of air flow around multiple objects

using the discrete vortex method', *J. Wind Engineering and Industrial Aerodynamics*, **56**, 213-234, (1995).

[23] C. Marchioro and M. Pulvirenti, 'Hydrodynamics in two dimensions and vortex theory', *Comm. Math. Phys.*, **84**, 483-503, (1982).

[24] J. Goodman, 'Convergence of the random vortex method', *Comm. Pure Appl. Math.*, *60*, 189-220, (1987).

[25] D.G. Long, 'Convergence of the random vortex method in two dimensions', *J. Amer. Math. Soc.*, **1**, 799-804, (1988).

[26] T. Kida and M. Kurita, 'High Reynolds number flow past an impulsively started circular cylinder', *Computational Fluid Dynamics JOURNAL*, **34**, 489-508, (1995).

[27] Y. Ogami and T. Akamatsu, 'Viscous flow simulation using the discrete vortex model — the diffusion velocity method', *Computers & Fluids*, **19**, 433-441, (1991).

[28] M. Shintani and T. Akamatsu, 'Investigation of two-dimensional discrete vortex method with viscous diffusion model', *Computational Fluid Dynamics JOURNAL*, **3**, 237-254, (1994).

[29] N.R. Clarke and O.R. Tutty, 'Construction and validation of a discrete vortex method for the two-dimensional incompressible Navier-Stokes equations', *Computers & Fluids*, **23**, 751-783, (1994).

[30] Y. Ogami and Y. Ayano, 'Flows around a circular cylinder simulated by the viscous vortex model — the diffusion velocity method', *Computational Fluid Dynamics JOURNAL*, **4**, 383-399, (1995).

[31] J.H. Strickland, S.N. Kempka, and W.P. Wolfe, *Viscous Diffusion of Vorticity using the Diffusion Velocity Method*, 121-122, Second International Workshop on Vortex Flows and Related Numerical Methods, Montreal, 1995

[32] J.T. Beale and A. Majda, 'High order accurate vortex methods with explicit velocity kernels', *J. Comput. Phys.*, **58**, 188-208, 1985.

[33] P. Degond and S. Mas-Gallic, 'The weighted particle method for convection-diffusion equations, part 1: the case of an isotropic viscosity', *Math. Comput.*, **53**, 485-507, (1989).

[34] P. Degond and S. Mas-Gallic, 'The weighted particle method for convection-diffusion equations, part 2: the anisotropic case, *Math. Comput.*, **53**, 509-525, (1989).

[35] A. Leonard and P. Koumoutsakos,'High resolution vortex simulation of bluff body flows', *J. Wind Engineering and Industrial Aerodynamics*, **46 & 47**, 315-325, (1993).

[36] P. Koumoutsakos and A. Leonard, 'High-resolution simulations of the flow around an impulsively started cylinder using vortex method', *J.Fluid Mech.*, **296**, 1-38, (1995).

[37] S. Subramaniam and L. van Dommelen, *A New Diffusion Scheme in Vortex Methods for Three Dimensional Incompressible Flows*, 140-141, Second International Workshop on Vortex Flows and Related Numerical Methods, Montreal, 1995

[38] D. Fishelov, 'A new vortex scheme for viscous flows', *J. Comp. Phys.*, **86**, 211-224, (1990).

[39] D. Fishelov, 'Vortex schemes for viscous flow', *Lectures in Applied Mathematics, Vortex Dynamics and Vortex Methods*, **28**, 153-164, (1991).

[40] Z.Y. Lu, 'Diffusing-vortex numerical scheme for solving incompressible Navier-Stokes equations', *J. Comp. Phys.*, **95**, 400-435, (1991).

[41] C.C. Chang and R.L. Chern, 'A numerical study of flow around an impulsively started circular cylinder by a deterministic vortex method', *J. Fluid Mech.*, **233**, 243-263, (1991).

[42] G. Russo, 'A deterministic vortex method for the Navier–Stokes equations', *J. Comp. Phys.*, **108**, 84-94, (1993).

[43] Y. Nakanishi and K. Kamemoto, 'Numerical simulation of flow around a sphere with vortex blobs', *J. Wind Engineering and Industrial Aerodynamics*, **46 & 47**, 363-369, (1993).

[44] H.N. Najm, *A Hybrid Vortex Method with Deterministic Diffusion*, 207-222, Vortex Flows and Related Numerical Methods, Kluwer Academic Pubkishers, Dordrecht, 1993.

[45] C. Greengard, 'The core spreading vortex method approximates the wrong equation', *J. Comput. Phys.*, **61**, 345-348, (1985).

[46] S.N. Kempka and J.H. Strickland, 'A method to simulate viscous diffusion of vorticity by convective transport of vortices at a non-solenoidal velocity', *SAND93-1763*, **UC-700**, 3-28, (1993).

[47] G.H. Cottet and S. Mas-Gallic, 'A particle method to solve transport-diffusion equations. Part II: the Navier-Stokes equations', *Numer. Math.*, *57*, 805-827, (1990).

[48] J. Barnes and P. Hut, 'A hierarchical O(NlogN) force-calculation algorithm', *Nature*, **324**, 446-449, (1986).

[49] G.K. Batchelor, *An Introduction to Fluid Dynamics*, Cambridge University Press, Cambridge, 1967.

[50] A.W. Appel, 'An efficient program for many-body simulation', *SIAM J. Sci. Statist.*

Comput., **6**, 85-103, (1985).

[51] J.G. Jernigan, 'Direct n-body simulations with a recursive center of mass reduction and regularization', *I.A.U. SYMP*, **113**, 275-284, (1985).

[52] L. Van Dommelen and E.A. Rundensteiner, 'Fast, adaptive summation of point forces in the two-dimensional Poisson equation', *J. Comput. Phys.*, **83**, 126-147, (1989).

[53] L. Greengard and V. Rokhlin, 'A fast algorithm for particle simulations', *J. Comput. Phys.*, **73**, 325-348, (1987).

[54] C.I. Draghicescu, 'An efficient implementation of particle methods for the incompressible Euler equations', *SIAM J. Numer. Anal.*, **31**, 1090-1108, (1994).

[55] R.W. Hockney and J.W. Eastwood, *Computer Simulation Using Particles*, McGraw-Hill, New York, 1981.

[56] S.B. Baden, 'Very large vortex calculations in two dimensions', *Lecture Note in Mathematics, Vortex Methods*, **1360**, 96-120, (1987).

[57] S.B. Baden and E.G. Puckett, 'A fast vortex method for computing 2d viscous flow', *J. Comput. Phys.*, **91**, 278-297, (1990).

[58] Y. Ogami and A.Y. Cheer, 'Simulations of Unsteady Compressible Fluid Motion by an Interactive Cored Particle Method', *SIAM Journal on Applied Mathematics*, **55**, 1204-1226, (1995).

[59] Y. Ogami, H. Inaba, and I. Arimoto, 'Particle method for isentropic compressible flow simulation', *Computational Fluid Dynamics JOURNAL*, **5**, 341-360, (1996).

[60] Y. Ogami, 'Numerical simulation of compressible viscous N–S equations by the particle method', *Proceedings of the 10-th Symposium of Computational Fluid Dynamics*, Tokyo, 104-105, (1996) (in Japanese).

[61] W.H. Hui, *Generalized Lagrangian Formulation of Computational Fluid Dynamics*, 382-398, Computational Fluid Dynamics Review 1995, Wiley, Chichester, 1995

[62] J.J. Monaghan, 'Smoothed particle hydrodynamics', *Annu. Rev. Astron. Astrophys.*, **30**, 543-574, (1992).

[63] J.J. Monaghan and R.A. Gingold, 'Shock simulation by the particle method SPH', *J. Comp. Physics*, **52**, 374-389, (1983).

[64] J.J. Monaghan, *Application of the Particle Method SPH to Hypersonic Flow*, 357-365, Computational Techniques and Applications: CTAC-85, Elsevier Science Publishers B.V., North-Holland, 1986.

[65] E.A.C. Henneken and V. Icke, 'SPH faces Emery's jump', *Computer Physics Communications*, **74**, 239-246,(1993).

[66] I.M. Katz and E.J. Shaughnessy, 'Computer aided analysis of 1-d compressible flow problems in a Lagrangian particle description using the α method', *Computers & Fluids*, **18**, 75-101, (1990).

[67] I.M. Katz and E.J. Shaughnessy, 'The equations of 1-d compressible flow with area change in a Lagrangian particle description', *Computers & Fluids*, **18**, 183-190, (1990).

[68] A.B. Konstantinov and S.A. Orszag, 'Extended Lagrangian particle-in-cell (ELPIC) code for inhomogeneous compressible flows', *J. Scientific Computing*, **10**, 191-231, (1995).

[69] G.A. Sod, 'A compressible vortex method with application to the interaction of an oblique shock wave with a boundary layer', *Appl. Numer. Math.*, **8**, 257-273, (1991).

METHOD OF DISCRETE VORTICES IN AERODYNAMIC PROBLEMS REDUCIBLE TO SINGULAR INTEGRAL EQUATIONS

S. M. BELOTSERKOVSKY   I. K. LIFANOV   L.N.POLTAVSKY

Dept. of Mathematics, N.E. Joukovski Airforce Engineering Academy; Russia

Abstract

Addressed in this article are only those problems of aerodynamics that can be reduced to linear singular or hypersingular equations and for which the mathematical foundation of the method of discrete vortices or the method of discrete closed vortex frames is given. Such problems include noncirculatory problems for piecewise smooth contours (surfaces), the linear stationary problem for the rectangular finite-span wing and the linear nonstationary problem for a thin airfoil.

1 Background

The intensive development of aviation and the rapid progress in computer technology have resulted in the emergence of a new scientific research method - the numerical experiment - and stimulated the creation of novel approaches to solving applied and fundamental problems of aerodynamics. This has given rise, in particular, to the creation and development of the method of discrete vortices (MDV) and the formation on its basis the computational vortex mechanics of fluids. In this survey article (and possibly in subsequent works on this subject), the characteristics are presented of this methodology formed by our creative team during the last three to four decades.

First of all, it should be stressed that the so-called vortex methods have found a new life having successfully brought together physical, mathematical and computational aspects of the problem. From the classical heritage of the past we should cite the works of N.Ye.Zhukovsky [1], L.Prandtl [2] and the book of V.V.Golubev [3].

The early 1950's saw the beginning of the creation and application of the MDV as a basic powerful tool for solving practical problems [4-7], these results were summarized at this phase in monographs [7-9]. Since the beginning of the 1970's, "pure" mathematicians have been joining this process, having provided a new impetus for it [9-15].

The formation of the new methodology have been favored by a permanent creative association between aerodynamicists, mathematicians and specialists in computational methods.It turned out that quite often a research physicist more quickly finds a radical way to solving a problem, being based on an understanding of its essence, than a mathematician treating it in a more formal way; the numerical experiment supported with certain exact relationships and solutions for verification allows one to perform without a strict proof a preliminary choice of a computational algorithm. After this, there is a broad field open to strict proofs, generalization and extention of the ideas into neighbouring areas.

The most promising proved to be natural approaches when the entire process of flow formation (nonstationary formulation of problems) and discretization in space and time is considered.

The following conclusion should be stressed: mathematical activity is given new powerful wings by the use of numerical simulation and becomes, as it were, a mathematical laboratory.

2 Physical Formulation of the Problem

We shall consider ideal incompressible fluid flow past a surface σ_1 (a contour in two-dimensional problems). It is supposed that the flow is smooth, i.e., a fluid particle having found itself on the surface can move only along this surface not penetrating it. The nonpenetration condition requires that the normal component of the relative flow velocity $\overline{V}_{rel}$ be equal to zero on

Received on July 1, 1997.

the surface in flow:

$$\overline{V}_{rel} \cdot \overline{n}_{M_0} = 0, \qquad M_0 \in \sigma_1 , \qquad (2.1)$$

where $\overline{n}_{M_0}$ is the unit vector of the normal to the surface σ_1 at a given point M_0.

Let us recall the following fundamental fact. Behind a body moving in flow the aerodynamic trailing track forms, where vortex motion can be observed, that is, fluid particles participate not only in translational and deformational motions, but also in rotational moves as well. Outside this region, the flow is irrotational. Thus, outside a body and its track, the flow perturbed by the body is potential and the continuity equation becomes the Laplace equation

$$\Delta U = 0, \qquad M \notin \sigma_1 \cup \sigma_2, \qquad (2.2)$$

where $U = U(x,y,z)$ is the potential of the velocity field perturbed by the body, σ_2 is the track behind the moving body. In this case, far from the surface σ_1 and the track σ_2 perturbed velocities must vanishing, therefore for the required solution $U(x,y,z)$ to Equation (2.2) the following condition must be satisfied:

$$\nabla U \to 0, \qquad \rho(M, \sigma_1 \cup \sigma_2) \to \infty. \qquad (2.3)$$

Then, the physical nature of the problem under study dictates the choice of a flow scheme. Among flow schemes, the major ones are the following.

a) Circulatory separated flow satisfying all physically evident conditions including the requirement of the velocity and pressure being finite throughout the entire flow domain. The aforesaid has a decisive influence on the choice of a flow scheme. For example, when considering flow around a thin wing with sharp edges (leading, trailing or side ones), one has to admit descending of free vortices from all of the edges and impose the Chaplygin-Zhukovsky-Kutta condition requiring the velocity finiteness. Otherwise, the flow velocities at the edges will tend to infinite values.

b) Some simplified schemes of circulatory flow around bodies, for which certain conditions are lifted. Most widely employed schemes are ones that do not require the finiteness of the velocities and pressures at the leading and trailing edges of the wings as well as at breaks of a body's surface. As a result, the problem may be also solved as a stationary one, with free vortices descending only from the trailing edge.

c) In some cases a further simplification of the problem is possible, when the motion of free vortices is assumed to be known. For example, in the linear nonstationary problem for a thin wing, it is assumed that free vortices move in the wing's plane at a specified velocity along straight lines parallel to a given one.

d) Finally, in some cases flow about a body can be considered without regard for the track at all. One of these cases is so-called noncirculatory problems. According to this scheme, the flow outside the body is taken to be potential. This fact substantially simplifies solving aerodynamic problems for bodies of arbitrary configuration.

As for the physical rightfulness of such an approach, it is based on the following facts.

In the case of stationary oscillating bodies or those under the action of periodically varying flow (gust), free vortices, having descend from a body, are not carried away, but concentrate near it. Free vortices will have circulations of opposite signs, their effects on the body are mutually cancelled and the vortices themselves are practically mutually annihilated.

The second case, related to the scheme indicated, corresponds to the motion of low-lifting bodies in a continuous medium when the strength of vortex tracks is small and their influence on the flow about a body may be ignored.

In the aforementioned cases, firstly, the solution of aerodynamic problems is considerably simplified: the action of the flow on a body is determined by calculating the so-called virtual masses. Secondly, studying the dynamics of a body's behavior in a continuous medium is also simplified. The motion is described by the same equations of mechanics that are used for analyzing motion in vacuum, but in this case virtual masses must be added to inertial and mass characteristics of a body [8].

Besides, the track may be also ignored in solving problems on two-dimensional flow about airfoils.

In this survey, the MDV application is considered for the two aforementioned classes of problems, because there is a rather full mathematical foundation for it. It is precisely to the MDV mathematical aspects that the present survey is dedicated.

Later on, the authors intend to present a survey of the MDV application to problem involving construction of vortex tracks.

3 Reduction of Boundary-Value Problems of Aerodynamics to Singular and Hypersingular Equations

As follows from the previous section, aerodynamic problems on flow of an ideal incompressible fluid around a surface (contour in two-dimensional problems) with no vortex sheets descending from the surface (contour) reduce to the solution of the Neumann problem for the Laplace equation for the potential of unperturbed flow's velocity field, i.e., to the solution of the problem

$$\Delta U = 0, \quad M \notin \sigma_1, \qquad (3.1)$$

$$\left.\frac{\partial U}{\partial \overline{n}_{M_0}}\right|_{\sigma_1} = -\overline{U}_0 \cdot \overline{n}_{M_0} = f(M_0), \quad M_0 \in \sigma_1, \qquad (3.2)$$

where $\overline{U}_0$ is the oncoming flow velocity. We shall solve Problem (3.1), (3.2) using the double-layer potential. This allows one to consider from a single standpoint flow around both closed and open surfaces (contours), as well as complicated surfaces composed of them. In the two-dimensional case, this will be the function

$$U(x_0, y_0) = \frac{1}{2\pi} \int\limits_{\sigma_1} \frac{\partial}{\partial \overline{n}_M} \ln\left(\frac{1}{r_{MM_0}}\right) g(M) dS_M,$$

$$M_0 \notin \sigma_1, \qquad (3.3)$$

where $r_{MM_0} = |\overline{MM_0}|$ is the distance between points M and M_0, $g(M)$ is the density of this potential, S_M is the arc length of the contour σ_1.

Using a differentiation formula [12, 16] for the double-layer potential and the continuity of its normal derivative, Condition (3.2) may be written in the form of the following singular integral equation (SIE) in the derivative of the the potential density

$$\frac{1}{2\pi} \int\limits_{\sigma_1} \frac{y'_{os}(y_0 - y) + x'_{os}(x_0 - x)}{r^2_{MM_0}} \gamma(s) ds_M = f(M_0),$$

$$M_0 \in \sigma_1, \qquad (3.4)$$

where $x = x(s), y = y(s), s \in [a, b]$, is the parametric equation of the curve σ_1, $x_0 = x(s_0)$, $y_0 = y(s_0)$, $\gamma(s) = g'(s)$. The contour σ_1 is assumed to be piecewise-smooth.

The physical sense of the unknown function $\gamma(s)$ in this equation is the density of the vortex layer by which the contour in flow is modeled.

If contour σ_1 is open and smooth and represented parametrically as $x = x(t)$, $y = y(t)$, $t \in [-1, 1]$, Equation (3.4) can be written in the form

$$\frac{1}{2\pi} \int\limits_{-1}^{1} \frac{\gamma(t)dt}{t_0 - t} + \int\limits_{-1}^{1} K(t_0, t)\gamma(t)dt = f(t_0), \quad t_0 \in (-1, 1).$$

If the contour σ_1 is a segment, $K(t_0, t) \equiv 0$ in Equation (3.5), which is called in this case the characteristic equation [17,18].

A solution to Equation (3.5) is sought in the class of absolutely integrable functions . In this class of solutions, Equation (3.5) can have three classes of solutions: those of *index* $\chi = 1$, depending on one arbitrary constant and able to turn into infinity of the type $\frac{1}{\sqrt{1-t^2}}$ at the ends of the segment; those of *index* $\chi = 0$, for which the solution is uniquely determined by setting the end of the seqment where the solution goes to zero; those of *index* $\chi = -1$, for which the solution vanishes at both ends of the segment and is existent and unique when and only when the following condition is met:

$$\int\limits_{-1}^{1} (1 - t_0^2)^{-\frac{1}{2}} [f(x_0) - \int\limits_{-1}^{1} K(t_0, t)\gamma(t)dt] dt_0 = 0. \qquad (3.6)$$

If the contour σ_1 is closed and smooth, Equation (3.4) takes the form (in that case it is assumed that the parameter t varies on the segment $[0, 2\pi]$):

$$\frac{1}{4\pi} \int\limits_{0}^{2\pi} \mathrm{ctg}\frac{t_0 - t}{2} \gamma(t)dt + \int\limits_{0}^{2\pi} K(t_0, t)\gamma(t)dt = f(t_0),$$

$$t_0 \in [0, 2\pi] \qquad (3.7)$$

If σ_1 is a circumference, $K(t_0, t) \equiv 0$ in Equation (3.7) and this equation is called the characteristic one.

In the case being considered, for the right-hand side of Equation (3.7) the following condition is satisfied:

$$\int\limits_{\sigma_1} f(s) = 0, \qquad (3.8)$$

and the solution in the class of absolutely integrable functions is determined up to a constant [12,17].

In [10], the problem on ejection is addressed, i.e., a contour in flow is considered with a device for air ejection on one of its sides. This device creates over the side with the ejection point M_q a velocity field analogous to that created by a source with the intencity Q. In this case, on the opposite side of the contour the velocity field in the neighborhood of the ejection point has no singularities.

For mathematical simulation, the ejection device is substituted for a source [10]. And again, the problem of the flow around the contour σ_1 with a ejection device reduces to solving integral equation (3.4), but now in a class of functions that have at the ejection point non-integrable singularity of the form $\frac{1}{(t_q - t)}$, where $M_q(x(t_q), y(t_q))$ is source's location point, which will be hereafter referred to as the singularity point. In [12], it has been shown that the presence of one

440

singularity point on the contour increases the index of Equation (3.4) by unity (correspondingly, k such points increase the index of the equation by k). Thus, the problem on ejection has lead to the necessity to consider a new class of solutions to Equation (3.4).

In the three-dimensional case, the double-layer potential has the form

$$ U = \frac{1}{4\pi} \int_{\sigma_1} \frac{\partial}{\partial \bar{n}_M} \left(\frac{1}{r_{MM_0}} \right) g(M) ds_M, \quad M_0 \notin \sigma_1, \quad (3.9) $$

where σ_1 is a surface in flow, s_M is an element of its area.

In this case, for meeting boundary dondition (3.2), it is better to take advantage of the continuity of the normal derivative and obtain the integral equation

$$ \frac{1}{4\pi} \int_{\sigma_1} \frac{\partial}{\partial \bar{n}_{M_0}} \frac{\partial}{\partial \bar{n}_M} \left(\frac{1}{r_{MM_0}} \right) g(M) d\sigma_M = f(M_0), $$

$$ M_0 \in \sigma_1, \quad (3.10) $$

in the potential density.

If the surface σ_1 is smooth and open, Equation (3.10) has a unique solution [14]. If σ_1 is a smooth closed surface, a solution exists and is determined up to a constant C, and its right-hand side must satisfy the condition specified in [14]:

$$ \int_{\sigma_1} f(M) d\sigma = 0. \quad (3.11) $$

The value of C can be determined (and thus the unique solution singled out) by setting a value of the integral taken over the solution or a value of the solution at a point on the surface σ_1.

If σ_1 is a part of the plane OXY, Equation (3.10) takes the form

$$ \frac{1}{4\pi} \int_{\sigma_1} \frac{g(M)}{r_{MM_0}^3} d\sigma_M = f(M_0), \quad M_0 \in \sigma_1, \quad (3.12) $$

i.e. the integral must be understood in the sense of Hadamard principal value [19], therefore we shall call it hypersingular [12,14].

4 Mathematical Well-Posedness of the Two-Dimensional MDV

At first we shall consider the original version of the MDV with the uniform decomposition of a contour in flow. A basic principle of the MDV [5,10] lies in the fact that quadrature formulas of quadrangular type

with two grids of points are applied to the singular integral in Equation (3.4). The values of the unknown function are taken at points of one of the grids (where discrete vortices are positioned), while Equality (3.4) is satisfied at the points of the other grid — the boundary nonpenetration condition (these points are usually referred to as reference points [11]). Then SIE (3.4) is substituted for a system of linear algebraic equestions (SLAE), whose form depends on a contour in flow and the class of solutions. A fundamental role in obtaining the SLAE is played by the so-called B-condition of the MDV [11]. At the grid point nearest to the edge where the solution is unlimited, a discrete vortex is placed, whereas the nearest point to the edge where the solution is limited is a reference point.

Let us denote the left-hand side of Equation (3.4) by $I(M_0)$. The points where discrete vortices are placed we denote by M_i, $i = 1, ..., n,$; they are uniformly distributed over the arc length of each smooth component of the contour σ_1, whereas reference points we denote M_{oj}, $j = 1, ..., m$, they are uniformly distributed over the arc length between neighboring discrete vortices. Then the approximate value of the integral $I(M_0)$ at the point M_{oj} is evaluated through the formula

$$ I_n(M_{oj}) = \sum_{i=1}^{n} \omega_{i,j} \Gamma_i, \quad (4.1) $$

$$ \omega_{i,j} = \frac{1}{2\pi} \frac{y'_{os,j}(y_{oj} - y_i) + x'_{os,j}(x_{oj} - x_i)}{r_{M_i M_{oj}}^2}, $$

$$ \Gamma_i = \gamma_i h = \gamma(s_i) h, $$

where h is the arc step between neighboring points of the grid.

If σ_1 is a closed smooth contour, the values $I_n(M_{oj})$ converge uniformly over all points M_{oj} to values $I(M_{oj})$.

If σ_1 is a smooth open contour,

1) values $I_n(M_{oj})$ converge uniformly over points M_{oj}, on any part of the curve σ_1, located at a finite distance from the ends of this curve;

2) over all points M_{oj} there is integral convergence, i.e.,

$$ \lim_{h \to 0} \sum_{j=1}^{m} I_n(M_{oj}) h = \int_{\sigma_1} I(M_0) ds_\sigma. \quad (4.2) $$

More accurate estimations are given in [11]. As shown in [11], if σ_1 is a piecewise smooth curve, the convergence of the quadrature formulas of the MDV type to the exact values of integral at the points has the same nature as for a smooth open curve.

Concider now the numerical solution of SIE (3.4). For demonstrating the ideas of the MDV, we shall use as an example the numerical solution of SIE's (3.5) and (3.7), i.e., we shall address the cases where the contour σ_1 is smooth and closed or open.

At first we take σ_1 smooth and open and therefore use Equation (3.5).

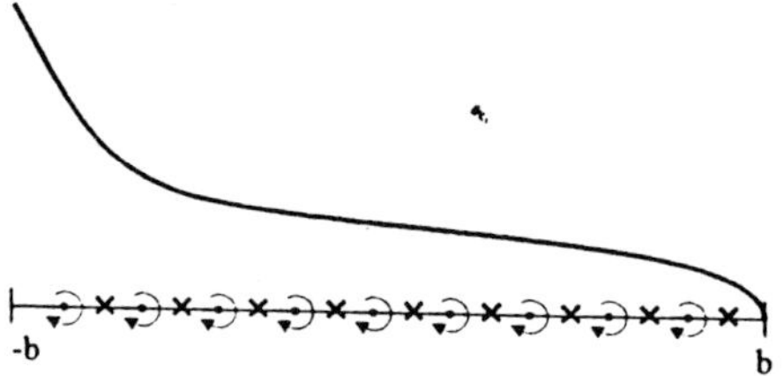

Fig.1:

Let us consider circulatory flow past a contour (profile). In this case the vortex strength $\gamma(t)$ becomes infinite at the leading edge (see Fig. 1), i.e., we take the solution of index $\chi = 0$. Therefore, in accordance with the B-condition of the MDV, the distribution pattern of the discrete vortices and reference points will be such as depicted in Fig. 1, i.e., the number of discrete vortices is equal to the number of reference points. Therefore, Equation (3.5) is substituted for the following SLAE

$$\sum_{i=1}^{n} \omega_{i,j}\Gamma_i = f_j, \qquad j = 1,...,n, \qquad (4.3)$$

where $\omega_{i,j}$ and Γ_j are defined by (4.1) and $f_j = f(M_{oj})$.

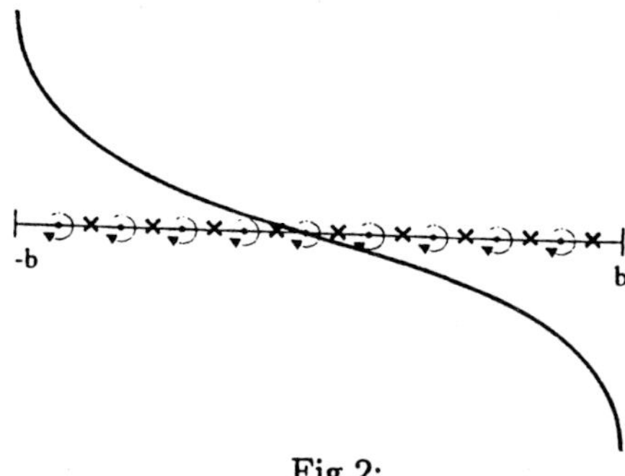

Fig.2:

If noncirculatory flow (solution of index $\chi = +1$) is considered, the number of discrete vortices is greater by unity than the number of reference points (see Fig. 2), the noncirculatory condition is discretized,

i.e., Equation (3.5) is substituted for the SLAE

$$\sum_{i=1}^{n} \omega_{i,j}\Gamma_i = f_j, \qquad j = 1,...,n-1,$$

$$\sum_{i=1}^{n} \Gamma_i = 0. \qquad (4.4)$$

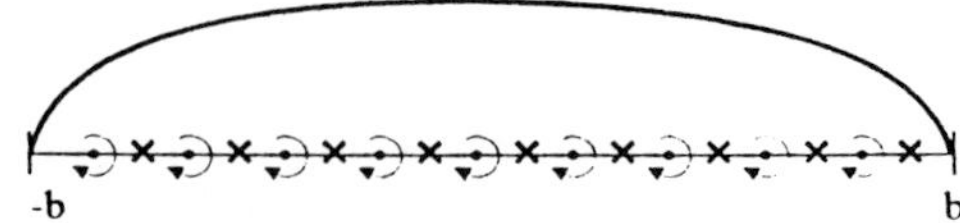

Fig.3:

And finally, consider flow with limited velocities (solution index $\chi = -1$). In this event, the solution is limited at both edges, the distribution pattern of discrete vortices and reference points is shown in Fig. 3 and, therefore, the number of discrete vortices is less by unity than the number of reference points. Now Equation (5) is substituted for the SLAE

$$\gamma_{on} + \sum_{i=1}^{n} \omega_{i,j}\Gamma_j = f_j, \qquad j = 1,...,n+1, \qquad (4.5)$$

where γ_{on} is the regularizing variable, i.e. without it System (4.5) is overdetermined and, as a rule, inconsistent. The variable γ_{on} tends to zero as $n \to \infty$ when and only when Condition (3.6) is met (if σ_1 is a segment) or an analogous condition [12,17] is satisfied in the general case.

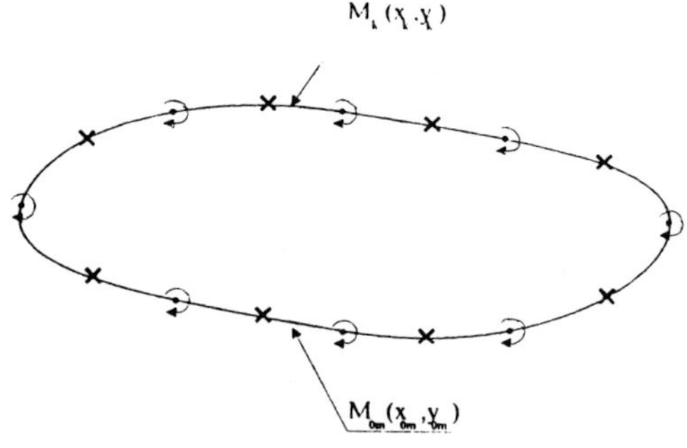

Fig.4:

If the contour σ_1 is smooth and closed, the numbers of discrete vortices and reference points are equal (Fig. 4). However, in this case, it is again necessary to discretize the noncircularity condition and, therefore, the SLAE that gives a numerical value to SIE (3.7) will have the form

$$\gamma_{on} + \sum_{i=1}^{n} \omega_{i,j}\Gamma_i = f_j, \qquad j = 1, ..., n, \qquad (4.6)$$

$$\sum_{i=1}^{n} \Gamma_i = 0.$$

Finally, let us consider an example of the numerical solution of a problem of flow past a contour with ejection. Let a source of intensity Q lies at a point M_q different from the angular and end points of the contour σ_1. Then, discrete vortices and reference points are selected so that the point M_q could be one of the reference points at $j = j_q$ and so that the nonpenetration condition could not be met. In this event, the discrete vortex nearest to that reference point M_q, may be considered as having a known strength, i.e., for example, instead of System (4.3) it is necessary to take the SLAE

$$\sum_{\substack{i=1 \\ i \neq j_q}}^{n} \omega_{i,j}\Gamma_i = f_j - \omega_{j_q,j}\Gamma_{j_q}, \quad j = 1, ..., n, \quad j \neq j_q, \quad (4.7)$$

where $\Gamma_{j_q} = \frac{2Q}{\pi}$.

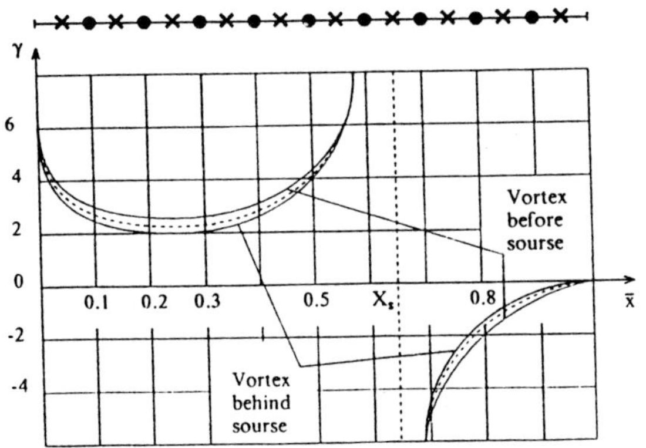

Fig.5:

For the case when σ_1 is a segment of the numerical axis OX, an example of the solution of System (4.7) is given in Fig. 5.

Systems (4.4)-(4.6) changes in an analogous way [11].

In all the cases (4.3)-(4.7), a mathematical foundation is obtained of the convergence of a numerical solution to the exact one using estimates of the same nature as in the case of the convergence of quadrature sums to the values of an integral [11,12].

Fig.6:

Analogous SLAE's must be considered also in the case if σ_1 is a simple piecewise smooth open or closed contour (profile) (see Fig. 6), discrete vortices and reference points being positioned to place discrete vortices at angular points.

Numerous cases of deriving SLAE's within the framework of the MDV for various curves and various problems of aerodynamics are described in monographs [11,12]. In these works, cases are considered where there is a mathematical foundation of numerical schemes, and problems are stated for which there is no such foundations.

Note about Convergence Improvement

In two-dimensional problems, for cases of smooth closed and open contours with the use of interpolation quadrature formulas for corresponding singular integrals of Cauchy type on a segment or with the Hibert-type kernel, References [11,12] show how to obtain the SLAE's of type (4.3)-(4.6) that provide convergence of a numerical solution to the exact one, the convergence being dependent on the smoothness of the contour and the on-coming flow.

5 Pseudo-Difference Operators

In many numerical schemes one can come across Toeplitz matrices, for example, in difference schemes, the MDV. In these problems, left-hand sides of equations are calculated in the same way as Fourier coefficients of the product of two functions or coefficients of Laurent's expansion of the product of two functions. One of these functions is completely determined by a numerical scheme of the equation being solved, while Fourier coefficients of other functions are unknown properties, which have to be found in the course of solution. Therefore, it is natural to try acting in the discrete case in the same way as one usually acts in the continuous case, and to introduce an analog of the pseudodifferential operator.

The operator referred to as the pseudo-difference operator (PDO) [13] is defined by the formula

$$A(\hat{x}) = \frac{1}{(2\pi)^n} \int\limits_{-\pi}^{\pi} ... \int\limits_{-\pi}^{\pi} \hat{A}(\varphi_1, ..., \varphi_n)\hat{x}(\varphi_1, ..., \varphi_n)$$

$$e^{-i(k_1\varphi_1+...+k_n\varphi_n)}d\varphi_1...d\varphi_n, \qquad (5.1)$$

where the function $\hat{A}(\varphi_1,...,\varphi_n)$ is called the symbol of the pseudo-difference operator; $k_m \in Z$ is the set of integers, i is the imaginary unit.

Let the operator defined by the formula

$$A(\hat{x}) = \frac{1}{(2\pi)^n} \int\limits_{-\pi}^{\pi} ... \int\limits_{-\pi}^{\pi} \hat{A}(\varphi_1,...,\varphi_n,k_1,...,k_n)$$

$$\hat{x}(\varphi_1,\varphi_2,...,\varphi_n)e^{-i(k_1\varphi_1+...+k_n\varphi_n)}d\varphi_1...d\varphi_n. \qquad (5.2)$$

be called the pseudo-difference operator with the variable symbol. It is assumed that the symbol of the operator defined by relationship (5.1) is a periodic function $\hat{A}(\varphi) \in L_\infty[-\pi,\pi]$ and its Fourier series converges to $\hat{A}(\varphi)$ in the norm $L_\infty[-\pi,\pi]$. Analogous restrictions are imposed on the function $\hat{A}(\varphi_1,...,\varphi_n,k_1,...,k_n)$ at fixed $k_1,...,k_n$, the function $\hat{x}(\varphi_1,...,\varphi_n) \in L_r, \quad r > 1$.

Then, the following relationships are valid

$$C_{k_1...k_n} = \sum_{j_1...j_n} a_{k_1-j_1,...,k_n-j_n} x_{j_1...j_n}, \qquad (5.3)$$

where $C_{k_1...k_n}$ are the values of the grid function $A(\hat{x})$: $a_{k_1...k_n}; x_{k_1...k_n}$ are the Fourier coefficients of the functions $\hat{A}(\varphi_1,...,\varphi_n); \hat{x}(\varphi_1,...,\varphi_n)$.

5.1 EXAMPLES.

1. Let us consider the second-order difference operator

$$B(U) = \frac{U_{k+1} - 2U_k + U_{k-1}}{h^2}, \quad h > 0,$$

which can be represented in a PDO form

$$A(\hat{x}) = \frac{1}{(2\pi)} \int\limits_{-\pi}^{\pi} \frac{(e^{i\varphi/2} - e^{-i\varphi/2})^2}{h^2} \hat{x}(\varphi)e^{-ik\varphi}d\varphi$$

$$= \frac{1}{2\pi} \int\limits_{-\pi}^{\pi} \frac{1}{h^2}(-4\sin^2\frac{\varphi}{2})\hat{x}(\varphi)e^{-ik\varphi}d\varphi,$$

where $\hat{x}(\varphi) = \sum_k U_k e^{ik\varphi}$.

2. Method of discrete vortices (one-dimensional case). Let us divide the numerical axis into intervals of the step h. Without the loss of generality, we can assume that one of the division points coincides with the origin point. The division points are determined by the formula $t_m = mh, \quad m \in Z$, where Z is the set of integers. Let us take $0 < \alpha < 1$ and choose another system of points $\{t_{oj}\}$,

where $t_{oj} = (j + \alpha)h, \quad j \in Z$. The grid function U is defined at grid nodes t_m, $U(t_m) = U_m$. The operator of the MDV is defined by the relation

$$M(U) = \sum_{m=-\infty}^{\infty} \frac{U_m h}{t_{oj} - t_m} = \sum_{m=-\infty}^{\infty} \frac{U_m}{(j-m) + \alpha}.$$

It is supposed that the values U_m of the grid function are such that

$$\hat{x}(\varphi) = \left(\sum_{k=-\infty}^{\infty} U_k e^{ik\varphi}\right) \in L_r[-\pi,\pi], \quad r > 1.$$

Let $\hat{A}(\varphi) = \sum_{j=-\infty}^{\infty} \frac{1}{j+\alpha}e^{ij\varphi}$, then the values f_k of its grid function $M(U)$ are found using relations

$$f(k) = A(\hat{x}) = \frac{1}{2\pi} \int\limits_{-\pi}^{\pi} \hat{A}(\varphi)\hat{x}(\varphi)e^{-ik\varphi}d\varphi,$$

i.e., the operator of the MDV can be represented in the form of the PDO.

The symbol of the PDO $\hat{A}(\varphi) = \sum_{j=-\infty}^{\infty} \frac{1}{j+\alpha}e^{ij\varphi} = \frac{\pi}{\sin\pi\alpha}e^{i(\pi-\varphi)\alpha}$. At $\alpha = \frac{1}{2}$ (aerodynamic case), we have $\hat{A}(\varphi) = i\pi e^{-i\frac{\varphi}{2}}$.

5.2 Restriction Problem for Pseudo-Difference Operators

Let us consider a set ε of indices $k = (k_1,...,k_n)$. The operator of the form

$$pA(x) = \frac{1}{(2\pi)^n} \int\limits_{-\pi}^{\pi} ... \int\limits_{-\pi}^{\pi} \hat{A}(\varphi_1,...,\varphi_n)\hat{x}(\varphi_1,...,\varphi_n)$$

$$e^{-(k_1\varphi_1+...+k_n\varphi_n)}d\varphi_1...d\varphi_n, \qquad (5.4)$$

$$(k_1,...,k_n) \in \varepsilon$$

is referred to as the restriction operator pA of PDO (1) into the set ε. The following problem is called the restriction problem for the PDO: given is a grid function $f(k_1,...,k_n)$, $k = (k_1,...,k_n) \in \varepsilon$ is a certain set of indices; it is required to find a grid function U, on the same set of indices for which the following is valid:

$$\begin{cases} pA(\hat{x}) = f \\ k \in \varepsilon, \end{cases} \qquad (5.5)$$

where $\hat{x}(\varphi) = \sum_{k\in\varepsilon} U_k e^{ik\varphi}, \quad k\varphi = k_1\varphi_1 + ... + k_n\varphi_n.$

To investigate certain issues, restriction problems into special sets of indices ε, can be useful: for example, let $k(k_1, k_2)$, $k_1, k_2 \in Z$, nd $\varepsilon_1 = \{k : k_1 \geq 0, k_2 \geq 0\}$, $\varepsilon_2 = \{k : k_1 \leq 0, k_2 \leq 0\}$. The restriction problems on ε_1 and ε_2 for a class of symbol $\hat{A}(\varphi_1, ..., \varphi_n)$ are solved using the Wiener-Hopf method.

5.3 Method of Discrete Vortices for the Prandtl Equation

To obtain parameters of non-separated flow around flat three-dimensional bodies, it is convenient to model the vortex layer on a body by pairs of discrete vortices in the two-dimensional case and by closed quadrangular vortex frames in the three-dimensional one. Strengths of these vortex formations equal the density of the double-layer potential distributed on a body, which gives the same potential as the perturbed flow. Thus, we come to the problem of finding the potential outside the body basing on its normal derivative on body's surface through the density of the double-layer potential.

For non-separated flow past a finite-span wing, we have the Prandtl integral equation (see (3.12))

$$\iint\limits_{\mathcal{G}} \frac{Q(x,z)}{r^3}\,dx\,dz = f(x_0, z_0), \qquad (5.6)$$

where $r = \sqrt{(x_0 - x)^2 + (z_0 - z)^2}$, $M_0(x_0, z_0) \in \mathcal{G}$, $\mathcal{G}$ is a two-dimensional domain; the integral in (5.6) is understood in the sense of Hadamard [12,19].

The numerical solution of Equation (5.6) is constructed as follows.

Let points $x_i = ih$, $x_{oj} = h(j + \frac{1}{2})$ form the canonical division of the OX axis while points $z_k = kh$, $z_{om} = (m + \frac{1}{2})h$, $h > 0$ form the canonical division of the axis OZ. Let us designate by M_h a set of rectangles $\prod_{ik}$ with vertices at points $M(x_i, z_k)$, $M(x_{i+1}, z_k)$, $M(x_{i+1}, z_{k+1})$, $M(x_i, z_{k+1})$ completely within $\mathcal{G}$. Then, an approximate solution to Equation (5.6) will be sought from the SLAE

$$\begin{cases} \sum\limits_{\prod_{ik} \in M_h} Q_h(x_{oi}, z_{ok}) \iint\limits_{\prod_{ik}} \frac{dx\,dz}{r_{jm}^3} = f(x_{oj}, z_{om}) \\ \prod_{jm} \in M_h, \end{cases} \qquad (5.7)$$

where $r_{jm} = \sqrt{(x_{oj} - x)^2 + (z_{om} - z)^2}$.

An analysis of System (5.7) had been carried out using the introduced concept of the pseudo-difference operator, its symbol and restriction problems. As a result, the following findings were obtained [14].

Theorem 1. Let $\mathcal{G}$ be a domain with an infinitely differentiable boundary. Then, if $f(x_0, z_0) \in C(\bar{\mathcal{G}})$, the following inequality is valid between the solution $Q_h(x_{oi}, z_{ok})$ of System (5.7) and the solution $Q(x, z)$ of Equestion (5.6):

$$\Delta_{ik} = |Q(x_{oi}, z_{ok}) - Q_h(x_{oi}, z_{ok})| < A|\ln n|^{\frac{3}{2}} h^{\frac{1}{4} - \varepsilon_1},$$

$0 < \varepsilon_1 < \frac{1}{4}$, for all points $M(x_{oi}, z_{ok})$, at an arbitrary distance δ from the boundary.

Theorem 2. Let $\mathcal{G}$ be a rectangle, $f(x_0, z_0) \in C(\bar{\mathcal{G}})$, then $\Delta_a \to 0$ as $h \to 0$, where

$$\Delta_a = \sum_{i=1}^{n} \sum_{k=1}^{n} |\Delta_{ik}|^a \cdot h^2, \qquad 1 \leq a \leq 2.$$

Theorem 3. If the conditions of Theorem 1 hold, the following estimate is true:

$$\sum_{\prod_{ik} \in M_h} |\Delta_{ik}| h^2 \leq A(\varepsilon) |\ln h|^{\frac{13}{4}} \cdot h^{\frac{1}{8} - \frac{7}{8}\varepsilon},$$

where the value of ε can be arbitrary one meeting the inequality $0 < \varepsilon < \frac{1}{7}$.

Let us introduce the following designations:

$$\tilde{Q}_{xh}(x_{oi}, z_{ok}) = \frac{Q_h(x_{oi+1}, z_{ok}) - Q_h(x_{oi}, z_{ok})}{h},$$

$$\tilde{Q}_{zh}(x_{oi}, z_{ok}) = \frac{Q_h(x_{oi}, z_{ok+1}) - Q_h(x_{oi}, z_{ok})}{h},$$

$$Q'_{xh}(x_{oi}, z_{ok}) = \frac{Q(x_{oi+1}, z_{ok}) - Q(x_{oi}, z_{ok})}{h},$$

$$Q'_{zh}(x_{oi}, z_{ok}) = \frac{Q(x_{oi}, z_{ok+1}) - Q(x_{oi}, z_{ok})}{h},$$

$$\tilde{\Delta}Q'_{xh}(x_{ol}, z_{op}) = \tilde{Q}_{xh}(x_{ol}, z_{op} - Q'_{xh}(x_{ol}, z_{op}),$$

$$\tilde{\Delta}Q'_{zh}(x_{ol}, z_{op}) = \tilde{Q}_{zh}(x_{ol}, z_{op} - Q'_{zh}(x_{ol}, z_{op}).$$

Then, the set $M_{1h} \subset M_h$ of rectangles $\prod_{ik}$, $S_{M_{1h}} \to S_{\mathcal{G}}$ as $h \to 0$ (S is the area) is constructively given, for which the following is true:

$$\sum_{\prod_{lp} \in M_{1h}} |\tilde{\Delta}Q'_{xh}(x_{ol}, z_{op})| h^2 \to 0 \quad as \quad h \to 0,$$

$$\sum_{\prod_{lp} \in M_{1h}} |\tilde{\Delta}Q'_{zh}(x_{ol}, z_{op})| h^2 \to 0 \quad as \quad h \to 0.$$

5.4 Multhopf Equation, Noncirculatory Flow

Le us consider the equation

$$\begin{cases} \int\limits_0^b dz \int\limits_0^l \frac{\Gamma(x,z)}{(z_0-z)^2}\left[1+\frac{x_0-x}{\sqrt{(x_0-x)^2+(z_0-z)^2}}\right] dx \\ \quad = f(x_0,z_0), x_0 \in (0,l), z_0 \in (0,b), \qquad (5.8) \\ \int\limits_0^l \Gamma(x,z)dx = 0, \quad \forall z \in (0,b). \end{cases}$$

The function $f(x_0,z_0) \in C[\bar{\mathcal{G}}]$, $\bar{\mathcal{G}} = [0,l] \times [0,b]$, $\Gamma(x,z) = \frac{\gamma(x,z)\sqrt{z(z-b)}}{\sqrt{x(l-x)}}$, $\gamma(x,z) \in C^1_\alpha[\bar{\mathcal{G}}]$. The integral in the first relationship of (5.8) is meant in the sense of Hadamard.

The SLAE for obtaining an approximate solution has the following form:

$$\begin{cases} \sum\limits_{i=1}^n \sum\limits_{k=1}^N \Gamma_h(x_i,z_{ok})\alpha(i,k,j,m)h = f(x_{oj},z_{om}), \\ j = 1,2,...,n-1; \; m = 1,...,N; \\ (n+1)h = l, \; b-h < (N+1)h \le b, \\ \sum\limits_{i=1}^n \Gamma_h(x_i,z_{ok}) = 0, \quad k = 1,...,N. \end{cases} \qquad (5.9)$$

The systems of points $\{x_i, x_{oj}\}$; $\{z_k, z_{om}\}$ correspondingly form the canonical divisions of $[0,l]$, $[0,b]$ with step h, i.e., $x_i = ih$, $x_{oj} = h(j+\frac{1}{2})$; $z_k = kh$, $z_{om} = h(m+\frac{1}{2})$.

The values $\alpha(i,k,j,m)$ of System (5.9) have the form

$$\alpha(i,k,j,m) =$$

$$= \left[\frac{x_{oj}-x_i+\sqrt{(x_{oj}-x_i)^2+(z_{om}-z_{k+1})^2}}{(x_{oj}-x_i)(z_{om}-z_{k+1})}\right.$$

$$\left. -\frac{x_{oj}-x_i+\sqrt{(x_{oj}-x_i)^2+(z_{om}-z_k)^2}}{(x_{oj}-x_i)(z_{om}-z_k)}\right].$$

Investigation performed yielded the following result (see [14]).

Theorem 4. Let the function $f(x_0,z_0)$ meet the requirements formulated for Equation (5.8). Then, for the solution of System (5.9) and the solution $\Gamma(x,z)$ of Equation (5.8), the inequality

$$|\Gamma_h(x_i,z_{ok})-\Gamma(x_i,z_{ok})| < C(\varepsilon)h^{\frac{1}{4}-\varepsilon}\cdot|\ln h|^{\frac{9}{2}}, \quad 0<\varepsilon<\frac{1}{4},$$

is valid for all points $M(x_i,z_{ok})$, whose coordinates satisfy the inequalities $\delta < x_i < l-\delta$, $\delta < z_{ok} < b-\delta$, $\delta > 0$ being arbitrary.

5.5 Method of Discrete Vortices for Numerically Solving the Linear Unsteady Problem for an Airfoil

Let a thin airfoil located on the segment $[-1,1]$ of the axis OX be in such unsteady flow at a low angle of attack α that from the trailing edge of the airfoil (point $x = 1$) a sheet of free vortices descends and these vortices move along the OX axis in its positive direction at an average velocity. To obtain some aerodynamic characteristics of the airfoil, the strength $\gamma(x,t)$ of the total vortex layer on it at the point x at the instant t and the strength $\delta(t)$ of the free vortex layer descending into flow at the same instant t. Using the nonpenetration condition at the point x_0 and the absence of the circulation along the fluid contour encompassing the airfoil and the track behind it, the following system of equations is obtained to find the functions $\gamma(x,t)$ and $\delta(t)$:

$$\begin{cases} \int\limits_{-1}^1 \frac{\gamma(x,t)dx}{x_0-x} + \int\limits_0^t \frac{\delta(\tau)d\tau}{x_0-1-(t-\tau)} = f(x_0,t), \\ \int\limits_{-1}^1 \gamma(x,t)dx + \int\limits_0^t \delta(\tau)d\tau = 0, \end{cases} \qquad (5.10)$$

where $x_0 \in (-1,1)$, $0 \le t \le T$, $f(x_o,t)$ satisfies the Holder condition on the set $[-1,1] \times [0,T]$. For the sake of simplisity, the average oncoming flow velocity is set to be equal to 1.

Because we consider non-separated flow about the airfoil with the vortex sheet descending from its trailing edge, the function $\gamma(x,t)$ must be unlimited at the leading edge (point $x = -1$) and limited at the trailing edge. Thus, the first equation in (5.10) we shall consider as a singular integral equation of the index $\kappa = 0$ with respect to the function $\gamma(x,t)$. The numerical solution of System (5.10) is constructed in the following way. Let us divide the set $[0,\infty)$ of the Ot axis by equally spaced points $t_k = k\Delta t$, $k = 1,2,...$. In each segment $[(i-1)\Delta t, i\Delta t]$ we take the point $\tau_i = (i-1)\Delta t + \Theta\Delta t$, $0 \le \Theta < 1$.

An approximate value of the function $\gamma(x,t_k)$ will be looked for in the form

$$\begin{cases} \gamma_n(x,t_k) = W(x)\varphi_n(x,t_k), \\ \varphi_n(x,t_k) = \sum\limits_{i=1}^n \frac{\varphi_n(x_i,t_k)P^{(\frac{1}{2},-\frac{1}{2})}(x)}{(x-x_i)P'^{(\frac{1}{2},-\frac{1}{2})}_n(x_i)}, \end{cases} \qquad (5.11)$$

where x_i, $i = 1,...,n$ are the roots of the Jacobi polinomial $P_n^{(\frac{1}{2},-\frac{1}{2})}(x)$ of the degree n, corresponding to the function $W(x) = \sqrt{\frac{1-x}{1+x}}$.

The values of $\varphi_n(x_i,t_k)$ and $\delta_{\Delta t}(t_k)$ will be found from the the system of equations

$$
\begin{cases}
\sum_{i=1}^{n} \dfrac{\varphi_n(x_i,t_k)a_i}{x_{oj}-x_i} + \sum_{r=1}^{k} \dfrac{\delta_{\Delta t}(t_r)\Delta t}{x_{oj}-1-(t_k-\tau_r)} = f(x_{oj},t_k), \\[2mm]
\sum_{i=1}^{n} \varphi_n(x_i,t_k)a_i + \sum_{r=1}^{k} \delta_{\Delta t}(t_r)\Delta t = 0, \\[2mm]
j = 1,...,n; \quad k = 1,...,p,
\end{cases}
\qquad (5.12)
$$

$$
a_i = \frac{1}{P'^{(\frac{1}{2},-\frac{1}{2})}_n(x_i)} \int_{-1}^{1} \frac{W(x)P_n^{(\frac{1}{2},-\frac{1}{2})}(x)dx}{x-x_i},
$$

where $x_{oj}, \quad j = 1,...,n$ are the roots of the Jacobi polinomial $P_n^{(-\frac{1}{2},+\frac{1}{2})}(x)$.

An analysis of System (5.12) yielded the following results.

Theorem 5. If for $f(x_0,t)$ on the right-hand side of System (5.12) the conditions $f''_{tt}(x_0,t) \in C[0,T]$, $f^{(r)}_{x_0}(x_0,t) \in H_\alpha[-1,1]$ are satisfied uniformly over t, the following estimate is true for the solution $\delta_{\Delta t}(t_k)$ of System (5.12) and solution $\delta(t)$ of System (5.10):

$$
\sum_{i=1}^{p} |\delta_{\Delta t}(t_i)-\delta(t_i)|h \leq C \left[\frac{1}{n(\Delta t)^{\frac{5}{2}}} + \frac{1}{n^{r+\alpha}\Delta t} + \sqrt{\Delta t} \right]
$$

at $\dfrac{1}{n(\Delta t)^{\frac{5}{2}}} < M$ and sufficiently small Δt.

Theorem 6. If the conditions of Theorem 5 are satisfied, for any $\delta > 0$ there exists such a number h that for all $\Delta t < h$ and all t_k, $\delta < t_k \leq T$, the following estimate is true:

$$
|\varphi(x_i,t_k) - \varphi_n(x_i,t_k)|
$$
$$
< \frac{N(\delta)}{1-x_i} \left[\frac{1}{n(\Delta t)^{\frac{5}{2}}} + \frac{1}{n^{r+\alpha}\Delta t} + \sqrt{\Delta t} \right],
$$

where $\varphi(x,t) = \sqrt{\frac{1+x}{1-x}}\gamma(x,t)$.

Similar results were obtained for airfoil moving with a varying velocity.

5.6 Neumann Problem

Dealing with problems on flow past surfaces, it is convenient to use closed quadrangular or triangular vortex frames corresponding to discrete vortex pairs in two-dimensional problems. The velocity field due to these vortex formations is equivalent to the velocity field from the double-layer potential of constant density yielding over the frame area the intensity equal to the strength of one frame.

With such a formulation of the problem on ideal incompressible flow past a lifting surface, we come to the solution of the Neumann problem for the Laplace equation outside the lifting surface through the double-layer potential, which leads to a strongly singular integral equation in the potential density.

The method of closed vortex frames can be used to numerically solve this strongly singular equation. Because of this, a need arose for construction of the theory of the Neumann problem with unclosed surfaces for the Laplace equation as well as for investigating strongly singular equations originating when solving the Neumann problem through the double-layer potential.

Let us consider the boundary-value problem

$$
\begin{cases}
\Delta u = 0, \quad x \notin \bigcup_{p=1}^{2} \bigcup_{i_p=1}^{n_p} S_{i_p}^{(p)}, \\[2mm]
\frac{\partial u}{\partial n}\big|_{S_{i_p}^{(p)}} = f_{i_p}^{(p)}(x), \quad p = 1,2; \quad i_p = 1,...,n_p,
\end{cases}
\qquad (5.13)
$$

where $S_1^{(1)}, ..., S_{n_1}^{(1)}$ and $S_1^{(2)}, ..., S_{n_2}^{(2)}$ are respectively closed and open (open manifold) compact surfaces of the C^∞ class, located at a positive distance one from another; $\frac{\partial}{\partial n}$ is the derivative with respect to normal to these surfaces.

The solution of Problem (5.13) will be sought through the double-layer potential. Let $\gamma_{i_p}^{(p)}(y)$ be the density of the double-layer potential on the surface $S_{i_p}^{(p)}$; using boundary conditions, we get the system of equations

$$
\sum_{p=1}^{2} \sum_{i_p=1}^{n_p} \frac{1}{4\pi} \frac{\partial}{\partial n_{l_m}^m} \int_{S_{i_p}^{(p)}} \nu_{i_p}^{(p)}(y) \frac{\cos\varphi_{xy}}{|x-y|^2} dS_{i_p y}^{(p)} = f_{l_m}^{(m)}(x),
$$

$$
(5.14)
$$

where $m = 1,2; \; l_m = 1,...,n_m, \quad x = (x_1,x_2,x_3) \in S_{l_m}^{(m)}, \quad y = (y_1,y_2,y_3) \in S_{i_p}^{(p)}, \quad \varphi_{xy}$ is the angle between the vector $x - y$ and the normal $\bar{n}_y$ to the surface $S_{i_p}^{(p)}$ at the point y.

Consider the Sobolev-Slobodetsky space $W_2^r(S_{i_1}^{(1)})$ of the functions $\nu_{i_1}^{(1)}(M)$ and the space $\overset{\circ}{W}{}^r_2(S_{i_2}^{(2)})$ of the functions $\nu_{i_2}^{(2)}(M)$. Let us designate by $W_2^r(S)$ the direct sum of all the spaces $W_2^r(S_{i_1}^{(1)})$ and $\overset{\circ}{W}{}^r_2(S_{i_2}^{(2)})$, and by $\hat{W}_2^r(S)$ the direct sum od the spaces $W_2^r(S_{i_1}^{(1)})$ and $W_2^r(S_{i_2}^{(2)})$.

Let $\prod(\nu) = \{\prod_{l_m}^{(m)}(\nu)\}$, $m = 1,2; \; l_m = 1,...,n_m$, where $\prod_{l_m}^{(m)}(\nu)$ is the operator on the left-hand side of (5.14) at fixed m end l.

The following theorems are proved.

Theorem 7. The operator $\prod(\nu)$ be the Fredholm operator (with index qual to zero) from $W_2^r(S)$ to $\hat{W}^{r-1}(S)$ at $0 < r < 1$, where the dimension of the kernel and cokernel is equal to the number of the closed surfaces.

Then the condition of solvability for System (5.14) has the form

$$\int_{S_{i_1}^{(1)}} f_{i_1}^{(1)}(y) dS_{i_1}^{(1)} = 0, \quad i_1 = 1, ..., n_1. \qquad (5.15)$$

Theorem 8. If $f \in \hat{W}^{r-1}(S)$, $\frac{1}{2} \le r < 1$, f being a vector function, the generalized solution of boundary-value problem (5.13) exists and is unique in the class of functions $u(x) \in W_2^1(\Omega)$, $\Omega = R_3 \setminus (\bigcup_{p=1}^{2} \bigcup_{i_p=1}^{n_p} S_{i_p}^{(p)})$, if conditions (5.15) are met.

REFERENCES

[1] N.E.Zhukovsky. Bound Vortices. Collected Works, Vol IV. Gostekhizdat, (1949), (in Russian)

[2] L.Prandtl. Beitrag zur Theorie der tragenden Flache, Zamm 16, No 6, 1939.

[3] V.V.Golubev. The Theory of Airplane Wings of Finite Span. Trudy Tsent. Aero. Gidr. Inst., No 108, 1931, (in Russian).

[4] S.M.Belotserkovsky. Three-deminsional non-steady motion of lifting surfaces, Prikl. Matem. i Mekh., XIX (4), 1955, (in Russian).

[5] S.M.Belotserkovsky. Studies in Aerodynamics of Modern Lifting Surfaces. Doctoral Thesis (in Russian), 1955.

[6] S.M.Belotserkovsky, E.M.Moiseyev and V.G.Tabachnicov. Atlas of Unsteady Characteristics for Wings With Various Planforms, Moscow, 1959, (in Russian).

[7] S.M.Belotserkovsky, A.S.Ginevsky and Ya.E.Polonsky. Force and Moment Aerodynamic Characteristics of Thin Profile Cascades, Oborongiz, 1962, (in Russian).

[8] S.M.Belotserkovsky. The Theory of Thin Wings in Subsonic Flow, USA, Plenum Press, 1967.

[9] S.M.Belotserkovsky, B.K.Skripachand, V.G.Tabachnikov. Wing in Nonstationary Gas Flow, Moscow, Nauca, 1971, (in Russian).

[10] S.M.Belotserkovsky and I.K.Lifanov. Numerical Methods in Singular Integral Equations, Nauka, Moscow, 1985, 256 p. (in Russian).

[11] S.M.Belotserkovsky and I.K.Lifanov. Method of Discrete Vortices, CRC Press, USA, 1993, 456 p.

[12] I.K.Lifanov. Singular Integral Equations and Discrete Vortices, VSP, the Netherlands, 1996, 478 p.

[13] I.K.Lifanov and L.N.Poltavsky. Generalized Fourier operators and their application to justification of some numerical methods in aerodynamics, Mathematical collection, V.183, N.5, 1992, p.79-114,(in Russian).

[14] L.N.Poltavsky. Mathematical justification of some numerical scheme in aerodynamics. Doctoral thesis, Moscow, 1993, (in Russian).

[15] L.N.Poltavsky. About one numerical scheme for Multhopf integral equation, Differention equations, V.30, N.9, 1994, (in Russian).

[16] D.Colton and R.Kress, Integral Equation Methods in Scattering Theory. Wiley, New York, 1984.

[17] N.I.Muskhelishvili, Singular Integral Equations, Noardhoff, Groningen, 1952.

[18] F.D.Gakhov, Boundary Value Problems, Nauka, Moscow, 1977, (in Russian).

[19] J.Hadamard, The Cauchy Problem for linear Partial Hyperbolic Equations, Nauka, Moscow, 1978, (in Russian).

REVIEW OF PRECONDITIONING FOR
THE COMPRESSIBLE FLUID DYNAMIC EQUATIONS

E. TURKEL

School of Mathematical Sciences, Sackler Faculty of Exact Sciences,
Tel-Aviv University; Tel-Aviv, Israel

1 Introduction

Over the past years several preconditioning methods have appeared with the aim to solve nearly incompressible flow problems with numerical algorithms designed for the compressible flows. The developments are motivated by several observations. First, there exist flow problems of mixed compressible/incompressible type. That is, part of the flow region is incompressible with locally low Mach numbers, whereas significant compressibility effects occur in other regions of the flow. A typical example in aerodynamics is the flow over a wing at a high angle of attack. Other examples include duct flow where the flow in most of the region is highly subsonic but sharp changes in the geometry of the duct cause the flow to become supersonic in a portion of the region. Moreover, low speed flows may be compressible, because of surface heat transfer or volumetric heat addition. Finally, engineers prefer to use existing compressible flow codes over the broadest range of flow conditions possible in order to avoid dealing with multiple flow codes.

The difficulty in solving the compressible equations for low Mach numbers are associated with the large disparity of the acoustic wave speed, $u + a$, and the waves convected at the fluid speed, u. The application of preconditioning changes the eigenvalues of the system of compressible flow equations in order to remove this large disparity of wave speeds. For example, the time derivatives are premultiplied by a matrix which slows the speed of the acoustic waves down towards the fluid speed. A review of preconditioning methods for convection dominated flow problems has been published by Turkel [25]. Successful applications to low Reynolds number viscous flows are reported by Choi and Merkle [5].

Received on July 1, 1997.

There are two basic views of preconditioning methods. The first, which arose in applications to elliptic equations, views preconditioning as a way to accelerate the convergence of the iterative method. This is usually done on the matrix level. Thus, once convergence is achieved the solution should be identical to the original solution. The second approach, which we advocate for low speed flow is to view preconditioning as changing the differential equations to be solved. Thus, as a consequence of preconditioning the basic scheme can be changed, e.g. the artificial viscosity or upwinding. Similarly, the boundary conditions satisfied are based on the preconditioned equations rather than the physical equations. The justification for this is presented in [27] is that the "standard" numerical schemes for the compressible equations do not converge to a scheme for the incompressible equations (using a pseudo-compressibility approach) as the Mach number goes to zero. However, a proper preconditioning leads to a numerical scheme that does behave appropriately for low Mach numbers.

We present a generalization of the preconditioners given by Turkel [25,26], Choi-Merkle [5] and van Leer et. al [30]. The characteristic variables needed to compute flow conditions at inflow or outflow boundaries are also derived. The accuracy of numerical solutions of the preconditioned flow equations is also investigated. We discuss nonconservative and conservative artificial dissipation models and the effects of the preconditioning matrix upon accuracy. The influence of dissipation coefficients and the choice of preconditioning matrix upon grid convergence of the numerical solutions are investigated. Applications to two and three dimensional inviscid and viscous flows are presented.

We demonstrate that preconditioning can be combined with well-known convergence acceleration techniques such as residual smoothing and multigrid. Indeed, it is no surprise that the clustering of eigenvalues with preconditioning improves the damping of transient high-frequency modes to an extent, that ef-

ficient multigrid computations of flows with very low
Mach numbers are now possible.

1.1 General requirements

For both the incompressible and compressible equations we consider systems of the form

$$w_t + f_x + g_y = 0.$$

This system is written in conservation form though for some applications this is not necessary. Our analysis will be based on the linearized equations so that the conservation form does not appear in the analysis though it does appear in the final numerical approximation. This system is now replaced by

$$\mathbf{P}^{-1}w_t + f_x + g_y = 0,$$

or in linearized form

$$\mathbf{P}^{-1}w_t + Aw_x + Bw_y = 0, \tag{1.1}$$

with A and B constant matrices.

In order for this system to be equivalent to the original system, in the steady state, we demand that $\mathbf{P}^{-1}$ have an inverse. This only need be true in the flow regime under consideration. We shall see later that frequently $\mathbf{P}$ is singular at stagnation points and also along sonic lines. Thus, we will consider strictly subsonic flow without a stagnation point. For transonic flow it is necessary to smooth out the singularity in a neighborhood of the sonic line. We also assume that the Jacobian matrices $A = \frac{\partial f}{\partial w}$ and $B = \frac{\partial g}{\partial w}$ are simultaneously symmetrizable. In terms of the 'symmetrizing' variables we also demand that $\mathbf{P}$ be positive definite.

Assuming the steady state has a unique solution, it does not matter which system we march to a steady state. We shall later see that for the finite difference approximations the steady state solutions are not necessarily the same and usually the preconditioned system leads to a better behaved steady state.

2 Incompressible Equations

We first consider the incompressible inviscid equations in primitive variables.

$$
\begin{aligned}
u_x + v_y &= 0 \\
u_t + uu_x + vu_y + p_x &= 0 \\
v_t + uv_x + vv_y + p_y &= 0
\end{aligned}
$$

We consider generalizations of Chorin's pseudo-compressibility method [6]. Using the preconditioning suggested in [25] (with $\alpha = 1$) we have

$$
\begin{aligned}
\frac{1}{\beta^2}p_t + u_x + v_y &= 0 \\
\frac{u}{\beta^2}p_t + u_t + uu_x + vu_y + p_x &= 0 \\
\frac{v}{\beta^2}p_t + v_t + uv_x + vv_y + p_y &= 0
\end{aligned}
\tag{2.1}
$$

We can also write this in matrix form using

$$\mathbf{P}^{-1} = \begin{pmatrix} 1/\beta^2 & 0 & 0 \\ u/\beta^2 & 1 & 0 \\ v/\beta^2 & 0 & 1 \end{pmatrix}, \mathbf{P} = \begin{pmatrix} \beta^2 & 0 & 0 \\ -u & 1 & 0 \\ -v & 0 & 1 \end{pmatrix}$$

Multiplying by $\mathbf{P}$ we rewrite this as

$$w_t + \mathbf{P}Aw_x + \mathbf{P}Bw_y = 0.$$

We also define

$$D = \omega_1 A + \omega_2 B \quad -1 \le \omega_1, \omega_2 \le 1$$

where ω_1, ω_2 are the Fourier transform variables in the x and y directions respectively. The speeds of the waves are now given by the roots of $det(\lambda I - \mathbf{P}A\omega_1 - \mathbf{P}B\omega_2) = 0$. Let $q = u\omega_1 + v\omega_2$. Then the eigenvalues of $\mathbf{P}D$ are

$$d_0 = q \quad , \quad d_{\pm} = \pm\beta \tag{2.2}$$

and so the 'acoustic' speed is isotropic.

The spatial derivatives involve symmetric matrices, i.e. D is a symmetric matrix but $\mathbf{P}$ is not symmetric. Thus, while the original system was symmetric hyperbolic the preconditioned system is no longer symmetric. In [25] it is shown that as long as

$$\beta^2 > (u^2 + v^2)$$

then the equations can be symmetrized. On the other hand the eigenvalues are most equalized if $\beta^2 = (u^2 + v^2)$ [25]. Hence, we wish to choose β^2 slightly larger than $u^2 + v^2$. However, numerous calculations verify, that in general, a constant β is the best for the convergence rate.

We wish to stress that β has the dimensions of a speed. Therefore, β cannot be a universal constant. There are papers that claim that $\beta = 1$ or $\beta = 2.5$ are optimal. Such claims cannot be true in general. It is simple to see that if one nondimensionalises the equation then β gets divided by a reference velocity. Hence, the optimal 'constant' β depends on the dimensionalization of the problem and in particular depends on the inflow conditions. In many calculations the inflow mass flux is equal to 1 or else $p + (u^2 + v^2)/2 = 1$. Such conditions will give an optimal β close to one. However, if one chooses the incoming mass flux as ten then the optimal β will be larger.

We next define the Bernoulli function

$$H = p + (u^2 + v^2)/2.$$

Bernoulli's theorem states that for steady inviscid flow H is constant along streamlines. We now multiply the second equation of (2.1) by u and the third equation of (2.1) by v and add these two equations. If $\beta^2 = u^2 + v^2$, the result is

$$H_t + uH_x + vH_y = 0. \qquad (2.3)$$

Thus, by altering the time dependence of the equations we have constructed a new equation in which H is convected along streamlines. Furthermore, if H is a uniform constant both initially and at inflow then H will remain constant for all time. On the numerical level this will usually not be true because of the introduction of an artificial viscosity or because of upwinding. For viscous flow, (2.3) is replaced by

$$H_t + uH_x + vH_y = \frac{1}{Re}(u\Delta u + v\Delta v)$$

We note that these relationships for H follow from the momentum equations and do not depend on the form of the continuity equation. Hence, we consider the following generalization of (2.1)

$$\frac{1}{\beta^2}p_t + aH_t + u_x + v_y = 0$$
$$\frac{\alpha u}{\beta^2}p_t + u_t + uu_x + vu_y + p_x = 0$$
$$\frac{\alpha v}{\beta^2}p_t + v_t + uv_x + vv_y + p_y = 0$$

where, a is a free parameter. The eigenvalues of PD are independent of the parameter a and are given by (2.2). For $a = 0, \alpha = 1$ we recover our original scheme. For $a = -1$ the time derivative of the pressure no longer appears in the continuity equation. For general β we have

$$\mathbf{P}^{-1} = \frac{1}{\beta^2}\begin{pmatrix} (a+1) & au & av \\ \alpha u & \beta^2 & 0 \\ \alpha v & 0 & \beta^2 \end{pmatrix},$$

$$\mathbf{P} = \frac{1}{d}\begin{pmatrix} \beta^2 & -au & -av \\ -\alpha u & 1 + a - \frac{a\alpha v^2}{\beta^2} & \frac{a\alpha uv}{\beta^2} \\ -\alpha v & \frac{a\alpha uv}{\beta^2} & 1 + a - \frac{a\alpha u^2}{\beta^2} \end{pmatrix}$$

where $d = 1 + a - a\alpha\frac{u^2+v^2}{\beta^2}$ and we require that $d \geq 0$. If $\beta^2 = u^2 + v^2$ and $\alpha = 1$ then

$$\mathbf{P} = \begin{pmatrix} u^2 + v^2 & -au & -av \\ -u & 1 + \frac{au^2}{u^2+v^2} & \frac{auv}{u^2+v^2} \\ -v & \frac{auv}{u^2+v^2} & 1 + \frac{av^2}{u^2+v^2} \end{pmatrix}$$

In [26] an analogy to the symmetric preconditioning of van Leer, Lee and Roe was constructed for the incompressible equations. If we choose $a = 1, \alpha = 1$ we get this preconditioning of van Leer et. al., i.e. **P** is symmetric .

These examples show that the preconditioning is not unique. If fact, since the determinant of the transpose of a matrix is equal to the determinant of the original matrix it follows that the transpose of **P** is also a preconditioner with the same eigenvalues for the preconditioned system. In general, these various systems will have similar eigenvalues but different eigenvectors for the preconditioned system. Calculations show that the system given by **P** in (2.1) is more robust and converges faster than that with the transpose preconditioner. This shows that it is not sufficient to consider just the eigenvalues but that the eigenvectors are also of importance. However, even when **P** is symmetric $\mathbf{P}D$ is not symmetric and so the eigenvectors of the preconditioned system do not form an orthogonal basis.

We next examine some general form that the preconditioner can have. For this analysis it is easier to use streamwise coordinates as suggested in [30] and so $v = 0$. Let u_* be some normalization of the velocity components, then

$$A = \begin{pmatrix} 0 & u_* & 0 \\ u_* & u & 0 \\ 0 & 0 & u \end{pmatrix}, \qquad B = \begin{pmatrix} 0 & 0 & u_* \\ 0 & 0 & 0 \\ u_* & 0 & 0 \end{pmatrix}$$

Then the "convective" eigenvector for the non-preconditioned system is

$$\begin{pmatrix} 0 \\ \omega_2 \\ -\omega_1 \end{pmatrix}.$$

The "acoustic" eigenvectors are given by

$$\begin{pmatrix} \frac{-u\omega_1 + \sqrt{(u\omega_1)^2 + 4u_*^2}}{2} \\ u_*\omega_1 \\ u_*\omega_2 \end{pmatrix}, \begin{pmatrix} \frac{-u\omega_1 - \sqrt{(u\omega_1)^2 + 4u_*^2}}{2} \\ u_*\omega_1 \\ u_*\omega_2 \end{pmatrix}$$

We now consider preconditioners of the form

$$\mathbf{P} = \begin{pmatrix} a & b & 0 \\ c & d & 0 \\ 0 & 0 & 1 \end{pmatrix}. \qquad (2.4)$$

We want the eigenvalues of $\mathbf{P}D$ to be $\omega_1 u$, $\pm u$. This gives us three relations for the four unknowns:

$$a = \frac{u^2}{u_*^2}$$
$$(b + c)u_* + du = 0$$
$$u^2 d - bcu_*^2 = u^2$$

The values suggested in [25] are $b = 0, c = -\frac{u}{u_*}, d = 1$ while the values suggested in [30] are $b = c = -\frac{u}{u_*}, d = 2$. We next present the eigenvectors of $\mathbf{P}D$ in terms of the elements of $\mathbf{P}$. We exclude the case $\omega_2 = 0, \omega_1 = \pm 1$ as in this case $\mathbf{P}D$ has a double eigenvalue u and the eigensystem completely changes. Then the "convective" eigenvector is

$$\begin{pmatrix} 0 \\ \omega_2 \\ -(1 + b\frac{u_*}{u})\omega_1 \end{pmatrix}.$$

The "acoustic" eigenvectors are given by

$$\begin{pmatrix} \frac{u^2}{u_*} - bu u\omega_1^2 - \frac{u^2\omega_1}{u_*} + bu\omega_1 \\ u_*(a + b\omega_1^2) - u_*(b + c)\omega_1 \\ (u_* b\omega_1 + u)\omega_2 \end{pmatrix},$$

$$\begin{pmatrix} \frac{u^2}{u_*} - bu u\omega_1^2 + \frac{u^2\omega_1}{u_*} - bu\omega_1 \\ u_*(a + b\omega_1^2) + u_*(b + c)\omega_1 \\ (u_* b\omega_1 - u)\omega_2 \end{pmatrix}.$$

We note that the convective eigenvector is the same as before the preconditioning for the choice $b = 0$. The two acoustic eigenvectors are orthogonal to each other if we choose $b = 0$ and $c^2 = \frac{u^2(1 - \frac{u^2}{u_*^2})}{u_*^2}$. This is similar, but not identical, to the choice suggested in [25]. There is no way to make the convective eigenvector normal to both acoustic eigenvectors for preconditioners of the form (2.4).

3   Algorithm I

We consider the preconditioned Euler equation written as

$$P^{-1}w_t + Aw_x + Bw_y = 0 \qquad (3.1)$$

The form of the matrices A, B depend on the choice of the variables w . We shall first consider the variables $w = w_0 = (p, u, v, S)$ with $dS = dp - a^2 d\rho$. We then have

$$A_0 = \begin{pmatrix} u & \rho a^2 & 0 & 0 \\ \frac{1}{\rho} & u & 0 & 0 \\ 0 & 0 & u & 0 \\ 0 & 0 & 0 & u \end{pmatrix} \quad B_0 = \begin{pmatrix} v & 0 & \rho a^2 & 0 \\ 0 & v & 0 & 0 \\ \frac{1}{\rho} & 0 & v & 0 \\ 0 & 0 & 0 & v \end{pmatrix}$$

The preconditioners we shall consider contain a free parameter β. Unfortunately, the papers that have dealt with this subject all use different definitions of this free variable. Thus, for Choi and Merkle $\beta \sim a^2$ while for Turkel $\beta^2 \sim u^2 + v^2$ (based on Chorin's work on pseudo-compressibility) while for van Leer $\beta = \sqrt{1 - M^2}$ (based on fluid dynamic notation). In this paper we shall concentrate on the Choi-Merkle and Turkel preconditioners. To avoid confusion we shall use the Choi-Merkle choice for β within this paper. In terms of these variables we consider the following preconditioner which is a generalization of previous preconditioners. For $\delta = 0$ we recover the preconditioner suggested by Turkel [25, 26] while for $\delta = 1, \alpha = 0$ we recover the preconditioner suggested by Choi and Merkle [5]. Note that δ is a new free parameter to generalize the Choi-Merkle and Turkel preconditioners. It does not correspond to the δ in the Choi-Merkle paper.

$$P_0^{-1} = \begin{pmatrix} \frac{a^2}{\beta M_r^2} & 0 & 0 & \delta \\ \frac{\alpha u}{\rho\beta M_r^2} & 1 & 0 & 0 \\ \frac{\alpha v}{\rho\beta M_r^2} & 0 & 1 & 0 \\ 0 & 0 & 0 & 1 \end{pmatrix} \qquad (3.2)$$

$$P_0 = \begin{pmatrix} \frac{\beta M_r^2}{a^2} & 0 & 0 & -\frac{\beta M_r^2}{a^2}\delta \\ -\frac{\alpha u}{\rho a^2} & 1 & 0 & \frac{\alpha u}{\rho a^2}\delta \\ -\frac{\alpha v}{\rho a^2} & 0 & 1 & \frac{\alpha v}{\rho a^2}\delta \\ 0 & 0 & 0 & 1 \end{pmatrix}$$

The guiding principle is that near stagnation points βM_r^2 should not be allowed to approach zero. We also desire that for supersonic flow that preconditioner be turned off, i.e. $\beta M_r^2 = a^2$, $\alpha = 0$, $\delta = 0$. We choose

$$\beta M_r^2 = K_2(u^2 + v^2) \qquad (3.3)$$

$$K_2 = K_1(1 + \frac{(1 - K_1 M_0^2)}{K_1 M_0^4} M^2)$$

In addition we make sure that βM_r^2 is never greater than a^2.

This returns βM_r^2 to its non-preconditioned value at M_0. K_1 typically varies between 1.0 and 1.1. $K-1$ also depends on α [25] and may depend on the local cell Reynolds number [5]. Thus,

$$P_0 A_0 = \begin{pmatrix} \frac{\beta M_r^2 u}{a^2} & \rho\beta M_r^2 & 0 & \frac{-\beta M_r^2 u\delta}{a^2} \\ \frac{1 - \frac{\alpha u^2}{a^2}}{\rho} & (1 - \alpha)u & 0 & \frac{\alpha u^2 \delta}{\rho a^2} \\ \frac{-\alpha uv}{\rho a^2} & -\alpha v & u & \frac{\alpha uv\delta}{\rho a^2} \\ 0 & 0 & 0 & u \end{pmatrix}$$

The eigenvalues of $P_0 A_0$ are

$$\lambda_0 = u \qquad (3.4)$$

$$\lambda_\pm = 0.5\left(zu \pm \sqrt{z^2 u^2 + 4\beta M_r^2(1 - \frac{u^2}{a^2})}\right)$$

$z = 1 - \alpha + \frac{\beta M_r^2}{a^2}$. In particular if $\alpha = 1 + \frac{\beta M_r^2}{a^2}$ then $z = 0$ and $\lambda_\pm = \pm\sqrt{\beta M_r^2(1 - M^2)}$. The right eigenvectors are given by the columns of

$$R^{-1} = \begin{pmatrix} \rho\beta M_r^2 & \rho\beta M_r^2 & 0 & 0 \\ \lambda_+ - \frac{\beta M_r^2}{a^2}u & \lambda_- - \frac{\beta M_r^2}{a^2}u & 0 & \frac{\delta u}{\rho a^2} \\ \frac{\alpha v \lambda_+}{u - \lambda_+} & \frac{\alpha v \lambda_-}{u - \lambda_-} & 1 & 0 \\ 0 & 0 & 0 & 1 \end{pmatrix}$$

We note that for $\delta = 0$ and $\alpha = 0$ the two eigenvectors for the convective speeds are orthogonal to each other and to the acoustic eigenvectors. Of course, orthogonality of eigenvectors depends on the basis chosen, i.e. on the variables used in constructing P_0 and A_0. Furthermore, as the Mach number goes to zero the determiant of R^{-1} depends only on βM_r^2. Thus, the condition number of the transformation is well behaved as long as βM_r^2 is bounded away from zero as the Mach number approaches zero.

For generalized coordinates replace this by

$$U = us_x + vs_y \quad |S_\xi|^2 = s_x^2 + s_y^2 \tag{3.5}$$

$$\lambda_\pm = \frac{zU \pm \sqrt{z^2U^2 + 4\beta M_r^2(|S_I|^2 - \frac{U^2}{a^2})}}{2}$$

where $S_\xi = (s_x, s_y)$ is the cell face vector in ξ-direction. The largest eigenvalue is then used to determine the (inviscid) time step. We note that these eigenvalues are independent of δ. For $M \sim 0$, $\lambda_\pm \sim 0.5u\left((1-\alpha) \pm \sqrt{(1-\alpha)^2 + 4}\right)$. So for $\alpha = 0$ the condition number is $\frac{1+\sqrt{5}}{1-\sqrt{5}} \sim 2.6$ while for $\alpha = 1$ the condition number is $1 + O(M)$. If we choose $\alpha = 1 + \frac{\beta M_r^2}{a^2}$ and $\beta M_r^2 = \frac{M^2 c^2}{1-M^2}$ then we get the same eigenvalues as those of the symmetric optimal preconditioner to be presented in the next section (see 5.1). For small Mach numbers this is similar to what we obtained with $\alpha = 1$ while at the sonic line α and β become infinite which reduces the condition number of the preconditioned system near the sonic line. We shall also express these matrices in terms of the variables $w_4 = (p, u, v, T)$. Let

$$g = 1 + (\gamma - 1)\delta = \begin{cases} 1, & \text{if } \delta = 0 \\ \gamma, & \text{if } \delta = 1 \end{cases}$$

Then,

$$A_4 = \begin{pmatrix} u & \rho a^2 & 0 & 0 \\ \frac{1}{\rho} & u & 0 & 0 \\ 0 & 0 & u & 0 \\ 0 & (\gamma-1)T & 0 & u \end{pmatrix}$$

$$B_4 = \begin{pmatrix} v & 0 & \rho a^2 & 0 \\ 0 & v & 0 & 0 \\ \frac{1}{\rho} & 0 & v & 0 \\ 0 & 0 & (\gamma-1)T & v \end{pmatrix}$$

Let $m^2 = \frac{\beta M^2}{a^2}$, then

$$P_4^{-1} = \begin{pmatrix} \frac{1}{m^2} - (\gamma-1)\delta & 0 & 0 & \frac{\gamma p}{T}\delta \\ \frac{\alpha u}{\rho\beta M^2} & 1 & 0 & 0 \\ \frac{\alpha v}{\rho\beta M^2} & 0 & 1 & 0 \\ \frac{1}{\rho c_p}\left(\frac{1}{m^2} - g\right) & 0 & 0 & g \end{pmatrix},$$

$$P_4 = \begin{pmatrix} m^2 g & 0 & 0 & -m^2\frac{\gamma p}{T}\delta \\ -\frac{\alpha u g}{\rho a^2} & 1 & 0 & \frac{\alpha u}{\rho a^2}\frac{\gamma p}{T}\delta \\ -\frac{\alpha v g}{\rho a^2} & 0 & 1 & \frac{\alpha v}{\rho a^2}\frac{\gamma p}{T}\delta \\ \frac{1}{\rho c_p}(m^2 g - 1) & 0 & 0 & 1 - (\gamma-1)m^2\delta \end{pmatrix}$$

The matrix presented in (3.2) while improving the condition in the neighborhood of $M = 0$ does not improve the condition number in the neighborhood of the sonic line. Indeed in (3.3) we explicitly removed the preconditioning as $M \to 1$. An alternative is to generalize (3.2) so that it is basically the same for low Mach numbers but reduces the condition number near the sonic line.

4 Algorithm II

In [7, 27] an alternate way of preconditioning the fluid dynamics system was introduced. In this approach we do not update the conservation variables, rather we update the variables

$$W = \begin{pmatrix} p' \\ \rho u \\ \rho v \\ E' \end{pmatrix}, \quad F = \begin{pmatrix} \rho u \\ \rho u^2 + p' \\ \rho uv \\ \rho H'u \end{pmatrix},$$

$$G = \begin{pmatrix} \rho v \\ \rho uv \\ \rho v^2 + p' \\ \rho H'v \end{pmatrix}, \quad Q = \begin{pmatrix} p' \\ \rho u \\ \rho v \\ H' \end{pmatrix}$$

$$\frac{\partial W}{\partial t} = P\left[\frac{\partial F}{\partial x} + \frac{\partial G}{\partial y}\right]$$

where $p' = p - p_\infty$, $E' = c_p\rho(T - T_\infty) - (p - p_\infty) + \frac{\rho(u^2 + v^2)}{2} = E + p_\infty - \frac{\gamma p_\infty}{(\gamma-1)\rho_\infty}\rho$ and $\rho H' = E' + p'$. Then

$$P = I + \Delta \cdot \begin{bmatrix} 1 - \frac{1}{\Delta} & -\frac{u}{G + h_\infty} & -\frac{v}{G + h_\infty} & \frac{1}{G + h_\infty} \\ -B_2 & \frac{uB_2}{G + h_\infty} & \frac{vB_2}{G + h_\infty} & -\frac{B_2}{G + h_\infty} \\ -B_3 & \frac{uB_3}{G + h_\infty} & \frac{vB_3}{G + h_\infty} & -\frac{B_3}{G + h_\infty} \\ -B_4 & \frac{uB_4}{G + h_\infty} & \frac{vB_4}{G + h_\infty} & -\frac{B_4}{G + h_\infty} \end{bmatrix}$$

$$B_1 = \frac{1}{\beta M_r^2} - \frac{1}{(\gamma - 1)h}$$

$$B_2 = (B_1 + \frac{\alpha}{\beta M_r^2})u$$

$$B_3 = (B_1 + \frac{\alpha}{\beta M_r^2})v$$

$$B_4 = B_1 H' + \frac{\alpha(u^2 + v^2)}{\beta M_r^2}$$

where βM_r^2 was previously defined and $h = c_p T$, $G = \frac{u^2 + v^2}{2}$, $\Delta = \frac{(G + h_\infty)\beta M_r^2}{h}$ and $\alpha = 1$.

With this approach all operations (e.g. residual smoothing, multigrid etc.) are done on the new variables and not the conservative variables. The rationale is that since the pressure is more appropriate than density at lower speeds we should interpolate the pressure between grids rather than the density. Similarly, since the density is almost constant it is more beneficial to smooth the residual of the pressure rather than the residual of the density. The eigenvalues of this system, and hence the time steps needed for stability, are the same as for Algorithm I, (3.5).

5 Symmetric Preconditioner

Until now we have stressed the importance of preconditioning for low Mach number flows. Indeed we have constructed the preconditioners by starting from the incompressible equations and analyzing what changes are needed to the compressible equations in order to recover the incompressible equations in the limit. Van Leer, Lee and Roe [30] began at the other end. They first noticed that the two dimensional supersonic equations can be diagonalized if an appropriate preconditioner is added to the system. They then constructed a preconditioner for the subsonic case that it some sense was the continuation of the supersonic case. This continuation can never be exact since there must be a singularity at the sonic line.

Furthermore, in contrast to the variables that we have used until now van Leer et al. expressed the inviscid equations in $(\frac{dp}{\rho a}, du, dv, dS)$ variables. This differs from what we have done above in the normalization of the pressure variable. With this new set of variables the Jacobian matrices for inviscid flow are symmetric. Adding a symmetric positive definite preconditioner yields a symmetric hyperbolic system:

$$P^{-1}\frac{\partial w}{\partial t} + A\frac{\partial w}{\partial x} + B\frac{\partial w}{\partial y} = 0$$

where P, A, B are symmetric and P is positive definite. Taking the inner product with we have by

integration by parts (ignoring boundary contributions) $(w, A\frac{\partial w}{\partial x}) = (w, B\frac{\partial w}{\partial x}) = 0$ Hence, we conclude that $\frac{\partial}{\partial t}(P^{-1}w, w) = 0$ and so $||w^2||_{P^{-1}}$ is constant in time. Since P^{-1} is positive definite, this norm is equivalent to the standard L^2 norm if there exists positive constants K_1 and K_2 such that

$$0 < K_1 \leq ||P|| \leq K_2$$

For the preconditioners that we shall consider the norm of P depends on the Mach number. Hence, we have well-posedness only as long as the Mach number is bounded away from zero. In the limit of the Mach number approaching zero, P becomes singular and the above analysis is no longer valid.

We now present the van Leer-Lee-Roe preconditioning for streamwise coordinates where the x axis is aligned with the flow and so $v = 0$. We express the preconditioner for the equations using $(\frac{dp}{\rho a}, du, dv, dS)$ variables. The coefficient matrices are given by

$$A = c\begin{pmatrix} M & 1 & 0 & 0 \\ 1 & M & 0 & 0 \\ 0 & 0 & M & 0 \\ 0 & 0 & 0 & M \end{pmatrix} \quad B = c\begin{pmatrix} 0 & 0 & 1 & 0 \\ 0 & 0 & 0 & 0 \\ 1 & 0 & 0 & 0 \\ 0 & 0 & 0 & 0 \end{pmatrix}$$

Define $D = A\omega_1 + B\omega_2$; then

$$P_V = \begin{pmatrix} \frac{\tau}{\hat{\beta}^2}M^2 & -\frac{\tau}{\hat{\beta}^2}M & 0 & 0 \\ -\frac{\tau}{\hat{\beta}^2}M & 1 + \frac{\tau}{\hat{\beta}^2} & 0 & 0 \\ 0 & 0 & \tau & 0 \\ 0 & 0 & 0 & 1 \end{pmatrix}$$

where

$$\hat{\beta} = \begin{cases} \sqrt{1 - M^2} & M < 1, \\ \sqrt{M^2 - 1} & M \geq 1 \end{cases}$$

$$\tau = \begin{cases} \sqrt{1 - M^2} = \hat{\beta} & M < 1, \\ \sqrt{1 - M^{-2}} = \frac{\hat{\beta}}{M} & M \geq 1 \end{cases}$$

β was introduced to change the envelope of the wave form while τ was introduced to change the amplitude of the waves. More details of the development of this preconditioner are given in [34]. Hence, we include both in P_V. The eigenvalues of $P_V D$, for subsonic flow, are

$$\lambda_0 = u\omega_1 \quad \text{twice}$$

$$\lambda_\pm = \pm\tau Mc\sqrt{\omega_1^2 + \frac{\omega_2^2}{1 - M^2}} \tag{5.1}$$

In order to compare this formula with the formulae of previous sections we wish to reformulate this preconditioner for the case where the flow is not aligned in the x direction. We denote the matrices in the streamwise and perpendicular directions as $A_\parallel$ and $A_\perp$ respectively. We next define the rotation matrices as

$$U = \begin{pmatrix} 1 & 0 & 0 & 0 \\ 0 & cos\theta & sin\theta & 0 \\ 0 & -sin\theta & cos\theta & 0 \\ 0 & 0 & 0 & 1 \end{pmatrix}$$

$$U^{-1} = \begin{pmatrix} 1 & 0 & 0 & 0 \\ 0 & cos\theta & -sin\theta & 0 \\ 0 & sin\theta & cos\theta & 0 \\ 0 & 0 & 0 & 1 \end{pmatrix}$$

To get the streamwise direction we shall choose

$$cos\theta = \frac{u}{\sqrt{u^2 + v^2}}$$
$$sin\theta = \frac{v}{\sqrt{u^2 + v^2}}$$

One can then verify that given the original matrices A, B.

$$A_\parallel = U(A cos\theta + B sin\theta)U^{-1}$$
$$A_\perp = U(-A sin\theta + B cos\theta)U^{-1}$$

Given numbers $\hat{\omega}_1, \hat{\omega}_2$ for $A_\parallel, A_\perp$ we define

$$\omega_1 = \hat{\omega}_1 cos\theta - \hat{\omega}_2 sin\theta$$
$$\omega_2 = \hat{\omega}_1 sin\theta + \hat{\omega}_2 cos\theta$$

note $\hat{\omega_1^2} + \hat{\omega_1^2} = \omega_1^2 + \omega_2^2$.

Also define $\mathbf{P} = U^{-1}\hat{P}U$. Then it is easy to verify that

$$\hat{\mathbf{P}}(A_\parallel \hat{\omega}_1 + A_\perp \hat{\omega}_2) = U\left[\mathbf{P}(A\omega_1 + B\omega_2)\right]U^{-1}$$

The van Leer-Lee-Roe preconditioning for general nonaligned flow in $\left(\frac{dp}{\rho a}, du, dv, dS\right)$ variables again decouples the last row and column. Hence we present only the 3×3 matrix:

$$\mathbf{P}_{VL} = \begin{pmatrix} \frac{\tau}{\beta^2}M^2 & \frac{-\tau}{\beta^2}u/a & \frac{-\tau}{\beta^2}v/a \\ \frac{-\tau}{\beta^2}u/a & b_{22} & b_{23} \\ \frac{-\tau}{\beta^2}v/a & b_{23} & b_{33} \end{pmatrix}$$

$$b_{22} = \left(\frac{\tau}{\hat{\beta}^2} + 1\right)\frac{u^2}{u^2 + v^2} + \tau\frac{v^2}{u^2 + v^2}$$

$$b_{33} = \left(\frac{\tau}{\hat{\beta}^2} + 1\right)\frac{v^2}{u^2 + v^2} + \tau\frac{u^2}{u^2 + v^2}$$

$$b_{23} = b_{32} = \left(\frac{\tau}{\hat{\beta}^2} + 1\right)\frac{uv}{u^2 + v^2}$$

The symmetric preconditioner has the nice property that it always leads to a well posed problem, at least for Mach numbers bounded away from zero. It is also useful in constructing canonical forms that separate the Euler equations into its elliptic and hyperbolic constitute parts (see section 14). However, this preconditioner seems to have difficulties near stagnation points. One possible reason is that the general preconditioner depends on the flow angle even near stagnation points. However, for very small Mach numbers the angle the flow makes with the xaxis is not well defined. This leads to a numerical indeterminacy near stagnation points. Possible cures for this are discussed in [31]. These cures depend on modifications that remove this dependency on the flow angle usually at the cost of a loss of optimality.

In general it is not clear that a condition number of one at $M = 0$ is a desirable trait. One purpose of preconditioning is to transform the infinite condition number at $M = 0$ to a finite and small condition number. However, in practice, choosing the optimal condition number of one results in a less robust scheme. Thus, for example, it is found that for Algorithm I and II $(3, 4)$ that $\alpha = 0$ leads to a more robust scheme even though $\alpha = 1$ is optimal at $M = 0$. One possible reason for this is that the preconditioning affects not only the time step but also the artificial viscosity. An optimal preconditioning lowers the artificial viscosity which improves the accuracy but deteriorates the robustness of the scheme.

6 Convergence to the Incompressible Equations

The time dependent Euler equations can be written as

$$\begin{aligned} p_t + up_x + vp_y + \rho a^2(u_x + v_y) &= 0 \\ u_t + uu_x + vu_y + \frac{p_x}{\rho} &= 0 \\ v_t + uv_x + vv_y + \frac{p_y}{\rho} &= 0 \\ S_t + uS_x + vS_y &= 0 \end{aligned} \qquad (6.1)$$

where a is the speed of sound given by $a^2 = \frac{\gamma p}{\rho}$.

The form of this system is unchanged if we nondimensionalize the equations. From now on we shall assume that u, v, p, ρ are nondimensional quantities where the dimensional variables are nondimensionalized by u_*, p_*, ρ_* with $p_* = \rho_* u_*^2$. Following [10] we define $\epsilon = \frac{u_*}{a_*} = M_*$. If the fluid is isentropic then

$$p = \frac{\rho^\gamma}{\gamma\epsilon^2} \quad , \quad a = \frac{\rho^{\frac{\gamma-1}{2}}}{\epsilon}. \qquad (6.2)$$

Hence, as ϵ goes to zero the speed of sound, a, goes to infinity and so the first equation in (6.1) reduces to $u_x + v_y = 0$.

It was pointed out in [25, 26] that these equations can be symmetrized by using $\frac{dp}{\rho a}$ as the independent variable rather than dp. Hence, we define ϕ by $d\phi = \frac{dp}{\rho a}$. For isentropic flow both p and a are functions only of the density and so using (6.2) this can be integrated explicitly. This gives $\phi = \frac{\rho^{\frac{\gamma-1}{2}}}{\frac{\gamma-1}{2}\epsilon}$. As the Mach number goes to zero ϕ tends to infinity and therefore, Gustafsson and Stoor [10] subtract a constant and define

$$\phi = \frac{\rho^{\frac{\gamma-1}{2}} - 1}{\frac{\gamma-1}{2}\epsilon}. \tag{6.3}$$

This amounts to specifying the constant in the integration of $d\phi$ from dp. They then prove, using energy methods, that for the linearized equations

$$a\phi_x \to \frac{\partial p_{incompressible}}{\partial x}.$$

Hence, ϕ and all its derivatives behave as $O(M)$ as $M \to 0$. Since $\rho \to 1$ and using the definition of $d\phi$ this is equivalent to

$$dp_{compressible} \to dp_{incompressible}.$$

Several papers [10, 12, 14], have proven the convergence of the solution to the compressible equations to the solution of the incompressible equations, for isentropic flow, as the Mach number goes to zero. For nonisentropic flow there are no formal proofs. However, it is clear that for viscous flows that the boundary condition on the temperature, adiabatic or isothermal is very important.

All these results refer to the time dependent physical equations. Once preconditioning is introduced time accuracy is lost and one can only discuss convergence of the steady state solutions. In this case one would hope that the time dependent preconditioned compressible equations converge to some time dependent preconditioning of the incompressible equations. In addition, one would also want this to be true on the numerical level. Thus, one would want to solve the preconditioned compressible equations by some numerical technique, on a fixed mesh and compare that with the solution of the preconditioned incompressible equations on the same mesh. Mathematically, we have two limit processes occurring: the Mach number going to zero and the mesh size going to zero. These two limits need not commute. If one first converges the mesh size and then the Mach number it is equivalent to the convergence proofs for the analytic case. The more interesting problem is to converge the Mach number and then converge the mesh, i.e. we use a fix mesh as the Mach number is reduced.

In particular this requires a careful study of the viscosities introduced by the scheme. We first consider an upwinding scheme. For the compressible case the Riemann solver should depend on the preconditioned problem. One would then need to show that this Riemann problem converges to a Riemann problem for some preconditioning of the incompressible equations. We next consider a central difference scheme with a scalar viscosity. In this case a high order even difference of some quantity is added separately to each equation, e.g. for the incompressible equations: pressure for the continuity equation, u and v for the momentum equations. For the compressible equations one normally adds a density difference to the continuity equation. In such a case it is obvious that the numerical scheme for the compressible equations cannot converge to the numerical scheme for the incompressible equations. Furthermore, for low Mach number flows the density is almost constant and so the higher order difference of the density does not add much of a viscosity to the continuity equation. As such, we conclude that the artificial viscosity for the compressible continuity equation should be based on pressure and not density (at least for low Mach numbers).

By convergence of the compressible equations to the incompressible equations we are merely verifying what happens to the difference equations as the Mach number goes to zero. This gives necessary but not sufficient conditions for the convergence of the solution of the numerical approximation to the preconditioned compressible equations to the numerical solution of the incompressible equations. We consider a central difference approximation together with a scalar artificial viscosity for the nondimensionalized preconditioned inviscid equations. For the incompressible equations in nonconservative form we consider the preconditioned system

$$p_t + \beta^2 (u_x + v_y) = h\left[(K_1 p_x)_x + (K_2 p_y)_y\right]$$
$$\frac{u}{\beta^2} p_t + u_t + uu_x + vu_y + p_x = h\left[(K_1 u_x)_x + (K_2 u_y)_y\right]$$
$$\frac{v}{\beta^2} p_t + v_t + uv_x + vv_y + p_y = h\left[(K_1 v_x)_x + (K_2 v_y)_y\right]$$

where each space derivative is approximated by a central difference with spacing h in each direction. We can also consider higher order accurate artificial viscosities but this doesn't change any of the properties that are of interest for the convergence to the incompressible limit. The time derivatives are replaced by a multistage scheme. K_1, K_2 are proportional to the largest eigenvalues of the coefficient matrix in the respective direction.

We next consider the same scheme for the preconditioned compressible inviscid equations, under the assumption that $p = p(\rho)$. It easier to analyze the convergence for the nonsymmetrical form since the pressure, p, converges and not $\frac{dp}{\rho a}$. Then for the compressible equations we have

$$p_t + \beta^2 [(\rho u)_x + (\rho v)_y)] = h[(K_1 p_x)_x + (K_2 p_y)_y]$$

$$\frac{u}{\beta^2} p_t + u_t + u u_x + v u_y + \frac{p_x}{\rho} = h[(K_1 u_x)_x + (K_2 u_y)_y]$$

$$\frac{v}{\beta^2} p_t + v_t + u v_x + v v_y + \frac{p_y}{\rho} = h[(K_1 v_x)_x + (K_2 v_y)_y]$$

Comparing these formulations it is obvious that if $\rho \to 1$ as $M \to 0$ then both converge to each other. It is crucial for both the time derivative and the artificial viscosity in the compressible continuity equation to be pressure based rather than density based. The preconditioning of the momentum equations is not important for this convergence.

We now consider how to construct a general artificial viscosity that will enable us to reach the incompressible limit without the difficulties previous presented. Consider

$$P^{-1} w_t + f_x + g_y = h[(Q_1 w_x)_x + (Q_2 w_y)_y].$$

We wish to find the dependence of $\mathbf{P}$ and Q_i on the Mach number as $M \to 0$ so that we get the proper convergence. We therefore consider the isentropic equations based on $w = (\phi, u, v)$ (see (6.3)). This has the symmetric form

$$\mathbf{P}^{-1} w_t + \begin{pmatrix} a_{11} & a_{12} & a_{13} \\ a_{12} & a_{22} & a_{23} \\ a_{11} & a_{23} & a_{33} \end{pmatrix} w_x +$$

$$\begin{pmatrix} b_{11} & b_{12} & b_{13} \\ b_{12} & b_{22} & b_{23} \\ b_{11} & b_{23} & b_{33} \end{pmatrix} w_y = h[(Q_1 w_x)_x + (Q_2 w_y)_y]$$

As $M \to 0$ $a_{12}, b_{13} = O(1/M)$ while $d\phi = O(M)$ while all other quantities are bounded. Hence, the leading terms in the first equation are all $O(1/M)$ while they are $O(1)$ for the second and third equations. Multiplying the first equation by M and taking the limit we get $u_x + v_y$ for the space derivatives on the left hand side. Using $d\phi = O(M)$, $du = O(1)$, $dv = O(1)$ we see that a necessary condition for convergence as $M \to 0$ is that P^{-1}, Q_1, Q_2 all have the form

$$\mathbf{P}^{-1}, \mathbf{Q_1}, \mathbf{Q_2} \sim \begin{bmatrix} O(\frac{1}{M^2}) & O(\frac{1}{M}) & O(\frac{1}{M}) \\ O(\frac{1}{M}) & O(1) & O(1) \\ O(\frac{1}{M}) & O(1) & O(1) \end{bmatrix} \quad (6.4)$$

If we would have considered the variables (p, u, v) instead of (ϕ, u, v) then all the terms would have been of order 1. However, the matrices A, B would not be symmetric. Hence, each set has its advantages and we shall consider both sets of variables.

As an example, consider the equations in $(\frac{dp}{\rho c}, du, dv)$ variables. Then

$$\mathbf{A_{comp}} = \begin{pmatrix} u & c & 0 \\ c & u & 0 \\ 0 & 0 & u \end{pmatrix}, \mathbf{B_{comp}} = \begin{pmatrix} v & 0 & c \\ 0 & v & 0 \\ c & 0 & v \end{pmatrix}$$

Since these equations are nondimensionalized $c = O(1/M)$. Let

$$\mathbf{D} = A\omega_1 + B\omega_2 = \begin{pmatrix} Rq\omega_1 & \omega_1 c & \omega_2 c \\ \omega_1 c & Rq\omega_1 & 0 \\ \omega_2 c & 0 & Rq\omega_1 \end{pmatrix},$$

where $q^2 = u^2 + v^2$, $R = \frac{u\omega_1 + v\omega_2}{q}$ and we define $\mathbf{P}^{-1}_{comp}$ and $\mathbf{P}_{comp}$ by

$$\mathbf{P}^{-1} = \begin{pmatrix} \frac{2}{M^2} - 1 & \frac{u}{qM} & \frac{v}{qM} \\ \frac{u}{qM} & 1 & 0 \\ \frac{v}{qM} & 0 & 1 \end{pmatrix},$$

$$\mathbf{P} = \frac{1}{1-M^2} \begin{bmatrix} M^2 & -\frac{Mu}{q} & -\frac{Mv}{q} \\ \frac{-Mu}{q} & 1-M^2+\frac{u^2}{q^2} & \frac{uv}{q} \\ -\frac{Mv}{q} & \frac{uv}{q} & 1-M^2+\frac{v^2}{q^2} \end{bmatrix}$$

By inspection $\mathbf{P}^{-1}_{comp}$ has the form desired of (6.4). To find the form of the artificial viscosity we return to the numerical approximation for

$$\mathbf{P}^{-1} w_t + f_x + g_y = h[(Q_1 w_x)_x + (Q_2 w_y)_y] \quad .$$

For constant $\mathbf{P}$ we want $\mathbf{P}Q_1 = |\mathbf{P}A|$. Hence, $Q_1 = \mathbf{P}^{-1}|\mathbf{P}A|$, $Q_2 = \mathbf{P}^{-1}|\mathbf{P}B|$. Then

$$(\mathbf{P}^{-1}|\mathbf{PD}|)_{comp} =$$

$$q \begin{pmatrix} \frac{S+R}{M^2} & \frac{|u|R}{qM} & \frac{|v|R}{qM} \\ \frac{|u|R}{qM} & \frac{u^2|R|}{q^2} + \frac{v^2 S}{q^2(1-M^2)} & \frac{|uv|R}{q^2} - \frac{|uv|S}{q^2(1-M^2)} \\ \frac{|v|R}{qM} & \frac{|uv|R}{q^2} - \frac{|uv|S}{q^2(1-M^2)} & \frac{v^2|R|}{q^2} + \frac{u^2 S}{q^2(1-M^2)} \end{pmatrix}$$

$$q^2 = u^2 + v^2, \quad R = \frac{u\omega_1 + v\omega_2}{q}, \quad S = \sqrt{(1-M^2)(1-M^2 R^2)}.$$

Letting $M \to 0$ we get $S \to 1$. To get the right limit of $(\mathbf{P}^{-1}|\mathbf{PD}|)_{comp}$ we must first multiply each element by the appropriate factor of M (or equivalently go to (p,u,v) variables). We then get

$$(\frac{dp}{\rho c}, du, dv)_{comp} \to (\frac{dp}{q}, du, dv)_{incomp}$$

and

$$(\mathbf{P}^{-1}|\mathbf{PD}|)_{comp} \to (\mathbf{P}^{-1}|\mathbf{PD}|)_{incomp} =$$

$$q \begin{pmatrix} |R| + 1 & \frac{|uR|}{q} & \frac{|vR|}{q} \\ \frac{|uR|}{q} & \frac{u^2|R|+v^2}{q^2} & \frac{|uv|(|R|-1)}{q^2} \\ \frac{|vR|}{q} & \frac{|uv|(|R|-1)}{q^2} & \frac{v^2|R|+u^2}{q^2} \end{pmatrix}.$$

Changing from the variables $(\frac{dp}{q}, du, dv)_{incomp}$ to $(dp, du, dv)_{incomp}$ we get the same viscosity matrix given by Arnone and Turkel [3].

We next point out that the conditions on the matrix $(\mathbf{P}^{-1}|\mathbf{PD}|)_{comp}$ are not satisfied by the non-preconditioned Roe matrices. Furthermore, even reasonable preconditioners need not satisfy these conditions. Consider, for example, the one dimensional system

$$\mathbf{P}^{-1}w_t + Aw_x = h(Qw_x)_x.$$

A reasonable choice is $\mathbf{P}^{-1} = |A|$ i.e. $\mathbf{P} = |A^{-1}|$. In this case all the wave speeds of $\mathbf{P}A$ are ± 1. Now

$$Q = \mathbf{P}^{-1}|\mathbf{P}A| = |A| \, ||A|^{-1}A| = |A| \sim$$

$$\begin{pmatrix} O(\frac{1}{M}) & O(1) & O(1) \\ O(1) & O(\frac{1}{M}) & O(1) \\ O(1) & O(1) & O(1) \end{pmatrix}$$

Q is the nonpreconditioned Roe matrix which does not have the desired property. We therefore conclude that for an upwind difference scheme the Riemann solver should be based on the preconditioned system and not the original scheme. For central difference schemes with a scalar artificial viscosity the viscosity in the continuity equation should include pressure terms. For a matrix viscosity $Q_1 = \mathbf{P}^{-1}|\mathbf{P}A|, Q_2 = \mathbf{P}^{-1}|\mathbf{P}B|$. Characteristics in the boundary conditions these should be based on the characteristics of the modified system and not the physical system. When using multigrid it is better to transfer the residuals based on the preconditioned system to the next grid since these residuals are more balanced than the physical residuals. However, conservation of the variables is now lost. Preconditioning is even more important when using multigrid than with an explicit scheme. With the original system the disparity of the eigenvalues greatly affects the smoothing rates of the slow components and so slows down the multigrid method, [16]. As indicated above there are accuracy difficulties at low Mach numbers [33]. Some of these can be alleviated by preconditioning the dissipation terms. For very small Mach numbers there is also a difficulty with round off errors as $\frac{p}{u^2+v^2} \to \infty$. Several people have suggested subtracting out a constant pressure from the dynamic pressure. A more

detailed analysis [9] suggests replacing the pressure p by $\tilde{p}$ where $p = \frac{p_0+\epsilon\tilde{p}}{\epsilon^2}$ and ϵ is a representative Mach number.

We conclude from the above remarks that the steady state solution of the preconditioned system may be different from that of the physical system. Thus, on the finite difference level the preconditioning can improve the accuracy as well as the convergence rate.

7 Artificial Viscosity

We have seen that the details of the artificial viscosity (or equivalently the Roe matrix) determines whether the approximation to the compressible equation converges to that of the incompressible equations as the Mach number approaches zero. The preconditionings discussed i.e. Algorithm I, Algorithm II and Van Leer et. al. all satisfy the necessary condition (6.4). We have previously shown that the "natural" one dimensional preconditioning $P^{-1} = |A|$ does not satisfy (6.4). Similarly the preconditioner proposed by Allmaras though it has nice high frequency properties provides no help for the low Mach limit. We now present the implementation of the artificial viscosity in more detail.

7.1 Basic Dissipation Scheme

For a central difference scheme it is necessary to add artificial dissipative terms to the central difference approximation of the spatial derivatives. One usually adds a nonlinear second difference to control oscillations near shocks and a linear fourth difference to damp high frequency oscillations [11]. We are interested in the functional form of these differences. Hence, our examples will be with a second-difference artificial dissipation. Extensions to fourth differences and nonlinearities are straightforward. Similarly, for one sided schemes the central difference plus artificial dissipation is replaced by a Roe matrix formulation. Extension to these schemes is straightforward.

A preconditioned finite difference scheme can be expressed in differential form as

$$\begin{aligned} \Delta w_c &= \Delta t P \left[\frac{\partial F}{\partial x} + \frac{\partial G}{\partial y} \right] \\ &= \Delta t P \left[A \frac{\partial w_c}{\partial x} + B \frac{\partial w_c}{\partial y} \right] \end{aligned} \tag{7.1}$$

where w_c represents the conservation variables and the partial spatial derivatives are replaced by central difference approximations, either cell centered or at nodal points. We shall consider both conservative and non-conservative ways of adding artificial dissipative terms. The dissipation need not be expressed in terms of the conservative variables w_c. Using another set of variables w_V we have $\Delta w_c = \frac{\Delta w_c}{\Delta w_V} \Delta w_V$. Using different variables we can introduce the appropriate preconditioner in two ways. One choice is to completely change variables in all places and the matrices A and B in (7.1) change as well as the preconditioner P. The new preconditioner P_i is then a function of P_0, see (3.2), given by $P_i = \frac{\partial w_i}{\partial w_0} P_0 \frac{\partial w_0}{\partial w_i}$. An alternative is to leave the spatial derivatives in conservation form and to transform only the time derivatives. Essentially following Choi and Merkle's notation we denote this preconditioner by Γ (our Γ is the inverse of that used by Choi and Merkle). Thus, in conservative variables we have $\Gamma_c = P_c$. We choose as our "basic" form that given in (p, u, v, S) coordinates (i.e. P_0). Let w_c be the conservation variables then for any other set of variables we have $\Gamma_i = \frac{\partial w_i}{\partial w_c} P_c = P_i \frac{\partial w_i}{\partial w_c} = \frac{\partial w_i}{\partial w_0} P_0 \frac{\partial w_0}{\partial w_c}$.

We now include an artificial dissipation and replace (7.1) by differences in a general coordinate system (ξ, η) yielding:

$$\frac{\Delta(w_V)_{i,j}}{\Delta t} = (\Gamma_V)_{i,j}.$$

$$\left[\frac{F_{i+1/2,j} - F_{i-1/2,j}}{\Delta\xi} + \frac{G_{i,j+1/2} - G_{i,j-1/2}}{\Delta\eta} \right] \\ - \left[\frac{d_{i+1/2,j} - d_{i-1/2,j}}{\Delta\xi} + \frac{d_{i,j+1/2} - d_{i,j-1/2}}{\Delta\eta} \right] \quad (7.2)$$

where

$$d_{i+1/2,j} = (|\sigma(PA)|)_{i+1/2,j} (w_{i+1,j} - w_{i,j})$$

$$d_{i,j+1/2} = (|\sigma(PB)|)_{i,j+1/2} (w_{i,j+1} - w_{i,j}).$$

All variables refer to to V coordinates and $A_V = \frac{\partial w_V}{\partial w_c} A \frac{\partial w_c}{\partial w_V}$, B_V is defined similarly, and σ is matrix function. For example, if σ is proportional to the spectral radius then we have a scalar artificial viscosity, in w_V variables, while if $\sigma(A) \sim A$ we have a matrix-valued artificial viscosity. Once we have found Δw_V we can either update this to find w_V or else we can convert back to w_c variables to calculate Δw_c. Thus, for example, residual smoothing and multigrid can be done on either the w_c variables or the w_V variables.

Formula 7.2 is easy to implement. For a scalar viscosity (i.e. σ is a scalar function) the viscosity remains scalar after preconditioning. The main difference is that now differences of the w_V variables are added to each of the conservation variables rather than differences of the conserved variables. We note that in the original paper of Jameson, Schmidt and Turkel [11] differences of ρH rather than E were added to the energy equation. However, the formulation (7.2) is nonconservative. Since we are also be interested in transonic and supersonic flows with shocks we present the addition of the artificial dissipation in a conservative formulation. Furthermore, for internal flow the conservative form is important even for subsonic flows, without shocks, in order to preserve global quantities, e.g. total mass flow. A conservative formulation is then given by:

$$\frac{\Delta(w_V)_{i,j}}{\Delta t} = (\Gamma_V)_{i,j}.$$

$$\left\{ \left[\frac{F_{i+1/2,j} - F_{i-1/2,j}}{\Delta\xi} + \frac{G_{i,j+1/2} - G_{i,j-1/2}}{\Delta\eta} \right. \right. \\ \left. \left. - \left[\frac{d_{i+1/2,j} - d_{i-1/2,j}}{\Delta\xi} + \frac{d_{i,j+1/2} - d_{i,j-1/2}}{\Delta\eta} \right] \right] \right\} \quad (7.3)$$

where

$$d_{i+1/2,j} = \Gamma^{-1}_{i+1/2,j} (|\sigma(P_V A_V)|)_{i+1/2,j} [(w_V)_{i+1,j} - (w_V)_{i,j}]$$

$$d_{i,j+1/2} = \Gamma^{-1}_{i,j+1/2} (|\sigma(P_V B_V)|)_{i,j+1/2} [(w_V)_{i,j+1} - (w_V)_{i,j}]$$

Even when σ is a scalar function we now have to evaluate matrix-vector product and so the numerical effort corresponds to a matrix-valued viscosity. We should mention that $A_V = \frac{\partial w_V}{\partial w_c} A \frac{\partial w_c}{\partial w_V}$, and similar expressions can be written for the other coordinate directions. Let $\hat{A}_V = \frac{\partial F}{\partial w_V}$ be the Jacobian matrix; then, $\hat{A}_V = \frac{\partial w_c}{\partial w_V} A_V$. Hence,

$$P_V A_V = \frac{\partial w_V}{\partial w_c} P_c A_c \frac{\partial w_c}{\partial w_V} = \Gamma_V A_c \frac{\partial w_c}{\partial w_V}$$

Because the spectral radius is invariant under a similarity transformation for a scalar viscosity, we can use either $P_V A_V$ or $P_c A_c$. Let $\Lambda = \mathrm{diag}(\lambda_1, \lambda_2, \lambda_3, \lambda_3, \lambda_3)$, where λ_1, λ_2, and λ_3 represent λ_+, λ_-, and λ_0 (i.e., the eigenvalues of PD), modified by cutoffs near the stagnation points. Define $|\Lambda| = \mathrm{diag}(|\lambda_1|, |\lambda_2|, |\lambda_3|, |\lambda_3|, |\lambda_3|)$. Then, $|PD| = R^{-1}|\Lambda|R$. For any vector $x = (x_1, x_2, x_3, x_4, x_5)^t$, $|PD|x = (R^{-1}|\Lambda|)z$, where $z = Rx$. Define $Y = \omega_1 x_2 + \omega_2 x_3 + \omega_3 x_4$. Then,

$$z = \begin{pmatrix} \frac{1}{\lambda_+ - \lambda_-} \left[\frac{\lambda_+ - (1-\alpha)q}{\rho\beta^2} x_1 + Y \right] \\ -z_1 + \frac{x_1}{\rho\beta^2} \\ z_3 \\ z_4 \\ (1-\gamma)x_1 + \frac{\gamma p}{T} x_5 \end{pmatrix}$$

where z_1 is the first element of the vector z. We note that z_3 and z_4 do not appear by themselves in subsequent formulas but rather in conjunction with other variables in the form

$$\omega_2 z_3 = \qquad -x_2 + \frac{1}{(\lambda_+ - q)(\lambda_- - q)}$$
$$\left[\frac{\alpha(u|\omega|^2 - q\omega_1)}{\rho} x_1 - (\beta^2\omega_1 - \alpha u q)Y \right]$$

$$\omega_2 z_4 = \qquad -x_4 + \frac{1}{(\lambda_+ - q)(\lambda_- - q)}$$
$$\left[\frac{\alpha(w|\omega|^2 - q\omega_3)}{\rho} x_1 - (\beta^2\omega_3 - \alpha w q)Y \right]$$

and

$$\omega_1 z_3 + \omega_3 z_4 = x_3 - \frac{1}{(\lambda_+ - q)(\lambda_- - q)}$$
$$\left[\frac{\alpha(v|\omega|^2 - q\omega_2)}{\rho} x_1 - (\beta^2\omega_2 - \alpha v q)Y \right]$$

Let

$$l_2 = \frac{\beta^2\omega_1 - \alpha u\lambda_+}{\lambda_+ - q}|\lambda_1|z_1$$
$$+ \frac{\beta^2\omega_1 - \alpha u\lambda_-}{\lambda_- - q}|\lambda_1|z_2 - (\omega_2 z_3)|\lambda_3|$$

$$l_3 = \frac{\beta^2\omega_2 - \alpha v\lambda_+}{\lambda_+ - q}|\lambda_1|z_1$$
$$+ \frac{\beta^2\omega_2 - \alpha v\lambda_-}{\lambda_- - q}|\lambda_1|z_2 + (\omega_1 z_3 + \omega_3 z_4)|\lambda_3|$$

$$l_4 = \frac{\beta^2\omega_3 - \alpha w\lambda_+}{\lambda_+ - q}|\lambda_1|z_1$$
$$+ \frac{\beta^2\omega_3 - \alpha w\lambda_-}{\lambda_- - q}|\lambda_1|z_2 - (\omega_2 z_4)|\lambda_3|$$

$$|PD|x = R^{-1}|\Lambda|z = \begin{bmatrix} \rho\beta^2(|\lambda_1|z_1 + |\lambda_2|z_2) \\ l_2 \\ l_3 \\ l_4 \\ \frac{T}{\gamma p}[(\gamma - 1)V_1 + |\lambda_3|z_5] \end{bmatrix}$$

where V_1 is the first element of this vector.

For a scalar viscosity (i.e., when σ is the spectral radius), the viscosity remains scalar after preconditioning. In this approach, differences of the Q_V variables, rather than those of the conserved variables, are added to each of the conservation equations.

The formulation used until now is nonconservative. Since we are also interested in solving transonic flows with shocks using this scheme, we present a conservative formulation of the artificial dissipation. By dropping the subscript V, we obtain

$$\Delta Q = \Delta t\Gamma \left\{ \frac{\partial F}{\partial x} + \frac{\partial G}{\partial y} + \frac{\partial H}{\partial z} + \left(\Gamma^{-1}\sigma(PA)\frac{\partial Q}{\partial x} \right)_x \right.$$
$$\left. + \left(\Gamma^{-1}\sigma(PB)\frac{\partial Q}{\partial y} \right)_y + \left(\Gamma^{-1}\sigma(PC)\frac{\partial Q}{\partial z} \right)_z \right\}$$

Even if σ is a scalar function, we must evaluate a matrix-vector product; as a result, the numerical effort is equivalent in complexity to a matrix-valued viscosity.

Following previous work on multigrid schemes for the Navier-Stokes equations, these artificial viscosity functions account for the ratio of the spectral radii in the different coordinate directions; that is,

$$\sigma(PA) = k^{(4)}\tilde{\sigma}(PA)\left[1 + \left(\frac{\rho(PB)}{\rho(PA)} \right)^\zeta + \left(\frac{\rho(PC)}{\rho(PA)} \right)^\zeta \right]$$

where $\tilde{\sigma}$ denotes the original viscosity function, ρ denotes the spectral radius, and $k^{(4)}$ is the artificial viscosity coefficient that corresponds to the fourth-difference terms. Similar relations are used to modify the artificial viscosity by cell aspect-ratio in the other coordinate directions. These scaling functions provide sufficient artificial dissipation for general curvilinear grids that contain high aspect-ratio cells.

In the previous section, a preconditioner was introduced that is dependent on the parameters α and δ. Because δ does not affect the eigenvalues of PA, it has no affect on the scalar artificial viscosity. Choosing $\alpha = 1$ reduces the largest eigenvalue, which also improves the condition number and decreases the artificial viscosity compared with $\alpha = 0$ case. We thus expect $\alpha = 1$ to slow the convergence compared with the case in which $\alpha = 0$, in spite of the fact that we have reduced the condition number. However, we expect that with $\alpha = 1$, the numerical accuracy will improve. For a matrix viscosity (or a Roe matrix), we expect similar but less pronounced dependence on α.

7.2 Preconditioned CUSP Scheme

In this section the use of specific dissipation schemes designed for shock capturing with a single-point shock structured is presented. More specifically, we describe the use of the CUSP scheme in combination with preconditioning.

We briefly review our previous description of CUSP for a Cartesian mesh. Extensions to a general mesh are straightforward.

$$\Delta w_c = \Delta t[\mathcal{L}_C + \mathcal{L}_{AD}], \qquad (7.4)$$

where w_c represents the conservative variables. $\mathcal{L}_C$ is the central difference approximation to the inviscid flux balance (possibly also including the physical viscous Navier-Stokes terms) while $\mathcal{L}_{AD}$ represents the artificial dissipation as a linear combination of second differences of the state vector w and F where F is the flux through a cell face.

$$\mathcal{L}_{AD} = - \left[\frac{d_{i+1/2,j} - d_{i-1/2,j}}{\Delta x} + \frac{d_{i,j+1/2} - d_{i,j-1/2}}{\Delta y} \right]$$

$$d_{i+1/2,j} = \frac{\nu c}{2}(w_{i+1,j} - w_{i,j}) + \frac{\beta}{2}(F_{i+1/2,j} - F_{i,j})|\vec{S}|$$

The dissipation coefficients, ν and β, can be defined such that a steady-state shock structure with a single interior point is obtained for shocks aligned with the cell face $(i+1/2,j)$. The particular form chosen for these parameters supports a clean shock capturing provided that the convective cell face Mach number is larger than 0.5. Moreover, it was shown that the eigenvalues of the artificial viscosity matrix are all equal to the convective cell face velocity for Mach numbers less than 0.5.

These observations yield sufficient guidance for the selection of appropriate preconditioning parameters for $CUSP$: First, the preconditioning matrix, P_c should be turned off within the shock structure so that the shock capturing properties of the $CUSP$ scheme are not perturbed. This is accomplished with the choice $\alpha = 0, M_0 = 0.5$ in (3.3). The nonconservative preconditioned $CUSP$ can then be written as

$$\Delta w_c = \Delta t \left[P_c \mathcal{L}_C + \mathcal{L}_{AD} \right], \tag{7.5}$$

Note, that we use conservative w_c variables and not (p, u, v, T) variables. The nonconservative form should give clean shocks if the post-shock Mach number is larger than 0.5. This is true, in particular, for transonic flows which create only moderate strength shocks. However, low post-shock Mach numbers can be observed for hypersonic flows. This is a partial motivation for considering a conservative form for the dissipation. This conservative is given by

$$\Delta w_c = \Delta t P_c \left[\mathcal{L}_C - ((P_c w)_{i+1/2,j} - (P_c w)_{i-1/2,j}) \right.$$
$$\left. - ((P_c w)_{i,j+1/2} - (P_c w)_{i,j-1/2}) \right]$$

This form is conservative for all Mach numbers. The shock capturing analysis holds for all cell faces where the local Mach number is larger than 0.5 and so $P_c = I$. Hence, this preconditioned scheme is consistent with the basic $CUSP$ scheme formulation. The disadvantage of the conservative formulation is that it is more expensive than the nonconservative form since P_c must be calculated at each of the neighbors $(i+1/2,j)$ etc. and a matrix-vector multiply is needed at each neighbor rather than just at the single point (i,j) for the nonconservative formulation.

8 Boundary Conditions

In many codes the boundary conditions in the far field are based on characteristic variables, even for viscous flow. Thus at inflow the incoming variables, corresponding to positive eigenvalues, are specified while the outgoing variables, corresponding to negative eigenvalues, are extrapolated. Once we change the time dependent equations we also change the characteristics of the system (though not the signs of the eigenvalues). Hence, it is necessary to also modify the boundary conditions for the preconditioned system. In addition, the original characteristic variables do not scale correctly as the Mach goes to zero and so we do not obtain well-posed boundary conditions in the limit.

Since the eigenvalues and eigenvectors are invariant under a similarity transform we shall use the (p, u, v, S) set of variables to determine the new eigenvalues and characteristic variables. We shall only consider the flow normal to the boundary which we take to be a line $x = const$. Thus, we need to use the characteristic variables we found above corresponding to the left eigenvectors of $P_0 A_0$. Flow variables at a boundary are then obtained by taking the values for R_1, R_2 and R_+ from upstream flow whereas downstream values are used for R_-. When we diagonalize $P_0 A_0 = RDR^{-1}$ the rows of R are the left eigenvectors of $P_0 A_0$ while the columns of R^{-1} are the right eigenvectors of $P_0 A_0$. Let variables with a subscript "b" denote the variables to be computed at the boundary. Variables with a subscript "a" denote the flow state approaching the boundary, i.e. specified at inflow, whereas variables with a subscript "l" are used for the flow leaving the boundary, e.g. extrapolated at inflow. Note, that this notation can be consistently used for inflow and outflow boundaries. The characteristic boundary conditions are expressed in (p, u, v, T) variables as:

$$p_b = \frac{\lambda_+ p_a - \lambda_- p_l}{\lambda_+ - \lambda_-} -$$
$$\frac{1}{\lambda_+ - \lambda_-} \left[u(1 - \alpha - \delta \frac{(\gamma - 1)\beta M_r^2}{a^2})(p_a - p_l) \right.$$
$$\left. - \rho\beta M_r^2(u_a - u_l) + \frac{\delta\rho\beta M_r^2 u}{T}(T_a - T_l) \right]$$

$$u_b = u_a - \frac{\lambda_+ - u(1 - \alpha)}{\rho\beta M^2}(p_b - p_a) \tag{8.1}$$

$$T_b = T_a + \frac{(\gamma - 1)T}{\gamma p}(p_b - p_a)$$

$$v_b = v_a - \frac{\alpha v}{\beta M^2 - \alpha u^2} \left[\frac{p_b - p_a}{\rho} + u(u_b - u_a) \right]$$

In generalized coordinates u and v are replaced by the normal and tangential components of the velocity relative to the boundary.

Computationally we observed that a simplified set of non-characteristic conditions yielded almost the same results as those of the characteristic boundary conditions. These are given by

$$v_b = v_a, \quad p_b = p_l, \quad u_b = u_a,$$

either $T_b = T_a$ or $\rho_b = \rho_a + \frac{p_b - p_a}{a^2}$ The simplified boundary condition has been used in all of the computations.

These boundary conditions are reasonable for exterior problems. However, for flow in interior regions one customarily specifies the total temperature, T_t, and total pressure, p_t, profiles together with the angle of the flow at inflow. At the exit a back pressure is usually given. Let

$$\frac{T_t}{T} = 1 + \frac{\gamma - 1}{2} M^2$$

$$\frac{p_t}{p} = \left\{ 1 + \frac{\gamma - 1}{2} M^2 \right\}^{\frac{\gamma - 1}{\gamma}}$$

$$\tan(\theta) = \frac{v}{u}$$

A reasonable boundary treatment for the inflow boundary is

$$p_b = p_l$$

$$T_b = T_t \left(\frac{p_b}{p_t} \right)^{\frac{\gamma - 1}{\gamma}}$$

$$u_b^2 = \frac{\frac{2\gamma}{\gamma - 1}(T_t - T_b)}{1 + \tan^2(\theta)}$$

$$v_b = u_b \tan(\theta)$$

At the exit we specify the back pressure and extrapolate u, v, T.

9   Implementation

We present the steps used to introduce preconditioning into a previously existing code based on Algorithm I. For simplicity we shall assume an explicit iteration scheme (e.g. Runge-Kutta) possibly augmented by implicit residual smoothing and multigrid.

1. Precondition the residual

   - Select a reasonable set of variables for the problem e.g. (p, u, v, T). We stress that other sets of variables are possible. This set is used for illustration only.

   - For nonconservative scalar artificial viscosity

     - multiply physical residual (inviscid and Navier-Stokes portions and physical forcing functions (if applicable)) by $\Gamma = P_T \frac{\partial w_4}{\partial w_1}$.
     - add scalar artificial viscosity in these new variables to the residual.

   - For conservative scalar viscosity

     - viscosity includes Γ^{-1} (see (7.3)). Hence, even a scalar viscosity requires a matrix-vector multiply.
     - add inviscid, viscous and artificial viscosity fluxes to obtain the total residual.
     - multiply this total residual by Γ.

2. Perform residual smoothing. At the end of the Runge-Kutta cycle restrict the residual to the next coarser grid. Thus both residual smoothing and restrictions are performed on the residuals in (p, u, v, T) variables.

3. Having calculated the total preconditioned residual in (p, u, v, T) multiply these by Δt. There are then two ways of updating the solution to the next Runge-Kutta stage.

   - One: multiply the above total residual by $\frac{\partial w_1}{\partial w_4}$ to get the residual in conservative variables. Add the new residual to the old conservative variables to get new conservative variables.
   - Two: Add the residuals in (p, u, v, T) variables to the old (p,u,v,T) variables to get (p, u, v, T) at the next Runge-Kutta stage. Calculate the conservative variables as nonlinear functions of (p, u, v, T).

4. Choose a new time step based on the preconditioning (3.4).

5. Modify the far field boundaries based on the new characteristics. If characteristic data is also used at physical boundaries (e.g. inflow to a nozzle) these should also be modified. Alternatively, specify or extrapolate (p, u, v, T) at the boundaries.

For Algorithm II (4) all operations including residual smoothing and interpolations between grids are done in terms of the $(p', \rho u, \rho v, E')$ variables. This scheme has the advantage that, in theory, setting

$M_\infty = 0$ would yield identical results to an incompressible code based on pseudo-compressibility. Algorithm I as presently implemented performs the residual smoothing and restriction in terms of (p, u, v, T) variables. However, restriction of the variables rather than residuals for the FAS algorithm is based on conservative variables. Similarly the prolongation from coarse to fine grids is based on conservative variables.

10   Jacobi Preconditioning

Consider the semi-discrete approximation

$$
w_t = \frac{f_{i+1/2,j} - f_{i-1/2,j}}{2\Delta x} + \frac{g_{i,j+1/2} - g_{i,j-1/2}}{2\Delta y}
$$
$$
+ \frac{1}{2}\left\{ \Delta_x(|A|\Delta_x w) + \Delta_y(|B|\Delta_y w) \right\}
$$

where Δ are appropriate finite difference operators. A and B are the Jacobians of f and g with respect to w respectively. Linearizing and taking Fourier transforms we get

$$
\begin{aligned}
G(\theta, \phi) &= i(A\sin(\theta) + B\sin(\phi)) \\
&- |A|(1 - \cos(\theta)) - |B|(1 - \cos(\phi))
\end{aligned}
$$

and so

$$
G(\pi, \pi) = -2|A| - 2|B| \tag{10.1}
$$

It is difficult to choose coefficients in a Runge-Kutta scheme that will damp all the eigenvalues of the matrix $G(\pi, \pi)$. Instead Allmaras [1, 2] and later Pierce-Giles [18] suggest replacing (10.1) by

$$
\begin{aligned}
\frac{(|A| + |B|)}{\rho(A) + \rho(B)} w_t &= \frac{f_{i+1/2,j} - f_{i-1/2,j}}{2\Delta x} \\
&+ \frac{g_{i,j+1/2} - g_{i,j-1/2}}{2\Delta y} \\
&+ \frac{1}{2}(\Delta_x(|A|\Delta_x w + \Delta_y(|B|\Delta_y w))
\end{aligned} \tag{10.2}
$$

where $\rho(A)$ is the spectral radius of A. (10.1) is now replaced by

$$
G(\pi, \pi) = -2\rho(A) - 2\rho(B)
$$

Hence, at the highest frequency G is essentially a scalar and so one can choose Runge-Kutta coefficients to damp this scalar quantity.

11   Preconditioning-squared

The Jacobi preconditioning (10.2) does nothing to improve the condition number of the system for low Mach numbers. On the other hand the low Mach number preconditioner (3.2) was not constructed to treat high frequencies errors. Hence, one can combine the two approaches and replace (10.2) by

$$
\begin{aligned}
2\frac{(|PA| + |PB|)}{\rho(PA) + \rho(PB)} w_t &= \\
P\left\{ \frac{f_{i+1/2,j} - f_{i-1/2,j}}{\Delta x} \right. &+ \frac{g_{i,j+1/2} - g_{i,j-1/2}}{\Delta y} \\
\left. + \Delta_x(P^{-1}|PA|\Delta_x w) \right. &\left. + \Delta_y(P^{-1}|PB|\Delta_y w) \right\}
\end{aligned}
$$

12   Time Dependent Problems

We consider two kinds of time dependent problems. For one, which is typified by acoustic problems, the requirements of accuracy and stability are approximately the same. Thus, the time restrictions required by stability of the scheme do not impose an unnecessary burden on the scheme as the same time steps are needed to resolve the changes in the time behavior of the scheme. For such problems explicit methods are the best. It is not clear how to combine this with preconditioning since the preconditioning that we have discussed until now destroys the time accuracy of the method.

A second time dependent problem is one where the time steps required by accuracy are much larger than those required by stability. One application is to forced motion where the time scale of the outside forces are much larger than the stability time step of an implicit scheme. For such problems it is useful to introduce an implicit method. However, to avoid time errors associated with large time steps it is advisable to use a fully implicit method and not an ADI type algorithm. However, this leads to the necessity of inverting large non-sparse matrices for a linear problem plus the linearization difficulties of a non-linear problem. One alternative is to consider the implicit method as a steady state problem for the next time step. Thus, any steady solver can be used to go from time t to time $t + \delta t$. It is only required that the steady state solver converge within a short enough time that this approach is more efficient than a straightforward explicit method. Thus, for example, if accuracy requirements lead to a CFL of 100 relative to an explicit method then this approach is efficient if we can solve for the steady state in 50 iterations or less. In order for this to be feasible it is necessary to start with a good initial guess which can be obtained by extrapolation from previous time data. In addition we only want the steady state solver to converge to the discretization level of the time dependent implicit scheme. Any attempt to solve the steady state problem to lower residuals is a waste of time.

As an example consider the simple time dependent problem

$$\frac{\partial w}{\partial t} + \frac{\partial f}{\partial x} = 0$$

We discretize this equation by the backward Euler method which is only first order in time.

$$w^{n+1} + \delta t \left(\frac{\partial f}{\partial x}\right)^{n+1} = w^n$$

To solve this problem we introduce a pseudo-time variable τ and solve

$$w_\tau + \frac{w}{\delta t} + \frac{\partial f}{\partial x} = \frac{w^n}{\delta t}$$

until a steady state in τ is reached. At this level we can now introduce preconditioning into the system and replace w_τ by $P^{-1}w_\tau$ as before.

13 Differential Preconditioners

We consider the linearized equations with constant coefficients. We first rewrite (6.1) in differential form. The Euler equations can be written as

$$w_t + \mathcal{L}w = 0 \tag{13.1}$$

with $w = (p, u, v, S)^t$. Define

$$Q = u\partial_x + v\partial_y$$

Since, all coefficients are assumed constant Q commutes with ∂_x and ∂_y and

$$\mathcal{L} = \begin{pmatrix} Q & \rho c^2 \partial_x & \rho c^2 \partial_y & 0 \\ \frac{1}{\rho}\partial_x & Q & 0 & 0 \\ \frac{1}{\rho}\partial_y & 0 & Q & 0 \\ 0 & 0 & 0 & Q \end{pmatrix} \tag{13.2}$$

Let

$$D = Q^2 - c^2(\partial_x^2 + \partial_y^2). \tag{13.3}$$

We now replace (13.1) by the preconditioned system

$$w_t + \mathbf{P_D}\mathcal{L}w = 0 \tag{13.4}$$

with

$$\mathbf{P_D} = \begin{pmatrix} Q^2 & -\rho c^2 \partial_x Q & -\rho c^2 \partial_y Q & 0 \\ -\frac{1}{\rho}\partial_x & Q^2 - c^2 \partial_y^2 & c^2 \partial_x \partial_y & 0 \\ -\frac{1}{\rho}\partial_y & c^2 \partial_x \partial_y & Q^2 - c^2 \partial_x^2 & 0 \\ 0 & 0 & 0 & D \end{pmatrix}$$

One can then verify that

$$\mathbf{P_D}\mathcal{L} = QDI, \qquad \mathbf{P}_D^{-1} = D^{-1}Q^{-1}\mathcal{L}$$

One can replace the D in the lower right corner of $\mathbf{P}_D$ by the identity matrix. Then $\mathbf{P_D}\mathcal{L}$ is not the identity matrix but is still a diagonal matrix. We can use simpler matrices than $\mathbf{P}_D$ by considering congruent transformations. Consider the symmetrizing variables $(\frac{dp}{\rho c}, u, v, S)$, then

$$\mathcal{L} = \begin{pmatrix} Q & c\partial_x & c\partial_y & 0 \\ c\partial_x & Q & 0 & 0 \\ c\partial y & 0 & Q & 0 \\ 0 & 0 & 0 & Q \end{pmatrix}$$

Let,

$$\mathbf{P_E} = \begin{pmatrix} Q & -c\partial_x & -c\partial_y & 0 \\ 0 & 1 & 0 & 0 \\ 0 & 0 & 1 & 0 \\ 0 & 0 & 0 & 1 \end{pmatrix}$$

$$\mathbf{P_E^t} = \begin{pmatrix} Q & 0 & 0 & 0 \\ -c\partial_x & 1 & 0 & 0 \\ -c\partial_y & 0 & 1 & 0 \\ 0 & 0 & 0 & 1 \end{pmatrix}$$

then

$$\mathbf{P_E}\mathcal{L}\mathbf{P_E^t} = \begin{pmatrix} DQ & 0 & 0 & 0 \\ 0 & Q & 0 & 0 \\ 0 & 0 & Q & 0 \\ 0 & 0 & 0 & Q \end{pmatrix}$$

so we have diagonalized $\mathcal{L}$ by a congruent transformation. But,

$$\mathbf{P_E}\mathcal{L}\mathbf{P_E}^t = \mathbf{P_E^{t^{-1}}}(\mathbf{P_E^t}\mathbf{P_E}\mathcal{L})\mathbf{P_E},$$

so the congruent transform is similar to a preconditioning with a positive definite matrix $\mathbf{P_E^t}\mathbf{P_E}$. Alternatively, $(\mathbf{P_E^t}\mathbf{P_E})\mathcal{L}$ is similar to a diagonal matrix.

$$\mathbf{P_E^t}\mathbf{P_E} = \begin{pmatrix} Q & -c\partial_x & -c\partial_y & 0 \\ -c\partial_x & 1 + c^2 \partial_x^2 & c^2 \partial_x \partial_y & 0 \\ -c\partial_y & c^2 \partial_x \partial_y & 1 + c^2 \partial_y^2 & 0 \\ 0 & 0 & 0 & 1 \end{pmatrix}$$

Note that $\mathbf{P_E^t}\mathbf{P_E}$ looks similar to $\mathbf{P_D}$ but is not identical. $\mathbf{P_E^t}\mathbf{P_E}$ has fewer derivatives along the identical but $\mathbf{P_E^t}\mathbf{P_E}\mathcal{L}$ is only similar to a diagonal matrix while $\mathbf{P_D}L$ is diagonal and even a scalar differential operator multiplying the identity matrix. These transformations are independent of the flow regime as long as the preconditioner is nonsingular.

It remains to show that $\mathbf{P}$ is nonsingular. We have four eigenvalues and corresponding eigenfunctions. As usual the entropy wave decouples. For this wave $\mathbf{P}$ has an eigenvalue D and an eigenfunction $(0,0,0,1)$. For the shear wave $\mathbf{P}$ has an eigenvalue D and the eigenvector is $(v_1, v_2, v_3, 0)$ where

$$Dv_1 = 0$$
$$D\left(\frac{\partial v_2}{\partial x} + \frac{\partial v_3}{\partial y}\right) = 0$$

The other two 'acoustic' eigenvalues of $\mathbf{P}$ are $Q^2 \pm cQ\sqrt{\partial_x^2 + \partial_y^2}$ and the eigenvectors satisfy the pseudo-differential equation

$$\left[c(\partial_x^2 + \partial_y^2) \pm Q\sqrt{\partial_x^2 + \partial_y^2}\;\right]\left(\frac{\partial v_3}{\partial x} - \frac{\partial v_2}{\partial y}\right) = 0$$
$$\mp\sqrt{\partial_x^2 + \partial_y^2} = \rho c\left(\frac{\partial v_2}{\partial x} + \frac{\partial v_3}{\partial y}\right)$$

We therefore have to show that the eigenvalues are all nonzero so that $\mathbf{P}$ is nonsingular. The operator D is just the potential operator i.e. for any variable w

$$Dw = (u^2 - c^2)w_{xx} + 2uvw_{xy} + (v^2 - c^2)w_{yy}$$

For subsonic flow this is an elliptic operator and so invertible. For supersonic flow D is a hyperbolic operator . Similarly, Q is a hyperbolic operator denoting convection along a streamline. Thus, given appropriate boundary conditions this too should be invertible. At a stagnation point Q is singular and so it is necessary to limit the values of u and v in the definition of Q so that they do not become too small in a neighborhood of the stagnation point. A similar smoothing is needed near the sonic line.

With residual smoothing and $\mathbf{P_D}$ we have increased the order of the system and so changed the number of boundary conditions needed for the equation to be well posed. To avoid this difficulty we do not solve the equation (13.4). Instead these preconditioners are used as a post processor for the usual Euler or Navier-Stokes equations. Thus, at each time step we calculate a residual based on one's favorite scheme. This gives a predicted value of the change in time, $\Delta w_{predicted}$. We also update the boundary conditions for the standard fluid dynamic equations. We then operate on Δw with $\mathbf{P}$ with the boundary condition that $\Delta w_{corrected} = 0$, i.e. we don't change the

boundary values calculated by the predictor. When we reach a steady state for the fluids equations we are solving $\mathbf{P}\Delta w_{corrected} = 0$ with zero boundary conditions. Since $\mathbf{P}$ is invertible $\Delta w = 0$, i.e. we preserve the steady state. Thus, in essence we are imposing the fluid dynamic boundary conditions between the $\mathbf{P}$ operator and the $\mathcal{L}$ operator.

We note that if the partial derivatives are all replaced by central differences then everything goes over since it is based only a formal multiplication and uses no properties of the derivative. Hence, beginning with a central difference approximation to the constant coefficient multidimensional Euler equations we can precondition it so that we have a set of scalar equations. This applied equally well for any aspect ratio. Hence, we need consider high aspect ratio problems only for the scalar equation. This use of the differential preconditioner is similar to the explicit/implicit residual averaging previously considered except that now the explicit portion is a matrix involving second differences rather than the scalar second differences considered before.

14   Canonical Decomposition

We shall restrict our analysis to the linear case with constant coefficients. Our typical system is either a steady state problem $\mathcal{L}u = 0$ or the unsteady problem $u_t + \mathcal{L}u = 0$. When diagonalizing such a system we distinguish between three types of diagonalization.

1. A similarity transformation is introduced so that $Q\mathcal{L}Q^{-1}$ is diagonal. In this case if we define $\mathbf{v} = Q\mathbf{u}$ then we have a diagonal system for the variables $\mathbf{v}$. Hence, this is equivalent to a change of variables.

2. We precondition the unsteady problem to get $\mathbf{P}^{-1}u_t + \mathcal{L}u = 0$ where $\mathbf{P}$ is positive definite and $\mathcal{L}$ is symmetric. We now assume a decomposition $\mathbf{P} = TS$ and define $v = T^{-1}u$. Then the equation becomes $v_t + (S\mathcal{L}T)v = 0$. A particular case is when $S = T^*$. This is convenient since every positive matrix $\mathbf{P}$ has the decomposition $\mathbf{P} = TT^*$. In this case we have a congruence transformation of $\mathcal{L}$. Hence, preconditioning is equivalent to all congruence transformation plus a similarity transformation.

3. In the above two cases we have implicitly assumed that the transformation Q contains only constant coefficients and not differential operators. Hence, we are not changing the order of the differential system. A different approach is to

allow the transformations to depend on derivatives. This is equivalent to replacing the first order system by a set of higher order equations. This again can be done via a change of variables, i.e. a similarity transform or else via a preconditioning i.e. a congruence transform.

In all these cases we demand that the steady state solution be independent of the transformations used so that either a similarity or congruence transformation may be allowed.

We first consider equations in one space direction plus time so that $\mathcal{L} = A\partial_x$. In this case any strongly hyperbolic $(m \times m)$ system of the form

$$\partial_t \mathbf{u} + A\partial_x \mathbf{u} = 0$$

is reducible to m scalar convection equations of the form

$$\partial_t(\ell_k \cdot \mathbf{u}) + \lambda_k \partial_x(\ell_k \cdot \mathbf{u}) = 0$$

where ℓ_k is a left eigenvector of A, and λ_k is the corresponding eigenvalue. This follows since for a strongly hyperbolic system A is diagonalizable. Thus, the canonical form is simply the familiar linear advection equation

$$\partial_t u + a\partial_x u = 0$$

In two dimensions we consider first the steady-state equations

$$A\partial_x u + B\partial_y u = 0 \tag{14.1}$$

When A and B are symmetric it is known that this equation can be diagonalized by a congruence transformation if and only if some real linear combination of A and B is positive definite. In that case the canonical form is again linear advection

$$a\partial_x u + b\partial_y u = 0. \tag{14.2}$$

If the above condition is not met, then we shall show that some of the eigenvalues of $A^{-1}B$ appear in conjugate pairs, and a second canonical form appears which is a (2×2) system, equivalent under scaling and axis transformations to the Cauchy-Riemann equations

$$\begin{pmatrix} 1 & 0 \\ 0 & -1 \end{pmatrix} \partial_x \begin{pmatrix} u \\ v \end{pmatrix} - \begin{pmatrix} 0 & 1 \\ 1 & 0 \end{pmatrix} \partial_y \begin{pmatrix} u \\ v \end{pmatrix} = 0.$$

The general two-dimensional system can then be written in a form that is block diagonal, with a mixture of scalar and (2×2) blocks. When the steady solution is sought by marching in a time-like variable, we show that a preconditioning matrix P can always

be found, such that this block structure is preserved in the transient solution of

$$\mathbf{P}^{-1}\partial_t \mathbf{u} + A\partial_x \mathbf{u} + B\partial_y \mathbf{u} = 0$$

The scalar blocks have then the canonical form

$$\partial_t u + a\partial_x u + b\partial_y u = 0. \tag{14.3}$$

The (2×2) blocks, although they behave elliptically in the steady state, behave hyperbolically in the transient phase, and the associated variables follow a wave equation

$$\frac{\partial^2 \psi}{\partial t^2} = \frac{\partial^2 \psi}{\partial x^2} + \frac{\partial^2 \psi}{\partial y^2} \tag{14.4}$$

We consider (14.3) and (14.4) as our basic canonical forms. Although both of these equations are hyperbolic, considered as time-dependent problems, only (14.3) is associated with any distinct direction (namely $dy/dx = b/a$). By contrast, (14.4), during the transient phase, is associated with characteristic surfaces, $(x - x_0)^2 + (y - y_0)^2 - (t - t_0)^2 = const$, rather than with individual directions. In the steady state, (14.3) is a hyperbolic equation while (14.4) reduces to an elliptic equation, Laplace's equation, and so has no uniquely distinguished directions.

In three dimensions, we have three canonical forms, which are

$$\frac{\partial \psi}{\partial t} + a\frac{\partial \psi}{\partial x} + b\frac{\partial \psi}{\partial y} + c\frac{\partial \psi}{\partial z} = 0,$$

and (3×3) systems equivalent to either

$$\frac{\partial^2 \psi}{\partial t^2} = \frac{\partial^2 \psi}{\partial x^2} + \frac{\partial^2 \psi}{\partial y^2} + \frac{\partial^2 \psi}{\partial z^2},$$

or

$$\left(\frac{\partial}{\partial t} - \frac{\partial}{\partial x} \right)^2 \psi = \frac{\partial^2 \psi}{\partial y^2} + \frac{\partial^2 \psi}{\partial z^2},$$

15 The Two dimensional Case

15.1 General Steady-state Problems

We consider

$$A\partial_x u + B\partial_y u = 0 \tag{15.1}$$

with A and B symmetric. This is a reasonable assumption since many physical systems can be described by symmetric hyperbolic equations.

According to a result of Au-Yeung [4] this equation can be diagonalized by a congruence transformation if a real linear combination of A and B is positive definite. In the present context, this condition has a physical interpretation. Suppose that we add a time-dependent term $\partial_t u$ to (15.1). Then the speeds with which waves propagate in a direction θ are the eigenvalues of $A\cos\theta + B\sin\theta$. If this matrix is positive definite for $\theta = \theta^*$, then the domain of influence of a disturbance at the origin lies in the half-plane $x\cos\theta^* + y\sin\theta^* > 0$. Therefore the domain of influence does not enclose the origin, and tangents to it from the origin may be drawn. These are the characteristic directions for the steady solution. This situation arises whenever the flow speed is greater than the speed of any signal relative to the medium (supersonic case). Thus, the two dimensional supersonic flow equations are always diagonalizable.

To deal with the subsonic case, we begin by attempting a formal diagonalization of (15.1). We premultiply the equation by a matrix S, and effect a change of variable $Qv = u$, so that

$$(SAQ)\partial_x v + (SBQ)\partial_y v = 0. \tag{15.2}$$

We require that SAQ, SBQ be representable as

$$SAQ = \Lambda_A^{\#},$$
$$SBQ = \Lambda_B^{\#},$$

where $\Lambda_A^{\#}, \Lambda_B^{\#}$ denote matrices having the same block diagonal structure. (Note that matrices sharing such a structure form a closed group under multiplication and inversion.) Then

$$(\Lambda_A^{\#})^{-1}\Lambda_B^{\#} = (Q^{-1}A^{-1}S^{-1})(SBQ) = Q^{-1}A^{-1}BQ$$

so that Q must represent a similarity transformation that block diagonalises $A^{-1}B$. (If A happens to be singular we may diagonalize $B^{-1}A$ instead). At the present stage, the matrix S is still arbitrary. We first attempt a complete diagonalization.

$$A^{-1}B = R\Lambda L \qquad \text{where} \qquad R = L^{-1}$$

If this succeeds, we can take $S = LA^{-1}$ and $Q = R$, to derive

$$\partial_x(Lu) + \Lambda\partial_y(Lu) = 0.$$

representing a set of characteristic equations for the characteristic variables $v_k = \ell_k \cdot u$ propagating along directions $dy/dx = \lambda_k$. This is a similarity transformation that reduces A to the identity matrix and diagonalizes B.

Suppose this fails in that

$$\det(A^{-1}B - \lambda I) = 0$$

has a pair of conjugate roots

$$\lambda_{1,2} = \lambda_R \pm i\lambda_I.$$

Then we will also find conjugate eigenvectors

$$\mathbf{r}_{1,2} = \mathbf{r}_R \pm i\mathbf{r}_I,$$
$$\ell_{1,2} = \ell_R \pm i\ell_I.$$

That is,

$$A^{-1}B = \begin{bmatrix} \cdot & \cdot & & & \\ \cdot & \cdot & & & \\ r_1 & r_2, & \cdot & \cdot & \cdot \\ \cdot & \cdot & & & \\ \cdot & \cdot & & & \end{bmatrix}$$

$$\begin{bmatrix} \lambda_1 & & & \\ & \lambda_2 & & \\ & & \cdot & \\ & & & \cdot \end{bmatrix}\begin{bmatrix} \cdot & \cdot & \ell_1 & \cdot & \cdot \\ \cdot & \cdot & \ell_2 & \cdot & \cdot \\ & & \cdot & & \\ & & \cdot & & \end{bmatrix}$$

Now introduce the complex matrices

$$J = \frac{1}{\sqrt{2}}\begin{bmatrix} 1 & 1 & & \\ i & -i & & \\ & & \sqrt{2} & \\ & & & \cdot \end{bmatrix},$$

$$J^T = \frac{1}{\sqrt{2}}\begin{bmatrix} 1 & i & & \\ 1 & -i & & \\ & & \sqrt{2} & \\ & & & \cdot \end{bmatrix}$$

and rewrite the factorization

$$A^{-1}B = (RJ^T)((J^T)^{-1}\Lambda J^{-1})(JL) = R^{\#}\Lambda^{\#}L^{\#}$$

Then it can be verified that $R^{\#}, L^{\#}$ are real matrices, and that $\Lambda^{\#}$ is the real symmetric matrix

$$\Lambda^{\#} = (J^T)^{-1}\Lambda J^{-1} = \begin{bmatrix} \lambda_R & \lambda_I & & \\ \lambda_I & -\lambda_R & & \\ & & \lambda_3 & \\ & & & \cdot \end{bmatrix}$$

Now we write the governing equations in the form (15.2) with $S = (R^{\#})^{-1}A^{-1}$ and $Q = (L^{\#})^{-1}$; the result, after partial simplification, is

$$(R^{\#})^{-1}(L^{\#})^{-1}\partial_x v + \Lambda^{\#}\partial_y v = 0$$

Here,

$$\begin{aligned}
(R^{\#})^{-1}(L^{\#})^{-1} &= (L^{\#}R^{\#})^{-1} \\
&= (JLRJ^T)^{-1} \\
&= (JJ^T)^{-1} \\
&= \begin{bmatrix} 1 & 0 & \\ 0 & -1 & \\ & & 1 \\ & & & \cdot \end{bmatrix}.
\end{aligned}$$

Thus, when expressed in terms of variables $\mathbf{v} = L^{\#}\mathbf{u}$, the equations decouple into $(m-2)$ advection equations, and the 2×2 symmetric elliptic system

$$\begin{pmatrix} 1 & 0 \\ 0 & -1 \end{pmatrix} \partial_x \begin{pmatrix} u \\ v \end{pmatrix} +$$

$$\begin{pmatrix} \lambda_R & \lambda_I \\ \lambda_I & -\lambda_R \end{pmatrix} \partial_y \begin{pmatrix} u \\ v \end{pmatrix} = 0. \tag{15.3}$$

If there is more than one conjugate pair of eigenvalues, the treatment above can easily be extended by adding additional (2×2) blocks to the matrix J, so we have

<u>THEOREM</u> Any $(m \times m)$ first order constant coefficient hyperbolic system in two dimensions decomposes into $(m - 2k)$ convection equations and k independent (2×2) symmetric elliptic systems.

<u>REMARK</u> The $(m \times 2)$ blocks that appear in $R^{\#}$ provide basis vectors for the subspaces inhabited by the elliptic subproblems; the $(2 \times m)$ blocks that appear in $L^{\#}$ are operators that project an arbitrary vector into these subspaces.

15.2 Steady Euler Equations

Take the vector of unknowns to be $d\mathbf{u} = (\frac{dp}{\rho c}, du, dv, dp - c^2 d\rho)^T$, and adopt coordinates aligned with the flow direction. Then A, B are the symmetric matrices

$$A = \begin{bmatrix} M & 1 & 0 & 0 \\ 1 & M & 0 & 0 \\ 0 & 0 & M & 0 \\ 0 & 0 & 0 & M \end{bmatrix}, B = \begin{bmatrix} 0 & 0 & 1 & 0 \\ 0 & 0 & 0 & 0 \\ 1 & 0 & 0 & 0 \\ 0 & 0 & 0 & 0 \end{bmatrix}$$

We then find that the equation $A^{-1}B = R\Lambda L$ becomes

$$A^{-1}B = \begin{bmatrix} 0 & 0 & \frac{M}{M^2-1} & 0 \\ 0 & 0 & \frac{-1}{M^2-1} & 0 \\ \frac{1}{M} & 0 & 0 & 0 \\ 0 & 0 & 0 & 0 \end{bmatrix} =$$

$$\begin{bmatrix} M & M & 0 & 0 \\ -1 & -1 & 1 & 0 \\ \sqrt{M^2-1} & -\sqrt{M^2-1} & 0 & 0 \\ 0 & 0 & 0 & 1 \end{bmatrix}$$

$$\begin{bmatrix} \frac{1}{\sqrt{M^2-1}} & 0 & 0 & 0 \\ 0 & \frac{-1}{\sqrt{M^2-1}} & 0 & 0 \\ 0 & 0 & 0 & 0 \\ 0 & 0 & 0 & 0 \end{bmatrix}$$

$$\begin{bmatrix} \frac{1}{2M} & 0 & \frac{1}{2\sqrt{M^2-1}} & 0 \\ \frac{1}{2M} & 0 & \frac{-1}{2\sqrt{M^2-1}} & 0 \\ \frac{1}{M} & 1 & 0 & 0 \\ 0 & 0 & 0 & 1 \end{bmatrix}$$

For supersonic flow, the transformation $\mathbf{v} = L\mathbf{u}$ produces four real characteristic equations, which are, in primitive variables,

$$\partial_x S = 0 \quad dS = dp - a^2 d\rho$$
$$\partial_x H = 0 \quad dH = dp + \frac{\rho}{2}(u^2)$$
$$(\beta\partial_x + \partial_y)(\beta dp + \rho u\, dv) = 0$$
$$(\beta\partial_x - \partial_y)(\beta dp - \rho u\, dv) = 0$$

where $\beta^2 = M^2 - 1$.

If $M^2 < 1$, two of these characteristic equations break down. We may then have recourse to the alternative factorization $R^{\#}\Lambda^{\#}L^{\#}$, which is

$$A^{-1}B = \begin{bmatrix} 0 & 0 & \frac{M}{M^2-1} & 0 \\ 0 & 0 & -\frac{1}{M^2-1} & 0 \\ \frac{1}{M} & 0 & 0 & 0 \\ 0 & 0 & 0 & 0 \end{bmatrix}$$

$$= \begin{bmatrix} \sqrt{2}M & 0 & 0 & 0 \\ -\sqrt{2} & 0 & 1 & 0 \\ 0 & -\sqrt{2(1-M^2)} & 0 & 0 \\ 0 & 0 & 0 & 1 \end{bmatrix}$$

$$\begin{bmatrix} 0 & \frac{-1}{\sqrt{1-M^2}} & 0 & 0 \\ \frac{-1}{\sqrt{1-M^2}} & 0 & 0 & 0 \\ 0 & 0 & 0 & 0 \\ 0 & 0 & 0 & 0 \end{bmatrix}$$

$$\begin{bmatrix} \frac{1}{\sqrt{2}M} & 0 & 0 & 0 \\ 0 & 0 & \frac{1}{\sqrt{2(1-M^2)}} & 0 \\ \frac{1}{M} & 1 & 0 & 0 \\ 0 & 0 & 0 & 1 \end{bmatrix}$$

From this factorization we recover the same streamwise characteristic equations as in the supersonic case. It is interesting to note that the entropy characteristic corresponds to a shared eigenvector of the matrices A, B, whereas the enthalpy characteristic does not. The two acoustic characteristics collapse into the elliptic system deriving from (15.3), which is

$$\begin{pmatrix} 1 & 0 \\ 0 & -1 \end{pmatrix} \partial_x \begin{pmatrix} \frac{\beta_*}{\rho u} dp \\ dv \end{pmatrix}$$

$$+ \begin{pmatrix} 0 & \frac{-1}{\beta_*} \\ \frac{-1}{\beta_*} & 0 \end{pmatrix} \partial_y \begin{pmatrix} \frac{\beta_*}{\rho u} dp \\ dv \end{pmatrix} = 0. \tag{15.4}$$

where $\beta_*^2 = 1 - M^2$.

These are the Cauchy-Riemann equations written in Prandtl-Glauert coordinates $(x, y/\beta_*)$. Hence, we can express this as two scalar second order elliptic equations.

This decomposition differs from that found by Ta'asan He employed, as canonical variables, (S, H, u, v) and his decomposition was not completely block diagonal, containing some additional coupling terms. The variables which emerge from the present treatment are (S, H, p, v) and there is no coupling between blocks.

15.3 Preconditioning

Suppose that the time-dependent two-dimensional problem obeys

$$\partial_t u + A \partial_x u + B \partial_y u = 0 \tag{15.5}$$

where A and B are symmetric. To find a decomposition of this equation as it stands would require a three-dimensional treatment. However, there are occasions when only the steady state is of interest, but it is convenient to compute it by following some time-like evolution. In such cases (15.5) may be replaced (sometimes advantageously [25, 26, 30]) by

$$\mathbf{P}^{-1} \partial_t u + A \partial_x u + B \partial_y u = 0 \tag{15.6}$$

where P is a 'preconditioning matrix'. To retain well-posedness, P must be positive definite. In this section we point out that P can always be chosen to preserve the canonical block-diagonal form of the steady state equations.

As in the steady case, premultiply (15.6) by $(R^\#)^{-1} A^{-1}$ and make the change of variable $\mathbf{v} = L^\# \mathbf{u}$; this gives

$$(R^\#)^{-1} A^{-1} \mathbf{P}^{-1} (L^\#)^{-1} \partial_t v$$
$$+ (J J^T)^{-1} \partial_x v + \Lambda^\# \partial_y v = 0.$$

For the decomposition to be preserved, it is necessary and sufficient that

$$(R^\#)^{-1} A^{-1} \mathbf{P}^{-1} (L^\#)^{-1} = D^\#$$

where $D^\#$ is block diagonal of the same form as the matrices that multiply the space derivatives. Solving for $\mathbf{P}$ gives

$$\mathbf{P} = (R^\# A D L^\#)^{-1}.$$

where the only constraint on $D^\#$ is that it shall be positive definite to ensure well-posedness.

To apply this analysis to the Euler equations we take

$$D^\# = \begin{pmatrix} P & Q & 0 & 0 \\ R & S & 0 & 0 \\ 0 & 0 & \frac{1}{M} & 0 \\ 0 & 0 & 0 & \frac{1}{M} \end{pmatrix}.$$

This assigns the fluid velocity u to each of the scalar waves. We require $\Delta = PS - QR > 0$ for positive definiteness. After carrying through the required operations we find that

$$\mathbf{P}^{-1} = \begin{pmatrix} \frac{1}{M^2} - P\frac{\beta_*^2}{M} & \frac{1}{M} & Q\beta_* & 0 \\ \frac{1}{M} & 1 & 0 & 0 \\ R\beta_* & 0 & -SM & 0 \\ 0 & 0 & 0 & 1 \end{pmatrix}.$$

and also

$$\mathbf{P} = \begin{pmatrix} -\frac{S}{\Delta}\frac{M}{\beta_*^2} & \frac{S}{\Delta}\frac{1}{\beta_*^2} & -\frac{Q}{\Delta}\frac{1}{\beta_*} & 0 \\ \frac{S}{\Delta}\frac{1}{\beta_*^2} & 1 - \frac{S}{\Delta}\frac{1}{M\beta_*^2} & \frac{Q}{\Delta}\frac{1}{M\beta_*} & 0 \\ -\frac{R}{\Delta}\frac{1}{\beta_*} & \frac{R}{\Delta}\frac{1}{M\beta_*} & -\frac{P}{\Delta}\frac{1}{M} & 0 \\ 0 & 0 & 0 & 1 \end{pmatrix}.$$

This is the most general preconditioner that preserves the canonical decomposition. It may be noted that $\det P = (\beta_*^2 \Delta)^{-1}$. The preconditioner of van Leer et. al. [30] corresponds to the special case $P = S = -1/(M\beta_*)$, $Q = R = 0$, $\Delta = 1/(M^2\beta_*^2)$.

It is still an open question which choice of $\mathbf{P}$ leads to the fastest and most robust convergence to a steady state for a numerical code. This is principally because it is not easy to translate these qualities into mathematical terms. Here, we will identify robustness with the behavior of the solutions as $M \to 0$. We identify the rate of convergence with the condition number i.e. the ratio of the largest and smallest propagation speeds in the transient solution. We wish this condition number to be chosen as small as possible. [1] We will also insist on symmetry requirements for well-posedness.

When the preconditioned equations are written in Cartesian coordinates not aligned with the flow, terms may appear which depend on the flow direction as the flow velocity vanishes. A necessary and sufficient condition to avoid this phenomenon is that $P_{22} = P_{33}$, which implies

$$\beta_*^2 (M\Delta + P) = S. \tag{15.7}$$

The wavespeeds in the direction θ are given by

$$\det[A \cos\theta + B \sin\theta - P^{-1}\lambda]$$

which gives a twice repeated root $\lambda = M\cos\theta$, together with the quadratic term

$$\Delta\beta^2\lambda^2 + [(P-S)\beta^2 \cos\theta + (Q+R)\beta \sin\theta]\lambda$$
$$= \beta^2 \cos^2\theta + \sin^2\theta.$$

---

[1] In practice life is not this simple. Choosing different preconditioners also changes the "artificial" viscosity associated with the scheme which in turn affects the rate of convergence. Hence, in applications it is not always the lowest condition number that gives the quickest convergence [27, 28].

To keep this symmetrical with respect to the streamline, we must take $Q + R = 0$, and we consider first the case $Q = R = 0 \Rightarrow \Delta = AD$. The wavespeeds in the x-direction are then $1/P$ and $-1/S$, and in the y-direction $/pm1/(\beta_* \sqrt{PS})$. It can be shown that the best condition number is obtained from the preconditioner of van Leer *et. al.*, but unfortunately this does not satisfy the condition (15.7) at low Mach numbers. We require, in fact, that as $M \to 0$, then

$$(S + 1/M)(P - 1/M) = -1/M^2.$$

This is the graph of a hyperbola in the SP plane. We are only concerned with the branch that passes through the origin (the other branch leads to imaginary wavespeeds), and only with the half of that branch for which P, S are both negative (the other half does not make P positive definite). The condition number based solely on the 'acoustic' waves is

$$K = \max(S/P, P/S)$$

which is unity at the origin and degrades away from it. Taking into account that the flow velocity is M, and that $|P| > |S|$, we actually have

$$K = \max\left(\frac{S}{P}, \frac{P}{S}, \frac{1}{MS}, MP\right).$$

The point on the hyperbola that minimizes this is

$$P = \frac{-1}{M} \quad S = \frac{-1}{2M}$$

leading to K=2.

Of course, the preconditioner does not have to satisfy (15.7) away from $M = 0$, and can be blended in with the van Leer *et.al.* preconditioner, by some such choice as

$$P = \frac{-1}{M\beta_*}, \; S = \frac{-1}{M * \beta_*(1 + \beta_*)}$$

which leads to the simple preconditioner

$$\mathbf{P} = \begin{pmatrix} \frac{M^2}{\beta_*} & \frac{-M}{\beta_*} & 0 & 0 \\ \frac{-M}{\beta_*} & \frac{1+\beta_*}{\beta_*} & 0 & 0 \\ 0 & 0 & \beta_*(1+\beta_*) & 0 \\ 0 & 0 & 0 & 1 \end{pmatrix},$$

$$\mathbf{P}^{-1} = \begin{pmatrix} \frac{1+\beta_*}{M^2} & \frac{1}{M} & 0 & 0 \\ \frac{1}{M} & 1 & 0 & 0 \\ 0 & 0 & \frac{1}{\beta_*(1+\beta_*)} & 0 \\ 0 & 0 & 0 & 1 \end{pmatrix}.$$

Rather strikingly, these are identical with those given by van Leer *et.al.*, except for the factors $(1 + \beta_*)$ in the $()_{33}$ elements. Further details can be found in [19].

16 Three Dimensions

16.1 General Steady-state Problems

The most common block diagonal forms in three-dimensional problems appear to have blocks that are either scalar or else (3×3). The possibility of (2×2) blocks is ruled out, apart from exceptional cases, by an anlysis of Gilquin, Laurens and Rosier [8], who considered the general (2×2) symmetric system in multi-dimensions. They only require that the system be strongly hyperbolic (i.e. combinations of the coefficient matrices can always be diagonalized). They then prove that such systems can always be decomposed into a convective part (possibly with speed zero) and a *two dimensional* elliptic part. Therefore, (2×2) blocks are incapable of representing three-dimensional elliptic systems.

Due to the complexity of the general three dimensional case we shall only consider the special case of the linearized Euler equations.

16.2 Subsonic

We consider the preconditioned subsonic equations for three dimensions in streamwise coordinates. We shall analyze the specific preconditioning suggested by van Leer, Lee and Roe [30]. By our previous discussion the decomposition also applies to the steady state equations without any preconditioning. Let $\beta_* = \sqrt{1 - M^2}$.

$$
\begin{aligned}
\frac{\beta_* + 1}{M^2}p_t + \frac{\rho u}{M^2}u_t + up_x + \rho c^2(u_x + v_y + w_z) &= 0 \\
\frac{1}{\rho u}p_t + u_t + uu_x + \frac{p_x}{\rho} &= 0 \\
\frac{1}{\beta_*}v_t + uv_x + \frac{p_y}{\rho} &= 0 \\
\frac{1}{\beta_*}w_t + uw_x + \frac{p_z}{\rho} &= 0 \\
S_t + uS_x &= 0
\end{aligned}
$$

S and H are convected along the stream line. We now have three equations for the unknowns (p,v,w). Let $(dq, dr, ds) = (\frac{\beta_*}{\rho u}dp, du, dv)$.

$$
\begin{pmatrix} q \\ r \\ s \end{pmatrix}_t + \begin{pmatrix} -\beta_* u & 0 & 0 \\ 0 & \beta_* u & 0 \\ 0 & 0 & \beta_* u \end{pmatrix} \begin{pmatrix} q \\ r \\ s \end{pmatrix}_x +
$$

$$
\begin{pmatrix} 0 & u & 0 \\ u & 0 & 0 \\ 0 & 0 & 0 \end{pmatrix} \begin{pmatrix} q \\ r \\ s \end{pmatrix}_y + \begin{pmatrix} 0 & 0 & u \\ 0 & 0 & 0 \\ u & 0 & 0 \end{pmatrix} \begin{pmatrix} q \\ r \\ s \end{pmatrix}_z = 0
$$

We next rescale x by defining $x = \beta_* x'$. We rewrite this equation in terms of x' and then drop the prime. The effect is to eliminate the β_* term from the x derivative. Then in this modified coordinate system we get

$$\begin{pmatrix} q \\ r \\ s \end{pmatrix}_t + \begin{pmatrix} -u & 0 & 0 \\ 0 & u & 0 \\ 0 & 0 & u \end{pmatrix} \begin{pmatrix} q \\ r \\ s \end{pmatrix}_x + \qquad (16.1)$$

$$\begin{pmatrix} 0 & u & 0 \\ u & 0 & 0 \\ 0 & 0 & 0 \end{pmatrix} \begin{pmatrix} q \\ r \\ s \end{pmatrix}_y + \begin{pmatrix} 0 & 0 & u \\ 0 & 0 & 0 \\ u & 0 & 0 \end{pmatrix} \begin{pmatrix} q \\ r \\ s \end{pmatrix}_z = 0$$

We wish to know when this first order system is equivalent to the Laplace equation. To do this we study the equivalence of the three dimensional Laplace equation with a first order system. Let

$$\frac{\partial^2 \psi}{\partial x^2} + \frac{\partial^2 \psi}{\partial y^2} + \frac{\partial^2 \psi}{\partial z^2} = 0.$$

Define $u = \psi_x$, $v = -\psi_y$ and $w = -\psi_z$. Then

$$\begin{pmatrix} -1 & 0 & 0 \\ 0 & 1 & 0 \\ 0 & 0 & 1 \end{pmatrix} \begin{pmatrix} u \\ v \\ w \end{pmatrix}_x + \begin{pmatrix} 0 & 1 & 0 \\ 1 & 0 & 0 \\ 0 & 0 & 0 \end{pmatrix} \begin{pmatrix} u \\ v \\ w \end{pmatrix}_y$$

$$+ \begin{pmatrix} 0 & 0 & 1 \\ 0 & 0 & 0 \\ 1 & 0 & 0 \end{pmatrix} \begin{pmatrix} u \\ v \\ w \end{pmatrix}_z = 0.$$

Hence, it is trivial that the second order three dimensional Laplace equation yields the first order system. However, one can not, in general, start with the (u, v, w) first order system and obtain a Laplace equation for each of the variables. By the existence of the potential ψ we also have $v_z - w_y = 0$. If this is added as an additional equation to the system then it is easily shown that (u, v, w) do satisfy a Laplace equation for each variable. Hence, to go from the system back to the second order Laplace equation for a potential ψ we have a compatibility requirement on the solution.

For the three dimensional Euler equations the "acoustic" portion corresponds to the Laplace equation (or time dependent wave equation) only if we satisfy a compatibilty constraint corresponding to the vanishing of the streamwise component of the vorticity. If this constraint is not met then we can only diagonalize the system by including higher derivatives.

Starting with (16.1) and differentiating we find that q satisfies the wave equation $\frac{\partial^2 q}{\partial t^2} = u^2 \Delta q$. However r and s do not obey a wave equation. Instead

$$\frac{\partial^2 r}{\partial t^2} = u^2 \left(\frac{\partial^2 r}{\partial x^2} + \frac{\partial^2 r}{\partial y^2} + \frac{\partial^2 s}{\partial y \partial z} \right)$$

$$= u^2 \left(\Delta r + \frac{\partial}{\partial z} \left(\frac{\partial s}{\partial y} - \frac{\partial r}{\partial z} \right) \right)$$

$$\frac{\partial^2 s}{\partial t^2} = u^2 \left(\frac{\partial^2 s}{\partial x^2} + \frac{\partial^2 s}{\partial z^2} + \frac{\partial^2 r}{\partial y \partial z} \right)$$

$$= u^2 \left(\Delta s - \frac{\partial}{\partial y} \left(\frac{\partial s}{\partial y} - \frac{\partial r}{\partial z} \right) \right)$$

Differentiate the equation for r by y and the equation for s by z and add we find that $\frac{\partial^2 (r_y + s_z)}{\partial t^2} = u^2 \Delta (r_y + s_z)$, i.e. $(r_y + s_z)$ also satisfies the wave equation. Since q satisfies a wave equation so does q_x and hence so does the divergence $q_x + r_y + s_z$. Define $\omega = r_z - s_y$ i.e. the z component of the vorticity. Differentiating the equation for r by z and the equation for s by y and subtracting we find that $\omega_t + u\omega_x = 0$. Thus, we have two "elliptic" variables, q and div(q,r,s) and one "hyperbolic" variable ω in addition to the previous two hyperbolic variables S and H. Hence, if we solve this three dimensional system with an initial condition that the vorticity is zero then the vorticity remains zero for all time. In this case we can go from the first order system back to the second order wave equation. In two dimensions the system is purely elliptic since ω is always identically zero.

16.3 Supersonic

In three dimensions we consider the van Leer et. al. preconditioned supersonic equations in streamwise coordinates. As before this also applies to the non-preconditioned steady state equations. Let $\beta = \sqrt{M^2 - 1}$.

$$\frac{\beta M + 1}{M^2} p_t + \frac{\rho c}{M} u_t + u p_x + \rho c^2 (u_x + v_y + w_z) = 0$$

$$\frac{1}{\rho u} p_t + u_t + u u_x + \frac{p_x}{\rho} = 0$$

$$\frac{M}{\beta} v_t + u v_x + \frac{p_y}{\rho} = 0$$

$$\frac{M}{\beta} w_t + u w_x + \frac{p_z}{\rho} = 0$$

$$S_t + u S_x = 0$$

Solving for (p, u, v, w, S) we get

471

$$p_t + \frac{\beta u}{M} p_x + \frac{\rho c^2 M}{\beta}(v_y + w_z) = 0$$

$$u_t + u u_x - \frac{1}{\rho}\left(\frac{\beta}{M} - 1\right) p_x - \frac{c}{\beta}(v_y + w_z) = 0$$

$$v_t + \frac{\beta}{M}\left(u v_x + \frac{p_y}{\rho}\right) = 0$$

$$w_t + \frac{\beta}{M}\left(u w_x + \frac{p_z}{\rho}\right) = 0$$

$$S_t + u S_x = 0$$

As before the system for (p,v,w) decouples from the rest. Let $\mathbf{w} = (\frac{\beta}{\rho u} dp, dv dw)^t$. Then

$$\mathbf{w}_t + \frac{\beta u}{M}\begin{pmatrix} 1 & 0 & 0 \\ 0 & 1 & 0 \\ 0 & 0 & 1 \end{pmatrix}\mathbf{w}_x \qquad (16.2)$$

$$+ \begin{pmatrix} 0 & c & 0 \\ c & 0 & 0 \\ 0 & 0 & 0 \end{pmatrix}\mathbf{w}_y + \begin{pmatrix} 0 & 0 & c \\ 0 & 0 & 0 \\ c & 0 & 0 \end{pmatrix}\mathbf{w}_z = 0$$

Normalizing (16.2) we have

$$\mathbf{w}_t + \mathbf{w}_x + B\mathbf{w}_y + C\mathbf{w}_z = 0$$

with

$$B = \begin{pmatrix} 0 & 1 & 0 \\ 1 & 0 & 0 \\ 0 & 0 & 0 \end{pmatrix} \qquad C = \begin{pmatrix} 0 & 0 & 1 \\ 0 & 0 & 0 \\ 1 & 0 & 0 \end{pmatrix}$$

This is a new canonical form which cannot be reduced to either a hyperbolic (convective) equation or to an elliptic (acoustic) equation. Instead since the coefficient matrix for the x direction reduces to the identity matrix we can introduce new variables $\tau = t$ and $\xi = x - t$. Then

$$\mathbf{w}_\tau + B\mathbf{w}_y + C\mathbf{w}_z = 0 \qquad \mathbf{w} = (q, r, s)^t$$

This is identical to the canonical form for the subsonic case considered above. Hence, q satisfies a wave equation and r, s satisfy

$$\frac{\partial^2 r}{\partial t^2} = \Delta r + \frac{\partial}{\partial z}\left(\frac{\partial s}{\partial y} - \frac{\partial r}{\partial z}\right)$$

$$\frac{\partial^2 s}{\partial t^2} = \Delta s - \frac{\partial}{\partial y}\left(\frac{\partial s}{\partial y} - \frac{\partial r}{\partial z}\right)$$

So, $r_y + s_z$ satisfies a two dimensional wave equation while $(s_y - r_z)_\tau = 0$ in the moving coordinate system. Hence, the vorticity component $s_y - r_z$ satisfies a one dimensional convection equation, in the steamwise direction, as we already have found.

Thus, the supersonic system consists of three convective equations for the entropy, enthalpy and vorticity that decouple. In addition we have a system of two equations that is a combined hyperbolic-elliptic system in the sense that it is an elliptic system in the y and z directions in a system moving in the x direction.

17 Comparison of Preconditioners

- Turkel and Van Leer preconditioners

    - For low speed flow: convergence and accuracy
    Gives large time step for low Mach numbers
    - Based on PDE i.e. low frequencies
    - Changes steady state solution

- Allmaras - Giles preconditioner

    $$(|A| + |B|)w_t = Aw_x + Bw_y + h(|A|w_{xx} + |B|w_{yy})$$

    - Works on high frequencies (high-high vs high-low)
    - Depends on individual scheme
    - At (π, π) converts matrix function to scalar
    - Does not change steady state solution
    - Does not give any benefit for low Mach numbers
    - Requires small time step - no residual smoothing allowed
    - Requires inversion of matrix (2D/3D)

    Preconditioning-squared

    Combine Turkel/van-Leer with Jacobi

    $$(|PA| + |PB|)w_t = PAw_x + PBw_y + h(|PA|w_{xx} + |PB|w_{yy})$$

18 Computational Results

18.1 Inviscid Flow

To assess the accuracy of the preconditioning versus the standard algorithm we compare the results for lift and drag for inviscid flow over NACA 0012 airfoil. The C-type grid has 224×40 cells clustered near the leading and trailing edges in order to allow accurate drag computations, for various Mach numbers, with the results from a panel method. As seen in table 1 the code using the preconditioning (both $\alpha = 0$ and $\alpha = 1$) gives lift and drag quite close to the panel method results with only a small dependence on the Mach number. The non-preconditioned code has large variations as the Mach number goes to zero and is not converging to the panel code results. In particular the drag, which should be zero for inviscid flow, is very large without preconditioning but close to zero when using preconditioning. In Figure 1 we compare the C_p for this case without and without the preconditioning. Other comparisons with incompressible solutions are presented in [29].

	Mach	panel	no prec	$\alpha=0$	$\alpha=1$
C_L	0.1	.241	.2448	.2434	.2429
	0.01		.2172	.2421	.2416
	0.001		.1097	.2421	.2416
C_D	0.1	0.0	.0008	.00027	.00023
	0.01		.0086	.00027	.00023
	0.001		.0731	.00027	.00023

Table 1: Comparison of lift and drag with/without preconditioning

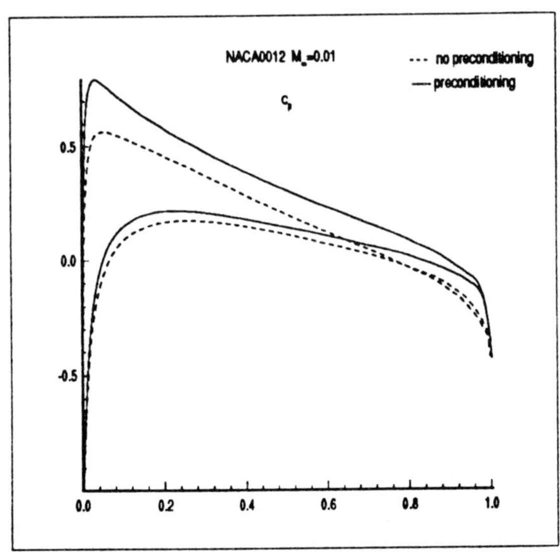

Figure 1: Comparison of C_p with and without preconditioning, $M_\infty = 0.01$

18.2   2-D Multi-element Airfoil

The 3-element airfoil configuration has been investigated both experimentally and theoretically [35]. Present computations are performed at a chord Reynolds number of 9 million and an angle of attack of 16.2 degrees using a variant of a cell-centered code [28]. The Mach number for this case is 0.2. A 20-block structured grid shown in Figure 2a, is used for the computations. This case has also been investigated with the original non-preconditioned version of the flow solver used in this work [32].

The convergence histories for this case are presented in Figure 2b, where the results from original non-preconditioned and current preconditioned (conservative) schemes are compared. The residual in the original code slow down considerably at approximately 4 orders; while the residual in the preconditioned scheme exhibits much better convergence. Note that the free-stream Mach number for this case (M=0.2) is generally not considered too low for compressible codes. However, several pockets of slow-moving flow exist in the cove regions of the slat and the main airfoil sections [32]. These pockets slow the convergence of standard compressible codes.

The computed pressure distributions for this case are compared in Figure 2c. As expected, very little difference is observed in the two sets of computed results. Furthermore these results compare quite favorably with the experimental data. Further results are presented in [28].

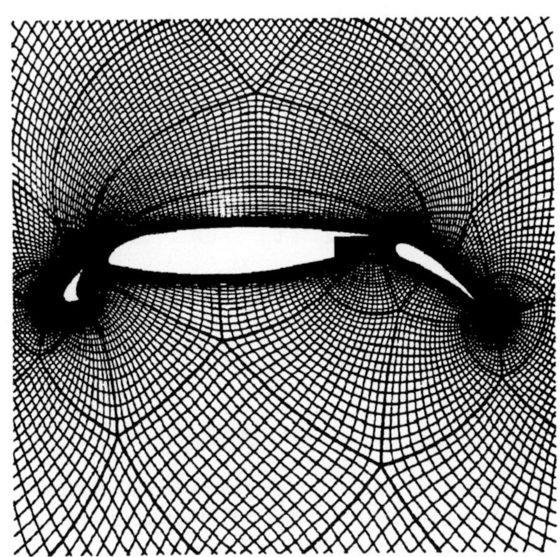

Figure 2a: Partial view of block-structured grid for 3-element high lift airfoil configuration

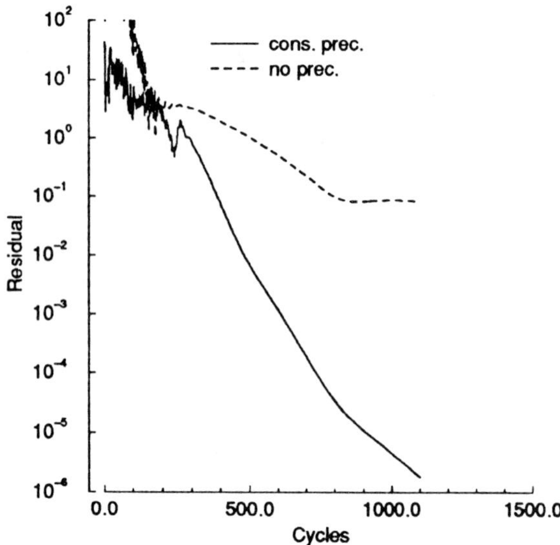

Figure 2b: Effect of preconditioning on convergence history for the the 3-element high-lift airfoil configuration

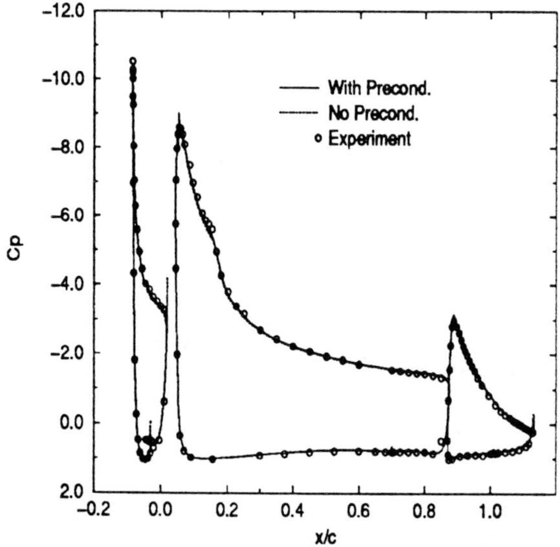

Figure 2c: Effect of preconditioning on pressure distributions for the 3-element high-lift airfoil configuration

18.3 Flow over 3D Wing

The three-dimensional calculations have been carried out with a cell-vertex method [15] which corresponds closely to *Code1* used for the 2D flows. The code uses central differencing with artificial viscosity, multistage time stepping with a five stage scheme, residual smoothing and a 3 level full multigrid with W-cycles.

Preconditioning is obtained using $\delta = 1, \alpha = 0$ and the dissipation according to (7.3).

We consider essentially incompressible viscous flow over an ONERA-M6 wing at an angle of attack of $\alpha = 2°$, Re=11 million. The effect of turbulence is modelled by the Baldwin and Lomax model. The grid used for the calculations has C-H topology and consists of 224 cells in the streamwise direction, 48 cells normal and 32 cells in the spanwise directions. Figure 3 shows the grid in the plane of symmetry. The residual convergence for computations at $M_\infty = 0.10, 0.01, 0.001$ is presented in Figure 4a. It is seen that the rate of convergence is independent of the free stream Mach number. Without preconditioning no convergence could be obtained for Mach numbers less than 0.1. Figure 4b compares the surface pressure distribution at spanwise section $\eta = 0.8$. There is no effect of the freestream Mach number visible. Mach number independence is further confirmed by Figure 4c where the total lift and pressure drag coefficients are plotted versus the Mach number. These values are constant if the Mach number approaches zero.

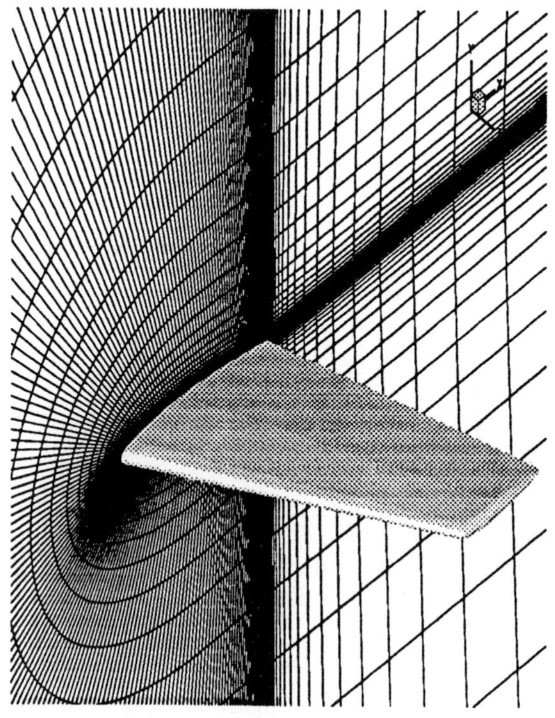

Figure 3: Wing surface and grid in symmetry plane

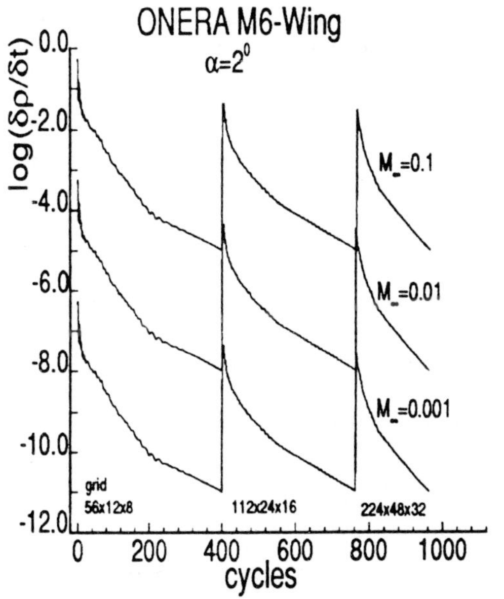

Figure 4a: Effect of freestream Mach number on residual convergence of 3D wing

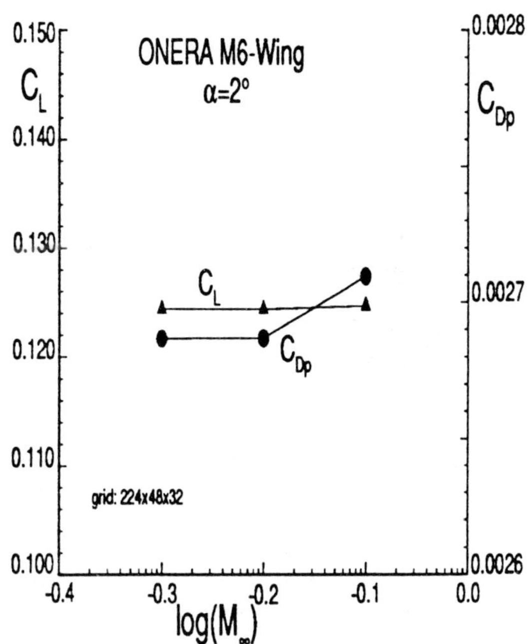

Figure 4c: Effect of freestream Mach number on force coefficients of 3D wing

References

[1] Allmaras, S., *Analysis of a Local Matrix Pre-conditioner for the 2-D Navier-Stokes Equations*, AIAA 11th Computational Fluid Dynamics Conference, 1995.

[2] Allmaras, S., *Analysis of Semi-Implicit Preconditioners for Multigrid Solution of the 2-D Compressible Navier-Stokes Equations*, AIAA paper 95-1651-CP, AIAA 12th Computational Fluid Dynamics Conference, 1995.

[3] Arnone A. and Turkel E., *Pseudo-compressibilty Methods for the Incompressible Flow Equations*, Proceedings 11th AIAA CFD Conference, pp. 349-357, AIAA paper 93-3329, 1993.

[4] Au-Yeung, Y.H. Proceedings Amer. Math. Soc., 20:545-548, 1969.

[5] Choi, Y.-H. and Merkle, C.L., *The Application of Preconditioning to Viscous Flows*, J. Comput. Phys., 105:207-223, 1993.

[6] Chorin, A.J., *A Numerical Method for Solving Incompressible Viscous Flow Problems*, J. Comput. Phys., 2:12-26, 1967.

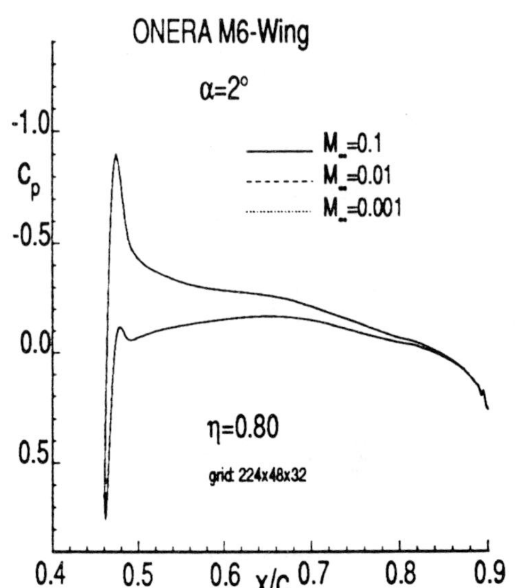

Figure 4b: Effect of freestream Mach number on pressure distribution of 3D wing

[7] Fiterman, A., Turkel, E. and Vatsa, V., *Pressure Updating Methods for the Steady-State Fluid Equations*, AIAA 95-1652-CP, AIAA 12th Computational Fluid Dynamics Conference, 1995.

[8] Gilquin, H., Laurens, J. and Rosier, C. *Multi-Dimensional Riemann Problems for Linear Hyperbolic Systems: Part I* in Nonlinear Hyperbolic Problems: Theoretical, Applied and Computational Aspects, eds A. Donato and F. Oliveri, pp. 276-279 Vieweg, 1993

[9] Gustafsson, B., *Unsymmetric Hyperbolic Systems and the Euler equations at Low Mach Numbers*, J. Scientific Computing, 2:123-136, 1987.

[10] Gustafsson, B. and Stoor, H., *Navier-Stokes Equations for Almost Incompressible Flow*, SIAM J. Numer. Anal., 28:1523-1547, 1991.

[11] Jameson, A., Schmidt, W. and Turkel, E., *Numerical Solutions of the Euler Equations by a Finite Volume Method using Runge-Kutta Time-Stepping Schemes*, AIAA paper 81-1259, 1981.

[12] Klainerman, S. and Majda, A., *Compressible and Incompressible Fluids*, Comm. Pure Appl. Math., 35:629-651, 1982.

[13] Kreiss, H.O. and Lorenz, J., **Initial-Boundary Value Problems and the Navier Stokes Equations**, Academic Press, Boston, 1989.

[14] Kreiss, H.O., Lorenz,J. and Naughton, M., *Convergence of the Solutions of the Compressible to the Solutions of the Incompressible Navier-Stokes Equations*, Advances in Appl. Math., 12:187-214, 1991.

[15] Kroll, N., Radespiel, R. and Rossow, C.-C., *Accurate and Efficient Flow Solvers for 3D Applications on Structured Meshes*, VKI Lecture Series, LS-04, 1994.

[16] Lee, D. and Van Leer, B., *Progress in Local Preconditioning of the Euler and Navier-Stokes Equations*, AIAA 11th Computational Fluid Dynamics Conference, pp. 338-348, 1993.

[17] Mesaros, L. and Roe, P., *Multidimensional Fluctuation Splitting Schemes Based on Decomposition Methods*, AIAA 12th Computational Fluid Dynamics Conference, pp. 582-591, 1995.

[18] Pierce, N.A. and Giles, M.B., *Preconditioning Compressible flow Calculations on Stretched Meshes* AIAA paper 96-0889, 1996.

[19] Roe, P. and Turkel, E. *The Quest for Diagonalization of Differential Systems*, Workshop on Barriers and Challenges in Computational Fluid Dynamics, Kluwer Publisher, 1996.

[20] Swanson, R.C. and Radespiel, R. *Cell Centered and Cell Vertex Multigrid Schemes for the Navier-Stokes Equations*, AIAA J., 29:697-703, 1991.

[21] Swanson, R.C. and Turkel, E., *Artificial Dissipation and Central Difference Schemes for the Euler and Navier-Stokes Equations*, AIAA paper 87-1107-CP, AIAA 8th Computational Fluid Dynamics Conference, 1987.

[22] Swanson, R.C. and Turkel, E., *On Central Difference and Upwind Schemes*,J. Comput. Phys., 101:292-306, 1992

[23] Ta'asan, S., *Canonical-Variables Multigrid Method for Steady-State Euler Equations. Fourteenth Inter. Conf. Numer. Meth. Fluid Dynamics*, Lecture Notes in Physics, Editor Deshpande et. al, Springer 1994.

[24] Ta'asan, S. *Essentially Optimal Multigrid Method for Steady State Euler Equations* AIAA Paper 95-0209, 1995.

[25] Turkel, E., *Preconditioned Methods for Solving the Incompressible and Low Speed Compressible Equations*, J. Comput. Phys., 72:277-298, 1987.

[26] Turkel, E., *A Review of Preconditioning Methods for Fluid Dynamics*, Applied Numerical Mathematics, 12:257-284, 1993.

[27] Turkel, E., Fiterman, A. and van Leer, B. *Preconditioning and the Limit to the Incompressible Flow Equations*, Computing the Future: **Frontiers of Computational Fluid Dynamics 1994**, editors D.A. Caughey and M.M. Hafez, Wiley Publishing, 215-234, 1994.

[28] Turkel, E., Vatsa, V.N. and Radespiel, R. *Preconditioning Methods for Low Speed Flow*, AIAA paper 96-2460, 1996.

[29] Turkel, E., Radespiel, R. and Kroll, N. *Assessment of Two Preconditioning Methods for Aerodynamic Problems* to appear Computers and Fluids.

[30] van Leer, B., Lee, W.T., Roe, P.L. , *Characteristic Time-Stepping or Local Preconditioning of the Euler Equations*, AIAA Paper 91-1552, 1991.

[31] van Leer, B., Mesaros, L., Tai, C-H and Turkel, E. *Local Preconditioning in a Stagnation Point*, AIAA paper 95-1654-CP, AIAA 12th Computational Fluid Dynamics Conference, 88-101, 1995.

[32] Vatsa, V.N., Sanetrik, M.D., Parlette, E.B., Eiseman, P. and Cheng, Z., Multi-block Structured Grid Approach for Solving Flows over Complex Aerodynamic Configurations, AIAA Paper 94-0655, 1994.

[33] Volpe, G., *Performance of Compressible Flow Codes at Low Mach Numbers*, AIAA J., 31:49-56, 1993.

[34] Wen-Tzong Lee, **Local preconditioning of the Euler equations**, Ph.D. dissertation, The University of Michigan, 1992.

[35] Workshop on high-lift flow physics for application to subsonic aircraft. A workshop held at NASA Langley Research Center, May 18-20, 1993.

ENHANCED ACCELERATION AND RECONDITIONING TECHNIQUES

Yousef SAAD

University of Minnesota, Department of Computer Science and Engineering,
200 Union St. SE, Minneapolis, MN 55455, USA

Abstract

The robustness and effectiveness if iterative solution methods in CFD habe improved significantly in recent years thanks to progress made in both accelerators and preconditioners. This paper highlights some of the idears which have been exploited to obtain these improvements.

1 Introduction

We consider the linear system

$$Ax = b \qquad (1)$$

in which A is nonsingular matrix, typically sparse and irregularly structured. Currently, iterative techniques based on preconditioned Krylov subspace projection offer the best compromise between efficiency and robustness. These techniques combine two ingredients which are both essential. First is the preconditioning transformation. An approximation M to the original linear system (1) must be found which is close enough to A in some sense but which if 'easily invertible' in the sense that it is inexpensive to compute $M^{-1}v$ for an arbitrary vector v. The original system may be transformed into the "left- preconditioned" equivalent system

$$M^{-1}Ax = M^{-1}b.$$

In this case the system $AM^{-1}z = b$ is solved for the unknown $z =\equiv Mx$, and the final x results is obtained through the post-transformation $x = M^{-1}z$. The main difference between these two approaches is that in the right preconditioned case, the actual residual norm is available at each step of the iterative process whereas in the second case only the preconditioned residual is available barring any additional computations. The second ingredient in the method is the "accelerator", typically a conjugate gradient-like method, based on projections on Krylov subspaces.

Let an initial guess x_0 to this system be given and denote by r_0 the initial residual vector $r_0 = b - Ax_0$.

Received on March 1, 1998.
This work was supported by NSF grant number CCR 9618827

General projection methods seek an approximate solution $\bar{x}$ from an affine subspace of approximants of the form $x + K$ by imposing a number of conditions, such as optimality of the approximate solution. The approximation is typically extracted from K by imposing a Petrov-Galerkin condition of the form:

$$b - a\bar{x} <> L \qquad (2)$$

where L is another subspace of the same dimension m as K. In this paper we will restrict our attention to Minimal Residual methods which minimize the residual norm of the approximate solution over all possible candidates in the subspace of approximants. In other words, the desired approximation x_m is such that

$$\|b - ax_m\|_2 = \min_{x \in x_0 + K} \|b - Ax\|_2 \qquad (3)$$

It is known that this minimum residual solution is the same as that obtained from a projection process onto K and orthogonally to $L = AK$, see. e.g., [25].

A Krylov subspace acceleration method is a projection method in which the subspace of approximants is the Krylov subspace:

$$K_m(A, r_0) = \mathrm{span}\left\{r_0, Ar_0, \ldots, A^{m-1}r_0\right\}$$

we will often simplify notation and denote by K_m. The residual vector $r_m = b - A\bar{x}$ of the approximate solution if of the form $r_m = \phi_m(A)r_0$, where ϕ_m is a certain polynomial of degree $\leq m$. The polynomial ϕ_m is referred to as the *residual polynomial* and methods which lead to residuals which have this form are referred as *polynomial acceleration methods*. A number of quite successful projection methods have been defined in this manner.

In order to calculate the desired approximate solution, the Arnold procedure is often used. This procedure computes an orthogonal basis of the Krylov subspace as follows.

ALGORITHM 1: Arnoldi

1.   Let $v_1 := r_0/\|r_0\|_2$.
2.   For $j = 1, 2, \ldots, m$ do:
3.   Compute $w_j := Av_j$
4.   For $i = 1, \ldots, j$ do
5. $h_{ij} = (w_j, v_i)$
6. $w_j := w_j - h_{ij}v_i$
7.   Enddo
8. $h_{j+1,j} = \|w_j\|_2$. If $h_{j+1,j} = 0$ Stop.
9. $v_{j+1} = w_j/h_{j+1,j}$.
10.   EndDo

This is essentially a modified Gram-Schmidt procedure in which at each step a new basis vector v_{j+1} is obtained by orthonormalizing Av_j against all previous v_i's at step j. The following notation will be used. The $n \times m$ matrix whose column vectors are the vectors $v_1, \ldots, v_m$ is denoted by V_m. The $(m+1) \times m$ upper Hessenberg (quasi-upper triangular) matrix whose coefficients h_{ij} are defined by the algorithm is denoted by $\bar{H}_m$. We also denote by H_m the $m \times m$ matrix obtained from $\bar{H}_m$ by deleting its last row. The important relation

$$AV_m = V_{m+1}\bar{H}_m, \tag{4}$$

shows easily that for an arbitrary vector in $x_0 + K_m$ which is written as $x_0 + V_m y$, the residual is

$$b - Ax = V_{m+1}[\beta e_1 - \bar{H}_m y] . \tag{5}$$

Using the orthonormality of the column-vectors of V_{m+1} we immediately obtain that y must minimize

$$\|\beta e_1 - \bar{H}_m y\|_2 \tag{6}$$

over all vectors in $\mathbf{R}^m$. This yields the GMRES algorithm [26].

Note that in practical situations, computing the approximate solution $\tilde{x}$ requires storing the m Arnoldi vectors $v_1, \ldots, v_m$ and this may become a big burden. One commonly used remedy is to restart the algorithm, i.e., if x_m is not accurate enough we restart GMRES with x_0 set to the most recent approximate solution obtained. However, this causes the algorithm to become sub-optimal and often to stagnate when the preconditioned matrix is not positive definite.

In the case when the system is preconditioned, then it is clear that the matrix A in the above descriptions is to be replaced by the preconditioned matrix $M^{-1}A$ or AM^{-1} depending on whether left or right preconditioning is used. There are applications where it is important to allow the preconditioner to vary from step to step in the inner Krylov subspace iteration. This is a situation which is gaining importance because of the impact of parallel processing in particular. Sometimes a preconditioner can be itself defined from an iterative process, such as a relaxation or a multigrid scheme. In this situation, M is not a constant matrix and can instead be viewed as an operator which changes at each step. When right preconditioning is used, it is possible to obtain a solution which is optimal in a certain sense. Flexible accelerators are a class of accelerators which respond to this need.

The GMRES algorithm applied to the system (2) will compute the approximation via the modified formula,

$$x_m = x_0 + M^{-1}V_m y_m .$$

where M is the preconditioner, y_m is the minimizer of (6) and $V_m = v_1, \ldots, v_m]$ is the orthonormal basis obtained from the Arnoldi algorithm. This forms a linear combination of the preconditioned vectors $z_i = M^{-1}v_i, i = 1, \ldots, m$, where v_i are the Arnoldi vectors. Because these vectors are all obtained by applying the same preconditioning matrix M^{-1} to the v's, we need not save them. We only need to apply M^{-1} to the linear combination of the $v's$, i.e., to $V_m y$. If we allowed the preconditioner to vary at every step, i.e., if z_j is now defined by

$$z_j = M_j^{-1}v_j$$

we must save these vectors and use them when computing x_m. This is the only difference between the standard and flexible variants of GMRES.

ALGORITHM 2: FGMRES

0.   Until convergence do :
1. $r_0 = b - Ax_0$; $\beta := \|r_0\|_2$; $v_1 = r_0/\beta$.
2.   For $j = 1, \ldots, m$ do:
3. $z_j := M_j^{-1}v_j$
4. $w := Az_j$
5.   Do $i = 1, \ldots, j$, $\begin{cases} h_{i,j} := (w, v_i) \\ w := w - h_{i,j}v_i \end{cases}$
6. $v_{j+1} = w/(h_{j+1,j} := \|w\|_2)$
7.   Define $Z_m := [z_1, \ldots, z_m]$.
8.   Compute $x_m = x_0 + Z_m y_m$ where
9. $y_m = \mathrm{argmin}_y\|\beta e_1 - \bar{H}_m y\|_2$.
10.   Set $x_0 \leftarrow x_m$.
11.   EndDo

For details see [22]. Similar techniques have also been developed by van der Vorst and Vuik [29] as well as by Axelsson and Vassilevski [1].

This algorithm can be used to enhance the robustness of GMRES in a number of ways. The key observation to make at this point is that there is no need for a formal preconditioning operation. Specifically, the new vector z_j in Line 3 of the algorithm can be an arbitrary new vector which is injected to the subspace. In fact, it may even be unrelated to the vector v_j. Examples of strategies that can be defined from Krylov subspace techniques by using these ideas are the following:

1. Introduce approximate eigenvectors associated with smallest eigenvalues.

2. Introduce approximate solutions to the last Krylov vector, possibly using a different solver [3].

3. Introduce a random new vector to the subspace.

The approximate eigenvectors in (1), typically numbering 4 to 10, are usually obtained from previous subspace projections and are improved as the iteration progresses, see e.g., [3, 19, 13, 9]. This strategy which will be discussed in the next section, has lead to significant improvements of the iterative solvers in some cases.

2 ENHANCED ACCELERATORS

In this section we describe just two ideas which have been used to develop enhanced Krylov accelerators. The first one is based on eigenvalue deflation and the second on a block version of GMRES.

2.1 Deflated Krylov solvers

The simplest way to exploit knowledge about the spectrum of A in order to enhance convergence of a Krylov solver is simply to add the desired eigenvectors directly to the Krylov subspace using the FGMRES framework. For eigenvalue problems similar techniques are commonplace and take the form of deflation [21] or Thick-Restarting [27]. Approaches of this type have also been proposed for linear systems by several authors, including Morgan [19], Karchenko and Yeremin [13], Burrage and Erhel [9]. The technique in [19] and [3] can be simply recast in terms of Flexible GMRES [22]. Approximate eigenvectors are injected as preconditioned quantities z_j for the last p steps of the algorithm. The subspace Z_m of flexible GMRES is therefore the subspace spanned by the vectors

$$Z_m = \{v_1, v_2, \ldots, v_{m-p}, u_1, u_2, \ldots, u_p\}$$

where the v_i's are the standard Arnoldi vectors associated with A or AM^{-1} when preconditioning is used, and $u_1, \ldots, u_p$ are approximate eigenvectors. The modification involves replacing Step 3 of Algorithm 2 by the following step:

3'. $\quad$ Set $z_j = \begin{cases} v_j & \text{if } j \leq m - p \\ u_{j-m+p} & \text{otherwise} \end{cases}$

When A is preconditioned from the right, the relation simlar to (4) which results from this is of the form

$$AM^{-1}Z_m = V_{m+1}\bar{H}_m \tag{7}$$

In the sequel we will use the notation

$$B = AM^{-1} . \tag{8}$$

The solution x then belongs to the subspace

$$x = x_0 + \text{span}\{v, Bv, \ldots, B^{m-p-1}v, u_1, \ldots, u_p\} ,$$

where $u_1, u_2, \ldots, u_p$ are estimated eigenvectors. It is computed exactly as is described in Lines 7-9 of Algorithm 2. After the approximate solution is obtained, and if further iterations are required, we need to restart the algorithm. Prior to this however, we must exploit the information cast in the relation (7) to improve the current set of eigenvectors $u_1, \ldots, u_p$ at the end of the Arnoldi loop, or to compute them for the first time for the first pass through GMRES. For the first pass through GMRES, no eigenvectors are available, and so $p = 0$. For problems with multiple right-hand sides, or in a non-linear problem, there may be estimates of eigenvalues available from a previous calculation, or a nearby problem. If these are used in the first pass through GMRES, convergence can be accelerated.

To estimate the eigenvectors of AM^{-1}, which are required in the procedure described above we can use a projection technique on the subspace available at the end of one outer loop of the process. To this end relation (7) must be exploited. The approximate eigenvector u to be extracted from the subspace $span\ \{Z_m\}$ can be expressed in the form

$$u = Z_m y$$

where y is an m-dimensional vector. Let θ be an associated eigenvalue. A number of different projection techniques can be invoked to extract the eigenvectors from the subspace spanned by Z_m. The one which was generally found to be the most effective in the context of deflated GMRES imposes the condition that the residual vector $(B - \tilde{\lambda}I)\tilde{u}$ of the approximate eigenpair $\tilde{\lambda}, \tilde{u}$ is orthogonal to the subspace $span\{BZ_m\}$. It is common to refer to this as a Galerkin procedure for approximating *harmonic* values, see Morgan [19, 18]. It is known to give, typically, better approximations for eigenvalues nearest zero than the standard Galerkin approximation.

The Galerkin condition for this approach yields,

$$(BZ_m)^H (BZ_m y - \theta Z_m y) = 0 \tag{9}$$

which leads to the Generalized eigenvalue problem,

$$Gy = \theta Fy \tag{10}$$

in which

$$F = (BZ_m)^H Z_m, \qquad G = (BZ_m)^H (BZ_m)$$

From (7), it is easily seen that

$$G = \bar{H}_m^H \bar{H}_m \quad \text{and} \quad F = \bar{H}_m^H V_{m+1}^H Z_m.$$

The generalized eigenvalue problem (10) can be solved by any technique for dense generalized eigenvalue problems.

However, a better alternative is to exploit the results of the FGMRES calculation. At the end of the FGMRES process the matrix $\bar{H}_m$ is factored as

$$Q_m \bar{H}_m = \bar{R}_m \qquad (11)$$

where Q_m is a unitary matrix of dimension $(m+1) \times (m+1)$ and $\bar{R}_m$ is an upper triangular matrix of size $(m+1) \times m$ whose last row is a zero row. This factorization is performed progressively in order to be able to compute the least-squares solution easily when the process is stopped, just as is done in the GMRES algorithm [25].

Define R_m to be the matrix obtained from $\bar{R}_m$ by deleting its last row. Then, clearly,

$$\bar{H}_m^H \bar{H}_m = \bar{R}_m^H \bar{R}_m = R_m^H R_m \ .$$

and, using the relation (11),

$$F = \bar{R}_m^H Q_m V_{m+1}^H Z_m = R_m^H \lfloor Q_m V_{m+1}^H Z_m \rfloor \qquad (12)$$

where $\lfloor X \rfloor$ is used to denote the matrix obtained by removing the last row of the matrix X. Therefore the generalized eigenvalue obtained is of the form

$$R_m^H R_m y = \theta R_m^H \lfloor Q_m V_{m+1}^H Z_m \rfloor \, y$$

which has the same solutions as

$$R_m y = \theta \lfloor Q_m V_{m+1}^H Z_m \rfloor \, y \ . \qquad (13)$$

To obtain the matrix on the right-hand side of (13) notice that $V_{m+1}^H Z_m$ must undergo the same rotations as those that transform $\bar{H}_m$ into R_m. These rotations are saved in the FGMRES algorithm.

This solution of this generalized eigenvalue problem will give us m eigenvalues θ_i, $i = 1, \ldots, m$ and associated eigenvectors y_i. Some of these eigenvalues will be good approximations to eigenvalues of B. The corresponding Ritz vectors given by $Z_m y$ will approximate eigenvectors of B. For additional details see, e.g., [3].

2.2 Block-GMRES

Many applications lead to the problem of solving linear systems with several right-hand-sides simultaneously. These linear systems will be denoted by

$$Ax^{(i)} = b^{(i)}, \quad i = 1, \ldots, p \qquad (14)$$

or in matrix form,

$$AX = B \qquad (15)$$

Here X and B are now dense $n \times p$ matrices in which p is small (e.g., less than 20). We can solve these p systems independently by handling each of the systems (14)

individually. An alternative is to solve these systems simultaneously by a block-Krylov method. These methods attempt to exploit information from the right-hand sides as a block and are usually more efficient than the separate right-hand side approach.

In block-Krylov subspace methods, the subspace of approximants is defined from a set or block of initial vectors. In order to simplify notation it is best to describe the method with a small block-size, e.g., $p = 3$. Consider then an initial block of the form

$$V_1 = [v^1, v^2, v^3]$$

The m-th Block-Krylov subspace is defined as

$$\text{span} \, \{V_1, AV_1, \ldots, A^{k-1}V_1\} \qquad (16)$$

The dimension of the subspace here is $k \times p$ where p is the block dimension. A block-Arnoldi generalization of the Arnoldi process can easily be derived, and a block GMRES algorithm can be defined from it, see, e.g., [25]. An implementation of the underlying block-Arnoldi process which is based on a variant developed by A. Ruhe [20] is given below. It is assumed that the initial block V_1 is orthonormal.

ALGORITHM 3: **Block-Arnoldi - Ruhe's variant**

1.   *Choose p initial orthonormal vectors $\{v_i\}_{i=1,\ldots,p}$.*
2.   *For $j = p, p+1, \ldots, m$ Do:*
3.   *Set $k := j - p + 1$;*
4.   *Compute $w_k := Av_k$;*
5.   *For $i = 1, 2, \ldots, j$ Do:*
6.   *$h_{i,k} := (w_k, v_i)$*
7.   *$w_k := w_k - h_{i,k}v_i$*
8.   *EndDo*
9.   *Compute $h_{j+1,k} := \|w_k\|_2$ and $v_{j+1} := w_k/h_{j+1,k}$.*
10.   *EndDo*

The particular case $p = 1$ coincides with the usual Arnoldi process. The dimension m of the subspace is not restricted to being a multiple of the block-size p as in other formulations of block Krylov Algorithms.

At the end of the j-loop of the algorithm we obtain a relation of the form

$$Av_k = \sum_{i=1}^{k+r} h_{ik} v_i \ .$$

As a consequence, the analogue of the relation (4) for Algorithm 3, is

$$AV_m = V_{m+p} \bar{H}_m \ . \qquad (17)$$

Here again the matrix V_j represents the $n \times j$ matrix with columns $v_1, \ldots v_j$. The matrix $\bar{H}_m$ is now of size $(m+p) \times m$.

481

A perceived disadvantage of the above algorithm relative to standard block formulations is that it gives up some potential parallelism. However, a slight modification in which p matrix-vector products are computed simultaneously every p steps at once will remedy this problem, In this modified implementation it suffices to replace Line 4 of the above algorithm by the following

4. if $(\mod(j-1,p) == 0))$ Then
    Compute $w_i = Av_i$ for $i = k, \ldots, k+p-1$

The p matrix-vector products are now computed simultaneously so there is no loss of parallelism whatsoever with this modified implementation. It is easy to see that this formulation is mathematically equivalent to the one described by Algorithm 3.

In the block generalization of GMRES for multiple right-hand sides the initial residuals $r_0^{(i)} = b^{(i)} - Ax_0^{(i)}$ are computed into a matrix R_0 whose columns are orthonormalized as follows

$$R_0 = V_1 C$$

in which C is an upper triangular $p \times p$ matrix. Each approximate solution $x^{(i)}$ is a vector of the form $x_0^{(i)} + V_m y^{(i)}$, or in Block form,

$$X = X_0 + V_m Y$$

where X, X_0 are of size $n \times p$, V_m is of dimension $n \times m$ and Y is of size $m \times p$. From (17) it is clear that:

$$
\begin{aligned}
B - AX &= R_0 - V_{m+p}\bar{H}_m Y \\
&= V_{m+1}[I_{m+p,p}C - \bar{H}_m Y] \quad (18)
\end{aligned}
$$

in which $I_{m+p,p}$ is the matrix consisting of the first $m+p$ columns of the $(m+p) \times (m+p)$ identity. The matrix $I_{m+p,p}C$ is $(m+p) \times p$ and its upper $p \times p$ block consists of the upper triangular matrix C. Denote by c_j the j-th column of this matrix. Since the column-vectors of V_{m+p} are orthonormal, in order to minimize the 2-norm of the individual columns of the block-residual (18) it is sufficient to minimize the 2-norm of

$$\bar{c}_j - \bar{H}_m y^{(j)}$$

over $y^{(j)}$.

This least-squares problem is similar to the one encountered for GMRES. However, the matrix $\bar{H}_m$ is band-Hessenberg. The usual way of solving the above least-squares problems is to use plane rotations [26]. Now p rotations are needed at each step j, instead of only one, to eliminate the elements below the main diagonal in column j of H_m. For additional details see [25].

A number of practical issues which arise must be addressed. We will consider only one of them here namely the issue of dealing with early convergence of the approximate solution to a particular system. It is quite unlikely that all systems will converge at the same time and so a mechanism must be provided for extracting these approximations, stopping the corresponding iterations, while the iterations with the other right-hand sides are continued.

Before addressing this issue, we begin by recalling some technical details on how convergence is monitored in practical block-GMRES algorithms. As was already mentioned, the initial block of residuals is orthogonalized with a Gram-Schmidt process. In the block-Krylov subspace of Krylov-dimension k (i.e., of dimension $m = kp$), the representation of this initial residual block is an $(k+1)p \times p$ matrix G which is filled with zeros except in the upper $p \times p$ submatrix which is upper triangular. For example these two matrices have the following shape for $p = 3, k = 2$, (therefore $m = 6$):

$$
\bar{H}_m =
\begin{pmatrix}
\star & \star & \star & \star & \star & \star \\
\star & \star & \star & \star & \star & \star \\
\star & \star & \star & \star & \star & \star \\
\star & \star & \star & \star & \star & \star \\
 & \star & \star & \star & \star & \star \\
 & & \star & \star & \star & \star \\
 & & & \star & \star & \star \\
 & & & & \star & \star \\
 & & & & & \star
\end{pmatrix}
\qquad
G_m =
\begin{pmatrix}
\star & \star & \star \\
 & \star & \star \\
 & & \star \\
 & & \\
 & & \\
 & & \\
 & & \\
 & & \\
 & &
\end{pmatrix}
$$

The Hessenberg matrix $\bar{H}_m$ obtained from the Block-Arnoldi algorithm is usually processed by plane rotations into an upper triangular matrix. The right-hand side block G representing the residuals is processed by the same rotations and results in an $(k+1)p \times p$ matrix whose bottom $p \times p$ block is upper triangular.

$$
\bar{H}_m =
\begin{pmatrix}
\star & \star & \star & \star & \star & \star \\
 & \star & \star & \star & \star & \star \\
 & & \star & \star & \star & \star \\
 & & & \star & \star & \star \\
 & & & & \star & \star \\
 & & & & & \star \\
\hline
 & & & & & \\
 & & & & & \\
 & & & & &
\end{pmatrix}
\qquad
G_m =
\begin{pmatrix}
\star & \star & \star \\
\star & \star & \star \\
\star & \star & \star \\
\star & \star & \star \\
\star & \star & \star \\
\star & \star & \star \\
\star & \star & \star \\
 & \star & \star \\
 & & \star
\end{pmatrix}
$$

The residual norm of the j-th linear system is the 2-norm of the j-th column of the $p \times p$ bottom block of G'_m. It is possible to determine when the solution to the j-th linear system has converged by computing this norm.

We now return to the problem of dealing with convergence of the approximations to the different linear systems being solved. There are several possible solutions. The simplest to implement, is to monitor convergence in a block fashion. All p approximate solutions are treated as a group and are considered as either converged or not converged simultaneously. This means that some residuals may be extremely small while others may still converging slowly toward zero. It is not uncommon to see situations in which residual norms differ by 10 orders of magnitude with this approach.

A better approach is to remove those approximations that have converged after each restart. Within each outer iteration the block-size is never reduced. Prior to restarting the approximate solutions are computed and those approximations which have converged to a solution are removed from the system along with the right-hand sides. Therefore, the block-size is reduced by a number equal to the number of converged solutions, at each restart of the outer loop. This approach is a good compromise between ease of implementation and efficiency.

Finally, it is also possible to reduce the block size *within each inner loop* and this approach may be preferable when m is large. The main idea of this approach can be easily explained by referring to the Block-Arnoldi algorithm 3. Assume that the stopping criterion outlined above indicates that a certain approximate solution has converged at step j of the algorithm (note that j starts at $j = p$). Then the vector v_{j+1} computed at this step as well as the vectors v_{j+1+p}, v_{j+1+2p}, etc, in future steps are no longer needed. Conceptually, one can think of filling in empty (e.g., zero) vectors for these columns which will not be used in the basis, while the computation will continue in the same way, except that computations related to these empty columns will be avoided. In reality we can achieve the same effect by simply reducing the block-size by one. Thus, at the next loop we can omit to increment j and then reduce p by one. The next value of k will be the same (hitting the same location in the array V) while p has been reduced for the rest of the computation. The Hessenberg matrices H_i will have a different number of columns from the V_i's. The remaining details are fairly straightforward.

We conclude this section by presenting an experiment to illustrate the behavior of block GMRES and deflated Krylov methods. In this experiment we use block GMRES with deflation (DBGMRES) to solve 12 linear systems with random right-hand sides. Initial guesses consisting of the zero vector are used. The matrix used is SHERMAN1 from the Harwell-Boeing collection, a matrix from petroleum reservoir modeling using $10 \times 10 \times 10$ regular mesh. A simple block SSOR preconditioning was used with the relation parameter $\omega = 0.5$ and a block-size of 2. The purpose of this experiment is to show the performance of a deflated block GMRES algorithm with various values of the block size (Nblock) and the number of eigenvector used in the deflation (Neig). We tested the algorithms with the block-sizes 1 (standard GMRES), 2, 3, and 4. The number of linear systems to be solved is always 12, so for example when Nblock = 3, DBGMRES

| | Total number of iterations | | |
Nblock	Neig = 0	Neig = 2	Neig = 3
1	2,731	1,182	1,088
2	4,508	1,239	1,181
3	8,015	1,312	1,291
4	11,633	1,391	1,339

Table 1: Iterations to converge with DBGMRES and block SSOR for Sherman1 and 12 right-hand sides using various block sizes. The subspace dimension is constant and equal to 40.

is called 4 times, to solve 3 linear systems in each call. In all cases when Neig > 0 the eigenvector information obtained for solving a group of right-hand sides was recycled and using as a starting set of eigenvectors for the next group. In the first experiment, the total dimension of the subspace used is always the same regardless of the block-size. It is set to $m = 40$. The storage requirement in this case is about the same for all tests.

The total number of iterations required to reduce the initial residual by a factor of 10^{-8} for all 12 linear systems is shown in Table 1. The iterations reported are tied to matrix-vector products, so each step consists of a matrix-vector product, a preconditioning operation and other operations related to the block-GMRES process.

One interesting observation to make is that while the number of iterations operations increases very rapidly with the block-size with Neig = 0 (no deflation), it increases only moderately for Neig ≥ 2. Another point is that it is most efficient to use a block-size of one in this case. Experiments reported in the literature usually indicate that for simultaneous right-hand sides, block GMRES performs better than standard GMRES. This is certainly true when the Krylov dimension is the same. Here Krylov dimension refers to the maximum degree of the polynomials in the Block-Krylov subspace, or the integer k in (16). Thus, for solving 12 linear systems with a block size of 4, a Krylov dimension of $40/4 = 10$ is used in the experiments. Block-GMRES with a total dimension of 40 and a block-size of 4 is expected to perform better than calling standard GMRES(10) 4 times for solving 1 linear systems in each call. A comparison of this type is made in the next table where the Krylov dimension is now kept constant and equal to $k = 20$. The memory requirement for each run is now dominated by the requirement to store the basis V_m which is $m = 20 * \text{Nblock}$, so they are substantially different.

The results in the table confirm that when memory availability is not an issue then it is more efficient to use block GMRES for solving several linear systems even in the case where Neig=0. As the block-size increases so does the total number of matrix-vector products required

	Total number of iterations		
Nblock	Neig = 0	Neig = 2	Neig = 3
1	5,460	1,337	1,296
2	4,508	1,239	1,181
3	3,712	1,182	1,147
4	2,662	1,171	1,119

Table 2: Iterations to converge with DBGMRES and block SSOR for Sherman1 and 12 right-hand sides using various block sizes. The Krylov dimension is equal to 20 for each case, so the subspace dimension used is 20*Nblock.

to solve all 12 linear systems. However, the most remarkable observation is that eigenvalue deflation is capable of reaching the same effect with a much more modest demand on memory. When Neig = 0, the best result in Table 2 which is achieved with Nblock=4, i.e., with $m = 80$, is still about twice as expensive as the result achieved with Neig=2, for a block-size of only one, i.e., with $m = 20$. This common situation indicates that block-GMRES is not as memory efficient as eigenvalue deflation.

3   PRECONDITIONERS

The most important component in the success of iterative methods is undoubtedly the preconditioning technique used. Preconditioning consists of transforming the original system, into one which will be easier to solve by a Krylov subspace method.

3.1   Block Preconditioners

The matrices which arise in CFD as well as other applications, are often block matrices, meaning that they can be considered as matrices the entries of which are themselves small dense matrices, typically of the same size. In such situations, it has been observed, see for example [4], that it is more effective to treat a the dense block as a single entity instead of considering the individual elements in the block in the incomplete factorization process. One method that is particularly successful in this context is the block level-based factorization, or block ILU(k). ILU(k) is an incomplete factorization strategy which uses the concept of 'level of fill'. A level of fill number is attributed to each fill-in element based on the levels of fill of the parents from which a fill-in is created during the factorization process. Those elements whose level of fill exceeds the number k are dropped. A block version of this called BILU(k) can easily be defined. It suffices to treat the dense blocks as elements and attribute a level of fill to each block during the factorization process, by a rule similar to that used for the scalar case. It is easy to see that the block version of ILU(k) is in fact mathematically equivalent to the point version. However, it is clear that its cost, both in storage and computations, is much lower. The computation and storage of the p^2 levels fill of a $p \times p$ block are now replaced by the computation and storage of a single level of fill.

A block version of ILUT [23] can also be defined. Recall that ILUT uses a dropping strategy based on the size of the fill-ins introduced during the factorization process. A threshold value is used to determine which elements are to be dropped. In addition, once a new row for L and a new row for U have been found only the largest k elements in L and the k largest elements in U are kept. One difficulty with the block generalization of this technique is that it is difficult to find a criterion for dropping a block. The most natural measure of the size of a block is its Frobenius norm and this has been used with some success, see, e.g., [4]. However, our experience is that often this is not enough. For example, we found that often the threshold based factorization drops blocks from the original pattern and this leads to poor incomplete factorization. Forcing the factorization to keep these initial blocks is an easy remedy. However, it seems that threshold based factorizations are not yet rigorously defined and they tend to underperform the ILU(k) factorizations for block matrices.

3.2   Stabilization and Regularization

Often the factors L and U which result from incomplete factorizations are 'unstable' in that the norm of $(LU)^{-1}$ can be huge. This is to be expected. For example a forward solve with L leads to a long linear recurrence for the solution vector being computed. This linear recurrence can clearly be unstable in case the matrix L is not diagonally dominant. Similarly, the backward recurrence resulting from a solve from U can be unstable when U is not diagonally dominant. In fact we found that a simple strategy for measuring this unstability is to solve the linear system

$$(LU)x = e$$

where e is the vector of all ones. Then the norm of x gives an upper bound of the condition number of $(LU)^{-1}$ which is a good estimate. One techniques which has been used to 'stabilize' the factors L and U is simply to add a diagonal to A prior to attempting the factorization. The simplest such strategy is to add a shift α to the diagonal elements. The criteria that should used to select α are still not too well understood though there has been some work on the issue in the past, see, e.g., [12, 17, 28, 24]. The issue can be understood as follows. Adding α to the diagonal elements will improve the stability of the factorization but the resulting factors are now for a different matrix. So increasing α will improve stability at the expense of accuracy. Finding an optimal compromise is not an easy task but heuristic strategies based on finding a small enough α which yields a good factorization have been proposed.

In addition to stabilization strategies, a number of authors have also developed regularization techniques, specifically in the context of block ILU methods. In block factorizations, it is required to invert small dense matrices explicitly. These matrices are the diagonal blocks of the U factor on the incomplete factorization. Often these matrices tend to be nearly singular in which case the use of their inverses may have a negative impact on the rest of the factorization process. A commonly used remedy for this problem is to exploit the singular value decomposition, see e.g., [6, 4]. Given the singular value decomposition of a block $D = U\Sigma V^T$, an approximation to its inverse is found in the form

$$D^{-1} \approx V\tilde{\Sigma}^{-1}U^T \tag{19}$$

where $\tilde{\Sigma}$ is the diagonal matrix whose diagonal entries are obtained from thresholding the singular values σ_i. Specifically, let σ_1 be the largest singular value of D and let α be a positive scalar ≤ 1. Then, each diagonal entry $\tilde{\sigma}_i$ is defined by

$$\tilde{\sigma}_i = \begin{cases} \sigma_i & \text{if} \quad \sigma_i \geq \alpha\sigma_1 \\ \alpha\sigma_1 & \text{Otherwise} \end{cases}$$

The approximate inverse defined in this fashion is guaranteed to have a condition number no worse than $1/\alpha$. Clearly, the larger α the more likely it is to obtain a stable factorization. However, the larger α the less accurate the resulting factorization.

3.3   Sparse approximate inverses

Sparse approximate inverse techniques have been recently advocated to remedy some of the difficulties that arise with standard factorizations [2, 10, 8, 11, 16, 15, 7, 5]. This approach consists of finding a matrix M which is sparse and which approximates the inverse of A directly. For example a number of strategies are based on find a sparse matrix M such that the matrix AM is close to the identity matrix. This can be achieved by attempting to minimize the Frobenius norm of $I - AM$, for example. Now the preconditioning operation will consist of a matrix-vector product with the matrix M instead of a solve.

These methods have been promoted for two distinct objectives in mind. The first objective is that of parallelism. These preconditioners are easy to compute in parallel and they are also easy to apply. The second objective

is that of robustness. Since there are no recurrences in applying the preconditioning operations, the resulting preconditioners are perfectly stable. However, our experience is that the success of approximate inverse techniques in solving difficult problems is mixed. These methods tend to be more successful for small matrices than for very large ones. For this reason, it seems that the most promising use of approximate inverse techniques is in combining them with other factorizations, such as block factorizations. For example, they can be used in developing block-incomplete factorizations for block-tridiagonal matrices, see e.g., [14]. Another idea described in [5] is to exploit blockings of the original linear system in the block form,

$$\begin{pmatrix} B & F \\ E & C \end{pmatrix} \begin{pmatrix} x \\ y \end{pmatrix} = \begin{pmatrix} f \\ g \end{pmatrix}. \tag{20}$$

This blocking may originate from a domain decomposition partitioning for example or may be the original structure of A, as in the case of the Navier-Srokes equations for incompressible flow.

A simple way to obtain an approximate factorization to A from the above blocking is to approximate the block LU factorization

$$M = LU$$

in which

$$L = \begin{pmatrix} B & 0 \\ E & S \end{pmatrix} \quad \text{and} \quad U = \begin{pmatrix} I & B^{-1}F \\ 0 & I \end{pmatrix}$$

where S is the Schur complement

$$S = C - EB^{-1}F.$$

Since F is a sparse block, the general approach of approximate inverse techniques can be effectively used to construct a sparse approximation Y to $B^{-1}F$. Once Y is obtained, then S is replaced by $M_S = C - EY$ in the L factor and $B^{-1}F$ is replaced by Y in the U factor. This yields an approximate block-factorization, Interestingly enough, this technique is remarkably robust, possibly due to the inherent decoupling of the block structure.

4   CONCLUSION

A variety of enhancements to preconditioned Krylov methods have resulted from several decades of research. Often it is difficult to predict in advance when these techniques will be helpful. Thus, eigenvalue deflation can be very useful in some cases, but may lead to little or no improvement in other cases. It is certain that some of these techniques should be implemented in any production code to enhance robustness since their additional cost is marginal. These include the techniques of eigenvalue deflation and regularization techniques for block factorization for example. Iterative methods based on preconditioned Krylov solvers require sophisticated drivers to decide which techniques are the most likely to succeed given some information on the problem. This 'black-box' approach to solving general linear systems is still elusive. Meanwhile, iterative solvers are most successful when information from the physical problem is used to extract a good preconditioner.

Acknowledgements. The Minnesota Supercomputer Institute provided the computer resources and an excellent research environment to conduct this research. This work would not have been possible without the help of Andrew Chapman in developing deflated GMRES and block GMRES codes as well as many other programs.

References

[1] O. Axelsson and P. S. Vassilevski. A block generalized conjugate gradient solver with inner iterations and variable step preconditioning. *SIAM Journal on Matrix Analysis and Applications*, 12, 1991.

[2] M. W. Benson and P. O. Frederickson. Iterative solution of large sparse linear systems arising in certain multidimensional approximation problems. *Utilitas Math.*, 22:127–140, 1982.

[3] A. Chapman and Y. Saad. Deflated and augmented Krylov subspace techniques. *Numerical Linear Algebra with Applications*, 1996. To appear.

[4] A. Chapman, Y. Saad, and L. Wigton. High-order ILU preconditioners for CFD problems. Technical Report UMSI 96/14, Minnesota Supercomputer Institute, 1996.

[5] E. Chow and Y. Saad. Approximate inverse techniques for block-partitioned matrices. *SIAM Journal on Scientific Computing*, 1997. To appear.

[6] E. Chow and Y. Saad. Experimental study of ilu preconditioners for indefinite matrices. Technical Report UMSI 97/95, Minnesota Supercomputer Institute, University of Minnesota, Minneapolis, MN, 1997.

[7] E. Chow and Y. Saad. Approximate inverse preconditioners for general sparse matrices. *SIAM Journal on Scientific Computing*, –, 1998. To appear.

[8] J. D. F. Cosgrove, J. C. Díaz, and A. Griewank. Approximate inverse preconditioning for sparse linear systems. *Intl. J. Comp. Math.*, 44:91–110, 1992.

[9] J. Erhel, K. Burrage, and B. Pohl. Restarted GMRES preconditioned by deflation. Technical Report –, IRISA, Rennes, France, 1994. to appear, Journal of Computational and Applied Mathematics, Dec. 1995.

[10] M. Grote and H. D. Simon. Parallel preconditioning and approximate inverses on the connection machine. In R. F. Sincovec, D. E. Keyes, L. R. Petzold, and D. A. Reed, editors, *Parallel Processing for Scientific Computing – vol. 2*, pages 519–523. SIAM, 1992.

[11] T. Huckle and M. Grote. A new approach to parallel preconditioning with sparse approximate inverses. Technical Report SCCM-94-03, Stanford University, Scientific Computing and Computational Mathematics Program, Stanford, California, 1994.

[12] D. S. Kershaw. The incomplete Choleski conjugate gradient method for the iterative solution of systems of linear equations. *Journal of Computational Physics*, 26:43–65, 1978.

[13] S. A. Kharchenko and A. Yu. Yeremin. Eigenvalue translation based preconditioners for the GMRES(k) method. *Numerical Linear Algebra with Applications*, 2:51–70, 1995.

[14] L. Yu Kolotilina and A. Yu Yeremin. On a family of two-level preconditionings of the incomplete block factorization type. *Soviet Journal of Numerical Analysis and Mathematical Modeling*, 1:293–320, 1986.

[15] L. Yu. Kolotilina and A. Yu. Yeremin. Factorized sparse approximate inverse preconditionings II. solution of 3-D FE systems on massively parallel computers. Technical Report EM-RR 3/92, Elegant Mathematics, Inc., Bothell, Washington, 1992.

[16] L. Yu. Kolotilina and A. Yu. Yeremin. Factorized sparse approximate inverse preconditionings I.Theory. *SIAM Journal on Matrix Analysis and Applications*, 14:45–58, 1993.

[17] T. A. Manteuffel. An incomplete factorization technique for positive definite linear systems. *Mathematics of Computations*, 34:473–497, 1980.

[18] R. B. Morgan. Computing interior eigenvalues of large matrices. *Linear Algebra and its Applications*, 154:289–309, 1991.

[19] R. B. Morgan. A restarted GMRES method augmented with eigenvectors. *SIAM J. Matrix Analysis and Applications*, 16:1154–1171, 1995.

[20] A. Ruhe. Implementation aspects of band Lanczos algorithms for computation of eigenvalues of large sparse symmetric matrices. *Mathematics of Computations*, 33:680–687, 1979.

[21] Y. Saad. *Numerical Methods for Large Eigenvalue Problems*. Halstead Press, New York, 1992.

[22] Y. Saad. A flexible inner-outer preconditioned GMRES algorithm. *SIAM Journal on Scientific and Statistical Computing*, 14:461–469, 1993.

[23] Y. Saad. ILUT: a dual threshold incomplete ILU factorization. *Numerical Linear Algebra with Applications*, 1:387–402, 1994.

[24] Y. Saad. Preconditioned Krylov subspace methods for CFD applications. In W. G. Habashi, editor, *Solution Techniques for large-scale CFD Problems*, pages 141–157. J. Wiley and sons, 1995.

[25] Y. Saad. *Iterative Methods for Sparse Linear Systems*. PWS publishing, New York, 1996.

[26] Y. Saad and M. H. Schultz. GMRES: a generalized minimal residual algorithm for solving nonsymmetric linear systems. *SIAM Journal on Scientific and Statistical Computing*, 7:856–869, 1986.

[27] A. Stathopoulos, Y. Saad, and K. Wu. Dynamic thick restarting of the davidson, and the implicitly restarted arnoldi methods. Technical Report UMSI 96/123, Minnesota Supercomputer Institute, University of Minnesota, Minneapolis, MN, 1996.

[28] H. A. van der Vorst. Stabilized incomplete lu-decompositions preconditionings for the tchebycheff iteration. In D. J. Evans, editor, *Preconditioning Methods: Analysis and Applications*, New-York, NY, 1983. Gordon and Breach.

[29] H. A. van der Vorst and C. Vuik. GMRESR: a family of nested GMRES methods. *Numerical Linear Algebra with Applications*, 1:369–386, 1994.

MULTILEVEL ELLIPTIC SOLVERS
ON UNSTRUCTURED GRIDS

Tony F. CHAN   Susie Go   Ludmil ZIKATANOV

Department of Mathematics, University of California at Los Angeles;
Los Angeles, CA 90095-1555, USA.
E-mail: chan@math.ucla.edu, sgo@math.ucla.edu, lzikatan@math.ucla.edu.

Abstract

An overview of multilevel methods on unstructured grids for elliptic problems will be given. The advantages which make such grids suitable for practical implementations are flexible approximation of the boundaries of complicated physical domains and the ability to adapt the mesh to resolve fine-scaled structures in the solution. Multilevel methods, which include multigrid methods and overlapping and non-overlapping domain decomposition methods, depend on proper splittings of appropriate finite element spaces: either by dividing the original problem into subproblems defined on smaller subdomains, or by generating a hierarchy of coarse spaces. The standard splittings used in structured grid case cannot be directly extended for unstructured grids because they require a hierarchical grid structure, which is not readily available in unstructured grids. We will discuss some of the issues which arise when applying multilevel methods on unstructured grids, such as how the coarse spaces and transfer operators are defined, and how different types of boundary conditions are treated. An obvious way to generate a coarse mesh is to re-grid the physical domain several times. We will propose and discuss different and possibly better alternatives: node nested coarse spaces and agglomerated coarse spaces.

1   Introduction

In this article, multilevel methods applied to elliptic problems on general unstructured grids will be discussed. We will describe various approaches for dealing with the solution of discrete equations arising from unstructured grids. Our interest will be in the performance of multilevel methods, including multigrid and domain decomposition methods.

The beauty of multilevel methods is that the convergence speed can often be proven to be independent of the problem size and they can be naturally parallelized. This makes them the most powerful and useful tool for a wide variety of applications. On the other hand, these methods require a hierarchical grid structure, which is not readily available in unstructured grids. In our context, we use them not as solvers on their own, but rather as preconditioners for Krylov subspace iterative methods.

Various approaches for dealing with these issues and their effect on the convergence properties of these methods will be covered. This article is organized as follows: Section 1 begins with an introduction to Krylov subspace methods and multilevel methods, followed by some two-level theory in Section 2. Specific examples of how to deal with node-nested multilevel methods are covered in Section 3. Section 4 concerns agglomerated multigrid methods.

Many of the topics described here represent previous and continuing joint work with Barry Smith and Jun Zou [2,3,4,5], (Section 3) and with Jinchao Xu [6] (Section 4).

1.1   Elliptic problems

Elliptic problems are one of the most extensively investigated problems in applied mathematics. Their relation to many physical models is well known and the theoretical and numerical results obtained in this area are very useful in practice. As a first approximation to more complicated physical and mathematical models

Received on September 1, 1997.
ONR under contract ONR-N00014-92-J-1890, and the Army Research Office under contract DAAL-03-91-C-0047 (Univ. of Tenn subcontract ORA4466.04 Amendment 1 and 2). The first two authors acknowledge support from RIACS under contract number NAS 2-13721 for visits to RIACS/NASA Ames. The third author is supported by Grant ONR-N00014-92-J-1890 and NSF Grant Int-95–06184.
This paper is an abridged version of the lecture notes which were prepared for the lecture course "28th Computational Fluid Dynamics", 3-7 March, 1997, at the von Karman Institute for Fluid Dynamics, Belgium

(such as those in computational fluid dynamics), elliptic problems are sometimes the only ones for which rigorous theoretical results are known. The design of numerical methods for such model problems can often be adapted and applied to more complicated situations. Elliptic problems are also important in their own right, for example in computational fluid dynamics in the solution of the pressure equation, implicit time integration schemes, etc.

In this section, we will state the model problems we consider. Our goal is to design effective solvers for the resulting systems of linear equations, and we will not pay much attention to the discretization techniques. Detailed discussions of the finite element element discretizations that we use can be found in [7, 8, 9, 10].

Let $\Omega \subset \mathbb{R}^d$ be a polygonal (polyhedral) domain, $d = 2, 3$. We consider the following variational (or Galerkin) formulation of an elliptic problem: Find $u \in H_0^1(\Omega; \Gamma_D)$ such that:

$$a(u, v) = F(v) \text{ for all } v \in H_0^1(\Omega; \Gamma_D), \tag{1.1}$$

where

$$\begin{aligned}
a(u, v) &= \int_\Omega \alpha(x) \nabla u \nabla v dx, \\
F(v) &= \int_\Omega F(x) v dx.
\end{aligned} \tag{1.2}$$

Here $H_0^1(\Omega; \Gamma_D)$ denotes the Sobolev space which contains functions which vanish on Γ_D with square integrable first derivatives. It is well known that (1.1) is uniquely solvable if $\alpha(x)$ is a strictly positive scalar function and F is square integrable.

We will use the simplest finite element discretization of the elliptic problem (1.1). First, we cover Ω with simplicial finite elements (triangles in $\mathbb{R}^2$ and tetrahedra in $\mathbb{R}^3$). Then the discrete problem can be formulated as follows:

Find $u_h \in V_h$ such that:

$$a(u_h, v_h) = F(v_h) \text{ for all } v_h \in V_h, \tag{1.3}$$

where V_h is the finite dimensional subspace of $H_0^1(\Omega; \Gamma_D)$ consisting of continuous functions linear on each of the simplexes forming the partition.

The values of the discrete solution on the grid nodes are then determined by solving the resulting system of linear equations:

$$Au = f, \tag{1.4}$$

where A is a symmetric and positive definite matrix, f is the right hand side and the nodal values of the discrete solution u_h will be obtained in u after solving the system (1.4). To obtain an accurate enough approximate solution of (1.1), one often has to solve huge discrete problems which are badly conditioned, with condition number growing like $O(h^{-2})$, where h is the characteristic mesh size. Our goal in the next sections will be to construct robust and effective methods for solving the discrete equations (1.4).

1.2 Unstructured grids

With the vast improvements in computational resources today, the motivating reasons for using structured grids over unstructured grids become less obvious. Cartesian or mapped Cartesian grids are popular because they are directional, so efficient methods can be used, such as the alternating direction implicit methods (ADI) and fast Fourier transforms (FFT). This structure, however, imposes limitations on the types of domains which can be considered. In addition, local refinement cannot be easily done without affecting large portions of the grid, so the ability to adapt the grids for resolving steep gradients in the solution is a source of difficulty.

One of the alternative approaches for dealing with complicated geometries is the composite grid method as proposed by Brown, Chesshire, Henshaw and Kreiss [11] and Chesshire and Henshaw [12].

Unstructured grids provide the flexibility needed to adapt to rapidly changing or dynamic solutions as well as complex geometries. These grids have irregular connectivity and so do not have to adhere to the strict structure of Cartesian-based grids, see Figure 1.1. The tradeoff is that, computations on unstructured grids require more complicated data structures and possibly modifications in some solvers, e.g. multilevel methods. We will discuss some such modifications in multilevel methods in this article.

1.3 Preconditioned iterative methods

As mentioned in the introduction, multilevel methods will be used in our framework as preconditioners in

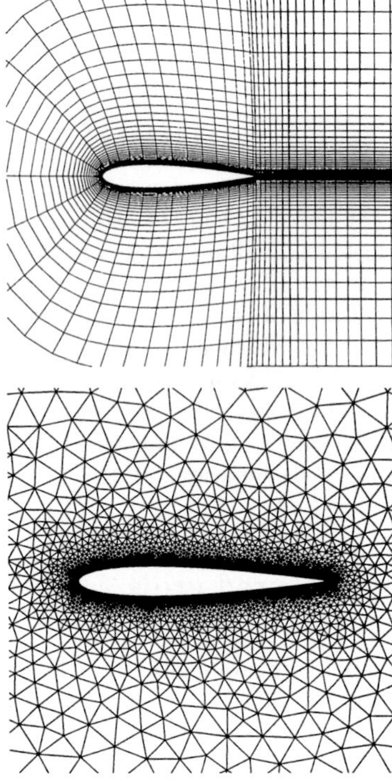

Figure 1.1: A structured grid (top) and unstructured grid (bottom).

Krylov subspace iterative methods, the most popular of which are: the Conjugate Gradient (CG) and the Generalized Minimum Residual (GMRES) method. In addition, there are many other Krylov subspace methods which can be used as alternatives, especially for non-symmetric A's. Some popular methods are BiCGSTAB, CGS, QMR, TFQMR, BCG and their many variants. At this point there is no widespread agreement on the relative merits of these methods. For more details concerning these methods we refer to the recent book by Saad [13].

It is often beneficial, before applying any iterative method, to write (1.4) in the following *preconditioned* form:

$$M^{-1}Au = M^{-1}f. \tag{1.5}$$

where M is called the preconditioner for A. The choice of M is very important because it can improve the convergence rate of the iterative method. A good preconditioner M for A should have the following properties:

- The action of $M^{-1}v$ for a given vector v should be less expensive to compute than $A^{-1}v$.

- The condition number $\kappa(M^{-1}A)$ should be as close to 1 as possible, preferably uniformly bounded above (with respect to the mesh size h).

- If A is SPD then M should be SPD.

When A and M are both SPD, it is more convenient to work with the symmetrized version of (1.5):

$$\begin{aligned}
\hat{A}\hat{u} &= \hat{f}, \quad \text{where} \quad \hat{A} = M^{-\frac{1}{2}}AM^{-\frac{1}{2}}, \\
\hat{u} &= M^{\frac{1}{2}}u \quad \text{and} \quad \hat{f} = M^{-\frac{1}{2}}f.
\end{aligned} \tag{1.6}$$

Oftentimes, it is useful to apply different preconditioners M_i at each step (e.g. inner inexact solvers). In this way, one obtains flexible solvers which can handle a wider class of problems. The standard Krylov subspace methods must be modified to handle such non-stationary preconditioners. One such method is known as *flexible* GMRES (see Saad [13]).

In the next sections, our particular interest will be focused on multilevel methods (such as domain decomposition methods and multigrid methods) used as preconditioners in PCG. The popularity of these methods as preconditioners is based on the fact that they exactly fit in the applications where finite element or finite difference method is used. In other words, the design of such preconditioners uses the properties of finite element spaces which allow precise optimal constructions and theoretical analysis to be done.

1.4   Multilevel methods

For many practical problems, the system of linear equations which arises from finite element or finite difference discretizations might be huge. A challenge is how to effectively solve such large systems of linear equations. Direct methods face the problem of excessive memory requirements and number of the floating point operations needed. In this connection, iterative methods, and especially multilevel methods such as multigrid and domain decomposition methods, are very attractive. These methods are popular because the amount of work required to solve a problem is on the order of the number of unknowns, the convergence rates are independent of the problem size and they can be easily parallelized.

1.4.1   Multigrid methods.

In this section, we briefly describe the multigrid methods for solving linear systems of discrete equations.

We will consider the case where these systems are obtained via finite element discretization of an elliptic partial differential equation. Detailed discussion on multigrid methods can be found in standard references, e.g. Briggs [14], Bramble [15], Hackbusch [16], and Xu [17, 18].

The idea behind multigrid methods is based on the fact that simple relaxation schemes such as Gauß-Seidel, Jacobi and Richardson possess a good smoothing property: they reduce the highly oscillatory part of the error very well in few iterations. This part of the error lies in the subspace spanned by the eigenvectors corresponding to large eigenvalues, i.e. the high frequencies. The global error, or the low frequencies unfortunately cannot be corrected well by such iterative schemes and this is where multigrid helps. The low frequencies from fine grid (say original one) are transfered to the coarse grid, where they behave like high frequencies, and are smoothed quickly by a simple relaxation scheme. Recursive application of this idea leads to the multigrid method.

We will denote the space which contains the solution u by V_J. We assume that the coarse grids are given and with each grid we associate a finite dimensional space (like V_J for the fine grid). We denote these spaces by $V_0, \ldots, V_{J-1}$. To unify the notation in this section we define $A_J := A$. We assume that the operators A_k, $k = 0, \ldots J - 1$, are given (these operators correspond to different approximations of A on the coarse grids). We also assume that the prolongation operator R_k^T and the smoothing operators S_k are also given. One can consider the action of the smoother on $g \in V_k$ as a fixed number of Gauß-Seidel or Jacobi iterations with right-hand side g and zero initial guess.

We view the multigrid method as a way of defining a preconditioner M_J. We will describe in matrix notation the action $M_J^{-1}g$ in the simplest case when one pre- and post-smoothing steps are applied.

The action of M_k^{-1} is then obtained through the following steps:

ALGORITHM 1.1 *(V-cycle multigrid preconditioner $M_k^{-1}g$)*

0. *If $k = 0$ then $M_0^{-1}g = A_0^{-1}g$*
1. Pre-smoothing: *Apply one transposed smoothing iteration with initial guess $x^0 = 0$ and right hand side g, i.e.*

$$x^1 = S_k^T g$$

2. Coarse grid correction:

(a) *Restrict the residual:*

$$q^0 = R_k(I - A_k S_k^T)g.$$

(b) *"Solve" on the coarse grid:*

$$\begin{aligned} q^1 &= M_{k-1}^{-1} q^0 \\ &= M_{k-1}^{-1} R_k (I - A_k S_k^T)g. \end{aligned}$$

(c) *Interpolate back and correct:*

$$\begin{aligned} x^2 &= x^1 + R_k^T q^1 \\ &= \left[S_k^T + R_k^T M_{k-1}^{-1} R_k (I - A_k S_k^T) \right] g. \end{aligned}$$

3. Post-smoothing: *Apply one smoothing iteration with initial guess x^2 and right hand side g, i.e.*

$$\begin{aligned} M_k^{-1}g &= x^2 + S_k(g - A_k x^2) \\ &= \big[S_k + S_k^T - S_k A_k S_k^T \\ &\quad + (I - S_k A_k) R_k^T M_{k-1}^{-1} R_k (I - A_k S_k^T) \big] g. \end{aligned}$$

Note that the above definition is recursive, the action of $M_k^{-1}g$ is defined in terms of $M_{k-1}^{-1}g$. Let us now consider the simplest case: a two-level method (when $J = 1$). For the sake of simplicity we omit the index 1 in the next equation. The preconditioner then is defined as follows:

$$\begin{aligned} M^{-1}g &= \big[S + S^T - SAS^T \\ &\quad + (I - SA) R^T A_0^{-1} R (I - AS^T) \big] g. \end{aligned}$$

1.4.2 Domain decomposition methods

Domain decomposition (DD) methods are divide-and-conquer methods which take a large problem defined on a physical domain, and appropriately decompose it into many smaller problems defined on subdomains. These smaller subdomain problems can then be solved quickly and independently of each other and their solution suitably combined, usually via an iterative process to obtain the solution to the original problem. Domain decomposition methods fall into two broad categories: overlapping DD (Schwarz methods) and nonoverlapping DD (substructuring or Schur complement methods). Our description here follows that in Chan-Mathew [19]; see also the recently published book by Smith, Bjørstad and Gropp [20]. We will not discuss the nonoverlapping domain decomposition methods here. For a detailed description and investigation of these methods we refer to [19, 20].

1.4.2.1 Overlapping DD In overlapping DD methods, a set of p overlapping subdomains are formed by taking a set of nonoverlapping subdomains $\{\Omega_i'\}_{i=1}^p$, and extending them to larger subdomains, $\{\Omega_i\}_{i=1}^p$ by some small distance. δ, see Fig. 1.2. The partitioning induced by such a decomposition amounts to an overlapping block decomposition of the system (1.4). Thus, the overlapping DD methods can be thought of as block iterative solvers, either overlapping block Jacobi or block Gauß-Seidel, depending on whether or not the the most updated iterates are used for boundary conditions.

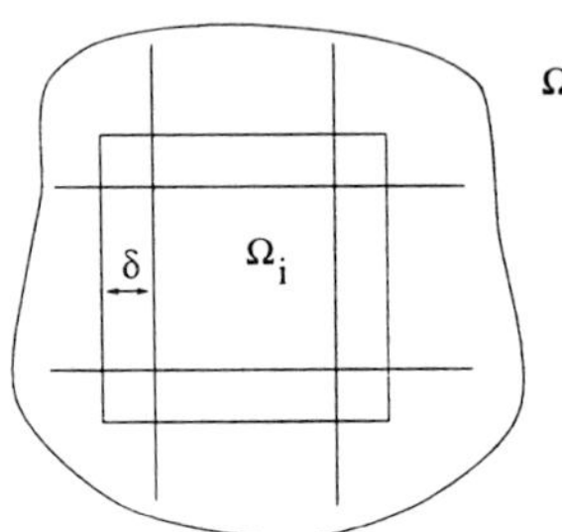

Figure 1.2: Generating a set of overlapping subdomains.

The main ingredients required in all DD methods are:

- *Restriction matrices*: Let R_i be the $n_i \times n$ restriction matrix of 1's and 0's which takes a full-length vector in $\mathbb{R}^n$ and maps it to a restricted vector in $\mathbb{R}^{n_i}$, where n_i denotes the number of unknowns in subdomain Ω_i. The effect on an n-vector is injection onto the subdomain, Ω_i.

- *Extension matrices*: Let R_i^T be the $n \times n_i$ extension matrix, which is defined as the transpose of the restriction matrix, R_i. The effect on an n_i-vector is identity on the subdomain, Ω_i, and zero extension outside the subdomain, i.e. on $\Omega \setminus \Omega_i$.

- *Subdomain matrices*: Define the local stiffness matrix on Ω_i to be $A_i = R_i A R_i^T$, where $A_i \in \mathbb{R}^{n_i \times n_i}$. Because the restriction and interpolation matrices consist only of 0's and 1's, the local stiffness matrices are simply principal submatrices of **A**.

- *Subdomain solvers*: Let A_i^{-1} symbolically denote the solver for the restricted operator. These can be either exact or inexact solvers (see Sec. 1.4.2).

The additive Schwarz (block Jacobi) method on p subdomains is given by:

$$u^{k+i/p} = u^{k+(i-1)/p} + R_i^T A_i^{-1} R_i(f - Au^k), \quad i = 1, ..., p.$$

In this form, it is seen that corrections are done simultaneously on p subdomains. Rewriting this as one equation reveals the preconditioned iterative method:

$$u^{k+1} = u^k + M_{as}^{-1}(f - Au^k)$$

where the preconditioner M_{as} is given by:

Additive Schwarz preconditioner.
(*block Jacobi on* A)

$$M_{as}^{-1} = \sum_{i=1}^p R_i^T A_i^{-1} R_i. \tag{1.7}$$

Instead of simultaneous corrections, the corrections can also be done successively, to yield the multiplicative Schwarz (block Gauß-Seidel) method, for $i = 1, ..., p$:

$$u^{k+i/p} = u^{k+(i-1)/p} + R_i^T A_i^{-1} R_i(f - Au^{k+(i-1)/p}).$$

Because the most currently updated information is used, this method will generally converge faster than additive Schwarz. The drawback is that it is less parallel (but this can be remedied by appropriate coloring of the subdomains).

For the multiplicative Schwarz method on p subdomains, the preconditioner can be written as:

Multiplicative Schwarz preconditioner.
(*block Gauß-Seidel on* A)

$$M_{ms}^{-1} = [I - \prod_{i=1}^p (I - R_i^T A_i^{-1} R_i A)]A^{-1}. \tag{1.8}$$

1.4.2.2 Coarse grid The domain of dependence for elliptic problems is the entire domain, but because Schwarz methods decompose the problem into smaller, independent problems, information from one subdomain must travel large distances to reach another subdomain. To avoid deterioration of the convergence rates of these methods, some sort of mechanism for the global transfer of data is needed. This is achieved, to some degree, by the overlapping of subdomains in the Schwarz methods. More overlap leads to more coupling between subdomains. However, this adds redundant work and communications overhead if too much overlap is introduced. Dryja and Widlund [21, 22] showed that the condition number for additive Schwarz (1.7) is given by:

$$\kappa(M_{as}^{-1}A) = O\left(H^{-2}\left(1 + \left(\frac{H}{\delta}\right)^2\right)\right).$$

The condition number is independent of h. For sufficient amount of overlap (choosing $\delta = O(H)$), the condition number is $O(H^{-2})$ and so will increase as H tends to zero. This means that the method will not be scalable to a large number of processors.

This deterioration can be remedied by introducing a coarse grid to achieve additional global coupling. In addition to the subdomain restriction, interpolation and stiffness matrices used in the one-level Schwarz methods, we need coarse versions of them: R_H, R_H^T, $A_H = R_H A R_H^T$, and A_H^{-1}. Here, R_H and R_H^T will instead be the full weighting restriction and linear interpolation matrices, respectively, which are commonly used in multigrid methods. The two-level additive Schwarz preconditioner can then be written as:

Additive Schwarz preconditioner with coarse grid.

$$M_{asc}^{-1} = R_H^T A_H^{-1} R_H + \sum_{i=1}^{p} R_i^T A_i^{-1} R_i.$$

It can be shown that the condition number for this two-level method is (see Section 2):

$$\kappa(M_{asc}^{-1}A) = O(1 + (H/\delta)^2),$$

and the method can be made independent of H, h with sufficient overlap by choosing $\delta = O(H)$.

1.4.2.3 Multilevel Schwarz Multilevel Schwarz is an extension of two-level Schwarz with L different coarse levels, each level being decomposed into p_l subdomains as previously described. We will denote the ith subdomain on the lth level as: Ω_i^l. Several different variants of multilevel Schwarz can be created, depending on when the most currently updated information is used:

- Fully additive multilevel methods would be additive among subdomains on the same level as well as additive between levels.

- Multilevel methods which are multiplicative among subdomains on the same level, but additive between levels can be viewed as "additive MG".

- Classical V-cycle MG can be viewed as a multilevel Schwarz method which is multiplicative both among subdomains on the same level as well as between levels.

The fully additive multilevel Schwarz preconditioner can be written as:

Fully additive multilevel Schwarz preconditioner.

$$M_{mlas}^{-1} = \sum_{l=1}^{L} \sum_{i=1}^{p_l} (R_i^l)^T (A_i^l)^{-1} (R_i^l).$$

1.4.2.4 Inexact subdomain solves In all of the domain decomposition methods described above, subdomain solves, A_i^{-1}, are required. These can be done either exactly or inexactly. Though the subdomain and coarse problems are much smaller than the original problem, it can still be quite expensive to attempt exact solves on these problems.

In the two-level additive Schwarz methods, we can simply replace the exact solves with inexact solves. Let $M_i \approx A_i$, $M_H \approx A_H$ represent the inexact solves. Then the preconditioner is given by:

$$\tilde{M}_{asc}^{-1} = R_H^T M_H^{-1} R_H + \sum_{i=1}^{p} R_i^T M_i^{-1} R_i.$$

1.5 Approaches for designing multilevel methods on unstructured grids

Multilevel methods require a hierarchical grid structure. For structured grids, the hierarchy can be recovered from the fine grid. For unstructured grids, however, there is no natural grid hierarchy. In addition, their lack of structure prevents these methods from exploiting regularity and using fast solvers as with structured grids. Difficulties exist in identifying coarse grid problems/spaces/boundary conditions which do not occur when using structured grids. The algorithms which are based on unstructured grids must be redesigned to handle these issues without sacrificing too much in terms of complexity and performance.

There are several approaches for constructing the coarse spaces for unstructured grids. One approach is to simply apply algebraic versions of multigrid methods [23]. However ignoring the geometric and differential nature of the underlying problem may lead to suboptimal performance.

Another approach (see Mavriplis [24]) is based on independently generated coarse grids and piecewise linear interpolation between the grids. The advantage of this approach is convenience: the coarse grids can be generated by using the same grid generator which produced the original fine grid. The disadvantage is that the interpolations can be expensive to apply since the set of nodes in the coarse grids are not related in any way to the nodes in the fine grid. Thus no fast search routines can be applied and the implementation will be $O(n^2)$.

An alternative approach is based on generating node-nested coarse grids, which are created by selecting subsets of a vertex set, retriangulating the subset, and using piecewise linear interpolation between the grids (see [25, 26]). This still provides an automatic way of generating coarse grids and now faster implementations of the interpolation (can be implemented in $O(n)$ time). The drawback is that in three dimensions, retetrahedralization can be problematic.

Another effective coarsening strategy proposed by Bank and Xu [27] uses the geometrical coordinates of the fine grid (which is available in most cases).

New coarsening strategies based on the algebraic approach recently were published by Hackbusch [28], Braess [29] and Reusken [30].

In many of these approaches, problems may occur in producing coarse grids which are valid and with boundaries which preserve the important features of the fine domain. One of most popular and promising new coarsening techniques which avoids this problem is based on the agglomeration technique (see Koobus, Lallemand and Dervieux [31]). Instead of constructing a proper coarse grid, neighboring fine grid elements are agglomerated together to form macroelements. Since these agglomerated regions are not standard finite elements, appropriate basis functions and interpolation operators must be constructed on them. Such algorithms have also been investigated by Mandel, Vaněk, Brezina [32]) and Vaněk, Křižková [33].

2 Introduction to convergence theory

As mentioned in Section 1.3, the estimate of the convergence rate of the PCG requires an estimate of the upper bound of $\kappa(M^{-1}A)$. In particular, estimates on the extreme eigenvalues of $M^{-1}A$ must be obtained. In this section, we first give a general framework for bounding $\kappa(M^{-1}A)$ and then we show how such an analysis can be carried out for the overlapping domain decomposition method. Such an analysis can show (or predict) the convergence rate and in most cases gives a good guess as to how the parameters and approximate operators should be chosen in order to get an optimal iterative method. For a similar approach in analyzing the convergence properties of iterative methods using general subspace splittings for structured meshes, we refer to [17].

We shall adopt a matrix approach for analyzing the domain decomposition methods, in the hope that it is more intuitive and easier to understand. In the analysis, we state the important theoretical results omitting their proofs. Detailed presentations of the analysis, rigorous proofs of the results quoted here, and more references can be found in [19], [20].

Although we will present the domain decomposition methods in matrix formulation, the use of Sobolev norms and semi-norms in $H^k(\Omega)$ cannot be avoided in a few places, so we will denote them by the conventional notation: $\|u\|_k$ (or $\|u\|_{k,\Omega}$) and $|u|_k$ (or $|u|_{k,\Omega}$), respectively (see [20]).

2.1 Subspace correction framework: matrix formulation

Our initial setting in matrix form is as follows: Let Ω be covered by p overlapping subdomains Ω_i, $i = 0, 1, \ldots, p$. Each subdomain Ω_i corresponds to a subspace $V_i \subset \mathrm{I\!R}^n$. The subspaces are defined through the restriction operators $R_i \in \mathrm{I\!R}^{n_i \times n}$, $i = 0, 1, \ldots, p$, and we set $V_i = \mathrm{Range}(R_i)$ Note here that we will interchangeably use the notation for the coarse grid versions denoted with subscript H in the previous section, with the subscript 0, when convenient. Here, V_0 denotes the coarse space.

We wish to construct a preconditioner for solving the following linear algebra problem

$$Au = f, \quad A \in \mathrm{I\!R}^{n \times n} \text{ is SPD.} \qquad (2.1)$$

Let us first explain the intuition behind the construction of a preconditioner based on this splitting of $\mathbb{R}^n$. It is natural to take the best approximation to the solution from each subspace, and then extend these different approximations to the whole $\mathbb{R}^n$ somehow in order to get a global solution. Thus the question is: What is the best correction to the k-th iterate u^k from V_i?

If we measured the error in the A-norm, $\|\cdot\|_A$, then this question can be reformulated as the following minimization problem:

$$\min_{y_i} \|(u^k + R_i^T y_i) - A^{-1}f\|_A. \tag{2.2}$$

The solution is given by:

$$\begin{aligned} y_i &= (R_i A R_i^T)^{-1} R_i (f - Au^k) \\ &= A_i^{-1} R_i (f - Au^k). \end{aligned} \tag{2.3}$$

The next iterate is then obtained via the equation (note that we correct here only in one subspace V_i)

$$u^{k+1} = u^k + R_i^T A_i^{-1} R_i (f - Au^k) \tag{2.4}$$

Example. The Jacobi iteration (see Section 1.3) corresponds to the splitting $V_i = span\{e_i\}$ where e_i is the i-th unit coordinate vector. The restrictions, R_i, in this case are defined as $R_i v = (v, e_i)e_i$.

Performing these subspace corrections simultaneously gives the additive subspace correction preconditioner:

$$M_{asc}^{-1} = \sum_{i=0}^{p} R_i^T A_i^{-1} R_i.$$

Defining now the projections $P_i \equiv R_i^T A_i^{-1} R_i A$, for $i = 0, \ldots p$, we get

$$M_{asc}^{-1} A = \sum_{i=0}^{p} P_i.$$

As we pointed out earlier, the convergence of the PCG method depends on the condition number of $M^{-1}A$. Thus our goal is to find an upper bound for $\kappa(M^{-1}A)$ which amounts to finding an upper bound for $\lambda_{\max}(M_{asc}^{-1}A)$ and a lower bound for $\lambda_{\min}(M_{asc}^{-1}A)$.

The estimate on the upper bound for $\lambda_{\max}(M_{asc}^{-1}A)$ is easier and it follows directly from the following simple lemmas.

Lemma 2.1 *P_i is a projection in $(\cdot, \cdot)_A$, i.e.*

$$AP_i = P_i^T A, \quad P_i^2 = P_i \quad \|P_i\|_A \leq 1.$$

Lemma 2.2 *The maximal eigenvalue of the preconditioned matrix satisfies the following inequality:*

$$\lambda_{\max}(M_{asc}^{-1}A) \leq p + 1.$$

REMARK. The bound given in the previous lemma can be easily improved to:

$$\lambda_{\max}(M_{asc}^{-1}A) \leq n_c + 1 \equiv c_1,$$

where n_c is the *number of colors to color Ω_i 's* in such a way that no two neighboring subdomains are colored the same color.

We next give an estimate on the lower bound for $\lambda_{\min}(M_{asc}^{-1}A)$. This estimate is based on the following *Partition Lemma* which plays a crucial role in the convergence analysis of the domain decomposition methods.

Lemma 2.3 (Partition Lemma). *(Matsokin-Nepomnyaschikh [34], Lions [35], and Dryja-Widlund [36, 21]). Assume that there exists a constant c_2 such that*

$$\min_{\substack{u = \sum u_i \\ u_i \in V_i}} \sum_{i=0}^{p} \|u_i\|_A^2 \leq c_2 \|u\|_A^2. \tag{2.5}$$

then

$$\lambda_{\min}(M_{asc}^{-1}A) \geq \frac{1}{c_2}.$$

The assumption made in the partition lemma (equation (2.5)) means that for any given u, a stable decomposition must exist in the sense that the sum of the "energy" of all the pieces u_i lying in V_i is bounded by the global energy norm of the decomposed vector. This assumption can be viewed as a condition on the V_i's, i.e. the subspaces must not introduce oscillations (high energy components) in u_i.

Combining lemmas 2.1– 2.3, we get our main theorem:

Theorem 2.1 *If assumption (2.5) holds, then the condition number $\kappa(M_{asc}^{-1}A)$ can be bounded by:*

$$\kappa(M_{asc}^{-1}A) \leq c_1 c_2. \tag{2.6}$$

This result suggests how to construct the decompositions in order to obtain optimal preconditioners: we want the constants c_1, c_2 to be independent of the problem parameters such as the number of subdomains, the characteristic mesh sizes h and H, jumps in the coefficients of the underlying PDE, etc. It is also desirable to make c_1 and c_2 as small as possible in order to get a condition number close to 1. But c_1 and c_2 depend on the size of overlaps in the subspaces V_i. More overlap will decrease c_2, but the number of colors c_1 will increase. On the other hand, small overlap will lead to large c_2 and small c_1. Thus the space decompositions have to be made in such a way to ensure that the product $c_1 c_2$ is as small as possible.

2.2 Application to two-level overlapping domain decomposition methods

As an example of the application of the above theory, we will present a detailed estimate for the condition number of the two-level Schwarz method.

2.2.1 The intuitive idea

As in the previous section, we first present the basic intuitive idea using a simple 1D version of (1.1).

We want a splitting which satisfies the partition assumption (2.5). Take Ω to be a fixed open interval on the real line and cover Ω with p overlapping subdomains Ω_i, $i = 1, \ldots, p$ (see fig 2.1). Consider the partition of unity θ_i corresponding to this covering. By construction the functions θ_i satisfy

$$\sum_{i=1}^{p} \theta_i = 1, \quad 0 \le \theta_i \le 1, \quad |\theta_i|_{1,\infty} \le \delta^{-1}, \qquad (2.7)$$

where δ is the size of the overlap and $|\theta|_{s,\infty}$ denotes the maximum norm, i.e. the maximum of the s-th derivative of θ_i. We define $u_i = \theta_i u$, so we have $u = \Sigma_{i=0}^{p} u_i$.

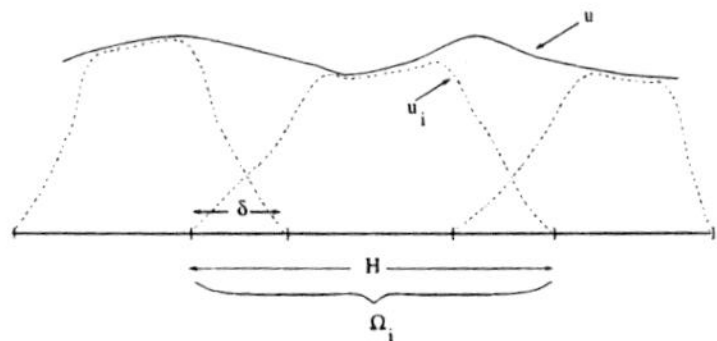

Figure 2.1: The partitioned u_i's without coarse grid.

Since A corresponds to a discretization of the second order elliptic operator, $\frac{d}{dx}\alpha(x)\frac{d}{dx}$, it is easy to see that the A-norm and the H^1-seminorm are equivalent in this case: $\|u\|_A \approx \|du/dx\|_0$. Our goal is to bound $\|u_i\|_A$ by $\|u\|_A$. Looking at Fig. 2.1, we see that the function u_i changes from $\|u\|_0$ to 0 over a distance δ and we get:

$$\|u_i\|_A^2 = \left\|\left(\frac{du}{dx}\right)\right\|_0^2 \le c\left(\frac{\|u\|_0}{\delta}\right)^2.$$

We still need to bound $\|u\|_0$ by $\|u\|_A$. But the function u, satisfying homogeneous Dirichlet boundary conditions, cannot change rapidly over the interval Ω if there is no significant change in the derivative. The well-known Poincaré inequality estimates the norm of the function with the norm of derivatives and its application leads to the following:

$$\left(\frac{\|u\|_0}{\delta}\right)^2 \le \tilde{c}\left(\frac{\|u\|_A}{\delta}\right)^2.$$

After summing over all subdomains, we get:

$$\sum_{i=1}^{p} \|u_i\|_A^2 \le O\left(\frac{1}{\delta^2}\right)\|u\|_A^2.$$

Therefore,

$$c_2 = O\left(\frac{1}{\delta^2}\right) = O\left(\left(\frac{H}{\delta}\right)^2 \frac{1}{H^2}\right).$$

From these inequalities, one may conclude that if the overlap is of size $O(H)$, then $\kappa(M^{-1}A) = O(H^{-2})$, which is an improvement over $O(h^{-2})$, but is still unsatisfactory. We can see that the overlapping subdomains alone cannot provide a stable partition of u.

It turns out that this dependence on H can be eliminated by using a global coarse space, V_H, which couples all the subdomains. The idea is to construct a coarse grid approximation u_H to u satisfying the following two important properties:

$$\|u_H\|_A \le c\|u\|_A \qquad (2.8)$$
$$\|u - u_H\|_0 \le cH\|u\|_A \qquad (2.9)$$

Define $w = u - u_H$ and the following partition of u:

$$u_i = \theta_i(u - u_H), \quad u = u_H + \sum_{i=1}^{p} u_i. \qquad (2.10)$$

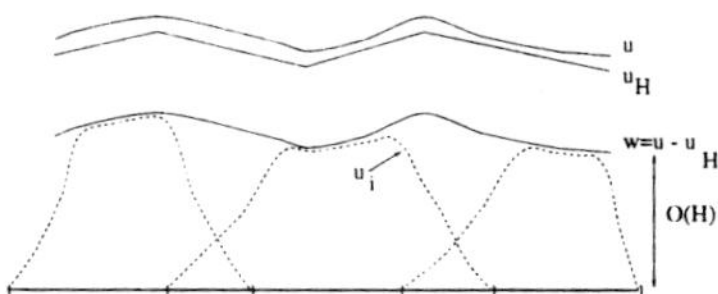

Figure 2.2: The partitioned u_i's with coarse grid.

Proceeding as before and taking into account that now the u_i's change from 0 to $O(H)$, we have

$$\|u_i\|_A^2 \le c \left(\frac{H\|u\|_A}{\delta} \right)^2 \le c \left(\frac{H}{\delta} \right)^2 \|u\|_A^2.$$

If we make the natural assumption that $\delta = O(H)$, the bound for c_2 now reads

$$c_2 = O\left(\left(\frac{H}{\delta} \right)^2 \right) = O(1).$$

Thus, we can see that the role of the coarse grid V_H is to make $\|u - u_H\|$ small enough ($O(H)$), so that it can be partitioned in a stable manner.

We would like to comment on the choice of u_H. As it can be seen, u_H is a purely theoretical construction and there is no need of its use in the algorithm. One possible choice is $u_H = \mathcal{R}_H u$, where $\mathcal{R}_H$ is a kind of interpolation or projection operator. Of course, $\mathcal{R}_H$ must satisfy properties similar to (2.8) and (2.9) namely:

$$
\begin{aligned}
|\mathcal{R}_H u|_1 &\le c|u|_1 \quad \text{(stability)} & (2.11) \\
\|u - \mathcal{R}_H u\|_0 &\le cH|u|_1. \quad \text{(approx.)} & (2.12)
\end{aligned}
$$

A natural candidate for such an operator is the nodal value interpolant on the coarse grid, $u_H = I_H u$. A drawback of such a choice is that in $3D$ this interpolation does not satisfy the stability property (2.11). To see this, we take w to be the basis function ψ_i^h associated with the grid node x_i. Then we have

$$
\begin{aligned}
|w|_1^2 &= \int_{\Delta_h} \left(\tfrac{1}{h} \right)^2 = O(h) \\
|I_H w|_1^2 &= \int_{\Delta_H} \left(\tfrac{1}{H} \right)^2 = O(H).
\end{aligned}
$$

The last estimate shows that the stability requirement is violated.

If the grid is structured, then a good and stable coarse grid approximation to the elements of V_h is the L_2-projection Q_H from $V_h \rightarrow V_H$ and we can define $\mathcal{R}_H = Q_H$, i.e. $u_H = Q_H u$. It is known that this

u_H satisfies the stability and approximation properties (2.11) and (2.12) (see Xu [17] or Dryja and Widlund [36]).

We now state the main result of this section.

Theorem 2.2 *Under the above assumptions (2.11) and (2.12) for the condition number $\kappa(M_{asc}^{-1}A)$, the following estimate holds*

$$\kappa(M_{asc}^{-1}A) = O(1 + (H/\delta)^2). \tag{2.13}$$

2.2.2 Some extensions

Inexact subdomain solves can easily be accommodated by using $\| \cdot \|_{M_i}$ satisfying:

$$\|u\|_A \le \omega \|u\|_{M_i} \qquad \forall u \in V_i.$$

Then the constant ω in the above inequality will be absorbed in the resulting bound for $\kappa(M^{-1}A)$.

The extension to *multilevel* Schwarz method is straightforward. An additional assumption, however is needed in this case:

$$(u_i, u_j)_A \le \varepsilon_{ij} \|u_i\|_A \|u_j\|_A \qquad \forall u_i \in V_i, \forall u_j \in V_j.$$

Such inequalities measure the abstract angles between the subspaces and are known in the literature as "strengthened Cauchy Schwarz inequalities". For a detailed discussion of the issues concerning multilevel theory we refer to Xu [17, 18], Chan-Mathew [19]. Note that $\rho(\{\varepsilon_{ij}\})$ will enter in the bound for $\kappa(M^{-1}A)$.

We now briefly comment on the convergence of the *multiplicative* Schwarz method which was described in Section 1.4.2. From (1.8), for the error e^{k+1}, we have:

$$e^{k+1} = (I - P_p) \cdots (I - P_0)e^k.$$

Since each $(I - P_i)$ is a projection in the A-norm, it immediately follows that $\|e^{k+1}\|_A \le \|e^k\|_A$. The following result gives a bound for the damping factor of the multiplicative iteration.

Theorem 2.3 *(Bramble, Pasciak, Wang, Xu [37]; and Xu [17]) Let V_i's satisfy the assumptions of the Partition Lemma. Then the following estimate is true:*

$$\|(I - P_p) \cdots (I - P_0)\|_A \le 1 - \frac{c}{c_2}.$$

where c depends on the number of colors for coloring the Ω_i's but is independent of p.

497

2.3 Convergence of multigrid methods

The convergence properties of *multigrid* methods (see Section 1.4.1) depend on many parameters. One can vary the number of smoothing steps, the smoothing operators, the interpolation and restriction operators, coarse grid operators, etc. There are two main approaches in constructing multigrid preconditioners. One of them uses nested subspace splittings of V_h and the other one uses non-nested spaces or specially interpolated bilinear forms (coarse grid matrices). The discussion of the convergence in both these cases is given in [15, 37, 17, 18]. Here we give the simplest convergence result in the case of *nested* spaces (i.e. $V_0 \subset V_1 \subset \ldots V_{J-1} \subset V_J = V_h$). and so-called full elliptic regularity assumption:

$$\|u\|_2 \leq c\|F\|_0,$$

where u is the solution, F is the right hand side of (1.1).

Again, the stability and approximation properties (2.11) and (2.12) are crucial in the convergence theory. The stability property (2.11) is automatically satisfied when the spaces are nested. It turns out that from the regularity assumption the following approximation property follows (see Xu [17, 18]):

There exists a constant, c_1, independent of the mesh parameters (i.e. of the mesh size h) such that

$$\|I - P_{k-1}v\|_A^2 \leq c_1 \frac{1}{\rho(A_k)} \|A_k v\|_2^2 \quad \forall v \in V_k, \,(2.14)$$

where P_k denotes the elliptic projection defined by $(AP_k u, v) = (Au, v) \ \forall v \in V_k$.

The other operator involved in the definition of M_J^{-1} is the smoother and we make the following assumption on it:

$$\frac{c_0}{\rho(A_k)}(v, v) \leq (S_{symm,k} v, v) \leq (A_k^{-1} v, v). \quad (2.15)$$

The smoother $S_{symm,k}$ is the symmetric version of S_k and is defined as: $S_{symm,k} := S^T + S - S^T A S$. Inequalities of the type (2.15) are satisfied by the Gauß-Seidel method.

An important thing to mention for the choice of the smoother is that we are trying to choose smoother which will quickly capture the high frequency components of the error, and we are not going to use it as a solver. For example if the matrix A_J corresponds to the five point finite difference stencil, it can be seen that the Jacobi method (with $\omega = 1$) is not good for a smoother. One must use the damped Jacobi method with $\omega < 1$.

The subspace correction framework applies to multigrid methods as well. As long as the stability and approximation properties are verified, the following convergence result holds:

Theorem 2.4 *Under the assumptions (2.14) and (2.15), the following estimate is true:*

$$\|I - M_J^{-1} A\|_A \leq 1 - \frac{c_0}{c_1}. \quad (2.16)$$

We want to point out that a convergence result similar to Theorem 2.4 is also true in the *non-nested* case. We refer to work by Bramble et.al. [37] or Chan-Zou [38, 39] for unstructured grids. The construction of interpolation operators satisfying the stability property might be an issue. Some possible stable definitions can be found in [38] and [39] and in Section 3.2.

In the next sections, we will define subspaces satisfying the above assumptions or directly satisfying the approximation property similar to (2.14). The verification of these assumptions is easy for structured grids and can be found in many papers. On unstructured grids, however, this might be rather complicated and tricky. Special attention should be paid to the construction of the spaces themselves, rather than using standard spaces and verifying the above inequalities.

3 Node-nested coarse spaces

Unstructured multilevel methods for solving linear systems like (1.4) require a hierarchy of coarse grids. Grids which are node-nested have the advantage that they can be automatically generated and that efficient methods can be used to create the interpolation and restriction operators needed to transfer information from one level to the other. Disadvantages are that for complicated geometries, particularly in three dimensions, special care must be taken to ensure that the coarse grids which are produced are valid and preserve the important geometric features of the fine domain. With unstructured meshes, the grid hierarchy can allow general grids which are non-quasiuniform and coarse grids whose boundaries may be non-matching to the boundary of the fine grid, so care must be applied when constructing intergrid transfer operators for various types of boundary conditions. In this section, we will discuss some possibilities.

3.1 Maximal independent set (MIS) coarsening

A maximal independent set of vertices in a graph is a subset of vertices which is *independent* in the sense that no two vertices in the subset are connected by an edge, and *maximal* if the addition of a vertex results in a dependent subset. An automatic approach to generating node-nested coarse grids is to take a maximal independent set (MIS) of the vertices and call this set, the set of coarse grid nodes, and then retriangulate it [25, 26]. A sequence of coarse grids can thus be created by repeated application of this technique.

A simple technique for finding a MIS of vertices is to first choose a MIS of the boundary vertices by choosing every other boundary vertex and eliminating all its nearest neighbors, and then find a MIS of the interior vertices by selecting a random interior vertex and eliminating all its nearest neighbors, and repeating the process until all vertices are either eliminated or selected. The resulting vertex subset is then retriangulated using for example, the same triangulation routine which generated the original fine grid.

3.2 Coarse-to-fine interpolations

In general, the resulting coarse grid domain will have boundaries which will not match the boundaries of the fine grid so the coarse space V_H is usually not a subspace of the fine space V_h. Indeed, even if $\Omega_H = \Omega_h$, V_H may still not be a subspace of V_h since the coarse elements are generally not the unions of some fine elements in unstructured grids. To construct a coarse-to-fine transfer operator, one can use the standard nodal value interpolant associated with the fine space, V_h.

ALGORITHM 3.1 *(Standard nodal value interpolation)*

1. **For** *each fine grid node,*
2.   Search *through all coarse grid elements until one which contains it is found.*
3.   **If** *the fine grid node is a coarse grid node, then*
4.   Set *the interpolant to be equal to that nodal value,*
5.   **Else**
6.   Set *it to be a linear interpolation of the 3 nodal values making up that coarse grid element (see Fig. 3.1).*

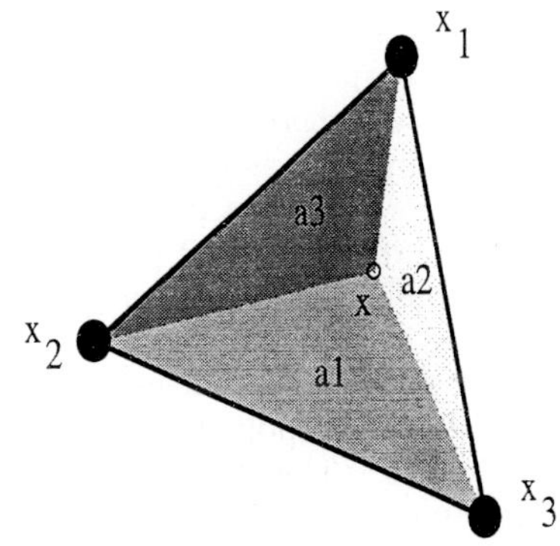

Figure 3.1: Barycentric (natural) coordinates: $\lambda_i(x) = \frac{\text{area}(a_i)}{\text{area}(\Delta_{123})}$, for $i = 1, 2, 3$, where Δ_{123} is the simplex with vertices x_1, x_2, x_3. The value of a function f at a point x is given by: $f(x) = \lambda_1(x)f(x_1) + \lambda_2(x)f(x_2) + \lambda_3(x)f(x_3)$.

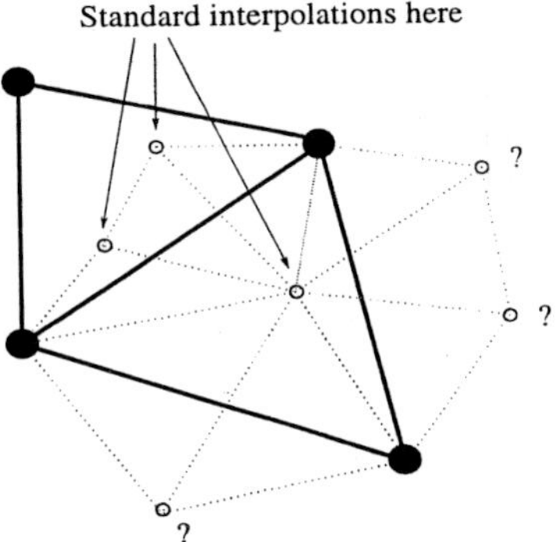

Figure 3.2: Use standard interpolation for fine grid nodes interior to a coarse element. What should be done for fine grid nodes which are not interior to any coarse grid element? One way is simply with extension by zero.

A naive implementation of this routine requires $O(n^2)$ time, but exploiting the node-nested property of the grids, one can implement this in $O(n)$ time, since only nearest coarse grid elements of a fine node need to be searched.

3.3 Interpolations on non-matching boundaries

Notice, however, that the standard nodal value interpolant is only well defined for those fine nodes lying also in the coarse domain $\bar{\Omega}_H$, but undefined for those fine nodes lying outside $\bar{\Omega}_H$. That is, in Step 2 of the standard nodal value interpolation (Algorithm 3.1), there is no provision for what to do if all the coarse grid elements have been searched, and none contains the fine grid node. A simple and natural way to remove this barrier is to assign those fine node values by zero. We denote this interpolant as the coarse-to-fine interpolant, $\mathcal{I}_h^0$. This zero extension interpolant works well for Dirichlet boundary conditions [26, 2] but will not be accurate nor stable for other boundary conditions.

We provide a simple one dimensional example to illustrate why better interpolants are needed at non-matching boundaries. This example has a Dirichlet boundary condition at the left boundary point and a homogeneous Neumann boundary condition at the right boundary point. The fine grid function, u, and the coarse grid approximation to it, U_H are shown. For Neumann boundary conditions, the elements from V_h which have to be interpolated are generally not zero at the Neumann part of the boundary. Recall from Section 1.1 that V_h is a subspace of $H_0^1(\Omega, \Gamma_D)$, whose elements are restricted to vanish only on Dirichlet boundary. Thus using a zero extension interpolant at a Neumann boundary will not be accurate enough and introduces a correction with high energy ($\|u - u_H\|$ is no longer $O(H)$), (see Fig. 3.3).

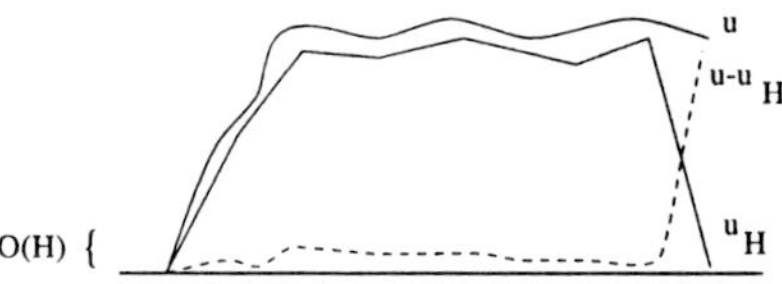

Figure 3.3: Non-matching boundaries: Zero extension interpolation is not accurate at the right end of the coarse domain.

To achieve better efficiency, we need to modify this intergrid operator to account for the Neumann condition. Two general ways to treat such boundaries are:

1. Modify the coarse grid domain to cover any fine grid boundaries of Neumann type and use standard nodal value interpolation.

2. Increase the accuracy of the interpolants by accounting for the Neumann condition for those fine nodes in $\Omega \backslash \Omega^H$.

The first approach is motivated by the fact that standard nodal value interpolants can still be used with efficiency, provided the coarse grid covers the Neumann boundary part of the fine grid (see Fig. 3.4). This was first proposed and justified in [2]. We shall denote this operator as the coarse-to-fine interpolant, $\mathcal{I}_h^1$. Let us still denote the modified coarse grid domain by $\bar{\Omega}_H$.

Then for all $v^H \in V^H$, the interpolant $\mathcal{I}_h^1$ is defined as:

$$
\mathcal{I}_h^1 v^H(x) \;=\; \begin{cases} v^H(x) & \text{for } x \in \Omega \cap \bar{\Omega}_H, \\ 0 & \text{for } x \in \Omega \setminus \bar{\Omega}_H. \end{cases}
$$

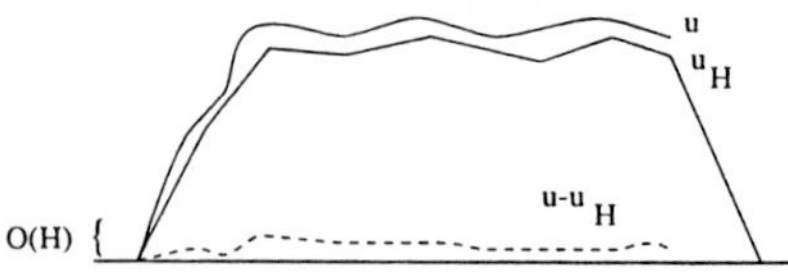

Figure 3.4: More accurate interpolation with $\mathcal{I}_h^1$ done at Neumann boundary.

This is a natural extension of v_H by zero outside the Dirichlet boundary part of the coarse grid domain. Similar zero extensions were used in Kornhuber-Yserentant [40] to embed an arbitrarily complicated domain into a square or cube in constructing multi-level methods on nested and quasi-uniform meshes for second order elliptic problems with purely Dirichlet boundary conditions.

Although the coarse-to-fine operator $\mathcal{I}_h^1$ works well for mixed boundary conditions, one has to modify the original coarse grid so that it covers the Neumann boundary part of the fine grid domain (see [4] for a description for modifying boundaries). This can be very difficult to do for complicated domains. To avoid modifying the original coarse grid, we now consider standard finite element interpolants which are modified only near Neumann boundaries. The idea is as follows: Let us consider a fine grid point, x, which lies outside the coarse grid domain. Find a nearby coarse grid triangle to x (say, τ_H with vertices x_1, x_2, x_3), and *extrapolate $u(x)$* using the values $u(x_1), u(x_2)$ and $u(x_3)$. Note that such an extrapolation should depend on the type of boundary condition at x.

We define the interpolant at x by using the nodes of the coarse boundary edge closest to x:

$$
\mathcal{I}_h^2 v^H(x) \;=\; \lambda(x) v^H(x_1^H) + (1 - \lambda(x)) v^H(x_2^H),
$$

where the coarse edge has endpoints x_1^H and x_2^H, and λ is the ratio of the lengths of two segments of this edge cut off by the normal line passing through x to the edge. This kind of interpolation was also used by Bank and Xu [41] in their construction of a hierarchical basis on a unstructured mesh.

We can also use a non-zero extension by extrapolation using barycentric functions:

$$
\mathcal{I}_h^3 v^H(x) \;=\; \lambda_1(x) v^H(x_1^H) + \lambda_2(x) v^H(x_2^H)
$$

$$+\lambda_3(x)v^H(x_3^H),$$

where $\lambda_1, \lambda_2, \lambda_3$ are the three barycentric coordinate functions corresponding to τ_H (see Figure 3.1). Note that the functions λ_1, λ_2 and λ_3 used in the definition of $\mathcal{I}_h^3$ satisfies $\lambda_1, \lambda_2, \lambda_3 \geq 0$ for $x \in \tau_H$, but not so for $x \notin \tau_H$. The barycentric coordinates may still be defined, provided we consider the area of a simplex to be orientation-dependent. That is. area is > 0 for "right-handed" triangles (clockwise) and area is < 0 for "left-handed" triangles (counter-clockwise).

We summarize the various interpolants:

$\mathcal{I}_h^0$: Zero extension with unmodified coarse boundaries,

$\mathcal{I}_h^1$: Zero extension with modified coarse Neumann boundaries,

$\mathcal{I}_h^2$: Nearest edge interpolation, and

$\mathcal{I}_h^3$: Nearest element interpolation.

3.4 Stability and approximation of the non-nested interpolation

The convergence theory for overlapping multilevel domain decomposition and multigrid methods require the coarse-to-fine grid transfer operator to possess the local optimal L^2-approximation and local H^1-stability properties as introduced in Sec. 2. The locality of these properties is essential to the effectiveness of these methods on highly non-quasi-uniform unstructured meshes.

Because the spaces are non-nested (they are *node-nested*, but still *non-nested*, as coarse grid elements are not unions of fine grid elements), in the theory discussed in Section 2, the u_H coarse space approximation to u should be defined as:

$$u_H = \mathcal{I}_h \mathcal{R}_H u.$$

Since $\mathcal{R}_H \in V_H \not\subset V_h$, we need to use the interpolation operator $\mathcal{I}_h$ to map $\mathcal{R}_H u$ back to V_h. The convergence theory now requires both $\mathcal{I}_h$ and $\mathcal{R}_H$ to possess the stability and approximation properties. When the mesh is quasi-uniform, the usual L^2-projection, Q_H, can be used for $\mathcal{R}_H$. But when the mesh is highly non-quasi-uniform, the constant in the approximation property (2.12) can deteriorate if we use Q_H. The trick then is to use a localized version of the L^2-projection, i.e. the so-called Clément's projection. It is known that this projection provides local stable and good approximations. We refer to Clément [42] for its definition and

Chan-Zou [38] and Chan-Smith-Zou [2] for its use in domain decomposition contexts.

For $\mathcal{I}_h$, we can take the coarse-to-fine interpolants introduced in Sec. 3.3 The key step then is proving the stability and approximation properties for $\mathcal{I}_h$. The proof that the non-nested standard interpolation used in the interior is stable and accurate can be found in Cai [43] and Chan-Smith-Zou [2]. The proof for the boundary-specific interpolations can be found in [5].

3.5 Numerical results

In this section, we provide some numerical results of domain decomposition and multigrid methods for elliptic problems on an unstructured airfoil mesh: see Figure 3.5. The well-known NASA airfoil mesh was provided by T. Barth and D. Jesperson of NASA Ames. Coarse grids were generated by the MIS approach as described above. All numerical experiments were performed using the Portable, Extensible Toolkit for Scientific Computation (PETSc) [44], running on a Sun SPARC 20. Piecewise linear finite elements were used for the discretizations and the resulting linear system was solved using either multilevel overlapping Schwarz or V-cycle multigrid as a preconditioner with full GMRES as an outer accelerator. In the experiments, the initial iterate is set to be zero and the iteration is stopped when the discrete norm of the residual is reduced by a factor of 10^{-5}.

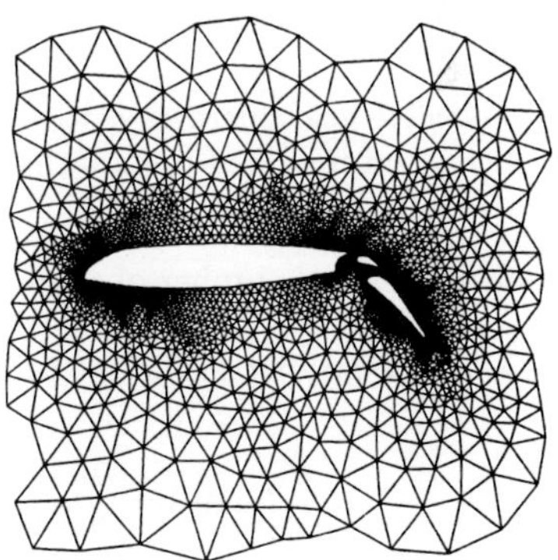

Figure 3.5: Unstructured NASA airfoil with 4253 nodes.

For partitioning, all the domains (except the coarsest) were partitioned using the recursive spectral bisection method [45], with exact solves for both the subdomain problems and the coarse grid problem. To generate overlapping subdomains, we first partition the domain into nonoverlapping subdomains and then extend

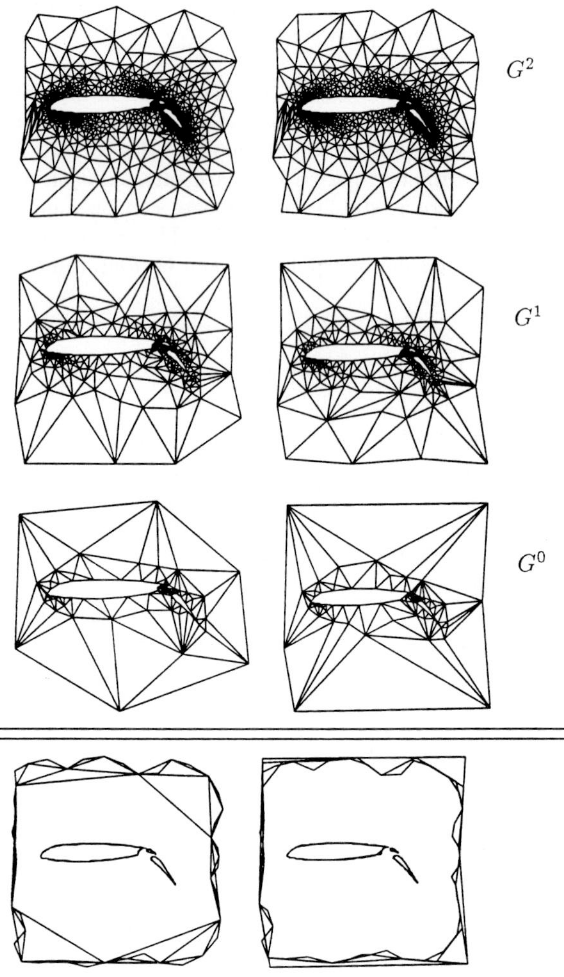

Figure 3.6: *Airfoil* grid hierarchy with unmodified boundaries (left) and modified boundaries (right).

each subdomain by some number of elements.

We solve a mildly varying coefficient problem on the airfoil:

$$\frac{\partial}{\partial x}((1 + xy)\frac{\partial u}{\partial x}) + \frac{\partial}{\partial y}((\sin(3y))\frac{\partial u}{\partial y}) = F(x, y),$$

where

$$F(x, y) = (4xy + 2)\sin(3y) + 9x^2 \cos(6y),$$

with either a purely Dirichlet boundary condition or a mixed boundary condition: Dirichlet for $x \leq 0.2$ and homogeneous Neumann for $x > 0.2$. For this problem, the non-homogeneous Dirichlet condition is $u = 2 + x^2 \sin(3y)$.

Fig. 3.7 shows results when a hybrid 4-level

multiplicative-additive Schwarz method is used (multiplicative between levels but additive among subdomains on the same level). As can be seen, deterioration of the method occurs with interpolant $\mathcal{I}_h^0$ when mixed boundary conditions are present. The next figure (Fig. 3.8) shows results for the multiplicative Schwarz method (both on the subdomains and between levels). This method behaves much like multigrid (see Table 1). In fact, this is nothing more than standard V-cycle multigrid with a block smoother used as a preconditioner. A V-cycle multigrid method with pointwise Gauss-Seidel smoothing and 2 pre- and 2 postsmoothings per level was used to produce the results in Table 1.

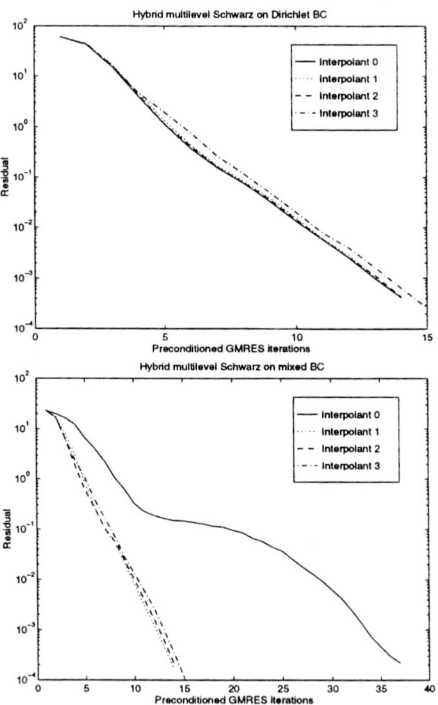

Figure 3.7: **Hybrid multiplicative-additive 4-level Schwarz**. Convergence history with purely Dirichet boundary conditions (top) or mixed boundary conditions (bottom).

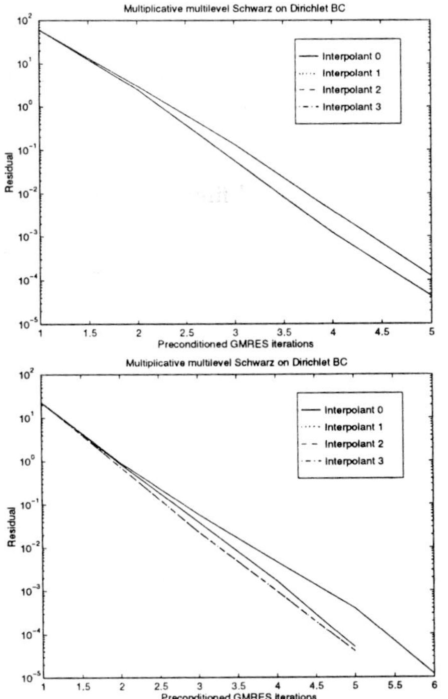

Figure 3.8: **Multiplicative 4-level Schwarz.** Convergence history with purely Dirichet boundary conditions (top) or mixed boundary conditions (bottom).

Table 1: **Multigrid.** Tables show the number of GMRES iterations to convergence with Dirichet or mixed boundary conditions.

Dirichlet boundary conditions

# of fine	MG	# of coarse	Interpolant Used			
grid nodes	levels	grid nodes	$\mathcal{I}_h^0$	$\mathcal{I}_h^1$	$\mathcal{I}_h^2$	$\mathcal{I}_h^3$
	2	1170	4	4	4	4
4253	3	340	4	4	4	4
	4	101	4	4	4	4

Mixed Dirichlet/Neumann boundary conditions

# of fine	MG	# of coarse	Interpolant Used			
grid nodes	levels	grid nodes	$\mathcal{I}_h^0$	$\mathcal{I}_h^1$	$\mathcal{I}_h^2$	$\mathcal{I}_h^3$
	2	1170	6	5	4	4
4253	3	340	6	4	5	5
	4	101	7	5	5	5

4 Agglomerated coarse spaces

4.1 Agglomerated multigrid methods on unstructured grids

In this section, we will consider a general agglomeration approach for constructing nested coarse spaces and transfer operators. The difference between this technique and the node-nested coarse spaces from the previous section is that here, we want to produce a nested sequence of spaces to be used in the multigrid method. The common point will be that our construction must satisfy the approximation and stability properties mentioned in the subspace correction framework in Section 2.

4.2 Coarse points and construction of macroelements

The agglomeration technique is based on the construction of a coarse grid with "macro-elements" consisting of unions of fine grid elements (triangles). An example of such a coarse grid is given on Fig. 4.1. Then, as in the standard finite element method, the basis functions in each coarse grid macroelement are appropriately defined. The coarse space V_H is then determined as the space spanned by these functions. If the coarse grid basis functions are defined as linear combinations of fine grid basis (i.e. the usual finite element basis), then V_H is a proper subspace of V_h, i.e. we obtain nested spaces by construction.

The construction of a basis in V_H is equivalent to the definition of the restriction matrix R_H, because the coordinates of these basis functions with respect to the fine grid basis form the rows of the restriction matrix. Thus, once the basis functions are defined, we have restriction R_H, the interpolation (or prolongation) R_H^T, the coarse grid operator $R_H A R_H^T$ and we can apply the V-cycle algorithm from section 1.4.1.

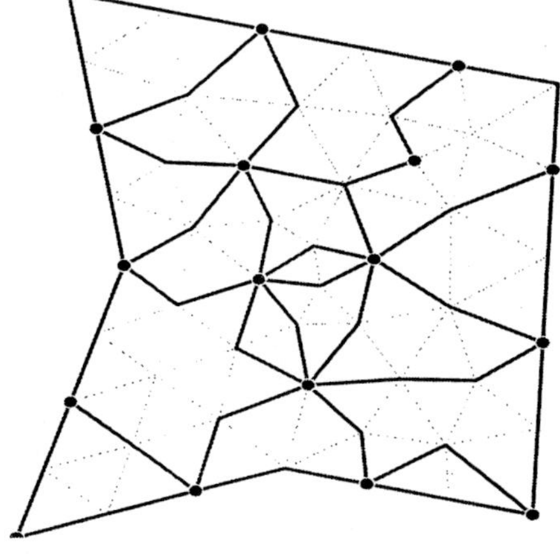

Figure 4.1: An example of macroelements

Although the nestedness is assured by construction, there are some important rules which should be followed in order to ensure good convergence rate of the resulting multilevel methods:

- *Smoothness:* The basis functions have to be smooth enough. This requirement is needed because the elements from V_H have to satisfy the stability property (2.11), which involves the A-norm of the coarse grid function.

- *Approximation:* The functions in V_H have to satisfy the approximation property (2.12). An implication of this is that the individual basis function in V_H cannot be independently chosen, and there must be some global relation which couples the basis functions.

- *Small supports:* The functions in V_H must have compact support. This requirement is based on the fact that once the basis $\{\Phi_i\}$ in V_H is given, then the coarse grid matrix elements are defined to be $a(\Phi_i, \Phi_j)$ (see (1.3)). Thus, if the basis functions have large supports, the coarse grid matrix will be dense, and the coarse grid problem is expensive to solve.

- *Conformity:* For finite element discretizations, it is desirable that the resulting coarse grid is formed by conformal macroelements, an analogue of conforming triangulations in finite element methods. This facilitates the analysis and construction of the algorithms.

- *Recursion:* The coarse grid should allow the recursive application of the algorithm to construct a multilevel method.

A careful look at these rules shows that it is difficult (even impossible) to satisfy all of them simultaneously. Usually some of them have to be weakened in order to satisfy others. For example, to have conforming macroelements of size $\approx 2h$ on an unstructured grid (which is a desirable choice in multigrid) is almost impossible. Taking smoother basis functions will increase the supports, and make the coarse grid operator denser.

A recent paper by Koobus, Lallemand and Dervieux [31] deals with agglomeration with the finite volume discretizations. The basis functions are piecewise constant on each cell, and the coarse grid cells are formed as unions of fine grid cells. A drawback of this algorithm is that the stability of the coarse grid basis functions is not easy to control.

An algebraic agglomeration algorithm can be found in the recent papers by Mandel, Vaněk, Brezina (see [32]) and Vaněk, Křížková [33]. Their approach uses an algebraically smoothed basis functions, and the coarse grid nodes are not explicitly defined. This allows the process of the basis construction to be more automatic, but it is more difficult to control the sparsity of the coarse grid operators.

Our approach is based first on the definition of the coarse grid points (using the MIS described in Section 3.1) and then using them to define the macroelements. The difference between the algorithm presented here and the agglomeration algorithms quoted above is in the more "geometrical" nature of our coarsening strategy. We will define our coarse grid space using macroelement edges and macroelement vertices (coarse points). In the numerical examples presented here, the definition the macroelement edges and coarse grid points is done using the dual graph of a given triangulation. Such an algorithm does not pretend to be computationally the best one, and we do not describe here this algorithm in detail. Our main concern will be the definition of a proper coarse space.

4.3 Coarse space basis functions.

Given the set of macroelements, we will now introduce three different ways to define the coarse grid basis functions. Let us first focus on meeting the first two rules given in the previous section: we need to define smooth basis functions so that the coarse space satisfy the approximation and stability properties.

To assure the approximation property, we should take a basis which preserves at least the constant function, i.e. the constant function must be always in the coarse space V_H. To do this, we first define basis functions possessing this property on the the macroelement boundaries, and after that we extend them into the interior of the macroelement as discrete harmonic functions. This extension obviously will not destroy the constant preserving property, because the constant function is harmonic.

We define the coarse basis functions on these edges as linear functions minimizing some quadratic functional. The $H^{1/2}$ norm on the boundary is one good choice for the quadratic functional, as it is the interface analogue of the A-norm.

Let the macroelement boundary be formed by ℓ edges from the fine grid, (see Fig 4.2) connecting two coarse grid points, x_0 and x_ℓ, and let this path contains the vertices $x_0, \ldots, x_\ell$. We define the basis function corresponding to the coarse grid node x_0 as follows:

Φ_0 is a linear function in $\mathbb{R}^2$:

$$\Phi_0 = a\xi + b\zeta + c, \quad (\xi, \zeta) \in \mathbb{R}^2.$$

We want : $\Phi_0(x_0) = 1$, $\Phi_0(x_\ell) = 0$. These are only two conditions and we have three parameters: a, b and c. To complete the set of conditions, we require that the function Φ_0 minimizes the functional (discrete $H^{1/2}$ norm):

$$F_{1/2}(\Phi_0) = \sum_{i=1}^{\ell} \sum_{j=i+1}^{\ell} \frac{h_i h_j}{h_{ij}^2}(\Phi_0(x_i) - \Phi_0(x_j))^2, \quad (4.1)$$

where h_i is the length of the edge (x_i, x_{i+1}) and $h_{ij} = |x_i - x_j|$. After using the first two conditions, this minimization is equivalent to minimization of a simple quadratic function of one variable which can be easily done analytically.

In the interior of the macroelements, we extend the basis functions by solving the equation:

$$a(\Phi_0, \phi) = 0, \qquad \text{for any} \quad \phi \in V_h. \qquad (4.2)$$

In this way, we define the basis in V_H and thus also defining the restriction operator R_H.

ALGORITHM 4.1 *(Geometric coarsening-Harmonic extension I)*

1. *The boundary values minimize the $H^{1/2}$-discrete norm.*
2. *For the nodes internal to the macroelements, the values of the basis functions are obtained via the harmonic extension, see Fig. 4.2.*

We will now give two simpler variants of this algorithm. The first one uses simpler boundary conditions: The function is defined on the boundary using the graph distance (see Fig 4.3). The graph distance $dist(i,j)$ is equal to the number of edges forming the shortest path connecting the vertices i and j, (in the case we consider these vertices are x_0 and x_ℓ).

ALGORITHM 4.2 *(Geometric coarsening-Harmonic extension II)*

1. *The boundary values are taken using the graph distance interpolation, see Fig. 4.3.*
2. *For the nodes internal to the macroelements, the values of the basis functions are obtained via the harmonic extension, see Fig. 4.3.*

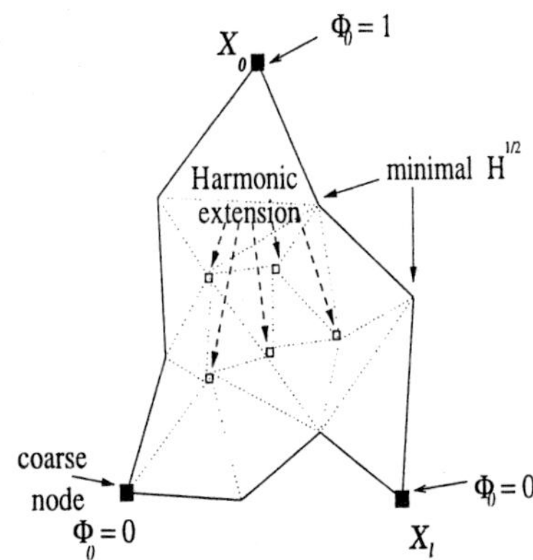

Figure 4.2: Minimize $H^{1/2}$ on the boundary and harmonic extension in the interior.

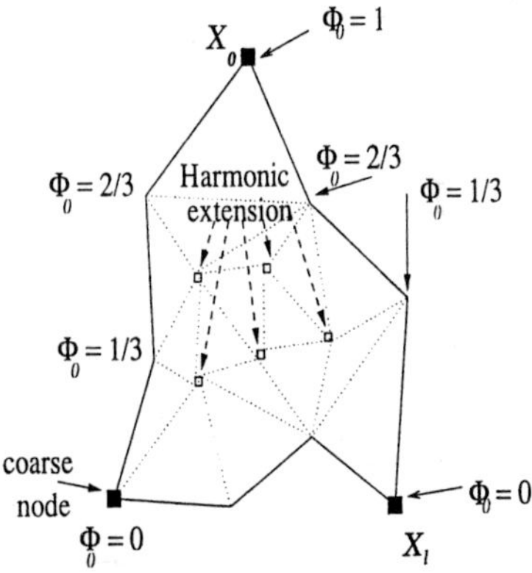

Figure 4.3: Graph distance on the boundary and smooth extension in the interior.

The second variant, which does not include harmonic extension is as follows:

ALGORITHM 4.3 *(Geometric coarsening)*
1. *On the macroelement boundaries the graph distance interpolation is used.*
2. *For the nodes internal to the macroelements and fictitious faces (see Step 4), the values of all basis functions whose supports form the macroelement take one and the same value: the reciprocal of the number of coarse points forming the macroelement, see Fig. 4.4.*

The approximation and stability properties of the agglomerated spaces given above are assured by the next lemma. A detailed proof will be included in the paper [6].

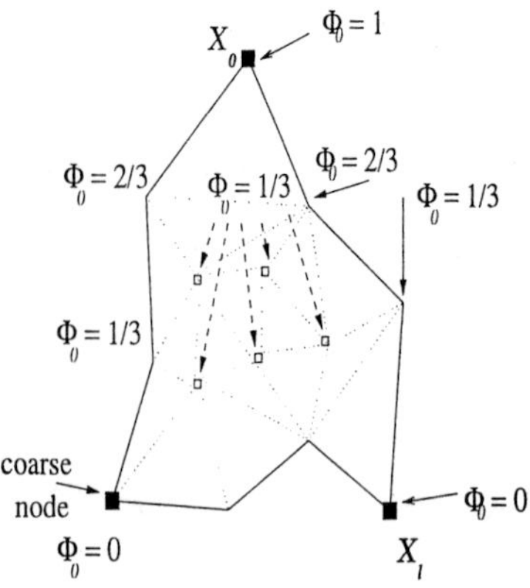

Figure 4.4: Graph distance on the boundary and in the interior.

Given a triangulation $\mathcal{T}_h$ and the corresponding linear finite element space $V_h \subset H^1(\Omega)$. Let $V_H \subset V_h$ be obtained by any of the three agglomeration algorithms described above. Let $Q_H : H^1(\Omega) \to V_H$ be the L^2-projection.

Lemma 4.1 *Assume that the constructed basis preserves the constant function. Then the following stability and approximation properties hold for the agglomerated subspaces:*

$$\|Q_H v\|_{1,\Omega} \leq c\|v\|_{1,\Omega}, \tag{4.3}$$
$$\|v - Q_H v\|_{0,\Omega} \leq cH|v|_{1,\Omega}, \quad \forall v \in H^1(\Omega). \tag{4.4}$$

4.4 Numerical examples

We consider an elliptic equation of following type:

$$\begin{cases} -\dfrac{\partial}{\partial x}\left(a(x,y)\dfrac{\partial u}{\partial x}\right) - \dfrac{\partial}{\partial x}\left(b(x,y)\dfrac{\partial u}{\partial x}\right) = 0, \\ u(x,y) = 1 \text{ on } \partial\Omega, \end{cases} \tag{4.5}$$

where $\Omega \subset \mathbb{R}^2$.

We use three types of coefficients for the equation (4.5) on three different grids. As a standard example we take $a = b = 1$, i.e. the Laplace operator.

In *Example 1*, the coefficients are mildly varying: $a(x,y) = (x^2 + y^2 + 1 + sin(x+y))$ and $b(x,y) = (x^2 + y^2 + 1 + cos(x+y))$. In *Example 2*, the coefficients are varying in the range $[10^{-3}, 30]$. For the grid

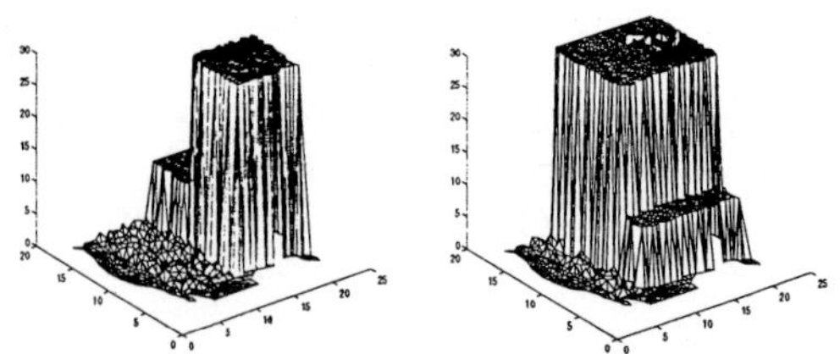

Figure 4.5: Surface plot of the coefficients for Example 2 — rapidly varying coefficients, $a(x,y)$ on the left, $b(x,y)$ on the right.

given in Fig. 4.6, the surface plot of the coefficients for *Example 2* are given in Fig. 4.5. For all the grids, the coefficients vary within the same range. In these experiments, we use the standard V-cycle preconditioner and the outer acceleration is done by the CG method. In the V-cycle , we use 1 pre- and 1 post-smoothing steps. The smoothing operator is forward Gauß-Seidel. The PCG iterations are terminated when the relative residual is less then 10^{-6}.

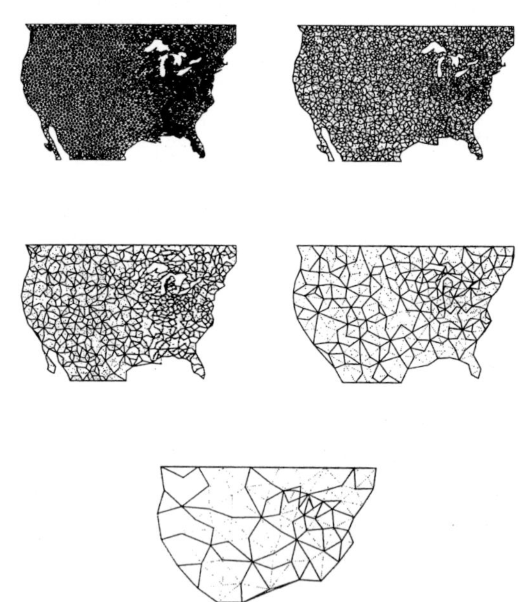

Figure 4.6: Macroelements for an unstructured grid level= 4 $N_h = 3422$; level= 3 $N_H^1 = 938$; level= 2 $N_H^2 = 268$; level= 1 $N_H^3 = 77$; level= 0 $N_H^4 = 21$.

In Figures 4.6–4.8, the macroelements are shown for different unstructured grids and different number of levels. Figure 4.9 shows the convergence histories for

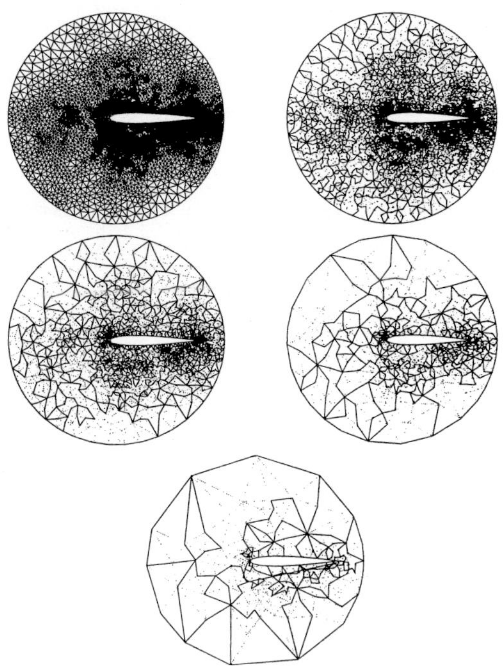

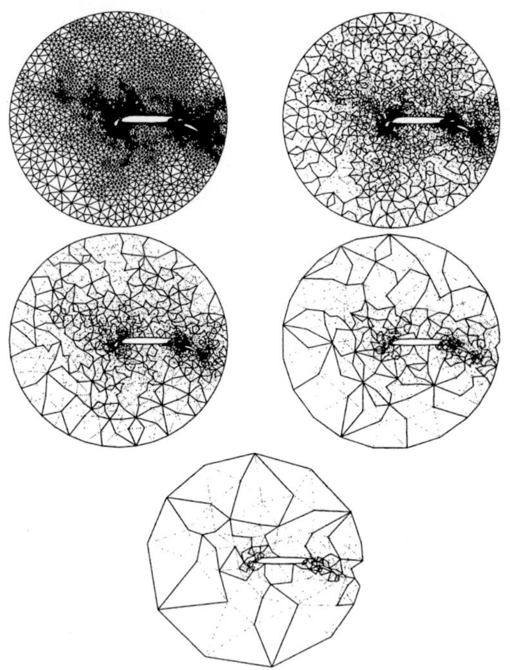

Figure 4.7: Macroelements for one element airfoil: level= 5 $N_h = 12665$; level= 4 $N_H^1 = 3404$; level= 3 $N_H^2 = 928$; level= 2 $N_H^3 = 257$; level= 1 $N_H^4 = 74$; level= 0 $N_H^5 = 24$.

Figure 4.8: Macroelements for four element airfoil: level= 5 $N_h = 12850$; level= 4 $N_H^1 = 3444$; level= 3 $N_H^2 = 949$; level= 2 $N_H^3 = 270$; level= 1 $N_H^4 = 80$; level= 0 $N_H^5 = 26$.

the different types of coefficients and different grids. All these experiments were done using the simplest interpolation algorithm, Alg. 4.3. Figure 4.10 shows the convergence histories for a varying number of unknowns on two different grids: a one-element airfoil with one internal boundary, and a four-element airfoil with four internal boundaries. The numerical experiments were done using Algorithm 4.2.

These computational results show that the convergence is uniform with respect to the mesh size h. The convergence in the experiments shown in Figure 4.10 is a little better because the aspect ratio of the grids is better and Algorithm 4.2 was used instead of Algorithm 4.3. The behavior for roughly varying coefficients is not as good, as seen in Fig. 4.9. The airfoil grids used for the experiments in this section were produced using Barth's SIMPLEX2D mesh generator. We note here that these grids were generated at random, with no special attention being paid to the quality of the meshes, and does not reflect any deficiencies in the mesh generator.

4.5 Extensions

Other interpolations can also be constructed in an algebraic way. They are known as matrix-dependent prolongations (see M. Griebel in [46]). In this case, the prolongation operator is defined by $R_H^T = [A_{11}^{-1} A_{12}, I]^T$. Here, A_{11} and A_{12} are the the blocks in A_J formed by the natural splitting of the unknowns into two non-overlapping subsets corresponding to fine and coarse grid unknowns respectively:

$$A_J = \begin{bmatrix} A_{11} & A_{12} \\ A_{21} & A_{22} \end{bmatrix}. \tag{4.6}$$

The block A_{11} corresponds to the contributions $fine - fine$. Then it is straightforward to see that the coarse grid matrix is equal to the Schur complement of A: $S = A_{J-1} = A_{22} - A_{21} A_{11}^{-1} A_{12}$. Unfortunately, A_{11}^{-1} is generally a dense matrix, which is a serious drawback of using this approach. There are ways of defining approximations to this type of matrix-dependent prolongations as proposed by A. Reusken [30] and M.

507

Grid-Fig. 4.6

Example	Reduction factor
Laplace	0.11754
Example 1	0.12298
Example 2	0.24887

Grid-Fig. 4.7

Example	Reduction factor
Laplace	0.20701
Example 1	0.20724
Example 2	0.42600

Grid-Fig. 4.8

Example	Reduction factor
Laplace	0.21144
Example 1	0.21454
Example 2	0.46994

1-element airfoil

nodes	Reduction factor
$N_h = 72139$	0.16834
$N_h = 18152$	0.14236
$N_h = 4683$	0.14911
$N_h = 1315$	0.14087

4-element airfoil

nodes	Reduction factor
$N_h = 72233$	0.15454
$N_h = 18328$	0.13836
$N_h = 4870$	0.15488
$N_h = 1502$	0.14727

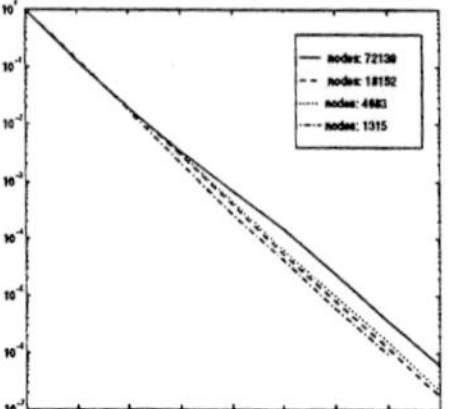
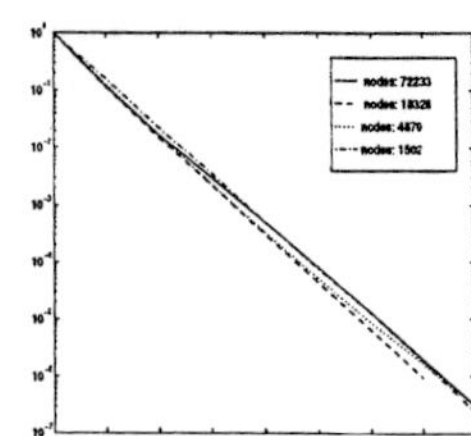

Figure 4.10: Convergence history and average reduction per iteration for varying number of unknowns

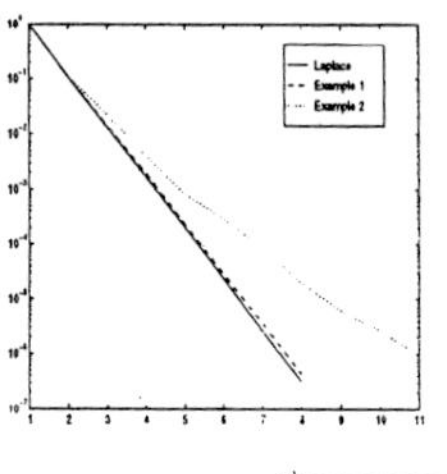
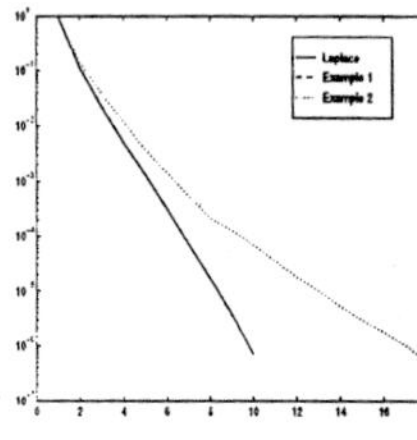
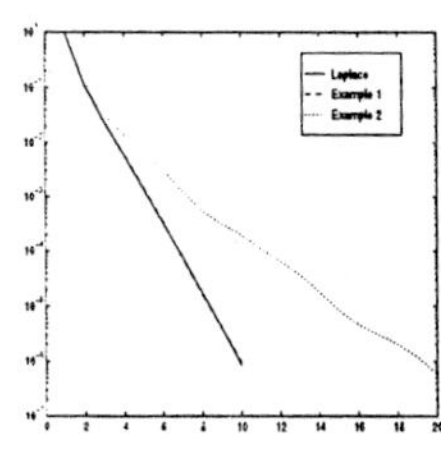

Figure 4.9: Convergence history and average reduction per iteration for Laplace equation, Example 1 and Example 2 on the different grids

Griebel in [46]. The agglomeration coarsening algorithm presented here and the aggregation given in [32] also might be viewed as ways of approximating the first entry $A_{11}^{-1} A_{12}$ in the prolongation with a sparse matrix.

4.5.1 Anisotropic problems

Another class of problems we have studied are the *anisotropic problems*. The problem in applying multigrid methods for such problems is that the smoother does not smooth the proper range of the high frequencies. A semi-coarsening (i.e. coarsening only in one direction) is often used to remedy this.

For anisotropic problems, the relevant changes in the agglomeration algorithm are straightforward. A dropping strategy can be used for the small off-diagonal

elements in A_k on each level. A new coarse grid operator $\tilde{A}_k$ is then obtained and this matrix corresponds to a new graph which is disconnected. Different dropping strategies can be applied (see [13]). Here we apply a simple one:
If

$$\sqrt{\frac{a_{ij} a_{ji}}{a_{ii} a_{jj}}} \leq 0.0001,$$

then set

$$a_{ii} := a_{ii} + a_{ij}; \quad a_{jj} := a_{jj} + a_{ji}; \quad a_{ij} := 0; \quad a_{ji} := 0.$$

Once this is done, we apply the usual *algebraic coarsening* algorithm [23]. In the next example, the algorithm which uses the dropping strategy is called *reduced graph algorithm*. Similar approaches for handling anisotropic problems can be found in [32].

The last numerical example in this section solves the Laplace equation with anisotropy introduced by the grid (see Fig. 4.11). The geometrical aspect ratio is of order 10^4. It can be seen that the algorithm which uses the anisotropic agglomeration is faster than the others.

4.6 Remarks

The agglomeration algorithms can provide a good approach for developing multilevel methods on unstructured grids. We have presented here a general technique for constructing basis for the coarse space satisfying stability and approximation properties. We have

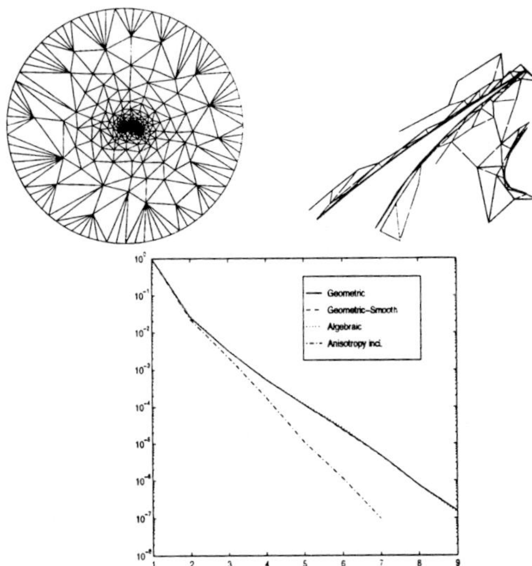

Figure 4.11: Macroelements (geometrical algorithm), second level geometric coarsening, zoomed 200 times near the airfoil. Convergence history for the stretched mesh: Geometrical coarsening; Geometrical coarsening-harmonic extension II: Algebraic coarsening; Reduced graph coarsening (anisotropy).

to point out that the general theory for the construction of agglomerated spaces on unstructured grids is still not fully developed. On the other hand, numerically these methods have good performance and can be applied to a large set of problems, including elliptic, anisotropic and convection dominated problems. For such experiments, we refer to Koobus, Lallemand and Dervieux [31], Mandel, Vaněk, Brezina (see [32]) and the experiments presented in this section.

We presented three different types of basis construction over agglomerated macroelements: Algorithm 4.1, Algorithm 4.2, Algorithm 4.3. We prefer to use Algorithm 4.3 for isotropic problems, because the convergence rate is as good as with the other two algorithms and this algorithm is simpler. The last numerical experiment we performed shows that for more complicated problems, such as anisotropic problems, the interpolation must be done depending of the direction of the anisotropy. In this case, the algorithms for constructing the coarse grid also need to be done very carefully, following the anisotropy direction.

5 Conclusions

There is no doubt that unstructured grids will be increasingly popular and the development of robust mul-

tilevel solvers on unstructured grids is important. We have presented two different approaches for constructing such solvers and they have their own areas of strength and weakness:

1. In some sense, the retriangulation algorithm is more natural and interfaces well with existing software (reusability). But problems may occur when solving 3D discrete equations because the retriangulation is not an easy task. The treatment of different types of boundary conditions must be done very carefully if one wishes to obtain a uniformly convergent iterative method.

2. The agglomeration algorithms offer many advantages: they are more algebraic, they produce nested coarse spaces, and are very robust. We have presented a framework for the design of agglomeration multilevel methods based on the stability and approximation properties of the underlying subspaces. Some of these methods have straightforward extension to 3D problems. The convergence of these types of methods is still not completely understood from a theoretical point of view.

Most importantly, we have shown that it is possible to design robust multilevel methods on unstructured grids which perform as efficiently as for structured grids.

References

[1] T. F. Chan, S. Go, and L. Zikatanov. Lecture notes on multilevel methods for elliptic problems on unstructured grids. Technical Report CAM 97–11, Department of Mathematics, University of California at Los Angeles, March 1997.

[2] T. F. Chan, B. Smith, and J. Zou. Overlapping Schwarz methods on unstructured meshes using non-matching coarse grids. *Numerische Mathematik*, 73(2):149–167, April 1996.

[3] T. F. Chan, B. Smith, and J. Zou. Multigrid and domain decomposition methods for unstructured meshes. In I.T. Dimov, Bl. Sendov, and P. Vassilevski, editors, *Proc. of the 3rd Int'l Conf. on Advances in Numerical Methods and Applications, Sofia, Bulgaria.*, pages 53–62. World Scientific, August 1994.

[4] T. F. Chan, S. Go, and J. Zou. Multilevel domain decomposition and multigrid methods for unstructured meshes: Algorithms and theory. Technical Report CAM-95–24, Department of Mathematics, University of California at Los Angeles, May 1995.

also in Proceedings of the Eighth International Conference on Domain Decomposition, May 1995, Beijing.

[5] T. F. Chan, S. Go, and J. Zou. Boundary treatments of multilevel methods on unstructured meshes. Technical Report CAM 96-30, Department of Mathematics, University of California at Los Angeles, September 1996.

[6] Tony F. Chan, J. Xu, and Ludmil Zikatanov. Agglomeration strategies in multigrid method. in preparation.

[7] G. Strang and G. Fix. *An Analysis of the Finite Element Method.* Prentice Hall, 1973.

[8] P. Ciarlet. *The finite element method for elliptic problems.* North-Holland, 1978.

[9] S. Brenner and L.R. Scott. *The mathematical theory of finite element methods.* Springer-Verlag, 1994.

[10] Claes Johnson. *Numerical solution of partial differential equations by the finite element method.* Cambridge University Press, Cambridge, 1990.

[11] D.L. Brown, G. Chesshire, W.D. Henshaw, and H.-O. Kreiss. On composite overlapping grids. In T.J. Chung and G.R. Karr, editors, *Finite Element Analysis in Fluids*, pages 544-559. UAH Press, 1989.

[12] G. Chesshire and W. D. Henshaw. Composite overlapping meshes for the solution of partial differential equations. *J. Comp. Phys.*, 90(1):1-64, 1990.

[13] Yousef Saad. *Iterative methods for sparse linear systems.* PWS Publishing company, 1996.

[14] William Briggs. *A multigrid tutorial.* Society for Industrial and Applied Mathematics, Philadelphia, 1987.

[15] J. Bramble. *Multigrid methods.* Pitman, Notes on Mathematics, 1994.

[16] W. Hackbusch. *Multi-grid methods and applications.* Springer Verlag, New York, 1985.

[17] J. Xu. Iterative methods by space decomposition and subspace correction. *SIAM Review*, 34:581-613, December 1992.

[18] J. Xu. *An introduction to multilevel methods.* To be published by Oxford University Press, 1996. Lecture notes: VIIth EPSRC Numerical analysis summer school, Leicester University, UK.

[19] T. F. Chan and T. Mathew. Domain decomposition algorithms. *Acta Numerica*, pages 61-143, 1994.

[20] B. Smith, P. Bjorstad, and W. Gropp. *Domain decomposition: parallel multilevel methods for elliptic partial differential equations.* Cambridge University Press, Cambridge, 1996.

[21] Maksymilian Dryja and Olof B. Widlund. Some domain decomposition algorithms for elliptic problems. In Linda Hayes and David Kincaid, editors, *Iterative Methods for Large Linear Systems*, pages 273-291, San Diego, California, 1989. Academic Press.

[22] Maksymilian Dryja and Olof B. Widlund. Additive Schwarz methods for elliptic finite element problems in three dimensions. In Tony F. Chan, David E. Keyes, Gérard A. Meurant, Jeffrey S. Scroggs, and Robert G. Voigt, editors, *Fifth Conference on Domain Decomposition Methods for Partial Differential Equations*, Philadelphia, PA, 1992. SIAM.

[23] J. W. Ruge and K. Stüben. Algebraic multigrid. In S. F. McCormick, editor, *Multigrid methods*, Frontiers in applied mathematics, pages 73-130, Philadelphia, Pennsylvania, 1987. SIAM.

[24] D.J. Mavriplis. Unstructured mesh algorithms for aerodynamic calculations. Technical Report 92-35, ICASE, NASA Langley, Virginia, July 1992.

[25] H. Guillard. Node-nested multi-grid method with Delaunay coarsening. Technical Report RR-1898, INRIA, Sophia Antipolis, France, March 1993.

[26] T. F. Chan and Barry Smith. Domain decomposition and multigrid methods for elliptic problems on unstructured meshes. In David Keyes and Jinchao Xu, editors, *Domain Decomposition Methods in Science and Engineering, Proceedings of the Seventh International Conference on Domain Decomposition, October 27-30, 1993, The Pennsylvania State University.* American Mathematical Society, Providence, 1994. also in Electronic Transactions on Numerical Analysis, v.2, (1994), pp. 171-182.

[27] R. E. Bank and J. Xu. An algorithm for coarsening unstructured meshes. Technical report, Dept. of Math., Univ. of Calif. at San Diego, 1994.

[28] W. Hackbusch. *Iterative solution of large sparse systems of equations.* Springer Verlag, Heidelberg, 1993. Applied Mathematical Sciences, Vol. 95.

[29] D. Braess. Towards algebraic multigrid for elliptic problems of second order. *Computing*, 55:379-393, 1995.

[30] A. A. Reusken. A multigrid method based on incomplete Gaussian elimination. Technical Report RANA 95-13, Eindhoven University of Technology, October 1995.

[31] B. Koobus, M. H. Lallemand, and A. Dervieux. Unstructured volume-agglomeration MG: solution of the Poisson equation. *International Journal for Numerical Methods in Fluids*, 18(1):27–42, 1994.

[32] P. Vaněk, J. Mandel, and M. Brezina. Algebraic multi-grid by smoothed aggregation for second and forth order elliptic problems. *Computing*, 1995. to appear.

[33] P. Vaněk and J. Křižková. Two-level method on unstructured meshes with convergence rate independent of the coarse space size. Technical Report 33, University of Colorado at Denver, January 1995.

[34] A. M. Matsokin and S. V. Nepomnyaschikh. A Schwarz alternating method in a subspace. *Soviet Mathematics*, 29(10):78–84, 1985.

[35] P. Lions. On the Schwarz alternating method. I. In Roland Glowinski, Gene H. Golub, Gérard A. Meurant, and Jacques Périaux, editors, *First International Symposium on Domain Decomposition Methods for Partial Differential Equations*, SIAM, Philadelphia, 1988.

[36] M. Dryja and O. Widlund. An additive variant of the Schwarz alternating method for the case of many subregions. Technical Report 339, also Ultracomputer Note 131, Department of Computer Science, Courant Institute, 1987.

[37] James H. Bramble, Joseph E. Pasciak, Junping Wang, and Jinchao Xu. Convergence estimates for multigrid algorithms without regularity assumptions. *Math. Comp.*, 57(195):23–45, 1991.

[38] T. F. Chan and J. Zou. Additive Schwarz domain decomposition methods for elliptic problems on unstructured meshes. *Numerical Algorithms*, 8:329–346, 1994.

[39] T. F. Chan and J. Zou. A convergence theory of multilevel additive Schwarz methods on unstructured meshes. Technical Report 95-16, Department of Mathematics, University of California at Los Angeles, March 1995. to appear in Numerical Algorithms.

[40] Ralf Kornhuber and Harry Yserentant. Multilevel methods for elliptic problems on domains not resolved by the coarse grid. In David Keyes and Jinchao Xu, editors, *Domain Decomposition Methods in Science and Engineering, Proceedings of the Seventh International Conference on Domain Decomposition, October 27-30, 1993, The Pennsylvania State University*, volume 180, pages 49–60. American Mathematical Society, Providence, 1994.

[41] R. Bank and J. Xu. An algorithm for coarsening unstructured meshes. *Numer. Math.*, 73:1–23, 1996.

[42] P. Clément. Approximation by finite element functions using local regularization. *R.A.I.R.O. Numer. Anal.*, R-2:77–84, 1975.

[43] Xiao-Chuan Cai. The use of pointwise interpolation in domain decomposition methods with nonnested meshes. *SIAM J. Sci. Comp.*, 16(1), 1995.

[44] William D. Gropp and Barry F. Smith. Portable Extensible Toolkit for Scientific Computation (PETSc). Available by anonymous ftp at **info.mcs.anl.gov** in the directory **pub/pdetools** or at **http://www.mcs.anl.gov/Projects/petsc/petsc.html**.

[45] A. Pothen, H. Simon, and K.P. Liou. Partitioning sparse matrices with eigenvector of graphs. *SIAM J. Mat. Anal. Appl.*, 11(3):430–452, 1990.

[46] T. Grauschopf, M. Griebel, and H. Regler. Additive multilevel preconditioners based on bilinear interpolation, matrix-dependent geometric coarsening and algebraic multigrid coarsening for second-order elliptic PDEs. *Applied Numerical Mathematics*, 23(1):63–97, february 1997. special issue on multilevel methods.

MULTIGRID METHODS FOR STEADY AND TIME-DEPENDENT FLOW

Chaoqun LIU

Numerical Simulation Group, Mathematics and Statistics
Louisiana Tech University, P.O. Box 3189, Ruston, LA 71272-0001, USA

Abstract

A recent development in application of multigrid and multilevel adaptive methods for fluid flow by the numerical simulation group at Louisiana Tech University is reviewed. The multigrid method has been successfully applied for a number of flow cases including both steady and time-dependent flows. The steady cases include laminar flow, turbulent flow, and turbulent combustion. The time-dependent cases include flow transition around airfoils at all speeds including incompressible, subsonic, and supersonic flows. The multigrid method has been found very efficient for both laminar and turbulent flows and also useful for complex geometries.

Received on June 24, 1997.

1 Introduction

Since Dr. Achi Brandt published his historic paper (Brandt, 1977), the multigrid method (MG) has been well understood (Hackbusch, 1985; McCormick, 1985; Briggs, 1987; Wesseling, 1992; Liu, 1995e) and widely used for computational fluid dynamics (CFD) (McCormick, 1983, 1985, 1987, 1989, 1991, 1993, 1995). The multigrid method was originally aimed at accelerating the convergence of linear elliptic problems, but is now widely applied for a variety of engineering problems. Moreover, the multilevel adaptive method (MAM, McCormick, 1989) further assists the local refinement and optimalization of grid usage. The distinctive advantage of MG over other numerical methods is its fast convergence rate which could be mesh-size or grid-number independent. The efficiency of MG for Poisson's equation is well demonstrated and recognized. An estimate of O(N) operations is given for the convergence of Poisson's equation via the full multigrid (FMG) method which shows MG is an optimal numerical method. However, the flow problem is much more complicated. It is a system with a number of differential equations, non-linear, and possibly singular. Further questions include the following.

- Can MG work well for turbulent flow when a turbulence model is used?
- Can MG improve the convergence for reacting flow?
- Can MG work efficiently for complex geometries?
- Can MG be used for time-dependent flow?

Recently significant progress has been made by the numerical simulation group (NSG) at Louisiana Tech University toward providing satisfying answers to these questions. This paper is arranged in such a way that we start with basic multigrid ideas and then study complex laminar and turbulent flow. Finally we discuss about the time-dependent flow.

2 Multigrid for Poisson's equation

2.1 Problems with iterative methods

Let us start with a 1-D Poisson's equation:

$$\begin{cases} -\frac{d^2 u}{dx^2} = f & x \in (0,1) \\ u = 0 & x = 0 \ \ or \ \ x = 1. \end{cases} \tag{1}$$

The central difference gives (Figure 1):

$$\begin{cases} -(u_{i-1} - 2u_i + u_{i+1}) = f_i \Delta x^2, & i = 1, 2, ..., N-1 \\ u_0 = 0 \\ u_N = 0 \end{cases}$$

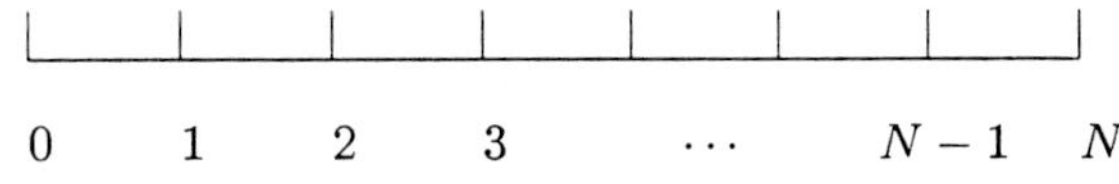

Figure 1. 1-D grids

We may write it in a matrix form:

$$\begin{bmatrix} 2 & -1 & & & \\ -1 & 2 & -1 & & \\ & & \ddots & & \\ & & & -1 & 2 \end{bmatrix} \begin{bmatrix} u_1 \\ u_2 \\ \cdot \\ \cdot \\ \cdot \\ u_{N-1} \end{bmatrix} = \begin{bmatrix} f_1 \\ f_2 \\ \cdot \\ \cdot \\ \cdot \\ f_{N-1} \end{bmatrix},$$

We can first give an initial guess u^0, which will introduce an initial error e^0, and then use iterative methods to reduce errors until certain criteria are satisfied:

$$\|e^n\| < \epsilon. \tag{2}$$

Here the superscript n represents the value associated with nth iteration, etc.

Let's take a look at Jacobi relaxation. Let

$$A = L + D + U, \tag{3}$$

where D is diagonal, and L and U are lower and upper triangle respectively.

$$Du^{\overline{n+1}} = F - (L+U)u^n, \tag{4}$$

$$u^{\overline{n+1}} = D^{-1}F - D^{-1}(L+U)u^n, \tag{5}$$

$$u^{n+1} = \omega u^{\overline{n+1}} + (1-\omega)u^n, \tag{6}$$

or

$$u^{n+1} = Mu^n + E, \tag{7}$$

where

$$M = [(1-\omega)I + \omega D^{-1}(L+U)]. \tag{8}$$

Here, ω is the parameter of relaxation and the superscript n represents the value associated with nth iteration, etc. Since

$$u \;=\; Mu + E \qquad (10)$$

where u is the exact solution of (2). Subtracting (8) from (10), we have

$$e^{n+1} \;=\; Me^n, \qquad (11)$$

or

$$e^n \;=\; M^{(n)}e^o, \qquad (12)$$

where

$$e^n \;=\; u^n - u.$$

The iteration matrix M is:

$$M = \begin{bmatrix} 1-\omega, & \frac{1}{2}\omega \\ \frac{1}{2}\omega, & 1-\omega, & \frac{1}{2}\omega \\ & & \ddots \\ & & \frac{1}{2}\omega & 1-\omega \end{bmatrix}.$$

The eigenfunction equation can be written as:

$$Mv \;=\; \lambda v, \qquad (13)$$

The kth eigenfunction is:

$$v_k \;=\; (\sin kx_1\pi, \,...\,, \sin kx_{N-1}\pi)^T, \qquad (14)$$

and the kth eigenvalue is:

$$\lambda_k = 1 - 2\omega \sin^2 \frac{k\pi}{2N}, \qquad k = 1, 2, \,...\,, N-1. \qquad (15)$$

Here, the eigenfunction v_k represents the kth error component and the eigenvalue λ_k is the amplification factor of v_k. Let us assume $\omega = 2/3$, then

$$\lambda_k = 1 - \frac{4}{3}\sin^2 \frac{k\pi}{2N}. \qquad (16)$$

The Jacobi relaxation has different amplification factors for different error components. The most important concept of multigrid methods is to divide the errors into two groups: high-frequency and low-frequency errors. For high-frequency error, $\frac{N}{2} \le k \le N-1$,

$$|\lambda_k| < \frac{1}{3}, \qquad (17)$$

which gives a fast reduction of high-frequency errors. For low-frequency error, e.g. $k = 1$ (when N is large),

$$\lambda_1 = 1 - \frac{4}{3}\sin^2 \frac{\pi}{2N} \approx 1 - \frac{\pi^2 h^2}{3}, \qquad (18)$$

so that $\lambda_1 \to 1$ as $h \to 0$, which is much too slow in convergence.

Assume f=0, then the exact solution is $u \equiv 0$. Let N=64 and $u^0 = v_1 + v_2 + v_{32} + v_{63}$, $e^0 = u^0 - u = u^0$, look at the change of these error components after 10 iterations:

$$\begin{aligned} e^{10} &= M^{(10)}v_1 + M^{(10)}v_2 + M^{(10)}v_{32} + M^{(10)}v_{63} \\ &= \lambda^{(10)}v_1 + \lambda^{(10)}v_2 + \lambda^{(10)}v_{32} + \lambda^{(10)}v_{63}. \qquad (19) \end{aligned}$$

We find

$$\begin{aligned} \lambda_1 = 1 - \frac{4}{3}sin^2\frac{\pi}{128} &= 0.99920, \\ \lambda_2 &= 0.99679, \\ \lambda_{32} &= 0.33333, \\ \lambda_{63} &= -0.33253. \end{aligned}$$

Then,

$$\begin{aligned} e^{10} &= 0.99203v_1 + 0.96836v_2 \\ &+ 0.0000169v_{32} + 0.0000165v_{63}. \qquad (20) \end{aligned}$$

From this simple example, we can easily find that the Jacobi relaxation is efficient for reducing high-frequency errors, but inefficient for low-frequency errors. In other words, Jacobi relaxation is a good smoother. The idea behind multigrid method is to use relaxation to reduce high-frequency errors and coarse grid correction to reduce low-frequency errors.

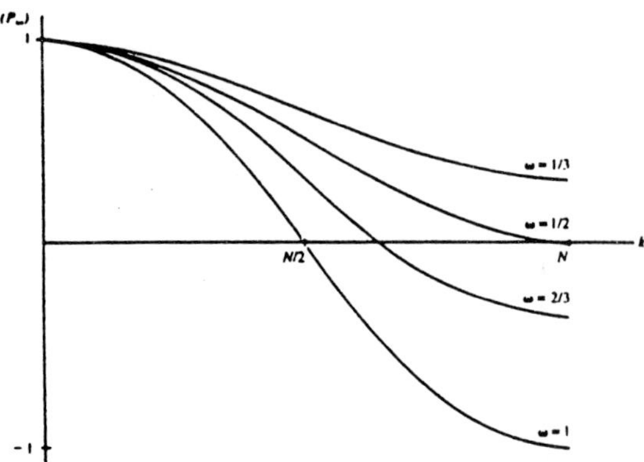

Figure 2. Jacobi relaxation

2.2 Multigrid V-cycle

The procedure for a two-level multigrid V-cycle (Figure 3) can be described as follows:
For a linear problem $L_h u_h = f_h$, where L_h is a linear operator,

1. Do fine grid relaxation 1 or 2 times,

$$u_h^{n+1} \;=\; Mu_h^n + E, \qquad (21)$$

   and calculate the residual:

$$R_h^{n+1} \;=\; f_h - L_h u_h^{n+1} . \qquad (22)$$

2. Solve the coarse grid equation:

$$\begin{aligned} L_{2h}v_{2h} &= I_h^{2h}R_h^{n+1} , \\ v_{2h} &= L_{2h}^{-1}I_h^{2h}(f_h - L_h u_h^{n+1}), \qquad (23) \end{aligned}$$

   where I_h^{2h} is a full-weighting restriction and L_{2h} is a coarse grid operator.

3. Do the coarse grid correction:

$$u_h^{n+2} \;=\; u_h^{n+1} + I_{2h}^h v_{2h} , \qquad (24)$$

   where I_{2h}^h is an interpolation operator.

This V-cycle represents a process:

$$\begin{aligned} u_h^{n+2} &= u_h^{n+1} + I_{2h}^h L_{2h}^{-1} I_h^{2h}(f_h - L_h u_h^{n+1}), \qquad (25) \\ e_h^{n+2} &= e_h^{n+1} + I_{2h}^h L_{2h}^{-1} I_h^{2h}(-L_h e_h^{n+1}) \\ &= [I - I_{2h}^h L_{2h}^{-1} I_h^{2h} L_h]e_h^{n+1} \\ &= CGe_h^{n+1}, \qquad (26) \end{aligned}$$

where

$$CG = I - I_{2h}^h L_{2h}^{-1} I_h^{2h} L_h \qquad (27)$$

is called the *coarse grid correction*.

Let us take

$$I_h^{2h} = c(I_{2h}^h)^T, \tag{28}$$

$$L_{2h} = I_h^{2h} L_h I_{2h}^h, \tag{29}$$

which is called the variational approach. Then,

$$CG = I - I_{2h}^h (I_h^{2h} L_h I_{2h}^h)^{-1} I_h^{2h} L_h. \tag{30}$$

In general, we can write the fine grid error as(Figure 4)

$$e_h^{n+1} = I_{2h}^h I_h^{\bar{2}h} e_h^{n+1} + d = I_{2h}^h e_{2h} + d, \tag{31}$$

where $e_{2h} = I_h^{\bar{2}h} e_h^{n+1}$, $d = e_h^{n+1} - I_{2h}^h I_h^{\bar{2}h} e_h^{n+1}$ and $I_h^{\bar{2}h}$ is an injection operator. Here, e_{2h} pretty much represents a smooth part of e_h^{n+1} and d is an oscillatory part.

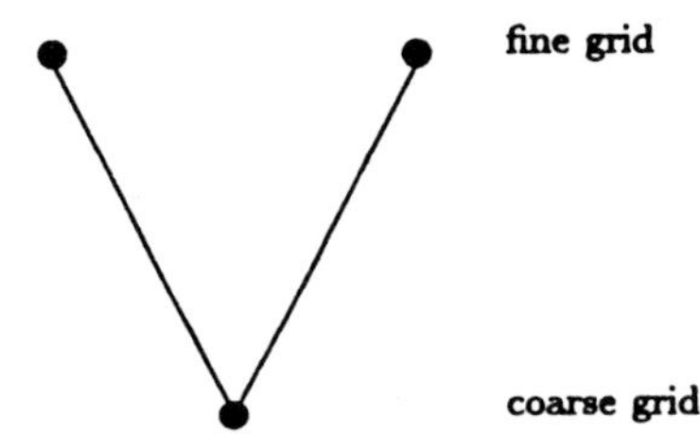

Figure 3. 2-level V-cycle

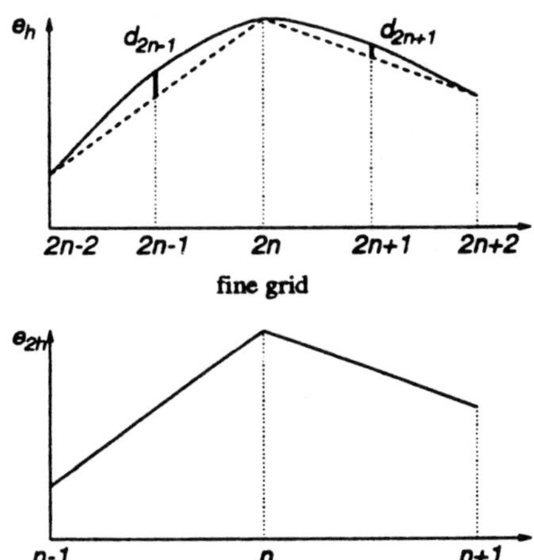

Figure 4. Errors on the fine and coarse grids

For a special case,

$$d = 0 \text{ and } e_h^{n+1} = I_{2h}^h e_{2h},$$

substituting this to (26),

$$\begin{aligned}
e_h^{n+2} &= (I - I_{2h}^h (I_h^{2h} L_h I_{2h}^h)^{-1} I_h^{2h} L_h) I_{2h}^h e_{2h} \\
&= I_{2h}^h e_{2h} - I_{2h}^h e_{2h} = 0. \tag{32}
\end{aligned}$$

We find that the entire e_{2h} is eliminated by coarse grid correction or the smooth part of the error is effectively reduced by CG. We can prove that, for the general case, the CG does not increase d and the iterative method can effectively reduce the part of d. Therefore, combined with iterative method, the MG V-cycle can reduce both low- and high-frequency errors efficiently. Because the CG can rapidly reduce the low-frequency errors, the convergence rate of MG V-cycle is pretty much governed by the convergence rate of relaxation for high-frequency errors, $\frac{N}{2} < k < N-1$, which is in general mesh-size independent.

2.3 Restriction, interpolation, and coarse grid operator

We usually use a full-weighting operator I_h^{2h} for the restriction of residual from the fine grid to the coarse grid, which can be constructed by a so-called area law (Figure 5, Liu, 1989d):

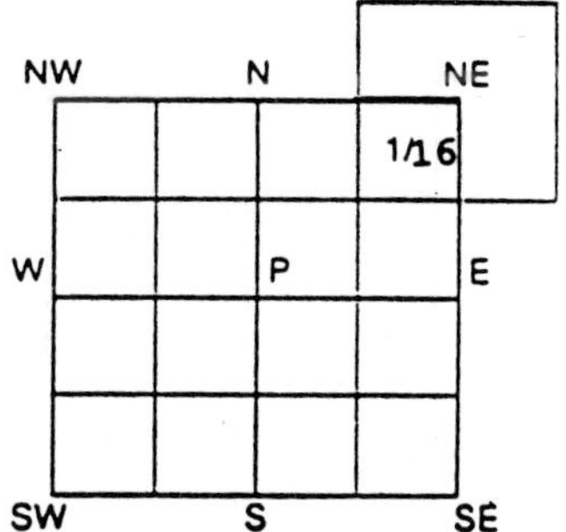

Figure 5. Restriction operator

The coarse grid volume, c, is divided by fine grid points, $P(\frac{1}{4})$, E, W, N, S ($\frac{1}{8}$ each), and NE, NW, SE, SW ($\frac{1}{16}$ each). Therefore, the stencil of the restriction is :

$$\begin{bmatrix} \frac{1}{16} & \frac{1}{8} & \frac{1}{16} \\ \frac{1}{8} & \frac{1}{4} & \frac{1}{8} \\ \frac{1}{16} & \frac{1}{8} & \frac{1}{16} \end{bmatrix}. \tag{33}$$

This law can be used for non-uniform grids and for interpolation which will give a linear or bi-linear interpolation. If we use the variational approach, $I_h^{2h} = c(I_{2h}^h)^T$ and $L_{2h} = I_h^{2h} L_h I_{2h}^h$, we will be led to a 9-point stencil for the coarse grid operator. We can find an approximate way to keep 5-point stencil (Figure 6). The coarse grid operator, L_{2h}, has three parts : restriction, fine grid operator, and interpolation. Let's define A_E^C as a contribution from E to C which only can be done through two paths : one is E shifted to $u_{i+1,j+1}$ and then contributed from $u_{i+1,j+1}$ to $u_{i,j+1}$ (or $A_{E\ i,j+1}^f$) and restricted to C ($\frac{1}{4}$); the other is E shifted to $u_{i+1,j}$ and then contributed from $u_{i+1,j}$ to $u_{i,j}$ (or $A_{E\ i,j}^f$) and restricted to C ($\frac{1}{4}$). Therefore,

$$A_E^c = \frac{1}{4} A_{E\ i,j+1}^f + \frac{1}{4} A_{E\ i,j}^f, \tag{34}$$

where the superscript C stands for the coarse grid and f for the fine grid, and the subscripts are indices of grid points.

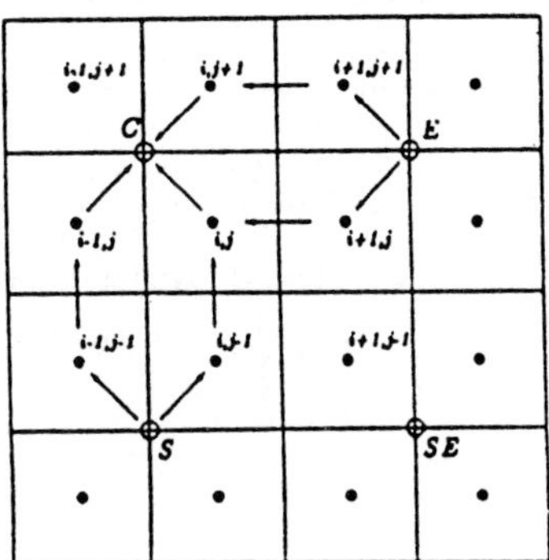

Figure 6. 5-point stencil on coarse grid

The other way to construct the coarse grid operator is to use the conservation principle (Figure 7, Liu et al, 1992a)) :

$$Flux_e^c = Flux_e^1 + Flux_e^3, \qquad (35)$$

where the subscript e stands for the east side of the control volume, superscript c for the coarse grid, and 1 and 3 are index of fine grid points.

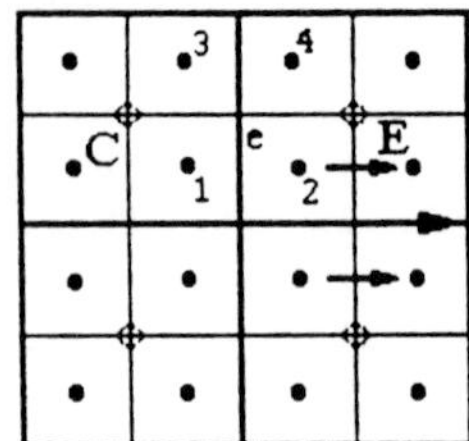

Figure 7. Conservation principle

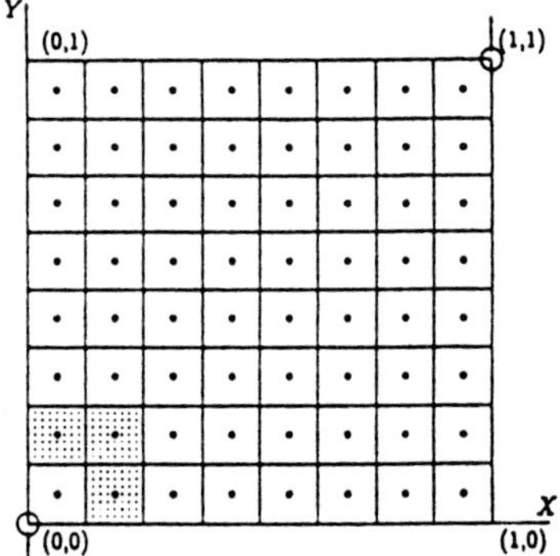

Figure 8. Jump interface

2.4 Elliptic equation with discontinuous coefficients (Liu et al, 1992a)

$$\nabla . \lambda \nabla P = f \qquad (36)$$

λ can jump to be several-orders higher at the interface of layers (Figure 8). We may use the following ways to solve the problem:

1. Harmonic averaging for diffusion coefficients:

$$\lambda_e = \frac{2\lambda^+ \lambda^-}{\lambda^+ + \lambda^-} \qquad (37)$$

2. Conservation principle for constructing the coarse grid operator (Figure 7)

3. Effective area law for constructing interpolation operator:

$$s^* = \frac{s}{\lambda} \qquad (38)$$

For a particular case (Figure 9), instead of using

$$I_{2h}^h = \begin{bmatrix} \frac{3}{16} & \frac{9}{16} \\ \frac{1}{16} & \frac{3}{16} \end{bmatrix}, \qquad (39)$$

we use

$$I_{2h}^h = \begin{bmatrix} \frac{3}{12} & \frac{5}{12} \\ \frac{1}{12} & \frac{3}{16} \end{bmatrix} \quad since \ \lambda^- >> \lambda^+.$$

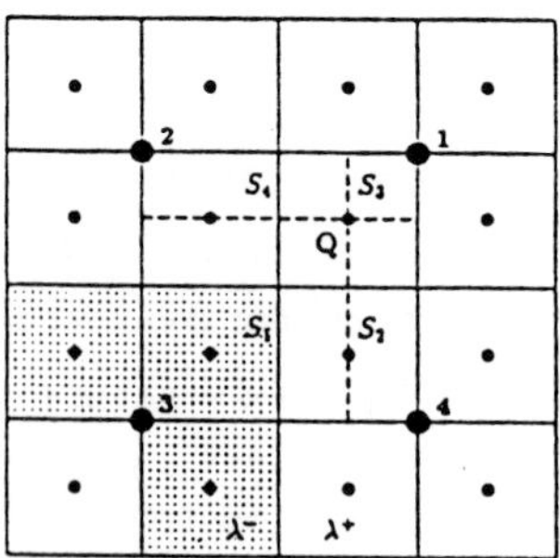

Figure 9. Effective area law.

2.5 Anisotropic grids

As shown by Brandt (1984), for a discrete Poisson's equation:

$$a(u_E^n - 2u_C^{n+1} + u_W^{n+1}) + c(u_N^n - 2u_C^{n+1} + u_S^{n+1}) = 0, \quad (40)$$

where $a << c$ for example, The point Gauss-Seidel relaxation gives an amplification factor:

$$\mu(\theta_1, \theta_2) = | \frac{A_\theta^{n+1}}{A_\theta^n} | = | \frac{ae^{i\theta_1} + ce^{i\theta_2}}{2a + 2c - ae^{-i\theta_1} - ce^{-i\theta_2}} | , \quad (41)$$

where (θ_1, θ_2) represents a Fourier component of errors, A_θ is the magnitude of the component, and the superscripts n and n+1 are associated with the values before and after the iteration. If a=c, for high-frequency errors, ($\frac{\pi}{2} \le |\theta| \le \pi$, $|\theta| = max\{|\theta_1|, |\theta_2|\}$)

$$\bar{\mu} = max\{\mu(\theta)\} = \mu\{\frac{\pi}{2}, arccos\frac{4}{5}\} = 0.5, \qquad (42)$$

which is acceptable, but if $c >> a$, when $\theta_1 = \frac{\pi}{2}$ and $\theta_2 = 0$,

$$\mu(\frac{\pi}{2}, 0) = [\frac{a^2 + c^2}{a^2 + (c + 2a)^2}]^{\frac{1}{2}} . \qquad (43)$$

The rate of smoothing is degenerated when μ approaches 1 as $a \to 0$. However, if we use line relaxation, we can obtain:

$$\mu(\theta_1, \theta_2) = | \frac{a}{2(a + c - a - c \cdot cos\theta_2) - ae^{-i\theta_1}} | , (44)$$

$$\mu(\frac{\pi}{2}, 0) = [\frac{a^2 + c^2}{a^2 + (c + 2a)^2}]^{\frac{1}{2}} . \qquad (45)$$

This gives a smoothing factor:

$$\bar{\mu} = max\{5^{-\frac{1}{2}}, \frac{a}{a + 2c}\} , \qquad (46)$$

which is still acceptable for high-frequency errors. The alternative is so called semi-coarsening. If we combine the line relaxation with the semi-coarsening, the solver will be very robust for high aspect-ratio grids.

3 Conservation principle (Liu, 1996e)

3.1 Boundary points

In order to achieve good coarse grid correction, we must find a nice coarse grid difference operator consistent with the fine grid difference operator. The reason is very simple: We must solve the same physical problems, using identical equations and boundary conditions on both the coarse grid and the fine grid. They cannot be different from each other. The variational or Galerkin method can keep the consistency, but we can also use the flux conservation to ensure the consistency which is more convenient even for complex problems. To guarantee the consistency between the coarse and fine grids, we should meet the following requirements :

1. Conservation for fine grid including interior and boundary points,

2. Conservation for coarse grid including interior and boundary points,

3. Same flux passing the fine and coarse grids.

Requirement 1 and 2 can be satisfied by good discretization of governing equations and proper specification of boundary conditions, but requirement 3 is frequently ignored, which leads to degeneration of multigrid convergence especially for non-smooth boundary condition and complex geometric boundaries. Let us look at following two cases as shown in Figure 10 and 11.

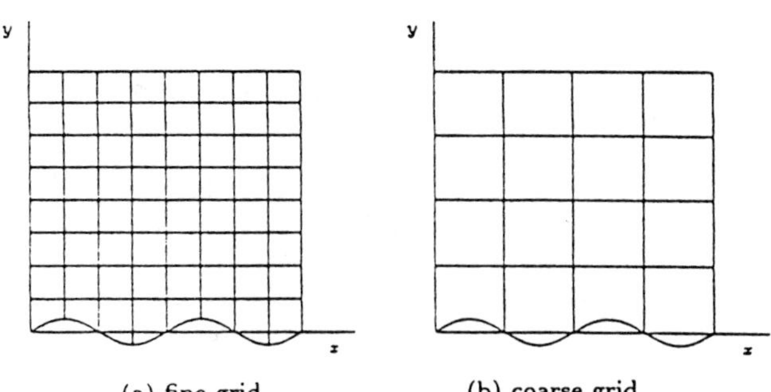

(a) fine grid (b) coarse grid

Figure 10. Boundary configuration is not smooth.

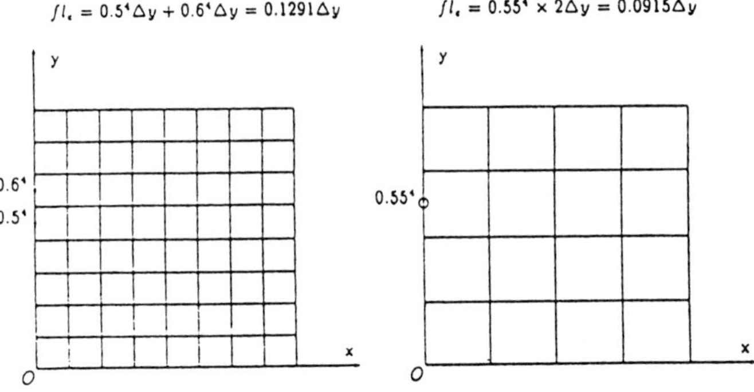

$fl_e = 0.5^4 \Delta y + 0.6^4 \Delta y = 0.1291 \Delta y$ $fl_e = 0.55^4 \times 2\Delta y = 0.0915 \Delta y$

Figure 11. Boundary value is not smooth

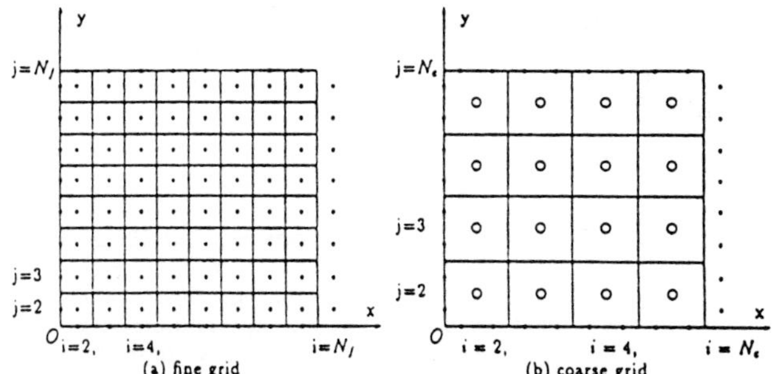

(a) fine grid (b) coarse grid

Figure 12. Same boundary for both fine and coarse grids

In the first case, the waved solid wall on the fine grid will become a flat plate as seen from the coarse grid. In the second case, the boundary value is not smooth and thus gives an inflow flux to the coarse grid which is different from the fine grid inflow flux. In order to avoid these mistreatments, we can use the same boundary points and boundary conditions for all levels of grids (Figure 12). In any case, we must guarantee that the flux passing the fine gird and coarse grid is the same or nearly the same for consistency. One easy way to do this is to use interpolation of the fine grid flux as the flux boundary condition for coarse grids.

3.2 Outflow and far-field boundary condition

The outflow and far-field boundary condition is often treated based on the direction of characteristics line (Figure 13).

For example, we can choose u as the characteristic value to decide the differencing direction, forward or backward, for incompressible flow. We may calculate:

$$fl_{e1} = \frac{\partial u}{\partial x}\Big|_1 \Delta y = \frac{u_{C1} - u_{W1}}{\Delta x}\Delta y,$$
$$\text{since } u_{C1} > 0 \qquad (47)$$

$$fl_{e2} = \frac{\partial u}{\partial x}\Big|_2 \Delta y = \frac{u_{E2} - u_{C2}}{\Delta x}\Delta y,$$
$$\text{since } u_{C2} < 0 \qquad (48)$$

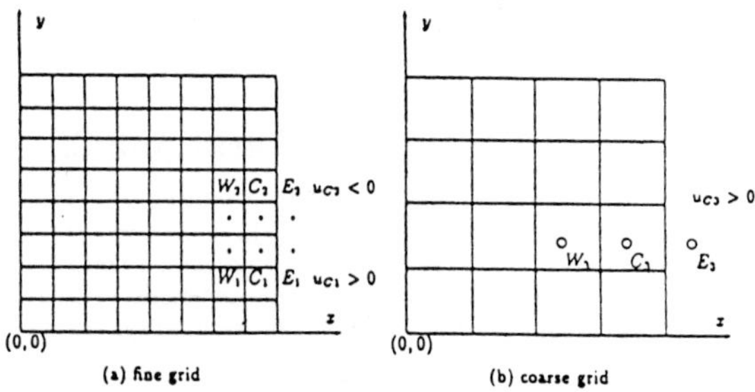

(a) fine grid (b) coarse grid

Figure 13. Outflow boundary condition for multigrid

However, in order to keep consistency with the fine grid, we cannot use this strategy on the coarse grid:

$$fl_{e3} = \frac{\partial u}{\partial x}\Big|_3 2\Delta y = \frac{u_{E3} - u_{C3}}{2\Delta x}2\Delta y,$$
$$\text{when } u_{C3} > 0, \qquad (49)$$

or

$$fl_{e3} = \frac{\partial u}{\partial x}\Big|_3 2\Delta y = \frac{u_{C3} - u_{W3}}{2\Delta x}2\Delta y,$$
$$\text{when } u_{C3} < 0, \qquad (50)$$

but have instead to use

$$fl_{e3} = fl_{e1} + fl_{e2}. \qquad (51)$$

For far field boundary, the principle is similar.

4 Conservation for multilevel adaptive grids (Liu, 1996e; Liao et al, 1997a)

This method uses local patch grids for refinement. This technique can be used to resolve shocks, stagnation points, boundary layers, recirculation zones, flame front, etc. The advantages of this technique include the following.

1. It does not need relocation of the existing grid points and thus will not distort the solution outside the refined areas.

2. The refined grid is also included in a multigrid cycle. Therefore, the convergence of multilevel adaptive grids is fast.

However, we have to treat the interface between the global coarse grid and local refined grid very carefully. Otherwise, the refined grid may provide a worse solution than the original non-refined grid. The boundary for refined grids may be the physical boundary, but, in most cases, is fine-coarse interface.

4.1 Composite grids (McCormick, 1986, Liu et al, 1987, 1989b)

A composite grid consists of coarse global grid, refined gird, and fine-coarse interface grid (Figure 14). Apparently, it is non-uniform grid and requires special schemes for interface grid points.

4.2 Flux interpolation for refined grid boundary

One convenient way to keep conservation for a composite grid is to do flux interpolation of the global grid on the interface and use it as the flux boundary condition for the refined grid.

4.3 FVE for interface grid points

The finite-volume-element (FVE, McCormick, 1989; Liu, 1989a, 1989b) method can solve the conservation problem for those interface grid points. To illustrate the basic FVE approach, consider the two rectangular grids in two dimensions as shown in Figure 14. For simplicity, consider the Poisson's equation

$$\nabla \cdot \nabla \phi = f. \tag{52}$$

The main idea here is to use the regular cell structures of the coarse and fine grids where possible, but adjust the volumes at the interface point to account for the irregularities there. As Figure 14 shows, this yields four types of grid points and volumes:

1. fine grids whose volumes are regular but smaller, denoted by Ω_f,

2. coarse grids whose volumes are regular but larger, denoted by Ω_c,

3. fine interface grids whose volumes are still regular, denoted by Ω_{fI},

4. coarse interface grids whose volumes are irregular, defined as Ω_{cI}.

Now we can discretize the differential equations on each of the four types of finite volumes using the Gauss divergence theorem (Figure 14):

$$\int_{\Omega_i} \nabla^2 \phi d\Omega_i = \int_{\Gamma_i} \nabla \phi \cdot \bar{n} d\Gamma_i = \sum_j f l_j, \tag{53}$$

where $f l_j$ is the flux through the jth side on surface of the finite volume Ω_i.

To calculate the fluxes, we need to evaluate $\nabla \phi$ at the volume surfaces located between node points. The basic idea

of FVE is to use triangular finite elements by linking immediate neighbor points, which cover the entire computational domain. In element k, we could use the relation

$$\nabla \phi_j = \sum_{l=1}^{3} (\nabla \phi)_l \cdot a_l(x_j, y_j), \tag{54}$$

where $\nabla \phi_j$ is the gradient evaluated along the volume interface within the kth element, $(\nabla \phi)_l$ is the gradient evaluated at the element's three node points (x_j, y_j), $j = 1, 2, 3$, and $a_l(x, y)$ is the nodal shape function in the kth element (assumed to be a linear function).

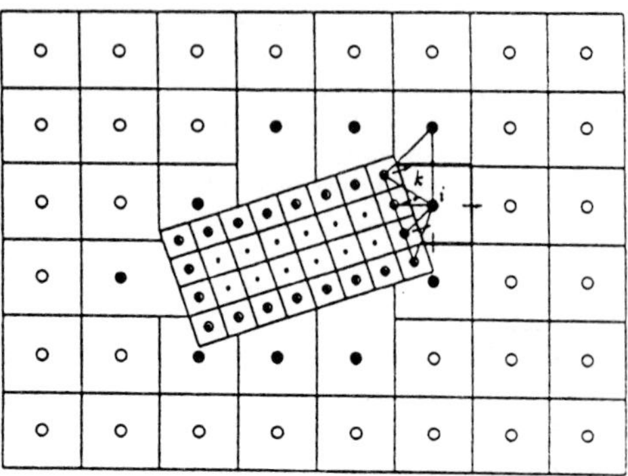

Figure 14. Finite volume-element discretization.

The above approach guarantees the conservation because:

- the computational domain is partitioned by finite volumes (covering with no overlap),

- the Gauss divergence theorem is used to integrate the differential equation in all four types of finite volumes, including regular fine and coarse grid volumes, regular fine interface grid volumes, and irregular coarse interface grid volumes,

- the triangular finite element method is used to provide an accurate estimate of the fluxes at volume interfaces.

This FVE approach finally gives us a conservative discretized system for the composite grid which can be written in the generic form

$$L_{h_i} \phi_i = f_i \qquad \text{in } \Omega \equiv \Omega_f + \Omega_c + \Omega_{fI} + \Omega_{cI}. \tag{55}$$

5 Line box relaxation for N-S equations (Liu et al, 1991a, 1991b)

5.1 Incompressible flow

The steady incompressible Navier-Stokes equations can be written as:

$$\nabla \cdot \vec{V} = 0, \tag{56}$$

$$\vec{V} \cdot \nabla \vec{V} - \frac{1}{Re} \nabla \cdot \nabla \vec{V} + \nabla P = 0, \tag{57}$$

where $\vec{V}$ is the velocity, P is the pressure, and Re is the Reynolds number.

Being different from Poisson's equation, this is a non-linear system. First, we should use full approximate scheme (FAS) for the non-linear problem. Second, we must find a systematic solver for the system. In order to achieve multigrid efficiency, a collective relaxation should be used.

5.1.1 Box relaxation

If the staggered grid is used, we can solve the following system of difference equations box by box (Figure 15):

$$A_E^C u_E + A_W^C u_W + A_N^C u_N + A_S^C u_S$$
$$-A_C^C u_C + \frac{P_W - P_C}{\Delta x} = S_{u_C}, \tag{58}$$

$$A_E^E u_{EE} + A_W^E u_C + A_N^E u_{NE} + A_S^E u_{SE}$$
$$-A_C^E u_E + \frac{P_C - P_E}{\Delta x} = S_{u_E}, \tag{59}$$

$$B_E^C v_E + B_W^C v_W + B_N^C v_N + B_S^C v_S$$
$$-B_C^C v_C + \frac{P_S - P_C}{\Delta y} = S_{v_C}, \tag{60}$$

$$B_E^N v_{NE} + B_W^N v_{NW} + B_N^N v_{NN} + B_S^N v_C$$
$$-B_C^N v_N + \frac{P_C - P_N}{\Delta y} = S_{v_N}, \tag{61}$$

$$\frac{u_E - u_C}{\Delta x} + \frac{v_N - v_C}{\Delta y} = 0, \tag{62}$$

where A and B are coefficients, the superscript represents the identification of the central point where we discretize, and the subscript represents the identification of neighboring points. We have four momentum and one continuity equations in each box for 5 unknowns, u_E, u_C, v_N, v_C, and P_C. The box relaxation requires solving a 5×5 system for 2-D and 7×7 system for 3-D and is therefore expensive.

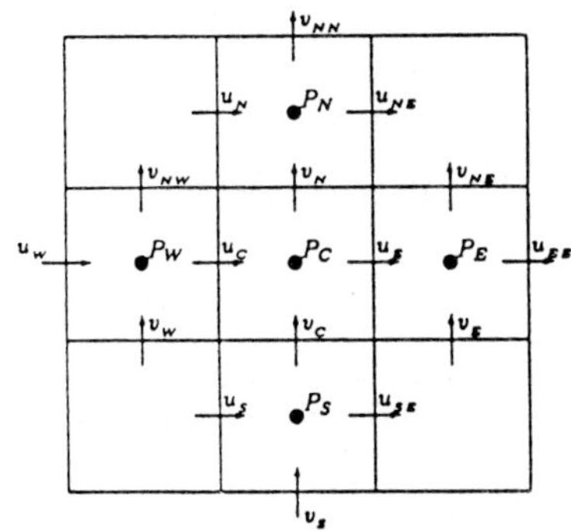

Figure 15. Box relaxation

5.1.2 Approximate box relaxation

To save computation cost, we develop a so-called approximate box relaxation (ABR) (Liu, 1995d). The idea behind ABR is to adjust u and v to satisfy the continuity equation in each box (Figure 16).

$$\frac{(u_E + \alpha_1 \delta) - (u_C - \alpha_2 \delta)}{\Delta x} +$$
$$\frac{(v_N + \alpha_3 \delta) - (v_C - \alpha_4 \delta)}{\Delta y} = 0, \tag{63}$$

$$\left(\frac{\alpha_1 + \alpha_2}{\Delta x} + \frac{\alpha_3 + \alpha_4}{\Delta y}\right)\delta =$$
$$-\left(\frac{u_E - u_C}{\Delta x} + \frac{v_N - v_C}{\Delta y}\right) = S_m, \tag{64}$$

$$\delta = S_m / \left(\frac{\alpha_1 + \alpha_2}{\Delta x} + \frac{\alpha_3 + \alpha_4}{\Delta y}\right). \tag{65}$$

Let

$$u_E \leftarrow u_E + \alpha_1 \delta,$$
$$u_C \leftarrow u_C - \alpha_2 \delta,$$
$$v_N \leftarrow v_N + \alpha_3 \delta,$$
$$v_C \leftarrow v_C - \alpha_4 \delta.$$

The continuity equation is then satisfied, but the question is how to determine $\alpha_1, \alpha_2, \alpha_3$, and α_4. This can be done by using 4 momentum equations in a box (Figure 16).

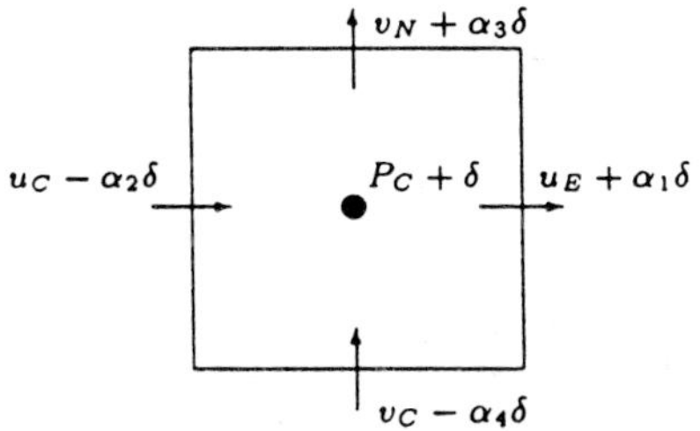

Figure 16. Approximate box relaxation

We may freeze P and do relaxation for each momentum equation. We can then assume momentum equations are satisfied or approximately satisfied. After that, we can adjust u and v to satisfy the continuity equation, but not let the adjustment contaminate those satisfactory momentum equations. This will provide a condition for determination of αs:

$$A_E^C \alpha_1 \delta + A_C^C \alpha_2 \delta - \frac{\delta}{\Delta x} = 0, \tag{66}$$

$$A_C^E \alpha_1 \delta + A_W^E \alpha_2 \delta - \frac{\delta}{\Delta x} = 0, \tag{67}$$

$$B_N^C \alpha_3 \delta + B_W^E \alpha_4 \delta - \frac{\delta}{\Delta y} = 0, \tag{68}$$

$$B_C^N \alpha_3 \delta + B_S^N \alpha_4 \delta - \frac{\delta}{\Delta y} = 0. \tag{69}$$

Solving the system, we obtain $\alpha_1 - \alpha_4$:

$$\frac{\alpha_2}{\alpha_1} = \frac{A_C^C - A_W^E}{A_C^E - A_E^C}, \quad \left(A_E^C + A_C^C \frac{A_C^C - A_W^E}{A_C^E - A_E^C}\right)\alpha_1 = \frac{1}{\Delta x};$$

$$\frac{\alpha_4}{\alpha_3} = \frac{B_C^C - B_S^N}{B_C^N - B_N^C}, \quad \left(B_N^C + B_C^C \frac{B_C^C - B_S^N}{B_C^N - B_N^C}\right)\alpha_3 = \frac{1}{\Delta y}.$$

This way provides a correction to u, v, and P to satisfy continuity and momentum equations as well.

We may approximately estimate $\alpha_1 - \alpha_4$. Assuming, for example,

$$A_C^C >> A_W^E, A_C^E >> A_E^C, \text{ and } A_C^C \approx A_C^E,$$

which is particularly true when a small time step is used, we have

$$\alpha_2 = \alpha_1, \text{ and } \alpha_1 = \frac{1}{A_C^C \Delta x}. \tag{70}$$

Similarly,

$$\alpha_4 = \alpha_3, \text{ and } \alpha_3 = \frac{1}{B_C^C \Delta y}. \tag{71}$$

This approximation will provide a significant saving in solving the box system since the system is decoupled and only one equation need be solved. For complex geometry, the approximation may be more complicated.

5.1.3 Line box relaxation (Liu et al, 1995d)

In general, we use high aspect ratio grids for Navier-Stokes equations. A line-box relaxation is thus preferred. Assume, for simplicity, $\alpha_1 = \alpha_2 = 1$, and $\alpha_3 = \alpha_4 = \alpha$, The adjustment for a 5-box line can be depicted by Figure 17.

From continuity equation, we can get 5 equations in 5 boxes:

$$2\frac{\delta_1}{\Delta x} + \frac{\alpha(\delta_1 - \delta_2)}{\Delta y} = S_{m1},$$

$$2\frac{\delta_2}{\Delta x} + \frac{\alpha(2\delta_2 - \delta_3 - \delta_1)}{\Delta y} = S_{m2},$$

$$\vdots$$

$$2\frac{\delta_5}{\Delta x} + \frac{\alpha(-\delta_4 + \delta_5)}{\Delta y} = S_{m5}. \qquad (72)$$

This is a tri-diagonal system and is very easy to solve for $\delta_1 - \delta_5$. We can then modify $u, v,$ and P to satisfy all continuity and momentum equations in a line of boxes.

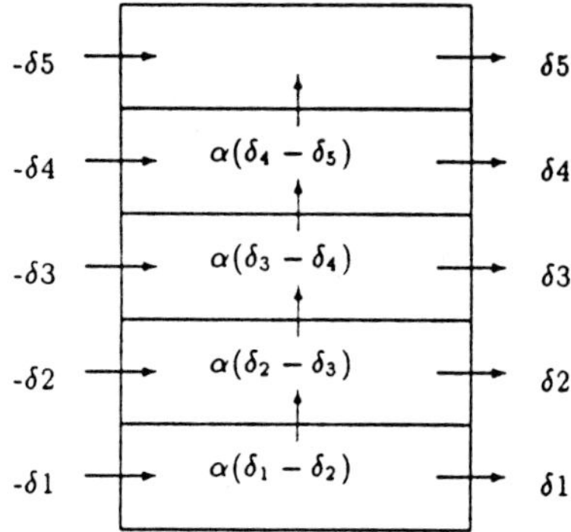

Figure 17. Line-box relaxation

5.2 Compressible flow (Liu et al, 1993b, 1993e)

The approximate line box relaxation for compressible flow is similar to those for incompressible flow. The difference is to correct ρu in stead of u (Figure 18):

$$\rho'_E u'_E \leftarrow \rho'_E u'_E + \beta\delta_1$$
$$\rho'_C u'_C \leftarrow \rho'_C u'_C - \beta\delta_1$$
$$\rho'_N u'_N \leftarrow \rho'_N u'_N + \delta_1 - \delta_2$$
$$\cdots \qquad (73)$$

Other steps are similar to those for incompressible flow solver, but we need to update the ρ.

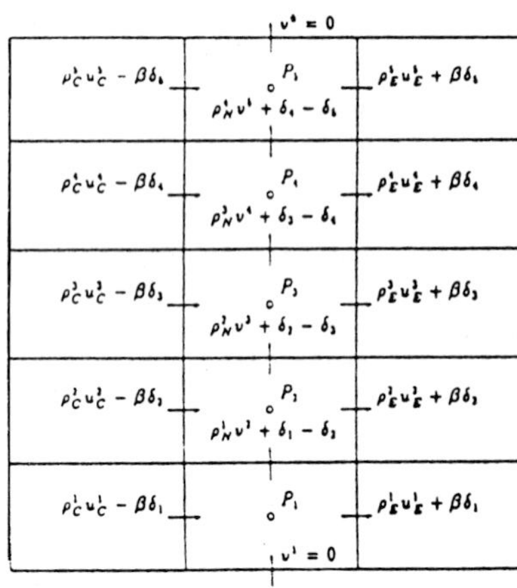

Figure 18. Line box relaxation for compressible flow

6 Turbulence model equation
(Zheng, et al, 1996c, 1997a, 1997b)

It is rare to find a successful report about multigrid method for turbulent flow. The main problem is caused by turbulence model equations. In order to achieve good MG efficiency, we must include the turbulence model equations in a multigrid cycle. We also need to consider the consistency principle for turbulence model equations. Therefore, we should find a model which is consistent for all levels of grid. Here, a $k - \omega$ model (Wilcox , 1988)is used for our calculation.

6.1 The Reynolds-averaged Navier-Stokes equations

The incompressible Reynolds-averaged Navier-Stokes equations can be summarized as:

$$\frac{\partial u_j}{\partial x_j} = 0, \qquad (74)$$

$$\frac{\partial u_i}{\partial t} + \frac{\partial u_i u_j}{\partial x_j} + \frac{1}{\rho}\frac{\partial p}{\partial x_j} + \frac{\partial}{\partial x_j}(\overline{u'_i u'_j})$$
$$- \frac{\partial}{\partial x_j}[\nu(\frac{\partial u_i}{\partial x_j} + \frac{\partial u_j}{\partial x_i})] = 0, \qquad (75)$$
$$i = 1, 2, 3,$$

where u_j is the velocity, p is the pressure, ρ is the constant density, ν is the molecular kinetic viscosity, u'_i, u'_j are fluctuation parts of velocity u_i and u_j, and $\overline{u'_i u'_j}$ is the Reynolds stress tensor which can be modeled by:

$$\tilde{\tau}_{i,j} = \frac{1}{\rho}\hat{\tau}_{i,j} = \overline{u'_i u'_j} = \frac{2}{3}\delta_{ij}k - \nu_t(\frac{\partial u_i}{\partial x_j} + \frac{\partial u_j}{\partial x_i}).$$

The turbulent kinetic energy k and eddy viscosity ν_t are determined from the turbulence model.

6.2 k-ω Turbulence Model

The k-ω model consists of two equations:
Turbulent mixing energy(k):

$$\frac{\partial k}{\partial t} + \frac{\partial}{\partial x_j}(u_j k) = \frac{\partial}{\partial x_j}\left[(\nu + \sigma_k \nu_T)\frac{\partial k}{\partial x_j}\right]$$
$$+ \tilde{\tau}_{i,j}\frac{\partial u_i}{\partial x_j} - \beta_k \omega k . \qquad (76)$$

Specific dissipation rate(ω):

$$\frac{\partial \omega}{\partial t} + \frac{\partial}{\partial x_j}(u_j \omega) = \frac{\partial}{\partial x_j}\left[(\nu + \sigma_\omega \nu_T)\frac{\partial \omega}{\partial x_j}\right]$$
$$+ (\gamma_\omega \omega/k)\tilde{\tau}_{i,j}\frac{\partial u_i}{\partial x_j} - \beta_\omega \omega^2 . \qquad (77)$$

The eddy viscosity is calculated from:

$$\nu_T = \gamma_k \frac{k}{\omega} . \qquad (78)$$

The closure coefficients are:

$$\beta_k = 9/100, \quad \gamma_k = 1, \quad \sigma_k = 1/2 ,$$
$$\beta_\omega = 3/40, \quad \gamma_\omega = 5/9, \quad \sigma_\omega = 1/2 . \qquad (79)$$

6.3 Point implicit techniques

For most two-equation models, the source terms are usually dominant and become stiff near the solid wall. The efficiency of solution then totally depends on the treatment of those source terms. Proper treatment of source terms will result in a robust and efficient solution of the turbulence model equations. In our work, a point-implicit technique is developed to improve the efficiency of the solution and, more importantly, to alleviate the stiffness of the governing equations exhibited in the near wall region.

The semi-discretized turbulence equation can be expressed as:

$$\frac{\Delta W}{\Delta t} = R(W) + S(W) \tag{80}$$

where $R(W)$ represents the convective and diffusive terms. The updating formula with source terms treated with point-implicit technique can be written as:

$$\left[\frac{1}{\Delta t}I - \frac{\partial S}{\partial W}\right]\Delta W = R(W) + S(W) . \tag{81}$$

From our computational experience, the following criteria should be considered in determining what should be kept in the matrix:

1. Generally, those terms contributing positively to the diagonal element of the coefficient matrix can be kept.

2. It is very important to keep the balance of the growth rate of the turbulent mixing energy (k) and dissipation (ω or ϵ). During the iteration, any one of them growing or decaying too fast will jeopardize the stability and deteriorate convergence.

3. It is necessary to maintain the combined matrix in the left side of equation (81) to be diagonally dominant. It is apparent that the combined matrix can be made diagonally dominant if the time step is chosen sufficiently small. However, in order to improve the computational efficiency, the matrix $\frac{\partial S}{\partial W}$ should be constructed to allow larger time step.

Based on the criteria described above, an optimized coefficient matrix is constructed:

$$-\frac{\partial S}{\partial W} = \begin{bmatrix} \beta^* \omega & 0 \\ 0 & \beta\omega \end{bmatrix} . \tag{82}$$

In the multigrid process, some techniques are found quite effective in acceleration of convergence.

1. The turbulence production term P_d, calculated from mean flow velocity strains, is better to be calculated only on the finest grid, and injected from the finest grid to the coarse grid. The reason for that is simple: the production term calculated on coarse grids is much smaller than the real values. It is not worth to calculate it again on the coarse grid.

2. Here it is suggested to update the formula for the k-ω model be put in a delta-form, because certain limiters are frequently imposed for the turbulence equation solution. The delta-form may avoid possible improper transfer of residuals of turbulence equations during the multigrid process, that might cause slow convergence or that cannot reach a fully converged solution at all.

6.4 Improvement to turbulence models

6.4.1 Total stress limitation

We can show that

$$\omega \geq \phi\gamma_k\sqrt{2(s_{i,j} - \frac{1}{3}\frac{\partial u_k}{\partial x_k}\delta_{i,j})^2}$$

$$\geq \phi\gamma_k\sqrt{(s_{i,j} - \frac{1}{3}\frac{\partial u_k}{\partial x_k}\delta_{i,j})\frac{\partial u_i}{\partial x_j}}$$

$$= \phi\gamma_k\sqrt{Production\ term} \tag{83}$$

where

$$\phi = \sqrt{3}/2, \text{ and } s_{i,j} = \frac{1}{2}(\frac{\partial u_i}{\partial x_j} + \frac{\partial u_j}{\partial x_i}).$$

For details of the derivation, please see Zheng et al (1997b). In our modification,

$$\mu_t = \gamma_k\frac{\rho k}{\omega} , \quad \omega = \max\left(\omega_o, \phi\gamma_k\sqrt{prod}\right) , \tag{84}$$

where ω_o is the solution from Wilcox original $k - \omega$ model.

6.4.2 Corrected formula for ω values at a solid wall

When the grid is fine enough to ensure that the normalized distance (y^+) from the first grid to the wall is less than 2, the ω value near the wall can be determined from the asymptotic solution:

$$\omega = \frac{6\nu}{\gamma_k y^2} \quad \text{as } y^+ \to 0 , \tag{85}$$

$$\omega_M = 2\omega_1 - \omega_W . \tag{86}$$

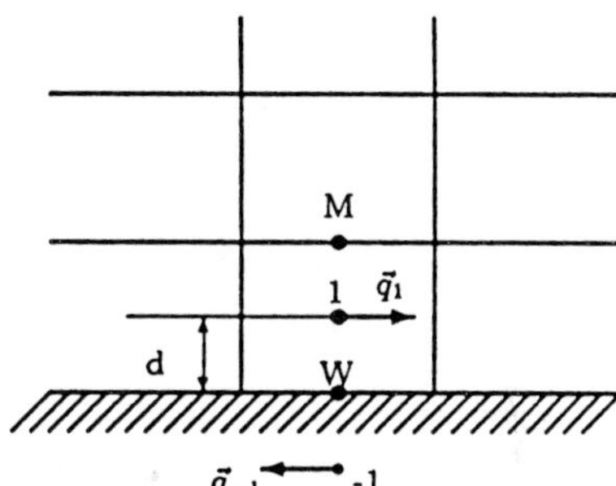

Figure 19. Wall Boundary Cell Sketch

It has been found that

$$\omega_W = 12\frac{\nu}{\gamma_k y^2} \tag{87}$$

usually gives good results (Figure 19).

If y^+ is larger than 2, equation (85) and equation (86) will give values of ω_W which are too small. For practical engineering problems, it is very hard to maintain the y^+ of the first grid point within 2 everywhere. Therefore a correction term is needed for reasons of both accuracy and grid-efficiency. The following formula is first proposed to correct equation (87):

$$\omega_A = 12\frac{\nu}{\gamma_k d^2} , \tag{88}$$

$$\omega_\Omega = Min(\omega_A, \Omega)C_e(y^+)^{\frac{2}{3}} , \tag{89}$$

$$\omega_W = Max(\omega_A, \omega_\Omega) , \tag{90}$$

where $C_e = 0.46$, a constant tuned for best results. Since the Min and Max are not continuous functions, they might slow down convergence in some cases. We found it is better to replace them with weighted interpolations. The final form of the empirical formula is:

$$\omega_A = 12 \frac{\nu}{\gamma_k d^2},\tag{91}$$

$$\omega_\Omega = \frac{\Omega \omega_A^2 + \omega_A \Omega^2}{\omega_A^2 + \Omega^2} C_e (y^+)^{\frac{3}{2}},\tag{92}$$

$$\omega_W = \frac{\omega_A^2 + \omega_\Omega^2}{\omega_A + \omega_\Omega}.\tag{93}$$

The above formulae have been tested for y^+ which ranges from 3 to 15. This is a range acceptable for engineering problems.

7 Multigrid for reacting flow (Liu et al, 1993b, 1993e, 1995c; Liao et al, 1995b, 1996c, 1996d; Zheng et al, 1995a)

Modeling of the finite-rate turbulent combustion becomes more and more important for engine design and pollutant prediction.

7.1 Governing system

For a 3-D turbulent reacting flow the Reynolds-averaged governing system can be written as follows:
Continuity equation:

$$\frac{\partial \rho}{\partial t} + \frac{\partial \rho u_j}{\partial x_j} = 0.\tag{94}$$

Momentum equation:

$$\begin{aligned}
\frac{\partial \rho u_i}{\partial t} + \frac{\partial \rho u_i u_j}{\partial x_j} + \frac{\partial P}{\partial x_j} + \frac{\partial}{\partial x_j}(\rho \overline{u_i' u_j'}) & \\
- \frac{\partial}{\partial x_j}[\mu(\frac{\partial u_i}{\partial x_j} + \frac{\partial u_j}{\partial x_i} - \frac{2}{3}\delta_{ij}\frac{\partial u_l}{\partial x_l})] & = 0, \\
i = 1,2,3, &
\end{aligned}\tag{95}$$

where μ is the molecular viscosity which is a function of temperature and species,

$$\rho \overline{u_i' u_j'} = \frac{2}{3}\delta_{ij}(\rho k + \mu_t \frac{\partial u_l}{\partial x_l}) - \mu_t(\frac{\partial u_i}{\partial x_j} + \frac{\partial u_j}{\partial x_i})$$

and

$$\mu_t = \rho C_\mu \frac{k^2}{\epsilon}.$$

Here, k is the turbulent kinetic energy, ϵ is its dissipation rate, μ is the molecular viscosity which is a function of temperature and species, and μ_t is called the turbulent eddy viscosity.
Energy equation:

$$\frac{\partial \rho h}{\partial t} + \frac{\partial \rho h u_j}{\partial x_j} - \frac{\partial P}{\partial t} - \frac{\partial}{\partial x_j}[(\frac{\mu + \mu_t}{\sigma_h})\frac{\partial h}{\partial x_j}] = 0.\tag{96}$$

Mixture fraction equation:

$$\frac{\partial \rho f}{\partial t} + \frac{\partial \rho f u_j}{\partial x_j} - \frac{\partial}{\partial x_j}[(\frac{\mu + \mu_t}{\sigma_m})\frac{\partial f}{\partial x_j}] = 0,\tag{97}$$

$$\alpha = 1,2,\cdots,N_S.$$

Chemical species equation:

$$\frac{\partial \rho Y_\alpha}{\partial t} + \frac{\partial \rho Y_\alpha u_j}{\partial x_j} - \frac{\partial}{\partial x_j}[(\frac{\mu + \mu_t}{\sigma_m})\frac{\partial Y_\alpha}{\partial x_j}] - R_\alpha = 0,\tag{98}$$

$$\alpha = 1,2,\cdots,N_S,$$

where R_α is the chemical reaction rate of species α.
Enthalpy expression:

$$h = h(Y_\alpha, T).\tag{99}$$

State equation:

$$\rho = \sum_\alpha \frac{Y_\alpha}{W_\alpha} \rho RT.\tag{100}$$

Here, R, Y_α, and W_α are the gas constant, the mass fraction and molecular weight of species α, respectively.

This is a nonlinear and coupled system since $\rho = \rho(Y_\alpha, T)$. It has all the problems associated with regular turbulent flow but, moreover, the chemical reaction cause more troubles. The reaction rate term R_α is a nonlinear function of species : $R_\alpha = R_\alpha(Y_1, Y_2, ...)$ which could be very large and dominant. Therefore, the species equations are closely coupled and have to be solved simultaneously. For laminar flames, the chemical reaction rate R_α for the αth species can be written as

$$R_\alpha = W_\alpha \sum_{j=1}^{N_R}[(\nu_{j\alpha}^P - \nu_{j\alpha}^R) \cdot (K_j^f \prod_{l=1}^{N_S} n_l^{\nu_{jl}^R} - K_j^b \prod_{l=1}^{N_S} n_l^{\nu_{jl}^P})],$$

where N_R is the total number of reaction steps, N_S is the total number of species, $\nu_{j\alpha}^P$ ($\nu_{j\alpha}^R$) refers to the stoichiometric coefficient of products (reactants), and n_l is the mole fraction of lth species. The function K_j^f (K_j^b) is the rate constant for the forward (backward) reaction step j. We assume K_j^f has the following Arrhenius temperature relationship,

$$K_j^f = A_j^f T_j^{\alpha_j^f} exp(-\frac{E_j^f}{RT}),\tag{101}$$

and K_j^b has a similar expression. However, it is very difficult to estimate R_α when turbulence is involved in the process and mixing becomes dominant. Moreover, the gradient of T and thus other variables is vary large around the flame front. All this makes it very difficult to obtain convergence of the solution. It is also difficult to obtain accurate results. It usually takes thousands of iterations for a 2-D problem to get fair convergence if you use conventional iterative methods. Although it is better to solve the whole system simultaneously, we solve them in groups considering the Navier-Stokes equations are coupled with species and turbulence model equations loosely.

7.1.1 Time-marching

A fully implicit time-stepping scheme is used for the calculation of turbulent combustion. For a methane-air diffusion flame, a system of 56 equations is solved at each time step. The system of equations are relaxed in groups as shown below:
(a) u_j, P
(b) k, ϵ, μ_t
(c) h, Y_α and , finally, updating
(d) ρ, μ_{eff} where $\mu_{eff} = \mu + \mu_t$.

7.1.2 Multigrid acceleration

A standard multigrid method is used for momentum and continuity equations for accelerating convergence. The process of a two-level FAS scheme used for the algebraic problem at each time step can be described by Chart 1.

The Gauss elimination method is used for the species equations which have 49 species and 229 chemical reaction steps. Again, the approximate line-box relaxation is used

522

for N-S equations. The MG is found to be much faster than the single grid solver.

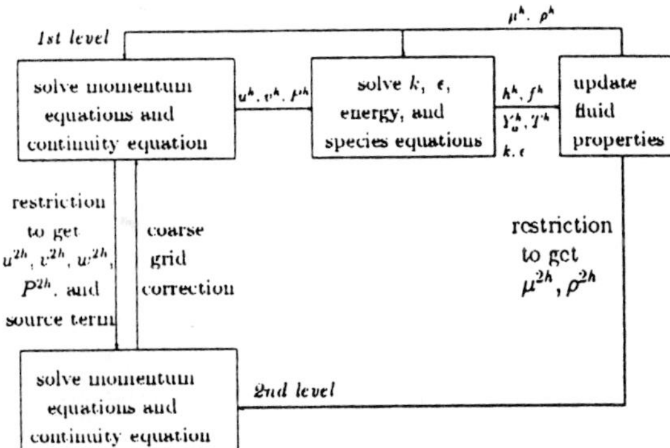

Chart 1. Process of a two-level
FAS scheme at each time step.

8 Direct numerical simulation for flow transition : incompressible flow (Liu et al, 1991a, 1991b, 1992b, 1993a, 1993c, 1993d, 1994b, 1994c, 1996a, 1996b, 1996f)

Flow transition and turbulence are most challenging problems in modern fluid mechanics. Strong interest in problems of flow transition around airfoils has developed recently. This is not only inspired by an effort towards understanding the mechanism of flow stability and transition, but is also required by advanced flight vehicle design. The laminar flow control, for example, which may delay transition which results in lower drag and therefore high fuel efficiency and the accurate prediction of aerodynamic forces and heating requirements.

While experimentation to provide sufficiently detailed data is still costly and difficult, direct numerical simulation (DNS) becomes more and more attractive. DNS was usually thought of as a tool to study basic flow with simple geometry when a temporal approach was used. However, as computer capability is dramatically increased and the cost is rapidly decreased, more and more research focus has been moved on the spatial approach.

The features of DNS research by the Louisiana Tech Team include the following.

- The whole process of flow transition including receptivity, linear growth, non-linear instability, breakdown, and transition to turbulence.

- General geometry including 2-d and 3-d airfoils.

- High Reynolds number up to 10^6.

- All speed including incompressible, subsonic, and supersonic flows.

8.1 Approaches

1. Grids: Conformal mapping and high-order grid generation are used to obtain high quality grids (continuous in curvature) and high-order Jacobien.

2. Time marching: Fully implicit time marching for incompressible flow and multigrid with line distributive relaxation for solving the discretized system at each time step. Explicit time marching for subsonic and supersonic flow. The multi-step Runge-Kutta method is used for time marching.

3. Discretization: A fourth-order central difference scheme is used for incompressible flow and a sixth-order compact central difference scheme is used for compressible flow.

4. Non-reflecting boundary conditions: A buffer domain with buffer functions is used for outflow boundary and sponges are used for far field and inflow boundaries.

5. Multilevel grids for local refinement to resolve small length scales.

6. Multilevel grid dissipation to stabilize the calculation to pass the breakdown stage.

8.2 High-order finite difference on staggered grid

8.2.1 Fully-implicit time-stepping

For a generic partial differential equation,

$$\frac{\partial \phi}{\partial t} + L\phi = S, \tag{102}$$

we use a second (or higher) order backward Euler difference in the time direction,

$$\frac{\partial \phi}{\partial t} = \frac{3\phi_{ijk}^{n+1} - 4\phi_{ijk}^{n} + \phi_{ijk}^{n-1}}{2\Delta t} + O(\Delta t^2). \tag{103}$$

Letting $L\phi = (L_h\phi)_{ijk}^{n+1}$, we then will have a fully implicit time-stepping. The fully implicit time-stepping has much better stability than the explicit one. It requires solving a large algebraic system at each time step arising by the fully implicit time-stepping. Fortunately, we only need to use one or two multigrid V-cycles to solve this large system and making the implicit scheme comparable to explicit schemes in CPU cost at each time step.

8.2.2 Fourth or higher order discretization on staggered grids

To minimize the numerical viscosity and phase error, fourth-order central differences in space are applied:

$$\frac{\partial \phi}{\partial x}\Big|_i \approx \frac{-\phi_{i+2} + 8\phi_{i+1} - 8\phi_{i-1} + \phi_{i-2}}{12\Delta x}$$

$$\frac{\partial^2 \phi}{\partial x^2}\Big|_i \approx \frac{-\phi_{i+2} + 16\phi_{i+1} - 30\phi_i + 16\phi_{i-1} - \phi_{i-2}}{12\Delta x^2}$$

$$\frac{\partial \phi}{\partial x}\Big|_{i+\frac{1}{2}} \approx \frac{-\phi_{i+2} + 27\phi_{i+1} - 27\phi_i + \phi_{i-1}}{24\Delta x} \tag{104}$$

Figure 20 depicts the stencil of U-equation in the (ξ, η) plane.

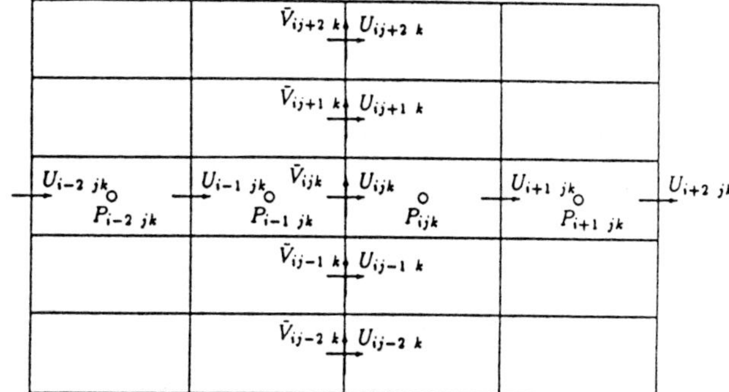

Figure 20. Stencil of U—equation in the (ξ, η) plane.

Since a staggered grid is used, we need to evaluate V at the points associated with U where we have no definition for V. We then have to evaluate V by high-order interpolation.

8.3 Discretized contravariant velocity for general geometry (Liu et al., 1996a, 1996b)

The contravariant velocity is defined as :

$$U = \frac{1}{J}(u\xi_x + v\xi_y + w\xi_z), \qquad (105)$$

$$V = \frac{1}{J}(u\eta_x + v\eta_y + w\eta_z), \qquad (106)$$

$$W = \frac{1}{J}(u\zeta_x + v\zeta_y + w\zeta_z), \qquad (107)$$

where (x, y, z) is the Cartesian coordinate, (ξ, η, ζ) is the curvilinear coordinate, (u, v, w) is the physical velocity, (U, V, W) is the contravariant velocity, and J is the transformation Jacobian from (x, y, z) to (ξ, η, ζ),

$$J = \frac{\partial(\xi, \eta, \zeta)}{\partial(x, y, z)}.$$

One of the advantages of using contravariant velocity is that it can maintain the simplicity of the continuity equation and compactness of the discretization stencil:

$$\frac{\partial U}{\partial \xi} + \frac{\partial V}{\partial \eta} + \frac{\partial W}{\partial \zeta} = 0. \qquad (108)$$

8.4 High order grid generation (Liu et al, 1996g)

The high-order discretization is important for DNS of flow transition and turbulence. It is crucial to acquire high-order accuracy in curvilinear coordinates for complex geometry. To achieve high-order accuracy, say fourth-order or higher, first, a high-order description of Jacobi metrics, e.g., ξ_x, ξ_y, etc. is required. Second, the high quality of grids, i.e., continuity in curvature must be guaranteed. As shown by numerical experiment, any discontinuity or disturbance in surface curvature will result in a totally different output for receptivity of inflow disturbance and transition. The second-order elliptic grid generation is no longer satisfactory.

8.4.1 Conformal mapping

One option for high-order grid generation is to use conformal mapping. This provides high-quality grids and analytic Jacobi metrics but is not convenient for complex geometry.

For a quarter cylinder, a three-step transformation is used to transfer the physical domain to the computational domain (see Figure 21):

1. $(\xi, \eta) \Rightarrow (\xi_1, \eta_1)$:

$$\xi_1 = \frac{1}{\sigma_A}\xi,$$

$$\eta_1 = \eta. \qquad (109)$$

This will ensure the points $(0,0)$ and $(a, 0)$ in the (ξ_1, η_1) space match the points $(-a, 0)$ and $(0, a)$ in the (x, y) space.

2. $(\xi_1, \eta_1) \Rightarrow (x_1, y_1)$:

$$x_1 = \frac{\sigma_B \xi}{\sigma_B + a - |\xi_1|} - a,$$

$$y_1 = \frac{\sigma_C y_{1max} \eta_1}{\sigma_C \eta_{max} + y_{1max}(\eta_{max} - \eta_1)}. \qquad (110)$$

This includes stretching in both directions. Here, σ is a parameter for stretching. The smaller the σ_B and σ_C, the stronger the stretching near the solid wall and the leading edge, respectively.

3. $(x_1, y_1) \Rightarrow (x, y)$: Joukowsky transformation

$$x_1 = \frac{x}{2}(1 + \frac{a^2}{x^2 + y^2}),$$

$$y_1 = \frac{y}{2}(1 - \frac{a^2}{x^2 + y^2}). \qquad (111)$$

For more complex geometry, more transformation steps are required.

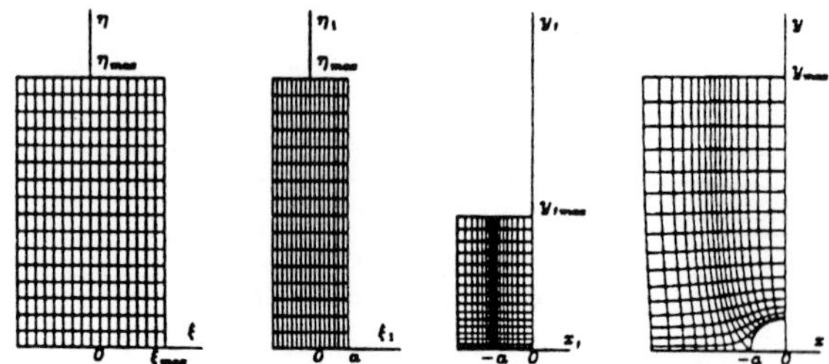

Figure 21. Transformation for a quarter of cylinder.

8.4.2 Fourth-order elliptic grid generation

For complex geometry, we usually use body-fitted grids. The grid is irregular in physical domain (x, y), but is uniform in computational domain (ξ, η).

This requires a coordinate transformation:

$$x = x(\xi, \eta),$$
$$y = y(\xi, \eta). \qquad (112)$$

In order to guarantee the one-to-one mapping and the existence of third-order derivatives, the fourth-order elliptic grid generation is required to solve a bi-harmonic system:

$$\nabla^4 x = P, \qquad (113)$$

$$\nabla^4 y = Q. \qquad (114)$$

For convenience, we introduce some intermediate variables, X and Y, and rewrite the grid generation system as

$$\begin{aligned} \nabla^2 x &= X, \\ \nabla^2 y &= Y, \\ \nabla^2 X &= P, \\ \nabla^2 Y &= Q. \end{aligned} \qquad (115)$$

where $\nabla^2 x = x_{\xi\xi} + x_{\eta\eta}, \nabla^2 y = y_{\xi\xi} + y_{\eta\eta}$.

The discretized system can be solved by a full multigrid (FMG) scheme. However, we use a higher order interpolation from the coarse grid to the fine grid, a third-order interpolation for a second-order difference scheme for the elliptic grid generation equation, for example (Liu et al, 1987), to achieve efficient FMG convergence.

8.5 Buffer out-flow boundary condition (Liu et al, 1994a)

Outflow boundary conditions have been the focus of study of many researchers. A buffer domain is appended to the end of the original outflow boundary to smear all possible reflections from the buffered outflow boundary (see Figure 22). The N-S equations are then modified to be good for both physical and buffer domains.

524

The new modified 2-D governing equations for perturbation in the computational (ξ, η) plane become:

$$\frac{\partial U}{\partial t} + \frac{1}{y_\eta}\frac{\partial(UU + 2U_0 U)}{\partial \xi} + \frac{\partial}{\partial \eta}\left(\frac{U_0 V + U V_0 + U V}{y_\eta}\right)$$

$$-\frac{b_{Re}}{Re_0^{\bullet}}\left(b \cdot \frac{\partial^2 U}{\partial \xi^2} + \frac{1}{y_\eta}\frac{\partial^2}{\partial \eta^2}\left(\frac{U}{y_\eta}\right) + y_\eta \eta_{yy}\frac{\partial}{\partial \eta}\left(\frac{U}{y_\eta}\right)\right)$$

$$+ y_\eta \frac{\partial P}{\partial \xi} = 0,$$

$$y_\eta \frac{\partial V}{\partial t} + \frac{\partial(U_0 V + U V_0 + U V)}{\partial \xi} + \frac{\partial(2V_0 V + V V)}{\partial \eta}$$

$$-\frac{b_{Re}}{Re_0^{\bullet}}\left(b \cdot y_\eta \frac{\partial^2 V}{\partial \xi^2} + \frac{1}{y_\eta}\frac{\partial^2 V}{\partial \eta^2} + y_\eta \eta_{yy}\frac{\partial V}{\partial \eta}\right) + \frac{\partial P}{\partial \eta} = 0,$$

$$\frac{\partial U}{\partial \xi} + \frac{\partial V}{\partial \eta} = 0.$$

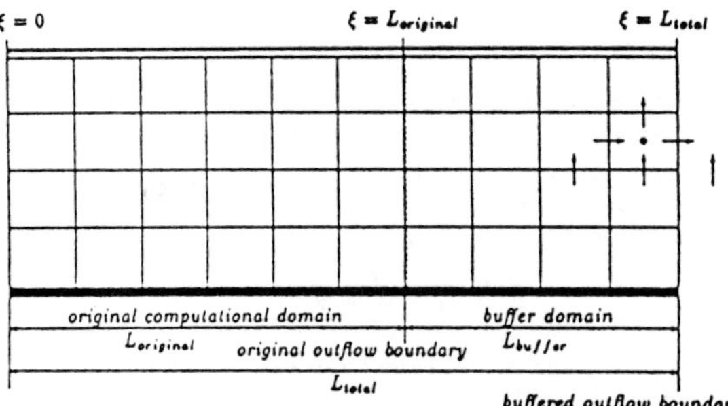

Figure 22. Extended computational domain.

The buffer functions are chosen as follows:

$$b(\xi) = \begin{cases} \frac{\tanh(L_{total} - \xi)}{\tanh(L_{buffer})} & L_{original} \le \xi \le L_{total}, \\ 1 & 0 \le \xi \le L_{original}, \end{cases}$$

$$b_{Re}(\xi) = \begin{cases} \frac{c}{L_{buffer}^2}(\xi - L_{original})^2 + 1 & \\ & L_{original} \le \xi \le L_{total}, \quad (116) \\ 1 & 0 \le \xi \le L_{original}. \end{cases}$$

It is clear that the first function decreases from one to zero very rapidly as one moves from the original outflow boundary to the buffered outflow boundary. The second function increasing from 1 to $c + 1$ is a quadratic function that is continuously differentiable at the original outflow boundary. Note that the total effect of these buffer functions is that toward the buffered outflow boundary:

- the momentum equations become increasingly convection dominated in the ξ−direction, while the equations generally become parabolic; and

- the momentum equations become highly diffusion dominated in the η−direction.

- the unstable modes will become stable in the buffer domain due to use of the buffer function b_{Re}.

This treatment makes the outgoing waves propagate outward without reflection in the ξ−direction, and any oscillation in the η−direction will be effectively smeared in the buffer domain.

8.6 Multilevel grid dissipation (Liu et al, 1995a)

The spatial DNS usually meets difficulties after the flow goes into the breakdown stage when the shear layer is developed and the large vortices break down to small scale vortices. The numerical simulation will thus have a huge energy burst.

To keep the numerical simulation going, we developed a *fine-coarse-fine grid dissipation technique*. To explain this technique, let us see what happens for a 1-D problem. We do the fine to coarse grid restriction and then do the coarse to fine grid interpolation at each time step:

$$u_c = I_h^{2h} u_f^{old},$$
$$u_f^{new} = I_{2h}^{h} u_c.$$

Here, we assume I_h^{2h} to be a linear restriction and I_{2h}^h a linear interpolation.

Define $L = \overline{AB}$ as a segment which has five grid points on the fine grid (see Figure 23), and ($u_A \equiv 0$ is assumed)

$$a = \frac{\int_0^L |u_f^{new}| dx}{\int_0^L |u_f^{old}| dx}$$

as the amplification factor of a fine-coarse-fine mapping.

Assume we have different frequency modes, e.g.,

$$\sin(K\frac{2\pi}{L}x), \quad K = \frac{2}{N-1}, \frac{4}{N-1}, \cdots, 1, \cdots,$$

where

$$L = l/\left(\frac{N-1}{4}\right),$$

N is the total number of grid points in the entire domain, l is the length of the entire domain, and $K > 1$ is either not visible or bad represented by the coarse grid. We can approximately get $a(K)$ by numerical integration for different modes.

$$a(K) = \frac{2|\sin(K\pi)| + |\sin(2K\pi)|}{|\sin(\frac{K\pi}{2})| + |\sin(K\pi)| + |\sin(\frac{3}{2}K\pi)| + |\frac{1}{2}\sin(2K\pi)|},$$

when $K = 1$, $a = 0$, when $K \to 0$, $a \to 1$. Figure 24 clearly shows that this kind of grid mapping, fine to coarse restriction and coarse to fine interpolation, significantly damps the high-frequency ($K \sim 1$), but has only a little effect on other modes ($K < 1$), and has almost no effect on low frequency modes ($K \to 0$).

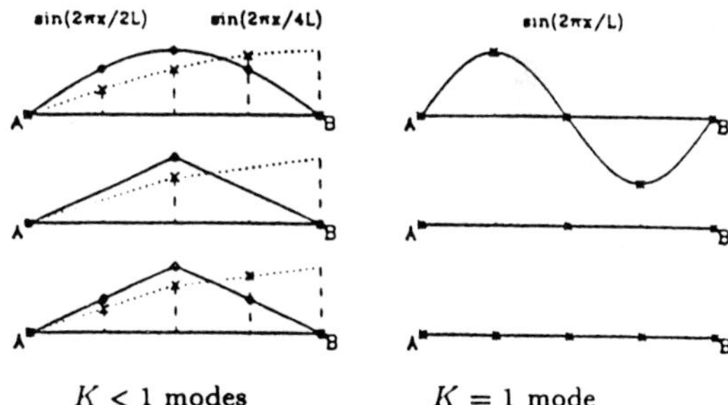

Figure 23. Fine-coarse-fine grid mapping.

Of course, we do not want to eliminate the high-frequency modes, but to restrict their energy growth. The actual procedure of this technique is

1. $u_c = I_h^{2h} u_f^{old}$,

2. $u_f^{new} = (1 - \sigma)u_f^{old} + \sigma I_{2h}^h u_c$.

Here, we choose

$$\sigma = c \cdot (u^2 + v^2 + w^2),$$

which is proportional to the perturbation energy, and $c = c(\Delta x, \Delta y, \Delta z, \Delta t)$. Therefore, there is very little damping to high-frequency modes when the perturbation is very small. In this way, we successfully keep the code running to simulate the whole process of transition: linear evolution, secondary instability, breakdown, and transition.

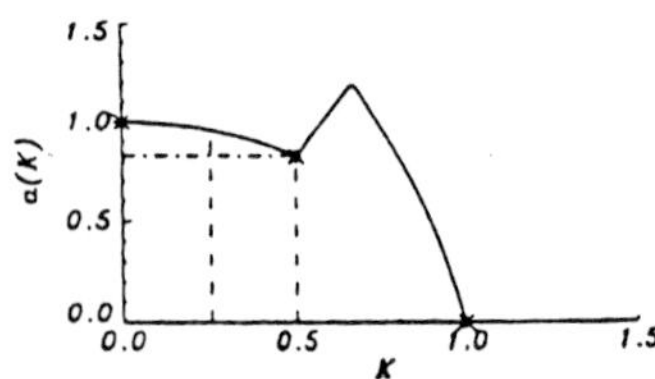

Figure 24. Amplification factor for different modes.

8.7 Multilevel multiscale simulation (Liu et al, 1995b; Chelyshkov et al, 1997)

The basic idea is that the flow is decomposed into several component groups according to spatial and temporal length scales. Each group has its own subdomain, governing system, mesh size, and discretization method. The simulation

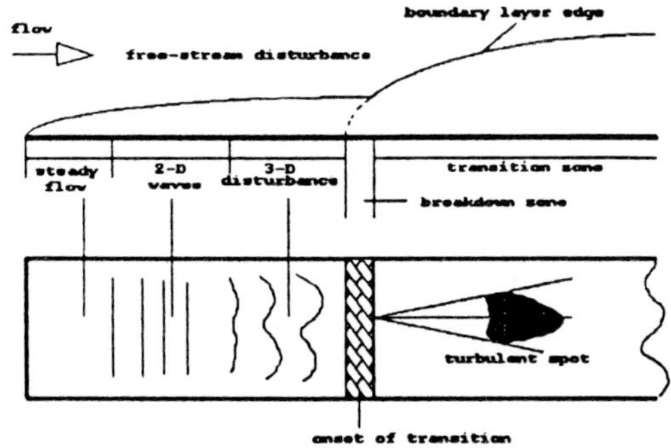

Figure 25. Idealized sketch of
transition process on a flat plate.

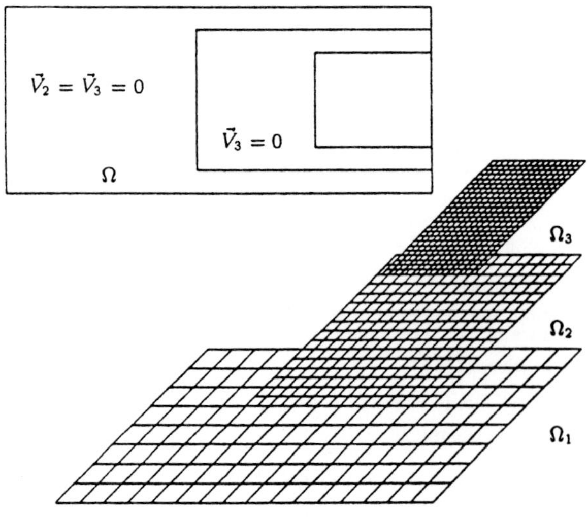

Figure 26. Multiple level grids.

Here we consider flat plate boundary layer flow as an example to describe the basic idea behind multilevel multiscale

simulation. Figure 25 depicts the natural flow transition process in a 3-D boundary layer, showing clearly the variations in flow regime scales.

Using the fact that the flow scale of interest is generally large in the free stream and the area before breakdown (Figure 25), we can consider the use of multiple levels of grid to resolve the flow. Figure 26 depicts the case of three levels used in our boundary layer example; Ω_j represents the domain that level j is used to resolve, with the whole computational domain given by

$$\Omega = \Omega_1 \cup \Omega_2 \cup \Omega_3.$$

To decompose the total flow according to those levels, suppose the physics is governed by the time-dependent Navier-Stokes equations, which we write as

$$\begin{aligned}
\frac{\partial \vec{V}}{\partial t} + L\vec{V} &= \vec{F} \qquad \text{in } \Omega, \\
\vec{V} &= \vec{U} + \vec{U}' \qquad \text{at inflow.}
\end{aligned} \tag{117}$$

Here, we also decompose the inflow vector into two components (usually, $\vec{U}$ is the base flow with large magnitude, and $\vec{U}'$ is the unsteady perturbation part with relatively small magnitude). We then decompose the total flow field into three components according to

$$\vec{V} = \vec{V}_1 + \vec{V}_2 + \vec{V}_3, \tag{118}$$

where $\vec{V}_1$, $\vec{V}_2$, and $\vec{V}_3$ represent increasingly more local and finer scales of the flow so that

$$\begin{aligned}
\vec{V}_2 &= 0 \qquad \text{in } \Omega - \Omega_2, \\
\vec{V}_3 &= 0 \qquad \text{in } \Omega - \Omega_3.
\end{aligned} \tag{119}$$

To define individual governing systems for each component, first consider the subdomain Ω_1, on which we impose the system

$$\begin{aligned}
\frac{\partial \vec{V}_1}{\partial t} + L^{\Omega_1}\vec{V}_1 &= \vec{F}_1 \qquad \text{in } \Omega_1, \\
\vec{V}_1 &= \vec{U} \qquad \text{at inflow.}
\end{aligned} \tag{120}$$

Here, L^{Ω_1} is the spatial difference operator in Ω_1. $\vec{V}_1$ represents the large scale flow without the inflow disturbance. Thus, (120) can generally be solved by low order schemes on a coarse grid. For subdomain Ω_2, we consider the governing system

$$\begin{aligned}
\frac{\partial \vec{V}_2}{\partial t} + L^{\Omega_2}(\vec{V}_1 + \vec{V}_2) &= I^{\Omega_2}_{\Omega_1}(L^{\Omega_1}\vec{V}_1 - \vec{F}_1) + \vec{F} \quad \text{in } \Omega_2, \\
\vec{V}_2 &= \vec{U}' \qquad \text{at inflow.}
\end{aligned} \tag{121}$$

Here, $I^{\Omega_2}_{\Omega_1}$ represents some interpolation operator to transfer between Ω_1 and Ω_2. Note that $\vec{V}_2$ represents the perturbation in the flow field due to the inflow disturbance $\vec{U}'$ and the presumably finer scale source term $\vec{F} - \vec{F}_1$. $\vec{V}_2$ has much smaller scale than does $\vec{V}_1$ and should be solved by a high-order scheme (L^{Ω_2}) on a fairly fine middle scale grid. For subdomain Ω_3, which we choose to be a small part of the flow domain, the governing system can be written as

$$\begin{aligned}
\frac{\partial \vec{V}_3}{\partial t} + L^{\Omega_3}(\vec{V}_1 + \vec{V}_2 + \vec{V}_3) &= I^{\Omega_3}_{\Omega_2}L^{\Omega_2}(\vec{V}_1 + \vec{V}_2) \quad \text{in } \Omega_3, \\
\vec{V}_3 &= 0 \qquad \text{on } \partial\Omega_3.
\end{aligned} \tag{122}$$

$\vec{V}_3$'s physical scale is considered to be very small so that (122) should be resolved on an extremely fine grid.

9 Direct numerical simulation for flow transition: compressible flow (Liu et al, 1997b,c,d,; Zhao et al, 1997; Sun et al, 1997)

As the power of today's computer increases, direct numerical simulation (DNS) for realistic flow at certain Reynolds numbers becomes feasible. It is generally agreed that compressibility does not change the fundamental physics for streamwise instabilities at subsonic Mach number, but it still exhibits many differences from incompressible instability, and the numerical simulation work is still very limited, especially for the complex geometries. Unlike the relatively comprehensive picture of transition scenarios in incompressible flows, nonlinear effects responsible for transition at high speeds are still very much a mystery.

9.1 Governing system

The 3-D, compressible, time-dependent Navier-Stokes equations on the general curvilinear coordinate system (ξ, η, ζ) can be expressed as follows:

$$\frac{\partial \mathbf{U}}{\partial t} + \frac{\partial(\mathbf{E} - \mathbf{E}_v)}{\partial \xi} + \frac{\partial(\mathbf{F} - \mathbf{F}_v)}{\partial \eta} + \frac{\partial(\mathbf{G} - \mathbf{G}_v)}{\partial \zeta} = 0, \quad (123)$$

where

$$\mathbf{U} = \frac{1}{J}\begin{pmatrix} \rho \\ \rho u \\ \rho v \\ \rho w \\ e_t \end{pmatrix}, \qquad \mathbf{E} = \frac{1}{J}\begin{pmatrix} \rho U \\ \rho U u + \xi_x p \\ \rho U v + \xi_y p \\ \rho U w + \xi_x p \\ (e_t + p)U \end{pmatrix}, \ldots$$

9.2 Numerical method

Because the boundary treatment for the compact difference scheme is sensitive, we use, for the subsonic cases, the standard sixth-order central difference for the convective terms, fourth-order central difference for the viscous terms. In the supersonic cases, we choose the sixth-order compact finite difference method (Lele, 1992) for both convection and diffusion terms.

For time-integration, we use the so-called compact low-storage third-order Runge-Kutta scheme (Wray, 1986). For the subsonic cases, we adopt the classic four-stage fourth-order Runge-Kutta scheme with sixth-order central difference in space directions.

We can rewrite the system into the semi-discrete form:

$$\frac{d}{dt}\mathbf{U}'_{i,j,k} + \mathbf{R}(\mathbf{U}'_{i,j,k}) = 0, \quad (124)$$

where $\mathbf{R}(\mathbf{U}'_{i,j,k})$ is the function defined by

$$\mathbf{R}(\mathbf{U}'_{i,j,k}) = [L_C + L_D]\mathbf{U}'_{i,j,k}, \quad (125)$$

where L_C and L_D are discretized operators for convection and diffusion, respectively. The fourth-order Runge-Kutta scheme used to solve a system of ODE corresponding to the discretized Navier-Stokes equations can be expressed as follows:

$$\begin{aligned}
\mathbf{U}'^{(0)} &= \mathbf{U}'^{(n)}, \\
\mathbf{U}'^{(1)} &= \mathbf{U}'^{(0)} - \frac{\Delta t}{2}\mathbf{R}^{(0)}, \\
\mathbf{U}'^{(2)} &= \mathbf{U}'^{(0)} - \frac{\Delta t}{2}\mathbf{R}^{(1)}, \\
\mathbf{U}'^{(3)} &= \mathbf{U}'^{(0)} - \Delta t\mathbf{R}^{(2)}, \\
\mathbf{U}'^{(4)} &= \mathbf{U}'^{(0)} - \frac{\Delta t}{6}(\mathbf{R}^{(0)} + 2\mathbf{R}^{(1)} + 2\mathbf{R}^{(2)} + \mathbf{R}^{(3)}), \\
\mathbf{U}'^{(n+1)} &= \mathbf{U}'^{(4)}.
\end{aligned} \quad (126)$$

Here, $\mathbf{R}^{(q)} = \mathbf{R}(\mathbf{U}'^{(q)})$, the superscript n denotes the time level at $n\Delta t$, and the mesh indices (i, j, k) associated with the solution vector $\mathbf{U}'$ and time step Δt are suppressed for convenience.

9.3 Boundary conditions

Figure 27 gives a schematic description of the computational domain. Supposing the blowing/suction strip is located relatively far away from the inflow boundary, we can assume

$$f'(x_0, y, z, t) = 0, \quad (127)$$

where f' stands for the variables u', v', w', ρ', and p', respectively.

For the boundary conditions at the wall, we use no-slip and isothermal conditions:

$$u' = w' = T' = 0. \quad (128)$$

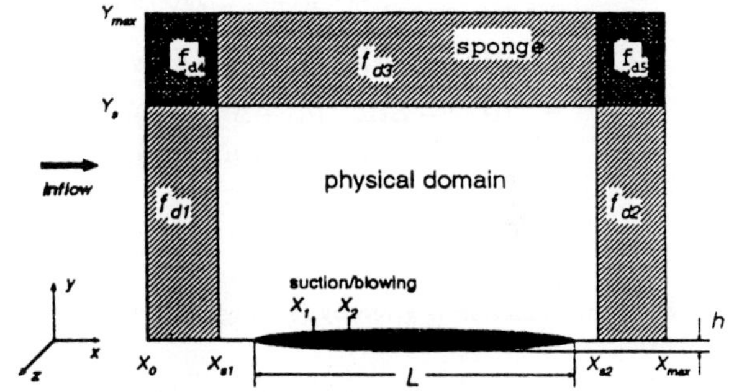

Figure 27 Sketch of the computational domain of a 2D airfoil with zero attack angle.

The wall normal velocity component is prescribed as a function of x, z, and t to generate the disturbances in the computational domain (Rist & Fasel, 1995):

$$v'(x, y_{wall}, z, t) = \begin{cases} f'_v(x, z, t) & x \in [x_1, x_2], \\ 0 & x < x_1 \text{ or } x > x_2. \end{cases}$$

In our work, we use

$$\begin{aligned}
f'_v(x, z, t) &= \varepsilon_{2d}\sin\left(\frac{2\pi(x - x_1)}{(x_2 - x_1)}\right)\sin(\omega_{2d}t) \\
&\quad + \varepsilon_{3d}\sin\left(\frac{2\pi(x - x_1)}{(x_2 - x_1)}\right)\cos(\beta z)\sin(\omega_{3d}t),
\end{aligned}$$

where ε_{2d} and ε_{3d} are the amplitudes for 2D and 3D disturbances, respectively; β is the controlled spanwise wave number; and ω_{2d} and ω_{3d} are the frequencies for 2D and 3D disturbances. This disturbance represents a combination of one 2D wave and a pair of oblique waves.

In the spanwise direction, a periodic boundary condition is imposed.

In the far field, an exponential decay condition (Fasel et al, 1990) together with the sponge layer treatment (Collis & Lele, 1996) are employed to satisfy the non-reflection requirement. The exponential decay condition yields

$$f_{NJ} = f_{NJ-1}e^{-\gamma(y_{NJ} - y_{NJ-1})}. \quad (129)$$

Here,

$$\gamma = \frac{\sqrt{\alpha_R^2 + \beta}}{\sqrt{Re}},$$

where α_R is the real part of the streamwise wave number.

The sponge layers are also embedded in the inflow and outflow sections (see Figure 27) to eliminate the reflection in the streamwise direction. For subsonic flow, the inflow sponge is necessary, while for supersonic flow, the inflow sponge can be removed.

For the Runge-Kutta type method, the sponge is loaded at each single stage, thus, (126) needs to be modified by adding an exponentially decaying source term in space directions to the right hand side:

$$\frac{d}{dt}\mathbf{U}'(x,y,z) + \mathbf{R}(\mathbf{U}'(x,y,z)) = -f_d(x,y)\mathbf{U}'(x,y,z).$$

This artificial term generates a so-called stiff differential equation which needs to be treated implicitly. This yields:

$$
\begin{aligned}
\mathbf{U}'^{(0)} &= \mathbf{U}'^{(n)}, \\
\mathbf{U}'^{(1)} &= [\mathbf{U}'^{(0)} - \frac{\Delta t}{2}\mathbf{R}^{(0)}]/[1 + \frac{\Delta t}{2}f_d(x,y)], \\
\mathbf{U}'^{(2)} &= [\mathbf{U}'^{(0)} - \frac{\Delta t}{2}\mathbf{R}^{(1)}]/[1 + \frac{\Delta t}{2}f_d(x,y)], \\
\mathbf{U}'^{(3)} &= [\mathbf{U}'^{(0)} - \Delta t\mathbf{R}^{(2)}]/[1 + \Delta t f_d(x,y)], \\
\mathbf{U}'^{(4)} &= [\mathbf{U}'^{(0)} - \frac{\Delta t}{6}(\mathbf{R}^{(0)} + 2\mathbf{R}^{(1)} + 2\mathbf{R}^{(2)} + \mathbf{R}^{(3)})] \\
&\quad /[1 + \Delta t f_d(x,y)], \\
\mathbf{U}'^{(n+1)} &= \mathbf{U}'^{(4)}.
\end{aligned}
\tag{130}
$$

The sponge function is given by (see Figure 27 for the location)

$$
f_d(x,y) = \begin{cases}
f_{d1}(x,y) & x \in [x_0, x_{s1}), & y \in [0, y_s) \\
f_{d2}(x,y) & x \in (x_{s2}, x_{max}], & y \in [0, y_s) \\
f_{d3}(x,y) & x \in [x_{s1}, x_{s2}], & y \in (y_s, y_{max}] \\
f_{d4}(x,y) & x \in [x_0, x_{s1}), & y \in (y_s, y_{max}] \\
f_{d5}(x,y) & x \in (x_{s2}, x_{max}], & y \in (y_s, y_{max}] \\
0 & & \text{elsewhere}
\end{cases}
$$

where

$$
\begin{aligned}
f_{d1}(x,y) &= A_{s1}\left(\frac{x_{s1} - x}{x_{s1} - x_0}\right)^{N_{s1}}, \\
f_{d2}(x,y) &= A_{s2}\left(\frac{x - x_{s2}}{x_{max} - x_{s2}}\right)^{N_{s2}}, \\
f_{d3}(x,y) &= A_{s3}\left(\frac{y - y_s}{y_{max} - y_s}\right)^{N_{s3}}, \\
f_{d4}(x,y) &= A_{s1}\left(\frac{x_{s1} - x}{x_{s1} - x_0}\right)^{N_{s1}} + A_{s3}\left(\frac{y - y_s}{y_{max} - y_s}\right)^{N_{s3}}, \\
f_{d5}(x,y) &= A_{s2}\left(\frac{x - x_{s2}}{x_{max} - x_{s2}}\right)^{N_{s2}} + A_{s3}\left(\frac{y - y_s}{y_{max} - y_s}\right)^{N_{s3}}
\end{aligned}
$$

Here, $A_{s1}, A_{s2}, A_{s3}, N_{s1}, N_{s2}$, and N_{s3} are prescribed constants. In our work, we choose $A_{s1} = A_{s2} = A_{s3} = 10$, and $N_{s1} = N_{s2} = N_{s3} = 3$.

9.4 Grid mapping

To insure that the numerical schemes of DNS have high accuracy, we need to make the accuracy of grid transformation schemes two orders higher than that of the flow scheme itself.

In subsonic and low supersonic Mach number flows, the primary mode dominates the transition process, while as Mach number goes to over $M_\infty = 3$, the second mode (Mach mode) becomes dominant. This requires use of different types of grid structure for the above two kinds of flows.

9.4.1 Grids for primary instability

For the problem of the flat plate boundary layer transition, the grid mapping function we used is

$$\eta(x,y) = \frac{\eta_{max}y(\sigma + y_{max})}{y_{max}(\sigma + y\sqrt{x_{in}/x})}\sqrt{\frac{x_{in}}{x}}. \tag{131}$$

Here, x_{in} is the location where Re^* is defined. Usually, we choose $x_{in} = x_2$ (see Figure 27). For the parallel base flow, the boundary layer thickness does not grow, so the above transformation becomes

$$\eta(y) = \frac{\eta_{max}y(\sigma + y_{max})}{y_{max}(\sigma + y)}. \tag{132}$$

9.4.2 Grids for second-mode instability

For high-speed ($M_\infty \geq 3$) flow transition, high gradients are usually concentrated near the wall and near the critical layer. With the aid of LST, we can estimate the location of critical layers, and then use an analytical mapping which clusters grid points in both wall and critical layer regions.

9.5 Filtering

Because of the use of the central difference scheme, the two-point "saw-tooth" oscillations will generally be generated, especially in the low viscous region. Those oscillations can induce some spurious physical waves. To avoid this phenomenon, we can either put some scheme viscosity through the biased upwinding scheme (Rai & Moin, 1993), or through the two-step MacCormack-like dissipative method (Gottlieb & Turkel, 1976), or through explicit filtering. Due to the simplicity of explicit filtering, we adopt it in our code. A sixth-order filter is applied every 10 time-steps in the wall-normal direction, and every 20 steps in the streamwise direction for the subsonic cases. During the study of supersonic transition, we find that only the wall-normal direction filtering is necessary. A compact sixth-order filtering is used every 10 time-steps in that direction.

9.6 Intergrid dissipation

As we did in the incompressible flow case, we use intergrid dissipation (Liu et al, 1995d) to introduce some grid viscosity to prevent the code from blowing up due to vortex breakdown. A brief description is as follows. Suppose f^h is the quantity obtained on the global grid G^h. We first restrict it to the next coarse level grid G^{2h}, and then obtain the new value of $f^{h(new)}$ on the global grid G^h. However, if we simply adopt this $f^{h(new)}$, the effects of small scales will all be ignored. Therefore, a dynamic weight function, σ, is introduced to maintain the effect of small eddies. This yields

$$
\begin{aligned}
f^{h(new)} &= \sigma I_{2h}^h I_h^{2h} f^{h(old)} + (1 - \sigma)f^{h(old)}, \\
\sigma &= \min\{2\kappa, 1\}, \\
\kappa &= \frac{1}{2}(\rho_0 + |\rho'|)(u'^2 + v'^2 + w'^2).
\end{aligned}
$$

Here, I_h^{2h} is a full-weighting restriction operator, and I_{2h}^h is a bi-linear interpolation operator.

For the detail of intergrid dissipation, see Liu & Liu (1995d).

10 Computational results

10.1 Incompressible laminar flow (Liu et al, 1993c, 1993d)

The multigrid method can substantially reduce the computation time for incompressible laminar flow. In order to reduce the residual of Navier-Stokes equations by three orders in magnitude, traditional iterative methods usually require hundreds or thousands of work units (one work unit is equivalent to one iteration on the finest grid), but the multigrid method only needs 25-50 work units (Figure 28). However, when the grid stretching or aspect ratio is large, we should use semi-coarsening with line distributive relaxation (or line box relaxation). Figure 29 gives a typical convergence history of semi-coarsening multigrid against single grid method.

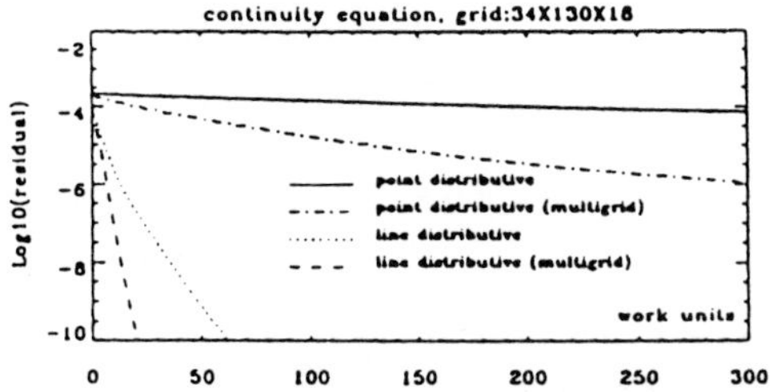

Figure 28. Comparison of convergence histories for relaxation and for relaxation used with multigrid for incompressible flow, Liu et al, 1993c

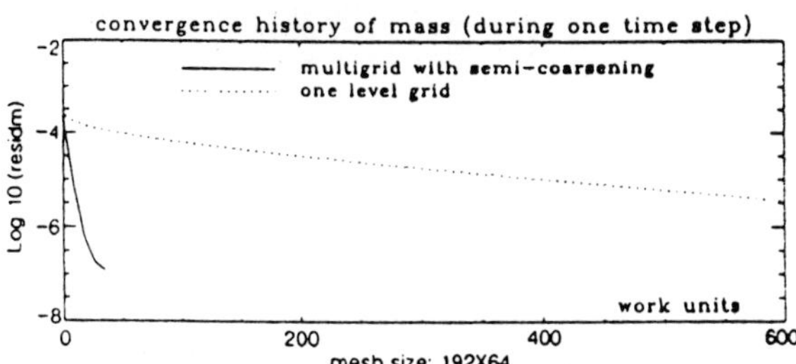

Figure 29. Convergence history of semi-coarsening multigrid method for incompressible flow, Liu et al, 1993d

10.2 Multiblock multigrid computation for submarines (Zheng et al, 1997b)

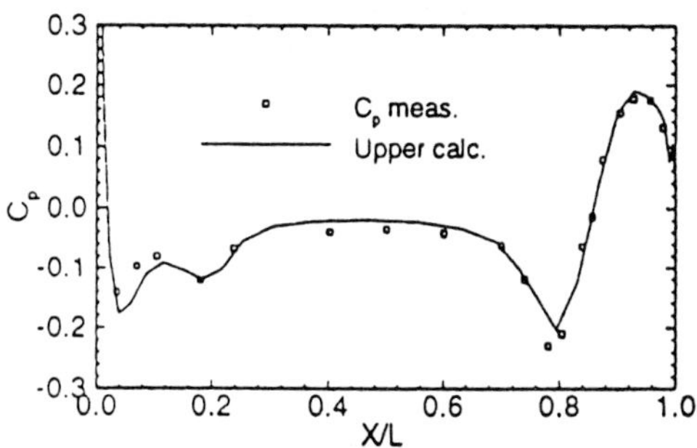

Figure 30. Bare hull pressure coefficient, $Re = 1.2 \times 10^7$, $49 \times 25 \times 33$ grid (Zheng et al, 1997b).

The test case is a turbulent flow about a submarine model, SUBOFF, which was developed by David Taylor Model Basin (Huang, 1992) to establish an experimental database for CFD code validation. The computations are carried out for the SUBOFF bare hull. Detailed comparison is made for the working condition of zero degree of incidence. The inflow Reynolds number is 1.2×10^7. Figures 30-31 show the computed pressure coefficients and skin friction coefficients at different locations with modified k-ω models. They are found in excellent agreement with experimental data.

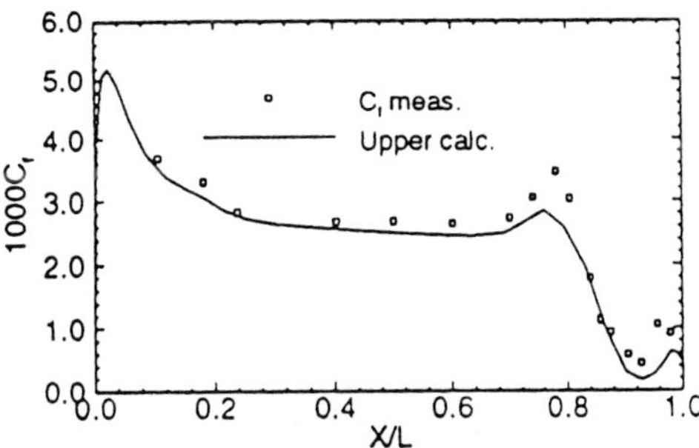

Figure 31. Bare hull Skin friction, $Re = 1.2 \times 10^7$, $49 \times 25 \times 33$ grid (Zheng et al, 1997b).

10.3 Laminar and turbulent diffusion flame (Liu et al 1993b; Liao et al, 1996d)

For combustion problem with detailed chemistry, due to the large governing system (species equations were added to the system) and large gradient of temperature and other variables around the flame front, the convergence of traditional iterative methods is much too slow. Ten thousands of iterations or more are normally required for a fair convergence. The multigrid can reduce the computation requirement to 100-500 work units which is around two orders faster. Figure 32 depicts a comparison of convergence history between single grid solver and multigrid methods. Because the multigrid can substantially reduce the computation time, we are able to do simulation for 3-D turbulent combustion with detailed chemistry in which 49 species and 229 reaction steps are considered for a methane-air reaction. The NO_x pollutant can be predicted because the detailed chemical mechanism is used. Figure 33 gives the distribution contours for NO in a channel combustor (Liao et al, 1996d).

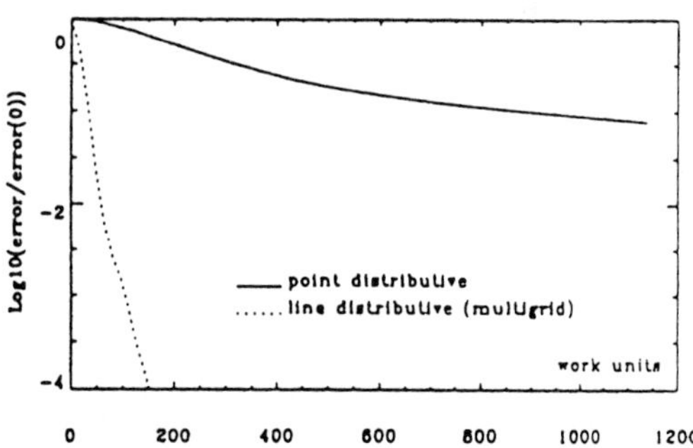

Figure 32. Comparison of convergence histories for traditional relaxation and line-distributive relaxation with multigrid for reacting flow Liu et al, 1993b

10.4 Multiple scale simulation for flow transition (Liu et all, 1995a)

To investigate the efficiency of our MMS approach, we choose to investigate the secondary instability case with $Re_0^* = 900$. We use only three levels to describe the flow. A $130 \times 18 \times 10$ grid is employed for both Ω_1 and Ω_2, which includes a 7 T-S wavelength physical domain and a 1 T-S wavelength buffer; a $42 \times 18 \times 18$ patch is used for Ω_3. The patch covers

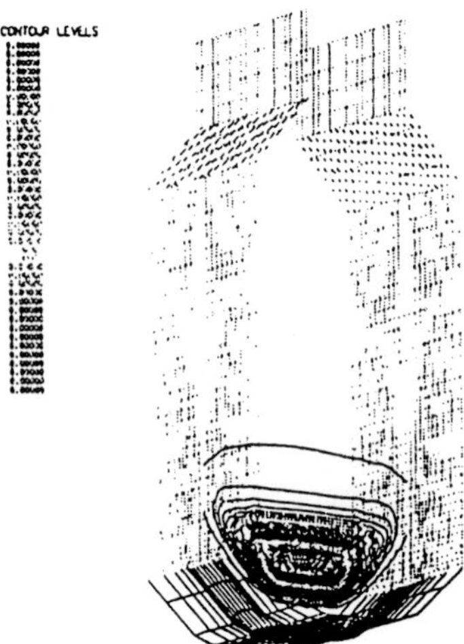

Figure 33. NO isopleths, contour from 0 to 4.1 x 10^{-6}, level is 1×10^{-7}, Liao et al, 1996b

the downstream half of the flat plate except for the buffer domain. The stretch parameter is $\sigma = 3.75$. As mentioned above, the Blasius similarity solution is employed as the base flow ($\vec{V}_1$), which is widely used as the base flow for flat plate transition. A Benney-Lin type disturbance,

$$U'^{(k)}(0,y,z,t) = \text{Real}\{\epsilon_{2d}\phi_{2d}^{(k)}(y)e^{-i\omega_{2d}t}$$
$$+ \ \epsilon_{3d+}\phi_{3d+}^{(k)}e^{-i\omega_{3d}t+i\beta z}$$
$$+ \ \epsilon_{3d-}\phi_{3d-}^{(k)}e^{-i\omega_{3d}t-i\beta z}\},$$

is imposed on the inflow to generate $\vec{V}_2$. Here, $\phi_{2d}(y)$ and $\phi_{3d\pm}(y)$ correspond, respectively, to 2-D and 3-D eigensolutions of the Orr-Sommerfeld equation, and the superscript (k) denotes different velocity components. Other parameters used in this work are as follows:

$$
\begin{aligned}
Re_0^{\bullet} &= 900, \quad Fr = 86 \ (\omega_{2d} = \omega_{3d} = 0.0774), \\
\beta &= 0.1, \quad y_{max} = \eta_{max} = 50, \\
\alpha_{2d} &= 0.2229 - 0.00451i, \\
\alpha_{3d} &= 0.2169 - 0.00419i, \\
\epsilon_{2d} &= 0.03, \quad \epsilon_{3d\pm} = 0.01, \\
x_0^{\bullet} &= 303.9, \quad x_{end}^{\bullet} = 529.4, \\
\Delta t &= T_{T-S}/240.
\end{aligned}
$$

Figure 34 depicts the contour plots of the spanwise perturbation vorticity ($\vec{V}_2$) in plane $y_0^{\bullet} = 0.1123$ at $t = 7T$, where T is the so-called T-S period. It is quite clear that within this level, the flow scale is still pretty large, and only large scale lambda waves can be resolved.

Figure 35 presents contour plots of spanwise vorticity produced by $\vec{V}_3$ in the same plane and at the same time as Figure 34. Though this level is still not fine enough to catch all of the scales in the flow field, some finer scales are resolved. We find that, in the patch (Ω_3), more vortices are generated on this level and they are amplified when they travel downstream. This is at least qualitatively correct. The final results produced by $\vec{V}_2 + \vec{V}_3$ are described in Figures 36 , showing clearly that more physical details can be found than in Figure 34.

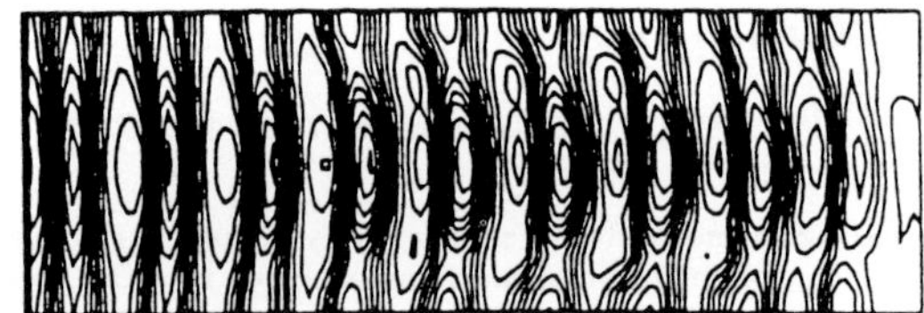

Figure 34. Contour plots of the spanwise vorticity produced by $\vec{V}_2$ in plane $y_0^{\bullet}=0.1123$, Liu et al, 1995a.

Figure 35. Contour plots of the spanwise vorticity produced by $\vec{V}_3$ in plane $y_0^{\bullet}=0.1123$, Liu et al, 1995a.

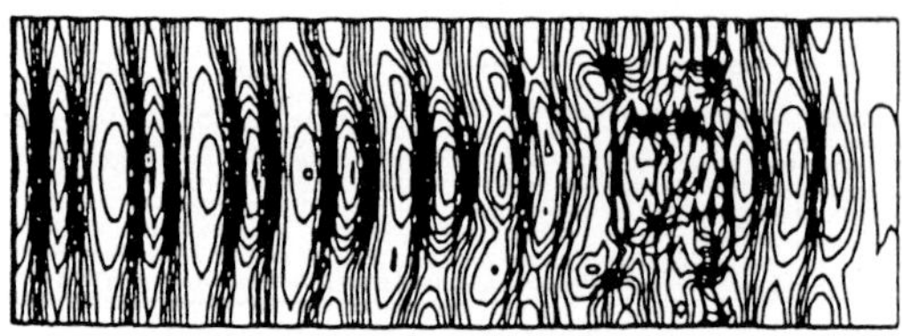

Figure 36. Contour plots of the spanwise vorticity produced by $\vec{V}_2 + \vec{V}_3$ in plane $y_0^{\bullet}=0.1123$, Liu et al, 1995a.

10.5 Direct numerical simulation for incompressible flow transition (Liu et al, 1995d, 1996a, 1996b)

Because the semi-coarsening multigrid associated with line distributive relaxation converges very fast and only 1 or 2 MG V-cycles are needed to solve the discretized system at each time step, we are able to use implicit time-stepping for real time dependent flow problem. Combined with high-order difference schemes, buffer outflow boundary conditions, and the intergrid dissipation, we are able to directly simulate the whole process of flow transition on a flat plate and airfoils at an acceptable computational cost. The direct numerical simulation cost for a flat plate is around 6 Cray-YMP hours.

10.5.1 Flat plate

The validation of the code is first examined in the linear instability stage over a smooth flat plate. The inflow perturbation for the 2D case is

$$u(0,y,z,t) = \text{Real}\{\epsilon_{2d}\phi_{2d}^{(k)}(y)e^{-i\omega_{2d}t}\}, \qquad (133)$$

for the 3D case is

$$u(0,y,z,t) = \text{Real}\{\epsilon_{3d}\phi_{3d}^{(k)}e^{-i\omega_{3d}t+i\beta z}\}, \qquad (134)$$

and for the 3D nonlinear breakdown case is

$$u(0,y,z,t) = \text{Real}\{\epsilon_{2d}\phi_{2d}^{(k)}(y)e^{-i\omega_{2d}t}$$

$$+ \quad \epsilon_{3d}\phi_{3d}^{(k)}e^{-i\omega_{3d}t+i\beta z}$$
$$+ \quad \epsilon_{3d}\phi_{3d}^{(k)}e^{-i\omega_{3d}t-i\beta z}\}. \tag{135}$$

Here, ϵ_{2d} and ϵ_{3d} are the amplitude of 2D and 3D disturbances. The grid transformation function in the wall-normal direction is given as

$$y(\eta) = \frac{y_{max}\sigma\eta}{\eta_{max}\sigma + y_{max}(\eta_{max} - \eta)}. \tag{136}$$

Here, we choose $\sigma = 3.5$, and $y_{max} = \eta_{max} = 50$ (based on δ^*). Both 2D ($\beta = 0$) and 3D ($\beta \neq 0$) cases are tested. For the 2D case, a $170 \times 34 \times 4$ grid is used (we use the 3D code do the 2D things), and for the 3D case, a $170 \times 34 \times 18$ grid is employed. We divide the whole computational domain into a physical domain and a buffer domain, which takes one T-S wavelength. Let $Re = 688$, $\omega_{2d} = \omega_{3d} = \omega = 0.075 (F = \omega/Re \times 10^6 = 109)$, and $\beta = 0.2$ (for the 3D case). The comparison of the maximum u and v amplitudes, the numerical profiles, and the amplitude distributions in both the wall-normal direction and in the streamwise direction coincide with the Linear Stability Theory (LST) results very well (Figure 37).

To verify the ability of this code, further tests on a smooth flat plate are performed. Using the experimental data of Saric et al (1984), we did both the fundamental breakdown and subharmonic breakdown simulation. The inflow boundary condition is assumed to have the form of formula (135). The computational results are shown in Figure 38. The numerical results match the experimental results very well. A contour plot of the relative helicity for a medium sized calculation is given in Figure 39, which shows clearly how the large scale vortices break into small ones.

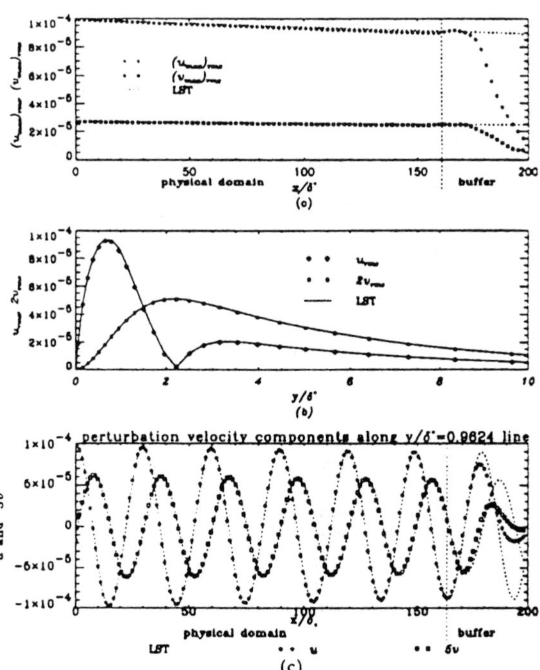

Figure 37. (a) Comparison of the maximum u and v amplitudes with LST. (b) Comparison of the numerical and LST profiles at $x/\delta^* = 114.91$ (2D case). (c) Amplitude of perturbation u and v distribution (2D case), Liu et al, 1996b.

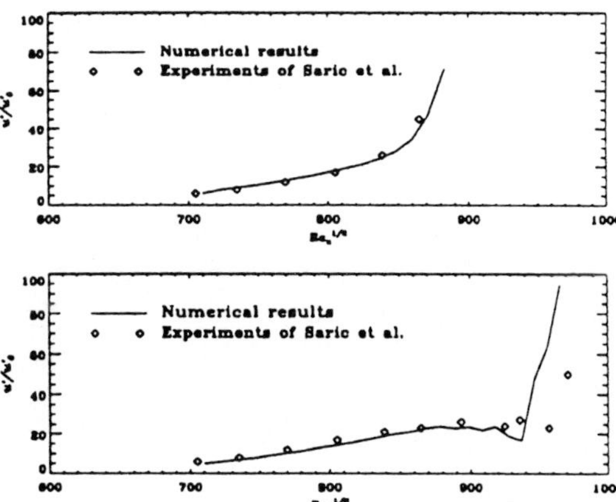

Figure 38. (a) Comparison of normalized perturbation velocity amplitude as a function of $Re_\delta (= Re_x^{1/2})$ with experimental results of Saric et al. (1984) for the fundamental breakdown case. (b) Comparison of normalized perturbation velocity amplitude obtained at $u = 0.4$ as a function of Re_δ with experimental results of Saric et al. (1984) for the subharmonic breakdown case. Liu et al, 1995d.

Figure 39. Relative helicity on the $j = 4$ (x, z)-plane. Spanwise periodic boundary condition is assumed. The second instability and breakdown process are shown clearly (Liu, 1996f).

10.5.2 Leading edge receptivity to free stream vorticity

The inflow disturbance is enforced very close to the symmetric axis with the form

$$u(0, y, t) = \text{Real}\{\epsilon_{enforce}A(|y| - y_c)e^{-\frac{(|y|-y_c)^2}{k}}e^{-i\omega t}\},$$

where A is a parameter for adjusting the amplitude of enforced disturbance, k is used to change the shape of u's profile, and y_c is the central point of the perturbation. Here, we use

$$k = 0.025, \quad y_c = 0.15,$$

other parameters are listed in Table 1.

Table 1. Parameters for 2D blunt body receptivity.

Re_h	2400
grids	$432 \times 34 \times 4$
domain height (based on h)	25
time-step/cycle of forcing	200
inflow disturbance amplitude	$\epsilon_{enforce} = 0.01$
Frequency	$\omega = 0.552(F = 230 \times 10^{-6})$
stretch (in formula (24))	$\sigma_C = 0.3, \quad \sigma_B = 50$
aspect ratio of ellipse	AR=6

Figure 40 depicts the contour plot of perturbation velocity u. It is very clear that energy of the long wavelength freestream disturbances has been converted into the short wavelength T-S wave after passing the leading edge regime, and the wavelength of the T-S wavelength is about 1/3 of the original freestream disturbance. This proves the theory of Goldstein (Goldstein, 1983, 1985).

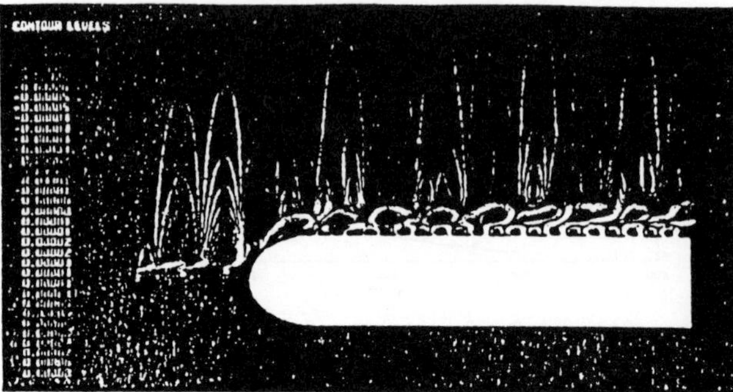

Figure 40. Streamwise perturbation velocity for the 2D ellipse-leading receptivity case, Liu, 1996f.

10.5.3 Whole process of transition over a 3D Joukowsky airfoil

In this case, a Joukowsky airfoil with infinite width and zero attack angle is considered, and the spanwise boundary condition is thus assumed to be periodic. The disturbance is imposed at free-stream. The pattern of disturbance is still the same: one 2D wave and two oblique waves. The parameters used in this simulation are listed in Table 2.

The simulation was carried out using an IBM RS 6000/590 workstation. It required approximately 140 hours of CPU time and 85 MB of memory.

The periodic disturbance is imposed at $x/h \approx -15$. Figure 41 depicts the instantaneous contour plots of perturbation $|\omega|$ on the $j = 6$ ($\approx 0.07h$ from the wing surface) grid surface after the 20th cycle of forcing. We can observe the nonlinear breakdown after some downstream location: the signs of randomness appear, the so-called spikes can then be observed, the flow field is no longer periodic in time, and the vortex scale becomes much smaller.

Figure 41. Instantaneous contour plots of perturbation $|\omega|$ on the $j = 6$ grid surface at $t = 17T$, Liu et al, 1996b.

Table 2. Parameters for simulation of 3D transition over a Joukowsky airfoil (natural transition).

Re_L	$200,000 (Re_h = 5000)$
L/h	40
grids	$432 \times 34 \times 18$
domain height (based on h)	25
time-step/cycle of forcing	200
disturbance amplitude	$\epsilon_{2d} = 0.02$, $\epsilon_{3d} = 0.02$
other parameters	$\omega = 1.2 (F = 240 \times 10^{-6})$
	$\beta = 1.0$
points on the airfoil	332 (streamwise)

10.6 Direct numerical simulation for compressible flow transition (Liu et al, 1997b, 1997d; Zhao et al, 1997a, 1997b)

We have developed a DNS code which can simulate the whole process of flow transition including the receptivity, linear growth, nonlinear instability, and transition for flat plate and airfoils at all speed, subsonic (e.g. $M_\infty = 0.5$) and supersonic (e.g. $M_\infty = 1.6$ and 4.5). Different types of the transition mechanism have been studied including first mode fundamental, subharmonic, and oblique transition. The second mode transition at high speed, e.g. $M_\infty = 4.5$, is also studied. Because of space limitations here we provide only a few figures describing these results.

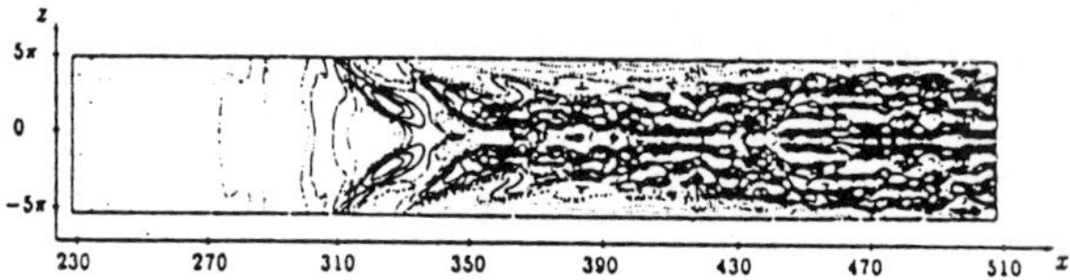

Figure 42. Instantaneous contour plots of perturbation spanwise vorticity (ω_z) on the $y^* = 0.85052$ (x, z) planes after 10 forcing periods, Liu et al, 1997b.

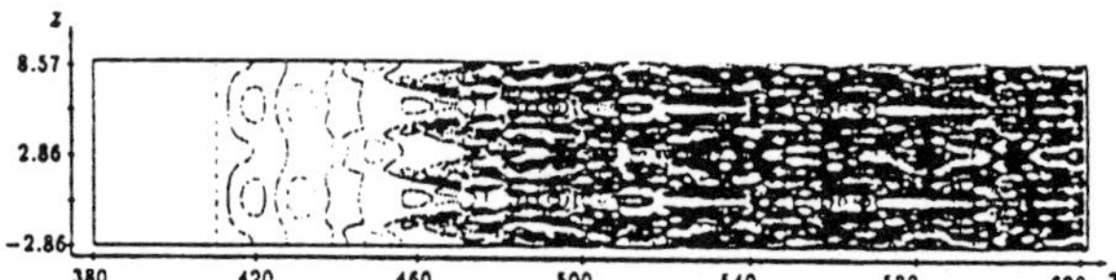

Figure 43. Instantaneous contour plots of perturbation spanwise vorticity (ω_z) on the $y^* = 0.6563$ (x, z) planes after 11 forcing periods, Liu et al, 1997b.

Figure 42 depicts the instantaneous contour plots of perturbation of spanwise vorticity (ω_z) on the $y^* = 0.85052$ (x, z) planes after 10 forcing periods in the case of K-type transition. For the H-type transition, Figure 43 shows the instantaneous contour plots of perturbation spanwise vorticity (ω_z) on the $y^* = 0.6563$ (x, z) planes after 11 forcing periods.

For supersonic flow, two breakdowns at Mach 1.6 are simulated in our work, **fundamental breakdown** and **oblique breakdown**.

As an example of fundamental breakdown, Figure 44 depicts the instantaneous contour plots of the streamwise perturbation velocity, u', after 14 forcing periods. It can be seen that even though we imposed only a pair of very weak 3-D disturbances (5% the magnitude of 2-D's), those disturbances are very unstable.

In order to present the oblique breakdown, the same parameters as for the simulation of fundamental breakdown are used. Figure 45 depicts the instantaneous contour plots of disturbance quantities, u'. Obviously, the structure of 3D disturbance development is different from that of the fundamental case. No lambda wave is observed in this case. Instead, a "zig-zag" type of vortices is observed, followed by a branching structure.

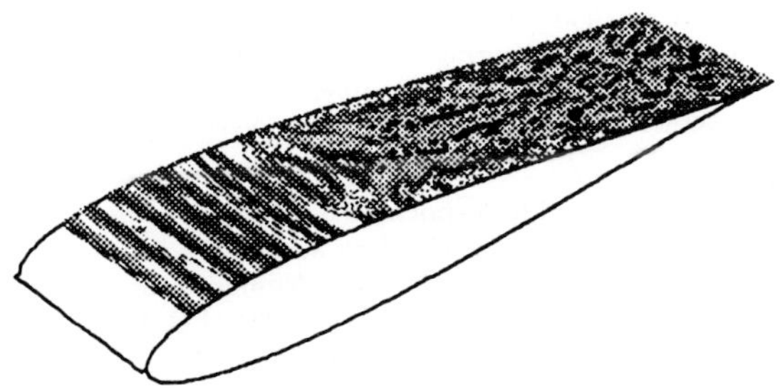

Figure 44 Instantaneous contour plots of u' for the Mach 1.6 Joukowsky boundary layer transition (fundamental breakdown), Liu et al, 1997c.

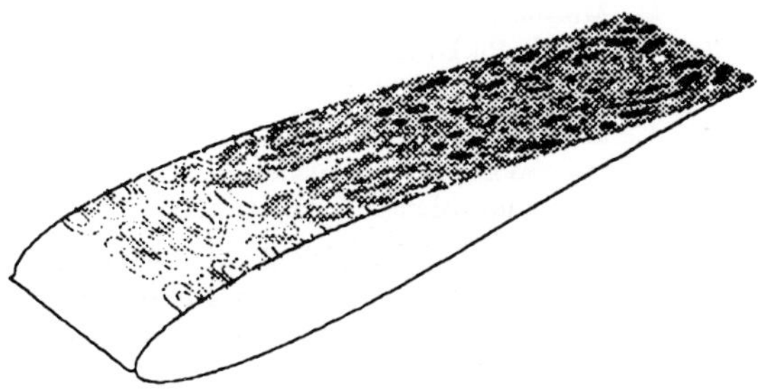

Figure 45 Instantaneous contour plots of u'. for the oblique breakdown case of a Mach 1.6 Joukowsky boundary layer transition on the j=10 grid surface ($\approx 0.05h$) at t=12T, Liu et al, 1997.

As the Mach number increases further, some new phenomena appear. For a Mach 4.5 flat plate, both first mode and second mode exist, and the transition mechanism becomes much more complicated. DNS is now thought to be a very important tool to investigate the mechanism for high speed flow. Figure 46 (a) depicts the mean streamwise velocity profiles obtained by our DNS code. The difference between the numerical solution and the laminar profile is obvious. Figure 46 (b) gives the mean friction coefficients for the same case, which clearly show the transition happened.

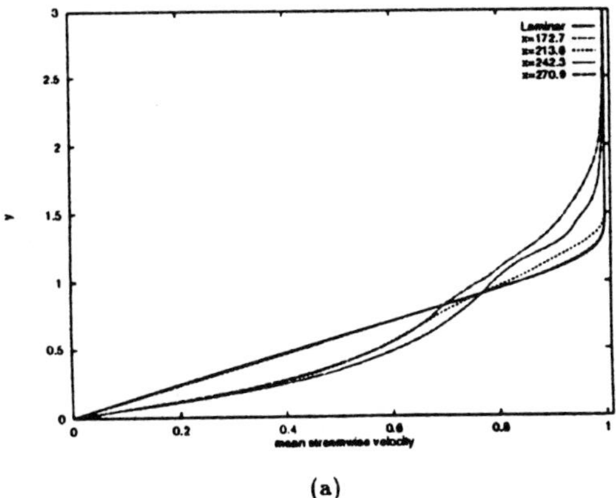

(a)

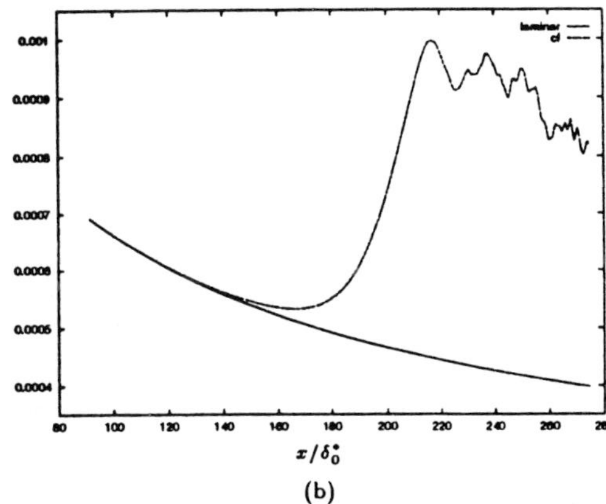

(b)

Figure 46 (a) Mean streamwise velocity profiles at different downstream location for a Mach 4.5 oblique breakdown case. (b) Mean friction coefficient of the transitional flow in a Mach 4.5 boundary layer.

11   Conclusion and discussion

After reviewing almost a decade of our work on the multigrid method, we would like to conclude this paper with the following observations.

1. The multigrid is an optimal method for linear elliptic problems.

2. For a Navier-Stokes system, we need use FAS and find a good systematic solver or collective solver, e.g. a box or distributive solver.

3. For stretched grids with high aspect ratio, we need to use a semi-coarsening with line or plane solvers.

4. We should use as many grid levels as possible until the coarsest grid has only a few points and the discretized equation can be solved exactly by direct or iterative methods on the coarsest grid.

5. For non-smooth boundary conditions and complex boundary geometries, we should use same boundary conditions for all levels of grids and, in general, follow the consistency principles.

6. The conservation principle can be also used to construct the coarse grid operator.

7. The multilevel adaptive method can use local refinement to resolve small scales, but needs to follow the conservation principle for interface. The finite-volume-element method for a composite grid can meet the conservation requirement.

8. The fast convergence of multigrid allows the use of implicit time-stepping for time-dependent problems with a cost comparable with explicit time-stepping but with much better stability.

9. The multilevel grid dissipation can overcome the non-physical energy burst and can simulate the whole process of flow transition.

10. The multigrid contribution to real time dependent problems includes use of different grid sizes to resolve different length scales or multiscale simulation.

12 Acknowledgments

The above work was accomplished by the numerical simulation group during the period of 1990-1997 under the support of NASA Langley Research Center (Grant No. NAS1-19016, NAS1-19312, NAG-1-1537, NAG-1-1651, NAG-1-1844, NAG-1-1891), NASA Lewis Research Center (Grant No. NAS3-2632, NAS3-27005), US Navy David Taylor Model Basin (Grant No. N00167-93-0009, N00024-95-C-4029), and US Air Force (Grant No. F08635-90-C-0368, F49620-95-1-0018, F49620-96-1-0122, F49620-97-1-0033). The author is grateful to all these sponsors. The author also wishes to thank Louisiana Tech University and the University of Colorado at Denver for the provision of facilities and for the technical support; he also wishes to thank DoD HPC for providing CRAY-C90 and SP2 time for this work.

References

[1] Brandt, A., Multi-level adaptive solutions to boundary value problems, Math. Comput., Vol 31, pp333-390, 1977.

[2] Brandt, A., *Multigrid techniques: Guide with application to fluid dynamics.* GMD Studie, GMD, St. Augustine, 1984.

[3] W.L. Briggs, 1987. A Multigrid Tutorial, SIAM Publications.

[4] Chelyshkov, VS, Grinchenko, VT, and Liu, C., Modeling of Near-Wall Quasi-Homogeneous Flow, Proceedings of First AFOSR International Conference on DNS/LES, Louisiana Tech University, Ruston, Louisiana, August 4-8, 1997.

[5] Collis, S.S. and Lele, S.K., A computational approach to swept leading-edge receptivity. *AIAA* 96-0180, 1996.

[6] Fasel, H., Rist, U., and Konzelmann, U., Numerical investigation of the three-dimensional development in boundary-layer transition. *AIAA J.* Vol 28, No.1, pp.29-37, 1990.

[7] Goldstein, M.E., 1983. The evolution of Tollmien-Schlichting waves near a leading edge, *J. Fluid Mech.* **127**, pp.959-981.

[8] Goldstein, M.E., 1985. Scattering of acoustic waves into Tollmien-Schlichting waves by small stream-wise variation in surface geometry, *J. Fluid Mech.* **154** pp.509-529.

[9] Gottlieb, D. and Turkel, E. Dissipative two-four methods for time-dependent problems. *Math. Comput.* Vol 30, No.136, pp.703-723, 1976.

[10] Hackbusch, W., Multi-grid methods and applications,, Springer, Berlin, 1985.

[11] Huang, T., Liu, H. L., Groves, N., Forlini, T., Blanton, J., and Gowing, S., Measurements of flows over an axisymmetric body with various appendages in a wind tunnel: the Darpa SUBOFF experimental program, *19th Symposium on Naval Hydrodynamics*, Seoul, Korea, August, 1992.

[12] Lele, S.K., Compact finite difference schemes with spectral-like resolution, *J. Comput. Phys.* **103**, pp.16-42, 1992.

[13] Liu, C. and McCormick, S., *Multigrid, elliptic grid generation, and fast adaptive composite grid method for solving transonic potential flow equations*, Lecture Note in Pure and Applied Mathematics, Marcel-Dekker, pp. 365-387, 1987.

[14] Liu, C. and McCormick, S., *The finite volume element method (FVE) for planar cavity flow*, 11th International Conference on Numerical Methods in Fluid Dynamics, Williamsberg, June 27 - July 1, 1988, and Lecture Note in Physics, Springer-Verlag, pp. 365-387, 1989a.

[15] Liu, C. and McCormick, S., *Multigrid, the rotated hybrid scheme and the fast adaptive composite grid method for planar cavity flow*, Numerical and Applied Mathematics, Edited by W. F. Ames, J.C. Baltzer AG, Scientific Publishing Co., IMACS, pp. 125-132, 1989b.

[16] Liu, C., *The finite volume element method (FVE) and fast adaptive composite grid method for the incompressible Navier-Stokes equations*, pp. 319-331, Proceedings of Fourth Copper Mountain Conference on Multigrid Method, Colorado, April, 1989c.

[17] Liu, C., *Multilevel adaptive methods in computational fluid dynamics*, Ph.D. thesis, University of Colorado at Denver, November, 1989d.

[18] Liu, C., Liu, Z., and McCormick, S., *Multigrid methods for flow transition in a planar channel*, Computer Physics Communications, 65, pp. 188-200, 1991a.

[19] Liu, C., Liu, Z., and McCormick, S., *Multilevel adaptive methods for incompressible flow in grooved channels*, J.of Computational and Applied Mathematics, vol. 38, pp. 283-295, 1991b.

[20] Liu, C., Liu, Z., and McCormick, S., *An efficient multigrid scheme for Poisson's equation with jump discontinuity*, Communications In Applied Numerical Methods, vol 8, pp. 621-631, 1992a.

[21] Liu, C. and Liu, Z., *Finite volume and multigrid methods for modeling instabilities in flat plate boundary layers*, J. of Wind Engineering, No. 52, pp. 412-417, August, 1992b.

[22] Liu, C. and Liu, Z., *High order finite difference and multigrid methods for spatially-evolving instability*, Journal of Computational Physics, Vol 106, pp. 92-100, 1993a.

[23] Liu, C., Liu, Z., and McCormick, S., *Multigrid methods for numerical simulation of laminar diffusion flames*, AIAA Paper 93-0236, Jan. 1993b.

[24] Liu, C. and Liu, Z., *Multigrid methods and high order finite difference for flow in transition*, AIAA Paper 93-3354, June, 1993c.

[25] Liu, C., Liu, Z., and McCormick, S., Multigrid methods for flow transition in three-dimensional boundary layers with surface roughness, NASA Contractor Report 4540, September, 1993d.

[26] Liu, C., Liu, Z., and McCormick, S., Multilevel adaptive methods for laminar diffusion flames, Journal of Scientific Computing, pp 341-355, Vol 8. No. 4, 1993e.

[27] Liu, Z. and Liu, C., Fourth order finite difference and multigrid methods for modeling instability in flat plate boundary layers - 2-D and 3-D approaches, J. Computers and Fluids, Vol 23, No.7, pp. 995-982, 1994a.

[28] Liu, Z., Liu, Z. X., Liu, C., and McCormick, S., Multi-level methods for temporal and spatial flow transition simulation in a rough channel, International Journal for Numerical Methods in Fluids, Vol 18, 1994b.

[29] Liu, C. and Liu, Z., Multiple scale simulation for transitional and turbulent flow, AIAA Paper 95-0777, Jan., 1995a.

[30] Liao, C., Liu, Z., and Liu, C., Implicit multigrid method for modeling 3-D turbulent diffusion flames with detailed chemistry, AIAA Paper 95-0801 , Jan., 1995b.

[31] Liu, Z., Liao, C.,, and Liu, C., Multigrid method for multi-step finite rate combustion, AIAA Paper 95-0205 , Jan., 1995c.

[32] Liu, C., and Liu, Z., Multigrid mapping and box relaxation for simulation of the whole process of flow transition in 3-D boundary layers, J. of Computational Physics, Vol 119, pp. 325-341, 1995d.

[33] Liu, C., Multigrid Method and Its Applications in Computational Fluid Dynamics, Book, Published by Tsinghua University Press, 1995e.

[34] Liu, Z., Xiong, G., Liu, C., and Joslin, R., Receptivity of free stream vortical disturbance of 2D and 3D airfoils, AIAA Paper 96-2084, June, 1996a.

[35] Liu, Z., Xiong, G., and Liu, C., Direct numerical simulation for the whole process of transition on 3-D airfoils, AIAA Paper 96-2081, June, 1996b.

[36] Liao, C., Liu, Z., Zheng, X., and Liu, C., Multilevel Adaptive Technique for Diffusion Flames with Complex Geometry, *AIAA 96-3127*, 1996c.

[37] Liao, C., Liu, Z., Zheng, X., and Liu, C., NOx Prediction in 3-D Turbulent Diffusion Flames By Using Implicit Multigrid Methods, Combust. Sci. and Tech. V. 119, 1-6, p. 219, 1996d.

[38] Liu, C., Principles in multilevel adaptive methods , Lecture notes, ACFD Short Course, Louisiana Tech University, pp. 239 - 264, Edited by Liu, C. and Liu, Z., June 24-28, 1996e.

[39] Liu, C., DNS for flow transition around complex geometry , Lecture notes, ACFD Short Course, Louisiana Tech University, pp. 343 - 376, Edited by Liu, C. and Liu, Z., June 24-28, 1996f.

[40] Liu, C., Guo, Y., High-order grid generation and its application in DNS, *Proceeding of 5th International Conference on Numerical Grid Generation in Computational Fluid Dynamics and Related Fields*, April 1–5, 1996g.

[41] Liao, C., Zheng, X., Liu, C., Sung, C., and Huang, T., Multilevel local refinement and multigrid methods for 3-D turbulent flows, AIAA paper 97-0911, 1997a.

[42] Liu, Z., Zhao, W., and Liu, C., Direct numerical simulation of transition in a subsonic airfoil boundary layer, AIAA Paper 97-0752, 1997b.

[43] Liu, Z., Zhao, W., Xiong, G., and Liu, C., Direct numerical simulation of transition in high-speed boundary layers around airfoil, AIAA Paper 97-0753, 1997c.

[44] Liu, C. and Liu, Z., Direct Numerical Simulation for Flow Transition Around Airfoils, Proceedings of First AFOSR International Conference on DNS/LES, Louisiana Tech University, Ruston, Louisiana, August 4-8, 1997d.

[45] Liu, Z., Zhao, W., and Liu, C., Direct Numerical Simulation of Flow Transition in a Compressible Swept-Wing Boundary Layer, Proceedings of First AFOSR International Conference on DNS/LES, Louisiana Tech University, Ruston, Louisiana, August 4-8, 1997e.

[46] McCormick, S. and Thomas, J., The fast adaptive composite grid (FAC) method for elliptic equations, Math. Comput., Vol 46, pp439-456, 1986.

[47] McCormick, S., Multigrid methods (Frontier in Applied Mathematics 3), SIAM, Philadelphia, 1987.

[48] McCormick, S., *Multilevel Adaptive Methods for Partial Differential Equations*, SIAM, Philadelphia., 1989.

[49] McCormick, S., Proceedings of Copper Mountain Conference on Multigrid Method, 1st, 1983; 2nd 1985; 3rd, 1987; 4th, 1989; 5th, 1991; 6th, 1993; 7th, 1995.

[50] Rai, M.M. and Moin, P. Direct simulations of transition and turbulence in a spatially evolving boundary layer. *J. Comput. Phys.* 109, pp.169-192, 1993.

[51] Rist, U. and Fasel, H. Direct numerical simulation of controlled transition in a flat-plate boundary layer, *J. Fluid Mech.* Vol 298, pp.211-248, 1995

[52] Saric, W. S., Kozlov, V. V. and Levchenko, C. Ya, Forced and Unforced Subharmonic Resonance in Boundary-Layer Transition, *AIAA 84-0007*, 1984.

[53] Sun, S., Lee, H., Liu, Z., and Liu, C., Parallel Direct Numerical Simulation of Compressible Flow Transition Using MPI, Proceedings of First AFOSR International Conference on DNS/LES, Louisiana Tech University, Ruston, Louisiana, August 4-8, 1997.

[54] Zhao, W., Liu, Z., and Liu, C., Numerical Investigation of Flow Transition in a Supersonic Airfoil Boundary Layer, Proceedings of First AFOSR International Conference on DNS/LES, Louisiana Tech University, Ruston, Louisiana, August 4-8, 1997.

[55] Zheng, X., Liao, C., Liu, Z., and Liu, C., A Mass Flux Based Implicit Multigrid Method for Modeling Multi-Dimensional Combustion, AIAA 95-2444, 31th AIAA/ASME/SAE/ASEE Joint Propulsion Conference and Exhibit, July 10, 1995, to be appeared on *AIAA Journal of Propulsion and Power*, 1995a.

[56] Zheng, X. , Liu, C., and Sung, C-H, Multigrid Solution of Incompressible Turbulent Flows by Using Two-Equation Turbulence Model, Copper Mountain Conference on Iterative Methods, April, 1996c, Copper Mountain, Colorado, USA.

[57] Zheng, X., Liao, C., Liu, C., Sung, C., and Huang, T., Preconditioned multigrid methods for unsteady incompressible flows, AIAA Paper 97-0445, 1997a.

[58] Zheng, X., Liao, C., Liu, C., Sung, C., and Huang, T., Multigrid, multiblock computation of incompressible flows using two-equation turbulent models, AIAA Paper 97-0626, 1997b.

[59] Wilcox, D. C., Reassessment of the scale-determining equation for advanced turbulence models, AIAA Journal, Vol 26, No. 11, pp1299-1310, 1988.

[60] Wray, A.A. Very low storage time-advancement schemes. *Internal Report*, NASA-Ames Research Center, Moffet Field, CA, 1986.

AIRFOIL SHAPE OPTIMIZATION USING THE IMPLICIT FUNCTION THEOREM

Yasuyoshi HORIBATA [1]

Research & Development Center, Toshiba Corporation; 4-1, Ukishima-cho, Kawasaki-ku, Kawasaki 210, Japan

Abstract

In this paper airfoil shape optimization is described. The shape is defined by a spline interpolating curve passing through data points, which are the design variables. An objective function, which depends on the design varibles, is minimized by using the sequential quadratic programming procedure. At each iteration of the sequential quadratic programming procedure, the gradient of the objective function with respect to the design vaiables is calculated by using the implicit function theorem. CPU time requirecd for the gradient calculation is greatly reduced by this method. An optimization example for a compressor cascade is presented.

1. Introduction

This paper deals with airfoil shape optimization. The shape is defined using n_D design variables $\boldsymbol{x}_D$. An objective function $F(\boldsymbol{x}_D)$ is introduced. The shape optimization is achieved by minimizimg the objective function. The objective function is minimized using the sequential quadratic programming (SQP) procedure. The SQP procedure iterates solving a quadratic programming subproblem. At each iteration, it requires the objective function $F(\boldsymbol{x}_D)$ and its gradient $\boldsymbol{g}$ with respect to the design variables $\boldsymbol{x}_D$.

The ith element of the gradient $\boldsymbol{g}$ can be calculated by using the finite difference

$$g_i \approx \frac{F(\boldsymbol{x}_D + h\boldsymbol{e}_i) - F(\boldsymbol{x}_D)}{h}$$

where $\boldsymbol{e}_i$ is the normalized ith coordinate vector and h is an increment. However, calculating all elements of the gradient requires executing as many flow simulations as the design variables, and consumes much CPU time. In this paper, the gradient $\boldsymbol{g}$ is calculated by using the implicit function theorem. The calculation requires only as much CPU time as one flow simulation, and hence CPU time is greatly reduced.

Similar gradient calculation methods have been reported for one-dimensional compressible flow inside a duct [1] and two-dimensional incompressible flow around an airfoil [2]. In this paper, two-dimensional compressible flow around a cascade is considered.

Received on February 28, 1997.

2. Optimization Algorithm

The shape optimization is formulated as a nonlinear optimization problem

$$\begin{aligned}
minimize \quad & F(\boldsymbol{x}_D) \\
subject\ to \quad & c_j(\boldsymbol{x}_D) = 0, \quad j = 1, \ldots, m_e \\
& c_j(\boldsymbol{x}_D) \geq 0, \quad j = m_e + 1, \ldots, m
\end{aligned}$$

The sequential quadratic programming procedure is used to solve this nonlinear optimization problem. The basic idea is the formulation of a quadratic programming subproblem. The SQP algorithm proceeds from a quadratic approximation of the Lagrange function

$$L(\boldsymbol{x}_D, \boldsymbol{v}) = F(\boldsymbol{x}_D) - \sum_{j=1}^{m} v_j c_j(\boldsymbol{x}_D)$$

where $\boldsymbol{v} = (v_1, \ldots, v_m)$ is a vector of Lagrange mutilpliers, and a linearization of the constraints. Let $\boldsymbol{x}_D^k$ be the kth estimate for the optimal design variables. The resulting quadratic programming subproblem is written in the form

$$\begin{aligned}
minimize \quad & \tfrac{1}{2}\boldsymbol{p}^T B_k \boldsymbol{p} + \nabla F(\boldsymbol{x}_D^k)^T \boldsymbol{p} \\
subject\ to &
\end{aligned}$$

$$\nabla c_j(\boldsymbol{x}_D^k)^T \boldsymbol{p} + c_j(\boldsymbol{x}_D^k) = 0, \quad j = 1, \ldots, m_e$$

$$\nabla c_j(\boldsymbol{x}_D^k)^T \boldsymbol{p} + c_j(\boldsymbol{x}_D^k) \geq 0, \quad j = m_e + 1, \ldots, m$$

where $\nabla F(\boldsymbol{x}_D^k)$ is the gradient of the objective function with respect to the design variables at $\boldsymbol{x}_D = \boldsymbol{x}_D^k$, B_k is a positive definite approximation of the Hessian matrix of the Lagrange function. Let $\boldsymbol{p}_k$ be the solution of the subproblem and $\boldsymbol{v}_k$ the corresponding vector of Lagrange multipliers. Then, the estimate for the optimal design variables is updated by

$$x_D^{k+1} = x_D^k + \alpha_k p_k$$

where α_k is a line search parameter, and is designed to produce a sufficient decrease of a merit function [3].

The SQP procedure, if properly applied, guarantees convergence towards a minimum value. The objective function $F(x_D)$ may, however, possess many minima. As opposed to genetic algorithms [4, 5], the SQP procedure is strongly attracted to all minima in $F(x_D)$, whether they be local or global, and there is no way to constraint the algorithm to converge only to the global minimum. Consequently, the only way to ensure that the global minimum point has been found is to repeat the minimization procedure starting from a different initial guess.

On the other hand, optimization using the SQP procedure usually reqires much less objective-function evaluations than genetic algorithms.

3. Calculation of the Gradient

The objective function is a function of the design variables x_D. Its gradient with respect to x_D is

$$g = \nabla F(x_D) \tag{1}$$

On the other hand, the objective function can be regarded as a function of the grid coordinates x_g and the flow variables x_f defined at the grid points in flow simulation. Hence, the ith element of g is written as

$$g_i = \frac{\partial F(x_f, x_g)}{\partial x_D^i}$$
$$= \sum_j \frac{\partial F(x_f, x_g)}{\partial x_g^j} \frac{\partial x_g^j}{\partial x_D^i} + \sum_j \frac{\partial F(x_f, x_g)}{\partial x_f^j} \frac{\partial x_f^j}{\partial x_D^i}$$
$$= h_i + (\nabla_f F)^T b_i$$

where

$$h_i = \sum_j \frac{\partial F(x_f, x_g)}{\partial x_g^j} \frac{\partial x_g^j}{\partial x_D^i}$$
$$\nabla_f F = \nabla_f F(x_f, x_g)$$
$$b_i = \frac{\partial x_f}{\partial x_D^i}$$

The vector $\nabla_f F$ is calculated by perturbing each flow variable with the grid coordinates fixed. The vector h is calculated by generating a new grid after each design variable is perturbed with the flow variables fixed.

The vector b_i is obtained by applying the implicit function theorem to the residual vector $R(x_f, x_g)$. The discretized Navier-Stokes equations are written as

$$R(x_f, x_g) = 0 \tag{2}$$

For the given grid coordinates x_g, flow simulation yields the flow variables x_f by solving equation (2). Hence, the flow variables x_f are an implicit function of the grid coordinates x_g. If the design variables x_D are perturbed by Δx_D, then the grid coordinates x_g and the flow variables x_f become $x_g + \Delta x_g$ and $x_f + \Delta x_f$, respectively. The new grid coordinates and flow varibales satisfy the discretized Navier-Stokes equations, too. Thus, one has

$$R(x_f + \Delta x_f, x_g + \Delta x_g) = 0$$

Developing a Taylor-series expansion for the left-hand side of this equation about (x_f, x_g) gives

$$R(x_f, x_g) + \sum_k \frac{\partial R}{\partial x_f^k} \Delta x_f^k + \sum_k \frac{\partial R}{\partial x_g^k} \Delta x_g^k = 0$$

From equation (1) this equation reduces to

$$\sum_k \frac{\partial R}{\partial x_f^k} \Delta x_f^k + \sum_k \frac{\partial R}{\partial x_g^k} \Delta x_g^k = 0 \tag{3}$$

The flow variables and grid coordinates are functions of the design varibles. Thus, Δx_f^k and Δx_g^k are expressed as

$$\Delta x_f^k = \sum_i \frac{\partial x_f^k}{\partial x_D^i} \Delta x_D^i$$
$$\Delta x_g^k = \sum_i \frac{\partial x_g^k}{\partial x_D^i} \Delta x_D^i$$

Substituting these into equation (3) gives

$$\sum_i \left(\sum_k \frac{\partial R}{\partial x_f^k} \frac{\partial x_f^k}{\partial x_D^i} + \sum_k \frac{\partial R}{\partial x_g^k} \frac{\partial x_g^k}{\partial x_D^i} \right) \Delta x_D^i = 0$$

This leads to

$$\sum_k \frac{\partial R}{\partial x_f^k} \frac{\partial x_f^k}{\partial x_D^i} + \sum_k \frac{\partial R}{\partial x_g^k} \frac{\partial x_g^k}{\partial x_D^i} = 0 \tag{4}$$

The Jacobian matrix is defined by

$$J_f = \begin{pmatrix} \frac{\partial R_1}{\partial x_f^1} & \frac{\partial R_1}{\partial x_f^2} & \cdots & \frac{\partial R_1}{\partial x_f^{n_f}} \\ \frac{\partial R_2}{\partial x_f^1} & \frac{\partial R_2}{\partial x_f^2} & \cdots & \frac{\partial R_2}{\partial x_f^{n_f}} \\ \vdots & \vdots & \ddots & \vdots \\ \frac{\partial R_{n_f}}{\partial x_f^1} & \frac{\partial R_{n_f}}{\partial x_f^2} & \cdots & \frac{\partial R_{n_f}}{\partial x_f^{n_f}} \end{pmatrix}$$

where n_f is the number of elements of the flow variables x_f. In terms of the Jacobian matrix, equation (4) is rewritten as

$$J_f b_i + d_i = 0 \tag{5}$$

where

$$d_i = \sum_k \frac{\partial R}{\partial x_g^k} \frac{\partial x_g^k}{\partial x_D^i}$$

If the determinant $|J_f|$ of the Jacobian matrix does not vanish, equation (5) has an unique solution. As a result, the perturbation Δx_f are uniquely determined by the perturbation Δx_g; this fact is referred to as the implicit function theorem.

From equation (5), b_i is rewritten as

$$b_i = -J_f^{-1} d_i$$

Hence, the ith element of g is rewritten as

$$\begin{aligned}
g_i &= h_i - (\nabla_f F)^T J_f^{-1} d_i \\
&= h_i - (J_f^{-1} d_i)^T \nabla_f F \\
&= h_i - d_i^T (J_f^{-1})^T \nabla_f F
\end{aligned} \tag{6}$$

Defining the matrix J_D by

$$J_D = (d_1\, d_2 \ldots d_{n_D})$$

equation (6) is rewritten in the vector form

$$g = h - J_D^T (J_f^{-1})^T \nabla_f F \tag{7}$$

The vector $(J_f^{-1})^T \nabla_f F$ in the second term on the right-hand side of equation (7) is the solution of the system of linear equations

$$J_f^T x = \nabla_f F \tag{8}$$

where x is a vector of unknowns.

The Jacobian matrix J_f is calculated by re-evaluating the residuals after each flow variable is perturbed while keeping x_g fixed. Te flow variables are always perturbed in such a way that the boundary conditions are never violated. The matrix J_D is calculated by re-evaluating the residuals after each design variable is perturbed and a new grid is generated while keeping x_f fixed. Equation (8) is solved by using an iterative procedure. Then, the gradient is calculated from equation (7)

4. Solution Method for the System of Linear Equations

Two-dimensional compressible flow around a cascade is considered. The flow variables x_f are written as

$$x_f = \begin{pmatrix} u_1 \\ u_2 \\ u_3 \\ u_4 \end{pmatrix}$$

where u_1, u_2, u_3, u_4 are $(n_f/4)$-vectors containing ρ, ρu, ρv, ρE, respectively, defined at the grid points.

Here, ρ, u, v, E are the density, the x, y components of the velocity, and the total energy per unit volume, respectively. The Jacobian matrix J_f is partitioned into 16 blocks as

$$J_f = \begin{pmatrix} J_{11} & J_{12} & J_{13} & J_{14} \\ J_{21} & J_{22} & J_{23} & J_{24} \\ J_{31} & J_{32} & J_{33} & J_{34} \\ J_{41} & J_{42} & J_{43} & J_{44} \end{pmatrix}$$

where J_{ij} is a $(n_f/4) \times (n_f/4)$ matrix, and has 21 non-zero diagonals. Similarly, the vector x of equation (8) is partitioned into 4 blocks as

$$x = \begin{pmatrix} x_1 \\ x_2 \\ x_3 \\ x_4 \end{pmatrix}$$

where x_1, x_2, x_3, x_4 are $(n_f/4)$-vectors. Thus, equation (8) is rewritten as

$$\begin{pmatrix} J_{11}^T & J_{21}^T & J_{31}^T & J_{41}^T \\ J_{12}^T & J_{22}^T & J_{32}^T & J_{42}^T \\ J_{13}^T & J_{23}^T & J_{33}^T & J_{43}^T \\ J_{14}^T & J_{24}^T & J_{34}^T & J_{44}^T \end{pmatrix} \begin{pmatrix} x_1 \\ x_2 \\ x_3 \\ x_4 \end{pmatrix} = \begin{pmatrix} \nabla_{u_1} F \\ \nabla_{u_2} F \\ \nabla_{u_3} F \\ \nabla_{u_4} F \end{pmatrix} \tag{9}$$

An iterative procedure is defined by rewriting equation (9) as

$$\begin{aligned}
(J_{11}^T + \lambda I)x_1^{n+1} = &-J_{21}^T x_2^n - J_{31}^T x_3^n \\
&- J_{41}^T x_4^n \\
&+ \lambda x_1^n + \nabla_{u_1} F
\end{aligned} \tag{10}$$

$$\begin{aligned}
(J_{22}^T + \lambda I)x_2^{n+1} = &-J_{12}^T x_1^{n+1} - J_{32}^T x_3^n \\
&- J_{42}^T x_4^n \\
&+ \lambda x_2^n + \nabla_{u_2} F
\end{aligned} \tag{11}$$

$$\begin{aligned}
(J_{33}^T + \lambda I)x_3^{n+1} = &-J_{13}^T x_1^{n+1} - J_{23}^T x_2^{n+1} \\
&- J_{43}^T x_4^n \\
&+ \lambda x_3^n + \nabla_{u_3} F
\end{aligned} \tag{12}$$

$$\begin{aligned}
(J_{44}^T + \lambda I)x_4^{n+1} = &-J_{14}^T x_1^{n+1} - J_{24}^T x_2^{n+1} \\
&- J_{34}^T x_3^{n+1} \\
&+ \lambda x_4^n + \nabla_{u_4} F
\end{aligned} \tag{13}$$

where I is the unit matrix and λ is a constant. Using equations (10) (11) (12) (13), the algorithm for solving equation (9) proceeds according to the following steps:

1. Guess initial values $x_1^0, x_2^0, x_3^0, x_4^0$.
2. Repeat the following calculations for $n = 0, 1, \ldots$ until x_1^{n+1}, x_2^{n+1}, x_3^{n+1}, x_4^{n+1} converge.
   - (a) Solve equation (10) for x_1^{n+1}. The conjugate residual (CR) method is used.
   - (b) Solve equation (11) for x_2^{n+1}. The CR method is used.
   - (c) Solve equation (12) for x_3^{n+1}. The CR method is used.
   - (d) Solve equation (13) for x_4^{n+1}. The CR method is used.

5. Numerical Test

The design method is tested for a compressor cascade. Figure 1 shows the initial compressor cascade. NACA 65 wing is chosen to be the initial airfoil shape. The inlet flow angle is 34.926°. The pitch is 36.8 mm. Inlet Mach number is around 0.78. The inlet total pressure and total temperature are 1.3148×10^5 Pa and 312.66 K, respectively. The outlet static pressure is 1.0052×10^5 Pa.

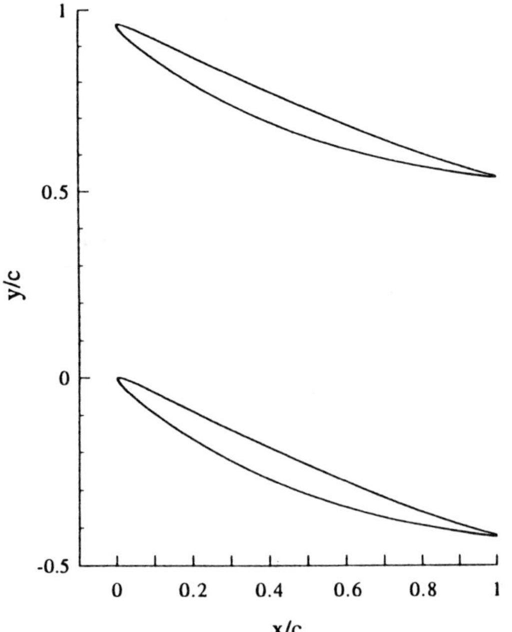

Fig.1: Initial compressor cascade.

5.1 Shape Definition and Design Variables

The shape is defined by a spline interpolating curve passing through data points. In Figure 2, the full and open symbols are spline data points used for the initial airfoil shape. The open symbols are fixed during optimization, whereas the full symbols are allowed to move in the direction normal to the local surface. These perturbations are the design variables. The number of the design variables is 10.

5.2 Validation of the Gradient Calculation Method

The objective function is taken to be the sum of the total pressure loss and the outlet flow angle deviation:

$$F(x_D) = w_1(P_{ti} - P_{t0}) + w_2(\theta_o - \bar{\theta}_o)^2$$

where P_{ti} and P_{t0} are the inlet and outlet total pressures, respectively, θ_o is the outlet flow angle and $\bar{\theta}_o$

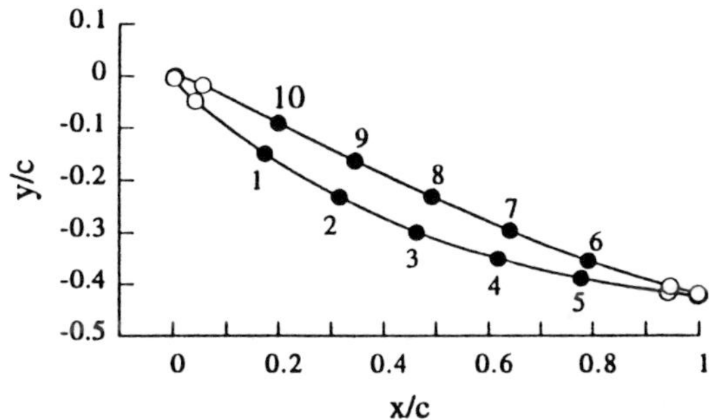

Fig.2: Spline data points. Open symbols are fixed; full symbols are allowed to move.

is the preferable value of the outlet flow angle; it is taken to be 17.30 degrees. The coefficients w_1 and w_2 are weights, which are taken to be $1/w_1 = 10^3$ Pa and $1/w_2 = 1$ degree, respectively. Constraints are applied so that the wing thickness can not be less than a given length. The leading edge and trailing edge are both fixed.

Grid is a 20 by 58 point H-mesh. A two-dimensional finite-difference time-marching solver is used.

In equations (10), (11), (12), and (13), λ is taken to be 6. About 8000 iterations of the outer loop are necessary for 3 significant figure accuracy. CPU time required for calculating the gradient is almost as much as that for one flow simulation.

Figure 3 shows the gradient calculated using the method proposed in this paper for the initial airfoil shape, and compares it with the gradient calculated by using finite differences. The largest difference is 54 % for the 4th design variable. The second largest difference is 23 % for the 5th design variable. The differences are less than 12 % for the other design variables. The absolute values of the 4th and 5th gradient elements are small relative to the other elements. Hence, the gradient calculated by the present method is accurate enough to be used for the optimization.

5.3 Shape Optimization

Figure 4 shows the history of the objective function, the total pressure loss, and the deviation of the outlet flow angle. The objective function and the total pressure loss are normalized by dividing by their respective initial values. The total pressure loss decreases about 10 % after 20 iterations. At that time, the outlet flow angle deviation remains around 0.1 degrees. Figure 5 shows the initial shape, and the shapes at the 5th, the 10th, the 15th, the 20th iteration. The position of the largest thickness is shifted toward the trailing edge. Figure 6 shows the corresponding pressure coefficients on the surfaces of the initial shape, and the shapes at the 5th, the 10th, the 15th, and the 20th iteration. On the lower surface, the position of the lowest pressure

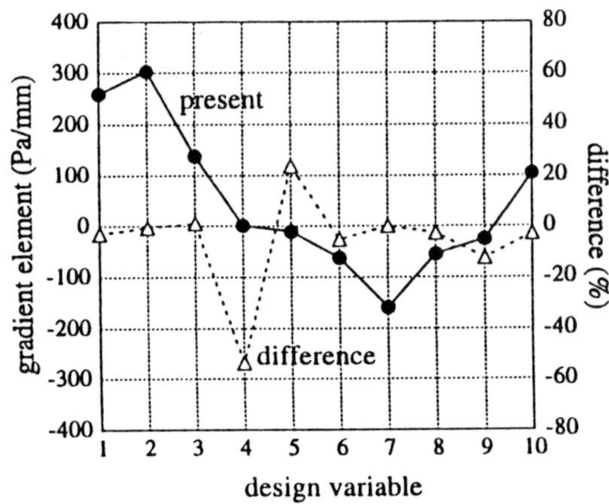

Fig.3: The gradient for the initial shape, and comparison with that calculated by finite differences.

is shifted close to the leading edge.

CPU time needed for the 20 iterations is about 90 min on CRAY Y-MP C90 (1 GFLOPS).

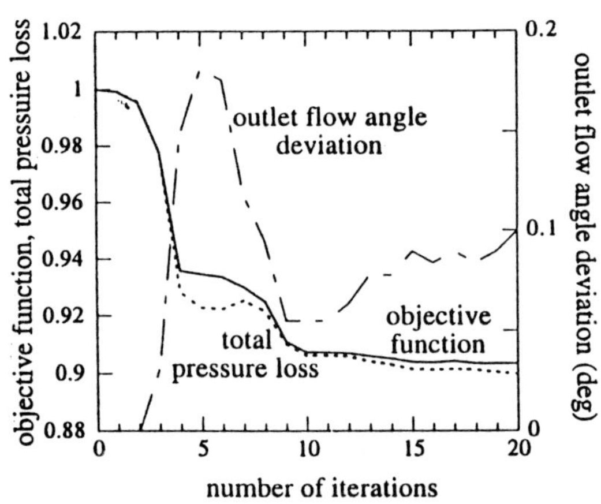

Fig.4: Objective function, total pressure loss, and outlet flow angle deviation vs. optimization iteration count

6. Conclusions

An optimization method for an airfoil shape is described. The sequential quadratic programming procedure is used. The gradient of the objective function is calculated by using the implicit function theorem. The calculation requires only as much CPU time as

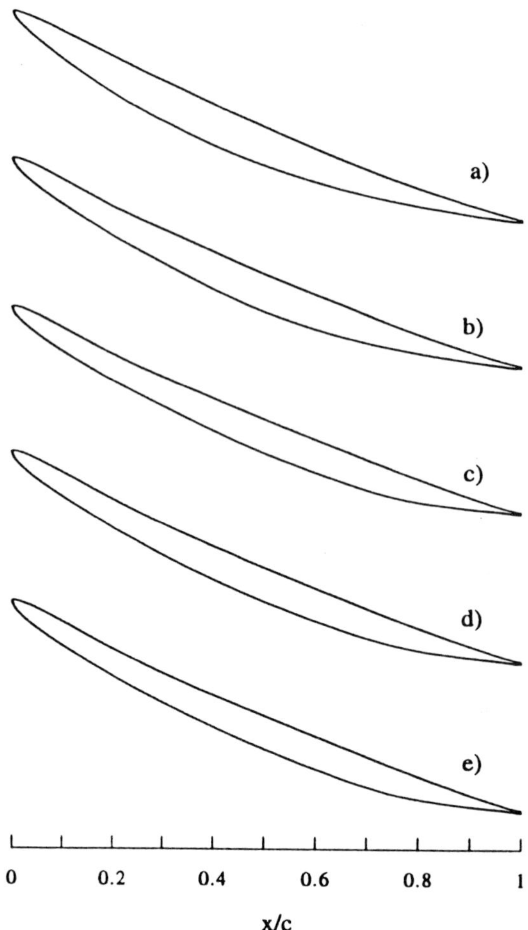

Fig.5: Airfoil shapes: a)NACA 65 wing b)shape at the 5th iteration c)shape at the 10th iteration d)shape at the 15th iteration e)shape at the 20th iteration

one flow simulation does. Thus, CPU time for the optimization is greatly reduced. A preliminary numerical test is done for a compressor cascade. The gradient calculated is almost the same as that obtained by using finite differences. The total pressure loss is decreased by about 10 % after 20 iterations.

REFERENCES

[1] P. D. Frank and G. A. Shubin; A comparison of optimization-based approaches for a model computational aerodynamics design problem, *J. Comput. Phys.*, **98**, 74-89, (1992).

[2] A. E. Dixon and C. A. J. Fletcher; Optimization applied to aerofoil design, *Proc. of 5th Int. Symp. Comput. Fluid Dynamics*, 161-165, (1993).

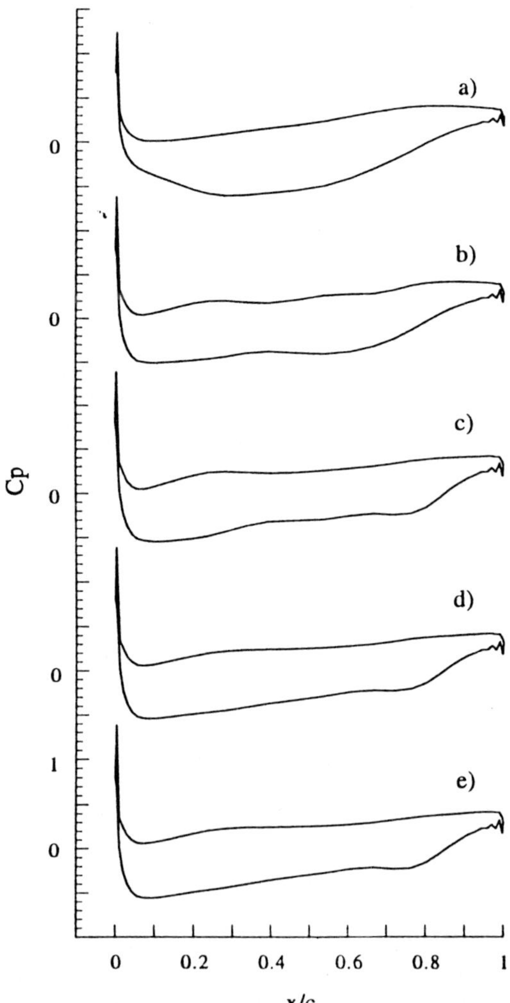

Fig.6: Pressure coefficients on the surfaces: a)NACA 65 wing b)shape at the 5th iteration c)shape at the 10th iteration d)shape at the 15th iteration e)shape at the 20th iteration

[3]   K. Schittkowski, NLPQL; A FORTRAN subroutine solving constrained nonlinear programming problems, *Annals of Operations Research*, **5**, 485-500, (1986).

[4]   H. Q. Chen, B. Mantel, J. Periaux, and M. Sefrioui; "Genetic method for optimization and industrial applications", Computational Fluid Dynamics Review, 541-547, (1995).

[5]   H. J. Kim and O. H. Rho; "Dual-point design of transonic airfoils using the hybrid inverse optimization method", AIAA Paper 97-0512, (1997).

AERODYNAMIC OPTIMIZATION ON UNSTRUCTURED MESHES WITH VISCOUS EFFECTS

Jonathan ELLIOTT [1]   Jaime PERAIRE [2]

1) Graduate Research Assistant, Member AIAA; Fluid Dynamics Research Laboratory MIT Department of Aeronautics and Astronautics
2) Associate Professor ;Fluid Dynamics Research Laboratory MIT

Abstract

A method is presented for performing lift-constrained drag minimization based on the 3D Euler equations and the 2D laminar Navier-Stokes equations. For the 3D Euler equations, the optimization system is based on a flow analysis scheme which uses gradients calculated on an elemental basis. For the 2D Navier-Stokes system, a change was made in the flow analysis scheme to one based on gradients calculated using a finite volume approach due to the fact that this scheme has drastically reduced memory costs associated with the storage of the residual jacobian, and allows direct extension to 3D Navier Stokes. Optimization exercises are presented which demonstrate the effectiveness of the optimization system in finding credible optimal geometries.

1. Introduction

Optimization based on the Euler equations is a useful capability for performing inverse pressure design of aerodynamic bodies especially for high Reynolds number flows where boundary layers are thin and the decambering associated with their displacement surfaces causes only small shock movements and lift changes. However almost any drag minimization exercise based on the Euler equations and applied to modern supercritical wings in cruise condition is doomed to failure. This is because the baseline wings have been designed to maximize L=D by pushing the upper surface shock as far aft as possible - at least over the highly-loaded outboard portion of the wing where the transonic effects are the most severe and where the wave drag is usually the highest. The beneficial lift improvement associated with this aft movement of the shock is counterbalanced by an increasingly severe adverse pressure gradient in the recovery region which typically, at shock positions of around 60 % a large role are those associated with high lift takeoff and landing configurations. In these cases, accurate modelling of the outer flow is often not possible even for fully attached flow, without inclusion of viscous effects . Typical regimes feature boundary layers whose displacement thickness form a significant proportion of the gaps (between elements) through which they pass. Hence these boundary layers have a large impact on the beneficial lift improvement often found by the interaction between elements. (This interelement interaction is heavily exploited by aerodynamic design engineers – but typically using unsatisfactory tools such as 2D viscous or 3D inviscid CFD tools, supported by copious expensive wind tunnel testing.) Furthermore local areas of separation are often present in these regimes, and indeed the maximum lift is often set by the onset of trailing edge separation on one or more elements. Further complications include off the surface slat wake flow reversal which can have a large impact on the lift achieved by landing configurations.

One effective way of including viscous effects is to solve a coupled viscous/inviscid system incorporating, for example, a viscous solver which solves the integral boundary layer equations [8]. This can result in a system which generates solutions more efficiently than Navier-Stokes solvers which typically require many points to resolve the boundary layer, resulting in both higher cost per iteration due to more mesh degrees of freedom and more iterations to convergence due to the requirement of smaller timesteps for stability, for both explicit and implicit time relaxation schemes. There are three factors which stack up against this option. Firstly, for separated flow, use of the direct coupling method (which most naturally fits into our current algorithm) results in an unstable scheme. Fully simultaneous [8] and semi-inverse schemes are possible alternatives which are stable but each would require significant amounts of work to incorporate into our current (explicit or point implicit) relaxation scheme. Secondly, while impressive work has been done on fully 3D integral boundary layer solvers, [22, 20], confidence in the aerospace industry in their results has not yet reached the levels found for their 2D counterparts and some theoretical issues remain to be resolved such as the imposition of hyperbolic boundary conditions. Application of the 2D results via strip theory may introduce a further source of error. Thirdly, use of displacement surfaces to implement the effect of the boundary layer on the inviscid flow leaves open the difficulty of applying this surface at geometric intersections such as wing/body intersections or wing/strut intersections.

On the other hand solving the Navier-Stokes equations throughout the flowfield is a more natural extension of the Euler optimization capability developed by the authors [9, 10]. Also, this capability can serve as a benchmark for future work towards coupled viscous/inviscid solvers.

AIAA-97-1849: presented at 13th AIAA CFD Conference, Snowmass, Colorado, June 1997

Although the algorithm presented herein has only been implemented for the 2D Navier-Stokes equations so far, it represents a significant preliminary step towards our ultimate goal of 3D Navier-Stokes optimization. The features of the 2D algorithm allow direct extension to a practical 3D algorithm. In particular, the residual Jacobian storage scheme implemented here leads to an efficient implementation of a 3D Navier-Stokes optimization algorithm.

2 Algorithms

In the course of extending the Euler optimization capability discussed in [9, 10, 11] to Navier-Stokes, it was found that switching to a spatial discretization scheme which can be described as mixed finite volume/finite element [24, 25] allowed a far more memory-efficient sensitivity calculation scheme than one based on the original scheme with viscous contributions. Since we also present 3D Euler results here and since we discuss the relative merits of either scheme vis-à-vis the sensitivity calculation, we find it appropriate to devote some space to discussion of the main components of either spatial discretization scheme.

Flow Analysis

The first scheme can be considered to be a finite volume scheme directly extended from the inviscid solver discussed in [9, 10]. In this case, exact gradients required by both viscous and artificial dissipation fluxes are found on an elemental basis. This scheme will henceforth be referred to as "Scheme I". The second scheme can be considered to be a mixed finite element/finite volume scheme directly extended from the fundamental algorithm underlying FELISA [25], in which the average gradients over each nodal control volume are used for the artificial dissipation and viscous fluxes. We will henceforth refer to this scheme as "Scheme II". The only difference between the two schemes is the way in which the gradients are calculated.

For illustration purposes, the algorithms are described based on the 2D algorithm. The 3D algorithm is a direct extension thereof. Both schemes begin with the integral form of the Navier-Stokes equations:-

$$\int_\Omega \frac{\partial \mathbf{U}}{\partial t} d\Omega = \int_\Omega \frac{\partial}{\partial x_j}(\mathbf{G}^j - \mathbf{F}^j)d\Omega = \oint_\Gamma (\mathbf{G}^j - \mathbf{F}^j)n_j d\Gamma \quad (1)$$

where $\mathbf{U} = (\rho, \rho u_1, \rho u_2, \rho E)^T$,

$$\mathbf{F}^j = \begin{bmatrix} \rho u_j \\ \rho u_1 u_j + p\delta_{1j} \\ \rho u_2 u_j + p\delta_{2j} \\ u_j(p + \rho E) \end{bmatrix}, \mathbf{G}^j = \begin{bmatrix} 0 \\ \tau_{1j} \\ \tau_{2j} \\ u_i \tau_{ij} + k\frac{\partial T}{\partial x_j} \end{bmatrix} \quad (2)$$

and Ω is a closed control volume with boundary, Γ. Here τ_{ij} is the tensor of viscous stresses and is given by

$$\tau_{ij} = \mu\left(\frac{\partial u_i}{\partial x_j} + \frac{\partial u_j}{\partial x_i}\right) - \frac{2}{3}\mu\delta_{ij}\frac{\partial u_k}{\partial x_k} \quad (3)$$

We first assume that $\mathbf{G}^j = 0$ and present the underlying inviscid scheme which is almost identical for both. The discretization of $\mathbf{U}$ and $\mathbf{F}^j$ on an unstructured triangulation of the domain is accomplished using piecewise linear polynomials. The spatial discretization is completed by using the finite volume formulation with control volumes associated with each node i consisting of all triangles having vertex i. The line integration given by Equation (1) is performed exactly around the outer boundary of this control volume. (It is noted in passing that an identical discretization can be achieved by forming a Galerkin weighted residual statement using the same piecewise linear functions as the weight functions, and lumping the mass matrix entries onto the diagonal.) Using this spatial discretization, the following set of semi-discrete equations results:-

$$\frac{d\mathbf{U}}{dt} + \mathbf{R}(\mathbf{U}) = 0 \quad (4)$$

where, through the use of an edge-based formulation, the residual at node i can be written as

$$\mathbf{R}_i = \sum_{ik=1}^{K}[\mathbf{R}]_{ik}. \quad (5)$$

Here ik represents the edge connecting nodes i and k and the residual increment for an interior edge is given by,

$$[\mathbf{R}]_{ik} = \sum_j \frac{\mathbf{F}_i^j + \mathbf{F}_k^j}{2}S_{ik}^j - \mathbf{D}_{ik} \quad (6)$$

where S_{ik}^j is the area vector associated with edge ik and is calculated as follows:-

$$(S_{ik}^1, S_{ik}^2) = \left(\frac{x_{k+1}^2 - x_{k-1}^2}{3}, \frac{x_{k-1}^1 - x_{k+1}^1}{3}\right) \quad (7)$$

where $k+1$ and $k-1$ are the 3rd nodes in either element containing edge ik. The second term on the right hand side in Equation (6) is an artificial dissipation flux added to stabilize the scheme and is given by:-

$$\mathbf{D}_{ik} = \frac{S}{2}\mathbf{P}|\mathbf{\Lambda}|(\mathbf{P}^{-1}\Delta\mathbf{U}_{ik} - L(\mathbf{P}^{-1}\Delta\mathbf{U}^-, \mathbf{P}^{-1}\Delta\mathbf{U}^+)) \quad (8)$$

Here $\Delta\mathbf{U}^-$ and $\Delta\mathbf{U}^+$ represent changes in $\mathbf{U}$ along an imaginary line segment extended past the positive and negative ends of the edge, of equal length and direction to those of the edge itself, $x_k^j - x_i^j$. $L(u, v)$ in Equation (8) is a nonlinear averaging function, constructed to retain coefficient positivity for the characteristic differences, $\Delta\mathbf{W}^\pm = \mathbf{P}^{-1}\Delta\mathbf{U}^\pm$, and therefore the LED character of the numerical scheme [19, 10]. It should be noted that if an exact average is returned by $L(u, v)$, then the combination of characteristic differences inside the paranthesis in Equation (8) can be seen to form a third order dissipative flux. (Note that $\mathbf{P}$ is the matrix of eigenvectors and $\mathbf{\Lambda}$ is the matrix of eigenvalues for the Roe matrix, $\mathbf{A}(\mathbf{U}_i, \mathbf{U}_k)$ based on the unit normal associated with S_{ik}^j and $\mathbf{A}$ incorporates the Harten correction [16] to avoid zero eigenvalues.)

The method of calculation of $\Delta\mathbf{U}^-$, $\Delta\mathbf{U}^+$ in Equation (8) is different for Scheme I and Scheme II. For Scheme I,

$\Delta \mathbf{U}^-$, $\Delta \mathbf{U}^+$ are formed exactly from the gradients in the elements that either end of the edge points into [3, 19]. For example,

$$\Delta \mathbf{U}^+ = \sum_j (\frac{\partial \mathbf{U}}{\partial x^j})^+ (x_k^j - x_i^j) = \epsilon_{ir}\Delta \mathbf{U}_{ir} + \epsilon_{is}\Delta \mathbf{U}_{is} \quad (9)$$

where the nodes in the adjacent element have been labelled i, r, s and $\epsilon_{ir}, \epsilon_{is}$ are, by construction, always positive. This feature is also critical in ensuring the LED nature of the scheme.

For Scheme II, $\Delta \mathbf{U}^-$, $\Delta \mathbf{U}^+$ are also calculated from the gradients, but here the gradients are calculated as averages over the control volumes:-

$$\frac{\partial \mathbf{U}_i}{\partial x^j} = \frac{1}{V_i}(\sum_{il=1}^{\mu_i} \frac{\mathbf{U}_i + \mathbf{U}_l}{2} S_{il}^j + \sum_{il=1}^{n_{bou}} \frac{5\mathbf{U}_i + \mathbf{U}_l}{4} S_{il}^l) \quad (10)$$

where μ_i is the number of nodes neighbouring node i, n_{bou} is the number of boundary edges contributing to the residual at node i and subscript il represents the edge connecting nodes i and l. The rigorous LED character of the scheme is lost but Scheme II tends to produce smoother and less dissipative solutions than Scheme I for transonic and low supersonic flows.

For both schemes, the means for inclusion of viscous terms recycles the nodal gradients used for the dissipative fluxes. For Scheme II, the edge-based data structure can be retained, and $\mathbf{F}_i^j$, for example, in Equation (6) is simply replaced by $\mathbf{F}_i^j - \mathbf{G}_i^j$. For Scheme I, on the other hand, inclusion of viscous effects demands that we return to an element-based data structure, whether the viscous terms are included via the Finite Volume or the Finite Element method [26]. This has a detrimental impact on the memory requirements for both flow analysis and sensitivity analysis (as pointed out in the next section).

Also note that for the Scheme II problems presented here, which are all Navier-Stokes problems, it was found desirable to use a smoother limiter function in Equation (8), to avoid the limiter switching to first order in regions of high curvature, such as the edge of the boundary layer. The limiter used was that employed in [15] and is

$$L(u, v) = minmod(2u, 2v, \frac{u + v}{2}). \quad (11)$$

Note that this limiter is more likely to return the third argument, and by examination of Equation (8), results in a scheme with a desirable third order dissipative flux. For the 3D Euler optimization problem, the standard minmod function was used.

Mesh Movement

For the 3D Euler calculations presented here, the interconnected springs algorithm was used to accomplish mesh movement. For the Navier-Stokes calculations, we reverted to a scheme in which the grid movement at a location x_j is simply given by the sum of the airfoil surface modal perturbations corresponding to that value of x_1. This eliminates

the potential problem of negative volume elements in the perturbed grid in the boundary layer due to the small mesh spacing in the normal direction. This will obviously have to be dealt with in a more satisfactory manner when extending to multi-element airfoils and 3D Navier-Stokes.

Sensitivity Analysis

Consider, for the purposes of simplicity of notation, that the design variable β contains only one component and so can be written as a scalar β. Furthermore, assume that for each condition j, the drag coefficient is given by

$$C_{D_j} = C_{D_j}(\mathbf{U}_j(\beta), \beta). \quad (12)$$

Differentiating the steady state version of Equation (4) and Equation (12) produce, respectively,

$$\frac{\partial \mathbf{R}}{\partial \mathbf{U}} \frac{\partial \mathbf{U}}{\partial \beta} = -\frac{\partial \mathbf{R}}{\partial \beta} \quad (13)$$

and

$$\frac{dC_{D_j}}{d\beta}\bigg| = \frac{\partial C_{D_j}}{\partial \beta} + \frac{\partial C_{D_j}}{\partial \mathbf{U}} \frac{\partial \mathbf{U}}{\partial \beta} \quad (14)$$

The first term on the right-hand side of (14) can be calculated relatively easily. The second term is far more difficult. Two approaches are the direct and the adjoint method.

The direct method involves *direct* solution of Equation (13) and substitution of the resulting vector $\partial \mathbf{U}/\partial \beta$ into Equation (14). The means by which Equation (13) is solved is discussed below. Note that Equation (13) must be solved once for each component of β.

The adjoint method is based on the recognition that Equations (13) and (14) can be combined to give:

$$\frac{dC_{D_j}}{d\beta}\bigg|_{\mathbf{R}=0} = \frac{\partial C_{D_j}}{\partial \beta} - \left[\frac{\partial C_{D_j}}{\partial \mathbf{U}}\right]^T \left[\frac{\partial \mathbf{R}}{\partial \mathbf{U}}\right]^{-1} \left[\frac{\partial \mathbf{R}}{\partial \beta}\right] \quad (15)$$

Due to associativity of matrix multiplication, the right-hand-side double product calculation may be performed either by first multiplying the Jacobian inverse by the term to its right – which gives the *direct* method – or by the term to its left – which gives the *adjoint* method. It is convenient to write the result of the intermediate calculation for the adjoint method as $-\psi$ and therefore ψ satisfies

$$\left[\frac{\partial \mathbf{R}}{\partial \mathbf{U}}\right]^T \psi = -\left\{\frac{\partial C_{D_j}}{\partial \mathbf{U}}\right\}^T \quad (16)$$

and

$$\frac{dC_{D_j}}{d\beta}\bigg|_{\mathbf{R}=0} = \frac{\partial C_{D_j}}{\partial \beta} + \psi^T \left\{\frac{d\mathbf{R}}{d\beta}\right\} \quad (17)$$

where ψ is called the *adjoint* variable.

Time Integration

It is possible to adopt exactly the same relaxation algorithm to solve equations (4), (13) and (16). This is done by introducing an artificial unsteady term into, for example, the adjoint equation.

$$\frac{d\psi}{dt} + \left[\frac{\partial \mathbf{R}}{\partial \mathbf{U}}\right]^T \psi - \left[\frac{\partial C_{D_j}}{\partial \mathbf{U}}\right]^T = 0 \qquad (18)$$

These schemes possess the same stability properties because the spectral radii corresponding to the direct and adjoint sensitivity analysis schemes are identical to that of the linearized flow analysis scheme which governs its asympotic convergence behaviour. The spectral radius of the error-mode amplification matrix implied by Equation (18) is obviously the same as that for the linearized flow analysis.

In the current research we have used a multi-stage explicit time-stepping scheme for the Euler calculations. For the Navier-Stokes calculations it was felt that an implicit scheme was needed due to the small sizes of the elements in the boundary layer in the normal direction. For preliminary studies a point implicit time-stepping algorithm scheme was used. It was found to allow at least a doubling of the asymptotic convergence rate for one fourth of the per-iteration CPU cost of the multistage scheme for the adjoint calculation. For a given problem we always use the same time marching algorithm for both flow and adjoint analysis.

The point implicit algorithm follows the ideas elucidated in [15, 14], but differs in that the exact block diagonal entries of the $\partial \mathbf{R}/\partial \mathbf{U}$ matrix are used on the left hand side as opposed to only the first order terms. In brief, the relaxation scheme is

$$\mathbf{L}_i \Delta \mathbf{U}_i = (\Omega_i + \Delta t_i \left.\frac{\partial \mathbf{R}}{\partial \mathbf{U}}\right|_{ii}) \Delta \mathbf{U}_i = \Delta t_i \mathbf{R}_i \qquad (19)$$

where

$$\mathbf{U}_i^{n+1} = \mathbf{U}_i^n + \Delta \mathbf{U}_i. \qquad (20)$$

To extend this timestepping scheme to the adjoint system, it is necessary to replace $\mathbf{L}$ with $\mathbf{L}^T$, since this ensures that the eigenvalues of the error amplification matrix remain the same as for linearized flow analysis. Justification for taking this approach was verified when it was observed that flow, direct sensitivity and adjoint analysis calculations exhibited identical asymptotic convergence behaviour, as can be seen in Figure 5.

The effect of the no-slip boundary conditions is to eliminate the momentum equation residuals at the wall from the system to be solved. Therefore the adjoint variables (which are in essence Lagrange multipliers for each nodal residual equation[11, 17]) corresponding to the momentum equations at the wall should have no impact on the final adjoint solution elsewhere. This was recently pointed out in reference [2]. This has been accomplished for our adjoint solver by performing an à posteriori sweep on the wall boundary nodes in which the momentum adjoint variables are zeroed out.

Finally, a viscous correction was made to the Δt_i in an effort to ensure that the viscous time step limit is not exceeded, thereby compromising stability. This correction is given by

$$\Delta t_i = \frac{\Delta t_i^{inv}}{1 + 4/Re_{\Delta s_i}} \qquad (21)$$

where Δt_i^{inv} is the time step limit found using linearized Fourier analysis for the inviscid scheme [12] and $Re_{\Delta s_i}$ is the Reynolds number based on the length of the smallest edge in the elements surrounding node i. It has been found by many researchers that this stability limit can have a large influence for well-resolved, low Reynolds number flows. For the 3D solutions presented here, Δt_i^{inv} also included a factor to ensure satisfaction of the TVD CFL-like condition [9].

Jacobian storage

One of the main obstacles to sensitivity calculation based on the Euler and Navier-Stokes equations using the direct and adjoint approaches is that associated with the large memory cost of storing the residual Jacobian matrix, $\partial \mathbf{R}/\partial \mathbf{U}$. The matrix is usually sparse due to the fact that the finite volume or finite element residual statements based on piecewise linear variation of the state vector, $\mathbf{U}$, result in only the immediate neighbours having an influence on $\mathbf{R}_i$, the residual at node i (an exception would be a spectral scheme). Dissipative fluxes required for stability and shock capturing can extend the stencil to include the neighbours of the neighbours depending on the formulation used. Nonetheless, it is obviously wasteful to store the whole sparse matrix, $\partial \mathbf{R}/\partial \mathbf{U}$.

One obvious improvement is to store the Jacobian entries on an edge basis. This results in the storage costs scaling as $O(N)$ rather than $O(N^2)$ for the full matrix. However the costs are still large and prohibitive for 3D cases unless a parallel architecture is used.

Several approaches have been used in the past to circumvent these memory cost problems. In some cases [5], schemes have been limited to first order accuracy in order that $\partial \mathbf{R}/\partial \mathbf{U}$ may be more easily stored due to the smaller stencil. An alternative option is to resort to schemes in which the entries in $\partial \mathbf{R}/\partial \mathbf{U}$ are recalculated "on-the-fly" [21, 9]. The authors found that although this obviously provides enormous memory savings, it results in large CPU cost increases [11] – about a factor of four increase for our 2D and 3D Euler schemes.

Use of the continuous sensitivity analysis approach [18, 27] avoids these problems since the relaxation scheme used for solving the flow analysis system can be recycled to solve the adjoint problem. Stability can be provided either by reverse biasing of the difference operators due to the reversed direction of the zone of dependence for the adjoint problem or simply by addition of a dissipative flux of the same form as that for the flow analysis scheme. Of course, one of the drawbacks of the continuous sensitivity analysis approach is that calculated sensitivities cannot be exact except at the limit of an infinitely fine grid.

The extension to Navier-Stokes introduces further complications. The rationale for using Scheme I [9, 10] originally was that a very small stencil results for the Euler equations with the resulting low memory costs for $\partial \mathbf{R}/\partial \mathbf{U}$. Unfortunately, when viscous stresses and heat fluxes are included when extending to the Navier-Stokes equations, Scheme I loses its small stencil advantage. (This is because exact integration of, for example, the viscous fluxes results in dependencies on the 3rd (and 4th in 3D) node in the 2 elements containing a given edge.) Switching to an element-based data structure because of this, causes an increase in memory for both flow analysis and a large increase in the size of $\partial \mathbf{R}/\partial \mathbf{U}$ for sensitivity analysis.

At first glance, reverting to the control-volume-average gradients in Scheme II would appear to have a still higher memory cost for $\partial \mathbf{R}/\partial \mathbf{U}$ storage especially in 3D. One might erroneously arrive at this conclusion because the residual contribution for a given edge is dependent on the nodal gradients at either end of the edge, and the gradient at node i is in turn dependent on its immediately adjacent neighbours as shown in Equation (10). Therefore the residual contribution for edge ik can be expressed as:-

$$[\mathbf{R}]_{ik} = [\mathbf{R}]_{ik}(\mathbf{U}_i, \mathbf{U}_k, \mathbf{U}_{i1}, \cdots, \mathbf{U}_{i\mu_i}, \mathbf{U}_{k1}, \cdots, \mathbf{U}_{k\mu_k}) \quad (22)$$

$[\mathbf{R}_{ik}]$ can be dependent on as many as $(2 \times 10 + 2)$ nodes in 2D and $(2 \times 40 + 2)$ nodes in 3D. Direct calculation of each dependency obviously results in a prohibitive memory cost for $\partial \mathbf{R}/\partial \mathbf{U}$.

This problem can be avoided by calculating $\partial \mathbf{R}/\partial \mathbf{U}$ in two stages. It can be seen from Equations (2), (6) and (8) that the residual contribution for a given edge can be written as

$$\mathbf{R}(\mathbf{U}) = \mathbf{R}(\mathbf{U}, \mathbf{U}_{x_i}; \mathbf{U}) \quad (23)$$

Following numerous previous researchers, e.g.[4], we apply the chain rule and obtain,

$$\begin{aligned}
\frac{d\mathbf{R}}{d\mathbf{U}} &= \frac{\partial \mathbf{R}}{\partial \mathbf{U}_I} + \frac{\partial \mathbf{R}}{\partial \mathbf{U}_{II}} \\
&= \frac{\partial \mathbf{R}}{\partial \mathbf{U}_I} + \frac{\partial \mathbf{R}}{\partial \mathbf{U}_{x_j}} \frac{\partial \mathbf{U}_{x_j}}{\partial \mathbf{U}}
\end{aligned}$$

where we have used the shorthand $\mathbf{U}_{x_j} = \partial \mathbf{U}/\partial x_j$. This doesn't appear to be much of a savings since $\partial \mathbf{U}_{x_j}/\partial \mathbf{U}$, at first glance, appears to be quite large. However, it is quickly found after examining the expression used to calculate $\mathbf{U}_{x_j}$, Equation (10), that each conservative variable component is only dependent on neighbouring nodal values of that same conservative variable component. Therefore $\partial \mathbf{U}_{x_j}/\partial \mathbf{U}$ is only nonzero if the same conservative variable components are being compared. Also $\partial \mathbf{U}_{x_j}/\partial \mathbf{U}$ is the same for all conservative variable components. Therefore $\partial \mathbf{U}_{x_j}/\partial \mathbf{U}$ is quite small, and $\partial \mathbf{R}/\partial \mathbf{U}$ can be stored using about the same amount of memory as was required to store the Euler residual jacobian based on the Scheme I discretization.

Optimization Strategy

A subspace BFGS algorithm is used to minimize drag with constraints on lift for all three examples discussed herein. The algorithm is elucidated in [11], but we will give a brief description. An increasingly accurate estimate to the Hessian (of the cost function with respect to the design variables) is developed in the subspace, $\mathbf{Z}$, orthogonal to the constraint gradients. This estimate is based on the usual BFGS formula, but uses variables which are all projected into the subspace [13]. The matrix $\mathbf{Z}_k$ which is used to perform the projection is found by performing an LQ decomposition of the constraint gradients, $\partial C_L/\partial \beta_j$. This formulation requires the solution of two adjoint problems per line search – one for C_L and one for C_D. The method appears to be quite robust since, as for the unconstrained BFGS algorithm, the Hessian approximation is guaranteed to remain positive definite provided the exact Hessian is positive definite and provided sufficient progress is made towards the minimum for each line search [13].

3 Validation

Flat plate flow analysis validation The grid used is shown in Figure 2. No slip boundary conditions are applied along the lower boundary from $x = 0$ to $x = 1.5$; free slip boundary conditions are applied on the lower boundary from $x = -1.5$ to $x = 0$; non-reflecting Riemann boundary conditions are applied at the left boundary; pressure is specified while the appropriate characteristic variables are extrapolated from the interior along the top and right boundaries. Flow is from left to right.

This test case or variations of it have been extensively tested by the second of the authors and other researchers [24, 1, 28]. At sufficiently large distances from the leading edge, the boundary layer profile is expected to closely approximate the Blasius profile – provided sufficient spatial resolution is used. The profiles found at $x = 1.215$ using Scheme II for successively coarser grids (coarsened by removing every other vertical and horizontal in multigrid fashion [1]) are shown in Figure 1. It can be seen that the solution spatially converges to a close approximation of the Blasius solution.

Airfoil flow validation A calculation was performed, once again using Scheme II, for the NACA0012 with freestream conditions of $M = 0.8, \alpha = 3.5°, Re = 2000$. The grid is shown in Figure 27. The C_p and M distributions are shown in Figures 28 and 29. There are 7081 points in the grid and 17 points in the boundary layer at the trailing edge, although only the first 9 elements are "structured" [23]. This small number of "structured" elements in the boundary layer is the cause of the wiggles observable in the C_p distribution. In spite of these low level errors, surface C_p and C_f distributions agree quite well with the corresponding distributions as found along the displacement surface by MSES, a coupled Euler/integral boundary layer code [6]. This can be seen in Figures 3-4. On the lower surface, agreement is quite good, while on the upper surface, the differences are due to the non-zero normal pressure gradient through the boundary layer. The C_p difference between displacement surface and the wall was sampled at several locations in the Navier-Stokes solution and found to agree quite closely with the differences observed in Figure

3. Further validation work at higher Reynolds numbers is currently in progress to confirm this hypothesis.

Airfoil sensitivity validation Sensitivity calculations were performed based on the solution described in the previous paragraph. As expected, and as found for the corresponding Euler calculations, asymptotic convergence rates for flow, adjoint and sensitivity analyses were very close as can be seen in Figure 5. Comparison of the resulting $\partial \rho u / \partial \alpha$ distribution (Figure 6) with that found using the finite difference method based on $\Delta \alpha = 0.01 \deg$ (Figure 8), reveals good agreement. Comparison of $\partial C_l / \partial \beta$ and $\partial C_d / \partial \beta$ as found by finite difference, adjoint and direct methods are shown in Table 1 below for design variables of α and a NACA 4-series meanline camber mode. Quite good agreement is found for α. The discrepancies for the camber mode are believed to be due to the finite difference step size used in making that estimate. $\Delta \beta_c = 0.001$ was used corresponding to a movement in the airfoil surface of 0.1% of chord.

To put a proper perspective on the source of this discrepancy, much can be learned by examining the sensitivities of the only flow quantities that contribute to the lift and drag. Surface pressure sensitivities are compared in Figure 7 revealing good agreement. Surface values of $\partial \tau_{11} / \partial \beta$ are shown in Figure 9 and also reveal good agreement. Similarly good agreement is observed for τ_{12} and τ_{22}. Since these 4 scalar values are the only ones that contribute to C_l and C_d, $\partial C_l / \partial \beta$ and $\partial C_d / \partial \beta$ are expected to show similarly good agreement. The larger than expected discrepancies shown in Table 1 are presumably due to the effect of small errors when the integration is performed. The difference of two large numbers is taken to obtain a small number, a scenario which is conducive to error magnification. Nonetheless it would probably be concluded at this juncture that the analytic derivatives are the ones with higher accuracy after examination of Figures 7 and 9. Further tests in the course of the optimization exercises by comparison of $\Delta \beta_j \partial C_l / \partial \beta_j$ with ΔC_l and $\Delta \beta_j \partial C_d / \partial \beta_j$ with ΔC_d provided further definitive evidence in this regard.

β	$\partial C_l / \partial \beta \vert_{FD}$	$\partial C_l / \partial \beta \vert_{adj}$	$\partial C_l / \partial \beta \vert_{dir}$
camber	-0.9086	-1.0075	-1.0075
α	0.04707	0.04811	0.04811

	$\partial C_d / \partial \beta \vert_{FD}$	$\partial C_d / \partial \beta \vert_{adj}$	$\partial C_d / \partial \beta \vert_{dir}$
camber	-0.0385	-0.04902	-0.04917
α	0.00627	0.00634	0.00634

Table 1: Comparison of sensitivities to α and NACA 4-series meanline mode

4 Results

3D Euler, Multipoint, lift-constrained drag minimization exercise For this case, Scheme I was used. This 3D case [11] is a double point drag-minimization exercise in which the cost function is given by $F = 0.5(C_{D_1} + C_{D_2})$. The two flight conditions are ($M_{\infty_1} = 0.9, C_{L_1} = 0.450$) and ($M_{\infty_2} = 1.6, C_{L_2} = 0.125$). The baseline geometry is a wing-body configuration with an area-ruled body. The wing has a low aspect ratio ($AR = 2.67$), with leading edge sweep, $\Lambda_{LE} = 45°$, trailing edge sweep, $\Lambda_{TE} = 0°$. It has uniform airfoil maximum thickness of 5% of chord, has uniform twist of $\phi = 0°$ and is uniformly uncambered across the span. The baseline grid contains 1.13 million elements and 211,000 points and is shown in Figure 10.

The optimization was performed using twelve design variables. Two of these were α_1 and α_2, the angles of attack at each condition. The other ten were combinations of four spanwise functions and three chordwise functions (twist and two camber modes). The adjoint method based on discrete sensitivity analysis was used to calculate sensitivities. These sensitivities and the more-easily calculated objective functions were fed into a BFGS optimization algorithm. After 7 BFGS iterations, the optimization process has practically converged, with a reduction of around 5.5% in the cost function.

Baseline and final C_p distributions are shown in Figures 11-14 for either condition while sectional geometry and C_p evolution are plotted in Figures 16-18. As expected, the nose has pitched down, the angle of attack has been reduced (resulting in a lower leading edge peak and a weakened or vanished leading edge shock), while the lift has been recovered by a camber increase along the aft part of the wing which is accompanied by an aft movement of the shock. It should be noted however, that the positioning of the shock this far aft would spell trouble in the physically realistic situation due to the presence of viscous effects. At normal Reynolds numbers, one would expect that either shock-induced boundary layer separation would occur or that separation would occur in the severe adverse pressure gradient in the recovery region.

The supersonic drag actually increases slightly but by a far smaller amount than the transonic drag decrease. The reader is referred to reference [11] for a more thorough exposition of the details of this optimization exercise, including design variable definitions, etc.

2D Attached Viscous Optimization Case Scheme II was used for flow analysis for this problem. The baseline geometry was a very thin airoil (with about 1% maximum thickness), with a thickness distribution that had already undergone considerable design and optimization for low Reynolds number viscous flow, albeit at a slightly higher Re=10000. The original airfoil is under consideration for use in micro unmanned aerial vehicles (μ-UAVs). It had significant camber, but it was decided that a well behaved zero-camber airfoil was required as an initial test for the optimization system, so the camber was removed from the airfoil by a simple geometric operation. At $\alpha = 3.5°$, $M = 0.8$, and $Re = 2000$ it was found that the boundary layer C_f distribution was well above zero everywhere. The high Mach number was chosen, not for its realism in representing the flight conditions of μ-UAVs, but because it allows quicker convergence of the flow analysis calculations. It was decided that drag minimization with a lift constraint was an appropriate initial test, even though not much of a drag reduction was expected. The design variables were chosen to be the modal amplitudes of Hicks-Henne camber functions of the form $\Delta y(x) = \sin(\pi x^p)$ or $\Delta y(x) = \sin(\pi(1 - x)^p)$ with p chosen to be such that peaks occur at 20%, 40%, 60% and 80% of chord. In addition, the angle of attack was chosen as a design variable.

The baseline grid, pressure distribution and Mach distribution are shown in Figures 19-21. The baseline grid contains 8506 nodes and at the trailing edge there are approximately 20 points in the boundary layer. However for this case, the elements with "structure" in the boundary layer number about 16, which is the probable cause of the reduced presence of wiggles in the solution compared to the NACA0012 described in the previous section. In spite of the low level error associated with these wiggles, the surface C_p distribution was found to match that found by MSES even more closely than the match shown in Figures 3-4. Note that although there is locally supersonic flow, no shocks are present, although the ability to capture these shocks is an inherent part of the flow analysis algorithm.

The final C_p distribution is shown in Figure 22 while the evolution of design variables, C_f distribution, C_p distribution and geometry are shown in Figures 23-26. It can be seen that the minimum appears to have been almost reached after about 4 line searches which is about what should be expected for a well behaved design surface. The cost function has only decreased by 18 counts or about 2.2% of the baseline C_d. This modest decrease was expected since the baseline solution exhibited quite healthy boundary layer behaviour. Note that the decrease has come about partially by a reduction of the leading edge pressure peak. The resulting reduction in the net pressure increase and adverse pressure gradient along the upper surface causes the momentum thickness to be lower at the upper surface trailing edge, indicating a more healthy boundary layer and lower drag.

Also it should be noted that this optimization exercise was reproduced using MSES and and LINDOP, its associated optimization driver [7]. A similar reduction in C_d was observed, although the final geometry was slightly different. It is believed that this is due to the relatively shallow minimum.

2D Separated Viscous Optimization Case Scheme II is also used for flow analysis for this problem. This case is a more challenging one due to the presence of separation over the aft 60% of the upper surface. However, more of a drag reduction is expected due to the poor initial health of the upper surface boundary layer. It was felt that thickness design variables would possess powerful leverage over setting the separation point on the upper surface, so two thickness design variables with peaks at 20% and 40% were chosen as well as one camber design variable with a peak at 50%, along with the angle of attack. In addition to the lift constraint, for this case, it was decided that an area constraint should also be imposed. Otherwise, the airfoil would be driven uselessly to zero thickness. The baseline grid, C_p and M distributions are shown in Figures 27-29.

It should also be noted that this optimization exercise did not proceed nearly as smoothly as the one discussed in the previous paragraph. The presence of the separated flow seemed to cause two major undesirable effects. Firstly, as soon as the flow separates on both lower and upper surface, the flow becomes unsteady due to the onset of vortex shedding, and no fully converged flow or adjoint solution can be found. The way this undesirable feature was dealt with in the optimization algorithm was to assume a very large value of the objective function at that point and thereafter to trace back up the line search direction until a steady solution could be found, and then to search for a new direction from there. The second undesirable effect was that the presence of separation caused larger deviations from quadratic behaviour than usual. This appears to have slowed down the convergence of the approximate Hessian to the value found at the minimum. It is not clear that a good approximation has been made for the final design point depicted. The optimization process was stopped after 3 line searches when successive iterations resulted in the search direction not changing significantly with the design poised to enter the part of design space in which the above-mentioned vortex-shedding unsteadiness occurs.

The final C_p distribution is shown in Figure 30 while the evolution of design variables, C_f distribution, C_p distribution and geometry are shown in Figures 31-34. The cost function for this case has decreased a far larger 120 counts or about 10.6% of the baseline C_d. This larger decrease was expected due to the presence of separation in the baseline solution. Note that the decrease has been accomplished partially by a reduction of the leading edge pressure peak. Also the thickness distribution has been redistributed such that the maximum is about 15% further aft. Like the early natural laminar flow airfoils [29], this delays the start of the adverse pressure gradient to aft of that maximum thickness point. Consequently the upper surface separation point moves from 45% to about 65%. This movement appears to be limited by the appearance of separation on the lower surface and the resulting above-mentioned unsteadiness as can be seen in Figure 32.

Agreement with a similar optimization exercise performed using LINDOP and MSES revealed similar initial behaviour although the LINDOP geometry evolved to one with negative camber – a configuration not allowed by the current Navier-Stokes optimizer due to the intervening onset of vortex shedding as the maximum camber passes through zero.

5　Conclusion

The problem addressed was the implementation of a capability to perform lift-constrained drag minimization on unstructured grids. To this end a previously described Euler optimization system was extended to perform this task in a multipoint fashion for 2 flight conditions. This study produced credible optimal geometries although it pointed to the need to include viscous effects in order to avoid unrealistic adverse pressure gradients.

Subsequently, a 2D laminar Navier-Stokes optimization capability was implemented and tested on two cases. One involved fully attached flow while the other contained an extensive region of separated flow. In spite of some complications unearthed in the course of the separated flow case – due to nonsmoothness of the design space surface and unsteadiness for some regions of design space – these studies confirm the effectiveness of the optimization system by also producing credible optimal geometries.

As a result of this study, it now appears that 2D laminar optmization is a capability that is well in hand. Extension of the capability to one based on 3D turbulent flow analysis is the problem area that should now be addressed to allow a practical industrial tool to evolve. It should be noted that most features of the overall approach extend directly as is to 3D Navier-Stokes. For example, the residual jacobian storage system developed here removes this as an obstacle to 3D Navier-Stokes optimization. However, grid generation, time integration and turbulence modelling are key areas on which attention will need to be focussed in the immediate future.

Acknowledgements

We would like to express our thanks to Professor Mark Drela for continuing invaluable assistance at various stages of this project and to Bill Herling and Fritz Roetman of the Boeing Company for many helpful discussions and for providing the 3D geometry. The authors would like to thank the Flight Test Information Systems Branch of the NASA Dryden Research Center (under grant NAG4-100) and the Aerothermodynamics Branch of the NASA Langley Research Center (under grant NAG-1-1587) for partially supporting this research.

References

[1] S. R. Allmaras. Contamination of Laminar Boundary Layers by Artificial Dissipation in Navier-Stokes Solutions. In *Proc. of the ICFD Conference on Numerical Methods for Fluid Dynamics*, Reading, England, UK, 1992.

[2] W. Anderson and V. Venkatakrishnan. Aerodynamic Design Optimization on Unstructured Grids with a Continuous Adjoint Formulation. AIAA-97-0643, 1997.

[3] P. Arminjon and A. Dervieux. Construction of TVD-like artificial viscosities on 2-dimensional arbitrary FEM grids. INRIA Report 1111, 1989.

[4] R. Beam and R. Warming. An Implicit Factored Scheme for the Compressible Navier-Stokes Equations. *AIAA Journal*, pages 393–402, April 1978.

[5] F. Beux and A. Dervieux. Exact-gradient shape optimization of a 2-D Euler flow. *Finite Elements in Analysis and Design*, 12:281–302, 1992.

[6] M. Drela. Newton Solution of Coupled Viscous/Inviscid Multielement Airfoil Flows. AIAA-90-1470, 1990.

[7] M. Drela. Design and Optimization Method for Multi-Element Airfoils. *AIAA*, 93-0969, 1993.

[8] M. Drela and M. Giles. Viscous-inviscid analysis of transonic and low Reynolds number airfoils. *AIAA Journal*, 25(10):1347–1355, 1987.

[9] J. Elliott and J. Peraire. Aerodynamic Design using Unstructured Meshes. AIAA-96-1941, 1996.

[10] J. Elliott and J. Peraire. Practical 3D Aerodynamic Design and Optimization using Unstructured Meshes. *AIAA Journal*, accepted for publication, 1997.

[11] J. Elliott and J. Peraire. Three-Dimensional, Multipoint, Lift-Constrained Drag Minimization using Unstructured Meshes. *Aeronautical Journal of the Royal Aeronautical Society*, submitted for publication, 1997.

[12] M. Giles. An Energy Stability Analysis of Multi-step Methods on Unstructured Meshes. Technical Report 87-1, MIT CFD Lab, 1987.

[13] P. E. Gill, W. Murray, and M. H. Wright, editors. *Practical Optimization*. Academic Press, 1981.

[14] P. Gnoffo, R. Candless, and H. Yee. Enhancements to Program Laura for Computation of Three-Dimensional Hypersonic Flow. AIAA Paper 87-0280, January 1987.

[15] P. A. Gnoffo. An Upwind-Biased Point-Implicit Relaxation Algorithm for Viscous, Compressible Perfect Gas Flows. TP 2953, NASA, February 1990.

[16] A. Harten and J. Hyman. Self adjusting grid methods for one-dimensional hyperbolic conservation laws. *Journal of Computational Physics*, 50:235–269, 1983.

[17] A. Iollo, G. Kuruvila, and S. Taasan. Pseudo-Time Method for Optimal Shape Design using the Euler Equations. Technical Report 95-59, ICASE, 1995.

[18] A. Jameson. Aerodynamic Design via Control Theory. *Journal of Scientific Computing*, 3:233–260, 1988.

[19] A. Jameson. Artificial Diffusion, Upwind Biasing, Limiters and their Effect on Accuracy and Multigrid Convergence in Transonic and Hypersonic Flows. AIAA-93-3359, 1993.

[20] B. H. Mughal. A Calculation Method for the Three-Dimensional Boundary-Layer Equations in Integral Form. Master's thesis, Dept. of Aeronautics and Astronautics, MIT, September 1992.

[21] J. Newman and A. Taylor. Three-dimensional Aerodynamic Shape Sensitivity Analysis and Design Optimization Using the Euler Equations on Unstructured Grids. AIAA-96-2464, 1996.

[22] B. A. Nishida. *Fully Simultaneous Coupling of the Full Potential Equation and the Integral Boundary Layer Equations in Three Dimensions*. PhD thesis, Dept. of Aeronautics and Astronautics, MIT, 1995.

[23] J. Peraire and K. Morgan. Unstructured Mesh Generation Including Directional Refinement for Aerodynamic Flow Simulation. *Finite Elements in Analysis and Design*, 25:343–356, 1997.

[24] J. Peraire, K. Morgan, M. Vahdati, and J. Peiro. The Construction and Behaviour of Some Unstructured Grid Algorithms for Compressible Flows. In *Proc. of the ICFD Conference on Numerical Methods for Fluid Dynamics*, Reading, England, UK, 1992.

[25] J. Peraire, J. Peiro, and K. Morgan. Finite Element Multigrid Solution of Euler Flows past Installed Aeroengines. *Computational Mechanics*, 11:433–451, 1993.

[26] J. Peraire, J. Peiro, and K. Morgan. Multigrid Solutions of the 3D Compressible Euler Equations on Unstructured Tetrahedral Grids. *International Journal for Numerical Methods in Engineering*, 36:1029–1044, 1993.

[27] J. Reuther and A. Jameson. Aerodynamic Shape Optimization of Wing and Wing-Body Configurations Using Control Theory. AIAA-95-0123, January 1995.

[28] S. Tatsumi, L. Martinelli, and A. Jameson. Flux-Limited Schemes for the Compressible Navier-Stokes Equations. *AIAA Journal*, 33:252–261, 1995.

[29] A. Von Doenhoff and A. Abbot. *Theory of Wing Sections*. Dover Publications, Inc, Mineola, NY, 1959.

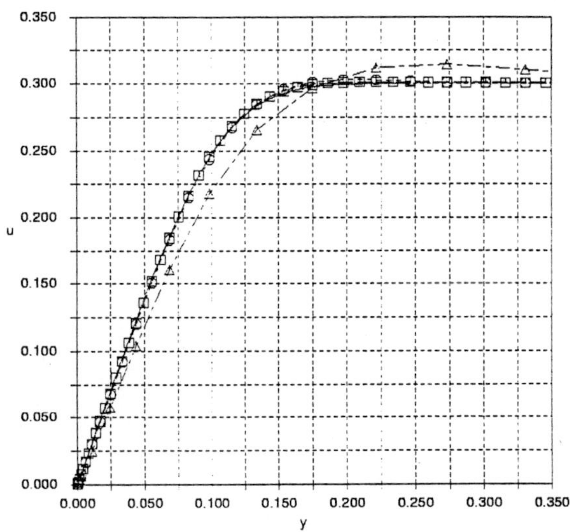

Figure 1: Spatial convergence of boundary layer profile at x=1.2 with local Blasius profile

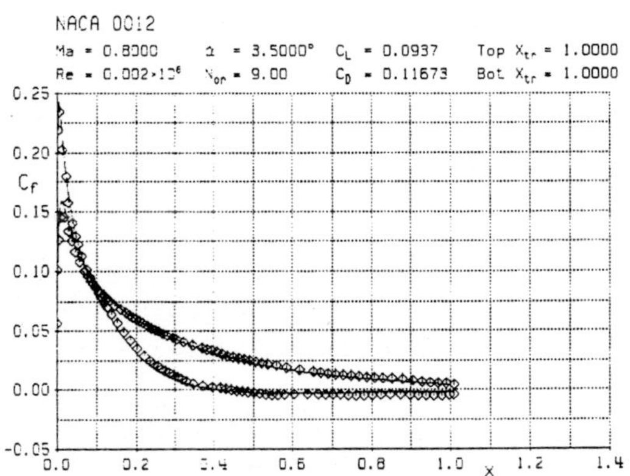

Figure 4: Surface C_f distribution for NACA0012, Re=2000, $M_\infty = 0.8$, $\alpha = 3.5$ deg: Scheme II and MSES solutions

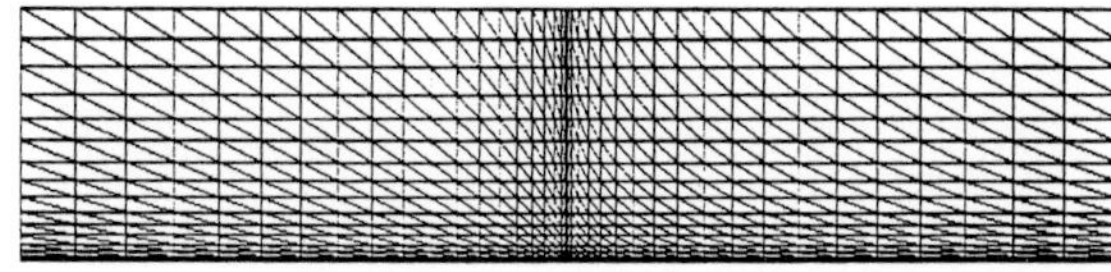

Figure 2: 41x17 grid used for flat plate calculations

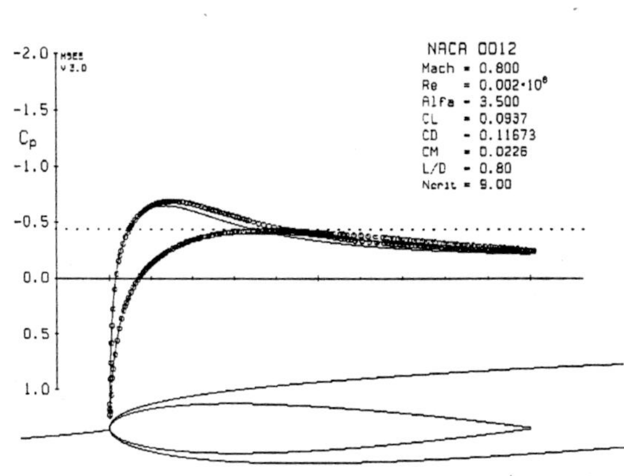

Figure 3: Surface C_p distribution for NACA0012, Re=2000, $M_\infty = 0.8$, $\alpha = 3.5$ deg: Scheme II and MSES solutions

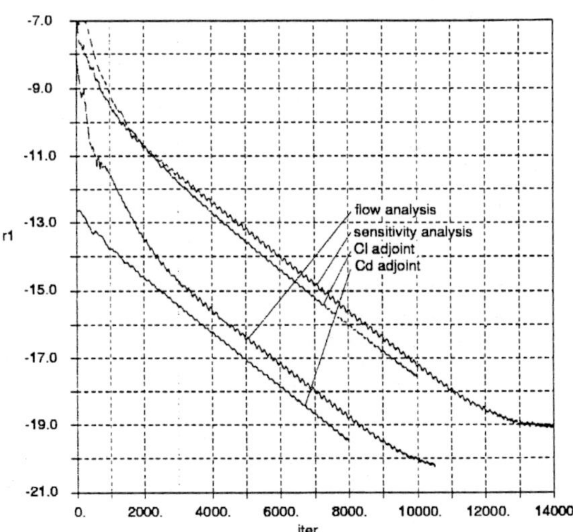

Figure 5: Convergence histories for NACA0012 airfoil, Re=2000, $M_\infty = 0.8$, $\alpha = 3.5$ deg

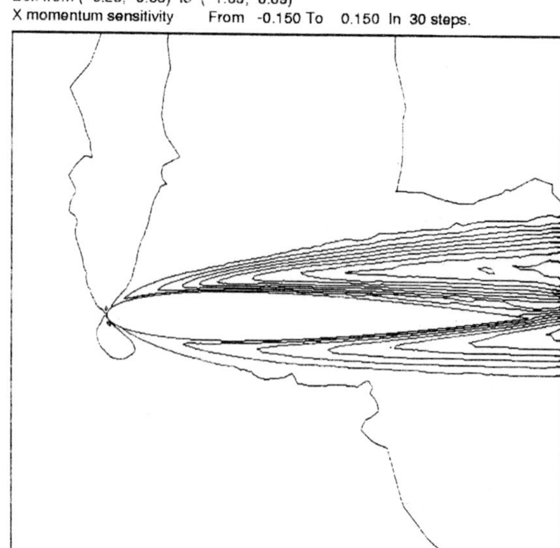

Figure 6: $\partial\rho u/\partial\alpha$ distribution for NACA0012, Re=2000, $M_\infty = 0.8$, $\alpha = 3.5$ deg (Scheme II) found by direct method

Figure 8: $\partial\rho u/\partial\alpha$ distribution for NACA0012, Re=2000, $M_\infty = 0.8$, $\alpha = 3.5$ deg (Scheme II)from finite difference with $\Delta\alpha = 0.01$

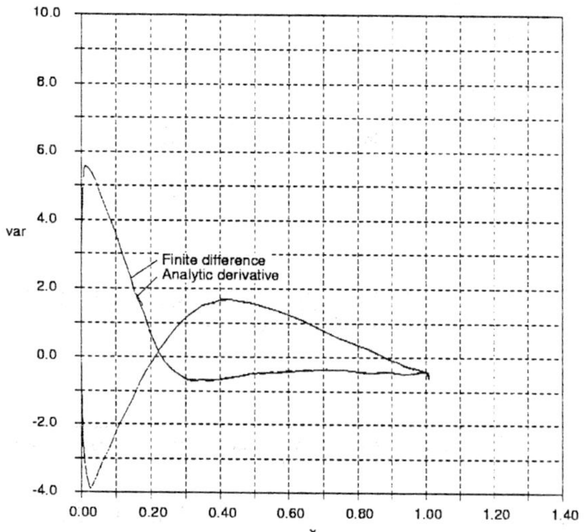

Figure 7: Surface $\partial p/\partial\beta_c$ distribution for NACA0012, Re=2000, $M_\infty = 0.8$, $\alpha = 3.5$ deg (Scheme II v. Finite Difference)

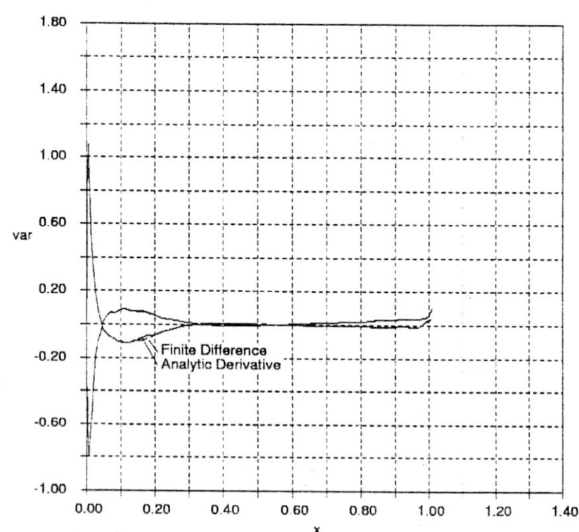

Figure 9: Surface $\partial\tau_{11}/\partial\beta_c$ distribution for NACA0012, Re=2000, $M_\infty = 0.8$, $\alpha = 3.5$ deg (Scheme II v. Finite Difference)

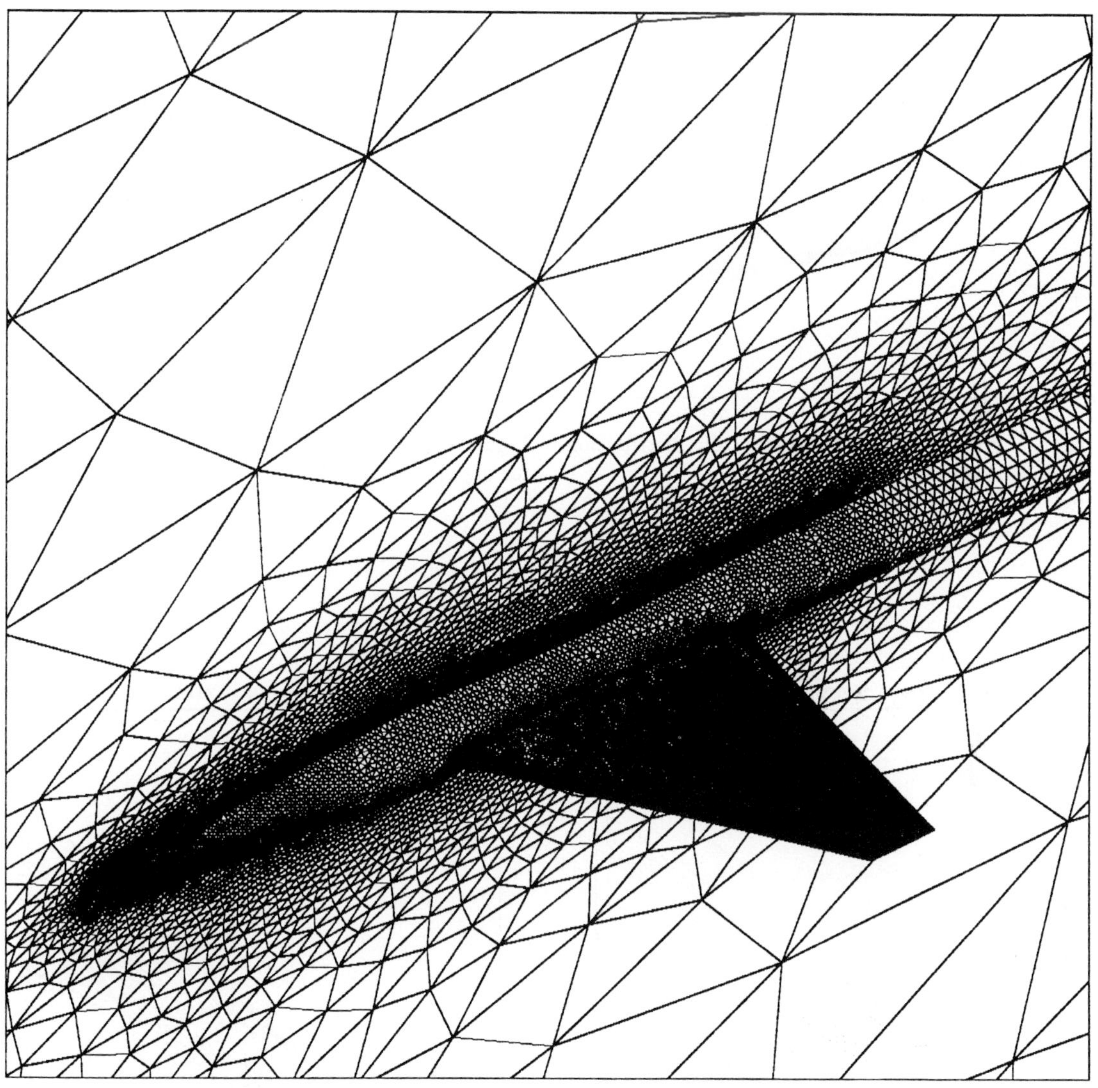

Figure 10: Baseline grid for 3D optimization exercise

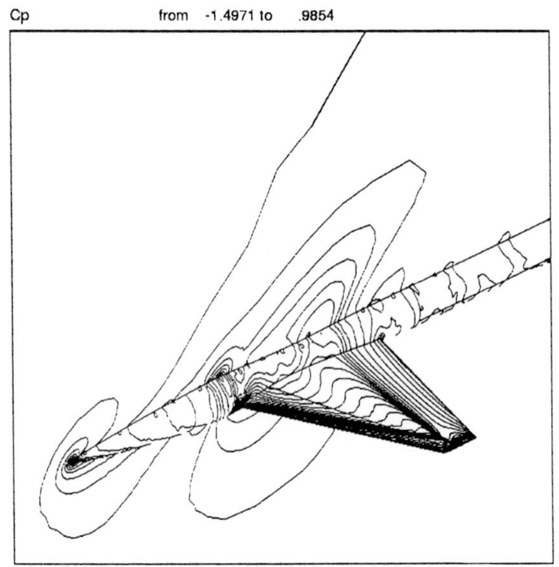

Figure 11: Condition 1 (M=0.9) baseline C_p distribution for 3D optimization exercise

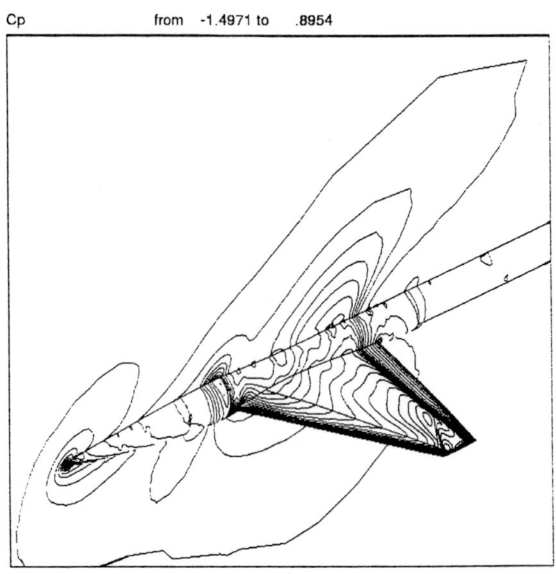

Figure 13: Condition 1 (M=0.9) "Design 7" C_p distribution for 3D optimization exercise

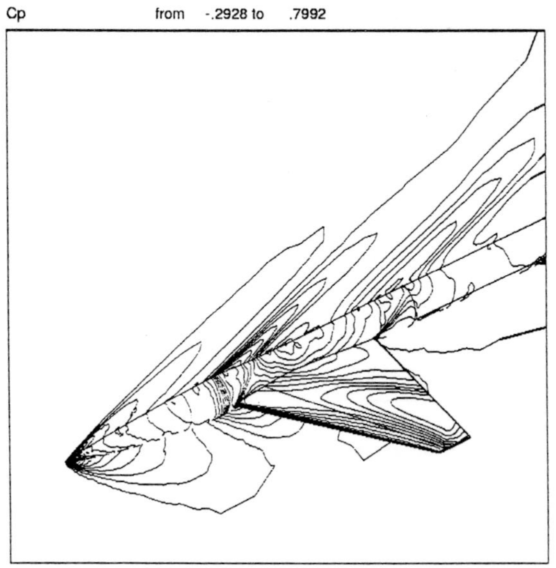

Figure 12: Condition 2 (M=1.6) baseline C_p distribution for 3D optimization exercise

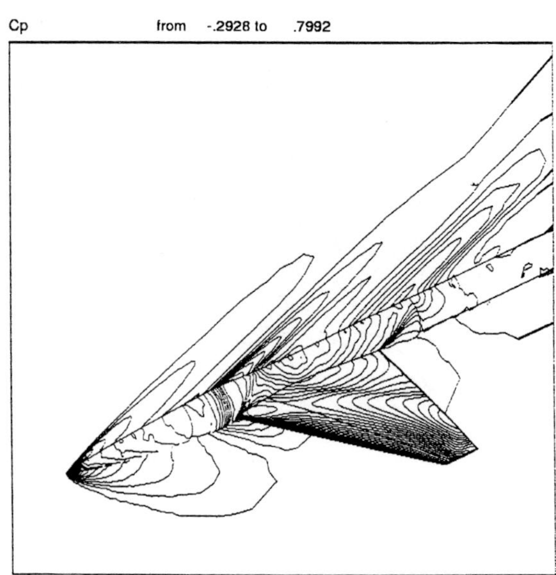

Figure 14: Condition 2 (M=1.6) "Design 7" C_p distribution for 3D optimization exercise

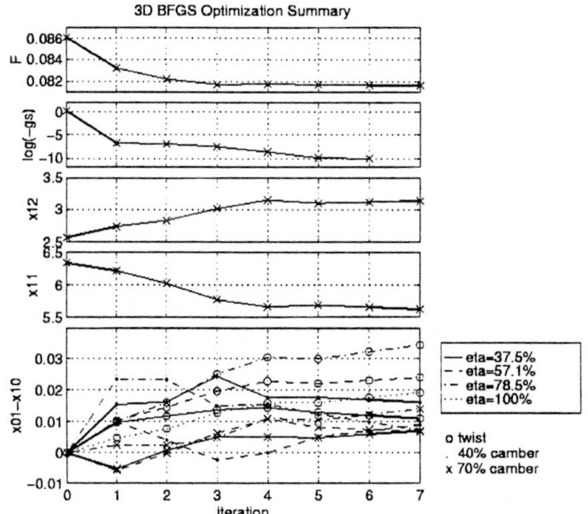

Figure 15: Cost function and design variable evolution for 2D BFGS optimization exercise

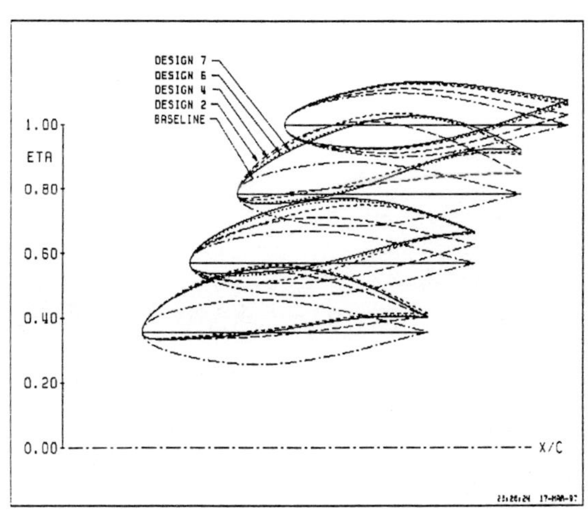

Figure 17: Geometry evolution for 3D BFGS optimization exercise

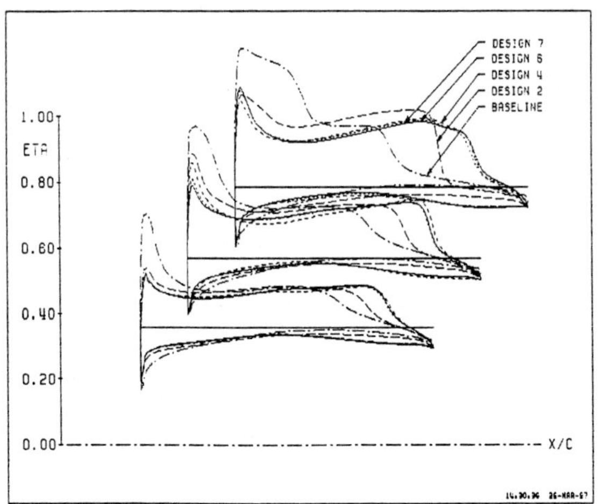

Figure 16: Condition 1 (M=0.9) C_p evolution for 3D optimization exercise

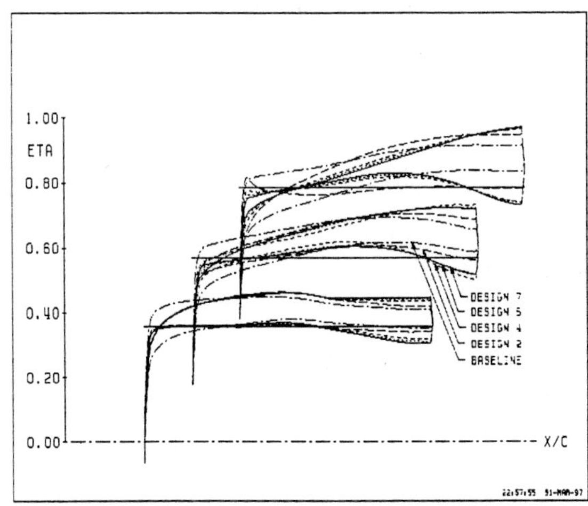

Figure 18: Condition 2 (M=1.6) C_p evolution for 3D optimization exercise

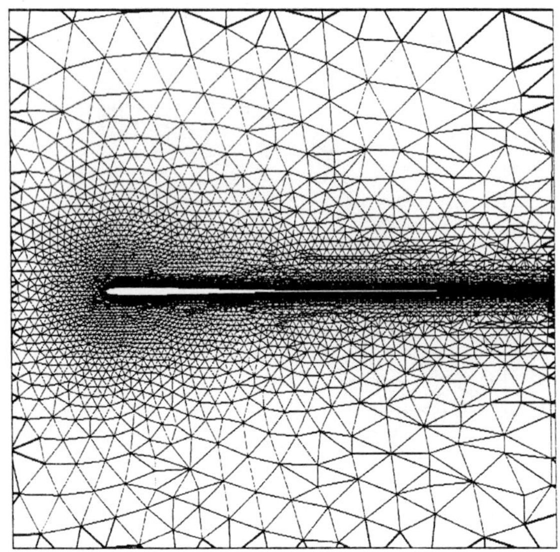

Figure 19: μ-UAV airfoil grid

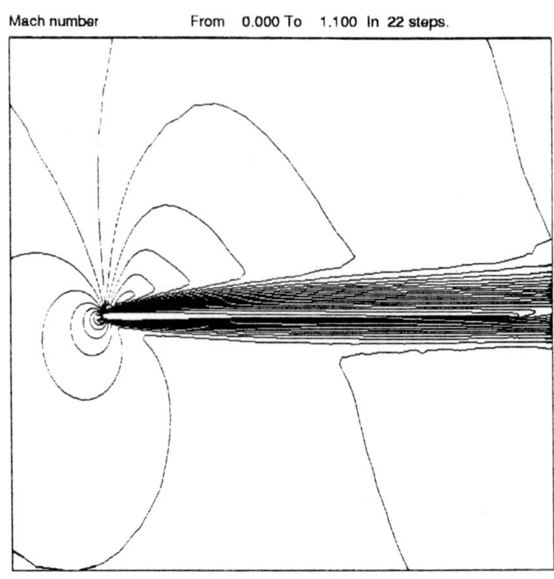

Figure 21: Baseline M distribution for μ-UAV airfoil, Re=2000, $M_\infty = 0.8$, $\alpha = 3.5\,\text{deg}$

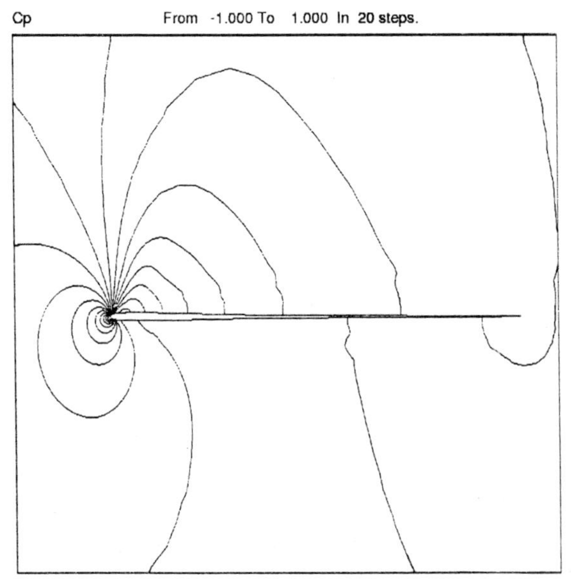

Figure 20: Baseline C_p distribution for μ-UAV airfoil, Re=2000, $M_\infty = 0.8$, $\alpha = 3.5\,\text{deg}$

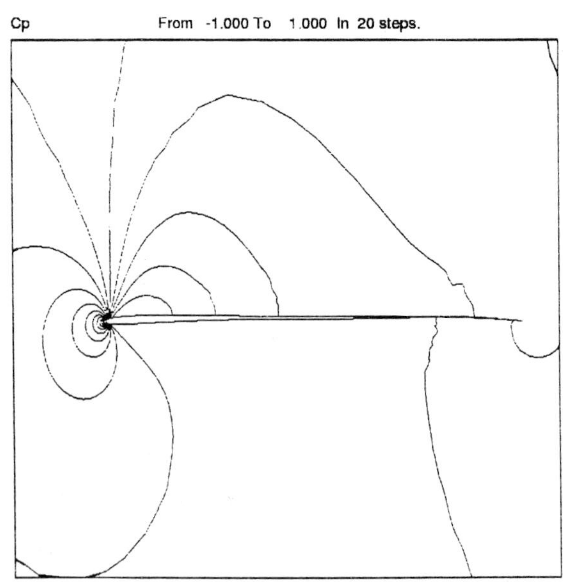

Figure 22: Final C_p distribution for μ-UAV airfoil, Re=2000, $M_\infty = 0.8$, $\alpha = 3.5\,\text{deg}$

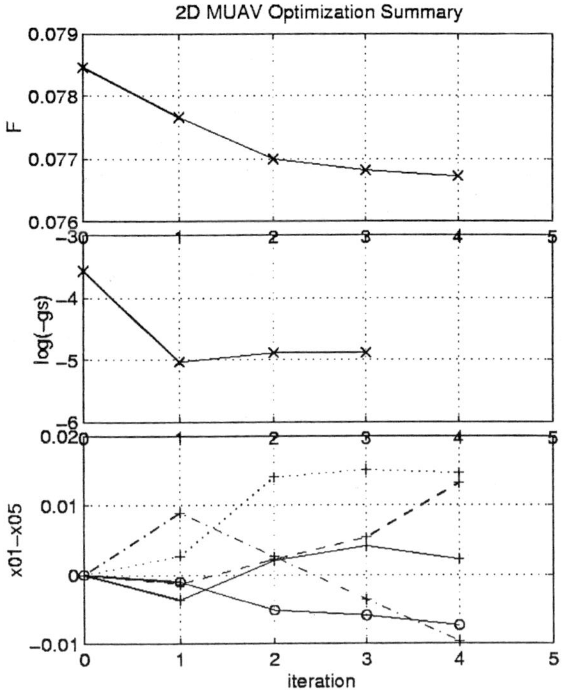

Figure 23: Design variable evolution for μ-UAV airfoil, Re=2000, $M_\infty = 0.8$, $\alpha = 3.5\,$deg

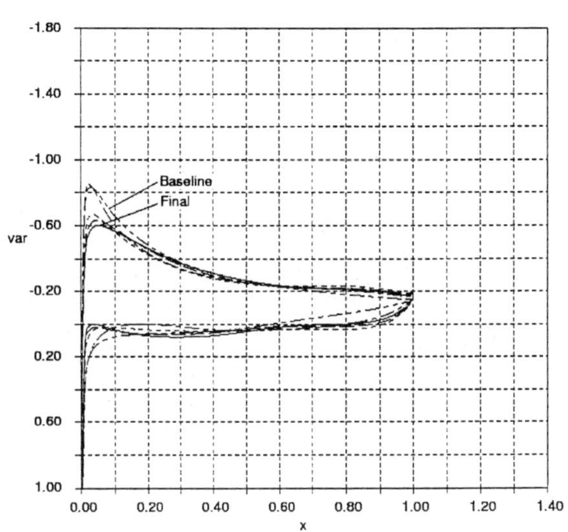

Figure 25: Surface C_p evolution for μ-UAV airfoil, Re=2000, $M_\infty = 0.8$, $\alpha = 3.5\,$deg

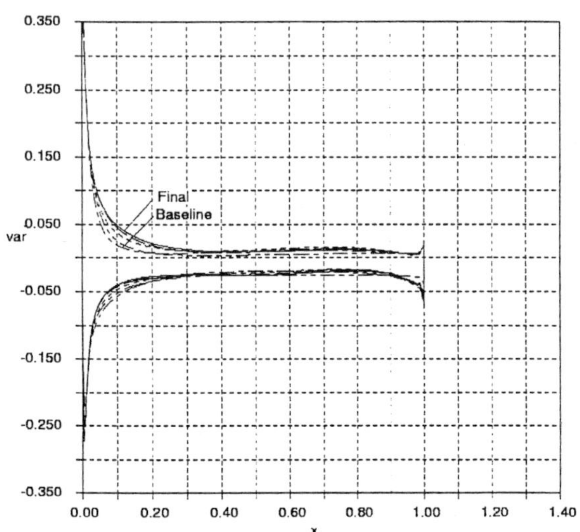

Figure 24: Surface C_f evolution for μ-UAV airfoil, Re=2000, $M_\infty = 0.8$, $\alpha = 3.5\,$deg

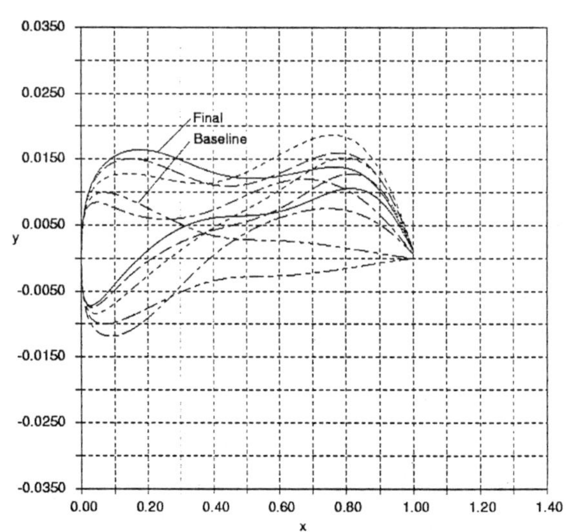

Figure 26: Geomety evolution for μ-UAV airfoil, Re=2000, $M_\infty = 0.8$, $\alpha = 3.5\,$deg

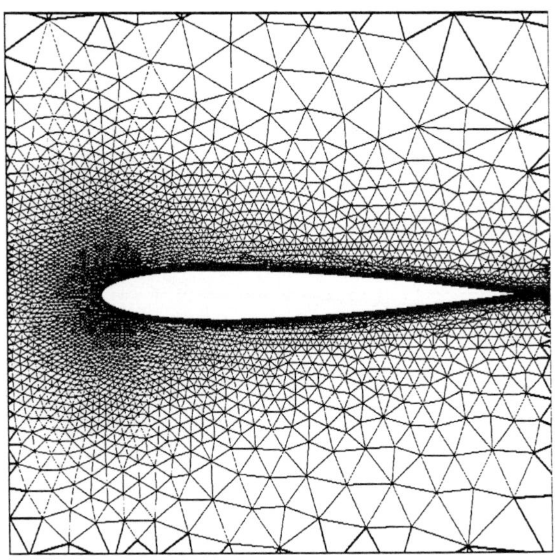

Figure 27: Re=2000 NACA0012 grid

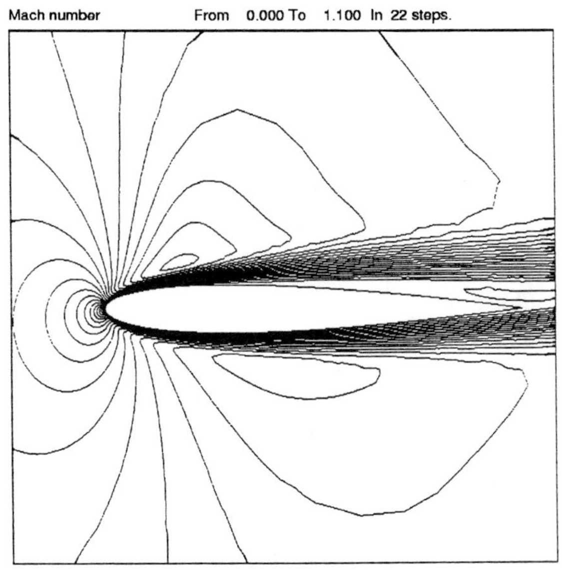

Figure 29: Baseline M distribution for NACA0012, Re=2000, $M_\infty = 0.8$, $\alpha = 3.5\,\mathrm{deg}$

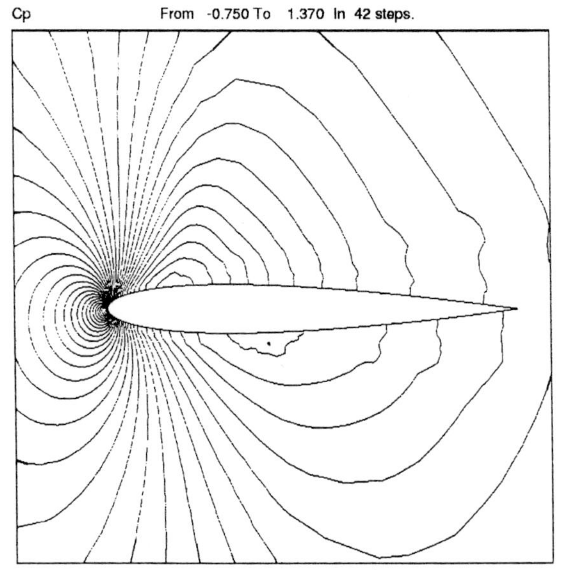

Figure 28: Baseline C_p distribution for NACA0012, Re=2000, $M_\infty = 0.8$, $\alpha = 3.5\,\mathrm{deg}$

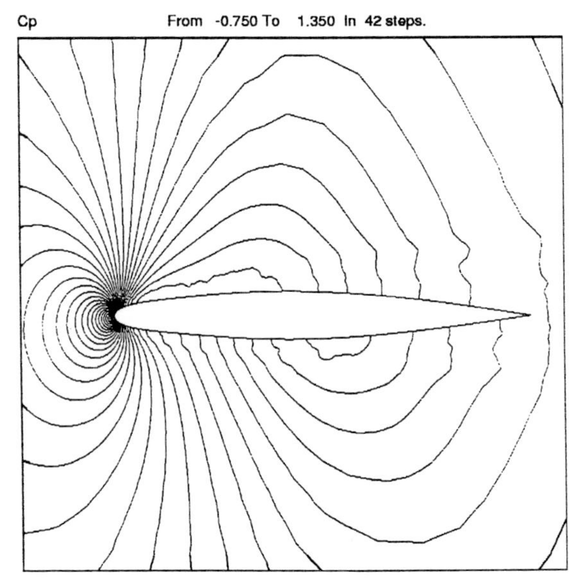

Figure 30: Final C_p distribution for NACA0012, Re=2000, $M_\infty = 0.8$, $\alpha = 3.5\,\mathrm{deg}$

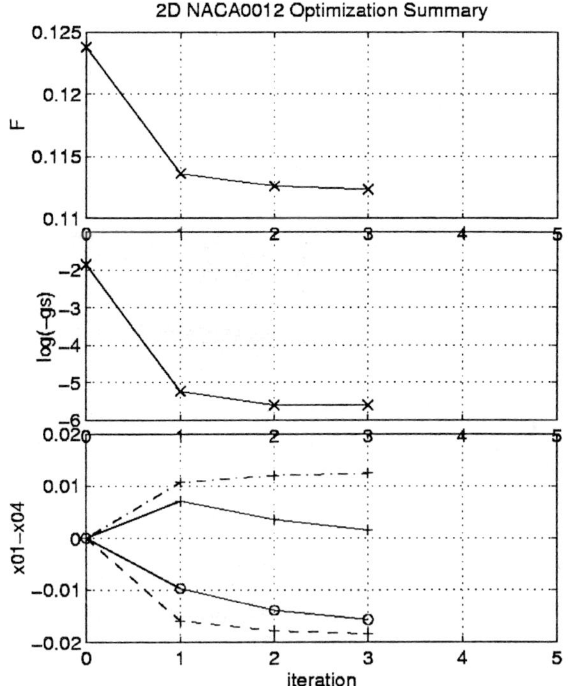

Figure 31: Design variable evolution for NACA0012, Re=2000, $M_\infty = 0.8$, $\alpha = 3.5$ deg

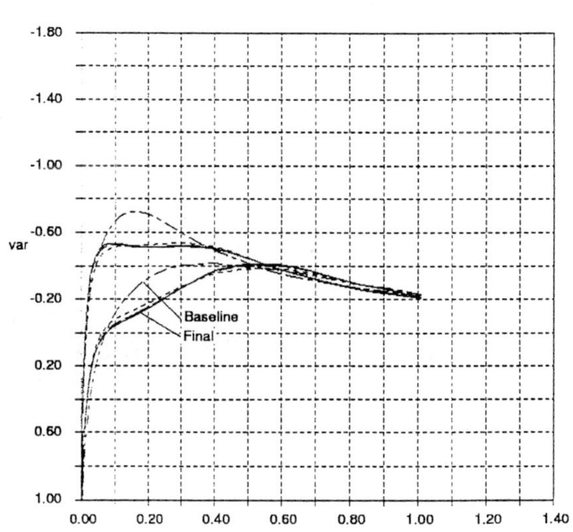

Figure 33: Surface C_p evolution for NACA0012, Re=2000, $M_\infty = 0.8$, $\alpha = 3.5$ deg

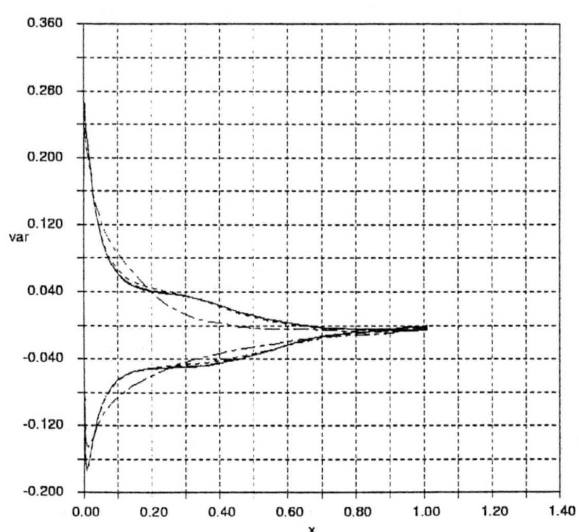

Figure 32: Surface C_f evolution for NACA0012, Re=2000, $M_\infty = 0.8$, $\alpha = 3.5$ deg

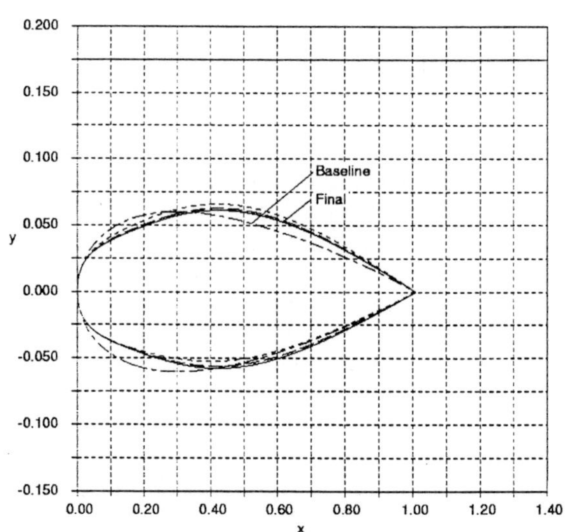

Figure 34: Geometry evolution for NACA0012, Re=2000, $M_\infty = 0.8$, $\alpha = 3.5$ deg

AERODYNAMIC DESIGN OPTIMIZATION USING ADVANCED CFD CODES, AUTOMATIC DIFFERENTIATION, AND PARALLEL COMPUTING

Amidu OLOSO [1]   Arthur C. TAYLOR, III [2]   James C. NEWMAN, III [3]

(1) Research Associate; Eagle Aeronautics; 123388, Warwick Blvd., Suite 301 Newport News, Virginia 23606

(2) Associate Professor; Dept. of Mechanical Engineering, Old Dominion University; Norfolk, Virginia 23529

(3) Assistant Professor; Dept. of Aerospace Engineering Mississippi State University, Starkville, Mississippi 39762

Abstract

This work presents a complete methodology for the use of an advanced CFD code in aerodynamic shape-design optimization of realistic aircraft configurations. The work consists of two parts. In the first part, the well-known, general-purpose, flow-analysis code CFL3D is accurately differentiated with respect to general geometric shape variations using an automatic-differentiation software tool ADIFOR. The result is a new flow-sensitivity-derivative code, CFL3D.ADII, which retains all of the capabilities of the original code (at least for steady flows), including the ability to model complex geometries with multiblock grids. Similar previous studies with advanced flow codes have yielded accurate sensitivity derivatives; however, computational efficiency was unacceptable. In the present study, however, computational efficiency of the differentiated flow code has been greatly improved, primarily through automatic differentiation in incremental iterative form and by a thorough restoration of the original code's vectorization for efficiency on Cray-type computers. In one example, a milestone result was achieved, in that the CPU-timing of the sensitivity-derivative code CFL3D.ADII was found to be more efficient than the original CFL3D code (on the basis of microseconds of CPU-time per multigrid-cycle per grid-point per design-variable). In the second part of the work, attention was focused on the use of CFL3D.ADII in the shape optimization studies of realistic geometries on parallel computers. A coarse-grain parallel implementation of CFL3D.ADII together with an alternative parallel 1-D line search scheme are used in shape improvement design studies of a realistic High-Speed Civil Transport wing/body geometry on an IBM-SP2 parallel computer. The geometry is represented by 60 design variables and over 200,000 grid points. The flow is supersonic and inviscid, represented by the 3-D Euler equations. In addition to making the handling of such a large problem possible, the use of parallel computation provided significantly reduced overall execution time and turnaround time.

1. Introduction

This work consists of two parts. In the first part, an efficient, general-purpose code for geometric-shape sensitivity analysis is developed and successfully tested; the code is called CFL3D.ADII (version 2.0). The code is fully compatible with the well-known, general-purpose, flow-analysis code CFL3D (version 4.1), and can be used to obtain accurate gradient information for subsequent use in aerodynamic-shape-design optimization.

The new flow-sensitivity code has been developed with the assistance of the automatic differentiation (AD) software tool ADIFOR (Automatic DIfferentiation of FORtran[1]). A straightforward, "black-box" application of AD to CFL3D yields a

Received on .

new source code which, when compiled and executed, will compute accurate sensitivity derivatives (SDs); unfortunately the resulting code is painfully inefficient computationally compared with the original flow-analysis code. This computational inefficiency is particularly marked for codes such as CFL3D when operating on Cray machines, where the computational efficiency of the original flow-analysis code depends heavily on a very high degree of vectorization.

In order to mitigate this problem of severe computational inefficiency, which results from a black-box (BB) application of AD to CFL3D, the flow code was differentiated (using AD) in what is known as incremental iterative form. Incremental iterative (II) form is described in detail in Korivi et al.[2]; the combined use of AD with II form was first proposed by Newman et al.[3], and first successfully tested with a 2D CFD code and documented in detail in Sherman et al.[4] Simply stated, AD in II form (i.e., ADII) means that only the flow-physics equations and boundary-condition equations are differentiated via AD. The flow-solution algorithm (i.e., the multi-grid, the approximate-factorization scheme, etc.) is not differentiated but is left unchanged to serve as the algorithm for solving (iteratively) the large systems of linear equations which are associated with computing the SDs. References [2] through [4] provide full details on the theory of computing discrete SDs from advanced CFD codes; in particular, on the ADII approach.

The first version of CFL3D.ADII (version 1.0) was successfully documented to yield accurate SDs with significantly greater computational efficiency than that obtained via a BB AD of CFL3D.[5,6] Nevertheless, the computational performance of version 1.0 was very disappointing, particularly on Cray machines where the vectorization of the AD code had some severe "bottlenecks". However, significant additional improvements in computational efficiency have now been realized in the newer version 2.0, as presented herein.

In the second part of the work, attention was focused on the use of CFL3D.ADII in the shape optimization studies of realistic geometries on parallel computers. Several efforts have been documented in the literature on the use of existing CFD analysis methods equipped with sensitivity analysis capability to perform aerodynamic shape-design optimization. The discrete sensitivity analysis approach, whether in the direct or adjoint form, has been used in several earlier studies within different CFD codes[7−12] for aerodynamic design optimization. With a few exceptions, (for example, the recent work of Reuther et al.[13] where a continuous adjoint formulation was applied to a complex multiblock 3-D geometry using the Euler equations and a

large number of design variables) most of the earlier efforts are limited to simplified flow physics, simple geometries, a small number of design variables, etc. The challenges posed by large-scale, industry-level aerodynamic design studies have largely not been met. These challenges include complicated 3-D geometries, complex flow physics, large number of design variables, large computational resource requirements that push the envelope of the capability of the currently available supercomputers resulting in large turnaround times. The present work is motivated by a desire to meet these challenges using discrete sensitivity analysis and parallel computing.

In this study, a coarse-grain parallel implementation of CFL3D.ADII was developed for distributed-memory parallel computers. The goal was to be able to perform sensitivity analysis for realistic aircraft configurations represented by a large number of design variables and a large computational grid. In addition, since the overall objective was to develop a complete methodology for aerodynamic design optimization that will significantly reduce the overall execution time and turnaround time, an alternative procedure employing parallel computation was developed to replace the time-consuming highly sequential one-dimensional (1-D) line search of the gradient based optimization techniques. The 1-D search may represent a "bottleneck" in aerodynamic design studies because it involves a series of CFD analyses performed sequentially (for each design cycle). A more complete presentation of the methods and results of the present study is given in Ref. [14].

CFL3D.ADII (Version 2.0)

The first part of this study, presented subsequently, is concerned with the development and validation of the CFL3D.ADII (v.2.0) code.

Code Capabilities

All features and capabilities of the original CFL3D (version 4.1) flow code have been included in the code CFL3D.ADII (version 2.0), with the exception of its unsteady flow capabilities; i.e., unsteady aerodynamic SDs cannot (yet) be calculated. However, all of the remaining powerful, user-friendly features of the original CFL3D code have been preserved. In particular, aerodynamic shape SDs can be accurately calculated for very complex geometries using the multiblock capabilities of this code; this includes its general, patched-grids capability and its overlapped-grids capability; even the embedded-grids feature of the original code has been included. In addition, the complete turbulence-modelling of the original code has been differentiated and included in CFL3D.ADII. Furthermore, for efficiently solving the linear sensitivity equations, the new code maintains the same algorithm capabilities that are featured in the original CFL3D code for solving the nonlinear flow equations. These algorithm features include the three-factor, spatially-spit, approximate-factorization procedure with a choice of either the efficient Roe-diagonalized scheme or the block-tridiagonal inversion scheme. In addition, multigrid and mesh-sequencing are retained for significantly accelerated solution of the flow-sensitivity equations.

Code Accuracy

The accuracy of the SDs produced by the CFL3D.ADII code has been validated through a comparison of its results with SDs obtained by the method of finite-differences. Sample results from one of the various validation studies are presented subsequently. The given example problem involves $M_\infty = 2.4$, inviscid flow over a wing-body, High-Speed Civil Transport (HSCT) type configuration (angle-of-attack, $\alpha = 1.9°$). The 3-D Euler equations are solved on a single-block grid having dimensions $193\times33\times33$ (a total of 210,177 grid points).

Table 1 shows a comparison (each in the form of a ratio) of SDs calculated from CFL3D.ADII with the SDs calculated by the method of central finite differences. In this table, sensitivity-derivative ratios are presented for derivatives of lift coefficient (C_L) and drag coefficient (C_D) with respect to five different geometric shape parameters which define the wing-section (airfoil) shapes at various points along the wing-span. As seen from the table, the agreement among the two methods is very good.

It is very significant to mention that the numerical error reduction for the nonlinear flow equations was ten orders-of-magnitude with the central finite-difference method, which was required for accuracy with the given step size. For the CFL3D.ADII code, however, the error reduction for the nonlinear flow equations and for the linear flow-sensitivity equations was only four orders-of-magnitude, to achieve the accuracy shown. Of course, in terms of the overall computational efficiency and accuracy of the two methods, this numerical phenomenon (also documented and explained in detail in earlier studies[7,14]) will serve to favor greatly the SDs computed via the CFL3D.ADII code (rather than via finite differences).

In another example problem[14], the CFL3D.ADII code accuracy was validated on a computational grid having two blocks, including a "patched-grid" interface between the blocks. The accuracy of the SDs computed in this more challenging multiblock example was comparable to that documented in Table 1. Thus the accuracy of the CFL3D.ADII code has been successfully documented in application to a complex-geometry, multiblock problem.

Code Efficiency Improvements

The CFL3D.ADII (version 2.0) code was evaluated with respect to its computational efficiency on Cray computers (i.e., a Cray-YMP in the subsequent results). In particular, a study was conducted to evaluate the computational cost of calculating SDs using the CFL3D.ADII (v.2.0) code compared to the cost of the same SDs using a code which was generated through a straightforward "black-box" (BB) application of AD. Henceforth, this latter code is referred to as CFL3D.ADBB. In addition, the computational cost of these two SD codes is evaluated in relationship to the computational cost of the original CFL3D flow code.

Subsequent results of CPU-timing comparisons are for the identical example flow problem which was described previously. In all cases, two-level multigrid is tested with a number of different combinations of code options. The Roe's upwind flux scheme with the Roe-diagonalized implicit solution algorithm was tested both with and without use of the min-mod limiter. In addition, the van Leer's upwind scheme with implicit, block-tridiagonal coefficient matrices was tested, again with and without the min-mod limiter.

Table 2a presents "baseline" CPU-timing results for the original CFL3D (v.4.1) flow code; timings are presented for the four different variations of code options which were tested. That is, results are presented for various combinations of upwind schemes, implicit operators, and flux limiting. (A short list of nomenclature for the abbreviations which are used in these tables is included immediately following the tables.) In Table 2a, timings are given in micro-seconds per multigrid-cycle per grid-point (μsecs/MGC/GP).

Table 2b presents a comparison of the CPU-timing results of the CFL3D.ADII (v.2.0) code with the same results for the CFL3D.ADBB code. Results are presented as <u>ratios</u> of timings in microseconds per multigrid-cycle per grid-point per design-variable (μsecs/MGC/GP/DV), normalized by the corresponding timing (from Table 2a) of the original CFL3D code (in μsecs/MGC/GP). Thus, with CPU-timing ratios, implicitly a comparison is made of the computational cost of computing the SDs with that of the original flow code. In addition, CPU-timing comparisons are given for both a single design variable (DV) as well as for a group of five DVs computed concurrently (with timings presented on a per DV basis). For multiple DVs, Sherman et al.[4] details the special computational implications of concurrently calculating these SDs on Cray machines.

As seen from Table 2b, dramatic improvements in computational efficiency are seen in the CFL3D.ADII (v.2.0) code over the CFL3D.ADBB code. Although not seen in the table, it is significant to note that the earlier version 1.0 of CFL3D.ADII, while more efficient than CFL3D.ADBB[5,6], was substantially less efficient than the new, improved version 2.0. Most of the gain of version 2.0 over version 1.0 was through vectorization improvements; i.e., by "unclogging" a few critical vectorization "bottlenecks" found in the earlier version 1.0.

One key result to note in Table 2b is that a few particular code options yield CPU-timing ratios that are less than one (for CFL3D.ADII). This has very significant computational implications with respect to computational efficiency comparisons of the present methodology for calculating aerodynamic SDs compared with the cost of obtaining these SDs via the method of finite differences. In fact, the ratios less-than-one seen in Table 2b can be considered to be a "milestone" in the development and application of the AD technology for efficiently calculating discrete aerodynamic SDs from advanced CFD codes.

Code Memory Requirements

The memory requirements of the CFL3D.ADII code are compared with the memory requirements of the original CFL3D code in Table 3, for the given example problem. (Note: Memory requirements of CFL3D.ADBB are similar to that of CFL3D.ADII.) The increase in memory requirements which is seen here is standard for the AD methodology. That is, for a given flow problem, the memory increase (over that of the original flow code) is approximately the memory requirement of the original code; this increase applies to each DV for which SDs are concurrently calculated.

Parallel 3-D Aerodynamic Shape Optimization

The second part of this study, presented subsequently, concentrates on the development of a procedure for shape-design optimization of realistic 3-D geometries.

Gradient Calculation "In Parallel"

In this section, the CFL3D.ADII code is specially adapted for use in a distributed-memory, parallel computing environment (i.e., either a cluster of workstations or an IBM-SP2-type machine.) The parallel implementation of the present study is similar to the "derivative strip-mining" technique of Bischof et al.[15] With this method, an identical copy of CFL3D.ADII is run on each node of the machine, but each node with different grid sensitivity data. The grid sensitivity data for each node (which is local to the node) can be either for a single design variable or for a group of design variables. The parallel implementation is very coarse because each node computes aerodynamic functions and the associated derivatives for the entire geometry. Also, no design variable is processed by more than one node. The internodal communication, which is minimal, is achieved via the standard Message Passing Interface (MPI). This parallel implementation is made possible because each node has enough memory to contain both the functions and their derivatives for at least one design variable for the entire geometry. If the problem size will not allow this, a finer grain implementation will have to be considered, in which the geometrical domain is partitioned as well. This latter implementation will have higher communication complexity because the geometrical domain is now processed by more than one node and internodal communication now includes data movement across grid-block boundaries during the function and derivative evaluation iterations.

In the subsequent design example, a total of 60 design variables were considered. Because of the problem size (over 200,000 grid points with the fluid flow modeled by the 3-D Euler equations), each regular node of the IBM-SP2 has sufficient memory to contain CFL3D.ADII (v.2.0) for only one design variable. This implies that, for each design (optimization) cycle, one node of the IBM-SP2 was dedicated to calculating the required SDs for each of the individual design variables; thus, it requires 60 nodes to calculate gradient information for 60 design variables in parallel.

Parallel 1-D Line Search

For the gradient-based optimization techniques employed in this work, a new design is obtained from the current design by this well-known equation

$$\mathbf{X}^{r+1} = \mathbf{X}^r + \alpha^\star \mathbf{S}^r \qquad (1)$$

where $\mathbf{X}^r$ and $\mathbf{X}^{r+1}$ are the current and the new vectors of shape variables respectively. $\alpha^\star > 0$ is the move parameter and $\mathbf{S}^r$ is the search-direction vector. In Eq. (1), to obtain the new design, $\alpha^\star$

and $\mathbf{S}^r$ are needed. $\mathbf{S}^r$ is computed using the values of the objective and constraint functions and their gradients at the current and, sometimes, previous design. The difficult part in computing $\mathbf{S}^r$ for aerodynamic design optimization is the gradient evaluation which is CPU-time and memory intensive. This difficulty is resolved by the parallel implementation of the CFL3D.ADII (v.2.0) discussed previously. The second unknown in Eq. (1) is the move parameter, α^*. In aerodynamic design optimization, evaluating α^* can be computationally intensive. Usually in practice, an initial α is proposed and a function evaluation is performed. If the design is improved, α is incremented and a new design is obtained. This process is continued until the optimum design for that particular search direction is trapped within some required tolerance of α^* (as in the golden-section method), or until a range of α is obtained which contains α^* (in which case a polynomial interpolation is then used to obtain α^*. Either way, the process of locating α^* is, by nature, highly sequential, since the current α is estimated based on the previous one. In aerodynamic design studies using CFD, this implies a series of CFD solutions computed one after the other, leading to a large amount of execution time. To mitigate the computational cost associated with the sequential 1-D line search, the so-called flow prediction or approximate analysis method[16,17] can be employed. In this method, a new solution at the next design point within the 1-D line search cycle is estimated, at a cheaper cost, by a truncated first-order-accurate Taylor series expansion of the solution algorithm about the current known design point. This approach is similar to the re-analysis technique usually employed in structural design optimization.[18] The approach is made possible because during the 1-D line search, the changes in design variable vector $\mathbf{X}$ are sometimes small and the new solution is expected to be close to the current "nearby" solution. The first problem with this approach is that, after using the approximate analysis for a number of iterations in the 1-D search loop, the predicted flow field solution begins to deteriorate, leading to very crude estimates for the values of the objective and the constraint functions, and also possibly poor convergence rates for the 1-D search process. This problem was overcome in Eleshaky[17] by performing an exact CFD analysis to update the flow field solution and to provide a new baseline solution if the 1-D search process does not converge after a predetermined number of approximate flow field analysis. The second problem is that the execution time for the approximate analysis can become quite high for large, 3-D problems.

In the present study, an alternative strategy is designed to reduce the execution time for the 1-D line search process; the approach exploits the multiprocessor environment offered by the IBM-SP2 or any similar architecture. The strategy proceeds as follows: The search direction vector $\mathbf{S}$ is computed the same way as in the sequential approach. However, after computing $\mathbf{S}$, before the optimizer starts the 1-D line search process, a number, say N, of CFD solutions are computed. Each CFD solution p, where p goes from 1 to N, is computed with a different estimated vector of design variable $_p\mathbf{X}_{estimate}$ obtained from

$$_p\mathbf{X}_{estimate} = \mathbf{X}^r +_p \alpha_{estimate}\mathbf{S} \qquad (2)$$

where, as mentioned earlier, $\mathbf{X}^r$ is the current design and $_p\alpha_{estimate}\mathbf{S}$ is the step size for CFD solution p. The quantity $_p\alpha_{estimate}$ is obtained as described subsequently. First, the maximum α, α_{max}, that will admit all possible α's which the optimizer may propose during the 1-D line search process is specified as

$$\alpha_{max} = min[\alpha_{max}^i], \quad i = 1 \quad \text{to NDV} \qquad (3)$$

where NDV is the number of design variables and α_{max}^i is the maximum α that will drive design variable i to its upper or lower bound and is obtained from

$$\alpha_{max}^i = \frac{X_i^u - X_i}{S_i} \qquad \text{if} \qquad S_i > 0 \qquad (4)$$

or

$$\alpha_{max}^i = \frac{X_i^l - X_i}{S_i} \qquad \text{if} \qquad S_i < 0 \qquad (5)$$

X_i is the current value of design variable i, while X_i^u and X_i^l are its upper and lower bounds respectively, and S_i is its component of the search direction vector. After α_{max} is evaluated, it is divided into N-1 equally spaced steps for the N CFD analyses. Thus, for the pth CFD analysis, $_p\alpha_{estimate} = (p - 1)\triangle\alpha$, where $\triangle\alpha = \alpha_{max}/(N - 1)$.

The N CFD analyses are completely decoupled and hence can be computed concurrently. This concurrency is exploited on the IBM-SP2 where the N solutions are computed on N nodes in parallel, a node for each solution. After the computations, there are a total of N different values of the objective function and N different values for each of the constraint functions which cover the entire range of α's the optimizer may need during the 1-D line search. Thereafter, a suitable polynomial is used to interpolate the values of the objective and the constraints as functions of α, one polynomial for each function. In this study, the natural cubic spline is employed and for the design studies, N is of O(10) and α_{max} is of O(1).

Once the cubic spline curves are generated, they are thereafter used to represent the behavior of the objective function and constraint functions (and to replace the computationally intensive CFD code) within the 1-D line-search procedure. The computational cost of evaluating the splines as well as the cost of obtaining the search direction (after the gradient information is made available) is negligible compared to the cost of a full CFD solution. The result is that the execution time for the 1-D line search is essentially reduced to the execution time for just one CFD analysis. The computation of the search direction as well as the overall coordination of the design process (checking for convergence, etc.) is achieved by using a general-purpose design optimization code ADS[19] which was modified as necessary for this study. The resulting code is a design package that incorporates the parallel implementation of the CFL3D.ADII (v.2.0) shape-sensitivity analysis code and the parallel 1-D line search procedure.

Design Optimization Study

To demonstrate the capabilities of the proposed methodology, design improvement studies were carried out on a realistic, proprietary aircraft geometry which is an HSCT-like wing-body configuration. This geometry is the same as the one used for the CFL3D.ADII code validation. The steady inviscid flow is modeled by solving the 3-D Euler equations. For simplicity, the present example involves wing-section shape design only, with the wing planform and the fuselage shapes being fixed.

A total of 60 wing-section shape design variables are selected; these 60 design variables are all camber and wing-twist parameters; no wing-section thickness variables were included in this particular design study. A total of 29 wing-section stations, distributed along the half-span of the wing, are established for this example design problem, with "lofting" of the airfoil (wing-section) shapes between these spanwise stations.

In addition to the preceding 60 geometric-shape design variables, the present example optimization problem is established as follows. The objective (function) is to minimize overall drag coefficient, C_D, subject to constraints: (1) the lift coefficient, C_L, can be no smaller than its initial value; and (2) the magnitude of the bending moment at the wing root, C_{M_x}, can be no larger than its initial value; and (3) there are physically appropriate "bounds" (i.e., limits) placed on each of the 60 design variables.

A total of five full design (optimization) cycles were successfully completed. A plot of C_D, C_L and C_{M_x}, normalized with their baseline values, versus the number of optimization cycles is shown in Fig. 1. As can be seen from this figure, it is not economical from a computational point-of-view to go beyond the fifth design cycle, because the relative change in objective function is just about 0.1% from the fourth to the fifth cycle. Also from Fig. 1, there is a reduction of 1.76% in C_D. The lift coefficient C_L is constrained to within 0.02% and C_{M_x} to within 0.2%. The drag reduction is achieved primarily by the changes in the mean camber line, Z, of the wing airfoil sections. This is illustrated in Fig. 2, where the percentage change in Z is plotted against normalized chordwise distance for the affected airfoil sections. The largest changes in Z (up to about 4%) occur in the inboard part of the wing. Z then gradually reduces as the wing tip is approached. Close to the wing tip, Z changes by as little as 0.5%. Moreover, camber perturbations are the largest in the aft regions of the inboard and mid-span wing sections. For the locations in the outboard region close to the wing tip, the wing sections tend to translate upward. The observed camber changes for all the airfoil sections indicate that the constraint at the wing trailing edge needs to be somewhat relaxed to allow possible upward movement, as the optimizer is suggesting. It shall be noted that an examination of the airfoil sections at each of the 29 spanwise wing stations reveals that the airfoil shapes which have resulted from the design are clearly physically realistic. This is due to the fact that the airfoil section thicknesses are held constant during the design process. Of course, since the objective was to minimize wave-drag, preventing the wing from becoming thinner is responsible for the relatively small drag reduction (only about 1.8%) obtained. In another design study for the same geometry but with over 100 design variables including wing-section thickness parameters (as well as camber and twist parameters), not only were the mean camber lines shifted significantly, but also the wing-section thicknesses were reduced substantially. As a result, there was a very substantial reduction in wave drag, although the optimized geometry was physically unrealistic (due to the large reduction in wing-section thicknesses). Hence the detailed results of the 100 design variable study are not presented here but are available elsewhere.[14]

The final values of the 60 design variables are within 15 to 85% of their upper or lower limits. Table 4 shows the comparison, in form of ratios, between the final values of 10 randomly selected design variables from the new parallel 1-D line-search scheme (described previously) and their counterparts from the traditional sequential 1-D line-search method. All ratios are essentially unity, indicating good accuracy agreement between the two methods. In terms of the execution time, the parallel approach is at least 2.5 times as fast as the sequential approach.

The average execution time for each design cycle on the IBM-SP2 is about 17 hours. Out of this, the gradient computation takes about 13 hours while the parallel 1-D line search takes about 4 hours. For the 1-D line search method used, the optimizer makes five calls to the cubic spline polynomials during each design cycle. This implies five function evaluations per cycle. This is confirmed by the sequential 1-D line search, which also goes through five function evaluations (in this case five CFD analyses) per cycle, with a total execution time of about 10 hours. As mentioned previously, the execution time for the sequential approach is about 2.5 times that of the parallel approach, not 5 times as would have been expected from five function evaluations. This is because in the sequential approach, solution restart files are employed. The potential for time saving using the parallel approach is even higher for 1-D line search options that involve larger number of function evaluations per design cycle, e.g., with the popular golden-section method. It should be noted that, regardless of the particular 1-D search method which is employed by the optimizer, the execution time will remain approximately constant (at just the execution time for one CFD analysis) for the parallel 1-D line-search implementation. On the other hand, the execution time for the sequential 1-D search implementation increases with the number of function evaluations. The turnaround time is about two days per design cycle. Thus, the entire five design cycles were completed in a total turnaround time of about 10 days, where all calculations were performed on the heavily-used IBM-SP2 of NASA's NAS computer systems.

Summary and Conclusions

In the first part of this study, a flow shape-sensitivity analysis code, CFL3D.ADII (v.2.0), is developed. The accuracy of the sensitivity derivatives (SDs) obtained from the code has been successfully validated. The computational resources required by the code are indeed very large, in terms of both CPU time and computer memory. However, very significant enhancements to code efficiency have been realized in the newer version 2.0 of CFL3D.ADII (in comparison with the straightforward, black-box approach for automatic differentiation of the CFD code). The results documented herein indicate that the new code will consistently generate the required geometric-shape SDs with significantly more reliability and less overall computa-

tional cost than that of the simple finite-difference approach.

In the second part of this study, the CFL3D.ADII (v.2.0) is incorporated in the formulation of an overall procedure targeted towards significant reduction in the execution time and the turnaround time for shape-design optimization of complex 3-D realistic aerodynamic configurations. This procedure includes (i) a coarse-grain parallel implementation of CFL3D.ADII for discrete sensitivity analysis, and (ii) the development of a parallel 1-D line search to replace the typical sequential 1-D line search in gradient-based design optimization. To demonstrate the procedure, design improvement studies of a realistic proprietary 3-D HSCT wing/body geometry were performed on an IBM-SP2 distributed-memory parallel computer. The geometry was represented by about 200,000 grid points and 60 design variables. The flow physics was inviscid, represented by the 3-D Euler equations; 60 SP2 nodes were employed in the computation of the sensitivity derivatives (gradients) of the required aerodynamic functions, using one node per design variable. The total execution time for the gradient evaluation for the 60 design variables is equivalent to that of a single design variable since communication among the nodes is negligible. The execution time for the parallel 1-D line search is effectively reduced to that of one full CFD analysis. For the 1-D line-search option selected in this study[19], the parallel 1-D line search approach is about 2.5 times as fast as the traditional sequential approach, despite the latter's judicious use of solution restart files. Both approaches gave final design variable vectors that are identical.

Acknowledgments

CFL3D.ADII (v.1.0) was developed with funding from the MDO branch of NASA-Langley Research Center, Hampton, VA, under contract NASI-19858, Task #77. The improved version 2.0 and the shape-design optimization procedure were developed with funding from the High-Speed Aerodynamics Technology Group of McDonnell Douglas Corporation, Long Beach, CA. Thanks are offered to Drs. S. Agrawal, P. Hartwich, S. Cheung, E. Unger and J. Hager of MDC; also acknowledgments to Drs. T. Zang, L. Green and P. Newman of the MDO branch of NASA-Langley Research Center, Hampton, VA.

References

1. Bischof, C. H., Carle, A., Corliss, G. F., Griewank, A., and Hovland, P., "ADIFOR: Generating Derivative Codes from FORTRAN Programs," *Scientific Programming*, Vol. 1, No. 1, 1992, pp. 11-29.

2. Korivi, V. M., Taylor, A. C. III, Newman, P. A., Hou, G. J.-W., and Jones, H. E., "An Approximately-Factored Incremental Strategy for Calculating Consistent Discrete Aerodynamic Sensitivity Derivatives," *Journal of Computational Physics*, Vol.113, No. 2, Aug. 1994, pp.336-346.

3. Newman, P. A., Hou, G. J.-W., Jones, H. E., Taylor, A. C. III, and Korivi, V. M., "Observations on Computational Methodologies for Use in Large-Scale Gradient-Based Multidisciplinary Design," AIAA Paper 92-4753-CP, Sep. 1992.

4. Sherman, L., Taylor III, A. C., Green L. L., Newman, P. A., Hou, G. J.-W., and Korivi, V. M., "First- and Second-Order Aerodynamic Sensitivity Derivatives via Automatic Differentiation with Incremental Iterative Methods," *Journal of Computational Physics*, Vol. 129, Dec. 1996, pp. 307-331.

5. Taylor III, A. C., "Automatic Differentiation of Advanced Flow-Analysis Codes in Incremental Iterative Form for Multidisciplinary Applications," Final Report, Master Contract NAS1-19858 (Task No. 77), Nov. 1995

6. Taylor, A. C. III; Green, L. L.; and Newman, P. A.; "Automatic Differentiation of the CFL3D Flow Code in Incremental Iterative Form," Poster session presented at the Second International SIAM Workshop on Computational Differentiation, Santa Fe, NM, Feb. 1215, 1996. (A short written abstract and copies of the poster session are available from the authors.)

7. Hou, G. J.-.W; Maroju, V.; Taylor, A. C. III; Korivi, V. M.; and Newman, P. A.; "Transonic Turbulent Airfoil Design Optimization Using Automatic Differentiation in Incremental Iterative Forms," AIAA Paper 95-1692-CP.

8. Korivi, V. M., Taylor III, A. C., Hou, G. J.-W., Newman, P. A., and Jones, H. E., "Sensitivity Derivatives for Three-Dimensional Supersonic Euler Code using Incremental Iterative Strategy," *AIAA Journal*, Vol. 32, No. 6, 1994, pp. 1319-1321.

9. Baysal, O., and Eleshaky, M. E., "Aerodynamic Design Optimization using Sensitivity Analysis and Computational Fluid Dynamics," AIAA Paper 91-0471, Jan. 1991.

10. Eleshaky, M. E., and Baysal, O., "Airfoil Shape Optimization using Sensitivity Analysis on Viscous Flow Equations," *Journal of Fluids Engineering*, Vol. 115, No. 1, Mar. 1993, pp. 75-84.

11. Burgreen, G. W., and Baysal, O., "Three-Dimensional Aerodynamic Shape Optimization of Wings using Sensitivity Analysis," AIAA Paper 94-0094, Jan. 1994.

12. Burgreen, G. W., and Baysal, O., "Three-Dimensional Aerodynamic Shape Optimization of Supersonic Delta Wings," AIAA Paper 94-0094, Jan. 1994.

13. Reuther, J., Jameson, A., Farmer, J., Martinelli, L., and Saunders, D., "Aerodynamic Shape Optimization of Complex Aircraft Configurations via Adjoint Formulation," AIAA Paper 96-0094, Jan. 1996

14. Oloso, A. O.; "Three-Dimensional Aerodynamic Design Optimization Using Discrete Sensitivity Analysis and Parallel Computing," Ph.D. Dissertation, Dept. of Mechanical Engineering, Old Dominion Universtiy, Norfolk VA, May 1997.

15. Bischof, C. H., Green, L. L., Haigler, K. J., Knuff, T. L., "Parallel Calculation of Sensitivity Derivatives for Aircraft Design Using Automatic Differentiation," AIAA Paper 94-4261, Sep. 1994.

16. Taylor III, A. C., Korivi, V. M., and Hou, G. J.-W., "Approximate Analysis and Sensitivity Analysis Methods for Viscous Flow involving Variation of Geometric Shape," AIAA Paper 91-1569, Jun. 1991.

17. Eleshaky, M. E., "A Computational Aerodynamic Design Optimization Method using Sensitivity Analysis," Ph.D. Dissertation, Dept.of Mechanical Engineering, Old Dominion University, Norfolk, VA, May 1992.

18. Noor, A. K., and Whitworth, S. L., "Sensitivity Analysis for Large Scale Problems," Sensitivity Analysis in Engineering, NASA CP-2457, NASA Langley Research Center, Hampton, VA, pp. 357-374, 1986.

19. Vanderplaats, G. N., "ADS - A Fortran Program for Automated Design Synthesis Version 1.10," NASA CR-177985, NASA-Langley Research Center, Hampton, VA, Sep. 1985.

Table 1: Accuracy Comparisons As Sensitivity Derivative Ratios ($\frac{CFL3D.ADII}{CentralFiniteDifference}$) of Aerodynamic Coefficients C_L (Lift) and C_D (Drag)

Design Variable (DV)	$\frac{\nabla CL_A DII}{\nabla CL_F D}$	$\frac{\nabla CL_A DII}{\nabla CL_F D}$
DV1	1.0008	1.0119
DV2	1.0000	1.0000
DV3	0.9928	1.9896
DV4	1.0023	1.0002
DV5	0.9990	1.0025

Nomenclature for Table 1

$\nabla C_L \equiv \frac{\partial C_L}{\partial DV} \equiv$ Sensitivity Derivative with respect to Design Variables (DVs 1-5); Similar Notation for C_D

ADII $\equiv$ Obtained from the CFL3D.ADII Code

FD $\equiv$ Obtained from the Central Finite-Difference Method

Table 2a: CPU Timings* for CFL3D, Version 4.1 with Different Code Options

Code Option	CPU Time (μsec/MGC/GP)
RD.NL	17.7
RD.MM	18.3
VL.NL	41.5
VL.MM	42.1

*CPU Time for Cray-YMP (Sabre)

Table 2b: CPU-Timing Ratios ($\frac{\mu sec/MGC/GP/DV \quad \text{for CFL3D.AD**}}{\mu sec/MGC/GP \quad \text{for CFL3D}}$) with Different Code Options

Code Option	With 1 DV		With 5 DVs (per DV)	
	CFL3D.ADBB	CFL3D.ADII (v.2.0)	CFL3D.ADBB	CFL3D.ADII (v.2.0)
RD.NL	66.79	2.73	13.82	2.04
RD.MM	90.40	2.96	18.52	2.06
VL.NL	77.16	1.82	15.62	0.878
VL.MM	87.28	1.92	17.64	0.905

Table 3: Memory Requirements

Code Tested	Memory Required (MW)
CFL3D	9.975
CFL3D.ADII with 1 DV	20.160
CFL3D.ADII with 5 DVs	58.784

Abbreviations used in Tables 2a, 2b and 3

GP = grid point

DV = design variable

MGC = multigrid cycle

RD = Roe's diagonalized upwind scheme

VL = van Leer's upwind scheme

NL = no limiters

MM = min-mod flux limiter

MW = mega-words (of memory)

μsec = micro-second (of CPU time)

Table 4: Comaparison (in form of <u>Ratios</u>) of Design Results between
the Parallel and the Sequential 1-D Line-Search Approaches

Design Variable (DV)	$\left(\dfrac{Value_{parallel}}{Value_{sequential}} \right)$
DV1	0.99997
DV6	0.99997
DV12	0.99998
DV18	1.0001
DV24	0.99997
DV30	0.99996
DV36	0.99997
DV42	0.99998
DV48	1.0000
DV60	1.0001

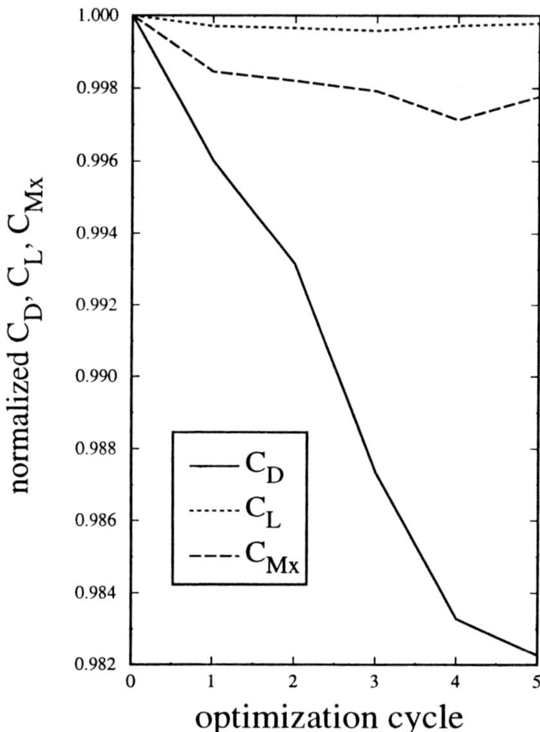

Fig. 1 Design history of the normalized drag (C_D), lift (C_L) and wing-root bending moment (C_{Mx}) coefficients.

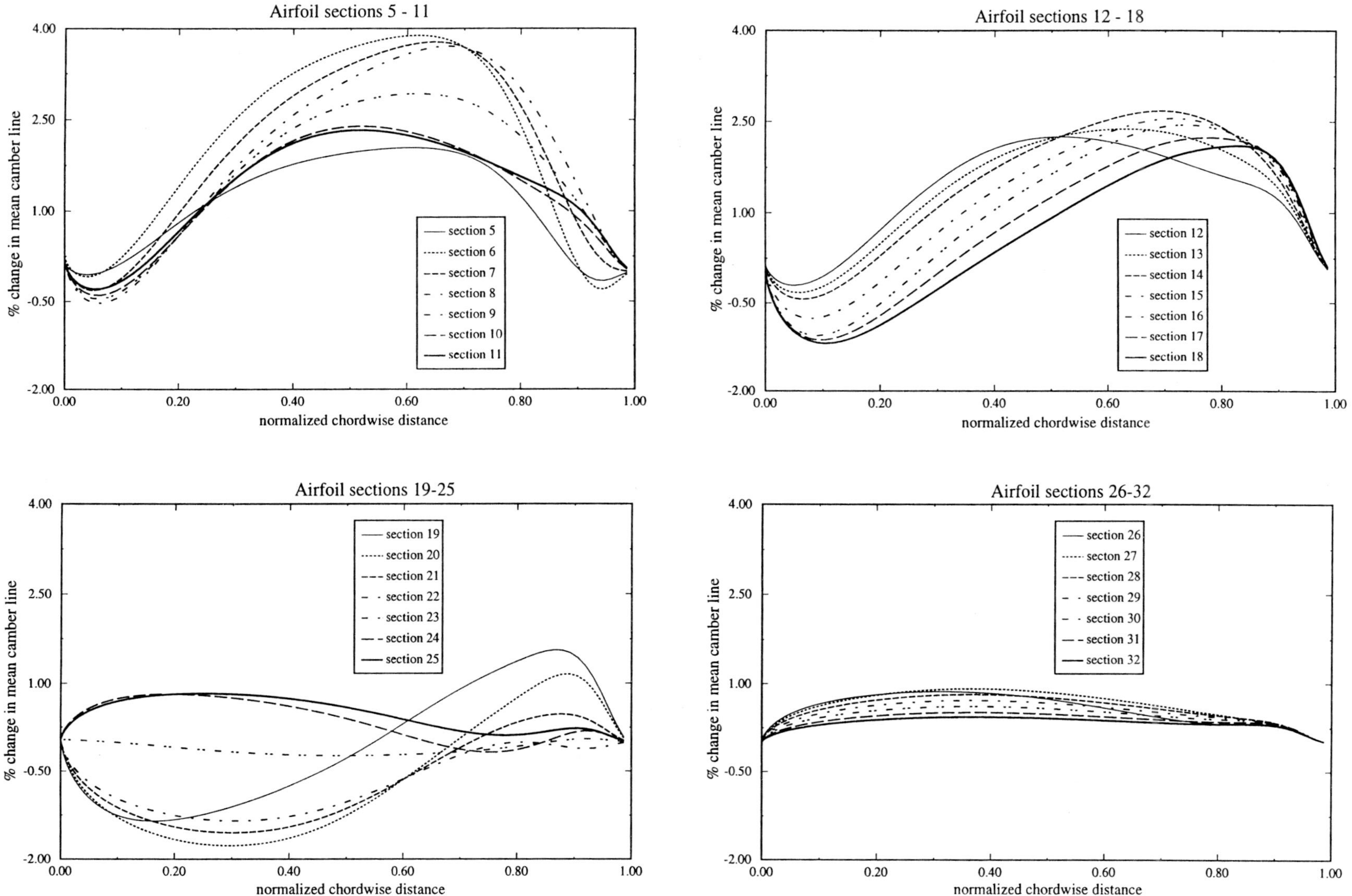

Fig. 2 Percentage change in the mean camber line for the affected wing airfoil sections.

Computational Fluid Dynamics Review 1998

Computational Fluid Dynamics Review 1998

Vol. II

Editors

M. Hafez
University of California, Davis

K. Oshima
University of Tokyo

World Scientific
Singapore • New Jersey • London • Hong Kong

Published by

World Scientific Publishing Co. Pte. Ltd.

P O Box 128, Farrer Road, Singapore 912805

USA office: Suite 1B, 1060 Main Street, River Edge, NJ 07661

UK office: 57 Shelton Street, Covent Garden, London WC2H 9HE

British Library Cataloguing-in-Publication Data
A catalogue record for this book is available from the British Library.

COMPUTATIONAL FLUID DYNAMICS REVIEW 1998

ISBN 981-02-3959-9 (Vol. I)
ISBN 981-02-3960-2 (Vol. II)
ISBN 981-02-3564-X (Set)

Printed in Singapore by Uto-Print

PREFACE

The first volume of *CFD Review* was published in 1995. The purpose of this new publication is to present comprehensive surveys and review articles which provide up-to-date information about recent progress in computational fluid dynamics, on a regular basis. Because of the multidisciplinary nature of CFD, it is difficult to cope with all the important developments in related areas. There are at least ten regular international conferences dealing with different aspects of CFD, including

* Grid Generation and Adaptation
* Artificial Viscosity and Upwind Schemes for Hyperbolic Systems
* Boundary Elements and Vortex Methods
* Finite Elements in Fluids
* Spectral Methods and High Order Schemes
* Iterative Algorithms and Preconditioning Techniques
* Domain Decomposition Methods
* Multigrid Methods
* Parallel Computation
* Flow Visualization

This list does not include optimization; a field bigger than all CFD areas combined.

Besides these specialized conferences, one should mention:

* International Conference on Numerical Methods in Fluid Dynamics
* International Symposium on CFD
* American CFD Meetings (AIAA, ASME, SIAM)
* European Conferences on CFD (ECCOMAS)
* Asian Conferences on CFD (ACFD)

as well as the British, Canadian, Australian, Japanese, Chinese conferences on CFD.

There are also the general conferences on computational methods and on fluid mechanics and aerodynamics in both engineering and applied mathematics circles.

It is a real challenge to keep up with all these activities and to be aware of essential and fundamental contributions in these areas. It is hoped that *CFD Review* will help in this regard by covering the state-of-the-art in this field.

The present book contains sixty-two articles written by authors from US, Europe Japan and China, covering the main aspects of CFD. There are five sections: general topics, numerical methods, flow physics, interdisciplinary applications, parallel computation and flow visualization. The section on numerical methods includes grids, schemes and solvers, while that on flow physics includes incompressible and compressible flows, hypersonics and gas kinetics as well as transition and turbulence. This book should be useful to all researchers in this fast-developing field.

Finally, we would like to thank Ben Ransom, Janeen Curtis, and Susan Torguson of UCD for their assistance with the computer work and the editing process.

Editors

CONTENTS OF VOLUME I

CONTENTS OF VOLUME II

INCOMPRESSIBLE NAVIER-STOKES SOLVERS USING PRIMITIVE VARIABLES

D. KWAK [1]   C. KIRIS [2]   J. DACLES-MARIANI [3]

S. ROGERS [1]   S. YOON [1]

(1) NASA Ames Research Center, Moffett Field, CA 94035

(2) MCAT Inc., Mountain View, CA 95127

(3) University of California, Davis, CA 95616

Abstract

This report reviews recent progress in incompressible Navier-Stokes solution methods and their application to problems of engineering interest. Discussions are focused on the methods designed for complex geometry applications in three dimensions, and thus are limited to primitive variable formulation. A summary of our recent progress in flow solver development is given followed by numerical studies of a few example problems of our current interest. Both steady and unsteady solution algorithms and their salient features are discussed. Solvers discussed here are based on a structured-grid approach using either a finite-difference or a finite-volume frame work.

1 Introduction

From an aerodynamicist's point of view, incompressible flow can be considered as a limiting case of compressible flow as the flow speed approaches a significantly low value compared to the speed of sound. There are a large number of flow problems of practical importance in aerospace and other fields which belong in this category. The incompressible Navier-Stokes equations, which represent these flows, pose a special problem of satisfying the mass conservation equation because it is not coupled to the momentum equations. Physically, these equations are characterized by the elliptic behavior of the pressure waves, the speed of which is infinite.

Various methods have been developed, which can be classified in numerous ways depending on the choice of formulations, variables, or algorithms. Since three-dimensional applications involving complex geometries are of our primary interest, the primitive variable formulation is chosen in the present study. The primitive variables, namely the pressure and the velocities, can easily be defined in real geometry compared to derived quantities like stream function or vorticity. Therefore, for convenience and flexibility, primitive variable formulations were used for developing incompressible Navier-Stokes codes (INS3D family of codes) at NASA Ames Research Center. The present article is intended to present our progress made since the review given by the first author in 1989 as a VKI Lecture note (Kwak, 1989). Some features will be repeated here for convenience to the reader and also

Received on July 30, 1997.

for discussing algorithmic characteristics and applications. The solution procedures presented here are mainly within a structured-grid framework. During the last several years, a large number of review articles and books on CFD discussed incompressible flow methods. For a more comprehensive review of computational methods for incompressible flow in general, readers are referred to these materials, i.e. Hirsch (1988), Hafez and Oshima (1995).

In Section II, solution methods are reviewed. Computed examples are then presented in Section III.

2 Solution Methods

In this section, two solution methods used in the development of INS3D, namely, an artificial compressibility method and a pressure projection method, are reviewed. The governing equations will be given first, followed by a discussion on the computational procedure related to the two methods.

2.1 Formulation

Three-dimensional incompressible flow with constant density is governed by the following Navier-Stokes equations:

$$\frac{\partial u_i}{\partial x_i} = 0 \tag{1}$$

$$\frac{\partial u_i}{\partial t} + \frac{\partial u_i u_j}{\partial x_j} = -\frac{\partial p}{\partial x_i} + \frac{\partial \tau_{ij}}{\partial x_j} \tag{2}$$

where t is the time, x_i the Carttesian coordinates, u_i the corresponding velocity components, p the pressure, and τ_{ij} ther viscoous-stress tensor. All the variables haave been nondimensionalized by a reference velocity and length scale. The viscous stress tensor can be written as

$$\tau_{ij} = 2\upsilon S_{ij} - R_{ij} \qquad (2.3)$$

$$S_{ij} = \frac{1}{2}\left(\frac{\partial u_i}{\partial x_j} + \frac{\partial u_j}{\partial x_i}\right) \qquad (2.4)$$

where, υ is the kinematic viscosity, S_{ij} is the strain-rate tensor, and R_{ij} are the Reynolds stresses. Various levels of closure models for R_{ij} are possible. In the present article, turbulence is simulated by an eddy viscosity model using a constitutive equation of the following form:

$$R_{ij} = \frac{1}{3}R_{kk}\delta_{ij} - 2\upsilon_t S_{ij} \qquad (2.5)$$

where υ_t is the turbulent eddy viscosity. By including the normal stress, R_{kk}, in the pressure, υ in equation (2.3) can be replaced by $\left(\upsilon + \upsilon_t\right)$ as follows.

$$\tau_{ij} = 2\left(\upsilon + \upsilon_t\right)S_{ij} = 2\upsilon_T S_{ij} \qquad (2.6)$$

In the remainder of this note the total viscosity, υ_T, will be represented simply by υ. The present formulations allow for spatially varying viscosity.

In general curvilinear coordinates, $\left(\xi, \eta, \zeta\right)$, the governing equations can be written as

$$\frac{\partial}{\partial t}\hat{u} = -\frac{\partial}{\partial \xi_i}(\hat{e}_i - \hat{e}_{vi}) + \hat{s} = -\hat{r} \qquad (2.7)$$

$$\frac{\partial}{\partial \xi_i}\left(\frac{U - (\xi)_t}{J}\right) = 0 \qquad (2.8)$$

where
$$\xi_i = \xi, \eta \text{ or } \zeta \text{ for i=1, 2, or 3}$$

$$\hat{u} = \frac{1}{J}\begin{bmatrix} u \\ v \\ w \end{bmatrix}$$

$$\hat{e}_i = \frac{1}{J}\begin{bmatrix} (\xi_i)_x p + uU_i \\ (\xi_i)_y p + vU_i \\ (\xi_i)_z p + wU_i \end{bmatrix}, \qquad (2.9)$$

$$\hat{e}_{vi} = \frac{\upsilon}{J}\nabla\xi_i \cdot \left(\nabla\xi_i\frac{\partial}{\partial \xi_i}\right)\begin{bmatrix} u \\ v \\ w \end{bmatrix}$$

$$U_i = \left(\xi_i\right)_t + \left(\xi_i\right)_x u + \left(\xi_i\right)_y v + \left(\xi_i\right)_z w$$

J = Jacobian of the transformation

$\hat{s}$ = source term

The source term $\hat{s}$ is used to represent centrifugal and Coriolis forces in a steady rotating reference frame, and will be discussed later in this section. For most flow applications, this term is set to zero.

2.2 Method Based on Compressible Flow Algorithm: Artificial Compressibility Method

Major advances in the state of the art in CFD have been made in conjunction with compressible flow computations. Therefore, it is of significant interest to be able to use some of these compressible flow algorithms for incompressible flows. To do this, the artificial compressibility method of Chorin (1967) can be used. In this formulation, the continuity equation is modified by adding a time-derivative of the pressure, resulting in

$$\frac{1}{\beta}\frac{\partial p}{\partial t} + \frac{\partial u_i}{\partial x_i} = 0 \qquad (2.10)$$

where β is an artificial compressibility parameter. Together with the unsteady momentum equations, this forms a hyperbolic-parabolic type of time-dependent system of equations. Thus, implicit schemes developed for compressible flows can be implemented. It is to be noted that the t no longer represents a true physical time in this formulation.

Physically, this means that waves of finite speed are introduced into the incompressible flow field as a medium to distribute the pressure. For a truly incompressible flow, the wave speed is infinite, whereas the speed of propagation of these pseudo waves depend on the magnitude of the artificial compressibility parameter. In a truly incompressible flow, the pressure field is affected instantaneously by a disturbance in the flow, but with artificial compressibility, there is a time lag between the flow disturbance and its effect on the pressure field. Ideally, the value of the artificial compressibility parameter is to be chosen as high as the particular choice of algorithm will allow so that the incompressibility is recovered quickly. This has to be done without lessening the accuracy and the stability property of the numerical method implemented. On the other hand, if the artificial compressibility is chosen such that these waves travel too slowly, then the variation of the pressure field accompanying these waves is very slow. This will interfere with the proper development of the viscous boundary layer. In viscous flows, the behavior of the boundary layer is very sensitive to the streamwise pressure gradient, especially when the boundary layer is separated. If separation is present, a pressure wave traveling with finite speed will cause a change in the local pressure gradient which will affect the location of the flow separation. This change in separated flow will feed back to the pressure field, possibly preventing convergence to a steady state. When the viscous effect is important for the entire flow field as in most internal flow problems, the interaction between the pseudo pressure-waves and the viscous flow field is especially important.

Artificial compressibility relaxes the strict requirement of satisfying mass conservation in each step. However, to utilize this convenient feature, it is essential to understand the nature of the artificial compressibility both physically and mathematically. Chang and Kwak (1984) reported details of the artificial compressibility, and suggested some guidelines for choosing the artificial compressibility parameter. Various applications which evolved from this concept have been reported for obtaining steady-state solutions (e.g., Steger and Kutler, 1977; Kwak et al. 1986; Chang et al., 1988; Choi and Merkle, 1985). To obtain time-dependent solutions using this method, an iterative procedure can be applied in each physical time step such that the continuity equation is satisfied (see, Merkle and Athavale, 1987, Rogers and Kwak ,1988 and 1991, Belov et. al., 1995). Further discussions on the artificial compressibility approach can be found in the literature (see, Temam, 1979, Rizzi and Eriksson, 1985).

2.2.1 Steady State Formulation

Combining equation (2.10) and the momentum equations gives the following system of equations:

$$\frac{\partial}{\partial \tau}\hat{D} = -\frac{\partial}{\partial \xi_i}\left(\hat{E}_i - \hat{E}_{vi}\right) + \hat{S} = -\hat{R} \quad (2.11)$$

where $\hat{R}$ is the right-hand-side of the momentum equation and can be defined as the residual for steady-state computations, and where

$$\hat{D} = \frac{D}{J} = \frac{1}{J}\begin{bmatrix} p \\ u \\ v \\ w \end{bmatrix}, \qquad \hat{E}_i = \begin{bmatrix} \beta\left(U_i - \left(\xi_i\right)_t\right)/J \\ \hat{e}_i \end{bmatrix},$$

$$\hat{E}_{vi} = \begin{bmatrix} 0 \\ \hat{e}_{vi} \end{bmatrix} \qquad (2.12)$$

When the governing equations are solved in a steadily rotating reference frames, the source term, $\hat{S}$, represents centrifugal and Coriolis terms. If the relative reference frame is rotating around the x-axis, the source term $\hat{S}$ is given by

$$\hat{S} = \begin{bmatrix} 0 \\ 0 \\ \Omega(\Omega y + 2w) \\ \Omega(\Omega z - 2v) \end{bmatrix}$$

where Ω is the rotational speed. In this report, the source term, $\hat{S}$, is set to zero other than for rotational steady solutions. Relative velocity components are written in terms of absolute velocity components u_a, v_a, and w_a as

$$u = u_a$$
$$v = v_a + \Omega z$$
$$w = w_a - \Omega y$$

In the steady-state formulation the equations are to be marched in a time-like fashion until the divergence of velocity in equation (2.10) converges to a specified tolerance. The time variable for this process no longer represents physical time, so in the momentum equations t is replaced with τ, which can be thought of as a pseudo-time or iteration parameter.

An unfactored implicit scheme can be obtained from the governing equations by linearizing the flux vectors about the previous time step.

$$\left[I + \alpha\Delta\tau J\left(\delta_{\xi_i}\left(\hat{A}_i^n - \Gamma_i\right) - \hat{H}\right)\right]\left(D^{n+1} - D^n\right)$$
$$= -\Delta\tau J\left[\delta_{\xi_i}\left(\hat{E}_i - \hat{E}_{vi}\right)^n - \hat{S}\right] \qquad (2.13)$$

where

$$\hat{A}_i = \text{Jacobian matrices of } \hat{E}_i$$

$\hat{H}$ = Jacobian matrix of the source term $\hat{S}$

$$\Gamma_1 D^{n+1} = \left(\frac{v}{J}\right)\nabla\xi \cdot \left(\nabla\xi_i \frac{\partial}{\partial\xi_i}\right)I_m \frac{\partial D}{\partial\xi}$$

Γ_2 and Γ_3 are similarly defined

$$\delta_\xi = \text{finite difference form of } \frac{\partial}{\partial\xi}$$

$\alpha = 1/2$ for trapezoidal, or 1 for Euler implicit
$$I_m = diag[0,1,1,1]$$

This equation is iterated in pseudo time until the solution converges to steady state, at which time the original incompressible Navier-Stokes equations are satisfied. A direct inversion of equation (2.13) would become a Newton iteration for a steady-state solution. In three dimensions, however, direct inversion of a large block banded matrix of the unfactored scheme would be impractical.

ADI scheme:

To overcome the difficulties of an unfactored implicit scheme, indirect methods have been devised by many researchers. The alternating direction implicit scheme by Beam and Warming (1978) and Briley and McDonald (1977) approximates the implicit operator in the unfactored scheme by a product of three one dimensional operators. It is difficult to apply the ADI scheme to equation (2.13) in its full matrix form. Noting that at steady-state the left-hand side of equation (2.13) approaches zero, a simplified expression for the viscous terms can be used on the left-hand side. To maintain the accuracy of the solution, the entire viscous terms are used on the right-hand side. Then equation (2.13) becomes

$$L_\xi L_\eta L_\varsigma L_H\left(D^{n+1} - D^n\right) = RHS \qquad (2.14)$$

where

$$L_{\xi_i} = \left[I + \frac{\Delta t}{2}J^{n+1}\delta_{\xi_i}\left(\hat{A}^n - \gamma_i\right)\right]$$

$$L_H = \left[I - \frac{\Delta t}{2}J^{n+1}\hat{H}\right]$$

γ_i = diagonal part of viscous terms
RHS = right-hand side of equation (2.13)

The inversion problems is reduced to a series of one-dimensional inversions

$$L_\xi \Delta\overline{D} = RHS$$
$$L_\eta \Delta\overline{\overline{D}} = \Delta\overline{D} \qquad (2.15)$$

$$L_\varsigma \Delta D = \Delta\overline{\overline{D}}$$

where

$$\Delta D = D^{n+1} - D^n$$

With the use of numerical dissipation, this scheme becomes conditionally stable. The ADI scheme introduces a factorization error which reduces the rate of convergence. However, this method introduced perhaps the first implicit scheme to CFD, thus viscous flow computations became more feasible. With the reduction of cost by diagonalizing the Jacobian (Pulliam and Chaussee, 1981), this scheme has been used in the development of many flow solvers. At NASA Ames, the first incompressible Navier-Stokes solver in generalized three-dimensional coordinate, INS3D code (Kwak et. al, 1984, 1986; Chang and Kwak, 1984, Rogers et al., 1987a), was developed (for user's guide, see Rogers et al. 1987b). The INS3D code was extensively used as a primary design tool for an upgraded design of the Space Shuttle main engine (SSME) hot gas manifold (see, Chang et. al, 1988; Yang et. al. 1987). This was probably the first major contribution by CFD in rocket propulsion system design.

LU-SGS scheme:

To improve the stability and convergence of the ADI scheme, several two-factor schemes have been developed. By combining the advantage of LU factorization and symmetric Gauss-Seidel (SGS) relaxation, Yoon and Jameson (1987) devised a new implicit algorithm called LU-SGS. By implementing the LU-SGS scheme to the artificial compressible formulation, the INS3D-LU code was developed (Yoon and Kwak, 1989). Applying the LU factorization to equation (2.13), the following LU scheme is obtained:

$$\left[I + \alpha\Delta\tau J\left(\delta_\xi^- \hat{A}^+ + \delta_\eta^- \hat{B}^+ + \delta_\varsigma^- \hat{C}^+ - \frac{1}{2}\hat{H}\right)\right]$$
$$\left[1 + \alpha\Delta\tau J\left(\delta_\xi^+ \hat{A}^- + \delta_\eta^+ \hat{B}^- + \delta_\varsigma^+ \hat{C}^- - \frac{1}{2}\hat{H}\right)\right]$$
$$\left(D^{n+1} - D^n\right) = -\Delta\tau J\left[\delta_{\xi_i}\left(\hat{E}_i - \hat{E}_{vi}\right)^n - \hat{S}\right] \qquad (2.16)$$

where δ_ξ^- and δ_ξ^+ are the backward- and forward-difference operators respectively, and $\hat{A}_i = \hat{A}$, $\hat{B}$, or $\hat{C}$ for $i = 1, 2,$ or 3 respectively.

The LU-SGS scheme can be written as

$$L_l L_d L_u \left(D^{n+1} - D^n \right) = RHS \qquad (2.17)$$

where

$$L_l = I + \alpha \Delta \tau J \left(\delta_\xi^- \hat{A}^+ + \delta_\eta^- \hat{B}^+ + \delta_\varsigma^- \hat{C}^+ \right.$$
$$\left. - \hat{A}^- - \hat{B}^- - \hat{C}^- - \hat{H} \right)$$

$$L_d = I + \alpha \Delta \tau J \left(\hat{A}^+ - \hat{A}^- + \hat{B}^+ - \hat{B}^- + \hat{C}^+ - \hat{C}^- \right)$$

$$L_u = I + \alpha \Delta \tau J \left(\delta_\xi^+ \hat{A}^- + \delta_\eta^+ \hat{B}^- + \delta_\varsigma^+ \hat{C} \right.$$
$$\left. + \hat{A}^+ + \hat{B}^+ + \hat{C}^+ \right)$$

The INS3D-LU code was based on the finite volume discretization while the original INS3D used a finite difference form. The derivatives of the viscous fluxes are approximated using second-order central differences. The convective flux terms are discretized using central differences, which require numerical dissipation terms for stability. By choosing different numerical dissipation models and Jacobian matrices, a variety of schemes can be developed. Further details can be found in the reference cited above.

<u>Line Relaxation</u>

The line-relaxation implicit scheme is formed not through factorization of the left-hand-side matrix, but through an iterative solution process. The discrete form of the matrix on the left-hand-side of equation (2.13) is a banded matrix composed of seven diagonals, where each entry of a diagonal consists of a 4 x 4 block. The discrete version of equation (2.13) is written as

$$[T,0,...,0,U,0,...,0,X,Y,Z,0,...,0,V,0,...,0,W]$$
$$\Delta D = RHS \qquad (2.18)$$

where T,U,V,W,X,Y and Z are the diagonals, with the Y vector being the main (center) diagonal. This matrix equation is approximately solved using an iterative approach. One of the three computational directions is chosen to be the implicit direction, and the sweeping through the domain proceeds in the other two directions. Using, for example, the ξ family, a tridiagonal matrix is formed by keeping the X,Y and Z diagonals on the left-hand-side, and multiplying the remaining diagonals by the latest known iteration of the ΔD solution vector, and shifting them to the right-hand-side. A forward sweep is composed of solving a block-tridiagonal system of the form:

$$[X,Y,Z]\Delta D^{l+1} = RHS - [T,0,...,0,U]\Delta D^{l+1}$$
$$-[V,0,...,0,W]\Delta D^l$$

and a backward sweep is similarly composed of solving:

$$[X,Y,Z]\Delta D^{l+1} = RHS - [T,0,...,0,U]\Delta D^l$$
$$-[V,0,...,0,W]\Delta D^{l+1}$$

where the l superscript denotes the sweep iteration number. The process is initialized by setting $\Delta D^0 = 0$.

The algorithm is implemented so that any or all of the three computational directions can be chosen for the sweep direction. The optimum direction and number of sweeps is very much problem dependent. Experience with this algorithm has shown that for most problems it is best to use the wall-normal direction as the implicit direction, and that on the order of 10 sweeps should be used.

2.2.2 Time-Accurate Formulation

Time-dependent calculation of incompressible flows are especially time consuming due to the elliptic nature of the governing equations. Physically, this means that any local change in the flow has to be felt by the entire flow field. Numerically, this means that in each time step, the pressure field has to go through one complete steady-state iteration cycle, for example, by Poisson-solver-type pressure iteration or artificial compressibility iteration method. In transient flow, the physical time step has to be small and consequently the change in the flow field may be small. In this situation, the number of iterations in each time step for getting a divergence-free flow field may not be as high as regular steady-state computations. However, the time-accurate computations are generally an order of magnitude more time-consuming than steady-state computation. Therefore, it is particularly desirable to develop computationally efficient methods either by implementing a fast algorithm and by utilizing computer characteristics such as vectorization and parallel processing.

A time-accurate method using artificial compressibility developed by Rogers and Kwak (1988, 1991) is summarized next. In this formulation the time derivatives in the momentum equations are differenced using a second-order, three-point, backward-difference formula

$$\frac{3\hat{u}^{n+1} - 4\hat{u}^n + \hat{u}^{n-1}}{2\Delta t} = -\hat{r}^{n+1} \qquad (2.19)$$

where the superscript n denotes the quantities at time $t = n\Delta t$ and $\hat{r}$ is the right-hand side given in equation (2.7). To solve equation (2.19) for a divergence free velocity field at the $(n+1)$ time level, a pseudo-time level is introduced and is denoted by a superscript m. The equations are iteratively solved such that $\hat{u}^{n+1,m+1}$ approaches the new velocity $\hat{u}^{n+1}$ as the divergence of $\hat{u}^{n+1,m+1}$ approaches zero. To drive the divergence of this velocity to zero, the following artificial compressibility relation is introduced:

$$\frac{p^{n+1,m+1} - p^{n+1,m}}{\Delta \tau} = -\beta \nabla \cdot \hat{u}^{n+1,m+1} \qquad (2.20)$$

where τ denotes pseudo-time and β is an artificial compressibility parameter. Combining equation (2.20) with the momentum equations gives

$$I_{tr}\left(\hat{D}^{n+1,m+1} - \hat{D}^{n+1,m}\right)$$
$$= -\hat{R}^{n+1,m+1} - \frac{I_m}{\Delta t}\left(1.5\hat{D}^{n+1,m} - 2\hat{D}^n + 0.5\hat{D}^{n-1}\right)$$

$$(2.21)$$

where $\hat{D}$ is the same vector defined in equation (2.13), $\hat{R}$ is the same residual vector defined in equation (2.11), and I_{tr} is a diagonal matrix given by

$$I_{tr} = diag\left[\frac{1}{\Delta t}, \frac{1.5}{\Delta t}, \frac{1.5}{\Delta t}, \frac{1.5}{\Delta t}\right]$$

Finally, the residual term at the $m+1$ pseudo-time level is linearized giving the following equation in delta form

$$\left[\frac{I_{tr}}{J} + \left(\frac{\partial \hat{R}}{\partial D}\right)^{n+1,m}\right]\left(\hat{D}^{n+1,m+1} - \hat{D}^{n+1,m}\right)$$
$$= -\hat{R}^{n+1,m} - \frac{I_m}{\Delta t}\left(1.5\hat{D}^{n+1,m} - 2\hat{D}^n + 0.5\hat{D}^{n-1}\right)$$

$$(2.22)$$

As can be seen, this equation is very similar to the steady-state formulation which can be rewritten for the Euler implicit case as

$$\left[\frac{1}{J\Delta \tau}I + \left(\frac{\partial \hat{R}}{\partial D}\right)^n\right]\left(\hat{D}^{n+1} - \hat{D}^n\right) = -\hat{R}^n$$

$$(2.23)$$

Both systems of equations will require the discretization of the same residual vector $\hat{R}$.

Since, in the artificial compressibility formulation, the governing equations are changed into a hyperbolic-parabolic type, some of the upwind differencing schemes which have been developed for the compressible Euler and Navier-Stokes equations can be utilized. The new version of the incompressible Navier-Stokes code, INS3D-UP code (see Rogers and Kwak, 1988), was then developed using an upwinding scheme. The method of Roe (1981) was used in differencing the convective terms by an upwind method that is biased by the signs of the eigenvalues of the local flux Jacobian. This is accomplished by casting the governing equations in their characteristic form and then forming the differencing stencil such that it accounts for the direction of wave propagation. In this code, the set of numerical equations are solved using a nonfactored line relaxation scheme similar to that employed by MacCormack (1985). The INS3D-UP has been used in a number of applications including steady and unsteady flow simulations.

Even though the time-accurate artificial compressibility formulation has been used successfully for unsteady flow calculations, it is desired to have an even more efficient solver for time-dependent problems. In the next section, a method based on a pressure projection approach will be discussed; this has been found to be an efficient method for obtaining time accurate solutions.

2.3 Method Based on Pressure Projection

In 1965, Harlow and Welch published the first primitive variable method using a Poisson equation for pressure. In this method, called the marker-and-cell (MAC) method, the pressure is used as a mapping parameter to satisfy the continuity equation. By taking the divergence of the momentum equation, the Poisson equation for pressure is obtained:

$$\nabla^2 p = \frac{\partial h_i}{\partial x_i} - \frac{\partial}{\partial t}\frac{\partial u_i}{\partial x_i} \qquad (2.24)$$

where

$$h_i = -\frac{\partial u_i u_j}{\partial x_j} + \frac{\partial \tau_{ij}}{\partial x_j}$$

The usual computational procedure involves choosing the pressure field at the current time step such that continuity is satisfied at the next time step. The original MAC method is based on a staggered arrangement on a 2-D Cartesian grid. The staggered grid conserves mass, momentum, and kinetic energy in a natural way and avoids odd-even point decoupling of the pressure encountered in a regular grid (Gresho and Sani, 1987). Even though the original method used an explicit Euler solver, various time advancing schemes can be implemented with this formulation. Ever since its introduction, numerous variations of the MAC method have been devised and successful computations have been made.

The MAC method can be viewed as a special case of the projection method (i.e. Chorin, 1968). In this method the strict requirement of obtaining the correct pressure for a divergence free velocity field in each step may significantly slow down the overall computational efficiency. To satisfy the mass conservation in grid space, the difference form of the second derivative in the Poisson equation has to be constructed consistent with the discretized momentum equation (see Kwak, 1989).

To solve for a steady-state solution, the correct pressure field is desired only when the solution is converged. In this case, the iteration procedure for the pressure can be simplified such that it requires only a few iteration at each time step. The best known method using this approach is the Semi-Implicit Method for Pressure-Linked Equations (SIMPLE) (Patankar, 1980 ; Chen et al., 1995). The unique feature of this method is the simple way of estimating the velocity and the pressure correction. This feature simplifies the computation but introduces empiricism into the method. Despite its empiricism, the method has been used successfully for many steady-state computations. It is not the intention of the present paper to evaluate this method, and readers interested in this approach are referred to the above cited references.

<u>Computational Procedure</u>

Here, the time evolution can be approximated by several steps where operator splitting can be accomplished by treating the momentum equations as a combination of convection, pressure, and viscous terms. The common application of this method, generally known as a fractional step method (see Yanenko, 1971, Marchuck. 1975), is done in two steps. The first step is to solve for an auxiliary velocity field using the momentum equation in which the pressure-gradient term can be computed from the pressure in the previous time step. In the second step, the pressure is computed which can map the auxiliary velocity onto a divergence-free velocity field.

This procedure is illustrated by the following example in Cartesian coordinates:

Step 1: Calculate auxiliary or intermediate velocity, $\hat{u}_i$, by

$$\frac{\hat{u} - u_i^n}{\Delta t} = \frac{1}{2}\left(3\tilde{H}_i^n - \tilde{H}_i^{n-1}\right) - \frac{\delta p^n}{\delta x_i}$$
$$+ \frac{1}{2}\frac{1}{\mathrm{Re}}\nabla^2\left(\hat{u}_i + u_i^n\right) \qquad (2.25)$$

where

$$\tilde{H}_i = -\frac{\delta}{\delta x_j}u_i u_j \ , \ \mathrm{Re=Reynolds\ number}$$

Here, the advection terms are advanced by a second-order Adams-Bashforth method. The pressure gradient term is added to the procedure, so as to minimize the pressure correction in each time step.

Step 2: Solve for the pressure correction.
In the second step, the momentum equation can be written as

$$\frac{u_i^{n+1} - \hat{u}_i}{\Delta t} = -\frac{1}{2}\frac{\delta}{\delta x_i}\left(\phi^{n+1} - \phi^n\right) \qquad (2.26)$$

where

$$p^n = \phi^n - \frac{\Delta t}{2\,\mathrm{Re}}\nabla^2\phi^n$$

This equation combined with continuity equation results in the following Poisson equation for the pressure correction.

$$\nabla^2\left(\phi^{n+1} - \phi^n\right) = \frac{2}{\Delta t}\frac{\delta}{\delta x_i}\hat{u}_i \qquad (2.27)$$

Once the pressure correction is computed, new pressure and velocities are calculated as follows:

$$p^{n+1} = p^n + \left(\phi^{n+1} - \phi^n\right) - \frac{\Delta t}{2\,\mathrm{Re}}\nabla^2\left(\phi^{n+1} - \phi^n\right)$$
$$\qquad (2.28)$$
$$u^{n+1} = u^n - \frac{\Delta t}{2}\frac{\delta}{\delta x_i}\left(\phi^{n+1} - \phi^n\right) \qquad (2.29)$$

Successful methods have been developed using fractional step approach in generalized coordinates (see Rosenfeld et al., 1991, 1993 and Wessling et al. 1992). One particular aspect of this approach requiring special care is the intermediate boundary conditions. Rosenfeld et al. (1991) devised a generalized scheme where physical boundary conditions can be used at intermediate steps. As with other pressure based methods, the efficiency of the fractional step method depends on the Poisson solver. A multigrid acceleration, which is physically consistent with the elliptic field, is one possible avenue to enhance the computational efficiency.

Rosenfeld et al. (1991) defined the dependent variables such that the pressure and the volume fluxes are at the center and on the faces of the primary cells, respectively. This selection is equivalent to a finite-difference formulation over a staggered grid with the choice of scaled contravariant velocity components as the unknowns. The viscous terms are treated by an approximate factorization. The resulting solver was successfully validated for time-dependent problems which require small physical time steps.

To relax the CFL-number restriction in three dimensions and to improve grid-dependent robustness of the solver, Kiris and Kwak (1996) implemented a relaxation scheme where both convective and viscous terms are treated implicitly. Here the first step is to solve the momentum equation for an auxiliary velocity field. Three-point backward differencing formula is used to discretize the momentum equations in time, as shown below:

$$\frac{1}{2\Delta t}\left(3\hat{u}_i - 4u_i^n + u_i^{n-1}\right) = -\frac{\delta p^n}{\delta x_i} + h_i(\hat{u}_i) \quad (2.30)$$

The operator splitting is formulated such that overall temporal accuracy is achieved. The resulting solver was validated under grid conditions with a metric discontinuity. It was found that a large CFL number could be used without causing an instability problem. This is a desirable feature for unsteady flow computations when the time step is restricted by the physics of the problem. A time-accurate version, INS3D-FS code, was developed using this approach (Kiris and Kwak, 1996). In order to reduce the linearization error in the implicit solution of the equation (2.30), a subiteration procedure is applied. In most cases, three subiterations are sufficient requiring about the same cost as a third order implicit Runga-Kutta scheme. Therefore, by implementing a second order implicit Runga-Kutta scheme, the computational cost for time integration is reduced to a lower value than that of using a three-point backward difference formula with three subiterations.

An implicit Runga-Kutta scheme of order N can be written as follows,

$$\left[\frac{I}{\alpha_k \Delta t} + \left(\frac{\partial h}{\partial u}\right)^{k-1}\right]\delta u_i =$$

$$h_i^{k-1} - \frac{\partial p^{k-1}}{\partial x_i} - \frac{1}{\alpha_k \Delta t}\left(u_i^{k-1} - u_i^n\right) \quad (2.31)$$

where
$$\alpha_k = \frac{1}{N+1-k}$$
$$\delta u_i = u_i^k - u_i^{k-1}$$

N and k represent the order and the step number of the Runga-Kutta scheme respectively.

For N = 2, equation (2.31) becomes,

Step 1:
$$\left[\frac{2I}{\Delta t} + \left(\frac{\partial h}{\partial u}\right)^n\right]\left(\hat{u}_i - u_i^n\right) = h_i^n - \frac{\partial p^n}{\partial x_i} \quad (2.32)$$

Step 2:
$$\left[\frac{I}{\Delta t} + \left(\frac{\partial \hat{h}}{\partial u}\right)\right]\left(u_i^{n+1} - \hat{u}_i\right) =$$

$$h_i(\hat{u}_i) - \frac{\partial p^{n+1}}{\partial x_i} - \frac{1}{\Delta t}\left(\hat{u}_i - u_i^n\right) \quad (2.33)$$

In the current version of INS3D-FS, the viscous fluxes are computed by simple averaging which results in second order central differencing. The convective flux terms are computed using a higher-order upwind-biased stencil. Flux-difference splitting is used here to structure the differencing stencil based on the sign of the eigenvalues of the convective flux Jacobian. The resulting linear algebraic equations are solved by using the Gauss-Seidel line relaxations scheme. The solver has an option to employ GMRES iterations with ILU(0) preconditioner. The solution of the Poisson equation uses GMRES iterations with a point Gauss-Seidel preconditioning procedure.

III. COMPUTED RESULTS

In the previous sections, solution methods and associated flow solvers based on primitive variable approaches have been reviewed. The performance characteristics of these solvers, such as the accuracy, computational efficiency, and robustness, have to be assessed. In this section, simple test cases are presented to show some of theses characteristics. Computed results are then presented for engineering problems simulated using two of the INS3D family of codes, INS3D-UP and INS3D-FS. These codes represent the artificial compressibility and the pressure projection methods, respectively.

3.1 Couette Flow

Grid quality, grid resolution, boundary conditions and physical modeling are some of the major factors affecting the accuracy and convergence of a flow solver. There is no rigorous guideline as to the "goodness of grid." However, there exists a varying degree of sensitivity to grid quality resulting from different discretization and algorithm approaches. Here the laminar Couette flow example is presented to show robustness of the two flow codes when the grid is not smooth.

The computational grid was generated with 63x63 points. As shown in Figure 1, the grid was intentionally distorted to introduce slope discontinuities and non-orthogonality in the computational domain. The exact solution of the Couette flow is a linear velocity profile. The flow was started with freestream velocity everywhere except the stationary wall. The lower wall is stationary and the upper wall is moving at a constant speed. Periodic conditions were imposed at the inflow and outflow boundaries.

This case was first computed using INS3D-FS. The CFL number was 100, where the CFL number is defined as

$$CFL = \max\left(\left|U^{\xi}\right| + \left|U^{\eta}\right| + \left|U^{\varsigma}\right|\right) * dt/Volume$$

The U-velocity profile is linear as shown in Figure 2. The result shows no indication of any nonlinear grid effects. The convergence history is plotted in Figure 3. The solid line shows the maximum residual of the momentum equations, and the dashed line represents the maximum divergence of velocity. The maximum divergence of velocity plot levels at about 10^{-5} because the solution of the Poisson equation is terminated after reaching $\varepsilon = 10^{-5}$ accuracy. For time-accurate cases, this error limit can be reduced further to machine accuracy.

The same case is computed with INS3D-UP with β value of 1 and 10. Maximum residuals and divergence of velocity values are plotted in Figure 4. The convergence behavior of the two methods is very similar.

3.2 Pulsitile Flow in a Constricted Channel

To discuss the characteristics of the two approaches in computing time-accurate flows, pulsitile flow in a constricted channel is chosen. The geometry is consistent with the experimental setup of Park (1989), and is shown in Figure 5. The height of the constriction is given by $a = 0.57$. This is the distance from the top wall of the channel to the lowest point in the constriction. The length of the channel upstream of the constriction is given by $L_u = 7$. The length of constriction is $L_c = 4.66$, and the downstream portion of the channel is given by $L_d = 15.34$.

The inflow boundary for the experiment was at 100 channel heights upstream of the constriction. However, the computational inflow boundary was placed at seven channel heights upstream of the constriction with a parabolic velocity profile such that the mass flow matches that of the experimental setup. The pulsitile inflow velocity is given by the shape function given in Figure 6. This problem was previously computed by Wiltberger et al. (1993), and repeated here for the purpose of comparing the two algorithms.

To obtain time-accurate solutions, the INS3D-UP is subiterated each time step until conservation of mass is satisfied, while INS3D-FS is time accurate at each time step. The resulting solutions are comparable as shown in Figures 7 and 8. In the experiment the location of the center of the vortex along the bottom wall, defined as B-vortex, was measured. The B-vortex grows immediately behind the constriction and is shed downstream. In Figure 7, streamline contours generated from these computations are shown at a time increment of 0.1 period. The results from the two codes show the same phenomena, where INS3D-FS produces larger vortical structure throughout the entire period. The location of B-vortex is then plotted against the experimental data in Figure 8. Both codes compare well with the experimental results.

To compare the time accurate procedure, the iteration process is studied in detail within one time step advancement. In INS3D-UP code, the time accuracy is achieved by subiterating within each step. Therefore, the upstream propagating pressure wave from the exit boundary has to have enough elapsed time to balance the viscous effect within the channel. The number of iterations required to accomplish this is directly related to the magnitude of the artificial compressibility parameter, defined as β earlier. The phenomena is shown in Figure 9a. On the other hand, INS3D-FS uses the Poisson equation for pressure to map the flow field into a divergence-free velocity field at the new time level. The Poisson iteration, which is done here using a point Gauss-Seidel iteration with GMRES convergence acceleration, does not exhibit wave-like phenomena as in the artificial compressibility case. This is shown in Figure 9b. Subiterating an implicit inversion as in INS3D-UP is usually more expensive compared to a Poisson iteration. For a steady state solution, however, the INS3D-UP can take a much larger time step than INS3D-FS. In sample computations, INS3D-FS required 1.5 to 2 times more CPU time than INS3D-UP for steady state solutions. In time dependent cases where a small physical time step is required, INS3D-FS runs faster than INS3D-UP by a factor of three or greater. The algorithmic characteristics of these two approaches will be discussed further in a future report.

3.3 Wingtip Vortex Formation and Propagation

The study of wingtip vortices has been of major importance in many areas of fluid engineering. Its significance can be seen in problems such as aircraft spacing during landing and take off, blade/vortex interaction on rotorcraft performance, and tip vortex cavitation on ship propeller performance. Although there has been a great deal of work done on the tip vortex in the form of theoretical, experimental and computational studies, very few actually address the near-field detail. The present study focused on the near field physics which will provide inflow conditions for simulating wake propagation in the intermediate and far field regions.

Formation on the Wing Surface

The computational studies on the formation and roll-up process were performed by Dacles-Mariani et al. (1993, 1996) using the INS3D-UP code in conjunction with an experimental study by Chow, et al. (1993). The initial roll-up process and the near field detail have been studied in great detail both experimentally and computationally.

In the computational study, the computational domain consists of the wing-wind tunnel wall geometry, as shown in Figure 10a, which is a close approximation to the experimental set-up in the 32 in. x 48 in. low speed wind tunnel at NASA Ames. The computational domain includes a rectangular half-wing with a NACA 0012 airfoil section, a rounded wing tip and the surrounding boundaries. The wing has an aspect ratio of 0.75 and was mounted inside a wind tunnel at 10 degrees angle of attack. The flow is turbulent with a Reynolds number of 4.6 million based on the chord length. The inflow boundary conditions for the velocity profiles were prescribed using experimental values and the inflow pressure was computed based on the method of characteristics using a one-dimensional Riemann

invariant. At the solid surface, (wind tunnel walls and the surface of the model), the velocity was specified to be zero.

A single grid approach (see Figure 10a) was used for this study. Resolving the physics of this problem requires a very dense grid. As a result, the CPU requirement was substantial. A total run time of 28 CPU hours on a Cray C-90 were required for a five order of magnitude drop in the residual for the 2.5 million grid point case. This study focused primarily on resolving the fundamental issues of the tip vortex computation without addressing the need to reduce the total number of grid points. However, it is recognized that for more practical implementations, the grid density requirement will have to be addressed.

As shown in Figure 10b, a vortex is formed at the tip of the wing fed by the vorticity from the tip boundary layer. A pressure differential between the upper and lower surfaces of a wing drives the fluid in the boundary layer around the tip and towards the suction side of the wing. The discrete vortex formed becomes highly three-dimensional and complex. As the vortex moves downstream, it rolls up more and more of the wing wake until its circulation is nominally equal to that of the wing. This roll-up distance is small compared to the separation of aircraft on the landing approach path, but may not be small compared to the distance between interacting lifting surfaces. The flow in the near-field is therefore important in its own right as well as providing possible means of controlling the far-field vortex.

As the flow progresses downstream, the maximum crossflow velocity increases in the vortex core. An axial pressure gradient in turn develops that accelerates the fluid in the vortex core. Phillips and Graham (1984) have shown that for a turbulent vortex, assuming that $\overline{v_\theta'^2} = \overline{v_r'^2}$, we get

$$\frac{\partial P}{\partial r} = \frac{v_\theta^2}{r} + \frac{\partial \overline{v_r'^2}}{\partial r}$$

This equation shows that the dominant term which contributes to the radial pressure gradient is a term which contains v_θ, which in turn affects the peak axial velocity. Note, however, that the second term on the right-hand side of the equation may not be neglected because of the high near field core turbulence intensities measured in the experiment by Zilliac et al. (1993). This shows that the resolution of the core properties are dependent not only on the gradient of the circumferential velocity, but also on one of the turbulent stress terms. If this term is not properly accounted for a very good comparison between measured and computed values may not be achieved.

Propagation in the Near Field

Once formed, the wake vortex almost preserves its strength for a long time due to its inviscid nature in the core region. This requires special turbulence modeling as well as high grid resolution in the core region. First the fractional step code, INS3D-FS, is used to investigate the level of error resulting from the spatial differencing and turbulence model. Then the wake velocity profiles from INS3D-UP and INS3D-FS are compared. In this study the Baldwin-Barth (1991) one-equation turbulence model was used.

For INS3D-FS computations, the computational domain includes the region from the trailing edge of the wing (x/c=1.0) to 0.673 chordlengths downstream of the wing with an H-H grid topology. Extensive experimental data (Garcia et. al., 1992) are available at x/c=1.0, 1.12, 1.24, 1.447, and 1.673. The experimental velocity profile at x/c=1.0 station is used as an inflow boundary condition. The pressure at boundaries is calculated from the compatibility condition. The computations are carried out using a relatively fine grid with dimensions of 36x82x82. The solution is converged to machine accuracy in 1000 iterations. The CPU time required for this computation is about 3.5 CRAY-C90 hours. Initially, the computations were carried out using a coarse grid. It was found that the vortex core velocity peak values were underpredicted. The grid resolution was increased by doubling the number of grid points in the k, and l directions, from grid dimension of 36x42x42 to 36x82x82. The prediction of the peak values at the vortex core was improved, but not substantially. The numerical results indicated that there is an excessive amount of numerical dissipation at the vortex core as it progresses downstream. This was consistent with the findings from the tip vortex study using artificial compressibility (see Dacles-Mariani et. al., 1993, 1996). The excessive numerical diffusion at the vortex core was reduced when the production term in the turbulence model was modified using the norm of the strain rate tensor (see, Kiris and Kwak, 1996).

In Figures 11a and 11b, the effects of the turbulence model as well as the third and fifth order convective differencing schemes are compared by showing the axial progression of flow quantities along vortex coreline. The amount of numerical dissipation is large when third-order flux difference splitting is used. In order to reduce this numerical dissipation, one can use a finer grid or can increase the stencil in the upwind-biased differencing. Since the cost of increasing the accuracy of the differencing is much less than that of increasing the grid size, the fifth order upwind differencing is used. It should be pointed out that the overall spatial accuracy of the method is second-order even though fifth order upwind differencing is used for the convective terms. That is because the volume and surface area vectors are evaluated to second-order accuracy. A simple averaging is used for the metric terms at half point locations, and a second order central differencing is used for the viscous terms. However, increasing the stencil size in upwind differencing has a significant effect in reducing the amount of numerical dissipation, compared to lower-order differencing. The results using third and fifth order schemes are compared in Figures 11a and 11b.

As the wake propagates downstream the core region still exhibits excessive dissipation, primarily due to high turbulent viscosity. To study the effects of the production term in the turbulence model, various combinations of the vorticity magnitude and the strain rate have been examined (Kiris and Kwak, 1996). The effect of the turbulence model in the core region can clearly be seen in Figures 11a and 11b. By modifying the production term in the core to incorporate the strain rate the difference from experiment in the core region was reduced to less than 2 %.

Finally, the velocity magnitude and the crossflow velocity at three different wake locations are plotted in Figure 12. As shown in the figure, the results from the two codes are reasonably close and both show the capability of capturing the vortex core quite well.

3.4 Application to Inducer and Impeller of a Liquid Rocket Engine

Until recently, the high performance pump design process was not significantly different from that of three decades ago. During this time, a vast amount of experimental and operational experience has revealed that there are many important features of pump flows that are not fully accounted for in the semi-empirical design process. During that same time span, huge strides have been made in computers, numerical algorithms, and physical modeling. After applying the original INS3D code to the redesign of the SSME hot gas manifold, enhancing the performance of the turbomachinery components in advanced rocket engines became of major importance. This prompted the development of a CFD procedure for rocket pump flow simulation. The liquid fuel and oxidizer pumps of interest are operating at constant speed. Therefore, rotational steady-state solutions are sought first, leading to the development of a pump simulation procedure using the INS3D-UP code, in a steady rotating reference frame.

Rocket pumps involve full and partial blades, tip leakage and exit boundary to diffuser. In addition to the geometric complexities, a variety of flow phenomena are encountered in turbopump flows. These include turbulent boundary layer separation, wakes, transition, tip vortex, three-dimensional effects, and Reynolds number effects. In order to use CFD in the design process, the computational flow analysis tools must be validated so that designers can define the accuracy and variations of these tools. The validation of the CFD procedure for pump applications has focused on a rocket inducer. Extensive computational validations were performed by the Pump Technology Team organized by NASA Marshal Space Flight Center (see, Garcia et. al, 1992).

An inducer provides a pressure rise to the impeller inflow to prevent cavitation on the impeller blades. The design flow of this Rocketdyne inducer is 2,236 gal/min with a design speed of 3,600 rpm. The tip diameter is 6 inches. In the computation, tip-leakage effects are included with a tip clearance of 0.008 inches. The problem was nondimensionalized with a reference length of one inch and the average inflow velocity of 339.6 in/sec. The Reynolds number of this calculation is 191,800. As shown in Figure 13, the computational upstream section is extended with a 10 inch long straight tube. The pump consists of 6 blade passages; only one of the blade passages is modeled. An H-H grid topology with grid dimensions of 187x27x35 was used. In Figure 14 and 15, the cross sections are shown where experimental measurements and computational results are compared. Computed results for planes B and C are presented in Figure 16. Both total velocity and the flow angles compare quite well inside the blade passage (Plane B) and near the exit of the impeller blades (Plane C). Further details can be found in the paper by Kiris et al. (1993).

The resulting computational procedure was applied to the flow through the SSME High Pressure Fuel Turbo-Pump impeller and to the development of an advanced pump impeller (Kiris and Kwak, 1994). The results from the advanced impeller flow analysis are presented next. Detailed comparisons between the computed results and experimental measurements were presented by Garcia et al. (1994).

In Figure 17, a cross sectional view of an advanced impeller is shown schematically. The computational model of an advanced pump includes the impeller and the exit cavity region. Figure 18 shows the computational grid near the hub region of the impeller. The impeller design flow rate is 1,205 gal/min with a design speed of 6,322 rpm. The Reynolds Number for this calculation was 181,273 per inch. In Figure 19, the meridional velocity, which is the circumferentially averaged axial velocity, is shown at the impeller discharge. A relative x-distance is measured from the shroud to the hub, where x=1.0 is the hub. The meridional velocities, Cm, were integrated along a radial strip for each constant x-position and they were nondimensionalized by the wheel tip speed of 249.5 ft/sec. The meridional velocity distribution for 5% and 10% recirculation from the exit shroud cavity were also plotted. When the exit shroud cavity has leakage to the impeller eye, the velocity peak at the impeller exit moves toward to the center of the b2 width, where b2 is defined as the blade height at the impeller exit (see Figure 17). However, the shroud leakage has only minor effects on the solution at r/rtip=1.0275 (Figures 19). In Figure 19, the symbols represent experimental data by Brozowski (1994), and the lines represent Cm distributions for the flow with vaneless space at the exit of the impeller. The test data shows that the peak is closer to the center of the b2 width. The discrepancy between the computed results and experimental data is partially due to the recirculation flow in the hub cavity. The leakage at the hub cavity leads to stronger recirculation region which shifts the velocity peak to the center of b2 width. Since the CFD analysis did not include the leakage at the hub cavity, the predicted recirculation region in the vaneless space is not as strong as in the experimental study. Figure 20 shows blade-to-blade velocity distributions at the impeller exit. Blade-to-blade velocity distribution illustrates the impeller exit flow distortion. Symbols represent the experimental data and the lines represent computed results. The jet-wake pattern, which produces an unsteady load on the diffuser vanes, was captured at both meridional locations. Overall, the numerical results compare reasonably well with the experimental data.

IV. CONCLUDING REMARKS

In this paper, incompressible Navier-Stokes solvers primarily designed for three-dimensional flow simulations are discussed. Since the primitive variable formulation causes fewer complications in setting the boundary conditions, the discussion has been limited to the primitive variable formulation. The computed results are presented to illustrate the numerical procedures. Even though computer speed and memory have been increased substantially in the recent past, the speed and the memory requirements of a flow solver are still major factors affecting the turnaround time. INS3D-UP, which is an upwind finite-difference code based on an artificial compressibility approach, is being applied to a wide variety of applications for steady-state, time-accurate and rotational-steady solutions. INS3D-FS, which is based on a pressure projection method using a finite volume discretization on staggered grids, is intended for solving time-dependent problems. For steady-state solutions, INS3D-FS requires about 1.5 to 2 times more CPU time than INS3D-UP. For unsteady computations, INS3D-FS is time-accurate while INS3D-UP requires subiteration at each time level. Thus, for time-accurate computations in general, INS3D-FS is more economical than INS3D-UP. These solvers have been utilized in many applications of major engineering significance. As these codes are applied to a wider variety of problems, it is hoped to quantify the advantages and shortcomings of these codes in more detail. Despite the limitations in speed, accuracy, and grid-dependency of the numerical solution procedures, these codes can be of significant value to developers of modern flow devices when the range of their applicability is properly understood.

REFERENCES

Baldwin, B.S. and Barth, T.J., "A One-Equation Turbulence Transport Model for High Reynolds Number Wall-Bounded Flows," AIAA Paper 91-0610, 1991.

Beam, R. M., and Warming, R. F., "An Implicit Factored Scheme for the Compressible Navier-Stokes Equations," AIAA J., Vol. 16, pp. 393-402, 1978.

Belov, A. Martinelli, L. and Jameson, A., "A New Implicit Algorithm with Multigrid for Unsteady Incompressible Flow Calculations," AIAA Paper 95-0049, 1995

Briley, W.R. and McDonald, H., "Solution of the Multidimensional Compressible Navier-Stokes Equations
by a Generalized Implicit Method," J. Comp. Phys, Vol 24, No. 4, pp.372-397, 1977.

Brozowski, L. A., Ferguson, T.V., and Rojas, L., "Impeller Flow Field Laser Velocimeter Measurements," Proceedings of the Fifth International Symposium on Transport Phenomena and Dynamics of Rotating Machinery, May 9-11, 1994, Kuanupulu, Maui.

Chang, J. L. C., and Kwak, D., "On the Method of Pseudo Compressibility for Numerically Solving Incompressible Flows," AIAA Paper 84-0252, AIAA 22nd Aerospace Sciences Meeting, Reno, NV, January 9-12, 1984.

Chang, J.L.C., Kwak, D., Rogers, S. E. and Yang, R-J, "Numerical Simulation Methods of Incompressible Flows and an Application to the Space Shuttle Maine Engine," Int. J. Numerical Method in Fluids, Vol. 8, pp. 1241-1268, 1988.

Chen, Y.S., Shang, H.M and Chen, C.P., "Unified CFD Algorithm with a Pressure Based Method," 6th Int'l. Symposium on Comp. Fluid Dyn., Sept 4-8, 1995, Lake Tahoe, NV.

Choi, D. and Merkle, C.L., "Application of Time-iterative Schemes to Incompressible Flow," AIAA J., Vol. 23, No. 10, 1518-1524, 1985.

Chorin, A. J., "A Numerical Method for Solving Incompressible Viscous Flow Problems," J. Comp. Phys.,
Vol. 2, pp.12-26, 1967.

Chorin, A.J., "Numerical solution of Navier-Stokes equations," Mathematics of Computation, Vol. 22, No. 104, 745-762, 1968.

Chow, J.S., Zilliac, G.G., and Bradshaw, P., " Near-Field Formation of a Turbulent Wingtip Vortex," AIAA 93-0551, 31st Aerospace Sciences Meeting, Reno, NV., Jan. 11-14, 1993.

Dacles-Mariani, J. S., S. Rogers, D. Kwak, G. Zilliac, and J. Chow, "A Computational Study of a Wingtip Vortex Flowfield," AIAA 24th Conference in Fluid Dynamics, Orlando, FL, July 1993. AIAA 93-3010.

Dacles-Mariani, J., Kwak, D., and Zilliac, G, "Accuracy Assessment of a Wingtip Vortex Flowfield in the Near-Field Region," 34th Aerospace Sciences Meeting and Exhibit, Jan. 15-18, 1996. Reno, NV.

Garcia, R., McConnaughey, P., and Eastland, A., "Activities of MSFC Pump Stage Technology Team," AIAA Paper No. 92-3232, 1992.

Garcia, R., McConnaughey, P., and Eastland, A., "Computational Fluid Dynamics Analysis for the Reduction of Impeller Discharge Flow Distortion," AIAA Paper No. 94-0749, 1994.

Gresho, M. P. and Sani, R. L., "On Pressure Boundary Conditions for the Incompressible Navier-Stokes Equations," Int. J. Numerical Methods in Fluids, Vol. 7, pp. 1111--1145, 1987.

Hafez, M. and Oshima, K., ed. Computational Fluid Dynamics 1985, John Wiley and Sons, 1995.

Harlow, F. H. and Welch, J. E., "Numerical Calculation of Time-Dependent Viscous Incompressible Flow with Free Surface," Phys. Fluids, Vol. 8, No. 12, pp. 2182-2189,1965.

Hirsch, C., "Numerical Computation of Internal and External Flows," John Wiley & Sons, 1988.

Kiris, C., Chang, L., Kwak, D., and Rogers, S. E., "Incompressible Navier-Stokes Computations of Rotating Flows," AIAA Paper No. 93-0678, 1993.

Kiris, C. and Kwak, D., "Progress in Incompressible Navier-Stokes Computations for the Analysis of Propulsion Flows,'" NASA CP 3282, Vol II, Advanced Earth-to-Orbit Propulsion Technology, 1994.

Kiris, C. and Kwak, D., "Numerical Solution of Incompressible Navier-Stokes Equations Using a Fractional-Step Approach," AIAA Paper 96-2089, AIAA 27th Fluid Dynamics New Orleans, LA, June 17-20, 1996.

Kwak, D., Chang, J. L. C., Shanks, S. P., and Chakravarthy, S., "An Incompressible Navier-Stokes Flow Solver in Three-Dimensional Curvilinear Coordinate System Using Primitive Variables," AIAA Paper 84-0253, AIAA 22nd Aerospace Sciences Meeting, Reno, NV, January 9-12, 1984.

Kwak, D., Chang, J. L. C., Shanks, S. P., and Chakravarthy, S., "A Three-Dimensional Incompressible Navier-Stokes Flow Solver Using Primitive Variables," AIAA J, Vol. 24, No. 3, pp 390-396, Mar. 1986.

Kwak, D., "Computation of Viscous Incompressible Flows," von Karman Institute for Fluid Dynamics, Lecture Series 1989-04. Also NASA TM 101090, March 1989.

MacCormack, R. W., "Current Status of Numerical Solutions of the Navier-Stokes Equations," AIAA Paper 85-0032, 1985.

Marchuk, G.M., Methods of Numerical Mathematics, Springer-Verlag, 1975.

Merkle, C. L. and Athavale, M., "Time-Accurate Unsteady Incompressible Flow Algorithms Based on Artificial Compressibility," AIAA Paper 87-1137, 1987.

Park, D.K., "The Biofluid mechanics of Arterial Stenoses," M.Sc. thesis, LeHigh University, Bethlehem, Pennsylvania, 1989.

Patankar, S.V., Numerical Heat Transfer and Fluid Flow, Hemisphere Publishing Co., New York, 1980.

Phillips, W.R.C. and Graham, J.A.H., "Reynolds-Stress Measurement in a Turbulent Trailing Vortex," J. Fluid Mech., Vol. 147, pp 353-371, 1984.

Pulliam, T. H., and Chaussee, D. S., "A Diagonal Form of an Implicit Approximate-Factorization Algorithm," J. Comp. Phys., Vol. 39, pp. 347-363, 1981.

Rizzi, A. and Eriksson, L.-E., "Computation of Inviscid Incompressible Flow with Rotation," J. Comp. Phys., Vol. 153, pp 275-312 , 1985.

Roe, P. L., "Approximate Riemann Solvers, Parameter Vectors, and Difference Schemes," J. Comp. Phys., Vol. 43, pp 357, 1981.

Rogers, S. E., Chang, J. L. C., and Kwak, D., "A Diagonal Algorithm for the Method of Pseudocompressibility," J. Comp. Phys. Vol. 73, No. 2, pp. 364-379, 1987.

Rogers, S. E., Kwak, D. and Chang, J. L. C., "INS3D-An Incompressible Navier-Stokes Code in Generalized Three-Dimensional Coordinates," NASA TM 100012, November 1987.

Rogers, S. E. and Kwak, D., "An Upwind Differncing Scheme for the Time-Accurate Incompressible Navier-Stokes Equations," AIAA Paper 88-2583, AIAA 6th Applied Aerodynamics Conference, Williamsburg, VA, June 6-8, 1988.

Rogers, S. E., Kwak, D. and Kiris, C., "Steady and Unsteady Solutions of the Incompressible Navier-Stokes Equations," AIAA J. Vol. 29, No. 4, 603-610, April 1991

Rosenfeld, M., Kwak, D. and Vinokur, M., "A Fractional Step Solution Method for the Unsteady Incompressible Navier-Stokes Equations in Generalized Coordinate Systems," J. Comp. Phys., Vol. 94, No.1, pp 102-137, May, 1991.

Rosenfeld, M., and Kwak, D., "Multigrid Acceleration of a Fractional-Step Solver in Generalized Curvilinear Coordinate Systems," AIAA J. Vol. 31, No. 10, pp 1792-1800, October, 1993.

Steger, J. L. and Kutler, P., "Implicit Finite-Difference Procedures for the Computation of Vortex Wakes," AIAA J., Vol. 15, No. 4, pp. 581-590, Apr. 1977.

Temam, R., _Navier Stokes Equations,_ Revised Edition., North Holland, 1979.

Wessling, P., Segal, A., van Kan, J.J.I.M., Oosterlee, C.W. and Kassels, C.G.M., "Finite Volume Discrtization of the Incompressible Navier-Stokes Equations in General Coordinates on Staggered Grids," Computational Fluid Dynamics Journal, vol. 1, pp 27-33, April 1992

Wiltberger, N.L., Rogers, S. E. and Kwak, D. "A Comparison of Two Incompressible Navier-Stokes Algorithms for Unsteady Internal Flow," NASA TM 108794, November 1993.

Yanenko, N.N., _The Method of Fractional Steps,_ Springer-Verlag, Berlin, 1971.

Yang, R-J, Chang, J. L. C., and Kwak, D., "A Navier-Stokes Simulation of the Space Shuttle Main Engine Hot Gas Manifold ," AIAA Paper 87-0368, 1987.

Yoon, S. and Jameson, A., "An LU-SSOR Scheme for the Euler and Navier-Stokes Equations," AIAA Paper 87-0600, 1987.

Yoon, S. and Kwak, D., "LU-SGS Implicit Algorithm for Three-Dimensional Incompressible Navier-Stokes Equations with Source Term," AIAA Paper 89-1964, 1989.

Zilliac, G.G., Chow, J.S., Dacles-Mariani, J. and Bradshaw, P., "Turbulent Structure of a Wingtip Vortex in the Near Field," AIAA Paper 93-3011, AIAA 24th Fluid Dynamics Conference, Orlando, FL, July 1993.

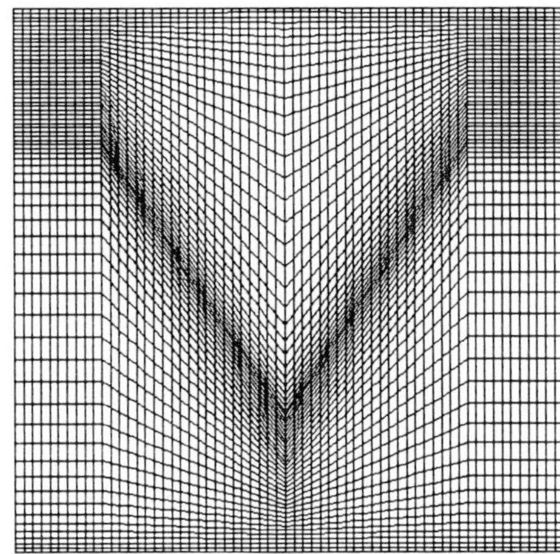

Figure 1 : Computational grid with 63x63 mesh points for Couette flow.

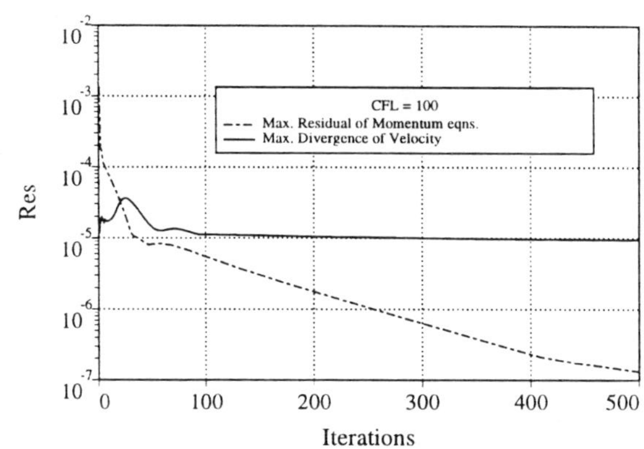

Figure 3 : Convergence history for Couette flow using the fractional-step code, INS3D-FS

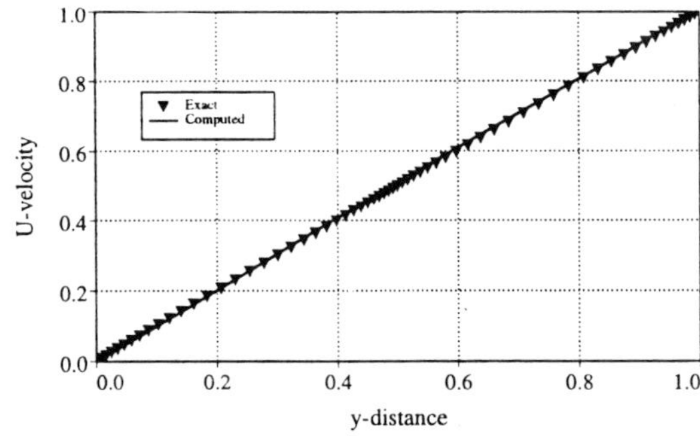

Figure 2 : U-velocity profile at for Couette flow.

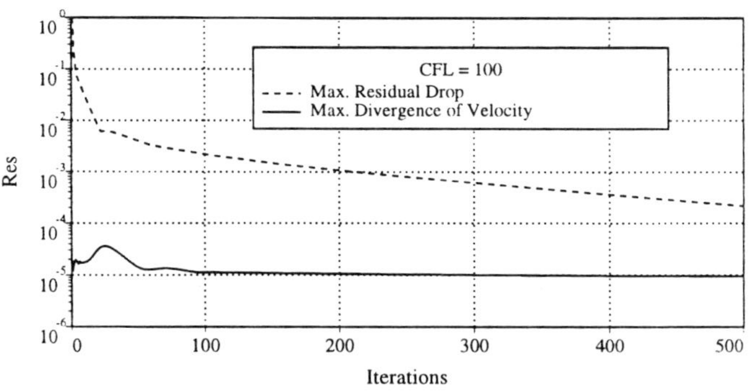

Figure 4 : Convergence history for Couette flow using the artificial compressibility code, INS3D-UP

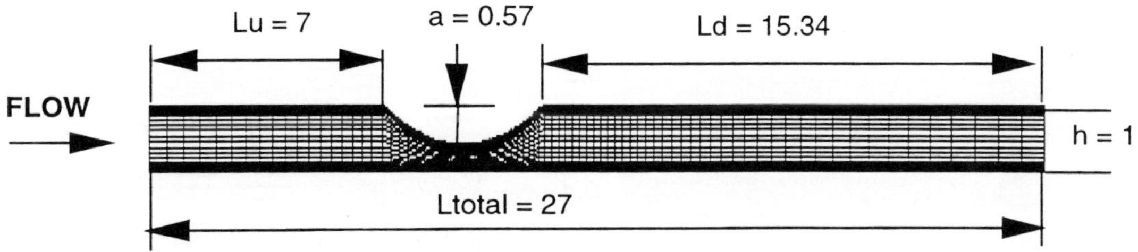

Figure 5 : Physical dimension of computational model of a constricted channel.Computation grid size : 241x63

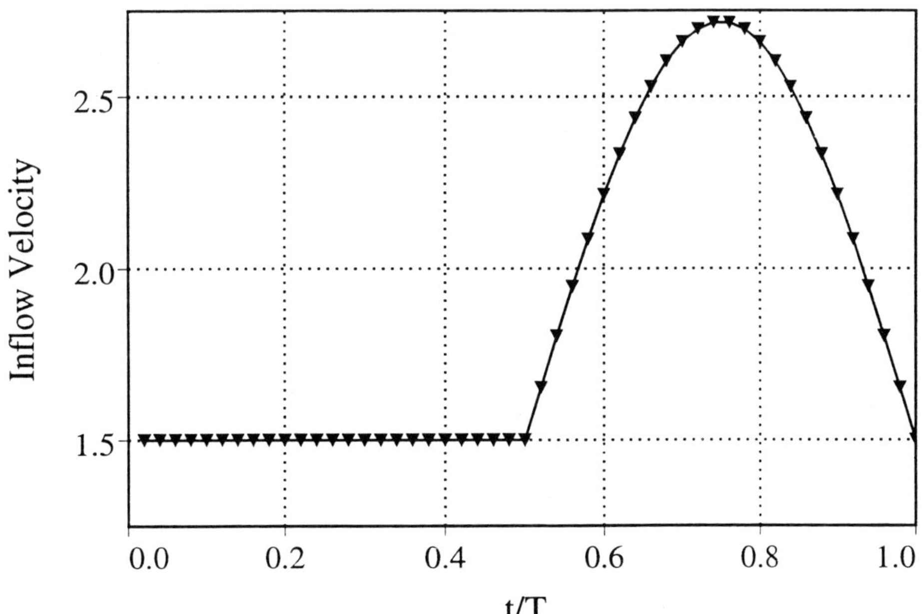

Figure 6 : Inflow velocity magnitude versus time for one period for constricted channel. Re=131.9, based on inflow velocity

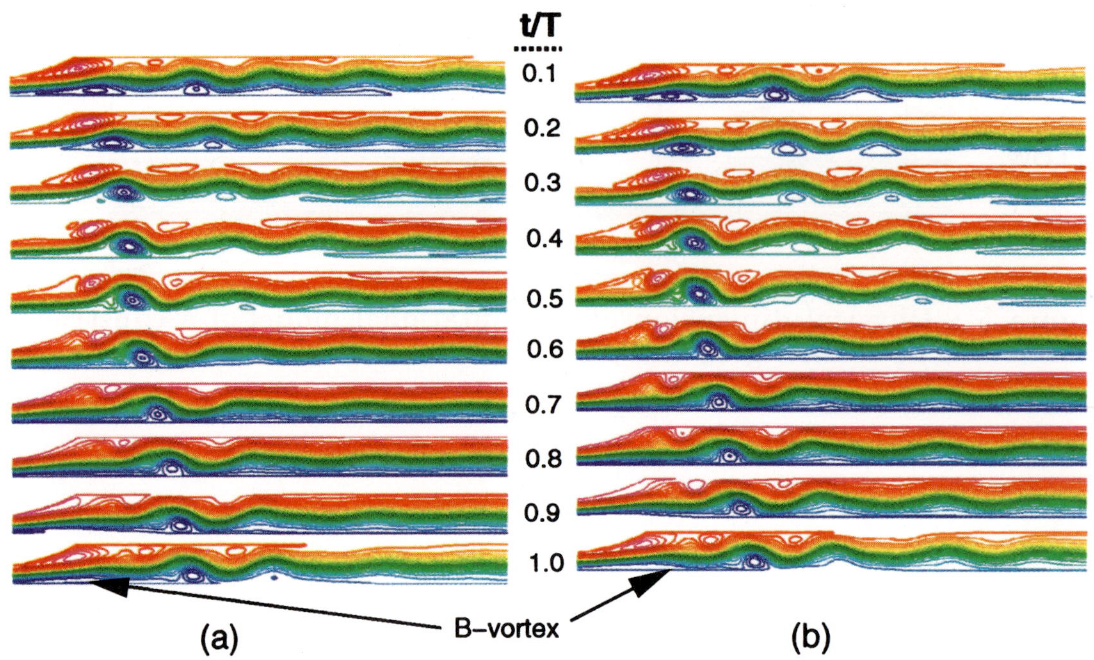

Figure 7 : Streamline at 0.1T to 1.0T : (a) INS3D-Up results, (b) INS3D-FS results.

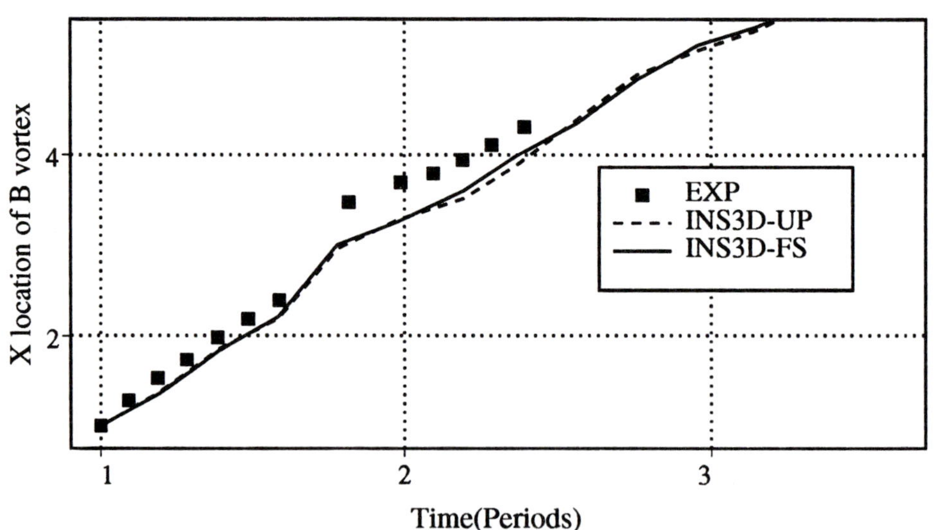

Figure 8 : Location of B-vortex versus period.

590

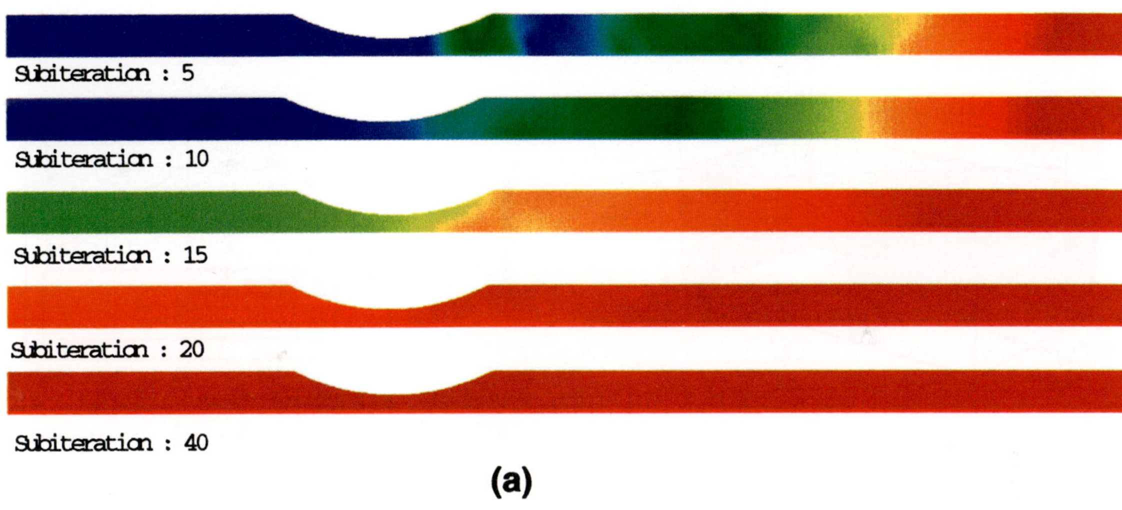

(a)

(b)

Figure 9 : Pressure correction within one time step: (a) INS3D-UP subiteration, (b) INS3D-FS Poisson iteration.

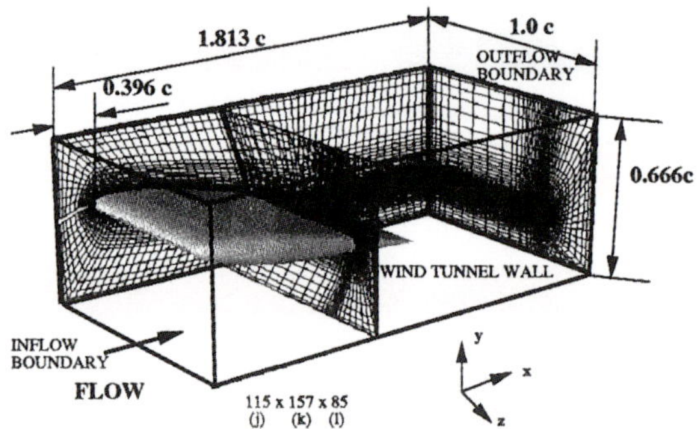

Figure 10-a : Schematic of the wind tunnel test section and the computational domain.

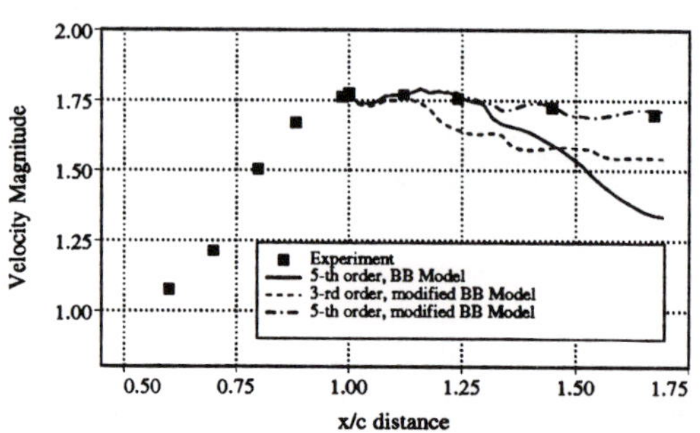

Figure 11-a : Axial progression of velocity magnitude along vortex coreline.

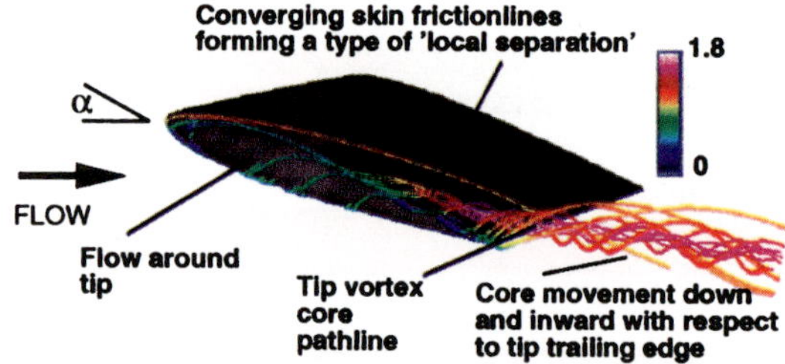

Figure 10-b : Particle traces showing tip vortex rollup and core movement: Re=4.6x10^6

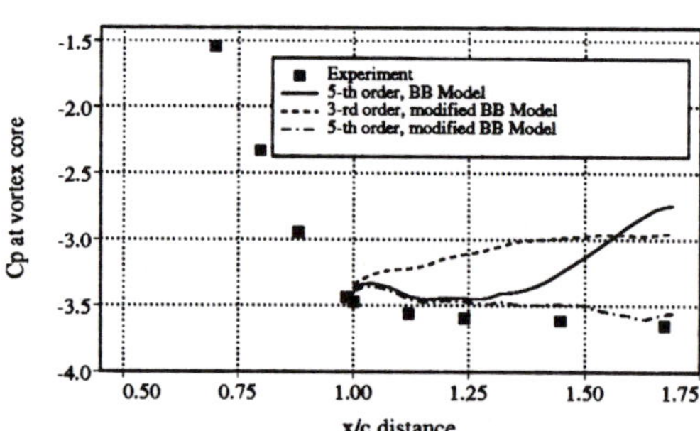

Figure 11-b : Axial progression of Cp magnitude along vortex coreline.

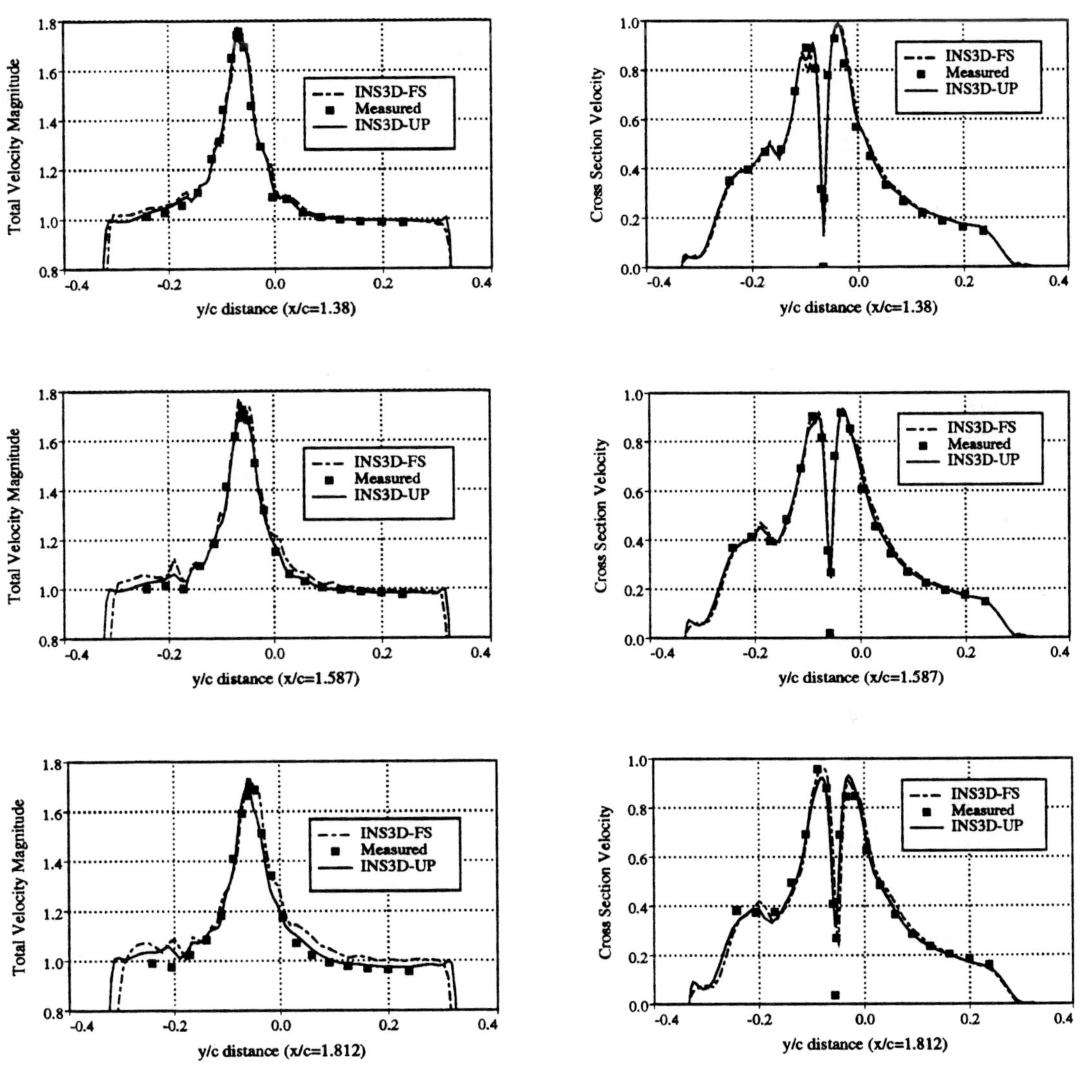

Figure 12 : Comparison of velocity magnitude and crossflow velocity across wake vortex at three different positions.

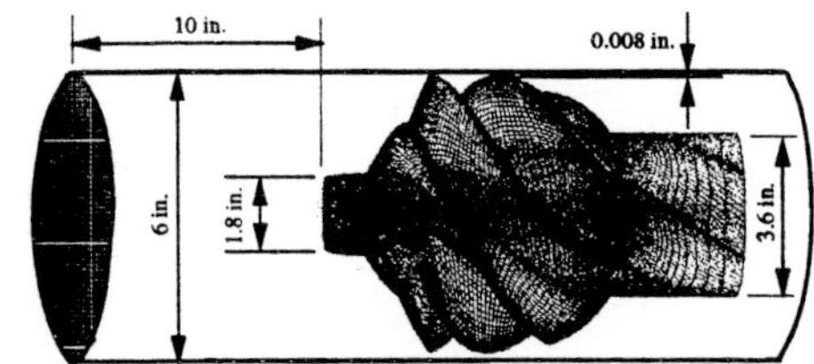

Figure 13 : Rocketdyne turbopump inducer configuration.

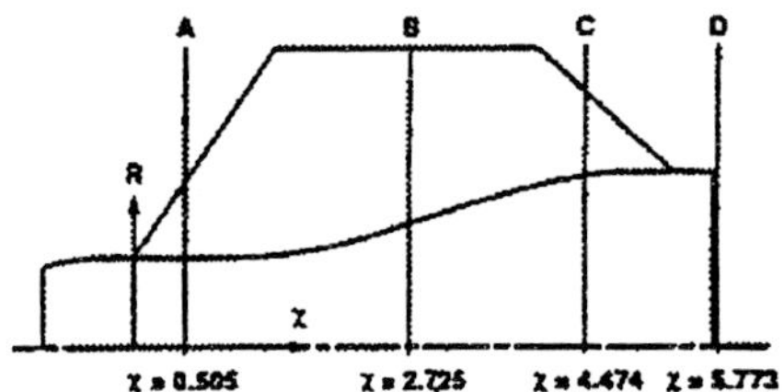

Figure 14 : Schematic representation of planes where experimental data are available.

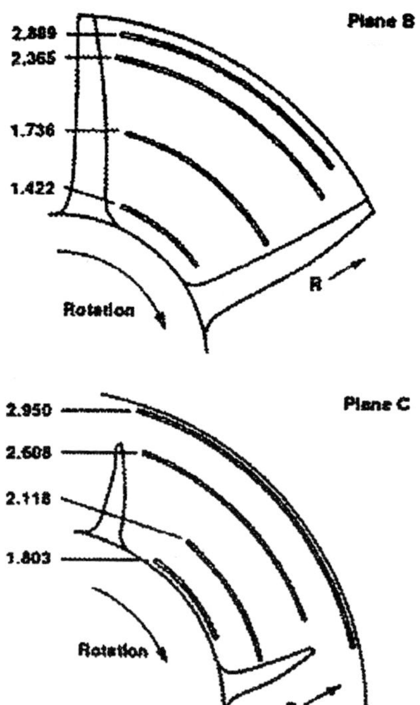

Figure 15 : Schematic representation of circular arcs where the results are compared with experimental data in plane B and C.

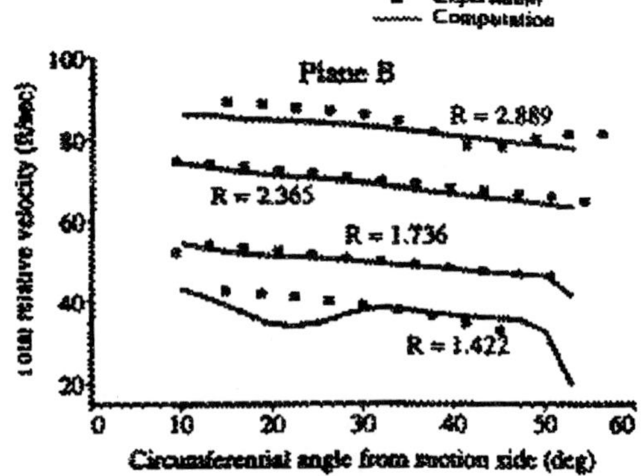

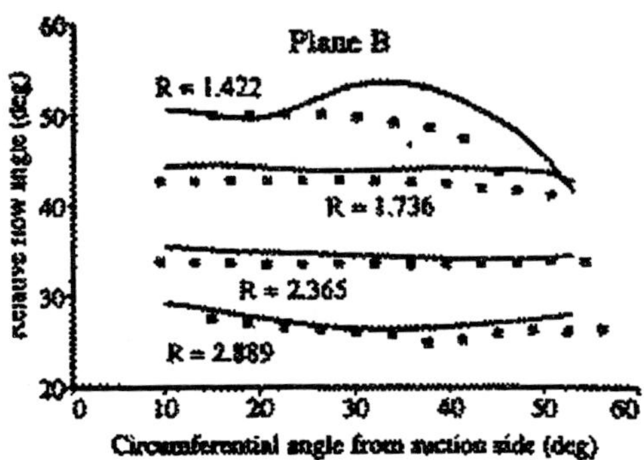

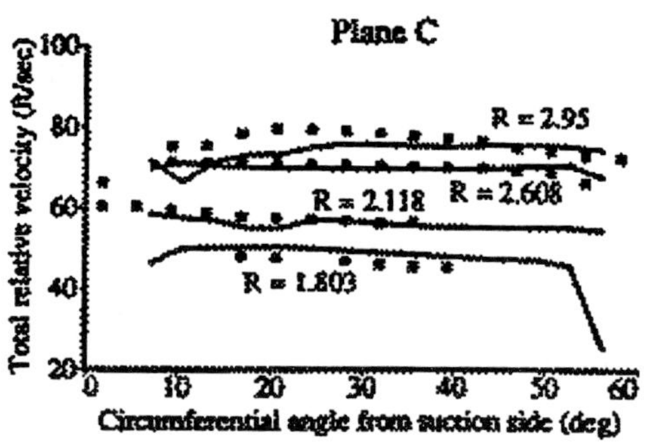

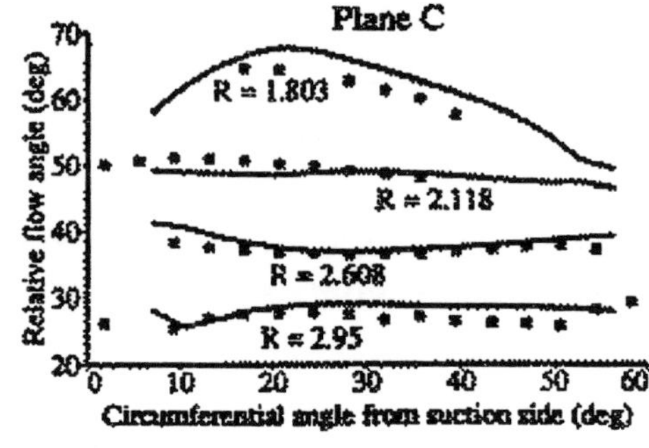

Figure 16 : Comparison of relative total velocity and relative flow angle in plane B and C.

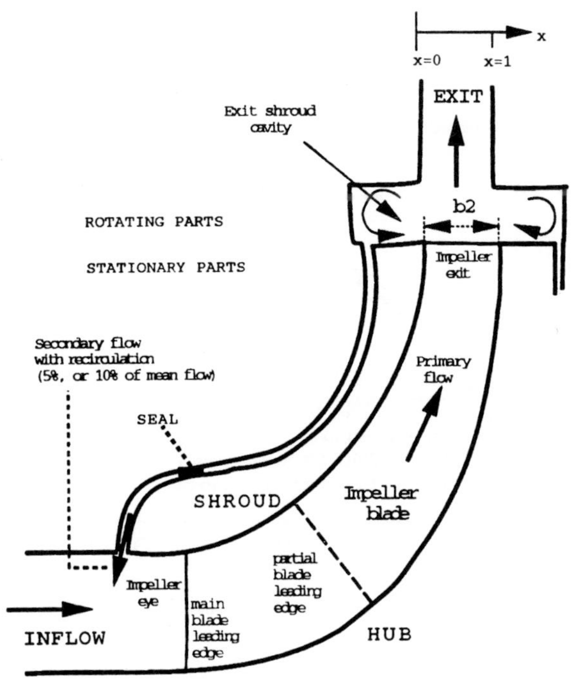

Figure 17. Schematic view of an advanced pump impeller cross-section.

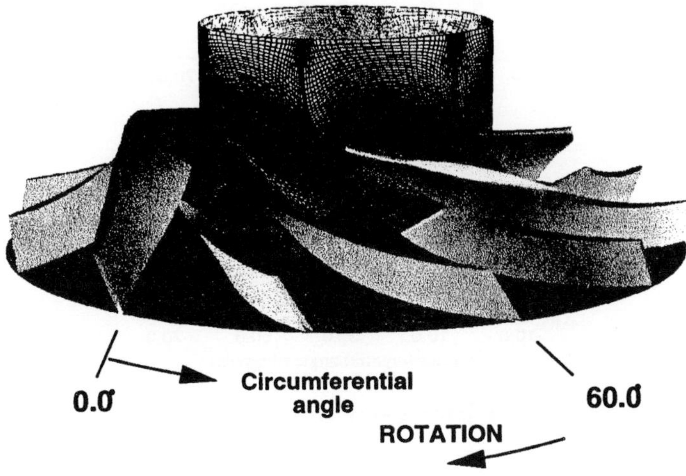

Figure 18. Advanced pump impeller computational grid on the hub surface.

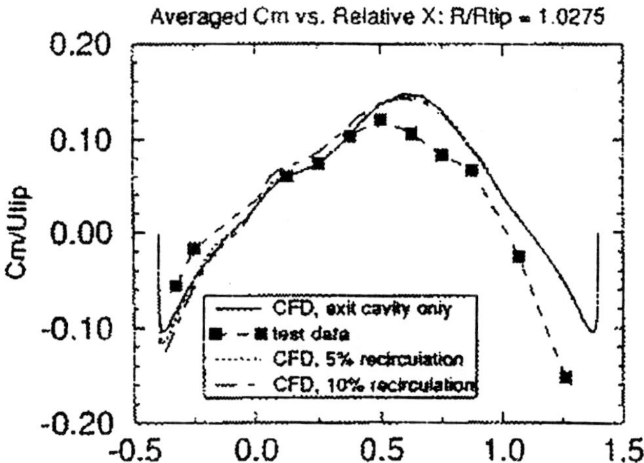

Figure 19. Comparison of circumferentially averaged meridional velocity at the impeller exit.

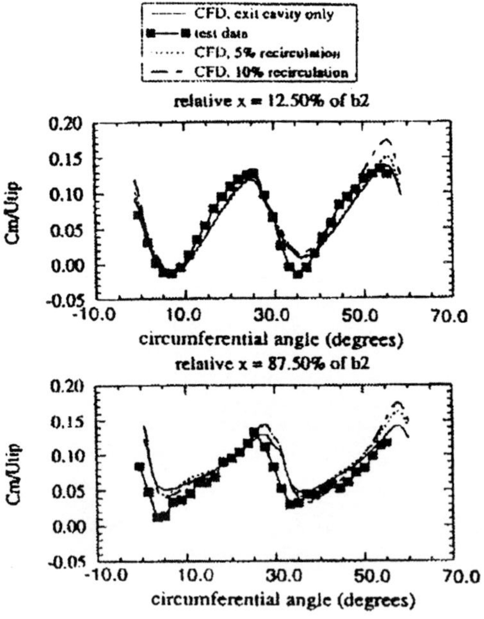

Figure 20. Comparison of blade-to-blade meridional velocity at the impeller exit.

SIMULATION OF UNSTEADY INCOMPRESSIBLE VISCOUS FLUID FLOW BY FRACTIONAL STEP FINITE ELEMENT METHOD: A REVIEW

Yan DING [1]   Mutsuto KAWAHARA [2]

(1)   Department of Civil Engineering, Chuo University; Kasuga 1-13-27,
      Bunkyo-ku, Tokyo 112, Japan   email: ding@sip.civil.chuo-u.ac.jp
(2)   The address is the same as that above.   email:kawa@sip.civil.chuo-u.ac.jp

Abstract

In this review, it is presented to survey of the recent development of the finite element method (FEM) which can be applied to the time-dependent problems of Navier-Stokes. This paper focuses on the fractional step method applied to the areas of unsteady incompressible fluid flow. Several frames about time discretization and the corresponding algorithms are analysed. With respect to practical problems in the simulation of incompressible fluid flow, we discuss some results of both steady and unsteady flow. The attention of application to free surface flows also is taken. Thus, the descriptions of fluid motion, i.e., Lagrangian method, Eulerian method, and Arbitrary Lagrangian-Eulerian method (ALE), are presented in this review.

1. Introduction

The FEM appears to be very convenient and most promising for obtaining spatial approximation to problems with complex geometries. Finite element theory provides an effective methodology for construction of discrete and/or semi-discrete approximations to variational boundary value problems in computational mechanics, including the reduced forms of the Navier-Stokes equations in fluid mechanics. The classic theory is found on the earliest fluid mechanics applications employed direct extension [1, 5, 6]. It was subsequently recognized that the Galerkin weak-statement, identical to the variationally-motivated formulation, was inadequate for problem applications in computational fluid dynamics (CFD) at large Mach and Reynolds numbers. This prompted development of various finite element methods. To overcome the disadvantage of Galerkin formulations in the computation of advection-diffusion equations, e.g., Navier-Stokes equations, the modified Galerkin methods such as the streamline upwind/Petrov-Galerkin formulations (SUPG), the Taylor-Galerkin formulations, etc., have been proposed [3, 10]. Those methods were applied successfully to the finite element solution of incompressible flow problems.

In the computation of unsteady incompressible viscous fluid flows, the fractional-step methods have been

Received on January 10, 1997.

proven to be effective semi-implicit FEM method, have some advantages of stability and simplicity. The fractional step (or projection) method was developed independently by Chorin [4] and Temam [43] . In his original description of the method, Chorin used a nonstaggered mesh and an implicit ADI time integration of the modified momentum equations. For finite-difference application, however, it has become more commonly implemented with an explicit time integration and a staggered mesh. Donea et al [10] have proposed a finite element method of the fractional step method. By diagonalizing the mass matrix using the standard row-sum lumping technique, they are able to use a purely explicit time integration. Simura and Kawahara [42] , using an equal-order finite element fractional step method, investigated a procedure for applying a computed nonhomogeneous Dirichlet outflow boundary condition for the pressure equation. Gresho and Chan [13] have proposed a fractional step method on the basis of a finite element method (i.e., Projection-2 method), in which the mass matrix is treated as a nearly consistent one. By the conjugate gradient (CG)-like solver, the projection-2 method has been applied to the simulation of large scale incompressible flow problems [7]. Williams and Baker [44] have pointed out that the key theoretical feature in the continuum theory development centered on the dual roles played by pressure as the mechanism instantaneously enforcing conservation of mass while also being a force in a mechanical balance law. They proposed a primitive variable CFD algorithm called the Continuity Constraint Method (CCM) on the basis of FEM. This method has been evaluated that it is equally applicable to development of a finite-volume or finite-difference form. Ding and Kawahara [8] have planed a nearly fractional step method associated with the finite element method and applied it to simulation of the transient flow past a circular cylinder subject to a nonreflective open boundary condition. The three-step explicit scheme of the fractional step method was proposed by Dr. Jiang which was applied to the time-dependent computation of incompressible viscous flows in order to sustain the temporally high accuracy [19, 20, 21]. In Perot's paper [37], the fractional step method for solving the incompressible Navier-Stokes equations in primitive variables is analyzed as a block LU decomposition. In this formulation, it is shown that poor temporal accuracy (first-order) is not due to boundary conditions, but due to the method itself. Perot [37] presented a generalized block LU decomposition that overcomes this difficulty of poor temporal accuracy.

In the analysis of unsteady incompressible viscous fluid flows, a nonlinear free surface motion is of practical importance in the areas of engineering. As is well known, the numerical solution of viscous fluid flow is complicated by the presence of free surface. There are two reasons for this. First, the position of free surface varies with time in a manner not known *a priori*, second the free surface boundary condition must be imposed. The problem is usually analysed on the basis of Eulerian/Lagrangian description. But the Eulerian description is rather inconvenient to express the complicated free surface configuration. Employing the Lagrangian description, fluid in the interior of a finite element always remains in that element and fluid boundaries always move with the element boundaries, the treatment of movement of free surface boundary is more simpler than that of Eulerian method. Several finite element methods with the Lagrangian description were presented by the research group of Prof. Kawahara in Chuo University. Those include the Lagrangian FEM [27, 38, 39], and arbitrary Lagrangian-Eulerian(ALE) finite element method [36, 40]. The ALE technique is suitable for the computation of highly nonlinear problems, but a large amount of computational time may be required.

Because of the coupling features between velocity and pressure in the governing equations of incompressible fluid flow, the complexity gives rise to more difficulties of the numerical computation. By applying the fractional step method, the unsteady incompressible Navier-Stokes equations can be transformed into the form in which velocity and pressure can be computed in an independent manner, then velocity and pressure are computed by means of some suitable numerical schemes. The velocity correction method is one of the fractional step method(semi-implicit method). Hayashi et al [17] have discussed in detail two kinds of the velocity correction method under the Lagrangian system (i.e., Method A and Method B), and the other

two kinds of iterative methods by which the pressure field is directly calculated on the basis of the pressure Poisson equation derived from the momentum equations (i.e., Method C and Method D). The results of numerical experiments showed that the computation by Method C is more effective in practice.

The remainder of the review is organized as follows. Section 2 describes the mathematical theory associated with incompressible fluid flow problems, the basic theory frame of fraction step method is constructed for the well-posed Navier-Stokes equations. In section 3, the finite element formulation and discritization are discussed briefly. In section 4, two kinds of descriptions in fluid dynamics, i.e., Lagrangian and Eulerian methods, are expressed in details, and four kinds of CFD algorithms associated with the simulation of free surface flows are discussed in details. In section 5, several numerical examples and applications of fraction step method will be summarized. Section 6 will concludes this review.

2. MATHEMATICAL THEORY

In this section, we will describe the governing equations of incompressible flow, boundary conditions, and the Pressure Poisson Equations (PPE) which is used widely in the computational fluid dynamics. The adequate boundary condition for the PPE will be presented.

2.1 Basic Equations and General Boundary Conditions

The partial differential equations set governing viscous, laminar flow of an incompressible fluid is a subject class of the familiar and nonlinear Navier-Stokes system. Let Ω be the fluid domain, which is surrounded by a piecewise smooth boundary Γ. By using indical notation and the summation convention for repeated indices, the corresponding basic constraint of conservation of mass, momentum are

Continuity

$$u_{i,i} = 0 \qquad in \ \Omega, \qquad (1)$$

Momentum

$$\frac{Du_i}{Dt} + \frac{p_{,i}}{\rho} - \nu(u_{i,j} + u_{j,i})_{,j} + f_i = 0 \qquad in \ \Omega, \ (2)$$

where ρ, ν, and f_i denote fluid density, kinematic viscosity and body forces respectively. The fields of velocity and pressure are defined as the functions of spatial/Eulerian coordinates x_i

$$u_i = u_i(x_i, t), \qquad (3)$$

$$p = p(x_i, t). \qquad (4)$$

The boundary Γ consists of two types of boundaries, i.e., the boundary Γ_1 is one on which the velocity is given, the boundary Γ_2 is the other on which the stress force is specified. The boundary conditions for the basic equations can be expressed in the following forms

$$u_i = \hat{u}_i \qquad on \ \ \Gamma_1, \qquad (5)$$

$$(-\frac{p}{\rho}\delta_{ij} + \nu(u_{i,j} + u_{j,i})) \cdot n_j = \hat{t}_i \qquad on \ \ \Gamma_2, \ (6)$$

where the superscript caret denotes a function which is given on the boundary, δ is Kronecker's delta. n_j means the direction cosine of the outward normal to the boundary with respect to coordinates x_j. If $\Gamma = \Gamma_1$, i.e., $\Gamma_2 = \emptyset$, $\hat{u}_i$ must satisfy

$$\int_\Gamma \hat{u}_i \cdot n_i d\Gamma = 0. \qquad (7)$$

This well-posed problem subjects to the initial condition

$$u_i(x_i, 0) = u_i^0(x_i) \qquad in \ \ \Omega, \qquad (8)$$

where the initial velocity field $u_i^0(x_i)$ also must satisfy the constraint of continuity in the computational domain.

For free surface flows which will be discussed in the next section, the free surface condition is

$$\frac{D\eta}{Dt} = 0 \qquad on \ \ \Gamma_f, \qquad (9)$$

where η denotes the height function of the free surface and Γ_f means the free surface (boundary).

2.2 Pressure Poisson Equation and Pressure Boundary Conditions

Time Discrete Form

Most of generating numerical schemes don't directly discrete and solve the mass conservation equation (1), since the solution of velocity and pressure from solving the equation (1) and (2) simultaneously is of inconvenient. According the strategy in the fractional step method, the numerical scheme in which indirectly solve the pressure Poisson equation, which satisfies the continuous constraint, has been extensively applied to the computation of flow field. Applying the divergence operator to (2) gives

$$\frac{p_{,ii}}{\rho} = -(\frac{\partial u_i}{\partial t})_{,i} - u_{j,i}u_{i,j} - u_j u_{i,ij}$$
$$+\nu(u_{i,j} + u_{j,i})_{,ij} + f_{i,i} \qquad (10)$$

as the pressure Poisson equation (PPE) associated with the Navier-stokes equations. Because of the unknown term $\frac{\partial u_i}{\partial t}$, several improved forms of (10) were used for the computation of practical problem. In the fractional step method, forward differencing from the level t_n to $t_{n+1} = t_n + \Delta t$, the time-dependent acceleration can be approximately expressed as follows

$$\frac{\partial u_i}{\partial t} = \frac{u_i^{n+1} - u_i^n}{\Delta t} + O(\Delta t). \qquad (11)$$

(11) implies that either the accuracy is the first order and the time increment must be restricted for stability. Assuming that the velocities in (n+1)th increment satisfy the incompressible constraint, i.e.

$$u_{i,i}^{n+1} = 0 \qquad (12)$$

The semi-discrete form of PPE can be expressed as [16, 20]

$$\frac{p_{,ii}^{n+1}}{\rho} = \frac{u_{i,i}^n}{\Delta t} - u_{j,i}^n u_{i,j}^n - u_j^n u_{i,ij}^n$$
$$+\nu(u_{i,j}^n + u_{j,i}^n)_{,ij} + f_{i,i}^n. \qquad (13)$$

To solve this elliptic equation, however (assuming for the moment that velocity field is known) requires that the boundary conditions be given on Γ, and these can apparently be of the Dirichlet , Neumann, or Robin type, or even one of them [16].

Analytic Form

The well-known form of PPE is obtained by assuming sufficient smoothness and operating on (1) with the divergence operator to give, considering the constraint of continuity (2), the analytic PPE is

$$p_{,ii} = -(u_j u_{i,j})_{,i} + f_{i,i}. \qquad (14)$$

The discussion of details for (14) can be found in the Gresho's review [12]. However, the attention to the implicit assumption in (14) should be paid, i.e., the acceleration is divergence free to obtain (14). The boundary condition of pressure in (14) is derived from the momentum equations, i.e.,

$$p_{,i} \cdot n_i = (-\frac{\partial u_i}{\partial t} - u_j u_{i,j} + \nu u_{i,ii} + f_i) \cdot n_i. \qquad (15)$$

Gresho [12] has pointed out that the additional boundary condition $p = 0$ specified on the open boundary sacrifices $u_{i,i} = 0$ on Γ, therefore this Dirichlet condition is an invalid boundary condition in the simulation of external flows(e.g., transient flow past a circular cylinder).

2.3 Projection of Divergence Error

Divergence Error Potential

In many of prediction-correction methods, the continuity error distribution can be projected on a scalar field [12]. In order to express the solenoidal velocity solution at time (n+1)th level, for the approximated velocity field $\tilde{u}^{n+1}$ computed from a discrete momentum equations, the divergence error can be expressed as the gradient of a scalar field, i.e.,

$$u_i^{n+1} - \tilde{u}_i^{n+1} = \phi_{,i}. \qquad (16)$$

Taking the divergence operator into (16), substituting (12), the Poisson equation of potential ϕ is derived as

$$\phi_{,ii} = -\tilde{u}_{i,i}^{n+1}. \qquad (17)$$

The additional boundary condition of ϕ can be found in Williams and Baker [44].

Pressure Correction Equation (PCE)

If the time discretization method is given by the *backward Euler method*, the semi-discrete momentum equations in the (n+1)th time level can be written as

$$\frac{u_i^{n+1} - u_i^n}{\Delta t} + u_j^n u_{i,j}^{n+1} = -\frac{p_{,i}^{n+1}}{\rho} + \nu u_{i,jj}^{n+1} + f_i, \quad (18)$$

and the equations of approximated velocities $\tilde{u}^{n+1}$ is

$$\frac{\tilde{u}_i^{n+1} - u_i^n}{\Delta t} + u_j^n \tilde{u}_{i,j}^{n+1} = -\frac{\tilde{p}_{,i}^{n+1}}{\rho} + \nu \tilde{u}_{i,jj}^{n+1} + f_i. \quad (19)$$

Subtracting $\tilde{u}_i^{n+1}$ from u_i^{n+1}, then

$$\frac{p_{,i}^{n+1}}{\rho} = \frac{\tilde{p}_{,i}^{n+1}}{\rho} - \frac{\phi_{,i}}{\Delta t} - u_j^n \phi_{,ij} + \nu \phi_{,ijj}. \quad (20)$$

For simplicity, the actual PCE can be reduced as the simpler form, i.e.,

$$p^{n+1} = \tilde{p}^{n+1} - \rho \frac{\phi}{\Delta t}. \quad (21)$$

Braza et al [2] have concluded that the results from (20) and (21) are similar for the case of flow around a circular cylinder by means of finite difference method, only the convergence is improved when using the complete equation(20). In practice, the formulation (21) can be computed explicitly point by point, instead of the computation of the Galerkin weak formulation.

3. FINITE ELEMENT FORMULATIONS

3.1 Single-Step Schemes

(1) Velocity Pressure Formulation (u-p mixed FEM)

The dependent variables in the continuity equation (1) and the momentum equations (2) are the velocities and pressure at all nodes. the arbitrary element velocity and pressure are approximated as

$$u_i = \Phi_N u_{iN} \qquad (N = 1, 2, \ldots\ldots N_v), \quad (22)$$

$$p = \Psi_M p_M \qquad (N = 1, 2, \ldots\ldots N p), \quad (23)$$

where Φ_N denotes the interpolation function of velocity, Ψ_M is that of pressure. N_v, N_p are the velocity nodes and the pressure nodes, respectively. Generally, Φ_N is chosen as a quadratic function due to the second order momentum transfer, and Ψ_M is considered as a linear function due to the one order pressure transfer. The weak formulations of (1) and (2) are obtained by using the standard Galerkin method, and considering the boundary conditions of velocity and pressure (5), (6),

$$\int_\Omega u_{j,j} \Psi_M d\Omega = 0, \quad (24)$$

$$\int_\Omega (\frac{\partial u_i}{\partial t} \Phi_N - u_i u_j \Phi_{N,j} + \frac{p \Phi_{N,j}}{\rho} - \nu(u_{i,j} + u_{j,i}) \Phi_{N,j}) d\Omega =$$

$$\int_{\Gamma_2} \hat{t}_i \Phi_N d\Gamma - \int_{\Gamma_1} u_i u_j \Phi_N^* n_j d\Gamma + \int_\Omega f_i \Phi_N d\Omega, (25)$$

where Φ_N^* represents the interpolation for the variation of boundary force. If it is put in the forward differential formulation (11), the scheme is explicit one with first-order accurate in time, if it is substituted by the Eulerian backward differentiation, the scheme becomes a fully implicit one which is unconditionally stable [15]. Due to the difficulty of solution to the simultaneous equation (24) and (25), it is usually to use some iteration methods, e.g. Newton-Raphson method, perturbation method [22].

(2) The Simplest Algorithm of Solving the Pressure Poisson Equation

It is difficult to directly solve the simultaneous equations as shown above. But the pressure Poisson equation (10) satisfies indirectly the incompressible constraint of equation (1), it is reasonable to construct a fractional step method for an unsteady flow by using the pressure Poisson equation (13). Generally, the numerical procedure can be simply expressed as follows:

1. Assuming the initial condition velocity and pressure;

2. Solving the pressure field p^{n+1} from Eq. (13);

3. Solving the velocity field u_i^{n+1} from the discrete equation of (2) ;

4. Solving other scalar fields;

5. Forward one time increment and iterating from (2) to (4).

This kind of scheme is semi-implicit and conditionally stable in which the time accuracy is first order. The analysis of the other single-step schemes can be found in the review of Gunzburger [15].

(3) Projection-2 Method

Gresho [11] has proposed a fractional step method named as Projection-2 method. Rather than solve the Navier-Stokes equations simultaneously, this fractional step method decouple and solve the continuity equation (1) and the momentum equations (2) independently. Daniels and Peters [7] showed that this semi-implicit fractional step method yields accurate results despite its approximate nature. We refer to Gresho [11] and Gresho and Chan [13] for further description and to Daniels and Peters [7] for further implementation details in the simulation of large scale incompressible flow problems.

(4) Iteration and Correction Methods

Although most of the fractional step schemes decouple and solve the Navier-Stokes equations independently, it is rather plausible for the enforcement of continuity constraint. In the respect of finite volume method, the SIMPLE-like schemes are typically semi-implicit iteration-correction method [44]. On the basis of the iteration-correction procedure, the constraint of continuity can be enforced to a desirable divergence error.

In terms of the θ-implicit time integration, Williams and Baker [44] have constructed the iterative method called the continuity constraint method (CCM), in which the solenoidal velocity field is obtained by means of the iterative procedure. In this scheme, the corrector in the iterative cycle plays a key role on the enforcement of continuity condition. After the convergent velocity field is obtained, the genuine pressure field is computed from the PPE (14). See Williams and Baker [44] for the further discussion.

In virtue of the Eulerian forward differentiation for the time discretization, Ding and Kawahara [8] proposed a simpler iteration-correction method which is similar to the CCM scheme. By specified the Neumann condition for the pressure, the results on the outlet boundary of external flow can be passed through smoothly [9]. The artificial Dirichlet boundary of pressure is avoidable in the simulation of external flow problems.

3.2 Streamline Upwind/Petrov-Galerkin Method (SUPG)

As is well known, it has been found that the standard Galerkin spatial discretization methods lead to central difference approximations of differential operators. Galerkin finite element solution to convective transport problem are often corrupted by spurious node to node oscillations or *wiggles*. To preclude such oscillation, various numerical procedures have been proposed, the most popular being the use of [?] differencing on the convective term. When the modified weighting functions are employed to all terms in the equations, the formulation is referred to as streamline upwind/Petrov-Galerkin weighted residual method [3]. The SUPG formulation is believed to be most accurate, and has been applied to solve the hyperbolic equation of convection-diffusion and the incompressible flows.

3.3 Taylor-Galerkin Method and Multiple-Step Method

For the computation of unsteady flow, a low-order time-step with the conventional Galerkin formulation often leads to produce the unstable numerical results. Donea et al [10] proposed the Taylor-Galerkin method which provides another accurate finite method for the hyperbolic problem governed the convection-diffusion equation. With respect to the advection equation, the starting point of this idea is the substitution of space derivation for the time derivation in a Taylor series expansion, as used in the derivation of the Lax-Wendroff method [19].

By means of a forward-time Taylor series expansion in time derivative, the variable value in the (n+1)th

time step can be given as

$$f^{n+1} = f^n + \Delta t \frac{\partial f}{\partial t} + \frac{\Delta t^2}{2} \frac{\partial^2 f}{\partial t^2} + \frac{\Delta t^3}{6} \frac{\partial^3 f}{\partial t^3} + O(\Delta t^4). \tag{26}$$

The following formulation is the approximation of Eq.(26) which constructed one step explicit Eulerian scheme

$$f^{n+1} = f^n + \Delta t \frac{\partial f}{\partial t}. \tag{27}$$

As the above mentioned reason, this is a central scheme which leads to instability in the convection dominated flows. By approximating Eq.(26) to the second-order accuracy, the two step explicit scheme can be expressed as follows [25, 30]

$$f^{n+1/2} = f^n + \frac{\Delta t}{2} \frac{\partial f^n}{\partial t}, \tag{28}$$

$$f^{n+1} = f^n + \Delta t \frac{\partial f^{n+1/2}}{\partial t}. \tag{29}$$

The spatial discretization of Eq.(28) and (29) can be obtained from using the standard Galerkin finite element method. According to this idea, Jiang and Kawahara [19] proposed the third-order accuracy, i.e., the three step scheme

$$f^{n+1/3} = f^n + \frac{\Delta t}{3} \frac{\partial f^n}{\partial t}, \tag{30}$$

$$f^{n+1/2} = f^n + \frac{\Delta t}{2} \frac{\partial f^{n+1/3}}{\partial t}, \tag{31}$$

$$f^{n+1} = f^n + \Delta t \frac{\partial f^{n+1/2}}{\partial t}. \tag{32}$$

The three step scheme retains the merits of Taylor-Galerkin method, and does not involve any other higher-order spatial derivative. In practice, it can be applied to solve the non-linear, multi-dimensional flow problems [19, 20, 21].

4. LAGRANGIAN OR/AND EULERIAN FINITE ELEMENT METHODS

4.1 Eulerian/Lagrangian Description

There are two kinds of description, that is, Eulerian and Lagrangian description in fluid mechanics. The former corresponds to the usual coordinates which the spatial domain of observator is fixed, but the later is one which the particle at varying times is tracked. The relation between the spatial and material domain is shown Fig. 1. The position vector of a material point P in this region is represented by $X_i = (X_1, X_2, X_3)$. The coordinates X_i are called the material coordinates. The particle originally at P is located at the point p and has the position vector $x_i = (x_1, x_2, x_3)$, which give the current position of the particle, is referred to as the spatial coordinates.

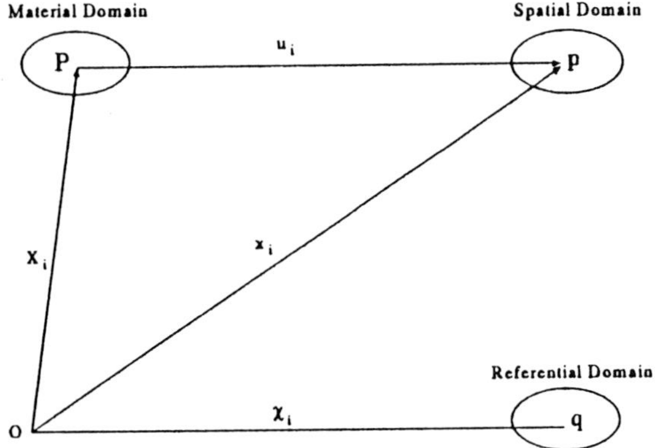

Fig.1: Coordinates System

The motion of body in the Lagrangian description carries various particles through various spatial position. It is mathematically expressed in the following relation

$$x_i = x_i(X_j, t). \qquad (33)$$

Therefore, the mappings (33) should be continuous, single-valued and possess a unique inverse for the Eulerian description as

$$X_i = X_i(x_j, t) \qquad (34)$$

The necessary and sufficient condition of the transformation between two kinds of coordinates is that the Jacobian determinant J ($J = det|\frac{\partial x_i}{\partial X_j}|$) should not vanish.

Because the velocity quantity in the Lagrangian description is given for a fixed particle at varying time, i.e., $U_i(X_i, t)$, the representation of acceleration is simply obtained from the Lagrangian coordinates, i.e., $\frac{\partial U_i(X_i, t)}{\partial t}$. But the velocities in the Eulerian coordinates are given at fixed point in space, $U_i(x_i, t)$, the acceleration formula can be expressed as

$$\frac{DU_i}{Dt} = \frac{\partial U_i}{\partial t} + U_j \frac{\partial U_i}{\partial X_j}. \qquad (35)$$

In general, the time rate of change of the continuous function f is given by the material derivative as follows

$$\frac{Df}{Dt} = \frac{\partial f}{\partial t} + U_j \frac{\partial f}{\partial X_j}. \qquad (36)$$

In the Eulerian finite element calculation, a coordinates system is stationary in the referential frame, in which fluid moves from element to element, the mass, momentum and energy of fluid are exchanged within every elements. The Eulerian method of numerical computation has some advantages: (i) the fluid can undergo arbitrary great distortion without loss of accuracy, and (ii) the outflow boundary are particularly easy to handle. However, the basic Eulerian methodology encounters some difficulties, e.g., (i) it is inconvenient for material interface to move through the mesh, and (ii) the local region of fine resolution is difficult to achieve. Therefore, the basic Eulerian method may not be directly applied to practical problem with respect to free surface flows. By combined with several tracing techniques, the Eulerian description can lead to many numerical computational procedures for fluid flows involving free surface, for example, marker and cell (MAC), the volume of fluid(VOF), the height function method and so on.

On the other hand, the Lagrangian finite element calculations are characterized by a coordinates system that move with the fluid, and each computational element always contains the same fluid element. The Lagrangian method has several convenient and valuable features, e.g., (i) the material interfaces can be specifically delineated and precisely followed, (ii) the free surface boundary conditions are easily applied and (iii) the curved rigid boundaries of arbitrary shape can be presented. The principal disadvantage is the inability to cope easily with strong distortion [40]. Thus the Lagrangian method is more recommendable from the computational point of view on the problems of free surface flows.

4.2 Lagrangian Finite Element Method (LFEM)

This section describes the finite element methods for the analysis of transient free surface flows in conjunction with the Lagrangian computational techniques, The basic equations and general boundary conditions as above are still valid for the Lagrangian description of free surface flows. The additional conditions are the elevation and the pressure on the surface(i.e., atmospheric pressure). By using the conventional Galerkin method, the discretization of governing equations can be also set up as above. So this section represents the computational scheme in connection with the Lagrangian finite element method of free surface flows only.

Method A

In the computational schemes, the material derivative in the Lagrangian coordinates can be expressed simply as

$$\frac{DU_i}{Dt} = \frac{U_i^{n+1} - U_i^n}{\Delta t} + O(\Delta t), \qquad (37)$$

The location of a nodal point at the (n+1)th level is linearly approximated by

$$X_i^{n+1} \equiv x_i^{n+1} = x_i^n + \frac{\Delta t}{2}(u_i^{n+1} + u_i^n). \qquad (38)$$

To obtain the velocities and the pressures at time t^{n+1}, the substitution of formula (37) in (2) can lead to the intermediate velocity $\tilde{U}_i^{n+1}$ as follows

$$\tilde{U}_i^{n+1} = U_i^n - \Delta t[\frac{p_{,i}^n}{\rho} - \nu(U_{i,j}^n + U_{j,i}^n) - f_i^n]. \quad (39)$$

The approximated velocities of flows can be derived from the above equation. As is well known, this equation does not satisfy the incompressible constraint (1). Assuming the exact velocities U_i^{n+1} that satisfy the incompressible continuity condition, the formula of the exact velocities can be also expressed as the above form. Because the curl of $p_{,i}$ is vanished, the following relation can be found

$$Curl\ U_i^{n+1} = Curl\ \tilde{U}_i^{n+1}. \quad (40)$$

Therefore, a scalar potential ϕ plays the role of correction in this method, can be derived from (40), i.e

$$U_i^{n+1} = \tilde{U}_i^{n+1} + \phi_{,i}. \quad (41)$$

Because the divergence of U_i^{n+1} is equal to zero, the resulting potential for ϕ such as (17) can be obtained by solving the Poisson equation of the scalar potential. Of course, the new value of pressure at the (n+1)th time increment can be directly calculated by

$$p^{n+1} = p^n - \rho\frac{\phi}{\Delta t} \quad (42)$$

The algorithm of the velocity correction method(i.e.Method A) [17] can be summarized as follows:

1. The approximate velocity $\tilde{U}_i^{n+1}$ is computed from Eq.(39);

2. The correction potential ϕ is calculated from the Poisson equation (17) of itself;

3. The exact velocity U_i^{n+1} which satisfies the incompressible constraint can be obtained from Eq.(41);

4. The new values of pressure are calculated directly from Eq.(42);

5. U_i^{n+1}, p^{n+1} are replaced with U_i^n, p^n, and proceed to the next time cycle.

Method B

Upon the substitute of a function $F(U_i)$ in Eq.(2), the momentum equations are expressed by

$$\frac{DU_i}{Dt} = -\frac{p_{,i}}{\rho} + F(U_i), \quad (43)$$

where the function of $F(U_i)$ consists of the terms of viscosity and body forces in Eq.(2). The following relation on $F(U_i)$ can be verified, i.e.,

$$DivF(U_i) = 0. \quad (44)$$

The above formula shows that the divergence of F is vanished, but the curl of F isn't zero. Meanwhile, as was stated above, the curl of pressure gradient is vanished, but the divergence of it can't become zero.

Because the pressure gradient has a role of transforming the divergence of velocity field, but not the curl, however the function F has another role of changing the curl of of velocity field, not the divergence, the motion of fluid flows can be divided into two subprocesses of flows, i.e. the first is the transformation of curl, the second is the change of divergence. So the equations of motion can be split into two terms as

$$U_i^{n+1} = \tilde{U}_i^{n+1} + \Delta U_i, \quad (45)$$

where ΔU_i is considered as the correction velocity which has the exact change of divergence. As was stated above, the intermediate velocity is defined as the following forms

$$\tilde{U}_i^{n+1} = U_i^n + \Delta t(F(U_i))^n, \quad (46)$$

$$\Delta U_i = U_i^{n+1} - \tilde{U}_i^{n+1} = -\Delta t\frac{p_{,i}^{n+1}}{\rho}. \quad (47)$$

Assuming the velocity field in the (n+1)th time increment satisfies the incompressible conditions, the following pressure Poisson equation can be obtained

$$\frac{p_{,ii}^{n+1}}{\rho} = \frac{\tilde{U}_{i,i}^{n+1}}{\Delta t}. \quad (48)$$

The algorithm of Method B can be summarized as follows:

1.The intermediate velocity $\tilde{U}_i^{n+1}$ is obtained from Eq.(46);

2.The pressure Poisson equation (48) is solved, p^{n+1} is obtained;

3.The exact velocity U_i^{n+1} is calculated from Eq.(47);

4. U_i^{n+1} is replaced with U_i^n and proceed to the next time cycle.

So Method B is a fractional step method by which the procedure of program is split into two subprocesses.

Method C

To overcome the difficulties of the treatment for the free surface condition, Method C employs the pressure Poisson equation which directly is derives from the momentum equations. This equation such as (13) can be expressed in the Lagrangian coordinates as follows

$$\frac{p_{,ii}^{n+1}}{\rho} = \frac{U_{i,i}^n}{\Delta t} + \nu(U_{i,j}^n + U_{j,i}^n)_{,ij} + f_{i,i}^n. \quad (49)$$

This equation implies the incompressible constraint upon the substitute of $U_{i,i}^{n+1} = 0$ into Eq.(10). By replacing Eq.(2) with Eq.(37), the equations of U_i^{n+1} are showed by

$$U_i^{n+1} = U_i^n - \Delta t(\frac{p_{,i}^{n+1}}{\rho} - \nu(U_{i,j}^n + U_{j,i}^n)_{,j} - f_i^n). \quad (50)$$

Because the values of U_i^{n+1} are required in the computation of pressure Poisson equation to compute the additional term Ω_i^{n+1} which have relation to the boundary condition,the iterative technique is employed to approach the exact value within one time step [17].

The algorithm of Method C can be simply expressed below:

1. Solving the pressure Poisson Eq.(49),

2. Computing the velocity field from Eq.(50).

Method D

This method is similar to Method C except for the construction of iteration. To circumvent the above-mentioned difficulty associated with the calculation of pressure equation, the iterative sequence in Method C is exchanged, i.e., after the velocity of U_i^{n+1} is obtained, the pressure Poisson equation can be solved at the (n+1)th time point. The algorithm of this method is omitted here(Referring to [17]).

Remarks

Hayashi, et al [17] discussed the above-mentioned four variations of the fractional step methods. By means of the computation of solitary wave propagation with moderate wave height, the results obtained showed that Method C is recommendable for practical problems. The Lagrangian FEM presented above is suitable to the moderate wave propagation problems.

4.3 Arbitrary Lagrangian-Eulerian Method (ALE method)

Although a purely Lagrangian method can be used for solving the problem of moderate distortions in fluid motions, this method still has a disadvantage of inability to cope easily with strong distortions, which so often characterize flows of interest. Therefore, the ALE method was proposed by Hirt et al [18], which combined the Eulerian description's advantages with the pure Lagrangian description's abilities to easily handle the free surface flows, and alleviate the shortcomings of both Eulerian and Lagrangian method.

In the ALE method, the material time derivatives can be defined by

$$\frac{D\mathbf{f}}{Dt} = \frac{\partial \mathbf{f}}{\partial t}|_{\chi_i} + (v_i - w_i)\frac{\partial \mathbf{f}}{\partial x_i}, \quad (51)$$

where $\mathbf{f}$ is any physical quantity which is a continuous function of the spatial x_i and time t. The velocity $w_i(= \frac{\partial x_i}{\partial t}|_{\chi_i})$ represents that of a representative grid point and $v_i(= \frac{\partial x_i}{\partial t}|_{x_i})$ denotes the velocity of a material point coinciding with the grid point at time t,

the coordinates X_i are material coordinates, x_i and χ_i are respectively the spatial and referential coordinates, as shown in Fig. 1. The Jacobian determinant for the transformation between referential and spatial coordinates is defined as

$$\overline{J} = det|\frac{\partial x_i}{\partial \chi_i}|. \quad (52)$$

The Reynolds transport theorem is applied to an arbitrary control volume moving through a deforming continuum that is then approached to a point. This yields the resulting field equations in the ALE description as

Continuity

$$\frac{\partial(\rho\overline{J})}{\partial t}|_{\chi_i} + \overline{J}(\rho(v_i - w_i))_{,j} = 0 \quad (53)$$

For incompressible fluids , this equation reduces to $v_{i,i} = 0$ in Ω.

Momentum

$$\frac{\partial v_i}{\partial t} + (v_j - w_j)v_{i,j} + \frac{p_{,i}}{\rho}$$
$$-\nu(v_{i,j} + v_{j,i})_{,j} - f_i = 0 \quad (54)$$

Equations (53) and (54) may be interpreted as material conservation laws with respect to arbitrary moving grid points. One advantage of the ALE description is that,in general, the velocity w_i of the referential frame may not be the same as the coincident material point velocity v_i, as in the Lagrangian description, or it may be vanished,as in the Eulerian description. The relative velocity term $(v_i - w_i)$ becomes zero, resulting in the vanishing of the convective terms, and consequently the set of equations become Lagrangian coordinates. By setting $w_i = 0$, a purely Eulerian description is obtained.

In the free surface boundary condition, the time evolution of the height function is governed by a kinematic equation expressing the fact that the free surface must move with the fluid

$$\frac{\partial \eta}{\partial t}|_{\chi_i} + (v_i^{(s)} - w_i^{(s)})\eta_{,i} = 0, \quad (55)$$

where $v_i^{(s)}$ and $w_i^{(s)}$ are the x_i-components of the fluid particle velocity and mesh velocity at the nodal points of free surface.

Rezoning Techniques

In practical computation of ALE method, the basic hydrodynamic part of the each cycle of the ALE method consists of two phases, the first phase is a typical Lagrangian calculation, the second is the rezoning techniques,in which the rezoning velocity is specified to reduce distortions in the fluid domain and performs all the convective flux calculations. The hydrodynamic problem would be run for a while with the pure Lagrangian code and then be stopped when the mesh becomes somewhat distorted. The rezoning code would then be implemented to smooth out the mesh, so that a desirable mesh configuration can be maintained. Then, the mesh would be passed back to the hydrodynamics code for next time step of calculations. By this use of rezoning technique, the mesh distortions may be held down enough to permit the problem to run satisfactorily. For the detailed description of the ALE scheme, one may be referred to Ramaswamy and Kawahara [40] and Palanisamy and Kawahara [36].

5. NUMERICAL EXAMPLES AND APPLICATIONS

The numerical examples and applications of unsteady incompressible viscous fluid flows by FEM are summarized as Table 1. The cases of computation for incompressible viscous fluid flows, by means of the Eulerian description, include as follows:

1. Laminar incompressible viscous fluid flow in a cavity,

2. Twin vortices behind a flat plate,

3. Vortex pairing in a mixing layer,

4. Transient flows past a circular cylinder,

5. High Reynolds number viscous fluid flows,

6. Tidal current flows.

In Figs. 2 and 3, the benchmark problem for flow in the lid-driven cavity is simulated by means of the iterative method proposed by Ding and Kawahara [8]. For $Re=1000$, we present the finite elemental mesh as fine as 81×81 (in which the total number of triangular elements is 12800 and the total nodal points are 6561) to obtain the grid-independent solution. The streamline, isobars, iso-vorticity, and comparison of u-velocity along vertical line through geometric center of cavity in Figs. 2 (a), (b), (c), and (d), respectively. For $Re=5000$, meshes 41×41, 81×81, and 161×161 are discretized spatially to pursue a convergence test. The results, i.e., streamline, isobar, iso-vorticity, and comparison of u-velocity such as those for $Re=1000$, are presented in Fig2. 3(a), (b), (c), and (d). Both of the two numerical results for $Re=1000$ and $Re=5000$ are in good agreement with those of other benchmark solution [14, 20]. In this case, the total mass source in the computational domain is less than 1.0×10^{-17} (in which all arithmetic was carried out in 64-bit precision). It means the law of mass conservation is enforced numerically in virtue of this kind of iterative procedure. We refer to Ding and Kawahara [8, 9] for the details.

<u>Remarks</u>

1. The Taylor-Galerkin method is capable of producing numerical schemes of high accuracy for the solution of convective transport problems .

2. The three-step finite element method has advantages of accuracy and simplicity, at the same, it is suitable for nonlinear, multi-dimensional convective flows and problems with complex outlet boundary conditions.

3. Recognition of the dual role of the pressure in the incompressible fluid flows leads to construct a new time-accuracy primitive variable CFD algorithm [44] In virtue of the fractional step method, these requirements are accommodated in the CFD algorithm, i.e., (i) applying a potential function to enforce the continuity constraint, (ii) using the derived PPE (13) or (14) to obtain an accurate and discretely continuous pressure solution, and (iii) employing mathematically well-posed and physically motivated Neumann boundary condition (15) for pressure [12, 9].

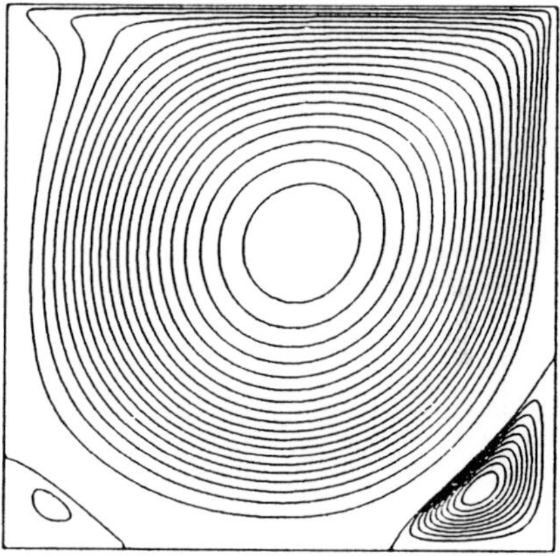

(a) Streamline for $Re=1000$
(Streamfunction $\Psi_{max} = 1.712 \times 10^{-3}$, $\psi_{min} = -0.119$)

(b) Pressure for $Re=1000$
($p_{max} = 0.708$, $p_{min} = -0.121$)

Fig.2: Simulation of flow in the lid-driven cavity for $Re=1000$ with mesh 81×81

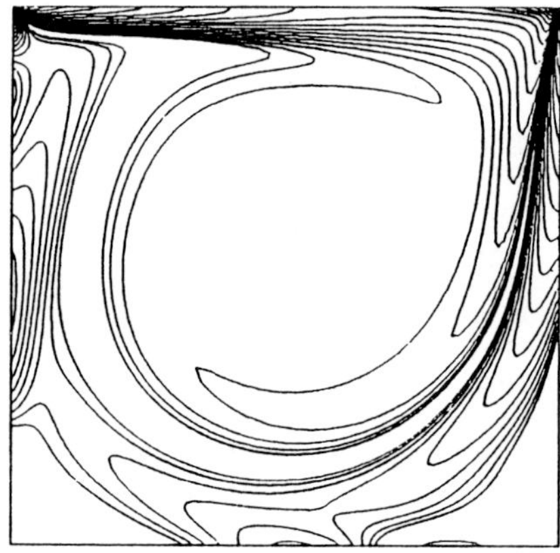

(c) Vorticity for *Re=1000*

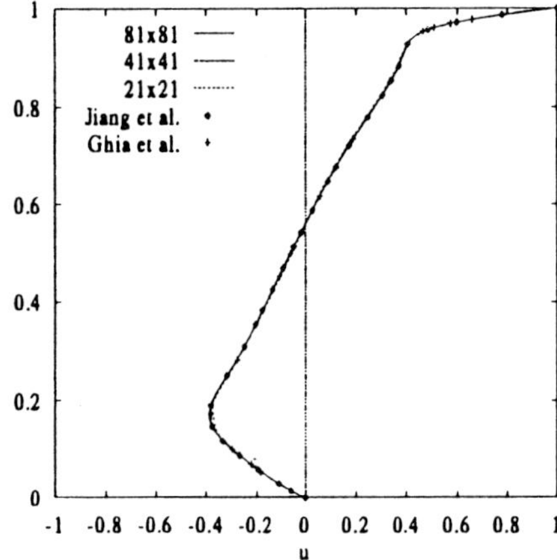

(d) Comparison of *u*-velocity along vertical
line through geometric center of cavity

Fig.2: Continued

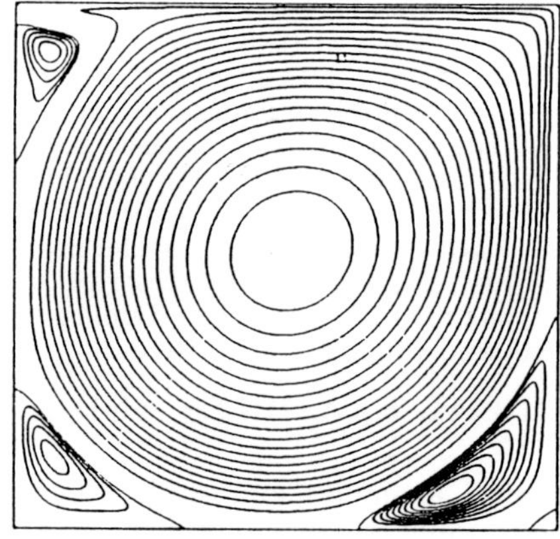

(a) Streamline for *Re=5000*
(Streamfunction $\Psi_{max} = 3.039 \times 10^{-3}$,
$\psi_{min} = -0.122$)

(b) Pressure for *Re=5000*
($p_{max} = 0.471$, $p_{min} = -0.123$)

Fig.3: Simulation of flow in the lid-driven cavity
for *Re=5000* with mesh 161×161

The cases of computation for free surface fluid flows
in Table 1 include as follows:

1. Sloshing problems,
2. Broken dam problems,
3. Forced oscillation flows in a container,
4. Solitary wave propagation,
5. Wave run-up of channel,
6. Tsunami wave propagation,
7. Density flows.

In Fig. 4, the numerical example is the simulation
of solitary wave propagation by ALE method in which
the solver for fluid flow is based on Method C as described above, where the initial conditions are similar
to those in [17]. The snapshots of flow patterns and
the two-dimensional free surface shapes shown in Fig.
4 describe the water wave oscillation very well. In
this Figure, two time step results at $t = 6s$ and $7s$ are
shown. The comparison of water elevation between
numerical simulation and experiment is depicted in
Fig. 5. The results of elevation between numerical
simulation and experiment are in very close agreement
[45]. See [17, 45] for the details.

Remarks:

1. The Lagrangian method is suitable for the computations of free surface flows. The numerical examples identify the method's flexibility in term of geom-

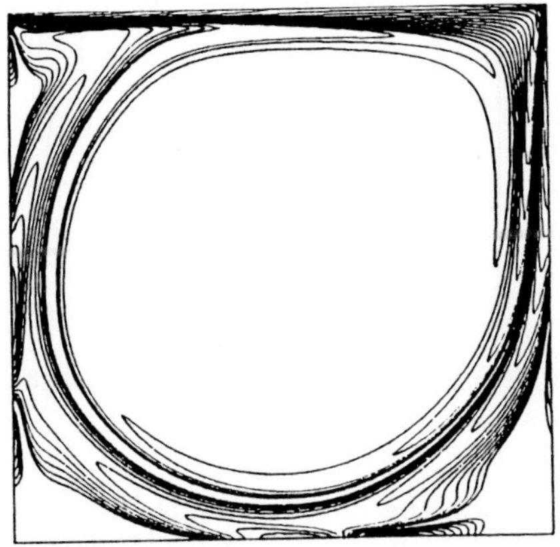

(c) Vorticity for *Re=5000*

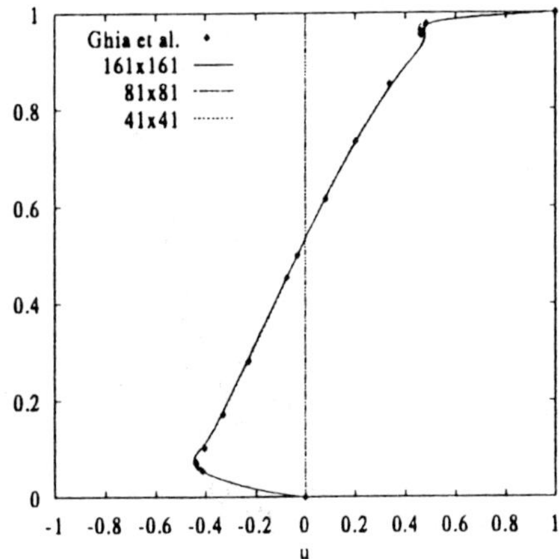

(d) Comparison of *u*-velocity along vertical
line through geometric center of cavity

Fig.3: Continued

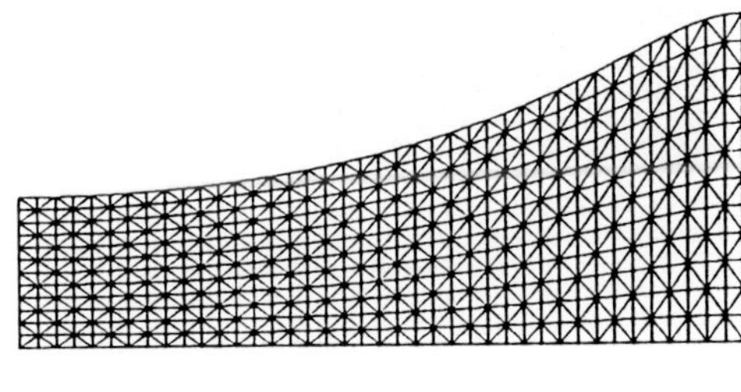

(a) Mesh Division

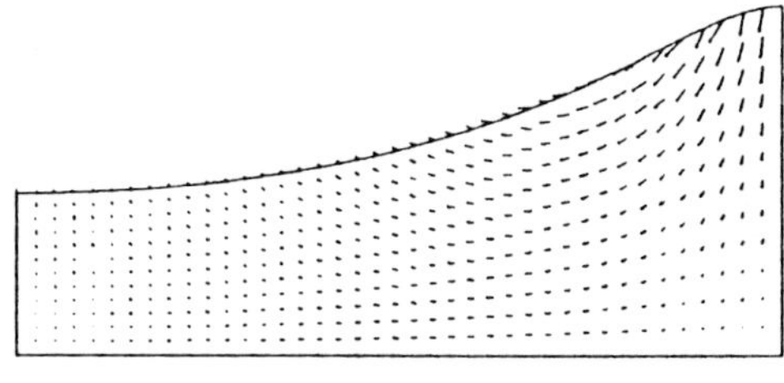

(b) Velocity

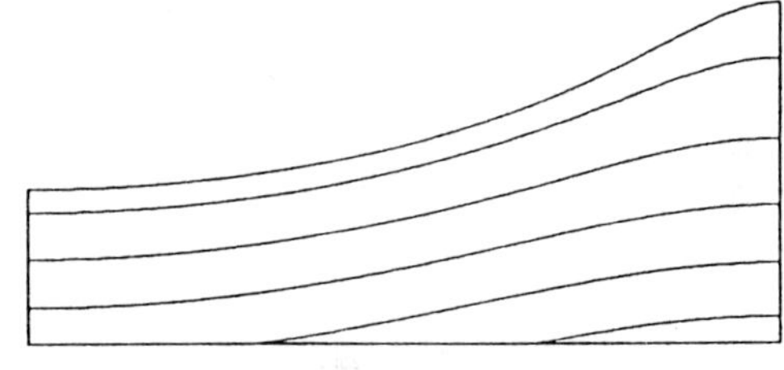

(c) Pressure

Fig.4.1: Computed results of solitary wave
propagation at *t=6.0s* by ALE and
Method C

etry and boundary conditions,as well as its robustness
in obtaining solutions to difficult problems.

2. The ALE method has abilities to resolve arbitrary confining boundaries, large distortions. It proves
to be a useful tool in the solution not only of unsteady
viscous free surface flows [41], but generally of time-
dependent fluid-structure interaction problems [33].

6. FINAL CONCLUSIONS

In this report, the fractional step finite element
methods of unsteady incompressible viscous fluid flows
are summarized. It emphatically discusses the single
and multiple step method for the calculation of convective flows, the pure Lagrangian method and ALE
method (the mixed method). The Lagrangian method

has successfully used for free surface flows, meanwhile,
the ALE method can be handle the large distortion
problems. The fractional step methods show some
advantages of accuracy, stability, simplicity, and the
ability to solve the complex convection dominated incompressible flows.

Acknowledgements

Y. D. wishes to thank K. Yamamoto for their assistance. Y. D. acknowledges the hospitality of the
Applied Mechanics Laboratory at the Chuo University where part of this work was conducted.

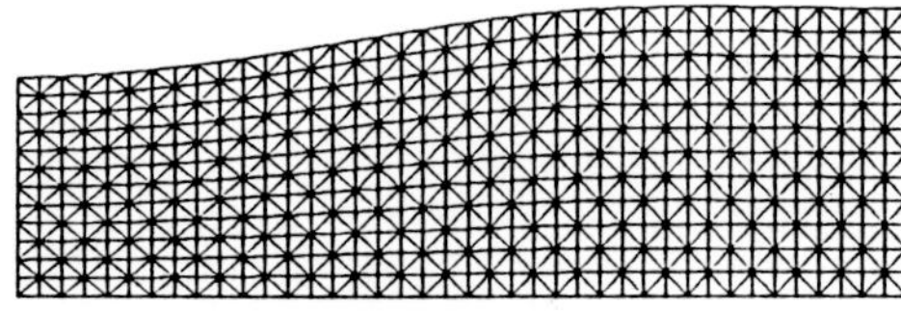

(a) Mesh Division

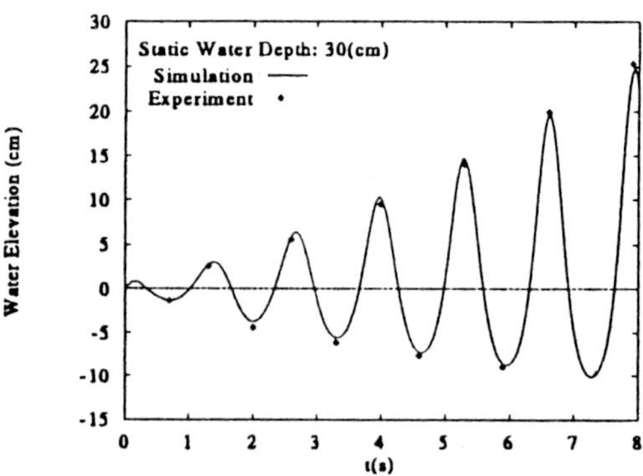

Fig.5: Water elevation on left-hand wall of tank versus time

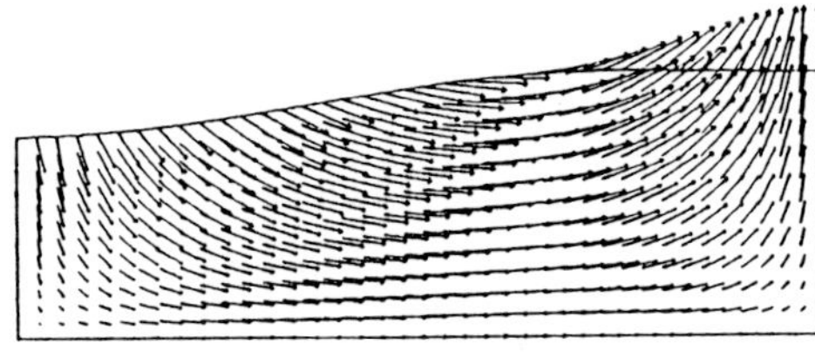

(b) Velocity

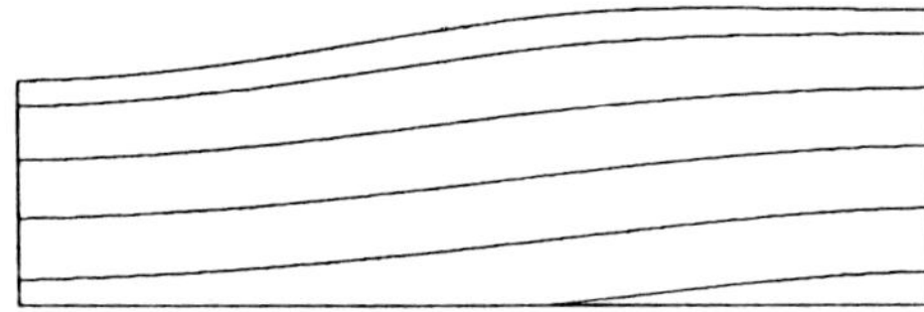

(c) Pressure

Fig.4.2: Computed results of solitary wave propagation t=7.0s by ALE and Method C

REFERENCES

[1] A.J. Baker, *Finite Element Computational Fluid Mechanics*, Hemisphere Pub. Co., McGraw-Hill Book Co., New York, (1983).

[2] M. Braza, P. Chassaing, and H. Ha Minh, 'Numerical study and physical analysis of the pressure and velocity fields in the near wake of a circular cylinder', *J. Fluid Mech.*, **165**, 79-130, (1986).

[3] A.N. Brooks and T.J.R. Hughes, 'Streamline upwind/Petrov-Galerkin formulations for convection dominated flows with particular emphasis on the incompressible Navier-Stokes equations', *Comp. Meth. App. Mech. Eng.*, **32**, 199-259, (1982)

[4] A.J. Chorin, 'Numerical solution of the Navier-Stokes equations', *Math. Comp.*, **22**, 745-762, (1968).

[5] T.J. Chung, *Finite Element Analysis in Fluid Dynamics*, Hemisphere Pub. Co., McGraw-Hill Book Co., New York, (1978).

[6] J.J. Connor and C.A. Brebbia, *Finite Element Techniques for Fluid Flow*, Butterworth & Co. Ltd, London, (1976).

[7] H. Daniels and A. Peters, 'Solving large incompressible time-dependent flow problems on scalable parallel systems', in *Solution Techniques for Large-scale CFD Problems*, edited by W. G. Habashi, John Wiley & Sons Ltd, 27-40, (1995).

[8] Y. Ding and M. Kawahara, 'Improved velocity correction method with constraint of continuity using finite element method', Proc. Conf. on Comp. Engrg. Sci., 1(1), (JSCES), 373-376, Tokyo, (1996).

[9] Y. Ding and M. Kawahara, 'Improved velocity correction method for incompressible viscous flows with outlet boundary', Proc. 2nd Asian CFD Conf.,Vol. 2, pp295-300, Tokyo, (1996).

[10] J. Donea, S. Giuliani, and H. Laval, 'Finite element solution of the unsteady Navier-Stokes equations by a fraction step method', *Comp. Meth. Appl. Mech. Eng.*, **30**, 53-73, (1982).

[11] P.M. Gresho, 'On the theory of semi-implicit projection methods for viscous incompressible flow and its implementation via a finite element method that also introduces a nearly consistent mass matrix, Part 1: Theory', *Int. J. Numer. Methods Eng.*, **11**, 587-620, (1990).

[12] P.M. Gresho, 'Incompressible fluid dynamics: some fundamental formulation issues', *Annu. Rev. Fluid Mech.*, **23**, 413-453, (1991).

[13] P.M. Gresho and S.T. Chan, ' On the theory of semi-implicit projection methods for viscous incompressible flow and its implementation via a

finite element method that also introduces a nearly consistent mass matrix: Part 2: implementation', *Int. J. Num. Meth. Eng.*, **11**, 621-659, (1990).

[14]   U. Ghia, K.N. Ghia, and C.T. Shin, 'High-Re solution for incompressible flow using the Navier-Stokes equations and a multigrid method', *J. Comp. Phy.*, **48**, 387-411 (1982).

[15]   M.D. Gunzburger, 'Navier-Stokes equations for incompressible flows: finite element methods', in *Handbook of Computational Fluid Mechanics*, edited by. Roger Peyret, et al., 99-157, Academic Press, (1996).

[16]   K. Hatanaka and M. Kawahara, 'A fractional step finite element method for conductive-convective heat transfer', *Int. J. Numer. Methods Heat Fluid Flow*, **1**, 77-94, (1991).

[17]   M. Hayashi, K. Hatanaka, and M. Kawahara, 'Lagrangian finite element method for free surface Navier-Stokes flow using fractional step methods', *Int. J. Numer. Methods Fluids*, **13**, 805-840, (1991).

[18]   C.W. Hirt, A.A. Amsden, and J.L. Cook, *J. Comp. Phy.*, **14**, 227, (1974).

[19]   C.B. Jiang and M. Kawahara, 'A three-step finite element method for convection dominated incompressible flow', *Comp. Fluid Dyn. J.*, **1**(4), 443-462, (1993).

[20]   C.B. Jiang and M. Kawahara, 'The analysis of unsteady incompressible flows by a three-step finite element method', *Int. J. Numer. Methods Fluids*, **16**, 793-811, (1993).

[21]   C.B. Jiang and M. Kawahara, 'A three-step finite element method for unsteady incompressible flows', *Comp. Mech.*, **11**, 355-370, (1993).

[22]   M. Kawahara, K. Yoshimura, and H. Ohsaka, 'Steady and unsteady finite element analysis of incompressible viscous fluid', *Int. J. Numer. Mechods Eng.*, **10**, 437-456, (1976).

[23]   M. Kawahara, M. Kobayashi and K. Nakata, 'Multiple level finite element analysis and its applications to tidal current flow in Tokyo Bay', *Appl. Math. Modelling*, **7**, 197-211, (1983).

[24]   M. Kawahara and K. Kashiyama, 'Selective lumping finite element method for near shore current', *Int. J. Numer. Methods Fluids*, **4**, 71-94, (1984).

[25]   M. Kawahara, H. Hirano, and Y. Iriye, 'A three-dimensional two-step explicit finite element method for high Reynolds number viscous fluid flow', *Finite Element in Fluids*, **6**, 199-217, (1985).

[26]   M. Kawahara, T. Kodama, and M. Kinoshita, 'Finite element method for Tsunami wave propagation analysis considering the open boundary condition', *Computer & Mathematics with Applications*, **16**(1/2), 139-152 (1988).

[27]   M. Kawahara and A. Anjyn, (1988) 'Lagrangian finite element method for solitary wave propagation', *Comp. Mech.*, **3**, 299-307, (1988).

[28]   M. Kawahara, H. Sakurai, and K. Kashiyama, 'Boundary-type element method for wave propagation analysis', *Int. J. Numer. Methods Fluids*, **8**, 559-578, (1988).

[29]   T. Kodama and M. Kawahara, 'Multiple level finite element analysis for tidal current flow with non-reflective boundary condition', *Structural Eng./Earthquake Eng.*, JSCE, **9**(1), 77s-87s, (1992).

[30]   H. Laval and L. Quartapelle, 'A fractional-step Taylor-Galerkin method for unsteady incompressible flows', *Int. J. Numer. Methods Fluids*, **12**, 501-513, (1990).

[31]   R. LΦohner, 'Design of incompressible flow solvers: practical aspects', in *Incompressible Compressible Computational Fluid Dynamics - Trends and Advances*, edited by Max. D. Gunzburger and Roy A. Nicolaides, Cambridge Univ. Press, 267-293, (1993).

[32]   A. Maruoka, S. Ohta, and M. Kawahara, 'A finite element analysis of incompressible flow using equal-order interpolation', 6th Int. Symp. on CFD, Sept. 4-8, Lake Tahoe, Navada, 762-767, (1995).

[33]   A. Maruoka, K. Nakakubo, H. Hirano, and M. Kawahara, 'ALE finite element analysis for flow around tandem circular cylinders', Proc. Int. Conf. on Finite Elements in Fluids - New trends and applications, Venezia, 15-21 Oct., 195-204, (1995).

[34]   T. Okamoto and M. Kawahara, 'Two-dimensional sloshing analysis by Lagrangian finite element method', *Int. J. Numer. Methods Fluids*, **11**, 453-477, (1990).

[35]   T. Okamoto, M. Kawahara, N. Ioki, and H. Nagaoka, 'Two-dimensional wave runup analysis by selective lumping finite element method', *Int. J. Numer. Methods Fluids*, **14**, 1219-1243, (1992).

[36]   V. Palanisamy and M. Kawahara, 'A fractional step arbitrary Lagrangian-Eulerian finite element method for free surface density flows', *Comp. Fluid Dyn. J.*, **1**, 57-77, (1993).

[37]   J.B. Perot, 'An analysis of the fractional step method', *J. Comp. Phy.*, **108**, 51-58, (1993).

[38]   B. Ramaswamy, M. Kawahara, and T. Nakayama, 'Lagrangian finite element method for the analysis of two-dimensional sloshing problems', *Int. J. Numer. Methods Fluids*, **6**, 659-670, (1986).

[39]   B. Ramaswamy and M. Kawahara 'Lagrangian finite element analysis applied to viscous free surface fluid flow', *Int. J. Numer. Methods Fluids*, **7**, 953-984, (1987).

[40]   B. Ramaswamy and M. Kawahara, 'Arbitrary Lagrangian-Eulerian finite element method for unsteady, convective, incompressible viscous free surface fluid flow', *Int. J. Numer. Methods Fluids*, **7**, 1053-1075, (1987).

[41] B. Ramaswamy, 'Numerical simulation of unsteady viscous free surface flow', *J. Comp. Phy.*, **90**, 396-430, (1990).

[42] M. Shumura and M. Kawahara, 'Two dimensional finite element flow analysis using the velocity correction method',*Structural Eng./Earthquake Eng.*, **5**(2), 255-263, (1988).

[43] R. Temam, 'Sur l'approximation de la solution des équations de Navier-Stokes par la méthode des pas fractionnaires (i)', *Arch. Rational Mech. Anal.*, Vol. 32, pp135-153, (1969).

[44] P.T. Williams and A.J. Baker, 'Incompressible computational fluid dynamics and the continuity constraint method for the three-dimensional Navier-Stokes equations', *Num. Heat Transfer B*, **29**, 137-273, (1996).

[45] K. Yamamoto and M. Kawahara, 'A structural oscillation control using tuned liquid damper based on finite element method', In *Proc. 1st Int. Conf. on Structural Dynamics - Dynamic Load Problem on Structure*, Nov. 28-29, Aula Barat ITB, Bandumg, Indonesia, 307-317, (1996).

TABLE 1 Summary of Numerical Simulations for Incompressible Fluid Flows

(1)	Flow (2)	(3)	Investigator (4)	Description (5)	Discretization (6)	Technique (7)	Conclusions (8)	Applications (9)
Incompressible Viscous Flow	Laminar Natural Convective Flow	2-D Vertical Structure	Hatanaka & Kawahara (1991) [16]	Eulerian (N-S eqs.)	Eulerian Backward Differentiation	Method C	• A simple algorithm structure. • The Neumann boundary condition of pressure field is involved in the time derivative term of velocity field. •The effect of temperature induced buoyancy in a 2-D cavity was presented.	Natural convection in 2-D Cavity
	Laminar Flow	2-D	Jiang & Kawahara (1993) [19]	ditto	3-step method	Modified Taylor-Galerkin Scheme	•The method retains the 3-order accuracy and uniform CFL property of the original Taylor-Galerkin method. •The method is suitable for nonlinear, multi-dimensional convective flows.	•Steady cavity flows •backward-facing step flows •Twin vortices behind a flat plate (steady)
	Density Flow	2-D	Jiang & Kawahara (1993) [20]	ditto	ditto	ditto	•The method is extended to unsteady incompressible flow.	
	Mixing Layer	2-D	Jiang & Kawahara (1993) [21]	ditto	ditto	ditto	•The results obtained shown the development of vortex pairing, and were in good agreement with those of observation.	
	Viscous Flows	2-D	Ding & Kawahara (1996) [8, 9]	ditto	Single Step Method	Improved velocity correction method	•The continuity constraint can be satisfied by the iterative strategy. •The artificial pressure open boundary is avoidable. The Neumann type boundary condition is specified.	•Steady cavity flow • Transient flow past a circular cylinder •Steady double-recirculations in backward-facing step
	Transient Flow	2-D & 3-D	Maruoka, Ohta, & Kawahara (1995) [32]	ditto	Eulerian Backward-differentiation (semi-implicit)	Modified PPE	•A stabilized PPE is computed. •The equal-order interpolation function is employed.	• Transient flows past a circular cylinder
	Viscous Flows	2-D	Gresho & Chan (1990) [13]	ditto	Eulerian Backward-differentiation (semi-implicit)	Project 2	•A semi-implicit projection method. •A nearly consistent mass matrix	
	Steady Viscous Flows	3-D	Williams & Baker (1996) [44]	Reynolds-averaged N-S eqs	θ-implicit time integration	Continuity constraint method (CCM)	•A equal-order interpolation. • Reynolds-averaged PPE. • Iterative strategy.	• Developing flow in a square duct. • Natural convection in a cavity • Step-wall diffuser
	Viscous Flows	2-D & 3-D	Löhner (1993) [31]	Eulerian with compressible assumption	θ-implicit time integration	Preconditioned Uzawa method	• Mixed element •Artificial Viscosity term.	• Driven Cavity • Circular Cylinder • Airfoil •Inlet Flow in a channel
	Viscous Flows	2-D	Perot (1993) [37]	Eulerian (N-S Eqs.)	Crank-Nicholson method	Single-step method	•An approximate block LU decomposition. •High accurate time discretization. •Finite Difference Method	

611

TABLE 1 Continued

(1)	Flow (2)	(3)	Investigator (4)	Description (5)	Discretization (6)	Technique (7)	Conclusions (8)	Applications (9)
Free Surface Flows	Free Oscillation	2-D	Hayashi, et al (1991) [17]	Lagrangian (N-S Eqs.)	Backward Differentiation Method	Lagrangian (Method A ~ D)	• Method C is recommendable for practical computation. •To solve the strong non-linear wave propagation problems, the ALE method must be introduced.	• Solitary wave Propagation
	Wave Run-up	2-D	Okamoto, et al (1992) [35]	Eulerian (Shallow Water Equation)	ditto	Two-Step Explicit Method	• The automatic mesh generation technique is effective for the 2-D analysis of moving boundary problems. •The detailed information about the land topography and incident wave is necessary.	•Run-up analysis of non uniform slope and a wedage-shaped cross section channel. •Tsunami Run-up caused by an earthquake
	Density Flow	2-D Vertical Structure	Palanisamy & Kawahara (1993) [36]	Lagrangian-Eulerian	ditto	ALE method + Rezoning	•The mixing of two different density fluids in a tank is analyzed. the results shown the method's capability and suitability to obtain useful and reliable results.	
	Wave Run-up	2-D	Kawahara, et al (1988) [26]	Eulerian (Shallow Water Equation)	ditto	Two-Step Explicit Method	• The two-step explicit scheme for treatment of open boundary is sucessfully formulated.	•Tokachi-Oki Tsunami
	Tidal Flow	Quasi-3-D	Kawahara, et al (1983) [23]	ditto	ditto	Multiple level FEM	•The arrangement of nodal points and finite elements needn't necessarily be coincident at each level. •The flow pattern and order of flow magnitude computed are suitable for design of practical port. •The specification of boundary conditions at each level has some difficulties.	•Tokyo Bay •Density Current
			Kodama & Kawahara (1992) [29]	ditto	Two-Step Explicit Scheme	ditto	•The open boundary condition does not cause the spurious reflective wave on the open boundary in the computation of multiple level flow.	• Tokyo Bay
	Near shore Current	2-D Depth-Averaged Structure	Kawahara & Kashiyama (1984) [24]	Eulerian	Two-Step Explicit Scheme	Standard Galerkin FEM	•The wave direction and height are determined. •The wave reflection and diffraction are ignorant.	•Fujisawa Coast
	Sloshing	2-D Vertical Structure	Ramaswamy, et al (1986) [38]	Lagrangian	Semi-Implicit Scheme	Method A	•This algorithm is rather simple. •It is particularly attractive for large scale problems.	•Oscillating tank

TABLE 1 Continued

(1)	Flow (2)	(3)	Investigator (4)	Description (5)	Discretization (6)	Technique (7)	Conclusions (8)	Applications (9)
Free Surface Flows	Viscous Flow	2-D Vertical Structure	Ramaswamy & Kawahara (1987) [39]	ditto	ditto	Method B	•It is effective for the use of Lagrangian FEM to simulate free surface viscous flows. •The iterative methods provide saving in computer execution times and storage requirement compared with other FEM. •Its success has been attributed to the strict enforcement of continuity constraint.	• Broken Dam • Nonlinear Oscillation •Rayleigh-Taylor Instability
			Ramaswamy & Kawahara (1987) [40]	Lagrangian-Eulerian	Iterative Strategy	ALE Method	•A minimum of stored information. *bullet*Highly stable and convergent feature. •This method is not affected by the spurious phenomenon of spatial oscillation of the pressure.	•Solitary wave propagation. •Large amplitude sloshing.
			Kawahara & Anjyu (1988) [27]	Lagrangian	ditto	Method A + Rezoning	•The deformation of solitary wave has been clearly computed using the rezoning technique, in which the slope of channel bed is variable.	•Solitary wave propagation (with a variety of channel bottom
	Water Oscillation	2-D	Kawahara, et al (1988) [26]	Helmhotz & Mid-slope Equations	ditto	Boundary-Type FEM	•Numerical results obtained by the 3- and 4-mode elements are in agreement with the observed data.	•Water oscillation of constant depth channel & inclined bottom channel. •Oscillation wave of Ohfunato Bay.
	Wave in a Container	2-D	Okamoto & Kawahara (1990) [34]	Lagrangian	Implicit method	Method B	•A double-valued free surface position can be presented by using Lagrangian FEM.	•Large-amplitude sloshing wave in a container of complicated shape with roofs.
Fluid-Structure Interaction	Transient Flow	2-D	Maruoka, et al (1995) [33]	Lagrangian-Eulerian	Trapezoidal Method	Semi-Implicit Method + ALE	•A stabilization technique (Balancing Tensor Diffusivity) •Equal-order interpolation for the finite elemental discretization •By means of ALE method the dynamic features of flow around the oscillating cylinder were obtained.	•Viscous flow around tandem circular cylinders

PROBLEMS AND SOLUTIONS RELATED TO RAPID AND IMPULSIVE CHANGES FOR INCOMPRESSIBLE FLOWS

P.M. GRESHO [1]   R.L. SANI [1]

(1) Lawrence Livermore National Laboratory
(2) University of Colorado

1 Introduction

We shall be interested in examining so-called *impulsive* changes in normal boundary conditions for the incompressible Navier-Stokes equations, both theoretically and numerically *via* the Galerkin finite element method (GFEM) and several time-marching methods. We begin by stating two facts:

1. In the *strictest* sense, impulsive (instantaneous / discontinuous) changes in the normal component of the velocity are illegal in that they cause violations of incompressibility, as well as concomitant unbounded pressure briefly.

2. For the very common case in the fluid dynamics literature, the misnomer impulsive start from rest is confusingly employed. It is a misnomer because the true initial condition for most of these analyses and/or simulations is potential flow far from a fluid at rest.

To help clarify the above issues, we shall also study rapid changes, via driving functions for normal velocity like $(1 - \exp(-\lambda t)$ for large λ — eventually permitting λ to become unbounded. We focus on but a single simple and common example: flow past a circular cylinder. The startup both rapid and impulsive was treated extensively in Gresho and Sani (1997) (hereafter referred to as GS). In this paper, we concentrate mostly on the opposite case: sudden shutdown (rapid and impulsive) of the flow past the same cylinder from an IC that was generated in the startup phase.

For a discussion of some of the previous work in the startup case, see GS. For the shutdown case, we cite the few with which we are familiar: Gresho (1991a) Wang and Dalton (1991), Chang and Maxey (1995), who state, An impulsively stopped free stream... is

much different than an impulsively started flow, and Mei and Lawrence (1996), who state, "When it is suddenly brought to rest, the wake behind the body will continue to move to the right with the wake origin traveling as $x \approx t$".

In the remainder of this paper, we shall analyze the problems associated with rapid/impulsive changes both in the continuum and in the finite h, finite Δt discrete world in which we are forced to do our computations. For the latter case, we shall demonstrate the performance of two popular elements ($Q_2 P_{-1}$ and $Q_1 Q_0$) and the manner in which they (or any other approximate method) behave; *viz.*, dismal failure at small time especially for the impulsive case with recovery fortunately occurring once the mesh near the cylinder has had time to recognize the situation and respond to the severe challenge. Another problem, somewhat surprising, is the discovery that the stable-for-Stokes-flow element ($Q_2 P_{-1}$) is less stable (bigger wiggles) for potential flow than the $Q_1 Q_0$ element that is deemed by some to be unstable (again for Stokes flow).

2 Theory

We shall sneak up on impulsive changes as follows: Find $\underline{u}$ and P from

$$\frac{\partial \underline{u}}{\partial t} + \underline{u} \cdot \nabla \underline{u} + \nabla P = \nu \nabla^2 \underline{u} \tag{1}$$

and

$$\nabla \cdot \underline{u} = 0 \quad \text{in} \quad \Omega \tag{2}$$

with

$$\underline{u} = \underline{w_0} \exp -\lambda t + \underline{w_1} (1 - \exp \{-\lambda t\}) \quad \text{on} \quad \Gamma \tag{3}$$

with
$$\underline{u} = \underline{u}_0 \quad \text{at} \quad t = 0$$

Received on July 31, 1997.

that satisfies both

$$\nabla \cdot \underline{u}_0 = 0 \quad \text{in} \quad \Omega \tag{4}$$

and

$$\underline{n} \cdot \underline{u}_0 = \underline{n} \cdot \underline{w}_0 \quad \text{on} \quad \Gamma; \tag{5}$$

where λ is a parameter that will be allowed to become arbitrarily large—with $\lambda \to \infty$ defining *our* impulsive change. (Actually, the Dirichlet BC of (2) will only be applied on a portion of Γ; the remaining BC's will be presented later.)

The pressure Poisson equation (PPE) implied by (1),

$$\nabla^2 P = \nabla \cdot \left(\nu \nabla^2 \underline{u} - \underline{u} \cdot \nabla \underline{u} \right) \quad \text{in} \quad \Omega \tag{6}$$

and, from (1) and (2), it's implied (Neumann) BC (see, e.g., Gresho and Sani [1987], GS),

$$\partial P / \partial n = \underline{n} \cdot \left[\nu \nabla^2 \underline{u} - \underline{u} \cdot \nabla \underline{u} - \partial \underline{u} / \partial t \right]$$
$$= \underline{n} \cdot \left(\nu \nabla^2 \underline{u} - \underline{u} \cdot \nabla \underline{u} \right)$$
$$- \underline{n} \cdot \left(\underline{w}_1 - \underline{w}_0 \right) \lambda e^{-\lambda t} \quad \text{on} \quad \Gamma, \tag{7}$$

are especially important to understand. Hopefully it is obvious (at least after due reflection) that for λ sufficiently large, and t sufficiently, small the pressure satisfies

$$\nabla^2 P \cong 0 \quad \text{in} \quad \Omega \tag{8}$$

with $\partial P / \partial n \cong -\underline{n} \cdot \left(\underline{w}_1 - \underline{w}_0 \right) \lambda e^{-\lambda t} \quad \text{on} \quad \Gamma; \tag{9}$

i.e., the acceleration BC *dominates* the problem.

Remarks:
1) For impulsive (or rapid) starts from rest, $\underline{u}_0 = \underline{0}$ and $\underline{w}_0 = \underline{0}$.

2) For impulsive (or rapid) stops, $\underline{w}_1 = \underline{0}$.

Noting that (8) and (9) are 'solved' by

$$P = -\phi \lambda e^{-\lambda t}, \tag{10}$$

where ϕ is the potential function satisfying

$$\nabla^2 \phi = 0 \quad \text{in} \quad \Omega \tag{11}$$

and

$$\partial \phi / \partial n = \underline{n} \cdot \left(\underline{w}_1 - \underline{w}_0 \right) \quad \text{on} \quad \Gamma, \tag{12}$$

permits the realization that an impulsive change via Dirichlet BC changes in the normal velocity is actually nothing more than a potential flow 'adjustment' + vortex sheet; i.e. for $\lambda \to \infty$ our 'new' IBVP is simply this: Find $\underline{u}$ and P with the *modified* IC and BC,

$$\underline{u} = \underline{u}_0 - \nabla \phi \quad \text{in} \quad \Omega \quad \text{and} \quad \underline{u} = \underline{w}_1 \quad \text{on} \quad \Gamma; \tag{13a, b}$$

and we note the presence of a vortex sheet on Γ since $\underline{\tau} \cdot \underline{u} = \underline{\tau} \cdot \underline{w}_1$ on Γ but $\underline{\tau} \cdot \underline{u} = \underline{\tau} \cdot \underline{u}_0 - \underline{\tau} \cdot \nabla \phi \neq \underline{\tau} \cdot \underline{w}_1$ just off of the surface. For λ large-but-finite, this is of course only an approximation (still for small time) to the true IBVP given by (1) through (5).

It is also worthwhile pointing out the following projection connection for $\lambda \to \infty$: The solution of (11)–(13) can also be derived as follows: (1) make a step change in the BC from $\underline{n} \cdot \underline{u} = \underline{n} \cdot \underline{w}_0$ to $\underline{n} \cdot \underline{u} = \underline{n} \cdot \underline{w}_1$; (2) call the resulting *non-solenoidal* velocity $\tilde{\underline{u}}$ (i.e. $\tilde{\underline{u}} = \underline{u}_0$ in Ω with $\underline{n} \cdot \tilde{\underline{u}} = \underline{n} \cdot \underline{w}_1$ ($\neq \underline{n} \cdot \underline{u}_0$) on Γ; (3) this is realized via (i) solve $\nabla^2 \phi = \nabla \cdot \tilde{\underline{u}}$ in Ω with $\partial \phi / \partial n = \underline{n} \cdot (\tilde{\underline{u}} - \underline{u}) = 0$ on Γ, where $\nabla \cdot \tilde{\underline{u}}$ is to be interpreted as a Dirac delta function on Γ; (4) compute $\underline{u} = \tilde{\underline{u}} - \nabla \phi$, which will give the same result as Eqn. (13). (Admittedly, the 'continuous' approach, with $\lambda \to \infty$ at the end, is somewhat more intuitively acceptable—but the *equivalence* is important.)

Note from (9) that $\lambda \to \infty$ puts a version of the Dirac delta function, a so-called generalized function, on the boundary:

$$\delta(t) \equiv \lambda e^{-\lambda t}, \tag{14}$$

which satisfies $\int_0^\infty \delta(t) dt = 1$ and $\int_0^t f(t) \delta(t) dt = f(0)$ for $\lambda \to \infty$, and causes $\partial P / \partial n$ to become unbounded.

Note too that this same limit causes both an infinite pressure and a violation of incompressibility— both briefly—the latter a consequence of a step change discontinuity in the normal velocity at the boundary ($\underline{n} \cdot \underline{u} = \underline{n} \cdot \underline{w}_0$ at $t = 0$ but $\underline{n} \cdot \underline{u} = \underline{n} \cdot \underline{w}_1$ for $t = 0^+$). It is this violation that has caused us to state that impulsive changes are "illegal" for incompressible flow (cf. Gresho 1991b, GS). Arbitrarily large λ's are quite permissible, of course. Here, however, we shall 'soften up' and take the broader view that impulsive changes are just as legitimate as are generalized solutions.

Consider now the application of the above theory to a particular case: rapid startup of flow past a circular cylinder (radius a) in an unbounded domain. We first treat the inviscid (potential flow) case, as the resulting analytical solutions provide some useful insight.

For this case, the following solution (in polar coordinates) is easily obtained:

$$\underline{u}_I = \left(1 - e^{-\lambda t}\right)\nabla\phi, \tag{15}$$

where

$$\phi = w_1\left(r + a^2/r\right)\cos\theta . \tag{16}$$

Also,

$$P_I = -\phi\lambda e^{-\lambda t} + P_{pot}\cdot\left(1 - e^{-\lambda t}\right)^2 \tag{17}$$

gives the concomitant pressure, where the potential pressure is

$$P_{pot} = \frac{w_1^2}{2}\frac{a^2}{r^2}\left(2\cos 2\theta - \frac{a^2}{r^2}\right). \tag{18}$$

For the limiting case of an impulsive start, $\lambda \to \infty$ and we get

$$\underline{u}_I = H_1(t)\nabla\phi \tag{19}$$

and

$$P_I = -\phi\delta(t) + H_2(t)P_{pot}, \tag{20}$$

where we 'interpret' $H_i(t)$ as Heaviside step functions— $H_i(0) = 0$ and $H_i(t) = 1$ for $t > 0$. Note that $\delta(t)$ *generates* the potential flow and $H_i(t)$ *maintains* it, which is our interpretation of an (ideal) impulsive start from rest.

For the viscous (no-slip) case, we do not know the exact solution, but in GS we have developed (with much help from A.C. Hindmarsh, to whom we remain indebted) the following useful approximation (model), valid within the boundary layer (BL) and close to the cylinder; i.e., for $r - a \ll \sqrt{4\nu t}$ and $r - a \ll a$:

$$u_\theta \cong -4w_1\ \sin\theta\cdot\frac{r-a}{\sqrt{\pi\nu\tau_{ac}}}\ D\!\left(\sqrt{\lambda t}\right), \tag{21}$$

where $\tau_{ac} \equiv 1/\lambda$ is the acceleration time constant and $D(\cdot)$ is Dawson's integral, which behaves like so: $D(y) \cong y$ for $y \ll 1$, $D(y) \approx \frac{1}{2y}$ for $y \gg 1$, with $D(y)$ attaining its only maximum of $D(\sim 0.92) \cong 0.54$. The corresponding pressure is given by

$$P = P_I - \frac{4\nu w_1}{r}\frac{\cos\theta}{}\cdot\frac{a}{\sqrt{\pi\nu\tau_{ac}}}D\!\left(\sqrt{\lambda t}\right). \tag{22}$$

For both the rapid start and the impulsive start, we emphasize that $D(0) = 0$—and also note that $D\!\left(\sqrt{\lambda t}\right)\!\big/\!\sqrt{\tau_{ac}} \to \frac{1}{2\sqrt{t}}$ for $\lambda \to \infty$ (and $t > 0$), which latter result makes our model agree with previous impulsive start theoretical analyses [e.g.,. Wang (1968), Collins and Dennis (1973), Bar-Lev and Yang (1975)] in that both viscous and pressure contributions to the drag coefficient, for small time, can be shown to be (see GS for details)

$$C_D^V = C_D^P \cong \frac{4}{w_1}\sqrt{\frac{\pi\nu}{\tau_{ac}}}D\!\left(\sqrt{\lambda t}\right) \tag{23}$$

giving, for $\lambda t \ll 1$,

$$C_D^V = C_D^P \cong 4\lambda\sqrt{\pi\nu t}\big/w_1 , \tag{24}$$

which corresponds to the acceleration phase of flow within the BL, while for $\lambda t \gg 1$ we get

$$C_D^V = C_D^P \cong \frac{2}{w_1}\sqrt{\frac{\pi\nu}{t}}, \tag{25}$$

which accounts for the deceleration within the BL—and is the result referred to above in the impulsive start literature (none of whom treat the accelerating phase), and we emphasize that even for the limiting case $(\lambda \to \infty)$, (25) *only* applies for $t > 0$; $C_D^V = C_D^P = 0$ at $t = 0$. The final contribution to the drag coefficient comes from the acceleration portion of the transient pressure field and can be derived as

$$C_D^A = \frac{2\pi a}{w_1}\lambda e^{-\lambda t} , \tag{26}$$

using (17) and (22).
The total drag coefficient is, of course,

$$C_D = C_D^A + C_D^P + C_D^V , \tag{27}$$

and we note that P_{pot} in (17) contributes naught to C_D, per d'Alembert. Thus, for λ arbitrarily large, C_D starts at an arbitrarily large value owing to C_D^A (the Dirac function, in the limit) and for $t > 0$ but small, goes like $\sqrt{\nu/t}$ in which both the viscous component *and* the pressure component are 'caused' by the no-slip BC, the pressure part needing to utilize $\nabla\cdot\underline{u} = 0$ to clearly see its viscous 'roots'; see GS for details.

Finally we note, for $\lambda \to \infty$, that there exists a near-vortex sheet on the cylinder at $t = 0^+$. For a 'classic' impulsive start, $(\lambda = \infty)$ if such a thing exists, the IC is potential flow for $r > a$ with a *true* vortex sheet on $r = a$ caused by suddenly imposing the no-slip BC. We are in fact somewhat 'bothered' by such a definition because it is quite misleading; the flow is not *at rest* at $t = 0$.

3. NUMERICS

So much for theory. How does a CFD code deal with the above issues, which clearly must lead to some serious numerical challenges for large λ and, as we shall show, to total failure at small t for $\lambda = \infty$? We begin by returning to the analytical pressure solution, (22), and note its satisfaction of the following BC on the cylinder (for λ sufficiently large that the contribution from P_{pot} is negligible):

$$\left.\frac{\partial P}{\partial r}\right|_a = \frac{4vw_1\cos\theta}{a\sqrt{\pi v\tau_{ac}}}\ D\!\left(\sqrt{\lambda t}\right), \qquad (28)$$

which behaves like (and causes the viscous part of the pressure to also behave like) $O\!\left(\lambda\sqrt{t}\right)$ for small $t\,(\lambda t \ll 1)$ and like $O\!\left(1/\sqrt{t}\right)$ for 'large' $t\,(\lambda t \gg 1)$. But the '*numerical*' version of this Neumann BC, derived in GS, is rather different:

$$\left.\frac{\partial P}{\partial r}\right|_a \cong \frac{vw_1\cos\theta}{a^2}(a/h - 4)\!\left(1 - e^{-\lambda t}\right), \qquad (29)$$

where the second term might be argued to at least *approximate* the physics (accelerating potential flow) *outside* the BL. (h is the distance from the cylinder to the first node point in the fluid.) But the first and spurious term, which *necessarily dominates* for all cases of interest ($h \ll a$) does not—and the physics *within* the BL is completely lost. Not only does this 'numerical BC' fail to describe reality, it clearly diverges and leads to unbounded pressure for $h \to 0$! By requiring that the far field PPE BC, $\partial P/\partial r = -w_1\cos\theta \cdot \lambda e^{-\lambda t}$, dominate the spurious one in (29) the following (approximate) "window of non-believability," can be derived (see GS for details):

$$\tau_{ac}\ln\frac{ah}{v\tau_{ac}} < t < \tau_{MTB}, \qquad (30)$$

where
$$\tau_{MTB} \equiv h^2/4v \qquad (31)$$

is called the Minimum Time of Believability—and actually came about while analyzing the limiting case $\lambda = \infty$, the impulsive start. For $\lambda = \infty$, the numerical solution can not be believed for $t < \tau_{MTB}$, whereas for the rapid startup (finite λ) the numerical solution is mostly believable for $t < \tau_{ac}\ln(ah/v\tau_{ac})$ because the

acceleration-dominated phase is not so difficult to compute successfully. It is not believable when (30) applies, and both types of startup are believable/useful when $t > \tau_{MTB}$, because the mesh can now 'follow' the viscous diffusion process because the BL now contains at least 1 node point. Of course, for the finite λ case, a good mesh design would preclude the left inequality in (30) via

$$\tau_{MTB} = h^2/4v = \tau_{ac}\ln(ah/v\tau_{ac}) \qquad (32)$$

which, for given v and τ_{ac}, can be solved for h—and we point out the bad news that $\tau_{ac} \to 0 \Rightarrow h \to 0$. Even when (32) is satisfied, however, the details of the solution within the BL cannot be captured by the finite mesh for $t < \tau_{MTB}$; e.g., in GS is shown the following for the 'numerical version' of C_D^V:

$$C_D^V \cong 2\pi v\lambda t/w_1 h \qquad (33)$$

for $t < \tau_{MTB}$, vis-a-vis the correct result, given in (23)–(25).

Enough on the 'theory' of the spatial numerics. Now we address the fact that, just as we cannot compute with $h = 0$, so too can we not time-integrate our resulting differential-algebraic equations (DAE's) exactly—we must also deal with finite Δt. The GFEM approximation to (1)–(5) is (cf., e.g., GS for details)

$$M\dot{u} + N(u)u + CP = -Ku + f(t) \qquad (34a)$$

and
$$C^T u = g(t) \qquad (34b)$$

with, when well-posed (in the 'strict' sense), an IC u_o that satisfies

$$C^T u_o = g(0) \equiv g_o. \qquad (35)$$

These equations also imply the following discrete PPE—complete with built in BC's:

$$\left(C^T M^{-1}C\right)P = C^T M^{-1}[f - Ku - N(u)u] - \dot{g}, \qquad (36)$$

and we briefly discuss how two simple ODE methods behave when applied to either the index 2 system of DAE's given by (34), or the index 1 system 'defined' by (34a) and (36); i.e., the PPE replaces (34b) in the index 1 formulation. Implicit time integration methods are 'natural for the index 2 formulation whereas the index 1 formulation is—after the mass is lumped (which is not always possible) to convert M (and thus M^{-1}) to a diagonal matrix—amenable to explicit methods; see GS for details on these issues.

But since we will also show some results using the penalty method, which method reduces the index on the DAE's to zero (i.e., to ODE's), we first briefly *summarize* the method: the fluid's incompressibility is 'relaxed' via the 'penalty' equation that relates pressure to incompressibility *violation*,

$$QP = \lambda\left(C^T u - g\right), \tag{37}$$

where λ is the "penalty parameter" and is very large (typically $10^6 - 10^{10}$), and Q is the pressure mass matrix (see, e.g., GS). Inserting (37) into (34a) yields the (very stiff) system of penalty ODE's:

$$M\dot{u} + N(u)u = -\left(K + \lambda CQ^{-1}C^T\right)u$$
$$+ f(t) + \lambda CQ^{-1}g(t). \tag{38}$$

See Engelman et al. (1982) for discussion related to the "penalty matrix," $B \equiv CQ^{-1}C^T$, and see GS for a discussion of the spurious penalty transient—whose behavior we shall later demonstrate. Note the implied ODE for the pressure [from (37) and (38)]:

$$\frac{1}{\lambda}Q\dot{P} + \left(C^T M^{-1} C\right)P$$
$$= C^T M^{-1}\left[f - (K + N(u))u\right] - \dot{g}, \tag{39}$$

vis-a-vis (36), to which (39) 'returns' when $\lambda t \gg 1$. Finally, we apologize for the introduction of a different parameter with the same symbol (λ)—but 'excuse ourselves' by stating that we will not (but could have) apply the penalty method to the $e^{-\lambda t}$ case.

We now turn to the simplest implicit ODE method; backward Euler applied to the index 2 DAE's gives

$$\frac{M\left(u_{n+1} - u_n\right)}{\Delta t} + N\left(u_{n+1}\right)u_{n+1}$$
$$+ CP_{n+1} = -Ku_{n+1} + f_{n+1} \tag{40}$$

and
$$C^T u_{n+1} = g_{n+1}, \tag{41}$$

whereas the simplest explicit method (forward Euler, FE) applied to the index 1 DAE's gives

$$\frac{M\left(u_{n+1} - u_n\right)}{\Delta t} + N\left(u_n\right)u_n$$
$$+ CP_n = -Ku_n + f_n, \tag{42}$$

wherein P_n is first computed from

$$\left(C^T M^{-1} C\right)P_n = C^T M^{-1} .$$
$$\left[f_n - Ku_n - N(u_n)u_n\right] - \frac{g_{n+1} - g_n}{\Delta t} . \tag{43}$$

We presented these two well-known methods in detail only because we need to address the question of how they 'perform' for a *step change* in velocity that violates discrete mass conservation; i.e., for our impulsive changes, we are employing IC's that do not satisfy $C^T u_o = g_o$—and we are facing an ill-posed system of DAE's, and one for which an honest/rigorous trapezoid rule (TR) for time integration would 'announce' the ill-posedness via WIGGLES (ringing via $2\Delta t$ oscillations; see GS for details).

Thus, starting with BE for $n = 0$ we obtain from (40) and (41) for the first time step, with $C^T u_1 = g_1$ but $C^T u_0 \neq g_0$,

$$\left(C^T M^{-1} C\right)P_1 = \frac{C^T u_0 - g_0}{\Delta t}$$
$$+ C^T M^{-1}\left[f_1 - Ku_1 - N(u_1)u_1\right] - \frac{g_1 - g_0}{\Delta t} \tag{44}$$

for the pressure—where in our case $g_1 = g_0$, for the impulsive start (or stop). Clearly, when $C^T u_0 \neq g_0$, it gives $P_1 \to \infty$ as $\Delta t \to 0$, thus reflecting the ill-posedness of impulsive changes. It turns out, however, that we can turn this apparent 'problem' into a *solution* simply by noting that for sufficiently small Δt (44) becomes

$$\left(C^T M^{-1} C\right)\phi \cong C^T u_0 - \dot{g}_0, \tag{45}$$

where $\phi \equiv \Delta t P_1$, which remains finite for all Δt, corresponds to the "potential flow adjustment" presented earlier for the continuous case. Thus, *for sufficiently small Δt, a single BE step performs the L^2-projection* to the div-free subspace that is associated with impulsive changes in normal velocity—a fact that we shall put to good use later.

Switching now to the explicit Euler method on the lowered-index DAE's, we examine the first step of FE applied to (42) and (43) by setting $n = 0$ there. Inserting P_0 from (43) into (42) and operating on the result with $C^T M^{-1}$ gives

$$C^T u_1 - g_1 = C^T u_0 - g_0, \qquad (46)$$

a 'general' result in the sense that *any* ODE method applied to the index 1 DAE's will 'preserve' the divergence (see Gresho 1991a, GS, for details). Thus, the 'PPE method' can *not* be used to obtain useful results when applied to impulsive changes.

Later, we shall demonstrate the above; i.e., we will show useful results from BE and discuss the useless ones from FE. We will also show good results for rapid, not impulsive, changes via the $e^{-\lambda t}$ BC. In GS we spent lots of time on both rapid and impulsive startups, both theoretical and via the GFEM and two time-marching methods, BE and TR, both in the 'smart' mode (variable Δt based on physics). Here we shall summarize *some* of that work, to set the stage for what is new herein—viz.,: (1) impulsive and fast shutdowns, (2) the behavior of several time integration methods applied to the 'illegal'/impulsive case, and (3) the solution to the sudden stop case via the penalty method. Thus we shall consider this paper to be an *extension* of the impulsive (and rapid) start case that is Section 3.19 in GS.

4. NUMERICAL RESULTS FOR STARTUPS

In this section we summarize and paraphrase some of the results in Section 3.19 of GS—partly because they are rather interesting and merit a second visit, but mainly to set the stage for the next section—rapid shutdowns. Fig. 1 shows both the domain and the mesh employed. The origin of the $x-y$ coordinate system is at the center of the unit radius cylinder, and the domain covers $-3 \le x \le 3$ and $0 \le y \le 2$.

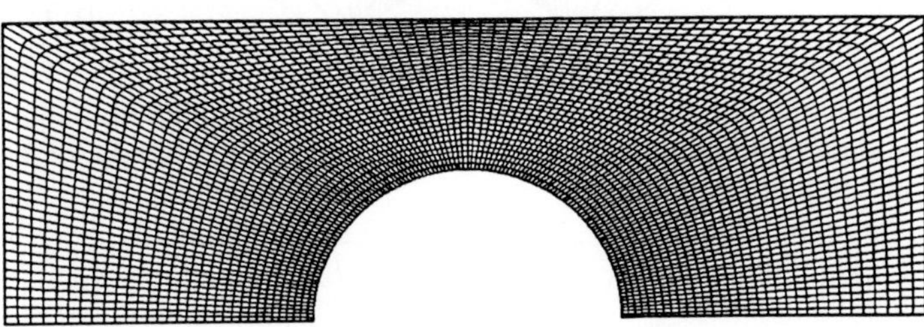

(a) The full domain.

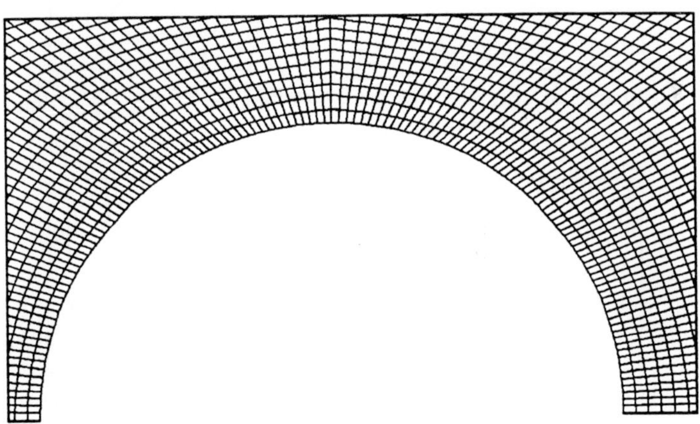

(b) Zoom near the cylinder.

Figure 1:   The mesh of 4290 Q_2P_{-1}(9/3) elements; 16,965 nodes.

The IC is $\underline{u}_0 = \underline{0}$ and the BC's at the inlet ($x = -3$) are $u = w_1\left(1 - e^{-\lambda t}\right)$, $v = 0$ where $w_1 = 0.1$ and $\lambda = 100$ or ∞—the latter case (impulsive) being 'solved' by the technique discussed above of taking one very small BE time step. Homogeneous natural boundary conditions (NBC's) were used at $x = 3$ ($v \frac{\partial u}{\partial x} = P$ and $v \frac{\partial v}{\partial x} = 0$) as outflow conditions; $\underline{u} = \underline{0}$ on the cylinder, and symmetry BC's were used elsewhere. We chose a Reynolds number, $\mathrm{Re} \equiv 2aw_1/v$, of 1000 which translates to $v = 0.0002$. The mesh shown in Fig 1(a) has 4290 9-node elements, with 16,965 nodes—and we used an equivalent mesh (same number of nodes) for the 4-node element.

To 'set the stage', we show first the pressure field for two steady-state results, in Figs. 2 and 3: potential flow and Stokes flow. Fig. 4 shows the velocity potential, which will be important for both startups and shutdowns. The potential flow pressure in Fig. 2, while looking quite good, is actually associated with a somewhat *bad* velocity—revealed by the wiggles in the line plots of Fig. 5.

The less 'stable' (for Stokes flow) element, Q_1Q_0 (4-node velocity, piecewise-constant pressure), was run on (virtually) the same mesh with the *less wiggly* results shown on the right side of the same figure. We do not understand the cause of these wiggles—which vanish later in time when viscosity has had a chance to 'help', and which also decrease with mesh refinement thus not precluding convergence—and we implore again (as we did in GS) the FEM mathematicians to study this second-order elliptic problem ($\underline{u} = \nabla\phi$ and $\nabla \cdot \underline{u} = 0$ with mixed methods) using 'Stokes elements'.

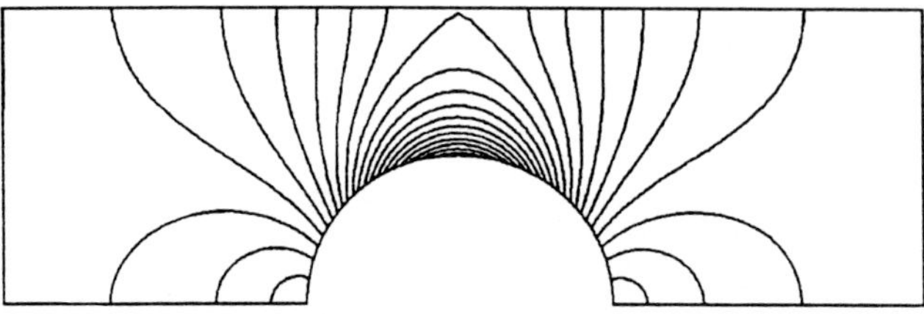

Figure 2: Potential flow pressure; $P_{max} = 0.00505$, $P_{min} = -0.0284$ ($\Delta P = 0.00167$)

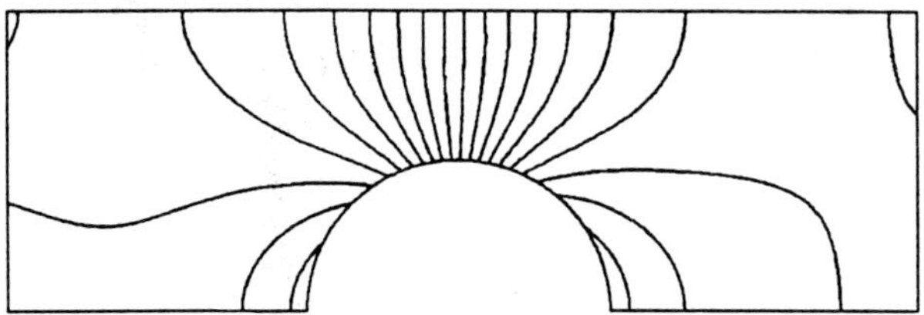

Figure 3: Stokes flow pressure; $P_{max} = 3.73 \times 10^{-4}$, $P_{min} = -0.5 \times 10^{-4}$ ($\Delta P = 2.14 \times 10^{-5}$).

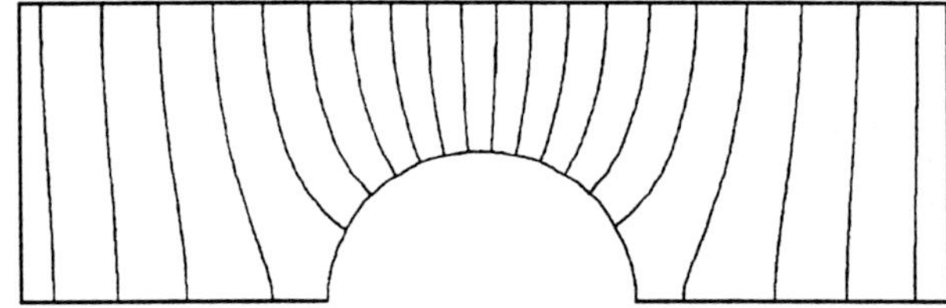

Figure 4: Velocity potential ; $\phi_{max} = 0.80$, $\phi_{min} = 0$ ($\Delta\phi = 0.04$)

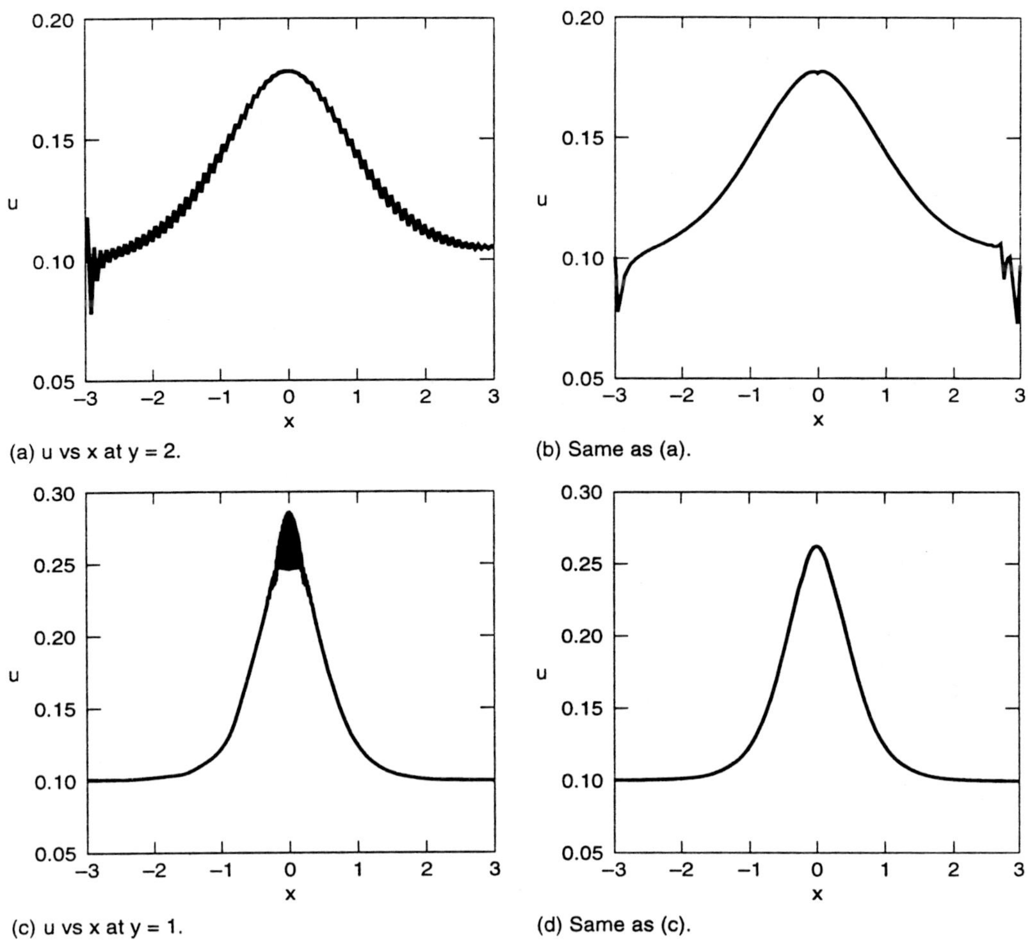

Figure 5: Potential flow: 9/3 results on the left, 4/1 on the right.

Moving on to the transient behavior, we point out that using either the potential flow plus no slip IC (realized via $\underline{u}_0 = \underline{0}$, $u = w_1 = 0.1$ at inlet, and the L^2-projection via the small Δt BE 'trick' described earlier), or the $w_1\left(1 - e^{-\lambda t}\right)$ inlet BC, gives the 'same' numerical results for $\lambda t >\sim 10$ (see GS for details, and for the differences when $\lambda t < 10$); we display here merely a sample of our ('long time')computed results—

in Figs. 6 through 8, just to give the reader a 'feel' for the startup results, *and* to provide an IC, that in Fig. 8, for the shutdown simulations to follow. Here, and in the sequel, all results are obtained using the $Q_2 P_{-1}$ element (also called 9/3). (The 4/1 element would deliver virtually the same results.) Flow separation, plus advection, began somewhat prior to $t = 5$, and visible eddy growth was present by $t = 5$.

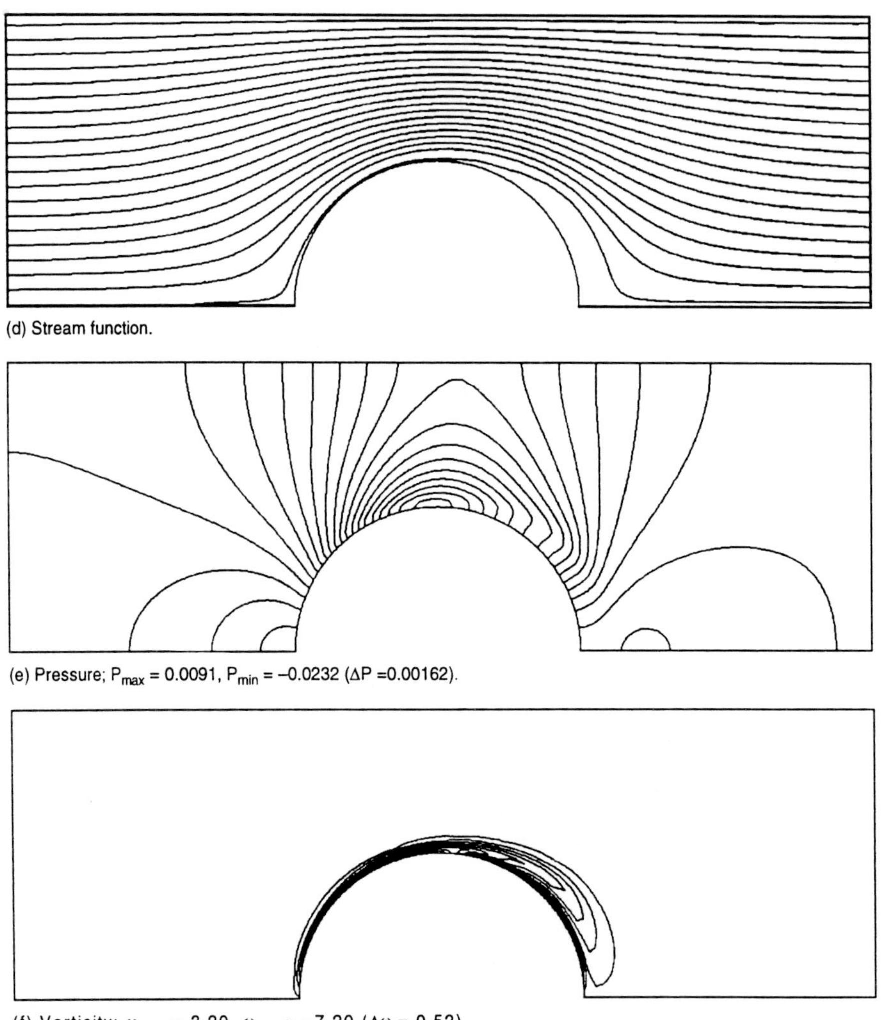

(d) Stream function.

(e) Pressure; $P_{max} = 0.0091$, $P_{min} = -0.0232$ ($\Delta P = 0.00162$).

(f) Vorticity; $\omega_{max} = 3.20$, $\omega_{min} = -7.20$ ($\Delta\omega = 0.52$).

Figure 6: Solution at $t = 10$.

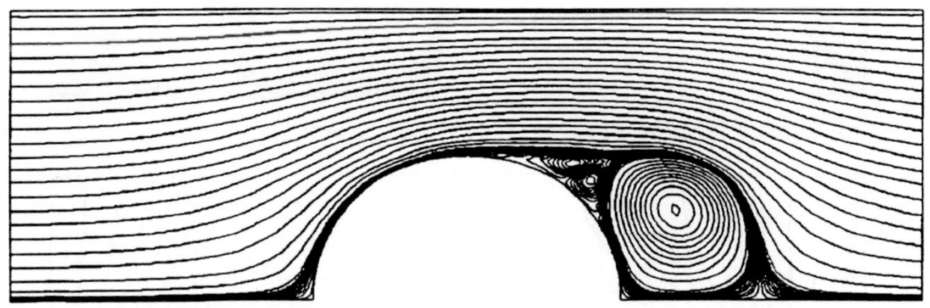

(a)   Stream function

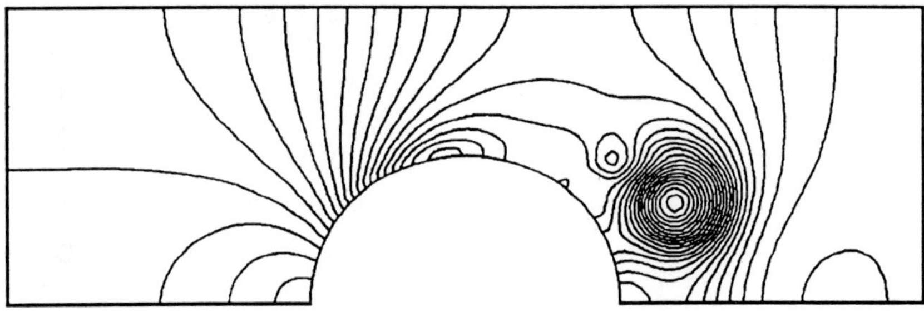

(b)   Pressure; $P_{max} = 0.014$, $P_{min} = -0.034$ ($\Delta P = 0.0032$).

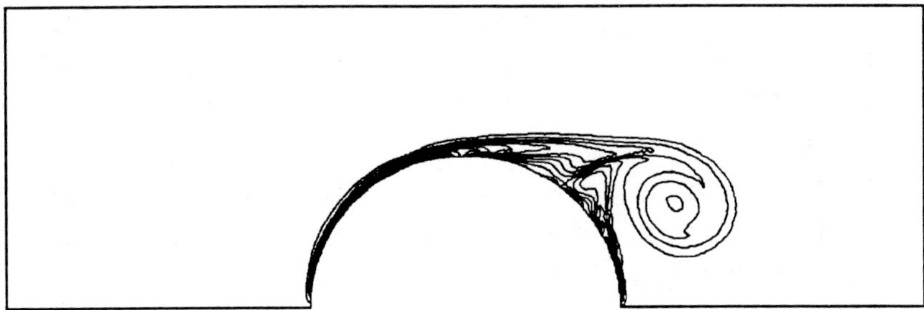

(c)   Vorticity; $\omega_{max} = 4.62$, $\omega_{min} = -7.12$ ($\Delta \omega = 0.78$).

Figure 7: Solution at $t = 25$.

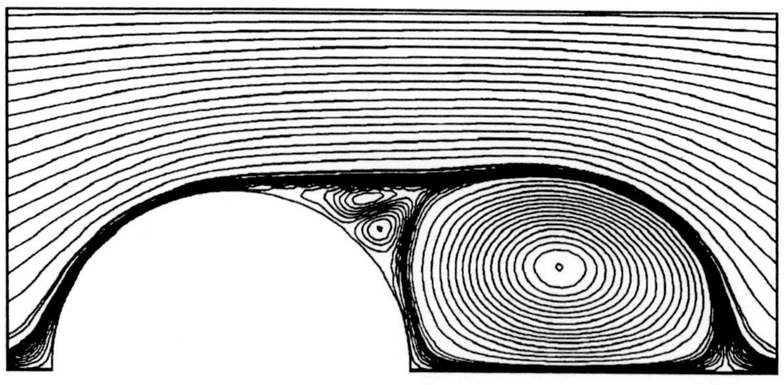

(a)   Stream function $\left(\Psi_{min} = -0.07325, \Psi_{max} = 0.2\right)$.

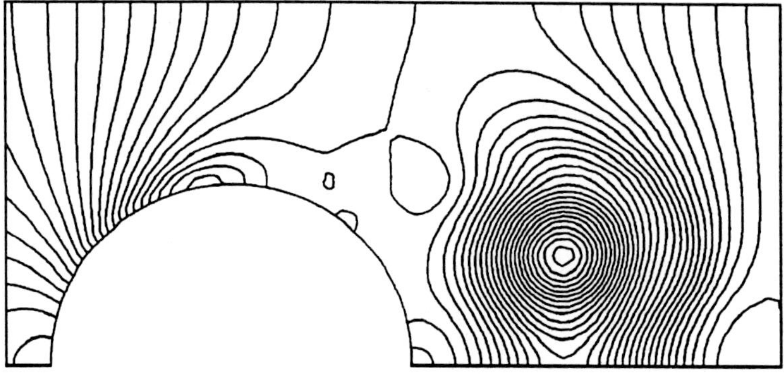

(b)   Pressure; $P_{max} = 0.014$, $P_{min} = -0.040$ ($\Delta P = 0.0027$).

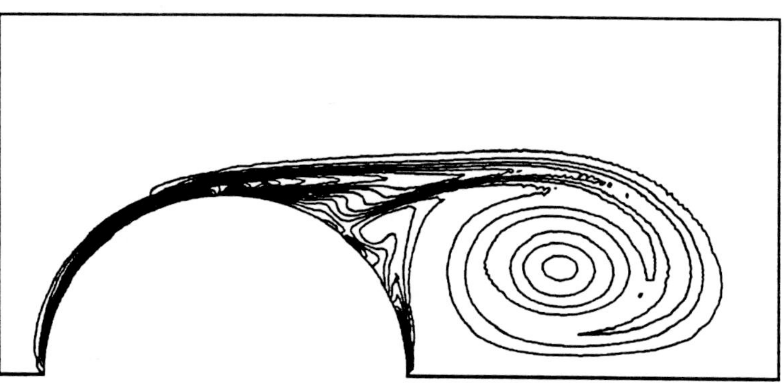

(c)   Vorticity; $\omega_{max} = 2.09$, $\omega_{min} = -7.01$ ($\Delta\omega = 0.35$).

Figure 8:   Solution at $t = 40$.

For further details regarding the startup cases, as well as drag coefficients *and* the demonstrated bad early time results—such as pressure and vorticity $\sim O(1/h)$ for $t < \tau_{MTB}$, et al.—see GS.

5. NUMERICAL RESULTS FOR SHUTDOWNS

We now switch gears and, starting from the solution in Fig. 8, present and discuss some shutdown results—performed both impulsively and quickly, the latter via $w_0 = 0.1$ and $\lambda = 100$ at the inlet in (2), with $w_1 = 0$. We shall also investigate the behavior of the penalty method. In all cases, we employed slippery BC's ($f_\tau = 0$, a zero *pseudo*-shear stress—see GS for details) except on the cylinder, which had $\underline{u} = \underline{0}$.

We begin by noting the somewhat remarkable fact that there are rather many seemingly different/disparate ways of getting to (virtually) the same place—a result that may surprise some. (We hope so!) The "place" they get to, starting from the IC's discussed above, is this: the vorticity-perserving potential flow adjustment plus a new vortex sheet on the cylinder. Here is the list of those that we have discovered thus far—all of which are virtually independent of Reynolds number, and most of which are discussed in GS:

1. An L^2-projection.

2. One very small backward Euler time step.

3. Exponential decay BC($e^{-\lambda t}$) for large λ.

4. The penalty method with accurate time integration through the *spurious* penalty transient.

5. The penalty method with inaccurate time integration.

6. One very small FE step on the index 2 DAE's.

7. The exact (analytical) generalized (and discontinuous) solution to the (ostensibly ill-posed) index 2 DAE's.

Another way to perhaps state these results is this: We have found six different ways to closely approximate the L^2-projection—some which we demonstrate below.

Also interesting—and somewhat surprising— is that each of these techniques can generate *nearly* the same result (at least for small t) for 3 *different* sets of BC's:

1. Close both ends (slam the door at inlet and outlet).

2. Close only the right end.

3. Close only the left (inlet) end.

In the latter two cases, the "standard" homogeneous ("do nothing") NBC's $v\,\partial u/\partial x - P = 0 = v\,\partial v/\partial x$, are applied at the open end.

Also noteworthy is the general observation that, since the impulsive shutdown is obtained (effectively if not directly) by subtracting the appropriate *potential* flow from the given flow, the closer the original flow *is* to a potential flow, the closer will be the sudden stop to a slow and 'boring' flow—with the limit being this: any potential flow IC will be stopped dead; $\underline{u} = \underline{0}$ everywhere. Another way to state this is this: *only* a potential flow can be impulsively *stopped*. [In the general case, (only) the vortical portion of the flow remains after the 'attempted' impulsive stop.] This fact, in fact, leads to a potentially useful way to debug portions of a code; viz (1) solve a (slippery) potential flow problem with the code, (2) slam the doors on this IC and take one small BE time step (still permitting slip, of course). Any non-zero flow that your code computes should be close to zero—and vary like $O(\Delta t)$. We in fact begin our shutdown presentation with such a flow; Fig. 9 shows the residual flow from the code used in this paper (FIDAP)—in which only the left end is closed off (the small inflow/outflow at the right in Fig. 9(a) would vanish if both ends had been closed). The upper left corner "noise" results from the use of large, distorted elements, and is further discussed in GS. But the main points are these: (1) the flow is nearly zero,

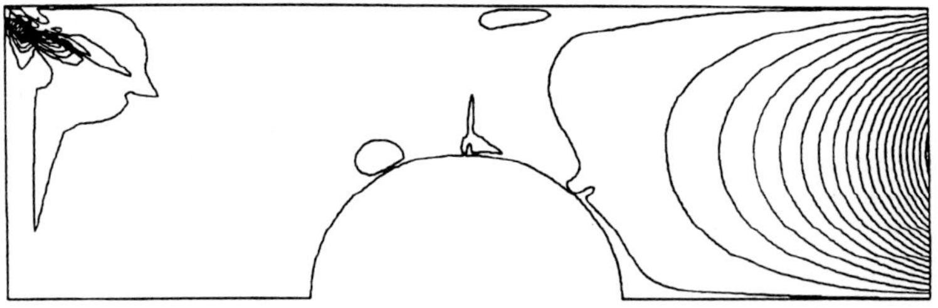

(a) Stream function ($\psi_{min} = -3.4 \times 10^{-9}$, $\psi_{max} = 5.6 \times 10^{-8}$)

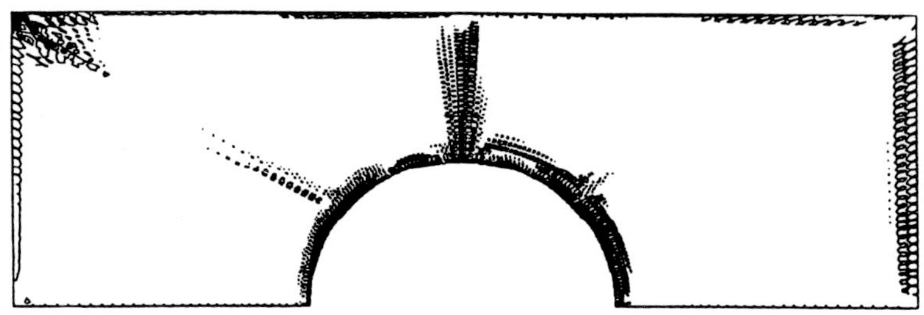

(b) vorticity ($\omega_{min} = -5.8 \times 10^{-5}$, $\omega_{max} = 5.5 \times 10^{-5}$)

Figure 9:   Residual flow upon closing the left end on a potential flow.

and (2) the error *does* scale with Δt (here $\Delta t = 5 \times 10^{-6}$) thus verifying both theory and code.

We now present some shutdown simulations starting from the Re = 1000 solution in Fig. 8—and, unless stated to the contrary, the results to follow are also at Re=1000. Fig. 10 shows the impulsive stop stream function after one small BE step ($\Delta t=10^{-5}$) for each of the 3 BC's discussed above—and Fig. 11 shows the corresponding vorticity and pressure for (virtually) all 3.

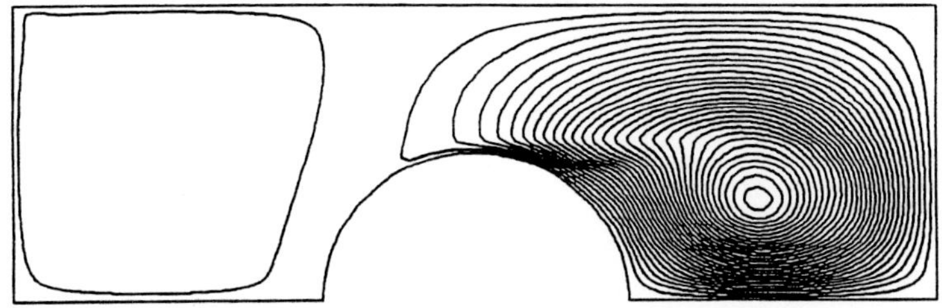

(a)   Both ends closed ($\psi_{min} = -0.12293,\ \psi_{max} = 0.00224$)

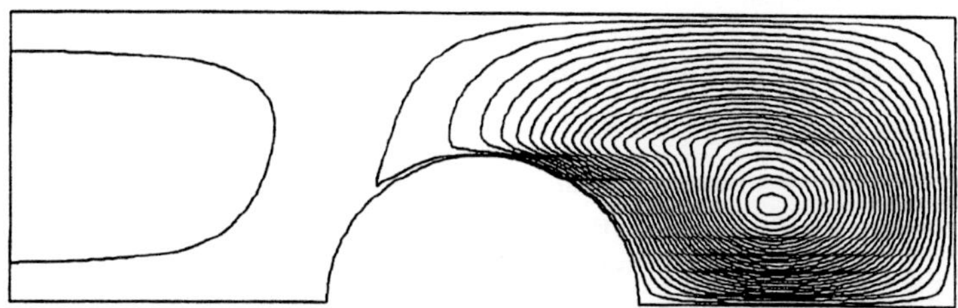

(b)   Right end closed, left end open ($\psi_{min} = -0.12293,\ \psi_{max} = 0.00356$)

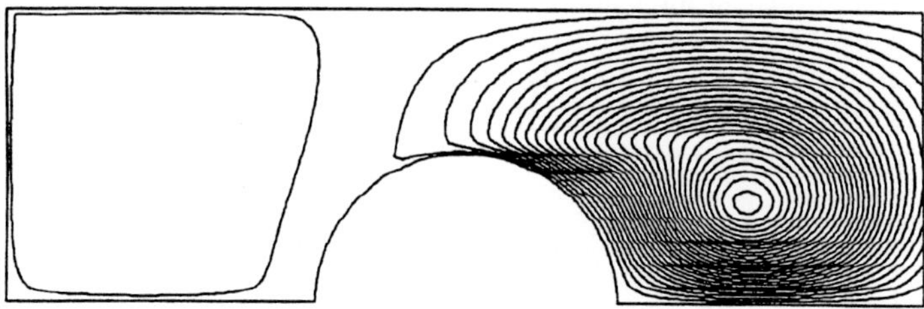

(c)   Left end closed, right end open ($\psi_{min} = -0.12589,\ \psi_{max} = 0.00224$)

Figure 10:   Flow field (stream function) after 1 small time step via backward Euler.

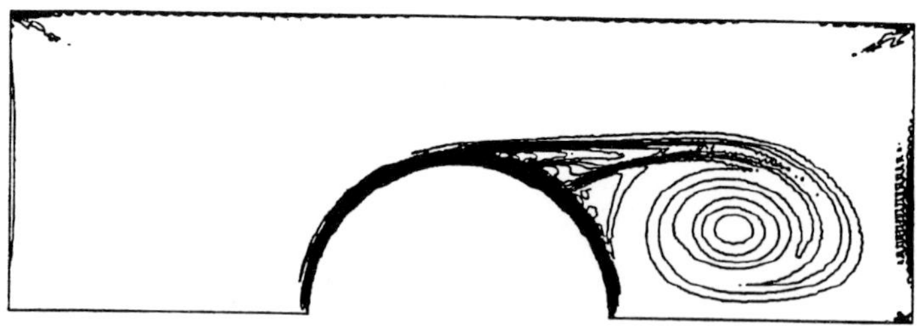

(a) Vorticity ($\omega_{min} = -7.687$, $\omega_{max} = 23.455$)

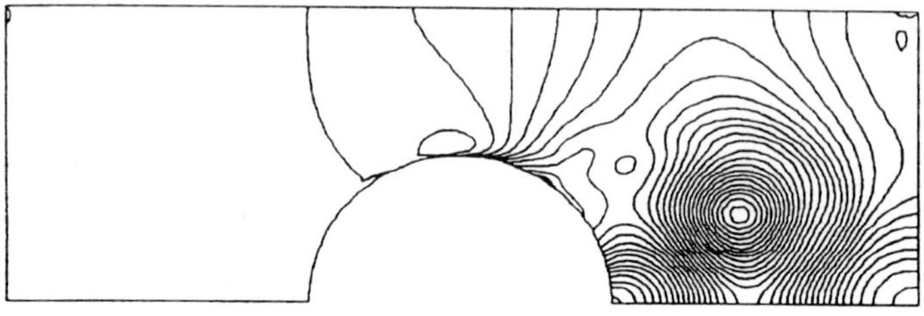

(b) Pressure ($P_{min} = -0.0459$, $P_{max} = 0.000025$)

Figure 11:   Vorticity and pressure corresponding to Figure 10.

It is obvious that the differences in ψ are quite small, showing once more the utility of the homogeneous NBC's as OBC's. Noteworthy also are the following: (1) The trapped eddy is spinning clockwise, as is the large eddy in Fig. 8, and the flow "upstream" is near-zero—both consequences of subtracting the potential flow from the viscous flow—and its intensity in the eddy is now greater ($\psi \cong -0.125$ at the center of the eddy, vs. -0.073 prior to the "projection"); (2) The vortex "sheet" has a peak vorticity of $+23.5$ on the cylinder (a rather far cry from infinity!), which corroborates almost perfectly with that from the

impulsive *start* case reported in GS: -23.6; (3) The vorticity away from the vortex sheet is virtually the same as that in Fig. 8—another consequence of the potential flow adjustment: conservation of vorticity; (4) The pressure is that after 2 time steps, since the 1st step is a potential field—per Fig. 4 and Eqn. (45).

Next, in Figs. 12 through 15 for the left-end-closed BC, we trace the evolution of the trapped flow—and point out that the only "brakes" on the system (besides viscous dissipation) are those from the no-slip BC on the cylinder; all other BC's are frictionless.

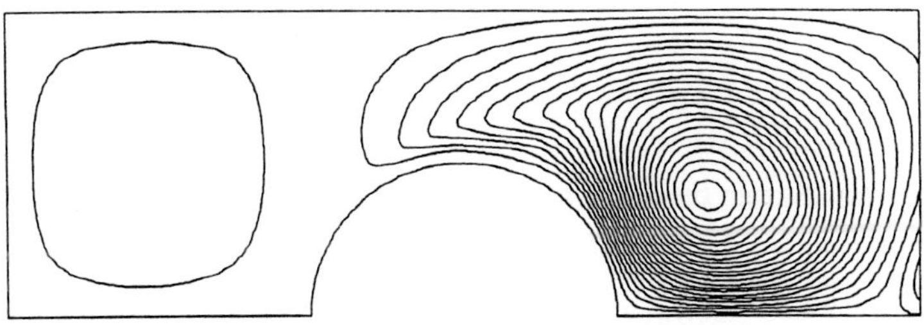

(a) Stream function ($\psi_{\min} = -0.12364,\ \psi_{\max} = 0.00859$)

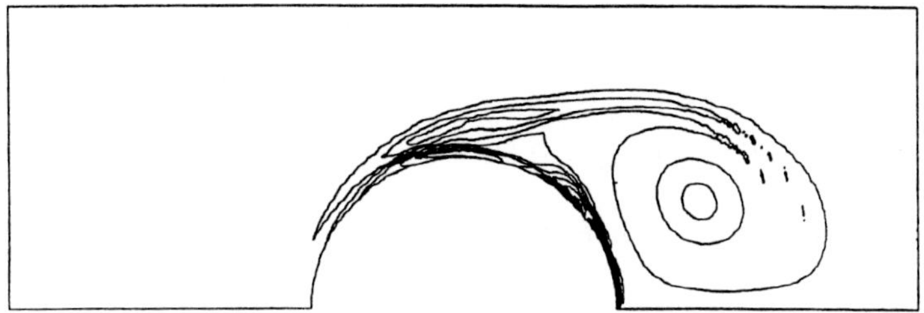

(b) Vorticity ($\omega_{\min} = -1.335,\ \omega_{\max} = 10.406$)

Figure 12:   Solution at $t = 5$ after impulsive stop; left end closed.

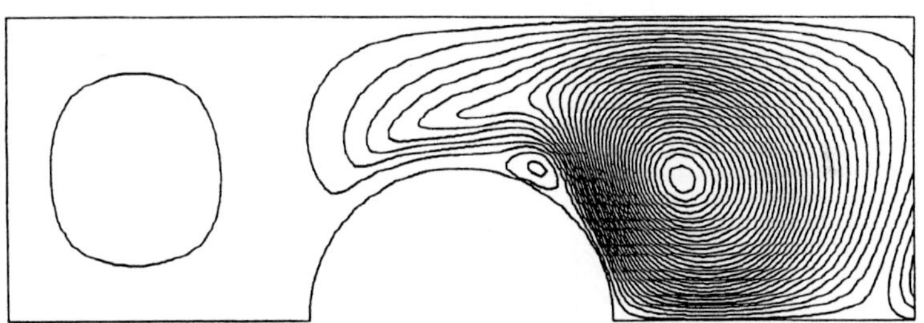

(a) Stream function ($\psi_{\min} = -0.12076,\ \psi_{\max} = 0.01038$)

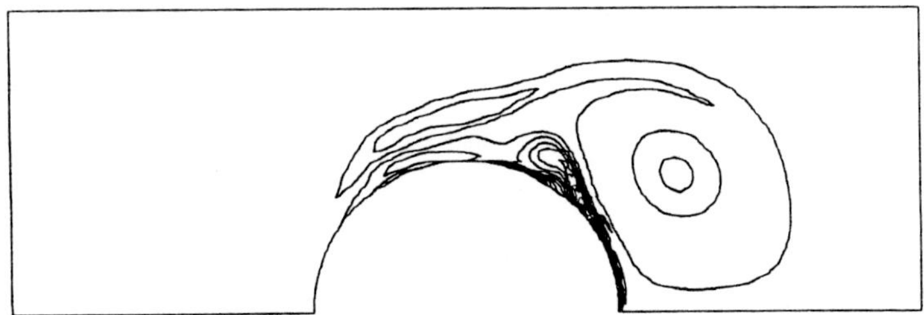

(b) Vorticity ($\omega_{\min} = -4.5260,\ \omega_{\max} = (11.571)$)

Figure 13:   As in Figure 12 except $t = 10$.

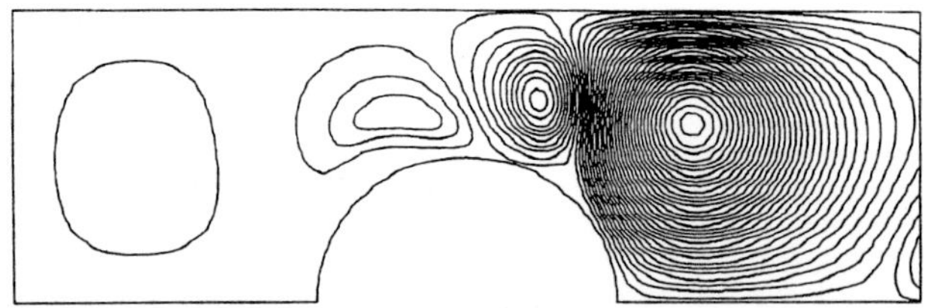

(a) Stream function ($\psi_{min} = -0.11109$, $\psi_{max} = 0.03460$)

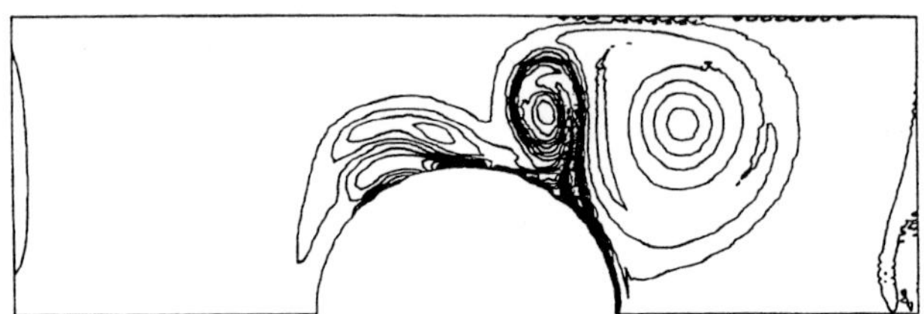

(b) Vorticity ($\omega_{min} = -1.9212$, $\omega_{max} = 6.2933$)

Figure 14: As in Figure 12 except $t = 20$.

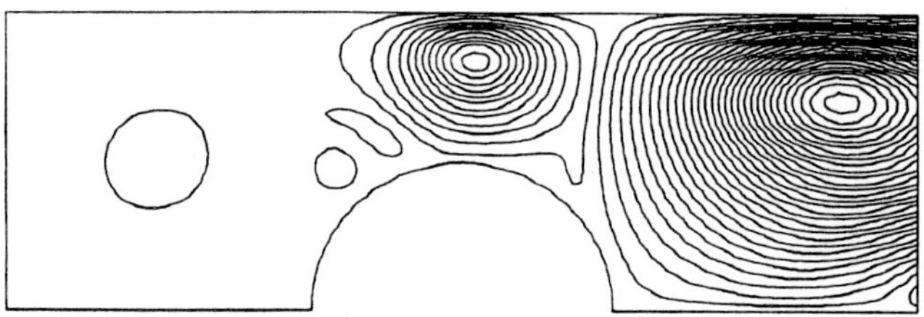

(a) Stream function ($\psi_{min} = 0 - 0.09867$, $\psi_{max} = 0.040937$)

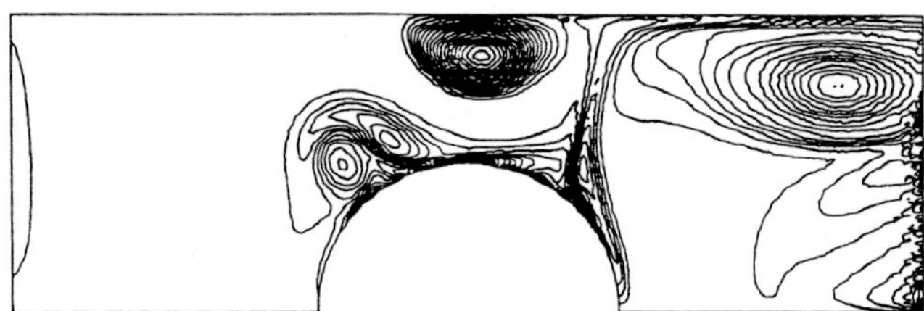

(b) Vorticity ($\omega_{min} = -1.333$, $\omega_{max} = 1.308$)

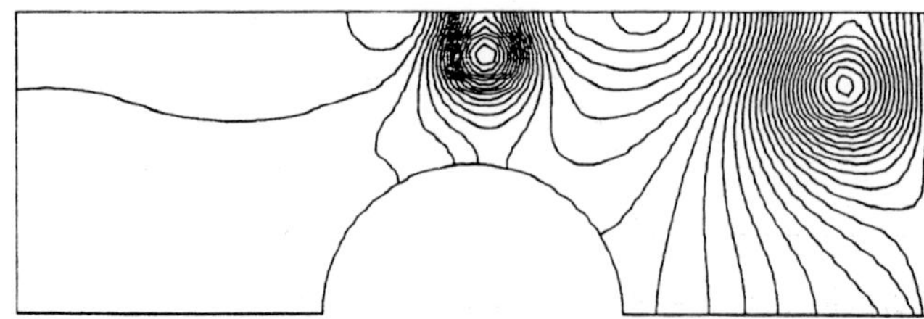

(c) Pressure ($P_{min} = -0.01383$, $P_{max} = 0.014525$)

Figure 15: As in Figure 12 except $t = 40$, and pressure is shown.

The combination of the initial large vortex and the vortex sheet evolves into a separated flow that spawns another eddy (Fig. 13) and yet another (Fig. 14), while the large eddy's center performs another sort of clockwise rotation on the right side of the cylinder—and these figures can also be 'interpreted' as an upstream motion of the original 'wake'. Although we went beyond t=40 (Fig. 15) in our simulations, we show no more because—in fact—the OBC at the right finally *did* cause some poor, nonphysical behavior. We could have closed both ends, but chose not to because

we wanted to compare the sudden stop with the $e^{-\lambda t}$ shutdown, which BC would be pretty awkward to apply at the right end. Thus, this and all remaining simulations are for the case of closing off only the left (inlet) boundary.

Figs. 16 through 19 show the analogous results for the somewhat unreal case of Stokes flow—starting from the same IC—and were generated just to see how the flow would evolve with advection absent (no wake "washing" over the cylinder).

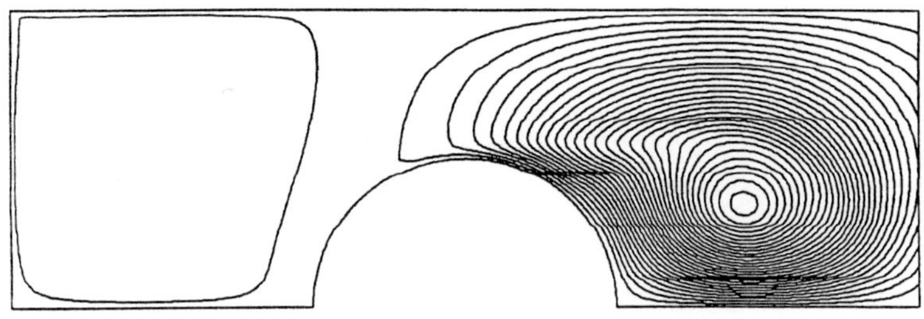

(a)   Stream function ($\psi_{min} = -0.12583$, $\psi_{max} = 0.002259$)

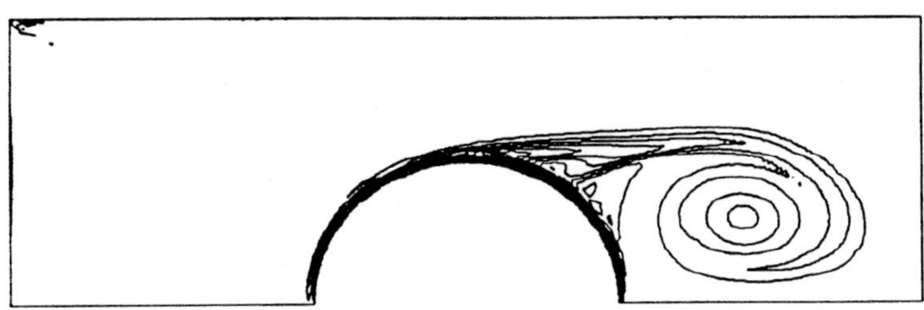

(b)   Vorticity ($\omega_{min} = -6.8677$, $\omega_{max} = 20.619$)

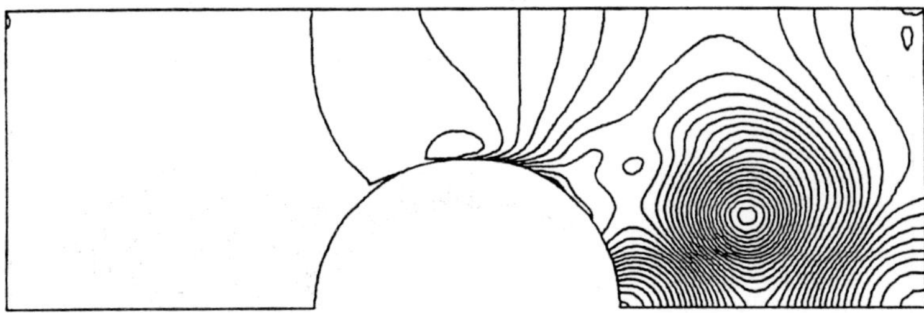

(c)   Pressure ($P_{min} = -0.007235$, $P_{max} = 0.002236$)

Figure 16:   Stokes flow solution at $t = 0.1$ after impulsive stop; left end closed.

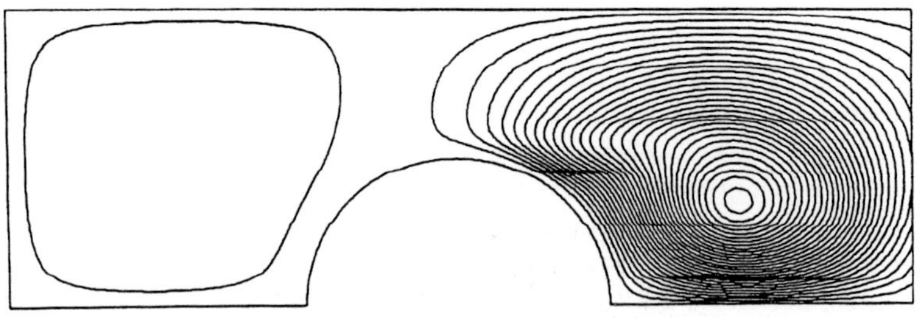

(a) Stream function ($\psi_{min} = -0.12538$, $\psi_{max} = 0.002336$)

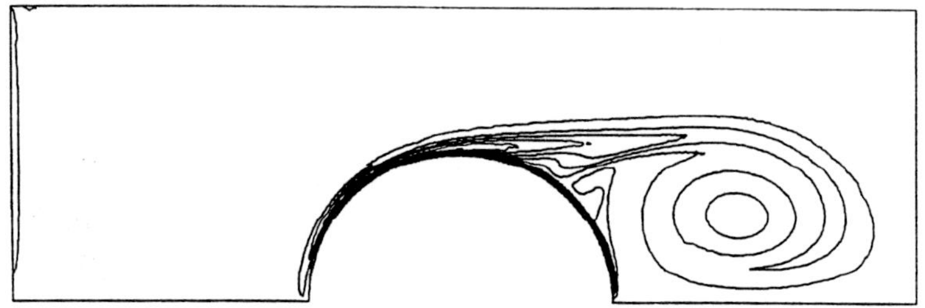

(b) Vorticity ($\omega_{min} = -2.5263$, $\omega_{max} = 10.993$)

Figure 17:   As in Figure 16 except $t = 1$.

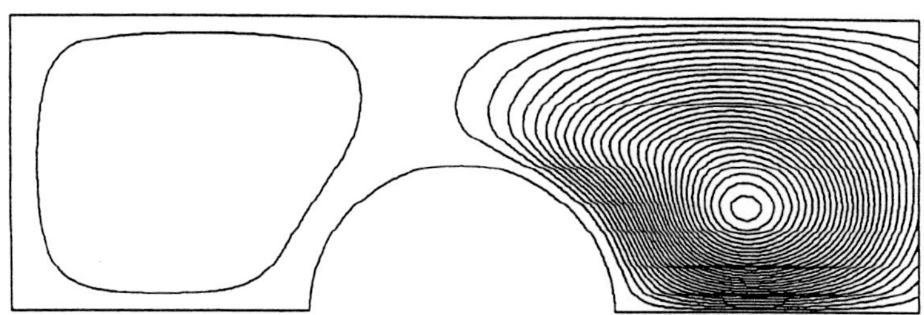

(a) Stream function ($\psi_{min} = -0.12246$, $\psi_{max} = 0.002432$)

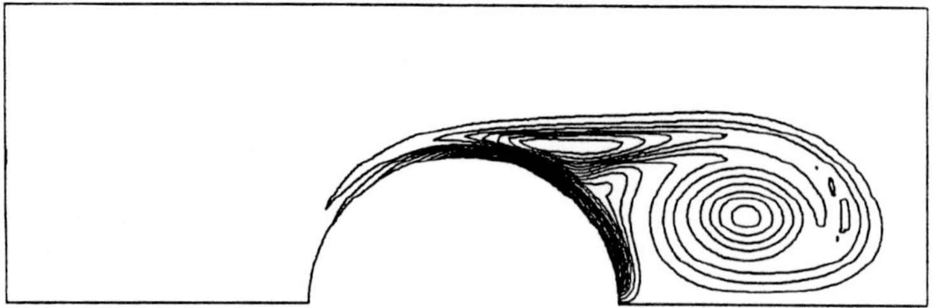

(b) Vorticity ($\omega_{min} = -1.2031$, $\omega_{max} = 3.2935$)

Figure 18:   As in Figure 16 except $t = 10$.

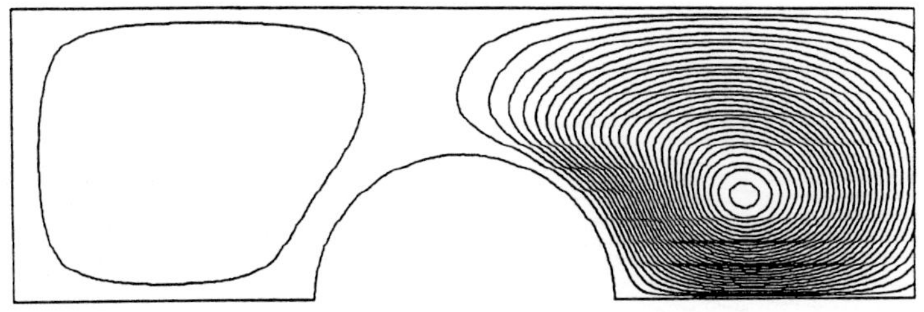

(a) Stream function ($\psi_{\min} = -0.11511,\ \psi_{\max} = 0.002372$)

(b) Vorticity ($\omega_{\min} = -0.99258,\ \omega_{\max} = 1.5285$)

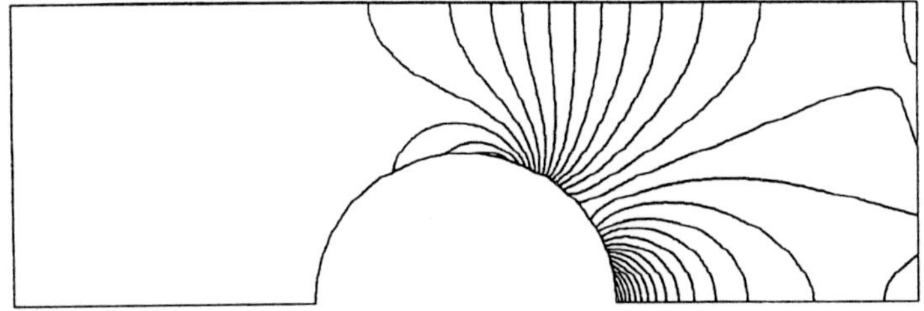

(c) Pressure $P_{\min} = -0.000329,\ P_{\max} = 0.000323$

Figure 19:   As in Figure 16 except $t = 40$.

Whereas the stream function decay is rather boring, the vorticity evolution via diffusion is interesting. Recall that vorticity flux through the cylinder surface is given by $v\,\partial\omega/\partial n = \partial P/\partial \tau$ (see Gresho 1991a); the flux is large when the tangential pressure gradient is. Finally we point out that the decay toward zero flow occurs on a *very* long time scale; $\tau_D \equiv D^2/v = 20{,}000$.

Now we turn to the "rapid" shutdown, via the $e^{-\lambda t}$ decay of the inlet BC. And we only show the part that is interesting and different from the impulsive stop—namely that for $t \le 10\tau_{ac} = 10/\lambda = 0.1$, because for times greater than this the two results are very close. Thus, Figs. 20 through 26 show the interesting evolution when one "shuts down the pump" via an exponential spindown.

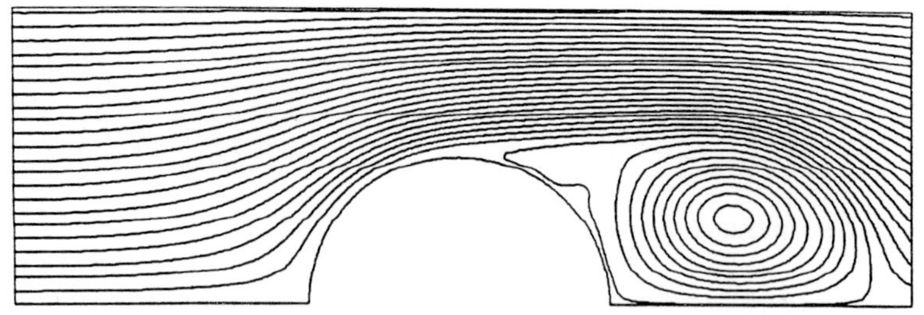

Figure 20: Stream function at $t = 0.001$ (inflow = 90.5% of initial) for $e^{-\lambda t}$ shutdown; ($\psi_{min} = -0.0773$, $\psi_{max} = 0.18097$).

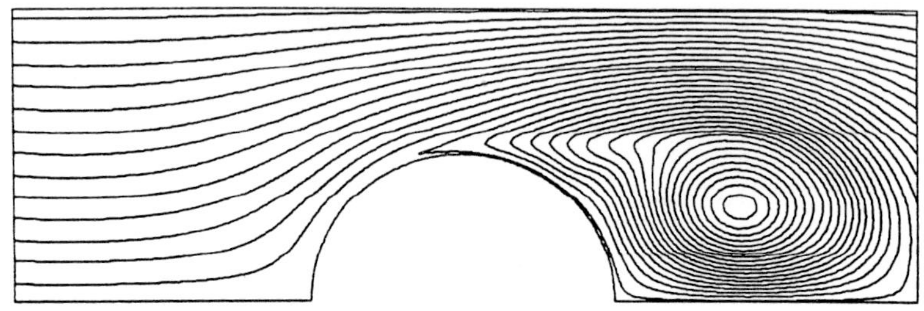

Figure 21: As in Figure 20 except $t = 0.01$ (inflow = 36.8% of initial); $\psi_{min} = -0.10485$, $\psi_{max} = 0.07390$.

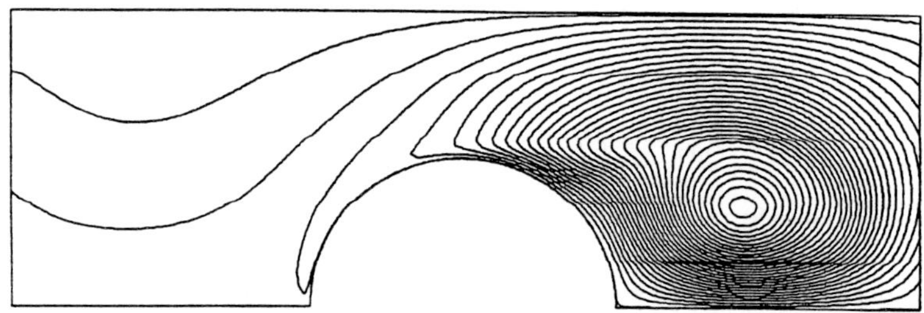

Figure 22: As in Figure 20 except $t = 0.03$ (inlet = 4.98% of initial); $\psi_{min} = -0.12283$, $\psi_{max} = 0.01022$.

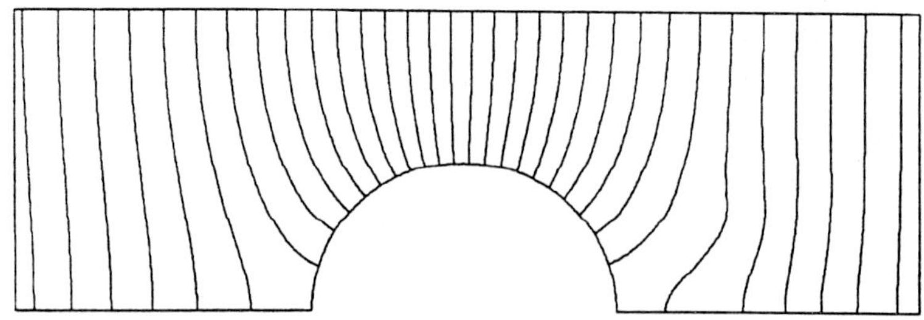

Figure 23: As in Figure 20 except pressure at $t = 0.04$ ($P_{min} = -1.5455$, $P_{max} = 0.00011$).

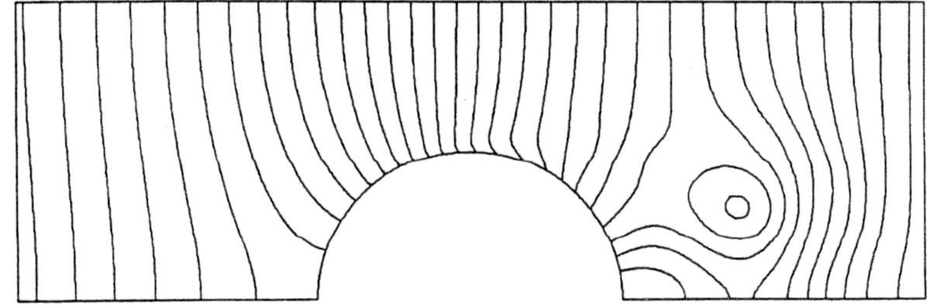

Figure 24:   As in Figure 23 except $t = 0.06$ ($P_{\min} = -0.2306$, $P_{\max} = 0.00011$).

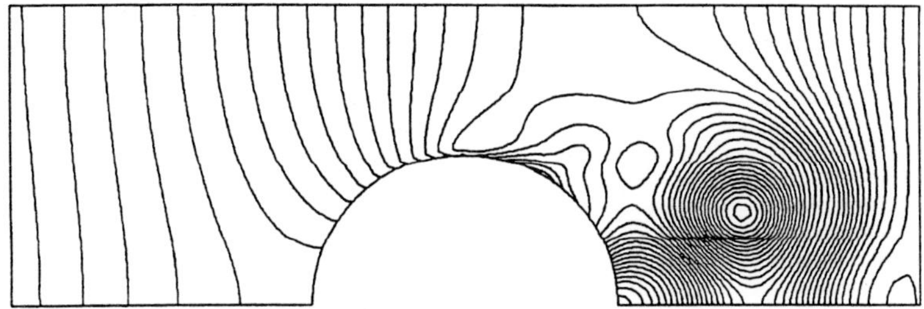

Figure 25:   As in Figure 23 except $t = 0.08$ ($P_{\min} = -0.04248$, $P_{\max} = 0.00240$).

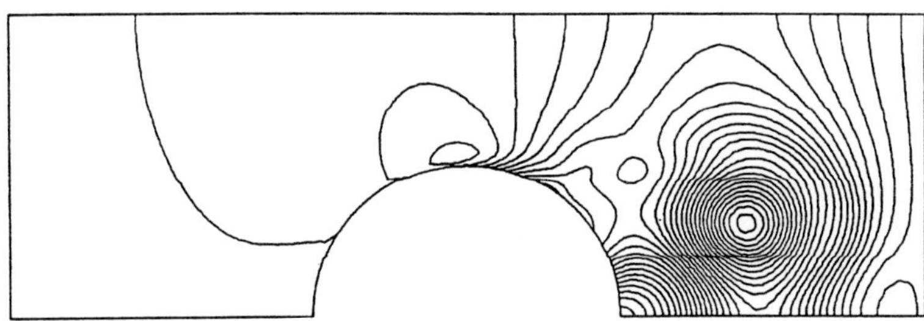

Figure 26:   As in Figure 23 except $t = 0.10$ ($P_{\min} = -0.03733$, $P_{\max} = 0.00918$).

The flow field seems to "evolve" somewhat more quickly than the pressure in that it has undergone most of its transition by $t \cong 3\tau_{ac}$ (Fig. 22) and the new, opposite-signed, vortex sheet is already nearly at full strength (not shown)—yet the pressure is still "dominated" by the deceleration transient (i.e., it's still mostly a potential field decaying like $e^{-\lambda t}$); see Fig. 23. But from $5\tau_{ac}$ to $10\tau_{ac}$ (Figs. 24 through 26), the pressure undergoes a *rapid* transition from acceleration-dominated to one reacting to the advective source term $-\left[\nabla \cdot (\underline{u} \cdot \nabla \underline{u}) \right]$ and the viscous Neumann BC; see (6) and (7). Beyond $t \cong 0.10$, this simulation closely mimics that in Figs. 12 through 15.

Turning now to the penalty method, we solved the impulsive stop problem (up to a time corresponding to the attainment of the equivalent L^2 - projection) in two ways: (1) using an initially very small time step and integrating the penalty ODE's with a smart, variable-step time-accurate integrator (see GS for how to do this) and (2) via a "large" Δt, 2-time-step procedure that by-passes the entire penalty transient yet, perhaps surprisingly, ends up at the same place. We chose the penalty parameter, λ, to be 10^7, and began our time-accurate simulation with $\Delta t_0 = 10^{-10}$. Conversely, for the 2-step run, we set $\Delta t_o = 10^{-2}$ and took just two BE steps. Defining the penalty time constant as $\tau_p \equiv D^2 / \lambda = 4 \times 10^{-7}$, gives $\Delta t_o / \tau_p = 0.00025$ for the time-accurate case, whereas $\Delta t / \tau_p = 25000$ for the 2 step case—rather a large range of step sizes.

Figs. 27 through 30 show the penalty pressure [from (39); actually, of course, from (37)] for the impulsive stop at the left end (slam the left door, leave the right one open)

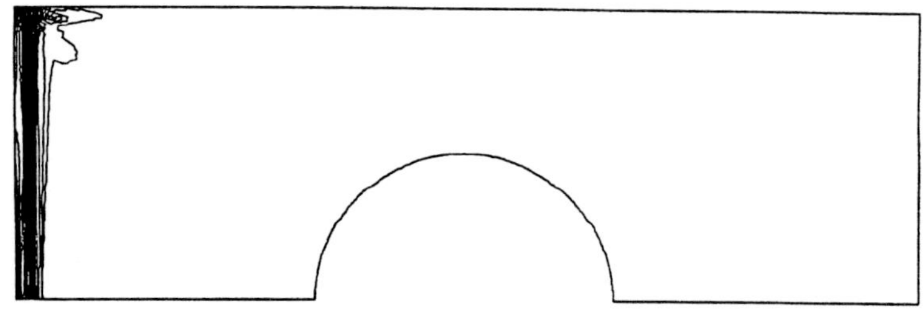

Figure 27: Pressure at $t = 0$ for penalty method (left end closed); $P_{min} = -1.69 \times 10^7$, $P_{max} = 1.33 \times 10^6$.

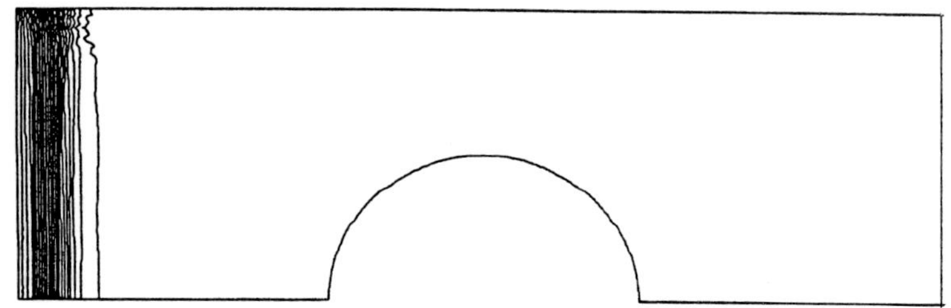

Figure 28: Penalty pressure at $t = 10^{-9}$ ($P_{min} = -5.44 \times 10^6$,. $P_{max} = 4.48 \times 10^4$)

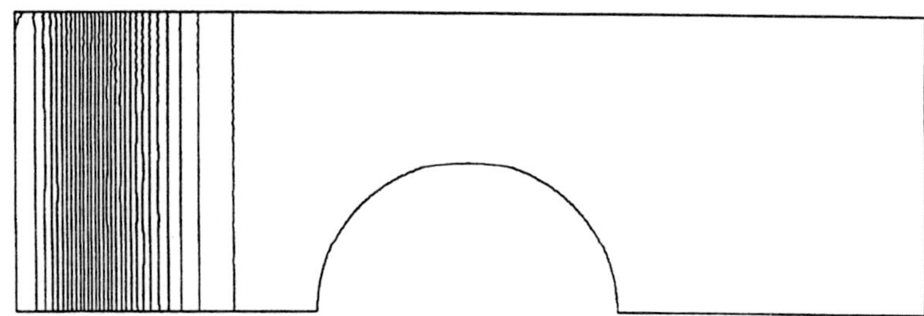

Figure 29: Penalty pressure at $t = 10^{-8}$ ($P_{min} = -1.68 \times 10^6$, $P_{max} = 7.53 \times 10^{-3}$).

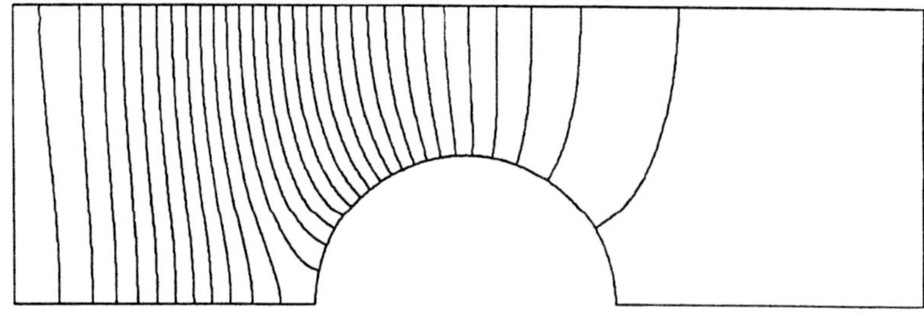

Figure 30: Penalty pressure at $t = 10^{-7}$ ($P_{min} = -7.26 \times 10^5$, $P_{max} = 1.97$).

During this phase of the penalty transient (the spurious, pressure 'shock wave,' portion), the bulk of the velocity field remains virtually unchanged from the IC; i.e., both ψ and ω look much like Fig. 8 and the magnitudes change little until t reaches $10^{-7}(=1/\lambda)$ or so. But they change *rapidly* for $t >\sim 10^{-7}$ and are 'finished' by $\sim t = 10^{-5}$. Fig. 31(a) shows the stream function at $t = 10^{-6}$—for which the pressure, in Fig. 31(b) still looks like a potential—which is clearly tending toward the desired result

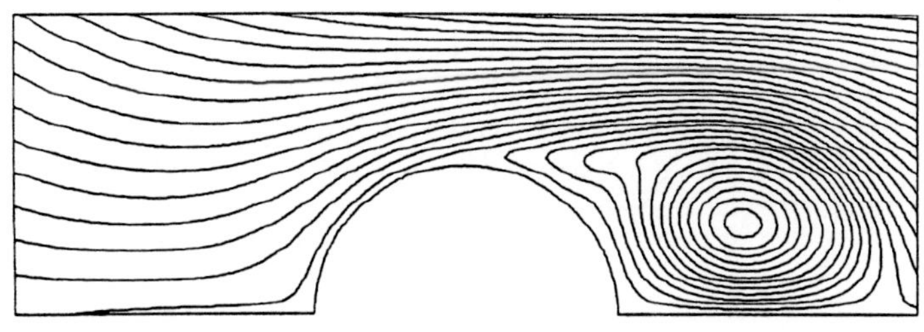

(a)   Stream function ($\psi_{min} = -0.09075,\ \psi_{max} = 0.13054$)

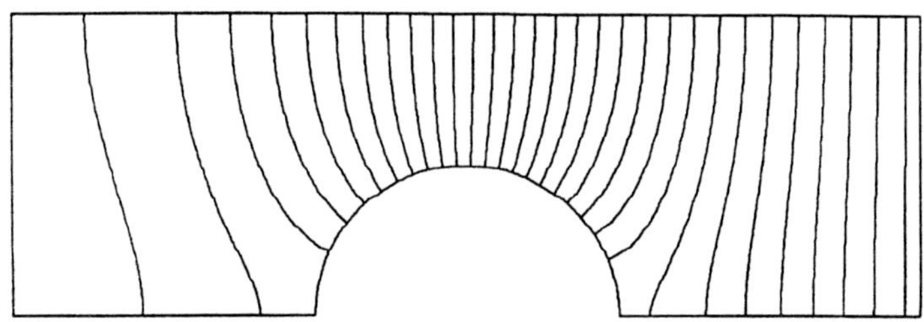

(b)   Pressure ($P_{min} = -2.16 \times 10^5,\ P_{max} = 0.034$)

Figure 31:   Penalty solution at $t = 10^{-6}$.

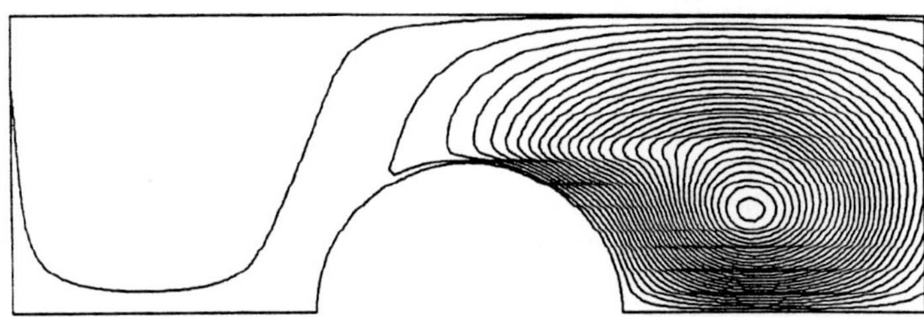

(a)   Stream function ($\psi_{min} = -0.12546,\ \psi_{max} = 0.00269$)

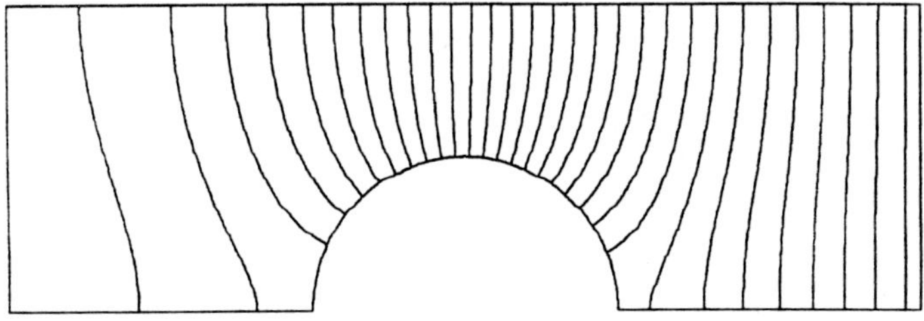

(b)   Pressure ($P_{min} = -872.1,\ P_{max} = 0.00022$)

Figure 32:   Penalty solution at $t = 10^{-5}$

The penalty transient for the velocity is over by $t \cong 10^{-5}$ [see Fig. 32(a), both ψ and ω now agree very closely with the projected solution], but certainly *not* for the pressure (Fig. 32(b), which is still more like a potential function—of the previous, $e^{-\lambda t}$, case. Even 10-fold farther in time ($t = 10^{-4} = 250\tau_p$) *still* shows a potential-like pressure—see Fig. 33. Finally, after one more decade, $t = 10^{-3}$, we see the pressure transient completed; the pressure field in Fig. 34 agrees well with the "projected" pressure field corresponding to Fig. 11b.

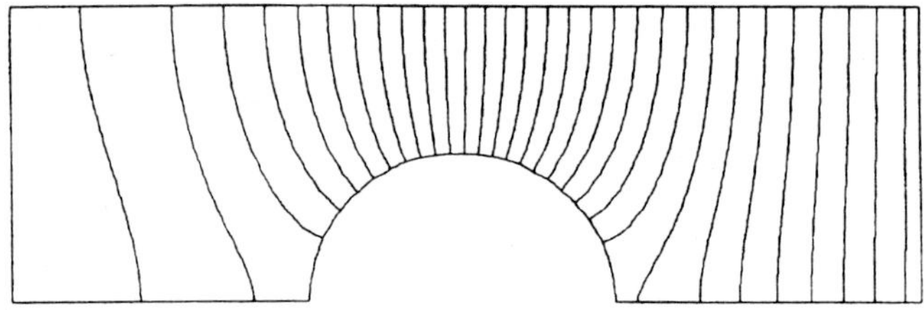

Figure 33:   Penalty pressure at $t = 10^{-4}$
($P_{\min} = -0.03673$, $P_{\min} = 2.26 \times 10^{-5}$, $P_{\max} = 13.38$)

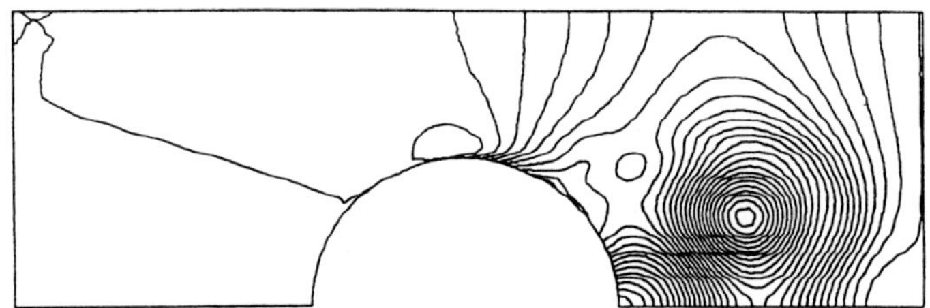

Figure 34:   Penalty pressure at $t = 10^{-3}$ ($P_{\min} = -0.03673$, $P_{\max} = 0.01002$)

Incidentally, the entire penalty transient was accurately tracked via the smart (variable-step) BE method described in GS; Δt grew to about 10^{-3} by a time of 10^{-3}, the entire simulation needing only 80 steps.

Having shown that a time-accurate penalty transient finds the correct "common" result at the conclusion of the penalty transient, we performed a much more important/relevant simulation that overlooks the spurious penalty transient. After all, if it were *necessary* to track this transient in order to "recover" properly, the penalty method would be much less attractive—especially for fixed-Δt integrators! Fortunately, this is not the case. With no need to show the results, we simply state that the 2-step run using $\Delta t = 10^{-2}$ (or 10^{-4}, in fact) achieved the proper result for the velocity after a single BE time step (as for the index 2 DAE formulation), while a 'proper' result for P was not obtained until step 2 because—as discussed in GS—the first 'large Δt' penalty pressure is, like BE on the index 2 problem, really $\phi/\Delta t$. So we have shown again that the penalty method is quite viable when $\lambda \Delta t$ is large, causing the spurious penalty transient to be gracefully by-passed via BE. (Do *not* try this trick using the more accurate trapezoid rule!)

To conclude this portion of the presentation, we present in Table 1 a summary of results. Although mostly self-explanatory, we offer a few additional remarks:

1. The near-invariance of ψ and ω between steps 1 and 2 of the 2-step runs 'justifies' our Δt selection.

2. The solutions at $t = 0.10$ have suffered some reduction in the vorticity; perhaps a larger λ is needed.

3. These *mathematical* simulations naturally preclude the two 'real-world' phenomena of cavitation and water hammer.

The final time integration method applied to the sudden shutdown was explicit Euler—applied to the index 1/PPE version of the DAE's. But the results are hardly worth presenting, let alone showing. The "div-preservation" equation, (46), holds true and causes the following behavior: (1) the only mass imbalance at

$t = 0$ occurs in the first column of elements at the inlet—via the change in BC from $u = w_0 = 0.1$ to $u = 0$; (2) the requirement to preserve the div *necessarily* causes the u-velocity at the first row of nodes to the right of the inlet nodes to hold a value very close to w_0; (3) the result of this is this: the flow is not shut-down—rather, it's much like discarding the first column of elements and applying the BC $u = w_0$ at the second column. And this is just what occurred: FE produced a continued integration of the startup run!!

To finish the presentation, we show in Fig. 35 the drag coefficients for three of the shutdown cases presented earlier.

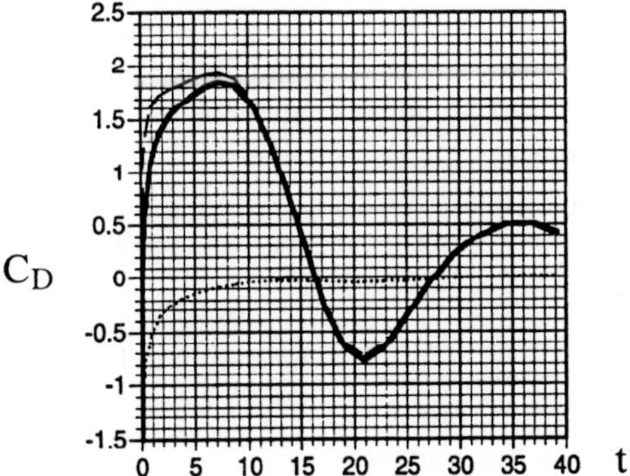

(a)   Exponential decay ($e^{-\lambda t}$) for Re = 1000; $C_D(0) \cong -8000$

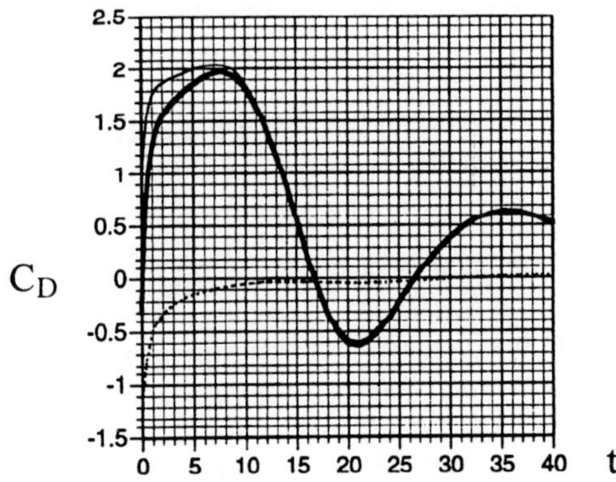

(b)   Impulsive stop for Re = 1000

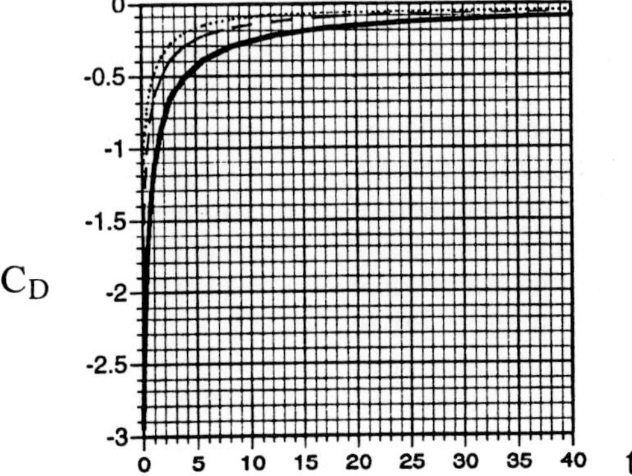

(c)   Impulsive stop for Re = 0

Figure 35:   Drag coefficients for three stopped flows.

The heavy lines are the total drag coefficients, the lighter solid lines are the pressure contributions, and the dotted lines are the viscous contributions. In all three cases, the initial value of C_D ($t = 40$ for the startup case) was about 3; see GS. Figs. 35(a) and (b) show the Re = 1000 shutdowns, about which we make four remarks:

1. Negative C_D ⇒ the fluid force on the cylinder is in the $-x$-direction.

2. $C_D(t)$ is truncated/clipped at -1.5 for the exponential case for plotting purposes; $C_D(0)$ is actually about -8000, and approximates the analytical result given by $C_D(t) = -2\pi a\lambda e^{-\lambda t}/w_o$ from the decelerating potential flow; see (26).

3. $C_D(0) \approx -0.3$ for the impulsive stop is a spurious result caused by our finite mesh; $C_D(0^+)$ should be $-\infty$ owing to the vortex sheet.

4. The small difference (for $t > \tau_{MTB} \cong 0.6$) between Figs. 35 (a) & (b), which are not even *visible* for the analogous and simpler startup case (see GS), are a reflection of the extra 'dynamics'; i.e., there is enough difference in the two cases for $t < \tau_{MTB}$ to be 'noticed,' even on our finite mesh.

Finally, Fig. 35(c) shows the impulsive stop drag coefficient for Stokes flow. Again the value of $C_D(0)$ is quite spurious; it is in fact not too far from the (negative of the) empirically-determined *mesh-dependent* result reported in GS— $C_D \cong 38v/w_o h \cong 3.45$ for the startup case, both representing the finite h spurious representation of a vortex sheet. Finally, we opine that all of our $C_D(t)$ results are reasonably accurate for $t > \tau_{MTB} \cong 0.6$, partly because 9/3 and 4/1 'agree'.

6. CONCLUDING COMMENTS

It might seem that our shutdown simulations are too much influenced by our tightly-bounded domain and—especially for those familiar with the problem of stopping a moving cylinder (or sphere, or other object)—perhaps not even 'correct' in that the original wake/eddy simply tends to mostly spin in place. While we plead guilty to a rather small domain (designed originally for studying only small-time results near the cylinder for fast startups), we believe and assert that a larger one (even much larger) would have only 'secondary' effects on the resulting shutdown flow (and M. Maxey agrees with us; personal communication) ; i.e., the eddy motion and behavior would be much the same as presented here. Evidence for this assertion can be seen in Fig. 13, and related discussion, in Chang and Maxey (1995).

Finally, in addition to the obvious conclusions from the results presented, we offer the following conclusion, based on our experience with both startups (see GS for details) and shutdowns (summarized herein): our $e^{-\lambda t}$ method is both more realistic than the 'impulse-method' *and* it sheds better understanding via analysis of the large λ situation (ultimately for $\lambda \to \infty$).

ACKNOWLEDGMENTS

We have profited considerably from discussions with Profs. R. Mei, M. Maxey, and J. Brady—and Dr. P. Lovalenti. Assistance from Drs. A. Hindmarsh and D. Veyret are also gratefully acknowledged, as is the expert document preparation by A. Henke. This work was sponsored by the U.S. Department of Energy Environmental Sciences Division and performed by the Lawrence Livermore National Laboratory under Contract No. W-7405-Eng-48.

REFERENCES

1. Bar-Lev, M. and Yang, H.T. (1996), Initial flow field over an impulsively started circular cylinder, *J. Fluid Mech.*, **72**, Pt. 4, pp. 625–647, (1975).

2. Chang, E.J. and Maxey, M.R. (1995), Unsteady flow about a sphere at low to moderate Reynolds number. Part 2. Accelerated motion, *J. Fluid Mech.*, **303**, pp. 133–153, (1995).

3. Collins, W.M. and Dennis, S.C.R. (1973b), The initial flow past an impulsively started circular cylindar, *Quart. Journ. Mech. and Applied Math.*, **XXVI**, Pt. 1, (1973).

4. Engelman, M.S. and Sani, R.L. (1982), Consistent vs. reduced integration penalty methods for incompressible media using several old and new elements, *Int. J. Num. Meth. Fluids*, **2**, pp. 25–42 (1982).

5. Gresho, P.M. (1991a), Some current CFD issues relevant to the incompressible Navier-Stokes Equations, *Computer Methods in Applied Mechanics and Engineering*, **87**, pp. 201–252, North-Holland, (1991).

6. Gresho, P.M. (1991b), Incompressible fluid dynamics: Some fundamental formulation issues, *Annu. Rev. Fluid Mech.*, **23**, pp. 413–53, (1991).

7. Gresho, P.M. and Sani, R.L. (1987), On pressure boundary conditions for the incompressible Navier-Stokes Equations, *Int. J. Num. Meth. Fluids*, **7**, pp. 1111–1145 (1987).

8. Gresho, P.M. and Sani, R.L. (1997), *Incompressible Flow and the Finite Element Method, Vol 1: Advection-Diffusion and Isothermal Laminar Flow*, John Wiley and Sons, Chichester (in Press).

9. Mei, R. and Lawrence, C.J. (1996), The flow field due to a body in impulsive motion, *J. Fluid Mech.*, **325**, pp. 79–111, (1996).

10. Wang, C.-Y. (1968), A note on the drag of an impulsively started circular cylinder, *J. Math. and Phys.*, **47**, p. 451.

11. Wang, X. and Dalton, C. (1991), Numerical solutions for impulsively started and decelerated viscous flow past a circular cylinder, *Int. J. Num. Meth. Fluids*, **12**, pp. 383–400 (1991).

	IC[1]	Both closed; step1/step2[2]	Right closed; step1/step2[2]	Left closed; step1/step2[2]	$e^{-\lambda t}$ at $t = 0.10$[3]	Impulsive at $t = 0.10$[3]	Penalty[4] at $t = 10^{-3}$	Penalty at $\Delta t \equiv .01$; step1/step2
$-\psi_{\min}$	0.07325	0.12293/0.12293	0.12293/0.12293	0.12589/0.12589	0.12579	0.12577	0.12589	0.12587/0.12587
$\psi_{\max}$	0.20000	0.00224/0.00224	0.00356/0.00356	0.00224/0.00224	0.00225	0.00225	0.00224	0.00256/0.00244
$-P_{\min}$	0.03997	$82070^{(5)}/0.04595$	$0.0837^{(5)}/0.03560$	$80356^{(5)}/0.03672$	0.03733	0.03682	0.03673	$4.4 \times 10^{-5}/0.03443$
$P_{\max}$	0.01431	$0^{(6)}/0^{(6)}$	$81739^{(5)}/0.01027$	$0.0577^{(5)}/0.0100$	0.00918	0.00986	0.01002	$80.319^{(5)}/0.01302$
$-\omega_{\min}$	7.012	7.690/7.690	7.686/7.686	7.690/7.690	7.164	7.118	7.681	7.599/7.510
$\omega_{\max}$	2.088	23.455/23.454	23.454/23.453	23.483/23.483	20.600	20.344	23.447	23.127/23.127

(1) $t = 40$ from startup run

(2)   BE 'projections' (2 steps at $\Delta t = 10^{-5}$)

(3)   Left end closed

(4)   Time-accurate

(5)   The "pressure" is actually $\phi/\Delta t$

(6)   Pressure 'pegged' at right edge (1 node) to set hydrostatic level

TABLE 1.   *Summary of Results*

PARALLEL FINITE ELEMENT COMPUTING METHODS FOR UNSTEADY FLOWS WITH INTERFACES

T. TEZDUYAR   S. ALIABADI   M. BEHR

Department of Aerospace Engineering and Mechanics, Army High Performance
Computing Research Center; University of Minnesota;
1100 Washington Avenue South; Minneapolis, MN 55455

Abstract

We provide an overview our finite element methods for parallel computation of unsteady flow problems with interfaces such as two-fluid and free-surface flows. The methods we discuss are the Deformable-Spatial-Domain/Stabilized Space-Time (DSD/SST) formulation, which is an interface-tracking technique, and the Enhanced-Discretization Interface-Capturing Technique (EDICT). Both methods are based on the stabilized finite element formulations which possess good stability and accuracy properties.

In the DSD/SST method, the finite element formulation of the problem is written over its associated space-time domain. This automatically takes into account the motion of the interfaces. The mesh update is achieved with strategies that focus on moving the mesh in an effective way and minimizing the frequency of remeshings. Although this interface-tracking method results in accurate representation of the interfaces, in cases with complex, rapidly-changing interfaces, especially in 3D simulations, reducing the frequency of remeshing becomes difficult, and sometimes not feasible.

With the EDICT, we solve, over a non-moving mesh, the Navier-Stokes equations together with an advection equation governing the evolution of an interface function with two distinct values identifying the two fluids. To increase the accuracy in modeling the interfaces, we use finite element functions corresponding to enhanced discretization at and near the interface. These functions are designed to have multiple components, with each component coming from a different level of mesh refinement over the same computational domain.

With the test problems presented here, we demonstrate that the EDICT can be used very effectively to increase the accuracy of the base finite element formulations. With parallel implementations of the DSD/SST method and EDICT, we have brought our simulation capability to a point where we can now address a wide-range of unsteady flow problems with interfaces.

1 Introduction

When we need to solve an unsteady flow problem with interfaces, such as fluid-fluid interfaces or free-surface flows, the location of the interface is also an unknown function of time and needs to be determined as part of the overall solution. Depending on the circumstances and the wave lengths at the interface, one can use a fixed mesh or moving mesh method. An interface-tracking method typically requires a moving mesh method, whereas an interface-capturing method can very efficiently be implemented with a fixed mesh approach.

In the category of moving mesh methods, we can mention moving finite element, arbitrary Lagrangian-Eulerian, and the space-time formulations. The Deformable-Spatial-Domain/Stabilized Space-Time (DSD/SST) finite element formulation for flows with moving boundaries and interfaces was first introduced, in the context incompressible flows, in Tezduyar et al.[1,2]. The examples in [2] included 2D simulation of a free-surface flow problem with long-time integration of the free surface. The stabilization methods with the streamline-upwind /Petrov-Galerkin (SUPG) [3]

Received on July 1, 1997.

643

and pressure-stabilizing/Petrov-Galerkin (PSPG) [4] terms provide stability and accuracy in computation of advection-dominated flows and permit usage of equal-order interpolation functions for velocity and pressure. The stabilized space-time formulation was also used earlier in [5] and [6] to solve problems with fixed domains. Considering the increased cost involved in using a space-time formulation, we do not see a sufficient advantage in using this method on problems with just fixed domains. From our experience with many simulations, we believe that equally satisfactory results can be obtained, with much less cost, by using semi-discrete methods with the SUPG and PSPG stabilization.

In the stabilized space-time formulation of a flow problem with moving boundaries and interfaces, because the formulation of the problem is written over its space- time domain, the motion of the boundaries and interfaces are taken into account automatically. The finite element functions are linear both in space and time, continuous in space, but discontinuous in time. The computations are carried out one space-time "slab" at a time, where the "slab" is a slice of the space-time domain between two consecutive time levels. The changes in the shape of the spatial domain is managed by updating the mesh with the combined approaches of moving the mesh (without changing the element connectivities), and remeshing it (i.e., generating a new set of nodes and element) as needed when the mesh distortion becomes to high. Moving the mesh is accomplished by special methods for specific problems and by automatic methods for more general cases. In 3D simulations, especially those based on parallel computing, reducing the frequency of automatic remeshing becomes essential, because the cost of frequent 3D automatic mesh generation could become prohibitive. Therefore, in designing our mesh update methods, we have always taken reducing the frequency of remeshing as a high priority issue. With parallel implementation of the DSD/SST formulation, we carried out simulations for a number of 3D problems, such as sloshing in a vertically vi-

brating container [7], a gas stream impinging on a liquid [8], and flow past hydraulic structures [9].

In flow problems with complex and very unsteady interfaces, especially in 3D simulations, reducing the frequency of remeshing becomes a very difficult task, and sometimes no longer feasible. In such cases, interface-capturing methods with fixed meshes can still be used. Typically these interface-capturing methods are more flexible but provide less accurate representation of the interface compared to the interface-tracking methods. These methods can be used as practical alternatives to carry out the simulations when compromising the accurate representation of the interfaces becomes less of a concern than facing major difficulties in moving the mesh to track such interfaces.

The Enhanced-Discretization Interface-Capturing Technique (EDICT) was first introduced in [10]. The objective was to enhance the spatial discretization around the interface to have higher accuracy in representing it. We start with the basic approach of a volume of fluid (VOF) method [11–13]. The Navier-Stokes equations are solved over a non-moving mesh. An interface function with two distinct values serves as a marker identifying the two fluids. The interface function is transported with a time-dependent advection equation. Our objective is to accurately represent and advect the front between the two distinct values of the interface function and to accurately model the interface between the two fluids. In EDICT, the spatial discretizations are based on stabilized finite element formulations with the SUPG, PSPG, and least-squares terms. These terms are added to the standard Galerkin formulation of the momentum equation, incompressibility constraint and the advection equation governing the interface function. We also incorporated into the method a front-sharpening algorithm [14].

To increase the accuracy in representing the interface, we use function spaces corresponding to enhanced discretization at and near the interface. To achieve this, we start with a base mesh that we call Mesh-1. A subset of these elements are identified as those at and near the interface. A more refined Mesh-2 is constructed by patching together second-level meshes generated over each element in this subset. Which elements will be in this subset will change from one time level to other, depending on which elements the interface is passing through. An element which is in this subset now, may be out of it some time later, and come back in again some time after that. For each element in this subset there will be a unique second-level mesh. If an automatic mesh generator is used to generate that, the mesh will be generated only once and stored somewhere, to be used later if that element needs a second-level mesh again. The trial and weighting functions for velocity and pressure will all have two components each: one coming from Mesh-1 and the second one coming from Mesh-2.

To further increase the accuracy in representing the interface, we construct a third-level Mesh-3 for the interface function. This is done by identifying a subset of the elements in Mesh-2 as those at and very near the interface. Mesh-3 is constructed by patching together third-level meshes generated over each element in this subset. Which elements in Mesh-2 will be in this subset will change from one time level to other, depending on where the interface is. The construction of Mesh-3 from Mesh-2 will be very similar to the construction of Mesh-2 from Mesh-1. The trial and weighting functions for the interface function will have three components, each coming from one of these three meshes.

Mesh-2 and Mesh-3 will not be re-defined every time step but frequently enough to keep the interface in zones covered with these higher level meshes. In parallel implementation, re-defining these meshes will involve some load balancing cost, but we do not expect this to be a major cost. We will of course have the option of not having a third-level mesh for the interface function. We also have the option of having zones covered by Mesh-2 and Mesh-3 to overlap. An advantage in keeping them non-overlapping, however, is that we can keep Mesh-2 wider than Mesh-3, target keeping the interface within Mesh-3, and solve for the interface function only over the part of the computational domain covered by Mesh-2.

Both in DSD/SST formulation and EDICT, at every time step of a simulation, we need to solve a large, coupled system of nonlinear equations. These systems are solved with Newton-Raphson iterations. Each step of the Newton-Raphson sequence requires the solution of a linear equation system. These systems are also solved iteratively, using diagonal preconditioners with the GMRES [15] update technique. At each step of these iterations, the residual of the linear equation system needs to be computed. To reduce the memory requirements associated with large-scale computations, a matrix-free iteration technique is employed to perform these residual computations. This vector-based technique [16, 17] totally eliminates the need to compute or store any coefficient matrices, even at the element level. For more information on the iterative computation strategies used here, see [9].

We have implemented the DSD/SST formulation on the Thinking Machines CM-5, CRAY T3D, and the SGI multi-processor systems: a 20-processor ONYX and a 12-processor POWER CHALLENGE. We refer the interested reader to [9] for more information. The EDICT has so far been implemented on the shared-memory parallel computing platform of the SGI multi-processor systems, and this was reported in [10].

In Section 2 we review the governing equations. The DSD/SST formulation and the mesh update strategies are reviewed, respectively, in Section 3 and Section 4. The finite element formulation for the EDICT is presented in Section 5, and the enhanced-discretization concept is described in Section 6. We present our numerical examples in Section 7, and end with concluding remarks in Section 8.

2. GOVERNING EQUATIONS

The flow computations are based on the solution of time-dependent Navier-Stokes equations of incompressible flows. In our notation, Ω_t and $(0,T)$ will denote the space and time domains, and Γ_t the boundary of Ω_t. In the DSD/SST formulation, the spatial domain will change with respect to time, and the subscript t indicates that time-dependence. In the EDICT, however, we will drop this subscript. The symbols $\rho(\mathbf{x},t)$, $\mathbf{u}(\mathbf{x},t)$ and $p(\mathbf{x},t)$ represent the density, velocity and pressure, respectively. The external forces (e.g. gravity) are represented by $\mathbf{f}(\mathbf{x},t)$. The momentum conservation equations and the incompressibility constraint can be written in the following vector form:

$$\rho\left(\frac{\partial \mathbf{u}}{\partial t} + \mathbf{u}\cdot\nabla\mathbf{u} - \mathbf{f}\right) - \nabla\cdot\boldsymbol{\sigma}$$
$$= \mathbf{0} \quad \text{on } \Omega_t \quad \forall t \in (0,T), \quad (1)$$

$$\nabla\cdot\mathbf{u} = 0 \quad \text{on } \Omega_t \quad \forall t \in (0,T), \quad (2)$$

where

$$\boldsymbol{\sigma}(\mathbf{u},p) = -p\mathbf{I} + \mathbf{T}, \quad (3)$$
$$\mathbf{T} = 2\mu\boldsymbol{\varepsilon}(\mathbf{u}). \quad (4)$$

Here $\mathbf{I}$ is the identity tensor, μ is the dynamic viscosity, and $\boldsymbol{\varepsilon}(\mathbf{u})$ is the strain rate tensor defined as

$$\boldsymbol{\varepsilon}(\mathbf{u}) = \frac{1}{2}\left(\nabla\mathbf{u} + (\nabla\mathbf{u})^T\right). \quad (5)$$

In the EDICT, to model fluid-fluid interfaces, we consider two immiscible fluids, A and B, with densities ρ_A and ρ_B and viscosities μ_A and μ_B.

Remark

(1) If we want to model a liquid-gas interaction, we can let Fluid A be the liquid and Fluid B the gas. If we want to model a free-surface problem where Fluid B is irrelevant, we can do that by assigning a sufficiently low density to Fluid B.

An interface function ϕ serves as a marker identifying Fluid A and B with the definition $\phi = \{1$ for Fluid A and 0 for Fluid B$\}$. The interface between the two fluids is approximated to be at $\phi = 0.5$. In this context, ρ and μ are defined as

$$\rho = \phi\rho_A + (1-\phi)\rho_B, \quad (6)$$
$$\mu = \phi\mu_A + (1-\phi)\mu_B. \quad (7)$$

The evolution of the interface function ϕ is governed by a time-dependent advection equation:

$$\frac{\partial\phi}{\partial t} + \mathbf{u}\cdot\nabla\phi = 0 \quad \text{on } \Omega \quad \forall t \in (0,T), \quad (8)$$

Remark

(2) One can also see Equations (6) - (8) as those representing the constitutive law of the fluid system. How accurately this law will be modeled will depend on how accurately the front between $\phi = 1$ and $\phi = 0$ will be represented and advected.

Remark

(3) We will not address here the surface tension effects at the interfaces.

The set of equations given so far in this section needs to be completed with an appropriate set of boundary conditions, a divergence-free initial condition on the velocity field:

$$\mathbf{u}(\mathbf{x},0) = \mathbf{u}_0, \quad \nabla\cdot\mathbf{u}_0 = 0 \quad \text{on } \Omega_0, \quad (9)$$

and the initial profile of the interface function.

3. DSD/SST FORMULATION

The DSD/SST formulation begins with the weak form of the governing equations being written over the associated space-time domain of the problem, by dividing this domain into a sequence of space-time slabs Q_n, where Q_n is the slice of the space-time domain between the time levels t_n and t_{n+1}. The integrations involved in the weak form are then performed over Q_n. The finite element interpolation functions used are continuous in space but discontinuous across time levels. To reflect this situation, we use the notation $(\cdot)_n^-$ and $(\cdot)_n^+$ to denote the function values at t_n as approached from below and above respectively. Each space-time slab Q_n is decomposed into space-time elements Q_n^e, where $e = 1, 2, \ldots, (n_{el})_n$. The subscript n used with n_{el} is to account for the general case in which the number of space-time elements may change from one space-time slab to other. In our computations, we use first-order polynomials as interpolation functions.

In writing the DSD/SST formulation, we first assume that for each slab Q_n, we have some appropriately-defined finite-dimensional space-time function spaces for trial and weighting functions corresponding to velocity and pressure. The trial function spaces will be denoted by $(\hat{\mathcal{S}}_u^h)_n$ and $(\hat{\mathcal{S}}_p^h)_n$, and the weighting function spaces by $(\hat{\mathcal{V}}_u^h)_n$ and $(\hat{\mathcal{V}}_p^h)_n$ $(= (\hat{\mathcal{S}}_p^h)_n)$. While the superscript h implies that these are finite-dimensional function spaces, the subscript n implies that corresponding to different space-time slabs we can have different spatial discretizations.

The stabilized space-time formulation of Equations (1) and (2) can then be written as follows: given $(\mathbf{u}^h)_n^-$, find $\mathbf{u}^h \in (\hat{\mathcal{S}}_u^h)_n$ and $p^h \in (\hat{\mathcal{S}}_p^h)_n$ such that $\forall \mathbf{w}^h \in (\hat{\mathcal{V}}_u^h)_n$ and $\forall q^h \in (\hat{\mathcal{V}}_p^h)_n$:

$$\int_{Q_n} \mathbf{w}^h \cdot \rho \left(\frac{\partial \mathbf{u}^h}{\partial t} + \mathbf{u}^h \cdot \boldsymbol{\nabla} \mathbf{u}^h - \mathbf{f} \right) dQ$$

$$+ \int_{Q_n} \boldsymbol{\varepsilon}(\mathbf{w}^h) : \boldsymbol{\sigma}(p^h, \mathbf{u}^h) dQ$$

$$+ \int_{Q_n} q^h \boldsymbol{\nabla} \cdot \mathbf{u}^h dQ$$

$$+ \int_{\Omega_n} (\mathbf{w}^h)_n^+ \cdot \rho \left((\mathbf{u}^h)_n^+ - (\mathbf{u}^h)_n^- \right) d\Omega$$

$$+ \sum_{e=1}^{(n_{el})_n} \int_{Q_n^e} \tau_{\text{MOM}} \frac{1}{\rho} \left[\rho \left(\frac{\partial \mathbf{w}^h}{\partial t} + \mathbf{u}^h \cdot \boldsymbol{\nabla} \mathbf{w}^h \right) \right.$$

$$\left. - \boldsymbol{\nabla} \cdot \boldsymbol{\sigma}(q^h, \mathbf{w}^h) \right]$$

$$\cdot \left[\rho \left(\frac{\partial \mathbf{u}^h}{\partial t} + \mathbf{u}^h \cdot \boldsymbol{\nabla} \mathbf{u}^h - \mathbf{f} \right) - \boldsymbol{\nabla} \cdot \boldsymbol{\sigma}(p^h, \mathbf{u}^h) \right] dQ$$

$$+ \sum_{e=1}^{(n_{el})_n} \int_{Q_n^e} \tau_{\text{CONT}} \boldsymbol{\nabla} \cdot \mathbf{w}^h \, \rho \boldsymbol{\nabla} \cdot \mathbf{u}^h dQ$$

$$= \int_{P_n} \mathbf{w}^h \cdot \mathbf{h}^h dP. \tag{10}$$

Here $\mathbf{h}^h$ represents the Neumann-type boundary condition associated with the momentum equation, P_n is lateral boundary of the space-time slab, and τ_{MOM} and τ_{CONT} are the stabilization parameters.

The solution to (10) is obtained sequentially for all of the space-time slabs $Q_1, Q_2, \ldots, Q_{N-1}$, and the computations start with

$$(\mathbf{u}^h)_0^- = \mathbf{u}_0^h. \tag{11}$$

In the formulation given by Equation (10), the first four integrals, together with the right-hand-side, represent the time-discontinuous Galerkin formulation of (1)–(2). The fourth integral enforces, weakly, the continuity of the velocity field in time. The two series of element-level integrals in the formulation are the least-squares stabilization terms.

4. MESH UPDATE METHODS FOR THE DSD/SST FORMULATION

In the DSD/SST formulation, as the shape of the spatial domain changes in time, the mesh needs to be updated to accommodate this change. The only rule the motion of the mesh needs to follow is that at the interface the normal velocity of the mesh has to match the normal velocity of the fluid. Beyond that, the mesh can be updated in any fashion desired, with the main objective being to reduce the frequency of remeshing (i.e., generating a new set of nodes and elements). In 3D simulations, if the remeshing requires calling an automatic mesh generator, the cost of automatic mesh generation in 3D becomes an overwhelming reason for trying to minimize the frequency of remeshing.

In some cases where the changes in the shape of the computational domain allow it, a special-purpose mesh moving method can be used in conjunction with a special-purpose mesh generator. In such cases, simulations can be carried out without calling an automatic mesh generator and without solving any additional equations to determine the motion of the mesh. An earlier example, 3D parallel computation of sloshing in a vertically vibrating container, can be found in [7].

In general, we can use an automatic mesh moving scheme [18] to move the nodal points. This is accomplished by solving the modified equations of linear elasticity to determine the motion of the internal nodes. The boundary conditions for these mesh motion equations are specified in such a way that they match the normal velocity of the fluid at the interface. The additional set of equations are also solved by using parallel computing methods. An earlier example of using this mesh update strategy, 3D parallel computation of flow over a hydraulic structure, can be found in [9].

5. FINITE ELEMENT FORMULATION FOR THE EDICT

To write the stabilized finite element formulation for the EDICT, we first assume that we have some appropriately-defined finite-dimensional function spaces for trial and weighting functions corresponding to velocity, pressure and interface function. The trial function spaces will be denoted by $(\mathcal{S}_u^h)_n$, $(\mathcal{S}_p^h)_n$, and $(\mathcal{S}_\phi^h)_n$, and the weighting function spaces by $(\mathcal{V}_u^h)_n$, $(\mathcal{V}_p^h)_n$ $(= (\mathcal{S}_p^h)_n)$, and $(\mathcal{V}_\phi^h)_n$. The subscript n in this case implies that corresponding to different time levels we can have different spatial discretizations. We will give more precise definition of these function spaces later.

The stabilized formulations of Equations (1), (2), and (8) can be written as follows: given $\mathbf{u}_n^h$ and ϕ_n^h, find $\mathbf{u}_{n+1}^h \in (\mathcal{S}_u^h)_{n+1}$, $p_{n+1}^h \in (\mathcal{S}_p^h)_{n+1}$, and $\phi_{n+1}^h \in (\mathcal{S}_\phi^h)_{n+1}$, such that, $\forall \mathbf{w}_{n+1}^h \in (\mathcal{V}_u^h)_{n+1}$, $\forall q_{n+1}^h \in (\mathcal{V}_p^h)_{n+1}$, and $\forall \psi_{n+1}^h \in (\mathcal{V}_\phi^h)_{n+1}$:

$$\int_\Omega \mathbf{w}_{n+1}^h \cdot \rho \left(\frac{\partial \mathbf{u}^h}{\partial t} + \mathbf{u}^h \cdot \nabla \mathbf{u}^h - \mathbf{f} \right) d\Omega$$

$$+ \int_\Omega \boldsymbol{\varepsilon}(\mathbf{w}_{n+1}^h) : \boldsymbol{\sigma}(p^h, \mathbf{u}^h) d\Omega$$

$$+ \int_\Omega q_{n+1}^h \nabla \cdot \mathbf{u}^h d\Omega$$

$$+ \sum_{e=1}^{n_{el}} \int_{\Omega^e} \left(\tau_{\mathrm{SUPG}} \mathbf{u}^h \cdot \nabla \mathbf{w}_{n+1}^h + \frac{\tau_{\mathrm{PSPG}}}{\rho} \nabla q_{n+1}^h \right)$$

$$\cdot \left[\rho \left(\frac{\partial \mathbf{u}^h}{\partial t} + \mathbf{u}^h \cdot \nabla \mathbf{u}^h - \mathbf{f} \right) - \nabla \cdot \boldsymbol{\sigma}(p^h, \mathbf{u}^h) \right] d\Omega$$

$$+ \sum_{e=1}^{n_{el}} \int_{\Omega^e} \tau_{\mathrm{CONT}} \nabla \cdot \mathbf{w}_{n+1}^h \rho \nabla \cdot \mathbf{u}^h d\Omega$$

$$= \int_\Gamma \mathbf{w}_{n+1}^h \cdot \mathbf{h}^h d\Gamma, \tag{12}$$

$$\int_\Omega \psi_{n+1}^h \left(\frac{\partial \phi^h}{\partial t} + \mathbf{u}^h \cdot \nabla \phi^h \right) d\Omega$$

$$+ \sum_{e=1}^{n_{el}} \int_{\Omega^e} \tau_\phi \mathbf{u}^h \cdot \nabla \psi_{n+1}^h \left(\frac{\partial \phi^h}{\partial t} + \mathbf{u}^h \cdot \nabla \phi^h \right) d\Omega$$

$$= 0, \tag{13}$$

where τ_{SUPG}, τ_{PSPG}, τ_{CONT} and τ_ϕ are the stabilization parameters.

Remark

(4) In Equation (12), the first three integrals, together with the right-hand-side, represent the Galerkin formulation of (1)-(2). The first series of element-level integrals in the formulation are the SUPG and PSPG stabilization terms. The second series of element-level integrals are the least-squares stabilization terms based on the incompressibility constraint. In Equation (13), the first integral represents the Galerkin formulation of (8), while the series of element-level integrals are the SUPG stabilization terms.

Remark

(5) In time discretization, the time derivatives, $\partial \mathbf{u}/\partial t$ and $\partial \phi/\partial t$ are represented as follows:

$$\frac{\partial \mathbf{u}}{\partial t} = \frac{\mathbf{u}_{n+1}^h - \mathbf{u}_n^h}{\Delta t}, \tag{14}$$

$$\frac{\partial \phi}{\partial t} = \frac{\phi_{n+1}^h - \phi_n^h}{\Delta t}, \tag{15}$$

where Δt is the time step size between time levels n and n+1. In this time discretization, the functions $\mathbf{u}^h$, p^h and ϕ^h are represented as follows:

$$\mathbf{u}^h \leftarrow (1 - \alpha)\mathbf{u}_n^h + \alpha\mathbf{u}_{n+1}^h, \qquad (16)$$

$$p^h \leftarrow p_{n+1}^h, \qquad (17)$$

$$\phi^h \leftarrow (1 - \alpha)\phi_n^h + \alpha\phi_{n+1}^h, \qquad (18)$$

where α is a time-integration parameter controlling the stability and accuracy of the integration. Normally we set $\alpha = 0.5$.

For a closer-to-discontinuous representation of the interface function ϕ, we incorporate a two-step interface-sharpening algorithm to the formulation. In the first step of this algorithm, the unacceptable values of ϕ are filtered out. Since we want $0 \leq \phi \leq 1$, any value of ϕ less than 0 or greater than 1 are clipped and set to 0 and 1 respectively. In the second step, $\phi = 0.5$ is assumed to be the interface and ϕ is sharpened around 0.5 as follows:

$$\text{for } |\phi_{n+1} - \phi_n| \leq b :$$
$$\phi \leftarrow 2^{a-1}\phi^a \qquad \text{for } 0.0 \leq \phi \leq 0.5,$$
$$\phi \leftarrow 1 - 2^{a-1}(1 - \phi)^a \quad \text{for } 0.5 < \phi \leq 1.0, (19)$$

where b and a are user-defined parameters. In the computations reported in this article, $b = 0.1$ and $a = 2.0$.

6. CONSTRUCTION OF FUNCTION SPACES – ENHANCED DISCRETIZATION

To construct the function spaces corresponding to time level n, we start with a base mesh which we will call Mesh-1. The set of elements and nodal points will be denoted by ϵ_n^1 and η_n^1. The subscript n implies that Mesh-1 might change from one time level to other, reflecting the possibility that occasionally we might replace the base mesh with a different one (see Figure 1).

A second-level and more refined mesh will be constructed over a subset $(\epsilon_n^1)_n^2$ of the these elements. This Mesh-2 will be generated by patching together the second-level meshes generated over each of the elements in $(\epsilon_n^1)_n^2$ (see Figures 1 and 2).

Remark

(6) The second subscript n implies that this subset might change from one time level to other, even though ϵ_n^1 might be the same. In other words, for a given Mesh-1, which elements of this mesh will be declared to be in $(\epsilon_n^1)_n^2$ might change from one time level to other. An element which might be declared to be in $(\epsilon_n^1)_n^2$ at some time level, might fall out of it at some other time, and yet come back in again at some time later.

Remark

(7) For each element in ϵ_n^1, there will be a unique second-level mesh. Therefore, if an element is declared to be in $(\epsilon_n^1)_n^2$ for a second time, the refined mesh generated over that element will be the same as the one generated at the earlier declaration. This way, if an automatic mesh generator is being used to generate these meshes, the cost for that mesh generation will be a one-time cost.

The set of elements and nodal points for Mesh-2 will be denoted by ϵ_n^2 and η_n^2.

A third-level and even more refined mesh will be constructed over a subset $(\epsilon_n^2)_n^3$ of the elements in Mesh-2. This Mesh-3 will be generated by patching together the third-level meshes generated over each of the elements in $(\epsilon_n^2)_n^3$ (see Figures 1, 2 and 3).

Remark

(8) The statements in Remarks 6-7 apply in this case too, with the mesh level numbers referred to in Remarks 6-7 shifted up by one.

The set of elements and nodal points for Mesh-3 will be denoted by ϵ_n^3 and η_n^3.

Remark

(9) We will limit ourselves to linear elements in 2D and 3D, as well as bilinear and trilinear elements, respectively, in 2D and 3D.

We decompose $\mathbf{u}_n^h$ as follows:

$$\mathbf{u}_n^h = \mathbf{u}_n^1 + \mathbf{u}_n^2. \tag{20}$$

The function $\mathbf{u}_n^1$ comes from a space of functions with the basis set consisting of the shape functions associated with all the nodes in η_n^1, excluding those "surrounded" by the elements in $(\epsilon_n^1)_n^2$. The function $\mathbf{u}_n^1$ also needs to satisfy the Dirichlet-type boundary conditions, except at those nodes that have been surrounded at the boundary of Ω.

The function $\mathbf{u}_n^2$ comes from a space of functions with the basis set consisting of the shape functions associated with all the nodes in η_n^2, excluding those at the boundaries of the zones covered by the elements in ϵ_n^2. However we do include the nodes at the boundary of Ω unless they coincide with the nodes in η_n^1 that have not been surrounded.

We decompose p_n^h in exactly the same way, except for recognizing the fact that the references to Dirichlet-type boundary conditions do not apply:

$$p_n^h = p_n^1 + p_n^2. \tag{21}$$

We decompose ϕ_n^h with more enhancement:

$$\phi_n^h = \phi_n^1 + \phi_n^2 + \phi_n^3. \tag{22}$$

The function ϕ_n^1 comes from a space of functions with the basis set consisting of the shape functions associated with all the nodes in η_n^1, excluding those "surrounded" by the elements in $(\epsilon_n^1)_n^2$. The function ϕ_n^1 also needs to satisfy the Dirichlet-type boundary conditions, except at those nodes that have been surrounded at the boundary of Ω.

The function ϕ_n^2 comes from a space of functions with the basis set consisting of the shape functions associated with all the nodes in η_n^2, excluding those at the boundaries of the zones covered by the elements in ϵ_n^2. We also exclude the nodes surrounded by the elements in $(\epsilon_n^2)_n^3$. However we do include the nodes at the boundary of Ω unless they coincide with the nodes in η_n^1 that have not been surrounded.

The function ϕ_n^3 comes from a space of functions with the basis set consisting of the shape functions associated with all the nodes in η_n^3, excluding those at the boundaries of the zones covered by the elements in ϵ_n^3. However we do include the nodes at the boundary of Ω unless they coincide with the nodes in η_n^2 that have not been surrounded.

The weighting functions are decomposed in a similar fashion:

$$\mathbf{w}_n^h = \mathbf{w}_n^1 + \mathbf{w}_n^2, \tag{23}$$

$$q_n^h = q_n^1 + q_n^2, \tag{24}$$

$$\psi_n^h = \psi_n^1 + \psi_n^2 + \psi_n^3. \tag{25}$$

The components of each weighting function are defined in the same way as we did for the trial functions, except that the weighting functions need to satisfy the homogeneous form of the Dirichlet-type boundary conditions.

Our objective with this enhanced discretization is to capture the interface as accurately as possible by using more refined meshes for the velocity, pressure, and interface function, and possibly even more refined meshes for the last. However we do this in a dynamic fashion by defining $(\epsilon_n^1)_n^2$ and $(\epsilon_n^2)_n^3$ depending on which elements in ϵ_n^1 and ϵ_n^2 the interface is passing through, and re-define these subsets occasionally to track the interface.

Remark

(10) We update $(\epsilon_n^1)_n^2$ and $(\epsilon_n^2)_n^3$ not every time step but with sufficient frequency to keep the interface within the zones covered by these subsets of elements. Although it would not be "illegal" for the interface to fall out of

these zones, we attempt to estimate or keep track of for how many time steps of the simulation the interface will remain inside these zones. How many time steps one can carry out the simulations without re-defining these subsets of elements will depend, among other things, on how "wide" we decide to keep these zones around the interface.

Remark

(11) Whenever we redefine these subsets, the mesh generation cost will not be a significant one. If we have to use an automatic mesh generator to generate the second- and third-level meshes, we will be able to use and re-use the meshes which have been generated at the very beginning of the simulations and stored.

Remark

(12) In parallel implementation, re-defining the subsets will require that load balancing is re-done. We do not anticipate this to be a major cost, since the METIS mesh partitioning [19] package performs extremely well, and its parallel version is now functional.

Remark

(13) It is possible to eliminate ϕ_n^3 by not choosing to go to a third-level of refinement. It is also possible to design the second- and third-level meshes in such a way that they overlap. One of the advantages in keeping them as non-overlapping meshes is that, by keeping Mesh-2 "wider" than Mesh-3, one can chose to limit the existence (as an unknown) of ϕ to Mesh-2, and therefore solving for it only over the part of the computational domain covered by Mesh-2. With this, we have to make sure that the interface remains in Mesh-2 zone. Since our objective will be to keep the interface in Mesh-3 zone, this would also keep it in Mesh-2 zone, even if the interface occasionally falls slightly out of the Mesh-3 zone.

7. EXAMPLES

2D simulation of surface waves in hydraulic structures. This simulation complements the earlier 3D simulation of flow in the spillway of the Olmsted Dam on the Ohio River reported in [9]. This dam design is currently under study by the U.S. Army Corps of Engineers. This simulation is an example application of the DSD/SST formulation coupled with the elasticity-based automatic mesh update method. The free-surface height is determined from an advection equation governing the height, and a stabilized formulation is used to solve this additional equation.

The geometrical model represents a 500 ft-long section of the navigation pass crest and stilling basin, and includes a long upstream channel, the spillway crest, and the stilling basin. The initial mesh consists of 7,966 space-time nodes and 7,571 triangular-based prismatic space-time elements. A flow rate of 271 cfs/ft is assumed, and the Reynolds number based on the inflow velocity and the spillway length is 7×10^6. The upstream Froude number based on the inflow velocity and depth is 0.217. Starting with a flat water surface, the flow is computed for 185 time steps with a time step size of 0.1 s until large mesh deformations force us to remesh. Frames from this initial part of the simulation are shown in Figures 4 and 5, displaying the deforming mesh and the velocity field. After remeshing, the flow is computed for another 2000 time steps with a time step size of 0.025 s and reaches a time-periodic state. The frames from this part of the simulation are shown in Figures 6 and 7.

This simulation is part of a collaborative project with R. Stockstill and C. Berger at the Waterways Experiment Station.

Sloshing in a tanker during braking. In this problem, sloshing in a partially-filled fuel tanker during braking is simulated by using the Interface-Capturing Technique with no enhanced discretization. The tanker suddenly brakes and reduces its speed at the rate of 0.2g. Here the tanker is assumed to be partitioned into six compartments, and only sloshing in one of the compartments is simulated. Figure 8 shows the geometry of the tanker. Also, due to the assumed symmetry of the problem with respect to the span-wise central plane, only half of the compartment is modeled.

A mesh with 1,250,260 hexahedral elements and 1,284,381 nodes is used in the simulation. At each time step, we need to solve a coupled system of nonlinear equations with 6,245,441 unknowns (4,961,060 unknowns for the Navier-Stokes equations and 1,284,381 unknowns for the interface function). At each time step four nonlinear iterations are performed. Iterations are carried out with matrix-free computations and the GMRES update technique with a Krylov space size of 60. Each nonlinear iteration takes approximately 4 minutes on a 512-node Thinking Machines CM-5. For 120 time steps the simulation took around 32 hours.

The graph in Figure 9 shows the time-history of the horizontal forces exerted by the sloshing fuel on the compartment. The images in Figure 10 show, at different instants, the sloshing and pressure distribution.

Sloshing in a tanker driving over a bump. This simulation too was carried out by using the Interface-Capturing Technique with no enhanced discretization. Here, a smaller tanker moving at 10 m/s drives over a bump 30 cm-high. In this case, the suspension system of the tanker absorbs the initial displacements due to the bump and transfers the generated forces to the structure of the tanker. The 3D rigid-body dynamics equations are coupled to the finite element formulation of the flow problem and solved simultaneously for the motion of the tanker as function of time. The fluid dynamics equations are written in a non-inertial frame. The pictures in Figure 11 show the dimensions of the tanker.

The mesh used consists of 343,560 hexahedral elements and 357,911 nodes. At each time step, a coupled system of nonlinear equations with 1,704,661 unknowns (1,346,750 unknowns for the Navier-Stokes equations and 357,911 unknowns for the interface function) is solved. The computation was carried out on a CRAY T3E with 32 processors.

The images in Figure 12 show, at different instants, the motion of the tanker, sloshing and pressure distribution.

2D sloshing in a container. A container with dimensions 10 cm × 7.5 cm (horizontal × vertical) is filled 2/3 with water and 1/3 with air. It is suddenly subjected to the gravitational acceleration ($g = 9.8$ m/s^2) and a horizontal acceleration of magnitude 0.2g.

We first compute this problem with DSD/SST formulation using a highly-refined moving mesh. The computational domain is discretized using 200 elements in the horizontal direction and 150 elements in the vertical direction, and this results in 30,000 quadrilateral elements and 30,351 nodes. The two images in Figure 13 show, at $t = 0.57$ s, the finite element mesh and the air-water interface. We will refer to this solution, obtained with the Interface-Tracking Technique, as Solution-IT. We will use it to evaluate the solutions obtained with the Interface-Capturing Technique with no enhanced discretization and the EDICT.

We use this structured mesh at its initial non-deformed configuration to compute the same problem with the Interface-Capturing Technique with no enhanced discretization. The two images in Figure 14 show, at $t = 0.57$ s, the finite element mesh and the air-water interface. We will refer to this solution, obtained with the Interface-Capturing Technique, as Solution-IC.

Time histories of the horizontal forces exerted on the container derived from Solution-IT and Solution-IC are compared in Figure 15. The agreement between the solutions is satisfactory.

Next, we compute the problem with the Interface-Capturing Technique using unstructured, triangular meshes. The base mesh, Mesh-1, consists of 13,739 elements and 7,021 nodes. Mesh-2 is obtained by subdividing each element in $(\epsilon_n^1)_n^2$ into four elements, in a style as shown in Figure 16. Mesh-3 is obtained by subdividing each element in $(\epsilon_n^2)_n^3$ into four elements, in a style as shown, again, in Figure 16. We set $(\epsilon_n^2)_n^3 = \epsilon_n^2$; this means that we have the Mesh-2 and Mesh-3 overlapping. We carry out the computations with a time step size of 0.005 s. We redefine $(\epsilon_n^1)_n^2$ at every five time steps, with all elements in ϵ_n^1 within a distance of 0.5 cm from the interface declared to be in $(\epsilon_n^1)_n^2$.

Solution-1 is obtained by using the base discretization, where all trial and weighting functions come only from Mesh-1. Solution-2 is obtained by using the EDICT, where all trial and weighting functions come from Mesh-1 $\oplus$ Mesh-2 (with no third-level discretization for the interface function). Solution-3 is obtained by using a more enhanced discretization, where the trial and weighting functions for velocity and pressure come from Mesh-1 $\oplus$ Mesh-2, and for the interface function from Mesh-1 $\oplus$ Mesh-2 $\oplus$ Mesh-3.

Figure 17 shows Mesh-1 together with Mesh-2 and Mesh-3 (both shown on top of Mesh-1) at $t = 0.5$ s. The graphs in Figure 18 show the time histories of the horizontal forces exerted on the container for all four solutions. We assume that the target solution is Solution-IT, obtained with the Interface-Tracking Technique, using a highly-refined mesh. It is clear from these graphs that Solution-1 has significant frequency and amplitude errors. Solution-2 is superior to Solution-1, but it is Solution-3 that is in best agreement with Solution-IT.

Axisymmetric filling/impact. This problem was simulated with the EDICT. We have a circular cylinder with both the radius and height 10 cm. The lower half of the cylinder is filled with a liquid with density equal to that of water but viscosity 100 times that of water. The upper half is filled with air. At $t = 0.0$ s we start injecting the same liquid through a circular section with radius 2.15 cm and positioned concentrically at the top of the cylinder. The injection stream has a uniform flow speed of 1.0 m/s. Figure 19 shows the problem geometry.

A finite element mesh with triangular elements is used in the computation. The base mesh, Mesh-1, consists of 30,000 elements and 15,251 nodes. Mesh-2 and Mesh-3 are generated in the same way as they were generated in the previous test problem. However, this time, at every redefinition of $(\epsilon_n^1)_n^2$, all elements in ϵ_n^1 within a distance of 0.7 cm from the interface are declared to be in $(\epsilon_n^1)_n^2$.

The solution is obtained by using an enhanced discretization, where the trial and weighting functions for velocity and pressure come from Mesh-1 $\oplus$ Mesh-2, and for the interface function from Mesh-1 $\oplus$ Mesh-2 $\oplus$ Mesh-3.

Figure 20 shows Mesh-2 (on top of Mesh-1) at $t = 0.15$ s. Figure 21 shows a sequence of air-liquid interactions seen at different instants during the simulation of this problem. The pictures show the injection stream impacting the still liquid, formation of surface waves, and entrapment of air in the liquid.

8.   CONCLUDING REMARKS

In this paper we have presented an overview of our parallel finite element computing methods for simulation of unsteady flow problems involving fluid-fluid interfaces and free surfaces. We discussed the Deformable-Spatial-Domain/Stabilized Space-Time (DSD/SST) formulation, which was developed earlier, and the Enhanced-Discretization Interface-Capturing Technique (EDICT), which was introduced very recently. The DSD/SST formulation is an interface-tracking method and works with meshes that need to be updated as the computations proceed. The EDICT, on the other hand, can work with non-moving meshes but requires the solution of an advection equation governing the evolution of an interface function. Both methods involve stabilized finite element formulations, and the stabilization results in good stability and accuracy properties for the methods.

For comparable mesh refinement, the DSD/SST method leads to more accurate representation of the interface. To reduce the cost associated with automatic mesh generation, the mesh update strategies used in the DSD/SST formulation are based on the objective of minimizing the frequency of remeshing. However, meeting this objective in 3D problems involving complex and very unsteady interfaces becomes very difficult, and sometimes not the realistic thing to try.

In the EDICT, to enhance the spatial discretization near the interfaces, we use finite element functions with multiple components. Each of these components corresponds to one of the multiple levels of mesh refinement used at and near the interface. Here we have used the EDICT to increase, in an efficient way, the accuracy of our stabilized formulations which function as the base finite element formulations for solution of the Navier-Stokes equations and the advection equation governing the interface function. Other finite element formulations can also be used as base formulations, including those well-tuned to accurately solve the scalar advection equation.

Both the DSD/SST formulation and EDICT have been implemented in parallel. The DSD/SST formulation, which is the more mature one, has been implemented on several parallel computing platforms and have been applied to a number of 3D problems. The EDICT so far has been implemented in a shared-memory parallel computing paradigm. The results obtained with the EDICT show that, compared to the base interface-capturing method, the Technique substantially increases the accuracy in representing the interface. We believe that these two methods give us the capability to solve a wide range of free-surface and two-fluid problems with varying degrees of interface complexities.

9. ACKNOWLEDGMENT

This work is sponsored by ARPA and by the Army High Performance Computing Research Center under the auspices of the Department of the Army, Army Research Laboratory cooperative agreement number DAAH04-95-2-0003/contract number DA-HH04-95-0008. The content does not necessarily reflect the position or the policy of the government, and no official endorsement should be inferred. The CRAY time was provided, in part, by the University of Minnesota Supercomputer Institute.

REFERENCES

[1] T.E. Tezduyar, M. Behr, and J. Liou, "A new strategy for finite element computations involving moving boundaries and interfaces – the deforming-spatial-domain/space-time procedure: I. The concept and the preliminary tests", *Computer Methods in Applied Mechanics and Engineering*, **94** (1992) 339–351.

[2] T.E. Tezduyar, M. Behr, S. Mittal, and J. Liou, "A new strategy for finite element computations involving moving boundaries and interfaces – the deforming-spatial-domain/space-time procedure: II. Computation of free-surface flows, two-liquid flows, and flows with drifting cylinders", *Computer Methods in Applied Mechanics and Engineering*, **94** (1992) 353–371.

[3] T.J.R. Hughes and A.N. Brooks, "A multi-dimensional upwind scheme with no crosswind diffusion", in T.J.R. Hughes, editor, *Finite Element Methods for Convection Dominated Flows*, AMD-Vol.34, 19–35, ASME, New York, 1979.

[4] T.E. Tezduyar, "Stabilized finite element formulations for incompressible flow computations", *Advances in Applied Mechanics*, **28** (1991) 1–44.

[5] T.J.R. Hughes and G.M. Hulbert, "Space-time finite element methods for elastodynamics: formulations and error estimates", *Computer Methods in Applied Mechanics and Engineering*, **66** (1988) 339–363.

[6] P. Hansbo and A. Szepessy, "A velocity-pressure streamline diffusion finite element method for the incompressible Navier-Stokes equations", *Computer Methods in Applied Mechanics and Engineering*, **84** (1990) 175–192.

[7] M. Behr and T.E. Tezduyar, "Finite element solution strategies for large-scale flow simulations", *Computer Methods in Applied Mechanics and Engineering*, **112** (1994) 3–24.

[8] G. Wren, S. Ray, S. Aliabadi, and T. Tezduyar, "Simulation of flow problems with moving mechanical components, fluid- structure interactions and two-fluid interfaces", to appear in *International Journal for Numerical Methods in Fluids*, 1997.

[9] T. Tezduyar, S. Aliabadi, M. Behr, A. Johnson V. Kalro, and M. Litke, "Flow simulation and high performance computing", *Computational Mechanics*, **18** (1996) 397–412.

[10] T.E. Tezduyar, S. Aliabadi, and M. Behr, "Enhanced-Discretization Interface-Capturing Technique", to appear in *Proceedings of the ISAC '97 High Performance Computing on Multiphase Flows*, Tokyo, Japan, 1997.

[11] C. W. Hirt and B. D. Nichols, "Volume of fluid (VOF) method for the dynamics of free boundaries", *Journal of Computational Physics*, **39** (1981) 201–225.

[12] D. L. Youngs, "Time-dependent multimaterial flow with large fluid distortion", in K. W. Morton and M. J. Baines, editors, *Numerical Methods in Fluid Dynamics*, Notes on Numerical Fluid Mechanics, 273–285, Academic Press, New York, 1984.

[13] C. M. Lemos, "Higher-order schemes for free-surface flows with arbitrary configurations", *International Journal for Numerical Methods in Fluids*, **23** (1996) 545–566.

[14] S. Aliabadi and T. Tezduyar, "3D simulation of free-surface flows with parallel finite element method", to appear in *Computational Mechanics '97, Proceedings of International Conference on Computational Engineering Science*, San Jose, Costa Rica, 1997.

[15] Y. Saad and M. Schultz, "GMRES: A generalized minimal residual algorithm for solving nonsymmetric linear systems", *SIAM Journal of Scientific and Statistical Computing*, **7** (1986) 856–869.

[16] Z. Johan, T.J.R. Hughes, and F. Shakib, "A globally convergent matrix-free algorithm for implicit time-marching schemes arising in finite element analysis in fluids", *Computer Methods in Applied Mechanics and Engineering*, **87** (1991) 281–304.

[17] S.K. Aliabadi and T.E. Tezduyar, "Parallel fluid dynamics computations in aerospace applications", *International Journal for Numerical Methods in Fluids*, **21** (1995) 783–805.

[18] A.A. Johnson and T.E. Tezduyar, "Mesh update strategies in parallel finite element computations of flow problems with moving boundaries and interfaces", *Computer Methods in Applied Mechanics and Engineering*, **119** (1994) 73–94.

[19] G. Karypis and V. Kumar, "Multilevel k-way partitioning scheme for irregular graphs", Technical Report 95-064, University of Minnesota, Department of Computer Science, 1995.

Figure 1. Mesh−1. The set of elements are denoted by $\in_n^1$. The elements marked by ○ are those declared to be in the subset $(\in_n^1)_n^2$.

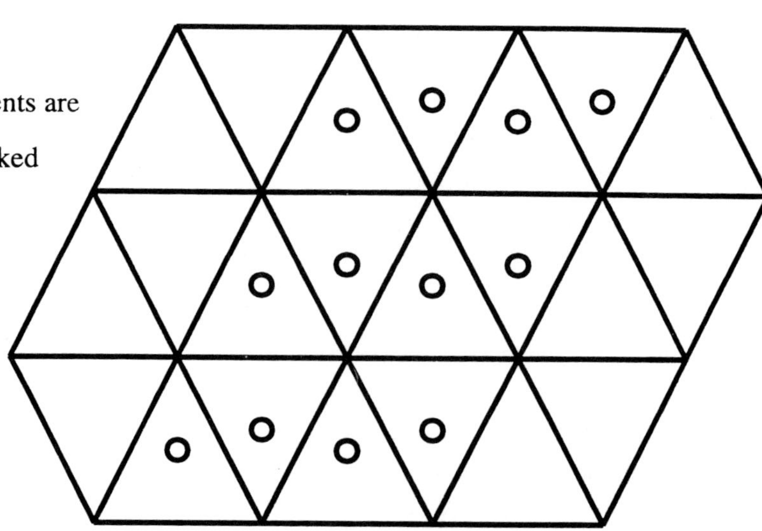

Figure 2. Mesh−2. The set of elements are denoted by $\in_n^2$. The elements marked by □ are those declared to be in the subset $(\in_n^2)_n^3$.

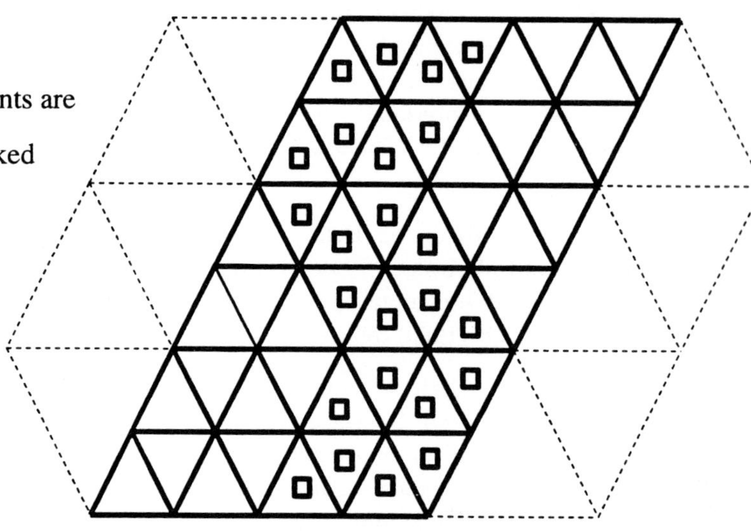

Figure 3. Mesh−3.

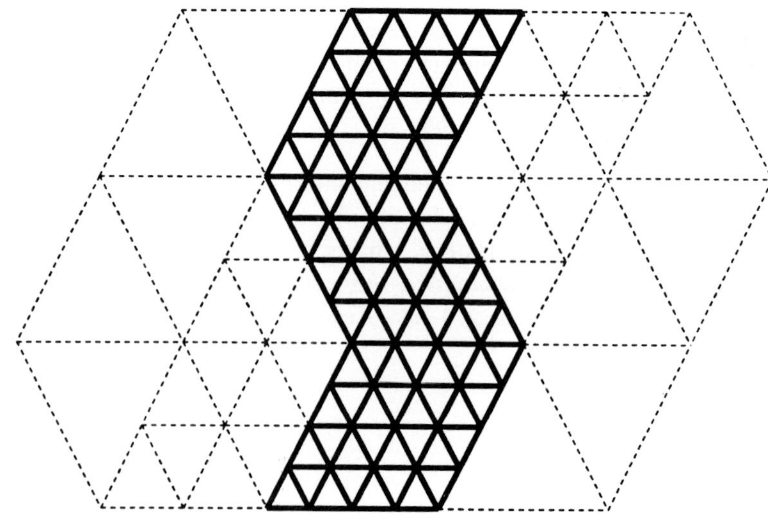

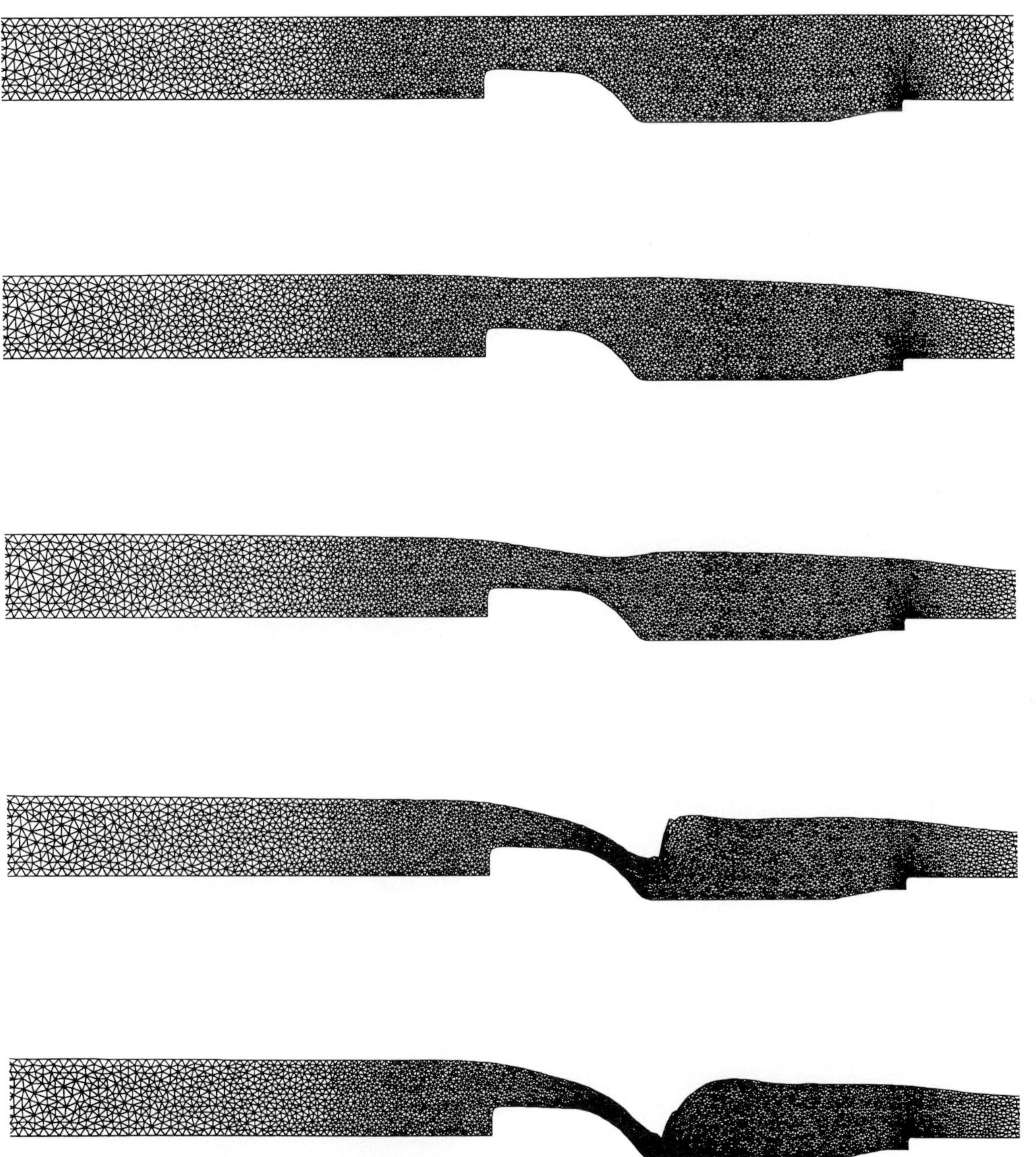

Figure 4. 2D simulation of surface waves in hydraulic structures. Deforming mesh at $t = 0$, 5, 10, 15 and 18 seconds.

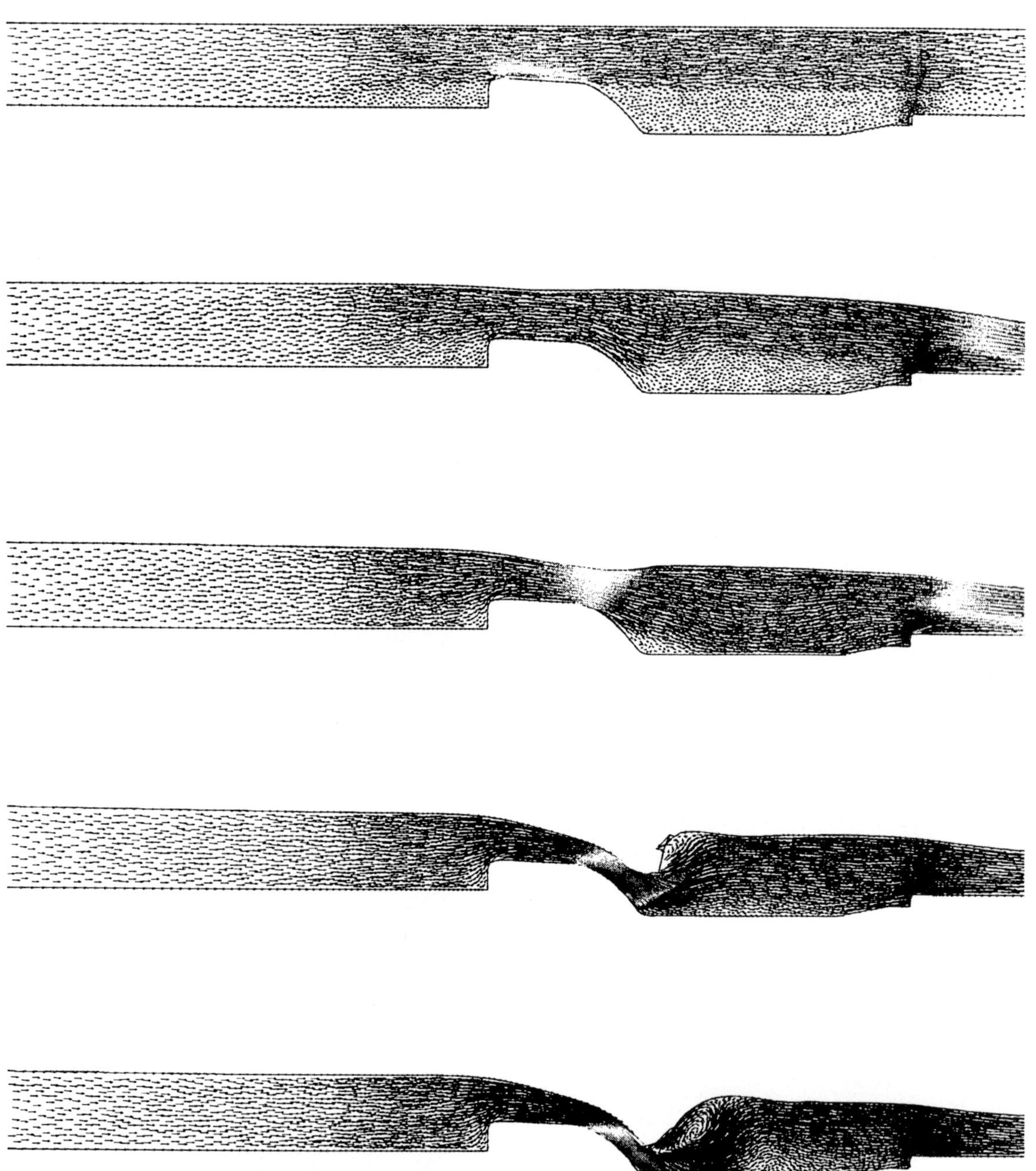

Figure 5. 2D simulation of surface waves in hydraulic structures. Velocity field at $t = 0$, 5, 10, 15 and 18 seconds.

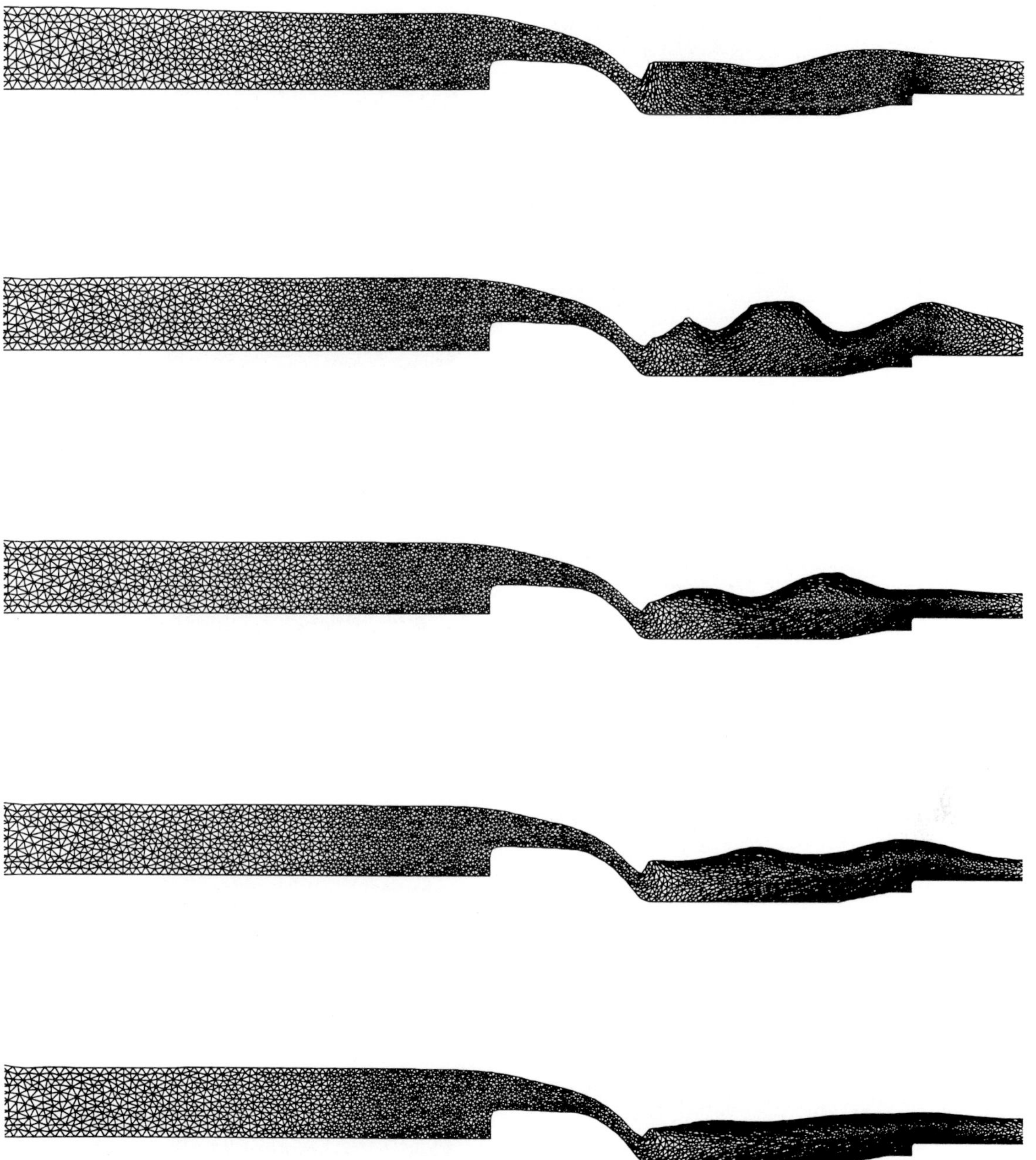

Figure 6. 2D simulation of surface waves in hydraulic structures. Deforming mesh at $t = 25$, 35, 45, 55 and 65 seconds.

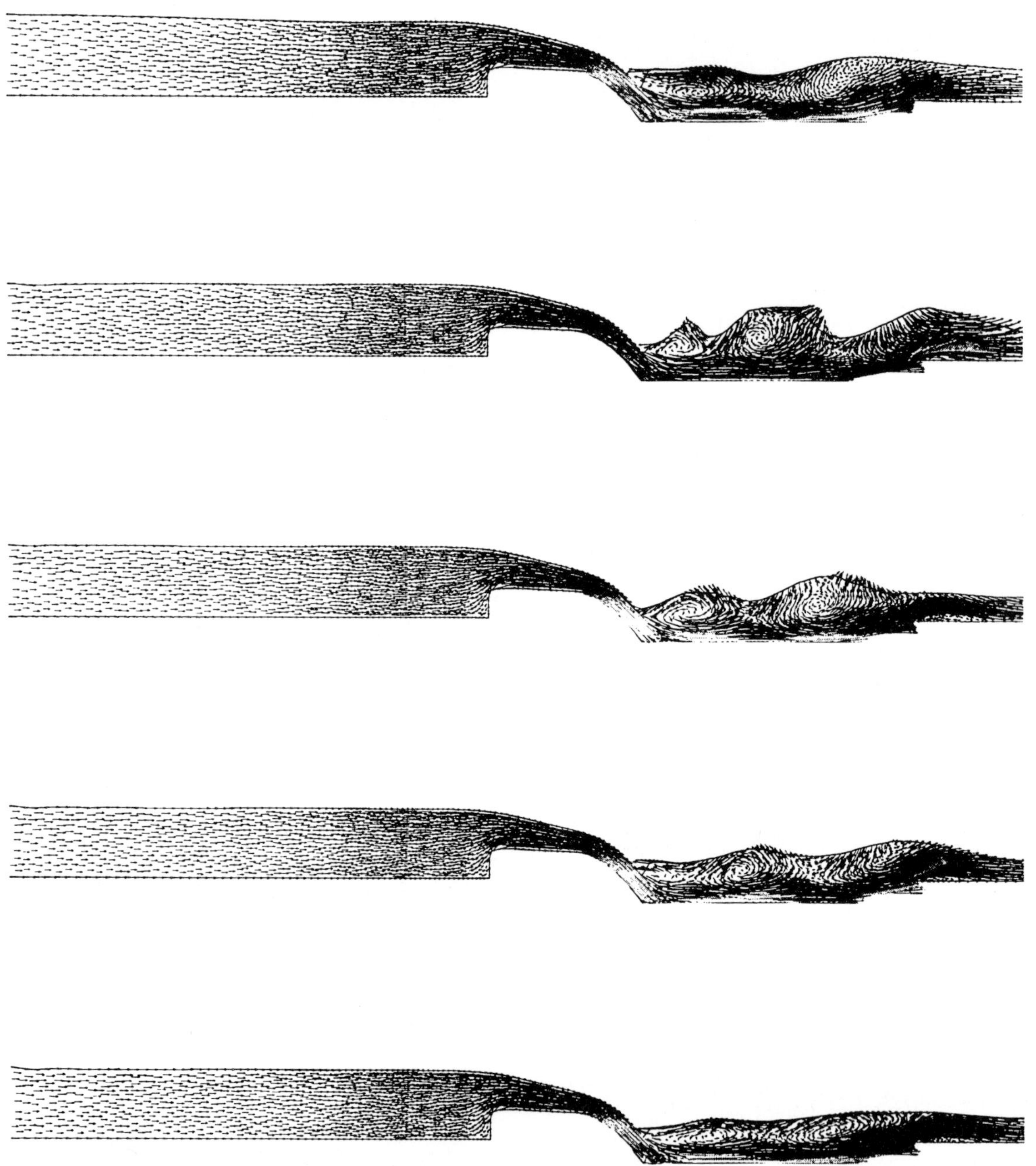

Figure 7. 2D simulation of surface waves in hydraulic structures. Velocity field at $t = 25, 35, 45, 55$ and 65 seconds.

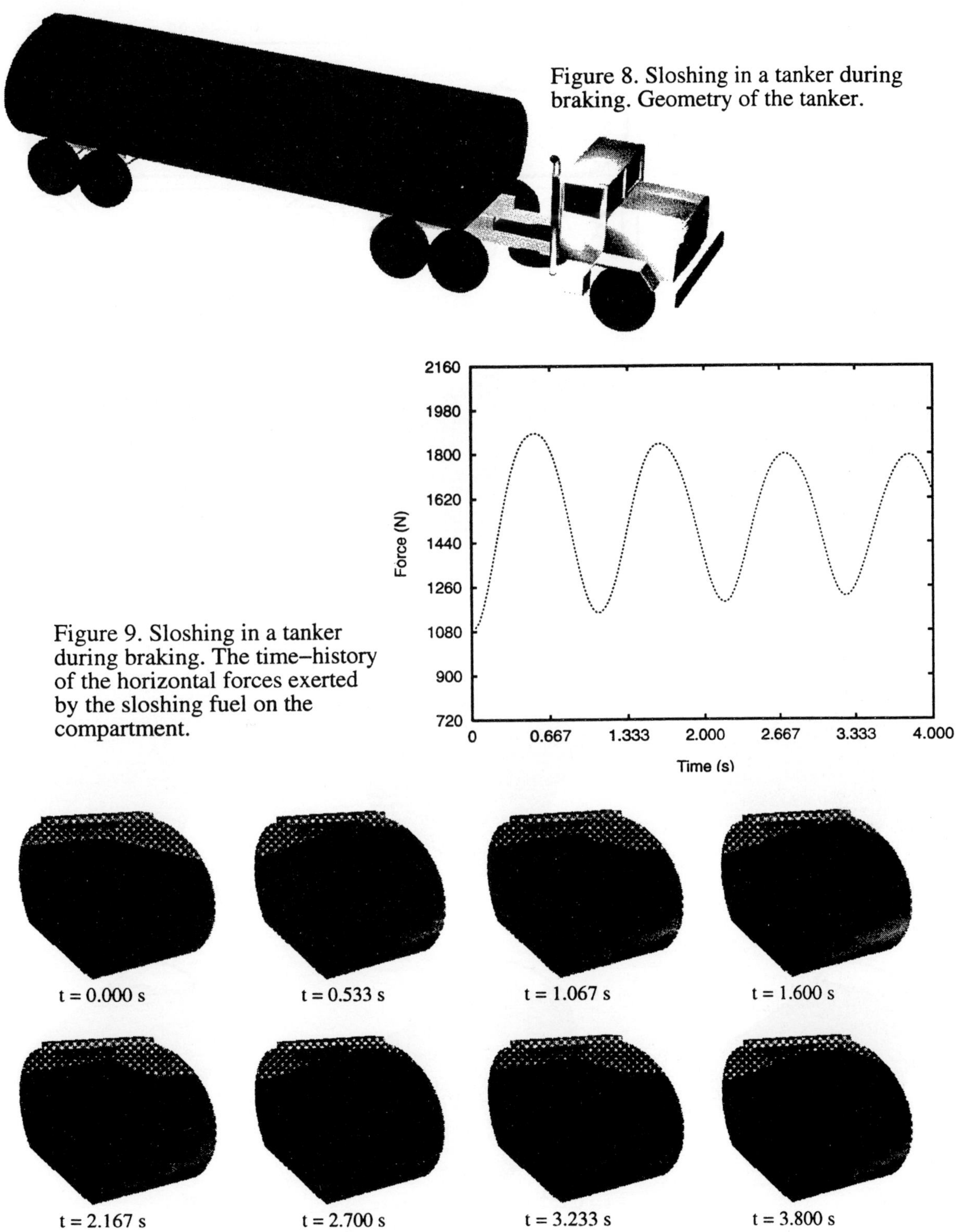

Figure 8. Sloshing in a tanker during braking. Geometry of the tanker.

Figure 9. Sloshing in a tanker during braking. The time–history of the horizontal forces exerted by the sloshing fuel on the compartment.

t = 0.000 s   t = 0.533 s   t = 1.067 s   t = 1.600 s

t = 2.167 s   t = 2.700 s   t = 3.233 s   t = 3.800 s

Figure 10. Sloshing in a tanker during braking. The images show, at different instants, the sloshing and pressure distribution.

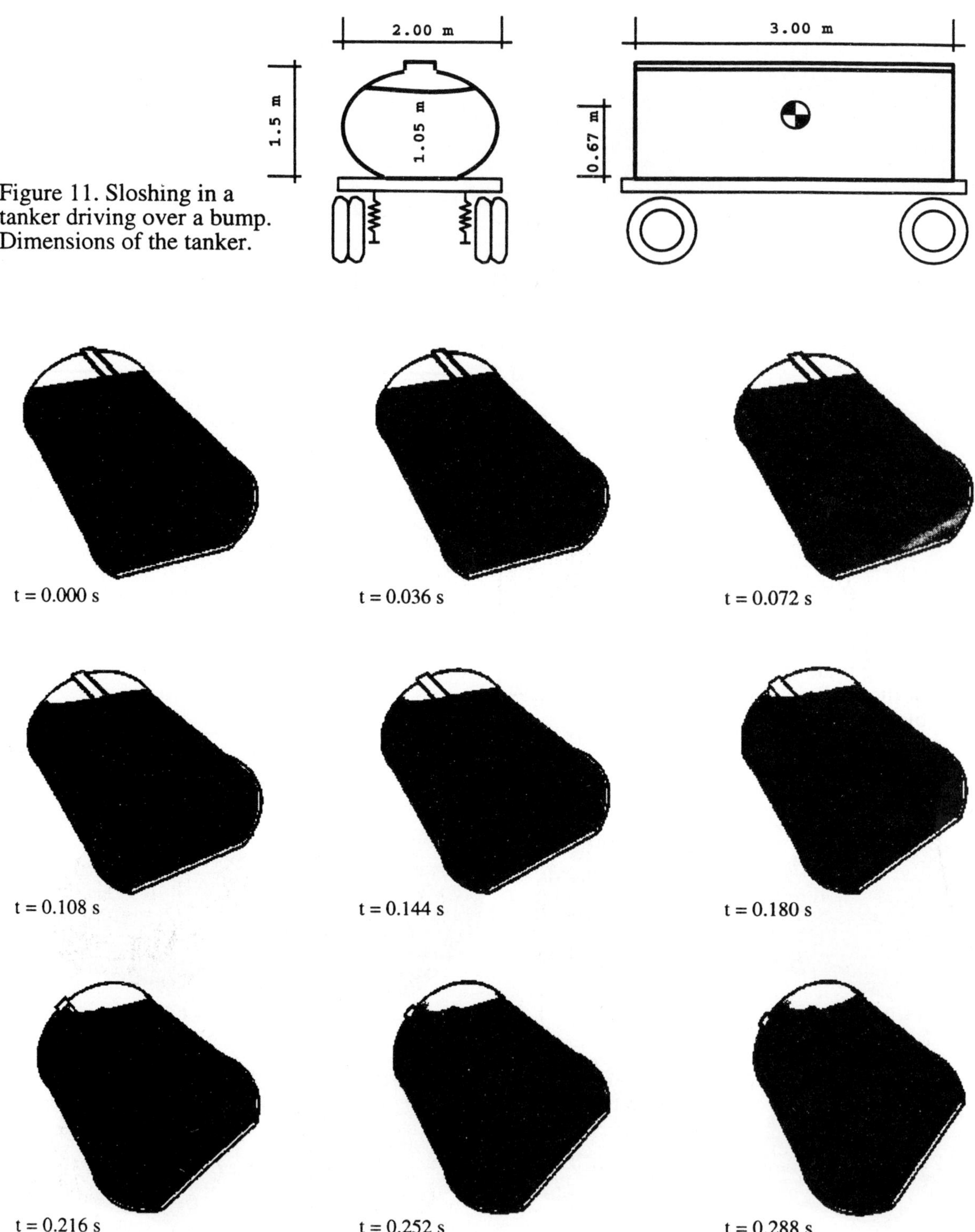

Figure 11. Sloshing in a tanker driving over a bump. Dimensions of the tanker.

t = 0.000 s

t = 0.036 s

t = 0.072 s

t = 0.108 s

t = 0.144 s

t = 0.180 s

t = 0.216 s

t = 0.252 s

t = 0.288 s

Figure 12. Sloshing in a tanker driving over a bump. The images show, at different instants, the motion of the tanker, sloshing and pressure distribution.

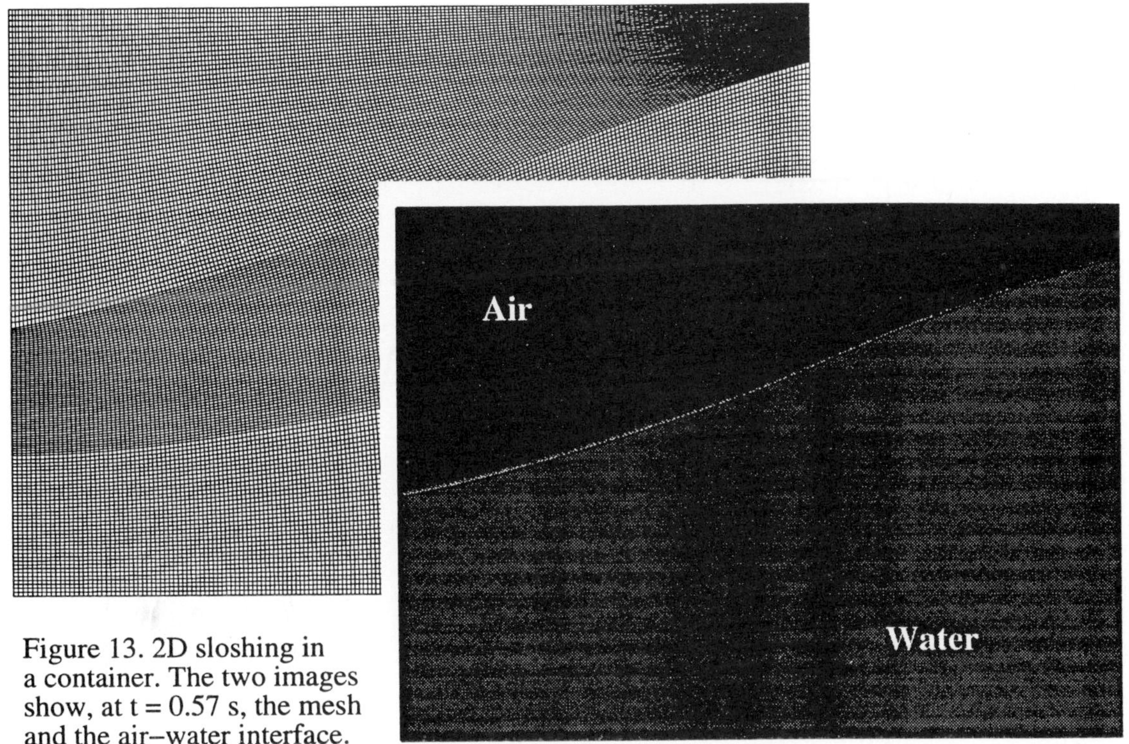

Figure 13. 2D sloshing in a container. The two images show, at t = 0.57 s, the mesh and the air–water interface.

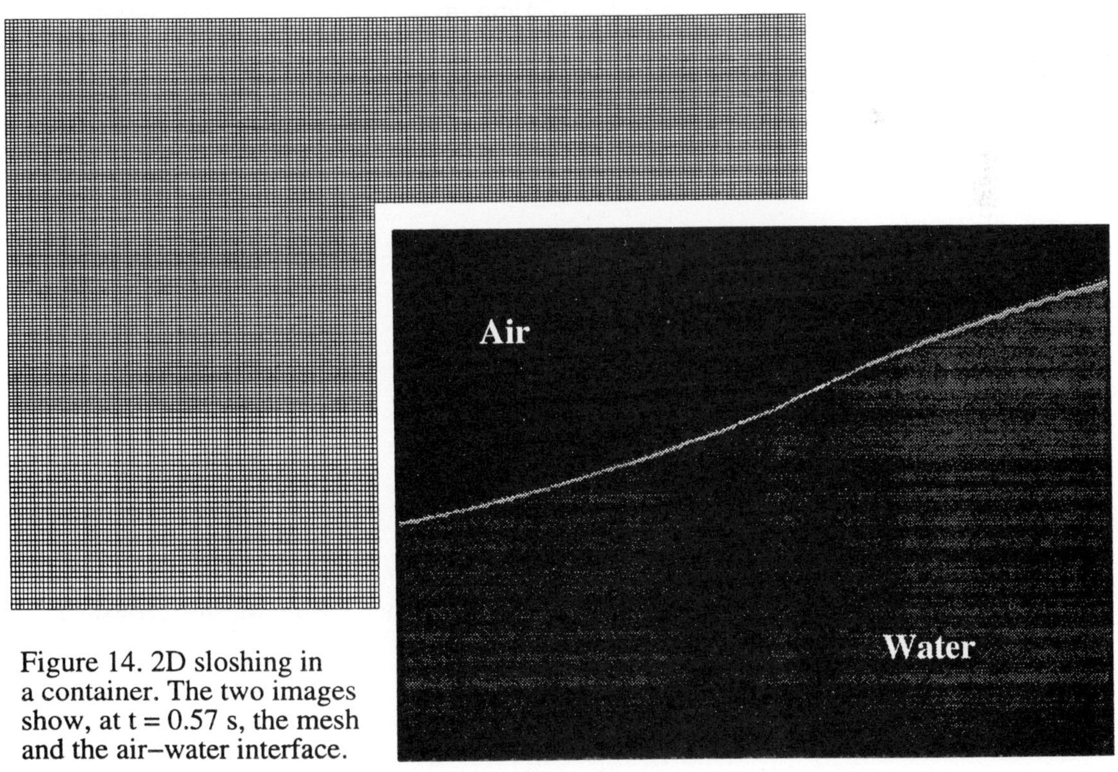

Figure 14. 2D sloshing in a container. The two images show, at t = 0.57 s, the mesh and the air–water interface.

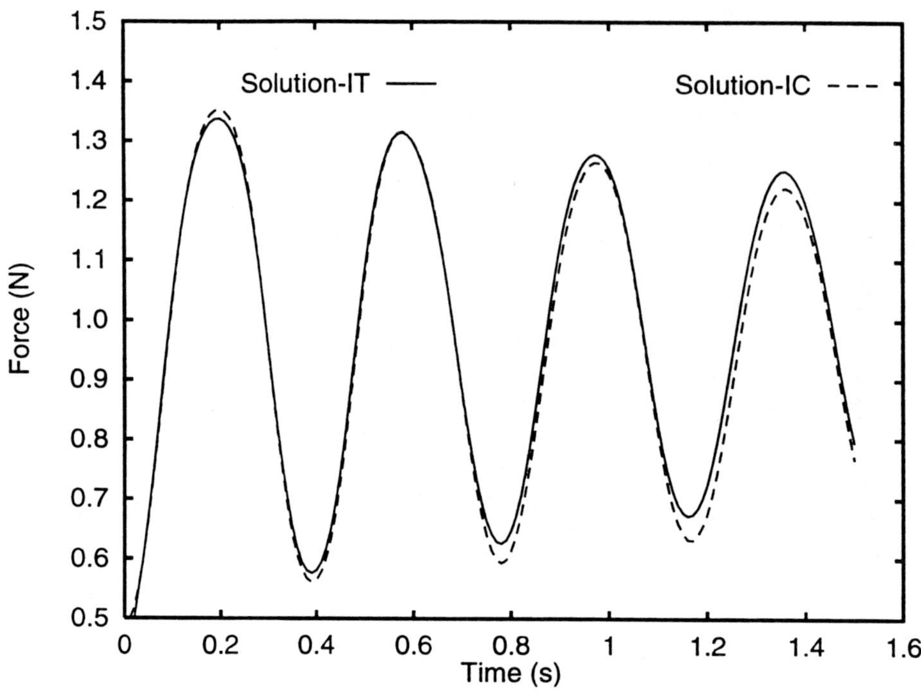

Figure 15. 2D sloshing in a container. Time histories of the horizental forces exerted on the container derived from Solution–IT and Solution–IC.

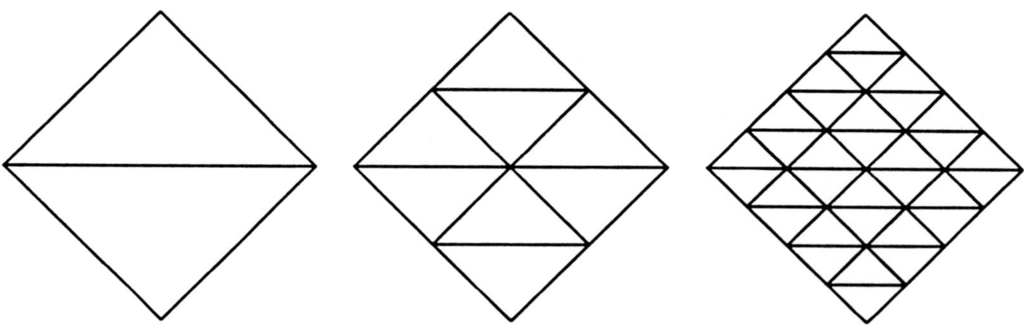

Figure 16. 2D sloshing in a container. Representive elements from Mesh–1, Mesh–2 and Mesh–3.

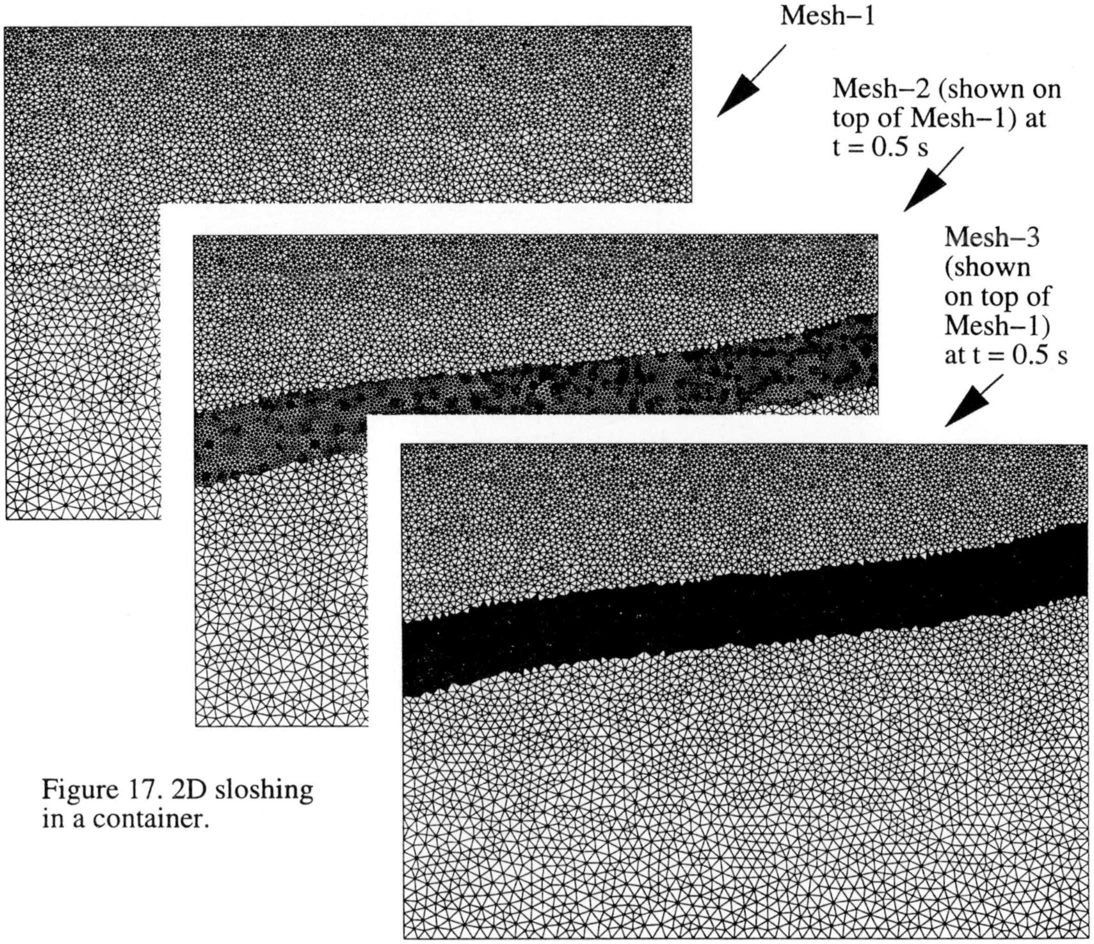

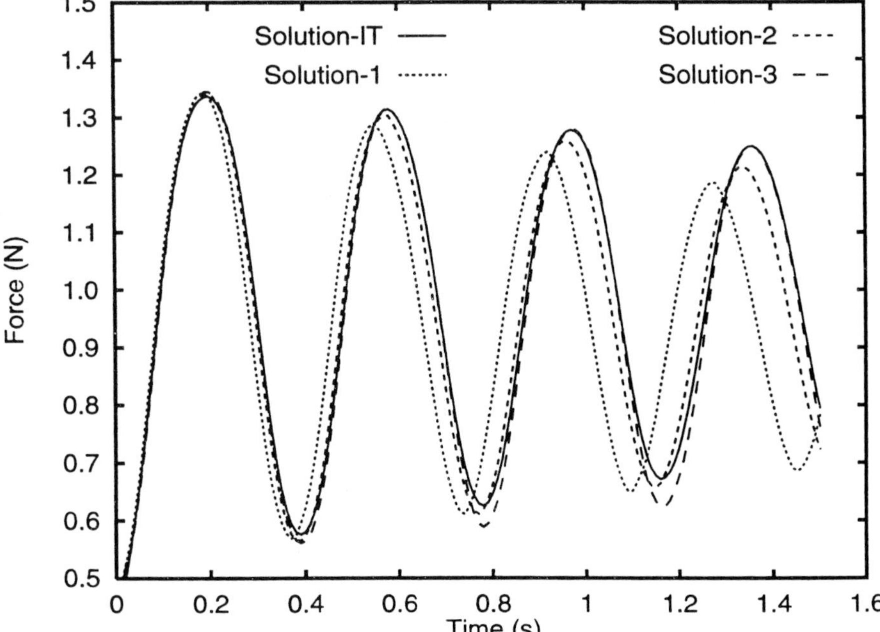

Figure 17. 2D sloshing in a container.

Figure 18. 2D sloshing in a container. Time histories of the horizontal forces exerted on the container for all four solutions.

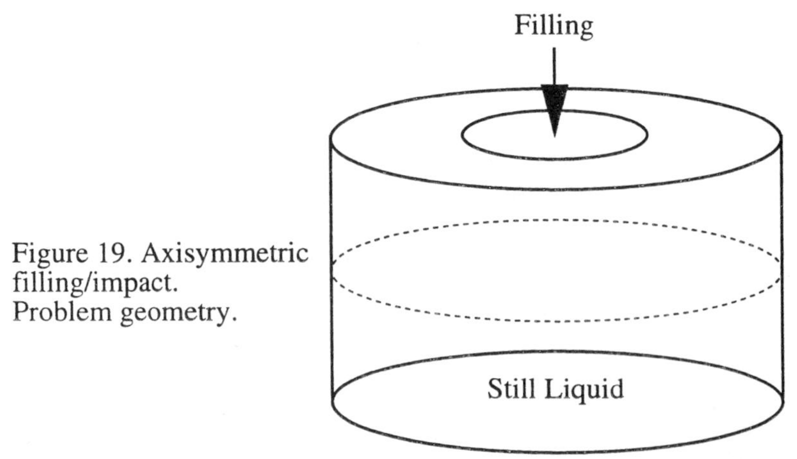

Figure 19. Axisymmetric filling/impact. Problem geometry.

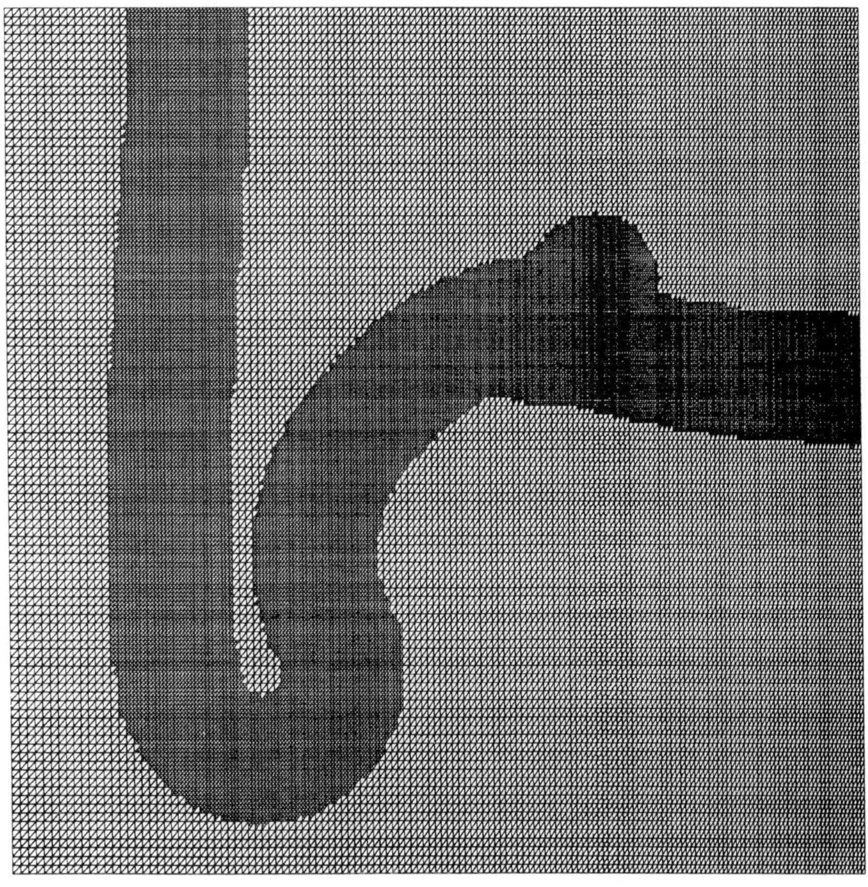

Figure 20. Axisymmetric filling/impact. Mesh–2 (shown on top of Mesh–1) at t = 0.15 s.

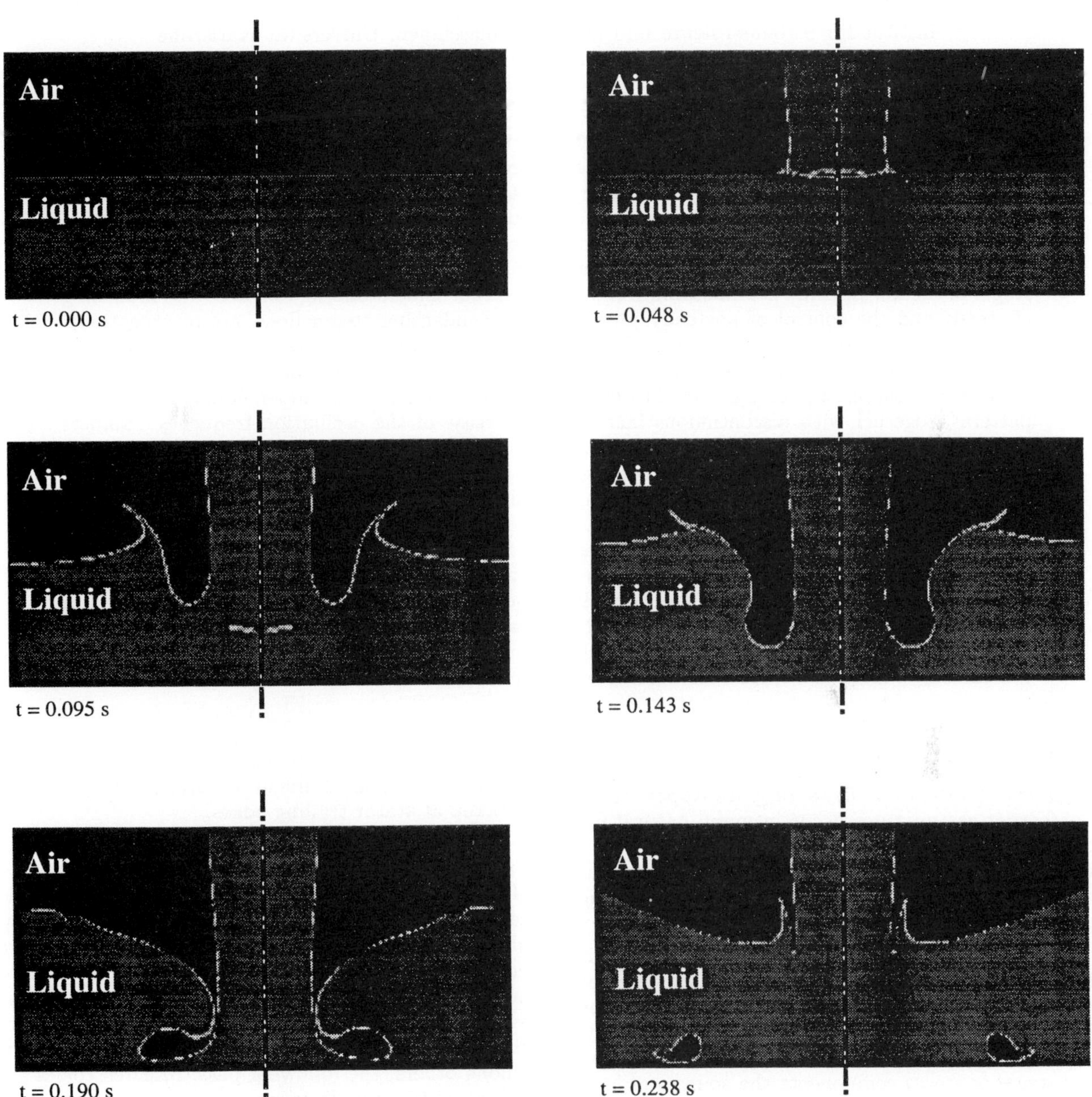

Figure 21. Axisymmetric filling/impact. The pictures show a sequence of air–liquid interactions.

TWO-PHASE FLOW INSTABILITIES
IN CHANNELS AND TURBINE CASCADES

Günter H. SCHNERR   Michael HEILER

Institut für Strömungslehre und Strömungsmaschinen, Universität Karlsruhe
(TH), Kaiserstrasse 12, D-76128 Karlsruhe, Germany

Abstract

The purpose of research on this subject is the fact that worldwide most of the plants of the electrical power industry have been in service for more than 25-30 years and recent advances in the development of CFD codes have made possible the detailed calculation of complex single phase flows in turbine stages. However no commercially available code contains an adequate representation of the steam phase transition mechanism, nor the corresponding condensation loss [?]. Because about 10% of the total electrical net power output of plants up to 1500 MW is produced in the last stages of the low pressure steam turbines it is obvious, that understanding and accurate modelling of steady and the control of unsteady phenomena in condensing steam flows are of exceptional great importance for the redesign of modern steam turbine bladings.

Performing experiments and numerical calculations of condensing transonic flows in nozzles with varying shapes we detected new instabilities and bifurcations with sudden changes of the flow patterns, together with discontinuous increase or decrease of the oscillation frequency (Schnerr et al. [?]). The amplitude of the pressure waves increases locally for more than 100 % and the dispersion of the liquid phase changes completely. The first aspect is certainly relevant for flutter excitation of large blades in low pressure stages and the second affects losses and erosion by the droplet impact on the blade surfaces.

In this paper we first investigate the formation of flow instabilities with bifurcations in arbitrary channel geometries with unsymmetric boundaries and we confirm the existence of bifurcation dynamics in axial cascades using vapor/carrier gas mixtures, which is the operating test fluid of the atmospheric indraft wind tunnel where these new instabilities were detected. We then present numerical results of steady and unsteady nucleating vapor flows using actual blading, operating at low pressure conditions, typical for observing condensation.

1 Introduction

High speed expansion flows of pure vapor or vapor/carrier gas mixtures occur in many natural processes and are important for technical applications in mechanical and chemical engineering, e.g. in steam turbines, jet engines and for safety control of large power plants.

Losses in turbine stages have been subject of intense study for many years. Such studies were conducted using nozzles and cascades of airfoil sections in atmospheric indraft wind tunnels or more recently in steam tunnels. Aside of homogeneously condensing steam or steam components the sources of losses in turbine bladings are manifold. E.g. from heterogeneous condensation caused by mechanical or chemical impurities of the steam or by the formation and movement of coarse water, by the movement of water films along bladings, entrainment of water films and droplet breakup at stator trailing edges.

From EPRI, the Electric Power Reserach Institute, Palo Alto, CA, USA it is reported that about 40% of the failures of turbine bladings have no identifiable cause, and on average 45% of blade failures are multiple blade failures [?]. Consequently, there is a definite need to improve turbine blade analysis, modelling and numerical simulation including two-phase phenomena, especially the control of instabilities in low pressure stages with high wetness fractions. In the same report, among others, the following basic research topics are suggested to be undertaken

- Excitation mechanism such as condensation shocks, stall and flutter, which are only partially understood at present.
- Possible flow instabilities and other flow excitations to gain improved understanding of other failure mechanisms.

Received on August 1, 1997.

Therefore, the purpose of the current research reported here, is to provide a more systematic structure and understanding of the mechanism of self-excited flow instabilities including higher order bifurcations which basically evolve from the interaction of the compressibility of the fluid with the embedded condensation process.

In the following we concentrate on the aerodynamic losses caused by the dynamics of self-excited instabilities in flows with homogeneously condensing vapor or vapor components of the fluid. During the expansion of steam in turbines the thermodynamic state path crosses the saturation line and the subsequent stages operate with a highly dispersed two-phase mixture of vapor and tiny droplets. Modern rotors of low pressure stages may come up to 3-4 meters in diameter. Together with the required rotational frequency corresponding to the electrical line frequency of 50 Hz (or 60 Hz depending on country), the circumferential speed reaches values of 500-600 m/s causing transonic/supersonic flow through the blade passages. Due to the high cooling rate of the expansion, the fluid nucleates to form a large number of growing tiny droplets that constitute the wetness (here about 5-6 %). For the stability of the flow it is important that the latent heat release to the flow happens near Mach number unity with the greatest sensitivity against smallest disturbances. The flow is then thermally choked (Delale, Schnerr, Zierep [4]) and shocks are formed which may become unstable. The interaction of the nucleating flow with compression shocks initiates self-excited oscillations with typical frequencies of 200 - 1500 Hz.

The phenomenon of self-excited oscillations in Laval nozzle flows caused by the interaction of shocks with nonequilibrium condensation was discovered by Schmidt [5]. For moist air and pure steam, Barschdorff [6, 7] conducted the first quantitative experiments. The first similarity formula for the oscillation frequency for a constant nozzle throat height and varying supply conditions was derived by Zierep and Lin [8]. Subsequently, experimental results of unsteady steam flows were presented by Barschdorff and Filippov [9]. The first numerical 1-D results of unsteady two-phase flow driven by nonequilibrium condensation of water vapor were presented by Saltanov and Tkalenko [10]. Recently Collignan [11] reproduced the characteristic frequency minimum of wet steam nozzle flows using a 1-D Euler code,

even though the calculated frequencies were higher compared with experiments. By approaching real flow conditions, White and Young [12] presented 2-D numerical results of inviscid unsteady condensing wet steam flow in a slender nozzle. Two-dimensional effects in unsteady nozzle flows of water vapor/carrier gas mixtures together with the frequency dependence on the nozzle shape and on the supply conditions were investigated by Schnerr et al. [13], Mundinger [14], and Adam [15]. Recently, numerical results of inviscid steady state 2-D and quasi 3-D condensing cascade flows were published by Young [16]. Bakhtar [17] developed a 2-D code for condensing cascade flows with an additional viscous model. Ishizaka et al. [18] provided the first Navier-Stokes solver to calculate 2-D cascade flows with homogeneous condensation.

Until now the existence of self-excited oscillatory instabilities in linear cascades has not yet been clarified, except for the investigation of Bakhtar [19] which indicates that periodic boundary conditions cannot prevent the onset of periodic instabilities with moving shocks. The aim of our current research is to prove the existence of these instabilities and to find the conditions under which the topology of instabilities including the bifurcation dynamics (Adam and Schnerr[20]), known from condensing nozzle flows, establishes in turbine cascades, too.

2 PHYSICAL MODEL

The homogeneous nucleation process is modelled with the classical nucleation theory of Volmer, Frenkel, and Zel'dovich. Dealing with water vapor/carrier gas mixtures at very low partial vapor pressures, the Hertz-Knudsen law is well suitable for modelling the droplet growth. Simulating steam flows in turbines requires droplet growth expressions which cover the whole Knudsen number range and take the temperature as a driving potential into account, e.g. see Gyarmathy [21], Young [22, 23], or Peters [24]. The theory of the droplet growth models of Young and Peters is based on the Langmuir model (three layer concept). Here the flow field around a droplet is divided into an outer region, where the equations of the continuum mechanics apply and an inner Knudsen layer (collison free zone), where transfer processes are kinetically controlled. Young's model which also includes the possibility to account for an inert gas is able to predict finite temperature and vapor pressure jumps across the Knudsen layer. Therefore it provides a physical interpretation for the origin of the reversed temperature gradient phenomenon. The model of Peters includes an uniform pressure hypothesis. This allows the determination of the droplet temperature from the ambient thermodynamic state saving the evaluation of the energy equation in the Knudsen layer. Gyarmathy's expression represents a droplet growth equation which models the kinetic controlled processes by applying a Knudsen number dependent correction factor to the continuum mechanics heat and mass transfer equation. We herein use the model given by Gyarmathy because calculations using this formula show good agreement with nozzle flow experiments, which are particularly relevant to LP steam turbine conditions. Furthermore this expression provides the lowest CPU-time requirements.

So far, heterogeneous condensation effects caused by particles or chemical impurities inside the mixture, are not yet taken into account. The surface tension and the mass accommodation coefficient are determined corresponding to our early steady state experiments, i.e. no correction is necessary for application to unsteady flows.

3   NUMERICAL METHOD

A comparison of computed nozzle flow results with experiments shows that viscosity effects are negligible, even for flows with moving shocks. Therefore, we solve the time dependent 2-D Euler-equations in conservation form, coupled to four additional conservation equations for the liquid phase. Air and water vapor are treated as perfect gases, whereas the properties of the mixture depend on the condensate mass fraction g. The hyperbolic system of eight conservation equations for the two-phase flow is then solved with a MUSCL-type finite volume method on structured body fitted grids [14]. The source terms are treated by applying the fractional step method in order to split the equations into a homogeneous and an inhomogeneous part [25]. This allows to account for different time scales of the flow and of the condensation process, respectively. Both sets of equations are solved by explicit second order accurate time integration. The fluxes at the cell interfaces are determined by the van Leer flux vector splitting for real gases using the frozen speed of sound [26]. The vector of unknowns at the cell interfaces is calculated using the κ-scheme with the van Albada limiter [27], which reduces the accuracy from third order in smooth regions to first order near shocks. More details of the numerical scheme can be found in Schnerr et al. [2, 28].

The internal structure of the code in conjunction with the time marching method guarantees an equal treatment of corresponding cells of symmetric grids, i.e. of corresponding cells of the upper and lower half plane of a symmetric 2-D Laval nozzle. Consequently, the numerical scheme is not responsible for the formation of unsymmetric disturbances in an initially symmetric flow having a perfect symmetric nozzle shape with a symmetric grid. However, even when a perfect symmetric numerical code (including perfect symmetric discretization and initialisation) is used, unsymmetric disturbances can grow. The reason is the limited accuracy of the computed results. For example, starting with a symmetric initialisation the individual calculations of the corresponding cells are identical, except for the sign (different direction of one flow vector component). However, on the SNI VPP300 where our calculations are performed, the different signs of the two numbers lead to different accuracies of these numbers because the negative sign allocates one digit. Consequently, for a great many of calculations, different absolute values in the corresponding cells will most probably appear. These differences are very small. When they appear, they are of the order of the precision of the variables (in our calculations always double precision is applied which means accuracy of $0.5 \cdot 10^{-16}$). If there is an inherent instability in the flow, these differences can grow exponentially as shown by Adam [15] and Adam and Schnerr [20] leading to bifurcation solutions with unsymmetric flow patterns.

4   RESULTS AND DISCUSSION

4.1   Channel Flows

Most of the fundamental research devoted to nonequilibrium condensation processes of vapor or vapor components is performed using simplified

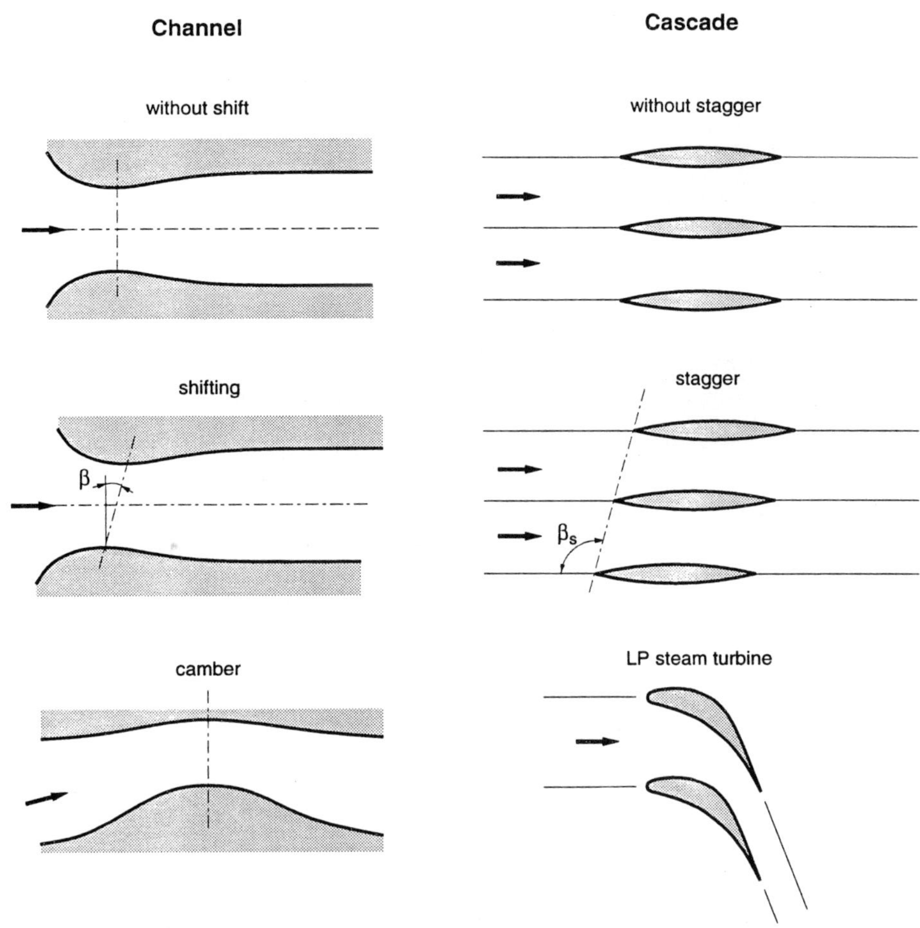

Fig. 1: Systematic variation of the geometry of channel and cascade flows

boundary conditions like quasi 1-D or 2-D nozzles and channels of perfect symmetry with respect to their center-lines.

Here, with reference upon important technical applications like steam turbine cascades, we regard the important question of the stability and existence of bifurcation solutions in flows where no symmetry exists at all. Unsymmetric boundary conditions are introduced and all possible effects on the dynamic response are analyzed by stepwise introducing shift and camber of the channel walls. Subsequently, the same systematic variation is applied to axial cascades (Fig. 1). The question to be clarified is, how can unsymmetric boundaries affect the existence of the frequency bifurcation and how significant are the qualitative and quantitative effects on the oscillation dynamics? In the following the 'symmetric', i.e. quasi-symmetric oscillation mode of a flow within un-

symmetric boundaries corresponds to the dynamics of symmetric flow patterns in perfect symmetric boundaries.

4.1.1 Shifted Nozzle A1

Without shift the shape of the plane 2-D nozzle A1 is designed to produce a shock-free parallel outflow with a supersonic exit Mach number $M_e = 1.2$. To create unsymmetry the upper and the lower wall are shifted by the angle 5° and 20°, respectively (Fig.1 - left). Figure 2 depicts the weak oblique shocks which already exist in steady adiabatic flow, caused by the displacement of the nozzle walls according to the different shift angles β. Obviously, with condensation the bifurcation phenomenon exists, too. The numerical results of Fig. 3 depict at the lower branch symmetric $(\beta = 0^o)$ or quasi symmetric $(\beta > 0^o)$ low fre-

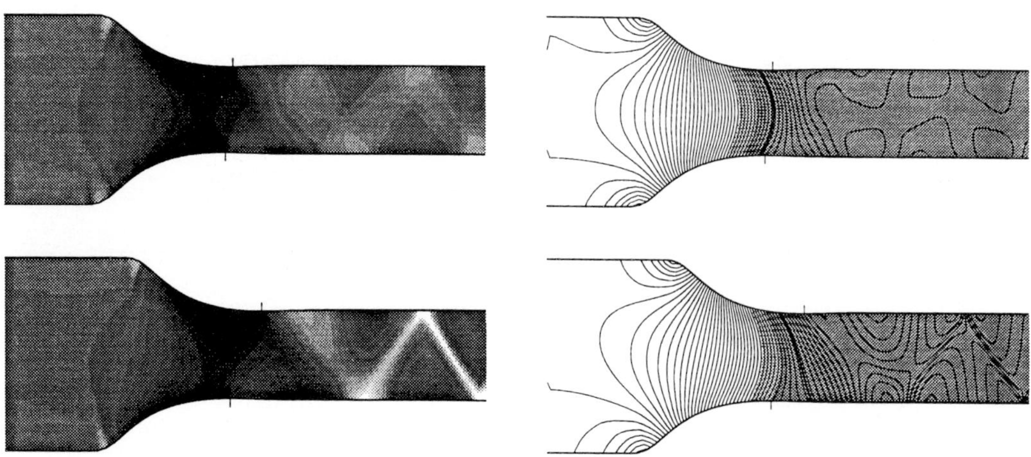

Fig. 2: Steady adiabatic flow in the shifted nozzle A1, reservoir conditions $T_{01} = 295K$, $p_{01} = 1bar$; left - numerical simulated schlieren pictures, right - frozen Mach number M_f (light grey $M_f > 0.8$, dark grey $M_f > 0.9$, increment $\Delta M_f = 0.02$); top: nozzle shift $\beta = 5°$, bottom: nozzle shift $\beta = 20°$.

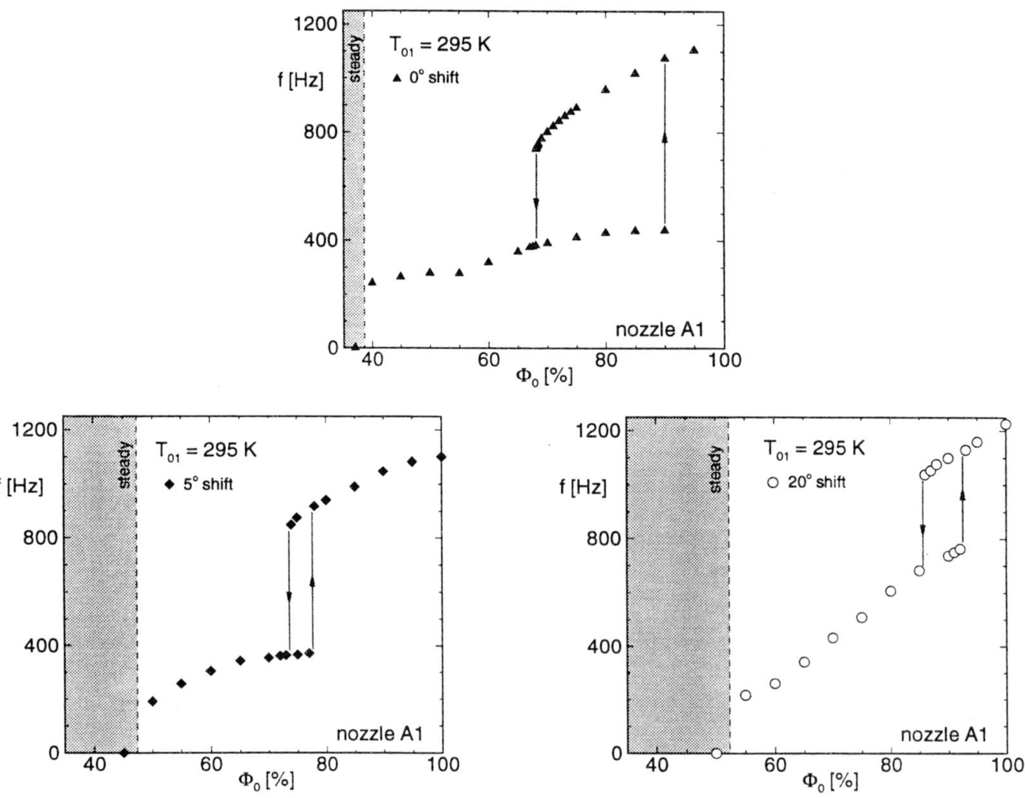

Fig. 3: Bifurcation – Frequency dependence on the water vapor pressure ratio Φ_0 in the shifted nozzle A1 for constant reservoir values of temperature $T_{01} = 295K$ and pressure $p_{01} = 1bar$; top: no shift, bottom - left: nozzle shift $\beta = 5°$, bottom - right: nozzle shift $\beta = 20°$.

quency oscillations and unsymmetric high frequency oscillations at the upper branch, i.e. the bifurcation dynamics is not prevented by unsymmetric boundaries. This supports the relevance of our experimental discovery of this new phenomenon to be independent of small dislocations of the nozzle blocks. However, the width of the hysteresis and the frequency difference between the two branches decreases with increasing unsymmetry of the boundaries, which is quite understandable.

4.1.2 Shifted Nozzle A1A

In order to prove that these phenomena are not restricted to a limited class of nozzle shapes, especially not to nozzles with very slender, and therefore very sensitive supersonic regions, we investigated a plane nozzle where the walls are formed by circular arcs, i.e. with constant radius of wall curvature of $R^* = 300mm$ and a very strong shift of $\beta = 40^o$. The numerical schlieren simulations of Fig. 4 show clearly the formation of the two typical wave patterns for identical conditions of the oncoming flow. Independent of whether the flow is quasi symmetric ('symmetric') or not, the flow accelerates to supersonic exit Mach numbers $M_e > 1$. Due to the supersonic boundary condition at the exit of nozzle A1A the bifurcation is clearly a local instability in the most sensitive transonic throat region, which again supports the validity and accuracy of our experimental findings. Figures 5 and 6 give more detailed insight into details of the interaction of the transonic flow with the nucleation process and the condensate formation. According to the iso-Mach lines (Figs. 5 and 6 - right) the maximum Mach number at the nozzle exit reaches $\overline{M_e} = 1.5$.

4.1.3 Higher Order Bifurcation in Nozzle A1H

Here, in order to create a different type of unsymmetry, only the lower half of the symmetric nozzle A1 is regarded, this half nozzle is called A1H. Because the inviscid boundary condition at the upper plane nozzle wall, i.e. the tangential flow condition, is equivalent to the symmetry condition at the center-line of the original nozzle A1, there is no difference between the symmetric oscillations in the nozzles A1 and A1H, respectively. It is interesting that for reservoir conditions $T_{01} = 295K$

and $p_{01} = 1bar$ the first order bifurcation (Fig. 3) does not exist in the half nozzle A1H. After increasing the reservoir temperature to $T_{01} = 305K$ a second order bifurcation establishes with a much higher oscillation frequency of 2047Hz, i.e. the frequency increases by a factor 4-5 from 440Hz to more than 2000Hz, which is about twice of the first order bifurcation in the symmetric nozzle A1. Due to the strong heat release the whole condensing nozzle flow is subsonic and the compression waves are weak (Fig. 7). With respect to similarity considerations it is interesting that the bifurcation dynamics is controlled by the time scale of the wave propagation perpendicular to the main flow direction (Adam and Schnerr [29]). Comparison of calculated frequencies at identical reservoir conditions for the nozzles A1 and A1H is made in Tab. I. Conditions according to the first and second line of Tab. I confirm the agreement of the symmetric modes, but only the first order bifurcation in nozzle A1. The data set at the third line corresponds to the schlieren simulation of Fig. 7 - right.

Table I: Comparison of oscillation frequencies in nozzles A1 and A1H

		symm. mode		unsymm. mode	
T_{01}	Φ_0	nozzle		nozzle	
$[K]$	$[\%]$	A1	A1H	A1	A1H
295	90	446	447	1068	–
305	75	440	440	1051	–
305	95	–	–	1334	2047

In contrast to nozzle A1 the amplitude of the pressure fluctuation at the throat decreases strongly (Fig. 8). The significantly higher mean value of the unsymmetric pressure oscillations ($\overline{p}/p_{01} \approx 0.60$ corresponding to $\overline{M} \approx 0.87$) shows that here the flow is not likely to be accelerated to supersonic velocities during the oscillation cycles.

In conclusion the existence of bifurcation phenomena is well confirmed. In tendency unsymmetric boundary conditions reduce the quantitative differences between the 'symmetric' and unsymmetric oscillation frequencies a bit, which is understandable, but they do not prevent one of the oscillation modes.

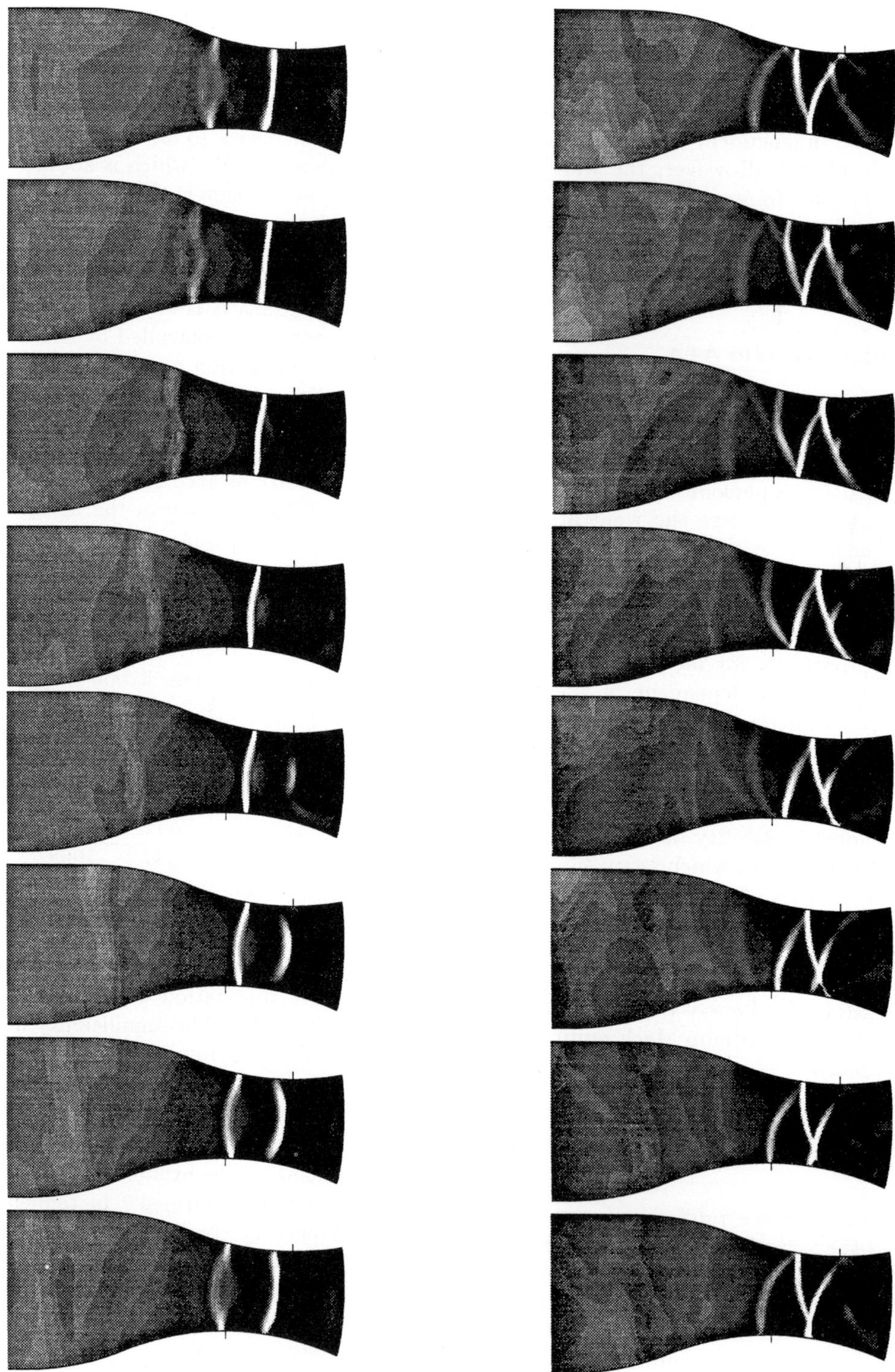

Fig. 4: One cycle of the 'symmetric' and unsymmetric flow oscillations in the shifted nozzle A1A, shift angle $\beta = 40^{\circ}$, for identical reservoir conditions $T_{01} = 295K$, $p_{01} = 1.0bar$, $\Phi_0 = 90\%$, numerically simulated schlieren pictures; left: 'symmetric' $f = 854Hz$, right: unsymmetric $f = 1082Hz$; flow from left to right, time increases from top.

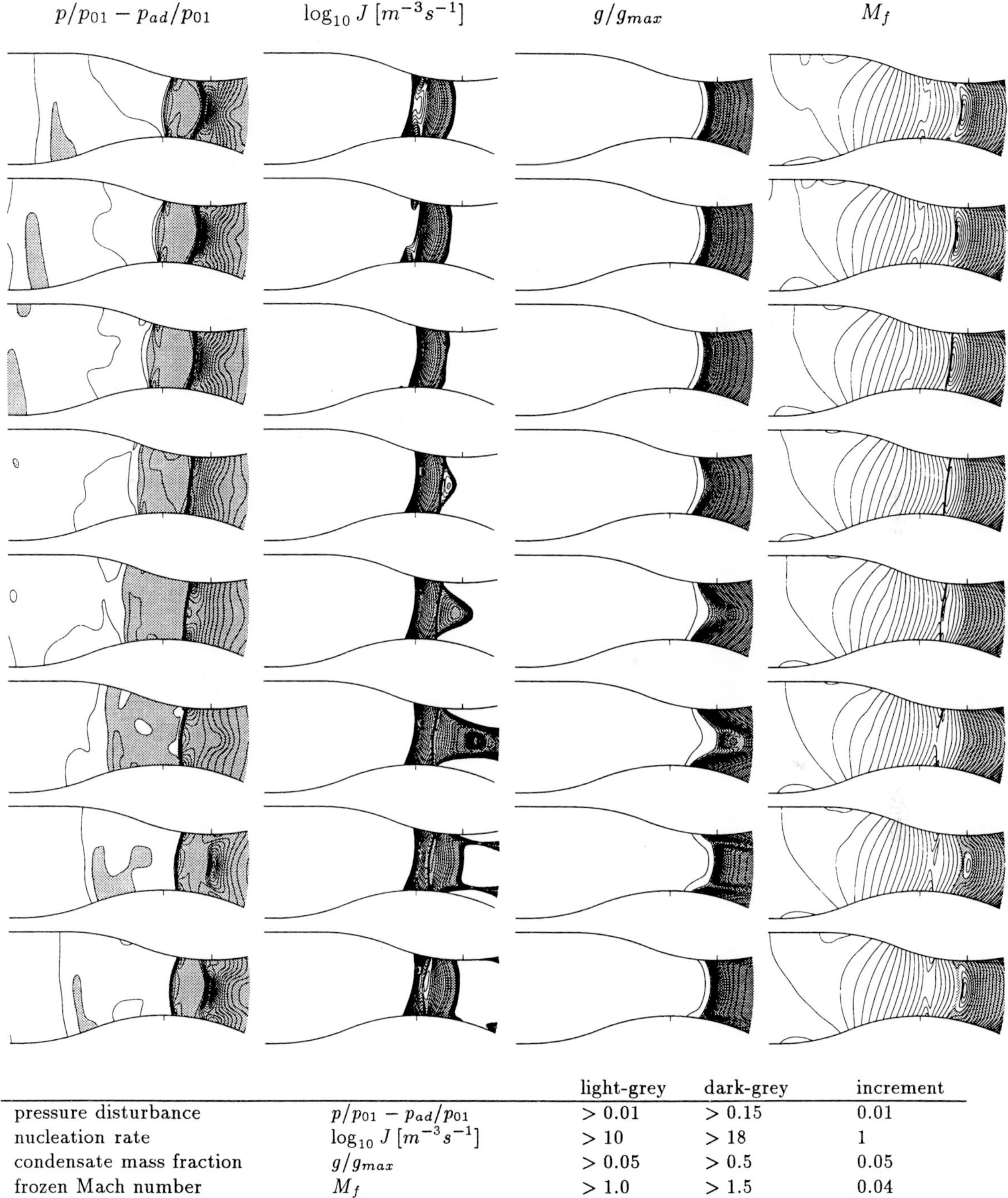

		light-grey	dark-grey	increment
pressure disturbance	$p/p_{01} - p_{ad}/p_{01}$	> 0.01	> 0.15	0.01
nucleation rate	$\log_{10} J \, [m^{-3}s^{-1}]$	> 10	> 18	1
condensate mass fraction	g/g_{max}	> 0.05	> 0.5	0.05
frozen Mach number	M_f	> 1.0	> 1.5	0.04

Fig. 5: One cycle of the 'symmetric' oscillation in the shifted nozzle A1A, shift angle $\beta = 40°$; reservoir conditions: $T_{01} = 295K$, $p_{01} = 1.0bar$, $\Phi_0 = 90\%$; frequency f=854 Hz; flow from left to right, time increases from top.

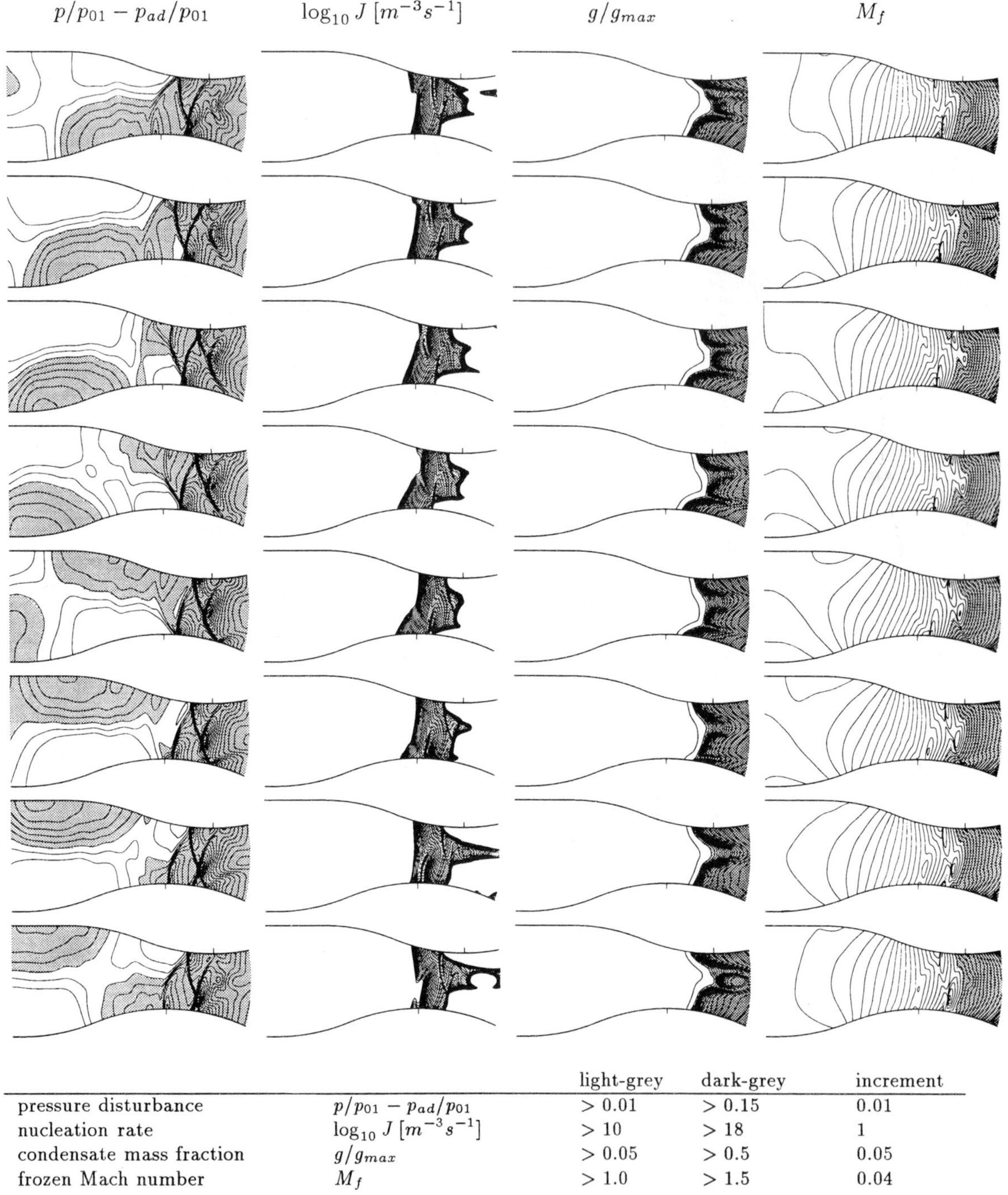

		light-grey	dark-grey	increment
pressure disturbance	$p/p_{01} - p_{ad}/p_{01}$	> 0.01	> 0.15	0.01
nucleation rate	$\log_{10} J\ [m^{-3}s^{-1}]$	> 10	> 18	1
condensate mass fraction	g/g_{max}	> 0.05	> 0.5	0.05
frozen Mach number	M_f	> 1.0	> 1.5	0.04

Fig. 6: One cycle of the unsymmetric oscillation in the shifted nozzle A1A, shift angle $\beta = 40^{\circ}$; reservoir conditions: $T_{01} = 295K$, $p_{01} = 1.0 bar$, $\Phi_0 = 90\%$; frequency f=1082 Hz; flow from left to right, time increases from top.

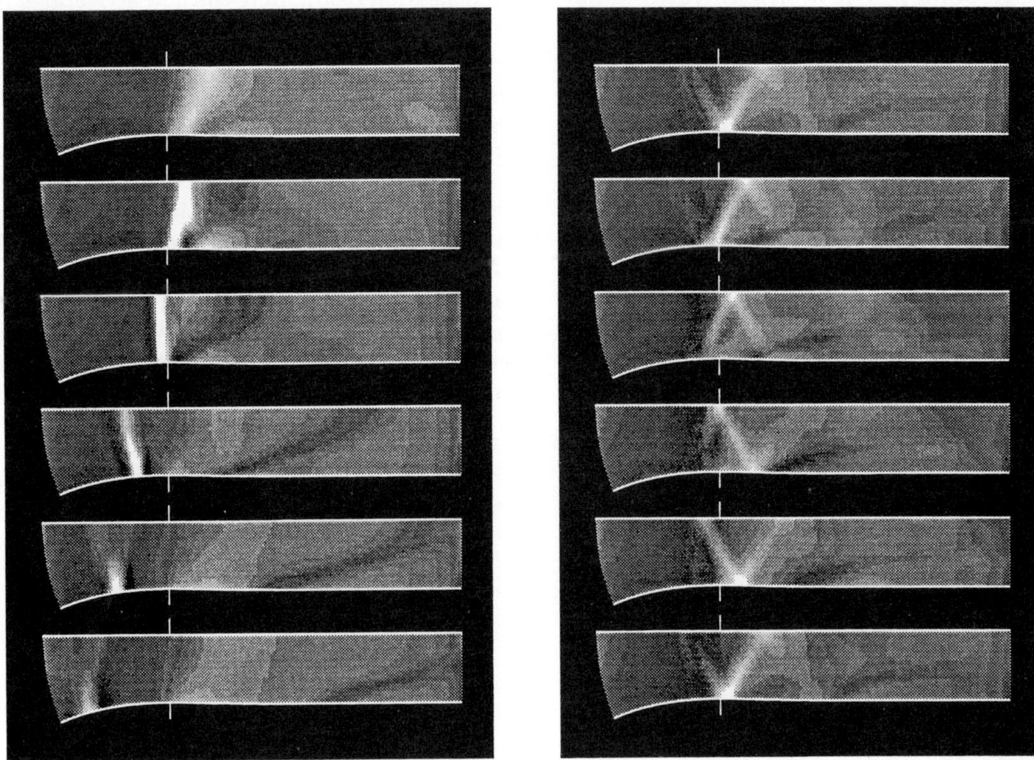

Fig. 7: 'Symmetric' and unsymmetric flow oscillations in the half nozzle A1H, reservoir conditions $T_{01} = 305K$, $p_{01} = 1bar$; numerically simulated schlieren pictures; left: 'symmetric' $\Phi_0 = 75\%$, $f = 440Hz$; right: unsymmetric $\Phi_0 = 95\%$, $f = 2047Hz$; flow from left to right, time increases from top.

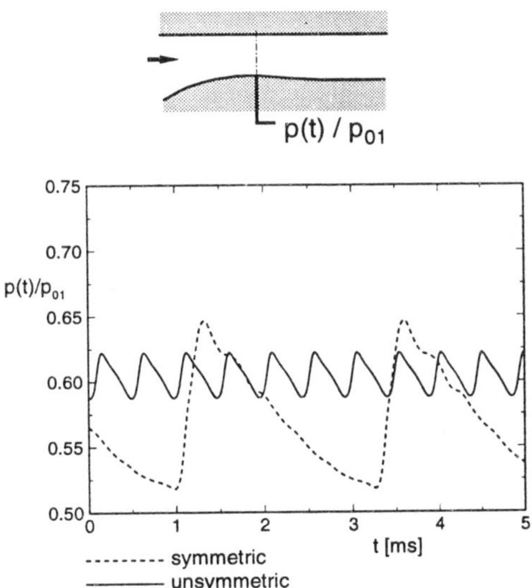

Fig. 8: Time dependent pressure ratio p/p_{01} at the nozzle throat (lower nozzle wall) for the 'symmetric and unsymmetric oscillations in the half nozzle A1H corresponding to Fig. 7.

4.2 Unsteady Flow in Linear Cascade A1A

In contrast to boundaries formed by nozzle walls, the periodicity of cascades ahead and behind the bladings accounts for the interaction of the pressure and suction sides of the blades, which is a quite different and more sensitive configuration in bifurcation dynamics. As a first approach, the blades are simply formed by circular arcs of 12.5 % thickness. Without stagger, the cooling rate at the equivalent throat position is the same as that of the nozzle A1 where the bifurcation was first detected experimentally and perfectly reproduced by numerical calculations. Currently we investigate the cascade angle $\beta_s = 95^o$ (Fig. 9).

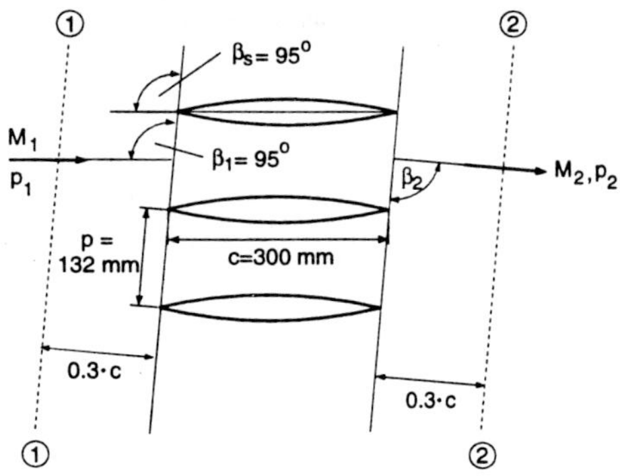

Fig. 9: Geometry and reference frames 1 and 2 of the staggered cascade A1A; p = blade pitch, c = blade chord.

Figure 10 shows numerical simulations of schlieren pictures for the adiabatic flow without

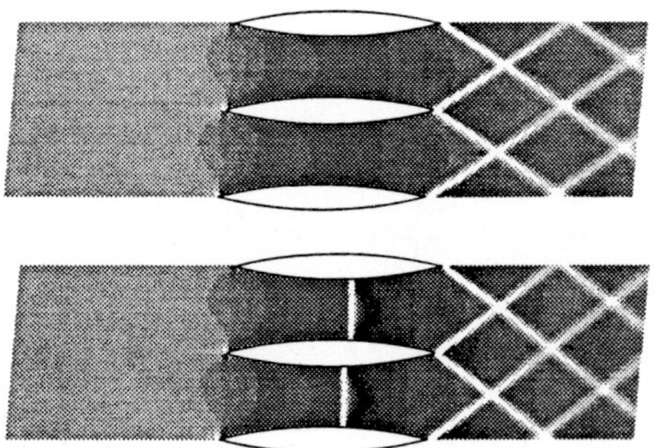

Fig. 10: Steady adiabatic and condensing flow in the staggered cascade A1A, water vapor/carrier gas mixture; $\beta_s = 95^o$; reservoir conditions: $T_{01} = 295\ K$, $p_{01} = 1\ bar$; numerically simulated schlieren pictures; top: adiabatic, vapor pressure ratio (relative. humidity) $\phi_0 = 0$ %, bottom: condensing, $\phi_0 = 55$ %; flow from left to right.

condensation (top). Here the flow accelerates from $M_1 = 0.46$ to $M_2 = 1.74$. The numerical schlieren picture below depicts a steady state condensing flow with a stationary normal shock in the nucleation region. Only a slight increase of the vapor pressure leads to the first mode with 'symmetric' oscillations. Figure 11 shows the surface pressure distribution over the blades. It clearly shows normal shocks with local subsonic flow behind. Due to condensation the Mach number behind the cascade is now reduced ($M_{f,2} = 1.62$) and the flow angle β_2 increases only slightly from $\beta_2 = 89.7^o$ without condensation to $\beta_2 = 89.8^o$ with condensation, respectively (Table II).

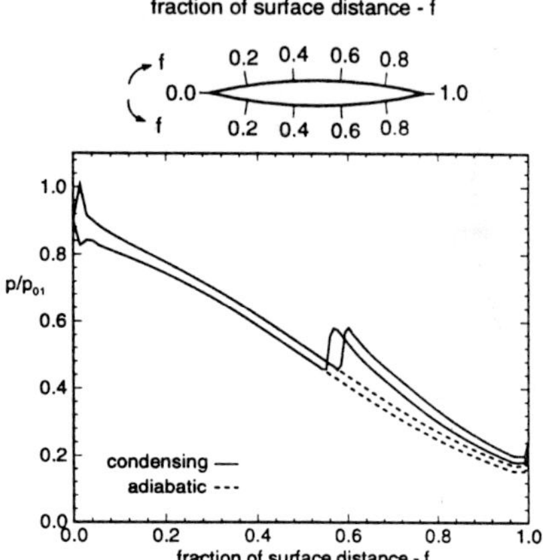

Fig. 11: Adiabatic and condensing blade surface pressure distribution for the staggered cascade A1A, $\beta_s = 95^o$; reservoir conditions according to Fig. 2.

Table II: Flow properties at the reference frames 1 and 2 for the staggered cascade A1A, $\beta_s = 95^o$, reference frames and reservoir conditions corresponding to Figs. 1 and 2.

Flow	$M_{f,1}$	$M_{f,2}$	$p_1[bar]$	$p_2[bar]$	$\beta_1[^o]$	$\beta_2[^o]$
adiabatic	0.46	1.74	0.866	0.187	95	89.7
condensing	0.46	1.62	0.866	0.212	95	89.8

Increasing the vapor pressure ratio at constant reservoir values of the temperature T_{01} and of the mixture pressure p_{01} (water vapor/carrier gas mixtures) initiates the periodic oscillations of Figs. 12 and 13. Obviously, for identical reservoir conditions two different flow patterns with significantly different frequencies exist. One period of the 'symmetric' oscillation mode with moving normal shocks at left is compared with one cycle of

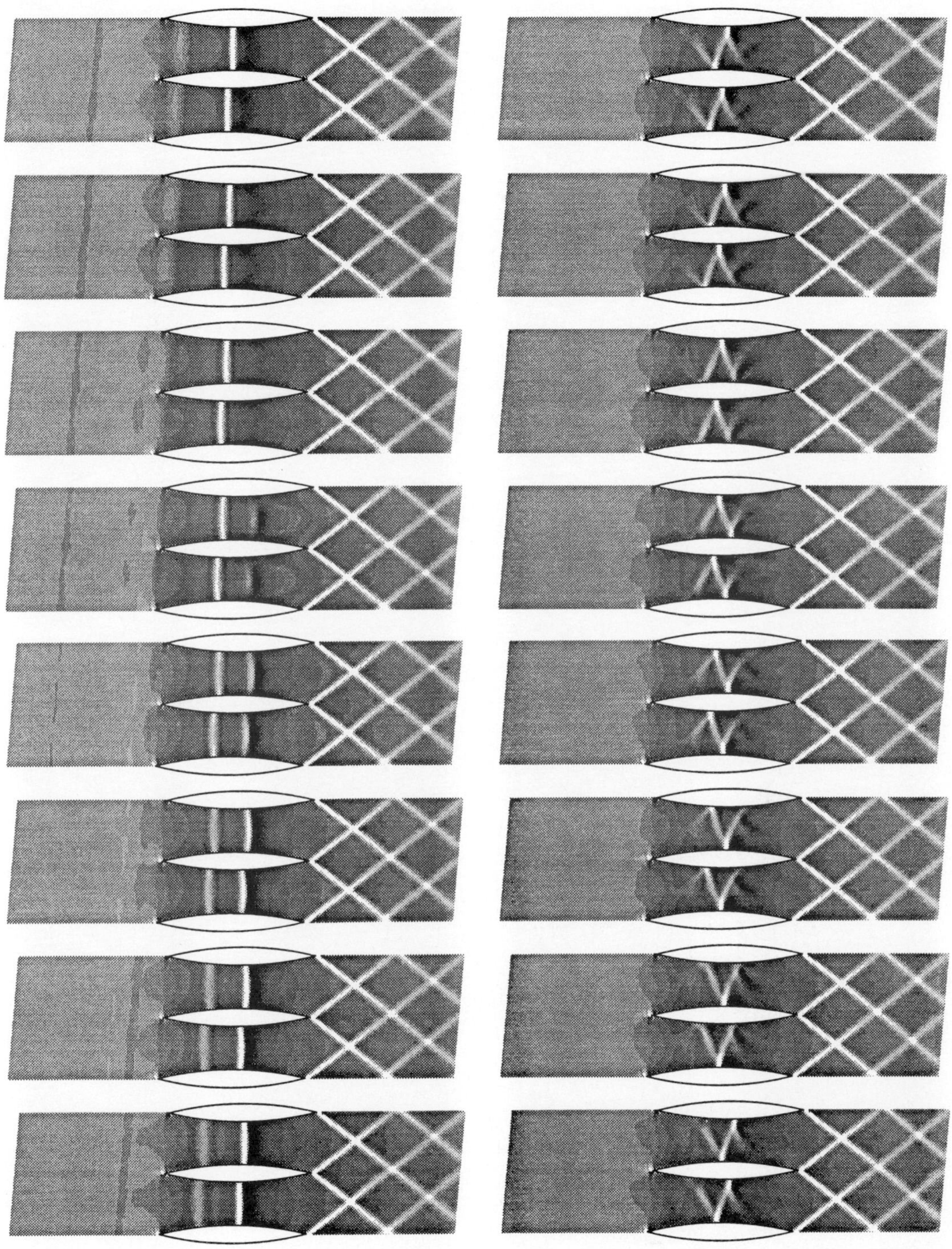

Fig. 12: One cycle of the 'symmetric' and unsymmetric flow oscillations in the staggered cascade A1A, $\beta_s = 95^o$, for identical reservoir conditions $T_{01} = 295\ K$, $p_{01} = 1\ bar$, $\Phi_0 = 100\ \%$, numerically simulated schlieren pictures; left: 'symmetric' $f = 866\ Hz$, right: unsymmetric $f = 1181\ Hz$; flow from left to right, time increases from top.

$(p-p_{ad.})/p_{01}$
0
0.1
0.2

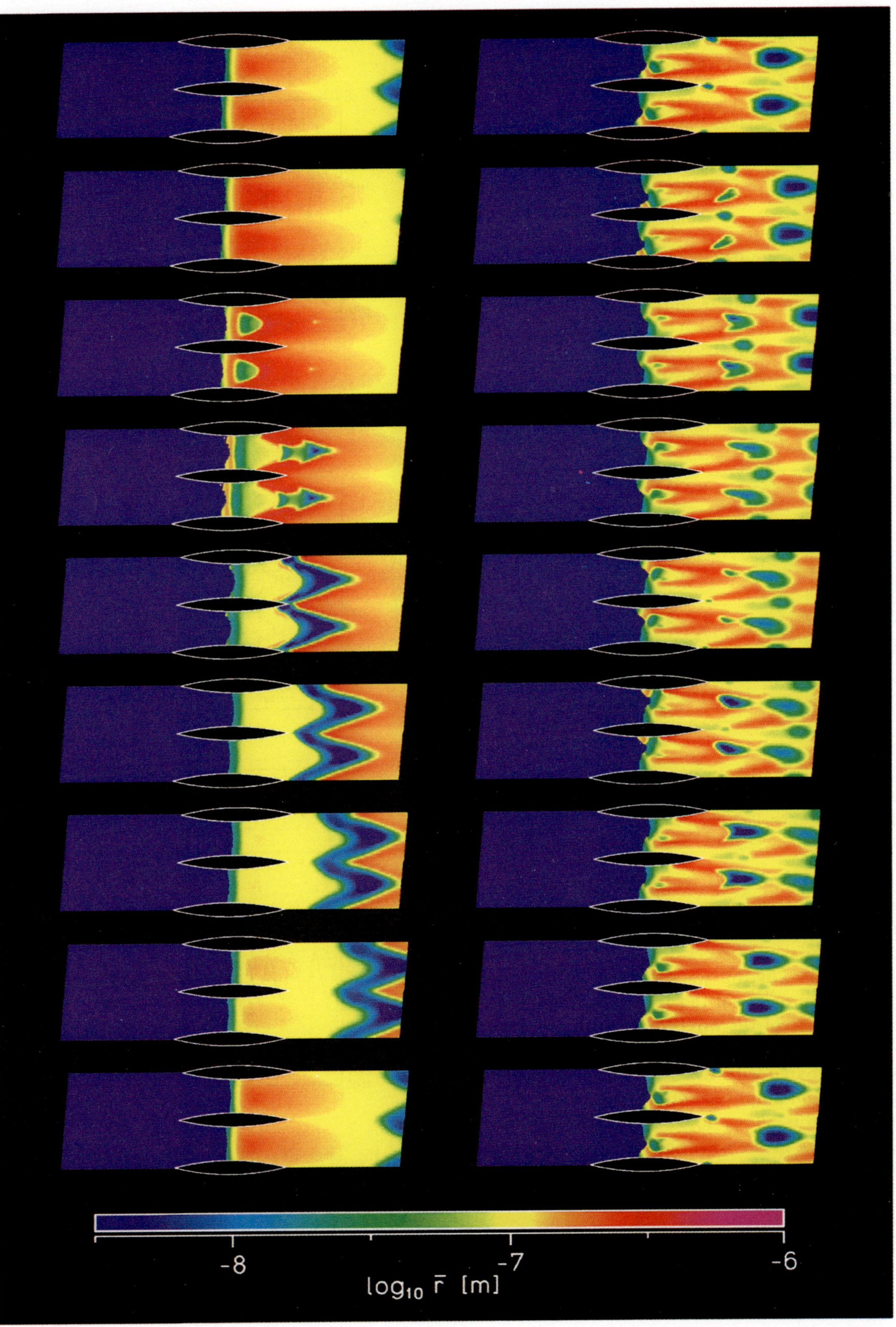

-8
-7
-6
log₁₀ r̄ [m]

Fig. 13: (colored pages) One cycle of the 'symmetric' and unsymmetric flow oscillations in the staggered cascade A1A, $\beta_s = 95°$, for identical reservoir conditions $T_{01} = 295\ K, p_{01} = 1\ bar, \Phi_0 = 100\ \%$; left: 'symmetric' $f = 866\ Hz$, right: unsymmetric $f = 1181\ Hz$; flow from left to right, time increases from top;

$(p - p_{ad})/p_{01}$ –   static pressure disturbance with respect to adiabatic flow,

$\log_{10}\bar{r}$ –   averaged droplet radius.

the unsymmetric oscillation with an additionally upward and downward moving oblique shock system (the time starts from top). In both cases the shocks are always formed in the supersonic region and they move and decay in the subsonic oncoming flow. Switching from the 'symmetric' to the unsymmetric mode, the frequency increases from 866 Hz to 1181 Hz (Fig. 14).

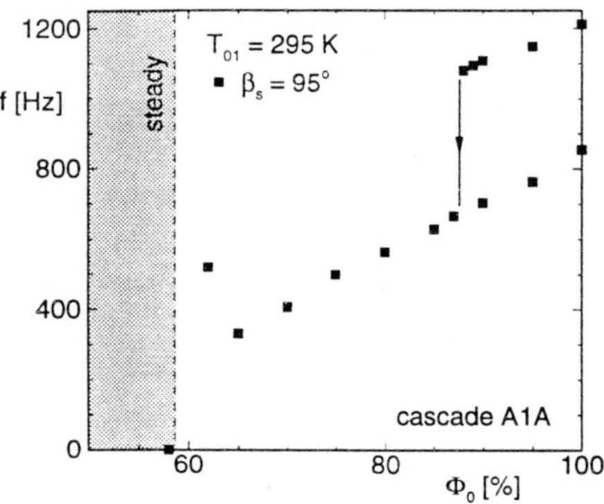

Fig. 14: Bifurcation – frequency dependence on the water vapor pressure ratio Φ_0 in the staggered cascade A1A, $\beta_s = 95°$, for constant values of reservoir temperature $T_{01} = 295\,K$ and pressure $p_{01} = 1\,bar$.

As the main result we can conclude, the existence of bifurcations is not prevented by periodic boundary conditions, i.e. in axial cascades. Figure 15 shows interesting details of the two different modes. On top we see the time dependent static pressure signal (normalized with the reservoir pressure of the mixture $p_{01} = 1$ bar) on the blade near the throat position, indicated by the dot. Compared with the 'symmetric' mode at this position, the amplitude of the unsymmetric pressure waves is about 40 % higher. How does the liquid phase in the wake change depending on the oscillation dynamics? Because we use the so called surface averaged droplet radius, our actual model cannot resolve individual droplet radius spectra. However, the area averaged condensate mass frac-

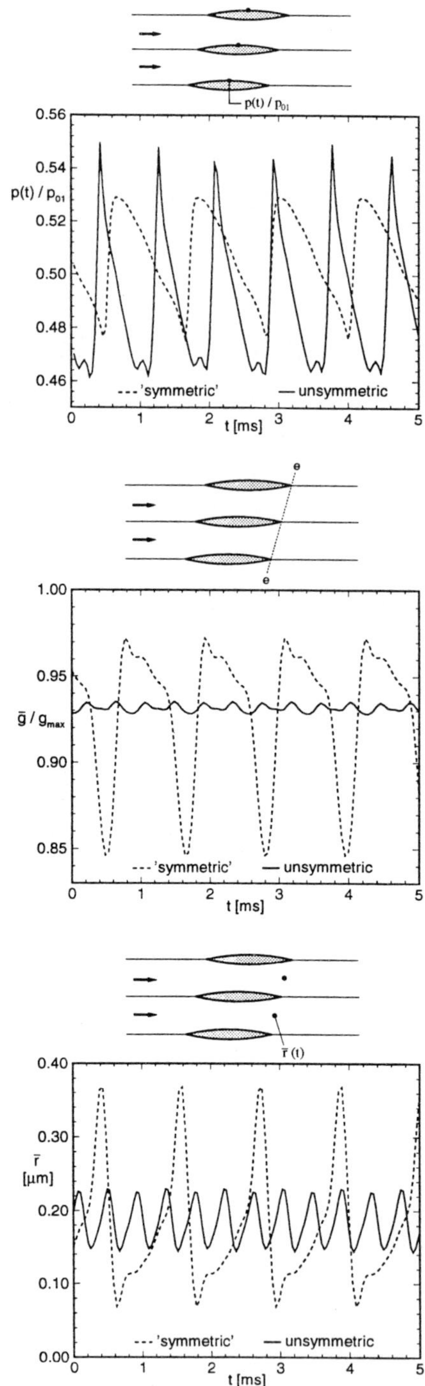

Fig. 15: Time dependent pressure ratio p/p_{01} (top), mean condensate mass fraction $\overline{g}/g_{max}$ (middle) and mean droplet radius $\overline{r}$ (bottom) at the indicated positions for the 'symmetric' and unsymmetric oscillations in the staggered cascade A1A corresponding to Fig. 4.

682

tion $\overline{g}/g_{max}$, calculated directly at the exit plane $\overline{e-e}$ of the cascade, shows interesting and quantitative important effects. The significant difference of the variation of $\overline{g}/g_{max}$, is caused by the different shock structures and shock strength of these modes. The strong variation of $\overline{g}/g_{max}$ within the 'symmetric' mode is caused by the relative strong normal shocks which instantaneously evaporate the condensate, whereas the weaker oblique shocks do not change the static temperature significantly. Therefore, within the unsymmetric higher frequency oscillation the condensate mass fraction remains approximately constant. Investigating the time variation of the surface averaged droplet radius $\overline{r}$ at the local position shown in Fig. 15 confirms this result. The minimum and maximum droplet radii formed during one period differ significantly. The 'symmetric' mode is more likely to produce large and tiny droplets depending on the phase relationship within one cycle.

4.3 Steady Flow in Low Pressure Steam Turbine Cascade L

For comparison, blading and flow conditions are the same as investigated by White et al.[30]. In addition, we included a systematic variation of the trailing edge geometry which is known to have a great influence on the efficiency (Fig. 16).

Figure 17 shows the discretization and the high resolution grid around the leading edge and in the region where condensation is expected, overall the H-grid consists of 326×40 cells. Figure 18 shows the adiabatic flow (left) and the steady condensing flow with sharp trailing edges. The high accuracy of the code and the high spatial resolution reproduce the local flow field in all details. Typically at sharp trailing edges no shocks appear at the pressure side. The oblique shock at the suction side starts as a strong shock with subsonic flow behind and then it weakens with $M > 1$ downstream of the shock (Fig. 21). It is interesting to note that the trailing edge is embedded in subsonic flow.

If the condensation process is included in the numerical simulation, a nearly normal shock forms in the wake flow. The small local subsonic bubble behind this shock can clearly be seen in Fig. 19 (left) by the frozen iso-Mach lines, together with the nucleation rate J (Fig. 19-middle) and the wetness fraction (Fig. 19-right). The maximum nuclation rate is about $J = 20 \ [m^{-3}s^{-1}]$ and the

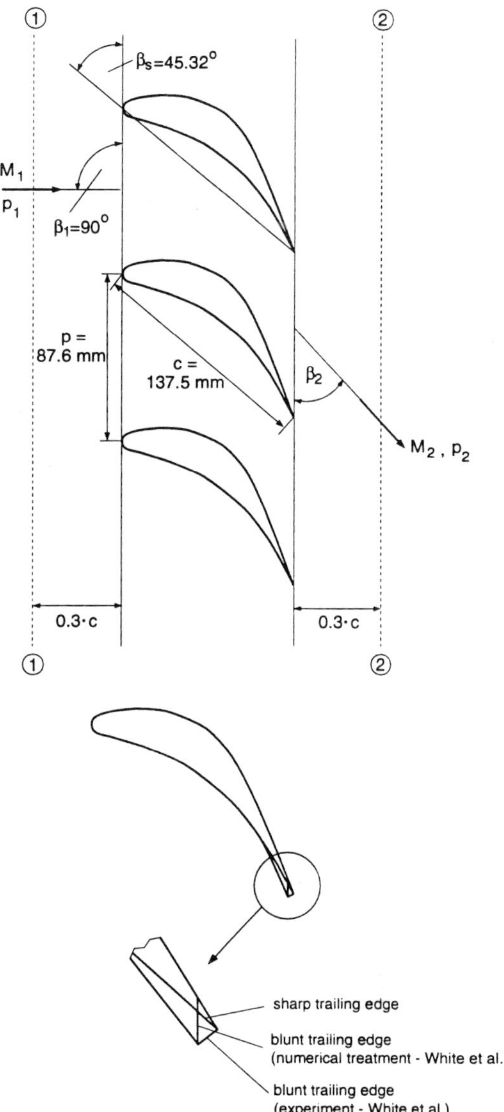

Fig. 16: Geometry and reference frames 1 and 2 of the low pressure steam turbine cascade L; top: definition of the reference frames, bottom: modifications of the trailing edge.

wetness fraction reaches nearly 5 %. The already weakly supercritical configuration of this condensing flow field indicates that only a slight intensification of the heat release is expected to create an unstable situation with moving shocks.

The equivalent numerical schlieren simulation for the blunt trailing edge is shown in Fig. 20. Without condensation (Fig. 20-left) there is no oblique shock reflection at the suction side and

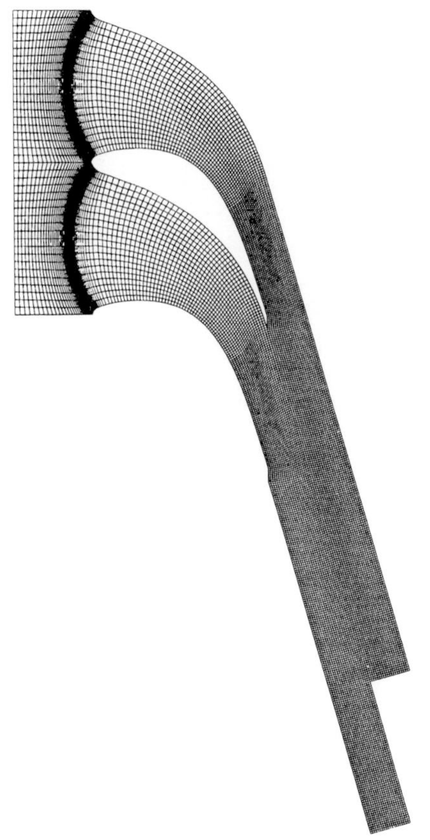

Fig. 17: Computational H-mesh for the LP steam turbine cascade, 326 × 40 cells (for better demonstration two meshes are plotted).

Table III: Flow properties at the reference frames 1 and 2 for the LP steam turbine cascade L, reference frames, reservoir and downstream conditions corresponding to Figs. 16 and 18.

Flow	Trailing edge	$M_{f,1}$	$M_{f,2}$	p_1 [bar]	p_2 [bar]	β_1 [^{o}]	β_2 [^{o}]
adia-batic	sharp	0.19	1.11	0.393	0.161	90	20.5
	blunt	0.18	1.05	0.394	0.165	90	19.0
cond-ensing	sharp	0.19	1.21	0.393	0.153	90	23.5
	blunt	0.18	1.16	0.394	0.157	90	22.0

4.4 Unsteady Flow in Low Preassure Steam Turbine Cascade L

From the existence of steady normal shocks in the cascade L for conditions investigated in the previous chapter, it can be concluded that only a slight increase of the wetness fraction or a small decrease of the inlet temperature T_1 would initiate oscillating instabilities. Because the conditions were given by the experiments of White et al. [30] we replaced the droplet growth law of Gyarmathy by the Hertz-Knudsen law, which is known to overestimate the condensate formation in regions of Knudsen numbers less than unity.

The result is very surprising. The flow immediately becomes unsteady with an unexpected high oscillation frequency of about 3.500 Hz. In addition the time averaged maximum of the wetness fraction is only slightly higher compared with steady flow, which confirms the previous conclusion that the test conditions of White et al. were critical and very near to the limit of stability. Detection and measuring of such high frequencies is a difficult task, usually the frequencies in condensing steam flows in Laval nozzles are about 400 Hz, i.e. one order below. From video reproductions of the unsteady flow in the cascade L another important effect could be detected. The variation of the flow angle β_2 during one cycle is much greater than the difference of $\Delta\beta_2 = 3^o$ of adiabatic and steady condensing flows.

the normal shock in the condensation region (Fig. 20-right) intensifies. At the suction side the supersonic flow accelerates at the blunt edge and reaches a maximum Mach number of $M_f = 1.75$. A strong oblique shock which decelerates to subsonic flow follows. Obviously this trailing edge variation enlarges the local subsonic flow region in the wake (Fig. 21). Figure 22 compares the static pressure distributions over the blade surface for both trailing edge configurations. Once again it shows the local strong acceleration passing through the additional Prandtl-Meyer expansion at the blunt edge. Table III shows the important effect that condensation increases the flow angle β_2 [31]. This means that the work performed by the blade decreases when operating in the two-phase region. It is interesting to note that the quantitative effect of $\Delta\beta_2 \approx 3^o$ is independent of the shape of the trailing edge.

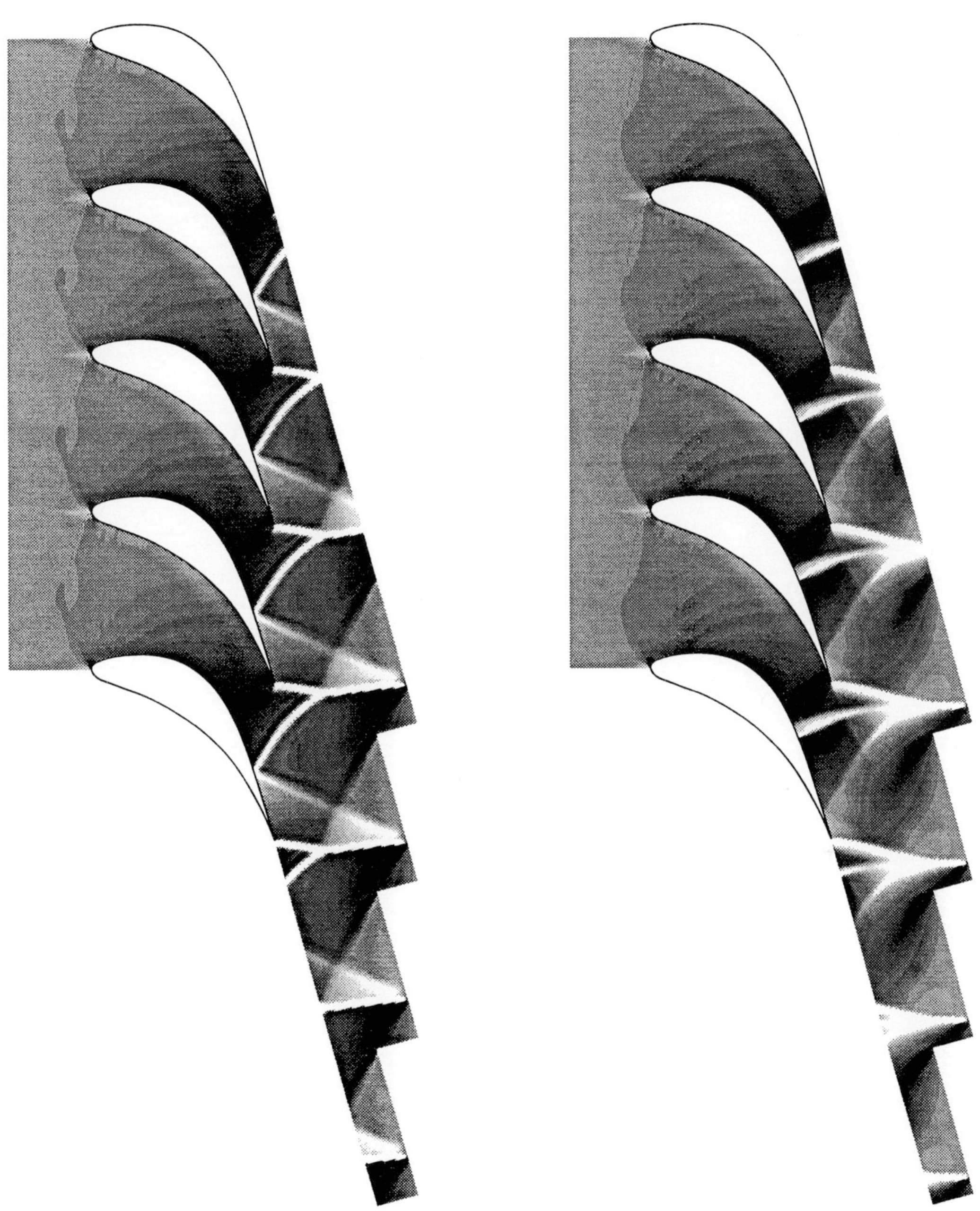

Fig. 18: Steady adiabatic and condensing flow in the LP steam turbine cascade L, sharp trailing edge, reservoir conditions $T_{01} = 354\ K$, $p_{01} = 0.403\ bar$, downstream condition $p_{ex} = 0.163\ bar$; numerically simulated schlieren pictures; definition of the reference frames according to Fig. 16; flow from left to right;

left   :   adiabatic flow (condensation process switched off) $M_1 = 0.19$, $\beta_1 = 90°$, $M_2 = 1.11$, $\beta_2 = 20.5°$,

right   :   condensing flow, $M_{f,1} = 0.19$, $\beta_1 = 90°$, $M_{f,2} = 1.21$, $\beta_2 = 23.5°$.

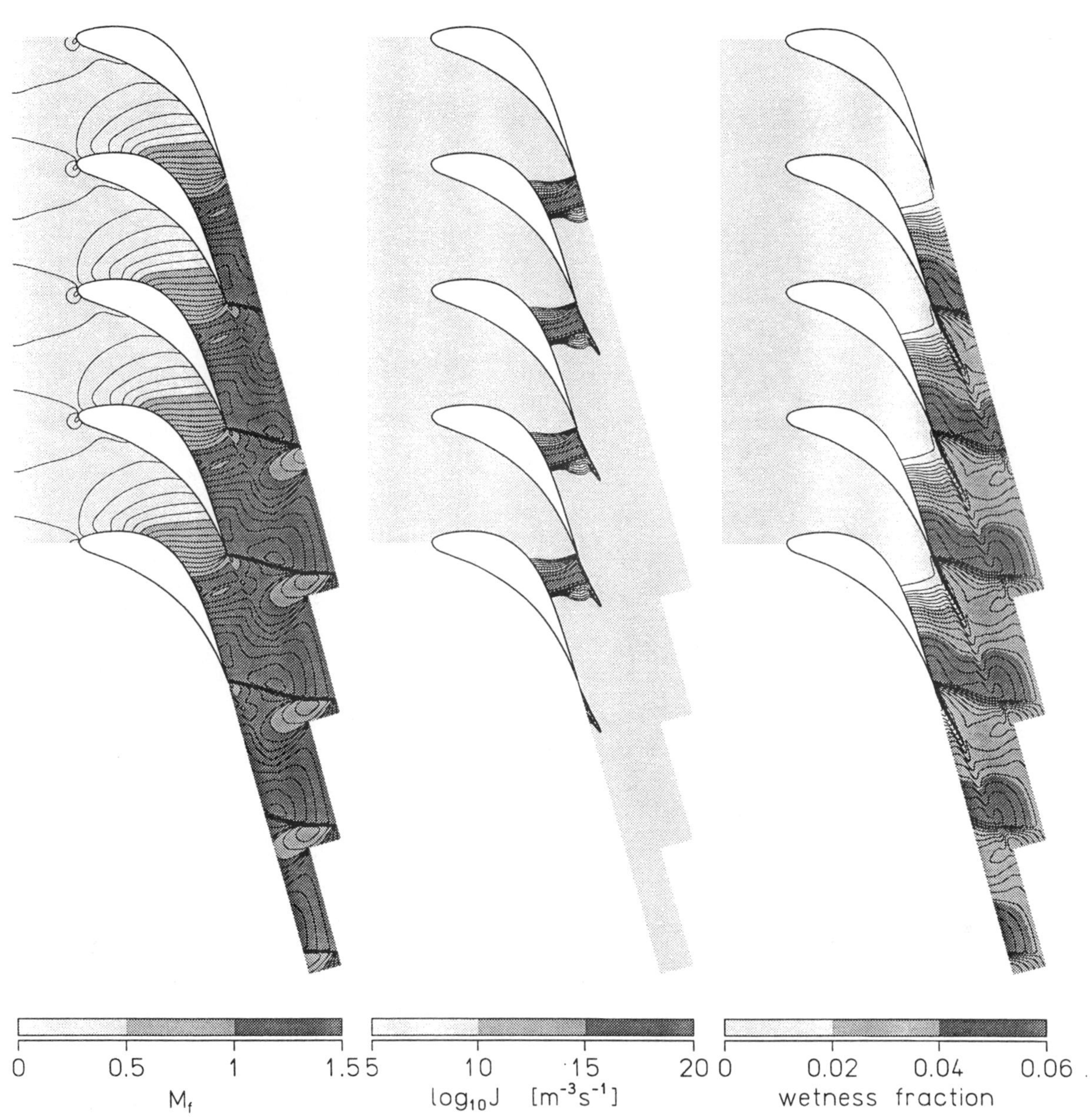

Fig. 19: Steady condensing flow in the LP steam turbine cascade L, sharp trailing edge, reservoir and downstream conditions and values at the reference frames according to Fig. 18; flow from left to right;

left	:	frozen Mach number M_f, increment $\Delta M_f = 0.05$,
middle	:	nucleation rate $\log J$, increment $\Delta(\log J) = 2$,
right	:	wetness fraction g, increment $\Delta g = 0.002$.

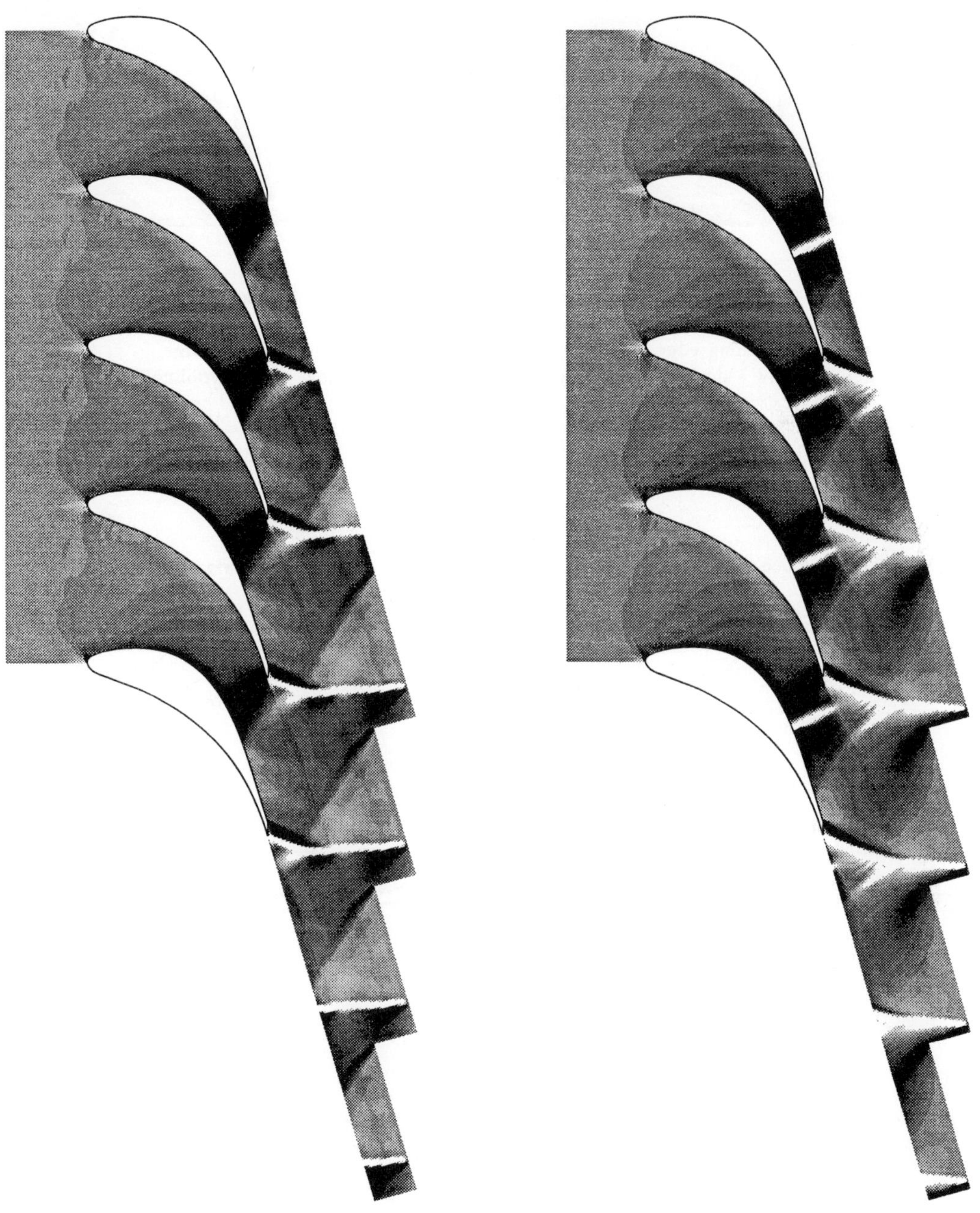

Fig. 20: Steady adiabatic and condensing flow in the LP steam turbine cascade L, blunt trailing edge, reservoir and downstream conditions and definition of the reference frames according to Figs. 16,18; numerically simulated schlieren pictures; flow from left to right;

left   :   adiabatic flow (condensation process switched off), $M_1 = 0.18$, $\beta_1 = 90°$, $M_2 = 1.05$, $\beta_2 = 19.0°$,

right   :   condensing flow, $M_{f,1} = 0.18$, $\beta_1 = 90°$, $M_{f,2} = 1.16$, $\beta_2 = 22.0°$.

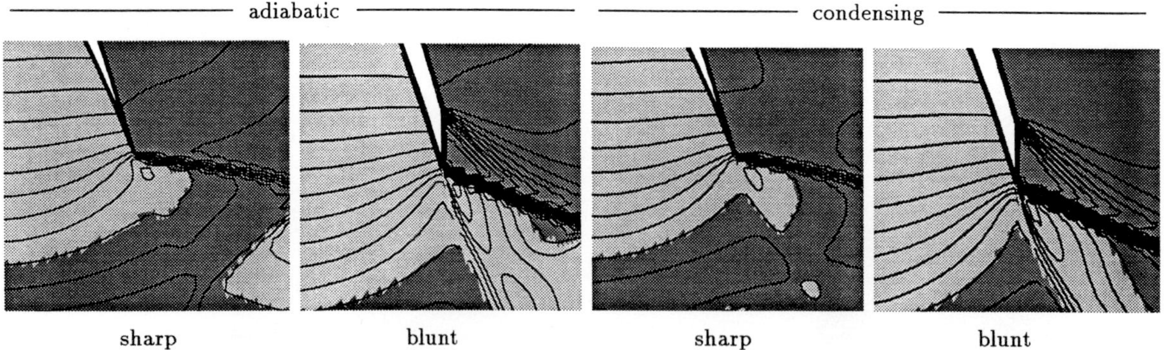

Fig. 21: Clipping of the flow around the trailing edge for the LP steam turbine cascade L; frozen Mach number M_f, light grey $M_f < 1$, dark grey $M_f > 1$, increment $\Delta M_f = 0.05$;
left: adiabatic flow - sharp trailing edge,   middle-left: adiabatic flow - blunt trailing edge,
middle-right: condensing flow - sharp trailing edge,   right: condensing flow - blunt trailing edge.

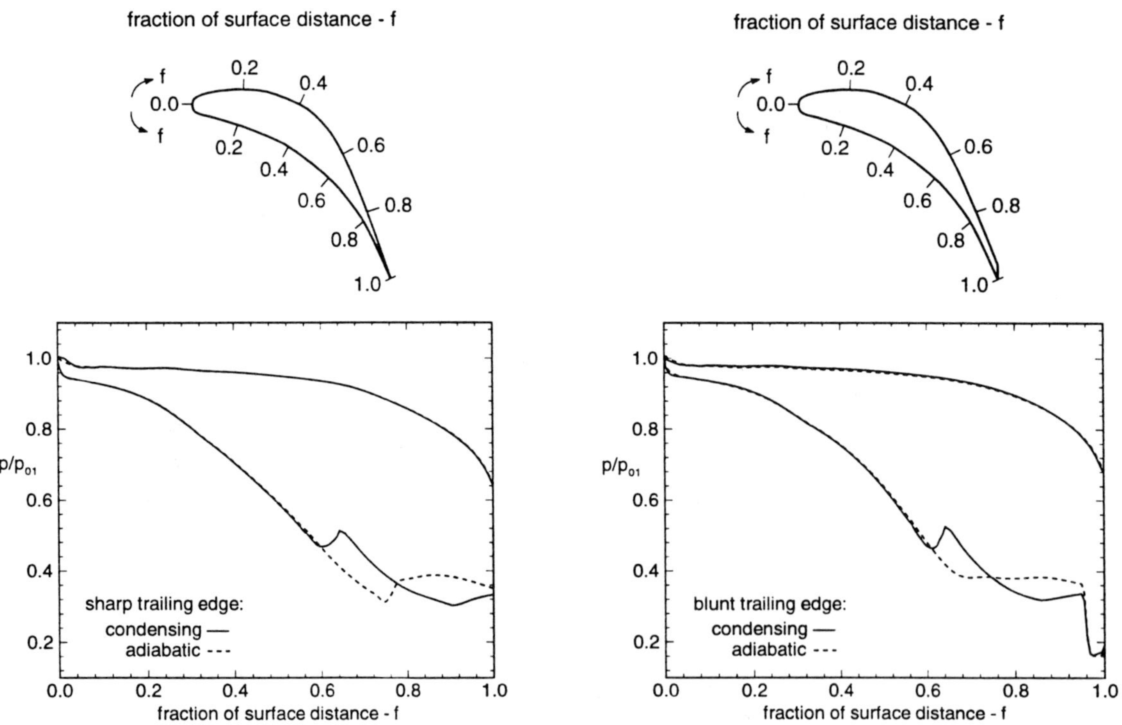

Fig. 22: Adiabatic and condensing blade surface pressure distribution for the LP steam turbine cascade L, different trailing edges; reservoir and downstream conditions according to Fig. 18.

5 CONCLUSIONS

With respect to the fact that about 40 % of failures of LP turbine bladings reported to EPRI/EEI[3] have no identifiable cause, there is a definite need to improve the understanding and analysis of turbine blade flows. Our current work is well suitable to perform this basic research and is, therefore, devoted to investigate excitation mechanisms such as supercritical shocks by condensation and flow instabilities, responsible for the additional dynamic load on the bladings and loss of efficiency. Based on our findings in simple geometries we confirmed that higher order instabilities in condensing flows can develop in axial cascades, too. Consequently, criteria for definition of the limit of stability of condensing steam flow in cascades are required. Furthermore, inclusion of heterogeneous [32] and binary nucleation models [33] is required to reproduce thermodynamics of real steam flow operating in big power plants.

Concerning excitation mechanisms, it is well-known that the interaction of adiabatic moving shocks, e.g. caused by the natural unsteadyness in turbines, with steady condensation regions, may initiate periodic pressure waves of much higher amplitude compared with self-excited oscillations. Finally, extension to efficient 3-D flow calculations is required including Euler [34] and Navier Stokes solvers.

Acknowledgement

The authors would like to express their gratitude to the Deutsche Forschungsgemeinschaft for supporting our research by the contracts Schn 352/13-4,5.

References

[1] Smith, D., Life Optimization is Critical to Extending Power Plant Life, Power Engineering, February, 1991.

[2] Schnerr, G.H., Adam, S., Lanzenberger, K and Schulz, R, Multiphase Flows: Condensation and Cavitation Problems. In: Computational Fluid Dynamics Review 1 (eds. M. Hafez, K. Oshima), John Wiley & Sons, New York, London, 1995.

[3] Dewey, R.P., McCloskey, T.H., Rieger, N.F., Analysis of Steam Turbine Blade Failures in the Utility of Industry, ASME paper 83-JPGC-Pwr-20.

[4] Delale, C.F., Schnerr, G.H., Zierep, J., The Mathematical Theory of Thermal Choking in Nozzle Flows, Z. Angew. Math. Phys. (ZAMP), Vol. 44, 1993, pp. 943-976.

[5] Schmidt, B., Beobachtungen über das Verhalten der durch Wasserdampfkondensation ausgelösten Störungen in einer Überschallwindkanaldüse, Dissertation, Fakultät für Maschinenbau, Universität Karlsruhe (TH), Germany, 1962.

[6] Barschdorff, D., Kurzzeitfeuchtemessung und ihre Anwendung bei Kondensationserscheinungen in Lavaldüsen, Strömungsmechanik und Strömungsmaschinen, 1967, 6, pp. 18-39.

[7] Barschdorff, D., Verlauf der Zustandsgrößen und gasdynamische Zusammenhänge bei der spontanen Kondensation reinen Wasserdampfes in Lavaldüsen, Forschung im Ingenieurwesen, 1971, 37 No. 5, pp. 146-157.

[8] Zierep, J., Lin, S., Ein Ähnlichkeitsgesetz für instationäre Kondensationvorgänge in Lavaldüsen, Forschung im Ingenieurwesen, 1968, 34 No. 4, pp. 97-132.

[9] Barschdorff, D., Filippov, G. A., Analysis of Special Conditions of the Work of Laval Nozzles with Local Heat Supply, Heat Transfer — Soviet Research, 1970, 2, pp. 76–87.

[10] Saltanov, G. A., Tkalenko, R. A. Investigation of Transonic Unsteady-State Flow in the Presence of Phase Transformations, Zh. Prikl. Mek. i Tek. Fiz. (UdSSR), 1975, 6, pp. 42–48.

[11] Collignan, B., Contribution à l'Etude de la Condensation Instationnaire en Ecoulement Transsonique. Ph.D. Thesis, Université Pierre & Marie Curie, Paris, 1994. France.

[12] White, A.J., Young, J.B., A Time Marching Method for the Prediction of Two-Dimensional, Unsteady Flows of Condensing Steam, J. of Propulsion and Power, 1993, 9, pp. 579-587.

[13] Schnerr, G.H., Adam, S., Mundinger, G., Frequency Control of Shock Oscillations in High Speed Two-Phase Flow, In Proc.: Fourth Triennal Intern. Symp. on Fluid Control, Measurement and Visualization (FLUCOME'94), CERT-ONERA, Aug. 29–Sept. 2, 1994, ed. P.Hebrard ENSAE Toulouse France, 1994, Vol. 2, pp. 957–962.

[14] Mundinger, G., Numerische Simulation instationärer Lavaldüsenströmungen mit Energiezufuhr durch homogene Kondensation. Dissertation, Fakultät für Maschinenbau, Universität Karlsruhe (TH), Germany, 1994.

[15] Adam, S., Numerische und experimentelle Untersuchung instationärer Düsenströmungen mit Energiezufuhr durch homogene Kondensation. Dissertation, Fakultät für Maschinenbau, Universität Karlsruhe (TH), Germany, 1996.

[16] Young, J.B., Two-dimensional, Nonequilibrium Wet-Steam Calculations for Nozzles and Turbine Cascades, J. of Turbomachinery, 1992, Vol 114, pp. 569-579.

[17] Bakthar, F., Mahpeykar, M.R., Abbas, K.K., An Investigation of Nucleating Flows of Steam in a Cascade of Turbine Blading - Theoretical Treatment, Transaction of the ASME, 1995, Vol. 117, pp. 138-144.

[18] Ishizaka, K., Ikohagi, T., Daiguji, H., A High-Resolution Numerical Method for Transonic Non-equilibrium Condensation Flows through a Steam Turbine Cascade, In: Proc. of the 6th Int. Symp. on Comp. Fluid. Dynamics (ISCFD), Lake Tahoe, USA, 4.-8. Sept. 1995.

[19] Bakthar, F., So, K.S., A Study of Nucleating Flow of Steam in a Cascade of Supersonic Blading by the Time-marching Method, Int. J. Heat and Fluid Flow, 1991, Vol. 12, No. 1, pp. 54-62.

[20] Adam, S., Schnerr, G.H., Instabilities and Bifurcation of Nonequilibrium Two-phase Flows, J. Fluid Mech, October, 1997.

[21] Gyarmathy, G., Two-phase Steam Flow in Turbines and Seperators, eds. Moore, M.J. and Sieverding, C.H., Hemisphere, 1976, capt.3, pp. 57-82.

[22] Young, J.B., The Condensation and Evaporation of Liquid Droplets in a Pure Vapour at Arbitrary Knudsen Number, Int. J. Heat Mass Transfer 34, 1991, pp. 1649-1661.

[23] Young, J.B., The Condensation and Evaporation of Liquid Droplets at Arbitrary Knudsen Number in the Presence of an Inert Gas, Int. J. Heat Mass Transfer 36, 1993, pp. 2941-2956.

[24] Peters, F., Meyer, K.A.J., Measurement and Interpretation of Growth of Monodispersed Water Droplets Suspended in Pure Vapour, Int. J. Heat Mass Transfer 38, 1995, pp. 3285-3293.

[25] Oran, E.S., Boris, J.P., Numerical Simulation of Reactive Flow, Elsevier, Amsterdam, 1987.

[26] Shuen, J.S., Liou, M.S., van Leer, B., Inviscid Flux-Splitting Algorithms for Real Gases with Nonequilibrium Chemistry, J. Comp. Phys. 90, 1990, pp. 371-395.

[27] Anderson, W.K., Thomas, J.L., van Leer, B., Comparison of Finite Volume Flux Vector Splittings for the Euler Equations, AIAA Journal 24, 1986, pp. 1453-1460.

[28] Schnerr, G.H., Adam, S., Mundinger, G., New Modes of Periodic Shock Formation in Compressible Two-phase Flows. In Proc.: IUTAM Symposium Waves in Liquid/Gas and Liquid/Vapor Two-phase Systems. Kyoto, May 9–13, 1994, eds. S. Morioka, L. van Wijngaarden, Kluwer Academic Publishers 1995, pp. 377–386.

[29] Adam, S., Schnerr, G.H. Similarity Considerations of Oscillatory Instabilities in Two-phase Flows. In: Proc. Japanese-German Symposium on Multiphase Flows, Tokyo, September 25-27, 1997, (eds. Saito, T., Müller, U.), 1997.

[30] White, A.J., Young, J.B., Walters, P.T., Experimental Validation of Condensing Flow Theory for a Stationary Cascade of Steam Turbine Blades, Phil. Trans. R. Soc. London A 354, 1996, pp. 59-88.

[31] Young, J.B., Snoeck, J., Aerothermodynamics of Low Pressure Steam Turbines and Condensers, eds. Moore, M.J. and Sieverding, C.H., Springer, 1987, capt. 4, pp. 87-133.

[32] Chirikhin, A.V., Numerical Study of Nonequilibrium Heterogeneous - Homogeneous Condensation of Streams in Supersonic Nozzles, J. Fluid Dynamics, Vol. 12, No. 1, 1977, pp. 114-120.

[33] Kalikmanov, V.I., van Dongen, M.E.H., Quasi-One-Component Theory of Homogeneous Binary Nucleation, Physical Review E, Vol. 51, Iss 5, 1995, pp. 4391-4399.

[34] Liberson, A., Kosolapov, Y., Rieger, N., Hesler, S., Calculation of 3-D Condensing Flows in Nozzles and Turbine Stages, EPRI Nucleation Workshop, Rochester, New York, October 24-26, 1995.

ADVANCED COMPUTATIONAL TECHNIQUES FOR DETAILED ANALYSIS OF FLOWS OVER FIXED AND ROTARY WING GEOMETRIES

Lakshmi N. SANKAR [1]   Mert E. BERKMAN [2]   Nathan HARIHARAN [1]

(1) School of Aerospace Engineering, Georgia Institute of Technology;
Atlanta, GA 30332-0150

(2) CFD Research Corporation, Huntsville, AL 35805

Abstract

First-principles based techniques for the prediction of fixed and rotary wing wake geometry are described. It is demonstrated that fifth order accuracy schemes do substantially better than third order spatial accuracy schemes in capturing the details of the vortex core structure. It is demonstrated that the use of embedded grids can further enhance the resolution of the tip vortex, particularly if the boundary conditions and the order of interpolation accuracy are carefully maintained to be fifth order. A hybrid approach where the costly Navier-Stokes analysis is confined to small viscous regions near the blade surface is also described. Sample applications of these methods to the vortex wake behind a fixed wing, and the lift and surface pressure distributions over various rotors are presented.

1 Introduction

The strong wake shed from rotary wings interacts with the blades, the fuselage and in some cases with the tail rotor and causes performance and noise problems for helicopters. Particularly, in hover and low speed forward flight the wake remains in the vicinity of the vehicle and dominates rotor acoustics and vibratory loads. Therefore, in any method attempting to solve rotor flows emphasis is given to the treatment of the wake, the inflow it produces and its effects on blade loading.

The earliest methods for modeling rotors were based on an extension of Prandtl's lifting line theory for wings. In these techniques, the individual blades were modeled as line vortices, and the wake was modeled as a deformed helix. During 1970s and 1980's, these methods were augmented by modern CFD techniques. Caradonna and Isom applied the transonic small disturbance theory to lifting rotors [1]. Chang [2] modified the full potential flow solver FLO22 for isolated wings to model rotors. Egolf and Sparks [3] modified Chang's work by embedding the vortex element associated with the tip vortex with the potential flow field. Their approaches solved either the steady or quasi-steady form of the potential flow equation. Sankar et al [4,5], Strawn [6], Bridgeman et al. [7], Strawn and Caradonna [8] developed unsteady full potential flow based rotor solvers. Ramachandran et al. [9,10] solved the full potential equation and included the rotor wake effects using a Lagrangean based approach

for tracking the vortex filaments.

During the late 1980's, Euler methods matured to a point where calculation of the rotor flow field in hover and forward flight was feasible. These methods solved the mass, momentum and energy conservation equations in a time dependent fashion using finite-difference or finite-volume methods. These solvers did not include viscous effects but could analyze the transonic flow with non-isentropic shocks. Sankar et al. [11], Agarwal and Deese [12], and Hassan et al. [13] developed Euler solvers for isolated rotors. Again, the wake effects were modeled as an inflow angle of attack table supplied from a separate comprehensive analysis. In Ref. 14, Wake and Sankar developed a Navier-Stokes code to analyze rotor flow fields in hover and forward flight.

During the 1980s, a new class of Euler/Navier-Stokes codes were developed in an effort to capture the rotor wake from first principles without any need for external wake models. Removing the need for external information that depends on rotor geometry is a big step in the true simulation of the rotor flow field. These first-principles based solvers are particularly useful in analyzing new or complex rotor blades where no experimental data are available. Strawn and Barth [15], Srinivasan and McCroskey [16], Srinivasan et al. [17-19], Duque [20-21] all solved the hovering rotor flow fields by capturing the rotor wake in an Eulerian fashion, and from first principles. Hariharan [22] developed high order accuracy schemes for capturing rotor wakes. Bangalore [23] studied the use of high lift devices such as slats to enhance the maneuverability of rotorcraft.

Received on July 1, 1997.

Recently, a new class of methods has become available, which combine the simplicity of Lagrangean wake methods for capturing trailing vortices, the efficiency of potential flow methods and the accuracy of Navier-Stokes methods. These methods are discussed in references [24-27].

Scope of the Present Work

The present work is organized as follows. A compressible Navier-Stokes solver that is second or third order accurate in time, and fifth order accurate in space is briefly described. An overset grid approach is next described, which allows sharp gradients in the flow field such as shocks and vortices to be captured on a locally embedded fine grid. Next, a hybrid formulation, where the time consuming calculations are confined only to small regions surrounding the rotor is discussed.

A number of results are presented for rotors in hover and in forward flight, and for a wing in forward flight. The section thrust and surface pressure distributions, and the velocity field in and around the tip vortices are presented and compared with measurements.

Because the mathematical formulation behind this work is well developed, and has been extensively documented [22-23, 28], the present formulation is very briefly discussed. The emphasis of this work is on the predictive capabilities of the approaches described here.

Mathematical and Numerical Formulation

<u>Navier-Stokes Analysis:</u>
The three-dimensional compressible Navier-Stokes equations are solved in an integral form in the present study. These equations may be formally written as:

$$\frac{\partial}{\partial \tau} \iiint_V q \cdot dV + \oiint_S \left[\overline{F} - q \cdot \overline{V}_G \right] \bullet \overline{n} dS = \iint_S \overline{R} \bullet \overline{n} dS$$

(1)

Here, the symbol V refers to the control volume on which the governing equations are applied in the above integral form. The symbol τ represents time; The quantities F and G are inviscid and viscous fluxes evaluated at the boundaries S of the control volume. Finally, V_G is the velocity of the grid surfaces in an inertial coordinate system. All the motions of the blade such as pitching, flapping and the rotations about

the rotor shaft enter the calculations through this term. This is in contrast to calculations in a rotational frame, where the centrifugal, angular acceleration and Coriolis terms are explicitly represented.

A fifth order accurate essentially non-oscillatory (ENO) scheme is used in the present work to compute the flow properties q to the left and right side of each of the six cell face S. An adaptive stencil is used, which uses only the smoothest part of the flow properties information in the cell, and its neighbors.

The semi-discrete form of the equations may be written as:

$$\frac{d(qV)}{dt} = R$$

(2)

Here the symbol V represents the cell volume. At any time step 'n+1', this equation is solved using an implicit iterative scheme of the form:

$$\left[\frac{1}{\Delta t} - \frac{\partial R}{\partial (qV)} \right] \Delta(qV) = \left[R - \frac{\partial (qV)}{\partial t} \right]^{n+1}$$

(3)

where the objective of the iteration is to drive $\Delta(qV)$ to zero, thereby satisfying Equation (2). The left side matrix is approximately factored into several smaller tridiagonal factors facilitating the inversion. Of course, when the iterations converge, the factorization errors disappear. If only a single iteration is done, then the present scheme reduces to a classical non-iterative ADI scheme.

It may also be noted that matrix Equation (3) may be solved using multigrid techniques. The left side operator as well as the right side "Residual" may be injected onto coarser grids and then inverted. This multigrid approach has been used in some wing-alone calculations, but were not used in this study.

<u>Hybrid Navier-Stokes/Full Potential Approach:</u>
In this approach, the above Navier-Stokes calculations are done only in a small region surrounding the rotor blade. At node points away from the viscous region, the following simplified form of the governing equations is used:

$$\frac{\partial}{\partial \tau} \iiint \rho dV + \oiint \rho \left[\overline{V} - \overline{V}_G \right] \bullet \overline{n} dS = 0$$

(4)

The flow velocity $\overline{V}$ is made up of three components: free stream velocity, disturbance potential based velocity, and induced velocity due to embedded vortices:

$$\overline{V} = \overline{V}_\infty + \overline{\nabla}\phi + \overline{q}_v \tag{5}$$

The density is assumed to be related to velocity potential through isentropic gas law, neglecting temporal variation of the induced velocity, $\overline{q}_v$.

A 3-D unsteady compressible potential flow solver developed by Sankar et al. [28] is used to solve Equation (4) along with isentropic energy equation. It uses a three factor ADI scheme to invert the matrix system for ϕ. The potential flow solver can capture embedded weak shock waves, and can model the convection of acoustic waves in a time accurate manner.

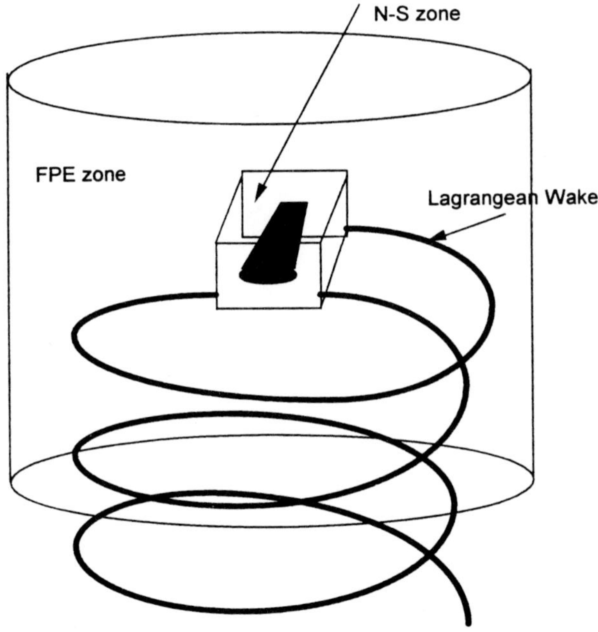

Figure 1. Flow Field Partitioning in Hybrid Method

For a lifting rotor, the presence of the rotor wake leads to pockets of rotational flow in the vicinity of the tip vortices trailing from the rotor. In the Navier-Stokes zone the wake is captured as a part of the solution. However, the potential solver cannot capture embedded vortex filaments such as embedded tip vortices. A mechanism must be therefore provided in the flow solver to model this effect.

A Lagrangean wake model has been developed to account for the wake induced rotational flow in the potential zone. The inboard vortex wake is neglected once it leaves the Navier-Stokes zone, a concentrated tip vortex is used to model far wake effects. The helical tip vortices up to 10 revolutions are modeled by a connected series of straight line elements. These vortical elements leaving the Navier-Stokes zone are tracked with markers. The vortical induced velocity field associated with these elements is found by applying the Biot-Savart law:

$$\overline{q}_v = \sum_{elements} \frac{\Gamma}{4\pi} \frac{\overline{dl} \times \overline{r}}{\left|r^3\right|} \tag{6}$$

Here $\overline{r}$ is the distance between the vortex element of length $\overline{dl}$ and the point where the induced velocity, $\overline{q}_v$ is calculated. The induced velocity field is basically a function of two parameters: the wake shape and the tip vortex strength, Γ. A classical wake or a prescribed wake shape is used to start the solution process and the wake shape is allowed change by letting wake markers move with the local flow speed. The strength of the tip vortex is taken to be the peak bound circulation and updated from one time level to the next as the loading on the blade changes.

Such a wake treatment enables carrying of the tip vortex for several revolutions without any diffusion. Figure 1 illustrates flow field decomposition and the modules involved in the technique.

<u>Overset Grid Scheme:</u>

In rotary wing in forward flight applications, it is difficult to know the vortex trajectory a priori. As a result, the grid in the vicinity of the vortices is not likely to be adequately clustered. In other instances, for example in rotor airframe interaction applications, a single body-fitted grid around the rotor and the airframe can not be generated. For these reasons, an overset capability has been implemented in the present method. In this approach, independent curvilinear refined grids are placed on top of existing grids. The base grid and the overset grid will not coincide as in conventional patched grids. Thus, information between the base grid and the overset grid must be exchanged by three-dimensional interpolation. Efficient 3-D interpolation techniques and search procedures that identify holes and fringe points have been developed.

Results and Discussion

In this section, a number of results are presented to demonstrate the capabilities of the present approach.

<u>Fixed Wing Tip Vortex Generation Process:</u>
A NACA0015 wing tested by McAlister et al. [29], has been studied. These experimental data are quite extensive including velocity profiles across the cross-section of the tip vortex at various stations downstream of the wing. A C-grid consisting of 121 points in the streamwise direction, 25 points in the spanwise direction and 31 points in the normal direction was constructed. The angle of attack of the wing was set at 12 degrees and the free stream Mach number was 0.18.

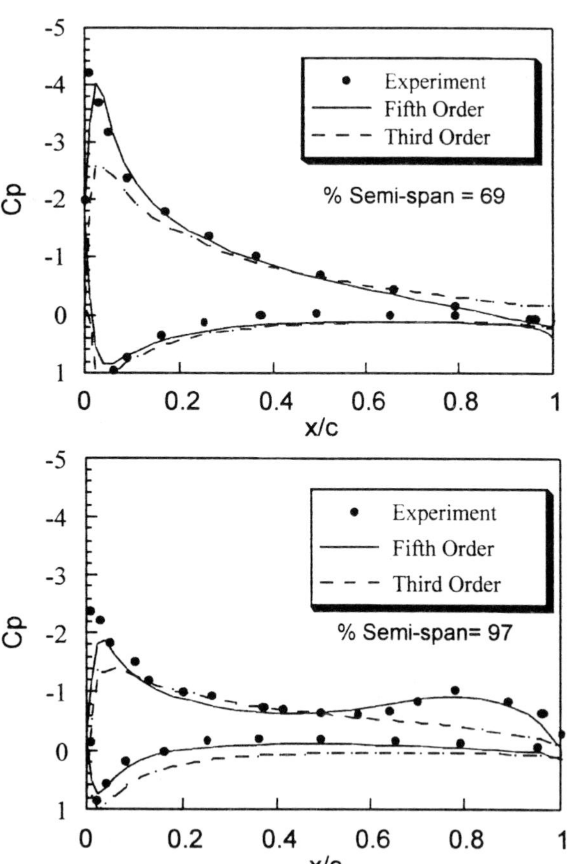

Figure 2. Surface Pressures over a Rectangular Wing Tested by McAlister, α=12°, M∞= 0.18.

Figure 2 presents the surface pressure distribution comparison between the two schemes, at

two spanwise stations. The solution computed by the fifth order scheme is found to be superior at all the stations, comparing better with the experimental values. At the inboard stations the suction peak is picked up better by the fifth order scheme, and the overall agreement with experiment is better. Very close to the tip (97% semi-span), the difference between the third and the fifth order scheme is even more marked. The fifth order scheme correctly predicts the rear bump in the suction side, caused by the formation of the tip vortex and the subsequent roll-up over the wing upper surface.

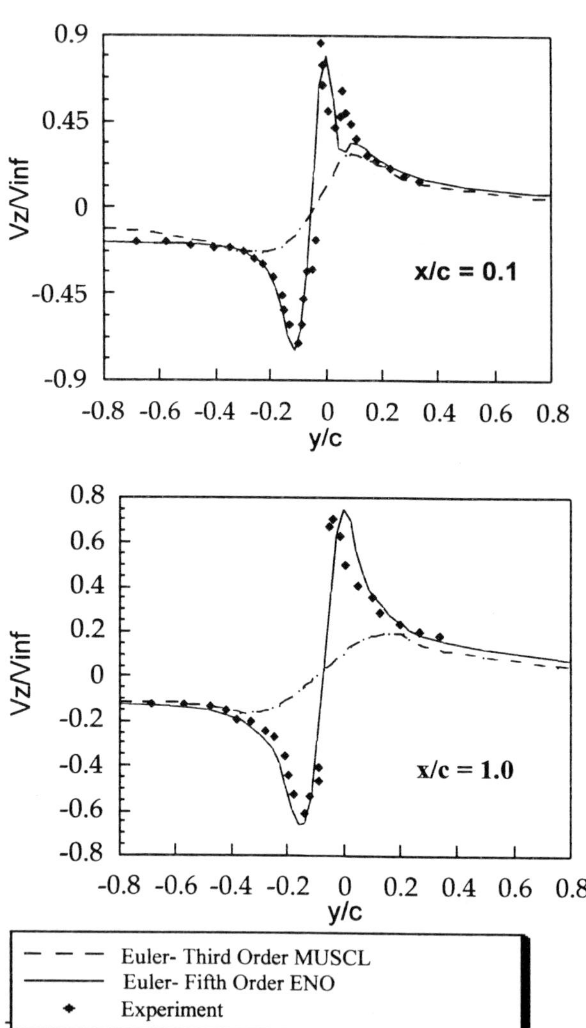

Figure 3. Variation of the Component of Velocity normal to the Wing Planform across the Vortex Core, Rectangular Wing

To study of tip vortex evolution in detail, the inviscid calculations were repeated with 40 points in the spanwise direction, retaining the same number of points in the other two directions. Figure 3 shows the velocity (V_z) profile normal to the wing plane across the tip vortex core at two streamwise locations behind the trailing edge. The fifth order solution is again found to be superior to the third order solution. Immediately behind the trailing edge (x/c =0.1, 0.2, 1.0) the fifth order solution agrees very well with the experiments.

Around x/c > 2 the velocity peaks captured by the fifth order scheme starts to diminish when compared to experiments. The grid stretching in the streamwise direction becomes too large to produce the exact peak.

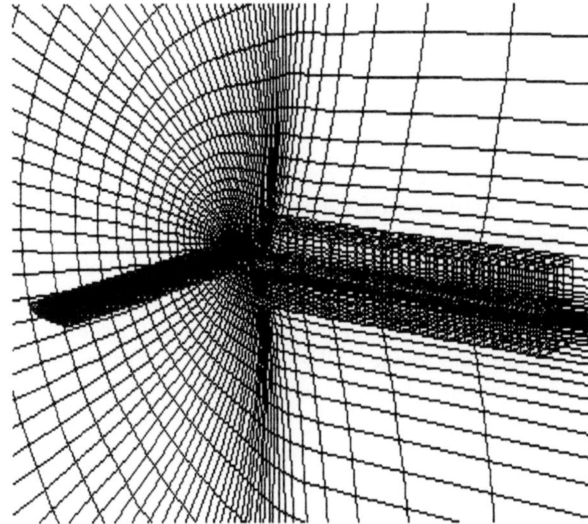

Figure 4. Embedded Grid for Improved Resolution of the Tip Vortex

An embedded grid shown on Figure 4 removes this difficulty. The normal velocity field was accurately computed up to six chords away from the rotor as shown in Figure 5. The fifth order scheme with an embedded grid was also able to capture the axial component of velocity inside the vortex core, as shown in Figure 6.

<u>Rotary Wing Analysis- Forward Flight</u>
The full Navier-Stokes analysis has been applied to wings in hover and forward flight. Sample results from the Ph. D. Dissertation of Bangalore [23] for a UH-60A rotor in forward flight is given in Figure 7. These calculations involving four blades required 1.2 Million grid points, and were done using a distributed computing strategy. Although the surface pressure distribution over most of the rotor is in good agreement, the integrated loads shown in Figure 8 do

not agree with the measurements well. Considerable additional work is needed in the areas of tip vortex modeling and 3-D dynamic stall modeling to improve agreement between the analyses and the measurements.

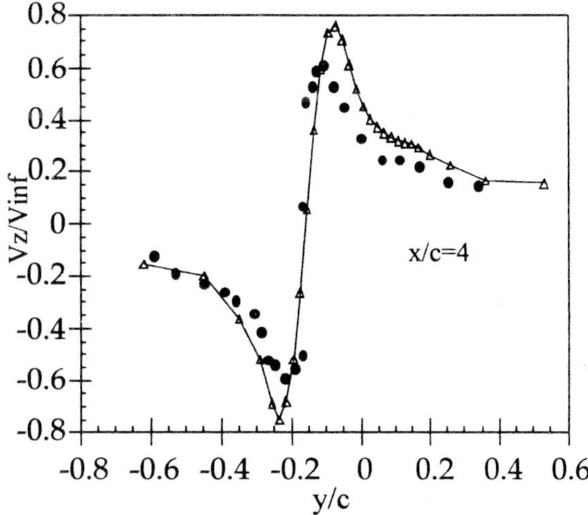

Figure 5. Variation of Normal Component of Velocity inside the Vortex Core with the Embedded Grid, Rectangular Wing

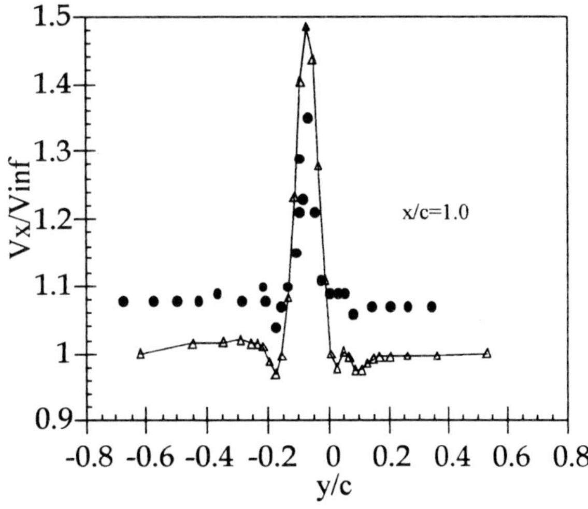

Figure 6. Variation of Axial Velocity within the Vortex Core with the Embedded Grid, Rectangular Wing

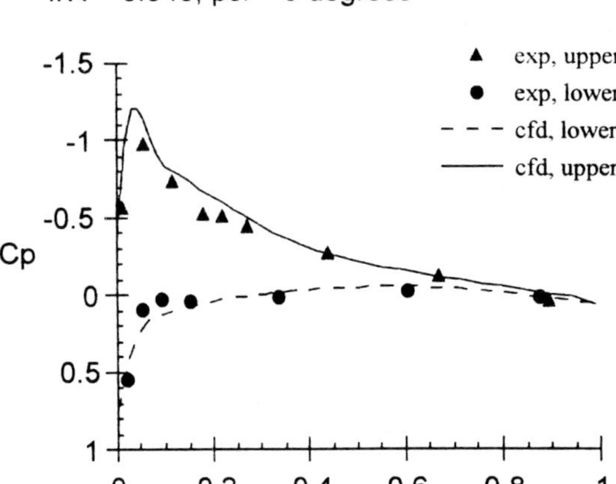

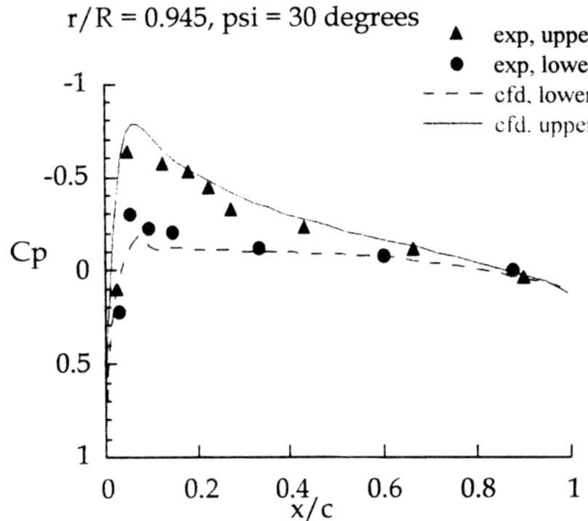

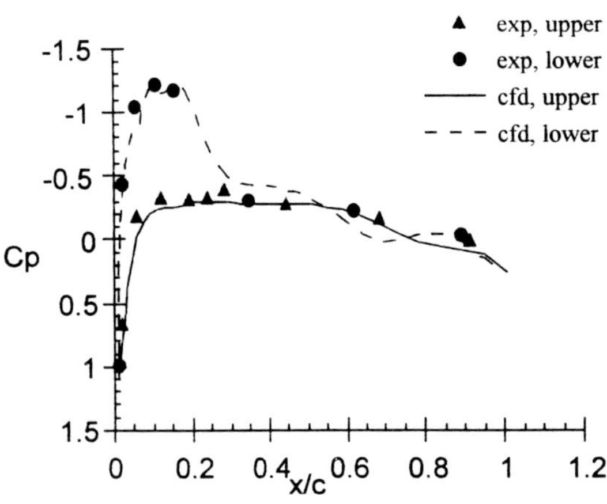

Figure 7: Surface Pressure Distribution, 95 % Radial Station at Selected Azimuthal Locations, M_{tip} = 0.628, advance ratio = 0.3, UH-60A Rotor

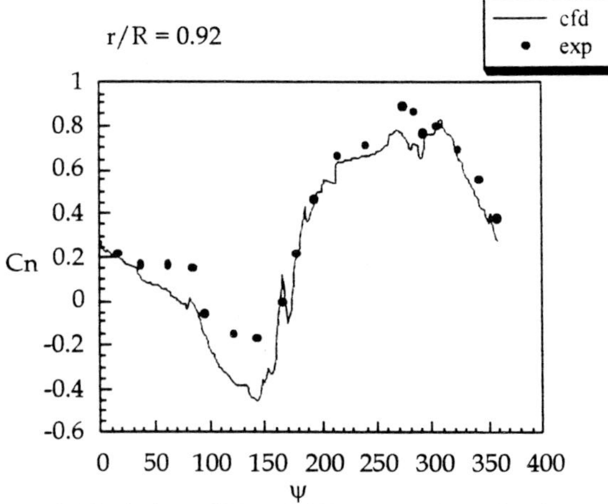

Figure 8. Variation of Normal Forces at 92% for the UH-60A Rotor, M_{tip} = 0.628, advance ratio = 0.3

Hybrid Analysis:

As discussed earlier, the CPU time for first principles based analysis can be reduced by a factor of 2 or more if most of the flow field is modeled using potential flow techniques. The resulting hybrid solver has been validated for a typical current generation rotor, the four bladed UH-60A rotor in hover. The blade planform has an aspect ratio of 15.3 and a maximum twist of 13°. The blade has a rearward sweep of 20° starting from a rotor radius of 93%. The blade is made up of two airfoil sections, SC1095 as the main airfoil and SC1095R8 section in the midspan [30]. Figure 9 shows the computational grid for UH-60A blade where the smaller Navier-Stokes is distinguished with darker grid lines. A H-O grid with 90 azimuthal, 43 radial, and 80 normal nodes are used for this case. The blade is discretized by 40 chordwise and 27 radial grid points. The Navier-Stokes zone extends half blade chord upstream and downstream of the blade. Typically only 40% of the grid nodes lie in the Navier-Stokes zone.

Figure 10 shows the radial variation of section thrust coefficient. Figure 11 gives surface pressures at four radial stations. The results are in excellent agreement with the experimental measurements. In these simulations the blade has 10° collective pitch and a tip Mach number of 0.628. In these studies, the wake marker locations were iteratively adjusted until the wake filaments were force free, as in conventional free wake analyses. These calculations required only 50% of CPU time required by a full blown Navier-Stokes analysis over the same grid.

Figure 9. Computational Grid for UH-60A Rotor, Hybrid Method

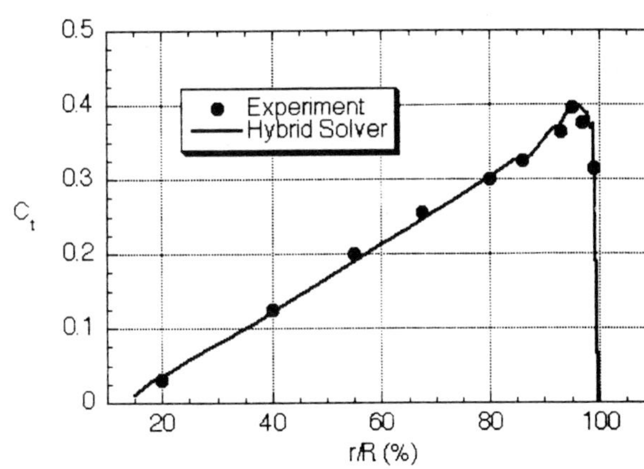

Figure 10. C_t over UH-60A Rotor in Hover, $M_{tip} = 0.628$, 10° Collective Pitch; Hybrid Method

The hybrid solver is also applicable to rotors in forward flight. Figure 12 shows the pressure contours around a two bladed rotor in non-lifting high speed forward flight. The blade has a NACA 0012 section and an aspect ratio of 7. It is seen that the pressure contours pass smoothly across the Navier-Stokes/Full Potential zone interfaces without any false reflections.

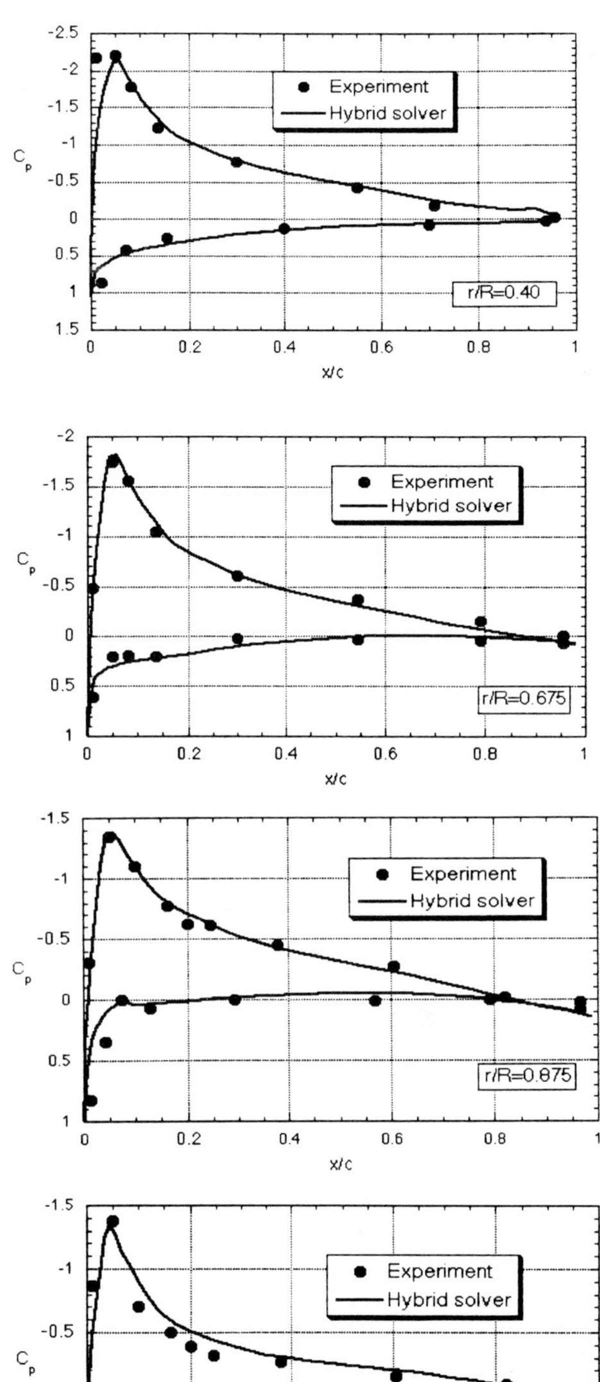

Figure 11. Surface Cp over UH-60A Rotor in hover; $M_{tip} = 0.628$, 10° Collective Pitch; Hybrid Method.

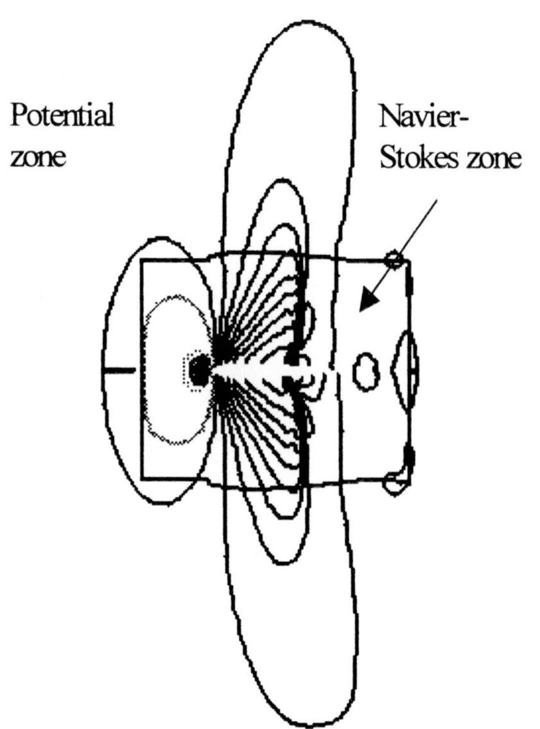

Figure 12. Pressure Contours, Non-lifting Rotor, M_{tip} = 0.8, advance ratio = 0.2; Hybrid Method,

The Navier-Stokes/Full Potential code is compared against the H-34 flight test data [Ref. 31]. The rotor has four blades with NACA 0012 sections. The blade has a linear twist of -8° and an aspect ratio of 20.48. The helicopter has a forward speed of 23 knots and advance ratio is 0.064. The blade pitching and flapping motion is defined by

$$\theta_{root} = \theta_c + \theta_{1c} \cos\psi + \theta_{1s} \sin\psi$$
$$\beta = \beta_0 + \beta_{1c} \cos\psi + \beta_{1s} \sin\psi$$

$$(7)$$

Higher order blade harmonics are neglected. Here ψ is the azimuthal position of the blade, and the pitching and flapping coefficients for the particular case are θ_c = 12.3°, θ_{1c} = 2.2°, θ_{1s} = -0.35°, and β_0 = 3.6°, B_{1c} = 0.01°, B_{1s} = -0.34°.

The spanwise lift distribution at 3 azimuthal stations are given in Figure 13. The lift is predicted well at ψ=0°. At the advancing position the lift is overestimated except near the tip. On the retreating side, the lift is predicted accurately up to 85% blade.

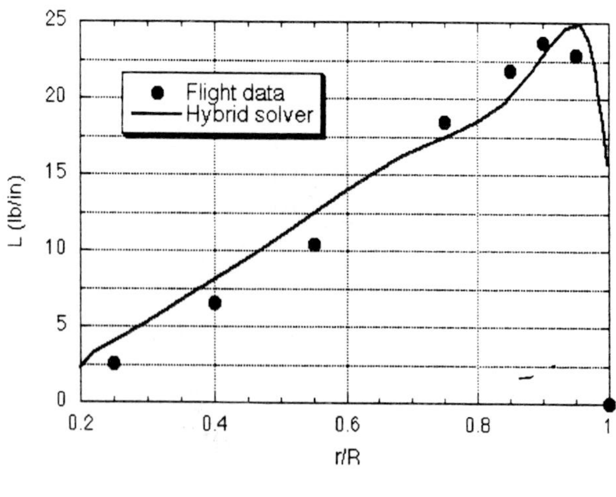

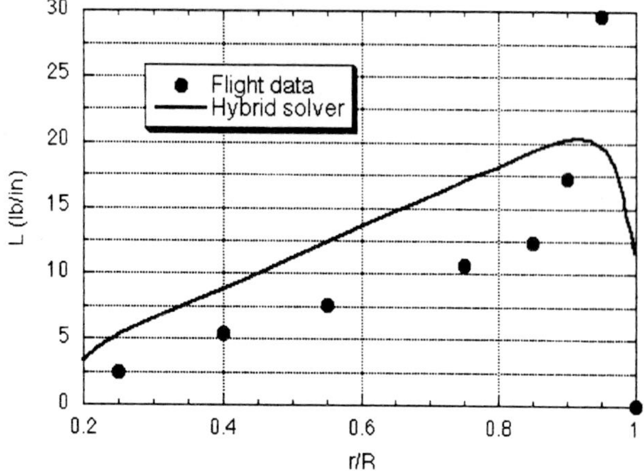

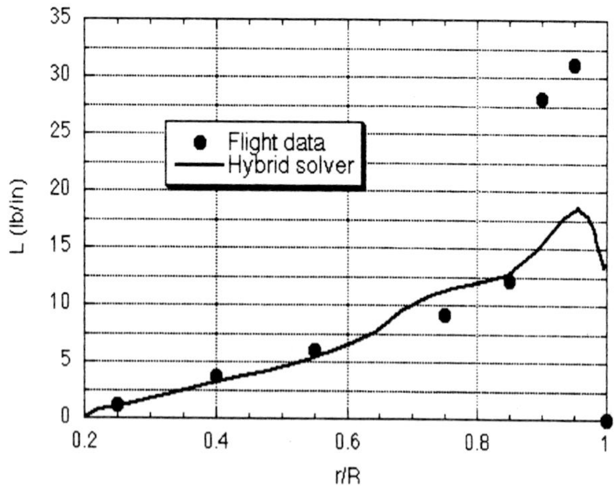

Figure 13. Lift Variation along the Blade at Backward, Advancing and Retreating Blade Positions, $\psi = 0°, 90°, 270°$; H-34 rotor, advance ratio = 0.064; Hybrid Method

Concluding Remarks

Advanced CFD techniques for detailed flow field analyses of fixed and rotary wings have been developed. It is shown that fifth order accuracy schemes are more effective in capturing the vortex core structure. Embedded grids have been shown to enhance the resolution of the tip vortex. A hybrid technique has been shown to predict accurate surface loads for isolated rotor flows. Such a technique reduces the computational time significantly and eliminates the tip vortex diffusion problem in the far field.

Acknowledgments

This work was supported by the National Rotorcraft Technology Center and by the U. S. Army Research Office under their Rotorcraft Center of Excellence program, and by the National Rotorcraft Technology Center. The second author was supported in part by a contract from Sikorsky Aircraft Corporation.

References

1. Caradonna, F. X. and Isom, M. P., "Subsonic and Transonic Potential Flow over Helicopter Rotor Blades, AIAA Journal, No. 12, Dec 1972, pp. 1606-1612.

2. Chang, I. C., " Transonic Flow Analysis for Rotors, " NASA TP 2375, July 1984.

3. Egolf, T. A. and Sparks, S. P., " A Full Potential Rotor Analysis with Wake Influence using an Inner-Outer Domain Technique, " 42nd Annual AHS Forum, June 1986.

4. Sankar, L. N. and Prichard, D., " Solution of Transonic Flow Past Rotor Blades using the Conservative Full Potential Equation, " AIAA Paper 85-5012, October 1985.

5. Sankar, L. N., Malone, J. B. and Tassa, Y., " A Strongly Implicit Procedure for Steady Three-dimensional Transonic Potential Flow, " AIAA Journal, Vol. 20, No. 5, 1982.

6. Strawn, R.C., " Numerical Modeling of Rotor Flows with a Conservative form of the Full Potential Equation, " AIAA Paper 86-0079, July 1986.

7. Bridgeman, J. O., Strawn, R. C. and Caradonna, F. X., " An Entropy and Viscosity Corrected Potential Method for Rotor Performance prediction, " 44th Annual AHS Forum, June 1988.

8. Strawn, R. C. and Caradonna, F. X., "Conservative full potential model for unsteady transonic rotor flows, " AIAA Journal, Vol. 25, February 1987, pp. 193-198.

9. Ramachandran, K., Tung, C. and Caradonna, F. X., "Rotor hover performance prediction using a free-wake computational fluid dynamics method, " Journal of Aircraft, Vol. 26, December 1989, pp. 1105-1110.

10. Ramachandran, K., Moffitt, R. C., Owen, S. J. and Caradonna, F. X., " Hover Performance Prediction using CFD, " 50th Annual AHS Forum, June 1994.

11. Sankar, L.N., Wake, B.E. and Lekoudis, S.G., "Solution of the Unsteady Euler Equations for Fixed and Rotor Wing Configurations," Journal of Aircraft, Vol. 23, No.4, April 1986, pp283-289.

12. Agarwal, R. K. and Deese, J. E., " An Euler Solver for Calculating the Flow Field of a Helicopter Rotor in Hover and Forward Flight, " AIAA 19th Fluid Dynamics, Plasma Dynamics and Laser Conference, June 1987.

13. Hassan, A. A., Tung, C. and Sankar, L. N., " An Assessment of Full Potential and Euler Solutions for Self generated Blade-Vortex Interactions, " 46th Annual AHS Forum, May 1990.

14. Wake, B.E. and Sankar, L.N., "Solution of Navier-Stokes Equations for the Flow over a Rotor Blade", AHS Journal, April 1989.

15. Strawn, R.J and Barth, J.T "A Finite-Volume Euler Solver for Computing Rotary-Wing Aerodynamics on Unstructured Meshes," Proceedings of the 48th Annual AHS Forum, Washington D.C, June 1992.

16. Srinivasan, G. R. and McCroskey, W. J., "Navier-Stokes Calculations of Hovering Rotor Flow fields, " Journal of Aircraft, Vol. 25, No. 10, October 1988, pp. 865-874.

17. Srinivasan, G.R., Baeder, J. D., Obayashi, S. and McCroskey, W. J., "Flow field of a Lifting Rotor in Hover : A Navier-Stokes Simulation, " AIAA Journal, Vol. 30, No. 10, October 1992.

18. Srinivasan, G.R and Baeder, J.D "TURNS: A free wake Euler/Navier-Stokes Numerical Method for Helicopter Rotors", AIAA Journal, Volume 31, Number 5 May 1993.

19. Srinivasan, G.R, Raghavan, V., Duque, E.P.N and McCroskey, W.J "Flow field analysis of Modern Helicopter rotors in Hover by Navier-Stokes Method", presented at the AHS International Technical Specialists meeting on Rotorcraft Acoustics and Rotor Fluid dynamics, Oct 1991, Philadelphia, PA.

20. Duque, E. P. N., " A Numerical Analysis of the British Experimental Rotor Program Blade, " 45th Annual AHS Forum, Boston, MA, May 1989.

21. Duque, E. P. N. and Srinivasan, G. R., "Numerical Simulation of a Hovering Rotor using Embedded

Grids, " 48th Annual AHS Forum, Washington D.C., June 1992.

22. Hariharan, N., Sankar, L. N., "Higher Order Numerical Simulation of Rotor Flow Field," AHS Forum and Technology Display, Washington, DC., May 1994.

23. Bangalore, A., Computational Fluid Dynamic Studies of High Lift Rotor Systems Using Distributed Computing, " Ph. D Thesis, Georgia Institute of Technology, Atlanta, GA, May 1994.

24. Mello, O. A. F., "An Improved Hybrid Navier-Stokes/Full Potential Method for Computation of Unsteady Compressible Flows, " Ph. D Thesis, Georgia Institute of Technology, Atlanta, GA, November 1994.

25. Berezin, C. R. and Sankar, L. N., "An Improved Navier-Stokes/Full Potential Coupled Analysis for Rotors, " <u>Mathematical Computational Modeling,</u> Vol. 19, No. 3/4, 1994, pp. 125-133.

26. Moulton, M. A., Hafez, M. M. and Caradonna, F. X., "Zonal Procedure for Predicting the Hovering Performance of a Helicopter, " <u>ASME Journal,</u> Vol. 184, 1994.

27. Berkman, M.E., Sankar, L. N., Berezin, C. R. and Torok, M. S., "A Navier-Stokes/Full Potential/Free Wake Method for Advancing Multi-Bladed Rotors," Proceedings of 53rd Annual forum of the AHS, VA, May 1997.

28. Sankar, L., Malone, J., Tassa, Y., "An Implicit, Conservative Algorithm for Steady and Unsteady Transonic Potential Flows," Proceedings of the AIAA 5th Computational Fluid Dynamics Conference, 1981.

29. McAlister, K. W., Takahashi, R. K., "NACA 0015 Wing Pressure and Trailing Vortex Measurements," USAAVSCOM TR-91-A-003, 1991.

30. Lorber, P. F., Statuter, R. C., Landgrebe, A. J. "A Comprehensive Hover Test of the Airloads and airflow of an Extensively Instrumented Model Helicopter Rotor," Proceedings of 45th Annual Forum of the AHS, MA, May 1989.

31. Scheiman, J., "A Tabulation of Helicopter Rotor-Blade Differential Pressures, Stresses and Motions as Measured in Flight," NASA TM X-952, Mar. 1964.

DEVELOPMENT AND APPLICATIONS OF AN UNSTRUCTURED GRID–BASED CFD SOLVER

S. V. RAMAKRISHNAN   Barna BIHARI   Kuo-Yen SZEMA

Vijaya SHANKAR   Dale OTA

Rockwell Science Center, Thousand Oaks, CA 91360

Abstract

The role of computational simulation in the design of advanced aerospace and commercial products is becoming prominent with advances in software technology (geometry modeling and grid generation, solution algorithms and visualization tools), as well as in computer hardware with increasingly powerful workstations and parallel platforms. While the computational technology has advanced to allow detailed simulations of very complex physical processes employing millions of grid cells running on parallel computing architectures, one of the pacing issues in the advancement of this technology for routine use as an engineering design tool is to make it cost–effective and quick–turnaround to meet the demands of shorter design cycle time with reduced budgets. This paper describes some of the recent advances at Rockwell in the development and application of a quick–turnaround, unstructured grid-based parallel computational technology addressing the bottlenecks in various stages of simulation.

Introduction

The development of next generation advanced aerospace systems and commercial products such as automobiles, will demand significantly improved overall performance and reduced life–cycle costs while facing drastic reduction in allocated budget and time for successful design–to–market. This requires vast improvements to be made to the entire design process from the conceptual design to the final design. Over the past thirty years, computational simulations have gradually transitioned from a research technology stage to the current role of serving as engineering tools supporting the design process. To be resposive to the demanding design schedules and performance requirements, the development, implementation and dissemination of computational tools must become efficient and cost-effective at all phases of the computational process starting from a CAD geometry model to the final solution. This paper addresses some of the computational issues related to pre– and postprocessing, implementation of physical models, accuracy of algorithms for unstructured meshes, and emerging computing environments (workstation cluster and massively parallel). The goal is to not only mature the computational technology for quick–turnaround

single discipline analysis/design problems, but also to advance the technology to address multidisciplinary analysis/optimization that supports the concurrent engineering design process.

In spite of the great strides made in maturing the CFD technology, gaps still exist. Some of them are 1) robustness of the process involved in going from a CAD geometry to surface and volume grid set up, 2) Accuracy and reliability of algorithms and their sensitivity to grid geometry, 3) ability to properly characterize complex physical processes such as turbulence, transition, high angle of attack vortex and separated flows, and 4) effective use of parallel computing power. However, opportunities exist for improvements in all these areas. We need to continue to invest in the development of robust algorithms that best represent the physics, understand how to achieve better accuracy even in stretched meshes to minimize the problem size in terms of execution and memeory requirements, create better and faster geometry/grid set up and repair tools, develop efficient parallel code architectures with improved domain decomposition and load balancing procedures for effective use of workstation clusters and parallel platforms.

Quick–Turnaround Metrics

The design of high performance aerospace platforms will greatly benefit from quick–turnaround simulation

Received on .

tools that can accurately model the complex physical processes that must be integrated together in the design process. Development of a computational environment that can provide such quick–turnaround simulations must address all aspects of the computational process listed here:

- **Pre–Processing** deals with the generation of surface and volume cells starting from a raw CAD geometry. In general, the initial CAD representation has a number of surface imperfections such as patch overlaps, missing patches and twisted patches that need to be fixed before setting up the surface and volume grids for the solver stage. Achieving proper grid resolution on the surface and in the volume to represent the physics properly is critical to obtain meaningful and accurate solutions. The preprocessing stage that goes from a CAD drawing to a computational mesh is a major bottleneck in the simulation process and usually consumes nearly 55% of total labor time. Automated procedures for CAD repair and grid generation can lead to substantial reduction in the total time for final solution.

- **Processing** involves running the code on a single node or parallel nodes (massively parallel or workstation clusters). Some of the issues are 1) implementing advanced algorithms and physics (material characterization, impedance layers, resistive cards,$\cdots$) addressing memory and execution efficiency and accuracy, 2) grid adaptation to solution, and 3) load balancing and scalable performance on parallel platforms.

- **Postprocessing** involves the visualization and animation of solutions. Some of the issues are 1) database management , 2) 3–D visualization, and 3) efficient data transfer to other disciplines (aeroloads to structures) for multidisciplinary optimization.

- **Validation** is critical for maturation of computational technology. Development of quick–turnaround methodologies must not overlook the effort involved in extensive validation of physics and numerics implemented in the computational process.

The Rockwell approach to developing a quick–turnaround computational addresses all stages of the computational effort. The central theme is a unification of methods that covers the following: 1) Unification of algorithms addressing high order accurate space and time discretization; 2) unification of gridding that allows both unstructured and structured or hybrid grid arrangements with dynamic adaptation to solution; 3) parallel computing including workstation clusters; and 4) unification of disciplines that allows the same computational framework to handle CFD, CEM, structures, acoustics, heat transfer and other disciplines.

Some of the metrics in measuring the effectiveness of the quick–turnaround process are 1) the time involved in going from a CAD geometry representation to a first complete solution, 2) the number of complete solution that can be performed in a day supporting design studies, 3) uncertainty reduction in the computational simulation, and 4) the speed of computation. For example, in the late '80s, it took about six months or more to go from a raw CAD geometry file to a first Euler solution employing a multizonal structured grid framework. With advancements in CAD geometry tools, unstructured grid–based efficient algorithms and parallel computing, the time to perform the same process has been reduced to a few days. Another quick–turnaround measure is the number of CFD solutions that can be performed in a day once the first solution is established. Today, we can routinely perform ten or more complete Euler solutions in a day to support the design of aerospace vehicles using workstation clusters or parallel platforms. This quick–turnaround computational framework is currently playing a critical role in the design of many aerospace configurations (reusable launch vehicles, Shuttle upgrades, linear aerospike engine, hypersonic weapons, Airborne Laser, etc.). Uncertainty management is a critical part of risk assessment in the design process. One of the goals is to reduce the level of uncertainty in the computational simulation from the current 5% level to within 1% level by year 2000. To achieve this, some of the steps to be taken are 1) improvements to the solution algorithm to allow better physics while reducing errors in computation, and 2) designing experimental data collection for code validation purposes.

Solution Procedure

Time-dependent, compressible Navier-Stokes equations in conservation-law form may be written as:

$$\mathbf{q}_t + \nabla \cdot \mathbf{F}(\mathbf{q}) = 0, \qquad (2.1a)$$

on $(\mathbf{x}, t) \in D \times (0, \infty), D \subset \mathbf{R}^3$, with initial conditions

$$\mathbf{q}(\mathbf{x}, 0) = \mathbf{q}_0(\mathbf{x}). \qquad (2.1b)$$

and appropriate boundary conditions. Here we denote $\mathbf{x} = (x, y, z)$, $\mathbf{q} = \{q_i\}_{i=1}^5 = (e, \rho, \rho u, \rho v, \rho w)$, and

$$\mathbf{F}(\mathbf{q}, \nabla \mathbf{q}) = [\mathbf{f_i}(\mathbf{q}) + \mathbf{f_v}(\mathbf{q}, \nabla \mathbf{q})]\hat{i} + [\mathbf{g_i} + \mathbf{g_v}]\hat{j} + [\mathbf{h_i} + \mathbf{h_v}]\hat{k} \qquad (2.1c)$$

where each component of the flux vector $\mathbf{F}$, namely $\mathbf{f}$, $\mathbf{g}$, and $\mathbf{h}$, has a convective (inviscid) and a diffusive (viscous) component: For completeness, their definitions are included below:

$$\mathbf{f_i} = [(e+p)u, \rho u, p + \rho u^2, \rho uv, \rho uw]^T, \qquad (2.1d)$$

$$\mathbf{g_i} = [(e+p)v, \rho v, \rho uv, p + \rho v^2, \rho vw]^T, \qquad (2.1e)$$

$$\mathbf{h_i} = [(e+p)w, \rho w, \rho uw, \rho uw, p + \rho w^2]^T. \qquad (2.1f)$$

and

$$\mathbf{f_v} = -[u\tau_{xx} + v\tau_{xy} + w\tau_{xz} - q_x, 0, \tau_{xx}, \tau_{xy}, \tau_{xz}]^T, \qquad (2.1g)$$

$$\mathbf{g_v} = -[u\tau_{xy} + v\tau_{yy} + w\tau_{yz} - q_y, 0, \tau_{xy}, \tau_{yy}, \tau_{yz}]^T, \qquad (2.1h)$$

$$\mathbf{h_v} = -[u\tau_{xz} + v\tau_{yz} + w\tau_{zz} - q_z, 0, \tau_{xz}, \tau_{yz}, \tau_{zz}]^T. \qquad (2.1i)$$

In the above, e is total energy per unit volume, p is pressure, ρ is density, u, v, w are the three components of the velocity, while the components of the shear-stress tensor τ and heat-flux vector q are given by:

$$\tau_{xx} = \frac{2}{3}\mu(2u_x - v_y - w_z), \qquad (2.1j)$$

$$\tau_{yy} = \frac{2}{3}\mu(2v_y - u_x - w_z), \qquad (2.1k)$$

$$\tau_{zz} = \frac{2}{3}\mu(2w_z - u_x - v_y), \qquad (2.1l)$$

$$\tau_{xy} = \mu(u_y + v_x), \qquad (2.1m)$$

$$\tau_{xz} = \mu(w_x + u_z), \qquad (2.1n)$$

$$\tau_{yz} = \mu(v_z + w_y), \qquad (2.1o)$$

$$(q_x, q_y, q_z) = -\kappa \nabla T, \qquad (2.1p)$$

where T denotes temperature and μ and κ are the coefficients of viscosity and thermal conductivity, respectively.

The computational domain D is discretized into cells $C_j, j = 1, ..., N$, which form an unstructured grid. In our cell-centered finite- volume scheme the numerical solution $\mathbf{v}$ approximates the cell average of the solution:

$$\mathbf{v}_j^n \approx \frac{1}{|C_j|} \int_{C_j} \mathbf{q}(\mathbf{x}, n\Delta t) dV \quad , \qquad (2.2)$$

where $|C_j| = \int_{C_j} dV$ and n is the time-step number. The finite-volume numerical scheme we use to solve (2.1) is then given by the following two-stage method:

$$\mathbf{v}_j^{n+\frac{1}{2}} = \mathbf{v}_j^n - \frac{\Delta t}{2|C_j|} \int_{\partial C_j} \hat{n} \cdot \mathbf{F}(\mathbf{v}_j^{\star n}) dS \quad , \qquad (2.3a)$$

$$\mathbf{v}_j^{n+1} = \mathbf{v}_j^n - \frac{\Delta t}{|C_j|} \int_{\partial C_j} \hat{n} \cdot \mathbf{F}(\mathbf{v}_j^{\star n+\frac{1}{2}}) dS \quad , \qquad (2.3b)$$

where the surface integrals are approximated by second order quadrature (midpoint rule). Here the superscript $\star$ denotes the solution's point value on the cell boundary an infinitesimally small time later ("solution in the small"). It is normally obtained by solving the one-dimensional Riemann problem on the cell face given the left and right point values of the solution : $\mathbf{v}^\star = E(\mathbf{v}_L, \mathbf{v}_R)$ (E here is also referred to as the *evolution* operator). We typically use Roe's approximate Riemann solver for E. To compute the $\mathbf{v}_{L,R}$ we need to *reconstruct* the point values of the solution from its cell averages $\mathbf{v}$. While in (2.3a) we use piecewise constant reconstruction in each cell, for the second stage (2.3b) piecewise linear reconstruction is used, which requires knowledge of the first derivative. The gradient necessary for linear reconstruction may be obtained by using the Riemann solution $\mathbf{v}^{\star n}$ of the first stage:

$$\nabla \mathbf{v}_j^n \approx \frac{1}{|C_j|} \int_{\partial C_j} \hat{n} \mathbf{v}_j^{\star n} dS \quad , \qquad (2.4)$$

The gradient necessary for computing the viscous flux should be based on a central, rather than an upwind stencil. This is accomplished by simply replacing the $\mathbf{v}^{\star n}$ in (2.4) by a volume-weighted average of the two neighboring cells, but leaving $\mathbf{F}$ in (2.3) as the usual (Roe's) Riemann flux.

To be able to reliably handle strong shocks we also employ a minmod limiter, in a face-by-face sense, where gradients of neighboring cells j and l are compared as follows:

$$\mathbf{G}_j^n = \text{minmod}(\nabla \mathbf{v}_j^n, \nabla \mathbf{v}_l^n) \quad , \qquad (2.5a)$$

where
$\text{minmod}(x, y) = \frac{1}{2}(\text{sgn}(x) + \text{sgn}(y)) \min(|x|, \beta|y|)$, and β is a compression constant. Now the linearly interpolated value at the cell face centroid will be given by the following reconstruction operation:

$$R(\mathbf{v}_j^{n+\frac{1}{2}}) = \mathbf{v}_j^{n+\frac{1}{2}} + (\mathbf{r} - \mathbf{r}_j) \cdot \mathbf{G}_j^n \quad , \qquad (2.5b)$$

where $\mathbf{r}$ and $\mathbf{r}_j$ are the position vectors of the face and cell centroids, respectively.

The resulting method is explicit and of second order accuracy in both space and time, although its first order version can also be used by simply using $\mathbf{v}^{n+\frac{1}{2}}$. The time update is similar to a two-stage Runge-Kutta method, where the spatial reconstruction is first order accurate for the first stage, and second order accurate for the second stage. It can also be thought of as a Lax-Wendroff type, but upwind predictor-corrector scheme(Ref. 1). Because of the TVD limiter in (2.5a), it also gets an adaptive stencil flavor, similar to that of an ENO (essentially nonoscillatory) reconstruction, but without its high computational cost.

Boundary Conditions

The implementation of boundary conditions ensures consistency in flux computations. That is, as in the case of any interior cell boundary, computation of fluxes for a cell face that lies on the boundary of the computational domain involves determination of "left" and "right" states and Roe's approximate Riemann solver. The state that corresponds to the "outside" of the domain should satisfy the appropriate boundary conditions. For instance, when computing fluxes for a cell on the left boundary of the domain where inviscid tangency condition is to be satisfied, the "left" state should be such that the corresponding velocity vector should be tangential to the surface. This manner of imposing boundary conditions ensures that only the information at a boundary that corresponds to waves propagating into the computational domain is actually used in the computation of fluxes.

Grid Generation

A tool named UNIVG has been developed at Rockwell Science Center for generating unstructured grids. UNIVG accepts specification of surface geometry in the form of a collection of patches. A patch geometry could be specified either in the IGES format or by specifying sufficient number of non-intersecting lines on the patch. Each line in turn is discretized by an ordered collection of sufficient number of points. Triangular elements are first generated on the boundary of the computational domain satisfying user specified clustering requirements. The computational domain is then discretized in the form of tetrahedral cells using the "advancing front" technique.

UNIVG has the capability to develop an unstructured grid that includes a specified "cloud" of points as nodes. This is sometimes useful in controlling the distribution of cells in the computational domain.

The Code Architecture

The Navier-Stokes solver developed using the algorithm described in section 2 is identified by the acronym ICAT-1. The solver requires following data as input:

- a) cells
  - 1) number of cells
  - 2) centroids
- 3) volumes

- b) interior faces
  - 1) number of interior faces
  - 2) centroids.
  - 3) two parent cells for each face
  - 4) normals

- c) boundary faces
  - 1) number of boundary faces
  - 2) centroids.
  - 3) parent cells
  - 4) normals

The output from the grid generator, which is in the form of (x, y, z) coordinates of the nodes and the connectivity between nodes and cells, is processed to generate the data listed above. Here, interior faces are cell surfaces that are common to two neighboring (parent) cells. Boundary faces, on the other hand, are cell faces that belong to only one cell.

The algorithm involves two time-steps. The solver consists of two loops for each time-step. The first loop is over the interior faces. For each interior face, the two neighboring cells are identified and the corresponding v^ns are used as the "left" and "right" states to compute the flux tensor $\mathbf{F}$ and $\mathbf{v}^*$. The contribution of $\mathbf{F}$ and $\mathbf{v}^*$ to $\Delta \mathbf{v}$ (the change in $\mathbf{v}$) and $\nabla \mathbf{v}^n$ are added for the two neighboring cells. The second loop in the solver performs similar computations for boundary faces. The only difference in this case is the fact that a boundary face contributes to only one cell. The steps involved in the second time-step are almost the same except for the fact that computation of $\nabla \mathbf{v}^n$ is not required.

Multigrid Implementation

The multigrid implementation described here (Ref. 2) employs only one set of dependent variables, namely, the fine-grid cell-averages. Even coarser-grid computations update only the fine-grid values. Therefore, only a minimal increase in computer storage is required. Adaptation of the scheme described earlier for multigrid involves three steps. The first step is the formation of coarser grids (Ref. 3). The second step identifies the interior faces of the coarser grid which are a subset of the fine-grid interior faces. The third step deals with some modifications to the solver.

The procedure for the formation of a coarser-grid from a finer-grid is as follows:

- 1) Create a data base to associate each node with its "parent" cells. Parent cells of a node are the cells that contain the given node as a vertex.

- 2) Choose any one of the available nodes as the first node. It may be possible to improve efficiency by employing certain criteria for the selection of the node, but for the present discussion that is not important.
- 3) Associate all the "parent" cells of the chosen node with group #1 and mark the node and the cells as "not available".
- 4) Choose any one of the available nodes as the second node. Again, recipies are available for optimizing the choice of the second node, but are not important for the current discussion.
- 5) Identify all available "parent" cells of the chosen node and check to ensure that agglomeration of these cells yields a connected cell. If not, agglomerate by starting with any one of these cells and adding its neighbors from the available "parent" cells of the chosen node to the group. This process may be optimized by enusring that the group so formed is the largest possible one.
- 6) Mark the node and the chosen cells as "not available" and associate the chosen cells with the next group.
- 7) Repeat steps 4, 5 and 6 until no more cells are available.

The steps described above result in creating a map between the fine-grid cells and groups (coarser-grid cells). Each fine-grid cell belongs to one and only one group. A group may have any number of fine-grid cells ≥ 1. Note that this process may result in some groups with number of cells as few as just 1. It is possible to prescribe a minimum volume for groups and eliminate groups that do not satisfy this requirement by agglomerating cells that belong to such groups with adjacent groups.

Optimization of the agglomeration process described above is a topic in itself and hence will not be discussed in detail here. It suffices to say that a substantial gain in convergence accelertion may be obtained using any reasonable recipe.

The second step that identifies the interior faces of a coarser grid is relatively straight forward. In this case, an interior face is a face of a cell whose "parent" cells belong to different groups. The "parent" cells of an interior face are the two fine-grid cells that the face belongs to.

The solver computes fluxes exactly the same way as it is done for a single grid. However, in this case, the fluxes are computed for only a subset of the interior faces corresponding to fine-grid. The flux-contribution from an interior face is accumulated in the first-cells of the two adjacent groups that the face belongs to. Note that the first-cell is used only for convenience and any one of the cells of a group may be used for this purpose. The boundary faces for a coarser grid are the same as those for the fine-grid and the flux contribution from a boundary face is added to the contribution from the interior faces for the corresponding group. At the end of the flux computations, the net-flux is stored at locations corresponding to the first-cells of the groups. This net-flux is then distributed among the members of a group using a "weight". The simplest "weight" that suffices corresponds to assuming that $\Delta \mathbf{v}$ is the same for all the cells that belong to a group.

The last paragraph clearly demonstrates the simplicity of the modifications involved in going from single-grid to multigrid. To summarize, the modifications are:

- 1) Flux computations are performed only for a subset of the fine-grid interior faces.
- 2) The net-flux for a group (coarser-grid cell) is then distributed among the members of the group.

The usual "restriction" and "prolongation" operations involved in a multigrid technique are hidden in the procedure described above. Here "restriction" corresponds to computing fluxes only for a subset of fine-grid interior faces and "prolongation" assumes $\Delta \mathbf{v}$ is the same for all the fine-grid cells that belong to a group. Note that provision exists for modifying the "prolongation" operator by prescribing a suitable "weighting" function.

Residual Smoothing

The agglomeration process described above has many elements of arbitrariness in the sense that depending on the order in which the nodes are selected, different grouping of the cells will result. In order to take this factor into account and also to improve convergence rate a simple smoothing procedure is employed. For each node, $\Delta \mathbf{v}$ is computed by volume-weighted-averaging of the $\Delta \mathbf{v}$'s of the parent cells. Then, for each cell, $\Delta \mathbf{v}$ is computed as an average of the $\Delta \mathbf{v}$'s of the nodes of the cell. This smoothing procedure necessitates allocation of storage for the $\Delta \mathbf{v}$'s of the nodes.

Parallel Implementation

The first-step involved in the parallel implementation is the "domain decomposition" which involves dividing of the computational domain into subregions of (almost) equal number of cells and renumbering of the cells to ensure continuity of the numbering of cells belonging to the same subdomain. For this purpose, we employ a recursive spectral bisection (RSB) method (Ref. 5). The number of subregions into which the computational domain is divided is determined by the number of CPU *nodes* over which parallel-processing is to be performed. In order to establish the communication links between any two given *nodes*, we identify those cells whose face-neighbors also belong to other *nodes*. These then formed the *communication buffer* for the particular *node*, since they will be the only cells that would need to send and receive information across nodal boundaries. Such information include the solution's cell average, its gradient, and cell centroid coordinates. Once these quantities are passed to the buffer, fluxes along subdomain boundaries are computed in exactly the same manner as for any other cell-face internal to the node.

Our parallel implementation employs the synchronous, *host-node* model where the host calls multiple parallel programs on multiple CPU's, or *nodes* (Ref. 4). Once the domain decomposition is completed, the host launches a process onto each *node*. The *host-node* and *node-node* communication channels are then established, which are used to transmit the data partitions to the respective nodes. During each stage of each time step, cells on *nodal* boundaries will communicate with those of adjacent *nodes* in order to obtain cell average and gradient data. At the end of the run, each *node* passes the solution array back to the host, which, in turn, assembles the global solution and prepares it for output. An outline of the algorithm is shown below, where ".h" and ".n" suffixes indicate host and node modules, respectively :

- (I.h) Read grid and solution
- (II.h) Decompose domain and renumber cells
- (III.h) Establish connectivity and set up communication buffer
- (IV.h) Send data to respective subdomains
- (I.n) Read data from host
- (II.n) First stage of solve
- (II.a.n) Write solution (from comm. buf.) to adjacent nodes
- (II.b.n) Compute inviscid fluxes and gradients – interior faces
- (II.c.n) Read solution (into comm. buf.) from adjacent nodes
- (II.d.n) Compute inviscid fluxes and gradients – nodal boundary faces
- (II.e.n) Compute inviscid fluxes and gradients – physical boundary faces
- (II.f.n) Write gradient and centroid (from comm. buf.) to adjacent nodes
- (II.g.n) Read gradient and centroid (into comm. buf.) from adjacent nodes
- (II.h.n) Compute viscous fluxes – interior, nodal, and physical boundary faces
- (II.i.n) First order time update of solution
- (III.n) Second stage of solve
- (III.a.n) Write solution (from comm. buf.) to adjacent nodes
- (III.b.n) Compute fluxes (inv. and vis.) – interior faces
- (III.c.n) Read solution (into comm. buf.) from adjacent nodes
- (III.d.n) Compute fluxes (inv. and vis.) – nodal boundary faces
- (III.e.n) Compute fluxes (inv. and vis.) – physical boundary faces
- (III.f.n) Second order time update of solution
- (IV.n) Write solution data to host
- (V.h) Assemble, postprocess and write solution to disk

The communication link between the *nodes* is established using a standard MPI (Message Passing Interface) implementation, which enables the same code to be run on virtually all massively parallel platforms, as well as workstation clusters. In the absence of a front end, the "host" is replaced by e.g. node 0, but the above flowchart should otherwise remain the same.

Results and Discussion

We have conducted a series of numerical experiments on laminar flow past a circular cylinder, with Reynolds numbers in the steady state regime. For $Re = 40$, for instance, we get the familiar velocity field pattern of Fig. 1a. The separation bubble lengths for six different Reynolds numbers show excellent comparison with experimental data (Fig. 1b). This series of validation runs was done using 15980 cells, the decomposed domain of Fig. 1c (RSB algorithm of [3]), and 32 nodes of our nCUBE2 parallel computer. We also obtained excellent load balancing (Fig. 1d), and scalability performance (Fig. 1e). For the latter, we ran the same problem on hypercube dimensions 3 to 7, and even on 128 nodes the actual speed-up was about 90% of linear scaling.

In order to test and validate the code on a moderate size 3-D problem, we converted a seven zone structured grid into our unstructured bookkeeping using 218400 hexahedral cells, and computed the Mach 10.19 flow, with Reynolds number (per meter) 16677. For zero angle of attack we get the symmetric surface pressure contours of Fig. 2a, and the very sharp hypersonic bow shock on the symmetry plane, as shown in Fig. 2c. Pressure coefficient comparisons with experimental data are given in Fig. 2b (on the intersection of the symmetry plane and Apollo surface). This case was run on 128 nodes of the nCUBE2, using the domain decomposition of Fig. 2d. The load balancing plot of Fig. 2e shows a fairly even load distribution, with a communications overhead amounting to about 10% of the total run time. The scalability plot (128 to 512 nodes) shows a performance which is about 85% of the linear speed-up (Fig. 2f).

Conclusions

Computational simulation is maturing and becoming an integral part of product development in both aerospace and commercial markets. We now face a real challenge in making these tools cost effective and quick–turn–around to be responsive to the market pressures faced by the product development teams. We will continue to seek advances in the numerical modeling of the physical processes as well as to better utilize the increasing power of workstation clusters and parallel architectures.

Acknowledgements

The work presented here represents the combined efforts of many years of research, development and implementation of computational tools at different divisions of Rockwell, namely, Rocketdyne, Space Systems Division, North American Aircraft Division and Science Center. In addition to funds provided through Rockwell IR&D projects, several government contracts from AFOSR, NASA, Army Research Laboratory, Wright Patterson AFB and others have contributed this computational development.

References

1. K-Y. Szema, S.V. Ramakrishnan, V.V. Shankar, and K. Rajagopal, "Application of a generalized Lax-Wendroff scheme for unstructured Euler computations," AIAA Paper No. 96-2401, 1996.

2. S.V. Ramakrishnan and K. Rajagopal, "A Multigrid Technique for unstructured grids based on a generalized Lax-Wendroff," AIAA Paper No. 97-0447, 1997.

3. V. Venkatakrishnan and D. J. Mavriplis, "Agglomeration multigrid for the three-dimensional Euler equations," AIAA Paper 94-0069, June 1994.

4. H.D. Simon, "Partitioning of unstructured problems for parallel processing," Computing Systems in Engineering, Vol. 2, No. 2/3, pp.135-148, 1991.

5. B. L. Bihari, S. V. Ramakrishnan, V. V. Shankar and S. Palaniswamy,"Massively Parallel Implementation of an Explicit Algorithm on Unstructured Grids," Proceedings of the 15th International Conference on Numerical Methods in Fluid Dynamics, June 24-28, 1996 in Monterey, california.

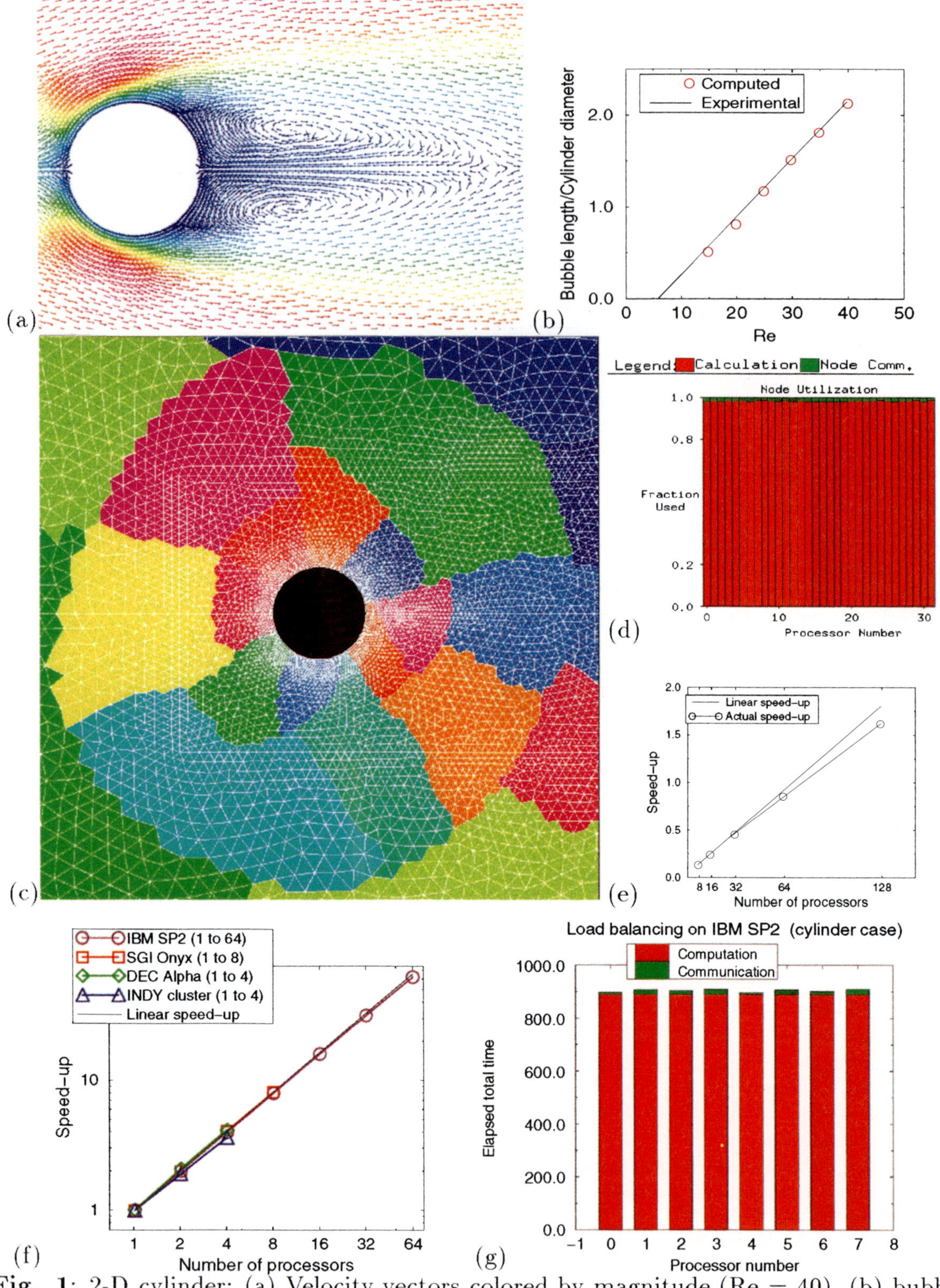

Fig. 1: 2-D cylinder: (a) Velocity vectors colored by magnitude (Re = 40), (b) bubble length, (c) subdomains (32 nodes), (d) loads (32 nCUBE2 nodes), (e) scaling – nCUBE2 (8 to 128 nodes), (f) scaling – MPI (1 to 64 nodes), (g) loads (8 IBM SP2 nodes).

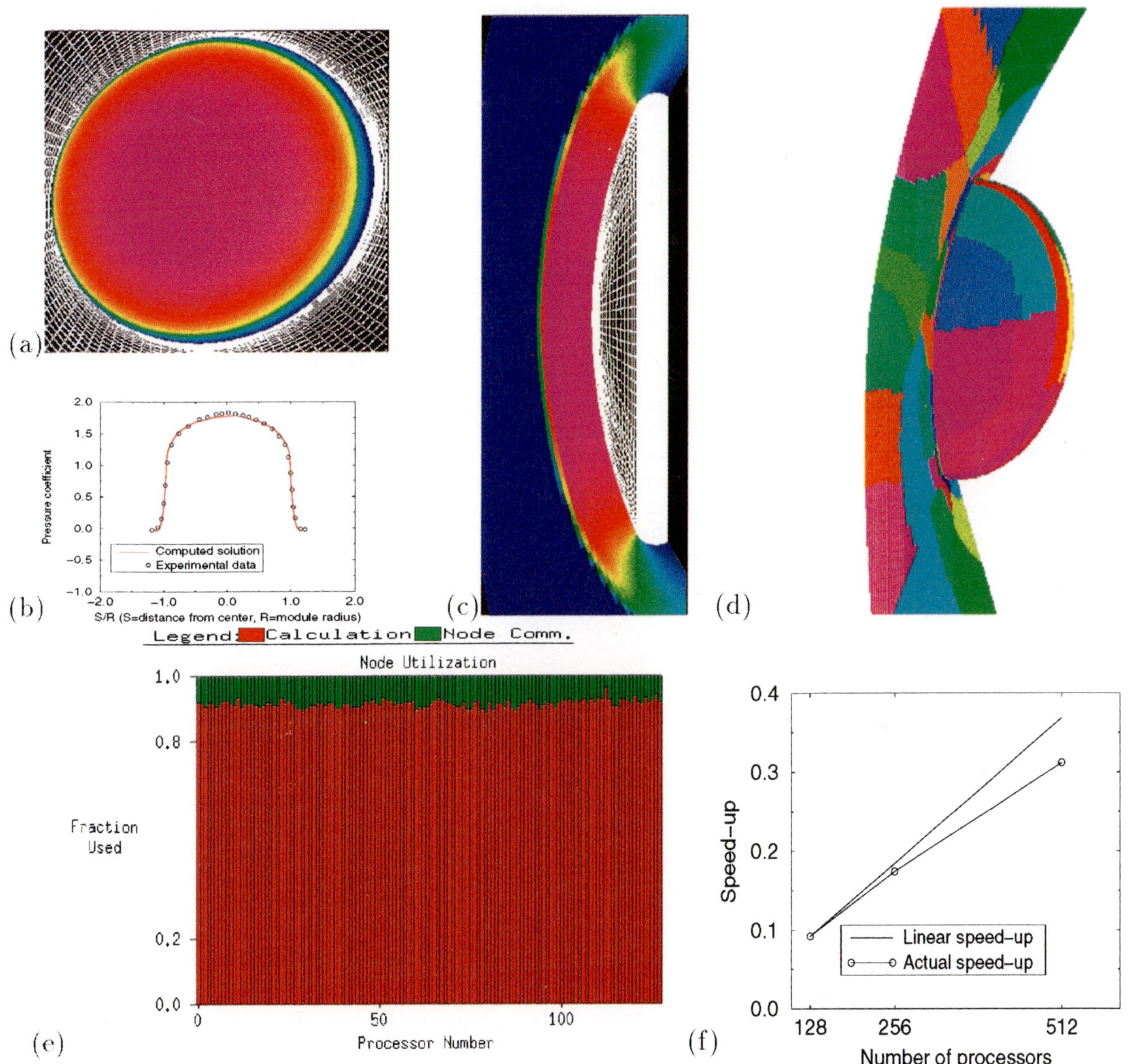

Fig. 2: Apollo re-entry module: (a) Surface pressure, (b) c_p profile along symmetry diagonal, (c) Pressure contours in symmetry plane, (d) domain decomposition (128 nodes), (e) load balancing (128 nodes), (f) scaling (128 to 512 nodes).

LARGE-SCALE SIMULATION RESULTS FOR COMPRESSIBLE AEROSPACE APPLICATIONS

Arthur RIZZI [1]   Jan Vos [2]   Michael RUDGYARD [3]

Pénélope LEYLAND [2]   Thilo SCHÖNFELD [4]

1) Aeronautical Engineering Dept., Royal Institute of Technology (KTH), S-10044 Stockholm, Sweden
2) y Hydraulic Machines & Fluid Mechanics Institute (IMHEF), Swiss Federal Institute of Technology (EPFL), CH-1015 Lausanne, Switzerland
3) Oxford University Computing Laboratory, Parks Rd., Oxford OX1 3QD, United Kingdom
4) Centre Europ'een de Recherche et de Formation Avance'e en Calcul Scientifique (CERFACS), 42 Avenue Gustave Coriolis, F-31057 Toulouse Cedex, France

Abstract

Large-scale simulation results for industrial aerospace applications obtained with the structured-mesh solver NSMB and with the unstructured-mesh solver AVBP are reviewed. A brief overview of the basic theory and a comparison of methodologies is given for these two types of solvers: the physical models for compressible turbulent flow are first described, and the different numerical formulations used to solve the governing equations are then examined, with a discussion of their implementation on vector and parallel computers. Finally, three large-scale simulation results are presented: two on transonic aircraft aerodynamics, and one on hypersonic spacecraft aerothermodynamics.

1. Introduction

The simulation of flow past aircraft or space vehicles requires the treatment of complex geometrical configurations. The two general approaches to handle the geometrical problem are either solvers based on a structured mesh or solvers based on an unstructured mesh. A structured-mesh solver treats the complexity by the so-called multi-block approach where the grid can be composed of different blocks of structured hexahedronal grids [1]–[3]. An unstructured solver on the other hand uses grids of polyhedronal cells that can be constructed automatically. Most commonly the cells are tetrahedal, but pentahedral and hexahedral cells are becoming more popular.

It is fair to say that either methodology possesses its strengths and weaknesses. The main advantages of structured methods can be briefly summarized as their higher computational efficiency and solution accuracy, whereas the disadvantages are their inflexibility for meshing extremely complex configurations and the problems that arise in conjunction with grid adaption. On the other hand unstructured methods are distinguished by their flexibility for handling complicated geometries, while the drawbacks are their higher computational cost per element and potentially lower accuracy due to skewed elements.

For the past two decades we have been developing computational aerodynamics tools to solve industrial problems in the aerospace field. During this time 3-D Euler solvers have become a standard tool used in industry on a daily basis today. Even 2D Navier Stokes solvers have almost entered that same status of a standard industrial tool. Although industry today uses 3D solvers on a limited basis due to the large memory requirements and high computer costs associated with Navier Stokes simulations that have enough grid points to resolve properly the viscous regions, their use is growing and it will not be long before they also join the same category, especially if the potential of massively parallel computers to bring the resolution of the Navier Stokes equations for aerodynamic design within acceptable costs is realized. However, obtaining good efficiency with a CFD code on these massively parallel computers is not a trivial task. For this reason it is quite common for industry to work with academics on the development and testing of 3D Navier Stokes codes.

Received on August 1, 1997.

We have developed a structured-mesh multi-block Navier-Stokes solver **NSMB** within an industry-academic consortium of European partners: SAAB Aircraft, Aerospatiale Avions, the research institute CERFACS, and the technical universities EPFL and KTH [4, 5]. The unstructured-mesh Navier-Stokes solver **AVBP** has been developed at CERFACS and Oxford University Computing Laboratory, with contributions from a European research network including EPFL and KTH among others [30]. Both efforts have been promoted and funded by various European and national agencies, and in many ways the work and the resulting codes are a pan-European enterprise.

The theme of this paper is to review the large-scale simulation results for aerospace applications that we have obtained with the structured-mesh solver **NSMB** as well as with the unstructured-mesh solver **AVBP**. We present an overview of the basic theory for these two types of solvers, including large-scale simulation results. Because space is limited we restrict ourselves to discussing some aerospace computations produced with our own codes.

The paper begins by presenting the physical models used to describe compressible turbulent flow, an overview is given of the numerical formulation to solve the governing equations, followed by a discussion of its implementation on vector and parallel computers. Three large-scale simulation results are then presented.

2 GOVERNING EQUATIONS

The governing equations are the 3D Reynolds-averaged compressible Navier–Stokes equations in conservative form. They are derived from the corresponding instantaneous equations by introducing conventional Reynolds-averaging on density ρ, pressure p, laminar stress tensor τ_ℓ and laminar heat flux vector $\mathbf{q}_\ell$ and by introducing Favre-averaging [6] on the velocity vector $\mathbf{v}$, stagnation enthalpy H and energy per unit mass E. If written in integral form over a control volume $\mathcal{V}$ with the boundary $\partial\mathcal{V}$ and the outer normal unit vector $\mathbf{n}$, the equations may be expressed as

$$\frac{\partial}{\partial t} \int_\mathcal{V} \mathbf{u}\, d\mathcal{V} + \int_{\partial\mathcal{V}} \mathcal{F}(\mathbf{u}, \nabla\mathbf{u}) \cdot \mathbf{n}\, dS = 0 \qquad (2.1)$$

where we have defined the vector of the conservative variables $\mathbf{u}$ and the flux tensor $\mathcal{F}$. In our description below, it is convenient to subdivide this tensor as $\mathcal{F} = \mathcal{F}^0 + \mathcal{F}^1$, were the zeroth order flux $\mathcal{F}^0 = \mathcal{F}^0(\mathbf{u})$ depends only on values of the solution vector, while the first order flux $\mathcal{F}^1 = \mathcal{F}^1(\mathbf{u}, \nabla\mathbf{u})$ depends additionally on gradients of the solution. For the conservation of mass, momentum and energy we have:

$$\mathbf{u} = \begin{pmatrix} \rho \\ \rho\mathbf{v} \\ \rho E \end{pmatrix} \qquad \mathcal{F}^0 = \begin{pmatrix} \rho\mathbf{v} \\ \rho\mathbf{v}\mathbf{v} + p\overline{I} \\ \rho H\mathbf{v} \end{pmatrix}$$

$$\mathcal{F}^1 = \begin{pmatrix} 0 \\ -\tau \\ -\tau \cdot \mathbf{v} + \mathbf{q} \end{pmatrix} \qquad (2.2)$$

with temperature T, total stress tensor $\tau = \tau_\ell + \tau_t$, total heat flux $\mathbf{q} = \mathbf{q}_\ell + \mathbf{q}_t$ and the unit tensor $\overline{I}$.

The laminar viscous terms are stated by the Newtonian formulation, Stokes' hypothesis and Fourier's law, respectively,

$$\tau_\ell = \mu_\ell\, [\, \nabla\mathbf{v} + (\nabla\mathbf{v})^T - \frac{2}{3}(\nabla\cdot\mathbf{v})\,\overline{I}\,] \qquad (2.3)$$

$$\mathbf{q}_\ell = -\frac{\mu_\ell C_p}{Pr_\ell} \nabla T \qquad (2.4)$$

where Pr_ℓ is the Prandtl number, C_p the specific heat at constant pressure, and μ_ℓ the laminar viscosity related to the local temperature by the Sutherland formula.

The closure problem, namely finding constitutive equations for the additional contribution due to turbulence, is solved by selecting a turbulence model. Two eddy-viscosity turbulence models are in widespread use, the algebraic Baldwin-Lomax model and the two transport-equation k - ε turbulence model, and we describe the latter here. These models relate the Reynolds stresses and the turbulent heat flux to the mean flow and thereby close the system of equations.

2.1 $k - \varepsilon$ Turbulence Model of Chien

The turbulence model that we describe in detail here is the low-Re-number k - ε model originally proposed by Chien [8] to predict incompressible turbulent flows and then extended by Coakley ([9] to the compressible case. This Chien model is a standard two-equation closure based on the modeled k and ε transport equations given by

$$\frac{\partial(\rho k)}{\partial t} + \nabla \cdot (\rho k\mathbf{v}) - \nabla \cdot \left[\left(\mu + \frac{\mu_t}{\sigma_k} \right) \nabla k \right] = \omega_k$$

$$(2.5)$$

$$\frac{\partial(\rho\varepsilon)}{\partial t} + \nabla\cdot(\rho\varepsilon\mathbf{v}) - \nabla\cdot\left[\left(\mu+\frac{\mu_t}{\sigma_\varepsilon}\right)\nabla\varepsilon\right] = \omega_\varepsilon' \tag{2.6}$$

where k is the turbulent kinetic energy per unit mass and ε represents its dissipation rate.

After absorbing these two additional transport equations, the governing system (2.1) can be rewritten as

$$\frac{\partial}{\partial t}\int_\mathcal{V}\mathbf{u}\,d\mathcal{V} + \int_{\partial\mathcal{V}}\mathcal{F}(\mathbf{u},\nabla\mathbf{u})\cdot\mathbf{n}\,dS = \int_\mathcal{V}\Omega\,d\mathcal{V} \tag{2.7}$$

where $\mathbf{u}$ has two more components and Ω is the vector of source terms:

$$\mathbf{u} = \begin{pmatrix}\rho\\\rho\mathbf{v}\\\rho E\\\rho k\\\rho\varepsilon\end{pmatrix}, \qquad \Omega = \begin{pmatrix}0\\0\\0\\\omega_k\\\omega_\varepsilon\end{pmatrix}. \tag{2.8}$$

The zeroth and first order flux tensors $\mathcal{F}^0$ and $\mathcal{F}^1$ are then

$$\mathcal{F}^0 = \begin{pmatrix}\rho\mathbf{v}\\\rho\mathbf{vv}+p\overline{I}\\\rho H\mathbf{v}\\\rho k\mathbf{v}\\\rho\varepsilon\mathbf{v}\end{pmatrix} \qquad \mathcal{F}^1 = \begin{pmatrix}0\\-\tau\\-\tau\cdot\mathbf{v}+\mathbf{q}\\-\left(\mu_\ell+\dfrac{\mu_t}{\sigma_k}\right)\nabla k\\-\left(\mu_\ell+\dfrac{\mu_t}{\sigma_\varepsilon}\right)\nabla\varepsilon\end{pmatrix} \tag{2.9}$$

In addition, the turbulent kinetic energy enters into

$$E = c_v T + \frac{1}{2}(\mathbf{v}\cdot\mathbf{v}) + k \tag{2.10}$$

$$\tau_t = \mu_t\,[\,\nabla\mathbf{v}+(\nabla\mathbf{v})^T - \frac{2}{3}(\nabla\cdot\mathbf{v})\,\overline{I}\,] - \frac{2}{3}\rho k\,\overline{I} \tag{2.11}$$

The turbulent viscosity is calculated from the Kolmogorov–Prandtl relation,

$$\mu_t = C_\mu\,f_\mu\,\rho\,\frac{k^2}{\varepsilon}, \tag{2.12}$$

and the two turbulent source terms read

$$\omega_k = P_k - \rho\varepsilon + S_k, \qquad \omega_\varepsilon = C_1 f_1\frac{\varepsilon}{k}P_k - C_2 f_2\rho\frac{\varepsilon^2}{k} + S_\varepsilon \tag{2.13}$$

where $P_k = (\tau_t\cdot\nabla)\cdot\mathbf{v}$ is the production rate of kinetic energy, k.

The model of Chien is a so called low-Re-number model which means that equations (2.5) and (2.6) must be integrated all the way to the wall to resolve the whole boundary layer including

the viscous sublayer. The various constants and damping functions used in this model are given by:

$$C_\mu = 0.09, \quad C_1 = 1.35, \quad C_2 = 1.8$$

$$\sigma_k = 1.0, \; \sigma_\varepsilon = 1.3$$

$$f_\mu = 1 - exp\,(-0.0115y^+) \tag{2.14}$$

$$f_1 = 1, \qquad f_2 = 1 - 0.22\,exp\,[-(R_t/6)^2].$$

In these functions, R_t is the turbulent Reynolds number, y^+ is the wall coordinate and u_τ the friction velocity, with the following definitions:

$$R_t = \frac{\rho k^2}{\mu\varepsilon}, \qquad y^+ = \frac{\rho u_\tau y}{\mu}, \qquad u_\tau = \sqrt{\frac{\tau_w}{\rho}} \tag{2.15}$$

where τ_w represents the wall shear stress and y is the normal distance from the wall. It is easy to see that for $y^+ \to 0$ the functions f_μ and f_2 go to zero in the laminar sublayer.

Lastly, the two source terms S_k and S_ε, one in each of the turbulent transport equations, need to be specified to complete the model. Chien constructed them to provide wall dissipation in order to handle properly the low-Re-number effects in the near-wall region. These two terms are given by

$$S_k = -\frac{2\mu k}{y^2}, \qquad S_\varepsilon = -\frac{2\mu\varepsilon}{y^2}\,exp\,(-0.5\,y^+). \tag{2.16}$$

The equations (2.5), (2.6) and (2.12) to (2.16) close the system of equations (2.7).

3 NUMERICAL FORMULATION

The natural discretisation of a system of conservation laws such as Eq.(2.1) requires the evaluation of flux-balance integrals within typical control volumes. The types of numerical schemes that can be constructed to approximate these integrals can be classified according to the the spatial approximation that is made, the geometrical shape of the volume element and to where the discrete data is located spatially around the volume. Two categories are of interest to us now :

- **Finite-volume schemes** for which the unknowns are stored at the centre of the control volumes and numerical fluxes are then evaluated across common interfaces. For cell centred methods, the control volumes are the cells or elements themselves, while for vertex centred methods the control volume is typically the dual cell. The underlying grids may be structured or unstructured. The NSMB code typifies this approach for *structured* meshes.

- **Weighted-residual schemes** where the unknowns are stored at the nodes of the mesh even though a flux integral is evaluated for each cell or element — neighbouring cell residuals are then averaged in order to obtain a nodal approximation. Cell vertex finite volume schemes as well as some simple finite element schemes fall into this category. The AVBP code typifies this approach for *unstructured* meshes.

3.1 Finite-Volume Approximation

We begin by considering the usual finite-volume method applied to Eq.(2.7). For a typical cell with n faces and volume $\mathcal{V}$, we make use of the approximation

$$\frac{d}{dt}\mathbf{u}_C \approx -\frac{1}{\mathcal{V}_C}\int_{\partial \mathcal{V}_C} \mathcal{F}(\mathbf{u}, \nabla \mathbf{u}) \cdot \mathbf{n}\, dS + \Omega_C$$

$$\approx -\frac{1}{\mathcal{V}_C}\sum_{f=1}^{n} \mathbf{F}_f + \Omega_C \qquad (3.1)$$

where $\mathbf{u}_C$ and Ω_C are discrete values of the dependent variables $\mathbf{u}$ and source terms Ω that are associated with the control volume C, and $\mathcal{F}_f$ is some approximation to the flux integration *over* the face f of the cell. For classical finite volume schemes the problem then becomes one of determining appropriate spatial schemes for $\mathbf{F}_f$ and time integration methods for the semidiscrete Eq.(3.1). In the present context, where we are most interested in steady–state solutions, the accuracy of the temporal scheme is of little interest to us since such schemes are only used as a means of iterating towards the steady state. For such problems it is convenient to define a spatial residual as:

$$\mathbf{Res}_C(\mathbf{u}) = \frac{1}{\mathcal{V}_C}\sum_{f=1}^{n} \mathbf{F}_f - \Omega_C, \qquad (3.2)$$

which expresses the balance of the numerical flux and the source term.

In the case of structured hexahedral meshes with data stored at cell centres, $n = 6$, and we can denote the grid cells and unknowns using the integer triplet $\{i, j, k\}$. If we only consider zeroth order fluxes, the *numerical flux function* $\mathbf{F}_f^0$ across the face $i + \frac{1}{2}$ can often be written as the sum of an average flux plus a dissipation term (which may be artificial or numerical) in the form

$$\mathbf{F}_{i+\frac{1}{2}}^0 = \frac{1}{2}[\mathcal{F}^0(u_i) + \mathcal{F}^0(u_{i+1})] \cdot \mathbf{S}_{i+\frac{1}{2}} + \mathbf{D}_{i+\frac{1}{2}}, \quad (3.3)$$

where $\mathbf{S}_{i+\frac{1}{2}}$ is an area–weighted vector associated with the face. The term $\mathbf{D}_{i+\frac{1}{2}}$ may be a classical artificial dissipation, used in combination with Jameson–type (centred) schemes, or the numerical dissipation associated with an upwind scheme. Upwind schemes involve the decomposition of the fluxes in the direction of $\mathbf{S}_{i+\frac{1}{2}}$ and the term $\mathbf{D}_{i+\frac{1}{2}}$ is a matrix function. In Jameson-type schemes consistent artificial viscosity terms are added which are scalar functions. The precision of the scheme may be increased by evaluating the flux function and/or the dissipation terms using extrapolated values rather than cell centre values; the extrapolated values may be calculated by recovering gradients (or higher order derivatives) in a manner similar to that described for cell vertex methods below.

3.1.1 Scalar Artificial Dissipation

The spatial discretization scheme employed in NSMB is a centered finite volume scheme using second and fourth order artificial dissipation terms to capture discontinuities and to damp odd/even oscillations permitted by centred schemes. It uses the standard nonlinear blend of second and fourth differences

$$\mathbf{D}_{i+\frac{1}{2}} = \epsilon_{i+\frac{1}{2}}^{(2)} \delta \mathbf{u}_{i+\frac{1}{2}} - \epsilon_{i+\frac{1}{2}}^{(4)} \delta^3 \mathbf{u}_{i+\frac{1}{2}}, \qquad (3.4)$$

where $\delta \mathbf{u}_{i+\frac{1}{2}} = \mathbf{u}_{i+1} - \mathbf{u}_i$ and $\delta^3 \mathbf{u}_{i+\frac{1}{2}} = \mathbf{u}_{i+2} - 3\mathbf{u}_{i+1} + 3\mathbf{u}_i - \mathbf{u}_{i-1}$. The coefficients ϵ are determined by

$$\epsilon_{i+\frac{1}{2}}^{(2)} = \kappa^{(2)} \hat{\lambda}_{i+\frac{1}{2}}^* \zeta_{i+\frac{1}{2}} \qquad (3.5)$$

$$\epsilon_{i+\frac{1}{2}}^{(4)} = \max(0, \kappa^{(4)} \hat{\lambda}_{i+\frac{1}{2}}^* - \epsilon_{i+\frac{1}{2}}^{(2)}) \qquad (3.6)$$

where the interface values are determined by averaging $\epsilon^{(4)}_{i+\frac{1}{2}} = \frac{1}{2}(\epsilon^{(4)}_i + \epsilon^{(4)}_{i+1})$. The variable ζ is a sensor designed to activate the second-difference dissipation in the regions of shocks and to de-activate it elsewhere. This is often defined as a function of pressure, such that $\zeta_{i+\frac{1}{2}} = \max(dp_i, dp_{i+1})$, where:

$$dp_i = \frac{|p_{i+1} - 2p_i + p_{i-1}|}{|p_{i+1} + 2p_i + p_{i-1}|} \qquad (3.7)$$

The term $\hat{\lambda}^*$ is a measure of the zeroth-order flux function $\mathbf{F}^0_f$ and is used to scale the artificial dissipation. It is commonly chosen as the spectral radius λ^* of the product of the flux Jacobian $\mathcal{A} = \frac{\partial \mathcal{F}^0}{\partial \mathbf{u}}$ and the face vector $\mathbf{S}_f$, eg.:

$$\hat{\lambda}^*_{i+\frac{1}{2}} = (\mathbf{v} \cdot \mathbf{S} + c|\mathbf{S}|)_{i+\frac{1}{2}}, \qquad (3.8)$$

where c is the local speed of sound. Equation (3.8) corresponds to a purely isotropic definition of λ. This is not a good choice for high aspect ratio cells, and 3D Vatsa and Wedan [10] propose

$$\hat{\lambda}^*_{i+\frac{1}{2}} := \hat{\lambda}^*_{i+\frac{1}{2}} \left[1 + (\hat{\lambda}^*_{j+\frac{1}{2}}/\hat{\lambda}^*_{i+\frac{1}{2}})^\alpha + (\hat{\lambda}^*_{k+\frac{1}{2}}/\hat{\lambda}^*_{i+\frac{1}{2}})^\alpha \right] \qquad (3.9)$$

where $0 < \alpha < 1$ ($\alpha = 0.5$ is often used). The dissipation coefficients $\kappa^{(2)}$ and $\kappa^{(4)}$ in Eqs.(3.5) and (3.6) are constants to be specified by the user.

Another scaling approach is that due to Martinelli [11]. He proposes to scale the artificial dissipation flux by a function of the cells aspect ratios. This method is less dissipative within the boundary layer, thus avoiding modification of the effective local Reynold's number by the artificial dissipation.

3.1.2 Matrix Artificial Dissipation

The coefficients ϵ given in Eqs.(3.5) and (3.6) are scalars. For a broad class of upwind schemes the dissipation is more complicated, but more in accordance with the physics of the flow. If we define $\hat{A}_f = \mathcal{A}_f \mathbf{S}_f$, with corresponding matrices of right and left eigenvectors $\hat{S}_f$ and $\hat{S}_f^{-1}$ respectively, many of these methods can be written (see [7]):

$$\mathbf{D}_{i+\frac{1}{2}} = \frac{1}{2} \hat{A}'_{i+\frac{1}{2}} \delta u_{i+\frac{1}{2}} \qquad (3.10)$$

where the coefficient $\hat{A}'$ is no longer a scalar but a matrix function of eigenvalues and limiter terms

$$\hat{A}'_{i+\frac{1}{2}} = \hat{S}_{i+\frac{1}{2}} \hat{\Lambda}'_{i+\frac{1}{2}} \hat{S}^{-1}_{i+\frac{1}{2}} \qquad (3.11)$$

where $\hat{\Lambda}'$ is a diagonal matrix.

For symmetric TVD for instance

$$\hat{\Lambda}'_{i+\frac{1}{2}} = diag \left(| \hat{\lambda}^{(k)}_{i+\frac{1}{2}} | \left[1 - Q^{(k)}_{i+\frac{1}{2}} (\mathbf{r}^-_{i+\frac{1}{2}}, \mathbf{r}^+_{i+\frac{1}{2}}) \right] \right) \qquad (3.12)$$

where $\hat{\lambda}^{(k)}$ is the kth eigenvalue of $\hat{A}$, $Q^{(k)}$ is a limiter function acting on the kth component of a vector, and the ratios $\mathbf{r}$ of scaled differences are defined as

$$\mathbf{r}^-_{i+\frac{1}{2}} = \frac{\hat{S}^{-1}_{i-\frac{1}{2}} \delta \mathbf{u}_{i-\frac{1}{2}}}{\hat{S}^{-1}_{i+\frac{1}{2}} \delta \mathbf{u}_{i+\frac{1}{2}}} \qquad \mathbf{r}^+_{i+\frac{1}{2}} = \frac{\hat{S}^{-1}_{i+\frac{3}{2}} \delta \mathbf{u}_{i+\frac{3}{2}}}{\hat{S}^{-1}_{i+\frac{1}{2}} \delta \mathbf{u}_{i+\frac{1}{2}}} \qquad (3.13)$$

The limiter functions therefore act on the variations of the 'characteristic' variables that are the elements of $\hat{S}_f^{-1} \mathbf{u}$.

Instead of limiting each of the characteristic fields individually, one can also limit directly using a single scalar function. This leads to upwind schemes that bear some resemblance to the classical central schemes with dissipation. For instance the dissipation of the symmetric TVD scheme using the minmod limiter becomes

$$\hat{A}'_{i+\frac{1}{2}} = | \hat{A}_{i+\frac{1}{2}} | \left[2 - \text{minmod}(1, r^+_{i+\frac{1}{2}}) - \text{minmod}(1, r^-_{i+\frac{1}{2}}) \right] \qquad (3.14)$$

where $|\hat{A}_f|$ is similar to $\hat{A}_f$ except that its eigenvalues have been replaced with their absolute values, and the ratios of differences are defined by Eq. (3.13) with differences of some appropriate scalar quantity.

3.1.3 Diffusive Terms

As yet we have only considered the approximation of the zeroth order flux integral and have ignored diffusive terms. Their inclusion is straightforward, although we shall see that several different approximations are possible. In NSMB, viscous terms are computed at cell interfaces using an auxiliary grid with control volumes C' with volumes $\mathcal{V}_{C'}$. These control volumes are defined by connecting the cell centres surrounding a given face. Gradients of the velocity and temperature at the center of the face are then computed using Gauss's theorem applied to C':

$$(\nabla v)_{C'} \approx \frac{1}{\mathcal{V}_{C'}} \int_{\partial \mathcal{V}_{C'}} v \, \mathbf{n} \, dS$$

$$\approx \frac{1}{\mathcal{V}_{C'}} \sum_{f'=1}^{6} v_{f'} \, \mathbf{S}_{f'} \qquad (3.15)$$

where $v_{f'}$ denotes some average value of v on the faces of the auxilary volume, and $\mathbf{S}_{f'}$ is the corresponding area–weighted normal vector. Once the solution gradients have been calculated at the faces of primary cells, a flux function $\mathbf{F}_f^1$ may be defined and used directly in Eq.(3.2).

3.2 Weighted Cell Residual Method

In the weighted cell residual approach the unknowns $\mathbf{u}$ are based at the nodes of the mesh although the residuals $\mathbf{Res}_C$ (3.2) are calculated within each cell C. Unlike the cell centre approach, were the flux function is calculated using weighted averages across the cell face, the cell vertex scheme averages nodal values of the flux tensors along the face. For tetrahedral elements, where the faces are triangular, this amounts to the use of the mid–point integration rule. Elements with (non–planar) quadrilateral faces may be treated by averaging both of the approximations that are obtained after subdividing the face into triangles. Because this approach is exact when the fluxes themselves vary linearly, the cell vertex method is said to be 'linearity preserving', or LP, on arbitrary meshes.

Unlike the cell centre method, where both the unknowns and the spatial residuals are calculated on the cells, the cell vertex method involves taking weighted combinations of the cell residuals in order to calculate a nodal iterate. In the most general case, and for unstructured or hybrid meshes, a cell–based *distribution matrix* d_C^j determines the form of this weighting from cell C to its vertices j. This allows us to calculate a *nodal* residual $\mathbf{Res}_i$, a point–wise (rather than cell–wise) approximation to the spatial terms of the equation:

$$\mathbf{Res}_i = \frac{1}{\mathcal{V}_i} \left(\sum_{\{C|i \in C\}} \mathcal{V}_C \, d_C^i \, \mathbf{Res}_C \right) \qquad (3.16)$$

Here, the notation $\sum_{\{C|i \in C\}}$ denotes the sum of all cells C meeting at the node i, and $\mathcal{V}_i$ is a control volume (dual cell) associated with each node such that $\sum_i \mathcal{V}_i$ is the total volume of the mesh.

Several forms of the distribution matrices may be defined [14], although most authors use either variants of Lax–Wendroff [15] or SUPG schemes, or a central-difference scheme [16]. An important property of the choice of weighting coefficients d_C^i is that these should obey the criterion

$$\sum_{\{i|i \in C\}} d_C^i = I, \quad \forall C \qquad (3.17)$$

since this ensures discrete conservation in the steady state [14, 17]. In the present discussion, and in order to make simple comparisons with the cell centre scheme, we shall only consider the central difference technique. We then have the simple coefficients

$$d_C^i = \frac{1}{n_{vert}} I, \quad \forall \, i \in C \qquad (3.18)$$

where n_{vert} is the number of vertices of the element. With this particular definition, the volumes associated with each node may be written as

$$\mathcal{V}_i = \sum_{\{C|i \in C\}} d_C^i \mathcal{V}_C \qquad (3.19)$$

3.2.1 Artificial Dissipation

For most simple distribution matrices, artificial dissipation must be added to the scheme in order to capture shock waves as well as to damp spurious high frequency oscillations [14, 19] that can seriously affect the convergence to steady state, as well as the final solution itself. The standard artificial dissipation model described above uses a combination of a fourth difference for background-smoothing, and a second difference for shock-capturing. On unstructured meshes the latter is usually a scaled approximation of a Laplacian–type operator, and is constructed by distributing edge differences [20, 23] or differences between nodal values and cell–averaged values [17]. The background smoothing is then calculated in a similar manner. However, this straightforward approach leads to a loss of precision when applied to the distorted or stretched elements [22] that are often found within unstructured grids.

To avoid this problem, Schönfeld and Rudgyard [18] use a technique that aims to reduce the influence of the grid on the converged solution.

Although the artificial viscosity they devised reduces to the standard model (3.4) – (3.6) for simple one dimensional problems on a mesh with constant spacing, it also preserves linear varying solutions on arbitrary meshes in three dimensions. It is based on the method described by Morgan *et al* [23] and makes use of a nodal approximation to the solution gradients, denoted $(\nabla \mathbf{u})_i$. It may be formulated for cells, edges or faces, depending on the chosen data structure. In the former case this becomes

$$D_i = \sum_{\{i|i \in C\}} \left\{ (\epsilon_C^{(2)} + \epsilon_C^{(4)})(\overline{\mathbf{u}}_C - \mathbf{u}_i) - \epsilon_C^{(4)} \overline{(\nabla \mathbf{u})}_C \cdot (\overline{\mathbf{x}}_k - \mathbf{x}_i) \right\} \quad (3.20)$$

where, in analogy to (3.5) – (3.6),

$$\epsilon_C^{(2)} = \kappa^{(2)} \hat{\lambda}_C^* \zeta_C \quad (3.21)$$

$$\epsilon_C^{(4)} = \max(0, \kappa^{(4)} \hat{\lambda}_C^* - \epsilon_C^{(2)}) \quad (3.22)$$

and $\kappa^{(2)}$ and $\kappa^{(4)}$ are user–defined parameters, $\hat{\lambda}_C^*$ is a suitable cell–wise scaling and the bar notation represents a cell average of nodal quantities. The term ζ_C should be O(1) near discontinuities but small elsewhere. This is ensured using modifications of the functions suggested by Swanson and Turkel [24] and Morgan *et al* [23], since the formulation given by the latter authors has been found to be extremely dissipative due to its limiter–like properties. It is given as a function of the pressure:

$$\zeta_C = \max_{i \in C} \frac{\|\overline{(\nabla p)}_C - (\nabla p)_i\|}{\|\overline{(\nabla p)}_C\| + \|(\nabla p)_i\| + \beta \frac{\overline{p}_C}{L_C}} \quad (3.23)$$

where β is an appropriate constant and L_C is proportional to a local length scale.

In order to recover nodal values of the gradients $(\nabla \mathbf{u})_i$ and $(\nabla p)_i$ we may calculate a cell approximation and then distribute this to the nodes. The cell gradients are given using Gauss's theorem, eg.

$$\left(\frac{\partial \mathbf{u}}{\partial x} \right)_C \approx \frac{1}{\mathcal{V}_C} \int_{\partial \Omega_C} (\mathbf{u}, 0, 0) \cdot \mathbf{n} dS \quad (3.24)$$

and this is approximated in exactly the same manner as the cell residuals themselves. A weighted average of neighbouring cell gradients then defines a nodal approximation. Although many other interpolation procedures are possible, the simplest technique is given by

$$(\nabla \mathbf{u})_i = \frac{1}{\mathcal{V}_i} \sum_{\{C|i \in C\}} \mathcal{V}_C d_C^i (\nabla \mathbf{u})_C \quad (3.25)$$

with d_C^i defined in (3.18).

With this artificial dissipation model the nodal residual (3.16) is now modified to be

$$\mathbf{Res}_i = \frac{1}{\mathcal{V}_i} \left(\sum_{C_i} \mathcal{V}_C \, d_C^i \, \mathbf{Res}_C - \mathcal{D}_i \right). \quad (3.26)$$

3.2.2 Diffusive Terms

In most discretisations of the Navier–Stokes equations using vertex data, viscous terms are added to individual nodes *after* the calculation of the inviscid contributions from neighbouring cells. However, it is also possible to include the first–order flux tensor directly in the calculation of the *cell residual*, and thus retain the weighted cell residual approach. This approach is straightforward to implement once we have recovered the nodal gradients (eg. using (3.25)) since we may use the chain rule to estimate the corresponding gradients of velocity, temperature and the turbulent quantities k and ϵ. These may then be used directly in order to calculate values of the flux functions of (3.2).

3.3 Boundary Conditions

The boundary conditions group into three general classes: 1) physical conditions, e.g. $\mathbf{v}_{wall} = 0$ or an isothermal wall, 2) farfield conditions where the mesh terminates, and, in the case of structured multiblock codes, 3) geometric conditions that relate to the mesh topology and connectivity, e.g. the condition of a face folding over onto itself along a coordinate cut.

For example at an adiabatic wall the gradient of the temperature normal to the wall is zero, the velocity vanishes, and the normal pressure gradient is determined from the momentum equation

$$\frac{\partial p}{\partial n} = \mathbf{n} \cdot (\mathbf{grad} \cdot \tau) \quad (3.27)$$

Inflow to or outflow from a farfield boundary is handled by the usual extrapolating or setting the appropriate Riemann invariant according to the slope of its associated characteristic.

For cell centre schemes, the imposition of physical and farfield boundary conditions involves in general the use of 'ghost' cells, and is well–documented [7].

For cell vertex schemes, the situation is slightly more complicated, since the boundary conditions must be imposed on the unknowns themselves. See [33] for a full discussion of this topic.

3.4 Convergence Acceleration

Multigrid is commonly used to accelerate the solution to a steady state. The multigrid method we outline here is based on the use of Full Approximation Storage (FAS) cycles where sawtooth or V-cycles as well as full multigrid can be chosen [26],[27]. The basic strategy is described briefly for structured meshes, since more complicated interpolation operators are required in the unstructured case, although the basic ideas remain the same [23].

Consider a sequence of grids $G_m, m = 1, ... M$ with M being the finest level and with cell sizes $h_m = 2h_{m+1} = 4h_{m+2} = \cdots$. Then on the finest grid G_M the semidiscrete problem (3.2) becomes after dropping subscripts

$$\frac{d}{dt}\mathbf{u}_M + \mathbf{Res}_M(\mathbf{u}_M) = 0 \qquad (3.28)$$

where $\mathbf{Res}_M(\mathbf{u}_M)$ is the spatial discretization of the steady-state flux balance (residual) on the finest mesh M.

3.4.1 Fine-to-Coarse Sweep

A smooth approximation to Eq.(3.28) may be obtained by iteratively solving (smoothing steps described below) a modified semi-discrete equation on a coarser grid level m

$$\frac{d}{dt}\mathbf{u}_m + \mathbf{Res}_m(\mathbf{u}_m) = \mathbf{Cor}_m^M \qquad (3.29)$$

where the right-hand-side $\mathbf{Cor}_m^M$ is the fine-to-coarse correction acting as a forcing function on discretization error. It maintains the truncation error of the fine grid G_M on the coarser grids G_m and is defined recursively as

$$\begin{aligned}
\mathbf{Cor}_m^M &= \mathbf{Res}_m(\mathcal{I}_{m+1}^m \mathbf{u}_{m+1}) + \mathcal{II}_{m+1}^m \left[\mathbf{Cor}_{m+1}^M \right. \\
&\quad \left. -\mathbf{Res}_{m+1}(\mathbf{u}_{m+1})\right]
\end{aligned} \qquad (3.30)$$

where the operators $\mathcal{I}_{m+1}^m$ and $\mathcal{II}_{m+1}^m$ restrict the state variables $\mathbf{u}_{m+1}$ and the discrete steady-state approximation $\mathbf{Res}_{m+1}$ as well as the defect corrections $\mathbf{Cor}_{m+1}^M$ from grid G_{m+1} to G_m. They are defined as full-weighting operators

$$\mathcal{I}_{m+1}^m \mathbf{u}_{m+1} = \frac{\sum \mathcal{V}_{m+1}\mathbf{u}_{m+1}}{\sum \mathcal{V}_{m+1}}$$

$$\mathcal{II}_{m+1}^m \mathcal{R}_{m+1} = \sum \mathcal{R}_{m+1} \qquad (3.31)$$

with the total residual of Eq.(3.29) $\mathcal{R}_{m+1}$ defined as

$$\mathcal{R}_{m+1} = \mathbf{Cor}_{m+1}^M - \mathbf{Res}_{m+1}(\mathbf{u}_{m+1}) \qquad (3.32)$$

The summations in Eqs.(3.31) are over the 8 fine cells contained within a coarse cell.

The fine-to-coarse sweep solves the temporally discretized Eq.(3.29) on the coarse grid G_m

$$\mathcal{S}\Delta\mathbf{u}_m + \mathbf{Res}_m(\mathbf{u}_m^{(0)}) = \mathbf{Cor}_m^M \qquad (3.33)$$

where $\mathbf{u}_m^{(0)}$ is the current solution on grid G_m around which Eq.(3.33) has been linearized for implicit time integration, or is the first estimate for an explicit time step. For its value restrict the next finer solution

$$\mathbf{u}_m^{(0)} = \mathcal{I}_{m+1}^m \mathbf{u}_{m+1} \qquad (3.34)$$

and obtain subsequent estimates $\mathbf{u}_m$ which are smoothed by solving Eq.(3.33) for $\Delta\mathbf{u}_m = \mathbf{u}_m - \mathbf{u}_m^{(0)}$. The operator $\mathcal{S}$ is designed to have good smoothing properties and depends on the time discretization that one chooses (see below). Once a smooth solution $\mathbf{u}_m$ is determined, Eqs.(3.34) and (3.30) transfer the results to the next coarser grid and the procedure is repeated until the coarsest grid is reached.

3.4.2 Coarse-to-Fine Sweep

After some smoothing iterations on the coarsest grid, the coarse-to-fine sweep of the multigrid cycle begins by updating the previous solution on the finer grid G_m with the current smoothed solution $\mathbf{u}_{m-1}$ on the next coarser grid G_{m-1} by prolongation. The FAS procedure transfers to the fine grid only the correction between the old fine grid solution $\mathbf{u}_m^{old}$ and the smoothed coarse-grid solution $\mathbf{u}_{m-1}$

$$\mathbf{u}_m^{new} = \mathbf{u}_m^{old} + \mathcal{P}_{m-1}^m(\mathbf{u}_{m-1} - \mathcal{I}_m^{m-1}\mathbf{u}_m^{old}) \qquad (3.35)$$

where $\mathcal{P}_{m-1}^m$ is the prolongation operator using piecewise constant interpolation. In each of the 8 fine cells contained within a coarse cell, the prolongated correction is taken to be the coarse cell correction, i.e.

$$\mathcal{P}_m^{m+1}\Delta\mathbf{u}_m = \Delta\mathbf{u}_m \qquad (3.36)$$

In a V-cycle the new solution $\mathbf{u}_m^{(0)}$ on the finer mesh is smoothed by iterating Eq.(3.33) before proceeding to the next finer level. In a sawtooth cycle this smoothing step is omitted and the solution is immediately prolongated to the next grid level.

For a steady-state solution the coarse-grid boundary conditions can be used without correction simply by transferring their change on the coarser grid back to the fine grid with the FAS prolongation (3.35). One can also freeze the boundary conditions on the coarse grid and update them only on the fine grid after each smoothing step.

3.4.3 Explicit Time Smoother

Source terms for the turbulence variables can be treated using a diagonalized point implicit scheme, or are treated explicitly. In all cases, the system of equations is coupled directly, and the turbulence variables are solved simultaneously with the mean flow.

The time-smoothing operator $\mathcal{S}$ in Eq.(3.33) is carried out by a standard multi–stage Runge–Kutta technique. For one stage of a Runge-Kutta scheme the smoothing operator simply becomes the diagonal matrix $\mathcal{S} = diag[\frac{1}{\alpha \Delta t}]$ where α is a constant. Using multiple stages of a general M-stage scheme greatly increases the damping effect of the scheme.

$$
\begin{aligned}
\mathbf{u}^{(0)} &= \mathbf{u}^n \\
&\vdots \\
\Delta \mathbf{u}^{(m)} &= -\alpha_m \Delta t\, \mathcal{R}(\mathbf{u}^{(m-1)}) \\
\mathbf{u}^{(m)} &= \mathbf{u}^{(0)} + \Delta \mathbf{u}^{(m)} \quad\quad (3.37) \\
&\vdots \\
\mathbf{u}^{n+1} &= \mathbf{u}^{(M)}
\end{aligned}
$$

for $m = 1, \ldots, M$, where $\mathcal{R}(\mathbf{u})$ is defined in Eq.(3.32), and the Runge-Kutta coefficients α_m are given by the user.

Integrating with local time steps accelerates the convergence to a steady-state solution because signals propagate faster. A second acceleration technique, which also increases the high-frequency damping, is implicit residual smoothing whereby one Runge-Kutta stage of Eq.(3.37) reads

$$(1 - \omega)\delta^2 \Delta \mathbf{u}^{(m)} = -\alpha_m \Delta t\, \mathcal{R}(\mathbf{u}^{(m-1)}) \quad (3.38)$$

where δ^2 denotes some discrete approximation to the scaled Laplacian operator. This may be solved iteratively, although we may simplify the procedure on structured meshes by replacing the smoothing operator by three uni–directional operators:

$$
\begin{aligned}
(1 - a\delta_i^2)(1 - b\delta_j^2)(1 - c\delta_k^2)\Delta \mathbf{u}^{(m)} &= \\
-\alpha_m \Delta t\, \mathcal{R}(\mathbf{u}^{(m-1)}) & \quad (3.39)
\end{aligned}
$$

where δ_i^2 denotes a second difference in the grid direction i, and the smoothing coefficients a, b, and c are $O(1)$. Such an approach allows the CFL number to be increased by a factor of two or three.

3.4.4 Locally Implicit Methods

As complete global matrix implicit methods are still too costly and impractical, most implicit CFD codes are based on point implicit schemes whereby at each point ones solves the system

$$\frac{1}{\Delta t}(\mathbf{u}_i^{n+1} - \mathbf{u}_i^n) + div\mathcal{F}(\mathbf{u}_i^{n+1}) = \Omega_i$$

by linearising around the state $\mathbf{u}_i^n$. This leads to a sub-problem with a $NU \times NU$ Matrix-Vector resolution, where NU denotes the number of unknowns in the state vector W^n, which is performed by some iterative scheme. These methods are extremely local, and can be used as sub parts of a multigrid cycle, or as a more precise residual smoothing by taking just the diagonal terms. Their incorporation within a Runge-Kutta outer loop allows unsteady calculations with higher CFL numbers. As such they are considered more as convergence accelerators than as methods in themselves. However, they allow direct treatment within a multi block structure, exactly as explicit methods. A more promising field are implicit methods based on residual minimisation, which are least–squared type minimisation techniques on the actual residuals coming from the governing differential equations. For finite volume schemes the implementation of these methods are more difficult than for the cell vertex ones, and for the multiblock structure several theoretical concepts need to be solved, [28].

4 STRUCTURE OF FLOW SOLVERS

4.1 Dataflow in a CFD Simulation

A complete flow simulation can be subdivided into the following four processes:

- Generation of the surface mesh

- Generation of the grid (structured/unstructured, multiblock/single block)

- Solving the Euler or Navier Stokes equations

- Visualization and analysis of results

Data is exchanged among these four stages, and if necessary, modifications are made to improve the solution or solution process. For example, the analysis of the calculated results may lead to adaptation of the grid so as to improve the resolution of particular flow features, or it may be decided that more time steps are required in order to converge the solution further.

In many CFD codes, data is transferred by means of files, for example there is a file for the surface definition, a file for the mesh, and a file for the calculated solution. For large calculations (i.e., calculations with more than 1 Million grid points), these files become rather large, and reading them in order to visualize a specific flow region can take considerable time, especially when this region is near the end of the file. However, in the case of multi–block structured meshes, or unstructured meshes that have been partitioned into blocks, it is possible to have a mesh and solution file for each block, which permits the solution to be read in a given block quite rapidly. Some codes therefore define files for each of the blocks, although this requires the management of a substantial number of files — with 64 blocks, this requires 128 files assuming a mesh file and solution file for each block.

With the development of the multi block solver **NSMB**, a different strategy was adopted based on the use of the MEM-COM data base system [29]. MEM-COM is a data management system for memory and memory-to-disk data handling. MEM-COM contains several data access modules, as for example the Data Base manager and the Relational Table Manager which are used by **NSMB**. With the MEM-COM data base system, all data used in the four processes listed above are saved in a single data base file, which appears as an ordinary UNIX-file on the file system. The data base file contains several data sets which are identified by their name and their type. Each data set can be accessed independent of the other data sets, hence accessing the last data set saved in the data

base file is as fast as accessing the first data set. Two different data type of data sets are used by **NSMB**, data sets to store variables (typically the coordinates and solution vector) and relational table datasets which contain entries in the form keyword + value. The relational tables are in **NSMB** used to specify the information on the grid and on the calculation made. Since all the information on the flow simulation is saved in the data base file, the data base file can be used for archiving.

Finally, MEMCOM is coupled to the graphics package BASPL, and the results of simulations can be directly visualized. In case the MEMCOM data base file needs to be transferred between computers with a different binary representation, tools are available to copy the data base with automatically converting the binaries.

Both **NSMB** and **AVBP** use Dynamic Memory Allocation of the variables at run time.

NSMB uses the Dynamic Memory Manager (DMM) supplied with the MEM-COM data base system. Variable storage is allocated block by block, as required by distributed memory parallel computers where each processor has its own memory, and blocks are assigned to a given processor. On shared memory (parallel) computers, the DMM automatically provides the pointer arrays to different processors. At run time, **NSMB** reads the number of blocks from the MEM-COM data base file, and for each block it reads the grid size and allocates the necessary memory for each variable.

AVBP uses Fortran extensions such as the POINTER statement in order to dynamically allocate the arrays. The local grid refinement method, developed within **AVBP** [18, 20] takes advantage of this dynamic memory allocation since the corresponding arrays grow as the grid is refined and more nodes and cells need to be stored.

4.2 Parallel Computing

Performing large scale computations requires large computational resources in terms of memory and of CPU-time. As an example, the calculation with **NSMB** for the HALIS geometry using 1.7 Million grid points presented in Section 5.3 required 85 Mword of memory on the Cray YMP M94/4-512 of EPFL, and the CPU time needed to reach convergence was more than 50 CPU hours. The present trend in Navier Stokes simulations is to use even finer grids (between 3 and 10 Million grid points), and these type of calculations are difficult to make on the present generation of vector computers. Distributed memory parallel computers, as for example the IBM SP2 and the Cray T3D, seem to have the potential to make these kind of calculations possible since they offer the required memory and CPU power. However, programming these computers is substantially more complex than programming for vector computers, and it is very difficult to get more than 10 - 20 % of the peak performance per node, this compared to the 50% of the peak performance a Cray YMP computer delivers. Moreover, optimizations made in a code to run efficiently on vector computers lead to a very poor performance on RISC architecture based parallel computers as IBM SP2 and Cray T3D, and vice-versa.

Both **NSMB** and **AVBP** were designed for using parallel computers by assigning blocks or partitions to different processors. Once per time step (or more often when needed) the information on the block interfaces is exchanged between the processors using standard message passing libraries such as PVM or PARMACS. Within **NSMB** and **AVBP** the communication routines are hidden by using a set of Macro's (called IPM [36]), which form the interface to PVM, PARMACS and, in the near future, MPI. In the CRAY-T3D version of **AVBP**, these macros are obsolete, and communication is performed using the Cray *shmem* directives.

A client-server approach was originally adopted by both codes as the parallelization strategy, where the client (typically a node of a parallel computers), asks and receives data from the server.

For **NSMB**, the server is located on the host computer (at EPFL the Cray YMP M94/4-512), and it reads the input file and the MEM-COM data base file and sends the data to the different nodes (clients) of the T3D. After the calculation has been finished, the clients send the data to the server which saves the data in the MEM-COM data base. The only difference from the user point of view between a serial and a parallel execution is that when running on a parallel computer a file needs to be read which specifies which block goes to which processor. When this file is not present a serial execution is made.

Two major problems appeared when porting **NSMB** to parallel computers. First, for turbulent calculations using the Baldwin Lomax turbulence model [34], it appeared that the initial implementation needed to be improved because the domain decomposition tool (see the next section) splitted blocks parallel to the wall in boundary layers. Since the Baldwin Lomax model requires information normal to a wall point, this information could be located in another block on another processor, and a genuine general parallel multi block implementation of the Baldwin Lomax model was developed to solve this problem.

The second problem concerned the optimization for RISC processors. The do-loops in the most time consuming routines were written as 1D loops with vectorization on the first array index for multi dimensional arrays. This yields very long vector loops, even for small blocks. RISC processors prefer to have the shortest array index as the first dimension, and ijk-versions of do-loops are preferable since the amount of work is reduced since no work is carried out in the ghost cells. To make **NSMB** running efficiently on both vector and RISC computers, the most time consuming routines are available as 1D-version and IJK-version. Moreover, a procedure was developed which flips automatically the array indices of multi dimensional arrays to get the shortest index in the leading dimension. When the code is installed, the user can choose between 1D and IJK version, and between flipped and non-flipped version, and the corresponding source code is generated using the C-preprocessor.

The **AVBP** code makes use of a modern *parallel library* that aims to free the non-specialist user from the need to consider aspects of high performance computing such as data partitioning, message passing or parallel I/O. An attempt to create such a general framework for parallel computations on arnitrary meshes has resulted in the prototype **COUPL** library (The CERFACS and Oxford University Parallel Library, previously named **CPULib** [31]).

For the present implementation of **AVBP** using **COUPL**, a client-server paradigm has been chosen. However, unlike most master-slave implementations, the **COUPL** master process is completely independent of the slave application, and may be viewed as a 'black' box, even though it is responsible for partitioning the mesh, dealing with the I/O and coordinating some of the actions of the slaves. The user is expected to write the code that corresponds to the slave application, and communicates with the generic master process, as well as other slaves, using subroutine calls from the library. He is not responsible for calculating any pointers either for exchanging data at partition interfaces, or for mapping data to or from the original mesh, and at no stage is he expected to deal with the underlying message-passing library itself. Although the master-slave paradigm is suitable for several HPC platforms, an SPMD implementation is arguably more appropriate for many newer machines. Fortunately, due to the use of a standard subroutine interface an SPMD version of **COUPL** may be implemented in a manner that is transparent to **AVBP** — this is achieved by allowing the root node to act as both the master and a slave.

4.3 Multi Block Implementation and Mesh Partitioning

Within **NSMB**, a multi block simulation is regarded as a multiple single block calculation (where each block can be executed in parallel on a different processor), hence it is only on the level of the incorporation of the boundary condition that there is a difference between single and multi block calculation due to the introduction of a block connectivity boundary condition. The boundary conditions are implemented such that a Runge Kutta stage can be executed in a block independent of its neighbouring blocks. This multi block implementation is very suitable when using explicit schemes.

To be able to use a parallel computer efficiently, it is necessary to split the (multi block) grid coming from the grid generator in such a way that a balanced calculation is obtained on the parallel computer. For structured grids, the multi block decomposition of the original grid is defined by the constraints of the mesh generation process, and not by requirements imposed by using a parallel computer.

At KTH, a domain decomposition tool, called MB-Split, was developed which takes a structured multi block grid coming from the mesh generator and splits it in such a way that a balanced calculation is obtained on a parallel computer using the number of nodes specified by the user. The number of processors can be on a single machine, or on a cluster of heterogeneous machines. The input to MB-Split is the MEM-COM data base containing the grid, and an input file specifying the number of nodes and the estimated performance of each node when using a heterogeneous cluster of nodes. The output of MB-Split is a new

MEM-COM data base for the decomposed grid, together with a file which specifies which block goes to which processor. Note that **NSMB** permits to have several blocks per processor, which greatly facilitates the load balancing of complex calculations.

A powerful feature of MB-Split is the possibility to merge a splitted grid back into the original multi block grid coming from the grid generator, and the possibility to split not only the grid, but also a solution stored in a MEM-COM data base. This feature was implemented for several reasons. First, a fine grid computation is never made from free stream conditions, but always from a solution obtained on a coarser grid. Coarse grid calculations can be made using less nodes on a parallel computer, hence they use a different block decomposition than the fine grid. The easiest way to interpolate a solution from a coarse grid to a fine grid is when the grids have the same multi block topology, which is provided by the original multi block decomposition coming from the grid generator. Another reason for merging the solution back to the original grid is that the post processing is facilitated since the original grid will contain less blocks than the grid used for the parallel computation. In case the grid is modified (without changing the topology) to better capture the important flow features, the solution in the merged grid is directly available as initial solution.

An unstructured code can handle large grids for complex geometries without the need for a multi-block mesh per se. However, a multiple domain approach helps to improve computational efficiency, reduces storage requirements and offers a natural framework for data parallelism using distributed or shared memory parallel computers. Although unstructured data sets are irregular, they have the advantage of being homogeneous (contrary to block-structured grids), which allows us to split a computational domain arbitrarily into any number of sub-domains. Several real-time partitioning algorithms are provided within **COUPL**, and are therefore integrated within **AVBP**. These include Greedy, Reverse Cuthill-McKee, Coordinate and Inertial Bisection techniques, applied to meshes of tetrahedral, prismatic, pyramidal and hexahedral elements. Such methods aim to load-balance the calculation whilst using heuristics to minimize interface lengths so as to reduce the communication between processors in a distributed memory environment.

4.4 Boundary Condition Implementation

In multi block simulations, two types of boundary conditions can be distinguished, boundary conditions which are local to a block (symmetry plane, solid wall, inflow/outflow boundary conditions), and boundary conditions which require information from neighbouring blocks (block connectivity boundary conditions).

Initially, **NSMB** used the concept of boundary arrays and ghost cells to impose the boundary conditions, which offered a large flexibility. More recently this concept was abandoned for two reasons, first the boundary arrays require additional storage, and second filling the boundary arrays is costly due to the indirect addressing. At present, **NSMB** uses still two layers of ghost cells at each side of the computational domain, however, these cells are now directly filled to satisfy the physical or numerical boundary conditions. For block connectivity boundary conditions the values of the state vector and pressure in the adjacent block are directly copied into the ghost cells.

NSMB offers the flexibility to have several types of boundary conditions on a face of a block using the concept of boundary condition windows, which are defined by their start and end indices. This concept is especially useful for parallel computations where a face of a block is often connected to several other blocks.

AVBP does not require the definition of ghost variables in order to impose far–field or physical boundary conditions. Nodes for which a particular boundary condition applies are stored as 'patches' — lists into the global node numbers. Indirect addressing is then used to impose values there.

At *internal* boundaries the cell-vertex formulation of our unstructured solver results in grid points that are located directly at the interface between two blocks. **COUPL** stores copies of the nodes that lie on these interfaces for each partition, although these are themselves distinguished as being owned or unowned by appropriate processors so that all sets are effectively partitioned. Since **COUPL** also allows us to choose between overlapping and non–overlapping partitions, a wide range of block interface treatments is available to **AVBP**, including the possibility of overlapping computations and communications.

5 COMPUTED RESULTS

5.1 NTF-Deltawing

Although the geometry of a deltawing is simple, the simulation of the vortex-dominated flow over the wing at high angle of attack is a challenging problem due to the interaction of multiple vortices, possible interaction of shock waves with the vortices, and most often the flow is turbulent.

Here results are presented of the flow over the $65°$ swept deltawing with a round leading edge that has been studied experimentally in the NASA Langley NTF cryogenic windtunnel. Experimental data are available for Reynolds numbers between 5 and 150 million, and here results are presented for a turbulent calculation with $M_\infty = 0.85$, $\alpha = 12°$ and $Re = 9 \times 10^6$ based on the root wing chord.

The finest mesh used contained 550'000 grid points, the maximum y^+ value near the wall was equal to 0.2, and there were 80 points located between wall and free stream boundary. The algebraic Baldwin Lomax turbulence model was used in the calculations. Calculations after 2000 and 2600 time steps were compared, and it showed that there are no difference in the results. Figure **??** shows the particle traces above the wing, and the vortex is clearly visible. Figure **??** shows the isobars in the plane at 40% chord station, the primary vortex is clearly visible, and a secondary vortex is visible near the leading edge. Comparing the calculated results with the experimental data at the same Reynolds number, see Fig. **??**, shows large discrepancies in the C_p: the maximum value is underpredicted, and the center of the vortex core in the calculations is shifted inboard. There are several possible explanations for this behaviour:

- In the Baldwin Lomax model the outer length scale is computed using the maximum of the function $F(y)$, where y is the distance from the wall. In simple wall flows, this function has one maximum, however this is not the case for the flow over a deltawing. The function $F(y)$ can have multiple maxima, and the selection of the *wrong* maximum may lead to an overprediction of the eddy viscosity.

- Another possible cause is that the grid is not fine enough in the region of the vortex and this causes excessive diffusion of the vorticity.

- A third possibility could be due to the uncertainty in the transition location. For the calculation we have specified that the entire flow is fully turbulent. What the conditions actually are in the experiment is unknown to us.

5.2 Falcon Jet Airplane

The Falcon Jet airplane used as an example for flow simulation with our unstructured flow solver **AVBP** is a generic twin-engine version of the real jet plane, often referred to as the FLACON-JJ. The free-stream flight conditions are the transonic flow at Mach=0.85 under a small angle of attack $\alpha = 2°$.

Two versions of AVBP were used: a standard version using PVMe on the IBM SP2 with 32 processors, and the special EPFL-CRAY T3D version using 16 processors. The mesh has been generated using the FELISA tetrahedral mesh generator of Peraire *et al* . The final mesh is composed of 155932 nodes and 847361 tetrahedra. A detail of the surface mesh is given in Fig ??. The surface mesh is extremely regular and adapted to the domains of flow features. The outer shell boundary is also superimposed showing the high concentration of grid points around the airplane.

The results for an Euler calculations are displayed in Figure ??. A fixed engine flow is assumed. the pressure contours on the surface of the plane illustrate the transonic regime flight constraints.

5.3 Halis

The HALIS geometry is a generic shuttle geometry constructed from the windward side of the US Orbiter, with a simplified leeward side. This geometry has been tested in "cold " windtunnels (ONERA S4), and is presently being tested in several "hot" high enthalpy wind tunnels (DLR HEG, ONERA F4, CALSPAN) as part of a project to find equivalent parametrisation for hypersonic Wind Tunnel to Flight extrapolation. The Orbiter flight data notably has brought insight into a pitching moment anomaly, due to real gas effects in the reentry phase.

The flowfield around the HALIS geometry is extremely complex, comprising large zones of viscous interactions, cross flow shocks, shock wave boundary layer interactions, wakes, and recirculation zones.

Very fine meshes are required to capture these physical effects with sufficient accuracy. The grid for the calculation presented here was composed of 17 blocks, and contained in total 1.7 Million grid points.

The calculation presented here was part of an ESA/ESTEC Workshop on Hypersonic Flows held in November 1994. Several groups in Europe made the same computation, and experimental data of the ONERA S4 windtunnel were available. Most of the calculations presented were made on a structured grid, and it appeared that 1.7 Million grid points was not sufficient to predict correctly the C_p and Stanton Number at the body flap. One group presented results using hybrid meshes (structured in viscous regions, unstructured outside), but their mesh was rather coarse, and they only studied a part of the geometry. Figure ?? shows the surface mesh of the HALIS, each block on the surface plotted in a different color. Note that the full calculation included the wake. The conditions of the calculation presented here correspond to the S4MA ONERA Conditions (Cold) with the so-called "HALIS Heat Flux/Pressure model(1/90)" with $15°$ half body-flap, placed within the test chamber with a $\alpha = 40°$ angle of attack, $P_i = 25$ bar and $T_i = 1100°K$ giving a freestream Mach number of approximately 9.8. The Reynolds number per meter was equal to $Re/m = 1.67\ 10^6$. Figure ?? show the isoMach contours in the symmetry plane. The bow show wave is clearly visible, including the shock wave coming from the body flap.

Figure ?? shows the stream lines in the symmetry plane in the region of the body flap. This figure shows clearly the complexity of the base flow region, including a recirculation zone.

Figure ?? shows the isoMach lines in 2 cuts along the coordinate x=constant, just before and just after the end of the body. The cross flow shock is clearly visible, including a large separation area on the leeward side of the body.

Figures ?? and ?? show the C_p in the symmetry plane for the calculation with 1.7 Million grid points, and for the calculation using a medium grid of 200000 points. As can be seen in Fig ??, the C_p is underpredicted compared to the experimental results, but the fine grid showed yielded better results compared to the medium grid. In the region of the bodyflap and at the body flap, the results show that the peak C_p value is largely underpredicted. Again the fine grid yields better results.

Figures ?? and ?? show the Stanton number in the Symmetry plane for the fine and medium grid together with experimental data. The calculated Stanton number between $\frac{x}{L} = 0.1$ and $\frac{x}{L} = 0.4$ is overpredicted compared to the experimental data, while further downstream the Stanton number is slightly underpredicted. In the bodyflap region (see Fig ??) the medium grid yields the best results just upstream of the body flap, but the grid is insufficient at the bodyflap and just downstream of it.

After the ESA/ESTEC Workshop, one of the contributors to the HALIS test case made a calculation using an extremely fine grid in the region of the bodyflap [37]. These results showed that with this grid it was possible to predict the C_p in the bodyflap region in agreement with experimental data. In the same report, several problems with the HALIS geometry definition were indicated, which probably have an influence on the results. In particular, the surface definition was not normal to the symmetry plane in the nose region, and the surface patches were not closed, hence gaps existed in the surface description.

6 CONCLUSIONS

We have reviewed two well–known approaches to solving the Navier-Stokes equations for steady compressible flow in aerospace applications, namely a cell-centered finite-volume code using structured meshes and a cell-vertex finite-volume code using unstructured meshes. A two-equation turbulence model has been described, and some details about the numerical methods have been presented. A discussion along with some remarks were also given about the implementation of these methods on vector and parallel computers, including modern concepts for data structure and modular programming techniques.

Three large-scale simulation results have been presented: the first treated transonic flow over an aircraft component, a delta wing, and the second two were simulations for complete vehicle configurations. One of these concerned the study of the aerothermodynamics around the hypersonic spacecraft HALIS using a grid with 1.7 M grid points. Altogether the three cases are fairly representative of the state of the art in computational aerodynamics. The best results have been obtained for the inviscid and laminar flow cases, and the poorest result for the turbulent flow over the delta wing. This points out the weak link in the chain, namely the turbulence modeling, and that there is much room for improvement in this area. We have not emphasized the convergence-acceleration procedure in this paper, but here is another area for improvement in the future. Multigrid can be very effective if the mesh is sufficiently smooth, but the effectivity drops off quickly as soon as cell-aspect ratios change too rapidly between neighboring cells. This puts into conflict the goals, on the one hand, to resolve accurately the boundary layers with an economy of grid points and, on the other hand, to obtain as rapid convergence to steady state as possible. The simulations presented here are also indicative of the large-scale simulations that can be done today, typically grid sizes of several million points. The large High- performance computers that soon will be available should allow the use of grids with 10 to 20 million points. It then remains to be seen whether the solution on such a grid can be driven to a steady state.

7 ACKNOWLEDGEMENTS

Roland Richter of Cray Research Inc. working on the Cray-EPFL Parallel Application Technology Programme is acknowledged for his help in porting the **AVBP** code to the Cray T3D.

The portable parallel version of the **NSMB** code was developed in the EC ESPRIT III Project Parallel Aero, the Swiss part in this project was funded by the Swiss Office for Education and Science under contract OFES 93.0214.

Collaborative work on **AVBP** was partly funded by the EC Human Capital and Mobility Programme within the Network Project 'A Modular Approach to CFD, Iterative Methods and High Performance Computing', Contract number CHRX-CT92-0042.

References

[1] A. Rizzi, P. Eliasson, I. Lindblad, Ch. Hirsch, C. Lacor and J. Hauser, The Engineering of Multiblock/Multigrid Software for Navier–Stokes Flows on Structured Meshes, *Comput-*

ers & Fluids, Vol. 22, No. 2/3, pp. 341–367, 1993

[2] J. Flores and N.M. Chaderijian, Zonal Navier Stokes Methodology for Flow Simulation about a Complete Aircraft. J. of Aircraft, 27, pp 583-590, (July 1990)

[3] J.L. Thomas, R.W. Walters, T. Reu, F.Ghaffari, R.P. Weston and J.M. Luckring, Application of a Patched Grid Algorithm to the F/A-18 Forebody Leading Edge Extension Configuration, J. of Aircraft, 27, pp 749-756, (Sept 1990)

[4] P. Leyland, F. Perrel and J.B. Vos A Family of Multiblock Codes for Computational Aerothermodynamics: Application to Complete Vehicle Hypersonic Flows. AIAA Paper 93-3042 .

[5] P. Leyland, J. Vos, V. van Kemenade, and A. Ytterstrom, NSMB: a Modular Navier Stokes Multiblock Code for CFD, AIAA Paper No 95-0568.

[6] A. Favre, Equations des Gas Turbulents Compressibles, Journal de Mecanique, Vol. 4, No. 3, pp. 361–390, 1965

[7] C. Hirsch, Numerical Computation of Internal and External Flows, Vol 2: Computational Methods for Inviscid and Viscous Flows John Wiley, New York, (1990)

[8] K.Y. Chien, Predictions of Channel and Boundary Layer Flows with A Low Reynolds Number Turbulence Model, AIAA Journal, Vol. 20, No. 1, pp. 33–38, 1982

[9] T. Coakley, "Turbulence Modeling Methods for the Compressible Navier–Stokes Equations", AIAA paper 83–1693, 1983

[10] V.N. Vatsa and B.W. Wedan, Development of an Efficient Multigrid Code for 3D Navier Stokes Equations, AIAA Paper No 89-1791, 1989.

[11] L. Martinelli, Ph.D Thesis, Princeton Univ 1987.

[12] R. Richter and P. Leyland, Entropy Correcting Schemes and Non-Hierarchical Auto-Adaptive Dynamic Finite Element Type Meshes: Applications to Unsteady Aerodynamics, Intl J. Num Meth Fluids, Vol. 20, No. 9, May 1995.

[13] P. Leyland, F. Benkhaldoun, N. Maman and B. Larrouturou, Dynamical mesh adaptation criteria for accurate capturing of stiff phenomena in combustion. To appear in Int.Journ. Num. Meth. in Heat and Mass Transfer 1993.

[14] M.A. Rudgyard, Cell Vertex Methods for Steady Inviscid Flow, VKI, Belgium, 1993.

[15] R.H. Ni, A Multiple Grid Method for Solving the Euler Equations, AIAA J., Vol 20, 1982, pp.1565-1571.

[16] R. Radespiel and R. Swanson, An Investigation of Cell-Centered and Cell-Vertex Multigrid Schemes for the Navier-Stokes Equations, AIAA Paper No 89-0548, 1989.

[17] Y. Kallinderis, Numerical Treatment of Grid Interfaces for Viscous Flows, JCP, Vol 98, 1992, pp. 129-144.

[18] T. Schönfeld and M. Rudgyard, A Cell-Vertex Approach to Local Mesh Refinement for the 3-D Euler Equations, AIAA Paper No 94-0318, 1994.

[19] P. Crumpton and G. Shaw, Cell Vertex Finite Volume Discretizations in Three Dimensions, Int. J. Num. Meth Fluids, Vol 14, No. 5, 1992, pp. 505-527

[20] T. Schönfeld, Adaptive Mesh Refinement Methods for Three-Dimensional Inviscid Flow Problems, Int. J. Comp. Fluid Dyn., Vol 4, pp. 363-391., 1995

[21] T. Schönfeld, M. Rudgyard and P. Crumpton, A Parallel Navier–Stokes Solver for Hybrid Meshes, accepted for Proceed. of 5th Int. Conf. on Grid Generation., Starkville/MS, April 1996.

[22] P. O'Brien and M. Hall, A Comparison of the Effects of Grid Distortion of Finite Volume Methods for Solving the Euler Equations, in Proc 12th Intl Conf Num Meth Fluid Dyn, Springer, Oxford, 1990.

[23] K. Morgan, J. Peraire and J. Peiro, Unstructured Grid Methods for Compressible Flows, AGARD-Proceed., France, 1992.

[24] R. Swanson, and E. Turkel, On Central-Difference and Upwind Schemes, ICASE Rep, 1990.

[25] R. Struijs, A multi-dimensional upwind discretization method for the Euler equations on unstructured grids Ph.D Thesis Delft, 1993.

[26] P. Wesseling, *An Introduction to Multigrid Methods*, John Wiley, New York, 1991.

[27] W. Schröder and D. Hänel, An Unfactored Implicit Scheme with Multigrid Acceleration for the Solution of the Navier Stokes Equations, *Comp & Fluids*, **15**, pp 313-336, 1987.

[28] T. Tefy, P. Leyland, Solveurs implicites avec structure multi-blocs pour les équations de Navier-Stokes Congrès National d'Analyse Numérique, Clermont Ferrand, France, May 1995.

[29] S. Merazzi, MEM-COM An Integrated Memory and Data Management System MEM-COM User Manual Version 6.2 SMR TR-5060, November 1994, SMR Corporation, P.O. Box 41, CH- 2500 Bienne

[30] M. Rudgyard, T. Schönfeld, G. Audemar, R. Struijs, P. Leyland, A Modular Approach for Computational Fluid Dynamics, Paper presented at the ECCOMAS Conference, Stuttgart, 1994.

[31] M. Rudgyard and T. Schönfeld, CPULib – A Software Library for Parallel Application on Arbitrary Meshes, *Proc. Parallel CFD 95*, Pasadena, 1995.

[32] R. Richter Distributed CFD Using Dynamically Allocated arrays within a Fortran77 Framework and Use of Dynamically allocated arrays within a F77 and Shmem Framework, CUG Denver, March 1995.

[33] M.B.Giles Nonreflecting Boundary Conditions for Euler Equation Calculations. AIAA Journal, Vol 28, No 12, pp 2050-2058

[34] Baldwin B.S. and Lomax H. Thin Layer Approximation and Algebraic Model for Separated Turbulent Flows AIAA-paper 78-257, 1978

[35] R. Richter and P. Leyland, Master-Slave Performance of Unstructured Flow Solvers on the CRAY T3D, to appear *Proc. Parallel CFD 95*, Caltech, June 1995.

[36] L. Giraud, P. Noyret, E. Sevault, V. Van Kemenade, IPM 2.3, User's Guide and Reference Manual, CERFACS, November 1994.

[37] J.M.A. Longo and H. Nickel, Navier Stokes Solution for the flow around the HALIS configuration -Cold Windtunnel Conditions-DLR IB 129-94/27, 1995

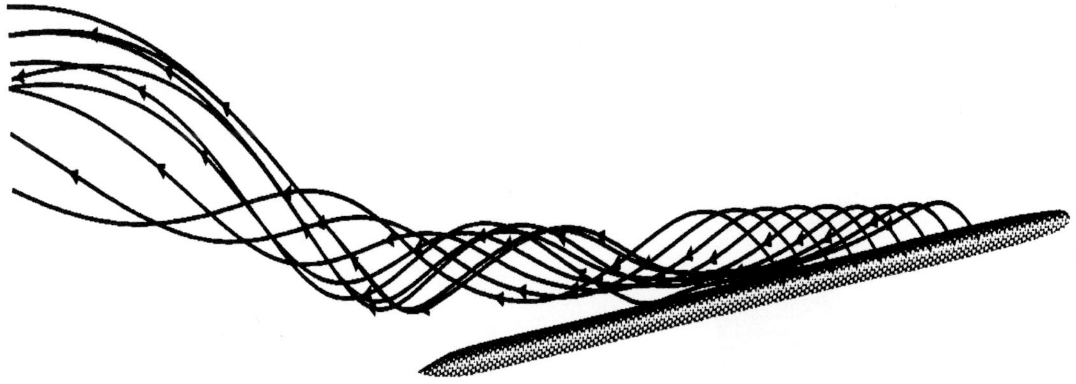

Figure 1: Particle traces depict the leading-edge vortex over the NTF delta wing.

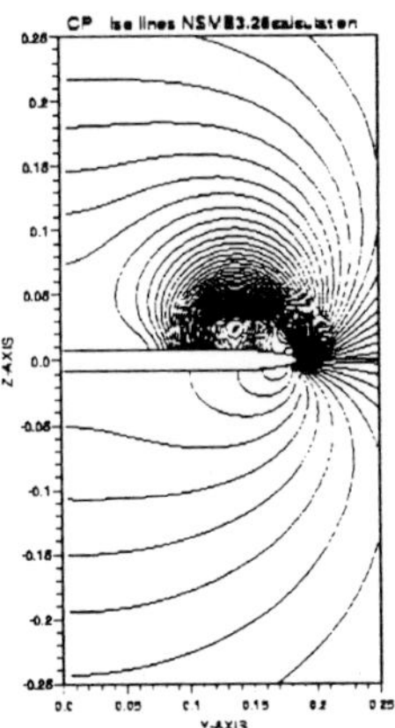

Figure 2: Isobars around the NTF wing at the 40% chord station.

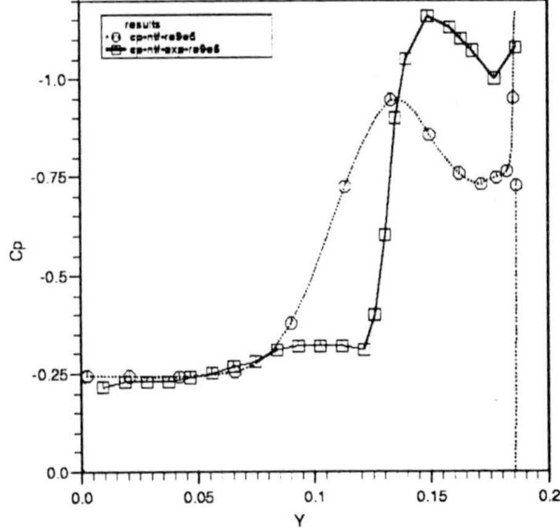

Figure 3: Comparison of the computed and measured spanwise C_p distribution at 40% chord .

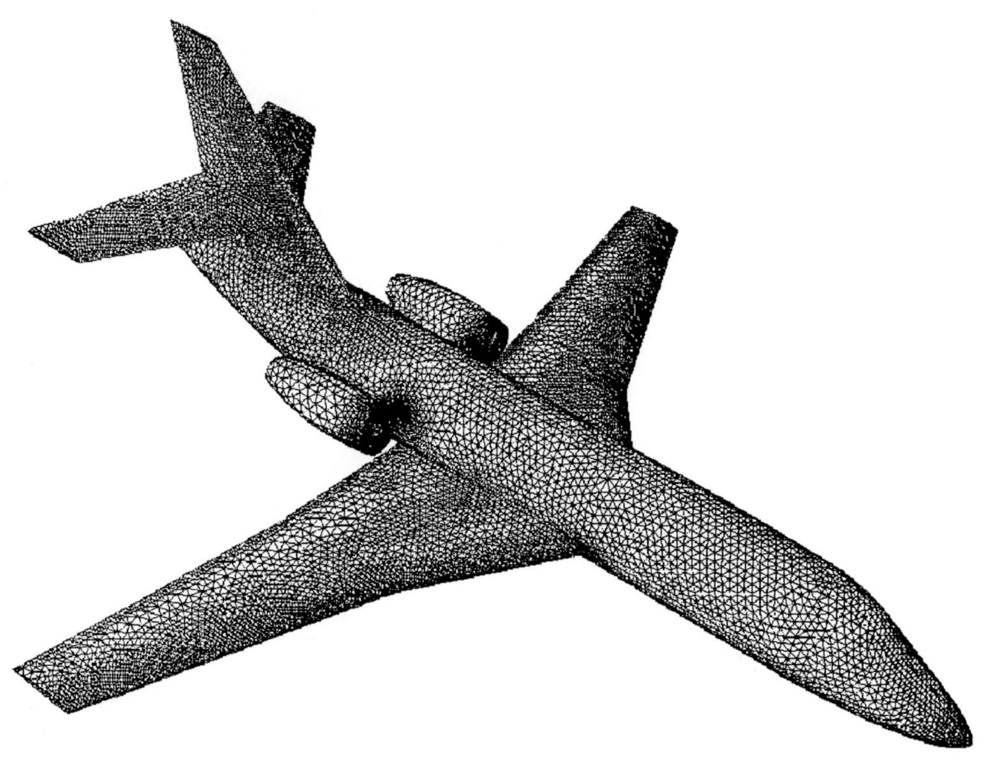

Figure 4: Surface mesh FALCON jet plane

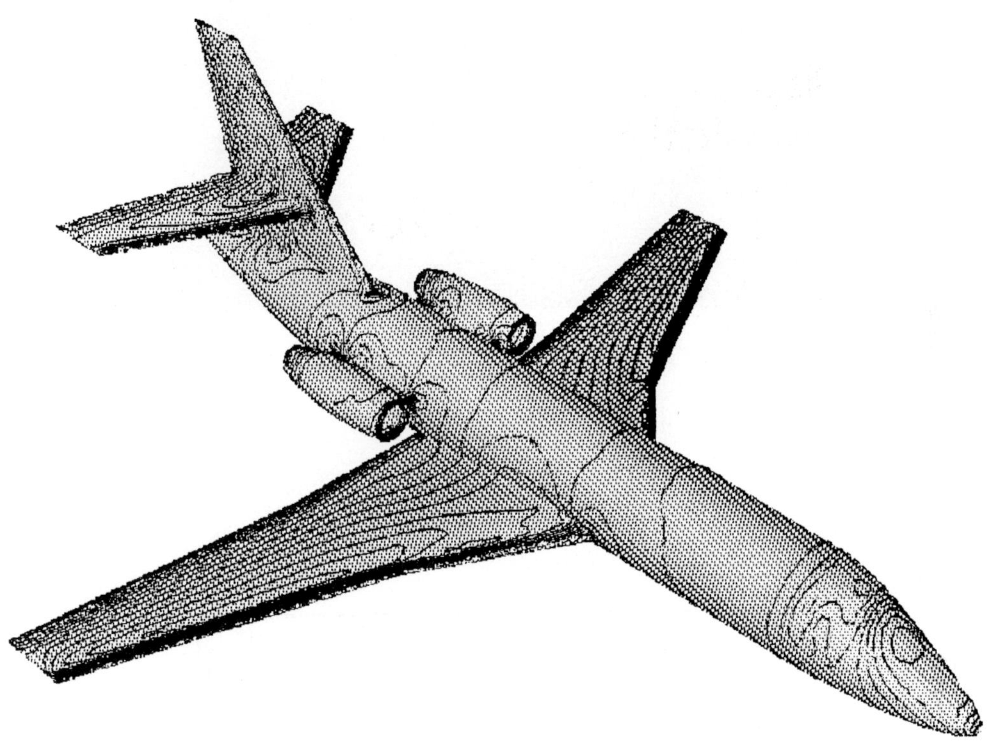

Figure 5: Pressure contours from an Euler solution using 155932 nodes and 847361 tetrahedral elements for a transonic flow over the FALCON jet plane - $M_\infty = 0.85$, $\alpha = 2.0°$.

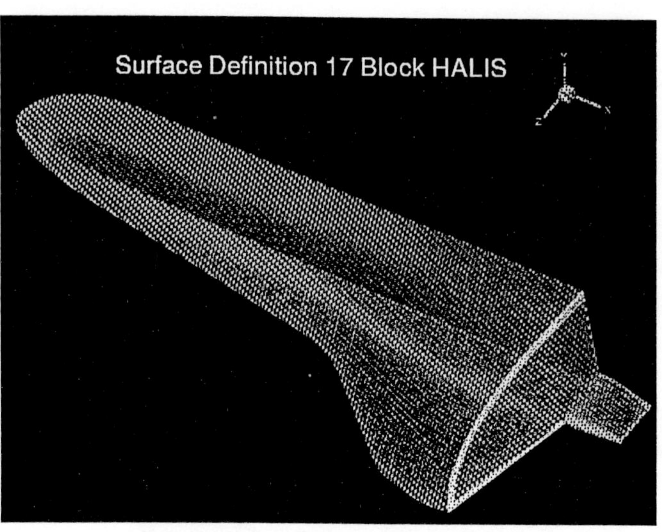

Figure 6: Surface grid around the HALIS hypersonic space vehicle.

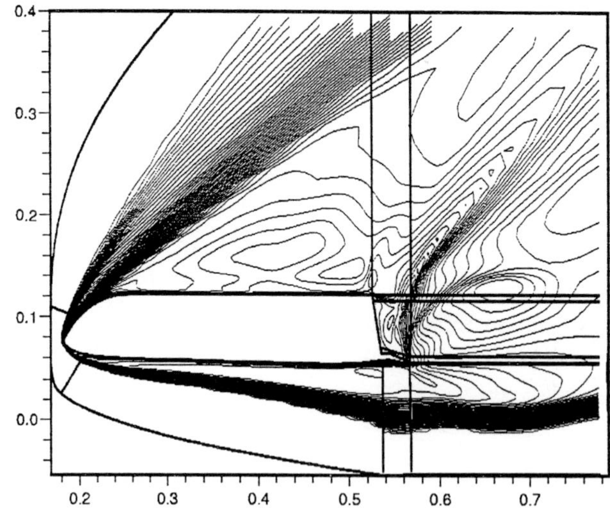

Figure 7: IsoMach contours in the symmetry plane of the HALIS hypersonic space vehicle.

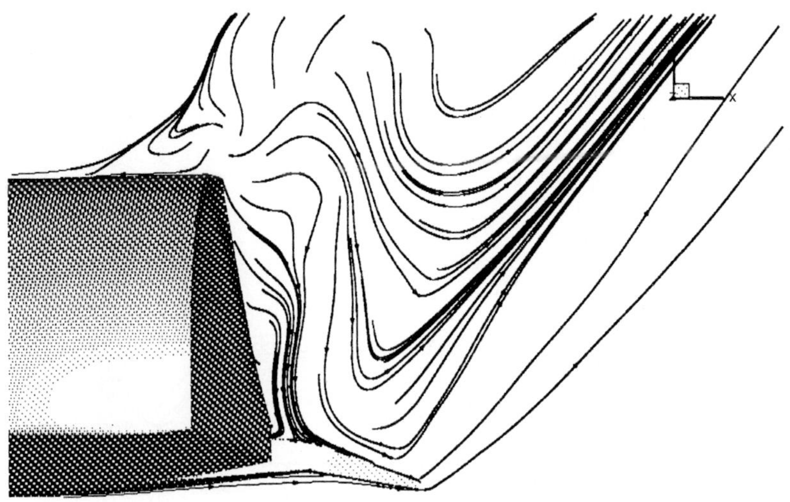

Figure 8: Streamlines in the symmetry plane in the region of the body flap, HALIS hypersonic space vehicle.

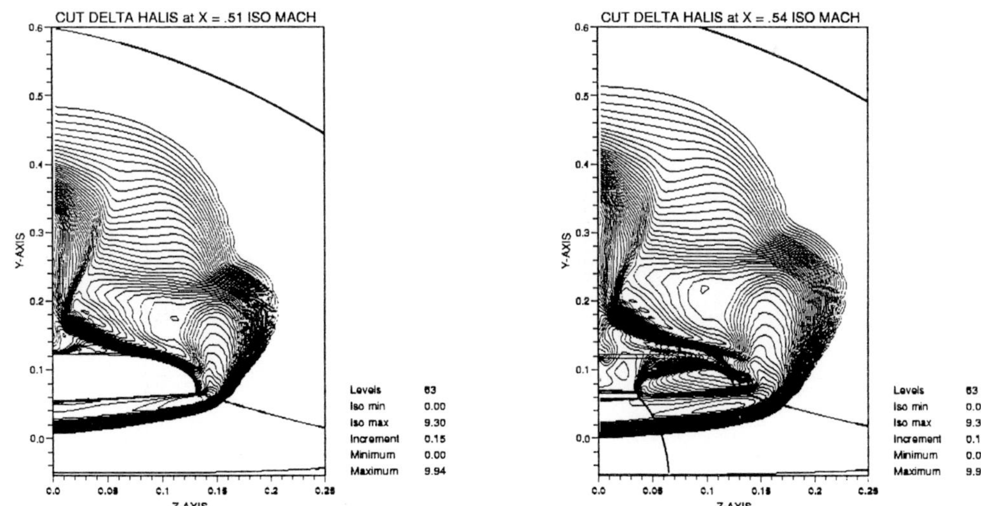

Figure 9: IsoMach contours in a plane x=constant just ahead of and just after the end of the body, HALIS hypersonic space vehicle.

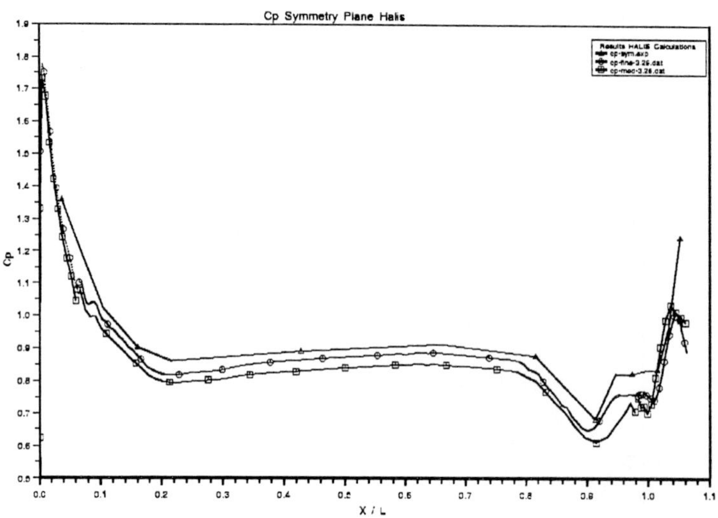

Figure 10: C_p distribution on the HALIS hypersonic space vehicle in the symmetry plane, 1.7 M grid points.

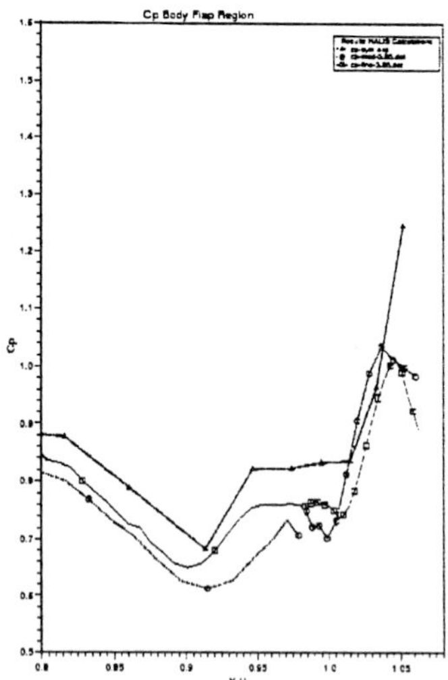

Figure 11: C_p distribution on the HALIS hypersonic space vehicle in the symmetry plane, medium grid of 200000 points.

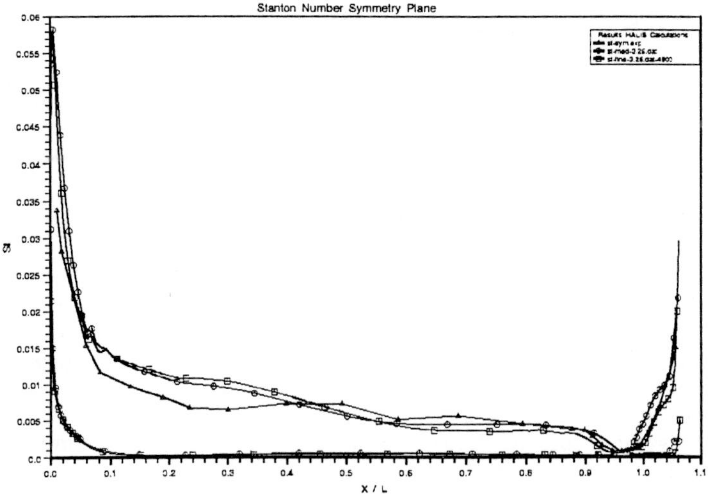

Figure 12: Fine-mesh Stanton number distribution compared with experiment.

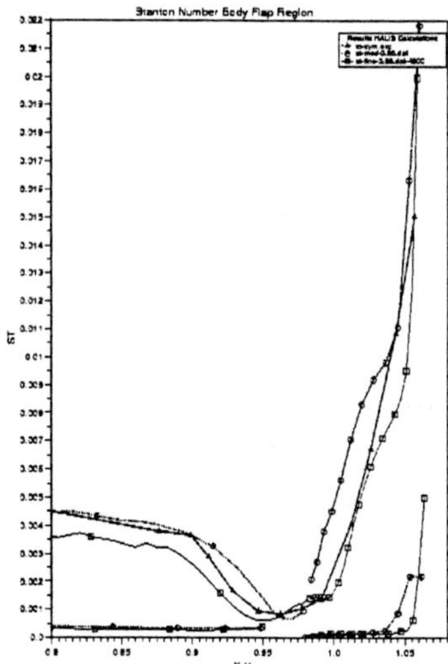

Figure 13: Medium-mesh Stanton number distribution compared with experiment.

NAVIER-STOKES PREDICTIONS OF MISSILE BODY SEPARATED FLOW FIELDS

Walter B. STUREK [1]   Duane FRIST [1]   Malcolm TAYLOR [1]

Hugh THORNBURG [2]   Bharat SONI [2]

(1)   U. S. Army Research Laboratory, Aberdeen Proving Ground, MD
(2)   Mississippi State University, Starkville, MS

Abstract

This paper reports a subset of results from a study carried out under the auspices of The Technical Cooperation Program (TTCP) with participants from Canada, the UK, and the US. The purpose of this study was to apply Navier-Stokes computational techniques to a complex flow field with highly separated flow for a missile shape to evaluate the predictive technology. The portion of the study reported here includes results for three Navier-Stokes computations for transonic and supersonic velocities at 8° and 14° angles of attack. The computational results were compared to experimental measurements for surface pressure, pitot surveys of the outer flow field, and strain gage force measurements. Computational results are reported for seven turbulence models and laminar flow. For the conditions of this study, no "best" turbulence model could be identified in the comparisons to experiment. Predictions of surface pressure and outer vortical flow in regions of highly separated flow indicate further development of the predictive technology is required.

Introduction

Computational Fluid Dynamics (CFD) has enjoyed considerable success for application to missile and aircraft configurations at small to moderate angles of attack. The ability of these techniques to accurately predict flows in which large regions of separated flow exist is less well known. In order for CFD to be regarded as a critical technology for the design and assessment of missile systems in the future, it is important that the ability of the technology to accurately predict effects of separated flow fields be defined. Influencing factors such as advancing computer technology; funding limitations which restrict the availability of experimental testing in the design phase of hardware development, and advances in computational modeling technology have served to increase interest in the development and validation of computational predictive technology.

The results reported here were part of an international cooperative study under the auspices of The Technical Cooperation Program (TTCP) with participants from Canada, the United Kingdom and the United States. The scope of the study involved the application of 3D Navier- Stokes computational techniques to predict the flow fields about long l/d bodies at moderate angles of attack for transonic and supersonic velocities. The computational results were compared to experimental data provided for this study by the Defense Research Agency, Bedford, UK, Birch(1). These data include surface pressure, flow field pitot pressure surveys and force measurements.

Computational Techniques

The computational techniques used for the portion of the study reported here are: OVERFLOW(2); NPARC(3,4); and CFL3D(5). These codes include finite-difference and finite volume implicit numerical techniques and have multi- block capabilities. The numerical algorithms use central and upwind differencing techniques. These codes are widely used, and their characteristics have been reported in open literature.

The question of grid resolution was addressed in this study by making comparisons for several grid resolutions. Since time-marching codes require substantial computer resources for even modest grid resolution, this effect has been addressed by comparing results for grid resolutions that ranged from 61x71x81 to 81x111x141. For this range of grid resolutions, the solutions were considered to be equivalent.

Experimental Data

The model configuration of interest for this study is a 3-caliber ogive with a 10-caliber cylindrical afterbody. A Schlieren photograph showing the model mounted in the wind tunnel for test conditions of Mach = 3.5 and alpha = 8° is shown in Figure 1. The test conditions for the six test cases used in this study are listed in Table 1. The surface pressure measurements include 32 axial stations for 41 circumferential locations from the wind to the lee-side at a 4.5° increment. The pitot pressure surveys provide a dense array of measurements at three axial stations, X/D = 5.5, 8.5, and 11.5. Strain gage force balance measurements were also available.

Results

Surface Pressure Distributions

Each test case presents significant physical modeling challenges. The following discussion attempts to provide an overview of the flow fields along with a "typical" comparison between computation and experiment. Due to space limitations, results for surface pressures and outer flow fields are shown only for Test Cases 2 and 3. The computed results shown here for Test Cases 2 and 3 were obtained using the OVERFLOW code. The turbulence model used is the Baldwin-Lomax(6) algebraic model with a Degani-Schiff(7) modification. Results were also obtained for laminar viscous effects.

Case 1. Mach = 1.45, α = 14°, Re/d = 0.667x10^6. The laminar results indicate strong separation on the wind-side prior to x/d = 3.5 whereas the turbulent results and experiment indicate separation on the lee-side for 3.5<x/d<4.5. The lee-side is strongly separated for x/d>4.5. The results at x/d=6.5, 7.5, and 9.5 show the experimental pressure greater than the prediction on the wind-side where close agreement between computation and experiment is expected. This behavior is, perhaps, evidence that a wind-side shock is triggered from a position upstream of x/d=6.5 and results in a pressure jump that the computed results do not show. The reason for the computation not picking up this behavior is not known. It is noted that the laminar predictions also do not predict the increase in surface pressure on the wind-side for x/d>5.5 even though the laminar predictions indicate early boundary-layer separation that could trigger a wind-side shock.

Case 2. Mach = 1.8, α = 14°, Re/d = 0.667x10^6, Figure 2.
The laminar results indicate strong separation for x/d=3.5, which is not supported by the experimental data or the turbulent calculations. The laminar viscous results consistently predict early flow field separation compared to the experiment and turbulent viscous results. The experiment indicates the flow is strongly separated for x/d > 4.5. A pressure jump is noted at x/d = 9.5 similar to that noted for Case 1.

Case 3. Mach = 2.5, α = 14 °, Re/d = 1.123x10^6, Figure 3.
The laminar viscous results indicate early separation of the viscous layer compared to the experiment and the turbulent viscous results. Good agreement is achieved on the wind-side for all results shown. This test case was selected as the top priority because the Mach number and Reynolds number of the flow provided the expectation for high-quality experimental measurements and fully turbulent viscous flow, which would yield the best conditions for agreement between computation and experiment.

Case 4. Mach = 3.5, α = 8°, Re/d = 1.123x10^6.
The laminar viscous results indicate separation for x/d = 4.5 and greater. Good agreement is shown on the wind-side for all stations except at x/d = 2.4 where the calculations predict a higher pressure than the experiment. No indication of a pressure jump resulting from an upstream shock is apparent as was seen in the lower Mach number flows. The turbulent viscous predictions for x/d >7.5 predict a lower first suction peak than is indicated by the experimental data.

Case 5. Mach = 3.5, α = 14 °, Re/d = 1.123x10^6.
The laminar and turbulent viscous results are in close agreement over the full missile body. The separated flow region on the lee-side is characterized by a relatively flat pressure distribution compared to the other test cases. For x/d > 5.5, the turbulent results indicate earlier separation than the laminar results with the experiment located in between.

Case 6. Mach = 0.7, α = 14 °, Re/d = 0.667x10^6.
Computational modeling required moving the outer boundary farther away from the body in order to lessen effects of the outer boundary interacting with the flow on the body. This reduced the density of the grid resolution and was cause for some concern. The turbulent viscous predictions are obviously in better agreement with the experiment than laminar. The lee-side suction peak is not predicted well until x/d > 7.5 is reached. Also, the turbulent results indicate some jitter in the surface pressure profiles, particularly on the wind-side for x/d > 3.5.

In summary, the test case results indicate that turbulent viscous modeling is required and that the flow fields are highly separated on the lee-side.

There is some indication that a wind-side shock occurs, resulting in a pressure jump on the wind-side at downstream stations for test cases with free stream Mach Number < 2.0. This effect is not predicted by the computations.

Statistical Analysis

The traditional method for evaluating the accuracy of flow field computations has been for the engineer to visually inspect the results, make qualitative comparisons, and draw conclusions. This process can be very satisfactory for limited data sets to evaluate. However, it was decided to explore the possibility that statistical analysis techniques could assist in the evaluation process.

The technique used here is to consider the difference between the computation and experiment(8). This difference has been used to obtain a mean value, standard deviation, and variance between the experiment and computation. An example of the results of this analysis technique for all six test cases in which the statistical quantities have been summed for forty-one circumferential positions at nine axial stations is shown in Figure 4. The visualization of the analysis shown in Figure 4 is in the form of a box and whisker plot. Depicted are OVERFLOW computations using the BLDS turbulence model. The summary shown consists of the median of the errors, the inner-quartiles (the 25th and 75th percentiles) which determine the vertical dimension of the box, and the maximum and minimum error to determine the whisker lengths. These results indicate that the accuracy of the solutions for M > 3 are clearly more accurate than those for M < 3.

This technique shows promise for providing a useful tool for use by the engineer when comparing numerous computational and experimental results; however, it does not diminish the value of examination of individual profile comparisons.

Turbulence Models

A variety of turbulence models have been utilized for the comparisons with experimental data. The models tested are: Baldwin-Lomax(BL); Baldwin-Lomax-Degani-Schiff(BLDS); k-epsilon(ke), k-omega(kw), Spalart-Allmaras(SA); and Baldwin-Barth(BB). An example of comparisons for Test Case 3 using the NPARC code for four turbulence models is shown in Figure 5. The BB model appears to provide the best results for the first suction peak. The BL model,

where the turbulent length scale is determined by a search to the flow field outer boundary, consistently overpredicts the location of the first suction peak. It is of interest to compare this result with that of Figure 3 where the BLDS model was used. None of the models predicts well the separated flow region for x/d = 6.5 and 7.5, where there is a distinct second suction peak. For x/d = 9.5 and 11.5, the models provide equivalent results in the lee-side separated flow. A similar comparison is shown in Figure 6 for results obtained using CFL3D. The results shown here for Test Case 3 are consistent with those for all six test cases.

Statistical analysis results for the various turbulence models are shown in Figure 7 for CFL3D and Figure 8 for NPARC. These results emphasize that comparable results are obtained for each test case for the various turbulence models and CFD codes. The message indicated here is that the more computer-intensive two-equation turbulence models do not provide a clear advantage over the less sophisticated algebraic or one-equation turbulence models.

Vortical Flow

The experimental data included measurements of pitot pressures in the outer flow fields using total head probes. These data provide the opportunity to evaluate the ability to predict the size, strength, and location of these vortices. These comparisons are made by direct comparisons of pitot pressures calculated from the flow field computational data to the experimental measurements.

An example of comparisons between the computations and experiment for the pitot pressure surveys is shown in Figure 9 for M= 1.8 for x/d = 5.5, 8.5, 11.5 for laminar and turbulent viscous modeling. This example indicates that turbulent modeling provides encouraging, although not fully satisfactory, accuracy for the axial positions shown.

Aerodynamic Forces and Moments

The experimental data available included strain gage balance measurements for normal force, pitching moment, and axial force. The results, not shown, indicate that predictions are within +/- 3% agreement with experiment for normal force and +/- 5% for pitching moment. This degree of accuracy is within acceptable requirements for initial design and optimization analysis for most projectile and missile systems.

Closing Remarks

Although flowfield details such as the definition of the separated vortex and the surface pressure in regions of separated flow are not predicted as well as desired, the aerodynamic force and moment predictions are accurate within +/- 5% for these test cases. This level of accuracy is usually adequate for weapon design requirements. Thus, CFD is a viable predictive tool for the class of body configurations considered in this study and should be of increasing importance as computer technology continues to advance, thus permitting the predicted results to be obtained in a timely fashion.

Acknowledgments

The authors would like to express their appreciation of TTCP Technical Panel WTP-2 for chartering this study.

Nomenclature

d	model cylinder diameter, 3.7 inches
C_m	pitching moment coefficient
C_n	Normal force coefficient
M	Mach Number
x	axial distance from nose of missile
y	vertical distance from axis
z	horizontal distance from axis
α	angle of attack, degrees
phi	circumferential angle about axis

References

1. Birch, T., private communication.

2. Buning, P., et al., "OVERFLOW User's Manual," NASA-Ames Research Center, June 1995.

3. Cooper, G. K. and Sirbaugh, J. R., "PARC Code: Theory and Usage," AEDC-TR-89-15, December 1989.

4. NPARC Alliance Technical Team, "A User's Guide to NPARC Version 2.0," Noverber 1994.

5. Thomas, J., Krist, S., and Anderson, W., "Navier-Stokes Computations of Vortical Flows Over Low Aspect-Ratio Wings," *AIAA Journal*, Vol. 28, No. 2, 1990, pp. 205-212.

6. Baldwin, B. S., and Lomax, H., "Thin Layer Approximation and Algebraic Model for Separated Turbulent Flow," AIAA Paper 78-257, Jan. 1978.

7. Degani, D., and Schiff, L. B., "Computation of Supersonic Viscous Flows Around Pointed Bodies at Large Incidence," AIAA Paper 83-0034, Jan. 1983.

8. Taylor, M. and Sturek, W., "Statistical Analysis of Surface Pressure Measurements vs. Computational Predictions," Army Research Laboratory Technical Report ARL-TR-1318, Aberdeen Proving Ground, MD, March 1997.

Table 1. Wind Tunnel Test Conditions

Case Number	Mach Number	Angle of Attack	Reynolds No, $\times 10^6$/ft
1	1.45	14	2.0
2	1.8	14	2.0
3	2.5	14	4.0
4	3.5	8	4.0
5	3.5	14	2.0
6	0.7	14	2.0

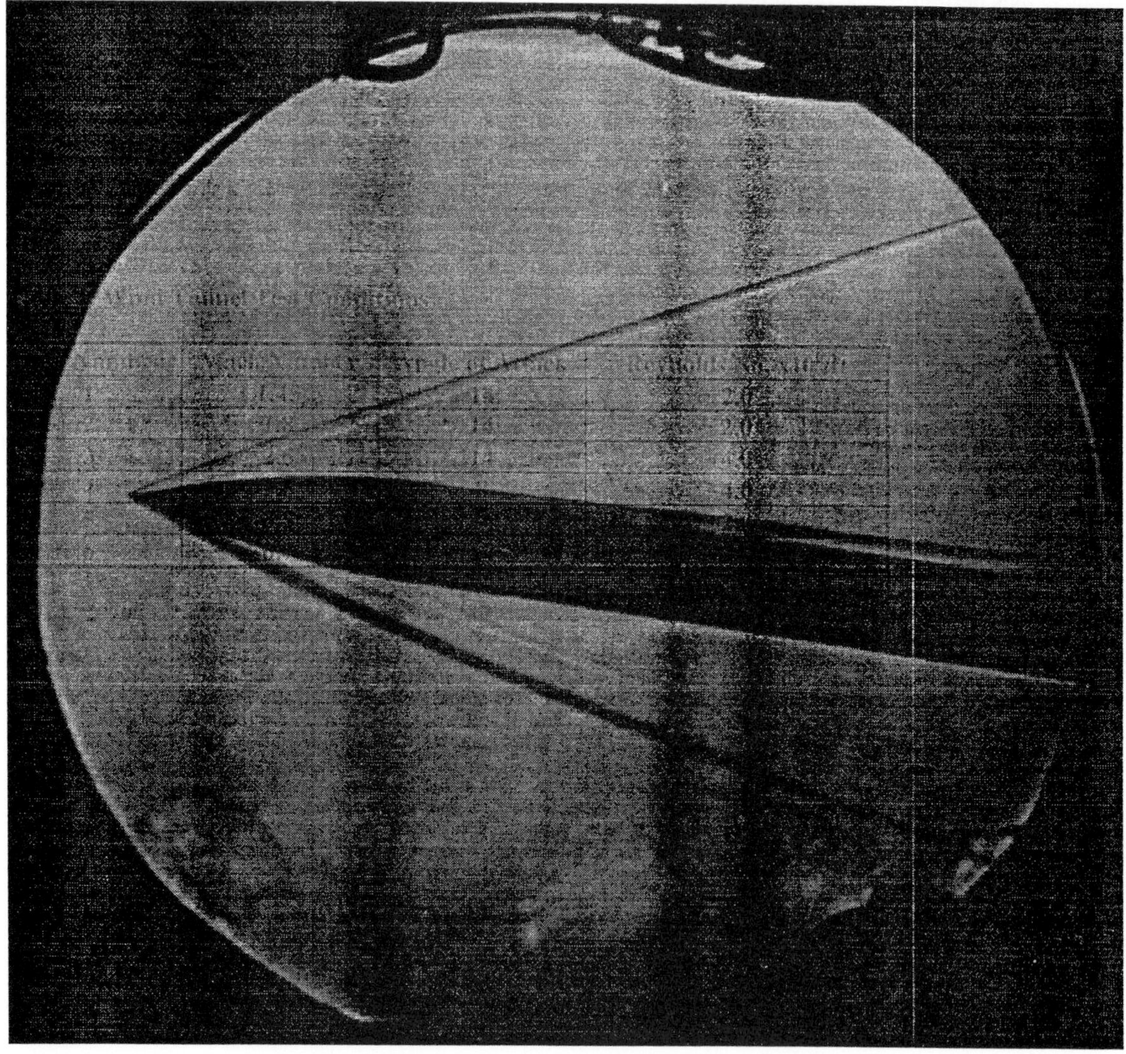

Figure1. Schlieren photograph of model mounted in wind tunnel, Mach =3.5, $\alpha = 8°$.

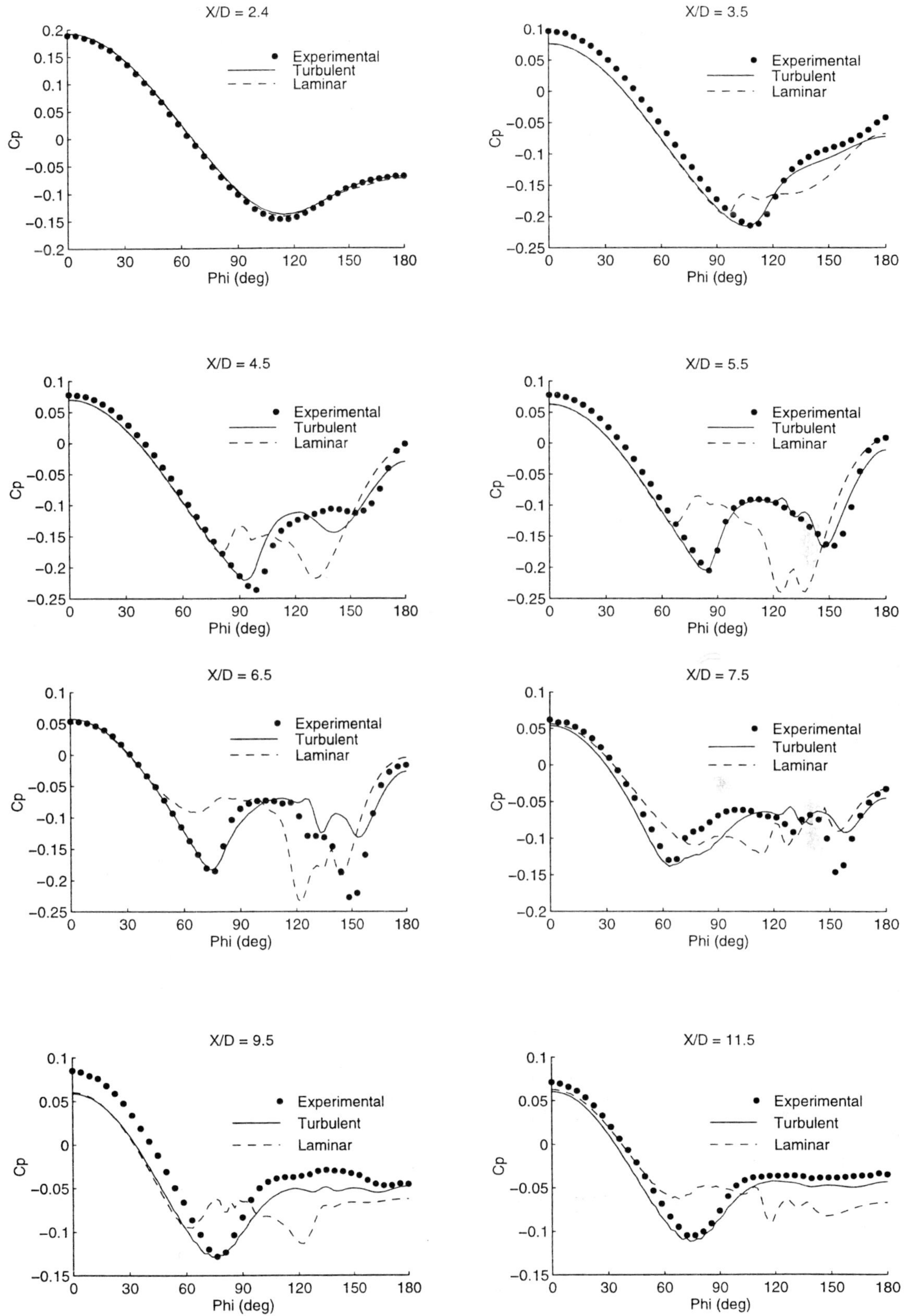

Figure 2. Test Case Overview, Case 2, Mach = 1.8, α = 14°, Re/D = 0.667x10^6 .

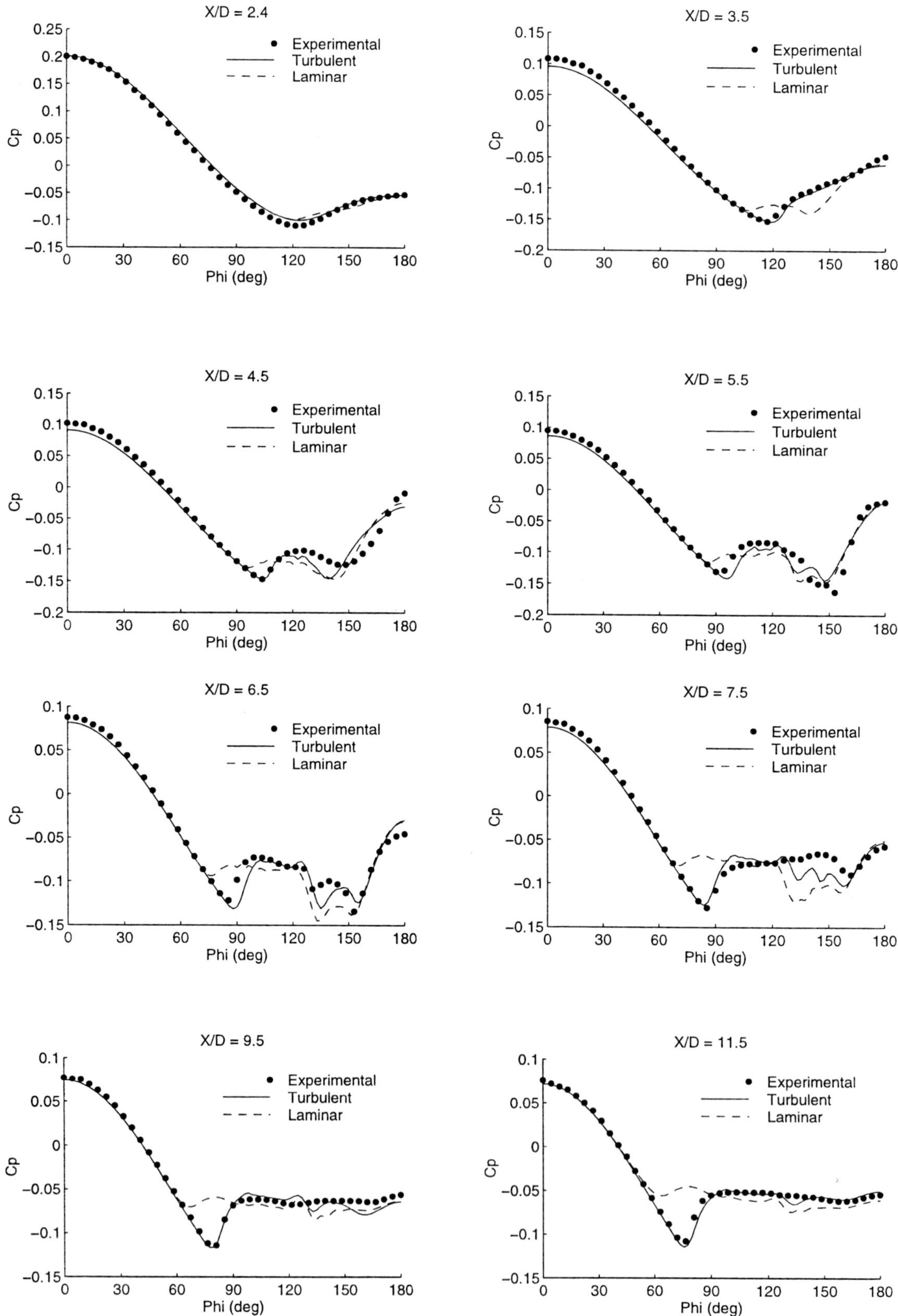

Figure 3. Test Case Overview, Case 3, Mach = 2.5, $\alpha = 14°$, Re/D = 1.123×10^6

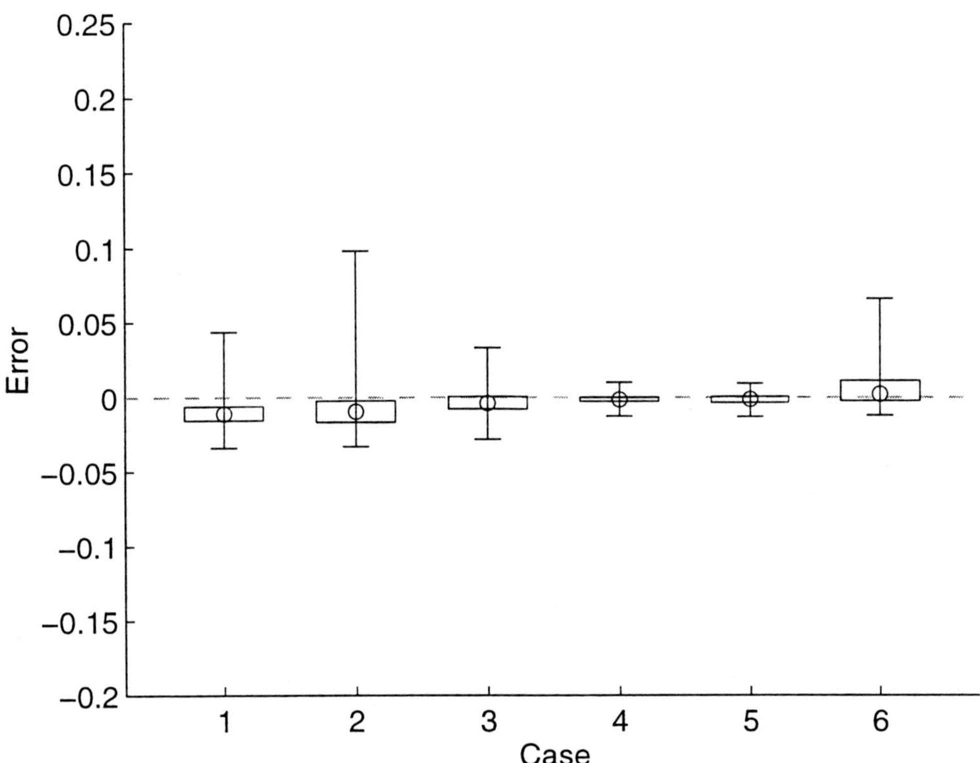

Figure 4. Statistical analysis of Overflow results for BLDS turbulence model, all test cases.

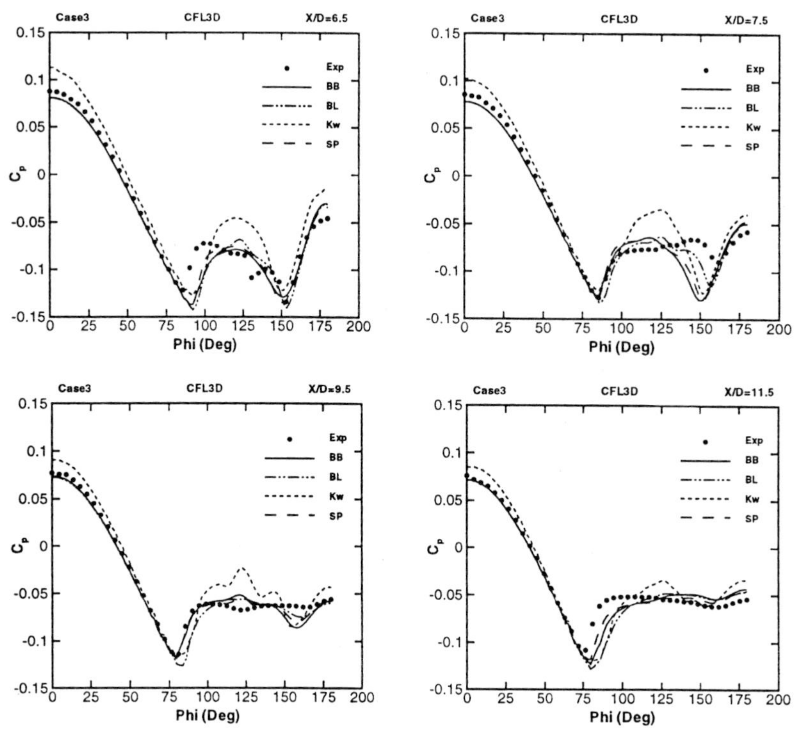

Figure 5. Turbulence modeling comparisons, CFL3D results, M=2.5, $\alpha=14°$

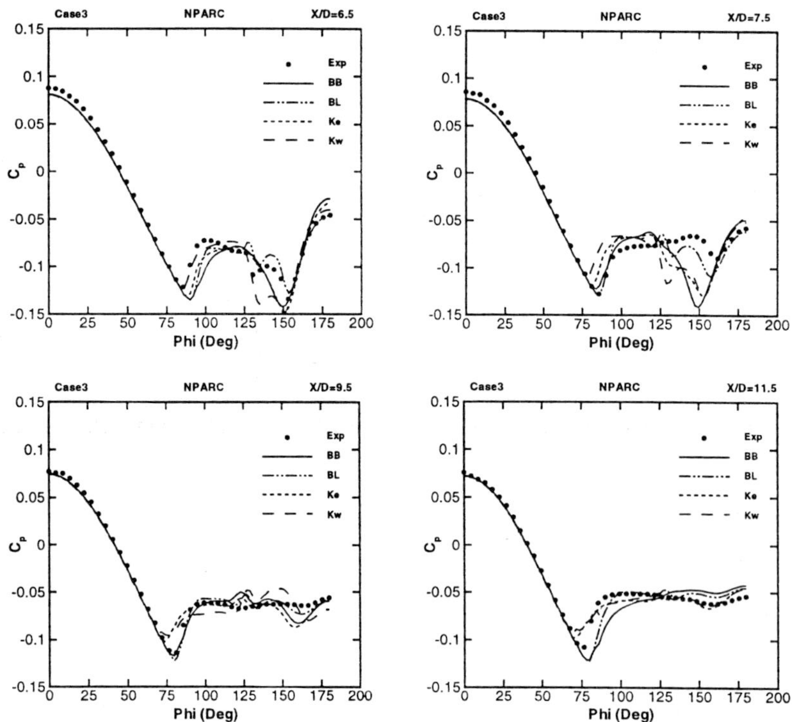

Figure 6. Turbulence modeling comparisons, NPARC results, M=2.5, $\alpha=14°$

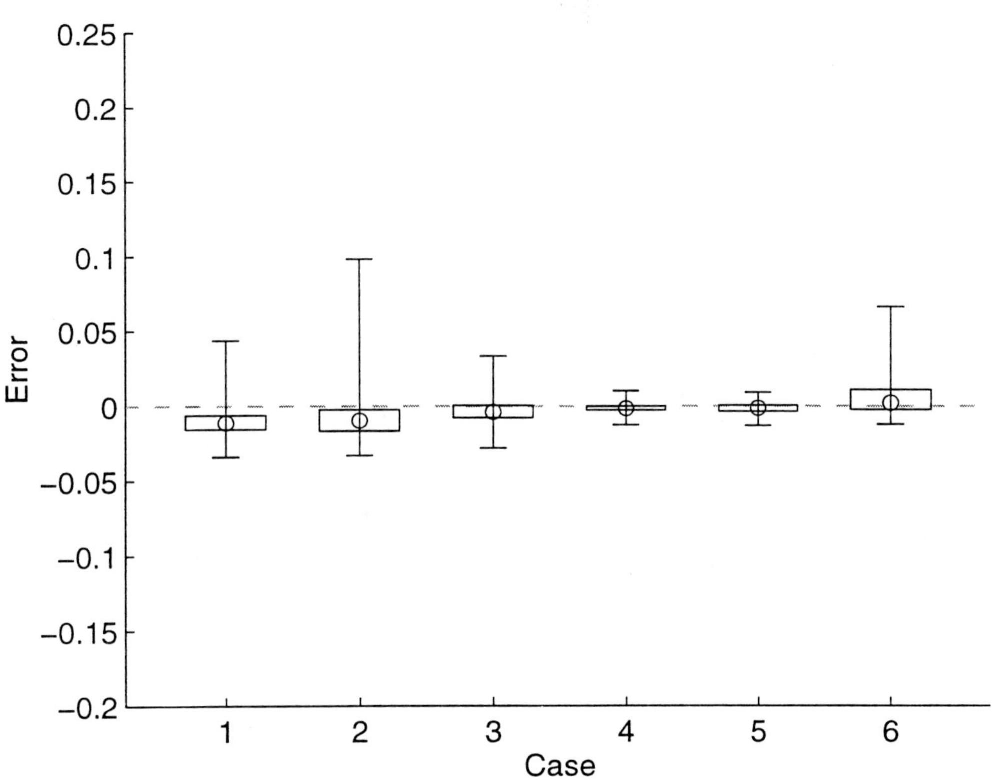

Figure 6. Statistical analysis of Overflow results for BLDS turbulence model, Test Cases 1-6.

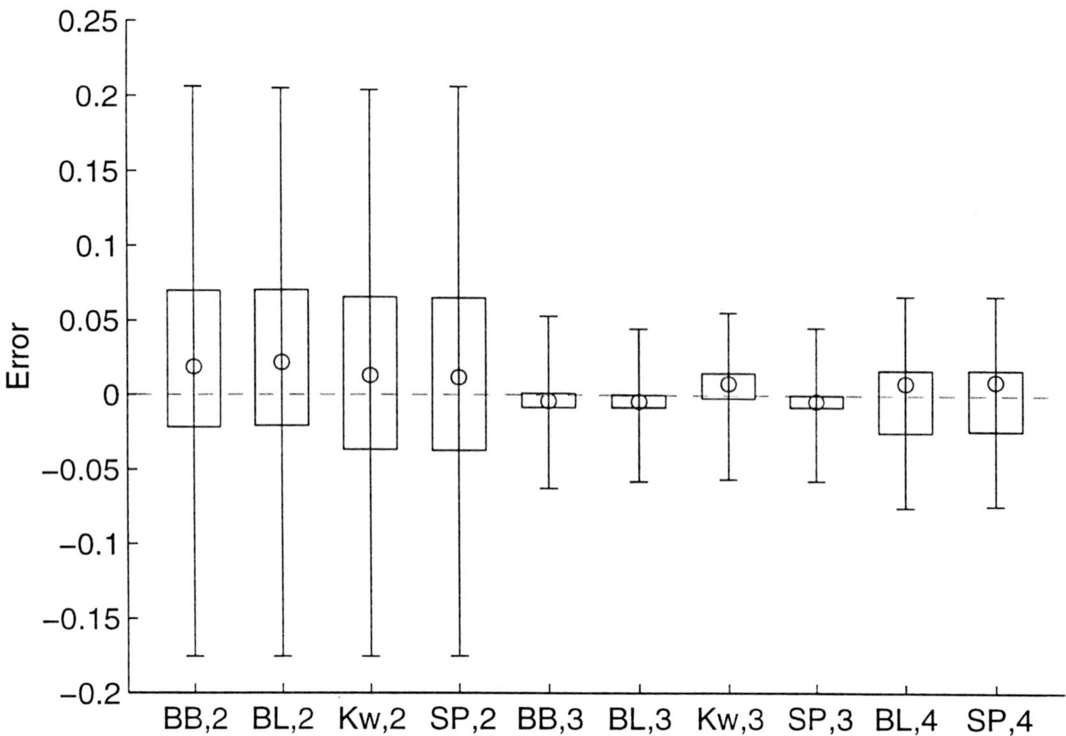

Figure 7. Statistical analysis turbulence model comparisons, CFL3D, Test Cases 2, 3, 4.

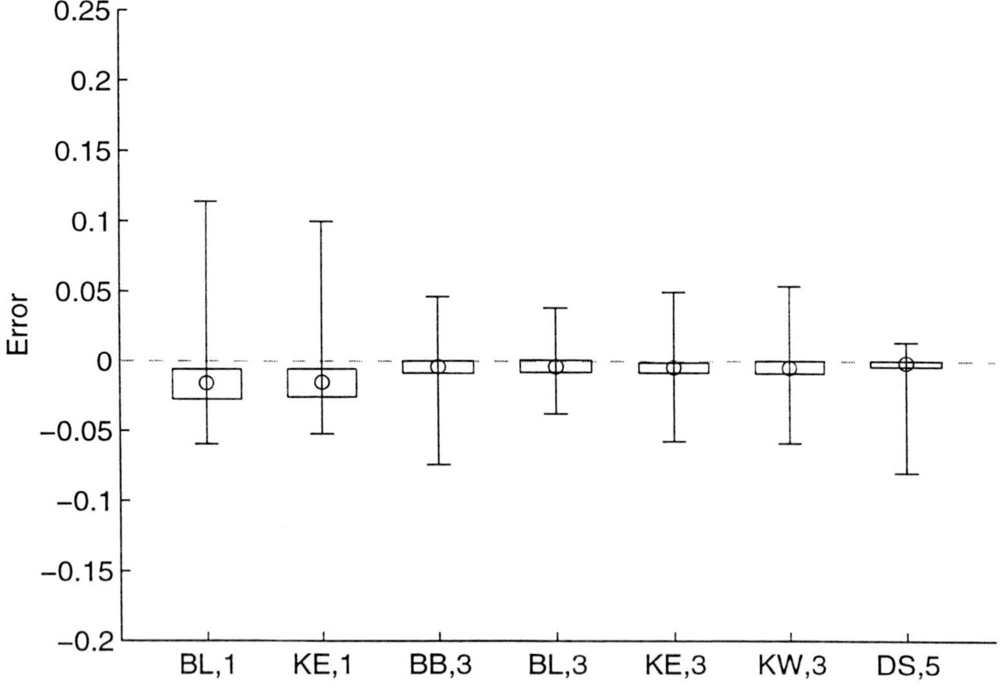

Figure 8. Statistical analysis turbulence model comparisons, NPARC, Test Cases 1, 3, 5.
(Note: DS refers to BLDS turbulence model)

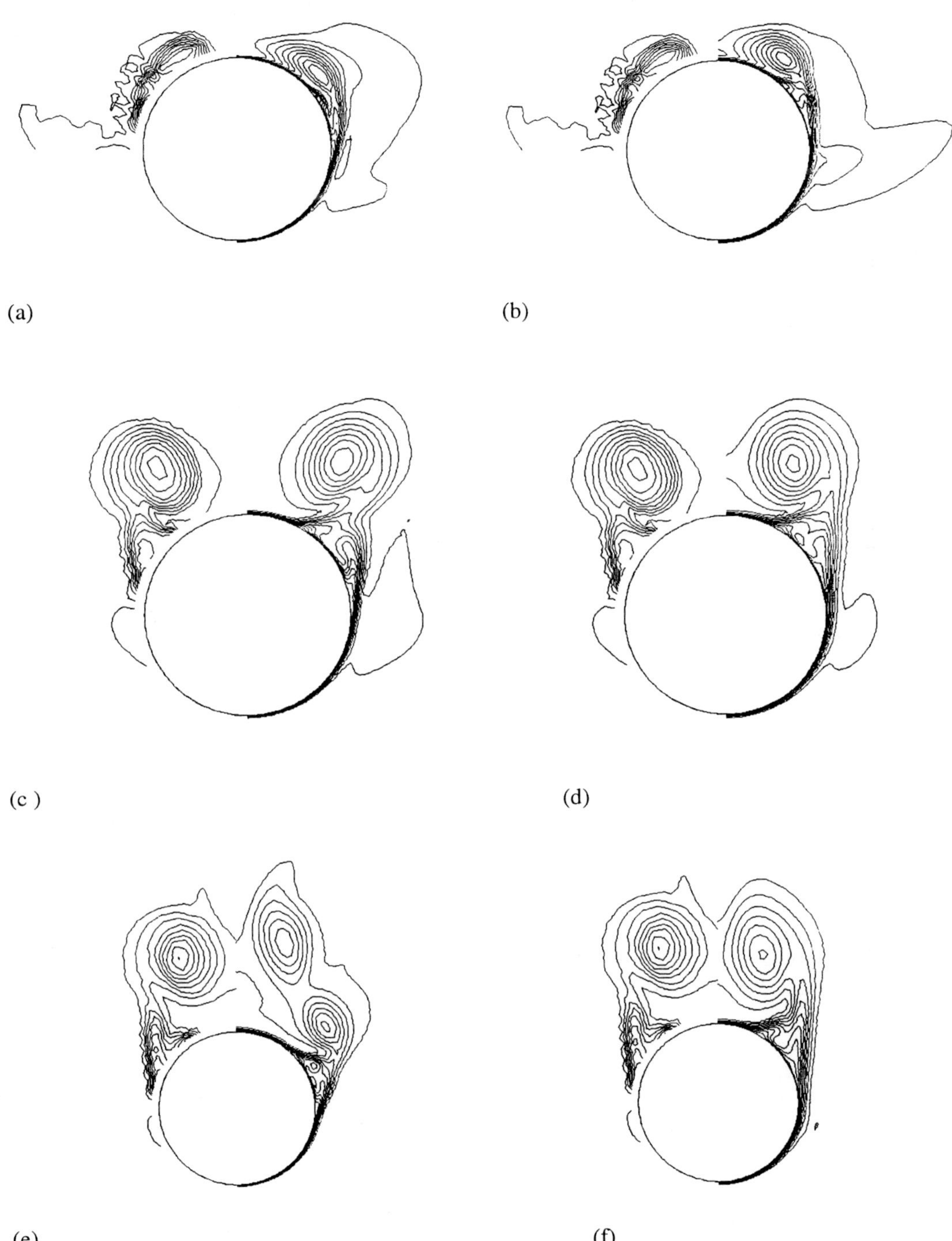

Figure 9. Pitot pressure contours for Case 2, Overflow Code, BLDS turbulence model, laminar (a,c,e), turbulent (b,d,f), x/d=5.5(a,b), x/d=8.5(c,d), x/d=11.5(e,f),left side experiment, right side computation.

COMPUTATIONS OF BLAST WAVE PROPAGATION USING AN OVERSET MOVING ZONAL METHOD

Kozo FUJII [1]   Fumio SHIMIZU [2]

(1) The Institute of Space and Astronautical Science 3-1-1, Yoshinodai, Sagamihara, Kanagawa, 229, JAPAN
(2) Department of Mechanical System Engineering, Kyushu Institute of Technology, Kawazu 680-4, Iizuka, Fukuoka, 820, JAPAN

Abstract

This paper describes computational efforts to develop an accurate computational tool for the purpose of establishing a better estimation method of the safety distance for the blast wave. Since this type of simulations require high level of grid resolution, two approaches are considered. One is an overset zonal method and the other is an unstructured solution adaptive method. Two-dimensional computations are carried out and the computed results indicate that the former approach is adequate for this problem. Based on the techniques developed for the two-dimensional simulations, three-dimensional simulations are carried out and the effect of the ground surface geometry is discussed. Sufficient grid resolutions are obtained within the available computer memory and speed. The computed result clearly shows that the ground surface has an important effect on the strength of the blast wave even far away from the point of explosion. To make the result useful for other practical problems, assessments of the computational aspect are given for the required resolution of the computational grid for this type of problems.

1. Introduction

Investigation of blast wave propagation is one of the important research topics of unsteady shock wave behavior[1,2]. How the blast wave behaves when propagating on the ground surface is of considerable importance for many practical applications such as an accidental explosion of chemical or nuclear plants or rocket propellants. Volcano eruptions may be another application. In these problems, accurate estimations of the blast wave strength are required because it is the key factor of the damage.

Figure 1 shows the schematic picture of the blast wave propagation. This figure shows the time history of the pressure observed at a certain distance from the point of explosion. The pressure stays at the atmospheric value until the frontal shock wave reaches the point of measurement. After an abrupt increase of the pressure due to the shock wave, the pressure decreases gradually and becomes lower than the atmospheric value due to the following expansions and eventually comes back to the atmospheric value. The

Received on February 15, 1997.

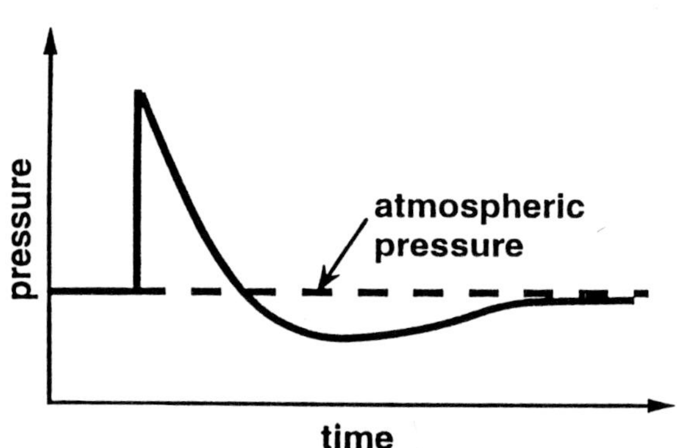

Fig.1: Pressure change due to the blast wave

rapid pressure rise is called "overpressure" and it defines the strength of the blast wave. Figure 2 shows the history of the instantaneous pressure profiles in space. As the blast wave propagates, the overpressure decays exponentially in general. For the estimation of the "safety distance" for an abrupt explosion, one of the criteria is the strength of the overpressure. Conventionally, the empirical method based on the one-dimensional theory and the experimental data by

Kingery and Pannill[3] is widely used for the evaluation of the "safety distance". In this theory, the overpressure simply depends on the amount of energy and the distance, and other factors such as the ground surface geometry and/or atmospheric conditions are not considered. When considering the diffraction, reflection and other interactions that may occur for the shock wave passing over some geometry, it is imagined that ground surface geometry may change the strength of the blast wave. Discretization methods easily handle such effects and therefore developing such methods may improve the accuracy of the estimation of "safety distance". Computational efforts using Lagrangean approach exist but they are restricted to the primitive simulations under the assumption of point symmetry or axisymmetry[4,5].

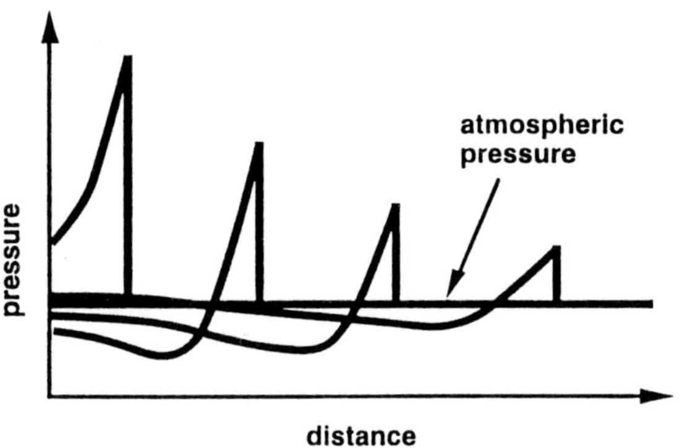

Fig.2: Instantaneous pressure profiles of the blast wave

In the present paper, a computational method using an overset zonal moving grid system is developed for an accurate evaluation of "safety distance". Since this type of simulations requires high level of grid resolution, two approaches are considered. One is an overset zonal method and the other is an unstructured solution-adaptive grid method. Two-dimensional computations are carried out and the computed results indicate that the former approach is adequate for this problem. Based on the techniques developed for two-dimensional simulations, three-dimensional simulations are carried out and the effect of ground surface geometry are discussed. To make the result useful for other practical problems, assessments of the computational aspect are also given for the required resolution of the computational grid for this type of problems.

2. Computational Approach

2.1 Choice of Approach

Although discretization methods are adequate for the simulation of blast wave propagation, it is impor-
tant to check the quantitative accuracy of the results as the discretization errors may change the strength of the blast wave. One-dimensional point explosion was considered and the simulations were carried out prior to the practical computations. Figure 3 shows the computed overpressure versus grid resolution. In this computation, the energy of explosion and the distance are defined where the overpressure becomes less than 1 % of the atmospheric pressure because the overpressure for the "safety distance" is conventionally considered to be 0.2 psi which is roughly 1.4 % of the atmospheric pressure. The horizontal axis is the number of grid points in the direction of the blast wave propagation. Even with 5000 grid points, the asymptotic value is not yet obtained and the result indicates more than 6000 grid points are required.

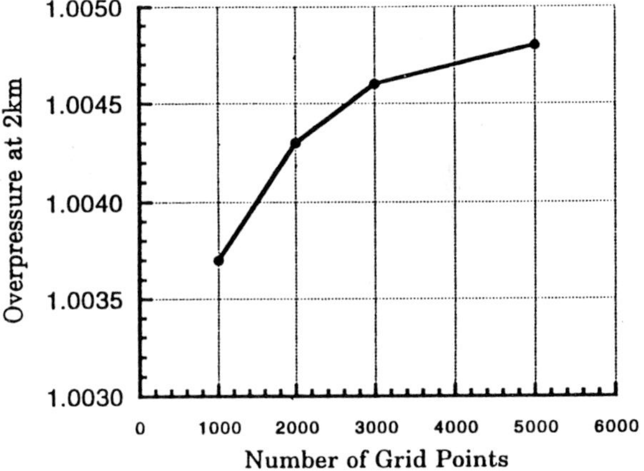

Fig.3: Overpressure vs grid resolution for one-dimensional point explosion

The one-dimensional simulations showed that some special techniques may be required for an accurate simulation. Two methods are considered. First one is an overset zonal moving grid. Since we are interested in the strength of the blast wave, the region that requires high accuracy is restricted in the certain area that moves with the speed of the blast wave. In the overset zonal method, a moving fine grid region can be locally defined to capture the blast wave without any difficulties. Second one is an unstructured grid with the solution-adaptive grid strategy. This method can automatically concentrate the grid cells to the region of blast wave. Both methods were tried[6-9]. Figure 4 shows the computed maximum overpressure distributions for these two methods in the two- dimensional problem. The moving grid solution seems slightly better (because it shows higher overpressure) than the unstructured grid solution, but they are almost identical in the region away from the explosion. So let's assume that these two results are as accurate each

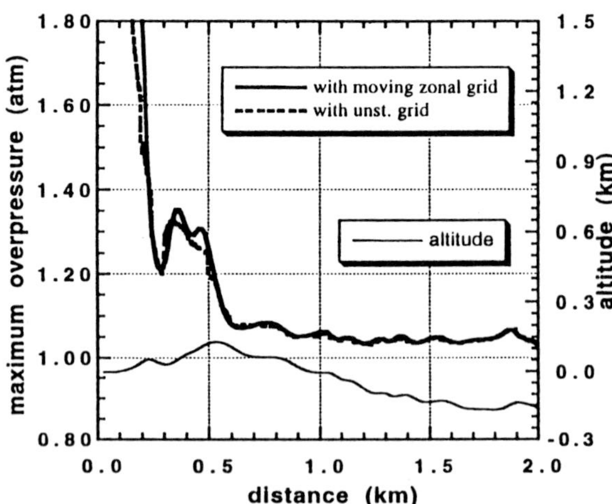

Fig.4: Computed maximum overpressure distributions - comparison of the zonal-grid solution and the unstructured-grid solution -

other. Although fair comparison is very difficult, we concluded that the former approach is more efficient for the present problem. There are two main' reasons. First, the unstructured solution- adaptive grid method requires large amount of memory. In this two-dimensional example, the moving overset grid computation required 1 hours on VPP500/1PE with 12,000 grid points in total. It required several MB of memory. On the other hand, the unstructured grid computation required less than 1 hour on the same computer with 100,000 grid points. It requires 20-25 MB of memory. The reason why the number of grid points are so different is that the shock wave is almost orthogonal to the ground surface, and therefore the unstructured grid (that requires the grid increase in both directions) need many more grid points. Shock wave interaction makes the configuration of the frontal shock wave complicated only near the ground surface. The unstructured grid with triangles spends large number of grid points for the region out of our interest. This leads to the fact that the unstructured grid approach requires large amount of memory. In three-dimensional problems using tetrahedra, this drawback would be more pronounced. Our estimation showed 5 to 10GB memory would be necessary for the unstructured- grid computations. On the other hand, overset zonal grid computations require 1 or 2 GB of memory in three dimensions. This is the first reason for the choice of overset zonal method. Second reason is another important reason for having adopted the overset zonal moving grid approach. During the course of the study, we have found that the local grid resolution is enhanced when using moving grids. When the grids move with the frontal shock wave, the resultant shock wave was more crisply captured than that would be obtained with the stationary grid having the same grid resolution. Appendix A shows the detailed analysis.

2.2 Computational Algorithm and FSA Zonal Method

Euler equations are considered here because the viscous effect on the ground surface would change the flow field left over by the frontal shock wave but would not change the strength of the frontal shock wave. The equations are written in the generalized coordinate system where ground surface is one of the boundaries. Here, only three dimensional equations are written but the similar formulation was derived for two dimensional equations.

$$\partial_\tau \widehat{Q} + \partial_\xi \widehat{E} + \partial_\eta \widehat{F} + \partial_\zeta \widehat{G} = \chi \left\{ \widehat{Q_f} - \widehat{Q} \right\} \qquad (1)$$

In Eq.(1), the original Euler equations are modified to include the source terms that are used for the data transfer from one zone to the other in the FSA (Fortified Solution Algorithm) zonal method [10]. The FSA zonal method is briefly explained here. In Eq.(1), $\widehat{Q_F}$ is a solution that is known in advance. In the zonal method, the physical variables obtained in the other zone having higher priority are interpolated to the variables at each grid point in the current zone and are given as $\widehat{Q_f}$. The switching parameter Ξ is set to be sufficiently large compared to all the other terms at the grid points on which the solution $\widehat{Q_f}$ should be enforced, and is set to be zero on the other grid points. When $\Xi >> 1$, the added source term simply forces the solution vector $\widehat{Q}$ to be $\widehat{Q_f}$, otherwise it blends $\widehat{Q_f}$ with $\widehat{Q}$. When $\Xi = 0$, the equations go back to the original Euler equations. They are solved with no effects of $\widehat{Q_f}$.

The convective terms are discretized using the flux difference splitting scheme with Roe's approximate Rieman Solver[11]. Third-order upwind biased differencing is achieved using the MUSCL interpolation with the van Albada's flux limiter[12]. Since the computation requires time accuracy, two-step Runge-Kutta time integration scheme is adopted. This can be written as

predictor step:

$$\{I + \chi\Delta\tau\} \left\{ \widehat{Q^*} - \widehat{Q^n} \right\} = -\frac{1}{2}\Delta\tau \left\{ \frac{\partial \widehat{E}}{\partial \xi} + \frac{\partial \widehat{F}}{\partial \eta} + \frac{\partial \widehat{G}}{\partial \zeta} \right\}^n$$

corrector step:

$$\{I + \chi\delta\tau\}$$

$$= -\Delta\tau \left[\left\{ \frac{\partial \widehat{E}}{\partial \xi} + \frac{\partial \widetilde{F}}{\partial \eta} + \frac{\partial \widehat{G}}{\partial \zeta} \right\}^{*} - \chi \left\{ \widehat{Q_f} - \widehat{Q^n} \right\} \right] \quad (2)$$

when considering the source terms for the FSA zonal method. The solution algorithm has been modified so that the method above is to be properly implemented in the solution process. Here the superscripts n indicate the level of the time step. Equation (2) shows that it reduces to the standard Runge-Kutta scheme when $\chi = 0$. If χ is sufficiently large, the algorithm reduces simply to $\widehat{Q^{n+1}} = \widehat{Q_f}$. Thus, in the region that χ is set to be large, Euler algorithm is turned off and the given solution $\widehat{Q_f}$ is " fortified" there. The details of the formulation as well as the merits of the FSA zonal method compared to the conventional CHIMERA approach is discussed in Ref.10.

At each time step, all the zones are solved sequentially and the time integration is continued until the blast wave reaches 2.5 km. In the overlapped region, the information of the solution for each zone is transferred by the FSA technique. Each grid point has a flag of χ that defines if the equations are solved in that particular zone $\{\chi = 0\}$ or interpolated from the solutions in the other zones $(\chi >> 1)$. The list of flags χ and the coefficients necessary for the interpolation is computed at each time step.

The grid system is composed of three zonal grid components. One is the base grid which covers the whole region. Second is the centered grid which is prepared to avoid the topological singularity in the center of the explosion. Third one is the locally-defined moving grid covering the blast wave region. This grid moves at the speed of sound. The grid could be moved with the exact speed of the frontal shock wave by monitoring the speed of the shock wave, but the speed is fixed to the speed of sound as the grid speed is not necessarily the same as the shock wave as indicated in the research shown in Appendix A. Note that the speed of the shock wave is close to the speed of sound except very near the point of explosion where the shock wave is very strong. In two-dimensional computations, the base grid also moves with the speed of sound as nothing happens outside the blast wave. The grid is initially small and becomes spread as the blast wave propagates. Figure 5 shows the time sequence of the computational grid in two- dimensional simulations. In two-dimensional problems, moving grid is prepared only for the blast wave propagating to the east direction. In three-dimensional problems, the base grid is fixed and local moving grids are defined for all the directions to see the effect of the ground surface geometry. To achieve this within the restriction of the memory size of 256 MB, the computational region is divided into 8 parts and eight computations are car-

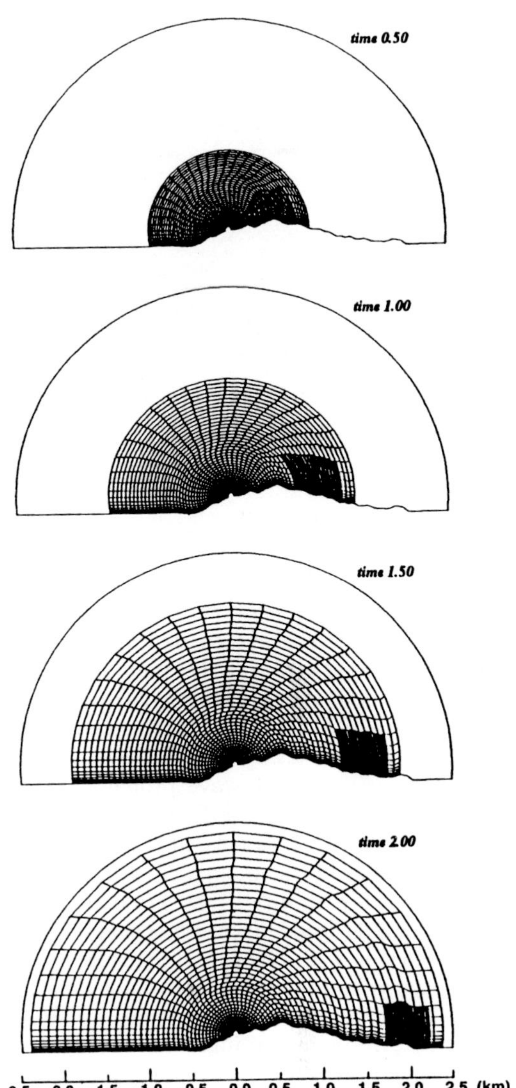

Fig.5: Time sequence of the computational grids for two-dimensional problem

ried out with the local moving fine-grid region located only in one part. Figure 6 shows one instantaneous shot of the base, centered, and all the eight moving grids together. Each computation uses roughly 0.5 million grid points in total. The computations were carried out on Fujitsu VPP500/1PE and the required computer time was roughly 20 hours.

One thing to note here is the consideration for the deformation of the computational grids. Geometric Conservation Law (GCL) should be satisfied when computational grids are deforming. In the present paper, all the metrics and Jacobians are computed considering the geometrical conservation[13] both in time

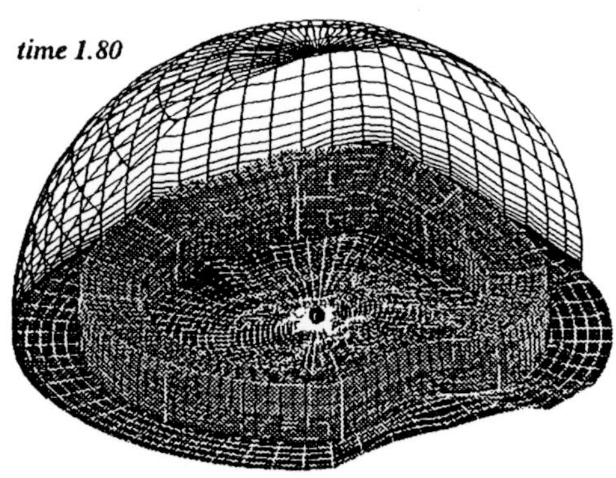

Fig.6: Computational grid for three-dimensional problem

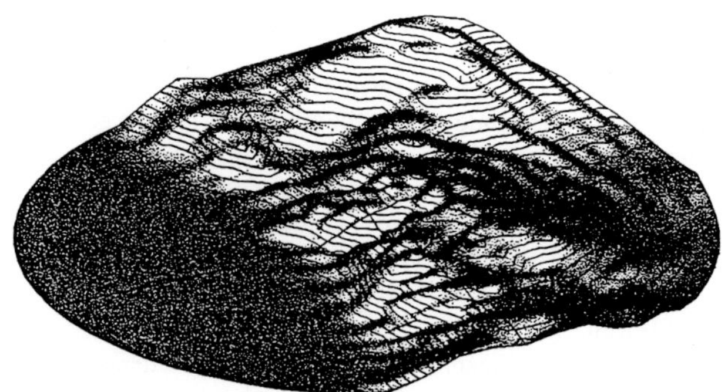

Fig.7: Ground surface geometry considered in the present study

and space to satisfy the GCL.

3. Computed Results

The ground surface geometry is taken from a real physical model which is shown in Fig.7. The west area (left side) is the sea. The point of explosion is the center and is located about 200 m above the sea level. There is a small hill in the east direction (right) from the point of explosion along the line from north to south. The blast wave propagating to the east direction goes over this hill of 300 m high. The energy core of 100 atm pressure and 2000K temperature having 6.2m radius is assumed in two dimensional problem. The core radius is set 17 m in the three dimensional simulations. Two dimensional computation is carried out with the ground surface to be cut from west-east plane in Fig.7 (See the geometry distributions in Fig.5). Two dimensional problem does not have any physical meanings since the blast wave propagates not in spherical but in cylindrical shape, but the three dimensional problem corresponds to the realistic case with the released energy of 55 tons TNT equivalence. Outside the core, atmospheric conditions are given. The pressure is 1 atm and the temperature is 300K there. Density gradient in the horizontal direction is not considered. The region to be simulated is about 2.5 km from the center of explosion because the strength of the blast wave is estimated to become less than 0.2 psi there.

Before conducting three-dimensional simulations, a lot of two-dimensional simulations only with the base and centered grids were conducted to see the effect of grid resolution and other factors. Figure 8 shows the observed overpressure versus the grid spacing in the direction of the blast wave propagation. From this figure, the solution becomes asymptotic value for the grid spacing to be 0.001. Since the distance is normalized by 1 km in the present simulations, it corresponds

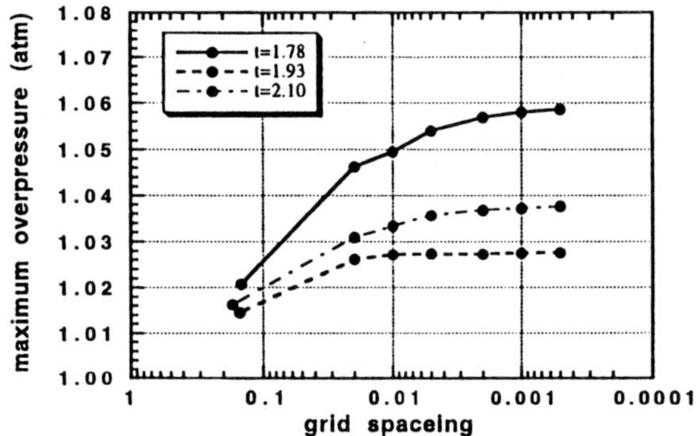

Fig.8: Effect of the grid resolution in the direction of blast wave propagation

to 1 m. Note that more computations were carried out with larger energy release or larger change of the surface geometry. From all these results, it was concluded that the grid size of about 1 m is necessary to capture the overpressure of a few percents of the atmospheric pressure.

Another observation to check the solution accuracy was the grid spacing ratio for the two adjacent grids normal to the ground surface. As the ground surface changes its geometry, and the boundary condition is given by the extrapolated value using the solutions on these grid points, we have to be careful for the grid spacing. Here, minimum spacing is set to be 0.005 from several preliminary computations. When the grid spacing ratios are taken 1.35 and 1.6 (it means the second grid spacing is 1.35 and 1.6 times larger than the first grid spacing from the ground, respectively), the results show the difference of 0.3is not negligible as is shown in Fig.9. On the other hand, the result for 1.4 is almost the same as the result for 1.35. The final grids are distributed based on the observation. An important conclusion here is that the grid distribution not only in the direction of the blast wave propagation but also in the lateral direction should be carefully examined.

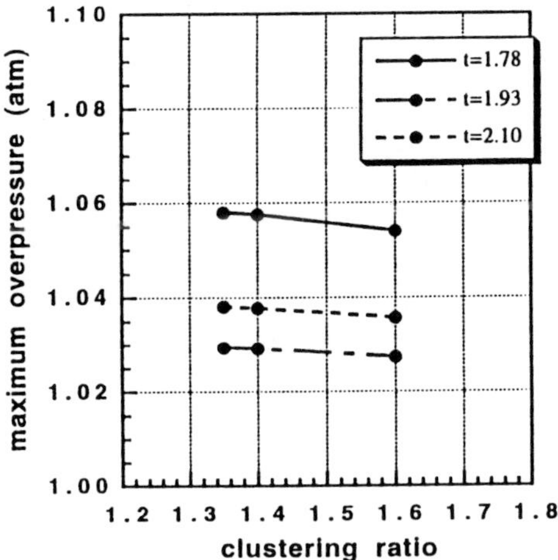

Fig.9: Effect of the ratio of the grid spacing normal to the ground surface

The solutions corresponding to the computational grids shown in Fig.5 are shown in Fig.10. The frontal shock wave and the following expansion are both captured well. The contours are not cylindrical due to the effect of the ground surface geometry. The zonal interface does not create any problems. In Fig.11, distributions of the maximum overpressure computed by the two- dimensional simulation are plotted. The result for the flat ground surface (dash line) and the ground surface height distribution are plotted for reference. The case having the realistic ground surface geometry is called REAL and the case with a flat surface ground is called FLAT. The overpressure essentially decreases with the distance similarly to the flat ground case, but local ground geometry influences to the result very much. There is a hill 300m high 500m east away from the point of explosion. The overpressure becomes much higher than that of the flat case in the frontal side of the hill, but becomes much lower in the rear side. This can intuitively be imagined from the fact that people try to hide behind some obstacles to avoid a strong pressure wave. The pressure increase in the frontal side occurs due to the shock reflection at the foot of the front corner of the hill. The pressure decrease in the rear side occurs due to the shock diffraction on the top of the hill. The behavior of the pressure when the blast wave passes over the hill can be predicted by the typical shock wave problem (See for instance, Ref.14). To see the effect of the ground surface geometry, the difference of the overpressure of the REAL case and the FLAT case normalized by the

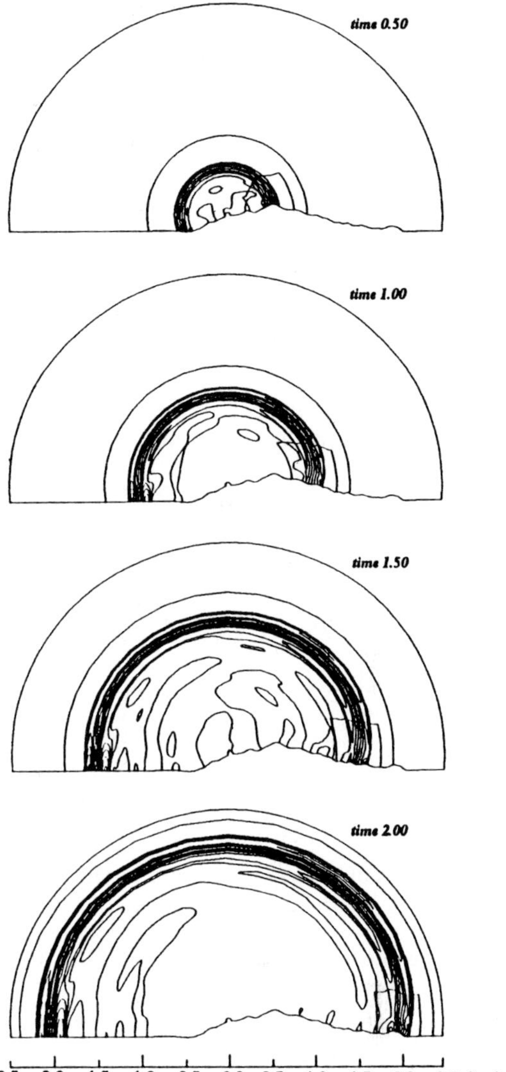

Fig.10: Computed pressure contour plots for two-dimensional problem

overpressure of the FLAT case is plotted in Fig.12. This can be considered, so to speak, the effect of the ground surface without the exponential decay. The gradient of the ground surface geometry in the direction of the blast wave propagation is computed and plotted in the same figure. These two plots obviously show the same tendency and it is clear that the behavior of the overpressure has strong correlation with the gradient of the ground surface geometry. It is, however, noted at the same time that quantitative difference exist between two plots in some area. Since the shock wave is weak except near the point of explosion, it may be possible to deduce some theoretical formula

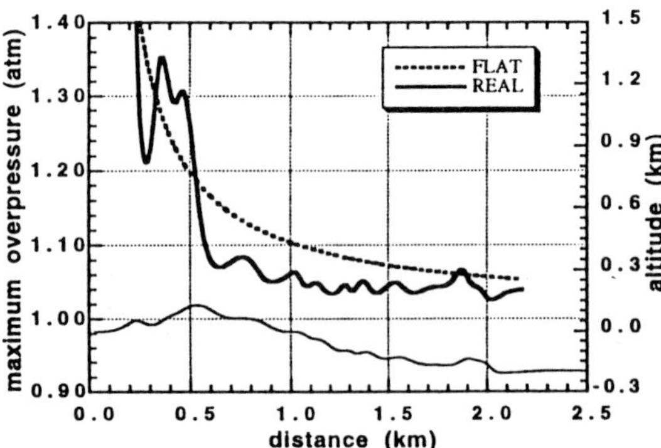

Fig.11: Computed maximum overpressure distributions - comparison of the practical ground surface and the flat ground surface -

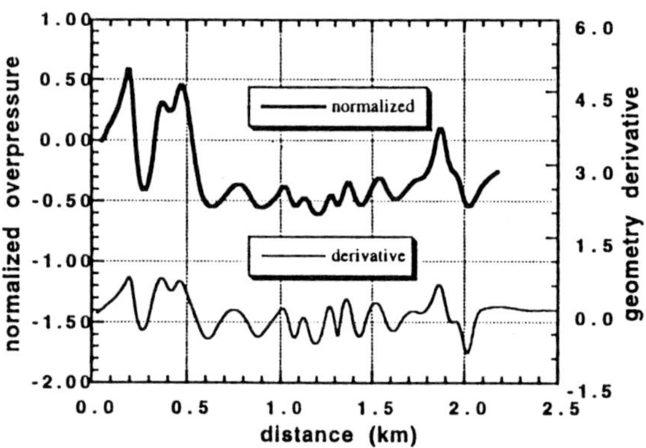

Fig.12: Normalized overpressure and the gradient of the ground surface

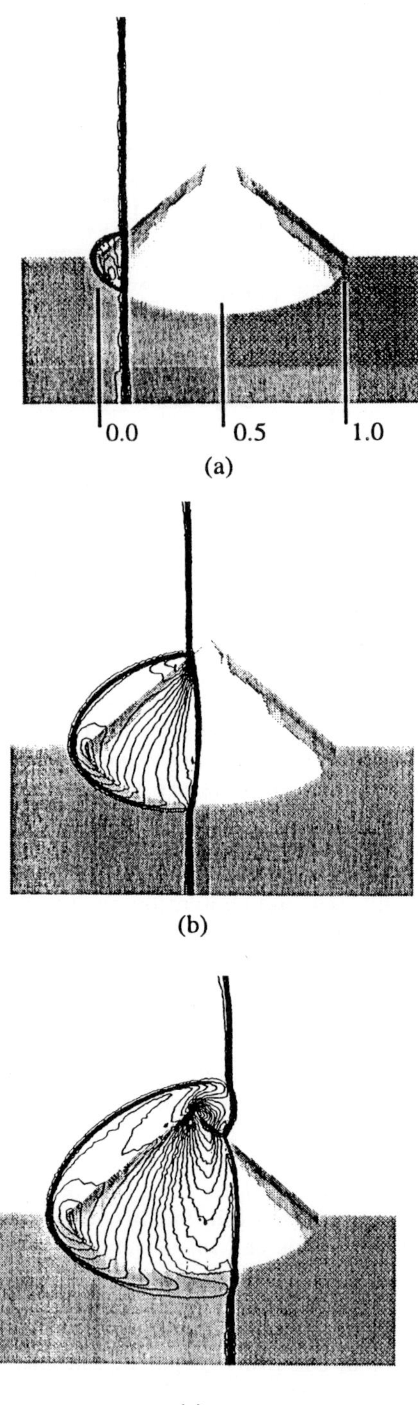

Fig.13: (a)-(c) Time sequence of the pressure contour plots for the shock wave passing over a cone

to describe the overpressure behavior, but it will be the future work.

The two dimensional computations showed that the ground surface geometry has strong influence on the strength of the blast wave and as a result on the overpressure. It was also shown that the gradient of the ground surface geometry is important for the overpressure distribution. Now, in three dimensions, the effect may be weaker because spherical expansion creates stronger decay of the blast wave than cylindrical expansion, and mountains or hills are not usually two dimensional. Another diffraction effect may exist from the side of these obstacles. Figures 13 (a)-(f) are the pressure contour plots that were taken from Ref.14 and it shows the flow field created near the cone when a plane shock wave passes over it. In this simulation, computations were carried out using a solution-adaptive unstructured grid method because the purpose was to find out the three-dimensional complicated diffraction effect. Although shock waves and blast waves have different nature, similar shock wave

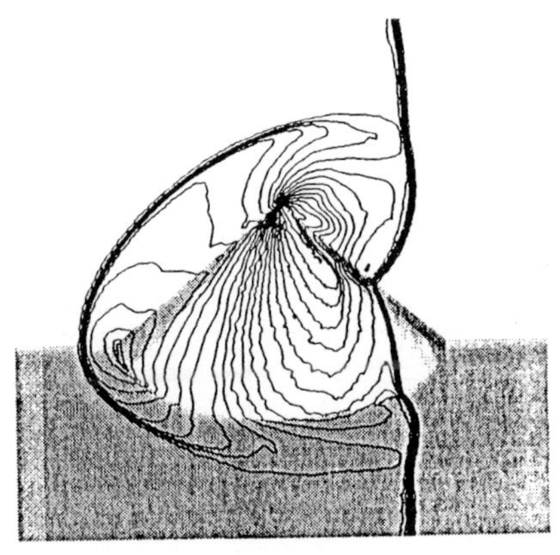

(d)

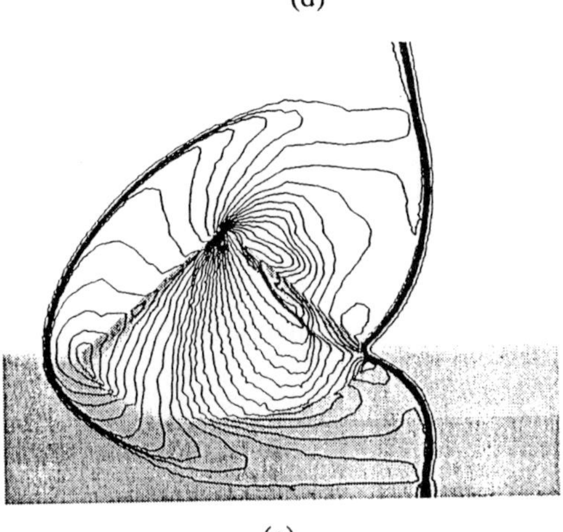

(e)

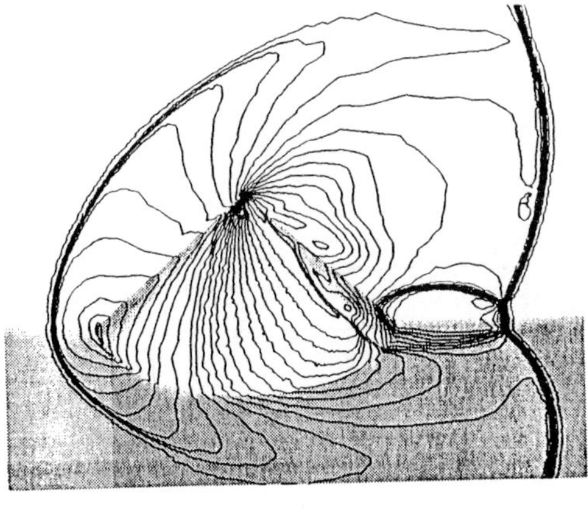

(f)

Fig.13: (d)-(f) Time sequence of the pressure
contour plots for the shock wave passing
over a cone

interaction may occur. The angle of the cone is different at each cutting surface and therefore different reflection patterns appear in each section as is pointed out in Ref.14. More specifically, the angle of incidence changes as the shock wave proceeds except the center line. After the incident shock wave passes the apex of the cone, the diffraction occurs. The shock speed becomes different and the shock wave tends to focus on the ground behind the cone as can be seen in Fig.13 (f). Three-dimensional vortical flow structure remains in the rear portion of the cone even after the shock wave has passed. This result indicates that two-dimensional understanding of the blast wave behavior may not be sufficient even qualitatively.

Three-dimensional computations are next carried out. Computational details were described in the previous section. The computed overpressure distributions in the WEST, EAST, NORTH, SOUTH are plotted in Fig.14. In this figure, some of the instantaneous pressure profiles are also plotted. The shock wave is crisply captured in all the directions. The overpressure distribution in the WEST direction shows that strong pressure increase occurs at 0.6 km. It is obviously due to the diffraction of the shock wave. The diffraction effect may be stronger in three dimensions due to the focus of the shock wave in this area as indicated in Fig.13 (f). The overpressure is very different in each direction and it indicates that the ground surface geometry effect is important and should be considered for an accurate estimation of the safety distance of the blast wave. To see the effect of the ground surface geometry, the computed overpressure normalized by the overpressure of the FLAT ground is plotted in Fig.15 to show the effect of the ground surface. The gradient of the ground surface geometry in the direction of the blast wave propagation is also plotted in the same figure. Although the gradient of the ground surface does not sufficiently explain the overpressure behavior, strong correlation exists. It is partly due to the three-dimensional effect but further research is also necessary to pick up better expressions of "normalized overpressure".

The regions in which the overpressure becomes larger than 2-10 % of the atmospheric pressure are plotted in Fig.16. The region strongly depends on the ground surface geometry as expected from Fig.14. Although the three dimensional simulations were conducted only for a specific example, the result shows that the present approach is adequate for the blast wave simulations and gives a lot of information to the estimation of "safety distance".

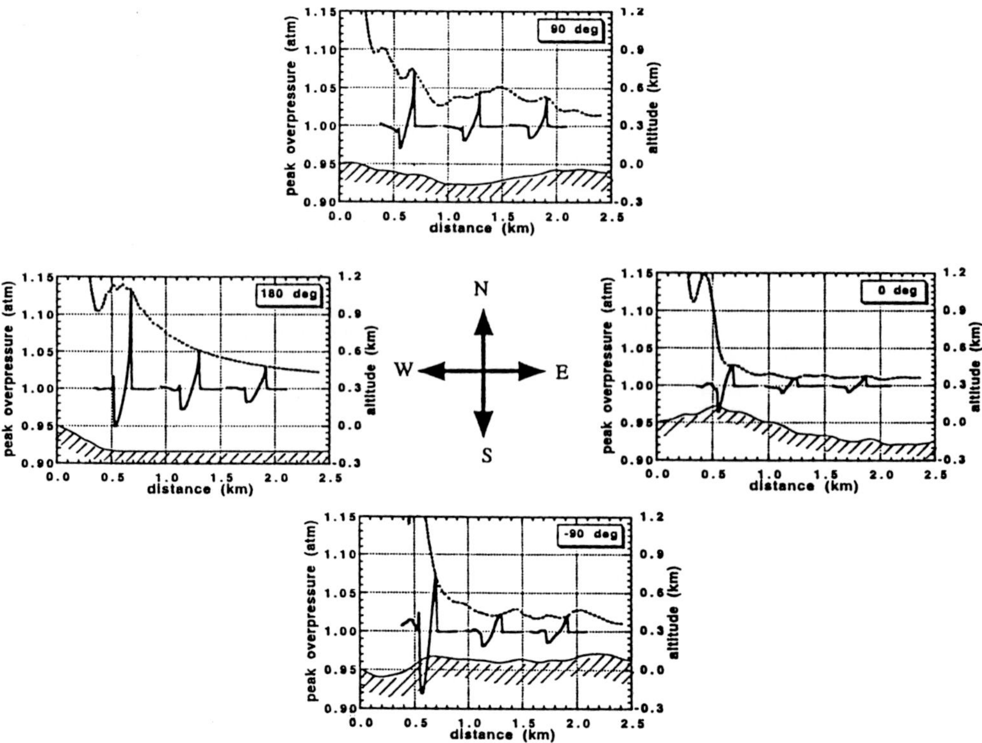

Fig.14: Computed maximum overpressure distributions in the four directions

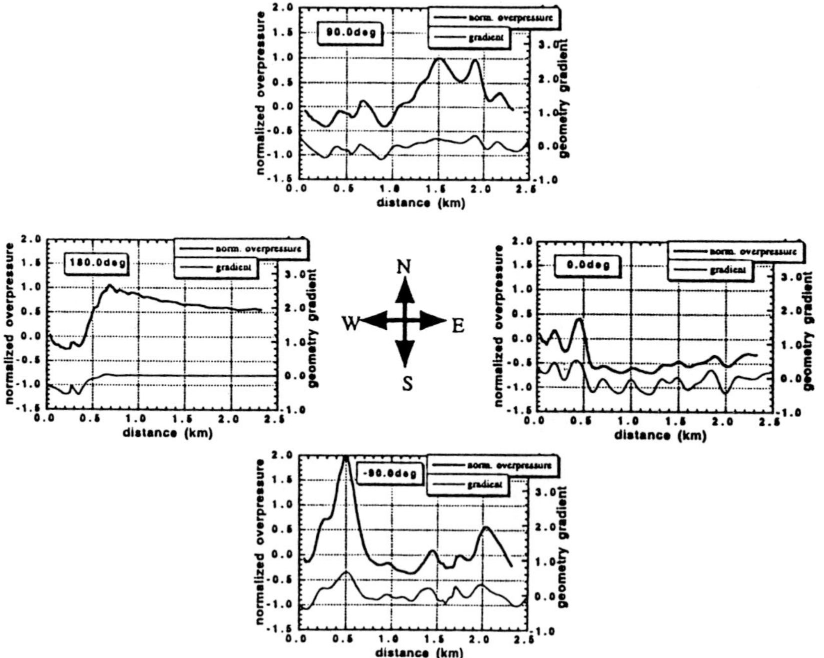

Fig.15: Normalized overpressure and the gradient of the ground surface in the four directions

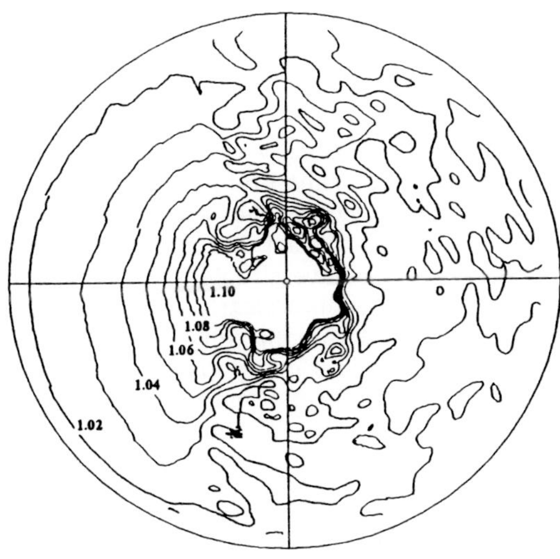

Fig.16: Overpressure distributions on the ground surface

4. Conclusions

An accurate computational tool has been developed for the purpose of establishing a better estimation method of the safety distance for the blast wave. Since this type of simulations require high level of grid resolution, two approaches were considered. One is an overset zonal method and the other is an unstructured solution-adaptive grid method. Two-dimensional computations showed the advantages of the overset zonal method. Many of the two dimensional computations were carried out to make the assessments of the effect of the computational grids to the solutions. Based on the techniques developed for two-dimensional simulations, three-dimensional simulations were carried out using the overset zonal method and the effect of the ground surface geometry was discussed. Sufficient grid resolutions were obtained within the available computer memory and speed. The computed result clearly showed that the ground surface has an important effect on the strength of the blast wave even far away from the point of explosion and the method developed here can be a good tool for the better estimation of the strength of the blast wave.

Acknowledgment

The two dimensional solutions for the unstructured solution-adaptive grid were obtained when Dr. Dimitri Sharov, currently a research engineer for Fujitsu Computational Mechanics Center of Fujitsu Corporation was a visiting scientist of the ISAS (Institute of Space and Astronautical Science). Computations shown in Appendix A for the accuracy of moving grid system were carried out by Mr. Erik A. Nijs, currently a graduate student at Delft University in Netherlands during his stay for three months under a traineeship at the ISAS. The authors would like to thank these two researchers for their contributions to the present study.

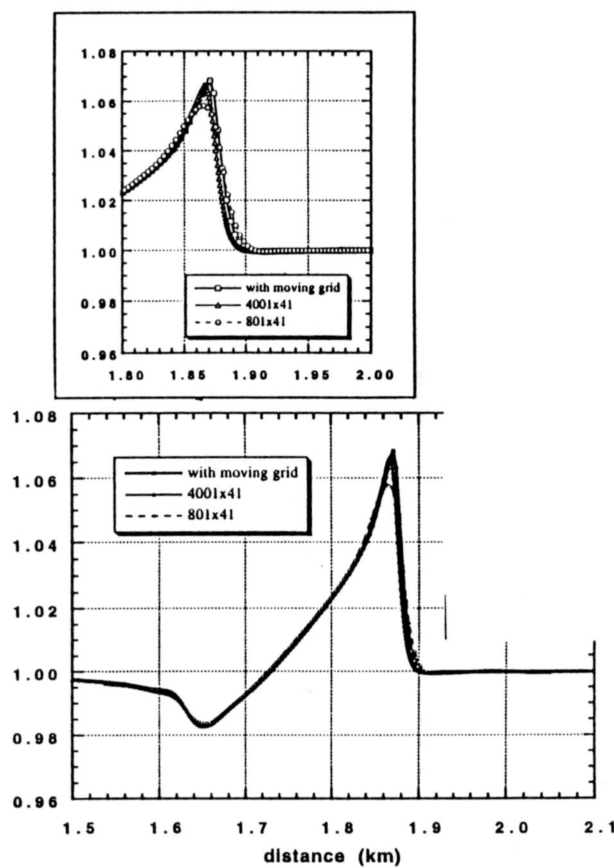

Fig.A1: Pressure distributions for three types of grids

Appendix A Accuracy Assessment of Moving Grids

Figure A1 shows the computed pressure profiles for the two-dimensional problem when the blast wave reaches 1.8 km from the point of explosion. There are three plots. Thick solid line is the result for the zonal computation with the local grid moving with the speed of sound. Note that the speed of the frontal shock wave is almost the same as the speed of sound because the shock wave is very weak in most of the flow field. Thin solid line is the result for the single-zone computation with 4001 grid points. Dash line shows

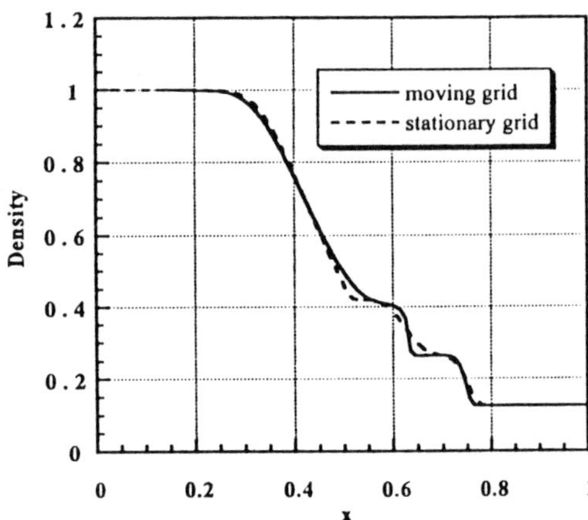

Fig.A2: Density distributions for the shock tube problem

the result for the single-zone computation with 801 grid points. Obviously, the shock wave with 801 grid points are smeared out due to the discretization error when compared to the other two results. In this example, the spatial resolution of the moving zonal grid is the same as the resolution of the 801 grid-points solution. However the result for the moving grid is much better and may be better than that for the 4001 grid points. The result indicates that movement of the grid somehow increases the grid resolution. To survey the reason, moving grid system is formulated for one-dimensional scalar and system of equations. When, the grid moves at the velocity u_g, the basic equation for a linear scalar equation becomes

$$\frac{\partial u}{\partial \tau} + \{c - u_g\}\frac{\partial u}{\partial \xi} = 0 \qquad (3)$$

Here c is constant. It is clearly noticed that the convective speed is changed from in the stationary grid to $c - u_g$ in the moving grid. All the truncation error terms for the space derivatives have the coefficient of this convective speed $c - u_g$ as shown in Eq.(2).

$$T.E. = \frac{\Delta t}{2}\frac{\partial^2 u}{\partial t^2} + \frac{\{c - u_g\}\Delta x}{2}\frac{\partial^2 u}{\partial t^2}$$
$$- \frac{\{\Delta t\}^2}{6}\frac{\partial}{\partial t}\frac{\partial^2 u}{\partial t^2} - \{c - u_g\}\frac{\partial\{\Delta x\}^2}{\partial 6}\frac{\partial^3 u}{\partial x^3} + h$$

The truncation errors from the time derivative terms can be translated to the space derivative errors using the relations in the original equations. Therefore all the truncation errors have $c - u_g$ as their coefficient and would decrease when u_g were set closer to c. In fact, when the grid moves at the velocity c, the solution does not practically change at all and correspondingly there are no discretization errors. The result indicates that the key reason for the improvement of the grid resolution in the moving grid system comes from the fact that the coefficients of the convective speed (eigenvalues for the system of equations) are reduced.

A series of computations were carried out for the one-dimensional shock tube problem to verify it. Figure A2, for instance, shows the result when the grid moves at the speed of contact discontinuity. Since the shock tube problems have simple wave regions, the decrease of the eigenvalues clearly improves the resolution of the contact discontinuity. The shock wave moves at $u + c$ in the stationary grid and becomes almost c in the moving grid. The coefficient becomes smaller and the resolution of the shock wave is improved although not as dramatically as the contact discontinuity. On the other hand, the modified speed of expansion fan has now the coefficient of $-c$ which is negatively larger. Therefore, the expansion fan is smeared out compared to the solution of the stationary grid. Eulerian computation with the moving grid system can be considered to have "Lagrangean" effect. The truncation error analysis shown here demonstrates that it can be explained as the change of the coefficients of the truncation errors. The results here indicate that the grid resolution can be improved even when the grid moves not at the exactly same speed as the phenomenon interested in.

REFERENCES

[1] Baker, W. E., "Explosion in Air," Univ. Texas Press, Austin and London, 1972.

[2] Kinnery, G. F. and Graham, K. J., "Explosirve Shocks in Air," Springer-Verlag, 1985.

[3] Kingery, C. N. and Pannill, B. F., "Peak Overpressure Vs Scaled Distance for TNT Surface Bursts (Hemispherical Charges)," Ballistic Research Laboratories, MR.-1518, 1964.

[4] Brode, H. L., "Blast Wave from a Spherical Charge," The Physics of Fluids, Vol. 2, No. 2, pp. 217-229, 1959.

[5] Plooster, M. N., "Shock Wave from Line Sources. Numerical Solutions and Experimental Measurements," The Physics of Fluids, Vol. 13, No. 11, pp. 2665-2675, 1970.

[6] Shimizu, F., Fujii, K. and Higashino, F., "Ground Surface Effect on the Blast Wave Propagation in Two Dimensions," Transaction of the Japan Society for Aeronautical and Space Sciences, Vol. 36, No. 111, pp. 36-46, 1993.

[7] Shimizu, F., Fujii, K. and Higashino, F., "Three-Dimensional Blast Wave Propagating

on the Realistic Ground Geometry," 6th International Symposium on Computational Fluid Dynamics, Collection of Technical Papers, pp.1148-1153, Sep., 1995.

[8] Sharov, D. and Fujii, K., "Unstructured Adaptive Mesh Method and Its Application to a Blast Wave Propagation Problem," The preprint of the 3rd World Congress on Computational Mechanics (WCCM III), May, 1994.

[9] Sharov, D. and Fujii, K., "Three-Dimensional Adaptive Bisection of Unstructured Grids for Transient Compressible Flow Computations," AIAA paper 95-1708, AIAA 12th Computational Fluid Dynamics Conference, June, 1995.

[10] Fujii, K., "Unified Zonal Method Based on the Fortified Solution Algorithm," Journal of Computational Physics, Vol. 118, pp. 92-108, 1995.

[11] Roe, P. L., "Characteristic-Based Schemes for the Euler Equations," Annual Review of Fluid Mechanics, pp. 337-365, 1986.

[12] Thomas, J. L., van Leer, B. and Walters, B. W., "Implicit Flux-Split Schemes for the Euler Equations," AIAA Paper 85-1680, 1985.

[13] Obayashi, S., "Free-Stream Capturing in Fluid Conservation Law for Moving Coordinates in Three Dimensions," NASA CR177572, 1991.

[14] Miyaji, K. and Fujii, K., "Unsteady Behavior of the Shock Wave Passing over Two and Three Dimensional Obstacles," AIAA Paper 96-2445, 14th AIAA Applied Aerodynamics Conference, June, 1996.

HIGH-ORDER NUMERICAL METHODS FOR UNSTEADY HYPERSONIC FLOW SIMULATIONS

Xiaolin ZHONG   Gregory H. FURUMOTO

Mechanical and Aerospace Engineering Department
University of California, Los Angeles, CA 90095

Abstract

Advances in numerical methods and rapid increases in memory capacity and computational speeds of available computers have made computational fluid dynamics (CFD) an essential tool for the development of hypersonic vehicles. This paper presents our recent work on developing high-order numerical methods for transient nonequilibrium hypersonic flow simulations and for the direct numerical simulation (DNS) of transitional and turbulent hypersonic boundary layers. The applications of these numerical techniques to unsteady hypersonic flow computations are demonstrated through several examples of results. A brief overview of general CFD methods as they are applied to hypersonic flow simulations in the continuum regime is also given.

1 Introduction

Recent interest in hypersonic flows stems from the proposed designs of high speed trans-atmospheric vehicles for both civil and military use [1]. An important aerodynamic phenomenon associated with hypersonic vehicles is that real gas effects, which include vibrational mode excitation, chemical reactions, ionization, and radiation, play a significant role in the aerodynamic and thermal loadings to the vehicles. Other important aerodynamic phenomena include shock-shock and shock-boundary-layer interactions, laminar-turbulent transition, turbulence, and rarefied gas effects. Early studies of hypersonic flows mainly encompassed experiments in ground based test facilities or in the form of flight tests such as the X-15 program [2,3], and analytical techniques such as those presented in [4]. An overview of the field of hypersonics spanning a period of nearly four decades can be found in [2,4-9].

In recent years, significant advances in numerical methods and rapid increases in memory capacity and computational power of available computers have made CFD an essential tool for the development of hypersonic vehicles. Numerical simulations offer the promise of combining sophisticated physical models and complex geometries for hypersonic flows at actual flight conditions to provide a valuable tool for both the study of hypersonic flow phenomena and the design of hypersonic vehicles. Currently, many computer codes have been developed to solve complex three-dimensional hypersonic flow problems including real gas effects in the continuum and rarefied gas flow regimes. Examples of these are, among others, the work of Candler [10-12], the TVD codes of Yee [13-16], the LAURA code developed at NASA Langley by Gnoffo, *et al.* [17-19], and the recent work of Brück at DLR [20,21]. The CFD methods developed for these codes are mainly second-order upwind schemes for computing steady or slightly unsteady hypersonic flow simulations using the shock-capturing approach.

Many advanced high-resolution upwind shock-capturing schemes have been developed for accurate computations of real-gas hypersonic flows with complex strong shock wave interactions. At the same time, both structured and unstructured grid methods have been used to handle complex three-dimensional geometries. The reviews for state-of-the-art of CFD methods for viscous real-gas hypersonic flows can be found in [22] and [23].

While steady flow simulations are essential for the design of hypersonic vehicles, the development of numerical methods for transient flow phenomena are also important for many hypersonic applications. One reason is due to the fact that many shock boundary-layer interactions have been found to be inherently unsteady, and the unsteadiness has strong effects on the aerodynamic parameters of the flows. Another application is in the study of the growth of instability waves in the boundary layers and the laminar-turbulent transition of hypersonic boundary layers. Hypersonic flows associated with such phenomena are inherently transient three-dimensional flows containing a wide range of time and length scales. The numerical simulations of such transient flows require high-order accurate CFD methods in order to capture all of the time and length scales of the flow transient. Standard shock-capturing CFD methods developed for steady or slightly unsteady flow computations are not appropriate for transient flow

Received on July 1, 1997.
1) e-mail: xiaolin@seas.ucla.edu

applications because their accuracy level is not high enough.

The purpose of this paper is to present our recent work on developing high-order numerical methods and simulating unsteady and transient nonequilibrium hypersonic flows in the continuum regime [24–31]. The application of these numerical techniques to unsteady hypersonic flow computations are demonstrated through sample results simulating transient hypersonic flows. Meanwhile, an brief overview is also given for general CFD methods as it is applied to nonequilibrium hypersonic simulations in the continuum regime. No efforts have been made to provide an exhaustive review on other CFD methods for hypersonic flows developed for steady flow simulations which have been reviewed in by many authors, such as those in [22, 23].

II. Hypersonic Flow Regime and Features

A Mach number of 5 is often used as a dividing line between supersonic and hypersonic flows. However, unlike the demarcation between supersonic and subsonic flows, a strict definition based on a Mach number value is insufficient to truly define the hypersonic flow regime. This is because the hypersonic regime is characterized not by a characteristic speed, but by certain thermophysical and fluid dynamic flow phenomena [2, 4]. In addition, rarefied gas dynamics effects are often part of the hypersonic phenomena because of the high altitudes at which hypersonic vehicles are designed to operate. Rarefied gas dynamics is beyond the scope of this paper. Further details can be found in [32–34].

From a thermophysical point of view, the high speeds create very strong shocks which produce large post shock temperatures on the order of thousands of Kelvin. At these elevated temperatures, several thermodynamic and chemical phenomena occur. First, the internal modes of polyatomic species, such as vibrational mode, can become excited. Second, for polyatomic species, the high temperatures can lead to species dissociation and the initiation of chemical reactions, such as the dissociation of oxygen and nitrogen into atomic species and the subsequent formation of various nitrogen oxides. Third, the actual internal structure of the atom can be disrupted by ionization. This adds the further complication of having charged particles in the flow field. Finally, at very high temperatures, the fluid itself may produce a significant amount of heat through thermal radiation.

From a fluid dynamical point of view, hypersonic flows are characterized by strong shock layers (the region between a shock wave and the body responsible for creating the shock wave) much thinner than those in supersonic flows, and the shock wave lies very close to the body. For flow in the viscous regime, the thin shock layer leads to viscous interaction whereby the boundary layer and the shock wave strongly interact. When this occurs, the boundary layer thickness becomes a significant fraction of the shock layer thickness and can provide a displacement effect, altering the shape of the shock wave. In both the inviscid and viscous regimes, the high shock curvature near the nose of the body creates strong entropy gradients which flow downstream in what is known as the entropy layer which plays an important role in the instability and transition of shear layers and boundary layers behind the shock. More in depth treatments of the thermophysical and fluid dynamical phenomena in hypersonics flows are given in [2, 7, 35, 36].

These phenomena which characterize hypersonic flow can be illustrated by examining representative hypersonic flowfields in which they occur. In particular, two broad classes of flows with high temperature and viscous effects are examined here: shock interactions and hypersonic boundary layer instability and laminar-turbulent transition.

Shock Interactions

There are two main types of viscous shock interactions that are of interest in the hypersonic flows about aerospace vehicles. The first is the shock-boundary layer interaction which is caused by direct shock impingement on a surface. The second is the shock-shock interference problem which is caused by the intersection of two or more shocks in the flow field.

A shock-boundary layer interaction occurs when a shock wave produced in the bulk flowfield impinges on the boundary layer near a body surface. In particular, when an oblique shock is reflected by a flat plate in a viscous flow, a shock-boundary layer interaction occurs. The large adverse pressure gradient due to the incident shock hitting the wall causes the boundary layer to separate. High local temperatures occur in the region of shock impingement which can bring about nonequilibrium phenomena [37, 38].

Shock-shock interference occurs when an oblique shock intersects the bow shock ahead of a body. In hypersonic applications, an example of a shock-shock interaction occurs when the body bow shock intersects the bow shock ahead of the engine intake cowl lip of a SCRAMJet engine. The complex system of shocks and shear layers produced by a shock-shock interaction can have a significant impact on the aerodynamic loading and thermal heating of the vehicle, depending on the flow geometry. Comprehensive studies of these phenomena can be found in works by Edney [39], Holden [40], and Wieting, *et al.* [41]. Of particular interest is the type IV shock-shock interaction because of the high surface heating rates it produces [39] and the potentially catastrophic effects associated with these heating rates [3,40,42]. The type IV shock-shock interaction, illustrated in Figure 2, occurs when the incident shock intersects the bow shock near the geometric stagnation line. This interaction creates a transmitted shock impinging upon the lower bow shock and a supersonic jet impinging on the body surface with a terminating strong shock. At the jet impingement point, extremely high surface pressures and heating rates are encountered [24,40,41]. This type of interaction has been shown to be inherently unsteady [24,26,43,44] and are affects by real-gas effects at high temperatures.

Boundary-Layer Laminar-Turbulent Transition

The prediction of laminar-turbulent transition location in hypersonic boundary layers is critical to the accurate calculations of drag and thermal loads for the aerodynamic design and control of hypersonic vehicles. In 1989, the Committee on Hypersonic Technology for Military Application of the National Research Council [45] identified the difficulty of accurately predicting the transition location as one of the main limitations on current CFD codes for computing hypersonic flows over the National Aero-Space Plane (NASP). The stability and transition of supersonic and hypersonic boundary layers was reviewed in [46–50].

The transition paths in boundary layers are the result of nonlinear response of the laminar boundary layers to forcing disturbances. The forcing disturbances, as discussed and reviewed by Bushnell [51], can originate from many difference sources: freestream acoustic disturbances, freestream turbulence, freestream vorticity disturbances, freestream entropy disturbances, surface roughness, surface vibrations, disturbances generated by other parts of the vehicles, etc. In an environment with small initial disturbances corresponding to those encountered in hypersonic flight, the paths to transition consist of three stages: 1) receptivity, 2) linear eigenmode growth or transient growth, and 3) nonlinear breakdown to turbulence. The receptivity process [52] is the process of the conversion of the environmental disturbances into instability waves, such as the Tollmien-Schlichting (T-S) waves, in the boundary layers. The linear eigenmode growth of boundary-layer instability waves can be solved as the eigen-solutions of the homogeneous linearized disturbance equations such as the Orr-Sommerfeld equation. The relevant instability waves developed in hypersonic boundary layers are the T-S wave and inviscid waves of higher

(Mack) modes discovered by Mack [46,53], the Görtler instability [54] over concave surfaces, and the three-dimensional cross flow instability [50]. Theoretical [53] and experimental [55,56] studies found that the first and second mode instabilities are simultaneously present in the hypersonic boundary layers where the second mode is most unstable at high Mach numbers. The breakdown of linear instability waves occurs after the growth of linear instability waves reach certain magnitudes. At this stage, the flow develops nonlinear secondary instabilities [57,58] and transition to turbulence. Compared with low-speed boundary layer transition, the transition of hypersonic boundary layers over practical hypersonic vehicles with blunt leading edges is affected by the additional effects of nose bluntness, shock interactions, and real-gas effects at high temperatures. Figure 3 shows a schematic of wave fields near the hypersonic leading edge generated by freestream disturbances.

Direct numerical simulation, which solves the full Navier-Stokes equations as an initial-boundary problem by numerical methods, has recently become an important tool in instability and transition studies [59,60]. We [30,61] have conducted the DNS of the receptivity of a hypersonic boundary layer to 2-D and 3-D freestream acoustic disturbances for a Mach 15 flow over a parabolic leading edge. It was found that the instability waves developed in the hypersonic boundary layer behind the bow shock contain both the first and second mode instabilities.

III. Physical Models and Governing Equations

Physical models used for the numerical simulations of nonequilibrium hypersonic flows must capture the complex thermophysical phenomena that characterize these flows. Detailed discussions of the relevant flow physics can be found in works by Vincenti and Kruger [36] and Clarke and McChensey [35]. There are many different ways to formulate the governing equations for nonequilibrium flow, depending on what approximations are made in the modeling process. In the context of continuum based (CFD) numerical simulations, some of the more well known are outlined in works by Park [62], Lee [63], Gupta, *et al.* [64], and Hauser [65]. The work presented here is a nonequilibrium air model based primarily on those of Park and Candler [10,62]. Governing equations are formulated for a two-temperature model with five species (nonionizing) finite rate air chemistry. The effects of ionization and radiation are not considered for the sake of simplicity.

Conservation Equations

In the sections that follow, sums are taken over indices ranging from 1 to either NS or nd. NS is the total number of species being considered, while nd is the number of diatomic species being considered. Only two-dimensional governing equations are presented for the sake of simplicity. In conservative form, the multicomponent Navier-Stokes Equations, along with the vibrational energy equation, are:

$$\frac{\partial \rho_i}{\partial t} + \frac{\partial}{\partial x}(\rho_i u + j_{ix}) + \frac{\partial}{\partial y}(\rho_i v + j_{iy}) = w_i \qquad (1)$$

$$\frac{\partial(\rho u)}{\partial t} + \frac{\partial}{\partial x}(\rho u^2 + p - \tau_{xx}) + \frac{\partial}{\partial y}(\rho u v - \tau_{xy}) = 0 \quad (2)$$

$$\frac{\partial(\rho v)}{\partial t} + \frac{\partial}{\partial x}(\rho u v - \tau_{xy}) + \frac{\partial}{\partial y}(\rho v^2 + p - \tau_{yy}) = 0 \quad (3)$$

$$\frac{\partial E_v}{\partial t} + \frac{\partial}{\partial x}(u E_v + q_{vx}) + \frac{\partial}{\partial y}(v E_v + q_{vy}) = w_v \quad (4)$$

$$\frac{\partial E}{\partial t} + \frac{\partial}{\partial x}[u(E + p) - u\tau_{xx} - v\tau_{xy} + q_x]$$
$$+ \frac{\partial}{\partial y}[v(E + p) - u\tau_{xy} - v\tau_{yy} + q_y] = 0 \qquad (5)$$

where E_v and E are the vibrational and total energies per unit volume, the w_i's are the chemical source terms for species i, and w_v is the source terms for the vibrational mode. For the thermodynamic state equations, the system is taken to be a mixture of thermally perfect gases, i.e.,

$$p = \sum_{i=1}^{NS} \rho_i R_i T_t \qquad (6)$$

where p is the bulk pressure, T_t is the translational temperature, ρ_i is the species density, and R_i is the species specific gas constant.

The energy equations for diatomic and monatomic species are:

$$e_i = \frac{5}{2}R_i T_t + R_i \frac{\theta_{vi}}{e^{\theta_{vi}/T_v} - 1} + h_i^{\circ} \qquad (7)$$

$$e_i = \frac{3}{2}R_i T_t + h_i^{\circ} \qquad (8)$$

where the vibrational energy equation for a diatomic species modeled by an independent vibrational temperature, T_v [10,36,66]. In the above equations, θ_{vi} is the characteristic vibrational temperature of species i and h_i° is the species heat of formation. The total energies per unit volume are given by:

$$E_v = \sum_{i=1}^{nd} \rho_i e_{vi} \qquad (9)$$

$$E = \sum_{i=1}^{NS} \rho_i e_i + \rho \frac{u^2 + v^2}{2} \qquad (10)$$

In the continuum regime, the viscous stresses are modeled by:

$$\tau_{xx} = \frac{2}{3}\mu \left[2\frac{\partial u}{\partial x} - \frac{\partial v}{\partial y} \right] \qquad (11)$$

$$\tau_{xy} = \mu \left[\frac{\partial u}{\partial y} + \frac{\partial v}{\partial x} \right] \qquad (12)$$

$$\tau_{yy} = \frac{2}{3}\mu \left[2\frac{\partial v}{\partial y} - \frac{\partial u}{\partial x} \right] \qquad (13)$$

For chemically reacting flow, species mass diffusion fluxes are given by full multicomponent diffusion models where the flux of species i is dependent upon its own concentration gradient, and the gradients of all the other species [67–69]. Such a multicomponent model can be computationally prohibitive for numerical simulations. Therefore, a common approximate model used is to assume each species follows Fick's Law of diffusion for a binary gas mixture. Thus, each species is treated as diffusing into the remaining bulk as if it were a binary mixture of the species in question and everything else [67]. Using this model, the mass diffusion fluxes for species i are given by:

$$j_{ix} = -\rho D_i \frac{\partial (\rho_i/\rho)}{\partial x} \qquad (14)$$

$$j_{iy} = -\rho D_i \frac{\partial (\rho_i/\rho)}{\partial y} \qquad (15)$$

The effects of diffusion due to thermal and pressure gradients are typically small and can be neglected for simplicity [70]. Heat diffusion is modeled using Fourier's Law:

$$q_x = -\kappa_t \frac{\partial T_t}{\partial x} - \kappa_v \frac{\partial T_v}{\partial x} + \sum_{i=1}^{NS} j_{ix} h_i \qquad (16)$$

$$q_y = -\kappa_t \frac{\partial T_t}{\partial y} - \kappa_v \frac{\partial T_v}{\partial y} + \sum_{i=1}^{NS} j_{iy} h_i \qquad (17)$$

with the total enthalpy, h_i, given by:

$$h_i = e_i + R_i T_t \qquad (18)$$

There are several transport coefficient models that can be used for flows with thermochemical nonequilibrium which are available in the literature [64, 67, 68]. The particular models used here use Wilke's mixture rule for viscosity [71], the Eucken relations for thermal conductivity [36], and curve fits for diffusivities [64] which are outlined in [72].

<u>Source Terms</u>

Vibration-translation interactions are modeled using the Landau-Teller model [36]:

$$Q_{T-V} = \sum_{j=1}^{nd} \frac{\rho_j R_j \theta_{vj} \left(\frac{1}{e^{\theta_{vj}/T_t} - 1} - \frac{1}{e^{\theta_{vj}/T_v} - 1} \right)}{\tau_{vj}} \qquad (19)$$

For the vibrational relaxation time of species j, τ_{vj}, the corrected Millikan and White formula [73] as proposed by Park [62] was used. This gives the vibrational time as:

$$\tau_{vj} = \frac{\sum_{i=1}^{NS} X_i}{\sum_{i=1}^{NS} X_i / \tau_{vij}^{MW}} + \tau_{cj} \qquad (20)$$

where X_i is the mole fraction of species i. The relaxation time τ_{vij}^{MW} is given by [73]:

$$\tau_{vij}^{MW} = \frac{1}{p} \exp\left[1.16 \times 10^{-3} \mu_{ij}^{-\frac{1}{2}} \theta_{vj}^{\frac{4}{3}} \right.$$
$$\left. \left(T_t^{-\frac{1}{3}} - 0.015 \mu_{ij}^{\frac{1}{4}} \right) - 18.42 \right] \qquad (21)$$

where p is in units of atmospheres, θ_{vj} is the characteristic vibrational temperature of species j, and μ_{ij} is

the reduced mass given by:

$$\mu_{ij} = \frac{\mathcal{M}_i \mathcal{M}_j}{\mathcal{M}_j + \mathcal{M}_i}, \qquad (22)$$

The correction factor is given by:

$$\tau_{cj} = \frac{1}{\bar{c}_j \sigma_v N} \qquad (23)$$

where $\bar{c}_j$ is the mean molecular speed given by: $\bar{c}_j = \sqrt{\frac{8 R_j T_t}{\pi}}$, σ_v is the limited collision cross section (in m^2) given by [74, 75]: $\sigma_v = 10^{-21} \left(\frac{50,000}{T_t} \right)^2$, and N is the total number density of the gas. The model used for vibration-dissociation coupling is the one employed by Candler in [10] and is given by

$$Q_{V-D} = \sum_{j=1}^{nd} w_j \frac{R_j \theta_{vj}}{e^{\theta_{vj}/T_v} - 1} \qquad (24)$$

The total vibrational source term is then:

$$w_v = Q_{T-V} + Q_{V-D} \qquad (25)$$

For the test cases considered in this paper, the temperatures are not expected to exceed $9000 K$ over most of the flowfield, which is the threshold for ionization [2]. Nonequilibrium chemistry is modeled using a five species air model:

$$N_2 + M_i \;\rightleftharpoons\; N + N + M_i \qquad (26)$$
$$O_2 + M_i \;\rightleftharpoons\; O + O + M_i \qquad (27)$$
$$NO + M_i \;\rightleftharpoons\; N + O + M_i \qquad (28)$$
$$N_2 + O \;\rightleftharpoons\; NO + N \qquad (29)$$
$$NO + O \;\rightleftharpoons\; O_2 + N \qquad (30)$$

where M_i denotes any of the i species. In general, the rate of formation of a reactant species in a general reaction of the form:

$$A + B \rightleftharpoons C + D \qquad (31)$$

can be written as:

$$\mathcal{R} = k_b [A][B] - k_f [C][D] \qquad (32)$$

In the Park chemistry model for air [62, 76], the forward rate coefficients are explicitly calculated by a modified Arrhenius equation. The backward rate coefficients are then deduced from the forward rates using the equilibrium constant:

$$k_b = k_f / K_{eq} \qquad (33)$$

The equilibrium constants, K_{eq}, are given by empirical curve fits with respect to temperature. Coefficient and curve fit values for a wide range of reactions of interest in hypersonics are tabulated in the various works of Park [62, 74, 76].

Boundary Conditions

For the boundary conditions on solid walls, a no-slip condition for velocity and temperature is used for flows within the continuum regime. Isothermal walls are also used as they are a common way to simulate cooled walls without solving the conjugate heat transfer problem. Catalytic effects can be handled by either assuming non-catalytic walls (frozen flow at the wall), fully catalytic walls (equilibrium composition assumed at the wall surface), or partially catalytic walls, which would then require a model for the surface chemistry effects. In the present study, non-catalytic wall were used. Studies using other boundary conditions such as catalytic walls and velocity and temperature slip for rarefied flows can be found in the literature [38, 77, 78].

IV. Numerical Methods

Once the physical model is defined, numerical methods are used to solve the governing system of partial differential equations in the flow field. Numerical methods are concerned with discretization of the equations in space and in time. Spatial discretization can be carried out using several approaches: finite volume, finite difference, finite element, and spectral methods. For hypersonic flow simulations with complex geometries, the finite volume method has been the method of choice because of its ability to enforce numerical flux conservation, which is essential for capturing shock waves, and to easily handle complex geometries. However, implementing high-resolution (third order or higher accuracy) algorithms in these finite-volume shock-capturing schemes can be difficult [79]. On the other hand, for flow fields with well defined geometries, finite difference methods are more efficient for implementation of third or higher order accurate algorithms. Shock fitting methods can also be used in high-order finite difference or spectral methods in the direct numerical simulations of transition and turbulent flows where the focus is on the study of the detailed fluid mechanical phenomena. Time discretization is used to integrate the semi-discrete equations after spatial discretization. Both explicit and implicit methods can be used. Care must be taken to ensure that the scheme chosen is both stable and accurate. In particular, the terms associated with the nonequilibrium thermophysics have a wide range of time scales which can add a high degree of numerical stiffness to the temporal discretization so special care must be taken when integrating in time.

Finite Volume Shock Capturing Methods

In the finite volume approach, the governing equations are formulated as integral equations in the conservation law form. This procedure automatically allows for discontinuities in the flow so no special tracking of singularities such as shocks is required. A more detailed description of the finite volume method can be found in many references such as [80] and [81]. In conservative form, the equations of motion can be written as:

$$\frac{\partial \mathbf{U}}{\partial t} + \frac{\partial}{\partial x}(\mathbf{F} + \mathbf{F_v}) + \frac{\partial}{\partial y}(\mathbf{G} + \mathbf{G_v}) = \mathbf{W} \qquad (34)$$

where the conserved quantity and source term vectors are:

$$\mathbf{U} = \begin{bmatrix} \rho_1 \\ \rho_2 \\ \rho_3 \\ \rho_4 \\ \rho_5 \\ \rho u \\ \rho v \\ E_v \\ E \end{bmatrix}, \qquad \mathbf{W} = \begin{bmatrix} w_1 \\ w_2 \\ w_3 \\ w_4 \\ w_5 \\ 0 \\ 0 \\ w_v \\ 0 \end{bmatrix} \qquad (35)$$

Inviscid fluxes are

$$\mathbf{F} = \begin{bmatrix} \rho_1 u \\ \rho_2 u \\ \rho_3 u \\ \rho_4 u \\ \rho_5 u \\ \rho u^2 + p \\ \rho vu \\ uE_v \\ u(E + p) \end{bmatrix}, \qquad \mathbf{G} = \begin{bmatrix} \rho_1 v \\ \rho_2 v \\ \rho_3 v \\ \rho_4 v \\ \rho_5 v \\ \rho uv \\ \rho v^2 + p \\ vE_v \\ v(E + p) \end{bmatrix} \qquad (36)$$

Viscous and diffusive fluxes are

$$
\mathbf{F_v} = \begin{bmatrix} j_{1x} \\ j_{2x} \\ j_{3x} \\ j_{4x} \\ j_{5x} \\ -\tau_{xx} \\ -\tau_{xy} \\ q_{vx} \\ Q_x \end{bmatrix}, \qquad \mathbf{G_v} = \begin{bmatrix} j_{1y} \\ j_{2y} \\ j_{3y} \\ j_{4y} \\ j_{5y} \\ -\tau_{xy} \\ -\tau_{yy} \\ q_{vy} \\ Q_y \end{bmatrix} \qquad (37)
$$

where $Q_x = -u\tau_{xx} - v\tau_{xy} + q_x$ and $Q_y = -u\tau_{xy} - v\tau_{yy} + q_y$.

The conservation laws are cast into an integral form in terms of grid-cell averages so that the finite volume technique could be applied. The system of conservation laws "integrates" for each ij cell to:

$$
\frac{\partial \mathbf{U}_{ij}}{\partial t} + \frac{1}{V_{ij}} \left[(\mathbf{E} \cdot \mathbf{S})_{i+\frac{1}{2},j} - (\mathbf{E} \cdot \mathbf{S})_{i-\frac{1}{2},j} \right.
$$
$$
\left. + (\mathbf{E} \cdot \mathbf{S})_{i,j+\frac{1}{2}} - (\mathbf{E} \cdot \mathbf{S})_{i,j-\frac{1}{2}} \right] = \mathbf{W}_{ij} \qquad (38)
$$

where

$$
\mathbf{E} = (\mathbf{F} + \mathbf{F_v})\hat{\imath} + (\mathbf{G} + \mathbf{G_v})\hat{\jmath} \qquad (39)
$$

and $\mathbf{S}$ is the cell face surface vector.

In shock capturing computations, the inviscid fluxes are evaluated by high-resolution upwind schemes in order to capture shock waves without numerical oscillations. High-resolution shock capturing schemes have received much attention in recent years and many algorithms have been developed, including flux vector splitting [82], Van Leer splitting [80], the Osher scheme [80], and Godunov [81,83] type schemes such as the Roe approximate solver [84]. The high-order extensions of these upwind schemes include those by Harten [85], the MUSCL scheme of Van Leer [80], the Piecewise Parabolic Method [86], the Flux Corrected Transport (FCT) method [87], TVD schemes [13,88], and ENO [89–91] schemes. These schemes have been extended to flows with real gas effects by several researchers, such as Liu and Vinokur [92], Shuen, Liou, and Van Leer [93,94], and Grossman and Cinella [95,96].

The use of a shock capturing method in aerodynamic calculations of complex hypersonic flows is presented here. The inviscid fluxes are solved via a second-order TVD formulation using characteristic variable extrapolation with the Roe flux difference splitting Riemann solver [84,96]. In the Roe schemes, the cell-face fluxes for the flux vector $\mathbf{F}$ at the cell face $i + 1/2$ may be expressed as [97]:

$$
\begin{aligned}
\mathbf{F}_{i+1/2} = {} & \frac{1}{2} \left[\mathbf{F}(\mathbf{U}^R) + \mathbf{F}(\mathbf{U}^L) \right] \\
& - \frac{1}{2} \hat{\mathbf{T}}^{-1} |\hat{\mathbf{\Lambda}}| \hat{\mathbf{T}} (\mathbf{U}^R - \mathbf{U}^L)
\end{aligned} \qquad (40)
$$

where the Jacobian of $\mathbf{F}$ ($\mathbf{A} = \frac{\partial \mathbf{F}}{\partial \mathbf{U}}$) is expressed in terms of $\mathbf{\Lambda}$, the diagonal matrix of its eigenvalues, and $\mathbf{T}^{-1}$ and $\mathbf{T}$, the eigenvector matrices, as $\mathbf{A} = \mathbf{T}^{-1}\mathbf{\Lambda}\mathbf{T}$. The ($\hat{\ }$) indicates the quantity is evaluated using the Roe averages at the $i + 1/2$ face, which are based on the formulations given by Grossman, et al. [95,96] for nonequilibrium gases. $\mathbf{U}^R$ and $\mathbf{U}^L$ are calculated from the characteristic variables, $\mathcal{W}^R, \mathcal{W}^L$, which are extrapolated to the cell faces using the minmod limiter [81] to control spurious oscillations. The characteristic variables are related to the conservative variables by:

$$
\mathcal{W} = \mathbf{T}\,\mathbf{U} \qquad (41)
$$

The left and right states at a given cell face are calculated using a slope limiter [81]. For cell face $i + 1/2$, the right state is:

$$
\mathcal{W}^R = \mathcal{W}_{i+1} - \frac{1}{2}\mathrm{minmod}[\Delta_{i+1}, \Delta_i] \qquad (42)
$$

where

$$
\begin{aligned}
\mathcal{W}_{i+1} &= \mathbf{T}_{i+1}\mathbf{U}_{i+1}, & (43) \\
\mathcal{W}_{i+2} &= \mathbf{T}_{i+1}\mathbf{U}_{i+2}, & (44) \\
\mathcal{W}_i &= \mathbf{T}_{i+1}\mathbf{U}_i, & (45) \\
\Delta_i &= \mathcal{W}_{i+1} - \mathcal{W}_i & (46)
\end{aligned}
$$

The left state is given by:

$$
\mathcal{W}^L = \mathcal{W}_i + \frac{1}{2}\mathrm{minmod}[\Delta_i, \Delta_{i-1}] \qquad (47)
$$

where

$$
\begin{aligned}
\mathcal{W}_i &= \mathbf{T}_i\mathbf{U}_i, & (48) \\
\mathcal{W}_{i+1} &= \mathbf{T}_i\mathbf{U}_{i+1}, & (49) \\
\mathcal{W}_{i-1} &= \mathbf{T}_i\mathbf{U}_{i-1}, & (50) \\
\Delta_i &= \mathcal{W}_{i+1} - \mathcal{W}_i & (51)
\end{aligned}
$$

From here, the left and right conservative variable vectors, $\mathbf{U}^R = \mathbf{T}_{i+1}^{-1}\mathcal{W}^R$ and $\mathbf{U}^L = \mathbf{T}_i^{-1}\mathcal{W}^L$ respectively, can be computed for the $i + 1/2$ cell surface.

ENO Shock Capturing Schemes for Transient Hypersonic Flow Simulations

The TVD upwind schemes, which are the most popular methods for hypersonic aerodynamic computations, are most suitable for steady and moderately unsteady flows because they reduce to first-order accuracy at local extrema of solutions. For numerical simulations where the focus is on transient flow phenomena instead of average aerodynamic parameters, the TVD schemes are less suitable because the loss of accuracy near the shock and at the extrema of the smooth solutions. Among high-resolution shock capturing schemes, the Essentially Non-Oscillatory (ENO) schemes [89,98,99] are most suitable for transient hypersonic flow computations because they are formally high-order accurate in smooth regions of the flow fields and are essentially non-oscillatory across shock waves. The ENO schemes achieve high-order accuracy in computing flux by non-oscillatory adaptive interpolation. The interpolating stencil of the ENO schemes is not fixed. It is chosen adaptively at every time step by comparing the magnitudes of two neighboring flux difference values to choose the one with smaller magnitude. The upwinding of the schemes are achieved by choosing the initial stencil according to the sign of the wave speed in the flux. The Roe average or a local Lax-Friedrichs schemes can be used as the building block as an entropy correction for the Roe-ENO scheme can be constructed. For the system of equations for nonequilibrium flows, the interpolation can be done in each characteristic field. The average eigenvalues and eigenvectors at the interface are evaluated using a Roe average procedure between U_i and U_{i+1}. The Roe average method for nonequilibrium air model derived by Grossman and Cinella [95] can be used. Since the original ENO schemes developed using finite volume approach are difficult and computationally expensive to extend to third-order or higher accuracy [98,100], Shu and Osher [99] developed efficient implementation of the ENO schemes using the finite difference approach. The ENO schemes were developed to improve the accuracy of the TVD schemes at the local extrema of smooth solutions. The main drawback of the ENO schemes is that they are less stable for steady-flow computations because the TVD condition is not strictly satisfied in the ENO schemes. Many modifications have been proposed to reduce the oscillations of the ENO schemes [101,102].

The ENO schemes have been used to compute transient hypersonic flows with shock-shock interference

hearting [24,25] and were used in the direct numerical simulations of shock/turbulence interactions [103,104], and shear layer instability [105].

High-Order Schemes for DNS of Transient Hypersonic Flows

For transient hypersonic flow simulations, such as the DNS of hypersonic boundary layer transition [59], it is necessary to resolve a wide range of time and length scales by high-order accurate schemes. Upwind shock capturing schemes do not have an adequate accuracy level for such simulations [106]. The development of high-order accuracy schemes has received much recent attention for applications ranging from computational aeroacoustics to the DNS of transition and turbulence. The traditional numerical methods for direct simulations have been the spectral methods because of their high accuracies [107]. But the applications of the spectral methods have been limited to flows in simple domains. Several alternative numerical methods have been developed for direct simulation of transitional and turbulent flows in more general geometries. Examples are the spectral element methods [108], high-order compact (Padé) finite difference methods [109,110], and high-order non-compact (explicit) finite difference methods [106]. Examples of using high-order compact and explicit finite difference methods in direct numerical simulations can be found in [110–112] and [113,114] respectively. For the DNS of stability and transition of hypersonic boundary layers over blunt bodies, it is difficult to apply the existing methods because of the stiffness of high Mach number flows with chemical reactions. Zhong [29] presented a new high-order (fifth and sixth order) upwind finite difference shock fitting method for the direct simulation of hypersonic flow with a strong bow shock and with stiff source terms. There are three main aspects of the new method for hypersonic flow DNS: new upwind high-order finite difference schemes, a high-order shock fitting formulation, and third-order semi-implicit Runge-Kutta schemes to treat stiff source terms [27].

The governing equations are discretized in the computational domain using the method of lines. For smooth flow behind the bow shock, high-order finite difference methods are used for spatial discretization of the equations, where the inviscid and viscous flux terms are discretized using different methods: central difference schemes for the viscous flux terms and upwind schemes for the inviscid flux terms. In [29], a family of finite-difference upwind schemes of third, fifth, and seventh orders have been presented for the direct numerical simulations of hypersonic boundary layers. Either compact or non-compact (explicit) schemes can be used. Each upwind scheme uses a central stencil with a free damping parameter α, which is chosen such that the inner scheme is stable when it is coupled with high-order numerical boundary schemes and the dissipation errors are smaller or comparable to phase errors. These high-order upwind schemes are used for the spatial discretization in the shock-fitting algorithm.

The general finite-difference approximation for $\partial u/\partial x$ located at i-th grid point can be written as [110,112]

$$\sum_{k=-M+M_0+1}^{M_0} b_{i+k}\, u'_{i+k}$$
$$= \frac{1}{h} \sum_{k=-N+N_0+1}^{N_0} a_{i+k}\, u_{i+k} \qquad (52)$$

where uniform grids with grid spacing of h are assumed, and u'_{i+k} is the numerical approximation of $\partial u/\partial x$ located at $(i+k)$-th grid point. On the right hand side of the equation, a total of N grid points are used for u_{i+k} with N_0 points bias with respect to the based point i. A similar grid combination of M and M_0 is used for u'_{i+k} on the left hand side of the equation. The schemes are compact finite difference schemes when $M \geq 2$, and they are explicit finite difference schemes when $M = 1$ and $M_0 = 0$. Ref. [29] considered a family of upwind compact and explicit high-order finite difference methods using central grid stencils, i.e.,

$$N = 2N_0 + 1 \qquad (53)$$
$$M = 2M_0 + 1 \qquad (54)$$

The coefficients a_{i+k} and b_{i+k} of the upwind schemes are determined such that the order of the schemes is one order lower than the maximum achievable order for the central stencil, i.e., the orders of the upwind schemes are always odd integers of $p = 2(N_0 + M_0) - 1$. As a result, there is a free parameter α in the coefficients a_{i+k} and b_{i+k}. The free parameter is set to be the coefficient of the leading truncation term which is a derivative of even order:

$$\sum_{k=-M_0}^{M_0} b_{i+k}\, u'_{i+k} = \frac{1}{h} \sum_{k=-N_0}^{N_0} a_{i+k}\, u_{i+k}$$
$$- \frac{\alpha}{(p+1)!} h^p \left(\frac{\partial u^{p+1}}{\partial^{p+1} x} \right)_i + \cdots \qquad (55)$$

where $p = 2(N_0 + M_0) - 1$, and α is the free parameter. All schemes with nonzero α are p-th order accurate, and they are central schemes of $(p+1)$-th order when

$\alpha = 0$. The choice of α is not unique, and it mainly affects the magnitudes of numerical dissipation. The specific value of α for an upwind scheme is chosen to be large enough to stabilize the high-order upwind inner scheme when it is coupled with stable boundary closure schemes, and to be small enough so that the dissipation errors are comparable to the dispersion errors of the inner scheme.

The detailed expressions of the upwind compact and explicit upwind schemes of fifth and seventh orders are given below. Ref. [29] chose a set of "recommended" values of α based on the accuracy and stability analysis. The specific formulas for a fifth-order upwind compact schemes are

$$b_{i-1}\, u'_{i-1} + b_i\, u'_i + b_{i+1}\, u'_{i+1}$$
$$= \frac{1}{h} \sum_{k=-2}^{2} a_{i+k}\, u_{i+k} - \frac{\alpha}{6!} h^5 \left(\frac{\partial u^6}{\partial^6 x} \right)_i + \cdots \qquad (56)$$

where

$$
\begin{aligned}
a_{i+2} &= \tfrac{5}{3} + \tfrac{5}{6}\alpha \\
a_{i+1} &= \tfrac{140}{3} + \tfrac{20}{3}\alpha & b_{i+1} &= 20 + 5\alpha \\
a_i &= 0 - 15\alpha & b_i &= 60 \\
a_{i-1} &= -\tfrac{140}{3} + \tfrac{20}{3}\alpha & b_{i-1} &= 20 - 5\alpha \\
a_{i-2} &= -\tfrac{5}{3} + \tfrac{5}{6}\alpha
\end{aligned}
$$

These schemes are fifth-order upwind compact schemes when $\alpha < 0$, and they reduce to the sixth-order central compact scheme when $\alpha = 0$. The recommended value for α is $\alpha = -1$, which corresponds to the following fifth-order upwind compact inner scheme:

$$25u'_{i-1} + 60u'_i + 15u'_{i+1} = \frac{1}{h}$$
$$\left(-\frac{5}{2}u_{i-2} - \frac{160}{3}u_{i-1} + 15u_i + 40u_{i+1} + \frac{5}{6}u_{i+2} \right) \qquad (57)$$

Similarly, the formulas for a fifth-order upwind explicit schemes are:

$$u'_i = \frac{1}{h\, b_i} \sum_{k=-3}^{3} a_{i+k}\, u_{i+k}$$
$$- \frac{\alpha}{6!\, b_i} h^5 \left(\frac{\partial u^6}{\partial^6 x} \right)_i + \cdots \qquad (58)$$

where

$$
\begin{aligned}
a_{i\pm3} &= \pm1 + \tfrac{1}{12}\alpha \\
a_{i\pm2} &= \mp9 - \tfrac{1}{2}\alpha \\
a_{i\pm1} &= \pm45 + \tfrac{5}{4}\alpha \\
a_i &= 0 - \tfrac{5}{3}\alpha \qquad\qquad b_i = 60
\end{aligned}
$$

These schemes are fifth-order upwind scheme when $\alpha < 0$, and they are sixth-order central scheme when $\alpha = 0$. The recommended value for α is $\alpha = -6$, and the corresponding fifth-order upwind explicit inner scheme is

$$
u_i' = \frac{1}{60h}\left(-\frac{3}{2}u_{i-3} + 12u_{i-2} - \frac{105}{2}u_{i-1} + \right.
$$
$$
\left. 10u_i + \frac{75}{2}u_{i+1} - 6u_{i+2} + \frac{1}{2}u_{i+3}\right) \qquad (59)
$$

High-order finite difference schemes require additional numerical boundary schemes at grid points near the boundaries of the computational domain. For a p-th order interior scheme, the accuracy of boundary schemes can be $(p-1)$-th order accurate without reducing the global accuracy of the interior scheme. Both one-sided compact and explicit finite difference schemes can be used as numerical boundary schemes. The expressions of the boundary schemes of up to sixth order can be found in [29]. The orders of accuracies for stable overall schemes when they are coupled with high-order interior schemes were analyzed in [29].

The high-order upwind schemes are applied to the inviscid flux vectors of the conservation equations according to the signs of the eigenvalues of the flux Jacobians. For the system of equations for hypersonic flows, the inviscid flux Jacobians generally contain both positive and negative eigenvalues. The upwind fifth-order schemes can be applied to the inviscid fluxes by either a simple local Lax-Friedrichs scheme or a flux splitting scheme to split the inviscid flux vectors into positive and negative wave fields. For example, the flux term E' can be split into two terms of pure positive and negative eigenvalues as follows

$$
E' = E'_+ + E'_- \qquad (60)
$$

where the flux E'_+ and E'_- contain only positive and negative eigenvalues respectively. Therefore, the spatial derivatives of the two flux terms on the right hand side of Eq. (60) are discretized differently, i.e., the first and second terms are discretized by an upwind and downwind high-order finite-difference schemes respectively. For the compressible Navier-Stokes equations, the derivatives of high-order terms related to viscous stress, heat conduction, and diffusion are discretized by standard high-order central finite difference schemes.

High-Order Semi-Implicit Runge-Kutta (SIRK)
Schemes for Stiff Equations

The spatial discretization of the governing equations leads to a system of first-order ordinary differential equations for the flow variables. For reacting hypersonic flow simulations, the thermo-chemical source term W is often stiff in temporal discretization. In [27], Zhong derived three kinds of high-order semi-implicit Runge-Kutta schemes for high-order temporal integration of the stiff governing equations. These schemes split the governing equations into stiff and non-stiff terms in the form of

$$
\frac{d\mathbf{u}}{dt} = \mathbf{f}(\mathbf{u}) + \mathbf{g}(\mathbf{u}) \qquad (61)
$$

where $\mathbf{u}$ is the vector of discretized flow field variables, $\mathbf{f}$ contains the non-stiff terms resulting from spatial discretization of the flux terms which can be computed explicitly, and $\mathbf{g}$ contains the stiff thermo-chemical source terms which need to be computed implicitly. The coefficients of the semi-implicit schemes were derived such that the schemes are high-order accurate with the simultaneous coupling between the implicit and explicit terms. In addition, the schemes are unconditionally stable for the stiff terms when a CFL condition is satisfied for the explicit terms.

Three versions of 3-stage third-order semi-implicit Runge-Kutta schemes for integrating Eq. (61) by simultaneously treating $\mathbf{f}$ explicitly and $\mathbf{g}$ implicitly are presented here:

ASIRK-3A Method:

$$
\left\{
\begin{aligned}
\mathbf{k}_1 &= h\{\mathbf{f}(\mathbf{u}^n) + \mathbf{g}(\mathbf{u}^n + a_1\mathbf{k}_1)\} \\
\mathbf{k}_2 &= h\{\mathbf{f}(\mathbf{u}^n + b_{21}\mathbf{k}_1) \\
&\quad + \mathbf{g}(\mathbf{u}^n + c_{21}\mathbf{k}_1 + a_2\mathbf{k}_2)\} \\
\mathbf{k}_3 &= h\{\mathbf{f}(\mathbf{u}^n + b_{31}\mathbf{k}_1 + b_{32}\mathbf{k}_2) \\
&\quad + \mathbf{g}(\mathbf{u}^n + c_{31}\mathbf{k}_1 + c_{31}\mathbf{k}_2 + a_3\mathbf{k}_3)\} \\
\mathbf{u}^{n+1} &= \mathbf{u}^n + \omega_1\mathbf{k}_1 + \omega_2\mathbf{k}_2 + \omega_3\mathbf{k}_3
\end{aligned}
\right.
$$

ASIRK-3B Method:

$$
\left\{
\begin{aligned}
&[\mathbf{I} - ha_1\mathbf{J}(\mathbf{u}^n)]\,\mathbf{k}_1 = h\{\mathbf{f}(\mathbf{u}^n) + \mathbf{g}(\mathbf{u}^n)\} \\
&[\mathbf{I} - ha_2\mathbf{J}(\mathbf{u}^n)]\,\mathbf{k}_2 = \\
&\qquad h\{\mathbf{f}(\mathbf{u}^n + b_{21}\mathbf{k}_1) \\
&\qquad + \mathbf{g}(\mathbf{u}^n + c_{21}\mathbf{k}_1)\} \\
&[\mathbf{I} - ha_3\mathbf{J}(\mathbf{u}^n)]\,\mathbf{k}_3 = \\
&\qquad h\{\mathbf{f}(\mathbf{u}^n + b_{31}\mathbf{k}_1 + b_{32}\mathbf{k}_2) \\
&\qquad + \mathbf{g}(\mathbf{u}^n + c_{31}\mathbf{k}_1 + c_{32}\mathbf{k}_2)\} \\
&\mathbf{u}^{n+1} = \mathbf{u}^n + \omega_1\mathbf{k}_1 + \omega_2\mathbf{k}_2 + \omega_3\mathbf{k}_3
\end{aligned}
\right.
$$

ASIRK-3C Method:

$$\begin{cases}
[\mathbf{I} - ha_1\mathbf{J}(\mathbf{u}^n)]\mathbf{k}_1 = h\{\mathbf{f}(\mathbf{u}^n) + \mathbf{g}(\mathbf{u}^n)\} \\
[\mathbf{I} - ha_2\mathbf{J}(\mathbf{u}^n + c_{21}\mathbf{k}_1)]\mathbf{k}_2 = \\
\qquad h\{\mathbf{f}(\mathbf{u}^n + b_{21}\mathbf{k}_1) + \\
\qquad\quad \mathbf{g}(\mathbf{u}^n + c_{21}\mathbf{k}_1)\} \\
[\mathbf{I} - ha_3\mathbf{J}(\mathbf{u}^n + c_{31}\mathbf{k}_1 + c_{32}\mathbf{k}_2)]\mathbf{k}_3 = \\
\qquad h\{\mathbf{f}(\mathbf{u}^n + b_{31}\mathbf{k}_1 + b_{32}\mathbf{k}_2) + \\
\qquad\quad \mathbf{g}(\mathbf{u}^n + c_{31}\mathbf{k}_1 + c_{32}\mathbf{k}_2)\} \\
\mathbf{u}^{n+1} = \mathbf{u}^n + \omega_1\mathbf{k}_1 + \omega_2\mathbf{k}_2 + \omega_3\mathbf{k}_3
\end{cases}$$

where $\mathbf{J}$ is the Jacobian of $\mathbf{g}$ vector. The parameters are:

ASIRK-3A, ASIRK-3B, and ASIRK-3C:

$\omega_1 = \frac{1}{8}$	$\omega_2 = \frac{1}{8}$	$\omega_3 = \frac{3}{4}$
$b_{21} = \frac{8}{7}$	$b_{31} = \frac{71}{252}$	$b_{32} = \frac{7}{36}$

ASIRK-3A:

$a_1 = .485561$	$a_2 = .951130$	$a_3 = .189208$
$c_{21} = .306727$	$c_{31} = .45$	$c_{32} = -.263111$

ASIRK-3B:

$a_1 = 1.40316$	$a_2 = .322295$	$a_3 = .315342$
$c_{21} = 1.56056$	$c_{31} = \frac{1}{2}$	$c_{32} = -.696345$

ASIRK-3C:

$a_1 = .797097$	$a_2 = .591381$	$a_3 = .134705$
$c_{21} = 1.05893$	$c_{31} = \frac{1}{2}$	$c_{32} = -.375939$

where a_1, a_2, a_3, c_{21}, and c_{32} are irrational numbers with six significant digits. The double-precision values of these parameters can be found in [27]. The first method above uses diagonally implicit Runge-Kutta methods for stiff term $\mathbf{g}$, which leads to a nonlinear equation at every stage of the implicit calculations if $\mathbf{g}$ is a nonlinear function of $\mathbf{u}$. The second and third methods use linearized implicit schemes for the stiff term $\mathbf{g}$. Methods B and C, which are similar to linearized implicit methods commonly used in computing reactive flows [115], are more efficient than the full implicit method A. However, for some stiff nonlinear problems, method A is necessary because it is more stable than the Rosenbrock semi-implicit Runge-Kutta method.

The SIRK methods have been developed further in [28,31] for non-autonomous systems with explicit time-dependent terms in the equations and low-storage version of the SIRK schemes to reduce the computer storage requirement to a minimum of two sets of memory locations for each variables. For example, a third-order

low-storage semi-implicit Runge-Kutta scheme is:

$$\begin{cases}
[\mathbf{I} - hc_1\mathbf{J}(\mathbf{u}^n)]\mathbf{k}_1 = h(\mathbf{f}(\mathbf{u}^n) + \mathbf{g}(\mathbf{u}^n)) \\
\mathbf{u}_1 = \mathbf{u}^n + b_1\mathbf{k}_1 \\
[\mathbf{I} - hc_2\mathbf{J}(\mathbf{u}_1 + \bar{c}_2\mathbf{k}_1)]\mathbf{k}_2 = \\
\qquad h(\mathbf{f}(\mathbf{u}_1) + \mathbf{g}(\mathbf{u}_1 + \bar{c}_2\mathbf{k}_1)) \\
\qquad + a_2[\mathbf{I} - hc_2\mathbf{J}(\mathbf{u}_1 + \bar{c}_2\mathbf{k}_1)]\mathbf{k}_1 \\
\mathbf{u}_2 = \mathbf{u}_1 + b_2\mathbf{k}_2 \\
[\mathbf{I} - hc_3\mathbf{J}(\mathbf{u}_2 + \bar{c}_3\mathbf{k}_2)]\mathbf{k}_3 = \\
\qquad h(\mathbf{f}(\mathbf{u}_2) + \mathbf{g}(\mathbf{u}_2 + \bar{c}_3\mathbf{k}_2)) \\
\qquad + a_3[\mathbf{I} - hc_3\mathbf{J}(\mathbf{u}_2 + \bar{c}_3\mathbf{k}_2)]\mathbf{k}_2 \\
\mathbf{u}^{n+1} = \mathbf{u}_2 + b_3\mathbf{k}_3
\end{cases} \qquad (62)$$

where the parameters obtained by the search to satisfy stability and accuracy requirement are listed below:

$$b_1 = \tfrac{1}{4} \quad b_2 = \tfrac{2}{9} \quad b_3 = 3 \quad a_2 = -\tfrac{1}{4} \quad a_3 = -\tfrac{29}{27}$$

$$c_1 = 2.267596813284564$$
$$c_3 = 2.309749357551431$$
$$\bar{c}_3 = -2.031219208388789$$
$$c_2 = 2.685297589634163$$
$$\bar{c}_2 = -1.143097033946135$$

These third-order Semi-Implicit Runge-Kutta methods are used for time-accurate computations in the direct simulation of transient hypersonic boundary layers. They are used to integrate hypersonic reacting flow equations simulations by treating the stiff source terms and the viscous terms in the wall-normal directions implicitly while the convective terms and the streamwise and spanwise directional viscous terms explicitly. Third-order accuracy in time can be achieved by using these SIRK methods.

High-Order Shock-Fitting Formulation for DNS of Hypersonic Flows

For the DNS of hypersonic boundary layers behind a well-defined but unsteady bow shock, a shock fitting formulation can be used to fit the bow shock in order to capture the shock wave geometry associated with the bow-shock physical oscillations. While shock capturing methods are suitable for aerodynamic calculations of time averaged pressure distributions, the shock-fitting method is more accurate in resolving physical short length-scale instability waves induced by shock interactions. The use of the shock fitting method makes it possible to use high-order linear schemes for spatial discretization of the flow equations behind the bow shock. Hussaini, Kopriva, Salas, and Zang [116] used the shock fitting spectral method to simulate shock/turbulent interaction. Recently, Cai [117] used a shock fitting method to compute two-dimensional detonation waves. Zhong[29] presented a high-order formulation for high-order shock fitting calculation of three-dimensional unsteady hypersonic flows. As a result, high-order schemes can be applied to the shock fitting calculations easily.

Figure 11 shows a schematic of the three-dimensional shock fitting computational mesh where the outer grid line is the bow shock. The shock fitting method treats the bow shock as a computational boundary at $\eta = \eta_{\max}$ as

$$\eta(x, y, z, t) = \eta_{\max} = \text{constant} \tag{63}$$

The flow variables behind the shock are determined by the Rankine-Hugoniot relation across the shock and a characteristic compatibility equation from behind the shock. As shown in Fig. 11, the position and velocity of the shock front are functions of $H(\xi, \zeta, \tau)$ and $H_\tau(\xi, \zeta, \tau)$, which are solved as unknown variables using high-order finite difference methods and advanced in time using the same time-stepping scheme as the conservations in the flow fields behind the shock.

The flow variables across the shock are governed by the Rankine-Hugoniot conditions. The Rankine-Hugoniot relations lead to jump conditions for flow variables behind the shock as functions of U_0 and the grid velocity v_n. In order to compute the flow variable behind the shock using the shock jump conditions, the velocity of the shock front v_n is determined by a characteristic compatibility equation at the grid point immediately behind the shock. It is found that the shock fitting computations and the shock geometry transformation relations are greatly simplified if we derive the characteristic compatibility equation in the conservation-law form, which can be derived directly from conservation equations in the direction along the η coordinates. Specifically, the governing equation in the computational domain at the point immediately behind the shock front can be written as:

$$\frac{1}{J}\frac{\partial U}{\partial \tau} + \frac{\partial F'}{\partial \eta} = \left(\frac{W}{J} - \frac{\partial E'}{\partial \xi} - \frac{\partial G'}{\partial \zeta} \right.$$
$$\left. - \frac{\partial E'_v}{\partial \xi} - \frac{\partial F'_v}{\partial \eta} - \frac{\partial G'_v}{\partial \zeta} - U\frac{\partial(\frac{1}{J})}{\partial \tau} \right) \tag{64}$$

where the equation is evaluated at point s behind the shock. In the equations, the Jacobian matrix, $B'_s = (\partial F'/\partial U)_s$, has the following eigenvalues:

$$\frac{|\nabla \eta|}{J}(u_n - v_n)_s, \quad \cdots, \quad \frac{|\nabla \eta|}{J}(u_n - v_n)_s,$$

$$\frac{|\nabla \eta|}{J}(u_n - v_n - c)_s, \quad \frac{|\nabla \eta|}{J}(u_n - v_n + c)_s \tag{65}$$

The corresponding left eigenvectors are

$$\mathbf{l_1}, \quad \mathbf{l_2}, \quad \cdots, \quad \mathbf{l_{N-1}}, \quad \mathbf{l_N} \tag{66}$$

where N is the number of independent variables in the equations.

The characteristic field approaching the shock from behind corresponds to the eigenvalue of $\frac{|\nabla \eta|}{J}(u_n - v_n + c)_s$. The compatibility relation for this characteristic field can be obtained by multiplying Eq. (64) by $\mathbf{l_N}$,

$$\mathbf{l_N} \cdot \left(\frac{\partial U}{\partial \tau} \right) = \mathbf{l_N} \cdot \left(\frac{W}{J} - \frac{\partial E'}{\partial \xi} - \frac{\partial F'}{\partial \eta} - \frac{\partial G'}{\partial \zeta} \right.$$
$$\left. - \frac{\partial F_v}{\partial \xi} - \frac{\partial F_v}{\partial \eta} - \frac{\partial F_v}{\partial \zeta} - U\frac{\partial(\frac{1}{J})}{\partial \tau} \right) J \tag{67}$$

On the other hand, the shock jump condition can be written as

$$[F'] = (\mathbf{F_s} - \mathbf{F_0}) \cdot \mathbf{a}$$
$$+ (U_s - U_0)\, b = 0 \tag{68}$$

where

$$\mathbf{a} = (\frac{\eta_x}{J})_s \mathbf{i} + (\frac{\eta_y}{J})_s \mathbf{j} + (\frac{\eta_z}{J})_s \mathbf{k} \tag{69}$$
$$b = (\frac{\eta_t}{J})_s . \tag{70}$$

Taking derivative of Eq. (68) with respective to τ in the computational space leads to

$$B'_s \frac{\partial U_s}{\partial \tau} - B'_0 \frac{\partial U_0}{\partial \tau} + (\mathbf{F_s} - \mathbf{F_0}) \cdot \frac{\partial \mathbf{a}}{\partial \tau}$$
$$+ (U_s - U_0)\frac{\partial b}{\partial \tau} = 0 \tag{71}$$

where the flux Jacobian

$$B' = \frac{\partial F'}{\partial U} \tag{72}$$

is the Jacobian of the η direction flux defined in the conservation equations, and

$$\frac{\partial \mathbf{a}}{\partial \tau} = \frac{\partial(\frac{\eta_x}{J})_s}{\partial \tau}\mathbf{i} + \frac{\partial(\frac{\eta_y}{J})_s}{\partial \tau}\mathbf{j} + \frac{\partial(\frac{\eta_z}{J})_s}{\partial \tau}\mathbf{k} \tag{73}$$

$$\frac{\partial b}{\partial \tau} = \frac{\partial (\eta_\tau / J)_s}{\partial \tau} \tag{74}$$

These time derivatives of the grid metrics can be derived by the same methods as those used in discretization of the interior equations.

Finally, the equation for the shock velocity can be obtained by multiplying both sides of Eq. (71) by $\mathbf{l_N}$, i.e.,

$$\frac{\partial b}{\partial \tau} = \frac{-1}{[\mathbf{l_N} \cdot (U_s - U_0)]}$$
$$\left[\frac{|\nabla \eta|}{J}(u_n - v_n + c)_s\, \mathbf{l_N} \cdot (\frac{\partial U}{\partial \tau})_s \right.$$
$$\left. + \mathbf{l_N} \cdot (\mathbf{F_s} - \mathbf{F_0}) \cdot \frac{\partial \mathbf{a}}{\partial \tau} - (\mathbf{l_N} \cdot \mathbf{B_0'})\frac{\partial U_0}{\partial \tau} \right]$$

where the term $\mathbf{l_N} \cdot \frac{\partial U_s}{\partial \tau}$ is computed using the characteristic relation (67), in which the spatial derivatives are discretized together with the discretization of the interior points for the governing conservation equations using the same schemes at the interior algorithm applied to boundary behind the shock. In the equation above, $\frac{\partial b}{\partial \tau}$ and $\frac{\partial a}{\partial \tau}$ can be expressed as functions of H and H_τ. Therefore, the equation for the shock acceleration can be obtained from Eq. (75) in the following form:

$$\frac{\partial H_\tau}{\partial \tau} = f\left(\xi,\ \zeta,\ U_s,\ \mathbf{l_{N-1}} \cdot (\frac{\partial U}{\partial \tau})_s, \right.$$
$$\left. U_0,\ \frac{\partial U_0}{\partial \tau},\ H,\ H_\tau \right) \tag{75}$$
$$\frac{\partial H}{\partial \tau} = H_\tau \tag{76}$$

The two equations above describe the shock normal velocity and shock shape, and they can be integrated in time simultaneously with the interior flow variables using Runge-Kutta methods. After the values of H and H_τ are determined, the flow variables behind the shock can be computed by the jump conditions across the shock. The grids and metrics are modified according to the new values of H and H_τ.

V. Results and Discussions

The high-order numerical methods presented in the preceding section have been applied for many transient and unsteady hypersonic flow computations. Three test cases are presented here: 1) steady shock-boundary-layer interaction with real gas effects; 2) unsteady shock-shock interference heating problem with

real gas effects using second-order shock-capturing TVD schemes with third-order semi-implicit Runge-Kutta (ASIRK-3C) schemes [26,118], 3) DNS of hypersonic boundary-layer receptivity to freestream acoustic disturbances [30,61] using fifth-order upwind schemes in the shock fitting formulation presented in the preceding section. The results of applications of the ENO schemes to transient hypersonic flows can be found in [24,25], and application of the SIRK schemes to reacting flow and detonation simulations can be found in [119,120].

<u>Steady Shock/Boundary Layer Interaction with Real Gas Effects</u>

The nonequilibrium hypersonic flows in a shock/boundary-layer interaction are simulated by a second-order TVD shock capturing scheme using a third-order SIRK time-stepping schemes for the source terms [26,72,118] to study an air freestream at a Mach number of 7.0 deflected by an angle of 22° by an incident shock. Results validating the 2-D TVD shock capturing code for nonequilibrium codes were can be found in [72], [26], and [118]. The code was able to accurately capture both the shock shape and stand off distance [26], parameters which have been demonstrated to be sensitive to the choice of thermochemical model [11].

The shock impinged upon a flat plate at a distance of 0.2134 m which resulted in a Reynolds Number based on the freestream and the shock impingement distance from the leading edge, of 3.571×10^5. No slip and isothermal wall boundary conditions were used on the plate surface. The wall temperature was $1200°K$. Numerical simulations were performed using three different thermo-chemical models. The first model assumed the a perfect (ideal) gas with frozen vibrational modes and frozen chemical modes (no dissociation/recombination). The second model assumed the diatomic molecules were vibrationally excited but chemically frozen. The third model assumed the gas was both vibrationally and chemically excited (reacting). All three cases used a Mach 7.0 freestream with $T_\infty = 1600\ K$ and $\rho_\infty = 0.0169\ kg/m^3$.

Figure 4 shows the translational temperature contours for the solutions of shock interaction using the three different gas models. The translational temperature contours are normalized by the freestream temperature and the y-axes in the contours have been magnified by a factor of 10 in order to clearly show the region around the shock impingement point. The contours show clearly the incident shock impinging on the wall, a separation bubble ahead of the shock impingement point, and the reflected shock where the separated flow reattaches behind the separation bubble. Immediately noticeable are the differences in the sizes of separation regions from model to model. The separation region is largest in Fig. 4(a) for the ideal gas model, decreases in size in Fig. 4(b) for the vibrational nonequilibrium model, and is smallest in Fig. 4(c) for the thermo-chemical nonequilibrium model.

The temperatures in the boundary layer behind the impinging shock were found to be about $9500\ K$ for the ideal case, $7700\ K$ for the chemically frozen case, and $6700\ K$ for the reacting case. There is a corresponding trend in the surface heat transfer profiles given in Fig. 5. The ideal case resulted in the highest heating rate to the wall in the shock impingement region while the lowest heating rate was given by the reacting case.

The decrease in temperature and bubble size is due to the endothermic nature of the nonequilibrium real gas effects. Between the ideal and vibrational nonequilibrium cases, the energy is distributed among more internal modes in the nonequilibrium case, resulting in a lower thermodynamic temperature in the latter case. The difference between the vibrational nonequilibrium case and the reacting flow case stems from the nature of the dominant chemical reactions. The post shock impingement area is dominated by the dissociation of N_2 into atomic nitrogen, which is an endothermic process, absorbing energy from the flow to break the chemical bonds. This reduces the temperature and pressure of the gas in the thermo-chemical nonequilibrium case. This pressure reduction results in a less severe adverse pressure gradient and a smaller separation bubble. The lower flow temperature near the isothermal wall for the reacting case results in smaller temperature gradients and lower heat transfers rates relative to the ideal and chemically frozen cases.

The present results indicate that real gas effects significantly influence the structure of hypervelocity shock/boundary layer interactions. In the test cases studied in this paper, the presence of real gas effects decreased the size of the shock induced separation region relative to the perfect gas case. Also, the post impingement surface heating rates were significantly reduced by the endothermic relaxation of the internal and chemical modes. The direction of the influence of real gas effects is dependent on the endothermic or exothermic nature of the nonequilibrium processes. Earlier work done by Ballaro and Anderson [37], and Grumet, et al. [38] indicated that real gas effects could increase the heating rates. However, in those studies, the dominating reactions near the wall were recombination reactions, which are exothermic, while in the present case, the dominating reactions were dissociation reactions, which are endothermic.

Unsteady Type IV Shock Interference Heating with Real Gas Effects

A simulation corresponding to an experimental type IV shock-shock interference heating problem was computed to test the ability of the code to model such complex flows. The shock interactions took place ahead of a 3 inch diameter cylinder in a Mach 8.03 flow with the following conditions: $p_\infty = 985.015\ Pa$, $\gamma_\infty = 1.4$, $T_\infty = 111.56\ K$, and $T_{wall} = 294.44\ K$. The flow deflection due to the impinging shock was 12°. This flow has been found to be inherently unsteady by both experimental [40, 121] and numerical [24, 122, 123] studies.

Figure 6 presents a comparison of the time averaged pressure coefficient along the body surface with experimental data given in [121]. Very good agreement between the computed and experimental results was obtained. Figure 7 shows a series of computed instantaneous nondimensional surface heating rate profiles compared with experimental data from [121]. Fair agreement between computation and experiment was obtained. The computations tended to underpredict the peak value of the surface heating. This result is consistent with other numerical studies of type IV flows [122, 124].

Numerical studies of perfect gas flows have reported a nondimensional frequency, or Strouhal number ($Sh = fD/u_\infty$), associated with the time variation of peak surface pressure of 1.3 [122] while experimental results report nondimensional frequencies of 0.13 to 0.45 [125] associated with the pressure variation at a fixed point on the body surface. A spectral analysis of the peak surface pressure and the pressure at a fixed point near the jet impingement point in the current simulations produced a dominant nondimensional frequency of 1.3 for the peak surface pressure, and 0.185 and 1.12 for the fixed surface points. These results are consistent with both existing numerical [24, 122] and experimental results [40, 125] for perfect gas flows.

To examine the real gas and unsteady effects, a higher enthalpy type IV shock-shock interaction was studied by computing two dimensional Navier-Stokes solutions. The freestream was an undissociated N_2 flow about a 0.0381 m radius cylinder with $M_\infty = 8.03$, $p_\infty = 985.015\ Pa$, $T_\infty = 800\ K$, and a Reynolds number of 2.57×10^5. The cylinder surface was a no-slip isothermal wall with a temperature of $1000\ K$. The flow deflection due to the impinging shock was 12°.

A dominant nondimensional frequency (Strouhal number, $Sh = fD/u_\infty$) associated with the time variation of peak surface pressure of 2.7 was observed in the computed nonequilibrium solution. This is over twice the nondimensional frequency of 1.2 computed for perfect gas flows [26,122]. The higher oscillation frequency exhibited by the nonequilibrium flow is due to the shorter length and time scales in the nonequilibrium case. For flows with similar freestream conditions, but with different thermochemical models, the upper bow shock standoff distances, nondimensionalized by the cylinder radius, are about 1.0 for the perfect gas [26] and 0.55 for the flow with nonequilibrium real gas effects.

Several researchers [24,122,126] have observed disturbances propagating from the shear layer toward the bow shock across the subsonic region of the flow. Disturbances from the shear layers near the body propagate towards the bow shocks. These perturbations disturb the bow shocks, causing oscillations which then affect the shock triple points and the transmitted shock. This creates perturbation waves within the jet which are convected back to the body and disturb the shear layer. A sequence of instantaneous translational temperature contours is presented in Figure 8. In Fig. 8-a, the jet impingement location is marked J, the point labeled A is a disturbance generated by the shear layer propagating toward the bow shock, and the point labeled B tracks a disturbance wave in the jet. The jet is nearly normal to the body surface and the surface pressure is at a local maximum. In Fig. 8-b, the jet, J, has reached a nearly type IVa configuration. This point corresponds to the minimum pressure point and the minimum surface heating peak. Disturbance A continues to move towards the bow shock, while disturbance B moves along the jet towards the body. Compression wave, C, is convected along the wall inside the shear layer. In Fig. 8-c, the jet, J, is once again nearly normal to the wall. Disturbance A has entered the relaxation zone behind the upper bow shock, and disturbance B is strongly interacting with the supersonic jet. The compression wave, C, has been convected off through the shear layer.

The results illustrate two major phenomena in type IV shock interactions. First, nonequilibrium real gas effects increase the dominant oscillation frequency when flow is inherently unstable. Real gas effects reduce the macroscopic length scales by reducing shock standoff distance and wall shear layer thickness. The relaxation phenomena introduce additional small time scales to the flow. Second, a feedback mechanism was observed whereby disturbance waves are created near the jet impingement point and propagate through the subsonic region to the bow shocks. The disturbed bow shocks create perturbations in the the supersonic jet. These perturbations feed back through the jet, inducing jet oscillations, which continue the unsteadiness.

Direct Numerical Simulation of Hypersonic Boundary Layer Instability

The receptivity of a hypersonic boundary layer to freestream acoustic waves is studied [30,61] by numerically solving the unsteady two and three-dimensional Navier-Stokes equations using a fifth-order shock fitting scheme [29] in the non-periodic streamwise and wall-normal directions and Fourier spectral collocation in the periodic spanwise direction. The spatially discretized equations are advanced in time using a Low-Storage Runge-Kutta scheme [127] for non-reacting flows or the semi-implicit Runge-Kutta schemes [27] for reacting flows. The numerical method has been tested and validated in [29]. For the current receptivity computations, grid refinements are used to ensure that both steady and unsteady solutions are grid independent.

In the simulations, the steady flow solutions are first obtained with no freestream disturbances. Subsequently, the freestream disturbances are superimposed on the basic flow solutions in front of the bow shock. The shock/disturbance interactions and generation of T-S waves in the boundary layer are solved without linearization using the full Rankine-Hugoniot relations at the shock and the Navier-Stokes equations in the flow fields. The freestream disturbances are assumed to be weak monochromatic planar acoustic waves with wave front oblique to the center line of the body in the $x - z$ plane at an angle of ψ. The perturbations of flow variable in the freestream introduced by the freestream acoustic wave before reaching the bow shock can be written in the following form:

$$q'_\infty = |q'|_\infty \, e^{i(k \cos \psi \, x + k \sin \psi \, z + \omega t)} \qquad (77)$$

where $|q'|$ represents one of the flow variables, $|u'|$, $|v'|$, $|w'|$, $|p'|$, and $|\rho'|$. The freestream perturbation amplitudes satisfy the following relations: $|u'|_\infty = \epsilon \cos \psi$, $|v'|_\infty = 0$, $|w'|_\infty = \epsilon \sin \psi$, $|p'|_\infty = \gamma M_\infty \epsilon$, and $|\rho'|_\infty = M_\infty \epsilon$. The parameter ϵ represents the freestream wave magnitude. The angle ψ is the angle of freestream wave with respect to the x axis in the x-z plane, where $\psi = 0$ corresponds to two-dimensional planar waves. The parameter k is the dimensionless freestream wave number which is related to the dimensionless circular frequency ω by: $\omega = k \left(\cos \psi + M_\infty^{-1} \right)$. The dimensionless frequency F is defined as: $F = \frac{\omega^* \nu^*}{U_\infty^{*2}} \times 10^6$. The body surface is a parabola given by $x^* = b^* y^{*2} - d^*$, where b^* a given constant and d^* is taken as the reference length. A flow variable with a superscript "*" is a dimensional variable, while dimensionless flow variables are denoted by the same dimensional notation but without the superscript "*".

The flow and geometric conditions are: $M_\infty = 15$, $Re_\infty = \frac{\rho_\infty^* U_\infty^* d^*}{\mu_\infty^*} = 6026.6$, $T_\infty^* = 192.989\,K$, $p_\infty^* = 10.3\,Pa$, $T_w^* = 1000\,K$, $\gamma = 1.4$, $R^* = 286.94\,Nm/kgK$, $Pr = 0.72$, $b^* = 40\,m^{-1}$, $d^* = 0.1\,m$, and the nose radius of curvature $= r^* = 0.0125\,m$. The body surface is assumed to be a non-slip wall with an isothermal wall temperature T_w^*. The perturbations imposed in the freestream are plane acoustic waves with a wave front angle ψ, where $\psi = 0$ corresponds to two-dimensional waves. Both two-dimensional and three-dimensional disturbances use the same two-dimensional basic-flow solution obtained numerically.

Steady two-dimensional solutions of the Navier-Stokes equations are obtained by advancing the solutions to a steady state without freestream perturbations. Figure 9 shows steady flow solutions for a set of 160×120 computational grid, velocity vectors, and steady entropy contours. The bow shock shape is obtained as the solution for the freestream grid line. The velocity vector plot in the figure shows the development of the boundary layer along the surface. The entropy contours show the entropy layer developing at the edge of the boundary layer. The swallowing of the entropy layer by the boundary layer has been shown to play an important role in the stability and transition of the boundary layer downstream [128]. The figure shows that the entropy layer begins to enter the outer region of the boundary layer when $x > 1$.

After obtaining the steady basic flow solutions, the unsteady flow solutions are solved by imposing acoustic disturbances to the basic flow in the freestream. Figure 10 shows the contours of instantaneous temperature perturbation T', the Fourier amplitude $|T'|$, and the phase angle φ_T (in degrees) after the flow field reach a periodic state for the case of freestream forcing frequencies of $F = 2655$ and $\epsilon = 5 \times 10^{-4}$. The instantaneous contours show the development of instability waves in the boundary layer on the surface. In general, the disturbance field between the shock and the body surface is a combination of the external forcing disturbance waves and the T-S waves generated in the boundary layer. Figure 10 shows very clearly the development of instability waves in the boundary layer along the surface because the instability waves are dominant in the boundary layer near the surface. Meanwhile the waves immediately behind the bow shock are mainly external forcing waves. On the body surface, both first and second modes are generated and propagated downstream along the wall. The first mode is generated near the leading edge ($x < 0.2$). The amplitudes of the first mode are amplified first and then decrease rapidly after reaching maximum values. After the first-mode decay, the second-mode disturbances become the dominant mode. The decay of first mode and the growth of the second mode at $x \approx 0.2$ are shown by a sudden change of phase angles near the body surface at that location. The phase structure change dramatically between the two modes, specifically the second mode has one more variation across the boundary layer. The contours of the Fourer amplitudes also show the switching of instability modes from region 1 to region 2 at about $x = 0.2$ on the surface.

Local parallel linear stability theory (LST) is used to identify the boundary layer eigenmodes and to analyze the instability mechanism and the bow-shock effects [129]. The LST is used to obtain instability modes based on the numerically obtained basic flow between the body and the bow shock. The first-mode and second-mode structure of the eigenfunctions compare well with the DNS results inside the boundary layer. An example of the comparison is shown in Fig. 12 on the first-mode eigenfunction amplitudes along the wall-normal direction at $x = 0.5521$ for the case of freestream forcing frequency $F = 1770$ and $\epsilon = 5 \times 10^{-4}$. The DNS results agree very well with LST results inside the boundary layer. The results also show that the agreement becomes better as the waves propagate further downstream. The disagreement in Fig. 12 outside the boundary layer near the bow shock is expected because of the effects of the forcing disturbances not included in the linear stability analysis. The LST also shows that the first mode is the least stable mode near the leading edge, while the second becomes the least stable mode further downstream. The results also shows that the receptivity always generate first-mode amplification near the leading edge followed by a first-mode decay and the second-mode dominance downstream. The maximum amplitudes and range of first-mode region depend on the value of the frequency. As F decreases, the first-mode amplification region expands and their peak values increase rapidly.

For weak monochromatic freestream forcing waves, the generation of instability waves are expected to be linear with respect to the forcing amplitudes for the disturbances with the same fundamental frequency. At the same time, receptivity and stability experiments [130,131] have shown the existence of high harmonics in addition to the waves of the fundamental frequency. Buter and Reed [132] found nonlinear supersonic harmonic in their incompressible receptivity simulations. Nonlinear higher harmonics are also found in the present DNS studies. Figure 13 shows the instantaneous entropy disturbances along the body surface for the case of $F = 1770$ and $\epsilon = 5 \times 10^{-4}$ for the fundamental frequency ($n = 1$) and their second harmonic ($n = 2$). The analysis of the results show that receptivity of the fundamental modes are governed by linear mechanism, while the second harmonic is nonlinear.

The receptivity of the hypersonic boundary layer to oblique three-dimensional freestream acoustic waves for the same two-dimensional basic flow has also been studied by DNS. Figure 14 shows the three-dimensional shock fitted grids using two computational zones resolved by $160 \times 120 \times 16$ and $200 \times 120 \times 16$ grids respectively. The unsteady flow fields are generated by imposing an oblique freestream disturbance at an angle of $\psi = 45°$ with wave amplitude of $\epsilon = 5 \times 10^{-3}$ and $F = 1770$. Figure 15 shows the contours of instantaneous perturbation v' after the flow field reaches a periodic state. The instantaneous contours of v' show the development of three-dimensional instability waves in the boundary layer on the surface, similar to the first-mode zone. The second mode region is also generated near the end of computational Zone 1. The figure shows similar trends as two-dimensional wave developments, i.e., the first-mode amplification near the leading edge and the second-mode dominant downstream. The nonlinear superharmonics in time and in the periodic spanwise direction is also found by a Fourier transform of the instantaneous perturbation solutions. The results show that the disturbance fields contain both fundamental mode, nonlinear second harmonic, and other nonlinear modes.

VI. Conclusions

High-order spatial and temporal schemes have been reviewed for transient hypersonic flow computations and for the direct numerical simulations of hypersonic reacting flows with strong bow shocks. These high-order methods are necessary when the emphasis of the studies is on the fundamental transient flow phenomena with a wide range of time and length scales. The new methods have been applied to a range of hypersonic reacting and non-reacting flow problems. It is concluded that for transient hypersonic flow simulations with complex shock structure, high-order ENO and TVD shock-capturing schemes are the method of choice, while for DNS of hypersonic flows with a well defined unsteady bow shock, the shock-fitting approach with high-order schemes, such as the fifth-order upwind schemes presented in this paper, is more accurate in resolving fine details of the flow physics. For the time integration, new third-order semi-implicit Runge-Kutta schemes have been shown to be accurate and robust for stiff hypersonic flow simulations.

Acknowledgments

This research was supported by the Air Force Office of Scientific Research under grant numbers F49620-94-1-0019 and F49620-95-1-0405 monitored by Dr. Len Sakell.

References

[1] G. F. Orton, L. F. Scuderi, P. W. Sanger, J. Artus, P. T. Harsha, G. Laurelle, and L. H. Shkadov. Airbreathing hypersonic aircraft and transatmospheric vehicles. In A. K. Noor and S. L. Venneri, editors, *Future Aeronautical and Space Systems*, volume 172 of *Progress in Aeronautics and Astronautics*. American Institute of Aeronautics and Astronautics, 1997.

[2] J. D. Anderson Jr. *Hypersonic and High Temperature Gas Dynamics*. McGraw-Hill, Inc., 1989.

[3] F. W. Burcham and J. Nugent. Local flow field around a pylon-mounted dummy ramjet engine on the x-15-2 airplane for mach numbers from 2.0 to 6.7. Technical Report NASA TN D-5638, Feb. 1970.

[4] W. D. Hayes and R. F. Probstein. *Hypersonic Flow Theory*. Academic Press, 1959.

[5] H. K. Cheng. Recent advances in hypersonic flow research. *AIAA Journal*, 1:295–310, 1963.

[6] H. K. Cheng. Perspectives on hypersonic viscous flow research. *Annual Review of Fluid Mechanics*, 25:455–484, 1993.

[7] M. Rasmussen. *Hypersonic Flow*. John Wiley and Sons, Inc., 1994.

[8] H. K. Cheng. Perspective on hypersonic nonequilibrium flow. *AIAA Journal*, 33:385–400, 1995.

[9] J. J. Bertin. *Hypersonic Aerothermodynamics*. AIAA, Washington, DC, 1994.

[10] G. V. Candler. *The Computation of Weakly Ionized Hypersonic Flows in Thermo-Chemical Nonequilibrium*. PhD thesis, Stanford University, 1988.

[11] G. V. Candler. *On the Computation of Shock Shapes in Nonequilibrium Hypersonic Flows*. AIAA Paper 89-0312, 1989.

[12] G. V. Candler. Chemistry of external flows. In *Aerothermochemistry for Hypersonic Technology*. Von Karman Institue for Fluid Dynamics Lecture Series 1995-04, 1995.

[13] H. C. Yee, G. H. Klopfer, and J. L. Montagné. High-resolution shock-capturing schemes for inviscid and viscous hypersonic flows. *Journal of Computational Physics*, 88:31–61, 1990.

[14] J. L. Montagné, H. C. Yee, G. H. Klopfer, and M. Vinokur. *Hypersonic Blunt Body Computations Including Real Gas Effects*. NASA TM-100074, 1988.

[15] G. H. Klopfer and H. C. Yee. *Viscous Hypersonic Shock-on-shock Interaction on Blunt Cowl Lips.* AIAA Paper 88-0233, 1988.

[16] G. H. Klopfer and H. C. Yee. *Numerical Study of Unsteady Viscous Hypersonic Blunt Body Flows with an Impinging Shock.* NASA TM-100096, 1988.

[17] P. A. Gnoffo. Point-implicit relaxation strategies for viscous hypersonic flows. In T. K. S. Murthy, editor, *Computational Methods in Hypersonic Aerodynamics.* Kluwer Academic Publishers, 1991.

[18] F. M. Cheatwood and P. A. Gnoffo. User's manual for the langley aerothermodynamic upwind relaxation algorith (laura). Technical Report NASA TM 4674, Apr. 1996.

[19] K. J. Weilmuenster, P. A. Gnoffo, F. A. Greene, H. H. Hamilton, and S. J. Alter. Hypersonic aerodynamic characteristics of a proposed single-stage-to-orbit vehicle. Technical Report AIAA Paper 95-1850, 1995.

[20] S. Brück. *Investigation of Shock-Shock Interactions in Hypersonic Reentry Flows.* Presented at the 20th International Symposium on Shock Waves, 1995.

[21] S. Brück, R. Radespiel, and J. M. A. Longo. Comparison of nonequilibrium flows past a simplified space-shuttle configuration. Technical Report AIAA Paper 97-0275, Jan. 1997.

[22] T.K.S. Murthy, Editor. *Computational Methods in Hypersonic Aerodynamics.* Kluwer Academic Publishers, The Netherlands, 1991.

[23] M. Capitelli, Editor. *Molecular Physics and Hypersonic Flows.* Kluwer Academic Publishers, The Netherlands, 1996.

[24] X. Zhong. Application of essentially nonoscillatory schemes to unsteady hypersonic shock-shock interference heating problems. *AIAA Journal*, 32(8):1606–1616, 1994.

[25] C.-E. Chiu and X. Zhong. Numerical Simulation of Transient Hypersonic Flow Using the Essentially Nonoscillatory Schemes. *AIAA Journal*, Vol. 34, No. 4, pp. 655–661, 1995. (Also AIAA paper 95-0469, 1995).

[26] G. H. Furumoto, X. Zhong, and J. C. Skiba. Numerical studies of real-gas effects on two-dimensional hypersonic shock-wave/boundary-layer interaction. *Physics of Fluids*, 9:191–210, Jan. 1997.

[27] X. Zhong. Additive Semi-Implicit Runge-Kutta Schemes for Computing High-Speed Nonequilibrium Reactive Flows. *Journal of Computational Physics*, Vol. 128, pp. 19-31. 1996.

[28] J. Shen and X. Zhong. Semi-Implicit Runge-Kutta Schemes for Non-Autonomous Differential Equations in Reactive Flow Computations. *AIAA paper 96-1868*, Also submitted to Journal of Computational Physics, June 1996.

[29] X. Zhong. Direct Numerical Simulation of Hypersonic Boundary-Layer Transition Over Blunt Leading Edges, Part I: New Numerical Methods and Validation (Invited). *AIAA paper 97-0755, 35th AIAA Aerospace Sciences Meeting and Exhibit, January 6-9, Reno, Nevada,* 1997.

[30] X. Zhong. Direct Numerical Simulation of Hypersonic Boundary-Layer Transition Over Blunt Leading Edges, Part II: Receptivity to Sound (Invited). *AIAA paper 97-0756, 35th AIAA Aerospace Sciences Meeting and Exhibit, January 6-9, Reno, Nevada,* 1997.

[31] J. J. Yoh and X. Zhong. Low-storage semi-implicit runge-kutta methods for reactive flow computations. *An extended abstract submitted to 36th AIAA Aerospace Sciences Meeting, Reno, NV, Jan. 12-15, 1998,* 1997.

[32] Samuel A. Schaaf and P. L. Chambre. Flow of rarefied gases. *Princeton Aeronautical Paperbacks*, 8, 1961.

[33] M. N. Kogan. *Rarefied Gas Dynamics.* New York, Plenum Press, 1969.

[34] E. P. Muntz. Rarefied gas dynamics. *Ann. Rev. Mech.*, 21:387–417, 1989.

[35] J. F. Clarke and M. McChesney. *The Dynamics of Real Gases.* Butterworths, 1964.

[36] W. G. Vincenti and C. H. Kruger. *Introduction to Physical Gas Dynamics.* John Wiley and Sons, Inc., 1965.

[37] C. A. Ballaro and J. D. Anderson Jr. *Shock Strength Effects on Separated Flows in Non-Equilibrium Chemically Reacting Air Shock Wave/Boundary Layer Interaction.* AIAA Paper 91-0250, 1991.

[38] A. A. Grumet, J. D. Anderson Jr., and M. J. Lewis. *A Numerical Study of Shock Wave/Boundary Layer Interaction in Nonequilibrium Chemically Reacting Air: The Effects of Catalytic Walls.* AIAA Paper 91-0245, 1991.

[39] B. Edney. *Anomalous Heat Transfer and Pressure Distributions on Blunt Bodies at Hypersonic Speeds in the Presence of an Impinging Shock.* Aeronautical Research Inst. of Sweden, FFA Rept. 115, Feb. 1968.

[40] M. S. Holden. *Shock-Shock Boundary Layer Interaction.* AGARD Rept. No. 764, July 1989.

[41] A. R. Wieting and M. S. Holden. Experimental shock wave interference heating on a cylinder at mach 6 and 8. *AIAA Journal*, 27:1557–1565, 1989.

[42] K. W. Iliff and M. F. Shafer. A comparison on hypersonic vehicle flight simulation and prediction results. Technical Report NASA TM 104313, Oct. 1995.

[43] G. Gaitonde and J. S. Shang. *The Performance of Flux-Split Algorithms in High-Speed Viscous Flows.* AIAA Paper 92-0186, 1992.

[44] C. A. Lind and M. J. Lewis. *A Numerical Study of the Unsteady Processes Associated with the Type IV Shock Interaction.* AIAA Paper 93-2479, 1993.

[45] National Research Council (U.S.). Committee on Hypersonic Technology for Military Application. Hypersonic Technology for Military Application. *Technical Report*, National Academy Press, Washington, DC., 1989.

[46] L. M. Mack. Boundary layer linear stability theory. In *AGARD report, No. 709*, 1984.

[47] M. V. Morkovin. Transition at Hypersonic Speeds. *ICASE Interim Report 1*, NASA CR 178315, May, 1987.

[48] D. Arnal. Laminar-Turbulent Transition Problems In Supersonic and Hypersonic Flows. *Special Course on Aerothermodynamics of Hypersonic Vehicles*, AGARD Report No. 761, 1988.

[49] E. Reshotko. Hypersonic stability and transition. *in Hypersonic Flows for Reentry Problems, Eds. J.-A. Desideri, R. Glowinski, and J. Periaux, Springer-Verlag*, 1:18–34, 1991.

[50] H. L. Reed and W. S. Saric. Stability of Three-Dimensional Boundary Layers. *Annual Review of Fluid Mechanics*, Vol. 21, pp. 235-284, 1989.

[51] D. Bushnell. Notes on Initial Disturbance Field for the Transition Problem. *Instability and Transition, Vol. I*, M. Y. Hussaini and R. G. Viogt, editors, pp. 217-232, Springer-Verlag, 1990.

[52] M. Morkovin. On the Many Faces of Transition. *Viscous Drag Reduction*, C.S. Wells, editor, Plenum, 1969.

[53] L. M. Mack. Linear Stability Theory and the Problem of Supersonic Boundary-Layer Transition. *AIAA Journal*, Vol. 13, No. 3, pp. 278-289, 1975.

[54] W. S. Saric. Götler Vortices. *Annual Review of Fluid Mechanics*, Vol. 26, pp. 379-409, 1994.

[55] J. M. Kendall. Wind Tunnel Experiments Relating to Supersonic and Hypersonic Boundary-Layer Transition. *AIAA Journal*, Vol. 13, No. 3, pp. 290–299, 1975.

[56] K. F. Stetson, E. R. Thompson, J. C. Donaldson, and L. G. Siler. Laminar Boundary Layer Stability Experiments on a Cone at Mach 8, Part 1: Sharp Cone. *AIAA paper 93-1761*, January 1983.

[57] Th. Herbert. Secondary Instability of Boundary Layers. *Annual Review of Fluid Mechanics*, Vol. 20, pp. 487-526, 1988.

[58] B. J. Bayly, S. T. Orszag, and Th. Herbert. Instability Mechanisms in Shear-Flow Transition. *Annual Review of Fluid Mechanics*, Vol. 20, pp. 359-391, 1988.

[59] L. Kleiser and T. A. Zang. Numerical Simulation of Transition in Wall-Bounded Shear Flows. *Ann. Rev. Fluid Mech.*, Vol. 23, pp. 495–535, 1991.

[60] H. L. Reed. Direct Numerical Simulation of Transition: the Spatial Approach. *Progress in Transition Modeling*, AGARD-Report-793 1994.

[61] X. Zhong. DNS of boundary-layer receptivity to freestream sound for 3-d hypersonic flows over blunt wedges and cones. *In proceeding of First AFOSR International Conference on Direct Numerical Simulation and Large Eddy Simulation, August 1997, Ruston, LA*, 1997.

[62] C. Park. *Nonequilibrium Hypersonic Aerothermodynamics.* Wiley Interscience, 1990.

[63] J. H. Lee. Basic governing equations for the flight regimes of aeroassisted orbital transfer vehicles. Technical Report AIAA Paper 84-1729, 1984.

[64] R.N. Gupta, J. M. Yos, R. A. Thompson, and K. P. Lee. *A Review of Reaction Rates and Thermodynamic Transport Properties for an 11-Species Air Model for Chemical and Thermal Nonequilibrium Calculations to 30000K.* NASA RP-1232, 1990.

[65] J. Häuser, J. Muylaret, H. Wong, and W. Berry. Computational aerothermodynamics for 2d and 3d space vehicles. In T. K. S. Murthy, editor, *Computational Methods in Hypersonic Aerodynamics*. Kluwer Academic Publishers, 1991.

[66] T. Gökçen. *The Computation of Hypersonic Low Density Flows with Thermochemical Nonequilibrium*. PhD thesis, Stanford University, 1989.

[67] R. B. Bird, W. E. Stewart, and E. N. Lightfoot. *Transport Phenomena*. John Wiley and Sons, 1960.

[68] J. O. Hirschfelder, C. F. Curtiss, and R. B. Bird. *Molecular Theory of Gases and Liquids*. John Wiley and Sons, 1954.

[69] F. G. Blottner. Finite difference methods of solution of the boundary-layer equations. *AIAA Journal*, 8:193–205, 1970.

[70] J. Cousteix, D. Arnal B. Aupoix, J. P. Brazier, and A. Lafon. Shock layers and boudnary layers in hypersonic flows. *Progress in Aerospace Sciences*, pages 95–212, 1994.

[71] C. R. Wilke. A viscosity equation for gas mixtures. *Journal of Chemical Physics*, 18(4):517–519, 1950.

[72] G. H. Furumoto, X. Zhong, and J. C. Skiba. *Unsteady Shock-Wave Reflection and Interaction in Viscous Flows with Thermal and Chemical Nonequilibrium*. AIAA Paper 96-0107, 1996.

[73] R. C. Millikan and D. R. White. Systematics of vibrational relaxation. *Journal of Chemical Physics*, 12:3209–3213, 1963.

[74] C. Park. Assessment of two-temperature kinetic model for ionizing air. *Journal of Thermophysics and Heat Transfer*, 3:233–244, 1989.

[75] L. C. Hartung, R. A. Mitcheltree, and P. A. Gnoffo. Coupled radiation effects in thermochemical nonequilibrium shock-capturing flowfield calculations. *Journal of Thermophysics and Heat Transfer*, 8:244–250, 1994.

[76] C. Park. Review of chemical-kinetic problems of future nasa missions, i: Earth entries. *Journal of Thermophysics and Heat Transfer*, 7:385–398, 1993.

[77] N. K. Makashev. On the boundary conditions for the equations of gas dynamics corresponding to the higher approximations in the Chapman-Enskog method of solution of the Boltzmann equation. *Fluid Dynamics (Translated From Russian)*, 14(3), May-June 1979.

[78] X. Zhong, R. W. MacCormack, and D. R. Chapman. Stabilization of the Burnett equations and application to hypersonic flows. *AIAA Journal*, 31(6), 1993.

[79] C. R. Mitchell A. G. Godfrey and R. W. Walters. Practical aspects of spatially high accurate methods. AIAA Paper 92-0054, 1992.

[80] C. Hirsch. *Numerical Computation of Internal and External Flows, Volume 2: Computational Methods for Inviscid and Viscous Flows*. John Wiley and Sons, 1990.

[81] R. J. Leveque. *Numerical Methods for Conservation Laws*. Birkhäuser, 1992.

[82] J. Steger and R. F. Warming. Flux vector splitting of the inviscid gasdynamics equations with application to finite difference methods. TM 78650, NASA, 1979.

[83] J. J. Quirk. A contribution to the great riemann solver debate. *International Journal for Numerical Methods in Fluids*, 18:555–574, 1994.

[84] P. L. Roe. Characteristic based schemes for the euler equations. *Annual Review of Fluid Mechanics*, 18:337–365, 1986.

[85] A. Harten. High resolution schemes for hyperbolic conservation laws. *Journal of Computational Physics*, 49:357–393, 1983.

[86] P. Collela and P. R. Woodward. The piecewise parabolic method (ppm) for gas-dynamical simulations. *Journal of Computational Physics*, pages 174–201, 1984.

[87] C. A. J. Fletcher. *Computational Techniques for Fluid Dynamics Volume II: Specific Techniques for Different Flow Categories*. Springer-Verlag, 1991.

[88] S. R. Chakravarthy, K-Y. Szema, U. C. Goldberg, J. J. Gorski, and S. Osher. Application of a new class of high accuracy tvd schemes to the navier stokes equations. Technical Report AIAA Paper 85-0165, 1985.

[89] A. Harten and S. Osher. Uniformly high order accurate essentially non-oscillatory schemes, I. MRC Technical Summary Report 2823, 1985.

[90] Ami Harten and Sukumar R. Chakravarthy. Multi-dimensional ENO schemes for general geometries. Report 91-16, CAM, August 1991. Department of Mathematics, University of California, Los Angeles.

[91] C. Shu and S. Osher. Efficient implementation of essentially non-oscillatory schemes. Contractor Report ICASE Report No. 87-33, NASA, May 1987.

[92] Y. Liu and M. Vinokur. Nonequilibirum flow computations i. an analysis of numerical formulations of conservation laws. *Journal of Computational Physics*, 83:373–397, 1989.

[93] J-S. Shuen, M-S. Liou, and B. Van Leer. Inviscid flux-splitting algorithms for real gases with non-equilibrium chemistry. *Journal of Computational Physics*, 90:371–395, 1990.

[94] M-S. Liou, B. Van Leer, and J-S. Shuen. Splitting of inviscid fluxes for real gases. *Journal of Computational Physics*, 87:1–24, 1990.

[95] B. Grossman and P. Cinella. Flux split algorithms with non-equilibrium chemistry and vibrational relaxation. *Journal of Computational Physics*, 88:131–168, 1990.

[96] B. Grossman and P. Cinella. Flux split algorithms for hypersonic flows. In T. K. S. Murthy, editor, *Computational Methods in Hypersonic Aerodynamics*. Kluwer Academic Publishers, 1991.

[97] D. Gaitonde and J. S. Shang. Accuracy of flux-split algorithms in high-speed viscous flows. *AIAA Journal*, 31:1215–1221, 1993.

[98] A. Harten, B. Engquist, S. Osher, and S. Chakravarthy. Uniformly High Order Accurate Essentially Non-oscillatory Schemes, III. *Journal of Computational Physics*, Vol. 71, No. 2, pp. 231-303, 1987.

[99] C.-W. Shu and S. Osher. Efficient implementation of essentially non-oscillatory schemes II. *Journal of Computational Physics*, 83:32–78, 1989.

[100] J. Casper. *Essentially Non-Oscillatory Shock Capturing Schemes to Multi-Dimensional Systems of Conservation Laws*. PhD thesis, Old Dominion University, December 1990.

[101] A. M. Rogerson and E. Meiburg. A numerical study of the convergence properties of ENO schemes. *Journal of Scientific Computing*, 5:151–167, June 1990.

[102] Chi-Wang Shu. Numerical experiments on the accuracy of ENO and modified ENO schemes. *Journal of Scientific Computing*, 5:127–149, June 1990.

[103] S. Lee, S. K. Lele, and P. Moin. Interaction of isotropic turbulence with a strong shock wave. *AIAA Paper 94-0311*, 1994.

[104] S.K. Lele K. Mahesh and P. Moin. The interaction of an isotropic field of acoustic waves with a shock wave. *Journal of Fluid Mechanics*, 300:383–407, 1995.

[105] Chi-Wang Shu, G. Erlebacher, T. A. Zang, D. Whitaker, and S. Osher. High-order ENO schemes applied to two- and three-dimensional compressible flow. Report 91-09, CAM, 1991. Department of Mathematics, University of California, Los Angeles.

[106] M. M. Rai and P. Moin. Direct Simulations of Turbulent Flow Using Finite Difference Schemes. *Journal of Computational Physics*, Vol. 96, pp. 15–53, 1991.

[107] C. Canuto, M. Y. Hussaini, A. Quarteroni, and T. A. Zang. *Spectral Methods in Fluid Dynamics*. Springer-Verlag, 1988.

[108] A. T. Patera. A Spectral Element Method for Fluid Dynamics: Laminar Flow in a Channel Expansion. *Journal of Computational Physics*, Vol. 54, pp. 468–488, 1984.

[109] S. A. Orszag and M. Israeli. Numerical Simulation of Viscous Incompressible Flows. *Annual Review of Fluid Mechanics*, Vol. 6, pp. 281–318, 1973.

[110] S. K. Lele. Compact Finite Difference Schemes with Spectral-like Resolution. *Journal of Computational Physics*, Vol. 103, pp. 16–42, 1992.

[111] C. D. Pruett, T. A. Zang, C.-L. Chang, and M. H. Carpenter. Spatial Direct Numerical Simulation of High-Speed Boundary-Layer Flows, Part I: Algorithmic Considerations and Validation. *Theoretical and Computational Fluid Dynamics*, Vol. 7, pp. 49–76, 1995.

[112] N. A. Adams and K. Shariff. A High-Resolution Hybrid Compact-ENO Scheme for Shock-Turbulence Interaction Problems. *Center for Turbulence Research, Stanford University, California*, Manuscript 155, March 1995.

[113] U. Rist and H. Fasel. Direct Numerical Simulation of Controlled Transition in Flat-Plate Boundary Layer. *Journal of Fluid Mechanics*, Vol. 298, pp. 211–248, 1995.

[114] M. M. Rai and P. Moin. Direct Numerical Simulation of Transition and Turbulence in a Spatially Evolving Boundary Layer. *Journal of Computational Physics*, Vol. 109, pp. 169–192, 1993.

[115] R. J. Leveque and H. C. Yee. A study of numerical methods for hyperbolic conservation laws with stiff source terms. *J. of Computational Physics*, 86:187–210, 1990.

[116] M. Y. Hussaini, D. A. Kopriva, M. D. Salas, and T. A. Zang. Spectral Methods for the Euler Equations: Part II–Chebyshev Methods and Shock Fitting. *AIAA Journal*, Vol. 23, No. 1, pp. 234–240, 1985.

[117] W. Cai. High-Order Hybrid Numerical Simulations of Two-Dimensional Detonation Waves. *AIAA Journal*, Vol. 33, No. 7, pp. 1248–1255, 1995.

[118] G. H. Furumoto. *Unsteady Shock-Wave Reflection and Interaction in Viscous Flow with Thermal and Chemical Nonequilibrium*. PhD thesis, UCLA, 1997.

[119] X. Zhong and T. Lee. Nonequilibrium Real-Gas Effects on Disturbance/Bow Shock Interaction in Hypersonic Flow Past a Cylinder. *AIAA paper 96-1856*, June 1996.

[120] J. J. Yoh and X. Zhong. Computations of Multi-Dimensional Detonation Using Semi-Implicit Runge-Kutta Schemes. *AIAA paper 97-0803, 35th AIAA Aerospace Sciences Meeting and Exhibit, January 6-9, Reno, Nevada*, 1997.

[121] M. S. Holden, J. R. Moselle, and J. Lee. *Studies of Aerothermal Loads Generated in Regions of Shock/Shock Interaction in Hypersonic Flow*. NASA CR-181893, October 1991.

[122] D. Gaitonde and J. S. Shang. On the structure of an unsteady type iv interaction at mach 8. *Computers and Fluids*, 24:469–485, 1995.

[123] C. A. Lind. *A Computational Analysis of the Unsteady Phenomena Associated with a Hypersonic Type IV Shock Interaction*. PhD thesis, University of Maryland, 1994.

[124] R. R. Thareja, J. R. Stewart, O. Hassan, K. Morgan, and J. Peraire. A point implicit unstructured grid solver for the euler and navier-stokes equations. *International Journal for Numerical Methods in Fluids*, 9:405–425, 1989.

[125] S. R. Sanderson. *Shock Wave Interaction in Hypervelocity Flow*. PhD thesis, California Institute of Technology, 1995.

[126] S. Yamamoto and S. Kano. *Sturcture of Bow Shock and Compression Wave Interactions in Unsteady Hypersonic Shock/Shock Interference Flow*. AIAA Paper 96-2152, 1996.

[127] J. H. Williamson. Low-Storage Runge-Kutta Schemes. *Journal of Computational Physics*, Vol. 35, pp. 48-56, 1980.

[128] E. Reshotko and N.M.S. Khan. Stability of the Laminar Boundary Layer on a Blunt Plate in Supersonic Flow. *IUTAM Symposium on Laminar-Turbulent Transition, R. Eppler and H. Fasel, editors*, Springer-Verlag, Berlin, pp. 186-190, 1980.

[129] S. H. Hu and X. Zhong. Linear Stability of Hypersonic Flow over a Parabolic Leading Edge. *AIAA paper 97-2015*, July 1997.

[130] K. F. Stetson and R. L. Kimmel. On Hypersonic Boundary Layer Stability. *AIAA paper 92-0737*, 1992.

[131] A. Dietz. Distributed Boundary Layer Receptivity to Convected Vorticity. *AIAA paper 96-2083*, June 1996.

[132] T. A. Buter and H. L. Reed. Boundary layer receptivity to free-stream vorticity. *Physics of Fluids*, 6(10):3368–3379, 1994.

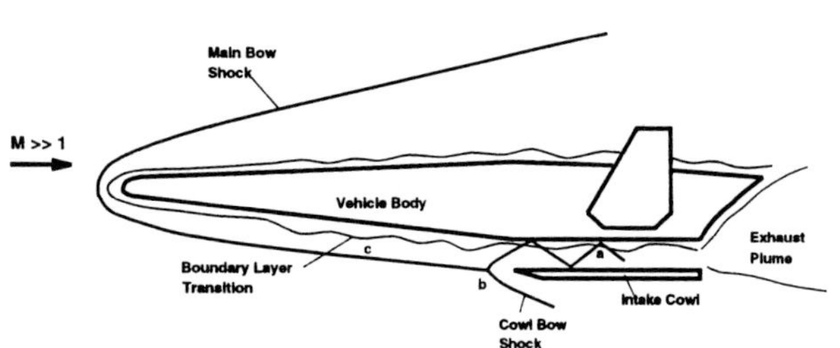

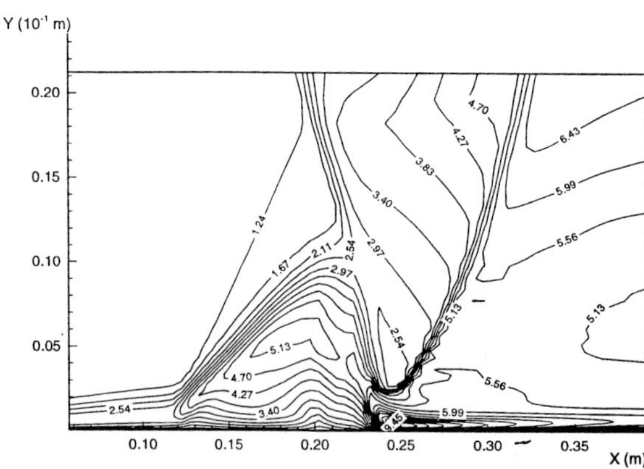

(a)

Figure 1: Typical hypersonic vehicle and associated shock interaction.

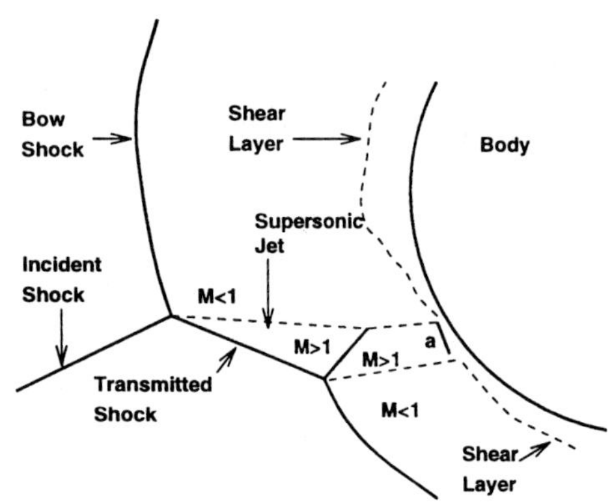

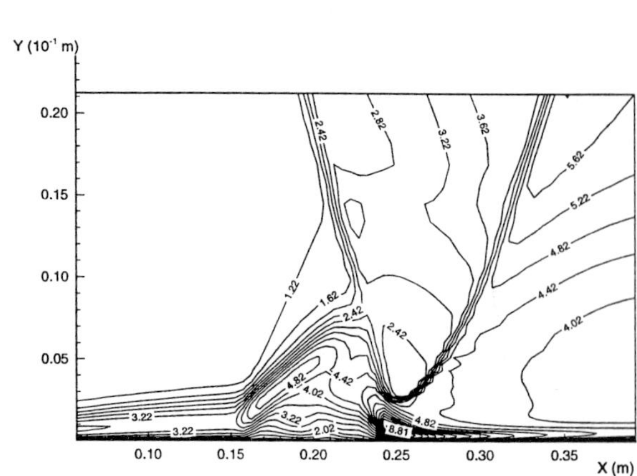

(b)

Figure 2: Schematic of type IV shock interference heating

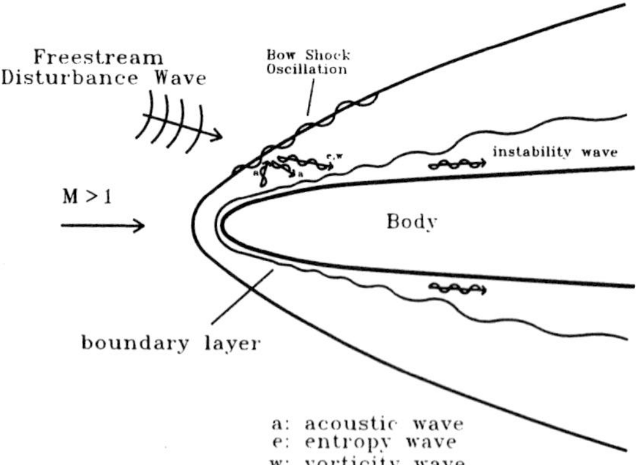

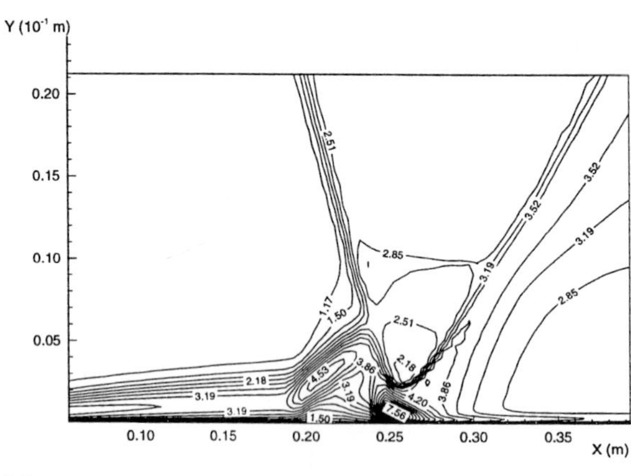

(c)

Figure 3: A schematic of the wave field of hypersonic boundary layer receptivity to free-stream disturbances.

Figure 4: Translational temperature (K) contours for shock/boundary layer interaction with different physical models: (a) ideal gas, (b) vibrational excitation and frozen chemistry,(c) full nonequilibrium gas ($T_\infty = 1600°K$, $M_\infty = 7$). The y axis has been magnified by a factor of 10 relative to the x axis for clarity.

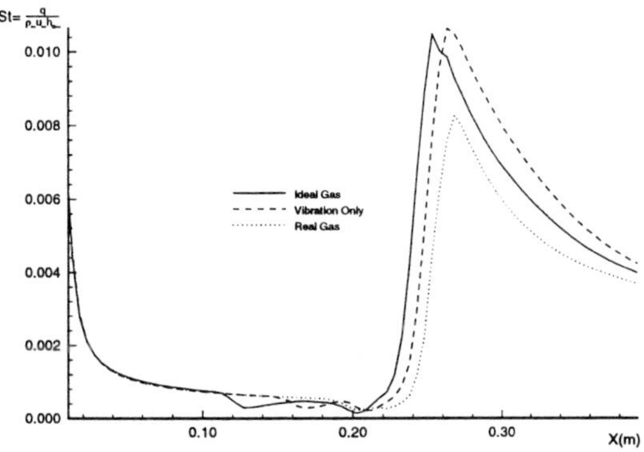

(a)

Figure 5: Normalized heat flux $\dot{q}/\rho_\infty u_\infty h_{o\infty}$ distribution along the flat plate for the three gas models.

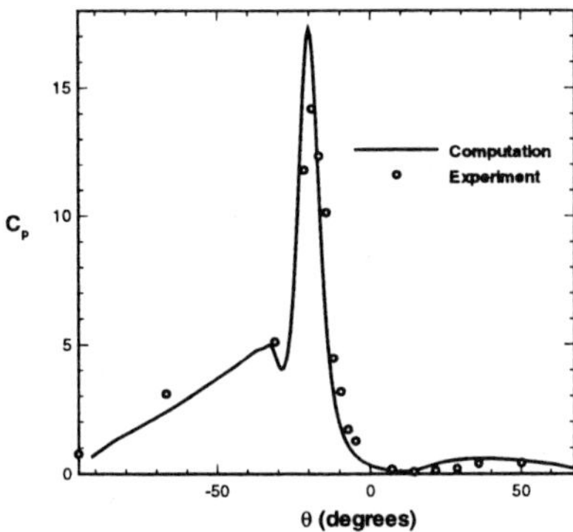

(b)

Figure 6: Computed surface pressure coefficient ($C_p = (p - p_\infty)/\frac{1}{2}\rho_\infty u_\infty^2$) profiles compared with experimental data of Holden.

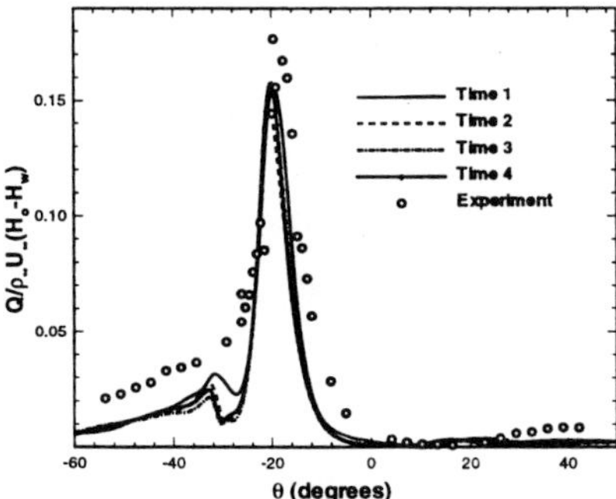

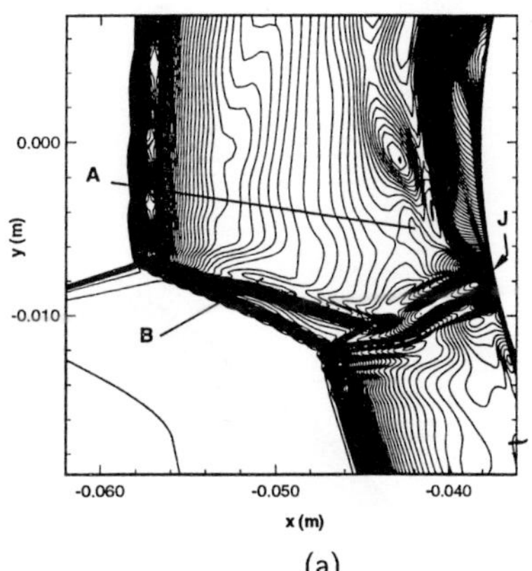

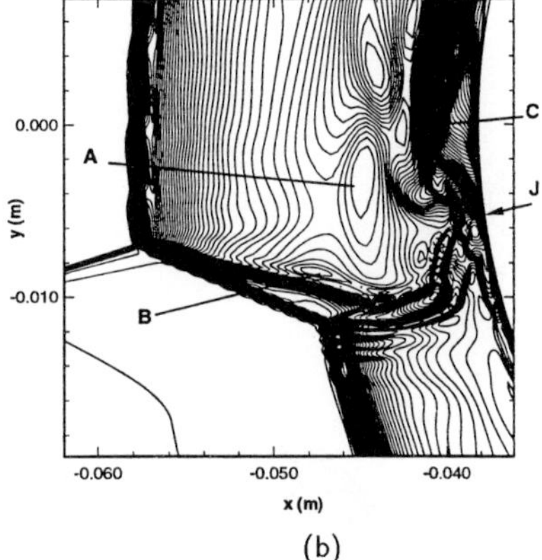

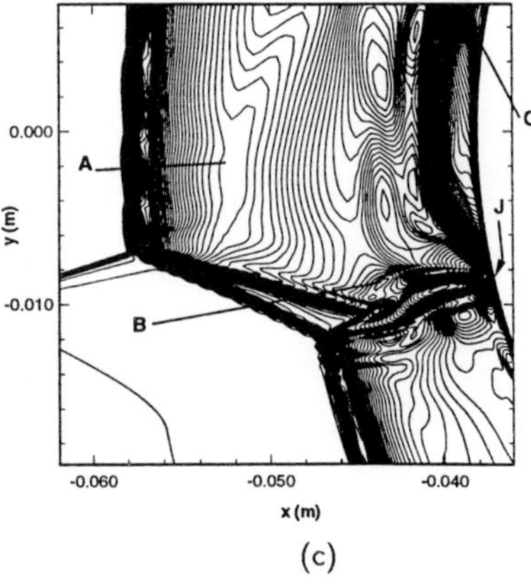

(c)

Figure 8: Translational temperature contours at three different times. J marks the location of the jet impingement point, A marks a disturbance generated by the shear layer, and B marks a disturbance generated at the bow shock.

Figure 7: Computed instantaneous surface Stanton number ($St = (\dot{q}/\rho_\infty u_\infty (H_o - H_w))$) profiles compared with experimental data of Holden.

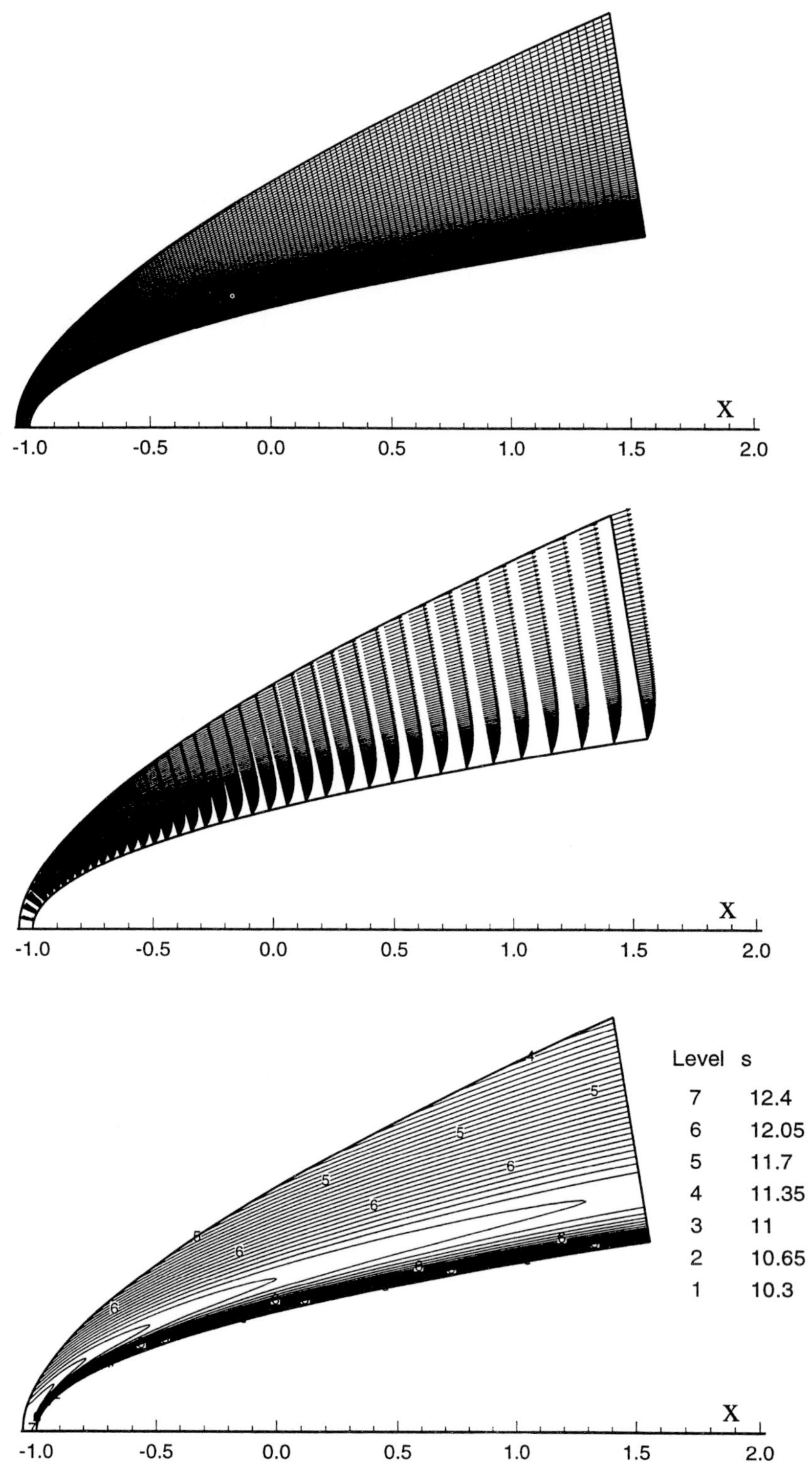

Figure 9: Base flow solutions for computational grid (upper figure) where the bow shock shape is obtained as the numerical solution for the freestream grid line, velocity vectors (middle figure), and entropy contours (lower figure).

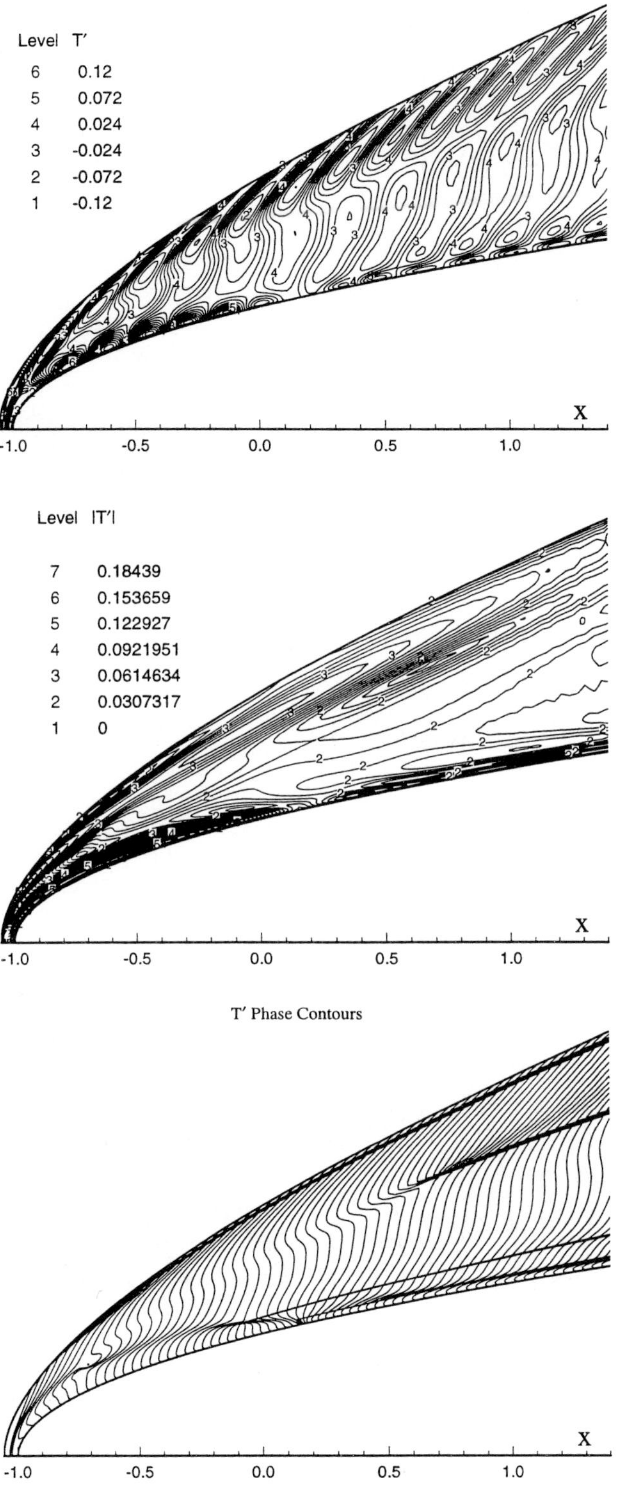

Figure 10: Unsteady temperature perturbation contours for the case of $F = 2655$: instantaneous T' (upper figure), Fourier amplitude $|T'|$ (middle figure), and Fourier phase angle φ_T (in degrees) of T' (lower figure).

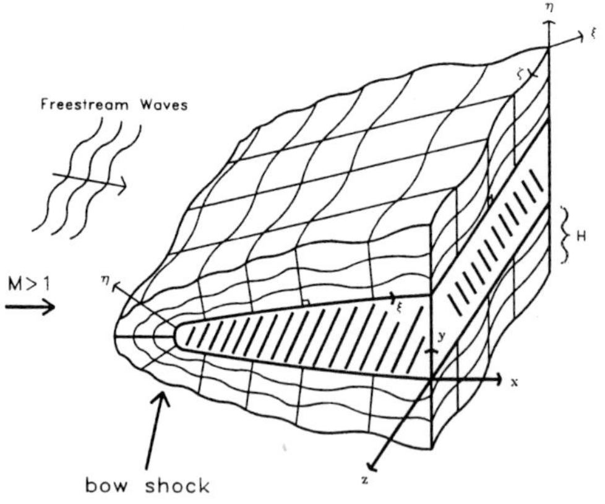

Figure 11: A schematic of three-dimensional shock fitted grids for the direct numerical simulation of hypersonic boundary-layer receptivity to freestream disturbances over a blunt leading edge.

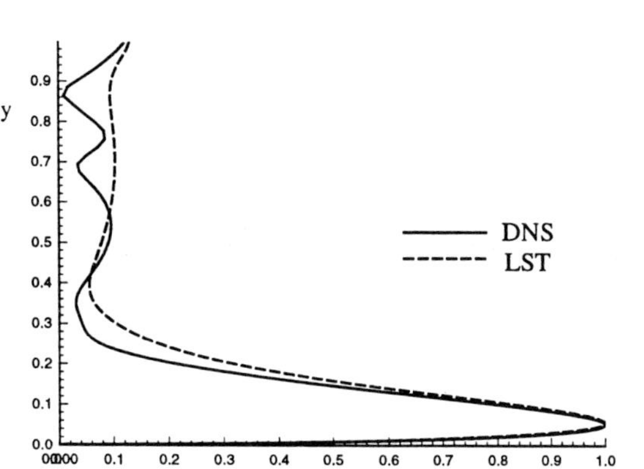

Figure 12: First-mode eigenfunction amplitudes along the wall-normal direction at $x = 0.5521$ ($F = 1770$).

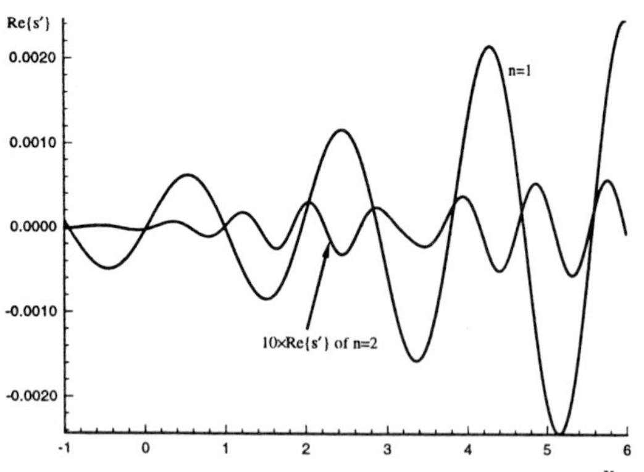

Figure 13: Instantaneous entropy disturbances along the body surface for the case of $F = 1770$ ($n = 1$: fundamental mode; $n = 2$: second harmonic).

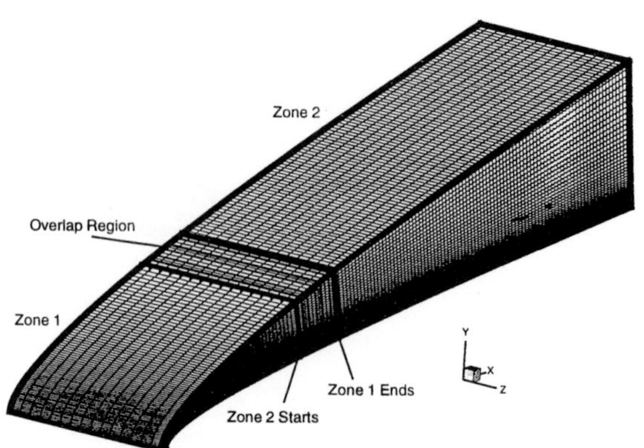

Figure 14: Computational grid for three-dimensional hypersonic boundary layer receptivity to freestream oblique disturbance waves using two overlap zones.

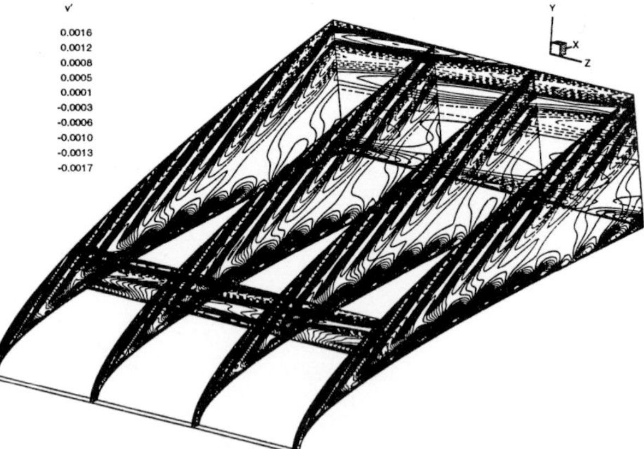

Figure 15: Instantaneous v' contours for the receptivity to freestream disturbances for three-dimensional hypersonic boundary-layer over a parabolic leading edge.

MODEL OF CONTINUUM MEDIUM WITH NONCLASSICAL KINEMATICS

V.P. MYASNIKOV   M.A. GUZEV

Institute of Automation and Control Processes, Far Eastern Division Russian
Academy of Sciences; 5, Radio St., Vladivostok, 690041, RUSSIA

Abstract

We propose the approach for generalizing the classical model of a continuum medium, whose local kinematics is not described by a diffeomorphic mapping. This approach is based on using the thermodynamics scheme in which distortions are the thermodynamics variables. We derived the complete system of equations for description of the medium.A special attention is payed to the consideration of the dissipative processes. We analyzed the Yang–Mills gauge approach from the thermodynamics point of view. Finally, the linear approximation of the gauge theory is considered.

1. Introduction

It is well known that classical continuum mechanics uses the dynamic Newton equations for elements of a medium and a kinematics hypothesis of continuity (see 2. Extention of kinematics). Different natural processes can be simulated in the framework of the classical theory. However there are physical phenomena demanding additional assumptions in the theory. To illustrate this situation we will present two hydrodynamics examples.

Propagating waves in an ideal fluid can make discontinuities inside the medium and form defect structures. It is well known that cavitation is a result of changing pressure in the fluid and it is appeared as a discontinuity of the fluid. In particular there is an acoustic cavitation in water when the amplitude of the acoustic pressure exceeds the acoustic cavitation threshold. The threshold is called as a strength of the fluid to a discontinuity.The strength is linked with as hydrophysical, hydrochemical factors and hydrostatic pressure as the sound field frequency. There are experimental data of measuring the acoustic cavitation threshold for sea water in [1–2].

Investigation of turbulent processes in ocean, atmosphere shows that principal feature of these processes is appearance and interaction of defect structures having different space scale. Simulation of turbulent phenomena in the framework of the classical model results in the parametrization problem. Its solution depends on the physical nature of defects. The study of the nature of these defects has shown that their presence disturbs the continuous structure of the medium.

While considering these physical phenomena we see the general problem: necessity of extending the kinematical hypothesis. This problem is known in the theory of plasticity. The investigators [3] proposed the Yang–Mills formalism for the macroscopic description of the plastic deformation of various materials. This formalism makes it possible to extend the set of kinematically permissible states of the medium as a result of the inroduction of additional (internal) degrees of freedom, whose description in terms of gauge fields characterizes the deviations from classical diffeomorphic kinematics. We analyzed [4–5] the characteristics of the Yang–Mills approach (the gauge theory) to the construction of new models of an ideal fluid. In the article [6] we formulated the Euler representation of the gauge theory of an ideal fluid and described dissipative (irreversible) processes in the fluid.

Importance of constructing models of reversible and irreversible processes for CFD was emphasized at The Second Asian CFD Conference [7], Tokyo,1996. In the gauge theory there remained the main difficulty : the physical interpretation of the gauge fields. We overcome it in this article by proposing the thermodynamics approach for constructing a model of continuum medium whose local kinematics is not described by the hypothesis of continuity. It allows us to link functions characterizing additional degrees of freedom with experimentally observable quantity — the dissipative function. The gauge theory follows from our approach on condition that there is an additional symmetry of the internal energy.

2. Extention of Kinematics

There is the fundamental kinematics assumption of the classical continuum mechanics — the continuity hypothesis. It implies [8] that the set of kinematically permissible states of the medium is described by the diffeomorphic mapping of the initial state $\vec{x}(\vec{\xi}, t_0) = \vec{\xi}$ into the current state $\vec{x}(\vec{\xi}, t)$. Then at each moment of time t the correspondence between the variables $\vec{x}$ and $\vec{\xi}$ is unambiguous, and we can consider the medium as

Received on January 25, 1997.

"

in the Lagrange representation as in the Euler representation.The quantities

$$\partial \xi^k(\vec{x}, t)/\partial x^i = p_i^k \tag{1}$$

characterize the medium deformation. They are the distortions. The transformation determinant $\vec{\xi} \to \vec{x}$ determines the density variation during motion: $\rho = \rho_0 det\|p_i^k\|$, where $\rho_0 = \rho(\vec{\xi}, t_0)$ is the density of the medium in the initial state. In the Euler representation the evolution of the distortions is described by the equation [8]:

$$\frac{dp_i^k}{dt} + p_s^k \frac{\partial v_s}{\partial x^i} = 0, p_i^k|_{t=t_0} = \delta_i^k, \tag{2}$$

in which $d/dt = \partial/\partial t + v_l \partial/\partial x^l$, v_l are the vector components. In (2) and what follows the summation convection applies.

On the foundation of the classical continuum mechanics we can analyze the behaviour of the medium with discontinuities. The general idea is to construct solutions containing singularities localized on sets whose dimension is less then the space dimension. We have pointed out that discontinuities disturb the diffeomorphic structure of the displacement field, and it is necessary to extend the set of kinematically permissible states of the medium. Introduction of additional degrees of freedom in the mechanics of a deformable solid was discussed in [3,9,10]. We extend the classical kinematics by introducing the general structure of distortions (1).

One considers the classical distortions p_i^k (1). These functions are determined through the displacement field $\vec{u}(\vec{x}, t) = \vec{x} - \vec{\xi}(\vec{x}, t)$. We introduce an arbitrary tensor set $P_i^k = P_i^k(\vec{x}, t)$ and formulate a condition at which the tensors are generated by the vector field in the threedimensional Euclidean space. Let us assume

$$\Gamma_{ij}^k = \partial P_j^k/\partial x^i - \partial P_i^k/\partial x^j = 0, \tag{3}$$

then there exists a vector field $\phi^k(\vec{x}, t) = \phi^k$, generating the tensors $P_i^k = \partial \phi^k/\partial x$. If ϕ^k doesn't contain singularities the medium has the classical kinematics: $\phi^k = \xi^k$ and P_i^k coinsides with the classical distortion p_i^k. If $\Gamma_{ij}^k \neq 0$ a generating field for P_i^k is absent and the kinematics isn't the classical one.

We define additive relation between P_i^k and the classical distortion p_i^k:

$$P_i^k = p_i^k + \varphi_i^k, \tag{4}$$

where $\varphi_i^k = \varphi_i^k(\vec{x}, t)$ is an additional field determining the field tensor $\Gamma_{ij}^k = \partial P_j^k/\partial x^i - \partial P_i^k/\partial x^j = \partial \varphi_j^k/\partial x^i - \partial \varphi_i^k/\partial x^j$. The tensor Γ_{ij}^k is the defect structure characteristic which doesn't change under the gradient transformation $\varphi_i^k \to \varphi_i^k + \partial \tau^k/\partial x^i$ with

an arbitrary function $\tau^k(\vec{x}, t)$. From the geometrical point of view this feature means that the defect structure possesses a group of internal symmetries.

Functions φ_i^k characterize the diviations from classical kinematics, and correspond to additional degrees of freedom. One writes the transport equation for φ_i^k by analogy with the equation (2). In accordance with (2) we have

$$\frac{d\varphi_i^k}{dt} + \varphi_s^k \frac{\partial v_s}{\partial x^i} = -I_i^k, \varphi_i^k|_{t=t_0} = 0. \tag{5}$$

There introduced a source I_i^k on the right-hand side of (5) which is responsible for emergence of nondiffeomorphic forms in the medium. The structure of I_i^k depends on the mechanical properties of the body and conditions of irreversible processes. From (2),(4),(5) it follows that the equation for the complete distortions (4) has the form:

$$\frac{dP_i^k}{dt} + P_s^k \frac{\partial v_s}{\partial x^i} = -I_i^k, P_i^k|_{t=t_0} = \delta_i^k. \tag{6}$$

By using (5) we get the equation for the field tensor Γ_{ij}^k:

$$\frac{D\Gamma_{ij}^k}{Dt} = \frac{d\Gamma_{ij}^k}{dt} + \frac{\partial v_l}{\partial x^i}\Gamma_{lj}^k + \Gamma_{il}^k \frac{\partial v_l}{\partial x^j}$$
$$= -\left(\frac{\partial I_j^k}{\partial x^i} - \frac{\partial I_i^k}{\partial x^j}\right), \tag{7}$$

$$\Gamma_{ij}^k|_{t=t_0} = 0.$$

If the right-hand side of the equation (7) isn't equal to zero, then the corresponding solution of the field tensor isn't trivial, and describes emergence of defect structures in the medium.

3. Thermodynamics of Continuum Mechanics with Nonclassical Kinematics

The distortions P_i^k and the field tensor Γ_{ij}^k are considered to be the thermodynamics parameteres of the medium, and the internal energy is $U = U(s, P_i^k, \Gamma_{ij}^k)$ with the entropy s. We follow the standard scheme of nonequilibrium thermodynamics [11]. One writes the laws conservation of mass, momentum, the first and the second law of thermodynamics:

$$\frac{\partial \rho}{\partial t} + \frac{\partial \rho v_k}{\partial x^k} = 0,$$

$$\rho \frac{dv_i}{dt} = \frac{\partial \sigma_{ij}}{\partial x^j}$$

$$\rho \frac{dU}{dt} = -\frac{\partial J_k^{(q)}}{\partial x^k} + \sigma_{ij}\frac{\partial v_i}{\partial x^j},$$

$$\rho \frac{ds}{dt} = -\frac{\partial J_k^{(s)}}{\partial x^k} + D, \quad D \geq 0. \tag{8}$$

Here $J_k^{(q)}$ $J_k^{(s)}$ are componets of the heat and entropy flow correspondingly, D is the dissipation function, σ_{ij} is the stress tensor, ρ is the density.

There is the Hibbs equality along the particle trajectory [11]:

$$\frac{dU}{dt} = T\frac{ds}{dt} + \frac{\partial U}{\partial P_i^k}\frac{dP_i^k}{dt} + \frac{\partial U}{\partial \Gamma_{ij}^k}\frac{d\Gamma_{ij}^k}{dt}. \qquad (9)$$

Substitution of the full time derivatives of the energy and the entropy (8) into (9) gives

$$-\frac{\partial J_k^{(s)}}{\partial x^k} + D$$
$$= -\frac{1}{T}\frac{\partial J_k^{(q)}}{\partial x^k}$$
$$+\frac{1}{T}\left(\sigma_{ij}\frac{\partial v_i}{\partial x^j} - \rho\frac{\partial U}{\partial P_i^k}\frac{dP_i^k}{dt} - \rho\frac{\partial U}{\partial \Gamma_{ij}^k}\frac{d\Gamma_{ij}^k}{dt}\right). \qquad (10)$$

By using the transport equation (6), (7) we exclude the time derivatives of P_i^k and Γ_{ij}^k in the right-hand side of (10), and get

$$\sigma_{ij}\frac{\partial v_i}{\partial x_j} - \rho\frac{\partial U}{\partial P_i^k}\frac{dP_i^k}{dt} - \rho\frac{\partial U}{\partial \Gamma_{ij}^k}\frac{d\Gamma_{ij}^k}{dt}$$
$$= \frac{\partial v_i}{\partial x_j}\left(\sigma_{ij} + \rho P_i^k\frac{\partial U}{\partial P_j^k} - 2\rho\Gamma_{il}^k\frac{\partial U}{\partial \Gamma_{lj}^k}\right)$$
$$+ I_j^k\left(\rho\frac{\partial U}{\partial P_j^k} - 2\frac{\partial}{\partial x^i}\rho\frac{\partial U}{\partial \Gamma_{ij}^k}\right)$$
$$+ 2\frac{\partial}{\partial x^i}\left(\rho I_j^k\frac{\partial U}{\partial \Gamma_{ij}^k}\right)$$
$$- \frac{J_k^{(q)}}{T^2}\frac{\partial T}{\partial x^k}. \qquad (11)$$

One substitutes (11) into (10) and has:

$$-\frac{\partial}{\partial x^i}\left[J_i^{(s)} - \frac{1}{T}J_i^{(q)} + \frac{2}{T}\rho I_j^k\frac{\partial U}{\partial \Gamma_{ij}^k}\right] + D$$
$$= \frac{1}{T}\frac{\partial v_i}{\partial x_j}\left(\sigma_{ij} + \rho P_i^k\frac{\partial U}{\partial P_j^k} - 2\rho\Gamma_{il}^k\frac{\partial U}{\partial \Gamma_{lj}^k}\right)$$
$$+ \frac{I_j^k}{T}\left(\rho\frac{\partial U}{\partial P_j^k} - 2\frac{\partial}{\partial x^i}\rho\frac{\partial U}{\partial \Gamma_{ij}^k}\right)$$
$$- \frac{1}{T^2}\left(J_i^{(q)} - \frac{2}{T}\rho I_j^k\frac{\partial U}{\partial \Gamma_{ij}^k}\right)\frac{\partial T}{\partial x^i}. \qquad (12)$$

The standard analysis in the framework of nonequilibrium thermodynamics [11] results in the following form of the entropy flow and the dissipation function:

$$J_i^{(s)} = \frac{J_i^{(q)}}{T} - \frac{2\rho}{T}I_j^k\frac{\partial U}{\partial \Gamma_{ij}^k}, \qquad (13)$$

$$DT = \frac{\partial v_i}{\partial x_j}\left(\sigma_{ij} + \rho P_i^k\frac{\partial U}{\partial P_j^k} - 2\rho\Gamma_{il}^k\frac{\partial U}{\partial \Gamma_{lj}^k}\right)$$
$$+ I_j^k\left(\rho\frac{\partial U}{\partial P_j^k} - 2\frac{\partial}{\partial x^i}\rho\frac{\partial U}{\partial \Gamma_{ij}^k}\right)$$
$$- \frac{1}{T}\left(J_i^{(q)} - \frac{2}{T}\rho I_j^k\frac{\partial U}{\partial \Gamma_{ij}^k}\right)\frac{\partial T}{\partial x^i}. \qquad (14)$$

The dissipative function is written in the bilinear form of the thermodynamics forces and flows:

$$DT = D_T(x) = X_i Y_i, \; D(x) \geq 0. \qquad (15)$$

From here it follows that $Y_i = \partial D/\partial x_i$. The complete thermomechanical description of the medium is founded on knowledge of independent functions: the internal energy and the dissipation function.

4. The Complete System of Equations

The internal energy and the dissipative function is supposed to be known then in accordance with (14), (15) we can write the state equation in the following form:

$$\sigma_{ij} = -\rho P_i^k\frac{\partial U}{\partial P_j^k} + 2\rho\Gamma_{il}^k\frac{\partial U}{\partial \Gamma_{lj}^k} + \frac{\partial D_T}{\partial e_{ij}},$$
$$\rho\frac{\partial U}{\partial P_j^k} - 2\frac{\partial}{\partial x^i}\rho\frac{\partial U}{\partial \Gamma_{ij}^k} = \frac{\partial D_T}{\partial I_j^k}, \qquad (16)$$

where $e_{ij} = \partial v_i/\partial x_j$. We take a linear approximation for the heat flow:

$$J_i^{(q)} = -\lambda\frac{\partial T}{\partial x^i} + 2\rho I_j^k\frac{\partial U}{\partial \Gamma_{ij}^k}. \qquad (17)$$

The phenomenological coefficient $\lambda \geq 0$ gives a positive value of the dissipation function. The relations (16),(17) and the equations (8),(5) (7) form the complete system for the medium with nonclassical kinematics:

$$U = U(s, P_i^k, \Gamma_{ij}^k), D_T = D_T(I_i^k, e_{ij}, T), T = \frac{\partial U}{\partial s},$$

$$\frac{dP_i^k}{dt} + P_s^k\frac{\partial v_s}{\partial x^i} = -I_i^k,$$
$$\frac{\mathcal{D}\Gamma_{ij}^k}{\mathcal{D}t} = -\left(\frac{\partial I_j^k}{\partial x^i} - \frac{\partial I_i^k}{\partial x^j}\right),$$

$$\rho \frac{dv_i}{dt} = \frac{\partial}{\partial x^j} \left(-\rho P_i^k \frac{\partial U}{\partial P_j^k} + 2\rho \Gamma_{il}^k \frac{\partial U}{\partial \Gamma_{lj}^k} + \frac{\partial D_T}{\partial e_{ij}} \right)$$

$$\frac{\partial \rho}{\partial t} + \frac{\partial \rho v_k}{\partial x^k} = 0, \tag{18}$$

$$\rho \frac{\partial U}{\partial P_j^k} - 2\frac{\partial}{\partial x^i} \rho \frac{\partial U}{\partial \Gamma_{ij}^k} = \frac{\partial D_T}{\partial I_k^i},$$

$$\rho T \frac{ds}{dt} = \frac{\partial}{\partial x_i} \left(\lambda \frac{\partial T}{\partial x^i} \right) + e_{ik} \frac{\partial D_T}{\partial e_{ik}} + I_k^i \frac{\partial D_T}{\partial I_k^i}.$$

5. The Dissipative Function

The dissipative function (14) characterizes the irreversible processes in the medium.It is supposed [12] to be a homogeneous function with respect to the thermodynamics forces. The degree of homogeneity is determined by the nature of a dissipative process in the medium. For the viscous mechanism of dissipation the function D_T has the degree of two:$D_T(\lambda e_{ij}) = \lambda^2 D_T(e_{ij})$. Then the stress (16) tensor has the form

$$\sigma_{ij} = -\rho P_i^k \frac{\partial U}{\partial P_j^k} + 2\rho \Gamma_{il}^k \frac{\partial U}{\partial \Gamma_{lj}^k} + 2\eta e_{ij} \tag{19}$$

where η is the coefficient of viscosity. In the case of $\Gamma_{ij}^k \to 0$ and $U = U(\rho)$ from (18), (19) we have the classical Navier-Stokes equation for $\vec{v}$.

The choice of a dissipative function with the degree of one corresponds to the threshold processes: they emerge in a medium with the stress tension of a critical value.Let us consider the threshold processes. The medium is supposed to be a homogeneous one. Then the dissipative function with the degree of one has the form $D_T = \tau \sqrt{I_\beta^\alpha I_\beta^\alpha}$ where τ is a dimensional parameter. We substitute this expression into (16) and get the representation for the flow:

$$\frac{\delta U}{\delta P_j^k} \equiv \rho \frac{\partial U}{\partial P_j^k} - 2\frac{\partial}{\partial x^i} \rho \frac{\partial U}{\partial \Gamma_{ij}^k} = \frac{\partial D_T}{\partial I_j^k} = \frac{\tau I_j^k}{\sqrt{I_\beta^\alpha I_\beta^\alpha}}. \tag{20}$$

The dissipative processes are described by the function φ_i^k which has the source I_i^k (5). In the terms of the source the threshold character of a dissipation means that the dissipation emerges on condition that $D_T/\tau = \sqrt{I_\beta^\alpha I_\beta^\alpha} \gg \xi$. The scalar parameter ξ characterizes a critical stress of a medium without emergence of dissipative properties. To take into account the last inequality we modify the formula (20) for all values of I_β^α

$$\frac{\delta U}{\delta P_j^k} = \frac{\tau I_j^k}{\sqrt{\xi^2 + I_\beta^\alpha I_\beta^\alpha}}. \tag{21}$$

There are known results in the limit cases of the expression (21) : if $\sqrt{I_\beta^\alpha I_\beta^\alpha} \ll \xi$ (we haven't the threshold intensity of the source) the function $\delta U/\delta P_j^k = 0$

— the dissipation is absent; if $\sqrt{I_\beta^\alpha I_\beta^\alpha} \gg \xi$, we have a situation with the intensive dissipative processes. For other values of I_β^α the formula (21) is continuously approximated with respect to the parameter ξ. We square (21) and take summation over all values of k, j:

$$I_\beta^\alpha I_\beta^\alpha = \frac{\xi^2 \delta U/\delta P_j^k \cdot \delta U/\delta P_j^k}{(\tau^2 - \delta U/\delta P_j^k \cdot \delta U/\delta P_j^k)}.$$

From here we get the representation of the source:

$$I_\beta^\alpha = \frac{\xi \delta U/\delta P_\beta^\alpha}{\sqrt{\tau^2 - \delta U/\delta P_j^k \delta U/\delta P_j^k}}. \tag{22}$$

The equations (5), (6), (22) shows that additional functons φ_j^k characterizing the diviations from classical kinematics are determined by the internal energy of the medium and its physical parameters. We can find the source directly by measuring the dissipative function D_T and supposing the homogeneity of D_T with respect to the source.

We consider a medium with intensive dissipation processes and the internal energy $U = U(P_j^k)$. From (20), (16) we have

$$\frac{\delta U}{\delta P_j^k} = \rho \frac{\partial U}{\partial P_j^k} = \frac{\tau I_j^k}{\sqrt{I_\beta^\alpha I_\beta^\alpha}}, \quad \sigma_{ij} = -\rho P_i^k \frac{\partial U}{\partial P_j^k}. \tag{23}$$

Combination of the formulae gives rise to the condition for the stress tensor σ_{ij} :

$$(P^{-1})_i^k \sigma_{ij} (P^{-1})_s^k \sigma_{sj} = \tau^2.$$

This relation shows the restriction for the stress tensor and the physical sense of the phenomenological parameter τ. For cavitation the parameter τ is linked with the the cavitation threshold.

It is noted that the geometrical structure of φ_j^k can change during the evolution deterninated by the equation (5). Let's suppose that the source I_j^k increases at infinity. We introduce the fields E_j^k and F_j^k defining potential and rotation componets correspondingly: $I_j^k = E_j^k + F_j^k$. If the functions $F_j^k \neq 0$ and $E_j^k = 0$ then the fields φ_j^k (7) are generated by the rotation componet of the source $\partial F_j^k/\partial x^i - \partial F_i^k/\partial x^j$. In this case the dissipation is linked with disturbance of classical kinematics which is described by φ_j^k. If $F_j^k = 0$ and $E_j^k \neq 0$ then the geometrical structure of the field of φ_j^k is a classical one: it is generated by E_j^k which is restored through the divergence $\partial E_j^k/\partial x^j$.

6. Gauge Model

We consider an ideal continuum medium with the Lagrangian $L = L(\partial_\mu \vec{x})$ where the function $\vec{x} =$

$\vec{x}\left(t,\vec{\xi}\right)$ characterizes the position of the medium points at each moment of time, ∂_μ denotes spatial and time differentiation at $\mu = i = 1,2,3$ and $\mu = t$ correspondingly. For an ideal medium the equations of motion can be obtained in the usual way from the condition of the extremality of the action functional $S = \int_{t_0}^{t_1} dt \int_v d\vec{\xi} L$. The Lagrangian L possesses a group of internal symmetries: it is invariant under transformations $\vec{x} \to \vec{x}' = \vec{x} + \vec{\tau}$, where $\vec{\tau}$ is an arbitrary constant vector. Sets of these $\vec{\tau}$ form a homogeneous group of translations $T(3)$. In order to extend the space of kinematic states which may be occupied by the medium, in accordance with the general Yang—Mills idea [3,13], we localize the group of internal symmetries of the Lagrangian, i.e., assume that the elements of the group are functions of the Lagrangian coordinates and time: $\vec{x} \to \vec{x}' = \vec{x} + \vec{\tau}\left(t,\vec{\xi}\right)$. Transformations of this type do not commute with the differentiation operators. In order to preserve the invariance of the Lagrangian under the action of the inhomogeneous group $T(3)$ we must replace the classical derivatives $\partial_\mu \vec{x}$ by the covariant derivatives $D_\mu \vec{x} = \partial_\mu \vec{x} + \vec{\psi}_\mu$ where $\vec{\psi}_\mu = \vec{\psi}_\mu\left(t,\vec{\xi}\right)$ are vectors, and introduce instead of L the Lagrangian $L_G = L\left(D_\mu\vec{x}\right)$. The rules of trasformations of the derivatives and the gauge fields are given by the expressions

$$D'_\mu \vec{x}' = D_\mu \vec{x}, \vec{\psi}'_\mu = \vec{\psi}_\mu - \partial_\mu \vec{\tau}. \tag{24}$$

The gauge fields characterize the deviation of the local structure of the displacement field from the classical structure described by a diffeomorfism. These functions correspond to additional (internal) degrees of freedom. The dynamical equations for determining $\vec{\psi}_\mu$ can be obtained from the principle of least action by introducing the field Lagrangian L_p. The structure of L_p can be found from the following analysis. The function L_p is equal to zero should we consider the classical kinematics. It means that $\vec{\psi}_\mu = 0$ with accuracy to a gradient transformation $\partial_\mu \vec{\tau}$. Then we have the equation of the classical continuity:

$$\vec{D}_{\mu\nu} = \partial_\mu D_\nu \vec{x} - \partial_\nu D_\mu \vec{x} = \partial_\mu \vec{\psi}_\nu - \partial_\nu \vec{\psi}_\mu \equiv 0. \tag{25}$$

The equation (25) is necessary and sufficient condition that the displacement field has the classical diffeomorphic kinematics and $D_\mu \vec{x} = \partial_\mu \vec{x}$. Hence L_p depends on the field tensor $\vec{D}_{\mu\nu}$.

The complete functional of the system is given by the expression

$$S_w = \int_{t_0}^{t_1} dt \int_v d\vec{\xi} \left[L\left(D_\mu\vec{x}\right) + L_p\left(\vec{D}_{\mu\nu}\right) \right]. \tag{26}$$

The variation of S_w with respect to variables $\vec{x}, \vec{\psi}_\mu$ leads to the equations of dynamics of an ideal medium and for the gauge fields correspondingly:

$$\frac{\partial L_G}{\partial q^i_\mu} = 2\partial_\nu \frac{L_p}{\partial D^i_{\nu\mu}}, \tag{27}$$

$$\partial_\mu \frac{\partial L_G}{\partial q^i_\mu} = 0, \tag{28}$$

The components q^i_μ denote $D_\mu x^i$. From (27) when the identity $\partial_\mu \partial_\nu \vec{D}_{\mu\nu} = 0$ is taken into account there follows a compatibility conditions that coincides with equation (28) for the medium. Because of the gauge invariance the solutions of the equations (27), (28) are related by the gauge transformation (24). In order for the fields to be determinated uniquely the system of equations (27), (28) must be supplemented by the gauge conditions which we will take in the form: $\vec{F}\left(\vec{\psi}_\mu\right) = 0$ with an arbitrary function $\vec{F}$. The choice of $\vec{F}$ is determined from the general physical consideration. The authors [3,9–10] use the Lorentz condition.

Gauge fields are nonclassical quantities so experimental interpretation of the corresponding solutions can be carried out when the problem formulation is correct. It is phenomenological hypothesis that is useful for this purpose.

Let us consider the problem formulation in the gauge theory with reference to the example of structures formation on the plane. Observation of the structures formation shows [4] that this process gives rise to division of the plane into regions with a piece-smooth boundary. Kinematics of the medium inside the regions is not classical in the general case. Then the boundary Γ is the classical continuum structure with respect to the internal region and we have the continuity equation (25) at the boundary:

$$\vec{D}_{\mu\nu}|_\Gamma = (\partial_\mu \vec{q}_\nu - \partial_\nu \vec{q}_\mu)|_\Gamma = 0. \tag{29}$$

From this point of view the processes of structures formation give rise to the emergence of a set Γ which dimension is less than dimension of the plane and the continuity equation is valid (29). Our problem formulation can be generalized for the case of the space. Then Γ is the set of plane structures for which the field tensor $\vec{D}_{\mu\nu}$ satisfies the equation (29).

7. Dissipation in the Gauge Model

Variational formulation of the gauge model means that there are laws of conservation in a medium so dissipation is not considered. Usually investigators tell about the energy redistribution between internal and external degrees of freedom: the displacement field corresponds to the first ones, the gauge field — to the second ones. However we introduce the gauge fields in the fourdimensional space of internal degrees of freedom: the transformation rules (24) contain space and time indeces. Then the whole functional (26) depends on gradients of velocities. It means, from the continuum mechanics point of view, that the gauge model is dissipative. We suppose that this difficulty of the gauge theory isn't taken into account by investigators.

In previous sections the thermodynamics discription of the dissipation in the Euler representation is proposed. Below we'll show link between energy redistribution processes of the gauge theory and the dissipative ones.

The classical Lagrangian of continuum mechanics is $L = \rho_0[|\partial \vec{x}/\partial t|^2/2 - U\left(\partial \vec{x}/\partial \xi^j\right)]$, where the internal energy of the continuum medium $U\left(\partial \vec{x}/\partial \xi^j\right)$ depends on its density ρ. In accordance with the Yang—Mills procedure we introduce the gauge Lagrangian $L_G = \rho_0\left[|D_t x^i|^2/2 - U\left(D_j x^i\right)\right]$.

For a homogeneous medium the field Lagrangian L_p depends on invariants $J_{\mu\nu} = |\vec{D}_{\mu\nu}|^2$. It is supposed to be a function $-L_p = \rho_0\left[U_1(J_{kj}) + U_2(J_{jt})\right]$. The internal energy $U_w = U_w(q_j^i, J_{kj}^i)$ of the medium—defects dynamical system is determined by the expression: $U_w = U + U_1$. Let's write the balance equation for U_w in the Lagrange variables:

$$\frac{\partial U_w}{\partial t} = \frac{\partial U_w}{\partial q_i^k}\frac{\partial q_i^k}{\partial t} + \frac{\partial U_w}{\partial D_{ij}^k}\frac{\partial D_{ij}^k}{\partial t}. \tag{30}$$

By using the definition of L_G, L_p, we rewrite equation of the system (27) for $\mu = j$ in the form

$$\frac{\partial U_w}{\partial q_j^i} = 2\partial_k \frac{\partial U_w}{\partial D_{kj}^i} + 2\partial_t \frac{\partial U_2}{\partial D_{tj}^i}.$$

From this and from (30) it follows that

$$\frac{\partial U_w}{\partial t} = 2\partial_k\left\{\frac{\partial U_w}{\partial D_{kj}^i}\partial_t q_j^i\right\} + 2\partial_t q_j^i\,\partial_t\frac{\partial U_2}{\partial D_{tj}^i}. \tag{31}$$

We will investigate the equilibrium states. For the classical model of a homogeneous medium from the equality of the velocity to zero in the equilibrium state there follows the constansy of the pressure throughout the volume occupied by the medium. A natural generalization of the classical equilibrium condition for the gauge model is the requirement that the total gauge momentum should be equal to zero: $D_t x^i = q_t^i = \partial_t x^i + \psi_t^i = 0$. Then the system of equations (27), (28) takes the form:

$$\frac{\partial U}{\partial q_j^i} = 2\partial_\mu\frac{\partial(U_1 + U_2)}{\partial D_{\mu j}^i}, \quad \partial_j\frac{\partial U_2}{\partial D_{jt}^i} = 0, \quad \partial_j\frac{\partial U}{\partial q_j^i} = 0. \tag{32}$$

Our additional assumption is a smalness of the displacement. It allows us to replace $\partial_t q_j^i \to D_{tj}^i$ in the formula (31), and identify Lagrange's and Euler's differentiation. Then the balance equation is written in following form:

$$\frac{dU_w}{dt} = -div\vec{J} + D_g, \quad D_g = 2D_{tj}^i\frac{\partial}{\partial t}\frac{\partial U_2}{\partial D_{tj}^i}. \tag{33}$$

Comparision of (33) with (8),(13) gives components $J_k = -2D_{tj}^i\partial U_w/\partial D_{ij}^k$ the internal energy the flow density $\vec{J}$, and the function D_g characterizing dissipation in the medium with a source D_{tj}^i.

In the nondissipative case the source is equal to zero: $D_{tj}^i = 0$. Solutions of the first system of equations (32)

$$\frac{\partial U}{\partial q_j^i} = 2\partial_k\frac{\partial U_1}{\partial D_{kj}^i}$$

define the thermodynamic equilibrium states of the medium—defects dynamical system.

The dissipative function D_g is supposed to be a function with the degree of one with respect to the source D_{tj}^i: $D_g = 2\gamma\sqrt{D_{tj}^i D_{tj}^i}$, where γ is a parameter.

From here and from (33) we have representation for the flow:

$$\frac{\partial}{\partial t}\frac{\partial U_2}{\partial D_{tj}^i} = \frac{\gamma D_{tj}^i}{\sqrt{I}}, \tag{34}$$

in which $I = D_{tk}^n D_{tk}^n = \sum_k J_{tk}$ is the source intensity. We modify the formula (34) by the analogy with (21). Then the system of equations (34) takes the form:

$$\frac{\partial}{\partial t}\frac{\partial U_2}{\partial D_{tj}^i} = \frac{\gamma D_{tj}^i}{\sqrt{\xi^2 + I}}. \tag{35}$$

The function $U_2(J_{jt})$ is $U_2(I)$. We multiply (35) by D_{tj}^i, and get the equation for the intensity I :

$$\frac{\partial I}{\partial t} = \frac{\gamma I}{\sqrt{\xi^2 + I}(U_2' + 2IU_2'')}, \tag{36}$$

where $'$ denotes differentiation with respect to the variable I. A solution for I depends on parametrization of U_2. We consider a minimal dynamical extension [11] for the gauge model by introducing $U_2 = I/2a$. The equation (36) is integrated and we have

$$s + \frac{\xi}{2}\ln\left|\frac{s - \xi}{s + \xi}\right| + const = \gamma a(t - t_0), \tag{37}$$

with $s = \sqrt{I + \xi^2}$. The intensive dissipative processes are characterized by the condition $I \gg \xi^2$, then from (37) it follows that the sorce intensity is a linear function with respect to time.

8. Linear Approximation in the Gauge Formalism (Acoustic Approximation)

In the previous section on the foundation of the gauge theory we considered the equilibrium state of a medium. Here our attention will be payed to small oscillations near the equilibrium state of an ideal fluid.

The ideal fluid is supposed to occupy a layer with a hard low boundary $x_2 \to -\infty$ and a free upper boundary $x_2 = \zeta(x_1,t)$. There introduce the potential function for the velocity $\vec{v} : \vec{v} = \nabla\varphi$. Because we consider small deviations from the equilibrium state one linerizes all functions near this state then the classical equation of motion for the potential coincides with the wave equation:

$$\Box\varphi = \frac{1}{c_a^2}\frac{\partial^2\varphi}{\partial t^2} - \frac{\partial^2\varphi}{\partial x^i \partial x^i} = 0 \qquad (38)$$

where the acoustic speed propagation c_a is supposed to be constant.

Below we will follow the standard scheme of the gauge formalism (see 6. Gauge model). The equation (38) is obtained from the condition of the extremality of the action fuctional $S = \int_{t_0}^{t_1} dt \int_v d\vec{x}\, L$ with the Lagrangian $L = L(\partial_\mu\varphi) = [|\partial\varphi/\partial t|^2/2 - c_a^2|\nabla\varphi|^2/2]$. The Lagrangian L is invariant with respect to internal transformations $\varphi \to \varphi' = \varphi + \tau$, where τ is an arbitrary constant fuction. The homogeneous group of translations $T(2)$ is formed from sets of these τ. We localize the group of symmetries of the Lagrangian,i.e., assume that the elements of the group are functions of the Euler coordinates and time $\varphi \to \varphi' = \varphi + \tau(t,\vec{x})$. Then we replace the classical derivatives ∂_μ by the covariant derivatives $D_\mu\varphi = \partial_\mu\varphi + \psi_\mu \equiv p_\mu$ where $\psi_\mu = \psi_\mu(t,\vec{x})$ are fuctions. And one introduces the gauge Lagrangian $L_G = L(D_\mu\varphi)$ instead of L. The rules of trasformations of the derivatives and the gauge fields are given by the expressions:

$$D'_\mu\varphi' = D_\mu\varphi,\; \psi'_\mu = \psi_\mu - \partial_\mu\tau. \qquad (39)$$

To obtain the dynamic equations for determination of ψ_μ we introduce the field Lagrangian L_p. We assume that the field Lagrangian L_p depends on invariants $J_{\mu\nu} = |D_{\mu\nu}|^2$ and it is analytical function of the second power of the homogeneity from $D_{\mu\nu}$ — minimal dynamical extension of the principle of least action : $L_p = -\varepsilon_\mu\varepsilon_\nu|D_{\mu\nu}|^2/4\alpha$, where $\varepsilon_t = 1/c^2, \varepsilon_j = -1; \alpha, c^2$ are phenomenological parameters. The complete functional of the system is given by the expression

$$S_w = \int_{t_0}^{t_1} dt \int_v d\vec{x}\Big\{\frac{1}{2}p_i^2 - \frac{c_a^2}{2}J$$
$$+\frac{1}{2\alpha c^2}(D_{1t}^2 + D_{2t}^2) - \frac{1}{2\alpha}D_{21}^2\Big\}. \qquad (40)$$

where the quantity $J = (p_1)^2 + (p_2)^2$. The variation of S_w with respect to variables φ, ψ_μ leads to the equations:

$$\partial_\mu D_{\mu t}\varepsilon_\mu = -\alpha c^2 p_t,$$
$$\partial_\mu D_{\mu j}\varepsilon_\mu = -\alpha c_a^2 p_j, \qquad (41)$$
$$\partial_t p_t = c_a^2 \partial_i p_i.$$

In accordance with the problem formulation in the gauge theory we are interested in the field tensors which are equal to zero at Γ. It was pointed out that boundary Γ is the classical continuum structure with respect to the internal region and there is the continuity equation (29):

$$D_{\mu\nu}|_\Gamma = (\partial_\mu D_\nu\varphi - \partial_\nu D_\mu\varphi)|_\Gamma = 0. \qquad (42)$$

First we consider the boundary with zero curvature. Then we write the equation for an arbitrary boundary règion in the form :$\Gamma(\vec{x}) = k_1 x_1 + k_2 x_2 = \gamma$ where γ is a constant.If the boundary changes with respect to time it is possible to write the simplest evolutional equation: $\Gamma(t,\vec{x}) = k_1 x_1 + k_2 x_2 - \omega t = \gamma$. In this case the natural "candidate" of a solution is $\cos\Gamma(t,\vec{x})$. The set of γ is $\gamma = \pi/2 + \pi k$ and we have a geometrical structure with the zero curvature. In accordance with our problem formulation we must write the system (41) through the field tensors and substitute representation $D_\mu\cos\Gamma(t,\vec{x})$ instead of D_{1t}, D_{2t}, D_{21}. A nontrivial solution exists on the condition that determinant of the system is equal to zero. It leads to the following relation between parameters:

$$\omega^4 - \omega^2\big\{|\vec{k}|^2(c^2 + c_a^2) + 2\text{æ}c_a^2\big\}$$
$$+(|\vec{k}|^2 + \text{æ})\big\{|\vec{k}|^2 c^2 + 2\text{æ}c_a^2\big\} = 0 \qquad (43)$$

with $\text{æ} = \alpha c^2$. There are roots of the equation (43):

$$\omega_\pm^2 = \frac{1}{2}\Big(|\vec{k}|^2(c^2 + c_a^2) + 2\text{æ}c_a^2 \pm \sqrt{D}\Big), \qquad (44)$$

$$D = |\vec{k}|^4(c^2 - c_a^2)^2 - 4\text{æ}c_a^4(\text{æ} + |\vec{k}|^2).$$

The expression (44) looks like the dispersion equation for waves in the medium with defects. This relation fixes the wavenumber $|\vec{k}|$. The parameter φ defines the direction of $\vec{k}$ which remains arbitrary. It allows us to describe different plane structures. The dispersion equation (44) with the condition of emergence of continuum medium $k_1 x_1 + k_2 x_2 - \omega t = \pi/2 + \pi n$ determines a net consisting of cells. The net is uniform for all frequencies. From (44) we see that there is a complex wavenumber $|\vec{k}|$ for a fixed defect parameter æ. From the physical point of view it corresponds to the existence of absorption linked with the defect structure. In the limit $\text{æ} \to 0$ the dispertion dependence $\omega_- = |\vec{k}|c_a$ is the classical one so ω_- is an acoustic frequency. The second solution $\omega_+ = |\vec{k}|c$ is determined by the defect structure, and it isn't equal to zero.

Let us consider the case of the nonzero curvature structures [6]. We suppose, for simplicity, $c = c_a$. Then from (41) there follows the system of equations for tensors:

$$\partial_{11}D_{1t} - \frac{1}{c^2}\partial_{tt}D_{1t} + \partial_{12}D_{2t} - \partial_{2t}D_{21} = \text{æ}D_{1t},$$

$$\partial_{22}D_{2t} - \frac{1}{c^2}\partial_{tt}D_{2t} + \partial_{12}D_{1t} + \partial_{1t}D_{21} = \text{æ}D_{2t},$$

$$(\partial_{11} + \partial_{22})D_{21} + \frac{1}{c^2}\partial_{2t}D_{1t} - \frac{1}{c^2}\partial_{1t}D_{2t} = \text{æ}D_{21}.$$
$$(45)$$

When $D_{1t} = 0$, $\partial_1 D_{2t} = \partial_t D_{21}$, the system (45) is reduced to the Gordon equation:

$$[\Box + \text{æ}]\begin{Bmatrix} D_{21} \\ D_{2t} \end{Bmatrix} = 0, \qquad (46)$$

where $\Box$ is the wave operator (38). Let us write the field tensor in the form $R(t, x_1, x_2)\cos\Gamma(t, x_1, x_2)$ and substitute this representation into (46). This leads to the following system of equations for determination of $R = R(t, x_1, x_2)$ ($\Gamma = \Gamma(t, x_1, x_2)$:

$$\Box R + \text{æ}R - \varepsilon_\mu(\partial_\mu\Gamma)^2 R = 0, \quad R\Box\Gamma + 2\varepsilon_\mu\partial_\mu R\partial_\mu\Gamma = 0. \qquad (47)$$

We suppose $\Gamma = \Gamma_0\big(x_2 - f(x_1 + vt)\big)$ where Γ_0 and v are the scale coefficients, a function $f = f(x_1 + vt)$ must be found.Factorization of $R(t, x_1, x_2) = R_1(x_1 + vt)R_2(x_2)$ allows to rewrite the system (47) in the form:

$$f'' + \frac{2R_1'}{R_1}f' + \Omega = 0, \qquad 2\gamma\frac{R_2'}{R_2} = -\Omega,$$
$$(48)$$
$$\frac{R_1''}{R_1} + \frac{\Omega^2}{4}\gamma^{-1} - (\Gamma_0^2 + \text{æ})\gamma - \Gamma_0^2(f')^2 = 0,$$

where primes denote differentiation with respect to variables, Ω is a constant, $\gamma^{-1} = 1 - v^2/c^2$. We choose $f' = q$ as an independent variable and introduce $f'' = F(q)$ then from (48) there follows an equation for $F = F(q)$:

$$\frac{FF'}{2q} - \frac{1}{4q^2}\big[3F^2 + 4\Omega F + \Omega^2\big] + \Gamma_0^2 q^2 + A = 0,$$

in which $A = (\Gamma_0^2 + \text{æ})\gamma + \Omega^2\gamma^{-1}/4$.

It is simple to find a solution in the case $\Omega = 0$: $F = q\sqrt{-4\Gamma_0^2 q^2 + F_0 q + 4A}$ with arbitrary constant F_0. Determination of $f = f(x_1 + vt)$ is reduced to calculation of two quadratures. Let's $A > 0$ (it corresponds to the case $v < c$) then

$$f(x_1 + vt) = -\frac{1}{\Gamma_0}$$
$$\text{arctan} \frac{\sqrt{F_0^2 + 64\Gamma_0^2 A}\,\exp(2\sqrt{A}(x_1 + vt)) + F_0}{\sqrt{64\Gamma_0^2 A}}.$$

When $F_0 = 0$ we have:

$$f(x_1 + vt) = \qquad (49)$$
$$-\frac{1}{\Gamma_0}\text{arctan}\exp\left(2\sqrt{\gamma(\Gamma_0^2 + \text{æ})}(x_1 + vt)\right).$$

The equation $\Gamma(t, x_1, x_2) = (2k + 1)\pi/2$ with an integer number k defines a continious structure in the fluid. This structure is the set of curves $x_2 = f(x_1 + vt) + \Gamma_0^{-1}(2k + 1)\pi/2$ and includes the free surface $x_2 = \zeta(x_1, t)$.This result shows that internal defect structure of the fluid can produce isolated waves (49) propagating on the free surface.It is simple to show that $\zeta = \zeta(x_1, t)$ determinated by the formula (49) satisfies $sin - Gordon$ equation

$$\frac{1}{c^2}\partial_{tt}\zeta - \partial_{x_1 x_1}\zeta + \frac{\Gamma_0^2 + \text{æ}}{\Gamma_0}\sin(4\Gamma_0\zeta) = 0. \qquad (50)$$

To satisfy the continuity condition (42) at $x_2 \to -\infty$ on Γ we must choose $\Omega < 0$.If the pertubation theory is correct for small values of Ω then a solution for ζ doesn't differ from (49) very much. Also we can suppose that solutions of the equation (50) define other possible position of the free surface different from (49).

Acknowlegments

This work was supported by the Russian Science Fundamental Foundation, Grant 96-01-00540

REFERENCES

[1] Akulichev V.A. and Ilyichev V.I. *Acoustic cavitation in sea water.* Proceedings of the International Symposium on Propeller and Cavitation, Wuxi, Editorial Office of Shipbuilding of China. Shanghai. 1986.P. 201–205.

[2] Akulichev V.A. *Cavitation nuclei and threshold of acoustic cavitation in sea water.* Proceedings of the 2-nd International Symposium on Cavitation. Edit. by H. Kato. The university of Tokyo. Japan. 1994. P. 343–348.

[3] Kadic A. and Edelen D.G. *A Gauge Theory of Dislocations and Disclinations.* Springler-Verlag. 1983.

[4] Guzev M. A., Myasnikov V. P. *Gauge-invariant hydrodynamics of an ideal fluid* (in Russian). Mekhanika Zhidkosti i Gaza. 1993. 4. P. 25–29.

[5] Myasnikov V. P., Guzev M. A. *Yang-Mills formalism in hydrodynamics of an ideal fluid.* Proceedings of The First Asian CFD Conference. Hong Kong. 1995. Vol. 2. P. 629–634.

[6] Myasnikov V. P., Guzev M. A. *Dissipative properties of the gauge model of an ideal fluid.* Proceedings of the Second Asian Computational Fluid Dynamics Conference.Tokyo.1996. Vol. 1. P. 261-263.

[7]   Koichi Oshima *ASIAN CFD:Where it comes, where it goes.* Proceedings of the Second Asian Computational Fluid Dynamics Conference.Tokyo.1996. Vol.1. P. 1-6.

[8]   Godynov C.K. *Elements of continuum mechanics*(in Russian). Nauka, Moscow. 1978.

[9]   Panin V.E., Grinyaev Yu.V., Danilov V.I.,et.al. *Structural Levels of Plastic Deformation and Fracture* (in Russian).Nauka, Novosibirsk. 1990.

[10]   Olemskoi A.N.,Sklyar I.A. *Evolution of solid defect structure in the process of a plastic deformation*(in Russian). Uspehi Fizicheskih Nauk. 1992. Vol. 162. P. 29–79.

[11]   De Groot S.R., Mazur P. *Non-equilibrium thermodynamics* North-Holland Publishing Company, Amsterdam. 1962.

[12]   Berdichevskii V.L., Sedov L.I.*Dynamical theory of continuosly distributed dislocations. Link with the theory of plasticity*(in Russian). Prikladnay matematika i mekhanika . 1967. Vol. 31. P. 980-1000.

[13]   Ryder L.H. *Quantum Field Theory* Cambridge University Press. 1985.

HIGHER-ORDER DISTRIBUTION FUNCTIONS, BGK-BURNETT EQUATIONS AND BOLTZMANN'S H-THEOREM

Ramesh BALAKRISHNAN (1) Ramesh K. AGARWAL (1)

(1) Graduate Research Assistant

(2) Bloomfield Distinguished Professor and Chair
Department of Aerospace Engineering, Wichita State University;
Wichita, Kansas 67260-0044

1 Introduction

A characteristic of hypersonic flows in the continuum-transition regime is that the shock thickness is of an order greater than or equal to that of the shock standoff distance. In this regime the linear stress-strain and linear heat conduction relationships break down as the Knudsen number increases. Consequently there is a need for higher-order constitutive relations to properly model the stress and heat transfer terms. One of the methods in this direction exploits the Chapman-Enskog asymptotic expansion to devise higher-order distribution functions. This expansion yields the Navier-Stokes equations for first-order departures from collisional equilibrium. In 1935, Burnett developed constitutive relationships for the stress and heat transfer terms by applying the Chapman-Enskog expansion to the Boltzmann equation, for second-order departures from collisional equilibrium. These equations will be referred to as the original Burnett equations. In 1939, Chapman and Cowling replaced the material derivatives in the original Burnett equations by spatial derivatives obtained from inviscid, isentropic approximations. This alternate form of the original Burnett equations will be referred to, henceforth, as the conventional Burnett equations. Expressing the material derivatives in terms of the spatial derivatives was considered acceptable as the Navier-Stokes and Burnett equations were perceived as first and second-order corrections to the Euler equations. The use of Euler equations (inviscid, isentropic approximation) to express the material derivatives retained the second-order accuracy of the Burnett equations. For reasons entirely unknown, the conventional Burnett equations and not the original Burnett equations became the set of higher-order constitutive relations studied during the past six decades.

In recent years Fiscko and Chapman and Zhong et al. (Refs. 1-2) have used the conventional Burnett equations to extend the numerical methods for continuum flow into the continuum-transition regime by incorporating the additional linear and non-linear stress and heat transfer terms in standard Navier-Stokes solvers. In one of the earliest attempts to solve the Burnett equations, Fiscko and Chapman (Ref. 1) solved the hypersonic shock structure problem by relaxing an initial solution to steady state. They obtained solutions for a variety of Mach numbers and concluded that the Burnett equations do indeed describe the normal shock structure better than the Navier-Stokes equations at high Mach numbers. They however experienced stability problems when they made the grids progressively finer. This was predicted by Bobylev (Ref. 3) who showed that the linearized conventional Burnett equations were unstable to small wavelength disturbances. In a subsequent attempt, Zhong (Ref. 2) showed that the Burnett equations could be stabilized by adding linear third order t om -Burne ns, in order to maintain second order accuracy, to th tress and heat transfer terms in the Burnett equations. This set of equations was termed the augmented Burnett equations. The coefficients (weights) of these linear third-order terms were determined by carrying out a linearized stability analysis of the augmented Burnett equation. The augmented Burnett equations did not present any stability problems when they were used to compute the flow parameters in the hypersonic shock structure and hypersonic blunt body flows. However attempts at computing the flowfields for blunt body wakes and flat plate boundary layers even with augmented Burnett equations have not been entirely successful. Further, the ad hoc addition of the linear super-Burnett terms and their necessity raises the question of whether the approximation used to create the conventional Burnett equations from the original Burnett equations introduces the small wavelength instabilities. Welder (Ref. 4) conjectured that an improved approximation for the material derivatives based on the Navier-Stokes equations rather than the

Received on July 1, 1997.

Euler equations might resolve the stability problems associated with the conventional Burnett equations. The linearized stability analysis of the resulting Burnett equations showed that they were unconditionally stable to small disturbances. However attempts to obtain numerical estimates of the shock structure gave rise to an interesting problem. It was observed that the total heat conduction (i.e. the sum of the Fourier and higher-order heat conduction terms) is positive at the leading edge of the shock thereby resulting in heat being conducted both upstream and downstream of the shock. In an attempt to reduce such unphysical heat conduction characteristics it was suggested that a set of modified Navier-Stokes based Burnett equations be used for shock structure computations. It has been noted by Chapman et. al (Refs. 5, 6) that linear stability analysis *alone* is not sufficient to explain the instability of Burnett equations with increasing Knudsen numbers as this analysis does not take into account many non-linear terms, products of first and higher order derivatives, that are present in the conventional Burnett equations. It was also conjectured that this instability may be due to the fact that the conventional Burnett equations violate the second law of thermodynamics at higher Knudsen numbers. In his work Comeaux (Ref. 7) proved that the conventional Burnett equations violate the H-theorem of Boltzmann and hence are not entropy consistent.

In order to gain a better understanding of the Burnett equations and the underlying assumptions it was decided that the equations be derived from the fundamentals. The primary objectives of this paper are:
a) To formulate a methodology, for deriving and integrating a new set of entropy consistent Burnett equations, that can be extended to higher dimensions.
b) To check if the second-order distribution function, that forms the basis for this new set of equations, violates the H-theorem.
c) To check if the constitutive relations for the Burnett stress and heat transfer correctly model the flow properties at high Knudsen numbers.
d) To computationally check if this new set of Burnett equations are stable to small wavelength disturbances.

The highly non-linear nature of the collision integral in the Boltzmann equation presents the biggest hurdle in devising a second order distribution function. This problem is circumvented by representing the collision integral in the Bhatnagar-Gross-Krook (BGK) form. This approximation assumes that any slight departure from the equilibrium distribution will eventually settle down to the equilibrium distribution exponentially. This approximation also assumes that the gas is dilute and hence the collision processes are predominantly binary in nature. Since only binary collisions are considered, the time taken for the non-equilibrium distribution to settle down to the equilibrium level is equal to the reciprocal of the collision frequency. The exact closed form analytical expression for the distribution function (Ref. 8) is derived by considering the first three terms of the Chapman-Enskog expansion. Moments of this distribution function with the collision invariant vector yield the BGK-Burnett equations. In this formulation we assume that the molecules are hard elastic spheres with no intermolecular forces acting between them.

In deriving the second order distribution function, an as yet unanswered question is the approximation for the material derivatives that appear in the second order terms. In a recent attempt (Ref. 9) the first and second order distribution functions were obtained iteratively by perturbation analysis of the 1-D Boltzmann equation. In this analysis the Euler equations were used to approximate the material derivatives in the first order distribution function. Moments of the first order distribution function with the collision invariant vector yielded the Navier-Stokes equations. In order to keep in step with the iterative refinement technique it was conjectured that the Navier-Stokes equations be used to approximate the material derivatives in the second order terms. Hence the second order distribution function has been derived by using the Navier-Stokes equations to express the material derivatives in terms of the spatial derivatives. It has been shown (Ref. 10) that this formulation ensures a positive entropy change. The BGK-Burnett equations are obtained by taking moments of this distribution function with the collision invariant vector. This set of Burnett equations contains all the stress and heat transfer terms reported by Fiscko and Chapman (Ref. 1) and has additional terms which are similar to the super-Burnett terms. Using linearized stability analysis it has been shown that these additional terms make the BGK-Burnett equations unconditionally stable.

In order to check if the entropy production is positive throughout the flowfield the Boltzmann H-theorem has been applied to the second-order distribution function. It has been shown that "H" is a monotonically decreasing function, thereby ensuring that the equations do not violate the second-law of thermodynamics. In order to ensure that the entropy gradient remains positive throughout the flow field, a set of boundary conditions has been derived using the Gibbs entropy equation. On applying this equation to the steady 1-D Navier-Stokes equations, it can be shown that the entropy production rate is always positive provided the stress and heat transfer terms become vanishingly small at stations far upstream and far downstream of the shock. On applying the same methodology to the steady 1-D BGK-Burnett equations, it was found that a positive entropy production can be ensured only by setting the heat transfer terms to zero at stations far upstream and far downstream of the shock. This necessitates setting the first, second and third derivatives of velocity and temperature to zero at the ends of the control volume enclosing the shock.

Kinetic upwind schemes were introduced by Deshpande (Ref. 11) to numerically integrate the Euler equations. In these class of solvers, flux splitting is performed at the Boltzmann level. Acheson et al. (Ref. 12) have shown that a Kinetic Wave/Particle Flux Split algorithm for the Euler equations is robust and computationally efficient. It has been shown by Jameson et al. (Ref. 13) and Perthame et al. (Refs. 14) that it is possible to design kinetic schemes for Euler and Navier-Stokes equations that are less dissipative when compared to the kinetic schemes of Deshpande (Ref. 11) that are based on the collisionless Boltzmann equation. Such kinetic schemes make use of the inherent upwinding property of the molecules. The kinetic theory based upwind schemes for Euler, Navier-Stokes and BGK-Burnett equations can be obtained by taking moments of the upwind differenced BGK-Boltzmann equation. An important feature of these schemes is the ease with which they can be extended to include higher order upwind differences. The novel BGK-Burnett equations are solved using the Kinetic Wave/Particle Split (KWPS) scheme. Numerical computations have been carried out for the 1-D hypersonic shock structure problem to assess the computational accuracy and efficiency of the KWPS upwind algorithm, when applied to the BGK-Burnett equations. The results of the hypersonic shock structure computations indicate that the present formulation ensures a positive entropy change for a range of Knudsen numbers.

2. FIRST-ORDER DISTRIBUTION FUNCTION AND N-S EQUATIONS

A general Maxwellian (equilibrium) distribution function has been derived by Deshpande (Ref. 11). This derivation takes into account the energy contributions due to the non-translational degrees of freedom of the gas molecules. Hence this formulation can in principle be used for any gas. Moments of the collisionless Boltzmann equation with the collision invariant vector give rise to the Euler equations. Owing to the highly non-linear nature of the collision integral a simplified Bhatnagar-Gross-Krook (BGK) model, that makes use of a single relaxation time for the various degrees of freedom, has been used to derive the first-order distribution function. It has been shown by Deshpande (Ref. 11) that moments of the BGK-Boltzmann equation, using the first order distribution function, with the collision invariant vector yield the Navier-Stokes equations.

2.1 BGK Model for the Collision Integral

The 1-D Boltzmann equation is represented as:

$$\frac{\partial f}{\partial t} + v \frac{\partial f}{\partial x} = \left\{ \frac{df}{d\tau} \right\}_{coll} \tag{1}$$

On non-dimensionalizing the above equation by defining the following non-dimensional variables,

$$\overline{x} = \frac{x}{L}, \quad \overline{f} = \frac{f}{f^{(0)}}, \quad \overline{v} = \frac{v}{C_{rms}}, \quad \overline{t} = \frac{t}{t_0} = \frac{C_{rms}t}{L} \quad \text{and}$$

$$\overline{\tau} = \frac{\tau}{\tau_0} = \frac{C_{rms}\tau}{\lambda}, \text{ we obtain}$$

$$\left(\frac{f^{(0)}}{t_0} \right) \frac{\partial \overline{f}}{\partial \overline{t}} + \left(\frac{C_{rms}f^{(0)}}{L} \right) \overline{v} \frac{\partial \overline{f}}{\partial \overline{x}} = \left(\frac{f^{(0)}}{\tau_0} \right) \left\{ \frac{d\overline{f}}{d\overline{\tau}} \right\}_{coll}$$

$$\left(\frac{\lambda}{L} \right) \left[\frac{\partial \overline{f}}{\partial \overline{t}} + \overline{v} \frac{\partial \overline{f}}{\partial \overline{x}} \right] = \left\{ \frac{d\overline{f}}{d\overline{\tau}} \right\}_{coll} \tag{2}$$

where,

$$f^{(0)} = \frac{\rho}{I_0} \left(\frac{\beta}{\pi} \right)^{\frac{D_f}{2}} \exp\left[-\beta(v - u)^2 - \frac{I}{I_0} \right] \tag{3}$$

denotes the Maxwellian distribution function, $C_{rms} = \sqrt{\overline{c^2}} = \sqrt{RT}$ denotes the root mean square molecular velocity, and λ denotes the mean free path.

Using the definition of the Knudsen number $\xi = \frac{\lambda}{L}$ and substituting for the non-dimensional distribution function, eq. (2) takes the form

$$\xi\left[\frac{\partial \bar{f}}{\partial \bar{t}} + \bar{v}\frac{\partial \bar{f}}{\partial \bar{x}}\right] = \left\{\frac{d\bar{f}}{d\bar{\tau}}\right\}_{coll} \tag{4}$$

From eq. (4) we note that $\xi = 0$ implies that $\left\{\dfrac{d\bar{f}}{d\bar{\tau}}\right\}_{coll} = 0$. Hence the number of replenishing collisions equals the number of depleting collisions. Hence $\xi = 0$ means that the kinetic processes are in collision equilibrium. On modeling the collision integral by the equation $\left\{\dfrac{d\bar{f}}{d\bar{\tau}}\right\}_{coll} = \dfrac{\bar{f}^{(0)} - \bar{f}}{\bar{\tau}}$, and

noting that $\bar{f}^{(0)} = 1$, the Boltzmann equation takes the form:

$$\xi\left[\frac{\partial \bar{f}}{\partial \bar{t}} + \bar{v}\frac{\partial \bar{f}}{\partial \bar{x}}\right] = \left\{\frac{d\bar{f}}{d\bar{\tau}}\right\}_{coll} = \frac{1 - \bar{f}}{\bar{\tau}} \tag{5}$$

The above representation for the collision integral in one of the most popular models, known as the BGK model. It assumes a single relaxation time to describe the evolution of the non-equilibrium distribution function. In order to gain some insight into the BGK model we represent the left hand side of the BGK-Boltzmann equation as a material derivative.

$$\xi\left[\frac{\partial \bar{f}}{\partial \bar{t}} + \bar{v}\frac{\partial \bar{f}}{\partial \bar{x}}\right] = \xi\frac{D}{D\bar{t}}\left(\bar{f}\right) = \frac{1 - \bar{f}}{\bar{\tau}} \tag{6}$$

On integrating equation eq. (6)

$$\left(\bar{f} - 1\right) = C_0 \exp\left(-\frac{\bar{t}}{\xi\bar{\tau}}\right) \tag{7}$$

Changing from non-dimensional to dimensional variables the above equation can be written as

$$f - f^{(0)} = C_0 f^{(0)} \exp\left(-\frac{t}{\tau}\right) \tag{8}$$

Defining the collision frequency as the reciprocal of the relaxation time

$$v = \frac{1}{\tau} \tag{9}$$

eq. (8) takes the form

$$f = f^{(0)} + C_0 f^{(0)} \exp\left(-\frac{t}{\tau}\right) \tag{10}$$

In order to fix the constant of integration we note that at t=0, $C_0 f^{(0)} = f - f^{(0)}$.

$$f = f^{(0)} + \left(f - f^{(0)}\right)\exp\left(-vt\right) \tag{11}$$

As the Knudsen number increases the flow remains in the non-equilibrium state for a much longer time. At standard atmospheric conditions the collision frequency is very high. Hence the non-equilibrium distribution function tends towards the equilibrium distribution function much faster. Since the BGK model uses only a single relaxation time it's validity is questionable. However it has been shown by Woods (Ref. 15) that this model is accurate up to the second-order, for Knudsen numbers less than unity.

2.2 First-Order Distribution Function

The first-order distribution function is obtained by considering the first two terms in the Chapman-Enskog expansion. The 1-D Boltzmann equation can be written as follows, using the BGK approximation for the collision integral J(f,f).

$$\frac{\partial f}{\partial t} + v\frac{\partial f}{\partial x} = J(f,f) = v\left(f^{(0)} - f\right) \tag{12}$$

The variables in eq. (12) are non-dimensionalized as:

$$\hat{x} = \frac{x}{L}, \quad \hat{v} = \frac{v}{C_{rms}}, \quad \hat{t} = \frac{tC_{rms}}{L}, \quad \hat{v} = \frac{v\lambda_\infty}{C_{rms}} \tag{13}$$

On substituting the non-dimensional variables, eq. (12) takes the following form:

$$\xi\left[\frac{\partial f}{\partial \hat{t}} + \hat{v}\frac{\partial f}{\partial \hat{x}}\right] = \hat{v}\left(f^{(0)} - f\right) \tag{14}$$

The Chapman-Enskog expansion for the first order distribution function is of the form $f = f^{(0)} + \xi f^{(1)}$, where $f^{(0)}$ denotes the equilibrium distribution function, $f^{(1)}$ denotes the first order term in the distribution function. Introducing this form of the distribution function into the non-dimensional 1-D BGK-Boltzmann equation and equating like powers of ξ yields:

$$\frac{\partial}{\partial \hat{t}}\left(f^{(0)}\right) + \hat{v}\frac{\partial}{\partial \hat{x}}\left(f^{(0)}\right) = -\hat{v}f^{(1)} \tag{15}$$

From the above equation the following equation for $f^{(1)}$ is obtained:

$$f^{(1)} = -\left(\frac{1}{\xi v}\right)\left[\frac{\partial}{\partial t}\left(f^{(0)}\right) + v\frac{\partial}{\partial x}\left(f^{(0)}\right)\right] \tag{16}$$

On expressing $f^{(1)} = f^{(0)}\Phi^{(1)}$ and substituting in eq. (16) the expression for $\Phi^{(1)}$ is obtained as shown:

$$\Phi^{(1)} = -\frac{1}{\xi v}\left[\frac{5c}{2\beta} - \frac{4(\gamma - 1)}{(3 - \gamma)}Ic - c^3\right]\frac{\partial \beta}{\partial x}$$

$$-\frac{1}{\xi v}\left[\beta(3 - \gamma)c^2 - 4\beta I\frac{(\gamma - 1)^2}{(3 - \gamma)} + \frac{(3\gamma - 5)}{2}\right]\frac{\partial u}{\partial x} \tag{17}$$

Details of the derivation are given in (Ref. 10). The moment of the distribution function 'f' is defined as

$$\langle \Psi, f \rangle = \int\limits_{0}^{\infty} \int\limits_{-\infty}^{\infty} \Psi f \, dv \, dI \qquad (18)$$

where, the collision invariant vector is defined as

$$\Psi = \left[1, v, \left(I + \frac{v^2}{2} \right) \right]^T \qquad (19)$$

The first order term satisfies the moment property $\left\langle \Psi, f^{(0)} \Phi^{(1)} \right\rangle = 0$. On taking moments of the BGK-Boltzmann equation, the following generic equation is obtained.

$$\frac{\partial}{\partial t} \langle \Psi, f \rangle + \frac{\partial}{\partial x} \langle \Psi, vf \rangle = \left\langle \Psi, v\left(f^{(0)} - f \right) \right\rangle = 0 \qquad (20)$$

On substituting the first-order distribution function the Navier-Stokes equations are obtained.

$$\frac{\partial}{\partial t} \left[\int\limits_{0}^{\infty} \int\limits_{-\infty}^{\infty} f\Psi \, dvdI \right] + \frac{\partial}{\partial x} \left[\int\limits_{0}^{\infty} \int\limits_{-\infty}^{\infty} vf\Psi \, dvdI \right] \qquad (21)$$
$$= v \int\limits_{0}^{\infty} \int\limits_{-\infty}^{\infty} \Psi \left(f^0 - f \right) dvdI = 0$$

Moments of the BGK-Boltzmann equation with Ψ using the first-order distribution function are given by the following equation

$$\frac{\partial}{\partial t} \left\langle \Psi, f^{(0)} + \xi f^{(1)} \right\rangle + \frac{\partial}{\partial x} \left\langle \Psi, v\left(f^{(0)} + \xi f^{(1)} \right) \right\rangle$$
$$= \left\langle \Psi, v\left(f^{(0)} - f \right) \right\rangle = 0 \qquad (22)$$

The 1-D Navier-Stokes equations are represented in conservation law form as:

$$\frac{\partial \mathbf{Q}}{\partial t} + \frac{\partial \mathbf{G}^i}{\partial x} + \frac{\partial \mathbf{G}^v}{\partial x} = 0 \qquad (23)$$

The elements of the field and flux vectors are:

$$\mathbf{Q} = \begin{bmatrix} \rho \\ \rho u \\ \rho e \end{bmatrix}, \qquad (24a)$$

$$\mathbf{G}^i = \begin{bmatrix} \rho u \\ p + \rho u^2 \\ pu + \rho ue \end{bmatrix}, \qquad (24b)$$

$$\mathbf{G}^v = \begin{bmatrix} 0 \\ -\tau_{xx}^{N-S} \\ -u\tau_{xx}^{N-S} + q_x^{N-S} \end{bmatrix} \qquad (24c)$$

The Navier-Stokes stress and heat transfer terms are given by the following expressions:

$$\tau_x^{N-S} = (3 - \gamma)\mu \frac{\partial u}{\partial x}, \text{ and } q^{N-S} = -k\frac{\partial T}{\partial x}$$

In obtaining the above relations the following equation has been used $\mu = \dfrac{P}{v}$. An interesting spin off from this formulation is a formal verification of Stokes hypothesis. A generalized equation relating the molecular and bulk viscosities is given by

$$\boxed{\widetilde{\mu} + (\gamma - 1)\mu = 0} \qquad (25)$$

From the above expression we note that the familiar form of the Stokes hypothesis

$$\boxed{\widetilde{\mu} + \frac{2}{3}\mu = 0} \qquad (26)$$

is valid only for mono-atomic gases !

3. SECOND-ORDER DISTRIBUTION FUNCTIONS AND BGK-BURNETT EQUATIONS

A closed form expression for the second-order distribution function is obtained by considering the first three terms of the Chapman-Enskog, asymptotic, expansion. This distribution function is used to obtain the 1-D BGK-Burnett equations. The problem of closure of the second-order distribution functions and the resulting non-uniqueness of the BGK-Burnett equations is explained. The linearized stability analysis has been applied to the BGK-Burnett equations to prove that the Navier-Stokes equations must be used to express the material derivatives in terms of the spatial derivatives in order to guarantee unconditional stability.

3.1 Second-Order Distribution Function

The second-order distribution function is obtained by considering the first three terms in the Chapman-Enskog expansion. The Chapman-Enskog expansion for the second order distribution function is of the form:

$f = f^{(0)} + \xi f^{(1)} + \xi^2 f^{(2)}$, where $f^{(0)}$ denotes the equilibrium distribution function, $f^{(1)}$ denotes the first order term in the distribution function and $f^{(2)}$ denotes the second order term. Introducing this form of the distribution function into the non-dimensional 1-D Boltzmann equation and equating like powers of ξ yields:

$$\frac{\partial}{\partial \hat{t}}\left(f^{(1)}\right) + \hat{v}\frac{\partial}{\partial \hat{x}}\left(f^{(1)}\right) = -\hat{v}f^{(2)} \qquad (27)$$

From the above equation the following equation for $f^{(2)}$ is obtained:

$$f^{(2)} = -\left(\frac{1}{\xi v}\right)\left[\frac{\partial}{\partial t}\left(f^{(1)}\right) + v\frac{\partial}{\partial x}\left(f^{(1)}\right)\right] \qquad (28)$$

The above equation forms the basis for obtaining an expression for the second-order distribution function.

Various methods have been devised to evaluate $f^{(2)}$. These methods make use of the physical premise that the field vector $\mathbf{Q}$ is the same, for the Euler equations, Navier-Stokes equations and Burnett equations. Accordingly the moments of the distribution function with the collision invariant vector Ψ must be the same irrespective of the form of the distribution function. Consequently, moments of the higher-order terms in the distribution function with the collision invariant vector must be zero. It has been shown that the moment of the first order term with the collision invariant vector is zero. Likewise, moments of the second order term in the distribution with the collision invariant vector **must** be equal to zero. Mathematically speaking this requires that the following equation be satisfied:

$$\left\langle \Psi, \xi^2 f^{(2)}\right\rangle$$
$$= \left\langle \Psi, -\left(\frac{\xi^2}{\xi v}\right)\left[\frac{\partial}{\partial t}\left(f^{(1)}\right) + v\frac{\partial}{\partial x}\left(f^{(1)}\right)\right]\right\rangle = 0 \qquad (29)$$

The above equation implies that

$$\left[\frac{\partial}{\partial t}\left(\left\langle \Psi, f^{(1)}\right\rangle\right) + \frac{\partial}{\partial x}\left(\left\langle \Psi, vf^{(1)}\right\rangle\right)\right] \equiv 0 \qquad (30)$$

It can be seen that the above equation cannot be satisfied as $\left\langle \Psi, vf^{(1)}\right\rangle$ equals the viscous flux terms in the Navier-Stokes equations. It is precisely this lack of closure[1] that necessitates the introduction of

---

[1] The problem of closure is present even in the original formulation of the Burnett equations. Burnett had used Sonine polynomials to evaluate the Burnett stress and heat transfer coefficients.

certain additional terms that force the second-order term to yield a zero moment with the collision invariant vector. The "weights" of these additional terms give rise to the BGK-Burnett coefficients. In the following sections two formulations have been considered to evaluate the second-order distribution function. It will be shown that the BGK-Burnett equations resulting from each of these formulations exhibit *similar* stability characteristics, although the weights and their distribution are very different. This non-uniqueness in the representation of the BGK-Burnett equations results from the lack of closure.

3.1.1 Method One

Expressing $f^{(2)}$ in terms of $f^{(1)}$, in a manner similar to the derivation of the first-order distribution function, results in the following expression.

$$f^{(2)} = \Phi^{(2)}f^{(1)} = \Phi^{(2)}\Phi^{(1)}f^{(0)} \qquad (31)$$

On substituting this expression for $f^{(2)}$, the following expression for $\Phi^{(2)}$ is obtained.

$$\Phi^{(2)}$$
$$= \Phi^{(1)} - \frac{1}{\xi v}\left[\frac{\partial}{\partial t}\left(\ln\left(\Phi^{(1)}\right)\right) + v\frac{\partial}{\partial x}\left(\ln\left(\Phi^{(1)}\right)\right)\right] \qquad (32)$$

On substituting for $\Phi^{(2)}$ in the Chapman-Enskog expansion for the distribution function, the following expression is obtained after some algebraic manipulation.

$$f = f^{(0)}\left[1 + \xi\Phi^{(1)} + \xi^2\left\{\Phi^{(1)}\right\}^2\right]$$
$$- \frac{\xi^2 f^{(0)}}{\xi v}\left[\frac{\partial}{\partial t}\left(\Phi^{(1)}\right) + v\frac{\partial}{\partial x}\left(\Phi^{(1)}\right)\right] \qquad (33)$$

In the above expression, the last term involving the partial differential equation in $\Phi^{(1)}$ needs to be evaluated. The following equation must be satisfied

$$\left\langle \Psi, f^{(0)}\left\{\left(\Phi^{(1)}\right)^2 + \Theta\right\}\right\rangle$$
$$= \int_0^\infty \int_{-\infty}^\infty \Psi f^{(0)}\left\{\left(\Phi^{(1)}\right)^2 + \Theta\right\}dvdI \qquad (34)$$
$$= 0$$

where,

$$\Theta = -\frac{1}{\xi v}\left\{\frac{\partial}{\partial t}\left(\Phi^{(1)}\right) + v\frac{\partial}{\partial x}\left(\Phi^{(1)}\right)\right\} \qquad (35)$$

In the following method, the term ' Θ ' is evaluated. From the first-order distribution function an expression for $\left(\Phi^{(1)}\right)^2$ is obtained in the form shown below:

$$\left(\Phi^{(1)}\right)^2 = \frac{1}{(\xi v)^2}\left[\begin{array}{l}\{Term_1\}\left(\frac{\partial\beta}{\partial x}\right)^2 \\ +\{Term_2\}\left(\frac{\partial u}{\partial x}\right)^2 \\ +\{Term_3\}\left(\frac{\partial u}{\partial x}\right)\left(\frac{\partial\beta}{\partial x}\right)\end{array}\right] \qquad (36)$$

where,

$$\{Term_1\} = \left\{\begin{array}{l}\left(\frac{25}{4}\right)\frac{c^2}{\beta^2} + 16\frac{(\gamma-1)^2}{(3-\gamma)^2}I^2c^2 \\ -20\frac{(\gamma-1)}{(3-\gamma)}I\frac{c^2}{\beta} + 8\frac{(\gamma-1)}{(3-\gamma)}Ic^4 \\ -5\frac{c^4}{\beta} + c^6\end{array}\right\}$$

$$\{Term_2\} = \left\{\begin{array}{l}(3-\gamma)^2\beta^2c^4 + (3\gamma-5)(3-\gamma)\beta c^2 \\ -8(\gamma-1)^2\beta^2Ic^2 + 16\frac{(\gamma-1)^4}{(3-\gamma)^2}\beta^2I^2 \\ -4\frac{(3\gamma-5)(\gamma-1)^2}{(3-\gamma)}\beta I + \frac{(3\gamma-5)^2}{4}\end{array}\right\}$$

$$\{Term_3\} = \left\{\begin{array}{l}\frac{5}{2}(3\gamma-5)\frac{c}{\beta} + 32\frac{(\gamma-1)^3}{(3-\gamma)^2}\beta I^2c \\ -8\frac{(\gamma-1)(4\gamma-5)}{(3-\gamma)}Ic \\ +16\frac{(\gamma-1)(\gamma-2)}{(3-\gamma)}\beta Ic^3 \\ +4(5-2\gamma)c^3 - 2(3-\gamma)\beta c^5\end{array}\right\}$$

On differentiating the above expression with respect to 't' and 'x' and writing the resulting expression in terms of the material derivative results in the following equations. The Euler or Navier-Stokes equations can be been used, subsequently, to express the material derivatives in terms of the spatial derivatives.

$$\frac{\partial\Phi^{(1)}}{\partial t} =$$

$$-\frac{1}{(\xi v)}\left\{\frac{5c}{2\beta} - 4\frac{(\gamma-1)}{(3-\gamma)}Ic - c^3\right\}\left(\frac{\partial^2\beta}{\partial t\partial x}\right)$$

$$-\frac{1}{(\xi v)}\left\{\begin{array}{l}(3-\gamma)\beta c^2 - 4\frac{(\gamma-1)^2}{(3-\gamma)}\beta I \\ +\frac{(3\gamma-5)}{2}\end{array}\right\}\left(\frac{\partial^2 u}{\partial t\partial x}\right)$$

$$+\frac{1}{(\xi v)}\left\{\frac{5c}{2\beta^2}\right\}\left(\frac{\partial\beta}{\partial x}\right)\left(\frac{\partial\beta}{\partial t}\right) \qquad (37)$$

$$-\frac{1}{(\xi v)}\left\{(3-\gamma)c^2 - 4\frac{(\gamma-1)^2}{(3-\gamma)}I\right\}\left(\frac{\partial u}{\partial x}\right)\left(\frac{\partial\beta}{\partial t}\right)$$

$$-\frac{1}{(\xi v)}\left\{-\frac{5}{2\beta} + 4\frac{(\gamma-1)}{(3-\gamma)}I + 3c^2\right\}\left(\frac{\partial\beta}{\partial x}\right)\left(\frac{\partial u}{\partial t}\right)$$

$$+\frac{1}{(\xi v)}\left\{2(3-\gamma)\beta c\right\}\left(\frac{\partial u}{\partial x}\right)\left(\frac{\partial u}{\partial t}\right)$$

On splitting the molecular velocity as the sum of the fluid and peculiar velocities (v=u+c):

$$v\frac{\partial\Phi^{(1)}}{\partial x} = (u+c)\frac{\partial\Phi^{(1)}}{\partial x}$$

$$-\frac{1}{(\xi v)}\left\{\frac{5c^2}{2\beta} - 4\frac{(\gamma-1)}{(3-\gamma)}Ic^2 - c^4\right\}\left(\frac{\partial^2\beta}{\partial x^2}\right)$$

$$-\frac{1}{(\xi v)}\left\{\begin{array}{l}(3-\gamma)\beta c^3 - 4\frac{(\gamma-1)^2}{(3-\gamma)}\beta Ic \\ +\frac{(3\gamma-5)}{2}c\end{array}\right\}\left(\frac{\partial^2 u}{\partial x^2}\right)$$

$$+\frac{1}{(\xi v)}\left\{\frac{5c^2}{2\beta^2}\right\}\left(\frac{\partial\beta}{\partial x}\right)^2 \qquad (38)$$

$$-\frac{1}{(\xi v)}\left\{(3-\gamma)c^3 - 4\frac{(\gamma-1)^2}{(3-\gamma)}Ic\right\}\left(\frac{\partial u}{\partial x}\right)\left(\frac{\partial\beta}{\partial x}\right)$$

$$-\frac{1}{(\xi v)}\left\{-\frac{5c}{2\beta} + 4\frac{(\gamma-1)}{(3-\gamma)}Ic + 3c^3\right\}\left(\frac{\partial\beta}{\partial x}\right)\left(\frac{\partial u}{\partial x}\right)$$

$$+\frac{1}{(\xi v)}\left\{2(3-\gamma)\beta c^2\right\}\left(\frac{\partial u}{\partial x}\right)^2$$

$$-\frac{u}{(\xi v)}\left\{\frac{5c}{2\beta} - 4\frac{(\gamma-1)}{(3-\gamma)}Ic - c^3\right\}\left(\frac{\partial^2\beta}{\partial x^2}\right)$$

$$-\frac{u}{(\xi v)}\left\{(3-\gamma)\beta c^2 - 4\frac{(\gamma-1)^2}{(3-\gamma)}\beta I + \frac{(3\gamma-5)}{2}\right\}\left(\frac{\partial^2 u}{\partial x^2}\right)$$

$$+\frac{u}{(\xi v)}\left\{\frac{5c}{2\beta^2}\right\}\left(\frac{\partial\beta}{\partial x}\right)^2 \qquad (38)$$

$$-\frac{u}{(\xi v)}\left\{(3-\gamma)c^2 - 4\frac{(\gamma-1)^2}{(3-\gamma)}I\right\}\left(\frac{\partial u}{\partial x}\right)\left(\frac{\partial\beta}{\partial x}\right)$$

$$-\frac{u}{(\xi v)}\left\{-\frac{5}{2\beta} + 4\frac{(\gamma-1)}{(3-\gamma)}I + 3c^2\right\}\left(\frac{\partial\beta}{\partial x}\right)\left(\frac{\partial u}{\partial x}\right)$$

$$+\frac{u}{(\xi v)}\left\{2(3-\gamma)\beta c\right\}\left(\frac{\partial u}{\partial x}\right)^2$$

$$\Theta = -\frac{1}{(\xi v)}\left\{\frac{\partial\Phi^{(1)}}{\partial t} + v\frac{\partial\Phi^{(1)}}{\partial x}\right\} =$$

$$\frac{1}{(\xi v)^2}\left\{\frac{5c}{2\beta} - 4\frac{(\gamma-1)}{(3-\gamma)}Ic - c^3\right\}\frac{D}{Dt}\left(\frac{\partial\beta}{\partial x}\right)$$

$$+\frac{1}{(\xi v)^2}\left\{(3-\gamma)\beta c^2 - 4\frac{(\gamma-1)^2}{(3-\gamma)}\beta I + \frac{(3\gamma-5)}{2}\right\}\frac{D}{Dt}\left(\frac{\partial u}{\partial x}\right)$$

$$-\frac{1}{(\xi v)^2}\left\{\frac{5c}{2\beta^2}\right\}\left(\frac{\partial\beta}{\partial x}\right)\left(\frac{D\beta}{Dt}\right) \qquad (39)$$

$$+\frac{1}{(\xi v)^2}\left\{(3-\gamma)c^2 - 4\frac{(\gamma-1)^2}{(3-\gamma)}I\right\}\left(\frac{\partial u}{\partial x}\right)\left(\frac{D\beta}{Dt}\right)$$

$$+\frac{1}{(\xi v)^2}\left\{-\frac{5}{2\beta} + 4\frac{(\gamma-1)}{(3-\gamma)}I + 3c^2\right\}\left(\frac{\partial\beta}{\partial x}\right)\left(\frac{Du}{Dt}\right)$$

$$-\frac{1}{(\xi v)^2}\left\{2(3-\gamma)\beta c\right\}\left(\frac{\partial u}{\partial x}\right)\left(\frac{Du}{Dt}\right)$$

$$+\frac{1}{(\xi v)^2}\left\{\frac{5c^2}{2\beta} - 4\frac{(\gamma-1)}{(3-\gamma)}Ic^2 - c^4\right\}\left(\frac{\partial^2\beta}{\partial x^2}\right)$$

$$+\frac{1}{(\xi v)^2}\left\{(3-\gamma)\beta c^3 - 4\frac{(\gamma-1)^2}{(3-\gamma)}\beta Ic + \frac{(3\gamma-5)}{2}c\right\}\left(\frac{\partial^2 u}{\partial x^2}\right)$$

$$-\frac{1}{(\xi v)^2}\left\{\frac{5c^2}{2\beta^2}\right\}\left(\frac{\partial\beta}{\partial x}\right)^2$$

$$+\frac{1}{(\xi v)^2}\left\{(3-\gamma)c^3 - 4\frac{(\gamma-1)^2}{(3-\gamma)}Ic\right\}\left(\frac{\partial u}{\partial x}\right)\left(\frac{\partial\beta}{\partial x}\right) \qquad (39)$$

$$+\frac{1}{(\xi v)^2}\left\{-\frac{5c}{2\beta} + 4\frac{(\gamma-1)}{(3-\gamma)}Ic + 3c^3\right\}\left(\frac{\partial\beta}{\partial x}\right)\left(\frac{\partial u}{\partial x}\right)$$

$$-\frac{1}{(\xi v)^2}\left\{2(3-\gamma)\beta c^2\right\}\left(\frac{\partial u}{\partial x}\right)^2$$

It can be verified that

$$\left\langle \Psi, f^{(0)}\left\{\left(\Phi^{(1)}\right)^2 + \Theta\right\}\right\rangle$$

$$= \int_0^\infty\int_{-\infty}^\infty \Psi f^{(0)}\left\{\left(\Phi^{(1)}\right)^2 + \Theta\right\}dvdI \neq 0 \qquad (40)$$

Hence we need to introduce additional terms with undetermined coefficients, so that eq.(40) is satisfied. The derivatives associated with these additional terms are assumed to have the same form as those in the expressions for $\left(\Phi^{(1)}\right)^2$ and Θ. On equating the moments of the resulting expression with the collision invariant vector to zero a set of simultaneous equations are obtained which on solving yields expressions for the undetermined coefficients that are functions of the specific heat ratio 'γ'. Hence these coefficients can be evaluated for any gas without the need for any kind of interpolation[2] ! These expressions are listed in (Ref. 18).

---

[2] This contrasts sharply with the earlier approaches of Zhong et. al where the Burnett coefficients for Argon were evaluated by linearly interpolating between the known coefficients for Hard Sphere and Maxwellian gases.

$$\left\{\left(\Phi^{(1)}\right)^2 + \Theta\right\} =$$

$$\frac{1}{(\xi v)^2}\left\{\frac{5c}{2\beta} - 4\frac{(\gamma-1)}{(3-\gamma)}Ic - c^3\right\}\frac{D}{Dt}\left(\frac{\partial\beta}{\partial x}\right)$$

$$+\frac{1}{(\xi v)^2}\left\{\begin{array}{l}(3-\gamma)\beta c^2 - 4\frac{(\gamma-1)^2}{(3-\gamma)}\beta I \\[4pt] +\frac{(3\gamma-5)}{2}\end{array}\right\}\frac{D}{Dt}\left(\frac{\partial u}{\partial x}\right)$$

$$+\frac{1}{(\xi v)^2}\left\{-\frac{5c}{2\beta^2} + \omega_3\frac{c^3}{\beta}\right\}\left(\frac{\partial\beta}{\partial x}\right)\left(\frac{D\beta}{Dt}\right)$$

$$+\frac{1}{(\xi v)^2}\left\{\begin{array}{l}(3-\gamma)(1-\omega_4)c^2 \\[4pt] +(\omega_5-4)\frac{(\gamma-1)^2}{(3-\gamma)}I\end{array}\right\}\left(\frac{\partial u}{\partial x}\right)\left(\frac{D\beta}{Dt}\right)$$

$$+\frac{1}{(\xi v)^2}\left\{\begin{array}{l}-\frac{5}{2\beta} + (4-\omega_2)\frac{(\gamma-1)}{(3-\gamma)}I \\[4pt] +(3-\omega_1)c^2\end{array}\right\}\left(\frac{\partial\beta}{\partial x}\right)\left(\frac{Du}{Dt}\right)$$

$$+\frac{1}{(\xi v)^2}\left\{-2(3-\gamma)\beta c + \omega_6\beta^2(3-\gamma)c^3\right\}\left(\frac{\partial u}{\partial x}\right)\left(\frac{Du}{Dt}\right)$$

$$+\frac{1}{(\xi v)^2}\left\{\begin{array}{l}\left(\frac{5}{2}-\alpha_1\right)\frac{c^2}{\beta} - 4\frac{(\gamma-1)}{(3-\gamma)}Ic^2 \\[4pt] +(\alpha_2-1)c^4\end{array}\right\}\left(\frac{\partial^2\beta}{\partial x^2}\right)$$

$$+\frac{1}{(\xi v)^2}\left\{\begin{array}{l}(3-\gamma)\beta c^3 + (\alpha_3-4)\frac{(\gamma-1)^2}{(3-\gamma)}\beta Ic \\[4pt] +\frac{(3\gamma-5)}{2}c\end{array}\right\}\left(\frac{\partial^2 u}{\partial x^2}\right)$$

$$+\frac{1}{(\xi v)^2}\left\{\begin{array}{l}\left(\frac{15}{4}-A_0\right)\frac{c^2}{\beta^2} + 16\frac{(\gamma-1)^2}{(3-\gamma)^2}I^2c^2 \\[4pt] -20\frac{(\gamma-1)}{(3-\gamma)}I\frac{c^2}{\beta} + 8\frac{(\gamma-1)}{(3-\gamma)}Ic^4 \\[4pt] +(A_1-5)\frac{c^4}{\beta} + c^6\end{array}\right\}\left(\frac{\partial\beta}{\partial x}\right)^2$$

$$+\frac{1}{(\xi v)^2}\left\{\begin{array}{l}(3-\gamma)^2\beta^2c^4 \\[4pt] +(1-B_0)(3\gamma-5)(3-\gamma)\beta c^2 \\[4pt] -8(\gamma-1)^2\beta^2Ic^2 - 2(3-\gamma)\beta c^2 \\[4pt] +16\frac{(\gamma-1)^4}{(3-\gamma)^2}\beta^2I^2 \\[4pt] +(B_1-4)\frac{(3\gamma-5)(\gamma-1)^2}{(3-\gamma)}\beta I \\[4pt] +\frac{(3\gamma-5)^2}{4}\end{array}\right\}\left(\frac{\partial u}{\partial x}\right)^2$$

$$+\frac{1}{(\xi v)^2}\left\{\begin{array}{l}\frac{15}{2}(\gamma-2)\frac{c}{\beta} \\[4pt] +(32-C_0)\frac{(\gamma-1)^3}{(3-\gamma)^2}\beta I^2c \\[4pt] +12\frac{(\gamma-1)(4-3\gamma)}{(3-\gamma)}Ic \\[4pt] +16\frac{(\gamma-1)(\gamma-2)}{(3-\gamma)}\beta Ic^3 \\[4pt] -(9\gamma-26)c^3 - 2(3-\gamma)\beta c^5\end{array}\right\}\left(\frac{\partial u}{\partial x}\right)\left(\frac{\partial\beta}{\partial x}\right)$$

$$(41)$$

The above expressions together with the first-order distribution function gives the second-order distribution function.

$$f = f^{(0)}\left[1 + \xi\Phi^{(1)} + \xi^2\left\{\Phi^{(1)}\right\}^2\right]$$
$$-\frac{\xi^2 f^{(0)}}{\xi v}\left[\frac{\partial}{\partial t}\left(\Phi^{(1)}\right) + v\frac{\partial}{\partial x}\left(\Phi^{(1)}\right)\right] \tag{42}$$

Moments of the BGK-Boltzmann equation, using the second-order distribution function, with the collision invariant vector yield the BGK-Burnett equations.

3.1.2 BGK-Burnett Equations (One)

The BGK-Burnett equations are derived by taking moments of the BGK-Boltzmann equation with the collision invariant vector. The second order distribution function derived in ᶓ 3.1.1 is substituted in the BGK-Boltzmann equation and moments with

with the collision invariant vector are evaluated as shown.

$$\frac{\partial}{\partial t}\left\langle \Psi, f^{(0)} + \xi f^{(1)} + \xi^2 f^{(2)} \right\rangle$$

$$+\frac{\partial}{\partial x}\left\langle \Psi, v\left(f^{(0)} + \xi f^{(1)} + \xi^2 f^{(2)} \right) \right\rangle \qquad (43)$$

$$=\left\langle \Psi, v\left(f^{(0)} - f \right) \right\rangle = 0$$

The 1-D BGK-Burnett equations are represented in conservation law form as:

$$\frac{\partial \mathbf{Q}}{\partial t} + \frac{\partial \mathbf{G}^i}{\partial x} + \frac{\partial \mathbf{G}^v}{\partial x} + \frac{\partial \mathbf{G}^B}{\partial x} = 0 \qquad (44)$$

The BGK-Burnett flux vector is given by:

$$\mathbf{G^B} = \begin{bmatrix} 0 \\[4pt] \begin{aligned} &\theta_0 \frac{\rho}{\beta^4 v^2}\left(\frac{\partial \beta}{\partial x}\right)^2 + \theta_1 \frac{\rho}{\beta v^2}\left(\frac{\partial u}{\partial x}\right)^2 \\ &+\theta_2 \frac{\rho}{\beta^2 v^2}\left(\frac{\partial \beta}{\partial x}\right)\frac{Du}{Dt} + \theta_3 \frac{\rho}{\beta v^2}\frac{D}{Dt}\left(\frac{\partial u}{\partial x}\right) \\ &+\theta_4 \frac{\rho}{\beta^2 v^2}\left(\frac{\partial u}{\partial x}\right)\frac{D\beta}{Dt} + \theta_5 \frac{\rho}{\beta^3 v^2}\left(\frac{\partial^2 \beta}{\partial x^2}\right) \end{aligned} \\[10pt] \begin{aligned} &\theta_0 \frac{\rho u}{\beta^4 v^2}\left(\frac{\partial \beta}{\partial x}\right)^2 + \theta_1 \frac{\rho u}{\beta v^2}\left(\frac{\partial u}{\partial x}\right)^2 \\ &+\theta_2 \frac{\rho u}{\beta^2 v^2}\left(\frac{\partial \beta}{\partial x}\right)\frac{Du}{Dt} + \theta_3 \frac{\rho u}{\beta v^2}\frac{D}{Dt}\left(\frac{\partial u}{\partial x}\right) \\ &+\theta_4 \frac{\rho u}{\beta^2 v^2}\left(\frac{\partial u}{\partial x}\right)\frac{D\beta}{Dt} + \theta_5 \frac{\rho u}{\beta^3 v^2}\left(\frac{\partial^2 \beta}{\partial x^2}\right) \\ &+\theta_6 \frac{\rho}{\beta^3 v^2}\frac{D}{Dt}\left(\frac{\partial \beta}{\partial x}\right) + \theta_7 \frac{\rho}{\beta^4 v^2}\left(\frac{\partial \beta}{\partial x}\right)\frac{D\beta}{Dt} \\ &+\theta_8 \frac{\rho}{\beta v^2}\left(\frac{\partial u}{\partial x}\right)\frac{Du}{Dt} + \theta_9 \frac{\rho}{\beta^2 v^2}\left(\frac{\partial^2 u}{\partial x^2}\right) \\ &+\theta_{10}\frac{\rho}{\beta^3 v^2}\left(\frac{\partial u}{\partial x}\right)\left(\frac{\partial \beta}{\partial x}\right) \end{aligned} \end{bmatrix} \qquad (45)$$

Generalized expressions for $\theta_i\,(\gamma)$ are given in the (Ref. 18).

3.1.3 BGK-BURNETT Stress and Heat Transfer Terms (One)

In order to gain some physical insight into the various terms in the BGK-Burnett flux vector they are split into the corresponding stress and heat transfer terms. Hence the BGK-Burnett flux vector can be written in terms of the stress and heat transfer as shown in eq. (46).

$$\mathbf{G^B} = \begin{bmatrix} 0 \\ -\tau_{xx}^{B} \\ -u\tau_{xx}^{B} + q_x^{B} \end{bmatrix} \qquad (46)$$

$$\begin{aligned} \tau_{xx}^{B} =&-\theta_0 \frac{\rho}{\beta^4 v^2}\left(\frac{\partial \beta}{\partial x}\right)^2 - \theta_1 \frac{\rho}{\beta v^2}\left(\frac{\partial u}{\partial x}\right)^2 \\ &-\theta_2 \frac{\rho}{\beta^2 v^2}\left(\frac{\partial \beta}{\partial x}\right)\frac{Du}{Dt} - \theta_3 \frac{\rho}{\beta v^2}\frac{D}{Dt}\left(\frac{\partial u}{\partial x}\right) \\ &-\theta_4 \frac{\rho}{\beta^2 v^2}\left(\frac{\partial u}{\partial x}\right)\frac{D\beta}{Dt} - \theta_5 \frac{\rho}{\beta^3 v^2}\left(\frac{\partial^2 \beta}{\partial x^2}\right) \end{aligned} \qquad (47)$$

$$\begin{aligned} q_x^{B} =&\,\theta_6 \frac{\rho}{\beta^3 v^2}\frac{D}{Dt}\left(\frac{\partial \beta}{\partial x}\right) + \theta_7 \frac{\rho}{\beta^4 v^2}\left(\frac{\partial \beta}{\partial x}\right)\frac{D\beta}{Dt} \\ &+\theta_8 \frac{\rho}{\beta v^2}\left(\frac{\partial u}{\partial x}\right)\frac{Du}{Dt} + \theta_9 \frac{\rho}{\beta^2 v^2}\left(\frac{\partial^2 u}{\partial x^2}\right) \\ &+\theta_{10}\frac{\rho}{\beta^3 v^2}\left(\frac{\partial u}{\partial x}\right)\left(\frac{\partial \beta}{\partial x}\right) \end{aligned} \qquad (48)$$

Expressing the material derivatives in eq.(47) and eq.(48) in terms of the spatial derivatives:

$$\begin{aligned} \tau_{xx}^{B} = -\frac{\mu^2}{p}&\begin{bmatrix} 2(\theta_1 - \theta_3)\left(\frac{\partial u}{\partial x}\right)^2 + 2\theta_3 \frac{RT}{\rho^2}\left(\frac{\partial \rho}{\partial x}\right)^2 \\ -2\theta_3 \frac{RT}{\rho}\left(\frac{\partial^2 \rho}{\partial x^2}\right) - 4\left(\frac{\theta_3}{2} + \theta_5\right)R\left(\frac{\partial^2 T}{\partial x^2}\right) \\ +8\left(\frac{\theta_0}{2} + \frac{\theta_2}{4} + \theta_5\right)\frac{R}{T}\left(\frac{\partial T}{\partial x}\right)^2 \\ +2(\theta_2 - \theta_3)\frac{R}{\rho}\left(\frac{\partial \rho}{\partial x}\right)\left(\frac{\partial T}{\partial x}\right) \end{bmatrix} \\ -\frac{\mu^3}{p^2}&\begin{bmatrix} -2\theta_2 R\left(\frac{\partial T}{\partial x}\right)\left(\frac{\partial^2 u}{\partial x^2}\right) - 2\theta_3 \frac{RT}{\rho}\left(\frac{\partial \rho}{\partial x}\right)\left(\frac{\partial^2 u}{\partial x^2}\right) \\ +2\theta_3 RT\left(\frac{\partial^3 u}{\partial x^3}\right) \end{bmatrix} \end{aligned} \qquad (49)$$

$$q_x^B = \frac{\mu^2}{\rho} \begin{bmatrix} 4\Big(\theta_{12}(\gamma-1)+\theta_{15}\Big)\left(\dfrac{\partial^2 u}{\partial x^2}\right) \\[4pt] -4\left(\begin{array}{l}\theta_{12}(\gamma-2)\\ +\theta_{13}(\gamma-1)\\ +\dfrac{\theta_{14}}{2}+\theta_{16}\end{array}\right)\dfrac{1}{T}\left(\dfrac{\partial T}{\partial x}\right)\left(\dfrac{\partial u}{\partial x}\right) \\[4pt] -2\dfrac{\theta_{14}}{\rho}\left(\dfrac{\partial \rho}{\partial x}\right)\left(\dfrac{\partial u}{\partial x}\right) \end{bmatrix}$$

$$+\frac{\mu^3}{p\rho}\begin{bmatrix} 8\Big((\gamma-1)\theta_{12}+\dfrac{\theta_{13}}{2}\Big)\dfrac{1}{T}\left(\dfrac{\partial T}{\partial x}\right)\left(\dfrac{\partial u}{\partial x}\right)^2 \\[4pt] +4(\gamma-1)\dfrac{\theta_{12}}{\rho}\left(\dfrac{\partial \rho}{\partial x}\right)\left(\dfrac{\partial u}{\partial x}\right)^2 \\[4pt] -8\Big((\gamma-1)\theta_{12}-\dfrac{\theta_{14}}{4}\Big)\left(\dfrac{\partial u}{\partial x}\right)\left(\dfrac{\partial^2 u}{\partial x^2}\right) \\[4pt] +4\dfrac{\gamma\theta_{12}}{Pr}\dfrac{R}{\rho}\left(\dfrac{\partial \rho}{\partial x}\right)\left(\dfrac{\partial^2 T}{\partial x^2}\right) \\[4pt] +\dfrac{8\gamma}{Pr}\Big(\theta_{12}+\dfrac{\theta_{13}}{2}\Big)\dfrac{R}{T}\left(\dfrac{\partial T}{\partial x}\right)\left(\dfrac{\partial^2 T}{\partial x^2}\right) \\[4pt] -\dfrac{4\gamma\theta_{12}}{Pr}R\left(\dfrac{\partial^3 T}{\partial x^3}\right) \end{bmatrix}$$

$$(50)$$

3.1.4 Method Two

An alternate method for obtaining the second-order distribution was needed in order to reduce the mathematical rigor, especially in two and three dimensions. Deriving the second-order distribution function by "Method One" calls for the expansion of $\left(\Phi^{(1)}\right)^2$, which is a formidable task in higher dimensions. Hence an easily extendible (to higher dimensions) method is presented which, *incidentally*, illustrates the non-uniqueness of the BGK-Burnett equations. From eq. (28):

$$f^{(2)} = -\left(\frac{1}{\xi v}\right)\left[\frac{\partial}{\partial t}\Big(f^{(1)}\Big) + v\frac{\partial}{\partial x}\Big(f^{(1)}\Big)\right] \quad (51)$$

$$f^{(2)} = -\left(\frac{1}{\xi v}\right)\left[\begin{array}{l}\dfrac{\partial}{\partial t}\Big(f^{(0)}\Phi^{(1)}\Big)+\dfrac{\partial}{\partial x}\Big(uf^{(0)}\Phi^{(1)}\Big)\\[4pt]+\dfrac{\partial}{\partial x}\Big(cf^{(0)}\Phi^{(1)}\Big)\end{array}\right] \quad (52)$$

The second-order distribution function **must** satisfy the property $\langle \Psi, f\rangle = \langle \Psi, f^{(0)}\rangle$.

This condition translates to the following equation:

$$\langle \Psi, \xi^2 f^{(2)}\rangle =$$
$$\left\langle \Psi, -\frac{\xi}{v}\left[\begin{array}{l}\dfrac{\partial}{\partial t}\Big(f^{(0)}\Phi^{(1)}\Big)+\dfrac{\partial}{\partial x}\Big(uf^{(0)}\Phi^{(1)}\Big)\\[4pt]+\dfrac{\partial}{\partial x}\Big(cf^{(0)}\Phi^{(1)}\Big)\end{array}\right]\right\rangle = 0$$

$$(53)$$

The above equation will not be satisfied unless some additional terms are added as $\langle \Psi, vf^{(0)}\Phi^{(1)}\rangle \neq 0$.

This moment in fact represents the viscous flux vector of the 1-D Navier-Stokes equations. The first-order term in the distribution function can be represented as shown below:

$$\Phi^{(1)} = -\frac{1}{\xi v}A^{(1)}(I,c)\frac{\partial\beta}{\partial x} - \frac{1}{\xi v}A^{(2)}(I,c)\frac{\partial u}{\partial x} \quad (54)$$

where,

$$A^{(1)}(I,c) = \left[\frac{5c}{2\beta} - \frac{4Ic(\gamma-1)}{(3-\gamma)} - c^3\right] \quad (55)$$

$$A^{(2)}(I,c) = \left[\beta c^2(3-\gamma) + \frac{(3\gamma-5)}{2} - 4I\beta\frac{(\gamma-1)^2}{(3-\gamma)}\right]$$

$$(56)$$

Define:

$$cA^{(i)}(I,c) = B^{(i)}(I,c) \text{ for } i=1,2.$$

$$\widetilde{\Phi}^{(1)} = -\frac{1}{(\xi v)}\widetilde{B}^{(1)}(I,c)\frac{\partial\beta}{\partial x} - \frac{1}{(\xi v)}\widetilde{B}^{(2)}(I,c)\frac{\partial u}{\partial x}$$

$$(57)$$

where,

$$\widetilde{B}^{(1)}(I,c) = B^{(1)}(I,c) + \omega_1\left(\frac{c^2}{\beta}-c^4\right) + \omega_2\left(\frac{I}{\beta I_0}c^2\right)$$

$$(58)$$

$$\widetilde{B}^{(2)}(I,c) = B^{(2)}(I,c) + \omega_3\left(\beta c^3 - \frac{I}{I_0}c + c\right) \quad (59)$$

It is now required that the moment $\langle \Psi, f^{(0)}\widetilde{B}^{(i)}(I,c)\rangle = 0$, in order to satisfy the property $\langle \Psi, f\rangle = \langle \Psi, f^{(0)}\rangle$.

Hence as outlined in § 3.1.3 a set of simultaneous equations are solved to evaluate the undetermined coefficients ω_1, ω_2 and ω_3. These undetermined coefficients are functions of the specific heat ratio 'γ'. Hence they can in principle be used for any gas. An expression for the second-order distribution function satisfying the above equation is given by:

$$f = f^{(0)} + \xi f^{(0)}\Phi^{(1)} - \frac{\xi}{v}\begin{bmatrix} \dfrac{\partial}{\partial t}\left(f^{(0)}\Phi^{(1)}\right) \\[6pt] +\dfrac{\partial}{\partial x}\left(uf^{(0)}\Phi^{(1)}\right) \\[6pt] +\dfrac{\partial}{\partial x}\left(cf^{(0)}\tilde{\Phi}^{(1)}\right) \end{bmatrix} \quad (60)$$

where,

$$\hat{\Phi}^{(1)} = -\frac{1}{\xi v}\tilde{A}^{(1)}\left(\tilde{I},c\right)\frac{\partial\beta}{\partial x} - \frac{1}{\xi v}\tilde{A}^{(2)}\left(\tilde{I},c\right)\frac{\partial u}{\partial x} \quad (61)$$

$$\tilde{A}^{(1)}\left(\tilde{I},c\right) = \left[\frac{\theta_1}{\beta}c + \frac{\theta_2}{\beta}\tilde{I}c + \theta_3 c^3\right] \quad (62)$$

$$\tilde{A}^{(2)}\left(\tilde{I},c\right) = \left[\beta\theta_4 c^2 + \theta_5\tilde{I} + \theta_6\right] \quad (63)$$

In the above expressions $\tilde{I} = \dfrac{I}{I_0}$ and c denotes the peculiar or thermal velocity, c = v-u. The coefficients θ_i, i = 1,2....6 are functions of the specific heat ratio 'γ'. The exact expressions for these coefficients are given in the (Ref. 19).

3.1.5 BGK-BURNETT Equations (Two)

On taking moments of the BGK-Boltzmann equation, using the second-order distribution function, with the collision invariant vector the BGK-Burnett equations are obtained. The BGK-Burnett equations can be cast in the conservation law form as in eq. (45). The elements of the BGK-Burnett flux vector are given in eq. (64). The flux vector can be cast in terms of the BGK-Burnett stress and heat transfer as in eq(s). (46)-(48). The stress and heat transfer terms are given by eq(s) (65)-(66). These expressions are in terms of the material derivatives. As in eq(s). (49)-(50) the 1-D Navier-Stokes equations have been used to express the material derivatives in terms of the spatial derivatives. These expressions are given by (67) and (68). The BGK-Burnett (formulations One and Two) stress and heat transfer terms have two sets of terms of orders μ^2 and μ^3. The former results when Euler equations are used to express the material derivatives in terms of the spatial derivatives and the latter is obtained when Navier-Stokes equations are used to express the material derivatives.

$$G^B = \begin{bmatrix} 0 \\[10pt] \begin{aligned} &\Omega_1\frac{\rho}{\beta v^2}\frac{D}{Dt}\left(\frac{\partial u}{\partial x}\right) - \Omega_1\frac{\rho}{\beta^2 v^2}\left(\frac{\partial u}{\partial x}\right)\frac{D\beta}{Dt} \\[6pt] &+\Omega_2\frac{1}{\beta^3 v^2}\left(\frac{\partial\rho}{\partial x}\right)\left(\frac{\partial\beta}{\partial x}\right) + \Omega_2\frac{\rho}{\beta^3 v^2}\left(\frac{\partial^2\beta}{\partial x^2}\right) \\[6pt] &\quad -\Omega_2\frac{3\rho}{\beta^4 v^2}\left(\frac{\partial\beta}{\partial x}\right)^2 \end{aligned} \\[20pt] \begin{aligned} &\Omega_1\frac{\rho u}{\beta v^2}\frac{D}{Dt}\left(\frac{\partial u}{\partial x}\right) - \Omega_1\frac{\rho u}{\beta^2 v^2}\left(\frac{\partial u}{\partial x}\right)\frac{D\beta}{Dt} \\[6pt] &+\Omega_2\frac{u}{\beta^3 v^2}\left(\frac{\partial\rho}{\partial x}\right)\left(\frac{\partial\beta}{\partial x}\right) + \Omega_2\frac{\rho u}{\beta^3 v^2}\left(\frac{\partial^2\beta}{\partial x^2}\right) \\[6pt] &-\Omega_2\frac{3\rho u}{\beta^4 v^2}\left(\frac{\partial\beta}{\partial x}\right)^2 + \Omega_1\frac{\rho}{\beta v^2}\left(\frac{\partial u}{\partial x}\right)\left(\frac{Du}{Dt}\right) \\[6pt] &-\Omega_3\frac{\rho}{\beta^3 v^2}\frac{D}{Dt}\left(\frac{\partial\beta}{\partial x}\right) + 3\Omega_3\frac{\rho}{\beta^4 v^2}\left(\frac{\partial\beta}{\partial x}\right)\left(\frac{\partial\beta}{\partial x}\right) \\[6pt] &+\Omega_4\frac{1}{\beta^2 v^2}\left(\frac{\partial\rho}{\partial x}\right)\left(\frac{\partial u}{\partial x}\right) + \Omega_4\frac{\rho}{\beta^2 v^2}\left(\frac{\partial^2 u}{\partial x^2}\right) \\[6pt] &\quad +(\Omega_2 - 2\Omega_4)\frac{\rho}{\beta^3 v^2}\left(\frac{\partial u}{\partial x}\right)\left(\frac{\partial\beta}{\partial x}\right) \end{aligned} \end{bmatrix}$$

$$(64)$$

3.1.6 BGK-BURNETT Stress and Heat Transfer Terms (Two)

On casting the BGK-Burnett flux vector in the form shown in eq. (46) the following expressions are obtained for the stress and heat transfer. On using the Navier-Stokes equations to express the material derivatives the following expressions are obtained:

$$\begin{aligned} \tau_{xx}^B = &-\Omega_1\frac{\rho}{\beta v^2}\frac{D}{Dt}\left(\frac{\partial u}{\partial x}\right) + \Omega_1\frac{\rho}{\beta^2 v^2}\left(\frac{\partial u}{\partial x}\right)\frac{D\beta}{Dt} \\[6pt] &-\Omega_2\frac{1}{\beta^3 v^2}\left(\frac{\partial\rho}{\partial x}\right)\left(\frac{\partial\beta}{\partial x}\right) - \Omega_2\frac{\rho}{\beta^3 v^2}\frac{\partial^2\beta}{\partial x^2} \\[6pt] &+3\Omega_2\frac{\rho}{\beta^4 v^2}\left(\frac{\partial\beta}{\partial x}\right)^2 \end{aligned}$$

$$(65)$$

$$
q_x^B = \Omega_1 \frac{\rho}{\beta v^2}\left(\frac{\partial u}{\partial x}\right)\frac{Du}{Dt} - \Omega_3 \frac{\rho}{\beta^3 v^2}\frac{D}{Dt}\left(\frac{\partial \beta}{\partial x}\right)
$$
$$
+3\Omega_3 \frac{\rho}{\beta^4 v^2}\left(\frac{\partial \beta}{\partial x}\right)\frac{D\beta}{Dt} + \Omega_4 \frac{1}{\beta^2 v^2}\left(\frac{\partial \rho}{\partial x}\right)\left(\frac{\partial u}{\partial x}\right)
$$
$$
+\left(\Omega_2 - 2\Omega_4\right)\frac{\rho}{\beta^3 v^2}\left(\frac{\partial u}{\partial x}\right)\left(\frac{\partial \beta}{\partial x}\right)
$$
$$
+\Omega_4 \frac{\rho}{\beta^2 v^2}\left(\frac{\partial^2 u}{\partial x^2}\right)
$$

$$(66)$$

$$
\tau_{xx}^B = -\frac{\mu^2}{p}\left[
\begin{array}{l}
a^{(1)}\left(\frac{\partial u}{\partial x}\right)^2 + a^{(2)}\frac{T}{\rho}\left(\frac{\partial^2 \rho}{\partial x^2}\right) \\[2mm]
+a^{(3)}\frac{T}{\rho^2}\left(\frac{\partial \rho}{\partial x}\right)^2 + a^{(4)}\frac{1}{\rho}\left(\frac{\partial \rho}{\partial x}\right)\left(\frac{\partial T}{\partial x}\right) \\[2mm]
+a^{(5)}\frac{1}{T}\left(\frac{\partial T}{\partial x}\right)^2 + a^{(6)}\left(\frac{\partial^2 T}{\partial x^2}\right)
\end{array}
\right]
$$
$$
-\frac{\mu^3}{p^2}\left[
\begin{array}{l}
a^{(7)}\frac{T}{\rho}\left(\frac{\partial \rho}{\partial x}\right)\left(\frac{\partial^2 u}{\partial x^2}\right) + a^{(8)}\frac{1}{Pr}\left(\frac{\partial u}{\partial x}\right)\left(\frac{\partial^2 T}{\partial x^2}\right) \\[2mm]
+a^{(9)}T\left(\frac{\partial^3 u}{\partial x^3}\right) + a^{(10)}\left(\frac{\partial u}{\partial x}\right)^3
\end{array}
\right]
$$

$$(67)$$

$$
q_x^B = \frac{\mu^2}{\rho}\left[
\begin{array}{l}
b^{(1)}\frac{1}{T}\left(\frac{\partial u}{\partial x}\right)\left(\frac{\partial T}{\partial x}\right) + b^{(2)}\left(\frac{\partial^2 u}{\partial x^2}\right) \\[2mm]
+b^{(3)}\frac{1}{\rho}\left(\frac{\partial \rho}{\partial x}\right)\left(\frac{\partial u}{\partial x}\right)
\end{array}
\right]
$$
$$
+\frac{\mu^3}{\rho\rho}\left[
\begin{array}{l}
b^{(4)}\left(\frac{\partial u}{\partial x}\right)\left(\frac{\partial^2 u}{\partial x^2}\right) + b^{(5)}\frac{1}{T\,Pr}\left(\frac{\partial T}{\partial x}\right)\left(\frac{\partial^2 T}{\partial x^2}\right) \\[2mm]
+b^{(6)}\frac{1}{\rho\,Pr}\left(\frac{\partial \rho}{\partial x}\right)\left(\frac{\partial^2 T}{\partial x^2}\right) + b^{(7)}\frac{1}{Pr}\left(\frac{\partial^3 T}{\partial x^3}\right) \\[2mm]
+b^{(8)}\frac{1}{T}\left(\frac{\partial T}{\partial x}\right)\left(\frac{\partial u}{\partial x}\right)^2 + b^{(9)}\frac{1}{\rho}\left(\frac{\partial \rho}{\partial x}\right)\left(\frac{\partial u}{\partial x}\right)^2 \\[2mm]
+b^{(10)}\frac{1}{T^2 Pr}\left(\frac{\partial T}{\partial x}\right)^3
\end{array}
\right]
$$

$$(68)$$

The coefficients $a^{(1)}$ - $a^{(10)}$ and $b^{(1)}$ - $b^{(10)}$ in the expressions for the BGK-Burnett stress and heat transfer terms are functions of 'γ' and are given in the (Ref. 19).

The expressions for the BGK-Burnett stress and heat transfer have all the derivatives present in the expressions for the Burnett stress and heat transfer as reported by Fiscko and Chapman when terms up to the order of μ^2 are considered. The coefficients, however, are very different. A comparison of the BGK-Burnett (up to order μ^2) and Burnett coefficients is shown in Table (1). Using the Navier-Stokes equations to express the material derivatives in terms of the spatial derivatives results in third order derivatives and products of first and second-order derivatives in the expressions for the BGK-Burnett stress and heat transfer. In must be noted that the Augmented Burnett equations have *only* the linear third-order terms. They do not have the nonlinear terms that are present in the BGK-Burnett formulation. The coefficients of the BGK-Burnett (up to order μ^3) stress and heat transfer terms and the Augmented Burnett terms are presented in Tables (2) and (3). In all these comparisons the BGK-Burnett coefficients correspond to the stress and heat transfer terms in eq(s) (67)-(68).

4. LINEARIZED STABILITY ANALYSIS AND BOLTZMANN'S H-THEOREM

An intuitive notion of stability requires that a mathematical model of a system when subjected to small perturbations will respond in a manner such that the effects of these perturbations will become vanishingly small over a finite period of time. A well known analytical tool that exploits this notion is the theory of small perturbations. Since the perturbations are assumed to be small, only linear terms need be considered for the purpose of analysis. In order to check if the BGK-Burnett equations are stable this simple analytical analysis is carried out by linearizing the equations. The response of these linearized BGK-Burnett equations is shown to be stabilizing for a wide range of Knudsen numbers. It is also shown by invoking Boltzmann's H-Theorem that the second-order distribution function, based on the BGK-Boltzmann equation, ensures that the irreversible entropy production rate is always positive.

4.1 Linearized Stability Analysis

It has been shown by Bobylev that the conventional Burnett equations are not stable to small wavelength disturbances. Hence the conventional Burnett equations tend to blow up when the mesh sizes are made progressively finer. In order to investigate the stability aspects of the BGK-Burnett equations two model problems are considered which study the response of a uniform gas to a 1-D periodic perturbation wave. In Case One the gas is assumed to be at rest and hence has a zero mean velocity. In Case Two the gas is assumed to have a uniform velocity specified by the free stream Mach number. Although Case One is a subset of Case Two, it was felt that both these cases be considered separately in order to identify the parameters on which the stability of the BGK-Burnett equations depend.

4.1.1 Case One (Zero Mean Velocity)

The initial density, temperature and velocity of the undisturbed gas at time $t = 0$ are ρ_0, T_0, and $u_0 = 0$ respectively. At $t = 0$ the gas is perturbed such that:

$$\rho = \rho_0 \left(1 + C_1 e^{\frac{i\omega x}{L_0}} \right) \tag{69}$$

$$T = T_0 \left(1 + C_2 e^{\frac{i\omega x}{L_0}} \right) \tag{70}$$

$$u = \sqrt{RT_0} \left(C_3 e^{\frac{i\omega x}{L_0}} \right) \tag{71}$$

Since the perturbations are assumed to be small the magnitudes of the coefficients in the expressions (69)-(71) are required to satisfy the inequality $|C_k| \ll 1, (k = 1,2,3)$. The characteristic length

$$L_0 = \frac{\mu_0}{\rho_0 \sqrt{RT_0}} = 0.783\lambda, \text{ where } \lambda \text{ denotes the}$$

mean free path. The non-dimensional circular frequency $\omega = \dfrac{2\pi}{\left(L / L_0 \right)} = 4.92 \dfrac{\lambda}{L} = 4.92 \text{Kn}$. On

introducing the perturbed quantities in the continuity, momentum and energy equations and neglecting products and powers of the same, the following equations are obtained.

$$\frac{\partial \rho}{\partial t} + \rho \frac{\partial u}{\partial x} = 0 \tag{72}$$

$$\frac{\partial u}{\partial t} + R \frac{\partial T}{\partial x} + \frac{RT}{\rho} \frac{\partial \rho}{\partial x} - \frac{1}{\rho} \frac{\partial \tau_x}{\partial x} = 0 \tag{73}$$

$$\frac{\partial T}{\partial t} + \frac{RT}{C_v} \frac{\partial u}{\partial x} + \frac{1}{\rho C_v} \frac{\partial q_x}{\partial x} = 0 \tag{74}$$

Equations (72)-(74) form the basis for deriving the linearized stability equations for both Case One and Case Two. Let us now define the following non-dimensional variables:

$$\rho' = \frac{\left(\rho - \rho_0 \right)}{\rho_0} \tag{75a}$$

$$T' = \frac{\left(T - T_0 \right)}{T_0} \tag{75b}$$

$$u' = \frac{u}{\sqrt{RT_0}} \tag{75c}$$

$$x' = \frac{x}{L_0} \tag{75d}$$

$$t' = \frac{tP_0}{\mu_0} \tag{75e}$$

where the subscript '$_0$' denotes unperturbed flow variables. On introducing these non-dimensional variables the continuity, momentum and energy equations take the form

$$\frac{\partial \rho'}{\partial t'} + \frac{\partial u'}{\partial x'} = 0 \tag{76}$$

$$\frac{\partial u'}{\partial t'} + \frac{\partial \rho'}{\partial x'} + \frac{\partial T'}{\partial x'} - \frac{1}{\rho_0 RT_0} \frac{\partial}{\partial x'} \left(\tau_{xx} \right) = 0 \tag{77}$$

$$\frac{\partial T'}{\partial t'} + \frac{R}{C_v} \frac{\partial u'}{\partial x'} + \frac{\sqrt{RT_0}}{C_v R \rho_0 T_0^2} \frac{\partial}{\partial x'} \left(q_x \right) = 0 \tag{78}$$

On substituting for τ_{xx} and q_x from the BGK-Burnett equations the following equations are obtained. It must be noted that the notation used corresponds to that in eq(s) (67)-(68) for the BGK-Burnett stress and heat transfer terms.

$$\frac{\partial \rho'}{\partial t'} + \frac{\partial u'}{\partial x'} = 0 \tag{79}$$

$$\frac{\partial u'}{\partial t'} + \frac{\partial \rho'}{\partial x'} + \frac{\partial T'}{\partial x'} - \frac{4}{3} \frac{\partial^2 u'}{\partial x'^2}$$
$$+ a'(2) \frac{\partial^3 \rho'}{\partial x'^3} + a'(6) \frac{\partial^3 T'}{\partial x'^3} + a'(9) \frac{\partial^4 u'}{\partial x'^4} = 0 \tag{80}$$

$$\frac{\partial T'}{\partial t'} + (\gamma - 1)\frac{\partial u'}{\partial x'} - \frac{\gamma}{Pr}\frac{\partial^2 T'}{\partial x'^2}$$
$$+ b'^{(2)}R(\gamma - 1)\frac{\partial^3 u'}{\partial x'^3} + b'^{(7)}\frac{(\gamma - 1)}{Pr}\frac{\partial^4 T'}{\partial x'^4} = 0 \tag{81}$$

where $a'(\ ) = \dfrac{a(\ )}{R}$, and $b'(\ ) = \dfrac{b(\ )}{R}$. These equations can be cast in the form of a vector equation as shown below:

$$\frac{\partial V'}{\partial t'} + M_1\frac{\partial V'}{\partial x'} + M_2\frac{\partial^2 V'}{\partial x'^2} + M_3\frac{\partial^3 V'}{\partial x'^3}$$
$$+ M_4\frac{\partial^4 V'}{\partial x'^4} = 0 \tag{82}$$

where,

$$V' = \begin{Bmatrix} \rho' \\ u' \\ T' \end{Bmatrix} \tag{83}$$

$$M_1 = \begin{bmatrix} 0 & 1 & 0 \\ 1 & 0 & 1 \\ 0 & (\gamma - 1) & 0 \end{bmatrix} \tag{84}$$

$$M_2 = \begin{bmatrix} 0 & 0 & 0 \\ 0 & -(3 - \gamma) & 0 \\ 0 & 0 & \dfrac{\gamma}{Pr} \end{bmatrix} \tag{85}$$

$$M_3 = \begin{bmatrix} 0 & 0 & 0 \\ \dfrac{a^{(2)}}{R} & 0 & \dfrac{a^{(6)}}{R} \\ 0 & b^{(2)}(\gamma - 1) & 0 \end{bmatrix} \tag{86}$$

$$M_4 = \begin{bmatrix} 0 & 0 & 0 \\ 0 & \dfrac{a^{(9)}}{R} & 0 \\ 0 & 0 & \dfrac{b^{(7)}(\gamma - 1)}{R\,Pr} \end{bmatrix} \tag{87}$$

Let us assume the solution of the above equation to be of the form

$$V' = \overline{V}e^{i\omega x'}e^{\phi t'} \tag{88}$$

where, $t' = \dfrac{t p_0}{\mu_0}$. The non-dimensional initial conditions for eq. (82) can be denoted in vector form

as $V'\big|_{t=0} = \overline{V}e^{i\omega x'}$, where $x' = \dfrac{x}{L_0}$, and $\phi = \alpha + i\beta$.

α denotes the attenuation coefficient and β denotes the dispersion coefficient. For stability $\alpha < 0$ as L decreases or in other words the flow must attenuate as the Knudsen number increases. Substituting eq. (88) in eq. (82) and simplifying yields eq. (89) when Euler equations are used to express the material derivatives.

$$\left[\phi I + i\omega M_1 - \omega^2 M_2 - i\omega^3 M_3\right]V_0 e^{i\omega x'}e^{\phi t'} = 0 \tag{89}$$

For a non-trivial solution eq. (90) must be satisfied.

$$\left|\phi I + i\omega M_1 - \omega^2 M_2 - i\omega^3 M_3\right| = 0 \tag{90}$$

When the Navier-Stokes equations are used to express the material derivatives in terms of the spatial derivatives eq. (91) is obtained.

$$\begin{bmatrix} \phi I + i\omega M_1 - \omega^2 M_2 - i\omega^3 M_3 \\ + \omega^4 M_4 \end{bmatrix}V_0 e^{i\omega x'}e^{\phi t'} = 0 \tag{91}$$

For non-trivial solutions eq. (92) must be satisfied.

$$\left|\phi I + i\omega M_1 - \omega^2 M_2 - i\omega^3 M_3 + \omega^4 M_4\right| = 0 \tag{92}$$

The trajectory of the roots of the characteristic eq(s). (90), (92) are plotted on the complex plane on which the real axis denotes the attenuation coefficient and the imaginary axis denotes the dispersion coefficient. For stability it is required that the roots lie to the left of the imaginary axis as the Knudsen number increases. Fig(s) 1-8 show the trajectory of the roots of the characteristic equation as the Knudsen number increases. From the plots it is observed that unconditional stability is guaranteed only when Navier-Stokes equations are used to express the material derivatives in terms of the spatial derivatives.

4.1.2 Case Two (Non-Zero Mean Velocity)

In order to check if high Mach numbers introduce any destabilizing effects with increasing Knudsen numbers, the BGK-Burnett equations were linearized under the assumption that the flow is uniform every where in the flow field. These linearized equations are checked for stability when perturbed. The methods of analysis is similar to that in Case One except for the velocity term. The unperturbed velocity is defined as

$$u_0 = M_\infty\sqrt{\gamma R T_0} \tag{93}$$

where M_∞ denotes the free stream Mach number. The non-dimensional velocity is defined as

$$u' = \frac{(u - u_0)}{\sqrt{RT_0}} \qquad (94)$$

On non-dimensionalizing eq(s) (72-74). the following equations are obtained.

$$\frac{\partial \rho'}{\partial t'} + \frac{\partial u'}{\partial x'} + M_\infty \sqrt{\gamma}\, \frac{\partial \rho'}{\partial x'} = 0 \qquad (95)$$

$$\frac{\partial u'}{\partial t'} + M_\infty \sqrt{\gamma}\, \frac{\partial u'}{\partial x'} + \frac{\partial \rho'}{\partial x'} + \frac{\partial T'}{\partial x'}$$
$$-\frac{1}{\rho_0 R T_0} \frac{\partial}{\partial x'}(\tau_{xx}) = 0 \qquad (96)$$

$$\frac{\partial T'}{\partial t'} + M_\infty \sqrt{\gamma}\, \frac{\partial T'}{\partial x'} + \frac{R}{C_v}\frac{\partial u'}{\partial x'}$$
$$+\frac{\sqrt{RT_0}}{C_v R \rho_0 T_0^2} \frac{\partial}{\partial x'}(q_x) = 0 \qquad (97)$$

On substituting the BGK-Burnett stress and heat transfer terms and simplifying, the following equation is obtained in vector form.

$$\frac{\partial V'}{\partial t'} + M_1 \frac{\partial V'}{\partial x'} + M_2 \frac{\partial^2 V'}{\partial x'^2} + M_3 \frac{\partial^3 V'}{\partial x'^3}$$
$$+M_4 \frac{\partial^4 V'}{\partial x'^4} = 0 \qquad (98)$$

where,

$$M_1 = \begin{bmatrix} M_\infty \sqrt{\gamma} & 1 & 0 \\ 1 & M_\infty \sqrt{\gamma} & 1 \\ 0 & (\gamma - 1) & M_\infty \sqrt{\gamma} \end{bmatrix} \qquad (99)$$

$$M_2 = \begin{bmatrix} 0 & 0 & 0 \\ 0 & -(3 - \gamma) & 0 \\ 0 & 0 & \dfrac{\gamma}{Pr} \end{bmatrix} \qquad (100)$$

$$M_3 = \begin{bmatrix} 0 & 0 & 0 \\ \dfrac{a^{(2)}}{R} & 0 & \dfrac{a^{(6)}}{R} \\ 0 & b^{(2)}(\gamma - 1) & 0 \end{bmatrix} \qquad (101)$$

$$M_4 = \begin{bmatrix} 0 & 0 & 0 \\ 0 & \dfrac{a^{(9)}}{R} & 0 \\ 0 & 0 & \dfrac{b^{(7)}(\gamma - 1)}{R\,Pr} \end{bmatrix} \qquad (102)$$

It can be seen that the difference between eq. (102) and eq.(82) is the M_1 matrix, which has non-zero diagonal terms. The trajectory of the roots of the characteristic equations for eq. (98) are plotted on the complex plane for increasing Knudsen numbers. These plots are shown in Fig(s) 9-16. It is noted that unconditional stability is guaranteed only when the Navier-Stokes equations are used to express the material derivatives in terms of the spatial derivatives. It is also noted that the parameters that tend to de-stabilize the linearized BGK-Burnett equations (when Euler equations are used to express the material derivatives) are the Knudsen number and the specific heat ratio. These plots also establish the fact that the free stream Mach number does not play any part in destabilizing the linearized BGK-Burnett equations.

4.1.3 Exact BGK-Burnett Equations (Non-zero Mean Velocity)

In order to check if the approximations introduced in expressing the material derivatives in terms of the spatial derivatives tend to destabilize the equations the original BGK-Burnett equations are considered for stability analysis. The material derivatives in the BGK-Burnett stress and heat transfer terms are considered without expressing them in terms of spatial derivatives. On following the procedure explained in § 4.1.1 the following equations are obtained.

$$\begin{bmatrix} \dfrac{\partial \rho'}{\partial t'} \\[2ex] \dfrac{\partial}{\partial t'}\left\{ u' + (3 - \gamma)\dfrac{\partial^2 u'}{\partial x'^2} \right\} \\[2ex] \dfrac{\partial}{\partial t'}\left\{ T' + \gamma \dfrac{\partial^2 T'}{\partial x'^2} \right\} \end{bmatrix} + M_1 \frac{\partial V'}{\partial x'}$$
$$+M_2 \frac{\partial^2 V'}{\partial x'^2} + M_3 \frac{\partial^3 V'}{\partial x'^3} = 0 \qquad (103)$$

where, M_1 and M_2 have the same elements as Case One and Case Two. The matrix M_3 is, however, different and is given below.

$$
M_3 = \begin{bmatrix}
0 & 0 & 0 \\[2ex]
0 & (3-\gamma)M_\infty\sqrt{\gamma} & -3\left(\theta_1 + \theta_2 + \dfrac{5}{3}\theta_3\right) \\[4ex]
0 & 4(\gamma-1)\left\{\theta_4\dfrac{6\gamma-3}{8(\gamma-1)} + \theta_5\dfrac{\gamma+3}{8(\gamma-1)} + \theta_6\dfrac{2\gamma}{8(\gamma-1)}\right\} & M_\infty\gamma\sqrt{\gamma}
\end{bmatrix}
$$

$$\text{(104)}$$

The elements in the above matrix correspond to the BGK-Burnett equations obtained from "Method One". A similar matrix can be obtained for the BGK-Burnett equations obtained by "Method Two". The trajectory of the roots of the characteristic equation of eq. (103) are plotted on the complex plane as in the earlier cases. These plots are shown in Fig(s) 17 - 20. These plots indicate that the original BGK-Burnett equations are stable only for Knudsen numbers below 0.1! Hence in order to guarantee unconditional stability, the Navier-Stokes equations **must** be used to express the material derivatives.

Linear stability analysis does not consider the many non-linear terms - powers and products of derivatives - that are present in the BGK-Burnett stress and heat transfer terms. Hence, this analysis, is at best only a necessary condition for the stability of these equations. A more rigorous proof of the stability of these equations involves verifying the Boltzmann's H-Theorem.

4.2 Boltzmann's H-Theorem

The BGK-Burnett equations must satisfy the second-law of thermodynamics. There, however, is no acceptable definition of entropy for a gas in a state of non-equilibrium. Physical intuition tells us that an isolated system will evolve from an arbitrary initial state to a state of equilibrium. Boltzmann's H theorem formalizes this notion, and also makes explicit the manner in which this evolution proceeds. A spatially homogenous gas is defined as one in which the density does not vary with position. Boltzmann's H theorem states that for a spatially homogenous gas the inequality, $\dfrac{\partial}{\partial t}(H) \leq 0$, must be satisfied when the gas approaches equilibrium. The quantity H which is shown to be the kinetic theory equivalent of entropy[1] is defined as $H = \displaystyle\int_{-\infty}^{\infty} f\ln f\, dv$.

In arriving at this definition of the H-function Boltzmann made the following assumptions:
a) The molecules comprising the gas do not have any internal energy. Hence the H function was defined *only* over the range of molecular velocities.
b) The gas was assumed to be monoatomic.

Since our definition of the Maxwellian and the first and second order distribution functions takes into account the energy contribution due to the various non-translational degrees of freedom and further does not assume the gas to be monoatomic, the definition of H must be modified (Refs. 20-21) to account for these differences. The modified definition of the H function can be shown to reduce to the classical (Boltzmann) definition of H for the specific case of a monoatomic gas.

4.2.1 Modified H-Function

The change in entropy in classical thermodynamics is given by the following expression:

$$(s_2 - s_1) = C_v \ln\frac{T_2}{T_1} - R\ln\frac{\rho_2}{\rho_1} \qquad (105)$$

where the subscripts '1' and '2' denote thermodynamic variables at equilibrium stations far upstream and far downstream respectively. The absolute entropy is given by the following expression up to an additive constant.

$$s = C_v \ln T - R\ln\rho + R\mathscr{C}_0 \qquad (106)$$

The above expression can be cast in the form

$$s = -R\left[\ln\rho + \frac{\ln\beta}{(\gamma-1)} - \mathscr{C}_0\right] \qquad (107)$$

---

[1] It must be noted that entropy according to classical thermodynamics is defined *only* for equilibrium systems. The quantity 'H' , however, is defined *even* for non-equilibrium systems.

where $\beta = \dfrac{1}{2RT}$. We now need to devise a method to arrive at the above expression from the Maxwellian distribution function which is also the equilibrium distribution function. The 1-D Maxwellian distribution function is given by the expression

$$f^{(0)} = F = \frac{\rho}{I_0} \sqrt{\frac{\beta}{\pi}} \exp\left[-\frac{I}{I_0} - \beta(v-u)^2 \right] \qquad (108)$$

where the average internal energy $I_0 = \dfrac{(3-\gamma)}{4\beta(\gamma-1)}$.

Equation (108) can be rewritten as:

$$\ln f^{(0)} = \ln F = \left[\begin{array}{c} \ln\rho + \dfrac{3}{2}\ln\beta \\[2mm] + \ln\left\{ \dfrac{4(\gamma-1)}{(3-\gamma)\sqrt{\pi}} \right\} - \beta u^2 \end{array} \right] \qquad (109)$$

$$-\frac{4\beta(\gamma-1)}{(3-\gamma)}I - 2\beta\left(\frac{v^2}{2}\right) + (2\beta u)v$$

On rearranging the terms in eq. (109)

$$\ln f^{(0)} - 2\beta I \frac{(5-3\gamma)}{(3-\gamma)}$$

$$= \left[\ln\rho + \frac{3}{2}\ln\beta + \ln\left\{ \frac{4(\gamma-1)}{(3-\gamma)\sqrt{\pi}} \right\} - \beta u^2 \right] \qquad (110)$$

$$-2\beta\left(I + \frac{v^2}{2} \right) + (2\beta u)v$$

Equation (4.35) is indeed a linear combination[2] of the collision invariants. Hence,

$$\left\langle \left\{ \begin{array}{c} \ln f^{(0)} \\[2mm] -2\beta I \dfrac{(5-3\gamma)}{(3-\gamma)} \end{array} \right\} \left(\frac{\partial f^{(0)}}{\partial t} + v\frac{\partial f^{(0)}}{\partial x} \right) \right\rangle$$

$$= \int\limits_{0}^{\infty}\int\limits_{-\infty}^{\infty} \left\{ \begin{array}{c} \ln f^{(0)} \\[2mm] -2\beta I \dfrac{(5-3\gamma)}{(3-\gamma)} \end{array} \right\} \left(\frac{\partial f^{(0)}}{\partial t} + v\frac{\partial f^{(0)}}{\partial x} \right) dv\, dI$$

$$= 0 \qquad (111)$$

The above equation can be recast in the following form by making use of the following identity:

---

[2] It can be shown that the moments of the BGK-Boltzmann equation with any linear combination of the collision invariants equals zero. Hence the need to express equation (109) as a linear combination of collision invariants.

$$\left\langle \Psi, \left(\frac{\partial f^{(0)}}{\partial t} + v\frac{\partial f^{(0)}}{\partial x} \right) \right\rangle = 0 \qquad (112)$$

$$\left\langle \left\{ \begin{array}{c} 1 + \ln f^{(0)} \\[2mm] -2\beta I \dfrac{(5-3\gamma)}{(3-\gamma)} \end{array} \right\} \left(\frac{\partial f^{(0)}}{\partial t} + v\frac{\partial f^{(0)}}{\partial x} \right) \right\rangle$$

$$= \int\limits_{0}^{\infty}\int\limits_{-\infty}^{\infty} \left\{ \begin{array}{c} 1 + \ln f^{(0)} \\[2mm] -2\beta I \dfrac{(5-3\gamma)}{(3-\gamma)} \end{array} \right\} \left(\frac{\partial f^{(0)}}{\partial t} + v\frac{\partial f^{(0)}}{\partial x} \right) dv\, dI$$

$$= 0 \qquad (113)$$

On simplifying, eq. (113) can be expressed as

$$\int\limits_{0}^{\infty}\int\limits_{-\infty}^{\infty} \left(\frac{\partial}{\partial t} + v\frac{\partial}{\partial x} \right) \left[\begin{array}{c} f^{(0)} \ln f^{(0)} \\[2mm] + \dfrac{(5-3\gamma)}{2(\gamma-1)} f^{(0)} \ln\beta \end{array} \right] dv\, dI = 0 \qquad (114)$$

The above equation can be expressed in the following compact form

$$\frac{\partial}{\partial t}\left(H^{(0)} \right) + \frac{\partial}{\partial x}\left(H_v^{(0)} \right) = 0 \qquad (115)$$

where the functionals $H^{(0)}$ and $H_v^{(0)}$ are defined in (116) and (117) as:

$$H^{(0)} = \int\limits_{0}^{\infty}\int\limits_{-\infty}^{\infty} \left[f^{(0)} \ln f^{(0)} + \frac{(5-3\gamma)}{2(\gamma-1)} f^{(0)} \ln\beta \right] dv\, dI \qquad (116)$$

$$H_v^{(0)} = \int\limits_{0}^{\infty}\int\limits_{-\infty}^{\infty} v\left[f^{(0)} \ln f^{(0)} + \frac{(5-3\gamma)}{2(\gamma-1)} f^{(0)} \ln\beta \right] dv\, dI \qquad (117)$$

On evaluating the moments in eq. (116) the following expression is obtained for $H^{(0)}$

$$H^{(0)} = \rho\left[\ln\rho + \frac{\ln\beta}{(\gamma-1)} - \mathcal{C}_o \right] \qquad (118)$$

On comparing eq(s). (118) and (106) we obtain:

$$\boxed{\rho s = -RH^{(0)}}^{[3]} \qquad (119)$$

The definition of the H-function can be extended to any distribution function. Accordingly the

---

[3] This relation establishes a link between Boltzmann's H-Theorem and the classical thermodynamics concept of entropy.

functionals H and H_v are defined as shown in eq(s). (120) and (121).

$$H = \int_0^\infty \int_{-\infty}^\infty \left[f \ln f + \frac{(5-3\gamma)}{2(\gamma-1)} f \ln \beta \right] dv\, dI \qquad (120)$$

$$H_v = \int_0^\infty \int_{-\infty}^\infty v \left[f \ln f + \frac{(5-3\gamma)}{2(\gamma-1)} f \ln \beta \right] dv\, dI \qquad (121)$$

It can be seen that the expression for H reduces to the classical definition $H = \int_0^\infty \int_{-\infty}^\infty f \ln f\, dv\, dI$, for the specific case of a monoatomic gas (i.e. $\gamma = \frac{5}{3}$).

For a spatially inhomogenous gas Grad has shown that the following inequality must be satisfied when the gas approaches equilibrium.

$$\boxed{\frac{\partial}{\partial t}(H) + \frac{\partial}{\partial x}(H_v) \le 0} \qquad (122)$$

The above inequality will be shown to be true for the first and second-order distribution functions. The first and second order distribution functions are given by expressions (123) and (124).

$$f = f^{(0)} + \xi f^{(0)} \Phi^{(1)} = f^{(0)}\left(1 + \xi \Phi^{(1)}\right) \qquad (123)$$

$$f = f^{(0)} + \xi f^{(0)} \Phi^{(1)} + \xi^2 f^{(0)} \Phi^{(2)}$$
$$= f^{(0)}\left(1 + \xi \Phi^{(1)} + \xi^2 \Phi^{(2)}\right) \qquad (124)$$

On expressing $\ln(f)$ as a Taylor series and considering only terms up to the second power in the Knudsen number the following approximations are obtained for the first and second order distribution functions.

$$\ln\left[f^{(0)}\left(1 + \xi \Phi^{(1)}\right)\right]$$
$$\cong \ln f^{(0)} + \xi \Phi^{(1)} - \frac{\xi^2 \left[\Phi^{(1)}\right]^2}{2} \qquad (125)$$

$$\ln\left[f^{(0)}\left(1 + \xi \Phi^{(1)} + \xi^2 \Phi^{(2)}\right)\right]$$
$$\cong \ln f^{(0)} + \left(\xi \Phi^{(1)} + \xi^2 \Phi^{(2)}\right) \qquad (126)$$
$$- \frac{\xi^2 \left[\Phi^{(1)} + \xi \Phi^{(2)}\right]^2}{2}$$

The H-Balance Equation for the Boltzmann equation is obtained by evaluating the moments in the following equation.

$$\frac{\partial}{\partial t}(H) + \frac{\partial}{\partial x}(H_v)$$
$$= \int_0^\infty \int_{-\infty}^\infty v\left(f^{(0)} - f\right)\left[\begin{array}{c} 1 + \ln f \\ -\dfrac{2(5-3\gamma)}{(3-\gamma)}\beta I \end{array}\right] dv\, dI \qquad (127)$$

On substituting eq. (123) in the above equation and retaining only terms up to the first power in ξ (Knudsen number) the right hand side (RHS) of the above equation takes the form:

$$\int_0^\infty \int_{-\infty}^\infty -v\xi f^{(0)} \Phi^{(1)} \left[\begin{array}{c} 1 + \ln f \\ -\dfrac{2(5-3\gamma)}{(3-\gamma)}\beta I \end{array}\right] dv\, dI \qquad (128)$$

The above integral equals zero as moments of the first-order distribution function with a linear combination[4] of the collision invariants are being evaluated. Since the first-order distribution function satisfies the property $\left\langle \Psi; \xi f^{(0)} \Phi^{(1)} \right\rangle = 0$, eq. (128) equals zero. Hence the first-order distribution function satisfies Boltzmann's H-Theorem. On evaluating eq. (127) up to the second power in ξ, the RHS of the equation takes the following form:

$$-\int_0^\infty \int_{-\infty}^\infty \left\{ \begin{array}{c} \xi f^{(0)} \Phi^{(1)} + \xi f^{(0)} \Phi^{(1)} \ln f^{(0)} \\ -\xi f^{(0)} \Phi^{(1)} \dfrac{2(5-3\gamma)}{(3-\gamma)}\beta I \\ +\xi^2 f^{(0)}\left[\Phi^{(1)}\right]^2 \end{array} \right\} dv\, dI \qquad (129)$$

$$-\int_0^\infty \int_{-\infty}^\infty \xi^2 f^{(0)} \Phi^{(2)} \left\{ \begin{array}{c} 1 + \ln f^{(0)} \\ -\dfrac{2(5-3\gamma)}{(3-\gamma)}\beta I \end{array} \right\} dv\, dI$$

The second integral equals zero as the second-order distribution function satisfies the moment property:

$$\left\langle \Psi; \xi^2 f^{(2)} \right\rangle = \left\langle \Psi; \xi^2 f^{(0)} \Phi^{(2)} \right\rangle = 0 \qquad (130)$$

For reasons given earlier moments of the second-order term with any linear combination of the collision invariants equals zero. Hence eq. (129) yields the following simplified expression:

---

[4] Moments of the first and second-order distribution function with any linear combination of the collision invariants equals zero.

$$-\int_0^\infty \int_{-\infty}^\infty \xi^2 f^{(0)}\left[\Phi^{(1)}\right]^2 dv\, dI \qquad (131)$$

On substituting for $\Phi^{(1)}$ and evaluating the moments

$$-\int_0^\infty \int_{-\infty}^\infty \xi^2 f^{(0)}\left[\Phi^{(1)}\right]^2 dv\, dI$$

$$= -\frac{1}{v^2}\left[\frac{5\rho}{4\beta^3}\left(\frac{\partial \beta}{\partial x}\right)^2 + \frac{\left(3\gamma^2 - 10\gamma + 11\right)}{2}\left(\frac{\partial u}{\partial x}\right)^2\right] \qquad (132)$$

It must be noted that eq. (132) is always less than zero. Hence the second-order distribution function satisfies Boltzmann's H-Theorem. It has been proved conclusively that both the first and second-order distribution functions satisfy the inequality

$$\frac{\partial}{\partial t}(H) + \frac{\partial}{\partial x}(H_v) \le 0 \qquad (133)$$

As a consequence the BGK-Burnett equations which are obtained from the BGK-Boltzmann equation by using the second-order distribution function are entropy consistent !

5. NUMERICAL ALGORITHM TO INTEGRATE THE BGK-BURNETT EQUATIONS AND COMPUTATIONAL TEST CASES

It has recently been shown by Jameson et. al (Ref. 13) and Perthame et. al (Ref. 14) that it is possible to construct less dissipative schemes for Euler and Navier-Stokes formulations from BGK-Boltzmann equations. Jameson has proved that these schemes uphold Boltzmann's H-Theorem and hence are Total -H -Diminishing (THD). Such a property eliminates the possibility of expansion shocks especially in the upstream regions hypersonic shock flows. In order to exploit the THD property of these kinetic schemes a KWPS algorithm has been derived for the BGK-Burnett equations.

5.1 KWPS Algorithm

An upwind algorithm for numerically integrating the BGK-Burnett equations is derived by taking moments of the discretized Boltzmann equation with the collision invariant vector. The molecular velocity is expressed as the sum of the fluid velocity and the peculiar velocity i.e. $v=u+c$. The upwind discretization of the BGK-Boltzmann equation can be written as:

$$\frac{f_j^{n+1} - f_j^n}{\Delta t} + \frac{1}{\Delta x}\left[\begin{array}{l}\left(\frac{u+|u|}{2}f\right)_j - \left(\frac{u+|u|}{2}f\right)_{j-1} \\ + \left(\frac{u-|u|}{2}f\right)_{j+1} - \left(\frac{u-|u|}{2}f\right)_j\end{array}\right]^n$$

$$+ \frac{1}{\Delta x}\left[\begin{array}{l}\left(\frac{c+|c|}{2}f\right)_j - \left(\frac{c+|c|}{2}f\right)_{j-1} + \left(\frac{c-|c|}{2}f\right)_{j+1} \\ - \left(\frac{c-|c|}{2}f\right)_j\end{array}\right]^n$$

$$= \left\{v\left(f^{(0)} - f\right)\right\}_j^n \qquad (134)$$

On taking moments of the above equation with the collision invariant vector, the first-order upwind scheme for the BGK-Burnett equations is obtained.

$$\frac{\mathbf{Q}_j^{n+1} - \mathbf{Q}_j^n}{\Delta t} + \left[\frac{u_{j+1}\mathbf{Q}_{j+1} - u_{j-1}\mathbf{Q}_{j-1}}{2\Delta x}\right]^n$$

$$- \left[\frac{|u_{j+1}|\mathbf{Q}_{j+1} - 2|u_j|\mathbf{Q}_j + |u_{j-1}|\mathbf{Q}_{j-1}}{2\Delta x}\right]^n$$

$$+ \frac{1}{\Delta x}\left[\mathbf{G}_j^{ai+} - \mathbf{G}_{j-1}^{ai+} + \mathbf{G}_{j+1}^{ai-} - \mathbf{G}_j^{ai-}\right]^n$$

$$+ \frac{1}{\Delta x}\left[\mathbf{G}_j^{v+} - \mathbf{G}_{j-1}^{v+} + \mathbf{G}_{j+1}^{v-} - \mathbf{G}_j^{v-}\right]^n$$

$$+ \frac{1}{\Delta x}\left[\mathbf{G}_j^{B+} - \mathbf{G}_{j-1}^{B+} + \mathbf{G}_{j+1}^{B-} - \mathbf{G}_j^{B-}\right]^n = 0 \qquad (135)$$

This methodology can be extended to higher order upwind schemes by using higher order upwind differences in the Boltzmann equation and taking moments of the resulting difference equation with the collision invariant vector. It is important to note that these moments have to be evaluated only once. The elements of the various split flux vectors are :

$$\mathbf{G}^{ai\pm} = \left[\begin{array}{c}\pm\dfrac{\rho}{2\sqrt{\pi\beta}} \\[2mm] \dfrac{p}{2} \pm \dfrac{\rho u}{2\sqrt{\pi\beta}} \\[2mm] \dfrac{pu}{2} \pm \dfrac{\rho}{2\sqrt{\pi\beta}}\left(\dfrac{p}{2}\dfrac{\gamma+1}{2(\gamma-1)} + \dfrac{u^2}{2}\right)\end{array}\right] \qquad (136)$$

$$\mathbf{G}^{v\pm} = \begin{bmatrix} \mp\dfrac{(3-\gamma)}{4\nu}\dfrac{\rho}{\sqrt{\beta\pi}}\dfrac{\partial u}{\partial x} \\[2ex] \hline -\dfrac{\rho(3-\gamma)}{\nu}\left\{\dfrac{1}{4\beta}\pm\dfrac{u}{4\sqrt{\beta\pi}}\right\}\dfrac{\partial u}{\partial x} \\[1ex] \mp\dfrac{\rho}{\nu}\left\{\dfrac{1}{4\beta^2\sqrt{\beta\pi}}\right\}\dfrac{\partial\beta}{\partial x} \\[2ex] \hline -\dfrac{\rho(3-\gamma)}{\nu}\left\{\dfrac{u}{4\beta}\pm\dfrac{u^2}{8\sqrt{\beta\pi}}\pm\dfrac{3\gamma-1}{16(\gamma-1)\beta\sqrt{\beta\pi}}\right\}\dfrac{\partial u}{\partial x} \\[1ex] +\dfrac{\rho}{\nu}\left\{\dfrac{\gamma}{8\beta^3(\gamma-1)}\pm\dfrac{u}{4\beta^2\sqrt{\beta\pi}}\right\}\dfrac{\partial\beta}{\partial x} \end{bmatrix} \tag{137}$$

$$\mathbf{G}^{B\pm} = \begin{bmatrix} \left\{\mathbf{G}^{B1\pm}\right\}\left(\dfrac{\partial\beta}{\partial x}\right)^2 + \left\{\mathbf{G}^{B2\pm}\right\}\left(\dfrac{\partial u}{\partial x}\right)^2 \\[1ex] +\left\{\mathbf{G}^{B3\pm}\right\}\left(\dfrac{\partial u}{\partial x}\right)\left(\dfrac{\partial\beta}{\partial x}\right) + \left\{\mathbf{G}^{B4\pm}\right\}\left(\dfrac{\partial\beta}{\partial x}\right)\left(\dfrac{Du}{Dt}\right) \\[1ex] +\left\{\mathbf{G}^{B5\pm}\right\}\left(\dfrac{\partial^2\beta}{\partial x^2}\right) + \left\{\mathbf{G}^{B6\pm}\right\}\left(\dfrac{\partial^2 u}{\partial x^2}\right) \\[1ex] +\left\{\mathbf{G}^{B7\pm}\right\}\left(\dfrac{\partial\beta}{\partial x}\right)\left(\dfrac{D\beta}{Dt}\right) + \left\{\mathbf{G}^{B8\pm}\right\}\left(\dfrac{\partial u}{\partial x}\right)\left(\dfrac{D\beta}{Dt}\right) \\[1ex] +\left\{\mathbf{G}^{B9\pm}\right\}\left(\dfrac{\partial u}{\partial x}\right)\left(\dfrac{Du}{Dt}\right) + \left\{\mathbf{G}^{B10\pm}\right\}\left(\dfrac{D}{Dt}\left(\dfrac{\partial\beta}{\partial x}\right)\right) \\[1ex] +\left\{\mathbf{G}^{B11\pm}\right\}\left(\dfrac{D}{Dt}\left(\dfrac{\partial u}{\partial x}\right)\right) \end{bmatrix} \tag{138}$$

The split BGK-Burnett flux vector has eleven constituent vectors. which are given below.

$$\mathbf{G}^{B1\pm} = \begin{bmatrix} \pm\dfrac{1}{v^2}\dfrac{\rho}{\beta^3\sqrt{\beta\pi}}\chi_1 \\[2ex] \dfrac{1}{v^2}\left\{\dfrac{\rho}{\beta^4}\chi_2\pm\dfrac{\rho u}{\beta^3\sqrt{\beta\pi}}\chi_1\right\} \\[2ex] \dfrac{1}{v^2}\left\{\pm\dfrac{\rho}{\beta^4\sqrt{\beta\pi}}\chi_3+\dfrac{\rho u}{\beta^4}\chi_2\pm\dfrac{\rho u^2}{\beta^3\sqrt{\beta\pi}}\left(\dfrac{\chi_1}{2}\right)\right\} \end{bmatrix} \tag{139a}$$

$$\mathbf{G}^{B2\pm} = \begin{bmatrix} \pm\dfrac{1}{v^2}\dfrac{\rho}{\sqrt{\beta\pi}}\chi_4 \\[2ex] \dfrac{1}{v^2}\left\{\dfrac{\rho}{\beta}\chi_5\pm\dfrac{\rho u}{\sqrt{\beta\pi}}\chi_4\right\} \\[2ex] \dfrac{1}{v^2}\left\{\pm\dfrac{\rho}{\beta\sqrt{\beta\pi}}\chi_6+\dfrac{\rho u}{\beta}\chi_5\pm\dfrac{\rho u^2}{\sqrt{\beta\pi}}\left(\dfrac{\chi_4}{2}\right)\right\} \end{bmatrix} \tag{139b}$$

$$\mathbf{G}^{B3\pm} = \begin{bmatrix} \dfrac{1}{v^2}\dfrac{\rho}{\beta^2}\chi_7 \\[2ex] \dfrac{1}{v^2}\left\{\pm\dfrac{\rho}{\beta^2\sqrt{\beta\pi}}\chi_8+\dfrac{\rho u}{\beta^2}\chi_7\right\} \\[2ex] \dfrac{1}{v^2}\left\{\dfrac{\rho}{\beta^3}\chi_9\pm\dfrac{\rho u}{\beta^2\sqrt{\beta\pi}}\chi_8+\dfrac{\rho u^2}{\beta^2}\left(\dfrac{\chi_7}{2}\right)\right\} \end{bmatrix} \tag{139c}$$

$$\mathbf{G}^{B4\pm} = \begin{bmatrix} \pm\dfrac{1}{v^2}\dfrac{\rho}{\beta\sqrt{\beta\pi}}\chi_{10} \\[2ex] \dfrac{1}{v^2}\left\{\dfrac{\rho}{\beta^2}\chi_{11}\pm\dfrac{\rho u}{\beta\sqrt{\beta\pi}}\chi_{10}\right\} \\[2ex] \dfrac{1}{v^2}\left\{\pm\dfrac{\rho}{\beta^2\sqrt{\beta\pi}}\chi_{12}+\dfrac{\rho u}{\beta^2}\chi_{11}\pm\dfrac{\rho u^2}{\beta\sqrt{\beta\pi}}\left(\dfrac{\chi_{10}}{2}\right)\right\} \end{bmatrix} \tag{139d}$$

$$\mathbf{G}^{B5\pm} = \begin{bmatrix} \pm\dfrac{1}{v^2}\dfrac{\rho}{\beta^2\sqrt{\beta\pi}}\chi_{13} \\[2ex] \dfrac{1}{v^2}\left\{\dfrac{\rho}{\beta^3}\chi_{14}\pm\dfrac{\rho u}{\beta^2\sqrt{\beta\pi}}\chi_{13}\right\} \\[2ex] \dfrac{1}{v^2}\left\{\pm\dfrac{\rho}{\beta^3\sqrt{\beta\pi}}\chi_{15}+\dfrac{\rho u}{\beta^3}\chi_{14}\pm\dfrac{\rho u^2}{\beta^2\sqrt{\beta\pi}}\left(\dfrac{\chi_{13}}{2}\right)\right\} \end{bmatrix} \tag{139e}$$

$$\mathbf{G}^{B6\pm} = \begin{bmatrix} \dfrac{1}{v^2}\dfrac{\rho}{\beta}\chi_{16} \\[2ex] \dfrac{1}{v^2}\left\{\pm\dfrac{\rho}{\beta\sqrt{\beta\pi}}\chi_{17}+\dfrac{\rho u}{\beta}\chi_{16}\right\} \\[2ex] \dfrac{1}{v^2}\left\{\dfrac{\rho}{\beta^2}\chi_{18}\pm\dfrac{\rho u}{\beta\sqrt{\beta\pi}}\chi_{17}+\dfrac{\rho u^2}{\beta}\left(\dfrac{\chi_{16}}{2}\right)\right\} \end{bmatrix} \tag{139f}$$

$$G^{B7\pm} = \begin{bmatrix} \dfrac{1}{v^2}\dfrac{\rho}{\beta^3}\chi_{19} \\[2ex] \dfrac{1}{v^2}\left\{\pm\dfrac{\rho}{\beta^3\sqrt{\beta\pi}}\chi_{20}+\dfrac{\rho u}{\beta^3}\chi_{19}\right\} \\[2ex] \dfrac{1}{v^2}\left\{\dfrac{\rho}{\beta^4}\chi_{21}\pm\dfrac{\rho u}{\beta^3\sqrt{\beta\pi}}\chi_{20}+\dfrac{\rho u^2}{\beta^3}\left(\dfrac{\chi_{19}}{2}\right)\right\} \end{bmatrix}$$

$$(139g)$$

$$G^{B8\pm} = \begin{bmatrix} \pm\dfrac{1}{v^2}\dfrac{\rho}{\beta\sqrt{\beta\pi}}\chi_{22} \\[2ex] \dfrac{1}{v^2}\left\{\dfrac{\rho}{\beta^2}\chi_{23}\pm\dfrac{\rho u}{\beta\sqrt{\beta\pi}}\chi_{22}\right\} \\[2ex] \dfrac{1}{v^2}\left\{\pm\dfrac{\rho}{\beta^2\sqrt{\beta\pi}}\chi_{24}+\dfrac{\rho u}{\beta^2}\chi_{23}\pm\dfrac{\rho u^2}{\beta\sqrt{\beta\pi}}\left(\dfrac{\chi_{22}}{2}\right)\right\} \end{bmatrix}$$

$$(139h)$$

$$G^{B9\pm} = \begin{bmatrix} \dfrac{1}{v^2}\rho\chi_{25} \\[2ex] \dfrac{1}{v^2}\left\{\pm\dfrac{\rho}{\sqrt{\beta\pi}}\chi_{26}+\rho u\chi_{25}\right\} \\[2ex] \dfrac{1}{v^2}\left\{\dfrac{\rho}{\beta}\chi_{27}\pm\dfrac{\rho u}{\sqrt{\beta\pi}}\chi_{26}+\rho u^2\left(\dfrac{\chi_{25}}{2}\right)\right\} \end{bmatrix}$$

$$(139i)$$

$$G^{B10\pm} = \begin{bmatrix} 0 \\[2ex] \dfrac{1}{v^2}\left\{\pm\dfrac{\rho}{\beta^2\sqrt{\beta\pi}}\chi_{28}\right\} \\[2ex] \dfrac{1}{v^2}\left\{\dfrac{\rho}{\beta^3}\chi_{29}\pm\dfrac{\rho u}{\beta^2\sqrt{\beta\pi}}\chi_{28}\right\} \end{bmatrix}$$

$$(139j)$$

$$G^{B11\pm} = \begin{bmatrix} \dfrac{1}{v^2}\left\{\pm\dfrac{\rho}{\sqrt{\beta\pi}}\chi_{30}\right\} \\[2ex] \dfrac{1}{v^2}\left\{\dfrac{\rho}{\beta}\chi_{31}\pm\dfrac{\rho u}{\sqrt{\beta\pi}}\chi_{30}\right\} \\[2ex] \dfrac{1}{v^2}\left\{\pm\dfrac{\rho}{\beta\sqrt{\beta\pi}}\chi_{32}+\dfrac{\rho u}{\beta}\chi_{31}\pm\dfrac{\rho u^2}{\sqrt{\beta\pi}}\left(\dfrac{\chi_{30}}{2}\right)\right\} \end{bmatrix}$$

$$(139k)$$

The variables χ_1 through χ_{32} are functions of γ and are given in (Ref. 18).

5.2. Entropy Considerations and Boundary Conditions

The BGK-Burnett equations are expressed in the conservation law form as:

$$\frac{\partial \mathbf{Q}}{\partial t} + \frac{\partial \mathbf{G}}{\partial x} = 0 \qquad (140)$$

The flux vector is given as:

$$\mathbf{G} = \begin{bmatrix} \rho u \\ \rho u^2 + p - \tau_{xx}^{N-S} - \tau_{xx}^{B} \\ \rho u e + p u + q_x^{N-S} + q_x^{B} \end{bmatrix} \qquad (141)$$

The boundary conditions for the BGK-Burnett equations are derived from the Gibbs entropy equation. The Gibbs relation can be expressed as:

$$T\nabla s = \nabla h - \frac{1}{\rho}\nabla p \qquad (142)$$

In one dimension the above equation takes the form:

$$\left(\frac{\partial s}{\partial x}\right) = \frac{1}{T}\left(\frac{\partial h}{\partial x}\right) - \frac{1}{\rho T}\left(\frac{\partial p}{\partial x}\right) \qquad (143)$$

The second law of thermodynamics states that the entropy production rate must be positive for a steady state process. Hence the difference in entropy between a station far downstream and a station far upstream of a shock must be positive. In order to derive a boundary condition based on the second law of thermodynamics two stations are chosen far upstream and far downstream of a normal shock. The subscripts '1' and '2' are used to identify the upstream and downstream locations respectively. On substituting for $\left(\dfrac{\partial h}{\partial x}\right)$ and $\left(\dfrac{\partial p}{\partial x}\right)$ from the flow equations, the following equation is obtained after simplification.

$$\overset{0}{m}(s_2 - s_1) = \int_1^2 \frac{\tau_x}{T}\left(\frac{\partial u}{\partial x}\right)dx - \int_1^2 \frac{1}{T}\frac{\partial}{\partial x}(q_x)dx \qquad (144)$$

Eq. (144) can be written in the following form:

$$\overset{0}{m}(s_2 - s_1) = \int_1^2 A\,dx + \left[-\frac{q_x}{T}\right]_1^2 + \int_1^2 B\,dx \qquad (145)$$

where

$$A = \frac{\tau_x}{T}\left(\frac{\partial u}{\partial x}\right) \quad \text{and} \quad B = -\frac{q_x}{T^2}\left(\frac{\partial T}{\partial x}\right) \qquad (146)$$

The above equation implies that setting the heat transfer q_x equal to zero at the ends of the control volume ensures a positive entropy gradient if 'A' and 'B' are positive. For the Navier-Stokes equations, a positive entropy gradient is ensured by setting the heat transfer q_x^{N-S} terms equal to zero at stations '1' and '2'. This is achieved by setting the derivative $\left(\dfrac{\partial T}{\partial x}\right)$ equal to zero. For the Navier-Stokes equations the integrands 'A' and 'B' take the form shown below. It can be seen that both 'A' and 'B' are positive.

$$A = \frac{\mu}{T}\left(\frac{\partial u}{\partial x}\right)^2 \text{ and } B = \frac{k}{T^2}\left(\frac{\partial T}{\partial x}\right)^2 \qquad (147)$$

Although a similar analytical analysis is difficult for the BGK-Burnett equations, it has been verified computationally (Ref. 18) that the integrands 'A' and 'B' are positive. This is ensured by setting all the derivatives in q_x^B equal to zero at stations '1' and '2' at the ends of the control volume enclosing the shock. Physically speaking this implies that the velocity and temperature gradients become vanishingly small at the ends of the control volume. The entropy consistent boundary conditions are obtained by setting the following derivatives equal to zero at stations '1' and '2' at the ends of the control. volume enclosing the shock.

$$\left(\frac{\partial u}{\partial x}\right) = 0; \ \left(\frac{\partial^2 u}{\partial x^2}\right) = 0; \ \left(\frac{\partial^3 u}{\partial x^3}\right) = 0 \ ;$$

$$\left(\frac{\partial T}{\partial x}\right) = 0; \ \left(\frac{\partial^2 T}{\partial x^2}\right) = 0; \text{ and } \left(\frac{\partial^3 T}{\partial x^3}\right) = 0 \qquad (148)$$

5.3 Numerical Experiments

The KWPS algorithm is applied to the 1-D BGK-Burnett equations for the Hypersonic Shock Structure Problem. The variables are finite differenced and they are treated explicitly. The objective of this numerical experiment is to test the computational stability of the entropy consistent BGK-Burnett equations by integrating them numerically on progressively finer grids. In order to test the stability of the algorithm, the scheme was applied initially to a coarse mesh of 101 grid points. The number of grid points was increased to 501.

The hypersonic shock structure for argon has been computed using the entropy consistent BGK-Burnett equations. The upstream flow conditions were specified and the downstream conditions were determined by the Rankine-Hugoniot relations. For purposes of comparison the same flow conditions as Zhong (Ref. 2) were used for computations. The parameters used are:

$$\boxed{\begin{aligned} T_\infty &= 300 \ ^\circ K \\ P_\infty &= 1.01325 \times 10^5 \ Nm^{-2} \\ \gamma_{Argon} &= 1.66 \\ \mu_{Argon} &= 22.7 \times 10^{-6} \ Nm^{-2} \end{aligned}} \qquad (149)$$

The Navier-Stokes solution was taken as the initial value for the BGK-Burnett equations. This smooth spatial distribution of variables is imposed on a mesh that corresponds to the expected shock thickness. The length of the control volume enclosing the shock is chosen to be equal to $1000\lambda_\infty$. For purposes of comparison the Knudsen number for the shock-structure problem is based on the length of the control volume. It must be noted, however, that a characteristic length cannot be defined for a 1-D problem. The mean free path based on the free stream parameters is obtained from the following relation:

$$\lambda_\infty = \frac{16\mu}{5\rho_\infty \sqrt{2\pi RT_\infty}} \qquad (150)$$

This is the mean free path that would exist in the unshocked region if the gas were composed of hard elastic spheres and had the same viscosity, density and temperature of the gas being considered. The solution is marched till the observed deviations are smaller than a preset convergence criterion. The solution is later checked to see if the fluxes are conserved at each point.

To begin with a set of computational experiments was carried out to compare the BGK-Burnett (with stress and heat transfer terms up to μ^2) solutions with the Burnett solutions of Fiscko and Chapman. For this purpose a hybrid scheme was used that used a first-order upwind KWPS scheme for the Euler flux vector and central differenced the viscous and BGK-Burnett/Burnett flux vectors. Tests were carried out for Mach 20 and Mach 35. Fig(s) 21 through 24 show the variations of specific entropy across the shock. It is observed that the BGK-Burnett equations show a positive change in entropy throughout the flow field. The conventional Burnett equations however break down due to a large negative entropy spike just ahead of the shock. Fig(s) 25 - 26 compare the KWPS solutions of the BGK-Burnett and Navier Stokes equations. The BGK-Burnett equations did not exhibit any kind of instability for the range of grid points and Knudsen numbers considered. The first order upwind computations of the BGK-Burnett equations takes about 20% to 40% more CPU time as compared to the first order upwind computations of the Navier-Stokes equations.

5.4 Conclusions

An entropy consistent set of BGK-Burnett equations has been derived from first principles. These equations have been numerically integrated to compute the hypersonic shock structure. The equations are computationally stable for the range of grid points and Knudsen numbers for which results are presented.

ACKNOWLEDGMENT

This work has been supported by AFOSR contract F49620-95-1-0125. The authors wish to acknowledge Dr. Len Sakell for his support and encouragement on this project.

REFERENCES

1. Fiscko, K. A., and Chapman, D. R., "Comparison of Burnett, Super-Burnett and Monte Carlo Solutions for Hypersonic Shock Structure", Proceedings 16th International Symposium on Rarefied Gas Dynamics, Pasadena, California, July 11-15, 1988.

2. Zhong, X., "Development and Computation of Continuum Higher Order Constitutive Relations for High-Altitude Hypersonic Flow", Ph.D. Thesis, Stanford University, 1991.

3. Bobylev, A. V., "The Chapman-Enskog and Grad Methods for Solving the Boltzmann Equation", Sov. Phys. Dokl., 27(1), January 1982.

4. Welder, W. T., Chapman, D. R., and MacCormack, R. W., "Evaluation of Various Forms of the Burnett Equations", AIAA-93-3094, AIAA 24th Fluid Dynamics Conference, July 6-9, 1993, Orlando, FL.

5. Zhong, X., MacCormack, and R. W., Chapman, D. R., "Stabilization of the Burnett Equations and Application to High Altitude Hypersonic Flows", AIAA 91-0770, 29th Aerospace Sciences Meeting. January 7-9, 1991, Reno, NV.

6. Fiscko, Kurt A., and Chapman, D. R., "Hypersonic Shock Structure with Burnett terms in the Viscous Stress and Heat Flux", AIAA 88-2733, AIAA Thermodynamics, Plasmadynamics, and Lasers Conference, June 27-29, 1988, San Antonio, Texas.

7. Comeaux, K. A., Chapman, D. R., and MacCormack, R. W., "An Analysis of the Burnett Equations Based on the Second Law of Thermodynamics", AIAA 95-0415, 33rd Aerospace Sciences Meeting. January 9-12, 1995, Reno, NV.

8. Balakrishnan R., and Agarwal, R. K., "Development of the First Order Distribution Function and a Kinetic Wave/Particle Flux Splitting Scheme for the Navier-Stokes Equations", report (unpublished), Department of Aerospace Engineering, Wichita State University, 1995.

9. Balakrishnan R., and Agarwal, R. K., "Formulation of a Second Order Distribution Function by Perturbation Analysis of the Boltzmann Equation", report (unpublished), Department of Aerospace Engineering, Wichita State University, 1995.

10. Balakrishnan R., and Agarwal, R. K., "Development of a Kinetic Wave/Particle Flux Splitting Scheme for the Burnett Equations", report (unpublished), Department of Aerospace Engineering, Wichita State University, 1995.

11. Deshpande, S. M., NASA Technical Paper No. 2613, 1986.

12. Acheson, K. E., and Agarwal, R. K., "A Kinetic Theory Based Wave/Particle Flux Splitting Scheme for Euler Equations", AIAA 95-2178, 266 AIAA Fluid Dynamic Conference, San Diego, CA, June 1995.

13. Xu, K., Martinelli, Luigi., and Jameson, A., "Gas Kinetic Finite Volume Methods, Flux-Vector Splitting, and Artificial Diffusion", Journal of Computational Physics 120, 48-65, 1995.

14. Perthame, B., and Tadmor, E., Commun. Math. Phys. 36, 501 (1991).

15. Woods, L. C., "Principles of Magnetoplasma Dynamics", Oxford Science Publications, 1987.

16. Grad, H., "On the Kinetic Theory of Rarefied Gases", Communications on Pure and Applied Mathematics. Vol. 2, Pages 331-407, 1949.

17. Balakrishnan, R., and Agarwal, R. K., "Entropy Consistent Formulation and Numerical Simulation of the BGK-Burnett Equations for Hypersonic Flows in the Continuum-Transition Regime", International Conference for Numerical Methods in Fluid Dynamics, 24-28 June, 1996, Monterey, California.

18. Balakrishnan, R., and Agarwal, R. K., "A Kinetic Theory Based Scheme for the Numerical Solution of the BGK-Burnett Equations for Hypersonic Flows in the Continuum-Transition Regime", AIAA 96-0602, 34th Aerospace Sciences Meeting & Exhibit, Jan 15-18, 1996, Reno, NV.

19. Agarwal, R. K., and Balakrishnan, R., "Numerical Simulation of BGK-Burnett Equations", Final Contract Report Submitted to AFOSR, Aug.' 15, 1996.

20. Deshpande, S. M., "Kinetic Flux Splitting Schemes", Computational Fluid Dynamics Review, 1995.

21. Ferziger, J. H., and Kaper, H. G., "Mathematical Theory of Transport Processes in Gases", Amsterdam, North-Holland Pub. Co. , 1972.

Table: 1

Stress and Heat Transfer Coefficients	Fiscko & Chapman (Air)	BGK-Burnett (Air)	Fiscko & Chapman (Argon)	BGK-Burnett (Argon)
$a^{(1)}$	1.749	0.96	1.749	0.446
$a^{(2)}$	-388.024	-459.2	-281.216	-277.472
$a^{(3)}$	388.024	459.2	281.216	277.472
$a^{(4)}$	-257.726	-5625.0	-186.784	-1216
$a^{(5)}$	403.522	-5166.0	292.448	-938.11
$a^{(6)}$	74.62	-5625	54.08	-1216.0
$b^{(1)}$	10.831	-21.633	10.831	-9.896
$b^{(2)}$	-2.269	0.183	-2.269	-0.194
$b^{(3)}$	-2.06	0.533	-2.06	-0.443

Table: 2

Stress Coefficients	Zhong (Air)	BGK-Burnett (Air)	Zhong (Ar)	BGK-Burnett (Ar)
$a^{(1)}$	1.749	0.96	1.749	0.446
$a^{(2)}$	-388.024	-459.2	-281.216	-277.472
$a^{(3)}$	388.024	459.2	281.216	277.472
$a^{(4)}$	-257.726	-5625	-186.784	-1216
$a^{(5)}$	403.522	-5166	292.448	-938.11
$a^{(6)}$	74.62	-5625	54.08	-1216
$a^{(7)}$	0	-459.2	0	-277.472
$a^{(8)}$	0	642.88	0	462.268
$a^{(9)}$	63.778	459.2	46.222	277.472
$a^{(10)}$	0	0.64	0	0.888

Table: 3

Heat Transfer Coef's	Zhong (Air)	BGK-Burnett (Air)	Zhong (Ar)	BGK-Burnett (Ar)
$b^{(1)}$	10.831	-21.633	10.831	-9.896
$b^{(2)}$	-2.269	0.183	-2.269	-0.194
$b^{(3)}$	-2.06	0.533	-2.06	-0.443
$b^{(4)}$	0	4.4	0	4.666
$b^{(5)}$	0	1406	0	866.84
$b^{(6)}$	0	-1406	0	-866.84
$b^{(7)}$	-2186	1406	-2041	866.84
$b^{(8)}$	0	0.7	0	0.833
$b^{(9)}$	0	-1.4	0	-1.666
$b^{(10)}$	0	-22500	0	-13870
$b^{(11)}$	-179.375	0	-130	0

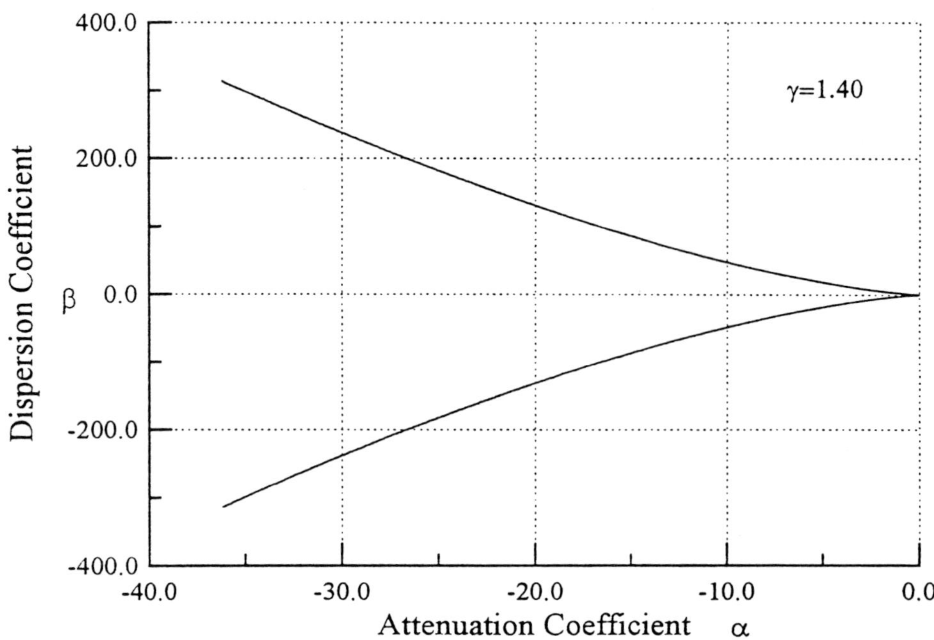

Fig. 1 Linearized Stability Analysis of the BGK-Burnett equations (One) with the Euler approximation for the material derivatives in the stress and heat transfer expressions.

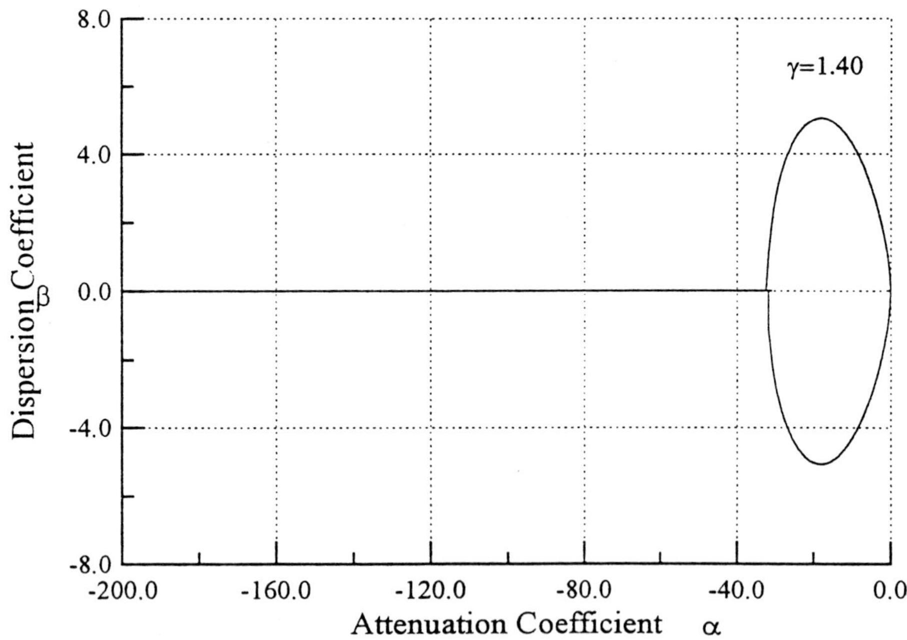

Fig. 2 Linearized Stability Analysis of the BGK-Burnett equations (One) with Navier-Stokes approximation for the material derivatives in the stress and heat transfer expressions.

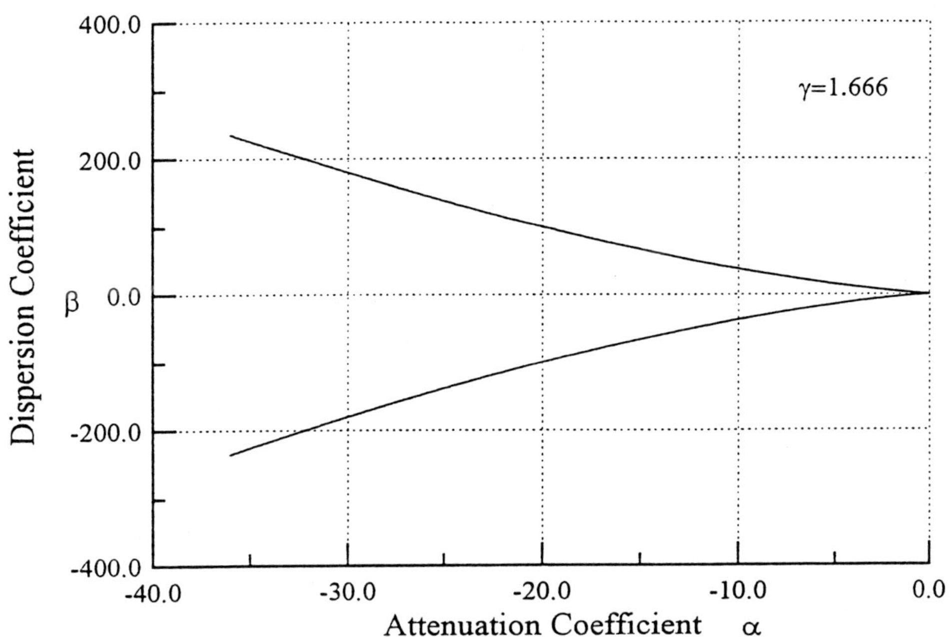

Fig. 3 Linearized Stability Analysis of the BGK-Burnett equations (One) with the Euler approximation for the material derivatives in the stress and heat transfer expressions.

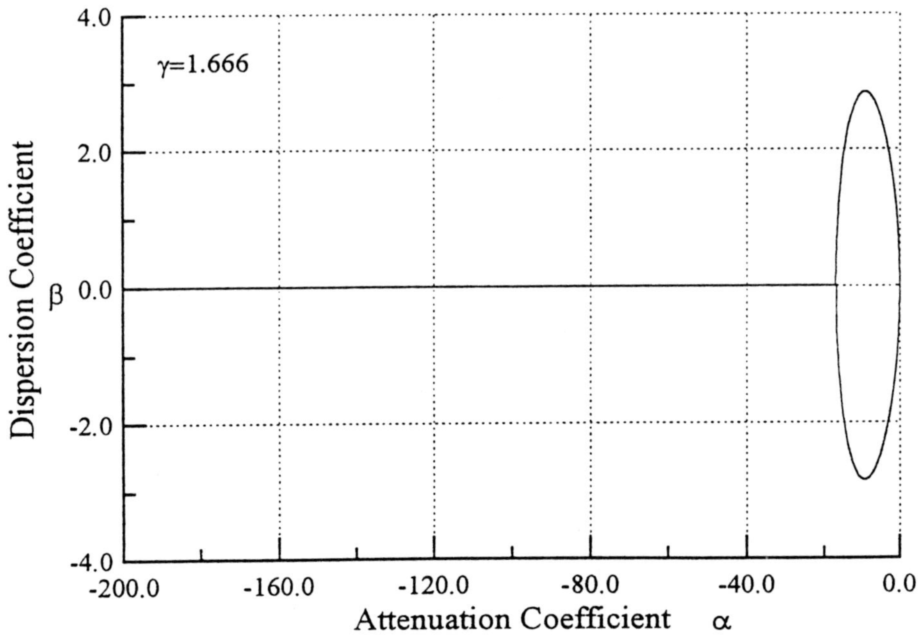

Fig. 4 Linearized Stability Analysis of the BGK-Burnett equations (One) with Navier-Stokes approximation for the material derivatives in the stress and heat transfer expressions

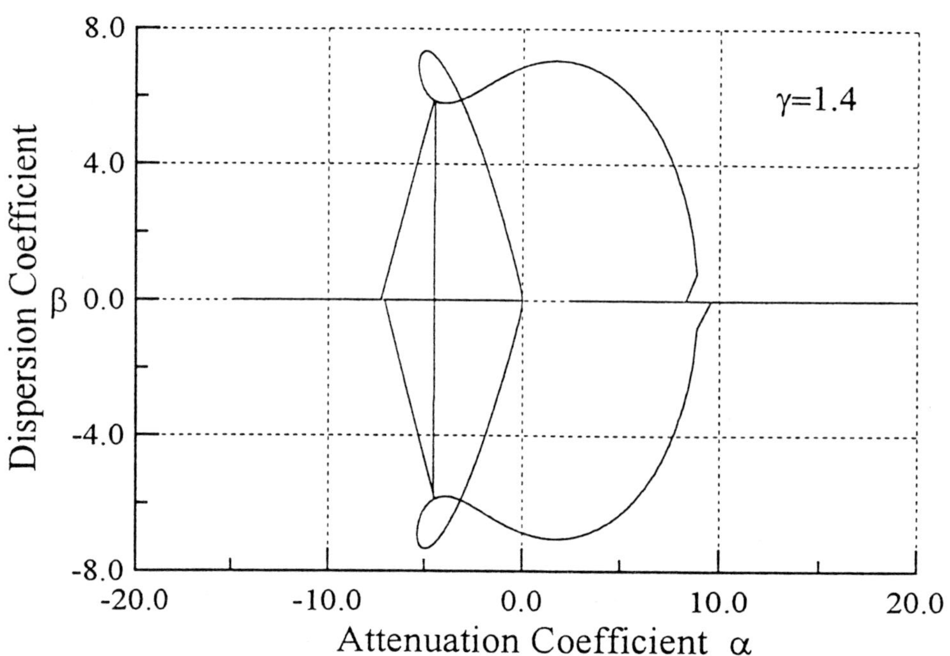

Fig. 5 Linearized Stability Analysis of the BGK-Burnett equations (Two) with the Euler approximation for the material derivatives in the stress and heat transfer expressions.

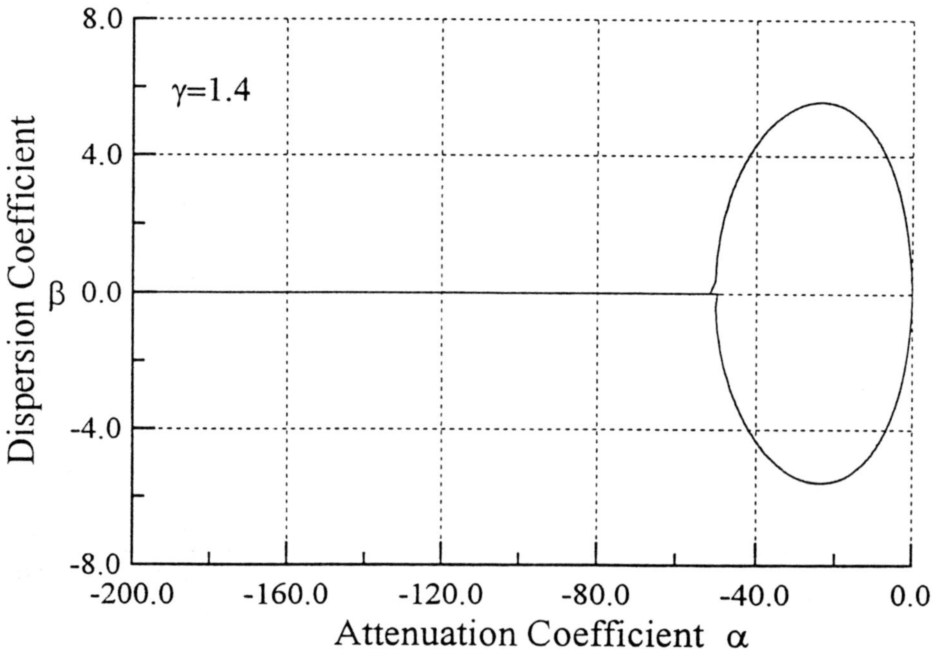

Fig. 6 Linearized Stability Analysis of the BGK-Burnett equations (Two) with Navier-Stokes approximation for the material derivatives in the stress and heat transfer expressions.

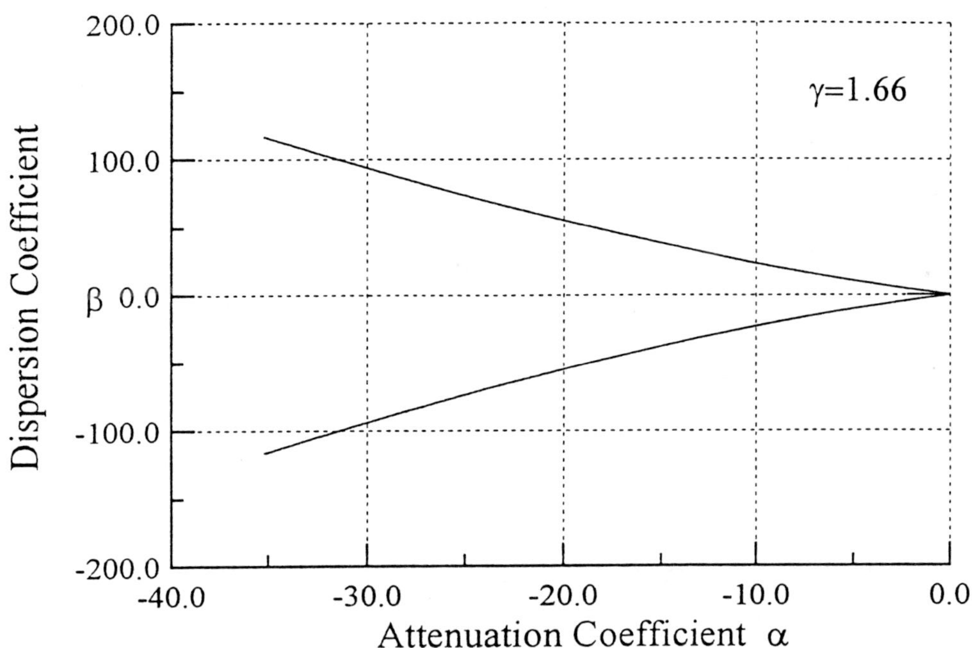

Fig. 7 Linearized Stability Analysis of the BGK-Burnett equations (Two) with the Euler approximation for the material derivatives in the stress and heat transfer expressions.

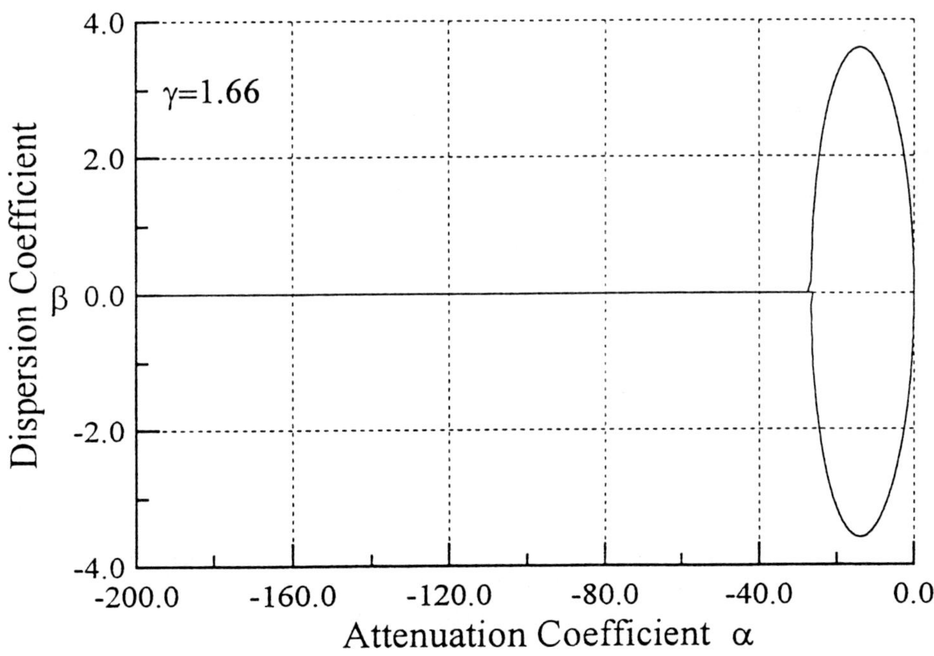

Fig. 8 Linearized Stability Analysis of the BGK-Burnett equations (Two) with Navier-Stokes approximation for the material derivatives in the stress and heat transfer expressions

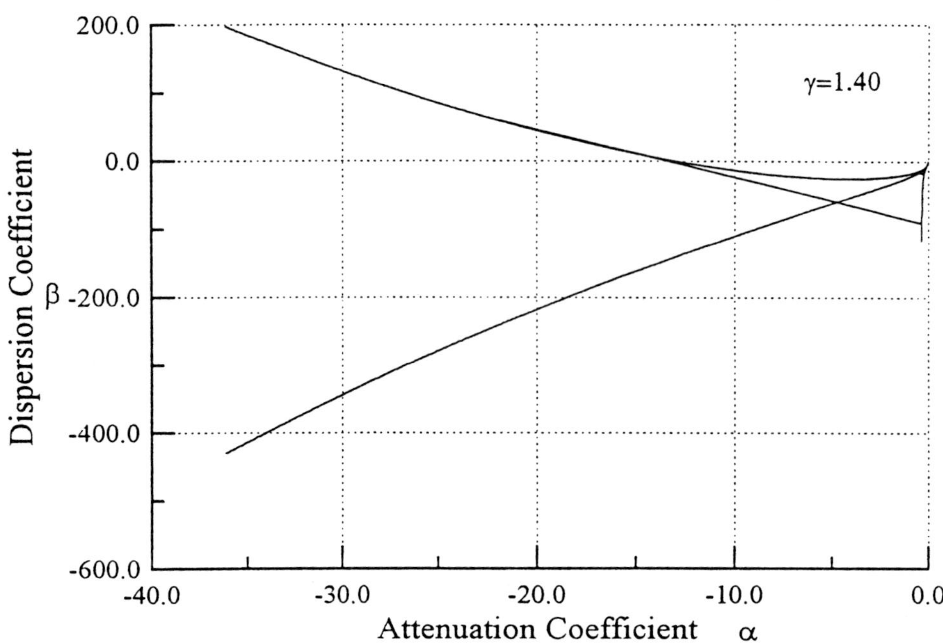

Fig. 9 Linearized Stability Analysis of the BGK-Burnett equations (One) with the Euler approximation for the material derivatives. Free stream Mach number, M=20.

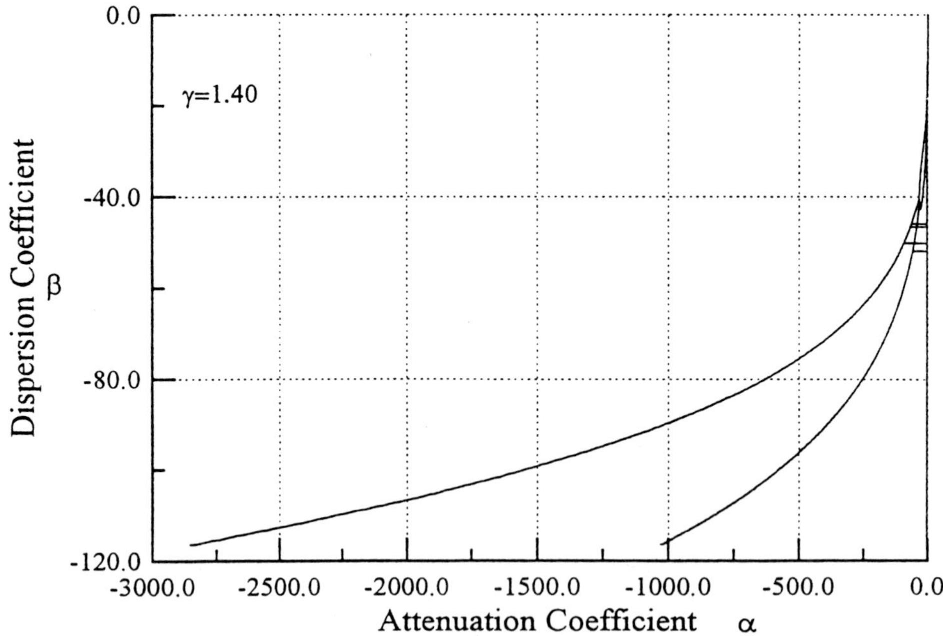

Fig. 10 Linearized Stability Analysis of the BGK-Burnett equations (One) with Navier-Stokes approximation for the material derivatives. Free stream Mach number M=20.

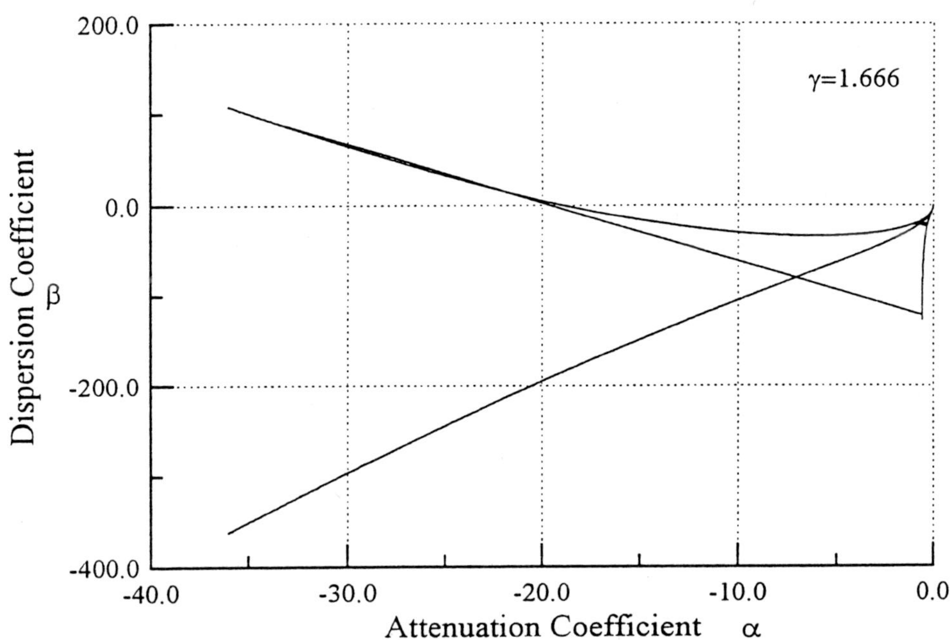

Fig. 11 Linearized Stability Analysis of the BGK-Burnett equations (One) with the Euler approximation for the material derivatives. Free stream Mach number, M=20.

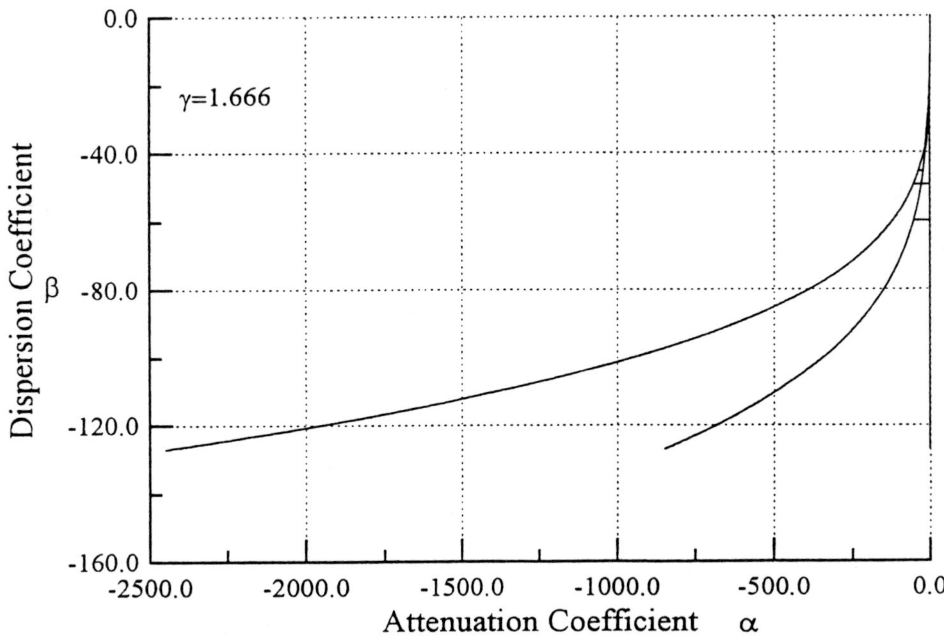

Fig. 12 Linearized Stability Analysis of the BGK-Burnett equations (One) with Navier-Stokes approximation for the material derivatives. Free stream Mach number, M=20.

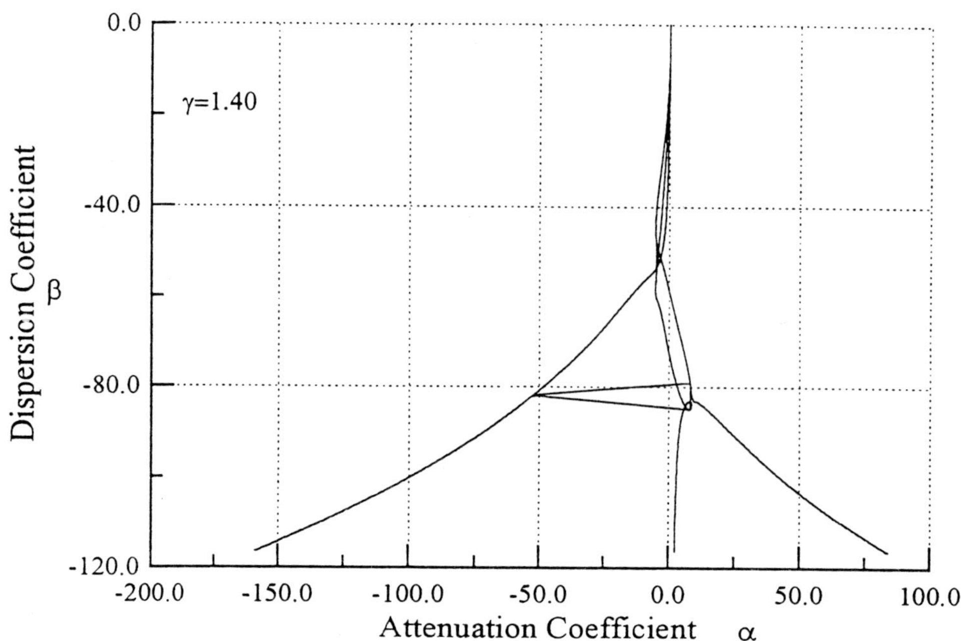

Fig. 13 Linearized Stability Analysis of the BGK-Burnett equations (Two) with the Euler approximation for the material derivatives. Free stream Mach number, M=20.

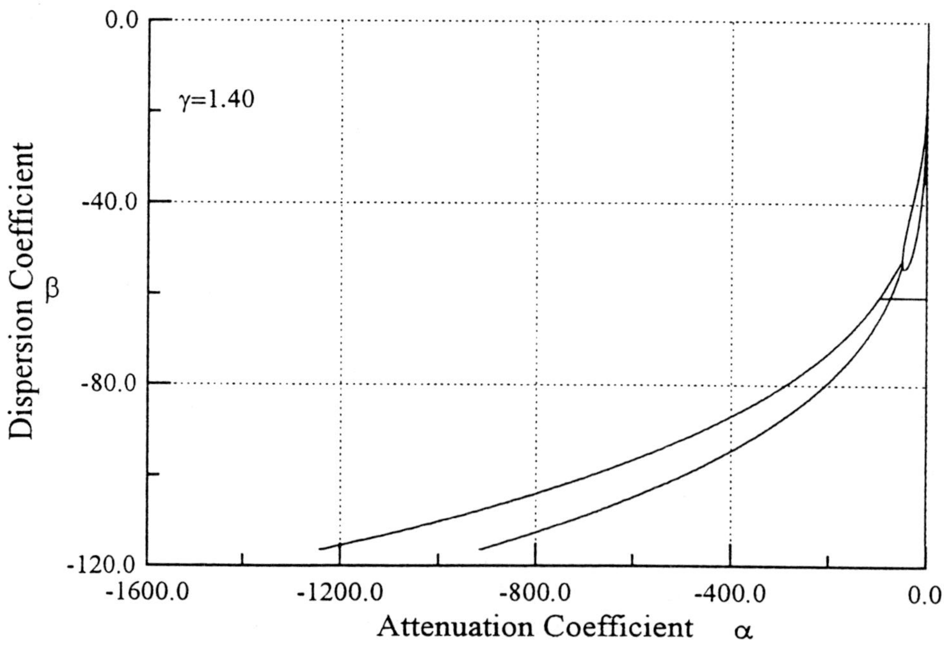

Fig. 14 Linearized Stability Analysis of the BGK-Burnett equations (Two) with Navier-Stokes approximation for the material derivatives. Free stream Mach number M=20.

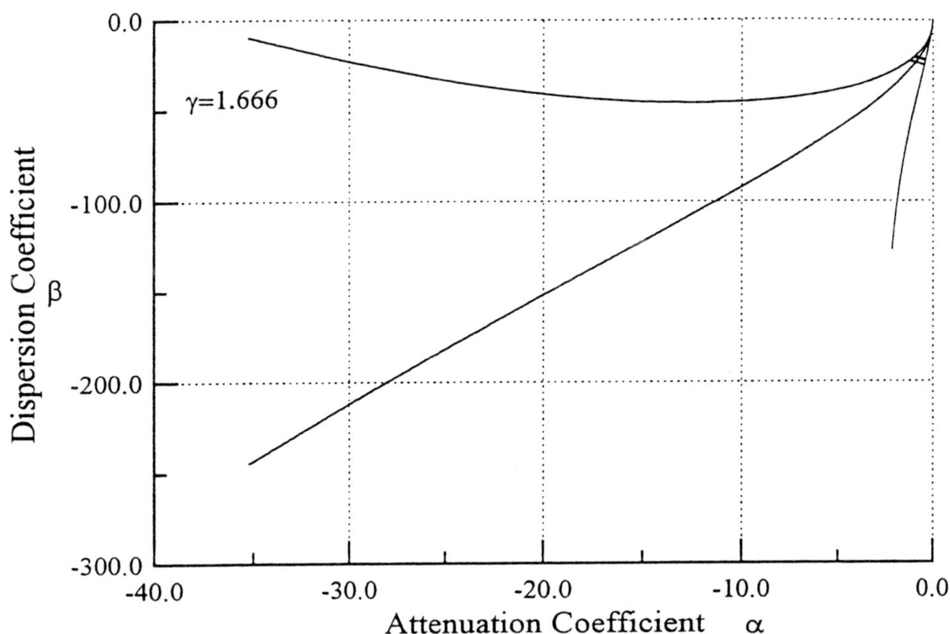

Fig. 15 Linearized Stability Analysis of the BGK-Burnett equations (Two) with the Euler approximation for the material derivatives. Free stream Mach number, M=20.

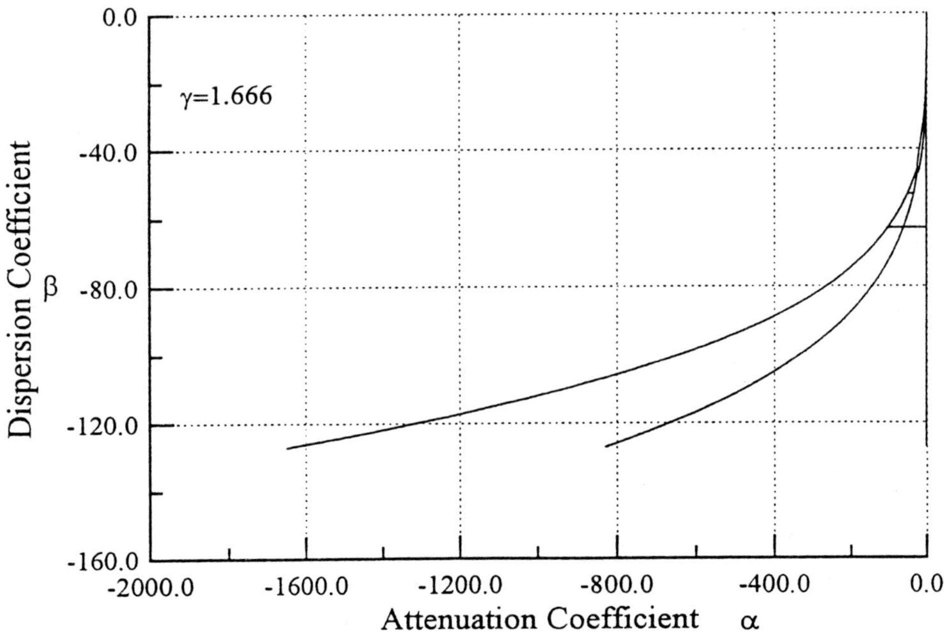

Fig. 16 Linearized Stability Analysis of the BGK-Burnett equations (Two) with Navier-Stokes approximation for the material derivatives. Free stream Mach number, M=20.

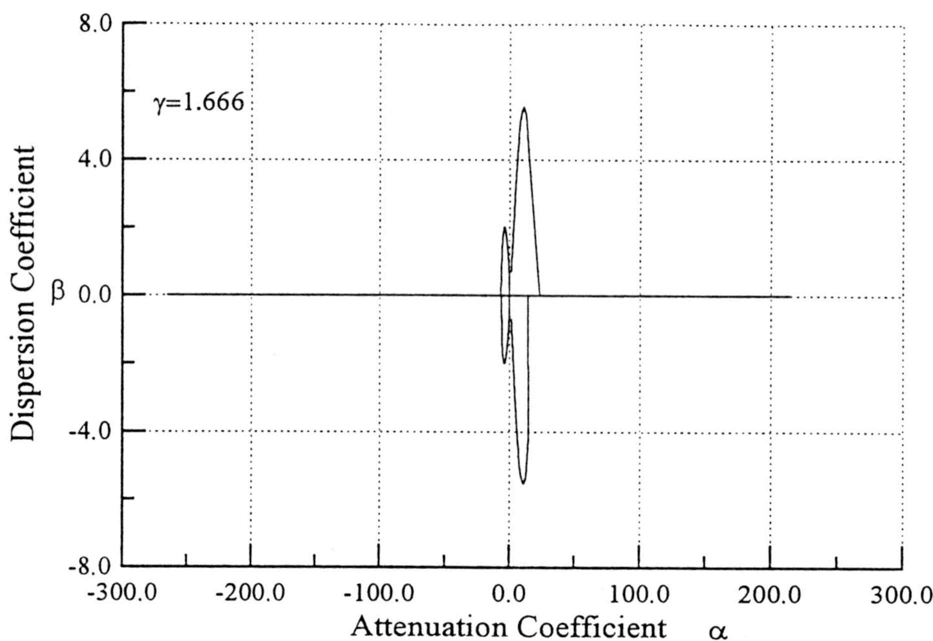

Fig. 17 Linearized Stability Analysis of the exact BGK-Burnett equations (One) Free stream Mach number, M=0.

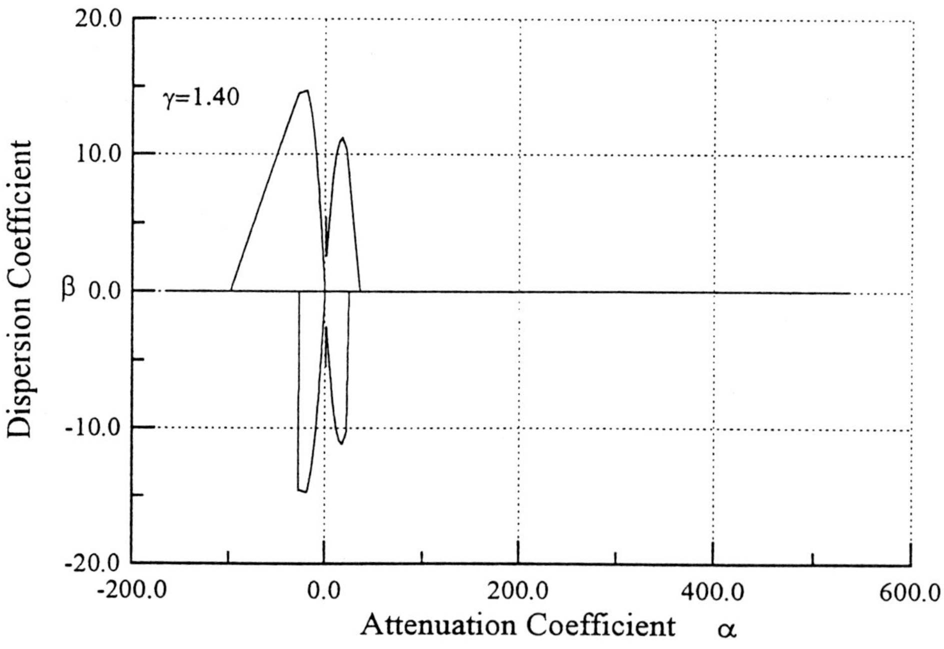

Fig. 18 Linearized Stability Analysis of the exact BGK-Burnett equations (One). Free stream Mach number M=0.

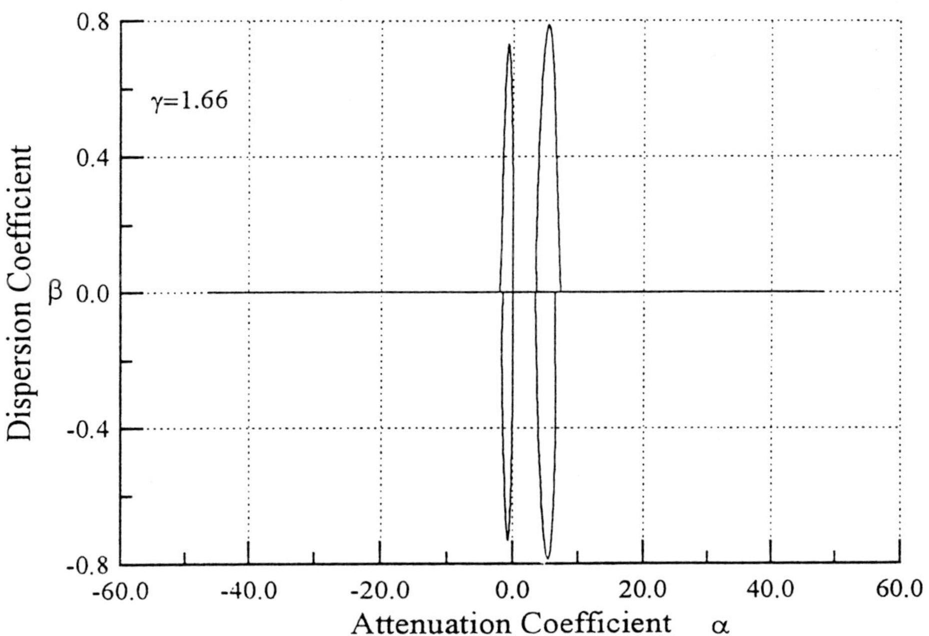

Fig. 19 Linearized Stability Analysis of the exact BGK-Burnett equations (Two). Free stream Mach number, M=0.

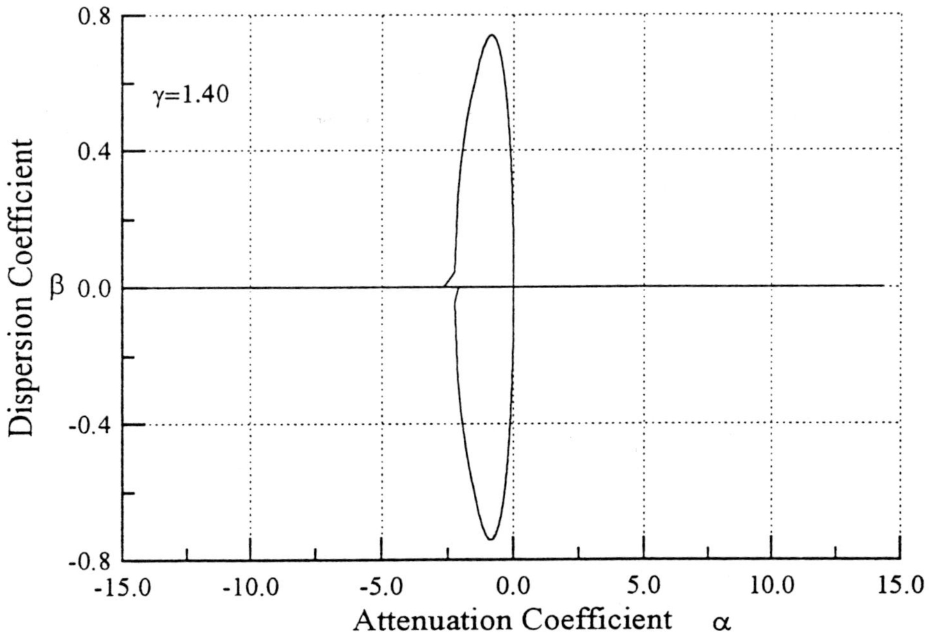

Fig. 20 Linearized Stability Analysis of the exact BGK-Burnett equations (Two). Free stream Mach number, M=0.

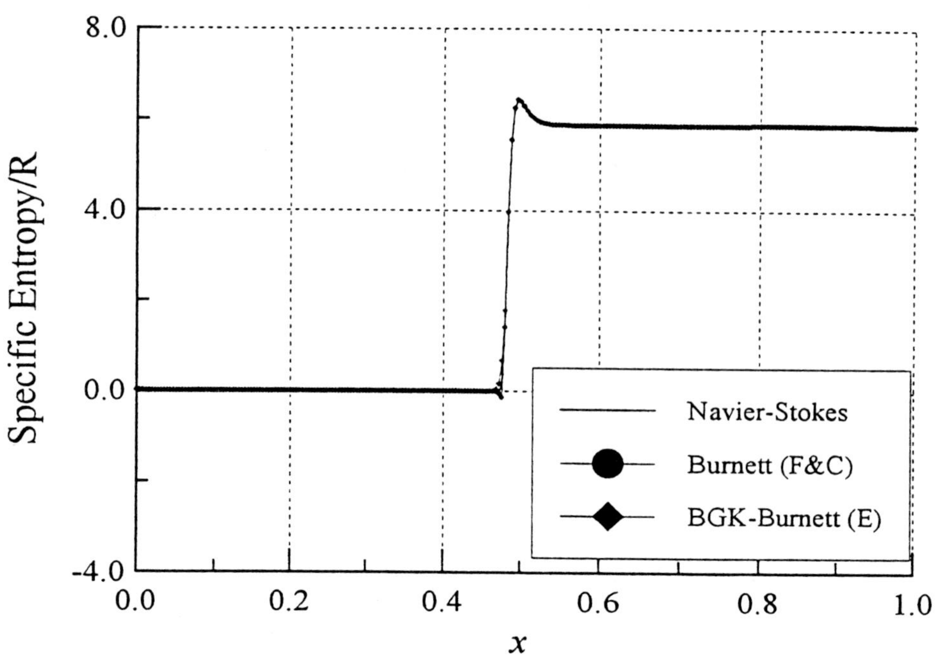

Fig. 21 Specific entropy variation across a Mach 20 (Argon) Normal shock. The free stream Knudsen number, Kn=0.0005

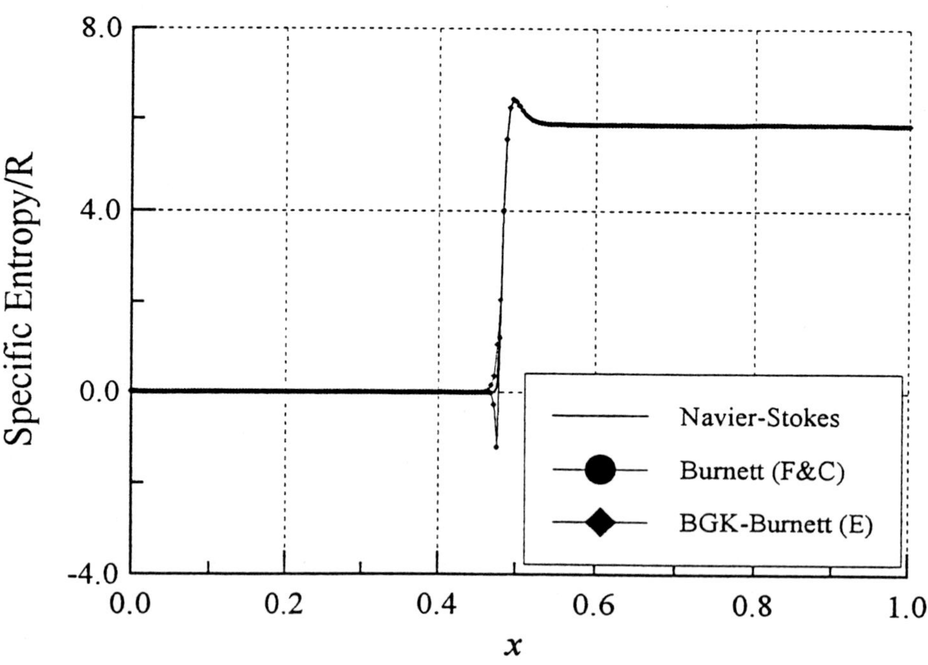

Fig. 22 Specific entropy variation across a Mach 20 (Argon) Normal Shock for a free stream Knudsen number of, Kn=0.001

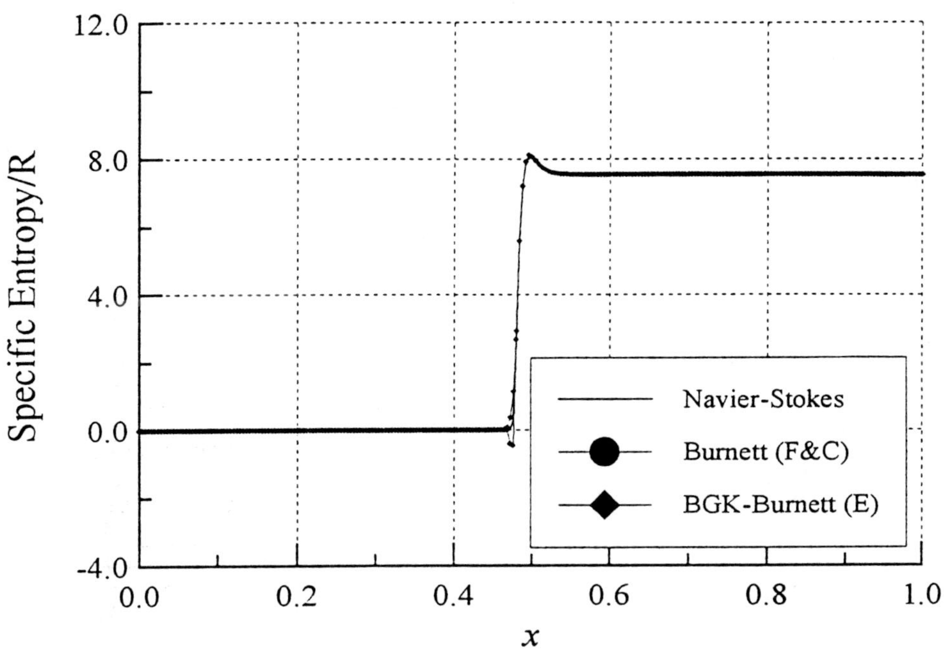

Fig. 23 Specific entropy variation across a Mach 35 (Argon) Normal Shock, for a free stream Knudsen number, Kn=0.0005

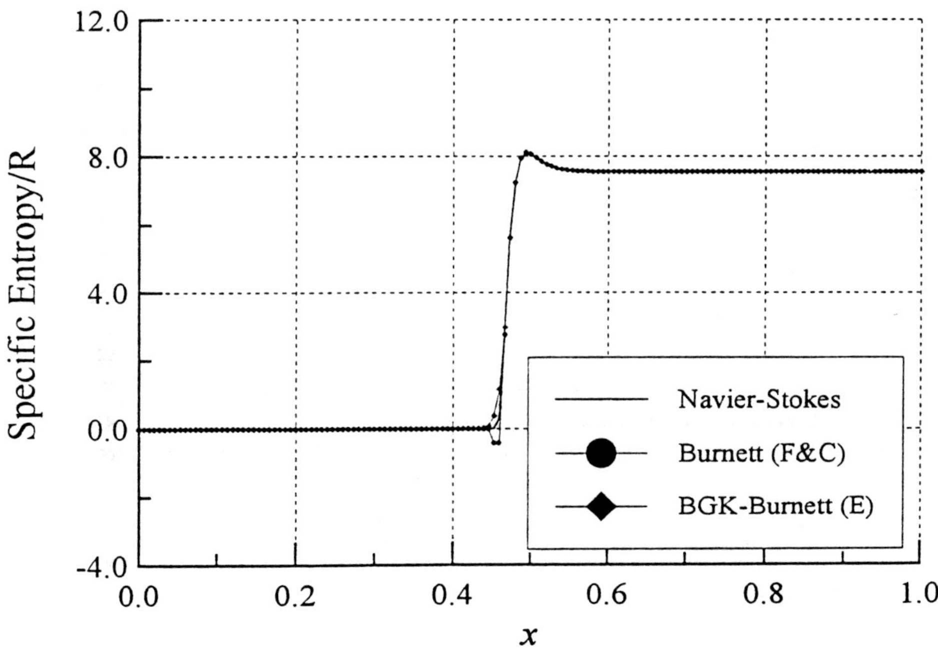

Fig. 24 Specific entropy variation across a Mach 35 (Argon) Normal Shock for a free stream Knudsen number, Kn=0.001

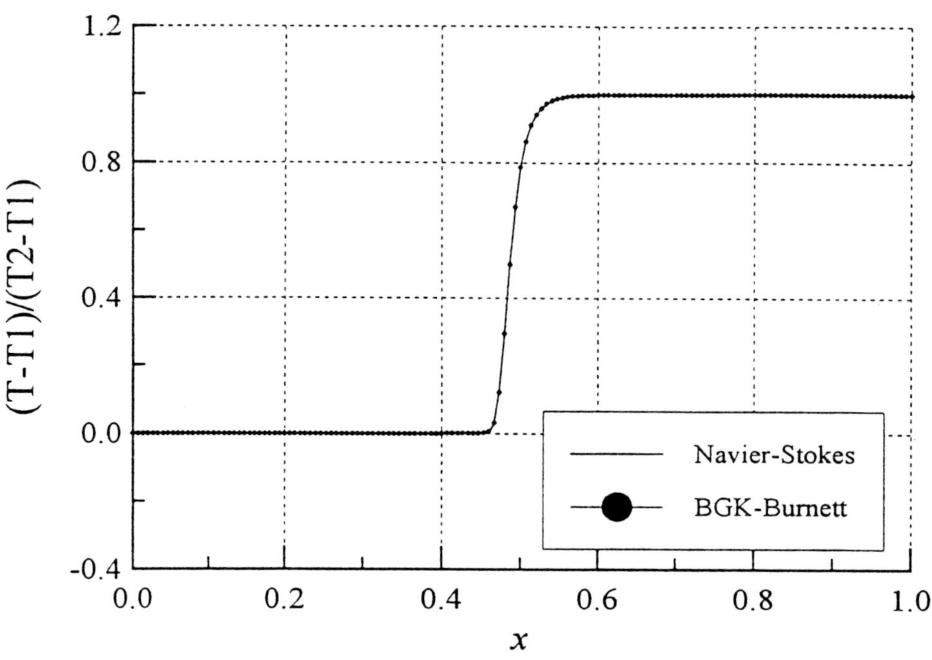

Fig. 25 Normalized temperature variation across a Mach 35 (Argon) Normal shock. The free stream Knudsen number, Kn=0.001. KWPS scheme with N-S equations for D()/Dt

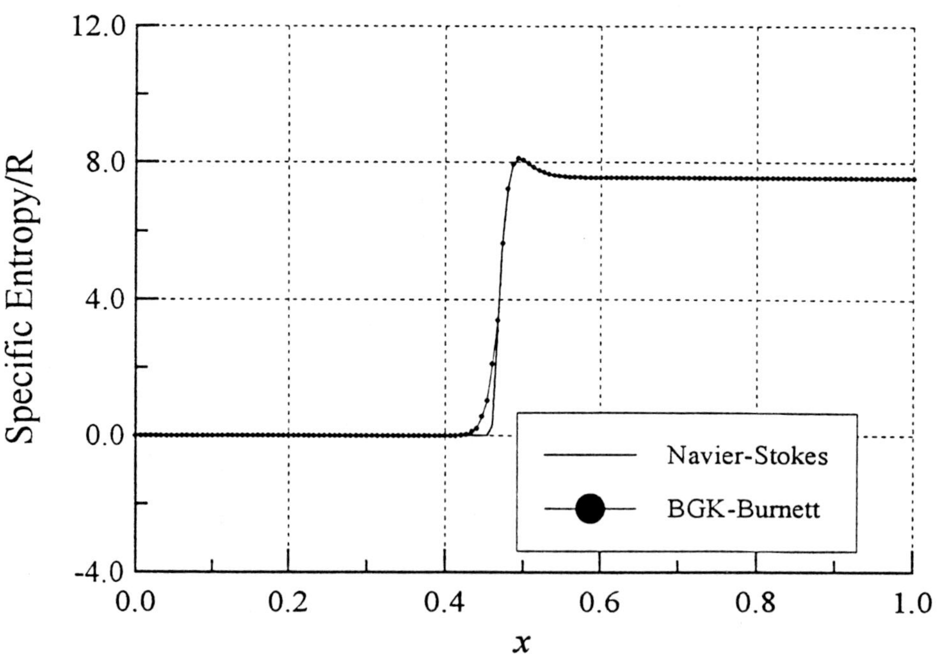

Fig. 26 Specific entropy variation across a Mach 35 (Argon) Normal Shock for a free stream Knudsen number of, Kn=0.001. KWPS scheme with N-S equations for D()/Dt.

COUPLING KINETIC MODELS
WITH NAVIER-STOKES EQUATIONS

P. LE TALLEC   J.P. PERLAT

INRIA, Domaine de Voluceau, B.P. 105, 78153 Le Chesnay Cedex, France.
email: patrick.letallec@inria.fr

1 Introduction

At high altitudes or in rarefied regimes, a gas is best modelled at a microscopic scale, as a collection of particles characterized by their velocity and position. The relevant mathematical model is then the Boltzmann equation, which governs the evolution of the space and velocity distribution of these particles and which can be solved numerically by Monte Carlo methods=[12].

This numerical solution requires a discretisation grid whose step is at most of the order of a particle mean free path (average distance covered by a particle between two collisions). For dense gases, this mean free path is very small compared to the macroscopic scale and the numerical solution of Boltzmann equations becomes impossible. Such regimes can then be modelled by the equations of Continuum Mechanics describing the gas by averaged thermodynamic quantities (typically mass density, average velocity, and temperature) whose evolution is governed either by the Euler equations or by the Navier-Stokes equations.

The transitional regime that lies between molecular flow and fluid dynamics, characterized by a Knudsen number (the ratio of the mean free path to the macroscopic scale) of order 10^{-3} to 10^{-1}, is much harder

Received on September 1, 1997.
This work has been largely supported by CEA-CESTA

to model. Because of the computational cost in both time and storage, Monte Carlo methods are not practical. On the other hand, because of strong desequilibrium effects, the equations of fluid dynamics are not sufficiently accurate in boundary layers or across shock fronts.

The first strategy used to describe such transitional regimes splits the gas flow in two regions: a low density molecular region, in which Monte Carlo methods must be used and a fluid region, in which the particle velocity distribution approaches the local thermodynamic equilibrium and where then fluid equations can be used. This strategy has been proposed in [5] for coupling Boltzmann and Euler equations and in [22], and [23] for coupling Boltzmann and Navier-Stokes.

The problem then is to overcome the complexity and noise of the local numerical solution of the Boltzmann equation. Another strategy uses averaged equations in moments obtained by a weighted velocity integration of the Boltzmann equations and closed by assuming a given a-priori expansion of the unknown velocity distribution in terms of these moments. The resulting system is then a combination of standard conservation laws and of additional relaxation equations, depending on the collision operator, and generalizing the usual constitutive laws of fluid dynamics. Such a strategy was used by H.Grad [3], with thirteen moment equations and a velocity distribution modeled by generalized

Hermite polynomials. However, the resulting system is not always hyperbolic and the predicted velocity distribution model can be negative, which is inconsistent with the underlying physical model. A more consistent approach has been recently proposed by D.Levermore [1] based on assumed velocity distributions of the form : $f(\underline{x}, \underline{v}, t) = exp(\underline{\alpha}(\underline{x}, t).\underline{m}(\underline{v}))$. In this framework, the purpose of the present paper is to propose a general strategy for solving flow problems in transitional regimes by coupling the Levermore moment equations used in the molecular region, and the Navier-Stokes equations used in the fluid regime. The paper begins by recalling how the Levermore's fourteen moments equations and associated boundary conditions can be properly derived from the Boltzmann kinetic model. The next step is then to obtain the Navier-Stokes equations by a Hilbert asymptotic expansion of these moment equations. This somewhat original derivation of the Navier-Stokes equations gives them a new kinetic interpretation in terms of the fourteen moments used in Levermore's expansion and is the basis of our coupling process. Indeed, following the ideas of [22], we then propose and implement a time explicit coupling strategy which acts as if we were solving the Levermore's equations everywhere, the Navier-Stokes equations being simply obtained locally by replacing the fourteen moments by their specific forms obtained through the above asymptotic expansion process. The final numerical scheme is then obtained by discretising the moment equation by a first-order kinetic numerical scheme in the molecular region and on the interface, and by using a standard finite volume discretisation of the Navier-Stokes equations elsewhere. Numerical simulations of shock problems and Couette flows are finally presented to validate this coupling strategy. In this paper, we will only treat the monoatomic case, but it can be easily extended to more general polyatomic gases by following the steps of [17], [20], [22].

2 Kinetic theories

2.1 Boltzmann equation

Let us consider a gas composed of identical monoatomic particles contained within a fixed spatial domain Ω. This gas is described at the kinetic level by the Boltzmann equation governing the evolution of the distribution function $f = f(\underline{x}, \underline{v}, t)$ over the particle phase space $\Omega \times IR^3$

$$\partial_t f(\underline{x}, \underline{v}, t) + \underline{v}.\partial_x f(\underline{x}, \underline{v}, t) = Q(f, f)(\underline{x}, \underline{v}, t). \quad (1)$$

The nonnegative distribution function f describes the local density of particles with position $\underline{x}$ and velocity $\underline{v}$. The operator $Q(f, f)(\underline{x}, \underline{v}, t)$ which describes the variation of distribution function due to binary collisions between particles conserves mass, velocity and energy

$$\int_{\underline{v} \in R^3} Q(f, f)(\underline{x}, \underline{v}, t)\phi(\underline{v})d\underline{v} = 0,$$

with $\phi(\underline{v}) = \{1 \,,\, \underline{v} \,,\, \frac{1}{2}||v||^2\}$, and satisfies the local dissipation relation

$$\int_{\underline{v} \in R^3} Q(f, f) \log f(\underline{x}, \underline{v}, t)d\underline{v} \leq 0. \quad (2)$$

The collision operator may have a rather complex form. Nevertheless, at the fluid limit, it can be replaced by a simpler one proposed by Bhatnagar, Gross and Krook, and which has the same transport properties as the original Boltzmann model. This simpler model describes the relaxation of f to the local equilibrium Maxwellian

$$\mathcal{M} = \frac{\rho}{(2\pi T)^{3/2}} exp \left(\frac{(v - u)^2}{2T} \right) \text{ by}$$

$$Q(f, f) = \frac{\rho RT}{\mu}(\mathcal{M}(f) - f) \quad (3)$$

and gives the correct Navier-Stokes viscosity μ at the hydrodynamic limit. Above, the Maxwellian $\mathcal{M}(f)$ is defined by using the moments $\rho, \underline{u}$ and T of the original distribution f. The associated heat conduction is given by $\kappa = \frac{5}{2} R\mu$ which corresponds to a gas with Prandtl number

$$Pr = \frac{5}{2} \frac{R\mu}{\kappa} = 1.$$

For most gases, we have $Pr \leq 1$. Thus, this simple approximation will generally not give the correct Navier-Stokes heat conduction. To obtain the right conduction, Levermore has introduced a generalized BGK collision operator of the form

$$Q(f, f) = \frac{\rho RT}{\mu}(\mathcal{M}(f) - \mathcal{G}(f)) + \frac{5}{2}\frac{\rho R^2 T}{\kappa}(\mathcal{G}(f) - f) \tag{4}$$

where $\mathcal{G}(f)$ is the quadratic Gaussian closure [14]

$$\mathcal{G}(f) = \exp\ (\alpha_0 + \alpha_i v_i + \alpha_{ij} v_i v_j)$$

with coefficients α obtained by equating the ten first moments $(\rho, \rho\underline{u}, \underline{\sigma})$ of $\mathcal{G}(f)$ to those of f.

2.2 Equations in Moments

The derivation of moment equations begins with the choice of a finite dimensional linear subspace $\mathcal{E}_m$ of functions of $\underline{v}$, with basis elements $m_i(\underline{v})$. Then, the moment system is obtained by integrating in velocity the Boltzmann equation once weighted by the vector $\underline{m}(\underline{v}) = (m_i(\underline{v}))_{i \in \mathcal{E}_m}$, yielding

$$\partial_t < \underline{m}f > + div(< \underline{m} \otimes \underline{v}f >) = < \underline{m}.Q(f) > . \tag{5}$$

Here $< g >$ denotes the integral of $g d\underline{v}$ over the velocity space and $< \underline{m}f >$ the vector of $\underline{m}$ moments of f. Since each equation introduces a flux $< \underline{m} \otimes \underline{v}f >$ which can only be evaluated by the next equation, the system is not closed. Moment closures must be found such that the resulting system respects conservation laws, physical symmetries (translation and rotation invariance), recovers the proper fluid dynamics approximations and such that the resulting moment can be realized by some non-negative distribution. These goals can be fulfilled if $\mathcal{E}_m$ satisfies the following conditions:

(I) span $\{1, \underline{v}, v^2\} \in \mathcal{E}_m$.

(II) $\mathcal{E}_m$ is invariant under the action of translations $T_{\underline{u}}$ and rotations T_o defined by

$$T_{\underline{u}}F = T_{\underline{u}}F(\underline{v}) = F(\underline{v} - \underline{u}) \quad \text{and} \quad T_oF = F(O^T\underline{v}).$$

(III) The convex cone

$$\mathcal{E}_m^c = \{\underline{m} \in \mathcal{E}_m; \int_{I\!R^d} \exp(-\underline{m}(\underline{v}))d\underline{v} < \infty\},$$

has an nonempty interior in $\mathcal{E}_m$.

Examples of such admissible spaces with maximal degree two are

$$\begin{aligned} \mathcal{E}_m &= span\ \{1, \underline{v}, v^2\}, \\ \mathcal{E}_m &= span\ \{1, \underline{v}, \underline{v} \otimes \underline{v}\}. \end{aligned}$$

To recover Navier-Stokes equations, one must use spaces containing at least thirteen moments, which leads to the fundamental choice of a polynomial space $\mathcal{E}_m$ with maximal degree four

$$\mathcal{E}_m = span\ \{1, \underline{v}, \underline{v} \otimes \underline{v}, v^2.\underline{v}, v^4\}.$$

2.3 Levermore's models

Once an admissible $\mathcal{E}_m$ has been defined, D.Levermore [1] closes the moment system (5) by assuming that the distribution function f is of the form

$$f(\underline{v}) = F(\underline{\alpha}, \underline{v}) = \exp(\underline{\alpha}.\underline{m}(\underline{v})), \tag{6}$$

with $\underline{m}(v) \in \mathcal{E}_m$ and

$$\underline{\alpha} = \underline{\alpha}(x, t) \in I\!R_c^M = \{\alpha_i \in I\!R, 1 \leq i \leq M,$$
$$\int_{v \in I\!R^d} \exp(\underline{\alpha}.\underline{m}(v))dv < +\infty\}. \tag{7}$$

The resulting system of moments

$$\partial_t < \underline{m} \exp(\underline{\alpha}.\underline{m}) >$$
$$+ div(< \underline{m} \otimes \underline{v} \exp(\underline{\alpha}.\underline{m}) >)$$
$$= < \underline{m} Q(\exp(\underline{\alpha}.\underline{m})) >, \qquad (8)$$

is the system proposed by Levermore. It contains M equations for M unknowns α. As proved in [1], Model (6) is nonnegative by construction, and the distribution F corresponds to the formal solution of the entropy constrained minimization problem:

$$J(F) = J(\exp(\underline{\alpha}.\underline{m}(v))) = \min_{f \in \mathcal{X}} \{ \int_{\underline{v}} (f \log(f) - f) d\underline{v} \}$$

over the space $\mathcal{X}$ defined by

$$\mathcal{X} = \{ f \in L^2(I\!R^d), f \log(f) \in L^2(I\!R^d),$$
$$f \geq 0, < \underline{m} f >= \underline{U}_\alpha \; in \; I\!R^M \}. \qquad (9)$$

Here, the α_i are the Lagrange multipliers of the minimization constraints

$$< \underline{m} f >= \underline{U}_\alpha,$$

and $J(f) = \int_{\underline{v}} (f \log(f) - f) d\underline{v}$ is the usual entropy function of the kinetic theory.

2.4 Differential form

Let us note the moments by

$$\underline{U}_\alpha(\underline{\alpha}) = < \underline{m} \exp(\underline{\alpha}.\underline{m}) >= \frac{\partial < \exp(\underline{\alpha}.\underline{m}) >}{\partial \underline{\alpha}} = \frac{\partial U}{\partial \underline{\alpha}},$$

the fluxes by

$$A_\alpha(\underline{\alpha}) = < \underline{m} \otimes \underline{v} \exp(\underline{\alpha}.\underline{m}) >= \frac{\partial < \underline{v} \exp(\underline{\alpha}.\underline{m}) >}{\partial \underline{\alpha}},$$

and the collision moments by

$$S(\underline{\alpha}) = < \underline{m}.Q(\exp(\underline{\alpha}.\underline{m})) > .$$

With this notation, the Levermore's system (8) reduces to the first order hyperbolic system :
Find $\underline{\alpha}(\underline{x}, t) \in I\!R_c^M$, such that

$$\partial_t \underline{U}_\alpha(\underline{\alpha}) + div A_\alpha(\underline{\alpha}) = S(\underline{\alpha}) . \qquad (10)$$

This system can be written either with respect to the unknown parameters $\underline{\alpha}$ of the distribution function, or with respect to the corresponding unknown moments $\underline{U}_\alpha$. To check the hyperbolicity, on has just to formally develop (10) into

$$U_{,\alpha\alpha}.\partial_t \underline{\alpha} + A_{,\alpha\alpha}. \nabla_x \underline{\alpha} = S(\underline{\alpha}).$$

By construction, $U_{,\alpha\alpha}$ is symmetric positive definite as the Hessian of the strictly convex function U and $A_{,\alpha\alpha}$ is symmetric as the Hessian of $A(\underline{\alpha}) = < \underline{v} \exp(\underline{\alpha}.\underline{m}) >$. So the system (10) is symmetric and thus is indeed hyperbolic.
We can also prove that the functions

$$\eta(\underline{V}) = \inf_\alpha (\underline{\alpha}.\underline{V} - U(\underline{\alpha})) = \underline{\alpha}(\underline{V}).\underline{V} - U(\underline{\alpha}(\underline{V})), \quad (11)$$

$$\Phi(\underline{V}) = (\underline{\alpha}.A_\alpha - A)(\underline{\alpha}(\underline{V})), \qquad (12)$$

with $\underline{\alpha}(\underline{V})$ the solution of $U_\alpha(\underline{\alpha}(\underline{V})) = \underline{V}$, define a pair of entropy functions for the system (10) : η is convex and satisfies

$$\nabla_{U_\alpha} \eta(\underline{U}_\alpha).\nabla_{U_\alpha} A_\alpha = \nabla_{\underline{U}_\alpha} \Phi(\underline{U}_\alpha); \qquad (13)$$

each solution of (10) satisfies the following entropy inequality

$$\partial_t \eta(U_\alpha) + div \Phi(U_\alpha) \leq 0. \qquad (14)$$

2.5 Boundary Conditions

The weak form in space of the first order hyperbolic system (10) is obtained by direct integration by parts and is given by

$$\int_\Omega \underline{\phi}.\partial_t \underline{U}_\alpha dx - \int_\Omega \partial_x \underline{\phi}.A_\alpha dx$$
$$+ \int_{\partial\Omega} \underline{\phi}.\{A_\alpha^+.\underline{n} + A_\alpha^-.\underline{n}\}d\gamma$$
$$= \int_\Omega \underline{\phi}.S(\underline{\alpha})dx, \ \forall \underline{\phi} \in C^1(\overline{\Omega}, R^M),$$

where $\underline{n}$ is the outward unit normal vector to $\partial\Omega$. The outgoing/incoming half-fluxes are defined here by

$$A_\alpha^{+/-}.\underline{n} = \int_{\underline{v}.\underline{n}\geq 0/\underline{v}.\underline{n}\leq 0} \underline{v}.\underline{n}\,\underline{m}F(\underline{\alpha}, \underline{v})d\underline{v}.$$

In order to have consistent and linearly well-posed kinetic boundary conditions on the boundary $\partial\Omega$ of the computational domain Ω, we have to characterize the incoming half-fluxes $A_\alpha^-.\underline{n}$ everywhere on this boundary

$$A_\alpha^-..\underline{n} = A_{imp}^- \text{ on } \partial\Omega. \tag{15}$$

At inflow, this value is imposed from inflow boundary conditions specifying the incoming distribution F.

On the solid boundaries, one uses the proper kinetic wall boundary condition specifying the reflected (or incoming) "half" of the distribution function f^- when the incident f^+ (or outgoing) half is given. Such a classical kinetic wall boundary condition is the one in which a certain fraction $k(\underline{x})$ of the incident particles, depending on its position, is specularly reflected, and the remaining particles are absorbed by the wall and re-emitted with a Maxwellian distribution associated to the temperature T_p and velocity $\underline{u}_p$ of the wall. This condition can be written

$$f^-(\underline{x}, \underline{v}) = k(\underline{x}).f^+(\underline{x}, \mathcal{R}\underline{v}) + (1 - k(\underline{x})).M(\rho, T_p, \underline{u}_p),$$
$$\tag{16}$$

where $\mathcal{R}$ is the reflection operator defined by

$$\mathcal{R}\underline{v} = \underline{v} - 2(\underline{v}.\underline{n})\underline{n}.$$

The "density ρ of the wall" is determined by the condition that the wall does not collect particles.

$$\int_{\underline{v}\in I\!R^d} \underline{v}.\underline{n}(f^- + f^+)(\underline{x}, \underline{v})d\underline{v} = 0. \tag{17}$$

In (16), the specular reflection corresponds to $k(\underline{x}) = 1$ and total accommodation corresponds to $k(\underline{x}) = 0$. In order to evaluate the incoming half-fluxes on the wall, we suppose that the kinetic boundary conditions (16) are still valid in flux average, if we replace f, solution of the full Boltzmann equation, by the Levermore's model F. So, we find

$$\int_{\underline{v}.\underline{n}<0} \underline{m}(\underline{v})F(\underline{x}, \underline{v}) - k(\underline{x})F(\underline{x}, \mathcal{R}\underline{v})$$
$$- (1 - k(\underline{x}))M(\rho, T_p, \underline{u}_p)\underline{v}.\underline{n}d\underline{v} = 0$$

with

$$\int_{\underline{v}} F(\underline{x}, \underline{v})\underline{v}.\underline{n}d\underline{v} = 0. \tag{18}$$

With this choice, the boundary conditions imposed to system (8) consist in imposing everywhere on the boundary the incoming half flux $A_\alpha^- \cdot \underline{n}$ by (15) with A_{imp}^- given at inflow from the known inflow distribution, and A_{imp}^- evaluated at any solid wall by

$$A_{imp}^- = \int_{\underline{v}.\underline{n}\leq 0} (k(\underline{x})F(\underline{x}, \mathcal{R}\underline{v})$$
$$+ (1 - k(\underline{x}))M(\rho, T_p, \underline{u}_p))\underline{m}(\underline{v})\underline{v}.\underline{n}d\underline{v}. \tag{19}$$

with density of $\mathcal{M}$ given by (18), that is $(A_\rho^+ + A_\rho^-).\underline{n} = 0$.

2.6   The 14-moments closure

As indicated earlier, the linear space $\mathcal{E}_m^{14} = \{1, \underline{v}.\underline{v}\otimes \underline{v}, v^2\underline{v}, v^4\}$ is the smallest admissible space generating a closure system which recovers formally the correct

Navier-Stokes approximation by asymptotic expansion. The resulting system is then given by

$$
\begin{cases}
\partial_t < F^{14} > & + \nabla_x . < \underline{v} F^{14} > \\
& = 0, \\[4pt]
\partial_t < \underline{v} F^{14} > & + \nabla_x . < \underline{v} \otimes \underline{v} F^{14} > \\
& = 0, \\[4pt]
\partial_t < \underline{v} \otimes \underline{v} F^{14} > & + \nabla_x . < \underline{v} \otimes \underline{v} \otimes \underline{v} F^{14} > \\
& = < \underline{v} \otimes \underline{v}\, Q(F^{14}) >, \\[4pt]
\partial_t < v^2 \underline{v} F^{14} > & + \nabla_x . < v^2 \underline{v} \otimes \underline{v} F^{14} > \\
& = < v^2 \underline{v}\, Q(F^{14}) >, \\[4pt]
\partial_t < v^4 F^{14} > & + \nabla_x . < v^4 \underline{v} F^{14} > \\
& = < v^4\, Q(F^{14}) > .
\end{cases}
$$

In order to write each equation in term of macroscopic quantities, we recall some definitions. As well known, the zeroth moment is the density

$$\rho(\underline{x},t) = < F^{14} > .$$

The average velocity corresponds to

$$\underline{u}(\underline{x},t) = \frac{1}{\rho} < \underline{v} F^{14} >,$$

and the intrinsic velocity $\underline{c} = \underline{v} - \underline{u}(\underline{x},t)$ and satisfies Clearly,

$$< \underline{c} F^{14} > = 0. \tag{20}$$

Second, third, and fourth c-moments are respectively

$$< \underline{c} \otimes \underline{c} F^{14} > = -\underline{\underline{\sigma}},\ \underline{\underline{\sigma}} = \text{ Cauchy stress tensor}, \tag{21}$$

$$\underline{\underline{Q}} = < \underline{c} \otimes \underline{c} \otimes \underline{c} F^{14} >, \tag{22}$$

and

$$\mathcal{R} = < \underline{c} \otimes \underline{c} \otimes \underline{c} \otimes \underline{c} F^{14} > . \tag{23}$$

These moments define the pressure

$$p = \frac{1}{3} < c^2 F^{14} > = -\frac{1}{3} tr(\underline{\underline{\sigma}}) = \rho R T, \tag{24}$$

the viscous stress tensor,

$$\underline{\underline{\tau}} = p \underline{\underline{Id}} + \underline{\underline{\sigma}} = \frac{1}{3} < c^2 F^{14} > - < \underline{c} \otimes \underline{c} F^{14} > , \tag{25}$$

and the heat flux,

$$\underline{q} = \frac{1}{2} < c^2 \underline{c} F^{14} > = \frac{1}{2} Tr(\underline{\underline{Q}}). \tag{26}$$

By contraction of (23), we introduce the auxiliary moments

$$\underline{\underline{R}} = < c^2 \underline{c} \otimes \underline{c} F^{14} >, \tag{27}$$

$$r = < c^4 F^{14} > = \mathcal{R}_{iijj}. \tag{28}$$

Finally, we introduce the last flux $\underline{s}$ defined by

$$\underline{s} = < c^4 \underline{c} F^{14} > . \tag{29}$$

With this notation, the moments and fluxes in the 14-moments closure write

$$
< \underline{m} F^{14} > = U_\alpha =
\begin{cases}
\rho \\
\rho \underline{u} \\
\rho \underline{u} \otimes \underline{u} - \underline{\underline{\sigma}} \\
\rho u^2 \underline{u} - tr(\underline{\underline{\sigma}}) \underline{u} - 2 \underline{\underline{\sigma}} . \underline{u} + 2 \underline{q} \\
\rho u^4 - 2 tr(\underline{\underline{\sigma}}) u^2 - 4 (\underline{\underline{\sigma}} . \underline{u}) . \underline{u} \\
\qquad + 8 \underline{q} . \underline{u} + r
\end{cases}
$$

$$
< v \otimes \underline{m} F^{14} > = A_\alpha =
\begin{cases}
\rho \underline{u} \\
\rho \underline{u} \otimes \underline{u} - \underline{\underline{\sigma}} \\
(\rho \underline{u} \otimes \underline{u} \otimes \underline{u}) - 3 \underline{\underline{\sigma}} \otimes \underline{u}_{sym} + \underline{\underline{Q}} \\
\rho u^2 \underline{u} \otimes \underline{u} - tr(\underline{\underline{\sigma}}) \underline{u} \otimes \underline{u} \\
\qquad - 4((\underline{\underline{\sigma}} . \underline{u}) \otimes \underline{u})_{sym} - \underline{\underline{\sigma}} u^2 \\
\qquad + 2 \underline{\underline{Q}} . \underline{u} + 4 (\underline{q} \otimes \underline{u})_{sym} + \underline{\underline{R}} \\
\rho u^4 \underline{u} - 2 tr(\underline{\underline{\sigma}}) u^2 \underline{u} - 4 \underline{u} . (\underline{\underline{\sigma}} . \underline{u}) \underline{u} \\
\qquad - 4 (\underline{\underline{\sigma}} . \underline{u}) u^2 + 8 (\underline{q} . \underline{u}) . \underline{u} \\
\qquad + 4 \underline{q} u^2 + 4 (\underline{\underline{Q}} . \underline{u}) . \underline{u} + r \underline{u} \\
\qquad + 4 \underline{\underline{R}} . \underline{u} + \underline{s}
\end{cases}
$$

and

$$\Xi_i = <m_i(\underline{v})Q(F^{14})>.$$

For the BGK collision operator, we have simply

$$\Xi = \left\{ \begin{array}{l} 0 \\ 0 \\ \dfrac{\rho RT}{\mu}(p\underline{\underline{Id}} + \underline{\underline{\sigma}}) \\ \dfrac{\rho RT}{\mu}2(p\underline{\underline{Id}} + \underline{\underline{\sigma}}).\underline{u} - 2P_r\dfrac{\rho RT}{\mu}\underline{q} \\ \dfrac{\rho RT}{\mu}(4p||u||^2 + 4\underline{u}.\underline{\underline{\sigma}}.\underline{u} - \dfrac{2}{\rho}\underline{\underline{\sigma}}:\underline{\underline{\sigma}} + 6\dfrac{p^2}{\rho}) \\ \quad -P_r\dfrac{\rho RT}{\mu}(8\underline{q}.\underline{u} + r - \dfrac{2}{\rho}\underline{\underline{\sigma}}:\underline{\underline{\sigma}} - 9\dfrac{p^2}{\rho}) \end{array} \right\}.$$

2.7 Non conservative formulation

The relation between Levermore's equations and more standard models of fluid dynamics is best understood when these equations are written in nonconservative form. If $(1C)$ (respectively $(2C)_i$, $(3C)_{ij}$, $(4C)_{0j}$, $(5C)$) respectively denote the original conservative form of these equations with respect to the moment $<F^{14}>$ (respectively $<m_iF^{14}>$, $<m_{ij}F^{14}>$, $<m_{0j}F^{14}>$, $<m_{14}F^{14}>$), the corresponding nonconservative form is systematically obtained by developing the gradients in time and space, and by elimination of the variations of mass, momentum and energy. First, a direct expansion of the space gradient in Equation (1C) directly gives the classical nonconservative form of the equation of mass

$$\frac{\partial\rho}{\partial t} + u_k\frac{\partial\rho}{\partial x_k} + \rho\frac{\partial u_k}{\partial x_k} = 0. \tag{30}$$

The equation in velocity is obtained by substracting this equation of mass (multiplied by u_i) from the equations $(2C)_i$, yielding the standard form

$$\frac{\partial u_i}{\partial t} + u_k\cdot\frac{\partial u_i}{\partial x_k} - \frac{1}{\rho}\frac{\partial\sigma_{ik}}{\partial x_k} = 0. \tag{31}$$

To characterize the evolution of the stress tensor $\underline{\underline{\sigma}}$, we first develop the time derivative of the moment $<m_{ij}F^{14}>$ into a sum of partial derivatives to be partially computed from the equation in velocity

$$\frac{\partial}{\partial t}(\rho u_i u_j - \sigma_{ij})$$
$$= -\frac{\partial\sigma_{ij}}{\partial t} + \rho u_i\frac{\partial u_j}{\partial t} + u_j\frac{\partial\rho u_i}{\partial t}$$
$$= -\frac{\partial\sigma_{ij}}{\partial t} - \rho u_i u_k\frac{\partial u_j}{\partial x_k} + u_i\frac{\partial\sigma_{jk}}{\partial x_k}$$
$$\quad -u_j\frac{\partial\rho u_i u_k}{\partial x_k} + u_j\frac{\partial\sigma_{ik}}{\partial x_k}.$$

Similarly the space derivatives of $<m_{ij}v_kF_{14}>$ can be developed into

$$\frac{\partial}{\partial x_k}(\rho u_i u_j u_k - \sigma_{ij}u_k - \sigma_{ik}u_j - \sigma_{jk}u_i + Q_{ijk}) =$$
$$u_j\frac{\partial\rho u_i u_k}{\partial x_k} + \rho u_i u_k\frac{\partial u_j}{\partial x_k} - \frac{\partial\sigma_{ij}u_k}{\partial x_k} - \sigma_{ik}\frac{\partial u_j}{\partial x_k}$$
$$-u_j\frac{\partial\sigma_{ik}}{\partial x_k} - \sigma_{jk}\frac{\partial u_i}{\partial x_k} - u_i\frac{\partial\sigma_{jk}}{\partial x_k} + \frac{\partial Q_{ijk}}{\partial x_k}.$$

Since in our generalized BGK model, the collision term takes the simple form

$$<m_{ij}Q(F^{14})> = \frac{\rho RT}{\mu}(p\delta_{ij} + \sigma_{ij}),$$

the equations $(3C)_{ij}$ take the final form

$$\frac{\partial\sigma_{ij}}{\partial t} + \frac{\partial u_k\sigma_{ij}}{\partial x_k} + \sigma_{jk}\cdot\frac{\partial u_i}{\partial x_k}$$
$$+\sigma_{ik}\cdot\frac{\partial u_j}{\partial x_k} - \frac{\partial Q_{ijk}}{\partial x_k}$$
$$= -\frac{\rho RT}{\mu}(p\delta_{ij} + \sigma_{ij}). \tag{32}$$

In particular, the equation in internal energy (or in pressure) corresponds to the trace of this set of equations (that is to the sum of the equations in σ_{ii})

$$\frac{\partial p}{\partial t} + \frac{\partial u_k p}{\partial x_k} - \frac{2}{3}\sigma_{jk}\cdot\frac{\partial u_j}{\partial x_k} + \frac{2}{3}\frac{\partial q_k}{\partial x_k} = 0, \tag{33}$$

which is the standard writing of the principle of conservation of energy. The equation governing the evolution of the heat flux q_i is obtained by a similar process. First, the time derivative of $<m_{0j}F^{14}>$ in equation $(4C)_{0j}$ can be developed into the sum

$$\frac{\partial <m_{0j}F^{14}>}{\partial t} = 2\frac{\partial q_i}{\partial t} + u^2 u_i \frac{\partial \rho}{\partial t} + \rho u^2 \frac{\partial u_i}{\partial t}$$
$$+2\rho u_i u_j \frac{\partial u_j}{\partial t} + 3p\frac{\partial u_i}{\partial t} + 3u_i \frac{\partial p}{\partial t}$$
$$-2\sigma_{ij}\frac{\partial u_j}{\partial t} - 2u_j \frac{\partial \sigma_{ij}}{\partial t}.$$

From (30), (31), (32), (33), this reduces to

$$\frac{\partial <m_{0j}F^{14}>}{\partial t} = 2\frac{\partial q_i}{\partial t} - u^2 u_i \frac{\partial \rho u_k}{\partial x_k} - \rho u^2 u_k \frac{\partial u_i}{\partial x_k}$$
$$+u^2 \frac{\partial \sigma_{ik}}{\partial x_k} - 2\rho u_i u_j u_k \frac{\partial u_j}{\partial x_k} + 2u_i u_j \frac{\partial \sigma_{jk}}{\partial x_k}$$
$$-3pu_k \frac{\partial u_i}{\partial x_k} + 3\frac{p}{\rho}\frac{\partial \sigma_{ik}}{\partial x_k} - 3u_i \frac{\partial pu_k}{\partial x_k} + 2\sigma_{jk}u_i \frac{\partial u_j}{\partial x_k}$$
$$-2u_i \frac{\partial q_k}{\partial x_k} + 2\sigma_{ij}u_k \frac{\partial u_j}{\partial x_k} - 2\frac{\sigma_{ij}}{\rho}\frac{\partial \sigma_{jk}}{\partial x_k}$$
$$+2u_j \frac{\partial \sigma_{ij}u_k}{\partial x_k} + 2\sigma_{jk}u_j \frac{\partial u_i}{\partial x_k} + 2\sigma_{ik}u_j \frac{\partial u_j}{\partial x_k}$$
$$-2u_j \frac{\partial Q_{ijk}}{\partial x_k} + \frac{\rho RT}{\mu}(2u_j(p\delta_{ij} + \sigma_{ij})).$$

On the other hand, the gradient of the flux vector $<v_k m_{0j}F^{14}>$ can be developed into

$$\frac{\partial}{\partial x_k}<v_k m_{0j}F^{14}> = u^2 u_i \frac{\partial \rho u_k}{\partial x_k} + \rho u_k u^2 \frac{\partial u_i}{\partial x_k}$$
$$+2\rho u_k u_i u_j \frac{\partial u_j}{\partial x_k} + 3u_i \frac{\partial pu_k}{\partial x_k} + 3pu_k \frac{\partial u_i}{\partial x_k}$$
$$-u^2 \frac{\partial \sigma_{ik}}{\partial x_k} - 2\sigma_{ik}u_j \frac{\partial u_j}{\partial x_k} - 2\sigma_{jk}u_j \frac{\partial u_i}{\partial x_k}$$
$$-2\sigma_{jk}u_i \frac{\partial u_j}{\partial x_k} - 2u_i u_j \frac{\partial \sigma_{jk}}{\partial x_k} - 2u_j \frac{\partial \sigma_{ij}u_k}{\partial x_k}$$
$$-2\sigma_{ij}u_k \frac{\partial u_j}{\partial x_k} + 2Q_{ijk}\frac{\partial u_j}{\partial x_k} + 2u_j \frac{\partial Q_{ijk}}{\partial x_k}$$

$$+2q_k \frac{\partial u_i}{\partial x_k} + 2u_i \frac{\partial q_k}{\partial x_k} + 2\frac{\partial q_i u_k}{\partial x_k} + \frac{\partial \mathcal{R}_{jjik}}{\partial x_k}.$$

From the particular form of the collision term

$$<m_{0j}Q(F^{14}> = \frac{\rho RT}{\mu}(2(p\delta_{ij} + \sigma_{ij})u_j) - \frac{5R\rho RT}{\kappa}(q_i),$$

Equation $(4C)_{0j}$ finally reduces to

$$2\frac{\partial q_i}{\partial t} + 2\frac{\partial q_i u_k}{\partial x_k} + 2Q_{ijk}\frac{\partial u_j}{\partial x_k} + 2q_k \frac{\partial u_i}{\partial x_k} \qquad (34)$$
$$+3\frac{p}{\rho}\frac{\partial \sigma_{ik}}{\partial x_k} - 2\frac{\sigma_{ij}}{\rho}\frac{\partial \sigma_{jk}}{\partial x_k} + \frac{\partial \mathcal{R}_{jjik}}{\partial x_k} = -\frac{5R\rho RT}{\kappa}(q_i).$$

The same strategy can finally be used to derive the equation governing the evolution of the fourteenth moment $\mathcal{R}_{iijj}$. The time derivatives (5C) are expressed in terms of

$$\partial_t \rho u^4, \quad 6\partial_t pu^2, \quad 4\partial_t(\sigma_{ij}.u_j)u_i \quad \text{and} \quad 8\partial_t q_i u_i$$

which are then eliminated by equations (30), (31), (32), (33), yielding

$$\frac{\partial <m_{14}F^{14}>}{\partial t} = -4u^2 \rho u_i u_k \frac{\partial u_i}{\partial x_k} + 4u^2 u_i \frac{\partial \sigma_{ik}}{\partial x_k}$$
$$-u^4 \frac{\partial \rho u_k}{\partial x_k} - 12pu_i u_k \frac{\partial u_i}{\partial x_k} + 12\frac{p}{\rho}u_i \frac{\partial \sigma_{ik}}{\partial x_k} - 6u^2 \frac{\partial pu_k}{\partial x_k}$$
$$+4u^2 \sigma_{jk}\frac{\partial u_j}{\partial x_k} - 4u^2 \frac{\partial q_k}{\partial x_k} + 4\sigma_{ij}u_i u_k \frac{\partial u_j}{\partial x_k}$$
$$-4\frac{\sigma_{ij}u_i}{\rho}\frac{\partial \sigma_{jk}}{\partial x_k} + 4\sigma_{ij}u_j u_k \frac{\partial u_i}{\partial x_k} - 4\frac{\sigma_{ij}u_j}{\rho}\frac{\partial \sigma_{ik}}{\partial x_k}$$
$$+4\sigma_{ij}u_i u_j \frac{\partial u_k}{\partial x_k} + 4u_i u_j u_k \frac{\partial \sigma_{ij}}{\partial x_k} + 4\sigma_{jk}u_i u_j \frac{\partial u_i}{\partial x_k}$$
$$+4\sigma_{ik}u_i u_j \frac{\partial u_j}{\partial x_k} - 4u_i u_j \frac{\partial Q_{ijk}}{\partial x_k}$$
$$+\frac{\rho RT}{\mu}(4(p\delta_{ij} + \sigma_{ij})u_i u_j)$$
$$-8Q_{ijk}u_i \frac{\partial u_j}{\partial x_k} - 8u_i \frac{\partial q_i u_k}{\partial x_k} - 8u_i q_k \frac{\partial u_i}{\partial x_k}$$

$$-12\frac{p}{\rho}u_i\frac{\partial\sigma_{ik}}{\partial x_k} + 8\frac{\sigma_{ij}u_i}{\rho}\frac{\partial\sigma_{jk}}{\partial x_k}$$

$$-4u_i\frac{\partial\mathcal{R}_{jjik}}{\partial x_k} - 4\frac{5R\rho RT}{\kappa}(u_iq_i)$$

$$-8q_iu_k\frac{\partial u_i}{\partial x_k} + 8\frac{q_i}{\rho}\frac{\partial\sigma_{ik}}{\partial x_k} + \frac{\partial\mathcal{R}_{iijj}}{\partial t}.$$

Similarly, the gradient of $< v_k m_{14}F^{14} >$ reduces to

$$\frac{\partial}{\partial x_k} < v_k m_{14}F^{14} > = u^4\frac{\partial\rho u_k}{\partial x_k} + 4u^2\rho u_iu_k\frac{\partial u_i}{\partial x_k}$$

$$+6u^2\frac{\partial pu_k}{\partial x_k} + 12pu_iu_k\frac{\partial u_i}{\partial x_k}$$

$$-4\sigma_{ij}u_iu_k\frac{\partial u_j}{\partial x_k} - 4\sigma_{ij}u_ju_k\frac{\partial u_i}{\partial x_k}$$

$$-4\sigma_{ij}u_iu_j\frac{\partial u_k}{\partial x_k} - 4u_iu_ju_k\frac{\partial\sigma_{ij}}{\partial x_k}$$

$$-4u^2u_i\frac{\partial\sigma_{ik}}{\partial x_k} - 8\sigma_{jk}u_iu_j\frac{\partial u_i}{\partial x_k}$$

$$-4u^2\sigma_{jk}\frac{\partial u_j}{\partial x_k} + 4u^2\frac{\partial q_k}{\partial x_k} + 8u_iq_k\frac{\partial u_i}{\partial x_k}$$

$$+8q_iu_k\frac{\partial u_i}{\partial x_k} + 8u_iq_i\frac{\partial u_k}{\partial x_k} + 8u_iu_k\frac{\partial q_i}{\partial x_k}$$

$$+4Q_{ijk}u_i\frac{\partial u_j}{\partial x_k} + 4Q_{ijk}u_j\frac{\partial u_i}{\partial x_k} + 4u_iu_j\frac{\partial Q_{ijk}}{\partial x_k}$$

$$+4\frac{\partial\mathcal{R}_{jjik}u_i}{\partial x_k} + \frac{\partial\mathcal{R}_{iijj}u_k}{\partial x_k} + \frac{\partial s_k}{\partial x_k}.$$

From the corresponding expression of the collision term

$$< m_{14}Q(F^{14}) > =$$
$$\frac{\rho RT}{\mu}(4pu^2 + 4\sigma_{ij}u_iu_j - \frac{2}{\rho}\sigma_{ij}\sigma_{ij} + 6\frac{p^2}{\rho})$$
$$-\frac{5R\rho RT}{\kappa}(4q_iu_i + \mathcal{R}_{iijj} - \frac{2}{\rho}\sigma_{ij}\sigma_{ij} - 9\frac{p^2}{\rho}),$$

the last equation (5C) finally reduces to

$$\frac{\partial\mathcal{R}_{iijj}}{\partial t} + \frac{\partial\mathcal{R}_{iijj}u_k}{\partial x_k} + 4\mathcal{R}_{jjik}\frac{\partial u_i}{\partial x_k}$$

$$+8\frac{q_i}{\rho}\frac{\partial\sigma_{ik}}{\partial x_k} + \frac{\partial s_k}{\partial x_k}$$

$$= \frac{\rho RT}{\mu}(-\frac{2}{\rho}\sigma_{ij}\sigma_{ij} + 6\frac{p^2}{\rho})$$

$$-\frac{5R}{2}\frac{\rho RT}{\kappa}(\mathcal{R}_{iijj} - \frac{2}{\rho}\sigma_{ij}\sigma_{ij} - 9\frac{p^2}{\rho}).$$

Regrouping this last equation with (30), (31),(32),(33) and (34) finally yields the nonconservative form of the Levermore's equations

$$\frac{\partial\rho}{\partial t} + u_k\cdot\frac{\partial\rho}{\partial x_k} + \rho\frac{\partial u_k}{\partial x_k} = 0, \qquad (35)$$

$$\frac{\partial u_i}{\partial t} + u_k\cdot\frac{\partial u_i}{\partial x_k} - \frac{1}{\rho}\frac{\partial\sigma_{ik}}{\partial x_k} = 0, \qquad (36)$$

$$\frac{\partial p}{\partial t} + \frac{\partial u_kp}{\partial x_k} - \frac{2}{3}\sigma_{jk}\cdot\frac{\partial u_j}{\partial x_k} + \frac{2}{3}\frac{\partial q_k}{\partial x_k} = 0, \qquad (37)$$

$$\frac{\partial\sigma_{ij}}{\partial t} + \frac{\partial u_k\sigma_{ij}}{\partial x_k} + \sigma_{jk}\cdot\frac{\partial u_i}{\partial x_k} + \sigma_{ik}\cdot\frac{\partial u_j}{\partial x_k}$$
$$-\frac{\partial Q_{ijk}}{\partial x_k} = -\frac{\rho RT}{\mu}(p\delta_{ij} + \sigma_{ij}), \qquad (38)$$

$$2\frac{\partial q_i}{\partial t} + 2\frac{\partial q_iu_k}{\partial x_k} + 2Q_{ijk}\frac{\partial u_j}{\partial x_k} + 2q_k\frac{\partial u_i}{\partial x_k} \qquad (39)$$
$$+3\frac{p}{\rho}\frac{\partial\sigma_{ik}}{\partial x_k} - 2\frac{\sigma_{ij}}{\rho}\frac{\partial\sigma_{jk}}{\partial x_k} + \frac{\partial\mathcal{R}_{jjik}}{\partial x_k} = -\frac{5R\rho RT}{\kappa}q_i,$$

$$\frac{\partial\mathcal{R}_{iijj}}{\partial t} + \frac{\partial\mathcal{R}_{iijj}u_k}{\partial x_k} + 4\mathcal{R}_{jjik}\frac{\partial u_i}{\partial x_k}$$

$$+8\frac{q_i}{\rho}\frac{\partial\sigma_{ik}}{\partial x_k} + \frac{\partial s_k}{\partial x_k}$$

$$= \frac{\rho RT}{\mu}(-\frac{2}{\rho}\sigma_{ij}\sigma_{ij} + 6\frac{p^2}{\rho})$$

$$-\frac{5R}{2}\frac{\rho RT}{\kappa}(\mathcal{R}_{iijj} - \frac{2}{\rho}\sigma_{ij}\sigma_{ij} - 9\frac{p^2}{\rho}), \qquad (40)$$

for i,j = 1,2,3.

The first three equations (35), (36), (37) correspond to the standard nonconservative formulation of the conservation laws of mass, momentum and energy. The last three ones (38), (39), (40) govern the evolution of the constitutive variables $\sigma_{ij}, q_i, \mathcal{R}_{iijj}$ and express therefore the constitutive properties of the fluid, which takes here the form of partial differential equations. These constitutive laws are not quite explicit since they involve auxiliary moments $\underline{\underline{Q}}$ and $\mathcal{R}_{jjik}$ which can only be obtained by computing explicitly the underlying kinetic distribution function F^{14}.

3 The Navier-Stokes limit

We now explain how to recover the Navier-Stokes equations (conservation and constitutive laws) from the Levermore fourteen moments system written in nonconservative form. This is based on an asymptotic expansion of these equations and of the corresponding unknowns σ_{ij}, Q_{ijk} (or q_i), $\mathcal{R}_{iijk}$ (or $\mathcal{R}_{iijj}$) in powers of the relaxation time ε

$$\sigma_{ij} = \sigma_{ij}^0 + \epsilon\sigma_{ij}^1 + \Theta(\epsilon^2),$$

$$Q_{ijk} = Q_{ijk}^0 + \epsilon Q_{ijk}^1 + \Theta(\epsilon^2),$$

$$\mathcal{R}_{iijk} = \mathcal{R}_{iijk}^0 + \epsilon\mathcal{R}_{iijk}^1 + \Theta(\epsilon^2).$$

By construction , the factors $\frac{\rho RT}{\mu}$ and $\frac{5R}{2}\frac{\rho RT}{\kappa}$ in front of the collision terms are proportional to $\frac{1}{\epsilon}$. Therefore, the highest order terms in the nonconservative form of the Levermore's equations are of order -1 in ε and correspond to the collision terms in the equations (38), (39) and (40). Solving then equations (38), (39) at order -1 then yields

$$\sigma_{ij}^0 = -p\delta_{ij}, \qquad (41)$$

$$q_i^0 = 0. \qquad (42)$$

Similarly, solving (40) at order -1 and substituting the above values of σ_{ij}^0 and q^0 yields

$$\mathcal{R}_{iijj}^0 = 15\frac{p^2}{\rho}. \qquad (43)$$

From these above calculations, we see that the first fourteen moments of F^{14} are those of a Maxwellian distribution. Since these moments characterize the distribution, we then deduce that at zero order F^{14} is a Maxwellian, and that therefore, its zero order moments are given by

$$Q_{ijk}^0 = < c_i c_j c_k \mathcal{M} > = 0,$$

$$\mathcal{R}_{iijk}^0 = < c_i c_i c_j c_k \mathcal{M} > = \frac{1}{3}r\delta_{jk} = 5\frac{p^2}{\rho}\delta_{jk}$$

$$s_k^0 = < c^4 c_k F^{14} > = < c^4 c_k \mathcal{M} > = 0.$$

By next solving (38) at order 0, and using the above expression of the zero order moments, we get

$$-\frac{\partial p\delta_{ij}}{\partial t} - \frac{\partial u_k p\delta_{ij}}{\partial x_k} - p\delta_{jk}\cdot\frac{\partial u_i}{\partial x_k} - p\delta_{ik}\cdot\frac{\partial u_j}{\partial x_k} = -\frac{\rho RT}{\mu}(\epsilon\sigma_{ij}^1)$$

Using the equation (37) to compute the time derivatives of p, this can be rewritten as

$$+\delta_{ij}\frac{\partial u_k p}{\partial x_k} - \delta_{ij}\frac{2}{3}p\delta_{jk}\frac{\partial u_j}{\partial x_k}$$
$$-\frac{\partial u_k p\delta_{ij}}{\partial x_k} - p\delta_{jk}\frac{\partial u_i}{\partial x_k} - p\delta_{ik}\frac{\partial u_j}{\partial x_k}$$
$$= -\frac{\rho RT}{\mu}(\epsilon\sigma_{ij}^1),$$

that is after simplification

$$\delta_{ij}\frac{2}{3}p\cdot\frac{\partial u_k}{\partial x_k} - p\frac{\partial u_i}{\partial x_j} - p\frac{\partial u_j}{\partial x_i} = -\frac{\rho RT}{\mu}(\epsilon\sigma_{ij}^1).$$

In other words, the zero order equation in $\underline{\sigma}$ yields the constitutive law

$$\epsilon \sigma_{ij}^1 = \mu \left(\frac{\partial u_j}{\partial x_i} + \frac{\partial u_i}{\partial x_j} - \frac{2}{3} \delta_{ij} \frac{\partial u_k}{\partial x_k} \right).$$

Similarly, the zero order expansion of Equation (39) yields

$$-3 \frac{p}{\rho} \frac{\partial p \delta_{ik}}{\partial x_k} - 2 \frac{p \delta_{ij}}{\rho} \frac{\partial p \delta_{jk}}{\partial x_k} + \frac{\partial 5 \frac{p^2}{\rho} \delta_{ik}}{\partial x_k} = -\frac{5 R \rho R T}{\kappa} (\epsilon q_i^1).$$

By a simple algebraic manipulation based on the thermodynamic definition of the pressure $p = \rho R T$, we can transform this equation into

$$+5 \rho R T \delta_{ik} \frac{\partial R T}{\partial x_k} = -\frac{5 R \rho R T}{\kappa} (\epsilon q_i^1),$$

which finally leads to the constitutive law

$$\epsilon q_i^1 = -\kappa \frac{\partial T}{\partial x_i}.$$

The first order expansion of the fourteenth moment is finally obtained by a zero order expansion of the last equation (40). Using the previously computed values of σ_{ij}^0, σ_{ii}^1, q_i^0, $\mathcal{R}_{iijj}^0 = 15 \frac{p^2}{\rho}$ and

$$\begin{aligned}
\sigma_{ij} \sigma_{ij} &= 3p^2 - 2\epsilon p \delta_{ij} \sigma_{ij}^1 + O(\epsilon^2), \\
&= 3p^2 - 2pTr\epsilon \sigma_{ij}^1 + O(\epsilon^2), \\
&= 3p^2 + O(\epsilon^2),
\end{aligned}$$

we get

$$\frac{\partial 15 \frac{p^2}{\rho}}{\partial t} + \frac{\partial 15 \frac{p^2}{\rho} u_k}{\partial x_k} + 20 \frac{p^2}{\rho} \delta_{ik} \frac{\partial u_i}{\partial x_k} = -\frac{5 R}{2} \frac{\rho R T}{\kappa} (\epsilon \mathcal{R}_{iijj}^1).$$

After expansion, and elimination of the time derivatives of ρ and p from the zero order expansion of the conservation equations (35) and (36), we get

$$30 \frac{p}{\rho} \left(-\frac{\partial u_k p}{\partial x_k} + \frac{2}{3} (-p \delta_{jk}) \frac{\partial u_j}{\partial x_k} \right)$$

$$-15 \frac{p^2}{\rho^2} \left(-u_k \cdot \frac{\partial \rho}{\partial x_k} - \rho \frac{\partial u_k}{\partial x_k} \right)$$

$$+ \frac{\partial \frac{15p^2}{\rho} u_k}{\partial x_k} + 20 \frac{p^2}{\rho} \frac{\partial u_k}{\partial x_k} = -\frac{5 R}{2} \frac{\rho R T}{\kappa} (\epsilon \mathcal{R}_{iijj}^1),$$

which can be simplified into

$$-\frac{5 R}{2} \frac{\rho R T}{\kappa} (\epsilon \mathcal{R}_{iijj}^1) = 0.$$

So finally, we have obtained the first order expression of the viscous fluxes :

$$\sigma_{ij} = -p \delta_{ij} + \mu \left(\frac{\partial u_j}{\partial x_i} + \frac{\partial u_i}{\partial x_j} - \frac{2}{3} \delta_{ij} \frac{\partial u_k}{\partial x_k} \right), \qquad (44)$$

$$q_i = -\kappa \frac{\partial T}{\partial x_i}, \qquad (45)$$

$$\mathcal{R}_{iijj} = 15 \frac{p^2}{\rho}. \qquad (46)$$

Together with the conservation laws (35)-(37), this defines the first order asymptotic expansion of the Levermore's equation. We recognize here the standard conservation laws and constitutive laws characterizing the evolution of a Newtonian viscous fluid, together with an additional law (46) defining the fourteenth moment of the associated Navier-Stokes distribution. In other words, the compressible Navier-Stokes equations correspond to a first order asymptotic expansion of the Levermore's equations. This means formally that a solution $\rho, \underline{u}, T, \underline{\sigma}, \underline{q}$ of the Navier-Stokes equation corresponds to an approximate solution

$$\underline{U}_\alpha^{NS} = \begin{cases} \rho \\ \rho \underline{u} \\ \rho \underline{u} \otimes \underline{u} - \underline{\sigma} \\ \rho u^2 \underline{u} - tr(\underline{\sigma}) \underline{u} - 2\underline{\sigma}.\underline{u} + 2\underline{q} \\ \rho u^4 - 2tr(\underline{\sigma}) u^2 - 4(\underline{\sigma}.\underline{u}).\underline{u} + 8\underline{q}.\underline{u} + \frac{15p^2}{\rho} \end{cases}$$

of the fourteen moment Levermore's equation, and conversely. In particular $\underline{U}^{NS}_\alpha$, and hence the corresponding Navier-Stokes solution, can be viewed as the moments of the positive "Navier-Stokes" distribution function $F_{NS}(\underline{v}) = exp(\underline{\alpha} \cdot \underline{v})$ where the coefficients $\underline{\alpha}$ are obtained by solving the moment compatibility equation

$$< \underline{m} exp(\underline{\alpha} \cdot \underline{v}) >= \underline{U}^{NS}_\alpha .$$

This new kinetic interpretation of the Navier-Stokes equation gives then a very natural way of coupling Navier-Stokes with more elaborate kinetic models. In contrast with the standard kinetic ansatz proposed by Chapman-Enskog or Grad [13], [19], this new distribution is positive and integrable. On the other hand, and although it is asymptotically equivalent to such standard ansatz for small values of $\underline{u} - \underline{v}$, it is no longer a polynomial perturbation of a Maxwellian, which forbids any explicit calculations of moments.

4　Coupling Strategy

4.1　The general approach

Following [22], we now explain how to couple the Navier-Stokes equations used far away from any solid boundary and more accurate kinetic models next to the obstacles. This approach is based on the following steps :

- approximate preliminary solution of the Navier-Stokes equations on the full computational domain using adequate slip boundary conditions at the solid walls [25];

- calculation of generalized residuals for detecting the regions where the proposed Navier-Stokes solution is inaccurate. For rarefied flows, the residuals are obtained by plugging the Navier-Stokes solution into a more general kinetic moment equation (such as Grad [3] or Levermore [1]);

- adaptive construction and meshing of the kinetic and hydrodynamic subdomains, finer models and finer grids being used in the regions of large residuals;

- development of adequate interface conditions matching inflow and outflow fluxes between the local kinetic and the global hydrodynamic models;

- solution of the coupled problem by a time marching algorithm.

4.2　Adaptive definition and validation of the kinetic region

The first problem in such a coupled numerical solution of transitional flows is to identify the kinetic zone Ω_{loc} where the Navier-Stokes solution F_{NS} which has been previously computed is not an adequate solution of the physical problem and where the Navier-Stokes model must be therefore replaced by a more detailed kinetic model. For this purpose, we use any available kinetic interpretation of the Navier-Stokes solution (Grad [3] or §3) and simply observe the quality of the associated Navier-Stokes distribution F^{NS} as solution of the finer Boltzmann equation. Plugging this solution into the Boltzmann equation leads to the following variational residual

$$\int_{\underline{v}} \varphi(\underline{v})[\frac{\partial F_{NS}}{\partial t} + \underline{v} \cdot \frac{\partial F_{NS}}{\partial \underline{x}} - Q(F_{NS}, F_{NS})]d\underline{v}$$
$$= R_\varphi(F_{NS})(\underline{x}, t), \qquad \forall \varphi, \forall \underline{x} \in \Omega.$$

As such, this residual is still too complex to compute. Therefore, we restrict ourselves to the test functions

$\varphi = \underline{m}$ which are used in the derivation of the relevant moments equation. With this choice, we can compute one local residual per finite element function $\psi_N(\underline{x})$ given by

$$
\begin{aligned}
R_N &= \frac{1}{\mathrm{Vol}\Omega_N} \left\| \int_\Omega < \underline{m} \left(\frac{\partial F_{NS}}{\partial t} - Q(F_{NS}, F_{NS}) \right) > \psi_N \right. \\
&\quad \left. - \frac{\partial \psi_N}{\partial \underline{x}} < \underline{m} F_{NS} > \psi_N \right\|.
\end{aligned}
$$

This residual is now simply the norm of the residual of the moment's equation when used with the available Navier-Stokes solution. Since the Navier-Stokes solution satisfies all conservation laws, the residuals must only be computed on the last constitutive equations (equations (38)-(40) when one uses Levermore's equations) which are better written in a nonconservative form.

In view of the above estimate of the quality of the Navier-Stokes solution, the corresponding kinetic region Ω_{loc} is therefore defined as the region of Ω where the residual computed above is larger than a given threshold.

We illustrate below this strategy on an example taken from [22], computing the hypersonic polyatomic flow around an ellipse at 30 degrees of angle of attack, Mach number at infinity $M_\infty = 20$, and Reynolds number at infinity $5000/m$.

The Navier-Stokes solution is calculated by a SUPG code provided by Dassault Aviation Company [21], using a mesh of 6882 nodes and noslip boundary conditions. The residual is based on a polyatomic Grad's interpretation of the Navier-Stokes equation. We present below the calculated residual isolines defining Boltzmann and Navier-Stokes regions, that we compare to the amount of rarefied desequilibrium predicted by a global Boltzmann calculation. We observe that apart from boundary problems induced by the noslip boundary condition in the Navier-Stokes

solution, both the Grad's residual and the global kinetic solution predict the same geometry for the rarefied domain (figures 1, 2, 3, 4). Similar results can be found for denser flows in ([9]). The use of Levermore's residuals is under development.

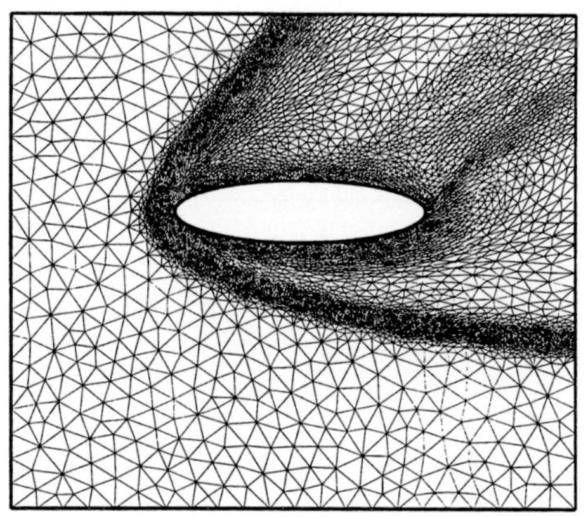

Figure 1: Adaptative Mesh

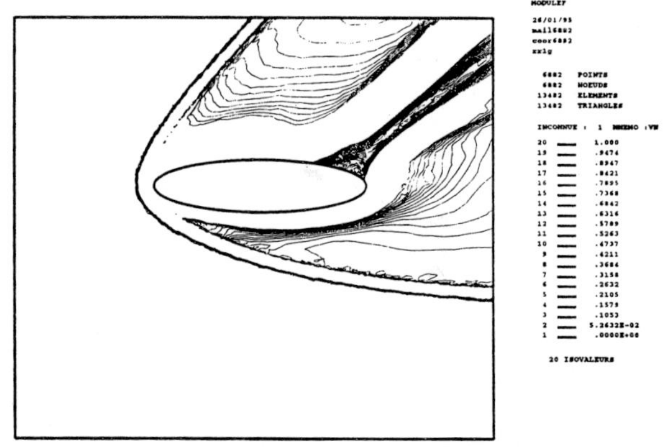

Figure 2: Grad's residual 0-5

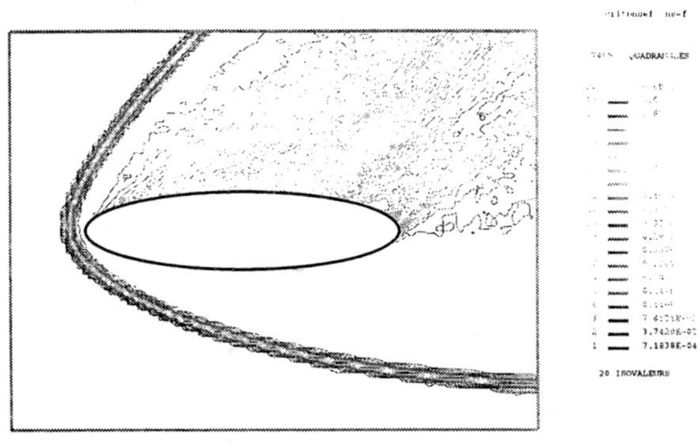

Figure 3: Boltzmann's residual

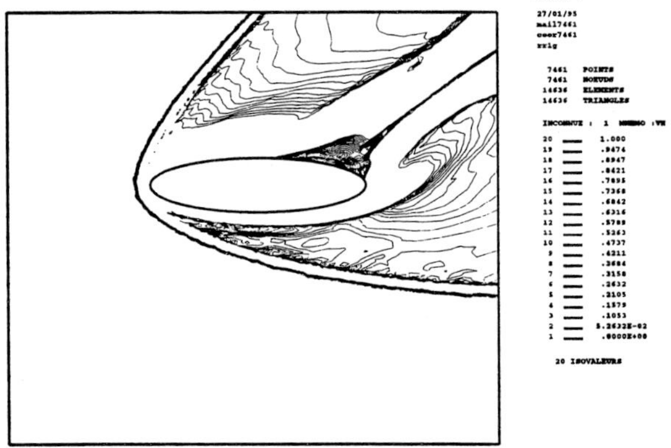

Figure 4: Grad's residual 0-1

4.3 Formulation of the coupled problem

Once knowing the adaptive partition of our original domain between a Navier-Stokes domain and a kinetic domain, we need to formulate a coupled problem combining the Navier-Stokes equation in the Navier-Stokes domain and the Levermore's fourteen moments system in the kinetic region. For this purpose, we will first introduce the coupling strategy in the general hyperbolic case of the multidomain formulation of Levermore's system and then degenerate one domain to the limiting Navier-Stokes model.

Let us therefore split the domain Ω into two nonoverlapping subdomains Ω_1 and Ω_2 with external unit normal vector $\underline{n}_1$ and $\underline{n}_2$, interface Γ_b, and consider the coupled system of well-posed subproblems obtained by solving the Levermore's system locally on each subdomain with appropriate inflow boundary conditions

$$\partial_t \underline{U}_\alpha^1 + div A_\alpha(\underline{U}_\alpha^1) = S(\underline{U}_\alpha^1) \text{ on } \Omega_1,$$
$$A_\alpha(\underline{U}_\alpha^1).\underline{n}_1 = A_{imp}^- + A_\alpha^+(\underline{U}_\alpha^1).\underline{n}_1 \text{ on } \partial\Omega_1,$$
$$A_{imp}^- = -A_\alpha^+(\underline{U}_\alpha^2).\underline{n}_2 \text{ on } \Gamma_b,$$
$$A_{imp}^- = \text{given on } \partial\Omega_1 \cap \partial\Omega_\infty,$$

$$\partial_t \underline{U}_\alpha^2 + div A_\alpha(\underline{U}_\alpha^2) = S(\underline{U}_\alpha^2) \text{ on } \Omega_2,$$
$$A_\alpha(\underline{U}_\alpha^2).\underline{n}_2 = A_{imp}^- + A_\alpha^+(\underline{U}_\alpha^2).\underline{n}_2 \text{ on } \partial\Omega_2,$$
$$A_{imp}^- = -A_\alpha^+(\underline{U}_\alpha^1).\underline{n}_1 \text{ on } \Gamma_b,$$
$$A_{imp}^- = \text{given on } \partial\Omega_2 \cap \partial\Omega_\infty.$$

By construction of the interface boundary conditions, any solution $(\underline{U}_\alpha^1, \underline{U}_\alpha^2)$ of this system is associated to a continuous interface flux

$$\begin{aligned} A_\alpha(\underline{U}_\alpha^1).\underline{n}_1 &= A_\alpha^-(\underline{U}_\alpha^1).\underline{n}_1 + A_\alpha^+(\underline{U}_\alpha^1).\underline{n}_1 \\ &= -A_\alpha^+(\underline{U}_\alpha^2).\underline{n}_2 - A_\alpha^-(\underline{U}_\alpha^2).\underline{n}_2 \\ &= A_\alpha(\underline{U}_\alpha^2).\underline{n}_1 \text{ on } \Gamma_b. \end{aligned}$$

Hence, it satifies the differential equation (10) everywhere on Ω (at least in a weak form). Since it also satifies the external boundary conditions (15) by construction, it is a solution of the original boundary value problem on the whole domain Ω.

The converse is also true, at least for the linearised hyperbolic problem ([18]) or for the discretised version of (10) to be presented in the next section : in other words, in these two particular situations, any solution of the global problem is also solution of the above coupled system. The advantage of the coupled form is that it gives the possibility of replacing one Levermore's subproblem by the Navier-Stokes asymptotic limit, yielding the final form of our coupled system

$$\partial_t \underline{U}^1_\alpha + div A_\alpha(\underline{U}^1_\alpha) = S(\underline{U}^1_\alpha) \text{ on } \Omega_1, \tag{47}$$

$$A_\alpha(\underline{U}^1_\alpha).\underline{n}_1 = A^-_{imp} + A^+_\alpha(\underline{U}^1_\alpha).\underline{n}_1 \text{ on } \partial\Omega_1, \tag{48}$$

$$A^-_{imp} = -A^+_\alpha(\underline{U}^{NS}_\alpha).\underline{n}_2 \text{ on } \Gamma_b, \tag{49}$$

$$A^-_{imp} = \text{ given on } \partial\Omega_1 \cap \partial\Omega_\infty, \tag{50}$$

$$\partial_t \begin{pmatrix} \rho \\ \rho\underline{u} \\ E \end{pmatrix} + div \begin{pmatrix} \rho\underline{u} \\ \rho\underline{u} \otimes \underline{u} - \underline{\sigma} \\ E\underline{u} - \underline{\sigma} \cdot \underline{u} + \underline{q} \end{pmatrix} = 0 \text{ on } \Omega_2, \tag{51}$$

$$A_5(\underline{U}^{NS}_\alpha).\underline{n}_2 = A^-_{imp} + A^+_5(\underline{U}^{NS}_\alpha).\underline{n}_2 \text{ on } \partial\Omega_2, \tag{52}$$

$$A^-_{imp} = -A^+_5(\underline{U}^1_\alpha).\underline{n}_1 \text{ on } \Gamma_b, \tag{53}$$

$$A^-_{imp} = \text{given on } \partial\Omega_2 \cap \partial\Omega_\infty. \tag{54}$$

Above, $E = \frac{1}{2}Tr(\underline{u}\otimes\underline{u} - \underline{\sigma})$ denotes the total energy density of the fluid, and A_5 denote the five components of the total flux A_α associated to the five conserved moments ρ, $\rho\underline{u}$ and $E = \frac{1}{2}Tr(\underline{u}\otimes\underline{u} - \underline{\sigma})$. Due to the specific structure of the Navier-Stokes equation, the boundary conditions (52) is only imposed on these five components A_5 of the flux A_α which actually correspond to the mass, momentum, and energy flux, respectively. This subset of boundary conditions is nevertheless sufficiently large to guarantee that the resulting coupled system perfectly conserves mass, momentum, and energy at the interface. More-

over, in the Navier-Stokes equations (51), the stress tensor $\underline{\sigma}$ and heat flux vector $\underline{q}$ are computed by the constitutive laws (44) and (45).

5 The Local Solvers

5.1 Kinetic solver for Levermore's model

We propose to take advantage of the kinetic structure of (10) to approach its exact solution by a first-order kinetic scheme with source term. This scheme is then written as a time integration formula with explicit flux splitting and implicit source term. It was introduced by B.Perthame [7] for the compressible Euler equations. This scheme builds a time approximation of (10) in the following way. Let Δt be a small time step. We are given the moment $\underline{U}^n_\alpha$ at time $t^n = n\Delta t$ for each point $\underline{x}$ and we obtain numerically the associated distribution

$$F^n(\underline{x}, \underline{v}) = exp(\alpha(\underline{x}, t^n).\underline{m}(\underline{v}))$$

by solving the entropy minimization problem subject to the constraint that

$$\int_{I\!R^d} \underline{m}(\underline{v}) F^n(\underline{x}, \underline{v}) d\underline{v} = \underline{U}^n_\alpha. \tag{55}$$

Then, we solve the underlying kinetic equation

$$\begin{cases} \partial_t f + \underline{v}.\partial_x f = Q(f) \\ f(\underline{x}, t^n, \underline{v}) = F^n(\underline{x}, \underline{v}) \end{cases} \tag{56}$$

with $t \in [n\Delta t, (n+1)\Delta t]$ in two steps:
(1) The linear transport equation

$$\begin{cases} \partial_t f + \underline{v}.\partial_x f = 0 \\ f(\underline{x}, t^n, \underline{v}) = F^n(\underline{x}, \underline{v}) \end{cases} \tag{57}$$

with solution

$$\tilde{f}(\underline{x}, t, \underline{v}) = F^n(\underline{x} - (t - t^n)\underline{v}, \underline{v}), \tag{58}$$

is first advanced from time t^n to time t^{n+1}.

(2) The collision equation

$$\begin{cases} \partial_t f = Q(f)(\underline{x}, \underline{v}, t) \\ f(\underline{x}, t^n, \underline{v}) = \tilde{f}(\underline{x}, t^{n+1}, \underline{v}), \end{cases}$$

is then solved between t^n and t^{n+1} by a first order implicit scheme in time, yielding

$$f(\underline{x}, t^{n+1}, \underline{v}) = \tilde{f}(\underline{x}, t^{n+1}, \underline{v}) + \Delta t Q(f)(\underline{x}, t^{n+1}, \underline{v}). \tag{59}$$

By combining (58) and (59), we obtain a consistent $O(\Delta t^2)$ approximation of the solution of (56) given by

$$f^{n+1}(\underline{x}, \underline{v}) = F^n(\underline{x} - (t^{n+1} - t^n)\underline{v}, \underline{v}) + \Delta t Q(f^{n+1})(\underline{x}, \underline{v}) \tag{60}$$

whose moments

$$U_\alpha^{n+1} = \int_{IR^d} m(\underline{v}).F^n(\underline{x} - \underline{v}\Delta t, \underline{v}) d\underline{v}$$

$$+ \Delta t \int_{IR^d} m(\underline{v}).Q(f^{n+1})(\underline{x}, \underline{v}) d\underline{v} \tag{61}$$

are first-order (in Δt) approximations of the solution to (10). In order to write the numerical scheme based on this approximation, we finally replace at each time step the moments U_α^{n+1} by their cell averages

$$U_{\alpha,j}^{n+1} = \frac{1}{\text{vol } \Omega_j} \int_{\Omega_j} U_\alpha^{n+1} dx, \tag{62}$$

which yields

$$\text{vol } (\Omega_j) U_{\alpha,j}^{n+1} = \int_{\Omega_j} \int_{IR^d} m(\underline{v}).F^n(\underline{x} - \underline{v}\Delta t, \underline{v}) d\underline{v} dx$$

$$+ \int_{\Omega_j} \int_{IR^d} m(\underline{v}).Q(f^{n+1}(\underline{x}, \underline{v})) d\underline{v} dx.$$

In general [24], the second term

$$\text{vol}(\Omega_j) S_{\alpha,j}^{n+1} = \int_{\Omega_j} \int_{IR^d} m(\underline{v}).Q(f^{n+1}(\underline{x}, \underline{v})) d\underline{v} dx$$

depends only on $U_{\alpha,j}^{n+1}$. For example, for the simplest BGK model, we have

$$S_{\alpha,j}^{n+1} = \frac{1}{\varepsilon} (\int \underline{m} \mathcal{M}(U_{\alpha,j}^{n+1}) - U_{\alpha,j}^{n+1}).$$

On the other hand, because $F^n(\underline{x} - \underline{v}(t - t^n), \underline{v})$ is the exact solution of the linear transport equation, the first integral can be written after time integration of (57)

$$\int_{\Omega_j} \int_{IR^d} m(\underline{v}).F^n(\underline{x} - \underline{v}\Delta t, \underline{v}) d\underline{v} dx$$

$$= \int_{\Omega_j} \int_{IR^d} m(\underline{v}).F^n(\underline{x}, \underline{v}) d\underline{v} dx$$

$$- \int_{t^n}^{t^{n+1}} \int_{\partial\Omega_j} \int_{IR^d} \underline{v}.\underline{n} m(\underline{v}) F^n(\underline{x} - \underline{v}(t - t^n), \underline{v}) d\underline{v} d\gamma dt.$$

If we now assume that due to appropriate CFL conditions, $\underline{x} - \underline{v}(t - t^n)$ remains almost always in one of the neighboring cell Ω_j or Ω_i for any $\underline{x} \in \partial\Omega_i \cap \partial\Omega_j$, if we introduce the associated local half-fluxes

$$A_{ji}^+(U_{\alpha,j}^n) = \int_{\partial\Omega_i \cap \partial\Omega_j} \int_{\underline{v}.\underline{n}_j > 0} \underline{v}.\underline{n}_j m(\underline{v}) F_j^n(\underline{v}) d\underline{v} d\gamma,$$

$$A_{ji}^-(U_{\alpha,i}^n) = \int_{\partial\Omega_i \cap \partial\Omega_j} \int_{\underline{v}.\underline{n}_j < 0} \underline{v}.\underline{n}_j m(\underline{v}) F_i^n(\underline{v}) d\underline{v} d\gamma,$$

$$A_{j\infty}^+(U_{\alpha,j}^n) = \int_{\partial\Omega \cap \partial\Omega_j} \int_{\underline{v}.\underline{n}_j > 0} \underline{v}.\underline{n}_j m(\underline{v}) F_j^n(\underline{v}) d\underline{v} d\gamma,$$

and if we take into account the kinetic boundary conditions (15) of section 2, we can write the proposed kinetic scheme under the final fully conservative form

$$U_{\alpha,j}^{n+1} = U_{\alpha,j}^n + \Delta t S_{\alpha,j}^{n+1}$$

$$- \frac{\Delta t}{\text{vol}(\Omega_j)} \sum_{\partial\Omega_i \cap \partial\Omega_j} (A_{ji}^+(U_{\alpha,j}^n) + A_{ji}^-(U_{\alpha,i}^n))$$

$$- \frac{\Delta t}{\text{vol}(\Omega_j)} (A_{j\infty}^+(U_{\alpha,j}^n) + \int_{\partial\Omega \cap \partial\Omega_j} A_{imp}^- d\gamma). \tag{63}$$

In this expression, the coupling conditions (49) at the interface are simply imposed explicitly in time by setting A_{imp}^- to the value of the outgoing half flux associated to the existing Navier-Stokes distribution

$$A_{imp}^- = -A_\alpha^+((\underline{U}_\alpha^{NS})^n).\underline{n}_2 \text{ on } \Gamma_b,$$

computed at this boundary at the previous time step. The validity of the above scheme is confirmed by the following theorem proved in [24]

Theorem 1 *Under an appropriate CFL condition, the one dimensional (in $\underline{x}$) version of the above numerical scheme (63) is stable, entropic and all the moments $U_{\alpha,j}^{n+1}$ are associated to a positive distribution function. In particular, this guarantees that all positive moments as pressure or density remain positive during the integration process.*

5.2 Navier-Stokes solver

Let us now consider the compressible Navier-Stokes equations which we formally write as

$$\frac{\partial W}{\partial t} + div[F(W)] = 0 \text{ on } \Omega, \tag{64}$$

with $W = (\rho, \rho v, E)$ the conservative variables, and $F = F^c + F^d$ the total flux (convective and viscous part). The problem consists in computing a solution of these equations, satisfying the boundary conditions imposed at infinity

$$W = W^\infty \text{ at infinity,}$$

and the coupling interface boundary conditions (52)-(53)

$$F(W) \cdot n = -(A_5^+(U_\alpha^{Lever}).\underline{n}_1 + A_5^+(U_\alpha^{NS}).\underline{n}_2,$$

on the interface. Here, the conservative flux A_5 must be computed by the numerical rule already used in the Levermore's region, otherwise, the discrete scheme will no longer be conservative at the interface. The global domain Ω is discretized using node centered cells (figure **??**) defined on an unstructured grid. Then, at each time step n and for each cell Ω_i, we solve

$$\int_{\Omega_i} \frac{W^{n+1} - W^n}{\Delta t} + \sum_j \int_{\partial\Omega_i \cap \partial\Omega_j} F^c(W^\infty) \cdot \underline{n}_i$$
$$+ \int_{\partial\Omega_i \cap \partial\Omega_\infty} F(W^n) \cdot \underline{n}_i + \int_\Omega F^d(W^n) \cdot \nabla\varphi_i$$
$$= -\int_{\partial\Omega_i \cap \Gamma_b} (-A_5^-((U_\alpha^{Lever})^n).\underline{n}_1 + A_5^+((U_\alpha^{NS})^n).\underline{n}_2) d\gamma.$$

Above φ_i is the nodal shape function associated to the node i, that is the continuous piecewise linear function with value 1 at node i and 0 at all other nodes. In all present numerical tests, in order to be robust in hypersonic regimes and to always return positive pressures and densities, we compute the convective flux

$$\int_{\partial\Omega_i \cap \partial\Omega_j} F^c(W^{n+1}) \cdot \underline{n}_i$$

by the hybrid upwind splitting (HUS) scheme of [26]. This scheme corrects the kinetic flux vector splitting associated of [7] by the formula

$$F(W_i, W_j) = F^+(W_i) + F^-(W_j) - F^-(W_R^*) + F^-(W_L^*)$$
$$\text{if } v^* \geq 0$$
$$= F^+(W_i) + F^-(W_j) + F^+(W_R^*) - F^+(W_L^*)$$
$$\text{if not.}$$

Above W_R^* and W_L^* are the left and right states calculated in the Osher-Solomon approximate Riemann solver (used in reverse order) and v^* is the speed of the contact discontinuity separating W_R^* and W_L^*.

By construction, this scheme returns positive pressures and densities, which makes it very robust in hypersonic situations. It also treats exactly contact discontinuities and therefore is less diffusive than the underlying kinetic scheme. Its second order implementation was realized by first computing nodal averages of gradients of W, and then constructing second order approximations W_{ij} and W_{ji} at $\partial C_i \cap \partial C_j$ using a MUSCL scheme with Van Albada limiters. In order to preserve its robustness, the scheme is automatically degraded to first order as soon as a minmod indicator detects gradient inversions.

6 Numerical Tests

We test below the coupling strategy in a transitional regime with Knudsen numbers ranging from 10^{-4} to 10^{-1}.

In all our tests, the convergence criteria for the coupled problem are

$$\|U_{NS}^{n+1} - U_{NS}^n\|_{L^2} \le 10^{-8}, \qquad (65)$$

in the Navier-Stokes region, and

$$\|U_\alpha^{n+1} - U_\alpha^n\|_{L^2} \le 10^{-8}, \qquad (66)$$

in the Kinetic region.

The solution is initialized by a Navier-Stokes solution with no slip boundary condition. The first application is a one-dimensional steady Couette flow. The solid walls are distant of 1 meter, the moving wall moves at $1000 m.s^{-1}$ (high speed Couette flow) and the corresponding internal domain is divided into 100 cells (h=.01) or 200 cells (h=.005). The considered monoatomic gas ($R = 287, \gamma = 5/3$) is initially at equilibrium with a temperature of $250K$ which is equal to the wall temperature. The initial density of the gas is equal to $5.19e^{-6} Kg.m^{-3}$, and its viscosity is taken to be proportional to the temperature square root (hard sphere model). Finally, the Prandtl number is taken at the standard value of Pr = 2/3.

In this case. the kinetic region is defined a priori (without using the Levermore residual) as the region closest to the wall at rest $0 \le x \le 0.3m$. The boundary conditions on the wall inside the kinetic region is derived from the total accommodation kinetic conditions. This is performed by setting $k(x) = 0$ in (18) and, temperature and velocity values to wall values in the Maxwellian (3). A noslip boundary condition is assumed at the other wall. It is not physically relevant, but it enables as to have a first validation of our coupling strategy.

Velocity and temperature profiles are shown in figures 5,6,7, 8,10 and 11, both for the Navier-Stokes and the coupled solutions. Velocity and temperature jumps at the kinetic surface are in the following table.

Mean Free Path λ	Space Discretisation step : h	Temperature slip T-Tw	Velocity slip $u - u_w$
10^{-5}	.01	$5.3^\circ K$	7.9m/s
	.005	$3.0^\circ k$	5.0m/s
10^{-3}	.01	$6.0^\circ K$	8.6 m/s
	.005	$4.0^\circ K$	5.3 m/s
10^{-2}	.01	$10.2^\circ K$	13.8 m/s
10^{-1}	.01	$31.0^\circ K$	43.5 m/s

There appears to be no kinetic effect on the solution, except the apparition of a velocity slip on the kinetic wall.

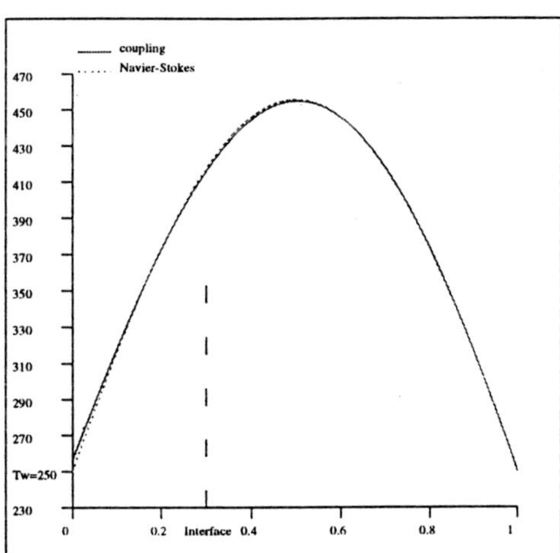

Figure 5: The temperature profile in the Couette flow at $(Kn)=10^{-5}$

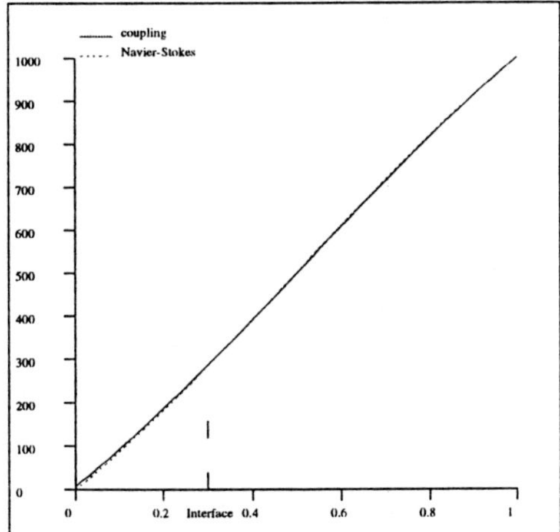

Figure 6: The Velocity profile in the Couette flow at $(Kn)=10^{-5}$

For a Knudsen number of 10^{-3}, the Navier-Stokes equations are still compatible with the Levermore equations. The difference between both profiles is limited and is due to the change of wall boundary

conditions between the full Navier-Stokes and the coupled simulation. The coupled simulation predicts a small jump at the wall and a smooth transition between the kinetic and the Navier-Stokes zone (figures 7,8).

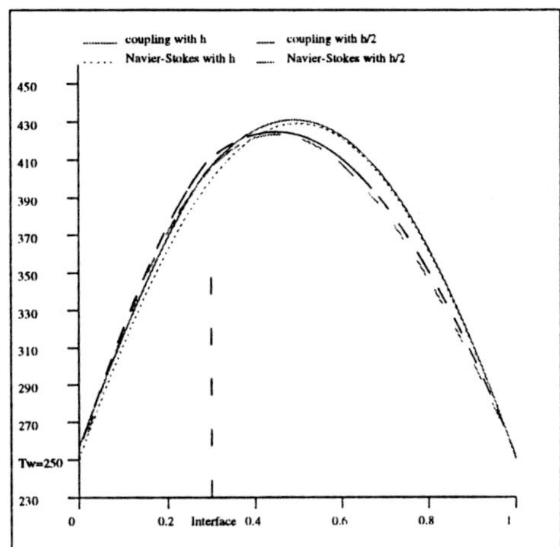

Figure 7: The temperature profile in the Couette flow at $(Kn)=0.001$

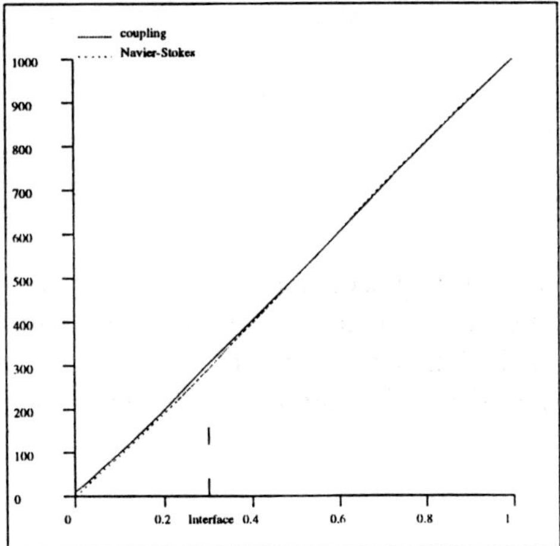

Figure 8: The Velocity profile in the Couette flow at $(Kn)=0.001$

To check the accuracy of the numerical code of the associated quadrature rules, we present above a plot of normal velocity (figure 9).It should be zero. The nonzero normal velocity obtained at the numerical level is a $o(h)$ numerical effect due to the fact that the integration rule used within the code is temperature dependent. The same effect can be observed in the variation of temperature profile in figure 7 as a function of h.

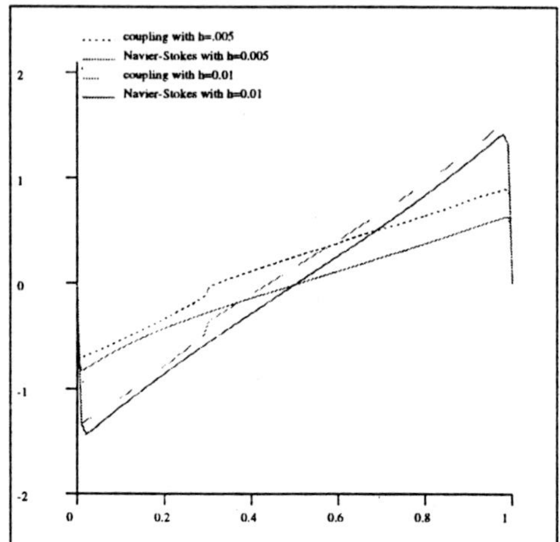

Figure 9: The Normal Velocity profile in the Couette flow at (Kn)=0.001

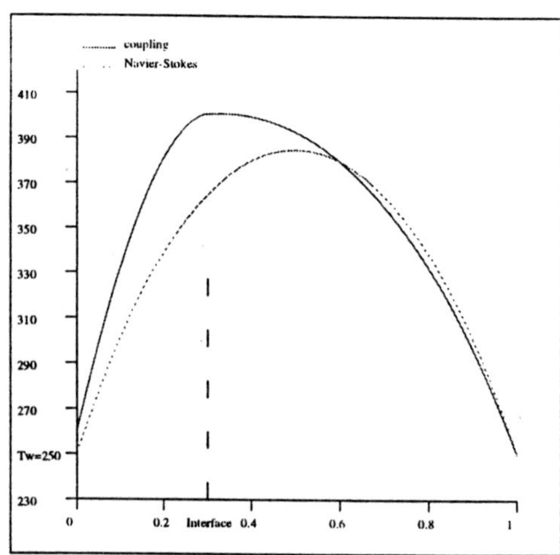

Figure 10: The temperature profile in the Couette flow at (Kn)=0.01

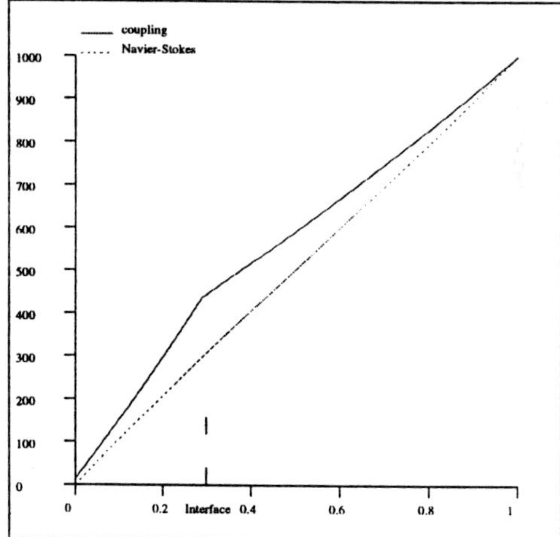

Figure 11: The Velocity profile in the Couette flow at (Kn)=0.01

For higher Knudsen numbers, the Navier-Stokes equations are nolonger compatible with the Levermore's solution, and therefore, we observe solution discontinuities at the interface (figures 10 and 11).

We now consider a preliminary test on a two-dimensional external hypersonic flow with an incidence of 10^0 over a flat plate of 1 meter. The gas $(R = 287, \gamma = 5/3, \rho = 5.19e^{-6})$ at infinity has a

Mach number of 18.62, a Reynolds number of 30687. $(Kn = 1.e^{-3})$ and a temperature of $194^{o}K$. The temperature of the plate is equal to $1000K$. The kinetic region is a domain given a priori surrounding the plate with interface defined as a function proportional to $\sqrt{x}$, where x is the horizontal distance from the tip of the plate. The boundary conditions on the plate is also derived from the total accommodation kinetic conditions. The same criterium for convergence is used (65,66). A course mesh of 3721 nodes is used.

We present iso-density lines (figure 12) and temperature, Mach values at 3 Cross-Sections x=.25,x=.55,x=.85. (figures 13)

Figure 12: hypersonic flow over a flat plate : Iso-density values

7 Conclusion

We have extended in this paper an existing general strategy for building a hierarchy of numerical models relating the Boltzmann to the Navier-Stokes equations. This strategy is based on a recent mathematically consistent ansatz proposed by D. Levermore, and briefly reviewed in the present paper. In this framework, we can adapt the general adaptive algorithm previously developed for coupling Navier-Stokes equations to local kinetic models.

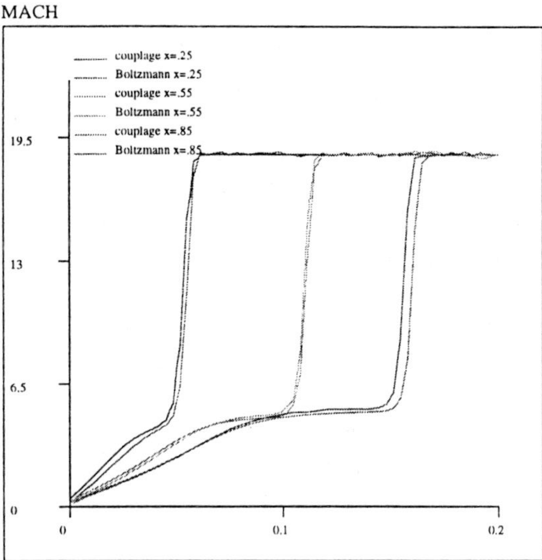

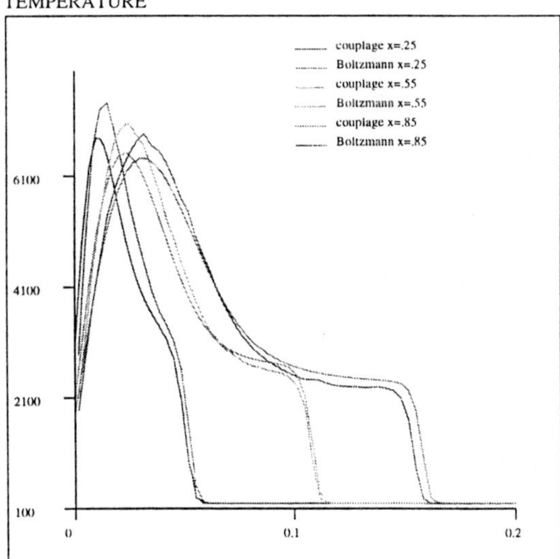

Figure 13: hypersonic flow over a flat plate : Mach and temperature values. Comparison with a Monte Carlo simulation solving Boltzmann equation.

The proposed multimodel coupling strategy has been illustrated by several numerical examples and appears to be operational for different type of gases in transitional regimes. It brings three major improvements:

- first, the kinetic model can be solved by a deterministic model, which removes all difficulties related to noise control in Monte Carlo simulations ;

- second, the new model is mathematically consistent, when treating either the entropy inequality or the kinetic boundary conditions ; it is therefore robust and physically relevant to situations with moderate to strong desequilibrium ;

- finally, the interface conditions in the coupling process are fully conservative, even at the discrete level.

The further work to be done concerns mainly the enrichment of the collision model, in order to be able to treat more general internal energy laws, that is more general collision operators.

References

[1] C.D. Levermore : Moment Closure Hierarchies for Kinetic Theories,*submitted to the journal of statistical physics, may 1995.*

[2] C. Cercignani : The Boltzmann Equation and its Applications ,*Applied Mathematical Sciences, Springer-Verlag, 1975.*

[3] H. Grad: On the Kinetic Theory of Rarefied Gazes, *Comm. Pure Appl. Math 2, pp 331-407, 1949.*

[4] C. Bardos : Une interprétation des relations existant entre les équations de Boltzmann, de Navier-Stokes et d'Euler à l'aide de l'entropie , *Math. Aplic Comp. V.6 n 1 pp 97-117, 1987.*

[5] P. Le Tallec, J.F. Bourgat, Mallinger F.,Perthame B.,Qiu Y.: Coupling Boltzmann and Navier-Stokes. *Rapport INRIA RR-2281, Aout 1994. Domain Decomposition methods in Science and Engineering, A.Quateroni. J.Periaux. Y.Kznetsov, O.Widlund eds, Contemporary Maths 157. AMS.1994. pp 377-398.*

[6] B. Khobalatte : Résolution numérique des équations de la mécanique des fluides par des méthodes cinétiques. *Thèse de l'Université d'Orléans 1994.*

[7] B. Perthame : Boltzmann type schemes for gas dynamics and the entropy property, *SIAM J.Numer. Anal., pp 1405-1421,Dec 1990.*

[8] B. Khobalatte, B. Perthame : Maximum Principle On The Entropy And Minimal Limitations For Kinetic Schemes, *Rapport INRIA RR-1628, 1994.*

[9] F.Mallinger : Couplage Boltzmann- Navier-Stokes, *Thèse de l'Université de Paris IX Dauphine et INRIA, septembre 1996.*

[10] P.Charrier, B.Dubroca, J.L.Feugeas: Etude numerique de modeles aux moments de Levermore en 2 dimensions, *communication personnelle, 1996.*

[11] R.J.LeVeque, H.C.Yee : A study of numerical Methods for hyperbolic conservation laws with stiff source terms,vol 86 pp 187-210, *Journal of Computational physics, 1990.*

[12] G.A.Bird : Molecular Gas Dynamics and the Direct Simulation of Gas Flow, *Oxford Engineering Science Series.42, Oxford Science Publications, 1994.*

[13] W.G.Vincenti, C.H.Kruger : Introduction to Physical Gas Dynamics, *Robert E.Krieger Publishing Company Malabar, Florida, 1986.*

[14] D.Levermore, W.J.Morokoff : The Gaussian Moment Closure for Gas Dynamics, *SIAM J. on Applied Mathematics, submitted 16 Feb.1996.*

[15] D.Levermore, Entropy Based Moment Closures for Kinetic Equations. *Transport Theory and Statistical Physics, submitted 29 April 1996.*

[16] P.Le Tallec, F.Mallinger : Coupling Boltzmann and Navier-Stokes Equations by Half Fluxes. *to appear in the Journal of Computational Physics.*

[17] J.F. Bourgat, L. Desvillettes, P. Le Tallec, B. Perthame, "Microreversible collisions for polyatomic gases and Boltzmann's theorem" ,*J. Mech., B/Fluids*, 13, pp 237-254, 1994.

[18] F. Gastaldi, L. Gastaldi, "On a domain decomposition approach for the transport equation: theory and finite element approximations", *IMA Journal of Numerical Analysis*, Vol. 14, pp , 1993.

[19] Hirschfelder, Curtiss, Bird ; *Molecular Theory of Gases and Liquids* J. Wiley and Sons (éd. 1954)

[20] P. Le Tallec, F. Mallinger, "Modélisation d'un gaz Polyatomique, Equations de Grad Généralisées et Validité de la Solution des Equations de Navier-Stokes" *Rapport de contrat final - Hermes* , April 1995.

[21] M. Mallet, "A finite element method for computational fluid dynamics", *Ph.D. Thesis*, Stanford University, 1985 .

[22] P. Le Tallec, F. Mallinger, *Adaptive Multimodel Decomposition in Fluid Mechanics*, Domain Decomposition Methods in Sciences and Engineering, R. Glowinski, J. Périaux, Z. C. Shi, O. Widlund eds., J. Wiley and Sons, 411, 426, 1997.

[23] J.F. Bourgat, P. LE Tallec, F. Mallinger, D. Tidriri, Y. Qiu : Numerical coupling of Boltzmann and Navier-Stokes, in Proceedings of the sixth IUTAM Conference on Rarefied Flows For Reentry Problems. Marseille, France, September 1992, R. Brun, A.A. Chikhaoui eds., 1992, 60-66.

[24] P. Le Tallec, J.P. Perlat, *Numerical Analysis of the Levermore's Moment System.* INRIA Research Report 3124, 1997.

[25] Gupta, Moss, Scott ; *Slip-Boundary Equations for Multicomponent Nonequilibrium Airflow* ; NASA report CR-181252 (1984).

[26] F.Coquel and M.S. Liou : Hybrid Upwind splitting (HUS) by a Field by Field Decomposition; NASA Technical Memorandum 106843, ICOMP-95-2, 1995.

APPLIED TRANSITION ANALYSIS

Thorwald HERBERT

DynaFlow, Inc., 3040 Riverside Dr. Columbus, Ohio 43221, USA   &

Dept. Mech. Eng., The Ohio State University; Columbus, Ohio 43210-1107, USA

Abstract

Transition analysis in the engineering environment serves to determine the border between the regions of laminar and turbulent flow in boundary layers. Knowledge of this border permits improved estimates of drag and heat transfer, and is key to the design of low-drag components and flow-control systems. We discuss problems of transition analysis, improvements to the current approach based on linear stability characteristics, and the more advanced nonlinear techniques enabled by the parabolized stability equations. Besides the issues of efficient computation crucial in engineering practice, we address improvements in software design and usability, data quality, and physical models of the transition process.

1 Introduction

Transition analysis is most often performed to obtain improved estimates of the drag of aircraft components or to improve the design of these components toward reduced drag. At supersonic speeds, or in gas turbines, the analysis may simultaneously or exclusively focus on the heat transfer to the surface. Knowledge of the "transition point" or transition front permits switching the viscous flow computation from laminar to turbulent, typically by activating the turbulence model in solvers for the Reynolds-averaged Navier-Stokes equations. Variation of the transition front with design variables is key to the selection of low-drag components and to the implementation of flow-control systems.

The aerodynamic bodies of interest for transition analysis are mostly wings, tail fins, nacelles, turbine vanes or blades, and in some cases the fuselage of smaller aircraft. Vehicles for flight at high supersonic and hypersonic speeds typically do not consist of the traditional components and form a special class of transition problems.

The tools for transition analysis vary in wide ranges that are different in the external aerodynamics and gas-turbine communities. The tools reach from pure guesses, through criteria based on the momentum-thickness Reynolds number Re_θ, criteria based on turbulent-spot production rates, increasingly elaborate turbulence models, to the e^N method, which has deep roots in the linear stability theory (LST) of boundary layers. A new approach to transition analysis on the basis of parabolized stability equations (PSE) has been proposed by Herbert (1991) and slowly encroaches the engineering environment. At the high turbulence levels in flows through high-

pressure turbines, turbulence models may be indeed the best option. At low turbulence levels of less than 2%, characteristic for flows in wind tunnels and atmospheric flight, methods based on the stability characteristics of the boundary layer are more appropriate, and we will limit the discussion to these methods.

Probably the most intense efforts in transition analysis have been spent on the e^N method in context with Boeing 757 flight tests to investigate concepts of natural laminar flow (NLF) and laminar-flow control (LFC) for aircraft wings. More recently, these efforts have been extended to the high-speed civil transport. Unfortunately, the extensive data are proprietary, and the agreement of the test results with the predictions cannot be evaluated. While success of the e^N method has been reported elsewhere (Malik 1990), scattered pieces of information indicate that the results are generally unreliable.

Two major series of flight tests have been performed in Europe to investigate NLF concepts and evaluate methods of transition analysis. The series of ATTAS flight tests were carried out on the laminar-flow glove of a VFW 614 aircraft for Deutsche Airbus GmbH. A second series of tests was performed on the glove of a Fokker F100 under the ELFIN (European Laminar Flow Investigation) program. Some information on the Fokker F100 glove design (Dressler et al. 1992), the approach to transition prediction (Schrauf et al. 1992), and the comparison of e^N predictions with test results (Schrauf 1994, Schrauf et al. 1997) has been published. Selected flight tests have been analyzed using the linear and nonlinear PSE (Schrauf et al. 1996, Herbert & Schrauf 1996). These studies as well as similar applications to transition problems under the high disturbance levels in experiments on turbines vanes (Herbert et al. 1993) have revealed the capabilities and limitations of traditional and advanced transition analysis.

Received on September 3, 1997.

Discrepancies between predictions and observations of transition points are not only caused by insufficient physical models of transition or different results for the stability characteristics, but also by inappropriate input data, inviscid-flow or boundary-layer computations, and weaknesses of the analysis software. Codes used in the commercial environment provide options and require choices that can mislead an engineer without special training to undesirable results. Often, the demand for higher efficiency of the computation has compromised the physics of the problem and reduced the benefit of the analysis.

The first generation of our codes for applied transition analysis implemented LST and linear and nonlinear PSE in general curvilinear coordinates for compressible flows up to high Mach numbers, and served well for investigating nonlinear aspects of transition and evaluating the PSE method. The current second software generation increases computational efficiency significantly, and is designed to provide reliable results generally without human interference. In the following, we give an overview over background and status of our efforts.

Transition in Aerodynamic Flows

Aerodynamic boundary layers are in general three-dimensional (3D), i.e. they have three velocity components that vary in all three spatial directions. Most references to the "stability of 3D boundary layers" concern flows such as Falkner-Skan-Cooke flow, flow over a rotating disk, boundary layers over an infinite swept wing, or the flow over a circular cone at zero angle of attack. In these flows, the three velocity components depend only on two spatial variables. We denote these flows as *quasi-2D*. This distinction is important for applications of the LST.

The three-dimensionality may be weak on a finite swept wing with moderate taper and twist over a span that is large in comparison with the chord, e.g. on the wing of a Fokker F100[3] with rear-mounted engines. Three-dimensionality is more pronounced on the short wings of fighter planes like the F-16XL[10] as shown by the isobars in Figure 1, or over the short span of tail fins and turbine blades. The three-dimensionality on the wings of commercial airplanes may be further enhanced by pylons and wing-mounted engines and the wing gloves used in flight tests as shown in Figure 2 for Boeing 757 flight experiments[11].

Careful attention to the spanwise pressure variations must be given when computing the boundary layer as the basic flow for stability analysis. Even if this task is successfully completed, the three-dimensionality of the boundary layer introduces uncertainty in the subsequent analysis of stability characteristics for transition evaluation caused by the lack of a rigorous linear

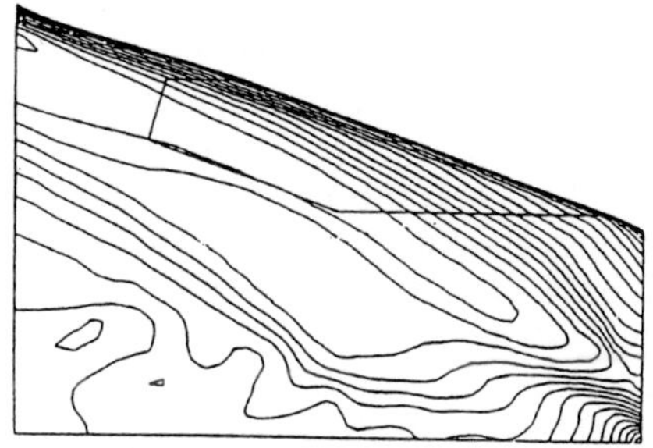

Figure 1: Upper surface isobars on the wing of an F-16XL.

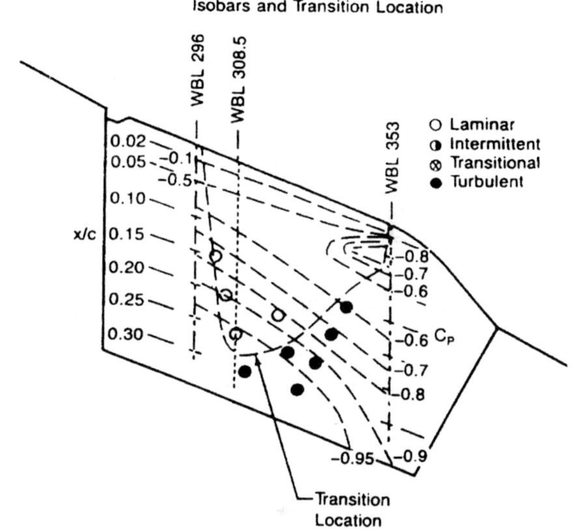

Figure 2: Upper surface isobars on a Boeing 757 NLF glove.

stability theory for 3D boundary layers.

The Transition Process

Transition is a physical process caused by disturbances to an otherwise steady laminar boundary layer. This process in essence involves three ingredients:

1. Receptivity, the conversion of the disturbances to modes internal to the boundary layer,

2. Amplification of the internal modes by algebraic transient growth and exponential growth owing to (linear) instability, and

3. Secondary instabilities and nonlinear interaction of the internal modes and breakdown of the laminar flow.

The last step leads to a modification of the mean flow and to changes in the skin-friction and heat-transfer coefficients. The amplification of internal modes may be absent or unnecessary at high disturbance levels (bypass transition). The process is *sensitive* to small changes in the mean flow, parameters, and disturbances. On the other hand, the drag estimate changes strongly with the transition location.

The transition process is governed by the Navier-Stokes equations. Although direct numerical simulation (DNS) has provided much insight in idealized situations, application to aircraft components is precluded by disparate scales. Resolving disturbances with wavelengths in the millimeter range at frequencies of some kHz on the wing or tail of an airliner would require many orders of magnitude more grid points than what research computations afford. The resolution might still be insufficient to capture important details of the surface roughness. Solving the Navier-Stokes equations for the steady laminar basic flow is possible yet expensive, because of the small scale of the boundary layer (displacement thickness ≈ 1 mm) which needs to be accurately resolved.

Steps of the Analysis

Since the transition process cannot be attacked as a whole, its analysis is usually decomposed into a sequence of steps:

1. Specification of input data (geometry, fluid properties, flow parameters)

2. Computation of the inviscid flow (Euler, potential, or panel code)

3. Computation of the viscous boundary layer

4. Stability and transition analysis

In this approach, transition analysis is not a stand-alone operation but part of a process. In the sense of total quality management, every step of this process must provide the necessary quality to achieve the full benefit of the analysis. While the quality of the input data is a separate issue, each of the three computational steps must be considered under the aspects of *verification* (solving the equations right) and *validation* (solving the right equations) as discussed by Roache[12]. Ideally, the independent verification of the computational codes should be supported by comparison of the results of competing codes for the same input data, because boundary-layer codes and stability codes for compressible flow are no trivial matter, and the results are sensitive to small errors in formulation or numerical treatment. Such comparisons are supported in Europe by the Eurotrans program.

Concerning the validation issue, the process should adhere as closely as possible to the physics of the problem, and therefore, should be treated with the best possible approximations to the governing Navier-Stokes equations. The choice of these approximations must consider the flow (including the boundary layer) over aerodynamic bodies as 3D. Solving the 3D Euler equations is generally preferable to using a potential or panel code. The reduced computational effort for potential solvers may favor their use in the design stage. Computing the 3D boundary layer is preferable to computing multiple 2D sections under the assumption of an infinite or conical wing. Analyzing stability and transition with the 3D extension of the linear and nonlinear PSE is preferable to the use of the original formulation for quasi-2D boundary layers (with three velocity components depending on two spatial variables). Use of the PSE is preferable to using the traditional LST which provides only the linear level of analysis and assumes locally parallel flow.

There remain, however, some inevitable difficulties. The external disturbances affecting transition (e.g. in the atmospheric environment) are insufficiently known. The receptivity mechanisms that selectively "import" disturbances of suitable scales into the boundary layer can be cast into quantitative relations in only a few cases. Many of the physical processes involved in transition are yet obscure. The intrinsic nonlinearity of the breakdown stage with mean-flow modifications and changes in skin-friction and heat transfer requires computations unacceptable to an engineering environment that wants quick answers at low cost. In summary, we want to compute intricate physical phenomena which we barely understand with incomplete input data at little cost to the "customer". CFD is often unable to solve a well-posed problem correctly. Transition analysis is not faced with a well-posed problem. The analysis can only attempt to provide the engineer with physically meaningful data at acceptable cost and outline their restricted utility where appropriate. In more advanced nonlinear analyses, the lack of precise input and receptivity information can be compensated by models of standard environments (similar to the hypothetical standard atmosphere) which may be improved or replaced as more insight and quantitative data become available.

The Traditional Approach

Given the swept, tapered, and twisted wing of an airplane, generating the pressure distribution over the surface is the first major task. For further analysis, this solution is sliced into numerous sections typically along cuts in free-stream direction along the spars of the wing. For every section, the viscous flow is computed using a 2D boundary-layer code, e.g. for tapered wings[13] with conical pressure distribution or locally infinite swept wings[14]. Horton & Stock[15] have pointed out various deficiencies of the Kaups-Cebeci code and derived an improved formulation for tapered wings.

The output files of this step provide the basis for the transition analysis along each section. The three levels of computation are often performed by different groups. Handling all sections requires performing many runs and keeping track of all the files, listings, and graphs produced, a procedure prone to error. For swept wings, the procedure also ignores that the boundary layer and disturbances traverse the sections and may evolve differently than suggested by the final results.

The transition analysis along the sections is based on the e^N method[16, 17] which in turn rests on the linear stability theory[18]. To understand the computational aspects of this approach and of more advanced methods, it is necessary to take a short look at this basis.

Linear Stability Theory

In Cartesian coordinates x, y, z and for incompressible flow, we decompose the total flow into the known basic flow $\mathbf{V}, P$ and disturbances $\mathbf{v}', p'$ where $\mathbf{v}' = (u', v', w')$. From the Navier-Stokes equations we obtain the non-linear disturbance equations which can be linearized if the disturbances are sufficiently small. Key to the LST is the assumption of a *locally parallel flow*, i.e. the assumption of a basic flow velocity in the form $\mathbf{V} = [U(y), 0, W(y)]$. In this case, the disturbances can be written as *normal modes*

$$\mathbf{q}'(x, y, z, t) = \mathbf{q}(y) \exp[i(\alpha x + \beta z - \omega t)] , \quad (1)$$

where $\mathbf{q}' = [u', v', w', p']^T$ is the vector of flow variables and $\mathbf{q} = [u, v, w, p]^T$ is the vector of the associated shape functions. The exponential factor with (complex) wavenumbers α and β in x and z, respectively, and (complex) frequency ω describes the wave nature of the solution. Introducing the normal modes (1) into the linear disturbance equations yields the normal-mode or *stability equations*, which in compact form can be written as an eigenvalue problem

$$L\mathbf{q} = 0 \quad (2)$$

with homogeneous boundary conditions on u, v, w. The operator L contains differentiation in y only. After proper manipulation, L can assume different forms, e.g. the form of the Orr-Sommerfeld and Squire equations.

For given U and W, the eigenvalue problem can be reduced to solving a complex characteristic equation of the form

$$\mathcal{F}(Re, \alpha, \beta, \omega) = 0 \quad (3)$$

which yields two quantities per eigenmode, provided all others are given. For boundary layers, it is most appropriate to specify the real quantities Re, β, and ω and

determine $\alpha = \alpha_r + i\alpha_i$ from equation (3) to obtain the spatial growth rate $-\alpha_i$ in x-direction. Because different powers of α appear in the equations and cause difficulties with many eigenvalue solvers, the problem is often solved for the eigenvalue ω with real Re, α, and β to obtain the temporal growth rate ω_i. The spatial growth rate is approximated by Gaster's transformation $\alpha_i \approx -\omega_i/c_g$ where c_g is the group velocity. In today's computational environment, the small savings by solving the temporal problem do not outweigh the uncertainty in the results and the lengthy discussions on procedures.

Although we deal with ordinary differential equations, solving e.g. the Orr-Sommerfeld problem numerically is not an easy task, as witnessed by a considerable body of literature on this matter. Spectral methods are well suited for solving the incompressible eigenvalue problem. For compressible flows which do not permit an Orr-Sommerfeld-type reformulation, the matrices of the spectral method become too large. We have successfully used three-point (7-th order) and two-point (4-th order) Hermitian methods. The compact method[19] requires higher derivatives of the basic-flow velocity that introduce unnecessary inaccuracies.

N Factors

The e^N method was originally developed for incompressible 2D flows over airfoils. Assuming a 2D TS wave with fixed real frequency ω and $\beta = 0$ has an amplitude A_0 at the point s_0 where instability begins, the growth of the wave amplitude along the arclength s is given by

$$N(s; \omega) = \ln \frac{A(s; \omega)}{A_0} = -\int_{s_0}^{s} \alpha_i(\xi; \omega) d\xi \quad (4)$$

where α_i has to be found by solving the eigenvalue problem (2) under the local flow conditions in the streamwise changing boundary layer[*]. Taking the envelope of these growth curves over the range of unstable frequencies, we obtain

$$N_\omega(s) = \max_\omega N(s; \omega) . \quad (5)$$

Correlation with empirical data shows that transition occurs at a point s_t where $N_\omega(s_t) = N_\omega^*$, and the "limiting value" N_ω^* was found in the range between 7 and 9. This correlation links transition to a certain degree of amplification, $A(s_t)/A(s_0) \approx \exp(N_\omega^*)$. The variation in the limiting values may be caused by empirical data taken at different disturbance levels. Even in the absence of empirical data, the envelope $N_\omega(s)$ can be used for trade-off analysis because lower values indicate a weaker tendency toward transition.

The simplicity of this method is lost in the compressible flows over aircraft components. It can be shown

---

[*]For wings, the arclength s measured normal to the leading edge replaces x in eq. (1), and z is oriented along the leading edge.

that a physical disturbance maintains a fixed frequency during its evolution. In 3D flows, however, it is unclear how the wavenumber vector $\mathbf{k} = (\alpha_r, \beta_r)$ varies from point to point along the surface and in which direction the disturbance grows. Only for quasi-2D flows on an infinite wing, Mack has shown[20] that the spanwise wavenumber $\beta = $ const. during the downstream evolution. In this case it is reasonable to assume uniform disturbances along the span, $\beta_i = 0$, and to base transition criteria on

$$N(s; \omega, \beta) = -\int_{s0}^{s} \alpha_i(\xi; \omega, \beta)d\xi \ . \tag{6}$$

Analysis of the amplitude-growth curves for different ω and β requires a tremendous number of solutions to the LST eigenvalue problem and considerable computational efforts.

To reduce the computational demand, many different e^N "strategies" have been developed[17]. In some cases, only special classes of normal modes are considered, e.g. N_{CF} for steady crossflow modes ($\omega = 0$) or N_{TS} for TS waves with the wavenumber vector locally aligned in edge-streamline direction. Other methods such as the envelope method integrate the local maxima of the growth rate, although no physical mode can grow with the integrated rates. Additional shortcuts and violations of physical constraints have been introduced in the strive for a convenient "universal" limiting value N^*. Options to include or exclude the effects of surface curvature or compressibility further enhance the variety of results. By choice of strategy, options, and numerical settings which are usually undocumented in the output, the same user of the same code can produce different results on different days.

Certainly, the neglect of the disturbance environment and of nonlinear processes is a major deficiency of the e^N method. How useful this method can be to an engineer is hard to evaluate, because many results have been generated with disregard for the physics of the linear instabilities, the key ingredient of the method.

Parabolized Stability Equations

In an attempt to obtain more accurate growth rates by accounting for the weak streamwise growth of the boundary layer, we observed that the WKBJ approximation leads to partial differential equations similar to the parabolized Navier-Stokes equations (PNS), except for additional terms containing frequency and wavenumbers: the *parabolized stability equations* or *PSE* for short. The successful application to linear and nonlinear stability problems in incompressible and compressible flat-plate boundary layers[21] and the prospects of solving inhomogeneous receptivity problems suggested the use of the PSE for practical applications[1].

Overviews over the PSE method and applications in 2D and quasi-2D flows have been given by Herbert[22, 23].

To provide the basis for a wider range of applications, we have recently extended the PSE to 3D boundary layers[24] to account for the small variations of the velocity components U and W in the streamwise and spanwise directions as well as the component V normal to the surface which is non-zero to provide the mass balance as the displacement thickness δ^* changes. We introduce the small parameter $\epsilon = \mathcal{O}(Re^{-1})$ and the slowly varying variables $\xi = \epsilon s$, $\zeta = \epsilon z$, where $Re = Re_s^{1/2}$, $Re_s = U_e s/\nu$, and U_e the edge velocity. The basic flow then takes the form $\mathbf{V} = (U(\xi, y, \zeta), \epsilon V(\xi, y, \zeta), W(\xi, y, \zeta))$ and is governed by Prandtl's boundary-layer equations.

For the disturbances $\mathbf{q}'(s, y, z, t)$ with components $\mathbf{q}' = (u', v', w', p')^T$ in incompressible flow we cannot enforce a similar form because $v' = \mathcal{O}(u')$ and the wavenumbers are not small. However, we can adapt the concept of normal modes such that

$$\mathbf{q}'(s, y, z, t) = \mathbf{q}(\xi, y, \zeta) \exp[\, i\theta(s, z, t)\,] \tag{7}$$

with a slowly varying shape function $\mathbf{q}$ and a well-defined phase function θ, provided oscillations and growth of the disturbance are absorbed into the phase function. This latter property can be enforced by imposing the integral norms

$$\int_0^\infty \mathbf{v}^\dagger \frac{\partial \mathbf{v}}{\partial \xi} dy = 0 \ , \quad \int_0^\infty \mathbf{v}^\dagger \frac{\partial \mathbf{v}}{\partial \zeta} dy = 0 \ , \tag{8a,b}$$

where the dagger denotes the complex conjugate. This choice makes the total kinetic energy $\int_0^\infty |\mathbf{v}|^2 dy$ of the shape functions independent of ξ and ζ. The growth $-2\theta_i$ directly refers to the total kinetic energy of the disturbance. Moreover, we obtain

$$\omega = -\frac{\partial \theta}{\partial t} \ , \quad \mathbf{k}(\xi, \zeta) = (\alpha, \beta) = \nabla\theta \tag{9}$$

with a slowly varying wavenumber vector $\mathbf{k}(\xi, \zeta)$. Since $\nabla \times \nabla\varphi = 0$ for any scalar function φ, the wavenumbers must satisfy the irrotationality condition[20]

$$\nabla \times \mathbf{k} = 0 \ , \quad \frac{\partial \alpha}{\partial \zeta} = \frac{\partial \beta}{\partial \xi} \ , \tag{10a,b}$$

a relation that guarantees the uniqueness of the phase function and its independence of the integration path between two points in the s, z plane. Substitution of equations (7) and (9) into the nonlinear disturbance equations yields up to order $\mathcal{O}(\epsilon)$ the parabolized stability equations

$$(L + \epsilon L')\mathbf{q} + \epsilon M_1 \frac{\partial \mathbf{q}}{\partial \xi} + \epsilon M_2 \frac{\partial \mathbf{q}}{\partial \zeta} =$$

$$\mathbf{r} - \epsilon \left[\frac{d\alpha}{d\xi} N_1 \mathbf{q} + \frac{d\beta}{d\zeta} N_2 \mathbf{q} \right] \tag{11}$$

where the operators $L, L', M_1, M_2, N_1, N_2$ act only in
y. The tedious detail of these operators is different
for incompressible and compressible boundary layers
and depends on the coordinate system. The term in
square brackets can be neglected, because N_1, N_2 originate from the viscous terms and are of order $\mathcal{O}(Re^{-1})$.
The operator L' contains the nonparallel contributions
$V, \partial U/\partial\xi, \partial U/\partial\zeta, \cdots$ of the basic flow. The function
$\mathbf{r} = \mathbf{r}(y)$ on the right hand side represents nonlinear
terms. For a linear analysis of a single mode (7), $\mathbf{r}$ can
be neglected. After linearization, the limit $\epsilon \to 0$ formally retrieves the governing equation $L\mathbf{q} = 0$ of the
LST.

In quasi-2D flows, $\mathbf{V}$ and $\mathbf{q}$ are independent of ζ
and the operators M_2, N_2 are absent. Equation (8b)
is identically satisfied, and the irrotationality condition
reduces to $\partial\beta/\partial\xi = 0$ or $\beta = \mathrm{const.}$.

The solution of eq. (11) under the conditions (8) and
(10b) requires both initial and boundary conditions.
The boundary conditions can be inhomogeneous. To
model the situation considered by the LST, we provide
an initial condition obtained from solving for an LST
mode with proper frequency and spanwise wavenumber
at some initial position s_0 and assume homogeneous
boundary conditions on $\mathbf{v}$ in y. These data are sufficient to solve the PSE for quasi-2D flow by a marching
procedure in s.

In 3D flows, it is necessary to specify initial conditions along the inflow boundaries of the domain in the
s, z plane shown by the dashed line in Figure 3. The
situation is similar to the initial-boundary-value problem for the 3D boundary-layer equations. Here we have
to prescribe the disturbances that enter the flow along
the inflow boundary. In some cases, it is possible to
provide initial conditions from a single LST eigenmode
at s_0, z_0 by applying the PSE along s under neglect
of spanwise variations and integrating the PSE along z
under neglect of streamwise variations as indicated in
Figure 3. With the known disturbance along the inflow
boundary, the PSE can be integrated in the domain of
dependence. In other cases, it is necessary to maintain
uniform initial disturbances along ζ by other means,
e.g. by inhomogeneous boundary conditions.

Similar to the PNS, the PSE are not strictly
parabolic. The residual ellipticity can lead to numerical
instabilities if the marching step is too small. Because
$\mathbf{q}$ varies weakly along the surface and oscillations and
growth are absorbed by the phase function, the integration can proceed in large steps. Satisfying equations
(8) requires iteration to adjust the wavenumbers to the
local conditions. This iteration converges faster than
the iteration on eigenvalues of the LST problem at a
given discretization in y.

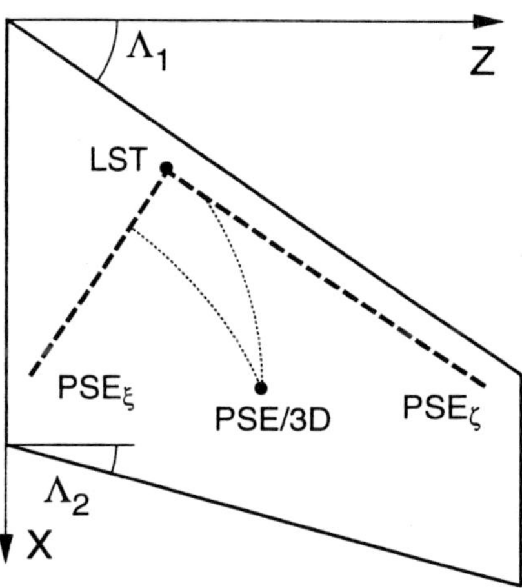

Figure 3: Initial conditions for the PSE/3D integration

Linear PSE for Quasi-2D Flows

The linear PSE are well suited to compute amplitude-
growth curves similar to $N(s; \omega, \beta)$ of eq. (6) by starting
from an LST mode just upstream of s_0. The integration does not only provide improved results but also
proceeds faster than the traditional procedure. The
amplitude, measured as the square root of the total kinetic energy, is a direct result of the computation. To
distinguish PSE and LST results, we denote the PSE
result with $n(s; \omega, \beta)$.

If the effects of wall curvature and of the nonparallel
terms (L' in eq. (11)) are neglected, the results are similar to the LST results. Curvature on convex surfaces
hardly changes the growth of TS waves, but strongly
stabilizes crossflow (CF) vortices on swept wings. In
the PSE analysis, the inherent streamwise changes in
the boundary-layer flow over a convex surface partly
compensate the effect of curvature alone, as shown in
Figure 4 for a 64A010 swept wing at $Ma = 1.5$. The
use of LST with surface curvature does not permit accounting for these streamwise changes and provides unexpectedly low values of N_{CF} or may suppress CF instability completely.

The marching solution of the PSE tracks the evolution of the physical modes according to the given initial
data, including ω and β. There is no need and no opportunity for strategies. This property is in favor of providing useful results, although the practicing engineer
may be disappointed that many of his former choices
disappear. The only exception is n_{CF} which is part of
the $N_{TS} - N_{CF}$ method used e.g. at Boeing and Airbus
in slightly different forms. The envelope n_{TS} is available in a modified form that correctly keeps $\beta = \mathrm{const.}$
during the integration. Differences from the traditional
result are minor, because curvature and nonparallelism
have only small effects on TS waves and the direction
of the edge streamline usually does not change much

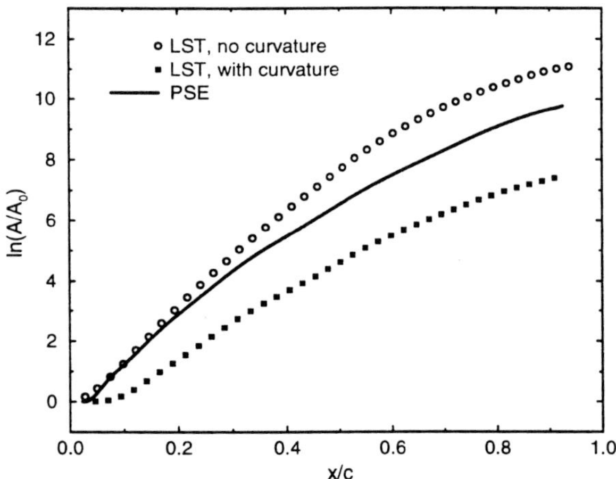

Figure 4: Amplitude growth of a steady crossflow vortex on a swept wing according to the LST with and without curvature in comparison with PSE results.

in the region where TS instability is prominent. Since the use of the PSE demands less computer time than solving the LST eigenvalue problem, the physically best justified approach may be ultimately accepted for commercial applications.

For a transition period, and to aid comparisons, we have also implemented a selection of LST strategies, since the LST is anyway needed to generate initial conditions and to search for suitable starting points s_0. The main use of the LST, however, is made in an innovative module we named "stability synopsis". A prerequisite for the computation of N factors is the knowledge of unstable modes. Given a new configuration, traditional codes require a tedious trial-and-error search for possible instabilities, that can be shortened by a skilled user and by analysis of boundary-layer characteristics such as crossflow Reynolds numbers, which may be produced by the boundary-layer code. To reduce the time-to-solution and increase the reliability of the results, we have developed the stability synopsis to automate the analysis. Using general criteria, small tables, and short calculations, this module checks for attachment line, TS, CF, and Görtler instability and estimates the regions of instability. The coarse estimates are refined using LST to provide starting values for the N factor calculations for every type and region of instability.

The code interfaces to various file formats of codes that provide boundary-layer profiles, including the 3D boundary-layer code BL3D[25] In addition, we have integrated a new boundary-layer code for tapered wings to compute all quantities necessary for a complete PSE analysis and currently integrate the capability to compute 3D boundary layers from given pressure distributions. This virtually redundant effort was necessary to obtain accurate transverse velocity components V and derivatives like $\partial U/\partial \xi$ which are not provided by traditional codes and cannot be retrieved from their results with sufficient accuracy. Although BL3D is supposed to produce these data, we found various weaknesses of this code, and the V component does not satisfy the correct conditions at the edge of the boundary layer. Though not yet optimized, our code is 6 times faster than BL3D and computes the boundary layer on a 40×64 surface grid in 55 seconds[†].

Given qualified input data, the optional use of the synopsis permits unattended operation of the n factor calculation. The synopsis module has been successfully tested over a large variety of test cases from turbine blades to supersonic wings. For a typical swept wing with CF and TS instability, the automated procedure from the pressure distribution to printed graphs for $n_{TS} - n_{CF}$ analysis currently takes less than 40 seconds per section, with about 1 second used for computing the boundary layer and 6 seconds for the stability synopsis.

The code operates under a graphical user interface on distributed computers. An initial setup can be specified by a privileged user. Normal operation allows only a few options such as the selection of input format and file, graphs or listings for printing or display, and the spanwise section for 3D boundary layers. Multiple sections can be simultaneously analyzed on a group of workstations to reduce the time-to-solution for large commercial projects. Additional windows can be invoked to perform standard tasks of linear stability analysis and to prepare input data for the nonlinear PSE analysis.

Linear PSE for 3D Flows

The PSE for 3D boundary layers have been implemented in a special version of our code that is restricted to body-fitted coordinates for locally infinite wings. Applications so far served to explore the concept and to establish basic results on the stability of 3D flows. The implementation for general curvilinear coordinates is in progress.

As a model wing for analyzing 3D effects, we have chosen the faired circular cylinder used by Poll[26] for detailed studies on the evolution and transition of crossflow vortices. The pressure distribution on this wing was measured by Poll and presented in analytical form. Because geometry and pressure distribution are nonproprietary and reproducible, various flows at different sweep angles and freestream speeds are common benchmarks for the evaluation of boundary-layer and stability codes. The specific case considered here is at laboratory condition, a sweep angle of 63 degrees, and a freestream speed of 50.5 m/s. The chord of the wing measured normal to the leading edge is $c = 0.4572$ m. To produce a spanwise variation of the flow over the wing, we consider a rectangular suction panel aligned

---

[†]Times are for an SGI Indigo 2, R10000, 195 MHz

with the coordinates s, z. The suction panel is operated with mass-flow rates typical of LFC systems.

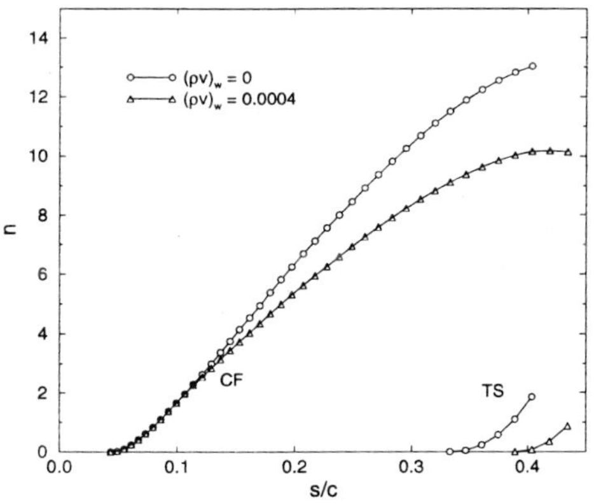

Figure 5: Envelopes n_{CF} and n_{TS} with and without suction.

With the PSE code for quasi-2D flow, we can calculate the envelope of amplitude growth curves for steady CF vortics and TS waves. The results are shown in Figure 5 for two cases with and without suction. The curves terminate at the last streamwise station before the boundary layer separates. This point is shifted downstream by suction. Suction stabilizes both CF and TS modes. The growth of TS waves is insignificant in this flow and we consider only one of the "most dangerous" CF modes with a spanwise wavenumber of 1400 m^{-1}. Results for other modes are qualitatively similar.

For 3D boundary layers, it is useful to change the graphical representation of the data from the curves in Figure 5 to contours of n in the $s/c, z/c$ plane. If the analysis of spanwise sections is performed by repeated runs for quasi-2D flow, the contours of n suggest that the disturbance amplitude jumps to a lower value as the flow enters the suction panel and the crossflow modes adapt their growth and propagation characteristics instantaneously to the modified flow conditions.

This picture changes remarkably, if we analyze the same configuration with the extended PSE approach for the 3D boundary layer, as demonstrated in Figure 6. As z/c increases and the crossflow mode enters the flow over the suction panel, the contours of constant n do not jump to a different location but slowly change over. For the higher values of n, adapting to the conditions of the flow with suction requires about 35% chord. The growth vector $\mathbf{k}_i = (\alpha_i, \beta_i)$ is oriented normal to the lines $n = $ const. The orientation varies with z/c, although the boundary-layer flow is uniform across the suction panel.

The interpretation of Figure 6 is simplified by analyzing the contours of the streamwise and spanwise growth rates α_i and β_i. Whereas we expect reduced streamwise growth over the suction panel, this growth is actually stronger in some areas but is combined with a spanwise

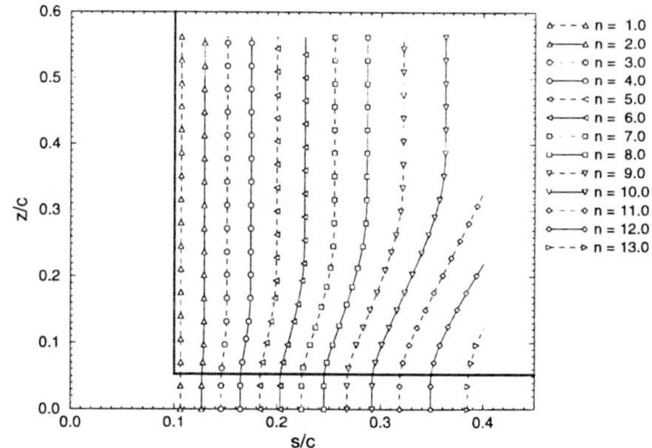

Figure 6: Contours of n in the $z/c, s/c$ plane for a steady crossflow vortex of wavelength 1400 m^{-1} in the presence of a suction panel (dotted line) obtained using the PSE/3D.

decay. The spanwise growth rates β_i shown in Figure 7 are positive over the panel and indicate the need for the amplitude of the CF mode to decrease as z/c increases until the amplitude ultimately reaches the conditions in the flow with suction. The irrotationality condition requires that the increase of β_i along s/c at the edge of the panel be accompanied by an increase in α_i in z/c direction.

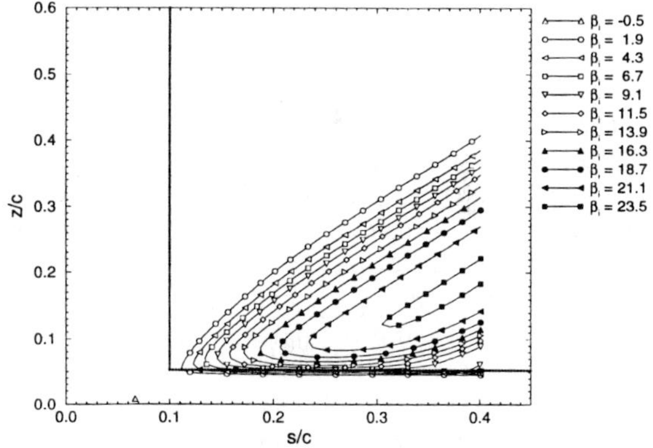

Figure 7: Contours of the spanwise growth rate β_i in the $z/c, s/c$ plane for a steady crossflow vortex of wavelength 1400 m^{-1} in the presence of a suction panel (dotted line) obtained using the PSE/3D.

The growth rates and direction of amplification in the flow over the suction panel vary although the boundary layer is uniform in the direction of z/c. This result leads to the important conclusion that the vector of growth is not determined by the local flow conditions alone but by their combination with the history of the mode. Therefore, neither of the suggested local procedures discussed by Arnal[17] can provide the correct amplification rate for the N-factor integration.

Figures 6 and 7 show a characteristic region over which the changes between the flows without and with suction occur. This region is characterized by $\beta_i \neq 0$. One boundary of this region is given by the edge of the panel at fixed z/c. By comparison with various lines associated with the basic flow and the CF mode we found that the other boundary is given by the group line emanating from the corner of the panel. The good agreement of the group line (LST) with the upper branch of the lines $\beta_i = $ const. (PSE) is a welcome validation and verification of the PSE analysis for 3D flows.

The results demonstrate that the growth characteristics and n factors for instabilities in 3D boundary layers may significantly deviate from the predictions of the local stability analysis and PSE predictions under the assumption of quasi-2D flow. Additional changes will be caused by computing the basic flow with a 3D boundary layer code. In this case, the effect of spanwise changes in the suction rate will smoothly affect the basic flow and further delay the adaptation of the stability characteristics to the conditions over the suction panel. The 3D version of the PSE can significantly improve the quality of stability and transition analysis in aerodynamic flows over wings with twist and engines, strongly swept and tapered tail fins, hypersonic bodies, turbine blades, and others.

The 3D approach also offers important operational advantages. Given the 3D inviscid flow e.g. in form of a Plot3D file, the complete boundary-layer and transition analysis can be automatically performed. The results in Figure 7 have been generated on a 64×40 grid which requires solving the 3D PSE in 2560 points. This run required about 2 minutes. The time can be further reduced by improvements of the numerical algorithms and by working on a coarser grid than what we choose for research applications. To obtain the envelope of n factors, the computation has to be repeated for selected modes in the unstable range of wave numbers, a standard procedure used for generating Figure 5. These computations can run simultaneously on a group of workstations or a multi-processor server. The total time-to-solution for a complete wing can then be reduced to less than 3 minutes instead of several days with traditional codes. On a single workstation, the sequential analysis of boundary layer and selected modes for n_{TS} and n_{CF} requires 30–40 minutes. This estimate may exceed the time for the (coarser and less reliable) section-by-section analysis by a factor of 2, but imposes less effort on the engineer who has to select an input file, hit "Run", and receives organized output of physically correct data after 3 or 40 minutes, depending on the computer environment.

Nonlinear PSE Studies

Besides the benefits of the PSE for linear analysis, they are capable of attacking inhomogeneous problems associated with receptivity and the nonlinear problems that arise in the transition process. In the simplest case, the initial conditions for nonlinear runs are given as a collection of LST modes of the form (7) and a coupled system of equations (11) is solved for the starting modes as well as for new modes created by forcing. Tests with the initial conditions taken from experiments show, that the PSE can be used for transition simulations at a small fraction of the cost of DNS for results of similar quality[23].

In practical applications, we face a lack of knowledge regarding the disturbance environment and receptivity mechanisms. We have attempted to bridge this gap by combining basic knowledge on receptivity, primary, and secondary instability with empirical parameters to construct "input models" for nonlinear PSE runs in quasi-2D flows[9]. A model for K-type transition was developed to analyze flows at the high disturbance levels typical of low-pressure turbines. The parameters for K-type transition were tuned using experiments on a heated flat plate[27] at a turbulence level $Tu = 0.4\%$ and the model was successfully tested at higher turbulence levels up to $Tu = 2.4\%$. The same input model generated results in good agreement with measurements on convex and concave heated plates at $Tu \approx 0.6\%$. A similar model was developed for subharmonic transition at low disturbance levels. These input models are flexible enough to incorporate new research findings on receptivity and the disturbance environment as they become available.

ERCOFTAC Test Case T3A

We have applied the K-type input model to analyze the ERCOFTAC test case T3A[28] for a flat plate at a turbulence level $Tu = 3\%$ and obtained the results shown in Figure 8. Results are given for two approximations with different Fourier truncations N and M in streamwise and spanwise direction, respectively. The agreement with the experimental data is comparable with DNS results for this test case. The lower minimum and steeper rise of C_f in the PSE (and DNS) result may be attributed to the neglect of intermittency. Our result indicates that instability mechanisms are still at work in this case considered as "bypass transition". The turbulence level of 3% is near the upper limit for PSE applications, as long as we have no better grasp on the receptivity to high free-stream turbulence. This case is governed by nonlinearity from the onset of instability and is outside the predictive capabilities of the e^N method.

The K-type input model was applied to the flow in the nozzle vane of an annular turbine stator for which

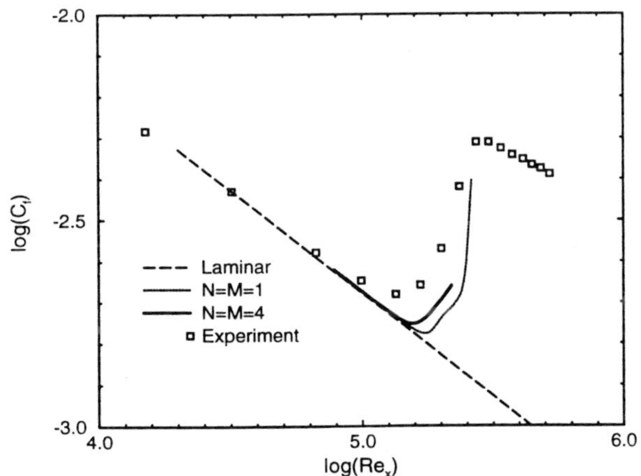

Figure 8: Transition for ERCOFTAC test case T3A at $Tu = 3\%$. Comparison of PSE result and experimental data[28].

experimental data are available. The pressure distribution at mid span was obtained using a 2D panel code. The boundary-layer was generated with a modified Kaups-Cebeci code. After a survey of the linear stability of the flow, the input model was adapted to the turbulence level $Tu = 0.5\%$. The PSE results for the variation of the Stanton number along the arclength s/c on the suction side provide a transition location upstream of the the experimental value. While some of this shift may be attributed to the simple input model, we have found the main reason in the deviation between computed and actual *laminar flow* in the experiments. Comparison with a Navier-Stokes solution produced with RVC3D[29] reveals that this deviation is partly caused by significant three-dimensionality of the basic flow. The computational results also suggest that the use of a standard boundary-layer code is inappropriate to obtain the correct flow in the narrow passage, because the edge conditions of the boundary layer disagree with the strong velocity gradients of the inviscid flow. The Navier-Stokes solution was too inaccurate to be used directly for the transition analysis. Continuation of these studies awaits improved codes for computing the viscous flow through turbines.

Transition in Flight Tests

The atmospheric disturbance environment experienced in flight tests and commercial flight is largely unknown. The transition mechanisms in flows on swept wings are insufficiently explored. It is known, however, that transition can be caused either by dominating TS or CF instability. To shed light on this problem area, we have performed a detailed PSE analysis of selected ATTAS flight tests. These tests were carried out on the laminar-flow glove of a VFW 614 aircraft in the range

of $0.35 \leq Ma \leq 0.7$ and $12 \cdot 10^6 \leq Re \leq 30 \cdot 10^6$. The sweep angle varied from 18 to 24 degrees by flying with sideslip such that the test envelope covered the range from dominating TS instability to dominating crossflow instability. The ATTAS flight tests have served for detailed evaluations of e^N strategies[5].

TS Dominated Transition

We have performed PSE analyses of selected TS-dominated cases to study transition mechanisms on swept wings and to extract information on the atmospheric environment needed in the input models[7]. The secondary instability mechanisms for 2D flows[30], subharmonic, fundamental, and combination resonance, were tracked for increasing sweep angles and found to carry over to quasi-2D flows with minor changes. Owing to the loss of symmetry across the edge streamline, the "left" and "right" subharmonic waves $(1,1)$ and $(1,-1)$ initially show different characteristics, but synchronize and grow simultaneously as the growing TS wave strengthens the parametric instability[‡]. Fundamental resonance occurs primarily in combination with weakly amplified crossflow vortices. The amplification characteristics of the secondary instabilities are sufficiently similar to the 2D case to obtain rough estimates from results for 2D flows.

Using the linear amplitude-growth curves and data on secondary instability, the most dangerous parameter combinations were estimated and refined in numerous PSE runs. The process was repeated for different transition mechanisms and two test cases A and B. In-flight infrared photographs exhibit transition near 17% chord for case A, 40% chord for case B. Figure 9 summarizes the PSE results for subharmonic transition in case B and shows the variation of the transition location with the initial amplitude A_{20} of the streamwise TS wave. The transition location observed in flight can be attributed to an amplitude level of waves in the range of $2 - 3 \cdot 10^{-6}$. Similar levels in the range of $1 - 3 \cdot 10^{-6}$ lead to fundamental or subharmonic transition in case A. The initial amplitudes of CF vortices have a weak influence on the location of the fundamental transition point.

It is currently unknown how the atmospheric environment varies with altitude and other flight conditions. So far, support for the broader validity of our estimates can only be derived from the e^N method. With the often cited limiting value $N^* = 9$ and $A_0 = 3 \cdot 10^{-6}$, this method predicts transition when the TS wave reaches an amplitude of 2.4%, shortly downstream of the onset of strong secondary instability. Our results show the usefulness of N factors for TS waves, provided these are correctly calculated under physical constraints. The

---

[‡]We denote with (n, m) a mode with frequency $n\omega_r$ and spanwise wavenumber $m\beta_r$.

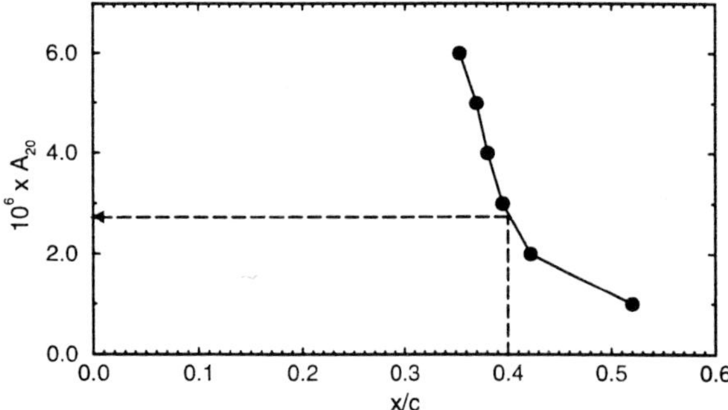

Figure 9: Variation of the transition location with the disturbance level A_{20} for subharmonic resonance in ATTAS test case B.

linear growth of TS waves is relevant to the transition location because nonlinear effects on TS waves are negligible at amplitudes below 1.5%. Nonlinearity is important, however, to cause the explosive rise of secondary instabilities.

CF Dominated Transition

Crossflow modes typically grow stronger than TS waves owing to the inviscid nature of inflectional instabilities. Earlier nonlinear PSE studies of CF vortices in swept Hiemenz flow[31, 22] have shown that the modes quickly attain amplitudes in excess of 10% and saturate at a still higher level. Saturation of single modes and the interesting interactions between steady and traveling modes are consistent with experiments on a flat plate with imprinted pressure variation[32].

Flows over swept wings are different, though, since the significant growth of the boundary layer causes the range of unstable steady and unsteady CF modes to change from high wavenumbers near the leading edge to much lower wavenumbers downstream, as shown by the diagram of growth curves in Figure 10 for ATTAS test case C[8].

The infrared record for case C exhibits the characteristic zig-zag pattern of crossflow-dominated transition between 15% and 28% chord. The steady pattern suggests that transition originates from distributed roughness and steady crossflow modes. Crouch found[33] that receptivity to roughness produces a broad band of steady modes especially in the leading-edge region. Details on the roughness level of the ATTAS glove are unknown.

Systematic PSE runs were performed to track the nonlinear evolution of single steady and traveling modes, the interaction of steady and unsteady modes, and interactions between steady modes of different spanwise wavenumbers. As in swept Hiemenz flow, single, continuously amplified crossflow modes of low wavenumbers can nonlinearly grow to large, slowly

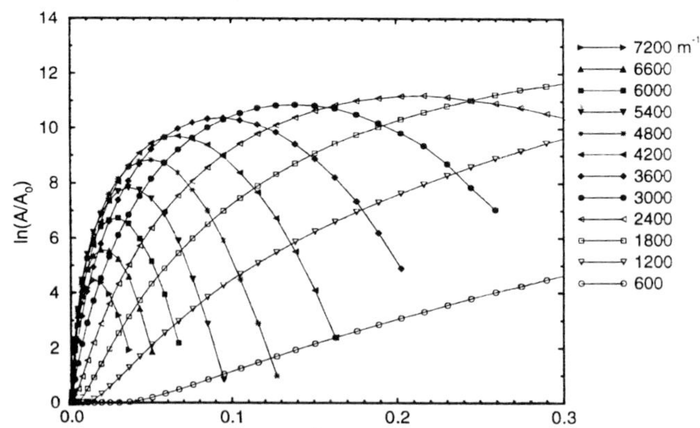

Figure 10: Growth curves for steady crossflow vortices of different wavenumbers along x/c in ATTAS case C. where c is the chord length.

varying saturation amplitudes that depend weakly on the initial amplitudes. This saturation phenomenon undermines the validity of N factors for CF modes. Single modes of high wavenumbers nonlinearly grow to a maximum amplitude and then decay because of the mismatch of their wavelength with the growing boundary-layer thickness, leaving behind a mean-flow distortion (MFD) that decays on its own viscous scale.

The results on the interaction of multiple steady modes are most illuminating for the usefulness of the N^*_{CF} values. Figure 11 shows the result of a PSE run that started with uniform initial amplitudes of $A_{0m} = 2 \cdot 10^{-6}$ for the eight modes with wavenumbers between 3000 m^{-1} to 7200 m^{-1} that are marked by the full symbols in Figures 10 and 11.

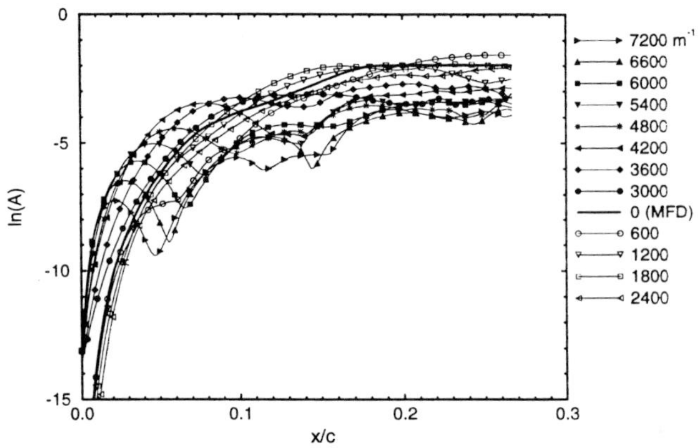

Figure 11: Interaction of steady crossflow vortices of different wavenumbers with $A_{0m} = 2 \cdot 10^{-6}$ for modes with $\beta \geq 3000$ m^{-1}. ATTAS case C.

Discrepancies between the linear and nonlinear "picture" are obvious: none of the modes decays in the nonlinear run, and for $x/c > 0.1$ the flow is dominated by low-wavenumber modes (open symbols) that were not introduced by the initial conditions and are expected from Figure 10 to exhibit modest growth. The mode with the smallest wavenumber of 600 m^{-1} is the lowest in Figure 10, but highest at the end of the nonlinear run. The highest amplitudes at $x/c = 0.3$ exceed 25%.

The mechanisms underlying this evolution are known. Initially, the high-wavenumber modes grow, create the mean-flow distortion, and force new modes at their wavenumber differences. Further downstream, the new modes are increasingly amplified, while the high-wavenumber modes tend to decay. Ultimately, the strengthened low-wavenumber modes dominate and maintain the modes with higher wavenumbers as harmonics and at the sum of their wavenumbers. The results in Figure 11 and the generic nature of the nonlinear mechanisms involved suggest that the creation of unexpectedly strong CF vortices can lead to CF-dominated transition at unusually small limiting N^* values and at unexpected locations. Strong growth to high amplitudes, strong nonlinear interactions, and saturation are likely causes for the failure of limiting N^*_{CF} values in predicting the transition location. These conclusions are supported by the "pathological" behavior of some Fokker F100 flight tests.

For practical applications, it will be necessary to account for the nonlinear effects in CF dominated transition without repeating the tedious computation and interpretation of the results. We have developed a relatively simple model to estimate saturation phenomena and nonlinear mode interactions on the basis of the known linear amplitude-growth curves. The parameters of this model can be determined by a few fast single-mode PSE runs and seem to vary little with flight conditions.

So far, we have not been able to determine the initial amplitudes of steady CF modes because the roughness conditions in the accessible flight tests have not been documented. The inverse method used in TS-dominated cases cannot yet be applied to CF modes because the high-frequency secondary instability of finite-amplitude CF vortices is not yet reproduced by the PSE computations. Therefore, we cannot make a direct link between computed and observed transition points. Closing this gap in our capabilities is the next goal in our studies of ATTAS test cases.

Conclusions

The use of the PSE for transition analysis has not only enhanced our insight into transition issues and led to a more selective application of the e^N method for aerodynamic flows, but also increased our understanding for the needs of the engineering environment. Many CFD codes used today in this environment need to be replaced by software that adheres to higher quality standards. Our new generation of boundary-layer and stability codes has responded to these needs. These codes provide the speed and ease-of-use desired for practical applications. Clarification of remaining issues such as the high-frequency breakdown of CF vortices and the full implementation of the PSE for 3D boundary layers rank high on our list of priorities.

Acknowledgment

The development and evaluation of the PSE method have been supported by the Air Force Office of Scientific Research. The PSE code for compressible problems in curvilinear coordinates has been developed in cooperation with G. K. Stuckert, N. Lin, S. Huang, and D. C. Hill of DynaFlow, Inc. This work has been supported by Wright Laboratories, NASA Ames Research Center, and NASA Lewis Research Center. The analysis of the ATTAS flight tests is conducted in cooperation with G. Schrauf, Daimler-Benz Aerospace Airbus GmbH, and has been supported by the German technology program RAWID.

References

[1] Th. Herbert. Boundary-layer transition – analysis and prediction revisited. AIAA Paper 91-0737, 1991.

[2] M. R. Malik. Stability for laminar flow control design. In *Viscous Drag Reduction in Boundary Layers*, volume 123 of *Progress in Astronautics and Aeronautics*, pages 3–46, 1990.

[3] U. Dressler, H. Hansen, S. Rill, K. H. Horstmann, C. H. Rohardt, and G. Wichmann. Design of the Fokker F100 natural laminar flow glove. In *First European Forum on Laminar Flow Technology*, volume DGLR-Bericht 92-06, pages 152–163, 1992.

[4] G. Schrauf, H. Bieler, and P. Thiede. Transition prediction – the Deutsche Airbus view. In *First European Forum on Laminar Flow Technology*. DGLR-Report 92-06, 1992.

[5] G. Schrauf. Transition prediction using different linear stability analysis strategies. AIAA Paper 94-1848, 1994. Submitted to *AIAA J. Aircraft*.

[6] G. Schrauf and Th. Herbert. A comparison of a 'regular' and a 'pathological' f100 flight test case using linear and nonlinear pse methods. In *Stability and Transition of Boundary Layers*, volume Abstracts, Euromech Colloquium 359, Stuttgart, Germany, March 10–13, 1997, 1997.

[7] G. Schrauf, Th. Herbert, and G. K. Stuckert. Evaluation of transition in flight tests using nonlinear PSE analysis. *AIAA J. Aircraft*, 33:554–560, 1996.

[8] Th. Herbert and G. Schrauf. Crossflow-dominated transition in flight tests. AIAA Paper 96-0185, 1996. Submitted to *AIAA J.*

[9] Th. Herbert, G. K. Stuckert, and V. Esfahanian. Effects of free-stream turbulence on boundary-layer transition. AIAA Paper 93-0488, 1993.

[10] C. J. Woan, P. B. Gingrich, and M. W. George. CFD validation of a supersonic laminar flow control concept. AIAA Paper 91-0188, 1991.

[11] L. J. Runyan, G. W. Bielak, R. Behbehani, A. W. Chen, and R. A. Rozendaal. 757 NLF glove flight test results. In *Research in Natural Laminar Flow and Laminar-Flow Control*, volume NASA CP 2487, Part 3, pages 795–818, 1987.

[12] P. J. Roache. Quantification of uncertainty in computational fluid dynamics. *Annu. Rev. Fluid Mech.*, 29:123–160, 1997.

[13] K. Kaups and T. Cebeci. Compressible laminar boundary layers with suction on swept and tapered wings. *J. Aircraft*, 14:661–667, 1977.

[14] E. Elsholz. Ein inverses LISW-Grenzschicht-Verfahren. MBB-Report TE-1681, 1988.

[15] H. P. Horton and H. W. Stock. Computation of compressible laminar boundary layers on swept, tapered wings. *J. Aircraft*, 32:1402–1405, 1995.

[16] D. Arnal. Description and prediction of transition in two-dimensional, incompressible flow. In *Special Course on Stability and Transition of Laminar Flow*, pages 2/1–71. AGARD Report 709, 1984.

[17] D. Arnal. Boundary layer transition: predictions based on linear theory. In *Special Course on Progress in Transition Modelling*, pages 2/1–63. AGARD Report 793, 1994.

[18] L. M. Mack. Boundary-layer linear stability theory. In *Special Course on Stability and Transition of Laminar Flow*, pages 3/1–81. AGARD Report 709, 1984.

[19] M. R. Malik. COSAL - A black box stability analysis code for transition prediction in three-dimensional boundary layers. NASA CR-165925, 1982.

[20] L. M. Mack. Transition prediction and linear stability theory. In *Laminar-Turbulent Transition*, pages 1/1–22. AGARD CP 224, 1977.

[21] F. P. Bertolotti. *Linear and nonlinear stability of boundary layers with streamwise varying properties.* PhD thesis, The Ohio State University, Columbus, Ohio, 1991.

[22] Th. Herbert. Parabolized stability equations. In *Special Course on Progress in Transition Modelling*, pages 4/1–34. AGARD Report 793, 1994.

[23] Th. Herbert. Parabolized stability equations. *Ann. Rev. Fluid Mech.*, 29:245–283, 1997.

[24] Th. Herbert. On the stability of 3D boundary layers. AIAA Paper 97-1961, 1997.

[25] V. Iyer. Three-dimensional boundary-layer program (BL3D) for swept subsonic or supersonic wings with applications to flow control. NASA Contractor Report 4531, 1993.

[26] D. I. A. Poll. Some observations of the transition process on the windward face of a long yawed cylinder. *J. Fluid Mech.*, 150:329–356, 1985.

[27] K. H. Sohn and E. Reshotko. Experimental study of boundary layer transition with elevated freestream turbulence on a heated flat plate. NASA TM 187068, 1991.

[28] A. M. Savill. A synthesis of T3 test case predictions. In O. Pironneau, W. Rodi, I. L. Ryhmig, A. M. Savill, and T. V. Truong, editors, *Numerical Simulation of Unsteady Flows and Transition to Turbulence*, pages 404–442. Cambridge University Press, 1992.

[29] R. V. Chima and J. W. Yokota. Numerical analysis of three-dimensional viscous flows in turbomachinery. *AIAA J.*, 28:798–806, 1990.

[30] Th. Herbert. Secondary instability of boundary layers. *Ann. Rev. Fluid Mech.*, 20:487–526, 1988.

[31] M. R. Malik and F. Li. Crossflow disturbances in three-dimensional boundary layers: nonlinear development, wave interaction and secondary instability. *J. Fluid Mech.*, 268:1–36, 1994.

[32] B. Müller and H. Bippes. Experimental study of instability modes in a three-dimensional boundary layer. In *Fluid Dynamics of Three-Dimensional Turbulent Shear Flows*. AGARD CP 438, 1988.

[33] J. D. Crouch. Theoretical studies on the receptivity of boundary layers. AIAA Paper 94-2224, 1994.

BOUNDARY-LAYER TRANSITION PREDICTION TOOLKIT

Mujeeb R. MALIK

normalsize High Technology Corporation; Hampton, VA

Abstract

Laminar-turbulent transition on aerodynamic surfaces occurs between often widely apart limits, namely the upper limit relevant in extremely low-disturbance environments and the lower limit given by the bypass transition. The e^N method only provides the upper bound on the transition location, while experimental information or direct Navier-Stokes simulations are required to determine the lower bound. Computational tools (receptivity theory/linear Navier-Stokes, nonlinear parabolized stability equations, etc.) are now available to estimate the departure from the upper bound due to factors such as surface roughness and acoustic disturbances but await validation against experiments. In this paper, progress made in boundary-layer stability and transition is reviewed and description is given for the computational tools developed to ease the determination of the upper limit via N-factor calculations. This includes a suction optimization code for the efficient design of laminar flow control wings. A brief discussion of some of the physical issues associated with boundary-layer transition is given and some key results, including those pertaining to the question of absolute instability, are presented.

Received on September 10, 1997.

1 Introduction

Boundary-layer transition is a subject which at first sight appears narrow but, in fact, is vast and offers many challenges for a fluid dynamicist. It has drawn a great deal of attention in the past two decades because of its relevance to laminar flow control (LFC) technology and aerothermal design of aerospace vehicles. Here, progress made in transition prediction methodology during the past two decades is briefly reviewed. The review is not meant to be exhaustive and mostly covers the work of the author, his colleagues, and that of other associates, but reference is made to other works where appropriate. Some recent more comprehensive reviews on this subject have been given by Reshotko (1994), Arnal et al. (1995) and Reed et al. (1996).

My involvement in the area of stability and transition began in 1979 and I presented my first paper on the subject (Malik and Orszag, 1980) in Snowmass at the 1980 Fluid and Plasma Dynamics Conference. The seventeen years since then have seen tremendous progress in various areas associated with laminar- turbulent transition in boundary-layers. This includes development of accurate and robust methods for the hydrodynamic stability problem with particular emphasis on compressible flow (Malik, 1982; Malik et al., 1982; Macaraeg et al., 1988; Malik, 1990), advances in the understanding of boundary-layer receptivity (Goldstein, 1983, 1985; Ruban, 1985; Kerschen et al., 1990; Choudhari and Streett, 1994; Crouch, 1994; Saric et al., 1994; Denier et al., 1991), discovery of secondary instability (Orszag and Patera, 1983; Herbert, 1983, 1988), application of boundary-layer stability computations to complex three-dimensional (3D) flows (cf. Iyer, 1996), development of parabolic and parabolized stability equations (PSE) methods (Hall, 1983; Herbert, 1991; Bertolotti et al., 1992; Chang, et al., 1991; Li and Malik, 1997), and direct numerical simulation (DNS) of transition (Wray and Hussaini, 1984; Kleiser and Zang, 1991; Pruett and Chang 1995).

These advances have encompassed a wide speed regime and have helped us understand some of the ways in which stream disturbances are internalized in the boundary-layers as instability modes [Tollmien-Schlichting (TS), Görtler, crossflow, Mack modes]; new numerical techniques have enabled accurate description of linear and nonlinear growth of these instabilities, as well as the process of their breakdown to turbulence. Some numerical simulations have been carried through the transitional regime all the way to turbulence (Rai and Moin, 1993). Experiments (Saric and Thomas, 1984; Kachanov and Levchenko, 1984; Wilkinson and Malik. 1985; Bippes, 1991; Wlezien, 1994) have helped a great deal in developing this understanding as well as validating the theoretical and computational tools.

Despite all this advancement and our ability to compute readily receptivity. linear and nonlinear growth of primary and the ensuing secondary instabilities, one thing has remained the same during these 17 years: the method of choice for boundary-layer transition prediction (to be more precise, estimation), at least for high Reynolds number engineering applications, remains to be the eN method (Smith, 1952; Smith and Gamberoni, 1956; Van Ingen, 1956). This is because, except for some carefully controlled experiments, we generally do not know the disturbance field or the 兎xternal* forcing(s) which is(are) the root cause of laminar-turbulent transition. Transition is influenced by a number of parameters (cf. Bushnell and Malik, 1987; Bushnell, 1990) including stream turbulence (vorticity fluctuations. temperature spottiness, noise), particulates, roughness, acoustic radiation from other vehicle components and, possibly, electrostatics. This influence can be either at the receptivity stage or the instability growth stage or both. Various stages of the laminar-turbulent transition have been well illustrated by Morkovin (1985). There are, at least, two known routes to transition. The one which is better understood, and occurs in low-disturbance environments, involves: receptivity (internalization of stream/wall disturbances in the boundary-layer), linear/nonlinear growth of boundary-layer instability modes, secondary instabilities, and breakdown to turbulence. The second, which occurs in high-disturbance environments, involves "bypass" of the above linear process. Morkovin (1993) gives a discussion of the various categories of bypasses and their underlying physical mechanisms. Of some relevance, in this context, is Trefethern et al.'s (1993) transient or nonmodal growth which is an extension of Landahl's (1980) algebraic instability concept.

Laminar-turbulent transition is a very complex and delicate phenomena due to the preponderance of parameters that influence the outcome (e.g., Reshotko, 1976); nonetheless, its prediction is a deterministic problem. In other words, transition location is computable if one knows all the necessary details about parameters which influence transition in a given case. This is a big if, however, and one is often forced to estimate the location of transition onset using approximate methods or empirical correlations. The e^N method only provides one bound on the transition location, i.e., the "extremely" low-disturbance limit. The other limit is the "bypass" transition limit ($Re_{\theta min}$, Morkovin, 1985) which we only empirically know for various flows, although DNS can play an important role in determining this limit (cf. Spalart, 1988). These two limits are often wide apart and transition can occur anywhere in between. Thus, in practice, transition prediction is a sorting out process where one first determines these two limits for a given flow configuration and then determines where it may actually occur depending upon the sensitivity to surface finish, steps/gaps/waviness, wind tunnel and flight environments, surface temperature distribution, pressure gradients, suction-hole distribution and hole Reynolds numbers (for LFC). It is well-known that while suction in general stabilizes boundary-layer instabilities, strong suction may trip the flow or a particular suction distribution may be more advantageous than others because of their effects not only on stability but also on receptivity (suction holes could constitute potential roughness sites adversely affecting transition). Parametric sensitivity to departure from the upper limit can then be obtained either by using empirical correlations [e.g., Mack's (1977) N-factor correlation for effect of "turbulence" on two-dimensional (2D) boundary-layer transition] or through more sophisticated tools such as receptivity theory, nonlinear PSE, and DNS.

In this paper, we first describe briefly various computational tools used in linear stability analysis (e^N) as well as for nonlinear evolution of disturbances. We then give some examples of the use of the e^N method for estimating transition onset in flows where primary instability modes constitute Görtler, TS (first mode) and crossflow disturbances and also point out instances were these predictions are rendered "useless" due to surface roughness or free-stream noise. The physical mechanisms of transition, particularly in crossflow dominated flows, are also discussed. Other topics of interest in boundary-layer transition or of potential concern to an LFC designer are briefly described.

2. Computational Tools for Studies in Stability and Transition

2.1 Graphical Transition Prediction Toolkit (GTPT™)[*]

Various boundary-layer and stability codes have been developed for computing 2D and three-dimensional (3D) boundary-layers and the associated instability modes. The usefulness (and the limitations) of the linear theory or PSE-based e^N method are well understood. For a designer using the e^N method, it is important that the computational tools are robust and packaged in an easy to use fashion. This is the goal of the Graphical Transition Prediction Toolkit. The graphical toolkit, GTPT™, is designed to provide a convenient platform for use of these codes by engineers and analysis of the results from each code. The boundary-layer and stability codes which are all based upon compressible governing equations are described below.

BLQ3D: This is a quasi-3D boundary-layer code which solves 2D/axisymmetric and quasi--3D (infinite swept-wing, infinite swept tapered wing with conical similarity) boundary-layer equations. Pressure distribution along one streamwise cut of the airfoil is required, along with suction distribution, if needed, and wall temperature distribution for non-adiabatic wall cases. The generated solution files are saved to be used as inputs to the stability codes.

BL3D: This is a fully three-dimensional boundary-layer code developed for finite swept-wings. It requires the prescription of inviscid flow-field and wall suction and temperature distribution if needed.

LST3D: Unlike the COSAL code (Malik, 1982), this is a *fully* three-dimensional linear stability theory (LST) code which solves the spatial eigenvalue problem and can treat both crossflow and TS disturbances under quasi-parallel approximation. In the quasi-parallel approximation, the effect of boundary-layer growth is assumed to be of a lower order and, hence, ignored. In other words, the mean flow is a function of the wall-normal coordinate y only. If x and z represent the coordinates in the plane parallel to the wall and t is time, then a disturbance ϕ can be represented as

---

[*] GTPT is a trademark of High Technology Corporation (HTC).

$$\phi(x,y,z,t) = \hat{\phi}(y)e^{i(\alpha x + \beta z - \omega t)} + \text{complex conjugate} \quad (1)$$

Where α, β are the x, z wavenumbers and ω is the disturbance frequency. Substituting (1) in linearized Navier-Stokes equations results in homogeneous equations which along with homogeneous boundary conditions, constitute an eigenvalue problem described by the complex *dispersion relation*

$$\alpha = \alpha(\omega, \beta) \quad (2)$$

which is solved numerically. For a general wave packet α, β, and ω can all be complex (e.g., see §5.0). For a convective monochromatic disturbance, however, one can use spatial theory where ω is real and the wavenumbers are complex. This option is used in LST3D. The following N-factor strategies can then be employed:

1. The saddle-point method: This option requires that the group velocity ratio be real, i.e., $\partial\alpha_i/\partial\beta_r = 0$. An additional condition is provided by either requiring that the disturbance growth rate be locally maximized or by fixing β_r (the real part of β). The former has been used by Cebeci and Stewartson (1980) and Cebeci et al. (1988). The latter more physical approach, suggested by Nayfeh (1980), requires repetitive calculations for various β_r. Both these approaches allow the growth rate direction to automatically switch from the attachment-line to the chordwise direction further downstream. It is the attachment-line region, however, where nonparallel effects are strong and are ignored in this quasi-parallel calculation.

2. Fixed wave-orientation: N-factor calculations are performed for a disturbance with fixed wave angle with respect to the streamline.

3. Fixed wavelength: The dimensional spanwise wavelength of the disturbance is kept constant during the N-factor computation. For an infinite swept-wing, the "spanwise" direction is clearly defined. For a finite swept tapered wing, however, this could either be the direction parallel to the leading-edge or some prescribed direction such as conical lines for a wing.

PSE3D: This is a parabolized stability equations (PSE) based code for three-dimensional boundary-layers over infinite or finite swept wings. Mean flow variation and surface curvature effects can be accounted for by using this code. Hence, this code allows consideration of all the "known" convective instability modes of two- and three-dimensional boundary-layer flows. The PSE code is also used to study nonlinear evolution of disturbances.

In order to demonstrate the capability of the three-dimensional PSE code, we consider the case of an infinite-swept wing boundary-layer subject to crossflow instability.

Wall suction is provided near the leading-edge for laminar flow control. In order to construct a fully 3D mean flow, we set the suction level to zero in a small area ($0.79 \times 1.18in$) near $x = 1ft$ in Fig. 1(a) where the chordwise velocity component of the 3D basic flow, computed by BL3D, is shown. Spanwise variation of the basic flow is evident and a wake region trailing the suction "blocker" is clearly visible. Note that the assumed C_p distribution does not vary in the spanwise direction and any spanwise mean flow variation is caused by the suction blocker only. Evolution of a stationary crossflow disturbance was computed using the 3D PSE approach as shown in Fig. 1(b). The figure shows the contour plot of N-factors over and in the wake of the suction blocker. The basic flow pattern is provided as a reference using the contour lines (dashed lines) of the chordwise velocity component at a height of 12% of the boundary-layer thickness. If transition were to occur at a constant N-factor, the transition front will appear as a wedge behind the suction blocker. In the regions away from the suction blocker, the infinite-swept wing solution is recovered.

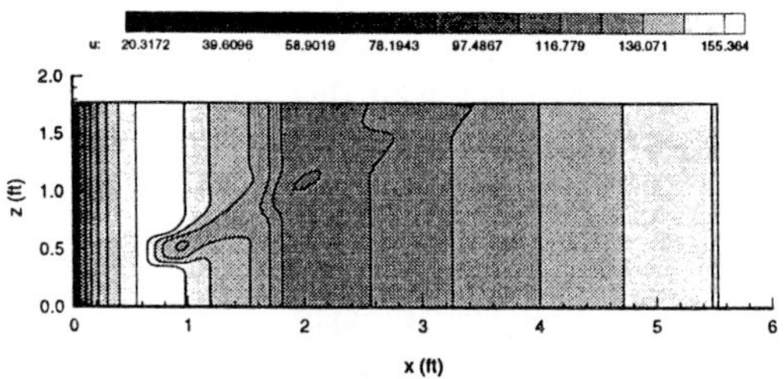

Fig. 1(a). Contour plot of the chordwise velocity component of the basic flow over a "suction blocker" in an infinite swept-wing.

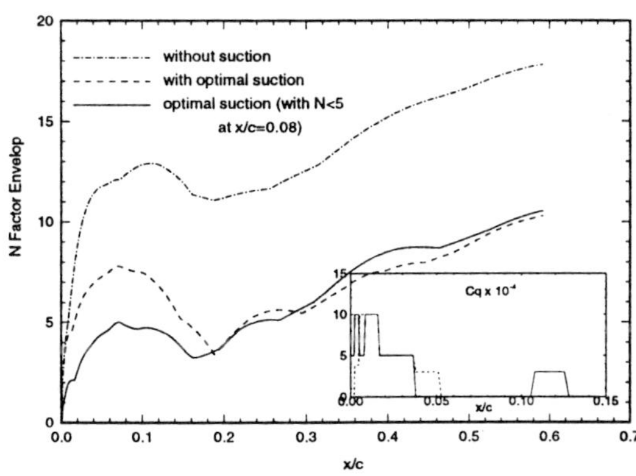

Fig. 1(b). Computed N-factors using 3D PSE for stationary crossflow disturbances. The background (dotted lines) show the basic flow variation.

2.2 Suction Optimization Code

For practical design, suction optimization is needed to minimize suction system requirements (both the total suction and the perforated area). Some early work reported in open literature is from Reed and Nayfeh (1981) who considered the TS problem. More recent work is reported by Balakumar and Hall (1996). The challenge in suction optimization for complex swept-wing flows, as with any other optimization, is to develop a code which is robust enough to be used by an aircraft designer. Suction optimization code developed at HTC can utilize either LST3D or PSE3D in the design process and will be incorporated in GTPT™. An example calculation for a swept-tapered wing is shown in Fig. 2 where the "envelope" of all the N-factor curves with and without suction are shown along with the predicted suction distribution. The objective function used in the optimization was to minimize N-factor at $x/c = .6$ for a given amount of total suction. Other options, e.g., minimization of suction for transition (as determined by a prescribed N-factor) at a fixed x/c, have also been incorporated. Figure 2 shows two solutions of the suction optimization. The first one (broken line) only minimizes the N-factor at the prescribed x/c. Nevertheless, it is seen that there is a peak in the N-factor of about 8 at $x/c \approx .06$. This early peak should be a reason for concern in view of the uncertainties associated with the e^N method of transition estimation. Therefore, a multipoint optimization scheme has been used to obtain the second solution (solid line) where an additional constraint requiring $N < 5$ for $x/c = .08$ has been imposed. This solution is better than the former from the viewpoint of LFC design.

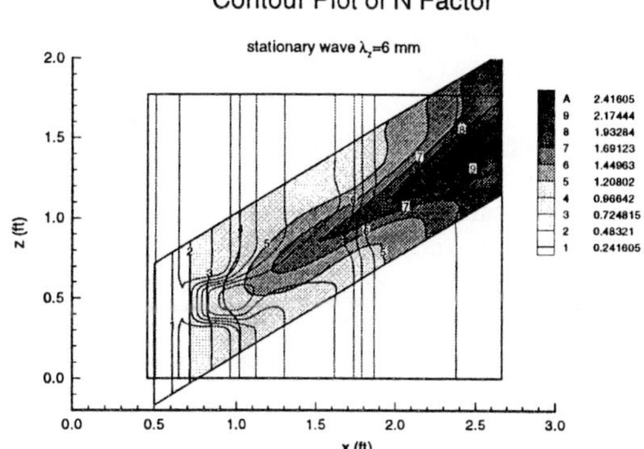

Fig. 2. Two solutions of suction optimization for a swept wing. The first solution (broken line) only minimizes the N-factor at $x/c \approx .6$ for a given total suction. An additional constraint ($N < 5$ at $x/c = .08$) is imposed for the second solution (solid line) using same total suction.

2.3 2D Eigenvalue Code

There is a class of problems which is governed by homogeneous governing equations and boundary conditions but, since the mean flow varies strongly in two dimensions, the resulting eigenvalue problem is governed by partial differential equations (PDE's). We call this a 2D eigenvalue problem, in contrast with the "1D" eigenvalue problem governed by a system of ordinary differential equations (e.g., the Orr-Sommerfeld equation). In the 2D eigenvalue problem, one must discretize the PDE's in two space variables resulting in a large matrix eigenvalue problem. Examples of physical problems which require the use of 2D eigenvalue techniques include corner layer, attachment-line boundary-layer and secondary instability studies. Two-dimensional eigenvalue techniques have been developed both for incompressible and compressible flows.

2.4 Linearized Navier-Stokes

The Linearized Navier-Stokes (LNS) approach has been used by Streett (1995), and more recently by Guo et al. (1997), for computation of instability waves in 2D and 3D boundary-layers. In this approach, the solution is assumed to have only one prescribed frequency and, thus, the governing equations are rendered time-independent which can be solved much more efficiently for problems where only a time-asymptotic solution is desired. Of course, it is more costly to solve this elliptic set of equations than the parabolized stability equations but the LNS approach offers various advantages over the PSE approach. For example, the LNS approach can be used in receptivity studies (Choudhari and Streett, 1994) and in flows which are rapidly varying or for tracking multiple instability modes.

PSE would tend to fail for such applications. LNS is of particular relevance in studying high Mach number flows at low Reynolds numbers.

2.5 A Strategy for Transition Prediction

Direct numerical simulations (DNS) of Navier-Stokes equations have been successfully used to compute some bypass transition cases (Rai and Moin, 1993; Spalart, 1988). Its use in high Reynolds number complex flows, however, is prohibitively expansive for present-day computers particularly when transition occurs in a low-disturbance environment. An alternative strategy can be constructed by considering various stages of the transition process depicted in Fig. 3.

Once the external disturbance field has been modeled/specified, appropriate computational tools can be used to carry through the various stages of transition as depicted in Fig. 4. This approach should prove more efficient than a brute-force application of DNS. If nonlinear PSE (NPSE) is used to capture secondary instabilities, the 2D eigenvalue calculations are not required. A good estimate of transition onset can be obtained by unsteady NPSE calculations and DNS is only required if one needs to compute through the transition zone. Pruett and Chang (1995) used combined NPSE and DNS to compute through the transition zone for a Mach 8 boundary-layer.

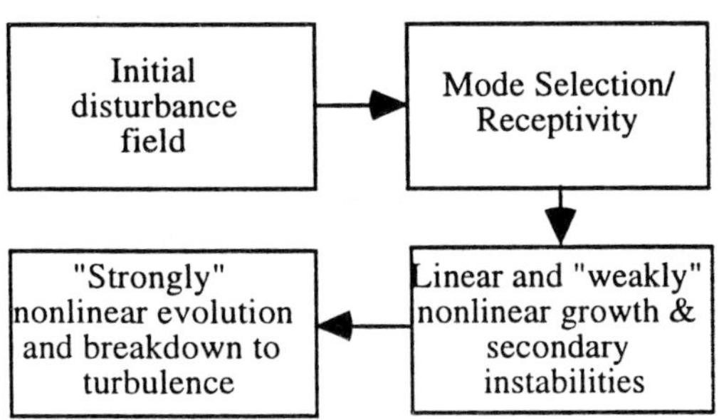

Fig. 3. Stages in the transition process in low-disturbance environments.

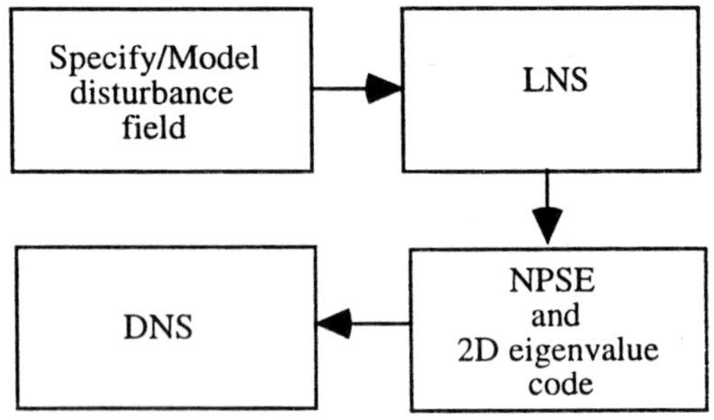

Fig. 4. Computational tools for transition prediction.

When nothing is known about the disturbance field and only the "upper" bound for the location of onset of transition is required, the e^N method can be employed. In "extremely" low-disturbance environments, this "upper" bound turns out to be quite close to the actual location of transition onset. Some such examples are provided in the next section.

3. Some Practical Applications

3.1 Görtler and First Mode Disturbances: The Case of the "Quiet" Tunnel

In order to appreciate the level of care needed to make transition "predictable" by methods such as the e^N method, we consider the Langley Mach 3.5 low-disturbance ("quiet") tunnel, Fig 5. Several filters have been installed in the air supply system and the settling chamber to remove moisture/particulates and to reduce acoustic disturbance level (Beckwith and Bushnell, 1988). As a result the rms pressure fluctuation in the settling chamber reduces to about .005 percent of stagnation pressure and particulate size is dropped to less than $1\mu m$. A honeycomb and several turbulence screens are used in the settling chamber to control vorticity fluctuations. These fluctuations are further reduced because of the large velocity expansion ratios in supersonic, as well as hypersonic, nozzles. Boundary-layer removal slots, which remove the turbulent boundary-layers developing along the settling chamber walls, are provided upstream of the nozzle throat. Therefore, new laminar boundary-layers develop along the four nozzle walls. The nozzle shown is of rectangular cross-section with two contoured walls and two flat walls (the side walls). The flat side walls become transitional in the throat region via crossflow instability due to the crossflow induced by the curvature of the inviscid streamlines which in turn is induced by the contoured walls. In addition, vortices present in the four corners become highly unstable and contribute another possible cause of transition in the corner layers. Noise radiation from the side walls to the center region of the test section is not a factor for rapid expansion nozzles with relatively large width-to-height ratio. Noise radiated from the turbulent boundary-layers along the contoured walls is the main limiting factor for the length of the "quiet" test core. Transition along the contoured walls results due to amplification of Görtler vortices and e^N calculations (Chen et al., 1985) indicated that the transition onset location could be well-correlated with an N-factor of about 9–11 for a range of unit Reynolds numbers (3×10^6/ft — 18×10^6/ft). Of course, this could only be achieved if nozzle surfaces were well polished and the roughness Reynolds number R_k ($R_k = U_k k/\nu_k$, where k is the height of the roughness, U_k the streamwise velocity at the height k in the absence of roughness and v_k the kinematic viscosity at that location) values in the throat region were maintained to be less than 12 (the critical value of R_k for low-speed zero-pressure gradient case is about 400) which corresponded to a maximum peak-to-valley defect of .5 micron (Beckwith and Bushnell, 1988). With such a care, it was possible to maintain laminar boundary-layers on the nozzle walls and as a result, the radiated normalized rms pressure fluctuations on the nozzle centerline were measured to be no more than .05 percent. A sharp cone model placed in this low-disturbance test core yields transition Reynolds numbers of 7×10^6 to 9×10^6 (Fig. 6), depending upon unit Reynolds number and surface finish, which compares well with adiabatic flight data (Beckwith and Bushnell, 1988; Beckwith et al., 1983).

When the nozzle wall boundary-layers are turbulent, the radiated noise level in the test section is increased by nearly two-orders of magnitude and the transition Reynolds number on the cone models is decreased and falls into the range of that found in conventional wind tunnels. A more dramatic shift in the transition Reynolds number due to noise occurs in the case of flat-plate boundary-layer (Fig. 7) such that the ratio of cone/flat-plate transition Reynolds number switches from about .65 to 2.5 due to the noise. The "low noise" transition Reynolds number are well correlated by the e^N method using N ~ 10 both for the case of flat-plate and cone. No method currently exists for predicting the effect of noise on transition; transition in conventional supersonic nozzles is not predictable by currently available and exercised computational techniques.

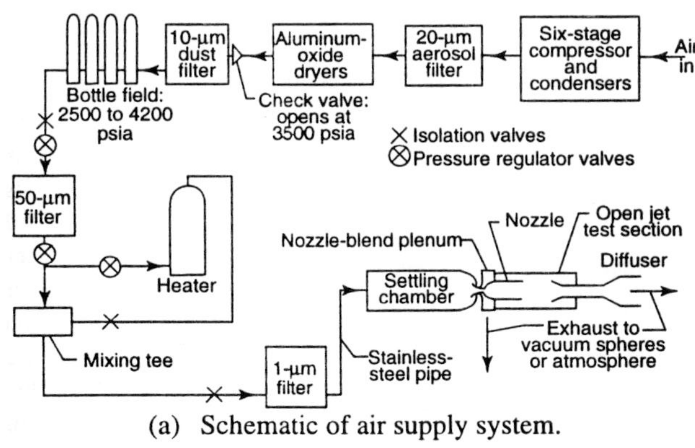

(a) Schematic of air supply system.

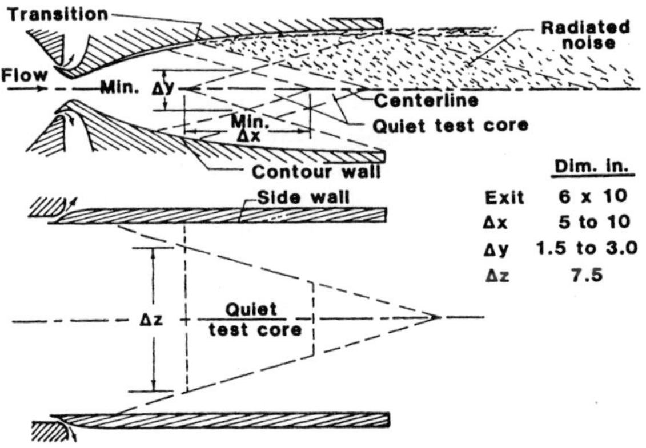

(b)   Quiet test core size and shape

Fig. 5. Langley Mach 3.5 "quiet" tunnel. The figure gives an idea about the care needed to yield boundary-layer transition on the nozzle wall predictable by the e^N method.

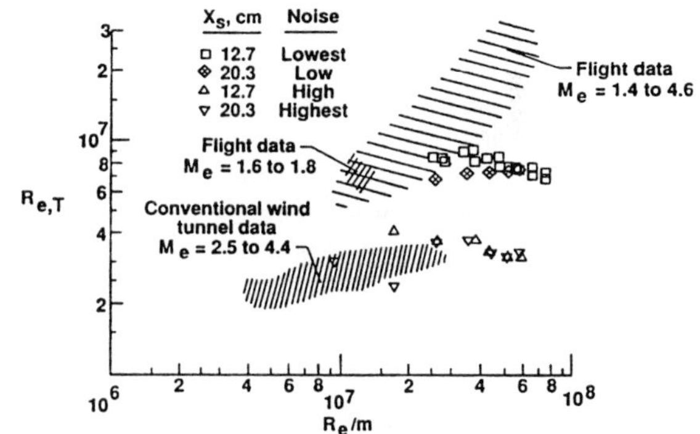

Fig. 6. Measured transition Reynolds numbers on sharp cones (quiet and conventional tunnels).

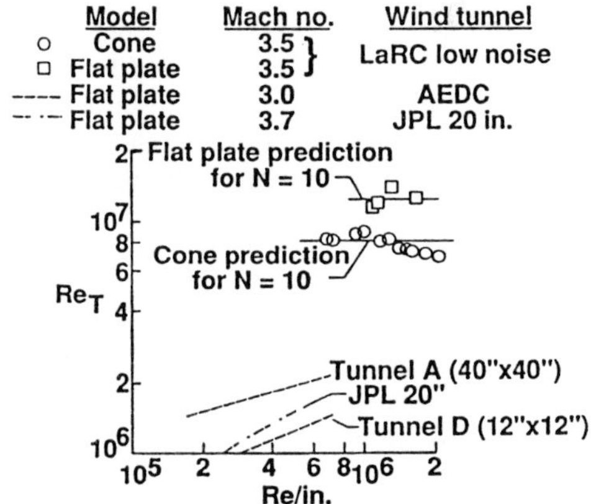

Fig. 7. Comparison of transition onset Reynolds numbers on cone and flat-plate in "quiet" and "noisy" tunnels.

In the above example, even though the forward parts of the cone and flat-plate models were in the low-noise region, transition actually occurs in the high-incident noise region. It was therefore of interest to extend the length of the "quiet" test core so that transition on aerodynamic models can be obtained in a completely low-noise environment. This required delaying transition on the nozzle walls, i.e., controlling Görtler instability. Görtler instability, being a curvature-induced instability, is difficult to control with wall suction or cooling. The latter control, in fact, tends to destabilize this instability and could also aggravate the roughness sensitivity due to the thinning of the boundary-layer. Crossflow (a swept-nozzle!) can tend to stabilize Görtler vortices (Hall, 1985), but then crossflow instability could introduce its own associated instability which is further enhanced by concave curvature (Zurigat and Malik, 1995). Therefore, the only effective way to control Görtler instability is to reduce the concave curvature. Beckwith et al. (1988) designed a new long slow expansion nozzle which could yield much longer quiet test core resulting in higher $Re_{\Delta x}$. The longer the nozzle, however, the less the width of the quiet test rhombus due to the noise radiated from the side walls and corner vortices. Hence, even though a long, quiet test section can be designed considering the contoured walls only, its width becomes so small that such a design becomes unattractive. Therefore, the concept was applied to design an axisymmetric nozzle and such a nozzle was fabricated and tested (Chen et al., 1990).

Figure 8 shows $R_{\Delta x}$ (Reynolds number based on the length of the quiet test case) from test data for two rapid expansion (R.E.) pilot quiet nozzles and the new Advanced Mach 3.5 axisymmetric quiet nozzle (Adv.) over the unit Reynolds number range tested. The predicted (using Görtler $N = 9$) values of $R_{\Delta x}$ are included for comparison. For each nozzle, the measured maximum surface roughness, k, in the throat region is also listed to show the effect of surface finish on the performance. The data for the pilot nozzle and the advanced nozzle indicate an increasing trend of $R_{\Delta x}$ with unit Reynolds number, R_∞, except for the large values of k. This unit Reynolds number effect is believed to be caused by the increasing local favorable pressure gradients, that suppresses the growth of Görtler vortices, as transition moves upstream along the nozzle wall with increasing R_∞. The maximum value of $R_{\Delta x}$ obtained from the advanced nozzle is about 1.4×10^7. Except for the problem of surface finish, values of $R_{\Delta x}$ for the advanced nozzle (designed using $N_G = 9$) are more than double the value of the pilot quiet nozzles. It is unfortunate that this nozzle was never used to test cone models to see if the transition Reynolds number remains to be 9×10^6 (the highest obtained in the pilot tunnels) or if higher numbers

are achieved. It may be noted that the Görtler N-factor calculations given in the figure were performed using the quasi-parallel approach. Recent calculations performed at HTC using PSE show that nonparallel effect is small for the conditions relevant for these nozzles (see Fig. 9).

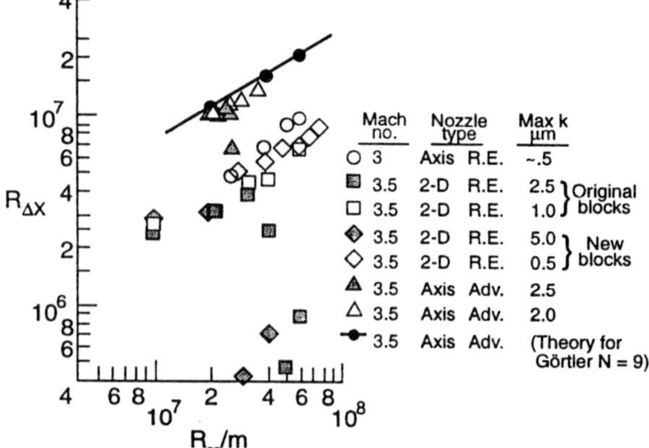

Fig. 8. Quiet test core Reynolds numbers in the advanced Mach 3.5 axisymmetric quiet nozzle (Adv.) and pilot quiet nozzles (Rapid Expansion, R.E.).

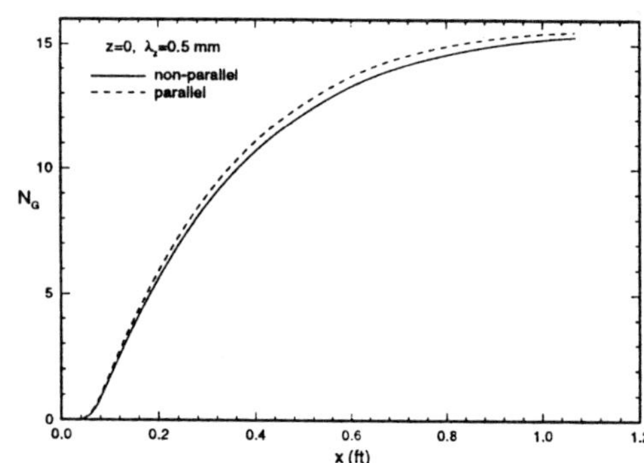

Fig. 9. Comparison of quasi-parallel and PSE results for Görtler instability in the Mach 3.5 nozzle.

The effect of surface finish (in the throat region) can be clearly seen (in Fig. 8). With $k = 2.5\mu m$, laminar flow is lost for $R_\infty/m > 2.2$; when $k = 2\mu m$, laminar flow can be extended up to about $3.5 \times 10^6/m$ and beyond that it suddenly disappears. The $R_k = 12$ criterion (Beckwith et al., 1988) indicates that the maximum value of k should be less than $1.2\mu m$ in order to achieve the design values of $R_{\Delta x}$ at higher values of R_∞. This example demonstrates how "lots" of laminar flow is entirely lost by very "little" roughness making the prediction of the e^N method completely useless. Similarly, the effect of noise on the first mode is evident in Fig. 7.

3.2 Crossflow Mode: The Case of ASU Experiment

The influence of micron-sized roughness on crossflow-induced transition has been clearly brought out in the Arizona State University (ASU) experiment (Radeztsky et al., 1993) where transition location is influenced by roughness as small as $.5\mu m$ ($R_k \sim .001$), provided roughness is located near the leading edge. There are, however, two differences between this case and the $R_k = 12$ criterion for the "quiet" nozzle. In the swept-wing case, roughness has a direct "input" into the stationary crossflow instability and the roughness with the quoted values of R_k of up to about 4.5 moves the location of transition but does not trip the boundary-layer. In the case of the nozzle, roughness is located in the throat region where boundary-layer is stable (linearly) to centrifugal instability and therefore any stationary perturbation has a large region of damping before they begin to amplify in the downstream concave curvature region. The boundary-layer in the throat region is also stable to TS instability owing to the favorable pressure gradients. Therefore, the loss of laminar flow with a roughness of $R_k \approx 12$ is a complete "bypass" of the usual transition process. This should serve as a challenge problem for DNS of transition or, perhaps, an opportunity to demonstrate the usefulness of the transient growth approach (Henningson et al., 1995).

Figure 10 shows N-factor calculations using nonlinear PSE for the ASU experiment ($\alpha = -4°$, $Re_c = 2.4 \times 10^0$). Calculations are shown for the most amplified stationary disturbance with spanwise wavelength of $12mm$, with three different initial amplitudes. When the leading edge was polished to yield surface finish of $.25\mu m$, transition occurred at $x/c \approx .7$ for a slightly higher Re_c of 2.6×10^6 (Radeztsky et al., 1993). At this location, the linear N-factor for stationary disturbances is about 8.8 (it will be somewhat higher for $Re_c = 2.6 \times 10^0$ and even higher if traveling disturbances are considered). Thus, linear theory does well in predicting this crossflow-induced transition as it does in the case of Görtler-induced transition (Fig. 8) and oblique first mode-induced transition (Fig. 7).

Fig. 10. Effect of initial amplitude on the growth of stationary crossflow disturbance with $\lambda_z = 12mm$ for the ASU wing (nonlinear PSE).

It is clear from the above example that the crossflow case is no exception to the usefulness of the linear theory based e^N method for estimation of transition. As noted above, the e^N method provides an "upper limit" on the transition Reynolds number which can be achieved with extreme care. It first appears quite surprising that a roughness of height $.5\mu m$ ($Re_k \approx .001$) begins to influence transition location, but the fact that a roughness directly "inputs" into the stationary crossflow disturbance explains this extreme sensitivity. If one assumes that amplitudes of $O(20\%)$ (see §4.2) are needed for crossflow-induced transition and that e^9 amplification precedes this stage, it results in the initial disturbance amplitude of about .0025 percent. While receptivity calculations are not yet available, it is not unlikely that a wall perturbation of $.5\mu m$ in a $.8mm$ thick boundary-layer would produce a mean flow disturbance of .0025 percent; a $9\mu m$ [the painted surface case of Dagenhart et al. (1989)] disturbance should produce a much bigger response. Therefore, the crossflow problem is much more sensitive to the roughness particularly when it is located near the critical point (onset of instability). The same should hold true for stationary Görtler vortices.

3.3 Crossflow in the Supersonic Regime: Mach 3.5 Delta-Wing Model

Experiments in the Langley Mach 3.5 "quiet" tunnel were performed on a 15-inch long symmetric delta wing model (leading-edge sweep = 77.1°) with (Cattafesta, 1996) and without (Cattafesta et al., 1994) wall suction. Figure 11 shows the transition front measured on the suction model using temperature sensitive paint (TSP) technique. Linear stability calculations were also performed and the values of the N-factor which cross the transition front lie between 8 to 10.5. Earlier calculations for the solid model (with no suction holes) indicated that an N-factor of about 14 ("envelope" approach) correlates the transition onset front quite well over a range of Reynolds numbers and angles of attack. Thus the N-factors at transition are lower in the case of suction due probably to the adverse effect of suction holes which induce roughness effect and enhanced receptivity.

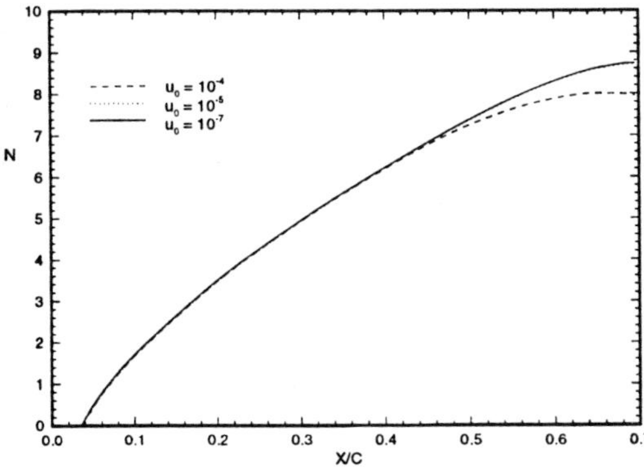

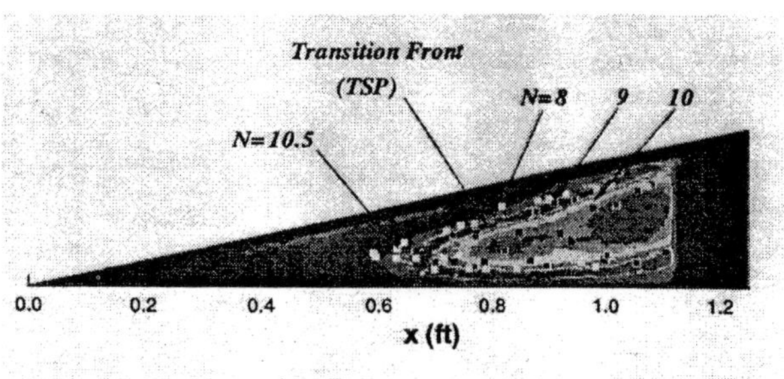

Fig. 11. Measured transition front using TSP on the suction model ($M_\infty = 3.5$), along with some N-factor results.

Constant N-factor contours (both with and without suction) indicated that the transition front is parallel to the leading-edge on one side and to the line of symmetry on the other and that it turns around in the middle. Earlier experiments (Cattafesta et al., 1994) using the thermocouple technique for transition detection did not support the calculations as it indicated that the transition front on the inside was not parallel to the line of symmetry, rather perpendicular to it. TSP results, however, turned out to be in agreement with the e^N calculations.

The observed and computed transition front can also be discerned from the contours of crossflow Reynolds numbers R_{cf} ($K_{cf} = |w_{max}|\delta_{0.1}/\nu$, the usual definition used in low-speed experiments); however, the values of R_{cf} are much higher (~600) than those found in the low-speed experiments. This was also found to be the case in the experiments of King (1991) on a cone at incidence in the same tunnel. Malik et al. (1991) performed computations for a Mach 8 cone at incidence and found that at the locations where the N-factor reaches 10, crossflow Reynolds number reaches values of about 2000. Malik et al. (1991) suggested that the higher values of crossflow Reynolds number result due to ~M^2 growth in the boundary-layer thickness (i.e., the compressibility effect) which can be scaled out by defining a new crossflow Reynolds number (see also Malik and Balakumar, 1992) R_{cf} where

$$\overline{R}_{cf} = \frac{R_{cf}}{1 + \dfrac{\gamma - 1}{2}\sqrt{\mathrm{Pr}}M_e^2} \ .$$

The value of $\overline{R}_{cf}$ at transition (measured or predicted using $N \approx 10$) was found to lie in the neighborhood of 200 (i.e., same range as the low speed case) for low supersonic as well as hypersonic flows. Reed and Haynes (1994) presented a more formal derivation of the compressibility effect and also accounted for nonadiabatic wall temperature.

3.4 <u>The Attachment-Line Boundary-Layer</u>

The attachment-line boundary-layer on a swept-wing requires careful consideration since, in general, when this boundary-layer becomes turbulent, laminar flow is lost on the entire wing. It was found by Hall et al. (1984) that the attachment-line is subject to small-amplitude traveling disturbances above a Reynolds number, $\overline{R}$, of 583. The frequencies of these viscous TS-like disturbances agreed well with the experimental results of Pfenninger and Bacon (1969) and Poll (1979,1980). More recently, Lin and Malik (1996) used the 2D eigenvalue approach to solve for the attachment-line instability. Their results agreed with the HMP mode found by Hall, Malik and Poll (1984), but this new approach allowed them to discover additional instability modes and to account for leading-edge curvature, ignored by Hall et al. This effect was found to be small (Lin and Malik, 1997) for practical subsonic wings. N-factor calculations performed for the HMP mode correlates well with the available experimental data (Fig. 12). Lin and Malik (1995) also considered the supersonic attachment-line problem using the 2D eigenvalue approach and found the nonparallel effect to be significantly destabilizing because the critical Reynolds number decreased from $\overline{R} \approx 573$ to $\overline{R} \approx 349$.

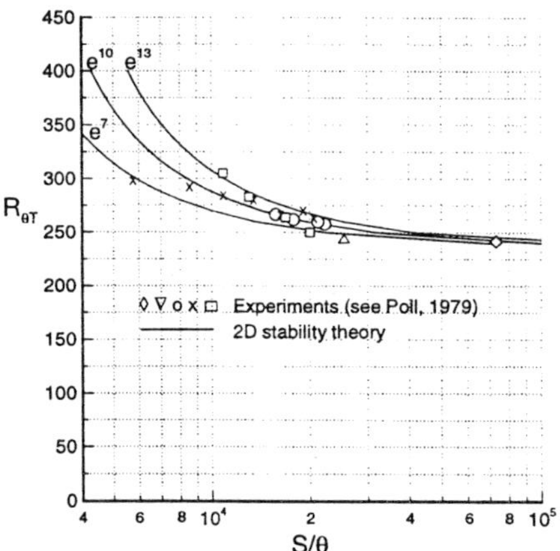

Fig. 12. Comparison of e^N results (HMP mode) and experiments for the attachment-line boundary-layer.

Experiments (Pfenninger and Bacon, 1969; Poll, 1979,1980) have shown that the attachment-line boundary-layer can sustain turbulent flow at Reynolds numbers much below the critical value of 583 quoted above. Large amplitude disturbances are known to damp out only at $\overline{R} < 245$ [e.g., Poll (1980)]. Hall and Malik (1986) used the weakly nonlinear theory, as well as direct numerical simulations, to show that the attachment-line boundary-layer is subject to subcritical instability which allows large amplitude disturbances to amplify above a critical Reynolds number. The value of this critical Reynolds number was found to be about 535 below which all large disturbances pertaining to the HMP mode were damped. Later calculations by Jiménez et al. (1990) failed to find the subcritical instability, but the recent direct Navier-Stokes simulations by Joslin (1994,1995) supported the conclusion drawn by Hall and Malik (1986). In any case, neither the results of Hall and Malik nor those of Joslin fill the wide gap between 583 and 245. DNS results of Spalart (1988) showed that the turbulence could not be sustained in the attachment-line boundary-layer below $\overline{R} < 245$, thus agreeing with the experiments. He was further able to show that the "turbulent" boundary-layer at $\overline{R} > 245$ could be "laminarized" with wall suction; this practically significant result has also been obtained experimentally by Poll and Danks (1995). Their experimental results showed that the boundary-layer could be laminarized up to $\overline{R}$ of 600 with sufficient wall suction.

3.5 Streamline Curvature Mode

Malik and Poll (1985) used orthogonal curvilinear coordinates to study the effect of (in-plane) streamline curvature on crossflow instability. Their results were later corrected in Malik and Balakumar (1993). Itoh (1994), using similar streamline-aligned coordinates, showed that in-plane streamline curvature introduced instability of its own. He later formulated (Itoh, 1996) the same problem in "natural" coordinates for the swept leading-edge flow and found the same mode of instability. Guo et al. (1997) used LNS to find instability in the parameter regime where Itoh found the "streamline curvature" mode. It seems that this instability cannot be captured by the method of multiple-scales as a perturbation solution to the Orr-Sommerfeld equation to account for nonparallel effects but PSE should be able to capture it. In any case, the growth rate for this mode is smaller than the usual crossflow mode, so an e^N calculation will consider this mode to be insignificant except that it has been found (Guo et al., 1997) that this curvature-induced instability is much less sensitive to wall suction as compared to the crossflow instability. Therefore, depending upon flow parameters, this mode deserves careful attention in LFC design using wall suction.

3.6 Corner Flow Instability

The corner viscous layer is relevant in various aerodynamic configurations such as wing body junction flows, engine inlets and wind tunnels of rectangular or square cross-section. Experiments by Zamir (1981) indicated that the viscous layer in a streamwise corner, formed by the intersection of two semi-infinite flat plates at 90° to each other, becomes unstable at a Reynolds number of about 10^4 as compared to the critical Reynolds number ($U_0 x / \nu$) of about 9×10^4 for the Blasius flow. Balachandar and Malik

(1995) performed an inviscid instability analysis of the similarity mean flow based on the formulation of Rubin and Grossman (1971). Their 2D eigenvalue analysis showed that the corner layer is inviscidly unstable. If this inviscid instability were to persist at small Reynolds numbers, this would corroborate with Zamir's experiments. Lin et al. (1996), however, performed viscous stability analysis of the mean flow computed by using steady Navier-Stokes equations. This analysis showed that the inviscid mode is stable at low Reynolds numbers. The corner layer was found to be unstable to viscous instability but the critical Reynolds number was no lower than that for the Blasius case. At present, no clear explanation is available for the discrepancy between the experiment and the computations. It is likely that the experimental observation was related to some sort of nonlinear phenomenon or that the experiments were not carefully performed and need to be repeated.

Lin et al. (1996a) and Wang et al. (1997) have also performed 2D eigenvalue calculations for corner flow in a supersonic nozzle of square cross-section. For this flow, it is found that the corner layer develops vortical structures which are highly unstable.

3.7 Wavy Wall

Although modern manufacturing techniques can provide smooth surfaces that are compatible with laminar flow, manufacturing tolerance criteria are needed for unavoidable surface imperfections. These imperfections include waviness, bulges, steps and gaps as well as three-dimensional roughness elements. Experimental studies which attempt to provide these criteria for humps and wavy walls are those of Fage (1943) and Carmichael and Pfenninger (1959).

Wie and Malik (1997) used interacting boundary-layer (IBL) and linear PSE to provide such a criterion for transition in 2D boundary-layers. They computed ΔN (increase in the N-factor due to waviness) for various parameters (e.g., see Fig. 13) and obtained a correlation for the change in N-factor due to waviness, ΔN, as

$$\Delta N = c\,\frac{nk^2 Re}{\lambda}$$

where n is the number of waves, k the total wave height, λ the wavelength and Re the unit Reynolds number. The constant c depends upon the stream pressure gradient and has a value of .14 for waviness on a flat plate. The results showed that the critical size of waviness below which waviness has no influence on TS instability does not exist, provided that the waviness is located on the right of the lower branch of the neutral curve.

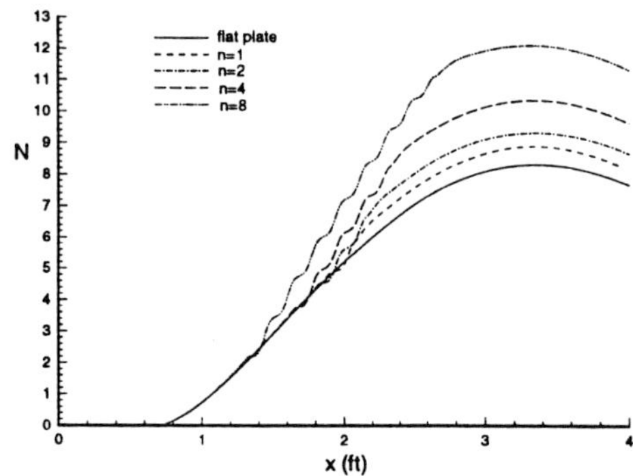

Fig. 13. Effect of number of waves on the N-factor in a wavy-wall boundary-layer.

Calculations based on IBL and quasi-parallel stability were earlier made by Nayfeh et al. (1988), Cebeci and Egan (1988) and Masad and Iyer (1994) for a hump and by Masad and Nayfeh (1994) for backward- and forward-facing steps.

3.8 Hypersonic Boundary-Layer Transition

The pioneering work of Mack (1969,1984) and Kendall (1967, 1975) laid out the foundation for hypersonic boundary-layer stability and transition. Both the theory and experiments showed the existence of second-mode disturbances. The acoustic second-mode disturbance, now called the Mack mode, was later verified in a series of experiments by Stetson et al. (1983, 1984). This mode which is destabilized by wall cooling, becomes dominant at a Mach number of about 4. Calculations by Malik (1987,1989), however, showed that, in an e^N sense, the switch between the first and second mode takes place at a Mach number of about 7 for adiabatic wall conditions. For cold walls, this switch over occurs at lower Mach numbers since wall cooling stabilizes the first mode and destabilizes the second mode. Recent experiments (Lachowicz et al., 1996) performed in a Mach 6 "quiet" tunnel, designed using the concept described in Beckwith et al. (1988), provided support for the predicted effect of adverse pressure-gradient on second-mode disturbances. Figure 14 [from Balakumar and Malik (1994)] shows that the N-factor at transition onset is about 8.5 for the experiment performed on a cone flare model in this tunnel. The predicted most amplified frequency (~220kHz) was also observed to be the most amplified frequency in the experiments of Lachowicz et al. (1996).

An important issue in hypersonic transition is the effect of nose bluntness. Calculations by Malik et al. (1990) agreed with the experimental findings of Stetson et al. (1984) in that small nose bluntness stabilizes the boundary-

layer. The critical Reynolds number increased by almost a factor of 15 due to bluntness (for the nose Reynolds number of 31250), although the predicted ($N = 10$) transition Reynolds number only increased by about 25%. The ratio of "blunt" to "sharp" transition Reynolds number increases with decreasing N-factor (i.e., increasing free-stream disturbance). Some experiments indicate that the trend switches (stabilizing to destabilizing) for large bluntness, but no satisfactory theoretical explanation is as yet available.

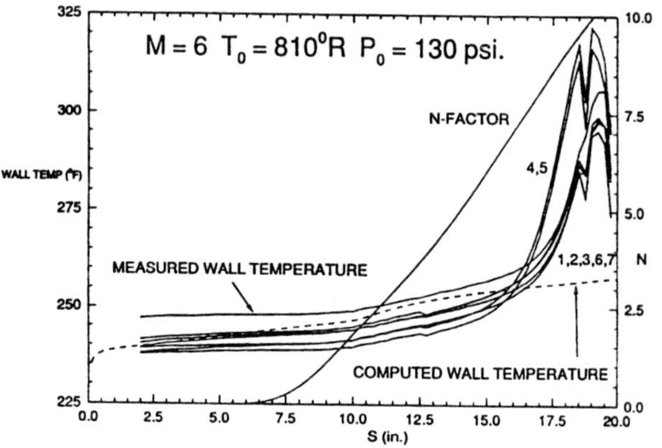

Fig. 14. Comparison between measured and computed (laminar) wall temperatures and the second mode N-factor calculation for the most amplified frequency ($f = 220kHz$) for a cone-flare model.

The effect of real gas has been studied by Malik (1989a,1990), Malik and Anderson (1991), Stuckert and Reed (1994) and recently by Hudson et al. (1996). Malik et al. (1989) performed stability calculations using equilibrium assumption for re-entry F cone (Wright and Zoby, 1977) and obtained an N-factor of 7.5 at the initiation of transition for this Mach 20 flight experiment. One issue that became prominent during the National Aerospace Plane (NASP) transition program was the role of supersonic modes in hypersonic transition (Malik and Macaraeg, 1993). The dominant second-mode disturbances in the perfect gas examples cited above are subsonic, i.e., their phase speed $c_r > 1 - 1/M_e$. Supersonic modes with $c_r < 1 - 1/M_e$ have been discussed by Mack (1969,1984), but are known to be either stable or only weakly unstable relative to the subsonic mode in wall-bounded flows (for supersonic mixing-layers and jets, the significance of the supersonic modes is well-known, cf. Tam and Hu, 1989; Macaraeg and Streett, 1989, 1991; Malik and Chang, 1997). Stability calculations using chemically reacting equilibrium gas assumption showed that supersonic modes gain a new significance in the e^N sense. Predicted ($N = 10$) transition location was several feet apart on a hypersonic body, depending upon the perfect and equilibrium gas assumption. Recently, Chang et al. (1997) have used PSE to study the effect of finite-rate chemistry and chemical equilibrium. For an example calculation (with M_∞ = 20), they find that N-factor results using finite-rate chemistry lie in between the results obtained by perfect gas and equilibrium gas assumptions. In such a situation, one must employ finite-rate chemistry for transition estimation since the error introduced by using equilibrium gas assumption (or the perfect gas) is rather large.

4. Transition Mechanisms

No attention is paid to actual transition mechanisms when e^N is used for estimation of location of transition. This is true of the low-speed TS case as well as the Görtler, first-mode, second-mode and crossflow cases discussed above. While significant progress has been made in understanding the breakdown mechanisms in two- and three-dimensional boundary-layers, the use of this capability to date in transition prediction has been limited due to the lack of precise knowledge of the initial conditions. This situation will change as progress is made towards developing appropriate models for the environmental disturbances and their imprint in the boundary-layer. Carefully performed receptivity experiments with detailed characterization of environmental disturbances are in dire need.

Here, we briefly review some of the transition mechanisms relevant in 2D and 3D boundary-layers. These, coupled with the yet to obtain receptivity information, would form the basis for a more rational approach for transition prediction.

4.1 TS and Mack Modes

There are various breakdown mechanisms for the low-speed TS case and the most obvious ones are:

1. Fundamental resonance or Klebanoff-type;
2. Subharmonic resonance or Craik- or Herbert-type;
3. Combination resonance;
4. Oblique mode interaction.

These mechanisms have been discussed by Craik (1971), Herbert (1988), Coarke and Mangano (1989), Kachanov et al. (1984) and Henningson et al. (1995).

Compressible stability calculations for the cone and flat-plate suggest that transition occurs due to oblique first-mode disturbances. No experiments have been performed in the "quiet" tunnel to actually understand the transition mechanism or even to see if first-mode disturbances are naturally present. Other experiments performed in noisy tunnels do show the presence of oblique modes but they were excited by surface glow discharge (Kosinov et al., 1990). Naturally occurring second-mode disturbances have

been observed in the Mach 6 "quiet" tunnel (Lachowicz et al., 1996).

The most likely scenario for the actual mechanism of transition for the flat-plate/cone case is the oblique-mode breakdown (Bestek et al., 1992; Chang and Malik, 1994). In this case, a pair of oblique modes interact to produce a streamwise vortex. The mutual and self-interaction of the streamwise vortex and the oblique modes results in the rapid growth of other harmonic waves and transition soon follows. The r.m.s. amplitude of the streamwise velocity component is found to be on the order of 4%–5% at the transition onset location marked by the rise in mean wall shear. Chang and Malik found that, for this mechanism, initial amplitude of the oblique modes has to be no more than ~.001% for transition Reynolds numbers observed in "quiet" tunnels. Chang and Malik (1993) found that the same oblique mode mechanism is operative in hypersonic boundary-layers.

Kosinov et al. (1994) found subharmonic resonance in their flat-plate experiment at Mach 2. Nonlinear PSE allows one to efficiently construct various scenarios for transition, provided initial conditions are prescribed.

4.2 Pseudo-Saturation of Crossflow Disturbances

Malik et al. (1994) used nonlinear PSE to study the evolution of crossflow disturbances in a model 3D boundary-layer. They observed that nonlinearity suppresses the fundamental whose growth rate eventually begins to oscillate about a small value. This saturation or pseudo-saturation was earlier observed by Malik (1986) for rotating-disk flow and by Meyer and Kleiser (1988) for the Falkner-Skan-Cooke boundary-layer. One of the solutions obtained by Gajjar (1996) also shows oscillatory behavior for the nonlinear crossflow vortex. Clearly secondary instabilities would soon destroy this pseudo-equilibrium state resulting in loss of laminar flow. For the model boundary-layer, Malik et al. (1994) found that the nonlinear growth rate begins to depart from the linear growth rate when the crossflow disturbance amplitude reaches about 4%. In Fig. 10, for the ASU wing, the amplitude of the disturbance at x/c = .45 (where nonlinear and linear calculations begin to depart for u_0 = .0001) is found to be 4.13%.

Figure 15 shows nonlinear PSE results with three different initial amplitudes to compare with the experiments of Reibert (1996). In all three cases, the amplitude of the fundamental reaches the range of 17–18% at $x/c \approx .5$ (where transition is observed in the experiment). Secondary instability calculations, of the type described in §4.3 below, are underway. It is hoped that these calculations will shed some light on the apparent insensitivity of transition location to the change in roughness height k which varied from $6\mu m$ to $48\mu m$.

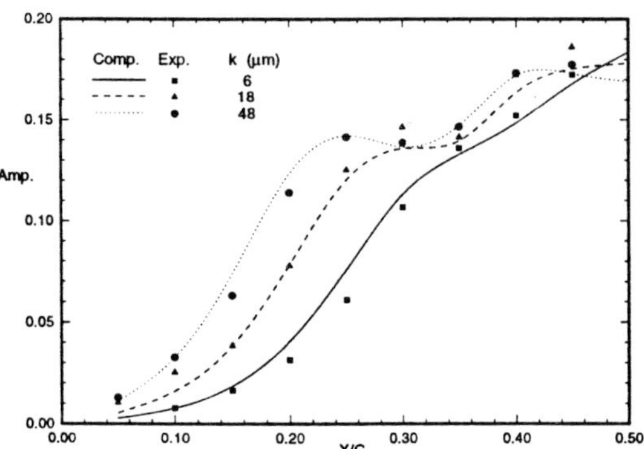

Fig. 15. Nonlinear evolution of crossflow disturbances (fundamental mode) and comparison with experiments (Reibert, 1996) using three different roughness heights.

4.3 High-Frequency Secondary Instability

There are several possible scenarios for transition in crossflow-dominated 3D boundary-layers, particularly on swept wings. The one in which the "quasi-saturated" stationary crossflow disturbances develop high-frequency secondary instability was observed by Kohama et al. (1991) in the ASU experiment and computed for a model boundary-layer by Malik et al. (1994). The more recent experimental work can be found in Kohama et al. (1996). Earlier, this same phenomena was found by Kohama (1984) and theoretically computed by Balachandar et al. (1992) for rotating disk flow. Poll (1985) observed a disturbance with $17.5kHz$ frequency in his swept-cylinder experiment. Secondary instability computations by Malik et al. (1996) showed that the disturbance with the highest growth rate has a frequency of about $17kHz$, thus agreeing with Poll's result.

Computational results by Malik et al. (1996) show that the modulated mean flow is subject to two types of secondary modes where one mode correlates with the spanwise shear and the other with the vertical shear. This is similar to the case in the Görtler vortex problem (Li and Malik, 1995) where these modes are referred to as sinuous and varicose (horse-show vortex) modes (Swearingen and Blackwelder, 1987; Hall and Horseman, 1991). Thus, apart from the role of traveling disturbances, mechanisms for breakdown of Görtler and crossflow vortices are quite similar (Malik and Li, 1993). For the supersonic nozzle where Görtler instability is present, transition likely occurs via a similar secondary instability mechanism.

4.4 Traveling and Stationary Disturbance Interactions

Although no direct measurements were made, Cattafesta et al. (1994) suggested that traveling disturbances must have been dominant in his Mach 3.5 delta-wing. The role of traveling vs. stationary disturbances has been a subject of some controversy in low-speed experiments but now it is widely accepted that this is a receptivity issue pertaining to the relative importance of surface finish and the environmental disturbances (turbulence, etc.). Thus, stationary disturbances are relevant on a wing with "roughness" but otherwise low-disturbance environment (e.g., Dagenhart et al., 1989) while traveling disturbances would dominate in an experiment performed in relatively high turbulence environment (cf. Muller and Bippes, 1988). While Poll (1985) observed both stationary and traveling disturbances, Takagi and Itoh (1994) only observed traveling disturbances when the experiment was performed on a "large" diameter swept cylinder, diminishing the roughness effect.

A simple case of the stationary/traveling wave interaction was studied by Malik et al. (1994). When the initial amplitude of stationary vortex is large compared to the traveling mode, the stationary vortex dominates most of the downstream development. Eventually, however, the traveling mode becomes of the same order as the stationary mode. The situation changes when the initial amplitudes of the traveling and the stationary modes are the same. Owing to its higher growth rate, the traveling mode dominates most of the downstream development and the growth of the stationary mode is suppressed. Some of the features observed by Mueller and Bippes (1988) were captured qualitatively by the results computed by Malik et al. (1994).

The rotating-disk experiment of Corke and Knasiak (1996) suggested a triad resonance between traveling modes of two frequencies and stationary modes. This resonance led to "difference" interactions providing energy to low wavenumber stationary modes which became dominant near transition to turbulence. Steady and traveling wave interactions have also been studied by Herbert and Schrauf (1996) in a swept-wing boundary-layer. Clearly, nonlinear PSE provides an efficient tool for studying such interactions provided initial amplitudes can be prescribed from receptivity studies.

5.0 Spatio-Temporal Growth and Absolute Instability

We have so far only considered convective instabilities, i.e., those for which an impulse response would decay to zero for large time since the induced instability is "swept-away" from the source with its group velocity. The spatial linear stability theory, PSE and the harmonic LNS of §2.4 can be used to describe the behavior of such instabilities at all points. On the other hand, if the impulse response is unbounded for large time, then the flow is called absolutely unstable. In this case, the energy is trapped in the neighborhood of the source and grows in time. A stability theory capable of describing space-time evolution of instabilities (or time-dependent LNS or DNS) must be used to describe it.

Using complex Fourier and Laplace transform techniques, Briggs (1964) and Bers (1975) derived general conditions for convective and absolute instabilities. If α and ω are the disturbance wavenumber and frequency, respectively, then it is required that a saddle point singularity between two spatial branches of the dispersion relation must exist in the α plane for $Im(\omega) > 0$. Such a singularity is known as a "pinch point." The "group velocity" $\partial\omega/\partial\alpha = 0$ at such a point. If the saddle point results due to the collision of two spatial branches originating in the same half of the α-plane, then it constitutes a second-order pole resulting only in transient growth. For the pinch point to result in absolute instability, the collision must occur between two spatial branches originating in two distinct halves of the α-plane.

Absolute instability has been studied, for example, by Huerre and Monkewitz (1985) in free-shear layers and by Koch (1985) for wake flows; Huerre and Monkewitz (1990) provide a review on the subject. The results of Balakumar and Malik (1990) for rotating-disk flow provided a hint that this flow might be absolutely unstable but this possibility was not explored because it appeared that the instability might occur at too high a Reynolds number. Recently, however, Lingwood (1995, 1996) showed that this boundary-layer becomes absolutely unstable at a Reynolds number of 507, i.e., just below the Reynolds numbers where disk flow becomes transitional. Lingwood (1996a) then performed an experiment in an attempt to show that this instability can be realized in an experimental set up. While wall roughness may dominate the crossflow problem resulting in the transition mechanism mentioned in §4.3, the result of Lingwood (1995) has opened a new dimension to the boundary-layer transition prediction issue particularly for ultra-smooth surfaces. Therefore, the possibility of absolute instability needs to be investigated for other boundary-layers.

Lin et al. (1997) recently studied the stability of the 3D boundary-layer formed on an infinite-swept cylinder. The problem they considered was of an impulse response of a spanwise periodic disturbance with real wavenumber β. They were able to find pinch points in the chordwise plane for this flow near the leading-edge at Reynolds numbers slightly below the Reynolds number at which the

attachment-line becomes unstable to infinitesimal disturbances.

Figure 16 shows some results from their calculation. Figure 16(a) shows the neutral curves for absolute instability for various $\overline{R}$. The critical $\overline{R}$ for the 60° swept cylinder is found to be 540.5. Figure 16(b) shows the two spatial branches at $\overline{R} = 632$ in the α-plane. At the pinch point (denoted by the intersection of curves C^+ and C^-), $\beta = .24$ and $\omega = (.08412, .00106)$. It should be noted that for a 3D wave packet, pinch points are required to occur simultaneously in the α and β planes for flow to be absolutely unstable. In the present calculation, the problem that is studied is the impulse response to a disturbance imposed all along the infinitely long cylinder. Under such a scenario, the disturbances are likely to become unbounded near the leading-edge provided $\overline{R} > 540.5$. In current LFC design, $\overline{R}$ is generally maintained at a much lower level to avoid attachment-line contamination. This instability, however, may be relevant for future applications of LFC to large aircraft where contamination may be suppressed by other means.

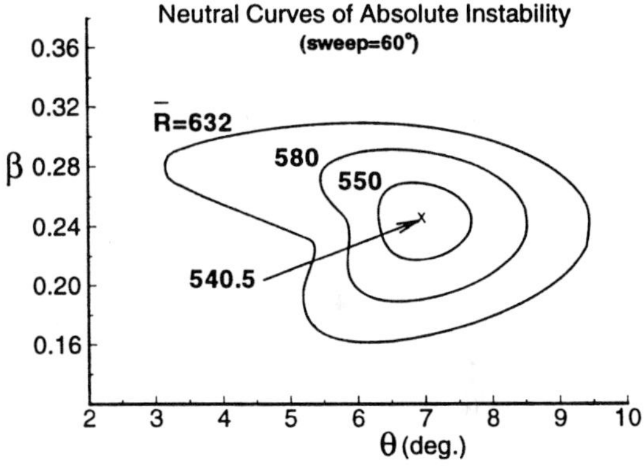

(a)   neutral curves; θ is the angle measured from the attachment-line.

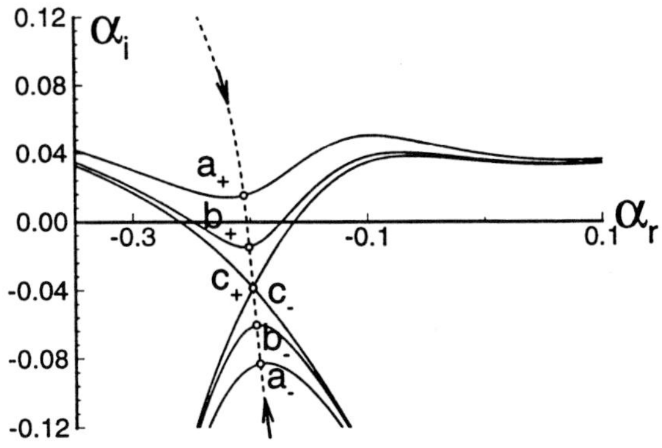

(b) pinch point in the α (chordwise) plane

Fig. 16. Spatio-temporal instability in the swept-cylinder boundary-layer (sweep angle 60°).

6. Concluding Remarks

This paper reviews recent progress made in the prediction (estimation) of boundary-layer transition. For a given flow geometry, transition onset constitutes two often widely apart limits: one for the "ultra-smooth" surface/low-disturbance case and the other for the "rough" surface/high-disturbance case. The former is determined by the e^N method while the latter requires experimental information or direct Navier-Stokes simulation. Transition prediction is a sorting out process where one first determines the upper limit and then estimates its onset by considering the effects of specific surface quality and free-stream environment. Computational tools for determining the upper limit for fully 3D boundary-layers have been developed, with particular emphasis for use by less-experienced users. This includes a suction optimization code for laminar flow control design. Computational tools have also been developed to account for surface quality effects but await validation against experiments.

A brief discussion of some of the physical issues associated with boundary-layer transition such as wall roughness, waviness and disturbance nonlinearity as well as transition breakdown mechanisms is given. The issue of absolute instability in a swept leading-edge boundary-layer is also addressed.

This review only considers application of stability theory relevant to laminar flow control technology or transition in low-disturbance environments. Transition in highly-disturbed environments (e.g., gas turbine blades) or other specific applications such as transition in high-lift systems is not considered here although the latter could be studied by a combination of computational tools described herein for linear and nonlinear evolution of disturbance.

References

Arnal, D., Casalis, G. & Juillen, J. C., 1995 "A Survey of the Transition Prediction Methods: from Analytical Criteria to PSE and DNS," *Laminar Turbulent Transition*, Sendai, Japan, Springer Verlag, R. Kobayashi, ed., pp. 3–14.

Balachandar, S. & Malik, M. R. 1995 "Inviscid Instability of Streamwise Corner Flow," *J. Fluid Mech.*, Vol. 282, pp. 187–201.

Balachandar, S., Streett, C. L. & Malik, M. R., 1992 "Secondary Instability in a Rotating Disk Flow," *J. Fluid Mech.*, Vol. 242, pp. 323–347.

Balakumar, P. & Hall, P., 1996 "Optimum Suction Distribution for Transition Control," AIAA 96–1950.

Balakumar, P. & Malik, M. R., 1990 "Traveling Disturbances in Rotating-Disk Flow," *TCFD*, Vol. 2, pp. 125–137.

Balakumar, P. & Malik, M. R., 1994 "Effect of Adverse Pressure Gradient and Wall Cooling on Instability of Hypersonic Boundary Layers," High Technology Corp. Report HTC–9404, Hampton, VA.

Beckwith, I. E. & Bushnell, D. M., 1988 "Wind-Tunnel Simulation of High-Speed Flight Transition," Seminar on Problems of Simulation in Wind Tunnels, USSR Academy of Sciences, Novosibirsk, USSR.

Beckwith, I. E., Chen, F.-J. & Malik, M. R., 1988 "Design and Fabrication Requirements for Low-Noise Supersonic/Hypersonic Wind Tunnels," AIAA Paper 88–0143.

Beckwith, I. E., Creel, T. R., Jr., Chen, F.-J. & Kendall, J. M., 1983 "Free-Stream Noise and Transition Measurements on a Cone in a Mach 3.5 Pilot Quiet Tunnel," NASA TP–2180.

Bertolotti, F. P., Herbert, Th. & Spalart, P. R., 1992 "Linear and Nonlinear Stability of the Blasius Boundary Layer," *J. Fluid Mech.*, Vol. 242, pp. 441–474.

Bers, A., 1975 "Linear Waves & Instabilities," *Physique des Plasmas* (ed. C. DeWitt & J. Peyraud), Gordon & Breach, pp. 117–215.

Bestek, H., Thumm, A. & Fasel, H., 1992 "Direct Numerical Simulation of Three-Dimensional Breakdown to Turbulence in Compressible Boundary Layers," 13th Int. Conf. Num. Meth. Fluid. Dyn., Rome, July 6–10.

Bippes, H., 1991 "Experiments on Transition in Three-Dimensional Accelerated Boundary Layer Flows," The Royal Aeronautical Soc. Conf. on Boundary Layer Transition & Control, Cambridge, UK.

Briggs, R. J., 1964 *Electron-Stream Interaction with Plasmas*, MIT Press.

Bushnell, D. M., 1990 "Notes on Initial Disturbance Fields for the Transition Problem," *Instability and Transition*, Vol. I, Hussaini & Voigt, eds., Springer-Verlag, pp. 217–232.

Bushnell, D. M. & Malik, M. R., 1987 "Application of Stability Theory to Laminar Flow Control — Progress and Requirements," *Stability of Time Dependent & Spatially Varying Flows*, D. L. Dwoyer & M. Y. Hussaini, eds., Springer-Verlag, pp. 1–17.

Carmichael, B. H. & Pfenninger, W., 1959 "Surface Imperfection Experiments on a Swept Laminar Suction Wing," Northrop Corp. Report NOR–59–454 (BLC–124).

Cattafesta, L. N., III, 1996 "Swept-Wing Suction Experiments in the Supersonic Low-Disturbance Tunnel," High Technology Corporation Report HTC–9605, Hampton, VA.

Cattafesta, L. N., III, Iyer, V., Masad, J. A., King, R. A. & Dagenhart, J. R., 1994 "Three-Dimensional Boundary-Layer Transition on a Swept Wing at Mach 3.5" AIAA 94–2375.

Cebeci, T. & Egan, D. A., 1988 "The Effect of Wave-Like Roughness on Transition," AIAA Paper 88–0139.

Cebeci, T. & Stewartson, K., 1980 "On Stability and Transition of Three-Dimensional Flows," *AIAA J.*, Vol. 18, pp. 398–405.

Cebeci, T., Chen, H. H. & Arnal, D., 1988 "A Three-Dimensional Linear Stability Approach to Transition on Wings at Incidence," AGARD CP–438, Fluid Dynamics of Three-Dimensional Turbulent Shear Flows & Transition, Cesme, Turkey, pp. 17–1 –17–13.

Chang, C.-L. & Malik, M. R., 1993 "Non-Parallel Stability of Compressible Boundary Layers," AIAA Paper 93–2912.

Chang, C.-L. & Malik, M. R., 1994 "Oblique-Mode Breakdown and Secondary Instability in Supersonic Boundary Layers," *J. Fluid Mech.*, Vol. 273, pp. 323–359.

Chang, C.-L., Vinh, H. & Malik, M. R., 1997 "Hypersonic Boundary-Layer Stability with Chemical Reactions using PSE," AIAA Paper 97–2012.

Chang, C.-L., Malik, M. R., Erlebacher, G. & Hussaini, M. Y., 1991 "Compressible Stability of Growing Boundary Layers Using Parabolized Stability Equations," AIAA Paper 91–1636.

Chen F. J., Malik, M. R. & Beckwith, I. E., 1985 "Instabilities and Transition in the Wall Boundary Layers of Low Disturbance Supersonic Nozzles," AIAA Paper 85–1573.

Chen, F-J., Malik, M. R. & Beckwith, I. E., 1989 "Boundary-Layer Transition on a Cone and Flat Plate at Mach 3.5," *AIAA J.*, Vol. 27, No. 6, pp. 687–693.

Chen, F-J., Malik, M. R. & Beckwith, I. E., 1990 "Advanced Mach 3.5 Axisymmetric Quiet Nozzle," AIAA Paper 90–1592.

Choudhari, M. & Streett, C. L., 1994 "Theoretical Prediction of Boundary-Layer Receptivity," AIAA Paper 94-2223.

Corke, T. C. & Knasiak, K. F., 1996 "Cross-Flow Instability with Periodic Distributed Roughness," IUTAM Symposium on Nonlinear Instability & Transition in Three-Dimensional Boundary Layers, P. W. Duck & P. Hall, eds., Kluwer, pp. 267–282.

Corke, T. C. & Mangano, R. A., 1989 "Resonant Growth of Three-Dimensional Modes in Transitioning Blasius Boundary Layers," *J. Fluid Mech.*, Vol. 209, pp. 93–150.

Craik, A. D. D., 1971 "Nonlinear Resonant Instability in Boundary Layers," *J. Fluid Mech.*, Vol. 50, pp. 393–413.

Crouch, J. D., 1994 "Receptivity of Boundary Layers," AIAA Paper 94–2224.

Dagenhart, J. R., Saric, W. S., Mousseaux, M. C. & Stack, J. P., 1989 "Crossflow Vortex Instability and Transition on a 45-Degree Swept Wing," AIAA Paper 89–1892.

Denier, J. P., Hall, P. & Seddougui, S. O., 1991 "On the Receptivity Problem for Görtler Vortices and Vortex Motion Induced by Wall Roughness," *Phil. Trans. R. Soc. Lond. A* **335**, pp. 51–85.

Fage, A., 1943, "The Smallest Size of Spanwise Surface Corrugation which Affect Boundary Layer Transition on an Airfoil," British Aeronautical Research Council, Report and Memoranda No. 2120.

Gajjar, J. S. B., 1996 "On the Nonlinear Evolution of a Stationary Cross-Flow Vortex in a Fully Three-Dimensional Boundary Layer Flow," IUTAM Symposium on Nonlinear Instability and Transition in Three-Dimensional Boundary Layers, P. W. Duck & P. Hall, eds., Kluwer, pp. 317–326.

Goldstein, M. E., 1983 "The Evolution of Tollmien-Schlichting Waves Near a Leading Edge," *J. Fluid Mech.*, Vol. 127, pp. 59–81.

Goldstein, M. E., 1985 "Scattering of Acoustic Waves into Tollmien-Schlichting Waves by Small Streamwise Variations in Surface Geometry," *J. Fluid Mech.*, Vol. 154, p. 485.

Guo, Y., Malik, M. R. & Chang, C.-L., 1997 "A Solution Adaptive Approach for Computation of Linear Waves," AIAA Paper 97–2072.

Hall, P., 1983 "The Linear Development of Gortler Vortices in Growing Boundary Layer," *J. Fluid Mech.*, Vol. 30, pp. 41–58.

Hall, P., 1985 "The Görtler Vortex Instability Mechanism in Three-Dimensional Boundary Layers," *Proc. R. Soc. Lond. A*, Vol. 399, p. 135.

Hall P. & Horseman, N. J., 1991 "The Linear Inviscid Secondary Instability of Longitudinal Vortex Structures in Boundary Layers" *J. Fluid Mech.*, Vol. 232, pp. 357–375.

Hall, P. & Malik, M. R., 1986 "On the Instability of a Three-Dimensional Attachment-Line Boundary Layer: Weakly Nonlinear Theory and a Numerical Approach," *J. Fluid Mech.*, Vol. 163, pp. 257.

Hall, P., Malik, M. R. & Poll, D. I. A., 1984 "On the Stability of an Infinite Swept Attachment Line Boundary Layer," *Proc. Roy. Soc. Lond. A*, 395, pp. 229–245.

Henningson, D. S., Berlin, S., Lundbladh, A., 1995 "Spatial Simulations of Bypass Transition in Boundary Layers," IUTAM Symp., Laminar Turbulent Transition, Sendai, Japan, Springer Verlag, R. Kobayashi (ed.), pp. 263–270.

Herbert, T., 1983 "Secondary Instability of Plane Channel Flow to Subharmonic Three-Dimensional Disturbances," *Phys. Fluids*, Vol. 26, No. 4, pp. 871874.

Herbert, T., 1988 "Secondary Instability of Boundary Layers," *Ann. Rev. Fluid Mech.*, Vol. 20, pp. 487–526.

Herbert , Th., 1991 "Boundary-Layer Transition — Analysis and Prediction Revisited," AIAA Paper 91–0737.

Herbert, Th. & Schrauf, G., 1996 "Crossflow-Dominated Transition in Flight Tests," AIAA Paper 96–0185.

Hudson, M. L., Chokani, N. & Candler, G. V., 1996 "Linear Stability of Hypersonic Flow in Thermochemical Nonequilibrium," AIAA Paper 96-0671.

Huerre, P. & Monkewitz, P. A., 1985 "Absolute and Convective Instabilities in Free Shear Layers," *J. Fluid Mech.*, Vol. 159, pp. 151–168.

Huerre, P. & Monkewitz, P. A., 1990 "Local and Global Instabilities in Spatially Developing Flows," *Ann. Rev. Fluid Mech.*, Vol. 22, pp. 473–537.

Itoh, N., 1994 "Effects of Wall and Streamline Curvatures on Instability of 3-D Boundary Layers," IUTAM Symp., Laminar-Turbulent Transition, Sendai, Japan, Springer Verlag, R. Kobayashi, (ed.), pp. 321–330.

Itoh, N., 1996 "Simple Case of the Streamline-Curvature Instability in Three-Dimensional Boundary Layers," *J. Fluid Mech.*, Vol. 317, pp. 129–154.

Iyer, V., 1996 "Stability Code Calibration from Preliminary F16XL Flight Data," High Technology Corporation Report HTC–9603, Hampton, VA.

Jiménez, J., Martel, C., Agüi, J. C. & Zufiria, J. A., 1990 "Direct Numerical Simulation of Transition in The Incompressible Leading Edge Boundary Layer," ETSIA, MF–903.

Joslin, R. D., 1994 "Direct Simulation of Evolution and Control of Nonlinear Instabilities in Attachment-Line Boundary Layers," AIAA Paper 94–0826.

Joslin, R. D., 1995 "Direct Simulation of Evolution and Control of Three-Dimensional Instabilities in Attachment-Line Boundary Layers," *J. Fluid Mech.*, Vol. 291, pp. 369–392.

Kachanov, Yu. S. & Levchenko, V. Ya., 1984 "The Resonant Interaction of Disturbances at Laminar-Turbulent Transition in a Boundary Layer," *J. Fluid Mech.*, Vol. 138, p. 209.

Kendall, J. M., 1967 "Supersonic Boundary Layer Stability Experiments," in *Boundary Layer Transition Study Group Meeting*, W. D. McCauley, ed., Report No. BSD-TR–67–213, Vol. II, U.S. Air Force, pp. 10–1–10–8 (available from DTIC as AD 820 364).

Kendall, J. M., 1975 "Wind Tunnel Experiments Relating to Supersonic and Hypersonic Boundary-Layer Transition," *AIAA J.*, Vol. 13, No. 3, pp. 290–299.

Kerschen, E. J., Choudhari, M. & Heinrich, R. A., 1990 "Generation of Boundary Instability Waves by Acoustic and Vortical Freestream Disturbances," *Laminar-Turbulent Transition*, Vol. IV, R. Kobayashi, ed., Springer.

King, R. A., 1991 "Mach 3.5 Boundary-Layer Transition on a Cone at Angle of Attack," AIAA Paper 91–1804.

Kleiser, L. & Zang, T. A., 1991 "Numerical Simulation of Transition in Wall-Bounded Shear Flows," *Ann. Rev. Fluid Mech.*, Vol. 23.

Koch, W., 1985 "Local Instability Characteristics and Frequency Determination of Self Excited Wake Flows," *J. Sound Vib.*, Vol. 99, pp. 53–83.

Kohama, Y., 1984 "Study on Boundary-Layer Transition of a Rotating Disk," *Acta Mechanica*, Vol. 50, pp. 193–199.

Kohama, Y, Saric, W. S. & Hoos, J. A., 1991 "A High-Frequency, Secondary Instability of Crossflow Vortices that Leads to Transition," Proc. The Royal Aeronautical Society Conf. on Boundary Layer Transition & Control, Cambridge, UK, pp. 4.1 – 4.13.

Kohama, Y., Onodera, T. & Egami, Y., 1996 "Design and Control of Crossflow Instability Field," IUTAM Symposium on Nonlinear Instability and Transition in Three-

Dimensional Boundary Layers, P. W. Duck & P. Hall, eds., Kluwer, pp. 147–156.

Kosinov, A. D., Maslov, A. A. & Shevelkov, S. G., 1990 "Experiments on the Stability of Supersonic Laminar Boundary Layers," *J. Fluid Mech.*, Vol. 219, pp. 621–633.

Kosinov, A. D., Semionov, N. V., Shevelkov, S. G. & Zinin, O. I., 1994 "Experiments on the Nonlinear Instability of Supersonic Boundary Layers," IUTAM Nonlinear Instability of Nonparallel Flows, D. T. Valentine, S. P. Lin & W.R.C. Phillips, eds., Springer, pp. 196–205.

Lachowicz, J. T., Chokani, N. & Wilkinson, S. P., 1996 "Boundary-Layer Stability Measurements in a Hypersonic Quiet Tunnel," *AIAA J.*, Vol. 34, No. 12, pp. 2496–2500.

Landahl, M. T., 1980 "A Note on Algebraic Instability of Inviscid Parallel Shear Flows," *J. Fluid Mech.*, Vol. 98, p. 243.

Li, F. & Malik M. R., 1995 "Fundamental and Subharmonic Secondary Instability of Görtler Vortices," *J. Fluid Mech.*, Vol. 297, pp. 77–100.

Li, F. & Malik, M. R., 1997 "Spectral Analysis of Parabolized Stability Equations," *Computers & Fluids*, Vol. 26, No. 1, pp. 279–297.

Lin, R.-S. & Malik, M. R., 1995 "Stability and Transition in Compressible Attachment-Line Boundary-Layer Flow," SAE Paper 952041, Aerotech '95, Los Angeles, CA.

Lin, R-S. & Malik, M. R., 1996 "On the Stability of Attachment-Line Boundary Layers: Part 1. The Incompressible Swept Hiemenz Flow," *J. Fluid Mech.*, Vol. 311, pp. 239–255.

Lin, R-S. & Malik, M. R., 1997 "On the Stability of Attachment-Line Boundary-Layers: Part 2. The Effect of Leading-Edge Curvature," *J. Fluid Mech.*, Vol. 333, pp. 125–137.

Lin, R.-S., Wang, W.-P. & Malik, M. R., 1996 "Linear Stability of Incompressible Viscous Flow Along a Corner," 1996 ASME Fluids Engineering Division Summer Meeting, San Diego, CA.

Lin, R.-S., Edwards, J. R., Wang. W.-P. & Malik, M. R., 1996a "Instabilities of a Mach 2.4 Slow-Expansion Square Nozzle Flow," AIAA Paper 96–0784.

Lin, R.-S., Li, F. & Malik, M. R., 1997 "Absolute Instability in Swept Leading-Edge Boundary-Layers," in preparation.

Lingwood, R. J., 1995 "Absolute Instability of the Boundary Layer on a Rotating Disk," *J. Fluid Mech.*, Vol. 299, pp. 17–33.

Lingwood, R. J., 1996 "On the Effects of Suction and Injection on the Absolute Instability of the Rotating-Disk Boundary Layer," *Phys. Fluids*, Vol. 9, No. 5, pp. 1317–1328.

Lingwood, R. J., 1996a "An Experimental Study of Absolute Instability of the Rotating-Disk Boundary-Layer Flow," *J. Fluid Mech.*, Vol. 314, p. 373.

Macaraeg, M. G. & Streett, C. L., 1989 "New Instability Modes for Bounded, Free Shear Flows," *Physics of Fluids*, A, Vol. 8, pp. 1305-1307.

Macaraeg, M. G. & Streett, C. L., 1991 "Investigation of Supersonic Modes and Three-Dimensionality in Bounded Free Shear Flows," *Computer Physics Communication*, Vol. 65, pp. 201-208.

Macaraeg, M. G., Streett, C. L. & Hussaini, M. Y., 1988 "A Spectral Multi-Domain Technique Applied to High-Speed Chemically Reacting Flows," in *Domain Decomposition Methods for PDE'S*, SIAM Publication, (eds., T. Chan et al.).

Mack, L. M., 1969 "Boundary Layer Stability Theory," Rept. 900-277 Rev. A, Jet Propulsion Lab., Pasadena, CA.

Mack, L. M., 1977 "Transition Prediction and Linear Stability Theory," AGARD CP–224, pp. 1-1 to 1-22.

Mack, L. M., 1984 "Boundary Layer Linear Stability Theory," AGARD Report No. 709, pp. 3/1 – 81.

Malik, M. R., 1982 "COSAL — A Black Box Compressible Stability Analysis Code for Transition Prediction in Three-Dimensional Boundary Layers," NASA CR–165925.

Malik, M. R., 1986 "Numerical Simulation of Transition in a Three-Dimensional Boundary Layer," *10th Intl. Conf. on Numerical Methods in Fluid Dynamics*, Proceedings, Beijing, F. G. Zhuang & Y. L. Zhu, eds., Springer-Verlag, pp. 455-461.

Malik, M. R., 1987 "Prediction and Control of Transition in Hypersonic Boundary Layers," AIAA Paper 87–1414.

Malik, M. R., 1989 "Prediction and Control of Transition in Supersonic and Hypersonic Boundary Layers," *AIAA J.*, Vol. 27, No. 11, pp. 1487–1493.

Malik, M. R., 1989a "Transition in Hypersonic Boundary Layers," presented at the 4th Symposium on Numerical & Physical Aspects of Aerodynamic Flows, Long Beach, CA.

Malik, M. R., 1990 "Numerical Methods for Hypersonic Boundary Layer Stability," *J. Comput Physics*, Vol. 86, No. 2, pp. 376–413.

Malik, M. R., 1990 "Stability Theory for Chemically Reacting Flows," *Laminar-Turbulent Transition*, , D. Arnal & R.

Michel, eds., Toulouse, France, Springer-Verlag, pp. 251–260.

Malik, M. R. & Anderson, E. C., 1991 "Real Gas Effects on Hypersonic Boundary-Layer Stability," *Phys. Fluids A*, Vol. 3(5), pp. 803–821.

Malik, M. R. & Balakumar, P., 1992 "Nonparallel Stability of Rotating Disk Flow Using PSE," *Instability, Transition & Turbulence* (Hussaini, Kumar & Streett, eds.), Springer-Verlag, pp. 168–180.

Malik, M. R. & Balakumar, P., 1993 "Linear Stability of Three–Dimensional Boundary Layers: Effect of Curvature and Non-Parallelism" AIAA Paper 93–0079.

Malik, M. R. & Chang C.-L., 1997 "PSE Applied to Supersonic Jet Instability," AIAA Paper 97–0758.

Malik, M. R. & Li, F. 1993 "Secondary Instability of Görtler and Crossflow Vortices," Proc. of the Internal Symposium on Aerospace and Fluid Science, Institute of Fluid Science, Tohoku University, Sendai, Japan, pp. 460–477.

Malik, M. R. & Macaraeg M. G., 1993 "Supersonic Modes in Hypersonic Boundary-Layers," NASP Transition Workshop — HYFLITE, Nov. 1–2, 1993, NASA Langley Research Ctr.

Malik, M. R. & Orszag, S. A., 1980 "Comparison of Methods for Prediction of Transition by Stability Analysis," AIAA Paper No 80-1430.

Malik, M. R. & Poll, D. I. A., 1985 "Effect of Curvature on Three-Dimensional Boundary-Layer Stability," *AIAA J.*, Vol. 23, No. 9, pp. 1362–1369.

Malik, M. R., Chuang, S. & Hussaini, M. Y., 1982 "Accurate Numerical Solution of Compressible Stability Equations," *ZAMP*, Vol. 33, pp. 189–201.

Malik, M. R., Spall, R. E. & Chang, C.-L., 1989 "Transition Prediction in Hypersonic Boundary Layers," Sixth National Aero-Space Plane Symposium, Monterey, CA.

Malik, M. R., Spall, R. E. & Chang, C.-L., 1990 "Effect of Nose Bluntness on Boundary Layer Stability and Transition," AIAA Paper 90–0112.

Malik, M. R., Spall, R. E. & Chang, C.-L., 1990a "Effect of Nose Bluntness on Boundary Layer Stability and Transition," AIAA Paper 90–0112.

Malik, M. R., Balakumar, P. & Chang, C.-L., 1991 "Linear Stability of Hypersonic Boundary Layers (U)," NASP Paper No. 189.

Malik, M. R., Li, F. & Chang, C.-L., 1994 "Crossflow Disturbances in Three-Dimensional Boundary Layers: Nonlinear Development, Wave Interaction and Secondary Instability," *J. Fluid Mech.*, Vol. 268, pp. 1–36.

Malik, M. R., Li, F. & Chang, C.-L., 1996 "Nonlinear Crossflow Disturbances and Secondary Instabilities in Swept-Wing Boundary Layers," IUTAM Symposium on Nonlinear Instability & Transition in Three-Dimensional Boundary Layers, P. W. Duck & P. Hall, eds., Kluwer, pp. 257–266.

Masad, J. A. & Iyer, V. 1994, "Transition Prediction and Control in Subsonic Flow Over Hump," *Physics of Fluids*, Vol. 6, No. 1, pp. 313–328.

Masad, J. A. & Nayfeh, A. H., 1994 "Stability of Separating Subsonic Boundary Layers," NASA CR–4638.

Meyer, F. & Kleiser, L., 1988 "Numerical Investigation of Transition in 3D Boundary Layers," AGARD CP–438, Fluid Dynamics of Three-Dimensional Turbulent Shear Flows and Transition, Cesme, Turkey, pp. 16–1 –16–17.

Morkovin, M. V., 1985 "Bypass Transition to Turbulence and Research Desiderata," *Transition in Turbines*, NASA CP–2386, pp. 161–199.

Morkovin, M., 1993 "Bypass-Transition Research: Issues and Philosophy," *Instabilities and Turbulence in Engineering Flows*, Ashpis, Gatski and Hirsh, eds., Kluwer, pp. 3–30.

Müller, B. & Bippes, H., 1988 "Experimental Study of Instability Modes in a Three-Dimensional Boundary Layer," AGARD CP–438, Paper 13, pp. 13–1 –13–15.

Nayfeh, A. H., 1980 "Stability of Three-Dimensional Boundary Layers," *AIAA J.*, Vol. 18, pp. 406–416.

Nayfeh, A. H., Ragab, S. A. & Al-Maaitah, A., 1988 "Effect of Bulges on the Stability of Boundary Layers," *Phys. Fluids*, Vol. 31, pp. 796–806.

Orszag, S. A. & Patera, A. T., 1983 "Secondary Instability of Wall-Bounded Shear Flows," *J. Fluid Mech.*, Vol. 128, 347–385.

Pfenninger, W. & Bacon, J. W., 1969 "Amplified Laminar Boundary Layer Oscillations and Transition at the Front Attachment Line of a 45° Flat-Nosed Wing With and Without Boundary Layer Suction," *Viscous Drag Reduction*, ed. C. S. Wells, Plenum, pp. 85–105.

Poll, D. I. A., 1979 "Transition in the Infinite Swept Attachment Line Boundary Layer," *The Aeronautical Quarterly*, Vol. XXX, pp. 607.

Poll, D. I. A., 1980 "Three-Dimensional Boundary Layer Transition via the Mechanisms of 'Attachment Line Contamination' and 'Cross Flow Instability'," IUTAM Symp., Laminar-Turbulent Transition, Stuttgart, Germany, Springer-Verlag, R. Eppler & H. Fasel, eds., pp. 253–262.

Poll, D. I. A., 1985 "Some Observations of the Transition Process on the Windward Face of a Long Yawed Cylinder," *J. Fluid Mech.*, Vol. 150, pp. 329–356.

Poll, D.I.A. & Danks, M., 1995 "Relaminarisation of the Swept Wing Attachment-Line by Surface Suction," IUTAM Symp., Laminar Turbulent Transition, Sendai, Japan, Springer Verlag, R. Kobayashi (ed.), pp. 137–144.

Pruett, C. D. & Chang, C.-L., 1995 "Spatial Direct Numerical Simulation of High-Speed Boundary-Layer Flows. Part II: Transition on a Cone in Mach 8 Flow," *TCFD*, Vol. 7, pp. 397–424.

Radeztsky, R. H., Reibert, M. S., Saric, W. S. & Takagi, S., 1993 "Effect of Micron-Sized Roughness on Transition in Swept-Wing Flows," AIAA Paper 93–0076.

Rai, M. M. & Moin, P., 1993 "Direct Numerical Simulation of Transition and Turbulence in a Spatially Evolving Boundary Layer," *J. Comput. Phys.*, Vol. 109, pp. 169–192.

Reed, H. L. & Haynes, T. S., 1994 "Transition Correlation in Three-Dimensional Boundary Layers," *AIAA J.*, Vol. 32, No. 5, pp. 923–929.

Reed, H. L. & Nayfeh, A. H., 1981 "Stability of Flow over Plates with Porous Suction Strips," AIAA Paper 81–1280.

Reed, H. L., Saric, W. S. & Arnal, D., 1996 "Linear Stability Theory Applied to Boundary Layers," *Annu. Rev. Fluid Mech.*, Vol. 28, pp. 389–428.

Reibert, M. S., 1996 "Nonlinear Stability, Saturation, and Transition in Crossflow-Dominated Boundary Layers," Ph.D. Dissertation, Arizona State University, May 1996.

Reshotko, E., 1976 "Boundary Layer Stability and Transition," *Ann. Rev. Fluid Mech.*, Vol. 8, pp. 311–349.

Reshotko, E., 1994 "Boundary Layer Instability Transition and Control" AIAA Paper 94–0001.

Ruban, A. I., 1985 "On Tollmien-Schlichting Wave Generation by Sound," *Laminar-Turbulent Transition*, V. V. Kozlov, ed., Springer-Verlag, pp. 313–320.

Rubin, S. G. & Grossman, B., 1971 "Viscous Flow Along a Corner: Numerical Solution of the Corner Layer Equations," *Q. Appl. Maths.*, Vol. 24. p. 169.

Saric, W. S. & Thomas, A. S. W., 1984 "Experiments on the Subharmonic Route to Turbulence in Boundary Layers," *Turbulence & Chaotic Phenomena in Fluids* (ed. T. Tatsumi), North-Holland, pp. 117-122, .

Saric, W. S., H. L. Reed & E. J. Kerschen, 1994 "Leading Edge Receptivity to Sound: Experiments, DNS, and Theory," AIAA 94–2222.

Smith, A. M. O., 1952 "Design of the DESA–2 Airfoil," Douglas Aircraft Co., ES17117, AD 143008.

Smith, A. M. O. & Gamberoni, N., 1956 "Transition, Pressure Gradient, and Stability Theory," Douglas Aircraft Company, Inc., Report ES 26388.

Spalart, P. R., 1988 "Direct Numerical Study of Leading-Edge Contamination," AGARD CP–438, Fluid Dynamics of Three-Dimensional Turbulent Shear Flows & Transition, Cesme, Turkey, pp. 5–1 –5–13.

Stetson, K. F., Thompson, E. R., Donaldson, J. C. & Siler, L. G., 1983 "Laminar Boundary Layer Stability Experiments on a Cone at Mach 8. Part 1: Sharp Cone," AIAA Paper 83–1761.

Stetson, K. F., Thompson, E. R., Donaldson, J. C. & Siler, L. G., 1984 "Laminar Boundary Layer Stability Experiments on a Cone at Mach 8, Part 2: Blunt Cone," AIAA Paper 84–0006.

Streett, C. L., 1995 "Computation of Wave Packets in Swept-Wing Flows," IUTAM Symposium on Nonlinear Instability & Transition in 3D Boundary Layers, June 1995.

Stuckert, G. K. & Reed, H. L., 1994 "Linear Disturbances in Hypersonic, Chemically Reacting Shock layers," *AIAA J.*, Vol. 32, No. 7, pp. 1384–1393.

Swearingen, J. D. & Blackwelder, R. F., 1987 "The Growth and Breakdown of Streamwise Vortices in the Presence of a Wall," *J. Fluid Mech.*, Vol. 182, pp. 255–290.

Takagi, S. & Itoh N., 1994 "Observation of Traveling Waves in the Three-Dimensional Boundary Layer Along Yawed Cylinder," *Fluid Dynamics Research*, Vol. 14, pp. 167–189.

Tam, C.K.W. & Hu, F. Q., 1989 "The Instability and Acoustic Wave Modes of Supersonic Mixing Layers Inside a Rectangular Channel," *J. Fluid Mech.*, Vol. 203, pp. 51–76.

Trefethern, L. N., Trefethen, A. E., Reddy, S. C. & Driscoll, T. A., 1993 "Hydrodynamic Stability without Eigenvalues," *Science*, Vol. 261.

Van Ingen, J. L., 1956 "A Suggested Semi-Empirical Method for the Calculation of the Boundary Layer Transition Region," Univ. of Techn., Dept. of Aero. Eng., Rept. UTH–74, Delft.

Wang, W.-P., Lin, R.-S., Malik, M. R. & Edwards, J. R., 1997 "Control of Corner Flow Vortices by Geometry Shaping in Mach 2.4 Rectangular Nozzles," AIAA Paper 97-2228.

Wilkinson, S. P. & Malik, M. R., 1985 "Stability Experiments in the Flow over a Rotating Disk," *AIAA J.*, Vol. 23, pp. 588–595.

Wie, Y.-S. & Malik, M. R., 1997 "Effect of Surface Waviness on Boundary-Layer Transition in Two-Dimensional Flow," to appear *Computers & Fluids*.

Wlezien, R. W., 1994 "Measurement of Acoustic Receptivity," AIAA Paper 94–2221.

Wray, A. & Hussaini, M. Y., 1984 "Numerical Experiments in Boundary Layer Stability," *Proc. Roy. Soc. Lond. A*, Vol. 392, pp. 373–389.

Wright, R. L. & Zoby, E. V., 1977 "Flight Boundary Layer Transition Measurements on a Slender Cone at Mach 20," AIAA Paper 77–719.

Zamir, M., 1981 "Similarity and Stability of the Laminar Boundary Layer in a Streamwise Corner," *Proc. R. Soc. Lond. A*, Vol. 337, p. 269.

Zurigat, Y. H. & Malik, M. R., 1995 "Effect of Crossflow on Görtler Instability in Incompressible Boundary Layers," *Phys. Fluids*, Vol. 7, No. 7, pp. 1616–1625.

TURBULENCE MODELING –
PROGRESS AND FUTURE OUTLOOK

Joseph G. MARVIN[1]   George P. HUANG[2]

(1) NASA-Ames Research Center, Moffett Field, CA.
(2) Department of Mechanical Engineering, University of Kentucky; Lexington, KY.

Abstract

Progress in the development of the hierarchy of turbulence models for Reynolds-averaged Navier-Stokes codes used in aerodynamic applications is reviewed. Steady progress is demonstrated, but transfer of the modeling technology has not kept pace with the development and demands of the CFD tools. An examination of the process of model development leads to recommendations for a mid-course correction involving close coordination between modelers, CFD developers and application engineers. In instances where the old process is changed and cooperation enhanced, timely transfer is realized. A turbulence modeling information data base is proposed to refine the process and open it to greater participation among modeling and CFD practitioners.

1. Introduction

Despite the significant advances of Computational Fluid Dynamics (CFD), turbulence modeling is still critical to its success for aerodynamic applications. New aircraft performance requirements are pushing the envelope of our experience and "time to market" considerations are requiring more sophisticated CFD early in the design cycle. A successful CFD tool would enhance our understanding of the effects of viscous, high Reynolds number flows associated with the design of these aircraft. The three-dimensional time-dependent solution of the Navier-Stokes equations could provide an exact description of the turbulent motion, but the range of time and length scales associated with turbulence are such that they cannot be resolved when computing complex aerodynamic flows. As a consequence, the Reynolds averaged form of the Navier-Stokes equations together with a turbulence model are the most practical means today of computing complex aerodynamic flows. Industry has started to use Reynolds-averaged Navier-Stokes (RANS) codes in their design cycle, even though the physical modeling is less than satisfactory. There is no doubt significant progress in the field of turbulence modeling has been made. However, in contrast to the revolutionary pace of CFD development, turbulence modeling development has been evolutionary and the resulting pace of improvement has frustrated the CFD community. Some of the factors leading to the frustration are: the number of modeling research and development studies has proliferated over the last decade, yet no clear choice for suitable models seems to be emerging; the test cases used to validate turbulence models are only weakly relevant to "real" aerodynamic flow applications; reported successes are fragmentary or inconsistent and numerical aspects and their impact on efficiency are rarely discussed; and, there is a general lack of a systematic effort to transfer "successful" models to CFD application codes.

The latter point is a crucial aspect too often ignored and results in CFD applications being performed with models that fail to represent the best available alternative. Modeling and CFD development typically proceed on independent paths. CFD developers are expected to pick models from a myriad of choices. And, there is general lack of feedback between the groups on modeling successes and failures. It is absolutely necessary to rectify the situation because the likelihood of having a single code with a single turbulence model that can solve the breadth of CFD applications encountered in aerodynamics is remote. A viable systematic effort requires attention to the complete model development process and close collaboration among the two disciplines.

In this paper, progress in modeling is described first. Next, the process of model development is examined to provide insight into the reasons for the perceived slow pace of progress. Ideas for a mid-course correction in the process that could lead to a more satisfactory pace are proposed and discussed. Examples are given throughout to provide foundation to the ideas being advocated.

Received on August 1, 1997.

2 Progress

2.1 Background

The formalism used to derive the Reynolds-averaged Navier Stokes equations leads to the well known closure problem wherein the fluid motion is described by its mean and suitably averaged fluctuating motion. The latter is accomplished through turbulence modeling. A review of early progress in modeling provides an informative prelude to our discussion on the current status.

In 1968, the AFOSR-IFP Stanford Conference[1] was held to determine whether emerging numerical techniques together with a turbulence model were sufficient to solve spatially developing boundary layers. Finite difference numerical procedures were judged to provide suitable means for solving the boundary layer form of the Reynolds averaged Navier-Stokes equations. The majority of comparisons with data were accomplished with incompressible codes using eddy viscosity models. No single modeling approach emerged as superior, but the potential of using RANS codes was demonstrated.

In 1969[2] and 1972[3] NASA conferences were held to assess early modeling progress on compressible boundary layers and shear layers. The available models, most of which were eddy-viscosity types and corrected for compressibility through Morkovin's hypothesis, were generally successful for boundary layers, but failed in some specific instances such as incompressible reattaching shear layers and growth rate predictions of compressible shear layers. Suggestions for further study were prevalent throughout the discussions, but optimism for expected improvements prevailed and follow-on research intensified.

During the next decade, considerable effort was undertaken to improve turbulence models and broaden the range of applications. Spurred by the arrival of the supercomputer, the effort focused on the closure problem. In particular, the development of two-equation and higher order Reynolds Stress transport closure models pioneered by Launder and his colleagues offered the hope for accurate predictions of complex turbulent flows and perhaps the development of a universal model.

The 1980-1981 AFOSR-HTTM-Stanford Conference[4] was organized to assess general progress in modeling complex turbulent flows during the decade of the '70's. The conclusions that emerged from the conference evaluation committee were mixed. On the one hand, it was concluded that complex flows (still simple compared to those associated with aeronautical applications), including separated flows, could indeed be calculated with the more sophisticated numerical methods and models. On the other hand, the accuracy of the computations and the performance of the higher order Reynolds stress models was disappointing. Including more physics did not necessarily lead to more accurate results and there was concern that computational errors (e.g. numerical convergence and grid resolution) affected the resulting conclusions. It was also apparent that the search for a universal model was not realized.

After a decade, Bradshaw, Launder and Lumley initiated a sequel to the 1980-81 Conference. Their purpose was to assess subsequent progress in RANS model developments for complex flows during the intervening years. They developed and provided a modified, narrower scope evaluation data base, and required contributors to provide the results, either by postmail or electronic mail. The format was interactive and allowed modelers to see their comparisons relative to others and make adjustments if they thought they could improve their own predictions.

Bradshaw *et al.*[5] presented their conclusions on modeling progress with the aid of a single figure. The conclusions that emerged from their study was disappointing in light of the latitude provided to the contributors. Figure 1, taken from their report, shows the skin friction coefficient on a flat plate at momentum-thickness Reynolds number of 10,000 plotted against I_{100}, the value of U/u_τ at $u_\tau y/\nu = 100$. The known value from experiment is depicted by the large "plus" symbol, which covers the range of experimental scatter. The other symbols represent various contributors' predictions and they scatter dramatically about a mean curve fit through them. They do not reproduce the experimental results. The flagged symbols are values predicted using the standard $k - \epsilon$ model, some with wall functions and some with integration to the wall. They do not agree with one another and show that different numerical schemes can lead to scatter in predictions even for the same model. Bradshaw *et al.* surmised that physical modeling progress was extremely difficult to evaluate in the absence of any consistent numerical assessment procedure even for the supposedly "universal" viscous wall region.

It is becoming clearer as more and more code assessment studies are conducted that errors are not attributable to modeling deficiencies alone. For example, at the recent workshop on CFD code assessment for turbomachinery[6], it was concluded that model deficiencies, numerical code errors and lack of training in the use of CFD codes were all factors contributing to differences between numerical solutions and their subsequent differences with data. The organizers also concluded that a single test case, although relevant to a practical flow, was probably too complicated to assess the accuracy of various turbulence models.

2.2 Current Status

The status of various modeling closures will be discussed in the order to which the Reynolds stresses are approximated.

The Reynolds stress transport models solve the transport equation for Reynolds stresses, $\overline{u_i u_j}$ directly:

$$\frac{D\overline{u_i u_j}}{Dt} = d_{ij} + P_{ij} + \phi_{ij} - \epsilon_{ij} \qquad (1)$$

The first term in the L.H.S. and the second term in the R.H.S. are convection and production terms, respectively, and require no modeling; the first term in the R.H.S. is the diffusion term and is in general modeled by a gradient diffusion approximation; the third term in the R.H.S. is the pressure-strain correlation. A majority of the modeling efforts have focused on this term. The last term is the dissipation term which is in general related to the scalar dissipation rate, governed by a transport equation. This type of model requires 7 equations, in addition to the mean flow equations. Currently, there are a least four major variants being developed: Launder *et. al.*[7], Shih and Lumley[8], Fu *et. al.*[9] and Speziale *et. al.*[10]. The main difference between these variants is the pressure strain correlation modeling. For a detailed review of these models, see ref. 11, 12, 13, 14 and 15. These models are just beginning to be evaluated for application to aerodynamic flows.

The Reynolds stress closure can be simplified by introducing a stress-strain relationship:

$$\overline{u_i u_j} = -2\nu_t(S_{ij} + H.O.T.) + \frac{2}{3}\delta_{ij}k \qquad (2)$$

where $S_{ij} = 1/2(\partial U_i/\partial x_j + \partial U_j/\partial x_i)$, ν_t is a turbulent eddy viscosity and the higher order terms (H.O.T.) contain non-linear products of the strain and vorticity tensors, S_{ij} and $\Omega_{ij} = 1/2(\partial U_i/\partial x_j - \partial U_j/\partial x_i)$, respectively. The concept of including the high order terms in the stress-strain relationship was introduced in 1975 by Pope[16], who used the algebraic stress assumption of Rodi[17] to derive an explicit form for the higher order terms. Currently, there are at least three variants for this type of model: Gatski and Speziale[18], Shih *et. al.*[19] and Craft *et. al.*[20]. In all these models, the

turbulent eddy viscosity, ν_t, is related to the turbulent kinetic energy, k and its dissipation rate, ϵ, by

$$\nu_t = c_\mu(S^*, \Omega^*)\frac{k^2}{\epsilon} \qquad (3)$$

where S^* and Ω^* are the dimensionless strain and vorticity invariants, $S^* = \sqrt{2S_{ij}^2 k/\epsilon}$ and $\Omega^* = \sqrt{2\Omega_{ij}^2 k/\epsilon}$, respectively, and the values of k and ϵ are obtained from their corresponding modeled transport equations.

Menter[21] observed that the ratio of production to dissipation of turbulent kinetic energy can be significantly larger than one in adverse pressure gradient flows. Under such conditions, employing a constant value of c_μ (≈ 0.09) leads to over-prediction of the turbulent shear stress. To remedy this problem, he proposed a modification to the eddy viscosity when the ratio of production and dissipation becomes large. His model provides significant improvement in predicting adverse pressure gradient flows, separated flows and transonic shock separated flows. See figure 2.

Interestingly, his modification, although intuitive and tested for linear eddy viscosity models, is in accord with (3), proposed for the non-linear models. Figure 3 shows the variation of c_μ with S^* compared with his. All models show a decrease in c_μ at high S^* values corresponding to those for adverse pressure gradient and separated flows. Based on this observation, it is believed that the non-linear models now under development will provide similar improvements. See for example ref. 22 and 23.

The majority of models employed in practical applications are linear eddy viscosity models. The value of eddy viscosity can be obtained by a family of two equation models, namely, $k^m\epsilon^n - k^p\epsilon^q$ models. For example, if one picks $m = 1/2$, $n = 0$, $p = -1$ and $q = 1$, one arrives at the $q-\omega$ model. In theory, one form of the model can be transformed exactly into another form. Investigators have made refinements that lead to a wide variety of formulations. Among these, the most frequently used are: the $q - \omega$ model[24], the $k - \epsilon$ model[25], the $k - \omega$ model[26], the $k - l$ model[27], the $k - \tau$ model[28] and the SST model[21]. The latter is a blend of the $k - \epsilon$ model in the outer region and the $k - \omega$ model in the near wall region. As mentioned above, it limits the ratio of production to dissipation. Results of computations with some of these models will be provided later.

In order to enhance numerical robustness and efficiency, Baldwin and Barth[29] proposed a single transport equation for ν_t. Spalart and Allmaras[30] proposed an improved version of a ν_t model and results using this model will be shown subsequently. Other variants of ν_t models were derived earlier. See for example ref. 31.

The most primitive form of eddy viscosity models is formed by relating its value directly to the mean velocity. These algebraic models are often employed in applications because they are numerically robust. The most often used models are: the Baldwin-Lomax[32] and the Cebeci-Smith[33] models. Johnson and King[34] improved the physical modeling for adverse pressure gradient and transonic flow applications by introducing an additional O.D.E. that accounts for turbulence history effects. The modification greatly improved predictions of transonic wing performance[35]. An example[36] showing calculations with this model of the flow field and pressures for a 747 wing-body combination is shown in Figure 4.

Several alternatives are available for modeling in the vicinity of solid surfaces: 1) the use of wall functions; 2) switching to a 1-eqn or a 2-eqn model near the wall; and 3) integrating the model equations to the wall. The wall function techniques make use of the law of the wall. The region between the first grid point and the wall is divided into a single[37] or several zones[38,39]. For compressible flows, the van Driest law of the the wall formally and correctly extends the wall functions to compressible flows [40,41]. The use of 1-equation[42,43,44,45] (or 2-equation[46]) models in the sublayer allows the transport of turbulent energy to be accounted for. Despite success with the latter approach many modelers prefer to use low Reynolds number damping that allows direct integration of the governing equations down to the wall. For a description of the more commonly applied methods used with 2-equation models see: the $q - \omega$ model of Coakley[24]; the $k - \epsilon$ models of Launder and Sharma[25], Chien[47] and Lien and Leschziner[48]; the $k - \omega$ model of Wilcox[26]; and the SST model of Menter[21]. The $k - \epsilon - \overline{v^2}$ model of Durbin[49,50] allows direct integration to the wall without recourse to a damping function by introducing an additional $\overline{v^2}$-equation, as does the Wilcox $k - \omega$ model.

The development of near wall modeling for Reynolds stress transport equations is still evolving [51,52,53,54,55,56]. Demonstration applications of these models are still limited to simple boundary layer flows.

Favre averaging is the most common technique for extending incompressible models into the compressible flow regime. A comprehensive investigation of the differences between Favre and Reynolds averaging may be found in Huang et. al.[57]. Generally, in addition to the compressible mean dilatation terms, additional terms such as the dilatation-dissipation and pressure-dilatation fluctuation terms arise and require modeling. Zeman et. al.[58] and Sarkar et. al.[59] independently proposed that dilatation-dissipation augments the solenoidal dissipation by a function of M_t^2, where M_t is the turbulent Mach number. Their models have been applied to compressible mixing layers and successfully predict the decrease in shear layer spreading at high Mach numbers[60]. Models for the pressure-dilatational fluctuation correlation have also been proposed[61,62,63]. These models were also focused on predictions of simple sheared mixing layers. Recent work by Huang et. al[57,64] suggests, however, that both of these models for dilatational-dissipation and the pressure-dilatation fluctuation correlations may not be useful for near-wall flows. Indeed, experience has shown that for prediction of subsonic and supersonic flows these two modifications degrade the results and are not recommended.

For the prediction of hypersonic near-wall flows, Coakley and Huang[65,66] proposed two modifications for current two-equation eddy viscosity models: one compression modification to cure the under-prediction of separation bubble size caused by shock wave/boundary layer interactions; and the other, a length scale modification to remedy the over-prediction of heat transfer rate near re-attachment points. These two modifications have been validated for a range of hypersonic flows. The modifications are formulated in a general manner so they can be applied in any existing two-equation model formulation. An example illustrating the effects of these modifications on the prediction of the hypersonic flow over a cylinder flare is shown in figure 5.

3 Process of Model Development

In the Spring of 1994, a NASA Turbulence Modeling Standards Committee was formed to undertake an independent evaluation of the process of turbulence model development and suggest changes that could enhance the quality of models used in CFD application codes. The Committee membership, comprised of experts from NASA, Industry and Academia, is listed in the Acknowledgement section.

The Committee established a set of turbulence modeling evaluation criteria or standards that turbulence modelers would be encouraged to apply. These criteria would enable CFD'ers to make intelligent choices for models to incorporate into their application code. In addition, they would provide modelers (and the aeronautical community at large) with relevant comparative metrics to assess progress on new or improved model development and to identify modeling shortcomings.

Four critical elements of the process where change is necessary will be discussed here. The first involves numerical procedures. The second involves the selection and use of standard validation data bases as a means of assessing model performance. The third involves establishing standard numerical solutions to insure accurate implementation of models into application codes. The fourth involves the development of a mechanism for insuring that "successful" models get transported into application codes. This element is the most challenging and important part of the total process. It requires renewed cooperation among modelers and code developers.

3.1 Numerical Procedures

Integration of the conservation and accompanying modeling equations requires careful consideration of errors resulting from the numerical procedures. As discussed above, inattention to such procedures clouds the interpretation of a turbulence model's performance.

Various alternative methods for implementing models in RANS codes may lead to numerical stability problems. In most cases, instabilities are caused by improper treatment of source terms. Often this is construed as "bad" modeling physics. Therefore it is essential that the modelers fully describe their procedures for handling source terms. A detailed discussion of the successful numerical strategies for implementing various types of models and corresponding computational times is provided by Gatski[12]. In addition, sensitivity of the solutions to the number of grid points, to boundary conditions (e.g. free-stream and inflow) and to y^+ at the first grid point adjacent to the wall must be taken into consideration.

The following solutions for an incompressible plate flat flow illustrate how sensitivity studies should be conducted and they serve to explain some of the factors for the scatter of the results shown in Figure 1. The following conditions were used in the calculations unless it is otherwise stated. The skin friction is reported at $Re_\theta = 10,000$, where the solution is not affected by the inflow conditions. The ratio of turbulent viscosity to molecular viscosity in the freestream, μ_t/μ_l, and the ratio of the square root of the freestream turbulent kinetic energy to the freestream velocity, $\sqrt{k}/U_\infty$, were both

taken to be 0.1%. The computational box was defined by $Re_L = 2 \times 10^7$ (resulting in Re_θ up to 2.5×10^4) with a height to length ratio of 0.02. The grid points were expanded exponentially from the wall to the top of the domain with an expansion ratio determined by the choice of the value of y^+ at the first grid point. The value of y^+ at the first grid point was kept approximately 0.1. A detailed discussion is given in Ref. 67.

Grid sensitivity. Computations using the several popular turbulence models were performed with 50, 100, 250, 500 and 1000 grid points in the y direction. This corresponds to approximately 35, 65, 145, 265 and 465 grid points inside the boundary layer at $Re_\theta = 10000$. The results, shown in Figure 6, are presented as the percent error with respect to the solution obtained using 1000 grid points. The zero- and one-equation models are less sensitive to the grid refinement. In general, errors can be controlled within less than 2% if 100 grid points are used in the calculations (corresponding to 60 grid points inside the boundary layer).

Sensitivity to y^+ at the first grid point, y_1^+. Computations were performed with $y_1^+ \approx 0.014, 0.14, 0.4, 0.7, 1.1$ and 1.4 using 500 points in the y-direction. The results, shown in Figure 7, are presented as the percent error with respect to the solution obtained using $y_1^+ \approx 0.014$. The $k - \epsilon$ model is the most sensitive. In general, one should limit the value of y_1^+ to be less than 0.3 to have accurate solutions.

Freestream boundary condition sensitivity. Computations were performed by fixing the freestream value of $\sqrt{k}/U_\infty$ at 0.1% while varying the value of freestream turbulent viscosity to molecular viscosity ratio according to $\mu_t/\mu_l = 10^n$. The values of n were chosen to be $-6, -3, -1$ and 0. Figure 8 shows the sensitivity as a percent of the value of the model solutions for $n = -6$. The $k - \omega$ model is very sensitive to freestream conditions. See Menter[68]. In general, one should maintain the value of n less than -3 (for the $k - \omega$ model).

Code invariant test. Three codes were used to establish a standard solution for the incompressible flat plate flow: a boundary layer code[69], an incompressible Navier-Stokes code (INS2D)[70] and a compressible Navier-Stokes code[71]. The solutions for each code were established following the guidelines recommended above. Figure 9 shows the comparison of the results obtained by the three codes using Menter's SST model. Differences in the region $Re_\theta < 5000$ were mainly due to the different inflow conditions used in the three solutions. Differences in skin friction coefficients are less 1% after

$Re_\theta > 5000$. Comparisons using the other models show similar results. The solutions are essentially code independent and the results provide a standard numerical solution for each of the models.

3.2 Standard Validation Data

Turbulence modeling to a large extent is an empirical science. Models mimic the real physics and confidence is established by comparing modeled solutions against experiment or direct numerical simulations. Since direct numerical simulations are limited, comparisons with experiment provide the best measure of a model's performance for engineering applications.

The NASA Standards Committee deliberated extensively on the type of data base needed to facilitate interaction between modelers and CFD developers. They concluded that a standard data base should be developed. They recommended the small set of experimental test cases shown in table 1 as a first step in evaluating models for external aerodynamic flows. The choices were based on personal experience and knowledge about the data base and prior computations of them. The motivation was to rule out possibilities of including inferior data and unforeseen difficulties in specifying the boundary conditions. Furthermore, the flows were considered challenging enough to sort out modeling differences and could comprise the metrics for assessing model improvement. The majority of the flows are 2-dimensional, mainly because there is still a dearth of well defined 3-dimensional experiments.

3.3 Standard Solutions

Standard solutions for a number of the test cases employing several turbulence models have been established by researchers at the Ames Research Center[67]. The group, Dr. T. J. Coakley, Dr. J. A. Bardina, and Dr P .G. Huang, maintained the rigorous standards for numerical procedures discussed previously. In most cases several codes were used to insure, to as great a degree as possible, that numerical stability issues were eliminated. These benchmark solutions enable modelers and code developers alike to gauge model development progress. Moreover they provide code developers with a means to validate model implementation.

Four models were selected in their study: the Launder-Sharma $k - \epsilon$ model[25], the Wilcox $k - \omega$ model[26], the Spalart-Allmaras ν_t model[30] and the Menter SST model[21]. Some example results follow.

Driver's adverse pressure gradient boundary layer. Figure 10 shows comparisons of the pressure and skin friction coefficients. With the exception of the $k - \epsilon$ model, all models predict flow separation. Overall the SST model provides the best solution.

Bachalo-Johnson axisymmetric bump. Figure 11 shows the comparison of the pressure coefficients along the surface of the axisymmetric bump. Both $k - \epsilon$ and $k - \omega$ models predict shock positions too far downstream as a result of underpredicting the size of the separation bubble. The SST and Spalart-Allmaras models show better agreement.

Transonic airfoil - RAE2822. The comparison of the pressure and skin friction coefficients are shown in figure 12. Again, both the $k - \epsilon$ and $k - \omega$ models predict the shock position too far downstream. All the models overpredict the pressure recovery toward the trailing edge.

From these few examples it is apparent that a code developer would probably not want to use either the $k - \epsilon$ or $k - \omega$ models in transonic flow applications.

3.4 Turbulence Modeling Technology Transfer

The last critical element of the process is the transfer of technology from modelers to code developers and application engineers. Historically, this has been the weak link in the process. The left diagram in figure 13 depicts the flow of modeling technology for the typical development paradigm. The lower box and triangle represent the suite of test cases used to demonstrate model performance. The upper box and inverted triangle represent the more complex engineering applications. The gap between causes a delay in the transfer because of the uncertainty in knowledge of how models will perform in more complex applications and the numerical impact they might have on the performance of application codes.

A new paradigm, depicted in the diagram on the right, is needed. An overlap must be created to affect timely transfer. It requires close cooperation between modelers, code developers and application engineers. For example, modelers will need to cooperate in extending evaluations to more complex flows and to adhere to the standards discussed above; code developers will need to participate to some degree in the evaluation by performing some of the test flows with their own codes to insure models have been implemented properly. Code developers and application engineers will need to work with modelers to break down their very complex applications conceptually into more manageable generic flow representations in order to create a broader modeling standard data base. Creating the overlap will also provide a better feedback mechanism on model performance in complex flows that can spur timely improvements.

Indeed, the pace at which "successful" models are implemented in application codes has accelerated in the few instances where this new paradigm is beginning. One such example was the formation of a Turbulence Modeling Integration Team at Ames Research Center whose focus was directed toward improving modeling for high lift applications using the INS-2D code. Menter[21] was instrumental in developing his own model and assessing its performance along with other models by using many of the standard test cases recommended above. It is noteworthy that he used the basic algorithm in the INS-2D code to develop his model and make his assessments. As a result, Rogers[84] was able to make use of all his work almost immediately by transferring the model subroutines to the full INS-2D code. Feedback on model performance was almost immediate. The first publication dates for Menter's model and Rogers' benchmark high lift computations were separated by only six months (Compare this with the difference of several years in dates for the development and application of the Johnson-King model in the benchmark wing calculation shown in figure 4).

Additional information was collected during the preparation of this paper to further show the importance of the cooperative aspect of model development. In table 2, a list of turbulence models available in several NASA application codes is provided. As a baseline in all these codes, the Baldwin-Lomax model is available. What is interesting is that the best of the newly developed 1- and 2-equation models were incorporated with the assistance of modelers. Some examples of benchmark calculations using the these models are presented next.

The first example involves the calculation of a McDonnell-Douglas multi-element airfoil using INS-2D[84]. The test data were the focus of a NASA CFD Challenge Workshop on high-lift hosted at NASA Langley. Lift curve variation with angle of attack to values near maximum lift were predicted quite well by the new models. Figure 14 shows a comparison of the calculations of pressure coefficient at a Reynolds number of 9 million and 21 degrees angle of attack using the two models that performed the best against the standard data base. Both the Spalart-Allmaras and SST models provide good predictions. The calculations are not as good at and beyond maximum lift and that information feedback is spurring work on model improvements.

Another example is the calculation of the flow over a military F/A-18 E/F wing section model[85]. A major difference between the commercial and the military airfoils is that the military airfoil stalls at a markedly lower angle of attack ($\approx$ 3 degrees) due to leading edge flow separation. Figure 15 shows a comparison of the calculations and experiment for the pressure coefficient at a Reynolds number of 16 million and 4 degrees angle of attack. Again, calculations with the two models match the experimental data.

Benchmark calculations[86] of a transport wing mounted on a circular cross section fuselage, have been made with the NASA TLNS 3-D code (a derivative of CFL3D) as part of a code validation program. Figure 16 shows comparisons with the measured pressure coefficients at two spanwise locations. The calculations using the Spalart-Allmaras and SST models show better agreement in the shock region compared with those using the Baldwin-Lomax model. Better predictions of lift and moment coefficients are realized as a consequence.

4 Future Outlook

The evolution of turbulence modeling improvement requires substantial and synergistic interaction among CFD developers, modelers and experimentalists. It was shown above that the pace of development could be accelerated through a coordinated effort. What needs to be done in the future is to expand the concept and open it to greater participation. An excellent vehicle for accomplishing this is to establish a turbulence model information system data base, accessible on the internet through the World Wide Web. The information system would provide streamlined access to the information needed to gage development of turbulence models, to correctly implement them into application codes and to provide feedback for initiating timely improvements.

Figure 17 presents a conceptual sketch of a possible information system. The inner shell is the heart. It would contain such information as model descriptions, numerical strategies, standard data bases, and standard numerical solutions showing comparisons with the data base. As a dynamic and evolving system it could be continually updated as developments warrant. The outer shell is comprised of CFD'ers, modelers and experimentalists. Individually or in teams they would have streamlined access and coordination channels through the Web. For example, modeler's could access the standard benchmark test cases to measure their improvements; experimentalists could determine gaps in the data base and propose and perform additional experiments from a more informed perspective; and CFD developers and application engineers would have the opportunity to make informed choices of models and validate their implementation. Teams could be formed, formally or informally, as a means of accelerating development for specific applications. Geographical and organizational boundaries would not have to be considered but access to some of the information could be controlled, if necessary.

An oversight filter for the heart of the system would have to be provided in order to avoid proliferation of unchecked information. This could be accomplished by forming an oversight committee comprised of members from academia, industry and research laboratories. The committee could meet at regular intervals to evaluate progress, to discuss application needs, to concur on updating the experimental and numerical data bases, and to make informed assessments on the current state-of-the-art.

With such a system in place, the future prospects for providing timely, accurate RANS codes for engineering applications are promising.

5   Concluding Remarks

Turbulence modeling is critical to the development of accurate RANS CFD codes. Progress in model improvement has been evolutionary over the past few decades. It is now possible to predict pressure distributions in aerodynamic applications with confidence using mixing length eddy viscosity models so long as the pressure gradients are small and shock waves are weak because these models give correct viscous displacement effects. Skin friction and heat transfer can also be predicted to engineering accuracy's (5%), but attention must be given to grid refinement and free stream boundary condition influence on the model choice. The situation is quite different for applications involving strong pressure gradient, strong shock waves and separation. When the flow is either subsonic or supersonic and any separation regions are small, eddy viscosity models, both mixing length and 1- and 2-equation models, can also provide adequate predictions of pressures because displacement effects are properly computed. Eddy viscosity models must be modified to properly account for shock location and attendant small separation for transonic flows

before pressures can be predicted because the displacement effects which have a first order influence are not correct. The model of Johnson and King is an example of an improvement to mixing length models to handle this situation. No definitive conclusions regarding the ability to predict skin friction and heat transfer in these applications, or for that matter even pressures and separation extent for applications involving the strong interactions, can be given. Progress toward this end is being made by improving eddy viscosity models and moving toward higher-order model closures. Indeed, prospects are good. However, these improvements are not always recognized by the CFD developers, their pace appears too slow, and perhaps more importantly, their timely introduction into application codes is not being realized.

Four critical elements of the process for model development were examined in detail and systematic procedures were proposed to remedy the situation. The first involved insurance that adequate numerical procedures were followed to establish such things as source term stability, grid refinement, and free stream boundary condition requirements. The second involved the creation of a standard data base to provide benchmarks for assessing progress. The third involved the creation of standard numerical solutions to insure accurate utilization of models in application codes. The fourth involved following a systematic mechanism for transfer of "successful" models into the application codes. All of these elements require renewed cooperation between model and code developers. In a small number of instances where these procedures are followed, model improvements were implemented in application codes in a timely fashion. Moreover, the feedback from benchmark application computations has spurred modelers to search for new improvements.

In the future, these procedures need to be refined and opened to greater participation among the modeling research and application communities. A logical path to this end can be realized using a turbulence information system data base. The heart of the system would contain such information as model descriptions, numerical strategies for solving the model equations, standard data bases, and standard numerical solutions. It would be a "living" system with filtered, continuous updating. Access to the system could be easily developed over the internet through the World Wide Web. Such an approach holds great promise for the future prospects of turbulence modeling.

Acknowledgements

The authors wish to thank members of NASA Turbulence Modeling Standards Committee for their valuable inputs: T. J. Coakley (ARC), D. A. Driver (ARC), T. Gatski (LaRC), N. Mansour (ARC), R. Mankbadi (LeRC), S. K. Robinson (LaRC), T.-H. Shih (CMOTT-ICMOP, LeRC), R. Cosner (MDA-E), M. Sindir (Rocketdyne-Rockwell), B. Smith (Lockheed), P. Spalart (Boeing), P. Bradshaw (Stanford U.) and C. Speziale (Boston U.). We also wish to thank D. A. Johnson for pointing out that Menter's SST and the non-linear eddy viscosity models can be usefully compared when plotting c_μ against S^*. Special thanks go to: Drs. S. E. Rogers, N. Georgiadis and C. L. Rumsey for providing us information on INS2D/3D, NPARC and CFL3D, respectively. Finally, we would like to thank Drs. T. J. Coakley, M. L. Merriam and Prof. P. Bradshaw for their comments and suggestions.

References

1. Kline, S. J., Markovin, M. V., Sovran, G. and Cockrell, D. J., editors, *Proceeding of Computation of Turbulent Boundary Layers - 1968 AFOSR-IFP-Stanford Conference*, Vol. 1 & 2, Thermoscience Division, Dept. of Mech. Engrg, Stanford University, 1969.

2. Bertram, M. H., editor, *Compressible Turbulent Boundary Layers*, NASA SP-216, 1969.

3. *Free Turbulent Shear Flows*, Vol. 1 & 2, NASA SP-321, 1973.

4. Kline, S. J., Cantwell, B. J. and Lilley, G. M., editors, *Proceeding of the 1980-1981 AFOSR-HTTM-Stanford Conference on Complex Turbulent Flows: Comparison of Computation and Experiment*, Vol. 1-3, Thermoscience Division, Dept. of Mech. Engrg, Stanford University, 1981.

5. Bradshaw, P., Launder B. E. and Lumley, J. L., JFE, in press, 1996.

6. Strazisar, A. J. and Denton, J. D., Global Gas Turbine News, IGTI, May/June, 1995, pp. 12-14.

7. Launder, B. E., Reece, G. J. and Rodi, W., J. Fluid Mech., vol 68, part 3, pp. 537-566, 1975.

8. Shih, T.-H. and Lumley, J. L., Tech. Rep. FDA-85-3, Cornell University, 1985.

9. Fu, S., Launder, B. E. and Tselepidakis, D. P., Rep TFD/87/5, Mech. Eng'rg Dept., UMIST, 1987.

10. Speziale, C. G., Sarkar, S. and Gatski, T. B., J. Fluid Mech., Vol. 227, pp. 245-272, 1991.

11. Speziale, C. G., Chapter 5, in *Simulation and Modeling of Turbulent Flow*, Ed. Gatski, T. B., Hussaini, M. Y. and Lumley, J. L, Oxford U. Press, 1996.

12. Gatski, T., Chapter 6, in *Handbook of Computational Fluid Mechanics*, ed. R. Peyret, Academic Press Ltd, 1996.

13. Leschziner, M. A., XXVI IAHR Congress "Hydra 200", London, Sept, 1995.

14. Shih, T.-H., NASA CR-198458, 1996.

15. Bradshaw, P., Engineering Foundation Conference on Turbulent Heat Transfer, San Diego, March 10-15, 1996.

16. Pope, S. B., J. Fluid Mech., 72, pp. 331-340, 1975.

17. Rodi, W., Z. Angew. Math Mech., 56, T219-T221, 1976.

18. Gatski, T. B. and Speziale, C. G., J. Fluid Mech., 254, pp. 59-78, 1993.

19. Shih, T.-H., Zhu, J. and Lumley, J. L., Comput. Methods Appl. Mech. Engrg, Vol 125, pp. 287-302, 1995.

20. Craft, T. J., Launder, B. E., and Suga, K, In *Proc. 5th Int. Symp. on Refined Flow Modelling and Turbulence Measurements*, p. 125, Presses Ponts et Chaussées, Paris, 1993.

21. Menter, F. R., AIAA J., Vol. 32, No. 8, pp.1598-1605, 1994.

22. Abid, R., Rumsey, C. and Gatski, T, AIAA J., Vol. 33, No. 11, pp. 2026-2031, 1995.

23. Abid, R., Rumsey, C. and Gatski, T, AIAA 96-0565, 1996.

24. Coakley, T. J., NASA TM-88333, 1986.

25. Launder, B. E. and Sharma, B. I., Letters in Heat and Mass Transfer, 1, pp. 131-138, 1974.

26. Wilcox, D. C., AIAA J., Vol. 26, No. 11, pp. 1299-1310, 1988.

27. Smith, B. R., AIAA-94-2386, 1994.

28. Speziale, C. G., Abid, R. and Anderson, E. C., AIAA J., Vol. 30, No. 2, pp 324-331, 1992.

29. Baldwin, B. S. and Barth, T. J., NASA TM 102847, 1990.

30. Spalart, P. R. and Allmaras, S. R., La Recherche Aèrospatiale, 1, pp. 5-21, 1994.

31. Shur, M., Strelets, M., Zaikov, L., Gulyaev, A., Kozlov, V. and Secundov, A., AIAA 95-0863, 1995.

32. Baldwin, B. S. and Lomax, H., AIAA-78-257, 1978.

33. Cebeci, T. and Smith, A. M. O., *Analysis of Turbulent Boundary Layer*, Series in Appl. Math. & Mech., Vol. XV, Academic Press, 1974.

34. Johnson, D. A. and King, L. S., AIAA J., Vol. 23, No. 11, pp. 1684-1692, 1985.

35. Johnson, D. A., AIAA J., Vol. 25, No. 2, pp. 252-259, 1987.

36. Yu, N., AIAA-92-2651, 1992.

37. Launder, B. E. and Spalding, D. B., Comput. Methods in Appl. Mech. Eng., Vol. 3, pp. 269-289, 1974.

38. Chieng, C. C. and Launder, B. E., Numerical Heat Transfer, Vol. 3, pp. 189-207, 1980.

39. Johnson, R. W. and Launder, B. E., Numerical Heat Transfer, Vol. 5 pp. 493-496.

40. Viegas, J. R. and Rubesin, M. W., AIAA 85-0180, 1985.

41. Huang, P. G. and Coakley, T. J., *Engineering Turbulence Modeling and Experiments 2*, W. Rodi and F. Martelli, editors, Elsevier Science Publishers B. V., pp. 731-739, 1993.

42. Iacovides, H. and Launder, B. E., ASME 90-GT-24, ASME Int. Gas Turb. Congress, Brussels, 1990.

43. Rodi, W., AIAA-91-0216, 1991.

44. Horstman, C. C., AIAA J., Vol. 30, No. 6, pp. 1480-1481, 1992.

45. Lien F. S. and Leschziner, M. A., The Aeronautical Journal, 99, pp. 125-144, 1995.

46. Abou Haidar, N. I., Iacovides, H. and Launder, B. E., in *77th Symp. of the Propulsion and Energetics Panel on CFD Techniques for Propulsion applications*, San Antonian , TX, 1991.

47. Chien, K. Y., AIAA J., Vol. 20, No. 1, pp. 33-38, 1982.

48. Lien F. S. and Leschziner, M. A., *Engineering Turbulence Modeling and Experiments 2*, W. Rodi and F. Martelli, editors, Elsevier Science Publishers B. V., pp. 217-228, 1993.

49. Durbin, P. A., Theoretical and Computational Fluid Dynamics, Vol. 3, No. 1, pp. 1-13, 1991.

50. Durbin, P. A., AIAA J., Vol. 33, No. 4, pp. 659-664, 1995.

51. Launder, B. E. and Shima, N., AIAA J., Vol. 27, No. 10, pp. 1319-1325, 1989.

52. Lai, Y. G. and So, R. M. C., J. Fluid Mechanics, Vol. 221, pp. 641-673, 1990.

53. So, R. M. C., Lai, Y. G. and Hwang, B. C., AIAA J., Vol. 29, No. 8, pp. 1202-1213, 1991.

54. Craft, T. J. and Launder, B. E., Turbulent Shear Flows 10, pp 20-25 – 20-30, 1995.

55. Jakirlič,S. and Hanjalič, K., Turbulent Shear Flows 10, pp 23-25 – 23-30, 1995.

56. Durbin, P., J. Fluid Mech., Vol. 249, pp. 465-498, 1993.

57. Huang, P. G., Coleman, G. N. and Bradshaw, P., J. Fluid Mech., Vol. 305, pp. 185-218, 1995.

58. Zeman, O., Phys. Fluids A 2(2), pp. 178-188, 1990.

59. Sarkar, S. Erlebacher G., Hussaini, M. Y. and Kreiss, H. O., J. Fluid Mech., Vol. 227, pp. 473-493, 1989.

60. Viegas, J. R. and Rubesin, M. W., AIAA J., Vol. 30, No. 10, pp. 2369-2379, 1992.

61. Baz, A. AM. EL and Launder, B. E., *Engineering Turbulence Modelling and Experiments* (ed W. Rodi and F. Martelli), Elsevier, 1993.

62. Sarkar, S. Erlebacher, G. and Hussaini, M. Y., *Turbulent Shear Flows* (ed. F. Durst *et. al.*), Springer, 1992.

63. Zeman, O., AIAA 93-0897, 1993.

64. Huang, P. G., Bradshaw, P. and Coakley, T. J., AIAA J., Vol. 31, pp. 1600-1604, 1993.

65. Coakley, T. J. and Huang, P. G., AIAA 92-0436, 1992.

66. Huang, P. G. and Coakley, T. J., AIAA 93-0200, 1993.

67. Bardina, J. E., Huang, P. G. and Coakley, T. J., to appear in a NASA TM, 1996.

68. Menter, F. R., AIAA J., Vol. 30, No. 6, pp. 1657-1659, 1992.

69. Huang, P. G., Schwarz, W. R. and Bradshaw, P., Internal Report, Dept. of Mech. Engrg., Stanford University, June, 1990.

70. Roger, S. E. and Kwak, D., NASA TM-103911, 1992.

71. Huang, P. G. and Coakley, T. J., AIAA 92-0547, 1992.

72. Driver, D. M., AIAA 91-1787, 1991.

73. Berg, B. van den, Elsenaar, A., Lindhout, J. P. F. and Wesseling, P., JFM 70, 127, 1975

74. Bradshaw, P. and Pontikos, N. S., JFM, 159, pp. 105-130, 1985.

75. Bachalo, W. D. and Johnson, D. A., AIAA J., Vol. 24, pp. 437-443, 1986.

76. Cook, P., Mcdonald, M. and Firmin, M., AGARD AR-138, 1979.

77. Bradshaw, P., in *1980-81 AFOSR-HTTM-Stanford Conference on Complex Turbulent Flows*, ed. by S. J. Kline, B. J. Cantwell and G. M. Lilley, Stanford University, Stanford, California, Vol. 1 pp. 364-368, 1981.

78. Davis, D. O. and Gessner, F. B., AIAA J., 30(2), pp. 367-375, 1992.

79. Johnson, P. L. and Johnston, J. P., Stanford Rept MD-53, 1989.

80. Alving, A.E., Smits, A. J. and Watmuff, J. H. JFM 211, pp. 529-556, 1990.

81. Driver, D. M. and Seegmiller, H. L., AIAA J., Vol. 23, No. 2, pp. 163-171, 1985.

82. Mateer, G. G., Seegmiller, H. L., Hand, L. A. and Szodruch, J., NASA TM 103933, 1992.

83. Wideman, J. K., Brown, J. L., Miles, J. B. and Özcan, O., NASA TM 108824, 1994.

84. Rogers, S. E., Menter, F. R., Durbin, P. A. and Mansour, N. N., AIAA-94-0291, 1994.

85. Kern, S. B., AIAA-96-0057, 1996.

86. Marconi, F., Siclari, M., Carpenter, G. and Chow, R., AIAA-94-2237, 1994.

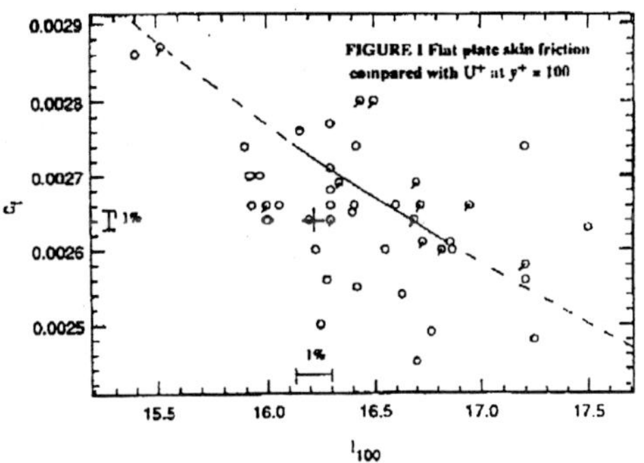

Figure 1: Flat plate skin friction compared with U^+ at $y=100$.

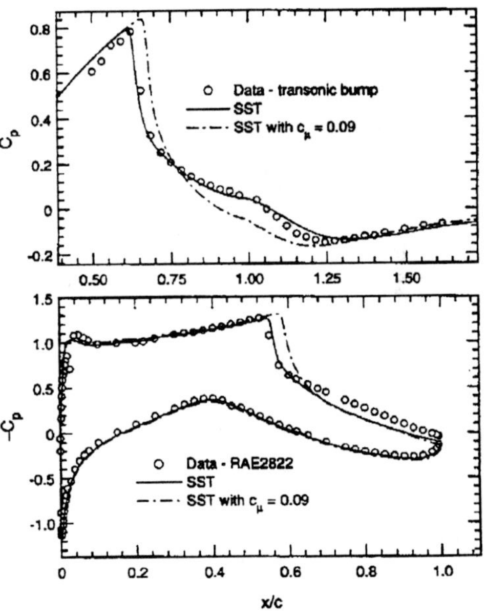

Figure 2: pressure coefficients along the surface of (a) the transonic bump and (b) the RAE 2822 CASE 10.

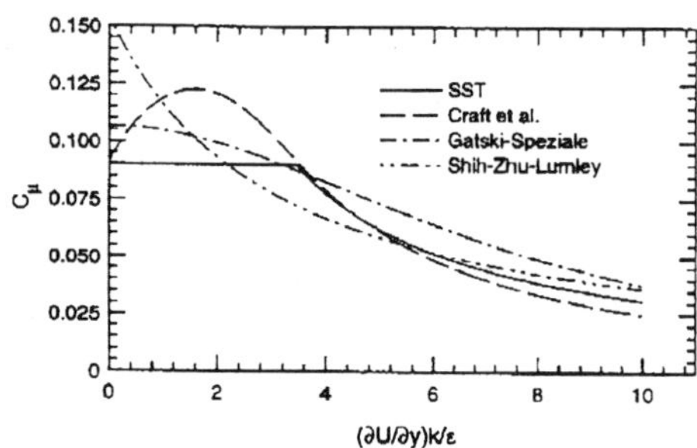

Figure 3: Variation of c_μ with S^*.

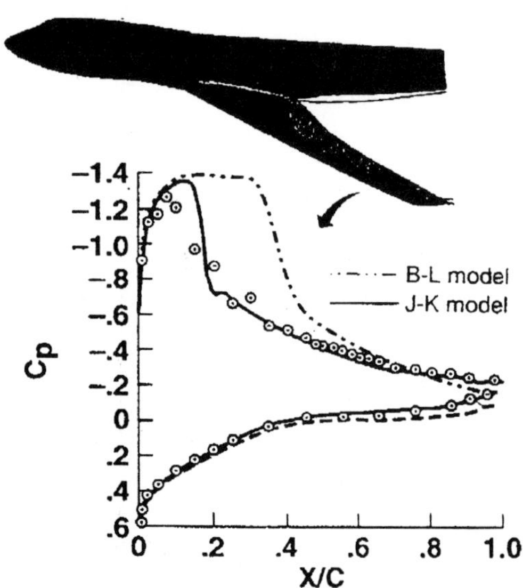

Figure 4: Improved turbulence model for a separated flow calculation of a 747-200 aircraft, $M_\infty = 0.85, \alpha = 5.7^\circ$.

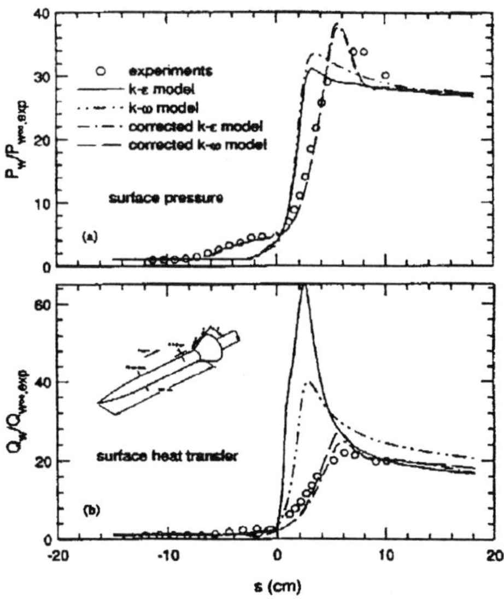

Figure 5: Compressible modifications - prediction of surface pressure and heat transfer of a hypersonic flow ($M = 7$) over 35° cylinder-flare.

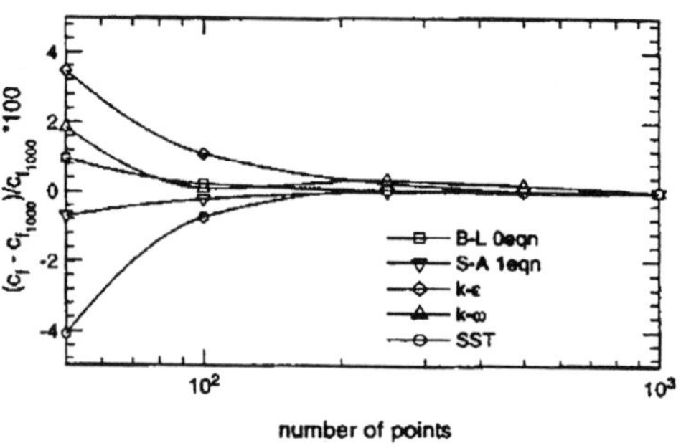

Figure 6: Sensitivity of predicted skin friction on a flat plate to grid refinement.

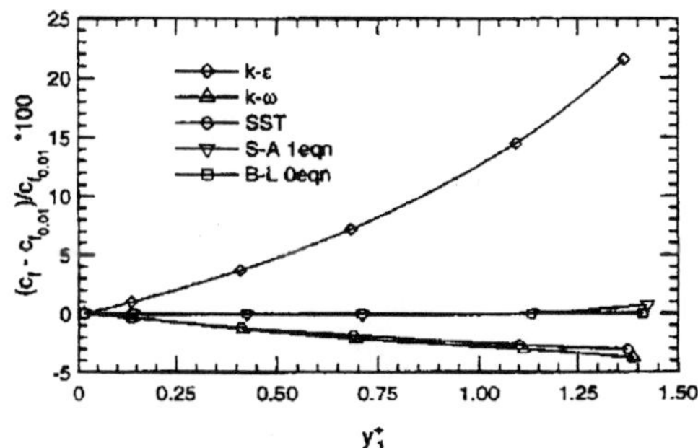

Figure 7: Sensitivity of predicted skin friction on a flat plate to the value of y_1^+.

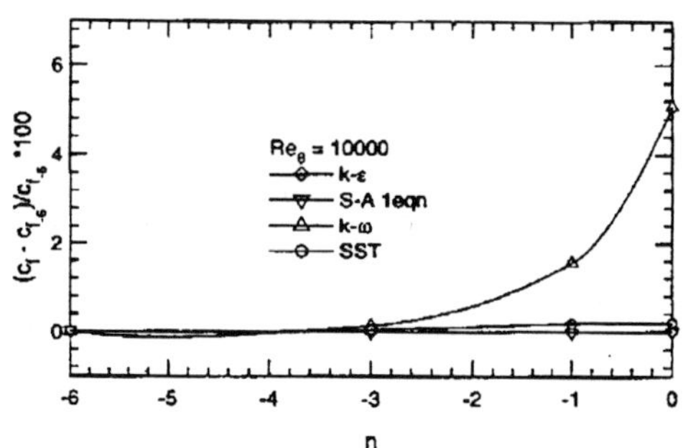

Figure 8: Sensitivity of predicted skin friction on a flat plate to free stream boundary conditions.

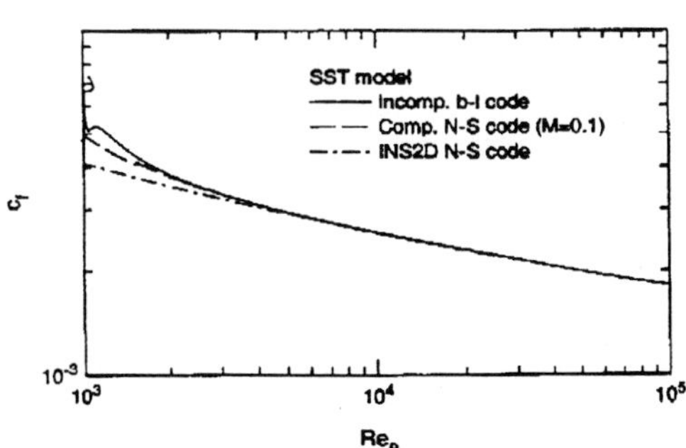

Figure 9: Code-to-code comparison of predicted skin friction on a flat plate.

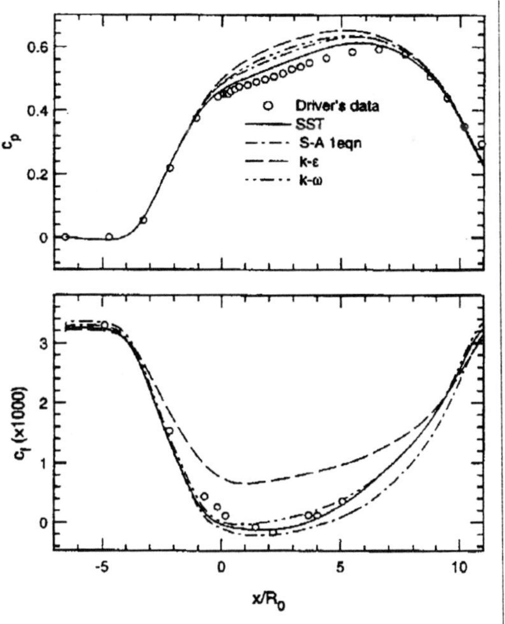

Figure 10: Comparison of surface pressure and skin friction coefficients: Driver's adverse pressure gradient flow.

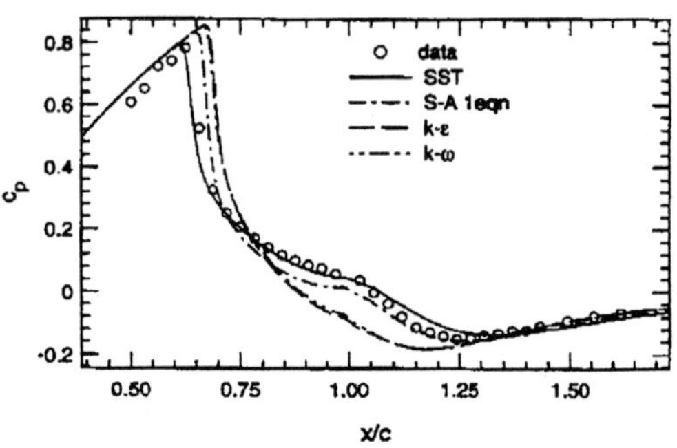

Figure 11: Comparison of surface pressure coefficient: Transonic bump flow: $M = 0.875$ and $Re/L = 13.1 \times 10^6/m$.

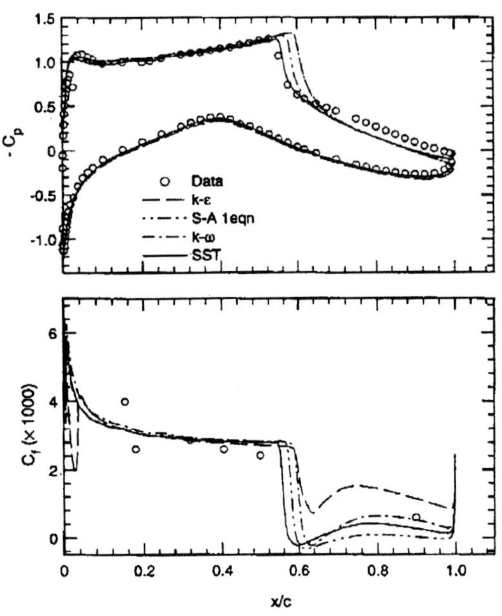

Figure 12: Comparison of surface pressure and skin friction coefficients: RAE2822, Case10.

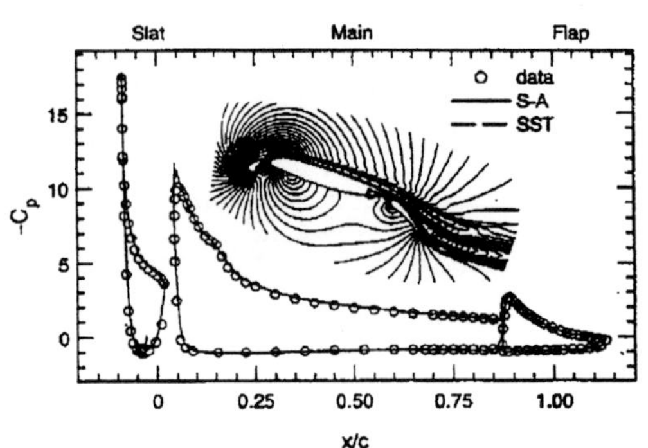

Figure 14: Prediction of surface pressure coefficient on a McDonnell-Douglas multi-element airfoil. $Re = 9 \times 10^6$ and $\alpha = 21^o$.

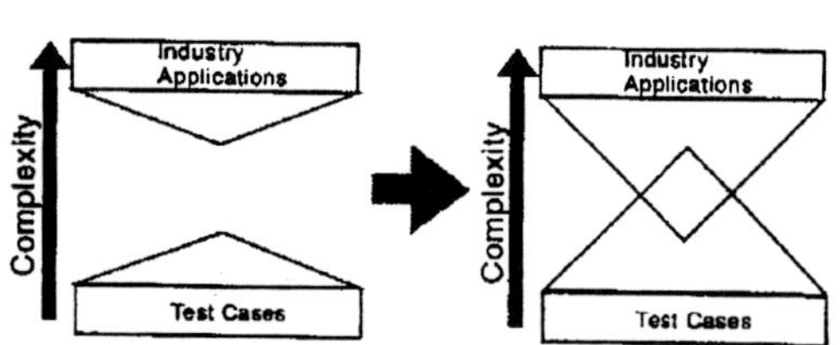

Figure 13: Old and new paradigms used in turbulence model development.

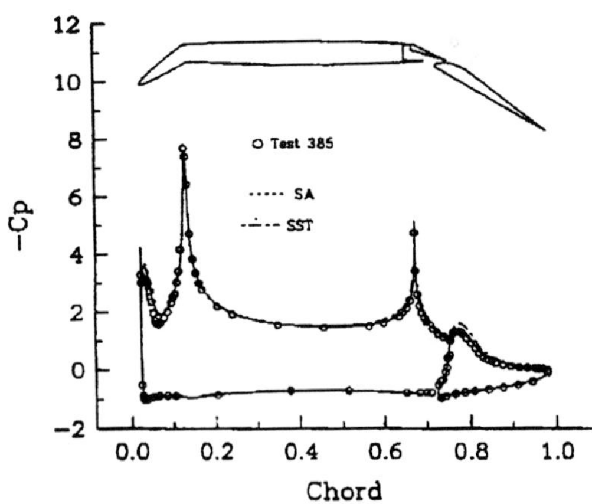

Figure 15: Prediction of surface pressure coefficient on a F/A-18E/F multi-element airfoil. $Re = 12 \times 10^6$ and $\alpha = 4^o$.

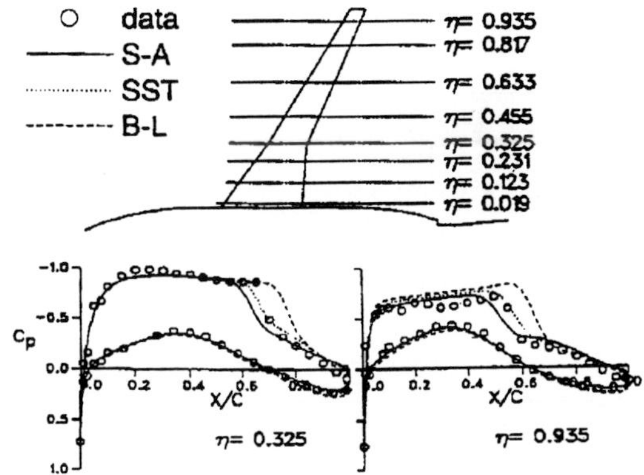

Figure 16: Prediction of surface pressure coefficient on a transport wing/simple body configuration. $M_\infty = .88$ and and $\alpha = .47^\circ$.

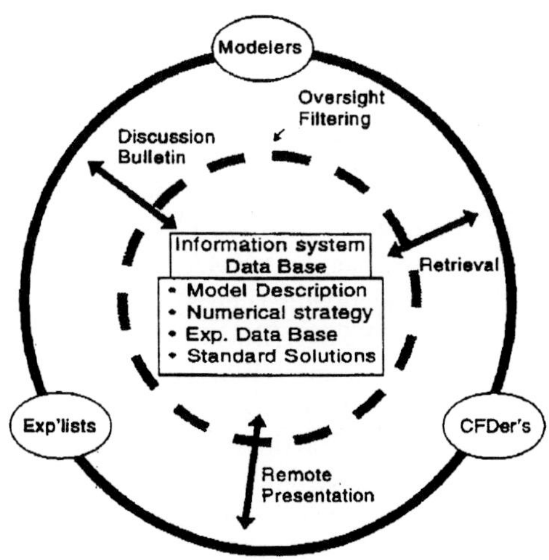

Figure 17: A conceptual information system for turbulence modeling.

	Subsonic, incompressible	*Transonic*	*Supersonic*
Priority cases	• Flat plate • APG boundary layer[72] • Self similar mixing layer • Self similar round jet • Self similar plane jet • Self similar plane wake • Infinite yawed wing[73,74]	• Axisymmetric bump[75] • RAE2822 airfoil[76]	• Flat plate $M = 5$ • Compressible mixing layer[77]
Optional cases	• 3-D boundary layer in transition duct[78] • Concave and convex curvatures[79,80] • Backward-facing step[81]	• MBB airfoil[82]	• $M = 2.9$ incipient separation[83]

Table 1: Recommended test cases for external aerodynamic flows.

Code	*Turbulence Models Available*			*Supporting Modelers*
	0-eqn	1-eqn	2-eqn	
NPARC (Lewis)	B-L P.D.Thomas Ahn's RNG	B-B S-A	Chien $k - \epsilon$ Shih-Lumley $k - \epsilon$ $k - \omega$	Shih, Yang & Zhu (CMOTT) Sirbaugh (NYMA) Georgiadis (Lewis)
INS2D/3D (Ames)	B-L	B-B S-A	SST	Spalart(Boeing) Menter(Eloret) Rogers(Ames)
CFL3D (Langley)	B-L	B-B S-A	$k - \omega$ SST	Johnson & Barth(Ames) Menter(Eloret) Gatski & Rumsey(Langley) Abid(HTC)

Table 2: Turbulence Models Available in NASA CFD Codes.

TAMING TURBULENCE

John KIM

Mechanical and Aerospace Engineering Department,
University of California; Los Angeles, CA 90095-1597

Abstract

A brief overview of the recent progress achieved by the author's research group on turbulence control for drag reduction in turbulent boundary layers is given. Control schemes are developed based on the premise that the most effective way to control turbulent boundary layers is through proper manipulation of the near-wall streamwise vortices. Two different methods — neural networks and suboptimal control theory — are utilized to achieve this goal. The resulting feedback laws are applied to a low Reynolds-number turbulent channel flow. Numerical experiments indicate that both approaches yield substantial drag reduction. The application of a system theory approach to the feedback stabilization to delay the transition in a laminar channel flow is also discussed. It is shown that the system theory approach is a valuable tool both for designing control systems to stabilize the flow as well as for understanding the physics of controlled transitional flows.

1 Introduction

Turbulence is the most common fluid motion encountered in daily life, but it was not until the historic demonstration by Osborne Reynolds reported in 1883 that investigators started looking carefully at the nature of turbulence. Since then, turbulence has attracted the attention of some of the best minds in engineering, physics, and mathematics because of its intellectual challenges as well as its importance in many practical engineering applications. Despite all of the attention, however, turbulence has been incredibly resistant to studies that would reveal its nature, and progress has been painfully slow. However, examples of the utilization of available knowledge of turbulence are abundant. From dimples in golf balls to vortex generators on commercial airplane wings, the designers' endeavors to control turbulence in order to achieve optimal results are evident. Most previous flow-control research, however, has been ad-hoc and based on the investigator's physical intuition. Only recently have systematic approaches to flow control appeared in the literature.

The potential payoffs of managing and controlling turbulent flows that occur in various engineering applications are significant. For example, it is estimated that a 10% reduction in fuel cost of commercial aircraft would yield a 40% increase in the profit margin of an airline [1]. Understanding the transition process from laminar to turbulent flow is also a critical element in designing aerodynamically acceptable shapes.

The purpose of this paper is to give a brief overview of the recent progress on boundary layer control that

has been made by the author's research group at UCLA. We have been using nontraditional approaches to explore new methods for turbulence control. Three examples are given in this paper, each of which is an excerpt of a full paper. The reader is referred to these papers for detailed descriptions of the particular approach and results. It should be also mentioned that many research efforts for turbulence control are currently being carried out, but this paper is not intended to review them. Two such efforts worth mentioning are those by the group at Stanford University, led by Professor Moin, and the other by the group at Cornell University, led by Professor Lumley. The reader is referred to Refs. [2] and [3], respectively, for alternative perspectives on the present topic.

This paper is organized as follows. A brief review of relevant turbulence structures in turbulent boundary layers is given in Sec. 2. Sections 3–5 each present a different example of boundary layer control, followed by concluding remarks in Sec. 6.

In this paper, I shall use (x, y, z) for the streamwise, wall-normal, and spanwise coordinates, respectively, and (u, v, w) for the corresponding velocity components.

2 Near-wall turbulence structures

Before presenting our recent work on boundary layer control, a brief review of near-wall turbulence structures, which motivates all of our control approaches described herein, is given.

Turbulence in wall-bounded shear flows is strongly affected by the near-wall region. Most production and dissipation of the kinetic energy take place in the buffer layer. The ubiquitous structural features in this region are low- and high-speed "streaks," which consist mostly of a spanwise modulation of the streamwise velocity, and streamwise vortices. Although we have learned a

Received on August 17, 1997.
Presented by invitation at the 28th AIAA Fluid Dynamics Conference, Snowmass Village, Colorado, June 29-July 2, 1997; AIAA Paper 97-1791.

great deal about their kinematics, very little is known or agreed upon about the *dynamics* of streaks and streamwise vortices in spite of much effort and attention they have attracted during the past three decades (*e.g.*, see Ref. [4] for a review). There is no consensus, for example, about how such structures are generated and sustained nor about what determines their well-known scales. Hamilton *et al.* [5] and Waleffe [6] have proposed a dynamical model, in which streaks and streamwise vortices are maintained by a self-sustaining process. Jiménez and Pinelli [7] propose a similar regeneration mechanism based on their clever numerical experiments. Many other position papers by active investigators will appear on this particular issue in a forthcoming research monograph [8].

At present, our knowledge of near-wall structures is primarily limited to kinematic aspects: the streaks have a characteristic average spacing of about 100 wall units [1] ([9], [10], [11]), while the streamwise vortices are observed to have a characteristic average diameter of about 20-40 wall units [11]. The streaks are often linked to a sequence of events called the "bursting process," in which they move away from the wall, oscillate, and break up. The bursting process is believed to be the essence of the turbulence production mechanism [12]. The streaks have also been linked to the genesis of "horseshoe" vortices through a Kelvin-Helmholtz type roll-up (*e.g.*, Ref. [13]). Another interpretation of the streaks, one to which we subscribe, is that they are the wakes left behind the streamwise vortices as the latter are advected downstream at a speed close to the local mean flow [14].

In line with this interpretation, Kim [14] has proposed the near-wall streamwise vortex as the most relevant turbulence structure from the perspective of drag reduction in turbulent boundary layers. This point of view is supported by the observation that streamwise vortices have been found to be responsible for both "ejection" and "sweep" events of the bursting process [4]. Recent studies have also shown that the high skin-friction regions in turbulent boundary layers are closely related to the near-wall streamwise vortices [15], [16]. These regions of high skin friction on the wall are created by the inrush of high-speed fluid toward the wall ("sweep"), which is induced by a strong streamwise vortex. Choi *et al.* [16] showed that a significant drag reduction is possible when the surface boundary condition is modified to suppress the near-wall streamwise vortices. They also relate the drag-reducing mechanism on a riblet surface to the restricted interaction between the riblet and streamwise vortices [17].

Our work on boundary-layer control is therefore based on the premise that the most effective way to control turbulent boundary layers for drag reduction is through proper manipulation of the near-wall streamwise vortices. In our numerical experiments, we use surface blowing and suction as the control input, but other approaches could be used as well. For instance, in other applications, we have also used a wall-normal electromagnetic force as the control input, which resulted in similar effects. All numerical results discussed in this paper have been obtained by direct numerical simulations of turbulent and transitional channel flows using the numerical methods reported in Kim *et al.* [11].

3 Application of a neural network to drag reduction

Choi *et al.* [16] used blowing and suction at the wall equal and opposite to the wall-normal component of velocity at $y^+ = 10$. They showed that this control effectively mitigated the streamwise vortices, giving approximately 25% drag reduction in a turbulent channel flow. Although the method employed in their work is impractical, since the information at $y^+ = 10$ is usually not available, it demonstrates that a control scheme that reduces the skin-friction by manipulating the near-wall streamwise vortices.

The objective of the work reported in Lee *et al.* [18] is to seek wall actuations, in the form of blowing and suction at the wall, which are dependent only on the wall-shear stress to achieve a substantial drag reduction. This requires knowledge of how the wall-shear stresses respond to wall actuations (*i.e.*, the correlation between the wall-shear stresses and the wall actuations). Because of the complexity of the solutions to the Navier-Stokes equations, however, it is not possible to find such a correlation in closed form or to approximate it in simple form. Instead, we used a neural network to approximate the correlation that then predicts the optimal wall actuation to achieve the minimum of the skin-friction drag. Lee *et al.* [18] describe how neural networks with a small number of shared weights were constructed and trained off-line, and then an on-line control scheme for drag reduction based on that neural network was implemented. A standard two-layer feedforward network with hyperbolic tangent hidden units and a linear output unit was used. The functional form of the final neural network was:

$$v_{jk} = W_a \tanh \left(\sum_{i=-(N-1)/2}^{(N-1)/2} W_i \left. \frac{\partial w}{\partial y} \right|_{j,k+i} - W_b \right) - W_c \, ,$$

$$1 \leq j \leq N_x \quad \text{and} \quad 1 \leq k \leq N_z \, , \tag{1}$$

where W denotes weight, N is the total number of input weights, and the subscripts j and k denote the numerical grid point at the wall in, respectively, the streamwise and spanwise directions. N_x and N_z are the number of computational grid points in each direction. The

---

[1]That is, $100\,\nu/u_\tau$, where ν is the kinematic velocity and $u_\tau = (\nu|dU/dy|_w)^{1/2}$ is the wall-shear velocity.

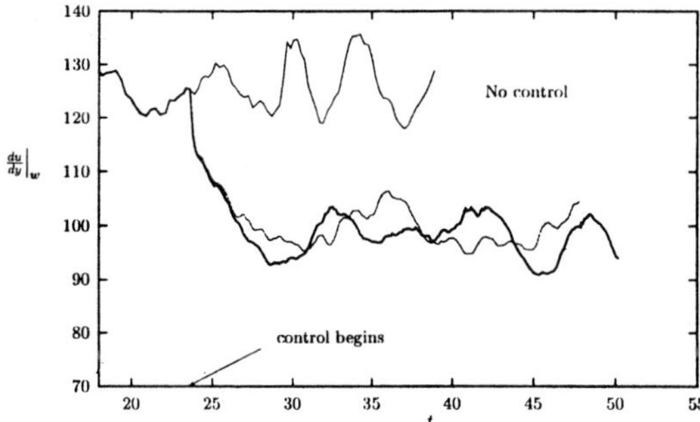

Figure 1: Mean wall-shear stress histories for various control schemes compared to the no-control case: ——— , no control; ——— , on-line control with neural network with 7x1 template; ········ , control with 7 fixed weights.

summation is done over the spanwise direction. Seven neighboring points ($N = 7$), including the point of interest, in the spanwise direction (corresponding to approximately 90 wall units) were found to provide enough information to adequately train and control the near-wall structures responsible for the high skin friction. Note that the input to the neural network is $\partial w/\partial y$ at the wall, not $\partial u/\partial y$. Initially $\partial u/\partial y$ and $\partial w/\partial y$ at the wall at several instances of time were used as input data fields and the actuation at the wall was used for the output data of the network. Experimentally we found that only $\partial w/\partial y$ at the wall from the current time was necessary for sufficient network performance. It should be noted that a brute force application of a neural network, whose architecture and input parameters were designed without consideration for relevant flow physics, led to either no convergence or excessively long training time.

Applying this control scheme to direct numerical simulations of a turbulent channel flow at low Reynolds number resulted in about 20% drag reduction (Fig. 1). The computed flow fields were examined to determine the mechanism by which the drag reduction was achieved. The most salient feature of the controlled case was that the strength of the near-wall streamwise vortices was substantially reduced (Fig. 2). This result further substantiates the notion that a successful suppression of the near-wall streamwise vortices leads to a significant reduction in drag.

An examination of the weight distribution from the on-line neural network led to a very simple control scheme that worked equally well while being computationally more efficient. This simple control scheme indicates that the optimum blowing and suction at the wall should be in the form,

$$v_w \sim \overline{\left.\frac{\partial}{\partial z}\frac{\partial w}{\partial y}\right|_w} , \qquad (2)$$

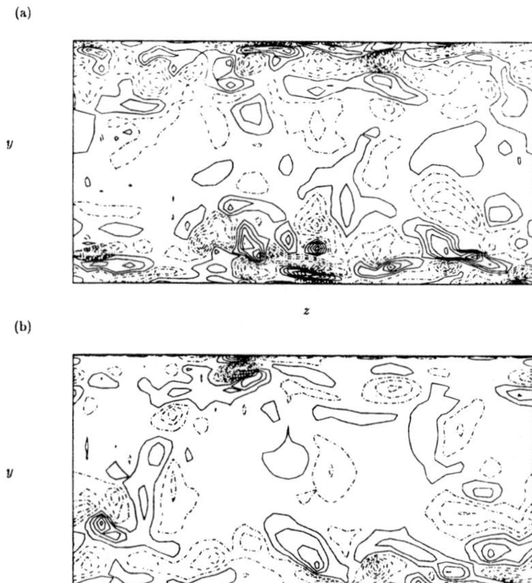

Figure 2: Contours of streamwise vorticity in a cross-flow plane: (a) no control; (b) control using 7 fixed weights. The contour level increment is the same for both figures. Negative contours are chain-dotted.

where the overbar represents a local spatial average with high wavenumber components reduced. The converged weight distribution can be expressed analytically, thus making the implementation of this control scheme relatively easy. This control scheme produces a distribution of wall actuations that are very similar to those produced by the v-control of Choi *et al.* [16].

The simple pattern of the weight distribution found from the nonlinear network suggested the possibility of using a linear network. Although a linear network produced a similar pattern, the pattern was not well preserved in time. The weights varied significantly, which increases the difficulty in circuit implementation since a larger dynamic range is required. The nonlinear network, however, produced bounded weights, simplifying circuit implementation. This suggests that a certain amount of nonlinearity is needed to capture a stable coherent pattern for the turbulent flow.

4 Application of suboptimal control theory to drag reduction

As already mentioned in the introduction, most previous control work has been rather ad-hoc, in that it was mainly based on the investigator's intuition. More systematic approaches using an optimal control theory have appeared recently. Choi *et al.* [19] proposed a "suboptimal" control procedure, in which the iterations required for a global optimal control were avoided by seeking an optimal condition over a short time period. The suboptimal control procedure was successfully applied to control of the Burgers equation. Bewley and Moin [20] were the first to apply the suboptimal control procedure to a turbulent flow and reported a drag reduction of about 17%. The procedure requires full velocity information throughout the flow in order to solve an adjoint problem, from which a feedback control input was then derived. In spite of this obvious drawback, the fact that a control theory applied to a turbulent flow resulted in a substantial drag reduction is encouraging, especially since their control procedure was derived rigorously from a control theory in which a pre-determined cost functional was minimized.

In Lee *et al.* [21] we demonstrated that a wise choice of the cost functional coupled with a variation of the formulation can lead to a more practical control law. We showed how to choose a cost functional and how to minimize it to yield simple feedback control laws that only require quantities at the wall as input. One of the laws requires spatial information on the wall pressure over the entire wall, and the other requires information, also over the entire wall, on one component of the wall shear stress. We then derived more practical control schemes that only require local wall pressure or local surface-shear stress information and showed that they work equally well.

In this approach, the choice of the cost functional to be minimized is critical to the performance of the control. Two different cost functionals were chosen based on our observation of the successful control of Choi *et al.* [16]. As shown in Fig. 3, Choi *et al.*'s blowing and suction, which are equal and opposite to the wall-normal velocity component at $y^+ = 10$, effectively suppress a streamwise vortex by the counteracting up/down motion induced by the vortex. This blowing and suction creates locally high pressure in the near-wall region marked with '+', and low pressure in the region marked with '−', in Fig. 3. A crucial aspect of the present analysis is the observation that this blowing and suction *increases* the pressure gradient in the spanwise direction under the streamwise vortex near the wall. This suggests that we should seek blowing and suction that increases the pressure gradient in the spanwise direction near the wall for a short time period (*i.e.*, in the suboptimal sense) in order to achieve a drag reduction similar to that achieved by Choi *et al.* [16]. The cost functional $\mathcal{J}(\phi)$ to be minimized is then

$$\mathcal{J}(\phi) = \frac{\ell}{2A\Delta t} \int_S \int_t^{t+\Delta t} \phi^2 \, dt \, dS$$
$$- \frac{1}{2A\Delta t} \int_S \int_t^{t+\Delta t} \left(\frac{\partial p}{\partial z}\right)^2_w dt \, dS , \quad (3)$$

where the integrations are over the wall, S, in space and over a short duration in time, Δt, which typically

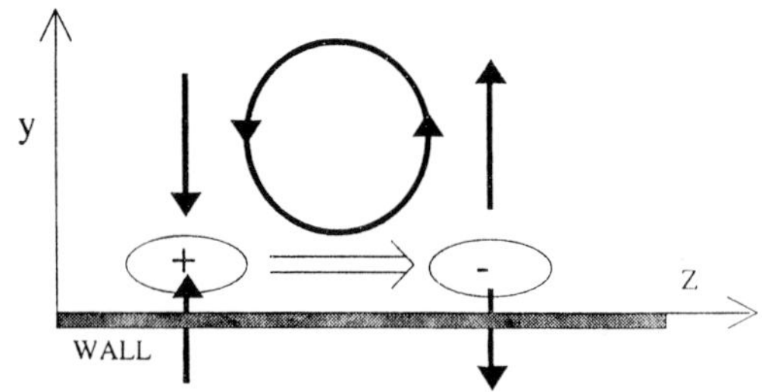

Figure 3: Schematic of a pressure field induced by a control based on $y^+ = 10$.

corresponds to the time step used in the numerical computation, and ℓ is the relative price of the control, since the first term on the right-hand side represents the cost of the actuation, ϕ (*i.e.*, the surface blowing and suction). Note that the minus sign in front of the second term, since we want to *maximize* the pressure gradient. It should be noted that the spanwise pressure gradient at the wall will be reduced eventually when the strength of the near-wall streamwise vortices are reduced through successful control. Here, blowing and suction that increase the spanwise pressure gradient for the next step are sought as a suboptimal control. The suboptimal control procedure described in Lee *et al.* [21] leads to

$$\phi = v_w \sim \overline{\frac{\partial^2 p_w}{\partial z^2}} , \quad (4)$$

where p_w denotes the wall pressure, and the overbar represents a local spatial average with high wavenumber components reduced.

Another wall quantity that indicates similar changes of the near-wall dynamics that are due to the altered pressure field is the spanwise shear at the wall, $\partial w/\partial y$. Because of the added pressure gradient in the spanwise direction below the streamwise vortex, spanwise motion near the wall is also induced, thus increasing the spanwise shear stress at the wall (see Fig. 3). Thus another valid choice of the cost functional to be minimized is

$$\mathcal{J}(\phi) = \frac{\ell}{2A\Delta t} \int_S \int_t^{t+\Delta t} \phi^2 \, dt \, dS$$
$$- \frac{1}{2A\Delta t} \int_S \int_t^{t+\Delta t} \left(\frac{\partial w}{\partial y}\right)^2_w dt \, dS . \quad (5)$$

Following the procedure that led to Eqn. (4) yields the optimum actuation:

$$\phi = v_w \sim \overline{\frac{\partial}{\partial z} \frac{\partial w}{\partial y}}\Big|_w . \quad (6)$$

Equation (6) indicates that the optimum wall actuation should be proportional to a locally averaged spanwise derivative of the spanwise shear at the wall. Note that this result is very similar to that found in the application of a neural network in Sec. 3. The only difference in the two expressions is how the local average should be performed; see Lee *et al.* [21] for details.

Applying Eqns. (4) and (6) to a turbulent channel flow produced, respectively, about 16% and 22% drag reduction [21]. Turbulence characteristics of the controlled flow field are very similar to those obtained in Sec. 3.

5 Systems control-theory approach to transition control

In this section, we consider the problem of stabilizing laminar boundary-layer flows. The goal is to design a feedback control system that stabilizes the flow so that no unstable modes exist in the new system. This is a fundamental change in the way transition control is approached. Instead of allowing unstable modes to appear and then "canceling" them with out-of-phase modes, we develop here a control system that changes the system from being inherently unstable to being inherently stable. In essence, this approach changes the philosophy of the problem from thinking about how inputs can mitigate an inherently unstable system to thinking about how sensors and actuators can be added to form an entirely new *stable* system.

In Joshi *et al.* [22], a system theory framework is presented for the linear stabilization of a laminar channel flow. The linearized Navier-Stokes equations are converted into the state variable representation of the system using a Galerkin method. This representation relies on the fact that the motion of any finite-dimensional dynamic system can be expressed as a set of first-order ordinary differential equations in matrix form:

$$\frac{d\mathbf{x}}{dt} = A\mathbf{x} + B\mathbf{q} \tag{7}$$

$$\mathbf{z} = C\mathbf{x} . \tag{8}$$

The vector $\mathbf{x}$ is the state vector of the system, $\mathbf{q}$ is the input variable (*i.e.*, control input), and $\mathbf{z}$ is the output variable (*i.e.*, sensor output). The matrix A is the dynamic matrix of the system, and B and C are, respectively, the input and output matrices. The state space and transfer function models are related as

$$H(s) = \frac{Z(s)}{Q(s)} = \frac{\prod_{j=1}^{J}(s-z_j)}{\prod_{i=1}^{I}(s-p_i)}$$
$$= C(sI - A)^{-1}B , \tag{9}$$

where s is the complex Laplace variable, z_j and p_i are, respectively, the zeros and poles of the system.

Joshi *et al.* [22] show that in addition to the well-studied system eigenvalues, the location of system zeros are important in linear stability control. The location of system zeros determines the effect of feedback control on both stable and unstable eigenvalues. In addition, system zeros can be used to determine sensor locations that lead to simple feedback control schemes. Joshi *et al.* showed that the channel flow can be stabilized using a simple, constant-gain feedback, integral-compensator controller. The goal of the control design was stability. By choosing proper sensor locations, they were able to achieve a stable, closed-loop system that was extremely robust to changing Reynolds numbers. It was also shown that a controller designed with linear theory also has a strong stabilizing effect on two-dimensional finite-amplitude disturbances. As a result, the secondary instabilities resulting from infinitesimal three-dimensional disturbances in the presence of a finite-amplitude two-dimensional disturbance cease to exist [22].

Joshi *et al.* [22] further illustrate that the system theory approach is a powerful tool both for designing control systems to stabilize flows, as well as understanding the physics of controlled transitional flows. Transfer-function models yield information on optimal sensor locations and sensor types. State variable models also show how each mode, both stable and unstable, is affected by feedback control. In addition, control theory concepts, such as observability and controllability, are used to explain possible pitfalls in flow control.

In our subsequent work ([23], [24]), linear quadratic Gaussian (LQG) optimal control theory is exploited to give faster perturbation energy dissipation. Controllers built using LQG methods for infinite-dimensional systems must, in theory, be infinite dimensional. Since it is impossible to implement infinite-dimensional controllers, we employed reduced-order controllers for the infinite order plant.

6 Conclusions

Successful applications of a neural network and a suboptimal control theory to a turbulent channel flow for drag reduction have been described. In the neural network approach, the network was able to identify a correlation between the wall-shear stress and the desired wall actuations. The optimum actuation, which is proportional to the spanwise shear stress at the wall, is not something one would normally consider for drag reduction. Apparently the properly trained neural network is capable of identifying the most relevant information from the given input and output. Although the present approach lacks the "ad-hoc" nature of earlier strategies, the design and application of the neural network nevertheless benefited from physical insight into the flow under consideration. In the suboptimal control-theory approach, the cost functional to be minimized was derived from an earlier control scheme that was designed to mitigate the interaction between streamwise vortices and the wall. The resulting feedback-control law was very similar to that found by the neural network. This is quite surprising, since the two approaches use totally different routes to get to a similar optimum actuation: the former uses a "black-box" approach to achieve the goal, whereas the latter tries to minimizes a quantity that is not directly related to the skin-friction drag. Both optimum actuations, which differ only slightly in the manner in which the input data are locally averaged, are related to the near-wall streamwise vortices. Thus our premise that proper manipulation of the near-wall streamwise vortices is the most effective way to control turbulent boundary layers is validated.

The system theory approach is a promising way to explore new flow control strategies, not only for the stablization of a system (as illustrated above), but also for control of turbulent boundary layers. In contrast to the neural network approach, which learns about a system that it treats as a black box, an estimator in the system theory approach is built based on the information that is representative of the system. However, our current system theory approach is, in principle, limited to linear problems, although we have shown that it can work surprisingly well when applied to a nonlinear disturbance. Many other modern control theories, such as disturbance attenuation design [25] and H_∞ [26], need to be explored. Such effort is currently underway at UCLA.

Although the control schemes presented in this paper are significant improvements over earlier approaches that require velocity information throughout the flow field, there are many technical issues that must be resolved before these control schemes can be implemented in real practice. Precise blowing and suction distribution over a surface, for example, is difficult to implement. Other approaches, such as surface movement by deformable walls, may prove to be more practical. Another issue worth mentioning is the time delay between sensing and actuation, which was not included in any of our numerical experiments. In a real situation, there will be a finite time between sensing and actuation, which must be accounted for.

In closing, we have explored new control strategies utilizing relatively new non-traditional technologies, and the results are promising. It is our belief that full exploitation of modern control theory should enhance our capability of taming turbulence for flow control.

Acknowledgments

This paper describes the work being carried out by the author's students and colleagues at UCLA. I thank them all. I am grateful to Dr. Gary Coleman for comments on a draft of this manuscript. The financial support from Air Force Office of Scientific Research (Drs. James McMichael and Mark Glauser) and Office of Naval Research (Dr. Patrick Purtell) during the course of this work is gratefully acknowledged. The computer time has been provided by the NAS Program of NASA Ames Research Center, San Diego Supercomputer Center, and DoD High Performance Computer Center.

References

[1] D. M. Bushnell, private communication, 1996.

[2] P. Moin and T. R. Bewley, "Application of control theory to turbulence," Twelfth Australian Fluid Mechanics Conference, Sydney, Australia, Dec. 10-15, 1995.

[3] J. L. Lumley, "Control of turbulence," AIAA Paper 96-0001, The 1996 Dryden Lecture in Research of the American Institute of Aeronautics and Astronautics, 34th AIAA Aerospace Sciences Meeting and Exhibit, January 15-18, Reno, Nevada, 1996.

[4] S. K. Robinson, "Coherent motions in the turbulent boundary layer," *Ann. Rev. Fluid Mech.* **23**, 601, 1991.

[5] J. M. Hamilton, J. Kim and F. Waleffe, "Regeneration mechanisms of near-wall turbulence structures," *J. Fluid Mech.* **287**, 317, 1995.

[6] F. Waleffe, "On a self-sustaining process in shear flows," *Phys. Fluids* **9**, 883, 1997.

[7] J. Jiménez and A. Pinelli, "Wall turbulence: How it works and how to damp it," AIAA PAper 97-2112, 4th AIAA Shear Flow Control Conference, June 29-July 2, 1997, Snowmass Village, Colorado.

[8] Self-Sustaining Mechanisms of Wall Turbulence, Panton (ed.), Computational Mechanics Publication, 1997.

[9] S. J. Kline, W. C. Reynolds, F. A. Schraub and P. W. Runstadler, "The structure of turbulent boundary layers," *J. Fluid Mech.* **30**, 741, 1967.

[10] C. R. Smith and S. P. Metzler, "The characteristics of low-speed streaks in the near-wall region of a turbulent boundary layer," *J. Fluid Mech.* **129**, 27, 1983.

[11] J. Kim, P. Moin and R. D. Moser, "Turbulence statistics in fully developed channel flow at low Reynolds number," *J. Fluid Mech.* **177**, 133, 1967.

[12] H. T. Kim, S. J. Kline and W. C. Reynolds, "The production of turbulence near a smooth wall in a turbulent boundary layer," *J. Fluid Mech.* **50**, 133, 1971.

[13] M. S. Acarlar and C. R. Smith, "A study of hairpin vortices in a laminar boundary layer. Part II: hairpin vortices generated by fluid injection," *J. Fluid Mech.* **175**, 43, 1987.

[14] J. Kim, "Study of turbulence structure through numerical simulations: the perspective of drag reduction," in AGARD Report (R-786), AGARD FDP/VKI Special Course on "Skin Friction Drag Reduction," March 2-6, 1992, VKI, Brussels, Belgium.

[15] A. G. Kravchenko, H. Choi and P. Moin, "On the relation of near-wall streamwise vortices to wall skin friction in turbulent boundary layers," *Phys. Fluids A* **5**, 3307, 1993.

[16] H. Choi, P. Moin and J. Kim, "Active turbulence control for drag reduction in wall-bounded flows," *J. Fluid Mech.* **262**, 75, 1994.

[17] H. Choi, P. Moin and J. Kim, "Direct numerical simulation of turbulent flow over riblets," *J. Fluid Mech.* **255**, 503, 1993.

[18] C. Lee, J. Kim, D. Babcock, and R. Goodman, "Application of neural networks to turbulence control for drag reduction," *Phys. Fluids* **9**, No. 6, 1740, 1997.

[19] H. Choi, R. Temam, P. Moin and J. Kim, "Feedback control for unsteady flow and its application to the stochastic Burgers equation," *J. Fluid Mech.* **253**, 1993.

[20] T. R. Bewley and P. Moin, "Optimal control of turbulent channel flows," *Active Control of Vibration and Noise*, ASME DE-Vol. 75, 1994.

[21] C. Lee, J. Kim and H. Choi, "Suboptimal control of turbulent channel flow for drag reduction," submitted for publication, 1997.

[22] S. S. Joshi, J. L. Speyer and J. Kim, "A systems theory approach to the feedback stabilization of infinitesimal and finite-amplitude disturbances in plane Poiseuille flow," *J. Fluid Mech.* **332**, 157, 1997.

[23] S. S. Joshi, J. L. Speyer and J. Kim, "Linear Stabilization of Plane Poiseuille Flow Using an Optimal Control Approach," submitted for publication, 1997.

[24] S. S. Joshi, J. L. Speyer, and J. Kim, "Modelling and control of two dimensional Poiseuille flow," Proceedings of the 34th IEEE Conference on Decision and Control (CDC), New Orleans, LA, December 13-15, 1995.

[25] I. Rhee and J. L. Speyer, "A game theoretic approach to a finite-disturbance attenuation problem," IEEE Transactions on Automatic Control, AC-**36**, No. 9, 1991.

[26] J. K. Doyle, K. Glover, P. P. Khargonekar, and B. A. Francis, "State space solutions to standard H_2 and H_∞ control problems," IEEE Transactions on Automatic Control, AC-**34**, No. 8, 1989.

PROGRESS IN SUBGRID-SCALE COMBUSTION MODELING

Andrew W. COOK [1]   James J. RILEY [2]

1) Nuclear and Hydrodynamic Applications Group, Applied Theoretical and
   Computational Physics Division, Los Alamos National Laboratory;
   Los Alamos, New Mexico 87545, USA

2) Department of Mechanical Engineering, University of Washington,
   Seattle, Washington 98195

Abstract

The technique of large eddy simulation (LES) has proven very useful for the prediction of turbulent flows. Its extension to turbulent, reacting flows has been rather slow, however, due to difficulties in modeling the chemistry in the subgrid scales. In this paper, the general technique of large eddy simulation is briefly described, as well as several recent approaches to the simulation of turbulent, reacting flows. One particular method, based upon the concept of flamelets, is discussed in detail. In this approach, finite-rate chemistry is accounted for by invoking the flamelet approximation, and an assumed form is employed for the subgrid-scale or 'large eddy' probability density function. The method requires the simulation of the filtered mixture fraction, its second moment, and its dissipation rate. Some *a priori* tests of the model are presented, giving optimism that large eddy simulation can be successfully applied to turbulent combustion.

Received on September 3, 1997.

Nomenclature

(Superscript $\star$ indicates variable is dimensional.)

D = nondimensional diffusivity, $\dfrac{D^\star}{D_\infty}$

D_∞ = reference diffusivity

Da_i = Damköhler number of ith reaction, $\dfrac{\rho_\infty K_i^\star L_\infty}{U_\infty}$

e = internal energy per unit mass, $\dfrac{T}{\gamma M^2(\gamma - 1)}$

H = total enthalpy, $\dfrac{\gamma}{(\gamma - 1)}T + \displaystyle\sum_{i=1}^{N} h_i Y_i$

h_i = nondimensional enthalpy of formation of species i, $\dfrac{h_i^\star}{R_\infty T_\infty}$

$K_i^\star$ = pre-exponential rate coefficient of ith reaction

$k^\star$ = thermal conductivity

L_∞ = reference length scale

Le = Lewis Number, $\dfrac{Sc}{Pr}$

M = Mach number, $\dfrac{U_\infty}{\sqrt{\gamma R_\infty T_\infty}}$

N = total number of chemical species in flow

Pe = Peclet Number, $ReSc$

Pr = Prandtl number (assumed constant), $\dfrac{c_p^\star \mu^\star}{k^\star}$

p = nondimensional pressure, $\dfrac{p^\star}{\rho_\infty U_\infty^2}$

R_∞ = mass-based ideal gas constant

Re = Reynolds number, $\dfrac{\rho_\infty U_\infty L_\infty}{\mu_\infty}$

Sc = Schmidt number, $\dfrac{\mu_\infty}{\rho_\infty D_\infty}$

T = nondimensional temperature, $\dfrac{T^\star}{T_\infty}$

T_{ai} = activation temperature of ith reaction

T_∞ = reference temperature

t = nondimensional time, $\dfrac{U_\infty t^\star}{L_\infty}$

U_∞ = reference velocity

u_i = ith component of nondimensional velocity vector, $\dfrac{u_i^\star}{U_\infty}$

$\dot{w}_i$ = reaction rate of the ith species, $\dot{w}_i(Da_i, \rho, Y_i, T_{ai})$

x_i = nondimensional distance in ith-direction, $\dfrac{x_i^\star}{L_\infty}$

Y_i = mass fraction of species i

γ = ratio of specific heats, $\dfrac{c_p^\star}{c_v^\star}$

μ = nondimensional dynamic viscosity, $\dfrac{\mu^\star}{\mu_\infty}$

μ_∞ = reference dynamic viscosity

ν = nondimensional kinematic viscosity, $\dfrac{\nu^\star}{\nu_\infty}$

ν_∞ = reference kinematic viscosity, $\dfrac{\mu_\infty}{\rho_\infty}$

ξ = conserved scalar mixture-fraction

ξ_{st} = stoichiometric value of mixture-fraction

ρ = nondimensional density, $\dfrac{\rho^\star}{\rho_\infty}$

ρ_∞ = reference density

τ_{ij} = viscous stress tensor, $\mu\left(\dfrac{\partial u_i}{\partial x_j} + \dfrac{\partial u_j}{\partial x_i} - \dfrac{2}{3}\delta_{ij}\dfrac{\partial u_k}{\partial x_k}\right)$

χ = scalar dissipation rate, $\dfrac{\mu}{Pe}\nabla\xi \cdot \nabla\xi$

II. Introduction

II..1 Large Eddy Simulation

Direct numerical simulation (DNS) of turbulence involves numerically solving the fully-nonlinear, three-dimensional, time-dependent Navier-Stokes equations, accompanying species conservation equations and relevant equations of state. All the significant space and time scales are resolved. Statistical results are obtained by spatial or temporal averaging over the computed flow fields. The advantage of this approach, when compared to other computational methods, is that no closure modeling is needed; the principal disadvantage is that the numerical resolution in space and time is very limited (e.g., to about 2 decades in each spatial direction), limiting the technique to somewhat low Reynolds and Damköhler numbers. This approach is presently used mainly as a research tool. analogous to laboratory experiments. For reviews of DNS applied to turbulent combustion, see, e.g., Givi [1] and Jou and Riley [2].

The methodology of large eddy simulation offers the potential of avoiding the limitations of low Reynolds and Damköhler numbers inherent in DNS. In this approach, the governing equations are spatially filtered, prior to numerical solution, to eliminate scales of motion not resolvable by the computational mesh. Filtering the momentum equation gives rise to a subgrid-scale (SGS) stress tensor which represents the effects of the unresolved scales on the resolved field. The subgrid stress tensor is sometimes combined with the viscous stress tensor so that the form of the equation remains the same. It is useful to think of the filtered momentum equation as governing a hypothetical 'LES fluid' which is non-Newtonian, i.e., a fluid in which the smallest scales of motion are smoothed through the action of an 'eddy viscosity' rather than by molecular viscosity. The model for the SGS stress tensor is often based upon the behavior of the resolvable velocity field at scales near to the grid spacing. This approach applied to non-reactive flows has been very useful as a research tool, and has significant potential to be used more in applications, especially as the methodology improves, and as computers become larger and faster.

The principal advantages of LES are: it eliminates the Reynolds and Damköhler number restrictions of DNS, at least in theory; it contains the physics of the larger-scale motions, which generally control most turbulent flow processes; and only the smaller scale motions need to be modeled. These small scales contain little energy (so that the results may not be too sensitive to the model employed), and are probably somewhat universal in character (so that one model may work in a variety of situations). The disadvantages of LES are that modeling is still needed, and that the methodology requires significant computer resources.

When LES is applied to chemically-reacting flows, additional modeling is required. Conservation equations for reacting chemical species contain nonlinear reaction rate terms as well as nonlinear advection terms. Developing models for filtered reaction rates has been thought to be very difficult since, for large Damköhler numbers, reaction zones occur at the subgrid scales [3]. If such terms can be accurately modeled, however, then the potential capabilities of LES can be extended to turbulent reacting flows.

II..2 Approaches to Combustion

A limited number of studies have addressed large eddy simulation of turbulent, reacting flows. Probably the first was by Schumann [4] who neglected the subgrid-scale effects of the chemistry altogether, assuming that the filtered reaction rate could be modeled as $\overline{w_i(Y_1, Y_2, \ldots, T)} = w_i(\overline{Y}_1, \overline{Y}_2, \ldots, \overline{T})$ where the overbar denotes a filtered quantity. For moderate to fast chemistry, this approximation has been known to introduce considerable error when applied to the Reynolds-averaged (or Favre-averaged) equations, and there is no reason to believe that this will not be the case when applied to the spatially-filtered equations used in large eddy simulations.

Since this work, two general lines of approach have been used in developing subgrid-scale models for turbulent, reacting flows: one based upon the mixture-fraction, and the other based upon the 'large-eddy' or 'filtered' probability density function. The approaches using the mixture-fraction are based upon the well-known flamelet approximation, whereby the concentration of chemical species, the reaction rates, and other aspects of the chemical reaction can all be directly related to the mixture-fraction and its dissipation rate. A prime example of this is the large eddy laminar flamelet model (LELFM), which will be discussed in detail in this paper. A similar model has been developed by Jiménez et al. [5], the differences being in the treatment of the joint 'large-eddy' probability density of the mixture-fraction and its dissipation rate.

Gao and O'Brien [6] have developed the equation for the large-eddy pdf, and discussed various theoretical issues in this regard. More recently Colucci et al. [7] have introduced the methodology to solve this equation, utilizing the approach which Pope developed for solving the ensemble-averaged pdf equation. Lagrangian-based Monte-Carlo methods are proposed, and tested against DNS data. Another Monte-Carlo method being applied as a subgrid-scale model is the linear-eddy model of Kerstein. McMurtry et al. [8] and Menon et al. [9] have utilized the methodology in two-dimensional, large eddy simulations with promising results. Both of these Monte-Carlo approaches have the potential advantage that, at least in theory, more physics can be added than in the mixture-fraction-based models. They have the disadvantage, however, of being very computationally intensive, since an additional computational dimension is introduced into the problem due to the use of the Monte-Carlo method at each grid point. Thus, to date, these models have only been applied to two-dimensional flows.

Frankel *et al.* [10] also formulated a pdf-based model for finite-rate chemistry, although their method relies on modeling the joint distribution of fuel and oxidizer, and hence modeling the subgrid-scale species covariance. Depending on the accuracy and generality of a model for this covariance, this method could have significant potential for future use.

III.   Problem Formulation

III..1   Filtering

Consider a two-feed combustion problem with fuel carried by feed 1 and oxidant carried by feed 2. The goal is to solve for the large-scale behavior of density, velocity, pressure, temperature and relevant chemical species. Equations governing the large-scale features of the flow are obtained by spatially filtering the conservation equations for mass, momentum and internal energy. In addition, a filtered equation of state and a filtered transport equation for a passive scalar are included in the set of equations to be solved. It will be shown how information about the conserved scalar can be used to predict all of the chemical species produced and consumed in the combustion process. A homogeneous filter operator is defined by the convolution integral

$$\overline{\phi}(\mathbf{x}) \equiv \int_{\mathcal{D}} G(|\mathbf{x} - \mathbf{x}'|; \Delta)\phi(\mathbf{x}')d\mathbf{x}' \qquad (1)$$

where the subscript $\mathcal{D}$ indicates that the integration is performed over the entire computational domain. The filter kernel G is normalized,

$$\int_{-\infty}^{\infty} G(r; \Delta)\, dr = 1 \;, \qquad (2)$$

and has a characteristic width Δ which is directly related to the grid spacing of the computational mesh. Due to its dependence on Δ, G is often referred to as the 'grid' filter. For an anisotropic grid, Δ is usually chosen as $\Delta = (\Delta x \Delta y \Delta z)^{1/3}$ [11], although other types of averages have also been used [12].

III..2   Governing Equations

Except for certain difficulties which may occur near solid boundaries, it is straightforward to show that Eq. (1) commutes with derivative operators. Applying Eq. (1) to the nondimensionalized governing equations yields:

$$\frac{\partial \overline{\rho}}{\partial t} + \frac{\partial \overline{\rho u_j}}{\partial x_j} = 0 \qquad (3)$$

$$\frac{\partial \overline{\rho u_i}}{\partial t} + \frac{\partial \overline{\rho u_i u_j}}{\partial x_j} = -\frac{\partial \overline{p}}{\partial x_i} + \frac{1}{Re}\frac{\partial \overline{\tau_{ij}}}{\partial x_j} \qquad (4)$$

$$\frac{\partial \overline{\rho e}}{\partial t} + \frac{\partial \overline{\rho e u_j}}{\partial x_j} = -\overline{p\frac{\partial u_j}{\partial x_j}} + \frac{1}{Re}\overline{\tau_{jk}\frac{\partial u_j}{\partial x_k}}$$

$$+\frac{1}{M^2 RePr}\frac{\partial}{\partial x_j}\overline{\left(\frac{\mu}{(\gamma - 1)}\frac{\partial T}{\partial x_j}\right)} + \frac{1}{\gamma M^2}\sum_{i=1}^{N} h_i \overline{\dot{w}_i} \qquad (5)$$

$$\frac{\partial \overline{\rho \xi}}{\partial t} + \frac{\partial \overline{\rho \xi u_j}}{\partial x_j} = \frac{1}{ReSc}\frac{\partial}{\partial x_j}\overline{\left(\mu\frac{\partial \xi}{\partial x_j}\right)} \qquad (6)$$

$$\overline{p} = \frac{\overline{\rho T}}{\gamma M^2} \qquad (7)$$

where the temperature dependence of viscosity is given by

$$\mu = 1.37\frac{T^{3/2}}{(T + 0.37)} \;. \qquad (8)$$

Here it has been assumed that all species have equal specific heats, molecular weights and diffusivities. Also neglected are radiant energy transfer and work due to body forces; although, such terms could be included in Eq. (5) if these effects were expected to be important in a particular problem. The assumption of equal diffusivities is more valid at higher Reynolds numbers where small-scale diffusion is not expected to directly affect on the overall combustion process. The assumption of equal molecular weights is not absolutely necessary, i.e., the gas constant for the mixture could be computed as a function of space and time, given knowledge of the species mass fractions [13]. However, it is a convenient assumption if the gas constant does not vary significantly, e.g., if much of the chemical system is composed of a dilutant. Equation (8) is Sutherland's formula for air at moderate temperatures; hence, it is assumed that air is the primary dilutant.

III..3   Scalar Based Chemistry

The mixture-fraction is defined as [14]

$$\xi \equiv \frac{(Z_i - Z_{i2})}{(Z_{i1} - Z_{i2})} = \frac{(\beta - \beta_2)}{(\beta_1 - \beta_2)} \qquad (9)$$

where Z_i is the mass fraction of element i and the subscripts 1 and 2 refer to the uniform composition in the fuel and oxidizer streams, respectively. It can be seen that $\xi = 1$ in the fuel stream and $\xi = 0$ in the oxidant stream. The quantities β, β_1 and β_2 refer to Shvab-Zeldovich functions of the species mass fractions, and will be defined later for a particular case. Without the assumption of equal diffusivities for all chemical species, a differential diffusion term would appear in

Eq. (6). It has also been assumed in Eq. (6) that the Schmidt number is not a function of temperature, i.e., $\nu^\star/D^\star = \nu_\infty/D_\infty$ so that, with the nondimensionalization, $D = \nu = \mu/\rho$.

III..4   Simplifications

The large-scale equations are simplified by recasting the variables as Favre-filtered quantities. A Favre-filtered, i.e., density-weighted, variable is denoted by a tilde and is defined as

$$\tilde{\phi} = \frac{\overline{\rho\phi}}{\overline{\rho}} \ . \tag{10}$$

The flow field is thus decomposed into Favre-filtered (resolved) and subgrid-scale (unresolved) components, i.e., $\phi = \tilde{\phi} + \phi''$. Note that typically $\overline{\tilde{\phi}} \neq \tilde{\phi}$ and therefore $\overline{\phi''} \neq 0$. Exceptions are spectral truncation filters and the discrete filtering employed in Schumann's control volume method [15].

In nonreacting flows, subgrid fluctuations in viscosity are typically neglected. However, in reacting flows, heat released at the small scales may have an important effect on the viscosity μ and its correlation with T, u_i and ξ. In order to examine this issue, T in Eq. (8) is replaced by $\tilde{T} + T''$ and the numerator and denominator are expanded in a binomial series, leading to

$$\mu = 1.37 \frac{\tilde{T}^{3/2}}{(\tilde{T} + 0.37)}$$
$$\left\{ 1 + \frac{3}{2}\frac{T''}{\tilde{T}} - \frac{T''}{(\tilde{T} + 0.37)} + \mathcal{O}\left[\left(\frac{T''}{\tilde{T}}\right)^2\right] \right\} \ . \tag{11}$$

The filtered viscosity is then

$$\overline{\mu} \approx 1.37 \overline{\left(\frac{\tilde{T}^{3/2}}{(\tilde{T} + 0.37)}\right)} \tag{12}$$

assuming that $T''/\tilde{T}$ is small. We will assume that Eq. (12) is valid; however, this has yet to be investigated for a reacting flow.

In addition to the expression for $\overline{\mu}$, there are five terms in the filtered equations that require special treatment: these are the viscous term in the momentum equation; the pressure-dilatation term, dissipation function, and conduction term in the energy equation; and the diffusive term in the scalar equation. These terms can be rewritten as:

$$\overline{\tau}_{ij} = \overline{\mu}\left(\frac{\partial \tilde{u}_i}{\partial x_j} + \frac{\partial \tilde{u}_j}{\partial x_i} - \frac{2}{3}\delta_{ij}\frac{\partial \tilde{u}_k}{\partial x_k}\right)$$
$$+ \left[\overline{\mu\left(\frac{\partial u_i}{\partial x_j} + \frac{\partial u_j}{\partial x_i} - \frac{2}{3}\delta_{ij}\frac{\partial u_k}{\partial x_k}\right)}\right.$$
$$\left. -\overline{\mu}\left(\frac{\partial \tilde{u}_i}{\partial x_j} + \frac{\partial \tilde{u}_j}{\partial x_i} - \frac{2}{3}\delta_{ij}\frac{\partial \tilde{u}_k}{\partial x_k}\right)\right] \tag{13}$$

$$\overline{p\frac{\partial u_j}{\partial x_j}} = \overline{p}\frac{\partial \tilde{u}_j}{\partial x_j} + \left[\overline{p\frac{\partial u_j}{\partial x_j}} - \overline{p}\frac{\partial \tilde{u}_j}{\partial x_j}\right] \tag{14}$$

$$\overline{\tau_{jk}\frac{\partial u_j}{\partial x_k}} = \overline{\tau}_{jk}\frac{\partial \tilde{u}_j}{\partial x_k} + \left[\overline{\tau_{jk}\frac{\partial u_j}{\partial x_k}} - \overline{\tau}_{jk}\frac{\partial \tilde{u}_j}{\partial x_k}\right] \tag{15}$$

$$\overline{\frac{\mu}{(\gamma - 1)}\frac{\partial T}{\partial x_j}} = \frac{\overline{\mu}}{(\gamma - 1)}\frac{\partial \tilde{T}}{\partial x_j}$$
$$+ \left[\overline{\frac{\mu}{(\gamma - 1)}\frac{\partial T}{\partial x_j}} - \frac{\overline{\mu}}{(\gamma - 1)}\frac{\partial \tilde{T}}{\partial x_j}\right] \tag{16}$$

$$\overline{\mu\frac{\partial \xi}{\partial x_j}} = \overline{\mu}\frac{\partial \tilde{\xi}}{\partial x_j} + \left[\overline{\mu\frac{\partial \xi}{\partial x_j}} - \overline{\mu}\frac{\partial \tilde{\xi}}{\partial x_j}\right] \ . \tag{17}$$

Until recently, LES of nonreacting compressible flows have typically neglected all the terms in brackets [16, 17]. Vreman $et\ al.$ [18] have demonstrated that terms arising from subgrid fluctuations in viscosity can be neglected but that the bracketed terms in (14) and (15) are important at high Mach numbers. They have proposed the following models for these terms, which appear to correlate well with exact terms computed from DNS:

$$P_d \equiv \left[\overline{p\frac{\partial u_j}{\partial x_j}} - \overline{p}\frac{\partial \tilde{u}_j}{\partial x_j}\right] \approx 2.2\left[\overline{\overline{p}\frac{\partial \tilde{u}_j}{\partial x_j}} - \overline{\overline{p}}\frac{\partial \tilde{\tilde{u}}_j}{\partial x_j}\right] \tag{18}$$

$$F_d \equiv \left[\overline{\tau_{jk}\frac{\partial u_j}{\partial x_k}} - \overline{\tau}_{jk}\frac{\partial \tilde{u}_j}{\partial x_k}\right]$$
$$\approx 8\left[\overline{\overline{\tau}_{jk}\frac{\partial \tilde{u}_j}{\partial x_k}} - \overline{\overline{\tau}}_{jk}\frac{\partial \tilde{\tilde{u}}_j}{\partial x_k}\right] \ . \tag{19}$$

For reacting flows, it is not yet clear whether the bracketed terms in Eqs. (13), (16) and (17) are important. It will here be assumed that they can be neglected based on the fact that μ is a slow function of T and on the assumption that differences between plain-filtered and Favre-filtered values are small.

With the above approximations, the Favre-filtered equations become:

$$\frac{\partial \overline{\rho}}{\partial t} + \frac{\partial \overline{\rho}\tilde{u}_j}{\partial x_j} = 0 \tag{20}$$

$$\frac{\partial \overline{\rho}\tilde{u}_i}{\partial t} + \frac{\partial \overline{\rho}\tilde{u}_i\tilde{u}_j}{\partial x_j} = -\frac{\partial \overline{p}}{\partial x_i} + \frac{1}{Re}\frac{\partial \overline{\tau}_{ij}}{\partial x_j} - \frac{\partial \sigma_{ij}}{\partial x_j} \tag{21}$$

$$\frac{\partial \overline{\rho}\tilde{e}}{\partial t} + \frac{\partial \overline{\rho}\tilde{e}\tilde{u}_j}{\partial x_j} = -\overline{p}\frac{\partial \tilde{u}_j}{\partial x_j} + \frac{1}{Re}\overline{\tau}_{jk}\frac{\partial \tilde{u}_j}{\partial x_k}$$
$$+ \frac{1}{M^2 RePr}\frac{\partial}{\partial x_j}\left(\frac{\overline{\mu}}{(\gamma - 1)}\frac{\partial \tilde{T}}{\partial x_j}\right)$$
$$+ \frac{1}{\gamma M^2}\sum_{i=1}^{N} h_i\overline{w}_i - \frac{\partial q_j}{\partial x_j} - P_d + F_d \tag{22}$$

$$\frac{\partial \overline{\rho}\tilde{\xi}}{\partial t} + \frac{\partial \overline{\rho}\tilde{\xi}\tilde{u}_j}{\partial x_j} = \frac{1}{ReSc}\frac{\partial}{\partial x_j}\left(\overline{\mu}\frac{\partial \tilde{\xi}}{\partial x_j}\right) - \frac{\partial \zeta_j}{\partial x_j} \tag{23}$$

$$\bar{p} = \frac{\bar{\rho}\widetilde{T}}{\gamma M^2} \qquad (24)$$

where:

$$\sigma_{ij} = \bar{\rho}\left(\widetilde{u_i u_j} - \widetilde{u}_i \widetilde{u}_j\right) \qquad (25)$$

$$q_i = \bar{\rho}\left(\widetilde{u_i e} - \widetilde{u}_i \widetilde{e}\right) \qquad (26)$$

$$\zeta_i = \bar{\rho}\left(\widetilde{u_i \xi} - \widetilde{u}_i \widetilde{\xi}\right) \ . \qquad (27)$$

In order to close this set of 'large-eddy' equations, models are needed for σ_{ij}, q_i, ζ_i and $\overline{w_i}$. In addition, a method for predicting filtered chemical species mass fractions, i.e., $\overline{Y_i}$, is required. The following section sumarizes a method proposed by Moin $et~al.$ [17] for modeling σ_{ij}, q_i and ζ_i. A closure for the chemistry is derived in subsequent sections.

IV.   Closure

In turbulent combustion, large temperatures, and hence large viscosities, may cause relaminarization in various flow regions. Models for σ_{ij}, q_i and ζ_i should vanish is such locations. The dynamic model of Germano $et~al.$ [19] utilizes spectral information in the large-scale field to extrapolate the small-scale stresses. A nice result of this is that the model gives 0 for laminar flows. Recently, a Lagrangian form of the dynamic model [20] was successfully used to predict the flux of turbulent kinetic energy toward the subgrid scales in DNS of compressible reacting flow [21].

IV..1   SGS Stress Tensor

Following Moin et al. [17], σ_{ij} is recast in terms of spatialy filtered variables, i.e.,

$$\sigma_{ij} = \overline{\rho u_i u_j} - \frac{\overline{\rho u_i}\,\overline{\rho u_j}}{\bar{\rho}} \qquad (28)$$

Applying a second 'test' filter to Eq. (4) gives rise to the test-filtered stresses,

$$\psi_{ij} = \widehat{\overline{\rho u_i u_j}} - \frac{\widehat{\overline{\rho u_i}}\,\widehat{\overline{\rho u_j}}}{\widehat{\bar{\rho}}} \ . \qquad (29)$$

An overbrace is here used to denote a quantity that has been filtered using a test filter of width $\widehat{\Delta} > \Delta$. Germano $et~al.$ [19] found an optimal ratio of test to grid scales to be $\widehat{\Delta}/\Delta = 2$. Applying the test filter to Eq. (28) and subtracting it from Eq. (29) leads to

$$L_{ij} = \psi_{ij} - \widehat{\sigma}_{ij} = \left(\frac{\widehat{\overline{\rho u_i}\,\overline{\rho u_j}}}{\bar{\rho}}\right) - \frac{\widehat{\overline{\rho u_i}}\,\widehat{\overline{\rho u_j}}}{\widehat{\bar{\rho}}}$$

$$= \widehat{\bar{\rho}\widetilde{u}_i\widetilde{u}_j} - \frac{\widehat{\bar{\rho}\widetilde{u}_i}\,\widehat{\bar{\rho}\widetilde{u}_j}}{\widehat{\bar{\rho}}} \qquad (30)$$

which may be directly calculated from the resolved fields, and as such, is sometimes called the 'resolved' or 'window' stress. The dynamic procedure assumes a prescribed functional form for the small scales, and furthermore assumes that this form is independent of filter width. Thus the model is expected to work best if the grid and test scales reside in the inertial range of the turbulent kinetic energy spectrum. The prescribed function is usually a Smagorinsky model, although recently it has been demonstrated that Kolmolgorov's model may be used, in conjunction with Smagorinsky's formula, to generate a family of dynamic models [22].

Using the trace-free Smagorinsky eddy viscosity model for both σ_{ij} and ψ_{ij} gives

$$\sigma_{ij} - \frac{1}{3}\delta_{ij}k_e = -C\bar{\rho}\Delta^2\left|\widetilde{S}\right|\left(\widetilde{S}_{ij} - \frac{1}{3}\delta_{ij}\widetilde{S}_{nn}\right) \qquad (31)$$

$$\psi_{ij} - \frac{1}{3}\delta_{ij}K_e = -C\,\widehat{\bar{\rho}}\,\widehat{\Delta}^2\left|\widehat{\widetilde{S}}\right|$$

$$\left(\widehat{\widetilde{S}}_{ij} - \frac{1}{3}\delta_{ij}\widehat{\widetilde{S}}_{nn}\right) \qquad (32)$$

where:

$k_e = \sigma_{kk}$ is twice the subgrid-scale kinetic energy

$K_e = \psi_{kk}$ is twice the test-scale kinetic energy

$$\widetilde{S}_{ij} = \frac{1}{2}\left(\frac{\partial\widetilde{u}_i}{\partial x_j} + \frac{\partial\widetilde{u}_j}{\partial x_i}\right)$$

$$\left|\widetilde{S}\right| = \left(2\widetilde{S}_{lm}\widetilde{S}_{lm}\right)^{1/2}$$

$$\widehat{\widetilde{S}}_{ij} = \frac{1}{2}\left(\frac{\partial\widehat{\widetilde{u}}_i}{\partial x_j} + \frac{\partial\widehat{\widetilde{u}}_j}{\partial x_i}\right)$$

$$\left|\widehat{\widetilde{S}}\right| = \left(2\widehat{\widetilde{S}}_{lm}\widehat{\widetilde{S}}_{lm}\right)^{1/2} \ .$$

The second term on the left hand side of Eq. (31) assures that the subgrid stress tensor is isotropic in the absence of shear. Note that $\widetilde{S}_{kk} = 0$ in an incompressible flow. The constant C is allowed to be a function of space and time and is determined by inserting Eqs. (31) and (32) into Eq. (30) to get

$$L_{ij} - \frac{1}{3}\delta_{ij}L_{kk} = \widehat{CM_{ij}} - CZ_{ij} \qquad (33)$$

where:

$$M_{ij} = \Delta^2\bar{\rho}\left|\widetilde{S}\right|\left(\widetilde{S}_{ij} - \frac{1}{3}\delta_{ij}\widetilde{S}_{nn}\right) \qquad (34)$$

$$Z_{ij} = \widehat{\Delta}^2\,\widehat{\bar{\rho}}\left|\widehat{\widetilde{S}}\right|\left(\widehat{\widetilde{S}}_{ij} - \frac{1}{3}\delta_{ij}\widehat{\widetilde{S}}_{nn}\right) \ . \qquad (35)$$

It is difficult to solve Eq. (33) for C, since it appears in the argument of the test-filter integral. Ghosal *et al.* [23] introduced a constrained variational problem for determining C which involves a Fredholm integral equation of the second kind. The integral equation is solved iteratively using an under-relaxation technique. A simpler approach was introduced by Piomelli and Liu [24] who noted that by filtering out the smallest scales of motion, the highest frequencies in the flow are also removed. Therefore, C is a slowly-varying function of time and may be estimated at timestep n via a first-order backward extrapolation, i.e.,

$$C \approx C_g = C^{n-1} + \Delta t \frac{\partial C}{\partial t}$$

$$\approx C^{n-1} + \frac{t^n - t^{n-1}}{t^{n-1} - t^{n-2}}(C^{n-1} - C^{n-2}) \ . \tag{36}$$

The C appearing under the overbrace in Eq. (33) is replaced by C_g, which is known. At this point (33) represents nine equations (five independent) in one unknown. This situation, in which C is over determined, is ironically opposite to the original closure problem. There are many ways in which Eq. (33) may be used to pick a value (or values) for C. Lilly [25] advocates a least squares fit that minimizes the error in Eq. (33). In this approach E_c is defined to be the square of the error in Eq. (33), i.e.,

$$E_c = (L_{ij} - \frac{1}{3}\delta_{ij}L_{kk} - \overbrace{C_g M_{ij}} + C Z_{ij})^2 \ . \tag{37}$$

Setting $\partial E_c / \partial C = 0$ and solving for C gives

$$C = -\frac{\left(L_{ij} - \frac{1}{3}\delta_{ij}L_{kk} - \overbrace{C_g M_{ij}}\right) Z_{ij}}{Z_{mn} Z_{mn}} \tag{38}$$

This represents the minimum of E_c, since $\partial^2 E_c / \partial C^2 = 2Z_{ij}^2 > 0$.

In incompressible flows, k_e may be absorbed into the pressure and need not be modeled. This may also be done for low Mach number flows by absorbing k_e into the second-order or dynamic pressure, the first-order or thermodynamic pressure being constant. In these cases, a modified pressure is computed while the real pressure remains unknown. In high Mach number flows however, the actual pressure must be computed in order to solve the energy equation. In this case, it is necessary to model k_e explicitly. The subgrid-scale kinetic energy may be parameterized using Yoshizawa's expression [26], i.e.,

$$k_e = C_I \, \overline{\rho} \Delta^2 \left| \widetilde{S} \right|^2 \ . \tag{39}$$

Similarly, the test scale kinetic energy may be written as

$$K_e = C_I \widehat{\overline{\rho}} \, \widehat{\Delta}^2 \left| \widehat{\widetilde{S}} \right|^2 \ . \tag{40}$$

Test filtering Eq. (39) and subtracting it from Eq. (40) yields

$$L_{ii} = K_e - \widehat{k_e} = \widehat{\overline{\rho} \widetilde{u}_i \widetilde{u}_i} - \frac{\widehat{\overline{\rho} \widetilde{u}_i} \, \widehat{\overline{\rho} \widetilde{u}_i}}{\widehat{\overline{\rho}}} \ . \tag{41}$$

Plugging Eqs. (39) and (40) into Eq. (41) and solving for C_I leads to

$$C_I = \frac{\widehat{\overline{\rho} \widetilde{u}_i \widetilde{u}_i} - (1/\widehat{\overline{\rho}}) \, \widehat{\overline{\rho} \widetilde{u}_i} \, \widehat{\overline{\rho} \widetilde{u}_i} + \Delta^2 \overbrace{C_G \overline{\rho} \left| \widetilde{S} \right|^2}}{\widehat{\Delta}^2 \, \widehat{\overline{\rho}} \left| \widehat{\widetilde{S}} \right|^2} \tag{42}$$

where

$$C_G = C_I^{n-1} + \frac{t^n - t^{n-1}}{t^{n-1} - t^{n-2}}(C_I^{n-1} - C_I^{n-2}) \ , \tag{43}$$

again employing the suggestion of Piomelli and Liu.

IV..2 Numerical Stability

In the current modeling, the eddy viscosity is defined as

$$\mu_T = \frac{1}{2} C \overline{\rho} \Delta^2 \left| \widetilde{S} \right| \ . \tag{44}$$

The total viscosity is defined as $\mu_{tot} = \overline{\mu} + \mu_T$. Since negative total viscosities correlated over long times can lead to numerical instabilities [27], μ_{tot} must be constrained in an LES to be non-negative. By placing the stability constraint on μ_{tot}, the model is capable of producing only small amounts of backscatter [24].

IV..3 SGS Heat Flux

In low Mach number flows, the second term on the left hand side of Eq. (5) is negligible and hence, no model is necessary for q_i [28]; however, in high Mach number flows q_i must be modeled. In order to derive a model for subgrid-scale heat flux, q_i is recast in terms of spatially filtered variables, i.e.,

$$q_i = \overline{\rho u_i e} - \frac{\overline{\rho u_i} \, \overline{\rho e}}{\overline{\rho}} \ . \tag{45}$$

Now let Q_i denote the heat flux at the test filter scale, i.e.,

$$Q_i = \widehat{\overline{\rho u_i e}} - \frac{\widehat{\overline{\rho u_i}} \, \widehat{\overline{\rho e}}}{\widehat{\overline{\rho}}} \ . \tag{46}$$

Each is assumed to obey an eddy diffusivity model, i.e.,

$$q_i = -\frac{C \overline{\rho} \Delta^2}{Pr_t} \left| \widetilde{S} \right| \frac{\partial \widetilde{e}}{\partial x_i} \tag{47}$$

$$Q_i = -\frac{C \, \widehat{\overline{\rho}} \, \widehat{\Delta}^2}{Pr_t} \left| \widehat{\widetilde{S}} \right| \frac{\partial \, \widehat{\widetilde{e}}}{\partial x_i} \ , \tag{48}$$

where Pr_t is a turbulent Prandtl number and C is defined in Eq. (38). Let

$$\mathcal{L}_i = Q_i - \widehat{\widetilde{q}}_i = \overline{\left(\frac{\overline{\rho u_i}\ \overline{\rho e}}{\overline{\rho}}\right)} - \frac{\widehat{\overline{\rho u_i}}\ \widehat{\overline{\rho e}}}{\widehat{\overline{\rho}}}$$

$$= \widehat{\overline{\rho \widetilde{u_i} \widetilde{e}}} - \frac{\widehat{\overline{\rho \widetilde{u_i}}}\ \widehat{\overline{\rho \widetilde{e}}}}{\widehat{\overline{\rho}}} \tag{49}$$

which is directly computable from the LES variables. Pr_t is determined by substituting Eqs. (47) and (48) into Eq. (49) to get

$$\mathcal{L}_i = \mathcal{R}_i - \frac{\Lambda_i}{Pr_t} \tag{50}$$

where:

$$\mathcal{R}_i = \Delta^2 \overline{\frac{C\overline{\rho}}{Pr_g} \left|\widetilde{S}\right| \frac{\partial \widetilde{e}}{\partial x_i}} \tag{51}$$

$$\Lambda_i = \widehat{\Delta}^2\, C\, \widehat{\overline{\rho}}\, \left|\widehat{\widetilde{S}}\right| \frac{\partial\, \widehat{\widetilde{e}}}{\partial x_i} \tag{52}$$

$$Pr_g = Pr_t^{n-1} + \frac{t^n - t^{n-1}}{t^{n-1} - t^{n-2}}(Pr_t^{n-1} - Pr_t^{n-2}) . \tag{53}$$

Let Υ be defined as the square of the error in Eq. (50), i.e.,

$$\Upsilon = \left(\mathcal{L}_i + \frac{\Lambda_i}{Pr_t} - \mathcal{R}_i\right)^2 . \tag{54}$$

Setting $\partial\Upsilon/\partial Pr_t = 0$ and solving for Pr_t gives

$$Pr_t = \frac{\Lambda_j \Lambda_j}{(\mathcal{R}_i - \mathcal{L}_i)\,\Lambda_i} . \tag{55}$$

IV..4 SGS Scalar Energy Flux

A model for ζ_i is derived in the same way as the model for q_i, leading to

$$\zeta_i = -\frac{C\overline{\rho}\Delta^2}{Sc_t} \left|\widetilde{S}\right| \frac{\partial\widetilde{\xi}}{\partial x_i} . \tag{56}$$

Here Sc_t is a turbulent Schmidt number, given by

$$Sc_t = \frac{\mathcal{J}_j \mathcal{J}_j}{(\mathcal{N}_i - \mathcal{L}'_i)\,\mathcal{J}_i} \tag{57}$$

where:

$$\mathcal{L}'_i == \widehat{\overline{\rho \widetilde{u_i}}\widetilde{\xi}} - \frac{\widehat{\overline{\rho \widetilde{u_i}}}\ \widehat{\overline{\rho \widetilde{\xi}}}}{\widehat{\overline{\rho}}}$$

$$\mathcal{N}_i = \Delta^2 \overline{\frac{C\overline{\rho}}{Sc_g} \left|\widetilde{S}\right| \frac{\partial\widetilde{\xi}}{\partial x_i}}$$

$$\mathcal{J}_i = \widehat{\Delta}^2\, C\, \widehat{\overline{\rho}}\, \left|\widehat{\widetilde{S}}\right| \frac{\partial\, \widehat{\widetilde{\xi}}}{\partial x_i}$$

$$Sc_g = Sc_t^{n-1} + \frac{t^n - t^{n-1}}{t^{n-1} - t^{n-2}}(Sc_t^{n-1} - Sc_t^{n-2}) .$$

V. SGS Chemistry Model

V..1 Laminar Flamelet Theory

In order to construct models for $\overline{\dot{w}_i}$ and $\overline{Y_i}$, it is first necessary to obtain an accurate physical description of the small-scale reaction zones. Consider an LES grid cell in which chemical reactions are possibly taking place. Assume that the reactions occur in thin flames that are locally one-dimensional, and that the flow field is locally counterflow [29]. This description of the reaction zones is appropriate if the Damköhler numbers and/or activation energies of the reactions are fairly large. The thin flames are located in the vicinity of the stoichiometric surface, defined by $\xi_{st}(\mathbf{x},t)$. Assuming a flame to be locally steady, and that ξ is a monotonic function of the local coordinate normal to the flame, the governing equations for Y_i can be transformed from physical space to ξ space [29]. The resulting set of laminar flamelet equations can be expressed as

$$\chi \frac{d^2 Y_i}{d\xi^2} = \dot{w}_i , \quad i = 1, .., N . \tag{58}$$

This equation set is coupled through the $\dot{w}_i$ terms, which are functions of ρ, T and Y_i. The effects of the local turbulent straining field are contained in ξ and χ. Four terms originally present in Eq. (58) have been neglected. The removal of these terms is based on the assumption that the reaction rate is fairly fast or that the activation energy is rather large, either case causing the reaction zone width in ξ space to be small. In such cases, the second derivative and reaction rate terms dominate the equations. The species mass fractions satisfy the boundary conditions:

$$Y_i(\xi = 0) = Y_{i2} \tag{59}$$

$$Y_i(\xi = 1) = Y_{i1} \tag{60}$$

where Y_{i1} and Y_{i2} are the uniform values of Y_i in feeds 1 and 2 respectively.

V..2 Counterflow

Equations (58), (59) and (60) constitute a coupled set of nonlinear boundary value problems. Before this set can be solved, it is first necessary to assume a form for the ξ dependence of χ. Due to the dynamics of the scalar gradient vector, there is some tendency for the surfaces of constant ξ to align orthogonal to the most compressive principle axis of the strain rate field [30]. Furthermore, experimental evidence suggests that mixing and reaction take place in local regions of steady, one-dimensional, laminar counterflow [31]. For this case, χ can be determined analytically and is given by

$$\chi = \chi_o F(\xi) \tag{61}$$

where

$$F(\xi) = \exp(-2[\text{erf}^{-1}(2(\xi - \xi^-)/(\xi^+ - \xi^-) - 1)]^2)$$

$$\text{for } \xi^- \leq \xi \leq \xi^+ . \tag{62}$$

Here χ_o is the local peak value of χ within the reaction layer, erf^{-1} is the inverse error function (not the reciprocal), and ξ^- and ξ^+ are the minimum and maximum values of ξ on either side of the layer, i.e., locations in the flow where $\nabla \xi = 0$ [32]. It has been argued that $\xi^- = 0$ and $\xi^+ = 1$ are acceptable modeling assumptions [33], even though the layer-like mixing structures are rarely bounded by pure fuel and oxidizer [34]. This is based on the observation that $\chi(\xi)$ need only be accurate where the subgrid PDF of ξ is large, and on the results of a number of model tests with and without the assumption being made.

V..3 Low Mach Number Combustion

In many devices, such as those that burn fossil fuels for energy, the combustion occurs at flow speeds much slower than the speed of sound. The Mach number of these flows is low, yet the density may vary due to heat release. In simulating these flows, the acoustic modes can be removed from the governing equations, resulting in significant computational savings [35, 28]. If a low Mach number approximation is applied to the governing equations, then the ideal gas equation becomes

$$p^{(0)} = \rho T \tag{63}$$

where $p^{(0)}$ is the first-order or thermodynamic pressure, which is constant in space [28]. If combustion takes place in an open domain, then $p^{(0)}$ is also constant in time, in which case ρ is known in terms of T alone. In such a regime, the number of parameters in Eq. (58) can be reduced by relating T, and thereby ρ, to ξ and Y_i. This is accomplished by making reference to the total enthalpy H which satisfies the equation [13],

$$\frac{\partial \rho H}{\partial t} + \frac{\partial \rho H u_j}{\partial x_j} = \frac{1}{RePr} \frac{\partial}{\partial x_j} \left(\mu \frac{\partial H}{\partial x_j} \right) . \tag{64}$$

Neglected in Eq. (64) are radiant heat transfer and work due to body forces. If $Pr = Sc$ then Eq. (64) is identical to the conservation equation for ξ and, therefore, H must be linearly related to ξ. This relationship is determined by ensuring consistency of the inflow conditions, i.e.,

$$H = \left[\frac{\gamma}{(\gamma - 1)}(T_1 - T_2) + \sum_{i=1}^{N} h_i(Y_{i1} - Y_{i2}) \right] \xi$$

$$+ \frac{\gamma}{(\gamma - 1)} T_2 + \sum_{i=1}^{N} h_i Y_{i2} \tag{65}$$

where T_1 is the temperature of the fuel feed and T_2 is the temperature of the oxidant feed. Temperature can then be expressed as a function of Y_i and ξ, i.e.,

$$T = \left[T_1 - T_2 + \frac{(\gamma - 1)}{\gamma} \sum_{i=1}^{N} h_i(Y_{i1} - Y_{i2}) \right] \xi$$

$$+ T_2 + \frac{(\gamma - 1)}{\gamma} \sum_{i=1}^{N} h_i(Y_{i2} - Y_i) . \tag{66}$$

With ρ and T known in terms of ξ and Y_i, Eq. (61) is inserted into Eq. (58) and the system, Eqs. (58), (59) and (60), is solved to obtain $Y_i(\xi, \chi_o)$. With the species mass fractions known in terms of ξ and χ_o, the reaction rates, i.e., $\dot{w}_i(\xi, \chi_o)$, can also be computed.

V..4 Large Eddy PDF

By assuming that reactions occur in thin regions of one-dimensional counterflow, the ξ dependence of χ is known through Eqs. (61) and (62). Furthermore, by assuming that $\xi^- = 0$ and $\xi^+ = 1$, χ_o then represents the value of χ where $\xi = 0.5$. The modeling thus implies that χ_o is independent of ξ; therefore, $\overline{Y_i}$ can be expressed as

$$\overline{Y_i} = \int_0^1 \int_{\chi_o^{min}}^{\chi_o^{max}} Y_i(\xi, \chi_o) P(\chi_o) P(\xi) d\chi_o d\xi \tag{67}$$

where χ_o^{min} and χ_o^{max} are the minimum and maximum values of χ_o within the grid volume. In Eq. (67), $P(\xi)$ is the large-eddy probability density function (LEPDF) [6], which gives the frequency distribution of ξ within the grid cell. Likewise, $P(\chi_o)$ gives the subgrid-scale probability density of χ_o. To simplify notation, no distinction has been made between the random variables and their probability space counterparts.

It has been shown that $Y_i(\xi, \chi_o)$ is a slow function of χ, and hence of χ_o, [36, 37, 38]. Therefore, if the interval $\chi_o^{max} - \chi_o^{min}$ is not too large then $Y_i(\xi, \chi_o)$ and $\dot{w}_i(\xi, \chi_o)$ can be approximated by the first two terms in the Taylor series expansion about the average of χ_o, e.g.,

$$Y_i(\xi, \chi_o) \approx Y_i(\xi, \overline{\chi_o}) + \left. \frac{\partial Y_i}{\partial \chi_o} \right|_{\overline{\chi_o}} (\chi_o - \overline{\chi_o}) . \tag{68}$$

Inserting Eq. (68) into Eq. (67) and integrating over χ_o gives

$$\overline{Y_i} = \int_0^1 Y_i(\xi, \overline{\chi_o}) P(\xi) d\xi . \tag{69}$$

The integral in Eq. (69) is carried out by assuming a Beta-distribution for $P(\xi)$ [39, 10], i.e.,

$$P(\xi) = \frac{\xi^{a-1}(1 - \xi)^{b-1}}{B(a, b)} , \quad a = \overline{\xi} \left[\frac{\overline{\xi}(1 - \overline{\xi})}{\xi_v^2} - 1 \right] ,$$

$$b = a/\overline{\xi} - a , \quad \xi_v^2 = \overline{\xi^2} - \overline{\xi}^{\,2} \tag{70}$$

where $B(a, b)$ is the Beta function and ξ_v^2 is the subgrid-scale variance of ξ.

Filtered reaction rates are computed in the same manner as the filtered species mass fractions, i.e., by expanding $\dot{w}_i(\xi, \chi_o)$ in a Taylor series about $\overline{\chi_o}$ and then integrating $\dot{w}_i(\xi, \overline{\chi_o})$ with $P(\xi)$. Finally, $\overline{\chi_o}$ must be related to $\overline{\chi}$ in order to make use of existing models. This is done by averaging Eq. (61) to obtain

$$\overline{\chi} = \overline{\chi_o} \int_o^1 F(\xi) P(\xi) d\xi . \tag{71}$$

V..5 Chemistry Tables

Prior to running an LES, tables may be constructed for $\overline{Y_i}(\tilde{\xi}, \tilde{\xi}^2, \overline{\chi})$ and $\overline{w_i}(\tilde{\xi}, \tilde{\xi}^2, \overline{\chi})$ which will depend on the following flow parameters: $p^{(0)}$, T_1, T_2, T_{ai}, h_i, Y_{i1}, Y_{i2}, Pe, and Da_i. This is accomplished in the following way. Firstly, $\tilde{\xi}$ and $\tilde{\xi}^2$ are chosen and $P(\xi)$ is computed from Eq. (70). Then $\overline{\chi_o}$ is chosen and $\overline{\chi}$ is computed using Eq. (71). The LFM solutions can then be specified in terms of $Y_i(\xi, \overline{\chi})$. Next, Eq. (66) is used, along with Eq. (63) and the LFM solutions $Y_i(\xi, \overline{\chi})$, to compute $\rho(\xi, \overline{\chi})$. With $P(\xi)$ and $\rho(\xi, \overline{\chi})$ known, $\tilde{\xi}$ and $\tilde{\xi}^2$ can then be computed. Finally, $\overline{Y_i}$ is computed from Eq. (69) and $\overline{w_i}$ is obtained similarly. Note that $\overline{Y_i}$ and $\overline{w_i}$ are initially obtained in terms of $\tilde{\xi}$, $\tilde{\xi}^2$ and $\overline{\chi_o}$, but may be tabulated as functions of $\tilde{\xi}$, $\tilde{\xi}^2$ and $\overline{\chi}$. Also, since ρ is a known function of ξ and $\overline{\chi}$, the Favre-filtered variables $\overline{Y_i}$ and $\overline{w_i}$ could also be tabulated.

V..6 Modeling ξ_v^2

In addition to $\tilde{\xi}$, the tables for $\overline{Y_i}$ and $\overline{w_i}$ require $\tilde{\xi}^2$ and $\overline{\chi}$ as inputs. Therefore, these quantities must be computed in an LES along with all the other variables of motion. An inexpensive method for estimating $\tilde{\xi}^2$ is to use the similarity model

$$\overline{\rho} \left(\tilde{\xi}^2 - \tilde{\xi}^2 \right) \approx \overline{\widetilde{\rho\xi\xi}} - \frac{\overline{\widetilde{\rho\xi}} \ \overline{\widetilde{\rho\xi}}}{\overline{\rho}} . \tag{72}$$

In the limit of constant density flow, Eq. (72) reduces to the similarity model discussed in Ref. [40]. Such a model has been successfully applied in large eddy simulations of decaying turbulence [21] and of a mixing layer [5].

VI. Modeling $\overline{\chi}$

VI..1 ξ-Energy

In deriving a model for $\overline{\chi}$, consider the equations governing ξ-energy in the resolved scales, i.e., $\tilde{\xi}^2$ and in the subgrid scales, i.e., $\tilde{\xi}^2$. The $\tilde{\xi}^2$ equation is obtained by multiplying the mixture-fraction equation by ξ and

Favre-filtering to get

$$\frac{\partial \overline{\rho} \tilde{\xi}^2}{\partial t} + \frac{\partial \overline{\rho} \tilde{u}_j \tilde{\xi}^2}{\partial x_j} = \frac{1}{Pe} \frac{\partial}{\partial x_j} \left(\overline{\mu} \frac{\partial \tilde{\xi}^2}{\partial x_j} \right) - 2\overline{\chi} - \frac{\partial \eta_j}{\partial x_j} , \tag{73}$$

where

$$\eta_j = \overline{\rho} \left(\widetilde{u_j \xi^2} - \tilde{u}_j \tilde{\xi}^2 \right) .$$

The $\tilde{\xi}^2$ equation is obtained by Favre-filtering the mixture-fraction equation then multiplying by $\tilde{\xi}$ to get

$$\frac{\partial \overline{\rho} \tilde{\xi}^2}{\partial t} + \frac{\partial \overline{\rho} \tilde{u}_j \tilde{\xi}^2}{\partial x_j} = \frac{1}{Pe} \frac{\partial}{\partial x_j} \left(\overline{\mu} \frac{\partial \tilde{\xi}^2}{\partial x_j} \right) - 2 \frac{\overline{\mu}}{Pe} \left(\frac{\partial \tilde{\xi}}{\partial x_j} \right)^2$$
$$- 2\tilde{\xi} \frac{\partial \zeta_j}{\partial x_j} . \tag{74}$$

At the larger scales, $\tilde{\xi}^2$ is approximately equal to $\tilde{\xi}^2$, the difference between the two being due to the filtering of ξ at the smaller scales. This implies, in particular, that the transfer rate of both quantities to the subgrid scales is nearly identical. Assuming in addition that the transfer rate of ξ-energy to the subgrid scales is equal to its dissipation rate at those scales, a comparison of Eqs. (73) and (74) suggests the model for $\overline{\chi}$:

$$\overline{\chi} \approx \frac{C_\chi \mu}{Pe} \left(\frac{\partial \tilde{\xi}}{\partial x_j} \right)^2 . \tag{75}$$

This is the first term in a model for $\overline{\chi}$ proposed by Girimaji and Zhou [41].

VI..2 Determining C_χ

The constant in Eq. (75) can be determined for the simple case of constant-density, isotropic turbulence such that $\tilde{\xi} = \overline{\xi}$. First assume that ξ is transformable, e.g., assume a periodic flow domain then let the period go to infinity. Forward and inverse Fourier transforms of ξ are defined as

$$\hat{\xi}(\mathbf{k}) \equiv \frac{1}{2\pi} \int_{-\infty}^{\infty} \exp(-i\mathbf{k} \cdot \mathbf{x}) \xi(\mathbf{x}) \, d\mathbf{x} , \tag{76}$$

$$\xi(\mathbf{x}) \equiv \int_{-\infty}^{\infty} \exp(i\mathbf{k} \cdot \mathbf{x}) \hat{\xi}(\mathbf{k}) \, d\mathbf{k} , \tag{77}$$

where the integrals are three-dimensional and $\mathbf{k}$ is a wavevector given in *radians per unit length*. The Fourier transform of $\partial \overline{\xi} / \partial x_j$ is

$$\widehat{\frac{\partial \overline{\xi}}{\partial x_j}} = -ik_j \hat{G}(k; \Delta) \hat{\xi}(\mathbf{k}) , \tag{78}$$

where $\hat{G}$ is a function only of the magnitude of $\mathbf{k}$, i.e., $k^2 \equiv \mathbf{k} \cdot \mathbf{k} = k_j k_j$. Writing $\partial \overline{\xi} / \partial x_j$ as the inverse transform of Eq. (78) and squaring both sides (and summing on j) leads to

$$\frac{\partial \overline{\xi}}{\partial x_j} \frac{\partial \overline{\xi}}{\partial x_j} = \int_{-\infty}^{\infty} \int_{-\infty}^{\infty} \exp(i(\mathbf{k} + \mathbf{m}) \cdot \mathbf{x})$$

$$(-k_j m_j)\widehat{G}(k;\Delta)\widehat{G}(m;\Delta)\widehat{\xi}(\mathbf{k})\widehat{\xi}(\mathbf{m})\,d\mathbf{k}\,d\mathbf{m} \ . \qquad (79)$$

Since $\xi(\mathbf{x})$ is real, $\widehat{\xi}(\mathbf{m}) = \widehat{\xi}^*(-\mathbf{m})$ where the asterisk denotes the complex conjugate. For homogeneous turbulence, the Fourier amplitudes $\widehat{\xi}(\mathbf{k})$ and $\widehat{\xi}^*(-\mathbf{m})$ are statistically orthogonal [42]; hence, the ensemble average $\left\langle \widehat{\xi}(\mathbf{k})\widehat{\xi}^*(-\mathbf{m}) \right\rangle$ is zero unless $-\mathbf{m} = \mathbf{k}$. The ensemble average of Eq. (79) thus becomes

$$\left\langle \frac{\partial \overline{\xi}}{\partial x_j} \frac{\partial \overline{\xi}}{\partial x_j} \right\rangle = \int_{-\infty}^{\infty} k^2 \widehat{G}^2(k;\Delta) \left\langle \widehat{\xi}^*(\mathbf{k})\widehat{\xi}(\mathbf{k}) \right\rangle \, d\mathbf{k} \ .$$
$$(80)$$

The integral in Eq. (80) is over all three-dimensional space and hence is equivalent to an integral over a spherical shell of radius k, followed by an integration over all shells. For example, if $d\sigma$ denotes a differential surface element of a shell, then Eq. (80) can be written

$$\int_{-\infty}^{\infty} k^2 \widehat{G}^2(k;\Delta) \left\langle \widehat{\xi}^*(\mathbf{k})\widehat{\xi}(\mathbf{k}) \right\rangle \, d\mathbf{k}$$

$$= \int_0^{\infty} k^2 \widehat{G}^2(k;\Delta) \oiint \left\langle \widehat{\xi}^*(\mathbf{k})\widehat{\xi}(\mathbf{k}) \right\rangle \, d\sigma\,dk \ . \qquad (81)$$

The shell integral in Eq. (81) is equal to twice the three-dimensional, scalar energy spectrum, i.e., $2E_\xi(k)$. Repeating the analysis for $\overline{\chi}$ leads to the result

$$\left\langle \overline{\frac{\partial \xi}{\partial x_j} \frac{\partial \xi}{\partial x_j}} \right\rangle = \left\langle \frac{\partial \xi}{\partial x_j} \frac{\partial \xi}{\partial x_j} \right\rangle = 2 \int_0^{\infty} k^2 E_\xi(k) \, dk \ .$$
$$(82)$$

Setting

$$\left\langle \frac{\partial \xi}{\partial x_j} \frac{\partial \xi}{\partial x_j} \right\rangle = C_\chi \left\langle \frac{\partial \overline{\xi}}{\partial x_j} \frac{\partial \overline{\xi}}{\partial x_j} \right\rangle$$

and solving for C gives

$$C_\chi = \frac{\int_0^{\infty} k^2 E_\xi(k) \, dk}{\int_0^{\infty} k^2 \widehat{G}^2(k;\Delta) E_\xi(k) \, dk} \ . \qquad (83)$$

Given the filter function G and the form of $E_\xi(k)$ (e.g., $k^{-5/3}$ in the inertial-convective range, k^{-1} in the viscous-convective range), C_χ is computable.

VII.   A Few Results

VII..1   Incompressible DNS Data

Data sets from $(256)^3$ point Direct Numerical Simulations of incompressible, isotropic, temporally-decaying, reacting turbulence were used to investigate the accuracy of the chemistry model. A key issue is the behavior of the model for a range of the activation temperature, T_a. Therefore, in the DNS, μ and ρ were kept constant in order to isolate the effects of T_a on the model performance. The chemical mechanism followed the single-step reaction: $Fuel + (r)Oxidizer \rightarrow (1+r)Product + Heat$, where r is the mass of oxidizer required to react with a unit mass of fuel. The ξ field

was initialized in a statistically homogeneous manner with regions of ones (fuel) and zeros (oxidizer) separated by thin mixing zones. The mean value of ξ was $\langle \xi \rangle = 0.25$, where the angle brackets denote an average over the entire $(256)^3$ domain. An equation for fuel was solved with the oxidizer and product obtained from the relations:

$$Y_o = Y_{o2}\,(1 - \xi) + r\,(Y_f - \xi Y_{f1}) \qquad (84)$$

$$Y_p = (r + 1)(\xi Y_{f1} - Y_f) \ . \qquad (85)$$

Here Y_f, Y_o and Y_p are the fuel, oxidizer and product mass fractions, respectively. For this case, the Shvab-Zeldovich functions in Eq. (9) are defined as:

$$\beta \equiv Y_f - Y_o/r$$

$$\beta_1 \equiv Y_{f1}$$

$$\beta_2 \equiv -Y_{o2}/r$$

and the stoichiometric value of the mixture-fraction is given by

$$\xi_{st} \equiv \frac{Y_{o2}}{(rY_{f1} + Y_{o2})} \ . \qquad (86)$$

In the DNS, the feed concentrations were set to $Y_{f1} = Y_{o2} = 1$ and r was assigned a value of 3, resulting in $\xi_{st} = 0.25$. The Y_f field was initialized according to the fast chemistry limit, i.e.,

$$Y_f = Y_{f1} \begin{cases} 0 & \text{if } \xi \leq \xi_{st} \\ (\xi - \xi_{st})/(1 - \xi_{st}) & \text{if } \xi > \xi_{st} \end{cases} . \qquad (87)$$

This provided initially high temperatures in the thin mixing regions, thus giving a 'spark' to initiate the reaction. The Schmidt number of the simulation was set to correspond to combustion in air, i.e., $Sc = 0.72$.

For single-step chemistry, it is only necessary to solve one laminar flamelet equation, which can be formulated in terms of Y_f, Y_o or Y_p. In terms of fuel, Eq. (58) is

$$\chi \frac{d^2 Y_f}{d\xi^2} = \dot{w}_f \ , \qquad (88)$$

where

$$\dot{w}_f = Da\rho Y_f \rho Y_o \exp(-T_a/T) \ . \qquad (89)$$

Note that Eqs. (61) and (84) are needed, in addition to Eq. (89), in order to solve Eq. (88).

Three simulations were performed, in which T_a took on values of 0, 4 and 8. The adiabatic flame temperature, defined as

$$T_f = \frac{Y_{o2} + rY_{f1} + Y_{f1}Y_{o2}q(\gamma - 1)/\gamma}{Y_{o2} + rY_{f1}} \ , \qquad (90)$$

was assigned a value of 8.14. Here q is the heat of combustion, defined as

$$q = h_f + rh_o - (r + 1)h_p \ , \qquad (91)$$

where h_f, h_o and h_p are the enthalpies of formation of the fuel, oxidizer and product respectively.

The DNS data fields were filtered onto a 16 X 16 X 16 point LES mesh. Then exact values for $\overline{Y_p}$, $\overline{w_f}$, $\overline{\xi}$, ξ_v^2, and $\overline{\chi}$ could be obtained by averaging over the $(16)^3$ DNS grid points in each LES grid cell. The latter three quantities were then used to obtain model values for $\overline{Y_p}$ and $\overline{w_f}$ using the equations of the previous section. The data were taken at one large-eddy turnover time after initialization of the ξ and Y_f fields. The eddy turnover time is defined as $u_{rms}^o t/l^o$, where $u_{rms} \equiv (\langle \vec{u} \cdot \vec{u} \rangle /3)^{1/2}$ and l is the integral scale of the turbulence [43]. The superscript $()^o$ indicates a value at $t = 0$.

Since the intent of LES is to resolve the large eddies, the filtered DNS fields represent an LES only if the filter width Δ is significantly smaller than the integral scale l. Furthermore, since the subgrid models appear to rely on inertial range behavior of the eddies near the grid-scale, Δ should be much greater than the Kolmogorov scale. At the time of filtering, the rato of filter width to the integral, Taylor and Kolmogorov scales of the DNS turbulence was $\Delta/l = 0.347$, $\Delta/\lambda = 1.59$ and $\Delta/\eta = 29.8$, respectively. The turbulent Damköhler number was $Da_t \equiv K^\star l^\star / u_{rms}^\star = 11.2$. The Reynolds number based on the integral scale was $Re_l \equiv u_{rms}^\star l^\star / \nu^\star = 417$ and the Taylor Reynolds number was $Re_\lambda \equiv u_{rms}^\star \lambda^\star / \nu^\star = 91$. The segregation of fuel and oxidizer was measured as $[\langle \xi^2 \rangle - \langle \xi \rangle^2]/[\langle \xi \rangle - \langle \xi \rangle^2] = 0.533$; initially this parameter had a value of 0.814.

VII..2   Effect of Arrhenius Kinetics

Figure 1 shows the correlation of exact and model values for $\overline{Y_p}$ for the three cases. Each point in the plots represents $\overline{Y_p}$ at a particular LES grid point. The plots on the left show the predictions of the large eddy laminar flamelet model (LELFM). The plots on the right show the results of assuming equilibrium (infinite Damköhler number) chemistry. In the plots on the right hand side ECL stands for Equilibrium Chemistry Limit. This limit provides a reference for judging the performance of LELFM. In the equilibrium chemistry limit

$$Y_p = (r+1)Y_{f1}\xi_{st} \begin{cases} \xi/\xi_{st} & \text{if } \xi \leq \xi_{st} \\ (1-\xi)/(1-\xi_{st}) & \text{if } \xi > \xi_{st} \end{cases},$$

$$(92)$$

so that, with the beta-pdf for ξ, the model prediction for $\overline{Y_p}$ becomes

$$\frac{\overline{Y_p}}{(r+1)Y_{f1}\xi_{st}} = \frac{I_{\xi_{st}}(a+1,b)}{\xi_{st}} \left[\frac{a}{a+b} \right]$$

$$+ \frac{I_{1-\xi_{st}}(b+1,a)}{1-\xi_{st}} \left[\frac{b}{a+b} \right]. \qquad (93)$$

where a and b are given in Eq. (70) and $I_\lambda(\alpha, \beta)$ is the Incomplete beta function [44, 40].

Figure 2 displays the correlation of exact and modeled values of $\overline{w_f}$ for the three cases. The LELFM prediction is shown in the plots on the left hand side and the Means Closure (MC) prediction is shown on the right hand side for comparison. The Means Closure is given by

$$\overline{w_f} = Da\rho\overline{Y_f}\rho\overline{Y_o} \exp(-T_a/\overline{T}).$$

A quantatative analysis is performed by applying a least squares fit to the data in Figures 1 and 2. The slope of the best-fit line, i.e., m, where $y = mx$, is given in Table 1 for $\overline{Y_p}$ and $\overline{w_f}$. The results for $\overline{Y_p}$

Table 1: Slope of linear, least-squares-fit line of the data in Figures 1 and 2.

	$T_a = 0$	$T_a = 4$	$T_a = 8$
$\overline{Y_p}$ LELFM	1.037	1.113	1.145
$\overline{Y_p}$ ECL	1.221	1.438	1.820
$\overline{w_f}$ LELFM	0.770	0.718	0.976
$\overline{w_f}$ MC	3.503	1.804	1.480

show that, as T_a is increased, the slope of the ECL prediction increases much more than the slope of the LELFM prediction. The results for $\overline{w_f}$ show that the LELFM and MC predictions improve with increasing T_a. In every case, the LELFM prediction for $\overline{w_f}$ is much better than the MC prediction.

The LELFM results for $\overline{Y_p}$ degrade somewhat with increasing T_a. This is due to the increase of chemical time scales relative to local flow time scales. However, as the chemical equilibrium departures increase, the accuracy of LELFM relative to ECL improves due to the fact that there are no mechanisms in ECL to account for these departures. At higher T_a the reaction zones are narrower, thus several of the assumptions of laminar flamelets should be more valid. The LELFM approach appears quite accurate for predicting $\overline{w_f}$ at high T_a. The same analysis was also applied to $(128)^3$ DNS fields at a Reynolds number of $Re_l = 135$ with virtually identical results.

VII..3   Sources of Error

Regarding the scatter in the LELFM values for $\overline{w_f}$, this may be partly due to neglection of subgrid-scale fluctuations in χ_o. Consider Eqs. (88) and (89) with $T_a = 0$. On the stoichiometric surface ξ_{st}, $Y_f = Y_o \sim \Delta\xi_R$, where $\Delta\xi_R$ denotes the width of the reaction zone. Furthermore $\partial^2 Y_f/\partial \xi^2 \sim 1/\Delta\xi_R$, from which it follows that $Y_f \sim \chi_o^{1/3}$ [37, 38]. This means that Eq. (68) is a good approximation for the species mass fractions. However, a similar approximation may be less accurate for the reaction rates, since (at least for isothermal chemistry) $\overline{w_f} \sim Y_f Y_o \sim \chi_o^{(2/3)}$. This conclusion seems justified by the DNS data, since the scatter in $\overline{w_f}$ is observed to be greater than the scatter in $\overline{Y_p}$.

Some additional modeling error occurs if ξ_{st} is either very high or very low. Extreme values of ξ_{st} mean that the integral in Eq. (69) is essentially determined over a very small range of ξ. In other words, for low ξ_{st}, the model becomes very sensitive to the shape of the assumed beta-pdf between 0 and ξ_{st}. In order to examine this effect, equilibrium values for Y_p were computed from the $(256)^3$ DNS, using Eq. (92), then filtered to obtain exact values for $\overline{Y_p}$. Modeled values were computed from the DNS using Eq. (93). Since the chemistry is exactly determined in terms of the equilibrium expression for Y_p, the only assumption being tested is that of a beta-pdf. Model versus exact values are plotted in Figure 3, for equilibrium chemistry, using various values for ξ_{st}. The value of $\xi_{st} = 0.0476$ corresponds to methane-air combustion and the value of $\xi_{st} = 0.0244$ corresponds to hydrogen-air combustion (considered as single-step reactions). The plots show that the correlation decreases as ξ_{st} becomes very low; however, the model results for the methane-air and hydrogen-air cases look quite reasonable. Therefore, the beta-pdf model should be applicable to common combustion problems, albeit with some degradation of accuracy for reactions with a very low ξ_{st}.

Some additional errors in LELFM can be expected if modeled values for ξ_v^2 and $\overline{\chi}$ are used instead of exact values. The effects of errors in ξ_v^2 and $\overline{\chi}$ on the overall model have been examined in Refs. [40] and [45]. It appears that errors in modeling ξ_v^2 and $\overline{\chi}$ have only a small effect on the overall accuracy of the chemistry model; however, further investigations are needed to better quantify these effects.

VIII. Conclusions

This paper has described how to apply LES to turbulent, reacting flows. The potential of LES for treating nonpremixed combustion problems is very good due to the fact that the rate of mixing is controlled by the large eddies. Certain simplifications to the Favre-filtered equations, commonly used for nonreacting flows [17], may be less valid for reacting flows with large heat release. In such cases, subgrid fluctuations in temperature may need to be taken into account.

Currently there are two main approaches to modeling the SGS chemistry: mixture-fraction based methods and Monte-Carlo based methods. The mixture-fraction methods require the least resources and can be used immediately in three-dimensional simulations. However, they are currently limited by the principal assumptions of fairly fast chemistry and equal diffusivities. Monte-Carlo methods have the potential to treat more physics, e.g., slower chemistry, differential diffusion, etc. However, they require roughly two orders of magnitude more computing resources, one for the Monte-Carlo simulation at each grid point and one for the detailed chemistry at each grid point.

The large eddy laminar flamelet model predicts filtered chemical species and reaction rates, given a Favre-filtered mixture-fraction, its subgrid-scale variance and dissipation rate. The model is employed by building tables for $\overline{Y_i}$ and $\overline{w_i}$ prior to running an LES. Since these quantities are obtained in an LES via table lookup, the chemistry model is very inexpensive to use. The LELFM predictions for $\overline{Y_i}$ are substantially more accurate than what would be obtained by assuming equilibrium chemistry. Increasing the activation temperature decreases the accuracy of the model by lowering the rate of reaction. However, at higher T_a, LELFM is much better than the equilibrium chemistry assumption. The values for $\overline{w_i}$ from LELFM are far more accurate than Means Closure predictions.

Finally, the beta-pdf was demostrated to be adequate for modeling reactions with very low values of ξ_{st}. The assumed beta-pdf approach should thus be applicable to many common reactions, e.g., methane-air combustion. However, errors in the beta-pdf model increase as ξ_{st} approaches zero or unity.

Acknowledgments

This work was supported by NSF under grant number CTS-9415280, AFOSR under contract number F49620-97-1-0092 and by DOE. The DNS code was written by V. Nilsen and executed on the Connection Machine Model 5 at the National Center for Supercomputing Applications. In addition, we wish to thank Professor George Kosály for much input in deriving the chemistry model.

References

[1] Givi P. (1989) Model Free Simulations of Turbulent Reactive Flows. *Prog. Energy Combust. Sci.* **15**, 1–107.

[2] Jou W. H. and Riley J. J. (1989) Progress in Direct Numerical Simulations of Turbulent, Reacting Flows. *AIAA J.* **27**, 1543–1556.

[3] Pope S. B. Computations of Turbulent Combustion: Progress and Challenges (1990) In *Proceedings of 23rd Sympoisium (Int.) on Combustion*, pages 591–612. The Combustion Institute, Pittsburgh, PA, 1990.

[4] Schumann U. (1989) Large-Eddy Simulation of Turbulent Diffusion with Chemical Reactings in the Convective Boundary Layer. *Atmospheric Environment.* **23**, 1713–1726.

[5] Jiménez J. , Liñán A. , Rogers M. M. , and Higuera F. J. (Submitted for publication) A-Priori Testing of Sub-grid Models for Chemicaly

Reacting Nonpremixed Turbulent Shear Flows. *J. Fluid Mech.*

[6] Gao F. and O'Brien E. E. (1993) A Large-Eddy Simulation Scheme for Turbulent Reacting Flows. *Phys. Fluids A.* **5**, 1282–1284.

[7] Colucci P. J. , Jaberi F. A. , Givi P. , and Pope S. B. The Filtered Density Function for Large Eddy Simulation of Turbulent Reacting Flows Bulletin of the American Physical Society, November, 1996.

[8] McMurtry P. A. , Menon S. , and Kerstein A. R. A Linear Eddy Subgrid Model for Turbulent Reacting Flows: Applications to Hydrogen-air Combustion (1992) In *Proceedings of 24th Symp. (int.) on Combustion*, pages 271–278. The Combustion Institute, Pittsburgh, PA, 1992.

[9] Menon S. . McMurtry P. A. , and Kerstein A. R. *A Linear-Eddy Mixing Model for Large Eddy Simulation of Reacting Homogeneous Turbulence* (1993), pages 287–314 Cambridge University Press.

[10] Frankel S. H. , Adumitroaie V. , Madnia C. K. , and Givi P. Large-Eddy Simulation of Turbulent Reacting Flows By Assumed Pdf Methods (1993) In *Engineering Applications of Large Eddy Simulations*, volume **162**, pages 81–101. Fluids Engineering Division, ASME, 1993.

[11] Ferziger J. H. *Subgrid-Scale Modeling* (1993), pages 37–54 Cambridge University Press.

[12] Ragab S. A. and Sheen S.-C. *Large Eddy Simulation of Mixing Layers* (1993), pages 255–285 Cambridge University Press.

[13] Libby P. A. and Williams F. A. *Turbulent Reacting Flows* (1980), volume **44**, chapter 1 Springer-Verlag, Berlin.

[14] Bilger R. W. *Turbulent Flows with Nonpremixed Reactants* (1980), volume **44**, chapter 3, pages 2–36 Springer-Verlag, Berlin.

[15] Schumann U. Results of a Numerical Simulation of Turbulent Channel Flows (1973) In Dalle-Donne M. , editor, *Int'l. Meeting of Reactor Heat Transfer*, pages 230–251, 1973.

[16] Erlebacher G. , Hussaini M. Y. , Speziale C. G. , and Zang T. A. ICASE Report 90-76, ICASE/NASA Langley Research Center, 1990.

[17] Moin P. , Squires K. , Cabot W. , and Lee S. (1991) A Dynamic Subgrid-Scale Model for Compressible Turbulence and Scalar Transport. *Phys. Fluids A.* **3**, 2746–2757.

[18] Vreman A. W. , Geurts B. J. , and Kuerten J. G. M. *Subgrid-Modelling in LES of Compressible Flow* (1994), pages 133–144 Kluwer Academic Publishers.

[19] Germano M. . Piomelli U. , Moin P. , and Cabot W. H. (1991) A Dynamic Subgrid-Scale Eddy Viscosity Model. *Phys. Fluids A.* **3**, 1760–1765.

[20] Meneveau C. , Lund T. S. , and Cabot W. A Lagrangian Dynamic Subgrid-Scale Model of Turbulence (1994) In *Proceedings of the 1994 Summer Program*, pages 271–299. Center for Turbulence Research, 1994.

[21] Réveillon J. and Vervisch L. (1996) Response of the Dynamic LES Model to Heat Release Induced Effects. *Phys. Fluids A.* **8**, 2248–2250.

[22] Carati D. . Jansen K. , and Lund T. A Family of Dynamic Models for Large-Eddy Simulation Center for Turbulence Research, Annual Research Briefs, 1995.

[23] Ghosal S. . Lund T. S. , Moin P. , and Akselvoll K. (1995) A Dynamic Localization Model for Large-Eddy Simulation of Turbulent Flows. *Phys. Fluids A.* **286**, 229–255.

[24] Piomelli U. and Liu J. (1995) Large-Eddy Simulation of Rotating Channel Flows using a Localized Dynamic Model. *Phys. Fluids A.* **7**, 839–848.

[25] Lilly D. K. (1992) A Proposed Modification of the Germano Subgrid-Scale Closure Method. *Phys. Fluids A.* **4**, 633–635.

[26] Yoshizawa A. (1986) Statistical Theory for Compressible Turbulent Shear Flows, with the Application to Subgrid Modeling. *Phys. Fluids A.* **29**, 2152–2164.

[27] Lund T. S. . Ghosal S. , and Moin P. Numerical Experiments with Highly-Variable Eddy Viscosity Models (1993) In Ragab S. A. and Piomelli U. , editors, *Engineering Applications of Large Eddy Simulations*, volume **162**, pages 7–11. Fluids Engineering Division, ASME, 1993.

[28] Cook A. W. and Riley J. J. (1996) Direct Numerical Simulation of a Turbulent Reactive Plume on a Parallel Computer. *J. Comput. Phys.* **129**, 263–283.

[29] Peters N. (1984) Laminar Diffusion Flamelet Models in Non-premixed Turbulent Combustion. *Prog. Energy Combust. Sci.* **10**, 319–339.

[30] Nomura K. K. and Elghobashi S. E. (1992) Mixing Characteristics of an Inhomogeneous Scalar in Isotropic and Homogeneous Sheared Turbulence. *Phys. Fluids A.* **4**, 606–625.

[31] Bish E. S. (1996) *A New Model for Nonequilibrium Mixing-Chemistry Coupling in Nonpremixed and Partially Premixed Turbulent Combustion* PhD thesis, The University of Michigan, Ann Arbor, MI.

[32] Bish E. S. and Dahm W. J. A. (1995) Strained Dissipation and Reaction Layer Analyses of Nonequilibrium Chemistry in Turbulent Reacting Flows. *Combust. Flame.* **100**, 457.

[33] Cook A. W. , Riley J. J. , and Kosály G. (1997) A Laminar Flamelet Approach to Subgrid-Scale Chemistry in Turbulent Flows. *Combust. Flame.* **109**, 332–341.

[34] Southerland K. B. and Dahm W. J. A. A Four-Dimensional Experimental Study of Conserved Scalar Mixing in Turbulent Flows Report No. 026779-12, The University of Michigan, Ann Arbor, MI, 1994.

[35] McMurtry P. A. , Jou W.-H. , Riley J. J. , and Metcalfe R. W. (1986) Direct Numerical Simulations of a Reacting Mixing Layer with Chemical Heat Release. *AIAA J.* **24**, 962.

[36] Kuznetsov V. R. and Sabel'nikov V. A. Turbulence and Combustion (1990) In Chigier N. , editor, *Combustion: An International Series.* Hemisphere Publishing Corp., 1990.

[37] Mell W. E. , Nilsen V. , Kosály G. , and Riley J. J. (1994) Investigation of Closure Models for Nonpremixed Turbulent Reacting Flows. *Phys. Fluids A.* **6**, 1331–1356.

[38] Buriko Y. Y. , Kuznetsov V. R. , Volkov D. V. , Zaitsev S. A. , and Uryvsky A. F. (1994) A Test of a Flamelet Model for Turbulent Nonpremixed Combustion. *Combust. Flame.* **96**, 104–120.

[39] Williams F. A. (1985) *Combustion Theory 2nd ed.* Addison-Wesley.

[40] Cook A. W. and Riley J. J. (1994) A Subgrid Model for Equilibrium Chemistry in Turbulent Flows. *Phys. Fluids A.* **6**, 2868–2870.

[41] Girimaji S. S. and Zhou Y. (1996) Analysis and Modeling of Subgrid Scalar Mixing using Numerical Data. *Phys. Fluids A.* **8**, 1224–1236.

[42] Batchelor G. K. (1953) *The Theory of Homogeneous Turbulence* Cambridge University Press, Cambridge.

[43] H. Tennekes and Lumley J. L. (1972) *A First Course in Turbulence* MIT Press, Cambridge.

[44] Madnia C. K. and Givi P. *Direct Numerical Simulation and Large Eddy Simulation of Reacting Homogeneous Turbulence* (1993), pages 315–346 Cambridge University Press.

[45] Cook A. W. , Riley J. J. , and deBruynKops S. M. A Subgrid-Scale Model for Nonpremixed Turbulent Combustion Presented at the Eleventh Annual Turbulent Shear Flow Conference in Grenoble, France, 1997.

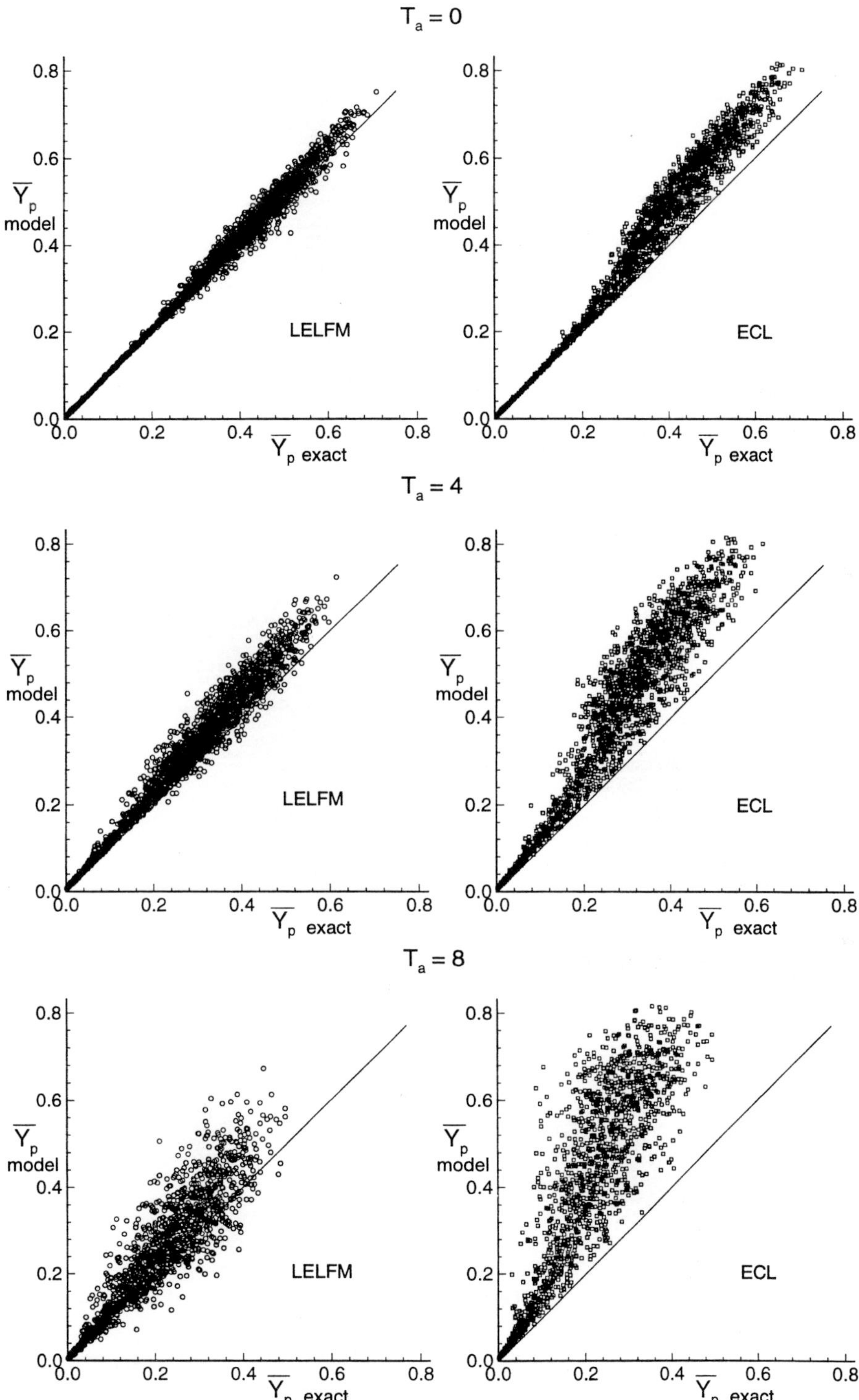

Figure 1: Correlation of exact and modeled product
mass fractions at various activation temperatures.

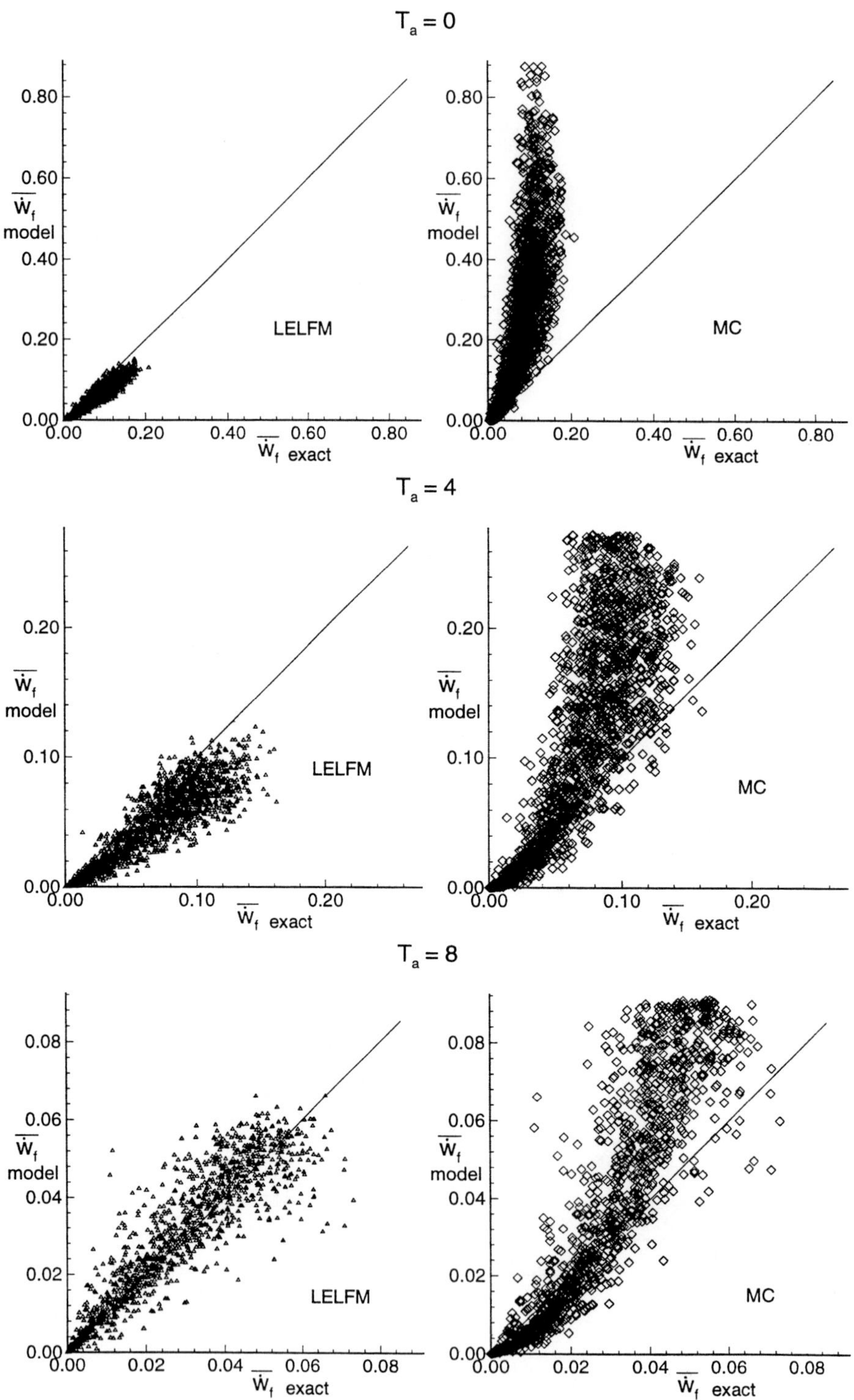

Figure 2: Correlation of exact and modeled reaction rates at various activation temperatures.

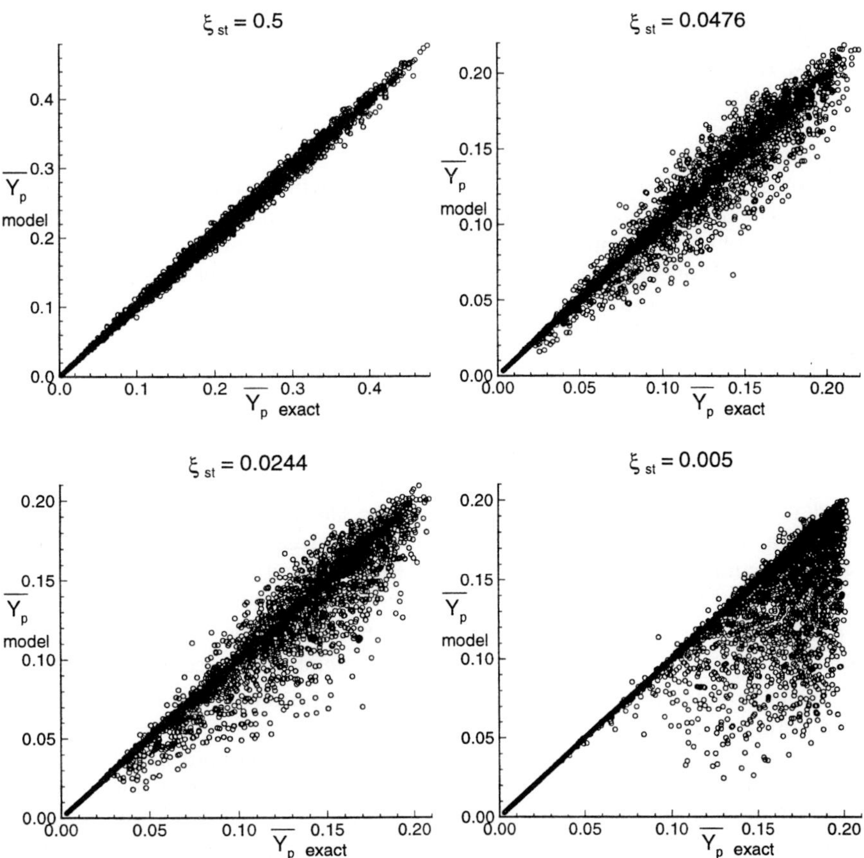

Figure 3: Correlation of exact equilibrium chemistry results with beta-pdf model at various stoichiometries.

APPLICATION OF NUMERICAL METHODS
TO MODELING THE STELLAR WIND
AND INTERSTELLAR MEDIUM INTERACTION

N. POGORELOV [2]   T. MATSUDA [1]

(1)   Department of Earth & Planetary Sciences, Kobe University;
1-1 Rokkodai-cho, Nada-ku, Kobe 657, JAPAN
(permanent address) Institute for Problems in Mechanics;
101 Vernadskii Ave., Moscow 117526, RUSSIA
(2)   Department of Earth & Planetary Sciences, Kobe University;
1-1 Rokkodai-cho, Nada-ku, Kobe 657, JAPAN

Abstract

Interaction between the stellar wind, including the solar wind, and the interstellar medium has long been the subject of investigation by both astrophysicists and fluid dynamicists. This is, first, due to the possibility of comparison of physical models for such interaction with the measurements performed by Voyager, Pioneer, and Ulysses spacecrafts. On the other hand, a complicated structure of the flow containing several discontinuities makes it a challenging problem for the application of modern numerical methods both in gasdynamic and magnetogasdynamic (MHD) cases.

In the solar wind case, the problem becomes even more complicated, since the charge-exchange processes between ions and neutral particles must be taken into account. The continuum equations are not applicable to the description of the neutral particle motion, for their mean free path is much larger than the characteristic length scale of the problem. In this case, either approximate coupling models or direct Monte-Carlo simulation are required. The spatial nonuniformity of the solar wind and its perturbations and periodicity make the problem three-dimensional and non-stationary. From a mechanical viewpoint the problem represents the interaction of the uniform interstellar medium and the spherically-symmetric (or asymmetric) solar wind flow. We consider various approaches used by different authors to solve this problem numerically.

The presence of the contact surface dividing the two flows rises the question of its stability. We discuss the reasons of such instabilities and parameters which influence it.

The presence of the interstellar magnetic field necessitates solution of the MHD equations for proper analysis of the obtained data. Although the system of governing equations remains hyperbolic in this case, the multivariance of the exact solution to the MHD Riemann problem makes inefficient its application for regular calculations. On the other hand, the solution to the linearized Riemann problem is nonunique. We discuss the possible ways of applying the Roe-type methods and some simplified approaches for numerical solution of the ideal MHD equations. One of the difficulties in the solution of the MHD system is the satisfaction of the magnetic field divergence-free condition. Different ways to solve this task are discussed. If the magnetic field vector in the uniform interstellar medium flow is not parallel to the velocity vector, the problem becomes three-dimensional. Both approximate and exact numerical solutions are considered which were applied in this case.

Far-field numerical boundary conditions play an essential role in astrophysical applications owing to very large length scales usual for these problems. We discuss several approaches that may be useful to solve problems similar to the stellar wind and interstellar medium interaction.

1. INTRODUCTION

The problem of the stellar wind, with the emphasis on the solar wind, interaction with the interstellar medium has long been the topic of interest for astrophysicists and specialists in the field of the solar–terrestrial physics [2], [3], [21], [28], and [69]. In [6],

Received on March 1, 1997.

the application of the continuum equations for this problem is systematically discussed. The first qualitative model for the interaction of the solar wind (SW) and the local interstellar medium (LISM) was proposed by E.N. Parker [69]. He assumed the interstellar medium to be a subsonic stream with the Mach number $M_\infty \ll 1$ (see Fig. 1).

Here HP is a heliopause dividing the SW and the LISM flow. Generally speaking the above assumption is not correct, since the velocity of the interstellar medium is $V_\infty \sim 20$ km/s and the scattering experiments for the solar radiation show that the temperature T_∞ of charged particles constituting the LISM is about $10^4 K$. Taking into account that the number density of charged particles is usually considered to be $n_\infty \sim 0.1\,\mathrm{cm}^{-3}$, the LISM flow can most likely be supposed supersonic than subsonic. The supersonic model of the interaction was first proposed in [5]. The important feature of this approach lies in the application of the continuum (Euler gasdynamic) equations only to the charged particles of both counteracting winds. Although the presence of turbulent pulsations of plasma is supposed to be insignificant for the mean flow structure, their influence is realized by a remarkable change of transport coefficients due to the possibility of scattering of charged particles on electromagnetic plasma fluctuations. This results in the substantial decrease of their mean free path compared with that calculated on the basis of the Coulomb collisions. The solar wind consists mainly of electrons and protons with the number density $n_e \sim 10\,\mathrm{cm}^{-3}$ and velocity $V_e \sim 400$–500 km/s and is also considered supersonic. The index "e" corresponds to values measured at $1\mathrm{AU} = 1.5 \times 10^{11}$ km, that is, at the Earth distance from the Sun. Thus, we can consider this problem, from a gasdynamic viewpoint, as a an interaction of the supersonic spherically-symmetric (or asymmetric) source flow of SW with the uniform supersonic LISM flow. This assumption gave rise to a so-called two-shock model. Generally speaking, this model can be easily obtained numerically if one choses as initial data for this interaction the arbitrary jump between the SW and the LISM parameters.

The source flow is essentially a combination of supersonic jet and blunt-body flows which are perfectly well studied and described in classical gas dynamics. The extension to the solar wind and space situation [31], [95] has been validated by numerous spacecraft observations of the plasma environment of several planets. Different flow regimes and shock-wave flow structure for the SW–LISM interaction was discussed in [106].

In [7] and [113] the axisymmetric problem of the interaction was considered on the basis of the shock-

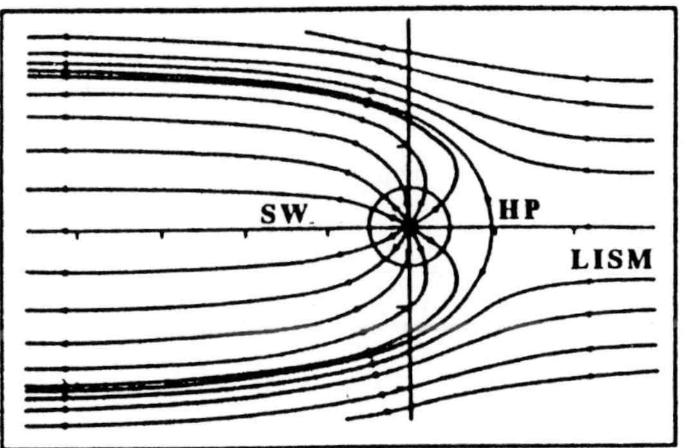

Fig.1: Schematic picture of the SW–LISM interaction [3]

fitting approach, but due to the limitations of the applied numerical method the authors calculated only the upwind part of the flow. In [55] and [92] the problem of the stellar wind interaction with the interstellar medium was analyzed in the closed region surrounding the star. These numerical results confirmed the scheme [106], but the bullet shape of the internal shock was obtained for a broader range of parameters. The general schematic picture is shown in Fig. 2.

Here TS is the inner shock terminating the solar wind within the heliopause (HP), BS is the bow shock, or the outer shock. TS at a certain point may turn to form the Mach disk (MD). At this triple point the reflected shock (RS) and the slip line (SL) originate.

This picture is similar to that suggested for the SW–comet interaction in [106]. The termination shock configuration is caused by its Mach-type reflection from the symmetry axis.

In [3], [18], [29], [32], [40], [104], and [105] the opinion was stated that the resonance recharge processes between the neutral and the charged particles should strongly influence the flow picture. The description of the neutral particle motion cannot be made on the basis of the continuum approach. For this reason in [8] an approximate method of taking into account this influence was suggested and, later, in [11] a self-consistent model was developed that takes into account the recharge processes. The Monte-Carlo method was used to calculate the trajectories of neutral particles. As was admitted in [104] and confirmed by [8] and [11], the charge-exchange effect effectively diminishes the Mach number of the LISM flow. This justifies development of alternative subsonic models of the interaction [48].

Of great importance are also nonstationary problems associated with the variable solar activity. In [15], [75], and [97] the problem was investigated of the time-dependent SW perturbation influence on the

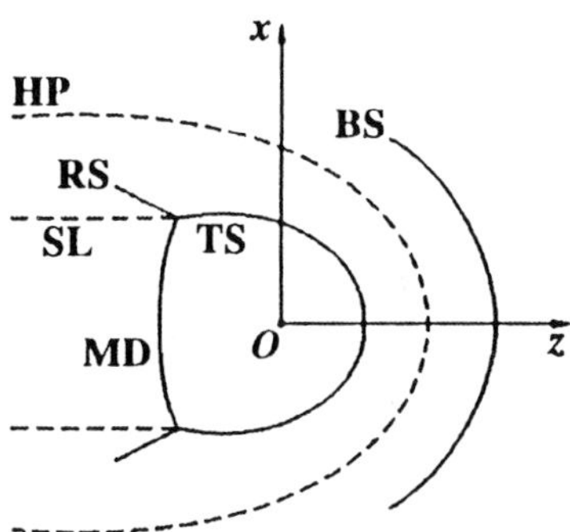

Fig.2: General scheme of supersonic interaction

whole flow structure. The nonstationary picture of the flow and the termination shock response to the 11-year variation of the solar wind were studied in [17], [47], and [76].

T. Matsuda et al. [55] discussed the instabilities of the contact discontinuity dividing the SW and the LISM flow. These instabilities originated in its lateral region and near the stagnation point of the flow. Unstable solutions were obtained in the parameter range which was not exactly suitable for the solar wind and the local interstellar medium flow and, therefore, were disputed in [96]. Recently in [52] a hydrodynamic instability of the heliopause driven by plasma–neutral charge-exchange processes was discussed.

According to the solar minimum observations by the Ulysses spacecraft, the solar wind properties depend on helioaltitude. The solar wind in this case is no longer spherically-symmetric (see also [99]). In [71] a 3D problem of the interaction was used and obtained results were compared with those for the isotropic solar wind.

The influence of the interstellar magnetic field can be important in the regions, for which the magnetic pressure is comparable by its value with the dynamic pressure of the flow. Magnetic field causes an increase of the maximum speed of small perturbations in the LISM flow, thus leading to the decrease of its effective Mach number [33], [34]. The ordinary Mach number of the LISM is often assumed to be $M_\infty = 2$. At the same time the magnitude of the LISM magnetic field, although not known very well, is estimated within 7×10^{-7} and 3×10^{-6} Gauss [3]. This means, as will be shown later, that the magnetic pressure can exceed the value of the thermal pressure and is the reason of including magnetic field into consideration. Its influence can be especially effective when the charge-exchange processes are included, since both this effects lead to the decrease of the effective Mach number. This means that the LISM flow can become subsonic

and the bow shock can disappear (see also [33] and [104]). The stellar wind–interstellar medium interaction with taking into account magnetic field was first studied in [59], although some of the results seem to be misinterpreted (see [12]). In the latter paper the problem was investigated by the shock-fitting method only in the upwind part of the flow due to the limitations of the numerical scheme. In [80] the solution was presented of the axisymmetric problem (the LISM magnetic field strength vector was assumed to be parallel to its velocity vector) of the SW–LISM interaction in the closed region surrounding the star. In [34] the shape of the heliopause was studied on the basis of the Newtonian approach in the 3D case of the arbitrary angle between the LISM magnetic field and velocity vectors. The authors found this shape approximately by equating the values of the total pressure on the both sides of the heliopause. In [81] this problem was first examined numerically.

The complicated pattern of the flow containing a number of interacting shocks, enhanced by various physical phenomena, makes it a challenging problem for the application of modern numerical methods invented for pure gasdynamic and magnetogasdynamic application. Numerical solution of this problem is often associated with the solution of such accompanying problems as non-reflecting boundary conditions, numerical implementation of the condition of the magnetic charge absence, etc. These problems are of general importance for the young, but quickly developing, field called computational fluid dynamics. In Section 2 of this review we present the mathematical statement of the problem, write out the system of governing equations and boundary conditions. The choice of initial conditions for the magnetogasdynamic (MHD) interaction is discussed. In Section 3 we discuss stationary solutions of the gasdynamic problem on the basis of shock-fitting and shock-capturing methods. In Section 4 different approaches are discussed which take into account the charge-exchange processes. In Section 5 we describe instabilities originating under certain circumstances in this problem and the reasons causing them. In Section 6 nonstationary solutions are considered resulting from the solar wind disturbances and its periodicity. In Section 7 we briefly discuss the effects of the solar wind spatial asymmetry. And, finally, Section 8 deals with numerical modeling of the solar wind interaction with the magnetized interstellar medium.

2. Mathematical statement of the problem

To make the paper more concise, we write out in this section the mathematical statement of the problem based on the MHD equations. The Euler gasdynamic equations can be easily obtained from the former one by omitting the terms containing the magnetic field strength and the equations describing the behavior of its components.

2.1 The system of governing equations

The system of governing equations for a MHD flow of an ideal, infinitely conducting, perfect plasma in the Cartesian coordinate system x, y, z, shown in Fig. 2 (y-axis is perpendicular to the picture plane), can be written as follows (one fluid approximation):

$$\frac{\partial \mathbf{U}}{\partial t} + \frac{\partial \mathbf{E}}{\partial x} + \frac{\partial \mathbf{F}}{\partial y} + \frac{\partial \mathbf{G}}{\partial z} + \mathbf{H} = \mathbf{0}, \qquad (1)$$

where

$$\mathbf{U} = \begin{bmatrix} \rho \\ \rho u \\ \rho v \\ \rho w \\ e \\ B_x \\ B_y \\ B_z \end{bmatrix}, \quad \mathbf{E} = \begin{bmatrix} \rho u \\ \rho u^2 + p_0 - \dfrac{B_x^2}{4\pi} \\ \rho u v - \dfrac{B_x B_y}{4\pi} \\ \rho u w - \dfrac{B_x B_z}{4\pi} \\ (e + p_0)u - \dfrac{B_x}{4\pi}(\mathbf{v} \cdot \mathbf{B}) \\ 0 \\ u B_y - v B_x \\ u B_z - w B_x \end{bmatrix},$$

$$\mathbf{F} = \begin{bmatrix} \rho v \\ \rho u v - \dfrac{B_x B_y}{4\pi} \\ \rho v^2 + p_0 - \dfrac{B_y^2}{4\pi} \\ \rho v w - \dfrac{B_y B_z}{4\pi} \\ (e + p_0)w - \dfrac{B_y}{4\pi}(\mathbf{v} \cdot \mathbf{B}) \\ v B_x - u B_y \\ 0 \\ v B_z - w B_y \end{bmatrix},$$

$$\mathbf{G} = \begin{bmatrix} \rho w \\ \rho u w - \dfrac{B_x B_z}{4\pi} \\ \rho v w - \dfrac{B_y B_z}{4\pi} \\ \rho w^2 + p_0 - \dfrac{B_z^2}{4\pi} \\ (e + p_0)w - \dfrac{B_z}{4\pi}(\mathbf{v} \cdot \mathbf{B}) \\ w B_x - u B_z \\ w B_y - v B_z \\ 0 \end{bmatrix}$$

In system (1) ρ, u, v, w, B_x, B_y, and B_z are the density and the components of the velocity $\mathbf{v}$ and of the magnetic field strength vector $\mathbf{B}$. We introduced here also the total pressure $p_0 = p + \mathbf{B}^2/8\pi$ (p is the thermal pressure) and the total energy per unit volume

$$e = \frac{p}{\gamma - 1} + \frac{\rho(u^2 + v^2 + w^2)}{2} + \frac{\mathbf{B}^2}{8\pi},$$

where $\gamma = 5/3$ is the specific heat ratio corresponding to the fully ionized plasma. The quantities of density, pressure, velocity, and magnetic field strength are normalized, respectively, by ρ_∞, $\rho_\infty V_\infty^2$, V_∞, and $V_\infty \sqrt{\rho_\infty}$, where the index "$\infty$" marks the values in the uniform LISM flow. Time and the linear dimension are respectively related to L/V_∞ and L, where L is equal to 1 AU. The above formulation implies that molecular and magnetic viscosities, heat conductivity, and anomalous transport effects are neglected. The source term $\mathbf{H}$ can be both of physical and of geometrical origin and will be specified separately for each problem. For example, if the flow is axisymmetric system (1) can be rewritten in the plane coordinate system x, z as follows:

$$\frac{\partial \mathbf{U}}{\partial t} + \frac{\partial \mathbf{E}}{\partial x} + \frac{\partial \mathbf{G}}{\partial z} + \mathbf{H} = \mathbf{0}, \qquad (2)$$

where

$$\mathbf{U} = \begin{bmatrix} \rho \\ \rho u \\ \rho w \\ e \\ B_x \\ B_z \end{bmatrix}, \quad \mathbf{H} = \frac{1}{x} \begin{bmatrix} \rho u \\ \rho u^2 - \dfrac{B_x^2}{4\pi} \\ \rho u w - \dfrac{B_x B_z}{4\pi} \\ (e + p_0)u - \dfrac{B_x}{4\pi}(\mathbf{v} \cdot \mathbf{B}) \\ 0 \\ u B_z - w B_x \end{bmatrix},$$

$$\mathbf{E} = \begin{bmatrix} \rho u \\ \rho u^2 + p_0 - \dfrac{B_x^2}{4\pi} \\ \rho u w - \dfrac{B_x B_z}{4\pi} \\ (e + p_0)u - \dfrac{B_x}{4\pi}(\mathbf{v} \cdot \mathbf{B}) \\ 0 \\ u B_z - w B_x \end{bmatrix},$$

$$
\mathbf{G} = \begin{bmatrix} \rho w \\ \rho u w - \dfrac{B_x B_y}{4\pi} \\ \rho w^2 + p_0 - \dfrac{B_z^2}{4\pi} \\ (e + p_0)w - \dfrac{B_z}{4\pi}(\mathbf{v}\cdot\mathbf{B}) \\ w B_x - u B_z \\ 0 \end{bmatrix}
$$

This system is valid in the half-plane $x0z$ and can be obtained from (1) in the assumption of cylindrical symmetry. Its another form is

$$
\frac{\partial x \mathbf{U}}{\partial t} + \frac{\partial x \mathbf{E}}{\partial x} + \frac{\partial x \mathbf{G}}{\partial z} + \tilde{\mathbf{H}} = \mathbf{0}, \tag{3}
$$

where

$$
\tilde{\mathbf{H}} = [0, -p_0, 0, 0, 0, 0]^{\mathrm{T}}
$$

Though both presentations of governing equations are mathematically equivalent, for numerical reasons it is often more convenient to use the former one, since it is supposed to give more stable results in the vicinity of the geometrical singularity $x = 0$. The Euler gasdynamic equations can be obtained in different forms from systems (1) and (2) by assuming $\mathbf{B} \equiv \mathbf{0}$.

2.2 Initial and boundary conditions

Calculations are usually performed in the computational region between the inner and the outer spherical surface (circular in the axisymmetric case). The flow from the Sun is supposed to be supersonic at the termination shock distance. For this reason we specify all parameter values at the inner boundary sphere. The uniform LISM flow is also supersonic and we can specify the parameter values at the inflow side of the outer boundary. The treatment of outflow boundary can be more complicated. In [96] all parameters were extrapolated with the zeroth order along the fluid particle trajectory. This approach cannot be expected suitable for deeply subsonic outflow boundary. Another approach was proposed in [92]. This method is based on introducing imaginary cells next to the boundary. These cells are filled with the LISM gas at infinity. To find the flux through the outer boundary the Riemann problem is solved between the imaginary and the adjacent cell values. At the boundary segments with the subsonic–supersonic transition the rarefaction wave relations are used. This results in the following interpretation (see [75] and [78]) of the method initially developed for a purely gasdynamic case and makes possible its extension to MHD problems [80].

Consider the method based on the two strictly nonreflecting conditions: the well-known extrapolation condition for a supersonic exit that provides a characteristically compatible approximation of the equations on the boundary and the procedure [75], [78], developed earlier for gasdynamic flows.

The idea of application of the relations in the rarefaction wave for the realization of the far-field boundary conditions lies in the artificial locating of the sonic point on the exit boundary. If the flow is supersonic at infinity such a procedure gives reasonable results and allows one to perform calculations in the cases for which other known approaches fail. The interpretation of our method is the following. Assume that parameters inside the chosen computational region fully define the flow behavior outside the boundary. In the case of subsonic exit the only possible elementary Riemann problem configuration for the above system is a rarefaction wave whose fan covers the boundary. In this case, if the self-similar variable value is known, we can locally continue the internal field to the boundary. That is why, an additional condition is that the flow velocity attains the sonic value there.

Consider the hyperbolic system for the vector $\mathbf{U}$ of unknown variables in the vicinity of the boundary (the right one for definiteness) in the form

$$
\frac{\partial \mathbf{U}}{\partial t} + A\frac{\partial \mathbf{U}}{\partial x} = \mathbf{0}, \tag{4}
$$

where $\mathbf{U} = \mathbf{U}(x,t)$, and x is the variable in the direction normal to the boundary Γ, t is time, and $A(\mathbf{U})$ is the coefficient matrix with a complete set of eigenvectors and only real eigenvalues. We seek the solution in the form of a simple wave $\mathbf{U} = \mathbf{U}(x,t) = \mathbf{U}(\xi)$, where $\xi = \frac{x}{t}$. By substituting this representation into Eq. (4), we obtain

$$
(A - \lambda I)\mathbf{U}_\xi = \mathbf{0}, \quad \lambda = \xi, \tag{5}
$$

where I is the identity matrix. Owing to Eq. (5), the vector $\mathbf{U}_\xi$ is the eigenvector of A for the eigenvalue $\lambda = \xi$. This means that we need to solve the following system of ordinary differential equations supplemented by the nondifference relation:

$$
\mathbf{U}_\xi = d(\mathbf{U}, \lambda)\mathbf{r}(\mathbf{U}, \lambda), \quad \lambda(\mathbf{U}) = \xi, \tag{6}
$$

where $\mathbf{r}$ is the right eigenvector (the vector-column) of A defined up to the scalar multiplier d. The eigenvalue in the rarefaction wave varies like $\lambda(\mathbf{U}) = \xi$. This condition completes the system for determining $\mathbf{U}$ and d. While realizing this boundary condition we must integrate Eq. (6) over ξ from $\xi_0 = \lambda(\mathbf{U}_0)$, where $\mathbf{U}_0$ represents the initial subsonic parameters inside the region, to $\xi = \xi_\Gamma = 0$, that is, to the sonic point.

Consider this approach, first, for pure gas dynamics. Let us choose the vector of unknowns in Eq. (4) in the form $\mathbf{U} = (\rho, u, v, w, a)^T$, where ρ is the density, u is the velocity vector component normal to Γ, v and w are its tangential components, and a is the speed of sound. The minimum eigenvalue in this case is $\lambda = u - a$ and the related eigenvector is

$$\mathbf{r} = \left[1, -\frac{a}{\rho}, 0, 0, \frac{(\gamma - 1)a}{2\rho} \right]^T, \tag{7}$$

where γ is the adiabatic index. System (6) in this case acquires the form

$$\rho_\xi = d, \ u_\xi = -\frac{ad}{\rho}, \ v_\xi = 0, \ w_\xi = 0,$$

$$a_\xi = \frac{(\gamma - 1)ad}{2\rho}, \ u - a = \xi. \tag{8}$$

This system can be exactly integrated, as its invariants are

$$\left(\frac{p}{\rho^\gamma} \right)_\xi = 0, \ \left(u + \frac{2a}{\gamma - 1} \right)_\xi = 0, \ v_\xi = 0, \ w_\xi = 0. \tag{9}$$

Thus, we obtain

$$a_\Gamma = \frac{\gamma - 1}{\gamma + 1} \left(u_0 + \frac{2}{\gamma - 1} a_0 \right), \ u_\Gamma = a_\Gamma, \ v_\Gamma = v_0, \tag{10}$$

$$w_\Gamma = w_0, \ \rho_\Gamma = \rho_0 \left(\frac{a_\Gamma}{a_0} \right)^{\frac{2}{\gamma - 1}}$$

The index "0" indicates the values belonging to the inner region. On the discreet mesh this means that they are taken from the center (or from the left side) of the cell adjacent to the boundary. Equations (10) must be supplemented by the condition at the supersonic exit if $(u/a)_0 \geq 1$: $\mathbf{U}_\Gamma = \mathbf{U}_0$.

These conditions are mutually consistent and coincide for $u_0 = a_0$. Note that in this case we did not need the explicit expression for d which can be easily found from the second and the fifth equations in (8). Besides the exact derivation based on the relations in the rarefaction wave, we give here the approximate relations keeping in mind such systems for which no exact expressions of this kind can be written out. For this purpose we first exclude d by substituting the first equation from (8) into the other ones. Then we obtain

$$u_\xi = -\frac{a\rho_\xi}{\rho}, \ v_\xi = 0, \ w_\xi = 0,$$

$$a_\xi = \frac{(\gamma - 1)a\rho_\xi}{2\rho}, \ u - a = \xi \tag{11}$$

Now, approximating Eqs. (11) by finite differences we arrive at the following relations:

$$a_\Gamma = \frac{\gamma - 1}{\gamma + 1} \left(u_0 + \frac{2}{\gamma - 1} a_0 \right), \ u_\Gamma = a_\Gamma, \ v_\Gamma = v_0,$$

$$w_\Gamma = w_0, \ \rho_\Gamma = \left[1 + \frac{2}{\gamma - 1} \left(\frac{a_\Gamma}{a_0} - 1 \right) \right]. \tag{12}$$

In this approximation solution (12) differs from (10) only in the entropy invariant and represents its linearization.

Now we proceed to the MHD equations. Let us choose the unknown vector in Eq. (4) in the form

$$\mathbf{U} = [\rho, \ u, \ v, \ w, \ a, \ B_x, \ B_y, \ B_z]^T. \tag{13}$$

where, in addition to the purely gasdynamic case, the components appear of the magnetic field strength vector normal (B_x) and tangential (B_y and B_z) to Γ. The minimum eigenvalue in this case is $\lambda = u - a_f$, where a_f is the largest of the two magnetosonic speeds a_f and a_s ($a_f > a > a_s$):

$$a_{f,s} = \frac{1}{2} \left[\left(a^2 + \frac{|\mathbf{B}|^2}{4\pi\rho} + \frac{a|B_x|}{\sqrt{\pi\rho}} \right)^{1/2} \right.$$

$$\left. \pm \left(a^2 + \frac{|\mathbf{B}|^2}{4\pi\rho} - \frac{a|B_x|}{\sqrt{\pi\rho}} \right)^{1/2} \right], \ |\mathbf{B}|^2 = B_x^2 + B_y^2 + B_z^2 \tag{14}$$

The eigenvector corresponding to this eigenvalue is

$$\mathbf{r} = \left[1, -\frac{a_f}{\rho}, \alpha B_y, \alpha B_z, \frac{(\gamma - 1)a}{2\rho}, 0, \beta B_y, \beta B_z \right]^T, \tag{15}$$

$$\alpha = \frac{aa_s}{2\rho\sqrt{\pi\rho}(a^2 - a_s^2)}, \ \beta = \frac{a^2}{\rho(a^2 - a_s^2)}$$

In this case system (6) acquires the form

$$\rho_\xi = d, \ u_\xi = -\frac{a_f d}{\rho}, \ v_\xi = \alpha B_y d, \ w_\xi = \alpha B_x d,$$

$$a_\xi = \frac{(\gamma - 1)ad}{2\rho}, \ (B_x)_\xi = 0, \ (B_y)_\xi = \beta B_y d, \tag{16}$$

$$(B_z)_\xi = \beta B_y d, \ u - a_f = \xi$$

Now in system (16) we substitute the equation for a by the equation for a_f, which can be easily obtained from Eq. (14) by direct differencing with respect to ξ and by using Eqs. (16):

$$(a_f)_\xi = \frac{\vartheta d}{\rho}, \quad \vartheta = \rho \mathbf{r} \frac{\partial a_f}{\partial \mathbf{U}}. \qquad (17)$$

It can be shown in this connection that for any admissible values of functions we have $\vartheta + a_f \geq 0$.

Then, by passing from d to ρ_ξ, we obtain the reduced system of equations that can be approximated similarly to Eq. (11)

$$\begin{aligned}
(a_f)_\Gamma &= \left(\frac{u\vartheta + a_f^2}{\vartheta + a_f}\right)_0, \quad \rho_\Gamma = \rho_0\left(1 + \frac{u - a_f}{\vartheta + a_f}\right)_0, \\
u_\Gamma &= (a_f)_\Gamma, \quad v_\Gamma = v_0 + (\alpha B_y)_0(\rho_\Gamma - \rho_0), \\
w_\Gamma &= w_0 + (\alpha B_z)_0(\rho_\Gamma - \rho_0), \quad (B_x)_\Gamma = (B_x)_0, \\
(B_y)_\Gamma &= (B_y)_0[1 + \beta_0(\rho_\Gamma - \rho_0)], \\
(B_z)_\Gamma &= (B_z)_0[1 + \beta_0(\rho_\Gamma - \rho_0)], \\
(a)_\Gamma &= (a)_0 + (\gamma - 1)(a/2\rho)_0(\rho_\Gamma - \rho_0)
\end{aligned} \qquad (18)$$

Note that in contrast to the purely gasdynamic case the velocity components tangential to the boundary, generally speaking, are different from their internal values in the presence of the magnetic field.

The case of the triple degeneration of eigenvalues ($B_y^2 + B_z^2 = 0$ and $a^2 \to B_x^2/4\pi\rho$), when $\vartheta + a_f \to \infty$, can be easily avoided by assuming $B_y^2 + B_z^2 = \varepsilon$, where ε is a small positive number.

The approach described above turned out to give stable results in contrast to the attempts of a straight application of the well-known non-reflecting boundary conditions, say, [102]. It is worth mentioning in this connection that a comprehensive review (more than 200 references) of various versions and modifications of non-reflecting boundary conditions can be found in [43] and [44] (see also [37] and [102]).

As initial values the jump can be chosen between the SW and the LISM parameters at a fixed distance R_f from the Sun smaller than the TS stand-off distance. For $R < R_f$ the SW parameter distribution is specified. The magnetic field pressure in this region is supposed to be negligibly small comparing with the SW hydrodynamic pressure, thus $\mathbf{B}_e = 0$. For $R > R_f$ the uniform distribution of the LISM pressure and density is assumed. If we solve an MHD problem, that is, the LISM flow is magnetized, it is wise to specify a magnetic field strength distribution satisfying the divergence-free condition. For this reason the magnetic and the velocity field in the LISM and in the SW flow are initially joined in the computational region so that $\mathbf{B}$ conserves a constant angle with $\mathbf{v}$ and

div $\mathbf{B} = 0$. This is done by assuming for $R > R_f$

$$U = -V_\infty\left[1 - \left(\frac{R_f}{R}\right)^3\right]\cos\theta$$

$$V = 0 \qquad (19)$$

$$W = V_\infty\left[1 + \frac{1}{2}\left(\frac{R_f}{R}\right)^3\right]\sin\theta$$

Here U, V, and W represent the spherical components of the velocity vector in the directions R, ϕ, and θ, respectively. This distribution corresponds to an incompressible fluid flow velocity distribution over a sphere, directed along the z-axis (θ-axis). The magnetic field is initialized in the same way, except that the field configuration is rotated about the y-axis so that the magnetic field vector is tilted with respect to the velocity vector by the desired angle.

If neutral particles are to be taken into account, we must specify their number density and velocity at the inflow.

3. Stationary gasdynamic calculations

A gasdynamic model for the SW interaction with the supersonic interstellar wind was first suggested in [7]. A cylindrical formulation was adopted which corresponds to a uniform LISM flow and a spherically-symmetric SW. The calculations were performed on the basis of a simplified thin-layer approximation. The solar wind was assumed to be decelerated mainly in the process of its interaction with the charged component, or plasma component, of the LISM. With the latter assumption calculations can also be made without simplifications using the Euler gasdynamic equations. Both shock-fitting and shock-capturing approaches can be found in publications and we describe them briefly in the following subsections.

3.1 Shock-fitting methods

Taking into account the adopted two-shock model with the contact discontinuity between the shocks, one can easily use shock-fitting methods in the upwind part of the interaction region. The idea of this approach lies in the subdivision of the computational region into subregions of smooth flow. In these smooth subregions any numerical scheme of sufficient order of accuracy can be used. Derivatives in this approach must never be approximated by finite differences across discontinuities. The latter are traced as boundary lines. Proper jump relations are used to determine the change of parameters across these boundaries and their new position in the course of time. This approach is very economical, since 1) you need not perform calculations in the regions of the uniform LISM and the spherically-symmetric SW, which are known beforehand, thus reducing the size of the computational domain and 2) you can avoid spurious oscillations around discontinuities inherent in application of shock-capturing methods. For this reason one can use linear high-order of accuracy numerical schemes not worrying about high non-oscillatory resolution of discontinuities. In the shock-fitting method their position and intensity are determined exactly. Certain difficulties originate if two or more discontinuities are interacting with each other. In this case, in principle, one can still use a shock-fitting approach using exact solutions for the problems of a discontinuity interaction. The algorithm, however, can become rather complicated (see [60], [61], [62], and [64]). In the shock-fitting approach a quasi-linear form of the system of governing equations is usually chosen rather than a conservation-law form.

On the basis of the shock-fitting approach the problem under consideration was first solved in [7] by the Babenko–Rusanov implicit scheme [4]. Due to the limitations of the numerical approach only upwind part of the interaction was calculated. A polar coordinate system R, θ was used and calculations were performed in the region restricted by the ray $\theta = \theta_{\max}$ at which the velocity normal to this boundary remained supersonic. The system of linear algebraic equations on the computational grid was solved together with the Rankine–Hugoniot conservation relations and the relations on the contact discontinuity. The computational region was located between the termination and the bow shock. The steady-state solution was obtained as $t \to \infty$ with the boundary conditions independent of time.

A physicist usually seeks dimensionless similarity parameters of the problem. For the considered problem they are represented by the Mach numbers M_e and M_∞ of the SW and the LISM flow, respectively, the ratios of their dynamic pressures $K = n_e V_e^2 / n_\infty V_\infty^2$ and stagnation temperatures $\chi = T_{e0}/T_{\infty 0}$. The specific heat ratio γ is usually adopted to be equal to 5/3 which corresponds to the fully ionized plasma. If neutral particles are not taken into account, the flow from the inner side of TS becomes hypersonic and, therefore, the results are only weakly dependent on the choice of M_e taken at 1 AU. Though results formally depend on χ (say, for $\chi = 1$ the density jump over CD is absent) an analysis of conservation relations on the discontinuities in the upwind part of the interaction region make us conclude that they can be recalculated using results for one particular value of χ. As K is the similarity parameter (see [5] and [7]), for the hypersonic solar wind and $K > 1$ the results are independent of K if the distances are

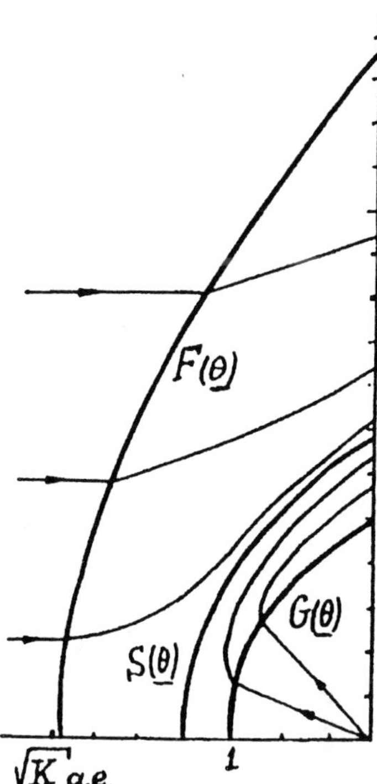

Fig.3: Self-similar picture of shock wave position [7]

measured in units of $1\mathrm{AU} \times \sqrt{K}$. In Fig. 3, the shapes of the outer (F) and of the inner (G) shock wave, the contact surface (S) and some streamlines are presented for the case $M_\infty = 2$.

In [113] a new shock-fitting numerical algorithm was described and applied to the problem under consideration that provided results with such a high accuracy that in the upwind part of the interaction it may represent a testing benchmark for all newly developed methods. This method is based on the composite explicit-implicit finite-difference scheme [86]. The well-known explicit Lax–Wendroff and the implicit Babenko scheme [4] represent its constituent parts. The algorithm is organized in such a way that depending on the CFL number value either explicit or implicit approximation is used. This approach makes the proposed method more economical than a purely implicit scheme. Among the schemes based on this approach we can also mention [53] and [74]. The latter method is the extension of MacCormack's explicit-implicit scheme for the steady Euler equations hyperbolic with respect to one of the space coordinates. A steady-state solution of the SW–LISM interaction was obtained in [113] by a quasi-marching method in which nonstationary problem was solved for $t \to \infty$ for each radial ray. Such an approach also allowed to save a

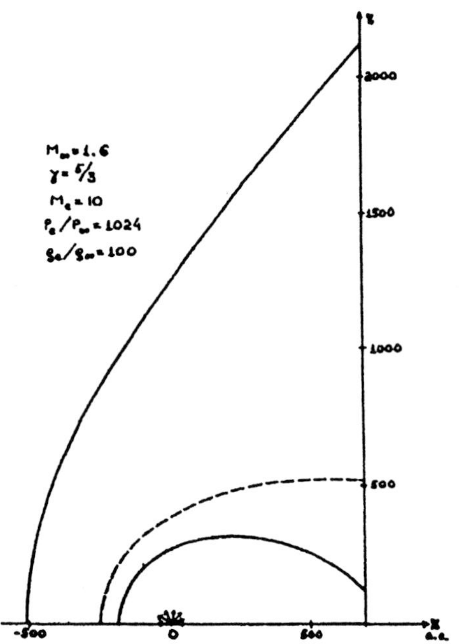

Fig.4: Discontinuity pattern: shock-fitting approach [113]

computational time comparing with a direct solution of the nonstationary system. One of the computational results is shown in Fig 4.

One must admit here that, in contrast to the results from [7], calculations were performed for a rather long distance in the downwind direction, actually until the flow along the z-axis shown in Fig. 4 remained supersonic. This resulted in a considerable unjustifiable elongation of the termination shock, as will be seen from the subsequent subsection. The discrepancy between the results is caused by the limitation of the quasi-marching approach that cannot take into account the Mach-type reflection of the inner shock from the symmetry axis. Nevertheless, the results are without any doubt quite reliable up to a certain distance in the wake region. No need to mention that the time necessary for obtaining the steady-state solution in this case is considerably less than that in the case of applying shock-capturing methods.

3.2 Shock-capturing methods

In shock-capturing methods we calculate finite differences across discontinuities. This may cause spurious oscillations of the solution if non-monotone numerical schemes are used. On the other hand, all linear schemes of the order of accuracy higher than one

are non-monotone [38]. For this reason one or another artificial viscosity must be used [85] or nonlinear numerical schemes ought to be applied. It is not our task to give a review of high-resolution TVD (total variation diminishing) schemes in this paper (see [42] and [112] for a regular mathematical background). Although both finite-difference and finite-volume methods can be equivalently applied in the latter schemes, we shall dwell mainly on the finite-volume formulation and monotonic upstream schemes for conservation laws (MUSCL) approach, since they are more descriptive.

To solve axisymmetric system (2), let us introduce a polar mesh

$$\begin{aligned}
f_{l,n}^k &= f(R_l, \theta_n, t^k), \quad t^k = k\Delta t; \\
R_l &= R_{\min} + (l-1)\Delta R, \quad l = 1, 2, \ldots, L; \\
\theta_n &= (n - 2.5)\Delta\theta, \quad n = 1, 2, \ldots, N; \\
\Delta R &= (R_{\max} - R_{\min})/(L-1), \\
\Delta\theta &= \pi/(N-4)
\end{aligned} \tag{20}$$

with the center in the star. Then for each cell system (2) in the finite-volume formulation can be rewritten as follows:

$$\begin{aligned}
&R_l\,\Delta R\,\Delta\theta\,\frac{\mathbf{U}_{l,n}^{k+1} - \mathbf{U}_{l,n}^k}{\Delta t} + \\
&(R_{l+1/2}\bar{\mathbf{E}}_{l+1/2,n} + R_{l-1/2}\bar{\mathbf{E}}_{l-1/2,n})\,\Delta\theta + \\
&(\bar{\mathbf{E}}_{l,n+1/2} + \bar{\mathbf{E}}_{l,n-1/2})\,\Delta R + R_l\,\Delta R\,\Delta\theta\,\mathbf{H}_{l,n} = 0.
\end{aligned} \tag{21}$$

Here $\bar{\mathbf{E}}$ is the flux normal to the boundary, defined as:

$$\bar{\mathbf{E}} = n_1\mathbf{E} + n_2\mathbf{G}, \tag{22}$$

where $\mathbf{n} = (n_1, n_2)$ is a unit outward vector normal to the cell surface.

Equation (21) has a time-discretized conservation-law form for an individual computational cell. Various numerical schemes are specified by the method chosen to calculate the numerical flux $\bar{\mathbf{E}}$ through the cell boundary surfaces. In [96] the two-step Lax–Wendroff scheme is used with the second order of accuracy. As usual for such schemes, an additional smoothing must be introduced to remove high-frequency oscillations and overshoots and undershoots originating near smeared shocks owing to the non-monotonicity of the scheme. Another possible flux calculation formulas which also include artificial viscosity were applied in [70]. It is based on the ZEUS fractional step code [98]. All methods using artificial viscosity for oscillation damping contain an empirical viscosity coefficient which must be adjusted in a way suitable for any particular problem. The compromise is between the effective monotonization of the solution and its deterioration. Nonlinear high-resolution numerical schemes are free from this drawback.

To attain the second order of accuracy in space, a piecewise-linear distribution of parameters inside computational cells can be adopted [112]. One can use the simplest "minmod" reconstruction procedure

$$\mathbf{U}^R_{l+1/2} = \mathbf{U}^k_{l+1} - \frac{1}{2}\min\mathrm{mod}(\Delta\mathbf{U}^k_{l+1/2}, \Delta\mathbf{U}^k_{l+3/2}) \quad (23)$$

$$\mathbf{U}^L_{l+1/2} = \mathbf{U}^k_{l} + \frac{1}{2}\min\mathrm{mod}(\Delta\mathbf{U}^k_{l-1/2}, \Delta\mathbf{U}^k_{l+1/2}), \quad (24)$$

$$\min\mathrm{mod}(x,y) = \mathrm{sgn}(x)\max\{0,\min[|x|, y\,\mathrm{sgn}(x)]\},$$

where $\Delta\mathbf{U}^k_{l+1/2} = \mathbf{U}^k_{l+1} - \mathbf{U}^k_{l}$, and $\mathbf{U}^R_{l+1/2}$ and $\mathbf{U}^L_{l+1/2}$ defined by Eqs. (23)–(24) represent parameter values on the right and on the left side of the cell surface with the index "$l+1/2$". The index "n" is omitted in these formulas. The reconstruction procedure in the angular direction is similar. To attain better resolution of the contact discontinuity, one can use more compressive slope limiting procedure for the density, e. g.,

$$\rho^R_{l+1/2} = \rho^k_{l+1} - \min\mathrm{mod}(\Delta\rho^k_{l+1/2}, \Delta\rho^k_{l+3/2}, \tilde{\Delta}), \quad (25)$$

$$\rho^L_{l+1/2} = \rho^k_{l} + \min\mathrm{mod}(\Delta\rho^k_{l-1/2}, \Delta\rho^k_{l+1/2}, \tilde{\tilde{\Delta}}), \quad (26)$$

$$\tilde{\Delta} = 0.25(\Delta\rho^k_{l+1/2} + \Delta\rho^k_{l+3/2}),$$

$$\tilde{\tilde{\Delta}} = 0.25(\Delta\rho^k_{l-1/2} + \Delta\rho^k_{l+1/2}),$$

$$\min\mathrm{mod}(x,y,z) = \mathrm{sgn}(x)\max\{0,\min[|x|, y\,\mathrm{sgn}(x), |z|]\}$$

The fluxes $\bar{\mathbf{E}}(\mathbf{U}^R, \mathbf{U}^L)$ through the cell surfaces can be found by different methods. Wide recognition acquired TVD shock-capturing methods based on the exact or some of the approximate solutions to the Riemann problem. The first steady-state solution of the SW–LISM interaction problem in the closed region surrounding the star was obtained in [92] on the basis of the Osher approximate Riemann problem solver [24]. In this approach an approximate solution to the Riemann problem is formed using elementary simple waves which correspond to definite eigenvectors of the Euler gasdynamic system and separate the regions of the constant flow. In the exact solution, a rarefaction wave, a contact discontinuity and a shock wave generally appear in various combinations. In the approximate solution used in [92] shock waves are approximated by compression waves. This approach connects parameter values on the right and on the left side of the computational cell by a number of algebraic relations. Such approach, however, seems to be more time-consuming than that based on Roe's solution of the linearized Riemann problem [88] (see also a comprehensive description of characteristic-based schemes

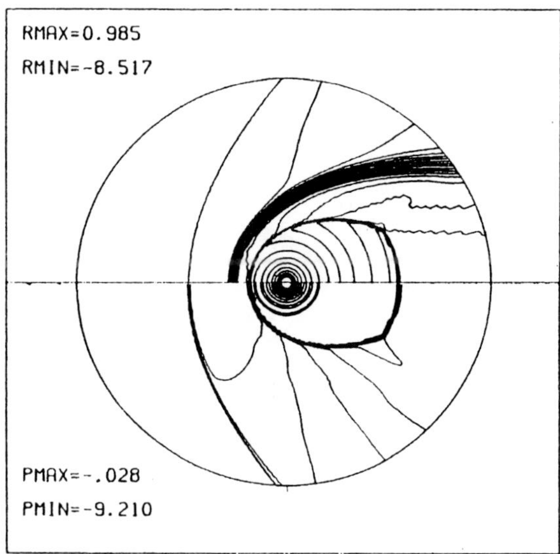

Fig.5: Discontinuity pattern: shock-capturing approach [75]

for the Euler equations [89]). A numerical flux in this case is calculated as follows:

$$\bar{\mathbf{E}}(\mathbf{U}^R, \mathbf{U}^L) = \frac{1}{2}\left[\mathbf{E}(\mathbf{U}^L) + \mathbf{E}(\mathbf{U}^R) - S|\Lambda|S^{-1}(\mathbf{U}^R - \mathbf{U}^L)\right] \quad (27)$$

Here $S(\bar{\mathbf{U}})$ and $S^{-1}(\bar{\mathbf{U}})$ are the matrices formed by the right and by the left eigenvectors, respectively, of the frozen Jacobian matrix

$$\bar{J} = \frac{\partial\bar{\mathbf{E}}(\bar{\mathbf{U}})}{\partial\mathbf{U}}$$

The value of $\bar{\mathbf{U}}(\mathbf{U}^L, \mathbf{U}^R)$ is chosen so that the conservation relations on shocks are exactly satisfied. The matrix $|\Lambda|$ is a diagonal matrix consisting of the frozen Jacobian matrix eigenvalue moduli.

The important peculiarity of the latter method is that, although it gives the solution of the linearized problem, the exact satisfaction of the Rankine–Hugoniot relations on shocks provides their more adequate and sharp resolution. This method was applied to the problem under consideration in [55] and [75].

Although all the mentioned methods give essentially similar pattern of discontinuities in the computational region, we present here, for the convenience of the further discussion, the picture from [75]. The interstellar plasma number density of protons is assumed to be $n_{H+} \approx 1\,\mathrm{cm}^{-3}$. The velocity of LISM relative to the solar system is about $20\,\mathrm{km/s}$, while the speed of sound of the LISM gas is about $10\,\mathrm{km/s}$. Thus, LISM

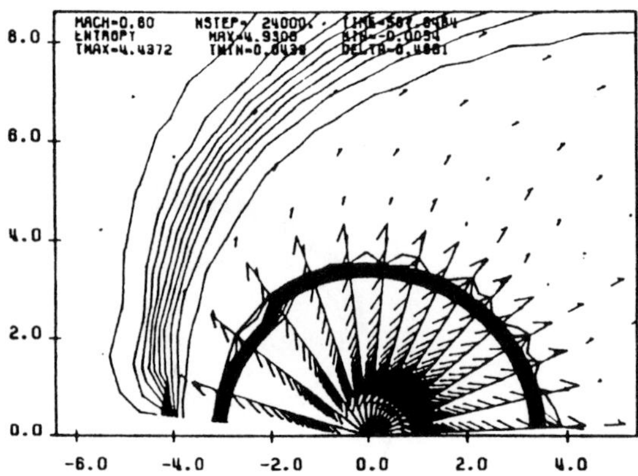

Fig.6: Entropy contours with velocity vectors for the model with the Mach number 0.6 [93]

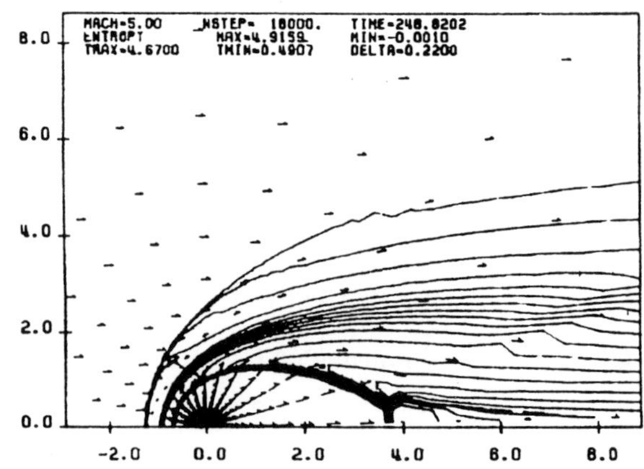

Fig.7: Entropy contours and velocity vectors for the Mach number 5 [93]

flow is supersonic. The SW charged particle number density is chosen $n_{H+} \approx 10\,\mathrm{cm}^{-3}$, its velocity is $V_e \approx 500\,\mathrm{km/s}$, the speed of sound $c_e \approx 100\,\mathrm{km/s}$ at the distance of the Earth's orbit (1 AU). The nondimensional parameters of the problem are Mach numbers of LISM and SW M_∞, M_e, the relations of dynamic pressures $K = \rho_e V_e^2 / \rho_\infty V_\infty^2$ and stagnation temperatures $\chi = T_{0e}/T_{0\infty}$ of SW and LISM. This corresponds to the following values of dimensionless parameters of the problem: $M_\infty = 2$, $M_e = 5$, $\chi = 400$, $K = 6250$. The specific heat ratios for SW and LISM are supposed to be 5/3. The calculation is performed in the ring region with the inner and outer circle radii being $R_{\min} = 10$ and $R_{\max} = 500$ AU with 99 and 116 cells in the radial and in the angular direction, respectively. Constant pressure (below the symmetry axis) and constant density natural logarithm contours of the steady-state solution are presented in Fig. 5. Results are given in the polar region with the inner and outer circle radii 10 and 400 AU. All peculiarities are seen of the shock wave pattern shown schematically in Fig. 2.

Calculations for a large variety of the stellar wind and interstellar medium parameters, including those very different from the parameters of the solar wind, were performed in [55] and [92]. These parametric studies may reflect various situations corresponding to ejecting stars and their environment. Although from the purely gasdynamic viewpoint the LISM flow is supersonic, the presence of the interstellar magnetic field and charge-exchange processes can decrease the effective Mach number of the uniform interstellar medium flow, thus making it subsonic. For this reason several authors [93], [97] performed calculations of a subsonic interaction on the basis of the Euler equations. It is quite clear that, unlike the two-shock Baranov's

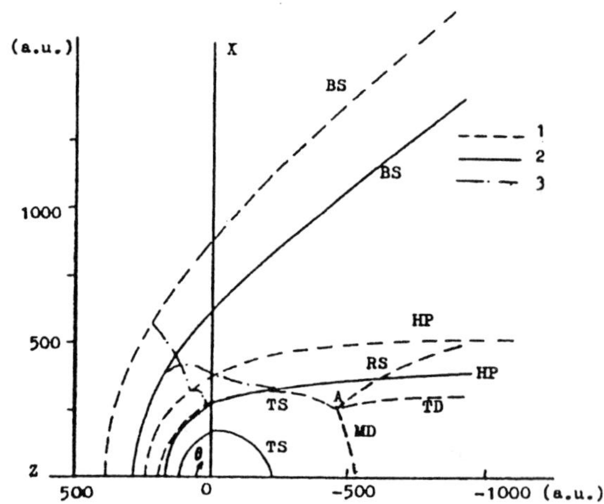

Fig.8: Geometrical pattern of the interface. Results of the numerical calculations for $n_{H\infty} = 0(1)$ and $n_{H\infty} = 0.14\,\mathrm{cm}^{-3}(2)$; curves (3) are the sonic lines. Positions of the bow shock (BS), termination shock (TS), heliopause (HP), reflected shock (RS), tangential discontinuity (TD), and Mach disk (MD) are shown.

model, no bow shock can appear in the stationary solution. As will be shown later, the termination shock in this case does not have a bullet shape and the difference between its stand-off distances in the upwind and in the downwind direction is not so pronounced. This can also be seen in Fig. 6, corresponding to $M_\infty = 0.6$. If the stellar wind Mach number is larger than that of the solar wind, the termination shock becomes elongated in the backward direction and the size of the Mach disk diminishes (see Fig 7, corresponding to $M_\infty = 5$). The latter two figures are taken from [93].

Summarizing the subject of this section, we would like to admit that, although application of shock-fitting methods can provide a solution with a very high precision at low computational costs, they can be used only in the flow regions with a simple shock-wave pattern which is known beforehand (say, the upwind part of the SW–LISM interaction). Attempts to promote calculations for larger distances into the downwind region results in the substantial distortion of the results, like it was in [113], in which the possibility of the Mach-type reflection of the termination shock was not taken into account. A shock-fitting method in its rigorous sense was not realized, since the exact relations connecting parameters in the triple point were not used. The best way out in this case is to combine shock-fitting and shock-capturing methods, as it was done in [11].

4. Stationary solutions including neutral particles

Since we restrict ourselves in this review to the description of numerical methods used to solve the SW–LISM interaction problem, more or less complete description of physical processes governing the charge exchange between the plasma and the neutral component of the flow lies beyond the scope of this work. It can be found in the corresponding references mentioned in Introduction. However, it is quite clear that the LISM is a partially ionized gas and, therefore, a consistent model must be developed accounting for the mutual influence of charged and neutral particles. Although both helium and hydrogen atoms are present, the authors usually disregard helium atoms, since their cosmic abundance is much less than that of hydrogen atoms.

The first attempts to estimate the influence of this process, say, [7] and [87], did not include the influence of the hydrogen atoms on the plasma component. The first self-consistent model of the heliospheric interface was suggested in [8]. In that paper the authors introduced the source terms accounting for the momentum and the energy transfer between the two components into the system of the Euler gasdynamic equations in the quasi-linear form. These source terms were presented in the form:

$$\rho \nu_c (\mathbf{v}_H - \mathbf{v}) \tag{28}$$

for the momentum equations (1) and

$$\rho(\gamma - 1)\nu_c \left[\frac{(\mathbf{v}_H - \mathbf{v})^2}{2} - \frac{3kT}{2m_H} + \frac{3kT_H}{2m_H} \right] \tag{29}$$

for the energy equation. Here k is the Boltzmann constant and m_H is the mass of the hydrogen atom.

For the collision frequency between protons and atoms the following formula was used:

$$\nu_c = \frac{\rho_H \sigma Q}{m_H}, \quad Q = \left[(\mathbf{v}_H - \mathbf{v})^2 + \frac{128k(T + T_H)}{9\pi m_H} \right]^{\frac{1}{2}},$$

where σ is the effective charge-exchange cross-section.

The hydrogen atoms were assumed to conserve their velocity and temperature in the process of their interaction with protons and the origin of the secondary hydrogen atoms was neglected. That allowed to describe the behavior of neutrals in the region between the two shocks by the system

$$\begin{aligned}
\mathbf{v}_H &= \mathbf{v}_{H_\infty} = \text{const}, \\
T_H &= T_{H_\infty} = \text{const}, \\
\text{div}(\rho_H \mathbf{v}_H) &= -\rho \nu_c
\end{aligned}$$

Incorporating this source term into implicit method [4] does not represent any difficulty and the solution was obtained for various values of dimensionless parameters. The major effect of charge exchange on the heliospheric interfaces is to decrease the distances to the TS, HP, and BS. The original developers of the approach [8], which considers the flow of neutral particles as a hydrodynamic flow, themselves admitted in [9] its main drawbacks: (1) such description is hardly justifiable, since the mean free-path of the hydrogen atoms is not smaller than the characteristic length of the problem; (2) the Maxwellian distribution of the atoms was adopted to calculate the plasma momentum and energy losses. This was the reason of applying the Monte-Carlo method for simulation of the hydrogen atom trajectories [54]. It gives a possibility to evaluate the source terms in the momentum and the energy equations for the plasma component on the basis of the kinetic description of the hydrogen atoms. An iteration method [10] was proposed to solve both systems of equations. The numerical results are discussed in [11] and [12]. This approach is nowadays the only one which can be considered physically consistent from the viewpoint of computational fluid dynamics. In Fig. 8, the geometrical interface between the two flows is presented [11] reflecting the influence of the charge exchange processes. The picture is presented in the $X0Z$ plane, where $0Z$ coincides with the axis of symmetry and is antiparallel to the vector on the LISM velocity (the Sun is in the coordinate system origin). The solid and the dotted lines in Fig. 8 correspond to $n_{H_\infty} = 0.14\,\text{cm}^{-3}$ and $n_{H_\infty} = 0$, respectively. The parameters of the plasma component were the following: $n_e = 7\,\text{cm}^{-3}$, $V_e = 450\,\text{km/s}$, $M_e = 10$, $n_\infty = 0.07\,\text{cm}^{-3}$, $V_\infty = 25\,\text{km/s}$, and $M_\infty = 2$. One can see a large influence of the described processes on the flow pattern and it is quite clear that any realistic calculation of the problem must take them into consideration.

Recently in [70] and [115] the simplified approach [8] was modified by developing the multifluid description for the neutrals which tried to take into account the highly non-Maxwellian nature of the neutral distribution. Since in these papers the plasma was assumed Maxwellian only in each region between the discontinuity surfaces, the neutral population produced by the charge exchange in these regions has the same basic characteristics as those of the plasma component. Although this approach still remains approximate, one must admit that it allowed its developers to obtain the solutions of the three-dimensional and nonstationary problems which have not been solved yet on the basis of the Monte-Carlo method.

5. Hydrodynamic instabilities in the SW–LISM interaction

As was admitted in [106], the contact discontinuity between the inner and the outer subsonic region is potentially the subject to the fluid Kelvin–Helmholtz or MHD flute instabilities which could produce some of the inhomogeneous structures of astropause tails. The study of instabilities accompanying the problem of the SW–LISM interaction with and without magnetic field is an important independent task which lies outside the scope of this review. We are going to clarify here only the effect of the choice of numerical methods on their origin and development in the problem in which two flows possessing substantially different entropies collide at a contact surface calculated by shock-capturing methods. It is more or less clear that discontinuity-fitting methods are hardly applicable for this task.

As was convincingly shown in [58] and [93], complicated nonstationary patterns can be obtained owing to the Rayleigh–Taylor instabilities if the interaction is accompanied by the gravitational effect from the star. In the case of the solar wind the gravity can be neglected, since the Hoyle–Littleton accretion radius is about 4.5 AU, which is much smaller than the expected location of the termination shock. In [55], however, the stellar wind–interstellar medium interaction was studied for parameters different from those corresponding to the solar wind and the results were obtained clearly indicating instability of the flow. The Osher numerical scheme [24] was used and an algebraically generated irregular grid rather than the polar one applied in the previous study of the problem [93]. That type of the grid is supposed to give a finer resolution at large distances from the ejecting center. The influence of the interstellar medium Mach num-

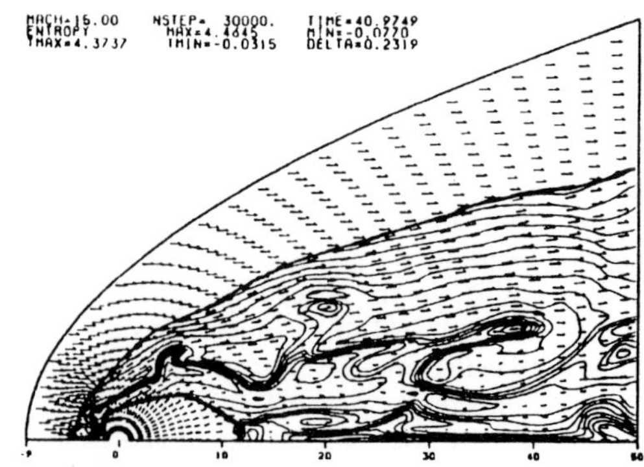

Fig.9: Entropy contours and velocity vectors [55] ($M_\infty = 15$)

ber was studied in a wide range from $M_\infty = 0.6$ to 15. A highly nonstationary solution was obtained in this case, instabilities originating both near the astropause stagnation point and in its lateral region (Fig. 9).

The Kelvin–Helmholtz instability in the latter region are quite possible as the gases with different entropy move along the both sides of a contact surface at different speeds. The origin of the Kelvin–Helmholtz instability near the stagnation point, however, is questionable, since velocities are very small there. Such stagnation point instabilities, however, were found earlier in the calculations of astrophysical jets [65], [66], and [95]. Similar instabilities were also found in [56] and [57] in the calculations of opposing and forward-facing jets. They were also observed in the experiments [36], [90], and [91]. It is worth mentioning, however, that the instabilities of the latter type were not obtained in the calculations for parameters close to the solar wind – interstellar medium interaction, as was reported in [96]. Moreover, N. Pogorelov in his calculations occasionally obtained instabilities near the stagnation point, even for a rather rough mesh, using the Roe-type MUSCL scheme under unfavorable selection of the parameter reconstruction method. Note that one or another parameter interpolation based on the assumption of the linear, parabolic [25], or higher-order distributions inside computational cells is usually used to increase the order of accuracy of the chosen numerical scheme. Say, in [75] and [76] the interpolation of characteristic variables was used which had been discovered to give much more stable results in the vicinity of the geometrical singularities at $\theta = 0$ and $\theta = \pi$. Note that both the polar grid used in [76] and the algebraically generated O-type grid incorporated into the algorithm [55] possess such a singularity. In [80] the choice of the form of governing equations (2) or (3) was claimed to have a certain effect on the stability of results.

There also exits another aspect of the problem. As was mentioned in the previous section, the interaction between the interstellar atoms and the solar plasma ions via resonant charge-exchange collisions greatly affects the steady-state structure of the global heliosphere. The flow of interstellar ions is diverted around the nose of the heliopause, whereas the neutral particles penetrate into the heliosphere impeded aside from the charge-exchange collisions. Thus, there is a velocity difference between them near the nose which performs like a drag force acting on the plasma (see Eq. 28). Since the LISM density is larger at the heliopause than that of the SW, the contact surface is potentially Rayleigh–Taylor unstable. That kind of instability was admitted in [52] and [115]. Its origin was confirmed by several numerical experiments, including the one applying the entirely different particle-in-cell code [20], and by comparing the instability linear growth rate with the theoretical one. However, as was admitted earlier, the multifluid hydrodynamic model in which all fliuds were supposed to be in a local thermal equilibrium was used in the above-mentioned papers to determine the motion of the neutral particles. This makes the obtained results questionable from the viewpoint of the spatial scale of instabilities. Paper [52] also neglects the effect of energetic solar wind neutrals created by the charge exchange inside the heliopause. The drag force caused by these neutrals tends to compensate the inward force described earlier and partially suppress the instability.

The study of hydrodynamic instabilities originating in the SW–LISM interaction problem is far from its conclusion also due to a number of physical phenomena that can affect it, such as, the interstellar magnetic field, cosmic rays, etc. It is important to admit in this review that, investigating instabilities, one must be very careful in order to distinguish those of physical and of numerical origin.

6. Nonstationary SW–LISM interaction

As was mentioned above and is widely accepted, the solar wind changes its speed from supersonic to subsonic through a termination shock. A number of reasons can cause temporal asymmetries of the termination shock and, therefore, the interaction pattern as a whole. Among them are disturbances of the solar wind and its 11-year periodicity. Due to these reasons the heliospheric shock will move in response to variation in upstream solar wind conditions. In [100], a kinematic analysis is made of the solar wind driven temporal variations in the heliospheric termination shock distance. In [15], [16], and [63] the motion of this shock was analyzed analytically on the basis of one-dimensional gasdynamic model. It was admitted that the termination shock would rather be nonstationary and would resemble a distorted asymmetric balloon with some part moving inward and others moving backward. In [97] the termination shock response to large-scale solar wind fluctuations were studied numerically using the Lax–Wendroff scheme. The LISM flow was assumed subsonic. In [47], on the basis of the similar numerical method 11-year solar wind variation influence on the inner shock was investigated. In [75] and [76] nonstationary problems were modeled by a numerical solution of the Euler gasdynamic equations (2) in the finite-volume formulation (21) using a MUSCL-type TVD high-resolution numerical scheme. The suggestion was made of a piecewise-linear distribution of the characteristic parameters inside the cells to determine the values at their boundaries and slope limiters were used to attain a TVD property. The results presented below were obtained using the formulas [1]:

$$\bar{\mathbf{E}}_{l+1/2,n} = \bar{\mathbf{E}}(\mathbf{U}^L, \mathbf{U}^R), \qquad (30)$$
$$\mathbf{U}^L = \mathbf{U}_{l,n} + \mathbf{U}'_{l,n}\Delta R/2,$$
$$\mathbf{U}^R = \mathbf{U}_{l+1,n} - \mathbf{U}'_{l+1,n}\Delta R/2,$$
$$\mathbf{U}'_{l,n} = S_{l,n}\mathbf{W}'_{l,n},$$
$$\mathbf{W}'_{l,n} = \frac{(b_m^2 + c)a_m + (a_m^2 + c)b_m}{a_m^2 + b_m^2 + 2c},$$
$$\mathbf{a} = S_{l,n}^{-1}(\mathbf{U}_{l+1,n} - \mathbf{U}_{l,n}),$$
$$\mathbf{b} = S_{l,n}^{-1}(\mathbf{U}_{l,n} - \mathbf{U}_{l-1,n})$$

Here c is a small positive value used to avoid division by zero. In these formulas S and S^{-1} are (4×4) matrices, constructed using right and left eigenvectors of the Jacobian matrix $\partial\bar{\mathbf{E}}/\partial\mathbf{U}$.

The fluxes presented by Eq. (30) were defined on the basis of Roe's approximate Riemann solver [88].

Fluxes through another pair of cell surfaces can be obtained similarly.

The promotion of the solution in time was performed in the following way:

$$\mathbf{U}_{l,n}^{(1)} = \mathbf{U}_{l,n}^k + \frac{\Delta t}{2}\frac{\partial\mathbf{U}_{l,n}^k}{\partial t};$$
$$\mathbf{U}_{l,n}^{k+1} = \mathbf{U}_{l,n}^k + \Delta t\frac{\partial\mathbf{U}_{l,n}^{(1)}}{\partial t},$$

where $t = k\,\Delta t$, $k = 0, 1, \ldots$, and Δt is defined by the time resolution and by the CFL condition.

The LISM proton number density was assumed to be $n_{H+} = 1\,\mathrm{cm}^{-3}$. The velocity of LISM relative to the solar system is about 20 km/s, while the speed of sound of the LISM gas is about 10 km/s. The SW protons number density was adopted $n_{H+} = 10\,\text{cm}^{-3}$, while its velocity is $V_e \approx 500\,\text{km/s}$, the speed of sound $c_e \approx 100\,\text{km/s}$ at the distance of the Earth's orbit (1 AU). This corresponds to the following values of dimensionless parameters chosen for the initial data: $M_\infty = 2$, $M_e = 5$, $\chi = 400$, $K = 6250$. The stationary initial flow was obtained by the same numerical method using a time-stabilization approach. The calculation was performed in the ring region with the inner and outer circle radii being $R_{\min} = 10$ and $R_{\max} = 500$ AU. On the inner surface all parameters were specified as functions of time by formulas:

$$U = 25, \quad \rho(t) = \rho(0)(1 + 4\exp[-(t-3)^2]),$$
$$p(t) = 3.2325\rho^\gamma(t),$$

since this boundary is supersonic. Here U is the radial velocity component. Although the choice of parameters is somewhat nonrealistic, as far as the solar wind is concerned, it still can be used for a qualitative analysis of a nonstationary picture of the interaction.

The initial distributions of pressure (below the symmetry axis) and density logarithms are shown in Fig. 5. The size of the outer circle in the figures is 400 AU. Similar isolines are presented in Figs. 10–13 at different moments of time (in the units $1\,\text{AU}/U_\infty$). The growing part of the disturbance, interacting with the inner shock, moves it, first, from the centre, Fig. 10, the intensity of IS increasing. Later, it tends to achieve the initial position.

The penetration of the disturbance in the region between IS and CD is seen on both charts of isolines at different moments of time. The disturbance, while crossing IS, increases greatly, and after some time a local pressure maximum originates between IS and CD (Fig. 11). This leads to the effect of suction of SW gas to the centre accompanying IS motion in the same direction. This pressure extremum line gradually moves towards CD, see Figs. 12–13.

The parts of the computational region more remote from the source position suffer the same changes later in time.

As soon as the disturbance meets different points of IS, the latter suffers substantial distortion. When the triple point on IS is reached, there appear two triple points with different reflected shocks and recirculation zone between them (Fig. 13). As soon as the source intensity becomes constant, the position of IS gradually, but rather slowly, moves towards its initial position. The maximum increase of MD stand-off distance is greater than that of IS along the ray

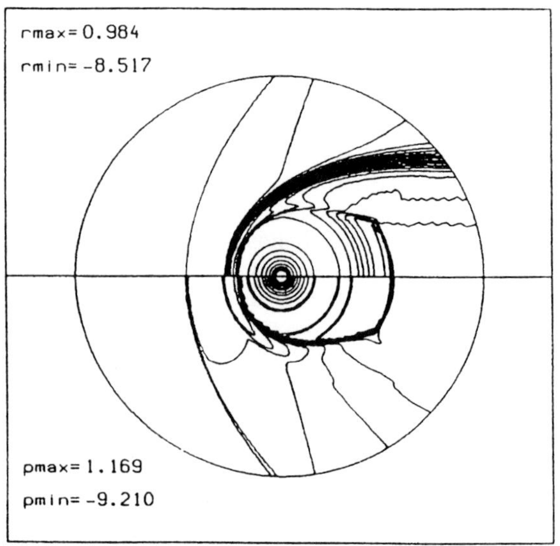

Fig.10: Pressure and density logarithm isolines [75], $t = 8$

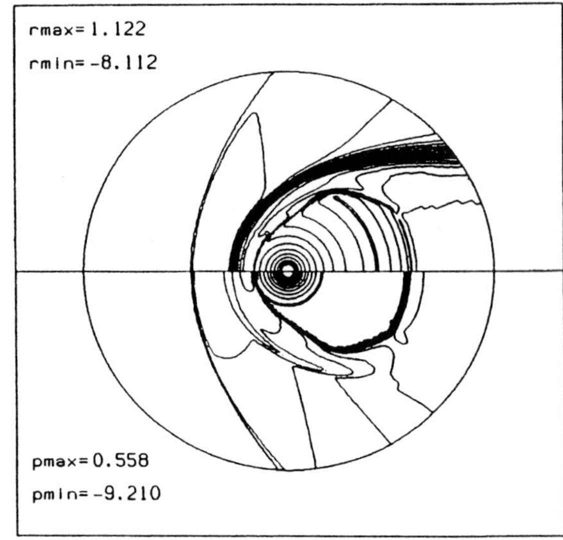

Fig.11: Pressure and density logarithm isolines [75], $t = 12$

$\theta = 0$.

The relaxation of the flow towards some stationary solution is very slow ($t > 220$). When $t > 20$, in the region near $\theta = 0$ additional local pressure extrema vanish. The extended vortex region originates behind MD and it is up to $t = 60$, when these vortices vanish, probably due to numerical viscosity.

Now we present some results [76] concerning the

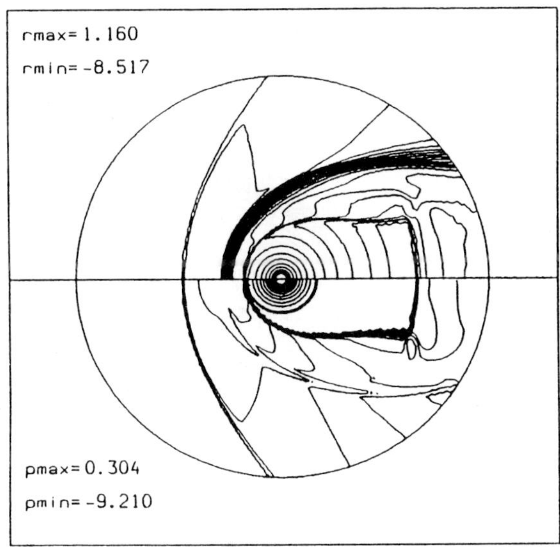

Fig.12: Pressure and density logarithm isolines [75], $t = 18$

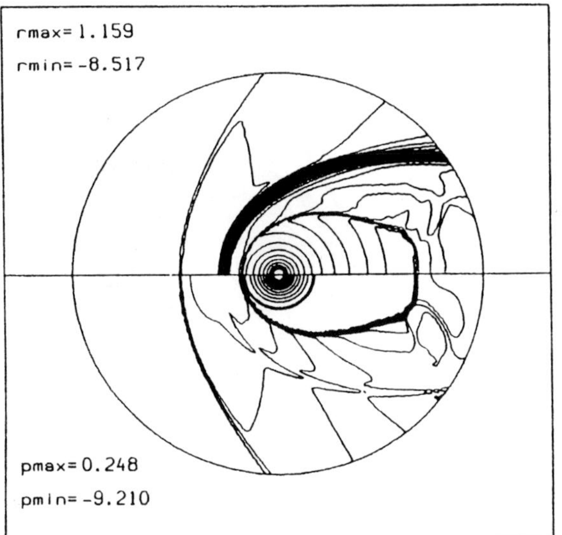

Fig.13: Pressure and density logarithm isolines [75], $t = 21$

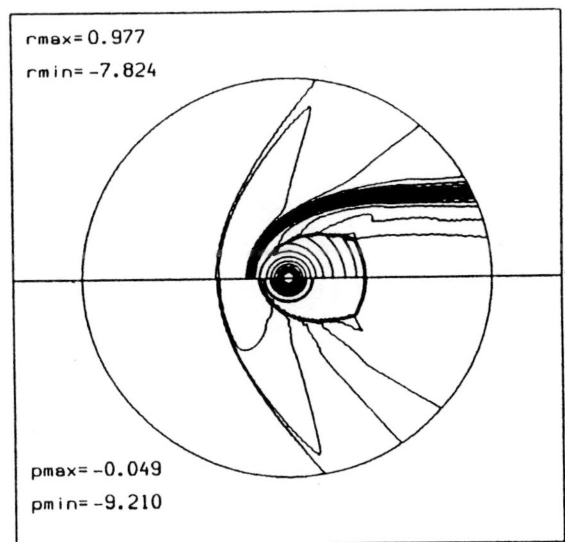

Fig.14: Pressure and density logarithm isolines [76], $t = 0$

periodic SW–LISM interaction. In this calculation the LISM proton number density, its velocity and the speed of sound are $n_\infty = 0.1\,\mathrm{cm}^{-3}$, $V_\infty = 20\,\mathrm{km/s}$ and $c_\infty = 10\,\mathrm{km/s}$, respectively. Parameters of the solar wind change substantially within the 11-year period of the solar activity. The concentration of charged particles $n_e = 1.56\,\mathrm{cm}^{-3}$ and the radial SW velocity $V_e = 400\,\mathrm{km/s}$ are chosen for the minimum of the so-

lar activity, while $n_e = 8\,\mathrm{cm}^{-3}$ and $V_e = 500\,\mathrm{km/s}$ correspond to its maximum (see [6] and [21]). Thus, for dimensionless parameters we have $K = 6250$, $\chi = 256$, $M_\infty = 2$, and $M_e = 5$ in the minimum and $K = 50000$, $\chi = 400$, $M_\infty = 2$, and $M_e = 5$ in the maximum of the solar activity. These values are chosen as the basic points for the sine function approximating the time dependence of K and χ within 11 years. The dimensionless time unit is 86.8 days. The time step is chosen to be $\Delta t \approx 0.4$ days. Parameter distribution corresponding to the stationary solution in the minimum solar activity is chosen as initial data. The results are obtained by the MUSCL numerical method described above. The calculation is performed in the ring region with the inner and the outer circle radii being $R_{\mathrm{min}} = 14$ and $R_{\mathrm{max}} = 700$, respectively.

The isolines of the pressure (below the symmetry axis) and the density logarithms corresponding to the initial data are shown in Fig. 14. They are presented for the outer circle size equal to 560 AU. The number of cells is 99 and 116 in the radial and in the angular direction, respectively. At the initial stage of the flow development, the increase of the parameter K results in the TS motion from the Sun. The motion of the compression wave through TS causes the origin of a new shock wave propagating from the center. This can be seen in Fig. 15 corresponding to $t = 20$. This shock penetrates through the contact discontinuity and moves towards the bow shock, see Fig. 16 ($t = 36$).

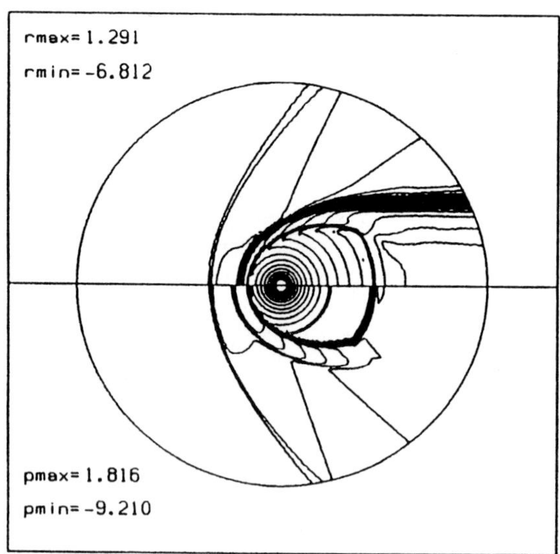

Fig.15: Pressure and density logarithm isolines [76], $t = 20$

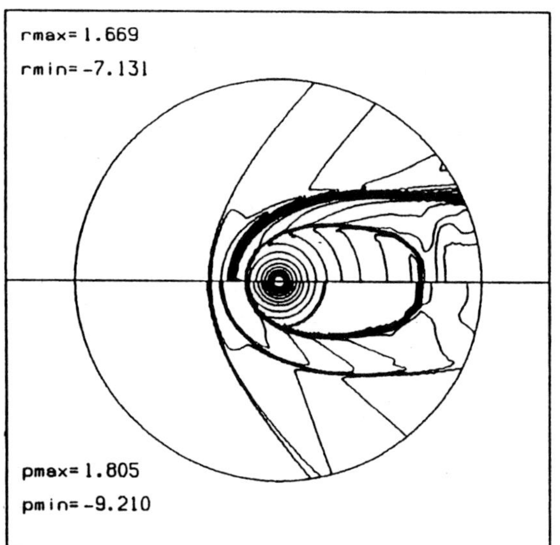

Fig.16: Pressure and density logarithm isolines [76], $t = 36$

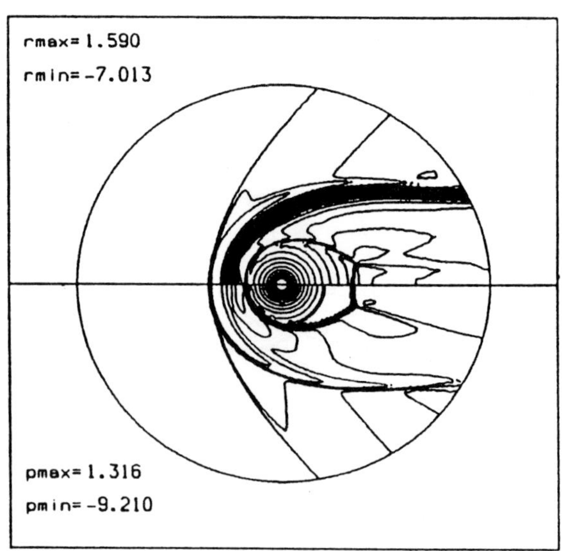

Fig.17: Pressure and density logarithm isolines [76], $t = 60$

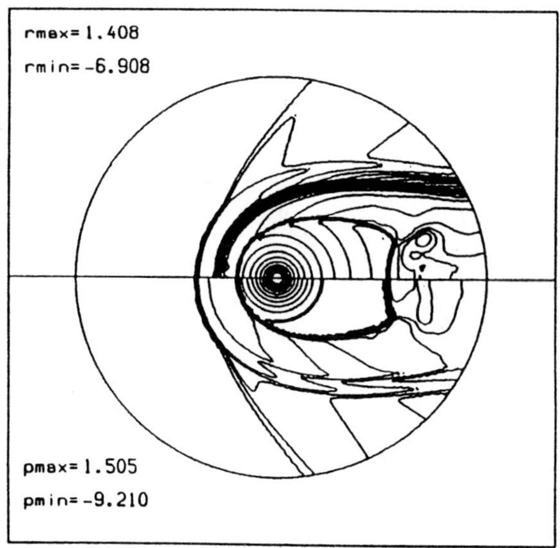

Fig.18: Pressure and density logarithm isolines [76], $t = 84$

The decreasing part of the periodic function causes a backward motion of the termination shock towards the Sun. This leads to the origin of new flow division surfaces, reverse flow zones, and vortices of variable size and intensity in the wake region. At $t = 60$, see Fig. 17, the size of the bullet shape termination shock becomes minimum again. Later on a next nonstationary shock wave appears, etc. A definite 11-year periodicity is developed in the shape of the termination shock. Parameter distribution between the inner and the bow shock is determined by the propagation of shock waves traveling one after another and interacting with the less intensive waves reflected from the bow shock (Fig. 18).

This modeling shows that the flow pattern is substantially nonstationary which must be taken into account when analyzing responses from the space vehicles crossing these discontinuities. Variable solar activity is usually accompanied by the asymmetry of the solar wind which will be considered in the next section.

Application of a high-resolution numerical methods allows one to avoid spurious oscillations near shocks and provides their sharper resolution using smaller number of computational cells than it is necessary in nonmonotone schemes with artificial viscosity.

7. Nonuniform solar wind – interstellar medium interaction

Distant solar wind deviations from spherical symmetry induced by the interaction with neutral interstellar hydrogen due to photoionization and charge-exchange processes were studied in [40] by a perturbation technique. The penetration of neutral particles deep inside the heliosphere results in a substantial increase of the distant solar wind temperature. In all models described in the previous sections we considered the stellar (solar) wind as spherically-symmetric. It can fairly easily become asymmetric, since charge-exchange processes are clearly more effective in the forward region of the interaction because their efficiency is proportional to the velocity difference between neutral and charged particles. The perturbation analysis has shown that a strongly asymmetric distribution of the neutral hydrogen within the heliosphere causes asymmetric deceleration and extremely nonuniform distribution of the distant solar wind temperature, thus leading to non-radial gradients and flows.

This is, however, only one (external) reason of the solar wind asymmetry. Another one is determined by a mere variation of the solar wind symmetry in time within its 11-year activity period (see, e.g., speculations in [76]). According to the solar minimum observations by Ulysses, the solar wind parameters depend on the helioaltitude. These data (see [72] and [73]) indicate that two large polar coronal holes, one in the northern and the other in the southern hemisphere, produce a hotter, lower-density, higher-speed wind comparing with the ecliptic wind. In [71] the calculations were performed of a nonuniform solar wind–interstellar medium interaction using the earlier mentioned three-dimensional time-dependent ZEUS method [98]. The solar wind boundary conditions were taken from the published Ulysses data. According to them, the ram pressure increases 1.5 times from the ecliptic plane to the solar pole. That is why, the termination shock in the calculations was found to be elongated along the solar axis which, in turn, resulted in an increased flow in the ecliptic plane compared with that over the solar poles. The authors reported

a pronounced effect of the solar wind asymmetry on the global structure of the termination shock and heliopause. Both a two-shock (supersonic LISM) and a one-shock (subsonic LISM) model were considered. It is worth mentioning once again in this connection that, if only the motion of charged particles is considered, the LISM flow is definitely supersonic. Pure gasdynamic subsonic models for the problem under consideration are sometimes used to account for the net effect of the charge-exchange processes, the influence of cosmic rays [35], and interstellar magnetic field. The latter subject will be discussed in the next section.

The paper [71] definitely indicates, from our viewpoint, that any realistic model for the SW–LISM interaction must include both neutral particles, the asymmetry of the solar wind, and, as a consequence, the solar wind periodicity, since the asymmetry varies within the solar activity period. This, of course, does not prevent an investigation of different physical effects separately.

8. Solar wind interaction with the magnetized interstellar medium

The presence of the interstellar magnetic field necessitates solution of the MHD equations for modeling of the SW–LISM interaction. The influence of magnetic field becomes important if a magnetic pressure $B^2/8\pi$ becomes comparable with a dynamic pressure. Magnetic field causes an increase of the maximum speed of small perturbations in the LISM flow, especially in the direction normal to the direction of the magnetic field, thus resulting in the decrease of the effective Mach number [33] and [34]. Though the magnitude and the direction of the interstellar magnetic field is not perfectly known, the estimates from [3] indicate the possible importance of its presence. The value and the direction of the LISM magnetic field can affect the global structure of the interaction not only directly, but also by modulating the distance between the bow shock and the heliopause which determines the transparency of this layer for the LISM neutrals. This, in turn, affects the observed line profiles of the solar Lyman-α backscattered emission and estimates of the LISM parameters (see the discussion in [12] and [50]). If the LISM magnetic field is not parallel to its velocity, the problem becomes three-dimensional. Later in the section we discuss the results of such MHD modeling.

It is worth mentioning, that although the value of the solar magnetic field is often neglected in numerical modeling of the problem under consideration at the distances of the termination shock, its influence is substantial at Earth's magnetosphere distances [68] and [109]. In [107] the global structure of the outer heliosphere was studied in the axisymmetric formulation for a subsonic interstellar medium with taking into account a time-varying poloidal magnetic field. In [108] and [67] the toroidal magnetic field in the heliosheath was found to increase with the distance from the Sun. The complicated three-dimensional nostationary behavior of the flow was studied which showed the importance of taking this component of the magnetic field into account. We restrict ourselves to noting this aspect of the problem and will not discuss it below.

8.1 On the eigenvector and eigenvalue systems of MHD equations

The system of ideal MHD equations in the conservation-law form is presented by Eq. (1). This system can be directly rewritten in the quasi-linear form which is more convenient for the characteristic analysis. It is well-known that the equation div $\mathbf{B} = 0$ expresses the absence of magnetic charge. It is also evident that, if magnetic charge is absent initially, it will not appear mathematically at any time instant. From this viewpoint the above equation is excessive. If we rewrite the system of MHD equations [51]

$$\frac{\partial \rho}{\partial t} + \operatorname{div} \rho \mathbf{v} = 0,$$

$$\frac{\partial \mathbf{v}}{\partial t} + (\mathbf{v} \cdot \nabla)\mathbf{v} = -\frac{\nabla p}{\rho} - \frac{\mathbf{B} \times \operatorname{rot} \mathbf{B}}{4\pi\rho},$$

$$\frac{dp}{dt} = a\frac{d\rho}{dt}$$

$$\frac{\partial \mathbf{B}}{\partial t} = \operatorname{rot}(\mathbf{v} \times \mathbf{B})$$

in quasilinear form not paying attention to the condition div $\mathbf{B} = 0$, the following system is obtained (a is the acoustic speed of sound):

$$\frac{\partial \mathbf{u}}{\partial t} + A_1 \frac{\partial \mathbf{u}}{\partial x} + A_2 \frac{\partial \mathbf{u}}{\partial y} + A_3 \frac{\partial \mathbf{u}}{\partial z} = 0, \qquad (31)$$

where

$$\mathbf{u} = (\rho,\, u,\, v,\, w,\, p,\, B_x,\, B_y,\, B_z)^T,$$

$$A_1 = \begin{pmatrix} u & \rho & 0 & 0 & 0 & 0 & 0 & 0 \\ 0 & u & 0 & 0 & \frac{1}{\rho} & 0 & \frac{B_y}{4\pi\rho} & \frac{B_z}{4\pi\rho} \\ 0 & 0 & u & 0 & 0 & 0 & -\frac{B_x}{4\pi\rho} & 0 \\ 0 & 0 & 0 & u & 0 & 0 & 0 & -\frac{B_x}{4\pi\rho} \\ 0 & \rho a^2 & 0 & 0 & u & 0 & 0 & 0 \\ 0 & 0 & 0 & 0 & 0 & u & 0 & 0 \\ 0 & B_y & -B_x & 0 & 0 & 0 & u & 0 \\ 0 & B_z & 0 & -B_x & 0 & 0 & 0 & u \end{pmatrix},$$

$$A_2 = \begin{pmatrix} v & 0 & \rho & 0 & 0 & 0 & 0 & 0 \\ 0 & v & 0 & 0 & 0 & -\frac{B_y}{4\pi\rho} & 0 & 0 \\ 0 & 0 & v & 0 & \frac{1}{\rho} & \frac{B_x}{4\pi\rho} & 0 & \frac{B_z}{4\pi\rho} \\ 0 & 0 & 0 & v & 0 & 0 & 0 & -\frac{B_y}{4\pi\rho} \\ 0 & 0 & \rho a^2 & 0 & v & 0 & 0 & 0 \\ 0 & -B_y & B_x & 0 & 0 & v & 0 & 0 \\ 0 & 0 & 0 & 0 & 0 & 0 & v & 0 \\ 0 & 0 & B_z & -B_y & 0 & 0 & 0 & v \end{pmatrix},$$

$$A_3 = \begin{pmatrix} w & 0 & 0 & \rho & 0 & 0 & 0 & 0 \\ 0 & w & 0 & 0 & 0 & -\frac{B_z}{4\pi\rho} & 0 & 0 \\ 0 & 0 & w & 0 & 0 & 0 & -\frac{B_z}{4\pi\rho} & 0 \\ 0 & 0 & 0 & w & \frac{1}{\rho} & \frac{B_x}{4\pi\rho} & \frac{B_y}{4\pi\rho} & 0 \\ 0 & 0 & 0 & \rho a^2 & w & 0 & 0 & 0 \\ 0 & -B_z & 0 & B_x & 0 & w & 0 & 0 \\ 0 & 0 & -B_z & B_y & 0 & 0 & w & 0 \\ 0 & 0 & 0 & 0 & 0 & 0 & 0 & w \end{pmatrix},$$

Solution of the characteristic equation

$$\det(A_1 - \lambda I) = 0$$

gives the following eigenvalues:

$$\lambda_{1,2} = u, \qquad \lambda_{3,4} = u \pm \frac{B_x}{\sqrt{4\pi\rho}}, \qquad (32)$$

$$\lambda_{5,6,7,8} = u \pm \frac{1}{2}\left[\left(a^2 + \frac{\mathbf{B}^2}{4\pi\rho} + \frac{B_x a}{\sqrt{\pi\rho}}\right)^{1/2}\right.$$

$$\left. \pm \left(a^2 + \frac{\mathbf{B}^2}{4\pi\rho} - \frac{B_x a}{\sqrt{\pi\rho}}\right)^{1/2}\right] \qquad (33)$$

Note that in one-dimensional treatment the equation for B_x reduces to

$$\frac{\partial B_x}{\partial t} + u\frac{\partial B_x}{\partial x} = 0, \qquad (34)$$

that is, to the one-dimensional convection equation for B_x. Of course, in the truly one-dimensional problem (all values depend only on the spatial variable x) one can simply assume $B_x \equiv$ const and omit the corresponding equation. On the contrary, if we are going to apply the solution of the one-dimensional MHD Riemann problem to determine the flux through the cell boundary, such an assumption is too excessive, since only the integral $\oint B_n d\sigma$ over the whole computational cell must be equal to zero. Among eigenvalues

(32)–(33) the first two correspond to the entropy and B_x convection waves, $\lambda_{3,4}$ corresponds to the Alfvén, or rotational, waves, and the other ones to the slow and to the fast magnetosonic wave. Omitting Eq. (34), we reduce the system to 7×7. Both the extended 8×8 system and the reduced one have real eigenvalues and a degenerate set of eigenvectors. One can easily derive expressions for them. Otherwise, one can refer to [22], [83], and [101]. As was admitted in [39] and [84], we can use the extended system to derive an approximate solution to the MHD Riemann problem. Another possibility is to derive formulas for the 7×7 system and use a convection equation to find B_x (x is normal to the cell boundary) on the cell surface. It is also useful to realize that by collecting the source term from system (31) to arrive at the conservative form similar to (1), we obtain the following system:

$$\frac{\partial \mathbf{U}}{\partial t} + \frac{\partial \mathbf{E}}{\partial x} + \frac{\partial \mathbf{F}}{\partial y} + \frac{\partial \mathbf{G}}{\partial z} + \mathbf{H}_{\mathrm{div}} = \mathbf{0}, \qquad (35)$$

where

$$\mathbf{H}_{\mathrm{div}} = \mathrm{div}\,\mathbf{B} \left(0, \frac{B_x}{4\pi}, \frac{B_y}{4\pi}, \frac{B_z}{4\pi}, \frac{\mathbf{v} \cdot \mathbf{B}}{4\pi}, u, v, w \right)^{\mathrm{T}}$$

This form of the system will be used later to satisfy the divergence-free condition.

Since the application of high-resolution numerical schemes to MHD flows and to the problem under consideration, in particular, has not yet become common (see some tests in [103]), we give their description in the next subsection.

8.2 High-resolution numerical schemes for MHD equation

TVD upwind and symmetric differencing schemes have recently become very efficient tool for solving complex multi-shocked gasdynamic flows. This is due to their robustness for strong shock wave calculations. A general discussion of the modern high-resolution shock-capturing methods and their application for a variety of gasdynamic problems can be found in [42] and [112]. The extension of these schemes to the equations of the ideal magnetohydrodynamics (MHD) is not straightforward. First, the exact solution [49] of the MHD Riemann problem is too multivariant to be used in regular calculations. Second, several different approximate solvers [22], [23], [26], [41], [77], [84], and [114] applied to MHD equations are now at the stage of investigation and comparison.

The schemes [22], [23], [41], and [77], [84] are based on the MHD extensions of Roe's linearization procedure [88]. In [22], the attempt of such extension was made and the second order upwind

scheme was constructed that demonstrated several advantages in comparison with the Lax–Friedrichs, the Lax–Wendroff, and the flux-corrected transport scheme [30]. Roe's procedure, however, turned out to be realizable only for the special case with the specific heat ratio $\gamma = 2$. The reason of such behavior of MHD equations is that there is not any single averaging procedure to find a frozen Jacobian matrix of the system. Another linearization approach is used in [23], [41], [77], and [84] in which the linearized Jacobian matrix is not a function of a single averaged set of variables, but depends in a complicated way on the variables on the right- and on the left-hand side of the computational cell surface. In [79] and [82] this procedure was shown to be nonunique. A multiparametric family of linearized MHD approximate Riemann problem solutions was presented that assured an exact satisfaction of the conservation relations on discontinuities. A proper choice of parameters is necessary to avoid physically inconsistent solutions.

Consider the one-dimensional system of MHD equations

$$\frac{\partial \mathbf{U}}{\partial t} + \frac{\partial \mathbf{F}}{\partial x} = \mathbf{0}, \qquad (36)$$

where $\mathbf{U} = (\rho, \rho u, \rho v, \rho w, e, B_y, B_z)^{\mathrm{T}}$ and

$$\mathbf{F(U)} = \begin{pmatrix} \rho u \\ \rho u^2 + p_0 - B_x^2/4\pi \\ \rho uv - B_x B_y/4\pi \\ \rho uw - B_x B_z/4\pi \\ (e + p_0)u - (uB_x + vB_y + wB_z)B_x/4\pi \\ uB_y - vB_x \\ uB_z - wB_x \end{pmatrix}$$

In these formulas $e = p/(\gamma-1)+\rho(u^2+v^2+w^2)/2+ (B_x^2+B_y^2+B_z^2)/8\pi$ is the total energy per unit volume, $p_0 = p+(B_x^2+B_y^2+B_z^2)/8\pi$ is the total pressure, p and ρ are pressure and density, $\mathbf{v} = (u, v, w)$ is the velocity vector, $\mathbf{B} = (B_x, B_y, B_z)$ is the magnetic field vector, and γ is the adiabatic index. We assume all functions to depend only on time t and on the linear coordinate x. Our aim is to construct a solution to Eq. (36) for $t > 0$ for the piecewise-constant initial distribution of $\mathbf{U}$: $\mathbf{U} = \mathbf{U}_1$ for $x < 0$ and $\mathbf{U} = \mathbf{U}_2$ for $x > 0$. It is assumed, owing to the divergence-free condition, that $B_x = B_{x1} = B_{x2} \equiv \mathrm{const}$.

Let us first find the exact expression for the matrix $A = \Delta\mathbf{F}/\Delta\mathbf{U}$, where $\Delta\mathbf{U} = \mathbf{U}_1 - \mathbf{U}_2$ and $\Delta\mathbf{F} = \mathbf{F}_1 - \mathbf{F}_2$. The function $\mathbf{F}$ being nonlinear, the expression for A is determined nonuniquely. In fact, if we choose some nondegenerate substitution $\mathbf{s} = \mathbf{s}(\mathbf{U})$, then from the exact equalities $\Delta\mathbf{F} = A_F\Delta\mathbf{s}$ and $\Delta\mathbf{U} = A_U\Delta\mathbf{s}$ it follows that $A = A_F(A_U)^{-1}$. The exact analytic expressions for A_F and A_U can be written out explicitly if $\mathbf{U}$ and $\mathbf{F}$ are fractional linear functions of $\mathbf{s}$ or polynomials with respect to its components. We use here the equivalent transforms of the type $\Delta(BC) = \frac{1}{2}(B_1 + B_2)\Delta C + \frac{1}{2}(C_1 + C_2)\Delta B$. The structure and the simplicity of A depends on the choice of $\mathbf{s}$. The matrix A is an approximation to the Jacobian matrix $J = \dfrac{\partial\mathbf{F}}{\partial\mathbf{U}}$ and must conserve the main hyperbolic properties of J. It must be representable in the form $A = \Omega_R\Lambda\Omega_L$, where Ω_L and Ω_R are the matrices of its left and right eigenvectors, respectively, $\Omega_R\Omega_L = I$, and I is the identity matrix; $\Lambda = ||\lambda_i\delta_{ij}||$ is the diagonal matrix of the real eigenvalues of A, and δ_{ij} is the Kronecker delta.

Then the sought solution to Eq. (36) for $t > 0$ acquires the form

$$\mathbf{U}(\xi) = \frac{1}{2}(\mathbf{U}_1 + \mathbf{U}_2 + \Omega_R S(\xi)\Omega_L\Delta\mathbf{U}), \quad (37)$$

$$\mathbf{F}(\xi) = \frac{1}{2}(\mathbf{F}_1 + \mathbf{F}_2 + \Omega_R|\Lambda(\xi)|\Omega_L\Delta\mathbf{U}) \quad (38)$$

where $\xi = \frac{x}{t}$, $S(\xi) = ||\operatorname{sgn}(\lambda_i - \xi)\delta_{ij}||$, and $|\Lambda(\xi)| = ||\lambda_i\operatorname{sgn}(\lambda_i - \xi)\delta_{ij}||$. Equations (37) and (38) determine the piecewise-constant functions $\mathbf{U}(\xi)$ and $\mathbf{F}(\xi)$ that glue the right and the left values of the initial distributions via the system of jumps. If the Hugoniot-type condition is valid $\Delta\mathbf{F} = \lambda\Delta\mathbf{U}$, where λ is the jump velocity, then λ is one of the eigenvalues of A, since $\Delta\mathbf{F} - \lambda\Delta\mathbf{U} = (A_F - \lambda A_U)\Delta\mathbf{s} = (A - \lambda E)A_U\Delta\mathbf{s} = (A - \lambda E)\Delta\mathbf{U} = 0$. Thus, $\det(A - \lambda E) = 0$, $\Delta\mathbf{U} \neq 0$ is the eigenvector of A, and relations (37)–(38) describe the jump exactly.

The solution of this kind was first constructed for the equations of ideal gas dynamics [88]. In [22] such solution was given for Eq. (36) in the case $\gamma = 2$. The approximate solution for an arbitrary adiabatic index was proposed in [41]. Here we present the procedure for obtaining the extension of Roe's linearization procedure for MHD equation and show that it is not unique. Thus, the solution from [41] is the particular case of the multiparametric family of approximate solutions to the MHD Riemann problem.

We choose the vector $\mathbf{s}$ as a generalization of that for pure gas dynamics:

$$\mathbf{s} = \begin{pmatrix} R \\ U \\ V \\ W \\ \mathcal{I} \\ Y \\ Z \end{pmatrix} = \begin{pmatrix} \sqrt{\rho} \\ \sqrt{\rho}u \\ \sqrt{\rho}v \\ \sqrt{\rho}w \\ \sqrt{\rho}H \\ B_y/\sqrt{\rho} \\ B_z/\sqrt{\rho} \end{pmatrix}$$

Then

$$\mathbf{U} = \begin{pmatrix} R^2 \\ RU \\ RV \\ RW \\ U_5 \\ RY \\ RZ \end{pmatrix},$$

where

$$U_5 = \frac{R\mathcal{I}}{\gamma} + \frac{(U^2 + V^2 + W^2)(\gamma - 1)}{2\gamma} - \frac{(Y^2 + Z^2)R^2(2 - \gamma)}{8\pi\gamma}$$

and $H = (e + p_0)/\rho$ is the total enthalpy. The vector $\mathbf{F}$ in the variabless acquires the form

$$\mathbf{F} = \begin{pmatrix} RU \\ F_2 \\ UV - RYB_x/4\pi \\ UW - RZB_x/4\pi \\ U\mathcal{I} - (UB_x/R + VY + WZ)B_x/4\pi \\ UY - VB_x/R \\ UZ - WB_x/R \end{pmatrix},$$

where $F_2 = U^2 + (\gamma - 1)R\mathcal{I}/\gamma - (U^2 + V^2 + W^2)(\gamma - 1)/2\gamma + (Y^2 + Z^2)R^2(2 - \gamma)/8\pi\gamma$.

Using the new expressions for $\mathbf{U}$ and $\mathbf{F}$, we can find the matrices A_U, A_F, and $A_\lambda = A_F - \lambda A_U$. We present only the expression for A_λ:

$$
A_\lambda = \begin{pmatrix}
u-2\lambda & 1 & 0 & 0 & 0 & 0 & 0 \\
A_{21} & \frac{\gamma+1}{\gamma}u-\lambda & \frac{1-\gamma}{\gamma}v & \frac{1-\gamma}{\gamma}w & \frac{\gamma-1}{\gamma} & q_y & q_z \\
A_{31} & v & u-\lambda & 0 & 0 & -\frac{B_x}{4\pi} & 0 \\
A_{41} & w & 0 & u-\lambda & 0 & 0 & -\frac{B_x}{4\pi} \\
A_{51} & A_{52} & A_{53} & A_{54} & u-\frac{\lambda}{\gamma} & A_{56} & A_{57} \\
A_{61} & h_y & -\frac{B_x}{\rho} & 0 & 0 & u-\lambda & 0 \\
A_{71} & h_z & 0 & -\frac{B_x}{\rho} & 0 & 0 & u-\lambda
\end{pmatrix},
$$

where

$$
A_{21} = \frac{\gamma-1}{\gamma}\mathcal{H} - \lambda u + q, \quad A_{31} = -\frac{B_x h_y}{4\pi} - \lambda v,
$$

$$
A_{41} = \frac{B_x h_z}{4\pi} - \lambda w, \quad A_{51} = \frac{uB_x^2}{4\pi\rho} + \lambda\left(q - \frac{\mathcal{H}}{\gamma}\right),
$$

$$
A_{52} = \mathcal{H} - \frac{B_x^2}{4\pi\rho} + \lambda u\frac{1-\gamma}{\gamma},
$$

$$
A_{53} = -\frac{B_x h_y}{4\pi} + \lambda v\frac{1-\gamma}{\gamma},
$$

$$
A_{54} = -\frac{B_x h_z}{4\pi} + \lambda w\frac{1-\gamma}{\gamma},
$$

$$
A_{56} = -\frac{vB_x}{4\pi} + \lambda q_y, \quad A_{57} = -\frac{wB_x}{4\pi} + \lambda q_z,
$$

$$
A_{61} = \frac{vB_x}{\rho} - \lambda h_y, \quad A_{71} = \frac{wB_x}{\rho} - \lambda h_z
$$

The following notions are adopted in the above relations: $\rho = \sqrt{\rho_1\rho_2}$, $u = \overline{\sqrt{\rho}u}/\sqrt{\rho}$, $v = \overline{\sqrt{\rho}v}/\sqrt{\rho}$, $w = \overline{\sqrt{\rho}w}/\sqrt{\rho}$, $\mathcal{H} = \overline{\sqrt{\rho}H}/\sqrt{\rho}$, $h_y = \overline{B_y/\sqrt{\rho}}/\sqrt{\rho}$, $h_z = \overline{B_z/\sqrt{\rho}}/\sqrt{\rho}$, where $\overline{f}$ means arithmetic averaging. Besides,

$$
q = \frac{2-\gamma}{4\pi\gamma}\left(\overline{Y}^2 + \overline{Z}^2 + \frac{\theta_1}{4}(\Delta Y)^2 + \frac{\theta_2}{4}(\Delta Z)^2\right.
$$

$$
\left. + \frac{\eta_1}{4\overline{R}}\overline{Y}\Delta Y\Delta R + \frac{\eta_2}{4\overline{R}}\overline{Z}\Delta Z\Delta R\right),
$$

$$
q_y = \frac{2-\gamma}{4\pi\gamma}\left(\overline{Y}\,\overline{R} + \frac{1-\theta_1}{4}\Delta Y\Delta R + \frac{1-\eta_1}{4\overline{R}}(\Delta R)^2\overline{Y}\right),
$$

$$
q_z = \frac{2-\gamma}{4\pi\gamma}\left(\overline{Z}\,\overline{R} + \frac{1-\theta_2}{4}\Delta Z\Delta R + \frac{1-\eta_2}{4\overline{R}}(\Delta R)^2\overline{Z}\right),
$$

where θ_1, θ_2, η_1, and η_2 are arbitrary parameters. Their origin is caused by the presence in the expressions for $\Delta\mathbf{F}$ of the terms containing the factors $\Delta R\Delta Y$ and $\Delta R\Delta Z$ which can be attributed both to the terms proportional to ΔR and ΔY or ΔZ. This results in an additional parametrization of the entries of the matrices A_F and A_U. It is not difficult to find that

$$
\det A_\lambda = \frac{2K}{\gamma}\left(K^2 - \frac{B_x^2}{4\pi\rho}\right)\left\{(K^2 - c^2 - \alpha)\right.
$$

$$
\left. \times \left(K^2 - \frac{B_x^2}{4\pi\rho}\right) - K^2\left[\frac{\rho}{4\pi}(h_y^2 + h_z^2) + \beta\right]\right\},
$$

where

$$
K = u - \lambda, \quad c^2 = (\gamma-1)\left[I - \frac{u^2 + v^2 + w^2}{2}\right.
$$

$$
\left. - \frac{B_x^2}{4\pi\rho} - \frac{(h_y^2 + h_z^2)\rho}{4\pi}\right],
$$

$$
\alpha = \gamma(\delta - \delta_y h_y - \delta_z h_z)/2 = \gamma(q - q_y h_y - q_z h_z),
$$

$$
\beta = \gamma(h_y\delta_y + h_z\delta_z), \quad \delta = q - \frac{2-\gamma}{4\pi\gamma}(h_y^2 + h_z^2)\rho,
$$

$$
\delta_y = q_y - \frac{2-\gamma}{4\pi\gamma}h_y\rho, \quad \delta_z = q_z - \frac{2-\gamma}{4\pi\gamma}h_z\rho
$$

The equation for α can be rewritten in form

$$
\alpha = \frac{2-\gamma}{32\pi}\left[\theta_1(\Delta Y)^2 + (\theta_1 + \eta_1 - 1)\frac{\overline{Y}\Delta Y\Delta R}{\overline{R}}\right.
$$

$$
+ (\eta_1 - 1)\left(\frac{\overline{Y}\Delta R}{\overline{R}}\right)^2 + \theta_2(\Delta Z)^2
$$

$$
\left. + (\theta_2 + \eta_2 - 1)\frac{\overline{Z}\Delta Z\Delta R}{\overline{R}} + (\eta_2 - 1)\left(\frac{\overline{Z}\Delta R}{\overline{R}}\right)^2\right]
$$

The eigenvalues of A are equal to u, $u\pm b$, where $b = |B_x|/\sqrt{4\pi\rho}$, and to the four roots of the biquadratic equation

$$
K^4 - 2pK^2 + Q = 0, \tag{39}
$$

where

$$
2p = c^2 + \alpha + b^2 + (h_y^2 + h_z^2)\rho/4\pi + \beta,
$$

$$
Q = (c^2 + \alpha)b^2
$$

If $c^2 + \alpha \geq 0$ and $(h_y^2 + h_z^2)\rho/4\pi + \beta \geq 0$, the roots of this equation are real and the diagonal matrix composed of the eigenvalues acquires the form

$$
\Lambda = \mathrm{diag}\,\|u+a_f,\ u+b,\ u+a_s,\ u,\ u-a_s,\ u-b,\ u-a_f\|
$$

The roots a_f and a_s are the largest and the least root of Eq. (39) (the fast and the slow magnetosonic waves) and b corresponds to the Alfvénic waves. The remaining eigenvalue corresponds to the entropy waves.

The peculiarity of our approach lies in the strict ordering of the eigenvalues. This provides the absence of their additional nonphysical degeneration which is not inherent in J. Note that the choice of other parameter vectors s can break this property. In particular, such a degeneration appears if q_y and q_z are not proportional to $\overline{Y}$ and $\overline{Z}$, respectively. This leads to the most simple admissible choice of θ: $\theta_1 = \theta_2 = 1$. In the MHD case, in contrast to pure gas dynamics, it is not possible to construct the matrix A depending on a single average vector.

Let us calculate Ω_R and Ω_L. It is convenient to introduce the matrix Ω_r instead, for which $\Omega_R = A_U\Omega_r$. This matrix consists of seven columns r. For the eigenvalues $\lambda = u + sa$, where $s = \pm 1$ and $a = a_f, a_s,$ or 0, the corresponding vector-columns $\mathbf{r} = \mathbf{r}(s, a)$ are the following:

$$\mathbf{r} = (1,\, u + 2sa,\, v - sh_y M,\, w - sh_z M,\, r_5,\, h_y N,\, h_z N)^T$$

where

$$r_5 = -\mathcal{H} + u^2 + v^2 + w^2 + 2sau - (vh_y + wh_z)sM$$
$$+ \left[2a^2 - q - (q_y h_y + q_z h_z)N\right]\gamma/(\gamma - 1),$$

$$M = M(a) = \frac{aB_x}{2\pi(a^2 - b^2)}, \quad N = N(a) = \frac{a^2 + b^2}{a^2 - b^2}$$

For the eigenvalues $\lambda = u + sb$ ($s = \pm 1$) the corresponding vector acquires the form

$$\mathbf{r} = \begin{pmatrix} 0 \\ 0 \\ h_z^* \\ -h_y^* \\ vh_z^* - wh_y^* \\ -sh_z^*\sqrt{4\pi/\rho}\,\mathrm{sgn}\,B_x \\ sh_y^*\sqrt{4\pi/\rho}\,\mathrm{sgn}\,B_x \end{pmatrix}$$

where $h_y^* = h_y/|\mathbf{h}|$ and $h_z^* = h_z/|\mathbf{h}|$.

When using the above formulas for $|\mathbf{h}| \to 0$, the indeterminacies of the type 0/0 must be resolved. This can be done, e. g., by the substitution $h_y = |\mathbf{h}|\sin\varphi$ and $h_z = |\mathbf{h}|\cos\varphi$.

The matrix Ω_L can be found similarly by introducing Ω_l such that $\Omega_L = D^{-1}\Omega_l$, where D is a diagonal matrix specified by the equality $A_U\Omega_R D^{-1}\Omega_l = I$. It consists of seven rows. For the eigenvalues $\lambda = u + sa$, where $s = \pm 1$ and $a = a_f, a_s,$ or 0, the corresponding vector-rows $\mathbf{l} = \mathbf{l}(s, a)$ are the following:

$$l_1 = \frac{c^2 - a^2 + sau}{\gamma - 1} - \frac{u^2 + v^2 + w^2}{2}$$
$$- 2s\pi(vr_y + wr_z)M + (h_y r_y + h_z r_z)L,$$

$$l_2 = u - \frac{sa}{\gamma - 1}, \quad l_3 = v + 2\pi M sr_y,$$

$$l_4 = w + 2\pi M sr_z, \quad l_5 = -1,$$

$$l_6 = -r_y L + \frac{\rho h_y}{4\pi}, \quad l_7 = -r_z L + \frac{\rho h_z}{4\pi}$$

$$r_y = \frac{h_y}{4\pi} + \frac{\gamma q_y}{\rho(\gamma - 1)},$$

$$r_z = \frac{h_z}{4\pi} + \frac{\gamma q_z}{\rho(\gamma - 1)}, \quad L(a) = \frac{[1 + N(a)]\rho}{2}$$

For $\lambda = u + sb$ ($s = \pm 1$) we obtain

$$l_1 = wh_y^* - vh_z^*, \quad l_2 = l_5 = 0, \quad l_3 = h_z^*, \quad l_4 = -h_y^*,$$
$$l_6 = -sh_z^*\sqrt{\rho/4\pi}\,\mathrm{sgn}\,B_x, \quad l_7 = sh_y^*\sqrt{\rho/4\pi}\,\mathrm{sgn}\,B_x$$

The matrix D has the form

$$D = \mathrm{diag}\,\|d(a_f),\, 2,\, d(a_s),\, -d(0),\, d(a_s),\, 2,\, d(a_f)\|,$$

where

$$d(a) = -\frac{2}{\gamma - 1}\Big\{a^2 + c^2 + \alpha$$
$$+ [(h_y^2 + h_z^2)\rho/4\pi + \beta](1 + N)N/2\Big\}$$

In practice, if $B_x \to 0$ or $|\mathbf{h}| \to 0$, the indeterminacy of the type 0/0 must be resolved in the above relations for $M = M(a)$ and $N = N(a)$ at $a = a_s$. This can be done by using the biquadratic equation for the roots. It is not difficult to find that

$$N(a_s) = -\frac{(a_f^2 - b^2)(c^2 + \alpha + 2p + b^2)}{\varepsilon|\mathbf{h}|^2(a_f^2 + b^2)},$$

$$M(a_s) = -\frac{(a_f^2 - b^2)\,\mathrm{sgn}\,B_x}{\varepsilon a_f|\mathbf{h}|^2}\sqrt{\frac{(c^2 + \alpha)\rho}{\pi}},$$

where

$$\varepsilon = \frac{1}{4\pi}\left\{\rho + \frac{(2 - \gamma)(\Delta R)^2}{4}\left[(2 - \eta_1)(h_y^*)^2 + (2 - \eta_2)(h_z^*)^2\right]\right\}$$

To resolve the indeterminacy, eigenvectors $\mathbf{r}(\pm 1,\, a_s)$ and $\mathbf{l}(\pm 1,\, a_s)$ are multiplied by $|\mathbf{h}|$, and the corresponding $d(a_s)$ by $|\mathbf{h}|^2$. Then, the substitution is made similar to that used for the regularization of the Alfvénic eigenvectors.

No nonphysical degeneration occurs for $\gamma = 2$, since in this case $\alpha \equiv 0$ and $\beta \equiv 0$ and the characteristic equation has only real roots. If we choose $\theta_1 = \theta_2 = 1$ and $\eta_1 = \eta_2 = 2$ for an arbitrary γ, see [41], one can easily show that $\alpha \geq 0$ and $\beta \equiv 0$, thus giving only real roots of the characteristic equation for any admissible right and left values.

Another choice is $\theta_1 = \theta_2 = 1$ and $\eta_1 = \eta_2 = 0$. In this case $\beta \geq 0$, whereas $\alpha \geq 0$ only for the right and the left values connected via the Hugoniot-type conditions. The possible degeneration of the formula can be avoided by the regularization. It is worth mentioning that the above manipulations related to the matrix A_λ were performed for arbitrary q, q_y, and q_z. This allows us, by substituting q^* for q, where

$$q^* = |q - h_y q_y - h_z q_z| + h_y q_y + h_z q_z,$$

and leaving q_y and q_z unchanged, to conserve the properties of J universally for any $0 \leq \eta_1 = \eta_2 \leq 2$. For small jumps this regularization introduces the error of the second order of smallness and does not prevent its application for constructing the solutions of the second order of accuracy. It does not distort the solution for the MHD jumps, since in this case $\alpha \geq 0$. We also have $\beta \geq 0$ for the above choice of η_1 and η_2. Such an approach preserves the jump relations and is intermediate between the techniques using the exact expression for $A = \Delta \mathbf{F}/\Delta \mathbf{U}$ [22], [41], and [88] and approaches using different approximations to A.

The family of approximate solutions to the MHD Riemann problem presented here generalizes the known approximate quasi-linearized and linearized solutions of this problem and preserves the Hugoniot-type relations on the jumps. By using proper reconstruction techniques, we can increase the order of accuracy of obtaining the fluxes in Eq. (38). In this case the indices "1" and "2" must be attributed to the parameter values on the right and on the left side of the computational cell.

Due to the complexity of formulas, this approach is much more cumbersome than Roe's linearization method in pure gas dynamics. Recently in [26], a nonlinear approximate Riemann problem solver is suggested in which all the waves emanating from the initial discontinuity are treated as discontinuous jumps. That is why, it is applicable only for weak rarefactions. Moreover, the solver proposed is somewhat time-consuming and sensitive to the initial approximation for the iteration process.

Taking into account the above-mentioned remarks some simplified approaches are welcome which should (1) satisfy TVD property and (2) be enough economical and robust.

In [14], the second order of accuracy in time and space TVD Lax–Friedrichs type scheme is suggested that gives a great simplification of the numerical algorithm in the finite-volume formulation comparing with the schemes which use the precise characteristic splitting of Jacobian matrices. The results obtained by this scheme were compared with those from [22] and [27] and a good agreement was observed. In this scheme we substitute the diagonal eigenvalue matrix in Eq. (38) by the diagonal matrix with the spectral radius (the maximum of eigenvalue magnitudes) of A on its diagonal. Using the proper parameter reconstruction to find their values on the computational cell surfaces, we obtain the second order of accuracy.

This scheme is less dissipative than the original Lax–Friedrichs scheme and can be applied to calculations of discontinuous MHD flows. Being much simpler than the scheme based on Roe's linearization method, it still gives numerical results with the reasonable accuracy.

Another important subject of discussion is that certain initial- and boundary-value problems can be solved nonuniquely using different shocks or different combinations of shocks, whereas physically one would expect only unique solutions. The situation differs from that in pure gas dynamics, where all entropy-increasing solutions are evolutionary and physically admissible. This means that the necessary conditions of the well-posedness for the linearized problem of their interaction with small disturbances are satisfied. In MHD case, on the contrary, the condition of the entropy increase is necessary, but not sufficient. Only slow and fast MHD shocks turned out to be evolutionary, while intermediate (or improper slow) shocks are to be excluded [45] and [49]. Nonevolutionary shocks are not simply unstable in ideal MHD. Their decay into evolutionary jumps occurs under infinitesimal perturbation within infinitesimal time. On the other hand, numerical viscosity (including the presence of finite conductivity) and numerical dispersion of a numerical scheme make such intermediate structures existent for a certain time interval before their destruction. It is very important to realize that this interval has nothing to do with the real interval of existence of intermediate waves in non-ideal plasma and is grid- and numerical scheme-dependent. If the viscosity and/or the conductivity are substantial, such compound structure can exist for a long time depending on the amount of viscosity. This was admitted and used for the explanation of certain physical phenomena in [111].

In [22], the ideal MHD Riemann problem was solved with initial data consisting of two constant states lying to the right and to the left of the centerline of the computational domain. Being solved as a strictly coplanar problem, it included a nonevolutionary compound shock. Such shocks must decay and are not realizable in physical problems. The peculiarity of MHD is that there exist discontinuities that are nonevolutionary only with respect to Alfvénic (rotational) disturbances. That is why, if a strictly coplanar problem is considered (velocity and magnetic field vectors lie in the same plane and the system of MHD equations includes only two vector components) the construction of the solution is possible both with evolutionary and nonevolutionary shock waves. The solution in this case is nonunique. If the full set of three-dimensional MHD equations is solved and a small tangential disturbance is added to the magnetic field vector, a rotational jump splits from the compound wave and it degrades into the slow shock. This means that the compound wave is unstable against tangential disturbances and is nonevolutionary in three dimensions (see [14]). That means that one must be very careful reducing the dimension of the system (say, in axisymmetric problems) to avoid the origin of nonadmissible solutions. The necessity of three-dimensional consideration of MHD Riemann problems has been admitted recently in [26]. One must take into account this feature of MHD equations in the construction of numerical algorithms. In other words, if we are going to solve an axisymmetric problem, in general, a full set of three-dimensional equations must be used.

8.3 Shock-fitting calculations

In [13] the interaction between the solar wind and the magnetized plasma component of the local interstellar medium were calculated by solving the axisymmetric MHD equations using the shock-fitting approach. The solar wind was assumed nonmagnetized. The magnetic field of the interstellar medium was assumed parallel to its velocity vector. The system of MHD equations in the quasi-linear form was solved by the iteration method. Each iteration consisted of the two steps. At the first step, the gasdynamic part of the system was solved assuming magnetic field known from the previous time step. At the second iteration step, the Maxwell equations were solved assuming velocities known. At the first step of each iteration magnetic field was assumed zero. The gasdynamic part was solved by the time-stabilization method [113] (see Section 3). For the known distribution of gasdynamic parameters the magnetic field in the considered case can be determined analytically as

$$\mathbf{B} = B_\infty \frac{\rho \mathbf{v}}{\rho_\infty V_\infty}$$

These equation follows from the induction equations and the MHD jump conservation relations [51] for $\mathbf{B} \parallel$

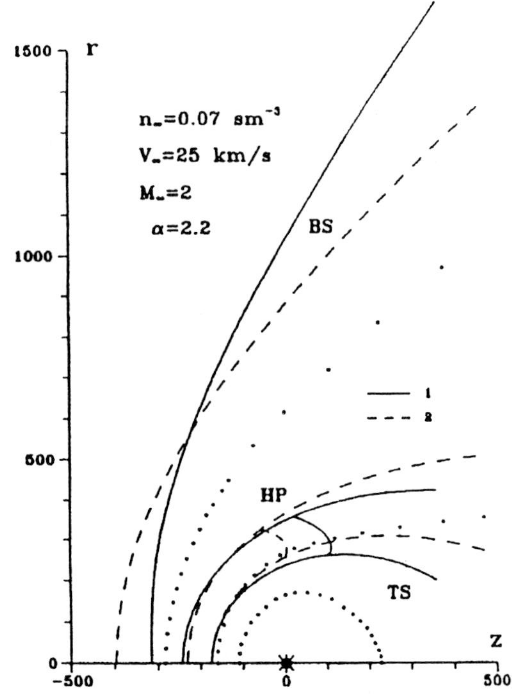

Fig.19: (1) Geometrical pattern of the interface for $\alpha = 2.2$ [13]. Positions of the bow shock (BS), the termination shock (TS), heliopause (HP), and sonic line in the solar wind. (2) The same lines for $\alpha = 0$. The discontinuities calculated in [11] in the presence of the neutral hydrogen are shown by dotted lines

v.

The following solar wind and LISM parameters were chosen:

$$n_e = 7\,\mathrm{cm}^{-3}, \ V_e = 450\,\mathrm{km/s}, \ M_e = 10,$$
$$n_\infty = 0.07\,\mathrm{cm}^{-3}, \ V_\infty = 25\,\mathrm{km/s}, \ M_\infty = 2$$

The dimensionless value of magnetic field was specified via parameter $\alpha = 2\sqrt{\pi}/A$, where $A = V_\infty/\sqrt{B_\infty/4\pi\rho}$ is the Alfvén number.

The geometrical pattern of the flow is shown in Fig. 19.

It is quite clear from this picture that both the LISM magnetic field and the charge-exchange processes play an important role in the interaction. One can notice that, although a shock-fitting approach perfectly well works in the forward part of the interaction, its limitations reveal themselves in the tail part of the flow. The applicability of the described approach is also limited to axisymmetric problems.

8.4 Shock-capturing calculations

All mentioned in the previous subsection necessitates the application of shock-capturing methods for modeling of the SW–LISM interaction. In [59] this problem was solved in the closed region surrounding an ejecting star on the basis of the flux-splitting method [46]. The choice of parameters was far from those adopted nowadays for the considered problem. Although, as a whole, a physically consistent results were obtained, the resolution of discontinuities was rather poor and some of the obtained data were misinterpreted and disputed in [13].

In [80] the solution of the axisymmetric problem (the LISM magnetic field strength vector is assumed to be parallel to its velocity vector) is presented for realistic SW and LISM parameters. The numerical method applied was developed by Pogorelov [14] and is the high-resolution second-order of accuracy version of the Lax–Friedrichs scheme. This method gives a drastic simplification of the numerical algorithm comparing with the methods based on the exact characteristic splitting of the Jacobian matrices in the MHD equations. Simple but very effective numerical boundary conditions at the far-field were suggested (see Section 2), which allowed one to avoid the influence of spurious reflections from the subsonic outer boundary. An effective procedure of satisfying the condition of source-free magnetic fields (divergence-free condition) was used, which is related to the approach [83].

The formulation of the problem was given is Section 2. Following [14], the numerical flux at the radial cell interface (27) was calculated by the formula

$$\bar{\mathbf{E}}_{l+1/2,n} = \frac{1}{2} \left[\bar{\mathbf{E}} \left(\mathbf{U}^R_{l+1/2,n} \right) + \bar{\mathbf{E}} \left(\mathbf{U}^L_{l+1/2,n} \right) + \mathbf{\Phi}_{l+1/2,n} \right]$$

$$\mathbf{\Phi}_{l+1/2,n} = -\hat{\mathbf{R}}_{l+1/2,n} \left(\mathbf{U}^R_{l+1/2,n} - \mathbf{U}^L_{l+1/2,n} \right)$$

(40)

Here $\hat{\mathbf{R}}_{l+1/2,n}$ is the diagonal matrix with the same elements on its diagonal equal to the spectral radius r (the maximum of eigenvalue magnitudes) of the Jacobian matrix $\frac{\partial \bar{\mathbf{E}}}{\partial \mathbf{U}}$:

$$r = |U| + a_f, \quad a_f^2 = \frac{1}{2} \left((a^*)^2 + \sqrt{(a^*)^4 - 4a^2 b_R^2} \right),$$

$$b_R = B_R/(4\pi\rho)^{1/2}, \quad B^2 = B_R^2 + B_\theta^2,$$

$$(a^*)^2 = (\gamma p + B^2/4\pi)/\rho, \quad a^2 = \gamma p/\rho,$$

where U is the radial velocity component and B_R and B_θ are the radial and the angular component of the magnetic field strength vector. Similar formulas can be written for the fluxes in the angular direction.

Having the second order of accuracy, the proposed scheme is much less dissipative than the original Lax–Friedrichs method and provides incomparably better shock resolution. Calculations were performed in the polar computational region with $R_{\min} = 24$ and $R_{\max} = 1200$. The mesh was $R \times \theta = 99 \times 116$.

Mathematically, if we choose the initial distribution of parameters with div $\mathbf{B} = 0$, the Maxwell equations will preserve this value in the steady solution. In fact, for problems solved numerically the regions of div $\mathbf{B} \neq 0$ can be accumulated, especially in the vicinity of discontinuities, see [19]. It is quite clear that the application of the one-dimensional Riemann problem solvers, implying that the component B_n of the magnetic field vector normal to the boundary is constant, contradicts to the condition $\oint B_n d\sigma = 0$ over the whole computational cell. In [83] an approximate Riemann problem solver for MHD equations was proposed on the basis of the modified system (35) which is conservative only in a steady state. Using the Lax–Friedrichs-type scheme, we do not need such modification, since we do not solve the Riemann problem to find the fluxes at the cell surfaces. The correction in our case can be made by adding to the source term of Eq. (2) the value proportional to div $\mathbf{B}$:

$$\mathbf{H}' = \text{div } \mathbf{B} \left[0, \frac{B_x}{4\pi}, \frac{B_z}{4\pi}, \frac{\mathbf{v} \cdot \mathbf{B}}{4\pi}, u, w \right]^T$$

This term acts to annihilate the error accumulated if system is solved in the conservation-law form (see [19]). It is worth noting that this approach lies in the framework of Powell's procedure and the correction term is not small only in the regions of comparably large errors in div $\mathbf{B}$. This term, in fact, is equal by the value and opposite by the sign to the appropriate terms proportional to div $\mathbf{B}$ appearing by differentiating $\mathbf{E}$ and $\mathbf{G}$ in Eq. (1). The divergence of magnetic field strength vector can be approximated over the finite volumes as

$$\text{div } \mathbf{B} = [R^2_{l+1/2}(B_n)_{l+1/2,n} + R^2_{l-1/2}(B_n)_{l-1/2,n}]/R_l^2 \Delta R +$$
$$[(B_n)_{l,n+1/2} \sin \theta_{n+1/2} + (B_n)_{l,n-1/2} \sin \theta_{n-1/2}]/R_l \sin \theta_n \Delta$$

This simple procedure gives a powerful tool for the realization of the divergence-free condition and necessitates only a slight modification of the existing codes for solving MHD equations. It seems limited, however, only to steady-state calculations.

Numerical results were obtained for the same flow parameters as in the previous subsection and for the three different values of the Alfvén number: $A = 10, 2, \sqrt{2}$. The first value corresponds to the very small magnetic field (see Fig. 20) and can be used for

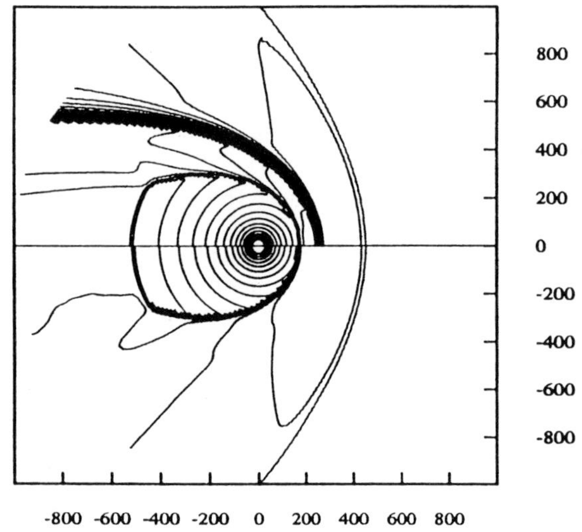

Fig.20: Pressure (below the symmetry axis) and density logarithm isolines, $A = 10$

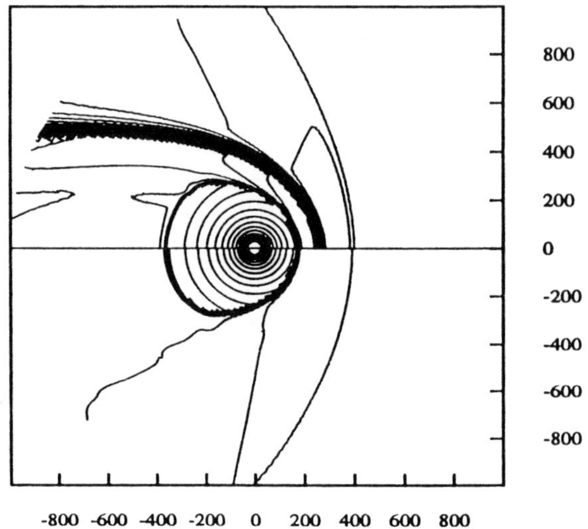

Fig.21: Pressure (below the symmetry axis) and density logarithm isolines, $A = 2$

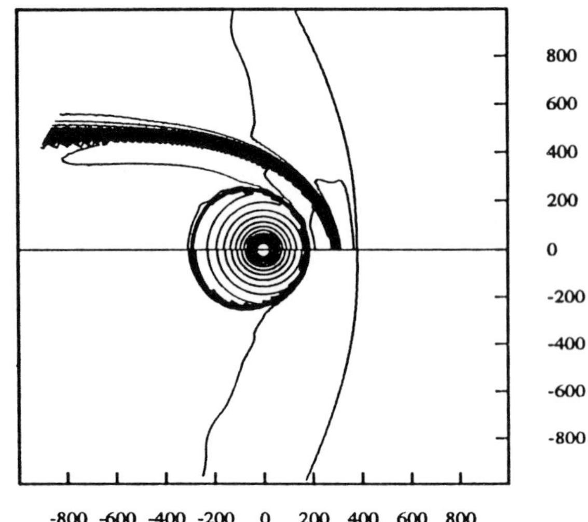

Fig.22: Pressure (below the symmetry axis) and density logarithm isolines, $A = 1.414$

comparison with purely gasdynamic results in which the Roe-type Riemann problem solver was used to obtain the steady state solution, see Subsection 3.2. In Fig. 20, the pressure (below the symmetry axis) and the density logarithm isolines are presented ($A = 10$). All features of the flow pattern (see Fig. 2) are sharply resolved, while the low time-consumption and the simplicity of the algorithm are quite clear. In Fig. 21 the same distributions are shown for the case $A = 2$.

In Figs. 22–23, the case with $A = \sqrt{2}$ is presented, which corresponds to $p_{\mathrm{magn}}/p_{\mathrm{thermal}} \approx 1.67$ at infinity, that is, $B_\infty \approx 2.5$ (dimensional value $\sim 2.3 \times 10^{-6}$ Gauss). In Fig. 23, the streamlines (lower half) and the magnetic field lines are shown.

Near the symmetry axis the heliocentric distance of the heliopause increases due to the magnetic field tension, whereas it decreases at the side parts due to the magnetic pressure. On the other hand, although the magnetic field is absent in the region inside the heliopause, its action is revealed by the increased value of the total pressure at infinity. This leads to substantial decrease of the termination shock stand-off distance in the downstream region. We can see the absence of the Mach disk structure in the backward direction. This means that the velocity along the termination shock becomes subsonic. The same occurs if the charge-exchange is taken into account [11]. The bow shock stand-off distance along the symmetry axis becomes smaller than in the absence of magnetic field.

As the magnetic field increases, the effective Mach number, generally speaking, diminishes. The similar effect is produced by the presence of neutral hydrogen

atoms and, if both of these factors exceed a definite value, the bow shock can disappear. In this case system (2) becomes elliptic and other methods must be applied for its solution. It is worth noting, however, that the speed of propagation of magnetosonic waves depends on the direction with respect to the magnetic field vector. That means that the effective Mach number varies along the bow shock. As will be shown later, this results in a highly asymmetric shape of the bow shock if magnetic field is not parallel to the velocity vector.

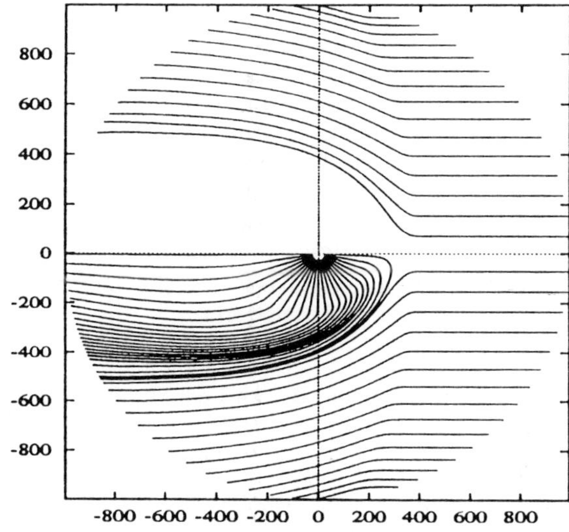

Fig.23: Streamlines (below the symmetry axis) and magnetic field lines, $A = 1.414$.

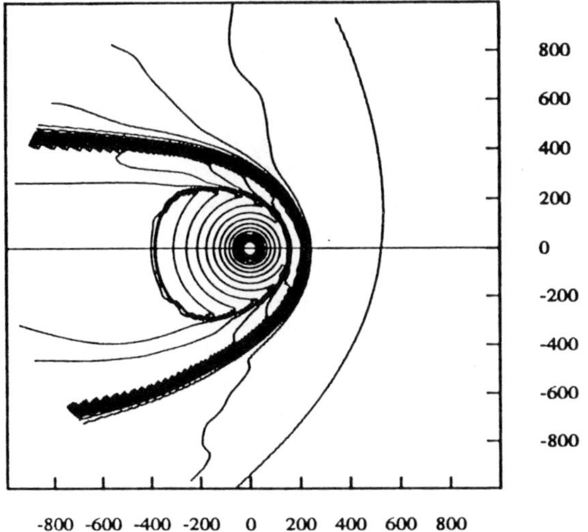

Fig.24: Density logarithm isolines in the symmetry plane

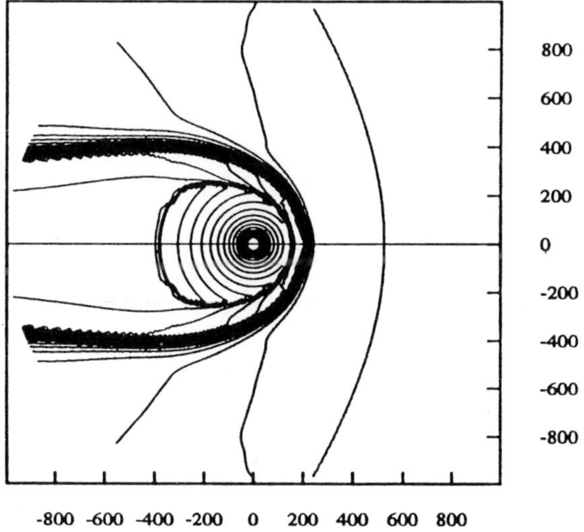

Fig.25: Density logarithm isolines in the plane $\varphi = 90° - 270°$

If the magnetic field strength at infinity is equal to its probable upper limit of 3×10^{-6} Gauss, the LISM flow becomes subsonic, even if the charge-exchange processes are neglected. This can be seen from the simple estimate. We can calculate the LISM effective Mach number as $M_{\text{eff}}^2 = \rho_\infty V_\infty^2/(\gamma p_\infty + B_\infty^2/4\pi)$ (this holds, e. g., at the symmetry axis for $\mathbf{V}_\infty \perp \mathbf{B}_\infty$. Thus, $1/M_{\text{eff}}^2 = 1/M_\infty^2 + 1/A^2$ and for $M_\infty^2 = 4$ the effective Mach number remains larger than unity only

for $A > 1.15$. Thus, both effects are of great importance for the interpretation of data obtained in the space experiment. The performance of the algorithm for the realization of the magnetic field source-free condition can be seen from Fig. 23. The magnetic field lines remain parallel to the streamlines. One can see that no magnetic field penetrates into the heliosphere. This result can not be achieved without a special treatment of the divergence-free condition.

The result of the magnetic field influence on the whole structure of the flow is not sufficiently investigated. It is clear that the flow pattern for $\mathbf{B}_\infty \not\parallel \mathbf{V}_\infty$ is three-dimensional. MHD modeling of the heliopause shape on the basis of the Newtonian approximation was performed in [34]. The results of the three-dimensional modeling of the solar wind interaction with the magnetized interstellar medium were presented in [81].

Calculations were performed in the spherical computational region with $R_{\text{min}} = 24$ and $R_{\text{max}} = 1200$. The mesh is $R \times \theta \times \varphi = 99 \times 116 \times 21$. All parameters are the same as in the previously described axisymmetric calculation and $A = 2$. In Figs. 24–25 the logarithm density isolines are shown for the case of the angle between $\mathbf{v}_\infty$ and $\mathbf{B}_\infty$ equal to $\alpha = 45°$ in the cross-sections $\varphi = 0$–$180°$ and $90°$–$270°$, respectively. The LISM influences in this case the shape of the bow shock as well as the location of the stagnation point at the heliopause surface in a way consistent with the simplified study [34]. In addition, the results show the existence of magnetosheath current layers providing proper rotation of the magnetic field with respect to the velocity vector from 45° to 0 or 180° at the heliopause surface. The bow shock wave stand-off distance is larger in the regions with a larger angle between the magnetic field and the shock normal. The contact surface is substantially contracted by the magnetic pressure in the xy-plane (Fig. 25) rather than in the symmetry xz-plane (Fig. 24). As was mentioned earlier, the size of the zone between the bow shock and the heliopause is very important in view of the charge-exchange processes in this zone.

9. Conclusions

In this work we presented a review of the application of numerical methods to modeling of the stellar wind interaction with the interstellar medium. This is only one among various domains of their application to space simulation problems. The environment of distant stars is not so well investigated as the solar system. That is a particular reason of an extremely intensive study of the solar wind and the local interstellar medium interaction. The problem is rather complicated even in gasdynamic and MHD formulations, since the flow pattern contains a number of intersecting discontinuities and is substantially three-dimensional. The study of this and other astrophysical and industrial problems has recently summoned the extension of high-resolution numerical methods to magnetohydrodynamic flows. The peculiarity of the problem considered in this paper is that a continuum approach is applicable only to the plasma component of the both flows. Trajectories of neutral particles must be calculated using a direct Monte-Carlo simulation. Although several authors (see brief discussion in [110]) argue that the application of a fluid approximation neutrals gives results close to those obtained on the basis of the self-consistent Euler–Boltzmann formulation, there is still a necessity to realize this possibility and to perform a critical comparison of numerical data for the same set of defining parameters.

The influence of the cosmic rays on the interaction is rather well known, but their inclusion into available numerical algorithms is at the initial stage.

We examined in this review mainly numerical aspects of the problem. This work by no means can be considered as an exhaustive description of physical phenomena which take place in the interaction region. Discussing the subject, we inevitably had our preferences and paid them more attention. The list of references can be extended by hundreds of publications on the considered problem, but we hope that even those mentioned in this review give an opportunity to realize the main processes defining the problem and approaches to their numerical modeling.

Acknowledgement

The authors are grateful to the Japanese Society of Computational Fluid Dynamics and, personally, to Prof. Koichi Oshima who invited them to write this review. The work of T.M. was supported by the Grant-in-Aid for scientific research C-08640375 of the Japanese Ministry of Education, Science, Sports, and Culture. N.P. was supported, in part, by the Russian Foundation for Basic Research Grant 95-01-00835.

Special thanks to K. Okuda for the help in the preparation of the graphic material.

REFERENCES

[1] G.D. van Albada, B. van Leer B., and W.W. Roberts, *Astron. & Astrophys.*, **108**, 76 (1982)

[2] H. Alfvén, *Cosmic Electrodynamics*, Clarendon Press, Oxford, 1950

[3] W.I. Axford, *Solar Wind, NASA SP-308*, 609 (1972)

[4] K.I. Babenko and V.V. Rusanov, *Proc. 2nd U.S.S.R. Meeting on Theoretical Mechanics. Review papers. 2.* Nauka, Moscow, 1965

[5] V.B. Baranov, K.V. Krasnobaev, and A.G. Kulikovskii, *Doklady AN SSSR*, **194**, 41 (1970) [transl. into English as Soviet Physics Doklady]

[6] V.B. Baranov and K.V. Krasnobaev, *Hydrodynamic Theory of Cosmic Plasma*, Nauka, Moscow, 1977 [in Russian]

[7] V.B. Baranov, M.G. Lebedev, and M.S. Ruderman, *Astrophys. and Space Sci.*, **66**, 441 (1979)

[8] V.B. Baranov, M.K. Ermakov, and M.G. Lebedev, *Fluid Dynamics*, No. 5, 754 (1982)

[9] V.B Baranov, *Interaction of the solar wind with the external plasma*, in: *Physics of the Outer Heliosphere*, Pergamon, New York, 287 (1990)

[10] V.B. Baranov, M.G. Lebedev, and Yu.G. Malama, *Astrophys. J.*, **375**, 347 (1991)

[11] V.B. Baranov and Yu.G. Malama, *J. Geophys. Res.*, **98**, 15157 (1993)

[12] V.B. Baranov and Yu.G. Malama, *J. Geophys. Res.*, **100**, 14755 (1995)

[13] V.B. Baranov and N.A. Zaitsev, *Astron. and Astrophys.*, **304**, 631 (1995)

[14] A.A. Barmin, A.G. Kulikovskiy, and N.V. Pogorelov, *J. Comput. Phys.* **126**, 77 (1996)

[15] A. Barnes, *J. Geophys. Res.*, **98**, 15137 (1993)

[16] A. Barnes, *J. Geophys. Res.*, **99**, No. A4, 6553 (1994)

[17] A. Barnes, *Space Sci. Rev.*, **72**, 233 (1995)

[18] P.W. Blum and H.J. Fahr, Nature, **223**, 936

(1969)

[19] J.U. Brackbill and D.C. Barnes, *J. Comput. Phys.*, **35**, 326 (1980)

[20] J.U. Brackbill, J. Comput. Phys., **96**, 163 (1991)

[21] J.C. Brandt, *Introduction to the Solar Wind*, W.H. Freeman, New York, 1970

[22] M. Brio and C.C. Wu, *J. Comput. Phys.*, **75**, 400 (1988)

[23] P. Cargo and G. Gallice, *ZAMM Special Issue I. Numerical Analysis, Scientific Computing, Computer Science*, 369 (1996)

[24] S. Chakravarthy and S. Osher, *AIAA J.*, **21**, 1241 (1983)

[25] P. Colella and P.R. Woodward, *J. Comput. Phys.*, **54**, 174 (1984)

[26] W. Dai and P.R. Woodward, *J. Comput. Phys.*, **111**, 354, (1994).

[27] W. Dai and P.R. Woodward, *J. Comput. Phys.*, **115**, 485, (1994).

[28] L.E. Davis, Jr., *Phys. Rev.*, **100**, 1440 (1955)

[29] A.J. Dessler, *Rev. Geophys.*, **5**, 1 (1967)

[30] C.R. DeVore, *J. Comput. Phys.*, **92** 142 (1991)

[31] M. Dryer, A.W. Rizzi, and W.-W. Shen, *Astrophys. and Space Sci.*, **22**, 329 (1973)

[32] H.J. Fahr, *Solar Physics*, **30**, 193 (1973)

[33] H.J. Fahr, *Adv. Space Res.*, **6**, 13 (1986)

[34] H.J. Fahr, et al., *Annales Geophysicae*, **6**, 337 (1988)

[35] H.J. Fahr, H. Fichtner, and S.Grzedzielski, *Solar Physics*, **137**, 355 (1992)

[36] P.J. Finley, *J. Fluid Mech.*, **26**, 337 (1966)

[37] D. Givoli, *J. Comput. Phys.*, **94**, 1 (1991)

[38] S.K. Godunov, *Mat. Sbornik*, **47**, 271 (1957) [in Russian] Transl. *US Joint Publ. Res. Service, JPRS 7226* (1969)

[39] T.I. Gombosi, et al., *J. Geophys. Res.*, **101**, No. A7, 15233 (1996)

[40] S. Grzedzielski and R. Ratkiewicz, *Acta Astron.*, **25**, 177 (1975)

[41] T. Hanawa, Y. Nakajima, and K. Kobuta, *Dept. of Astrophysics Nagoya University Preprint No. 94-34* (1994).

[42] Ch. Hirsch, *Numerical Computation of External and Internal Flows*, John Wiley & Sons, Chichester, 1990

[43] M.A. Il'gamov, *Nonreflecting boundary conditions on the calculation domain boundaries*, in: *Dynamics of Shells in Flows, Publ. 4-75*, Kazanskii Fiz.-Tekh. Inst KF Akad. Nauk S.S.S.R., Kazan, 1985 [in Russian]

[44] M.A. Il'gamov, *Review of research on nonreflecting conditions on the boundaries of the calculational domain*, in: *Numerical Boundary Conditions, Publ. 6-54*, Kazanskii Fiz.-Tekh. Inst KF Akad. Nauk S.S.S.R., Kazan, 1990 [in Russian]

[45] A. Jeffrey and T. Taniuti, *Nonlinear Wave Propagation*, Academic Press, New York, 1964.

[46] T. Jyounouchi, N. Tsujimura, and M. Yasuhara, in: *Proc. 4th International Conference in Computational Fluid Dynamics*, Nagoya, 964 (1989)

[47] S.R. Karmesin, P.C. Liewer, and J.U. Brackbill, *Geophys. Res. Lett.*, **22**, 1153 (1995)

[48] I.K. Khabibrakhmanov and D. Summers, *J. Geophys. Res.*, **101**, 7609 (1996)

[49] A. Kulikovskiy and G. Lyubimov, *Magneto-hydrodynamics*, Addison–Wesley, Reading, MA, 1965.

[50] R. Lallemant, et al., *Astrophys. J.*, **396**, 696 (1992)

[51] L.D. Landau and E. Lifshits, *Electrodynamics of Continuous Media*, Pergamon, New York, 1960

[52] P.C. Liewer, S.R. Karmesin, and J.U. Brackbill, *J. Geophys. Res.*, **101**, 17119 (1996)

[53] R.W. MacCormack, *AIAA Paper 81-0110* (1981)

[54] Yu.G. Malama, *Astrophys. Space Sci.*, **176**, 21 (1991)

[55] T. Matsuda, et al., *Progr. Theor. Phys.*, **81**, 810 (1989)

[56] T. Matsuda, et al., *Progr. Theor. Phys.*, **84**, 837 (1990)

[57] T. Matsuda, et al., *Progr. Theor. Phys.*, **84**, 856 (1990)

[58] T. Matsuda, et al., *CFD J.*, **1**, No. 2, 115 (1992)

[59] T. Matsuda and Y. Fujimoto, in: *Proc. 5th Int. Symp. on Comput. Fluid Dyn., Sendai, August 31 – September 3, 1993*, **2**, 186 (1993)

[60] G. Moretti, *Computers and Fluids*, **7**, 191 (1979)

[61] G. Moretti and L. Zanetti, *AIAA J.*, **22**, 758 (1984)

[62] G. Moretti, *Computers and Fluids*, **15**, 59 (1987)

[63] K. Naidu and A. Barnes, *J. Geophys. Res.*, **99**, No. A6, 11553 (1994)

[64] M. Napolitano and A. Dadone, *AIAA J.*, **23**, 1343 (1985)

[65] M.L. Norman, et. al., *Astron. & Astrophys.*, **113**, 285 (1982)

[66] M.L. Norman, K.-H.A. Winkler, and L. Smarr, in: *Atrophysical Jets* (eds. A. Ferrari and A.G. Pacholczyk), D. Reidel, 227 (1982)

[67] S. Nozawa and H. Washimi, *Publ. Astron. Soc. Japan,* **49**, No. 3 (1997)

[68] T. Ogino, *J. Geophys. Res.,* **91**, No. A6, 6791 (1986)

[69] E.N. Parker, *Astrophys. J.,* **134**, 20 (1961)

[70] H.L. Pauls, G.P. Zank, and L.L. Williams, *J. Geophys. Res.,* **100**, No. A11, 21595 (1995)

[71] H.L. Pauls and G.P. Zank, *J. Geophys. Res.,* **101**, 17081 (1996)

[72] J.L Phillips, et al., *Science,* **268**, 1030 (1995)

[73] J.L Phillips, et al., in: *Solar Wind,* **8**, in press (1996)

[74] N.V. Pogorelov and Yu.D. Shevelev, *U.S.S.R. Comput. Math. Math. Phys.,* **25**, 1391 (1985)

[75] N.V. Pogorelov, in: *Proc. 5th Int. Symp. on Comput. Fluid Dyn., Sendai, August 31 – September 3, 1993,* **3**, 7 (1993) [see also *CFD J.,* **6**, No. 2, 213 (1997)]

[76] N.V. Pogorelov, *Astron. and Astrophys.,* **297**, 835 (1995)

[77] N.V. Pogorelov, A.A. Barmin, A.G. Kulikovskiy, and A.Yu. Semenov, *Collection of Papers, 6th International Conference on Computational Fluid Dynamics (Lake Tahoe, September 5–9, 1995),* **2**, 952, University of California, Davis, 1995.

[78] N.V. Pogorelov and A.Yu. Semenov, *Comput. Math. Math. Phys.,* **36**, No. 3, 395 (1996) [translated from Zh. Vychisl. Mat. Mat. Fiz.]

[79] N.V. Pogorelov and A.Yu. Semenov, in: *Numerical Methods in Engineering,* John Wiley & Sons, Chichester, 1022 (1996)

[80] N.V. Pogorelov and A.Yu. Semenov, *Astron. and Astrophys.,* **321**, 330 (1997)

[81] N.V. Pogorelov and T. Matsuda, in: *Proc. 5th Int. School/Symposium for Space Simulations, March 13–19, Kyoto, 1997* (1997) [*J. Geophys. Res.,* in press (1997)]

[82] N.V. Pogorelov and A.Yu. Semenov, *Comp. Math. Math. Phys.,* **37**, 320 (1997)

[83] K.G. Powell, *Report NM-R9407,* Centrum voor Wiskunde en Informatica, Amsterdam (1994)

[84] Powell K.G., et al., in: *Proc. AIAA 12 Comput. Fluid Dyn. Meeting* (1995)

[85] T.H. Pulliam, *AIAA Paper 85-0438* (1985)

[86] Yu.B. Radvogin, *Preprint No. 8,* Keldysh Inst. Appl. Math., Moscow, 1987 [in Russian]

[87] H. Ripken and H.J. Fahr, *Astron. & Astrophys.,* **122**, 121 (1983)

[88] P.L. Roe, *J. Comput. Phys.,* **43**, 357 (1981)

[89] P.L. Roe, *Characteristic-based Schemes for the Euler equations,* in: *Ann. Rev. Fluid Mech.,* **18**, 337 (1986)

[90] D.J. Romeo and J.R. Sterrett, *NASA TN D-1605* (1963)

[91] D.J. Romeo and J.R. Sterrett, *AIAA J.* **3**, 544 (1965)

[92] K. Sawada, E. Shima, T. Matsuda, and T. Inaguchi, *Mem. Fac. Engg. Kyoto Univ.,* **48**, 240 (1986)

[93] E.Shima, T. Matsuda, and T. Inaguchi, *Monthly Notices Royal Astron. Soc.,* **221**, 687 (1986)

[94] L. Smarr, M.L. Norman, and K.-H.A. Winkler, *Phisica,* **12D**, 83 (1984)

[95] J.R. Spreiter, A.L. Summers, and A.W. Rizzi, *Planet. Space Sci.,* **18**, 1281 (1970)

[96] R.S. Steinolfson, V.J. Pizzo, and T. Holzer, *Geophys. Res. Lett.,* **21**, No. 4, 245 (1994)

[97] R.S. Steinolfson, *J. Geophys. Res.,* **99**, No. A7, 13307 (1994)

[98] J.L. Stone and M.L. Norman, *Astrophys. J. Suppl.,* **80**, 753 (1992)

[99] S.T. Suess, D.H. Hathaway, and A.J. Dessler, *Geophys. Res. Lett.,* **14**, No. 9, 977 (1987)

[100] S.T. Suess, *J. Geophys. Res.,* **98**, No. A9, 15147 (1993)

[101] M.T. Sun, S.T. Wu, and M. Dryer, *J. Comput. Phys.,* **116**, 330 (1995)

[102] K.W. Thompson, *J. Comput. Phys.,* **68**, 1 (1987)

[103] G. Tóth and D. Odstrĉil, *J. Comput. Phys.,* **128**, 82 (1996)

[104] M.K. Wallis, *Nature Phys. Sci.,* **233**, 23 (1971)

[105] M.K. Wallis, *Nature,* **254**, 202 (1975)

[106] M.K. Wallis and M. Dryer, *Astrophys. J.,* **205**, 895 (1976)

[107] H. Washimi, *Adv. Space Res.,* **13**, No. 6, 227 (1993)

[108] H. Washimi and T. Tanaka, *Space Sci. Rev.,* **78**, 85 (1996)

[109] K. Watanabe and T. Sato, *J. Geophys. Res.,* **95**, No. A1, 75 (1990)

[110] L.L. Williams, D.T. Hall, H.L. Pauls, and G.P. Zank, *Astrophys. J.,* **476**, 366 (1997)

[111] C.C. Wu and C.F. Kennel, *Geophys. Res. Lett.,* **19**, 2087 (1992)

[112] H.C. Yee, *NASA TM-101088* (1989)

[113] N.A. Zaitsev and Yu.B. Radvogin, *Preprint No. 86,* Keldysh Inst. Appl. Math., Moscow, 1990 [in Russian]

[114] A.L. Zachary and P. Colella, *J. Comput. Phys.,* **99**, 341, (1992)

[115] G.P. Zank, H.L. Pauls, L.L. Williams, and D.T. Hall, *J. Geophys. Res.,* **101**, No. A10, 21639 (1996)

APPROXIMATE SOLUTION OF AN OCEAN DYNAMICS PROBLEM

E.M.VIKHTENKO   A.G.ZARUBIN

Khabarovsk State University of Technology, Center of Fundamental Research;
136 Tikhookeanskaya st., Khabarovsk 680035 Russia

Abstract

In the article a number of approximate methods of the solution of ocean dynamics is offered. Evaluation of rate of convergence of the approximate solutions to the exact one, is carried out numerically.

1. Introduction

Mathematical problems of dynamics of the ocean, development and substantation of numerical methods of the solution of problems of dynamics of the ocean, study of existence and uniqueness of the solutions of such problems are studied extensively by many authors. The detailed bibliography and exposition of received results can be found in monography [1].

In the present work a problem of ocean dynamics is investigated. For the solution of iterative, projective-iterative methods and Rothe-Galerkin method are considered. With the help of these methods one can construct the approximate solutions of an initial value problem, for which its convergence to the exact one is established, and rate of convergence is obtained.

2. Statement of a Problem

Let Ω be a limited pool, i.e. cylindrical area with forming, parallel axis OZ and directed downwards; $\partial\Omega$ is a lateral cylindrical surface; R is a depth of pool. We denote by (u, v, w) a vector of speeds, where u, v are horizontal directing on directions of axes OX, OY, w is vertical component.

In reference [1] the definition of flow components to investigate the following initial boundary value problems is given:

A problem 1. Find the solution of a system of equations

$$u'_t - \mu_1 \Delta u - \nu_1 u''_{zz} - lv$$

$$+\frac{g}{\rho_0}\left(\int_0^z \frac{\partial\rho}{\partial x}dz_0 - \frac{1}{R}\int_0^R\int_0^z \frac{\partial\rho}{\partial x}dz_0\,dz\right) = f_1,$$

$$v'_t - \mu_1 \Delta v - \nu_1 v''_{zz} + lu$$

Received on December 26, 1996.

$$+\frac{g}{\rho_0}\left(\int_0^z \frac{\partial\rho}{\partial y}dz_0 - \frac{1}{R}\int_0^R\int_0^z \frac{\partial\rho}{\partial y}dz_0\,dz\right) = f_2,$$

$$\rho'_t - \mu_2 \Delta\rho - \nu_2\rho''_{zz}$$

$$+\Gamma\int_z^R \left(\frac{\partial u}{\partial x} + \frac{\partial v}{\partial y}\right) dz_0 = f_3, \tag{1}$$

satisfying to initial and boundary conditions

$$u(x,y,z,0) = v(x,y,z,0) = \rho(x,y,z,0) = 0, \tag{2}$$

$$\frac{\partial u}{\partial z} = \frac{\partial v}{\partial z} = \frac{\partial\rho}{\partial z} = 0, \tag{3}$$

where $(x, y, z) \in \partial\Omega_0$ or $\partial\Omega_R$,

$$u = v = \rho = 0, \text{where } (x, y, z) \in \partial\Omega. \tag{4}$$

A problem 2. Find the solution of a system (1) under following initial and boundary conditions:

$$u(x,y,z,0) = v(x,y,z,0) = \rho(x,y,z,0) = 0, \tag{5}$$

$$\frac{\partial u}{\partial z} = \frac{\partial v}{\partial z} = \frac{\partial\rho}{\partial z} = 0, \quad \text{where } (x,y,z) \in \partial\Omega_0, \tag{6}$$

$$u = v = \frac{\partial\rho}{\partial z} = 0, \quad \text{where } (x,y,z) \in \partial\Omega_R,$$

$$u = v = \rho = 0, \quad \text{where } (x,y,z) \in \partial\Omega. \tag{7}$$

Here $\partial\Omega_R$ and $\partial\Omega_0$ are the bottom and top basis of cylinder Ω respectively, ρ is a deviation of density from the average value ρ_0, l is a Coriolis parameter, ν_1, ν_2 and μ_1, μ_2 are the coefficients of vertical and horizontal turbulent viscosity, g is a acceleration of gravity, Γ is some average (for considered region of the ocean) distribution of temperature. Δ is a plane Laplace operator.

We define the necessary function spaces. We denote by $W_2^2(\Omega)$ as the Sobolev's space [2]. Set of functions, belonging $W_2^2(\Omega)$ and satisfying to conditions (3)-(4) ((6)-(7)), we denote by $\hat{W}_2^2(\Omega)$ ($\tilde{W}_2^2(\Omega)$). Put $H = [L_2(\Omega)]^3$, $H_1 = [\hat{W}_2^2(\Omega)]^3$ ($H_1 = [\tilde{W}_2^2(\Omega)]^3$). By $B_2 = B_2(0, T; H)$ we denote the Hilbert space all strongly measurable on $[0, T]$ with values in H of functions, for which the norm

$$\|\psi(t)\|_{B_2} = \left(\int_0^T \|\psi(t)\|_H^2 \, dt \right)^{1/2}, \quad 0 < T < \infty$$

is finite.

Let us consider the function $\psi(t)$ with values in the Hilbert space H_1. Let $\psi(t)$ has continuous derivative $\psi'(t)$ in the space H. In set of such functions we define the norm

$$\|\psi(t)\|_{B_2^1} = \left(\int_0^T \left(\|\psi'(t)\|_H^2 + \|\psi(t)\|_{H_1}^2 \right) dt \right)^{1/2}.$$

The supplement of given set under this norm gives the Hilbert space B_2^1.

On the space H_1 the operators

$$A = \begin{pmatrix} -\Delta_1 & 0 & 0 \\ 0 & -\Delta_1 & 0 \\ 0 & 0 & -\Delta_2 \end{pmatrix}, \quad K = \begin{pmatrix} 0 & -lI & F_1 \\ lI & 0 & F_2 \\ F_3 & F_4 & 0 \end{pmatrix},$$

where

$$\Delta_j = \mu_j \Delta - \nu_j \frac{\partial^2}{\partial z^2}, \quad j = 1, 2,$$

$$F_1 = \frac{g}{\rho_0} \left(\int_0^z \frac{\partial}{\partial x} dz_0 - \frac{1}{R} \int_0^R \int_0^z \frac{\partial}{\partial x} dz_0 \, dz \right),$$

$$F_2 = \frac{g}{\rho_0} \left(\int_0^z \frac{\partial}{\partial y} dz_0 - \frac{1}{R} \int_0^R \int_0^z \frac{\partial}{\partial y} dz_0 \, dz \right),$$

$$F_3 = \Gamma \int_z^R \frac{\partial}{\partial x} dz_0, \quad F_4 = \Gamma \int_z^R \frac{\partial}{\partial y} dz_0$$

are defined.

We put

$$\varphi = \begin{pmatrix} u \\ v \\ \rho \end{pmatrix}, \quad f = \begin{pmatrix} f_1 \\ f_2 \\ f_3 \end{pmatrix}.$$

In accepted definitions the system (1) can be recorded in a form

$$\frac{d\varphi}{dt} + A\varphi(t) + K\varphi(t) = f(t), \quad \varphi(0) = 0, \qquad (8)$$

where $\varphi(t)$ and $f(t)$ are unknown and given function, determined on $[0, T]$ with values in H.

For a problem (8) we consider following approximate methods:

— **iterative**

$$\frac{d\varphi_n(t)}{dt} + A\varphi_n(t) + K\varphi_{n-1}(t) = f(t), \quad \varphi_n(0) = 0, \, (9)$$

where starting point $\varphi_0(t)$ is arbitrary from the space B_2^1;

— **projective-iterative**

$$\frac{d\varphi_n^m}{dt} + P_m A\varphi_n^m + P_m K\varphi_{n-1}^m = P_m f, \quad \varphi_n^m(0) = 0, \, (10)$$

where $\varphi_n^m(t)$ for each n and m belongs to the space H^m, H^m is a linear hull of elements

$$\phi_{ijk} = \begin{pmatrix} \sin \frac{i\pi x}{a} \sin \frac{j\pi y}{b} \cos \frac{k\pi z}{R} \\ \sin \frac{i\pi x}{a} \sin \frac{j\pi y}{b} \cos \frac{k\pi z}{R} \\ \sin \frac{i\pi x}{a} \sin \frac{j\pi y}{b} \cos \frac{k\pi z}{R} \end{pmatrix},$$

$$i = 1, 2, \ldots, m$$
$$j = 1, 2, \ldots, m$$
$$k = 0, 1, 2, \ldots, m$$

for a problem (1)-(4) or linear hull of elements

$$\phi_{ijk} = \begin{pmatrix} \sin \frac{i\pi x}{a} \sin \frac{j\pi y}{b} \cos \frac{(2k+1)\pi z}{2R} \\ \sin \frac{i\pi x}{a} \sin \frac{j\pi y}{b} \cos \frac{(2k+1)\pi z}{2R} \\ \sin \frac{i\pi x}{a} \sin \frac{j\pi y}{b} \cos \frac{k\pi z}{R} \end{pmatrix}$$

for a problem (1), (5)-(7), and P_m is orthoprojector of H onto H^m;

— **Rothe-Galerkin method**

$$\frac{\phi_s^m - \phi_{s-1}^m}{\tau} + P_m A\phi_s^m + P_m K\phi_{s-1}^m = P_m h^s,$$

$$\phi_0^m = 0, \quad s = 1, 2, \ldots, N, \qquad (11)$$

where N is a positive integer number, $\tau = T/N$. The function $\psi_\tau = \{\psi_j\}_0^N$ is the net function from $H(N)$, where $H(N)$ is the cartesian product N of the Hilbert spaces of net functions with the norm

$$\|\psi_\tau\|_{H(N)} = \left(\tau \sum_{s=0}^N \|\psi_s\|_H^2 \right)^{1/2}.$$

Let $\psi_\tau^m = \{P_m \psi_j\}_0^N = \{\psi_j^m\}_0^N$.

Theorem 1. *Let $f(t) \in B_2(0, T; H)$. Then the problem (8) has a unique solution $\varphi(t)$ from B_2^1. At* each n the problem (9) has a unique solution $\varphi_n(t)$ from B_2^1 and for the approximate solutions $\varphi_n(t)$ the estimates

$$\sup_t \|\varphi_n(t) - \varphi(t)\|_H^2 \le d_1 \frac{(d_2 T)^n}{n!},$$

$$\|\varphi_n(t) - \varphi(t)\|_{B_2^1}^2 \le d_3 \left(\frac{d_2 T}{n!}\right)^{1/2}$$

are fulfilled, where the positive constants d_1, d_2, d_3 are independent of n.

Theorem 2. *Let a condition of the theorem 1 is satisfied. Then the problem (10) has a uneque solution for every n and m. The approximate solutions $\varphi_n^m(t)$ of problems (10) converge to the exact solution $\varphi(t)$ in the space B_2^1 with rate*

$$\|\varphi(t) - \varphi_n^m(t)\|_H^2 \le d_4 \left(\frac{(d_5 T)^n}{n!} + m^{-1}\right), \quad (12)$$

where the positive constants d_4 and d_5 are independent of n and m.

Theorem 3. *Let a condition of the theorem 1 is satisfied. Let $f(t)$ be from the space $C^1([0,T]; H)$, $f(0)$ belongs $D(A^{1/2})$. Then the problem (11) has a unique solution ϕ_s^m for every n, m and the estimate*

$$\|\varphi(t_s) - \phi_s^m\|_H \le d_6 \left(\tau + m^{-1/2}\right),$$

take place, where the positive constant d_6 is independent of s and m.

3. PROOF OF THEOREMS

Before the proofs of the theorems 1 - 3 we shall establish some properties of operators A and K.

Lemma 1 *The operators A and K have the following properties:*

1) operator A is the selfadjoint positive-defined operator, the domain of definition $D(A) = H_1$;

2) for any $\varphi \in H$ the inequality

$$\|K\varphi\|_H \le M_1 \|A\varphi\|_H^{1/2} \|\varphi\|_H^{1/2}, \quad (13)$$

take place, the constant $M_1 > 0$ is independent of φ;

3) for any φ_1 and $\varphi_2 \in H_1$ such correlation

$$|(K\varphi_1, \varphi_2)_H| \le M_2 \|\varphi_1\|_H \|A^{1/2}\varphi_2\|_H, \quad (14)$$

is true, the constant $M_2 > 0$ is independent of choice of elements φ_1 and φ_2.

Proof. It is easy to show, that the operators $-\Delta_i$ are selfadjoint positive-defined operators. Hense, the operator A is selfadjoint positive-defined operator too.

We estimate the norm of element $K\varphi$ in the space H, we have

$$\|K\varphi\|_H^2$$
$$= \| - lv + F_1\rho\|_{L_2(\Omega)}^2 + \|lu + F_2\rho\|_{L_2(\Omega)}^2$$
$$+ \|F_3 u + F_4 v\|_{L_2(\Omega)}^2$$
$$\le M_3 \left(\|v\|_{L_2(\Omega)}^2 + \|u\|_{L_2(\Omega)}^2 + \|F_1\rho\|_{L_2(\Omega)}^2\right.$$
$$\left. + \|F_2\rho\|_{L_2(\Omega)}^2 + \|F_3 u\|_{L_2(\Omega)}^2 + \|F_4 v\|_{L_2(\Omega)}^2\right).$$
$$(15)$$

For summand $\|F_1\rho\|_{L_2(\Omega)}^2$ we can write

$$\|F_1\rho\|_{L_2(\Omega)}^2$$
$$= \left(\frac{g}{\rho_0}\right)^2 \int_\Omega \left(\int_0^z \frac{\partial\rho}{\partial x} dz_0 - \frac{1}{R}\int_0^R \int_0^z \frac{\partial\rho}{\partial x} dz_0\, dz\right)^2 d\Omega$$
$$\le 2\left(\frac{g}{\rho_0}\right)^2 \left(\int_\Omega \left(\int_0^z \left|\frac{\partial\rho}{\partial x}\right| dz_0\right)^2 d\Omega\right.$$
$$\left. + \int_\Omega \left(\int_0^R \int_0^z \left|\frac{\partial\rho}{\partial x}\right| dz_0\, dz\right)^2 d\Omega\right). \quad (16)$$

We consider first integral in (16). Applying the Holder inequality, and then multiplicate inequality [2], we receive

$$\int_\Omega \left(\int_0^z \left|\frac{\partial\rho}{\partial x}\right| dz_0\right)^2 d\Omega$$
$$\le \int_\Omega \left(\int_0^R \left|\frac{\partial\rho}{\partial x}\right| dz_0\right)^2 d\Omega$$
$$\le M_4 \int_\Omega \left|\frac{\partial\rho}{\partial x}\right|^2 d\Omega$$
$$\le M_5 \|\Delta\rho\|_{L_2(\Omega)} \|\rho\|_{L_2(\Omega)}.$$

Similarly, for second integral in (16) we obtain

$$\int_\Omega \left(\int_0^R \int_0^z \left|\frac{\partial\rho}{\partial x}\right| dz_0\, dz\right)^2 d\Omega \le M_6 \|\Delta\rho\|_{L_2(\Omega)} \|\rho\|_{L_2(\Omega)}.$$

The similar inequalities can be obtained for other summands in (15). So, inequality (13) take place.

From the form of operators A and K we have

$$(K\varphi_1, \varphi_2)_H = \int_\Omega (-lv_1 + F_1\rho_1)u_2\, d\Omega$$

$$+ \int_\Omega (lu_1 + F_2\rho_1)v_2 \, d\Omega + \int_\Omega (F_3 u_1 + F_4 v_1)\rho_2 \, d\Omega$$

$$(17)$$

for any $\varphi_1 = (u_1, v_1, \rho_1)$ and $\varphi_2 = (u_2, v_2, \rho_2)$.

Let us consider, for example, $\int_\Omega F_3 u_1 \rho_2 \, d\Omega$. Integrating by parts, we obtain

$$\int_\Omega F_3 u_1 \rho_2 \, d\Omega = \Gamma \int_\Omega \frac{\partial}{\partial x}\left(\int_0^z u_1 \, dz\right) \rho_2 \, d\Omega$$

$$= -\Gamma \int_\Omega \left(\int_0^z u_1 \, dz\right) \frac{\partial \rho_2}{\partial x} \, d\Omega.$$

From here we have

$$\left| \int_\Omega F_3 u_1 \rho_2 \, d\Omega \right| \leq \Gamma \int_\Omega \left| \int_0^R u_1 \, dz \frac{\partial \rho_2}{\partial x} \right| d\Omega$$

$$\leq M_7 \left(\int_\Omega u_1^2 \, d\Omega\right)^{1/2} \left(\int_\Omega \left(\frac{\partial \rho_2}{\partial x}\right)^2 d\Omega\right)^{1/2}$$

$$\leq M_8 \|u_1\|_{L_2(\Omega)} \|\nabla \rho_2\|_{L_2(\Omega)}. \tag{18}$$

Similarly (18) it is possible to estimate all right part (17). Therefore inequality (14) is proved.

Proof of Theorem 1

The uniqueness solvability of a problem (8) is shown in [3]. Multiplying the equality (9) to $2\varphi_n(t_1)$ and integrating on t_1 from 0 to $t \leq T$, we obtain

$$\|\varphi_n(t)\|_H^2 + 2\int_0^t \|A^{1/2}\varphi_n(t_1)\|_H^2 \, dt_1$$

$$\leq 2\int_0^t |(h(t_1), \varphi_n(t_1))_H| \, dt_1$$

$$+ 2\int_0^t |(K\varphi_{n-1}(t_1), \varphi_n(t_1))_H| \, dt_1. \tag{19}$$

We apply to a right part (19) the elementary inequality $2|ab| \leq \varepsilon a^2 + \varepsilon^{-1} b^2$, where $\varepsilon > 0$. As the operator A is positive-defined, $\|A^{1/2}\psi\|_H^2 \geq \gamma^2 \|\psi\|_H^2$ for any element $\psi \in H_1$. Then for sufficiently small ε, accoding to (14) we come to the inequality

$$\|\varphi_n(t)\|_H^2 \leq M_9 + M_{10} \int_0^t \|\varphi_{n-1}(t_1)\|_H^2 \, dt_1, \tag{20}$$

where M_9 and M_{10} are independent of n and t.

Applying (20), beginning with $n = 1$, we have

$$\|\varphi_1(t)\|_H^2 \leq M_9 + M_{10} M_{11} t,$$

$$\|\varphi_2(t)\|_H^2 \leq M_9 + M_{10} M_{11} t + M_{11}\frac{M_{10}^2 t^2}{2}$$

and, finally,

$$\|\varphi_n(t)\|_H^2 \leq M_{12} \sum_{k=0}^n \frac{(M_{10} t)^k}{k!} \leq M_{13}, \tag{21}$$

where the constant M_{13} is independent of n and t.

In [4] the following inequality is established:

$$\int_0^T \left(\|\varphi_n'(t)\|_H^2 + \|A\varphi_n(t)\|_H^2\right) dt \leq$$

$$\leq M_1 \left(\int_0^T \|h\|_H^2 \, dt + \int_0^T \|K\varphi_{n-1}(t)\|_H^2 \, dt\right), \tag{22}$$

where the constant $M_{14} = 2\exp\left(T\gamma^{-2}\right)$.

From (22) and subordination of operator K (the inequality (13) is true) we have

$$\int_0^T \left(\|\varphi_n'(t)\|_H^2 + \|A\varphi_n(t)\|_H^2\right) dt \leq$$

$$\leq M_{1y} + M_{16}\int_0^T \|A\varphi_{n-1}(t)\|_H \|\varphi_{n-1}(t)\|_H \, dt, \tag{23}$$

where the constant M_{15} and M_{16} are independent of n. We transform the (23) by analogy with (19), and we choose small ε so that $0 < M_{16}/2\varepsilon = q < 1$, then from (23) and (21) it is easy to obtain, that

$$\int_0^T \|A\varphi_n(t)\|_H^2 \, dt \leq M_{17} + q\int_0^T \|A\varphi_{n-1}(t)\|_H^2 \, dt. \tag{24}$$

From (24), accoding to choice q, we have

$$\int_0^T \|A\varphi_n(t)\|_H^2 \, dt \leq M_{18}, \tag{25}$$

where constant M_{18} is independent of n.

Let $\varphi(t)$ be a solution of the problem (8). Then function $\psi_n(t) = \varphi_n(t) - \varphi(t)$ satisfies to following correlation:

$$\frac{d\psi_n(t)}{dt} + A\psi_n(t) + K\psi_{n-1}(t) = 0.$$

Multiplying last identity to $2\psi_n(t)$ and integrating on t_1 from 0 to $t \leq T$, we obtain

$$\|\psi_n(t)\|_H^2 + 2 \int_0^t \left\| A^{1/2}\psi_n(t_1) \right\|_H^2 dt_1$$

$$\leq 2 \int_0^t |(K\psi_{n-1}(t_1), \psi_n(t_1))_H| \, dt_1.$$

From here, similarly (20), we come to the inequality

$$\|\psi_n(t)\|_H^2 \leq M_{19} \int_0^t \|\psi_{n-1}(t_1)\|_H^2 \, dt_1, \qquad (26)$$

where $M_{19} > 0$ is independent of n and t.

Applying consecuently (26), we receive

$$\sup_t \|\psi_n(t)\|_H^2 \leq M_{20} \frac{(M_{21}T^n)}{n!}. \qquad (27)$$

So, the first estimate of the theorem 1 is proven.

Further, analogously (22) we can write the inequality

$$\|\psi_n(t)\|_{B_2^1}^2 \leq M_{22} \int_0^T \|K\psi_{n-1}(t)\|_H^2 \, dt.$$

From here and from (13) we have the inequality

$$\|\psi_n(t)\|_{B_2^1}^2 \leq M_{22} \int_0^T \|A\psi_{n-1}(t)\|_H \, \|\psi_{n-1}(t)\|_H \, dt$$

$$\leq M_{23} \left(\int_0^T \|A\psi_{n-1}(t)\|_H^2 \, dt \right)^{1/2}$$

$$\times \left(\int_0^T \|\psi_{n-1}(t)\|_H^2 \, dt \right)^{1/2}.$$

From last inequality, from (25) and (27) second estimate of the theorem 1 follows.

Proof of Theorem 2.

We consider the equation

$$\frac{d\varphi^m}{dt} + P_m A \varphi^m + P_m K \varphi^m = P_m f(t). \qquad (28)$$

In [4] is shown, that Galerkin's equation (28) at the initial condition $\varphi^m(0) = 0$ has a unique solution, and the estimate

$$\|\varphi(t) - \varphi^m(t)\|_H \leq M_{24} m^{-1/2} \qquad (29)$$

is true, the constant M_{24} is independent of m.

It is easy to see, that the projective-iterative process (10) for a problem (8) is iterative for a problem (28). Therefore it is necessary to establish for the operators $P_m A$ and $P_m K$ the same properties, which were established in Lemma 1 for the operators A and K.

Clearly, the operators $P_m A$ and $P_m K$ are necessary to consider on H^m, and the domains of definition of operators $P_m A$ and $P_m K$ coincide with H^m. It is easy to see, that the operator $P_m A$ is selfadjoint on H^m. From positive definiteness of operator A in H positive definiteness of operator $P_m A$ on H^m follows, i.e. $(P_m A\psi, \psi)_H \geq \gamma^2 \|\psi\|_H^2$.

The operator $P_m A$ has the inverse bounded operator $(P_m A)^{-1}$ on H^m. From said above follows, that

$$\|P_m K\psi\|_H \leq \|K\psi\|_H$$
$$\leq M_1 \|A\psi\|_H^{1/2} \|\psi\|_H^{1/2}$$
$$\leq M_1 \|P_m A\psi\|_H^{1/2} \|\psi\|_H^{1/2}.$$

Let ψ_1 and ψ_2 be from H^m, then

$$|(P_m K\psi_1, \psi_2)_H| = |(K\psi_1, \psi_2)_H|$$
$$\leq M_2 \|\psi_1\|_H \left\| A^{1/2}\psi_2 \right\|_H$$
$$\leq M_2 \|\psi_1\|_H \left\| (P_m A)^{1/2}\psi_2 \right\|_H.$$

So, the properties, similar (13) and (14) are established. Hence, according to the theorem 1, estimate

$$\sup_t \|\varphi_n^m - \varphi^m\|_H^2 \leq M_{25} \frac{(M_{26}T^n)}{n!}$$

takes place. From here and from (29) we obtain

$$\sup_t \|\varphi_n^m(t) - \varphi(t)\|_H^2 \leq M_{25} \frac{(M_{26}T^n)}{n!} + M_{27} m^{-1},$$

where constant M_{25}, M_{26} and M_{27} are independent of m and n.

The convergence $\{\varphi_n^m\}$ to the exact solution $\varphi(t)$ of the problem (8) in B_2^1 follows from the convergence $\{\varphi^m(t)\}$ to $\varphi(t)$ under the norm of space B_2^1 [3] and convergence in B_2^1 a sequence $\{\varphi_n^m(t)\}$ to $\varphi^m(t)$ at any m, the last follows from the theorem 1.

Proof of Theorem 3

Let the net function $f^\tau = \{f^s\}_0^N$ approximate the right part (1) with the order τ, i.e. $\max_s \|h^s - f(t_s)\|_H \leq M_{28}\tau$.

From positive definiteness and selfadjointness of the operator A in H follows (see, for example, [5]), that the operator $I + \tau P_m A$ has the inverse operator and $\|(I + \tau P_m A)^{-1}\| \leq 1$ for any m and s. Therefore the equation (11) has a unique solution ϕ_s^m and the function ϕ_s^m belongs the space H^m.

Scalar product of the equation (11) to $2\tau\phi_s^m$ after some transformations gives

$$\|\phi_s^m\|_H^2 - \|\phi_{s-1}^m\|_H^2 + \tau^2 \left\|\frac{\phi_s^m - \phi_{s-1}^m}{\tau}\right\|_H^2$$
$$+2\tau \left\|A^{1/2}\phi_s^m\right\|_H^2$$
$$\leq \tau \left(\varepsilon \|\phi_s^m\|_H^2 + \frac{1}{\varepsilon}\|h^s - f(t_s)\|_H^2\right.$$
$$\left.+\frac{1}{\varepsilon}\|f(t_s)\|_H^2 + 2\left|\left(K\phi_{s-1}^m, \phi_s^m\right)_H\right|\right).$$

From here and from (14) the inequality

$$\|\phi_s^m\|_H^2 - \|\phi_{s-1}^m\|_H^2 + \tau\left(2 - \frac{\varepsilon^2}{\gamma} - \varepsilon M_2\right)\left\|A^{1/2}\phi_s^m\right\|_H^2$$
$$\leq \tau \left(\frac{M_{28}}{\varepsilon}\tau^2 + \frac{1}{\varepsilon}\|f(t_s)\|_H^2 + \frac{M_2}{\varepsilon}\|\phi_{s-1}^m\|_H^2\right)$$

follows.

We take here $\varepsilon > 0$ so, that $2 - \varepsilon\gamma^{-2} - \varepsilon M_2 > 0$, then

$$\|\phi_s^m\|_H^2 - \|\phi_{s-1}^m\|_H^2$$
$$\leq \tau \left(M_{28}\tau^2 + M_{29}\|f(t_s)\|_H^2 + M_{30}\|\phi_{s-1}^m\|_H^2\right),$$

where constant M_{28}, M_{29} and M_{30} are independent of m and s.

Sum the last inequality on s from 1 to $k \leq N$, we obtain

$$\|\phi_k^m\|_H^2 \leq M_{31}\left(\tau^2 + \|f(t)\|_{B_2}^2 + \sum_{s=1}^{k}\tau\|\phi_{s-1}^m\|_H^2\right).$$

From here and from Gronwall-Bellman inequality [6] the estimate

$$\|\phi_s^m\|_H^2 \leq M_{32} \qquad (30)$$

follows, the constant M_{32} is independent of $s = 1, 2, \ldots, N$.

Multiply scalarly equation (11) by $2\tau A\phi_s^m$, we obtain

$$\left\|A^{1/2}\phi_s^m\right\|_H^2 - \left\|A^{1/2}\phi_{s-1}^m\right\|_H^2$$
$$+\tau^2 \left\|\frac{A^{1/2}\left(\phi_s^m - \phi_{s-1}^m\right)}{\tau}\right\|_H^2 + 2\tau\|A\phi_s^m\|_H^2$$
$$\leq \tau \left(\varepsilon\|A\phi_s^m\|_H^2 + \frac{1}{\varepsilon}\|h^s - f(t_s)\|_H^2\right.$$
$$\left.+\frac{1}{\varepsilon}\|f(t_s)\|_H^2 + 2\left|\left(K\phi_{s-1}^m, A\phi_s^m\right)_H\right|\right).$$

As the operator K subordinated to the operator A (the inequality (13) is true), from the last corralation at $\varepsilon = 1$, we have

$$\left\|A^{1/2}\phi_s^m\right\|_H^2 - \left\|A^{1/2}\phi_{s-1}^m\right\|_H^2 + \tau M_{33}\|A\phi_s^m\|_H^2$$
$$\leq \tau \left(M_{34}\|h^s - f(t_s)\|_H^2 + M_{35}\|f(t_s)\|_H^2\right.$$
$$\left.+M_{36}\|A\phi_{s-1}^m\|_H\|\phi_{s-1}^m\|_H\right). \qquad (31)$$

Sum (31) over s from 1 to $k \leq N$, then

$$\tau\sum_{s=1}^{k}\|A\phi_s^m\|_H^2$$
$$\leq M_{37} + \tau M_{36}\sum_{s=1}^{k}\|A\phi_{s-1}^m\|_H\|\phi_{s-1}^m\|_H. \qquad (32)$$

From (32) we have the correlation

$$\tau\sum_{s=1}^{k}\|A\phi_s^m\|_H^2$$
$$\leq M_{37} + \tau M_{36}\sum_{s=1}^{k}\frac{1}{2}\varepsilon\|A\phi_{s-1}^m\|_H^2$$
$$+\tau M_{36}\sum_{s=1}^{k}\frac{1}{2}\varepsilon^{-1}\|\phi_{s-1}^m\|_H^2. \qquad (33)$$

From (33) and (30) by the choice $\varepsilon = 2q/m$, $q < 1$ it follows

$$\tau\sum_{s=1}^{k}\|A\phi_s^m\|_H^2 \leq M_{37} + \tau q\sum_{s=2}^{k}\|A\phi_s^m\|_H^2 + \tau q\|A\phi_0^m\|_H^2,$$

where the constant M_{37} is independent of k and m. From here

$$\tau\sum_{s=1}^{k}\|A\phi_s^m\|_H^2 \leq (1-q)^{-1}M_{37} = M_{38}, \qquad (34)$$

where M_{38} is independent of k and m.

The method (11) is the Rothe method for Galerkin's equations (28), therefore in the beginning we shall estimate the norm $\|\phi_s^m - \varphi^m(t_s)\|$.

We denote $\psi_s^m = \phi_s^m - \varphi^m(t_s)$. The elements ψ_s^m are the solution of the equation

$$\frac{\psi_s^m - \psi_{s-1}^m}{\tau}$$
$$+P_mA\psi_s^m + P_mK(\phi_{s-1}^m - \varphi^m(t_{s-1}))$$
$$+P_mK(\varphi^m(t_{s-1}) - \varphi^m(t_s))$$
$$= P_m(h^s - f(t_s))$$
$$-\left(\frac{\varphi^m(t_s) - \varphi^m(t_{s-1})}{\tau} - \frac{d\varphi^m(t_s)}{dt}\right). \qquad (35)$$

Multiply scalarly (35) by $2\tau\psi_s^m$, after some transformations we obtain

$$\|\psi_s^m\|_H^2 - \|\psi_{s-1}^m\|_H^2 + 2\tau\left\|A^{1/2}\psi_s^m\right\|_H^2$$

$$\leq 2\tau\Bigg(\|P_m(h^s - f(t_s))\|_H \, \|\psi_s^m\|_H$$

$$+ M_1\left\|A^{1/2}\psi_s^m\right\|_H \|\psi_{s-1}^m\|_H$$

$$+ M_1\|\varphi^m(t_s) - \varphi^m(t_{s-1})\|_H \left\|A^{1/2}\psi_s^m\right\|_H$$

$$+ \left\|\frac{\varphi^m(t_s) - \varphi^m(t_{s-1})}{\tau} - \frac{d\varphi^m(t_s)}{dt}\right\|_H \|\psi_s^m\|_H \Bigg).$$

Using the similar transformation as above we can show, that

$$\|\psi_s^m\|_H^2 - \|\psi_{s-1}^m\|_H^2$$

$$\leq \tau\Bigg(M_{39}\|h^s - f(t_s)\|_H^2 + M_{40}\|\psi_{s-1}^m\|_H^2$$

$$+ M_{41}\|\varphi^m(t_s) - \varphi^m(t_{s-1})\|_H^2$$

$$+ M_{42}\left\|\frac{\varphi^m(t_s) - \varphi^m(t_{s-1})}{\tau} - \frac{d\varphi^m(t_s)}{dt}\right\|_H^2 \Bigg).$$

Summarizing the last correlation over s from 1 to $k \leq N$, we obtain

$$\|\psi_k^m\|_H^2$$

$$\leq \tau\Bigg(M_{39}\sum_{s=1}^{k}\|h^s - f(t_s)\|_H^2 + M_{40}\sum_{s=1}^{k}\|\psi_{s-1}^m\|_H^2$$

$$+ M_{41}\sum_{s=1}^{k}\|\varphi^m(t_s) - \varphi^m(t_{s-1})\|_H^2$$

$$+ M_{42}\sum_{s=1}^{k}\left\|\frac{\varphi^m(t_s) - \varphi^m(t_{s-1})}{\tau} - \frac{d\varphi^m(t_s)}{dt}\right\|_H^2 \Bigg).$$

$$(36)$$

We estimate each summand of the right part (36) separately.

According to the condition of approximation of function $h(t)$, we have

$$\tau\sum_{s=1}^{k}\|h^s - f(t_s)\|_H^2 \leq M_{43}\tau^2. \qquad (37)$$

In [7] is shown, that under to fulfilment of conditions of the theorem 3, corralation

$$\tau\sum_{s=1}^{k}\left\|\frac{\varphi^m(t_s) - \varphi^m(t_{s-1})}{\tau} - \frac{d\varphi^m(t_s)}{dt}\right\|_H^2 \leq M_{44}\tau^2 \quad (38)$$

is correct.

From the Lagrange's theorem follows, that

$$\|\varphi^m(t_s) - \varphi^m(t_{s-1})\|_H \leq \tau\left\|\frac{d\varphi^m(\theta_s)}{dt}\right\|_H,$$

and as $\sup_t \|d\varphi^m(t)/dt\|_H \leq M_{45}$ [8], then

$$\|\varphi^m(t_s) - \varphi^m(t_{s-1})\|_H \leq \tau M_{45},$$

therefore

$$\tau\sum_{s=1}^{k}\|\varphi^m(t_s) - \varphi^m(t_{s-1})\|_H^2 \leq \tau^2 M_{46}. \qquad (39)$$

Connecting (36), (37), (38) and (39), we obtain

$$\|\psi_k^m\|_H^2 \leq \tau^2 M_{47} + \tau M_{48}\sum_{s=1}^{k}\|\psi_{s-1}^m\|_H^2.$$

From here and from difference analogue of Gronwall-Bellman inequality [6] estimate

$$\sup_k \|\psi_k^m\|_H \leq \tau M_{48}. \qquad (40)$$

follows.

From (40) and [3] it is easy to obtain

$$\|\varphi(t_s) - \phi_s^m\|_H$$
$$\leq \|\varphi(t_s) - \varphi^m(t_s)\|_H + \|\varphi^m(t_s) - \phi_s^m\|_H$$
$$\leq M_{49}\left(\tau + m^{-1/2}\right).$$

4. NUMERICAL EXPERIMENTS

In this section we consider a practical finding of the approximate solutions of problems (1)–(4) and (1),(5)–(7), constructed on Rothe-Galerkin method. We devide the segment $[0, T]$ on N parts with the step τ, $\tau = T/N$. On each time step $t_q = q\tau$ the approximate solutions of a problem (1)–(4) are under construction in the form of finite sums

$$u_s^m = \sum_{ijk=0}^{m} A_{ijk}^q \sin\frac{i\pi x}{a}\sin\frac{j\pi y}{b}\cos\frac{k\pi z}{R}, \qquad (41)$$

$$v_s^m = \sum_{ijk=0}^{m} B_{ijk}^q \sin\frac{i\pi x}{a}\sin\frac{j\pi y}{b}\cos\frac{k\pi z}{R},$$

$$\rho_s^m = \sum_{ijk=0}^{m} C_{ijk}^q \sin\frac{i\pi x}{a}\sin\frac{j\pi y}{b}\cos\frac{k\pi z}{R},$$

the approximate solutions of a problem (1), (5)–(7) — in the form of sums

$$u_s^m = \sum_{ijk=0}^{m} A_{ijk}^q \sin\frac{i\pi x}{a} \sin\frac{j\pi y}{b} \cos\frac{(2k+1)\pi z}{2R}, \quad (42)$$

$$v_s^m = \sum_{ijk=0}^{m} B_{ijk}^q \sin\frac{i\pi x}{a} \sin\frac{j\pi y}{b} \cos\frac{(2k+1)\pi z}{2R},$$

$$\rho_s^m = \sum_{ijk=0}^{m} C_{ijk}^q \sin\frac{i\pi x}{a} \sin\frac{j\pi y}{b} \cos\frac{k\pi z}{R}.$$

The unknown coefficients A_{ijk}^q, B_{ijk}^q and C_{ijk}^q in representation (41) are the solutions of the system of algebraic equations, which is equivalent to operator equation (8).

Solving this system of equations under known on $(q-1)$ time step on a coefficients A_{ijk}^{q-1}, B_{ijk}^{q-1} and C_{ijk}^{q-1} we define unknown coefficients on q time step. Thus becomes possible to observe the evolution m".+nf((of the initial condition, prescribed by coefficients A_{ijk}^0, B_{ijk}^0 and C_{ijk}^0 under influence of external forces, described by functions $f_1(x,y,z,t)$, $f_2(x,y,z,t)$, $f_3(x,y,z,t)$.

Let $f_2 \equiv 0$, $f_3 \equiv 0$, and

$$f_1 = \begin{cases} f_0(z-z_0)^2, & 0 \le z \le z_0, \\ 0, & z_0 < z \le R. \end{cases}$$

At such representations of functions f_1, f_2, f_3 under $z_0 \ll R$ a situation is simulated, when on a surface of pool Ω in a direction from west for east wind blows with constant force (f_0 characterizes a component of wind pressure over a surface of the ocean along the axis Ox). Let the coefficients $A_{ijk}^0 = 0$, $B_{ijk}^0 = 0$ and $C_{ijk}^0 = 0$. Such values of coefficients describe a condition of rest in a initial moment of a time, $u = 0$, $v = 0$, $\rho = 0$.

In numerical experiment pool in the ocean to the sizes $1000 \times 1000\,\mathrm{m}^2$ and depth $R = 2000\,\mathrm{m}$ was considered. The values of constants were chosen following: $\mu_1 = 1.0 \cdot 10^3$, $\mu_2 = 1.0 \cdot 10^4$, $\nu_1 = 1.5 \cdot 10^{-1}$, $\nu_2 = 1.5 \cdot 10^{-1}$, $\ell = 7.0 \cdot 10^{-5}$, $g = 9.8$, $\rho = 1.0$, $\Gamma = 1.0$, $z_0 = 0.05R$, $\tau = 0.01$. In representation (41) at the accounts N was set equal 5.

On drawings 1–5 results of numerical accounts for a described situation are submitted. Under influence of force f_1 the surface layers of a water of ocean pool begin to move in a direction of wind from west for east (fig.1), simultaneously arises counter bottom flow (fig. 2). In intermediate layers of a liquid there are flows, arrangement and kind of which vary with a depth (fig.3 and fig.4). With current of a time at preservation of character of external effect the speed surface and bottom flows grows: the coefficient $A_{110}^{25} = 6.35$, and $A_{110}^{100} = 22.19$, system of internal flows is reconstructed (fig.5).

The density of a water also varies. At east border of pool the density of a water grows, deviation from the average value $\rho > 0$, and at western - decreases ($\rho < 0$) (fig.6).

REFERENCES

[1] Mathematical models of circulation in the ocean / Under edit. Marchuk G.I., Sarkisjan S.A. Moscow, 1980

[2] O.A.Ladyshenskaya, B.A.Solonnikov, N.N.Uraltseva Linear and Quasi-linear Equations of Parabolic Type, Moscow, 1967

[3] A.G.Zarubin On a Convergence Rate of Faego-Galerkin method for Linear non-stationary Equations; Diff.Equations, V.18 N.4 pp.639–645

[4] P.Oja On Convergence and Stability of the Galerkin Method for the Parabolic Equations with a differentiable operator; Labour till the mathematician and mechanic, Tartu, V.17 pp.184–210.

[5] A.G.Zarubin, M.F.Tiunchik On a Rothe-Galerkin Method for one Class of Linear non-stationary Equations; Diff.Equations. V.19 N.12 pp.2141–2148

[6] V.B.Demidovich On asymptotic Behavior of Solutions of Finite-Difference Equations. 1. General aspects; Diff.Equations, V.10 N.12 pp.2267–2278.

[7] J.-L.Lions, E.Magnes Problems aux limits non homogenes et applications, Paris, V.1 1968

[8] A.G.Zarubin On a Convergance Rate of Rothe-Galerkin Method for Operator-Differential Equations; Diff.Equations, V.22 N.12 pp.2135–2144.

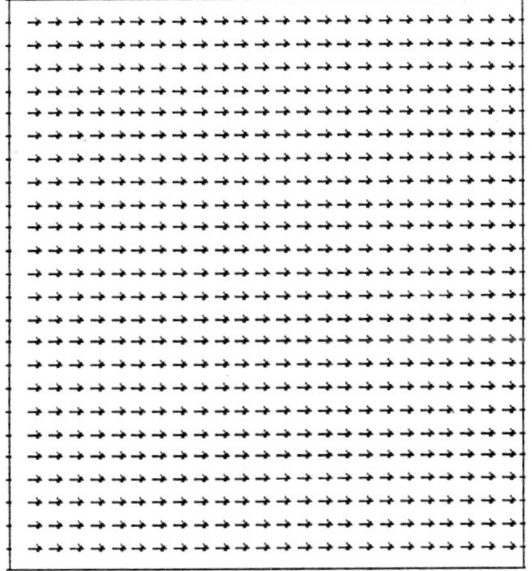

Fig.1: Field of the speeds *u-v*
$z=0$, $t=50\,\tau$

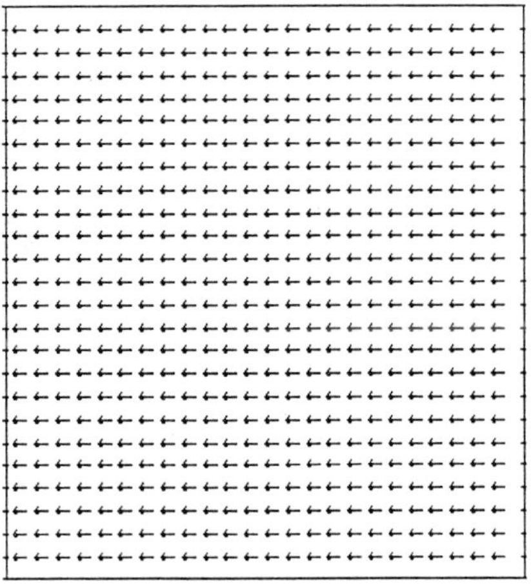

Fig.2: Field of the speeds *u-v*
$z=2000$ m, $t=50\,\tau$

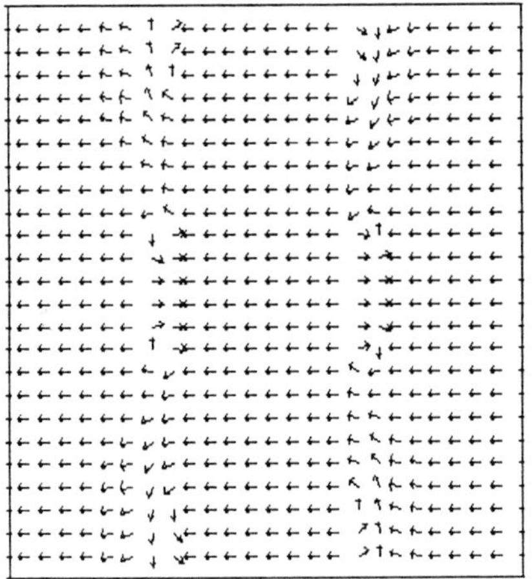

Fig.3: Field of the speeds *u-v*
$z=400$ m, $t=50\,\tau$

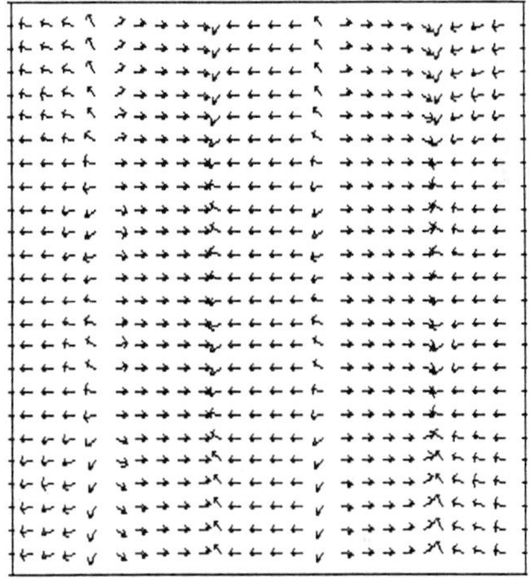

Fig.4: Field of the speeds *u-v*
$z=1200$ m, $t=50\,\tau$

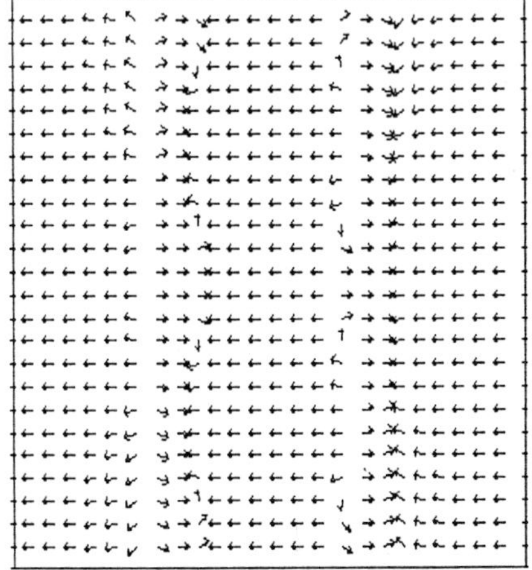

Fig.5: Field of the speeds *u-v*
$z=400$ m, $t=75\,\tau$

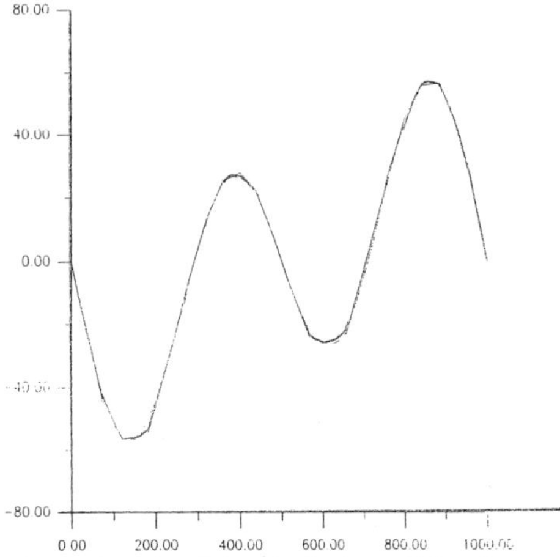

Fig.6: Graph of the deviation of density
$x=500$ m, $z=400$ m, $t=75\,\tau$

THE SHALLOW WATER EQUATIONS: STATE OF THE ART AND NEW TRENDS

Maria Morandi CECCHI   Luca SALASNICH

Dipartimento di Matematica Pura ed Applicata, Università di Padova;
Via Belzoni 7, I–35131 Padova, Italy

Abstract

In this paper we review the state of art for shallow water equations using various approches by finite element methods. Such methods are very effective and powerfull and can be still more efficient by the use of parallel computers. Numerical results are given of an implementation made for the Venice Lagoon.

Contents

1   Introduction

Fluid dynamics concerns itself with the investigation of the motion and equilibrium of fluids and is one of the oldest branches of physics [1]. Nowadays computational fluid dynamics is an important engineering tool like wind tunnels and the practical applications to its study are numerous. For instance, it is used in aeronautical sciences, in meteorology, in thermo–hydraulics, in the petroleum industry, in plasma physics, etc [2].

In this paper we concentrate on the computational methods for the shallow water equations. In Section 2, after a brief review of functional analysis, we discuss the weak formulation of a differential problem and its discretization by the finite element method. In Section 3 we derive and discuss the Navier–Stokes equations. In Section 4 we analyze the numerical methods used to solve the Navier–Stokes equations both with coupled and decoupled velocity and pressure. In Section 5 we derive the shallow water equations from the Navier–Stokes equations and describe the finite element methods used to solve them. In Section 6 we introduce the

Received on November 25, 1996.

domain decomposition methods which are mostly used nowadays. Finally in Section 7 we present a finite element model to solve the motion problem of the shallow water due to the tides into the Venice lagoon.

2   Functional Spaces and Finite Element Method

The L^p spaces, the Hilbert spaces and the Sobolev spaces are the functional spaces used in the finite element theory [3,4,5].

Definition 2.1 *Let Ω be a regular and compact domain of $\mathbf{R}^d$ and $\partial\Omega$ its boundary. Let be $p \in \mathbf{R}$ and $1 \le p < \infty$.*

$$L^p(\Omega) = \{f : \Omega \to \mathbf{R} : f \text{ is measurable and}$$

$$\int_\Omega |f|^p d\eta < \infty\},$$

where η is the Lebesgue measure. This space has the norm

$$\|f\|_{L^p} = \|f\|_p = \left\{\int_\Omega |f(x)|^p d\eta\right\}^{1/p}.$$

Definition 2.2 *Let be $p = \infty$, then*

$$L^\infty(\Omega) = \{f : \Omega \to R : f \text{ is measurable and}$$

$$\exists\, C : |f(x)| \le C \text{ q.e. on } \Omega\}$$

with the norm

$$\|f\|_{L^\infty} = \|f\|_\infty = \inf\{C : |f(x)| \le C \text{ q.e. on } \Omega\}.$$

It is possible to proof that for any $p, 1 \le p \le \infty$, L^p is a Banach space, i.e. a normed and complete vectorial space [3].

Remark 2.3 *Complete means that any Cauchy succession is convergent. A succession $\{v_n\}_{n \in \mathbf{N}} \in L^P$ is a Cauchy succession if $\forall \epsilon > 0 \ \exists \nu$ such as $n.m > \nu$ implies $\|v_n - v_m\|_{L^p} < \epsilon$.*

Definition 2.4 *An Hilbert space is a vectorial space with the scalar product (u, v) and complete respect the norm $\|u\| = (u, u)^{1/2}$.*

Remark 2.5 *$L^2(\Omega)$ with the scalar product $(u, v) = \int_\Omega u(x)v(x)dx$ is an Hilbert space.*

Definition 2.6 *Let V be a normed space with norm $\|\cdot\|$. The dual space of V, indicated with V'. is the space of all the linear and continuous functions $L : V \longrightarrow \mathbf{R}$. The norm of V' is given by*

$$\|L\|_{V'} = \sup_{v \in V} \frac{L(v)}{\|v\|} \ .$$

Definition 2.7 *Let $D(\Omega)$ (or $C_0^\infty(\Omega)$) the space of continuously differentiable functions $f : \Omega \longrightarrow \mathbf{R}$ with a compact support K in Ω such as $f(\Omega - K) = 0$. The space of distributions $D'(\Omega)$ is the dual space of $D(\Omega)$.*

The finite element method is based on a variational formulation of the differential problem and the functional spaces which must be used are the Sobolev spaces [3].

Definition 2.8 *For $k \in \mathbf{N}$ and $1 \leq p \leq \infty$ the Sobolev spaces $W^{k,p}$ are given by*

$$W^{k,p}(\Omega) = \{v \in L^p(\Omega) : D^\alpha v \in L^p(\Omega) \text{ for } |\alpha| \leq k\} \ .$$

Here $\alpha = (\alpha_1, ..., \alpha_N)$ is a multi-index with $|\alpha| = \sum_{i=1}^N \alpha_i$ and

$$D^\alpha = \frac{\partial^{|\alpha|} v}{\partial x_1^{\alpha_1} ... \partial x_N^{\alpha_N}} \ .$$

In such spaces the norm is defined as

$$1 \leq p \leq \infty \quad \|v\|_{W^{k,p}(\Omega)} = \|v\|_{k,p.\Omega} = \left(\sum_{|\alpha| \leq k} \|D^\alpha v\|_{L^p}^p \right)^{1/p} ,$$

$$p = \infty \quad \|v\|_{W^{k,\infty}(\Omega)} = \|v\|_{k,\infty,\Omega} = \max_{|\alpha| \leq k} \|D^\alpha v\|_{L^\infty} \ .$$

Definition 2.9 *The $H^k(\Omega) = W^{k,2}(\Omega)$ spaces are given by*

$$H^k(\Omega) = \{v \in L^2(\Omega) : D^\alpha v \in L^2(\Omega) \text{ for } |\alpha| \leq k\} \ .$$

These spaces are Hilbert spaces for the scalar product

$$(u, v)_{H^k(\Omega)} = \sum_{|\alpha| \leq k} (D^\alpha u, D^\alpha v)_{L^2(\Omega)} \ ,$$

with norm

$$\|u\|_{H^k(\Omega)} = \left(\sum_{|\alpha| \leq k} \|D^\alpha u\|_{L^2(\Omega)} \right)^{1/2} \ .$$

Definition 2.10 *The $W_0^{k,p}(\Omega)$ spaces are defined for $k \in N$ and $1 \leq p < \infty$ as*

$$W_0^{k,p}(\Omega) = \text{ closing of } C_0^\infty(\Omega) \text{ on } W^{k,p}(\Omega) \ ,$$

where $C_0^\infty(\Omega) = \{v \in C^\infty(\Omega) : v \text{ is a function with compact support}\}.$

Obviously if $p = 2$ then $H_0^k(\Omega) = W_0^{k,2}(\Omega)$.

For the functions of $H^2(\Omega)$ is useful the Green formula: $\forall u \in H^2(\Omega), v \in H^1(\Omega)$ we have

$$\int_\Omega \Delta u \, v \, dx = \int_{\partial\Omega} \frac{\partial u}{\partial \nu} v \, d\gamma - \int_\Omega \nabla u \cdot \nabla v \, dx \ ,$$

where Δu is the Laplacian of u and $\frac{\partial u}{\partial \nu} = \nabla u \cdot n$, n is the unit vector of the external normal to $\partial\Omega$. This formula can be used also when Δ is substituted with a more general elliptic operator [3].

It is useful to introduce also the Sobolev spaces of real index.

Definition 2.11 *Let s be a real number. The Sobolev space $H^s(\Omega)$ is given by*

$$H^s(\Omega) = \{v \in L^2(\Omega) : (1 + |\psi|^2)^{s/2} \bar{v}(\psi) \in L^2(\Omega)\} \ ,$$

where $\bar{v}(\psi)$ is the Fourier transform of v.

It is possible to prove (see [3]) that for all $s > 1/2$ the space $H^{s-1/2}(\Sigma)$, with $\Sigma \subset \partial\Omega$, can be characterized as follows:

$$H^{s-1/2}(\Sigma) = \{\gamma_\Sigma v : v \in H^s(\Omega)\} \ ,$$

where γ_Σ is a linear continuous map such that $\gamma_\Sigma v = v|_\Sigma$. In particular we have

$$H^{1/2}(\partial\Omega) = \{\gamma_0 v : v \in H^1(\Omega)\} \ ,$$

where $\gamma_0 v = v|_{\partial\Omega}$.

Because in fluid dynamics is present also the time T, it is necessary to introduce the $L^p(0, T; V)$ spaces.

Definition 2.12 *For $1 \leq p < \infty$ the $L^p(0, T; V)$ spaces are given by*

$$L^p(0, T; V) = \{v :]0, T[\longrightarrow V : v \text{ is measurable and}$$

$$\int_0^T \|v(t)\|_V^p \, dt < \infty\}$$

$L^p(0, T; V)$ is a Banach space for the norm:

$$\|v\|_{L^p(0,T;V)} = \left(\int_0^T \|v(t)\|_V^p \, dt \right)^{1/p} \ .$$

For $p = 2$ and V Hilbert space, $L^2(0, T; V)$ is an Hilbert space for the scalar product

$$(u, v) = \int_0^T (u(t), v(t))_V \, dt \ .$$

2.1 Finite Element Method

The finite element method is a numerical method to approximate the solution of a differential problem that is written in variational form on a space V. From a differential problem (D) we can obtain a variational problem (P) by multiplying the equation for a test function v, by integrating over the space domain Ω and then by applying the Green formula and the boundary conditions. The variational problem (P) obtained in this way is called weak formulation of the problem (D) or weak problem [4].

Definition 2.13 *The general form of the weak problem is defined as:*
(P) find $u \in V$ such as $a(u,v) = L(v)$ $\forall v \in V$,
where:
i) V is an Hilbert space with scalar product $(\cdot,\cdot)_V$ and norm $\|\cdot\|_V$;
ii) $a(\cdot,\cdot) : V \times V \to \mathbf{R}$ is a bilinear continuous form;
iii) $L : V \to \mathbf{R}$ is a linear continuous form.

For example, let us consider two differential problems:

$$(D_1) \begin{cases} -\Delta u + au = f & \text{in } \Omega \\ u = 0 & \text{on } \partial\Omega \end{cases}$$

which is elliptic with homogenous Dirichlet boundary conditions, and

$$(D_2) \begin{cases} -\Delta u + au = f & \text{in } \Omega \\ \frac{\partial u}{\partial \nu} = g & \text{on } \partial\Omega . \end{cases}$$

which is elliptic with non–homogenous Neumann conditions to the boundary. Here Δu is the Laplacian of u and $\frac{\partial u}{\partial \nu} = \nabla u \cdot n$, n is the unit vector of the external normal to $\partial\Omega$. The *weak formulation* of the differential problems (D_1) and (D_2) is given by:

$$(P_1) \begin{cases} V = H_0^1(\Omega) \\ a(u,v) = \int_\Omega \left(\sum_{i=1}^d \frac{\partial u}{\partial x_i} \frac{\partial v}{\partial x_i} + auv \right) dx \\ L(v) = \int_\Omega fv dx, \text{ where } a \in L^\infty(\Omega), a \geq 0 \text{ a.e. in } \Omega, \\ \qquad f \in L^2(\Omega); \end{cases}$$

$$(P_2) \begin{cases} V = H^1(\Omega) \\ a(u,v) = \int_\Omega \left(\sum_{i=1}^d \frac{\partial u}{\partial x_i} \frac{\partial v}{\partial x_i} + auv \right) dx \\ L(v) = \int_\Omega fv dx + \int_{\partial\Omega} gv d\gamma, \\ \quad \text{where } a \in L^\infty(\Omega), a \geq a_0 > 0 \text{ a.e. in } \Omega, \\ \qquad f \in L^2(\Omega), g \in L^2(\partial\Omega). \end{cases}$$

The V space for differential problems like (D_2) is $H^1(\Omega)$, while for problems like (D_1) is $H_0^1(\Omega)$. In general we have $H_0^1(\Omega) \subseteq V \subseteq H^1(\Omega)$.

Definition 2.14 *The bilinear form $a(\cdot,\cdot) : V \times V \to \mathbf{R}$ is called coercive (or V-elliptic) if $\exists \alpha > 0$ such that*

$$a(v,v) \geq \alpha \|v\|_V^2 \qquad \forall v \in V .$$

The coercivity is necessary to prove the following lemma, known as the Lax–Milgram lemma [3].

Lemma 2.15 *Consider the general weak problem (P) with a coercive bilinear form $a(\cdot,\cdot)$. Then there exists one and only one solution of (P), and the solution is stable because*

$$\|u\|_V \leq \frac{1}{\alpha} \|L\|_X .$$

where $\|L\|_X = \sup_{v \in V} \frac{L(v)}{\|v\|_V}$.

If, in addition to the previous assumption, the bilinear form $a(\cdot,\cdot)$ is also *symmetric*, i.e. $a(u.v) = a(v,u)$ $\forall u,v \in V$, then the solution u of (P) can be regarded as the unique solution of the following *minimization problem*:
(M) find $u \in V$ such as $F(u) = \min_{v \in V} F(v)$,
where $F(v) = \frac{1}{2} a(v,v) - L(v)$.

Moreover, it is possible to prove that the two problems (M) and (P) are equivalent [5]. Obviously if the bilinear form $a(\cdot,\cdot)$ is not symmetric the minimization problem can not be formulated.

We have seen that the continuity and the coercivity of the bilinear form $a(\cdot,\cdot)$ implies the existence of a unique solution of the weak problem (P). However, there are important examples (such as the Navier–Stokes equations) in which the associated bilinear form is not coercive but it satisfies a weaker condition known as *the inf–sup condition*.

Definition 2.16 *We say that the bilinear form $a(\cdot,\cdot)$ satisfies the inf–sup conditions on V if there exists $\beta > 0$ such that*

$$\sup_{v \in V} \frac{a(u,v)}{\|v\|_V} \geq \beta \|u\|_V \qquad \forall u \in V$$

and

$$\sup_{u \in V} \frac{a(u,v)}{\|u\|_V} \geq \beta \|v\|_V \qquad \forall v \in V$$

Remark 2.17 *Clearly, if $a(\cdot,\cdot)$ is symmetric both conditions are the same. Note that both conditions can be written as*

$$\inf_{u \in V} \sup_{v \in V} \frac{a(u,v)}{\|u\|_V \|v\|_V} \geq 0$$

which justifies the usual terminology.

Remark 2.18 *If $a(\cdot,\cdot)$ is coercive it satisfies the inf–sup conditions. In fact,*

$$\sup_{v \in V} \frac{a(u,v)}{\|v\|_V} \geq \frac{a(u,u)}{\|u\|_V} \geq \alpha \|u\|_V$$

and $\alpha = \beta$.

The inf–sup conditions ensure the existence of a unique solution of (P), and in fact they also necessary (if the bilinear form is symmetric it has to satisfy one inf–sup condition).

Theorem 2.19 *The continuous bilinear form $a(\cdot,\cdot)$ satisfies the inf–sup conditions if and only if the weak problem (P) has a unique solution u for any L and $||u||_V \leq ||L||_{V'}$.*

Now we introduce the Galerkin approximation to the solution of the weak problem (P).

Definition 2.20 *Let $V_N \subset V$ be a finite dimensional space. The Galerkin approximation is the following discrete problem:*
(G) find $u_N \in V_N$ such as $a(u_N, v_N) = L(v_N)\ \forall v_N \in V_N$.

It is important to remark a fundamental difference between coercive forms on V and forms which satisfy the inf–sup on V but are not coercive.

Remark 2.21 *If the bilinear form $a(\cdot,\cdot)$ is coercive on V then it is also coercive on any subspace, in particular on V_N, and the Galerkin approximation u_N is well defined. Instead, the inf–sup condition on V is not inherited by subspaces, and so, when the form is not coercive, the inf–sup condition has to be verified on V_N in order to have u_N well defined.*

A natural question is whether $lim_{N\to\infty} u_N = u$ provided the spaces V_N are chosen in an appropriate way. Clearly, if the Galerkin approximation converges to u we have that

$$d(u, V_N) = \inf_{v \in V_N} ||u - v_N|| \to \infty \text{ when } N \to \infty . \qquad (1)$$

This limit is a natural property to ask on the subspaces (it means that they approximate u), and we would like to know if it is also a sufficient condition for convergence. An answer follows from the following lemma, known as Cea's lemma.

Lemma 2.22 *If the bilinear form $a(\cdot,\cdot)$ is continuous and coercive then,*

$$||u - u_N||_V \leq \frac{M}{\alpha} \inf_{v_N \in V_N} ||u - v_N||_V .$$

The lemma says that the Galerkin approximation u_N is like the best approximation in V_N to u up a constant depending on the form $a(\cdot,\cdot)$. In particular we have the following convergence result.

Theorem 2.23 *If the bilinear form $a(\cdot,\cdot)$ is continuous and coercive and the spaces V_N are such that (1) holds then $lim_{N\to\infty} u_N = u$.*

Suppose now that the form $a(\cdot,\cdot)$ is not coercive but it satisfies the inf–sup conditions on V. In order to have the Galerkin approximation (G) well defined we have to assume (an in concrete cases is has to be proved) that $a(\cdot,\cdot)$ satisfies the inf–sup condition also on V_N, i.e. that there exists $\beta > 0$ such that

$$\sup_{v_N \in V_N} \frac{a(u_N, v_N)}{||v_N||_V} \leq \beta ||u_N||_V \qquad \forall u_N \in V_N \qquad (2)$$

Note that, since V_N is finite dimensional the second inf–sup condition follows from this one.

In order to prove the convergence, we will ask that the constant β be independent of N. Under this assumption we have the following generalization of Cea's lemma due to Babuska [6].

Lemma 2.24 *If the bilinear form $a(\cdot,\cdot)$ is continuous and satisfies the inf–sup condition (2) then*

$$||u - u_N||_V \leq (1 + \frac{M}{\beta}) \inf_{v_N \in V_N} ||u - v_N||_V .$$

In particular, if β is independent of N, the constant in this error estimate is independent of N.

As an immediate consequence we have the following convergence result.

Theorem 2.25 *If the bilinear form $a(\cdot,\cdot)$ is continuous and satisfies the inf–sup condition (2) with β independent of N, and the spaces V_N are such that (1) holds, then $lim_{N\to\infty} u_N = u$.*

2.2 Finite Element Spaces

One of the most important and widely used Galerkin approximation are those based on spaces of picewise polynomial functions. The question is how to construct good approximation subspaces. The finite element method provides a systematic way of constructing this kind of subspaces.

We assume that $\partial\Omega$ is a polygonal curve, so Ω is a polygonal domain. The finite element method is a Galerkin Method where the V_N spaces, which we now call V_h, are finite element spaces which satisfy the following conditions.

(1) With the triangulation T_h we divide $\overline{\Omega}$ in a finite number of subsets K, called *finite elements*, so that

$$\overline{\Omega} = \cup_{K \in T_h} K, \qquad \forall K_1, K_2 \qquad Int(K_1) \cap Int(K_2) = \emptyset .$$

For adjacent triangles on the plane it is imposed that no vertex can stay on the side of another triangle. The parameter of the triangulation is defined as

$$h = \max_{K \in T_h} \text{diam}(K) ,$$

$$\text{diam}(K) = \text{diameter of } K = \text{longest side of } K .$$

On the basis of the triangulation of $\overline{\Omega}$ we define a finite element space $V_h \subset V$.

The functions of the V_h space are characterized by the following conditions.

2) Let $P_k = \{v_h|_K : v_h \in V_h\}$ be the space of the restrictions of the functions of V_h to the element K. P_K is polynomial space on K and a function of V_h is a polynomial function.

For example, in the case of triangles on the plane we can define V_h as $V_h v : v$ is continuous in $\Omega, v|_K$ is linear $\forall K \in T_h$. We have $V_h \subset H^1(\Omega)$ and, adding the condition $v|_{\partial\Omega} = 0$, $V_h \subset H_0^1(\Omega)$. It is possible to proof [5] that $V_h \subset H^1(\Omega) \iff V_h \subset C^0(\Omega)$.

3) There must be a basis of V_h, $\{w_k\}_{k=1}^{N_h}$, given by functions with compatible domain.

In the case of triangles on the plane, given a function $v \in V_h$ and called $v(N_i)$, $i = 1, \ldots . N_h$, its values in the nodes of the triangulation, the functions of the basis $\varphi_j \in V_h, j = 1, \ldots, N_h$, are defined as:

$$\varphi_j(N_i) = \delta_{ij} \, , \qquad \text{with } \delta_{ij} \text{ Kronecker symbol .}$$

The domain of φ_j is given by the triangles which have in common the node N_j. A function $v \in V_h$ can be written

$$v(x) = \sum_{j=1}^{N_h} v(N_j)\varphi_j(x) \, , \qquad x \in \overline{\Omega}.$$

If the discrete solution is written respect the basis $\{w_k\}_{k=1}^{N_h}$, $u_h = \sum_{k=1}^{N_h} u_k w_k$, then the discrete problem is reduced to the solution of the linear system

$$\sum_{k=1}^{N_h} a(w_k, w_l)u_k = L(w_l) \, , \qquad l = 1, \ldots, N_h,$$

where the matrix of coefficients, $A = [a(w_k, v_l)]_{k,l=1,\ldots,N_h}$, is invertible and positive definite if the bilinear form $a(\cdot, \cdot)$ is coercive. The matrix A is called rigidity matrix and the vector $b = [b_l] = [L(w_l)]_{l=1}^{N_h}$ is called weight vector. The construction of the system $Au = b$ is obtained adding the contributions of any element to A and b (see [4] and [5]).

For example, one way of approaching the problem is the following:
if $a(u, v)$ is the bilinear form of the problems (P_1) and (P_2) then

$$a(w_k, w_l) = \sum_{K \in T_h} a_K(w_k, w_l) =$$

$$= \sum_{K \in \mathrm{supp}w_k \cap \mathrm{supp}w_l} a_K(w_k, w_l) \, ,$$

where $a_K(w_k, w_l) = \int_k (\nabla w_k \cdot \nabla w_l + a w_k w_l)dx$.

Let us define the space $P_r(K)$ as

$$P_r(K) = \{p : p \text{ is a polynomial of degree } \leq r \text{ on } K\}$$

then the V_h space of the triangles on the plane reads

$$V_h = \{v \in C^0(\overline{\Omega}) : v|_K \in P_1(K), \forall K \in T_h\} \subset V = H^1(\Omega).$$

If $\Omega \subset \mathbf{R}^2$ then the finite element space defined on the triangulation T_h is given by

$$V_h = \{v \in C^0(\overline{\Omega}) : v|_K \in P_2(K), \forall K \in T_h\} \subset V = H^1(\Omega).$$

Because a function $p \in P_2(K)$ is determined by the values that it assumes on the vertices of the triangle K and on the mean points of the three sides, we have that a function $v \in V_h$ is completely determined by its values on the nodes of T_h and on the mean points of all the sides of the triangles of T_h. These values are the degrees of freedom of a function on V_h.

3 Fluid Dynamics and Navier–Stokes Equations

In fluid dynamics one assumes that the macroscopic fluid properties, for example mean density, mean pressure and mean viscosity, vary continuously with the position in the fluid system and with the time. Thus fluid properties are expressed as continuous functions of position and time only. On this basis, it is possible to establish equations governing the motion of a fluid, which are independent, in their form, from the nature of the particle structure [1].

Definition 3.1 *The Eulerian velocity* $\mathbf{u}(\mathbf{x}(t), t)$ *of the fluid is a vectorial field* $\mathbf{u} : \Omega \times]0, T[\to \mathbf{R}^3$, *where* $\Omega \subset \mathbf{R}^3$ *is the bounded domain of the fluid, with boundary* $\partial\Omega$, *and* $]0, T[\subset \mathbf{R}^+$ *is the time interval, with* T *a prefixed time.*

Remark 3.2 *The Eulerian velocity* $\mathbf{u}(\mathbf{x}(t), t)$ *satisfies the equation*

$$\frac{d}{dt}\mathbf{u} = \frac{\partial\mathbf{u}}{\partial t} + (\mathbf{u} \cdot \nabla)\mathbf{u} \, .$$

This equation is known as the d'Alembert–Euler acceleration formula. We call

$$\frac{D}{Dt} = \frac{\partial}{\partial t} + \mathbf{u} \cdot \nabla$$

the material derivative. The $\partial\mathbf{u}/\partial t$ *is the local acceleration and represents the acceleration felt by an observer fixed at a location in space. The term* $(\mathbf{u} \cdot \nabla)\mathbf{u}$ *is the acceleration felt by an observer at a given instant due to a change in position in space. It is the convective acceleration. Indeed, if* $f(\mathbf{x}(t), t)$ *is any function of position and time (scalar of vector), than we have*

$$\frac{d}{dt}f(\mathbf{x}(t), t) = \frac{\partial f}{\partial t} + (\mathbf{u} \cdot \nabla)f = \frac{Df}{Dt}(\mathbf{x}(t), t) \, .$$

Definition 3.3 *The matter density $\rho(\mathbf{x}, t)$ of the fluid is a scalar field $\rho : \Omega \times]0, T[\longrightarrow \mathbf{R}$, where $\Omega \subset \mathbf{R}^3$ is the bounded domain of the fluid and $]0, T[\subset \mathbf{R}$ is the time interval.*

Definition 3.4 *The fluid is called incompressible if $\forall(\mathbf{x}, t) \in \Omega \times]0, T[$*

$$\frac{d}{dt}\rho(\mathbf{x}, t) = 0 \ .$$

Definition 3.5 *The fluid is called isotropic if $\forall(\mathbf{x}.t) \in \Omega \times]0, T[$*

$$\nabla \rho(\mathbf{x}, t) = 0 \ .$$

The conservation of the fluid mass implies that the quantity of fluid entering a certain volume in space must be balanced by the quantity leaving it. From it the following theorem is derived.

Theorem 3.6 *The fluid satisfies the equation of continuity*

$$\frac{\partial \rho}{\partial t} + \nabla \cdot (\rho\, \mathbf{u}) = 0 \ . \tag{3}$$

Proof. Let $D \subset \Omega$ a regular subdomain of Ω. To conserve the mass, the rate of change of mass of the fluid in D has to be equal to the mass flux across the boundary ∂D of D

$$\frac{\partial}{\partial t} \int_D \rho \, d\mathbf{x} = - \int_{\partial D} (\rho\, \mathbf{u}) \cdot \mathbf{n} \, d\gamma \ ,$$

where $\mathbf{n}$ denotes the exterior normal at ∂D. Using the divergence theorem:

$$\int_D \nabla \cdot \mathbf{v} \, d\mathbf{x} = \int_{\partial D} \mathbf{v} \cdot \mathbf{n} \, d\gamma$$

and the fact that D is arbitrary, we get immediately the thesis. $\square$

Lemma 3.7 *An incompressible fluid satisfies the equation $\nabla \cdot \mathbf{u} = 0$.*

Proof. The equation of continuity, using the cartesian components of the vector position $\mathbf{x} = [x_1, x_2, x_3]^T$ and velocity $\mathbf{u} = [u_1, u_2, u_3]^T$, can be written

$$\frac{\partial \rho}{\partial t} + \frac{\partial}{\partial x_j}(\rho u_j) = \frac{\partial \rho}{\partial t} + u_j\frac{\partial \rho}{\partial x_j} + \rho\frac{\partial u_j}{\partial x_j} = \frac{d\rho}{dt} + \rho\frac{\partial u_j}{\partial x_j} = 0,$$

where the repeated index means summation, and by using the definition of incompressibility we have the thesis. $\square$

The conservation of mass and its consequence, the equation of continuity, have been examined. In fluid dynamics is also conserved the momentum of the fluid. This leads to the equations of motion for the fluid. In their most elementary form these equate the rate of change of momentum of a fluid with the sum of all forces acting on that fluid.

Definition 3.8 *The inertial force $\mathbf{F}^{(i)}$ of the fluid is defined as*

$$\mathbf{F}^{(i)} = \int_\Omega \rho\frac{d\mathbf{u}}{dt} \, d\mathbf{x} \ .$$

This force equals the rate of change of linear momentum of the fluid.

Definition 3.9 *The external force $\mathbf{F}^{(e)}$ acting on the fluid is given by*

$$\mathbf{F}^{(e)} = \int_\Omega \rho\, \mathbf{f} \, d\mathbf{x} \ ,$$

where $\mathbf{f} \in \mathbf{R}^3$ is the external force per unit mass.

Definition 3.10 *The contact force $\mathbf{F}^{(c)}$ is defined as*

$$\mathbf{F}^{(c)} = \int_{\partial\Omega} \phi^{\mathbf{n}} \, d\gamma \ ,$$

where $\mathbf{n} \in \mathbf{R}^3$ is the unit vector of the internal normal to the boundary $\partial\Omega$. The vector $\phi^{\mathbf{n}}$ is generated by the stress tensor ϕ, a linear application $\phi : \mathbf{R}^3 \longrightarrow \mathbf{R}^3$, so that $\forall \omega \in \mathbf{R}^3 \ \phi^\omega = \phi\, \omega$ is the specific stress of the fluid in the direction ω.

Remark 3.11 *The specific stress vector in the direction ω can be written in cartesian coordinates $\phi^\omega = [\phi_1^\omega, \phi_2^\omega, \phi_3^\omega]^T$ as $\phi_i^\omega = \phi_{ij}\, \omega_j$, where ϕ_{ij} are the components of the stress tensor and the repeated index means summation.*

Lemma 3.12 *The contact force $\mathbf{F}^{(c)} = [F_1^{(c)}, F_2^{(c)}, F_3^{(c)}]^T$ can be written*

$$F_i^{(c)} = - \int_\Omega \frac{\partial \phi_{ij}}{\partial x_j} \, d\mathbf{x} \ . \qquad i, j = 1, 2, 3 \ .$$

Proof. For the i–component of the contact force we have:

$$F_i^{(c)} = \int_{\partial\Omega} \phi_i^{\mathbf{n}} \, d\gamma = \int_{\partial\Omega} \phi_{ij}n_j \, d\gamma \ ,$$

and from the divergence theorem

$$\int_{\partial\Omega} \phi_{ij}n_j \, d\gamma = - \int_\Omega \frac{\partial \phi_{ij}}{\partial x_j} \, d\mathbf{x} \ ,$$

which completes the proof. $\square$

Theorem 3.13 *The indefinite equations of the dynamics of the fluid are given by*

$$\frac{\partial u_i}{\partial t} + u_j\frac{\partial u_i}{\partial x_j} + \frac{1}{\rho}\frac{\partial \phi_{ij}}{\partial x_j} = f_i \qquad i, j = 1, 2, 3 \tag{4}$$

Proof. For the fluid the second law of dynamics of Newton is $\mathbf{F}^{(i)} = \mathbf{F}^{(e)} + \mathbf{F}^{(c)}$. Then by using the lemma 2.12 and the remark 2.2 we get the thesis. $\square$

Definition 3.14 *The fluid is inviscid (non–viscous or Eulerian) if the stress tensor is given by*

$$\phi_{ij} = P\delta_{ij} \, ,$$

where $P = P(\mathbf{x})$ is the pressure and δ_{ij} is the Kronecker symbol.

Theorem 3.15 *The equations of motion of an inviscid fluid are given by*

$$\frac{\partial \mathbf{u}}{\partial t} + (\mathbf{u} \cdot \nabla)\mathbf{u} + \frac{1}{\rho}\nabla P = \mathbf{f} \, . \tag{5}$$

Proof. It is sufficient to insert the stress tensor of the inviscid fluid into the indefinite equations of the dynamics of the fluid. $\square$

These equations are called *Euler equations* and were obtained by Euler in 1755. Such equations were generalized including viscosity by Navier in 1822 and Stokes in 1845 [1].

Definition 3.16 *The fluid is Newtonian (or Newtonian viscous) if the stress tensor satisfies the constitutive equations*

$$\phi_{ij} = P\delta_{ij} - 2\mu D_{ij} - \lambda D_{ij}\delta_{ij} \, ,$$

where P is the pressure, μ and λ are the first and second viscosity, δ_{ij} is the Kronecker symbol and the deformation rate tensor is given by

$$D_{ij} = \frac{1}{2}\Big(\frac{\partial u_i}{\partial x_j} + \frac{\partial u_j}{\partial x_i}\Big) \, .$$

Experimental results on thick liquids show a considerable departure from the constitutive equations of the Newtonian fluid. Examples are lubrificants, colloids, polymer solutions and pastes. They are called non–Newtonian liquids. Several nonlinear constitutive equations have been proposed by introducing nonlinearity and elasticity or memory to explain the behavior of such highly viscous fluids.

The equations of a Newtonian fluid are called *Navier–Stokes equations* and in the case of a inviscid fluid they give the Euler equations. As an example, we introduce the Navier–Stokes equations for an isotropic and incompressible fluid.

Theorem 3.17 *The equations of the isotropic incompressible Newtonian fluid are given by*

$$\frac{\partial \mathbf{u}}{\partial t} + (\mathbf{u} \cdot \nabla)\mathbf{u} + \nabla p - \nu\nabla^2\mathbf{u} = \mathbf{f} \, ,$$

$$\nabla \cdot \mathbf{u} = 0 \, , \tag{6}$$

where $p = P/\rho$ is the reduced pressure and $\nu = \mu/\rho$ is the reduced or kinematic viscosity.

Proof. For a incompressible fluid we have $\nabla \cdot \mathbf{u} = \frac{\partial u_j}{\partial x_j} = 0$. The stress tensor of the incompressible Newtonian viscous fluid becomes

$$\phi_{ij} = P\delta_{ij} - 2\mu D_{ij} \, .$$

Now it is sufficient to insert this tensor in the indefinite equations of the dynamics of the fluid with the isotropic condition $\nabla\rho = 0$. $\square$

The Navier–Stokes equations are to be integrated in the space–time domain $\Omega \times]0, T[\subset \mathbf{R}^3 \times \mathbf{R}^+$ once an appropriate set of initial and boundary conditions has been defined. While the former conditions, in general, only need to reproduce a feasible shape of the current solution, the latter are to be set and treated with much care due to the effect they can have on the evolution process [7,8].

Let us call $\mathbf{b}$ the boundary velocity, than the boundary condition is given by

$$\mathbf{u}|_{\partial\Omega} = \mathbf{b}(\mathbf{x}_{\partial\Omega}, t) \tag{7}$$

with $\mathbf{x}_{\partial\Omega} \in \partial\Omega$. The condition (7) is not always appropriate for the numerical integration of the system (5). Therefore most usually the boundary condition is decomposed in the normal $\mathbf{n} \cdot \mathbf{u}|_{\partial\Omega}$ and tangent $\mathbf{u} - \mathbf{n} \cdot \mathbf{u}|_{\partial\Omega}$ components. where $\mathbf{n}$ is the outer normal to the boundary $\partial\Omega$. The normal condition reads

$$\mathbf{n} \cdot \mathbf{u}|_{\partial\Omega} = \mathbf{n} \cdot \mathbf{b}(\mathbf{x}_{\partial\Omega}, t) \tag{8}$$

while for the tangent we have

$$\mathbf{u} - \mathbf{n} \cdot \mathbf{u}|_{\partial\Omega} = \mathbf{b} - \mathbf{n} \cdot \mathbf{b}(\mathbf{x}_{\partial\Omega}, t) \tag{9}$$

which is called *no–slip condition.*

In the case of closed boundary it is always valid the condition (8), while the no–slip condition is imposed only when the viscous terms of the fluid on the boundary are important. But we observe that it is not possible to impose boundary conditions for the pressure because the system should be over. In the case of open boundary it is instead useful to impose boundary conditions for the pressure and only the direction of the flows.

In general we impose that the fluid velocity satisfies the initial condition

$$\mathbf{u}|_{t=0} = \mathbf{u}_0(\mathbf{x}) \, . \tag{10}$$

and the global boundary condition, for any $t > 0$

$$\int_\Omega \mathbf{n} \cdot \mathbf{u}|_{\partial\Omega} = 0 \, , \tag{11}$$

which follows from the integration of equation (5) on Ω and the divergence theorem. This boundary condition is weaker than (8).

4 Numerical Models for Coupled and Decoupled Variables

In the numerical solution of the Navier–Stokes equations the time evolution is usually treated in the framework of finite difference theory, in particular by using a θ–scheme. Instead the spatial problem is studied by using the finite element method [8,9]. An important role is played by the velocity and the pressure. Models with coupled or decoupled velocity and pressure imply a different treatment of the diffusive term and of the incompressibility condition.

4.1 Model for Coupled Variables

Here we introduce a model of coupled variables (velocity and pressure) due to Glowinski [10]. The Navier–Stokes equations can be written as

$$\frac{\partial \mathbf{u}}{\partial \mathbf{t}} + A\mathbf{u} + B\mathbf{u} = \mathbf{f} ,$$

where A is the Stokes operator:

$$A\mathbf{u} = -\nu\nabla^2\mathbf{u} + \nabla p \quad \text{with} \quad \nabla\cdot\mathbf{u} = 0 ,$$

and B is the convection operator:

$$B\mathbf{u} = \mathbf{u}\cdot\nabla\mathbf{u} .$$

The θ-scheme can be written (for $n > 0$):

$$\frac{1}{k\theta}(\mathbf{u}^{n+\theta} - \mathbf{u}^n) + A\mathbf{u}^{n+\theta} + B\mathbf{u}^n = \mathbf{f}^{n+\theta} , \quad (12)$$

$$\frac{1}{(1-2\theta)k}(\mathbf{u}^{n+1-\theta} - \mathbf{u}^{n+\theta}) + A\mathbf{u}^{n+\theta} + B\mathbf{u}^{n+1-\theta} = \mathbf{f}^{n+1-\theta} , \quad (13)$$

$$\frac{1}{k\theta}(\mathbf{u}^{n+1} - \mathbf{u}^{n+1-\theta}) + A\mathbf{u}^{n+1} + B\mathbf{u}^{n+1-\theta} = \mathbf{f}^{n+1} , \quad (14)$$

where k is the time step and $0 < \theta < 1$. This method divides the time step in three parts by using the θ parameter. In this way we have to solve two stationary Stokes problems, for (12) and (13), and a nonlinear problem for the convective term $B = (\mathbf{u}\cdot\nabla)$.

4.1.1 The Stokes Problem

Finite element methods in which two spaces are used to approximate two different variables receive the general denomination of *mixed methods*. In the Stokes problem the velocity $\mathbf{u}$ and the pressure p are two natural independent variables and so the mixed formulation is the natural one.

Definition 4.1 *The generalized Stokes problem is:* (D_{SP}) *find* $\mathbf{u}(\mathbf{x}) \in \mathbf{R}^n$ *and* $p(\mathbf{x}) \in \mathbf{R}$ *such as*

$$\alpha\mathbf{u} - \nu\nabla^2\mathbf{u} + \nabla p = \mathbf{f} \qquad in\ \Omega$$

$$\nabla\cdot\mathbf{u} = 0 \qquad in\ \Omega$$

$$\mathbf{u} = \mathbf{u}_{\partial\Omega} \qquad on\ \partial\Omega$$

Here α and ν are positive constants and $\mathbf{f}$ is a function of $\Omega \subset \mathbf{R}^n$.

The Stokes problem can be considered as a time–discretization of the Navier-Stokes equations if α is the inverse the time step and f is an approximation of the convective term $-\mathbf{u}\cdot\nabla\mathbf{u}$.

Definition 4.2 *We define the space*

$$J(\Omega) = \{\mathbf{u} \in H^1(\Omega)^n : \nabla\cdot\mathbf{u} = 0\} ,$$

its subspace

$$J_0(\Omega) = \{\mathbf{u} \in J(\Omega)^n : \mathbf{u}|_{\partial\Omega} = 0\} ,$$

and the scalar products

$$(\mathbf{a},\mathbf{b}) = \int_\Omega a_i b_i \qquad \mathbf{a},\mathbf{b} \in L^2(\Omega)^n ,$$

$$(A, B) = \int_\Omega A_{ij} B_{ij} \qquad A, B \in L^2(\Omega)^{n\times n} .$$

Definition 4.3 *The weak Stokes problem is given by:* (P_{SP}) *find* $\mathbf{u} \in J_0(\Omega)$ *such that*

$$\alpha\,(\mathbf{u},\mathbf{v}) - \nu\,(\nabla\mathbf{u},\nabla\mathbf{v}) = (\mathbf{f},\mathbf{v}) \quad \forall\mathbf{v} \in J_0(\Omega)$$
$$\mathbf{u} - \mathbf{u}'_{\partial\Omega} \in J_0(\Omega)$$

where $\mathbf{u}'_{\partial\Omega}$ is an extension of $\mathbf{u}_{\partial\Omega}$ in $J_0(\Omega)$.

The following theorems can be proved.

Theorem 4.4 *If $\mathbf{u}_{\partial\Omega} \in H^{1/2}(\partial\Omega)^n$ and if $\int_{\partial\Omega} \mathbf{u}_{\partial\Omega}\cdot\mathbf{n}\,d\mathbf{x} = 0$ then it exist an extension $\mathbf{u}'_{\partial\Omega} \in J(\Omega)$ of $\mathbf{u}_{\partial\Omega}$.*

Theorem 4.5 *If $\mathbf{f} \in L^2(\Omega)^n$ and $\mathbf{u}'_{\partial\Omega} \in J(\Omega)$, the weak Stokes problem (P_{SP}) has a unique solution.*

Theorem 4.6 *If the solution of the weak Stokes problem (P_{SP}) is C^2 then it is also solution of the generalized Stokes problem (D_{SP}). Instead if $\{\mathbf{u}, p\} \in H^1(\Omega)^n \times L^2(\Omega)$ is solution of the generalized Stokes problem (D_{SP}) then $\mathbf{u}$ is also solution of the weak Stokes problem (P_{SP}).*

Definition 4.7 *We define the following subspaces of V_h*

$$J_h = \{\mathbf{v}_h \in V_h : (\nabla\cdot\mathbf{v}_h, q)_h = 0\ \forall q \in Q_h\} ,$$
$$J_{0h} = \{\mathbf{v}_h \in J_h : \mathbf{v}_h|_{\partial\Omega} = 0\} .$$

J_{0h} is determined by the choice of the space V_h, approximation of $H^1(\Omega)$, the space Q_h, approximation of $L^2(\Omega)$, and by the choice of the quadrature formula $(\cdot,\cdot)_h$. Let us consider $\{J_h\}_h$, a succession in the finite dimensional space J_{0h} such as

$$\forall\mathbf{v} \in J_0(\Omega)\ \exists\ \mathbf{v}_h \in J_{0h} \ :\ \|\mathbf{v}_h - \mathbf{v}\| \to 0 \text{ when } h \to 0 .$$

Definition 4.8 *The Galerkin approximation to the Stokes Problem is:*
(G_{SP}) *find* $\mathbf{u}_h \in J_{0h}$ *such as:*

$$\alpha\,(\mathbf{u}_h, \mathbf{v}_h) - \nu\,(\nabla \mathbf{u}_h, \nabla \mathbf{v}_h) = (\mathbf{f}, \mathbf{v}_h)$$

$$\forall \mathbf{v}_h \in J_{0h} \tag{15}$$

$$\mathbf{u}_h - \mathbf{u}'_{\partial \Omega_h} \in J_{0h}$$

where $\mathbf{u}'_{\partial \Omega_h}$ *is an approximation of* $\mathbf{u}'_{\partial \Omega}$ *in* J_h.

Then it is possible to prove the following theorem.

Theorem 4.9 *If* J_{0h} *is not empty and* $V_{0h} \subset H_0^1(\Omega)^n$ *(Ω polygonal), then the Galerkin problem* (G_{SP}) *has a unique solution and it is equivalent to*

$$\alpha\,(\mathbf{u}_h, \mathbf{v}_h) - \nu\,(\nabla \mathbf{u}_h, \nabla \mathbf{v}_h) + (\nabla p_h, \mathbf{v}_h) = (\mathbf{f}, \mathbf{v}_h) \quad \forall \mathbf{v}_h \in J_{0h}$$

$$(\nabla \cdot \mathbf{u}_h, q_h) = 0 \quad \forall q_h \in Q_h$$

$$\mathbf{u}_h - \mathbf{u}'_{\partial \Omega_h} \in V_{0h}, \quad p_h \in Q_h \ .$$

Moreover it is possible to prove the following important theorem [11].

Theorem 4.10 *The problem* (G_{SP}) *has a a stable solution if*

$$\inf_{q_h \in Q_h} \sup_{\mathbf{v}_h \in V_h} \frac{b(\mathbf{v}_h, q_h)}{\|\mathbf{v}_h\|_V \|q_h\|_Q} \geq \beta' > 0 \tag{16}$$

with $b(\mathbf{v}_h, q_h) = (\nabla \mathbf{v}_h, q_h)$.

This result is the inf–sup condition for the Galerkin approximation to the Stokes problem. In the finite element method it is necessary to choose the elements so that the functional spaces V_h e Q_h satisfy the inf–sup condition [4.11].

4.2 Model for Decoupled Variables

We analyze now a model of decoupled velocity and pressure, called also *splitting method* [7]. This model is based essentially on the following theorem due to Ladyzhenskaya (see [2]).

Theorem 4.11 *Any vectorial field* $\mathbf{v}$ *defined on* Ω *has the following unique orthogonal decomposition*

$$\mathbf{v} = \mathbf{w} + \nabla \varphi, \tag{17}$$

where $\mathbf{w}$ *is a zero–divergence vector which has also a zero normal to the boundary* $\partial \Omega$ *of* Ω.

Let us suppose that the boundary condition for the velocity $\mathbf{u}$, solution of the Navier–Stokes equation. is given by

$$\mathbf{n} \cdot \mathbf{u}|_{\partial \Omega} = 0$$

Then it is natural to introduce the operator $\mathcal{P}_{J_0^0}$ of orthogonal projection on the space

$$J_0^0(\Omega) = \{\mathbf{w} \in L^2(\Omega), \quad \nabla \cdot \mathbf{w} = 0. \quad \mathbf{n} \cdot \mathbf{w}|_{\partial \Omega} = 0\} \ .$$

The space $J_0^0(\Omega)$ is a closed subset of $L^2(\Omega)$.

It is easy to show that the projection of the Navier–Stokes equation on $J_0^0(\Omega)$ eliminates the pressure. In fact, by using the scalar product defined in $L^2(\Omega)$ we see that $\forall \mathbf{w} \in J_0^0(\Omega)$ $(\mathbf{w}, \nabla p) = 0$ so that

$$(\mathbf{w}, \nabla p) = -(\nabla \cdot \mathbf{w}. p) = 0 \ .$$

The project Navier–Stokes equation is then given by:

$$\frac{\partial \mathbf{u}}{\partial \mathbf{t}} = \mathcal{P}_{J_0^0}[-(\mathbf{u} \cdot \nabla)\mathbf{u} + \nu \nabla^2 \mathbf{u}]. \tag{18}$$

The importance of the projector $\mathcal{P}_{J_0^0}$ is that it separates the diffusive term from the incompressibility condition. In fact, by using this projective method the time-discretization is obtained in to successive steps. The first step is to calculate the flow velocity $\mathbf{u}^{n+1/2}$, which does not satisfy the incompressibility condition:

$$\begin{aligned}
\frac{\mathbf{u}^{n+1/2} - \mathbf{u}^n}{\Delta t} &= -(\mathbf{u}^n \cdot \nabla)\mathbf{u}^n + \nu \nabla^2 \mathbf{u}^n, \\
\mathbf{u}^{n+1/2}|_{\partial \Omega} &= \mathbf{b}^{n+1} \ .
\end{aligned} \tag{19}$$

The second step is the solution of the system:

$$\begin{aligned}
\frac{\mathbf{u}^{n+1} - \mathbf{u}^{n+1/2}}{\Delta t} &= -\nabla p^{n+1}, \\
\nabla \cdot \mathbf{u}^{n+1} &= 0, \\
\mathbf{n} \cdot \mathbf{u}^{n+1}|_{\partial \Omega} &= \mathbf{n} \cdot \mathbf{b}^{n+1} \ ,
\end{aligned} \tag{20}$$

which completes the time step by obtaining $\mathbf{u}^{n+1}$ and p^{n+1}.

5 Shallow Water Equations

In this section we show how to obtain the shallow water equations from the Navier–Stokes equations. In the shallow water hypothesis we assume that the characteristic horizontal scale L for the motion (the wave length) is longer than the average height $\bar{h}$ of the fluid. i.e. $L >> \bar{h}$. In such hypothesis vertical acceleration and velocity are negligible and the flux becomes almost horizontal [12,13].

Definition 5.1 *The total height of the water is given by*

$$h(x_1, x_2, t) = H(x_1, x_2) + \eta(x_1, x_2, t) \ ,$$

where H is the depth of the water in the stationary condition above the reference level $x_3 = 0$. and η is the time-dependent difference, i.e. the height of the free surface.

Remark 5.2 *The total height h is a scalar field, $h : \Omega \times]0, T[\longrightarrow \mathbf{R}$, where $\Omega \subset \mathbf{R}^2$ is the bounded domain of the water and $]0, T[\subset \mathbf{R}$ is the time interval, with T a prefixed time.*

The external forces acting on the water are of extreme importance to determine the equations of motion of the system.

Definition 5.3 *The force of gravity is defined as*

$$\mathbf{f}_g = \int_\Omega \rho \, \mathbf{g} \, dx \,,$$

where $\mathbf{g} = [0, 0, -g]^T$ is the vector acceleration of gravity.

Theorem 5.4 *The inviscid shallow water equations in presence of the force of gravity are given by:*

$$\frac{\partial \mathbf{U}}{\partial t} + \frac{\partial \mathbf{F}_1}{\partial x_1} + \frac{\partial \mathbf{F}_2}{\partial x_2} = 0 \qquad (21)$$

where $\mathbf{U} = [\eta, u_1, u_2]^T$ is the vector of the conservative variables. The flow vectors $\mathbf{F}_1$ and $\mathbf{F}_2$ are

$$\mathbf{F}_1 = [H u_1, g\eta, 0]^T \,, \quad \mathbf{F}_2 = [H u_2, 0, g\eta]^T \,,$$

where g is the acceleration of gravity.

Proof. Let us consider the inviscid Navier–Stokes equations. The external force is given by $\mathbf{f} = [0, 0, -g]^T$. The Navier–Stokes equations for inviscid shallow water projected on the x_3 axis are reduced to

$$\frac{\partial p}{\partial x_3} = -g \,,$$

because for a inviscid fluid we have $\nu = 0$ and the term

$$\frac{du_3}{dt} = \frac{\partial u_3}{\partial t} + u_1 \frac{\partial u_3}{\partial x_1} + u_2 \frac{\partial u_3}{\partial x_2} + u_3 \frac{\partial u_3}{\partial x_3}$$

is supposed to be negligible.

By integrating between x_3 and η we obtain

$$\int_{x_3}^\eta \frac{\partial p}{\partial x_3} dx_3 = - \int_{x_3}^\eta g \, dx_3$$

and so

$$p = p_a + g(\eta - x_3) \,,$$

where $p_a = p(x_3 = \eta)$ is the reduced atmospheric pressure, which is supposed to be constant. From this equation we obtain

$$\frac{\partial p}{\partial x_1} = g \frac{\partial \eta}{\partial x_1} \,, \quad \frac{\partial p}{\partial x_2} = g \frac{\partial \eta}{\partial x_2} \,.$$

Instead on the x_1 and x_2 axis we have

$$\frac{\partial u_1}{\partial t} + u_1 \frac{\partial u_1}{\partial x_1} + u_2 \frac{\partial u_1}{\partial x_2} + \frac{\partial p}{\partial x_1} = 0 \,,$$

$$\frac{\partial u_2}{\partial t} + u_1 \frac{\partial u_2}{\partial x_1} + u_2 \frac{\partial u_2}{\partial x_2} + \frac{\partial p}{\partial x_2} = 0 \,,$$

because also term u_3 is negligible.

The equation of continuity for incompressible fluids

$$\frac{\partial u_1}{\partial x_1} + \frac{\partial u_2}{\partial x_2} + \frac{\partial u_3}{\partial x_3} = 0 \,,$$

can be written as

$$\frac{\partial u_1}{\partial x_1} + \frac{\partial u_2}{\partial x_2} = -\frac{\partial u_3}{\partial x_3} \,,$$

and integrating between $-H$ and η becomes:

$$\int_{-H}^\eta \left(\frac{\partial u_1}{\partial x_1} + \frac{\partial u_2}{\partial x_2} \right) dx_3 = - \int_{-H}^\eta \frac{\partial u_3}{\partial x_3} dx_3 = -u_3|_{x_3=\eta} \,.$$

where $u_3|_{x_3=-H} = 0$. Because $u_3 = \frac{d\eta}{dt}$ we have

$$\int_{-H}^\eta \left(\frac{\partial u_1}{\partial x_1} + \frac{\partial u_2}{\partial x_2} \right) dx_3 = -u_3|_{x_3=\eta} =$$

$$= -\left(\frac{\partial \eta}{\partial t} + u_1|_{x_3=\eta} \frac{\partial \eta}{\partial x_1} + u_2|_{x_3=\eta} \frac{\partial \eta}{\partial x_2} \right) \,.$$

so that

$$\frac{\partial u_1}{\partial t} + u_1 \frac{\partial u_1}{\partial x_1} + u_2 \frac{\partial u_1}{\partial x_2} + g \frac{\partial \eta}{\partial x_1} = 0,$$

$$\frac{\partial u_2}{\partial t} + u_1 \frac{\partial u_2}{\partial x_1} + u_2 \frac{\partial u_2}{\partial x_2} + g \frac{\partial \eta}{\partial x_2} = 0 \,.$$

In conclusion

$$\frac{\partial \eta}{\partial t} + \frac{\partial H u_1}{\partial x_1} + \frac{\partial H u_2}{\partial x_2} = 0 \,,$$

$$\frac{\partial u_1}{\partial t} + g \frac{\partial \eta}{\partial x_1} = 0,$$

$$\frac{\partial u_2}{\partial t} + g \frac{\partial \eta}{\partial x_2} = 0,$$

which are the equations of the thesis in explicit form. $\square$

If we consider large regions of water it is necessary to include other forces, like the Coriolis force and the Chezy force, which models the friction of the water at the bottom [9,12,13].

Definition 5.5 *The Coriolis force is defined as*

$$\mathbf{f}_{cor} = \int_\Omega \omega \wedge \mathbf{u} \, dx \,,$$

where $\omega = [0, 0, w_3]^T$ is the rotation vector of the earth and $\omega_3 = k_0$ is called Coriolis coefficient.

Definition 5.6 *The Chezy force is defined as*

$$\mathbf{f}_{ch} = \int_{\Omega} \frac{g\mathbf{u}|u|}{k_1^2 h}\, d\mathbf{x}\ ,$$

where $|u| = \sqrt{u_1^2 + u_2^2}$, g *is the scalar acceleration of gravity,* h *is the total height of the water and* k_1 *is the Chezy coefficient.*

Theorem 5.7 *Let us consider the inviscid shallow water with external forces given by the gravity force, the Coriolis force and the Chezy force. The shallow water equations read*

$$\frac{\partial \mathbf{U}}{\partial t} + \frac{\partial \mathbf{F}_1}{\partial x_1} + \frac{\partial \mathbf{F}_2}{\partial x_2} = \mathbf{R}_s \qquad (22)$$

where $\mathbf{U} = [\eta, u_1, u_2]^T$ *is the vector of the variables, the flow vectors* $\mathbf{F}_1$ *and* $\mathbf{F}_2$ *are*

$$\mathbf{F}_1 = [Hu_1, g\eta, 0]^T\ , \quad \mathbf{F}_2 = [Hu_2, 0, g\eta]^T\ ,$$

and the source vector $\mathbf{R}_s$ *is given by*

$$\mathbf{R}_s = [0, k_0 u_2 - \frac{gu_1|u|}{k_1^2 h}, -k_0 u_1 - \frac{gu_2|u|}{k_1^2 h}]^T\ ,$$

where g *is the acceleration of gravity,* k_0 *and* k_1 *are the coefficients of Coriolis and Chezy and* $H(x_1, x_2)$ *is the depth of the water in the stationary condition.*

Proof. In the shallow water equations (21) the external force was $\mathbf{f} = [0, 0, -g]$. In this case the total external force is given by

$$\mathbf{f} = [k_0 u_2 - \frac{gu_1|u|}{k_1^2 h}, -k_0 u_1 - \frac{gu_2|u|}{k_1^2 h}, -g]^T\ ,$$

where g is the acceleration of gravity and k_0 and k_1 are the coefficients of Coriolis and Chezy forces. Substituting the new f in the equations (21) we obtain the thesis. $\square$

Here the Chezy force has been introduced phenomenologically but we observe that it can be obtained explicitly in the shallow water equations starting from the Newtonian (viscid) Navier–Stokes equations.

Theorem 5.8 *Let us consider the viscid shallow water with external forces given by the gravity force and the Coriolis force. The 3–dimensional viscid shallow water equations are*

$$\frac{\partial u_1}{\partial t} + u_1 \frac{\partial u_1}{\partial x_1} + u_2 \frac{\partial u_1}{\partial x_2} + g\frac{\partial \eta}{\partial x_1} - \nu\nabla^2 u_1 = k_0 u_2\ , \quad (23)$$

$$\frac{\partial u_2}{\partial t} + u_1 \frac{\partial u_2}{\partial x_1} + u_2 \frac{\partial u_2}{\partial x_2} + g\frac{\partial \eta}{\partial x_2} - \nu\nabla^2 u_2 = -k_0 u_1\ , \quad (24)$$

$$\int_{-H}^{\eta} \left(\frac{\partial u_1}{\partial x_1} + \frac{\partial u_2}{\partial x_2} \right) dx_3 = -\left(\frac{\partial \eta}{\partial t} + u_1|_{x_3=\eta}\frac{\partial \eta}{\partial x_1} + u_2|_{x_3=\eta}\frac{\partial \eta}{\partial x_2} \right) \qquad (25)$$

where g *is the acceleration of gravity,* ν *is the viscosity,* k_0 *is the coefficient of Coriolis and* $h(x_1, x_2, t) = H(x_1, x_2) + \eta(x_1, x_2, t)$ *is the total height of the water.*

Proof. Let us consider the viscid Navier–Stokes equations. The external force is given by $\mathbf{f} = [k_0 u_2, -k_0 u_1, -g]^T$. On the x_1 and x_2 axis we have

$$\frac{\partial u_1}{\partial t} + u_1 \frac{\partial u_1}{\partial x_1} + u_2 \frac{\partial u_1}{\partial x_2} + \frac{\partial p}{\partial x_1} - \nu\nabla^2 u_1 = k_0 u_2\ ,$$

$$\frac{\partial u_2}{\partial t} + u_1 \frac{\partial u_2}{\partial x_1} + u_2 \frac{\partial u_2}{\partial x_2} + \frac{\partial p}{\partial x_2} - \nu\nabla^2 u_2 = -k_0 u_1\ .$$

On the x_3 axis the viscid shallow water equation is the same of the inviscid one:

$$p = p_a + g(\eta - x_3)\ ,$$

where $p_a = p(x_3 = \eta)$ is the reduced atmospheric pressure, which is supposed to be constant. From this equation we obtain

$$\frac{\partial p}{\partial x_1} = g\frac{\partial \eta}{\partial x_1}\ , \quad \frac{\partial p}{\partial x_2} = g\frac{\partial \eta}{\partial x_2}\ .$$

In conclusion the equations on the x_1 and x_2 axis are given by

$$\frac{\partial u_1}{\partial t} + u_1 \frac{\partial u_1}{\partial x_1} + u_2 \frac{\partial u_1}{\partial x_2} + g\frac{\partial \eta}{\partial x_1} - \nu\nabla^2 u_1 = k_0 u_2\ ,$$

$$\frac{\partial u_2}{\partial t} + u_1 \frac{\partial u_2}{\partial x_1} + u_2 \frac{\partial u_2}{\partial x_2} + g\frac{\partial \eta}{\partial x_2} - \nu\nabla^2 u_2 = -k_0 u_1\ .$$

The equation of continuity for incompressible fluids

$$\frac{\partial u_1}{\partial x_1} + \frac{\partial u_2}{\partial x_2} + \frac{\partial u_3}{\partial x_3} = 0\ ,$$

can be written as

$$\frac{\partial u_1}{\partial x_1} + \frac{\partial u_2}{\partial x_2} = -\frac{\partial u_3}{\partial x_3}\ ,$$

and integrating between $-H$ and η becomes:

$$\int_{-H}^{\eta} \left(\frac{\partial u_1}{\partial x_1} + \frac{\partial u_2}{\partial x_2} \right) dx_3 = -\int_{-H}^{\eta} \frac{\partial u_3}{\partial x_3}\, dx_3 = -u_3|_{x_3=\eta}\ ,$$

where $u_3|_{x_3=-H} = 0$. Finally, because $u_3 = \frac{d\eta}{dt}$, we obtain

$$\int_{-H}^{\eta} \left(\frac{\partial u_1}{\partial x_1} + \frac{\partial u_2}{\partial x_2} \right) dx_3 = -u_3|_{x_3=\eta} =$$

$$= -\left(\frac{\partial \eta}{\partial t} + u_1|_{x_3=\eta}\frac{\partial \eta}{\partial x_1} + u_2|_{x_3=\eta}\frac{\partial \eta}{\partial x_2} \right)\ .$$

This completes the proof. $\square$

We observe that the velocities $u_1 = u_1(x_1, x_2, x_3, t)$ and $u_2 = u_2(x_1, x_2, x_3, t)$ are functions also of the x_3 variable, so the system is 3–dimensional. It is possible to obtain a 2–dimensional system by performing the substitution

$$u_1(x_1, x_2, x_3, t) \rightarrow a(\psi)u_1(x_1, x_2, t) \ ,$$

$$u_2(x_1, x_2, x_3, t) \rightarrow a(\psi)u_2(x_1, x_2, t) \ ,$$

where

$$\psi = \frac{x_3 + H}{\eta + H} \qquad \text{with} \qquad \int_0^1 a(\psi)d\psi = 1 \ ,$$

and then by integrating the new equations over the x_3 variable. In this way one can prove the following theorem [12,13].

Theorem 5.9 *Let us consider the viscid shallow water with external forces given by the gravity force and the Coriolis force. The 2–dimensional viscid shallow water equations are*

$$\frac{\partial \mathbf{U}}{\partial t} + \frac{\partial \mathbf{F}_1}{\partial x_1} + \frac{\partial \mathbf{F}_2}{\partial x_2} = \mathbf{R} \qquad (26)$$

where $\mathbf{U} = [h, hu_1, hu_2]^T$ *is the vector of the variables, the flow vectors* $\mathbf{F}_1$ *and* $\mathbf{F}_2$ *are*

$$\mathbf{F}_1 = [hu_1, hu_1^2 + \frac{1}{2}g(h^2 - H^2), hu_1u_2]^T \ ,$$

$$\mathbf{F}_2 = [hu_2, hu_1u_2, hu_2^2 + \frac{1}{2}g(h^2 - H^2)]^T \ ,$$

and the source vector

$$\mathbf{R} = \mathbf{R}_s + \frac{\partial \mathbf{R}_{d1}}{\partial x_1} + \frac{\partial \mathbf{R}_{d2}}{\partial x_2}$$

is given by

$$\mathbf{R}_s = [0, k_0 hu_2 + g(h - H)\frac{\partial H}{\partial x_1} - \frac{gu_1|u|}{k_1^2 h}, -k_0 hu_1 +$$

$$+ g(h - H)\frac{\partial H}{\partial x_2} - \frac{gu_2|u|}{k_1^2 h}]^T \ ,$$

$$\mathbf{R}_{d1} = [0, 2\nu h\frac{\partial u_1}{\partial x_1}, \nu h(\frac{\partial u_2}{\partial x_1} + \frac{\partial u_1}{\partial x_2})]^T \ ,$$

$$\mathbf{R}_{d2} = [0, \nu h(\frac{\partial u_2}{\partial x_1} + \frac{\partial u_1}{\partial x_2}), 2\nu h\frac{\partial u_2}{\partial x_2}]^T \ ,$$

where g *is the acceleration of gravity,* ν *is the viscosity,* k_0 *and* k_1 *are the coefficients of Coriolis and Chezy and* $h(x_1, x_2, t) = H(x_1, x_2) + \eta(x_1, x_2, t)$ *is the total height of the water.*

For the flow vectors $\mathbf{F}_i = \mathbf{f}_i(x_1, x_2, \mathbf{U}(x_1, x_2, t))$, we have

$$\frac{\partial \mathbf{F}_i}{\partial x_i} = \frac{\partial \mathbf{f}_i}{\partial x_i} + \frac{\partial \mathbf{f}_i}{\partial \mathbf{U}}\frac{\partial \mathbf{U}}{\partial x_i} = \frac{\partial \mathbf{f}_i}{\partial x_i} + A_i\frac{\partial \mathbf{U}}{\partial x_i} \qquad i = 1, 2 \ .$$

The shallow water equations can be generalized including also the wind effect. We shall discuss these aspects in the last section.

The shallow water equations are to be integrated in the space–time domain $\Omega \times]0, T[\subset \mathbf{R}^2 \times \mathbf{R}^+$ once an appropriate set of initial and boundary conditions has been defined

$$\mathbf{U}(\mathbf{x}, t = 0) = \mathbf{U}_0(\mathbf{x}) \qquad \forall \mathbf{x} \in \Omega \ . \qquad (27)$$

$$\mathbf{U}(\mathbf{x}, t) = \mathbf{U}_{\partial\Omega}(\mathbf{x}, t) \qquad \forall \mathbf{x} \in \partial\Omega \ , \ \forall t \in]0, T[\ . \qquad (28)$$

While the initial conditions, in general, only need to reproduce a feasible shape of the current solution, the boundary conditions are to be set and treated with much care due to the effect they can have on the evolution process [14,15].

6 Domain Decomposition Methods

To solve the shallow water equations it is necessary to use sofisticated numerical methods because of the large number of nodes that are involved. Different approaches have been used by iterative methods like conjugate gradient and Cholewski decomposition. Also the domain decomposition method has shown great interest according to different ways followed by different authors. We observe that recently there has been an increase of research activities in the area of parallel computing due to the advent and growth of various parallel processing architectures. The domain decomposition approach can achieve the highest level of parallelism in the numerical solution of partial differential equations. The idea of domain decomposition goes beck to Schwarz [16,17], however only with last years the researchers developed and extended these ideas to use them with parallel architectures.

The domain decomposition technique can be employed for the solution of problems defined on irregular domain, using the same equations to be solved on all domains [18,19]. This technique allows the use of different numerical schemes and different resolution or different types of elements for the finite element method [20]. The domain decomposition technique can be used also if the problem requires different mathematical models for different subdomains as often happens in fluid dynamics using a viscous model near the boundary and a inviscid model in the far field [21,22].

For the domain decomposition we have the overlapping and non–overlapping approaches. For the non–overlapping method most of the applications, up to now, have been developed for the interfaces or more specifically to find good

preconditioners of the conjugate gradient algorithm [23] for
the linear system arising from the discretization of elliptic
partial differential equations in two or three dimensional
domains [24,25].

Sometimes it is convenient to follow another approach
which constructs a decomposed–domain preconditioner for
the simultaneous iterative procedure on the whole domain.
One of the advantages of this approach is that only approximate subdomain solvers are required. This alternative
approach was first proposed in [26.27] and further studied
in [28,29]. An other work on this line on shallow water is
[30]. Another formulation is given in [14], in this paper
the set of first order nonlinear hyperbolic partial differential equations are written for the shallow water problems
which have very important applications in meteorology and
oceanography. These equations can be used to study ty-
des and surface water run–off. They can also be used to
study large–scale waves in the atmosphere and ocean if
terms representing the effects of the earth's rotation (Coriolis terms) are included. The equations used in such case
are 2–dimensional shallow water equations where the Chezy
term is neglected. These equations are solved by using a
Galerkin finite element method for the primitive equations
of numerical weather prediction [31]. Such method gives
a good numerical solution in meteorological problems. For
the solution it is used the Shur complement [14].

It is well known that in the finite element method the internal degree of freedom can be condensed at the element
level before the assembly process and if the same procedure is applied at a subregion it can be considered as a
new structure at the subregion level. In such procedure
two classes of variables are usually identified, namely the
internal variables of the nodes within subdomains, and the
interfaces variables relevant to nodes belonging to two or
more subdomains. When the discretization for finite elements is made the internal variables can be numbered either before or after the interface ones.

The numerical integration method of the shallow water
equations for the Venice lagoon will be discussed in details
in the next section. Here we discuss domain decomposition of the lagoon because we are particularly interested
in the computational speed–up resulting when using the
domain decomposition technique. We consider the subdivision of the domain Ω under consideration in the following
way: Ω is subdivided into two subdomains Ω_1 and Ω_2 where
$\Omega = \Omega_1 \cup \Omega_2 \cup \Gamma$ and $\Omega_1 \cap \Omega_2 = 0$ and where Γ is the boundary between the two regions Ω_1 and Ω_2 [32]. Let us denote
the original nodal numbers as the old numbering, while the
nodal numbers after renumbering (namely the substructuring numbering) will be denoted as the new numbering.
This becomes cumbersome if we think to the finite element
numbering method in which the relationship among global
nodes, local nodes and element numbering turns out to be

very different if we try to formulate the problem using the
new numbering systems for various mesh resolutions.

Therefore we use the following rule: the node indices
are renumbered in sequence starting from the region Ω_1
following with the region Ω_2 and finally with the boundary
Γ. As a consequence of such rule the matrix A of the linear
system to be solved

$$Ax = f$$

has the form

$$A = \begin{pmatrix} A_{11} & 0 & A_{13} \\ 0 & A_{22} & A_{23} \\ A_{31} & A_{32} & A_{33} \end{pmatrix}$$

incomparable with the topology of the central lagoon.
Therefore it would have been too difficult to think of a
method with overlapping regions and almost inextensible
to a such complicated area.

The nature of the method itself makes it possible to apply any order of decomposition [14,35] necessary to extend
such methodology to the entire lagoon. This opens the way
to the use of parallel computers, in fact the times obtained
making sequential calculations are comparable to those obtained on the same machine with Conjugate Gradient solver
that has given in this case from 2 to 3 seconds for ini-
tializations and 351 seconds for a complete solution of the
same simulation time with identical time step. Therefore
if such advantages are obtained with sequential calculation
a greater advantage is expected on the global evaluation
time due to the use of a parallel calculation. Also topological properties of the domain suggest to use the domain decomposition to obtain the advantage of evaluating together
regions of comparable physical characteristic. Nevertheless
the domain decomposition have to be realized conveniently,
namely each subdomain should have almost the same number of nodes and the internal boundaries should have the
number of nodes smaller as possible. Two cases are considered, fox example. in the case of subdivision in only two
subdomains:

	CASE 1	CASE 2
External Nodes	508	350
Internal Nodes	251	381
Boundary Nodes	131	159
Total Time	635 sec	581 sec
Initialization Time	170 sec	190 sec
Iteration Time	464 sec	390 sec

As can be seen the case 1 is unbalanced and the case 2
is balanced.

The time step for the implicit scheme applied to the central lagoon of Venice has been evaluated every 5 minutes for
a space discretization of 890 nodes for the central lagoon.

To be able to make a faster model the domain decomposition idea has been applied to such region. The choice of the global time–step is then based on the triangles of smaller size along the canals of greater depth, although a greater time–step would be required in those areas where a coarser mesh would be sufficient. Therefore the possibility of using the domain decomposition is extremely important, allowing to subdivide the region studied so to treat with a Courant number enough homogeneous and to save CPU–time.

In the case of the total Venice lagoon we decompose it into 10 subdomains and the nodes of the interfaces collected into a unique set (see Figure 1). The number of nodes of the entire discretization is 1967 and the elements are 3423. The results of the calculation obtained using the domain decomposition method without parallelization have approximately a speed–up of 25 per cent over a Conjugate Gradient solver used on a not decomposed domain.

7 Case Study: the Venice Lagoon

In this section we present a model to calculate the motion of the water into the Venice lagoon. The equation considered are the shallow water equations. All simplification are obtained considering different depth of the lagoon and in particular a finer discretization is considered where the water runs with greater speed namely along the more important canals. A numerical method based on a finite element discretization is presented and numerical results are given. The model has shown to be very stable and capable to preserve the water quantity of the system. The wind effect is considered and shows to be very effective on the behaviour of the model [33–39].

As shown in the previous section the inviscid shallow water equations are given by

$$\frac{\partial \mathbf{U}}{\partial t} + \frac{\partial \mathbf{F}_1}{\partial x_1} + \frac{\partial \mathbf{F}_2}{\partial x_2} = \mathbf{R}_s \tag{29}$$

where $\mathbf{U} = [\eta, u_1, u_2]^T$ is the vector of the conservative variables, the flow vectors $\mathbf{F}_1$ and $\mathbf{F}_2$ are

$$\mathbf{F}_1 = [Hu_1, g\eta, 0]^T, \quad \mathbf{F}_2 = [Hu_2, 0, g\eta]^T,$$

and the source vector $\mathbf{R}_s$ is given by

$$\mathbf{R}_s = [0, k_0 u_2 - \frac{gu_1|u|}{k_1^2 H}, -k_0 u_1 - \frac{gu_2|u|}{k_1^2 H}]^T,$$

where g is the acceleration of gravity, k_0 and k_1 are the coefficients of Coriolis and Chezy and $H(x_1, x_2)$ is the depth of the water in the stationary condition. Here we use the hypothesis of long waves, i.e. $\eta << H$, and the hypothesis that the viscous terms are negligible.

Now we apply the splitting method putting

$$\mathbf{F}_i = \mathbf{F}_i^* + \mathbf{F}_i^{**},$$

$$\mathbf{U}^{(n+1)} = \mathbf{U}^{(n)} + (\Delta \mathbf{U}^*)^{(n+1)} + (\Delta \mathbf{U}^{**})^{(n+1)},$$

with $\mathbf{F}_i^* = 0$, $\mathbf{F}_i^{**} = \mathbf{F}_i$, and $\Delta \mathbf{U}^*$ and $\Delta \mathbf{U}^{**}$ the increments of the solution vector. The system (29) can be divided in:

$$\frac{\partial \Delta \mathbf{U}^*}{\partial t} = \mathbf{R}_s. \tag{30}$$

and

$$\frac{\partial \Delta \mathbf{U}^{**}}{\partial t} + \frac{\partial \mathbf{F}_1^{**}}{\partial x_1} + \frac{\partial \mathbf{F}_2^{**}}{\partial x_2} = \mathbf{0}. \tag{31}$$

The equation (30) is discretized in time by using a Taylor expansion to the second order

$$(\Delta \mathbf{U}^*)^{(n+1)} = \Delta t \left(\frac{\partial \Delta \mathbf{U}^*}{\partial t}\right)^{(n)} + \frac{\Delta t^2}{2} \left(\frac{\partial^2 \Delta \mathbf{U}^*}{\partial t^2}\right)^{(n)}, \tag{32}$$

and from

$$\frac{\partial \Delta \mathbf{U}^*}{\partial t} = \mathbf{R}_s, \quad \frac{\partial^2 \Delta \mathbf{U}^*}{\partial t^2} = \frac{\partial \mathbf{R}_s}{\partial t} = \frac{\partial \mathbf{R}_s}{\partial \Delta \mathbf{U}^*}\frac{\partial \Delta \mathbf{U}^*}{\partial t}, \tag{33}$$

the equation (32) can be written

$$(\Delta \mathbf{U}^*)^{(n+1)} = \Delta t(\mathbf{R}_s)^{(n)} + \frac{\Delta t^2}{2}(\mathbf{GR}_s)^{(n)}, \tag{34}$$

where

$$\mathbf{G} = \frac{\partial \mathbf{R}_s}{\partial \Delta \mathbf{U}^*}.$$

Because of the computational complexity in the evaluation of the right term of equation (34), we use a two–step version of the Taylor-Galerkin algorithm [38–40]. This is given by

a) approximation of $(\mathbf{U}^*)^{(n+1/2)}$ with the Taylor expansion

$$\mathbf{U}^{(n+1/2)} = \mathbf{U}^{(n)} + \frac{\Delta t}{2}(\mathbf{R}_s)^{(n)} \tag{35}$$

from which we can evaluate $\mathbf{R}_s^{(n)}$;

b) approximation of $\mathbf{R}_s^{(n+1/2)}$ with the Taylor expansion

$$(\mathbf{R}_s)^{(n+1/2)} = (\mathbf{R}_s)^{(n)} + \frac{\Delta t}{2}\left(\frac{\partial \mathbf{R}_s}{\partial t}\right)^{(n)} = \tag{36}$$

$$(\mathbf{R}_s)^{(n)} + \frac{\Delta t}{2}(\mathbf{GR}_s)^{(n)},$$

from which we obtain $(\mathbf{GR}_s)^{(n)}$ as

$$(\mathbf{GR}_s)^{(n)} = \frac{2}{\Delta t}\left[(\mathbf{R}_s)^{(n+1/2)} - (\mathbf{R}_s)^{(n)}\right]. \tag{37}$$

The spatial approximation with the finite elements method is obtained by using linear form functions for the integer step $(n, n+1, \ldots)$ and constant functions for the half step $(n - 1/2. n + 1/2, \ldots)$. In this way the equation (34) reads

$$(\mathbf{M}\Delta\mathbf{U}^*)_j^{(n+1)} = \Delta t \int_\Omega (\mathbf{R}_s)^{(n)} \phi_j \, d\mathbf{x} +$$

$$+ \frac{\Delta t^2}{2} \int_\Omega \frac{2}{\Delta t} \left[(\mathbf{R}_s)^{(n+1/2)} - (\overline{\mathbf{R}}_s)^{(n)} \right] \phi_j \, d\mathbf{x} = \quad (38)$$

$$= \Delta t \int_\Omega \left\{ (\mathbf{R}_s)^{(n+1/2)} + \left[(\mathbf{R}_s)^{(n)} - (\overline{\mathbf{R}}_s)^{(n)} \right] \right\} \phi_j \, d\mathbf{x} ,$$

where $j = 1, 2, \cdots, N$, N is the total number of nodes, ϕ_j is the weight linear function of the node j, the bar denotes mean values calculated on the element and

$$\mathbf{M} = [M]_{ij} = \int_\Omega \phi_i \phi_j \, d\mathbf{x} , \quad (39)$$

is the mass matrix.

The equation (31) is given, in explicit form, by

$$\begin{cases} \frac{\partial}{\partial t}(\Delta\eta^{**}) + \frac{\partial}{\partial x_1}(Hu_1) + \frac{\partial}{\partial x_2}(Hu_2) = 0 \\ \frac{\partial}{\partial t}(\Delta u_1^{**}) + \frac{\partial}{\partial x_1}(g\eta) = 0 \\ \frac{\partial}{\partial t}(\Delta u_2^{**}) + \frac{\partial}{\partial x_2}(g\eta) = 0 \end{cases} \quad (40)$$

The time discretization is obtained by using the θ-method proposed in [41]

$$\begin{array}{l} (\Delta\eta^{**})^{(n+1)} + \Delta t[\frac{\partial}{\partial x_1}(Hu_1^{(n+\theta_1)}) + \frac{\partial}{\partial x_2}(Hu_2^{(n+\theta_1)})] = 0 \\ (\Delta u_1^{**})^{(n+1)} + \Delta t \frac{\partial}{\partial x_1} p^{(n+\theta_2)} = 0 \\ (\Delta u_2^{**})^{(n+1)} + \Delta t \frac{\partial}{\partial x_2} p^{(n+\theta_2)} = 0 \end{array}$$
$$(41)$$

where $p = g\eta$, θ_1 and θ_2 are real parameters between 0 and 1 and

$$\begin{array}{l} u_i^{(n+\theta_1)} = u_i^{(n)} + \theta_1[(\Delta u_i^*)^{(n+1)} + (\Delta u_i^{**})^{(n+1)}] \\ p^{(n+\theta_2)} = p^{(n)} + \theta_2(\Delta p^{**})^{(n+1)} \end{array} \quad (42)$$

Note that the term $(\Delta p^*)^{(n+1)}$ does not appear because the splitting method (see paragraph 4.2) does not include variations in the water elevation due to (34). Substituting (42) into (41), we can write

$$\begin{array}{l} \frac{1}{g}(\Delta p^{**})^{(n+1)} + \Delta t\theta_1 \sum_{i=1}^2 \frac{\partial}{\partial x_i}[H(\Delta u_1^{**})^{(n+1)}] = \\ -\Delta t \sum_{i=1}^2 \frac{\partial}{\partial x_i}(Hu_1^{(n)}) + \Delta t\theta_1 \sum_{i=1}^2 \frac{\partial}{\partial x_i}[H(\Delta u_1^*)^{(n+1)}] \\ (\Delta u_1^{**})^{(n+1)} + \Delta t\theta_2 \frac{\partial}{\partial x_1}[(\Delta p^{**})^{(n+1)}] = -\Delta t \frac{\partial}{\partial x_1} p^{(n)} \\ (\Delta u_2^{**})^{(n+1)} + \Delta t\theta_2 \frac{\partial}{\partial x_2}[(\Delta p^{**})^{(n+1)}] = -\Delta t \frac{\partial}{\partial x_2} p^{(n)} \end{array}$$
$$(43)$$

Here $(\Delta u_1^{**})^{(n+1)}$ and $(\Delta u_2^{**})^{(n+1)}$ can be obtained from the second and third equation as a function of $(\Delta p^{**})^{(n+1)}$ and then substituted in the first equation which becomes

$$\frac{1}{g}(\Delta p^{**})^{(n+1)} - \Delta t^2 \theta_1 \theta_2 \sum_{i=1}^2 \frac{\partial}{\partial x_i}[H\frac{\partial}{\partial x_i}(\Delta p^{**})^{(n+1)}] =$$

$$= -\Delta t \{ \sum_{i=1}^2 \frac{\partial}{\partial x_i}[H(u_1^{(n)} + \theta_1 (\Delta u_i^*)^{(n+1)})] - \quad (44)$$

$$-\Delta t\theta_1 \sum_{i=1}^2 \frac{\partial}{\partial x_i}(H \frac{\partial}{\partial x_i} p^{(n)}) \} .$$

By using again linear triangular elements for the spatial discretization with the finite elements method, and by using the Green formulas to the terms which include the second derivatives. the variational formulation (44) gives the following system of discrete equations

$$\left(\tfrac{1}{g}\mathbf{M} + \Delta t^2\theta_1\theta_2\mathbf{S} \right) (\Delta p^{**})^{(n+1)} =$$
$$-\Delta t \left\{ \sum_{i=1}^2 \mathbf{Q}_i \left[H \left(u_1^{(n)} + \theta_1 (\Delta u_i^*)^{(n+1)} \right) \right] + \Delta t\theta_1 \mathbf{S} p^{(n)} \right\}$$
$$(45)$$

where

$$\mathbf{S} = \sum_{i=1}^2 \left(\int_\Omega \frac{\partial[\phi]}{\partial x_i} H \frac{\partial[\phi]^T}{\partial x_i} d\mathbf{x} \right) \quad \mathbf{Q}_i = \int_\Omega [\phi] \frac{\partial[\phi]^T}{\partial x_i} d\mathbf{x}.$$
$$(46)$$

When $(\Delta p^{**})^{(n+1)}$ is evaluated by (45), the second and third equations of (43) can be solved in $(\Delta u_1^{**})^{(n+1)}$ and $(\Delta u_2^{**})^{(n+1)}$; their discretization to finite elements is given by

$$\mathbf{M}(\Delta u_i^{**})^{(n+1)} = -\Delta t\mathbf{Q}_i \left[p^{(n)} + \theta_2(\Delta p^{**})^{(n+1)} \right] \quad i = 1, 2.$$
$$(47)$$

It is easy to see that this set of equations is similar to (38); as a consequence it can be solved in the same way.

In summary, the solution at each time step implies the following operations:

a)   solve (38) $(\Delta u^*)^{(n+1)}$ and $(\Delta u^*)^{(n+1)}$ (here $(\Delta p^*)^{(n+1)} = 0$ because the first component of $\mathbf{R}_s$ is zero);

b)   calculate $(\Delta p^{**})^{(n+1)}$ from (45);

c)   solve (47) for $(\Delta u_1^{**})^{(n+1)}$ and $(\Delta u_2^{**})^{(n+1)}$;

d)   calculate $\mathbf{U}_{(n+1)} = \mathbf{U}_{(n)} + (\Delta\mathbf{U}^*)^{(n+1)} + (\Delta\mathbf{U}^{**})^{(n+1)}$.

We have added a convective term. The new system is given by

$$\frac{\partial\eta}{\partial t} + \frac{\partial(Hu_1)}{\partial x_1} + \frac{\partial(Hu_2)}{\partial x_2} = 0 ,$$

$$\frac{\partial u_1}{\partial t} + g\frac{\partial\eta}{\partial x_1} = k_0 u_2 - \frac{g|u|u_1}{k_1^2 H} - (2u_1\frac{\partial u_1}{\partial x_1} + \frac{u_1^2}{H}\frac{\partial H}{\partial x_1} +$$

$$+ u_1\frac{\partial u_2}{\partial x_2} + u_2\frac{\partial u_1}{\partial x_2} + \frac{u_1 u_2}{H}\frac{\partial H}{\partial x_2}) ,$$

$$\frac{\partial u_2}{\partial t} + g\frac{\partial\eta}{\partial x_2} = -k_0 u_1 - \frac{g|u|u_2}{k_1^2 H} - (2u_2\frac{\partial u_2}{\partial x_2} + \frac{u_2^2}{H}\frac{\partial H}{\partial x_2} +$$

$$+u_1\frac{\partial u_2}{\partial x_1} + u_2\frac{\partial u_1}{\partial x_1} + \frac{u_1 u_2}{H}\frac{\partial H}{\partial x_1}\Big)\ .$$

The splitting method is the same with a new source vector

$$\mathbf{R}_s = [0, k_0 u_2 - \frac{g|u|u_1}{k_1^2 H} - R_1, -k_0 u_1 - \frac{g|u|u_2}{k_1^2 H} - R_2]$$

where R_1 and R_2 are

$$R_1 = (2u_1\frac{\partial u_1}{\partial x_1} + \frac{u_1^2}{H}\frac{\partial H}{\partial x_1} + u_1\frac{\partial u_2}{\partial x_2} + u_2\frac{\partial u_1}{\partial x_2} + \frac{u_1 u_2}{H}\frac{\partial H}{\partial x_2})$$

$$R_2 = (2u_2\frac{\partial u_2}{\partial x_2} + \frac{u_2^2}{H}\frac{\partial H}{\partial x_2} + u_1\frac{\partial u_2}{\partial x_1} + u_2\frac{\partial u_1}{\partial x_1} + \frac{u_1 u_2}{H}\frac{\partial H}{\partial x_1}).$$

The convective term is included in the right side and it is treated in the explicit part of the integration method.

To describe the Venice Lagoon we must add the wind effect. Let v the wind velocity respect the water and ξ an adimensional constant, the wind term is given by [42]

$$\tau = \xi\frac{|v|v}{H}\ .$$

and the original system becomes

$$\frac{\partial \eta}{\partial t} + \frac{\partial(Hu_1)}{\partial x_1} + \frac{\partial(Hu_2)}{\partial x_2} = 0$$

$$\frac{\partial u_1}{\partial t} + g\frac{\partial \eta}{\partial x_1} = k_0 u_2 - \frac{g|u|u_1}{k_1^2 H} + \xi\frac{|v|v_1}{H}$$

$$\frac{\partial u_2}{\partial t} + g\frac{\partial \eta}{\partial x_2} = -k_0 u_1 - \frac{g|u|u_2}{k_1^2 H} + \xi\frac{|v|v_2}{H}$$

In the approximation that the water velocity is negligible respect the wind velocity, the wind term is an external parameter to the system, i.e. the wind velocity respect the soil. We take a time–dependent wind velocity but space-independent. The Venice Lagoon has been discretized with 1967 nodes and 3423 elements as shown in Figure 1. In Figure 2 we plot the height of the water in the Venice Lagoon obtained by using the splitting method.

The model has been verified by using the experimental data of the Ufficio Idrografico and Mareografico di Venezia (month of Fabruary 1994) [43]. The data have been introduced at the mouth of the lagoon as boundary conditions on the open boundary. Therefore a comparison has been made between the simulation and the real data on the internal measurement stations of the lagoon [44]. The experimental and simulated data present a good agreement in all the internal measurement stations [45,46,47]. In Figure 3 we plot the results of the station of Burano island. The results of such station are more significative then others because it is farther from the mouths of inlet.

References

[1] A. J. Chorin and J. E. Marsden, *A Mathematical Introduction to Fluid Mechanics* (Springer, Berlin, 1980).

[2] L. Quartapelle, *Numerical Solution of the Incompressible Navier-Stokes Equations* (Birkhouse, Bessel, 1993).

[3] H. Brezis, *Analisi Funzionale* (Liguori Editore, Napoli, 1986).

[4] F. Brezzi and M. Fortin, *Mixed and Hybrid finite element methods* (Springer, Berlin, 1991).

[5] C. Johnson, *Numerical Solution of Partial Differential Equations by Finite Element Method* (Cambridge University Press, Cambridge, 1987).

[6] I. Babuska, 'Error Bound for Finite Element Method', Numer. Math. **16**, 322–333 (1971).

[7] R. Temam, *Navier-Stokes Equations* (North Holland, Amsterdam, 1984).

[8] G.F. Carey and J.T. Oden, *Finite Elements: Fluid Mechanics* (Prentice–Hall, New Jersey, 1986).

[9] O. Pironneau, *Finite Element Methods for Fluids* (John Wiley – Masson, New York, 1989).

[10] R. Glowinski, 'Finite Element Methods for the Numerical Simulation of Incompressible Viscous Flow. Introduction to The Control of The Navier-Stokes Equation', Lect. Appl. Math. **28**, 219–301 (1991).

[11] F. Brezzi and K.J. Bathe, 'A discours on the stability conditions for mixed finite element formulations', Comp. Meth. Appl. Mech. Eng. **82**, 27–57 (1990).

[12] O.C. Zienkiewicz and J.C. Heinrich, 'A unified treatment of steady-state shallow water and two dimensional Navier-Stokes equations', Comp. Meth. Appl. Mech. Eng., 673–698 (1979).

[13] J. Peraire, O.C. Zienkiewicz and K. Morgan, 'Shallows water problems: a general explicit formulation', Int. J. Numer. Meth. Eng. **22**, 547–574 (1986).

[14] I.M. Navon and Y. Cai, 'Domain decomposition and parallel processing of a element model of the shallow water equations', Comp. Meth. Appl. Mech. Eng. **106**, 179–212 (1993).

[15] M. Morandi Cecchi and A. Pica, 'General Simulation Modelling in Shallow Water Problems with Data Assimilation Computer Techniques and Applications', Ed. by W.R. Blai and E. Cabrera, pp. 249–262, (Springer-Verlag, Berlin, 1992).

[16] H.A. Schwarz, 'Uber einige Abbildungasaufgaben', Ges. Math. Abh. **11**. 65–83 (1869).

[17] J.S. Przemienniechi, 'Matrix structural analysis of substructures', AIAA J. **1**, 138–147 (1963).

[18] D.A. Kopriva, 'Computation of hyperbolic equations on camplicated domains with patched and overset Chebyshev grids', SIAM J. Sci. Statist. Comput. **10**, 120–132 (1989).

[19] M.M. Rai, 'A conservative treatment of zonal boundaries for Euler equation calculations', J. Comput. Phys. **62**, 472–503 (1986).

[20] D.A. Kopriva, 'Domain decomposition with both spectral and finite difference methods for the accurate computation of flows with shoks', Appl. Numer. Math. **6**, 141–151 (1989).

[21] Q.V. Dinh, R. Glowinski, J. Periaux and G. Terrasson, 'On the coupling of viscous and inviscid models for incompressible fluid flows via domain decomposition', in the 1st International Symposium on Domain Decomposition Methods for Partial Differential Equations, Ed. by R. Glowinski, G.H. Golub, G.A. Meurant and J. Periaux, pp. 350–369 (SIAM. Philadelphia, 1988).

[22] R. Glowinski, J. Periaux and G. Terrasson, 'On the coupling of viscous and inviscid models for compressible fluid flows via domain decomposition', in the 3rd International Symposium on Domain Decomposition Methods for Partial Differential Equations, Ed. by T.F. Chan, R. Glowinski, J. Periaux, and O.B. Widlund, pp. 64–97 (SIAM. Philadelphia, 1990).

[23] P. Concus, G.H. Golub and D.P. O'Leary, 'A generalized conjugate gradient method for the numerical solution of elliptic partial differential equations', in *Sparse Matrix Computations*, Ed. by J.R. Bunch and D.J. Rose, pp. 309–332 (Academic Press, New York, 1976).

[24] M. Dryja, 'A capacitance matrix method for Dirichlet problem on polygon regions', Numer. Math. **39**, 51–64 (1982).

[25] M. Dryja, 'A finite–element capacitance method for elliptic problems on regions partioned into subregions', Numer. Math. **44**, 153–168 (1984).

[26] J.H. Bramble, J.E. Pasciak and A.H. Schatz, 'An iterative method for elliptic problems on regions partitioned into substructures', Math. Comput. **46**, 361–369 (1986).

[27] J.H. Bramble, J.E. Pasciak and A.H. Schatz, 'The construction of preconditioners for elliptic problems by substructures', Math. Comput. **47**, 103–134 (1986).

[28] T.F. Chan and D. Goovaerts, 'A note on the efficiency of domain decomposed incomplete factorizations', SIAM J. Sci. Statist. Comput. **11**, 794–803 (1990).

[29] C. Borgers, 'The Neumann–Dirichlet domain decomposition method with inexact solvers on the subdomains', Numer. Math. **55**, 123–136 (1989).

[30] Y. Cai and I.M. Navon, 'Parallel Block Preconditioning techniques for the numerical simulation of the shallow water flow using finite element methods', J. Comput. Phys. **122**, 39–50 (1995).

[31] J. Steppeler, 'Energy conserving Galerkin finite element scheme for the primitive equations of numerical weather prediction', J. Comput. Phys. **69**, 258–264 (1987).

[32] M. Morandi Cecchi, L. Padoan and G. Scenna, 'A shallow water model by domain decomposition method', in *Parallel Computing: Problems, Methods and Applications*, Ed. by P. Messina and A. Murli, pp. 361–371 (Elsevier, New York, 1992).

[33] M.Morandi Cecchi, 'Study of the behaviour of oscillatory waves in a lagoon', Int. J. Numer. Meth. Eng. **27**, 103–112 (1989).

[34] M. Morandi Cecchi and L. Padoan, 'A shallow water model for a lagoon by a finite element method', in the 8th International Conference on Computational Methods in Water Resources, Ed. by G. Gambolati *et al.*, pp. 157–162 (Springer-Verlag, Berlin, 1990).

[35] O.C. Zinkiewicz and J. Wu, 'A new semi–implicit or explicit algorithm for shallow water equations', Proc. Int. Symp. on Environmental Hydraulics. Hong Kong, 16–18 December, p. 971–976 (Balchema Press, Rotterdam, 1991).

[36] R. Lohner, K. Morgan and O.C. Zinkiewicz, 'The use of Domain Splitting with an explicit hyperbolic solver', Comp. Meth. Appl. Mech. Eng. **45**, 313–329 (1984).

[37] M. Morandi Cecchi, A. Pica and E. Secco, 'Tidal Flow Analisis With Core Minimization', in the 8th International Conference on Finite Elements in Fluids: New Trends and Applications, Finite Elements in Fluids. Ed. by K. Morgan, E. Onate, J. Periaux, J. Peraire. D.C. Zienkiewicz, pp. 1026–1036 (Pineridge Press, U.K., 1993).

[38] M. Morandi Cecchi, A. Pica and E. Secco, 'Domain Decomposition of a Finite Element Model of a Lagoon', in the 8th International Conference on Finite Elements in Fluids: Ed. by K. Morgan *et al.*, pp. 990–998 (Pineridge Press, U.K., 1993).

[39] M. Morandi Cecchi, A. Pica and E. Secco, 'A Study of The Venice Lagoon by Domain Decomposition of a Finite Element Model', in the 9th International Conference on Finite Elements in Fluids: New Trends and Applications, Venezia 1995, Ed. by M. Morandi Cecchi, pp. 1417–1426 (SM, Padova, 1995).

[40] J. Donea, 'A Taylor–Galerkin method for convective transport problems', Int. J. Numer. Meth. Eng. **19** 101–119 (1984).

[41] W. L. Wood and O. C. Zienkiewicz, 'A unified set of single step algorithms. Part 1: General formulation and applications', Int. J. Numer. Meth. Eng. **20**, 1529–1552 (1984).

[42] G. Di Silvio, 'Modelli matematici per lo studio della propagazione di onde lunghe e del trasporto di materia nei corsi d'acqua e nelle zone costiere' in Equazioni Differenziali dell'Idrologia e dell'Idraulica, Ed. by G. Casadei, G. Gambolati, C. Minnaja, pp. 11–41 (Patron, Bologna, 1979).

[43] L. D'Alpaos and G. Di Silvio, 'Le correnti di marea nella Laguna di Venezia, parte I: Misure di velcità col metodo aerofotogrammetrico, parte II: Utilizzazione dei risultati per la taratura dei modelli matematici' (Ministero dei Lavori Pubblici, 1979).

[44] C.N.R. Grandi Masse, 'Foglio accompagnatorio dei

rilevamenti digitalizzati delle profondità nella laguna di Venezia' (C.N.R. Grandi Masse, Venezia, 1970).

[45] M. Morandi Cecchi, A. Pica and E. Secco, 'Projection Method for Shallow Water Equations', Int. J. Numer. Meth. in Fluids, **25**, 1–15 (1997).

[46] M. Morandi Cecchi, L. Salasnich, 'Shallow Water Theory and its Application to the Venice Lagoon', Comp. Meth. in Appl. Mech. and Eng. (1997), to appear.

[47] M. Morandi Cecchi, S. De Marchi, E. Secco, 'Un Modello Idrodinamico per lo Studio del Moto della Laguna di Venezia', Atti del Convegno per la Conclusione del Progetto Sistema Lagunare Veneziano (1997), to appear.

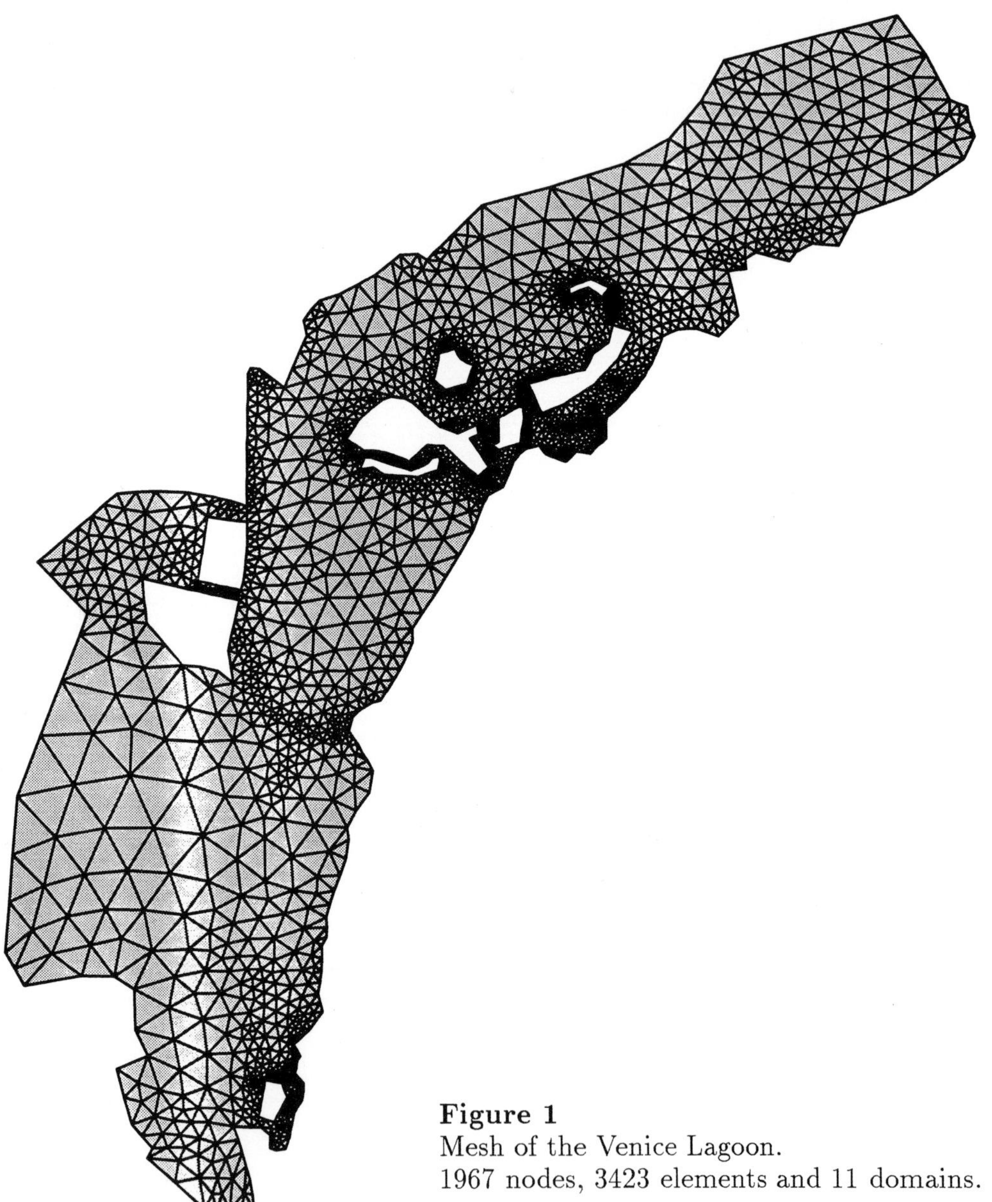

Figure 1
Mesh of the Venice Lagoon.
1967 nodes, 3423 elements and 11 domains.

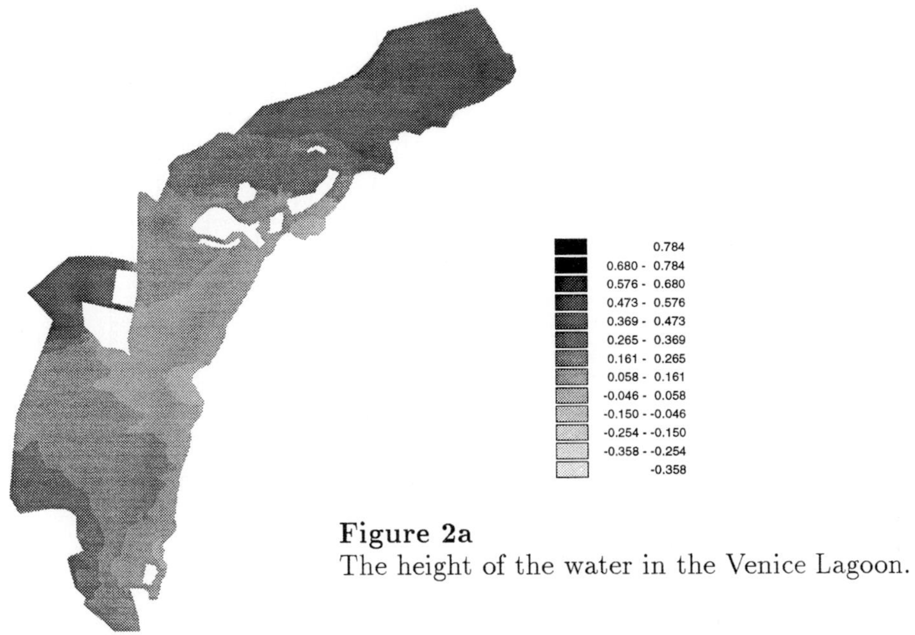

Figure 2a
The height of the water in the Venice Lagoon.

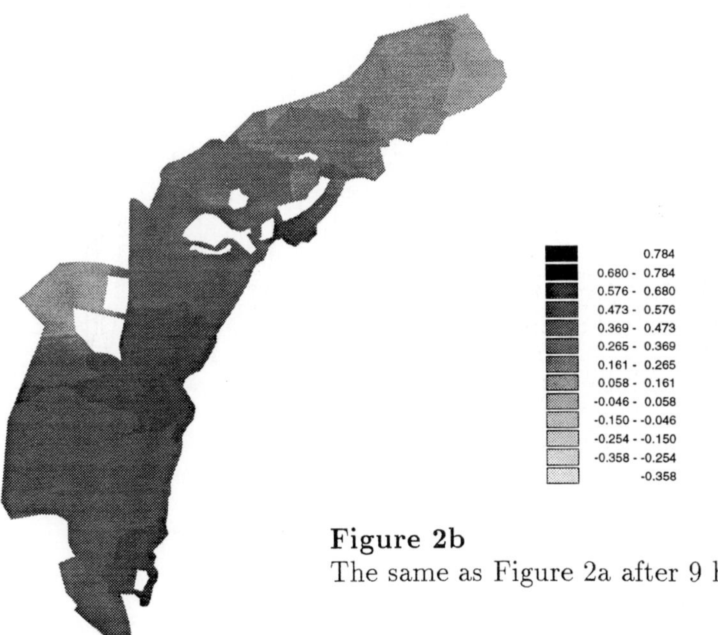

Figure 2b
The same as Figure 2a after 9 hours.

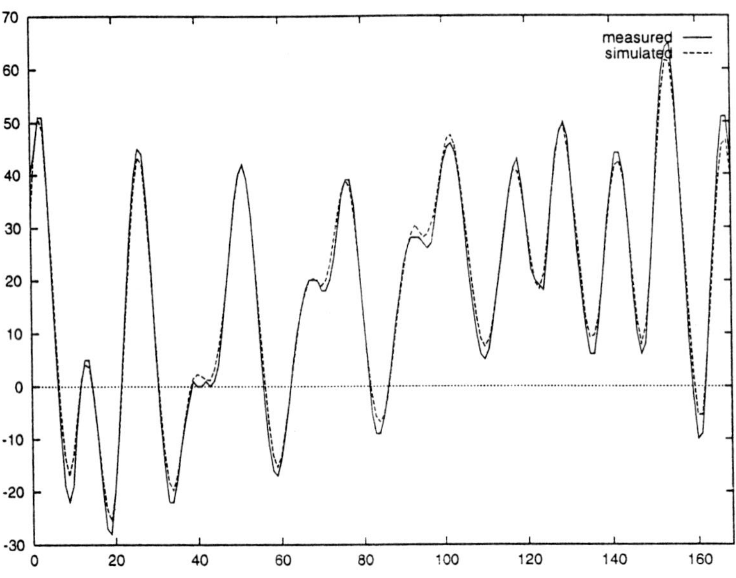

Figure 3a
Field data of Burano island for the first week.
Measured (continuous line) *vs* simulated (dotted line).

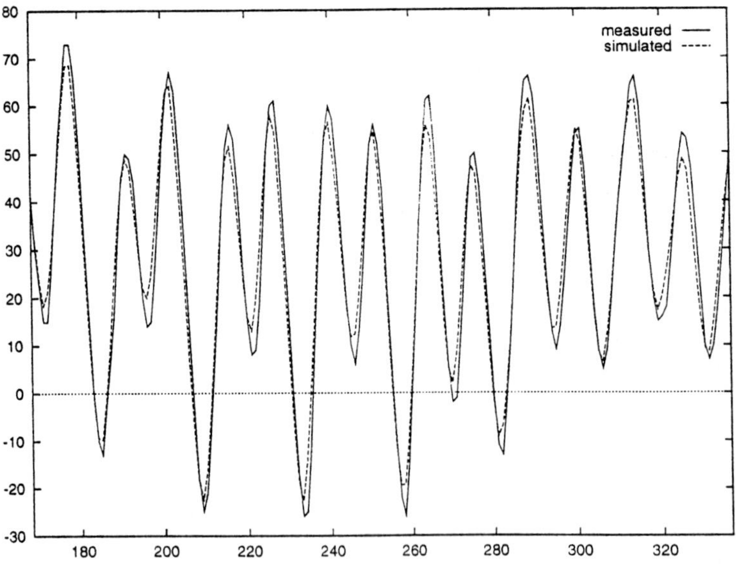

Figure 3b
Field data of Burano island for the second week.
Measured (continuous line) *vs* simulated (dotted line).

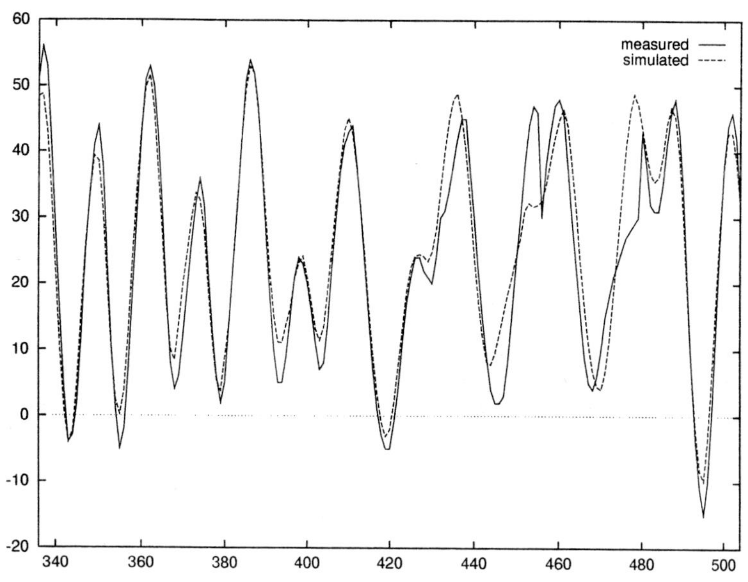

Figure 3c

Field data of Burano island for the third week.
Measured (continuous line) *vs* simulated (dotted line).

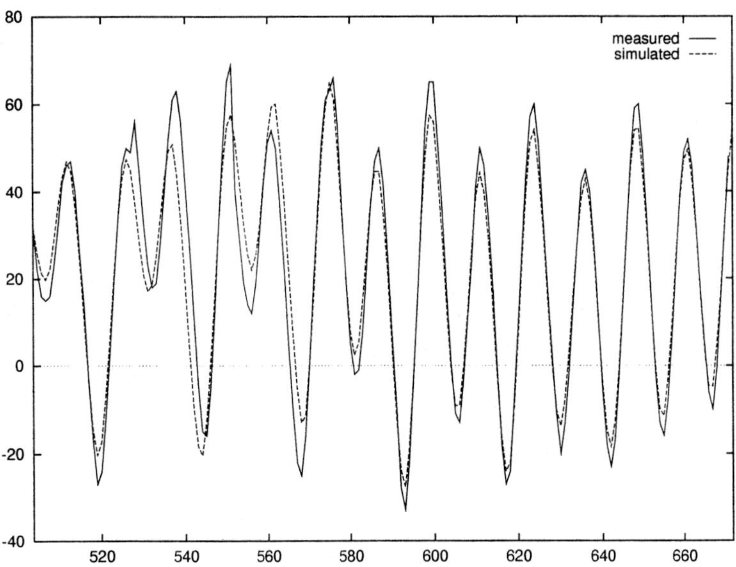

Figure 3d

Field data of Burano island for the forth week.
Measured (continuous line) *vs* simulated (dotted line).

NUMERICAL SIMULATIONS OF MULTIPHASE FLOWS

Yoichiro MATSUMOTO [1]   **Shu TAKAGI** [2]

(1) Department of Mechanical Engineering, The University of Tokyo,
7-3-1 Hongo, Bunkyo-ku Tokyo 113, Japan
(2) Department of Mechano-Aerospace Engineering, Tokyo Institute of
Technology, 2-12-1 Ohkayama, Meguro-ku Tokyo 152, Japan

Abstract

Numerical simulations of multiphase flows are generally classified into two categories. One is direct numerical simulation(DNS) of all phases and the other is simulation with the multiphase flow models like two fluid model. Recent progress of computer performance and numerical algorithms enable us to conduct DNS of relatively complicated multiphase flows. In the present paper, some of the techniques recently developed for tracking the interfaces are reviewed and examples of numerical simulations are shown. These results show quantitatively good agreements with experimental ones. It is also shown that macro-scale model of bubbly flows, which is constructed precisely taking micro scale phenomena like the motion of bubbles into account, can predict the transient phenomena and instabilities of the multiphase flows in a good accuracy.

1 Introduction

Multiphase flows are consisted of two or more phases of gas, liquid and solid. There are many kinds of combination of phases, such as gas-liquid, gas-solid, liquid-solid, gas-liquid-solid, etc. All of these are called multiphase flow if each phases interacts through some hydrodynamical effects. In the present paper, gas-liquid two-phase flow is considered. Even in the gas-liquid two phase flow, the flow patterns are very complicated and they are classified into the following types; bubbly flow, slug flow, froth flow, annular flow and mist flow. When the volumetric fraction of gas, which is called void fraction, is small, gas phases are dispersed such as bubbles in liquid continuum phase. On the other hand, when the void faction is large, liquid phases are dispersed such as droplets in gas continuum phase. These are called bubbly flow and mist flow respectively. Between these two extreme types, slug, froth and annular flows are observed. In a slug flow, bubbles that have larger size than the characteristic length of the flow, like the diameter of pipe, is observed and those bubbles are separated by the liquid slug that contains many small bubbles. In a froth flow, the interface of large bubble becomes very turbulent. In an annular flow, the liquid phase flows along the pipe wall and the gas phase flows in the center area separately.

Received on February 20, 1998.

Since the flow patterns in gas-liquid two-phase flow are such complicated, it is nearly impossible to analyze the "detail" structure of the flows with one numerical scheme. There are some numerical codes(Liles et al.[1985], Ransom et al.[1985]) which are developed in connection with nuclear reactor safety and can treat the changes of these flow patterns. However, the governing equations of these models contain some empirical relations whose physical meanings are not clear and they are not suitable to analyze the detailed structure of multiphase flow. Here, our interest is focused on the detail structures related to bubbly flows and the characteristics of bubbly flows are mentioned more in detail.

When the pressure rise, gas changes its volume considerably, however, liquid hardly changes its volume. The gas density is only order of one thousandth of liquid one. The liquid has much larger heat capacity than the gas. Those characteristics cause the complicated features to multiphase flows. For example, the speed of sound in bubbly liquid changes very largely depending on the void fraction and the speed can be very low. This is because the bulk modulus of bubbly liquid becomes close to that of gas phase and its density becomes close to that of liquid phase. Due to this effect, there is a wide distribution of Mach number in bubbly flows which have wide distribution of void fraction. For the simulation of these types of flows, a robust numerical method for wide distributions of Mach number must be applied. This is just one example of characteristic behaviors of bubble flows.

In general, in bubbly flows, there are the micro scale phenomena which are governed by each bubble motion, the mezzo scale phenomena which are governed by the time and space averaged structure in the local area, and the macro scale phenomena like large scale flow structures which are governed by the micro and mezzo scale phenomena. If the characteristic time of a local phenomenon is longer than that of main phenomenon, a relaxation phenomenon is observed when the state changes from one to the other. For example, since the characteristic time of thermo-fluid phenomena in bubble motion is rather long, long relaxation phenomena are observed in the wave propagation in bubbly liquid.

These multi-time and -length scale structure are one of the significant feature of the multiphase flows. It is not difficult to imagine that numerical simulations of the flows with such large variations of the scales is very difficult. Therefore, it is often the case that bubbly flow models are used for the numerical simulations with some assumptions for the simplification. However, recent progress of computer performance and numerical algorithms enable us to conduct direct numerical simulations of relatively complicated multiphase flows. These results give a lot of information such as forces acting on each bubble, detail structures of the flow field etc. In the present paper, some of the techniques recently developed for tracking the interfaces are reviewed and examples of numerical simulations are shown. It is also shown that the well-established bubbly flow model can predict the transient phenomena and instabilities of the flow in a good accuracy.

2 Modeling of Multiphase Flow

Considering the numerical simulations of multi-phase flow, one of the essential difficulties comes from the jump of some physical quantities at the interface such as viscosity, density, pressure etc. Although a single phase flow is differentiable in everywhere as far as fluid is assumed to be a continuum, it is not the case of multi-phase flow and there exists a discontinuity of physical properties at the interface. Due to the existence of this discontinuity, basic equations for entire flow field are not differentiable in a classical sense. However, if each phase is considered individually and the interfaces are treated as boundaries of each phase, the governing equations are expressed in differential form. In this case, the conservation equations of mass, momentum and energy for each phases are expressed as follows.

Mass Conservation Equation:

$$\frac{\partial \rho_k}{\partial t} + \frac{\partial}{\partial x_i} \rho_k u_{ki} = 0 \qquad (1)$$

Momentum Conservation Equation:

$$\frac{\partial \rho_k u_{ki}}{\partial t} + \frac{\partial}{\partial x_j} \rho_k u_{kj} u_{ki} = -\frac{\partial p_k}{\partial x_i} + \frac{\partial \tau_{kij}}{\partial x_j} + \rho_k f_{ki} \qquad (2)$$

Energy Conservation Equation:

$$\frac{\partial}{\partial t} \rho_k \left(e_k + \frac{1}{2} u_{ki} u_{ki} \right)$$

$$+ \frac{\partial}{\partial x_j} \rho_k \left(e_k + \frac{1}{2} u_{ki} u_{ki} \right) u_{kj}$$

$$= \frac{\partial}{\partial x_j} \left(-p_k \delta_{ij} + \tau_{kij} \right) u_{ki} + \rho_k f_{ki} u_{ki} - \frac{\partial q_{kj}}{\partial x_j}$$

$$(3)$$

Boundary conditions on a discontinuous interface are obtained from the limit operation of conservation equations in each phase on the interface. They are expressed as follows.

B.C. for Mass:

$$[\rho (u_i - u_{si}) n_i] = 0 \qquad (4)$$

B.C. for Momentum:

$$[\rho u_i (u_j - u_{sj}) n_j + p n_i - \tau_{ij} n_j] = F_{si} \qquad (5)$$

B.C. for Energy:

$$\left[\rho \left(e + \frac{1}{2} u_i u_i \right) (u_j - u_{sj}) n_j + p n_j u_j \right.$$

$$\left. - \tau_{ij} n_j u_i + q_j n_j \right] = F_{si} u_i \qquad (6)$$

The exact solutions of multi-phase flows can be obtained if the set of equations mentioned above are solved with proper initial and boundary conditions. However, in general, it is extremely difficult to obtain the numerical results of these equations, because the motion of interface is very complicated and it sometimes includes a topological change of the flow field by breaking up or coalescence of free surfaces. Therefore, as is the same as turbulent flow, it is often the case that so-called averaged equations are used to simulate these types of complicated flow. There are several types of modeling for gas-liquid two phase flows. The brief explanation of these models are shown below.

(A) Two Fluid Model:

Gas and liquid phases are treated individually and the governing equations are expressed for each phase.

A set of averaged equations is closed by the constitutive equations which express the relations of mass, momentum and energy transfer at the interface.

(B) Mixture Model:

The governing equations are constructed assuming multiphase flows as one mixture. The averaged equations are closed by the correlation expressions for velocity and temperature differences between the phases. Mixture model is classified into the following three types related to a treatment of slip velocity between gas and liquid phases.

Homogeneous Flow Model:

Non-slip velocity is assumed between gas and liquid phases. This is the simplest model and is frequently used for numerical analysis.

Slip Model:

Velocity ratio of gas and liquid phases are expressed by a slip ratio and it requires a proper correlation expression for the slip velocity.

Drift Flux Model:

Drift velocity is defined as the difference between each phase flux and mixture's flux. The model is constructed using this drift velocity and it requires one of the equations for drift velocities as a constitutive equations.

The detail procedures of averaging for two-fluid model and mixture model are shown in Drew[1983] and Kataoka[1990]. Here, two fluid model which is considered to be suitable to analyze detail behaviors of multi-phase flows are shown. Conducting an ensemble average to mass, momentum and energy equations for each phases, and expressing the averaged value by $\bar{}$, conservation equations for phase k are expressed as follows.

Mass Conservation Equation:

$$\frac{\partial}{\partial t}\left(\alpha_k \overline{\rho_k}\right) + \frac{\partial}{\partial x_i}\left(\alpha_k \overline{\rho_k u_{ki}}\right) = \Gamma_k \quad (k = g, l) \quad (7)$$

Momentum Conservation Equation:

$$\frac{\partial}{\partial t}\left(\alpha_k \overline{\rho_k u_{ki}}\right) + \frac{\partial}{\partial x_j}\left(\alpha_k \overline{\rho_k u_{kj} u_{ki}}\right)$$

$$= -\alpha_k \frac{\partial \overline{p}}{\partial x_i} + \frac{\partial}{\partial x_j}\left(\alpha_k \overline{\tau_{kij}}\right) + \alpha_k \overline{\rho_k f_i}$$

$$+ M_{ki} \quad (k = g, l) \quad (8)$$

Energy Conservation Equation:

$$\frac{\partial}{\partial t}\overline{\alpha_k \rho_k \left(e_k + \frac{1}{2}u_{ki}u_{ki}\right)}$$

$$+ \frac{\partial}{\partial x_j}\overline{\alpha_k \rho_k \left(e_k + \frac{1}{2}u_{ki}u_{ki}\right) u_{kj}}$$

$$= \frac{\partial}{\partial x_j}\overline{\alpha_k(-p\delta_{ij} + \tau_{kij})u_{ki}} + \alpha_k \overline{\rho_k f_{ki} u_{ki}}$$

$$- \frac{\partial}{\partial x_j}\left(\alpha_k \overline{q_{ki}}\right) + E_k \quad (k = g, l) \quad (9)$$

Boundary conditions at the interface are expressed as follows.

B.C. for Mass:

$$\Gamma_k = \overline{-\rho_{ks}\left(u_{ksi} - u_{si}\right)n_{ki}a_s} \quad (10)$$

B.C. for Momentum:

$$M_{ki} = \overline{-\left\{\rho_{ks}u_{ksi}\left(u_{ksj} - u_{sj}\right)n_{kj}\right.}$$

$$\overline{\left. + p_{ks}n_{ki} - \tau_{ksij}n_{kj}\right\}a_s} \quad (11)$$

B.C. for Energy:

$$E_k = \overline{-\left\{\rho_{ks}\left(e_{ks} + \frac{1}{2}u_{ksi}u_{ksi}\right)\left(u_{ksj} - u_{sj}\right)n_{kj}\right.}$$

$$\overline{\left. + p_{ks}n_{kj}u_{ksj} - \tau_{ksij}n_{kj}u_{ksi} + q_{kj}n_{nj}\right\}a_s}$$

$$(12)$$

In each model, these quantities are expressed as functions of some averaged physical quantities. Farther, in addition to the fluxes transferred between phases, shear stress and heat transfer quantity on the wall must be given as boundary conditions for the averaged equations. These constitutive equations have large effects to the numerical results and lots of correlation expressions are proposed for them. However, satisfactory results are not always obtained and the researches are still going on.

As is mentioned above, in many cases, motions of gas-liquid interface are very complicated and cannot be solved by direct numerical simulations. However, recent development of numerical schemes for free surface flow makes it possible to perform a direct numerical simulations of these flows in some cases. These numerical methods show very high possibility of much wider applications in the near future. They are briefly reviewed and the simulation examples are shown in the following sections.

3 Numeracal Simulations of Free Surface Flow

3.1 Brief review of the numerical method

Flow with a change of boundary shape is often observed in many industrial situations, such as flow around a rotor blade, in an elastic tube or multi-phase flow. It is generally difficult to obtain a stable and accurate solutions of these flows due to the existence of moving boundary. In particular, in case of free surface flow, this difficulty becomes more serious than that in other moving boundary flows where the boundary shape and velocity are given as boundary conditions. This is because boundary shapes and velocity are decided as a part of solution in free surface flows and this generally makes the solutions unstable.

One of the popular methods to treat a free surface flow is Boundary Element Method (BEM). This method is very powerful if governing equations are linear (potential flow, Stokes flow). For a gas-liquid interface, since vorticity generation on the interface are much smaller than that on a rigid wall, potential flow solutions often show good agreement with experimental results. In this method, original differential equations are transformed to boundary integral equations by using the Green functions and motions of the free surface are obtained by solving these integral equations numerically. One of the most important advantages of this method is a reduction of dimensions for governing equations. That is, two dimensional problems become one dimensional and three dimensional problems become two dimensional. Therefore, as far as governing equations are linear, this method affords the higher accuracy and less time consuming with less computational memory, compared with solving differential equations in the same problem. In case of Stokes flow, numerical treatment becomes a little complicated and is discussed in Pozrikidis[1992]. For the simulation of a bubble in liquid, most of the numerical studies were based on the potential flow theory in the past decades (Oguz & Prosperetti[1993], Boulton-Stone & Blake[1993], etc.).

Talking about a rising bubble in a liquid, not so many studies have been conducted with Navier-Stokes(N.S) equation. The simulation of a rising bubble is more difficult to obtain a numerically stable solution than that of a horizontal free surface. This is because a bubble surface constitutes a closed surface in a liquid and it deforms under the existence of static pressure gradient. One of the most famous studies with N.S. eq. is the one done by Ryskin & Leal[1984]. They used a boundary-fitted coordinate system to express the bubble surface. Although their results show very good agreements with the former experimental data and give a lot of knowledge about a separation behind a bubble, the method is restricted only for the steady state. Takagi & Matsumoto[1995] developed a similar method for unsteady three dimensional motions and give a knowledge about an bubble behavior in the simple shear flow and the path instability of a rising bubble(spiral or zigzag motion). Although this method is good for the analysis of a single bubble, it is not easy to implement for bubble-bubble or bubble-wall interaction problems.

Recent progress of the computer performance and numerical algorithms produce some new techniques to treat free surface flows with a rectangular grid system. Here in this paper, theses methods are reviewed briefly and examples of numerical simulations using one of these methods are shown later.

3.2 Free surface solver with rectangular grid system

The first numerical solver for free surface is the famous MAC (Marker And Cell) method developed by Harlow & Welch[1965]. In this method, free surface position at new time step is found by tracking the motions of marker particles which is spread out on the surface at the former time step. After some modified versions of MAC method, Hirt & Nichols[1981] developed different types of method which is called VOF (Volume of Fluid) method, and modified versions of this method are frequently used these days. Recently lots of new methods are proposed and some of these methods are very powerful to be able to treat not only the two phase flow but also many types of moving boundary problems. In these methods, motions of gas-liquid interfaces are expressed by an advection of some indicator function in a fixed rectangular grid system. This function takes different values in each phases, for example, 0 in liquid phase and 1 in gas phase. The methods using these indicator functions are generally called front capturing method. In the sense that volume fraction of gas phase plays a role of indicator function in VOF method, VOF method can be considered as one of the front capturing methods. In this case, since the indicator function is directly related to physical property of volume fraction, the accuracy of mass conservation is excellent from the numerical point of view.

In general, mass conservation is not guaranteed in front capturing methods and the accuracy of mass conservation is decided by an accuracy of solving an advection equation for indicator function. Hence, it is desired that stable and accurate schemes with small numerical viscosity are used for solving the advection equation. Adoption of CIP (Takewaki & Yabe[1987]) or TVD methods for solving the advection equation comes from this requirement.

There is a method called Level Set Method(Sussman et al.[1994], Chang et al.[1996] and Zhao et al.[1996]). This method is originally developed by Osher & Sethien[1988] in UCLA and is getting very popular in many field. This method can handle not only the free surface flow but also many physical and manufacturing process such as solidification problems, etching processes etc. This method is similar to general front capturing methods except that a level set function which is smooth distance function is used to express the interface instead of the indicator functions which have very sharp gradient at the interface. The advantage of level set method is that the levels set function doesn't require too much attention for the advection of interfaces. This is because level set function takes a value of distance from the interface. It takes the value 0 at the interface and linearly increase with distance from the interface. On the other hand, as is mentioned already, the general indicator functions have very large gradient at the interface, where the functions change their value from 0 to 1. They require accurate scheme to advect the interface with less numerical viscosity. In this sense, the level set method is better than other front capturing methods. However, a level set function loses a characteristic of distance function after the advection of the function. Hence, it requires so-called reinitialization procedure at each time step to keep it as a distance function. The original reinitialization procedure by Sussman et al.[1994] is only for keeping a level set function as a distance function. This reinitialization procedure has an accumulation of small numerical error for mass conservation. Later, Chang et al.[1996] developed new reinitilization procedure to satisfy mass conservation exactly.

There is another type of methods which is developed by Unverdi & Tryggvason[1992]. This is called a front tracking method and it has a mixed feature of front capturing and MAC methods. In this method, gas-liquid interfaces are expressed explicitly by a group of segments in 2D and that of triangle elements in 3D. Since the interfaces are expressed explicitly, they are tracked in Lagrangian way and the method basically doesn't have problems of numerical viscosity which arises from the advection of indicator function. However, the method has an idea of indicator function to decide the viscosity and density field. That is, the jumps of viscosity, density and surface tension at the interface are described by smoothed Heaviside and Delta functions by a grid scale and numerical stability is accomplished by introducing them.

In these methods that employ indicator functions in rectangular grid systems, interface forces such as a surface tension are expressed as source terms in Navier-Stokes equation. In case of VOF or general front capturing methods, gas-liquid interface is expressed by a steep function and this gives a difficulty to obtain an accurate solutions of surface shape and surface curvature. Therefore, many studies (Brackbill et al.[1992], Ashgriz & Poo[1991], Lafaurie et al.[1994], etc.) have been done on this problem.

In the following sub-sections, two examples of numerical simulations using the methods mentioned above are shown. One is the simulations using boundary-fitted coordinate system and the other is by the front tracking method(Unverdi & Tryggvason[1992]) which employs rectangular grid system. In both cases, rising bubble problems are treated as ex- amples of simulations.

3.3 Simulations using boundary-fitted coordinate system

3.3.1 Numerical Method

Assumptions:

The following assumptions are used for the simulations.

(1) The surface tension coefficient of the gas-liquid interface is constant. Therefore, the liquid is assumed to be free of impurities and surfactants.

(2) The liquid phase is the incompressible Newtonian fluid.

(3) The density and viscosity of the liquid phase is sufficiently large compared with those of the gas. Hence, the dynamic pressure and viscous stress of the gas phase is negligible compared with those of the liquid.

(4) The volume of the rising bubble is constant. In real situations, the rising bubble changes its volume slightly because of the existence of the static pressure gradient. But this will be quite small for the gas bubble rising under atmospheric pressure.

Since our purpose is to simulate the unsteady motion of the gas bubble rising through the liquid which has the high density ratio of O(1000), these are the reasonable assumptions .With the similar assumptions, Ryskin & Leal[1984] and Takagi et al.[1995] got very good agreement with the experiment, although their calculations are restricted for the axisymmetric steady cases.

Solution Algorithm:

Generalized curvilinear coordinate system is used with the unsteady grid movement. Therefore, the grid is regenerated at each time step. The solution algorithm for axisymmetric simulation is shown here first.

(1) Assign a certain initial condition for the bubble shape and rising velocity. In the present study, a spherical bubble at rest is placed in a quiescent liquid.

(2) Generate an orthogonal grid such that a bubble surface coincides with one of the boundaries in the computational space.

(3) Full Navier-Stokes and continuity equations, which are expressed in the orthogonal curvilinear coordinate system, are solved numerically by the SIMPLER algorithm with the proper boundary conditions under the generated grid.

(4) Move the grid point on the bubble surface in a Lagrangian way, using the calculated velocity in process(3), and decide the shape of bubble at a new time step.

(5) Repeat processes (2)-(4).

For three dimensional unsteady motions, following modifications are made.

(M-1) 3-D grid is generated by expanding 2-D grid around the axis. The generated grids in 2-D and in 3-D are shown in Fig.1(a),(b).

(M-2) Basic equations are formulated for non-orthogonal three-dimensional grid systems.

Grid Generation:

Three dimensional grid is generated by the connection of two dimensional grids as is explained above. The way of generating two dimensional grid is the one developed by Duraiswami & Prosperetti[1992]. They developed the method to generate the orthogonal grid system using a Laplace equation expressed in curvilinear space. The basic advantage of their method is capability of grid density control by introducing a distortion function

Basic Equations:

Since the liquid phase is assumed to be Newtonian and incompressible, full Navier-Stokes equation is solved by SIMPLER method. Basic equations for axisymmetric flow are shown below.

Continuity Equation

$$\frac{1}{h_\xi h_\eta r} \left[\frac{\partial}{\partial \xi}(h_\eta r u_\xi) + \frac{\partial}{\partial \eta}(h_\xi r u_\eta) \right] = 0 \qquad (13)$$

where,

$$h_\xi = \sqrt{\left(\frac{\partial x}{\partial \xi}\right)^2 + \left(\frac{\partial r}{\partial \xi}\right)^2}, \; h_\eta = \sqrt{\left(\frac{\partial x}{\partial \eta}\right)^2 + \left(\frac{\partial r}{\partial \eta}\right)^2} \qquad (14)$$

Momentum Equation in ξ-direction

$$\frac{\partial u_\xi}{\partial t} + \frac{1}{h_\xi h_\eta r} \left[\frac{\partial}{\partial \xi}(h_\eta r u_\xi^2) + \frac{\partial}{\partial \eta}(h_\xi r u_{xi} u_\eta) \right]$$

$$= -\frac{1}{\rho_l}\frac{1}{h_\xi}\frac{\partial p}{\partial \xi} + \frac{1}{h_\xi}\frac{\partial x}{\partial \xi}g$$

$$+\nu \frac{1}{h_\xi h_\eta r} \left[\frac{\partial}{\partial \xi}\left(\frac{h_\eta r}{h_\xi}\frac{\partial u_\xi}{\partial \xi}\right) \right.$$

$$\left. + \frac{\partial}{\partial \eta}\left(\frac{h_\xi r}{h_\eta}\frac{\partial u_\xi}{\partial \eta}\right) \right] + S_\xi \qquad (15)$$

where, S_ξ comes by employing contravariant velocity components and is expressed as follows.

$$S_\xi = \frac{u_\eta^2}{h_\xi h_\eta}\frac{\partial h_\eta}{\partial \xi} - \frac{u_\xi u_\eta}{h_\xi h_\eta}\frac{\partial h_\xi}{\partial \eta}$$

$$-\nu \frac{1}{r^2 h_\xi}\frac{\partial r}{\partial \xi}\left(\frac{u_\xi}{h_\xi}\frac{\partial r}{\partial \xi} + \frac{u_\eta}{h_\eta}\frac{\partial r}{\partial \eta}\right)$$

$$+\nu \left[\frac{1}{h_\xi}\frac{\partial}{\partial \xi}\left(\frac{u_\eta}{h_\xi h_\eta}\frac{\partial h_\xi}{\partial \eta}\right) \right.$$

$$\left. -\frac{1}{h_\eta}\frac{\partial}{\partial \eta}\left(\frac{u_\eta}{h_\xi h_\eta}\frac{\partial h_\eta}{\partial \xi}\right) \right]$$

$$-\nu \frac{1}{h_\xi h_\eta}\left[\frac{\partial h_\xi}{\partial \eta}\frac{1}{h_\xi}\frac{\partial u_\eta}{\partial \xi} - \frac{\partial h_\eta}{\partial \xi}\frac{1}{h_\eta}\frac{\partial u_\eta}{\partial \eta} \right]$$

$$-\nu \frac{1}{h_\xi^2 h_\eta^2}\left[u_\xi\left(\frac{\partial h_\xi}{\partial \eta}\right)^2 + u_\xi\left(\frac{\partial h_\eta}{\partial \xi}\right)^2 \right] \qquad (16)$$

Momentum equation in η-direction is obtained by interchanging ξ and η.

Time derivative term:

Since the grid movement is treated in the present study, the time derivative term in physical space can be separated into two parts. One is the time derivative term in computational space, the other is the term due to grid movement. Compared to the case using Cartesian components, additional terms exist for the present variables because of the basis change in time and space due to grid movement.

Boundary Conditions:

The deformed bubble is treated here with the surface tension effect. Therefore, the stress free condition is imposed in the tangential direction along the surface and the condition of the stress balance among the pressure, normal viscous stress and the surface tension force is imposed in the normal direction.

3.3.2 Results and discussions

The following problems about a translating bubble are treated with this numerical technique.

(1) Axisymmetric motions

(2) Zigzag and spiral motions

(3) Effect of mean shear

(4) History force effect

(5) Effect due to an acceleration of a surrounding liquid

The results of theses problems are briefly shown below.

(1) Axisymmetric Motions:

Numerical simulations of a bubble rising steadily in a quiescent liquid are often used to examine the accuracy and stability of the numerical scheme. The results obtained by the present method are shown in

Fig.2 with the comparison of experimental and numerical results obtained by other's. Simulation conditions are shown in Table 1. It is shown that present numerical results show very good agreements with other results in the wide range of Reynolds($0 < Re < 200$) and Weber ($1 < We < 20$) number.

(2) Zigzag and Spiral Motions:

It is known experimentally that a bubble starts showing an unsteady three-dimensional (spiral or zigzag) motion beyond a certain size, although a small size bubble rises straightforward in a quiescent liquid. This phenomena can be explained by vortex sheddings from a bubble surface with increasing Reynolds number. However, in case of a bubble on which a tangential stress free condition is imposed, vortex shedding from a bubble surface, which is observed in rigid particle cases, never occurs if bubble keeps spherical shape. This makes the problem much more complicated. In Fig. 3,4, numerical results of axisymmetric and three dimensional simulations are shown in the condition that zigzag or spiral motions are observed experimentally. Numerical results of axisymmetric simulations show that a shape oscillation occurs in these conditions and three dimensional simulations show that zigzag motion occurs after a shape oscillation growth. Therefore, our numerical results suggest that the growth of axisymmetric shape oscillation causes three-dimensional motions of a rising bubble.

(3) Effect of mean shear:

A deformable bubble behavior in the existence of mean shear is discussed here. Schematic figure of the present problem is show in Fig.5. It is theoretically predicted in low Reynolds number that a rigid spherical particle in a simple shear flow moves in the region of higher slip velocity (left direction in Fig.5). However, experimental observation by Kariyazaki[1987] showed that deformed bubble moves in the other direction (right direction in Fig.5). Numerical simulation was performed on this problem in a wide range of Reynolds number ($2 < Re < 100$). The results reveal that bubble starts moving in the other direction with increasing the deformation as shown in Fig.6(a). However, in case of low Reynolds number, this tendency is quite small and bubble moves to smaller slip velocity region mainly due to the deformation effect (Fig.6(a)). This is because lift force comes form the inertia effect and a spherical particle or bubble in a stokes flow cannot feel the lift force because of the kinematic reversibility.

(4)History force effect:

History force is one of the topics many researchers are working on recently. Takagi & Matsumoto[1996] conducted numerical simulation for a spherical bubble rising unsteadily in a quiescent liquid and investigated the history force effect. The numerical method have a minor change of algorithm for keeping a spherical shape. One of the main conclusions we obtained is that history force effect is negligible for a bubble rising beyond $Re > 50$.

(5) Effect due to an acceleration of a surrounding liquid:

A bubble can feel the force by an acceleration of a surrounding liquid. This effect is investigated by introducing a periodic body force in momentum equations. Numerical results reveal that bubble behaviors are classified into four types as shown in Fig.7, depending on the ratio of deformation time scale to that of translational and forcing.

3.4 Simulations by the front tracking method

3.4.1 Numerical Method

This method treats a multi-phase flow as one fluid which has jumps of density, viscosity and pressure at the interface. The momentum equations are expressed as follows.

$$\frac{\partial \rho \boldsymbol{u}}{\partial t} + \nabla \cdot (\rho \boldsymbol{u}\boldsymbol{u}) = -\nabla p + \nabla \cdot \mu(\nabla \boldsymbol{u} + \nabla \boldsymbol{u}^T)$$

$$+ \int \delta(x - x_f)\sigma \kappa \boldsymbol{n} da. \quad (17)$$

where, the velocity is denoted by $\boldsymbol{u}$, density by ρ, viscosity by μ, pressure by p, gravity by $\boldsymbol{g}$, surface tension by σ, and twice the mean curvature κ.

The surface tension term is included in the momentum equation and the gas and liquid phase are solved simultaneously. The basic difference of the present method compared to the other similar type methods(VOF, Level Set, CIP...) is the advection of the density and viscosity field. The present method explicitly track the interface. The physical properties at the interface are transported with the interface, instead of solving Eq.(18). This prevents the numerical diffusion coming in for the advection of the density and viscosity. Because of this explicit tracking of the interface, this method also enables treating the two nearly parallel interfaces in the same cell, which cannot be treated in VOF type method.

$$\frac{\partial \phi}{\partial t} + \boldsymbol{u} \cdot \nabla \phi = 0 \quad (18)$$

Assumptions:

The following assumptions are used for the simulations.

(1) Both gas and liquid phases are incompressible Newtonian.

(2) Surface tension coefficient is constant.

<u>Solution Algorithm:</u>

(1) Decide a density and viscosity field by solving a Poisson equation for Indicator function.

(2) Solving Navier-Stokes equation in a rectangular grid system. (Surface tension effect is included in N.S. equation.

(3) Surface elements are transferred in Lagrangian way and a reconstruction of the elements are given to have nearly the same size.

(4) Repeat the process (1)-(3) with time development.

In the process(1), density and viscosity at each grid point is decided as follows.

$$\rho(x) = \rho_0(x) + (\rho_b(x) - \rho_0(x))I(x) \qquad (19)$$

$$\mu(x) = \mu_0(x) + (\mu_b(x) - \mu_0(x))I(x) \qquad (20)$$

where, $I(x)$ is called Indicator function which take the value 0 in a liquid phase and 1 in a gas phase. Poisson equation for Indicator function is derived as follows.

First, the gradient of Indicator function is expressed as follows.

$$G(x) = \nabla I = \sum_j D(x - x^{(l)})n^{(l)}\Delta s^{(l)} \qquad (21)$$

Where, $D(x)$ is a distribution function which is corresponding to the smoothed delta function. Taking a divergence of this equation, the following Poisson equation for the indicator function is obtained.

$$\nabla^2 I = \nabla \cdot G \qquad (22)$$

This equation is solved numerically and the value of indicator function at each grid point is decided.

3.4.2 Results and discussions

From the numerical simulations using boundary-fitted coordinate system, it was found that a bubble rising through an uncontaminated liquid behaves very much like a bubble(particle) in potential flow beyond $Re = 50$. That is, potential flow approximation is valid as far as a single bubble behavior is analyzed. It is noted that this characteristic is quite different from the rigid particle case in which vortex shedding occurs in high Reynolds number. Therefore, it is often the case that potential flow approximations are used for analysis of bubble cloud behavior. Under this approximation, Sangani & Didwania[1993] conducted a direct numerical simulation of rising bubble clouds in potential flow. It was found from this simulation that bubble clouds in potential flow form a plane clustering structure in horizontal direction. However, such a clustering structure is not observed in real experiments and lots of discussions arises on this phenomenon. Here, to obtain information on this phenomenon, bubble-bubble interaction problems are investigated using Tryggvason's front tracking method. The simulations are conducted for a bubble rising in a moderate Reynolds numbers ($1 < Re < 100$) and the factors which can break up the plane clustering structure are discussed.

Two bubbles are initially spherical and start rising due to a buoyancy force with interacting each other. The initial configuration of two bubbles are shown in Fig.8 and simulation conditions are shown in Table 2. As shown in Fig.8, four patterns of initial configurations are selected for the simulations. Although the flow field is three dimensional and we adopted three dimensional Navier-Stokes equation for the simulation, initial positions of two bubble are set on the same plane perpendicular to horizontal plane

Relative trajectories of geometric center of bubble2 to that of bubble1 in CaesA is shown in Fig.9. Although bubble-bubble interaction can cause a three dimensional motion of bubbles, the geometric centers of bubbles are moving in the same plane as initial bubbles are set. It is shown in the figure that two bubbles set in a side-by-side position(CaseA-1) migrate mainly in horizontal direction and keep a stable position in the perpendicular direction. In CaseA-2, which is the case that two bubbles are set in the direction inclined 30 degree from horizontal plane, bubbles keeps very stable positions and relative distance doesn't change such as no interaction exists. On the other hand, once a bubble comes to underneath of the upper bubble like CaseA-4, the bubble are much affected by the upper bubble. This is due to a strong wake effect of the upper bubble. Simulation results of CaseB and CaseC are shown in Fig.10 and Fig.11 respectively. CaseB is corresponding to the simulation of $Re = 50 - 70$ with large deformation and CaseC is that of $Re = 3$ with spherical shape. It is shown that the lower bubble is attracted to the upper one in high Reynolds number case, which is against the prediction of potential flow theory. This shows that wake effect is not negligible in high Reynolds number case because the strong wake region is constructed with the increase of bubble deformation.

Time history of two bubble interactions in Case A-4 is shown in Fig.12. Since the interface is tracked in an explicit manner, the present method can treat two parallel interfaces in one cell. Therefore, as is shown in Fig.12, it is possible to simulate the situation that bubbles repel each other once after bubbles come to nearly attaching.

4 Bubbly Flow Simulations by Macro-Scale Model

Bubbly flows are observed in various industrial processes and sometimes the flow is a turbulent one. The particle-laden flows were simulated by considerable number of researchers (Shuen et al.[1985], Squires et al.[1991], Wang et al.[1993], Elghobashi et al.[1993], etc.) , where the influence of a turbulence on dispersed particles were treated by Lagrangian-tracking method. In these papers, the effect of two-way interaction was considered only for the momentum. However, bubbly flows have usually large volume fraction of dispersed phase compared to the particle-laden flows, so that it is required to take into account both mass and momentum interactions between two phases.

General descriptions of turbulent energy and dissipation for gas-liquid two phase flow were proposed by Kataoka & Serizawa[1989]. The set of equations for the Two-Fluid model was proposed by Liles et al.[1986] based on the local volume average. Zhang & Prosperetti[1994] proposed an equation system of bubbly flow based on the ensemble averaging technique. However those system equations have many unknown correlation terms not only as turbulence flow but also as two phase flow. Also the phenomenological modeling and experimental measurements are much more difficult than those for single phase turbulent flows.

Here, we employ the volume averaged conservation equations for mass and momentum of bubbly media. Each bubble motion is tracked by using the equation of translational motion of bubble in an infinite liquid. The HSMAC method (Hirt et al.[1972]) and the CIP scheme (Takewaki & Yabe[1987]) are adopted as the numerical solver for the system equations and the finite difference scheme for the advection terms respectively. In the present paper, the typical characteristics of three dimensional turbulent structure in bubbly two-phase flow driven by bubble plume are shown (Murai & Matsumoto[1995]).

4.1 Assumptions

Governing equations for bubbly flow used in the present study are formulated by using the following assumptions:

(1) Bubble size is smaller than the characteristic length of the flow.

(2) The volume of the bubble is in equilibrium with the surrounding pressure.

(3) Liquid phase is an incompressible Newtonian. Compressibility of bubbly flow is due to the volume change of the bubbles.

(4) Coalescence and fragmentation of bubbles are neglected since the void fraction of bubbly flow is less than 0.1 for the present simulation.

(5) Gas inside the bubble is non-condensable and obeys the perfect gas law. Also there is no mass diffusion of the gas at the bubble interface. Therefore the amount of gas inside the bubble is constant.

4.2 Governing equations

Conservation Equation of Liquid Phase:

$$\frac{\partial f_L \rho_L}{\partial t} + \frac{\partial f_L \rho_L u_{Li}}{\partial x_i} = 0. \tag{23}$$

Conservation Equation of Gas Phase:

$$\frac{\partial f_G \rho_G}{\partial t} + \frac{\partial f_G \rho_G u_{Gi}}{\partial x_i} = 0. \tag{24}$$

In this study, a bubble is treated as a point source and its volume is much smaller than the computational grid. In a general volume averaging method, the high-frequency noise is induced when a bubble passes over a cell boundary because the bubble motion in a grid is not able to be captured. Here, weighted averaging method is employed to calculate a void fraction in order to avoid such a numerical oscillation. A linear effective function ϕ, which indicates 1 at the grid center and 0 at the neighbor in an each direction, is adopted as the weighted function. The volume fractions are expressed as

$$f_G = \frac{1}{V_{cell}} \sum_i^{x_{Gi} \in V_{eff}} \phi(x_{Gi}) V_{Gi}, \qquad f_L = 1 - f_G, \tag{25}$$

where V_{cell} is a computational grid volume and V_{eff} is an effective volume of a bubble.

Conservation of Momentum in bubbly flow:

$$\frac{\partial f_L \rho_L u_{Li}}{\partial t} + \frac{\partial f_L \rho_L u_{Li} u_{Lj}}{\partial x_j}$$

$$+ \frac{\partial f_G \rho_G u_{Gi}}{\partial t} + \frac{\partial f_G \rho_G u_{Gi} u_{Gj}}{\partial x_j}$$

$$= -\frac{\partial p}{\partial x_i} + \frac{\partial}{\partial x_j}\left(\mu \frac{\partial u_{Li}}{\partial x_j}\right) + \frac{1}{3}\frac{\partial}{\partial x_i}\left(\mu \frac{\partial u_{Lj}}{\partial x_j}\right)$$

$$+ (f_L \rho_L + f_G \rho_G) g_i, \tag{26}$$

where μ is an effective viscosity due to disperse bubbles (Batchelor[1967]) and it is expressed as

$$\mu = \left[1 + f_G \left\{\frac{\mu_L + (5/2)\mu_G}{\mu_L + \mu_G}\right\}\right] \mu_L. \tag{27}$$

Equation of Translational Motion of a Bubble:

The various forces on a bubble which correspond to added mass, pressure, drag, lift and buoyancy are considered to formulate.

$$\frac{d}{dt}(\rho_G V_G u_{Gi}) + \frac{d}{dt}(\beta \rho_L V_G u_{Gi})$$

$$-\frac{D_L}{Dt}(\beta \rho_L V_G u_{Li}) = -V_G \frac{\partial p}{\partial x_i}$$

$$+V_G \left\{ \frac{\partial}{\partial x_j}\left(\mu \frac{\partial u_{Li}}{\partial x_j}\right) + \frac{1}{3}\frac{\partial}{\partial x_i}\left(\mu \frac{\partial u_{Lj}}{\partial x_j}\right) \right\}$$

$$-\frac{1}{2}\rho_L \pi r_G^2 C_D u_s (u_{Gi} - u_{Li})$$

$$-\rho_L V_G C_L \varepsilon_{ijk}(u_{Gj} - u_{Lj})\left(\varepsilon_{klm}\frac{\partial u_{Lm}}{\partial x_l}\right)$$

$$+\rho_G V_G g_i, \tag{28}$$

where $u_s = |u_{Gi} - u_{Li}|$ and $V_G = 4\pi r_G^3/3$. the added mass coefficient β, the substantial time derivative of the liquid phase D_L/Dt, the drag coefficient C_D, the lift coefficient C_L and the bubble Reynolds number Re_s are written in following equations.

$$\beta = \frac{1}{2},$$

$$\frac{D_L}{Dt} = \frac{\partial}{\partial t} + u_{Li}\frac{\partial}{\partial x_i},$$

$$C_D = \max\left\{\frac{16}{Re_s}, \frac{48}{Re_s}\left(1 - \frac{2.21}{\sqrt{Re_s}}\right)\right\}, \quad Re_s < 200,$$

$$C_L = \frac{1}{2},$$

$$Re_s = \frac{2r_G u_s}{\nu},$$

where drag and lift forces on a bubble are employed the theoretical formulas in a pure liquid. The C_D is expressed by the mixed formula of Stokes and potential flows (Moore[1959]). The C_L is expressed by Auton[1987]'s formula.

Position of Geometric Center of a Bubble:

$$X_{Gi}(t) = X_{Gi}(0) + \int_0^t u_{Gi}(t)dt. \tag{29}$$

Equilibrium Equation for Bubble Volume:

$$p_{G0}V_{G0} = \left(p_G - \frac{1}{4}\rho_L u_s^2\right)V_G. \tag{30}$$

Kinematic Condition of the Free Surface:

$$\frac{\partial h}{\partial t} + u_{fi}\frac{\partial h}{\partial x_i} = u_{ki}, \tag{31}$$

where h is the height at a free surface, u_f is liquid velocity on the free surface and the suffix k is a perpendicular component.

Normal Stress Balance on the Free Surface:

$$p_f = p_o - \sigma \nabla_f^2 h + 2\mu_L \frac{\partial u_{ki}}{\partial x_{ki}}, \tag{32}$$

where x_{f1} and x_{f2} are horizontal components and ∇_f^2 is expressed as

$$\nabla_f^2 = \frac{\partial^2}{\partial x_{f1}^2} + \frac{\partial^2}{\partial x_{f2}^2}.$$

4.3 Results and discussions

The flow driven by buoyancy force acting on bubbles are called bubble plume. In the present study, turbulent structures of bubble plume in a rectangular tank are numerically analyzed using the bubbly flow model.

Plume behavior in bubble plume:

The bubble plume has various flow patterns depending on tank shape, void fraction, and bubble size. The experimental observations have been reported by several researchers (Hussain et al.[1976], McDouall [1978], Alam et al. [1993], etc.). Here, the rectangular tank in a vertically long shape is used for the present simulation. The size of tank is 400mm in height and 100mm in width. Bubbles are injected from the square plate at the center of the bottom of tank. The size of bubble injection region is 20x20 mm^2. Density and kinematic viscosity of liquid are 10^3kg/m^3 and 5×10^{-6}m^2/s respectively. It is noted that bubble plumes in this condition have wide variations of characteristic length scales such as tank size, plume size and bubble size.

The developing bubble plume structures at 30 seconds after bubble injections are shown in Fig.13 . It is recognized from the figure that the structures of bubble plume are strongly affected by bubble radius and void fractions(=gas volume fractions). Spiral structure of bubble plume occurs under the condition of void fraction; f_G=0.05 and bubble radius; R=1.0mm. This spiral structure is caused by the circulating component of the ambient flow that comes into the neighborhood of bubble injection area. The spiral structure grows with flowing upwards, because bubbles are widely distributed in radial direction due to the lift force acting on bubbles.

When bubble radius is small (R=0.20mm), the smooth spiral structure vanishes. Instead, the core of bubble plume shrinks around the center axis and shows a strong meandering behavior due to the local shear stress. In the case of $R=0.5$mm, transient structures between the spiral and the meandering are observed.

Increasing the void fraction, bubble clouds drive an ambient liquid to flow strongly and the bubble plume becomes to have large fluctuations. Under the condition of $f_G=0.1$ and $R=1.0$mm, wave lengths of the spiral structure become shorter and bubbles spread out widely in radial direction. It is considered that this tendency comes from the following two effects. One is the diffusion of bubble number density by the spiral motion and the other is the lift force effect in a strongly rotating flow field. In the case of $R=0.2$mm, the bubble plume is largely distorted by shear stress and it shows three-dimensional complicated behaviors. In the upper region of the flow field, since smaller bubbles have less slip velocity, many tiny bubbles remains stagnantly near the free surface and the flow field itself also becomes stagnant.

Although no former reports are available on these phenomena, we observed these structures in our experiments in which bubble injector was carefully designed to keep an uniformity of bubble size. The observed results show quantitatively good agreement with the present numerical results. These complicated behaviors of bubble plume come from the effect of mass and momentum interactions between both phases. These phenomena can not be simulated by one-way method in which a liquid phase is not affected by gas phase. They can be simulated only by using two-way coupling methods which we employed.

Turbulence Fluctuation:

Fig.14(a), (b) shows the horizontal velocity fluctuations of liquid phase inside/outside the bubble plume at the three different locations in vertical direction, which are at 0.2, 0.45 and 0.7 of the tank height from the bottom. The results are corresponding to the condition of $f_G=0.05$ and $R=0.5$mm and 1.0mm. The measuring period is from $T=30$ to 40 sec. It is seen from the figure that fluctuations are lager inside the plume and that fluctuating frequency is higher at the higher location. This is because pseudo-turbulence generated by bubble motions plays an important role on the fluctuation inside a plume and the global structure of flow field dominates the phenomena outside a plume. In the case of $R=1.0$mm, fluctuation period becomes longer than the case of $R=0.5$mm. This is because the spiral structure has a larger fluctuation period than the meandering structure does. The pseudo-turbulence becomes more distinguished in the case of $R=1.0$mm and the fluctuation is no longer negligible compared with the entire fluctuations.

About the pseudo-turbulence, Lance & Bataille [1991] experimentally measured liquid jets containing small bubbles by Laser Doppler Velocimetry. Although their experiment is not same as the present flow field, their results show similar velocity fluctuations to our numerical results. These pseudo-turbulence are the essential characteristics of dispersed flows in which volumes of individual dispersed phase are not negligible.

5 Concluding Remark

Numerical simulations of multiphase flows are discussed, especially concerning about gas-liquid two phase flows. The numerical schemes recently developed for free surface flows are briefly reviewed and examples of the direct numerical simulations are shown. It is revealed that the results of direct numerical simulations show quantitatively good agreement with those of experiments in many cases. The numerical model of bubbly flow is also constructed taking micro scale phenomena like the motion of bubbles into account. It is revealed that the numerical model can predict the transient phenomena and instabilities of the multiphase flows in a good accuracy.

Acknowledgment

Authors thank to Mr.K.Sugiyama for his help in preparation of the manuscript.

REFERENCES

[1] Alam, M. and Arakeri, V.H., *J. Fluid Mech.*, **254**, 363, (1993)

[2] Ashgriz, N. and Poo,J.Y. "FLAIR: Flux Line-Segment Model for Advection and Interface Reconstruction", *J. Comp. Phys.*, **93**, 449-468, (1991)

[3] Auton, T.R., *J. Fluid Mech.*, **183**, 199, (1987)

[4] Batchelor, G.K., *An Introduction to Fluid Dy-namics*, Cambridge University Press, (1967)

[5] Boulton-Stone, J.M. and Blake,J.R., "Gas bubbles bursting at a free surface", *J. Fluid Mech.*, **254**, 437-466, (1993)

[6] Brackbill, J.U., Kothe, D.B. and Zemach, C., "A Continuuum Method for Modeling Surface Tension", *J. Comp. Phys.*, **100**, 335-354, (1992)

[7] Chang, Y.C., Hou, T.Y., Merriman, B. and Osher, S., "A Level Set Formulation of Eulerian Interface Capturing Methods for Incompressible Fluid Flows", *J. Comp. Phys.*, **124**, 449-464, (1996)

[8] Drew, D.A., "Mathematical modeling of two-phase flow", *Annu. Rew. Fluid Mech.*, **15**, 261-291, (1983).

[9] Duraiswami, R. and Prosperetti, A., "Orthogonal Mapping in Two Dimensions", *J. Comp. Phys.*, **98**, 254-268, (1992)

[10] Elghobashi, S. and Truesdell, G.C., *Phys. Fluids*,

A5(7), 1790, (1993)

[11] Harlow, F.H. and Welch, J.E., "Numerical Calculation of Time-Dependent Viscous Incompressible Flow of Fluid with Free Surface", *Phys. Fluids*, **8**, 2182-2189, (1965)

[12] Hirt, C.W. and Cook, J.L., *J. Comp. Phys.*, **10**, 324, (1972)

[13] Hirt, C.W. and Nichols, B.D. "Volume of Fluid (VOF) Method for the Dynamics of Free Boundaries," *J. Comp. Phys.*, **39**, 201-225, (1981)

[14] Hnat, J.G. and Buckmaster, J.D., "Spherical cap bubbles and skirt formation", *Phys. Fluids*, **19**, 182-194, (1976)

[15] Hussain, N.A. and Siegel, R., *J. Fluids Eng.*, March, 49, (1976)

[16] Kariyazaki, A. "The behavior of an isolated bubble in a uniform shear flow", *Trans. JSME.*, **53-487**, 744-749, (1986)

[17] Kataoka, I. and Serizawa, A., *Int. J. Multiphase Flow*, **15**, N0.5, 843, (1989)

[18] Kataoka, I., "The modeling and basic equations of gas-liquid two phase flow", *Japanese J. Multiphase Flow*, **4-4**, 275-283, (1990)

[19] Lafaurie, B. et al., "Modeling Merging and Fragmentation in Multiphase Flows with SURFER", *J. Comp. Phys.*, **113**, 134-147, (1994)

[20] Lance, M. and Bataille, J., *J. Fluid Mech.*, **222**, 95, (1991)

[21] Liles, D. et al. "TRAC-PF1/MOD1, An Advanced Best-Estimate Computer Program for Pressurized Water Reactor Thermal-Hydraulic Analysis", NUREG/CR-3658 LA-10157-MS, (1985).

[22] Liles, D. et al. "TRAC-PF1/MOD1, An Advanced Best-Estimate Computer Program for Pressurized Water Reactor Analysis", NUREG/CP-3858 LA-10157-MS, (1986).

[23] McDouall, T.J., *J. Fluid Mech.*, **85**, 655, (1978)

[24] Moore, D.W., *J. Fluid Mech.*, **6**, 113, (1959)

[25] Murai, Y. and Matsumoto, Y., "Turbulent Structure of a Bubble Plume", *Proc. of the Int. Symp. on Mathematical Modeling of Turbulent Flows*, 233-238, Tokyo, Japan, (1995)

[26] Oguz, H.N. and Prosperetti, A., "Dynamics of bubble growth and detachment from a needle," *J. Fluid Mech.*, **257**, 111-145, (1993)

[27] Osher, S. and Sethian, J.A., "Fronts Propagating with Curvature-Dependent Speed: Algorithms based on Hammilton-Jacobi formulations", *J. Comp. Phys.*, **89**, 12-49, (1988)

[28] Pozrikidis, C., *Boundary integral and singularity methods for linearized viscous flow*, Cambridge Univ. Press, (1992)

[29] Ransom, V.H. et al. "RELAP5/MOD2 Code Manual, Vol.1, Code Structure, System Models, and Solution Methods", NUREG/CR-4312,

EGG-2796, (1985).

[30] Ryskin, G. and Leal, L.G., "Numerical Solution of Free Boundary Problems in Fluid Mechanics. Part.2: Buoyancy Driven Motion of a Gas Bubble through a Quiescent Liquid," *J. Fluid Mech.*, **148**, 19-35, (1984).

[31] Sangani, A.S. and Didwania, A.K., "Dynamic simulations of flows of bubbly liquids at large Reynolds numbers", *J. Fluid Mech.*, **250**, 307-337, (1993)

[32] Shuen, J-S., et al., *AIAA Journal*, **23**, N0.3, 396, (1985)

[33] Squire, J.S. and Eaton, J.K., *J. Fluid Mech.*, **226**, 1, (1985)

[34] Sussman, M., Smereka, P. and Osher, S., "A Level Set Approach for Computing Solutions to Incompressible Two-Phase Flow", *J. Comp. Pays.*, **114**, 146-159, (1994)

[35] Takagi, S., Prosperetti, A. and Matsumoto, Y., "Drag Coefficient of a Gas Bubble in an Axisymmetric Shear Flow", *Phys. Fluids* **6**, 3186 (1994)

[36] Takagi, S. and Matsumoto, Y., "Three Dimensional Calculation of a Rising Bubble", *Proc. of the 2nd Int. Conf. on Multiphase Flow '95*, **1**, PD2-9 Kyoto, Japan. (1995)

[37] Takagi, S. and Matsumoto, Y., "Force Acting on a Rising Bubble in a Quiescent Fluid", *Proc. of the ASME Summer Meeting on Numerical Method for Multiphase Flow*, Crowe et al., ed., San Diego, CA. (1996)

[38] Takewaki, H. and Yabe, T., *J. Comp. Phys.*, **70**, (1987)

[39] Unverdi, S.O. and Tryggvason, G., "A Front-Tracking Method for Viscous, Incompressible Multi-fluid Flows", *J. Comp. Phys.*, **100**, 25-37, (1992)

[40] Wang, L-P., and Maxey, M.R., *J. Fluid Mech.*, **256**, 27, (1993)

[41] Zhang, D.Z., Prosperetti, A., *Phys. Fluids*, **A6(9)**, 2956, (1994)

[42] Zhao, H.K., Chan, T., Merriman, B. and Osher, S., "A Variational Level Set Approach to Multiphase Motion", *J. Comp. Phys.*, **127**, 179-195, (1996)

Table 1. Calculation conditions and comparison of terminalvelocity with previously reported results

Case	Eo	log Mo	Re	We	Compared Data	Velocity normalized by Compared Data
1	39.4	−1.19	20	15	H&B	1.005
2	116	2.93	2.5	16	B&W	0.962
3	17.0	−4.77	100	10	R&L	0.944
4	1.77	−7.75	200	4	TP&M	1.014

H&B	:	Hnat & Buckmaster[1976].
B&W	:	Bhaga & Weber[1981].
R&L	:	Ryskin & Leal[1984].
TP&M	:	Takagi et al.[1994].

Table 2. Calculation conditions $(\rho_l/\rho_g = 11,\ \mu_l/\mu_g = 10)$

Case	Morton #	Eotvos #
A	0.026	10.0
B	3.3×10^{-5}	10.0
C	0.026	2.0

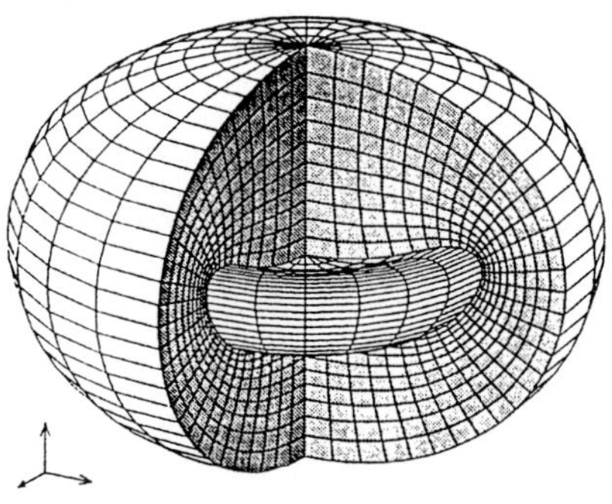

(a) 2-D grid around axis

(b) 3-D grid aound a bubble

Figure 1. Grid and coordinate system

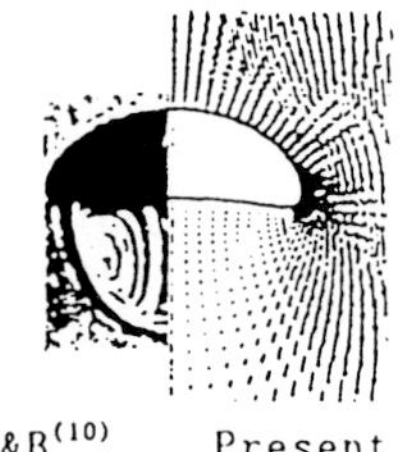

H&B[10]   Present   B&W[11]   Present
(a) Case 1   (b) Case 2

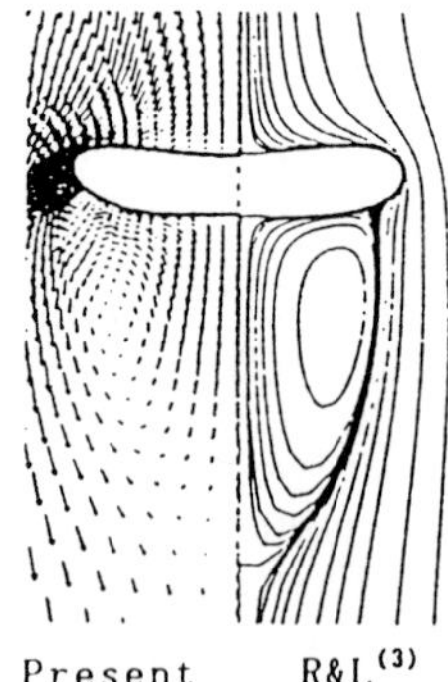

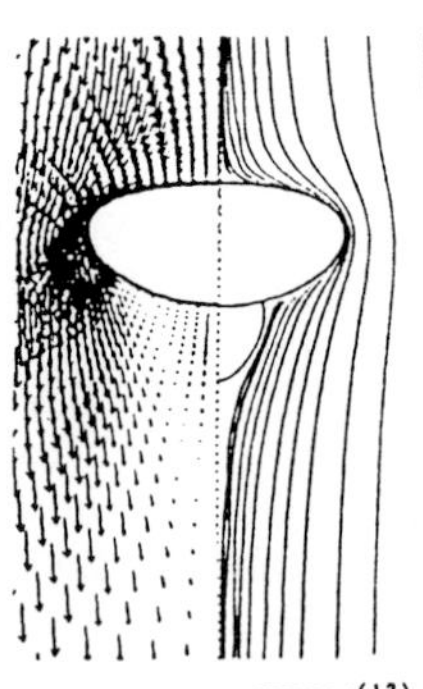

Present   R&L[3]   Present   TP&M[12]
(c) Case 3   (d) Case 4

Figure 2. Comparison of the bubble shape and flow field

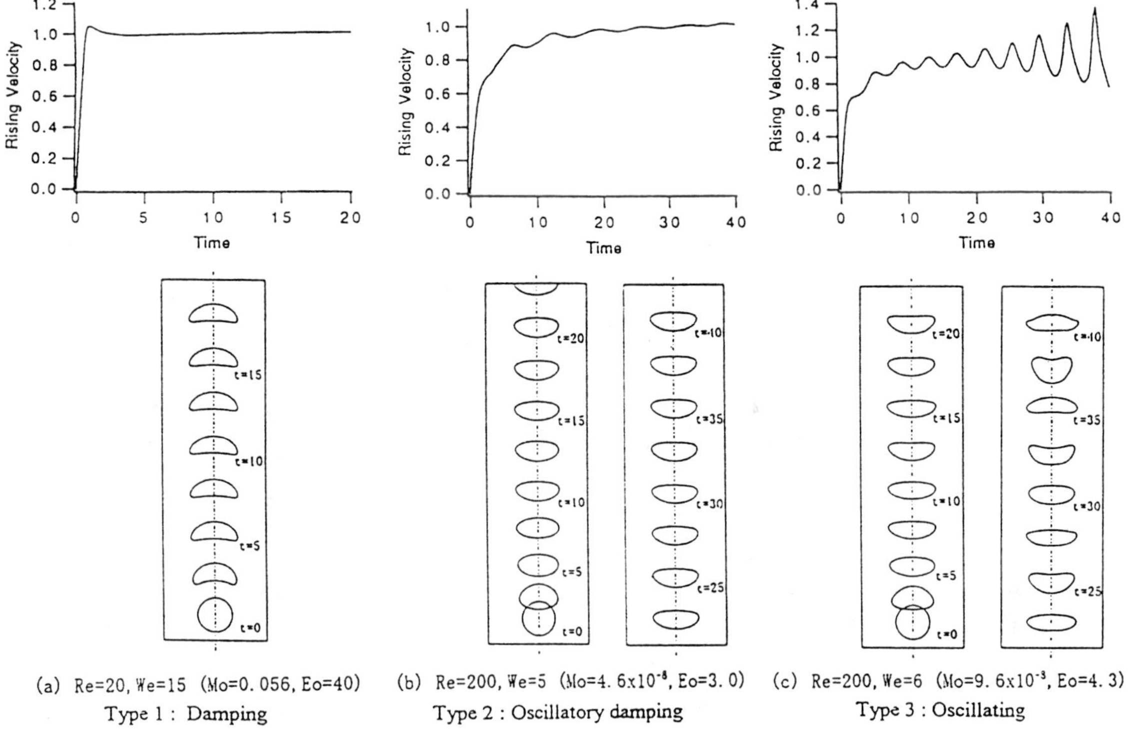

(a) Re=20, We=15 (Mo=0.056, Eo=40)   (b) Re=200, We=5 (Mo=4.6x10⁻⁴, Eo=3.0)   (c) Re=200, We=6 (Mo=9.6x10⁻³, Eo=4.3)

Type 1 : Damping   Type 2 : Oscillatory damping   Type 3 : Oscillating

Figure 3. Unsteady change of the bubble shape and rising velocity

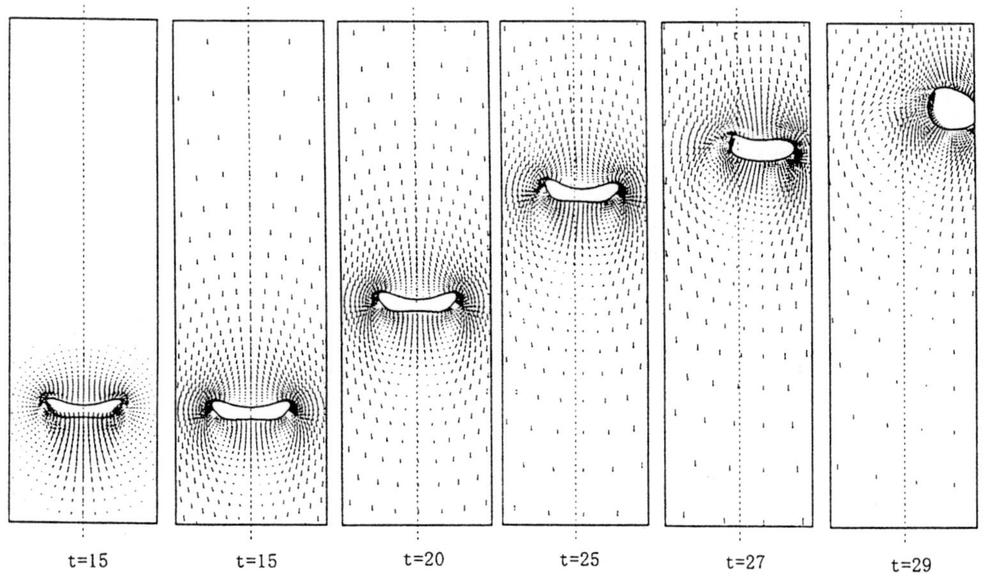

Figure 4. 3-D simulation of the rising bubble

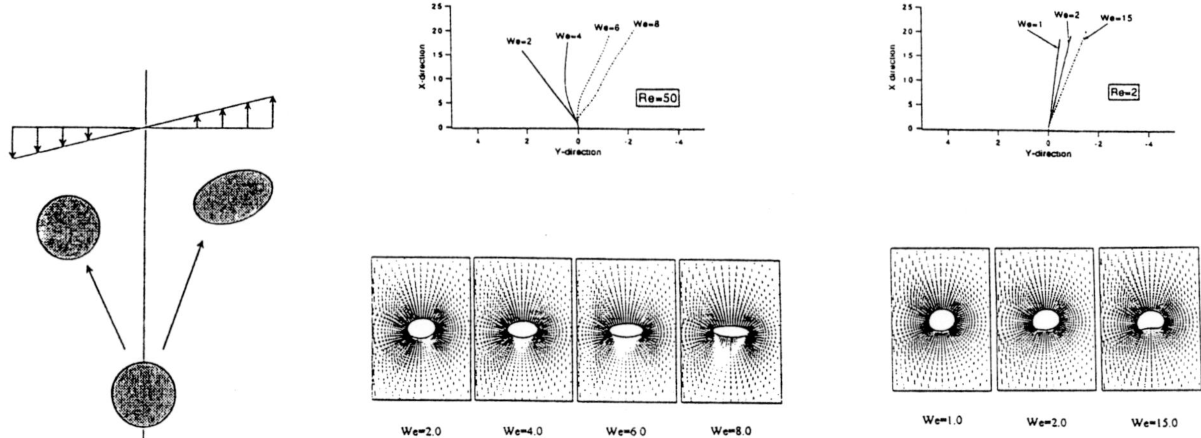

Figure 5. Rising bubble in a simple shear flow

Figure 6. Trajectories and shapes of a bubble in a simple shear flow

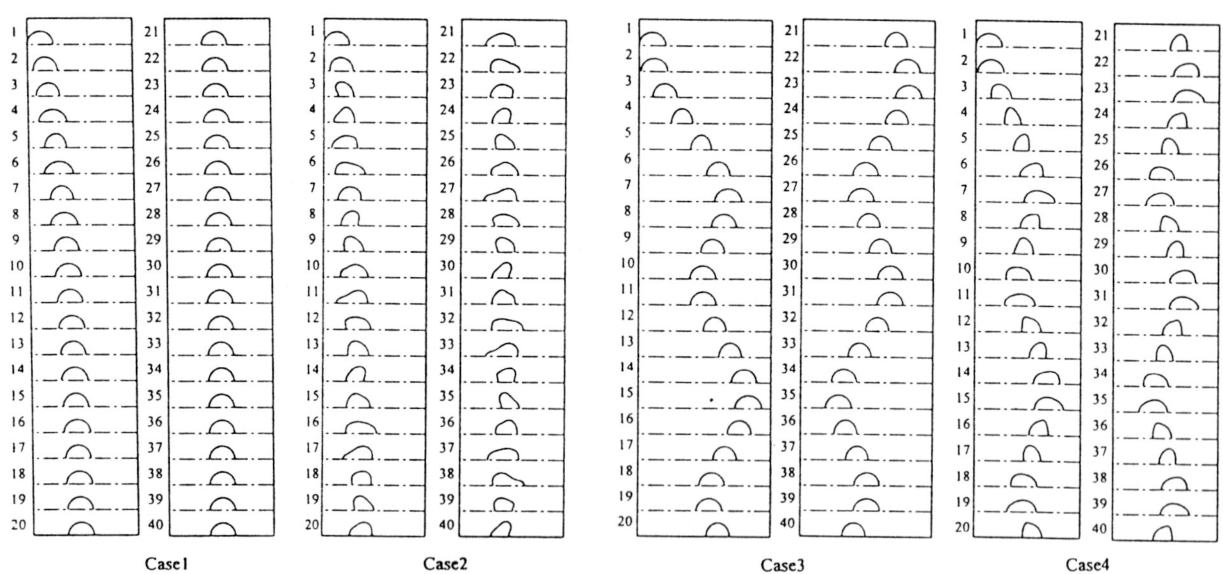

Figure 7. Unsteady change of a bubble shape
(Case 1,2: Small translational motion, Case 3,4: Large Trans. motion)

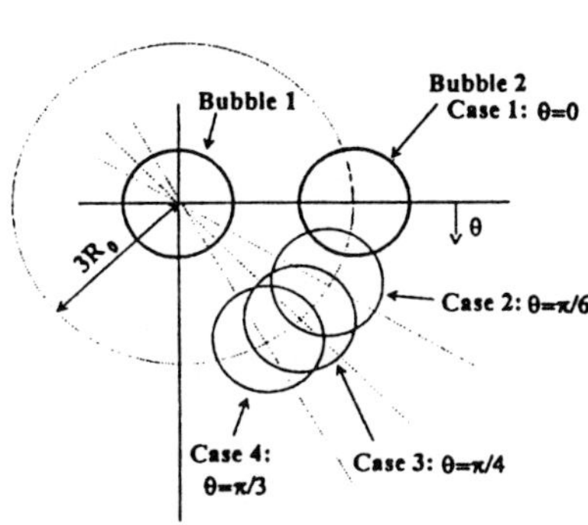

Figure 8.   Initial configulations of two bubbles

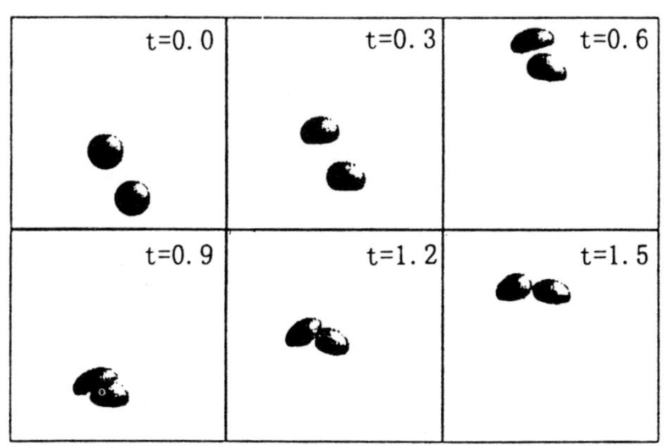

Figure 12 Time sequences of two bubble
interaction

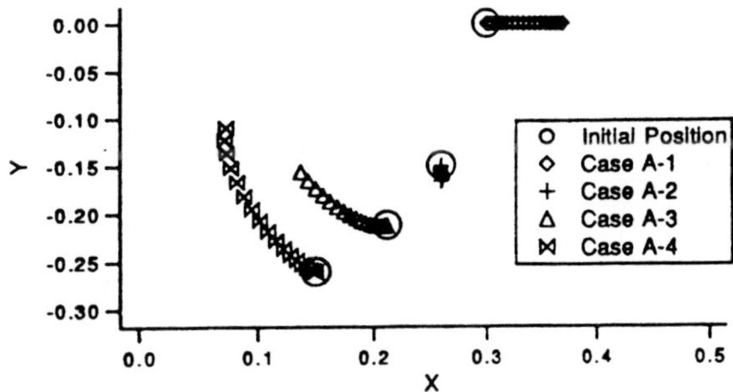

Figure 9.   Trajectories of geometric center
(Case A)

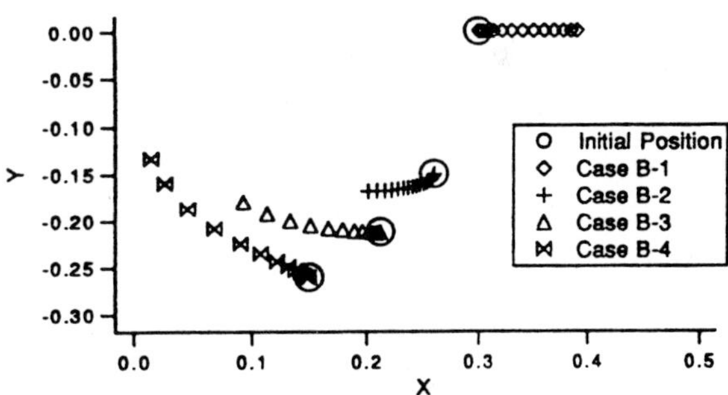

Figure 10.   Trajectories of geometric center
(Case B)

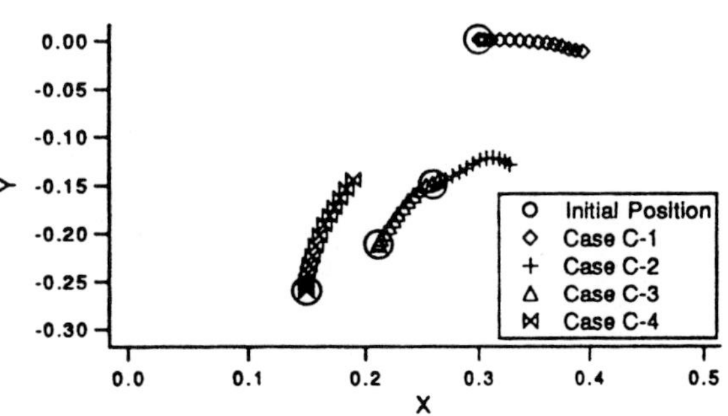

Figure 11.   Trajectories of geometric center
(Case C)

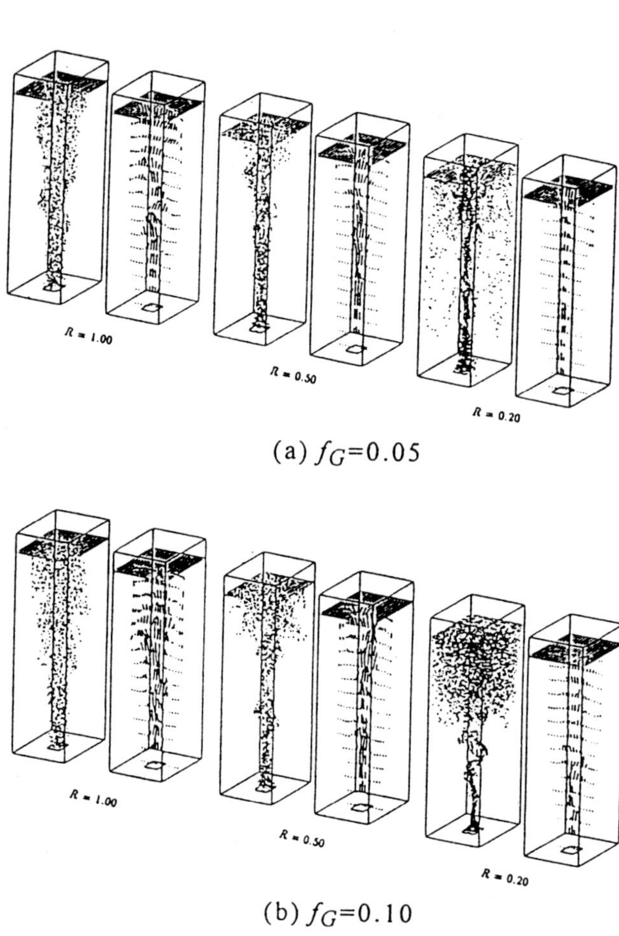

(a) f_G=0.05

(b) f_G=0.10

Figure 13. 3-D structure of bubble plume
(Bubble and velocity distributions)

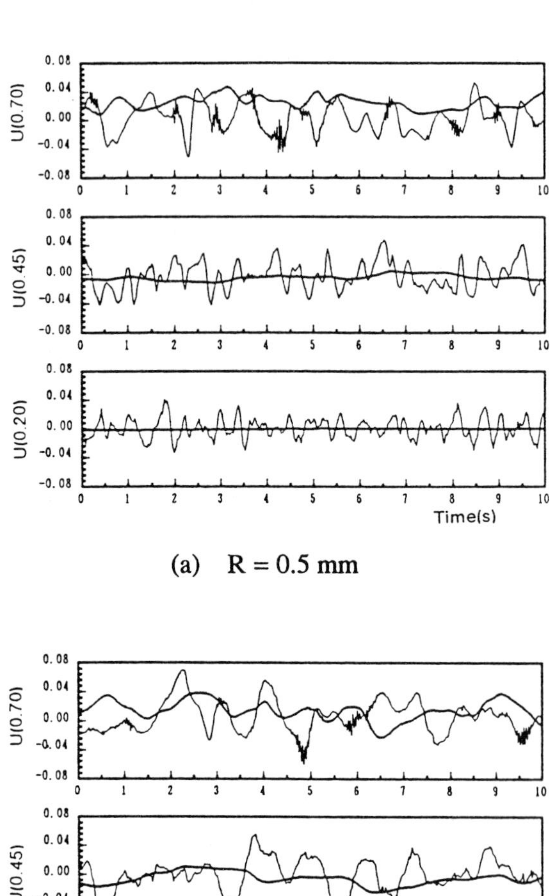

(a)   R = 0.5 mm

(b)   R = 1.0 mm

Figure 14.   Velocity fluctuations at different
height
(L(z=0.20   ),M(z=0.45   ),H(z=0.70   )
Thick(inside), Thin(outside))

UNIFIED SOLVER CIP FOR SOLID, LIQUID AND GAS

T. YABE

Tokyo Institute of Technology; Nagatsuda, Midori-ku, Yokohama 227, Japan
Tel: (81)45 924 5535   Fax: (81) 45 924 5536   e-mail: yabe@es.titech.ac.jp

Abstract

We have succeeded for the first time to simulate dynamic phase transition from metal to vapor. This success is due to the CIP (Cubic-Interporated Pseudoparticle/Propagation) method that can treat solid, liquid and gas together and can trace a sharp interface with almost one grid. The strategy of the CIP method and its various modifications are reviewed including application to laser- induced evaporation, slamming of ship and formation of milk crown in three dimensions.

1. INTRODUCTION

The developing speed of computer hardware is quite fast compared to software technology because of various innovations in basic technology. We need similar innovation to accelerate the developing speed of software. One of them will be a universal solver for computational engineering problems. Recent high technology requires new tools for combined analysis of materials in different phase state, e.g., solid, liquid and gas. A universal treatment of all phases by one simple algorithm is essential and we are at the turning point of attacking this goal .

For these types of problems such as welding and cutting processes, we need to treat topology and phase changes of the structure simultaneously. In freezing, condensation, melting and evaporation, the grid system aligned to the solid or liquid surface has no meaning and sometimes the mesh is distorted and even broken up. To solve these problems with Lagrangean representation in finite difference, finite element and boundary element methods will be quite a challenging task.

Toward this goal, we take Eulerian-approach based on CIP[1-3] method developed by the author which does not need adaptive grid system and therefore removes the problems of grid distortion caused by structural break up and topology change. The material surface can be captured almost by one grid throughout the computation. Furthermore, the code can treat all the phases of materials from solid state through liquid and two phase state to gas without restriction on the time step from high-sound speed.

In this paper, we will give a historical review of the CIP method and its strategy, then give some examples in various research fields.

2 CIP METHOD

2.1 Basic Scheme

In order to attack the problems mentioned above, we must first find a method to solve the interaction of compressible gas with incompressible liquid or solid. For compressible fluid, elaborate schemes like TVD or

---

Received on March 3, 1997.

ENO proved to be quite effective in capturing shock waves. However, since these schemes employ a conservative form of fluid equations, divergence of velocity which becomes zero in the incompressible limit cannot be treated independently of the advection part. On the contrary, incompressible schemes like QUICK or higher-order upwind schemes combined with improved MAC (Marker and Cell) procedure can treat divergence-free fluid vorticity and turbulence. However, these schemes cannot always treat a shock wave as a sharp discontinuity.

We need a scheme to treat both compressible and incompressible fluids simultaneously in one program to simulate the interaction of gas with liquid or solid. Fully implicit solvers can treat this procedure, but the convergence of iteration in highly distorted state is still a problem. Recently, we have proposed a new type of scheme CIP[1,2] to treat shock waves by a non-conservative scheme. By a simple extention, the CIP can be used for both compressible and incompressbile fluids simultaneously. This code is called the CCUP(CIP Combined and Unified Procedure) [3] and can treat incompressible fluid with full hydrodynamic equations. In combination with the surface capturing scheme presented (which will be called "digitizer" hereafter) in a previous paper[4], this scheme provides a useful tool to describe various physical processes which were never attacked.

In this section, we review the CIP method and illustrate its principle. The key issues of the CIP method are in the representation of advection term and splitted treatment of other terms. By this separation, the code can be extensible to compressible and incompressible fluids. Let us first start with a one-dimensional linear advection equation.

$$\frac{\partial f}{\partial t} + u \frac{\partial f}{\partial x} = 0 \tag{1}$$

The solution of Eq.(1) gives a simple translational motion of wave with a velocity u. The initial profile (solid line of Fig.1(a)) moves like a dashed line in a continuous representation. At this time, the solution at grid points is denoted by closed circles and is the same as the exact solution. However, if we eliminate the dashed line as in

Fig.1(b), it is hard to imagine the original profile and it is natural to retrieve the original profile like that shown by solid line in (c). Thus, numerical diffusion arises when we construct the profile by the linear interpolation even with the

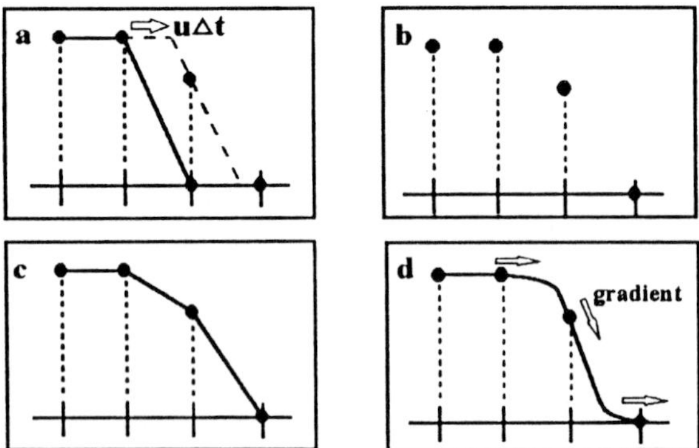

Figure 1. The principle of the CIP method. (a) solid line is initial profile and dashed line is an exact solution after advection, whose solution (b) at discretized points. (c) When (b) is linearly interpolated , numerical diffusion appears. (d) In the CIP, spatial derivative also propagates and the profile inside a grid cell is retrieved.

exact solution as shown in Fig.1(c). This process is the first-order upwind scheme. On the other hand, if we use quadratic polynomial for interpolation, it suffers from overshooting. This process is the Lax-Wendroff scheme or Leith scheme.

What made this solution worse ? This is why we neglect the behavior of the solution inside grid cell and merely follow after the smoothness of the solution. Therefore we should consider how to incorporate the real solution into the profile within a grid cell. We propose to approximate the profile as shown below. Let us differentiate Eq.(1) with spatial variable x, then we get

$$\frac{\partial g}{\partial t} + u\frac{\partial g}{\partial x} = -\frac{\partial u}{\partial x}g \qquad (2)$$

where g stands for the spatial derivative of f, $\partial f/\partial x$. In the simplest case where the velocity u is constant, Eq.(2) coincides with Eq.(1) and represents the propagation of spatial derivative with a velocity u. By this equation, we can trace the time evolution of f and g on the basis of Eq.(1). If g propagates as shown by the arrows in Fig.1(d), the profile after one step is limited to a specific profile. It is easy to imagine that by this limitation, the solution becomes very closer to the initial profile.

If two values of f and g are given at two grid points, the profile between these points can be described by cubic polynomial $F(x)=ax^3+bx^2+cx+d$. Thus, the profile at n+1 step can be obtained shifting the profile by $u\triangle t$ like $f^{n+1}=F(x-u\triangle t)$, $g^{n+1}=dF(x-u\triangle t)/dx$.

$$f_i^{n+1} = a\xi^3 + b\xi^2 + g_i\xi + f_i$$
$$g_i^{n+1} = 3a\xi^2 + 2b\xi + g_i \qquad (3)$$
$$a = \frac{g_i + g_{iup}}{D^2} + \frac{2(f_i - f_{iup})}{D^3}$$
$$b = \frac{3(f_{iup} - f_i)}{D^2} - \frac{2g_i + g_{iup}}{D}$$

where we define $\xi=-u\triangle t$ and $D=-sgn(u)\triangle x$, iup=i-sgn(u), since the upward direction depends on the sign of the velocity(=sgn(u)). An interpolation with f and its derivative is called Hermite spline. However, the key issue of the CIP scheme is in the way of determining the time evolution of spatial derivative. We proposed to determine them from spatial derivatives of Eq.(1) also. Therefore, the profile even within a grid cell   is determined so as to be consistent with the equation.

By repeating Eq.(3), we can get the solution of Eq.(1). One example is shown in Fig.2(a). A small overshooting is unavoidable but does not grow in time. In most of practical calculation, this quality is sufficient. However, for the description of the sharp interface, we need more accurate solution. We proposed two methods to improve the solution.

2.2 Digitizer

In some special cases, one need to treat sharp interface

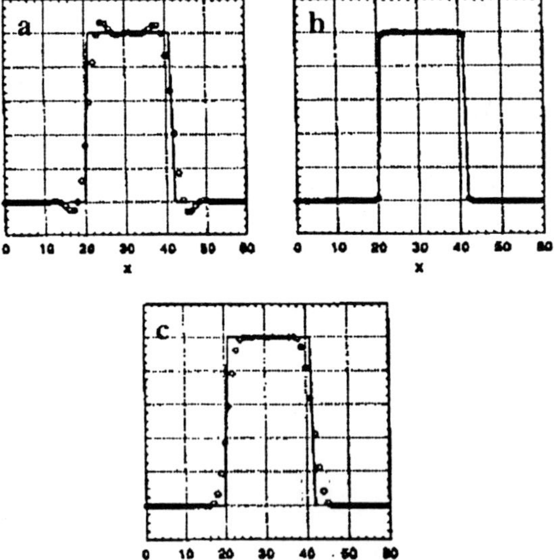

Figure 2. A square wave profile after 1000 time steps with CFL=$u\triangle t/\triangle x$=0.2. The solid line is an analytical solution and symbols are numerical solution. (a) original CIP, (b) transformation with tangent function and (c) rational CIP.

with exactly one grid.   There have been numerous methods proposed for treating interface between two different materials. These methods are divided into two groups. In one group the interface is described by a surface function, while in the other group the interface is defined as surface of a density function [5] such as VOF (Volume of Fluid Method). In the former case, main problem arises from a multi-valued function when the surface is strongly distorted or even broken up. Although this shortcoming will not arise in the latter case, the numerical scheme to describe an evolution of the density function without numerical diffusion is a problem which needs further investigation. The level set function[6] is another interesting example of the latter case and is worthy of further investigation.

Recently, we proposed a simple method [4,7] to treat the density function ϕ with high accuracy in multi-dimensions. For this purpose, we slightly modified the CIP method described above. We proposed to transform ϕ into $F(\phi)$. It is obvious that a new function $F(\phi)$ also obeys the same equation as the density function if it is monotone function of ϕ. We can choose an appropriate function to ensure monotone and sharp discontinuity. In the previous paper[4], we chose the tangent function:

$$F(\phi) = \tan\left[\pi\left(\phi - \frac{1}{2}\right)\right] \qquad (4)$$

and the equation for F

$$\frac{\partial F}{\partial t} + u\frac{\partial F}{\partial x} = 0 \qquad (5)$$

is solved. It should be noticed that the equation to be solved for the spatial derivative is the derivative of Eq.(5) but is not of Eq.(1) because the target equation is now Eq.(5). In all the time steps, only F and its derivative $\partial F/\partial x$ are calculated and then if it becomes necessary, ϕ is obtained by the inverse transformation of Eq.(4). The result with this scheme is shown in Fig.2(b), where no overshooting appears and very small diffusion is attained. This is clear from the definition of tangent function.

This tangent function is useful to represent the interface by the density function in which only digitized values 0 and 1 appear. This scheme is applied to the injection of heavier fluid into lighter fluid. In this case, we set the sound speed to quite a large value and hence the process treated here is almost incompressible although full hydrodynamic equations are used with the method described in section 3. The result is shown in Fig.3. The interface of two fluids has successfully been treated by one grid throughout the computation.

2.3 Rational Function CIP

Although the digitizer is useful for density function or for incompressible fluid, we meet the problem where density can change in time in each region separated by sharp interface. Although we can use the same technique even in this case[4], we had better consider more

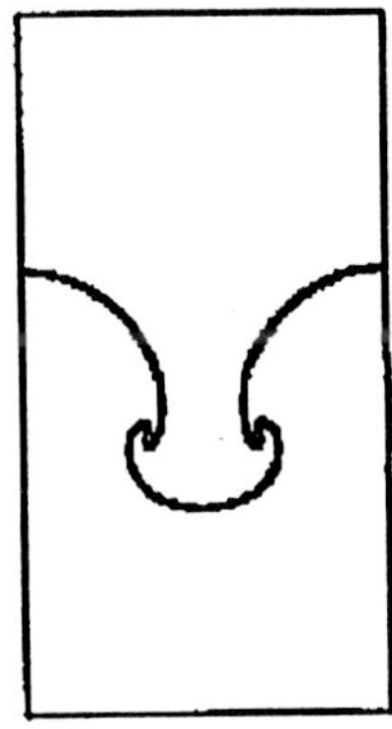

Figure 3. Injection of heavier liquid into lighter liquid. Equally-spaced 180×90 Cartesian fixed grids are used.

elaborate method to eliminate overshooting. We may use MmB scheme or limiter function. However, we prefer to use the rational function :

$$F(x) = \frac{ax^3 + bx^2 + cx + d}{1 + \alpha Bx} \qquad (6)$$

where α is the switching parameter. If α is 0, Eq.(6) is merely the CIP method. When α is 1, it can be convex-concave preserving scheme. The coefficients are determined as follows. At first, b, c, d and B are determined from the continuity of f and g at i and iup=i$\pm$1 for the function

$$F(x) = \frac{bx^2 + cx + d}{1 + Bx}$$

Nextly, we determine a, b, c and d from the same continuity requirement for the function (6), adopting the expression of B already given before. Thus we obtain

$$a = \left[g_i - S + (g_{iup} - S)(1 + \alpha BD)\right] / D^2$$
$$b = S\alpha B + (S - g_i) / D - aD$$
$$c = g + f\alpha B, \quad d = f_i$$
$$B = \left[\left|(S - g_i) / (g_{iup} - S)\right| - 1\right] / D \qquad (7)$$
$$S = (f_{iup} - f_i) / D$$

where D and iup are already defined in Eq.(3) (see Ref.[8,9] for details). As shown in Fig.2(c), monotone and convex-concave preserving scheme is attained.

2.4 Implicit CIP

Another interesting modification of the CIP method is its implicit version[10]. Although it is implicit, it is directly solved in non-iterative way. In obtaining Eq.(3), we shift the profile from the past to the present. On the contrary, the profile is shifted from future to the present in the implicit scheme. Therefore, we can replace f and g by f^{n+1} and g^{n+1}, and change x to be -x in Eq.(3). Rearranging this result, we obtain the expression:

$$
\begin{aligned}
f_i^{n+1} &= (\kappa + 1)^{-3}[-\kappa(\kappa + 1)(\kappa g_{iup}^{n+1} - g_i^n)D \\
&\quad + \kappa^2(\kappa + 3)f_{iup}^{n+1} + (3\kappa + 1)f_i^n] \\
g_i^{n+1} &= (\kappa + 1)^{-3}[(\kappa + 1)\{(\kappa - 2)g_{iup}^{n+1} \\
&\quad - (2\kappa - 1)g_i^n\} + 6\kappa(f_{iup}^{n+1} - f_i^n)/D] \\
\kappa &= -u\Delta t/D \geq 0
\end{aligned}
\tag{8}
$$

Fortunately, only two points are connected in the CIP method. For example, j and j-1 are related in case of u>0. Therefore, even in the implicit solution, we can directly solve it from the upwind direction. It is easy to extend it to the case where velocity changes its sign[10]. Figure 4 shows the result and demonstrates its stability and correctness. Surprisingly, combination of digitizer with implicit scheme gives a very accurate result even for CFL=$u\Delta t/\Delta x$=4.

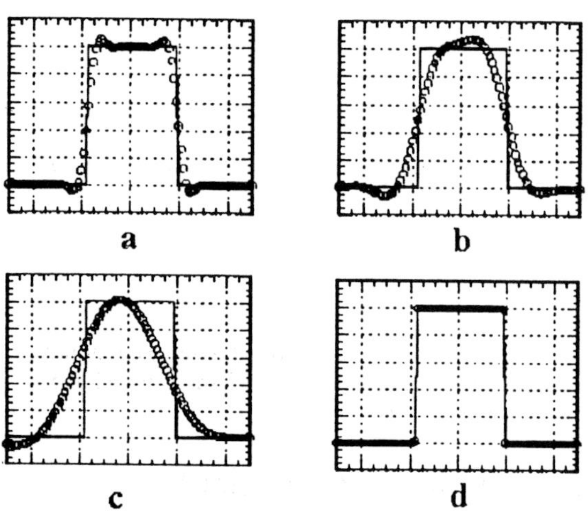

a   b
c   d

Figure 4. The result with implicit CIP. (a) CFL=0.5, (b) CFL=4.0, (c) CFL=10 and (d) CFL=4 with digitizer.

2.5 Multi-Dimensions

We extend here the CIP method to higher dimensions. The method described in the following can be used in any scheme proposed in section 2. Let us first consider a two-dimensional hyperbolic equation in Cartesian coordinates x,y :

$$
\frac{\partial f}{\partial t} + u\frac{\partial f}{\partial x} + v\frac{\partial f}{\partial y} = g,
\tag{9}
$$

As will be shown later, equations in quite large research fields can be put into this form. The equations for the first spatial derivatives are derived from Eq.(9) to be

$$
\frac{\partial f_x}{\partial t} + u\frac{\partial f_x}{\partial x} + v\frac{\partial f_x}{\partial y} = g - \frac{\partial u}{\partial x}f_x - \frac{\partial v}{\partial x}f_y \equiv R_x
\tag{10}
$$

$$
\frac{\partial f_y}{\partial t} + u\frac{\partial f_y}{\partial x} + v\frac{\partial f_y}{\partial y} = g - \frac{\partial u}{\partial y}f_x - \frac{\partial v}{\partial y}f_y \equiv R_y
\tag{11}
$$

For simplicity we use a rectangular grid with constant spacing Δx, Δy but it can be easily extended to curvilinear system as given in section 5.1.

The spatial profile can be constructed with cubic polynomial in two directions like

$$
F(X, Y) = \sum_{l=0}^{3} \sum_{m=0}^{3} C_{l,m} X^l Y^m
$$

Among various families of polynomial written in this form, in the first paper of multi-dimensional CIP[11], we have proposed to use

$$
\begin{aligned}
F_{i,j}(x,y) &= C_{3,0}X^3 + C_{2,0}X^2 + f_{x_{i,j}}X + f_{i,j} \\
&\quad + C_{0,3}Y^3 + C_{0,2}Y^2 + f_{y_{i,j}}Y \\
&\quad + C_{2,1}X^2Y + C_{1,1}XY + C_{1,2}XY^2,
\end{aligned}
\tag{12}
$$

to be consistent with one-dimensional CIP in one-direction. Here we define X=x-x_i , Y=y-y_j. Thus 10 unknowns are determined from the continuity of f, $\partial f/\partial x$, $\partial f/\partial y$ at (i,j), (i+1,j), (i,j+1) and the continuity of f at (i+1,j+1) [11]. As in the one-dimensional case, we split the advection and non-advection terms which are sequencially solved as will be shown in section 3.2.

Although this choice is not the best one, it is sufficient for small time step $u\Delta t/\Delta x$<0.5 as will be imagined from the definition: $\partial f/\partial x$, $\partial f/\partial y$ at (i+1,j+1) are not continuous and are not accurate. In Fig.5, we show a result of two-dimensional solid-body revolution. Figure 5(a) shows an initial profile and Fig.5(b) the profile after one revolution when Eq.(12) is applied. Although the scheme is quite simple, it can produce a result comparable to existing higher order schemes. If we adopt digitizer to this problem, the result is much improved as in Fig.5(c).

It is important to note that Eq.(12) gives a correct result for $u\Delta t/\Delta x$<0.5. If we want to extend the scheme to implicit one proposed in section 2.4, we had better improve the profile. In order to keep the continuity of $\partial f/\partial x$, $\partial f/\partial y$ at (i+1,j+1), we must increase the polynomials. One choice is proposed by Aoki[12] as

$$F_{i,j}(x,y) = C_{3,0}X^3 + C_{2,0}X^2 + f_{x_{i,j}}X + f_{i,j}$$
$$+ C_{0,3}Y^3 + C_{0,2}Y^2 + f_{y_{i,j}}Y + C_{3,1}X^3Y \qquad (13)$$
$$+ C_{2,1}X^2Y + C_{1,1}XY + C_{1,2}XY^2 + C_{1,3}XY^3 ,$$

Thus, $C_{3,1}$ and $C_{1,3}$ are additional unknowns and can be determined by the increased continuity conditions at $(i+1,j+1)$. This can improve the profile in the implicit solution for $u\triangle t/\triangle x \gg 1$. Surprisingly, however, Eq.(12) gives a similar result although slight deformation is unavoidable.

This scheme can be applied to Vlasov equation

(a)

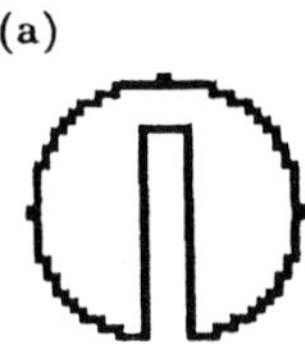

(b)

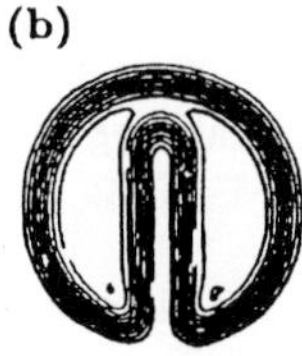

(c)

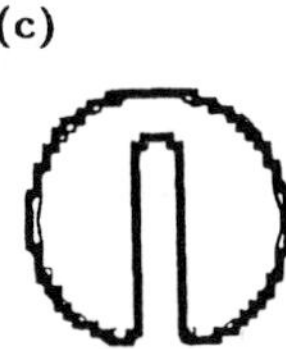

Figure 5. Rotation of a solid body. Inside a $100\times$ 100 grid system, a sphere of radius 15 having a slit of 5 width located at $(x,y)=(0,75)$. This shape rotates around the center at $(50,50)$ in 628 time steps. (a) Initial profile, (b) after one rotation with Eq.(12), and with (c) Eq.(12) plus digitizer.

$$\frac{\partial f}{\partial t} + u\frac{\partial f}{\partial x} + \frac{F}{m}\frac{\partial f}{\partial u} = 0$$

in two-dimensional x,u space. As will be shown in section 3.1, the CIP method is exactly conservative for constant velocity field. This is also true if u is independent of x and F/m is independent of u which is also valid even for $F\propto u\times$ B when magnetic fields B exist. This property is quite useful for Vlasov simulation. Furthermore, at high energy tail of the distribution function, time step is limited by CFL condition there although the tail needs not be accurately solved.. Therefore, implicit CIP method in multi-dimensions will be useful for this project [13].

3 TIME SPLITTING AND CONSERVATION

3.1 Constant Velocity Field

We should note that the CIP method described in the previous section is exactly conservative for constant velocity field. Let us prove this in one dimension. The CIP procedure Eq.(3) gives

$$\sum_i f_i^{n+1} = (2\chi^2 - 3\chi + 1)\chi\Delta x\sum_i g_i^n + \sum_i f_i^n$$
$$\sum_i g_i^{n+1} = (6\chi^2 - 6\chi + 1)\chi\Delta x\sum_i g_i^n \qquad (14)$$

where $\chi = -u\Delta t/\Delta x$. If $\Sigma\, g$ vanishes initially, it will do so throughout the calculation. Therefore, Eq.(14) gives the exact conservation $\Sigma\, f^{n+1} = \Sigma\, f^n$. Even if $\Sigma\, g^n$ is set to a finite value initially, the factor $|\, 1-6\,\chi +6\,\chi^2\,|$ in Eq.(14) is less than unity for $|\,\chi\,| \leqq 1$ hence $\Sigma\, g^n$ quickly converges to zero.

3.2 General Case

In the CIP method, a one-dimensional hyperbolic equation :

$$\frac{\partial f}{\partial t} + \frac{\partial(fu)}{\partial x} = h \qquad (15)$$

is modified by using $H \equiv h - f\partial u/\partial x$ to be

$$\frac{\partial f}{\partial t} + u\frac{\partial f}{\partial x} = H \qquad (16)$$

Since the CIP method uses $g \equiv \partial f/\partial x$ as well as f, we derive an equation for g by taking a spatial derivative of Eq.(16) as

$$\frac{\partial g}{\partial t} + u\frac{\partial g}{\partial x} = H' - \frac{\partial u}{\partial x}g \qquad (17)$$

where ' in H' means the spatial derivative. Equations (16) and (17) are split into two phases : advection phase

$$\frac{\partial f}{\partial t} + u\frac{\partial f}{\partial x} = 0 ,$$
$$\frac{\partial g}{\partial t} + u\frac{\partial g}{\partial x} = 0 , \qquad (18)$$

which are solved by simply shifting profile as given in section 2, and non-advection phase

$$\frac{\partial f}{\partial t} = H$$
$$\frac{\partial g}{\partial t} = H' - \frac{\partial u}{\partial x}g \qquad (19)$$

are solved by finite difference scheme or finite element scheme. The latter has been attacked already by Tanahashi and Tosaka[14]. The former can be given simply as

$$f_i^{n+1} = f_i^{\bullet} + H_i \Delta t$$

$$\frac{g_i^{n+1} - g_i^{\bullet}}{\Delta t} = \frac{f_{i+1}^{n+1} - f_{i-1}^{n+1}}{2\Delta x \Delta t} - \frac{f_{i+1}^{\bullet} - f_{i-1}^{\bullet}}{2\Delta x \Delta t} - \left(\frac{\partial u}{\partial x}\right)_i g_i^{\bullet}$$

We should note that in the last equation we do not use the explicit form of derivative H'. Instead we use spatial derivative of $(f^{n+1}-f^*)/\triangle t$ which is equivalent to H'. As described in the previous paper, this treatment makes scheme free from complication of H' and can be used for implicit treatment of H like thermal conduction.

Thus in the advection phase f^n, g^n are advanced to $f^{\bullet}, g^{\bullet}$, and in the non-advection phase $f^{\bullet}, g^{\bullet}$ are advanced to f^{n+1}, g^{n+1}.

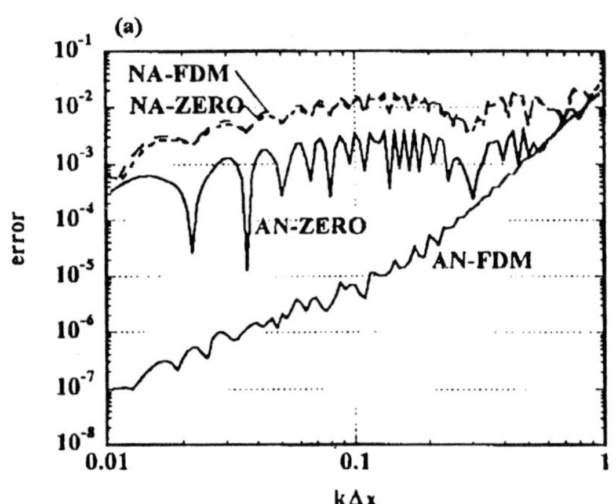

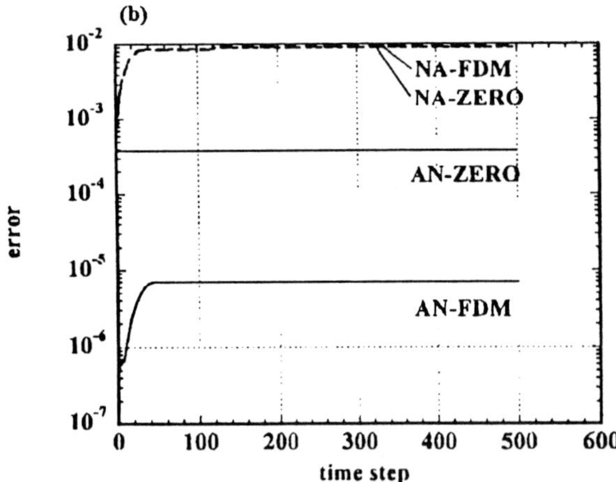

Figure 6. (a) The conservation error when the velocity profile u(x)=1+0.25sin(kx) is used. The error =(sum of f)/(sum of initial f). NA: non-advection first and advection second. AN : advection first . (b) Time evolution of error for k△x=0.1.

Here we show the conservative property of the explicit CIP method. The grid size is $\triangle x$=1.0, and 700 grids are used. The periodic boundary condition is assumed. The initial condition is f=1.0 at $206 \leqq x \leqq 226$, otherwise f=0. $\triangle t = 0.4$. Figure 6(a) shows an error occurred when the equation,

$$\frac{\partial f}{\partial t} + u \frac{\partial f}{\partial x} = -f \frac{\partial u}{\partial x} \qquad (20)$$

is solved under the given oscillating velocity field $u = 1 + 0.25\sin(kx)$ after 500 time steps. We should note that conservative property is improved when the order of calculation is changed. In early papers, we calculated non-advection terms at first and then advection terms. We call this order of calculation 'NA-method' temporarily . However, we found recently that the conservation becomes much better by several orders of magnitude when the advection terms are calculated at first. We call this order of calculation 'AN-method' temporarily. Furthermore, it is interesting to observe the effect of initial condition of the derivatives. Figure 6(b) shows the time evolution of errors. When we adopt $g = 0$ initially, the conservation is broken during one time step and then quite accurate solution follows. We call this procedure "ZERO". For example, AN-ZERO starts with 4×10^{-4} error after one step.

On the other hand, if we adopt $g = (f_{i+1} - f_{i-1})/2\Delta x$ as initial derivatives, serious error at initial one time step disappears. We call this procedure FDM (finite difference method.)

For highly accurate solution, this initial error becomes important. As shown in Fig.6(a), large difference between AN-ZERO and AN-FDM is due to this initial error. The solutions by NA-ZERO and NA-FDM are similar because much larger error occurs later on although initial error of 4×10^{-4} is the same as that in AN-ZERO.

4   UNIVERSAL TREATMENT

4.1 Hydrodynamic Equations

In order to solve all the materials in a universal form, we must find an appropriate equation for solid, liquid and gas. We use full hydrodynamic equations for these materials, which can be written in a form :

$$\frac{df}{dt} \equiv \frac{\partial f}{\partial t} + (u \cdot \nabla)f = g, \qquad (21)$$

where $f=(\rho, u, T)$ and $g =(-\rho \nabla \cdot u, -\nabla p/\rho +Q_u, -P_{TH}\nabla \cdot u/\rho C_V +Q_E)$, where ρ is the density, u the velocity, p the pressure, T the temperature, Q_u represents viscous, stress terms and surface tension etc., and Q_E includes viscous heating, heat conduction and heat source. Here C_V is the specific heat for constant volume and we defined $P_{TH}=T(\partial p/\partial T)_{\rho}$. The equation of energy is derived from the first principle of thermodynamics

$$TdS = dU + pdV = \left(\frac{\partial U}{\partial T}\right)_v dT + \left(\frac{\partial U}{\partial V}\right)_T dV + pdV \qquad (22)$$

$$= C_v dT + T\left(\frac{\partial p}{\partial T}\right)_v dV$$

where S is the entropy, U the internal energy and $V = 1/\rho$ the volume. The last expression in Eq.(22) is obtained from the Helmholtz free energy F=U-TS and thermodynamic consistency

$$\left(\frac{\partial p}{\partial T}\right)_v = -\left(\frac{\partial^2 F}{\partial V \partial T}\right) = \left(\frac{\partial S}{\partial V}\right)_T = \frac{1}{T}\left[p + \left(\frac{\partial U}{\partial V}\right)_T\right] \qquad (23)$$

It is illustrative to give an explicit expression of P_{TH} in special case. For ideal fluid, it becomes pressure since p is linearly proportional to T, while for two phase flow if the pressure is given by the Clausius-Clapeyron relation,

$$p = p_0 \exp\left(-\frac{L}{RT}\right), \quad P_{TH} = T\left(\frac{\partial p}{\partial T}\right) \propto L \qquad (24)$$

where R is the gas constant and therefore P_{TH} is proportional to latent heat L and P_{TH} represents the latent heat release owing to the increase of volume followed by the increase of fraction of gas in two phase flow. For more general equation of state given in section 4.2, C_v and P_{TH}

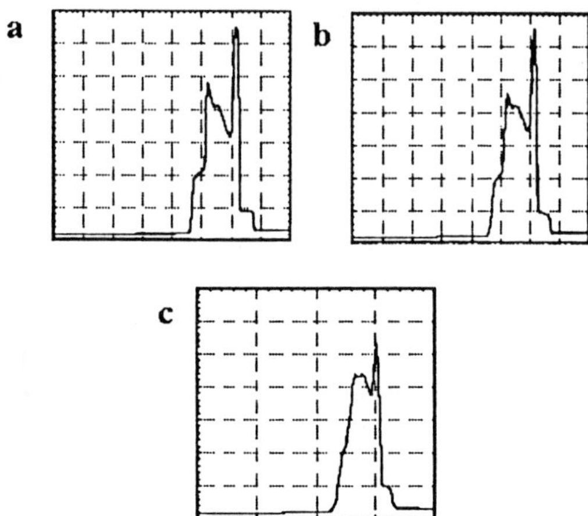

Figure 7. Density profile at t=0.038 where (a) 800 , (b) 400 and (c) 200 equally space grids are used.

can be calculated analytically or numerically throughout all phases.

The CIP method solves the equations like Eq.(21) by dividing those into non-advection and advection phases as given in section 3.2. A cubic-interpolated profile is shifted in space in the advection phase as shown in the previous sections and then nonadvection phase is calculated by finite difference method.

As shown in the previous papers, we can trace

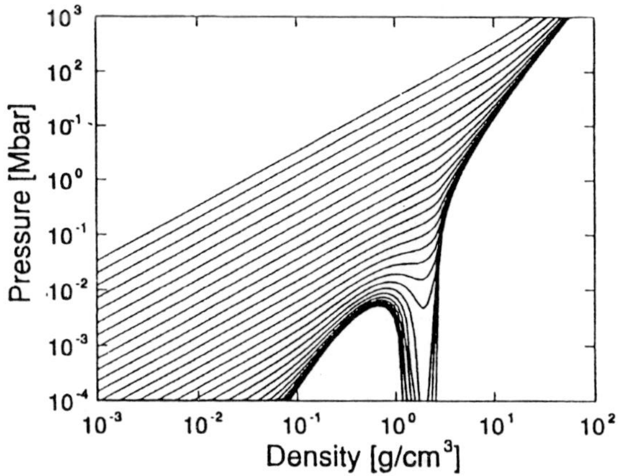

Figure 8. Equation of state of aluminum. Each line represents isotherms.

shock waves correctly with the CIP method although it uses fluid equations written in a non-conservative form or in primitive Euler representation.

This scheme has been tested by an example of two interacting blast waves given by Woodward and Colella[15]. Here, initial pressure p is p=1000 for x<0.1, p=0.01 for 0.1<x<0.9 and p=100 for 0.9<x<1. Although Woodward and Colella used minimum grid size of $\triangle$x=1/9600, we have succeeded to reproduce the result even with $\triangle$x=1/400 (Fig.7b).

4.2 Universal Solver

Since we are treating hydrodynamic equations in a non-conservative form, it is easy to extend it to include both incompressible and compressible fluids. Let us consider again the origin of the difficulty. In the Fig.8, we plot the iso-temperature contour of aluminum automatically generated from the semi-analytical formula. In the gas phase where density is sufficiently low, the pressure is in proportion to the density. Therefore, we may solve the density first in Eq.(21) and then after temperature is obtained, we use EOS(equation of state) in Fig.8. However, near the solid density of 2.7 g/cm^3, the pressure rises very sharply. If we use the same procedure there, the pressure can change easily by 3-4 orders of magnitude even with small error of density around few tens of percent. Therefore, the strategy to solve the density first is broken in this area. This is the reason why the universal treatment of solid, liquid and gas is a difficult task. In attacking this problem, the physicist in incompressible fluid invented an interesting technique. We will translate the strategy they used from a different view point and reconsider the technique. If the pressure is very sensitive to the density, we had better solve pressure at first. This means that we should rotate Fig.8 by 90 degree. If we have a way to solve pressure at first, then we get density very accurately at the solid density. Since the pressure is proportional to density in the gas phase, this strategy does not harm the solution there either. Then how to realize this strategy ? Our method starts with the thermodynamic relation :

$$\Delta p = \left(\frac{\partial p}{\partial \rho}\right)_T \Delta\rho + \left(\frac{\partial p}{\partial T}\right)_\rho \Delta T \tag{25}$$

where $\Delta p = p^{n+1} - p^*$ and $*$ represents a profile after advection. The same expression is used for ρ and T. Therefore, if $\Delta\rho$ and ΔT are predicted, Δp can be

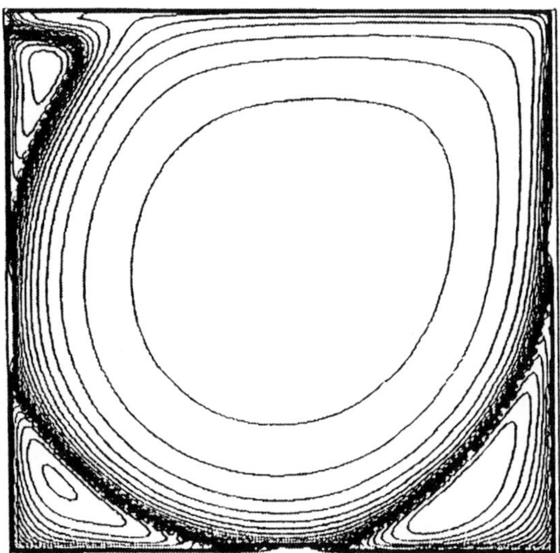

Figure 9. Shear-driven cavity flow in two dimensions for Re=5000. Contours of stream function ϕ at t=75 are plotted from ϕ =-1000 to 1000 with increments of 30, where ϕ =100log(- ϕ) for ϕ <0 (large circle near the center) and ϕ =100log(ϕ)+350 for ϕ >0 (secondary flows at corners), ϕ being the stream function calculated from u. The system size is 1×1 with 100×100 rectangular grids.

obtained since $\partial p/\partial\rho$, $\partial p/\partial T$ are already given from EOS. The advantage of the CIP is the separate treatment of the non-advection term. Therefore, we can limit our discussion here to the non-advection term only. We should note that this merit is quite important to get the final result.

Then we get

$$\Delta\rho = -\rho * \nabla\cdot\mathbf{u}^{n+1}\Delta t \tag{26}$$

$$\rho * C_v\Delta T = -P_{TH}\nabla\cdot\mathbf{u}^{n+1}\Delta t$$

where C_V is the specific heat for constant volume. The velocity $\mathbf{u}^{n+1}$ in above equations can be eliminated by using the equation of motion:

$$\Delta\mathbf{u} = -\frac{\nabla p^{n+1}}{\rho *}\Delta t \tag{27}$$

$(\Delta\mathbf{u}=\mathbf{u}^{n+1}-\mathbf{u}*)$ we obtain the equation for p^{n+1}[3,16]

$$\nabla\left(\frac{1}{\rho *}\nabla p^{n+1}\right) = \frac{p^{n+1} - p *}{\Delta t^2(\rho C_s^2 + \frac{P_{TH}^2}{\rho C_v T})} + \frac{\nabla\cdot\mathbf{u} *}{\Delta t} \tag{28}$$

where $P_{TH}=T(\partial p/\partial T)_\rho$ 、 $C_s^2=(\partial p/\partial\rho)_T$.

Thus velocity $\mathbf{u}^{n+1}$ can be calculated by Eq.(27) and then the density ρ^{n+1} is by Eq.(26). It is very important to note that in Eq.(28), ρ is inside the derivative on the lefthand side. At the interface between materials having large density difference, the continuity of acceleration $\nabla p/\rho$ is very important because the denominator ρ can change by several orders of magnitude in one grid. Equation (28) guarantees the continuity of $\nabla p/\rho$ at the discontinuity. By this procedure, we can treat all the material at once by simply changing its equation of state. We note again that this property is a consequence of the separate treatment of advection and non-advection terms, otherwise the continuity of $\nabla p/\rho$ is not guaranteed and a large density can not be traced.

This scheme is applied to two-dimensional shear-driven flow as in Fig.9. The system size is 1×1 in non-dimensional units described by 100×100 rectangular grids. The top wall moves with the speed of unity, and the nonslip boundary condition is imposed on all the walls. The real kinematic viscosity coefficient is 1/5000 and therefore the Reynolds number Re is 5000.

5 MISCELLANEOUS NUMERICAL MODEL

5.1 Curvilinear Coordinate

Let us consider the following equation described in the Cartesian coordinate.

$$\frac{Df}{Dt} \equiv \frac{\partial f}{\partial t} + u\frac{\partial f}{\partial x} + v\frac{\partial f}{\partial y} + w\frac{\partial f}{\partial z} = \mathbf{g} \tag{29}$$

Here, the vector $\mathbf{f}$ includes scalar quantities like density and pressure, and vector quantities like velocity and electromagnetic fields.

For general purpose, it is useful to keep the form of **f** in the Cartesian coordinate even for the curvilinear coordinate. If we use the general transformation from the Cartesian coordinate (x,y,z) to the curvilinear coordinate (ξ,η,ζ) and define

$$\begin{pmatrix} \partial/\partial x \\ \partial/\partial y \\ \partial/\partial z \end{pmatrix} = \begin{pmatrix} a_{11} & a_{12} & a_{13} \\ a_{21} & a_{22} & a_{23} \\ a_{31} & a_{32} & a_{33} \end{pmatrix} \begin{pmatrix} \partial/\partial\xi \\ \partial/\partial\eta \\ \partial/\partial\zeta \end{pmatrix}$$

Equation (29) can be rewritten as

$$\frac{D\mathbf{f}}{Dt} \equiv \frac{\partial\mathbf{f}}{\partial t} + U\frac{\partial\mathbf{f}}{\partial\xi} + V\frac{\partial\mathbf{f}}{\partial\eta} + W\frac{\partial\mathbf{f}}{\partial\zeta} \tag{30}$$

using contravariant velocities:

$$\begin{cases} U = a_{11}u + a_{21}v + a_{31}w \\ V = a_{12}u + a_{22}v + a_{32}w \\ W = a_{13}u + a_{23}v + a_{33}w \end{cases}$$

It is interesting to note that the numerical procedure of the advection phase in a curvilinear coordinate system is exactly the same as that in equally-spaced Cartesian coordinate system when the contravariant advection velocity U, V, W are used, while the vector **f** is not transformed into imaginary space as mentioned above. Note that the derivative g here should be translated as $\partial\mathbf{f}/\partial\xi$, $\partial\mathbf{f}/\partial\eta$, $\partial\mathbf{f}/\partial\zeta$ instead of $\partial\mathbf{f}/\partial x$, $\partial\mathbf{f}/\partial y$, $\partial\mathbf{f}/\partial z$.

The present scheme is applied to natural convection heat transfer. In this simulation, we did not use the assumption of incompressible fluid. Instead, we directly solved Eq.(21) with quite large sound speed. Figure 10 shows a cross section of the heat transfer system. Between inner and outer cylinders, atmospheric air is filled and gravitational acceleration is directed downward. Details of the calculation parameters are seen in Ref.[17].

5.2 Surface Tension

Many phenomena can be included in Q_u in Eq.(21). One of them is a surface tension. It's form is given by

$$Q_u = \frac{\sigma}{\rho}\kappa\nabla\phi$$

as proposed by Brackbill[18]. The curvature κ is calculated by the normal vector of the surface **n**.

$$\kappa = -\nabla\cdot\hat{\mathbf{n}} \quad , \quad \hat{\mathbf{n}} = \frac{\nabla\phi}{|\nabla\phi|}$$

Since the density function is accurately calculated by Eqs.(4) and (5), the surface tension can be calculated with a few grids, for example, a liquid drop with a diameter of 4 meshes can be traced even with Cartesian grids.

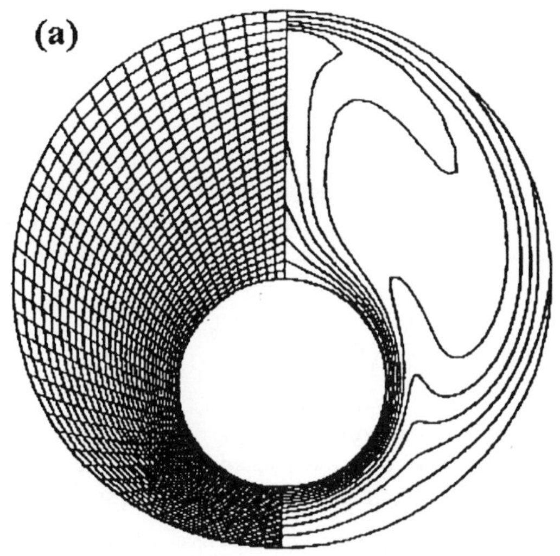
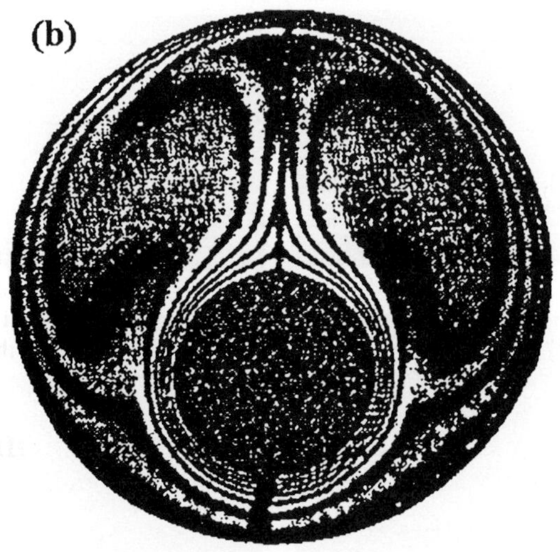

Figure 10. (a) The temperature contour(right) and the grids (left) used for a calculation. (b) The photograph of an experimental study (See Ref.[17]).

5.3 Elastic-Plastic Model

Elastic motion can be included also in Eq.(21) as

$$Q_u = \frac{1}{\rho} \frac{\partial \sigma_{ij}}{\partial x_j}$$

where σ is given by $\sigma_{ij} = s_{ij} - p\delta_{ij}$. The time development of stress is calculated to be

$$\frac{ds_{ij}}{dt} \equiv \dot{s}_{ij} = 2G\dot{\xi}_{ij}$$

$$\dot{\xi}_{ij} = \frac{1}{2}\dot{\varepsilon}_{ij} - \frac{1}{3}\dot{\varepsilon}_{kk}\delta_{ij}$$

$$\dot{\varepsilon}_{ij} \equiv \frac{\partial u_i}{\partial x_j} + \frac{\partial u_j}{\partial x_i}$$

where u is the fluid velocity. When the stress exceeds the von-Mises yield condition, the elastic material changes into plastic material , that is fluid dominated by viscous stress.

Most important problem for numerical solution is the treatment of the stress term when the term becomes quite large and explicit treatment of the term becomes unstable. In the followings, we shall briefly explain how we treat this problem numerically.

In many physical problems, we meet an equation for the time evolution of nonlinear dynamical system like

$$\frac{\partial U}{\partial t} = G(U) \tag{31}$$

where G(U) consists of spatial derivative or spatial integral of U in general, therefore the partial time derivative is used on the left-hand side of Eq.(31). When the characteristic time of G is quite short compared to the time scale which we are concerned with, we need to solve it implicitly. Ordinary procedure will be to linearlize G(U) in terms of U and solve it by matrix solver. However, when G includes highly nonlinear terms and integral of U, the linearlization does not promise the steady solution G(U)=0 for very large time step.

We here propose a very simple way to solve this implicitly. For this purpose we introduce residual function F(U) as

$$F(U) = \frac{\partial U}{\partial t} - G(U) \tag{32}$$

The procedure we propose is as follows;

1) Assume $U_0 = U^n$ and define $F(U_0) = -G(U_0)$.
2) Then obtain a solution U^{n+1} that satisfies F(U)=0. That is

$$U^{n+1} = U^n - G(U_0)\Delta t.$$

3) Then estimate the residual again

$$F(U^{n+1}) = \frac{U^{n+1} - U^n}{\Delta t.} - G(U^{n+1})$$

4) Get an improved estimation to minimize the residual.

$$U = U_0 + \frac{F(U_0)}{F(U_0) - F(U^{n+1})}(U^{n+1} - U_0) \tag{33}$$

5) Setting $U_0 = U$ and repeat from (2) to (5).

Let us first explain how Eq.(33) is derived. When $F(U_0)$ and $F(U_{n+1})$ are given in the grids points U_0, U_{n+1}, the value of U that makes F(U) being zero is obtained by linear extraporation and obtain U. It is easy to get this new value as

$$U = \frac{F(U_0)U^{n+1} - F(U^{n+1})U_0}{F(U_0) - F(U^{n+1})}$$

which is modified to a form in Eq.(33).

Next we show that the procedure given here is the same as the fully implicit solution of Eq.(31) in simple linear case:

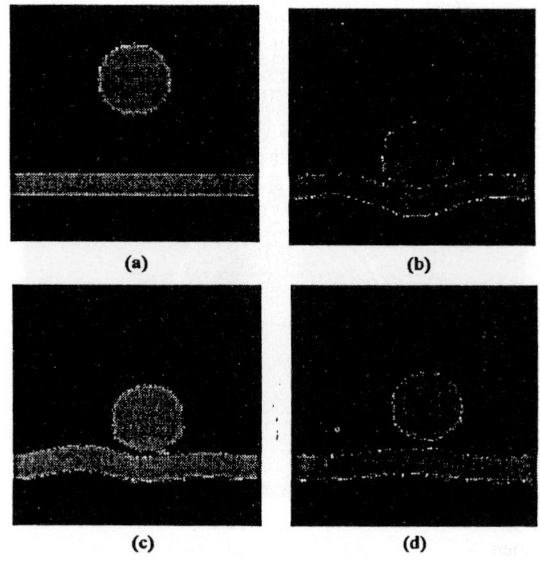

Figure 11. Example of elastic-plastic flow. The system is described by 100×100 Cartesian fixed grid and only 10 grids are used for the width of plate. Plate and ball are moving through fixed grid system. Black part is filled with air and also calculated.

$$\frac{dU}{dt} = -\gamma U$$

Then

1) $\quad F(U_0) = \gamma U_0$

2) $\quad U^{n+1} = U^n - \gamma \Delta t U^n$

3) $\quad F(U^{n+1}) = \dfrac{U^{n+1} - U^n}{\Delta t.} + \gamma U^{n+1} = -\gamma^2 \Delta t U^n$

4) $\quad U = \dfrac{1}{1 + \gamma \Delta t} U^n$

Thus final result is exactly the same as the fully implicit solution by a finite difference scheme :

$$\frac{U^{n+1} - U^n}{\Delta t} = -\gamma U^{n+1}$$

Figure 11 shows an example of elastic-plastic flow in which both ball and plate are moving through air in the fixed grid system.

5.4 Rigid Body Treatment

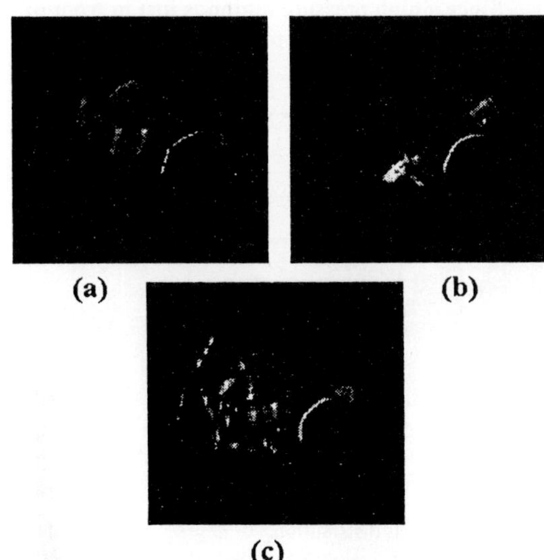

(a)   (b)

(c)

Figure 12. The solid structure is rotating just above water surface. Even with Cartesian fixed grid system, the CIP can trace the movement of the structure without numerical diffusion and topology change of water surface.

When a rigid body is suspended inside flow, it is expensive to use elastic-plastic model to realize the rigid motion by increasing the yield strength of matter. Instead, we may use economical and faster method[19] to realize this motion. This uses the density function ϕ as follows :

$$\frac{dU}{dt} = \frac{1}{M} \int \frac{du}{dt} \rho \phi dV, \quad M = \int \rho \phi dV$$

$$\frac{d\Omega}{dt} = \frac{1}{I} \int \mathbf{R} \times \frac{du}{dt} \rho \phi dV, \quad I = \int \mathbf{R} \cdot \mathbf{R} \rho \phi dV$$

$$\mathbf{R} = \mathbf{x} - \mathbf{X}, \quad \mathbf{X} = \frac{1}{M} \int \mathbf{x} \rho \phi dV$$

where du/dt inside the volume integral $\int dV$ is given by the fluid equation (21), since Eq.(21) is calculated for all area including rigid body. Thus the density function for the rigid body moves with the average translational velocity $\mathbf{U}$. The rotational motion of the body W is calculated by the moment of inertia I and the location of mass center $\mathbf{X}$. Therefore the combined velocity of the rigid body $\mathbf{V}$ is given by

$$\mathbf{V} = \mathbf{U} + \mathbf{R} \times \Omega$$

There are many applications of this scheme. We will show here only one typical example. One of the features of the code is the high ability to trace an interface. In all the simulations given in this paper have been performed in a Cartesian fixed grid system. Figure 12 shows a solid rotar rotating just above the water surface. As shown here, rigid body motion is traced without numerical diffusion for a long time and topological change of water surface is successfully treated.

6 APPLICATIONS

We have developed a numerical scheme to solve dynamics of solid, liquid and gas together. By this scheme, we are able to attack the elastic-plastic behavior of solid interacting with liquid and gas. Simplest example will be the problem of slamming of ship on water surface where moving solid body interact with water and air. Since the density ratio between water and air is 1000, conventional algorithm needs a special treatment at the interface of these two fluids. For complex boundary occurring in splashing, connecting two different solutions at the interface is quite a difficult task, because the grid system cannot be fitted to the interface and then one cell inside water has multiple neighbors which are difficult to be connected. Our scheme does not require such boundary condition to connect two different solutions given in gas and liquid, instead we solve simultaneously two phases and get one self consistent solution. In this section, we shall give some typical examples demonstrating the potentiality of the scheme.

6.1 Laser-Induced Evaporation

Figure 13 shows melting and evaporation of aluminum under the illumination of laser light, where the density changes from 2.7 to 10^{-4} g/cm^3. Aluminum solid is treated as an elastic material initially and then changes to liquid and vapor during phase transition. This change is simply realized by the equation of state. This example shows the high ability of the code to describe a sharp interface and to be robust enough to treat both compressible and incompressible fluid simultaneously.

The experiment was performed at the Institute of Laser Engineering, Osaka University regarding the x-ray source

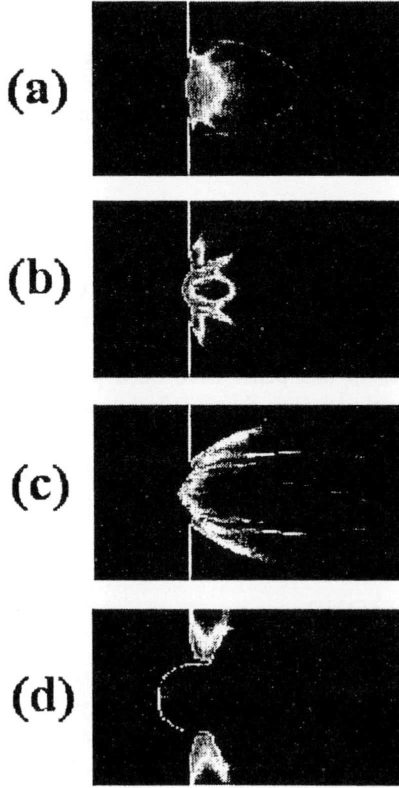

Figure 13. Density contour of aluminum (on the left) illuminated by laser light. Time sequence is 50, 100, 300 nsec. Filamentary structure explains the experimental results. (a) 40 ns, (b) 90 ns, (c) 290 ns with elastic-plastic effect. The crater does not grow after this. However, if elastic-plastic effect is switched off, it continues to grow like the contour at (d) 500 ns.

development[20] : a YAG laser of 650 mJ in 8nsec is used to obliquely illuminate an aluminum slab target with an angle of 45 degree to the target normal. Final crater depth and shape agree quite well with the simulation[20] and seems to

be anomalous because the cutting speed is 100 μ m/ 8nsec~ 10^6 cm/sec if this crater should have been created during laser pulse. Since the speeds of sound wave and elastic wave inside aluminum are order of 10^5 cm/sec, the cutting speed is much larger than these speeds. Is this speed physically possible? It is interesting to note that the crater is not formed during the laser pulse, but it develops gradually in the time scale of several 100 ns well after the laser pulse ended as shown in Fig.13. The very high temperature plasma more than a few tens of eV produced by the 8 ns laser pulse and most of them expands from the target. However, some of them still stay near the target for long time after the laser pulse because of recoil force from expanded plasmas and act as heat source to melt aluminum metal in the time scale of several 100 ns.

When the plasma temperature becomes less than melting temperature around 290 nsec (the time is measured from the laser peak) , the stress of aluminum whose strength is 0.248 Mbar and yield strength is 2.2976 kbar is recovered and no distortion occurs after that time. This yield stress is quite important to determine the final crater size. Without yield strength, the crater develops further even after 490 ns and the depth becomes more than 300 μ m as shown in Fig.13(d) although less difference is seen at the beginning around 90 ns.

The plasma heated crater formation leads to other interesting phenomena. Since the plasma acts not only as heat source but also as pressure source, the dynamic expansion of evaporated material at later time is strongly modified. Since a high pressure region is just in front of the evaporation surface, the vapor is forced to flow bypassing

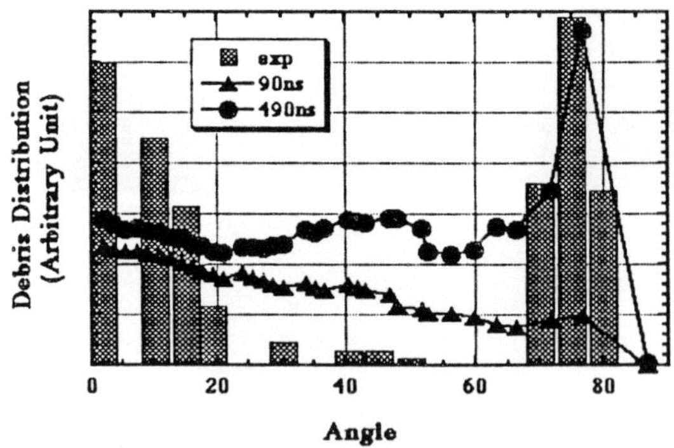

Figure 14. Debris distribution. 0 degree corresponds to the target normal. The histogram shows the experimental result, while circles and triangles show the accumulated mass from the simulation at 490 ns and 90 ns, respectively.

through a narrow channel between the metal surface and this pressure source. Therefore, the vapor preferentially flows toward a circumference with a large angle to the target normal. This effect is the exactly the same as that obtained in the experiment.   Figure 14 shows a distribution of debris from the targets.   The histogram is the experimental result and it was drawn from 2000 shots accumulated. Clearly there exist two peaks around 0 and 75 degree.   As in Fig.13, the plasma expands directed normally to the target at early time t<90 ns and this expansion is a bulk part of the laser-heated plasma.   As already stated, this expansion causes recoil force to the hot plasma surrounding aluminum surface. This main part of the expansion creates a peak at 0 degree. The triangles in Fig.14 show the distribution calculated from the time integration of mass flow ρ u up to t=90 ns. At an early stage t<90 ns, no peak appears around 75 degree. On the other hand, the expansion at later stage t>90nsec is limited to the sideward direction as stated before and creates the peak at 75 degree. Therefore accumulated distribution up to 490 ns shown by circles in Fig.14 increases mainly at 75 degree.

Simulation also predicts further interesting behavior. The expansion at t<40nsec is quite uniform because its temperature is quite high $\sim$ a few tens of eV.   The experiment supports this result and the debris around 0 degree is very fine and indistinguishable with an optical microscope.   On the other hand, the simulation result at t=290 nsec shows some filamentary streams flowing from the surface.   The experiment also supports this result and the debris at 75 degree consists of 1 to 20 μ m sized particles. Since the simulation is two-dimensional axisymmetric, we cannot estimate the particle size but we can suggest the origin of the filaments.

6.2 Slamming of Ship

In this section, we apply the method described in the previous section to the slamming problem.   For the simulation, we treat the solid part of ship body as rigid body. In this case, the surface is captured by the density function inside of which is forced to move with a constant velocity, while pressure distribution is solved by Eq.(28) even inside the solid.   The equation of state is ideal for air and Tait equation is used for water.   When the Tait equation is used, P_{TH} vanishes and the temperature decouples from all other equations.

In calculations shown here, we used fixed Cartesian grid system with 200 grids in horizontal (x) direction and 100-700 grids in vertical (y) direction along which the ship falls down. The ship width is 100 cm horizontally and attack angle is varied. The horizontal grid size Δ x is 1 cm. Vertical grid size Δ y changes depends on the case. The densities of water, air and ship are taken to be 1.0, 1.293 $\times 10^{-3}$, 0.5 g/cm^3, respectively.

After the impact upon water surface, pressure wave propagates side-ward whose peaks are located at the contact point between ship and water.   If we plot the pressure distribution along the bottom surface of the ship with distance x normalized by the distance c of the contact point

as shown in Fig.15, all the profiles converges to one line as suggested by Wagner's self-similar theory. When the impact velocity is varied, this maximum pressure is proportional to the square of the velocity as predicted by the theory at this angle.

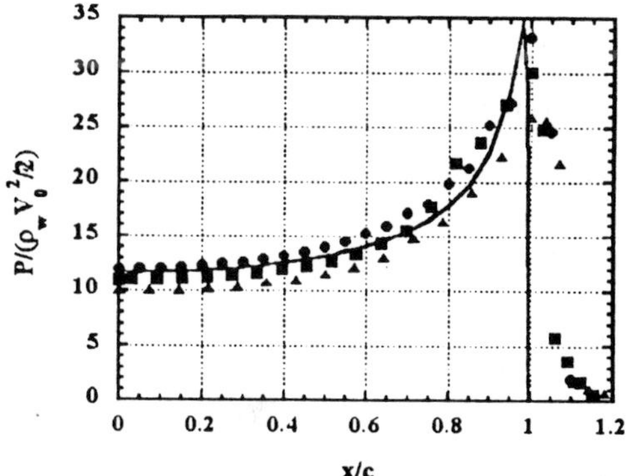

Figure 15. Comparison between the Wagner's self-similar theory (solid line) and simulation resuts (symbols).   When the profile is drawn with a distance x along the bottom of the ship scaled by a distance of the leading edge c, all the profiles at 57 (▲) , 66(■), 77(●) ms converge to the analytical result.

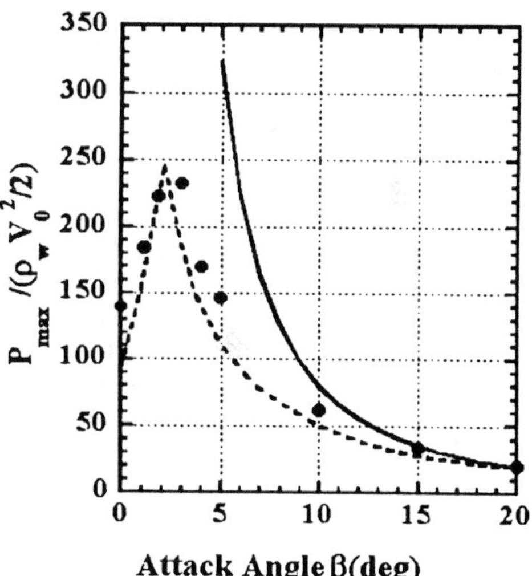

Figure 16. Attack angle dependence of peak pressure. Solid line : Wagner's theory. Dashed line : Chuang's experiment. Symbols : present simulation.

Then what happens when the attack angle is varied. The

Wagner's theory predicts the monotonous increase with decreasing angle and the infinite pressure at β =0. There was a contradicting experiment by Chuang in which pressure peaks at β =2.5. Since the Wagner's theory gives a very sharp pressure pulse for small β , this difference has been attributed to the crude resolution of the pressure gauge and estimated pressure with correction based on simulation pushed up the pressure to that of Wagner's.

We have repeated the calculations by changing attack angles and carefully checking the results with various grid size, and then finally obtaind the results shown in Fig.16. Surprisingly our results are fitted very well to the experimental results by Chuang. We shall analyze why such decrease occurs at low angle. The simulation shows that the pressure peak for β =0 appears before the ship arrives the water surface. This is because the air between the ship and water is compressed and works as a cussion. Also in small angle β <2, entrapped air is still observed and this acts as a cussion like in β =0. For small β , the pressure profile is far from the Wagner's theory and it gradually reaches the Wagner's when β increases.

Treating the solid, water and air simultaneously is quite a difficult task but we are now in a level to attack the problem of this kind. Since no other schemes are employed to solve these problems yet, we hope experimentalists to consider again the problem for low β to test the present results. As shown in the previous section, we are able to calculate the ship as elastic-plastic material. This problem will be studies in near future.

6.3 Milk Crown

Simulation of the coronet or "milk-crown" has long been a dream in computational physics. Several interface capturing schemes have been proposed to attack this problem. Although the present-day technique is already close to this goal, nobody reported the three-dimensional coronet formation before except for several two-dimensional works[21] pioneered by Harlow & Shannon[22], in which instability and therefore break-up of ring leading to coronet was not simulated. This is because the coronet is not merely a consequence of a free surface problem but we must solve surrounding gas as well. This requires a special treatment at the complex boundary of the coronet in which density ρ changes by 1000 times. In solving pressure p, it is quite a difficult task to guarantee the continuity of force $(\nabla p)/\rho$ across the complex interface having very large density ratio without special boundary condition there.

As shown in the previous sections, the CIP method can treat the interface of liquid and gas with almost one grid even in the fixed Cartesian grid system. The VOF[5] (Volume Of Fluid) and LSA[6] (Level Set Approach) may have a similar capability. The CIP has further advantage that it can solve all the materials from solid through liquid to gas together because it can treat very large density change keeping the continuity of $(\nabla p)/\rho$ at the interface without any special

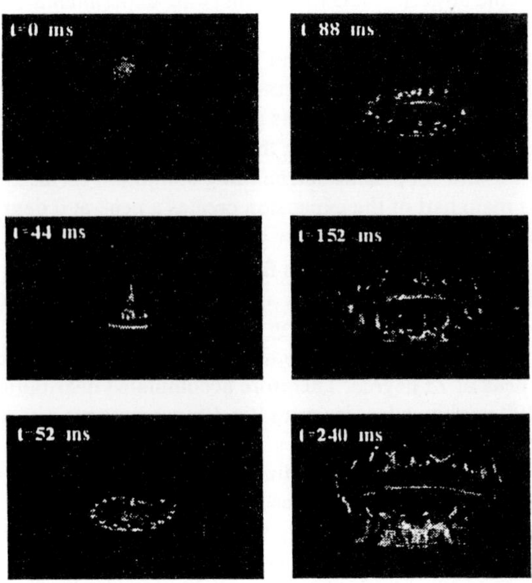

Figure 17. Water surface plot in the coronet formation. $100 \times 100 \times 32$ Cartesian grid is used.

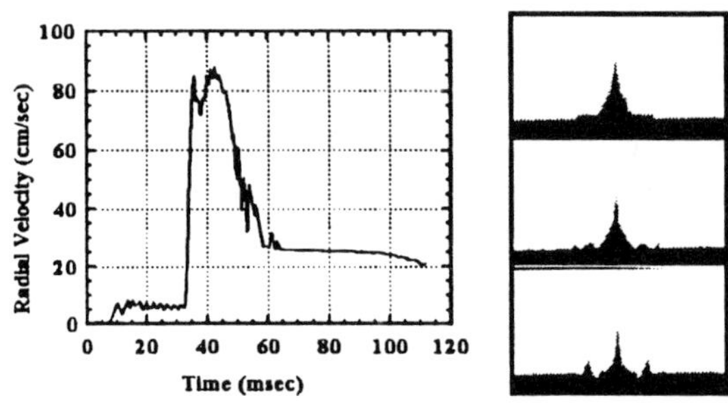

Figure 18. (Left) Time evolution of radial velocity. At the time of impact t=32 ms, large acceleration occurs and then deceleration after 44 ms until 60 ms. In the latter phase, the Rayleigh-Taylor instability occurs at the hump of thin film driven by a snow plough. (Right) cross sectional view of density at 48, 52, 62 ms.

boundary condition. This capability is essential for the application to the instability during the dynamics of coronet formation[23].

Figure 17 shows an iso-surface contour of density in which 100×100(horizontal) $\times 32$(vertical) fixed equally spaced grids are used with 0.1 cm grid spacing. A thin water film of 0.4 cm thick is placed on a solid plate and a water drop of 0.8 cm radius impacts from the top at a speed of 50 cm/sec. We solved air as well as water and the density ratio at the interface is almost 1000 as already mentioned. As shown in Fig.17 at t=52 and 88 ms, the irregular ring, which we call "ornament" of the coronet, appears at first and then laminar belt develops below the ornament later on at 152 and 240 ms in Fig.17.

The time evolution of the radial expansion velocity of the coronet can shed light on the formation mechanism of these structures. In the simulation of Fig.18 we performed two-dimensional axisymmetric simulation to separate the instability from its origin, that is, to avoid the influence of the instability on the main flow driving instability. Upon impact of the drop onto the liquid surface, liquid film is pushed radially outwards like a snow plough. The radial motion is initially accelerated by the high pressure of impact and then decelerated as shown in Fig.18. After impact until t=44 ms, flat ring of thin film like a "washer" appears on the original film as shown in Fig.17 because of raking up by a snow plough. The outer radius of the washer is 1.5 cm at t=44 ms and inner radius in contact with the drop is 0.8 cm, between these radii the film thickness is extremely uniform. At t=44 ms, deceleration starts and then outer leading edge of the washer moves apart from the main ring, while the latter gradually shrinking to the same radius and giving rise to a large hump which eventually leads to the coronet (see cross sectional view in Fig.18). This hump becomes unstable under the deceleration because of the Rayleigh-Taylor instability.

From Fig.18, we can estimate the deceleration of the motion to be $a=2.3 \times 10^3 cm/sec^2$ and from typical wavelength 0.3 cm (k=20) of the irregularity along the ring during deceleration we get the growth rate $\gamma =(ak)^{1/2}=214$ sec^{-1} for the Rayleigh-Taylor instability. This growth rate is sufficient to account for the evolution of the irregularity because of $\gamma \triangle t=5.56$ during the deceleration time $\triangle t=26$ ms. At t=60 ms, the deceleration stops and from this time on, because of the lack of the instability, laminar belt of the coronet begins to develop just below the ornament. Thus the well-known double structure consisting of the ornament on the top of laminar belt is generated as observed in typical experiment of the coronet[21]. We did not include the surface tension here, however, we believe that the surface tension plays a similar role to determine the wavelength of maximum growth and most important property observed here like the time evolution of the coronet formation may be similarly reproduced. Perhaps, the surface tension finally determines the formation of drops at the tip of irregularities and this effect will be studied in future.

7.   CONCLUSION

7.1 CIP Family

Recently there have been growing interest in the CIP method in Japan and many modified and advanced schemes appeared. One is the Differential Algebraic (DA) CIP method. In the scheme that we have introduced in this paper,

we treat the non-advection term by a finite difference method. Since the profile inside grid cell is already given, it is possible to estimate ∇p by taking direct differentiation of the cubic polynomial. Doing this, some care must be taken because for polynomial used in the CIP method, spatial derivative is not continuous at the grid point. Then for ∇p in the equation of motion is not continuous at the point where du/dt is calculated. In order to avoid this, space covering polynomial of u had better be shifted by a half of grid spacing relative to the space covering polynomial of p. This idea is similar to the staggerd grid system. Thus Utsumi[24] succeeded to capture shock wave propagation by algebraic procedure manipulating all the physical variables and those derivatives symbolically. In addition to the hyperbolic system, they extended the method to more general system including parabolic type equations.

In order to avoid staggerd space, Aoki extended the DA-CIP to higher-order polynomial by IDO (Interpolated Differential Operator)[25].

7.2 Summary

We have proposed a new tool to attack the simultaneous solution of all the materials. The success of the code is due to a high ability of tracing sharp interface even with fixed grid and flexibility of extension to various materials and physics. The region where the code can be applied is quite large. For example, the code is applied to the break-up[26] of a comet Shoemaker-Levy 9. Other applications are bubbly flow, combustion and chemical reaction, volcanic explosion, cavitation[27] and so on. Although some of them are not at present available in open literature, they will appear soon elsewhere.

ACKNOWLEDGEMENT

The author would like to thank T.Aoki, P.Y.Wang, F.Xiao, T.Utsumi for their help in developing CIP method and preparing this article.

REFERENCES

[1]   H.Takewaki, A.Nishiguchi and T.Yabe , 'The Cubic-Interpolated Pseudo-Particle (CIP) Method for Solving Hyperbolic-Type Equations'..*J. Comput. Phys.* **61**, 261-268, (1985).

[2]   T.Yabe *et al.*, 'A Universal Solver for Hyperbolc Equaitons by Cubic-Polynomial Interpolation', *Comput.Phys.Commun.* **66**,219-242, (1991).

[3]   T.Yabe and P.Y.Wang, 'Unified Numerical Procedure for Compressible and Incompressible Fluid', *J.Phys.Soc.Japan* ,**60**, 2105-2108 (1991) .

[4]   T.Yabe and F.Xiao, 'Description of Complex and Sharp Interface during Shock Wave Interaction with Liquid Drop', *J.Phys.Soc.Japan* **62**, 2537-2540, (1993) .

[5]   C.W.Hirt and B.D.Nichols, 'Volume of Fluid (VOF)

Method for the Dynamics of Free Boundaries', *J.Comput.Phys.* **39** , 201-225,(1981).

[6] S.Osher and J.A.Sethian, 'Fronts Propagating with Curvature-Dependent Speed : Algorithm based on Hamilton-Jacobi Formulation', *J.Comput.Phys.* **79**,12-49(1988).

[7] F.Xiao and T.Yabe, 'A Method to Trace Sharp Interface of Two Fluids in Calculation Involving Shocks', *Shock Waves* **4**, 101-108, (1994).

[8] F.Xiao, T.Yabe and T.Ito, 'Constructing Oscillation Preventing Scheme for the Advection Equation by a Rational Function': *Comput.Phys.Commn.* **93**, 1-12 (1996) .

[9] F.Xiao *et al.*, 'Constructing a Multi-Dimensional Oscillation Preventing Scheme for the Advection Equation by a Rational Function', *Comput. Phys .Commn.* **94**, 103-118 ,(1996).

[10] M.Ida and T.Yabe, 'Implicit CIP (Cubic-Interpolated Propagation) Method in One Dimension', *Comput.Phys.Comm.* **92**, 21-26 (1995).

[11] T.Yabe *et al.*, 'A Universal Solver for Hyperbolc Equaitons by Cubic-Polynomial Interpolation II.Two- and Three-Dimensional Solvers', *Comput.Phys. Commun.* **66**,233-242, (1991).

[12] T.Aoki, 'Multi-Dimensional Advection of CIP (Cubic-Interporate Propagation) Scheme', *CFD Journal,* **4**, 279-291, (1995).

[13] H.Makuuchi, T.Aoki and T.Yabe, 'Implicit CIP (Cubic-Interpolated Propagation) Method in Two Dimensions', submitted for publication (1997).

[14] T.Tanahashi and N.Tosaka, private communicatios.

[15] P.R.Woodward and P.Colella, 'The Numerical Simulation of Two-Dimensional Fluid Flow with Strong Shocks', *J.Comput.Phys.* **54**, 115-173, (1984).

[16] F.Xiao *et al*, 'An Efficient Model for Driven Flow and Application to GCB', *Comput. Model. & Sim. Eng.* **1** , 235-249,(1996) p.235.

[17] P.Y.Wang, T.Yabe and T.Aoki, 'A General Hyperbolic Solver - the CIP Method - Applied to Curvilinear Coordinate', *J.Phys.Soc.Japan,* **62**, 1865-1871, (1993).

[18] J.U.Brackbill, D.B.Kothe and C.Zemach, 'A Continuum Method for Modeling Surface Tension' *J.Comput.Phys.,* **100** , 335-354, (1992).

[19] F.Xiao, 'Numerical Schemes for Advection Equation and Multi-Layered Fluid Dynamics', Ph.D Thesis, Tokyo Institute of Technology (1996).

[20] T.Yabe *et al.*, 'Anomalous Crater Formation in Pulsed-Laser-Illuminated Aluminum Slab and Debris Distribution', *Research Report NIFS (National Institute for Fusion Science) Series, NIFS-417,* May (1996).

[21] A.Prosperetti and H.N.Oguz, ' The Impact of Drops on Liquid Surfaces and the Underwater Noise of Rain', *Annu. Rev. Fluid Mech.* **25**,577-602(1993).

[22] F.H.Harlow and J.P.Shannon,, 'The Splash of a Liquid Drop', *J.Appl.Phys.* **38**,3855-3866(1967).

[23] T.Yabe, P.Y.Wang ang Y.Zhang, 'Simulation Technique at Last Reached a Level that Reveals the Dynamics of Coronet Formation', submitted for publication (1997).

[24] T.Utsumi, 'Differential Algebraic Hydrodynamic Solver with Cubic-Polynomial Interpolation', CFD Journal, 4, 225-238, (1995).

[25] T.Aoki, 'IDO (Interpolated Differential Operator', to be published in *Comput.Phys.Commun.* (1997).

[26] T.Yabe *et al.*, 'Possible Explanation of the Secondary Flash and Strong Flare on IR Lightcurves upon Impact of Shoemaker-Levy 9". *Geophys. Res. Lett. ,* **22**, 2429-2432, (1995).

[27] T.Yabe and F.Xiao, 'Simulation Technique for Dynamical Evaporation Processes', *Nucl Eng & Design* **155**, 45-53 (1995).

THE NUMERICAL SIMULATION OF STRONGLY UNSTEADY FLOW WITH HUNDREDS OF MOVING BODIES

Rainald LÖHNER [1]   Joseph D. BAUM [2]
Chi YANG [1]   Hong LUO [2]

(1) GMU/CSI, The George Mason University; Fairfax, VA 22030-4444, USA

(2) Science Applications International Corporation;
1710 Goodridge Drive, MS 2-3-1, McLean, VA 22102, USA

Abstract

A methodology for the simulation of strongly unsteady flows with hundreds of moving bodies has been developed. An unstructured-grid, high-order, monotonicity preserving, ALF solver with automatic refinement and remeshing capabilities was enhanced by adding: equations of state for high explosives, deactivation techniques and optimal data structures to minimize CPU overheads, two new remeshing options, and a number of visualization tools for the pre- processing phase of large runs. The combination of these improvements has enabled the simulation of strongly unsteady flows with hundreds of moving bodies. Several examples demonstrate the effectiveness of the proposed methodology.

Received on September 10, 1997.

1 Introduction

A number of engineering applications require the prediction of strongly unsteady flowfields interacting with many (possibly hundreds) of moving bodies. Examples include flare and submunition deployment, fragmentation, and debris impact. In order to carry out simulations of this kind, the following requirements must be placed on the methodology used:

A accurate simulation of strongly unsteady flows with discontinuities;

Ability to handle several complex equations of state simultaneously;

Tracking and update of many independently moving bodies;

Correct treatment of body-body interactions, such as contact, spalation, etc.;

Rapid and error-free problem set-up and grid generation; and

Meaningful data reduction and visualization.

Over the years, we have developed and applied (see, e.g. [Ba91, Ba93, Ba95, Ba96]) a methodology to meet most of these requirements. The numerical techniques used for the Computational Fluid Dynamics (CFD) aspects are based on unstructured finite element techniques, using tetrahedral meshes, to treat complex geometries and /or physics. Extensive use is made of FEM-FCT [L*87] or other monotonicity preserving schemes [Lu94] to handle transient discontinuities. An arbitrary Lagrangean-Eulerian (ALE) frame of reference is employed for all equations, enabling the use of moving grids. Adaptive mesh refinement [L*92] is used extensively to track shocks and other discontinuities. Regions of elements distorted due to mesh motion are regridded automatically [L*90] using an advancing front technique [L*88, Sh95, L*96]. For the Computational Structural Dynamics (CSD) aspects, rigid bodies are treated through either 6 degree of freedom (6DOF) integrators, or a loose coupling [L*95] to impact bodies.

The current thrust is directed towards complex equations of state, better data structures for many-body applications, speed via deactivations of edges, better remeshing strategies for moving bodies, improved pre-processing and visualization tools, and validation through comparison to experimental data. This paper reports on progress made in each one of these areas, which has, synergistically, *resulted in the ability to simulate, on a farily routine basis*, flows that interact with hundreds of moving bodies.

2 Equations of State

Most high explosives are well modelled by the Jones-Wilkins-Lee (JWL) equation of state, given by:

$$p = A \left\{ 1 - \frac{\omega}{R_1 v} \right\} \exp^{-R_1 v} + \left\{ 1 - \frac{\omega}{R_2 v} \right\} \exp^{-R_2 v} + \omega \rho e , \tag{1}$$

where v denotes the relative volume of the gas:

$$v = \frac{V}{V_0} = \frac{\rho_0}{\rho} \tag{2}$$

Afterburni9ng is modelled by adding energy via a burn coefficient λ that is obtained from

$$\lambda_{,t} = a p^{1/6} \sqrt{1 - \lambda} , \tag{3}$$

where $\lambda = 0$ for unburned state, and $\lambda = 1$ for the fully burned material. After updating λ, the enrgy released is added as follows:

$$(\rho e)|^{n+1} = (\rho e)|^{n} + \rho Q(\lambda^{n+1} - \lambda^{n}) \ , \qquad (4)$$

where Q is the afterburn energy. Compared to the five unknowns required for the Euler equations with an ideal air equation of state, we require an additional two: the burn fraction b to determine with part of the material has ignited, and the afterburn coefficient λ. Observe that in the expanded state ($v \to \infty$), the JWL equation of state reduces to

$$p = \omega \rho e = (\gamma - 1)\rho e \ , \qquad (5)$$

where the correlation of ω and γ becomes apparent. The transition to air is made by comparing the density of air to the density of the high explosive. Given that $A >> B$, the decay of the first term in Eqn.(1) with increasing v is much faster. This implies that as v increases, we have

$$p \to Be^{-R_2 v} + \omega \rho e \ . \qquad (6)$$

The mixture of high explosive and air is considered as air when the effect of the B-term may be neglected, i.e.

$$\frac{p}{p_{cj}} = \epsilon = Be^{-R_2 v} \ , \qquad (7)$$

where p_{cj} denotes the Chapman-Jouget pressure and $\epsilon = O(10^{-3})$.

3. DEACTIVATION ZONES

Consider a typical explosion simulation. The major portion of CPU time is required to simulate the burning material. This is because the pressures are very high, and so are the velocities of the fluid particles. Once the material has burned out, one observes a drastic reduction of pressures and velocities, which implies a dramatic increase in the allowable timestep. Even though shocks travel much larger distances, this post-burn 'diffraction phase' takes less CPU-time than the burn phase. In order to speed up the simulation, the portions of the grid outside the detonation region are deactivated. The detonation velocity provides a natural speed beyond which no information can travel. Given that the major loops in an unstructured-grid flow solver are processed in groups (elements, faces, edges, etc.) for vectorization, it seemed natural to deactivate not the individual edge, but the edge-group. In this way, all inner loops can be left untouched, and the test for deactivation is carried out at the group-level. The number of elements in each edge-, face-, or element-group is kept reasonably small ($O(128)$) in order to obtain the highest percentage of deactivated edges without compromising performance. The points are renumbered

according to their ignition time assuming a constant detonation velocity. In this way, the work required for point-loops is minimized as much as possible. The edges and points are checked every 5-10 timesteps and activated accordingly. This deactivation technique leads to considerable savings in CPU at the beginning of run, where the timestep is very small and the zone affected by the explosion only comprises a small percentage of the mesh.

4. DYNAMICS AND INTERACTION OF MANY BODIES

The dynamics and interaction of many bodies pose a number of challenging problems. Two areas may be singled out as particularly critical:
- Contact algorithms; and
- Optimal data structures for the handling of body motion.

4.1 Contact Algorithms

For simple store separation problems (see [Sa96] for a cross-section of current capabilities), contact seldomly appears as an issue. However, for applications with many moving bodies, contact between the bodies is highly likely, and must therefore be accounted for. In order to treat contact, we employ a loose coupling algorithm [Lö95], linking the CFD code to an impact CSD code. CSD codes for impact have a long tradition of efficient contact algorithms, and it seemed thefore prudent to follow this path. The bodies in the flowfield are discretized using unstructured 8-noded brick finite elements, and are treated as either rigid, elastic, or elasto-plastic materials. The contact algorithm intrinsic in these codes then accounts for body-body interactions. Another option to treat this problem is via a potential field that has a repelling force as bodies come close to each other. This option is being explored at present.

4.2 Optimal Data Structures

Traditional applications of fluids interacting with moving bodies only considered a limited number of moving bodies. For example, store separation applications seldomly include more than 1-5 moving bodies [Sa96]. Computing with such a low number of moving bodies implies that when doing force evaluations, point movement, etc., the CPU penalty incurred by careless coding is barely noticeable. The situation reverses when one faces problems with hundreds of moving bodies. As an example, consider the evaluation of body forces and moments from surface pressures. A simplistic way to evaluate these would be to loop over all the faces, filter the ones belonging to a given body, and then sum up forces and moments for this reduced list of faces. This implies nbody-loops over the faces, where nbody is the number of bodies in the flowfield. It is clear that for a large number of bodies, the CPU penalty incurred by such a procedure is severe. A sample run with more than 200 bodies and several million elements indicated that body motion required 40% of the overall CPU time. For this reason, a number of data structures were implemented to arrive at optimal speeds for the handling of body motion. Among these, the following are of particular relevance:

- Linked lists to gain rapid access to the faces comprising a body (force evaluation);
- Linked lists to gain rapid access to the points comprising a body (movement of points on a body);
- Coloured list of edges in the moving mesh portion (mesh movement).

With these data structures the overhead for force, moment and rigid body point movement calculations could be reduced to less that 5% of the overall CPU time.

5. REMESHING STRATEGIES

Any field simulation with boundaries that undergo severe movement will need some form of mesh adjustment to cope with the change of spatial resolution dictated by the geometry and the physics. Nonconforming grids (e.g. Cartesian grids) do this by refining and coarsening the grid, with a subsequent adjustment of boundary conditions at the surface. Overlapping grids change the interpolation information in the overlap zones. Unstructured, conforming grids typically remove deformed elements and remesh the voids thus created. To date, we have used a combination of local and global remeshing to solve this class of problems. Over the last year, it has become apparent that two other ways of remeshing are also very useful:

a) Remeshing only the ALE region, and
b) Excluding the highly distorted Navier-Stokes region grids from remeshing.

The first option is particularly attractive if the number of deforming (moving) mesh layers surrounding bodies in motion comprises only a fraction of the total volume. Global remeshing is comparatively expensive in this case, with no additional advantage. The second option is essential for RANS simulations. For this class of problems, the regions of highly stretched elements close to the bodies in motion are moved in a rigid fashion, just as the surface points. These two remeshing options have improved dramatically the ability to simulate problems with many moving bodies.

6. PREPROCESSING TOOLS

The definition of boundary conditions, surface data, and desired elements size and shape in space are tedious enough for problems with a few moving bodies.

They become onerous for hundreds of moving bodies. Errors in the input data become impossible to discern without a graphical, intuitive pre-processing tool. We have therefore implemented in our FECAD preprocessing tool a number of features to define and check such items as surface geometry, boundary conditions, body-number, background sources, size attached to CAD-data, etc. It is hard to underestimate the benefit of this graphical pre-processing toolkit. Suffice it to say that without it, it would been impossible to conduct the calculations performed.

7. NUMERICAL EXAMPLES

7.1 <u>Weapon Fragmentation</u>: As a first example, we consider a weapon fragmentation experiment conducted recently. This case had approximately 200 fragments. Each of the fragments is treated as a separate, freely flying body. The weapon case was serrated in order to achieve reproducability of experiments. The high explosive was modelled using a JWL equation of state. The flow solver option employed is edge-based FEM-FCT. At $t = 0.0$, the minimum distance between the fragments is of 0.5 mm, while the average size of each fragment is 30 mm × 20 mm × 15 mm. In contrast, the room size is several meters. Typical meshes for this simulation were of the order of 2.5 Mtets, and the simulations required of the order of 8 hours on the CRAY C-90. Figure 1 shows the fragments at a certain time, together with their projected impact points on the walls, and the comparison to experiments. Given the complexity of the physical phenomena being modelled, the correlation with experimental data is surprisingly good.

7.2 <u>Weapon Fragmentation</u>: In this second example, a fragmenting weapon with approximately 500 fragments is simulated. As before, each of the fragments is treated as a separate, freely flying body. Figure 2 shows the fragments at a certain time, together with some of the flow quantities.

8. CONCLUSIONS AND OUTLOOK

A methodology for the simulation of strongly unsteady flows with hundreds of moving bodies has been developed. The numerical techniques used for the Computational Fluid Dynamics (CFD) aspects are based on unstructured finite element techniques, using tetrahedral meshes, to treat complex geometries and/or physics. Extensive use is made of FEM-FCT or other monotonicity preserving schemes to handle transient discontinuities. An arbitrary Lagrangean-Eulerian (ALE) frame of reference is employed for all equations, enabling the use of moving grids. Adaptive mesh refinement is used extensively to track shocks

and other discontinuities, and regions of elements distorted due to mesh motion are regridded automatically. High explosives are modelled using the Jone-Wilkins-Lee (JWL) equation of state. A deactivation technique has been implemented to minimize the CPU requirements during the burning phase of the explosive. Optimal data structures are included to minimize CPU overheads for problems with many bodies. For the Computational Structural Dynamics (CSD) aspects, rigid or bodies are treated through either 6 degree of freedom (6DOF) integrators, or a loose coupling to impact codes. Two new remeshing options for the class of problems considered here have decreased significantly CPU requirements. In addition, several pre-processing tools have been developed to define and check input data. This, as it turns out, is one of the keys to successfully conducting simulations with hundreds of moving bodies.
The combination of these different improvements has **resulted in the ability to simulate flows that interact with hundreds of moving bodies.**
Future developments will center on more sophisticated equations of state, improved diagnostics and simplified data reduction.

9. ACKNOWLEDGMENTS

This work was partially supported by DSWA, with Dr. Michael Giltrud as the technical monitor, as well as AFOSR, with Dr. Leonidas Sakell as the technical monitor.

10. REFERENCES

[Ba91] J.D. Baum and R. Löhner - Numerical Simulation of Shock Interaction with a Modern Main Battlefield Tank; *AIAA-91-1666* (1991).

[Ba93] J.D. Baum, H. Luo and R. Löhner - Numerical Simulation of a Blast Inside a Boeing 747; *AIAA--93-3091* (1993).

[Ba95] J.D. Baum, H. Luo and R. Löhner - Numerical Simulation of Blast in the World Trade Center; *AIAA-95-0085* (1995).

[Ba96] J.D. Baum, H. Luo, R. Löhner, C. Yang, D. Pelessone and C. Charman - A Coupled Fluid/Structure Modeling of Shock Interaction with a Truck; *AIAA-96-0795* (1996).

[Lö87] R. Löhner, K. Morgan, J. Peraire and M. Vahdati - Finite Element Flux-Corrected Transport (FEM-FCT) for the Euler and Navier-Stokes Equations; *Int. J. Num. Meth. Fluids* 7, 1093-1109 (1987).

[Lö88] R. Löhner and P. Parikh - Three-Dimensional Grid Generation by the Advancing Front Method; *Int. J. Num. Meth. Fluids* 8, 1135-1149 (1988).

[Lö90] R. Löhner - Three-Dimensional Fluid-Structure Interaction Using a Finite Element Solver and Adaptive Remeshing; *Computer Systems in Engineering* 1, 2-4, 257-272 (1990).

[Lö92] R. Löhner and J.D. Baum - Adaptive H-Refinement on 3-D Unstructured Grids for Transient Problems; *Int. J. Num. Meth. Fluids* 14, 1407-1419 (1992).

[Lö95] R. Löhner, C. Yang, J. Cebral, J.D. Baum, H. Luo, D. Pelessone and C. Charman - Fluid-Structure Interaction Using a Loose Coupling Algorithm and Adaptive Unstructured Grids; *AIAA-95-2259* (1995).

[Lö96] R. Löhner - Progress in Grid Generation via the Advancing Front Technique; *Engineering with Computers* 12, 186-210 (1996).

[Lu94] H. Luo, J.D. Baum and R. Löhner - Edge-Based Finite Element Scheme for the Euler Equations; *AIAA J.* 32, 6, 1183-1190 (1994).

[Sa96] L. Sakell and D.D. Knight (eds.) - *Proc. 1st AFOSR Conf. on Dynamic Motion CFD*, Rutgers University, New Brunswick, New Jersey, June (1996).

[Sh95] A. Shostko and R. Löhner - Three-Dimensional Parallel Unstructured Grid Generation; *Int. J. Num. Meth. Eng.* 38, 905-925 (1995).

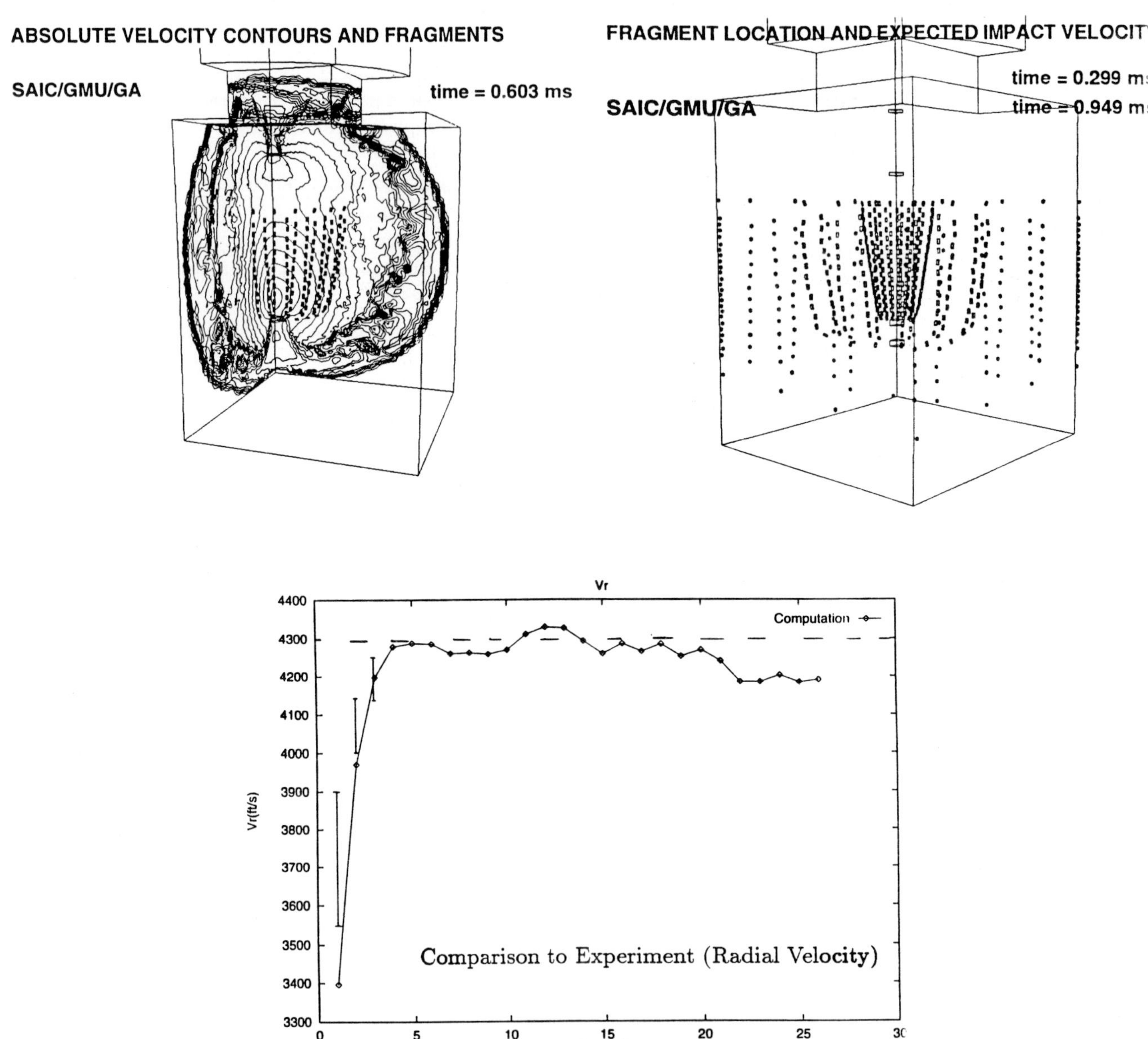

Fragmenting Object (JWL EOS, **ALE**, **200** Objects, Moving Meshes, Automatic Regridding)

NUMERICAL SIMULATION OF BLAST AND FRAGMENT IMPACT

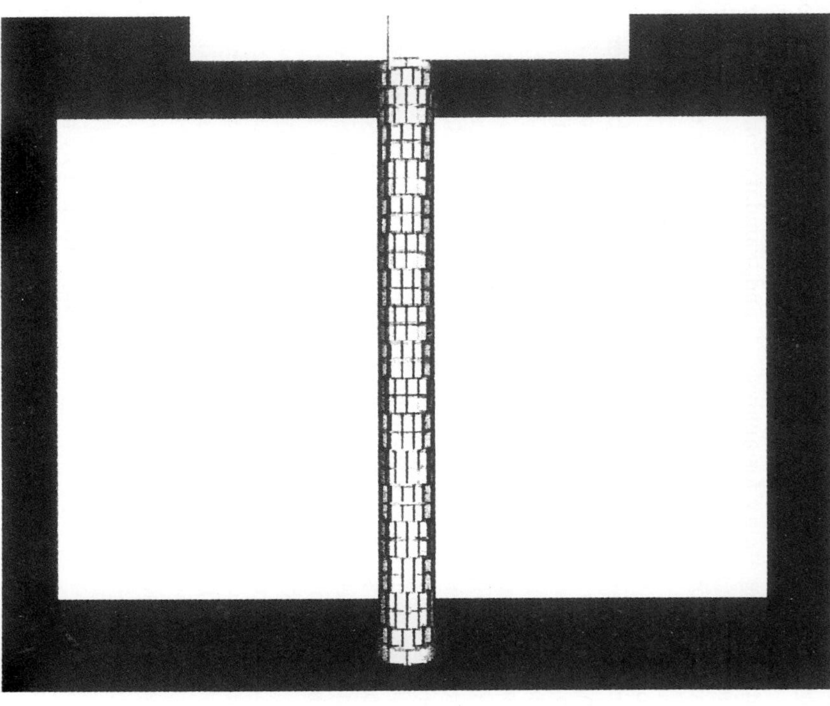

time = 0.0

time = 0.88 ms

ACCURATE FINITE DIFFERENCE ALGORITHMS FOR COMPUTATIONAL AEROACOUSTICS

John W. GOODRICH [1]   Jay HARDIN [2]

1) NASA Lewis Research Center; leveland, OH 44135
2) NASA Langley Research Center; Hampton, VA 23681

Abstract

A new method is presented for developing finite difference algorithms that are high order in both space and time, and several example algorithms are compared. Differences in efficiency of several orders of magnitude are shown, even for propagation through a few periods. Accurate and efficient propagation to $O[10^5]$ periods is shown to be possible with high order and high resolution finite difference algorithms. High resolution examples have spectral like properties. Two dimensional examples correctly propagate information along characteristic surfaces. Well posed, stable, and convergent artificial boundary conditions are reported.

1 Introduction

Computational Aeroacoustics (CAA) will be defined for the purposes of this paper as the use of Computational Fluid Dynamics (CFD) techniques for the direct calculation of any aspect of sound generation and propagation in air. CAA is a relatively new field of research, and has improved the ability for analysis of aeroacoustic phenomena while maintaining typical CFD capabilities for handling nonlinear equations in complex geometries. The most important challenge for CAA from the aeroacoustic viewpoint is the direct simulation of acoustic sources from the underlying fluid dynamics. CAA is also useful for the simulation of acoustic propagation, especially in the cases of a nonuniform mean flow, of a medium with nonuniform properties, and of scattering from multiple or complex objects. In order to faithfully generate and propagate high frequency acoustic waves, algorithms for CAA must have high accuracy in both space and time. Much recent work has been done to introduce improved algorithms into CAA simulations. Accurate versions of well established techniques such as the MacCormack method have been adapted for aeroacoustic applications [1]. New techniques such as the Dispersion Relation Preserving methods [2] have been developed in order to improve accuracy, with associated boundary conditions [3]. The accuracy and efficiency of time stepping for Dispersion Relation Preserving methods has been increased [4]. New Perfectly Matched Layer absorbing boundary conditions have been adapted to CAA [5]. Discontinuous Galerkin

Methods are being used to develop accurate algorithms for aeroacoustic simulations [6]. A fuller review of the current state of CAA has been done by Lele [7], with many recent references.

This paper reports on a particular line of active research that is intended to produce accurate algorithms and boundary conditions for CAA. A brief overview of CAA problems is given in Section 2 in order to establish a context for the algorithm developments that are presented in this paper. The Linearized Euler Equations in one space dimension are considered in Section 3 as a model problem for algorithm development and comparison. In Section 3.1 a general technique for algorithm development is presented, with examples of single step central methods that are up to eighth order accurate in both space and time. Accurate and stable boundary algorithms have been developed for these methods. Section 3.2 considers algorithms with Hermitian interpolation, and introduces single step explicit algorithms on alternating grids that are up to eleventh order accurate in both space and time. The algorithms from Sections 3.1 and 3.2 are compared by means of results from numerical experiments in Section 3.3. Particular attention is given to comparing accuracy and efficiency, and high order algorithms with high resolution are seen to be able to efficiently and accurately propagate to a true far field. An example is shown of propagation with eight grid points per wavelength through 50,000 periods with a maximum absolute error of only $O[10^{-6}]$. A normal mode analysis of high resolution algorithms is reported in Section 4, showing truly spectral like quantities for the

Received on July 1, 1997.

algorithms that are considered, with phase speeds that are virtually one for normal mode frequencies $\theta \in [0, \pi]$, and CFL numbers $\lambda \in [0, 1]$. The Linearized Euler Equations in two space dimensions are discussed in Section 5 as an example of nondiagonalizable hyperbolic systems in multiple space dimensions. In Section 5.1 a method for algorithm development is shown that correctly incorporates the propagation of information along characteristic surfaces. Section 5.2 presents grid refinement data from numerical experiments with algorithms for two space dimensions. Section 6 briefly presents new nonreflecting artificial boundary conditions for the Linearized Euler Equations in two space dimensions. The boundary conditions are formulated as a series of conditions that use an increasing number of auxiliary functions. It has been shown that each of the approximate artificially bounded problems is strongly well posed, and that the solutions to the sequence of problems with artificial boundaries converge exponentially to the solution of the original problem on an unbounded domain [11]. Fuller details for the results reported in this paper can be found in [8, 9, 10, 11].

2   Computational Aeroacoustics

In order to directly calculate sound generation and propagation from a fluid flow, CAA must deal with the equations for flows in a compressible medium. Batchelor [2] gives these equations in air as:

Conservation of Mass

$$\frac{\partial \rho}{\partial t} + \frac{\partial \rho u_i}{\partial x_i} = \dot{m}, \tag{1}$$

Conservation of Momentum

$$\frac{\partial \rho u_i}{\partial t} + \frac{\partial \rho u_i u_j + p_{ij}}{\partial x_i} = \dot{F}_i, \tag{2}$$

Equation of State

$$p = p(\rho, S) = \rho R T, \tag{3}$$

Energy Equation

$$T\frac{DS}{Dt} = c_p \frac{DT}{Dt} - \frac{\beta T}{\rho}\frac{Dp}{Dt} = \phi + \frac{1}{\rho}\frac{\partial}{\partial x_i}(k\frac{\partial T}{\partial x_i}), \tag{4}$$

with

$$p_{ij} = p\delta_{ij} + \mu[-\frac{\partial u_i}{\partial x_j} - \frac{\partial u_j}{\partial x_i} + \frac{2}{3}(\frac{\partial u_k}{\partial x_k})\delta_{ij}],$$

where p is pressure, ρ is density, the u_i are the velocity components, T is temperature, S is entropy,

μ is the viscosity coefficient, c_p is the specific heat at constant pressure, $\beta = -\frac{1}{\rho}(\frac{\partial \rho}{\partial T})_p$ is the thermal expansion coefficient, ϕ is the dissipation function proportional to the viscosity coefficient, k is the thermal conductivity of the fluid, R is the gas constant, $\dot{m}$ is the rate of mass introduction per unit volume, and the F_i are the body force components per unit volume.

Since $\beta = \frac{1}{T}$, the Equation of State (3) can be used for integrating the Energy Equation (4), producing the relation

$$p(\rho, S) = \rho^\gamma \exp(\frac{S - S_0}{c_v}),$$

where $\gamma = 1.4$, S_0 is the initial entropy, and c_v is the specific heat at constant volume. If the flow is isentropic, then $\frac{p}{\rho^\gamma}$ is constant. This set of equations supports the propagation of acoustic waves at speed c given by

$$c^2(\rho, p) = (\frac{\partial p}{\partial \rho})_S = \frac{\gamma p}{\rho} = \gamma R T.$$

Sound may be produced by variable mass introduction into a medium, or by fluctuating forces acting on a medium. The production of sound by these classical sources is well understood in cases such as musical instruments. Sound is also frequently produced by flows in which these two classical sources are absent. Lighthill [21] has analysed this case and shown that Equations (1) and (2) can be manipulated to yield the inhomogeneous linear wave equation

$$\frac{\partial^2 \rho}{\partial t^2} - c_0^2 \nabla^2 \rho = \ddot{m} - \frac{\partial F_i}{\partial x_i} + \frac{\partial^2 T_{ij}}{\partial x_i \partial x_j}, \tag{5}$$

where

$$T_{ij} = \rho u_i u_j + (p - c_0^2 \rho)\delta_{ij} + \mu[-\frac{\partial u_i}{\partial x_j} - \frac{\partial u_j}{\partial x_j} + \frac{2}{3}(\frac{\partial u_k}{\partial x_k})\delta_{ij}]$$

is the Lighthill stress tensor, and c_0 is the speed of sound in the ambient medium surrounding the sources. This equation suggests that in addition to the two classical sources, there is a third sound source due to the action of the Lighthill stresses on the medium. If these stresses were known, then the resulting acoustic field could readily be calculated from Equation (5) in the manner of a forced oscillation. Unfortunately, an examination of the Lighthill stress tensor shows that these stresses depend upon the acoustic field and cannot be known independently. This dilemma led to the development of approximations to the Lighthill stress tensor in terms of known quantities, including the case where surfaces are present. This research has led to useful results in the cases of supersonic jets [23], and of rotating blades [5]. The approximation of the source terms in these cases is not independent of some calculation of the basic flow. The CAA approach is to solve the governing equations directly, rather than to manipulate the governing equations to obtain Equation (5), with subsequent approximation of the source terms. In this CAA approach, the source terms arise naturally from the flow solution and do not require modelling. This idea was initially met with much trepidation because acoustic waves have relatively small amplitude and a wide frequency range. The direct calculation of sound generation and propagation from the fluid flow raises algorithmic issues because of the significant dissipation and dispersion errors of standard CFD methods, and because of the need for nonreflecting artificial boundary conditions for time dependent problems.

If a steady flow field is known, then the Linearized Euler Equations can be used for the simulation of the propagation of acoustic signals. The underlying steady flow can be either constant or varying in space. In the isentropic case with a constant mean convection field, and in cartesian coordinates, the equations for the acoustic disturbance can be written as:

Linearized Euler Equations

$$\frac{\partial u_j}{\partial t} + U_i \frac{\partial u_j}{\partial x_i} + \frac{\partial p}{\partial x_j} = 0,$$

$$\frac{\partial p}{\partial t} + U_i \frac{\partial p}{\partial x_i} + \frac{\partial u_i}{\partial x_i} = 0, \tag{6}$$

where p is the pressure and u_i is the ith velocity component of the acoustic disturbance, and U_i is the ith velocity component of the mean convection field. A discussion of this system can be found in Kreiss and Lorenz [18]. The performance of standard CFD techniques is readily analyzed when they are applied to the Linearized Euler Equations, and their properties stand out in sharp relief. It is actually quite difficult to accurately and efficiently simulate wave propagation over any considerable distance. Standard finite difference techniques have difficulty accurately propagating a signal beyond $O[10]$ periods without many grid points per wavelength. The algorithmic issues of dissipation and dispersion, resolution, accuracy, efficiency, artificial boundary conditions, and geometric complexity are all significant if propagation to a relatively far field with a broadband signal is to be attempted. Sound propagation is only simpler than source characterization because the nonlinear fluid dynamics do not have to be directly calculated if an acceptable flow field is provided.

Two CAA workshops have recently been held in order to validate the results of CAA, and to compare various CAA algorithms. Contributors were challenged to solve benchmark problems with whatever methods they preferred, and the results were compared with independently obtained solutions. Both workshops included linear and nonlinear problems. The first CAA workshop involved classical problems, and the published results [13] are generally good. The second CAA workshop included four categories of more realistic problems that are emblematic of technological applications, with results that will appear [12]. The first category of problem in the second CAA workshop is for sound from a point (line) source that is scattered by a sphere (cylinder) in three (two) space dimensions. The second category of problem is for an incoming duct mode propagating through a duct and radiating into open space. This problem was posed in cylindrical coordinates, and a schematic is shown in Figure 1. The third cat-

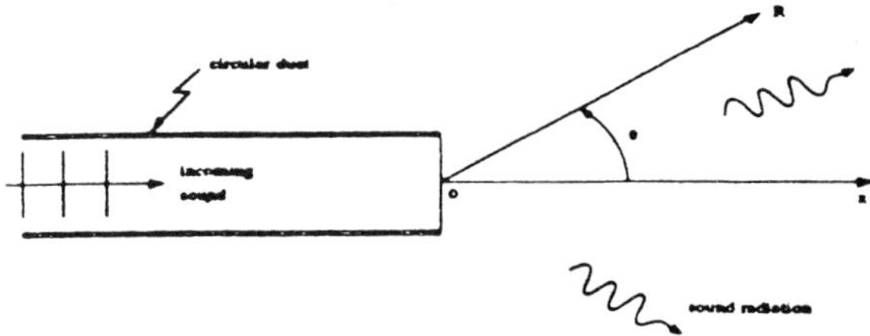

Figure 1: Duct Mode Propagation

egory problem is for an incoming gust scattered from a stationary y-periodic cascade of flat plates. Figure 2 shows a schematic for this problem. One of the cases proposed for the third category was propagation through a sliding downstream interface to the location of a second rotating cascade. The workshop results for the problems from these three categories are in good agreement with semi-analytic solutions. It should be noted that the acoustic sources

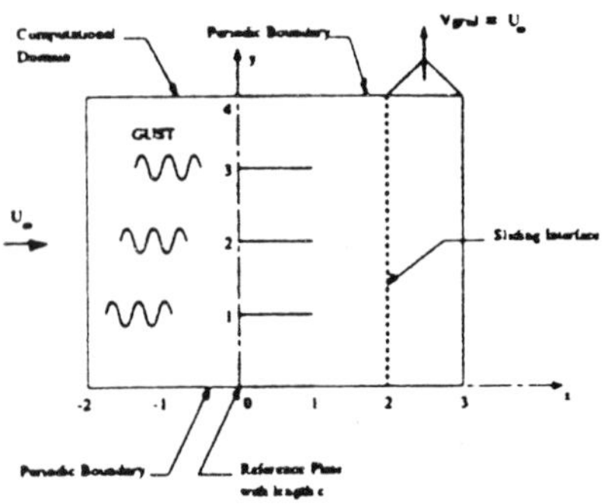

Figure 2: Flat Plate Cascade

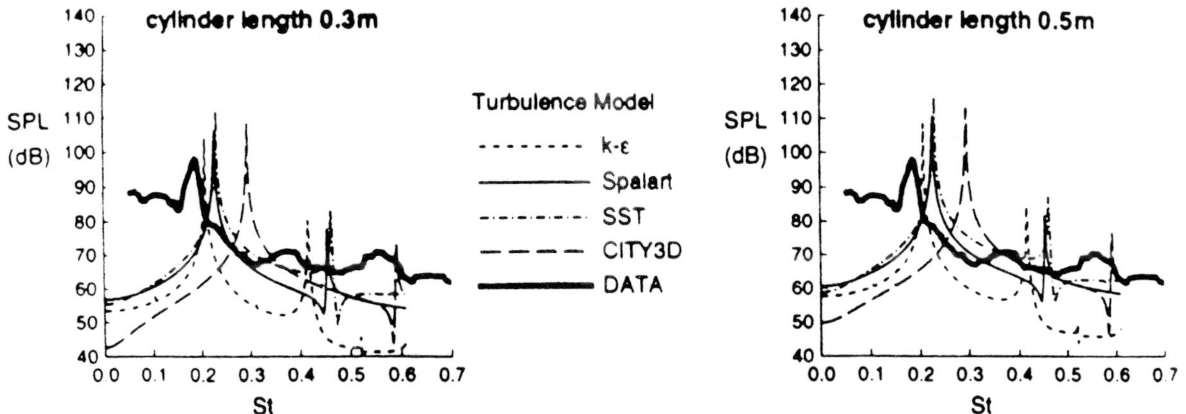

Figure 3: Aeolian Tone Spectra

are specified for these first three linear problems, and that sound propagation is with the Linearized Euler Equations in various coordinate systems. The fourth category of problem for the second CAA workshop represents the Aeolian tone generated by a cylinder in uniform flow, and is emblematic of noise generation by both automobiles and aircraft. The flow in this case is fundamentally viscous and massively separated, and the sound is generated by the flow. Solutions were requested at a Mach number of 0.2 and a Reynolds number of 90,000. These conditions create a flow which cannot be fully resolved by any current computer capability, with significant flow information at scales that range from the acoustic wavelengths down to the viscous flow scales. No semi-analytic solution is possible for this problem, so the numerical results were compared with experimental data. The comparisons were somewhat disapointing for the participants. Most of the solutions were reasonably close to the proper Strouhal number for the peak sound generation, and also demonstrated the correct dipole directivity pattern, but the spectral amplitudes varied dramatically. Typical results are shown in Figure 3, where the spectra from experimental data are compared with results from direct CAA calculations for two different cylinder lengths with several turbulence models. The variation between turbulence models is particularly noticable, but perhaps should not be suprising, since the turbulence models were developed to predict time averaged properties of turbulent flows, and in this situation they are being used to simulate time dependent phenomena. The results of these two workshops show that good progress has been made in CAA, but that there is still a need for improved methods in order to efficiently and accurately conduct CAA simulations. Progress is needed in both propagation algorithms and artificial boundary conditions, with a general requirement for greater accuracy in computed CAA results, and for greater efficiency in obtaining them.

3 Linearized Euler Equations in 1D

We will consider the Linearized Euler Equations (6) in one space dimension as a vehicle for introducing a new method for the development of finite difference algorithms that are accurate in both space and time. The nondimensionalized system for the isentropic case can be written as:

$$\frac{\partial u}{\partial t} + M\frac{\partial u}{\partial x} + \frac{\partial p}{\partial x} = 0,$$

$$\frac{\partial p}{\partial t} + M\frac{\partial p}{\partial x} + \frac{\partial u}{\partial x} = 0, \tag{7}$$

where u is the velocity, p is the pressure, and M is the constant mean convection Mach number. The Linearized Euler Equations in one space dimension (7) can be diagonalized and written in the equivalent decoupled form

$$\frac{\partial \omega_1}{\partial t} + (M-1)\frac{\partial \omega_1}{\partial x} = 0,$$

$$\frac{\partial \omega_2}{\partial t} + (M+1)\frac{\partial \omega_2}{\partial x} = 0,$$

where $\omega_1 = \frac{1}{2}(u-p)$ and $\omega_2 = \frac{1}{2}(u+p)$. The initial value problem for u and p with $u(x,0) = u_o(x)$ and $p(x,0) = p_o(x)$ for $x \in (-\infty, +\infty)$ is equivalent to decoupled inital value problems for the Riemann variables ω_1 and ω_2 with $\omega_1(x,0) = \frac{1}{2}(u_o(x) - p_o(x))$ and $\omega_2(x,0) = \frac{1}{2}(u_o(x) + p_o(x))$. The decoupled problems for ω_1 and ω_2 can be solved by the Method of Characteristics, and these solutions can be combined to provide a general solution for u and p, with

$$
\begin{aligned}
u(x,t) = \ &+\tfrac{1}{2}u_o(x-(M+1)t) \\
&+\tfrac{1}{2}p_o(x-(M+1)t) \\
&+\tfrac{1}{2}u_o(x-(M-1)t) \\
&-\tfrac{1}{2}p_o(x-(M-1)t), \\[2mm]
p(x,t) = \ &+\tfrac{1}{2}u_o(x-(M+1)t) \\
&+\tfrac{1}{2}p_o(x-(M+1)t) \\
&-\tfrac{1}{2}u_o(x-(M-1)t) \\
&+\tfrac{1}{2}p_o(x-(M-1)t).
\end{aligned}
\tag{8}
$$

This system has two characteristics, and separate solutions traveling with the characteristic velocities $M \pm 1$, respectively.

3.1 Algorithm Development

We will present a general approach for developing explicit finite difference algorithms by applying it to the Linearized Euler Equations (7). The first step in our general algorithm development is to interpolate the known data at t_n about the grid point x_i with order D polynomials in x,

$$
\begin{aligned}
u(x_i+x,t_n) \approx u_a(x) &= \textstyle\sum_{\delta=0}^{D} u_\delta x^\delta, \\
p(x_i+x,t_n) \approx p_a(x) &= \textstyle\sum_{\delta=0}^{D} p_\delta x^\delta.
\end{aligned}
\tag{9}
$$

The Method of Undetermined Coefficients is used with the known data on the grid to determine the expansion coefficients u_δ and p_δ. Note that particular choices of mesh and data are not specified, and that approximation of separate derivatives is not considered. Local polynomial approximation is equivalent to using a Taylor series expansion, and the expansion coefficients can be interpreted as spatial derivatives, with

$$
u_\delta = \frac{1}{\delta!}\frac{\partial^\delta u_a}{\partial x^\delta} \approx \frac{1}{\delta!}\frac{\partial^\delta u}{\partial x^\delta},
$$

$$
p_\delta = \frac{1}{\delta!}\frac{\partial^\delta p_a}{\partial x^\delta} \approx \frac{1}{\delta!}\frac{\partial^\delta p}{\partial x^\delta}.
\tag{10}
$$

For the sake of simplicity we will assume a uniform grid spacing, with space mesh size $h = \Delta x$, time step size $k = \Delta t$, and grid ratio $\lambda = \frac{k}{h}$. For the case of solution data specified on a uniform grid, the formulas for the expansion coefficients are easily recognized as multiples of familiar finite difference forms. A key feature of our approach to algorithm development is that overall accuracy is determined just by the order D of the spatial interpolation polynomial (9), and that the order of accuracy is independent

from the grid geometry or the type of data. This feature is independent from spatial dimension.

The second step in our general algorithm development is to propagate the local approximation to known solution data on a given grid, in this case by the general solution (8) which is an exact local propagator. With the local spatial interpolant as initial data, the local solution is approximated in space and time by

$$
\begin{aligned}
u(x_i+x,t_n+t) \approx \ &+\tfrac{1}{2}u_a(x-(M+1)t) \\
&+\tfrac{1}{2}p_a(x-(M+1)t) \\
&+\tfrac{1}{2}u_a(x-(M-1)t) \\
&-\tfrac{1}{2}p_a(x-(M-1)t), \\[2mm]
p(x_i+x,t_n+t) \approx \ &+\tfrac{1}{2}u_a(x-(M+1)t) \\
&+\tfrac{1}{2}p_a(x-(M+1)t) \\
&-\tfrac{1}{2}u_a(x-(M-1)t) \\
&+\tfrac{1}{2}p_a(x-(M-1)t).
\end{aligned}
\tag{11}
$$

The local solution approximation (11) is an exact solution of the Linearized Euler Equations (7), and correctly incorporates its wave dynamics for the initial data (9). The solution at the grid point x_i for the new time t_{n+1} is

$$
\begin{aligned}
u_i^{n+1} =\ &+\tfrac{1}{2}u_a(-(M+1)k) \\
&+\tfrac{1}{2}p_a(-(M+1)k) \\
&+\tfrac{1}{2}u_a(-(M-1)k) \\
&-\tfrac{1}{2}p_a(-(M-1)k) \\
\approx\ &u(x_i,t_n+k), \\[2mm]
p_i^{n+1} =\ &+\tfrac{1}{2}u_a(-(M+1)k) \\
&+\tfrac{1}{2}p_a(-(M+1)k) \\
&-\tfrac{1}{2}u_a(-(M-1)k) \\
&+\tfrac{1}{2}p_a(-(M-1)k). \\
\approx\ &p(x_i,t_n+k).
\end{aligned}
\tag{12}
$$

The algorithm form (12) represents a family of algorithms with various properties that depend upon a specification of the interpolant (9).

There are four related interpretations of the general solution approximation (11) and algorithm form (12). The most direct interpretation is that both forms represent a locally exact solution derived from the Method of Characteristics with a local truncated Taylor series approximation in space for the initial data. If the expansion coefficients from the interpolant (9) are interpreted as spatial derivatives (10), then the solution approximation (11) can be seen as a locally exact polynomial solution derived from a truncated Cauchy-Kovalevskaya series expansion in space and time. If the solution approximation (11) is restricted to the grid point x_i, then it can be reformulated as a truncated Taylor series expansion in time, and the algorithm form (12) can be recast as an expansion in k,

$$
\begin{aligned}
u_i^{n+1} = \ & u_0 - k(p_1 + Mu_1) \\
& + k^2(2Mp_2 + (M^2 + 1)u_2) \\
& - k^3((1 + 3M^2)p_3 + M(3 + M^2)u_3) \\
& + \ldots,
\end{aligned}
$$

$$
\begin{aligned}
p_i^{n+1} = \ & p_0 - k(u_1 + Mp_1) \\
& + k^2(2Mu_2 + (M^2 + 1)p_2) \\
& - k^3((1 + 3M^2)u_3 + M(3 + M^2)p_3) \\
& + \ldots,
\end{aligned}
\tag{13}
$$

where the grid ratio $\lambda = \frac{k}{h}$ is implicit, since the expansion coefficients u_δ and p_δ have h^δ in their denominators. If the formulas for the expansion coefficients from the interpolant (9) are used, then algebraic manipulation of the algorithm form (12) can produce a conventional explicit finite difference method in the form of a weighted sum of the stencil values of the solution at time t_n. The first interpretation ensures that the numerical method properly represents the wave dynamics of the partial differential equation, since the characteristic behavior of the solution is built into the numerical algorithm. The second interpretation provides an avenue for generalization and extension, since local analytic solutions are possible for hyperbolic systems in higher space dimensions. The third and fourth interpretations ground this family of algorithms in the tradition of finite difference methods. A key shift in perspective is toward the approximation of a solution of a system of partial differential equations, instead of focusing on the details of approximating separate terms in an equation.

We will consider four algorithms with uniform grids, central stencils, and values for u and p as data at each grid point. These four algorithms are on stencils with three, five, seven and nine grid points, with interpolants that are of second, fourth, sixth, and eighth order, and are labeled the c3o0ex, c5o0ex, c7o0ex, and c9o0ex algorithms, respectively. The c3o0ex method is the second order Lax-Wendroff method [19]. All of the algorithms are single step explicit methods with dispersive truncation errors, and each is stable if $\frac{k}{h} \leq \frac{1}{1+|M|}$, for any M. The order of accuracy in both space and time for each of the methods is the same as the order of interpolation for the method. Stable boundary algorithms for these methods have been developed, with the same orders of accuracy as the propagation algorithms, up to eighth order accuracy in space and time. These high order boundary algorithms are based on the characteristics of Equation (7), and depend upon

a consistent interpolation and coherent propagation of the solution over a half stencil boundary interval. Further details on this family of algorithms and their boundary conditions can be found in [8].

3.2 Hermitian Interpolation

We will introduce an additional family of algorithms based upon exact propagators with Hermitian spatial interpolation. In addition to the solution data for u and p, this family of algorithms may use and compute their spatial derivatives at each grid point. Particular members of this family of algorithms are distinguished by the stencil and the number of derivatives that are used.

If the solution data at t_n is approximated with local polynomial interpolants (9), then the first spatial derivatives of u and p are approximated by

$$
\begin{aligned}
\frac{\partial u}{\partial x}(x_i + x, t_n) &\approx \frac{\partial u_a}{\partial x}(x) = \sum_{\delta=1}^{D} \delta u_\delta x^{\delta-1}, \\
\frac{\partial p}{\partial x}(x_i + x, t_n) &\approx \frac{\partial p_a}{\partial x}(x) = \sum_{\delta=1}^{D} \delta p_\delta x^{\delta-1}.
\end{aligned}
\tag{14}
$$

Local approximations for other derivatives are obtained similarly. For algorithms that use the solution and its first derivative, interpolation is by the Method of Undetermined Coefficients with constraint equations from (9) for the solution data, and from (14) for the derivative data. This interpolation directly produces local approximations only for the solution components u and p, and approximations for their spatial derivatives are obtained from (14). As an example, if a third order interpolant is used on a two point stencil with data for a function and its x derivative at both points, then the x derivative of this interpolant yields a local second order approximation of the x derivative of the function. Local exact solution forms for spatial derivatives that are consistent with (11) are obtained by spatial differentiation of (11), with

$$
\begin{aligned}
\frac{\partial u}{\partial x}(x_i + x, t_n + t) \approx \ & \frac{1}{2}\frac{\partial u_a}{\partial x}(x - (M+1)t) \\
& + \frac{1}{2}\frac{\partial p_a}{\partial x}(x - (M+1)t) \\
& + \frac{1}{2}\frac{\partial u_a}{\partial x}(x - (M-1)t) \\
& - \frac{1}{2}\frac{\partial p_a}{\partial x}(x - (M-1)t),
\end{aligned}
$$

$$
\begin{aligned}
\frac{\partial p}{\partial x}(x_i + x, t_n + t) \approx \ & \frac{1}{2}\frac{\partial u_a}{\partial x}(x - (M+1)t) \\
& + \frac{1}{2}\frac{\partial p_a}{\partial x}(x - (M+1)t) \\
& - \frac{1}{2}\frac{\partial u_a}{\partial x}(x - (M-1)t) \\
& + \frac{1}{2}\frac{\partial p_a}{\partial x}(x - (M-1)t).
\end{aligned}
$$

The algorithms for the x derivatives of u and p which use these exact solution forms can be written in the form of truncated Taylor series expansions in time at the grid point x_i, with

$$
\begin{aligned}
u_{x_i}^{n+1} =\ & u_1 - 2k(p_2 + Mu_2) \\
& + 3k^2(2Mp_3 + (M^2+1)u_3) \\
& - 4k^3((1+3M^2)p_4 + M(3+M^2)u_4) \\
& + \ldots,
\end{aligned}
$$

$$
\begin{aligned}
p_{x_i}^{n+1} =\ & p_1 - 2k(u_2 + Mp_2) \\
& + 3k^2(2Mu_3 + (M^2+1)p_3) \\
& - 4k^3((1+3M^2)u_4 + M(3+M^2)p_4) \\
& + \ldots.
\end{aligned}
\tag{15}
$$

The algorithm forms (15) for u_x and p_x are directly comparable to the forms (13) for u and p, and provide time evolutions for the x derivatives that are consitent with the time evolutions for u and p. Algorithms for other derivatives have similar derivations.

The use of Hermitian interpolation with exact propagators on central stencils produces unstable algorithms. Stable, diffusive, symmetric algorithms with Hermitian interpolation and exact local propagators can be developed on alternating grids. The alternating grids are offset from each other by a half mesh width $\frac{h}{2}$. We will introduce two subfamilies of algorithms, using Hermitian interpolation on alternating two or four point stencils, and using exact propagators to advance through two half steps of size $\frac{k}{2}$ in time. The half time steps for these algorithms can be composed to produce a whole step algorithm on a single central stencil. The algorithms on alternating two point grids will be labeled the c3o0s2, c3o1s2, c3o2s2 and c3o3s2 methods, where the s2 refers to the use of two alternating grids, and the c3 refers to the spatial stencil. The o0 method uses only function data and is first order accurate in both space and time. The o1 method uses up to first derivative data and is third order in both space and time. For every grid point at each time step, the o1 algorithm uses two cubic interpolations (for u and p) to obtain expansion coefficients that are used to compute four distinct time evolutions (for u, p, u_x, and p_x). The o2 method uses up to second derivative data, is fifth order, and uses two fifth order interpolations for six time evolutions. The o3 method uses up to third derivative data, is seventh order accurate in both space and time, and uses two seventh order interpolations for eight time evolutions for every grid point at each time step. The algorithms on alternating four point stencils will be labeled the c5o0s2, c5o1s2, and c5o2s2 methods, and they are fourth, seventh, and eleventh order accurate in both space and time, respectively. The use and computation of derivative data clearly adds complexity, and requires more floating point operations (FLOPs) per

grid point per time level. The relative efficiency of algorithms in terms of total FLOPs that are required to achieve a preset error level at a specified time can be a more practical comparison than algorithm complexity, and it is useful to include this comparison in algorithm evaluation. This class of algorithms is also considered in [9].

3.3 Numerical Comparisons

Grid refinement data from simple numerical experiments will be used to compare the four explicit methods from Section 2.1, and the seven methods from Section 2.2 with Hermitian interpolation on staggered grids. The numerical experiments will be for the Linearized Euler Equations (7) with initial conditions

$$
p(x,0) = \sin(\pi x), \quad u(x,0) = 0,
$$

for $x \in (-1,1]$, and with periodic boundary conditions

$$
p(-1,t) = p(1,t), \quad u(-1,t) = u(1,t),
$$

for $0 \le t$. In order to minimize the effect of the CFL constraint $\lambda \le \frac{1}{1+|M|}$, we will take $M = 0$, with $\lambda = \frac{8}{10}$. Figure 4 and Figure 5 present the maximum absolute error in either u or p at $t = 10$, or after five periods of propagation. We have chosen this short time for these initial comparisons because the lowest order methods cannot be used at long times with reasonable grid resolutions.

In Figure 4, the horizontal axis is the $\log_2$ of the number of grid points per wavelength, from 2^2 to 2^{13}, and the vertical axis is the $\log_{10}$ of the maximum absolute error, from 10^1 to 10^{-15}. Each line in Figure 4 represents the data from one algorithm, and is labeled with the algorithm name. Figure 4 is from grid refinement data, and the slope of each line corroborates the order of accuracy of the corresponding method. The lowest ranges of the maximum errors for these algorithms are limited by accumulated roundoff error. The most interesting feature in Figure 4 is the data at the coarsest grid resolution with four grid points per wavelength. Note that this data is taken after twenty five time steps, and that the maximum error ranges from $O[1]$ to $O[10^{-6}]$. The $O[1]$ error is typical for standard finite difference methods, but the $O[10^{-2}]$ to $O[10^{-6}]$ errors are exceptional at this grid density. The low errors on this coarse grid show exceptionally high resolution for the c3o2s2, c5o1s2, c3o3s2, and c5o2s2 algorithms. It is clear that the total error for each method as a function of grid density is a combined effect of the initial

1040

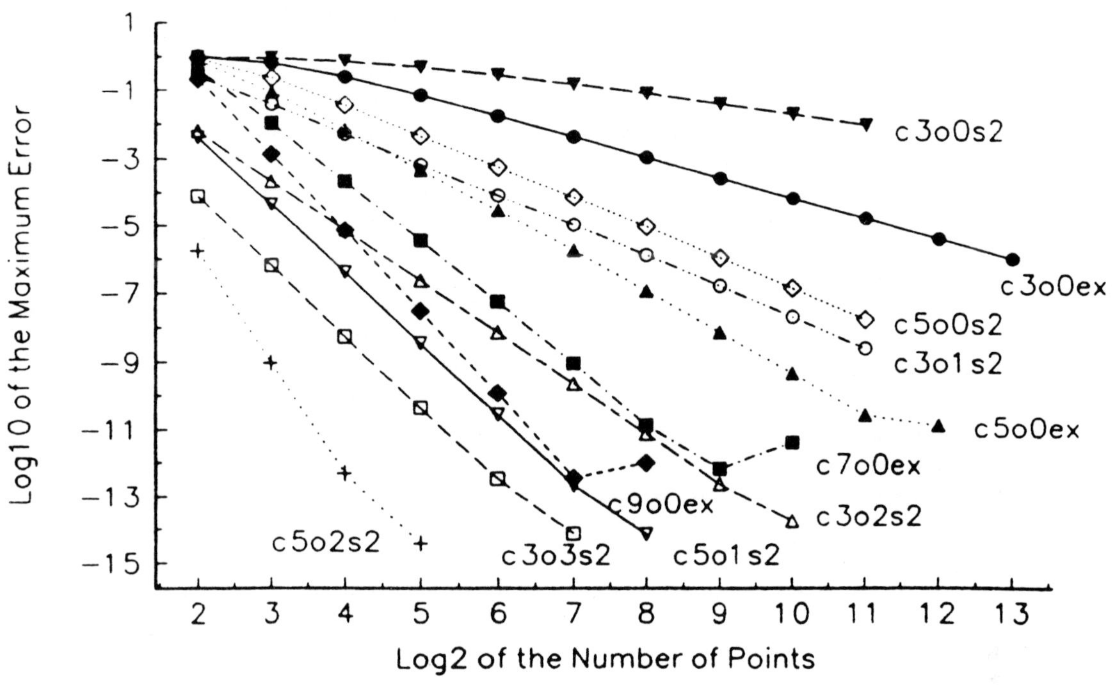

Figure 4: Maximum Absolute Error by Grid Points per Wavelength

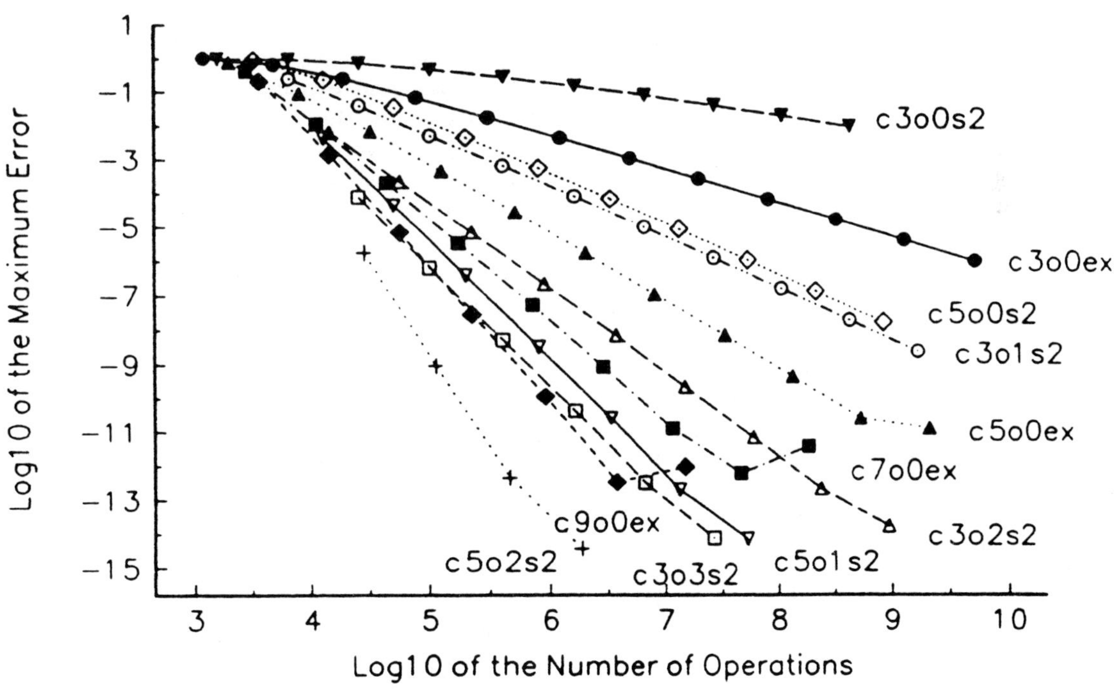

Figure 5: Maximum Absolute Error by Total FLOPs

coarse grid error and the rate of decrease of the error with grid refinement, or the level of resolution and the order of accuracy of each method. As an example, notice at each grid resolution that the errors from the sixth order c7o0ex method are greater than those from the fifth order c302s2 method, and that the errors from the eighth order c9o0ex method are greater than those from the seventh order c5o1s2 method. In both cases, there is a difference of more than an order of magnitude in the coarse grid errors between the standard algorithms and the lower order method that uses Hermitian interpolation. In these two cases, the greater accuracy of the relatively lower order methods is due to its higher resolution. Notice finally that the conventional c7o0ex and c9o0ex algorithms show a greater sensitivity to accumulated roundoff errors than the algorithms which use Hermitian interpolation.

Figure 5 presents the maximum absolute error in either u or p at $t = 10$, as a function of the total number of floating point operations required to complete the computation. The data in Figure 5 is from the same simulations that produced Figure 4, but now the horizontal axis is the $\log_{10}$ of the number of FLOPs, ranging from 10^3 to 10^{10}. Because modern computer systems have a combined multiply and add instruction, only the number of multiplications is counted. The first observation from this data is that for any given error level, the highest order method is always the most efficient. It is known that the relative efficiency of higher order methods increases as the error bound is lowered and as the simulation time is extended [18]. The algorithms from Section 3.2 that use Hermitian interpolation in effect change the system (7) by adding evolution equations for the derivative data. The data in Figure 5 shows that for our methods, the higher order algorithms will be more efficient even if they are more complex with a greater number of variables and equations than a simpler but lower order algorithm. In fact, note the comparisons between the data in Figure 5 from the two third order methods (c5o0s2, c3o1s2), and from the two seventh order methods (c5o1s2,c3o3s2). In both cases, the relatively more efficient method uses higher order derivative data and requires more evolution equations. The c5o0s2 method uses just u and p, with two evolution equations, while the c3o1s2 method also uses their derivatives, with four evolution equations, but the c3o1s2 method is more efficient at every error level. Similarly, the c5o1s2 method has four evolution equations, and the c3o3s2 method has eight, but the c3o3s2 method is more efficient. The relative efficiency of the methods with Hermitian interpolation

is due to their relatively high resolution. A second observation from Figure 5 is that the number of FLOPs required by the various algorithms differs by a factor of up to $O[10^5]$, even to compute out to just five periods with an order $O[10^{-5}]$ error. In one month there are $O[10^3]$ hours and $O[10^5]$ minutes. A final observation from Figure 5 is that the most complex methods use a significant number of FLOPs even at the coarsest resolution. At four grid points per wavelength the eleventh order c5o2s2 method uses $O[10^{4.5}]$ FLOPs to produce an $O[10^{-6}]$ error at five periods. If that level of accuracy is not required over five periods of propagation, then a less accurate algorithm may in fact be more efficient.

Table 1 contains data from all of the algorithms with eight gridpoints per wavelength at $t = 10$, $t = 1,000$, and $t = 100,000$. This grid resolution was chosen as being reasonably moderate but challenging. The data in Table 1 is for the same periodic problem that generated the data for Figure 4 and Figure 5, with $M = 0$ and $\lambda = \frac{8}{10}$. The algorithms are listed in descending order of accuracy in the column on the left. The criterion for including data at any of the three times was that the numerical solution has to contain some meaningful information and at least grossly correspond to the exact solution, with an error that is at least as low as $O[10^{-1}]$. By this criterion of utility, the first order c3o0s2 and second order c3o0ex methods are not useful even for propagation over five periods or wavelengths with a moderate grid density. With eight grid points per wavelength, the third order c5o0s2 and c3o1s2 methods, and the fourth order c5o0ex method are not useful except at $t = O[10]$, or $O[5]$ periods of propagation. The only methods that have either less than $O[10^{-2}]$ error at $t = 1,000$, or no more than $O[10^{-1}]$ error at $t = 100,000$, are the seventh order c5o1s2 and c3o3s2 methods, and the eleventh order

Table 1: Maximum Absolute Error in u or p

Method	$t = 10$	$t = 1,000$	$t = 100,000$
c3o0s2	$O[1]$	$O[1]$	$O[1]$
c3o0ex	$O[1]$	$O[1]$	$O[1]$
c5o0s2	2.42D-01	$O[1]$	$O[1]$
c3o1s2	3.97D-02	$O[1]$	$O[1]$
c5o0ex	9.14D-02	$O[1]$	$O[1]$
c3o2s2	2.27D-04	2.26D-02	$O[1]$
c7o0ex	1.11D-02	1.04D-01	$O[1]$
c5o1s2	4.52D-05	4.55D-03	3.75D-01
c3o3s2	6.74D-07	6.78D-05	6.75D-03
c9o0ex	1.39D-03	1.38D-02	$O[1]$
c5o2s2	9.33D-10	9.37D-08	9.37D-06

c5o2s2 method. Notice at each time level, that the higher order methods do not necessarily have the lowest errors. At $t = 10$ for example, the fifth order c3o2s2 method has lower errors than either the sixth order c7o0ex or the eighth order c9o0ex methods. Notice also at $t = 100,000$, or after 50,000 periods of propagation, that with eight grid points per wavelength the c3o3s2 algorithm has a maximum error of $O[10^{-3}]$, and the c5o2s2 of $O[10^{-6}]$. In fact, for the the eleventh order c5o2s2 algorithm, at $t = 1,000$ with with four grid points per wavelength the maximum error is $O[10^{-4}]$, and at $t = 1,000,000$ with with eight grid points per wavelength the maximum error is $O[10^{-5}]$, and with sixteen it is $O[10^{-8}]$. These levels of accuracy should redefine "propagation to the far field" for finite difference methods. Note that efficient far field propagation requires both high order accuracy and high resolution.

4   Normal Mode Analysis

The exceptional resolving power of the algorithms with Hermitian interpolation may be seen by a Normal Mode Analysis of their full discretization algorithmic forms [26]. For this purpose, we will consider the linear first order wave equation

$$\frac{\partial u}{\partial t} + M \frac{\partial u}{\partial x} = 0, \tag{16}$$

where M is the constant convection speed. In local coordinates, a normal mode for Equation (16) may be written as

$$u(x,t) = \alpha \exp[i\theta \frac{(x - Mt)}{h}], \tag{17}$$

with amplitude α, and frequency $\theta \in [0, \pi]$, where h is the spatial grid size. The symbol ρ for an algorithm is obtained from

$$\rho(\lambda, \theta) = \frac{u_i^{n+1}}{u_i^n}, \tag{18}$$

where u_i^{n+1} is the numerical solution at t_{n+1} and x_i produced by the algorithm, with the known solution at t_n on a stencil obtained from the normal mode (17). Recall that $\lambda = \frac{Mk}{h}$ is the CFL number. The algorithms which use Hermitian interpolation need spatial derivative data, which will be supplied on a stencil by the derivatives of (17). The symbol (18) is complex valued, and the amplification factor for an algorithm is just its norm, or

$$\|\rho(\lambda, \theta)\| = (Re[\rho]^2 + Im[\rho]^2)^{\frac{1}{2}}. \tag{19}$$

An algorithm that is entirely free from dissipation error, or numerical damping, will have $\|\rho(\lambda, \theta)\| \equiv 1$. The amplification factor (19) can be used to define the phase change per time step as

$$pc(\lambda, \theta) = \cos^{-1}[\frac{Re[\rho]}{\|\rho(\lambda, \theta)\|}], \tag{20}$$

where we have departed from Vichnevetsky and Bowles [26] by using a definition in terms of $\cos^{-1}$ insted of $\tan^{-1}$. In order to see the dispersion or phase speed qualities of an algorithm, we will normalize the phase change per time step (20), and define the relative phase change per time step as

$$rpc(\lambda, \theta) = \frac{1}{\lambda\theta} \cos^{-1}[\frac{Re[\rho]}{\|\rho(\lambda, \theta)\|}]. \tag{21}$$

An algorithm with no dispersion error, or that perfectly preserves the dispersion relationship for solutions of (16), will have $rpc(\lambda, \theta) \equiv 1$.

Recall from Figure 4, that the maximum absolute error at five periods of propagation with four grid points per wavelength ranged from $O[1]$ to $O[10^{-6}]$, and that four algorithms had exceptional coarse grid resolution. The maximum error with this grid density is $O[10^{-4}]$ for the c3o3s2 algorithm. Recall from Table 1 that only two algorithms had small errors after propagation over 50,000 periods with eight grid points per wavelength. The maximum error for this computation with the c3o3s2 algorithm is $O[10^{-3}]$. We will consider in detail the seventh order c3o3s2 algorithm as representative of algorithms which use Hermitian interpolation, and which have exceptional resolution and accuracy. The c3o3s2 algorithm for Equation (16) is easy to develop, since the exact local propagator solution form is readily obtained from the Method of Characteristics. The amplification factor and relative phase change per time step for the c3o3s2 algorithm have been obtained analytically, and data for plotting has been obtained numerically from these analytical expressions.

The amplification factor (19) for the c3o3s2 algorithm is plotted in Figure 6 as a function of CFL number $\lambda \in [0, 1]$, and of normal mode wave number $\theta \in [0, \pi]$, and in Figure 7 as a function of $\lambda \in [0, 1]$ in the limit at $\theta = \pi$. The data in Figure 6 shows no values for the amplification factor of the c3o3s2 algorithm that are greater than one, and the greatest departures from one occur in the limit at $\theta = \pi$. Note from Figure 7 that the most extreme range for the amplification factor is in $[0.9988, 1]$ at $\theta = \pi$. The other algorithms with Hermitian interpolation also have their most extreme properties in the limit at $\theta = \pi$, with the range of the amplification factor in $[0.65, 1]$ for the third order c3o1s2 method,

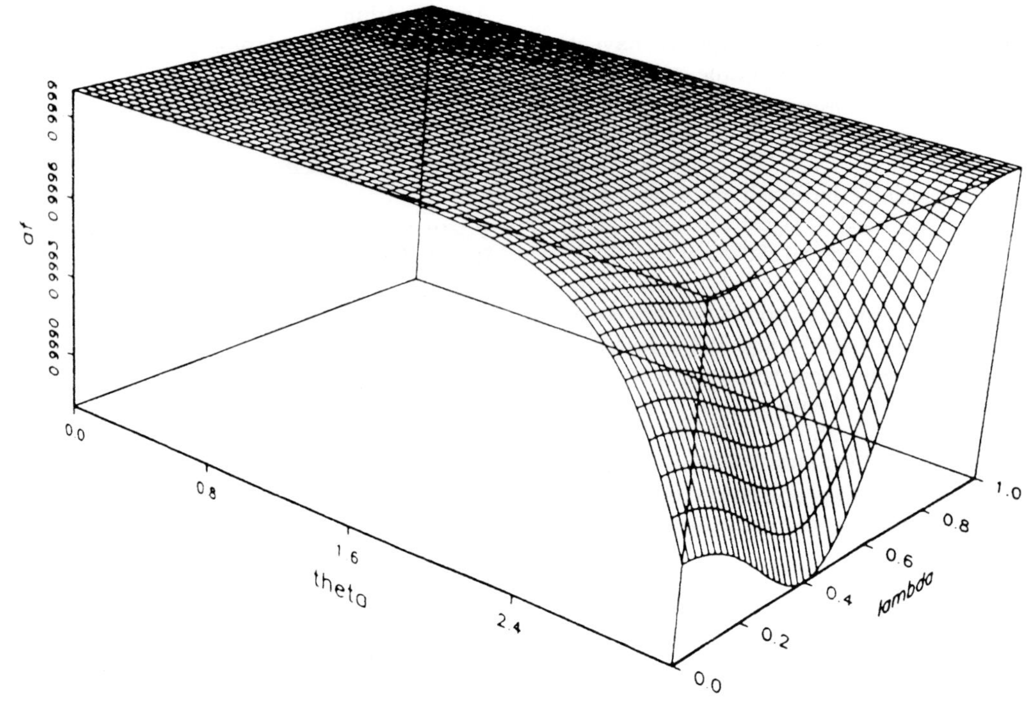

Figure 6: Amplification Factor for Algorithm c3o3s2

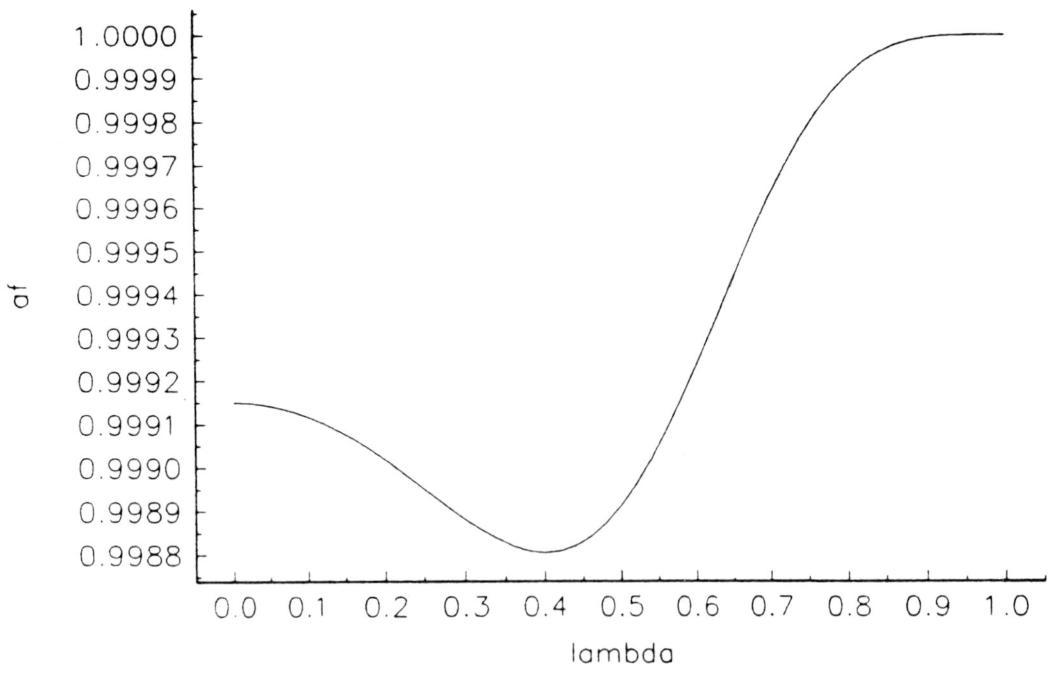

Figure 7: Amplification Factor for Algorithm c3o3s2 at $\theta = \pi$

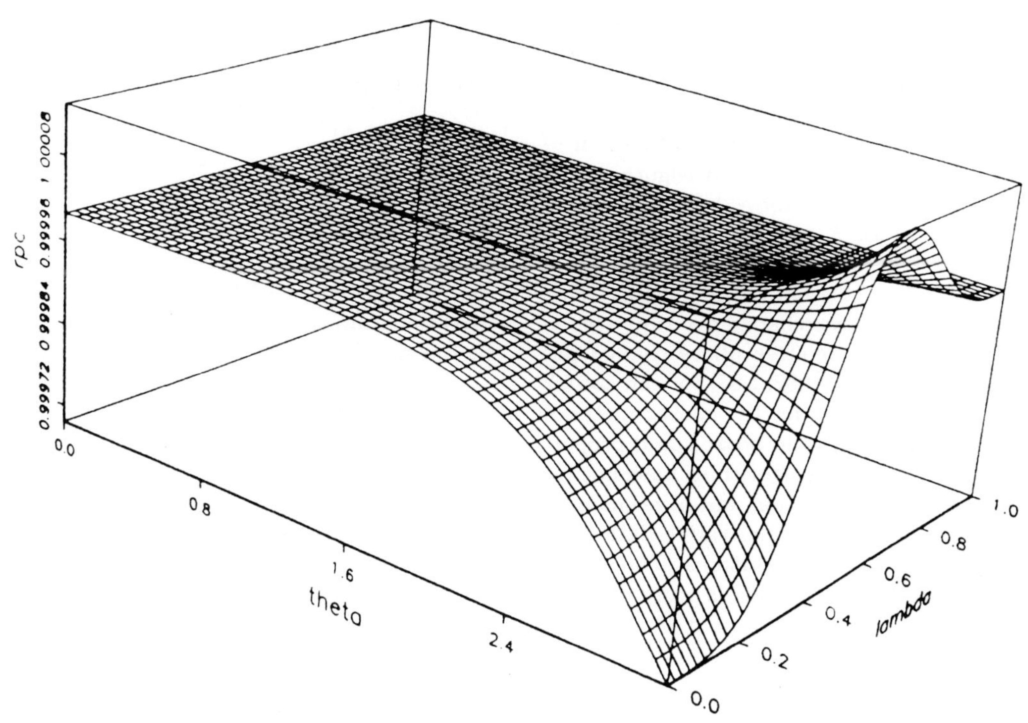

Figure 8: Relative Phase Change for Algorithm c3o3s2

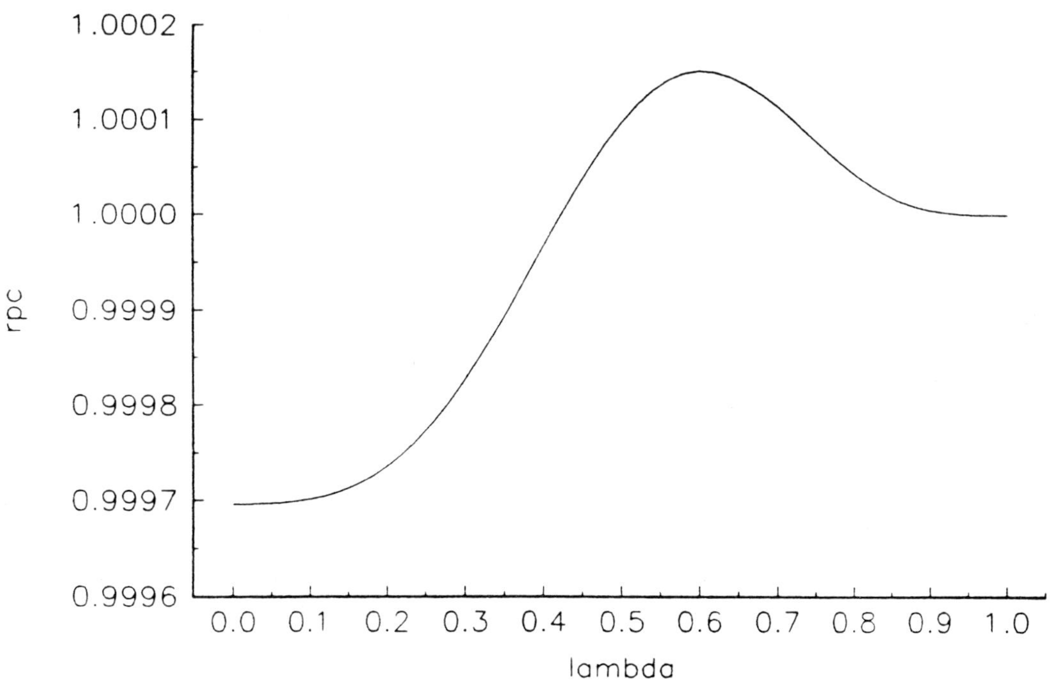

Figure 9: Relative Phase Change for Algorithm c3o3s2 at $\theta = \pi$

and in $[0.97, 1]$ for the c3o2s2 method. The relative phase change per time step (21) for the c3o3s2 algorithm is plotted in Figure 8 as a function of CFL number $\lambda \in [0, 1]$, and of normal mode wave number $\theta \in [0, \pi]$, and in Figure 9 as a function of $\lambda \in [0, 1]$ in the limit at $\theta = \pi$. In Figure 8, the relative phase change ranges above 1 for $\frac{1}{2} \leq \lambda$, and below 1 for $\lambda \leq \frac{1}{2}$ for $\theta \in [0, \pi]$. Here also the extreme behaviour is in the limit at $\theta = \pi$. Note from Figure 9, that in this limit, the most extreme range for the relative phase change per time step is in $[0.9997, 1.0002]$. The relative phase change per time step has its range in $[0.82, 1.07]$ for the c3o1s2 method, and in $[0.98, 1.004]$ for the c3o2s2 method. Note that the amplification and phase speed properties of these finite difference algorithms have not required any tuning or optimization, but have emerged from the use and evolution of derivative data with exact propagator Hermitian algorithms. The exceptional resolution of these methods is truly spectral like.

5   Linearized Euler Equations in 2D

Solutions to the Linearized Euler Equations in one space dimension propagate information along characteristic curves, and in two or more dimensions along characteristic surfaces. This is a significant change in the wave dynamics. In terms of the algorithm development presented above, the Method of Characteristics can no longer be used to obtain general exact solutions for the time propagation of a local spatial interpolant. In order to address this issue, we will consider the example of the Linearized Euler Equations in two space dimensions. For the nondimensionalized isentropic case, the Linearized Euler Equations (6) may be written in two dimensions as

$$\frac{\partial u}{\partial t} + U\frac{\partial u}{\partial x} + V\frac{\partial u}{\partial y} + \frac{\partial p}{\partial x} = 0,$$

$$\frac{\partial v}{\partial t} + U\frac{\partial v}{\partial x} + V\frac{\partial v}{\partial y} + \frac{\partial p}{\partial y} = 0, \qquad (22)$$

$$\frac{\partial p}{\partial t} + U\frac{\partial p}{\partial x} + V\frac{\partial p}{\partial y} + \frac{\partial u}{\partial x} + \frac{\partial v}{\partial y} = 0,$$

where p is the pressure disturbance, (u, v) is the velocity disturbance, and (U, V) is the constant mean convection velocity in Mach number. The system (22) is non-diagonalizable, with wave propagation along characteristic surfaces.

5.1   Algorithm Development

The first step of algorithm development for Equation (22) is to interpolate p, u and v. For a second order algorithm in two space dimensions, we will use a biquadratic interpolant. In local coordinates, the form of the interpolant for u about (x_i, y_j) at t_n is

$$u_a(x, y) = \sum_{\alpha,\beta=0}^{2} u_{\alpha,\beta} x^\alpha y^\beta$$

$$\approx u(x_i + x, y_j + y, t_n). \qquad (23)$$

Similar interpolants are used for v and p. Note that there are cross derivative terms up to the fourth order. The Method of Undetermined Coefficients is used to obtain the expansion coefficients, which may be interpreted in terms of spatial derivatives as

$$u_{\alpha,\beta} = \frac{1}{\alpha!\beta!}\frac{\partial^{\alpha+\beta}u_a}{\partial x^\alpha \partial y^\beta} \approx \frac{1}{\alpha!\beta!}\frac{\partial^{\alpha+\beta}u}{\partial x^\alpha \partial y^\beta}. \qquad (24)$$

On a symmetric 3×3 stencil with nine grid points, familiar finite difference approximations appear. The use of a symmetric multidimensional stencil for interpolation is consistent with the wave dynamics of Equation (22), since information is propagated to a stencil center from all directions, and not just along coordinate lines. Similar multidimensional interpolants are used for higher order algorithms. As an example, a biquartic fourth order interpolant can be defined on a symmetric 5×5 stencil with twenty-five grid points, and up to eighth order cross derivative terms. Symmetric high order interpolants do not need to use a square stencil, but may clip some points from the stencil corners. As an example, a modified fourth order interpolant can be used on the 5×5 square stencil without its four corner points. This stencil has twenty-one grid points, and the interpolant has up to sixth order cross derivative terms. The form of the interpolant in Equation (23) does not require a uniform, symmetric, or regular grid, but just sufficient data located so that the linear equations from the Method of Undetermined Coefficients can be solved.

The second step of algorithm development for Equation (22) is to propagate the local data interpolants. Exact local solutions for Equation (22) cannot be obtained in general, but useful solutions can be found if the initial data are finite degree polynomials such as Equation (23). If the local interpolant is viewed as a function of t as well as of x and y, then the solution may be approximated locally in both space and time. In the second order case for u, with the local biquadratic spatial interpolant in Equation (23), a local solution approximation in time and space may be written in the form

$$u_a(x, y, t) = \sum_{\alpha,\beta=0}^{2} \sum_{\gamma=0}^{4} u_{\alpha,\beta,\gamma} x^\alpha y^\beta t^\gamma$$

$$\approx u(x_i + x, y_j + y, t_n + t), \qquad (25)$$

with similar approximations for v and p. These three local solution approximations are substituted into Equation (22), and all of the coefficients with $\gamma \neq 0$ are obtained by requiring that Equation (22) is to be satisfied for all x, y, and t. Coefficients with $\gamma \neq 0$ are equivalent to mixed partial derivatives which include some order of time differentiation, and they may all be expressed entirely in terms of purely spatial derivatives, or coefficients for which $\gamma = 0$. Forcing the polynomial solution approximations with the form of Equation (25) to be an exact solution of Equation (22) is equivalent to deriving forms for the pure and mixed time derivatives directly from Equation (22). This is exactly the process that is used in proofs of the Cauchy-Kovalevskaya Theorem, such as in [6]. The Cauchy-Kovalevskaya method has been used to obtain expressions for various derivatives by Harten, et al. [13]. With a local exact polynomial solution for Equation (22), the multidimensional algorithm development follows the same course as in the one dimensional case. The local exact solution with the polynomial spatial interpolant as initial data is evaluated at the grid point (x_i, y_j) and at $t = t_{n+1}$ to obtain the new solution values, which can be written as

$$
\begin{aligned}
u_{i,j}^{n+1} &= u_a(0,0,k) \approx u(x_i, y_j, t_n + k), \\
v_{i,j}^{n+1} &= v_a(0,0,k) \approx v(x_i, y_j, t_n + k), \qquad (26) \\
p_{i,j}^{n+1} &= p_a(0,0,k) \approx p(x_i, y_j, t_n + k).
\end{aligned}
$$

The order of accuracy of these algorithms is the same in space and in time, and is the order of the spatial interpolant. In fact, since a local exact propagator is used for the time evolution, it can be said that this type of algorithm introduces error only in the spatial interpolation.

The algorithm forms for the two dimensional Linearized Euler Equations are derived with a local exact propagator, so that the numerical solution automatically incorporates the correct multidimensional wave dynamics. In this sense, these algorithms can be viewed as an extension of the Method of Characteristics to nondiagonalizable hyperbolic systems in multiple space dimensions, with propagation of solutions along characteristic surfaces. The interpretation of the expansion coefficients in terms of mixed spatial derivatives given in Equation (24) allows this type or algorithm to be viewed as a Cauchy-Kovalevskaya expansion which locally approximates the exact solution in both space and time. This viewpoint is particularly useful next to boundaries. The numerical solution forms given by Equation (26) may be interpreted as Taylor series expansions in

time at the grid point (x_i, y_j). Note that the time expansion coefficients in the biquadratic case given in Equation (25) are indexed up to $\gamma = 4$. Recall that the symmetric biquadratic spatial interpolant with the form of Equation (23) has spatial cross derivatives up to the fourth order. In order to accomodate all of the spatial derivative data from the interpolant, a local exact propagator solution will in general have higher order time derivatives than the order of the method. The time expansion at the grid point (x_i, y_j) may be truncated at any order, such as the desired order of the algorithm. This approach may be viewed as a Taylor Series or high order Lax-Wendroff [19] method in multiple space dimensions, but in general it is no longer an exact propagator method, and by neglecting higher order terms it does not correctly represent the multidimensional wave dynamics.

The algorithm class that is obtained in the way that we have sketched has many possible members. All of the possible algorithms will be explicit methods that use data from only one time level, and that have the same order of accuracy in both space and time. The accuracy of the methods is determined by the spatial interpolant, and can be of virtually any order. Further details are in [8]. As in the one dimensional case, Hermitian interpolation may be used with alternating grids to obtain high order and high resolution algorithms on small stencils.

5.2 Numerical Comparisons

We will briefly present data from numerical experiments with three algorithms that have been developed as described above for the two dimensional Linearized Euler Equations (22). The three algorithms that we consider are the second order c09o0ex method on a nine point 3×3 square stencil, with time terms up to the fourth order, the fourth order c21o0ex method on a twenty–one point stencil made from the 5×5 square by dropping the four corner points, with time terms up to the sixth order, and the sixth order c37o0ex method on a thirty–seven point stencil made from the 7×7 square by dropping three points from each corner, with time terms up to the eighth order. All three of these methods are single step explicit algorithms, they have the same order of accuracy in space as in time, and they all have time expansion terms that are two orders higher than their order of accuracy. Each of these methods is an exact propagator algorithm, and incorporates the correct multidimensional wave propagation for the interpolated data on its stencil. Numerical experiments with these three algorithms are for the initial data

$$
\begin{aligned}
p(x,y,0) &= \sin(\pi x)\sin(\pi y), \\
u(x,y,0) &= 0, \\
v(x,y,0) &= 0,
\end{aligned}
\tag{27}
$$

where $(x,y) \in [-1,1] \times [-1,1]$, with periodic boundaries in both x and y. The exact solution of the two dimensional Linearized Euler Equations (22) with the initial data (27) and biperiodic boundaries is easily obtained. The initial data for the numerical experiments in two dimensions is analogous to the data for the experiments in Section 2.3 in one space dimension. In one dimension we considered the case with no mean flow, but for the two dimensional comparisons we will consider the case where the mean flow is $U = 1$ and $V = 1$, with convection in the direction of a mesh diagonal.

Grid refinement data from the three algorithms that we are considering is presented in Table 2. The left hand column with heading $\frac{2}{h}$ gives the number of grid points per wavelength in x for $\sin(\pi x)$, and in y for $\sin(\pi y)$. The three columns on the right of Table 2 contain data from the three algorithms. The data in Table 2 is the maximum absolute error in p at $t = 10$, with grid ratio $\frac{k}{h} = \frac{2}{5}$, and for a series of grids with successively halved mesh sizes. The data in Table 2 confirms the order of accuracy of each of the methods. In order to asses the effect of spatial dimension on the accuracy of the algorithms that are produced by our methods, it is useful to compare the data from the one dimensional experiments in Section 2.3 with the data in Table 2. Recall that the one dimensional simulations have $M = 0$, with no convection, and $\frac{k}{h} = \frac{4}{5}$, so that with similar space resolution, the two dimensional simulations require twice as many time steps. From the one dimensional simulations, if $\frac{2}{h} = 16$, then the maximum absolute error in u or p is 2.59D-01 for the second order c3o0ex method, 7.12D-03 for the fourth order c5o0ex method, and 2.16D-04 for the sixth order c7o0ex method. If $\frac{2}{h} = 512$, then the

Table 2: Maximum Error in $|p|$

$\frac{2}{t}$	c09o0ex	c21o0ex	c37o0ex
4	8.85D-01	9.31D-01	7.35D-01
8	8.88D-01	1.68D-01	2.22D-02
16	3.44D-01	9.86D-03	3.01D-04
32	1.02D-01	6.89D-04	5.32D-06
64	2.70D-02	4.54D-04	8.82D-08
128	6.88D-03	2.90D-06	1.42D-09
256	1.74D-03	1.84D-07	2.29D-11
512	4.36D-04	1.15D-08	5.10D-12

maximum absolute error in u or p is 2.84D-04 for the second order c3o0ex method, 7.18D-09 for the fourth order c5o0ex method, and 6.30D-13 for the sixth order c7o0ex method. For pairs of algorithms with the same order of accuracy, the error levels from the one and two dimensional calculations are very close at each grid density, differing by a factor of approximately 2. This factor appears to be due primarily to the nonzero mean flow field $(U,V) = (1,1)$, which requires a lower CFL constraint and twice as many time steps at each grid resolution for the two dimensional simulations. If this difference is considered, then the two dimensional algorithms do not show any significant loss of accuracy as a function of grid density when compared to similar one dimensional algorithms. In particular, the truly multidimensional wave dynamics of Equation (22) are correctly incorporated in our algorithms, with propagation of information along characteristic surfaces. Also, the mean convection velocity for the two dimensional simulations is at an angle of $\frac{\pi}{4}$ to the grid lines, showing that the accuracy of our multidimensional algorithms is independent from the direction of convection, so that in this sense they can be said to be isotropic.

The effect of spatial dimension on relative efficiency is significant independently from its effect on accuracy. Consider the problem of ensuring an error at $t = 10$ of less than 5×10^{-4} with the initial data for the numerical experiments in one and two space dimensions. The grid refinement data can be interpolated to obtain the grid density that is required for any of the methods in either dimension, and this information can be used to calculate the total FLOPs required to attain the specifiec error level at the required time. In one space dimension, with no mean convection, the second order c3o0ex method requires $\frac{2}{h} \geq 420$, the fourth order c5o0ex method requires 32, and the sixth order c7o0ex method requires 16. In one dimension, the ratio between the number of multiplications required by the second and fourth order methods is approximately 108, and between the fourth and sixth is approximately 2.86. In two space dimensions, with mean convection $(U,V) = (1,1)$, the second order c09o0ex method requires $\frac{2}{h} \geq 496$, the fourth order c21o0ex method requires 40, and the sixth order c37o0ex method requires 16. In two dimensions, the ratio between the number of multiplications required by the second and fourth order methods is approximately 817, and between the fourth and sixth is approximately 9. The relative advantage of higher order methods appears to increase with spatial dimension, as well as with simulation time and a decrease in error bounds. The simple central methods for which we already have two dimension extensions are not as powerful and efficient as the methods that use and propagate derivative data. The multidimensional extension of the high order and high resolution algorithms is underway.

6 Artificial Boundary Conditions

The Linearized Euler Equations in one space dimension (7) can easily be provided with boundary conditions because the solution propagates along characteristic curves. The Linearized Euler Equations in two space dimensions (22) are representative of multidimensional systems which require special artificial boundary treatments. We will briefly present artificial inflow and outflow boundary conditions that have recently been developed by Hagstrom [11], and implemented by Goodrich and Hagstrom [10]. The development of the artificial boundary conditions follows Engquist and Majda [4], Giles [7], and Collino [3]. In the case where the constant mean convection U is positive, there will be an outflow at the $+x$ artificial boundary, and an inflow at the $-x$ boundary. If the Linearized Euler Equations (22) are diagonalized with respect to x, then two variables $r = u + p$ and $l = u - p$ may be defined which are like Riemann variables with convection normal to the $\pm x$ boundaries, but with coupling to multidimensional wave propagation through various y derivatives and v. The variable r can be viewed as "right going", and l as "left going". The essential idea that is used in deriving the boundary conditions is to apply a Fourier–Laplace transform to the Linearized Euler Equations (22), and then to develop a Padé approximation of the operator symbol. The approximation is implemented with auxiliary functions, and a sequence of local boundary conditions is obtained, depending upon the number of auxiliary functions that are used. The approximate problems with the resulting artificial boundary conditions are all well posed, and the solutions to the approximate problems converge exponentially with the number of auxiliary functions to the solution of the underlying problem on the open domain [11]. The essential constraint on the use of these boundary conditions is that the initial data must lie entirely within the numerical domain. There is no need for geometric assumptions about the location of a theoretical acoustic source, and there is no tuning for the frequency content of an acoustic signal.

At the $+x$ artificial outflow boundary, the "outgoing" variables $r = u + p$ and v are obtained by using the Cauchy-Kovalevskaya form of the solution approximation in space and time with a consistent interpolation and propagation over a half stencil width next to the boundary, which essentially amounts to a uniform and consistent interior differencing near the boundary. The "incoming" variable $l = u - p$ is obtained from the solution to the system

$$\frac{\partial l}{\partial t} + M_y \frac{\partial l}{\partial y} - M_x \frac{\partial v}{\partial y} + \sum_{j=1}^{m}(f_j + g_j) = 0,$$

$$\frac{\partial f_j}{\partial t} + \alpha_j \frac{\partial f_j}{\partial y} = \frac{\beta_j}{2}\frac{\partial^2(l-r)}{\partial y^2},$$

$$\frac{\partial g_j}{\partial t} - \alpha_j \frac{\partial f_j}{\partial y} = \frac{\beta_j}{2}\frac{\partial^2(l-r)}{\partial y^2},$$

$$(28)$$

where

$$\alpha_j = (\sqrt{1 - M_x^2}\cos\frac{j\pi}{2m+1} + M_y),$$

$$\beta_j = -(1 - M_x^2)\frac{\sin^2\frac{j\pi}{2m+1}}{2m+1}.$$

$$(29)$$

Note that l is defined only on the outflow boundary surface, and that Equation (28) has derivatives only with respect to t, or to y in the boundary surface. The variables p and u are obtained in the outflow boundary surface by combinations of r and l. The solutions of Equation (28) are forced by the $M_x\frac{\partial v}{\partial y}$ term in the evolution equation for l, since v is solved for from the interior, and by the $\frac{\partial^2(l-r)}{\partial y^2}$ term in the equations for the adjustment factors f_j and g_j, since r comes from the interior. The $\frac{\partial^2(l-r)}{\partial y^2}$ derivative is an approximation in the boundary surface to the $\frac{\partial^2 p}{\partial y^2}$ derivative.

At the $-x$ artificial inflow boundary, $l = u - p$ is "outgoing" from the interior of the numerical domain, and is obtained by solving from the interior, while $r = u + p$ and v are "incoming" and must be solved for on the boundary. The system for r and v on the inflow boundary surface is

$$\frac{\partial v}{\partial t} + M_y \frac{\partial v}{\partial y} + \frac{(1+M_x)}{2}\frac{\partial r}{\partial y} - \frac{(1-M_x)}{2}\frac{\partial l}{\partial y} = 0,$$

$$\frac{\partial r}{\partial t} + M_y \frac{\partial r}{\partial y} + \frac{1-M_x}{2}\frac{\partial v}{\partial y} + \sum_{j=1}^{m}(F_j + G_j) = 0,$$

$$\frac{\partial F_j}{\partial t} + \alpha_j \frac{\partial F_j}{\partial y} = \frac{\beta_j}{2}\frac{\partial^2 r}{\partial y^2},$$

$$\frac{\partial G_j}{\partial t} - \alpha_j \frac{\partial G_j}{\partial y} = \frac{\beta_j}{2}\frac{\partial^2 r}{\partial y^2},$$

where α_j and β_j are as in Equation (29). The equation for v in the inflow boundary surface is forced by the $\frac{\partial l}{\partial y}$ derivative. The evolution equation for r is coupled to the equation for v by the $\frac{\partial v}{\partial y}$ derivative, and this coupling is the entire forcing of the system for r and the adjustment factors F_j and G_j.

As an example of the use of these artificial boundary conditions, consider the propagation of a pressure pulse in an infinite duct, with the Linearized

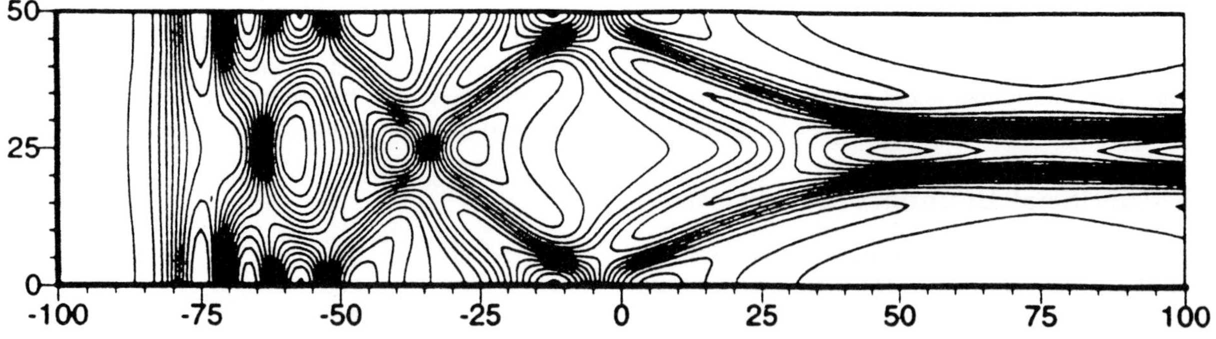

Figure 10: Pressure Contours at $t = 150$

Euler Equations (22). Figure 10 presents the pressure contours at $t = 150$. The initial data is $u = 0$, $v = 0$, and

$$p(x, y, 0) = \exp[-\frac{\ln(2)}{25}(x^2 + (y - 25)^2)],$$

with mean convection $U = 0.5$ and $V = 0$. The numerical domain is $(x, y) \in [-100, 100] \times [0, 50]$, with artificial inflow boundary at $x = -100$, and outflow at $x = +100$, and with walls at $y = 0$ and $y = 50$. The fourth order c21o0ex algorithm will be used for propagation, with a uniform mesh size $h = 1$ in both x and y, and $k = 0.25$ as the time step size. The artificial boundary conditions given above are implemented with fourth order accurate finite difference algorithms that are developed in the same way as the propagation algorithms, using exact polynomial solutions to the boundary equations. We will use $m = 2$ auxiliary functions. The wall boundary condition for the normal velocity component is $v = 0$. Since there is no vorticity in this disturbance, we will take $\frac{\partial u}{\partial y} = 0$ and $\frac{\partial p}{\partial y} = 0$ along the walls. Note in Figure 10 that the center of the disturbance has convected to $x = 75$. The complex interference patterns in the numerical domain represent the structure of the disturbance upstream from the convected center, and the mirror image structure has passed downstream past the artificial outflow boundary at $x = 100$. Since the nondimensionalized convection velocity is $U = 0.5$ as a Mach number, and the wave fronts expand at Mach 1, solution structures will propagate effects upstream. The solution in the numerical domain is still subject to the effect of the complex interference patterns that are downstream from the artificial outflow boundary. The symmetry in x of the plotted pressure contours in Figure 10 shows that the downstream effect of the interference patterns upstream from the disturbance center are mirrored by the upstream effect of the artifical boundary treatment, which replaces the entire downstream domain. In other words, the artificial boundary treatment is nonreflecting, and is capable of accurately capturing and transmitting the effect of a complex disturbance as it passes through the boundary. Further details are in [10].

Conclusions

Explicit, single step, high order, finite difference methods have been shown in one and two space dimensions, with the same order of accuracy in both space and time. Examples of up to the eleventh order accuracy are included. High order, high resolution methods that are truly spectral like have also been shown, with virtually correct amplification factor and phase speed properties for normal mode phase angles in $[0, \pi]$. Several orders of magnitude improvement in relative efficiency is shown, even for near field propagation with moderate error requirements. Accurate and efficient far field wave propagation requires both high order and high resolution, and accurate propagation to $O[10^5]$ periods with eight grid points per wavelength is shown. Two dimensional algorithms correctly propagate information along characteristic surfaces for nondiagonalizable hyperbolic systems in multiple space dimensions. Accurate artificial boundary conditions have been presented, with high order finite difference implementations.

References

[1] H.L. Atkins, "Quadrature Free Implementations of the Discontinuous Galerkin Method for Hyperbolic Equations," AIAA 96-1683, *The 2nd AIAA/CEAS Aeroacoustics Conference*, State College, PA, May, 1996.

[2] G. K. Batchelor, *An Introduction to Fluid Dynamics*, Cambridge University Press, Cambridge, 1967.

[3] F. Collino, "Conditions d'ordre élevé pour des modèles de propagation d'ondes dans des domaines rectangulaires," INRIA report 1790, 1993.

[4] B. Engquist and A. Majda, "Absorbing Boundary Conditions for the Numerical Simulation of Waves," *Math. Comp.*, **31**, p629, 1977.

[5] F. Farassat, "The Acoustic Analogy and the Prediction of the Noise of Rotating Machinery," *Int. Symp. on Theor. and Comp. Fluid Dynamics*, Tallahassee, Nov. 6-8, 1996.

[6] P. Garabedian, *Partial Differential Equations*, Wiley, New York, 1964.

[7] M. Giles, " Nonreflecting Boundary Conditions for Euler Equation Calculations," *AIAA J.*, **28**, p2050, 1990.

[8] J. W. Goodrich, "An Approach to the Development of Numerical Algorithms for first Order Linear Hyperbolic Systems in Multiple Space Dimensions: The Constant Coefficient Case," NASA TM 106928, 1995.

[9] J. W. Goodrich, "Accurate Finite Difference Algorithms," NASA TM 107377, 1996.

[10] J. W. Goodrich and T. Hagstrom, "Accurate Algorithms and Radiation Boundary Conditions for Linearized Euler Equations," AIAA 96-1660, *The 2nd AIAA/CEAS Aeroacoustics Conference*, State College, PA, May, 1996.

[11] T. Hagstrom, "On High Order Radiation Boundary Conditions," to appear in the proceedings of *The IMA Workshop on Computational Wave Propagation*, Minneapolis, Minn., September, 1994.

[12] J. Hardin, J.R. Ristorcelli, and C.K.W. Tam, eds., "ICASE/LaRC Workshop on Benchmark Problems in Computational Aeroacoustics," NASA CP-3300, 1995.

[13] A. Harten, B, Engquist, S. Osher and S. R. Chakravarthy, "Uniformly High Order Accurate Essentially Non-oscillatory Schemes, III," *J. Comput. Phys.*, **71**, p231, 1987.

[14] R. Hixon, "On Increasing the Accuracy of MacCormack Schemes for Aeroacoustic Applications," NASA CR 202311, ICOMP 96-11, 1996.

[15] F.Q. Hu, "On Absorbing Boundary Conditions for Linearized Euler Equations in a Perfectly Matched Layer," *J. Comput. Phys.*, **129**, p201, 1996.

[16] F.Q. Hu, M.Y. Hussaini, and J.L. Manthey, "Low Dissipation and Low Dispersion Runge-Kutta Schemes for Computational Acoustics," *J. Comput. Phys.*, **124**, p177, 1996.

[17] H. O. Kreiss and J. Lorenz, *Initial Boundary Value Problems and the Navier Stokes Equations*, Academic Press, New York, 1989.

[18] H. O. Kreiss and J. Oliger, "Comparison of Accurate Methods for the Integration of Hyperbolic Equations," *Tellus*, **24**, p199, 1972.

[19] P. D. Lax and B. Wendroff, "Systems of Conservation Laws," *Comm. Pure Appl. Math*, **13**, p217, 1960.

[20] S. K. Lele, "Computational Aeroacoustics: A Review," AIAA 97-0018, *The 35th Aerospace Sciences Meeting*, Reno, NV, January, 1997.

[21] M. J. Lighthill, "On Sound Generated Aerodynamically: I. General Theory," *Proc. Roy. Soc., Ser. A,* **211**, p564, 1962.

[22] C.K.W. Tam, "Supersonic Jet Noise," *Annual Rev. Fluid Mechanics* **27**, p17, 1995.

[23] C.K.W. Tam and J. Hardin, eds., "FSU/LaRC Workshop on Benchmark Problems in Computational Aeroacoustics," to appear, 1997.

[24] C.K.W. Tam and J.C. Webb, "Dispersion Relation Preserving Finite Difference Schemes for Computational Acoustics," *J. Comput. Phys.*, **107**, p262, 1993.

[25] C.K.W. Tam and J.C. Webb, "Radiation Boundary Conditions and Anisotropy Correction for Finite Difference Solutions of the Helmholtz Equation," *J. Comput. Phys.*, **113**, p122, 1994.

[26] R. Vichnevetsky and J. Bowles, *Fourier Analysis of Numerical Approximations of Hyperbolic Equations*, SIAM, Philadelphia, 1982.

PROGRESS IN COMPUTATIONAL ELECTROMAGNETICS

J.S. SHANG

Senior Scientist; Air Force Research Laboratory, Wright-Patterson Air Force
Base, Ohio 45433, USA

Nomenclature

B	Magnetic flux density
C	Coefficient matrix of flux-vector formulation
D	Electric displacement
E	Electric field strength
F	Flux vector component
H	Magnetic flux intensity
i, j, k	Index of discretization
J	Electric current density
n	Index of temporal level of solution
S	Similar matrix of diagonalization
t	Time
U	Dependent variables
V	Elementary cell volume
x, y, z	Cartesian coordinates
ξ, η, ζ	Transformed coordinates
λ	Eigenvalue
∇	Gradient, backward difference operator
Δ	Forward difference operator

1 Introduction

Computational electromagnetics (CEM) in the
present context is focused on numerical methods for
solving the time-dependent Maxwell equations. The
first-order divergence-curl equations together with associated initial/boundary conditions constitute the
hyperbolic partial differential equation system. The
solution of this type of differential equation system
is not necessarily analytical and has a distinctive domain of dependence in which all the data propagate
invaryingly along characteristics [1,2]. A series of numerical schemes has been devised to duplicate the
physics which is dominated by directional information
propagation. These numerical procedures are collec-
tively designated as characteristic-based methods and
in the most elementary form are the Riemann problem [3,4,5]. Characteristic-Based methods when applied to solve the time-dependent Maxwell equations
have exhibited many attractive attributes. In particular, this formulation can alleviate reflected waves
from the truncated computational domain easily and
can construct piecewise continuous solutions across
media interface. The former requirement is a fundamental dilemma of solving the initial-value problem on any finite memory size computer. The latter is always encountered when the electromagnetic
wave is propagating through different media . Equally
important, characteristic-based methods are derived
from the eigenvector and eigenvalue structure of the
Maxwell equations, the numerical stability and accuracy are superior than conventional methods.

In general, differential equations in the time domain CEM consist of two categories: the first-order
divergent-curl equations and the second-order curl-
curl equations [6-8]. In applications, further simplifications into frequency domain or the Helmholtz equations and the potential formulation have been accomplished. Poor numerical approximations to physical
phenomena can result from solving overly simplified
governing equations. Under these circumstances, no
meaningful quantification of errors for the numerical procedure can be achieved. Equally important, a
physically incorrect value and an inappropriate implementation of initial and/or boundary conditions are
another major source of error. The placement of the
farfield boundary and type of initial or boundary conditions have also played an important role. These concerns are easily appreciated in the light of the fact that
the governing equations are identical, only the different initial/boundary conditions generate different solutions.

Received on September 3, 1997.

The Maxwell equations in the time domain are difficult to solve by conventional numerical methods. Nevertheless, the pioneering efforts by Yee and others have attained impressive achievements [9-11]. Recently, numerical techniques in CEM have been further enriched by the computational fluid dynamics (CFD) community. A basic approach to enhance the accuracy of the computation can be derived from high resolution schemes or spectral methods. Substantial progress is being made in the compact difference method, optimized algorithm research, and unstructured grid formulation [12-15]. All these numerical techniques are devised to increase the numerical resolution of simulations over a wider range of the frequency spectrum. On the other hand, for an electromagnetic simulation associated with a large-scale configuration, the required number of mesh points to meet an accuracy specification is often beyond the reach of a conventional computing system.

In the last decade, through remarkable progress in micro chip and interconnect data link technology, a host of multiple address, message passing computers have became available for data processing. These scalable multi-processors or multi-computers, in theory, are capable of providing essentially unlimited computing resources for scientific simulations. However, the effective use of the distributed memory, message passing homogeneous multi-computer still requires a judicious trade off between a balanced work load and inter-processor communication. These requirements are intrinsically related to the numerical algorithms and hardware architectures. A synergism of the relatively new numerical procedures and scalable parallel computing capability will open up a new frontier in electromagnetics research. For this reason, a major portion of the present effort will be focused on introducing the relatively new characteristic-based finite-volume and finite-difference algorithms [4,5].

3   Governing Equations

The time dependent Maxwell equations for the electromagnetic field can be written as [6,7]:

$$\frac{\partial B}{\partial t} + \nabla \times E = 0 \tag{1}$$

$$\frac{\partial D}{\partial t} - \nabla \times H = -J \tag{2}$$

$$\begin{aligned}
\nabla \cdot B &= 0 \\
\nabla \cdot D &= \rho
\end{aligned} \tag{3}$$

where the ρ and J are the charge and current density respectively, and represent the source of the field. The constitutive relations between the magnetic flux density and intensity, as well as the electric displacement and field strength are $B = \mu H$ and $D = \epsilon E$. Since equations (1-2) contain information of the propagation

information of the electromagnetic field, they constitute the basic equations of CEM.

In order to complete the description of the differential system, initial and/or boundary values are required. For Maxwell equations, only the source of the field and a few physical boundary conditions at the media interfaces are pertinent [6,7]:

$$\begin{aligned}
n \times (E_1 - E_2) &= 0 \\
n \times (H_1 - H_2) &= J_s \\
n \cdot (D_1 - D_2) &= \rho_s \\
n \cdot (B_1 - B_2) &= 0
\end{aligned} \tag{4}$$

where the subscripts 1 and 2 refer to media on two sides of the interface. J_s and ρ_s are the surface current and charge densities of a perfect electrical conductor respectively.

Since all computing systems have finite memory, all CEM computations in the time domain must be conducted on a truncated computational domain. This intrinsic constraint requires a numerical farfield condition at the truncated boundary to mimic the behavior of an unbounded field. This numerical boundary unavoidably induces a reflected wave to contaminate the simulated field. In the past, absorbing boundary conditions at the farfield boundary have been developed from the radiation condition[1,16-18]. In general, a progressive order of accuracy procedure can be used to implement the numerical boundary conditions with increasing accuracy [16,17]. On the other hand, the characteristic-based methods which satisfy the physical domain of dependence requirement can specify the numerical boundary condition readily. For this formulation, the reflected wave can be suppressed by eliminating the undesirable incoming numerical data. Although the accuracy of the numerical farfield boundary condition is local coordinate system dependent, in principle this formulation under an ideal circumstance can effectively suppress artificial wave reflections.

4   Maxwell Equations on Curvilinear Frame

In order to develop a versatile numerical tool for computational electromagnetics for a wide range of applications, the Maxwell equations can be cast on a general curvilinear frame of reference [4,5,19]. The system of equations on general curvilinear coordinates can be derived by a coordinate transformation from the Cartesian frame[20,21]. The mesh system in the transformed space can be obtained by numerous grid generation procedures [21]. For a body-oriented coordinate system, the interface between two different media is easily defined by one of the coordinate surfaces. Along this coordinate parametric plane, all discretized nodes on the interface are precisely prescribed without the need for an interpolating procedure. In the transformed space, computations are performed on a uniform mesh space but the corresponding physical spacing can be highly clustered to enhance the numerical resolution. As an illustration of the numerical advantage for solving the Maxwell equations on non-orthogonal curvilinear, body-oriented coordinates, the scattered electromagnetic field simulation of a re-entry vehicle has been performed [22].

The most general coordinate transformation of the Maxwell equations in the time domain is definable by a one-to-one relationship between two sets of temporal and spatial independent variables. However for most practical applications, only the spatial coordinate transformation is sufficient.

$$\begin{aligned} \xi &= \xi(x,y,z) \\ \eta &= \eta(x,y,z) \\ \zeta &= \zeta(x,y,z) \end{aligned} \qquad (5)$$

The governing equation in the strong conservation form is obtained by dividing the chain-rule differentiated equations with the Jacobian of coordinate transformation and by invoking metric identities [20,21]. The time-dependent Maxwell equations on a general curvilinear frame of reference and in the strong conservative form are;

$$\frac{\partial U}{\partial t} + \frac{\partial F_\xi}{\partial \xi} + \frac{\partial F_\eta}{\partial \eta} + \frac{\partial F_\zeta}{\partial \zeta} = -J \qquad (6)$$

where the dependent variables are now defined as

$$U = U\left(B_x V, B_y V, B_z V, D_x V, D_y V, D_z V\right) \qquad (7)$$

V is the Jacobian of coordinate transformation and is also the inverse local cell volume. If the Jacobian has nonzero values in the computational domain, the correspondence between the physical and the transformed space is uniquely defined.

$$V = det \begin{bmatrix} \xi_x & \eta_x & \zeta_x \\ \xi_y & \eta_y & \zeta_y \\ \xi_z & \eta_z & \zeta_z \end{bmatrix} \qquad (8)$$

and ξ_x, η_x, ζ_x, etc are the metrics of coordinate transformation and can be computed easily from the definition given by equation (5). The flux vector components in the transformed space have the following form: After introducing the coordinate transformation, all coefficient matrices now contain metrics which are position dependent. This added complexity to the characteristic formulation of the Maxwell equations no longer permits the system of equations to be decoupled to acquire the true Riemann problem like that on the Cartesian frame [4,5,23].

5 Eigenvalues and Eigenvectors

The fundamental idea of the characteristic-based method for solving the hyperbolic system of equations is derived from the eigenvalue and eigenvector analyses of the governing equations. In a time-space plane, the eigenvalue which relates to the phase velocity of the wave actually defines the slope of the characteristic. All dependent variables within the time-space domain bounded by two intersecting characteristics are completely determined by the values along these characteristics and by their compatibility relationship. The direction of information propagation is also clearly described by these two characteristics [1,2]. In numerical simulation, the well-posedness requirement of initial or boundary conditions and the stability of a numerical approximation are also ultimately linked to the eigenvalues of governing equation [20]. Therefore, characteristic-based methods have demonstrated superior numerical stability and accuracy properties over others schemes [22,23]. However, characteristic-based algorithms also have an inherent limitation in that the governing equation can be diagonalized only in one space-time plane at a time. The multi-dimensional equations are required to split into multiple one-dimensional formulations. This limitation is not unusual for most numerical algorithms such as the approximate factored and the fractional-step schemes [20,23]. A consequence of this restriction is that solutions of the characteristic-based procedure may exhibit some degree of sensitivity to the orientation of the coordinate selected. This numerical behavior is consistent with the concept of optimal coordinates.

In the characteristic formulation on the Cartesian frame, data of wave motion are first split according to the direction of phase velocity and then transmitted in each respective orientation [23]. In each time-space plane, the direction of the phase velocity degenerates into either positive or negative orientation. They are commonly referred to as the right-running or the left-running wave components [1,2]. In the transformed coordinate space, signs of the eigenvalue are now determined the directions of data transmission [4,5]. The corresponding eigenvectors are the essential elements for diagonalizing the coefficient matrices and for formulating the approximated Riemann problem [3]. In essence, knowledge of eigenvalues and eigenvectors of Maxwell equations in the time domain becomes the first prerequisite of the present formulation.

The analytic process to obtain the eigenvalues and the corresponding eigenvectors of the Maxwell equations on general curvilinear coordinates is identical to that on the Cartesian frame. In each temporal-spatial planes t-ξ, t-η, and t-ζ, the eigenvalues are easily found by solving the six-degree characteristic equation associated

$$
F_\xi = \begin{bmatrix}
0 & 0 & 0 & 0 & -\frac{\xi_z}{\epsilon V} & \frac{\xi_y}{\epsilon V} \\
0 & 0 & 0 & \frac{\xi_z}{\epsilon V} & 0 & -\frac{\xi_x}{\epsilon V} \\
0 & 0 & 0 & -\frac{\xi_y}{\epsilon V} & \frac{\xi_x}{\epsilon V} & 0 \\
0 & \frac{\xi_z}{V\mu} & -\frac{\xi_y}{V\mu} & 0 & 0 & 0 \\
-\frac{\xi_z}{V\mu} & 0 & \frac{\xi_x}{V\mu} & 0 & 0 & 0 \\
\frac{\xi_y}{V\mu} & -\frac{\xi_x}{V\mu} & 0 & 0 & 0 & 0
\end{bmatrix}
\begin{Bmatrix} B_x \\ B_y \\ B_z \\ D_x \\ D_y \\ D_z \end{Bmatrix}
\tag{9}
$$

$$
F_\eta = \begin{bmatrix}
0 & 0 & 0 & 0 & -\frac{\eta_z}{\epsilon V} & \frac{\eta_y}{\epsilon V} \\
0 & 0 & 0 & \frac{\eta_z}{\epsilon V} & 0 & -\frac{\eta_x}{\epsilon V} \\
0 & 0 & 0 & -\frac{\eta_y}{\epsilon V} & \frac{\eta_x}{\epsilon V} & 0 \\
0 & \frac{\eta_z}{V\mu} & -\frac{\eta_y}{V\mu} & 0 & 0 & 0 \\
-\frac{\eta_z}{V\mu} & 0 & \frac{\eta_x}{V\mu} & 0 & 0 & 0 \\
\frac{\eta_y}{V\mu} & -\frac{\eta_x}{V\mu} & 0 & 0 & 0 & 0
\end{bmatrix}
\begin{Bmatrix} B_x \\ B_y \\ B_z \\ D_x \\ D_y \\ D_z \end{Bmatrix}
\tag{10}
$$

$$
F_\zeta = \begin{bmatrix}
0 & 0 & 0 & 0 & -\frac{\zeta_z}{\epsilon V} & \frac{\zeta_y}{\epsilon V} \\
0 & 0 & 0 & \frac{\zeta_z}{\epsilon V} & 0 & -\frac{\zeta_x}{\epsilon V} \\
0 & 0 & 0 & -\frac{\zeta_y}{\epsilon V} & \frac{\zeta_x}{\epsilon V} & 0 \\
0 & \frac{\zeta_z}{V\mu} & -\frac{\zeta_y}{V\mu} & 0 & 0 & 0 \\
-\frac{\zeta_z}{V\mu} & 0 & \frac{\zeta_x}{V\mu} & 0 & 0 & 0 \\
\frac{\zeta_y}{V\mu} & -\frac{\zeta_x}{V\mu} & 0 & 0 & 0 & 0
\end{bmatrix}
\begin{Bmatrix} B_x \\ B_y \\ B_z \\ D_x \\ D_y \\ D_z \end{Bmatrix}
\tag{11}
$$

with the coefficient matrices.

$$
\lambda_\xi = \left\{ -\frac{\alpha}{V\sqrt{\epsilon\mu}}, -\frac{\alpha}{V\sqrt{\epsilon\mu}}, \frac{\alpha}{V\sqrt{\epsilon\mu}}, \frac{\alpha}{V\sqrt{\epsilon\mu}}, 0, 0 \right\}
\tag{12}
$$

$$
\lambda_\eta = \left\{ -\frac{\beta}{V\sqrt{\epsilon\mu}}, -\frac{\beta}{V\sqrt{\epsilon\mu}}, \frac{\beta}{V\sqrt{\epsilon\mu}}, \frac{\beta}{V\sqrt{\epsilon\mu}}, 0, 0 \right\}
\tag{13}
$$

$$
\lambda_\zeta = \left\{ -\frac{\gamma}{V\sqrt{\epsilon\mu}}, -\frac{\gamma}{V\sqrt{\epsilon\mu}}, \frac{\gamma}{V\sqrt{\epsilon\mu}}, \frac{\gamma}{V\sqrt{\epsilon\mu}}, 0, 0 \right\}
\tag{14}
$$

where $\alpha = \sqrt{\xi_z{}^2 + \xi_y{}^2 + \xi_x{}^2}$, $\beta = \sqrt{\eta_z{}^2 + \eta_y{}^2 + \eta_x{}^2}$, and $\gamma = \sqrt{\zeta_z{}^2 + \zeta_y{}^2 + \zeta_x{}^2}$

One recognizes that the eigenvalues in each time-space plane contain multiplicities, and hence the eigenvectors do not have unique elements [4,5]. Nevertheless, linearly independent eigenvectors associated with each eigenvalue still have been found by reducing the coefficient matrix to the Jordan normal form. The eigenvectors are selected in such a fashion that the similar matrices of diagonalization will automatically degenerate to the identical form of the Cartesian frame as required.

From the eigenvector analysis, the similarity transformation matrices of diagonalization in each time-space plane are formed by using eigenvectors as the column arrays as shown in the following equations. For an example, the first column of the similar matrix of diagonalization, $[-\frac{\sqrt{\mu}\xi_y}{\sqrt{\epsilon}\alpha}, \frac{\sqrt{\mu}(\xi_x{}^2+\xi_z{}^2)}{\sqrt{\epsilon}\xi_x\alpha}, \frac{\sqrt{\mu}\xi_y\xi_z}{\sqrt{\epsilon}\xi_x\alpha}, -\frac{\xi_y}{\xi_x}, 0, 1]$ in the

t-ξ plane is the eigenvector corresponding to the eigenvalue $\lambda_\xi = -\frac{\alpha}{V\sqrt{\epsilon\mu}}$. Since the similar matrices of diagonalization, S_ξ, S_η, and S_ζ are non-singular, the left-hand inverse matrices, S_ξ^{-1}, $S^{-1}{}_\eta$, and $S^{-1}{}_\zeta$ are easily found. Although these left-hand inverse matrices are essential to the diagonalization process, they provide little insight for the following flux vector splitting procedure. The rather involved results are omitted here, but they can be found in references [4,5].

6 Flux-Vector Splitting

An efficient flux vector splitting algorithm for solving the Euler equations was developed by Steger and Warming [24]. The basic concept is equally applicable to any hyperbolic differential system. In most computational electromagnetics applications, the discontinuous behavior in solution is associated only with the wave across the interface of different media. The salient feature of the piecewise continuous solution domains of the hyperbolic partial differential equation stands out. The coefficient matrices of the time-dependent, three-dimensional Maxwell equations cast in the general curvilinear frame of reference contain metrics of coordinate transformation. Therefore, the equation system no longer has constant coefficients even in an isotropic and homogeneous medium. Under this circumstance, eigenvalues can change sign at any given field location due to the metric variations of coordinate transformation. Numerical oscillations have appeared in calculated results using the flux vector splitting technique

$$S_\xi = \begin{bmatrix}
-\dfrac{\sqrt{\mu}\,\xi_y}{\sqrt{\epsilon}\,\alpha} & \dfrac{\sqrt{\mu}\,\xi_z}{\sqrt{\epsilon}\,\alpha} & \dfrac{\sqrt{\mu}\,\xi_y}{\sqrt{\epsilon}\,\alpha} & -\dfrac{\sqrt{\mu}\,\xi_z}{\sqrt{\epsilon}\,\alpha} & 1 & 0 \\[2mm]
\dfrac{\sqrt{\mu}\left(\xi_x{}^2+\xi_z{}^2\right)}{\sqrt{\epsilon}\,\xi_x\,\alpha} & \dfrac{\sqrt{\mu}\,\xi_y\,\xi_z}{\sqrt{\epsilon}\,\xi_x\,\alpha} & -\dfrac{\sqrt{\mu}\left(\xi_x{}^2+\xi_z{}^2\right)}{\sqrt{\epsilon}\,\xi_x\,\alpha} & -\dfrac{\sqrt{\mu}\,\xi_y\,\xi_z}{\sqrt{\epsilon}\,\xi_x\,\alpha} & \dfrac{\xi_y}{\xi_x} & 0 \\[2mm]
-\dfrac{\sqrt{\mu}\,\xi_y\,\xi_z}{\sqrt{\epsilon}\,\xi_x\,\alpha} & -\dfrac{\sqrt{\mu}\left(\xi_x{}^2+\xi_y{}^2\right)}{\sqrt{\epsilon}\,\xi_x\,\alpha} & \dfrac{\sqrt{\mu}\,\xi_y\,\xi_z}{\sqrt{\epsilon}\,\xi_x\,\alpha} & \dfrac{\sqrt{\mu}\left(\xi_x{}^2+\xi_y{}^2\right)}{\sqrt{\epsilon}\,\xi_x\,\alpha} & \dfrac{\xi_z}{\xi_x} & 0 \\[2mm]
-\dfrac{\xi_z}{\xi_x} & -\dfrac{\xi_y}{\xi_x} & -\dfrac{\xi_z}{\xi_x} & -\dfrac{\xi_y}{\xi_x} & 0 & 1 \\[2mm]
0 & 1 & 0 & 1 & 0 & \dfrac{\xi_y}{\xi_x} \\[2mm]
1 & 0 & 1 & 0 & 0 & \dfrac{\xi_z}{\xi_x}
\end{bmatrix} \tag{15}$$

$$S_\eta = \begin{bmatrix}
-\dfrac{\left(\eta_y{}^2+\eta_z{}^2\right)\sqrt{\mu}}{\sqrt{\epsilon}\,\eta_y\,\beta} & -\dfrac{\eta_x\,\eta_z\,\sqrt{\mu}}{\sqrt{\epsilon}\,\eta_y\,\beta} & \dfrac{\left(\eta_y{}^2+\eta_z{}^2\right)\sqrt{\mu}}{\sqrt{\epsilon}\,\eta_y\,\beta} & \dfrac{\eta_x\,\eta_z\,\sqrt{\mu}}{\sqrt{\epsilon}\,\eta_y\,\beta} & \dfrac{\eta_x}{\eta_y} & 0 \\[2mm]
\dfrac{\eta_x\,\sqrt{\mu}}{\sqrt{\epsilon}\,\beta} & -\dfrac{\eta_z\,\sqrt{\mu}}{\sqrt{\epsilon}\,\beta} & -\dfrac{\eta_x\,\sqrt{\mu}}{\sqrt{\epsilon}\,\beta} & \dfrac{\eta_z\,\sqrt{\mu}}{\sqrt{\epsilon}\,\beta} & 1 & 0 \\[2mm]
\dfrac{\eta_x\,\eta_z\,\sqrt{\mu}}{\sqrt{\epsilon}\,\eta_y\,\beta} & \dfrac{\left(\eta_x{}^2+\eta_y{}^2\right)\sqrt{\mu}}{\sqrt{\epsilon}\,\eta_y\,\beta} & -\dfrac{\eta_x\,\eta_z\,\sqrt{\mu}}{\sqrt{\epsilon}\,\eta_y\,\beta} & -\dfrac{\left(\eta_x{}^2+\eta_y{}^2\right)\sqrt{\mu}}{\sqrt{\epsilon}\,\eta_y\,\beta} & \dfrac{\eta_z}{\eta_y} & 0 \\[2mm]
0 & 1 & 0 & 1 & 0 & \dfrac{\eta_x}{\eta_y} \\[2mm]
-\dfrac{\eta_z}{\eta_y} & -\dfrac{\eta_x}{\eta_y} & -\dfrac{\eta_z}{\eta_y} & -\dfrac{\eta_x}{\eta_y} & 0 & 1 \\[2mm]
1 & 0 & 1 & 0 & 0 & \dfrac{\eta_z}{\eta_y}
\end{bmatrix} \tag{16}$$

$$S_\zeta = \begin{bmatrix}
\dfrac{\sqrt{\mu}\left(\zeta_y{}^2+\zeta_z{}^2\right)}{\sqrt{\epsilon}\,\zeta_z\,\gamma} & \dfrac{\sqrt{\mu}\,\zeta_x\,\zeta_y}{\sqrt{\epsilon}\,\zeta_z\,\gamma} & -\dfrac{\sqrt{\mu}\left(\zeta_y{}^2+\zeta_z{}^2\right)}{\sqrt{\epsilon}\,\zeta_z\,\gamma} & -\dfrac{\sqrt{\mu}\,\zeta_x\,\zeta_y}{\sqrt{\epsilon}\,\zeta_z\,\gamma} & \dfrac{\zeta_x}{\zeta_z} & 0 \\[2mm]
-\dfrac{\sqrt{\mu}\,\zeta_x\,\zeta_y}{\sqrt{\epsilon}\,\zeta_z\,\gamma} & -\dfrac{\sqrt{\mu}\left(\zeta_x{}^2+\zeta_z{}^2\right)}{\sqrt{\epsilon}\,\zeta_z\,\gamma} & \dfrac{\sqrt{\mu}\,\zeta_x\,\zeta_y}{\sqrt{\epsilon}\,\zeta_z\,\gamma} & \dfrac{\sqrt{\mu}\left(\zeta_x{}^2+\zeta_z{}^2\right)}{\sqrt{\epsilon}\,\zeta_z\,\gamma} & \dfrac{\zeta_y}{\zeta_z} & 0 \\[2mm]
-\dfrac{\sqrt{\mu}\,\zeta_x}{\sqrt{\epsilon}\,\gamma} & \dfrac{\sqrt{\mu}\,\zeta_y}{\sqrt{\epsilon}\,\gamma} & \dfrac{\sqrt{\mu}\,\zeta_x}{\sqrt{\epsilon}\,\gamma} & -\dfrac{\sqrt{\mu}\,\zeta_y}{\sqrt{\epsilon}\,\gamma} & 1 & 0 \\[2mm]
0 & 1 & 0 & 1 & 0 & \dfrac{\zeta_x}{\zeta_z} \\[2mm]
1 & 0 & 1 & 0 & 0 & \dfrac{\zeta_y}{\zeta_z} \\[2mm]
-\dfrac{\zeta_y}{\zeta_z} & -\dfrac{\zeta_x}{\zeta_z} & -\dfrac{\zeta_y}{\zeta_z} & -\dfrac{\zeta_x}{\zeta_z} & 0 & 1
\end{bmatrix} \tag{17}$$

when eigenvalues change sign. A refined flux difference splitting algorithm has been developed to resolve fields with jump conditions [25,26]. The newer flux difference splitting algorithm is particularly effective at locations where the eigenvalues vanish. In general the governing equations are linear; at most the coefficients of the differential system are dependent of physical location and phase velocity. Therefore, the difference between the flux vector splitting [24] and flux difference splitting [25,26] schemes, when applied to the time-dependent Maxwell equations should not be overly significant.

The characteristic-based algorithms have a deep rooted theoretical base in describing the wave dynamics which is highly directional. The flux vectors F_ξ, F_η, and F_ζ will be split according to the sign of their corresponding eigenvalues. The split fluxes are then differenced by an upwind algorithm to honor the zone of dependence of an initial-value problem [3,4,5].

$$\begin{aligned}
F_\xi &= F_\xi^+ + F_\xi^- \\
F_\eta &= F_\eta^+ + F_\eta^- \\
F_\zeta &= F_\zeta^+ + F_\zeta^-
\end{aligned} \tag{18}$$

The flux vector components associated with the positive and negative eigenvalues are obtainable by a straightfor-ward matrix multiplication.

$$\begin{aligned}
F_\xi^+ &= S_\xi \lambda_\xi^+ S_\xi^{-1} U \\
F_\xi^- &= S_\xi \lambda_\xi^- S_\xi^{-1} U \\
F_\eta^+ &= S_\eta \lambda_\eta^+ S_\eta^{-1} U \\
F_\eta^- &= S_\eta \lambda_\eta^- S_\eta^{-1} U \\
F_\zeta^+ &= S_\zeta \lambda_\zeta^+ S_\zeta^{-1} U \\
F_\zeta^- &= S_\zeta \lambda_\zeta^- S_\zeta^{-1} U
\end{aligned} \tag{19}$$

The detailed expression of split flux vectors in the curvilinear frame can be found in references [4,5].

7 Spatial Discretization

Once the detailed split fluxes are known, formulation of the finite-difference approximation is straightforward. From the sign of an eigenvalue, the stencil of a spatially second or higher order accurate windward differencing can be easily constructed to form multiple one-dimensional difference operators (8,15,16). In this regard, the forward difference and the backward difference approximations are used for the negative and the positive eigenvalues respectively. The split flux vectors are evaluated at each discretized point of the field according to the signs of the eigenvalues. A windward second-order accurate procedure is best in illustrating this point.

$$
\begin{aligned}
If \quad \lambda < 0, \quad \Delta U_i &= \left[-3U_i + 4U_{i+1} - U_{i+2}\right]/2 \\
If \quad \lambda > 0, \quad \nabla U_i &= \left[\ 3U_i - 4U_{i-1} + U_{i-2}\right]/2
\end{aligned}
\tag{20}
$$

A higher numerical resolution over a frequency spectrum can be obtained by using a compact difference scheme and yet requires no greater grid-point stencil dimension. This approach is based on the Hermite's generation of Taylor series [27]. The accuracy gain is derived from additional derivative data at the boundary, and the solution is required to be smooth on a uniformly spaced mesh [12-15,27]. A five-point compact stencil formula for the approximation of a first-order derivative has the following expression [14,15]:

$$
\alpha U'_{i+1} + U'_i + \beta U'_{i-1} = b(U_{i+2} - U_{i-2})/2 \\
+ a(U_{i+1} - U_{i-1})/4
\tag{21}
$$

The desired high resolution characteristics of compact differencing schemes can be further optimized to achieve numerical behavior approaching that of a spectral method. Although the most popular compact differencing method has a tridiagonal stencil structure, there still is a need of a transitional scheme from boundary to interior domain. The transition scheme is not only required to transmit data from the boundary but must also preserve the stability and accuracy of interior calculations for the global order of resolution. Therefore, the development of a numerical boundary scheme is emerging as the pacing item for the high resolution method [12-15].

The finite-volume approximation solves the governing equation by discretizing the physical space into contiguous cells and balancing the flux vectors on the cell surfaces [19,20]. Thus in discretized form, the integration procedure degenerates into evaluation of the sum of all fluxes aligned with surface area vectors. This feature is distinctive from the point values derived from adjacent nodes as in the finite difference approximation. The van Leer's Kappa scheme (MUSCL) is frequently adopted in CEM in which solution vectors are reconstructed on the cell surface from the piecewise data of neighboring cells [4,5,19]. The reconstruction scheme spans a range from first-order to third-order upwind biased approximations by manipulating the parameters ϕ and κ [19-21].

$$
\begin{aligned}
U^+_{i+\frac{1}{2}} &= U_i + \phi/4\left[(1-\kappa)\nabla + (1+\kappa)\Delta\right]U_i \\
U^-_{1+\frac{1}{2}} &= U_i - \phi/4\left[(1+\kappa)\nabla + (1-\kappa)\Delta\right]U_{i+1}
\end{aligned}
\tag{22}
$$

where $\Delta U_i = U_i - U_{i-1}$ and $\nabla U_i = U_{i+1} - U_i$ are the forward and backward differencing discretization.

The compact differencing recently has also been applied to the finite-volume approximation [15]. For this relatively new advancement in CEM, the distinction between cell-averaged and pointwise variables is critical. In the reconstruction process to obtain the flux balance on the cell surface, the mid-point formula has to be used to interpolate the cell-averaged values to achieve the desired accuracy.

$$
\alpha U'_{i+3/2} + U'_{i+1/2} + \beta U'_{i-1/2} = b(U_{i+5/2} - U_{i-3/2})/2 \\
+ a(U_{i+3/2} - U_{i-1/2})/4
\tag{23}
$$

The compact differencing approximation also introduces a point of departure from the traditional characteristic-based method. In which most of the compact formulations are derived from the spatially central scheme, the flux vector splitting in the interior computing domain can be completely circumvented. The flux vector splitting is only required to be implement at the boundary. The grid-point stencil will now limit the dimension of a thin layer that can be simulated.

The most significant feature of the flux vector splitting scheme lies in its ability to easily suppress reflected waves from the truncated computational domain. In wave motion, the compatibility condition at any point in space is described by the split flux vector in the direction of the wave motion (1,2). However, if the grid system does not align with the motion, the compatibility will degenerate. A general approximated no-reflection condition can be given by setting the incoming flux component to zero.

$$
Either \quad \lim_{r \to \infty} F^+(\xi, \eta, \zeta) = 0 \quad or \quad \lim_{r \to \infty} F^-(\xi, \eta, \zeta) = 0
\tag{24}
$$

This unique attribute of the characteristic-based numerical procedure in alleviating a fundamental dilemma in CEM should be further refined.

8 Temporal Discretization

Although the fractional-step or the time-splitting algorithm has demonstrated greater efficiency in data storage and a higher data processing rate than predictor-corrector time integration procedures, it is limited to second-order accuracy in time [20,23]. For the fractional step method, the temporal second-order result is obtained by a symmetrically cyclic operators sequence [23].

$$
U^{n+2} = L_\xi L_\eta L_\zeta L_\zeta L_\eta L_\xi U^n
\tag{25}
$$

where L_ξ, L_η, and L_ζ are the difference operators for one-dimensional equations in the ξ, η, and ζ coordinates respectively.

In general, second-order and higher temporal resolution is achievable through multiple time step schemes [20]. However, one-step schemes are more attractive because they have less memory requirements and don't need special start-up procedures [15,16]. For future higher order accurate solution development potential, the Runge-Kutta family of single-step, multi-stage procedure is recommended. Although the temporal integration scheme is known to introduce dissipation when applied to the model wave equation in conjunction with a spatially central scheme, the numerical advantage is still significant [12,15]. This choice is equally effective for both the finite-difference and the finite-volume method. The four-stage, formally fourth order accurate scheme is presented as follows:

$$
\begin{aligned}
U^{n+1} &= U^n + (\Delta t/6)(U_1' + 2U_2' + 2U_3' + U_4') \\
U_1' &= U_1'(t, U^n) \\
U_2' &= U_2'(t + \Delta t/2, U^n + U_1'\Delta t/2) \\
U_3' &= U_3'(t + \Delta t/2, U^n + U_2'\Delta t/2) \\
U_4' &= U_4'(t + \Delta t, U^n + U_3'\Delta t)
\end{aligned}
$$

$$(26)$$

In Figure 1, the dispersive error of of several selected algorithms is presented. All numerical solutions were generated by solving the simple model wave equation. The windward difference schemes generally have a predominant leading phase error. The third-order bidiagonal compact difference procedure even outperforms fourth-order scheme in the wavenumber range beyond $\frac{\pi}{4}$. For the mid-point leapfrog and the upwind schemes, their one-dimensional solutions on uniform mesh have the perfect-shift property, but exhibit significant phase errors beyond the wavenumber range of $\frac{\pi}{8}$. Therefore in a long range wave propagating simulation, an alternative numerical algorithm may be required. The suppression of dispersive error by the optimized compact difference schemes over a greater wavenumber range has been demonstrated [15].

The superior isotropy characteristics of the compact and optimized compact difference schemes over the MUSCL algorithm is also illustrated by two-dimensional semi-discretized simulations in Figure 2. The comparative results are compressed into four separated quadrants with different normalized wavenumber. At the lowest normalized wave number $\omega = \pi/8$, corresponding to 16 nodes per wavelength, all numerical schemes contained negligible anisotropic error. The error increases and becomes unacceptable as the normalized wavenumber reaches the value of $\omega = 3\pi/4$. However, at the normalized wavenumber of $\pi/2$ where each wavelength is only supported by 4 nodes, the superior numerical resolution of the compact difference and optimized schemes exhibit developable potential [14,15]

9 Numerical Results

The technical merits of the characteristic-based methods for solving the time-dependent, three-dimensional Maxwell equations can best be illustrated by the two following illustrations. In Figure 3, the exact electrical field of a traveling wave is compared with numerical results. The numerical results of a single-step upwind explicit scheme (SUE) were generated at the maximum allowable time step size defined by the Courant-Friedrichs-Lewy (CFL) number of 2, $(\lambda \Delta x / \Delta t) = 2$ [20,23]. The numerical solutions are presented at instants when a right-running wave reaches the mid-point of the computational domain and exits the numerical boundary respectively. For this one-dimensional simulation, the characteristic-based scheme using the single-step upwind explicit algorithm exhibits the shift property which indicates a perfect translation of the initial value in space [23]. As the impulse wave moves through the initially quiescent environment, the numerical result duplicates the exact solution at each and every discretized point, including the discontinuous incoming wave front. Although this highly desirable property of a numerical solution is only achievable under very restrictive conditions and is not preserved for multi-dimensional problem, [19] the ability to simulate the non-analytic solution behavior in the limit is clearly illustrated.

In Figure 4, another outstanding feature of the characteristic-based method is highlighted by simulating the oscillating electric dipole. For the radiating electric dipole, the depicted temporal calculations are sampled at the instant when the initial pulse has traveled a distance of 2.24 wavelengths from the dipole. The numerical results are generated on a $(48 \times 48 \times 96)$ mesh system with the second order MUSCL scheme. Under the condition each wavelength is resolved by 15 mesh points and the difference between numerical results by the finite volume and the finite difference method is negligible. Under the present computational conditions, both numerical procedures uniformly yield excellent comparison with the theoretical result. The most interesting numerical behavior, however, is revealed at the truncated farfield boundary. The no-reflection condition at the numerical boundary is observed to be satisfied within the order of truncation error. For the spherically symmetric radiating field, the orientation of the wave is aligned with the radial coordinate, the suppression of the reflected wave within the numerical domain is the best achievable by the characteristic formulation.

The corresponding computed magnetic field intensity by both the second order accurate finite difference and the finite volume procedure is given in Figure 5. Again the difference in solution between the two distinct numerical procedures is indiscernible. For the oscillating electric dipole, only the x and the y components of the magnetic field exist. Numerical results attain an excellent agreement with theoretical values [4,5]. The third order accurate finite volume scheme also produces a similar result on the same mesh but at a greater allowable time step size (a CFL value of 0.87 is used vs 0.5 for the second order method). However, the third order windward biased MUSCL algorithm cannot reinforce rigorously the zone of dependence requirement, therefore the reflected wave suppression is incomplete at the truncated numerical domain and the calculation is not included here.

The numerical efficiency of CEM can be enhanced substantially by using scalable multicomputers. The characteristic-based finite volume computer program has been successfully mapped onto distributed memory systems by a rudimentary domain decomposition strategy [24]. For example a square waveguide, at five different frequencies until the cut-off occurred, is simulated. Figure 6 displays the x-component of the magnetic field intensity within the waveguide. The simulated transverse electric mode, $TE_{(1,1)}$, $E_z = 0$ which has a half-period of π along the x and y coordinates, is generated on a $(24 \times 24 \times 128)$ mesh system. Since the entire field is described by simple harmonic functions, the remaining field components are similar and only half the solution domain along the z coordinate is presented to minimize repetition. In short, the agreement between the closed-form and numerical solutions is excellent at each frequency. In addition, the numerical simulations also duplicate the physical phenomenon at the cut-off frequency below which there is no phase shift along the waveguide and the wave motion then ceases. At a grid-point density of 12 nodes per wavelength, the L2 norm [28] has a nearly uniform magnitude of $O(10^{-4})$. The improvement to the parallel and scalable numerical efficiency is one of the most promising area in CEM research.

The pioneering efforts in CEM usually employed the total-field formulation on staggered mesh systems [9,10]. The particular combination of numerical algorithm and procedure has been proven to be very effective. In the total-field formulation, the calculation must contain the residual of partial cancelations of the incident and the diffracted waves, and the incident wave must also traverse the entire computation domain. Both requirements impose severe demand on numerical accuracy of simulation. The nearfield electromagnetic energy distribution becomes a secular problem - a small difference between two variables of large magnitude. An alternative approach via the scattered-field formulation for RCS calculations appears to be very attractive. Particularly in this formulation, the numerical dissipation of the incident wave that propagate from the farfield boundary to the scatterer is completely eliminated from the computations. In essence, the incident field is directly specified on the scatterer's surface. A numerical advantage over that of the total field formulation is substantial.

The comparison of horizontal polarized RCS of a perfect electric conducting (PEC) sphere, $\sigma(\theta, 0.0)$, of the total-field and the scattered-field formulations at ka = 4.7 is presented in Figure 7. The validating data is the exact solution for the scattering of a plane electromagnetic wave by a PEC sphere which is commonly referred to as the Mie series [6,7]. Both numerical results are generated under identical computational conditions. Numerical results of the total-field formulation reveal far greater error than the scattered-field formulation. In the scattered-field formulation, the incident field data are described precisely by the boundary condition on the scatterer surface. Since the farfield electromagnetic energy distribution is derived from the nearfield parameters [6,7], the advantage of describing the incident data without error on a scatterer is tremendous. Numerical error of the total-field calculations are evident in the exaggerated peaks and troughs over the entire viewing angle displacement.

In Figure 8, the vertical polarized RCS $\sigma(\theta, 90.0)$ of the Ka = 4.7 case substantiates the previous observation. In fact, the numerical error of the total-field calculation is excessive in comparison with the result by the scattered-field formulation. The deviation of the total-field result from the theory is excessive and becomes unacceptable. In addition, computations by the total-field formulation exhibit a strong sensitivity to placement of the farfield boundary. A small perturbation of the farfield boundary placement leads to a drastic change in RCS prediction: a feature resembling the ill-posedness condition which is highly undesirable for numerical simulation. Since there is very little difference in computer coding for the two formulations, the difference in computing time required for an identical simulation is insignificant.

Once the numerical algorithmic and computational accuracy criteria were established, more complex three-dimensional scatterer configurations can be systematically simulated. In Figure 9, the HH polarization bistatic RCS of an ogive at 4 Ghz and by a Gaussian pulse is depicted. Two numerical results used the same grid dimension $(60 \times 79 \times 51)$, but different grid spacing distributions were obtained. The uniform grid system is constructed by 21 grid points per wavelength over the entire length of the ogive. The clustered grid system has a grid density per wavelength distribution from 10 at middle of the ogive to 55 near the apexes. The L2 norm discrepancy of present results to a method-of-moments solution [29] is 0.40 dB and 0.62 dB respectively.

Even more complex configurations with piecewise

continuous surfaces which required multiple-grid-block description have also been achieved. The bistatic RCS of an ogive cylinder when illuminated by a head-on incident wave in horizontal polarization at 2 GHz is presented in Figure 10. The numerical result is generated by a two-block grid system with the grid dimension of $(159 \times 109 \times 79)$ and $(57 \times 109 \times 93)$. The excellent agreement with a Method-of-Moments result [29] is clearly demonstrated.

In spite of rapid progress in the high-resolution numerical algorithm improvement, additional reduction of data processing time must be derived from concurrent computing. As a demonstration, the speed up of a 4.56 Gigabyte computation using MPI on the IBM SP2 wide node system is given in Figure 11. For this timing result, all PEC sphere scattering calculations are supported by a $(91 \times 241 \times 385)$ grid and excited by a harmonic incident wave at a angular frequency of 100. A minimum of 6 computational nodes is used to accommodate this memory size. The timing data therefore, are collected from 6 to 128 nodes. The speed-up of the present preliminary timing test is a factor of 13.7. The parallel efficiency drops from a value of 98.4 to 63.7 percent. Sustained research efforts in concurrent computing are maintained to achieve the potentially greater numerical efficiency.

10 Summary

In summary, recent progress in solving the three-dimensional Maxwell equations in the time domain has opened a new frontier in electromagnetics, plasmadynamics, optics, as well as the interphase between electrodynamics and quantum mechanics. The progress in microchip and interconnect network technology has led to a host of high performance distributive memory, message passing parallel computer systems. The synergism of efficient and accurate numerical algorithms for solving the Maxwell equations in the time domain with high performance multicomputers will propel the relatively new interdisciplinary simulation technique to practical and productive applications.

11 Acknowledgment

Author is grateful for the AFOSR sponsorship by Dr. A. Nachman. The computing support from the Department of Defense HPC Shared Resource Center at WPAFB is deeply appreciated

12 References

1. Sommerfeld, A., PARTIAL DIFFERENTIAL EQUATIONS IN PHYSICS, Academic Press, New York, 1949, Chapter 2.

2. Courant, R and Hilbert, D., METHODS OF MATHEMATICAL PHYSICS, Vol II, Interscience, New York, 1965.

3. Roe, P.L., Characteristic-Based Schemes for the Euler Equations, Ann Rev. Fluid Mech., Vol 18, 1986, pp.337-365.

4. Shang, J.S., Characteristic-Based Algorithms for Solving the Maxwell Equations in the Time Domain, IEEE Antennas and Propagation Magazine, vol 37, No 3, June 1995, pp 15-25.

5. Shang, J.S. and Fithen, R. M., A Comparative Study of Characteristic-Based Algorithms for the Maxwell Equations, J. Comp Phys, Vol 125, 1966, pp 378-394.

6. Elliott, R. A., ELECTROMAGNETICS, McGraw-Hill, New York, 1966, Chapter 5.

7. Harrington, R.F., TIME-HARMONIC ELECTROMAGNETIC FIELDS, McGraw-Hill, 1961.

8. Harrington, R.F., FIELD COMPUTATION BY MOMENT METHODS, Robert E Krieger Pub. Co, Malabar, FL, 1968, 4th edition.

9. Yee, K.S., Numerical Solution of Initial Boundary Value Problems Involving Maxwell's Equations, in Isotropic Media, IEEE Trans. Ant. Prop. Vol 14, No 3, 1966, pp 302-307

10. Taflove, A., Re-inventing Electromagnetics: Supercomputing Solution of Maxwell's Equations Via Direct Time Integration on Space Grids, Computing Sys Engineering, Vol 3, no 1-4, 1992, pp.153-168.

11. Shankar, V., Research to Application Supercomputing Trends for the 90's Opportunities for Interdisciplinary Computations, AIAA Preprint 91-0002, 29th Aerospace Science Meeting, Reno NV, January, 1991.

12. Lele, S.K., Compact Finite Difference Schemes with Spectral-like Resolution, J. Comp.Physics, Vol 103, 1992, pp 16-42.

13. Carpenter, M.K., Gottlieb, D., and Arbarbanel, S., Time-Stable Boundary Conditions for Finite-Difference Schemes Solving Hyperbolic Systems: Methodology and Application to High-Order Compact Schemes, J. Comp. Phys. Vol 111, No 2, April 1994, pp 220-236.

14. Shang, J.S. and Gaitonde, D., On High Resolution Schemes for Time-Dependent Maxwell Equations, AIAA Preprint 96-0832, 34th Aerospace Science Meeting, Reno NV, January 1996.

15. Gaitonde, D. and Shang, J.S., High-Order Finite-Volume Schemes in Wave Propagation Phenomena, AIAA Preprint 96-2335, 27th Plasmadynamics and Lasers Conf., New Orleans LA, June 1996.

16. Enquist, B. and Majda, A., Absorbing Bound-

ary Conditions for the Numerical Simulation of Waves, Math of Comp., Vol 31, July 1977, pp 629-651.

17. Higdon, R., Absorbing Boundary Conditions for Difference Approximation to Multi-dimensional Wave Equation, Math of Comp., Vol 47, No 175, 1986, pp 437-459.

18. Berenger, J., A Perfectly Matched Layer for the Absorption of Electromagnetic Waves, J. Comp Phys, Vol 114, 1994, pp 185-200.

19. Shang, J.S. and Gaitonde, D., Characteristic-Based, Time- Dependent Maxwell Equation Solvers on a General Curvilinear Frame, AIAA J Vol 33, No 3, March 1995, pp 491-498.

20. Anderson, D.A., Tannehill, J.C., and Pletcher, R.H., COMPUTATIONAL FLUID MECHANICS AND HEAT TRANSFER, Hemisphere Publishing Corp, New York, 1984.

21. Thompson, J. F., NUMERICAL GRID GENERATION, Elsevier Science Publishing Co., New York, 1982

22. Shang, J.S. and Gaitonde, D., Scattered Electromagnetic Field of a Reentry Vehicle, J. Spacecraft and Rockets, Vol 32, No 2, March-April, 1995, pp. 294-301.

23. Shang, J.S., A Fractional-Step Method for Solving 3-D, Time-Domain Maxwell Equations, J comp Phys Vol 118, No 1, April 1995, pp 109-119.

24. Steger, J.L. and Warming, R.F., Flux Vector Splitting of the Inviscid Gasdynamics Equations With Application to Finite Difference Methods, J. Comp. Phys., Vol 20, No 2, Feb. 1987, pp 263-293.

25. Van Leer, B., Flux-Vector Splitting for the Euler Equations, TR 82-30, ICASE, Sept. 1982, Also Lecture Notes in Physics, Vol 170, 1982, pp 507-512.

26. Anderson, W.K., Thomas, J.L., and van Leer, B., A Comparison of Finite Volume Flux Splittings for the Euler Equations, AIAA Paper 85-0122, AIAA 23rd Aerospace Science Meeting, Reno NV, January 1985.

27. Colatz, L., The Numerical Treatment of Differential Equations, Springer-Verlag, New York, 1966, p. 538.

28. Shang, J.S., Calahan, D.A., and Vikstrom, B., Performance of a Finite-Volume CEM Code on Multicomputers, Comp. Systems in Engineering, Vol 6, No 3, 1995, pp 241-250.

29. Putnam, J.N. and Gedera, M.B., A General Purpose 3-D Method of Moment Scattering Code, IEEE Antennas and Propagation Mag., Vol 35, No.2, April 1993, PP. 69-71.

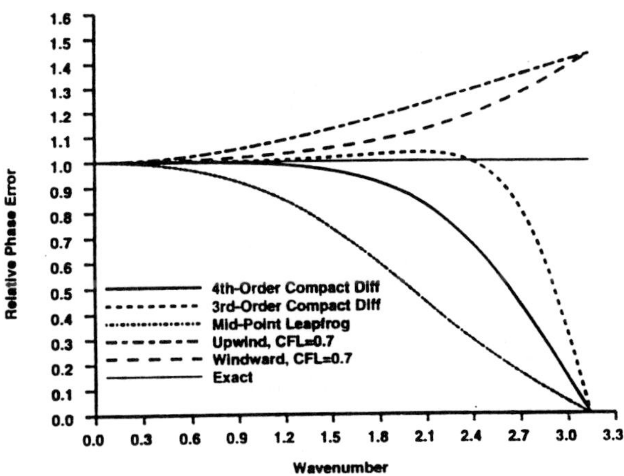

Figure 1: Dispersive Error in Solving Simple Model Wave Equation

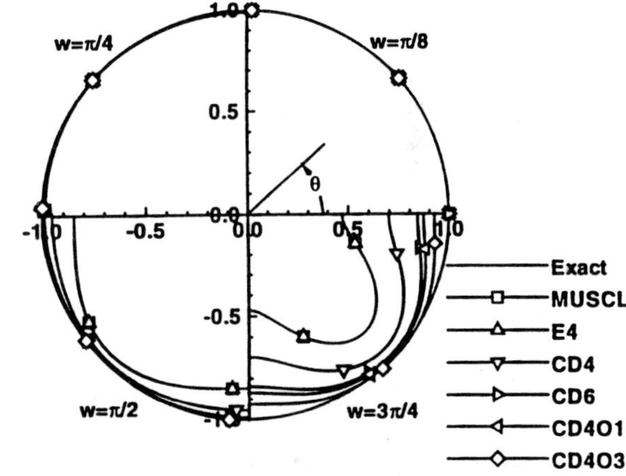

Figure 2: Anisotropic Error in Solving Simple Model Wave Equation

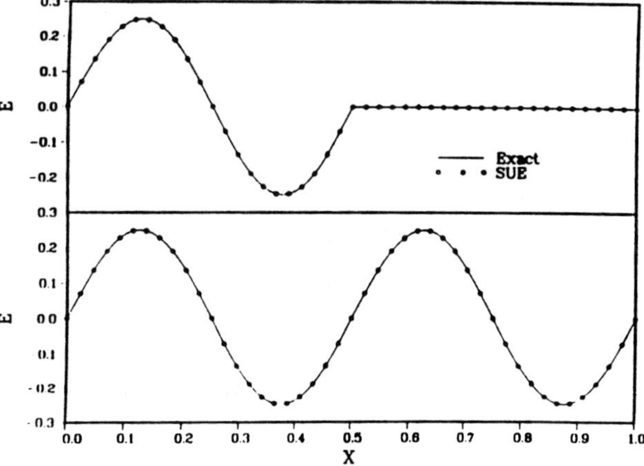

Figure 3: Perfect Shift Property of a One-Dimensional Waves Computation, CFL = 2

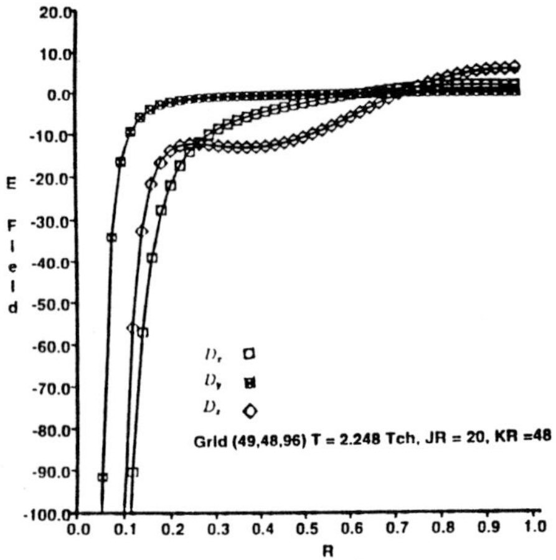

Figure 4: Instantaneous Oscillating Dipole Electric Field

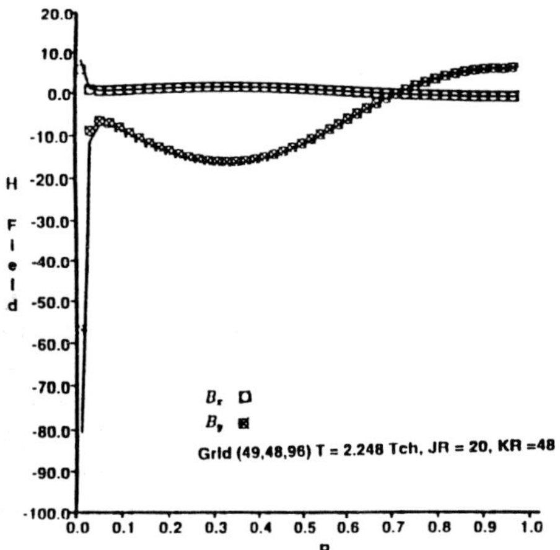

Figure 5: Instantaneous Oscillating Dipole Magnetic Field

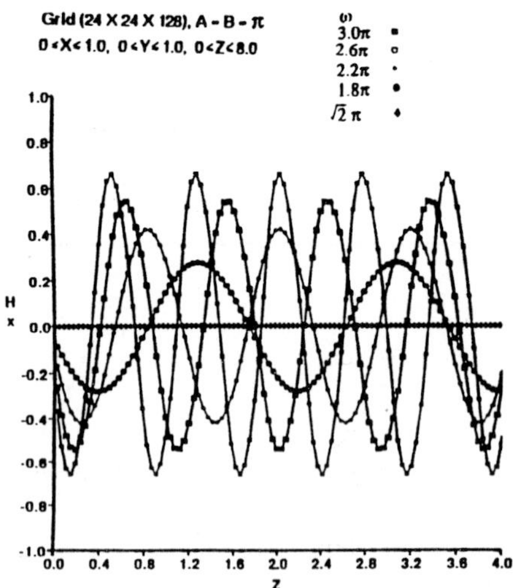

Figure 6: Cut-Off Frequency of a Rectangular Waveguide, TE(1,1)

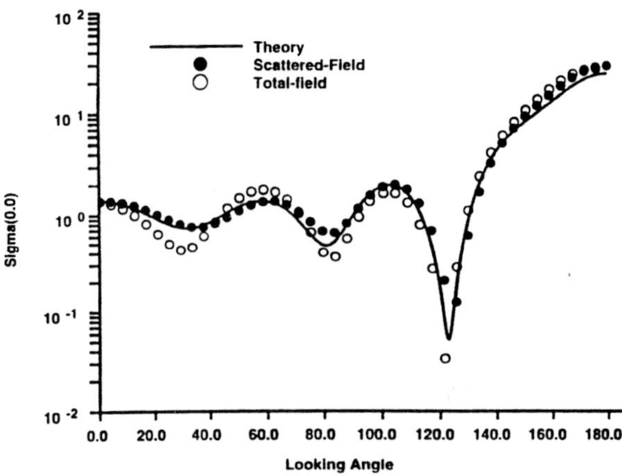

Figure 7: Total-Field and Scattered-Field RCS Calculations, HH Polarization

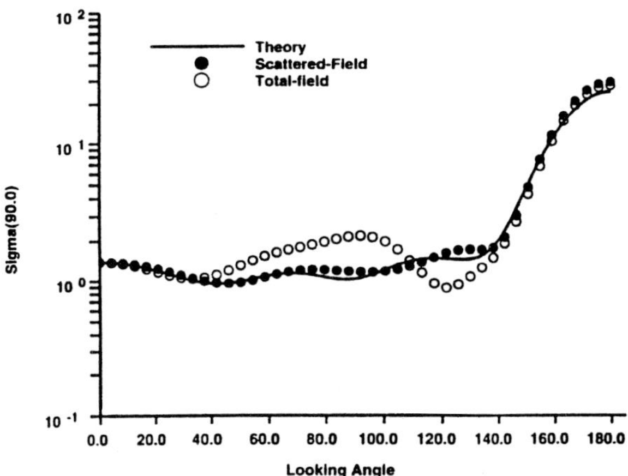

Figure 8: Total-Field and Scattered-Field RCS Calculations, VV Polarization

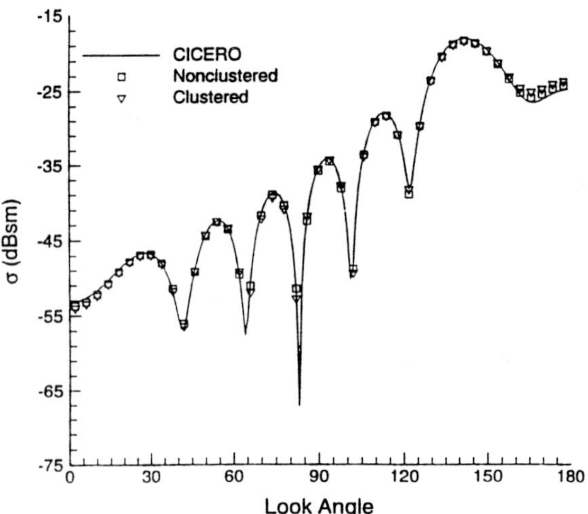

Figure 9: RCS Calculations of an Ogive, 4 Ghz

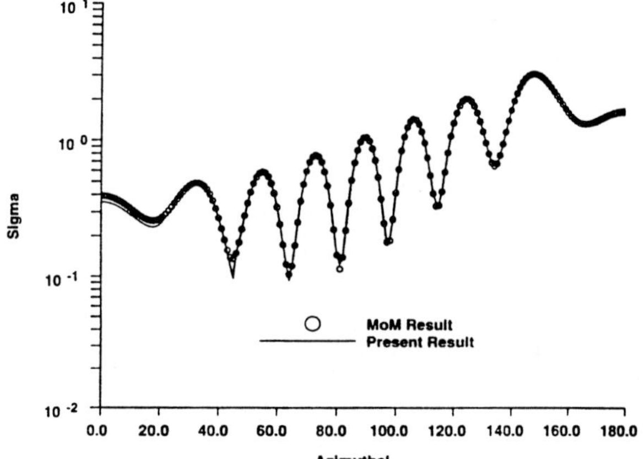

Figure 10: RCS Calculations of an Ogive Cylinder, 2 Ghz

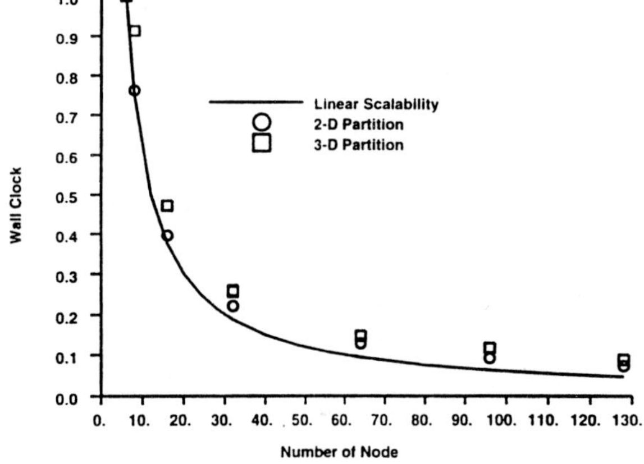

Figure 11: Speed-Up of Scattering Simulations by Concurrent Computing

HOMOGENEOUS AND HETEROGENEOUS DOMAIN DECOMPOSITION FOR COMPRESSIBLE FLUID FLOWS AT HIGH REYNOLDS NUMBERS

Alfio QUARTERONI [1][2] **Luca STOLCIS** [3]

(1) Department of Mathematics, Technical University of Milan; Piazza L. Da Vinci 32, 20133 Milano (Italy)

(2) CRS4, Via Sauro 10, 09123 Cagliari (Italy)

(3) FIAT Research Center, Str. Torino 50, 10043 Orbassano (Italy)

1 Introduction

Numerical methods for the simulation of compressible flows have became very popular during the last few decades. The enormous growth in computer power has allowed to simulate complicated problems (e.g. unsteady three-dimensional flowfields) by means of more and more sophisticated physical models, which now range from the Euler equations for inviscid non-heat conducting flows to the Navier-Stokes equations for viscous flows with heat transfer. These developments have affected all the disciplines involved with scientific computing, including, for instance, numerical algorithms and their implementation on modern supercomputers. Domain decomposition techniques have been also exploited especially after the introduction of multiprocessor machines, which have allowed the solution of certain types of large scale problems which are feasible with standard single-processor machines.

In its most classical version, domain decomposition methods provide a strategy for an efficient solution of the algebraic system arising from a given space-discretization method. As such, domain decomposition techniques can be regarded as an alternative algebraic approach (e.g., as a preconditioner) which is especially suitable for implementations on parallel architectures. In other cases, however, domain decomposition is a way for introducing novel numerical solutions to the given differential problem, allowing the partition of the computational domain into subdomains where various kind of numerical discretization schemes can be combined together. In this respect, domain decomposition provides a flexible environment for the design of novel numerical approximations to the (initial) boundary-value problems.

Such a flexibility can be pushed further on to allow the treatment of different kind of partial differential equations in different subregions of the given domain. This approach, which is referred to as *heterogeneous* domain decomposition (as opposed to the standard, *homogeneous* one), can be regarded as a new way for the mathematical modeling of physical problems, with the aim of reducing the complexity of the arising computational method and/or to enhance its accuracy.

In particular, for several cases of practical interest in fluid dynamics, it is physically sound to try to distinguish between different contributions and to try to employ the most appropriate physical models in different parts of the domain. The range of validity of such approach is rather wide.

A classical situation concerns boundary-layer simulations for viscous flows (e.g. flows over a flat plate, or around an airfoil). The full system of Navier-Stokes equations is used to describe the flowfield in a region embodying the boundary layer, whereas in the complementary (farfield) domain, where the inertial forces are dominant over the viscous forces, the reduced Euler equations can be employed (see [16]).

Another case of practical interest where this approach is valid is that of a high speed rarefied flow over a bluff body, e.g. the canopy of space vehicles. The most correct physical model to be employed for the numerical simulations is that of the Boltzmann equations. However, the Boltzmann equations can be replaced with lower order models (Navier-Stokes or Euler equations) in regions where such models are in rather good agreement with the physics, e.g. for this case in regions far from the body surfaces (see e.g. [8]).

A third situation is that of compressible viscous

Received on July 1, 1997.

flows with fixed transition between laminar- and turbulent regimes, e.g. wings or airfoil in wind tunnels, where the transition points are fixed through suitable mechanical devices on the surfaces of the objects. For such cases the flow-field can be split in three regions: the outer region, where a suitable physical model is represented by the Euler equations for inviscid flows; a laminar flow region upstream the transition points, described by the Navier-Stokes equations for laminar flows; a turbulent flow region, described by the Reynolds-averaged Navier-Stokes equations together with additional transport equations for some turbulent quantities.

In this paper, some of the heterogeneous methods devised for simpler advection-diffusion problems will be extended to the class of problems described by the Navier-Stokes equations, and a general approach which allow the use of different flow models in each sub-domain will be devised. In Section 2 the governing equations for compressible viscous and inviscid flows will be presented, and some consideration regarding one-dimensional flows will be introduced. Section 3 will be devoted to the presentation of the mathematical analysis of the conditions to be imposed on the interfaces between sub-domains. Three different cases will be considered: the fully viscous case, where in both domains the Navier-Stokes equations are to be solved; the inviscid case, where the Euler equations for inviscid flows are to be solved; the heterogeneous case where the Euler equations are solved in one domain and the Navier-Stokes equations are solved in the other. For sake of clarity, the discussion of the interface matching mechanism will be based on a partition of the computational region in two subregions only. Clearly, the extension to the case involving more than two subdomains (a situation which is more favourable for parallel computations) is straightforward. In Section 4 we will present the numerical framework adopted, which is based on finite volumes, and the details for the implementation of the interface conditions. Section 5 will present some selected applications. Conclusions will be drawn in Section 6.

2 Mathematical Equations for Compressible Flows

The Navier-Stokes equations for compressible fluid flows can be written as (see, e.g. [5], [14])

$$\frac{\partial \mathbf{W}}{\partial t} + \mathrm{div}\mathbf{F}(\mathbf{W}) = \frac{1}{Re}\mathrm{div}\mathbf{G}(\mathbf{W}) \quad in \ \Omega \times (0, T) \tag{2.1}$$

$$\mathbf{W}(\mathbf{X}, t = 0) = \mathbf{W}_0(\mathbf{X}) \quad \mathbf{X} \in \Omega \tag{2.2}$$

where $\Omega \subset \mathcal{R}^d$, $d = 1, ..., 3$, and the i-components

of the operator $\mathbf{div}(.)$ are defined as $\sum_j \partial(.)_{ij}$, with $\partial_i = \partial/\partial x_i$. The conserved variables are $\mathbf{W} = (\rho, \rho\mathbf{u}, \rho E)^T$, with ρ being the density of the fluid, $\mathbf{u}$ its velocity vector with components u_i, and E the total energy per unit mass. The convective and diffusive terms, $\mathbf{F}$ and $\mathbf{G}$, are given by the $(2+d) \times d$ matrices

$$\mathbf{F}(\mathbf{W}) = \begin{pmatrix} \rho\mathbf{u}^T \\ (\rho\mathbf{u} \otimes \mathbf{u}) + p\bar{\bar{I}} \\ \rho\mathbf{u}^T H \end{pmatrix},$$

$$\mathbf{G}(\mathbf{W}) = \begin{pmatrix} 0 \\ \bar{\bar{\tau}} \\ \bar{\bar{\tau}} \cdot \mathbf{u} + \mathbf{q}^T \end{pmatrix} \tag{2.3}$$

where $\mathbf{u} \otimes \mathbf{u}$ is the tensor with components $u_i u_j$, $\bar{\bar{I}}$ is the unit tensor, $\mathbf{q}$ is the heat flux, $(\bar{\bar{\tau}} \cdot \mathbf{u})_i := \sum_j \bar{\bar{\tau}}_{ij} u_j$. The components of the viscous stress tensor, $\bar{\bar{\tau}}$, are

$$\tau_{ij} = \mu[(\partial_i u_j + \partial_j u_i) - \frac{2}{3}div\mathbf{u}\delta_{ij}] \tag{2.4}$$

μ is the molecular viscosity, δ_{ij} the Kronecker symbol, H is the total enthalpy, $\rho H = \rho E + p$. The equation of state,

$$\rho(\gamma - 1)E = p + \rho\frac{|\mathbf{u}|^2}{2} \tag{2.5}$$

expresses the relation between pressure, p, the conserved variables, and the ratio of specific heats, γ. Eqs. (2.1) are written in non-dimensional form, where the non-dimensional quantities are obtained from the dimensional quantities, superscript $*$, by employing reference values, subscript o, as shown below

$$\mathbf{X} = \frac{\mathbf{X}^*}{L_o^*}, \ \rho = \frac{\rho^*}{\rho_o^*}, \ p = \frac{p^*}{\rho_o^*(|\mathbf{u}|_o^*)^2},$$

$$\mathbf{u} = \frac{\mathbf{u}^*}{|\mathbf{u}|_o^*}, \ E = \frac{E^*}{(|\mathbf{u}|_o^*)^2}, \ \mu = \frac{\mu^*}{\mu_o^*},$$

$$t = \frac{t^*}{L_o^*/|\mathbf{u}|_o^*}. \tag{2.6}$$

The viscous fluxes on the right-hand side of Eq. (2.1) are multiplied by the non-dimensional quantity $1/Re$, where $Re = \rho_o^*|\mathbf{u}_o^*|L_o^*/\mu_o^*$ is the Reynolds number, which is based upon a reference length L_o^* and is supposed to be a constant of the problem. With this particular notation, when $Re \to \infty$ its inverse tends to zero, and the viscous terms reduce their importance. The Euler equations for inviscid flows can be obtained by dropping the viscous terms, i.e. by taking $1/Re = 0$.

2.1 One-dimensional flows

The main characters of the Navier-Stokes equations for compressible flows can be enlighten by looking at one-dimensional flows. The study of the one-dimension inviscid counterpart of system (2.1) is of great interest when convection-dominated flows have to be computed.

$$\frac{\partial \mathbf{W}}{\partial t} + \frac{\partial \mathbf{F}}{\partial x} = \frac{\partial \mathbf{G}}{\partial x} \quad in \; \Omega \times (0,T) \qquad (2.7)$$

This time, the unknown quantities are $\mathbf{W} = (\rho, \rho u, \rho E)^T$, while the inviscid- and viscous fluxes read

$$\mathbf{F} = \begin{pmatrix} \rho u \\ \rho u^2 + p \\ \rho u H \end{pmatrix} , \quad \mathbf{G} = \begin{pmatrix} 0 \\ \frac{4}{3}\mu\frac{\partial u}{\partial x} \\ \frac{4}{3}\mu u\frac{\partial u}{\partial x} + q \end{pmatrix} \qquad (2.8)$$

In this Section we recall some known facts about the characteristic form of the one-dimensional Euler equations, which will be at the basis of our developments of heterogeneous methods for multidimensional flows.

The one-dimensional Euler equations in quasi-linear form become

$$\frac{\partial \mathbf{W}}{\partial t} + A\frac{\partial \mathbf{W}}{\partial x} = 0 \qquad (2.9)$$

with A being the Jacobian matrix of the system

$$A = \partial \mathbf{F}/\partial \mathbf{W}$$
$$= \begin{pmatrix} 0 & 1 & 0 \\ -(3-\gamma)u^2/2 & (3-\gamma)u & \gamma - 1 \\ (\gamma-1)u^3 - \gamma u E & \gamma E - 3(\gamma-1)u^2/2 & \gamma u \end{pmatrix}$$
$$(2.10)$$

In order to identify some relevant properties associated with the Euler equations for one-dimensional flows, it is more convenient to transform the equations from conservative- to primitive variables $\mathbf{V} = (\rho, u, p)^T$. Eq. (2.9) then becomes

$$\frac{\partial \tilde{\mathbf{V}}}{\partial t} + \tilde{A}\frac{\partial \tilde{\mathbf{V}}}{\partial x} = 0 \qquad (2.11)$$

with $\tilde{A}$ being the Jacobian matrix of the governing equations in primitive variables

$$\tilde{A} = \begin{pmatrix} u & \rho & 0 \\ 0 & u & 1/\rho \\ 0 & c^2\rho & u \end{pmatrix} \qquad (2.12)$$

Here, the speed of sound, $c = \sqrt{\gamma p/\rho}$, has been introduced for simplicity and also because of its physical and mathematical role. The eigenvalues of the above Jacobian matrix are

$$\lambda_1 = u , \quad \lambda_{2/3} = u \pm c \qquad (2.13)$$

and the left- and right eigenvectors matrices, $\tilde{L}$ and $\tilde{R}$ respectively, are

$$\tilde{L} = \begin{pmatrix} 1 & 0 & -\frac{1}{c^2} \\ 0 & 1 & \frac{1}{\rho c} \\ 0 & 1 & -\frac{1}{\rho c} \end{pmatrix} , \quad \tilde{R} = \begin{pmatrix} 1 & -\frac{\rho}{2c} & -\frac{\rho}{2c} \\ 0 & \frac{1}{2} & \frac{1}{2} \\ 0 & \frac{\rho c}{2} & -\frac{\rho c}{2} \end{pmatrix} \qquad (2.14)$$

The system can be diagonalized by using the eigenvectors of $\tilde{A}$, leading to a fully decoupled problem. A suitable set of variables, the *characteristic variables* $\tilde{\mathbf{W}}$, is introduced by means of the following definition

$$\delta\tilde{\mathbf{W}} = \tilde{L}\delta\tilde{\mathbf{V}} = \begin{pmatrix} \delta\rho - \frac{1}{c^2}\delta p \\ \delta u + \frac{1}{\rho c}\delta p \\ \delta u - \frac{1}{\rho c}\delta p \end{pmatrix} \qquad (2.15)$$

where with δ we have indicated an arbitrary variation of the variables, which for the present case could be either a space- or a time-derivative. The introduction of the above relations based on variations and not upon the quantities themselves, is motivated by the fact that the coefficients of the system, i.e. the elements of $\tilde{A}$, are not constant and depend from the solution itself. Since $\tilde{L}\tilde{A}\tilde{L}^{-1} = \Lambda = diag(\lambda_1, \lambda_2, \lambda_3)$, owing to Eq. (2.15), Eq. (2.11) becomes

$$\frac{\partial \tilde{\mathbf{W}}}{\partial t} + \Lambda\frac{\partial \tilde{\mathbf{W}}}{\partial x} = 0 \qquad (2.16)$$

Since Eq. (2.16) splits the system into three scalar independent equations

$$\frac{\partial \tilde{\mathbf{W}}_i}{\partial t} + \lambda_i\frac{\partial \tilde{\mathbf{W}}_i}{\partial x} = 0 , \quad i = 1, 2, 3 \qquad (2.17)$$

it follows that $\tilde{\mathbf{W}}_i$ is constant along the *characteristic curve* $S(t, x^{(i)}(t)) : \frac{dx^i}{dt} = \lambda_i$ for $i = 1, 2, 3$. By using the thermodynamic relations for isentropic flows on Eq. (2.15) and by recalling the definition of speed of sound, it is found that the characteristic variables become $\tilde{W}_1 = s, \tilde{W}_2 = R^+, \tilde{W}_3 = R^-$ where s is the entropy (which is defined as $s = c_v ln(p/\rho)$, with c_v being the specific heat at constant volume), and $R^\pm$ are the two Riemann invariants, which for isentropic flows take the form

$$R^\pm = u \pm \frac{2c}{\gamma - 1} \qquad (2.18)$$

In particular, entropy is constant along the characteristic curve $C^o : \frac{dx}{dt} = u$, and the two Riemann invariants $R^\pm$ are constant with $C^\pm : \frac{dx}{dt} = u \pm c$, the two incoming or outgoing characteristics curves.

3 Mathematical Analysis of Interface Conditions

As from the present section, we will consider a two-dimensional computational domain $\Omega \subset \mathcal{R}^2$ to be split into two disjoint sub-domains Ω_1 and Ω_2 with a common boundary Γ, with $\mathbf{n}$ denoting the unit outward normal vector to Ω_1 on Γ (see Fig.1).

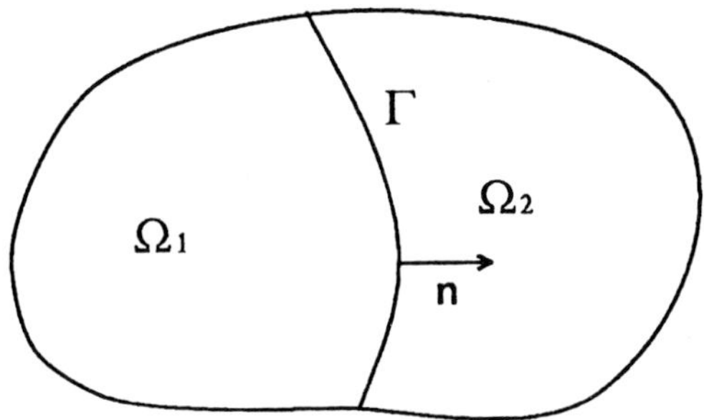

Fig.1: Computational domain $\overline{\Omega} = \overline{\Omega}_1 \cup \overline{\Omega}_2$

We will consider three cases: the inviscid problem, in which the Euler equations hold both in Ω_1 and Ω_2; the viscous problem, in which the Navier-Stokes equations are satisfied in both domains, and, finally, the heterogeneous case, in which the Euler equations are satisfied in one sub-domain, say Ω_1, whereas the Navier-Stokes equations hold in the other sub-domain Ω_2. In all cases, we will discuss the correct interface conditions that allow the proper matching between the solutions in Ω_1 and Ω_2 across Γ. Although these interface conditions can often be derived from a rigourous mathematical analysis on the partial differential equations underlining the various physical models considered, it is beyond the scope of this note to present the full details of this analysis. We will simply draw the conclusions and comments on their significance.

3.1 The viscous case

Let us consider the Navier-Stokes equations in the whole spatial domain Ω (see Section 2 for symbols' explanation)

$$\frac{\partial \mathbf{W}}{\partial t} + \mathrm{div}\mathbf{F}(\mathbf{W}) = \frac{1}{Re}\mathrm{div}\mathbf{G}(\mathbf{W}) \quad in \ \Omega \times (0, T)$$
$$(3.1)$$

supplemented by the initial condition on $\mathbf{W}$ at time $t = 0$, and by suitable boundary conditions. Denoting with $\mathbf{W}^{(i)}$ the restriction of $\mathbf{W}$ upon Ω_i, for $i = 1, 2$, and assuming enough regularity on the solution $\mathbf{W}$, Eq. (3.1) can be reformulated for each $0 < t < T$ in the following split form

$$\frac{\partial \mathbf{W}^{(i)}}{\partial t} + \mathrm{div}\mathbf{F}(\mathbf{W}^{(i)}) = \frac{1}{Re}\mathrm{div}\mathbf{G}(\mathbf{W}^{(i)})$$
$$in \ \Omega_i, \ i = 1, 2 \quad (3.2)$$

$$\left[\mathbf{F}(\mathbf{W}^{(1)}) - \frac{1}{Re}\mathbf{G}(\mathbf{W}^{(1)})\right] \cdot \mathbf{n}$$
$$= \left[\mathbf{F}(\mathbf{W}^{(2)}) - \frac{1}{Re}\mathbf{G}(\mathbf{W}^{(2)})\right] \cdot \mathbf{n} \quad on \ \Gamma \quad (3.3)$$

$$u^{(1)} = u^{(2)}, \ v^{(1)} = v^{(2)}, \ E^{(1)} = E^{(2)} \quad on \ \Gamma \quad (3.4)$$

In other words the sub-domain restrictions $\mathbf{W}^{(1)}$ and $\mathbf{W}^{(2)}$ of the overall Navier-Stokes equations satisfy the Navier-Stokes equations in Ω_1 and Ω_2 respectively. Let us note that some of the physical variables are required to be continuous on Γ, precisely, those which appear with their second derivatives in the conservation equations, namely u, v, and E (see Eqs. (2.1) and (2.3)). We point out that the continuity requirement, Eq. (3.4) follows from the property of $\mathbf{W}$ to be continuous on the whole domain Ω (this property is inherently enjoyed by the solutions of the Navier-Stokes equations). Clearly, Eqs. (3.2)-(3.4) inherit the same boundary conditions prescribed for $\mathbf{W}$ on $\partial\Omega$, and the same initial conditions on Ω at $t = 0$. Moreover, Eq. (3.3) entails the continuity of the normal fluxes. If n_x and n_y denote the x- and y-components of $\mathbf{n}$, respectively, we point out that the normal flux has the four components

$$\rho u n_x + \rho v n_y$$

$$[(\rho u^2 + p) + \tau_{xx}]n_x + [\rho uv + \tau_{xy}]n_y$$

$$[\rho uv + \tau_{xx}]n_x + [(\rho v^2 + p) + \tau_{yy}]n_y \qquad (3.5)$$

$$[\rho uH + (u\tau_{xx} + v\tau_{xy} + q_x)n_x + [\rho vH + (u\tau_{xy} + v\tau_{yy} + q_y)n_y$$

which result therefore to be continuous across Γ. In particular, since u and v are continuous (see Eq. (3.4)), from the continuity of $\rho u n_x + \rho v n_y$ it follows that

$$\rho \text{ is continuous at all the points of } \Gamma \qquad (3.6)$$

where $\rho u n_x + \rho v n_y \neq 0$ i.e. the density is continuous everywhere on Γ unless at those points where the flow is tangential. This latter condition is in agreement with the physical properties of compressible fluid flows, which basically admit two types of discontinuities: shock-waves or contact discontinuities. In particular, on contact discontinuities the normal velocity is zero, the pressure is continuous, and density and tangential velocity allow non-zero jumps.

The flux matching property, Eq. (3.5), is a natural consequence of the fact that the variables $\mathbf{W}^{(i)}$ which are the 'weak' solution to Eq. (3.1), i.e. they satisfy the weak form for all $0 < t < T$

$$\sum_{i=1}^{2} \left\{ \int_{\Omega_i} \frac{\partial \mathbf{W}_i}{\partial t} \Phi d\Omega_i \right.$$

$$\left. - \int_{\Omega_i} \left[\mathbf{F}(\mathbf{W}_i) - \frac{1}{Re} \mathbf{G}(\mathbf{W}_i) \right] \nabla \Phi d\Omega_i \right\} = 0 \quad (3.7)$$

for all sufficiently smooth test function Φ defined in Ω (e.g. $\Phi \in [L^2(\Omega)]^4$ with $\frac{\partial \Phi}{\partial x_j} \in [L^2(\Omega_i)]^4$ for j=1,...,2 possibly vanishing on $\partial\Omega \cap \partial\Omega_i$ or on a subset of it, depending upon the boundary conditions prescribed for $\mathbf{W}^{(i)}$). Indeed, passing from Eq. (3.1) to Eq. (3.4) one disregards the interface terms:

$$\int_{\Gamma} \left[\mathbf{F}(\mathbf{W}^{(1)}) - \mathbf{F}(\mathbf{W}^{(2)}) \right] \cdot \mathbf{n}\Phi d\Gamma$$

$$-1/Re \int_{\Gamma} \left[\mathbf{G}(\mathbf{W}^{(1)}) - \mathbf{G}(\mathbf{W}^{(2)}) \right] \cdot \mathbf{n}\Phi d\Gamma.$$

The fact that Φ is arbitrary, yields to condition (3.3). It is worth noticing that the energy does not appear explicitly in Eq. (3.5). However, if the heat fluxes are modelled with the standard Fourier's law, i.e. $\mathbf{q} = -k\nabla\theta$, with k being the heat conductivity and θ the temperature, since $\theta = E - |\mathbf{u}|/[2(\gamma - 1)]$, it is possible to express the last line in Eq. (3.5) in terms of total energy.

3.2 The inviscid case

Assume now that the inviscid Euler equations are to be solved

$$\frac{\partial \mathbf{W}}{\partial t} + \mathbf{div}\mathbf{F}(\mathbf{W}) = 0 \quad in \ \Omega \times (0, T) \qquad (3.8)$$

Still denoting with $\mathbf{W}^{(i)}$ the restriction of $\mathbf{W}$ to Ω_i, we have the equivalent formulation, for all $0 < t < T$

$$\frac{\partial \mathbf{W}^{(i)}}{\partial t} + \mathbf{div}\mathbf{F}(\mathbf{W}^{(i)}) = 0 \quad in \ \Omega_i, \ i = 1, 2 \quad (3.9)$$

$$\mathbf{F}(\mathbf{W}^{(1)}) \cdot \mathbf{n} = \mathbf{F}(\mathbf{W}^{(2)}) \cdot \mathbf{n} \quad on \ \Gamma \qquad (3.10)$$

This time the only matching conditions on Γ are those prescribing the continuity of the normal inviscid flux, whose components are

$$\rho u n_x + \rho v n_y$$

$$(\rho u^2 + p)n_x + \rho u v n_y$$

$$\rho u v n_x + (\rho v^2 + p)n_y \qquad (3.11)$$

$$\rho u H n_x + \rho v H n_y$$

If the interface Γ is not kept fixed but moves along the time, then denoting by $\sigma(t)$ its velocity at time t along the normal direction $\mathbf{n} = \mathbf{n}(t)$ the matching condition, Eq. (3.10) has to be replaced by

$$\left(\mathbf{W}^{(1)}(t) - \mathbf{W}^{(2)}(t) \right) \sigma(t)$$

$$- \left[\mathbf{F}(\mathbf{W}^{(1)}(t)) - \mathbf{F}(\mathbf{W}^{(2)}(t)) \right] \mathbf{n}(t) = 0$$

$$on \ \Gamma(t) \qquad (3.12)$$

If $\Gamma(t)$ coincides with a shock front $\delta(t)$, then Eq. (3.12) can be easily recognized as the Rankine-Hugoniot jump conditions holding across $\delta(t)$ (see, e.g., [7]).

3.3 The heterogeneous case

In several flow problems of physical interest that are described by the Navier-Stokes equations, the viscous terms are indeed negligible in some parts of the computational domain. If we identify with Ω_1 this subregion, we can decide to drop drastically the viscous stresses from the governing equations therein. The resulting problem is an heterogeneous model that entails the coupling between the Navier-Stokes equations, Eq. (3.1), in Ω_2 and the Euler equations, Eq. (3.8) in Ω_1, i.e. for all $0 < t < T$

$$\frac{\partial \mathbf{W}^{(1)}}{\partial t} + \mathrm{div}\mathbf{F}(\mathbf{W}^{(1)}) = 0 \quad in \ \Omega_1 \ , \qquad (3.13)$$

$$\frac{\partial \mathbf{W}^{(2)}}{\partial t} + \mathrm{div}\mathbf{F}(\mathbf{W}^{(2)}) = \frac{1}{Re}\mathrm{div}\mathbf{G}(\mathbf{W}^{(2)}) \quad in \ \Omega_2$$
$$(3.14)$$

A rigorous analysis on the correct interface matching conditions for multi-dimensional systems such as Eqs. (3.13)- (3.14) does not exist. However, as shown in [2], [3], [12], such theory exists for scalar equations and, under certain circumstances, for one-dimensional systems as well. Therefore, the interface matching conditions are obtained by extrapolation of the rigorous results determined for model problems. Following [12] [3] the matching conditions for Eq. (3.13)- (3.14) at the interface Γ are the following

$$\mathbf{F}(\mathbf{W}^{(1)}) \cdot \mathbf{n} = \mathbf{F}(\mathbf{W}^{(2)}) \cdot \mathbf{n} - \frac{1}{Re}\mathbf{G}(\mathbf{W}^{(2)}) \cdot \mathbf{n} \quad on \ \Gamma$$
$$(3.15)$$

$$\sum_{j=1}^{4}(R_n)_{ij}^{-1}\mathbf{W}_j^{(1)} = \sum_{j=1}^{4}(R_n)_{ij}^{-1}\mathbf{W}_j^{(2)}$$
$$at \ P \ \in \ \Gamma, \ i = 1, ..., N(P), \ 0 < t < T \quad (3.16)$$

The former relation, Eq. (3.15), expresses the continuity of the normal fluxes at the interface. This relation states that the inviscid normal flux associated with $\mathbf{W}^{(1)}$ has to be set equal to the total flux associated with $\mathbf{W}^{(2)}$, which contains the inviscid as well as the viscous contribution. The latter relation, Eq. (3.16), consitutes a less-straightforward extension of the relation for scalar equations, [12]. Owing to the hyperbolic nature of the system of the Euler equations, on the interface it is required to ensure the continuity of the characteristic variables 'entering' into the hyperbolic subdomain. Unfortunately, such conditions can be enforced in an exact way for one-dimensional systems only. For multi-dimensional problems, one has to rely to a locally one-dimensional approach, such as Eq. (3.16), where R_n are the matrices of the right eigenvectors of the Jacobian matrices, Λ_n, obtained by employing the normal velocity. Under these assumptions, Eq. (3.16) enforces the continuity of the invariants associated with the $N(P)$ characteristic curves $\{(t, \mathbf{x}(t)) : d\mathbf{x}^{(i)}/dt = \lambda_i\}$ defined by the *negative* eigenvalues λ_i of Λ_n. (For simplicity of exposition we have ordered the N(P) negative eigenvalues from i=1 to i=N(P), whilst λ_i, $i = N(P) + 1, ..., q$ denote the positive eigenvalues, with $q \leq d + 2$).

For the one-dimensional case, Eqs. (3.16) are equivalent to the requirement that the Riemann invariants associated to the negative eigenvalues (i.e. to the characteristic curves that are pointing into the "Euler region" Ω_1) are continuos. In the two-dimensional case, the same interpretation holds true provided that the Riemann invariants are replaced by their 1D-counterparts which refer to the straight line normal to Γ at point P. For simplicity of notations, in all cases we can call $\{R_i\}$ these functions.

At this stage it is very important to point out that the interface treatment for fully viscous, viscous/inviscid, or fully inviscid undergoes to some sort of scalability property. More precisely, the heterogeneous case can be seen as an *intermediate* case, which lies in between the inviscid and the viscous ones. Therefore, the interface conditions can be obtained by either 'upgrading' the interface conditions for the inviscid case, or by 'downgrading' the ones for the fully viscous case. For the particular case under investigation this scalability concept is summarized in Table 1, where the relations which are supposed to hold on the interface Γ are reported:

FULL NAVIER – STOKES :

$$\left\{ \begin{array}{l} \mathbf{F}(W^1)\cdot\mathbf{n} - \boxed{\frac{1}{Re}\mathbf{G}(W^1)\cdot\mathbf{n}} = \mathbf{F}(W^2)\cdot\mathbf{n} - \frac{1}{Re}\mathbf{G}(W^2)\cdot\mathbf{n} \quad on \ \Gamma \\[2ex] R_i^1 = R_i^2 \quad i = 1,...,N(P) \quad at \ P, \ P \in \Gamma \\[2ex] \boxed{R_i^1 = R_i^2 \quad i = N(P) + 1,...,q} \quad at \ P, \ P \in \Gamma \end{array} \right.$$

EULER/NAVIER – STOKES :

$$\left\{ \begin{array}{l} \mathbf{F}(W^1)\cdot\mathbf{n} = \mathbf{F}(W^2)\cdot\mathbf{n} - \boxed{\frac{1}{Re}\mathbf{G}(W^2)\cdot\mathbf{n}} \quad on \ \Gamma \\[2ex] \boxed{R_i^1 = R_i^2 \quad i = 1,...,N(P)} \quad at \ P, \ P \in \Gamma \end{array} \right.$$

EULER/EULER :

$$\mathbf{F}(W^1)\cdot\mathbf{n} = \mathbf{F}(W^2)\cdot\mathbf{n} \quad on \ \Gamma$$

Table 1: Scalability of interface conditions

We have put in the boxes the terms that have to be added when the hyerachically higher order model is employed (i.e. moving bottom-up). On the other hand, such terms can be seen as those that have to be dropped when going from the higher order model to the hyerarchically lower model (i.e. moving top-down). We notice also, that the continuity of all $\{R_i, i = 1, ...q\}$ implies that of $\mathbf{u}$ and E, see Eq. (3.4).

4 Numerical Approximation

The theory presented so far is not restricted to any particular kind of numerical discretization methods, and, since it has been developed directly from the differential form of the governing flow equations, it can be adapted to all types of space approximations. In the present work we will show how it can be implemented on a particular kind of finite volume approach, where the space discretization is kept separated from the time discretization, and the resulting equations in discrete form are integrated with an explicit time-marching algorithm.

Methods based upon the finite-volume spatial discretisation have become very popular during the last decade, mainly because they present some advantages of both finite difference and finite element schemes. In particular, they are very suitable for equations in conservation form and are able to solve the equations directly in the physical space without any need to transform them into a computational space. The advantage is an increase in flexibility of the method because all the problems associated with the evaluation of the Jacobians of the transformation matrix, which arise from finite-difference schemes, are removed. The governing equations in conservation form, can be written in the following integral form, which is obtained after the integration over a control volume Ω_K, whose measure $|\Omega_K|$ is assumed to be constant,

$$\frac{\partial \overline{\mathbf{W}}_K}{\partial t} + \frac{1}{|\Omega_K|} \oint_{\partial \Omega_K} \mathbf{F} \cdot \mathbf{n} ds = \frac{1}{|\Omega_K|} \oint_{\partial \Omega_K} \mathbf{G} \cdot \mathbf{n} ds \quad (4.1)$$

Obviously, the control volumes $\{\Omega_K\}$ provide a partition of the computational domain Ω. In this manner, Eq. (4.1) contains the cell-averaged values, $\overline{\mathbf{W}}_K$, of the conserved variables as unknown. Usually, the discretisation of Eq. (4.1) is performed in two separate steps: the spatial discretisation, which is performed first, and the time discretisation. This approach, also called "method of lines", reduces the integro-differential equations to a system of non-linear ordinary differential equations in time. The two dimensional flux vectors can be decomposed into the following Cartesian components: $\mathbf{F} = \mathbf{F}_x \mathbf{i} + \mathbf{F}_y \mathbf{j}$, and $\mathbf{G} = \mathbf{G}_x \mathbf{i} + \mathbf{G}_y \mathbf{j}$, with $\mathbf{i}$ and $\mathbf{j}$ being the normal unity vector components. Thus, the corresponding discrete scheme for Eq. (4.1) is given by

$$\frac{d\mathbf{W}_K}{dt} + \frac{1}{|\Omega_K|} \sum_{i=1}^{kedges} (\mathbf{F}_{x,i}\Delta y_i - \mathbf{F}_{y,i}\Delta x_i)$$

$$= \frac{1}{Re|\Omega_K|} \sum_{i=1}^{kedges} (\mathbf{G}_{x,i}\Delta y_i - \mathbf{G}_{y,i}\Delta x_i) \quad (4.2)$$

where the summation over the edges, *kedges*, forming the control volume represents the discretised form of the contour integral, e.g. the edges forming each triangle. $\Delta x_i = x_B - x_A$ and $\Delta y_i = y_B - y_A$ are the components of the edge outward normal vector $\mathbf{n}ds$, A and B being the indices of the two nodes belonging to the edge (see Figure 2). It is worth noticing that once the connectivity information are provided, such a scheme can be applied to any kind of control volumes of polygonal shapes.

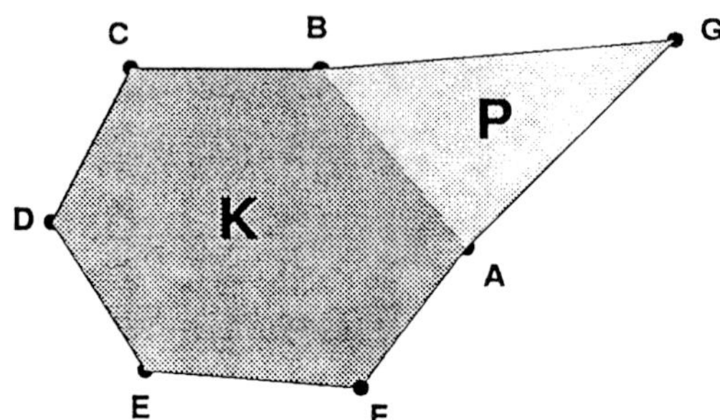

Fig.2: Nomenclature for cells, eddges and nodes

The flux components at the i-th edge can be evaluated either using the values of the fluxes at the centres of the cells or by using the values of the conserved variables. In the present method, the fluxes on each edge are evaluated using the values of the conserved variables on the edge, which are defined as a simple average of the two neighbouring cell centre values. For instance, if the i-th edge is the one shared by cells K and P in Figure 2, then the variables $\mathbf{W}_i$ are defined as

$$\mathbf{W}_i = \frac{1}{2}(\mathbf{W}_K + \mathbf{W}_P) \quad (4.3)$$

and the flux vector components on the same edge are computed as

$$\mathbf{F}_{x,i} = \mathbf{F}_x(\mathbf{W}_i), \quad \mathbf{F}_{y,i} = \mathbf{F}_y(\mathbf{W}_i),$$
$$\mathbf{G}_{x,i} = \mathbf{G}_x(\mathbf{W}_i), \quad \mathbf{G}_{y,i} = \mathbf{G}_y(\mathbf{W}_i) \quad (4.4)$$

The resulting scheme becomes identically equal to the second order three-point scheme for Cartesian orthogonal grids. However, finite-volume schemes depart from standard finite-difference schemes when applied to curvilinear reference systems, because they do not involve any coordinate transformation. Finite-volume schemes resemble also the Galerkin finite-element method when triangular cells and suitable shape functions are employed (see, e.g., [14]).

Both inviscid and viscous fluxes are discretized with standard centered discretization and artificial dissipation terms of the type proposed by Jameson, adapted for the use within an unstructured grid framework, are employed to damp odd-even decoupling due to the central discretisation of the convective terms and to ensure clean capture of shock waves. The artificial dissipation consists in a blending of discrete pseudo-Laplacian and biharmonic operators, controlled by a suitable non-linear shock sensor plus two additional input parameters [9], [17]. The steady-state solution is reached by marching in time the unsteady form of the equations by means of a standard explicit four stage time marching scheme. In order to enhance convergence towards the steady state solutions, local time stepping is employed together with an implicit residual averaging procedure. The resulting time integration scheme for the discrete equation can be written as follows

$$\mathbf{W}^{(m+1)} = \mathbf{W}^{(m)} + \alpha_m \Delta t (\mathbf{R}^{(m)} + \mathbf{D}^{(0)}) \qquad (4.5)$$

where $\mathbf{R}$ is the residual computed according to Eq. (4.2), $\mathbf{D}$ is the artificial dissipation term computed at the first stage, and the superscript m indicates the generic intermediate stage. At the first stage, one takes $\mathbf{W}^{(0)} = \mathbf{W}^{(n)}$. The following "standard" set of coefficients have been employed: $\alpha_0 = 1/4, \alpha_1 = 1/3, \alpha_2 = 1/2, \alpha_3 = 1$, [9].

4.1 Interface treatment

The boundary conditions to be set at the interface between two sub-domains are determined according to the type of physical models employed within the sub-domains. As shown in the previous section, we can have three types of interfaces: Euler/Euler, where in both sub-domains we solve the Euler equations for inviscid flows; Navier-Stokes/Navier-Stokes, where in both sub-domains we solve the Navier-Stokes equations for viscous flows; Euler/Navier-Stokes, where in one sub-domain we solve the Euler equations for inviscid flows and in the other we solve the Navier-Stokes equations for viscous flows.

The inclusion of the interface treatment within a numerical method can be performed in different ways. The simplest one, which we have adopted in the present implementation, uses an explicit solver for the time-discretization. In this case, at each time step the interface conditions are computed using the values at the previous time level, which are then used for the independent evaluation of the residuals in both domains, and the solution is updated.

When an implicit solver is used for the solution of the discretized system, it is more convenient the so-called iteration-by-sub-domains approach. At the new time-level, the two problems are coupled through their interface conditions, see Table 1. The decoupling is achieved by an iterative procedure that splits the interface conditions in two subsets: one to be associated with the domain Ω_1, the other with Ω_2. Either set provides the proper boundary conditions on Γ for the corresponding subdomain. At each stage of the iterative process there are therefore two independent problems to be solved within each subdomain. For more details on this aspect see, e.g., [3], [10], [15].

4.1.1 The viscous case

When the physical model employed in both sub-domains is that of a viscous flow, and the Navier-Stokes equations are solved in both Ω_1 and Ω_2, the interface conditions to be applied are those of Eqs. (3.3)-(3.4). In the present implementation we have assumed that the interface Γ is never tangential to the flow. Then, according to the discussion made in Section 3.1, the above conditions can be reduced to the continuity condition of all the conserved variables across the interface, i.e.

$$\mathbf{W}^{(1)} = \mathbf{W}^{(2)} \qquad on \ \Gamma \qquad (4.6)$$

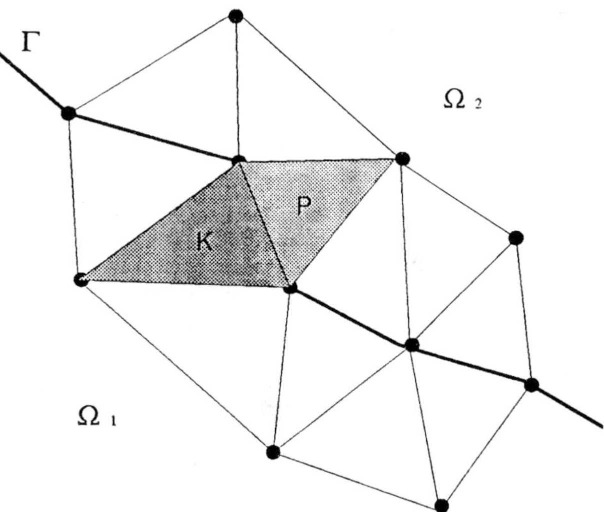

Fig.3: Interface and control volume

In the framework of the present numerical scheme based on a finite-volume spatial discretization, the above relation can be easily implemented by assuming that the interface Γ is coincident with the edges of the control-volumes used for the evaluation of the residuals, as shown in Figure 3. Thus, the fluxes on the interface, which are required for the computation of the terms in Eq. (4.2), can be determined in the usual manner by taking the values of the conserved variables on the interface as the average of the two neighbouring cells (which in this case belong to two different subdomains), i.e.

$$\mathbf{W}_\Gamma = \mathbf{W}_i = \frac{1}{2}(\mathbf{W}_K + \mathbf{W}_P) \qquad (4.7)$$

where with $\mathbf{W}_\Gamma$ we have denoted the values of $\mathbf{W}$ on the interface Γ. In this case, the interface condition, Eq. (4.6), is satisfied since we will have from Eq. (4.7) that $\mathbf{W}_\Gamma = \mathbf{W}^{(1)} = \mathbf{W}^{(2)}$. Furthermore, since the values on the edges are defined in a unique way, the resulting discretization of the fluxes is conservative and consistent with the global finite volume discretization adopted.

4.1.2 The inviscid case

The condition at the interface is based on the continuity of the characteristics entering in the domain. We have seen in Subsection 3.2, that on Γ we should prescribe the continuity of the fluxes, which for the two-dimensional Euler equations becomes an interface condition involving the normal fluxes, Eq. (3.10). If we assume that Γ doesn't cross any discontinuity, this condition on the normal fluxes can be satisfied by a condition on the conserved variables by explicitly setting $\mathbf{W}^{(1)} = \mathbf{W}^{(2)}$ on Γ. As previously discussed for the viscous case, this approach is consistent with our finite volume framework.

From a practical point of view, one is left with the problem to determine in the most suitable way the values on the interface.

If the approach used for viscous flows, Eq. (4.7), is adopted, the correct propagation mechanism through the interface would not be ensured. In other words, the straightforward application of Eq. (4.7) would be equivalent to having a set of non-reflecting boundary conditions on Γ, which is something that should be avoided. Therefore, a suitable way of evaluating $\mathbf{W}_\Gamma$ have to be devised in order to satisfy Eq. (3.10) and to allow the waves exiting from one domain to enter in the other one and vice-versa. In order to achieve such a goal, the values of the conserved variables on the interface are set according to the correct propagation of the characteristic variables. In particular, the present method is based on a simplified approach which assumes a locally one-dimensional flow, and therefore the relations devised for one-dimensional flows can be employed [12].

For two-dimensional flows, using as reference frame the outward normal to Ω_1, see Figure 1, we will have that the incoming characteristics carry the information belonging to Ω_2, R^-, to be evaluated as in Eq. (2.18) with the velocity normal to the interface instead

of u. The outgoing characteristics carry the information belonging to Ω_1, R^+, so that we can determine the values of the velocity normal to Γ, $(q_n)_\Gamma$, and the associated speed of sound, c_Γ, as follows

$$(q_n)_\Gamma = \frac{1}{2}\left(R^+ + R^-\right) ,$$

$$c_\Gamma = \frac{\gamma - 1}{4}\left(R^+ - R^-\right) \quad on \quad \Gamma \qquad (4.8)$$

The other two conditions on the tangential velocity and on the entropy are set according to the direction of the flow (see, e.g. [4]). From the above relations it is possible to get the values of the conserved variables at the interface, $\mathbf{W}_\Gamma$, to be used for the computation of the terms in Eq. (4.2) (Notice however that the right hand side of Eq. (4.2) is equal to sero for the inviscid case).

Because in the present method we employ an explicit time marching algorithm, the interface values evaluated using Eq. (4.8) involve only quantities at the previous time step which are also used to compute the residuals in both domains Ω_1 and Ω_2. It is important to note that the above procedure is used in conjunction with an explicit solver in each sub-domain, and cannot be employed when implicit solvers are used and a different approach, based on iterations-by-subdomains, must be adopted [10].

4.1.3 The heterogeneous case

The interface conditions for the heterogeneous case described in Section 3 can be also accommodated within the present finite volume method in a way similar to that for the viscous and inviscid case.

Let's assume that the Navier -Stokes equations are used in Ω_2 and that the Euler equations are used in Ω_1, and that interface is coincident with the edges forming the control volumes . Then, the interface conditions to be used on Γ are those of equations (3.15)-(3.16), which are to be evaluated by using the values of the conserved variables at the previous time step. At first glance these conditions are not too different from those for the homogeneous problems. However, it is easly verified that they are not even satisfied by simply setting $\mathbf{W}^{(1)} = \mathbf{W}^{(2)}$. In order to be consistent with our numerical method, which is based on the conservation of physical quantities at the discrete level, we have found convenient to split the procedure into two different steps: the evaluation of $\mathbf{W}$ on the interface, and the computation of the normal fluxes. The evaluation of $\mathbf{W}$ on the interface is performed by considering the direction of propagation of the characteristic variables as described in Subsection 4.1.2. The continuity of the invariants, Eq. (3.16) is implemented on a locally one-dimensional approximation. The values of the normal velocity and the speed of sound on the interface are computed as in Eq. (4.8), where 'viscous' and 'inviscid' quantities are involved. Then, depending upon the direction of the flow, the conserved quantities are determined accordingly. Once $\mathbf{W}_\Gamma$ has been determined, the fluxes on the interface can be computed.

Following [11], the matching condition will be implemented by evaluating the fluxes on the interface as $\mathbf{F}(\mathbf{W}_\Gamma) \cdot \mathbf{n}$ when the flow is moving from the *inviscid* domain Ω_1 to the *viscous* domain Ω_2, i.e. when $\mathbf{u} \cdot \mathbf{n} > 0$ on Γ. Whereas they will be evaluated as $\mathbf{F}(\mathbf{W}_\Gamma) \cdot \mathbf{n} - \frac{1}{Re}\mathbf{G}(\mathbf{W}_\Gamma) \cdot \mathbf{n}$ when $\mathbf{u} \cdot \mathbf{n} < 0$. It is worth noticing that this approach is consistent with the finite volume discretization since the fluxes are uniquely defined. On the other hand, as discussed in [11], this approach introduces an error on the solution in the region across the interface. However, for convection-dominated flows this error is neglegible since it is proportional to $1/Re$.

5 Applications

The methodologies presented in the previous sections have been employed for the computation of several standard CFD test cases for compressible viscous flows. All the simulations were performed with standard

The method presented has been validated with the aid of several standard test cases for viscous compressible flows. Here we present some of the results obtained.

5.1 Quasi-one-dimensional nozzle

The compressible flow through a nozzle with variable cross-sectional area, $A = A(x)$, can be simulated by using a set of equations derived from 2.7, with ad-hoc terms involving the area as a function of the coordinate. The corresponding modified equations read

$$\frac{\partial \mathbf{W}}{\partial t} + \frac{\partial \mathbf{F}}{\partial x} = \frac{\partial}{\partial x}(\mathbf{G}A) + \mathbf{J} \quad in \; \Omega \times (0,T) \quad (5.1)$$

where the influence of the cross-sectional area is contained in the additional source terms

$$\mathbf{J} = -\begin{pmatrix} \rho u \\ \rho u^2 \\ \rho u H \end{pmatrix} \frac{1}{A}\frac{dA}{dx} \quad (5.2)$$

whilst the unknowns remain unchanged. In the present work, the computations were performed for a nozzle length of 10 and a variable nozzle-section defined as $A(x) = 1.398 + 0.347 tanh(0.8x - 4)$. For an inlet Mach number of 1.26, the corresponding inviscid flow solution shows a shock located at a distance

of approximately 0.5 of the channel length from the inlet section.

This configuration has been used to test the heterogeneous interface treatment presented in Section 3.3. The Navier-Stokes equations for viscous compressible flows have been solved for $X < 6$, whilst the Euler equations (obtained from 5.1 by setting $\mathbf{G}A(x)$ to zero) have been solved for $X > 6$. In order to investigate the influence of this interface treatment upon the solution, five different viscosity coefficients, ranging from 0.001 up to 0.5, have been employed. The computations have been performed with a domain of 100 grid points equally spaced along the channel, which non-dimensional length has been taken equal to 10.

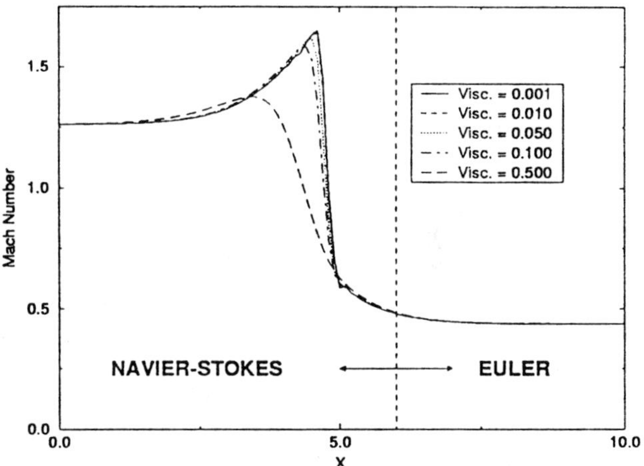

Fig.4: Quasi-one-dimensional nozzle -Mach number distribution-

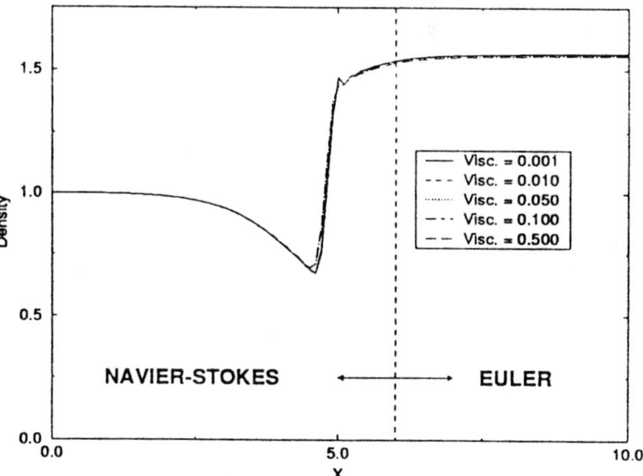

Fig.5: Quasi-one-dimensional nozzle -Density distribution-

Figures 4 to 7 show respectively the Mach number, density, velocity and energy distributions along the

1073

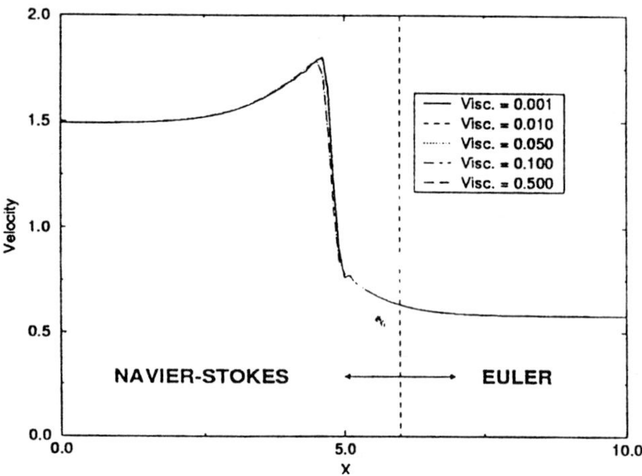

Fig.6: Quasi-ine-dimensional nozzle -Velocity distribution-

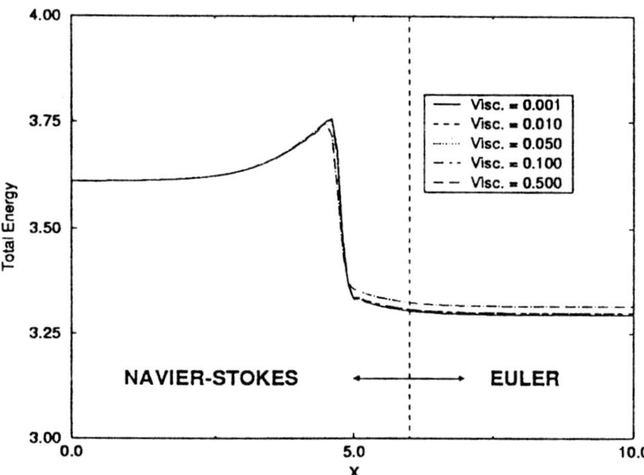

Fig.7: Quasi-one-dimensional nozzle -Total energy distribtion

channel for the various viscosity coefficients. For low viscosity (i.e. $\mu < 0.05$) the influence of the viscous terms is limited, so that the shock in proximity of the nozzle throat is still pronounced and the solution is very similar to an inviscid one. When the viscosity is further increased, the influence of the viscous terms is more pronounced and the shock is smeared. Apparently, there is no influence upon the solutions which shows a smooth matching at the interface for all the viscosity considered.

It is also worth noticing that the level of artificial dissipation has been kept at minimum to limit its influence on the solution. This is the reason why, for the low viscosity cases, just downstream of the shock there are small oscillations in the computed solutions.

5.2 Laminar flow over a flat plate

The validation was performed by computing the compressible laminar flow over a flat plate which represents an important test case because of the existence of an analytical solution. The resulting velocity profiles become function of a transformed coordinate normal to the plate surface, η, which is a function of the distance from the plate leading-edge and of the normal distance from the plate surface. The velocity profiles are strictly valid only for incompressible flows. However, their validity can be extended to compressible boundary layers if the normal coordinate is scaled to account for density variations.

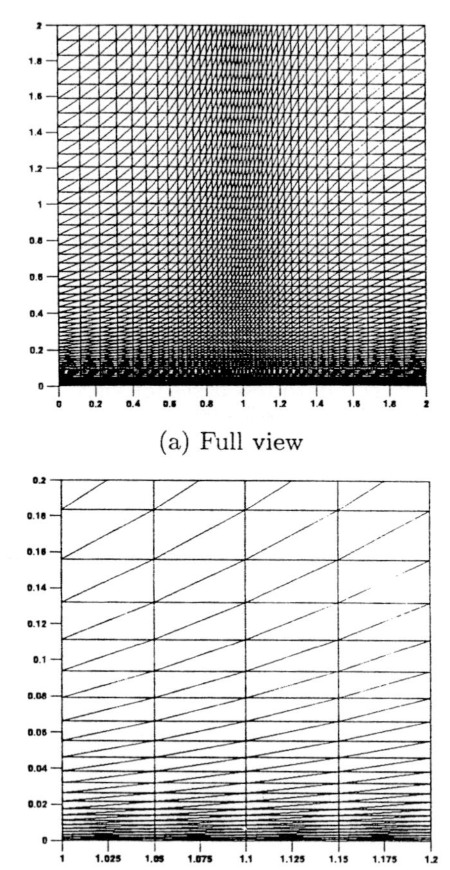

(a) Full view

(b) Enlarged view near the leading edge

Fig.8: Laminar flow over a flat plate -Computational grid-

The results presented in this section have been computed for a compressible flow with a freestream Mach number of 0.5 and a Reynolds number based on the length of the plate, of 10000. The computational grid employed for the final validation (see Figure 8) has been obtained by direct triangulation of a stretched 41x81 structured grid, with the outer boundary located at one chord upstream of the leading edge, two chords downstream, and six chords in the normal direction.

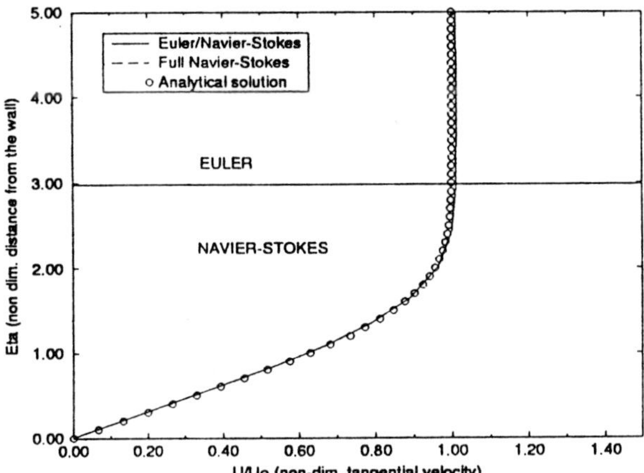

Fig.9: Laminar flow over a flat plate ($M = 0.5$, $Re = 10000$) -Tangential velocity profile at $X = 1.48$

For this case, the Navier-Stokes equations have been solved in the proximity of the plate, whereas the Euler equations have been employed in the other region. In all the computations, the interface between the two heterogeneous domains has been kept at a constant distance (equal to 0.04 the plate length) from the plate surface. The velocity profiles for the station at X=1.48 are presented in Figure 9, where there is no apparent difference between the results obtained with the heterogenous approach, the full viscous problem and the analytical solution.

A more detailed indication of the performance of the heterogenous approach is presented in Figure 10, where the absolute values of the differences between the two solutions (Euler/Navier-Stokes and full Navier-Stokes) are plotted against the distance from the plate surface. The Figure shows the results obtained for three different Reynolds numbers based on the plate length. The *error*, defined as above, decreases in proximity of the interface as the Reynolds number increases from 5000 up to 40000. This result are in line with the theory since the heterogenoeus conditions of Eqs. (3.15) - (3.16) do not prevent jumps on the interface. It is also worth noticing that the general behaviour of results of the present simulation of compressible fluid flows agree with those obtained by other authors for convection-diffusion problems (see [11]).

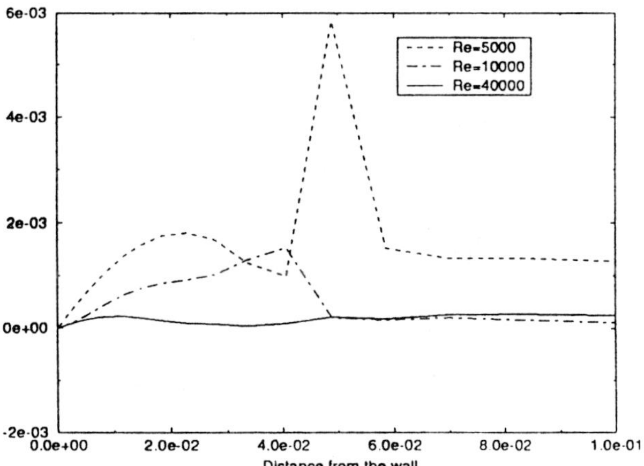

Fig.10: Laminar flow over a flat plate -Error distribution at $X = 1.48$ (Eerror = $\|u_{fullNS} - u_{coupled}\|$)

5.3 Compressible flow over a NACA 0012 airfoil

The NACA 0012 single airfoil was also employed to test the method for the computation of compressible flows around more realistic geometries.

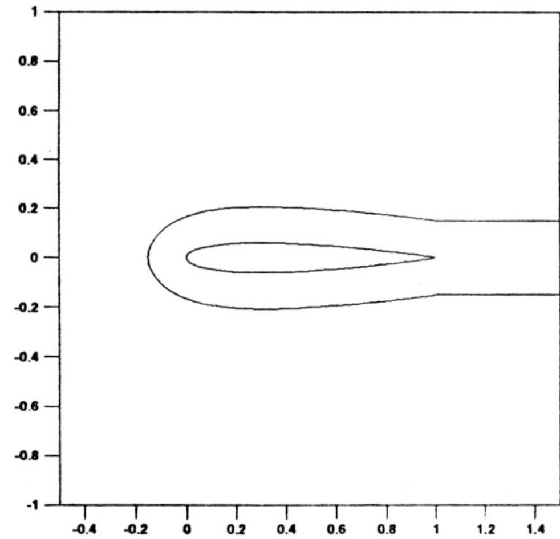

Fig.11: NACA 0012 airfoil -Computational domains

A well known workshop test case has been chosen for the validation. The flow is compressible but fully subcritical with a freestream Mach number of 0.5. This case has been and a chord-based Reynolds number of 10000 . The grid employed is shown in Figure 12, and has been obtained using a 161x41 structured C-type grid. The computed flow-field is presented in Figure 13, where the iso-Mach contours are depicted.

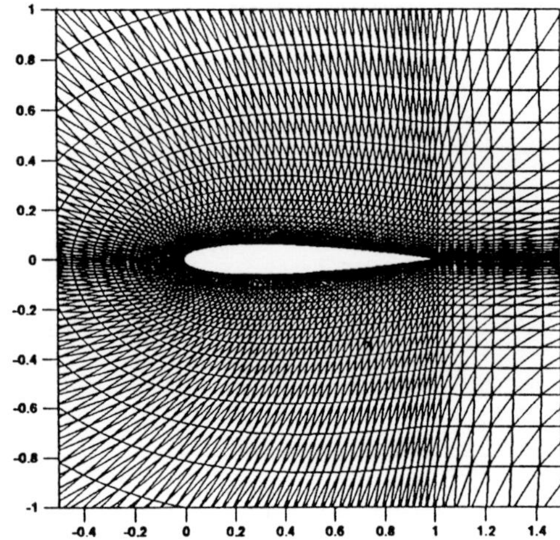

Fig.12: NACA 0012 airfoil -Computational grid (200×40)

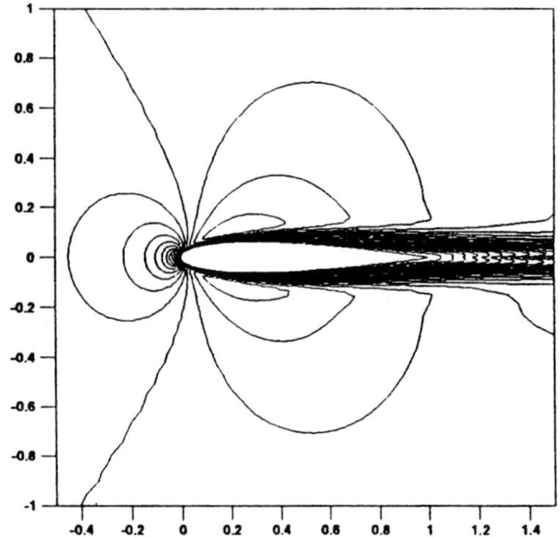

Fig.13: NACA 0012 airfoil -Iso-Mach contours (M_∞, $Re_\infty = 5000$)

Separation occur at about 80% chord. Downstream of the trailing edge the flow gradually reattaches, yielding to a well defined separated region which is visible in the iso-Mach plots. The present test case has been employed for testing both homogeneous- and heterogeneous domain decomposition method. The computational domain has been divided in two subdoimains, as shown in Figure 11, with one domain enclosing the airfoil and the wake developing downstream of its trailing edge. When the homogeneous approach is adopted and the Navier-Stokes equations are solved in both domains, whilst the Euler and Navier-Stokes equations are solved in the two different do-

mains when the homogeneouus approach is used. Conditions (3.3) to (3.4) , or (3.15) to (3.16) described in Section 3.1 has been applied on the interface between the domains. The results obtained agree very well with those obtained with a single domain. Figures 14 and 15 show the pressure and iso-Mach fields for both homogeneous- and heterogeneous approach which results indistinguishible from each other and from those for a single domain (which have been not shown for brevity).

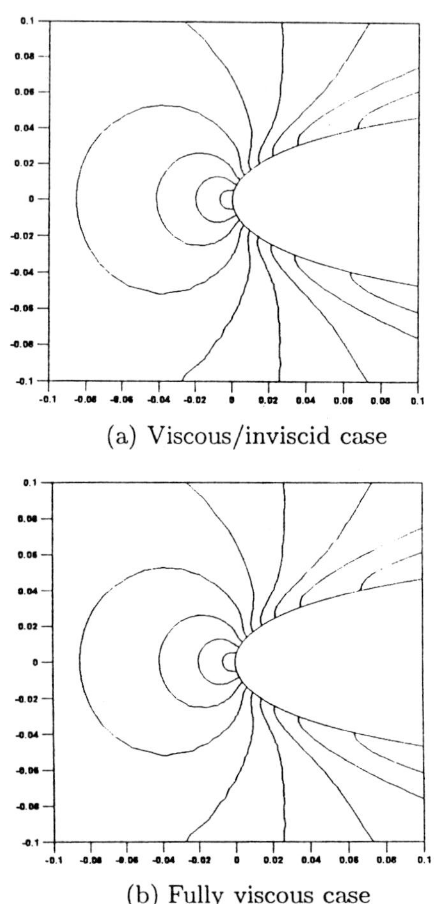

(a) Viscous/inviscid case

(b) Fully viscous case

Fig.14: NACA 0012 airfoil -Iso-Mach contours (Enlarged view)

5.4 Turbulent flow over a RAE 2822 airfoil

The heterogeneous method illustrated in this paper has been appied to turbulent compressible flows. The RAE 2822 airfoil has been chosen and the well known case 9 (Mach=0.734, $\alpha = 2.79$, Re = 6.5 million) has been simulated. The grid employed for the simulation was generated by direct triangulation of a 200x40 C-type structured grid, and the domain

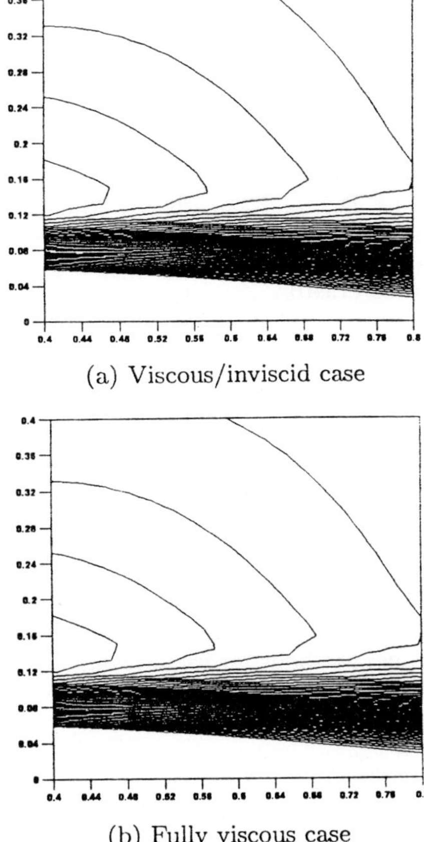

(a) Viscous/inviscid case

(b) Fully viscous case

Fig.15: NACA 0012 airfoil -Iso-Mach contours (Enlarged view)

the *turbulent* region downstream. Since the equations for the turbulent quantities refer to scalar quantities only, the interface conditions can be set as described in [12] where the coupling between advection equations and advection-diffusion equations is described. For simplicity, the present simulation was performed by retaining the full Navier-Stokes equations and by employing the heterogeneous coupling for the turbulent quantities as well.

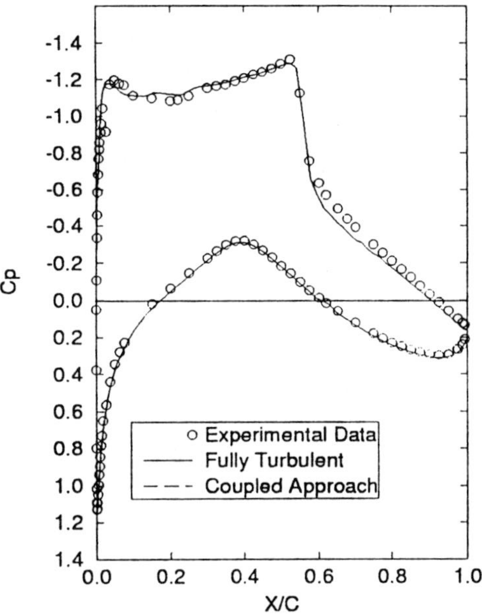

Fig.16: RAE 2822 airfoil -Surface pressure distribution

has been split into two sub-domains: $\Omega_1 : -11.0 < x < 0.03, -11.0 < y < 11.0$ and $\Omega_2 : -0.03 < x < 15.0, -11.0 < y < 11.0$.

The Navier-Stokes equations in mass-weighted averaged form have been solved in conjunction with a standard $k - \epsilon$ turbulence model. Transition from laminar to turbulent flow has been imposed according to the experiments at x = 0.03. Because the difficulties associated with a properly resolved simulation of transitional flows, CFD practitioners overcome the problem by simply solving the turbulence equations only downstream of the transition point, and by taking the turbulence quantities upstream equal to the free-stream values. This simple but effective method could be made more consistent with the numerical problem by just solving the transport part of the turbulence equations in the *laminar* part (i.e. upstream of the transition point), and the complete equations in

In Fig. 16, the surface pressure distributions obtained with the this heterogenous approach and those by a more standard method are compared with the experiments. The results are in very good agreement with each other and with the experimental data. The iso-Mach contours are shown in Fig. 17, were a smooth transition between the two sub-domains can be detected.

6   Acknowledgement

The first author is supported in part by "Fondi M.U.R.S.T. 40 %" and by the C.N.R. Strategic Project "Mathematics for Technology and Society". The present research work has been carried out with the financial support of the Sardinian Regional Authorities.

REFERENCES

[1] Frati, A., Pasquarelli, F., and Quarteroni, A. 'Spectral approximation to advection-diffusion problems by the fictitious interface method', *J.*

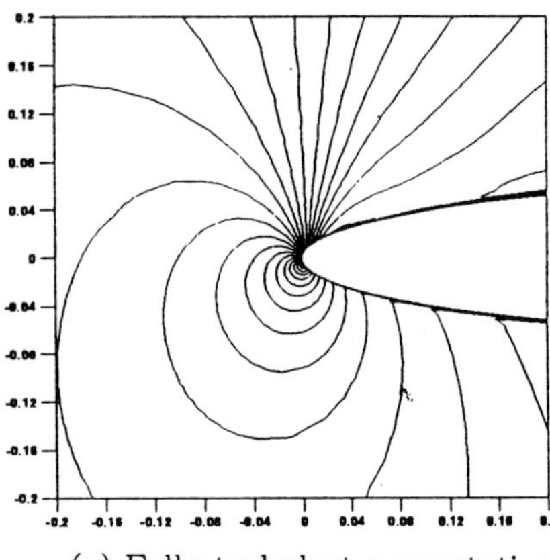

(a) Fully turbulent computation

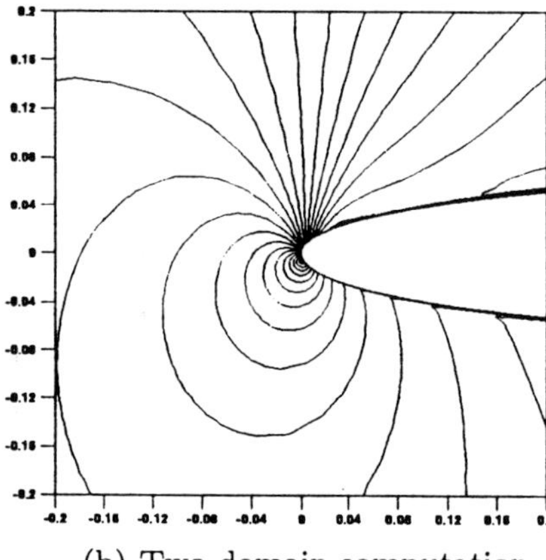

(b) Two-domain computation

Fig.17: RAE 2822 airfoil

of Comput. Phys., **107**, 201-212 (1993)

[2] Gastaldi, F. and Quarteroni, A. 'On the coupling of hyperbolic and parabolic systems: analytical and numerical approach', *Appl. Num. Math.*, **6**, pp. 3-31 (1988)

[3] Gastaldi, F., Quarteroni, A., and Sacchi-Landriani, G. 'On the coupling of two-dimensional hyperbolic and elliptic equations: analytical and numerical approach', in *Proceedings, Domain Decomposition Methods for Partial Differential Equations III*, edited by T. Chan *et al.* (SIAM, Philadelphia, 1990), pp. 22-63

[4] Hirsch, C. *Numerical Computation of Internal and Eternal Flows*, (J. Wiley, Chichester, 1990)

[5] Landau, L.D. and Lifshitz, E.M. *Fluid Mechanics* (Pergamon Press, Oxford, 1959)

[6] Launder, B.E. and Spalding, D.B. *Mathematical Models of Turbulence* (Academic Press, New York, 1972)

[7] Lax, P.D. *Hyperbolic Systems of Conservation Laws and The Mathematical Theory of SHock Waves*, (SIAM, Philadelphia, 1973)

[8] Bourgat, J.F., Le Tallec P., Perthame B. and Qiu, Y., 'Coupling Boltzmann and Euler Equations without overlapping', Domain Decomposition Methods in Science and Engineering, A. Quarteroni et Al. Eds., AMS, Rhode Island, 1994,, pp. 377-398

[9] Mavriplis, D.J. and Jameson A. 'Multigrid solution of the Navier-Stokes equations on triangular meshes', *AIAA Journal*, **28**, pp. 1415-1425 (1990)

[10] Quarteroni, A. 'Domain decomposition and parallel processing for the numerical solution of partial differential equations', *Surv. Math. Ind.*, 1, pp. 75-118 (1991)

[11] Quarteroni, A., Pasquarelli, F. and Valli, A. 'Heterogeneous Domain Decomposition: Principles, Algorithms, Applications', Domain Decomposition Methods for Partial Differential Equations, V, D.Keyes et Al., Eds., SIAM, Philadelphia, 1992, pp. 129-150

[12] Quarteroni, A. and Stolcis, L. 'Heterogeneous domain decomposition for compressible flows', Numerical Methods for Fluid Dynamics V, edited by Morton, K.W. and Baines, M.J., Oxford University Press, Oxford, 1995, pp. 113-128

[13] Quarteroni, A. and Valli, A. 'Theory and application of the Steklov-Poincare' operators for boundary-value problems: the heterogeneous operator case', in *Proceedings, Domain Decomposition Methods for Partial Differential Equations IV*, edited by Glowinski, R. *et al.* (SIAM, Philadelphia, 1991) pp.58-81

[14] Quarteroni, A. and Valli, A. *Numerical Approximation of Partial Differential Equations*, (Springer Verlag, Heidelberg, 1994)

[15] Quarteroni, A. and Valli, A. *Domain Decomposition Methods for Partial Differential Equations*, V.K.I. Lecture Series, March 1996

[16] Schlichting, H., *Boundary Layer Theory*, (McGraw Hill, New York, 1968)

[17] Stolcis, L. and Johnston, L.J. 'Solution of the Euler equations on unstructured grids for two-dimensional compressible flow', *The Aeronautical Journal*, **94**, pp.181-195 (1990)

[18] Stolcis, L. and Johnston, L.J. 'Compressible flow calculations using a two-equation turbulence model and unstructured grids', in *Proceeedings, Numerical Methods in Laminar and Turbulent Flow Vol.7*, edited by Taylor C. *et al.* (Pinerdge Press, Swansea, 1991) pp. 922-931

PROSPECTS FOR CFD ON PETAFLOPS SYSTEMS

David E. KEYES [1]   Dinesh K. KAUSHIK [2]   Barry F. SMITH [3]

[1]   Keyes is Associate Professor, Computer Science Department, Old Dominion University, Norfolk, VA 23529-0162,
      and Associate Research Fellow, ICASE, NASA Langley Research Center, Hampton, VA 23681-0001,
      keyes@icase.edu.

[2]   Kaushik is a Graduate Research Assistant, Computer Science Department, Old Dominion University,
      and an ICASE intern, kaushik@cs.odu.edu.

[3]   Smith is a Computer Scientist in the Mathematics and Computer Science Division at Argonne National
      Laboratory, Argonne, IL 60439-4844, bsmith@mcs.anl.gov.

Abstract

With teraflops-scale computational modeling expected to be routine by 2003–04, under the terms of the Accelerated Strategic Computing Initiative (ASCI) of the U.S. Department of Energy, and with teraflops-capable platforms already available to a small group of users, attention naturally focuses on the next symbolically important milestone, computing at rates of 10 15 floating point operations per second, or "petaflop/s". For architectural designs that are in any sense extrapolations of today's, petaflops-scale computing will require approximately one-million-fold instruction-level concurrency. Given that cost-effective one-thousand-fold concurrency is challenging in practical computational fluid dynamics simulations today, algorithms are among the many possible bottlenecks to CFD on petaflops systems. After a general outline of the problems and prospects of petaflops computing, we examine the issue of algorithms for PDE computations in particular. A back-of-the-envelope parallel complexity analysis focuses on the latency of global synchronization steps in the implicit algorithm. We argue that the latency of synchronization steps is a fundamental, but addressable, challenge for PDE computations with static data structures, which are primarily determined by grids. We provide recent results with encouraging scalability for parallel implicit Euler simulations using the Newton-Krylov-Schwarz solver in the PETSc software library. The prospects for PDE simulations with dynamically evolving data structures are far less clear.

Keywords— Parallel scientific computing, computational fluid dynamics, petaflops architectures.

1. Introduction

Future computing technology in general, and scientific computing technology in particular, will be characterized by highly parallel, hierarchical designs. This trend in design is a fairly straightforward consequence of two other trends: a desire to work with increasingly large data sets at increasing speeds and the imperative of cost-effectiveness. A system possessing large memory without a correspondingly large number of processors to act concurrently upon it is expensively out-of-balance. Fortunately, data use in most real programs has sufficient temporal and spatial locality to allow a distributed and hierarchical memory system, and this locality must be exploited at some level (by a combination of the applications programmer at the algorithmic level, the system software at the compiler and runtime levels, and the hardware). Research on petaflops[1] systems can be seen as paving the way for

exploiting hierarchical parallelism at all levels. Indeed, "petaflops" has come to refer to body of research dealing with very highly parallel computing, since petaflops computers are likely to have between 10^4 and 10^6 processors, with deep memory hierarchies.

A. Petaflops Numerology

Casting petaflops-scale computing into popular terms is a worthwhile exercise even for the quantitatively elite, if for no other reason than that this staggering (and staggeringly expensive) capability must be explained to others. With apologies for drawing significance to any number with an arbitrary dimension attached (i.e.. the second) except for its mnemonic value, we note that mainstream production scientific computing on workstations is carried out at approximately the *square-root* of 1 Pflop/s today: $\sqrt{10^{15}} \approx 31.5 \times 10^6$. The following commodity workstations perform the LINPACK- 100 benchmark at a rate within a few percent of 31.5 Mflop/s [11]:

<hr>

[1] In order to distinguish the plural of "floating point operations" from the rate "floating point operations per second," the rate is customarily abbreviated "flop/s", with an explicit "/" for the "per". We retain this distinction when quoting measurements, but we do not distinguish between "petaflops"

and "petaflop/s" when using the term as an adjective of scale. "Petaflops" will also be used in its general adjectival form to include the term "peta-ops," reflecting requirements to per- form integer and logical computation at comparable rates, in- dependently of or (often) in conjunction with floating point computation.

Received on August 8, 1997.

- SGI Indigo2 (200 MHz)
- IBM RS 6000-560 (50 MHz)
- DEC 3000-500 Alpha AXP (150 MHz)
- Sun Sparc 20 (90 MHz)

A typical sparse PDE computation performs somewhat below the dense LINPACK-100 rates, but with attention to cache residency through variable interleaving and subdomain blocking, it can come close.

There are also 31.5×10^6 seconds in a year, to within one-tenth of a percent. Therefore, a 1 Pflop/s computer could compute in one second what one of these workstations can compute in one year.

There are also 31.5×10^6 people presently living in the state of California, to within a few percent, based on an extrapolation from the 1990 federal census. Therefore, the processing power of a 1 Pflop/s computer (but not the requisite connectivity!) could be realized if everyone in California pooled a commodity scientific workstation to the task. This particular bit of numerology calls to mind that the electrical power consumption of a 1 Pflop/s computer built from commercial, off-the-shelf (COTS) components would be impressive.

As a final point of perspective, we note that the human brain has approximately 10^{12} neurons capable of firing at approximately 1 KHz, and is therefore a specialized peta-op/s "machine" weighing just three pounds and requiring far less power.

B. Interagency Petaflops Workshops

Since February 1994, there has been a systematic effort to explore the feasibility of and encourage the development of petaflops-scale computing by an informal interdisciplinary, interagency working group, subsets of which have met, typically for a week at a time, to consider:

- petaflops applications — what problems appear to require 1 Pflop/s or beyond for important benefits not achievable at smaller scales?
- petaflops architectures — how can balanced systems that store, transfer, and process the data of petaflops applications be supported with conceivable technologies?
- petaflops software — how can the gap between the complex hardware and the application community be spanned with tools that automate program preparation and execution?
- petaflops algorithms — how much concurrency can be exposed at various levels in a computational model and what fundamental requirements on capacity, bandwidth, latency, and processing arise from the underlying physics and mathematics?

The main contents of this chapter were originally created for, and have been informed by, the most recent of these meetings, the Petaflops Algorithms workshop in Williamsburg, VA, April 13-18, 1997. Fifty-five participants from federal agencies, universities, computer vendors, and other private computational organizations attempted to address the algorithmic research questions presented by potential of "affordable" petaflops systems by the year 2010.

The principal findings and recommendations have been outlined in [2], which concludes that petaflops computing is algorithmically feasible, in that at least some of today's key algorithms appear to be scalable to petaflops. Issues of interest to algorithmicists include the following, many of which are shared with the software and hardware communities:

- Concurrency
- Data locality
- Latency and synchronization
- Floating point accuracy (extended wordlength)
- Dynamic (data-adaptive) redistribution of workload
- Detailed performance analysis
- Algorithm improvement metrics
- New languages and constructs
- Role of numerical libraries
- Algorithmic adaptation to hardware failure

Participants made preliminary assessments of algorithm scalability, from as many diverse areas of high-performance computing as were represented, and applied a "triage"-style categorization: Class 1 – appearing to be scalable to petaflops systems, given appropriate effort; Class 2 – appearing scalable, provided certain significant research challenges are overcome; and Class 3 – appearing to possess major impediments to scalability, from our present perspective.

Many core algorithms from scientific computing were placed in Class 1 (scalable with appropriate effort), including: dense linear algebra algorithms; FFT algorithms (given sufficient global bandwidth); PDE solvers, based on static grids, including explicit and implicit schemes; sparse symmetric direct solvers, including positive definite and indefinite cases; sparse iterative solvers (given parallelizable preconditioners); "tree-code" algorithms for n-body problems and multipole or multiresolution methods; Monte Carlo algorithms for quantum chromodynamics; radiation transport algorithms; and certain highly concurrent classified (in the sense of national security) algorithms with *a priori* specifiable memory accesses.

Class 2 algorithms (scalable if significant challenges overcome) included a category of principal interest to CFD practitioners — namely, dynamic unstructured grid methods, including mesh generation, mesh adaptation and load balancing — along with several others: molecular dynamics algorithms; interior point-based linear programming methods; data mining, including associativity, clustering, and similarity search; sampling-based optimization, search, and genetic algorithms; branch and bound search algorithms; boundary element algorithms; symbolic algorithms, including Gröbner basis methods; discrete event simulation; certain further classified algorithms involving random memory accesses.

Into Class 3 (possessing major impediments to scalability) the participants placed: sparse unsymmetric Gaussian elimination, theorem-proving algorithms; sparse simplex linear programming algorithms; and integer relation and integer programming algorithms.

From these lists one may abstract the following contraindications for petaflops:

- Data dependencies that are random in characterization and determinable only at runtime (input-dependent dependencies);
- Insufficient speculative concurrency;
- Frequent uncoverable global synchronization;
- Multiphase algorithmic structure with disparate mappings of data to memories within alternating load-balanced phases; and
- Requirement of fast access to huge data sets by all processors.

Computational fluid dynamics as practiced at the contemporary state-of-the-art for problem with complex physics is sometimes characterized by this list. Adaptive methods cannot be statically balanced and mapped across processors, making incremental dynamic balancing and mapping necessary, together with performance monitoring and performance estimation to make cost-benefit analyses. Hybrid particle-field techniques often have unbalanced sequential phases when either the particle or the field computation is given priority over the other in data distribution. Lookup tables for complex state equations, constitutive relations, and cross-sections or reaction coefficients are often too large to replicate on each processor, but too nonlocally accessed to partition without sacrifice of efficiency.

In addition to these readily apparent contraindications, there is another complementary pair, of relevance to fluid dynamics simulations:

- Work requirements that scale faster than than $M^{4/3}$, where M is the main memory capacity; and
- Memory requirements that scale faster than $W^{3/4}$, where W is the (arithmetic) work complexity.

This constraint between memory and work scaling (or, alternatively, between memory and execution time scaling) is *not* likely to be as painful an issue for PDE-based computations as it may be for some others, since it reflects an architectural decision that is largely influenced to accommodate stencil-type computations on three-dimensional space-time grids (as we discuss further below). It is however, a new constraint, as applied in a two-sided manner. CFD practitioners are accustomed to *either* a memory or a time constraint, which they play up against — running the largest job that fits in memory for as much time as required on a dedicated system or running a job up against a temporal deadline with as much resolution as can be afforded. A tightly-coupled petaflops-capable system will be delicately balanced in its hardware configuration for a specific memory/processing rate model. Such systems will be too rare and too expensive to turn over in a dedicated fashion for an indefinite amount of time. They will also be too expensive to use without employing the full amount of memory most of the time. Algorithms that can trade space for time (such as methods that can vary discretization order, and thus the number of operations per grid vertex) will therefore extend more gracefully to an architecturally and economically constrained machine than algorithms that can only be run at a specific operation-count-to-memory ratio.

C. Technology Outlook

We conclude our introduction with a glimpse at a baseline COTS petaflops machine, and at a couple of nontraditional architectural directions. As we quote the educated guesses of others in this section, we begin with a caveat from Yogi Berra, philosopher in Baseball's Hall of Fame:

> *"Prediction is hard. Especially the future..."*

In its projections for the year 2007 (the target year of its current ten-year window, as of this writing) the Semiconductor Industry Association (SIA) anticipates that individual clock rates will continue their historically gratifying ascent as far as approximately 2GHz and then level off. This implies that at least 500,000-fold instruction concurrency is required (to achieve a product of 10^{15} operations per second), some of which will be found at the subprocessor level. Based on this number, and informed by other technology extrapolations, Stevens [24] has projected a COTS design. He envisions a 2,000-node system, with 32 processors per node, totaling 64,000 processors. This leaves approximately 8-fold concurrency to be found within a processor's own pipelined instruction stream (e.g., through multiple functional units). With 65 GB of shared memory per node, the system would have an aggregate of 130 TB. Approximately 80,000 disks (failing at the rate of approximately one every hour) would back this memory. The overall memory hierarchy (from processor registers to disks) would have 8 levels. The 2 GHz-clock multifunctional unit processors would be fed by approximately 240 GB/s of loads and 120 GB/s stores apiece (assuming dominantly triadic operations, $a \leftarrow b$ op c). This requires 180 data Bytes per cycle in and out of Level-1 cache, which would take up about 70% of an overall 2,048-bit wide path from L1 to CPU. Extrapolating from present pricing trends and practices, such a machine would cost approximately $32M for the CPUs and $174M for the overall system. Power consumption would be 11.5 MW and the annual power bill would be approximately $12M.

Sterling has led a design team that is looking well beyond COTS technology. The Hybrid Technology, Multi-threaded (HTMT) architecture [25] is looking towards a 100 GHz clock from quantum logic processors. At this rate, there will be a latency to DRAM of approximately 10,000 clocks. The 7-layer memory hierarchy of HTMT traverses the temperature spectrum from non-uniform random access (NURA) registers, cryogenic RAM (CRAM), at liquid helium temperatures, SRAM at liquid nitrogen temperatures, conventional DRAM, and high density holographic RAM, (HRAM), backed by disk. Programmer-specified "thread affinity" will reduce data hazards.

The Processor-in-Memory (PIM) design of Kogge et al. [20] will feature 100 TB of memory in 10,000 to 20,000 chips, each of which contains about 50 embedded "CPUs." The memory system will be like a live file with filters attached.

All designs are subject to the so-called "Tyranny of DRAM," which states that bandwidth between memory and the processors must be proportional to processor consumption of operands, even if latency is covered (through prefetching or some other technique). Many kernels, like the DAXPY and the FFT, do work that is a small constant (or at most a logarithmic) multiple of the size of the data set. The tyranny implies that progressively remote and slower levels of the memory system must provide proportionally wider pathways of data towards the CPU, so that the bandwidth product can be maintained during computational phases that cycle through the entire data set and do little work with each element.

II. PARTIAL DIFFERENTIAL EQUATION ARCHETYPES AND PARALLEL COMPLEXITY

Partial differential equations come in a wide variety, which explains why we have national laboratories instead of general purpose PDE libraries. Evolution equations come in time-hyperbolic and time-parabolic flavors, and equilibrium equations come in elliptic and spatially hyperbolic or parabolic flavors. Generally, hyperbolic equations are challenging to discretize since they support discontinuities, but easy to solve when addressed in characteristic form. Conversely, elliptic equations are easy to discretize, but challenging to solve, since their Green's functions are global: the solution at each point depends upon the data at all other points. The algorithms naturally employed for "pure" problems of these types vary considerably. CFD spans all of these regimes. Its problems can be of mixed type, varying by region, or of mixed type by virtue of being multicomponent in a single region (e.g., a parabolic system with an elliptic constraint). In a prospective chapter such as this one, we cannot afford to be algorithmically comprehensive, and fortunately, we do not need to be in order to accomplish some computational complexity estimates of generic value, since PDE computations have a great deal of complexity regularity within their algorithmic variety, due to their field nature. The resource requirements of a PDE problem can usually be characterized by the following parameters, for which typical values are suggested for problems in the ASCI class:

- N_x, spatial grid points (10^4–10^9)
- N_t, temporal grid points (1–...)
- N_c, components per point (1–10^2)
- N_a, auxiliary storage per point (0–25)
- N_s, grid points in "stencil" (7–30)

In terms of these parameters, typical memory requirements would be some small number of copies of the fields (successive iterates, overwritten in a shifted or moving-windowed manner) together with a copy of the current Jacobian: $N_x \cdot (N_c + N_a) + N_x \cdot N_c^2 \cdot N_s$. (We assume with the N_c^2 term in the Jacobian that all components depend upon all other components). The work for an explicit code, or for an implicit code in which the linear system is solved through a sparse iterative means, is a small multiple of: $N_x \cdot N_t \cdot (N_a + N_c^2 \cdot N_s)$.

For equilibrium problems solved by "good" implicit methods, work W scales slightly superlinearly in the problem size (or main memory M); hence the Amdahl-Case Rule applies: $M \propto W$ For evolutionary problems, work scales with with problem size times the number of timesteps. CFL-type arguments place the latter on the order of the resolution of each spatial dimension. For 3D problems, therefore, $M \propto W^{3/4}$, which leads to the conventional petaflops "memory-thin" scaling rule. The actual constant of proportionality between M and W can be adjusted over a very wide range by both discretization order (high-order im-

plies more work per point and per memory transfer) and by algorithmic tuning. If frequent time frames are to be captured, other resources — disk capacity and I/O rates — must both scale linearly with W. This is a more stringent scaling than for memory. For reasons of scope, we do not further address the scaling of peripherals; however, we note that significant research remains to be done with archiving data and I/O to support petaflops computing.

A. PDE Archetypes and Software Toolchain

The Computational Archetypes project at Caltech [9] has identified PDE archetypes according to the following classification:

- Local mesh computations
  - Concurrent
    * Explicit update schemes, diagonal relaxation schemes
    * Sparse matrix-vector multiplications
  - Sequential
    * Triangular relaxation schemes
    * Sparse approximate factorization schemes
- Global dimensionally-split computations
  - spectral schemes
  - ADI-like schemes
- Direct linear algebraic computations
  - Gaussian elimination in various orderings

With due respect to the importance of the latter, we concentrate on the prime archetypes for parallel CFD: concurrent local mesh computations, explicit and iterative implicit.

Before confining our attention to a few quantitative aspects of the solution algorithm, we note that solvers are just one link in a "toolchain" [19] for PDE computations worth doing at petaflops scales. This toolchain involves:

- Geometric modeling and grid generation
- Discretization (and automated code generation)
- Error estimation and adaptive refinement (h- and/or p-type)
- Task assignment
  - Domain partitioning
  - Subdomain-to-processor mapping
- Solution
  - Grid and operator "coarsening"
  - Automated or interactive steering
- Visualization, postprocessing, and application

interfacing

- Parallel performance analysis

The toolchain metaphor is useful in reminding that the solver is not all there is to a parallel computation, and may not be the most difficult part. Furthermore, the difficulty of one link may be affected by decisions in another, e.g., a solver may have to work harder in conjunction with a poor grid generator. The overall outcome of a computation may be limited by any weak link, making it difficult to attach relative merits to individual components. The toolchain metaphor is possibly misleading in that not all links are important in all problems, and not all important relationships are between links adjacent in list.

We make a few additional remarks on the toolchain, abstracting CFD-relevant remarks from [19]. Software components of the chain tend to be modular, with well-defined interfaces, because of both good design principles and the impossibility of any one individual or team being expert in all components. A few full, vertically integrated parallel toolchain environments exist today. Amdahl's "rake" eventually forces parallelization of all components; certainly, at least, for petaflops. As one tool is perfected, the parallel bottleneck shifts to another. Significant sharing and reuse of components occurs horizontally (across groups) at the "low" end of the toolchain. For instance grid generators and partitioners are easy to share since they interface to the rest of the environment through intermediate disk files. At higher levels, the compatibility of inner data structures becomes an issue, which limits sharing. Some reuse of software between components occurs vertically, such as between mesh generation and improvement algorithms, and between these and the solver. Though data-structure-specific, common operations are sufficiently generic to become candidates for vertical software reuse within a group (e.g., intermesh transfer operators, error estimators, and solvers for error estimators and for actual solution updates). The parallel scalability requirement discourages the use of graph algorithms that make frequent use of global information, such as eigenvectors. Instead, heavy use is made of maximal independent sets, which can be constructed primarily by a local, greedy algorithm, with local mediation at subdomain interfaces. Trees are

generally avoided as primary data structures in important inner-loop nearest-neighbor operations of PDE-based codes. Crucial trade-offs exist between time to access grid and geometry information and total memory usage: redundant data structures can reduce indirection at the price of extra storage.

B. Algorithms for PDEs

An explicit PDE solution algorithm has the following algebraic structure in moving from iterate $\ell - 1$ to iterate ℓ:

$$\mathbf{u}^\ell = \mathbf{u}^{\ell-1} - \Delta t^\ell \cdot \mathbf{f}(\mathbf{u}^{\ell-1}),$$

or, for higher temporal order schemes, a more general, fully known right-hand side:

$$\mathbf{u}^\ell = \mathbf{F}(\mathbf{u}^{\ell-1}, \mathbf{u}^{\ell-2}, \ldots).$$

Let N be the discrete dimension of a 3D problem and P the number of processors. Assume that the domain is of unit aspect ratio so that the number of degrees of freedom along an edge is $N^{1/3}$, and that the subdomain-to-processor assignment is isotropic, as well. The concurrency is pointwise, $\mathcal{O}(N)$. Since the stencil is localized, the communication-to-computation ratio enjoys surface-to-volume scaling: $\mathcal{O}\left((\frac{N}{P})^{-1/3}\right)$. The communication range is nearest-neighbor, except for timestep selection, which typically involves a global CFL stability check. The synchronization frequency is therefore once per timestep, $\mathcal{O}\left((\frac{N}{P})^{-1}\right)$. Storage per point is low — just a small multiple of N, itself. The data locality in the stencil update operations can be exploited both "horizontally" (across processors) and "vertically" (in cache). Load balancing is a straightforward matter of equipartitioning gridpoints while cutting the minimal number of edges, for static quasi-uniform meshes. Load balance becomes nontrivial when grid adaptivity is combined with the synchronization step of timestep selection.

The discrete framework for an implicit PDE solution algorithm has the form:

$$\frac{\mathbf{u}^\ell}{\Delta t^\ell} + \mathbf{f}(\mathbf{u}^\ell) = \frac{\mathbf{u}^{\ell-1}}{\Delta t^\ell},$$

with $\Delta t^\ell \to \infty$ as $\ell \to \infty$. We assume that pseudo-timestepping is used to advance towards a steady state. An implicit method may also be time-accurate, which generally leads to an easier problem than the steady-state problem, since the Jacobian matrix for the left-hand side is more diagonally dominant when the timestep is small. The sequence of nonlinear problems, $\ell = 1, 2, \ldots,$ is solved with an inexact Newton method. The resulting Jacobian systems for the Newton corrections are solved with a Krylov method, relying only on matrix-vector multiplications, so the stencil-based sparsity is not destroyed by fill-in. The Krylov method needs to be preconditioned for acceptable inner iteration convergence rates, and the preconditioning is the "make-or-break" aspect of an implicit code. The other phases parallelize well already, being made up of DAXPYs, DDOTs, and sparse MATVECs.

The job of the preconditioner is to approximate the action of the Jacobian inverse in a way that does not make it the dominant consumer of memory or cycles in the overall algorithm. The true inverse A^{-1} is usually dense, reflecting the global Green's function of the continuous PDE operator approximated by A. Given $Ax = b$, we want B approximating A^{-1} and a rescaled system $BAx = Bb$ (left preconditioning) or $ABy = b,\ x = By$ (right preconditioning). Though formally expressible as a matrix, the preconditioner is usually implemented as a vector-in, vector-out subroutine. A good preconditioner saves both time and space by permitting fewer iterations in the innermost loop, smaller storage for Krylov subspace. An Additive Schwarz preconditioner [6] accomplishes this in a localized manner, with an approximate solve in each subdomain of a partitioning of the global PDE domain. Optimal Schwarz methods also require solution of a global problem of small discrete dimension. Applying a preconditioner in an Additive Schwarz manner increases flop rates over a global preconditioner, since the smaller subdomain blocks maintain better cache residency.

Newton   Krylov   Schwarz

The pioneers of NKS methods.

Combining a Schwarz preconditioner with a Krylov iteration method inside an inexact Newton method leads to a recently assembled synergistic parallelizable nonlinear boundary value problem solver with a classical name: Newton-Krylov-Schwarz (NKS).

When nested within a pseudo-transient continuation scheme to globalize the Newton method [18], the implicit framework has four levels:

```
do l = 1, n_time
  SELECT TIME-STEP
  do k = 1, n_Newton
    compute nonlinear residual and Jacobian
    do j = 1, n_Krylov
      do i = 1, n_Precon
            solve subdomain problems concurrently
      enddo
      perform Jacobian-vector product
      ENFORCE KRYLOV BASIS CONDITIONS
      update optimal coefficients
      CHECK LINEAR CONVERGENCE
    enddo
    perform DAXPY update
    CHECK NONLINEAR CONVERGENCE
  enddo
enddo
```

The operations written in uppercase customarily involve global synchronizations.

The concurrency is pointwise, $\mathcal{O}(N)$, in most algorithmic phases *but only subdomainwise*, $\mathcal{O}(P)$, *in the preconditioner phase.* The communication-to-computation ratio is still mainly surface-to-volume, $\mathcal{O}\left(\left(\frac{N}{P}\right)^{-1/3}\right)$. Communication is still mainly nearest-neighbor in range, but convergence checking, orthogonalization/conjugation steps in the Krylov method, and the optional global problems add nonlocal communication. The synchronization frequency is often more than once per mesh-sweep, *up to the Krylov dimension (K),* $\mathcal{O}\left(K(\frac{N}{P})^{-1}\right)$. Similarly, *storage per point is higher by a factor of* $\mathcal{O}(K)$. Locality can still be fully exploited horizontally and vertically, and load balance is still straightforward for any static mesh.

C. Parallel Complexity Analysis

Given complexity estimates of the leading terms of:

- the concurrent computation,
- the communication-to-computation ratio, and
- the synchronization frequency,

and a model of the architecture including:

- internode communication (network topology and protocol reflecting horizontal memory structure), and
- on-node computation (effective performance parameters including vertical memory structure),

one can formulate optimal concurrency and optimal execution time estimates for parallel PDE computations, on per-iteration basis or overall (by taking into account any granularity dependence in the convergence rate).

For an algebraically simple example that is sufficient to elucidate the main issues in algorithm design, we consider a 2D stencil-based PDE simulation and construct a model for its parallel performance based on computation and communication costs. The basic parameters are as follows:

- n grid points in each direction, total memory $N = \mathcal{O}(n^2)$,
- p processors in each direction, total processors $P = p^2$,
- memory per node requirements $\mathcal{O}(n^2/p^2)$,
- execution time per iteration An^2/p^2 (A includes factors like number of components at each point, number of points in stencil, number of auxiliary arrays, amount of subdomain overlap),
- n/p grid points on a side of a single processor's subdomain,
- neighbor communication per iteration (neglecting latency) Bn/p, and
- cost of an individual reduction per iteration (assumed to be logarithmic in p with the frequency of global reductions included in the coefficient) $C \log p$.

A, B, and C are all expressed in the same dimensionless units, for instance, multiples of the scalar floating point multiply-add.

Putting the components together, the total wall-clock time per iteration is

$$T(n,p) = A\frac{n^2}{p^2} + B\frac{n}{p} + C \log p.$$

The first two terms fall as p increases; the last term rises slowly. An optimal p is found where $\frac{\partial T}{\partial p} = 0$, or

$$-2A\frac{n^2}{p^3} - B\frac{n}{p^2} + \frac{C}{p} = 0,$$

or

$$p_{opt} = \frac{B}{2C}\left[1 + \sqrt{1 + 8AC/B^2}\right] \cdot n.$$

Observe that p can usefully grow proportionally to n without limitation. The larger the problem size, the more processors that can be employed with the effect of reducing the execution time. In this limited sense, stencil-based PDE computations are scalable to arbitrary problem sizes and numbers of processors. The optimal running time is

$$T(n, p_{opt}(n)) = \frac{A}{\rho^2} + \frac{B}{\rho} + C \log(\rho n),$$

where $\rho = \frac{B}{2C}\left[1 + \sqrt{1 + 8AC/B^2}\right]$. This optimal time is *not* constant as the problem size (and number of processors) increases, but it degrades only logarithmically.

To simplify, consider the limit of infinite bandwidth so that the (asynchronous) nearest-neighbor exchanges take no time. Then,

$$p_{opt} = \sqrt{2A/C} \cdot n,$$

and

$$T(n, p_{opt}(n)) = C\left[\frac{1}{2} + \log(\sqrt{2A/C} \cdot n)\right].$$

This simple analysis is on a per iteration basis; a fuller analysis would multiply this cost by an iteration count estimate that generally depends upon n and p. We observe that although an algorithm made up of this mix of operations is formally scalable, the number of processors amongst which the problem should be divided varies inversely with C, the coefficient of the global synchronization term, and running time varies proportionally. Recall that the main difference in complexity per iteration between explicit and implicit methods in this context is the much greater frequency of synchronization for implicit methods. One of the main benefits provided in return for this synchronization is freedom from CFL limitations, and hence the prospect of an iteration count that is not constrained by the resolution of the grid.

The synchronization cost is made of two parts: the hardware and software latency of accessing remote data when the data is, in fact, ready, and the synchronization delay when the data is not ready. Since they are difficult to distinguish in practice, we lump them together under the term "latency" and consider strategies for latency tolerance.

D. Latency Tolerance

From an architect's perspective [10], there are two classes of strategies for tolerating latency: amortization (block data transfers) and hiding or covering (precommunication, proceeding past an outstanding communication in the same thread, and multithreading). The requirements for tolerating latency are excess concurrency in the program (beyond the number of processors being used) and excess capacity in the memory and communication architecture, in order to stage operands near the processors.

Any architectural strategy has an algorithmic counterpart, which can be expressed in a sufficiently rich high-level language. For instance, prefetching is partially under programmer control in some recent commercially available language extensions. In addition, however, algorithmicists have a unique strategy, not available to architects by definition: reformulation of the problem to create concurrency. Algorithmicists may note that not all nonzeros are created equal, and can create additional concurrency by neglecting nonzero couplings in a system matrix when they stand in the way. Algorithmicists may also accept a (sufficiently rapidly converging) outer iteration that restores the coupling in a less synchronous way, if it improves the concurrency of the iteration body. The reduction in the cost per iteration must more than offset the cost of the restorative outer iterations. An understanding of the convergence behavior of the problem, especially the dependence of the convergence behavior on special exploitable structure, such as heterogeneity (region-dependent variation) and anisotropy (direction-dependent variation), is required in order to intelligently suppress nonzero data dependencies. We briefly mention some ideas for latency-tolerant preconditioners, latency-tolerant accelerators, and latency-tolerant formulations.

The Additive Schwarz method (ASM) named above as the innermost component of the implicit NKS method is a perfect illustration of latency-tolerant preconditioner. We take a closer look at the construction of this method.

The operator B is formed out of (approximate) local solves on overlapping subdomains. The figure below shows a domain Ω decomposed into nine subdomains Ω_i, which are extended into overlapping subdomains Ω_i' that are cut off at the original boundary. The fine mesh spacing is indicated in one of the overlapping subdomains. This example is for a matching discretization in the overlapping subdomains, but nonmatching discretizations can be accommodated.

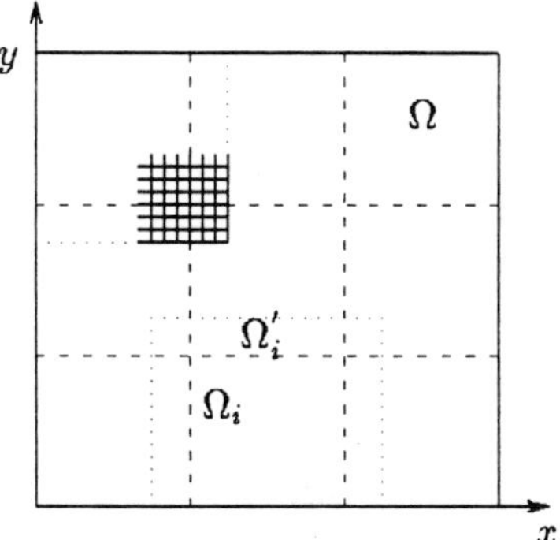

Let R_i and R_i^T be Boolean gather and scatter operations, mapping between a global vector discretized on the fine mesh and its ith subdomain support, and let

$$B = \sum_i R_i^T \tilde{A}_i^{-1} R_i.$$

The concurrency thus created is proportional to the number of subdomains. Part of the action of the R_i is indicated schematically in the figure below. The bold right segment of Ω_i and the bold left segment of Ω_j are the same physical points. The overlapping subdomains are shown pulled apart, and the padding of each with interior data of the other is indicated by the arrows and dashed rectangles. (The width of the overlap is exaggerated for clarity in this illustration.)

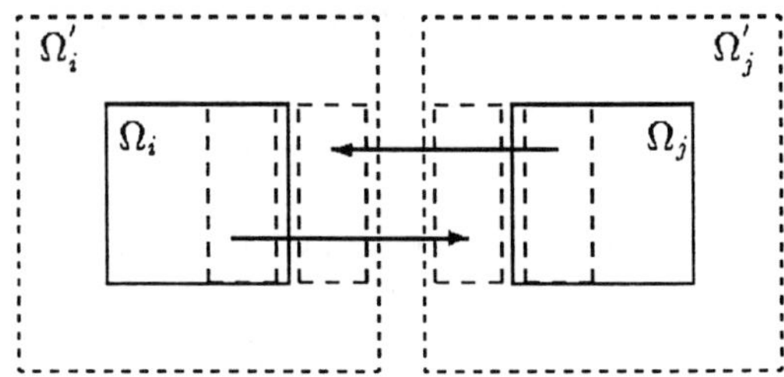

The amount of overlap obviously determines the amount of communication and the amount of redundant computation (on non-owned, buffered points).

A two-level form of Additive Schwarz is provably optimal in convergence rate for some problems [23], but requires an exact solve on a coarsened grid. Convergence theorems for scalar 3D elliptically dominated systems may be summarized as follows, where I estimates the number of iterations as a function of problem size N and number of subdomains (and processors) P:

- No preconditioning: $I \propto N^{1/3}$
- Zero-overlap Schwarz preconditioning: $I \propto (NP)^{1/6}$
- Generous-overlap Schwarz preconditioning: $I \propto (P)^{1/3}$
- Two-level, generous overlap Schwarz preconditioning: $I = \mathcal{O}(1)$

The PETSc library [3], [4] includes portable parallel parameterized implementations of Schwarz preconditioners, including the new, more communication efficient, Restricted Additive Schwarz (RAS) method [8].

Another example of a latency-tolerant preconditioner is the form of the Sparse Approximate Inverse (SPAI) recently developed in [14]. Here B is formed in explicit, forward-multiply form by performing a sparsity-constrained norm minimization of $\|AB-I\|_F$. The minimization decouples into N independent least squares problems, one for each row of B. An adaptively chosen sparsity pattern, such that $\|Ab_k - e_k\|_2 < \epsilon$ leads to $\kappa(AB) \leq \sqrt{\frac{1+\delta}{1-\delta}}$, where $\delta \propto N\epsilon^2$. ϵ chosen as a compromise between storage and convergence rate. The requirement on the smallness of ϵ appears pessimistic (in that B becomes denser as ϵ becomes smaller), but SPAI is worthwhile beyond the hypotheses of the theorem, just as Additive Schwarz is worthwhile with overlaps much smaller than required by the theory for optimality.

The concurrency created by SPAI is pointwise, in both the construction and the application of B. A parallel implementation of SPAI is described in [5]. (The next public release of PETSc will contain an interface to this package.) The sparsity profiles of an original matrix A and its SPAI, with a comparable number of differently positioned nonzeros are shown below (from [14]):

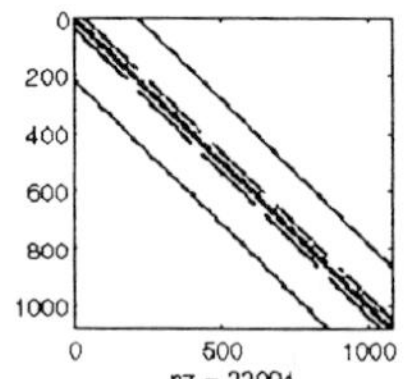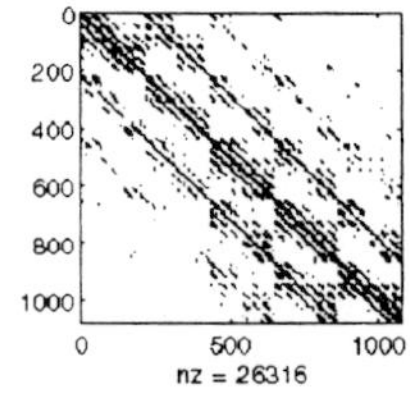

Modified forms of the classical Krylov accelerators of conjugate gradients (CG) and generalized minimal residuals (GMRES) can provide latency-tolerant accelerators. Krylov methods find the best solution to an N-dimensional problem in a K-dimensional Krylov space ($K \ll N$). Conventional Krylov methods orthogonalize (or conjugate) at every step to build up a well conditioned Krylov basis and to update the expansion coefficients of the solution in the enlarged basis. In infinite precision, this orthogonalization can be delayed for many steps at a time and "made up" in one multicomponent global reduction [12], some options for which are available in PETSc. In finite precision, delayed orthogonalization may be destabilizing, but for the low-accuracy requirements of an inner loop of a Newton method it may be tolerable, since the basis is flushed before it gets large. Furthermore, the requirement of performing all pairwise orthogonalizations may be avoided by construction during part of the iteration if the bases are generated from sparse seed vectors with sparse system matrices A. Many other tradeoffs of stability for reduced synchronization frequency have yet to be carefully investigated on realistic problems. Petaflops scale CFD will require a systematic assault on the synchronicity of Krylov basis generation.

The formulations of PDE algorithms, themselves may be made more latency-tolerant in ways that do not compromise the ultimate accuracy of the result, but only the minimal number of iterations required to achieve it. Many synchronization steps in conventional algorithms (e.g., convergence tests, global timestep selection) can be hidden by speculative computation of the next step based on a conservative prediction of the outcome. Such conservative predictions (that an iteration has not converged, or that a timestep cannot be increased) allow by-passing tests that would be recommended for minimal computational complexity if communication were free; but their communication costs may not justify the resulting instant adaptation.

Much work in PDE codes with complex physical models is related to updating auxiliary quantities used in Jacobian assembly, such as flop-intensive constitutive laws or communication-intensive table lookups. These can be "lagged" to slightly stale (or very stale) values with latency savings and acceptable convergence rate consequences.

Message-number versus message-volume trade-offs can be resolved in architecturally optimal ways, given latency and bandwidth models.

Furthermore, a "neighbor-computes" paradigm may sometimes be better than an "owner-computes" in cases in which the output of the computation is small but the inputs (residing on the neighbors) are large.

III. Case Study in the Parallel Port of an NKS-based CFD Code

Discussions of petaflops-scale computing ring hollow if not accompanied by experiences on contemporary parallel platforms that demonstrate that the currently provided technology has been absorbed. We therefore include in this chapter some parallel performance results for a NASA unstructured grid CFD code that is used to study the high-lift, low-speed behavior of aircraft in take-off and landing configurations. Our test cases, possessing only 1.4 million degrees of freedom, are miniscule on the petaflops scale, but we will show scalability of algorithmic convergence rate and per iteration performance over a wide range of numbers of processors, which we have every reason to believe can be extended as the hardware becomes available.

The demonstration code, FUN3D [1], is a tetrahedral vertex-centered unstructured grid code developed by W. K. Anderson of the NASA Langley Research Center for compressible and incompressible Euler and Navier-Stokes equations. FUN3D uses a control volume discretization with variable-order Roe schemes for approximating the convective fluxes and and a Galerkin discretization for the viscous terms. Our parallel experience with FUN3D is with the incompressible Euler subset thus far, but nothing in the solution algorithms or software changes for the other cases. Of course, convergence rate will vary with conditioning, as determined by Mach and Reynolds numbers and the correspondingly induced grid adaptivity. Furthermore, robustness becomes more of an issue in problems admitting shocks or making use of turbulence models. The lack of nonlinear robustness is a fact of life that is largely outside of the domain of parallel scalability. In fact, when nonlinear robustness is restored in the usual manner, through pseudo-transient continuation, the conditioning of the linear inner iterations is enhanced, and parallel scalability may be improved. In some sense, the Euler code, with its smaller number of flops per point per iteration and its aggressive trajectory towards the steady state limit may be a *more*, not less, severe test of scalability.

The solution algorithm we employ is pseudo-transient Newton-Krylov-Schwarz (ΨNKS), with point-block ILU(0) on the subdomains for the action of $\tilde{A}_i^{-1}$ (in the customary Schwarz notation; see above). The original code possesses a pseudo-transient Newton-Krylov solver already. Our reformulation of the global point-block ILU(0) of the original FUN3D into the Schwarz framework of the PETSc version is the primary source of additional concurrency. The timestep grows from an initial CFL of 10 towards infinity according to the switched evolution/relaxation (SER) heuristic of Van Leer & Mulder [21]. Our ΨNKS solver operates in a matrix-free, split-discretization mode, whereby the Jacobian-vector MATVEC operations required by the GMRES method are approximated by finite-differenced Fréchet derivatives of the nonlinear residual vector. The action of the Jacobian is therefore always "fresh." However, the submatrices used to construct the point-block ILU(0) factors on the subdomains as part of the Schwarz preconditioning are based on a lower-order discretization than the one used in the residual vector, itself. This is a common approach in practical codes, and the requisite distinctions within the residual and Jacobian subroutine calling sequences were available already in the FUN3D legacy version.

Conversion of the legacy FUN3D into the distributed memory PETSc version was begun in October 1996 and first demonstrated in March 1997. It has been undergoing continual enhancement since, largely with respect to single-node aspects, namely blocking, variable interlacing, and edge-reordering for higher cache efficiency. The original five-month, part-time effort included: learning about FUN3D and its mesh preprocessor, learning the MeTiS unstructured grid partitioning tool, adding and testing new functionality in PETSc (which had heretofore been used with structured grid codes; see, e.g. [13]), and restructuring FUN3D from a vector to a cache orientation. Porting a legacy unstructured code into the PETSc framework would take considerably less time today. Approximately 3,300 of the original 14,400 lines (primarily in FORTRAN77) of FUN3D are retained in the PETSc version. The retained lines are primarily SPMD "node code" for flux and Jacobian evaluations, plus some file I/O routines. PETSc solvers replace the rest. Parallel I/O and post-processing are challenges that remain.

A. Summary of Results on the Cray T3E and the IBM SP

We excerpt from a fuller report to appear elsewhere a pair of tables for a 1.4-million degree-of-freedom problem converged to machine precision in approximately 6.5 minutes, using approximately 1600 global fine-grid flux balance operations (or "work units" in the multigrid sense) on 128 processors of a T3E or 80 processors of an SP. Relative efficiencies of 75% to 85% are obtained over this range The physical configuration is a three-dimensional ONERA M6 wing up against a symmetry plane. This configuration has been extensively studied by our colleagues at NASA and ICASE, and throughout the international aerospace industry generally, as a standard case. Our tetrahedral Euler grids were generated by D. Mavriplis of ICASE. The largest contains 357,900 vertices, which implies that vector of four unknowns per vertex has dimension 1,431,600. We used a maximum Krylov dimension of 20 vectors per pseudo-timestep. The maximum CFL used in the SER pseudo-timestepping strategy is 10,000. The pseudo-timestepping is a nontrivial feature of the algorithm, since the norm of the steady state residual does not decrease monotonically in the largest grid case. (In practice, we might employ mesh sequencing so that the largest grid case is initialized from the converged solution on a coarser grid. In the limit, such sequencing permits the finer grid simulation to be initialized within the domain of convergence of Newton's method.)

The first table, for the Cray T3E, shows a relative efficiency in going from the smallest processor number for which the problem fits (16 nodes) to the largest available (128 nodes), of 85%. Each iteration represents one pseudo-timestep, including one Newton correction, and up to 20 Schwarz-preconditioned GMRES steps.

Cray T3E Performance (357,900 vertices)

procs	its	exe	speedup	η_{alg}	η_{impl}	$\eta_{overall}$
16	77	2587.95s	1.00	1.00	1.00	1.00
24	78	1792.34s	1.44	0.99	0.97	0.96
32	75	1262.01s	2.05	1.03	1.00	1.03
40	75	1043.55s	2.48	1.03	0.97	0.99
48	76	885.91s	2.92	1.01	0.96	0.97
64	75	662.06s	3.91	1.03	0.95	0.98
80	78	559.93s	4.62	0.99	0.94	0.92
96	79	491.40s	5.27	0.97	0.90	0.88
128	82	382.30s	6.77	0.94	0.90	0.85

The second table, for the IBM SP[2], shows a relative efficiency of 75% in going from 8 to 80 nodes. The SP has 32-bit integers, rather than the 64-bit integers of the T3E, so the integer-intensive unstructured-grid problem fits on just eight nodes. The average per node computation rate of the SP is about 50% greater than that of the T3E for the current cache-optimized version of the code.

IBM SP Performance (357,900 vertices)

procs	its	exe	speedup	η_{alg}	η_{impl}	$\eta_{overall}$
8	70	2897.46s	1.00	1.00	1.00	1.00
10	73	2405.66s	1.20	0.96	1.00	0.96
16	78	1670.67s	1.73	0.90	0.97	0.87
20	73	1233.06s	2.35	0.96	0.98	0.94
32	74	797.46s	3.63	0.95	0.96	0.91
40	75	672.90s	4.31	0.93	0.92	0.86
48	75	569.94s	5.08	0.93	0.91	0.85
64	74	437.72s	6.62	0.95	0.87	0.83
80	77	386.83s	7.49	0.91	0.82	0.75

Convergence is defined as a relative reduction in the norm of the steady-state nonlinear residual of the conservation laws by a factor of 10^{-12}. The convergence rate typically degrades slightly as number of processors is increased, due to introduction of increased concurrency in the preconditioner, which is partition-dependent, in general. We briefly explain the efficiency metrics in the last three columns of the tables.

Conflicting definitions of parallel efficiency abound, depending upon two choices:

[2]The configuration consists, more precisely, of 80 120MHz P2SC nodes with two 128 MB memory cards each connected by a TB3 switch, and is available at Argonne National Lab.

- What scaling is to be used as the number of
  processors is varied?
  - overall fixed-size problem
  - varying size problem with fixed memory per
    processor
  - varying size problem with fixed work per pro-
    cessor
- What form of the algorithm is to be used as
  number of processor is varied?
  - reproduce the sequential arithmetic exactly
  - adjust parameters to perform best on each
    given number of processors

In our implementations of NKS, we always adjust the subdomain blocking parameter to match
the number of processors, one subdomain per processor; this causes the number of iterations to vary,
especially since our subdomain partitionings are
not nested. The effect of the changing-strength
preconditioner should be examined independently
of the general effect of parallel overhead, by considering separate algorithmic and implementation
efficiency factors.

The customary definition of relative efficiency in
going from q to p processors $(p > q)$ is

$$\eta(p|q) = \frac{q \cdot T(q)}{p \cdot T(p)},$$

where $T(p)$ is the overall execution time on p processors (directly measurable). Factoring $T(p)$ into
$I(p)$, the number of iterations, and $C(p)$, the average cost per iteration, the algorithmic efficiency
is an indicator of preconditioning quality (directly
measurable):

$$\eta_{alg}(p|q) = \frac{I(q)}{I(p)}.$$

Implementation efficiency is the remaining (inferred) factor:

$$\eta_{impl}(p|q) = \frac{q \cdot C(q)}{p \cdot C(p)}.$$

Algorithmic efficiency (ratio of iteration count
of the less decomposed domain to the more decomposed domain – using the "best" algorithm
for each processor granularity) is in excess of 90%
over this range. The main reason that the iteration count is only weakly dependent upon granularity is that the pseudo-timestepping over the
early part of the iteration provides some parabol-
icity.

Implementation efficiency is in excess of 82%
over the experimental range, and maintains uni-
tarity over the early part of the range. Implementation efficiency is a balance of two opposing effects in modern distributed memory architectures. It may improve slightly as processors
are added, due to smaller workingsets on each
processor, with resulting better cache residency.
Implementation efficiency ultimately degrades as
communication-to-computation ratio increases for
a fixed-size problem after the benefits of cache residency saturate.

The low (82%) implementation efficiency for the
80-processor SP can be accounted for almost completely by communication overhead. PETSc provides detailed profiling capabilities that provide
the communication timings. The percentage of
wallclock time spent in communication on 80 processors of the SP is:

- 4% on nearest-neighbor communication to set
  ghostpoint values needed in function and Jacobian stencil computation (implemented using
  PETSc's vector scatter operations);
- 13% on globally synchronized reduction operations, further subdivided into:
  - 5% on norms, required in convergence tests,
    in vector normalizations in GMRES, and in
    differencing parameter selection in matrix-
    free MATVECs, and
  - 8% on groups of inner products, required
    in the classical Gram-Schmidt orthogonal-
    ization in GMRES. (Note that the percent-
    age lost to inner products would be much
    higher if the modified Gram-Schmidt (rec-
    ommended in [22] for numerical stability
    reasons) were used, since the modified ver-
    sion synchronizes on each individual inner
    product.)

The effect of the neighbor and global communications required in implicit methods for the parallel solution of PDEs is clearly seen from this profiling. We would expect an explicit code, tuned to
synchronize globally only infrequently on timestep
updates, to obtain approximately 95% fixed-size
efficiency on the SP, instead of 82%. The IBM SP
has communications performance (in both bandwidth and latency) that is particularly poor in re-

lation to its excellent computational performance. However, on any parallel computer with thousands of processors, algorithms requiring frequent global reductions will be of major concern.

Since we possess a sequence of unstructured Euler grids, we can perform a Gustafson-style scalability study by varying the number of processors and the discrete problem dimension in proportion. We note that the concept of Gustafson-style scalability does not extend perfectly cleanly to nonlinear PDEs, since added resolution brings out added physics and (generally) poorer conditioning, which may cause a shift in the "market basket" of kernel operations as the work in the nonlinear and linear phases varies. However, our shockless Euler simulation is a reasonably clean setting for this study, if corrected for iteration count. The table below shows three computations on the T3E over a range of 40 in problem and processor size, while maintaining approximately 4500 vertices per processor.

Cray T3E Performance — Gustafson scaling

vert	procs	vert/proc	its	exe	exe/it
357,900	80	4474	78	559.93s	7.18s
53,961	12	4497	36	265.72s	7.38s
9,428	2	4714	19	131.07s	6.89s

The good news in this experiment is contained in the final column, which shows the average time per parallelized pseudo-time NKS outer iteration for problems with similarly sized local workingsets. Less than a 7% variation in performance occurs over a factor of nearly 40 in scale. Provided that synchronization latency can be controlled as the number of processors is increased, via the ideas discussed in the previous section and many others not yet invented, we expect that indefinite scaling is possible. We insert the caveat that most petaflops-scale PDE computations will not be homogeneous, but will consist of interacting tasks with different types of physics and algorithmics. Predictions of scalability are invariably problem-dependent when such interactions need to be taken into account. Furthermore, most petaflops-scale PDE computations will require dynamically adaptive gridding, and the adaptivity phase may not scale anywhere near as gracefully as the solution phase exhibits here.

IV. PARALLEL IMPLEMENTATION USING PETSc

The parallelization paradigm we illustrate above in approaching a legacy code is a compromise between the "compiler does all" and the "hand-coded by expert" approaches. We employ the "Portable, Extensible Toolkit for Scientific Computing" (PETSc) [3], [4], a library that attempts to handle in a highly efficient way, through a uniform interface, the low-level details of the distributed memory hierarchy. Examples of such details include striking the right balance between buffering messages and minimizing buffer copies, overlapping communication and computation, organizing node code for strong cache locality, preallocating memory in sizable chunks rather than incrementally, and separating tasks into one-time and every-time subtasks using the inspector/executor paradigm. The benefits to be gained from these and from other numerically neutral but architecturally sensitive techniques are so significant that it is efficient in both the programmer-time and execution-time senses to express them in general purpose code.

PETSc is a large and versatile package integrating distributed vectors, distributed matrices in several sparse storage formats, Krylov subspace methods, preconditioners, and Newton-like nonlinear methods with built-in trust region or line-search strategies and continuation for robustness. It has been designed to provide the numerical infrastructure for application codes involving the implicit numerical solution of PDEs, and it sits atop MPI for portability to most parallel machines. The PETSc library is written in C, but may be accessed from user codes written in C, FORTRAN, and C++. PETSc version 2, first released in June 1995, has been downloaded thousands of times by users worldwide. PETSc has features relevant to computational fluid dynamicists, including matrix-free Krylov methods, blocked forms of parallel preconditioners, and various types of time-stepping.

A diagram of the calling tree of a typical ΨNKS application appears below. The arrows represent calls that cross the boundary between application-specific code and PETSc library code: all other details are suppressed. The top-level user routine performs I/O related to initialization, restart, and post-processing and calls PETSc subroutines to create data structures for vectors and matrices and to initiate the nonlinear solver. PETSc calls user routines for function evaluations $\mathbf{f}(\mathbf{u})$ and (approximate) Jacobian evaluations $\mathbf{f}'(\mathbf{u})$ at given state vectors. Auxiliary information required for the evaluation of $\mathbf{f}$ and $\mathbf{f}'(\mathbf{u})$ that is not carried as part of $\mathbf{u}$ is communicated through PETSc via a user-defined "context" that encapsulates application-specific data. (Such information typically includes dimensioning data, grid data, physical parameters, and quantities that could be derived from the state $\mathbf{u}$, but are most conveniently stored instead of recalculated, such as constitutive quantities.)

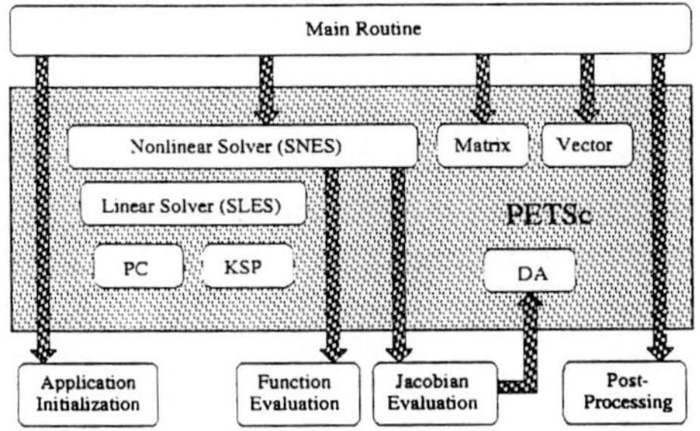

From our experience in writing and rewriting PDE codes for distributed memory machines, we have the following recommendations, which will undoubtedly continue to be relevant as codes are written in anticipation of an ultimate petaflops port.

- Abandon vector-based disk-striped data structures (e.g., node colorings) for cache-based data structures (e.g., subblocks).
- Interlace unknown fields so that most rapid ordering is within a point, not between points.
- Use the most convenient naming (global or local) for each given task, maintaining translation capability:
  - Physical boundary conditions rely on global names.
  - Many interior operations can be carried over from the uniprocessor code to SPMD node code by a simple "1-to-n" loop, with remapped entity relations (e.g., "vertices of edges", "edges of cells").
- Apply memory conservation aggressively; consider recomputation in cache rather than storage in memory.
- Micromanage storage based on knowledge of

horizontal (e.g., network node) and vertical (e.g., cache) boundaries.

V. NONTRADITIONAL SOURCES OF CONCURRENCY

We step back briefly from our narrow focus on data parallelism through spatial decomposition of a PDE grid to consider less traditional means of discovering the million-fold concurrency that will be required for petaflops-scale computing.

Time-parallelism is a counterintuitive but demonstrably interesting source of concurrency, even in evolutionary, causal simulations. A key idea of time-parallelism is that not all of the work that goes into producing a converged solution at time level ℓ is sequentially captive to a converged solution at time level $\ell - 1$. When an iterative method is employed, different components of the error may converge at different stages, and useful work may conceivably begin at level ℓ before the solution at $\ell - 1$ is completely globally converged. This is particularly true in nonlinear problems. The direction, volume, and granularity of interprocessor communications in temporal parallelism are different from those of spatial parallelism, as are the memory scalings, since multiple time-frames of the problem proportional to the temporary concurrency must be kept in fast memory. For reasons of scope, we do not pursue the corresponding parallel complexities here, but refer to [16], [17].

In addition to the data parallelism within an individual PDE analysis, there is data parallelism *between* PDE analyses when the analyses are evaluations of objective functions or enforcements of state variable constraints within a computational optimization context. Computational fluid dynamics is not about individual large-scale analyses, done fast and well-resolved and "thrown over the wall." Both the results *and* their sensitivities are desired. Often multiple forcings (right-hand sides) are available *a priori*, rather than sequentially, which permits concurrent evaluation. Petaflops-scale computing for CFD will mean 100 analyses running on 10,000 1Gflop/s processors earlier than it means 1 analysis running on 1,000,000 1Gflop/s processors.

Finally, we recall that computational fluid dynamics is not bound to a PDE formulation. The continuum approach is convenient, but not fundamental. In a flat, global memory system, it is natural to solve Poisson equations; in a hierarchical, distributed memory system, it is less natural. Nature is statistical, and enforces elliptic constraints like incompressibility through fast local collision processes. Among major phenomena in CFD only radiation is fundamentally "action at a distance." Lattice gas models have had a discouraging history, perhaps because they are too highly quantized, requiring massive statistics, and because their fundamental operations cannot exploit floating point hardware. Lattice Boltzmann models, on the other hand, seem highly promising. They are still quantized in space and time, but not in particle number, as quantized particles are replaced with continuous probability distribution functions. Lattice Boltzmann models possess ideal petaflops-scale concurrency properties: their two phases or relaxation and advection are alternatively completely local and nearest-neighbor in nature. There is no inherent global synchronization, except for assembling a visualization.

VI. SUMMARY OBSERVATIONS

The PDE-based algorithms for general purpose CFD simulations that we use today will in theory[3] scale to petaflops, particularly as the equilibrium simulations that are prevalent today go over to evolutionary simulations, with the superior linear conditioning properties of the latter in implicit contexts. The pressure to find latency tolerant algorithms intensifies. Longer word lengths (e.g., 128-bit floats) anticipated for petaflops-scale architectures, for more finely resolved — and typically worse-conditioned — problems, can assist in those forms of latency toleration, such as delayed orthogonalization, that are destabilizing. Solution algorithms are, in some sense, the "easy" part of highly parallel computing, and thornier issues such as parallel I/O and parallel dynamic redistribution schemes may ultimately determine the practical limits to scalability.

Summarizing the "state-of-the-art" of architectures and programming environments, as they affect parallel CFD, we believe that:

- Vector-awareness is *out*; cache-awareness is *in*; but vector-awareness *will return* in subtle ways

having to do with highly multiple-issue processors.

- Except for the Tera machine and the presently installed vector base, near-term large-scale computer acquisitions will be based on commodity cache-based processors.
- Driven by ASCI, large-scale systems will be of distributed-shared memory (DSM) type: shared in local clusters on a node, with the nodes connected by a fast network.
- Codes written for the Message Passing Interface (MPI) are considered "legacy" already and will therefore continue to be supported in the DSM environment; MPI-2 will gracefully extend MPI to effective use of DSM and to parallel I/O.
- High-performance Fortran (HPF) and parallel compilers are not yet up to the performance of message-passing codes, except in limited settings with lots of structure to the memory addressing [15]. Hybrid HPF/MPI codes are possible steps along the evolutionary process, with high-level languages automating the expression and compiler detection of structured-address concurrency at lower levels of the PDE modeling.
- Computational steering will be an important aspect of petaflops-scale simulations and will appear in the form of interpreted scripts that control SPMD compiled executables.

With respect to algorithms, we believe that:

- Explicit time integration is solved problem, except for dynamic mesh adaptivity.
- Implicit methods remain a major challenge, since:
  - Today's algorithms leave something to be desired in convergence rate, and
  - All "good" implicit algorithms have *some* global synchronization.
- Data parallelism from domain decomposition is unquestionably the main source of locality-preserving concurrency, but optimal smoothers and preconditioners violate strict data locality.
- New forms of *algorithmic* latency tolerance must be found.
- Exotic methods should be considered at petaflops scales.

With respect to the interaction of algorithms

---

[3] We are warned by Philosopher Berra: *"In theory there is no difference between theory and practice. In practice, there is."*

with applications we believe that the ripest remaining advances are interdisciplinary:

- Ordering, partitioning, and coarsening must adapt to coefficients (grid spacing and flow magnitude and direction) for convergence rate improvement.
- Trade-offs between pseudo-time iteration, nonlinear iteration, linear iteration, and preconditioner iteration must be understood and exploited.

With respect to the interaction of algorithms with architecture, we believe that:

- Algorithmicists must learn to think natively in parallel and avoid introducing unnecessary sequential constraints.
- Algorithmicists should inform their choices with a detailed knowledge of the memory hierarchy and interconnection network of their target architecture. It should be possible to develop very portable software, but that software will have tuning parameters that are determined by hardware thresholds.

ACKNOWLEDGEMENTS

The authors owe a large debt of gratitude to W. Kyle Anderson of the Fluid Mechanics and Acoustics Division of the NASA Langley Research Center for providing FUN3D as the legacy code that drove several aspects of this work. Satish Balay, Bill Gropp, and Lois McInnes of Argonne National Laboratory co-developed the PETSc software employed in this paper, together with Smith, under the the Mathematical, Information, and Computational Sciences Division subprogram of the Office of Computational and Technology Research, U.S. Department of Energy, under Contract W-31-109-Eng-38. Kaushik has been supported by NASA contract NAG1-1692 and by a GAANN fellowship from the U. S. Department of Education. Dr. George Lea of The National Science Foundation supported the work of Keyes at Old Dominion University under ECS-9527169, and all three authors have collaborated while in residence at ICASE under NASA contracts NAS1-19480 and NAS1-97046. Keyes and Smith also collaborated on this chapter while in residence at the Institute for Mathematics and its Applications at the University of Minnesota during June, 1997. Program development time on the NASA SP2s was provided through the Ames/Langley NAS Metacenter, under the Computational Aero Sciences section of the High Performance Computing and Communication Program.

REFERENCES

[1] W. K. Anderson (1997), FUN2D/3D (homepage).
http://fmad-www.larc.nasa.gov/~wanderso/Fun/fun.html

[2] D. H. Bailey, et al. (1997). *The 1997 Petaflops Algorithms Workshop Summary Report.*
http://www.ccic.gov/cicrd/pca-wg/pal97.html

[3] S. Balay, W. D. Gropp, L. C. McInnes, and B. F. Smith (1996). *The Portable, Extensible Toolkit for Scientific Computing, version 2.0.13* (code and documentation).
http://www.mcs.anl.gov/petsc

[4] S. Balay, W. D. Gropp, L. C. McInnes, and B. F. Smith (1997), *Efficient Management of Parallelism in Object-Oriented Numerical Software Libraries.* Modern Software Tools in Scientific Computing. E. Arge, A. M. Bruaset and H. P. Langtangen, eds., Birkhauser Press (to appear).
ftp://info.mcs.anl.gov/pub/petsc/scitools96.ps.gz

[5] S. T. Barnard and Robert L. Clay (1997). *A Portable MPI Implementation of the SPAI Preconditioner in ISIS++.* Technical Report NAS-97-002, NASA Ames Research Center.
http://science.nas.nasa.gov/Pubs/TechReports/NASreports/NAS-97-002/NAS-97-002.html

[6] X.-C. Cai (1989), *Some Domain Decomposition Algorithms for Nonselfadjoint Elliptic and Parabolic Partial Differential Equations.* Technical Report 461, Courant Institute, NYU.

[7] X.-C. Cai, D. E. Keyes, and V. Venkatakrishnan (1995), *Newton-Krylov-Schwarz: An Implicit Solver for CFD,* in "Proceedings of the Eighth International Conference on Domain Decomposition Methods" (R. Glowinski et al., eds.), Wiley, New York, pp. 387-400; also ICASE TR 95-87.
ftp://ftp.icase.edu/pub/techreports/95/95-87.ps

[8] X.-C. Cai and M. Sarkis (1997). *A Restricted Additive Schwarz Preconditioner for Nonsymmetric Linear Systems.* Technical Report, Computer Sci. Dept., Univ. of Colorado, Boulder.

[9] M. Chandy, et al. (1996). *The Caltech Archetypes/eText Project.*
http://www.etext.caltech.edu/

[10] D. E. Culler, J. P. Singh, and A. Gupta (1997). *Parallel Computer Architecture.* Morgan-Kaufman Press (to appear).
http://HTTP.CS.Berkeley.EDU/~culler/book.alpha/

[11] J. J. Dongarra (1997), *Performance of Various Computers Using Standard Linear Equations Software.* Technical Report CS-89-85 Computer Science Dept., Univ. of Tennessee, Knoxville.
http://www.netlib.org/benchmark/performance.ps

[12] J. Erhel (1995), *A parallel GMRES version for general sparse matrices,* ETNA 3: 160-176.
http://etna.mcs.kent.edu/vol.3.1995/pp160-176.dir/pp160-176.ps

[13] W. D. Gropp, L. C. McInnes, M. D. Tidriri, and D. E. Keyes (1997), *Parallel Implicit PDE Computations: Algorithms and Software,* Proceedings of Parallel CFD'97, A. Ecer, et al., eds., Elsevier (to appear).
http://www.icase.edu/~keyes/papers/pcfd97.ps

[14] M. J. Grote and T. Huckle (1997), *Parallel Preconditioning with Sparse Approximate Inverses.* SIAM J. Sci. Comput. (to appear).
http://www-sccm.stanford.edu/Students/grote/grote/spai.ps.gz

[15] M. E. Hayder, D. E. Keyes, and P. Mehrotra (1997), *A Comparison of PETSc Library and HPF Implementations of an Archetypal PDE Computation*, Proceedings of 4th National Symposium on Large-Scale Analysis and Design on High Performance Computers and Workstations (to appear).

[16] G. Horton (1994), *Time-parallelism for the massively parallel solution of parabolic PDEs*, in "Applications of High Performance Computers in Science and Engineering", D. Power, ed., Computational Mechanics Publications, Southampton (UK).

[17] G. Horton, S. Vandewalle, and P. H. Worley (1995), *An algorithm with polylog parallel complexity for solving parabolic partial differential equations*, SIAM J. Sci. Stat. Comput., **16**: 531–541.

[18] C. T. Kelley and D. E. Keyes (1996), *Convergence Analysis of Pseudo-Transient Continuation*, SIAM J. Num. Anal. (to appear); also ICASE TR 96-46.
`ftp://ftp.icase.edu/pub/techreports/96/96-46.ps`

[19] D. E. Keyes and B. F. Smith (1997), *Final Report on "A Workshop on Parallel Unstructured Grid Computations"*, NASA Contract Report NAS1-19858-92.
`http://www.icase.edu/~keyes/unstr.ps`

[20] P. M. Kogge, J. B. Brockman, V. Freeh, and S. C. Bass (1997). *Petaflops, Algorithms, and PIMs*, Proc. of the 1997 Petaflops Algorithms Workshop.

[21] W. Mulder and B. Van Leer (1985), *Experiments with Implicit Upwind Methods for the Euler Equations*, J. Comp. Phys., **59**: 232–246.

[22] Y. Saad and M. H. Schultz (1986), *GMRES: A Generalized Minimal Residual Algorithm for Solving Nonsymmetric Linear Systems*, SIAM J. Sci. Stat. Comp. **7**: 865–869.

[23] B. F. Smith, P. E. Bjørstad, and W. D. Gropp (1996). *Domain Decomposition: Parallel Multilevel Algorithms for Elliptic Partial Differential Equations*, Cambridge Univ. Press.

[24] R. W. Stevens (1997). *Hardware Projects for COTS-based Designs*, Proc. of the 1997 Petaflops Algorithms Workshop.

[25] T. Sterling (1997), *Hybrid Technology Multithreaded Architecture: Issues for Algorithms and Programming*, Proc. of the 1997 Petaflops Algorithms Workshop.

[26] T. Sterling, P. Messina, and P. H. Smith (1995), *Enabling Technologies for Petaflops Computing*, MIT Press.

DISTRIBUTED VISUALIZATION
IN COMPUTATIONAL FLUID DYNAMICS

Juan Raúl CEBRAL   **Rainald LÖHNER**

George Mason University, Fairfax, VA 22030, USA

Abstract

The use of parallel computers and in particular distributed memory architectures imposes a number of requirements on the visualization systems for Computational Fluid Dynamics (CFD). We describe the design and capabilities of our distributed visualization software which can handle parallel multidisciplinary applications, can connect to a running solver for on-line display and steering, supports collaborative visualization and can be used as a tool for the automatic making of movies and animations. This discussion serves as an overview of the current trends and developments in visualization as applied to CFD and other computational sciences.

1   Introduction

Large scale computing is becoming very common in both industry as well as in research institutions. In many instances, the simulation of complex 3D configurations is being carried out either on parallel computers or using networks of workstations. This last option is becoming very popular, and tremendous effort is being devoted to the development of efficient solvers for such architectures [1]. For these cases it is very convenient to have a distributed visualization system that does not require to transfer the data produced in a parallel super-computer to a local workstation for visualization [2],[3].

The simulations of complex transient problems generate a very large amount of data to be analyzed and are time consuming. Typically these calcula-

tions are part of a design loop, therefore fast visualization tools that allow the analyst to interpret the data while it is being produced are of fundamental importance [4],[8],[9],[10]. Any visualization system to be effectively integrated to the solvers must be general, interactive, able to deal with distributed data and perform computations in parallel. Furthermore, it is highly desirable to have a system that does not require the introduction of major modifications to the solvers or interferes with their performance.

In many instances, specially in the industry, different stages of a design or analysis loop are carried out by different groups. The total time of a simulation is greatly influenced by the set-up and post-processing times. A collaborative environment, where engineers of different groups can visualize and interact

---

Received on July 1, 1997.

with the data simultaneously [11], [12], can be a very helpful tool to accelerate the total time of the design loop.

The solution of multidisciplinary problems were two or more disciplines are solved simultaneously is becoming more and more widely used for the analysis of complex situations. The most common example is the interaction between fluids and structures. These problems are mostly solved by coupling independent solvers together. This kind of simulations require visualization software that can also deal with two or more disciplines at the same time.

The purpose of this paper is to give an overview of the current developments in visualization as applied to Computational Fluid Dynamics. We review some of the areas of research in the context of our own visualization system, hoping that it covers most of the current efforts.

2. Visualization Software

Visualization software can be classified into three major groups:

a) **Libraries**: a set of functions are provided to the developers in order to program their own visualization system. Libraries range from low level graphic libraries to high level visualization libraries. Some examples are OpenGL [13], Open Inventor [14], PGPLOT [15] and IDL [16]. High level parallel visualization libraries are beginning to be available, such as PGL [18].

b) **Application Builders**: consist in a set of basic modules that the users need to connect together in order to build an application for their particular needs. Usu-ally it is possible to add new modules to extend the capabilities of the system. Application builders include AVS [19], SGI Explorer [20], SuperGlue [21], etc.

c) **Turnkey Applications**: provides the user with a set of fixed visualization operations. The most severe drawback of the majority of this type of visualization system is that they do not support functional extension. However, for specific disciplines they usually satisfy most of the user's needs, requiring no extra work. Applications of this kind are PLOT3D [22], FAST [23], UFAT [24], pV3 [4], EnSight [5], FieldView [6] TecPlot [7] and our own *ZFEM* [17].

In what follows we concentrate on the design, capabilities and development of our distributed visualization system (*ZFEM*) originated on the needs for parallel multidisciplinary applications. The major goals of this software are: deal with distributed applications in parallel, multidisciplinary applications, both unstructured and structured grids as well as particles, on-line visualization, collaborative visualization, automatic movie making capabilities, fully interactive, user friendly and portable.

3. Data Visualization

We are primarily interested in data based on **unstructured grids** since they provide the most general framework for fluid flows in complex geometries. The basic data set consist of a list of points, a set of fields (scalars, vectors, etc.) defined on the points and a set of element lists connecting the points. Different element types are considered: triangles and quads for surface meshes and tetrahedra and hexahedra for volume grids. For visualization purposes, we always split quads into triangles or hexahedra into tetrahedra, being these our basic elements.

We also support visualization based on **structured grids** and **particles**. In the former case we automatically generate quad or hexahedral elements to connect the points. In the latter, we only keep the list of points (particles) and fields (their properties), no elements are considered. These two situations are particular cases of our general data set type. Even though this is not the best approach for structured grids, this methodology is general and allows us to consider both unstructured and structured grids as well as particles in a unified manner.

The visualization of **transient** or **time-dependent** problems is done with the help of a script system, as explained later on. The main idea is to process automatically a sequence of data files representing the state of the simulation at different times.

In the case of **multidisciplinary** applications, two or more disciplines are coupled together and solved simultaneously. The visualization of such situations is carried out by working in parallel on each discipline simultaneously.

Most of the large scale problems are solved on parallel machines, with both shared and distributed memory architectures. In the former case it is not necessary to do anything special for visualization. However, in the latter the output consist in a set of data files that may be **distributed** across different file systems and computers. A distributed visualization system avoids the need of shipping the data over to the local workstation where the rendering is performed.

In the sections that follows we give more details about the methodology we use to deal with each of these situations. However, we can now characterize the data we are working on and the operations we apply to it: we deal with distributed, time-dependent, multidisciplinary applications (fluids, structures, heat, electrodynamics, particles, etc.) based on three dimensional unstructured grids (structured grids are also considered).

A number of **data processing** options are available before any graphics: get the list of **boundary** elements of a grid (this results in a new set of elements connecting the points), computing **iso-surfaces** of a given field or a **cutting plane** through the domain (these yield a new set of points, fields and elements allowing the user to visualize them as if they were independent data sets), **filter** out points according to the values of a given field (in order to visualize only parts of the domain) and compute **derived quantities** for specific disciplines such as compressible and incompressible flows (new fields are generated, e.g. Mach numbers, vorticity, etc.).

Different **visualization techniques** can be applied to the data, including: **scalar glyphs** to display mesh points and particles, **mesh lines** to display grid elements, **surface shading** to paint a surface according to a given scalar value, **contour lines** to display scalar values on a surface, **vector glyphs** and **ribbons** to display vector fields. Research is still being done on different ways of visualizing vector [29], [30] and tensor [31], fields as well as identifying critical points of such fields [32], [?].

Interpolation on unstructured grids is a very important issue for some of the calculations mentioned above (iso-surfaces, cut-planes and ribbons). We use the available fast searching algorithms for unstructured grids [35].

The final images of a particular visualization are composed by applying a variety of **presentation** techniques to the scene. Our software supports: **viewing transformations** (rotation, translation, zoom and translation of the origin of rotation), arbitrary **clipping plane** to cut the objects in the scene and see inside, **lights** to enhance the 3D appearance of the objects, arbitrary **mirrors** for symmetry planes and to see objects "from behind", **text annotations** to include titles and **legends** to relate colors with field values. Also, the user has control over some scene components such as **coordinate axes**, **bounding box**, **floor** and other properties such as the **background color** and the **z-buffer** shifting. Many of the graphical attributes of the objects in the scene can be interactively modified without having to recompute the object itself (e.g. colors, line widths and styles, materials, uniform length of vectors, etc.).

4. Software Design

The design of *ZFEM* was based not only on the requirements imposed by the data to be visualized described in the previous section, but also on the desired **modes of operation**. The system can run in three different modes: **post-processing** where the user interactively works on the results of simulations stored in data files, **on-line display** where the visualizer is "connected" to a running solver in order to see the results as they are being produced, and **batch** where a script file is executed without user intervention in order to reproduce an earlier visualization session or to create movies.

With these ideas in mind the software was implemented as a set of independent modules that can run on the local workstation or on a remote computer and communicate with each other via message passing. Every module has an interpreter that can decode commands that are received from another module and execute the corresponding action. The events generated by the user are also transformed into commands that can be interpreted, stored or sent to other modules.

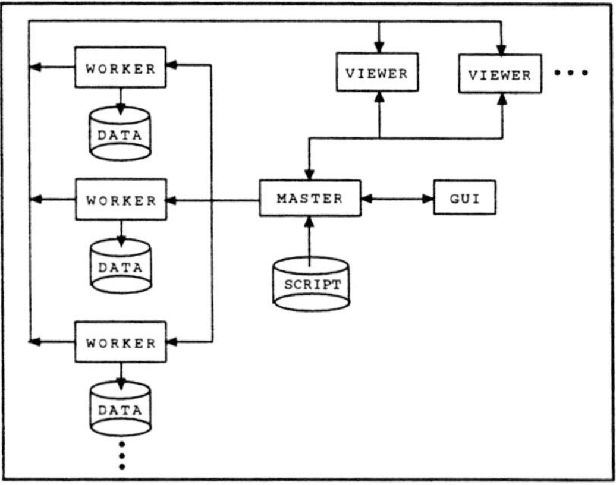

Figure 1: *ZFEM* architecture.

The basic modules and their main tasks are (see figure 1):

MASTER:

-control communications between other modules
-launch other modules
-execute scripts
WORKER:
-read data files in parallel on remote machines
-perform computations on the data in parallel
-create graphical objects for display
VIEWER:
-display graphical objects
-apply transformations desired by the user
-edit scene attributes and components
-dump image files
GUI:
-interact with the user to create plots

In order to be able to port *ZFEM* to different computers with a minimum of effort, it was programmed in C, the graphical interface in Motif, the graphics in OpenGL and the communications using PVM. Since these libraries are available for most of the computer systems, there should be little portability restrictions.

5. Distribution and Parallelism

As the size of the simulations increase, parallel environments are becoming very popular for the solution of large problems. Different architectures are being used: massively parallel distributed memory supercomputers, shared memory supercomputers and networks of heterogeneous computers. However the visualizations are usually carried out on graphical workstations of less computational power. This situation calls for the use of distributed visualization systems that allow to work with physically distributed data. Furthermore, with such a system one can take advantage of the different capabilities of different computers of an heterogeneous network. For example it is possible to combine the high computational power and large memory of supercomputers with the high speed graphics of the workstations. Various issues such as network performance, network bandwidth and synchronization between processes become important during the design stages of this kind of visualization software.

The basic visualization process can be subdivided into four steps: 1) load the data and perform some calculations, 2) create geometrical objects for display, 3) create an image from the geometrical objects and 4) display the image. There are four basic ways of distributing these tasks among processes of a distributed visualization software [26]:

a) **Distributed file system**: the data files residing on a remote computer can be accessed from the graphics workstation.

b) **Geometry partition**: the processes on the remote computers calculate the graphical objects to display on the local workstation and ship them over.

c) **Distributed graphics library**: the remote processes work directly on the parallel rendering of the image on the workstation.

d) **Frame buffer**: the image is completely computed on the remote machines and transferred to the workstation for display.

We have chosen to work with a geometry partition since it allows us to: access remote data files, perform parallel computations on the raw data that do not involve any graphics, and create graphical objects for display in parallel. With this approach the rendering process is completely local to the workstation, avoiding large network communications. This strategy also allows us to use the most appropriate data structures to hold the data on the remote computers and to store the graphical objects in the local workstation.

Our code can work with a single data file or with a "parallel project". In this last case, we assume that the parallel calculation was done using a domain decomposition and that a set of files was produced, each containing a subdomain. Thus, our parallel project consists in a set of data files located on a set of remote machines. A worker module is launched for each file to be read on the machine containing that file. The data is read by the workers on the remote machine without having to ship it to the local machine where the graphics are rendered. Once the data is read, the workers can perform different computations and generate graphical objects that are sent to the viewers for display. When a particular calculation is requested by the user (e.g. a cutting plane), each worker operates on its corresponding part of the domain. In some instances one of the workers acts as a "foreman" collecting the results of a parallel calculation (for example when finding the min or max of a field across the whole computational domain which yields a single scalar value). As we can see the parallelism is at the level of the data, with each worker performing a computation on its own part of the parallel data set. Other systems use different levels of parallelism, such as image decomposition and parallel rendering [27], [28].

As an example of a distributed visualization, consider the compressible flow about an F117 aircraft, shown in figure 3. In this case the solution was obtained by subdividing the domain into 10 subdomains, and running on 10 processors of a parallel computer. The visualization was done by launching 10 workers on 10 processors of a remote parallel computer (it can also be done by running all the workers on the same machine, or a different set of computers, all what we need is a worker for each data file). The figure shows how the surface of the aircraft was subdivided across processors (colors are according to processor number). A cut-plane was computed and pressure contour lines were displayed on this plane. A set of three ribbons was computed and displayed with colors according to the pressure in order to track the fluid velocity. Only half of the aircraft was considered for the solution because of its symmetry, thus a mirror was used to display the whole aircraft.

6. Exploration and Multidisciplinary Visualization

As indicated in figure 1, any number of viewer modules can be opened simultaneously. A set of these viewers is interactively "selected" by the user. These are the viewers that receive the graphical objects generated by the workers. The list of selected viewers can be changed at any time, thus directing new plots to the desired viewers. This capability allows the user to have any number of viewers displaying the same, different or a combination of plots in each of them.

When two or more viewers are selected, they become "linked", in the sense that if an action is executed in any of them it is also executed by the other linked viewers. Thus, if for example a rotation is performed in one viewer, a short message is sent to the other linked viewers so that the same rotation is also applied. In this way the view-point is "shared" by linked viewers. On the other hand, "unlinked" viewers keep their viewing transformations local.

This kind of **multiview** capability is very useful for exploration of data sets by performing **simultaneous visualization** of different features of the results of a simulation [34]. Different plots can be displayed on different viewers at the same time, sharing the view-point, in order to identify correlations between different quantities.

Consider for example the incompressible flow around a tuna fish. In figure 4 three linked viewers are used to display velocity vectors, pressure surface shading and boundary mesh lines on the tail of the fish. Note that the window border of the viewers show the viewer name (its number), it host (machine name) and a "$< - - -$" which acts as an indicator of whether the viewer is selected or not (all selected viewers are linked). Any action performed in any of the three viewers is simultaneously executed by the other viewers, allowing the user to inspect different features at the same time.

As a second example of simultaneous visualization, we show in figure 5 two unlinked viewers displaying surface pressure distribution from two independent viewpoints. The viewer on the left shows a pressure shading and pressure line contours, while the one on le right, pressure shading and mesh lines on the outer boundary.

The use of multiple viewers can also be used for **multidisciplinary** visualization. The results of a multidisciplinary application consist in a set data files, corresponding to the solutions of each discipline. For example a fluid-structure interaction simulation yields a fluid and a structural solution. Correlations between them can be found by displaying each of them in a different viewer with the same view-point. This situation is illustrated in figure 6, where two linked viewers are used to display simultaneously the results of the interaction between a shock wave impacting on a truck. One viewer shows the deformation of the solid model colored according to the structural velocity (structural solution) while the other shows the pressure distribution on the surface of the fluid mesh (fluid solution).

This kind of visualization is possible thanks to the ability to work with **multiple projects**. Any number of projects, either a single data file or a parallel project, can be opened at the same time. Therefore the user can work on different data sets simultaneously. In the case of multidisciplinary applications each discipline is treated as a different project. All the projects are read at the same time. Each project has its own set of worker modules working in parallel on the corresponding data files. They recognize the type of data (fluid, structure, etc.) and perform different actions accordingly (for example compute derived fields). Different plots created from the data of each projects can be directed to the same viewer or to different viewers, as already seen in figure 6.

Some applications involve not only continuous fields like in the case of fluid dynamics and structural dynamics but also particles. They can be visualized by using standard techniques. An example is provided in figure 7 where some scene components like a bounding box and a floor were added in order to increase the three-dimensional appearance of the view.

7. Collaboration

In a typical visualization session a single analyst interacts with a scene on his workstation. Communication of the results to other colleagues becomes very slow and inefficient since it is necessary to transfer the data to the others and have them produce the same plots again, or send them hardcopies or videos, which are absolutely not interactive.

A new approach that is becoming very interesting is the use of collaboration software. The basic idea is that several scientists interact with the same scene simultaneously and communicate through video-conferencing tools, which are already available. Different users can be located in different offices, different buildings or even different cities. This strategy can help increase the speed of design loops substantially.

In our geometry partition approach several modules can run locally on different workstations. In particular multiple viewers can be launched on different hosts. As explained before, the graphical objects are produced remotely and trans-

mitted to all the selected or linked viewers, thus a "shared" scene is obtained. When any user performs an action (for example a viewing transformation) a short message is sent to the other modules in order to perform the same action. This collaboration approach does not require a high bandwidth network since only small messages are sent from module to module. For example, when a rotation is performed, a message of the form "rotate rx ry rz" is sent to the other linked viewers. The shipping time of graphical objects to the viewers for display, can be thought of as part of the time required to create a given plot. Usually this transfer time is much smaller than the time required to actually create the requested graphical object (e.g., cut planes or iso-surfaces). Besides these operations are not done as frequently as for example viewing transformations. Note however that once the objects are created, they become local to the viewers and only small network traffic is necessary for collaboration. Thus "long distance" collaboration is possible with this strategy.

A collaborative visualization session is illustrated in figure 8, where two linked viewers display the same scene to two different users located far away. They both interact with the scene at the same time, and use a video-conference software to communicate. A text object is also used to point to different places in the scene (text objects can interactively be moved around the scene).

8. On-Line Visualization and Steering

In many instances, specially for time-dependent simulations it is very desirable to be able to interactively visualize data as it is being produced by a solver.

Usually graphics are either **fully coupled** with solvers or completely **decoupled**. In the former case the graphics routines are called from the solver directly. This approach has the drawback that the interactivity is lost. The user typically specifies some plotting parameters (such as the viewing point) at the beginning of the simulation and they remain fixed for the rest of the run. Even more, the user is not allowed to turn on or off the graphics at any time during the simulation. In the second case the solver just dumps the computed data into files that can be read with a post-processing software. This strategy has several drawbacks: many data files are produced for a time-dependent simulation, this takes time and storage, the visualization is not updated as new data is computed, and if the solver started with a parameter that specifies no data-dumping it remains that way through out the whole simulation and it is not possible to visualize anything.

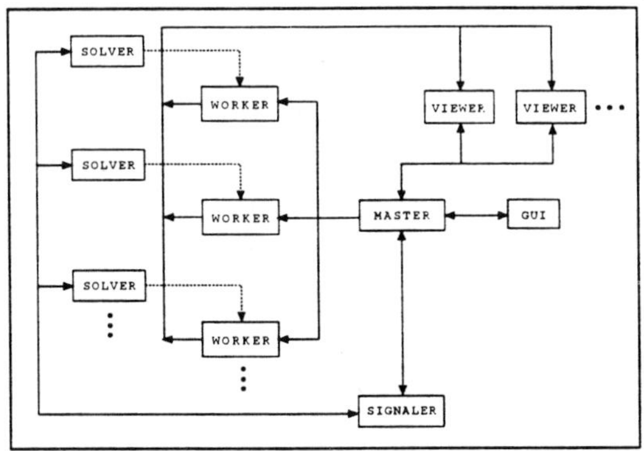

Figure 2: Architecture in on-line mode.

However, a visualization system can also be **loosely coupled** to a solver. The main idea is that the solver and the graphics remain two independent processes (which can run on different computers) as in the decoupled case, but the data is communicated every timestep from the solver to the plotter for display. In this way the user has complete interactivity with the visualization software and at the same time the data gets automatically updated when it is calculated. Different mechanisms can be used for the data communication between the solver and the display. For example, the codes could communicate through temporary files. Besides of being slow this method introduces difficult problems related to synchronization and blocking. Another alternative is to use a message passing for the communications. In this case the solver needs minor modifications in order to send the data to the visualizer. We have chosen to work with this last approach.

The connection to the solver is established via a signal file. A new module caller the SIGNALER writes a signal file (thus its name) which can be read by the solver. The solver checks every time step if the signal file exists. In such a case it reads from this file the necessary information to establish a communication with the SIGNALER. Once the communication is established the signaler informs the master of the number of worker modules needed, and once these are launched instructs the solver how to link each subdomain of the computation with a corresponding worker. Thus, for the on-line mode of operation we use the architecture illustrated in figure 2.

When the solver and the visualizer are connected, the solver sends the current data to the workers every time-step (or at a specified frequency) and continues its execution.

If the user requests an operation using the GUI (perform a calculation on the current data or create a plot of a certain quantity) this request is cast in the form of a command. This command is executed immediately and stored in a queue. When a new time-step is finished and the data has been updated, all the stored commands are re-executed. Therefore the same computations and plots are created for the new data set.

This methodology gives full interactivity to manipulate the scene during an on-line visualization. The user can also decide to "disconnect" from the solver at any time. In this case the signal file is removed, thus the solver will not attempt any communication and continues its computation. The user is free to connect again later to check the run.

It is interesting to note that this methodology opens a very easy way of **steering** the solver once the communication is established. It is only a matter of sending commands to the solver to carry out the desired actions. These actions include: stopping the calculation, dumping data or restart files, or changing certain parameters.

The series of pictures shown in figure 9 illustrate the on-line display for the calculation of the compressible flow about a pitching wing. They were taken while an implicit flow solver was updating the flow field. Note how the shocks move as the angle of attack changes. During the first time-step a pressure contour plot, a uniform shading of the surface and a clipping plane were created, the view-point was set and a snapshot taken. All these actions are subsequently re-executed every time-step after the flow field is updated.

Note that the same result could have been obtained by having the solver dump a file every time-step and then using *ZFEM* in batch mode in order to read one file after another, create the scene and dump the image to a file. Consecutively, the images can be combined into a movie.

9. Animations and Movies

The visualization of transient simulations is most naturally carried out by composing animations and movies, where the analyst can see changes in the solution in time. For these purposes we can use the on-line capabilities. However, the actual solution of the problem can be very time consuming, specially for large scale problems and complex geometries with moving bodies. Therefore, another alternative is to operate in batch mode.

Every time the user performs an action (through the graphical user interface), this action is cast into a command and stored in a script file. This file can later be executed in order to reproduce the visualization session (batch mode). However, repeatability is only one of the uses of this **script system**. It is also very useful to create demos or execute the same set of operations on different data sets. The second most interesting application is the production of movies. If the same set of operations are applied to a sequence of data files representing the solution at different time levels, one can obtain a series of pictures to create a movie. Our script system supports the addition of **do-loops** in order to loop through a series of data files (or parallel projects).

Besides the possibility of making movies by running scripts which translate, rotate and zoom the scene, it is possible to use a **navigation** camera. The view point of the camera is specified by the coordinates of the eye, the direction in which it is looking at and a vector specifying the vertical direction. The user can interactively create a "fly by" path that the camera follows. The camera moves from one point along this path to the next in an arbitrary number of steps, interpolating the eye, direction and vertical coordinates. If various "linked" viewers are used, all of them set the view-point to the current camera position. At each step, each viewer dumps into a file its current image. Thus, the necessary pictures for making a movie are obtained. Note that by using two (or more) viewers displaying different features, for example the mesh and a shading plot, one can get two (or more) sets of images to make a movie that shows the evolution of both quantities at the same time. By combining this navigation method with the scripts, quite complex movies can be produced. In this case, the script is in charge of updating the scene (for time-dependent data for example) while the camera provides a navigation through the current scene and creates the image files. Thus, with this methodology the making of a movie is completely automatic.

Figure 10 shows an example of a navigation camera being used to create a movie. It corresponds to a simulation of the incompressible flow around a tuna fish. In the first of the images we can see the path that was interactively specified for the camera. The following images contain the different views obtained as the camera moves through the points of this path. Obviously more complex path can be created in order to produce a more elaborated navigation.

10. Discussion

A new trend in visualization systems for CFD and other computational disciplines is beginning to emerge. The key idea is to be able to interact with a remotely running parallel solver in order to visualize the data as it is being generated. At this stage, we have developed a software that can deal with remote/distributed data in parallel. Different data sets (may be from different computational disciplines) can be visualized simultaneously and various analysts can interact with a common scene at the same time. This last feature can speed up the communications between different groups working concurrently on the same problem.

The use of a network of heterogeneous computers offers the possibility of taking full advantage of the capabilities of particular architectures, for example some modules could be optimized for shared memory machines or massively parallel supercomputers.

Most turnkey applications suffer from the problem of poor extensibility. Therefore, more research is needed in order to be able to consider arbitrary data formats, user defined data processing procedures and visualization techniques.

Collaborative environments are just coming into existence. Porting the technology described here to other stages of the engineering analysis is needed.

Future research will also concentrate on the development of integrated environments which provide the analyst with a single environment for all the stages of the analysis: pre-processing and problem set-up, automatic grid generation, problem solution, visualization, design loops and reporting capabilities (including web posting tools). These tools interact with each other and allows the user to make use of all the available computer resources in a transparent and friendly way. These environments could also include data-bases for storage of results and artificial intelligence algorithms to help the engineers select the most appropriate set of parameters for a particular application.

Acknowledgments

This work was partially supported by the CHSSI project, with Dr. William Sandberg as technical monitor and AFOSR with Dr. Leonidas Sakell as the technical monitor.

References

[1] Satofuka, N., Periaux, J., and Ecer, A. (eds.), "Parallel Computational Fluid Dynamics", Elsevier, North Holland, 1995.

[2] Haber, R., Bliss, B.. Jablonowski, D. and Jog, C., "A Distributed Environment for Run-Time Visualization and Application Steering in Computational Mechanics". *Computing Systems in Engineering*, Vol. 3, Nos. 1-4, 501-515, 1992.

[3] Forslung, D.W., Hansen, C., Hinker, P., John, W. St., Tenbrink, S., and Brewton, J., "High-Speed Networks, Visualization, and Massive Parallelism in the Advanced Computing Laboratory", *Computing Systems in Engineering*, Vol. 3, Nos. 1-4, 521-524, 1992.

[4] Haimes, R., "Visualization in a Parallel Processing Environment", *AIAA* 97-0348, 1997.

[5] see http: //www.ceintl.com/ news/ post/ N294/ 5.2.Rel.html

[6] see http: //www.merak.com/ software/ fieldview.html

[7] see http: //www.abcworks.co.uk/ tecplot/ tecpics.html

[8] Kuzuu, K. and Kaizaki, H., "Real Time Visualization of Flow Fields using Open GL System", *AIAA* 97-0235, 1997.

[9] Bryson, S., "Real-Time Exploratory Scientific Visualization and Virtual Reality", in *Scientific Visualization Advances and Challenges*, Academic Press, 1994.

[10] Earnshow, R.A. and Jem, M., "Fundamental Approaches to Interactive Real-Time Visualization Systems", in *Scientific Visualization Advances*

and Challenges, Academic Press, 1994.

[11] Anupam, V., Bajaj, C., Schikore, D. and Schikore, M., "Distributed and Collaborative Visualization", *Computer*, 37-43, July 1994.

[12] Chen, H., "Collaborative Systems: Solving the Vocabulary Problem", *Computer*, 58-66, May 1994.

[13] Neider, J., Davis, T. and Woo M., "OpenGL Programming Guide", Addison Wesley, 1993.

[14] Wernecke, J., "The Inventor Mentor", Addison Wesley, 1994.

[15] Pearson, T. J., "PGPLOT Graphics Sub-routine Library", version 5.2, 1995 (http://astro.caltech.edu/~tjp/ pgplot).

[16] "IDL User's Guide", Version 2.0, Research Systems, Inc., 1990.

[17] Cebral, J.R., "ZFEM: Collaborative Visualization for Parallel Multidisciplinary Applications", in *Proceedings of Parallel CFD'97*, Manchester, U.K., May 19-21, 1997.

[18] Crockett, T.W., "Beyond The Renderer: Software Architecture for Parallel Graphics and Visualization", *ICASE Report No. 96-75*, 1996.

[19] Upson, C., "The Application Visualization System (AVS): A Computational Environment for Scientific Visualization", *IEEE Computer Graphics and Applications*, Vol. 9, No. 4, 30-42, 1989.

[20] Silicon Graphics Inc. "IRIS Explorer User's Guide", document Number 007-1371-010, January 1992.

[21] Hulquist, J.P.M. and Raible, E.L., "SuperGlue: A Programming Environment for Scientific Visualization", *NAS Report RNR-92-014*, 1992.

[22] Walatka, P.P., Buning, P.G., Pierce, L. and Elson, P.A., "PLOT3D User's Manual", *NASA Technical Memorandum 101067*, March 1990.

[23] Walatka, P.P., Clucas, J., Plessel, T. and McCabe, K.R., "FAST User Guide", November 1994.

[24] Lane, D., "UFAT - A Particle Tracer for Time-Dependent Flow Fields", in *Proceedings of Visualization '94*, Washington D.C., 257-64, 1994.

[25] Earnshow, R.A. and Wiseman, N., *An Introductory Guide to Scientific Visualization*, Springer-verlag, 1992.

[26] Gerald-Yamasaki, M., Hulquist, J., Bryson, J., Kenwright, D., Lane, D., Walatka, P., Clucas, J. and Watson, V., "Visualization of Computational Fluid Dynamics", in *Computational Fluid Dynamics Review 1995*, eds. Hafez, M. and Oshima, K., John Wiley, 904-930, 1995.

[27] Whitman, S., Hansen, C.D. and Crockett, T.W., "Recent Developments in Parallel Rendering", *IEEE Computer Graphics and Applications*, 21-22, July 1994.

[28] Singh, J.P., Gupta, A. and Levoy, M., "Parallel Visualization Algorithms: Performance and Architec-

tural Implications", *Computer*, 45-55, July 1994.

[29] Hulquist,J.P.M, "Interactive Numerical Flow Visualization Using Stream Surfaces", *Computing Systems in Engineering*, Vol. 1, Nos. 2-4, 349-353, 1990.

[30] Globus, A., "Optimizing Particle Tracing in Unsteady Vector Fields", NASA RNR-94-001, January 1994.

[31] Hesselink, L., "Visualization of 3-D Tensor Fields", NASA-CR-201805, July 31, 1996.

[32] Helman, J.L. and Hesselink, L., "Visualizing Vector Field Topology in Fluid Flows", *IEEE Computer Graphics and Applications*, Vol. 11, 36-46, 1991.

[33] Ma, K.L. and Interrante V., "Extracting Features From 3D Unstructured Meshes for Interactive Visualization", NASA CR-201610, ICASE Report No. 96-60, pp. 13, November 1996.

[34] "Visiometrics, Juxtaposition and Modeling", *Physics Today*, Vol. 46, No. 3, 24-31, 1993.

[35] Löhner, R., "Robust, Vectorized Search Algorithms for Interpolation on Unstructured Grids", *J. Comp. Phys.*, Vol. 118, 380-387, 1995.

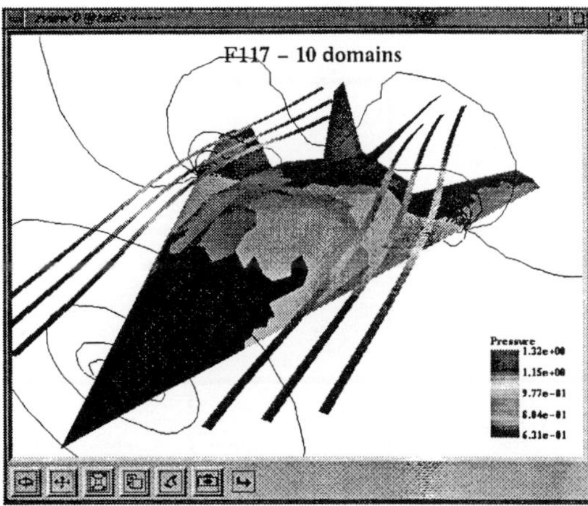

Figure 3: Parallel distributed visualization.

Figure 4: Simultaneous visualization with linked viewers.

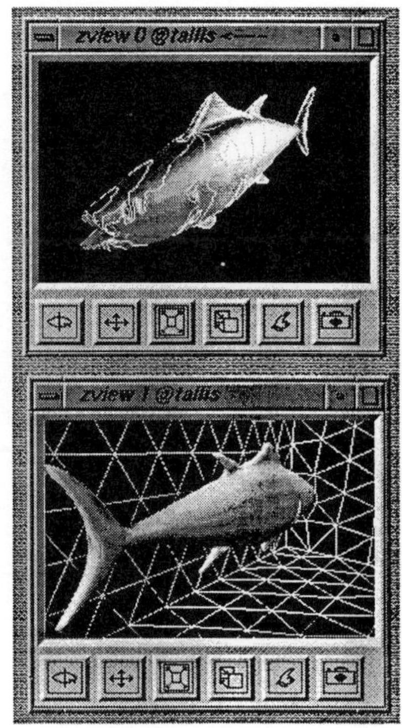

Figure 5: Simultaneous visualization with unlinked viewers.

Figure 6: Miltidisciplinary visualization.

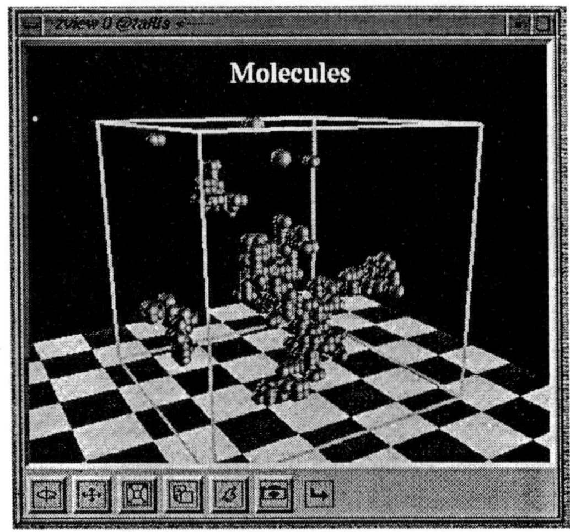

Figure 7: Particle visualization.

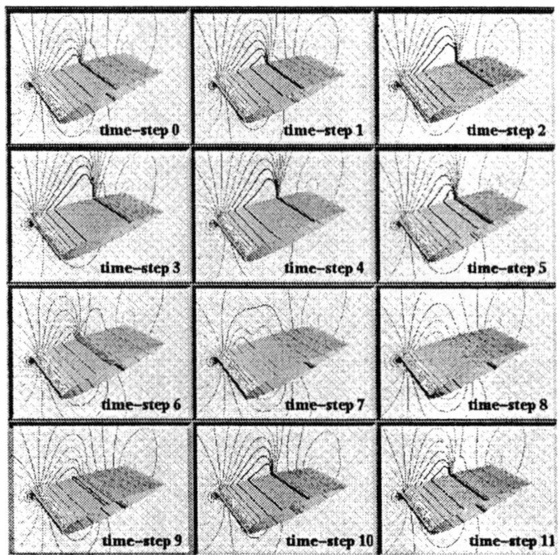

Figure 9: Interactive on-line visualization.

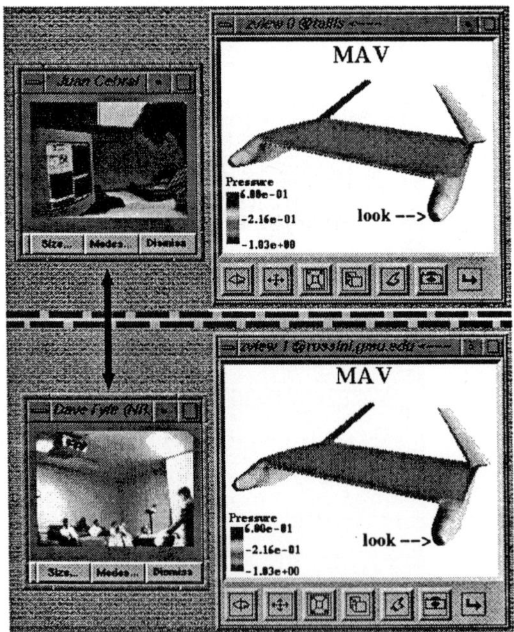

Figure 8: Collaborative visualization.

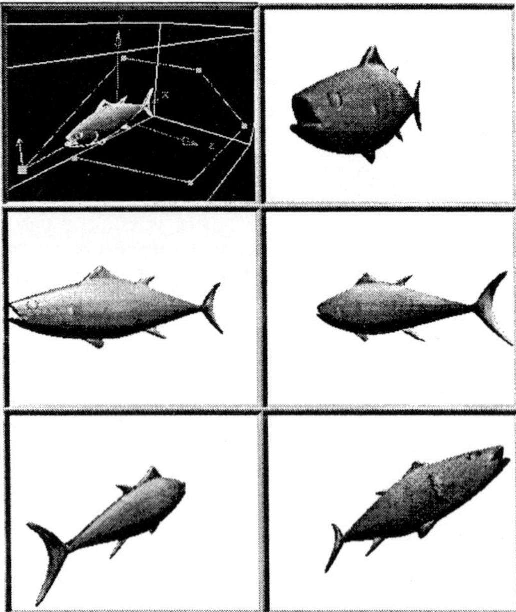

Figure 10: Navigation and movies.

THE VIRTUAL WINDTUNNEL

Steve BRYSON [2]   Sandy JOHAN [1]   Leslie SCHLECHT [2]

Brian GREEN [2]   David KENWRIGHT [2]   Michael GERALD-YAMASAKI [1]

(1) Data Analysis Branch, Numerical Aerodynamic Simulation Systems
Division, NASA Ames Research Center, Moffett Field, Ca. 94035-1000

(2) MRJ Technology Solutions

Abstract

This paper describes the virtual windtunnel, a virtual reality-based, near-real-time interactive system for CFD visualization. The virtual windtunnel supports several visualization techniques and meets several requirements, including extensibility, guaranteed performance while being very computationally intensive, time management including the maintenance of simultaneity in the virtual environment, and interface independence. Creating a framework which meets all of these requirements presented a major challenge. We describe the interaction techniques, object-oriented structure and process architecture, including interprocess communications and control. Subtle issues regarding the flow of time in the visualization environment are described and addressed. The problem of time-critical computation is discussed, and a solution is presented. Device independence of both the command and display structures are developed, providing the ability to use a wide variety of interface hardware options. The resulting framework supports a high-performance visualization environment which can be easily extended to new capabilities as desired.

1 Introduction

The virtual wind tunnel is the application of virtual reality interface techniques to the visualization of the results of computational fluid dynamics (CFD) simulations. The numerical data resulting from these simulations are typically time-dependent vector and scalar fields in three-dimensional space. CFD datasets typically contain extremely complex phenomena, involving time- and space-varying structures such as vortices, recirculation, and oscillation.

Virtual reality (sometimes referred to as virtual environments) is the use of various computer technologies including graphics, computation and interfaces to produce the effect of a three-dimensional computer-generated environment. In this environment a user can interact with objects that have a strong sense of three-dimensional spatial presence. This sense of spatial presence provides both an enhanced sense of three-dimensional spatial structure, and facilitates the ability to directly manipulate objects in three dimensions. The virtual windtunnel is an example of an immersive virtual environment, in which the user feels surrounded by the computer-generated environment.

Simultaneously meeting the requirements of general-purpose fluid flow visualization and virtual reality has proven to be very challenging. Issues of extensibility, scaling of code complexity, con-

figuration of the environment at run-time, and specified computation rates and response times all had to be addressed. These issues often result in conflicting requirements. The soft-ware frameworks that provide solutions to these problems is described in this paper. While these structures are designed for scientific visualization, they can also be used for other computationally intensive near-real-time interactive applications.

Several early prototypes of the virtual windtunnel [1][2] were developed which had very limited visualization capability. These prototypes proved the concept of virtual-reality-based visualization of simulated fluid flow, and demonstrated the advantages of three-dimensional display and interaction provided by virtual reality. Particle integration visualization techniques were implemented in both single-workstation and distributed modes of operation. Interaction was built around the VPL Dataglove Model II and display around the FakeSpace Labs BOOM family of displays. The problems addressed by these prototypes included time-varying visualization, manipulation techniques, and collaborative operation though shared distributed environments. Because these prototypes were dedicated to a single class of visualization techniques, they were very limited in their scope. These prototypes also did not address the critical issues of specified computation rate

---

Received on July 1, 1997.

and strict simultaneity of displayed phenomena.

Two fundamental weaknesses in these prototypes were identified: lack of versatility in terms of visualization options, and the requirement of the researcher to use a particular interaction hardware. Both of these weaknesses are addressed in the version of the virtual windtunnel described in this paper. Versatility in terms of visualization techniques was addressed by implementing an object-oriented structure that made it easy to add both new visualization capabilities and new visualization and environment control tools. Visualization and interface techniques that have been implemented using this framework are described in this paper and in [3]. The lack of versatility in terms of interface hardware options was addressed through the implementation of a structure that abstracts interaction and display to a layer in which it is easy to add new hardware options. Both of these types of versatility have been demonstrated by the rapid implementation of new features by individuals with no prior knowledge of the virtual windtunnel.

The current version of the virtual windtunnel is implemented in C++ on Silicon Graphics platforms and supports a variety of visualization techniques and interface hardware configurations. The virtual windtunnel has been released for evaluation purposes to two sites: NASA Langley Research Center, NASA Lewis Research Center, and NASA Goddard Space Flight Center. The response has been enthusiastic [4] and the virtual windtunnel has been used by CFD researchers to investigate their simulations. A public release expected in mid to late 1997.

This paper describes the underlying framework of the virtual windtunnel. In the next section several requirements for the virtual windtunnel are outlined. Section 3 briefly describes related work. In section 6 the object class structure of the virtual windtunnel framework is motivated and described. Section 7 describes the run-time software process structure, with particular attention being paid to process control. In section 8 the management of time in the virtual windtunnel is discussed. In section 9 time-critical computational techniques are discussed. Section 10 presents the structure that simplifies the addition of commands to the virtual windtunnel in a way that is independent of the source of that command. Section 11 describes the *device* structure, that implements interface hardware devices in a "plug and play" modular manner.

More information about the virtual windtunnel can be found at http://science.nas.nasa.gov/Software/VWT/vwt.html.

2 REQUIREMENTS FOR THE VIRTUAL WINDTUNNEL

The virtual windtunnel is at the intersection of two highly demanding applications of computer graphics: near-real-time interactive virtual environment systems and time-varying fluid flow visualization. In this section we outline the requirements of the underlying software framework described in this paper.

2.1 Requirements of Computational Fluid Dynamics Visualization

The numerical data resulting from CFD simulations are typically vector and scalar fields in three-dimensional space that change over time. Time-varying datasets are provided on computational grids, with the time evolution of the data being encoded as a series of data files. Thus there is a discrete sense of time built into the data. A given visualization task can involve the simultaneous examination of several data fields, such as pressure, density and velocity.

Fluid flow visualization typically involves the two-stage process shown in Fig. 1. Data (usually a precomputed file on disk) is processed into visualization geometry called **extracts** which are displayed using three-dimensional computer graphics. Extracts are specified by user input. While some extracts such as arrows for a vector field involve little computation, others can involve significant computation. Isosurfaces, for example, involve interpolating values on the computational grid to compute surfaces reflecting a value specified by the user. Streamlines involve the integration of a vector field starting from a point in space specified by the user. Once the extracts are computed they may be displayed with a variety of user options. In a complex flow simulation several visualization extracts may be required to exhibit interesting phenomena, as shown in Fig. 2.

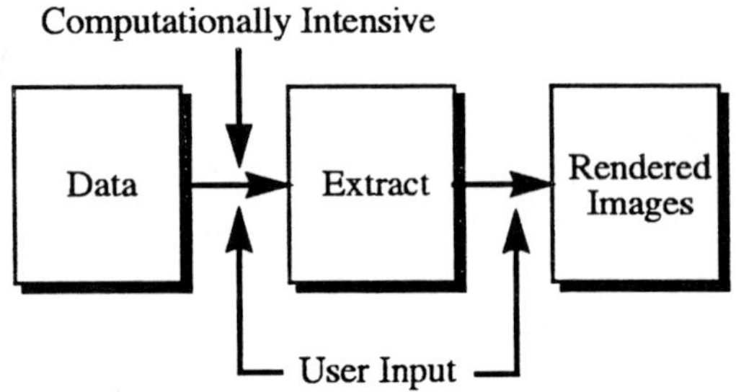

Fig. 1 The visualization process: data is processed to produce visualization extracts, that are rendered on the computer screen.

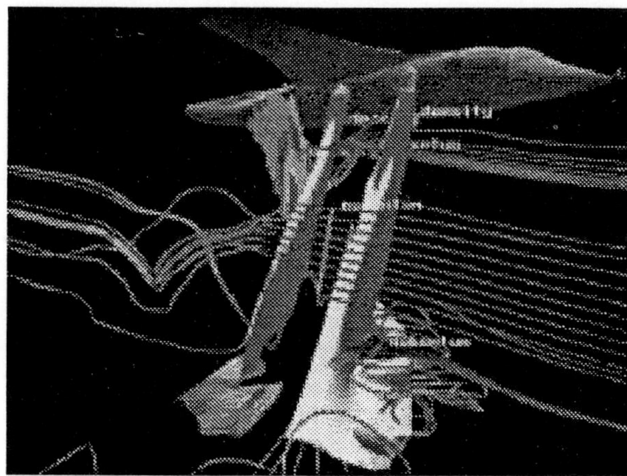

Fig. 2 An example of a complex visualization environment, showing streamlines, isosurfaces and cutting planes displaying the velocity vector field and density scalar field around a harrier aircraft in hover [6].

2.2 Requirements of Virtual Reality

Virtual reality (sometimes referred to as virtual environments) is the use of various computer technologies including graphics, computation and interfaces to produce the effect of a three-dimensional computer-generated environment. This effect is attained primarily through the use of a head-tracked display system. In the virtual environment objects have a strong sense of a location in three-dimensional space relative to the user, which we call **object spatial presence,** or simply spatial presence. Spatial presence provides enhanced perception of three-dimensional spatial structure as well as the enhanced ability to directly manipulate objects in three dimensions. These three-dimensional capabilities combine to make the exploration of complex three-dimensional structures significantly easier and faster than conventional visualization systems [5].

One of the lessons learned from the virtual windtunnel prototypes described in section 1 is that near-real-time three-dimensional interaction is the primary advantage of a virtual reality interface. Virtual reality interfaces require high performance in terms of high frame rates and low delays between user input and system response. The specific requirements will depend on the application. For applications in which interactive objects do not move except when manipulated by the user, experience has shown that a minimum frame rates of 10 frames per second and maximum latency of 0.1 seconds is tolerable. At lower frame rates, the virtual reality effect of object spatial presence is lost. At higher latencies user control of the environment becomes signifi-

cantly impaired [7]. When interactive objects themselves move in the virtual environment, higher frame rates and lower latencies are required [8].

The computation of visualization extracts easily saturates this 10 frames per second time constraint, even when the resulting visualizations can be displayed at 10 frames per second. For this reason the virtual windtunnel separates the computation and rendering of the visualization extracts into two asynchronous processes. This introduces two frame rates into the virtual windtunnel: the **graphics frame rate** and the **computation frame rate**. The graphics frame rate supports the virtual reality effect of spatial presence and the resulting interactivity. The computation frame rate produces new geometry either in response to a change in data or the change in the location of a visualization.

Experience has shown that the computational frame rate and latency requirements are somewhat more relaxed than those for the graphics. Latency in the response of a visualization to user motion can be as much as 0.2 to 0.3 seconds and still be useful, though longer latencies significantly impair the usefulness of the direct manipulation capability. The computation frame rate can be as low as 5 frames per second and still by useful so long as the 10 frames per second graphics animation rate is maintained.

As mentioned above, low frame rates impair the ability to directly manipulate objects in the environment. This is particularly true if the objects to be manipulated are moving. When the data varies with time the visualizations will in general be moving, updated at the computation frame rate. Because these moving visualizations would be difficult to manipulate, one of the design decisions of the virtual windtunnel is that user interaction be via tools that are stationary except when they are moved by the user. These tools are primarily graphics objects and so update at the graphics frame rate, providing fast user manipulation feedback. These tools, described in section 6.1, are easier to manipulate and control than the visualizations.

2.3 User-Driven Requirements

There are several higher level requirements that a general purpose visualization system must meet in order to be useful to the scientific visualization community.

- **Extensibility**: The visualization environment should be extensible in order to accommodate new visualization and interaction techniques. This extensibility should, whenever possible, be consistent with existing visualization and interaction techniques. As we will discuss throughout this paper, this extensibility is complicated by the complex list management, process communication, and user interaction logic structures in the virtual windtunnel. One of the primary motivations for the virtual windtunnel object structure is to hide these complications in high levels of the object hierarchy. In this way programmers can add new objects by following a template without having to understand the entire virtual windtunnel structure.

- **Versatility:** The user must be able to configure the environment at run-time, adding or deleting visualization and control objects at will. In addition, the user must be allowed to access any portion of the data set, at any timestep and control the flow of data timesteps.

- **Cooperative visualization:** The fluid flow community, like any scientific community, operates by cooperative investigation of phenomena. Thus a flow visualization system should support shared, cooperative visualization. This requirement is met in the virtual windtunnel either through a shared distributed environment option, or through displays options which allow several viewers at once.

- **User acceptance:** Flow researchers will use a system when the difficulties and training investment are outweighed by the advantages of the visualization system. To reduce the burden of use, the system should run with a variety of interface options, to match the user's needs, available hardware, and budget.

2.4 The Requirements of Direct Manipulation

Direct manipulation in the virtual windtunnel is through abstract static gestures (sometimes called poses) at a position and orientation in space determined by the tracking device. These gestures may be the result of button pushes or gesture recognition based on a glove device. There are three static gestures defined in the virtual windtunnel: **grab**, **point**, and **null**. The action of each gesture is dependent on the context in which the gesture is made.

Direct manipulation is based on mapping data at a position in space, usually the position of the user's hand or arrow pointer, to an action in the virtual environment. This position and orientation data must be mapped to visualizations in order to specify their extracts. Visualizations that are specified using data at a point in space are called **local visualizations**. For visualization techniques such as vectors, streamlines, and cutting planes

this is straightforward. Isosurfaces, however, are usually specified by value, without a spatial manipulation metaphor. For the virtual windtunnel, the concept of local isosurface was developed [9]. This isosurface is specified by sampling the value of a scalar field at a point in space and constructing the isosurface around that point. Using local isosurfaces the user can interactively explore the geometry of the scalar field.

3 RELATED WORK

Scientific visualization systems fall into roughly two classes: modular data-flow systems such as AVS [10] and dedicated visualization systems such as FAST [11]. The virtual windtunnel falls into the class of dedicated visualization systems.

Data flow systems allow the user to reconfigure visualization capabilities through the use of modules that act on the data. These modules typically transform the entire dataset, compute extract geometry, or render that geometry. Virtual reality interfaces have been implemented for rendering modules on these systems by some individuals, but this additional virtual reality capability only allows non-interactive viewing of the visualization results. Because data flow systems transform the entire data set, they are not well suited to the near-real-time visualization of time-varying data because of the very large amounts of data involved (though this issue is being addressed in more recent versions of data flow). For this reason, the data flow approach was not used in the virtual windtunnel.

There have been many scientific visualization systems produced that are designed for work in a particular scientific discipline. Many software architectures have been implemented in these systems, ranging from single process batch-based to distributed multi-platform visualization environments. Those systems that were designed for easy extensibility, however, were not designed for performance, while those designed for performance were programmed for a particular scientific problem rather than as a flexible system that can be used for a wide array of problems. The prototype virtual windtunnel systems were in this latter class. None of the frameworks of these systems served as a useful starting point for the production virtual windtunnel framework. Some of these dedicated systems have been implemented with an interactive virtual reality interface, such as those developed at the University of North Carolina [12][13], and particularly with the CAVE environment [14].

A system which has many of the features described in this paper was developed at Brown University [15]. This system is based on an interpreted, object oriented, multiprocessing system developed at Brown. While this system meets the extensibility requirements described in this paper, it does not have the required performance due to its interpreted nature.

Several virtual reality systems have been developed [13][16] which separate the graphics and computation process, usually by distributing these functions among several platforms.

Work related to time-critical graphics will be discussed in section 9.

4 THE USER INTERFACE

The user interface of the virtual windtunnel is based on the overall metaphor of the user being immersed in an environment which contains several objects. Some of these objects are interactive and can be created and destroyed. Other objects are created at start-up time and are not interactive. There are no objects in the virtual windtunnel system which are not part of the three-dimensional environment, though control of the system can occur via keyboard and text input.

Most operations in the virtual windtunnel are performed with a direct manipulation interface via three-dimensional trackers. There are three classes of operations:

- **Direct manipulation of objects in the environment**
- **Indirect control of the environment or objects in the environment via direct manipulation of virtual menus and sliders**
- **External commands via keyboard input or voice**

All operations are performed with the glove using only two gestures: grab (fist) and point.

4.1 Visualization Control Tools

Local visualizations (called "visualizations" in this section) are controlled through the use of **vtools**. Each vtool contains one or more **emitters**. An emitter is an object which contains a list of visualization objects and the data required to generate those visualizations. The position of each emitter is determined by the vtool containing that emitter. How an emitter controls a visualization will be described below in the visualization section.

Each vtool is assigned to a particular data field, and all visualizations on that vtool will visualize that data field. All emitters on a vtool contain the same set of visualizations. All visualizations of the same type (i.e. all streamlines or all isosurfaces) associated with a vtool share the same properties such as length, coloring and so on.

Each vtool may be grabbed and moved about in a

way special to the vtool type, described below. In all cases, the vtool can be grabbed in only specific places called **handles**. When the user's hand is near enough to one of these places to grab it with a fist gesture, a skeletal tessellated sphere appears and the vtool is said to be touched. When grabbed and moved about, the position of each emitter in the vtool is continuously updated, which in turn updates the visualizations attached to that emitter. The visualizations are continually updated by the computation process, resulting in the dynamic display of phenomena at the current vtool position.

Each vtool can display its associated data field, its position in physical coordinates and its position in computational coordinates.

At any particular time one vtool is distinguished as the "current vtool", which is drawn differently from the other vtools. All menu commands affecting vtools or visualizations apply only to the current vtool and its visualizations. A vtool can also be selected as the current vtool by pointing when touching one of its handles.

The most basic vtool type is the **sample point** (so-called to distinguish it from the point gesture). A sample point is a zero-dimensional object and is defined by a single location in space (Fig. 3). The sample point contains one emitter whose position is the position of the sample point. The sample point can grabbed and translated in space.

Fig. 3 A sample point emitting a streamline.

Another vtool type is the **rake**. A rake is a one-dimensional object defined as a straight line between two points in physical space (Fig. 4). A rake contains a variable number of emitters, which are equally spaced along the rake.

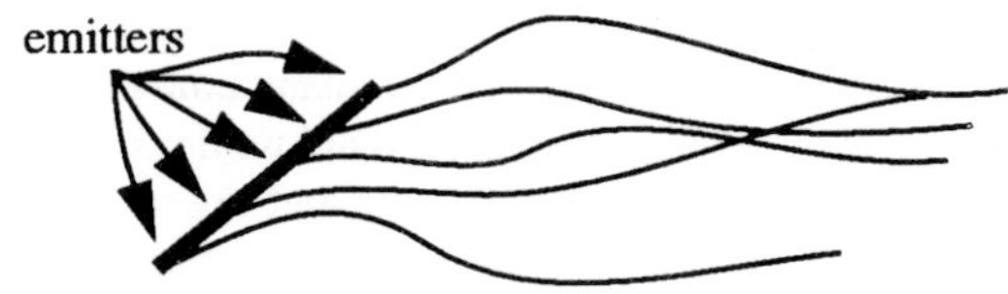

Fig. 4 A rake of 5 emitters containing streamlines.

The rake is a complex object, and may be grabbed in the middle or at one of the two ends (Fig. 5). Grabbing the rake in the middle causes the rake to rigidly translate with the user's hand. Controlling the orientation of the rake while grabbing its middle is a user-selectable option, but we find this method of orientation control to be somewhat limited. Grabbing one end of the rake causes that end of the rake to move with the user's hand while the other end remains motionless, providing control over the length and orientation of the rake.

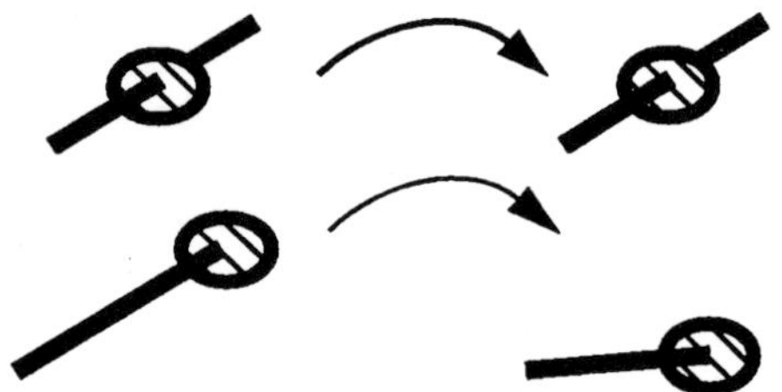

Fig. 5 Manipulating a rake. *Above:* grabbing the rake in the middle. *Below:* grabbing one end of the rake.

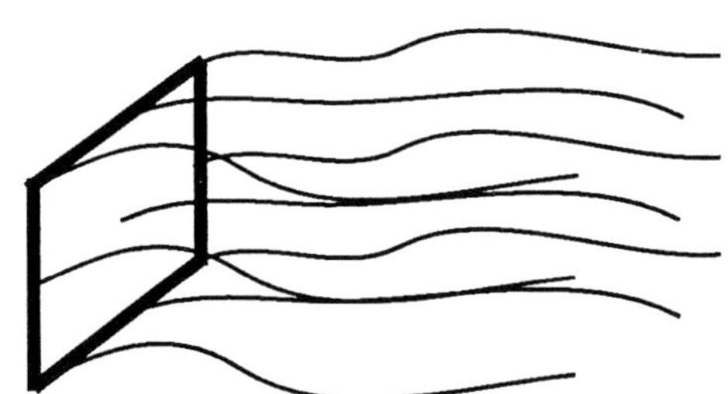

Fig. 6 A plane of 3x3 emitters containing streamlines.

Another vtool type is the **plane**. A plane is a two-dimensional object defined as the rectangle bounded by four vertices (Fig. 6). The plane contains a variable number of emitters equally spaced as an *nxn* array. A special case is made for the cutting and contour plane visualizations (see below), which have a single emitter at the center of the plane.

A plane is a complex object and may be grabbed in a variety of ways (Fig. 7): grabbing the center of the plane causes the plane to rigidly move with the user's hand. Grabbing one edge of the plane causes that edge to rigidly move with the user's hand while the opposing edge stays still, allowing the size and orientation of the plane to be controlled. Grabbing one corner of the plane allows that corner to be moved while constrained to the previous plane and while the opposing corner stays still. This allows the plane to be resized without changing its orientation.

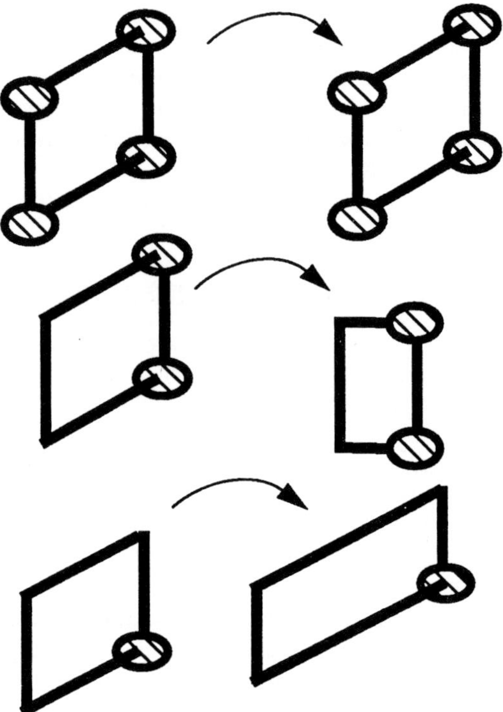

Fig. 7 Manipulating a plane. *Top:* grabbing the plane in the middle. *Middle:* grabbing one edge. *Bottom:* grabbing one corner of the plane.

The fourth vtool type is the **grid plane**. The grid plane is a face of a computational grid. It has emitters at every *n*th node of the grid. The grid plane is can be colored by the values of a scalar field as well. The grid plane is manipulated via a handle which appears as a three-dimensional crosshair on the grid plane. The choice of grid zone and grid face is determined via a menu and slider interface. When the grid plane is grabbed by its handle the plane is moved to the node nearest to the hand position with the currently selected orientation.

4.2 Other Tools

There are many aspects of the virtual windtunnel besides the placement of visualizations which are under user control. The properties of the vtools, navigation in the environment, properties of the visualizations and other aspects of the virtual environment are under user control and thus require a user interface. The tools which are used to control these properties are **menus, sliders,** and **markers.**

The virtual windtunnel uses hierarchical pop-up menus with the pointing interface developed by Jacoby and Ellis [17]. Making a point gesture in empty space in the virtual environment causes the top level menu to pop up in three-dimensional space within the user's field of view. The menu remains as long as the point gesture is held by the user. While the menu is up, the user's hand orientation information is used to point at various menu items. A line appears from the user's hand position to a cursor on the menu indicating where the user is pointing. The menu item which is pointed at it is highlighted. Releasing the point gesture selects that menu item, executing an associated callback routine.

Continuous environment parameters such as the number of emitters on a rake or plane are controlled via the manipulation of virtual **sliders**. These sliders exist in the three-dimensional environment and output values determined by the slider's thumb. The sliders can be moved by making a grab gesture in the region of the slider thumb. While the slider thumb is moved a callback routine associated with that thumb is executed with the value of the slider as argument. Sliders can contain a variable number of thumbs. Sliders come in one, two and three dimensional form, outputting one, two or three numbers respectively. Sliders are toggled on and off via the virtual menus. Sliders may be moved by grabbing their bodies away from the thumb and resized by grabbing resize handles on the slider.

The **marker** tool delineates a location in physical space. When a marker is grabbed and moved, a callback routine associated with that marker is executed with the position of the marker as the input argument.

The primary method of navigation in the virtual windtunnel is via the head-tracked display. This method of navigation is very limited in extent, however. Navigation over large distances within the environment uses a paradigm in which the user stays in one place and moves the environment about. This is accomplished by making a grab gesture in empty space and moving the entire environment with the motion of the user's hand. For large environments, scale is used to have the effect of moving the user: when the user wishes to move to another location far away in the environment, first the scale slider is used to make environment small enough so that the desired location is within reach. The "scale origin" marker, which sets the origin of the scale operation to its location when it is moved, is placed at the desired location. The scale slider is then used to make the environment large again. As the scale origin marker does not move during a scale operation the user finds himself at the desired location. As a concrete example, say that the user is at the tail of a large aircraft and wishes to examine the nose of the aircraft at the same scale. The scale slider is used to make the aircraft small so that the nose is within reach, the scale center marker is place at the nose of the aircraft, and the scale slider is

used to return the aircraft to the original scale. After these operations the user will find himself at the nose of the aircraft. This interface allows rapid and high-precision maneuvering in the environment. Tools other than vtools such as menus and sliders are not effected by this motion. The scale and grab paradigm is significantly easier to control than the point and fly navigation paradigm.

<u>Gestures</u>

The extensive use of hand gestures possibly read by the glove in the virtual windtunnel requires a robust and reliable gesture recognition algorithm. This is accomplished using only the middle joints of the four fingers. The values measured at the knuckle joints and the thumb are ignored. First the raw values output by the Dataglove are calibrated to actual finger bend angles. Then the angles read by the glove are compared with a lookup table, to identify if the gesture is either a fist or a point. Recognizing only three gestures (fist, point, no gesture) allows forgiving and tolerant gesture recognition. With this gesture recognition algorithm, new users require almost no training to make gestures reliably. The glove is calibrated to each user before the system is started.

5 VISUALIZATIONS

The virtual windtunnel currently supports the following visualizations:

- streamlines
- streaklines
- particle paths
- isosurfaces
- cutting planes
- contour planes
- tufts
- numerical display
- precomputed geometry in the ARCGRAPH format

Most of these visualizations are local in nature in the sense that they emanate from a point in space called a **seed point** and propagate outwards in some manner. This property of locality allows for simple user control metaphors (direct placement of the seed location) and for well-defined methods for limiting the time taken up be each visualization (through limiting the extent of propagation). Local visualizations are seeded via a seed-data object, which contains the data defining the visualization. The values in the seeddata for a given visualization are set by the emitter controlling that visualization.

Streamlines, streaklines and **particle paths** are particle advection techniques which were implemented in the original prototype of the virtual windtunnel [1].

For these three visualizations the seeddata contains the position of the emitter controlling this visualization. In an unsteady flow the vector field is taken at the vector field at the current timestep, so these streamlines are also known as "instantaneous streamlines". Streamlines are bidirectional or unidirectional at the user's choice. Streaklines are the single integration of particles along a vector field per frame, with new particles continually introduced at the seedpoint. Particle paths are the path of a particle introduced at the seed location in a time-varying vector field. For all particle advection visualizations, the amount of propagation, which determines the time required for computation, is determined by either a time allotment or a limitation on the number of integration steps in the case of streamlines or particle paths and particles in the case of streaklines. All particle advection is based on an adaptive integration method developed in [18]

Isosurfaces, cutting planes and **contour planes** are different manifestations of local isosurfaces [9]. A local isosurface for a scalar field is defined by seeddata which contains a location in space. The value of the scalar field is sampled at that point. We shall call the resulting scalar value the defining value for that isosurface. For the current computational cell a polygon is computed as a linear approximation to the isosurface of that value through linear interpolation on the cell edges, much as in the marching cubes algorithm. This algorithm is iterated for every computational cell which shares a face containing the defining scalar value. In this way the isosurface is propagated out from the seed point. This algorithm will generate only the connected component of the isosurface through the seed point, plus those components which happen to share a computational cell with the component through the seedpoint. This algorithm allows for the intuitive use of isosurfaces to interrogate the geometry of a scalar field. This algorithm also facilitates control over the time of computation and rendering by stopping the iteration when either a specified amount of time has passed or a specified number of polygons have been generated.

Cutting and contour planes are special cases of a local isosurface: in both cases the geometry of the plane is defined as a zero-value isosurface of the scalar field defined as the distance from the plane. Thus for cutting and contour planes the seeddata contain both the location of the seed point and a normal defined in different ways by different vtools. For the sample point vtool the normal is determined by an explicit user choice of one of the coordinate planes via the menu interface. For rake vtool the normal is given as a unit vector in the direction of the rake, so that the resulting plane is perpendicular to the rake. For the plane tool the normal is simply the nor-mal to the plane. Once the geometry of the plane is generated it is rendered either via a color mapping of the selected scalar field in the case of the cutting plane or via contour lines of the selected scalar field in the case of the contour plane.

Tufts are lines drawn from the seed position representing the value of a vector field. The direction and length of the line are determined by the vector value at the seed point. The scale of all tufts is controlled via a slider interface.

Numerical visualizations are numbers drawn near the seedpoint representing the value of the data at that point. How many numbers are drawn is determined by the type of data being visualized: one number for scalar values, three numbers for (three-dimensional) vector values.

All of the above visualizations except the cutting plane can be colored by a scalar field which may be the same as or different from the field defining that visualization. Thus one visualization can display two data fields.

Some of the above visualization techniques such as isosurfaces and cutting and contour planes work only for scalar fields, while others such as tufts and the particle advection techniques work only for vector fields. When a scalar visualization technique is applied to a vector field, the scalar technique visualizes the magnitude of that vector field by default. When a vector technique is applied to a scalar field the gradient of that scalar field is visualized by default. While these are the only derived field choices currently available we anticipate other derived choices such as vorticity and gradient magnitude in the future.

There are visualization techniques which are not local and do not follow the above discussion. One of these techniques is the grid plane described above in the visualization tools section, which is really a vtool color mapped via a scalar field. The other is the ability to import three-dimensional geometry in the NASA Ames ARCGRAPH format. This capability has been used to combine such feature detection techniques as vortex core identification [19] with interactive visualization. In an unsteady data set there is one set of ARCGRAPH geometry for each time step.

6 CLASS STRUCTURE

The primary motivations for the class structure in the virtual windtunnel are the ability to add new interface and visualization objects without either effecting or necessarily understanding the entire virtual windtunnel system, and to scale, that is to allow an unlimited number of new objects to be inserted into the system without having the software collapse from excessive complexity. Several individuals, ranging from sophomore summer students to experienced professionals, who have no previous knowledge of the virtual windtunnel have added subclasses of both vtool and visualization after only a week's worth of effort, proving the extensibility of this class structure.

6.1 Environment Objects and the Environment List

The virtual windtunnel is conceptualized as an environment that contains objects. All of these environment objects must know how to render themselves and may have computational tasks. This motivates a class hierarchy, shown in Fig. 8, with the class **envobj** for environment objects at the highest level. An envobj contains identifier information, *draw* and *compute* member functions, as well as information to support the time-critical functions described in section 9. The envobj class also contains the *find* member function, as well as the *grab* and *point* member functions that respond to user gestures as described in section 6.3. Even though there are environment objects, such as visualizations, whose interactivity is not currently used, the *find*, *grab* and *point* functions are defined at the envobj level because other applications of this framework may allow all environment objects to be interactive.

The envobj class is the parent of two subclasses: **tool** and **visualization.** The tool class is the parent of such object classes as menus, sliders, markers, and the visualization control tools described in section 6.2. Unlike conventional graphical user interfaces, interface tools such as menus and sliders appear inside the virtual environment. These tools are described more fully in [3] and will not be discussed in this paper.

Envobjs in the environment are managed through the **envlist** object, that contains lists, implemented as arrays, of all environment objects. It is the envlist object which iterates the environment objects, causing them to be computed, drawn, and, in the case of tools, to be found by the user. There are two primary lists in the envlist class: a list of all environment objects and a list of all tool objects. The list of environment objects is used to cause each object to compute its state in a compute traversal and draw itself in a draw traversal. The list of tool objects is used for the user search traversal to determine if the user is interacting with that tool, as described in section 6.3. The reason for maintaining a separate tools list is that there are typically many more visualization objects than tool objects. Restricting the search to the tools objects improves the search time. Insertion into the envobj list is handled by the envobj constructor calling an envlist member function. Insertion into the tool list is handled similarly by the tool constructor. In this way when a new subclass of visualization or tool is implemented in the virtual windtunnel it will be properly inserted into the appropriate lists.

6.2 Visualizations and Visualization Tools

As described in section 2.2, the virtual windtunnel class structure is designed with the philosophy that visualizations are not manipulated directly, but rather through visualization tools, or **vtools.** Vtools can be spa-

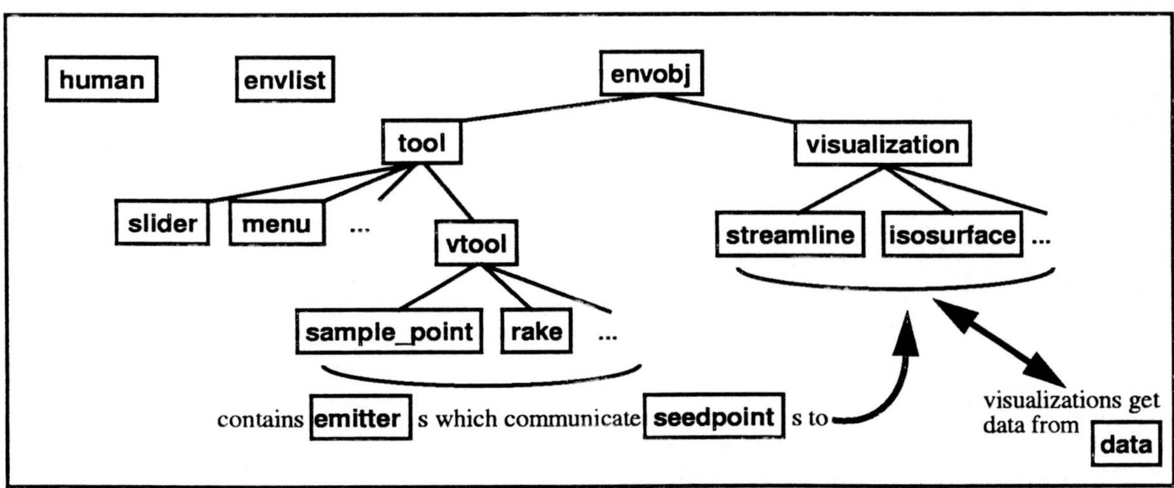

Fig. 8 The object hierarchy of the virtual windtunnel. Sample_point and rake are example subclasses of vtool, while streamline and isosurface are example subclasses of visualization. The envlist object maintains lists of the envobjs in the environment, and the human object manipulates objects through the envlist.

tially extended objects, containing groups of visualizations which are manipulated at one time.

There are two types of objects defined for the communication of data from vtools to visualizations. An **emitter** is a class which specifies a set of visualizations at a point in space. The emitter class contains a list of visualization objects, and a **seedpoint**. The seedpoint class contains all the data at a point in space necessary to specify a local visualization. A vtool contains one or more emitters. When a vtool is moved, the seedpoints contained in each of that vtool's emitters is set to the new position of that emitter. The emitters then inform all of their visualizations of the new seedpoint. Several visualizations can be displayed on the same vtool. Local visualizations contain an identifier of the data field to be visualized, a seedpoint, and a function which takes that data field and seedpoint as input and outputs visualization extract geometry.

The emitter structures within a vtool allows one vtool to contain emitters with different sets of visualizations, as well as having these visualizations display different data fields. A design decision was made to specify the visualization content and data field at the level of the vtool. Thus all emitters in a vtool have the same visualizations displaying the same data.

With this vtool/local visualization structure, new local visualization techniques can be implemented using a template of move, compute and draw functions. When implemented in this way the new visualization subclass will automatically work with all existing vtools. Similarly, a new vtool can be defined containing a set of emitters and will automatically work with all implemented local visualizations. In this way easy extensibility is achieved for any visualization technique that uses local spatial data as input. No knowledge of the rest of the virtual windtunnel structure such as the nature of the process structure, list management, or interaction logic is required.

6.3 User Interaction

User data, including head and hand tracker position and orientation data, gesture, and environment transformations such as scale are encapsulated in a **human** object. The human object controls the interaction with tools in the environment. Inside the human object there is a function pointer called *doit* which is executed once per graphics frame. The default content of *doit* is a pointer to a search function (contained in the envlist class) which traverses the list of tools and passes the human object to each tool. If the tool returns the message that the user is interacting with it, *doit* is set to a pointer to either the *point* or *grab* function of that tool as appropriate to the current human gesture data, and the

pointer to the tool is stored. The *grab* or *point* function of the tool is then executed once per graphics frame, with the human object as input data.

The C++ syntax of this operation is sufficiently obtuse to warrant explicit mention. Using the *grab* function of an object as an example, the function *get_grab_pointer* takes the function pointer *doit* as an argument and sets it to a pointer to the object's *grab* function:

Inside the human class doit and the gesture variables are declared:

```
void (envobj::*doit) (human *being)
int old_gesture;
int current_gesture;
```

Inside the envobj class the grab functions are defined

```
virtual void grab(human *);
void get_grab_pointer(int
                (envobj::*(&f)) (human *)
{ f = &envobj::grab; }
```

In the envlist class, when envlist determines that an object has been grabbed, get_grab_pointer is called a pointer to the object is stored in curobj.

```
human *being;
envobj *toollist[];

toollist[found_object]->
        get_grab_pointer(being->doit);
being->curobj = toollist[found_object];
being->old_gesture = being-
>current_gesture;
```

In the human class doit is executed by:

```
(curobj->*doit) (this);
```

The result of this example is that a pointer to the tool's *grab* function is placed into *doit*. This grab function is executed once per graphics frame, moving the tool in a way determined by the tool and the human object's hand tracker data. *Doit* is set back to the default function pointer whenever there is a change in the gesture data in the human object. Using this structure there is no need to maintain state information about which object is being manipulated beyond the contents of *doit* and *curobj*.

7 RUN-TIME SOFTWARE ARCHITECTURE

The run-time software architecture of the virtual windtunnel is designed to support both consistent high rendering rates and large amounts of computation. This architecture consists of two groups of processes, reflecting the difference in times scales between the rendering and computation tasks described in section 2. There is a graphics process group executing the draw functions of the environment objects, and a computation process group executing the compute functions, both executing asynchronously from each other. Both of these groups access environment objects through the envlist class as described in section 6.1, leading to the requirement of process locking particularly during object creation or deletion. In addition, the results of the computational process must be communicated to the rendering processes respecting the requirements outlined in section 8. These processes are outlined in Fig. 9.

This structure fits with a client-server distributed architecture, which allows shared environments for collaboration. The computational process group is on a server system and the graphics process group acts as the client. This is essentially the architecture implemented in the distributed virtual windtunnel [2].

Graphics Process Group

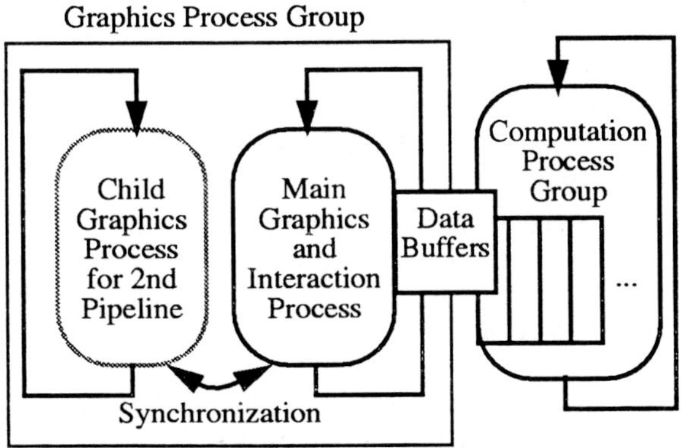

Fig. 9 The computational processes of the virtual windtunnel, including the optional child graphics process created when there is a second graphics pipeline used for stereoscopic display. The computation process creates several parallel subprocesses that call the compute functions of the environment objects.

7.1 The Graphics Process Group

When supported by the graphics hardware, the virtual windtunnel uses two graphics pipelines to produce visual stereoscopic images, one pipeline for each eye. The virtual windtunnel uses the Silicon Graphics OpenGL library, which allows lightweight shared-memory processes.

The graphics processes are synchronized through the use of a pair of flags in shared memory. At the end of a draw cycle, each process does not proceed unless the other process has indicated it has also completed its draw cycle.

The parent graphics process (which is also the parent process of the virtual windtunnel) handles user interaction, polling the interface devices and calling the human's interact function. Changes in the environment state, including the creation and deletion of objects are executed by this function.

7.2 The Computation Process Group

The computation process group is designed to take advantage of available multiple processors. It is comprised of one lightweight process which executes the environment objects' compute function in parallel using the SGI *m_fork* family of functions. The environment object list is traversed in parallel with each processor taking charge of an object in turn. Because the object list is implemented as an array, the parallel execution of this list is straightforward.

Environment objects only recompute themselves when their seedpoint has changed or when the data to be displayed has changed due to, for example, time variation. The objects themselves determine whether or not they require computation at the start of their compute functions, so all objects' compute functions are called by the computation process.

7.3 Process Locking

When objects are created and deleted in the parent graphics process these objects must be locked so that they are not being accessed by the computation process. Due to the implementation of the environment object lists as arrays in the envlist object, locking of the entire array is required. When this lock is requested by the graphics process, a flag is set which causes the computation process to stop. There are two reasons for stopping the computation process. The first is that some user interactions involve continuous parameter changes which create and destroy objects. An example is the selection of the number of emitters on a vtool, which is continuously controlled by a slider and results in the creation and destruction of many objects in several successive frames. If the computation process is allowed to reacquire the array lock between graphics frames, the continuous control becomes very jerky as the graphics process waits to reacquire the array lock. The second reason for stopping computation is that the array lock request by the graphics process is an interrogation which returns for processing as described in section 8. The very small time window between when the computation process releases and reacquires the array lock would often cause the graphics process' lock request to be missed it the computation process were not stopped. When the interaction is completed the computational process it told to proceed.

8 TIME MANAGEMENT

This section describes the time management structure of the virtual windtunnel, which control the flow of data time. After outlining the requirements of the time management structure, we give the time flow algorithm which satisfies these requirements. These algorithms are deeply related to the process structure described in section 7.

As mentioned in section 2, the way in which time-varying data is represented implies that time in the visualization environment has a discrete nature. This sense of time, which we shall call **data time,** is distinct from the user's sense of time, which is continuous. The naive way to implement data time is to map each timestep to a single pass through the computation process, i.e. a single traversal of the environment object's computation functions. Then the rate at which data time passes is clocked by the speed at which the environment object's computation functions can be executed. A somewhat richer time structure is implemented in the virtual windtunnel, which allows time flow to be manipulated by the user and allows exploration even when time is stopped.

The asynchronous nature of the virtual windtunnel process structure raises several issues about how the results of the computations should be provided to the graphics process. There are several requirements that must be met:

- **Minimal Delay:** The most recent computation data should be available to the graphics process.

- **Avoid Blocking the Computation Process:** The computation process should not have to wait for available buffers when consistent with other requirements.

- **Completeness:** Visualization extract geometry from any timestep which was computed in the computation process must be displayed.

- **Directed Flow:** The timesteps must be displayed in the order determined by the user specified time flow.

- **Correct Simultaneity:** All visualizations displayed at one time must be from the same data time value.

- **Time Flow Control:** The user should be able to change the flow of time, making it speed up (= allow the computation process to skip time steps), slow down (allow the same timestep to be addressed several times by the computation process), stopped, or run backwards.

- **Allow Exploration within a Timestep:** When the user has slowed or stopped the flow of time, visualizations should be recomputed by the computation process when they are moved.

The first two requirements, avoiding delays and blocking of the computational process, is accomplished via a triple-buffer structure called a **geometry buffer.** When a geometry buffer is filled by the computation process it is marked computed, so that it may be drawn. The other possible states are: available and busy. The usual operation of the triple-buffer structure, that would allow the faster process simply alternate between the two buffers not used by the slower process, contradicts the other requirements as described below.

The requirement of allowing exploration within a timestep implies the ability to compute a subset of all the visualization objects without computing the ones that have not changed. This in turn implies that the geometry buffers must be defined on the object level. The buffers of some objects may be changing while the buffers of others are static.

The requirements of completeness, directed time flow, and correct simultaneity are implemented by defining data time counters for each of the computation and graphics processes:

- **Computation Data Time:** The time value of the current data timestep being used for the computation process.

- **Graphics Data Time:** The time value of the data timestep that was used for the computation of the visualizations currently being displayed.

The computation processes contain a global computation time value and the graphics processes contain a global graphics data time value. Each buffer of the triple buffer structure mentioned above contains a data time value set by the computation process. Simultaneity at display time is then enforced by the draw process selecting a buffer with a data time value equal to the global graphics data time value.

Time flow control is implemented as a proportional scaling: current computation data time is incremented by the time rate. The time rate is a floating point number and may be positive, negative, or zero. User selected time flow control implies that we cannot base the requirement of directed time flow on an increasing data time counter. For this reason a new counter is introduced, which is always increasing.

- **Data Frame Time:** A counter that is incremented whenever the computation data time changes.

Data frame time essentially encodes the timeliness of the data. The computation and graphics processes contain global values of the data frame time. The data frame time is also contained in the geometry buffers.

None of the above counters, however, are incremented if the data time flow is slowed or stopped. When data time flow is slowed, for example, several executions of the computation process may take place before the computation data time and the data frame time change. Another counter is required to maintain synchronization within a computation frame, which may be thought of as a "fractional part" of the data frame time.

- **In Frame Time:** A counter that is incremented every time all of the environment objects are computed. This counter is set to zero every time the data frame time is incremented.

The problem of meeting the time management requirements is the implemented by the following algorithms. Let b_drawing = the buffer currently being drawn, and b1 and b2 be the other two buffers.

First we present the algorithm for finding the next buffer for computation. This algorithm is essentially looking for a buffer that can be overwritten without violating the completeness requirement.

First return any available buffer:

```
if (any buffer is marked AVAILABLE)
        return that buffer
```

Second, check to see if any buffers are from data frame times smaller than the data frame time that is currently being drawn. Such buffers correspond to data frame times that have already been drawn:

```
if (b1's data frame time < b_drawing's data frame time)
        return b1
if (b2's data frame time < b_drawing's data frame time)
        return b2
```

If we have not yet returned from this algorithm, the data frame times of b1, and b2 are greater than or equal to that of b_drawing. The requirement of completeness demands that we do not skip any values of the data frame time. If b1 and b2's data frame times are equal, pick the one with the smallest in time frame as the other will draw that data time frame:

```
if (b1's data frame time = b2's data frame time)
{
        if (b1's in frame time < b2's in frame time)
                return b1
        else
                return b2
}
```

If b1's or b2's in frame times are equal to b_drawing's in frame time, pick it, as that data time frame is being drawn:

```
if (b1's data in frame time = b_drawing's data in frame time)
        return b1
if (b2's data in frame time = b_drawing's data in frame time)
        return b2
```

If we have reached this point in the algorithm returning a buffer may cause the overwrite of a data frame time, violating the requirement of completeness. So:

```
return NO_BUFFER_AVAILABLE
```

The computational process waits until a buffer becomes available. Buffers become available after they have been drawn and released by the graphics process. This introduces a potential deadlock condition with the array lock described in section 7.3: the computation process has acquired the array lock and will not release it until a geometry buffer becomes available. The graphics process, in the meantime, may have completed its drawing and may have requested the array lock in response to a user command as described in section 7.3. At this point there is no drawing taking place in the graphics process, so no buffers will become available to the computation process. This deadlock condition is addressed by having the graphics process interrogate the lock using the IRIX *uscsetlock* call which returns a value indicating whether the lock is available. The graphics process repeatedly requests this lock for a short period of time. If the lock does not become available in that time, the command is archived for execution in the next graphics frame and drawing is resumed.

Selection of the draw buffer is much more straightforward. Return the buffer that has been computed, has data frame time = to the global graphics data frame time, and the largest value of in frame time. The largest value of in frame time indicates the most recently computed buffer, and is chosen to minimize frame latency. If no new buffer is available for drawing, then the current draw buffer is returned to be drawn again.

The time management structure described in this section is implemented by defining the **geom_buffer** template class which contains the buffers and buffer selection functions. The geom_buffer class contains an array of the template class specified at declaration. Then for each visualization class, a **geom_data** class is defined which encapsulates the extract geometry of that visualization. A geom_buffer object is then declared in the visualization class which takes the geom_data class as its template instantiation. The C++ syntax for this operation is:

```
class geom_data
{
        /* geometry data structures */
};


class visualization_subclass : visualiza-
tion
{
        geom_buffer<geom_data> geometry;
        /* everything else */
} ;
```

In the computation function for this visualization subclass, the computational buffer must be acquired for use:

```
/* declare a geom_data reference pointer
*/
geom_data &comp_geom_buf;

/* set to the current computational buffer
*/
comp_geom_buf = geometry.comp_buffer();

/* fill it in */
comp_geom_buf.data = value;
```

The draw buffer in the object's draw function is acquired in a similar way.

Using this template structure hides the operation of the geom_buf object. This allows a programmer to implement visualizations using the time management structure without having to understand its details.

9 TIME-CRITICAL COMPUTATION

In this section we address the issue of controlling the time required to compute and display the visualization objects. This problem is at the heart of the conflict between the requirement of accurate visualization on the one hand and high performance for virtual reality on the other. Many researchers have indicated that when interactively exploring a data set they would rather have a fast, less accurate answer than a slow accurate answer. Once interesting phenomena has been identified using a fast but inaccurate method, that phenomena can be verified with a slower but more accurate method. The degradation of accuracy must be implemented very delicately, respecting the needs of the researcher.

We first argue that the time-critical graphics problem is less severe than the time-critical computation problem. We then turn our attention to the computational problem, discussing ways to determine how much time a computation is allowed to take. Once a computational budget is established for an object we give examples of the parameterization of object computation that allow that budget to be automatically met. Finally, the results of a simple test of this time-critical structure is presented.

9.1 Time Critical Graphics vs. Time Critical Computation

Because scientific visualization geometry represents abstract data, photorealistic rendering is not required. This observation allows the choice of simple, fast rendering techniques for the visualization extracts: streamlines may be rendered as lines and isosurfaces may be rendered via Gouraud-shaded triangles. Further, almost all of the graphics in a visualization environment arise due to computation. For many visualization techniques the computational time cost per generated vertex (= time required to compute the visualization extract divided by the number of vertices in that extract) is lower than the graphical time cost of that vertex (= time required to draw the visualization extract divided by the number of vertices in that extract). Thus if the computation of an extract is optimized for a particular computation frame time, the resulting extract will be able to be rendered within a shorter graphics frame time. Further, once a visualization extract is computed it is extremely risky to graphically optimize the extract geometry: such an optimization may introduce misleading artifacts into the visualization, distorting the analysis of the data.

The only exception to these observations is the visualization geometry based on faces of the computational grid. These visualization techniques require little computational cost per vertex. If it is determined that the graphics frame rate is too long, it is a simple matter to subsample these visualization geometries. This subsampling must be designed with great care, however, as many computational grids are designed with the minimally tolerable spatial subsampling already built in.

Thus an upper limit on the amount of subsampling allowed must be implemented to avoid losing interesting phenomena. This limit should be user selectable.

9.2 Time Critical Computation

Controlling the time required for computation is a much more difficult problem. Once a total time for a computational frame has been specified by the user, there are two issues that must be addressed: determination of the how much of that total time each individual visualization computation is allowed, called the **time budget**, and modifying the computation to meet that time budget.

9.2.1 Determining the Time Budget of a Visualization

The determination of the computational time budget is the most difficult. Funkhouser and Sequin [20] developed a method that maximized benefit/cost ratio for the graphical rendering of all objects in an environment. The cost of an object was the graphical rendering time for that object at a given level of detail. Each object in the environment has several pre-computed representations at varying levels of detail. The benefit of the object is more complicated, involving the apparent size of the object on the screen, the accuracy of the representation of the object at its current level of detail, and how close the object is to the center of the screen, among other measures. The benefit/cost ratio is maximized through an iterative algorithm.

For time-varying scientific visualization, the estimate of the benefit of a visualization is more difficult. Measures like screen size and location in the viewer's visual field require the computation of the entire visualization: many visualization objects, such as streamlines and isosurfaces, may be objects that extend well away from their seedpoints. The location of a visualization in the user's field of view cannot be inferred from its seedpoint. Coherency from frame to frame cannot be used for time-varying data or when the visualization is moved, because the visualization may change drastically. A streamline, for example, may completely change direction when moved near a critical point in the vector field. This issue does not arise in non-time-varying flows for visualizations that are not moved, because these visualizations will not be recomputed.

For these reasons we compute the time budget for each visualization object through simple arithmetic. The entire computational frame is given a frame time budget, which is distributed across the visualizations. The actual time of the computational frame is measured. The ratio of the frame time budget to the actual time is computed. This ratio then multiplies the time budgets of the indi-

vidual visualization objects. This scenario is somewhat complicated by the possibility of radical changes in the computational time: the computation time can effectively drop to zero when time flow is slowed down or stopped and no visualizations are moved, or when computation is stopped during a user command. The computational time may also be very long for user-selected events such as the computation of a very large isosurface. These situations are handled by filtering the actual times, ignoring times that are too long or short (we currently ignore times that are a factor of 10 too long or short), and by clamping times between a factor of 2 and 10 too long or short to a factor of 2.

The user may specify that the visualization objects on a particular vtool be given a longer time budget at the expense of other visualizations in the environment. This is specified by a simple scale factor controlled by a slider which multiplies the time budgets of all the visualizations on that vtool. In this way the user can specify that a vtool is of more interest than other vtools.

The way in which the computational time budget is distributed among the visualization objects requires some care. If all visualization objects were of the same type, then each object's budget would simply be the computational time budget multiplied by the number of processors divided by the number of objects. Some visualizations, however, inherently take longer to compute interesting geometries than others. Isosurfaces require more time than streamlines, for example, and there are typically many more streamlines than isosurfaces in an environment. Thus an individual isosurfaces should be given a longer time budget than an individual streamline. Exactly how much longer depends on the visualization and is the subject of ongoing experimentation: we find that an increase of about a factor of three seems to give good results.

9.2.2 Meeting the Computational Time Budget

Once the time budget of a visualization has been set, that time budget is met in ways that depend on how that visualization is computed. A simple example is that of local isosurfaces, where there is a fixed cost per polygon so the only way to control the time required to compute the isosurface is to control the number of polygons. Because local isosurfaces are computed by growing out from the seedpoint, one simply computes isosurface triangles until the time budget is used up.

Particle integration techniques like streamlines offer a richer set of possibilities. The virtual windtunnel uses an adaptive particle integration technique [18] which offers three parameters that affect the computation speed. The simplest is the length of the streamline: compute the streamline until the time budget is used up.

The second parameter is to switch the integration from fourth-order Runge-Kutta, which requires four interpolations of the vector field, to second-order Runge-Kutta, which requires two interpolations of the vector field. This switch results in about an 80% speedup in the integration at a slight cost of accuracy. The third parameter is to switch off the adaptive computation, which results in an almost three-fold speedup but at the cost of a dramatic loss of accuracy. These two latter parameters are used when the resulting streamline has too few vertices. Thus the algorithm for parameterizing the streamline computations is:

```
if (last computation's vertex count < minimum number)
{
        if (using 4th-order Runge-Kutta)
                switch to 2nd-order Runge-Kutta
        else if (adaptive)
                turn off adaptive
}
while (time < time_budget)
        compute vertices
```

When the time budget increases, the adaptivity or 4th-order Runge-Kutta are switched back on when the number of vertices in the streamline reaches more than three times the minimum number of vertices.

9.3 Time-Critical Computation Results

The time-critical structure of the virtual windtunnel was tested by averaging the computational frame rate of several streamlines over 400 timesteps of a time-varying flow. The response to a shorter computation budget was to reduce the number of vertices in the streamline. The graphics frame rate for this measurement was 15 frames/second, so only lower computational frame rates were specified. The results are shown in table 1. We believe that the failure to meet the specified frame rate of 12.5 frames per second is due to the graphics frame rate overhead.

Table 1: Time Critical Results

Specified Frame Rate (frames/second)	Actual Frame Rate (frames/second)
5	5.6
6.7	6.89

Table 1: Time Critical Results

Specified Frame Rate (frames/second)	Actual Frame Rate (frames/second)
10	9.76
12.5	10.5

10 THE COMMAND OBJECT CLASS

Commands in the virtual windtunnel can come from many disparate sources including user gestures, menu commands, slider callbacks, start text scripts, keyboard input, and network messages. In the future we envision commands from voice recognition. In order to simplify the addition of new commands, the **command** object was created. The command object contains a name field and two overloaded functions: *SetValue* which executes the command and *GetValue* which returns information about the state determined by the command. These functions are overloaded by argument so that the same command can be executed with several different argument protocols. An example is the command which sets the scale of the environment. There are two versions of *SetValue* implemented for this command with two types of arguments: one which takes the scale value directly, and one which takes a pointer to a string of text from which the scale value is to be read. The *GetValue* function for the scale command returns the current scale value.

Commands are accessed by the actuator class, that contains a command pointer, set at creation time using the command's name. Actuators are contained in menus, sliders, or other command-generating objects.

Several commands determine the state of a vtool or the visualizations on that vtool. These commands access the selected tool through a global vtool pointer *current_vtool*. The value of current_vtool is set by the user through a variety of vtool selection methods.

11 INTERFACE HARDWARE INDEPENDENCE

As described in section 2, the virtual windtunnel is required to support several user interfaces, to allow the user to use the interface hardware that is available. A variety of interface hardware is currently supported by the virtual windtunnel, including the FakeSpace family of displays, stereoscopic projection screens, and conventional workstations, and several types of gloves and the conventional mouse for input. The ability to rapidly add new devices is key to the user acceptance of the virtual windtunnel. The effectiveness of the approach described in this section was demonstrated by the addition of new display and interaction hardware by the graphics group at Stanford University with no prior knowledge of the virtual windtunnel.

This versatility is implemented for user input by abstracting input devices as providers of tracker data: a head or hand tracker is defined in the virtual windtunnel as a piece of software that provides a position as a two- or three- dimensional vector and an orientation. A gesture tracker is defined as software what returns the current user gesture. Software is then written for each hardware interface that provides that data using a standard protocol. The hand tracker may be three dimensional for position and orientation trackers, or it may be two dimensional as in the case of the conventional mouse. The tool subclasses then implement their *find* and *grab* member functions (see section 6) in both two- and three- dimensional versions. The conventional mouse may also be used as a logical head tracker to provide view control when a three-dimensional head-tracked display is not available. In this case mouse motions are converted into orientation data that is passed to the virtual windtunnel.

Display independence is implemented by providing a single display function which takes a pointer to a graphics function as an argument. The display function then calls the rendering function in a way appropriate to the display hardware.

These display and input functions are implemented as switch statements in a support library function. Adding an interface device involves adding the code that handles the device hardware, and converts its data into the required format, to the library, then adding that device to the switch statements as appropriate. The virtual windtunnel code itself knows nothing about the devices being used except the dimensionality of the hand tracker device.

The devices that are used in a particular user session of the virtual windtunnel are specified at start-up time by an ascii file.

12 ACKNOWLEDGMENTS

We would also like to thank the NAS management for their continued support of the virtual windtunnel.

Brook Connor of Brown University and Al Globus of NASA Ames provided many helpful insights in the design of the object structure for the virtual windtunnel. Andries van Dam of Brown University has been a continual source of excellent summer students.

The virtual windtunnel has had many participants.

Creon Levit and the first author designed the initial concept and prototype. The distributed architecture was originally developed by Michael Gerald-Yamasaki and the first author. Jeff Hultquist contributed computational code to the original prototype. Rick Jacoby, Diglio Simoni, Ken Lao, and Currier McEwen contributed interface code. Tom Meyer, Han-Wei Shen, and David Kao have contributed visualization techniques.

13 REFERENCES

[1] Bryson, S. and Levit, C.," The Virtual Wind Tunnel: An Environment for the Exploration of Three Dimensional Unsteady Flows", *Proceedings of Visualization '91 San* Diego, Ca, Oct. 1991, also *Computer Graphics and Applications* July 1992

[2] Bryson, S. and Gerald-Yamasaki, M., "The Distributed Virtual Wind Tunnel", *Proceedings of Supercomputing '92 Minneapolis*, Minn., Nov. 1992

[3] The Virtual Windtunnel Users Manual, available at http://www.nas.nasa.gov/NAS/VWT

[4] Bryson, S., Johan, S., Globus, A., Meyer, T., and McEwen, C. "Initial User Reaction to the Virtual Windtunnel", AIAA 95-0114, 33rd AIAA Aerospace Sciences Meeting and Exhibit, Reno., Nevada. Jan. 1995

[5] Bryson, S., "Virtual Environments in Scientific Visualization", Proceedings of 1994 Virtual Reality Software and Technology, Singapore, Aug. 1994 also to appear in *Communications of the ACM.*

[6] Smith, M., Chawla, K., and Van Dalsem, W., "Numerical Simulation of a Complete STOVL Aircraft in Ground Effect", paper AIAA-91-3293, American Institute of Aeronautics 9th Aerodynamics Conference, Baltimore Md. 1991

[7] Sheridan, T. and Ferrell, W, *Man-Machine Systems*, MIT Press, Cambridge, Ma. 1974

[8] Bryson, S., "Impact of Lag and Frame Rate on Various Tracking Tasks", *Proceedings of SPIE Conference on Stereoscopic Displays and Applications*, San Jose, Ca., Feb. 1993

[9] Meyer, T. and Globus, A., "Direct Manipula-

tion of Isosurfaces and Cutting Planes in Virtual Environments", RNR Technical Report RNR-93-019, NASA Ames Research Center

[10] Upson, C., Faulhaber, T., Kamins, D., Laidlaw, D., Schlegel, D., Vroom, J., Gurwitz, R., and van Dam, A., "The Application Visualization System: A Computational Environment for Scientific Visualization", IEEE Computer Graphics and Applications, Volume 9, Number 4, July 1989

[11] Bancroft, G. V., Merritt, F. J., Plessel, T. C., Kelaita, R. K., McCabe, R., K., and Globus, A., "FAST: A Multi-Processed Environment for Visualization of Computational Fluid Dynamics", *Proceedings of Visualization '90,* IEEE Computer Society Press, October 1990

[12] Brooks, F. P. Jr., Ouh-Young, J., Blatter, J. J., and Kilpatrick, P. J., "Project GROPE - Haptic Displays for Scientific Visualization", *Computer Graphics: Proceedings of SIGGRAPH 90,* July 1990

[13] Taylor, R. M., Robinett, W., Chi, V. L., Brooks, F. P. Jr., and Wright, W., "The Nanomanipulator: A Virtual Reality Interface for a Scanning Tunneling Microscope", *Computer Graphics: Proceedings of SIGGRAPH 93,* Aug 1993

[14] Cruz-Neira, C., Leigh, J., Barnes, C., Cohen, S., Das, S., Englemann, R., Hudson, R., Papka, M., Siegel, L., Vasilakis, C., Sandin, D. J., and DeFanti, T. A., "Scientists in Wonderland: A Report on Visualization Applications in the CAVE Virtual Reality Environment", *Proceedings of the IEEE Symposium on Research Frontiers in Virtual Reality,* October 1993

[15] Herndon, K.P. and Meyer, T., "3D Widgets for Exploratory Scientific Visualization", *Proceedings of UIST '94*, ACM SIGGRAPH, November, 1994,

[16] Pausch R., Burnette, T., Capeheart, A.C., Conway, M., Cosgrove, D., DeLine, R., Durbin, J., Gossweiler, R., Koga, S., and White, J., "A Brief Architectural Overview of Alice, a Rapid Prototyping System for Vitrual Reality", *IEEE Computer Graphics and Applications,* May 1995.

[17] Jacoby, R. H., "Using Virtual Menus in a Virtual Environment". *Proceedings of the Symposium on Electronic Imaging Science & Technology,* International Society for Optical Engineering/Society for Imaging Science & Technology, Volume 1668.

[18] D. Kenwright and D. Lane, "Interactive Time-Dependent Particle Tracing Using Tetrahedral Decomposition", *IEEE Transactions on Visualization and Computer Graphics*, Vol 2(2), June 1996, pp. 120-129

[19] D. Kenwright and R. Haimes, "Vortex Identification - Applications in Aerodynamics", *Proceedings Visualization '97,* IEEE Computer Society Press, October 1997

[20] 1997.Funkhouser, T. A. and Sequin, C. H., "Adaptive Display Algorithm for Interactive Frame Rates Duing Visualization of Complex Virtual Environments", *Computer Graphics: Proceedings of SIGGRAPH 93,* Aug 1993